MEDIA UPDATE

7TH EDITION

Mathematical Reasoning FOR ELEMENTARY TEACHERS

Calvin T. **Long**

Duane W. **DeTemple**

Richard S. **Millman**

Pearson

Director, Portfolio Management: Anne Kelly
Courseware Portfolio Manager: Marnie Greenhut
Courseware Portfolio Management Assistant: Stacey Miller
Content Producer: Tara Corpuz
Managing Producer: Scott Disanno
Producer: Nicholas Sweeny
Manager, Courseware QA: Mary Durnwald
Manager, Content Development: Robert Carroll
Product Marketing Manager: Kyle DiGiannantonio
Field Marketing Manager: Andrew Noble
Marketing Assistant: Hanna Lafferty
Senior Author Support/Technology Specialist: Joe Vetere
Manager, Rights and Permissions: Gina Cheselka
Manufacturing Buyer: Carol Melville, LSC Communications
Production Management, Composition, and Illustrations: Aptara®, Inc.
Cover and Interior Design: Nancy Goulet/Studio Wink
Cover Image: Michael Ross/Flickr Open/Getty Images

The Library of Congress has cataloged the seventh edition as follows:

Long, Calvin T.
Mathematical reasoning for elementary teachers/Calvin T. Long, Washington State University, Duane W. DeTemple, Washington State University, Richard S. Millman, Georgia Institute of Technology.—7th edition.
pages cm
ISBN 978-0-321-90099-9
1. Mathematics—Study and teaching (Elementary) I. DeTemple, Duane W. II. Millman, Richard S., 1945- III. Title.
QA135.6.L66 2014
510—dc23

2013023848

3 18

Pearson

ISBN 13: 978-0-13-475882-4
ISBN 10: 0-13-475882-X

Dedication

To the memory of my good wife and constant helpmate, Jean. ▸ C.T.L.

To my wife, Janet, and my daughters, Jill and Rachel. ▸ D.W.D.

To Sandy, for her loving support. ▸ R.S.M.

Contents

*Available in MyMathLab or www.pearsonhighered.com/mathstatsresources

Preface

TO THE FUTURE TEACHER

You may be wondering what to expect from a college course in mathematics for prospective elementary or middle school teachers. Will this course simply repeat arithmetic and other material that you already know, or will the concepts be new and interesting? In this preface, we will give a positive answer to that question and at the same time provide a useful orientation to the text.

This book is designed to help you, as a future teacher, add to the depth of your knowledge about the mathematics of elementary and middle school. Most institutions structure their teacher education curriculum to start with a sequence of mathematics content courses. The content courses serve as a prerequisite for the teaching methods course, which deals with, among other ideas, how school children learn mathematics as they grow and develop. This text is filled with activities, investigations, and a host of problems with results and answers that are attractive, surprising, and unexpected, yet are designed to engage you thoughtfully doing mathematics.

The content needed for future teachers is covered fully and does so with an eye toward giving you a deep background into why things work (and why some things don't). Emphasis is placed on **the mathematical knowledge needed for teaching**, a topic very much a part of research in mathematics education today. This depth, called **conceptual understanding,** is a very important part of being a teacher. To decide whether the methods or ideas of your students are right or wrong and be able to explain to the students why is one of the most important aspects of teaching. In addition, the depth of your confidence and basic skills will be increased during this course as you participate in solving problems and performing operations in a number of different ways.

The **Common Core State Standards for Mathematics** (CCSS-M), or **Common Core**, is a new approach to the curriculum in mathematics and has been widely adopted. It has influenced this text significantly as you will soon see. In addition to content, the CCSS-M also advocates *pedagogy* (the practice of teaching). As of 2013, 46 states plus the District of Columbia are now working with the **Common Core** in mathematics.

Problem Solving and Mathematical Reasoning

Problem solving (or, said another way, "mathematical reasoning") is stressed throughout as a major theme of this book.

At first, problem solving may seem daunting, but don't be afraid to try and perhaps not succeed, because you will succeed as you keep trying. As you gain experience and begin to acquire an arsenal of strategies, you will become increasingly successful and will even begin to find the challenge of solving a problem stimulating and enjoyable. Quite often, with much surprise, this has been the experience of students in our classes as they successfully match wits with problems and gain insights and confidence that together lead to even more success.

You should not expect to see instantly into the heart of a problem or to immediately know how it can be solved. The text contains many problems that check your understanding of basic concepts and build basic skills, but you will also continually encounter problems requiring multiple steps and reflection. These problems are not unreasonably hard. (Indeed, many would be suitable for use in classes you will subsequently teach with only minor modifications.) However, they do require thought. Expect to try a variety of approaches, be willing to discuss possibilities with your classmates, and form a study group to engage in cooperative problem solving. This is the way mathematics is done, even by professionals, and as you gain experience, you will increasingly feel the real pleasure of success and the beginning development of a mathematical habit of your mind.

You will greatly improve your thinking and problem-solving skills if you take the time to write carefully worded solutions that explain your method and reasoning. Similarly, it will help you engage in mathematical conversations with your instructor and with other students. Research shows that mechanical skills learned by rote without understanding are soon forgotten and guarantee failure, both for you now and for your students later. By contrast, the ability to think creatively makes it more likely that the task can be successfully completed. *Conceptual understanding of the material is the key to your success and the future success of your students.*

How to Read This Book

Learning mathematics is not a spectator sport.

No mathematics textbook can be read passively. To understand the concepts and definitions and to benefit from examples, you must be an active participant in a conversation with the text. Often, this means that you need to check a calculation, make a drawing, take a measurement, construct a model, or use a calculator or computer. If you first attempt to answer questions raised in the examples on your own, the solutions written in the text will be more meaningful and useful than they would be without your personal involvement.

The odd problems are fully or partially answered in the back of the book. These answers give you an additional source of worked examples. But again, you will benefit most fully by attempting to solve the problems on your own or in a study group before you check your reasoning by looking up the answer provided in the text.

Guiding Philosophy and Approach

The content and processes of mathematics are presented in an appealing and logically sound way with these major goals in mind:

- to develop positive attitudes toward mathematics and the teaching of mathematics,
- to develop mathematical knowledge and skills, with particular emphasis on problem solving and mathematical reasoning,
- to develop a conceptual understanding of the mathematics of elementary and middle school,
- to develop excellent teachers of mathematics, and
- to understand the three major themes below of the mathematics and the ability to teach.

Three Major Themes

This text responds to three overarching themes that shape the content (conceptual development) and pedagogical skills (teaching excellence) required for the successful elementary or middle school teacher. The first of these themes is recognition of the *Principles and Standards for School Mathematics,* first set forth in 1989 by the National Council of Teachers of Mathematics (NCTM) and revised in 2000 to its current form. The second theme is problem solving, exemplified by Principles of Problem Solving, set forth in George Pólya's classic book *How to Solve It,* first published in 1945 and rewritten beautifully in 1988. Finally, the third and most recent theme is the recognition of the content and teaching standards found the *Common Core State Standards—Mathematics.*

The *Principles and Standards* of the NCTM, *Pólya's Problem Solving Principles,* and the *Common Core State Standards* at first may look different, but they really are not. They share a common vision of teaching skills and conceptual understandings of effective teachers of mathematics and they provide different ways to look at mathematical reasoning.

You will find the three themes infused in this text in the following ways:

- *The Principles and Standards of the NCTM*

  The goals of NCTM *Principles and Standards* for school mathematics are to "ensure quality, indicate goals, and promote positive changes in mathematics education in grades preK-12."

  An overview of the *Principles and Standards* is contained at the back of the book and should be reviewed at different times during the semesters as you work to look forward to "big picture" ideas.

- *Pólya's Principles of Problem Solving*

  The four principles of Pólya will be of tremendous help to you throughout this text. They will be used frequently in the early chapters, in particular, and are listed as

  Pólya's First Principle: Understand the Problem
  Pólya's Second Principle: Devise a Plan
  Pólya's Third Principle: Carry Out the Plan
  Pólya's Fourth Principle: Look Back

  Special attention is given to the fourth principle, "Look Back."
- *The Common Core State Standards—Mathematics*

  The mathematics topics included in this text thoroughly cover the content standards set forth in the CCSS-M. In addition to the content standards, the CCSS-M sets forth eight **Standards for Mathematical Practice**. (A complete statement of the standards can be found at the end of this book.) The goal of these standards is to ensure that teachers instill the following skills and approaches to reasoning in their students:

  1. Make sense of problems and persevere in solving them.
  2. Reason abstractly and quantitatively.
  3. Construct viable arguments and critique the reasoning of others.
  4. Model with mathematics.
  5. Use appropriate tools strategically.
  6. Attend to precision.
  7. Look for and make use of structure.
  8. Look for and express regularity in repeated reasoning.

We encourage you to read and compare throughout the ideas of the NCTM *Principles and Standards,* Pólya's Principles, and the *Common Core State Standards for Mathematics*. Excerpts and examples illustrating the standards and problem solving strategies are provided in every chapter. For example, when we are working an example, describing a concept, examining a definition, or solving an insightful problem, there will be times when you (and sometimes your students) will work in-depth to solve problems. Notions like exploring, explaining or expanding blend ideas from these three parts of the text to help readers recognize the mathematics that increases their mathematical habits of the mind.

Mathematical habits of mind are studied in mathematics education and have been used in earlier editions of this text. Since the mathematical habits of mind are very much a part of NCTM *Principles and Standards,* Pólya's Principles, and the *Common Core State Standards for Mathematics,* we will not continue to formally use that notion in this edition.

This text models effective teaching by emphasizing:

- manipulatives
- investigations
- activities for discovery
- written projects
- discussion questions
- appropriate use of technology

And, above all else,

- **problem solving**, **mathematical reasoning**, and **conceptual understanding.**

New to This Edition

The Common Core State Standards—Mathematics. The eight goals of the **Standards for Mathematical Practice** (SMP) are designed to teach our students to combine the mathematical practice (to see what is happening!) and to understand the mathematical content. The eight SMP principles describe ways in which future teachers "increasingly ought to engage with the subject matter as they grow in mathematical maturity and expertise." The SMP also places importance on the five NCTM process standards, which are found in the back of this text together with the five content standards. An important new idea is to make sure that the SMP can be "seen" in the combination of teaching and content. We have provided examples of the SMP goals in this text giving a way for you, as future teachers, to see what will happen when you are teaching.

Different Bases. Chapter 3 is reorganized in this edition in response to suggestions by instructors who felt that arithmetic in bases other than ten needed to be grouped completely in one section. To follow that approach, the authors have included all of base ten and its arithmetic (addition, subtraction, multiplication, and division) into the first three sections. The fifth section contains the nondecimal positional system with an emphasis on bases of five or six. However, while addition, subtraction and multiplication will be done, division in bases other than ten is not covered in the text as it increases significantly both the length and the complexity of long or short division. Some instructors may want to omit bases other than ten, which is certainly appropriate, and so Section 3.5 is optional and may be skipped. The other section, Section 3.4, is important, however, as it emphasizes estimation and mental arithmetic for elementary students (in base ten, of course.)

Into the Classroom. Completely new "Into the Classroom" features provide insights from active teachers, providing a window into their classroom via activities, projects, discussions, and ideas to engage children in the mathematics being covered in that particular chapter of the text. These features help students make connections between what they are currently learning in this textbook and what they will be teaching in their future classrooms.

Integrating Mathematics and Pedagogy (IMAP). IMAP video references have been added to the Annotated Instructor's Edition and allow future teachers to see elementary and middle school students working out numerical concepts. These videos provide an opportunity for valuable classroom discussion of the mathematics and knowledge of student understanding needed to teach concepts. The IMAP videos are available in MyMathLab.

Responding to Students Problems (RTS). Throughout the problem sets there are many new problems that give examples of the ways in which children try to use mathematical techniques. A really important part of being an excellent teacher is to be able to analyze what the children are doing and then give them help at a conceptual level or show them why their method works. The RTS problems show that future teachers will need a thorough understanding of mathematical content in order to answer students' questions. We want to thank Jean Anderson, who has 25 years of experience teaching in elementary and middle school in DeKalb, Georgia, for her contributions to the RTS problems and to Cameron Schriner for his help in constructing figures and tables in some problem sets.

Probability. The chapter has been completely revised to provide a comprehensive introduction to elementary probability for future K–8 teachers. With this new approach, the basics of probability are introduced quickly by their placement in the opening section. The next two sections develop counting principles with immediate applications to the calculation of probabilities; in this way, the importance of counting principles is readily apparent. The final section takes up selected additional topics—odds, expected values, geometric probability, and simulations, so that the chapter as a whole provides the background required of future teachers to meet the content standards of the NCTM and the **Common Core**.

Overview of Content

- **Problem Solving** We begin the text with an extensive introduction to problem solving in Chapter 1. This theme continues throughout the text in special problem-solving examples and is featured in the problems grouped under the headings "Thinking Critically," "Into the Classroom," and "Making Connections." New in this seventh edition is the use of the recently added Section 1.4, "Algebra as a Problem-Solving Strategy," as a platform for the expanded Chapter 8, which applies algebra to geometry.
- **Number Systems** Chapters 2, 3, 5, 6, and 7 focus on the various number systems and make use of discussion, pictorial and graphical representations, and manipulatives to promote an understanding of the systems, their properties, and the various modes of computation. There is plenty of opportunity for drill and practice, as well as for individual and cooperative problem solving, reasoning, and communication.
- **Number Theory** Chapter 4 contains much material that is new, interesting, and relevant to students' careers as future teachers. Notions of divisibility, divisors, multiples, greatest common divisors, greatest common factors, and least common multiples are developed first via informative diagrams and then through the use of manipulatives, sets, prime-factor representations, and the Euclidean algorithm.

- **Algebraic Reasoning and Representation** Although algebraic notions are used earlier in the text, Chapter 8 gives a careful and readable discussion of algebraic ideas needed in elementary and middle school. Included in the discussion are variables; algebraic expressions and equations; linear, quadratic, and exponential functions; simple graphing in the Cartesian plane; and especially the intimate relationship between algebra and geometry, in the last section of the chapter. All of these concepts are increasingly appearing in texts for elementary and middle school students. Schoolteachers must therefore understand algebraic and geometric ideas to be comfortable teaching from current texts. Chapter 8, however, is not meant as a comprehensive review of algebra. Its focus is on the algebra that is a part of the elementary and middle school curriculum.
- **Geometry** The creative and intuitive nature of geometric discovery is emphasized in Chapters 9, 10, 11, and 12. These chapters will help students view geometry in an exciting new way that is much less formal than they have seen before. The text's approach to geometry is constructive and visual. Students are often asked to draw, cut, fold, paste, count, and so on, making geometry an experimental science.

  Problem solving and applications permeate the geometry chapters, and sections on tiling and symmetry provide an opportunity to highlight the aesthetic and artistic aspects of geometry. Examples are taken from culturally diverse sources. Though it is optional, many of the concepts and construction of geometry are enhanced by their exploration with dynamic geometry software such as GeoGebra, Geometer's Sketchpad, and the like.
- **Statistics** Chapter 13, on statistics, is designed to give students an appreciation of the basic measures and graphical representations of data. The examples and problem sets use updated data and are also relevant for the children and the future teachers as the problems and examples are focused on education. This section has been modified in this edition to include "Responding to Students" problems and many new State Assessment problems. To show that statistics really is a part of the elementary school curriculum, 18 State Assessment problems are now included in Sections 13.1 and 13.2. There is also a discussion of the standardized normal distribution, as well as of z scores and percentiles.
- **Probability** Chapter 14 has been completely rewritten, with the first section introducing both *experimental probability*—probability based on experiences and repeated trials—as well as *theoretical probability*—that based on counting and other *a priori* considerations. Many of the fundamental terms and notations are covered in this introductory section, including outcome of a trial, sample space, event, and probability function. Abundant examples are given to clarify the concepts of equally likely outcomes, mutually exclusive events, complementary events, and independent events. The following two sections introduce the principles of counting—the addition and multiplication principles, combinations and permutations. These demonstrate their importance in the determination of theoretical probabilities. The concluding section completes the chapter's introduction to basic probability by discussing odds, expected values, geometric probability, and simulations.

Topics of Special Interest

The text includes several topics that many students will find especially interesting. These topics provide stimulating opportunities to hone such mathematical reasoning skills as problem solving, pattern recognition, algebraic representation, and use of calculators. The following topics are threaded into several chapters and problem sets;

- **The Fibonacci Number and the Golden Ratio.** The Fibonacci numbers (1, 1, 2, 3, 5, 8, . . .) and the Golden Ratio have surprised and fascinated people over the ages and continue to serve as an unlimited source for mathematical and pedagogical examples. It is not always obvious that there is a connection to the Fibonacci numbers. Much of the charm of such exercises consists in the surprise of discovery in unexpected places.
- **Pascal's Triangle.** This well-known triangular pattern that has roots in ancient China has unexpected applications to counting the number of paths through a square lattice and is replete with patterns awaiting discovery.
- **Triangular Numbers.** The numbers in the third column of Pascal's triangle (1, 3, 6, 10, 15, 21, . . .) appear in almost countless unexpected contexts.
- **Magic Squares and Other Magic Patterns.** These topics provide interesting practice in basic number patterns and number facts.

Features for the Future Classroom

A teacher of mathematics should be aware of both the current and historical development of mathematics, have some knowledge of the principal contributors to mathematics, and realize that mathematics continues to be a lively area of research. The text contains a number of features that future teachers will find to be valuable in the classroom:

- **Pólya Principles** have been used in an increasing number of examples, with solutions written to highlight his four-step approach to problem solving—an approach that will be quite useful.

New!

- **Common Core State Standards—Standards of Mathematical Practice**: It's important for future teachers to have a comfort level with what will be expected of them when they are in the classroom. The authors provide opportunities in context for you to become more familiar with the Standards of Mathematical Practice and how they relate to the content.
- **NCTM Principles and Standards for School Mathematics**: Classroom teachers appreciate the guidance offered by this document of continuing importance. The six principles address equity, curriculum, teaching, learning, assessment, and technology. The five content standards cover number and operations, algebra, geometry, measurement, and data and probability. The five process standards speak to problem solving, reasoning and proof, communications, connections, and representation.

New & Improved!

- **Into the Classroom** provides insights from active teachers, providing a window into their classroom via activities, projects, discussions, and ideas to engage children in the mathematics being covered in that particular chapter of the text. These features help students make connections between what they are currently learning in this textbook and what they will be teaching in their future classrooms.

New & Improved!

- **Cooperative Investigations** are activities within the body of the chapters that use small groups to explore the concepts under discussion. Working together is an important skill for future teachers as well as their future students. Additional activities can be found in MyMathLab and the corresponding Activities Manual by Dolan et al. References to the Activity Manual are found throughout the Annotated Instructor's Edition. Blue margin annotations indicate which activity from the manual would be useful when teaching specific content, making it much easier to integrate activities into the course.
- **Integrating Mathematics and Pedagogy (IMAP) videos** provide an opportunity to see children solve real problems and explain their problem solving process. These videos provide a glimpse of what a future classroom may be like and reinforce why a deeper conceptual understanding of mathematics is important for teachers.
- **Highlights from History** illustrate the contributions individuals have made to mathematics and provide a cultural, historical, and personal perspective on the development of mathematical concepts and thought.

Chapter Elements

The chapters following Chapter 1 are consistently and meaningfully structured according to the following pattern:

New & Improved!

- **Chapter Opener:** Each chapter opens with an introductory activity that introduces some of the principal topics of the chapter by means of *cooperative learning*. They are followed by a "Key Ideas" feature that shows the interconnections among the various parts of mathematics previously discussed and between mathematics and the real world. Beyond, there are, within the body of the chapters, small groups to explore the concepts under discussion.
- **Examples** are often presented in a *problem-solving mode*, asking students to independently obtain a solution that can be compared with the solution presented in the text. Solutions are frequently structured in the Pólya four-step format.
- **Think Clouds.** These notes serve as quick reminders and clarify key points in discussions.
- **Cooperative Investigations:** Each chapter includes a number of games, puzzles, and explorations to be completed in small groups. Most of these can be adapted for future elementary and middle school classrooms.
- **Common Core State Standards—Standards for Mathematical Practice:** In addition to content, the Common Core advocates Standards for Mathematical Practice (SMP), which will help elementary and middle school students develop a deeper conceptual understanding of the math

they are taught. Nearly all chapters have two "SMP" symbols in the margin noting a particular standard along with highlighted text. This is to help you make connections between the standards and the content and eventual implementation.

- **From the NCTM Principles and Standards.** Extensive excerpts from the *NCTM Principles and Standards* help you understand the relevance of topics and what students will be expected to teach.
- **Problem Sets** are organized according to the following categories:
  - **Understanding Concepts** reinforce basic concepts and provide ample practice opportunities.
  - New & Improved! **Into the Classroom** problems pose questions that cause you to carefully consider how you might go about clarifying subtle and often misunderstood points for your future students. Answering these questions often forces one to think more deeply and come to a better understanding of the subtleties involved, especially in a student classroom. Group or cooperative problems are included in this section. The number of such problems has increased significantly in this edition.
  - New & Improved! **Responding to Students** exercises provide future teachers the opportunity to see what mathematical questions and procedures children will come up with on their own and ways to respond to them. Many more have been added to this edition including more from middle school.
  - **Thinking Critically** problems offer problem-solving practice related to the section topic. Many of these problems can be used as classroom activities or with small groups.
  - **Making Connections** problems apply the section concepts to solving real-life problems and to other parts of mathematics.
  - **State Assessment** exercises are problems and problem types from various state exams providing insight into the standardized testing based on state standards in effect prior to adoption of the Common Core standards. Common Core assessment is under development at this time.
  - **Writing** exercises are interspersed throughout the problem types providing opportunities to convey ideas through written words and not just numbers and symbols.
- **Chapter in Relation to Future Teachers** is a brief essay that discusses the importance of the material just covered in the context of future teaching and helps place the chapter in relation to the remainder of the book.
- **End-of-Chapter Material** Each chapter closes with the following features:
  - **Chapter Summary** is in a table format, with more complete information, to make it more helpful for reviewing the content. The summary includes *Key Concepts, Vocabulary, Definitions,* and *Notation,* and may also include *Theorems, Properties, Formulas, Procedures,* and *Strategy.*
  - **Chapter Review Exercises** help students self-check their understanding of the concepts discussed in the chapter.

Note for the Instructor

The principal goals of this text are to impart mathematical reasoning skills, a deep conceptual understanding, and a positive attitude to those who aspire to be elementary or middle school teachers. To help meet these goals, we have made a concerted effort to involve students in mathematical learning experiences that are intrinsically interesting, often surprising, aesthetically pleasing, and focused on **mathematical knowledge for teaching**. With enhanced skill at mathematical reasoning and a positive attitude toward mathematics come confidence and an increased willingness to learn the mathematical content, skills, and effective teaching techniques necessary to become a fine teacher of mathematics.

In our own classes, we have found it extremely worthwhile to spend considerable time on Chapter 1. Problem solving has gone a long way toward changing student attitudes and promoting their ability to reason mathematically. A course that begins and continues with an extensive study of the number systems and algorithms of arithmetic is not attractive or interesting to students who feel that they already know these things and have found them dull. By contrast, the material in Chapter 1 and the many problems in the problem sets are new, stimulating, and not what students have previously experienced. We have found that, aside from increasing interest, the extensive time spent on Chapter 1 develops positive attitudes, an increasing mathematical knowledge for teaching, and skills that make it possible to deal much more quickly with the usual material on number systems, algorithms, and all the subsequent ideas that are important to the teaching of mathematics in elementary schools.

There are a number of different ways to use the text. Some instructors prefer to intersperse topics from Chapter 1 throughout their courses as they cover subsequent chapters. Another approach is to begin with Chapters 2 and 3, and present Chapter 1, and then continue with additional chapters.

We have also found that it is important to answer the frequently asked question, "Why are we here?" by going beyond the discussions of conceptual understanding and showing the kinds of questions that children may ask. The Integrating Mathematics and Pedagogy (IMAP) videos in MyMathLab are an especially useful tool. In some of the videos, the children understand the material well and in others they are confused; both serve a valuable purpose. One hour spent early in the course with a few well-chosen video clips is a tremendous help in answering the "Why?" question. There are also assignable IMAP video homework problems in MyMathLab.

Prerequisite Mathematical Background

This text is for use in mathematics content courses for prospective elementary and middle school teachers. We assume that the students enrolled in these courses have completed two years of high school algebra and one year of high school geometry. We do not assume that the students will be highly proficient in algebra and geometry, but rather that they have a basic knowledge of those subjects and reasonable arithmetic skills. Typically, students bring widely varying backgrounds to these courses, and this text is written to accommodate that diversity.

Course Flexibility

The text contains ample material for either two or three semester-length courses at Washington State University, at which elementary education majors are required to take two three-semester hour courses, with the option for an elective third course that is particularly suited to the needs of upper elementary and middle school teachers. Our text is used in all three courses. The following suggestions are for single semester-length courses, but instructors should have little difficulty selecting material that fits the coverage needed for courses in a quarter system:

- A first course, *Problem Solving and Number Systems*, covers Chapters 1 through 7. Our own first course devotes at least five weeks to Chapter 1. The problem-solving skills and enthusiasm developed in this chapter make it possible to move through most of the topics in Chapters 2 through 7 more quickly than usual. However, as noted earlier, some instructors prefer to intersperse topics from Chapter 1 among topics covered later in their courses. There is considerable latitude in which topics an instructor might choose to give a lighter or heavier emphasis. One section, Section 3.5, isolates bases other than ten. This approach allows students to focus on various kinds of positional systems that are not decimal.

- A second course, *Algebra, Basic Geometry, Statistics, and Probability*, covers Chapters 8 through 14, with the optional inclusion of computer geometry software.
- An alternative approach, *Informal Geometry*, covers Chapters 9 through 12, with an instructor deciding on software if needed. Many instructors may want to have their students become familiar with dynamic geometry software such as GeoGebra, which is now available as a free download.
- Once the basic notions and symbolism of geometry have been covered in Sections 9.1 and 9.2, the remaining chapters in geometry can be taken up in any order. Section 9.3, on figures in space, should be covered before the instructor takes up surface area and volume in Sections 10.4 and 10.5
- Many universities use the text for a three-course sequence: "Problem Solving and Number Systems" (Chapters 1–7), "Algebra and Geometry" (Chapters 8–12), and "Probability and Statistics" (Chapters 13 and 14).

Acknowledgments

We would like to thank the following individuals who reviewed either the current or previous editions of our text:

Khadija Ahmed
Monroe County Community College

Richard Anderson-Sprecher
University of Wyoming

James E. Arnold
University of Wisconsin–Milwaukee

Bill Aslan
Texas A & M University–Commerce

Scott Barnett
Henry Ford Community College

James K. Bidwell
Central Michigan University

Patty Bonesteel*
Wayne State University

Martin V. Bonsangue
California State University–Fullerton

James R. Boone
Texas A & M University

Peter Braunfeld
University of Illinois–Urbana

Tricia Muldoon Brown
University of Kentucky

Jane Buerger
Concordia College

Louis J. Chatterly
Brigham Young University

Phyllis Chinn
Humboldt State University

Lynn Cleary
San Juan College

Max Coleman
Sam Houston State University

Dr. Cherlyn Converse
California State University–Fullerton

Dana S. Craig
University of Central Oklahoma

Lynn D. Darragh
San Juan College

Addie Davis*
Olive-Harvey College

Allen Davis
Eastern Illinois University

Gary A. Deatsman
West Chester University

Sheila Doran
Xavier University

Arlene Dowshen
Widener University

Stephen Drake
Northwestern Michigan College

Joseph C. Ferrar
Ohio State University

Marjorie A. Fitting
San Jose State University

Gina Foletta
Northern Kentucky University

Grace Peterson Foster
Beaufort County Community College

Sonja L. Goerdt
St. Cloud State University

Suzie Goss*
Lone Star College, Kingwood

Tamela D. Hanebrink
Southeast Missouri State University

Lisa Hansen*
Western New England College

Ward Heilman
Bridgewater State College

Fay Jester
Pennsylvania State University

Wilburn C. Jones
Western Kentucky University

Carol Juncker
Delgado Community College

Eric B. Kahn
Bloomsburg University

Virginia Keen*
University of Dayton, Ohio

Jane Keiser
Miami University

Greg Klein
Texas A & M University

Mark Klespis
Sam Houston State University

Randa Kress
Idaho State University

Martha Ann Larkin
Southern Utah University

Verne Leininger
Bridgewater College

Charlotte K. Lewis
University of New Orleans

**Denotes reviewers of the seventh edition.*

Jim Loats
Metropolitan State College of Denver

Catherine Louchart
Northern Arizona University

Carol Lucas
University of Central Oklahoma

Jennifer Luebeck
Sheridan College

Kanchan Mathur*
Clark College

Vikki McNair*
East Central Community College

Dr. Dixie Metheny
Montana State University

David Anthony Milazzo
Niagara County Community College

Eldon L. Miller
University of Mississippi

Carla Moldavan
Berry College

Marlene M. Naquin
University of Southern Mississippi–Gulf Coast

Beth Noblitt
Northern Kentucky University

F. A. Norman
University of North Carolina–Charlotte

Jon Odell
Richland Community College

Bonnie Oppenheimer
Mississippi University for Women

Anthony Piccolino
Montclair State College

Buddy Pierce
Southeastern Oklahoma University

Jane Pinnow
University of Wisconsin–Parkside

Robert Powers
University of Northern Colorado

Tamela D. Randolph
Southeast Missouri State University

Craig Roberts
Southeast Missouri State University

Michael Roitman
Kansas State University

Jane M. Rood
Eastern Illinois University

Lisa M. Scheuerman
Eastern Illinois University

Darcy Schroeder
Arizona State University

Julie Sliva
San Jose State University

Carol J. Steiner
Kent State University

Richard H. Stout
Gordon College

Elizabeth Turner Smith
University of Louisiana–Monroe

Christine Wetzel-Ulrich
North Hampton Community College

Kimberly Vincent
Washington State University

Thanks also go to the following teachers who contributed their knowledge and experience for our Into the Classroom feature:

Kathryn Busbey
Kelly Lane Intermediate School

Nancy Campbell
Oxford School District

Robbin Crowell
Fort Worth Independent School District

Debra Goodman
Durango School District 9-R

Kristin Hanley
Clarkstown Central School District

Ann Hlabangana-Clay
Tower Hill School

Melissa Hosten
Maricopa County Education Service Agency

Jenifer G. Martin, M.A.
St. Ambrose Catholic School: University of Notre Dame ACE Academy

Tara Morey
Bayfield School District

Ralph Pantozzi
Kent Place School

Teri Rodriguez
Modesto City Schools

Marianne Strayton
Clarkstown Central School District

Simone Wells-Heard
Rockdale County Public Schools

April White
Westside Global Awareness Magnet School

Supplements

Student Supplements

Mathematics Activities for Elementary Teachers, Seventh Edition

Dan Dolan, Jim Williamson, and Mari Muri.
ISBN-10: 0-321-91511-9
ISBN-13: 978-0-321-91511-5

- Provides hands-on, manipulative-based activities keyed to the text that involve future elementary school teachers discovering concepts, solving problems, and exploring mathematical ideas.
- Colorful, perforated paper manipulatives in a convenient storage pack.
- Activities can be adapted for use with elementary students at a later time.

Student's Solutions Manual

Beverly Fusfield
ISBN-10: 0-321-90102-9
ISBN-13: 978-0-321-90102-6

- Provides detailed, worked-out solutions to all odd-numbered exercises.

Video Resources

Video lectures in MyMathLab make it easy and convenient for students to watch the videos from anywhere and are ideal for distance learning or supplemental instruction. Videos include optional subtitles.

Instructor Supplements

Annotated Instructor's Edition

ISBN-10: 0-321-91505-4
ISBN-13: 978-0-321-91505-4

- All answers included, with answers to most exercises on the page where they occur. Longer answers are in the back of the book.

The following supplements are ONLINE ONLY and are available inside your MyMathLab course or available for download in the Pearson Higher Education catalog at www.pearsonhighered.com/irc.

Instructor's Solutions Manual

Beverly Fusfield
ISBN-10: 0-321-91507-0
ISBN-13: 978-0-321-91507-8

- Provides complete solutions to all problems in the text.

Instructor's Testing Manual

- Contains prepared tests with answer keys for each chapter.

Instructor's Guide to *Mathematics Activities for Elementary Teachers*, Seventh Edition

- Contains answers for all activities, as well as additional teaching suggestions for some activities.

PowerPoint Lecture Presentation

- Fully editable lecture slides include definitions, key concepts, and examples for every section of the text.

TestGen®

- Enables instructors to build, edit, print, and administer a test, using a computerized bank of questions developed to cover all the objectives of the text.
- Algorithmically based, allowing instructors to create multiple, but equivalent, versions of the same question or test with the click of a button.
- Tests can be printed or administered online.

Media Supplements

MyMathLab® Online Course (access code required)

MyMathLab from Pearson is the world's leading online resource in mathematics, integrating interactive homework, assessment, and media in a flexible, easy to use format.

MyMathLab delivers **proven results** in helping individual students succeed.

- MyMathLab has a consistently positive impact on the quality of learning in higher education math instruction. MyMathLab can be successfully implemented in any environment–lab-based, hybrid, fully online, traditional–and demonstrates the quantifiable difference that integrated usage has on student retention, subsequent success, and overall achievement.
- MyMathLab's comprehensive online gradebook automatically tracks your students' results on tests, quizzes, homework, and in the study plan. You can use the gradebook to quickly intervene if your students have trouble, or to provide positive feedback on a job well done. The data within MyMathLab is easily exported to a variety of spreadsheet programs, such as Microsoft Excel. You can determine which points of data you want to export, and then analyze the results to determine success.

MyMathLab provides **engaging experiences** that personalize, stimulate, and measure learning for each student.

- **Personalized Learning:** MyMathLab offers several features that support adaptive learning: personalized homework and the adaptive study plan. These features allow your students to work on what they need to learn when it makes the most sense, maximizing their potential for understanding and success.
- **Exercises:** The homework and practice exercises in MyMathLab are correlated to the exercises in the textbook, and they regenerate algorithmically to give students unlimited opportunity for practice and mastery. The software offers immediate, helpful feedback when students enter incorrect answers.
- **Multimedia Learning Aids:** Exercises include guided solutions, sample problems, animations, videos, and eText access for extra help at point-of-use.

And, MyMathLab comes from an **experienced partner** with educational expertise and an eye on the future.

- Knowing that you are using a Pearson product means knowing that you are using quality content. That means that our eTexts are accurate and our assessment tools work. It means we are committed to making MyMathLab as accessible as possible. MyMathLab exercises are compatible with the JAWS 12/13 screen reader, and enables multiple-choice and free-response problem types to be read and interacted with via keyboard controls and math notation input. More information on this functionality is available at http://mymathlab.com/accessibility.
- Whether you are just getting started with MyMathLab, or have a question along the way, we're here to help you learn about our technologies and how to incorporate them into your course.

To learn more about how MyMathLab combines proven learning applications with powerful assessment, visit **www.mymathlab.com** or contact your Pearson representative.

Specific to This MyMathLab Course:

- New! "Show Work" questions enable professors to assign questions that require more detailed solutions to prove conceptual understanding, which is highly emphasized in the new Common Core State Standards.
- Problems requiring students to watch Integrating Mathematics and Pedagogy (IMAP) videos test students' understanding of concepts and content in the context of children's reasoning.
- New! Flashcards reinforce key vocabulary from the text in an interactive online format.
- New! Getting Ready section provides an opportunity for remediation in areas where students may need a little more support.

MathXL® Online Course (access code required)

MathXL® is the homework and assessment engine that runs MyMathLab. (MyMathLab is MathXL plus a learning management system.)

With MathXL, instructors can:

- Create, edit, and assign online homework and tests using algorithmically generated exercises correlated at the objective level to the textbook.
- Create and assign their own online exercises and import TestGen tests for added flexibility.
- Maintain records of all student work tracked in MathXL's online gradebook.

With MathXL, students can:

- Take chapter tests in MathXL and receive personalized study plans and/or personalized homework assignments based on their test results.
- Use the study plan and/or the homework to link directly to tutorial exercises for the objectives they need to study.
- Access supplemental animations and video clips directly from selected exercises.

MathXL is available to qualified adopters. For more information, visit our website at www.mathxl.com, or contact your Pearson representative.

About the Authors

Calvin Long received his B.S. in mathematics from the University of Idaho. Following M.S. and Ph.D. degrees in mathematics from the University of Oregon, he worked briefly as an analyst for the National Security Agency and then joined the faculty at Washington State University. His teaching ran the gamut from elementary algebra through graduate courses and frequently included teaching the content courses for prospective elementary school teachers.

His other professional activities include serving on numerous committees of the National Council of Teachers of Mathematics and the Mathematical Association of America, and holding various leadership positions in those organizations. Professor Long has also been heavily engaged in directing and instructing in-service workshops and institutes for teachers at all levels, has given more than 100 presentations at national and regional meetings of NCTM and its affiliated groups, and has presented invited lectures on mathematics education abroad.

Professor Long has coauthored two books and is the sole author of a text in number theory. In addition, he has authored over 90 articles on mathematics and mathematics education and also served as a frequent reviewer for a variety of mathematics journals, including *The Arithmetic Teacher* and *The Mathematics Teacher*. In 1986, he received the Faculty Excellence Award in Teaching from Washington State University, and in 1991, he received a Certificate for Meritorious Service to the Mathematical Association of America.

Aside from carrying out his professional activities, Cal enjoys listening to, singing, and directing classical music; reading; fly fishing; camping; and backpacking.

Duane DeTemple received his B.S. with majors in applied science and mathematics from Portland State College. Following his Ph.D. in mathematics from Stanford University, he was a faculty member at Washington State University, where he is now a professor emeritus of mathematics. He has been extensively involved with teacher preparation and professional development at both the elementary and secondary levels. Professor DeTemple has been a frequent consultant to projects sponsored by the Washington State Office of the Superintendent of Public Instruction, the Higher Education Coordinating Board, and other boards and agencies.

Dr. DeTemple has coauthored four other books and over 100 articles on mathematics or mathematics materials for the classroom. He was a member of the Washington State University President's Teaching Academy and, in 2007, was the recipient of the WSU Sahlin Faculty Excellence Award for Instruction and the Distinguished Teaching Award of the Pacific Northwest Section of the Mathematical Association of America.

In addition to teaching and researching mathematics, Duane enjoys reading, listening to and playing music, hiking, biking, canoeing, traveling, and playing tennis.

Richard Millman received a B.S. from the Massachusetts Institute of Technology and a Ph.D. from Cornell University both in mathematics. He is a professor of mathematics and was director of the Center for Education Integrating Science, Mathematics, and Computing at the Georgia Institute of Technology which supports STEM outreach in K–12. He was formerly the Outreach Professor of Mathematics at the University of Kentucky, where he was involved in both preservice and in-service teacher training for mathematics teachers.

Dr. Millman has coauthored four books in mathematics, coedited three others, and received ten peer-reviewed grants. He has published over 50 articles about mathematics or mathematics education and has taught a wide variety of mathematics and mathematics education courses throughout the undergraduate and graduate curriculum, including those for preservice teachers. He received, with a former student, an Excel Prize for Expository Writing for an article in *The Mathematics Teacher* and was a Member-at-Large of the Council of the American Mathematical Society. He was principal investigator and project director for ALGEBRA CUBED, a grant from the National Science Foundation to improve algebra education in rural Kentucky. He was the principal investigator of a Race to the Top grant form the Georgia Department of Education and another NSF grant, SLIDER, in which students use a curriculum based on engineering design in the context of building robots to learn eighth-grade physical science and math.

Rich enjoys traveling, writing about mathematics, losing golf balls, listening to music, and going to plays and movies. He also loves and is enormously proud of his grandchildren, with whom he enjoys discussing the conceptual basis of mathematics, among other topics.

1 Thinking Critically

COOPERATIVE INVESTIGATIONS
The Gold Coin Game

Material Needed

15 markers (preferably circular and yellow, if possible) for each pair of students.

Directions

This is a two-person game. Each pair of players is given 15 gold coins (markers) on the desktop. Taking alternate turns, each player removes one, two, or three coins from the desktop. The player who takes the last coin wins the game. Play several games, with each player alternately playing first. Try to devise a winning strategy, first individually as you play and then thinking jointly about how either the first or second player can play so as to force a win.

Questions to Consider

1. To discover a winning strategy, it might be helpful to begin with fewer coins. Start with just 7 coins, and see if it is possible for one player or the other to play in such a way as to guarantee a win. Try this several times, and do not move on to question 2 until the answer is clear from what happened with 7 coins.
2. This time, start with 11 coins on the desktop. Is it now possible for one player or the other to force a win? Play several games until both you and your partner agree that there is a winning strategy, and then see how the player using that strategy should play.
3. Extend the strategy you developed in step 2 to the original set of 15 coins.
4. Would the strategy work if you began with 51 markers? Explain carefully and clearly.

Variation

Devise a similar game in which the player taking the last coin *loses* the game, and explain how one player or the other can force a win for your new game.

This first chapter is dedicated to how one goes about solving a mathematical problem and how one learns to reason mathematically. Each problem to be solved needs some thought. In order to help the reader answer the questions, we present a large number of strategies for problem solving. The key question then is, "Which of the strategies should I use?" The answer is to do many problems for practice and you will ultimately instinctively go to the appropriate strategy for answering the problem. Of course, there may be many different ways to attack a problem, so it is important to try a number of strategies until you find one that works.

Why should a text devoted to future teachers focus on problem solving and mathematical reasoning? One of the most prominent features of current efforts to reform and revitalize mathematics instruction in American schools has been the recommendation that such instruction should stress problem solving and quantitative reasoning. That this emphasis continues is borne out by the fact that it appears as the first of the process standards in the National Council of Teachers of Mathematics' (NCTM's) *Principles and Standards for School Mathematics*, published in 2000. (See the Problem-Solving Standard on the next page and page 000 of the preface.) Children need to learn to *think* about quantitative situations in insightful and imaginative ways—just memorizing seemingly arbitrary rules for computation is unproductive.

Of course, if children are to learn problem solving, their teachers must themselves be good teachers of problem solving. Thus, the purpose of this chapter, and indeed of this entire book, is to help you to think more critically, analytically, and thoughtfully, in order to be more comfortable with mathematical reasoning and discourse and to bring those mathematical habits of the mind to your classroom.

These traits are a part of the Common Core State Standards for Mathematics, which we will call **"Common Core"** throughout and will be described more completely at the end of the next section.

KEY IDEAS

- The need to look for patterns: using inductive reasoning to form a conjecture
- Beginning to understand the Standards for Mathematical Practice of the Common Core State Standards for Mathematics (also known as CCSS-M) and the *Principles and Standards for School Mathematics* of the NCTM
- The four problem-solving principles of George Pólya
- A variety of available problem-solving strategies (12 are highlighted in this chapter, with more to come later)
- The idea of algebra as a problem-solving strategy
- The use of the Pigeonhole Principle
- Deductive reasoning
- The rule of indirect reasoning

FROM THE NCTM PRINCIPLES AND STANDARDS

Problem-Solving Standard

Instructional programs from prekindergarten through grade 12 should enable all students to—

- *build new mathematical knowledge through problem solving;*
- *solve problems that arise in mathematics and in other contexts;*
- *apply and adapt a variety of appropriate strategies to solve problems;*
- *monitor and reflect on the process of mathematical problem solving.*

Problem solving is the cornerstone of school mathematics. Without the ability to solve problems, the usefulness and power of mathematical ideas, knowledge, and skills are severely limited. Students who can efficiently and accurately multiply but who cannot identify situations that call for multiplication are not well prepared. Students who can both develop *and* carry out a plan to solve a mathematical problem are exhibiting knowledge that is much deeper and more useful than simply carrying out a computation. Unless students can solve problems, the facts, concepts, and procedures they know are of little use. The goal of school mathematics should be for all students to become increasingly able and willing to engage with and solve problems.

Problem solving is also important because it can serve as a vehicle for learning new mathematical ideas and skills (Schroeder and Lester 1989). A problem-centered approach to teaching mathematics uses interesting and well-selected problems to launch mathematical lessons and engage students. In this way, new ideas, techniques, and mathematical relationships emerge and become the focus of discussion. Good problems can inspire the exploration of important mathematical ideas, nurture persistence, and reinforce the need to understand and use various strategies, mathematical properties, and relationships.

1.1 An Introduction to Problem Solving

The problem that follows is an excellent and realistic example of problem solving that works well with fifth-grade students. Look for how many different ways there are to solve this problem and how many mathematical discussions there can be in a classroom.

When the children arrived in Frank Capek's fifth-grade class one day, this "special" problem was on the blackboard:

> *Old MacDonald had a total of 37 chickens and pigs on his farm. All together, they had 98 feet. How many chickens were there and how many pigs?*

After organizing the children into problem-solving teams, Mr. Capek asked them to solve the problem. "Special" problems were always fun and the children got right to work. Let's listen in on the group with Mary, Joe, Carlos, and Sue:

"I'll bet there were 20 chickens and 17 pigs," said Mary.

"Let's see," said Joe. "If you're right there are 2 × 20, or 40, chicken feet and 4 × 17, or 68, pig feet. This gives 108 feet. That's too many feet."

"Let's try 30 chickens and 7 pigs," said Sue. "That should give us fewer feet."

"Hey," said Carlos. "With Mary's guess we got 108 feet, and Sue's guess gives us 88 feet. Since 108 is 10 too much and 88 is 10 too few, I'll bet we should guess 25 chickens—just halfway between Mary's and Sue's guesses!"

These children are using a **guess and check** strategy. If their guess gives an answer that is too large or too small, they adjust the guess to get a smaller or larger answer as needed. This can be a very effective strategy. By the way, is Carlos's guess right?

Let's look in on another group:

"Let's make a table," said Nandita. "We've had good luck that way before."

"Right, Nani," responded Ann. "Let's see. If we start with 20 chickens and 17 pigs, we have 2 × 20, or 40, chicken feet and 4 × 17, or 68, pig feet. If we have 21 chickens, ..."

This is a powerful refinement of guess and check.

Chickens	Pigs	Chicken Feet	Pig Feet	Total
20	17	40	68	108
21	16	42	64	106
22	15	44	60	104
.	.	.	.	.
.	.	.	.	.
.	.	.	.	.

Making a table to look for a pattern is often an excellent strategy. Do you think that the group with Nandita and Ann will soon find a solution? How many more rows of the table will they have to fill in? Can you think of a shortcut?

Mike said, "Let's draw a picture. We can draw 37 circles for heads and put two lines under each circle to represent feet. Then we can add two extra feet under enough circles to make 98. That should do it."

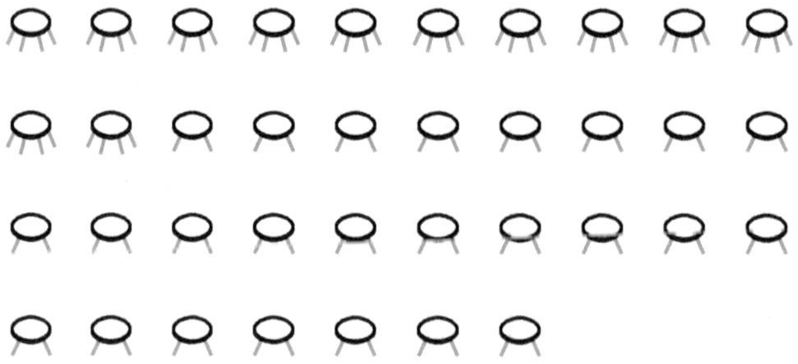

Drawing a picture is often a good strategy. Does it work in this case?

"Oh! The problem is easy," said Jennifer. "If we have all the pigs stand on their hind legs, then there are 2 × 37, or 74, feet touching the ground. That means that the pigs must be holding 24 front feet up in the air. This means that there must be 12 pigs and 25 chickens!"

98
−74
24

37
−12
25

It helps if you can be ingenious like Jennifer, but it is not essential, and children *can* be taught strategies like the following:

Guess and Check
Make a Table
Look for a Pattern
Draw a Picture

These and other useful strategies will be discussed later (see page 000 and Sections 1.4 and 1.5), but for now, let's try some problems on our own.

EXAMPLE 1.1 Guessing Toni's Number

Toni is thinking of a number. If you double the number and add 11, the result is 39. What number is Toni thinking of?

Solution 1

Guessing and checking

Guess 10.	$2 \cdot 10 + 11 = 20 + 11 = 31.$	This is too small.
Guess 20.	$2 \cdot 20 + 11 = 40 + 11 = 51.$	This is too large.
Guess 15.	$2 \cdot 15 + 11 = 30 + 11 = 41.$	This is a bit large.
Guess 14.	$2 \cdot 14 + 11 = 28 + 11 = 39.$	This checks!

Toni's number must be 14.

Solution 2

Making a table and looking for a pattern

Trial Number	Result Using Toni's Rule	
5	$2 \cdot 5 + 11 = 21$	
		2 larger
6	$2 \cdot 6 + 11 = 23$	
		2 larger
7	$2 \cdot 7 + 11 = 25$	
		2 larger
8	$2 \cdot 8 + 11 = 27$	
.	.	.
.	.	.
.	.	.

We need to get to 39, and we jump by 2 each time we take a step of 1. Therefore, we need to take

$$\frac{39 - 27}{2} = \frac{12}{2} = 6$$

more steps; we should guess $8 + 6 = 14$ as Toni's number, as before.

EXAMPLE 1.2 Guessing and Checking

(a) Place the digits 1, 2, 3, 4, and 5 in these circles so that the sums across and vertically are the same. Is there more than one solution?

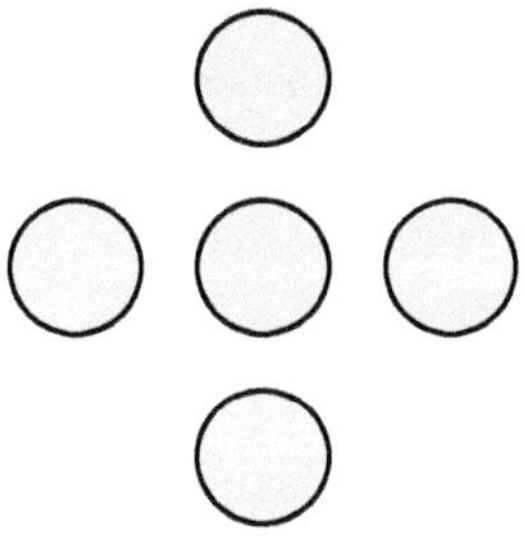

(b) Can part (a) be accomplished if 2 is placed in the center? Why or why not?

Solution

(a) Using the guess and check strategy, suppose we put the 3 in the center circle. Since the sums across and down must be the same, we must pair the remaining numbers so that they have equal sums. But this is easy, because $1 + 5 = 2 + 4$. Thus, one solution to the problem is as shown here:

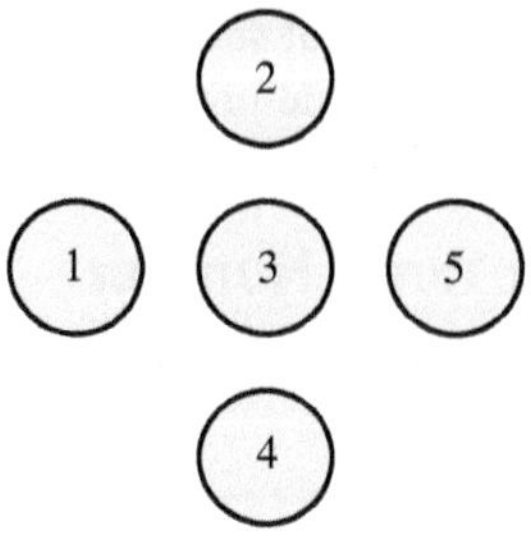

Checking further, we find other solutions, such as these:

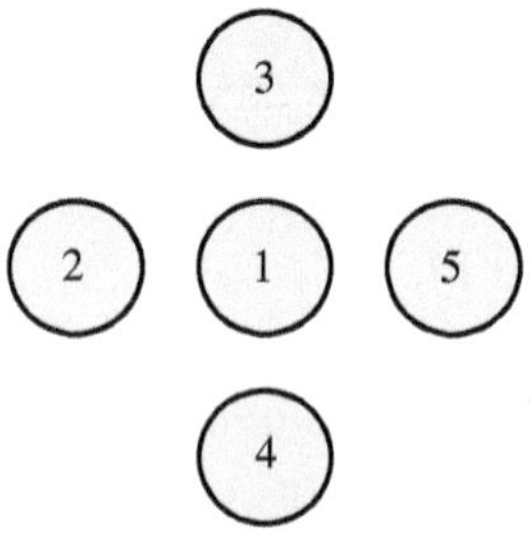

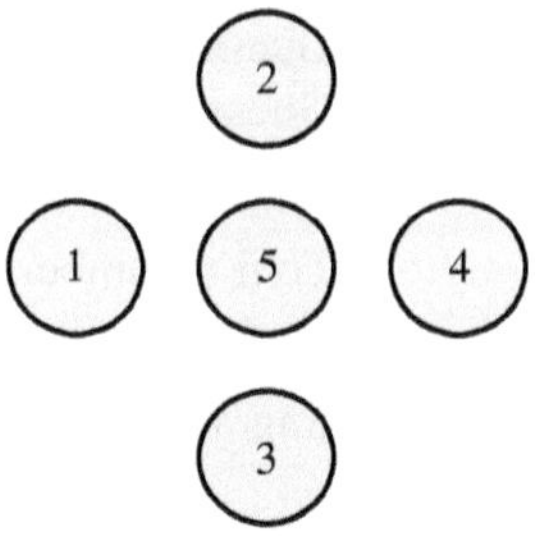

(b) What about placing 2 in the center? The remaining digits are 1, 3, 4, and 5, and these cannot be grouped into two pairs with equal sums, since one sum is necessarily odd and the other even. Therefore, there is no solution with 2 in the center circle.

1.1 Problem Set

Understanding Concepts

1. Levinson's Hardware has a number of bikes and trikes for sale. There are 27 seats and 60 wheels, all told. Determine how many bikes and how many trikes there are.

Bikes	Trikes	Bike Wheels	Trike Wheels	Total
17	10	34	30	64
18	9	36	27	63
·	·	·	·	·
·	·	·	·	·
·	·	·	·	·

(a) Use the guess and check strategy to find a solution.

(b) Complete the table to find a solution.

(c) Find a solution by completing this diagram.

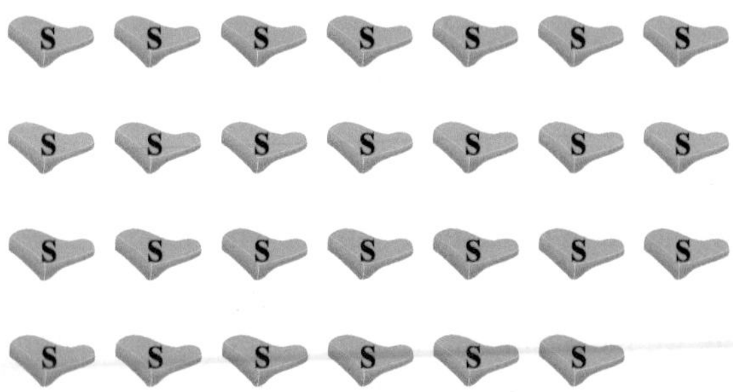

(d) Would Jennifer's method work for this problem? Explain briefly.

2. The spring concert at Port Angeles High School sold 145 tickets. Students were charged $3 each and adults $5 each. The income from the sale of tickets was $601. How many students and how many adults bought tickets?

3. **(a)** Mr. Akika has 32 18-cent and 29-cent stamps, all told. The stamps are worth $8.07. How many of each kind of stamp does he have?

(b) Summarize your solution method in one or two *carefully* written sentences.

4. Toni of Example 1.1 thinks of another number. She then triples it and subtracts 11, her result is 28.

(a) Using Guess and Check, what is Toni's number?

(b) Using Make a Table, what is Toni's number?

(c) Are there other methods to find out Toni's number? (*Hint:* Try algebra, which we will later use in Section 1.4.)

5. Xin has nine coins with a total value of 48 cents. What coins does Xin have? (*Hint:* Make an orderly list of the nickels, dimes, pennies, and quarters.)

6. Make up a problem similar to problems 1 and 2.

7. Who am I? If you multiply me by 5 and subtract 8, the result is 52.

8. Who am I? If you multiply me by -2 and add 12, the result is -4.

9. Make up a problem like problems 5 and 7.

10. **(a)** Place the digits 4, 6, 7, 8, and 9 in the circles to make the sums horizontally and vertically equal 19.

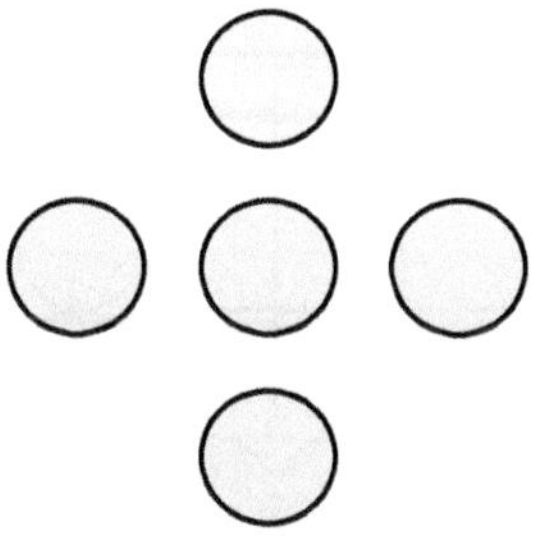

(b) Is there more than one answer to part (a)? Explain briefly. Digits should be used once and only once.

11. **(a)** Using the digits 1, 2, 3, 4, 5, and 6 once and only once, fill in the circles so that the sums of the numbers on each of the three sides of the triangle are equal.

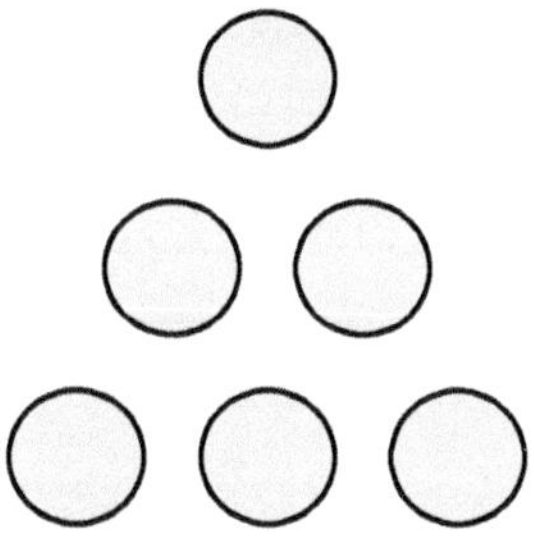

(b) Does part (a) have more than one solution?

(c) Write up a brief but careful description of the thought process you used in solving this problem.

12. In this diagram, the sum of any two horizontally adjacent numbers is the number immediately below and between them:

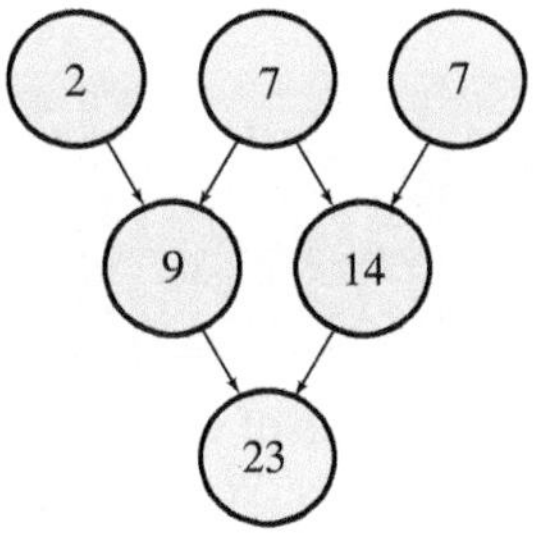

Using the same rule of formation, complete these arrays:

(a)

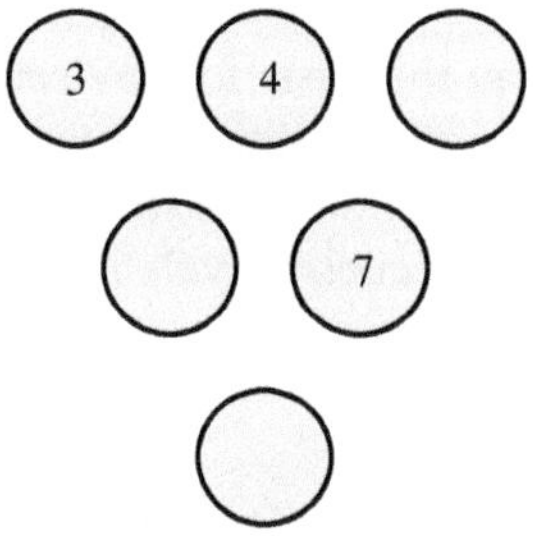

(b)

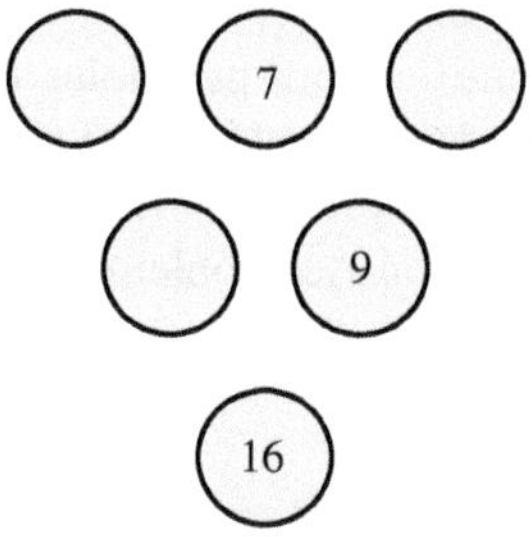

(c)

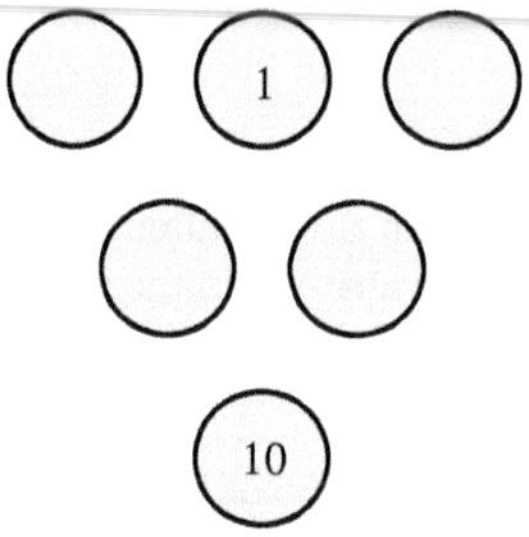

(d) Is there more than one solution to part (a)? part (b)? part (c)?

13. Study the sample diagram. Note that

$$2 + 8 = 10,$$
$$5 + 3 = 8,$$
$$2 + 5 = 7, \text{ and}$$
$$3 + 8 = 11.$$

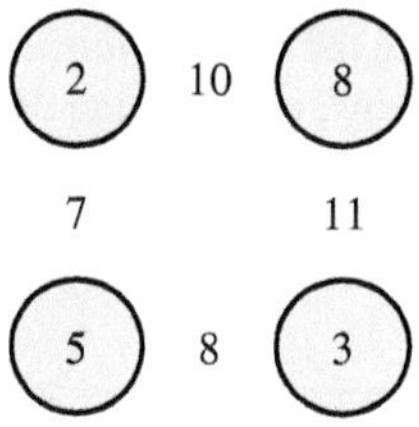

If possible, complete each of these diagrams so that the same pattern holds:

(a)

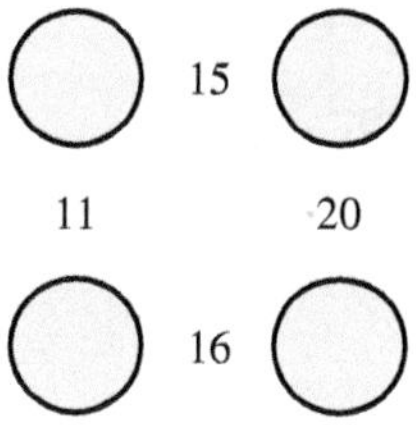

(b)

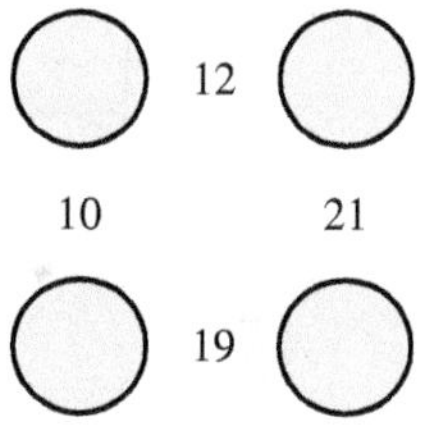

(c)

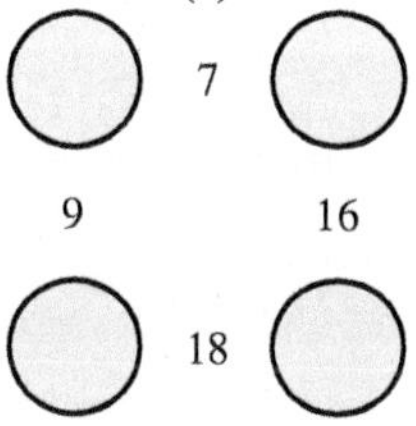

(d)

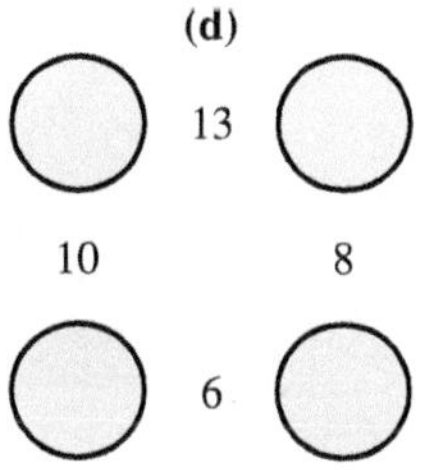

14. **(a)** Use each of the numbers 2, 3, 4, 5, and 6 once and only once to fill in the circles so that the sum of the numbers in the three horizontal circles equals the sum of the numbers in the three vertical circles.

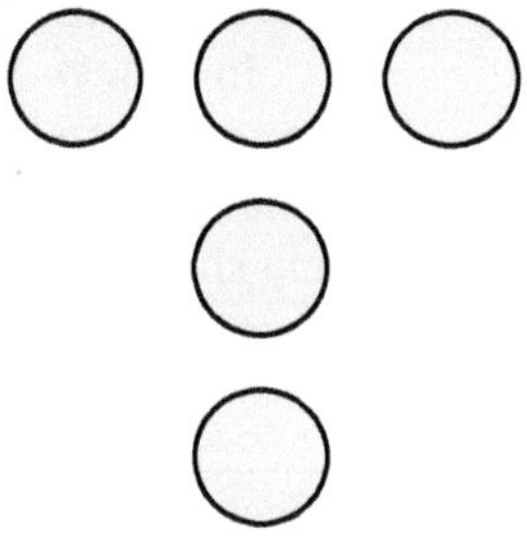

(b) Can you find more than one solution?

(c) (**Writing**) Can you have a solution with 3 in the middle of the top row? Explain in two *carefully* written sentences.

15. **(a)** In the following magic square, compute the sums of the numbers in each row, column, and diagonal of the square and write your answers in the appropriate circles:

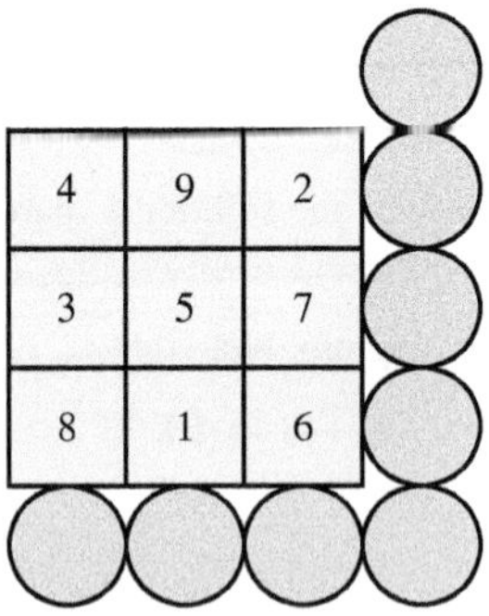

(b) Interchange the 2 and 8 and the 4 and 6 in the array in part (a) to create the magic *subtraction* square shown next. For each row, column, and diagonal, add the two end entries and subtract the middle entry from this sum.

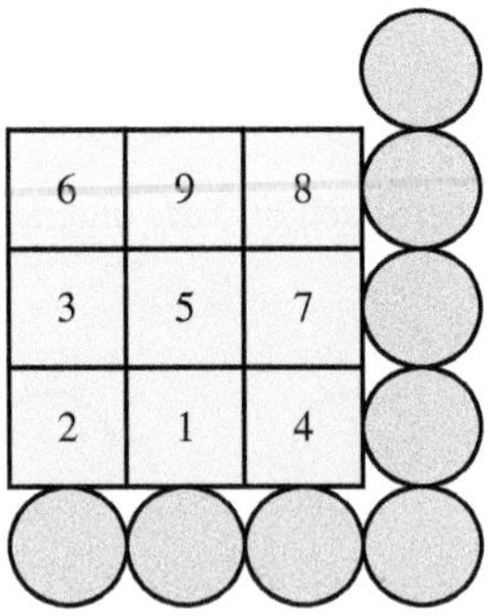

16. **(a)** Write the digits 0, 1, 2, 3, 4, 5, 6, 7, and 8 in the small squares to create another magic square. (*Hint:* Relate this to problem 15. Also, you may want to write these digits on nine small squares of paper that you can move around easily to check various possibilities.)

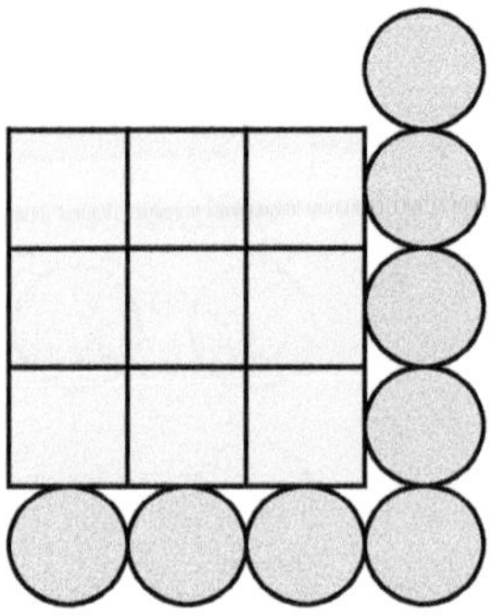

(b) Make a magic subtraction square using the numbers 0, 1, 2, 3, 4, 5, 6, 7, and 8. Digits should be used once and only once.

17. Study this sequence of numbers: 3, 4, 7, 11, 18, 29, 47, 76. Note that $3 + 4 = 7, 4 + 7 = 11, 7 + 11 = 18$, and so on. Use the same rule to complete these sequences:

(a) 1, 2, 3, ______, ______, ______, ______

(b) 2, ______, 8, ______, ______, ______, ______

(c) 3, ______, ______, 13, ______, ______, ______

(d) 2, ______, ______, ______, ______, 26

(e) 2, ______, ______, ______, ______, 11

1.2 Pólya's Problem-Solving Principles and the Standards for Mathematical Practice of the Common Core State Standards for Mathematics

Strategies

- Guess and check.
- Make an orderly list.
- Draw a diagram.

In this section, we will first talk about the marvelous four principles of George Pólya. These principles will help enormously to provide strategies for solving problems. We will next use the problem-solving standard of the National Council of Teachers of Mathematics (NCTM) from the introduction to this chapter (p. 3). We will then look to the Common Core State Standards for Mathematics (**Common Core**), especially its "Standards for Mathematical Practice (SMP)." Both the **Common Core** and NCTM are discussed in the preface.

In *How to Solve It*,* George Pólya identifies four principles that form the basis for any serious attempt at problem solving. He then proceeds to develop an extensive list of questions that teachers should ask students who need help in solving a problem. These are also questions that students can and should ask themselves as they seek solutions to problems. (The NCTM cites Pólya's insightful approach, as can be seen on page 10.)

*George Pólya, *How to Solve It* (Princeton, NJ: Princeton University Press, 1988).

Pólya's First Principle: Understand the Problem Although this principle seems too obvious to mention, students are often stymied by their efforts to solve a problem because they don't understand it fully or even in part. Teachers should ask students such questions as the following:

- Do you understand all the words used in stating the problem? If not, look them up in the index, in a dictionary, or wherever they can be found.
- What are you asked to find or show?
- Can you restate the problem in your own words?
- Is there yet another way to state the problem?
- What does *(key word)* really mean, and what is its definition?
- Can you work out some numerical examples that would help make the problem clear?
- Can you think of a picture or diagram that might help you understand the problem?
- Is there enough information to enable you to find a solution?
- Is there extraneous information and does that matter?
- What do you really need to know to find a solution?

Highlight from History: George Pólya (1887–1985)

How does one most efficiently proceed to solve a problem? Can the art of problem solving be taught, or is it a talent possessed by only a select few? Over the years, many have thought about these questions, but none so effectively and definitively as the late George Pólya, and he maintained that the skill of problem solving can be taught.

Born in Hungary in 1887, Pólya received his Ph.D. in mathematics from the University of Budapest. He taught for many years at the Swiss Federal Institute of Technology in Zurich and would no doubt have continued to do so but for the advent of Nazism in Germany. Deeply concerned by this threat to civilization, Pólya moved to the United States in 1940 and taught briefly at Brown University and then, for the remainder of his life, at Stanford University.

He was extraordinarily capable both as a mathematician and as a teacher. He also maintained a lifelong interest in studying the thought processes that are productive in both learning and doing mathematics. Among the numerous books that he wrote, he seemed most proud of *How to Solve It* (1945), which has sold over a million copies and has been translated into at least 21 languages. This book forms the definitive basis for much of the current thinking in mathematics education and is as timely and important today as when it was written.

Pólya's Second Principle: Devise a Plan Devising a plan for solving a problem once it is fully understood may still require substantial effort. But don't be afraid to make a start—you may be on the right track. There are often many reasonable ways to try to solve a problem, and the successful idea may emerge only gradually after several unsuccessful trials. A partial list of strategies includes the following, some of which we'll see later:

- use algebra
- guess and check
- make an orderly list or table
- think of the problem as partially solved
- eliminate possibilities
- solve an equivalent problem
- use the symmetry of a graph or picture
- consider special cases (experiment)
- use direct reasoning
- solve an equation
- look for a pattern
- use the pigeonhole principle
- draw a picture
- think of a similar problem already solved
- solve a simpler problem (experiment)
- use a model
- work backward
- use a formula
- be ingenious!

Skill at choosing an appropriate strategy is best learned by solving many problems. As you gain experience, you will find choosing a strategy increasingly easy—and the satisfaction of making the right choice and having it work is considerable! Again, teachers can turn the preceding list of strategies into appropriate questions to ask students in helping them learn the art of problem solving.

FROM THE NCTM PRINCIPLES AND STANDARDS

Apply and Adapt a Variety of Appropriate Strategies to Solve Problems

Of the many descriptions of problem-solving strategies, some of the best known can be found in the work of Pólya (1957). Frequently cited strategies include using diagrams, looking for patterns, listing all possibilities, trying special values or cases, working backward, guessing and checking, creating an equivalent problem, and creating a simpler problem. An obvious question is, How should these strategies be taught? Should they receive explicit attention, and how should they be integrated with the mathematics curriculum? As with any other component of the mathematical tool kit, strategies must receive instructional attention if students are expected to learn them. In the lower grades, teachers can help children express, categorize, and compare their strategies. Opportunities to use strategies must be embedded naturally in the curriculum across the content areas. By the time students reach the middle grades, they should be skilled at recognizing when various strategies are appropriate to use and should be capable of deciding when and how to use them.

SOURCE: Principles and Standards for School Mathematics by NCTM, *pp. 53–54. Copyright © 2000 by the National Council of Teachers of Mathematics. Reproduced with permission of the National Council of Teachers of Mathematics via Copyright Clearance Center. NCTM does not endorse the content or validity of these alignments.*

Pólya's Third Principle: Carry Out the Plan Carrying out the plan is usually easier than devising the plan. In general, all you need is care and patience, given that you have the necessary skills. If a plan does not work immediately, be persistent. If it still doesn't work, discard it and try a new strategy. Don't be discouraged; this is the way mathematics is done, even by professionals.

Pólya's Fourth Principle: Look Back Much can be gained by looking back at a completed solution to analyze your thinking and ascertain just what the key was to solving the problem. This is how we gain "mathematical power," the ability to come up with good ideas for solving problems never encountered before. In working on a problem, something may be lurking in the back of your mind from a previous effort that says, "I'll bet if . . . ," and the plan does indeed work! This notion of looking back is a part of a "mathematical habit of the mind," as the student will see what else may have happened while solving the problem. (See p. xv of the preface.)

Questions to ask yourself in looking back after you have successfully solved a problem include the following:

- What was the key factor that allowed me to devise an effective plan for solving this problem?
- Can I think of a simpler strategy for solving the problem?
- Can I think of a more effective or powerful strategy for solving the problem?
- Can I think of *any* alternative strategy for solving the problem?
- Can I think of any other problem or class of problems for which this plan of attack would be effective?

Looking back is an often overlooked but extremely important step in developing problem-solving skills.

Let's now look at some examples of problems and some strategies for solving them. We will then finish the section with the relationship between this problem and the Standards for Mathematical Practice (SMP).

Guess and Check

PROBLEM-SOLVING STRATEGY 1 Guess and Check

Make a guess and check to see if it satisfies the demands of the problem. If it doesn't, alter the guess appropriately and check again. When the guess finally checks, a solution has been found.

Guessing is like experimenting, giving us insight into what the next guess should be. A process of guessing, checking, altering the guess if it does not check, guessing again in light of the preceding check, and so on is a legitimate and effective strategy. When a guess finally checks, there can be no doubt that a solution has been found. If we can be sure that there is only one solution, then *the* solution has been found. Moreover, the process is often quite efficient and may be the only approach available, although students often feel that it is not "proper" to solve a problem by guessing.

EXAMPLE 1.3

Using Guess and Check

In the first diagram, the numbers in the big circles are found by adding the numbers in the two adjacent smaller circles as shown. Complete the second diagram so that the same pattern holds.

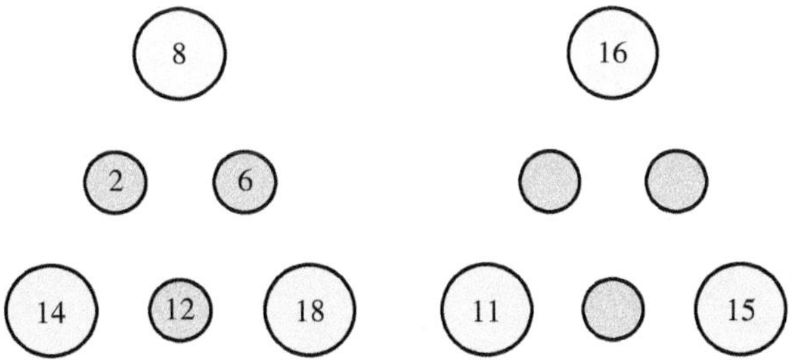

Solution

Understand the Problem

Considering the example, it is pretty clear that we must find three numbers—a, b, and c—such that

$$a + b = 16,$$
$$a + c = 11, \quad \text{and}$$
$$b + c = 15.$$

How should we proceed?*

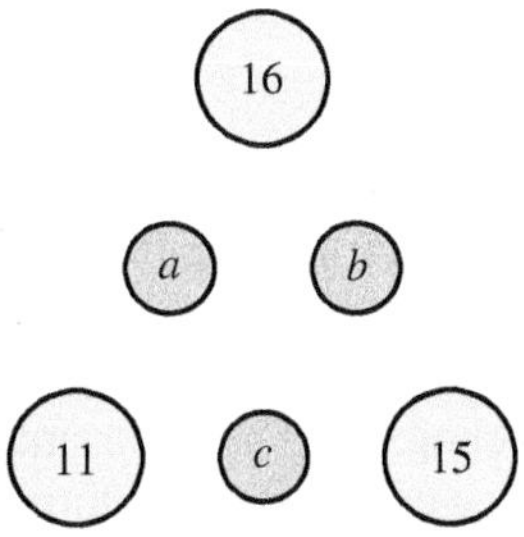

Devise a Plan

Let's try the *guess and check* strategy. It worked on several problems somewhat like this one in the last problem set. Also, even if the strategy fails, it may at least suggest an approach that will work.

Carry Out the Plan

We start by guessing a value for a. Suppose we guess that a is 10. Then, since $a + b$ must be 16, b must be 6. Similarly, since $b + c$ must be 15, c must be 9. But then $a + c$ is 19, instead of 11 as it is supposed to be. This does not check, so we guess again.

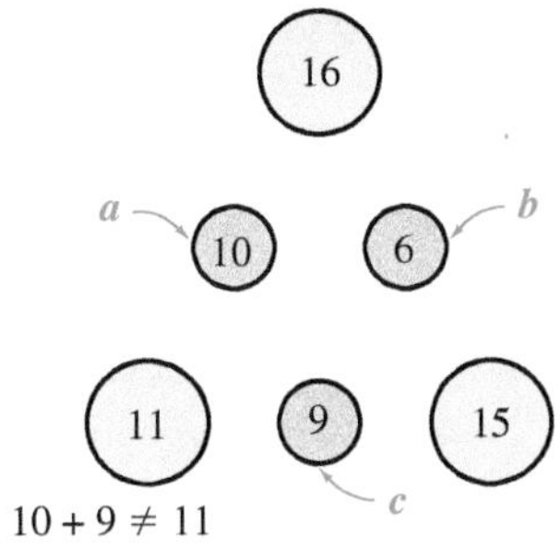

*Students who know algebra could solve this system of simultaneous equations, but elementary school students don't know algebra at this level.

Since 19 is too large, we try again with a smaller guess for a. Guess that a is 5. Then b is 11 and c is 4. But then $a + c$ is 9, which is too small, but by just a little bit. We should guess that a is just a bit larger than 5.

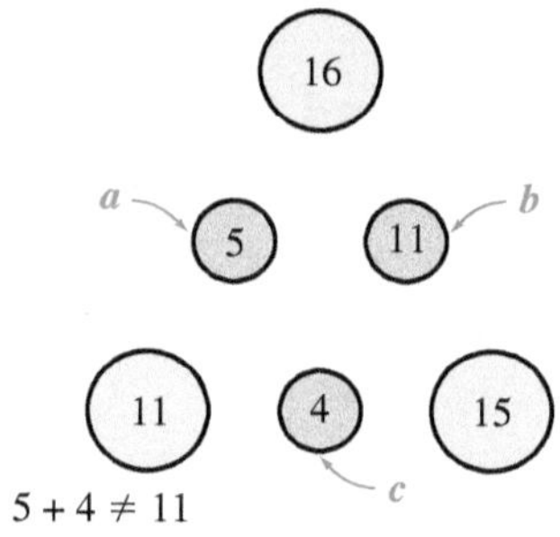
16
a
5
11
b
11
4
15
c
5 + 4 ≠ 11

Guess that $a = 6$. This implies that b is 10 and c is 5. Now $a + c$ is 11 as desired, and we have the solution.

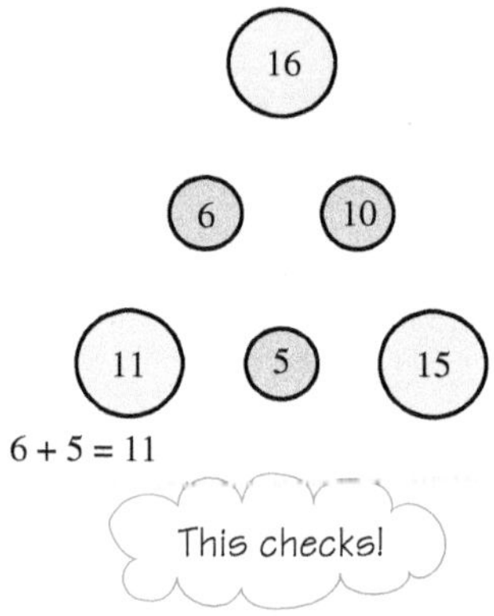
16
6
10
11
5
15
6 + 5 = 11
This checks!

Look Back

Guess and check worked fine. Our first choice of 10 for a was too large, so we chose a smaller value. Our second choice of 5 was too small but quite close. Choosing $a = 6$, which is between 10 and 5, but quite near 5, we obtained a solution that checked. Each check led us closer to the solution. Surely, this approach would work equally well on other similar problems.

But wait. Have we fully understood this problem? Might there be a way to "expand" the problem to find an easier solution?

Look back at the initial example and at the completed solution to the problem. Do you see any special relationship between the numbers in the large circles and those in the small circles?

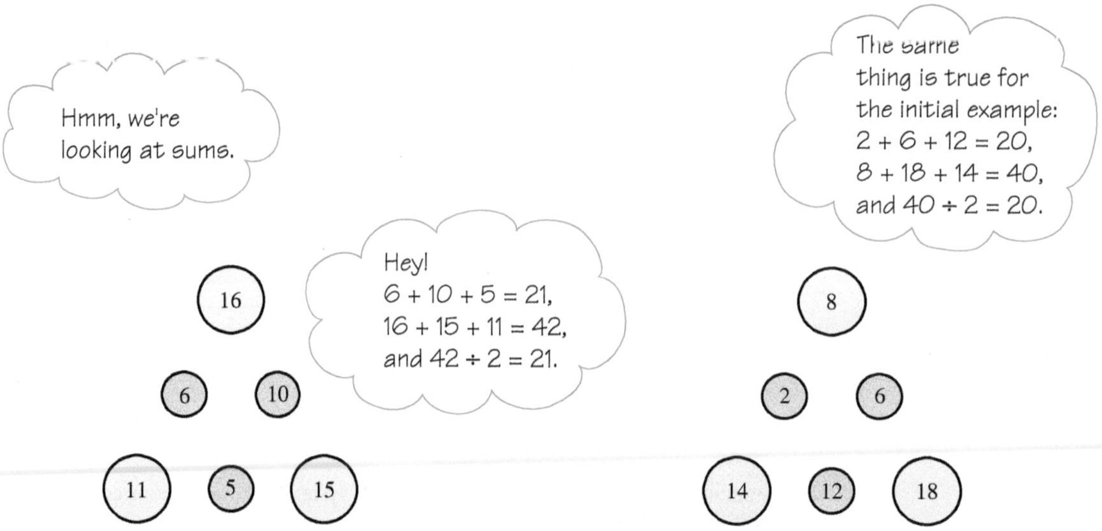
Hmm, we're looking at sums.
Hey! 6 + 10 + 5 = 21, 16 + 15 + 11 = 42, and 42 ÷ 2 = 21.
The same thing is true for the initial example: 2 + 6 + 12 = 20, 8 + 18 + 14 = 40, and 40 ÷ 2 = 20.
16
6
10
11
5
15
8
2
6
14
12
18

That's interesting: The sum of the numbers in the small circles in each case is just half the sum of the numbers in the large circles. Could we use this strategy to find another solution method?

Sure! Since $16 + 15 + 11 = 42$ and $a + b + c$ is half as much, $a + b + c = 21$. But $a + b = 16$, so c must equal 5; that is,

$$c = 21 - 16 = 5,$$
$$b = 21 - 11 = 10, \quad \text{and}$$
$$a = 21 - 15 = 6.$$

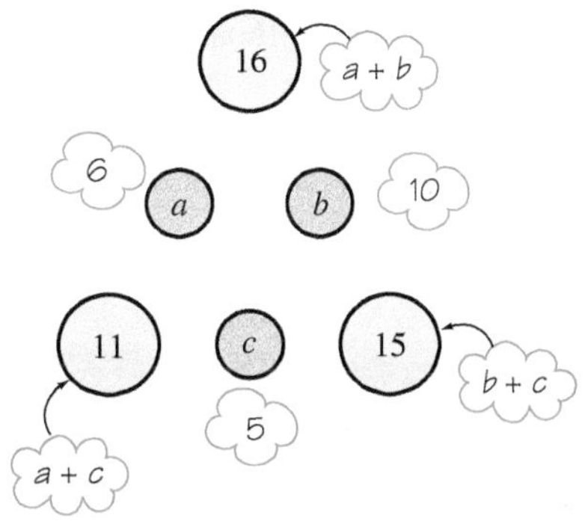

This is much easier than our first solution and, for that matter, the algebraic solution. Quickly now, does it work on this diagram? Try it.

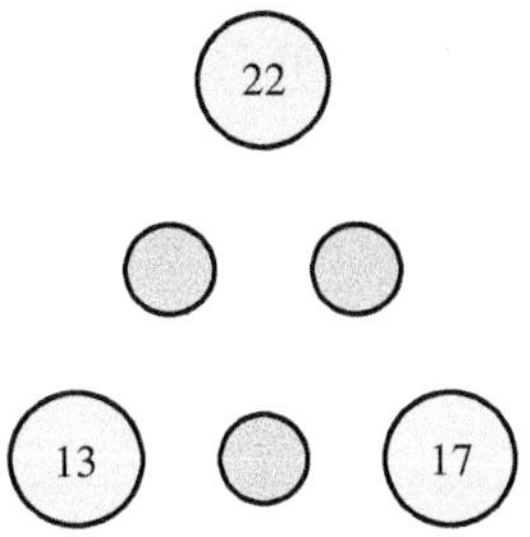

But there's one more thing: Do you understand *why* the sum of the numbers in the little circles equals half the sum of the numbers in the big circles? This diagram might help:

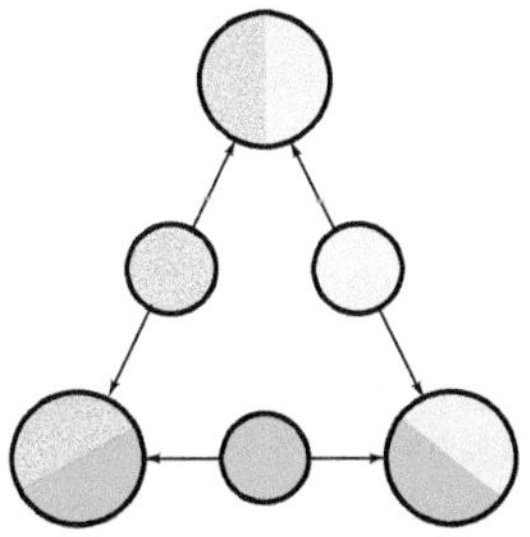

Make an Orderly List

PROBLEM-SOLVING STRATEGY 2 Make an Orderly List

For problems that require a consideration of many possibilities, make an orderly list or a table to ensure that no possibilities are missed.

Sometimes a problem may be sufficiently involved so that the task of sorting out all the possibilities seems quite forbidding. Often these problems can be solved by making a carefully structured list so that you can be sure that all of the data and all of the cases have been considered, as in the next example.

EXAMPLE 1.4

Making an Orderly List

How many different total scores could you make if you hit the dartboard shown with three darts?

Solution

Understand the Problem

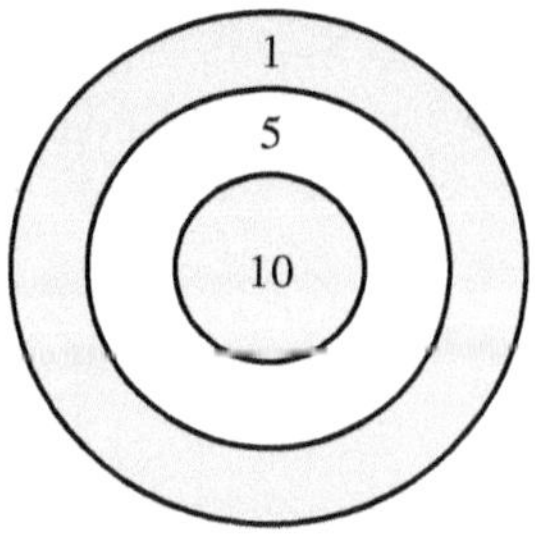

Three darts hit the dartboard and each scores a 1, 5, or 10. The total score is the sum of the scores for the three darts. There could be three 1s, two 1s and a 5, one 5 and two 10s, and so on. The fact that we are told to find the total score when throwing three darts at a dartboard is just a way of asking what sums can be made by using three numbers, each of which is either 1, 5, or 10.

It's often helpful to restate the problem in a different way.

Devise a Plan

If we just write down sums hit or miss, we will almost surely overlook some of the possibilities. Using an orderly scheme instead, we can make sure that we obtain all possible scores. Let's make such a list. We first list the score if we have three 1s, then if we have two 1s and one 5, then two 1s and no 5s, and so on. In this way, we can be sure that no score is missed.

Carry Out the Plan

Number of 1s	Number of 5s	Number of 10s	Total Score
3	0	0	3
2	1	0	7
2	0	1	12
1	2	0	11
1	1	1	16
1	0	2	21
0	3	0	15
0	2	1	20
0	1	2	25
0	0	3	30

The possible total scores are listed. There are ten different total scores.

Look Back

Here, the key to the solution was in being systematic. We were careful first to obtain all possible scores with three 1s, then two 1s, then no 1s. With two 1s, there could be either a 5 or a 10, as shown. For one 1, the only possibilities are two 5s and no 10s, one 5 and one 10, or no 5s and two 10s. Constructing the table in this orderly way makes it clear that we have not missed any possibilities.

Draw a Diagram

PROBLEM-SOLVING STRATEGY 3 Draw a Diagram

Draw a diagram or picture that represents the data of the problem as accurately as possible.

The aphorism "A picture is worth a thousand words" is certainly applicable to solving many problems. Language used to describe situations and state problems often can be clarified by drawing a suitable diagram, and unforeseen relationships and properties often become clear. As with the problem of the pigs and chickens on Old MacDonald's farm, even problems that do not appear to have pictorial relationships can sometimes be solved with this technique. Would you immediately draw a picture in attempting to solve the problem in the next example? Some would and some wouldn't, but it's surely the most efficient approach. Visual learners usually profit from making a diagram or a picture.

EXAMPLE 1.5

Using a Diagram

In a stock car race, the first five finishers in some order were a Ford, a Pontiac, a Chevrolet, a Buick, and a Dodge.

(a) The Ford finished 7 seconds before the Chevrolet.
(b) The Pontiac finished 6 seconds after the Buick.
(c) The Dodge finished 8 seconds after the Buick.
(d) The Chevrolet finished 2 seconds before the Pontiac.

In what order did the cars finish the race?

Solution

Understand the Problem

We are told how each of the cars finished the race relative to one other car. The question is, "Can we use just this information to determine the order in which the five cars finished the race?"

Devise a Plan

Imagine the cars in a line as they race toward the finish. If they do not pass one another, this is the order in which they will finish the race. We can draw a line to represent the track at the finish of the race and place the cars on it according to the conditions of the problem. We mark the line off in time intervals of one second. Then, using the first letter of each car's name to represent the car, we see if we can line up *B*, *C*, *D*, *F*, and *P* according to the given information.

Carry Out the Plan

In the following line with equally spaced points that represent 1-second time intervals, pick some point and label it *C* to represent the Chevrolet's finishing position:

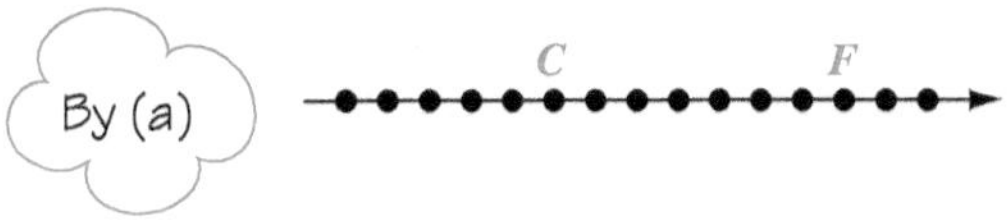

Then *F* is 7 seconds ahead of *C* by condition (a), as shown. Conditions (b) and (c) cannot yet be used, since they do not relate to the position of either *C* or *F*. However, (d) allows us to place *P* 2 seconds behind (to the left of) *C*, as follows:

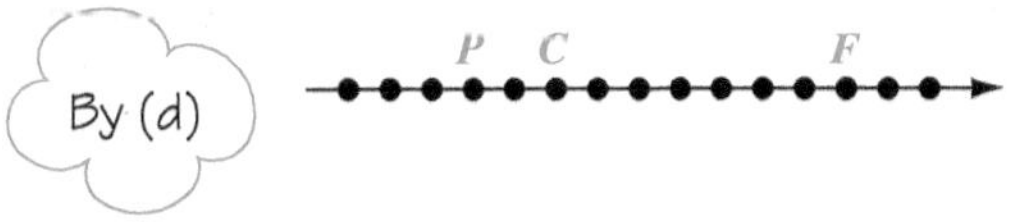

Since (b) relates the finishing position of *B* to *P*, we place *B* 6 seconds ahead of *P*:

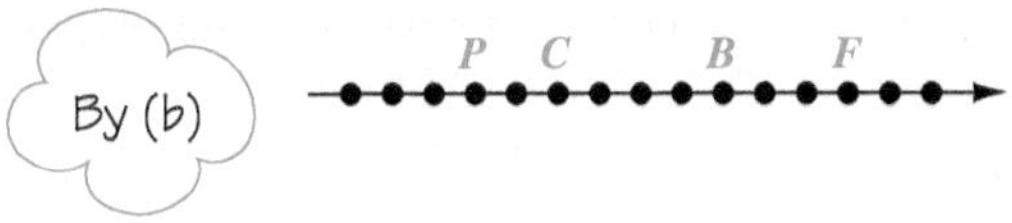

Similarly, (c) relates the finishing positions of *D* and *B* and allows us to place *D* 8 seconds behind (to the left of) *B*. Since this accounts for all the cars, a glance at the following diagram reveals the order in which the cars finished the race:

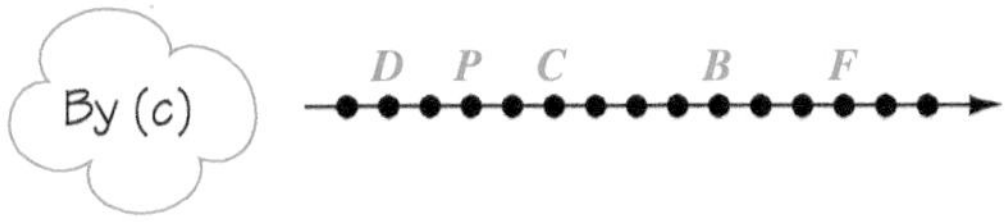

Note that we repeatedly drew the line for pedagogical purposes in order to show the placement of the cars as each new condition was used. Ordinarily, all the work would be done on a single line because it is not necessary to show what happens at each stage as we did here.

Look Back

Like the problem of the pigs and chickens on Old MacDonald's farm, this problem may not immediately suggest drawing a picture. However, having seen pictures used to solve these problems will help you to see how pictures can be used to solve other, even vaguely related, problems.

Pólya's Problem-Solving Principles and the Standards for Mathematical Practice of the Common Core Standards for Mathematics

As of 2013, in the United States, 45 states plus the District of Columbia have adopted the **Common Core State Standards for Mathematics.** This book covers its content through elementary and middle school in depth, and overlaps significantly with the content standards of the **Common Core**. In addition to content, the CCSS-M has SMP which will help very much.

The first descriptions of the Standards for Mathematical Practice (SMP) include this introductory paragraph:

> The Standards for Mathematical Practice describe varieties of expertise that mathematics educators at all levels should seek to develop in their students. These practices rest on important "processes and proficiencies" with longstanding importance in mathematics education. The first of these are the NCTM process standards of problem solving, reasoning and proof, communication, representation, and connections. The second are the strands of mathematical proficiency specified in the National Research Council's report *Adding It Up*: adaptive reasoning, strategic competence, conceptual understanding (comprehension of mathematical concepts, operations and relations), procedural fluency (skill in carrying out procedures flexibly, accurately, efficiently and appropriately), and productive disposition (habitual inclination to see mathematics as sensible, useful, and worthwhile, . . .[1]

Continuing from the Common Core, the following eight Standards for Mathematical Practice are designed to teach students to

1. Make sense of problems and persevere in solving them.
2. Reason abstractly and quantitatively.
3. Construct viable arguments and critique the reasoning of others.
4. Model with mathematics.
5. Use appropriate tools strategically.
6. Attend to precision.
7. Look for and make use of structure.
8. Look for and express regularity in repeated reasoning.

The Standards for Mathematical Practice elaborate and reinforce the importance of Pólya's four principles of problems solving. In particular, special attention is given to the fourth principle, to Look Back. As the preceding quoted paragraph makes clear, the SMP also places importance on the five NCTM process standards, which are found in the back of this text together with the five content standards. We encourage you to read and compare the Common Core State Standards, the NCTM *Principles and Standards,* and the problem solving principles of Pólya. Throughout this text, there will be excerpts from the NCTM standards and examples illustrating Pólya's principles. Moreover, when we are working an example, or describing a concept, or solving an insightful problem, there will occasionally be an "SMP" symbol in the margin that will encourage you (and, ultimately, your students) to think deeply about mathematics.

As an example, let's return to the Old MacDonald's farm problem from the beginning of the first section of this chapter. We will see how the first principle from the SMP helps us better explain this problem-solving exercise.

[1]Common Core State Standards for Mathematics

SMP 1 Make sense of problems and persevere in solving them.

"Mathematically proficient students . . . make conjectures about the form and meaning of the solution and plan a solution pathway rather than simply jumping into a solution attempt . . . . Mathematically proficient students . . . can understand the approaches of others to solving complex problems and identify correspondences between different approaches."

SMP 1 makes it clear there is far more to problem solving than simply getting the answer.

Mary, Joe, Carlos, and Sue discussed the problem and started with guessing first. However, the conversation showed that they were guessing *and then checking*. In fact, they then realized, upon looking back, that they had "invented" the notion of Guess and Check. Their classmates (Natalie and Ann) found other ways to do the problem (in other words, a different "solution pathway") and "discovered" the notion of the Make a Table strategy. Although their problem wasn't complex, it wasn't elementary either. They did recognize the various different approaches as there are many ways to do mathematics!

1.2 Problem Set

Understanding Concepts

1. (a) Nancy is thinking of a number. If you multiply it by 5 and add 13, you get 48. Could Nancy's number be 10? Why or why not?

   (b) What techniques could you use to find Nancy's number?

   (c) What is Nancy's number?

2. Lisa is thinking of a number. If you multiply it by 7 and subtract 4, you get 17. What is the number?

3. Vicky is thinking of a number. Twice the number increased by 1 is 5 less than 3 times the number. What is the number? (*Hint:* For each guess, compute two numbers and compare.)

4. In Mrs. Garcia's class, they sometimes play a game called *Guess My Rule*. The student who is It makes up a rule for changing one number into another. The other students then call out numbers, and the person who is It tells what number the rule gives back. The first person in the class to guess the rule then becomes It and gets to make up a new rule.

   (a) For Juan's rule, the results were

Numbers chosen	2	5	4	0	8
Numbers Juan gave back	7	22	17	−3	37

   Could Juan's rule have been "Multiply the chosen number by 5 and subtract 3"? Could it have been "Reduce the chosen number by 1, multiply the result by 5, and then add 2"? Are these rules really different? Discuss briefly.

   (b) For Mary's rule, the results were

Numbers chosen	3	7	1	0	9
Numbers Mary gave back	10	50	2	1	82

   What is Mary's rule?

   (c) For Peter's rule, the results were

Numbers chosen	0	1	2	3	4
Numbers Peter gave back	7	10	13	16	19

   Observe that the students began to choose the numbers in order, starting with 0. Why is that a good idea? What is Peter's rule?

5. How many different amounts of money can you pay if you use four coins including only nickels, dimes, and quarters?

6. How many different ways can you make change for a 50-cent coin by using quarters, nickels, dimes, and pennies?

7. List the three-digit numbers that use each of the digits 2, 5, and 8 once and only once.

8. List the three-digit numbers that use each of 0, 3, and 5 once and only once.

9. When Anita made a purchase, she gave the clerk a dollar and received 21 cents in change. Complete this table to show what Anita's change could have been:

Number of Dimes	Number of Nickels	Number of Pennies
2	0	1

10. Julie has 25 pearls. She put them in three velvet bags, with an odd number of pearls in each bag. What are the possibilities?

11. A rectangle has an area of 120 cm^2. Its length and width are whole numbers.

    (a) What are all the possibilities for the two numbers?

    (b) Which possibility gives the smallest perimeter?

12. The product of two whole numbers is 96 and their sum is less than 30. What are the possibilities for the two numbers?

13. Peter and Jill each worked a different number of days but earned the same amount of money. Use these clues to determine how many days each worked:

    Peter earned $20 a day.
    Jill earned $30 a day.
    Peter worked 5 more days than Jill.

14. A frog is in a well 12 feet deep. Each day the frog climbs up 3 feet, and each night it slips back 2 feet. How many days will it take the frog to get out of the well?

15. Bob can cut through a log in 1 minute. How long will it take Bob to cut a 20-foot log into 2-foot sections? (*Hint:* Draw a diagram.)

16. How many posts does it take to support a straight fence 200 feet long if a post is placed every 20 feet?

17. How many posts does it take to support a fence around a square field measuring 200 feet on a side if posts are placed every 20 feet?

18. Albright, Badgett, Chalmers, Dawkins, and Ertl all entered the primary to seek election to the city council. Albright received 2000 more votes than Badgett and 4000 fewer than Chalmers. Ertl received 2000 votes fewer than Dawkins and 5000 votes more than Badgett. In what order did each person finish in the balloting?

19. Nine square tiles are laid out on a table so that they make a solid pattern. Each tile must touch at least one other tile along an entire edge. The squares all have sides of length 1.

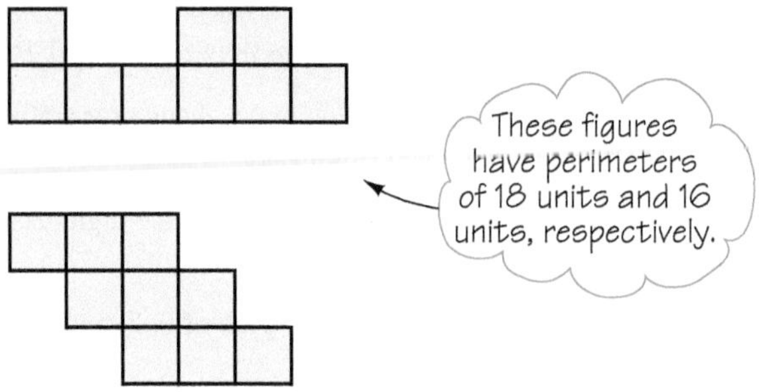

(a) What are the possible perimeters of the figures that can be formed? (The *perimeter* is the distance around the figure.)

(b) Which figure has the least perimeter?

20. For each of the strategies that follow, write a word problem that would use the method. Show the solution you have in mind.

(a) Guess and check

(b) Make an orderly list

(c) Draw a diagram

Into the Classroom

21. As in Example 1.3, the numbers in the big circles are the sums of the numbers in the two small adjacent circles. Working with small groups in grades 3, 4, or 5, place numbers in the empty circles in each of these arrays so that the same scheme holds.

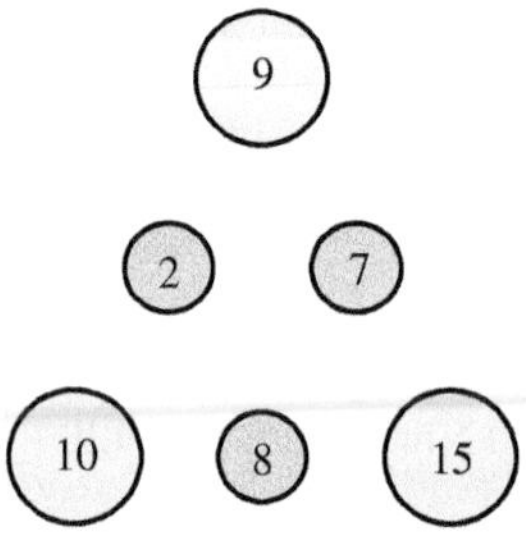

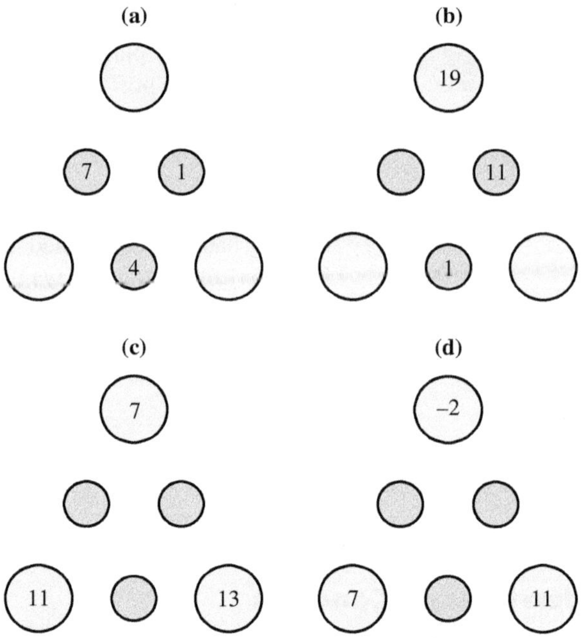

22. (**Writing**) Read one of the following from NCTM's *Principles and Standards for School Mathematics*:

(a) Problem-Solving Standard for Grades Pre-K–2, pages 116–121

(b) Problem-Solving Standard for Grades 3–5, pages 182–187

(c) Problem-Solving Standard for Grades 6–8, pages 256–261

Write a critique of the standard you read, emphasizing your own reaction. How do the recommendations compare with your own school experience?

State Assessments

23. (Grade 5)
As a fifth grader, Cathy has an excellent taste in clothing. She has decided how she will dress today. She is certain that she wants a skirt or a pair of shorts as a choice. After thinking about tee shirts, she decides that she chooses a red, blue, or lime green tee shirt. Cathy then wonders about the number of possibilities there are for her clothes today. She then decides to form a table.

Choice 1—bottoms	Choice 2—tee shirts
skirt	red
shorts	blue
	lime green

(a) From the choices in the table above, what are all of the various combinations of a pair of shorts or a skirt (Choice 1) and one of the solid tee shirts (Choice 2)? You will now help Cathy by showing her options in the form of an organized list.

(b) In addition to her previous clothes, her parents gave her the option of which kind of shoes she would like to pick. She can pick one of two different shoe categories: sneakers or sandals. How many different combinations can be found by Cathy of three pieces of clothing made out of skirts or shorts, tee shirts, and shoes? Explain how you got your answer.

24. (Washington State, Grade 4)
Four students create their own Good Fitness Games. The students run their fastest and do as many sit-ups and pull-ups as they can. Their results follow. The students want to pick an overall winner. They decide that all the events are equally important. Tell who you think the overall winner is. Explain your thinking using words, numbers, or pictures.

	50-Meter Dash	Sit-Ups	600-Meter Run	Pull-Ups
Sarah	10 seconds	42	3 minutes, 15 seconds	4
Jan	7 seconds	37	3 minutes, 50 seconds	2
Angel	8 seconds	38	3 minutes, 20 seconds	6
Mike	9 seconds	27	3 minutes, 30 seconds	8

25. (Grade 6)
Liz has a large collection of white chairs and also of black chairs. At first, she puts down a white chair. She then makes a second arrangement—a white chair surrounded by 4 black chairs as below. For the third arrangement, she has added to the second area with four white chairs. In the fourth position, Liz copied the third area and added four more black chairs. Assuming that the pattern continues, how many black chairs will Liz need for the tenth arrangement?

Black and White Chair Arrangement

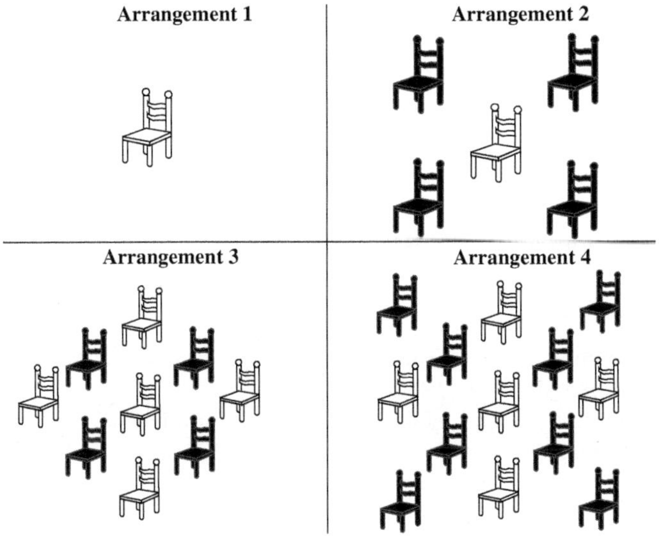

A. 17 B. 20
C. 24 D. 37

26. (Grade 6)
Sandy has 24 coins and Judith Ann has 18 coins. Each knows how many coins the other has. The sisters put their coins into groups with the same number of coins, but the number of coins in each of Sandy's groups may be different from the number of coins in Judith Ann's groups. They will win if the number of groups Sandy has matches the number of groups that Judith Ann has. What is the greatest number of groups Sandy could make and still have a chance at winning the prize?

Sandy

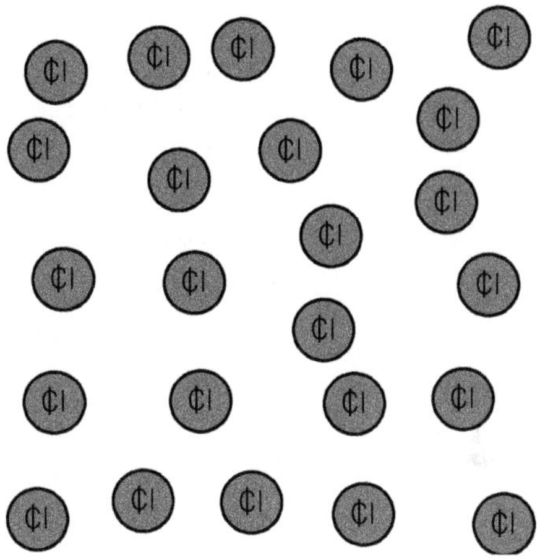

Judith Ann

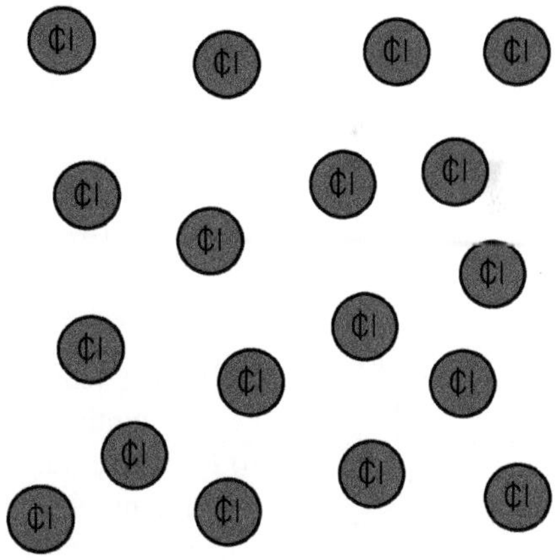

A. 2 B. 3 C. 6 D. 12

1.3 More Problem-Solving Strategies

Look for a Pattern

Strategies

- Look for a pattern.
- Make a table.
- Consider special cases.

PROBLEM-SOLVING STRATEGY 4 Look for a Pattern

Consider an ordered sequence of particular examples of the general situation described in the problem. Then carefully scrutinize these results, looking for a pattern that may be the key to the problem.

"Look for a pattern" is the most important of all problem-solving strategies. In fact, mathematics is often characterized as the study of patterns, and patterns occur in some form in almost all problem-solving situations. Think about the problems we have already considered, and you will see patterns everywhere—numerical patterns, geometrical patterns, counting patterns, listing patterns, rhetorical patterns—patterns of all kinds.

Some problems, like those in the next examples, are plainly pattern problems, but looking for a pattern is rarely a bad way to start solving a problem.

EXAMPLE 1.6

Looking for Patterns in Numerical Sequences

For each of the following numerical sequences, fill in the next three blanks:

(a) 1, 4, 7, 10, 13, _____, _____, _____
(b) 19, 20, 22, 25, 29, _____, _____, _____
(c) 1, 1, 2, 3, 5, _____, _____, _____
(d) 1, 4, 9, 16, 25, _____, _____, _____

Solution

Understand the Problem

In each case, we are asked to discover a reasonable pattern suggested by the first five numbers and then to continue the pattern for three more terms.

Devise a Plan

Questions we might ask ourselves and answer in search of a pattern include the following: Are the numbers growing steadily larger? Steadily smaller? How is each number related to its predecessor? Is it perhaps the case that a particular term depends on its two predecessors? On its three predecessors? Perhaps each term depends in a special way on the number of the term in the sequence; can we notice any such dependence? Are the numbers in the sequence somehow special numbers that we recognize? This is rather like playing *Guess My Rule.* Let's see how successful we can be.* This example also introduces the Fibonacci numbers, which are used throughout the book and shown in the Fibonacci Highlight from History.

Carry Out the Plan

In each of (a), (b), (c), and (d), the numbers grow steadily larger. How is each term related to the preceding term or terms in each case? Are the terms related to their numbered place in the sequence? Do they have a special form we can recognize?

(a) For the sequence in part (a), each number listed is 3 greater than its predecessor. If this pattern continues, the next three numbers will be 16, 19, and 22.

(b) Here, the numbers increase by 1, by 2, by 3, and by 4. If we continue this scheme, the next three numbers will be 5 more, 6 more, and 7 more than their predecessors. This would give 34, 40, and 47.

(c) If we use the ideas of (a) and (b) for this sequence, we should check how much greater each entry is than its predecessor. The numbers that must be added are *0*, *1*, *1*, and *2*; that is, $\mathit{0} + 1 = 1$, $\mathit{1} + 1 = 2$, $\mathit{1} + 2 = 3$, and $\mathit{2} + 3 = 5$. This just amounts to adding any two consecutive terms of the sequence to obtain the next term. Continuing this scheme, we obtain 8, 13, and 21.

The sequence 1, 1, 2, 3, 5, 8, 13, 21, . . . , where we start with 1 and 1 and add any two consecutive terms to obtain the next, is called the **Fibonacci sequence,** and the numbers

*Actually, there is a touchy point here. To be strictly accurate, any three numbers you choose in each case can be considered correct. There are actually infinitely many different rules that will give you any first five numbers followed by any three other numbers. What we seek here are relatively simple rules that apply to the given numbers and tell how to obtain the next three in each case.

are called the **Fibonacci numbers.** We denote the Fibonacci numbers by $F_1 = 1$, $F_2 = 1$, $F_3 = 2$, $F_4 = 3, \ldots, F_n =$ the nth Fibonacci number, and so on. In particular, observe that

$$F_3 = 2 = 1 + 1 = F_2 + F_1,$$
$$F_4 = 3 = 2 + 1 = F_3 + F_2,$$
$$F_5 = 5 = 3 + 2 = F_4 + F_3,$$

and so on. In general, any particular entry in the sequence is the sum of its two predecessors. The Fibonacci numbers first appeared in A.D. 1202 in the book *Liber Abaci* by Leonardo of Pisa (Fibonacci), the leading mathematician of the thirteenth century.

(d) Here, 4 is 3 larger than 1, 9 is 5 larger than 4, 16 is 7 larger than 9, and 25 is 9 larger than 16. The terms seem to be increasing by the next largest odd number each time. Thus, the next three numbers should probably be $25 + 11 = 36$, $36 + 13 = 49$, and $49 + 15 = 64$. Alternatively, in this case we may recognize that the numbers 1, 4, 9, 16, and 25 are special numbers. Thus, $1 = 1^2, 4 = 2^2, 9 = 3^2, 16 = 4^2$, and $25 = 5^2$. The sequence appears to be just the sequence of square numbers. The sixth, seventh, and eighth terms are just $6^2 = 36, 7^2 = 49$, and $8^2 = 64$, as before.

Look Back

In all four sequences, we checked to see how much larger each number was than its predecessor. In each case, we were able to discover a pattern that allowed us to write the next three terms of the sequence. In part (d), we also noted that the first term was 1^2, the second term was 2^2, the third term was 3^2, and so on. Thus, it was reasonable to guess that each term was the square of the number of its position in the sequence. This allowed us to write the next few terms with the same result as before.

Highlight from History: Fibonacci

The most talented mathematician of the Middle Ages was Leonardo of Pisa (ca. 1170–1250), the son of a Pisan merchant and customs manager named Bonaccio. The name of Leonardo, son of Bonaccio, was soon shortened simply to Fibonacci, which is still popularly used today. The young Fibonacci was brought up in Bougie—still an active port in modern Algeria. It was here and on numerous trips throughout the Mediterranean region with his father that Fibonacci became acquainted with the Indo–Arabic numerals and the algorithms for computing with them that we still use today. Fibonacci did outstanding original work in geometry and number theory and is best known today for the remarkable sequence

$$1, 1, 2, 3, 5, 8, 13, 21, 34, 55, 89, \ldots,$$

which bears his name. However, his most important contribution to Western civilization remains his popularization of the Indo–Arabic numeration system in his book *Liber Abaci*, written in 1202. This book so effectively illustrated the vast superiority of that system over the other systems then in use that it soon was widely adopted not only in commerce but also in serious mathematical studies. His use so simplified computational procedures that its effect on the rapid growth of mathematics during the Renaissance and beyond can only be characterized as profound.

We have already seen in earlier examples how making a table is often an excellent strategy, particularly when combined with the strategy of looking for a pattern. Like drawing a picture or making a diagram, making a table often reveals unexpected patterns and relationships that help to solve a problem.

Make a Table

PROBLEM-SOLVING STRATEGY 5 Make a Table

Make a table reflecting the data in the problem. If done in an orderly way, such a table may reveal patterns and relationships that suggest how the problem can be solved.

EXAMPLE 1.7

Applying Make a Table

(a) Draw the next two diagrams to continue this sequence of dots:

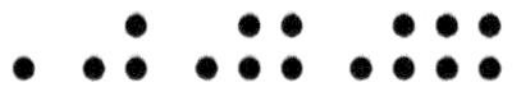 ______, ______,

(b) How many dots are in each figure?

______, ______, ______, ______, ______, ______

(c) How many dots would be in the one-hundredth figure?
(d) How many dots would be in the one-millionth figure?

Solution

Understand the Problem

What is given?

In part (a), we are given an ordered sequence of arrays of dots. We are asked to recognize how the arrays are being formed and to continue the pattern for two more diagrams. In part (b), we are asked to record the number of dots in each array in part (a). In parts (c) and (d), we are asked to determine specific numerical terms in the sequence of part (b).

Devise a Plan

In part (a), we are asked to continue the pattern of a sequence of arrays of dots. As with numerical sequences, our strategy will be to see how each array relates to its predecessor or predecessors, hoping to discern a pattern that we can extend two more times.

In part (b), we will simply count and record the numbers of dots in the successive arrays in part (a).

In parts (c) and (d), we will study the numerical sequence of part (b) just as we did in Example 1.6, hoping to discern a pattern and understand it sufficiently to determine its one-hundredth and one-millionth terms.

Carry Out the Plan

In part (a), we observe that the arrays of dots are similar but that each array has one more two-dot column than its predecessor. Thus, the next two arrays are

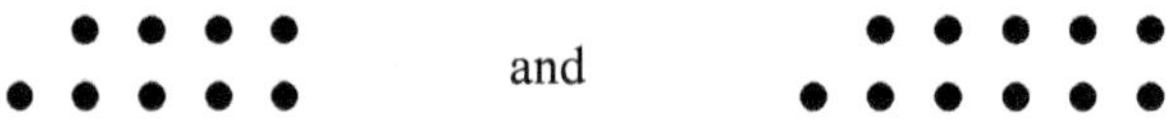

For part (b), we count the dots in each array of part (a) to obtain

$$1, 3, 5, 7, 9, 11, \ldots.$$

These are just the odd numbers, and we could tediously write out the first one million odd numbers and so answer parts (c) and (d). But surely there's an easier way.

Let's review how the successive terms were obtained. A table may help.

Reviewing this table carefully, we finally experience an Aha! moment:

Number of Entry	Entry
1	$1 = 1$
2	$3 = 1 + 2$
3	$5 = 1 + 2 + 2 = 1 + 2 \times 2$
4	$7 = 1 + 2 + 2 + 2 = 1 + 3 \times 2$
5	$9 = 1 + 2 + 2 + 2 + 2 = 1 + 4 \times 2$

The *second* term is $1 + 1 \times 2$.

The *third* term is $1 + 2 \times 2$.

The *fourth* term is $1 + 3 \times 2$.

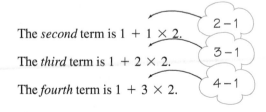

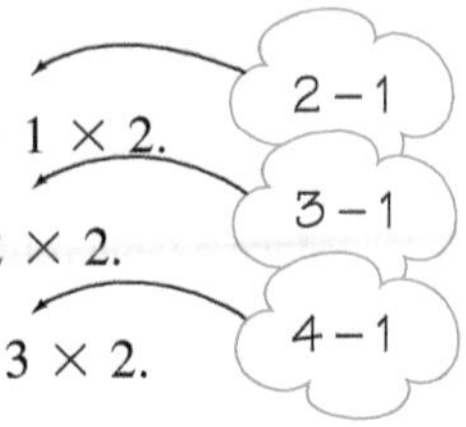

The number of 2s added is one less than the number of the term. Therefore, the one-hundredth term is

$$1 + 99 \times 2 = 199$$

100 − 1

and the one-millionth term is

$$1 + (1{,}000{,}000 - 1) \times 2 = 1 + 999{,}999 \times 2 = 1{,}999{,}999.$$

Look Back

The basic observation was that each diagram could be obtained by adding a column of two dots to its predecessor. Hence, the successive terms in the numerical sequence were obtained by adding 2 to each entry to get the next entry. Using this notion, we examined the successive terms and discovered that any entry could be found by subtracting 1 from the number of the entry, doubling the result, and adding 1. But this last sentence is rather cumbersome, and we have already seen that using *symbols* can make it easier to write mathematical statements. If we use n for the number of the term, the sentence in question can be translated into this mathematical sentence:

$$q_n = (n - 1) \times 2 + 1 = 2n - 2 + 1 = 2n - 1.^*$$

$2n - 1$ is the nth odd number.

Here q_n, read "q sub n," is the formula that gives the nth entry in the sequence of part (b). All we have to do is replace n by 1, 2, 100, and so on, to find the first entry, the second entry, the one-hundredth entry, and so on. Thus,

$$q_1 = 2 \times 1 - 1 = 1,$$
$$q_2 = 2 \times 2 - 1 = 3,$$
$$q_{100} = 2 \times 100 - 1 = 199,$$

and so on.

Use a Variable

In the preceding example and even earlier, we saw how using symbols or **variables** often makes it easier to express mathematical ideas and so to solve problems. We will look carefully at using algebra as a strategy in the next section.

The equation $q_n = 2n - 1$ used in the preceding example should remind you of your algebra courses in middle school or high school. This equation assigns, to each stage n in the process, a number given by $q_n = 2n - 1$, in other words, the nth odd number. If we prefer, we can go in the other direction; that is, if we are given an odd number q_n, we can tell what the stage (the value for n) is for the odd number.

Sometimes we want to use a variable in representing the general term in a sequence. Thus, as we saw in Example 1.7, $2n - 1$ is the nth odd number. In the expression $2n - 1$, n is the variable and it can be replaced by any natural number. For example, if we want to know which odd number 85 is, we need to determine n such that

$$2n - 1 = 85.$$

This implies that $2n = 86$, so $n = 43$. Hence, 85 is the 43rd odd number.[†]

[*]Symbols like this are often called *variables,* and the formula $q_n = 2n - 1$ is an example of a function. These notions are of considerable importance in mathematics and will be discussed in Chapter 8.
[†]These notions will be discussed in greater detail in the next section and in Chapter 8.

Consider Special Cases

PROBLEM-SOLVING STRATEGY 6 Consider Special Cases

In trying to solve a complex problem, consider a sequence of special cases. This will often show how to proceed naturally from case to case until arriving at the case in question. Alternatively, the special cases may reveal a pattern that makes it possible to solve the problem.

Pascal's Triangle

One of the most interesting and useful patterns in all of mathematics is the numerical array called "Pascal's triangle."

Consider the problem of finding how many different paths there are from A to P on the grid shown in Figure 1.1 if you can only move *down* along edges in the grid. If we start to trace out paths without care, our chances of finding all possibilities are not good.

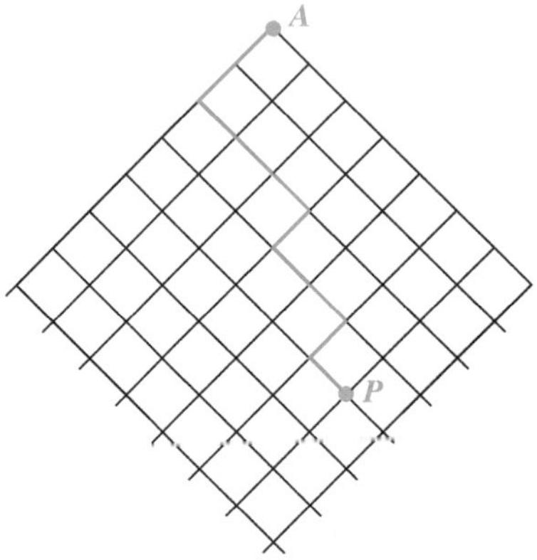

Figure 1.1
A path from A to P

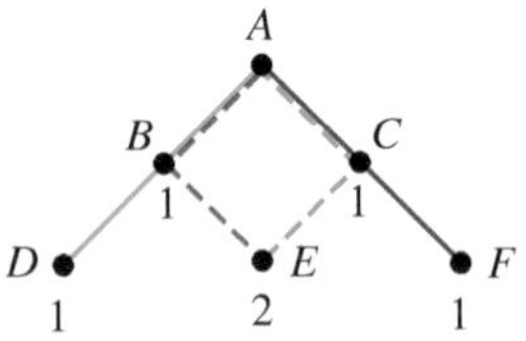

Figure 1.2
Solving an easier, similar problem

A strategy that is often helpful is to **solve an easier, similar problem** (special cases). It would certainly be easier if P were not so far down in the grid. Consider the easier, similar problem of finding the number of paths from A to E in Figure 1.2. This process begins our **experimenting** with this problem.

Or consider solving a similar *set* of easier problems all at once. How many different paths are there from A to each of B, C, D, E, and F? (Note that this is an example of considering a series of special cases.) Clearly, there is only one way to go from A to each of B and C, and we indicate this fact by the 1s under B and C in Figure 1.2. Also, the only route to D is through B, so there is only one path from A to D, as indicated. For the same reason, there is one path from A to F. By contrast, there are two ways to go from A to E—one route through B and one through C. We indicate this fact by placing a 2 under E on the diagram. This approach certainly doesn't solve the original problem, but it gives us a start and even suggests how we might proceed. Consider the diagram in Figure 1.3.

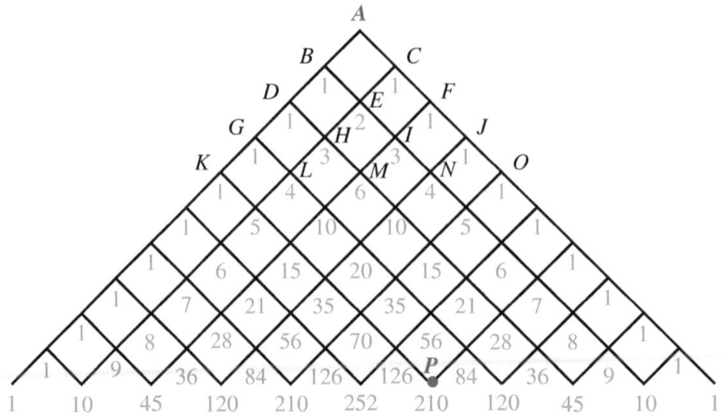

Figure 1.3
The number of paths from A to P

Having determined the number of paths from A to each of B, C, D, E, and F, could we perhaps determine the number of paths to G, H, I, and J and then continue on down the grid to eventually solve the original problem?

1. Always moving downward, we see that the only way to get to G from A is via D. But there is only one path to D and only one path from D to G. Thus, there is only one path from A to G, and we enter a 1 under G on the diagram, as shown.
2. The only way to get to H from A is via D or E. Since there is only one path from A to D and one from D to H, there is only one path from A to H via D. However, since there are two paths from A to E and one path from E to H, there are two paths from A to H via E. The number of paths from A to H is the number via D plus the number via E—that is, $1 + 2 = 3$ paths—and we enter 3 under H on the diagram as shown.
3. The arguments for I and J are the same as for H and G, so we enter 3 and 1 under I and J, respectively, on the diagram.
4. The first three steps reveal a very nice pattern that enables us to solve the original problem with ease. There can be only one path to any edge vertex on the grid, since we have to go straight down the edge to get to such a vertex. For any interior point on the grid, however, we can always reach that point by paths through the points immediately above and to the left and right of such a point. Since there is only one path from each of these points to the point in question, the total number of paths to this point is the *sum* of the number of paths to these preceding two points. Thus, we easily generate the number of paths from A to any given point in the grid by simple addition. In particular, there are 210 different paths from A to P, as we initially set out to determine.

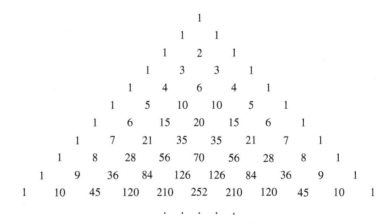

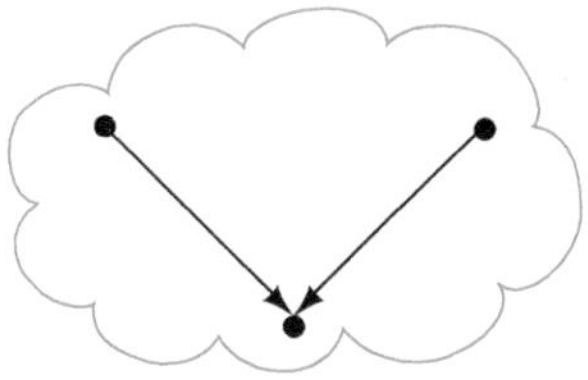

Without the grid and with an additional 1 at the top to complete a triangle, the number array in Figure 1.4 is called **Pascal's triangle** and comes up frequently in mathematics and this text. It also provides elementary school students with ways to experiment with numbers.

Figure 1.4
Pascal's triangle

1
1 1
1 2 1
1 3 3 1
1 4 6 4 1
1 5 10 10 5 1
1 6 15 20 15 6 1
1 7 21 35 35 21 7 1
1 8 28 56 70 56 28 8 1
1 9 36 84 126 126 84 36 9 1
1 10 45 120 210 252 210 120 45 10 1
. . . . .

The array is named after the French mathematician Blaise Pascal (1623–1662), who showed that these numbers play an important role in the theory of probability. However, the triangle was certainly known in China as early as 1303 by Chu Shih-Chieh.

Pascal's triangle is rich with remarkable patterns and is also extremely useful. Before discussing the patterns, we observe that it is customary to call the single 1 at the top of the triangle the *zeroth row* (since, for example, in the path-counting problem just discussed, this 1 would represent a path of length 0). For consistency, we will also call the initial 1 in any row the zeroth element in the row, and the initial diagonal of 1s the zeroth diagonal. (See Figure 1.5.) Thus, 1 is the zeroth element in the fourth row, 4 is the first element, 6 the second element, and so on.

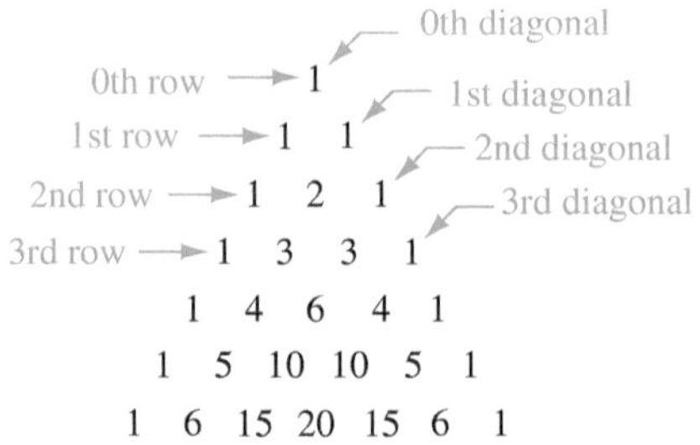

Figure 1.5 Numbered rows and diagonals in Pascal's triangle

EXAMPLE 1.8 Finding a Pattern in the Row Sums of Pascal's Triangle

Consider special cases. Look for a pattern.

(a) Compute the sum of the elements in each of rows zero through four of Pascal's triangle.
(b) Look for a pattern in the results of part (a) and guess a general rule.
(c) Use Figure 1.4 to check your guess for rows five through eight.
(d) Give a convincing argument that your guess in part (b) is correct.

Solution

Understand the Problem

In part (a), we must add the elements in the indicated rows. In part (b), we are asked to discover a pattern in the numbers generated in part (a). In part (c), we must compute the sums for four more rows and see if the results obtained continue the pattern guessed in part (b). In part (d), we are asked to argue convincingly that our guess in part (b) is correct.

Devise a Plan

Part (a) is certainly straightforward: We must compute the desired sums. To find the pattern requested in part (b), we should ask the question, "Have we ever seen a similar problem before?" The answer, of course, is a resounding "yes": All the problems in this section, but particularly Example 1.6, have involved looking for patterns. Surely, the techniques that succeeded earlier should be tried here. Appropriate questions to ask and answer include the following: "How are the successive numbers related to their predecessors?" "Are the numbers special numbers that we can easily recognize?" "Can we relate the successive numbers to their numbered locations in the sequence of numbers being generated?" Answering these questions should help us make the desired guess.

For part (c), we will compute the sums of the elements in rows five through eight to see if these numbers agree with our guess in part (b). If they *don't* agree, we will go back and modify our guess. If they *do* agree, we will proceed to part (d) and try to make a convincing argument that our guess is correct. About all we have to go on is the fact that the initial and terminal elements in each row are 1s and that the sum of any two consecutive elements in a row is the element between these two elements but in the next row down.

Carry Out the Plan

(a) $1 = 1$
$1 + 1 = 2$
$1 + 2 + 1 = 4$
$1 + 3 + 3 + 1 = 8$
$1 + 4 + 6 + 4 + 1 = 16$

(b) It appears that each number in part (a) is just twice its predecessor. The numbers are

$$1, \quad \overbrace{2 \cdot 1 = 2^1}^{2}, \quad \overbrace{2 \cdot 2 = 2^2}^{4}, \quad \overbrace{2 \cdot 2^2 = 2^3}^{8}, \quad \overbrace{2 \cdot 2^3 = 2^4}^{16}.$$

It appears that the sum of the elements in

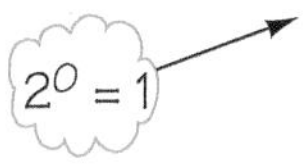

the zeroth row is 2^0,
the first row is 2^1,
the second row is 2^2,
and so on.

Our guess is that the sum of the elements in the nth row is 2^n.

(c) Computing these sums for the next four rows, we have

fifth row	$1 + 5 + 10 + 10 + 5 + 1 = 32 = 2^5$
sixth row	$1 + 6 + 15 + 20 + 15 + 6 + 1 = 64 = 2^6$
seventh row	$1 + 7 + 21 + 35 + 35 + 21 + 7 + 1 = 128 = 2^7$
eighth row	$1 + 8 + 28 + 56 + 70 + 56 + 28 + 8 + 1 = 256 = 2^8$

Since these results do not contradict our guess, we proceed to try to make a convincing argument that our guess is correct.

(d) What happens as we go from one row to the next? How is the next row obtained? Consider the third and fourth rows, shown in the diagram that follows. The arrows show how the fourth row is obtained from the third, and we see that each of 1, 3, 3, and 1 in the third row appears *twice* in the sum of the elements in the fourth row. Since this argument would hold for any two consecutive rows, the sum of the numbers in any row is just twice the sum of the numbers in the preceding row. Hence, from above, the sum of the numbers in the ninth row must be $2 \times 2^8 = 2^9$, in the tenth row it must be $2 \cdot 2^9 = 2^{10}$, and so on. Thus, the result is true in general, as claimed.

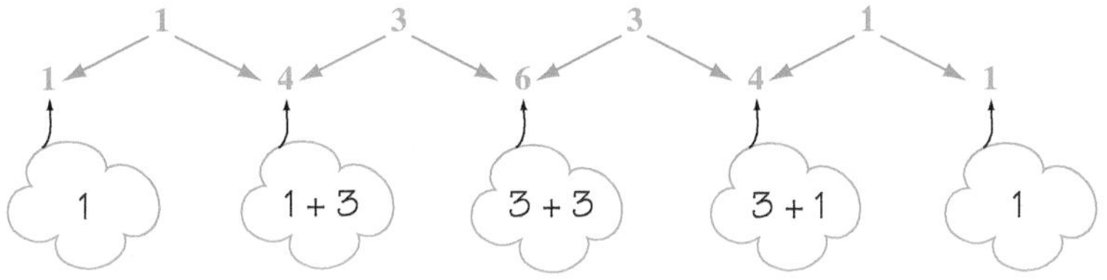

Look Back

Several aspects of our solution merit special comment. Beginning with the strategies of considering special cases and looking for a pattern, we were led to guess that the sum of the elements in the nth row of the triangle is 2^n. In an attempt to argue that this guess was correct, we considered the special case of obtaining the fourth row from the third. This showed that the sum of the elements in the fourth row was twice the sum of the elements in the third row. *Since the argument did not depend on the actual numbers that appeared in the third and fourth rows, but only on the general rule of formation of the triangle,* it would hold for any two consecutive rows and so actually proves that our conjecture is correct. This type of argument is called **arguing from a special case** and is an important problem-solving strategy.

The **arithmetic** (a-rith-met'-ic) **progression** is a very useful pattern, such as 2, 5, 8, 11, ... The arithmetic progression is defined by the property that each term is a **constant difference** greater (or less) than its predecessor by the same amount. For example, the difference in the preceding progression 2, 5, 8, 11, . . . , is that each time, the difference of consecutive terms is 3. Looking at the following consecutive terms, we see the clouds showing the constant difference.

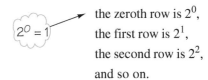

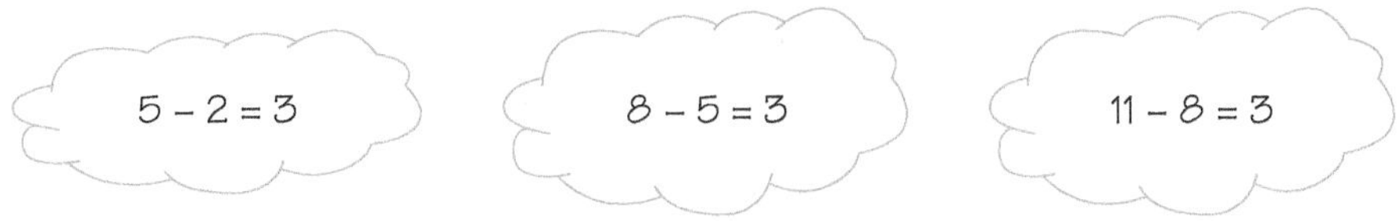

Do you think that an arithmetic progression can diminish? Is 9, 6, 3, 0, −3, −6, . . . , an arithmetic progression, and if so, what is its constant difference?

EXAMPLE 1.9 Working with Arithmetic Progressions

Is the sequence −2, 3, 8, 13, 18, . . . , 68 an arithmetic progression ending at 68? If so, find the number of terms.

Solution

To see if this sequence is an arithmetic progression, we must subtract any two consecutive numbers, starting with −2 and 3, to see if we always get the same number, a constant difference. The difference of the first two terms is 5, since $3 - (-2) = 5$. The other terms at the beginning of the sequence are

$$8 - 3 = 5, \qquad 13 - 8 = 5, \qquad 18 - 13 = 5, \ldots.$$

Yes, it does look like an arithmetic progression, because 5 is the difference for all consecutive numbers; 5 is the constant difference in this problem.

To find the number of terms, we look at the *second* term, 3, and see that it is 5 more than −2; the *third* term, 8, is 5 more than 3; the *fourth* term, 13, is 5 more than 8, and so on. If there are n numbers, then the pattern shows

$$-2 + (n - 1)5 \text{ is the } n\text{th term.}$$

Since the last term in our problem is 68, we need to solve

$$-2 + 5(n - 1) = 68.$$

We finish this exercise using algebra. We see that $5(n - 1) = 70$. Dividing both sides of the equation by 5, we have $n - 1 = 14$. The number of terms is, therefore, $n = 15$.

1.3 Problem Set

Understanding Concepts

1. Look for a pattern and fill in the next three blanks with the most likely choices for each sequence:

(a) 2, 5, 8, 11, ____ ____ ____

(b) −5, −3, −1, 1, ____ ____ ____

(c) 1, 1, 3, 3, 6, 6, 10, ____ ____ ____

2. Look for a pattern and fill in the blanks with the most likely choices for each sequence.

(a) 1, 3, 4, 7, 11, ____, ____, ____.

(b) 2, 8, 32, 128, ____, ____, ____.

(c) 1, 9, 25, 49, ____, ____, ____.

3. Find the number of terms in each of these arithmetic progressions:

(a) 5, 7, 9, . . . , 35

(b) −4, 1, 6, . . . , 46

(c) 3, 7, 11, . . . , 67

4. Consider the sequence 8, 5, 2, −1, . . . , −52.

(a) Is the sequence an arithmetic progression? Why or why not?

(b) How many terms are there in the sequence?

5. (a) Fill in the blanks to continue this sequence of equations:

$$1 = 1$$
$$1 + 2 + 1 = 4$$
$$1 + 2 + 3 + 2 + 1 = 9$$
$$1 + 2 + 3 + 4 + 3 + 2 + 1 = 16$$
$$\underline{\qquad\qquad} = \underline{\qquad}$$
$$\underline{\qquad\qquad} = \underline{\qquad}$$

(b) Compute this sum:

$$1 + 2 + 3 + \cdots + 99 + 100 + 99 + \cdots + 3 + 2 + 1 = \underline{\qquad}$$

(c) Fill in the blank to complete this equation:

$$1 + 2 + 3 + \cdots + (n - 1) + n + (n - 1) + \cdots + 3 + 2 + 1 = \underline{\qquad}$$

6. (a) Fill in the blanks to continue this sequence of equations:

$$1 = 0 + 1$$
$$1 + 3 + 1 = 1 + 4$$
$$1 + 3 + 5 + 3 + 1 = 4 + 9$$
$$\underline{\qquad\qquad} = \underline{\qquad}$$
$$\underline{\qquad\qquad} = \underline{\qquad}$$

(b) What expression, suggested by part (a), should be placed in the blank to complete this equation?

$$1 + 3 + 5 + \cdots + (2n - 3) + (2n - 1)$$
$$+ (2n - 3) + \cdots + 5 + 3 + 1 = \_\_\_\_\_\_$$

(*Hint:* The number preceding n is $n - 1$.)

7. **(a)** Fill in the blanks to continue this dot sequence in the most likely way:

______, ______

(b) What number sequence corresponds to the sequence of dot patterns of part (a)?

(c) What is the 10th term in the sequence of part (b)? the 100th term?

(d) Which term in the sequence is 101? (*Hint:* How many 3s must be added to 2 to get 101?)

8. (**Writing**) Writers of standardized tests often pose questions such as, "What is the next term in the sequence 2, 4, 8, . . . ?"

(a) How would you answer this question?

(b) Evaluate the expressions 2^n, $n^2 - n + 2$, and $n^3 - 5n^2 + 10n - 4$ in the following chart by replacing n successively by 1, 2, 3, and 4:

n	1	2	3	4
2^n				
$n^2 - n + 2$				
$n^3 - 5n^2 + 10n - 4$				

(c) In light of the results in part (b), what criticism would you make of the test writer who would write a test question like that above? Compare the wording above with that in problem 1 of this problem set.

9. Here is the start of a hundreds chart:

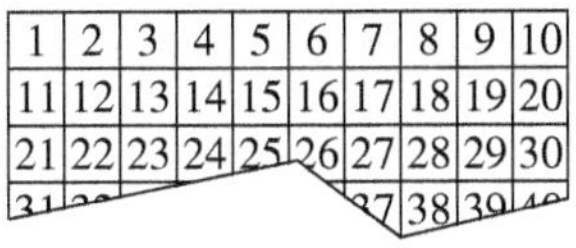

Shown next are parts of the chart. Without extending the chart, determine which numbers should go in the lavender squares.

(a)

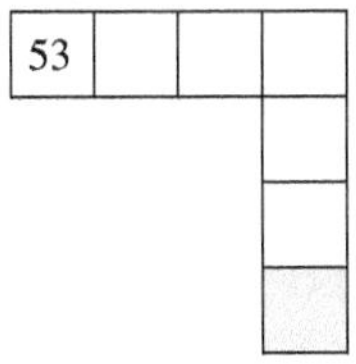

(b)

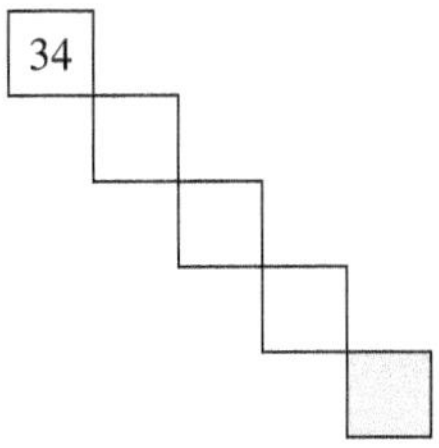

(c)

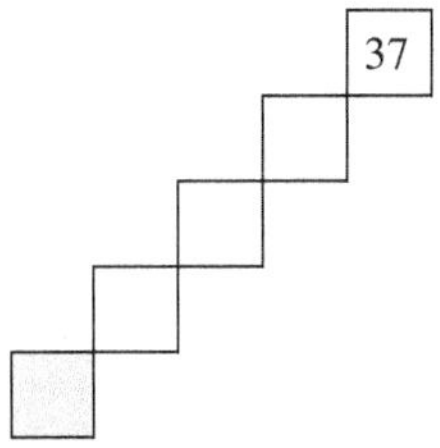

10. Refer to the chart in the preceding problem. Shown next are parts of the chart. Without extending the chart, determine which numbers should go in the lavender squares.

(a)

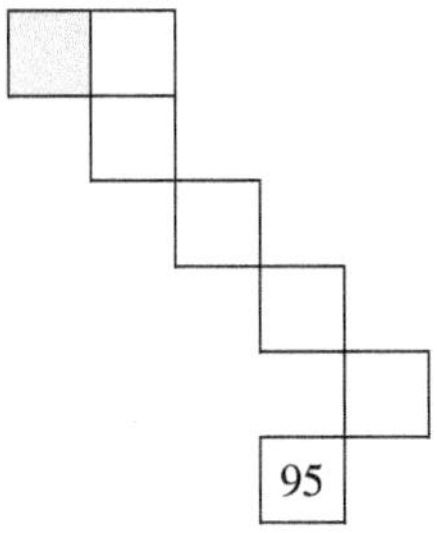

(b)

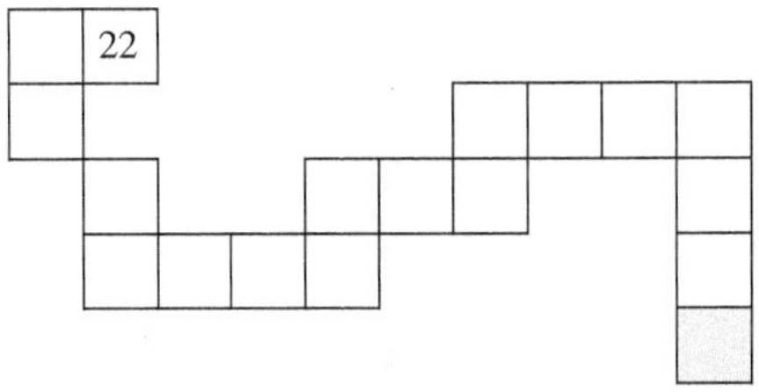

11. Five blue (B) and five red (R) discs are lined up in the arrangement B B B B B R R R R R:

Switching just two adjacent discs at a time, what is the least number of moves you can make to achieve the arrangement B R B R B R B R B R shown here?

(*Hint:* How many moves are required to rearrange B B R R to B R B R? B B B R R R to B B R B R R? and so on.)

12. **(a)** Complete the next two equations of this sequence:

$$1 = 1$$
$$1 - 4 = -3$$
$$1 - 4 + 9 = 6$$
$$1 - 4 + 9 - 16 = -10$$
$$\_\_\_\_\_\_ = \_\_\_\_\_\_$$
$$\_\_\_\_\_\_ = \_\_\_\_\_\_$$

(b) Write the seventh and eighth equations in the sequence of equations of part (a).

(c) Write general equations suggested by parts (a) and (b) for even n and for odd n, where n is the number of the equation.

13. **(a)** How many rectangles are there in each of these figures? (*Note:* Rectangles may measure 1 by 1, 1 by 2, 1 by 3, and so on.)

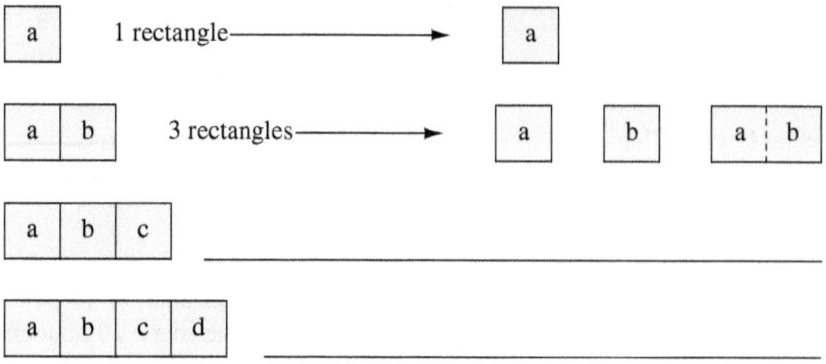

(b) How many rectangles are in this figure? ___________

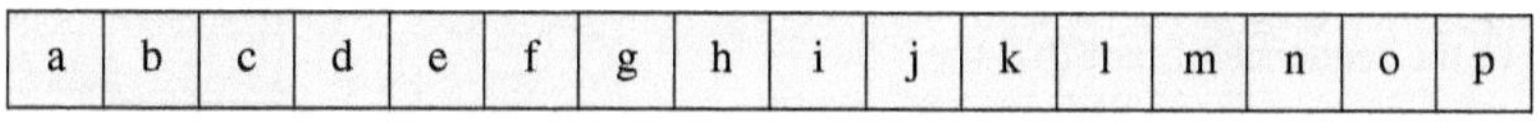

(c) How many rectangles are in a $1 \times n$ strip like that in part (b)?

(d) Argue carefully and lucidly that your guess in part (c) is correct. (*Hint:* How many of each type of rectangle begin with each small square?)

---

14. If one must always move downward along the lines of the grid shown, how many different paths are there from point *A* to each of these points?

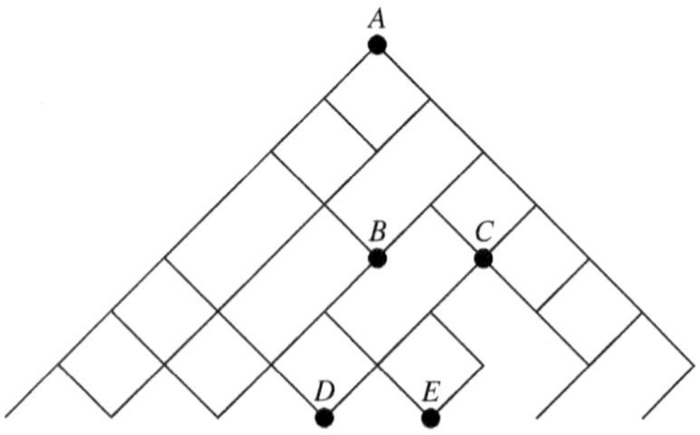

15. If one must always move upward or to the right on each of the grids shown, how many paths are there from *A* to *B*?

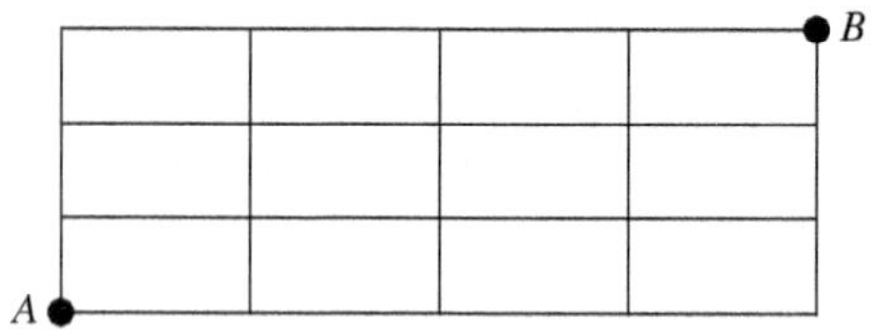

16. If one must always move upward or to the right on the gird shown, how many paths are there from *C* to *D*?

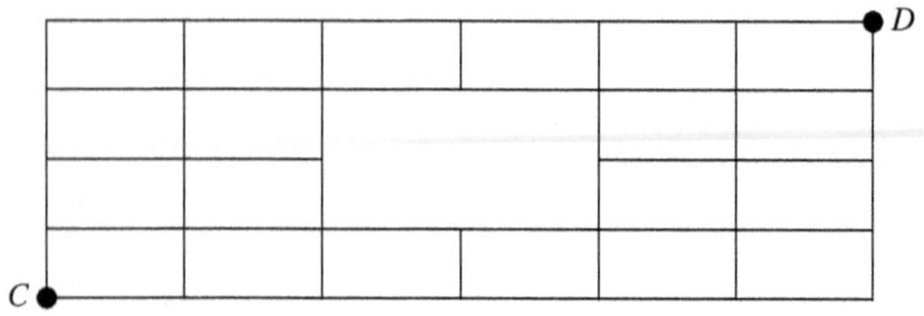

17. If one must follow along the paths of the following diagram in the direction of the arrows, how many paths are there from *C* to *E*?

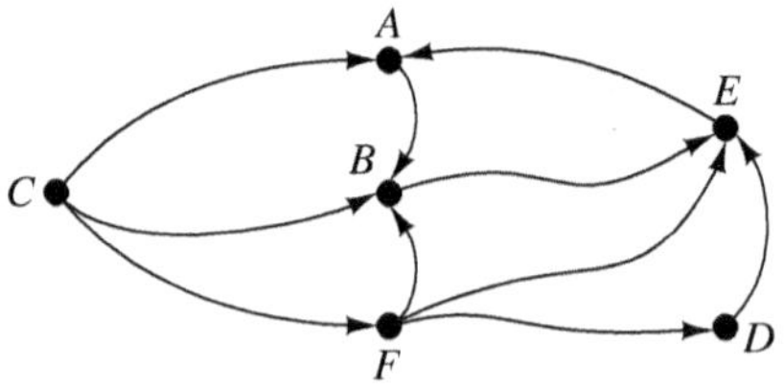

18. How many chords are determined by joining dots on a circle if there are

(a) 4 dots?

(b) 10 dots?

(c) 100 dots?

(d) *n* dots?

(e) Argue that your solution to part (d) is correct.

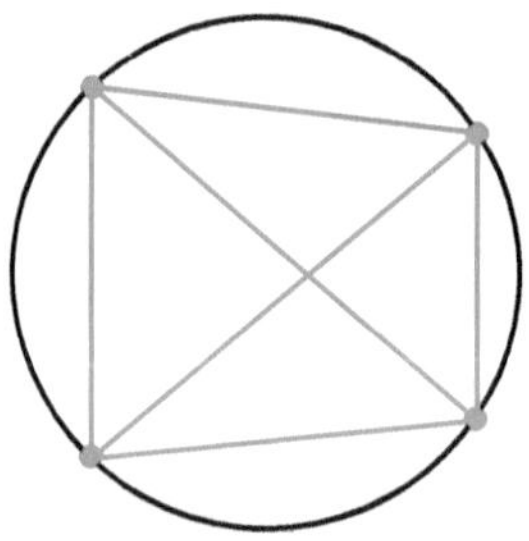

19. **(a)** How many games are played in a round-robin tournament with 10 teams if every team plays every other team once?

(b) How many games are played if there are 11 teams?

(c) (**Writing**) Is this problem related to problem 18 of this problem set? If so, how?

20. Here is an addition table:

+	0	1	2	3	4	5	6	7	8	9
0	0	1	2	3	4	5	6	7	8	9
1	1	2	3	4	5	6	7	8	9	10
2	2	3	4	5	6	7	8	9	10	11
3	3	4	5	6	7	8	9	10	11	12
4	4	5	6	7	8	9	10	11	12	13
5	5	6	7	8	9	10	11	12	13	14
6	6	7	8	9	10	11	12	13	14	15
7	7	8	9	10	11	12	13	14	15	16
8	8	9	10	11	12	13	14	15	16	17
9	9	10	11	12	13	14	15	16	17	18

(a) Find the sum of the entries in these squares of entries from the addition table:

2	3
3	4

5	6
6	7

11	12
12	13

15	16
16	17

Look for a pattern and write a clear and simple rule for finding such sums almost at a glance.

(b) Find the sum of the entries in these squares of entries from the table:

4	5	6
5	6	7
6	7	8

10	11	12
11	12	13
12	13	14

14	15	16
15	16	17
16	17	18

(c) Write a clear and simple rule for computing these sums.

(d) Write a clear and simple rule for computing the sum of the entries in any square of entries from the addition table.

21. Write a problem whose solution involves

(a) looking for a pattern.

(b) making up a table.

22. Write a problem whose solution involves

(a) using a variable.

(b) considering special cases.

23. (Read carefully before responding.) How much dirt is there in a hole 10 meters long, 5 meters wide, and 3 meters deep?

24. (Read carefully before responding.) If pencils are 5 cents each, how many are there in a dozen?

25. (Read carefully before responding.) Joni has two U.S. coins in her pocket. One of the coins is not a quarter. If the total value of the coins is 26 cents, what are the two coins?

State Assessments

26. (Washington State, Grade 4)
Look at the following list of numbers. Describe two different patterns you see in these numbers.

9 18 27 36 45 54 63 72 81 90

27. (Grade 4)
Duane came up with the following pattern of numbers:

0, 2, 10, 42, 170, ______.

What rule did Duane use to find the next number?

A. Add 2 to the last number.

B. Add 8 to the last number.

C. Quadruple (which means multiple by 4) the last number and add 2.

D. Double the last number and add 3.

1.4 Algebra as a Problem-Solving Strategy

Strategies

- Use a variable.
- Use two variables.

Use a Variable

In Example 1.7, we saw a use of the language of algebra when we wrote the solution of the example as $q_n = 2n - 1$. But algebra is more—much more—than just a language. Algebra can also solve problems that look very hard or appear to be a mathematical trick. In fact, we believe in **algebra as a great explainer** in much of mathematics at all levels because it clarifies so much. Algebra is now appearing informally in elementary school and is used by students in grade 7 or even earlier.

In this section, we shall explore the algebra-based problem-solving strategy "Use a Variable," in which we first use a variable name for what we are looking at and then do some algebra to solve the problem. Using symbols or **variables** often makes it easier to express mathematical ideas and so to solve problems with the algebra that you already know from school. In Chapter 8, we'll describe, in detail, algebra as it is needed for teaching in elementary or middle school, but because the use of algebra is so pervasive and so much a part of problem solving, we introduce algebra as the great facilitator early and continue it as a thread in the book.

DEFINITION Variable

A **variable** is a symbol (usually a letter) that can represent any of the numbers in some set of numbers.

EXAMPLE 1.10

Using a Variable—Gauss' Insight

Carl Gauss (1777–1855) is generally acknowledged as one of the greatest mathematicians of all time. When Gauss was just 10 years old, the teacher instructed the students in his class to add all the numbers from 1 to 100, expecting this to take a long time. To the teacher's surprise, young Gauss completed the task very quickly.

Solution

Gauss' strategy was to use a variable, and his insight was that numbers can be added in any order! If

$$s = 1 + 2 + 3 + \cdots + 100,$$

then, also,

$$s = 100 + 99 + 98 + \cdots + 1.$$

$1 + 100 = 101$
$2 + 99 = 101$
$3 + 98 = 101$
$\vdots$
$100 + 1 = 101$

Therefore, adding these two expressions for s, we obtain

$$2s = 101 + 101 + 101 + \cdots + 101,$$

a sum with 100 terms

$$2s = 100 \times 101,$$

and

$$s = \frac{100 \times 101}{2} = 5050.$$

The key in this example is to define a variable in such a way that we can finish the problem by using an algebraic insight or reasoning (like Gauss') or by recognizing the equation as one that we already know how to solve.

Highlight from History: Carl Friedrich Gauss (1777–1855)

Carl Friedrich Gauss was born of poor parents in Braunschweig, Germany, in 1777 and was both a child prodigy (see "Example 1.10, Using a Variable—Gauss' Insight") and an incredibly productive mathematician. Gauss contributed an enormous amount of new mathematics that not only solved a number of already existing problems but also opened many fresh paths of research into a variety of aspects of both applied and theoretical mathematics. He did seminal work in number theory (about how fast the number of primes grows), statistics (method of least squares; the Gaussian, or bell, curve), and non-Euclidean geometry, all before he was 17. He also produced groundbreaking research in establishing complex numbers, proved (when he was 20) that every polynomial of degree n can be solved and has n roots (called the "fundamental theorem of algebra") and gave three other proofs of this theorem later in his life. Gauss made substantive contributions to astronomy and physics, the geometry of curved surfaces, and the theory of functions. At the age of 63, he decided to learn Russian and spoke and read it well. His depth and reach across mathematics are truly amazing.

PROBLEM-SOLVING STRATEGY 7 Use a Variable

Often, a problem requires that a number be determined. Represent the number by a variable, and use the conditions of the problem to set up an equation that can be solved to ascertain the desired number.

EXAMPLE 1.11

Using a Variable to Determine a General Formula

Look at these corresponding geometrical and numerical sequences:

$$
\begin{array}{cccc}
\cdot & \begin{array}{c}\cdot\\\cdot\,\cdot\end{array} & \begin{array}{c}\cdot\\\cdot\,\cdot\\\cdot\,\cdot\,\cdot\end{array} & \begin{array}{c}\cdot\\\cdot\,\cdot\\\cdot\,\cdot\,\cdot\\\cdot\,\cdot\,\cdot\,\cdot\end{array}\\
1 & 3 & 6 & 10
\end{array}
$$

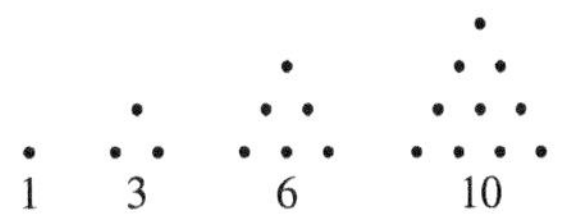

For fairly obvious reasons, the numbers 1, 3, 6, and 10 are called **triangular numbers.** (The diagram with a single dot is considered a *degenerate triangle.*) The numbers 1, 3, 6, and 10 are the first four triangular numbers. Find a formula for the nth triangular number.

Solution

Understand the Problem

Having just gone through a similar problem in Example 1.7, we understand that we are to find a formula for t_n, the nth triangular number.

Devise a Plan

The geometrical and numerical sequences in the statement of the problem suggest that we look for a pattern. How is each diagram related to its predecessor? How is each triangular number related to its predecessor? We'll try to guess patterns.

Carry Out the Plan

We add a diagonal of two dots to the first diagram to obtain the second, a diagonal of three dots to the second diagram to obtain the third, and so on. Thus, the next two diagrams should be as shown here (we can think of this process as experimenting):

$$
\begin{array}{cc}
\begin{array}{c}\cdot\\\cdot\,\cdot\\\cdot\,\cdot\,\cdot\\\cdot\,\cdot\,\cdot\,\cdot\\\cdot\,\cdot\,\cdot\,\cdot\,\cdot\end{array} &
\begin{array}{c}\cdot\\\cdot\,\cdot\\\cdot\,\cdot\,\cdot\\\cdot\,\cdot\,\cdot\,\cdot\\\cdot\,\cdot\,\cdot\,\cdot\,\cdot\\\cdot\,\cdot\,\cdot\,\cdot\,\cdot\,\cdot\end{array}
\end{array}
$$

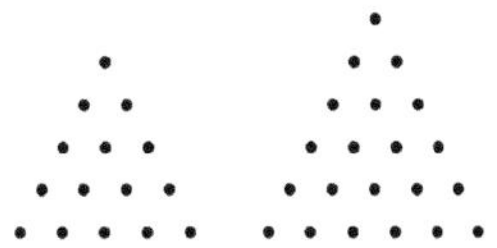

Numerically, we add 2 to the first triangular number to obtain the second, 3 to the second triangular number to obtain the third, and so on. To make this rule even more clear, we can construct the following table:

Number of Entry	Entry
1	$t_1 = 1$
2	$t_2 = 1 + 2 = 3$
3	$t_3 = 1 + 2 + 3 = 6$
4	$t_4 = 1 + 2 + 3 + 4 = 10$
5	$t_5 = 1 + 2 + 3 + 4 + 5 = 15$

Indeed, it appears that

$$t_n = 1 + 2 + 3 + \cdots + n.$$

Then, using Gauss' insight, we obtain

$$t_n = n + (n - 1) + (n - 2) + \cdots + 1.$$

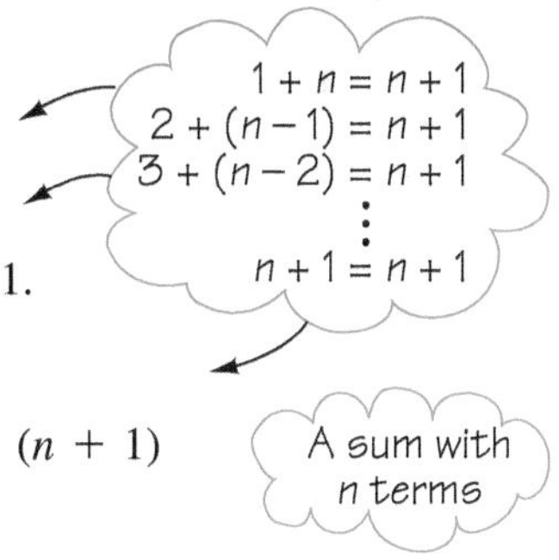

So,

$$2t_n = (n + 1) + (n + 1) + (n + 1) + \cdots + (n + 1)$$
$$= n(n + 1)$$

and

$$t_n = \frac{n(n + 1)}{2},$$

as required.

$$1 + n = n + 1$$
$$2 + (n - 1) = n + 1$$
$$3 + (n - 2) = n + 1$$
$$\vdots$$
$$n + 1 = n + 1$$

A sum with n terms

Look Back

Looking back, we note that the key to our solution lay in considering the sequence of special cases t_1, t_2, t_3, t_4, and t_5 and in looking for a pattern. This approach is an example of Problem-Solving Strategy 6, considering special cases or experimenting.

It is helpful to view algebraic thinking for problem solving as a circle of connected ideas, beginning and ending with the pattern or problem that is being investigated. The essential steps are shown in Figure 1.6 and are developed in more detail throughout the remainder of this section.

Figure 1.6
The steps in algebraic reasoning

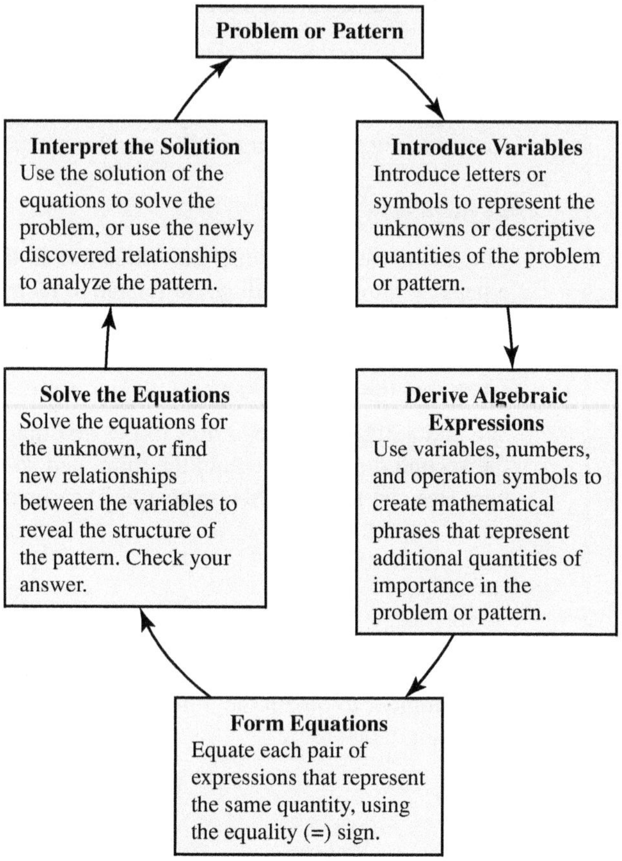
Problem or Pattern
Introduce Variables
Introduce letters or symbols to represent the unknowns or descriptive quantities of the problem or pattern.
Derive Algebraic Expressions
Use variables, numbers, and operation symbols to create mathematical phrases that represent additional quantities of importance in the problem or pattern.
Form Equations
Equate each pair of expressions that represent the same quantity, using the equality (=) sign.
Solve the Equations
Solve the equations for the unknown, or find new relationships between the variables to reveal the structure of the pattern. Check your answer.
Interpret the Solution
Use the solution of the equations to solve the problem, or use the newly discovered relationships to analyze the pattern.

EXAMPLE 1.12

Forming, Checking, and Evaluating Algebraic Expressions

Toothpicks can be used to form cross patterns, such as the three that follow. Additional crosses are formed by following the same pattern of construction. What is the number of squares in the nth cross? How many toothpicks are required to construct the nth cross?

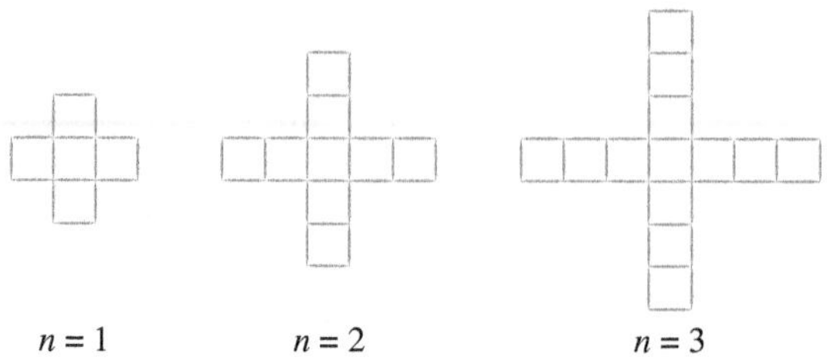
n = 1
n = 2
n = 3

Solution

Understand the Problem

We seek two expressions that each involve the variable n. One is to give the number of squares in the nth cross, and the other expression must give the number of toothpicks required to form the pattern. We see that n also gives the number of squares in each of the four "arms" of the cross.

Devise a Plan

Each new pattern is created by adding a new square to each of the four arms of the preceding pattern. That is, the number of squares is increased by four in each new pattern. This strategy suggests that we look for an expression of the form $a + 4n$, where we will need to determine the value of the constant a. Similarly, adding one square to an arm of the cross requires three additional toothpicks, so adding a square to each of the four arms requires 12 additional toothpicks. This pattern suggests that we look for an expression of the form $b + 12n$ to give the number of toothpicks in the nth cross, where the constant b will need to be determined.

Carry Out the Plan

In the first cross pattern, $n = 1$. Evaluating $a + 4n$ at $n = 1$ gives $a + 4$, and since there are 5 squares in the pattern, we get the equation $a + 4 = 5$. That is, $a = 1$, and the number of squares in the nth cross is given by $1 + 4n$. The constant b in the expression $b + 12n$ can be determined similarly: There are 16 toothpicks in the $n = 1$ pattern, so we must have $b + 12 = 16$. That is, $b = 4$, and the number of toothpicks in the nth pattern is given by the expression $4 + 12n$.

Look Back

The plan that has been followed will apply to any sequence in which each new term is a fixed number larger than the preceding term. That is, we have an arithmetic sequence. For example, to find an expression that gives the nth term of the sequence 3, 9, 15, 21, . . . , we would look for an expression of the form $c + 6n$, since 6 is the common difference of the sequence. Choosing $n = 1$, we note that the constant c must satisfy the condition $c + 6 = 3$. That is, $c = 3 - 6 = -3$. Thus, the nth term is given by the expression $6n - 3$.

EXAMPLE 1.13 Setting Up and Solving an Equation: Can I Get a C?

Larry has exam scores of 59, 77, 48, and 67. What score does he need on the next exam to bring his average for all five of the exams to 70?

Solution

We first need to designate a variable—one that gives the answer. A bit more formally, let the variable s denote Larry's minimum score needed on the fifth exam, which is what we are looking for. The domain of s is the set of numbers between 0 and 100. Using the variable s, we find that the expression giving Larry's average over all five tests is the sum of his scores divided by 5, or

$$\frac{59 + 77 + 48 + 67 + s}{5}.$$

Since Larry is looking for the lowest score s that will give him a 70 average, we form the equation

$$\frac{59 + 77 + 48 + 67 + s}{5} = 70.$$

To solve this equation, we multiply each side of the equation by 5 and add the sum of the first four test scores to get

$$251 + s = 350.$$

Finally, we subtract 251 from both sides of the equation to solve for s:

$$s = 350 - 251 = 99.$$

Larry must hope for a 99 or 100 on the last test. It is easy to check that our solution is correct by taking the average of 59, 77, 48, 67, and 99.

This section concludes with two more examples that illustrate the steps in algebraic reasoning shown in Figure 1.6.

EXAMPLE 1.14

Solving a Rate Problem: Tom and Huck Whitewash a Fence

Two years ago, it took Tom 8 hours to whitewash a fence. Last year, Huck took just 6 hours to whitewash the fence. This year, Tom and Huck have decided to work together so that they'll have time left in the afternoon to angle for catfish. How long will the job take the two boys?

Solution

Let T denote the time, in hours, that Tom and Huck together need to whitewash the fence. That is, T is the unknown. Since Tom can whitewash the fence in 8 hours, he can whitewash $\frac{1}{8}$ of the fence per hour. In T hours, Tom will have whitewashed $\frac{T}{8}$ of the fence. Similarly, Huck whitewashes $\frac{1}{6}$ of the fence per hour, and he therefore can whitewash $\frac{T}{6}$ of the fence in T hours. Working together, the boys will whitewash the entire fence when

$$\frac{T}{8} + \frac{T}{6} = 1.$$

Multiplying both sides by 48 gives the equivalent equation

$$6T + 8T = 48.$$

That is, $14T = 48$, so $T = \frac{48}{14} = 3\frac{3}{7}$. Thus, working together, Tom and Huck can whitewash the fence in just less than $3\frac{1}{2}$ hours and will enjoy an afternoon of fishing.

Many people who don't take time to think in depth about the Tom and Huck problem will answer 7 hours (because the average of 6 and 8 is 7). The "check your answer" part of our "Steps in Algebraic Reasoning" (Figure 1.6) shows that 7 hours can't be correct. After all, for Tom and Huck to take 7 hours means that it would take them longer together than it would if Huck did it himself!

Use Two Variables

There are many math tricks, puzzles, and jokes that circulate around the Internet. They often look quite mysterious and give the impression of magic. In fact, many can be understood with Pólya's problem-solving principles, and a number involve the use of algebra. Looking at math problems through the lens of algebra (**"algebra as the great explainer"**) can take the mystery away from a number of processes and provides a conceptual explanation. The next example is an actual e-mail that we received from our friend Amanda, who found the math trick amazing and now realizes that it is "just" algebra.

EXAMPLE 1.15

Using Variables (Amanda's Telephone Number E-Mail)

"Here is the math trick so unbelievable that it will stump you.

Grab a calculator. (You won't be able to do this one in your head.)
Key in the first three digits of your phone number (not the area code).
Multiply by 80.
Add 1.
Multiply by 250.
Add the last four digits of your phone number.
Add the last four digits of your phone number again.
Subtract 250.
Divide the number by 2.

Do you recognize the number? Does it always turn out to be your number?" Amanda asked in an e-mail.

Solution

Understand the Problem

To make sure that we understand what's going on, the first step is to see if the procedure works with our own phone number. Let's first try Amanda's phone number of 257-6821.

Step 6 has some ambiguity because the instruction "add the last four digits" could mean that we should add 6, 8, 2, and 1 or add the number 6821. If you try the first way, you'll see that the process fails, so we took the second interpretation.

The process does work for Amanda's number, because 2,576,821 of step 9 is 257 (10,000) + 6821. (See Table 1.1.) Thus, expressed as a phone number, 2,576,821 is 257-6821. Since there is nothing special about the number used, it may well be that the "math trick" always works. However, we can't try all the phone numbers, so just guessing and checking is not feasible. We should go to the next Pólya principle to see if Amanda's trick always works—that is, to see if we always get the original phone number back.

TABLE 1.1 VERIFICATION OF AMANDA'S MATH TRICK BY USING HER PHONE NUMBER, 257-6821

Step	Result
2. Start with 257	257
3/4. Multiply by 80, add 1	$80(257) + 1 = 20{,}561$
5. Multiply by 250	$(250)(20{,}561) = 5{,}140{,}250$
6/7. Add 6821 twice	5,153,892
8. Subtract 250	5,153,642
9. Divide by 2	2,576,821

Devise a Plan

The question "Does it always work?" in the problem gives a direction for our plan. Example 1.14 is solved by looking at all possibilities at once by letting the time needed to whitewash the fence be the variable T. Let's see whether that approach will work for Amanda's problem.

The first step in planning is to decide what variable to use for which quantity. It is tempting to let x be a phone number. If you try this approach, however, you will see that we can't do step 2 or 3 in terms of x because we must have only the first three digits of the seven-digit phone number. Looking at the procedure again, we see that, although there are seven digits, only the first three and the last four are used, none of them individually. Our plan, then, is to name the first three digits with the variable y and the last four with the variable z and go through the Pólya principles by using algebra.

Carry Out the Plan

Let the telephone number be given and let y be the first three digits of the number and z the last four. (Thus, in the case of Amanda's number, 257-6821, $y = 257$ and $z = 6821$.) We will now follow steps 2 through 9 exactly as in Table 1.1 and see what happens.

What is the final answer after all nine steps? The result at step 9 is $10{,}000y + z$. (See Table 1.2.) However, $10{,}000y$ takes the three-digit number y (257 in our example) and adds four zeros (to get 2,570,000). Thus, the first three digits of the seven-digit number $10{,}000y + z$ are the digits of the number y.

What are the last four digits of $10{,}000y + z$? They are exactly the digits of z (for example, $2{,}570{,}000 + 6821 = 2{,}576{,}821$). The "math trick" works because $10{,}000y + z$ is a way to express in algebra the phone number (without area code) y-z. (That's y hyphen z, not subtraction.) Note that we didn't need a calculator to do the problem!

TABLE 1.2 VERIFICATION OF AMANDA'S MATH TRICK BY USING THE PHONE NUMBER WHOSE FIRST THREE DIGITS ARE y AND LAST FOUR ARE z

Step	Results
2. Start with y	y
3/4. Multiply by 80, add 1	$80y + 1$
5. Multiply by 250	$250(80y + 1) = 20{,}000y + 250$
6/7. Add z twice	$20{,}000y + 250 + 2z$
8. Subtract 250	$20{,}000y + 2z$
9. Divide by 2	$(20{,}000y + 2z)/2 = 10{,}000y + z$

Look Back

In the explanation of Amanda's phone number problem, we used "algebra as the great explainer" because we were trying to show that something involving numbers was always correct. At first, we thought of solving the problem by using only one variable, but we recognized that there were really two different numbers involved; that is, a phone number is made up of two parts (plus an area code), and we needed different names for the two of them.

Algebraic steps came next and finished the problem. Since we proved the result for all seven-digit numbers, not just some specific ones, we have given a carefully reasoned argument that shows that the result is always true. This approach is also an example of the "Proof and Reasoning Standard" of the NCTM.

Amanda's problem asks whether your solution turns out to be your telephone number. Amanda realizes that, since any phone number will work, there must be a variable (or two) that will show the generality of the solution. In other words, an algebra strategy would be a good one to try.

SMP 2 Reason abstractly and quantitatively.

"Mathematically proficient students make sense of quantities and their relationships in problem situations. They bring . . . the ability to *decontextualize*—to abstract a given situation and represent it symbolically and manipulate the representing symbols as if they have a life of their own. . . ."

Amanda's insight that two variables are needed rather than one is valid because she recognizes that the three digits (257, or more abstractly, y) of her phone number, when multiplied by 10,000, will give the first three digits of her phone number. The use of y and z allows her phone number to be represented symbolically. All that remained was for Amanda to follow up by doing some algebraic manipulations as shown in Table 1.2. The SMP 2 (reason abstractly and quantitatively) makes it clear that there is far more to problem solving than simply getting the answer.

1.4 Problem Set

Understanding Concepts

1. (a) Draw three diagrams to continue this dot sequence:

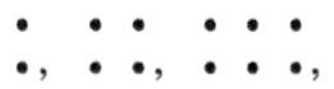

(b) What number sequence corresponds to the pattern of part (a)?

(c) What is the 10th term in the sequence of part (b)? The 100th term?

(d) Which even number is $2n$?

(e) What term in the sequence is 2402?

2. Example 1.1 (Guessing Toni's Number, in Section 1.1) was solved by using guess and check. Do the example again, applying algebra this time. (*Suggestion:* Use a variable.)

3. (a) Jackson is thinking of an integer, which, if you triple it and subtract 13, ends with a result of 2. What number is Jackson thinking of? (*Suggestion:* Use a variable.)

(b) Jackson decides to be more clever this time and he says he ends up with a result of 4 instead of 2. What whole number is Jackson thinking of?

4. (a) Maria picks an integer, divides it by 2, and then adds 12 to what she has. When she tells you that she now has 10, can you tell her what integer she started with? (*Suggestion:* Use a variable.)

(b) This time Maria wants to trick her teachers so she picks an integer, multiplies it by 2, and then adds 12. When she tells you that she now has 15, can you tell her what integer she started with?

5. The first three trains in a sequence of trapezoid trains are shown below. Verify that the number t of toothpicks required to form a train with c trapezoidal cars is given by the formula $t = 1 + 4c$.

6. A rectangular table seats 6 people: 1 person on each end and 2 on each of the longer sides. Thus, two tables placed end to end seat 10 people.

(a) How many people can be seated if n tables are placed in a line end to end?

(b) How many tables, set end to end, are required to seat 24 people?

7. (Handshake Problem) There are n people in a room, and each of them will shake hands with every other person once and only once. The general question of how many handshakes take place is best done through the insights gained by experimenting (Problem-Solving Strategy 6).

(a) If there are 3 people in the room, how many handshakes are there?

(b) If there are 6 people, how many handshakes are made?

(c) If there are 200 people, how many handshakes are there?

(d) If there are n people, how many handshakes are there?

Imagination can generate many variants of the Handshake Problem. This exercise gives rise to a number of other problems through the use of the Mathematical Habit of the Mind.

8. (Feuding Handshake Problem) There are 100 people in a room, half of whom don't speak to the other half. Assume that if they won't speak to each other, they won't shake hands. How many handshakes are there if everyone shakes hands only once (if they shake at all)?

9. (Marital Handshake Problem) There are 100 people consisting of 50 married couples in a room. Assuming that no husband or wife shakes the other's hand but everyone else shakes hands exactly once, how many handshakes are there?

10. (Your Handshake Problem) Make up a handshake problem similar to problems 7, 8, and 9.

11. Old MacDonald has 100 chickens and goats altogether in the barnyard. In all, there are 286 feet. How many chickens and how many goats are in the barnyard? (*Suggestion:* Use Problem-Solving Strategy 7.)

12. Consider the following sequence of equations:

$$1 = 1$$
$$3 + 5 = 8$$
$$7 + 9 + 11 = 27$$
$$13 + 15 + 17 + 19 = 64$$
$$\_\_\_ + \_\_\_ + \_\_\_ + \_\_\_ + \_\_\_ = \_\_\_$$

(a) Fill in the blanks to continue the sequence of equations.

(b) Guess a formula for the number on the right of the nth equation.

(c) Check that the expression $n^2 - n + 1$ generates the first number in the sum on the left of each equation and that $n^2 + n - 1$ generates the last number in the sum.

(d) Use the result of part (c) to prove that your guess to part (b) is correct. (*Hint:* How many terms are in the sum on the left of the nth equation?)

13. We have already considered the triangular numbers from page 33.

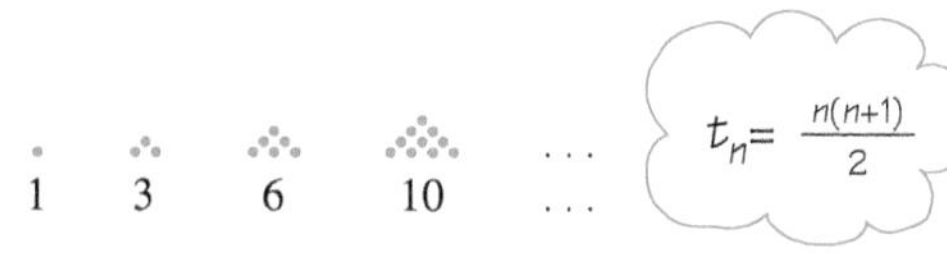

and the square numbers,

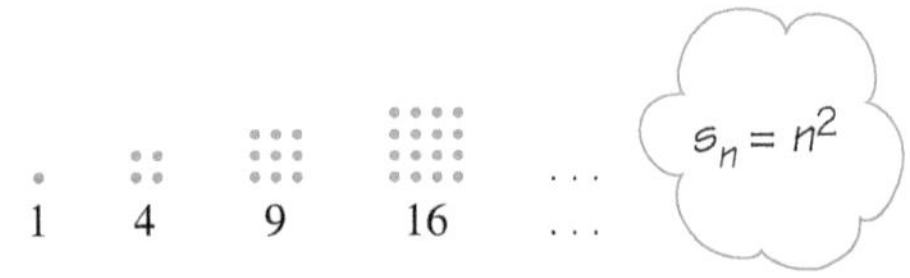

(a) Draw the next two figures to continue this sequence of dot patterns:

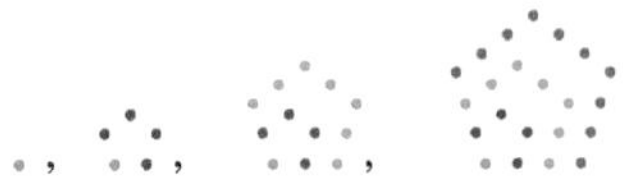

(b) List the sequence of numbers that corresponds to the sequence of part (a). These are called **pentagonal numbers.**

(c) Complete this list of equations suggested by parts (a) and (b):

$$1 = 1$$
$$1 + 4 = 5$$
$$1 + 4 + 7 = 12$$
$$1 + 4 + 7 + 10 = 22$$
$$\_\_\_\_\_\_\_\_ = \_\_\_\_\_\_\_\_$$
$$\_\_\_\_\_\_\_\_ = \_\_\_\_\_\_\_\_$$

Observe that each pentagonal number is the sum of an arithmetic progression.

(d) Compute the 10th term in the arithmetic progression 1, 4, 7, 10, ...

(e) Compute the 10th pentagonal number.

(f) Determine the nth term in the arithmetic progression 1, 4, 7, 10, ...

(g) Compute the nth pentagonal number, p_n.

14. **(a)** The **hexagonal numbers** are associated with this sequence of dot patterns:

Complete the next two diagrams in the sequence.

(b) Write the first five hexagonal numbers.

(c) What is the 10th hexagonal number?

(d) Compute a formula for h_n, the nth hexagonal number.

15. Example 1.3 of Section 1.2 was solved by using guess and check. Do the example again but this time with algebra. (Use three variables.)

16. The following figure shows an addition pyramid, in which each number is the sum of the two numbers immediately below it:

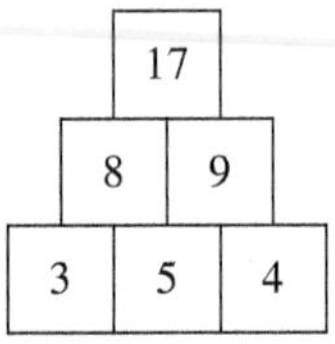

For each of the following incomplete addition pyramids, show that the values in each square can be determined by forming and solving an equation in the unknown variable shown:

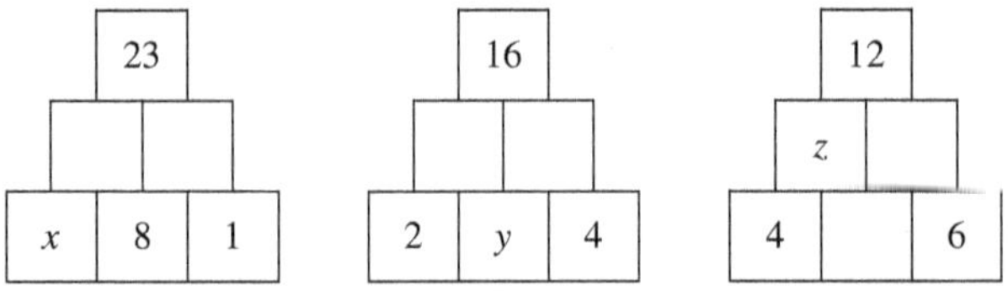

(a) **(b)** **(c)**

17. In the circle patterns shown, each number in a large circle is the sum of the numbers in the two adjacent small circles. The pattern on the left is complete, but the pattern on the right needs to be completed. Do so by letting x denote the value in one of the small circles. Now obtain expressions and equations that let you solve for x, and determine the values in all three small circles.

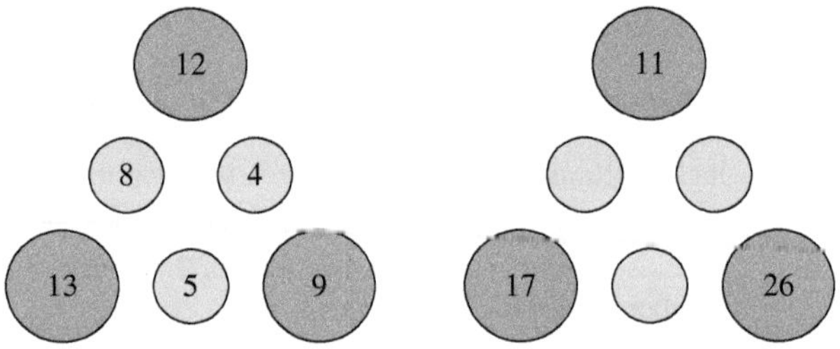

18. Find the numbers to place in each of the five small circles shown so that each number given in a large circle is the sum of the numbers in the two adjacent small circles. Do so by letting one of the values in a small circle be denoted by an unknown and then obtaining and solving an equation for the unknown.

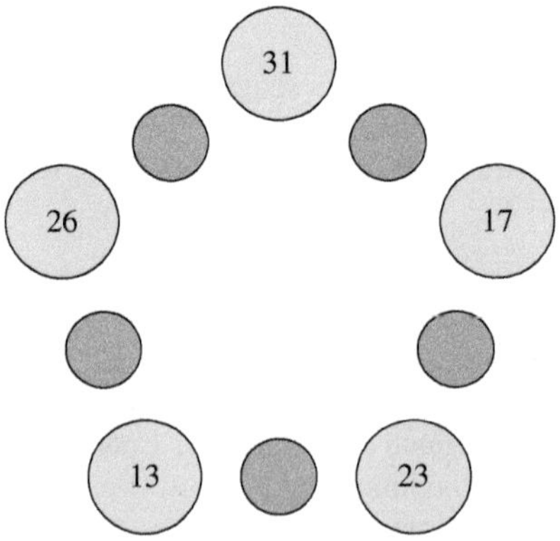

19. Use algebra to solve the following problems from Problem Set 1.1:

(a) Problem 13a.

(b) Problem 13d.

20. Dana received grades of 78, 86, and 94 on the first three exams. For an A, the average on four exams must be at least 90. What can Dana do to get an A? (*Hint:* Look at Example 1.13.)

21. Find a formula for the (positive) difference of the squares of consecutive integers by doing parts (a) and (b).

(a) By way of an experiment, fill in the following table of integers:

n	1	2	3	4	5	6
n^2	1	4				
$(n+1)^2$	4	9				
difference	3	5				

Do you see a pattern?

(b) Using algebra, write a formula for the difference of the squares of consecutive numbers (larger minus smaller), and give a reason that it works.

22. In Section 1.1, Mr. Capek's fifth-grade class found multiple methods of solving "Old MacDonald's problem." How would you solve it by using algebra?

Thinking Critically

23. Consider each of the following addition problems in which the digits are reversed:

$23 + 32 \quad 42 + 24 \quad 17 + 71 \quad 51 + 15 \quad 67 + 76$

(a) What do the answers of all of these have in common?

(b) Using algebra, show that if a and b are digits, then the sum of the numbers represented by ab and ba is always divisible by 11. (*Hint:* Since a is the first digit and b is the second, ab represents the number $10a + b$. What number does ba represent, and what is the sum of ab and ba?)

24. Consider each of the following subtraction problems in which the digits are reversed:

$32 - 23 \quad 42 - 24 \quad 71 - 17 \quad 51 - 15 \quad 76 - 67$

(a) What do the answers of all of these computations have in common?

(b) Using algebra, show that if a and b are digits, then the difference of the numbers represented by ab and ba is always divisible by 9. (*Hint:* Since a is the first digit and b is the second, ab represents the number $10a + b$. What number does ba represent, and what is the difference of ab and ba?)

25. A rectangle is three times as long as it is wide and has the same perimeter as a square whose area is 4 square feet larger than that of the rectangle. What are the dimensions of both the rectangle and the square?

26. How would you generalize the formula of problem 21?

(a) If x and y differ by 2, show that the difference of their squares is twice their sum.

(b) To formulate a generalization of part (a), what condition on the difference of x and y would you suggest? Write a problem that generalizes part (a) and prove your result by using algebra.

27. Using algebra, solve the Port Angeles High School exercise, which is problem 2 of Section 1.1.

28. Algebra can be used to prove results, as we will see in Chapter 4. Although the intent of this text is not to emphasize formal proofs, we do want to show the power of algebra. Here's a beginning problem: Show that the square of an even integer is a multiple of 4. (*Hint:* Start with the definition! We know that if x is an even number, then there is another integer n with $x = 2n$. What is x^2?)

State Assessments

29. (Grade 5)

✧ + ♡ = 14

♡ − ✧ = 6

Each ✧ has the same value. Each ♡ has the same value. What is the value of ♡?

30. (Grade 5)

Roberto had 90 baseball cards. He gave 10 cards to his sister. After that, he used the remaining cards for his five friends and demanded that each friend had the same amount of cards. How many cards did each of his friends get?

(a) 13 cards

(b) 16 cards

(c) 14 cards

(d) 17 cards

31. (Grade 8)

Two small children, Daniel and Christine, were playing a game to see who would be the first to reach the door. The children started the game by standing 20 meters away from the door, and then they each alternated doing the following:

- Daniel moved **one-half** the distance between himself and the door on each move.
- Christine moved **1 meter** toward the door on each move.

(a) How far was each child from the door after the **first** move?

(b) After **four** moves, was Daniel or Christine closer to the door? Show your work.

(c) Daniel thought that the game was unfair because he would never reach the door. Explain why his statement is correct or incorrect.

32. (Grade 7)

Monday and Tuesday taxi drivers went from Georgia Tech to Emory University the same number of times. On Wednesday, business wasn't as good and there were only 16 taxi rides from Georgia Tech to Emory. If the total of these rides on Monday, Tuesday, and Wednesday is 62 rides, how often did the taxi drivers go from Ga Tech to Emory on Tuesday?

A. 16

B. 24

C. 31

D. 23

1.5 Additional Problem-Solving Strategies

Strategies

- Work backward.
- Eliminate possibilities.
- Use the Pigeonhole Principle.

Working Backward

PROBLEM-SOLVING STRATEGY 8 Work Backward

Start from the desired result and work backward step-by-step until the initial conditions of the problem are achieved.

Many problems require that a sequence of events occur that results in a desired final outcome. These problems at first seem obscure and intractable, and you may be tempted to try a guess and check approach. However, it is often easier to work backward from the end result to see how the process would have to start to achieve the desired end. To make this idea more clear, consider the next example.

EXAMPLE 1.16

Working Backward—The Gold Coin Game

This is a two-person game. Place 15 gold coins (markers) on a desktop. The players play in turn and, on each play, can remove one, two, or three coins from the desktop. The player who takes the last coin wins the game. Can one player or the other devise a strategy that guarantees a win? (*Note:* This is the problem of the Cooperative Investigation presented at the beginning of the chapter, where you were expected to arrive at the solution by repeatedly trying simpler cases. We think that you will find the approach here much more effective and insightful.)

Solution

Understand the Problem

The assertion that the game is played with gold coins is just so much window dressing. The rules are that the players start with 15 objects; they can remove 1, 2, or 3 objects on each play; and the player who takes the last object wins the game. The question is, How can one play in such a way that he or she is sure of winning?

Devise a Plan

Since it is not clear how to begin to play or how to continue as the play proceeds, we turn the problem around to see how the game must end. We then work backward step-by-step to see how we can guarantee that the game ends as we desire.

Carry Out the Plan

We carry out the plan by presenting an imaginary dialogue that you could have with yourself to arrive finally at the solution to the problem. This dialogue (or conversation with a friend) is a sign that shows a thoughtful mathematical conversation.

Q. What must be the case just before the last person wins the game?
A. There must be 1, 2, or 3 markers on the desk.
Q. So how can I avoid leaving this arrangement for my opponent?
A. Clearly, I must leave at least 4 markers on the desk in my next-to-last move. Indeed, if I leave precisely 4 markers, my opponent must take 1, 2, or 3, leaving me with 3, 2, or 1. I can remove all of these on my last play to win the game.
Q. So how can I be sure to leave precisely 4 markers on my next-to-last play?
A. If I leave 5, 6, or 7 markers on my previous play, my opponent can leave *me* with 4 markers and he or she can then win. Thus, I must be sure to leave my opponent 8 markers on the previous play.
Q. All right. So how can I be sure to leave 8 markers on the previous play?
A. Well, I can't leave 9, 10, or 11 markers on the previous play, or my opponent can take 1, 2, or 3 markers as necessary and so leave me with 8 markers. But then, as just seen, my opponent can be sure to win the game. Therefore, at this point, I must leave 12 markers on the desk.

Q. Can I be sure of doing this?

A. Only if I play first and remove 3 markers the first time. Otherwise, I have to be lucky and hope that my opponent will make a mistake and still allow me to leave 12, 8, or 4 markers at the end of one of my plays. The following strategy outlines the play if I get to play first:

- I take 3 markers, leaving 12.
- My opponent takes 1, 2, or 3 markers, leaving 11, 10, or 9.
- I take 3, 2, or 1 marker as needed to make sure that I leave 8.
- My opponent takes 1, 2, or 3 markers, leaving 7, 6, or 5.
- I take 3, 2, or 1 marker as needed to ensure that 4 markers are left on the desk.
- My opponent takes 1, 2, or 3 markers, leaving 3, 2, or 1.
- I take the remaining markers and win the game!

Look Back

In looking back, it is important to ask such questions as these:

- What was the key feature that led me to eventually solve this problem?
- Could I use this strategy to solve variations of the problem? For example, suppose the game started with 21 coins or 37 coins or, in general, with n coins.
- Suppose that each player could take up to 5 coins at a time. How would that affect the strategy?
- Could I use the successful strategy I just employed to solve other similar (or not so similar) problems?
- Could I devise other problems for which this strategy would lead to a solution?

Working backward is crucial in many problem-solving situations, particularly with a problem like this one. The desired strategy to win the game is not at all clear. Here the strategy of working backward is somewhat similar to considering special cases and looking for a pattern. It is as if we started with just a few markers so that the strategy was more apparent and gradually increased the number of markers until we reached the given number of 15. Working backward is a powerful strategy that ought to be a major part of your portfolio of strategies.

Eliminate Possibilities

One way of determining what must happen in a given situation is to determine what the possibilities are and then to eliminate them one by one. If you can eliminate all but one possibility in this way, then that possibility must prevail. Suppose that either John, Jim, or Yuri is singing in the shower. Suppose also that you are able to recognize both John's voice and Yuri's voice but that you do not recognize the voice of the person singing in the shower. Then the person in the shower must be Jim. This is another important problem-solving strategy that should not be overlooked.

PROBLEM-SOLVING STRATEGY 9 Eliminate Possibilities

Suppose you are guaranteed that a problem has a solution. Use the data of the problem to decide which outcomes are impossible. Then at least one of the possibilities not ruled out must prevail. If all but one possibility can be ruled out, then it must prevail.

Of course, if you use this strategy on a problem and *all* possibilities can be correctly ruled out, the problem has no solution. Don't be misled. It is certainly possible to have problems with no solution! Consider the problem of finding a number such that 3 more than twice the number is 15 and 6 more than 4 times the number is 34. This problem has no solution, since the first condition is satisfied only by 6 and the second is satisfied only by 7. Yet $6 \neq 7$.

However, if we know that a problem has a solution, it is sometimes easier to determine what can't be true than what must be true. In this approach to problem solving, one eliminates possibilities until the only possibility left must yield the desired solution.

Consider the next example.

EXAMPLE 1.17

Eliminating Possibilities

Beth, Jane, and Mitzi play on the basketball team. Their positions are forward, center, and guard. Given the following information, determine who plays each position:

(a) Beth and the guard bought a milkshake for Mitzi.
(b) Beth is not a forward.

Solution

Understand the Problem

Given clues (a) and (b), we are to determine which girl plays each position.

Devise a Plan

The problem is confusing. Perhaps, if we make a table of possibilities, we can use the clues to eliminate some of them and so arrive at a conclusion.

Carry Out the Plan

The table showing all possibilities is as follows:

	Beth	**Jane**	**Mitzi**
forward			
center			
guard			

Using the clues in the order given, we see from (a) that neither Beth nor Mitzi is the guard, and we put Xs in the table to indicate this. But then it is clear that Jane is the guard, and we put an O in the appropriate cell to indicate this as well. The table now looks like this:

	Beth	**Jane**	**Mitzi**
forward			
center			
guard	X	O	X

But if Jane is the guard, she is not the center or forward, so we also X out these cells:

	Beth	**Jane**	**Mitzi**
forward		X	
center		X	
guard	X	O	X

This appears to be all the information we can get from condition (a), so we turn to condition (b), which tells us that Beth is not the forward. After we X out this cell, it becomes clear that Beth is the center. So we place an O in the Beth–center cell and an X in the Mitzi–center cell. This leaves the forward position as the only possibility for Mitzi, so we place an O in the Mitzi–forward cell. The completed table is as shown here:

	Beth	**Jane**	**Mitzi**
forward	X	X	O
center	O	X	X
guard	X	O	X

We conclude that Mitzi plays forward, Beth plays center, and Jane plays guard. Note that, in order to make the step-by-step process of eliminating possibilities clearer, we repeatedly redrew the table as the solution progressed. However, all the work could have been done in a single table and normally would have been.

Look Back

In this problem, we were confronted with data not easily analyzed. To bring order into this chaos, it seemed reasonable to make a table allowing for all possibilities and then to use the given statements to decide which possibilities could be ruled out. In this way, we were able to delete possibilities systematically until the only remaining possibilities completed the solution.

Eliminating possibilities is often a successful approach to solving a problem.

The Pigeonhole Principle

If 101 guests are staying at a hotel with 100 rooms, can we make any conclusion about how many people there are in a room? It is likely that a number of the rooms are empty, since some of the guests probably include married couples, families with children, and friends staying together to save money. But suppose most of the hotel's guests desire single rooms. How many such persons could the hotel accommodate? If there were just one person per room, all 100 rooms would be occupied, with one person left over. Thus, if *all* 101 guests are to be accommodated, there must be at least two persons in one of the rooms. To summarize,

If 101 guests are staying in a hotel with 100 guest rooms, then at least one of the rooms must be occupied by at least two guests.

This reasoning is essentially trivial, but it is also surprisingly powerful. Indeed, it is so often useful that it is called the **Pigeonhole Principle,** which is stated next.

PROBLEM-SOLVING STRATEGY 10 The Pigeonhole Principle

If m pigeons are placed into n pigeonholes and $m > n$, then there must be at least two pigeons in one pigeonhole.

For example, if we place three pigeons into two pigeonholes, then there must be at least two pigeons in one pigeonhole. To make this quite clear, consider all possibilities as shown here:

Pigeonhole Number 1	Pigeonhole Number 2
3 pigeons	0 pigeons
2 pigeons	1 pigeon
1 pigeon	2 pigeons
0 pigeons	3 pigeons

In every case, there are at least two pigeons in one of two pigeonholes.

A second useful way to understand this reasoning is to try to avoid the conclusion by spreading out the pigeons as much as possible. Suppose we start by placing one pigeon into each pigeonhole as indicated in the table that follows. Then we have one more pigeon to put into a pigeonhole, and it must go in either hole number 1 or hole number 2. In either case, one of the holes must contain a second pigeon and the conclusion follows.

Pigeonhole Number 1	Pigeonhole Number 2
1	1

EXAMPLE 1.18

Using the Pigeonhole Principle

A student working in a tight space can barely reach a box containing 12 rock CDs and 12 classical CDs. Her position is such that she cannot see into the box. How many CDs must she select to be sure that she has at least two of the same type of CDs?

Solution

Understand the Problem

The box contains 12 rock CDs and 12 classical CDs. The student is in a tight spot and can barely reach the box into which she cannot see. In one attempt, she wants to select just enough to be sure that she has at least two CDs of the same type. We must determine how many CDs she must select to ensure the desired result.

Devise a Plan

Let's consider the possibilities. To make sure that we don't miss one, we make an orderly list.

Carry Out the Plan

If the student chooses two CDs, she must have one of the following:

Two rock CDs	and	zero classical CDs
One rock CD	and	one classical CD
Zero rock CDs	and	two classical CDs

Two CDs are *not* enough; she might get one of each kind. But if she selects a third CD, she will end up with a third rock CD, a second rock CD, a second classical CD, or a third classical CD. In any case, she will have two CDs of the same type and the condition of the problem will be satisfied. Therefore, she needs to select only three CDs from the box.

Look Back

We certainly solved the problem by considering possibilities. But might there be an easier solution? Choosing CDs of two kinds is much like putting pigeons into two pigeonholes. Thus, if we select three CDs, then, by the Pigeonhole Principle, at least two must be the same kind, and we have the same result as before. Also, be sure to read the "Into the Classroom" feature shown next.

INTO THE CLASSROOM

Make Use of Incorrect Responses

Observe that the number 12 in the statement of the preceding problem is misleading (only two CDs of each kind in the box are really needed), and this causes many students to respond that the answer is 13. Often, students make incorrect responses that serve as good springboards to useful classroom discussions. Rather than just saying that the response of 13 *is* incorrect, a good teacher may say something like "Well, Pete, that's not quite right, but could you think of a question related to this problem for which 13 *is* the correct answer?" This response not only corrects Pete but also gives him an immediate opportunity to redeem himself in the eyes of the class. The ensuing discussion will both inform Pete and enhance the understanding of the entire class. Other questions that might be discussed are as follows: "What questions might be asked for which 14 is the correct answer?" "How many CDs must the student select to be sure that she has 12 rock CDs?" Also, one might repeat the problem with 12 rock CDs, 12 classical CDs, and 12 country and western CDs and ask similar questions. The possibilities are almost limitless. The idea of using a student's incorrect response (a thought process) is a wonderful way to explore the concepts that you are talking about.

1.5 Problem Set

Understanding Concepts

1. Play this game with a partner. The first player marks down 1, 2, 3, or 4 dots on a sheet of paper. The second player then adds to this by marking down 1, 2, 3, or 4 more dots. The first player to exceed a total of 30 loses the game. Can the first player or the second player devise a surefire winning strategy? Explain carefully.

2. As in the previous problem, play this game with a partner. The first player (Andy) marks down 1, 2, or 3 dots on a sheet of paper. The second player (Annabelle) then adds to this by marking another 1, 2, or 3 dots. The winning player is the first one to equal a total of 12 dots. Can one player or the other devise a surefire winning strategy? Explain carefully.

3. Consider this mathematical machine:

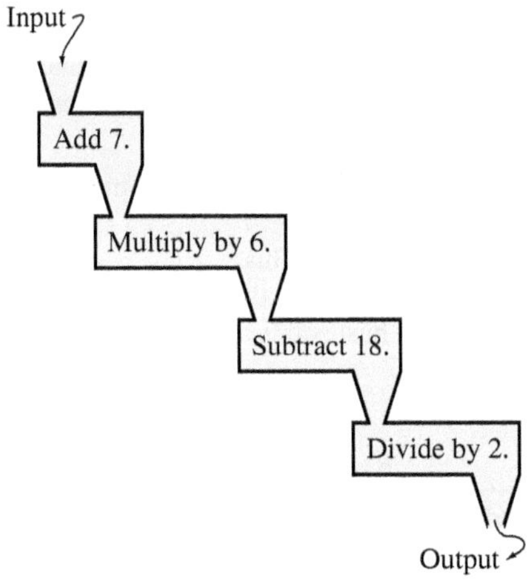

   (a) What number would you have to use as input if you wanted 39 as the output?
   (b) What would you have to input to obtain an output of 48?
   (c) Describe a different strategy for attacking this problem than the one you used for parts (a) and (b).
   (d) How would you solve parts (a) and (b) by the "use a variable" problem-solving strategy?

4. Consider this mathematical machine:

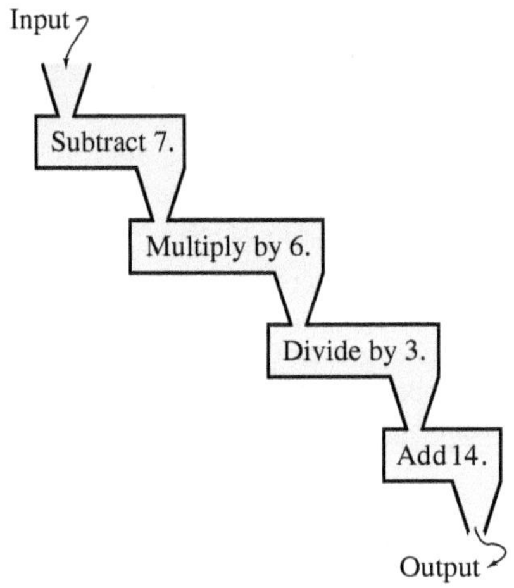

   (a) What number would you have to use as input if you wanted 39 as the output?
   (b) What would you have to input to obtain an output of 48?
   (c) Describe a different strategy for attacking this problem than the ones you used for parts (a) and (b).
   (d) How would you solve parts (a) and (b) by the "use a variable" problem-solving strategy?

5. Consider this mathematical machine:

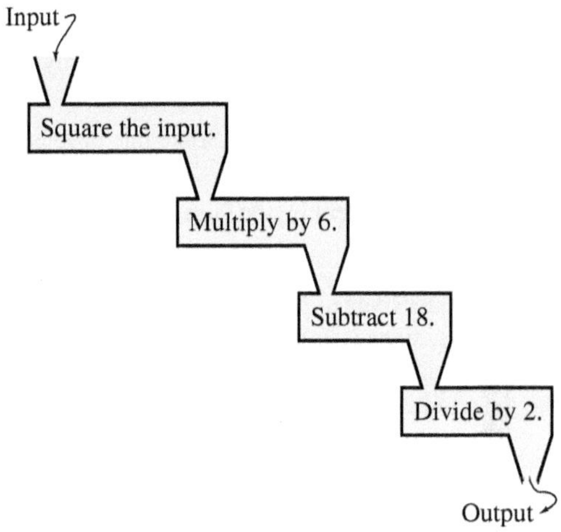

   (a) If the input is 3, what is the output?
   (b) What is the relationship between the output and the input using variables?

6. Consider this mathematical machine:

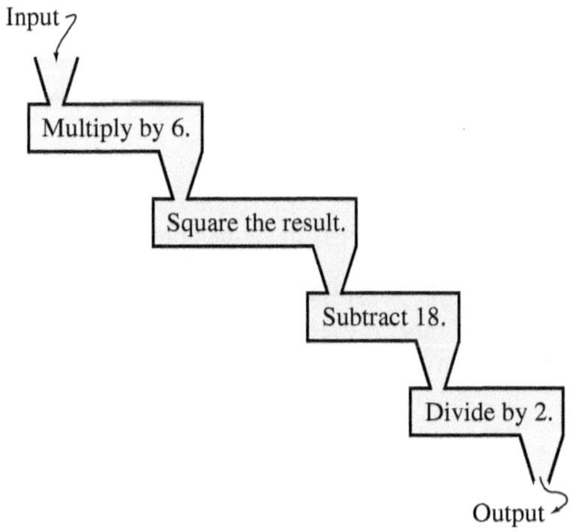

   (a) If the input is 3, what is the output?
   (b) What is the relationship between the output and the input using variables?
   (c) Are the machines of problems 5 and 6 the same? (That is, is the output of the two machines the same?)

7. Josh wanted to buy a bicycle but didn't have enough money. After Josh told his troubles to Sam, Sam said, "I can fix that.

See that fence? Each time you jump that fence, I'll double your money. There's one small thing though: You must give me $32 each time for the privilege of jumping." Josh agreed, jumped the fence, received his payment from Sam and paid Sam $32. Repeating the routine twice more, Josh was distressed to find that, on the last jump, after Sam had made his payment to Josh, Josh had only $32 with which to pay Sam and so had nothing left. Sam, of course, went merrily on his way, leaving Josh wishing that he had known a little more about mathematics.

(a) How much money did Josh have before he made his deal with Sam?

(b) Suppose the problem is the same, but this time Josh jumps the fence 5 times before running out of money. How much money did Josh start with this time?

8. Liping is thinking of a number. If you multiply her number by 71, add 29, and divide by 2, you obtain 263.

(a) What is Liping's number? Solve this problem by working backward.

(b) Solve the problem by the "use a variable" problem-solving strategy.

9. Alice Gardner likes to play number games with the students in her class, since it improves their skill at both mental arithmetic and critical thinking. Solve each of these number riddles she gave to her class. She is thinking of a number.

The number is odd.
It is more than 1 but less than 100.
It is greater than 20.
It is less than $5 \cdot 7$.
The sum of its digits is 7.
It is evenly divisible by 5.

What is the number? Was all the information needed?

10. Ms. Gardner is thinking of a number.

The number is not even.
The sum of its digits is divisible by 2.
The number is a multiple of 11.
It is greater than $4 \cdot 5$.
It is a multiple of 3.
It is less than $7 \cdot 8 + 23$.

What is the number? Is more than one answer possible?

11. Ms. Gardner is thinking of a number.

The number is even.
It is not divisible by 3.
It is not divisible by 4.
It is not greater than 9^2.
It is not less than 8^2.

What is the number? Is more than one answer possible?

12. Moe, Joe, and Hiram are brothers. One day, in some haste, they left home with each one wearing the hat and coat of one of the others. Joe was wearing Moe's coat and Hiram's hat. Whose hat and coat was each one wearing?

13. A stack of 10 cards numbered 0, 1, 2, . . . , 9 in some order lies face up on a desk. Form a new stack as follows: Place the top card face up in your hand, place the second card face up **under** the first card, place the third card face up **on top** of the new stack, place the fourth card face up **on the bottom** of the new stack, and so on. What should be the arrangement of the original stack so that the cards in the new stack are numbered in increasing order from the top down?

14. Saturday afternoon, Aaron, Boyd, Carol, and Donna stopped by the soda fountain for treats. Altogether, they ordered a chocolate malt, a strawberry milkshake, a banana split, and a double-dip walnut ice-cream cone. Given the following information, who had which treat?

(a) Both boys dislike chocolate.

(b) Boyd is allergic to nuts.

(c) Carol bought a malt and a milkshake for Donna and herself.

(d) Donna shared her treat with Boyd.

15. Four married couples belong to a bridge club. The wives' names are Kitty, Sarah, Josie, and Taneisha. Their husbands' names (in some order) are David, Will, Gus, and Floyd.

Will is Josie's brother.
Josie and Floyd dated some, but then Floyd met his present wife.
Kitty is married to Gus.
Taneisha has two brothers.
Taneisha's husband is an only child.

Use this table to sort out who is married to whom:

	Kitty	Sarah	Josie	Taneisha
David				
Will				
Floyd				
Gus				

16. (a) Jorge chose one of the numbers 1, 2, 3, . . . , 1024 and challenged Sherrie to determine the number by asking no more than ten questions to which Jorge would respond truthfully either "yes" or "no." Determine the number he chose if the questions and answers are as follows:

Questions	***Answers***
Is the number greater than 512?	no
Is the number greater than 256?	no
Is the number greater than 128?	yes
Is the number greater than 192?	yes
Is the number greater than 224?	no
Is the number greater than 208?	no
Is the number greater than 200?	yes
Is the number greater than 204?	no
Is the number greater than 202?	no
Is the number 202?	no

(b) In part (a), Sherrie was able to dispose of 1023 possibilities by asking just ten questions. How many questions would Sherrie have to ask to determine Jorge's number if it is one of 1, 2, 3, . . . , 8192? If it is one of 1, 2, 3, . . . , 8000? Explain briefly but clearly. (*Hint:* Determine the differences between 512, 256, 128, 192, and so on.)

(c) How many possibilities might be disposed of with 20 questions?

17. If it takes 867 digits to number the pages of a book starting with page 1, how many pages are in the book?

18. (a) How many students must be in a room to be sure that at least two are of the same sex?

(b) How many students must be in a room to be sure that at least six are boys or at least six are girls?

19. (a) How many people must be in a room to be sure that at least two people in the room have the same birthday (not birth date)? Assume that there are 365 days in a year.

(b) How many people must be in a room to be sure that at least three have the same birthday?

Into the Classroom

20. (**Writing**) Read one of the following from NCTM's *Principles and Standards for School Mathematics*:

(a) Reasoning and Proof Standard for Grades Pre-K–2, pages 122–126

(b) Reasoning and Proof Standard for Grades 3–5, pages 188–192

(c) Reasoning and Proof Standard for Grades 6–8, pages 262–267

Write a critique of the standard you read, emphasizing your own reaction. How do the recommendations compare with your own school experience?

Thinking Critically

21. Show that, in any collection of 11 natural numbers, there must be at least two whose difference is evenly divisible by 10. (*Helpful question:* When is the difference of two natural numbers divisible by 10?)

22. (a) In any collection of seven natural numbers, show that there must be two whose sum or difference is divisible by 10. (*Hint:* Try a number of particular cases. Try to choose numbers that show that the conclusion is false. What must be the case if the sum of two natural numbers is divisible by 10?)

(b) Find six numbers for which the conclusion of part (a) is false.

23. Show that, if five points are chosen in or on the boundary of a square with a diagonal of length $\sqrt{2}$ inches, at least two of them must be no more than $\frac{\sqrt{2}}{2}$ inches apart. (*Hint:* Consider the figure shown and use the Pigeonhole Principle.)

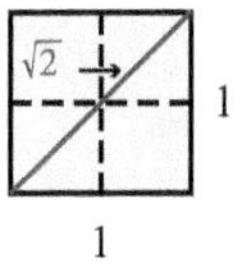

24. Show that if five points are chosen in or on the boundary of an equilateral triangle with sides 1 meter long, at least two of them must be no more than $\frac{1}{2}$ meter apart.

25. Think of 10 cups, with 1 marble in the first cup, 2 marbles in the second cup, 3 marbles in the third cup, and so on. Show that, if the cups are arranged in a circle in any order whatsoever, then some three adjacent cups in the circle must contain a total of at least 17 marbles.

26. A fruit grower packs apples in boxes. Each box contains at least 240 apples and at most 250 apples. How many boxes must be selected to be certain that at least three boxes contain the same number of apples?

27. Show that, at a party of 20 people, there are at least two people with the same number of friends at the party. Assume that the friendship is mutual. (*Hint:* Consider the following three cases: (i) Everyone has at least one friend at the party; (ii) precisely one person has no friends at the party; (iii) at least two people have no friends at the party.)

28. Argue convincingly that at least two people in New York have precisely the same number of hairs on their heads. (*Hint:* You may need to determine a reasonable figure for the number of hairs on a human head.)

State Assessments

29. (Washington State, Grade 4)
Dan baked some cookies. Sam took half of the cookies. Then Sue took half of the remaining cookies. Later, Lisa took half of the cookies that were left. When Dan came home, he saw only three cookies. Tell how you could figure out how many cookies Dan baked altogether. Explain your thinking using words, numbers, or pictures.

30. (Washington State, Grade 4)
Emily, Mei, and Andrew go to the same camp. They each like different games. Use the information in the figure shown on page 50 to find out which game Mei likes best. Their favorites are tug-of-war, rope skipping, and relay race.

Emily's favorite game does *not* use a rope.
Andrew does *not* like tug-of-war.

Which is Mei's favorite game?

A. Tug-of-war
B. Relay race
C. Rope skipping

Figure for Problem 30

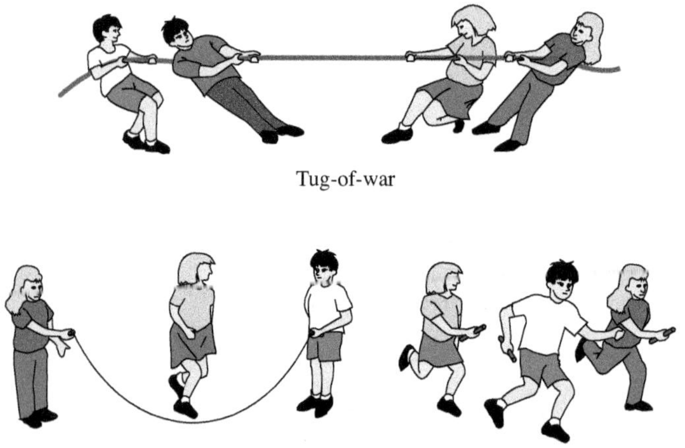

Tug-of-war

Rope skipping

Relay race

31. (California, Grade 5)
Which equation could have been used to create this function table?

x	*y*
−9	−5
−2	2
4	8
11	15

A. $y = \frac{x}{2}$

B. $y = 2x$

C. $y = x - 4$

D. $y = x + 4$

1.6 Reasoning Mathematically

In this concluding section on critical thinking and mathematical reasoning, we will extend our strategies of problem solving. In particular, we will discuss the following topics, each of which will be helpful in the chapters to come:

- inductive reasoning
- representational reasoning
- mathematical statements
- deductive reasoning

Reading the NCTM *Reasoning and Proof Standard* (p. 52) gives us an overview of the major role that mathematical reasoning plays in K–12 education. The point is that if we are mathematically curious enough to want to know why, how, or how general a statement can be made, then we are under the umbrella of the Standards of Mathematical Practice.

Inductive Reasoning

We use **inductive reasoning** to draw a general conclusion based on information obtained from specific examples. For example, think about the bears you've seen, maybe in zoos, in pictures in magazines, or perhaps even in the wild. On the basis of these experiences, you would likely draw the conclusion that bears are brown or black, or even white if you've seen a polar bear. Here's a mathematical example: Consider the square numbers 4, 9, 16, 25, and 36. Notice that they are either multiples of 4 or one more than a multiple of 4. We can check that this property of squares is also true for other squares—say, 49, 64, 81, and 100. Even $1^2 = 1$ passes our check, since 1 is one more than 4 times 0. Thus, inductive reasoning leads us to the generalization that the square of any whole number is either a multiple of 4 or one more than a multiple of 4.

DESCRIPTION Inductive Reasoning
Inductive reasoning is drawing a conclusion based on evidence obtained from specific examples. The conclusion drawn is called a **generalization.**

Inductive reasoning is a powerful way to create and organize information. However, it only suggests what *seems* to be true, since we have not yet checked that the property holds for *all* examples. For example, there is a rare type of bear in southeastern Alaska called the blue bear. Although it is a genetic variant of the black bear, its fur is dark blue. Thus, the existence of the blue bear tells us that the statement "All bears are black, brown, or white" is false. Such an example, one that disproves a statement, is called a **counterexample.** It is interesting to notice that a proof of a generalization requires us to demonstrate that a certain property holds for every possible case, but a generalization can be proved to be false by finding just *one* counterexample.

EXAMPLE 1.19

Using Inductive Reasoning in Mathematics

Examine the generalizations that follow. Test the validity of each generalization with additional evidence. If you believe that the generalization is valid, try to offer additional reasons that this is so. If you believe that the generalization may be false, search for a counterexample.

(a) Consider three consecutive integers, such as 8, 9, and 10. Exactly one of these three numbers is a multiple of 3. Similarly, each of the consecutive triples 33, 34, and 35 and 121, 122, and 123 includes precisely one multiple of 3. Thus, in any string of three consecutive integers, it is probably true that exactly one is a multiple of 3.

(b) Place n points on a circle. Next, join each pair of these points with a line segment (that is, with a chord of the circle) such that no more than two chords intersect at a single point. As shown in the following diagrams, the number of regions in the circle doubles with each additional point placed on the circle, giving the sequence 1, 2, 4, and 8:

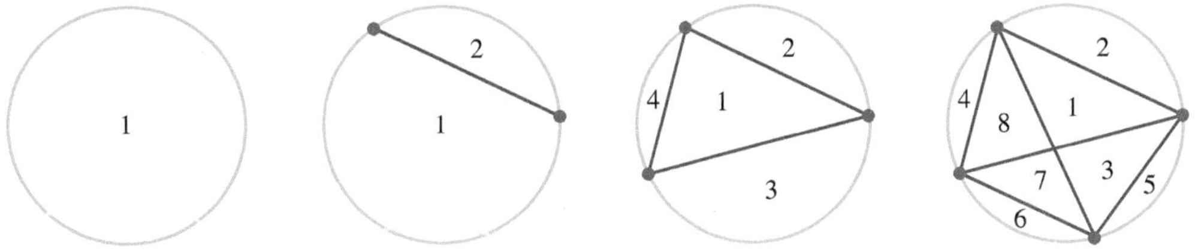

In conclusion, it is probably true that there are 2^{n-1} regions created by drawing chords between n points on a circle.

(c) $3 \times 439 = 1317$, and the sum of the digits of the product is $1 + 3 + 1 + 7 = 12$, a multiple of 3. Similarly, $3 \times 2687 = 8061$, and the sum of digits of this product is $8 + 0 + 6 + 1 = 15$, again a multiple of 3. Thus, it is probably true that the sum of the digits of any whole-number multiple of 3 is also a multiple of 3.

Solution

(a) Additional examples support the general conclusion. If the whole numbers are written in the form **0** 1 2 **3** 4 5 **6** 7 8 **9** 10 11 **12** 13 . . . , with the multiples of 3 shown in bold, we see that any three numbers in succession include exactly one of the bold numbers. Thus, the assertion *appears* to be true.

(b) Let's draw a sketch of the next two cases, with $n = 5$ and 6 points on the circle:

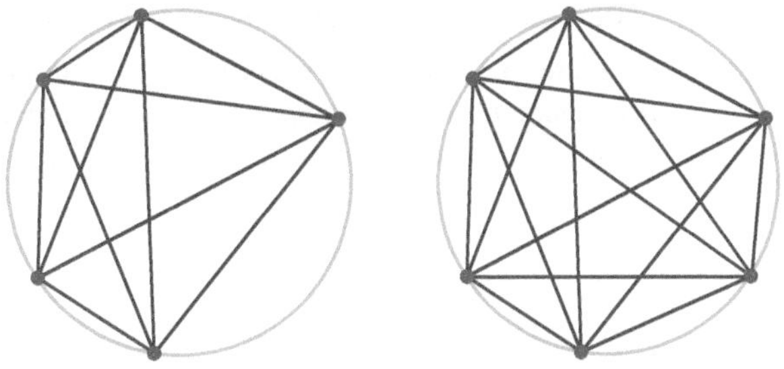

When $n = 5$ points are placed on the circle, we see that there are $16 = 2^{5-1}$ regions, which supports the generalization. However, the doubling pattern breaks down with $n = 6$ points on the circle, since we do not get 32 regions as we had expected. How many regions do the chords joining 6 points on a circle create?

(c) Additional examples all exhibit the same property. In Chapter 4, we will prove that the sum of the digits of any whole number divisible by 3 is also divisible by 3.

FROM THE NCTM PRINCIPLES AND STANDARDS

Reasoning and Proof Standard

Instructional programs from prekindergarten through grade 12 should enable all students to

- *recognize reasoning and proof as fundamental aspects of mathematics;*
- *make and investigate mathematical conjectures;*
- *develop and evaluate mathematical arguments and proofs;*
- *select and use various types of reasoning and methods of proof.*

During grades 3–5, students should be involved in an important transition in their mathematical reasoning. Many students begin this grade band believing that something is true because it has occurred before, because they have seen several examples of it, or because their experience to date seems to confirm it. During these grades, formulating conjectures and assessing them on the basis of evidence should become the norm. Students should learn that several examples are not sufficient to establish the truth of a conjecture and that counterexamples can be used to disprove a conjecture. They should learn that by considering a range of examples, they can reason about the general properties and relationships they find.

Mathematical reasoning develops in classrooms where students are encouraged to put forth their own ideas for examination. Teachers and students should be open to questions, reactions, and elaborations from others in the classroom. . . . There is clear evidence that in classrooms where reasoning is emphasized, students do engage in reasoning and, in the process, learn what constitutes acceptable mathematical explanation (Lampert 1990; Yackel and Cobb 1994, 1996).

SOURCE: Principles and Standards for School Mathematics by NCTM, *pp. 56, 188. Copyright © 2000 by the National Council of Teachers of Mathematics. Reproduced with permission of the National Council of Teachers of Mathematics via Copyright Clearance Center. NCTM does not endorse the content or validity of these alignments.*

PROBLEM-SOLVING STRATEGY 11 Use Inductive Reasoning

- Observe a property that holds in several examples.
- Check that the property holds in other examples. In particular, attempt to find an example in which the property does not hold (that is, try to find a counterexample).
- If the property holds in every example, state a generalization that the property is probably true in general.

A generalization that seems to be true, but has yet to be proved, is called a **conjecture.** Once a conjecture is given a proof, it is called a **theorem.** The logical thought process that brings us to a conjecture or, even better, the proof of a theorem shows a deep understanding of math.

Representational Reasoning

In mathematics, a *representation* or a manipulative is an object that captures the essential information needed for understanding and communicating mathematical properties and relationships. Often, the representation conveys information visually, as in a diagram, a graph, a map, or a table. At other times, the representation is symbolic, such as a letter denoting a variable or an algebraic expression or equation. The manipulative can also be a physical object, such as a paper model of a cube or an arrangement of pebbles to represent a whole number. The following excerpt from the NCTM *Principles and Standards for School Mathematics* points out that a representation can even be simply a mental image.

Here is an example to illustrate how a representation can reveal and explain a property. Consider the "pyramidal sums" shown in this list:

$$1 = 1$$
$$1 + 2 + 1 = 4$$
$$1 + 2 + 3 + 2 + 1 = 9$$
$$1 + 2 + 3 + 4 + 3 + 2 + 1 = 16$$

FROM THE NCTM PRINCIPLES AND STANDARDS

Representation Standard for Grades 3–5

Instructional programs from prekindergarten through grade 12 should enable all students to

- *create and use representations to organize, record, and communicate mathematical ideas;*
- *select, apply, and translate among mathematical representations to solve problems;*
- *use representations to model and interpret physical, social, and mathematical phenomena.*

In grades 3–5, students need to develop and use a variety of representations of mathematical ideas to model problem situations, to investigate mathematical relationships, and to justify or disprove conjectures. They should use informal representations, such as drawings, to highlight various features of problems; they should use physical models to represent and understand ideas such as multiplication and place value. They should also learn to use equations, charts, and graphs to model and solve problems.

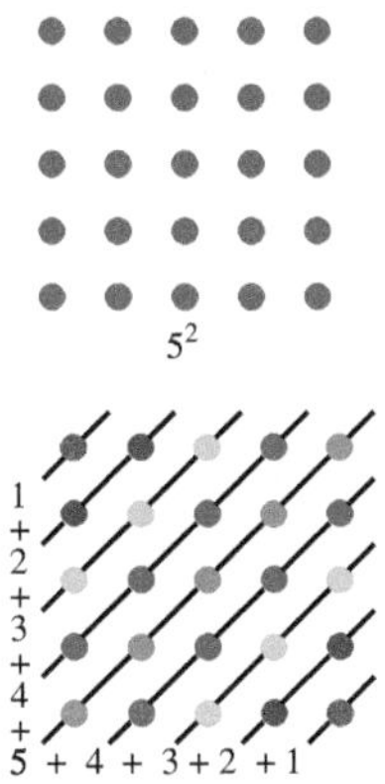

Figure 1.7
Any square number is a pyramidal sum

Using inductive reasoning, we would likely draw the conclusion that the nth pyramidal sum is the square number n^2. That is,

$$1 + 2 + 3 + \cdots + (n - 1) + n + (n - 1) + \cdots + 3 + 2 + 1 = n^2.$$

To see why this is so, recall that any square number can be represented as a square pattern of dots. For example, 5^2 is represented by the 5×5 array shown in Figure 1.7. If we sum the numbers of dots along each diagonal of the square, as seen at the right of the figure, we see very clearly why $1 + 2 + 3 + 4 + 5 + 4 + 3 + 2 + 1$ is equal to the square number 5^2. In our "mind's eye," where we can create a mental image of the most general case, we see why any pyramidal sum is a square number.

In the next example, we use dot drawings with colored pencils on squared paper in order to make some more discoveries about number patterns through the use of representations.

EXAMPLE 1.20

Using Dot Representations to Discover Number Patterns

Recall that the triangular and square numbers can be represented with dot patterns as shown here:

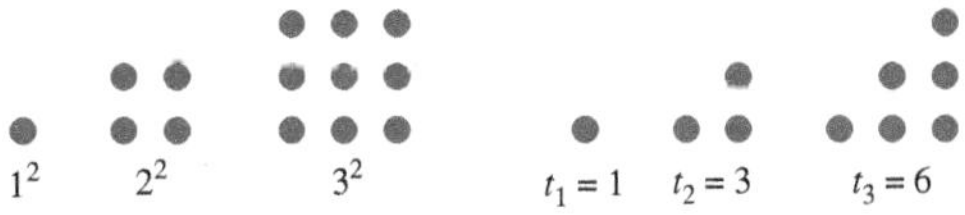

(a) Draw dot figures to show that $t_3 + t_3 = 3 \cdot 4$ and $t_4 + t_4 = 4 \cdot 5$. By inductive reasoning, why can you conclude that $t_n = \dfrac{n(n+1)}{2}$?

(b) Draw dot figures to show that $t_3 + t_4 = 4^2$ and $t_4 + t_5 = 5^2$. What generalization can you make?

(c) Draw dot figures to show that $7^2 = 1 + 8t_3$ and $9^2 = 1 + 8t_4$. (*Hint:* The 1 is the center dot in your square.) By inductive reasoning, what do you think is the value of $1 + 8t_{50}$? of $1 + 8t_n$?

Solution

(a) We generalize that two dot patterns, each representing t_n, can form an n-by-$(n + 1)$ rectangle. Therefore,

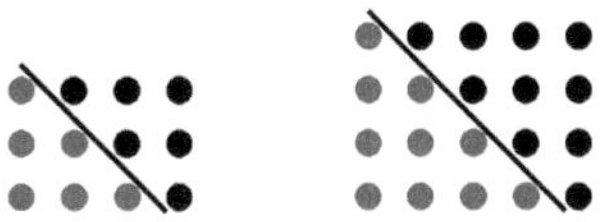

$$t_n + t_n = n(n + 1), \quad \text{or} \quad t_n = \frac{n(n+1)}{2}.$$

(b) The patterns shown suggest that the patterns representing t_{n-1} and t_n can be arranged to form an n-by-n square. We conclude that $t_{n-1} + t_n = n^2$.

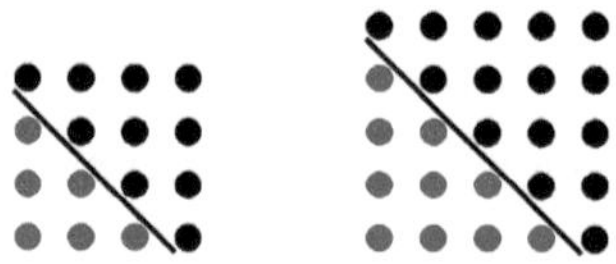

(c) Eight triangular dot patterns together with a single dot can be arranged to form a square with an odd number of dots in each row. Inductive reasoning suggests that, $1 + 8t_{50} = 101^2 = 10{,}201$ and, in general, that $1 + 8t_n = (2n + 1)^2$.

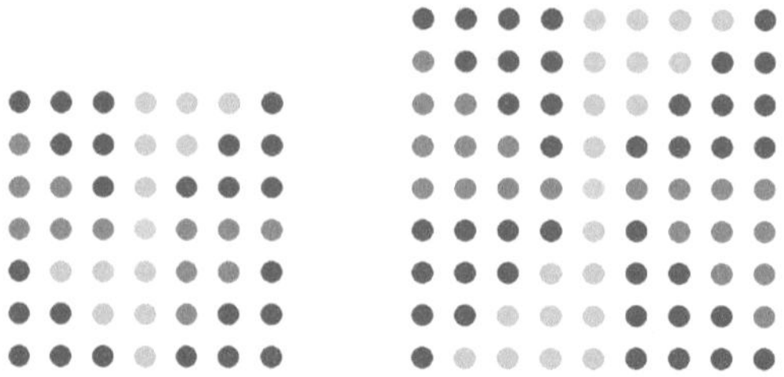

Mathematical Statements

In mathematics, a **statement** is a declarative sentence that is either true or false, but not both. Thus, "2 + 3 = 5" is a true statement, and "Pigs can fly" is a false statement. "Shut the door" is not a statement, since it is not a declarative sentence. The sentence "This sentence is a false statement" is declarative, but not a mathematical statement, since it is neither true nor false. (If it is true, then it is false, and if it is false, then it is true—so it is neither true nor false!)

Two statements—say, p and q—are often combined to form another statement by using **and** and **or** operations.

- **And:** "p and q" is true when statement p and statement q are both true, and false when either or both of p and q are false.
- **Or:** "p or q" is true when either statement p or statement q (or even both) is true, and false only when both p and q are false.

This description can also be given a representation known as a **truth table,** as shown in Figure 1.8.

Figure 1.8
The truth table for "p and q" and "p or q"

p	q	p and q	p or q
T	T	T	T
T	F	F	T
F	T	F	T
F	F	F	F

Yet another way to combine two statements is to form a **conditional statement,** or an **"if . . . then" statement.**

- **If . . . then:** "If p, then q" is true when the truth of statement p guarantees the truth of statement q.

For example, you have probably encountered the conditional statement "If a whole number has 0 as its last digit, then it is divisible by 10." Often, "If p then q" is read as "p implies q" and is written $p \rightarrow q$. The truth table for $p \rightarrow q$ is shown in Figure 1.9.

Figure 1.9
The truth table for $p \rightarrow q$, which shows that $p \rightarrow q$ is false when, and only when, p is true and q is false

p	q	$p \rightarrow q$
T	T	T
T	F	F
F	T	T
F	F	T

The big surprise in Figure 1.9 is that $p \rightarrow q$ is true when p is false, independently of whether q is true or false. Thus, "If the moon is made of green cheese, then pigs can fly" is a true conditional statement! Fortunately, the moon is not made of green cheese, so we don't need to be on the lookout for flying pigs.

Many theorems of mathematics have the "If p, then q" form.

EXAMPLE   1.21

Proving an "If . . . Then" Statement

Prove the following theorem:

If n *is any whole number, then* n^2 *is either a multiple of 4 or 1 larger than a multiple of 4.*

Solution

Understand the Problem

We must show that *any* whole number n, when squared, has one of two forms: Either it is a multiple of 4, so that $n^2 = 4j$ for some whole number j, or else $n^2 = 4k + 1$ for some whole number k. Earlier, we used inductive reasoning that supported the truth of the theorem.

Devise a Plan

The whole numbers are 0, 1, 2, 3, 4, 5, 6, . . . , alternating between even and odd. Their squares are 0, 1, 4, 9, 16, 25, 36, . . . , alternating between multiples of 4 and numbers that are 1 larger than a multiple of 4. These two mathematical facts suggest that we consider two cases: when n is even and when n is odd. We also know that when n is even, it can be written as $n = 2r$ for some whole number r, and when n is odd, it can be written as $n = 2s + 1$ for some whole number s. Thus, we need to consider the two cases $(2r)^2$ and $(2s + 1)^2$ and use algebra.

Carry Out the Plan

If $n = 2r$, it follows that $n^2 = (2r)^2 = 4r^2$. That is, $n^2 = 4j$, where $j = r^2$. Similarly, when $n = 2s + 1$, we have $n^2 = (2s + 1)^2 = 4s^2 + 4s + 1 = 4(s^2 + s) + 1$. That is, $n^2 = 4k + 1$ for $k = s^2 + s$.

$$(2s + 1)^2 = (2s + 1)(2s + 1)$$
$$= 4s^2 + 2s + 2s + 1$$
$$= 4s^2 + 4s + 1$$

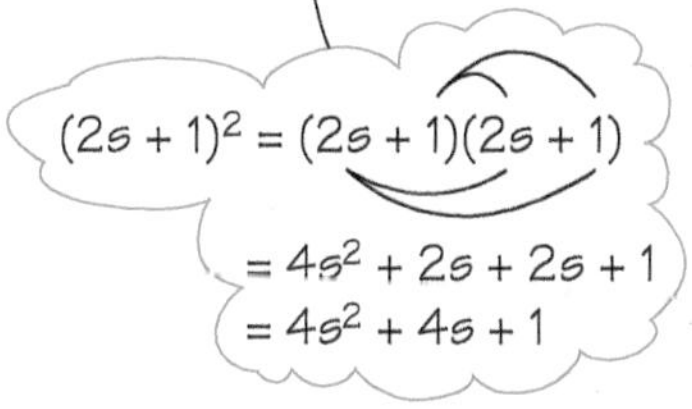

Look Back

Let's take a closer look at the squares of the odd numbers—that is, the numbers 1, 9, 25, 49, 81, ... These numbers are not only 1 larger than a multiple of 4; they're even 1 more than a multiple of 8. This was actually shown earlier in Example 1.20 (c): Using dot pattern representations, we discovered that $(2n + 1)^2 = 8t_n + 1$, where t_n is the nth triangular number. We can also use dot pattern representations to visualize why the square of an even number is four times another square. The following figure shows that $10^2 = 4 \times 5^2$:

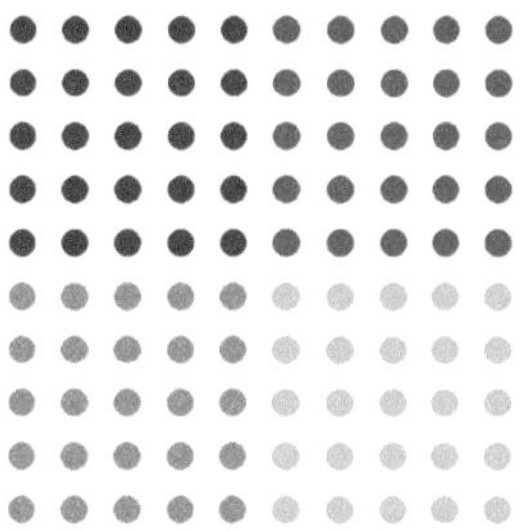

This proof involves algebra, once again showing how algebra can be used to explain many things.

Deductive Reasoning

Suppose that we have a collection of true statements. If we can argue on the basis of these statements that another statement must also be true, then we are using **deductive reasoning.** Example 1.21 is an example of deductive reasoning, since we showed that if a number is a square whole number, then it is either a multiple of 4 or 1 larger than a multiple of 4.

The list of true statements we begin with are known as the **premises,** or **hypotheses,** of the argument. The new true statement that we obtain is called the **conclusion** of the argument.

Here is another example of deductive reasoning:

Hypothesis:	***Statement 1.***	If you wish to become a successful elementary school teacher, then you must become proficient in mathematical reasoning.
	Statement 2.	You wish to become a successful elementary school teacher.
Conclusion:	***Statement 3.***	You must become proficient in mathematical reasoning.

In symbols, if we let p denote the statement "You wish to become a successful elementary school teacher" and q denote the statement "You need to become proficient in mathematical reasoning," then we have reached the conclusion by using the **rule of direct reasoning.**

RULE OF DIRECT REASONING

Hypotheses: $\begin{cases} \text{If } p, \text{ then } q \\ p \text{ is true} \end{cases}$

Conclusion: Therefore, q is true

The rule of direct reasoning may seem straightforward, but it is sometimes used incorrectly. Here is an example of invalid reasoning called the "fallacy of the converse."

If I am a good person, then nothing bad will happen to me.
Nothing bad has happened to me.
Therefore, I am a good person.

} INVALID REASONING

This argument is invalid, since even if $p \rightarrow q$ is true and q is true, nothing can be said about whether p is true or not. This concept is clearly shown in the truth table of Figure 1.9.

On the other hand, the truth table does show that if $p \rightarrow q$ is true and q is *false*, then p is also necessarily false. Thus, we have the very useful **rule of indirect reasoning.**

RULE OF INDIRECT REASONING

Hypotheses: $\begin{cases} \text{If } p, \text{ then } q \\ q \text{ is false} \end{cases}$

Conclusion: Therefore, p is false

The rule of indirect reasoning is often used to give a **proof by contradiction.** That is, if we take a statement p and derive a false statement q from it, then we know that p is false. The checkerboard tiling problem presented in Example 1.22 is a classic example of indirect reasoning and proof by contradiction. In Chapter 4, we will use this method to show that the $\sqrt{2}$ is not a rational number.

EXAMPLE 1.22

Tiling a Checkerboard with Dominos

It is easy to cover the 64 squares of an 8-by-8 checkerboard with 32 dominos, where each domino covers two adjacent squares, one red and one black, of the checkerboard. But what if two diagonally opposite squares are removed? Can the 62 remaining squares be covered with 31 dominos?

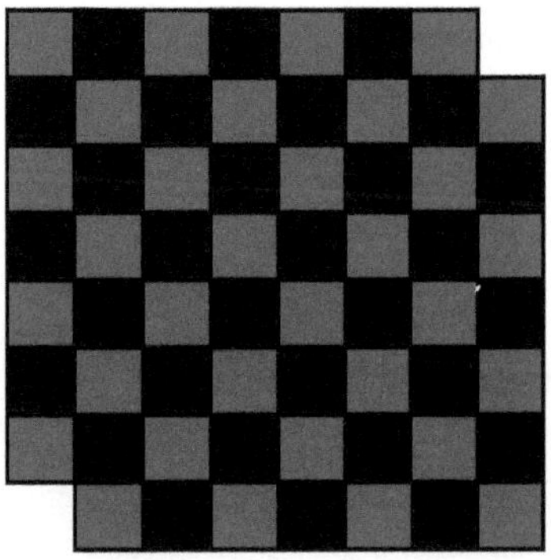

Solution

If you believed that the modified board could be tiled, you would search for an arrangement of 31 dominos that covers all 62 squares. After a few unsuccessful trials, you might begin to have doubts. Let's use indirect reasoning, or reasoning by contradiction, and assume (though you haven't found it) that there is an arrangement of 31 dominos that covers the modified board. Notice that each domino, whether vertical or horizontal, covers one red and one black square, so that an equal number of red and black squares must be covered. However, the two squares we've removed have the same color (black in the diagram). This means that we have left behind 30 squares of one color and 32 squares of the other color. This situation contradicts our observation that any arrangement of the dominos covers an equal number of squares of the two colors. Thus, the assumption that a domino tiling exists leads to a falsehood, so our assumption must also be false. That is, there is no tiling of the modified checkerboard with dominos.

The rule of indirect reasoning is also misused with some frequency. Here is an example of what is called the "fallacy of the inverse":

If I am wealthy, then I am happy.
I am not wealthy.
Therefore, I am not happy. } INVALID REASONING

An examination of the truth table in Figure 1.9 shows that, if $p \rightarrow q$ is true and p is false, q can be either true or false.

Deductive reasoning provides another important strategy for problem solving.

PROBLEM-SOLVING STRATEGY 12 Use Deductive Reasoning

- To show that a statement q is true, look for a statement p such that $p \rightarrow q$ is true. If p is true, you can deduce that q is also true by direct reasoning.
- To show that a statement p is false, show that $p \rightarrow q$, where q is false. That is, show that assuming p leads to a contradiction. Then p is false by indirect reasoning.

In this section, we have highlighted inductive, representational, and deductive reasoning. These modes of mathematical thought will be used throughout the remainder of this book and will be supplemented with some additional methods, such as proportional reasoning in Chapter 7, algebraic reasoning in Chapter 8, statistical reasoning in Chapter 13, and geometrical reasoning in Chapters 9–12.

1.6 Problem Set

Understanding Concepts

1. (a) Compute the value of these expressions:
$1 \times 8 + 1$, $12 \times 8 + 2$, and $123 \times 8 + 3$.

(b) Use inductive reasoning (don't calculate yet) to describe in words what you expect are the values of these expressions: $1234 \times 8 + 4$, $12345 \times 8 + 5$, $123456 \times 8 + 6$, $1234567 \times 8 + 7$, $12345678 \times 8 + 8$, and $123456789 \times 8 + 9$.

(c) Use a calculator to check that your inductive reasoning was correct in part (b).

2. Consider a three-digit number *abc*, where the digit *a*, in the hundreds position, is larger than the units digit *c*. Now reverse the order of the digits to get the number *cba*. Subtract your two three-digit numbers, and let the difference be *def*. Finally, reverse the digits of *def* to form the number *fed*, and add this number to *def*. Altogether, you should do the following addition and subtraction steps with your beginning number *abc*:

$$\begin{array}{r} abc \\ -\underline{cba} \\ def \end{array} \qquad \begin{array}{r} def \\ +\ \underline{fed} \\ ???? \end{array}$$

Choose several examples of three-digit numbers *abc*, and carry out the subtraction and addition steps. Use inductive reasoning to make a generalization about this process.

3. (a) Use a calculator to compute these products:
$1 \times 1089 =$ ________
$2 \times 1089 =$ ________
$3 \times 1089 =$ ________
$4 \times 1089 =$ ________
$5 \times 1089 =$ ________
$6 \times 1089 =$ ________
$7 \times 1089 =$ ________
$8 \times 1089 =$ ________
$9 \times 1089 =$ ________

(b) Did you have to compute all the products in part (a) to be pretty sure that you knew what all the answers would be? Explain briefly.

(c) Do you see any other interesting patterns in part (a)? Explain briefly.

4. (a) Use inductive reasoning on the function $f(x) = x^6 - 14x^4 + 49x^2 - 36$ to see what the value of $f(x)$ might be for all x. Start with $x = -1, 1, -2, 2, -3$, and 3, and substitute into the equation. What might you conclude from these substitutions?

(b) Is the guess that you would make from inductive reasoning using just the values for x above correct? Why?

Figure for Problem 3

5. Use a ruler to draw two line segments and label three points on each line as follows:

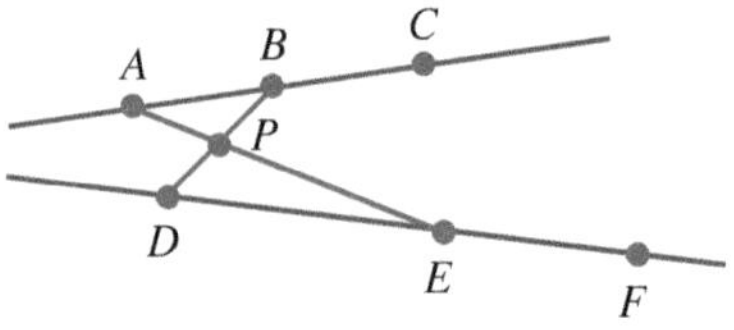

The segments $\overline{AE}$ and $\overline{BD}$ (shown in red) intersect to define point P. In the same way, draw the segments $\overline{AF}$ and $\overline{CD}$ to define the intersection point Q. Finally, draw the segments $\overline{BF}$ and $\overline{CE}$ to define the intersection point R. What seems to be special about the relative positions of the three points P, Q, and R? Use inductive reasoning to make a generalization, and test your conclusion by drawing a new pair of lines and points.

6. The given figure suggests that the number of pieces into which a pie is divided is doubled with each additional cut across the pie. Do you believe this is a valid generalization, or can you find a counterexample?

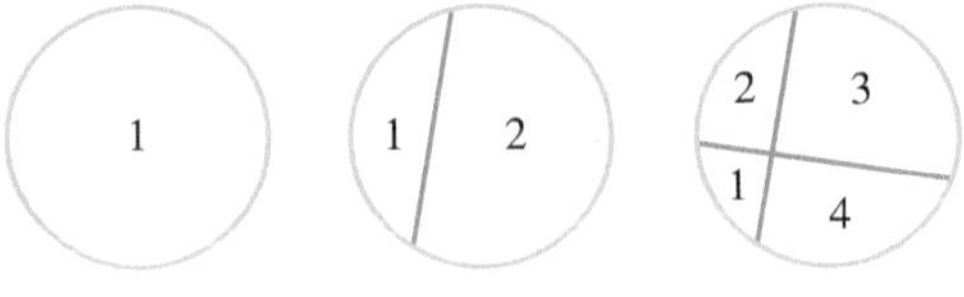

7. Suppose that logs are stacked on level ground. All of the logs in the bottom row must be side by side, and any log in an upper row must touch two lower logs. The ways to stack n circular logs are as follows for $n = 1, 2, 3$, and 4 logs, showing that the respective numbers of stacking arrangements are 1, 1, 2, and 3:

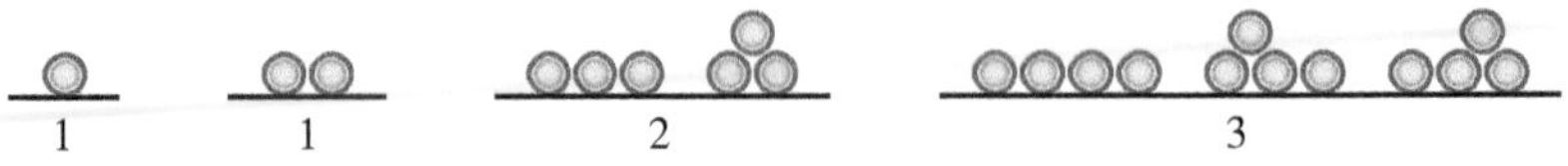

(a) Do the Fibonacci numbers $F_1 = 1$, $F_2 = 1$, $F_3 = 2$, $F_4 = 3$, $F_5 = 5$, $F_6 = 8, \ldots$, first encountered in Example 1.6 (e), correctly count the number of ways to stack n logs? Investigate this generalization by drawing arrangements of 5 and 6 logs.

(b) Use inductive reasoning to investigate the number of ways to stack logs at most two layers high, where, as before, the bottom row of logs must be side by side and logs in the second row must rest on two logs of the bottom row.

8. Suppose a flagpole is erected on one of n blocks, with all of the blocks to the right (if any) used to attach guy wires and an equal number of blocks to the left of the pole also used as points of attachment of guy wires. Any block can be used to attach at most one guy wire. Here are the three permissible arrangements of a flagpole and guy wires on a row of four blocks:

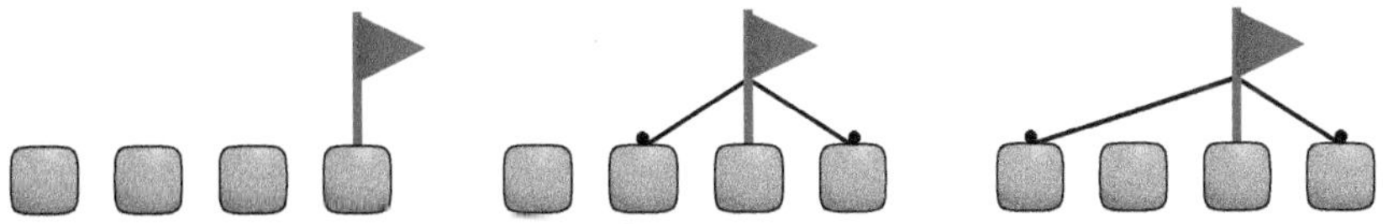

Use inductive reasoning to discover the number of arrangements of a flagpole and guy wires on a row of n blocks.

9. Consider the three triangles of pennies shown, each pointing upward. It is easy to see that the triangle of 3 pennies on the left can be inverted (that is, made to point downward) by moving one penny.

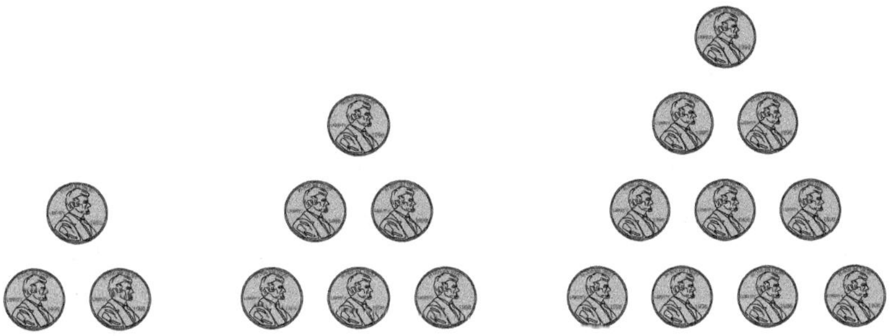

(a) What is the smallest number of pennies that must be moved in order to invert the 6-penny triangle?

(b) Show that the 10-penny triangle can be inverted by moving just 3 pennies.

(c) What do you think is the minimum number of pennies that must be moved to invert the 15-penny triangle? Experiment to see if you are justified in drawing this conclusion.

---

10. (a) The fifth pentagonal number P_5 is the sum of the arithmetic progression 1, 4, 7, 10, 13. (See problem 13, Problem Set 1.4.) Why does the dot pattern representation displayed here show that $P_5 = 1 + 4 + 7 + 10 + 13$? Give your answer in words and in a diagram drawn on triangular dot paper.

(b) Extend your dot representation of part (a) to show that P_6 is 51.

(c) Color a triangular dot pattern to show that $P_5 = 5 + 3t_4$, where $t_4 = 10$ is the fourth triangular number.

(d) Use triangular dot paper to show that $P_6 = 6 + 3t_5$.

(e) By inductive reasoning, and recalling that $t_{n-1} = \frac{1}{2}(n - 1)n$, obtain a formula for the nth pentagonal number $P_n = 1 + 4 + 7 + \cdots + (3n - 2)$.

11. Consider the **trapezoidal numbers** $1 + 1 + 1, 1 + 2 + 2 + 2 + 1, 1 + 2 + 3 + 3 + 3 + 2 + 1, \ldots$ The nth trapezoidal number is $1 + 2 + \cdots + n + n + n + \cdots + 2 + 1$ where n is a positive integer.

(a) Create a dot pattern representation of the trapezoidal numbers.

(b) Notice that $1 + 1 + 1 = 1 \times 3$, $1 + 2 + 2 + 2 + 1 = 2 \times 4$, and $1 + 2 + 3 + 3 + 3 + 2 + 1 = 3 \times 5$. Show how the dots in a representation of a trapezoidal number can be rearranged into a rectangle with two more columns than rows.

(c) Using inductive reasoning, show what general formula gives the nth trapezoidal number.

12. A log in the woods is just long enough to have space for seven frogs on top. Suppose there are three green frogs on the left and three red frogs on the right, with one empty space between. The frogs wish to change ends of the log, so that the three green frogs are on the right and the three red frogs are on the left. A frog can either move into an empty adjacent space or hop over one adjacent frog to an empty space on the opposite side of the frog being jumped over.

(a) Represent the frog problem with a physical model made from colored counters and a row of squares drawn on paper, as shown here:

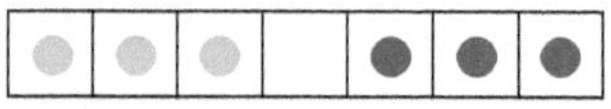

Use your model to show that the frogs can exchange ends of the log in 15 moves. This exercise may take several trials to find an efficient order in which to move the frogs. (*Suggestion:* Work in pairs, with one partner moving the counters and the other partner counting the number of moves.)

(b) On a three-space log that has space for one frog of each color and a single space between, the frogs can trade ends in three moves. What is the minimum number of moves to trade ends on a five-space log, where two green frogs start on the left and two red frogs start on the right?

(c) Examine your answers for the minimum number of moves to have $n = 1, 2,$ or 3 green frogs trade ends with the same number of red frogs on the other end of the log and with just one extra space between them. What conjecture can you make for the case of n frogs of each color separated by one extra space? (*Hint:* There is a connection between the frog-jumping problem and the trapezoidal numbers of problem 11.)

13. Use deductive reasoning to show that if n is a multiple of 3, then n^2 is a multiple of 9. (*Note:* A multiple of k is a number of the form ks, where s is a whole number.)

14. Use indirect reasoning to show that if n^2 is odd, then n is also odd. (This problem shows the value of algebra in a proof.)

15. Use indirect reasoning to show that there is no whole number, n, which solves the equation $2n + 16 = 35$.

The Chapter in Relation to Future Teachers

What is problem solving really about? Problem solving concerns itself with techniques to help analyze a situation, give insights, and, in many cases, suggest a course of action or even provide a solution regardless of whether the problem is one of mathematics or life. Your personal and work worlds are full of situations that need carefully thought out solutions. Our main point is that clear, analytical thinking aided by various techniques such as the ones in this chapter will help you and the children you will teach arrive at good solutions. Problem solving runs through the educational process. Encouraging it in your students will help prepare them for their future.

Is problem solving a part of the mathematical education of elementary and middle school students? Absolutely! The State Student Assessment questions are ample evidence of the clear need for your students to study problem-solving techniques in this chapter and others.

More importantly, problem solving runs throughout mathematics. Every homework exercise, test, or in-class assignment is a form of problem solving. Insights are obtained by looking for solutions by means of different problem-solving strategies, whether those strategies lead to a solution to the original problem or not. Mathematics is learned through the process of solving problems, not by obtaining a correct answer alone.

How can you learn to teach problem solving? During your next course in mathematics for teachers, a “methods” course, you will spend quite a bit of time answering that question. What follows are some bits of advice that come from this content course. Another fine source is the NCTM standards on problem solving on p. 3 and the Common Core State Standards for Mathematics.

- Set out now to develop a store of interesting and challenging problems appropriate to the skill levels of the students in your class. This process should continue over your professional lifetime. A good start is to keep your present textbook and to use many of the problems you find here.
- Really listen to your students. They often have good ideas. Listening carefully and helping your students to communicate clearly are important traits.
- Don’t be afraid of problems you don’t already know how to solve. Say, “Well, I don’t know. Let’s work together and see what we can come up with.” Indeed, many students will be excited and motivated by the prospect of collaborating with their friends or “working with their teacher.” They also learn not to be afraid to tackle the unknown, to make mistakes, and yet to persevere until a solution is finally achieved.
- Develop a long list of leading questions you can ask of your students and that they can ask themselves to help clarify their thinking and eventually arrive at a solution.

Chapter 1 Summary

Section 1.1 An Introduction to Problem Solving	Page Reference
CONCEPTS	
• **General introduction:** a classroom discussion.	3
STRATEGIES	
• **Guess and check strategy:** Make a guess and check it to see if it satisfies the demands of the problem. If it doesn't, alter the guess appropriately and check again. When the guess finally checks, a solution has been found.	4
• **Make a table:** Make a table reflecting the data in the problem. If done in an orderly way, such a table may reveal patterns and relationships that suggest how the problem can be solved.	4
• **Look for a pattern:** Consider an ordered sequence of particular examples of the general situation described in the problem. Then carefully scrutinize results, looking for a pattern that may be the key to the problem.	4
• **Draw a picture:** Draw a picture that represents the data of the problem as accurately as possible.	4

Section 1.2 Pólya's Problem-Solving Principles and the Common Core State Standards for Mathematics	Page Reference
CONCEPTS	
• **Understand the problem:** Make sure that the words, conditions, and what is to be found are completely understood.	9
• **Devise a plan:** Carefully consider the problem and think of a possible approach to finding a solution.	9
• **Carry out your plan:** If it doesn't work out, try again. If it still doesn't solve the problem, modify the plan and try yet again.	10
• **Look back:** It is important to reexamine your thinking to see what led you to the solution. This is how you gain "mathematical power."	10
• **Problem-solving strategy of "guess and check":** Guessing is a good way to start to gain understanding of a problem, but guessing gives a solution or solutions only if the guess checks.	10
• **Problem-solving strategy of "make an orderly list":** This strategy helps avoid the omission of other possibilities.	13
• **Problem-solving strategy of "draw a diagram":** A diagram often clarifies a problem.	14
STRATEGIES	
• **Pólya's principles:** The four principles that form the basis for any serious attempt at problem solving—understand the problem, devise a plan, carry out the plan, and look back.	8
• **Problem-solving strategies**	
• **Make an orderly list:** For problems that require a consideration of many possibilities, make an orderly list or a table to ensure that no possibilities are missed.	13
• **Draw a diagram:** Draw a diagram or picture that represents the data of the problem as accurately as possible.	14

• **Guess My Rule:** Game where one student makes up a rule for changing one number into another. The other students then call out numbers and the person who made up the rule tells what numbers the rule gives back. The first person in the class to guess the rule then gets to be the person to make up a new rule.	17
• **Standards for Mathematical Practice:** The eight SMP principles of the **Common Core** describe ways that future teachers ". . . increasingly ought to engage with the subject matter as they grow in mathematical maturity and expertise . . ."*	16

Common Core State Standards for Mathematics, p. 8.

Section 1.3 More Problem-Solving Strategies	Page Reference
CONCEPTS	
• **Look for a pattern:** Patterns often suggest what the answer to a problem should be.	19
• **Make a table:** A table helps you search for patterns and eliminate possibilities that do not meet all of the criteria required of a solution.	21
• **Consider special cases (experiment):** What is true in the general case must also be true in a special case. What is true in a special case *suggests* what is true in the general case.	24
• **Solve an easier, similar problem:** Solving an easier, but similar, problem often suggests how to solve the problem at hand.	24
• **Argue from special cases:** An argument made from a special case or similar problem, but depending on general principles (not properties applying only to the special case), often serves to provide a path to prove the general case.	24
DEFINITIONS	
• **Fibonacci number, F_n:** Any number in the Fibonacci sequence.	21
• **Fibonacci sequence:** The sequence 1, 1, 2, 3, 5, 8, 13, 21, . . . , where it starts with 1 and 1 and add any two consecutive terms to obtain the next.	20
• **Pascal's triangle:** A triangular array of numbers in which each number is the sum of the two directly above it. Its patterns are useful in probability and in binomial expansions, as well as other areas of math.	24
STRATEGIES	
• **Look for a pattern:** Consider an ordered sequence of particular examples of the general situation described in the problem. Then carefully scrutinize results, looking for a pattern that may be the key to the problem.	19
• **Make a table:** Make a table reflecting the data in the problem. If done in an orderly way, such a table may reveal patterns and relationships that suggest how the problem can be solved.	21
• **Experiment:** Use the strategy of considering easier, similar problems (special cases) in order to solve a given problem or describe a given pattern.	24
• **Consider special cases:** Consider a sequence of special cases when trying to solve a complex problem. This will often show how to proceed naturally from case to case until arriving at the case in question. Alternatively, the special cases may reveal a pattern that makes it possible to solve the problem.	24

Section 1.4 Algebra as a Problem-Solving Strategy	Page Reference
CONCEPTS	
• **Use a variable:** A symbol (or more than one) can often be used to represent and determine a number that is the answer to a problem. Variables represent quantities that are unknown or can change.	31

• **Algebra as a great explainer:** Algebra is a unifying concept.	31
• **Algebraic expressions:** An algebraic expression is a mathematical expression involving variables, numbers, and operation symbols.	34
• **Algebraic reasoning:** Algebraic reasoning is used to solve problems and understand patterns by following these steps: introduce variables, derive algebraic expressions, form equations, solve equations, and interpret the solution of the equations in the context of the original problem or pattern.	32
• **Equation:** Setting two algebraic expressions that represent the same quantity equal to one another creates an equation.	35
• **Triangular numbers, t_n:** t_n is the total number of dots in a triangle of n rows whose first row has 1 dot, second has 2 dots, third has 3, and so on, until the last row, which has n dots.	33
FORMULA	
• **Gauss' insight:** $2t_n = n(n + 1)$ where the sum of the first n integers is t_n.	32
STRATEGIES	
• **Use a variable or variables:** Use a variable, or variables, when a problem requires a number be determined. Represent the number by a variable, and use the conditions of the problem to set up an equation that can be solved to ascertain the desired number.	31
• **Algebra as a great explainer:** Algebra can solve problems that appear to be quite hard or mathematical tricks. It is used in many levels of mathematics because it clarifies so much.	31

Section 1.5 Additional Problem-Solving Strategies	Page Reference
CONCEPTS	
• **Work backward:** If you can't see how to start a solution, perhaps you can start from the desired conclusion and work backward to the beginning of the problem.	42
• **Eliminate possibilities:** If all possibilities but one can be ruled out, that possibility must be checked. If it works, then it is the answer.	43
• **Use the Pigeonhole Principle:** If you have items to consider that can be placed in different categories, but you have more items than categories, then at least two items must be in some one category. This fact may lead to a solution to your problem.	45
STRATEGIES	
• **Work backward:** Start from the desired result and work backward step-by-step until the initial conditions of the problem are achieved.	42
• **Eliminate possibilities:** Suppose you are guaranteed that a problem has a solution. Use the data of the problem to decide which outcomes are impossible. Then at least one of the possibilities not ruled out must prevail. If all but one possibility can be ruled out, then it must prevail.	43
• **Pigeonhole Principle:** If m pigeons are placed into n pigeonholes and $m > n$, then there must be at least two pigeons in one pigeonhole.	45

Section 1.6 Reasoning Mathematically	Page Reference
CONCEPTS	
• **Inductive reasoning:** This type of reasoning entails drawing a general conclusion on the basis of a consideration of special cases. It does not necessarily yield truth but suggests what may be true. It gives rise to conjectures.	50
• **Representational reasoning:** Various physical, pictorial, or even mental representations often make a problem clearer and make it possible to find a solution.	52

- **Mathematical statements:** These are declarative statements that are either true or false. 54
- **If . . . then:** A statement of the form "If p is true, then q is true." 54
- **Deductive reasoning:** This type of reasoning entails drawing necessary conclusions from given information to arrive at a proof or a solution to a problem. 56
- **Rule of direct reasoning:** If p implies q, then q can be shown to be true by showing that p is true. 56
- **Rule of indirect reasoning:** If p implies q and q is false, then p is false. This is the basis for proof by contradiction. 56

DEFINITIONS

- A **counterexample** is an example that disproves a statement. 51
- A **conjecture** is a generalization that seems to be true but has yet to be proved. 52
- A **theorem** is a conjecture that is given a proof. 52
- A **statement** is a declarative sentence that is either true or false but not both. 54
- A **conditional statement,** or "*if . . . then*" *statement*, is a way to combine two statements. "If p, then q" is true when the truth of statement p guarantees the truth of statement q. 54
- A **premise,** or **hypothesis,** is a list of true statements that we know at the beginning of an argument. 56
- The **conclusion** is the new true statement that we obtain for the argument. 56

STRATEGIES

- **Inductive reasoning:** First, observe a property that holds in several examples. Next, check that the property holds in other examples. In particular, attempt to find an example in which the property does not hold. Then, if the property holds in every example, state a generalization that the property is probably true in general (but there is no guarantee that it is true yet). 50
- **Proof by contradiction:** If we take a statement p and derive a false statement from it, then we know p is false. 56
- **Rule of indirect reasoning:** If a truth table shows $p \rightarrow q$ and q is false, then p is also necessarily false. 56
- **Rule of deductive reasoning:** If there is a collection of true statements, we can argue on the basis of these statements that another statement must also be true. 56

Chapter Review Exercises

Sections 1.1 and 1.2

1. Standard Lumber has 8-foot and 10-foot two-by-fours. If Mr. Zimmermann bought 90 two-by-fours with a total length of 844 feet, how many were 8 feet long? Do the problem twice, using different strategies.

2. **(a)** Using each of 1, 2, 3, 4, 5, 6, 7, 8, and 9 once and only once, fill in the circles in this diagram so that the sum of the three-digit numbers formed is 999:

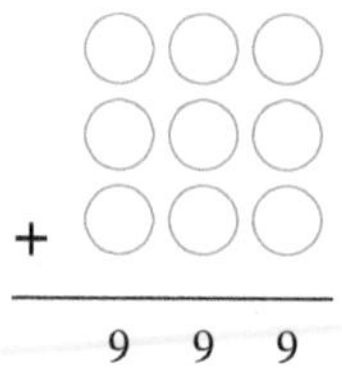

(b) Is there more than one solution to this problem? Explain briefly.

(c) Is there a solution to this problem with the digit 1 not in the hundreds column? Explain briefly.

3. Bill's purchases at the store cost \$4.79. In how many ways can Bill receive change if he pays with a 5-dollar bill?

4. How many three-letter code words can be made by using the letters *a*, *e*, *i*, *o*, and *u* at most once each time?

5. A flower bed measuring 8 feet by 10 feet is bordered by a concrete walk 2 feet wide. What is the area of the concrete walk?

6. Karen is thinking of a number. If you double it and subtract 7, you obtain 11. What is Karen's number?

7. **(a)** Chanty is It in a game of *Guess My Rule*. If you give her a number, she uses her rule to determine another number. The numbers the other students gave Chanty and her responses are as shown. Can you guess her rule?

Student Input	Chanty's Responses
2	8
7	33
4	18
0	−2
3	13
⋮	⋮

(b) Can you suggest a better strategy for the students to use in attempting to determine Chanty's rule? Explain briefly.

Section 1.3

8. Study this sequence: 2, 6, 18, 54, 162, ... Each number is obtained by multiplying the preceding number by 3. The sequences in (a) through (e) are formed in the same way but with a different multiplier. Complete each sequence.

(a) 3, 6, 12, _____, _____, _____

(b) 4, _____, 16, _____, _____, _____

(c) 1, _____, _____, 216, _____, _____

(d) 2, _____, _____, _____, 1250, _____

(e) 7, _____, _____, _____, _____, 7

9. Because of the high cost of living, Kimberly, Terry, and Otis each hold down two jobs, but no two have the same occupation. The occupations are doctor, engineer, teacher, lawyer, writer, and painter. Given the following information, determine the occupations of each individual:

(a) The doctor had lunch with the teacher.

(b) The teacher went fishing with Kimberly, who is not the writer.

(c) The painter is related to the engineer.

(d) The doctor hired the painter to do a job.

(e) Terry lives next door to the writer.

(f) Otis beat Terry and the painter at tennis.

(g) Otis is not the doctor.

10. **(a)** Write down the next three rows to continue this sequence of equations.

$$2 = 1^3 + 1$$
$$4 + 6 = 2^3 + 2$$
$$8 + 10 + 12 = 3^3 + 3$$

____________ = _____

____________ = _____

____________ = _____

(b) Write down the 10th row in the sequence in part (a).

11. Consider a circle divided by n chords in such a way that every chord intersects every other chord interior to the circle and no three chords intersect in a common point. Complete the table that follows and answer these questions:

(a) Into how many regions is the circle divided by the chords?

(b) How many points of intersection are there?

(c) Into how many segments do the chords divide one another?

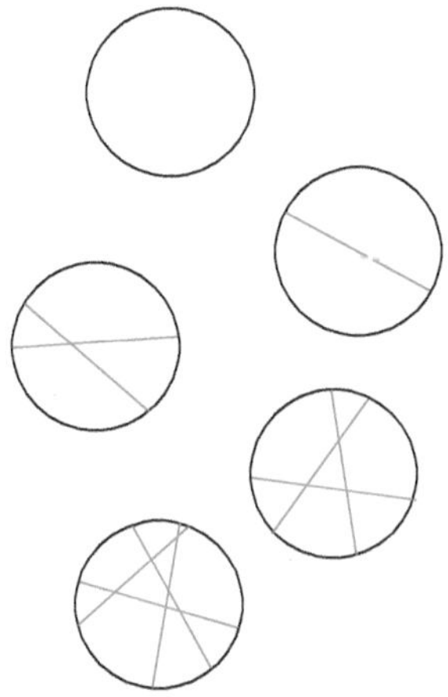

Number of Chords	Number of Regions	Number of Intersections	Number of Segments
0	1	0	0
1	2	0	1
2	4	1	4
3			
4			
5			
6			
⋮			
n			

Section 1.4

12. Bernie weighs 90 pounds plus half his own weight. How much does Bernie weigh?

13. A "hex-square train" is made with toothpicks, alternating between hexagonal and square "cars" in the train. For example, following are three-car and four-car hex-square trains:

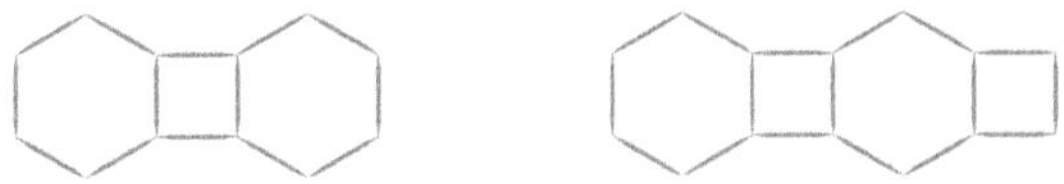

(a) Evidently, 14 toothpicks are required to form a three-car train. Give a formula for the number of toothpicks required to form an n-car train. (*Suggestion:* Take separate cases for trains with an odd or an even number of cars.)

(b) A certain hex-square train requires 102 toothpicks. How many hexagons and how many squares are in the train?

14. The sum of two real numbers is 7 and their difference is 9. Use the problem-solving strategy to solve for the numbers.

(a) Guess and check.

(b) Use a variable.

Section 1.5

15. How many cards must be drawn from a standard deck of 52 playing cards to be sure that

(a) at least two are of the same suit?

(b) at least three are of the same suit?

(c) at least two are aces?

16. How many books must you choose from among a collection of 7 mathematics books, 18 books of short stories, 12 chemistry books, and 11 physics books to be certain that you have at least 5 books of the same type?

Section 1.6

17. (a) Compute these products:

$67 \times 67 =$ _______

$667 \times 667 =$ _______

$6667 \times 6667 =$ _______

(b) Guess the result of multiplying 6,666,667 by itself. Are you sure your guess is correct? Explain in one *carefully* written sentence.

18. (a) Compute these products:

$1 \times 142{,}857 =$ _______

$2 \times 142{,}857 =$ _______

$3 \times 142{,}857 =$ _______

$4 \times 142{,}857 =$ _______

$5 \times 142{,}857 =$ _______

(b) Predict the product of 6 and 142,857. Now calculate the product and see if your prediction was correct.

(c) Predict the result of multiplying 7 times 142,857, and then compute this product.

(d) What does part (c) suggest about apparent patterns? Explain.

19. In Example 1.20, we used representations to show that the square of an odd number is one more than a multiple of 8. Use the fact from Example 1.20 that, if n is odd, then $n = 2s + 1$ for some whole number s and $n^2 = 4(s^2 + s) + 1$. Note that $s^2 + s = s(s + 1) = 2\frac{s(s + 1)}{2}$. Argue that $\frac{s(s + 1)}{2}$ is a whole number and, hence, that $n^2 = 8q + 1$ for some whole number q. That is, again, n^2 is one more than a multiple of 8.

20. In this section, we saw that the sum of the digits of a multiple of 3 is also a multiple of 3. Is the same true of multiples of 6? Why or why not?

2 Sets and Whole Numbers

COOPERATIVE INVESTIGATION
Counting Cars and Trains

Materials Needed

A set of Cuisenaire® rods (or colored number strips) for each cooperative group of three or four students. The rods, which have lengths from 1 to 10 centimeters, are color coded as follows:

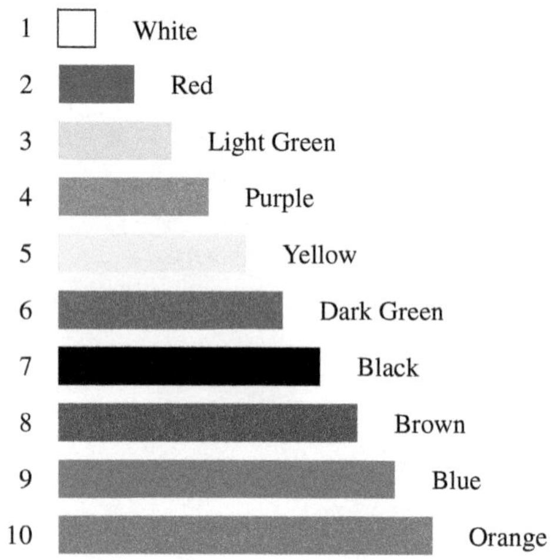

Directions

Form a train by placing one or more rods end to end; each rod in a train is a car. For example, in the figure to the right column there are four ways to form trains that have the same overall length as the light-green (LG) rod.

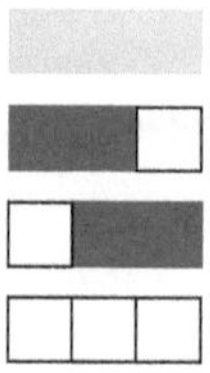

The order in which the cars appear is taken into account, so we consider the red–white and white–red trains as two different trains. (Imagine that the engine is the rightmost car and the caboose is the leftmost car.)

Questions to Consider

1. How many trains can you form that have the same length as the purple rod?
2. How many trains can you form that have the same length as the yellow rod?
3. What pattern do you observe in the number of trains of length 3, 4, and 5? On the basis of the pattern, predict the number of trains of length 8.
4. How many trains can be formed of length n, where n is any whole number and you use cars of lengths l up to n? Can you justify your conjecture? Can you give more than one justification?
5. A train made up of only red and white cars is called an RW-train. Answer questions 1 through 4 for RW-trains.
6. Call a train with no white cars a $\overline{\text{W}}$-train. Answer questions 1 through 4 for $\overline{\text{W}}$-trains.
7. (**Writing**) Write two additional questions that could be asked and then give solutions to them.

In the Cooperative Investigation activity above, you will start by trying a few examples (in questions 1, 2, and 3) and then decide (in question 4) whether a pattern emerges to solve the general case. You will see that experimenting with $n = 3, 4, 5$, and 8 gives insights into the solution of question 4 for any train of length n. What about others?

CHAPTER PREVIEW

The notion of a **set** (a collection of objects) is introduced in this chapter as the primary way of describing **whole numbers** such as 0, 1, 2, 3, ... We will see that operations on sets form the basis of fundamental arithmetic operations on whole numbers, such as addition, subtraction, multiplication, and division. Because each child learns differently, we give a number of models of addition, subtraction, multiplication, and division of whole numbers and look at their properties through these models.

KEY IDEAS

- Sets and operations on sets, such as union and intersection
- The relationship between sets and whole numbers
- Modeling whole-number addition through the use of sets or through measurement on the number line
- Subtraction and its models
- Various models of multiplication and division of whole numbers
- Properties of whole-number multiplication and division

2.1 Sets and Operations on Sets

The notion of a set originated with the German mathematician Georg Cantor (1845–1918) in the last half of the nineteenth century. Sets have now become indispensable to nearly every branch of mathematics. Sets make it possible to organize, classify, describe, and communicate. For example, each number system—the whole numbers, the integers, the rational numbers, and the real numbers—is best viewed as a set together with a list of the operations on the numbers and the properties that these operations obey.

Intuitively, a set is a collection of objects. An object that belongs to the collection is called an **element** or **member** of the set. Words like *collection, family,* and *class* are frequently used interchangeably with "set." In fact, a set of dishes, a collection of stamps, and the class of 2014 at your high school are all examples of sets.

Cantor requires that a set be *well defined.* This means two things. First, there is a **universe** of objects that are allowed into consideration. Second, there are only two choices for each object. Any object in the universe either is an element of the set or is not an element of the set. For example, "the first few presidents of the United States" is not a well-defined set, since "few" is a matter of varying opinion. By contrast, "the first three presidents of the United States" does provide an adequate verbal description of a set, with the understanding that the universe is all people who have ever lived.

There are three ways to define a set:

Word Description:	The set of the first three presidents of the United States
Listing in Braces:	{George Washington, Thomas Jefferson, John Adams}
Set-Builder Notation:	$\{x \mid x$ is one of the first three presidents of the United States$\}$

The last expression is read "the set of all x such that x is one of the first three presidents of the United States." More generally, set-builder notation, $\{x \mid x \ldots\}$, is read "the set of all x such that x is . . . ."

The order in which elements in a set are listed is arbitrary, so listing Jefferson, the third president, before Adams is permissible. However, each element should be listed just once. The letter x used in set-builder notation can be replaced with any convenient letter. The letter x is a variable representing any member of the universe, and the set contains exactly those members of the universe that meet the defining criteria listed to the right of the vertical bar. In some books, a colon replaces the vertical bar.

Capital letters $A, B, C, \ldots$ are generally used to denote sets. Membership is symbolized by $\in$, so that if P designates the preceding set of three U.S. presidents, then John Adams $\in P$, where $\in$ is read "is a member of" or "is an element of." The symbol $\notin$ is read "is not a member of," so James Monroe $\notin P$. It is useful to choose letters that suggest the set being designated.

One of the very useful notions in this chapter is that of the positive integers, or natural numbers. We formalize the concept with a definition using set notation.

DEFINITION Natural Numbers

A **natural number,** or **counting number,** is a member of the set $N = \{1, 2, 3, \ldots\}$,

where the ellipsis ". . ." indicates "and so on."

EXAMPLE 2.1 Describing Sets

Each set that follows is taken from the universe N of the natural numbers and is described in words, by listing the set in braces, or with set-builder notation. Provide the two remaining types of description for each set.

(a) The set of natural numbers greater than 12 and less than 17
(b) $\{x \mid x = 2n \text{ and } n = 1, 2, 3, 4, 5\}$
(c) $\{3, 6, 9, 12, \ldots\}$

(d) The set of the first ten odd natural numbers
(e) $\{1, 3, 5, 7, \ldots\}$
(f) $\{x \mid x = n^2 \text{ and } n \in N\}$

Solution

(a) $\{13, 14, 15, 16\}$; listing
$\{n \mid n \in N \text{ and } 12 < n < 17\}$; set builder
(b) $\{2, 4, 6, 8, 10\}$; listing
The set of the first five even natural numbers; word description
(c) The set of all natural numbers that are multiples of 3; word description
$\{x \mid x = 3n \text{ and } n \in N\}$; set builder
(d) $\{1, 3, 5, 7, 9, 11, 13, 15, 17, 19\}$
$\{x \mid x = 2n - 1 \text{ and } n = 1, 2, \ldots, 10\}$; set builder
(e) The set of the odd natural numbers; word description
$\{x \mid x = 2n - 1 \text{ and } n \in N\}$; set builder
(f) $\{1, 4, 9, 16, 25, \ldots\}$; listing
The set of the squares of the natural numbers; word description

Venn Diagrams

A Venn diagram is an example of how a representation can aid the teaching and understanding of a concept. The Representation Standard is discussed in detail in the NCTM's *Principles and Standards for School Mathematics.*

Sets can be represented pictorially by **Venn diagrams,** named for the English logician John Venn (1834–1923). The universal set, which we denote by U, is usually represented by a rectangle. Any set within the universe is represented by a closed loop lying within the rectangle. The region inside the loop is associated with the elements in the set. An example is given in Figure 2.1, which shows the Venn diagram for the set of vowels $V = \{a, e, i, o, u\}$ in the universe $U = \{a, b, c, \ldots, z\}$.

Figure 2.1
The Venn diagram showing the set of vowels in the universe of the 26-letter alphabet

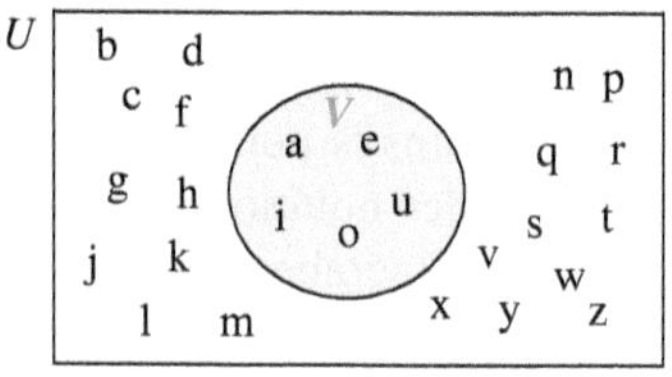

Coloring, shading, and cross-hatching are also useful devices to distinguish the sets in a Venn diagram. For example, set A is blue in Figure 2.2, and the set of elements of the universe that do not belong to A is colored red.

Figure 2.2
The Venn diagram of set A (shown in blue), its complement $\overline{A}$ (shown in red), and the universal set U (the region within the rectangle)

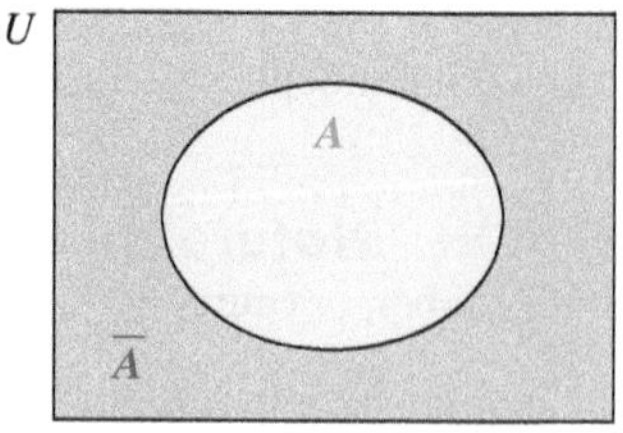

The elements in the universe that are not in set A form a set called the **complement** of A, which is written $\overline{A}$.

DEFINITION The Complement of Set *A*
The **complement of set *A*,** written $\overline{A}$, is the set of elements in the universal set U that are not elements of A. That is,

$$\overline{A} = \{x \mid x \in U \text{ and } x \notin A\}.$$

EXAMPLE 2.2

Finding Set Complements

Let $U = N = \{1, 2, 3, \ldots\}$ be the set of natural numbers. For the following sets E and F, find the complementary sets $\overline{E}$ and $\overline{F}$:

(a) $E = \{2, 4, 6, \ldots\}$
(b) $F = \{n \mid n > 10\}$

Solution

(a) $\overline{E} = \{1, 3, 5, \ldots\}$. That is, the complement of the set E of even natural numbers is the set $\overline{E}$ of odd natural numbers.
(b) $\overline{F} = \{1, 2, 3, 4, 5, 6, 7, 8, 9, 10\}$

Relationships and Operations on Sets

Consider several sets, labeled $A, B, C, D, \ldots$, whose members all belong to the same universal set U. It is useful to understand how sets may be related to one another and how two or more sets can be used to define new sets.

DEFINITION Subset
The set A is a **subset** of B, written $A \subseteq B$, if, and only if, every element of A is also an element of B.

If A is a subset of B, then every element of A also belongs to B. In this case, it is useful in the Venn diagram to place the loop representing set A within the loop representing set B, as shown in Figure 2.3.

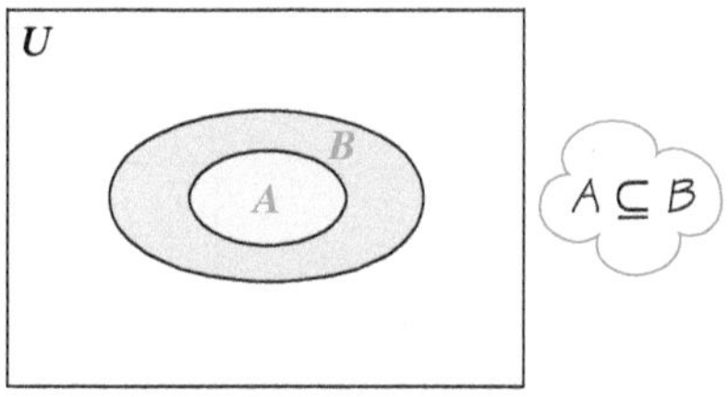

Figure 2.3
A Venn diagram when it is known that A is a subset of B

DEFINITION Equal Sets and Proper Subset
If two sets A and B have precisely the same elements, then they are **equal** and we write $A = B$. If $A \subseteq B$, but $A \neq B$, we say that A is a **proper subset** of B and write $A \subset B$.

If $A \subset B$, there must be some element of B that is not also an element of A; that is, there is some x for which $x \in B$ and $x \notin A$. The set of all women senators is a proper subset of the set of all senators.

It is useful to have some terminology and notation for a set with no elements in it. For example, the set of all ten-year-olds in the United States who vote in presidential elections is the empty set (or, more informally, empty).

DEFINITION Empty Set
A set that has no elements in it is called the **empty set** and is written $\varnothing$, similar to (but not the same as) the Greek letter phi.

DEFINITION Intersection of Sets
The **intersection** of two sets A and B, written $A \cap B$, is the set of elements common to both A and B. That is,

$$A \cap B = \{x \mid x \in A \text{ and } x \in B\}.$$

For example, {a, b, c, d} $\cap$ {a, d, e, f} = {a, d}. The symbol $\cap$ is a special mathematical symbol called a **cap.** (See Figure 2.4a.)

DEFINITION Disjoint

Two sets C and D are **disjoint** if C and D have no elements in common. That is, "C and D are disjoint" means that $C \cap D = \varnothing$. (See Figure 2.4b.)

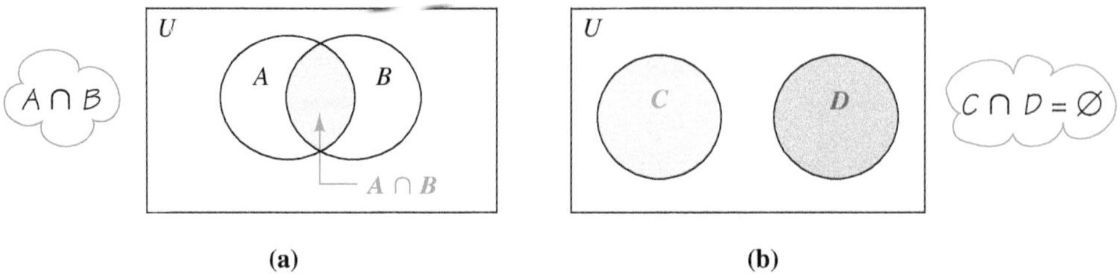

Figure 2.4
(a) A Venn diagram whose shaded region shows the intersection $A \cap B$ **(b)** A Venn diagram for sets C and D that are disjoint. That is, $C \cap D = \varnothing$

DEFINITION Union of Sets

The **union** of sets A and B, written $A \cup B$, is the set of all elements that are in A or B. That is,

$$A \cup B = \{x \mid x \in A \text{ or } x \in B\}.$$

For example, if A = {a, e, i, o, u} and B = {a, b, c, d, e}, then $A \cup B$ = {a, b, c, d, e, i, o, u}. Elements such as "a" and "e" that belong to both A and B are listed just once in $A \cup B$. The word *or* in the definition of union is used in the inclusive sense of "and/or." The symbol for union is the **cup,** $\cup$. The cup symbol must be carefully distinguished from the letter U used to denote the universal set. (See Figure 2.5.)

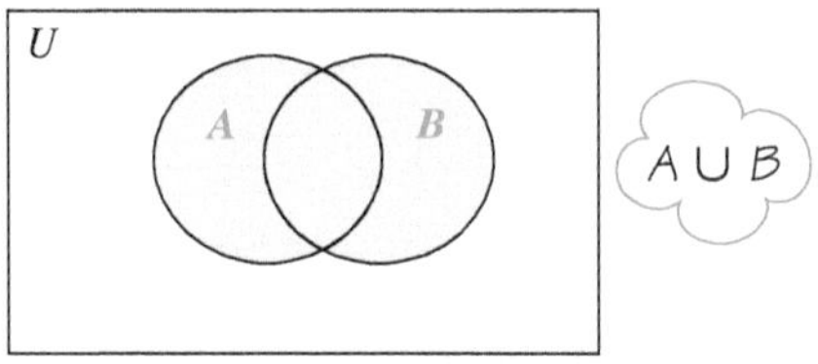

Figure 2.5
The shaded region corresponds to the union, $A \cup B$, of sets A and B

EXAMPLE 2.3

Performing Operations on Sets

Let U = {p, q, r, s, t, u, v, w, x, y} be the universe, and let A = {p, q, r}, B = {q, r, s, t, u}, and C = {r, u, w, y}. Locate all ten elements of U in a three-loop Venn diagram, and then find the following sets:

(a) $A \cup C$ **(b)** $A \cap C$ **(c)** $A \cup B$ **(d)** $A \cap B$
(e) $\overline{B}$ **(f)** $\overline{C}$ **(g)** $A \cup \overline{B}$ **(h)** $A \cap \overline{C}$

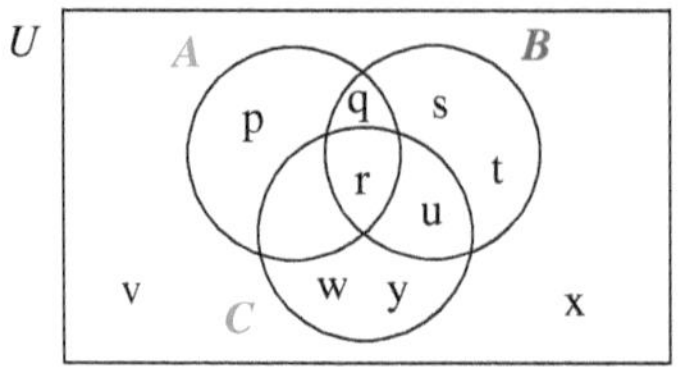

Solution

(a) $A \cup C$ = {p, q, r, u, w, y}
(b) $A \cap C$ = {r}
(c) $A \cup B$ = {p, q, r, s, t, u}
(d) $A \cap B$ = {q, r}
(e) $\overline{B}$ = {p, v, w, x, y}
(f) $\overline{C}$ = {p, q, s, t, v, x}
(g) $A \cup \overline{B}$ = {p, q, r, v, w, x, y}
(h) $A \cap \overline{C}$ = {p, q}

Using Sets for Problem Solving

The notions of sets and their operations are often used to understand a problem and communicate its solution. Moreover, Venn diagrams provide a visual representation that is useful for understanding and communication.

EXAMPLE 2.4

Using Sets to Solve a Problem in Color Graphics

The RGB color model has red, green, and blue light added together in various ways to reproduce a broad array of colors. The name of the model comes from the initials of the three additive primary colors: red, green, and blue.

The main purpose of the RGB color model is for the sensing, representing, and displaying of images in electronic systems, such as televisions and computers, though it has also been used in conventional photography. Use a Venn diagram to show which colors result.

Solution

Introduce three loops in a Venn diagram—one for each of the component colors. As shown here, eight colors can be achieved. Most computers also allow the intensity of each color to be specified. In this way, many more colors can be obtained.

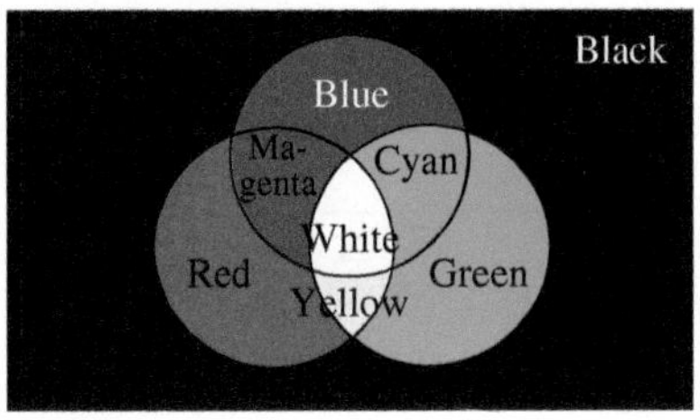

Notice that if red and green are excited, including blue will produce white. Similarly, if red is excited, including both green and blue will again produce white. In symbols, this relationship becomes

$$(R \cap G) \cap B = R \cap (G \cap B).$$

Since it does not matter where the parentheses are placed, this equation illustrates what is called the **associative property** of the intersection operation. Because the order in which the intersections are performed has no effect on the outcome, we do not need parentheses, and it is meaningful to write $R \cap G \cap B$. Similarly, union is associative, and it is meaningful to write $R \cup G \cup B$, without parentheses.

The associative property is just one example of a number of useful properties that hold for set operations and relations. The properties listed in the theorem that follows can be proved by reasoning directly from the definitions given earlier. They can also be justified by considering appropriately shaded Venn diagrams.

THEOREM Properties of Set Operations and Relations

1. **Transitivity** of inclusion

$$\text{If } A \subseteq B \text{ and } B \subseteq C, \text{ then } A \subseteq C$$

2. **Commutativity** of union and intersection

$$A \cup B = B \cup A$$
$$A \cap B = B \cap A$$

3. **Associativity** of union and intersection

$$A \cup (B \cup C) = (A \cup B) \cup C$$
$$A \cap (B \cap C) = (A \cap B) \cap C$$

4. Properties of the **empty set**

$$A \cup \varnothing = \varnothing \cup A = A$$
$$A \cap \varnothing = \varnothing \cap A = \varnothing$$

5. **Distributive property** of union and intersection

$$A \cap (B \cup C) = (A \cap B) \cup (A \cap C)$$
$$A \cup (B \cap C) = (A \cup B) \cap (A \cup C)$$

EXAMPLE 2.5

Verifying Properties with Venn Diagrams

(a) Verify the distributive property $A \cap (B \cup C) = (A \cap B) \cup (A \cap C)$ (first part of number 5 in preceding theorem).

(b) Show that $A \cup B \cap C$ is not meaningful without parentheses.

Solution

(a) To shade the region corresponding to $A \cap (B \cup C)$, we intersect the A loop with the loop formed by the overlapping circles of $B \cup C$:

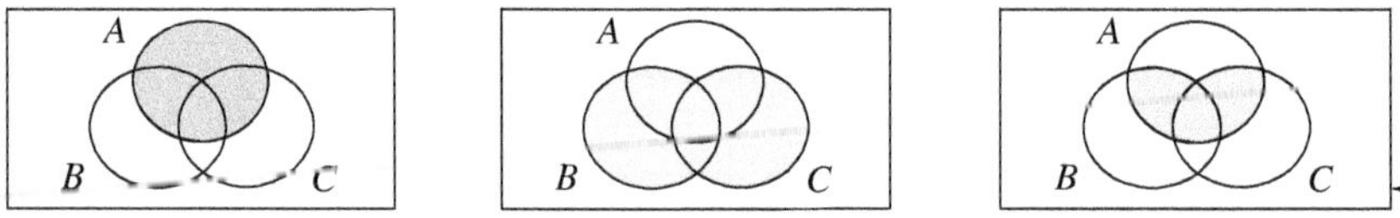

$A \quad \cap \quad (B \cup C) \quad = \quad A \cap (B \cup C)$

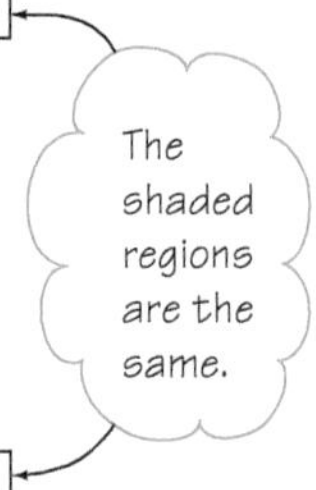

Next, we combine the two almond-shaped regions $A \cap B$ and $A \cap C$ to form $(A \cap B) \cup (A \cap C)$:

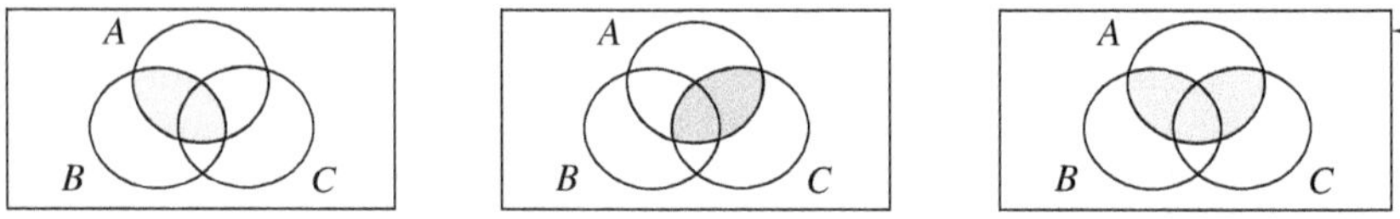

$(A \cap B) \quad \cup \quad (A \cap C) \quad = \quad (A \cap B) \cup (A \cap C)$

The shaded regions at the right agree, which verifies that intersection distributes over union; that is,

$$A \cap (B \cup C) = (A \cap B) \cup (A \cap C).$$

(b) The request in Example 2.5(b) means that, without specifying the order of the operations $\cap$ and $\cup$, the symbol $A \cup B \cap C$ makes no sense. If the union is taken first, we have the expression $(A \cup B) \cap C$, corresponding to the region shaded on the left in the following figure. If the intersection is taken first, we have the expression $A \cup (B \cap C)$, corresponding to the region shaded in the Venn diagram on the right. The shaded regions are different, so in general $(A \cup B) \cap C \neq A \cup (B \cap C)$.

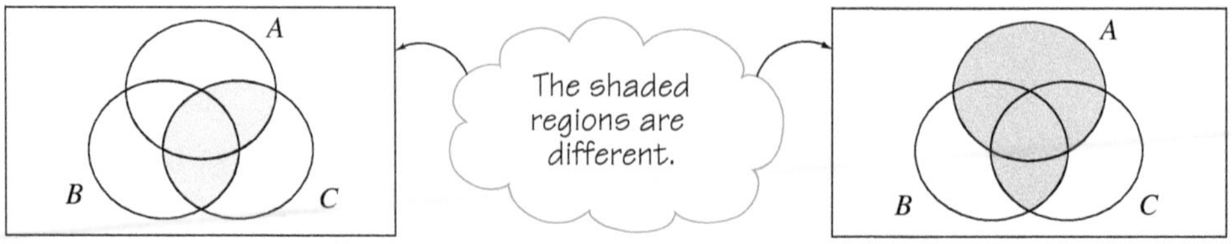

$(A \cup B) \cap C \qquad\qquad A \cup (B \cap C)$

We can also consider the particular case $A = \{a\}, B = \{a, b\}, C = \{b, c\}$, for which $(A \cup B) \cap C = \{b\}$ and $A \cup (B \cap C) = \{a, b\}$. We see that $(A \cup B) \cap C \neq A \cup (B \cap C)$, since $\{b\} \neq \{a, b\}$. An example showing that a statement is false is called a **counterexample.** Examples may shed some light as to why a statement *might* be true. A proof would show that it is always true. A counterexample shows that a statement is false.

Venn diagrams are part of the mathematics of elementary school, as we see from the State Student Assessment problems in this section and the next. More generally, sets appear in such diverse areas as digital photography, visual arts, and color graphics (Example 2.4). In art, the nineteenth-century technique of oil painting called pointillism consists of constructing the painting through the use of points of different colors; that is, objects, which are really subsets of the painting, can be identified as a collection of points, all of the same color. Seurat's painting *Sunday Afternoon on the Island of La Grande Jatte* is a beautiful example of this technique.

2.1 Problem Set

Understanding Concepts

Problem numbers in color indicate that the answer to that problem can be found at the back of the book.

1. Write the following sets by listing their elements:
- **(a)** The set of states in the United States that border Nevada
- **(b)** The set of states in the United States whose names begin with the letter M
- **(c)** The set of states in the United States whose names contain the letter Z

2. Write the following sets:
- **(a)** The set of letters used in the sentence "List the elements in a set only once."
- **(b)** The set of letters that are needed to spell these words: *team, meat, mate,* and *tame*
- **(c)** The set of states in the United States that contain the letter Q

3. Let $U = \{1, 2, 3, \ldots, 20\}$. Write the following sets by listing the elements in braces:
- **(a)** $\{x \in U \mid 6 < x \leq 13\}$
- **(b)** $\{x \in U \mid x \text{ is odd and } 9 \leq x \leq 13\}$
- **(c)** $\{x \in U \mid x = 2n \text{ for some } n \in N\}$

4. Let $U = \{1, 2, 3, \ldots, 20\}$. Write the following sets by listing the elements in braces:
- **(a)** $\{x \in U \mid x \text{ is divisible by } 4\}$
- **(b)** $\{x \in U \mid x = 2n + 1 \text{ for some } n \in N\}$
- **(c)** $\{x \in U \mid x = n^2 \text{ for some } n \in N\}$

5. Write these sets in set-builder notation, where $U = \{1, 2, \ldots, 20\}$.
- **(a)** $\{11, 12, 13, 14\}$
- **(b)** $\{6, 8, 10, 12, 16\}$
- **(c)** $\{4, 8, 12, 16, 20\}$
- **(d)** $\{2, 5, 10, 17\}$

6. Use set-builder notation to write the following subsets of the natural numbers:
- **(a)** The even natural numbers larger than 12
- **(b)** The squares of the odd numbers larger than or equal to 25
- **(c)** The natural numbers divisible by 3

7. Decide whether the following statements asserting set relationships are *true* or *false:*
- **(a)** $\{s, c, r, a, m, b, l, e, d\} = \{a, b, c, d, e, l, m, r, s\}$
- **(b)** $\{6\} \subseteq \{6, 7, 8\}$
- **(c)** $\{6, 7, 23\} = \{7, 23, 6\}$

8. Decide whether the following statements asserting set relationships are *true* or *false*:
- **(a)** $\{7\} \subset \{6, 7, 23\}$
- **(b)** $\{6, 7, 23\} \subset \{7, 23, 6\}$
- **(c)** $\{6\} \subseteq \{7\}$

9. Let $U = \{a, b, c, d, e, f, g, h\}$, $A = \{a, b, c, d, e\}$, $B = \{a, b, c\}$, and $C = \{a, b, h\}$. Locate all eight elements of U in a three-loop Venn diagram, and then list the elements in the following sets:
- **(a)** $B \cup C$
- **(b)** $A \cap B$
- **(c)** $B \cap C$
- **(d)** $A \cup B$
- **(e)** $\overline{A}$
- **(f)** $A \cap C$
- **(g)** $A \cup (B \cap C)$

10. Let $L = \{6, 12, 18, 24, \ldots\}$ be the set of multiples of 6, and let $M = \{45, 90, 135, \ldots\}$ be the set of multiples of 45.
- **(a)** Find four more elements of set M.
- **(b)** Describe $L \cap M$.
- **(c)** What is the smallest element in $L \cap M$?

11. Let $G = \{n \mid n \text{ divides } 90\}$ and $D = \{n \mid n \text{ divides } 144\}$. In listed form, $G = \{1, 2, 3, 5, 6, 9, 10, 15, 18, 30, 45, 90\}$.

(a) Find the listed form of the set D.

(b) Find $G \cap D$.

(c) Which element of $G \cap D$ is largest?

12. Draw and shade Venn diagrams that correspond to the following sets:

(a) $A \cap B \cap C$ **(b)** $A \cup (B \cap \overline{C})$

(c) $(A \cap B) \cup C$ **(d)** $\overline{A} \cup (B \cap C)$

(e) $A \cup B \cup C$ **(f)** $\overline{A} \cap B \cap C$

13. For each part, draw a Venn diagram whose loops for sets A, B, and C show that the conditions listed must hold.

(a) $A \subseteq C, B \subseteq C, A \cap B = \varnothing$

(b) $C \subseteq (A \cap B)$

(c) $(A \cap B) \subseteq C$

14. If $A \cup B = A \cup C$, is it necessarily true that $B = C$? Give a proof or provide a counterexample.

15. If $D \cap F = D \cap G$, is it necessarily true that $F = G$? Explain why or provide a counterexample.

16. Let U be the set of natural numbers 1 through 20, let A be the set of even numbers in U, and let B be the set of numbers in U that are divisible by 3.

(a) Find $\overline{A \cap B}, \overline{A} \cup \overline{B}, \overline{A \cup B}$, and $\overline{A} \cap \overline{B}$.

(b) What two pairs of the four sets of part (a) are equal?

17. The 12 shapes in the Venn diagram shown are described by the following attributes:

Shape: circle, hexagon, or triangle

Size: small or large

Color: red or blue

Let C, H, and T denote the respective sets of circular, hexagonal, and triangular shapes. Similarly, let S, L, R, and B denote the sets of small, large, red, and blue shapes, respectively. The set A contains the shapes that are small and not a triangle, so $A = S \cap \overline{T}$.

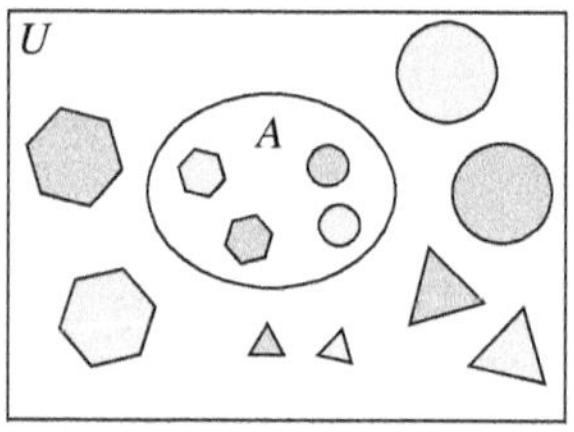

Describe which shapes are found in these sets:

(a) $R \cap C$ **(b)** $L \cap H$

(c) $T \cup H$ **(d)** $L \cap T$

(e) $B \cap \overline{C}$ **(f)** $H \cap S \cap R$

18. Let C, H, T, S, L, R, and B denote the sets of attribute shapes described in problem 17. The set of small hexagons (six-sided figures) can be written $S \cap H$ in symbolic form. Express the following sets in symbolic form:

(a) The large triangles

(b) The blue polygonal (i.e., noncircular) shapes

(c) The small or triangular shapes

(d) The shapes that either are blue circles or are red

Into the Classroom

19. A large cooperative activity can effectively teach set concepts to children. For example, let the classroom be the universe of the students. Use lengths of rope (about 30 feet, adjusting for the size of the class) to form Venn diagram loops on the floor. Place signs in the loops to identify the sets being considered. Once the signs are placed, have the children stand in the appropriate loop or in overlapping regions of two or more loops, depending on the defining property of the set. For example, if loop A corresponds to the girls in the class, then the girls should stand inside the loop and the boys should stand outside the loop. Now create other interesting classroom Venn diagram activities.

(a) Define sets B and C for which B is a subset of C.

(b) Define sets D and E that are disjoint.

(c) Define sets F, G, and H in a way that you believe there is a good likelihood that students will be found in all eight regions of the classroom Venn diagram.

20. Attribute cards can be made by drawing different shapes on cards, varying the figure shown, the color used, and so on. Here are some examples of a few attribute cards:

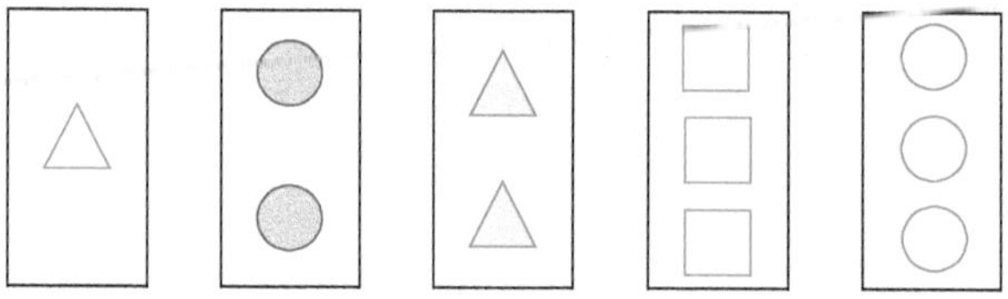

(a) Conjecture how many cards make a full deck. Explain how you got your answer.

(b) Form groups of three to five students. Each cooperative group is to design and make a set of attribute cards, drawing simple figures on small rectangles of card stock with colored pens.

(c) Exchange decks of attribute cards among the cooperative groups. Shuffle each deck and turn over just a few of the cards. Can each group predict the number of cards in the complete deck? Carefully explain the reasoning used.

21. **(Writing)** Many of the terms introduced in this section also have meaning in nonmathematical contexts. For example, *complement*, *union*, and *intersection* are used in ordinary speech. Other terms, such as *transitive*, *commutative*, *associative*, and *distributive*, are rare in everyday usage, but there are closely related words that are quite common, such as *transit*, *commute*, *associate*, and *distribute*. Discuss the differences and similarities in the meanings of the words and word roots of the terms introduced in this section.

Responding to Students

22. A student is told that $A \cup B$ is the set of all elements that belong to A or to B. However, when asked to find the union of $A = \{a, b, c, d\}$, and $B = \{c, d, e, f, g, h\}$, the student says that the answer is $\{a, b, e, f, g, h\}$. How is the word *or* in the definition of set union being misunderstood? Give the student an everyday example where *or* is used in the mathematical sense.

23. Students are told that the set of natural numbers (N) is a subset of the set of whole numbers (W). Have students represent this relationship with a Venn diagram and then explain it in words and symbols. Would your students think that $0 \in N$? Zero will be continue to be a source of confusion as children begin to develop mathematical number sense.

24. Joe, a fifth grader, asked why the union of any two sets (such as two chairs and three umbrellas meaning five, or six girls and four boys meaning ten) means addition. Joe says that he always gets the answer easily. How would you explain to Joe why his approach does or does not work?

Thinking Critically

25. Circular loops in a Venn diagram divide the universe U into distinct regions.

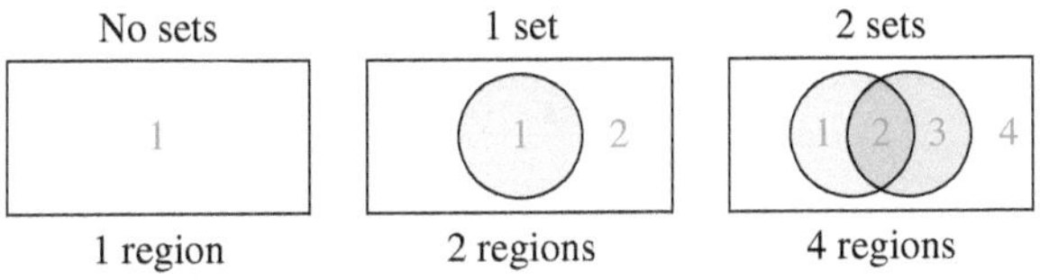

(a) Draw a diagram with three circles that gives the largest number of regions.

(b) How many regions do the four circles define in this figure?

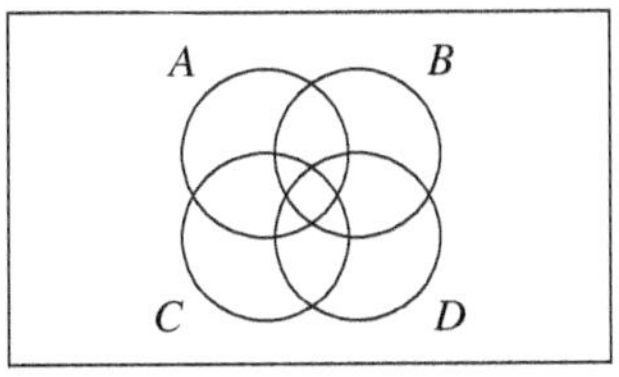

(c) Verify that the four-circle diagram is missing a region corresponding to $A \cap \overline{B} \cap \overline{C} \cap D$. What other set has no corresponding region?

(d) Will this Venn diagram allow for all possible combinations of four sets? Explain briefly but carefully.

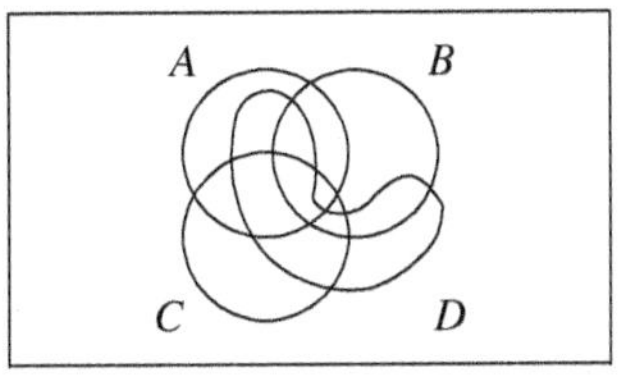

26. The English mathematician Augustus DeMorgan (1806–1871) showed that **(a)** $\overline{A \cap B} = \overline{A} \cup \overline{B}$ and **(b)** $\overline{A \cup B} = \overline{A} \cap \overline{B}$. These identities are now called DeMorgan's laws. Use a sequence of Venn diagrams to explain, in words and pictures, how to shade the regions corresponding to $\overline{A \cap B}$ and $\overline{A} \cup \overline{B}$. The final shaded regions obtained should be identical, justifying DeMorgan's first law.

27. Carefully explain how DeMorgan's second law (from problem 26b) could be understood by a fifth grader.

28. Each of the objects depicted in problem 17 is described by its attributes: shape (three choices), size (two choices), and color (two choices). The number of pieces in a full set of shapes can be varied by changing the number of attributes and the number of possible choices of an attribute. How many pieces are contained in the following attribute sets?

(a) Shape: circle, hexagon, equilateral triangle, isosceles right triangle, rectangle, square
Size: large, small
Color: red, blue

(b) Use the attributes in part **(a)** and also include this attribute:
Thickness: thick, thin

Making Connections

29. If a penny (P) and a nickel (N) are flipped, the possible outcomes can be listed by forming all of the subsets that could represent heads. There are four possible outcomes. $\varnothing$, {P}, {N}, and {P, N}. For example, $\varnothing$ is the outcome for which both coins land on tails.

(a) List the possible outcomes if a penny, nickel, and dime (D) are flipped. How many outcomes (i.e., subsets of {P, N, D}) did you find?

(b) List the subsets of {P, N, D, Q}, where Q represents a quarter. How many subsets did you find?

(c) How many of the subsets {P, N, D, Q} contain Q? How many do not contain Q?

(d) Use inductive reasoning to describe the number of subsets of a set with n elements.

30. **The ABO System of Blood Typing.** Until the beginning of the twentieth century, it was assumed that all human blood was identical. About 1900, however, the Austrian-American pathologist Karl Landsteiner discovered that blood could be classified into four groups according to the presence of proteins called antigens. This discovery made it possible to transfuse blood safely. A person with the antigens A, B, or both A and B has the respective blood type A, B, or AB. If neither antigen is present, the type is O. Draw and label a Venn diagram that illustrates the ABO system.

31. **The Rh System of Blood Typing.** In 1940, Karl Landsteiner (see problem 30) and the American pathologist Alexander Wiener discovered another protein that coats the red blood cells of some persons. Since the initial research was on rhesus monkeys, a person with the protein is classified as Rh positive (Rh+) and a person whose blood cells lack the protein is Rh negative (Rh−). Draw and label a Venn diagram illustrating the classification of blood into the eight major types: A+, A−, B+, B−, AB+, AB−, O+, and O−.

State Assessments

32. (Washington State, Grade 4)
Ms. Yonan took a survey in her class to see how many students like hamburgers, pizza, or hot dogs. The results of the survey are shown in the chart.

Number of Students Who Like		
Only 1 of These Foods	**Only 2 of These Foods**	**All 3 of These Foods**
14	10	4

Look at the following three Venn diagrams. Choose the diagram that best represents the results of this survey.

A.

B.

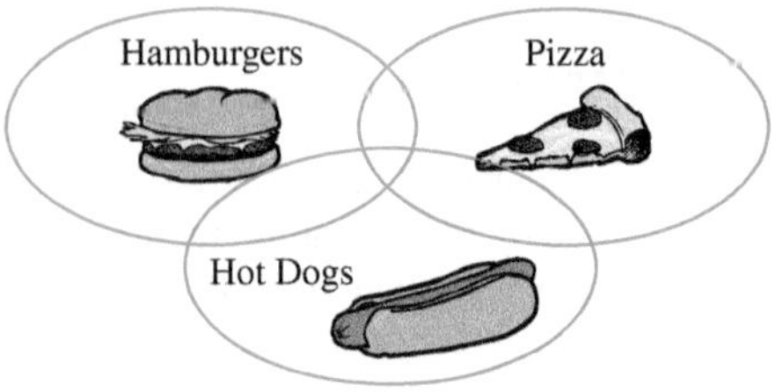

C.

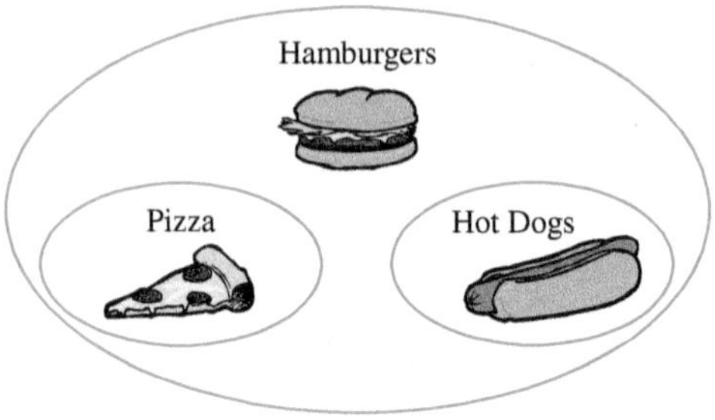

33. (Grade 5)
Look at the Venn diagram:

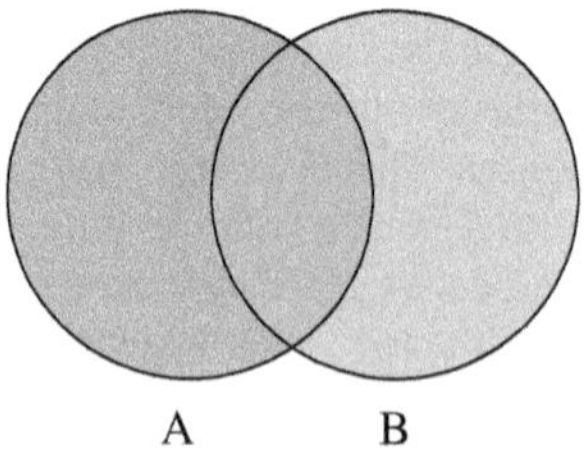

The set, A, contains all integers which are divisible by 3 and are from 3 to 21 including both 3 and 21. B contains all even positive integers from 2 to 24 inclusive. Describe what the elements of $A \cap B$ are by filling in the overlap of A or B.

34. (Grade 6)
Look at set V and set W shown below.

$$\text{Set } V = \{7, 14, 21, 28, 35, \ldots\}$$
$$\text{Set } W = \{3, 6, 9, 12, 15, 18, \ldots\}$$

Could any of the following numbers belong to both set V and set W?

F. 33 G. 49
H. 42 J. 31

35. (Grade 8)
Sarah is filling out the numbers in the Venn diagram. No number is to be entered more than once.

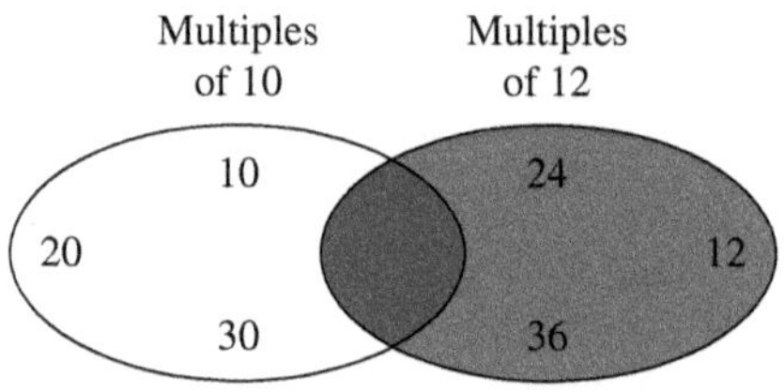

What is the smallest number that can be appropriately placed in the shaded (overlapped) area of the diagram?

A. 24 B. 240
C. 120 D. 60

36. (Grade 6)
There is a week of 54 school children who play music. There are 28 who play the cello and 23 who play the violin. There are 12 children who play both the cello and the violin. How many musicians did not use either a cello or violin?

2.2 Sets, Counting, and the Whole Numbers

If you attend a student raffle, you might hear the following announcement when the entry forms are drawn:

The student with ticket 50768-973 has just won second prize—four tickets to the big game this Saturday.

There are three types of numbers.

This sentence contains three numbers, each of a different type, and each serving a different purpose.

First, the number of the ticket identifies the ticket. Such a number is called a **nominal number** or an **identification.** A nominal number is a sequence of digits used as a name or label. Telephone numbers, social security numbers, account numbers with stores and banks, serial numbers, and driver's license numbers are just a few examples of the use of numbers for identification and naming. The role such numbers play in contemporary society has expanded rapidly with the advent of computers.

The next type of number used by the raffle announcer is one that represents a place in a sequence. The words *first, second, third, fourth,* and so on are used to describe the relative positions of the objects. A number that describes where an object is in an ordered sequence is called an **ordinal number.** The key word is "order." Thus, ordinal numbers communicate location in a collection. "First class," "second rate," "third base," "fifth page," "sixth volume," and "21st century" are all familiar examples of ordinal numbers.

The final use of a number by the raffle announcer is to tell how many tickets had been won. That is, the prize is a set of tickets, and *four* tells us *how many* tickets are in the set. More specifically, the **cardinal number** of a set is the number of objects in the set. Thus, a cardinal number, or **cardinality,** helps communicate the basic notion of "how many."

Notice that numbers, of whatever type, can be expressed verbally (in a language) or symbolically (in a numeration system). For example, the number of moons of Mars is "two" in English, "zwei" in German, and "dos" in Spanish. Symbolically, we could write II in Roman numerals. Numeration systems, both historic and the numbers used today, are described in Chapter 3.

In the remainder of this section, we explore the notion of "how many" (cardinal numbers) more fully. Of special importance is the set of whole numbers, which can be viewed as the set of cardinal numbers of finite sets.

One-to-One Correspondence and Equivalent Sets

Suppose that when you come to your classroom, each student is seated at his or her desk. You would immediately know, *without counting,* that the number of children and the number of occupied desks are the same. This example illustrates the concept of a **one-to-one correspondence** between sets, in which each element of one set is paired with exactly one element of the second set and each element of either set belongs to exactly one of the pairs.

DEFINITION One-to-One Correspondence

A **one-to-one correspondence** between sets A and B is an assignment, for each element of A, of exactly one element of B in such a way that all elements of B are used. It can also be thought of as a pairing of elements between A and B such that each element of A is matched with one and only one element of B and every element of B has an element of A assigned to it.

Figure 2.6 illustrates a one-to-one correspondence between the sets of board members and offices of the Math Club.

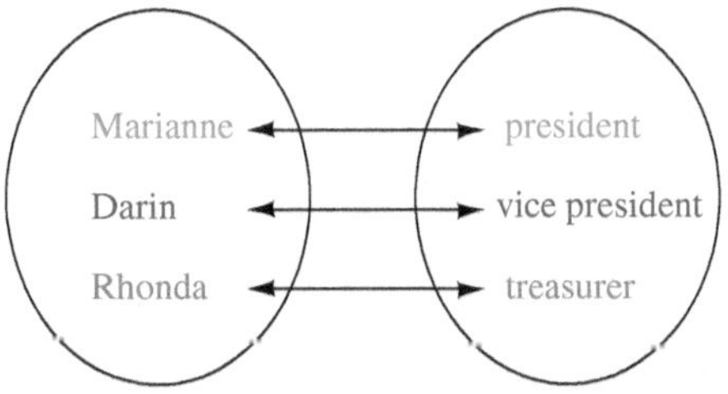

Figure 2.6
A one-to-one correspondence between the sets of board members {Marianne, Darin, Rhonda} and offices {president, vice president, treasurer} of the Math Club

Because there is a one-to-one correspondence between the sets {m, d, r} and {p, v, t} (abbreviating the names of the elements of each set), we say that they are **equivalent sets** and write {m, d, r} $\sim$ {p, v, t}. By contrast, the set {m, d, r} is neither equal nor equivalent to {m, d}.

More generally, we have the following definition:

DEFINITION Equivalent, or Matching, Sets

Sets A and B are **equivalent** if there is a one-to-one correspondence between A and B. When A and B are equivalent, we write $A \sim B$. We also say that equivalent sets **match.** If A and B are not equivalent, we write $A \nsim B$.

It is easy to see that equal sets match. To see why, suppose that $A = B$. Then, since each x in A is also in B, the natural matching $x \leftrightarrow x$ is a one-to-one correspondence between A and B. Thus, $A \sim B$. Conversely, there is no reason to believe that equivalent sets must be equal. For example, $\{\square, \bigstar\}$ is equivalent to $\{1, 2\}$, but $\{\square, \bigstar\} \neq \{1, 2\}$.

EXAMPLE 2.6

Investigating Sets for Equivalence

Let $A = \{x \mid x \text{ is a moon of Mars}\}$
$B = \{x \mid x \text{ is a former U.S. president whose last name is Adams}\}$
$C = \{x \mid x \text{ is one of the Brontë sisters of nineteenth-century literary fame}\}$
$D = \{x \mid x \text{ is a satellite of the fourth-closest planet to the sun}\}$

Which of these relationships, equal (=), not equal (≠), equivalent (~), or not equivalent (≁), holds between distinct pairs of the four sets?

Solution

It is useful to write the sets in listed form:

$A = D = \{\text{Deimos, Phobos}\}, B = \{\text{John Adams, John Quincy Adams}\}$, and
$C = \{\text{Anne, Charlotte, Emily}\}$.

Therefore, looking at the four sets, we see that

$$A \neq B, A \neq C, A = D, B \neq C, B \neq D, C \neq D,$$
$$A \sim B, A \nsim C, A \sim D, B \nsim C, B \sim D, \text{ and } C \nsim D.$$

The Whole Numbers

The sets {a, b, c}, {□, ○, △}, {Mercury, Venus, Earth}, and {Larry, Moe, Curly} are different, but they do share the property of "threeness." The English word *three* and the Indo-Arabic numeral 3 are used to identify this common property of all sets that are equivalent to {1, 2, 3}. In a similar way, we use the word *two,* and the symbol 2, to convey the idea that all of the sets that are equivalent to the set {1, 2} have the same cardinality.

Some sets are quite large, and it is difficult to know how large n must be for the set to be equivalent to the set $\{1, 2, 3, \ldots, n\}$. For example, the set of people alive in the world in the year 2012 would require n to be around 6.96×10^9 (i.e., n is about 6.96 billion). Even so, this is an example of a finite set. In general, a set is said to be **finite** if it either is the empty set or is equivalent to a finite set $\{1, 2, 3, \ldots, n\}$ for some natural number n. Sets that are not finite are called **infinite.** For example, the set N of all of the natural numbers is an infinite set. It is usually easy to know whether a set is finite or infinite, but not always. In Chapter 4, we'll see Euclid's clever proof that the set of prime numbers {2, 3, 5, 7, 11, 13, 17, 19, 23, 29, 31, . . .} is an infinite set.

Use $n(A)$ to denote the number of elements in set A.

The whole numbers, as defined shortly, allow us to classify any finite set according to how many elements the set contains. It is useful to adopt the symbol $n(A)$ to represent the cardinality of the finite set A. If A is not the empty set, then $n(A)$ is a counting number. The cardinality of the empty set, however, requires a new name and symbol: Informally, the set with no elements (the empty set) has nothing in it, so we write $n(\emptyset) = 0$. More formally, **zero** designates the cardinality of the empty set.

A common error is made by omitting the $n(\ )$ symbol: Be sure *not* to write "$A = 4$" when your intention is to state that $n(A) = 4$.

DEFINITION The Whole Numbers

Whole numbers are the cardinal numbers of finite sets—that is, the numbers of elements in finite sets. If $A \sim \{1, 2, 3, \ldots, m\}$, then $n(A) = m$ and $n(\emptyset) = 0$, where $n(A)$ denotes the cardinality of set A. The set of whole numbers is written $W = \{0, 1, 2, 3, \ldots\}$.

EXAMPLE 2.7

Determining Whole Numbers

For each set, find the whole number that gives the number of elements in the set:

(a) $M = \{x \mid x \text{ is a month of the year}\}$
(b) $A = \{\text{a, b, c}, \ldots, \text{z}\}$
(c) $B = \{n \in N \mid n \text{ is a square number smaller than 200}\}$
(d) $Z = \{n \in N \mid n \text{ is a square number between 70 and 80}\}$
(e) $S = \{0\}$

Solution

(a) $n(M) = 12$, since $M \sim \{1, 2, 3, \ldots, 12\}$, and this amounts to counting the elements in M.
(b) $n(A) = 26$
(c) $n(B) = 14$, since $14^2 = 196 \in B$ but $15^2 = 225 \notin B$. Note that we don't have to write out all of the members of B to solve this problem. Still, it is easy to do:
$B = \{1, 4, 9, 16, 25, 36, 49, 64, 81, 100, 121, 144, 169, 196,\}$, and B contains 14 elements.
(d) $n(Z) = 0$, since $8^2 = 64 < 70$ but $9^2 = 81 > 80$. Therefore, there are no square numbers between 70 and 80 and $Z = \varnothing$.
(e) $n(S) = 1$, since the set $\{0\}$ contains one element. This shows that zero is *not* the same as "nothing"!

Representing the Whole Numbers Pictorially and with Manipulatives

SMP 5 Use appropriate tools strategically.

"Mathematically proficient students consider the available tools when solving a mathematical problem. These tools might include pencil and paper, concrete models, a ruler, a protractor, a calculator, . . ."

This chapter and the rest of the book show the value of tools such as physical and pictorial representations for whole numbers. These manipulatives give a visual and material way to understand better the mathematics of elementary school. Tiles, cubes, number strips, and rods are all manipulatives.

Tiles Tiles are congruent squares, each about 2 centimeters ($\frac{3}{4}$ inch) on a side. They should be sufficiently thick to be easily picked up and moved about. Colored plastic tiles are available from suppliers, but tiles are easily handmade from vinyl tile or cardboard. Of course, beans, circular discs, and other objects can be used as well. However, square tiles can be arranged into rectangular patterns, and such patterns reveal many of the fundamental properties of the whole numbers. (See Figure 2.7.)

Figure 2.7
Some representations of "six" with square tiles

Cubes Cubes are much like tiles, but they can form both three-dimensional and two-dimensional patterns. Several attractive manipulatives are commercially available. Unifix™ Cubes can be snapped together to form linear groupings. MathLink® Cubes permit planar and spatial patterns, as we see in Figure 2.8.

Figure 2.8
Some representations of "12" with cubes

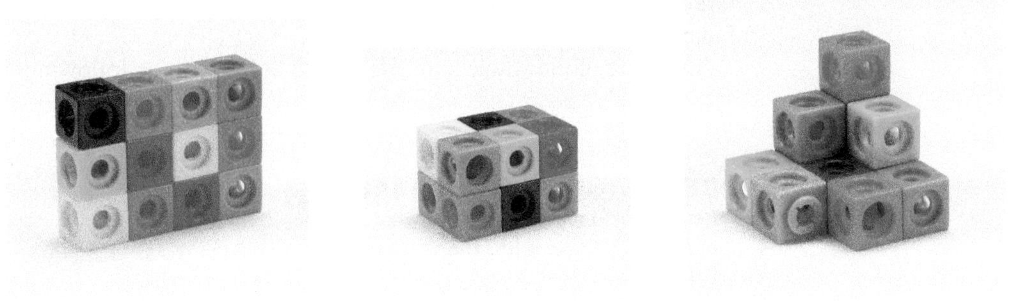

Number Strips and Rods Colored strips of cardboard or heavy paper, divided into squares, can be used to demonstrate and reinforce whole-number properties and operations. The squares should be ruled off, and colors can be used to visually identify the number of squares in a strip, as indicated in Figure 2.9. These number strips are nearly interchangeable with Cuisenaire® rods, which have been used effectively for many years. The colors shown in the figure correspond to those of Cuisenaire® rods. Whole numbers larger than 10 are illustrated by placing strips, or rods, end to end to form a "train."

Figure 2.9
Number strips for the natural numbers 1 through 10

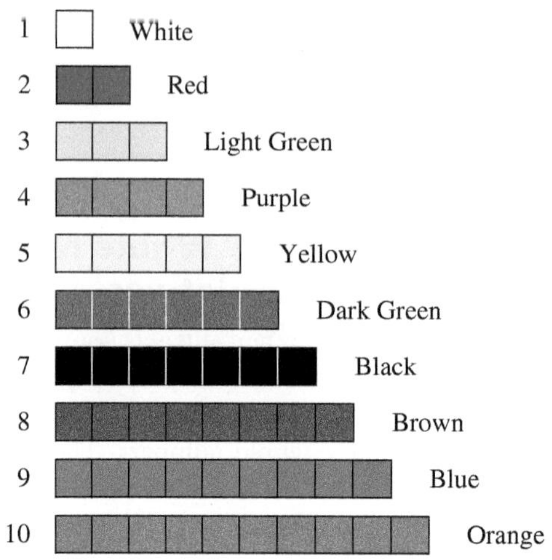

Number Line The number line is a picture, that is, a model in which two distinct points on a line are labeled 0 and 1 and then the remaining whole numbers are marked off in succession with the same spacing. (See Figure 2.10.) Any whole number is interpreted as a distance from 0, which can be shown by an arrow. The number-line model is particularly important because it can also be used to visualize the number systems that are developed later in this book.

The number line begins at 0, not at 1.

Figure 2.10
Illustrating "two" and "five" on the number-line model of the whole numbers

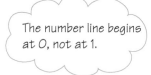

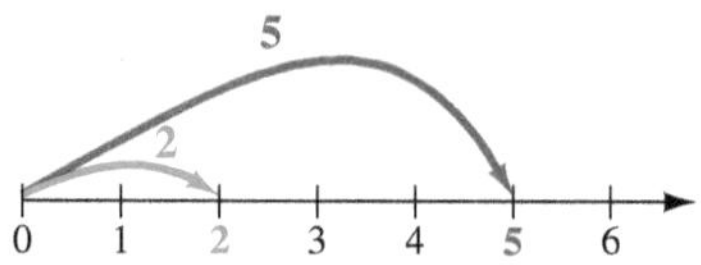

The number line will also help us represent operations involving whole numbers and the ordering of the whole numbers.

Ordering the Whole Numbers

We often wish to relate the number of elements in two given sets. For example, if each child in the class is given one cupcake, and there are some cupcakes left over, we would know that there are more cupcakes than children. Notice that children have been matched to a *proper* subset of the set of cupcakes.

The order of the whole numbers can be defined in the following way:

DEFINITION Ordering the Whole Numbers

Let $a = n(A)$ and $b = n(B)$ be whole numbers, where A and B are finite sets. If A matches a *proper* subset of B, we say that ***a* is less than *b*** and write $a < b$.

The expression $b > a$ is read "b is greater than a" and is equivalent to $a < b$. Also, $a \le b$ means "a is less than or equal to b." The use of rods or the number line shows that $a < b$ if, and only if, $a + c = b$ for some $c > 0$.

EXAMPLE 2.8

Showing the Order of Whole Numbers

Use **(a)** sets, **(b)** tiles, **(c)** rods, and **(d)** the number line to show that $4 < 7$.

Solution

(a) The following diagram shows that a set with 4 elements matches a proper subset of a set with 7 elements. Therefore, $4 < 7$. Note that 3 elements of the second set have no matches.

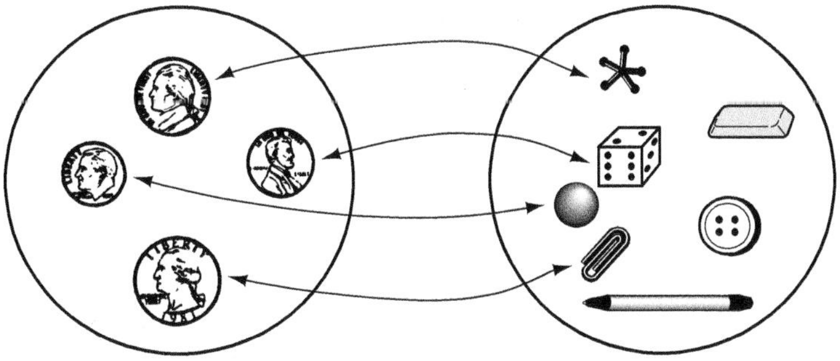

(b) By setting tiles side by side, it is seen that the 4 colored tiles match a proper subset of the 7 uncolored tiles. Here, 3 uncolored tiles are left unmatched.

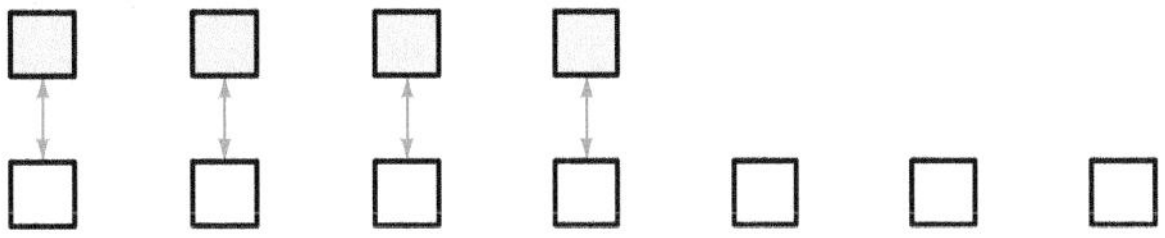

(c) With rods, the order of whole numbers is interpreted by comparing the lengths of the rods:

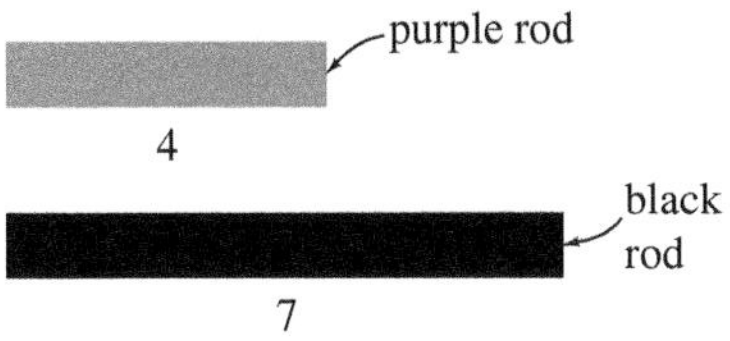

(d) On the number line, $4 < 7$ because 4 is to the left of 7.

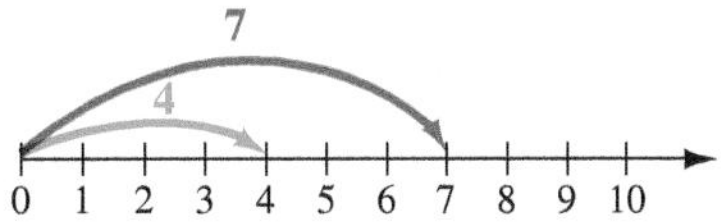

Problem Solving with Whole Numbers and Venn Diagrams

Venn diagrams and the associated concept of a whole number can often be used as a practical problem-solving tool, as we illustrate in the next example. These problems appear often in the State Student Assessments. (See problems 43 and 45 in Problem Set 2.2.)

EXAMPLE 2.9

Solving a Classification Problem

In a recent survey, the 60 students living in Harris Hall were asked about their enrollments in science, engineering, and humanities classes. The results were as follows:

24 are taking a science class.
22 are taking an engineering class.
17 are taking a humanities class.
5 are taking both science and engineering classes.
4 are taking both science and humanities classes.
3 are taking both engineering and humanities classes.
2 are taking classes in all three areas.

How many students are not taking classes in any of the three areas? How many students are taking a class in just one area? Using a Venn diagram, indicate the number of students in each region of the diagram.

Solution

Let S, E, and H denote the set of students in science, engineering, and humanities classes, respectively. Since $n(S \cap E \cap H) = 2$, begin by placing the 2 in the region of the Venn diagram corresponding to the subset $S \cap E \cap H$. Then, by comparing $n(S \cap E \cap H) = 2$ and $n(E \cap H) = 3$, we conclude that there is one student who is taking both engineering and humanities but not a science class.

This conclusion allows us to fill in the 1 in the Venn diagram. Analogous reasoning leads to the entries within the loops of the following Venn diagram:

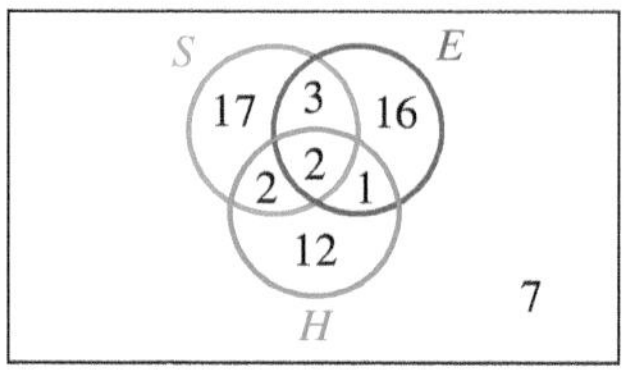

The values within the loops account for 53 of the 60 students, so it follows that 7 students are not taking classes in any of the three areas. Also, since $17 + 16 + 12 = 45$, we conclude that 45 students are taking a class in just one area.

2.2 Problem Set

Understanding Concepts

1. Classify by type—cardinal, ordinal, or nominal—the numbers that appear in these sentences:
   (a) On June 13, Chang was promoted to first vice president.
   (b) In the fourth week of class, Alonzo received a 93 on the second exam.

2. Classify by type—cardinal, ordinal, or nominal—the numbers that appear in these sentences.
   (a) The Atlanta Braves won the fourth game of the World Series on October 13, 2012.
   (b) My passport is 2213762789 and was the 23rd passport approved on June 18, 2013.

3. For each pair of sets, decide whether the sets are equivalent to one another:
   (a) $\{1, 2, 3, 4, 5\}$ and $\{x \mid x$ is a letter in the phrase "PANAMA BANANA MAN"$\}$
   (b) {a, b, c} and {w, x, y, z}
   (c) {o, n, e} and {t, w, o}
   (d) $\{0\}$ and $\varnothing$

4. For each pair of sets, decide whether the sets are equivalent to one another:
   (a) $\{-3, 5, -7, 9\}$ and $\{2, -4, 6, -8\}$
   (b) The set of all integers between 0 and 23 (inclusive) and the set of all integers between –7 and 16 (inclusive)
   (c) $\{0, 1\}$ and all fractions between 0 and 1 (inclusive)

5. Let A, B, and C be finite sets, with $A \subset B \subseteq C$ and $n(B) = 5$.
   (a) What are the possible values of $n(A)$?
   (b) What are the possible values of $n(C)$?

6. Let
   $A = \{n \mid n$ is a cube of a natural number and $1 \le n \le 100\}$;
   $B = \{s \mid s$ is a state in the United States that borders Mexico$\}$.
   Is $A \sim B$?

7. Use counting to determine the whole number that corresponds to the cardinality of these sets:
   (a) $A = \{x \mid x \in N \text{ and } 20 < x \le 27\}$
   (b) $B = \{x \mid x \in N \text{ and } x + 1 = x\}$

8. Use counting to determine the whole number that corresponds to the cardinality of these sets:
   (a) $C = \{x \mid x \in N \text{ and } (x - 1)(x - 9) = 0\}$
   (b) $D = \{x \mid x \in N, 1 \le x \le 100$, and x is divisible by both 5 and 8$\}$

9. Let $N = \{1, 2, 3, 4, \ldots\}$ be the set of natural numbers and $S = \{1, 4, 9, 16, \ldots\}$ be the set of squares of the natural numbers. Then $N \sim S$, since we have the one-to-one correspondence $1 \leftrightarrow 1, 2 \leftrightarrow 4, 3 \leftrightarrow 9, 4 \leftrightarrow 16, \ldots n \leftrightarrow n^2$. (This example is interesting, since it shows that an infinite set can be equivalent to a proper subset of itself.) Show that each of the following pairs of sets are equivalent by carefully describing a one-to-one correspondence between the sets:

(a) The whole numbers and natural numbers:
$W = \{0, 1, 2, 3, \ldots\}$ and $N = \{1, 2, 3, 4, \ldots\}$

(b) The sets of odd and even natural numbers:
$D = \{1, 3, 5, 7, \ldots\}$ and $E = \{2, 4, 6, 8, \ldots\}$

(c) The set of natural numbers and the set of powers of 10:
$N = \{1, 2, 3, 4, \ldots\}$ and $\{10, 100, 1000, \ldots\}$

10. Decide which of the following sets are finite:

(a) {grains of sand on all the world's beaches}

(b) {whole numbers divisible by the number 46, 182, 970, 138}

(c) {points on a line segment that is 1 inch long}

11. The figure that follows shows two line segments, L_1 and L_2. The rays through point P give, geometrically, a one-to-one correspondence between the points on L_1 and the points on L_2—for example, $Q_1 \leftrightarrow Q_2$.

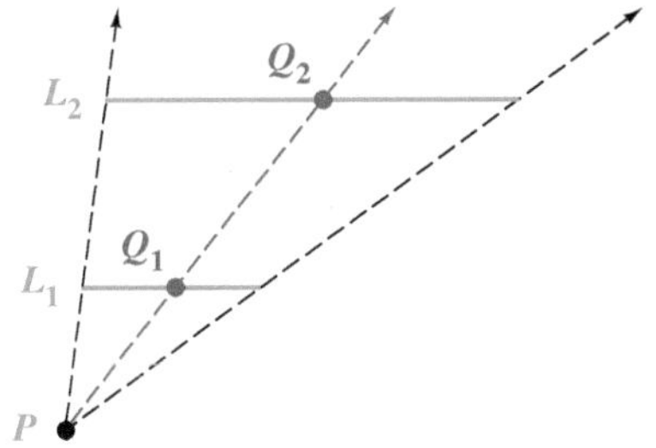

Use similar geometric diagrams to show that the figures that follow are equivalent sets of points. Describe all possible locations of point P.

(a) Two concentric circles

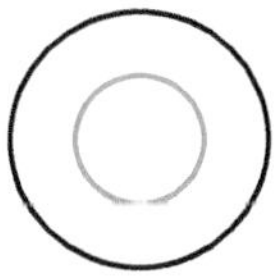

(b) A circle and an inscribed square

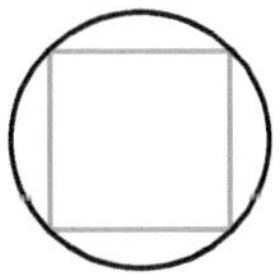

12. Look at the figure of problem 11 and describe all possible locations of the point P.

(a) A triangle and its circumscribing circle

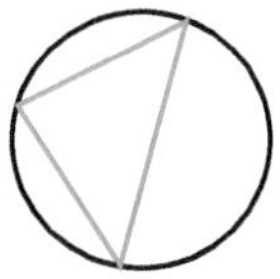

(b) A semicircle and its diameter

13. Decide whether the statements that follow are true or false. If false, give a counterexample; that is, give an example of two finite sets A and B that satisfy the hypothesis (the "if part"), but not the conclusion, of the statement.

(a) If $A \subseteq B$, then $n(A) \leq n(B)$.

(b) If $n(A) < n(B)$, then $A \subset B$.

(c) If $n(A \cup B) = n(A)$, then $B \subseteq A$.

(d) If $n(A \cap B) = n(A)$, then $A \subseteq B$.

14. Decide whether the statements that follow are true or false. If false, give a counterexample; that is, other than part (c), give an example of two finite sets A and B that satisfy the hypothesis (the "if" part), but not the conclusion, of the statement.

(a) If $A \subseteq B$, then $n(A \cup \varnothing) \leq n(B \cup \varnothing)$.

(b) If S has 15 elements and $A \subseteq S$, then $n(\overline{A}) = 15 - n(A)$.

(c) If N (the positive integers) has infinitely many elements and $B \subseteq N$, then the complement of B is infinite.

(d) $n(A \cup B) = n(A) + n(B)$.

15. Let A and B be finite sets.

(a) Explain why $n(A \cap B) \leq n(A)$.

(b) Explain why $n(A) \leq n(A \cup B)$.

(c) Suppose $n(A \cap B) = n(A \cup B)$. What more can be said about A and B?

16. Let $U = \{1, 2, 3, \ldots, 1000\}$. Also, let F be the subset of numbers in U that are multiples of 5, and let S be the subset of numbers in U that are multiples of 6. Since $1000 \div 5 = 200$, it follows that $n(F) = n(\{5 \cdot 1, 5 \cdot 2, \ldots, 5 \cdot 200\}) = 200$.

(a) Find $n(S)$ by using a method similar to the one that showed that $n(F) = 200$.

(b) Find $n(F \cap S)$.

(c) Label the number of elements in each region of a two-loop Venn diagram with universe U and subsets F and S.

17. A survey of 700 households revealed that 300 had only a TV, 100 had only a computer, and 100 had neither a TV nor a computer. How many households have both a TV and a computer? Use a two-loop Venn diagram to find your answer.

18. Finish labeling the number of elements in the regions in the Venn diagram shown, where the subsets A, B, and C of the universe U satisfy the conditions listed.

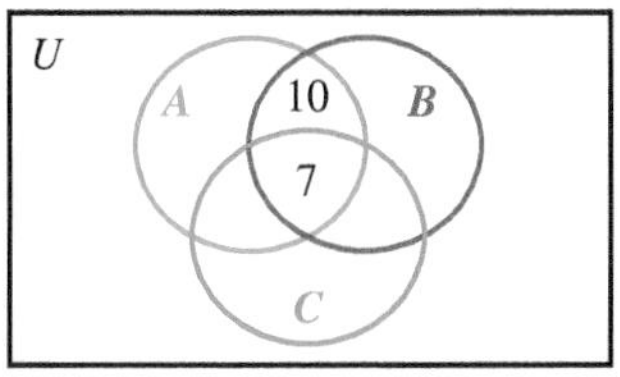

$n(U) = 100$
$n(A) = 40$
$n(B) = 50$
$n(C) = 30$
$n(A \cap B) = 17$
$n(B \cap C) = 12$
$n(A \cap C) = 15$
$n(A \cap B \cap C) = 7$

19. A poll of students showed that 55 percent liked basketball, 40 percent liked soccer, 55 percent liked football, 25 percent liked both basketball and soccer, 20 percent liked both soccer and football, 20 percent liked both basketball and football, and 10 percent liked all three sports. Use a Venn diagram to answer these questions: What percentage of students likes only one sport? What percentage does not like any of the three sports?

20. In a college mathematics class, all the students are also taking anthropology, history, or psychology, and some of the students are taking two, or even all three, of these courses. If (i) 40 students are taking anthropology, (ii) 11 students are taking history, (iii) 12 students are taking psychology, (iv) 3 students are taking all three courses, (v) 6 students are taking anthropology and history, and (vi) 6 students are taking psychology and anthropology,

(a) how many students are taking only anthropology?

(b) how many students are taking anthropology or history?

(c) how many students are taking history and anthropology but not psychology?

21. Number tiles can be arranged to form patterns that relate to properties of numbers.

(a) Arrange number tiles to show why 1, 4, 9, 16, 25, . . . are called the "square" numbers.

(b) Arrange number tiles to show why 1, 3, 6, 10, 15, . . . are called "triangular" numbers.

Into the Classroom

22. Describe how you might teach a very young child the idea of color and the words such as *blue* and *red* that describe particular colors. Do you see any similarities in teaching the idea of color and the idea of number, including such words as *two* and *five*?

23. Suggest a method to convey the notion of zero to a young child.

24. Imagine yourself teaching a third grader the transitive property of "less than," whereby $a < b$ and $b < c$ allow you to conclude that $a < c$. Write an imagined dialogue with the student, using number strips (or Cuisinaire® rods) as a manipulative.

25. Which of the numbers that follow is a square number (meaning the square of an integer)? Use a diagram to support your answer.

200	300	400	500	600

Responding to Students

26. Annabelle's parents told her that she could push the sixth-floor button of an elevator. A math professor in the elevator said to her, "How about hitting the third-floor button twice? Will that get you to the sixth floor?" What do you think Annabelle did?

27. Zack has trouble working a problem from the Fayette County (Kentucky) Public Schools Grade 4 Learning Check. He is not sure of where 27 or 14 should go. First do the Learning Check yourself, and then explain to Zack where 27 and 14 go and why. (This problem was given in October 2005 as a Mathematics Open-Response Question.)

The diagram that follows shows how Zack is grouping some numbers:

- One circle has multiples of 3.
- The other circle has multiples of 4.
- The shaded space is for numbers that are multiples of both 3 and 4.
- Numbers that are not a multiple of 3 or 4 go in the space outside the circles.

Zack has already placed the numbers 1, 5, 8, 12, and 30 in the correct spaces in the diagram.

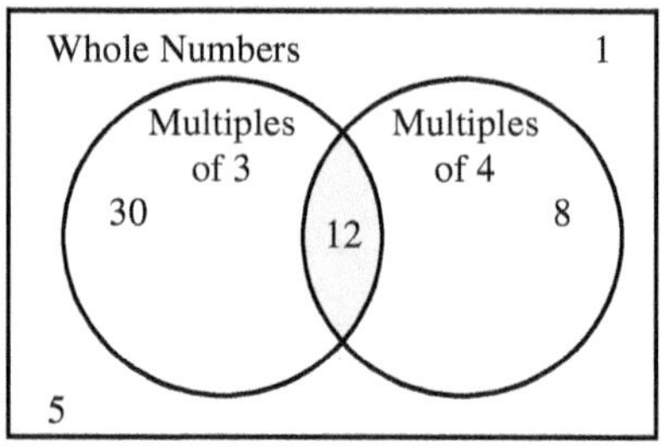

(a) Make a copy of the diagram on your paper. Write each of the following numbers in the correct space in your diagram:

24, 16, 27, 14, 32

(b) Identify one more number *less than 40* that goes in *each* of the four spaces in the diagram. Be sure to tell in which space in the diagram each number belongs. Explain how you know that the numbers you used are placed correctly.

28. Richard is working with square numbers and is confused. He says that 0^2 equals 0 and 2^2 equals 4, so $5^2 = 10$.

(a) What does Richard think that the rule is?

(b) Using centimeter grid paper, show Richard what the representations (manipulatives) of the first five square numbers are.

29. Jeff is working with numbers that are cubed and he is confused. He says that $0^3 = 0$ and 2^3 equals 6, so 5^3 must be 15.

(a) What does Jeff think the rule for exponents is?

(b) What manipulative (a physical object) can you draw to show Jeff a representation of each of the first five cube numbers?

Thinking Critically

30. (a) How many ways can three balls of different colors—yellow, red, and green—be put in a ball can?

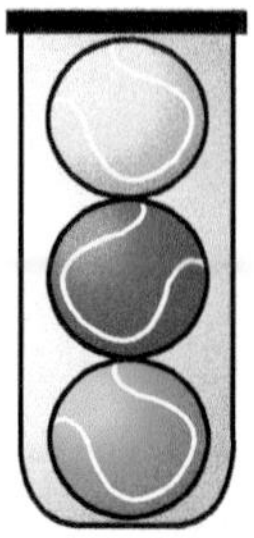

(b) How many different one-to-one correspondences are there between the two sets {1, 2, 3} and {y, r, g}?

(c) Explain why a dozen dyed eggs, all of different colors, can be placed in a carton in $12 \times 11 \times 10 \times 9 \times 8 \times 7 \times 6 \times 5 \times 4 \times 3 \times 2 \times 1$ ways. (*Suggestion:* How many ways can you put the first egg into the carton? How many ways can you put the second egg into the carton? . . . How many ways are left to put the last egg into the carton?)

(d) How many different one-to-one correspondences are there between the two sets {1, 2, 3, . . . , 12} and {a, b, c, d, e, f, g, h, i, j, k, l}?

31. Suppose your class has the set $C = \{a, b, c\}$ in mind. This set has one subset with no elements—namely, $\varnothing$; three subsets with one element—{a}, {b}, and {c}; three subsets with two elements—{a, b}, {b, c}, and {a, c}; and one subset with three elements—{a, b, c}. This fills in row $n = 3$ of the table below, where the entry in row n and column r gives the number of ways to choose different subsets of r elements from a set of n elements.

(a) Find all of the subsets of each of the four sets $\varnothing$, $A = \{a\}$, $B = \{a, b\}$, and $D = \{a, b, c, d\}$, and fill in the table through row $n = 4$.

(b) What pattern of numbers do you see in the table? Use the pattern to fill in the next two rows.

	$r = 0$	$r = 1$	$r = 2$	$r = 3$	$r = 4$	$r = 5$	$r = 6$
$n = 0$							
$n = 1$							
$n = 2$							
$n = 3$	1	3	3	1			
$n = 4$							
$n = 5$							
$n = 6$							

32. Evelyn's Electronics Emporium hired Steven's Survey Services (SSS) to poll 100 households at random. Evelyn's report from SSS contained the following data on ownership of a TV, VCR, or stereo:

TV only	8
TV and VCR (at least)	70
TV, VCR, and stereo	65
Stereo only	3
Stereo and TV (at least)	74
No TV, VCR, or stereo	4

Evelyn, who assumes that anyone with a VCR also has a TV, is wondering if she should believe the figures. Should she?

33. At a school with 100 students, 35 students were taking Arabic, 32 Bulgarian, and 30 Chinese. Twenty students take only Arabic, 20 take only Bulgarian, and 14 take only Chinese. In addition, 7 students are taking both Arabic and Bulgarian, some of whom also take Chinese. How many students are taking all three languages? None of these three languages?

34. Use the ideas of sets and the definition of the order relation for the whole numbers to justify the transitive property of "less than." That is, if k, l, and m are whole numbers satisfying $k < l$ and $l < m$, then $k < m$. (*Suggestion:* Consider sets satisfying $K \subset L \subset M$.)

35. A political polling organization sent out a questionnaire that asked the following question: "Which taxes—income, sales, or excise—would you be willing to have raised?" Sixty voters' opinions were tallied by the office clerk:

Tax	Number Willing to Raise the Tax
Income	20
Sales	28
Excise	29
Income and sales	7
Income and excise	8
Sales and excise	10
Unwilling to raise any tax	5

The clerk neglected to count how many, if any, of the 60 voters are willing to raise all three of the taxes. Can you help the pollsters? Explain how.

36. There are 40 students in the Travel Club. They discovered that 17 members have visited Mexico, 28 have visited Canada, 10 have been to England, 12 have visited both Mexico and Canada, 3 have been only to England, and 4 have been only to Mexico. Some club members have not been to any of the three foreign countries, and, curiously, an equal number have been to all three countries.

(a) How many students have been to all three countries?

(b) How many students have been only to Canada?

37. Letitia, Brianne, and Jake met at the mall on December 31. Letitia said that she intends to come to the mall every third day throughout the next year. Brianne said that she intends to be there every fourth day, and Jake said he would be there every fifth day. Letitia said that she knew she would be at the mall a total of 121 days, since 365 ÷ 3 is 121 with a remainder of 2. Brianne said that she'd be at the mall 91 days, since 365 ÷ 4 is 91 with a remainder of 1. Moreover, Brianne said that of those 91 days, she would expect to see Letitia 30 times, since they will both be coming every 12 days and 365 ÷ 12 is 30 with a remainder of 5.

(a) How many days will all three friends meet at the mall in the next year?

(b) Use reasoning similar to that of Letitia and Brianne to construct a three-loop Venn diagram that shows the number of times the three friends go to the mall in all of the possible combinations.

(c) How many days in the year will Jake be at the mall by himself?

(d) How many days in the year will none of the three be at the mall?

Making Connections

38. In the early 21st century, modern society entered "the digital information age." People are now accustomed to being identified by number as often as by name. Make a list of your own identification numbers—social security, credit card, telephone, and so on. It may be helpful to look through your wallet, cell phones, and so on!

39. Blood tests of 100 people showed that 45 had the A antigen and 14 had the B antigen. (See problem 30 of Section 2.1.) Another 45 had neither antigen and so are of type O. How many people are of type AB, having both the A and B antigens? Draw and label a Venn diagram that shows the number of people with blood types A, B, AB, and O.

State Assessments

40. (Washington State, Grade 4)
What numbers do W, X, and Y probably represent on the number line?

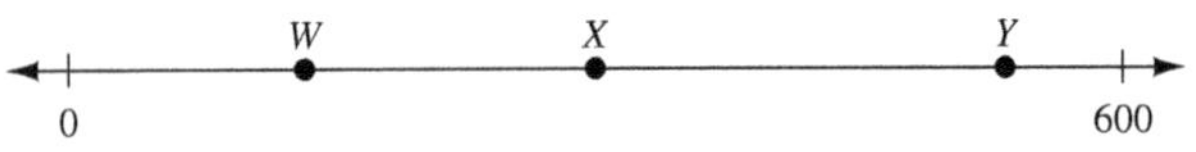

A. $W = 100, X = 200, Y = 500$
B. $W = 150, X = 300, Y = 400$
C. $W = 150, X = 300, Y = 525$

41. (Massachusetts, Grade 3)
Seth read 5 chapter books and 9 picture books. Anna read 16 chapter books.
Which number sentence correctly compares the total number of books Seth read with the number of books Anna read?

A. $5 + 9 < 16$
B. $5 + 9 > 16$

C. $16 = 5 + 9$
D. $9 = 16 - 5$

42. (Massachusetts, Grade 4)
Max and Sam wrote a number sentence to show that Max is older than Sam. In their number sentence,

- M represents Max's age in years, and
- S represents Sam's age in years.

Which number sentence shows that Max is older than Sam?

A. $M < S$
B. $M > S$
C. $M = S$
D. $M + S = 10$

43. (Grade 8)

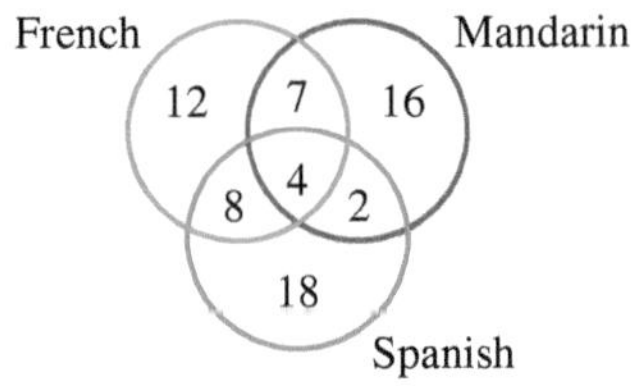

In middle school, students may take up to three language classes. The Venn diagram shows the number of students in Mrs. Lawton's homeroom class who are taking the electives of Mandarin, French, or Spanish. What is the *total* number of students who are taking *both* French and Spanish? *Use the Venn diagram above to answer this question.*

A. 19
B. 12
C. 4
D. 8

44. (Grade 7)
The numbers in Set N are multiples of 5.

$$N = \begin{array}{ccc} 5 & 15 & 20 \\ & 25 & 10 \end{array}$$

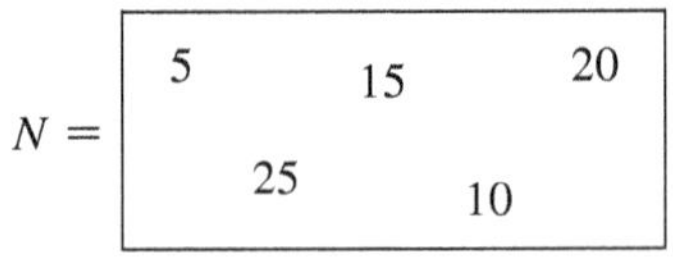

The numbers in Set P are multiples of 9.

$$P = \begin{array}{ccc} 9 & 27 & 36 \\ & 45 & 18 \end{array}$$

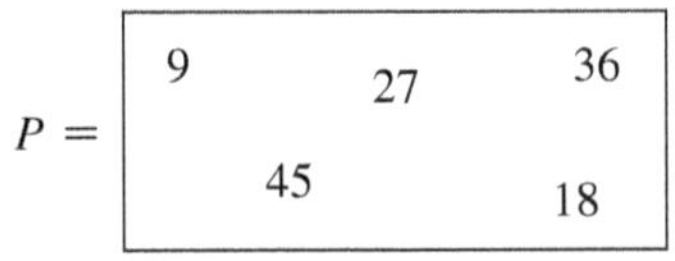

Which number of the three possibilities could be a number of both Set N and Set P?

A. 27
B. 25
C. 90

45. (Grade 7)
There are seven students at Buzz Middle School each of whom is either in mathematics or theater courses. They are Mary, Alice, Marilyn, Janice, Rui, Delay, and Taneisha. Mary and Alice are in mathematics only, Delay and Taneisha are in theater only, and Marilyn, Janice, and Rui are in both mathematics and theater. Which Venn diagram correctly represents the data?

A.
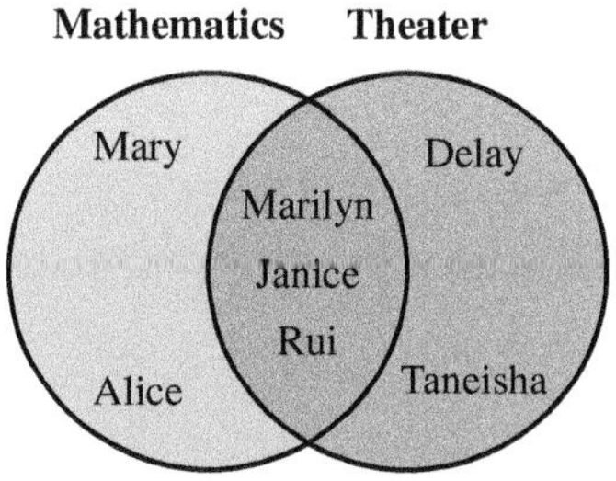

B.
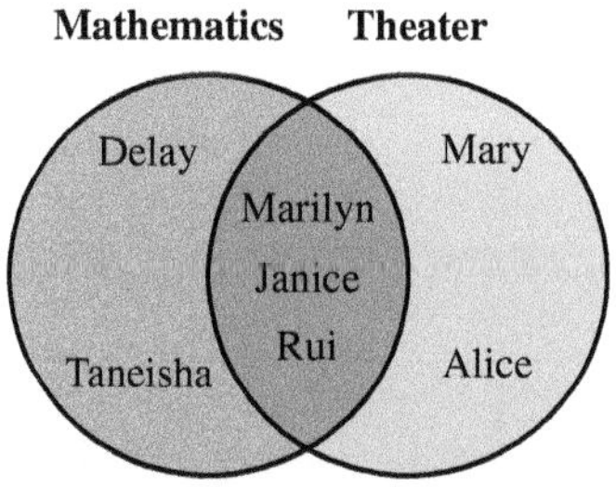

C.
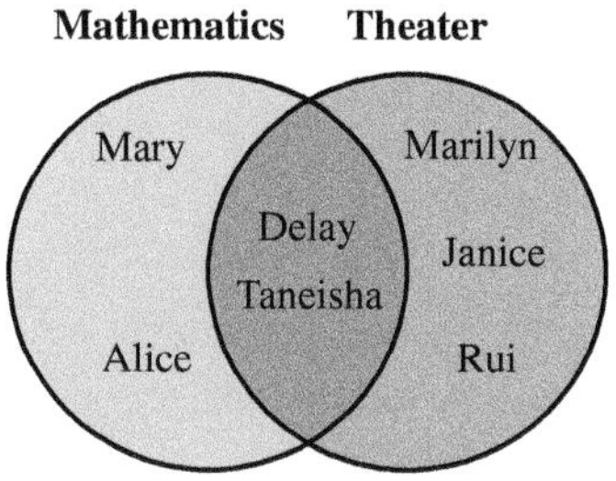

D.
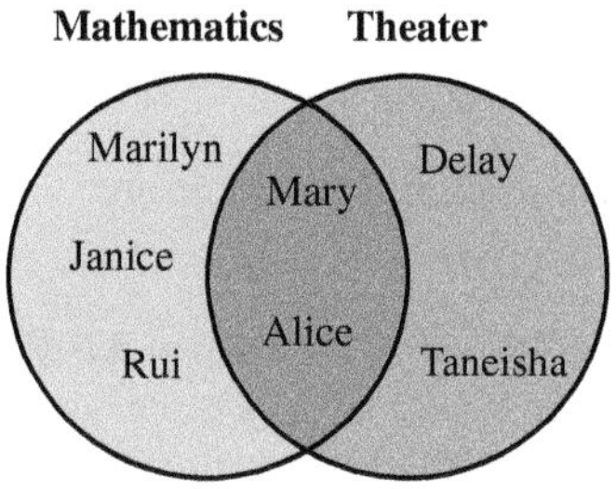

2.3 Addition and Subtraction of Whole Numbers

In this section, we introduce the operations of addition and subtraction on the set of whole numbers $W = \{0, 1, 2, 3, \ldots\}$. In each operation, two whole numbers are combined to form another whole number. Because *two* whole numbers are added to form the sum, addition is called a **binary operation.** Similarly, subtraction is defined on a pair of numbers, so subtraction is also a binary operation.

The definitions of addition and subtraction are accompanied by a variety of conceptual models that give the operations both intuitive and practical meaning. It is vitally important that children be able to interpret and express, in a variety of ways, operations and their properties through manipulatives and visualization so that they will have a conceptual understanding of arithmetic. Activities with these representations prepare them to understand computation and build confidence in their ability to select the appropriate operations for problem solving.

The Set Model of Whole-Number Addition

The whole numbers answer the basic question "How many?" For example, if Alok collects baseball cards and A is the set of cards in his collection, then $a = n(A)$ is the number of cards he owns. Suppose his friend Barbara has a collection of different baseball cards, forming a set B with $b = n(B)$ cards. If Alok and Barbara decide to combine their collections, the new collection would be the set $A \cup B$ and would contain $n(A \cup B)$ cards. The **addition,** or **sum,** of a and b can be defined as the number of cards in the combined collection. That is, the sum of two whole numbers a and b is given by $a + b = n(A \cup B)$, where A and B are disjoint sets, $a = n(A)$, and $b = n(B)$.

Addition answers the question "How many elements are in the union of two disjoint sets?" Figure 2.11 illustrates how the set model is used to show that $3 + 5 = 8$. First, disjoint sets A and B are found, with $n(A) = 3$ and $n(B) = 5$. Since $n(A \cup B) = 8$, we have shown that $3 + 5 = 8$. By using physical objects such as beans to form the sets, there is no question about disjointness.

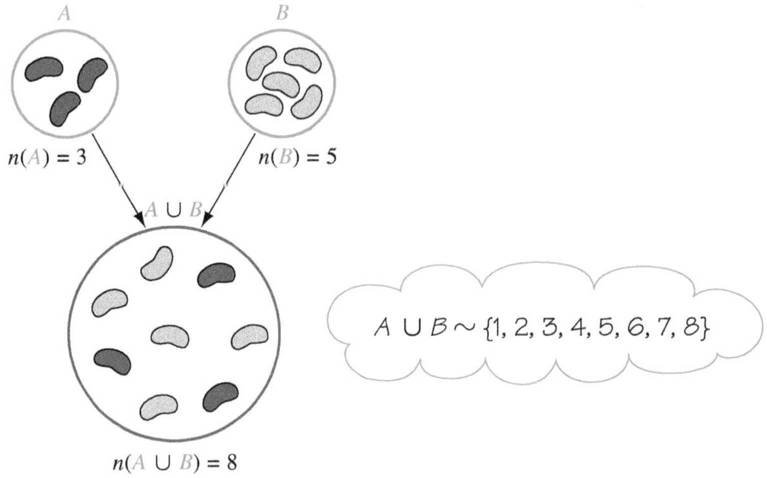

Figure 2.11
Showing $3 + 5 = 8$ with the set model of addition

Here is the exact definition:

> **DEFINITION The Addition of Whole Numbers**
> Let a and b be any two whole numbers. If A and B are any two disjoint sets for which $a = n(A)$ and $b = n(B)$, then the **sum of a and b,** written $a + b$, is given by
>
> $$a + b = n(A \cup B).$$

The expression $a + b$ is read "a plus b," where a and b are called the **addends** or **summands.**

$a + b$
↑ ↑ addends

EXAMPLE 2.10 Using the Set Model of Addition

The University Math Club membership includes 14 women and 11 men. The club members major in math, in physics, or in both of these areas.

(a) How many students belong to the Math Club?
(b) If 21 club members major at least in math, and 6 at least in physics, how many have majors in both?

Solution

(a) If C denotes the set of members of the club, then $C = M \cup W$. First, we name the sets M and W where M is the set of all men and W is the set of all women in the club. Since M and W are disjoint sets, the number of club members is

$$n(C) = n(M \cup W) = n(M) + n(W) = 11 + 14 = 25.$$

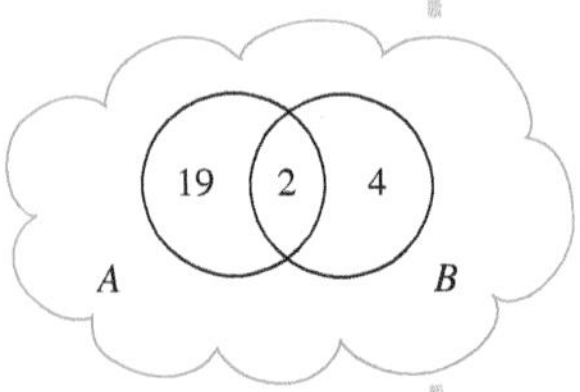

(b) Let A denote the set of club members with a math major and B the set of club members with a physics major. Since $n(A) + n(B) = 21 + 6 = 27$ is 2 more than the 25 members of the club, 2 club members belong to both sets A and B. That is, $n(A \cap B) = 2$ club members have a double major. Notice that the equation $n(A \cup B) = n(A) + n(B) - n(A \cap B)$ is valid in this case. (See problem 34 of the problem set for Section 2.3).

The set model of addition can be illustrated with manipulatives such as the tiles, cubes, strips, and rods described in the previous section. Many addition facts and patterns become evident when they are discovered and visualized by means of concrete representations. An example is the **triangular numbers** $t_1 = 1, t_2 = 3, t_3 = 6, \ldots$ introduced in Chapter 1. Recall that t_n denotes the nth triangular number, where the name refers to the triangular pattern that can be formed with t_n objects.

By using number-tile patterns as manipulatives, as shown in Figure 2.12, we see that the sum of any two successive triangular numbers forms a square number. Indeed, we have the general formula

$$t_{n-1} + t_n = n^2, \qquad n = 2, 3, \ldots.$$

Figure 2.12
The sum of two successive triangular numbers is a square number

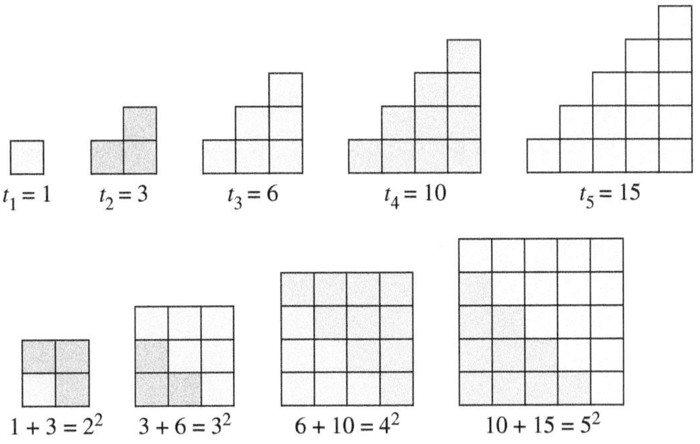

The Measurement (Number-Line) Model of Addition

On the number line, whole numbers are geometrically interpreted as distances. Addition can be visualized as combining two distances to get a total distance. If we wish to add 3 and 5, we can draw 3 as usual, starting at 0. The number 5 can be thought of, in this case, as starting at any point and going five whole numbers to the right. Combining the two distances 3 and 5 can be represented by starting the arrow for 5 at 3 and counting 5 to the right for its finish, as shown in Figure 2.13. It is important to notice that the two distances are not overlapping, and the tail of the arrow representing 5 is placed at the head of the arrow representing 3. The result of the addition is 8, which is indicated on the number line by circling the 8. Figure 2.13 shows a visual picture of addition.

Figure 2.13
Illustrating 3 + 5 on the number line

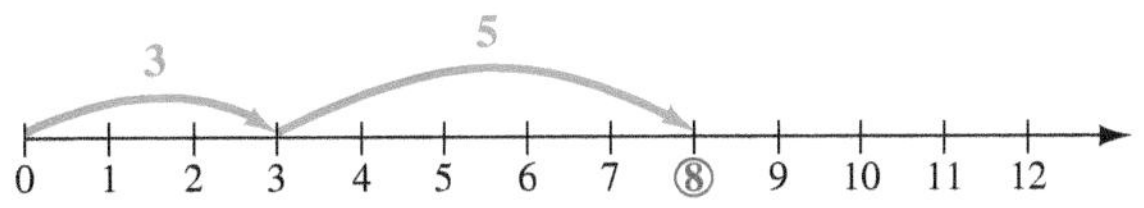

Properties of Whole-Number Addition

Using models or going back to definitions for addition is not a good way, of course, to actually do computation. In the next few pages, we give some properties of whole-number addition and subtraction that we already use automatically to help simplify the way in which we add and subtract. Proceeding in this manner gives a conceptual background for why the procedures that we have been accustomed to actually work.

The sum of any two whole numbers is also a whole number, so we say that the set of whole numbers has the **closure property** under addition. This property is so obvious for addition that it may seem unnecessary to mention it. However, the set of whole numbers is not closed under subtraction or division. For example, is $3 - 5$ a whole number? Even under addition, many subsets of whole numbers do not have the closure property. For example, if $D = \{1, 3, 5, \ldots\}$ denotes the subset of the odd whole numbers, then D does *not* have the closure property under addition. (For instance, the sum of 1 and 3 is even and hence is not in D.) By contrast, the set of even whole numbers, $E = \{0, 2, 4, 6, \ldots\}$, *is* closed under addition; this is so because the sum of any two even whole numbers is also an even whole number. Young children frequently can show why the sum of even numbers is an even number.

Some other important properties of whole-number addition correspond to properties of operations on finite sets. For example, the **commutative property** of union, $A \cup B = B \cup A$, proves that $a + b = b + a$ for all whole numbers a and b. Similarly, the **associative property** $A \cup (B \cup C) = (A \cup B) \cup C$ tells us that $a + (b + c) = (a + b) + c$. Since $A \cup \varnothing = \varnothing \cup A = A$, we obtain the additive-identity property of zero, $a + 0 = 0 + a = a$. Zero is said to be the **additive identity** because of this property.

The properties in the next theorem can all be proven from the definition of whole-number addition and what we already know about operations on sets. We will, however, use reasoning through models, instead of formal mathematical proof, to demonstrate the theorem.

THEOREM Properties of Whole-Number Addition

Closure Property	If a and b are any two whole numbers, then $a + b$ is a unique whole number.
Commutative Property	If a and b are any two whole numbers, then $a + b = b + a$.
Associative Property	If a, b, and c are any three whole numbers, then $a + (b + c) = (a + b) + c$.
Additive-Identity Property of Zero	If a is any whole number, then $a + 0 = 0 + a = a$.

The last three of these properties can be described informally as "order of addition doesn't matter," "grouping of addition doesn't matter," and "adding zero doesn't matter," respectively. (See problem 35 of the Problem Set for Section 2.4 for similar statements about multiplication.)

The properties of whole-number addition can also be illustrated with physical and pictorial models of the whole numbers. Figure 2.14 shows how the associative property can be depicted with number strips, which are another example of manipulate.

The addition properties are very useful when we add several whole numbers, since we are permitted to rearrange the order of the addends and the order in which pairs of addends are summed.

Figure 2.14 Illustrating the associative property $(3 + 5) + 2 = 3 + (5 + 2)$ with number strips or Cuisenaire® rods

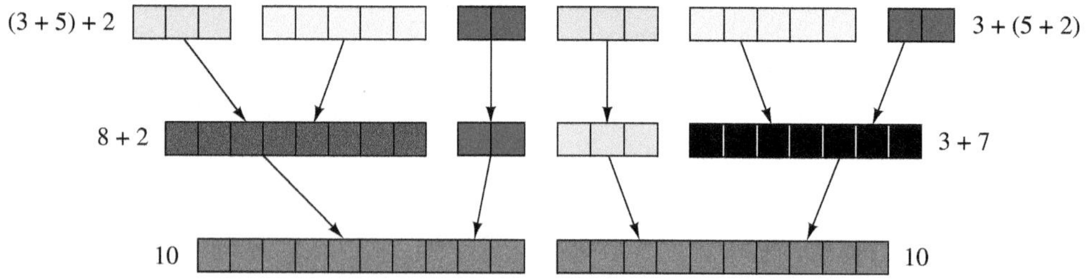

EXAMPLE 2.11

Using the Properties of Whole-Number Addition

(a) Which property justifies each of the following statements?

(i) $8 + 3 = 3 + 8$

(ii) $(7 + 5) + 8 = 7 + (5 + 8)$

(iii) A million plus a billion is not infinite.

(b) Justify each equality:

(i) $(20 + 2) + (30 + 8) = 20 + [2 + (30 + 8)]$

(ii) $= 20 + [(30 + 8) + 2]$

(iii) $= 20 + [30 + (8 + 2)]$

(iv) $= (20 + 30) + (8 + 2)$

Solution

(a) (i) Commutative property, (ii) associative property, (iii) the sum is a whole number by the closure property and is therefore a finite value.

(b) (i) Associative property, (ii) commutative property, (iii) associative property, (iv) associative property.

The number line is a very useful way for children to discover the properties of arithmetic by starting with a number line and putting the head of one arrow followed by the tail of the other. This is a picture of addition.

EXAMPLE 2.12

Illustrating Properties of Arithmetic on the Number Line via the Measurement Model

What properties of whole-number addition are shown on the following number lines?

(a)

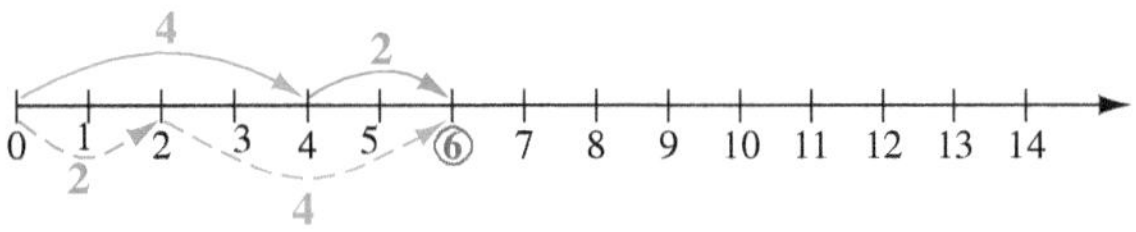

(b)

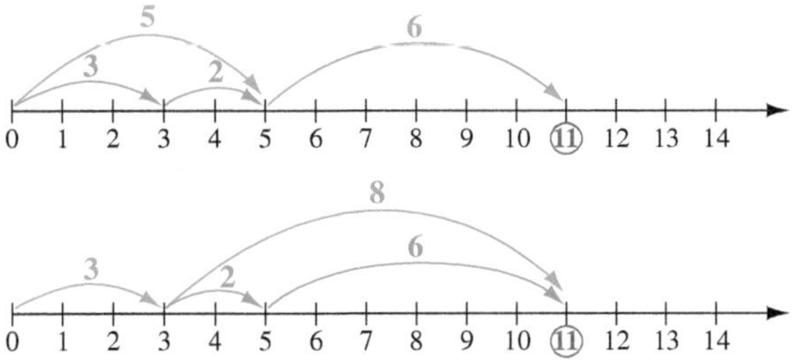

(c)

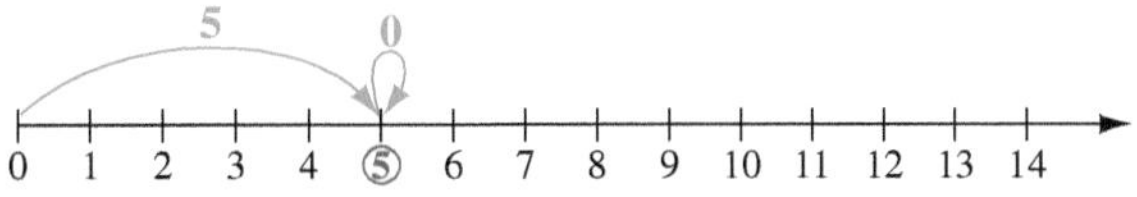

Solution

(a) The commutative property: $4 + 2 = 2 + 4$.
(b) The associative property: $(3 + 2) + 6 = 3 + (2 + 6)$.
(c) The additive-identity property: $5 + 0 = 5$.

INTO THE CLASSROOM

Using Addition Properties to Learn Addition Facts

As children learn addition, the properties of whole-number addition should become a habit of thought from the very beginning. Indeed, the properties are useful even for learning and recalling the sums of one-digit numbers, as found in the addition table shown. The commutative property $a + b = b + a$ means that the entries above the diagonal repeat the entries below the diagonal. Also, the first column of addition with 0 is easy by the additive-identity property. The next four columns, giving addition with 1, 2, 3, and 4, can be learned by "counting on." For example, $8 + 3$ is viewed as "8 plus 1 makes 9, plus 1 more makes 10, plus 1 more makes 11." The diagonal entries $1 + 1 = 2, 2 + 2 = 4, \ldots$ are the "doubles," which are readily learned by knowing how to count by twos: 2, 4, . . . , 18. The remaining entries in the table can be obtained by combining "doubles" and "counting on." For example, $6 + 8$ is viewed as $6 + (6 + 2)$, which is $(6 + 6) + 2$; knowing that 6 doubled is 12 and counting on 2 gives the answer of 14.

Other effective strategies include the following:

Making tens: For example, $8 + 6 = (8 + 2) + 4 = 10 + 4 = 14$.

Counting back: For example, 9 is 1 less than 10, so $6 + 9$ is 1 less than $6 + 10 = 16$, giving 15. In symbols, this amounts to

$$6 + 9 = (6 + 10) - 1 = 16 - 1 = 15.$$

Look for patterns. They also help you learn the facts!

+	**0**	**1**	**2**	**3**	**4**	**5**	**6**	**7**	**8**	**9**
0	0	1	2	3	4	5	6	7	8	9
1	1	2	3	4	5	6	7	8	9	10
2	2	3	4	5	6	7	8	9	10	11
3	3	4	5	6	7	8	9	10	11	12
4	4	5	6	7	8	9	10	11	12	13
5	5	6	7	8	9	10	11	12	13	14
6	6	7	8	9	10	11	12	13	14	15
7	7	8	9	10	11	12	13	14	15	16
8	8	9	10	11	12	13	14	15	16	17
9	9	10	11	12	13	14	15	16	17	18

Subtraction of Whole Numbers

A common way to introduce subtraction is to define subtraction in terms of a related addition problem. In other words, $8 - 3 = 5$ because $5 + 3 = 8$. More formally, we have the following definition:

DEFINITION Subtraction of Whole Numbers

Let a and b be whole numbers. The **difference of a and b,** written $a - b$, is the unique whole number c such that $a = b + c$. That is, $a - b = c$ if, and only if, there is a whole number c such that $a = b + c$.

Let's look for a minute at the definition of subtraction but use the letter x to be the difference of a and b. This notation would mean that $x = a - b$ when x is a whole number such that $a = b + x$. As an example, subtracting 5 from 8 is the same as solving $8 = 5 + x$. For x viewed this way, algebra becomes a way of describing subtraction. The last two sentences give an example of what's called "prealgebra" and is now a part of mathematics in grades 3–5.

The expression $a - b$ is read "a minus b," where a is the **minuend** and b is the **subtrahend.**

$$a - b = c$$

c is the difference of a and b

b is the subtrahend

a is the minuend

This definition relates subtraction to addition and is the definition most easily extended to the integers, rational numbers, and real numbers.

Since $8 = 5 + 3$, the definition tells us that $8 - 5 = 3$. However, the practical value of subtraction is not revealed in the definition. To understand the nature and value of subtraction as it is often introduced to children, we will introduce four conceptual models: **take away, missing addend, comparison,** and **number line** (or **measurement**).

The following four problems respectively illustrate each of the four conceptual models, and each has a manipulative to show physically what subtraction is. The manipulatives for the four models in this case are money, the chapters of a book, comparison of number of objects in a set (e.g., mice), and visualization of walking up a mountain trail.

Take away:
Eroll has \$8 and spends \$5 for a ticket to the movies. How much money does Eroll have left?

Missing addend:
Alice has read 5 chapters of her book. If there are 8 chapters in all, how many more chapters must she read to finish the book?

Comparison:
Georgia has 8 mice and Tonya has 5 mice. How many more mice does Georgia have than Tonya?

Number line (measurement):
Mike hiked up the mountain trail 8 miles. Five of these miles were hiked after lunch. How many miles did Mike hike before lunch?

In all four problems, the answer is 3, because we know the addition fact $8 = 5 + 3$. For example, Eroll's \$8, when written as \$5 + \$3, shows that a \$5 movie ticket and the \$3 still in his pocket account for all of the \$8 he had originally. The cashier at the box office "took away" 5 of Eroll's 8 dollars, leaving him with \$3. Thus, the first problem is an example of the **take-away model** of subtraction. Alice, having read 5 chapters, wants to know how many more chapters she must read; the emphasis now is on what number is to be added to 5 to get 8, so the second problem illustrates the **missing-addend model** of subtraction. Similarly, the remaining problems illustrate the **comparison model** and the **number-line** (or **measurement**) **model** of subtraction.

The four basic conceptual models of the subtraction $8 - 5$ are visualized as follows:

Take-Away Model

1. Start with 8 objects (dollars).
2. Take away 5 objects.
3. How many objects are left?

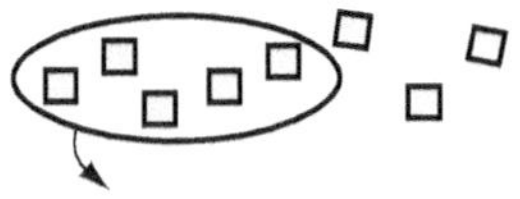

Missing-Addend Model

1. Start with 5 objects (chapters).
2. How many more objects are needed to give a total of 8 objects?

Comparison Model

1. Start with two collections, with 8 objects in one collection and 5 in the other (mice).

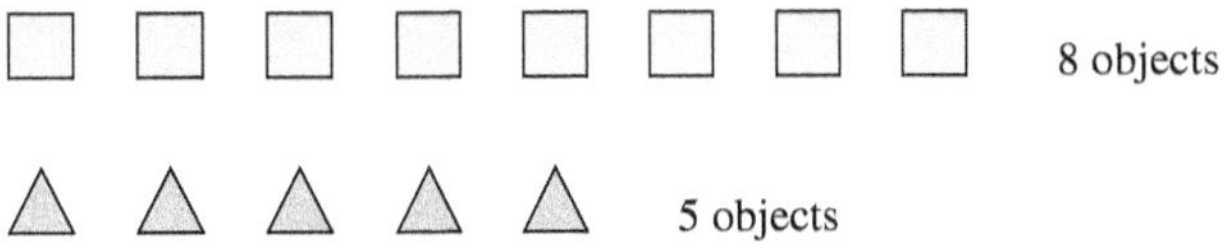

2. How many more objects (in this case, mice) are in the larger collection?

Number-Line Model

1. Move forward (to the right) 8 units (miles).

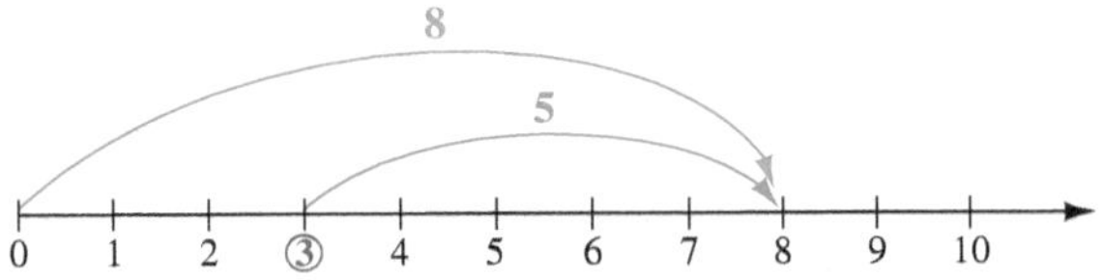

Put arrowheads together for subtracting and draw the subtrahend arrow backwards, from head to tail.

2. Remove a jump to the right of 5 units.
3. What is the distance from 0?

INTO THE CLASSROOM

Robin Crowell Discusses Physical and Pictorial Representations for Whole Numbers

Teachers are inspired when they see students excited about learning, helping to alleviate any hesitations in making changes to their instruction. The implementation of manipulatives and pictorial representations such as ten-frame mats provide a window into student thinking. Children naturally represent numbers using their fingers. Helping them to transfer their physical understanding to math manipulatives and then to representation is necessary for their continued understanding of concepts of our number system.

A teacher observed that the new second graders had difficulty counting a set of 20 counters to use for their lessons. She implemented the double ten-frame mat, first discussing with students how the tool would be useful. The students quickly went from being unable to gather the correct number of manipulatives to making statements such as "I need 4 more counters because $16 + 4 = 20$." Students demonstrated confidence in working with and making numbers to 20. The teacher subsequently observed the students drawing ten-frames to solve problems on assessments.

Next, second graders were learning to represent and compare three-digit numbers using base-ten manipulatives. After practicing building numbers, they moved to representing them pictorially. When comparing numbers, pictorial representations provided students an immediate visual of the place value where two numbers differ.

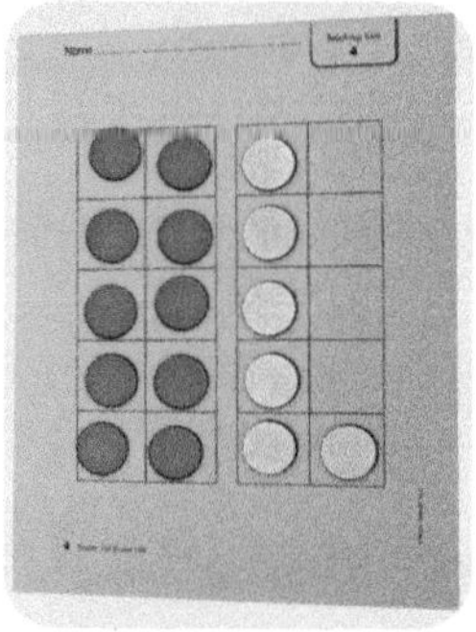

Notice that the head of the arrow representing 5 is positioned at the head of the arrow representing 8 and is drawn backwards, since it is subtracted. The result of the subtraction, 3, is shown by circling the 3.

It is important to notice that the whole numbers 8 and 5 are each represented by a right-pointing arrow. Since the outcome of the subtraction $8 - 5$ is 3, the arrow from 0 to 3 on the number line is also right pointing.

More generally, any whole number is represented on a number line as a right-pointing arrow. In Chapter 5, the whole numbers are enlarged to include the positive and negative integers. There, right-pointing arrows represent the positive integers (natural numbers) and left-pointing arrows represent the negative integers.

EXAMPLE 2.13

Identifying Conceptual Models of Subtraction

In each case, identify the conceptual model of subtraction that best fits the problem:

(a) Mary got 43 pieces of candy trick-or-treating on Halloween. Karen got 36 pieces. How many more pieces of candy does Mary have than Karen?

(b) Mary gave 20 pieces of her 43 pieces of candy to her sick brother, Jon. How many pieces of candy does Mary have left?

(c) Karen's older brother, Ken, collected 53 pieces of candy. How many more pieces of candy would Karen need to have as many as Ken?

(d) Ken left home and walked 10 blocks east along Grand Avenue, trick-or-treating. The last 4 blocks were after crossing Main Street. How far is Main Street from Ken's house?

Solution

(a) Comparison model
(b) Take-away model
(c) Missing-addend model
(d) Number-line model

The set of whole numbers is not closed under subtraction. For example, $2 - 5$ is not defined yet, since there is no whole number n that satisfies $2 = 5 + n$. Similar reasoning shows that subtraction is not commutative, so the order in which a and b are taken is important. Neither is subtraction associative, which means that parentheses must be placed with care in expressions involving subtractions. For example,

$$5 - (3 - 1) = 5 - 2 = 3, \quad \text{but} \quad (5 - 3) - 1 = 2 - 1 = 1.$$

COOPERATIVE INVESTIGATION
Diffy

The process in this activity is sometimes called "Diffy." The name comes from the process of taking successive differences of whole numbers, and the activity provides an interesting setting for practicing skills in subtraction. Work in pairs to check each other's calculations.

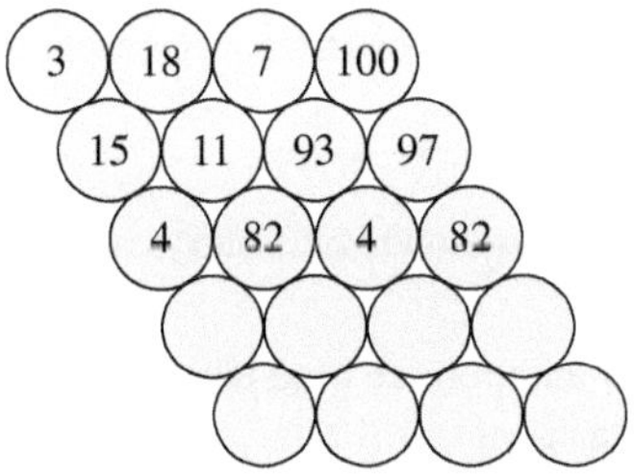

Directions

Step 1. Make an array of circles as shown, and choose four whole numbers to place in the top four circles.

Step 2. In the first three circles of the second row, write the differences of the numbers above and to the right and left of the circle in question, always being careful to subtract the smaller of these two numbers from the larger. In the fourth circle of the second row, place the difference of the numbers in the first and fourth circles in the preceding row, again subtracting the smaller number from the larger.

Step 3. Repeat step 2 to fill in successive rows of circles in the diagram. You may stop if you obtain a row of zeros.

Step 4. Replace the four numbers in the top row, and then repeat steps 1, 2, and 3 several times, each time replacing the four numbers in the top row with different numbers.

Questions

1. Do you think the process will always stop?
2. Can you find four numbers such that the process terminates at the first step? The second step? The third step? Try several sets of starting numbers.
3. On the basis of your work so far, what do you guess is the largest number of steps needed for the process to stop?
4. Can you find four starting numbers such that the process requires eight steps to reach termination?
5. Try Diffy with the starting numbers 17, 32, 58, and 107.

2.3 Problem Set

Understanding Concepts

1. Let $A = \{\text{apple, berry, peach}\}$, $B = \{\text{lemon, lime}\}$, and $C = \{\text{lemon, berry, prune}\}$.
   (a) Find (i) $n(A \cup B)$, (ii) $n(A \cup C)$, and (iii) $n(B \cup C)$.
   (b) In which case is the number of elements in the union *not* the sum of the number of elements in the individual sets?

2. Let $n(A) = 4$ and $n(A \cup B) = 8$.
   (a) What are the possible values of $n(B)$?
   (b) If $A \cap B = \varnothing$, what is the only possible value of $n(B)$?

3. Illustrate each of the given additions with a number-line diagram. Circle the value on the number line that corresponds to the sum.
   (a) $3 + 5$
   (b) $4 + 2$
   (c) $3 + (5 + 7)$

4. Illustrate each of the given additions with a number-line diagram. Circle the value on the number line that corresponds to the sum.
   (a) $5 + 3$
   (b) $0 + 6$
   (c) $(3 + 5) + 7$

5. Draw number strips (or rods) to illustrate the use of manipulatives.
   (a) $4 + 6 = 10$;
   (b) $2 + 8 = 8 + 2$; and
   (c) $3 + (2 + 5) = (3 + 2) + 5$.

6. Make up a word problem that uses the set model of addition to illustrate $30 + 28$.

7. Make up a word problem that uses the measurement model of addition to illustrate $18 + 25$.

8. Which of the given sets of whole numbers is closed under addition? If the set is not closed, give an example of two elements from the set whose sum is not in the set.

(a) $\{10, 15, 20, 25, 30, 35, 40, \ldots\}$
(b) $\{1, 2, 3, \ldots, 1000\}$
(c) $\{0\}$
(d) $\{1, 5, 6, 11, 17, 28, \ldots\}$
(e) $\{n \in N \mid n \geq 19\}$
(f) $\{0, 3, 6, 9, 12, 15, 18, \ldots\}$

9. What properties of addition are used in these equalities?

(a) $14 + 18 = 18 + 14$
(b) $12{,}345{,}678 + 97{,}865{,}342$ is a whole number.
(c) $18 + 0 = 18$
(d) $(17 + 14) + 13 = 30 + 14$
(e) $(12 + 15) + (5 + 38) = 50 + 20$

10. An easy way to add $1 + 2 + 3 + 4 + 5 + 6 + 7 + 8 + 9 + 10$ is to write the sum as $(1 + 10) + (2 + 9) + (3 + 8) + (4 + 7) + (5 + 6) = 11 + 11 + 11 + 11 + 11 = 55$.

(a) Compute $1 + 2 + 3 + \cdots + 20$. Describe your procedure.
(b) What properties of addition are you using to justify why your procedure works?

11. Illustrate each of the given subtractions with a number-line diagram. Remember to put the heads of the minuend and subtrahend arrows together.

(a) $7 - 3$ (b) $7 - 4$

12. Illustrate each of the given subtractions with a number-line diagram. Remember to put the heads of the minuend and subtrahend arrows together.

(a) $7 - 7$ (b) $7 - 0$

13. If $5 + 9 = 14$, then $9 + 5 = 14$, $14 - 9 = 5$, and $14 - 5 = 9$. Many elementary texts call such a group of four basic facts a "fact family."

(a) What is the fact family that contains $5 + 7 = 12$?
(b) What is the fact family that contains $12 - 4 = 8$?

14. (a) Draw four number-line diagrams to illustrate each of the number facts in the fact family (see the previous problem) $5 + 9 = 14$, $9 + 5 = 14$, $14 - 9 = 5$, and $14 - 5 = 9$.

(b) Repeat part (a) for the fact family that contains $11 - 4 = 7$.

15. For each of the following problems, which subtraction model—take away, missing addend, comparison, or measurement—corresponds best to that problem?

(a) Ivan solved 18 problems and Andreas solved 13 problems. How many more problems has Ivan solved than Andreas?
(b) On Malea's 12-mile hike to Mirror Lake, she came to a sign that said she had 5 miles left to reach the lake. How far has Malea walked so far?

16. For each of the following problems, which subtraction model—take away, missing addend, comparison, or measurement—corresponds best to that problem?

(a) Jake has saved \$45 toward the \$60 CD player he wants to buy. How much money does he have to add to his savings to purchase the player?
(b) Joshua has a book of 10 pizza coupons. If he used 3 coupons to buy pizza for Friday's party, how many coupons does he have left?

17. (**Writing**) For each of the following four conceptual models for subtraction, write a word problem that corresponds well to that problem.

(a) Take away
(b) Missing addend
(c) Comparison
(d) Measurement (number line)

18. Alejandro must read the last chapter of his book. It begins on the top of page 241 and ends at the bottom of page 257. How many pages must he read?

19. Notice that $(6 + (8 - 5)) - 2 = 7$, but a different placement of parentheses on the left side would give the statement $(6 + 8) - (5 - 2) = 11$. Place parentheses to turn these equalities into *true* statements:

(a) $8 - 5 - 2 - 1 = 2$
(b) $8 - 5 - 2 - 1 = 4$

20. Notice that $(6 + (8 - 5)) - 2 = 7$, but a different placement of parentheses on the left side would give the statement $(6 + 8) - (5 - 2) = 11$. Place parentheses to turn these equalities into *true* statements:

(a) $8 - 5 - 2 - 1 = 0$
(b) $8 + 5 - 2 + 1 = 12$
(c) $8 + 5 - 2 + 1 = 10$

21. Each circled number is the sum of the adjacent row, column, or diagonal of the numbers in the square array:

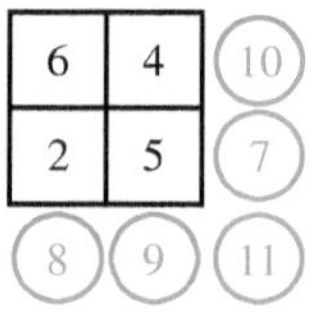

Fill in the missing entries in these patterns:

(a)
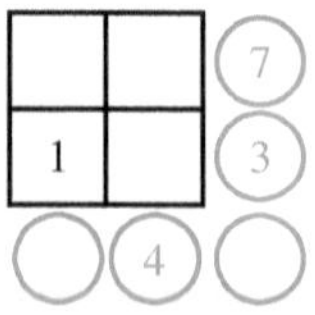

(b)
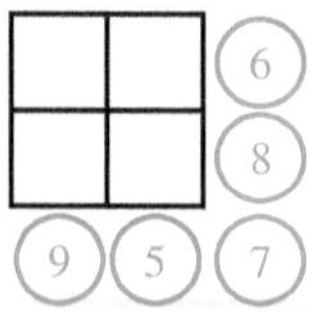

22. The first figure shown illustrates that the numbers 1, 2, 3, 4, 5, 6 can be placed around a triangle in such a way that the three numbers along any side sum to 9, which is shown circled.

Arrange the numbers 1, 2, 3, 4, 5, 6 to give the sums circled in the next three figures.

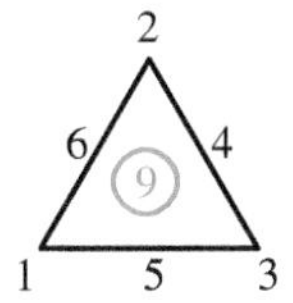

(a)

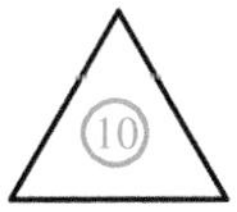

(b)

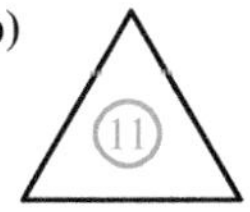

(c)

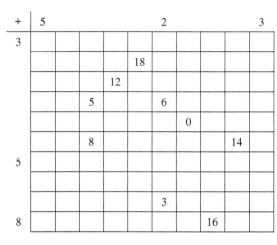

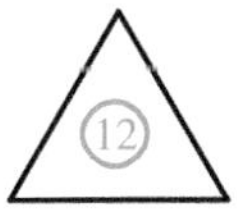

Into the Classroom

23. Suppose that Andrea told you that "Blake and I started out with the same number of marbles, but I gave him one of mine. Now Blake has one more marble than me." Why is Andrea's reasoning incorrect? How can you address her misunderstanding?

24. The addition operation can be represented with beans and loops of string, as illustrated in Figure 2.11. In words and pictures, describe how you would use this representation to illustrate the properties of whole-number addition to a youngster. For example, your figure illustrating the associative property will be something like the number-strip diagram in Figure 2.14.

25. Invent ways to use a double-six set of dominoes to teach children basic addition facts and addition properties. Notice that a blank half of the domino can represent zero.

26. How would you show that, for any whole number a, $a + 0 = a$ if

(a) you were starting from our definition of addition and the audience is your university class?

(b) you were talking to your class of fourth graders?

27. How would you explain to a student that filling in the boxes in (a) and (b) are related?

(a) $\square - 3 = 26$

(b) $\square = 29 - 3$

28. (**Writing**) Addition and subtraction problems arise frequently in everyday life. Consider such activities as scheduling time, making budgets, sewing clothes, making home repairs, planning finances, modifying recipes, and making purchases. Make a list of five addition and five subtraction problems you have encountered at home or in your work. Which conceptual model of addition or subtraction corresponds best to each problem?

29. **(a)** The table shown represents a partially complete scrambled addition table. The row and column headings are not set up in numerical order, and many entries are blank. See if the cooperative teams in your class can complete the table.

+	5					2				3
3										
					18					
				12						
			5			6				
							0			
			8						14	
5										
						3				
8								16		

(b) Have your cooperative team make up its own incomplete scrambled addition-table puzzle, similar to the one in part (a). It may be helpful to use two colors to separate the entries that are given from the entries that are determined by what is given. Trade puzzles among groups and see which presents the most challenge.

Responding to Students

30. Jane and Angel are twin sisters in the second grade who share a room and have a pet mouse collection of 8 mice. Angel says she adores 6 of them, but Jane says that that is impossible, since she loves 7 and there aren't 13 mice. Is Jane right? How would you explain to the sisters the conceptual reason for your answer?

31. In an attempt to do a three-digit subtraction, Yuan writes

$$\begin{array}{r} 706 \\ -\ 327 \\ \hline 421 \end{array}$$

(a) Identify what she is doing incorrectly.

(b) How will you help correct her error?

32. Phillip was asked to subtract 279 from 386 and writes

$$\begin{array}{r} 77 \\ 386 \\ -\ 279 \\ \hline 102 \end{array}$$

Identify what Phillip is doing incorrectly.

33. Carmen has learned an algorithm for subtraction. She uses it to find $52 - 17$ as follows:

$$52 + 10 = 62, \quad \text{so} \quad 52 + 10 = 50 + 12;$$
$$-(\underline{17} + \underline{10}) = -\underline{27}, \quad \text{so} \quad -(\underline{17 + 10}) = -(\underline{20 + 7})$$

So her answer is $30 + 5 = 35$.

(a) What properties does Carmen's algorithm use? Is her method correct?

(b) Now solve another subtraction problem, $71 - 38$, with Carmen's algorithm. Use the traditional algorithm to check your work.

Thinking Critically

34. If $A = \{a, b, c, d\}$ and $B = \{c, d, e, f, g\}$, then $n(A \cup B) = n(\{a, b, c, d, e, f, g\}) = 7, n(A) = 4, n(B) = 5,$ and $n(A \cap B) = n(\{c, d\}) = 2$. Since $7 = 4 + 5 - 2$, this suggests that

$$n(A \cup B) = n(A) + n(B) - n(A \cap B).$$

Use Venn diagrams to justify this formula for arbitrary finite sets A and B.

35. Since $12 \times 16 = 192$ and $5 \times 40 = 200$, it follows that among the first 200 natural numbers $\{1, 2, \ldots, 200\}$ there are 16 that are multiples of 12 and 40 that are multiples of 5. Just 3, namely, 60, 120, and 180, are multiples of *both* 5 and 12. Use the formula of problem 34 to find how many natural numbers in the set $\{1, 2, \ldots, 200\}$ are divisible by *either* 12 or 5 or both.

36. Adams's Magic Hexagon. In 1957, Clifford W. Adams discovered a magic hexagon in which the sum of the numbers in any "row" is 38. Fill in the empty cells of the partially completed hexagon shown, using the whole numbers $6, 7, \ldots, 15$, to re-create Adams's discovery. When completed, each cell will contain one of the numbers $1, 2, \ldots, 19$.

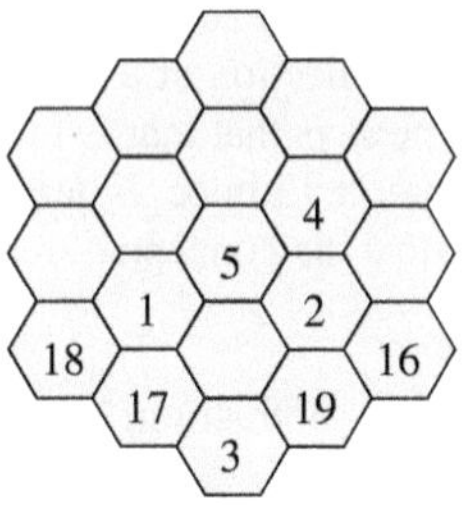

$18 + 17 + 3 = 38$
$3 + 19 + 16 = 38$

37. A Magic Hexagram. The numbers $1, 2, \ldots, 12$ can be entered into the 12 regions of the diagram that follows so that each of the six rows of five triangles (shown by the arrows) contains numbers that sum to the magic constant 33. The first six numbers have been put into place, so you are challenged to enter the last six numbers, 7 through 12, to finish this magic hexagram discovered in 1991 by the mathematicians Brian Bolt, Roger Eggleton, and Joe Gilks.

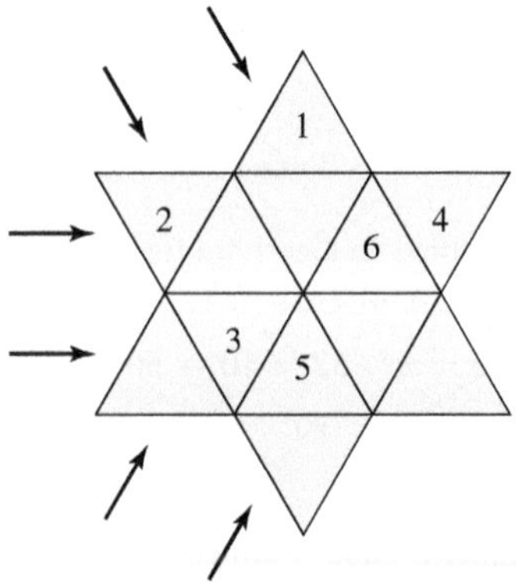

38. Other than all whole numbers, show that there is a nonempty subset of the whole numbers that is closed under subtraction. Is there more than one subset?

39. The set C contains 2 and 3 and is closed under addition.

(a) What whole numbers must be in C?

(b) What whole numbers may not be in C?

(c) Are there any whole numbers definitely not in C?

(d) How would your answers to (a), (b), and (c) change if 2 and 4 instead of 2 and 3 were contained in C?

40. The **triangular numbers** $t_1 = 1, t_2 = 3, t_3 = 6, \ldots$ are shown in Figure 2.12.

(a) Complete the following table of the first 15 triangular numbers:

n	1	2	3	4	5	6	7	8	9	10	11	12	13	14	15
t_n	1	3	6	10											

(b) The first 10 natural numbers can be expressed as sums of triangular numbers. For example, $1 = 1, 2 = 1 + 1, 3 = 3, 4 = 1 + 3, 5 = 1 + 1 + 3, 6 = 6, 7 = 1 + 6, 8 = 1 + 1 + 6, 9 = 3 + 6$, and $10 = 10$. Show that the natural numbers 11 through 25 can be written as a sum of triangular numbers. Use as few triangular numbers as possible each time.

(c) Choose 5 more numbers at random (don't look at your table!) between 26 and 120, and write each of them as a sum of as few triangular numbers as possible. What is the largest number of triangular numbers needed?

(d) On July 10, 1796, the 19-year-old Carl Friedrich Gauss wrote in his notebook, "EUREKA! NUM $= \triangle + \triangle + \triangle$." What theorem do you think Gauss had proved?

41. Fibonacci Sums. Recall the **Fibonacci numbers** (of Example 1.6) are 1, 1, 2, 3, 5, 8, 13, 21, 34, 55, 89, 144, ... Sameer claims that any natural number can be written as a sum of two or more distinct Fibonacci numbers. For example, $29 = 21 + 5 + 3$ and $55 = 34 + 21$. Do you think Sameer might be right? Investigate this possibility yourself, and offer your opinion to support or refute Sameer's contention.

State Assessments

42. (Massachusetts, Grade 4)
Corey worked these problems.

$2 + 8 = 10$	$6 + 6 = 12$	$10 + 4 = 14$
$12 + 2 = 14$	$8 + 4 = 12$	$2 + 6 = 8$

When Corey finished, he said, "I think that the sum of ANY two even numbers is always an even number." Maya looked at Corey's work. She thought a bit and said, "I think that the sum of any two ODD numbers is always an odd number."

(a) Is Corey correct? Explain the reasons for your answer by using pictures, numbers, or words.

(b) Is Maya correct? Explain the reasons for your answer by using pictures, numbers, or words.

43. (Massachusetts, Grade 4)
Which of the following problems CANNOT be solved using the number sentence below?

$$15 - 8 = \square$$

A. Siu Ping had 15 trading cards. She gave 8 to Jim. How many does she have now?
B. Siu Ping needs 15 more trading cards than Jim has. Jim has 8. How many does Siu Ping need?
C. Siu Ping has 15 trading cards. Jim has 8. How many more does Siu Ping have than Jim?
D. Siu Ping needs 15 trading cards. She has 8. How many more does she need?

44. (Grade 7)
Which situation is best represented by the equation $x - 4 = 16$?

A. Alisa picked 16 apples and ate $\frac{1}{4}$ of them. What is x, the number of apples she had left?
B. Jose ran for 16 minutes and walked for 4 minutes. What is x, the difference between the time he spent running and the time he spent walking?
C. James spent \$4 of his allowance and had \$16 left. What is x, the total amount of Jordan's allowance?
D. Annette has hit 4 of the last 16 balls pitched. What is x, the total number of balls pitched?

45. (Grade 3)
When you subtract one of these numbers from 800, the answer is less than 300. Which number is it?

A. 250   B. 152
C. 175   D. 525

46. (Grade 3)

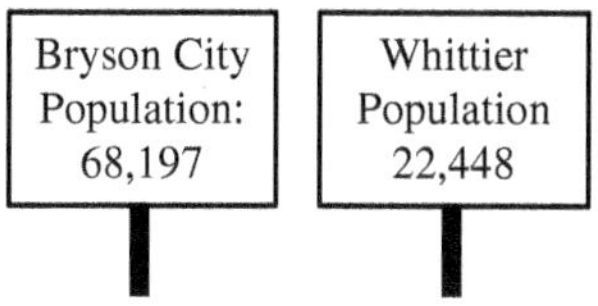

How much *greater* is the population of Bryson City than Whittier?

A. 45,749   B. 27,651
C. 87,745   D. 90,645

47. (Grade 3)
Barbara was born in 1975. When she was 16 years old, she went to university. Four years later she started her own company. How can you find out what year Barbara started her business?

A. $1975 - 16 + 4$
B. $1975 - 16 - 4$
C. $1975 + 16 + 4$
D. $1975 + 16 - 4$

48. (California, Grade 4)
What number goes in the box to make this number sentence true?

$$54 + \square = 71$$

A. 7   B. 17
C. 19   D. 27

2.4 Multiplication and Division of Whole Numbers

Multiplication of Whole Numbers

There are six conceptual models for the multiplication of two numbers (as we will see below): **multiplication as repeated addition,** the **array model for multiplication,** the **rectangular area model for multiplication,** the **skip-count model for multiplication,** the **multiplication tree model,** and the **Cartesian product model.** Each of these models provides a useful conceptual and visual representation of the multiplication operation. For a given context in a problem, it is frequently clear which model provides the most natural representation, but often there are several models that are each a reasonable choice. We will discuss now what they are and the properties that they have.

Multiplication as Repeated Addition

Misha has an after-school job at a local bike factory. Each day, he has a 3-mile round-trip walk to the factory. At his job, he assembles 4 hubs and wheels. How many hubs and wheels does he assemble in 5 afternoons? How many miles does he walk to and from his job each week?

These two problems can be answered by repeated addition. Misha assembles

$$4 + 4 + 4 + 4 + 4 = 20$$

Sum of 5 fours, written $5 \cdot 4$

hubs and wheels. Repeated addition is illustrated by the set diagram shown in Figure 2.15.

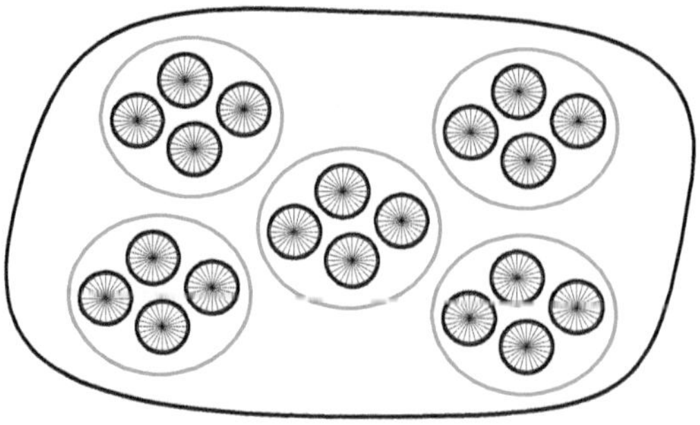

Figure 2.15
A set model shows that 5 times 4 is 20, since $4 + 4 + 4 + 4 + 4 = 20$ by repeated addition

Misha walks

$$3 + 3 + 3 + 3 + 3 = 15$$ Sum of 5 threes, written $5 \cdot 3$

miles each week. This is illustrated by the number-line model shown in Figure 2.16. It is also called **skip counting.** As a manipulatives, it gives a visualization of multiplication in the number line.

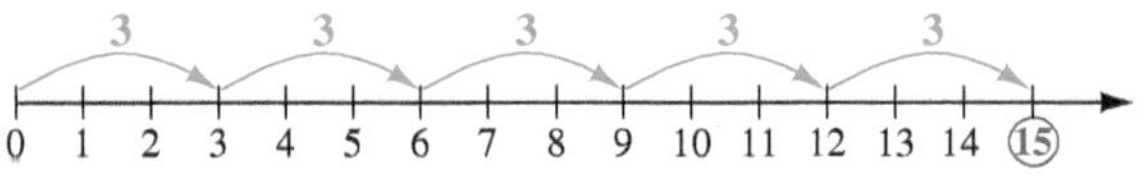

Figure 2.16
Number-line model to show that 5 times 3 is 15

If Misha worked only one day in the week, he would assemble $1 \cdot 4 = 4$ wheels. If he were sick all week and missed work entirely, he would not assemble any wheels; therefore $0 \cdot 4 = 0$.

Since multiplication is defined for all pairs of whole numbers, and the outcome is also a whole number, we see that multiplication is a binary operation that is closed on the set of whole numbers.

DEFINITION Multiplication of Whole Numbers as Repeated Addition

Let a and b be any two whole numbers. Then the **product** of a and b, written $a \cdot b$, is defined by

$$a \cdot b = \underbrace{b + b + \cdots + b}_{a \text{ addends}} \text{ when } a \neq 0$$

and by

$$0 \cdot b = 0.$$

The dot symbol for multiplication is often replaced by a cross, $\times$ (not to be mistaken for the letter x), or by a star, * (asterisk), the symbol computers use most often. Sometimes no symbol at all is used, or parentheses are placed around the factors. Thus, the expressions

$$a \cdot b, \quad a \times b, \quad a * b, \quad ab, \quad \text{and} \quad (a)(b)$$

all denote the multiplication of a and b. Each whole number, a and b, is a **factor** of the product $a \cdot b$, and often $a \cdot b$ is read "a times b."

The Array Model for Multiplication

Suppose Lida, as part of her biology research, planted 5 rows of bean seeds and each row contained 8 seeds. How many seeds did she plant in her rectangular plot?

The 40 seeds Lida planted form a 5-by-8 rectangular array, as shown in Figure 2.17. In the array model, the numbers of rows and columns in the rectangular array easily identify the factors in the multiplication. Also, the result of the product is found by counting the number of individual objects in the array. Notice that this picture (a manipulative) shows that the two ways of looking at the 40 seeds are $5 \cdot 8$ and $8 \cdot 5$, both of which are 40.

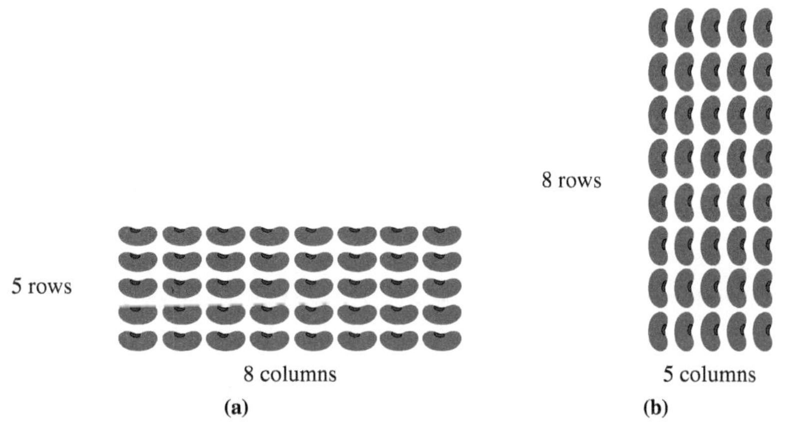

Figure 2.17
Array model showing that (a) $5 \cdot 8 = 40$ (b) $8 \cdot 5 = 40$

The Rectangular Area Model for Multiplication

Janet wants to order square ceramic tiles to cover the floor of her 4-foot-by-6-foot hallway. If the tiles are each 1 square foot, how many will she need to order? The rectangular area model, as shown in Figure 2.18, shows that 24 tiles are required.

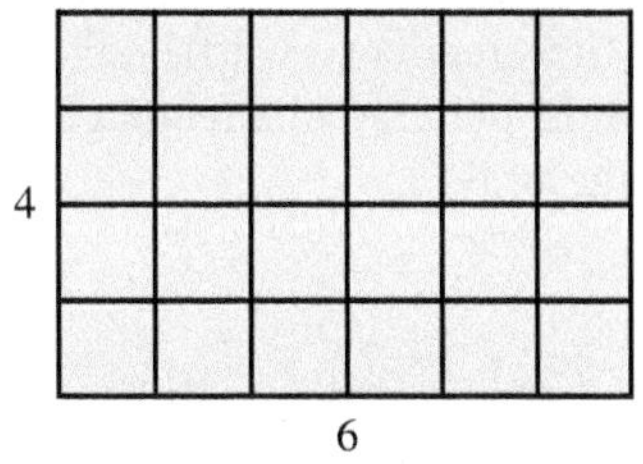

Figure 2.18
Rectangular area model showing that $4 \times 6 = 24$

In this model, the dimensions of the rectangle correspond to the factors and the area of the rectangle corresponds to the value of the product. In later chapters, we will show that the rectangular area model for multiplication is especially important because it extends from whole-number multiplication to the multiplication of rational numbers, integers, and even real numbers. For example, if Janet's floor measured $4\frac{1}{2}$ by 6 feet, the array would show that Janet needs $4\frac{1}{2} \times 6 = 27$ tiles, with some of the tiles needing to be cut.

The Skip-Count Model for Multiplication

In the early primary grades, examples similar to Figure 2.16 are a popular way to introduce multiplication. One other way of interpreting 5 times 3 is to start with 0 and skip to 3 as our first position, then skip 3 more to 6 as our second position (or, said another way, skip from 3 to 6 and count of 2). We then skip from 6 to 9 (count of 3), 9 to 12 (count of 4), and, finally, 12 to 15 (count of 5). We have skipped by 3 five times to get 15. This method for $a \cdot b$ is called "skip counting" because we skip by the number b exactly a times.

The Multiplication Tree Model

Melissa has a box of 4 flags, colored red, yellow, green, and blue. How many ways can she display 2 of the flags on a flagpole? It is useful to think of the two decisions Melissa must make. First, she must select the upper flag, and next, she must choose the lower flag from those remaining in the box. There are 4 choices for the upper flag, and for *each* of these choices there are 3 choices for the lower flag. The multiplication tree shown in Figure 2.19 represents the product 4×3.

The multiplication tree in Figure 2.19 looks more like a tree if you turn it upside down. Multiplication trees can also grow from the right or the left.

The Cartesian Product Model of Multiplication

At Jackson's Ice Cream Shop, a customer can order either a sugar or a waffle cone and one of four flavors of ice cream: vanilla, chocolate, mint, and raspberry. Any ice cream cone purchase can be

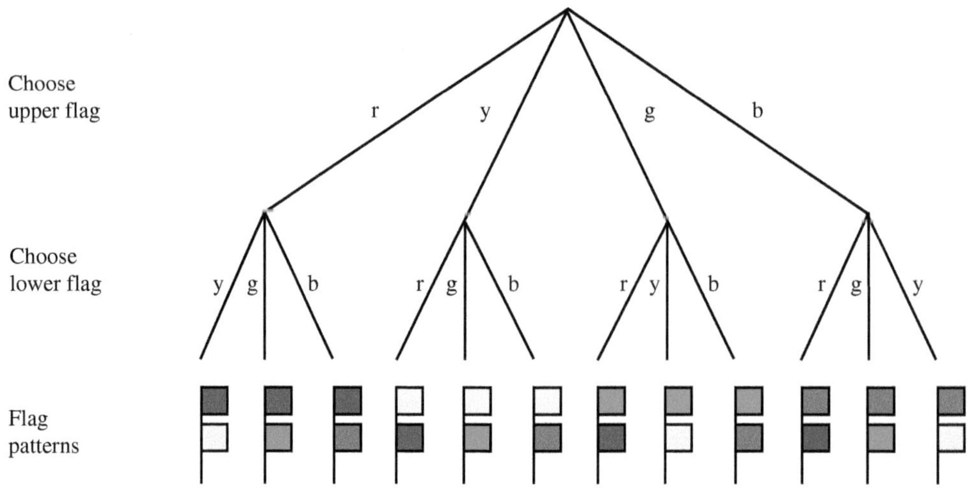

Figure 2.19
A multiplication tree showing that $4 \times 3 = 12$

written as an **ordered pair (a, b),** where the first component a of the ordered pair indicates the type of cone and the second component b of the ordered pair indicates the flavor. For example, if $C = \{s, w\}$ and $F = \{v, c, m, r\}$ are the respective sets of cone types and flavors, then an order for raspberry ice cream in a waffle cone corresponds to the ordered pair (w, r). The total set of ice cream orders can be pictured in a rectangular array much like the figures used to model multiplication as a rectangular array:

	Flavor			
Type of Cone	**v**	**c**	**m**	**r**
s	(s, v)	(s, c)	(s, m)	(s, r)
w	(w, v)	(w, c)	(w, m)	(w, r)

The set of ordered pairs shown in the table is called the **Cartesian product** of C and F, written $C \times F$, where the cross symbol $\times$ should not be confused with the letter x. The name "Cartesian" honors the French mathematician and philosopher René Descartes (1596–1650). Here is the general definition of the Cartesian product of two sets:

DEFINITION Cartesian Product of Sets

The **Cartesian product** of sets A and B, written $A \times B$, is the set of all ordered pairs whose first component is an element of set A and whose second component is an element of set B. That is,

$$A \times B = \{(a, b) \mid a \in A \text{ and } b \in B\}.$$

At Jackson's, there are $n(C) = 2$ ways to choose the type of cone and $n(F) = 4$ ways to choose the flavor of ice cream. This gives $n(C \times F) = 2 \cdot 4 = 8$ ways to order an ice cream cone. More generally, the Cartesian product of sets gives us an alternative way to define the multiplication of whole numbers:

ALTERNATIVE DEFINITION Multiplication of Whole Numbers via the Cartesian Product

Let a and b be whole numbers, and suppose that A and B are any sets for which $a = n(A)$ and $b = n(B)$. Then $a \cdot b = n(A \times B)$.

Since $\varnothing \times B = \varnothing$ (there are no ordered pairs in $\varnothing \times B$, since no first component can be chosen from $\varnothing$) and $0 = n(\varnothing)$, this definition of multiplication is consistent with the earlier definition of multiplication by 0—that is, $0 \cdot b = b \cdot 0 = 0$.

EXAMPLE 2.14

Using the Cartesian Product Model of Multiplication

To get to work, Juan either walks, rides the bus, or takes a cab from his house to downtown. From downtown, he either continues on the bus the rest of the way to work or catches the train to his place of business. How many ways can Juan get to work?

Solution

If $A = \{w, b, c\}$ is the set of possibilities for the first leg of his trip and $B = \{b, t\}$ is the next set of choices, then Juan has, altogether,

$$3 \cdot 2 = n(A \times B) = n(\{(w, b), (w, t), (b, b), (b, t), (c, b), (c, t)\}) = 6$$

ways to commute to work. We notice that it doesn't matter whether or not A and B have an element in common.

Properties of Whole-Number Multiplication

It follows from the definition that the set of whole numbers is **closed under multiplication:** The product of any two whole numbers is a unique whole number. We also observed earlier that $0 \cdot b = 0$ and $1 \cdot b = b$ for all whole numbers b.

By rotating a rectangular array through 90°, we interchange the number of rows and the number of columns in the array, but we do not change the total number of objects in the array. Thus, $a \cdot b = b \cdot a$, which demonstrates the **commutative property of multiplication.** An example is shown in Figure 2.20. The manipulatives are two rectangles of the same size, allowing us to look at the areas as numbers of little squares. One rectangle's area is $4 \cdot 7 = 28$. The other rectangle's area is $7 \cdot 4 = 28$. Clearly, from these manipulatives, $a \cdot b = b \cdot a$.

Figure 2.20
Multiplication is commutative: $4 \cdot 7 = 7 \cdot 4$

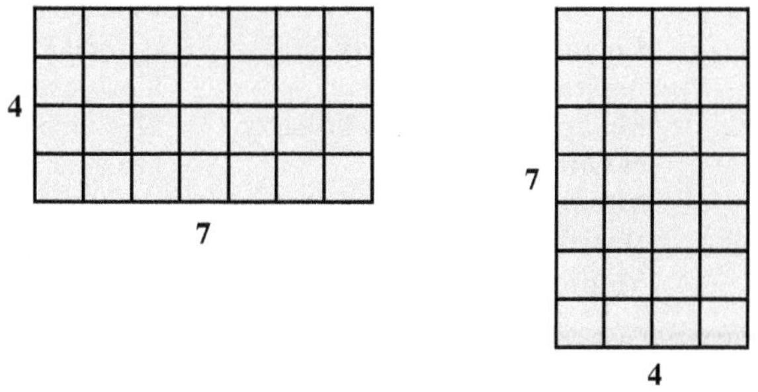

Since, as we already have seen, $1 \cdot b = b$ for any whole number b, it follows by the commutative property that $1 \cdot b = b \cdot 1 = b$. This makes 1 a **multiplicative identity.**

The reason that $a \cdot (b \cdot c) = (a \cdot b) \cdot c$ (the **associative property**) is that both sides of this equation can be represented as the volume of a box (rectangular prism, which is a three-dimensional manipulative) whose sides are a, b, and c. Only the length, width, and height are interchanged. (See Figure 2.21 for the case in which $a = 4$, $b = 3$, and $c = 5$.) This means that an expression such as $4 \cdot 3 \cdot 5$ is meaningful without parentheses: The product is the same for both ways parentheses can be placed.

Figure 2.21
Multiplication is associative:
$4 \cdot (3 \cdot 5) = (4 \cdot 3) \cdot 5$

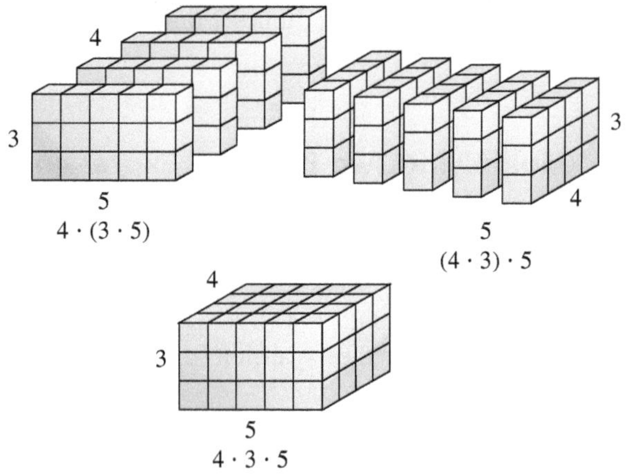

There is one more important property, the **distributive property,** that relates multiplication and addition. This property is the basis for the multiplication algorithm discussed in the next chapter. The distributive property can be nicely visualized by the rectangular area model. Figure 2.22 illustrates that $4 \cdot (6 + 3) = (4 \cdot 6) + (4 \cdot 3)$; that is, the factor 4 *distributes* itself over each term in the sum $6 + 3$. We can visualize Figure 2.22 as a swimming pool of size 4 yds by $(6 + 3)$ yds or 4 yds $\times$ 6 yds + 4 yds $\times$ 3 yds.

Figure 2.22
Multiplication is distributive over addition:
$4 \cdot (6 + 3) = (4 \cdot 6) + (4 \cdot 3)$

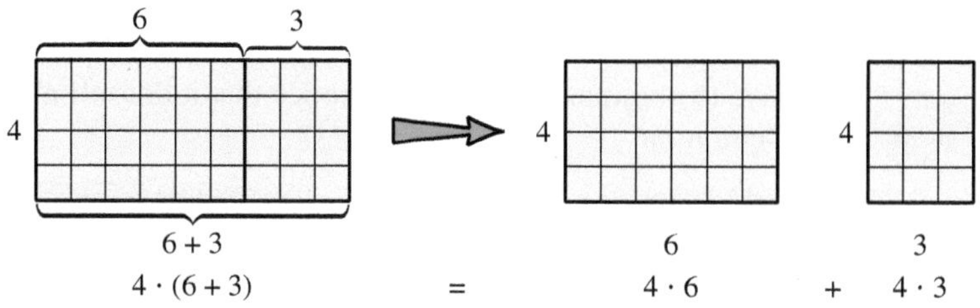

Shown next is a summary of the properties of multiplication on the whole numbers. Each property can be proved from the definition of multiplication of whole numbers.

THEOREM Properties of Whole-Number Multiplication

Closure Property	If a and b are any two whole numbers, then $a \cdot b$ is a unique whole number.
Commutative Property	If a and b are any two whole numbers, then $a \cdot b = b \cdot a$.
Associative Property	If a, b, and c are any three whole numbers, then $a \cdot (b \cdot c) = (a \cdot b) \cdot c$.
Multiplicative Identity Property of One	The number 1 is the unique whole number for which $b \cdot 1 = 1 \cdot b = b$ holds for all whole numbers b.
Multiplication-by-Zero Property	For all whole numbers b, $0 \cdot b = b \cdot 0 = 0$.
Distributive Property of Multiplication over Addition	If a, b, and c are any three whole numbers, then $a \cdot (b + c) = (a \cdot b) + (a \cdot c)$ and $(a + b) \cdot c = (a \cdot c) + (b \cdot c)$.

SMP 7 Look for and make use of structure.

"Mathematically proficient students look closely to discern a pattern or structure. Young students, for example, might notice that three and seven more is the same amount as seven and three more, or . . . . Later, students will see 7×8 equals the well-remembered $7 \times 5 + 7 \times 3$, in preparation for learning about the distributive property."

At the elementary and middle school levels, patterns become algebra, and the rules to work with these patterns are evident in the preceding theorem, *Properties of Whole-Number Multiplication.* These properties simplify mathematics a great deal. For example, how would you multiply 10 times 9 times 8 times 7, . . . , times 2 times 1, times 0? Is there a much faster way to get an answer?

EXAMPLE 2.15 **Multiplying Two Binomial Expressions**

(a) Use the properties of multiplication to justify the formula $(a + b)(c + d) = ac + ad + bc + bd$. Each of the factors $(a + b)$ and $(c + d)$ are called **binomials,** since each factor has two terms.
(b) Visualize $(2 + 4)(5 + 3)$ with an area model drawn on squared paper.
(c) Illustrate the expansion $(a + b)(c + d) = ac + ad + bc + bd$ with the area model of multiplication.
(d) Visualize $16 \cdot 28 = (10 + 6) \cdot (20 + 8) = 200 + 80 + 120 + 48$ with an area diagram.
(e) Show that $(x + 7) \cdot (2x + 5) = 2x^2 + 5x + 14x + 35$ with an area diagram.

Solution

(a) $(a + b)(c + d) = (a + b)c + (a + b)d$ Distributive property
$= ac + bc + ad + bd$ Distributive property
$= ac + ad + bc + bd$ Commutative property of addition

(b)

=

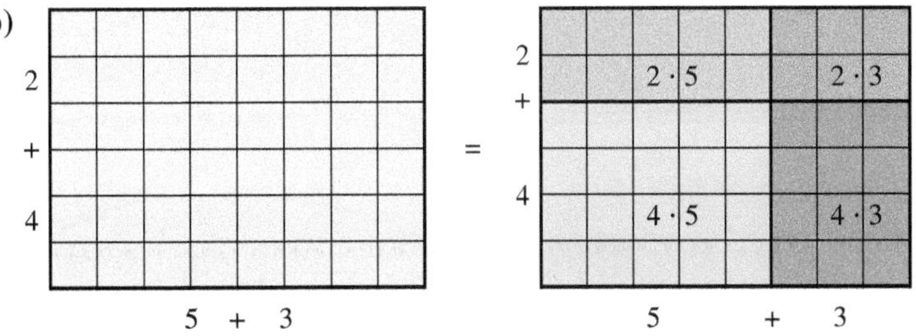

(c)

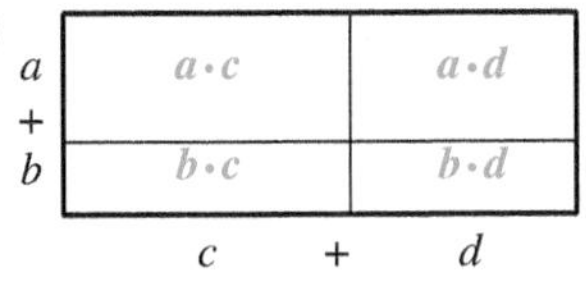

	c	$+$	d
a	$a \cdot c$		$a \cdot d$
$+$			
b	$b \cdot c$		$b \cdot d$

(d)

	20	$+$	8
6	120		48
$+$			
10	200		80

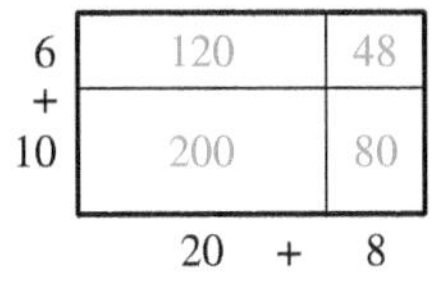

(e)

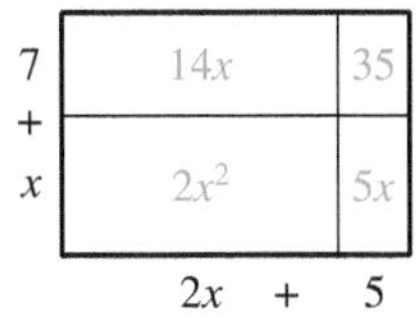

Division of Whole Numbers

The division of whole numbers is more difficult than multiplication. After all given a and b, which are whole numbers, $a \cdot b$ is always a whole number whereas $a \div b$ may not be. There are three conceptual models for the division $a \div b$ of a whole number a by a nonzero whole number b: the **repeated-subtraction** model, the **partition** model, and the **missing-factor** model. We will now discuss what these models or manipulatives are and the properties they have.

INTO THE CLASSROOM
Katie Busbey: The Partition Model of Division

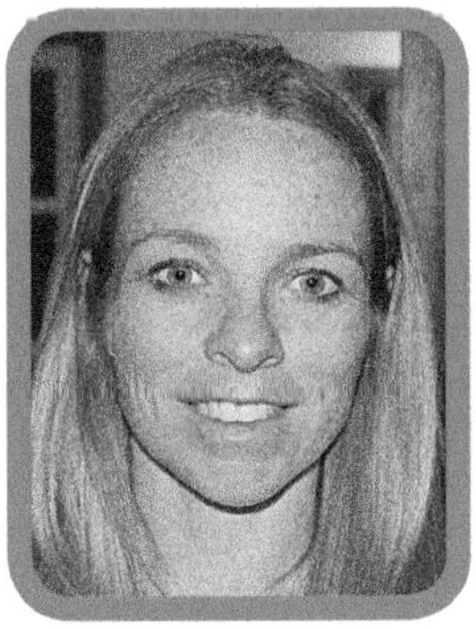

To build student interest, when introducing the partition model of division, I have the students plan party gift bags. At various stations throughout the classroom, students separate items into their allotted number of gift bags. For example, students may have 4 gift bags they need to fill. One station may include 28 pieces of candy, another station 24 stickers, a third station 16 unfilled balloons, and so forth. Students move from station to station using the real items to divide and then filling their bags with pictures of the items, leaving the real items for the next group. Groups use facilitation questions at each station to talk about the process of partitioning and to make connections to multiplication. Upon completion of the activity, the class discusses how many of each item should be included in the 4 gift bags and the relationships between the total number of items and number of items in the bag.

Students relate to the process of filling gift bags, especially with items that interest them. This allows students to make a personal connection to division and to apply their previous knowledge to the new concept.

In future grade levels, a similar activity can help introduce remainders and open a discussion with students about what they should do with the gift bag items that remain after filling all of the required gift bags. This activity provides a connection to interpreting the remainder and what it means in various situations.

The Repeated-Subtraction Model of Division

Ms. Rislov has 28 students in her class whom she wishes to divide into cooperative learning groups of 4 students per group. If each group requires a set of Cuisenaire® rods, how many sets of rods must Ms. Rislov have available? The answer, 7, is pictured in Figure 2.23 and is obtained by counting how many times groups of 4 can be formed, starting with 28. Thus, $28 \div 4 = 7$. The repeated-subtraction model can be realized easily with physical objects; the process is called **division by grouping.** Since groups of 4 are being "measured out" of the class of 28, repeated subtraction is also called **measurement division.**

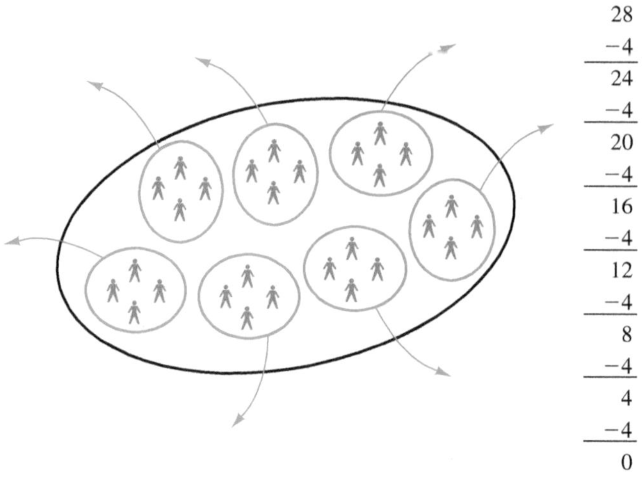

Figure 2.23
Division as repeated subtraction:
$28 \div 4 = 7$ because 4s can be subtracted from 28 seven times

The Partition Model of Division

When Ms. Rislov checked her supply cupboard, she discovered she had only 4 sets of Cuisenaire® rods to use with the 28 students in her class. How many students must she assign to each set of rods? The answer, 7 students in each group, is depicted in Figure 2.24. The partition model is also realized easily with physical objects, in which case the process is called **division by sharing** or **partitive division** (named after the word *partition*).

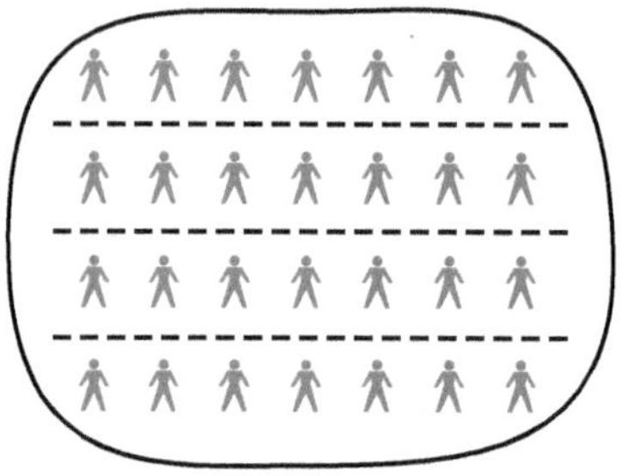

Figure 2.24
Division as a partition:
$28 \div 4 = 7$ because when 28 objects are partitioned into 4 equal-sized sections, there are 7 objects in each partition

The Missing-Factor Model of Division

In the repeated-subtraction model, $28 \div 4 = 7$ because, when grouped by 4s, seven groups will be formed. That is, $28 = 4 + 4 + 4 + 4 + 4 + 4 + 4 = 7 \cdot 4$. However, in the partition model, $28 \div 4 = 7$ because $28 = 7 + 7 + 7 + 7 = 4 \cdot 7$. In both cases, the division $28 \div 4$ can be viewed as finding the factor c for which $28 = 4 \cdot c$ or $28 = c \cdot 4$. The missing-factor model is usually the concept adopted to define division formally:

DEFINITION Division in Whole Numbers
Let a and b be whole numbers with $b \neq 0$. Then $a \div b = c$ if, and only if, $a = b \cdot c$ for a unique whole number c.

When we defined subtraction in the last section, we interpreted it in terms of algebra. What would an algebraic equation look like to describe division in whole numbers? By analogy, it would be as follows: Let a and b be whole numbers with $b \neq 0$. Then b divides a if, and only if, there is a unique whole number x for which $a = bx$. As an illustration, we can solve the equation $3 = 2x$

for x, but we won't get a whole number for an answer. That means that 2 does not divide 3. Said another way, the integers are not closed under division. Chapter 4, "Number Theory," will go into depth about this subtle difference between multiplication and division.

The symbol $a \div b$ is read "a divided by b," where a is the **dividend** and b is the **divisor.** If $a \div b = c$, then we say that ***b* divides *a*** or ***b*** is a **divisor** of a, and c is called the **quotient:**

$$a \div b = c$$

dividend (↑ a) — divisor (↑ b) — quotient (↑ c)

Division is also symbolized by a/b or $\frac{a}{b}$; the slash notation is used by most computers.

Any multiplication fact with nonzero factors is related to three other equivalent facts that form a **multiple-division fact family.** For example,

$$3 \cdot 5 = 15, \qquad 15 \div 3 = 5,$$
$$5 \cdot 3 = 15, \quad \text{and} \quad 15 \div 5 = 3$$

form a fact family. Note that a fact family can be represented with manipulatives such as the rectangular array model, as shown in Figure 2.25 with arrays of colored discs.

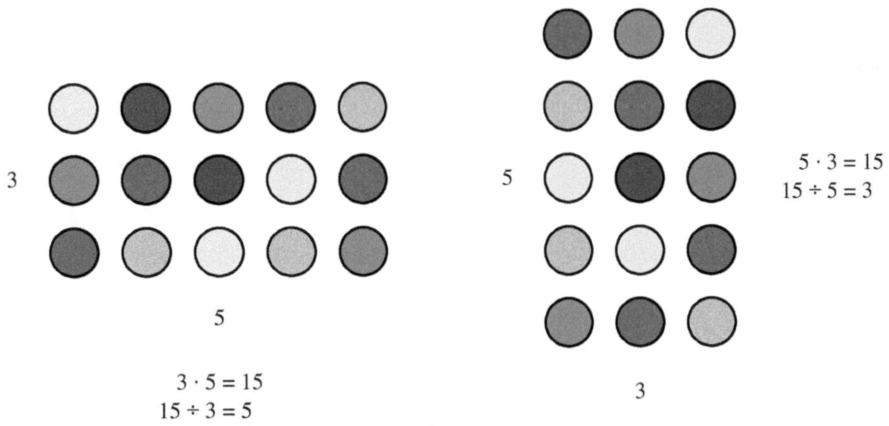

Figure 2.25 Rectangular arrays, formed with colored discs, illustrate the fact family $3 \cdot 5 = 15$, $15 \div 3 = 5$, $5 \cdot 3 = 15$, and $15 \div 5 = 3$

EXAMPLE 2.16

Computing Quotients with Manipulatives

Suppose you have 78 number tiles. Describe how to illustrate $78 \div 13$ with the tiles, using each of the three basic conceptual models for division.

Solution

(a) Repeated subtraction. Remove groups of 13 tiles each. Since 6 groups are formed, $78 \div 13 = 6$.

(b) Partition. Partition the tiles into 13 equal-sized parts. Since each part contains exactly 6 tiles, $78 \div 13 = 6$.

(c) Missing factor. Use the 78 tiles to form a rectangle with 13 rows. Since it turns out there are 6 columns in the rectangle, $78 \div 13 = 6$.

Division by Zero Is Undefined

The definition of division tells us that $12 \div 3 = 4$, since 4 is the unique whole number for which $12 = 3 \times 4$. Division *into zero* is also defined. For example, $c = 0$ is the unique whole number for which $0 = 5 \times c$, so $0 \div 5 = 0$. However, division *by zero* (or "$a \div 0$") is not defined for any whole number a. The reason for this can best be explained by taking the cases $a \neq 0$ and $a = 0$ separately.

Case 1: $a \neq 0$. In this case, "$a \div 0$" would be equivalent to finding the missing factor c that makes $a = c \cdot 0$. But $c \cdot 0 = 0$ for all whole numbers c, so there is no solution when $a \neq 0$. Thus, $a \div 0$ is undefined for $a \neq 0$.

Case 2: $a = 0$. In this case, "$0 \div 0$" is equivalent to finding a *unique* whole number c for which $0 = 0 \cdot c$. But that equation is satisfied by every choice for the factor c. Since no *unique* factor c exists, the division of 0 by 0 is also undefined.

In the division $a \div b$, there is no restriction on the dividend a. For all $b \neq 0$, we have $0 \div b = 0$, since $0 = b \cdot 0$.

The multiplicative identity 1 has two simple but useful relationships to division:

$$\frac{b}{b} = 1 \quad \text{for all } b, \quad \text{where } b \neq 0;$$

$$\frac{a}{1} = a \quad \text{for all } a.$$

Division with Remainders

Consider the division problem $27 \div 6$. There is no whole number c that satisfies $27 = c \cdot 6$, so $27 \div 6$ is not defined in the whole numbers. That is, the set of whole numbers is *not closed* under division.

By allowing the possibility of a **remainder,** we can extend the division operation. Consider $27 \div 6$, where division is viewed as repeated subtraction. Four groups of 6 can be removed from 27. This leaves 3, which are too few to form another group of 6. Thus, we can write

$$27 = 4 \cdot 6 + 3.$$

Here, 4 is called the **quotient** and 3 is the **remainder.** This information is also written

$$27 \div 6 = 4 \text{ R } 3,$$

where R separates the quotient from the remainder. A visualization of division with a remainder is shown in Figure 2.26. We are pulling away 4 groups of 6 each and one left with 3 individuals in Figure 2.26b.

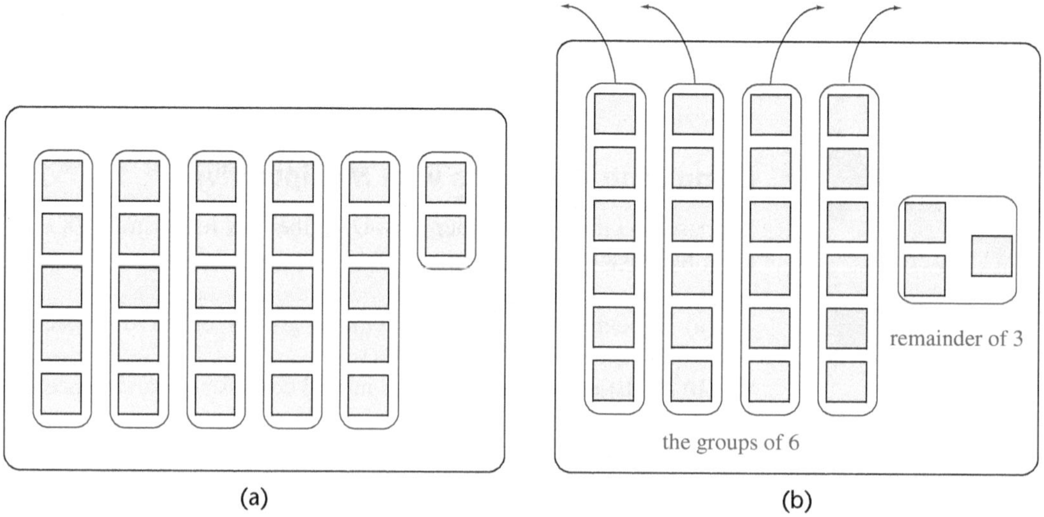

Figure 2.26
(a) collection of 27 squares
(b) Representing division with a remainder: $27 \div 6 = 4 \text{ R } 3$

In general, we have the following important result:

THEOREM The Division Algorithm

Let a and b be whole numbers with $b \neq 0$. Then there is a unique whole number q called the **quotient** and a unique whole number r called the **remainder** such that

$$a = q \cdot b + r, \qquad 0 \leq r < b.$$

It is common to write

$$a \div b = q \text{ R } r$$

if $a = q \cdot b + r, 0 \leq r < b$. The quotient q is the largest whole number of groups of b objects that can be formed from a objects, and the remainder r is the number of objects that are left over; hence, r is called the "remainder." The remainder is 0 if, and only if, b divides a according to the definition of division in whole numbers and $b \neq 0$. This important case is explored fully in Chapter 4.

Why Does Quotient with Remainder (the Division Algorithm) Work?

When we were in elementary school, we were all taught how to find the quotient and remainder but not why it works. In the example we discussed in the previous section, with $a = 27$ and $b = 6$, the quotient q is 4 and the remainder is $r = 3$.

Conceptually, by using the repeated subtraction of b, we have a process that gives the quotient and remainder of a by b. Here's how the process works in the case of the example we just did in which $b < a$, where $a = 27$ and $b = 6$. Starting with a and subtracting b, there are only two possibilities (Figure 2.27).

Figure 2.27
Algorithm for division as repeated subtraction

$27 \rightarrow 27 - 6 = 21$ → (if 21 is larger than b) → subtract b again

$27 \rightarrow 27 - 6 = 21$ → (if 21 is less than b) → finished

The subtraction process is repeated until we get to the finished stage. For our example, the picture of the algorithm (process) is shown in Figure 2.28.

Figure 2.28
Example of division algorithm as repeated subtraction

$$a = 27 \xrightarrow{-6} 27 - 6 = 21 \xrightarrow{-6} 15 \xrightarrow{-6} 9 \xrightarrow{-6} 3 \rightarrow \text{finished}$$

The algorithm stopped because, after subtracting b enough times (4 in this example), we ended with an integer, $r = 3$, so that $0 \leq r < b = 6$. How many times did we subtract b from a? We did it $q = 4$ times. Thus, $a - q\,b = r$, which is the same as $a = q\,b + r$. From this one example, and others from Problem Set 2.4, it becomes clear that this repeated-subtraction process will always stop. We certainly feel that the division algorithm always works, although having many examples are not a proof.

EXAMPLE 2.17

Using the Division Algorithm to Solve the Marching-Band Problem

Mr. Garza was happy to see that so many students in the school band had turned out for the parade. He had them form into rows of 6, but it turned out that just 1 tuba player was in the back row. To his dismay, when he re-formed the band into rows of 5, there was still a lone tuba player in the back row. In desperation, Mr. Garza had the band reassemble into rows of 7. To his relief, every row was filled! What is the smallest size possible for the parade band?

Solution

If we let n denote the number of students marching in the band, then we know that when n is divided by 6 (the number of rows) the remainder is 1. Thus, $n = 6q + 1$ for some whole number $q = 0, 1, 2, \ldots$. This means that n is somewhere in the list of numbers that are 1 larger than a multiple of 6:

$$1, 7, 13, 19, 25, 31, 37, 43, 49, 55, 61, 67, 73, 79, 85, 91, 97, 103, \ldots.$$

Similarly, $n \div 5$ has a remainder of 1, so n is also a number in this list:

$$1, 6, 11, 16, 21, 26, 31, 36, 41, 46, 51, 56, 61, 66, 71, 76, 81, 86, 91, 96, 101, \ldots.$$

Finally, 7 is a divisor of n, so n is one of these numbers:

$$7, 14, 21, 28, 35, 42, 49, 56, 63, 70, 77, 84, 91, 98, 105, \ldots.$$

Comparing the three lists, we find that 91 is the smallest number common to all three lists, so it is possible that there are 91 band members. Mr. Garza has arranged them in 13 rows, since $91 \div 7 = 13$.

FROM THE NCTM PRINCIPLES AND STANDARDS

Understanding Meanings of Operations and How They Relate to One Another

In grades 3–5, students should focus on the meanings of, and relationship between, multiplication and division. It is important that students understand what each number in a multiplication or division expression represents. For example, in multiplication, unlike addition, the factors in the problem can refer to different units. If students are solving the problem 29 × 4 to find out how many legs there are on 29 cats, 29 is the number of cats (or number of groups), 4 is the number of legs on each cat (or number of items in each group), and 116 is the total number of legs on all the cats. Modeling multiplication problems with pictures, diagrams, or concrete materials helps students learn what the factors and their product represent in various contexts.

Students should consider and discuss different types of problems that can be solved using multiplication and division. For example, if there are 112 people traveling by bus and each bus can hold 28 people, how many buses are needed? In this case, 112 ÷ 28 indicates the number of groups (buses), where the total number of people (112) and the size of each group (28 people in each bus) are known. In a different problem, students might know the number of groups and need to find how many items are in each group. They should learn the meaning of a remainder by modeling division problems and exploring the size of remainders given a particular divisor.

Exponents and the Power Operation

Instead of writing $3 \cdot 3 \cdot 3 \cdot 3 \cdot 3$, we can follow a notation introduced by René Descartes and write 3^5. This operation is called "taking 3 to the fifth power." The general definition is described as follows:

DEFINITION The Power Operation for Whole Numbers

Let a and m be whole numbers, where $m \neq 0$. Then ***a* to the *m*th power,** written a^m, is defined by

$$a^1 = a, \quad \text{if } m = 1,$$

and

$$a^m = \overbrace{a \cdot a \cdot \cdots \cdot a}^{m \text{ times}}, \quad \text{if } m > 1.$$

The number a is called the **base,** m is called the **exponent** or **power,** and a^m is called an **exponential expression.** Special cases include squares and cubes. For example, 7^2 is read "7 squared," and 10^3 is read "10 cubed." On most computers, 7^2 and 10^3 would be typed in as 7^2 and 10^3, where the circumflex ^ separates the base from the exponent. Calculators often have a [^] or [y^x] key to compute powers. For example, 3 [y^x] 2 [=] will give the answer 9. The reason that multiplication rules of exponentials are true is because we see how the exponents add when the base is the same.

EXAMPLE 2.18

Working with Exponents

Compute the following products and powers, expressing your answers in the form of a single exponential expression a^m:

(a) $7^4 \cdot 7^2$ **(b)** $6^3 \cdot 6^5$ **(c)** $2^3 \cdot 5^3$
(d) $3^2 \cdot 5^2 \cdot 4^2$ **(e)** $(3^2)^5$ **(f)** $(4^2)^3$

Solution

(a) $7^4 \cdot 7^2 = (7 \cdot 7 \cdot 7 \cdot 7) \cdot (7 \cdot 7) = 7 \cdot 7 \cdot 7 \cdot 7 \cdot 7 \cdot 7 = 7^6$

(b) $6^3 \cdot 6^5 = (6 \cdot 6 \cdot 6) \cdot (6 \cdot 6 \cdot 6 \cdot 6 \cdot 6)$
$= 6 \cdot 6 \cdot 6 \cdot 6 \cdot 6 \cdot 6 \cdot 6 \cdot 6 = 6^8$

(c) $2^3 \cdot 5^3 = (2 \cdot 2 \cdot 2) \cdot (5 \cdot 5 \cdot 5) = (2 \cdot 5) \cdot (2 \cdot 5) \cdot (2 \cdot 5)$
$= (2 \cdot 5)^3 = 10^3$

(d) $3^2 \cdot 5^2 \cdot 4^2 = (3 \cdot 3) \cdot (5 \cdot 5) \cdot (4 \cdot 4) = (3 \cdot 5 \cdot 4) \cdot (3 \cdot 5 \cdot 4)$
$= (3 \cdot 5 \cdot 4)^2 = 60^2$

(e) $(3^2)^5 = (3)^2 \cdot (3)^2 \cdot (3)^2 \cdot (3)^2 \cdot (3)^2$
$= (3 \cdot 3) \cdot (3 \cdot 3) \cdot (3 \cdot 3) \cdot (3 \cdot 3) \cdot (3 \cdot 3)$
$= 3 \cdot 3 \cdot 3 \cdot 3 \cdot 3 \cdot 3 \cdot 3 \cdot 3 \cdot 3 \cdot 3 = 3^{10}$

(f) $(4^2)^3 = (4^2) \cdot (4^2) \cdot (4^2) = (4 \cdot 4) \cdot (4 \cdot 4) \cdot (4 \cdot 4)$
$= 4 \cdot 4 \cdot 4 \cdot 4 \cdot 4 \cdot 4 = 4^6$

Example 2.18 reveals that the multiplication of exponentials follows useful patterns that can be used to shorten calculations. For example, $7^4 \cdot 7^2 = 7^{4+2}$ and $6^3 \cdot 6^5 = 6^{3+5}$ are two examples of the general rule $a^m \cdot a^n = a^{m+n}$ for multiplying exponentials with the same base. Similarly, $2^3 \cdot 5^3 = (2 \cdot 5)^3$ is a case of $a^n \cdot b^n = (a \cdot b)^n$, and $(3^2)^5 = 3^{2 \cdot 5}$ is an example of $(a^m)^n = a^{m \cdot n}$. The proof of the first rule of exponentials comes from counting how many times x repeats itself.

THEOREM Multiplication Rules of Exponentials

Let a, b, m, and n be whole numbers, where $m \neq 0$ and $n \neq 0$. Then

(i) $a^m \cdot a^n = a^{m+n}$;
(ii) $a^m \cdot b^m = (a \cdot b)^m$;
(iii) $(a^m)^n = a^{m \cdot n}$.

Proof of (i): $a^m \cdot a^n = a^{m+n}$

$$a^m \cdot a^n = \underbrace{a \cdot a \cdot \cdots \cdot a}_{m \text{ factors}} \cdot \underbrace{a \cdot a \cdot \cdots \cdot a}_{n \text{ factors}}$$
$$= \underbrace{a \cdot a \cdot \cdots \cdot a}_{m + n \text{ factors}}$$
$$= a^{m+n}$$

The proofs of (ii) and (iii) are similar.

If the formula $a^m \cdot a^n = a^{m+n}$ were extended to allow $m = 0$, it would state that $a^0 \cdot a^n = a^{0+n} = a^n$. This suggests that it is reasonable to define $a^0 = 1$ when $a \neq 0$.

DEFINITION Zero as an Exponent

Let a be any whole number, $a \neq 0$. Then a^0 is defined to be 1, or $a^0 = 1$.

Another way to see that $a^0 = 1$ makes sense is to look at various powers of a. Although we have not yet formally defined a to a negative power, we know from our previous experience that $a^{-n} = 1/a^n$. Thus, we can look at a sequence for $a = 2$ and guess what the question mark denotes:

$$2^4 = 16, 2^3 = 8, 2^2 = 4, 2^1 = 2, 2^0 = ?, 2^{-1} = \tfrac{1}{2}, 2^{-2} = \tfrac{1}{4}, 2^{-3} = \tfrac{1}{8}, 2^{-4} = \tfrac{1}{16}.$$

Isn't it natural to think that $2^0 = 1$ or, for that matter, that $a^0 = 1$, for any positive integer?

To see why 0^0 is *not* defined, notice that there are two conflicting patterns:

$$3^0 = 1, 2^0 = 1, 1^0 = 1, 0^0 = ?$$
$$0^3 = 0, 0^2 = 0, 0^1 = 0, 0^0 = ?$$

The multiplication formula $a^m \cdot a^n = a^{m+n}$ can also be converted to a corresponding division fact. For example,

$$a^{5-3} \cdot a^3 = a^{(5-3)+3} = a^5$$

so

$$a^5/a^3 = a^{5-3}.$$

In general, we have the following theorem:

THEOREM Rules for Division of Exponentials

Let a, b, m, and n be whole numbers, where $m \geq n > 0$, $b \neq 0$, and $a \div b$ is defined. Then

$$\textbf{(i)}\ b^m/b^n = b^{m-n}$$

and

$$\textbf{(ii)}\ (a^m/b^m) = (a/b)^m.$$

EXAMPLE 2.19

Working with Exponents

Rewrite these expressions in exponential form a^m:

(a) $5^{12} \cdot 5^8$ **(b)** $7^{14}/7^5$ **(c)** $3^2 \cdot 3^5 \cdot 3^8$
(d) $8^7/4^7$ **(e)** 2^{5-5} **(f)** $(3^5)^2/3^4$

Solution

(a) $5^{12+8} = 5^{20}$ **(b)** $7^{14-5} = 7^9$
(c) $3^{2+5+8} = 3^{15}$ **(d)** $(8/4)^7 = 2^7$
(e) $2^0 = 1$ **(f)** $3^{5\cdot 2-4} = 3^6$

2.4 Problem Set

Understanding Concepts

1. What multiplication fact is illustrated in each of the diagrams shown? Name the multiplication model that is illustrated.

(a)

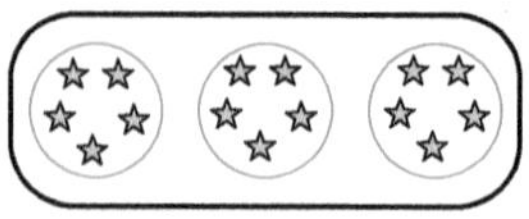

(b)

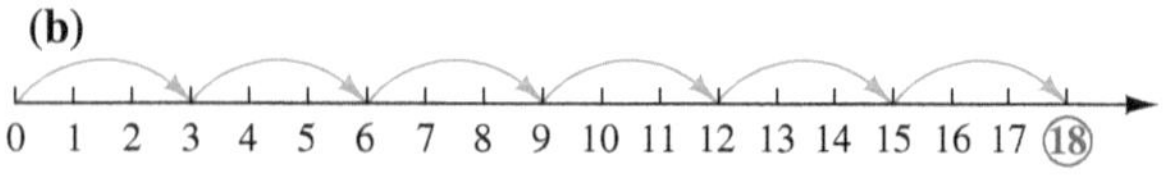

(c)

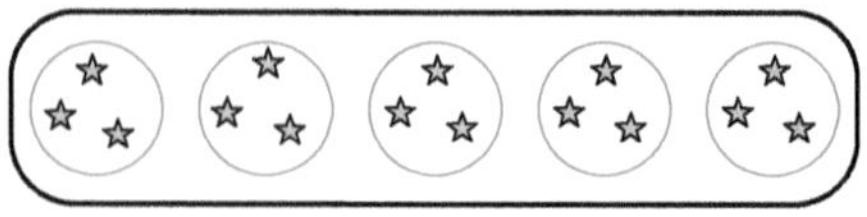

(d)

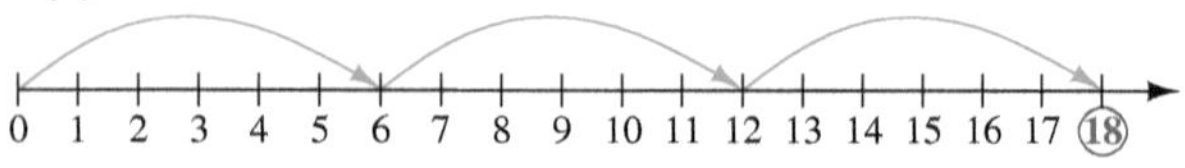

(e)

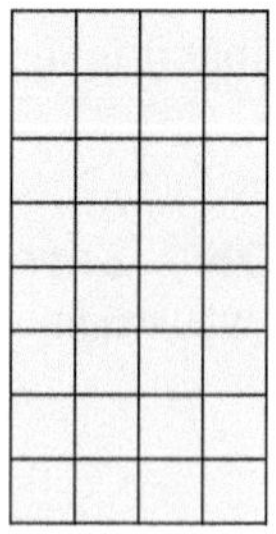

(f)

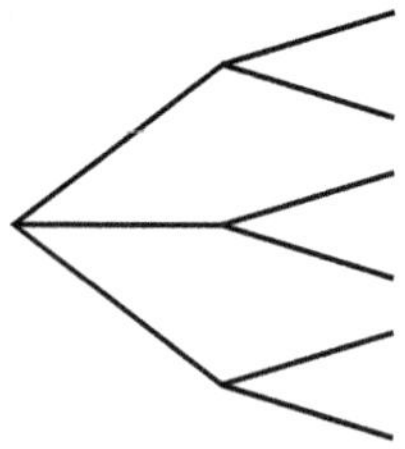

2. Discuss which model of multiplication—set model (repeated addition), number-line (measurement), array, rectangular area, multiplication tree, or Cartesian product—best fits the following problems:

(a) A set of dominoes came in a box containing 11 stacks of 5 dominoes each. How many dominoes are in the set?

(b) Marja has 3 skirts that she can "mix or match" with 6 blouses. How many outfits does she have to wear?

(c) Harold hiked 10 miles each day until he crossed the mountains after 5 days. How many miles was his hike?

(d) Ace Widget Company makes 35 widgets a day. How many widgets are made in a 5-day workweek?

(e) Janet's sunroom is 9 by 18 feet. How many 1-square-foot tiles does she need to cover the floor?

(f) Domingo rolls a die and flips a coin. How many outcomes are there?

3. Multiplication as repeated addition can be illustrated on most calculators. For example, $4 \cdot 7$ is computed by 7 ⊞ 7 ⊞ 7 ⊞ 7 ⊟. Each press of the ⊞ key completes any pending addition and sets up the next one, so the intermediate products $2 \cdot 7 = 14$ and $3 \cdot 7 = 21$ are displayed along the way. Many calculators have a "constant" feature, which enables the user to avoid having to reenter the same addend over and over. For example, ⊞ 734 ⊟ ⊟ ⊟, or 734 ⊞ ⊞ ⊞ may compute $3 \cdot 734$; it all depends on how your particular calculator operates.

(a) Explain carefully how repeated addition is best accomplished on your calculator.

(b) Use repeated addition on your calculator to compute the given products. Check your result by using the ⊠ key.

(i) $4 \cdot 9$

(ii) 7×536

(iii) $6 \times 47{,}819$

(iv) $56{,}108 \times 6$ (What property may help?)

4. The Cartesian product of finite and nonempty sets can be illustrated by the intersections of a crossing-line pattern, as follows:

{f, g, h} × {x, y}

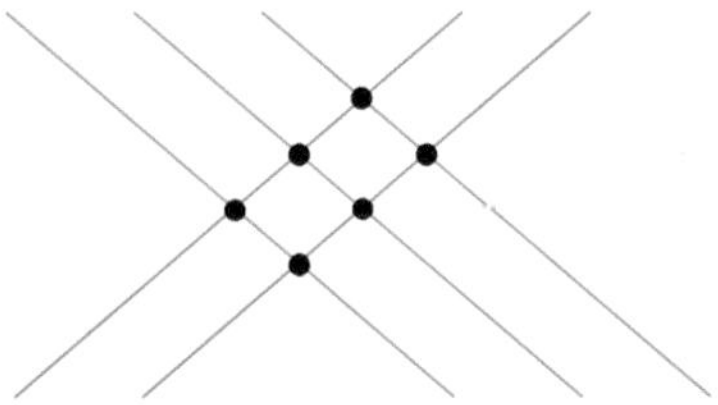

(a) Explain why the number of intersection points in the crossing-line pattern for $A \times B$ is $a \cdot b$, where $a = n(A)$ and $b = n(B)$.

(b) Draw the crossing-line pattern for {□, △} × {♥, ♦, ♣, ♠}.

5. Which of the given sets of whole numbers are closed under multiplication? Explain your reasoning.

(a) $\{1, 2\}$

(b) $\{0, 1\}$

(c) $\{0, 2, 4\}$

(d) $\{0, 2, 4, \ldots\}$ (the even whole numbers)

(e) $\{1, 3, 5, \ldots\}$ (the odd whole numbers)

(f) $\{1, 2, 2^2, 2^3\}$

(g) $\{1, 2, 2^2, 2^3, \ldots\}$

(h) $\{1, 7, 7^2, 7^3, \ldots\}$

6. Which of the given subsets of the whole numbers $W = \{0, 1, 2, \ldots\}$ are closed under multiplication? Explain carefully.

(a) $\{0, 1, 2, 3, 4, 6, 7, \ldots\}$ (i.e., the whole numbers except for 5)

(b) $\{0, 1, 2, 3, 4, 5, 7, 8, \ldots\}$

(c) $\{0, 1, 4, 5, 6, 7, 8, \ldots\}$

7. What properties of whole-number multiplication justify these equalities?

(a) $4 \cdot 9 = 9 \cdot 4$

(b) $4 \cdot (6 + 2) = 4 \cdot 6 + 4 \cdot 2$

(c) $0 \cdot 439 = 0$

8. What properties of whole-number multiplication justify these equalities?

(a) $7 \cdot 3 + 7 \cdot 8 = 7 \cdot (3 + 8)$

(b) $5 \cdot (9 \cdot 11) = (5 \cdot 9) \cdot 11$

(c) $1 \cdot 12 = 12$

9. What property of multiplication is illustrated by the manipulatives of the accompanying diagrams?

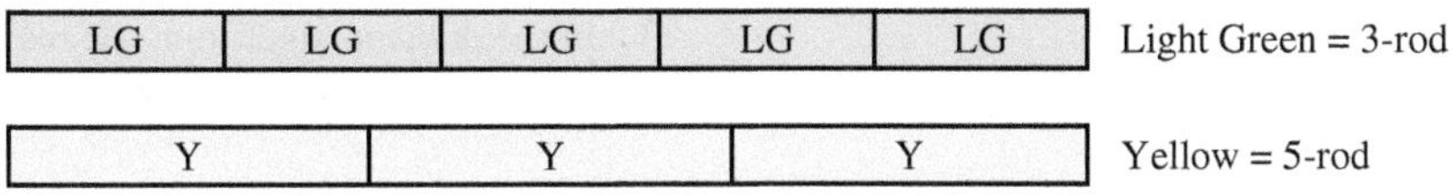

10. What property of multiplication is shown in the diagram that follows?

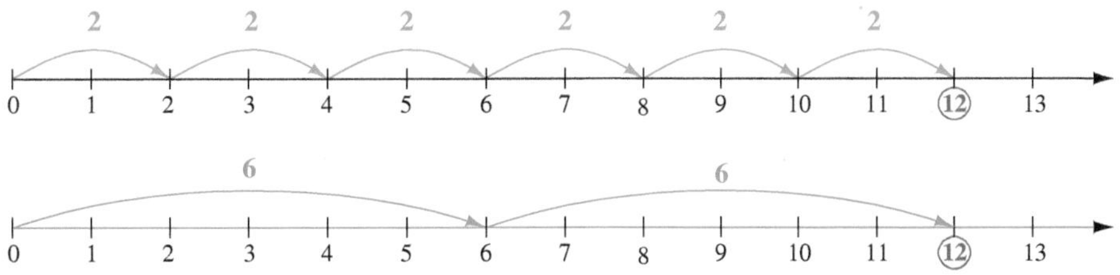

11. Use the rectangular area model to illustrate each of the given statements. Make drawings similar to Figure 2.22.

(a) $(2 + 5) \cdot 3 = 2 \cdot 3 + 5 \cdot 3$

(b) $3 \cdot (2 + 5 + 1) = 3 \cdot 2 + 3 \cdot 5 + 3 \cdot 1$

(c) $(3 + 2) \cdot (4 + 3) = 3 \cdot 4 + 3 \cdot 3 + 2 \cdot 4 + 2 \cdot 3$

12. Use the area of a swimming pool as a rectangular area model to show the distributive property of multiplication over addition by trying the example $2 \cdot (5 + 4) = 2 \cdot 5 + 2 \cdot 4$.

13. The **FOIL** method is a common way to recall and to visualize how to expand the product of two binomials $(a + b) \times (c + d)$: Multiply the **F**irst terms, the **O**uter terms, the **I**nner terms, and the **L**ast terms of the binomials, and sum the four products to obtain the formula $(a + b) \times (c + d) = ac + ad + bc + bd$. Carefully explain how the rectangular area diagram shown justifies the FOIL method. The important point of this problem is for you to understand the reason FOIL actually works rather than being able to "just use it." Again, it is the conceptual understanding that is crucial for teaching.

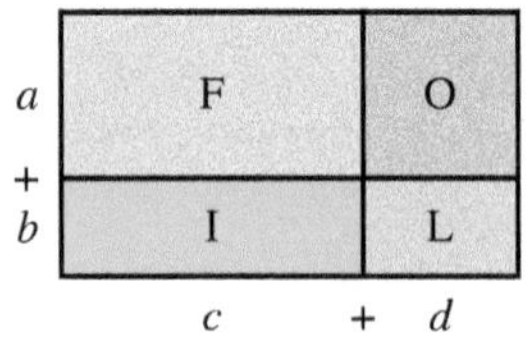

14. Use the following figure as a model to show that the product of trinomials, $(a + b + c) \cdot (d + e + f)$, can be written as a sum of nine products:

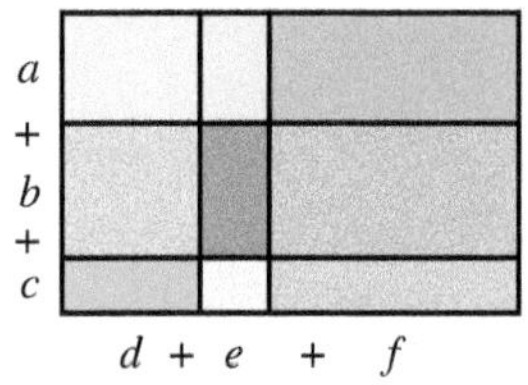

15. What properties of multiplication make it easy to compute these values mentally?

(a) $7 \cdot 19 + 3 \cdot 19$

(b) $24 \cdot 17 + 24 \cdot 3$

(c) $36 \cdot 15 - 12 \cdot 45$

16. What properties of multiplication make it easy to compute these values mentally?

(a) $7 \cdot 16 + 24 \cdot 7$

(b) $23 \cdot (25 \cdot 35 \cdot 24 \cdot 0)$

(c) $36 \cdot 15 + 12 \cdot 55$

17. Modify Figure 2.21 to show how the associative property $3 \cdot (2 \cdot 4)$ can be illustrated with arrays of cubes in three-dimensional space as a manipulative.

18. What division facts are illustrated in the accompanying diagrams?

(a)

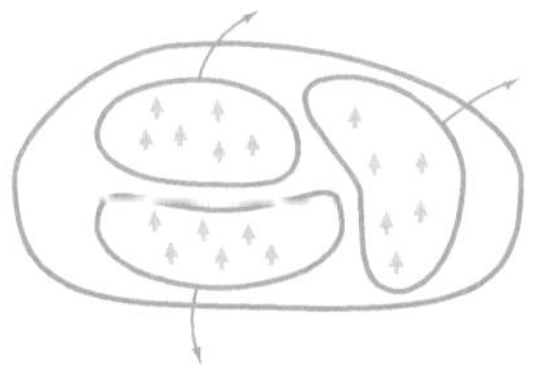

(b)

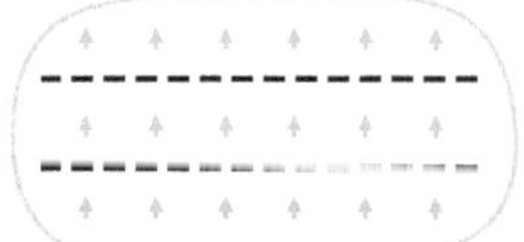

19. A 2-by-3 rectangular array is associated with the fact family $2 \cdot 3 = 6, 3 \cdot 2 = 6, 6 \div 2 = 3$, and $6 \div 3 = 2$. What fact family is associated with each of these rectangular arrays?

(a) 4 by 8 **(b)** 6 by 5

20. A 2-by-3 rectangular array is associated with the fact family $2 \cdot 3 = 6, 3 \cdot 2 = 6, 6 \div 2 = 3$, and $6 \div 3 = 2$. What fact family is associated with each of these rectangular arrays?

(a) 7 by 9

(b) 11 by 5

21. Discuss which of three conceptual models of division—repeated subtraction, partition, and missing factor—best corresponds to the problems that follow. More than one model may fit.

(a) Preston owes \$3200 on his car. If his payments are \$200 a month, how many months will Preston make car payments?

(b) An estate of \$76,000 is to be split among four heirs. How much can each heir expect to inherit?

(c) Anita was given a grant of \$375 to cover expenses on her trip. She expects that it will cost her \$75 a day. How many days can she plan to be gone?

22. In the division algorithm, use the method of repeated subtraction to solve the given division problems. Include a diagram that shows the algorithm with arrows, similar to the example of Figure 2.28.

(a) The remainder of 19 on division by 5

(b) The remainder of 18 on division by 9

(c) The remainder of 25 on division by 8

(d) What is the quotient in each of these cases?

23. In the division algorithm, use the method of repeated subtraction to solve the given division problems. Include a diagram that shows the algorithm with arrows, similar to the example of Figure 2.28.

(a) The remainder of 14 on division by 7

(b) The remainder of 7 on division by 14

(c) What is the quotient in each of these cases?

24. Use repeated subtraction on your calculator to compute the given division problems where remainders are possible. Be sure to take advantage of the "constant" feature of your calculator.

(a) $78 \div 13$ **(b)** $832 \div 52$

(c) $96 \div 14$ **(d)** $548{,}245 \div 45{,}687$

25. Solve for the unknown whole number in the following expressions:

(a) $y \div 5 = 5 \text{ R } 4$

(b) $20 \div x = 3 \text{ R } 2$

26. Solve for the unknown whole number in the following expressions:

(a) $z \div 15 = 3 \text{ R } 4$

(b) $20 \div (2x) = 3 \text{ R } 2$

(c) $(w + 2) \div 6 = 4 \text{ R } 5$

27. Rewrite each of the following in the form of a single exponential:

(a) $3^{20} \cdot 3^{15}$

(b) $(3^2)^5$

(c) $y^3 \cdot z^3$

28. Rewrite each of the following in the form of a single exponential:

(a) $4^8 \cdot 7^8$

(b) $x^7 \cdot x^9$

(c) $(t^3)^4$

29. Write each of the following as 2^m for some whole number m:

(a) 8 **(b)** $4 \cdot 8$

(c) 1024 **(d)** 8^4

30. Find the exponents that make the following equations true:

(a) $3^m = 81$ **(b)** $3^n = 531{,}441$

(c) $4^p = 1{,}048{,}576$ **(d)** $2^q = 1{,}048{,}576$

31. In the Division Algorithm, why is there no alternative in Figure 2.27 indicating what to do if one of the stages comes up with a negative remainder?

Into the Classroom

32. Large-group kinesthetic activities are often an effective way to deepen one's understanding of basic concepts. For example, have the children hold hands in groups of three and count the number of groups to illustrate the concept of division by 3 with the repeated-subtraction (grouping) model of division. Carefully describe analogous large-group activities that illustrate

(a) multiplication as repeated addition;

(b) multiplication as an array;

(c) partitive division;

(d) missing-factor division.

33. When the class was asked to pose a meaningful problem that corresponded to the division $14 \div 3$, Peter's problem was answered by 4, Tina's by 5, and Andrea's by 4 with a remainder of 2. Carefully explain how all three children may be correct by posing three problems of your own that give their answers.

34. **(Writing)** Discuss how division can be modeled with number strips as a manipulative. Write a brief essay that includes several examples, each illustrated with carefully drawn figures.

35. **(Writing)** The theorem that describes the property of whole numbers in addition in Section 2.3, page 89, is followed by an informal way to describe three of the four properties for addition. What would be your description of the middle four properties of whole-number multiplication in the theorem on page 000 of this section?

36. Each team is given a game sheet as shown. The teacher rolls three dice (or a die 3 times) to determine three numbers x, y, and z. Each team uses whole-number operations, including raising a number to a power, involving all three numbers exactly once, to obtain expressions equal to as many of the numbers 0, 1, 2, . . . , 18 as possible. The team filling all regions of the game sheet first is the winner. If no team has won at the end of the first roll, the numbers x, y, and z are replaced by rolling the dice again. Each team now fills in remaining empty positions on the game sheet, using the new set of three numbers. The dice are rolled again until there is a winning team. Suppose the dice turn up the numbers 2, 3, 3. Show that at least 14 of the regions on the board can be filled.

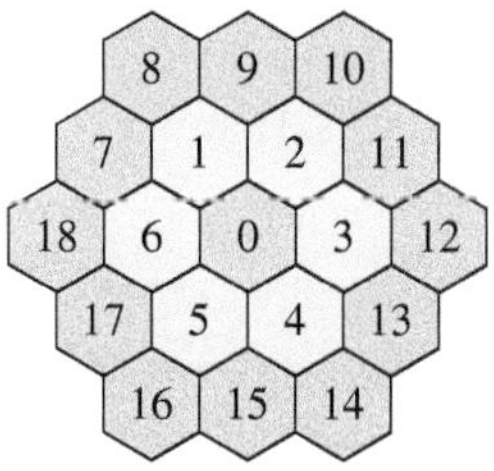

Responding to Students

37. A student is given two problems and asked to fill in the blanks. He responds to them as given in the second column:

Problem	$7 \times 7 =$ ___	Student Response	$7 \times 7 = 49$
Problem	___ $\div 7 = 7$	Student Response	$1 \div 7 = 7$

What is this student doing incorrectly?

38. Your student, Shannon, bought 18 nuts and 18 bolts. The bolts were 86 cents each and the nuts were 14 cents each. She figured out the cost in her head and handed the clerk \$18. She asked you why the clerk did it the way that he did (which follows). How would you respond to her?

$$\begin{array}{r} {}^{4} \\ 86 \\ \times 18 \\ \hline 688 \\ 86 \\ \hline 1548 \end{array} \qquad \begin{array}{r} {}^{3} \\ 14 \\ \times 18 \\ \hline 112 \\ 14 \\ \hline 252 \end{array} \qquad \begin{array}{r} {}^{1\,1} \\ 15.48 \\ +\ 2.52 \\ \hline 18.00 \end{array}$$

39. Nelson baked 36 cupcakes. In each pan, he baked 6 cupcakes. Students were asked to write an equation that tells how to find p, the number of *pans* of cupcakes he baked. One student answered that it was $36 - 6 = p$. What error did this student make and what would you suggest?

40. Emily is having trouble multiplying two-digit numbers. She is a wonderful art student, so you are trying to teach her a visual method for multiplication. How would you help her by drawing a rectangular area model for 32 × 23 after you've shown the partial products as follows?

$$\begin{array}{r} 32 \\ \underline{\times\ 23} \\ 6 \\ 90 \\ 40 \\ \underline{600} \\ 736 \end{array}$$

41. Jing, who is a fine middle school student, is challenged to find a whole number x such that $(x^2 - 3) \div 4 = 8$ R 1. Her answer is $x = \pm 6$. What subtle point is she missing?

42. Andrea was given a problem about the area of a rectangle. Her teacher asked her to see what the area would be if the width of the rectangle is a 2-foot log and the length is made up of two logs pushed together, a 5-foot log and a 1-foot log. She couldn't see how to do the computation because she knew the area had to be in square feet. She got 2 times 5 to be 10 square feet and 2 times 1 to be 2 feet. Andrea knew that they couldn't be the same units. How would you help her?

Thinking Critically

43. The binary operation ★ is defined for the set of shapes {○, □, △} according to the table that follows. For example, △ ★ □ = ○.

★	○	□	△
○	○	□	△
□	□	△	○
△	△	○	□

Is the operation closed? commutative? associative? Is there an identity shape?

44. **(a)** Verify that the three numbers in every row, column, and diagonal in the following square have the same product, making the square a *magic multiplication square*:

8	256	2
4	16	64
128	1	32

(b) Write each number in the magic multiplication square as a power of 2, and then explain how the multiplication square can be obtained from a related magic addition square.

(c) Create a magic multiplication square with the numbers 1, 3, 9, 27, 81, 243, 729, 2187, and 6561.

45. **(a)** The Math Club wants to gross \$500 on its raffle. Raffle tickets are \$2 each and 185 tickets have been sold so far. The answer is 65. What is the question?

(b) A total of 67 eggs was collected from the henhouse and placed in standard egg cartons. If the answer is 5, what is the question? If the answer is 7, what is the question?

46. There are 318 folding chairs being set up in rows in an auditorium, with 14 chairs being put in each row, starting at the front of the auditorium.

(a) If the answer is 22, what is the question?

(b) If the answer is 10, what is the question?

47. A certain whole number less than 100 leaves the remainders 1, 2, 3, and 4 when divided, respectively, by 2, 3, 4, and 5. What is the whole number?

State Assessments

48. (Grade 5)
Bobby is trying to decide which outfit to wear to a party. His choices are shown in the diagram.

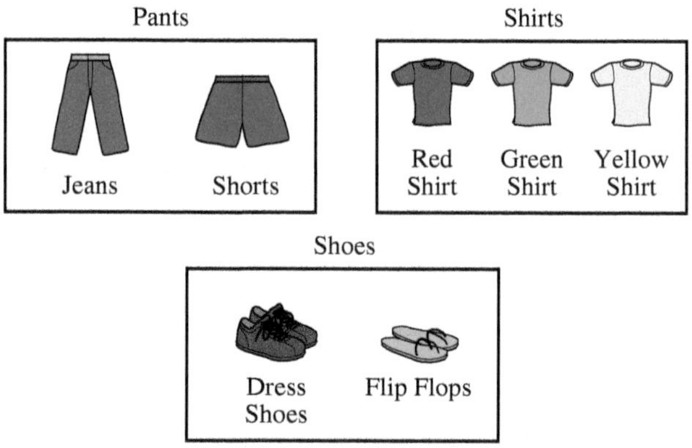

How many possible outfits can Bobby create if chooses 1 pair of pants, 1 shirt, and 1 pair of shoes?

A. 7 B. 8 C. 10 D. 12

49. (Grade 5)
Which expression fits the diagram below?

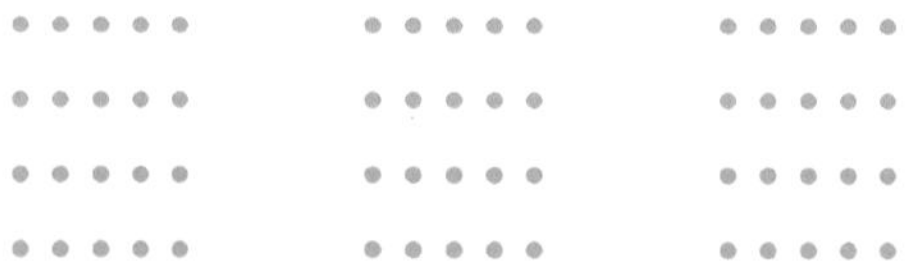

A. 4 × (5 × 5) B. 3 × (4 × 5) C. 3 + (3 × 4)
D. 3 + (3 + 4) E. 3 × (3 + 4)

50. (Grade 4)
Alonzo has a team of 16 baseball players. Today, he is going to have 4 pieces of each apple. Each of his

teammates will get 2 pieces. How many apples does Alonzo need to cut?

A. 10 B. 8 C. 12 D. 4

51. (California, Grade 4)

Justin solved the problem below. Which expression could be used to check his answer?

$$3\overline{)1364}\quad 454r2$$

A. $(454 \times 3) + 2$ B. $(454 \times 2) + 3$
C. $(454 + 3) \times 2$ D. $(454 + 2) \times 3$

52. (Grade 6)
A math sentence represents the total number of dog legs in a kennel of 13 dogs. If you know that □ represents the number of dogs, what does the △ represent? The equation is □ × △ = 52.

A. The amount of dog food that the dogs eat.
B. The number of male dogs in the kennel.
C. The number of legs on each dog.
D. The total number of legs in the kennel.

53. (Grade 6)
Steven writes 3^5 on the whiteboard. Which of the following expressions are another way to use repeating multiplication?

A. 3×5
B. 5×5
C. $3 \times 3 \times 3 \times 3 \times 3$
D. $5 \times 5 \times 5$

The Chapter in Relation to Future Teachers

What must a future teacher know in order to help children learn counting and the operations of the basic arithmetic of whole numbers? The simplest mathematical idea of all is counting. In Chapter 2, sets were introduced and were used in counting, addition, subtraction, multiplication, and division. A variety of models and manipulatives to perform these operations was discussed. The background of this chapter will help you respond to questions from students and work with them to understand the concepts of whole-number arithmetic. The goal toward which elementary school teachers strive is the combination of conceptual understanding and procedural fluency in students.

Chapter 2 Summary

Section 2.1 Sets and Operations on Sets	Page Reference
CONCEPTS	
• **Set:** A set is a collection of objects from a specified universe. A set can be described verbally, by a list, or with set-builder notation.	68
• **Venn diagram:** Sets can be visualized with Venn diagrams, in which the universe is a rectangle and closed loops inside the universe correspond to sets. Elements of a set are associated with points within the loop corresponding to that set.	70
• **Set operations and relations.** Sets are combined and related in the following ways: set complement ($\overline{A}$), subset ($A \subseteq B$), proper subset ($A \subset B$), intersection ($A \cap B$), and union ($A \cup B$).	71
DEFINITIONS	
• A **universe** (U) consists of the objects allowed into consideration for a set.	69
• An **element,** or member, is an object that belongs to the collection, or set.	69
• A **natural number,** or **counting number,** is a member of the set $N = \{1, 2, 3, \ldots\}$, where the ellipsis "..." indicates "and so on."	69
• The **complement of set,** written **($\overline{A}$),** is the set of elements in the universal set U that are not elements of A.	70
• The set A is a **subset** of B, written $A \subseteq B$, if, and only if, every element of A is also an element of B.	71

- A set A is an **equal set** of a set B, written $A = B$, if, and only if, A and B have precisely the same elements. 71
- A **proper subset, $A \subset B$,** is when $A \subseteq B$, but there must be some element of B that is not also an element of A, $A \neq B$. 71
- An **empty set** is a set that has no elements in it and is written $\varnothing$. 71
- The **intersection** of two sets A and B, written $A \cap B$, is the set of elements common to both A and B. 71
- **Disjoint** sets are two sets that have no elements in common, written $A \cap B = \varnothing$. 72
- The **union** of sets A and B, written $A \cup B$, is the set of all elements that are in A or B. 72
- A **counterexample** is a case that shows that a statement is false. 72

PROPERTIES

- **Transitive property of set inclusion:** If $A \subseteq B$ and $B \subseteq C$, then $A \subseteq C$ 73
- **Commutative property of union and intersection:** $A \cup B = B \cup A$, $A \cap B = B \cap A$ 73
- **Associative property of union and intersection:** $A \cup (B \cup C) = (A \cup B) \cup C$, $A \cap (B \cap C) = (A \cap B) \cap C$ 73
- **Distributive property of union and intersection:** $A \cap (B \cup C) = (A \cap B) \cup (A \cap C)$, $A \cup (B \cap C) = (A \cup B) \cap (A \cup C)$ 74

NOTATION

- **A set as a list in braces:** {a, b, . . .} 69
- **Set-builder notation:** $\{x \mid x \text{ is } (\textit{condition})\}$ 69

Section 2.2 Sets, Counting, and the Whole Numbers	**Page Reference**

CONCEPTS

- **Types of number:** Nominal numbers name objects (e.g., an ID number), ordinal numbers indicate position (e.g., 5th place), and cardinal numbers indicate the number of elements in a set (e.g., there are 9 justices of the U.S. Supreme Court). 78
- **Equivalence of sets:** Two sets are equivalent if, and only if, there is a one-to-one correspondence (or matching) of the elements of the two sets. 79
- **Manipulatives:** Physical, pictorial, and concrete representations of mathematics. 81
- **Whole numbers:** The whole numbers are the cardinal numbers of finite sets, with zero being the cardinal number of the empty set. Whole numbers can be represented and visualized by a variety of manipulatives and diagrams, including **tiles, cubes, number strips, rods,** and the **number line.** 80
- **Order of the whole numbers:** The whole numbers are ordered, so that m is less than n if a set with m elements is a proper subset of a set with n elements. 82

DEFINITIONS

- A **nominal number,** or **identification,** is a sequence of digits used as a name or label. 78
- An **ordinal number** describes location in an ordered sequence with the words *first, second, third, fourth,* and so on, communicating the basic notion of "where." 79
- A **cardinal number** of a set is the number of objects in the set, communicating the basic notion of "how many." 79

- **One-to-one correspondence** between sets is the concept that each element of one set is paired with exactly one element of another set, and each element of either set belongs to exactly one of the pairs. 79
- Sets A and B are **equivalent,** or **matching,** if there is a one-to-one correspondence between A and B, written $\boldsymbol{A \sim B}$. 79
- A **finite** set is a set that is either the empty set or a set equivalent to $\{1, 2, 3, \ldots, n\}$, for some natural number n. 80
- An **infinite** set is a set that is not finite. One way to think of an infinite set is that, if you were to list all members of the set, the list would go on forever. 80
- **Whole numbers** are the cardinal numbers of finite sets, which means they are the numbers of elements in finite sets. 80
- **Less than, $\boldsymbol{a < b}$:** A way to relate the number of elements in two given sets, A and B where $n(A) = a$ and $n(B) = b$. There are less elements in set A than there are in set B, or A is a proper subset of B. 82
- **Greater than, $\boldsymbol{a > b}$:** A way to relate the number of elements in two given sets, A and B where $n(A) = a$ and $n(B) = b$. There are more elements in set A than there are in set B, or B is a proper subset of A. 82

NOTATION

- **Zero, $\boldsymbol{0 = n(\varnothing)}$:** Zero represents the empty set, a set that has no members. 80

Section 2.3 Addition and Subtraction of Whole Numbers	Page Reference

CONCEPTS

- **Addition of whole numbers:** Addition of whole numbers is defined by $a + b = n(A \cup B)$, where $a = n(A)$, $b = n(B)$, and A and B are disjoint finite sets. Addition can be visualized on the number line with the **measurement model** or using the **set model.** 89
- **Closure property:** The sum of two whole numbers is a whole number. 91
- **Commutative property:** For all whole numbers a and b, $a + b = b + a$. 91
- **Associative property:** For all whole numbers a, b, and c, $a + (b + c) = (a + b) + c$. 91
- **Zero property of addition:** Zero is an additive identity, so $a + 0 = 0 + a = a$ for all whole numbers a. 91
- **Conceptual models of subtraction:** These models include the **take-away model,** the **missing-addend model,** the **comparison model,** and the **number-line (or measurement) model.** 94

DEFINITIONS

- A **binary operation** is an operation in which two whole numbers are combined to form another whole number. 89
- The **addition,** or **sum,** of a and b is the total number of the combined collection, written $\boldsymbol{a + b}$. 89
- The **addends,** or **summands,** of the expression $a + b$ are a and b. 90
- The **difference,** written $\boldsymbol{a - b}$, is the unique whole number c such that $a = b + c$. 94
- In the expression $a - b$, the **minuend** is a. 94
- In the expression $a - b$, the **subtrahend** is b. 94

PROPERTIES	
• **Properties of addition**	
• **Closure property:** If a and b are any two whole numbers, then $a + b$ is a unique whole number.	91
• **Commutative property:** If a and b are any two whole numbers, then $a + b = b + a$.	91
• **Associative property:** If a, b, and c are any three whole numbers, then $a + (b + c) = (a + b) + c$.	91
• **Additive-identity property of zero:** If a is any whole number, then $a + 0 = 0 + a = a$.	91

Section 2.4 Multiplication and Division of Whole Numbers	Page Reference
CONCEPTS	
• **Multiplication:** Multiplication is defined as a repeated addition, so that $a \cdot b = b + b + \cdots + b$, where there are a addends.	101
• **Conceptual models of multiplication:** These models include the **array model,** the **rectangular area model,** the **multiplication tree model,** the **skip-count model,** and the **Cartesian product** $[a \cdot b = n(A \times B)]$. Here, $A \times B$ denotes the Cartesian product of sets A and B.	101
• **Properties of multiplication:** For all whole numbers a, b, and c, $a \cdot b$ is a whole number (the closure property), $a \cdot b = b \cdot a$ (the commutative property), $a \cdot (b \cdot c) = (a \cdot b) \cdot c$ (the associative property), $a \cdot (b + c) = a \cdot b + a \cdot c$ (the distributive property), $1 \cdot a = a \cdot 1 = a$ (one is a multiplicative identity), and $0 \cdot a = a \cdot 0 = 0$ (multiplication-by-zero property)	105
• **Division:** $a \div b, b \neq 0$, is defined if, and only if, there is a unique whole number c such that $a = b \cdot c$.	107
• **Conceptual models of division:** These models include the **repeated-subtraction (grouping) model,** the **partition (sharing) model,** and the **missing-factor model,** where the missing factor c in the multiplication equation $a = b \cdot c$ is calculated.	107
DEFINITIONS	
• **Multiplication (product),** $\boldsymbol{a \cdot b}$**:** The operation of adding a to itself b times; also denoted $a \times b$, $a*b$, ab and $(a)(b)$.	102
• A **factor** is each whole number a and b of the product $a \cdot b$.	102
• **Repeated addition model:** A way to represent the multiplication operation: Where a and b are any two whole numbers, a multiplied by b, written $a \cdot b$, is defined by $a \cdot b = b + b + \cdots + b$ with a addends, when a is not zero and by $0 \cdot b = 0$.	102
• **Ordered pair** $(\boldsymbol{a}, \boldsymbol{b})$ is the representation of the first component, a, from one set and a second component, b, from another set. It also indicates the Cartesian coordinates of a point.	104
• The **Cartesian product** of sets A and B, written $\boldsymbol{A \times B}$, is the set of all ordered pairs whose first component is an element of set A and whose second component is an element of set B.	103
• **Division,** written $\boldsymbol{a \div b}$, or $\boldsymbol{a/b}$: Let a and b be whole numbers with $b \neq 0$. Then $a \div b = c$ if, and only if, $a = b \cdot c$ for a unique whole number c.	107
• In the expression $a \div b$, a is the **dividend.**	109
• In the expression $a \div b$, b is the **divisor.**	109
• For any positive whole number a (the base) and any whole number m (the **exponent,** or **power**), $\boldsymbol{a}$ **to the** $\boldsymbol{m}$**th power,** a^m, is defined by $a^m = a \cdot a \cdot \cdots \cdot a$, where there are m factors of a. Also, $a^0 = 1$.	112

THEOREM

- **The division algorithm:** Let a and b be whole numbers with $b \neq 0$. Then there is a unique whole number q called the **quotient** and a unique whole number r called the **remainder** such that $a = q \cdot b + r, 0 \leq r < b$. It is common to write $a \div b = q \text{ R } r$ if $a = q \cdot b + r, 0 \leq r < b$. 110

PROPERTIES

- **Properties of multiplication**
  - **Closure property:** If $a \cdot b$ are any two whole numbers, then $a \cdot b$ is a unique whole number. 105
  - **Commutative property:** If a and b are any two whole numbers, then $a \cdot b = b \cdot a$. 105
  - **Associative property:** If a, b, and c are any three whole numbers, then $a \cdot (b \cdot c) = (a \cdot b) \cdot c$. 105
  - **Multiplicative identity:** The number 1 is the unique whole number for which $b \cdot 1 = 1 \cdot b = b$ holds for all whole numbers b. 105
  - **Multiplication-by-zero property:** For all whole numbers b, $0 \cdot b = b \cdot 0 = 0$. 106
  - **Distributive property of multiplication over addition:** If a, b, and c are any three whole numbers, then $a \cdot (b + c) = (a \cdot b) + (a \cdot c)$ and $(a + b) \cdot c = (a \cdot c) + (b \cdot c)$. 106

FORMULA

- **Multiplication–division fact family:** The set of four equivalent facts that may be assumed when one of the four facts is given. In the family, there are two multiplication statements, for example, $a \cdot b = c$ and $b \cdot a = c$ (provided that neither a nor b is zero) and two division statements, for example, $c \div a = b$ and $c \div b = a$ (provided that neither a or nor b is zero). 109
- **Multiplication rules of exponentials:** Let a, b, m, and n be whole numbers, where $m \neq 0$ and $n \neq 0$. Then (i) $a^m \cdot a^n = a^{m+n}$ (ii) $a^m \cdot b^m = (a \cdot b)^m$ (iii) $(a^m)^n = a^{m \cdot n}$. 113

NOTATION

- **Exponential expression:** a^m, where a is the base, $a \neq 0$, and m is the exponent. 112

Chapter Review Exercises

Section 2.1

1. Let $U = \{n \mid n \text{ is a whole number and } 2 \leq n \leq 25\}$, $S = \{n \mid n \in U \text{ and } n \text{ is a square number}\}$, and $P = \{n \mid n \in U \text{ and } n \text{ is a prime number}\}$ ($n > 1$ is *prime* when it is divisible by just two different whole numbers: itself and 1). Let $T = \{n \mid n \in U \text{ and } n \text{ is a power of } 2\}$.
   (a) Write S, P, and T in listed form.
   (b) Find the following sets: $\overline{P}$, $S \cap T$, $S \cup T$, and $S \cap \overline{T}$.
2. Draw a Venn diagram of the sets S, P, and T in problem 1.
3. Replace each box $\square$ with one of the symbols $\cap$, $\cup$, $\subset$, $\subseteq$, or $=$ to give a correct statement for general sets A, B, and C.
   (a) $A \square A \cup B$
   (b) If $A \subseteq B$ and A is not equal to B, then $A \square B$.
   (c) $A \square (B \cup C) = (A \cap B) \cup (A \cap C)$
   (d) $A \square \emptyset = A$

Section 2.2

4. Let $S = \{\text{s, e, t}\}$ and $T = \{\text{t, h, e, o, r, y}\}$. Find $n(S)$, $n(T)$, $n(S \cup T)$, $n(S \cap T)$, $n(S \cap \overline{T})$, and $n(T \cap \overline{S})$.
5. Show that the set of square natural numbers less than 101 is in one-to-one correspondence with the set {a, b, c, d, e, f, g, h, i, j}.
6. Show that the set of cubes, $\{1, 8, 27, \ldots\}$, is equivalent to the set of natural numbers.
7. Label the number of elements in each region of a two-loop Venn diagram for which $n(U) = 20$, $n(A) = 7$, $n(B) = 9$, and $n(A \cup B) = 11$.

Section 2.3

8. Explain how to illustrate $5 + 2$ with
   (a) the set model of addition;
   (b) the number-line (measurement) model of addition.

9. What properties of whole-number addition are illustrated on the following number lines?

(a)

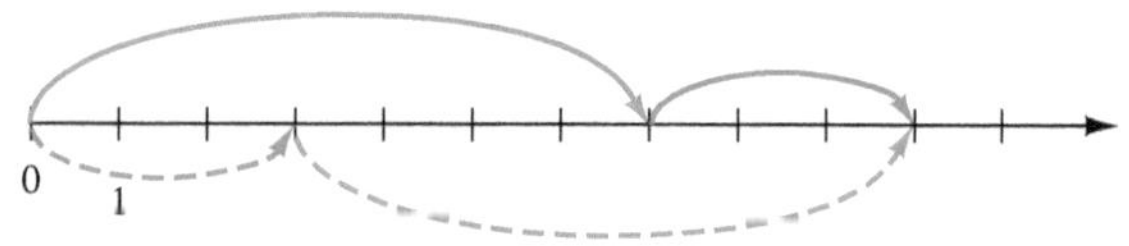

(b)

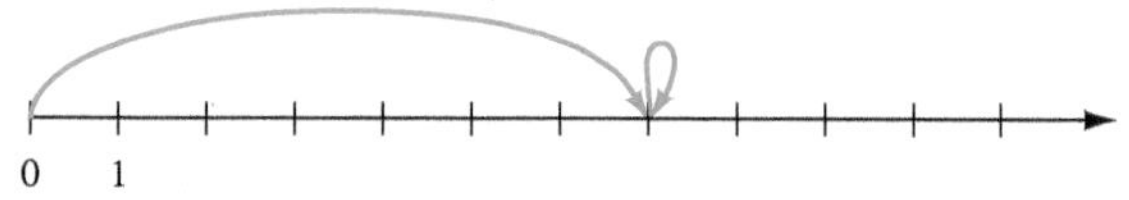

10. Draw figures that illustrate $6 - 4$,

(a) using sets. **(b)** using the number line.

Section 2.4

11. Draw representations of the product 4×2, using the given representation:

(a) A set diagram

(b) An array of discrete objects

(c) A rectangular area

(d) A multiplication tree

(e) A number-line diagram

12. Let $A = \{p, q, r, s\}$ and $B = \{x, y\}$.

(a) Find $A \times B$.

(b) What product is modeled with $A \times B$?

13. Whiffle balls 2 inches in diameter are packed individually in cubical boxes, and then the boxes are packed in cartons of 3 dozen balls. What are the dimensions of a suitable rectangular carton?

14. A drill sergeant lines up 92 soldiers in rows of 12, except for a partial row in the back. How many rows are formed? How many soldiers are in the back row?

15. Draw figures that illustrate the division problem $15 \div 3$, using these models:

(a) Repeated subtraction (grouping objects in a set)

(b) Partition (sharing objects)

(c) Missing factor (rectangular array)

3 Numeration and Computation

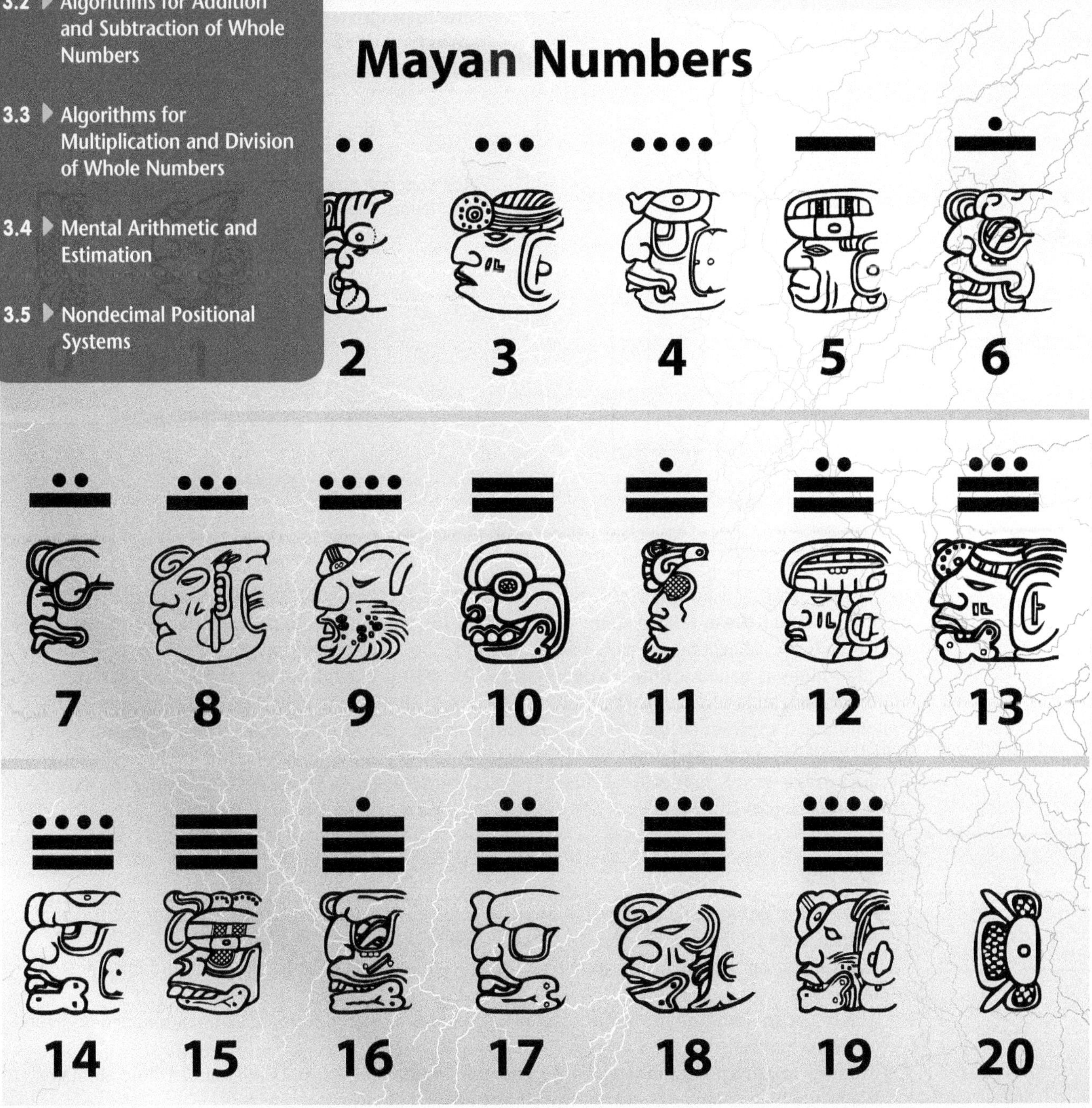

COOPERATIVE INVESTIGATION

Numbers from Rectangles

Material Needed

1. One rectangle of each of these shapes for each student:

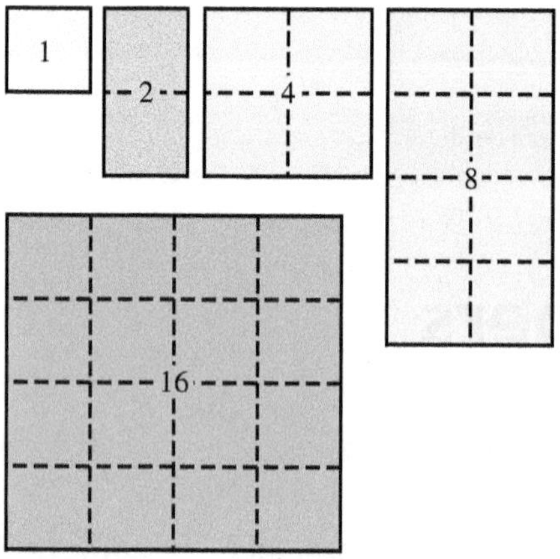

2. One record sheet like this for each student:

	16	8	4	2	1
0	0	0	0	0	0
1	0	0	0	0	1
18	1	0	0	1	0
19	1	0	0	1	1

	16	8	4	2	1
20	1	0	1	0	0
21					
38					
39					

Directions

Step 1. Use the rectangles to determine whether or not there are representations of each of the numbers 0, 1, 2, . . . , 39 as a sum of the numbers 1, 2, 4, 8, or 16, with each of the latter group of numbers used at most once.

Step 2. For each representation determined in step 1, record the numbers (rectangles) used by placing a 0 or a 1 in the appropriate columns of the record sheet. The rows for 0, 1, 18, 19, and 20 have been done for you.

(a) Do all the numbers from 0 through 39 have such a representation?

(b) What additional numbers could be represented if you had a 32 rectangle?

(c) Describe any interesting patterns you see on your record sheet.

CHAPTER PREVIEW

In Chapter 3, we will consider how numbers have been represented historically—especially in the modern **decimal (base-ten)** system. The symbols for writing numbers are called **numerals,** and the methods for calculating are called **algorithms.** Taken together, any particular system of numerals and algorithms is called a **numeration system.** We will give algorithms for addition, subtraction, multiplication, and division of whole numbers and will show some physical representations of these systems. Such representations depend crucially on the notion of **exchange** (what you may have called "borrowing" or "carrying" when you were in school) in the representation of a whole number. Because of the need for procedural fluency, the chapter finishes with methods of estimating answers and a discussion of some practices associated with mental arithmetic.

KEY IDEAS

- Numeration systems in the past and present, in both base ten and other bases, including the Egyptian, Roman, Babylonian, Mayan, and Indo-Arabic **(decimal)** systems
- Representation of numeration systems by physical objects such as units, strips, and mats or sticks in bundles
- The use of **exchanging** to simplify the tasks of addition, subtraction, multiplication, and division of whole numbers
- Methods **(algorithms)** for addition, subtraction, multiplication, and division of whole numbers
- Ways in which to make careful **estimations** of a final answer as a check
- Ways in which we might do mental arithmetic faster

3.1 Numeration Systems Past and Present

To appreciate the power of the Indo-Arabic (or Hindu-Arabic) numeration system, which is the present base-ten system or decimal system, as we now call it, it is important to know something about numeration systems of the past. Just as the idea of number historically arose from the need to determine "how many," the demands of commerce in an increasingly sophisticated society stimulated the development of convenient symbolism for writing numbers and methods for calculating.

The earliest means of recording numbers consisted of creating a set of tallies—marks on stone, stones in a bag, notches in a stick—one for one, for each item being counted. Indeed, the original meaning of the word *tally* was "a stick with notches cut into it to record debts owed or paid." Often such a stick was split in half, with one half going to the debtor and the other half to the creditor. It is still common practice today to keep count by making tallies, or marks, with the minor but useful refinement of marking off the tallies in groups of five. Thus,

is much easier to read as 23 than is

However, such systems for recording numbers were much too simplistic for large numbers and for calculating.

Of the various systems used in the past, we will consider only four here: the Egyptian, the Roman, the Babylonian, and the Mayan systems.

The Egyptian System

As early as 3400 B.C., the Egyptians developed a system for recording numbers on stone tablets using **hieroglyphics** (Table 3.1). This system was based on the number 10, as is our modern system, and probably for the same reason. That is, we humans come with a built-in "digital" calculator, as it were, with 10 convenient "keys" (fingers).

TABLE 3.1 EGYPTIAN SYMBOLS FOR POWERS OF 10

Power of 10	$10^0 = 1$	$10^1 = 10$	$10^2 = 100$	$10^3 = 1000$	$10^4 = 10{,}000$	$10^5 = 100{,}000$	$10^6 = 1{,}000{,}000$
Egyptian Symbol Description	a staff	 a yoke	 a scroll	 a lotus flower	 a pointing finger	 a fish	 an amazed person

The Egyptians had symbols for the first few powers of 10 and then combined symbols additively to represent other numbers. (See Table 3.1.) The system was cumbersome, not only because it required numerous symbols to represent even relatively small numbers but also because computation was awkward. We will now write 336 in the Egyptian system and then add two numbers.

EXAMPLE 3.1

Using Egyptian Notation to Write and Add Numbers

(a) What is the decimal number 336 in Egyptian notation?
(b) Add 336 and 125 in the Egyptian system.

(a) From Table 3.1, the number 336 is $300 + 30 + 6 =$ three scrolls + three yokes + 6 staffs.

$$\text{or } 336 = \text{999} \cap \cap \cap \begin{matrix} ||| \\ ||| \end{matrix}.$$

- three hundreds
- three tens
- six ones

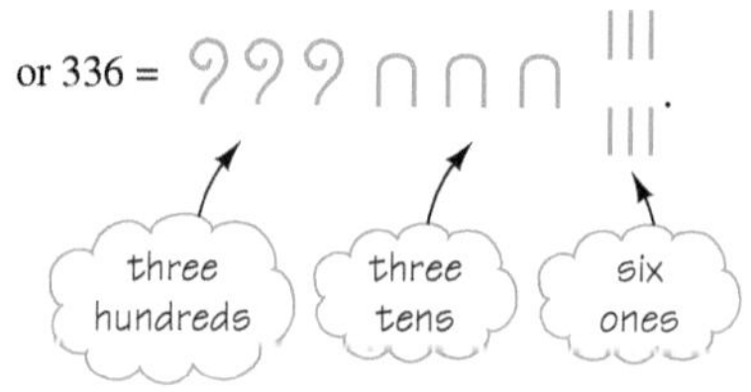

(b) Since 336 appears in part (a) in the Egyptian system, we note that 125 is written as

$$125 = \text{9} \cap \cap \begin{matrix} ||| \\ || \end{matrix}$$

- one hundred
- two tens
- five ones

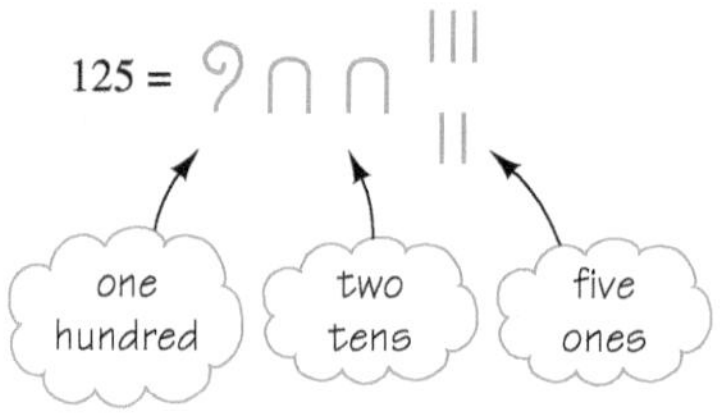

Then $336 + 125$ was found by combining the Egyptian symbols to obtain

$$\begin{matrix} ||||| \\ ||||| \end{matrix} = \cap$$

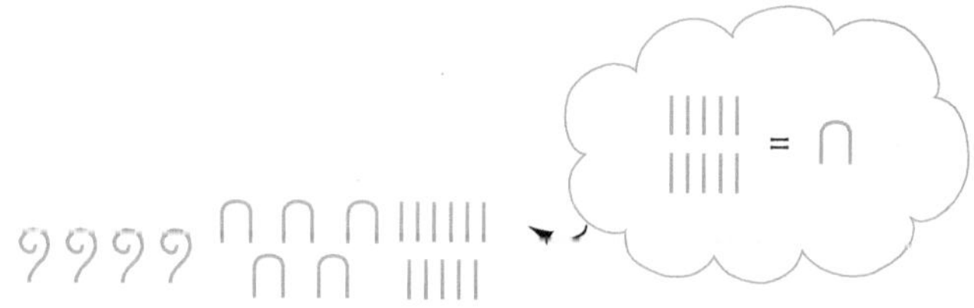

and then replacing 10 of the symbols for 1 by one symbol for 10 to finally obtain

$$\begin{array}{r} 336 \\ + 125 \\ \hline 461 \end{array}$$

representing 461. You can easily imagine how cumbersome multiplication (as repeated addition) and division (as repeated subtraction) was in this system.

The Roman System

The **Roman system of numeration** is already somewhat familiar from its current usage on the faces of analog watches and clocks, on cornerstones, and on the façades of buildings to record when they were built. Originally, like the Egyptian system, the Roman system was completely additive, with the familiar symbols shown in Table 3.2.

TABLE 3.2   ROMAN NUMERALS AND THEIR MODERN EQUIVALENTS

Roman Symbol	I	V	X	L	C	D	M
Modern Equivalent	1	5	10	50	100	500	1000

With these symbols, 1959 originally was written as

$$\text{M} \quad \text{D} \quad \text{CCCC} \quad \text{L} \quad \text{V} \quad \text{IIII}.$$

- one thousand
- five hundred
- four hundreds
- five
- fifty
- four ones

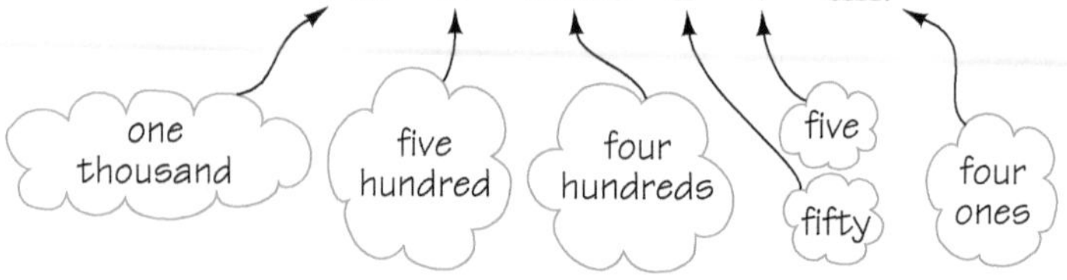

Later, a subtractive principle was introduced to shorten the notation: If a single symbol for a lesser number was written to the left of a symbol for a greater number, then the lesser number was to be *subtracted* from the greater number. The common representations of this principle were as shown here:

IIII was written as IV. $4 = 5 - 1$

VIIII was written as IX. $9 = 10 - 1$

XXXX was written as XL. $40 = 50 - 10$

CCCC was written as CD. $400 = 500 - 100$

DCCCC was written as CM. $900 = 1000 - 100$

With this subtractive principle, 1959 could be written more succinctly as

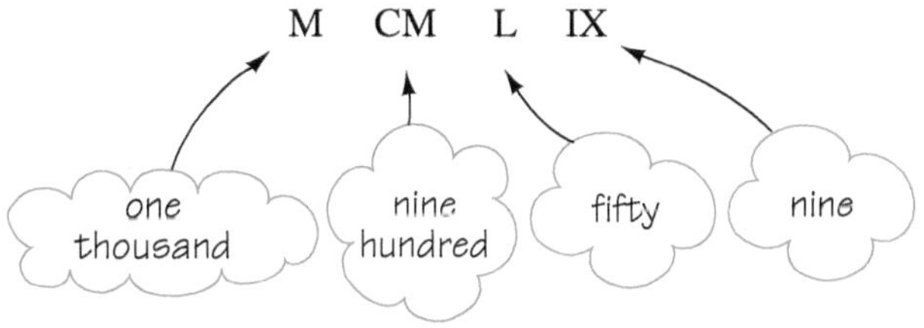

As with the Egyptian system, arithmetic* calculations using Roman numerals were quite cumbersome. For this reason, much of the commercial calculation of the time was performed on devices such as abacuses, counting boards, sand trays, and the like.

The Babylonian System

Developed about the same time as the Egyptian system, and much earlier than the Roman system, the **Babylonian system** was more sophisticated than either in that it introduced the notion of **place value,** in which the position of a symbol in a numeral determined the value of the numeral. In particular, this made it possible to write numerals for even very large numbers by using very few symbols. Indeed, the system utilized only two symbols, ▼ for 1 and ❮ for 10, and combined these additively to form the "digits" 1 through 59. Thus,

represented 21 and 34, respectively. Beyond 59, the system was positional to base 60 (a *sexagesimal* system), where the positions from right to left represented multiples of successive powers of 60 and the multipliers (digits) were the composite symbols for 1 through 59.

EXAMPLE 3.2

Using Babylonian Notation to Write a Number

Using base sixty in the Babylonian system, what base-ten numbers are given by the following?

(a)

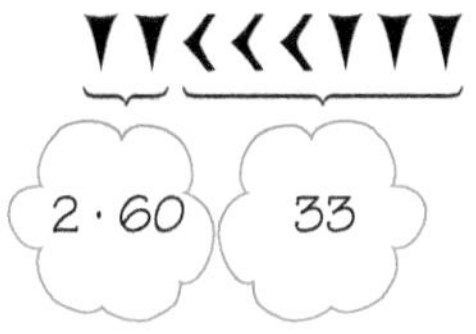

(b)

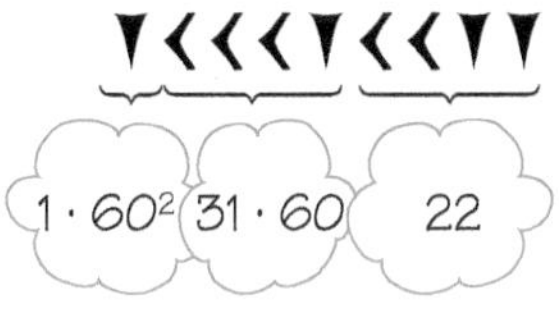

*When used as a noun, "arithmetic" is pronounced ə-rith′mə-tik. When used as an adjective, it is pronounced ar′ith-met′ik.

Solution

(a) Starting at the right are three ▼ symbols (each of which equals 1) and three ❮ symbols (each of which equals 10) for a total of 33. Then follow two ▼ symbols that, since the base is sixty, equal $2 \cdot 60^1$. The total is computed as $2 \cdot 60^1 + 33 = 153$.

(b) Using the place values of base sixty, this Babylonian system number is $1 \cdot 60^2 + 31 \cdot 60^1 + 22 = 5482$, which is clear from the think clouds.

A difficulty with the Babylonian system was the lack of a symbol for zero. In writing numerals, scribes simply left a space if a certain position value was not to be used; since spacing was not always uniform, this practice often made it necessary to infer values from context. For example, the Babylonian notation for 83 and 3623 originally were, respectively,

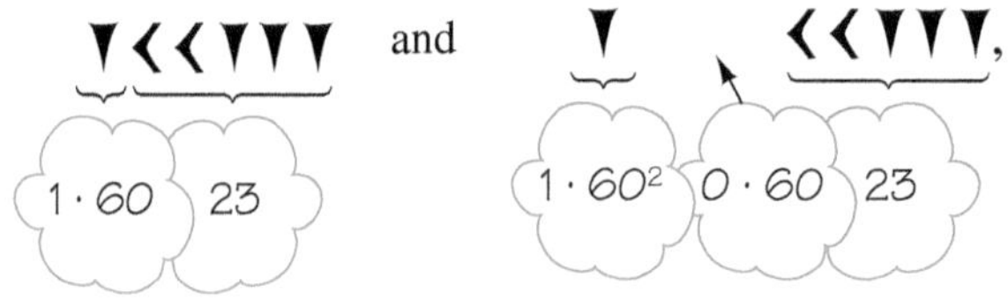

and these could easily be confused if the spacing were not clear. Indeed, there was no way even to indicate missing position values on the extreme right of a numeral, so ▼ could represent 1 or $1 \cdot 60^1 = 60$ or $1 \cdot 60^2 = 3600$, and so on. Eventually, the Babylonians employed the symbol ▲ as a placeholder to indicate missing position values, though they never developed the notion of zero as a number. With this symbol, ▼❮❮▼▼▼ was clearly understood as $1 \cdot 60 + 23 = 83$ and ▼▲❮❮▼▼▼ was unmistakably

$$1 \cdot 60^2 + 23 = 3623.$$

The Mayan System

One of the most interesting of the ancient systems of numeration was developed by the Mayans in the region now known as the Yucatán Peninsula, in southeastern Mexico. As early as A.D. 200, these resourceful people had developed a remarkably advanced society. They were the first Native Americans to develop a system of writing and to manufacture paper and books. Their learned scholars knew more about astronomy than was known at that time anywhere else in the world. Their calendar was very accurate, with a 365-day year and a leap year every fourth year. In short, many scholars believe that the Mayans developed the most sophisticated society ever attained by early residents of the Western Hemisphere.

Like their other achievements, the **Mayan system of numeration** was unusually advanced. Actually, there were *two* systems. Both systems were positional systems like the Babylonian system and our present base-ten, or decimal, system, and both contained a symbol for zero. One system was a base-twenty, or *vigesimal*, system based on powers of 20 and utilizing rather involved hieroglyphics for the numerals 1 through 19. Since the vigesimal system has not survived in written form, we describe the second system, which was a modification of the first, devised to facilitate computations related to the calendar. It has the advantage that $18 \cdot 20 = 360$ is much closer to the length of a year than is $20 \cdot 20 = 400.$[*] Thus, the positions were for 1, 20, $18 \cdot 20$, $18 \cdot 20^2$, $18 \cdot 20^3$, and so on, rather than $20^2, 20^3, 20^4, \ldots$ as with the first system. Also, the second system used a simple set of symbols for the numerals 0 through 19, as shown in Table 3.3.

TABLE 3.3   MAYAN NUMERALS FOR 0 THROUGH 19

Mayan Symbol	Modern Equivalent	Mayan Symbol	Modern Equivalent
	0		10
	1		11
	2		12
	3		13
	4		14
	5		15
	6		16
	7		17
	8		18
	9		19

The Mayans wrote their numerals in a vertical style with the unit's position on the bottom. For example, the number 43,487 would appear as

$$6 \cdot 18 \cdot 20^2 = 43{,}200$$
$$0 \cdot 18 \cdot 20 = 0$$
$$14 \cdot 20 = 280$$
$$7 \cdot 1 = \underline{\quad 7\quad}$$
$$43{,}487$$

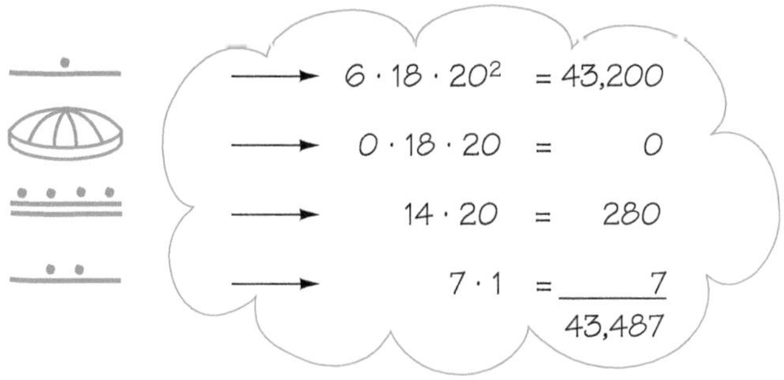

All that is needed to convert a Mayan numeral to modern notation is a knowledge of the digits and the value of the position each digit occupies, as shown in Table 3.4.

---

[*]The Mayan calendar was composed of 18 months of 20 days each, with 5 additional days not associated with any month and with the provision for a leap year.

TABLE 3.4 POSITION VALUES IN THE MAYAN SYSTEM

Position Level	Position Value
·	·
·	·
·	·
Fifth	$18 \cdot 20^3 = 144{,}000$
Fourth	$18 \cdot 20^2 = 7{,}200$
Third	$18 \cdot 20 = 360$
Second	20
First	1

EXAMPLE 3.3

Using Mayan Notation to Write a Number

Write 27,408 in Mayan notation.

Solution

We will not need the fifth, or any higher, position, since $18 \cdot 20^3 = 144{,}000$ is already greater than 27,408. How many 7200s, 360s, 20s, and 1s are needed? This question can be answered by repeated subtraction, or, more simply, by division. From the arithmetic

$$\begin{array}{r} 3 \\ 7200\overline{)27{,}408} \\ \underline{21{,}600} \\ 5\,808 \end{array} \qquad \begin{array}{r} 16 \\ 360\overline{)5808} \\ \underline{360} \\ 2208 \\ \underline{2160} \\ 48 \end{array} \qquad \begin{array}{r} 2 \\ 20\overline{)48} \\ \underline{40} \\ 8 \end{array} \qquad \begin{array}{r} 8 \\ 1\overline{)8} \\ \underline{8} \\ 0 \end{array}$$

it follows that we need three 7200s, sixteen 360s, two 20s, and eight 1s. Thus, in Mayan notation, 27,408 appears as

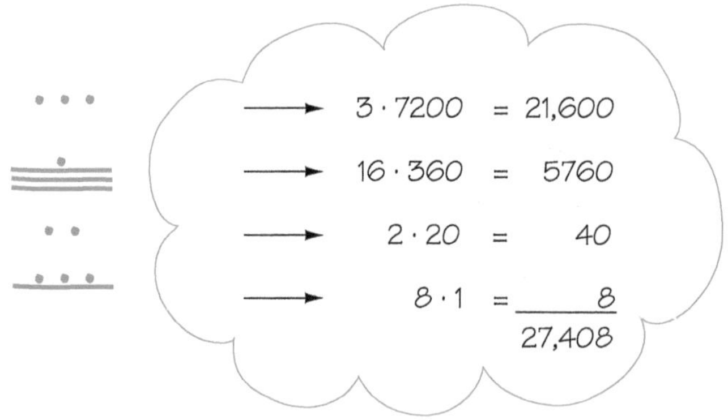

The ingenuity of this system, particularly since it involves the 18s, is easy to overlook. It turns out that such a positional system *will not work* unless the value of each position is a number that evenly divides the value of the next-higher position. This fact is not at all obvious!

The Indo-Arabic System

Today, the most universally used system of numeration is the **Indo-Arabic,** or **decimal, system.** The system was named jointly for the East Indian scholars who invented it at least as early as 800 B.C. and for the Arabs who transmitted it to the Western world. Like the Mayan system, it is a positional system. Since its *base* is ten, it requires special symbols for the numbers 0 through 9. Over the years, various notational choices have been made, as shown in Table 3.5.

TABLE 3.5 SYMBOLS FOR ZERO THROUGH NINE, ANCIENT AND MODERN

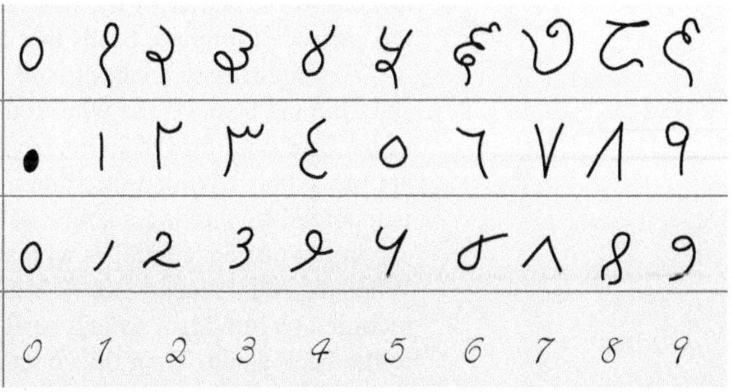

Tenth-century East Indian	
Current Arabic	٠ ١ ٢ ٣ ٤ ٥ ٦ ٧ ٨ ٩
Fifteenth-century European	
Modern cursive	0 1 2 3 4 5 6 7 8 9

The common symbols 0, 1, 2, 3, 4, 5, 6, 7, 8, and 9 are called **digits,** as are our fingers and toes, and it is easy to imagine the historical significance of this terminology. With these 10 symbols and the idea of positional notation, all that is needed to write the numeral for any whole number is the value of each digit and the value of the position the digit occupies in the numeral. In the decimal system, the positional values, as shown in Table 3.6, are well known.

TABLE 3.6 POSITIONAL VALUES IN BASE TEN

Position Names	**Hundred Thousands**	**Ten Thousands**	**Thousands**	**Hundreds**	**Tens**	**Units**
Decimal Form	100,000	10,000	1000	100	10	1
Powers of 10	10^5	10^4	10^3	10^2	10^1	10^0

As our number words suggest, the symbol 2572 means 2 thousands plus 5 hundreds plus 7 tens plus 2 ones. In so-called **expanded notation,** we write

2000 + 500 + 70 + 2

$$\begin{aligned} 2572 &= 2 \cdot 1000 + 5 \cdot 100 + 7 \cdot 10 + 2 \cdot 1 \\ &= 2 \cdot 10^3 + 5 \cdot 10^2 + 7 \cdot 10^1 + 2 \cdot 10^0. \end{aligned}$$

The amount each digit contributes to the number is the value of the digit times the value of the position the digit occupies in the representation.

Physical Models for Positional Systems

The Classroom Abacus as a Manipulative We now look at five models that show physically what a **manipulative** looks like and how it can aid our understanding. These examples of manipulatives include the classroom abacus; sticks in bundles; Unifix™ cubes; units, strips, and mats; and base-ten blocks.

Concrete physical materials offer different ways of learning mathematics and so appeal to the mind in various ways, depending on the student.

The Indo-Arabic scheme for representing numbers derived historically from the use of counting boards and abacuses of various types to facilitate computations in commercial transactions. Such devices are still useful today in helping children to understand the basic concepts. One device, commercially available for classroom use, is shown in Figure 3.1.

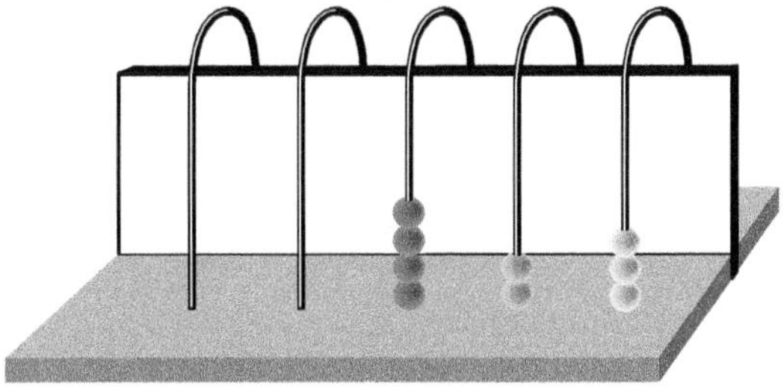

Figure 3.1
A classroom demonstration abacus

The device consists of a series of wire loops fixed into a wooden base and with a vertical shield affixed to the center of the base under the wire loops so that the back is hidden from the students. On each wire loop are beads that can be moved from behind the shield to the front and vice versa.

To demonstrate counting and positional notation, one begins by moving beads from behind the shield to the front on the wire to the students' right (the instructor's left), counting 1, 2, 3, and so on as the beads are moved to the front of the shield. Once 10 beads on the first wire are counted, all 10 are moved back behind the shield, and the fact that 10 beads on this wire have all been counted once is recorded by moving a single bead to the front of the shield on the second wire. The count 11, 12, 13, and so on then continues with a single bead on the first wire again moved forward with each count. When the count reaches 20, 10 beads on the first wire will have been counted a second time, and this is recorded by moving a second bead forward on the second wire and moving the beads on the first wire to the back again. Then the count continues. When the count reaches, for example, 34, the children will see the arrangement of beads illustrated schematically in Figure 3.2, with three beads showing on the second wire and four beads showing on the first wire. A natural way to record the result is to write 34—that is, 3 tens and 4 ones. Note that we never leave 10 beads up on any wire.

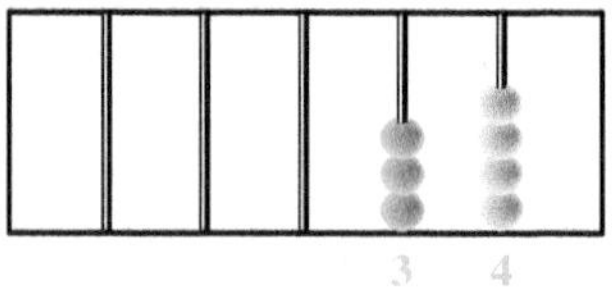

Figure 3.2
Thirty-four on the classroom abacus

What happens if we count to 100—that is, if we count 10 beads on the first wire 10 times? Since we move one bead on the second wire forward each time we count 10 beads on the first wire, we will have 10 beads showing on the second wire. But we record this on the abacus by moving one bead forward on the third wire and moving all beads on the second wire back behind the shield. Thus, each bead showing on the first wire counts for 1, each bead showing on the second wire counts for 10, each bead showing on the third wire counts for 100, and so on. If we count, for example, to 423, the arrangement of beads is as illustrated earlier in Figure 3.1, and the count is naturally recorded as 423.

Approached in this way, the notion of place value becomes much more concrete. One can actually experience the fact that each bead on the second wire counts for 10, each bead on the third wire counts for 100, and so on—particularly if one is allowed to handle the device and move the beads while counting. One can also see that the counting process on the abacus can proceed as far as desired, though more and more wire loops would have to be added to the device. It follows that any whole number can be represented in one, and only one, way in our modern system by using only the digits 0, 1, 2, 3, 4, 5, 6, 7, 8, and 9.

Other physical devices can also be used to illustrate positional notation, and many are even more concrete than the historical abacus just mimicked.

Sticks in Bundles One simple idea for introducing positional notation is to use bundles of small sticks, which can be purchased inexpensively at almost any craft store. Single sticks are 1s. Ten sticks can be bound together in a bundle to represent 10. Ten bundles can be banded together to represent 100, and so on. Thus, 34 would be represented as shown in Figure 3.3.

Unifix™ Cubes Here, single cubes represent 1s. Ten cubes snapped together to form a stick represent 10. Ten sticks bound together represent 100, and so on. Again, 34 would be represented as shown in Figure 3.4.

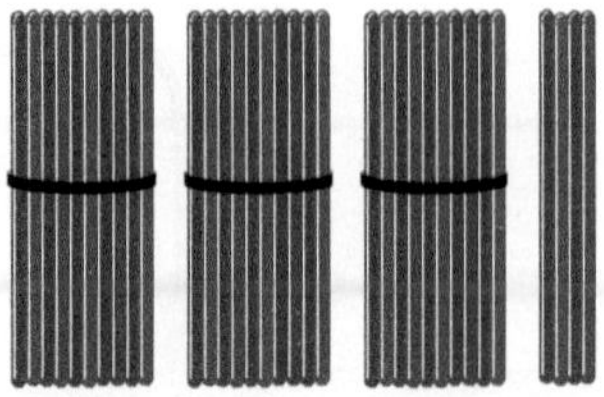

Figure 3.3
Representing 34 with sticks and bundles of sticks

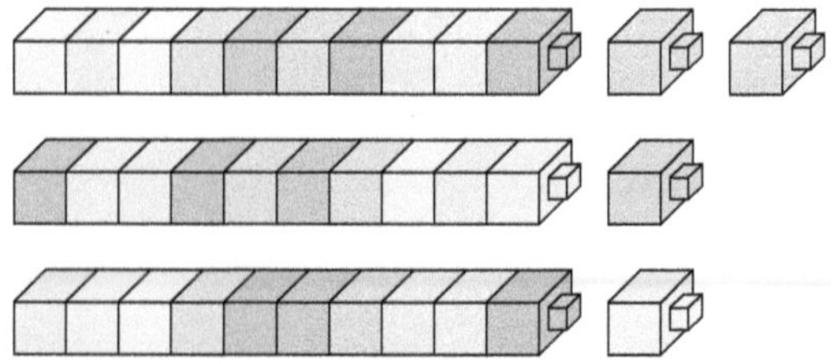

Figure 3.4
Representing 34 with Unifix™ cubes

Units, Strips, and Mats Pieces for this concrete realization of the decimal system are easily cut from graph paper with reasonably large squares. A unit is a single square, a strip is a block of 10 squares, and a mat is a square 10 units on a side, as illustrated in Figure 3.5. To make the units, strips, and mats more substantial and easier to manipulate, the graph paper can be pasted to a substantial tagboard or copied on reasonably heavy stock. Using units, strips, and mats, we would represent 254 as shown in Figure 3.6.

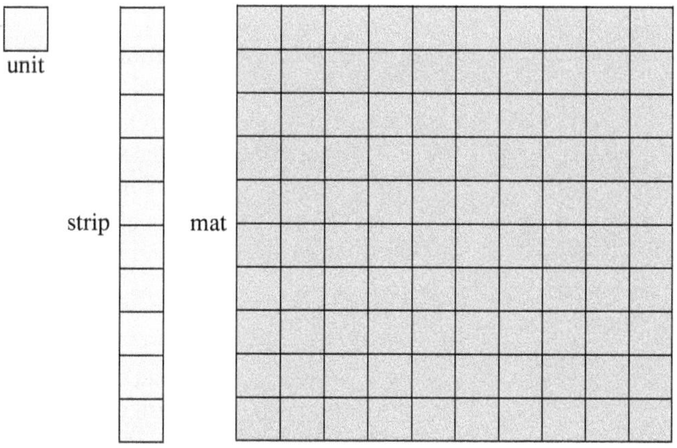

Figure 3.5
A unit, a strip, and a mat

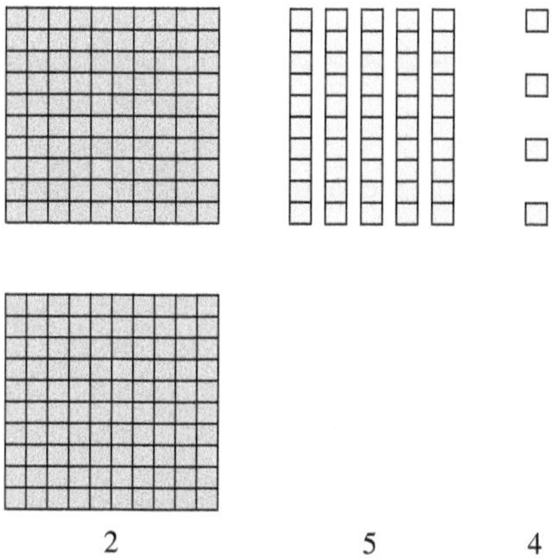

Figure 3.6
Representing 254 with units, strips, and mats

Base-Ten Blocks These commercially prepared materials include single cubes that represent 1s, sticks called "longs" made up of 10 cubes that represent 10, "flats" made up of 10 longs that represent 100, and "blocks" made up of 10 flats that represent 1000, as shown in Figure 3.7.

Figure 3.7
Base-ten blocks for 1, 10, 100, and 1000

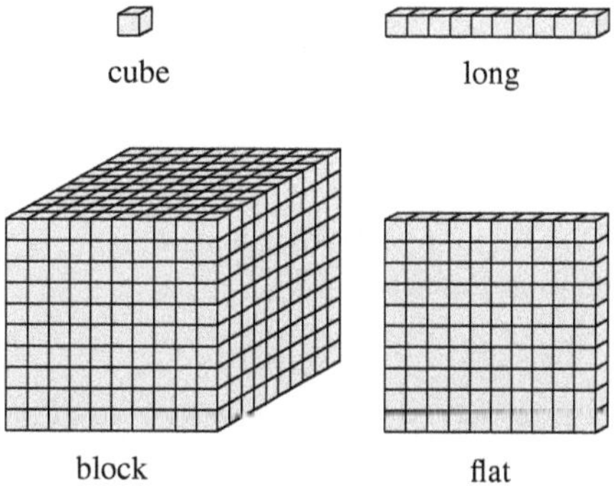

3.1 Problem Set

Understanding Concepts

1. Write the Indo-Arabic equivalent of each of the following:

(a) 

(b) MDCCXXIX

(c) DCXCVII

(d) 

(e) 

(f) 

2. Write the Indo-Arabic equivalent of each of the following:

(a) 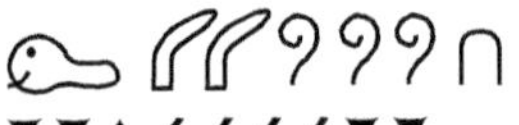 (b) CMLXXXIV

(c)  (d) 

(e) 

3. Write the following numbers in Egyptian notation:

(a) 11 (b) 597 (c) 1949

4. Write the following numbers in Roman notation, using the subtraction principle as appropriate:

(a) 9 (b) 1996

5. Write the following numbers in Roman notation, using the subtraction principle as appropriate.

(a) 974 (b) 2009

6. Write these numbers in Babylonian notation:

(a) 251 (b) 3022 (c) 18,741

7. Write these numbers in Mayan notation:

(a) 12 (b) 584 (c) 12,473

8. Add 

and ,

using only Egyptian notation. Write your answer as simply as possible.

9. Add MDCCCXVI and MCCCLXIV, using only Roman notation. Write your answer as simply as possible.

10. Add 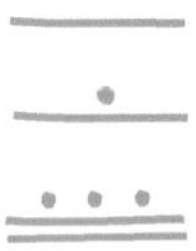 and 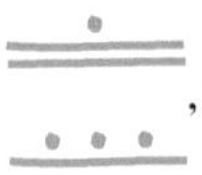, using only Mayan notation.

Write your answer as simply as possible.

11. Write 2002, 2003, and 2004 in Roman numerals.

12. To date, what year has had the Roman numeral with the most symbols? Give your answer in both Roman and Indo-Arabic numerals.

13. Draw a sketch to illustrate how the number 452 is represented with units, strips, and mats. (See Figures 3.5 and 3.6.) Use dots to represent units, vertical line segments to represent strips, and squares to represent mats.

14. Draw a sketch to illustrate the number 234, using base-ten blocks.

15. Draw a sketch of the exposed side of a classroom abacus to illustrate the number 2475.

16. Write 24,872 and 3071 in expanded notation.

17. Suppose you have 3 mats, 24 strips, and 13 units, for a total count of 553. Briefly describe the exchanges you would make to keep the same total count but have the smallest possible number of manipulative pieces.

18. Suppose you have 2 mats, 7 strips, and 6 units in one hand and 4 mats, 5 strips, and 9 units in the other.

(a) Put all these pieces together and describe the exchanges required to keep the same total count but with the smallest possible number of manipulative pieces.

(b) Explain briefly but clearly what kind of mathematics the manipulation in part (a) represents.

19. Suppose you have 3 mats and 6 units on your desk and want to remove a count represented by 3 strips and 8 units.

(a) Describe briefly but clearly the exchange that must take place to accomplish the desired task.

(b) After removing the 3 strips and 8 units, what manipulative pieces are left on your desk?

(c) What kind of mathematics does the manipulation in part (a) represent?

Into the Classroom

20. Recall that units, strips, and mats are typically made from graph paper (see Figure 3.5). How would you augment this simple set of manipulatives to make one representing the number 1000 for a lesson you might teach to a class of third graders?

21. Triana has 12 dollars and 33 dimes in her piggy bank, while Leona has 9 dollars and 11 dimes.

(a) If the girls pool their money, how many dollars and dimes will they have? Note that this question can be answered without adding; one can simply count on, instead.

(b) If the girls buy their mother a bottle of perfume for $25, how much money will they have left over?

(c) Describe any exchanges you made in determining the answer to part (b).

22. Jon works at odd jobs for Mr. Taylor. On four successive Saturdays, he worked 6 hours and 35 minutes, 4 hours and 15 minutes, 7 hours and 30 minutes, and 5 hours and 20 minutes, respectively. All told, how many 8-hour days, how many hours, and how many minutes was this amount of time equivalent to? (Make sure that the number of hours in your answer is less than 8 and that the number of minutes is less than 60.)

23. Make up a problem along the lines of problem 22 but using hours, days, and weeks.

24. (a) How does a problem like problem 21 help students to understand positional notation?

(b) Would understanding the notion of exchanging help students who are studying calculation with fractions to deal with a problem like this?

$$\begin{array}{r} 2\frac{3}{7} \\ +\ 2\frac{5}{7} \\ \hline \end{array}$$

Explain.

Responding to Students

25. How does a problem like problem 22 help students to understand positional notation? Explain.

26. Tom and Marsha are playing a game with 10-sided dice. The object of the game is to make the largest 5-digit number possible when 5 dice are rolled. The numbers rolled are 8, 0, 9, 0, and 2. Tom is losing most rounds. What can you do to help him win?

27. Sara, a first grader, has a paper version of a rod and is starting to cut it up. About halfway through, Sara's teacher asks why she is cutting up the rod. Sara says that she isn't doing anything bad. It is so she can learn about how to build. What do you think she might mean?

State Assessments

28. (Grade 3)
Find the missing number that makes the sentence true.
300 + ____ + 6 = 386

A. 70 B. 86 C. 80 D. 90

29. (Grade 4)
Look at the units, strips and mats below. What is the correct the number?

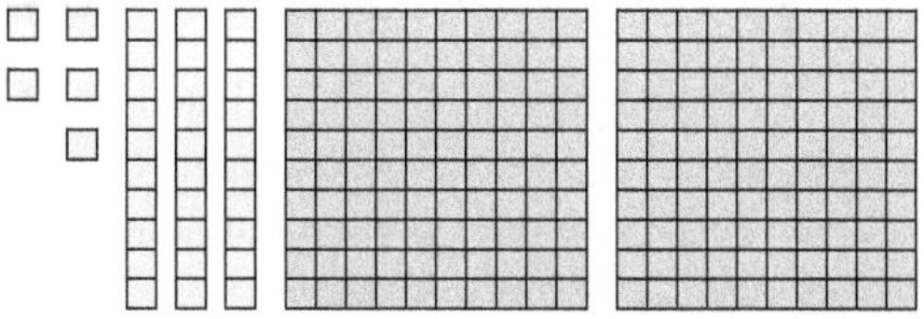

A. 235 B. 532
C. 425 D. 253

30. (Grade 4)
What does the same number as 278 mean?

A. 20 + 700 + 8
B. 200 + 70 + 8
C. 200 + 70 + 8
D. 2 + 7 + 8

31. (Grade 5)
The area of each of the four lakes is listed in the table below, measured in square miles.

Lake	Area (miles^2)
S	82,234
T	59,343
U	21,276
V	35,232
W	?

Lake W is larger in area than Lake T and Lake W is smaller than Lake S. Which could be the area of Lake W?

A. 87,234 sq. miles
B. 59,343 sq. miles
C. 57,789 sq. miles
D. 67,345 sq. miles

32. (Grade 4)
There were six million, two hundred and seven thousand, one hundred and thirty one students in California public schools in 2011-2012. How do you write this number in standard form?

A. 6,000,207,131
B. 600,207,131
C. 6,207,131
D. 6, 200,13

3.2 Algorithms for Addition and Subtraction of Whole Numbers

In this section, we will discuss <u>how</u> we add and subtract integers and <u>why</u> the methods actually work. The customary approaches are called the **addition algorithm** and the **subtraction algorithm.** The problems include other methods such as the **scratch method** for addition. (See problem 25 in Problem Set 3.2.) During your teaching career, you will find that children have a marvelous sense of invention. When a child comes up with a different way to add or subtract numbers, you, as a teacher, must decide whether the child has found a correct way to do the problem. If the child makes a mistake, you will need to correct that child's approach. The point here is, of course, that teachers must have a conceptual understanding of the material in order to be able to help their students, regardless of whether the child is correct or incorrect. You will practice helping students with their approaches in the "Responding to Students" problems throughout the text.

The first step in the construction of algorithms is to understand **positional notation.** We have seen in Section 3.1 that, for a positional system with, for example, base ten, the two-digit number 35 means 3 tens plus 5. In addition, there are many physical models for positional systems. We will first discuss making exchanges in base ten and then develop the addition and subtraction algorithms in a conceptual manner. **Place value,** or positional notation, is the foundation on which the operations of arithmetic are built. (See the NCTM's view of place value that follows.)

FROM THE NCTM PRINCIPLES AND STANDARDS

Place-Value Concepts and Calculators

Students also develop understanding of place value through the strategies they invent to compute. . . . Thus, it is not necessary to wait for students to fully develop place-value understandings before giving them opportunities to solve problems with two- and three-digit numbers. When such problems arise in interesting contexts, students can often invent ways to solve them that incorporate and deepen their understanding of place value, especially when students have opportunities to discuss and explain their invented strategies and approaches.

. . . . .

As students encounter problem situations in which computations are more cumbersome or tedious, they should be encouraged to use calculators to aid in problem solving. In this way, even students who are slow to gain fluency with computation skills will not be deprived of worthwhile opportunities to solve complex mathematics problems and to develop and deepen their understanding of other aspects of number.

SOURCE: Principles and Standards for School Mathematics by NCTM, *pp. 82, 86, 87. Copyright © 2000 by the National Council of Teachers of Mathematics. Reproduced with permission of the National Council of Teachers of Mathematics via Copyright Clearance Center. NCTM does not endorse the content or validity of these alignments.*

The Addition Algorithm

Consider the following addition:

$$\begin{array}{r} \overset{1}{2}8 \\ +45 \\ \hline 73 \end{array}$$

This process depends entirely on our positional system of notation. But why does it work that way? Why "carry the 1"? Why add by columns? To many people, these procedures remain a great mystery—you add this way because you were told to—and it's simply done by rote with no understanding.

As noted earlier, children learn abstract notions by first experiencing them concretely with devices they can actually see, touch, and manipulate. Thus, one should introduce the **addition algorithm** with manipulatives like base-ten blocks; sticks and bundles of sticks; units, strips, and mats; and abacuses—or, better yet, by using a number of these devices. For illustrative purposes here, we use units, strips, and mats.

After cutting out their strips and mats, students will be aware that there are 10 units on a strip and 10 strips, or 100 units, on a mat, and they will be aware that, in manipulating their materials, they can make these exchanges back and forth as needed.

EXAMPLE 3.4

Making Exchanges with Units, Strips, and Mats

Suppose a number is represented by 15 units, 11 strips, and 2 mats. What exchanges must be made in order to represent the same number but with the smallest number of manipulative pieces?

Solution

Understand the Problem

We are given the mats, strips, and units indicated in the statement of the problem and are asked to make exchanges that reduce the number of loose pieces of apparatus while keeping the same number of units in all.

Devise a Plan

Since we know that 10 units form a strip and 10 strips form a mat, we can reduce the number of loose pieces by making these exchanges.

Carry Out the Plan

If we actually had in hand the pieces described, we would physically make the desired exchanges. Here in the text, we illustrate the exchanges pictorially.

We reduce the number of pieces by replacing 10 units by 1 strip and 10 strips by 1 mat. This gives 3 mats, 2 strips, and 5 units, for 325 units as shown.

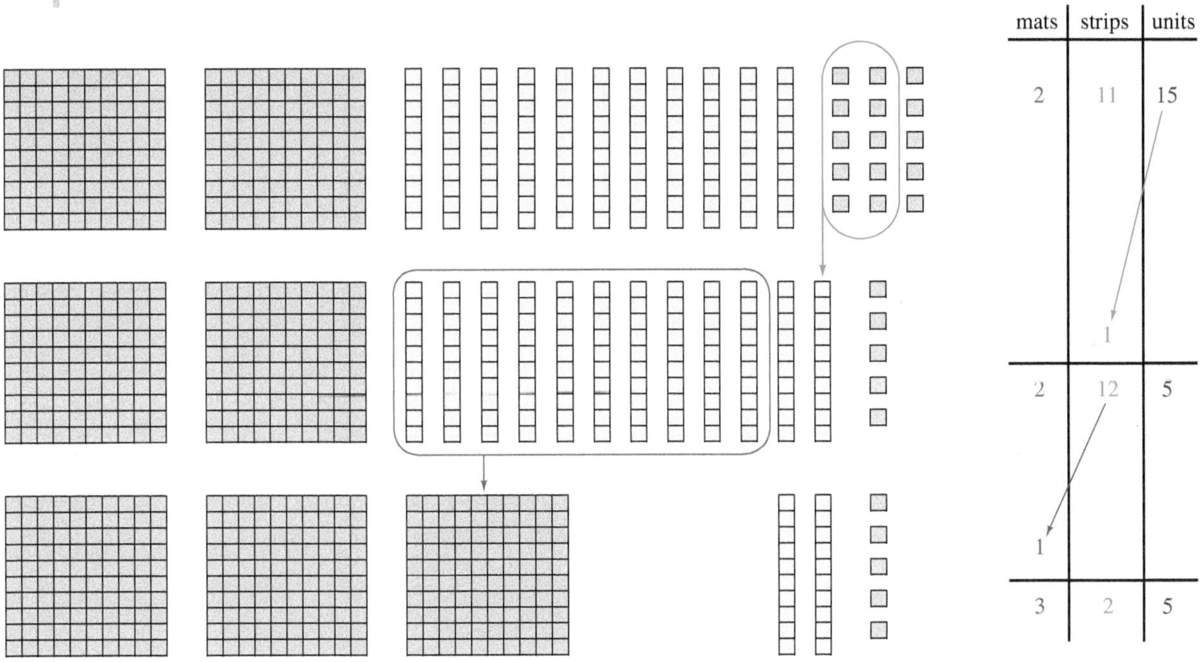

Look Back

No more combining can take place, because it takes 10 units to make a strip and 10 strips to make a mat. Thus, we finish with 3 mats, 2 strips, and 5 units, for a total of 325 units. Moreover, we have not changed the total number of units, since

$$2 \cdot 100 + 11 \cdot 10 + 15 = 200 + 110 + 15 = 325.$$

EXAMPLE 3.5 Developing the Addition Algorithm

Find the sum of 135 and 243.

Solution

With Units, Strips, and Mats

One hundred thirty-five is represented by 1 mat, 3 strips, and 5 units, and 243 is represented by 2 mats, 4 strips, and 3 units, as shown. All told, this gives a total of 3 mats, 7 strips, and 8 units. Therefore, since no exchanges are possible, the sum is 378. Note how this illustrates the column-by-column addition algorithm typically used in pencil-and-paper calculation.

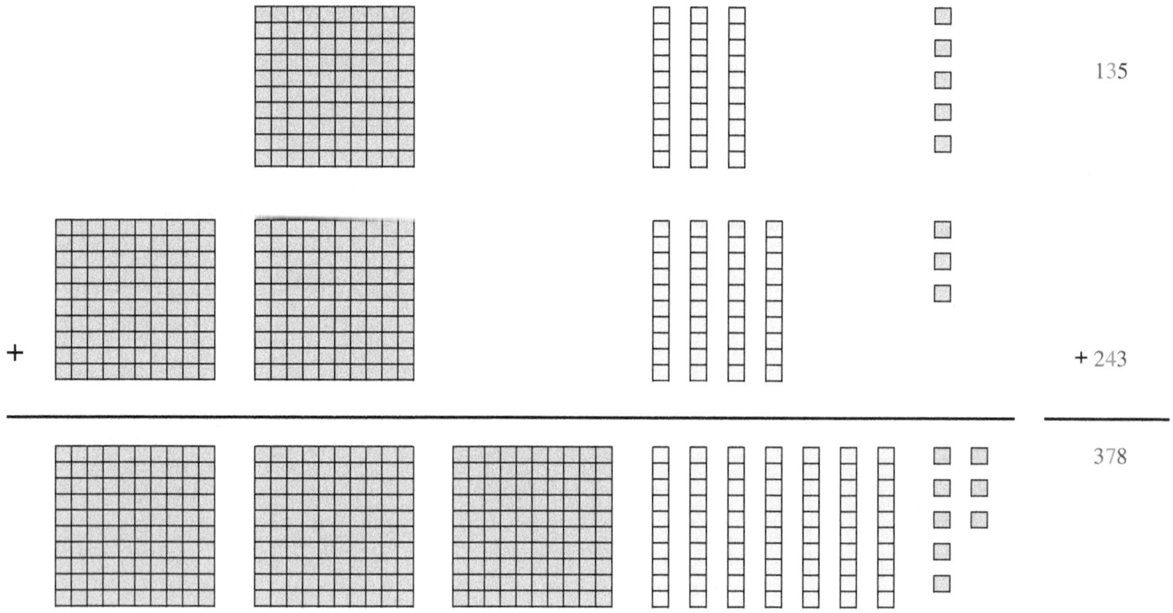

With Place-Value Cards

A somewhat more abstract approach to this problem is to use **place-value cards**—that is, cards marked off in squares labeled 1s, 10s, and 100s from right to left and with the appropriate number of markers placed in each square to represent the desired number. A marker on the second square is worth 10 markers on the first square, a marker on the third square is worth 10 markers on the second square, and so on. The addition of the two numbers in this example is illustrated as follows:

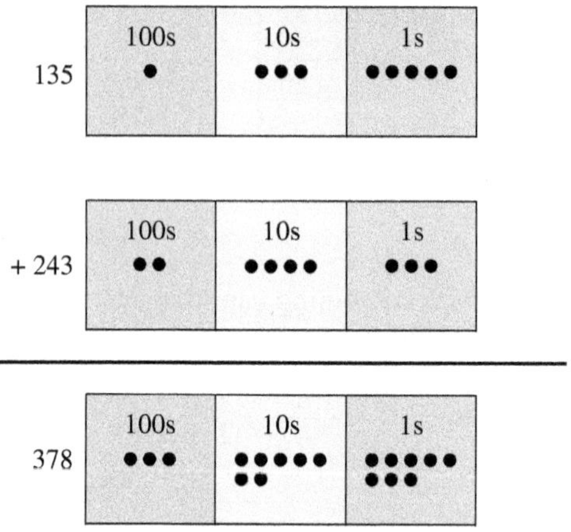

With Place-Value Diagrams and Instructional Algorithms

An even more abstract approach that leads finally to the usual algorithm is provided by **place-value diagrams** and **instructional algorithms.** The instructional algorithm is also called the **partial-sum algorithm.** The final algorithm is also known as the **traditional,** or **final, algorithm.** Here they are, applied to this example:

Place-Value Diagram			Instructional Algorithm		Final Algorithm
100s	10s	1s	135		135
1	3	5	+ 243		+ 243
+ 2	4	3	8	5 + 3	378
3	7	8	70	30 + 40	Compress the result to a single line.
			300	100 + 200	
			378		

Notice how the degree of abstraction steadily increases as we move through the various solutions. If the elementary school teacher goes directly to the final algorithm, many students will be lost along the way. The approach from the concrete gradually moving toward the abstract is more likely to impart the desired understanding.

EXAMPLE 3.6

Adding with Exchanging

Find the sum of 357 and 274.

Solution

With Units, Strips, and Mats

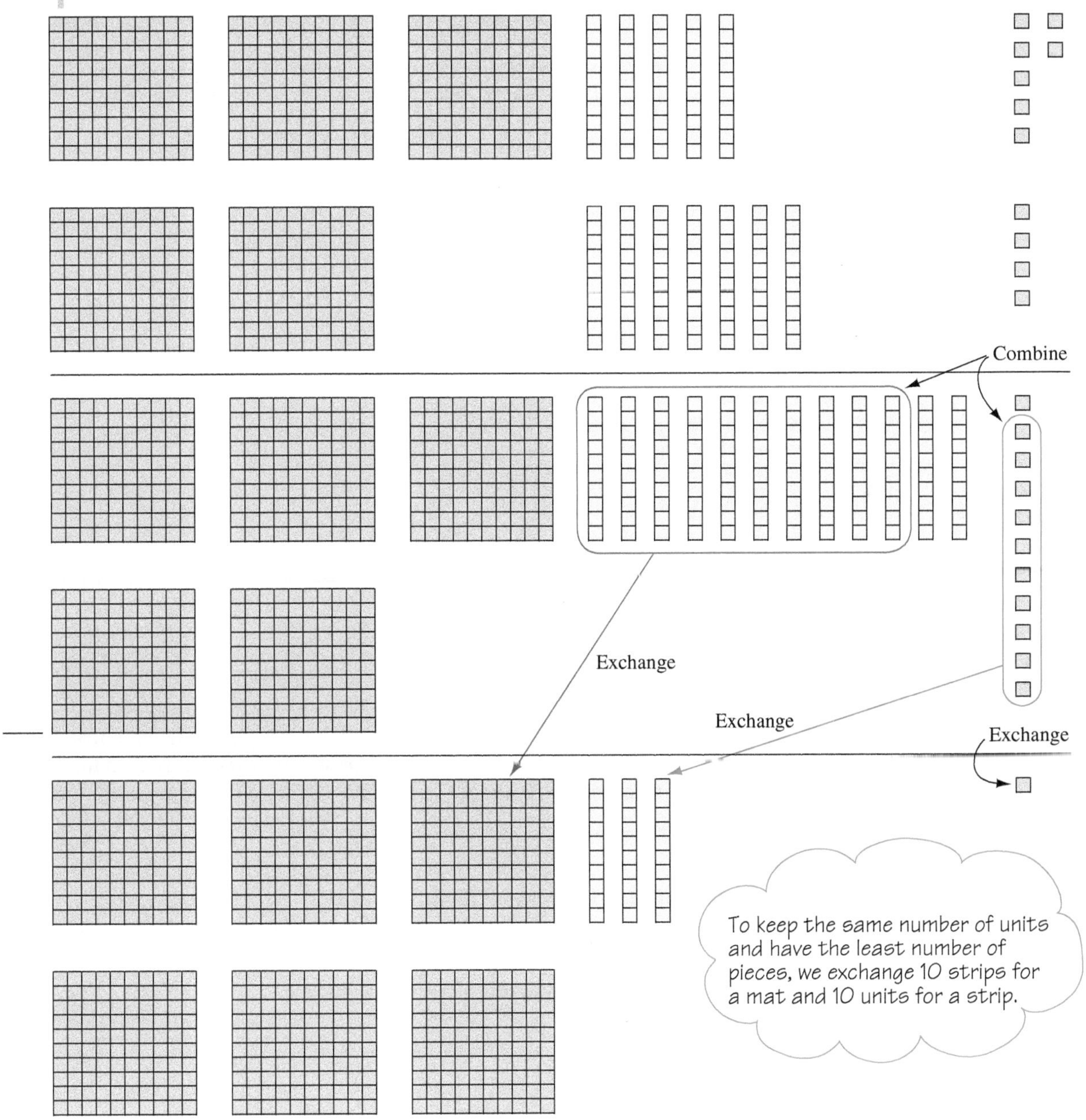
Combine
Exchange
Exchange
Exchange
To keep the same number of units and have the least number of pieces, we exchange 10 strips for a mat and 10 units for a strip.

Note how the preceding diagram illustrates not only the usual column-by-column addition of pencil-and-paper arithmetic but also "carrying," or, more appropriately stated, "exchanging." The 1 "carried" from the first column to the second indicates the exchange of 10 units for one 10 (represented by 1 strip), and the 1 carried from the second column to the third indicates the exchange of ten 10s

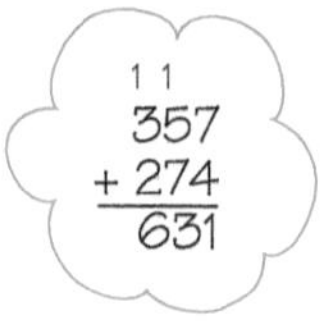
1 1
357
+ 274
631

(that is, 10 strips) for 100 (represented by 1 mat). Many teachers try to avoid the traditional word *carry,* since it encourages students to perform the algorithms by rote, with no understanding of what they are actually doing. Words like *exchange, trade,* and *regroup* are much more descriptive and actually describe what is being done.

With Place-Value Cards

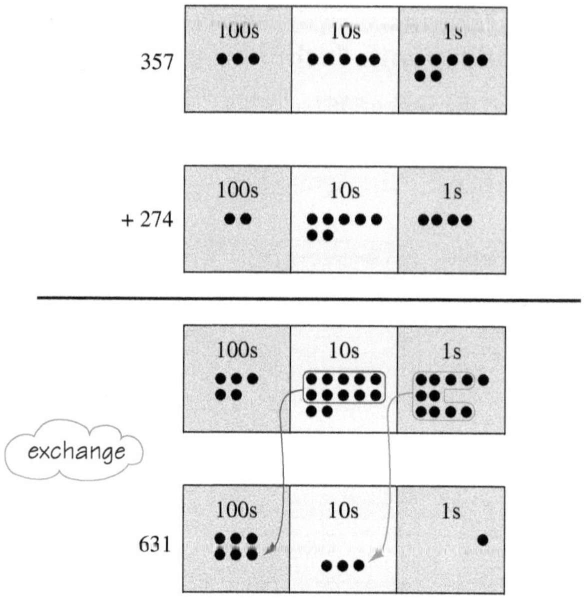

With Place-Value Diagrams and Instructional Algorithms

exchange

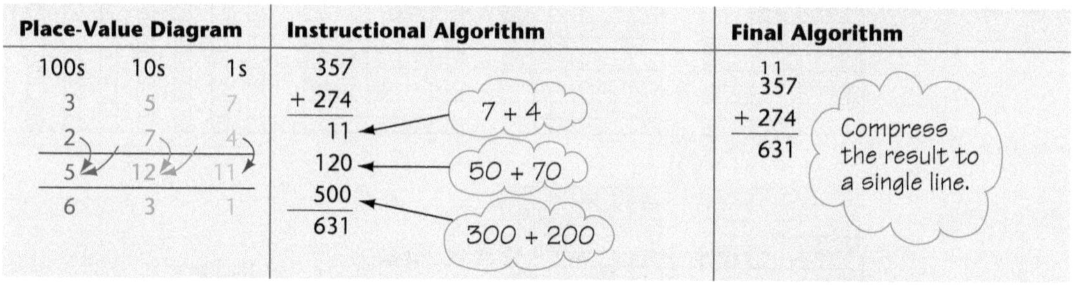
Place-Value Diagram
100s 10s 1s
3 5 7
2 7 4
5 12 11
6 3 1
Instructional Algorithm
357
+ 274
11
120
500
631
7 + 4
50 + 70
300 + 200
Final Algorithm
1 1
357
+ 274
631
Compress the result to a single line.

Notice again how the level of abstraction steadily increases as we move through these solutions. This example and the various solutions should make it clear to students that at no point are they "carrying a 1." It is always the case of exchanging ten 1s from the 1s column for one 10 in the 10s column, ten 10s from the 10s column for one 100 in the 100s column, and so on.

The Subtraction Algorithm

We can illustrate the **subtraction algorithm** in much the same way as we did the addition algorithm. For primary school children, the idea of subtraction is often understood in terms of "take away." Thus, if you have 9 apples and I take away 5, you have 4 left, as shown in Figure 3.8. The algorithm called **instructional** in Example 3.7 is also known as the **partial-difference algorithm.**

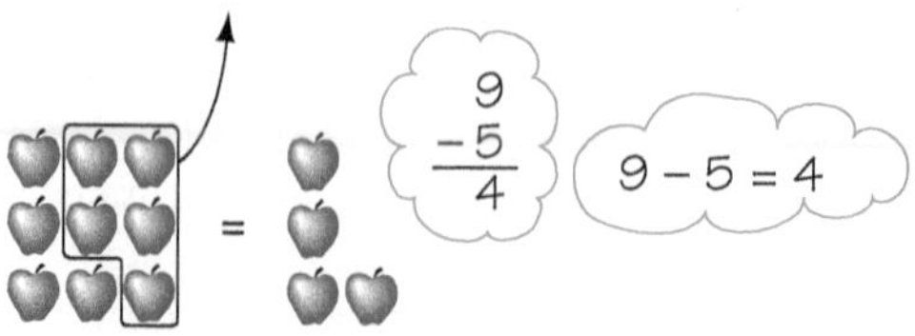

Figure 3.8
Subtraction as "take away"

EXAMPLE 3.7

Subtracting Without Exchanging

Subtract 243 from 375.

Solution

With Units, Strips, and Mats

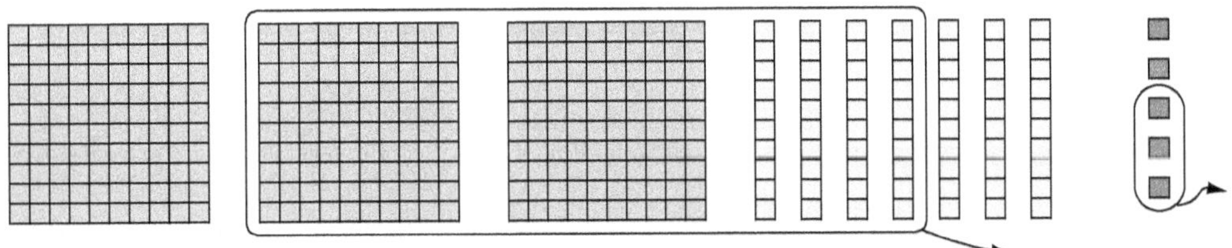

375 – 243 = 132

With Place-Value Cards

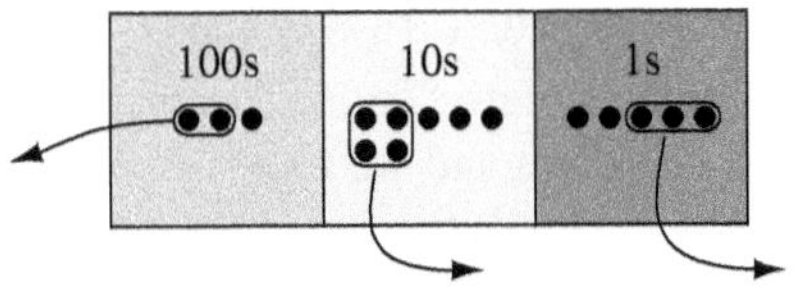

375 – 243 = 132

With Place-Value Diagrams and Instructional Algorithms

Place-Value Diagram	Instructional Algorithm	Final Algorithm
100s 10s 1s 3 7 5 – 2 4 3 1 3 2	375 – 243 2 ← 5 – 3 30 ← 70 – 40 100 ← 300 – 200 132	375 – 243 132 Compress the result to a single line.

EXAMPLE 3.8

Subtracting with Exchanging

Subtract 185 from 362.

Solution

With Units, Strips, and Mats

We start with 3 mats, 6 strips, and 2 units:

362

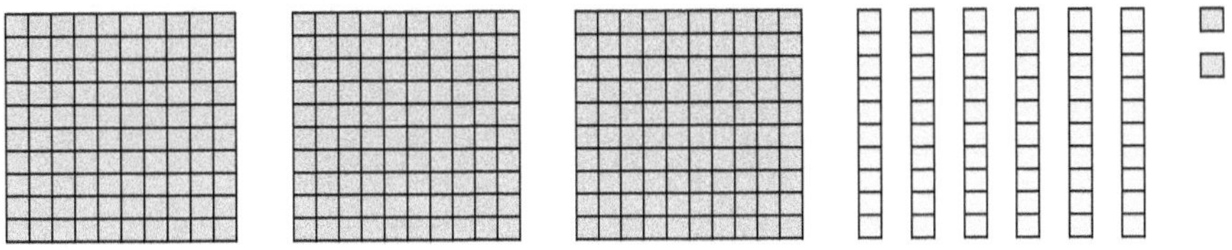

We want to take away 1 mat, 8 strips, and 5 units. Since we cannot pick up 5 units from our present arrangement, we exchange a strip for 10 units to obtain 3 mats, 5 strips, and 12 units:

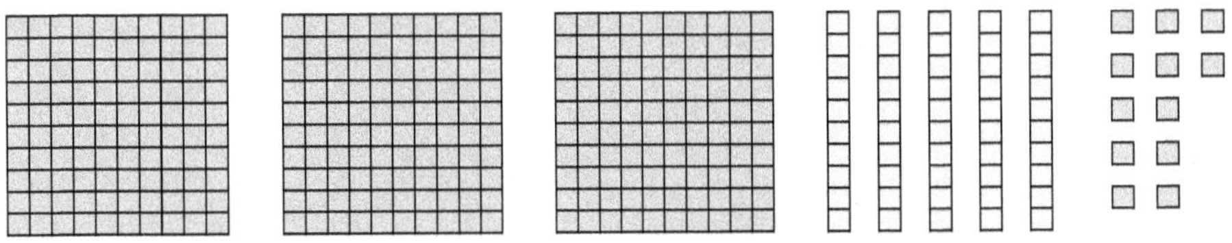

We can now take away 5 units, but we still cannot pick up 8 strips. Therefore, we exchange a mat for 10 strips to obtain 2 mats, 15 strips, and 12 units. Finally, we are able to take away 1 mat, 8 strips, and 5 units (i.e., 185 units), as shown here:

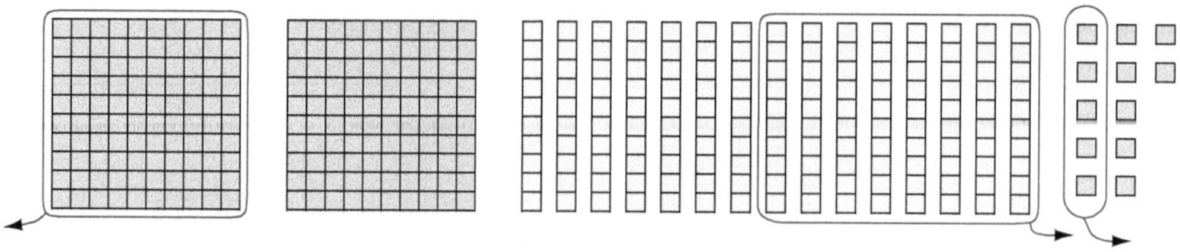

This leaves 1 mat, 7 strips, and 7 units. So

$$362 - 185 = 177.$$

With Place-Value Cards

We must take away 1 marker from the 100s square, 8 markers from the 10s square, and 5 markers from the 1s square. To make this possible, we trade 1 marker from the 10s square for 10 markers on the 1s square and we trade 1 marker on the 100s square for 10 markers on the 10s square. Now, taking away the desired markers, we have 1 marker left on the 100s square, 7 markers left on the 10s square, and 7 markers left on the 1s square, for 177:

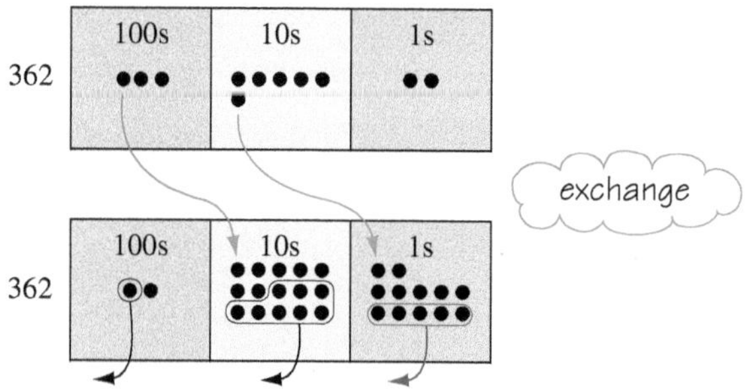

362 – 185 = 177

With Place-Value Diagrams and Instructional Algorithms

	Place-Value Diagram			Instructional Algorithm				Final Algorithm
	100s	10s	1s					
	3	6	2	3	6	2		
	− 1	8	5	− 1	8	5		
	3	5	12	3	5	(12)		$2\,{}^{1}5$
exchange	− 1	8	5	− 1	8	5	exchange	$\cancel{3}\,\cancel{6}\,{}^{1}2$
	2	15	12	2	(15)	(12)		−1 8 5
	− 1	8	5	− 1	8	5		1 7 7
	1	7	7	1	7	7		

3.2 Problem Set

Understanding Concepts

1. Sketch the solution to 36 + 75, using

(a) mats, strips, and units. Draw a square for a mat, a vertical line segment for a strip, and a dot for a unit.

(b) place-value cards marked 1s, 10s, and 100s from right to left.

2. Sketch the solution to 275 − 136, using

(a) mats, strips, and units.

(b) place-value cards marked 1s, 10s, and 100s from right to left.

3. Use the instructional algorithm for addition to perform the following additions:

(a) 23 + 44

(b) 57 + 84

4. Use the instructional algorithm for addition to perform the following additions:

(a) 324 + 78

(b) 608 + 513

5. Use the instructional algorithm for subtraction to perform these subtractions:

(a) 78 − 35

(b) 75 − 38

6. Use the instructional algorithm for subtraction to perform the following subtractions:

(a) 582 − 245

(b) 414 − 175

7. Perform these additions and subtractions, being careful not to leave more than 59 seconds or 59 minutes in your answers:

(a) 3 hours, 24 minutes, 54 seconds
+2 hours, 47 minutes, 38 seconds

(b) 7 hours, 56 minutes, 29 seconds
+3 hours, 27 minutes, 52 seconds

(c) 5 hours, 24 minutes, 54 seconds
−2 hours, 47 minutes, 38 seconds

(d) 7 hours, 46 minutes, 29 seconds
−3 hours, 27 minutes, 52 seconds

8. The column-by-column addition of the numbers 36 and 52 is shown next. State the property of the whole numbers that justifies each of these steps. We begin with expanded notation as a justification of the first line:

$36 + 52$	$= (3 \cdot 10 + 6) + (5 \cdot 10 + 2)$	expanded notation
	$= 3 \cdot 10 + [6 + (5 \cdot 10 + 2)]$	(a) ________
	$= 3 \cdot 10 + [(6 + 5 \cdot 10) + 2]$	(b) ________
	$= 3 \cdot 10 + [(5 \cdot 10 + 6) + 2]$	(c) ________
	$= 3 \cdot 10 + [5 \cdot 10 + (6 + 2)]$	(d) ________
	$= (3 \cdot 10 + 5 \cdot 10) + (6 + 2)$	(e) ________
	$= (3 + 5) \cdot 10 + (6 + 2)$	(f) ________
	$= 8 \cdot 10 + 8$	addition facts
	$= 88$	expanded notation

Into the Classroom

9. The hand calculation of the sum of 279 and 84 involves two exchanges and might appear as follows:

$$\begin{array}{r} {}^{1\,1}\\ 279 \\ 84 \\ \hline 363 \end{array}$$

Carefully describe each of the exchanges. How would you explain these to a third grader?

10. The calculation of the difference 30,007 − 1098 might look like this to your class.

$$\begin{array}{r} \overset{2}{\cancel{3}}\,{}^{1}\overset{9}{\cancel{0}}\,{}^{1}\overset{9}{\cancel{0}}\,{}^{1}\overset{9}{\cancel{0}}\,{}^{1}7 \\ -\ 1\,0\,9\,8 \\ \hline 2\,8,9\,0\,9 \end{array}$$

Carefully discuss the exchanges indicated.

11. While Sylvia was trying to balance her checkbook, her calculator battery went dead. When she added up the outstanding checks by hand, her work looked like this:

$$\begin{array}{r} {}^{1\,1}\\ 2109 \\ 308 \\ 19 \\ 207 \\ 129 \\ 208 \\ 219 \\ 307 \\ 29 \\ 108 \\ 17 \\ 209 \\ 118 \\ \hline 3987 \end{array}$$

Note the exchange from the units column to the hundreds column.

(a) Is the addition correct?

(b) Discuss each of the exchanges shown.

(c) Would you have proceeded as Sylvia did?

(d) How else might have the addition been carried out?

(e) What would you say to a student of yours who did the computation as Sylvia did?

12. In a long-distance relay race of 50 miles, the four runners on one team had the following times: 1 hour, 2 minutes, 23 seconds; 51 minutes, 31 seconds; 1 hour, 47 seconds; and 48 minutes, 27 seconds. What was the total time for the team? Comment on the exchanges involved in solving this problem. (*Note:* The final answer should be stated with the least possible number of minutes and seconds.)

13. **(Writing)** Read the Number and Operations Standard for Grades Pre-K–2, pages 78–88 in NCTM's *Principles and Standards for School Mathematics,* and write a critique of the standard, emphasizing your own reaction to the recommendations. How do the recommendations compare with your own school experience?

14. Give each pair of students a set of the 10 digits, each printed on a square of heavy paper.

(a) Have the students choose and place the digits on the desktop to form the greatest four-digit number that is a multiple of 6.

(b) Can any odd number be a multiple of 6? Why or why not?

(c) Have the students choose and place the digits on the desktop to form the least six-digit number for which the sum of the digits is at least 20.

(d) Have the students choose and place the digits on the desktop to form the greatest six-digit number for which the sum of the digits is less than 20.

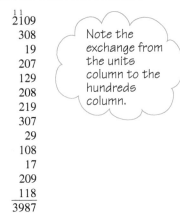

15. Form two four-digit numbers using each of 1, 2, 3, 4, 5, 6, 7, and 8 once, and only once, so that

(a) the sum of the two numbers is as large as possible.

(b) the sum of the two numbers is as small as possible.

(c) You can do better than guess and check on parts (a) and (b). Explain your solution strategy briefly.

(d) Is there only one answer to each of parts (a) and (b)? Explain in two sentences.

Responding to Students

16. How would you respond to one of your students who subtracted 229 from 2003 as

$$\begin{array}{r} {}^{1}\,{}^{9}\,{}^{9}\quad \\ \not{2}\not{0}\not{0}{}^{1}3 \\ -\quad 229 \\ \hline 1774 \end{array}$$

saying that he or she had exchanged one of the 200 tens for 10 units?

17. Examine the given student responses to the same addition problem below. What possible reason(s) can you give for their errors? How would you help each student correct the work?

$$\begin{array}{r} 426 \\ +\ 397 \\ \hline \end{array}$$

(a) Thomas, who sometimes isn't careful with routine single-digit arithmetic, writes the solution as

$$\begin{array}{r} 426 \\ +\ 397 \\ \hline 818 \end{array}$$

(b) Annabelle wrote down

$$\begin{array}{r} 426 \\ +\ 397 \\ \hline 713 \end{array}$$

(c) Xiao gives the solution as

$$\begin{array}{r} 426 \\ +\ 397 \\ \hline 803 \end{array}$$

18. Examine the given responses of fourth-grade students to the subtraction problem. Generally, subtraction errors are more common than addition errors.

$$\begin{array}{r} 722 \\ -\ 558 \\ \hline \end{array}$$

What errors is each student making, and why is the student making them? What steps would you take to help the students correct their mistakes?

(a) Richard writes the solution as

$$\begin{array}{r} {}^{11}\,{}^{12} \\ 7\not{2}\not{2} \\ -558 \\ \hline 264 \end{array}$$

(b) Jeff's computation is

$$\begin{array}{r} {}^{8}\,{}^{11}\,{}^{11} \\ \not{7}\not{2}\not{2} \\ -558 \\ \hline 363 \end{array}$$

(c) Sandy used the following steps:

$$\begin{array}{r} 10\ \ \\ \downarrow\ \ \\ {}^{8}\,{}^{1}\,{}^{12} \\ \not{7}\not{2}\not{2} \\ -558 \\ \hline 354 \end{array}$$

19. Xavier, a third grader, looks at finding the answer to 523 − 247. After he looks at what the teacher wrote, he asks, "Is the answer 411?" The calculation of the difference 523 − 247 might look like this:

$$\begin{array}{r} {}^{4}\,{}^{11}\ \ \\ \not{5}\not{2}{}^{1}3 \\ -247 \\ \hline 276 \end{array}$$

Carefully discuss the exchanges indicated. How would you explain these to Xavier?

20. Ms. Leung asks her fourth-grade class what 273 + 124 is and gets the correct answer of 397. She then wants them to consider the problem 274 + 124, but they must think before they add. What is the quickest way for the students to do the problem (with understanding)?

21. (**Writing**) Mrs. Giles writes on the board the equation $273 - 152 = 121$. She then asks her students if they would use that equation to write an explanation as to which of (a) and (b) is correct and why.

(a) $273 - 153 = 120$

(b) $273 - 153 = 122$

Thinking Critically

22. On his way to school, Peter dropped his arithmetic paper in a puddle of water, blotting out some of his work. What digits should go under the blots in these problems?

(a) $\begin{array}{r} 6■3 \\ +■51■ \\ \hline ■2282 \end{array}$ (b) $\begin{array}{r} 77■ \\ +■■2 \\ \hline 871 \end{array}$ (c) $\begin{array}{r} 8■■ \\ +362 \\ \hline ■■43 \end{array}$

(d) $\begin{array}{r} 248■ \\ -■22 \\ \hline ■1■9 \end{array}$ (e) $\begin{array}{r} 4■2 \\ -1843 \\ \hline ■15■ \end{array}$ (f) $\begin{array}{r} 34■5 \\ -■748 \\ \hline ■2■ \end{array}$

23. Find the missing digits in each of these addition problems:

(a) $\begin{array}{r} \_437 \\ 2\_1 \\ +\ 347\_ \\ \hline 6\_94 \end{array}$ (b) $\begin{array}{r} \_721 \\ 901\_ \\ +\ 71\_3 \\ \hline \_0\_26 \end{array}$

(c) $\begin{array}{r} 38\_1 \\ 24\_3 \\ +\ 512\_ \\ \hline \_5\_9 \end{array}$ (d) $\begin{array}{r} 5\_4 \\ 612\_ \\ +\ 8\_1 \\ \hline 76\_6 \end{array}$

24. Fill in the missing digits in each of these subtraction problems:

(a) $\begin{array}{r} \_3\_ \\ -\ 2\_1 \\ \hline 594 \end{array}$ (b) $\begin{array}{r} 3\_\_4 \\ -\ \_346 \\ \hline 175\_ \end{array}$

(c)
```
  7_4_
− _5_4
  808
```

(d)
```
  63__4
− 2_12_
  _6209
```

25. There is a rather interesting addition algorithm called the **scratch method** that proceeds as follows: Consider the sum

$$\begin{array}{r} 2\,834 \\ \not{5}\not{7}\not{6} \\ 4\,\not{8}35 \\ \not{2}\,743 \\ \hline 10{,}988 \end{array}$$

Begin by adding from the top down in the units column. When you add a digit that makes your sum 10 or more, scratch out the digit as shown and make a mental note of the units digit of your present sum. Start with the digit noted and continue adding and scratching until you have completed the units column, writing down the units digit of the last sum as the units digit of the answer, as shown. Now count the number of scratches in the units column and, starting with this number, add on down the tens column, repeating the scratch process as you go. Continue the entire process until all the columns have been added. This gives the desired answer. Explain why the algorithm works.

Making Connections

26. Julien spent 1 hour and 45 minutes mowing the lawn and 2 hours and 35 minutes trimming the hedge and some shrubs. How long did he work altogether?

27. (a) Mr. Kobayashi has four pieces of oak flooring left over from a job he just completed. If they are 3′ 8″, 4′ 2″, 6′ 10″, and 5′ 11″ long, what total length of flooring does he have left over? Make sure that the number of inches in your answer is less than 12.

(b) If Mr. Kobayashi uses 9′10″ of the flooring left over in part (a) to make a picture frame, how much flooring does he have left over then?

28. It was 25 minutes after 5 in the morning when Ari began his paper route. If it took him an hour and three-quarters to deliver the papers, at what time did he finish?

State Assessments

29. (Grade 5)
The movie, *Math Teachers Have Fun*, is scheduled on March 14. The schedule of the movie starts and ends at the following times.

Movie Times

Start	End
12:30 P.M.	3:00 P.M.
2:45 P.M.	5:15 P.M.
6:45 P.M.	9:15 P.M.
9:05 P.M.	11:35 P.M.

From the information in the table, which of the following statements is true?

F. The end time is exactly 2 hours and 45 minutes after the start time.
G. The end time is exactly 2 hours and 15 minutes after the start time.
H. The end time is exactly 2 hours and 30 minutes after the start time.
J. The end time is exactly 3 hours and 45 minutes after the start time.

30. (Grade 4)
If you change the digit 2 to 4 in the number 32,314, what will be the difference?

A. Three hundred B. Two hundred
C. Four thousand D. Two thousand

31. (Grade 4)
Ann picked pears on her family's farm. She picked 46 pears on Monday, 56 on Tuesday, and 39 on Wednesday. What is the total of the pears she picked after the three days?

A. 84 B. 96
C. 141 D. 102

32. (Grade 4)
A skier reached the level of 3776 feet above sea level. She is going up the next part of the mountain and will at the top which is 7248 feet high. Using the ski lift, how many more feet must she go up to get to the top so that she can ski back?

A. 4572 ft B. 4472 ft
C. 3776 ft D. 3472 ft

33. (California, Grade 4)
Solve: 619,581 − 23,183 = ?
Solve: 6747 + 321,105 = ?

3.3 Algorithms for Multiplication and Division of Whole Numbers

In most ancient numeration systems, multiplication and division were quite complicated. The decimal system makes these processes much easier, but many people still find them confusing. At least part of the difficulty has been that the ideas are often presented as a collection of rules to be learned by rote, with little or no effort made to impart understanding. In this section, we will see the conceptual basis of a number of different algorithms for multiplication and division; that is, you will now see why they work!

Multiplication Algorithms

Multiplication is repeated addition. Thus, $2 \cdot 9$ means $9 + 9$, $3 \cdot 9$ means $9 + 9 + 9$, and so on. But repeated addition is slow and tedious, and easier algorithms exist. As with addition and subtraction, these algorithms should be introduced by starting with concrete approaches and gradually becoming more and more abstract. The development should proceed through units, strips, and mats; place-value cards; classroom abacuses; and so on. Let's consider the product $9 \cdot 3$.

FROM THE NCTM PRINCIPLES AND STANDARDS

Research suggests that by solving problems that require calculation, students develop methods for computing and also learn more about operations and properties (McClain, Cobb, and Bowers 1998; Schifter 1999). As students develop methods to solve multidigit computation problems, they should be encouraged to record and share their methods. As they do so, they can learn from one another, analyze the efficiency and generalizability of various approaches, and try one another's methods. In the past, common school practice has been to present a single algorithm for each operation. However, more than one efficient and accurate computational algorithm exists for each arithmetic operation. In addition, if given the opportunity, students naturally invent methods to compute that make sense to them (Fuson forthcoming; Madell 1985). The following episode, drawn from unpublished classroom observation notes, illustrates how one teacher helped students analyze and compare their computational procedures for division:

Students in Ms. Sparks' fifth-grade class were sharing their solutions to a homework problem, 728 ÷ 34. Ms. Sparks asked several students to put their work on the board to be discussed. She deliberately chose students who had approached the problem in several different ways. As the students put their work on the board, Ms. Sparks circulated among the other students, checking their homework.

Henry had written his solution:

34×10=340
34×20=680

$$\begin{array}{r} 680 \\ +\ 34 \\ \hline 714 \end{array} \qquad \begin{array}{r} 728 \\ -714 \\ \hline 14 \end{array}$$

Henry explained to the class, "Twenty 34s plus one more is 21. I knew I was pretty close. I didn't think I could add any more 34s, so I subtracted 714 from 728 and got 14. Then I had 21 remainder 14."

EXAMPLE 3.9

Developing the Multiplication Algorithm

Compute the product of 9 and 3.

Solution

Using Units, Strips, and Mats

Since $9 \cdot 3 = 3 + 3 + 3 + 3 + 3 + 3 + 3 + 3 + 3$, we can illustrate this as shown with 9 rows of 3 units each. Simplifying the original array by appropriately exchanging units for strips, we eventually have 2 strips and 7 units, which is recorded as 27. Of course, elementary school children should actually handle the materials, making the necessary exchanges of units for strips.

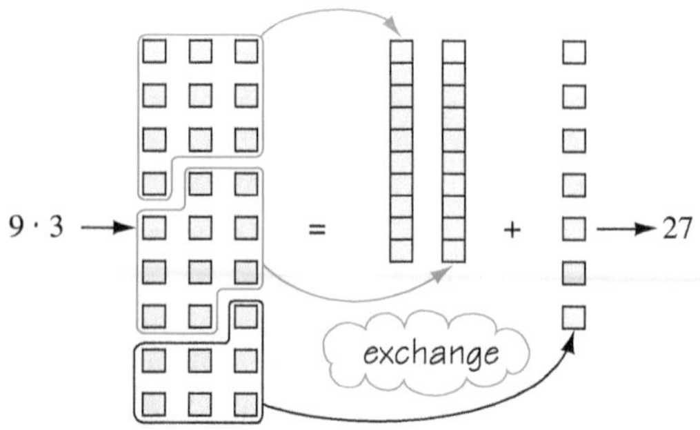

Using Place-Value Cards

With place-value cards, students should start with 9 rows of 3 markers each on the 1s square of their place-value cards, as shown here. They should then exchange 10 markers on the 1s square for 1 marker on the 10s square as many times as possible and record the fact that this gives 2 tens plus 7 units, or 27.

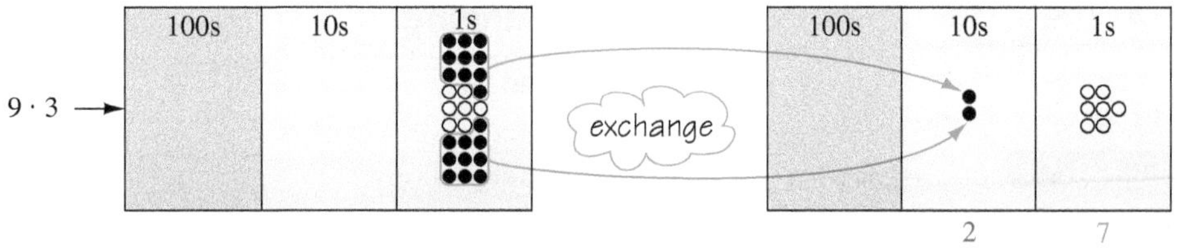

Using manipulatives as illustrated, children can experience, learn, and actually *understand* all the one-digit multiplication facts. It should not be the case that $9 \cdot 3 = 27$ is a string of meaningless symbols memorized by rote. This basic fact must be understood conceptually.

Once the one-digit facts are thoroughly understood *and memorized,* as they must be, one can move on to more complicated problems. Consider, for example, $3 \cdot 213$. This multiplication could be illustrated with units, strips, and mats, but, for brevity, we will go directly to place-value cards, expanded notation, and, finally, to an algorithm, as illustrated in Example 3.10. Again, we make use of the understanding that $3 \cdot 213 = 213 + 213 + 213$. For multiplication, the **instructional algorithm** is also known as the **partial-products algorithm.**

Multiplication can be done in a number of ways. The key to establishing methods to actually perform a multiplication is to have a deep understanding of positional systems and notation. Using this knowledge and the distributive law for whole numbers leads to a number of different algorithms. We will illustrate **expanded notation,** the instructional (or partial-products) algorithm, and the final algorithm. There are a number of other algorithms for multiplication, three of which are included in problems 17, 26, and 27 in Problem Set 3.3 (the **lattice algorithm, Egyptian algorithm,** and **Russian peasant algorithm,** respectively). The lattice algorithm is a method that is also taught at a number of schools in the United States.

In addition to the algorithms that are presented in this text, students will invent their own ways of multiplying or doing other operations. Their ideas should be encouraged, as we see from the preceding quote from NCTM.

EXAMPLE 3.10 Computing a Product with a Multidigit Number

Compute the product of 3 and 213.

Solution

Using Place-Value Cards, Expanded Notation, and Algorithms

Place-value cards

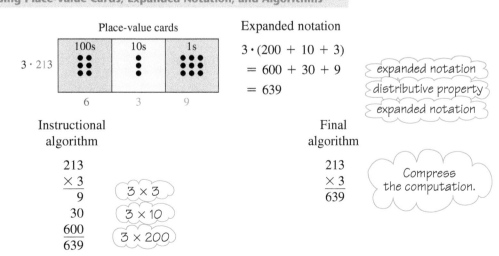

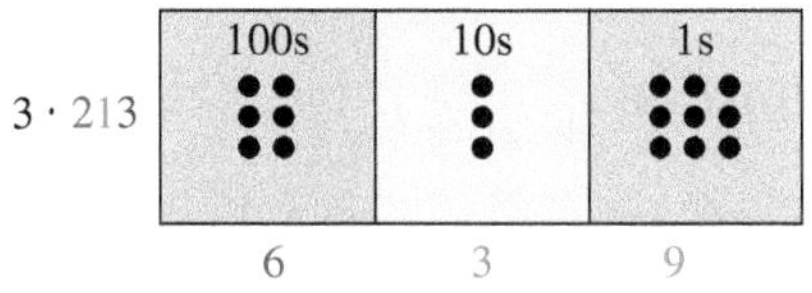

Expanded notation

$$3 \cdot (200 + 10 + 3)$$
$$= 600 + 30 + 9$$
$$= 639$$

expanded notation
distributive property
expanded notation

Instructional algorithm

$$\begin{array}{r} 213 \\ \times\, 3 \\ \hline 9 \\ 30 \\ 600 \\ \hline 639 \end{array}$$

3 × 3
3 × 10
3 × 200

Final algorithm

$$\begin{array}{r} 213 \\ \times\, 3 \\ \hline 639 \end{array}$$

Compress the computation.

The product $3 \cdot 213$ did not require exchanging. Consider the product $4 \cdot 243$. The presentation of this product proceeds from the concrete representation to the final algorithm.

EXAMPLE 3.11 Multiplying with Exchanging

Compute the product of 4 and 243.

Solution — Using Place-Value Cards, Expanded Notation, and Algorithms

Place-value cards

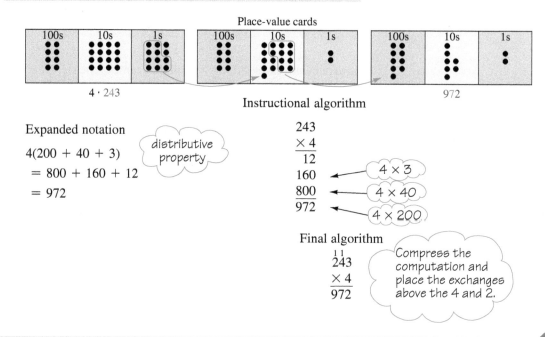

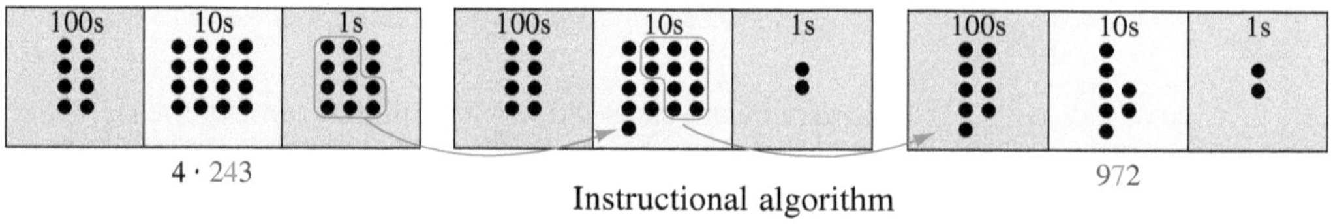

Instructional algorithm

Expanded notation

$4(200 + 40 + 3)$ (distributive property)

$= 800 + 160 + 12$

$= 972$

$$\begin{array}{r} 243 \\ \times\ 4 \\ \hline 12 \\ 160 \\ 800 \\ \hline 972 \end{array}$$

12 ← 4 × 3

160 ← 4 × 40

800 ← 4 × 200

Final algorithm

$$\begin{array}{r} {}^{1\,1} \\ 243 \\ \times\ 4 \\ \hline 972 \end{array}$$

Compress the computation and place the exchanges above the 4 and 2.

The preceding development represents many lessons. However, the various demonstrations should be clearly tied together, and enough time should be spent to ensure that each level of the chain of reasoning is understood before proceeding to the next.

Finally, consider the product $15 \cdot 324$. Unquestionably, the most efficient algorithm is that provided by the calculator. The pencil-and-paper algorithm, however, is not difficult, as the series of calculations in Example 3.12 shows.

EXAMPLE 3.12 Multiplying Multidigit Numbers

Compute the product of 15 and 324.

Solution — Using Expanded Notation, an Instructional Algorithm, and the Final Algorithm

Expanded notation

$15 \cdot 324 = (10 + 5) \cdot 324$ (distributive property)

$= 10 \cdot 324 + 5 \cdot 324$

$= 10(300 + 20 + 4) + 5(300 + 20 + 4)$ (expanded notation)

$= 3000 + 200 + 40 + 1500 + 100 + 20$ (distributive property)

$= 4860$

Instructional algorithm

	324	
	× 15	
	20	5 · 4
5 · 20	100	
	1500	5 · 300
10 · 4	40	
	200	10 · 20
10 · 300	3000	
	4860	

×	300 +	20 +	4
10 +	3000	200	40
5	1500	100	20

Sum = 4860

Final algorithm

$$\begin{array}{r} \overset{1\,2}{324} \\ \times\ 15 \\ \hline 1620 \\ 324 \\ \hline 4860 \end{array}$$

20
100
+ 1500
1620

40
200
+ 3000
3240

Division Algorithms

An approach to division discussed in Chapter 2 was repeated subtraction. This ultimately led to the so-called **division algorithm,** which we restate here for easy reference: The division algorithm is the foundation for the ways in which we actually do the division of one whole number by another. This foundation leads to a number of different algorithms, including the **long-division, short-division,** and **scaffold** (also called **pyramid**) algorithms, which are discussed next. We now return to base ten.

$$\begin{array}{r} 8 \\ 7\overline{)57} \\ \underline{56} \\ 1 \end{array}$$

THEOREM The Division Algorithm

If a and b are whole numbers with b not zero, there exists exactly one pair of whole numbers q and r with $0 \leq r < b$ such that $a = bq + r$. q is the **quotient,** and r is the **remainder,** of a divided by b.

$57 = 7 \cdot 8 + 1$

The Long-Division Algorithm Suppose we want to divide 941 by 7. Doing this by repeated subtraction will take a long time, even with a calculator. But suppose we subtract several 7s at a time and keep track of the number we subtract each time. Indeed, since it is so easy to multiply a number by 10, 100, 1000, and so on, let's subtract hundreds of 7s, tens of 7s, and so on. The work might be organized like this:

7)941			
700	Subtract	100	7s
241			
70	Subtract	10	7s
171			
70	Subtract	10	7s
101			
70	Subtract	10	7s
31			
28	Subtract	4	7s
3		134	number of 7s subtracted

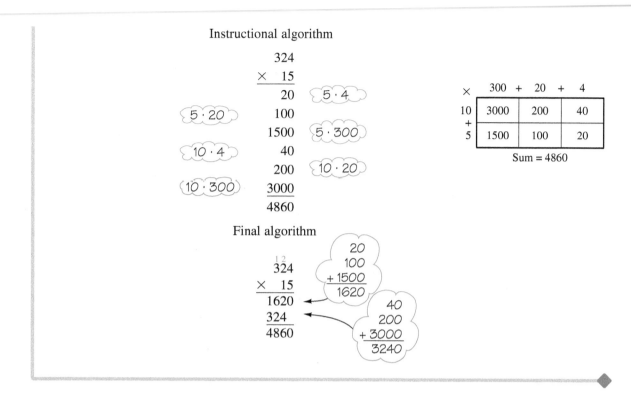

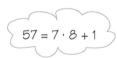

Since $3 < 7$, the process stops and we see that 941 divided by 7 gives a quotient of 134 and a remainder of 3. As a check, we note that $941 = 7 \cdot 134 + 3$. The preceding work could have been shortened if we had subtracted the three tens of 7s all at once and then the four 7s all at once, like this:

$$\begin{array}{r r l}
7\overline{)941} & & \\
\underline{700} & 100 & 7\text{s} \\
241 & & \\
\underline{210} & 30 & 7\text{s} \\
31 & & \\
\underline{28} & \underline{4} & \underline{7\text{s}} \\
3 & 134 &
\end{array}$$

A slightly different form of this algorithm, which we call the **scaffold algorithm,** is obtained by writing the 100, 30, and 4 above the division symbol. This algorithm is also known as the **pyramid algorithm** and appears as follows:

$$\begin{array}{r l}
\underline{134} & \\
4 & \\
30 & \\
100 & \\
7\overline{)941} & \\
\underline{700} & 7 \cdot 100 \\
241 & \\
\underline{210} & 7 \cdot 30 \\
31 & \\
\underline{28} & 7 \cdot 4 \\
3 &
\end{array}$$

Note that 1, 3, and 4 are the quotients when 9, 24, and 31, respectively, are divided by 7.

Many teachers prefer the scaffold algorithm as the final algorithm for division, since it fully displays the mathematics being done. Others still cling to the following form, which is easily obtained from the scaffold algorithm, even though it seriously masks the mathematics and forces many students to rely on rote memorization rather than understanding:

$$\begin{array}{r}
1\ 3\ 4 \\
7\overline{)9^{2}4^{3}1} \\
\underline{7} \\
24 \\
\underline{21} \\
31 \\
\underline{28} \\
3
\end{array}$$

Look at how compressed the notation is in this long-division algorithm or in the final algorithm compared with the scaffold algorithm. It may be that the notation is simplified, but the concepts can be obscured in the long-division or final algorithm.

A further example with a larger divisor may be helpful.

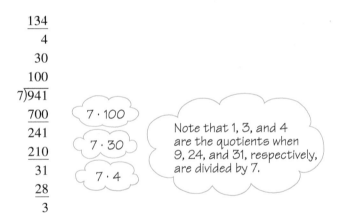

EXAMPLE 3.13

Using the Long-Division Algorithm

Divide 28,762 by 307.

Solution

Scaffold algorithm

```
          93
           3
          90
   307)28762
       27630
        1132
         921
         211
```

Standard algorithm

```
          93
   307)28762
       2763
        1132
         921
         211
```

Since 307 > 287, we ask how many 307s are in 28,762. Since 28 ÷ 3 ≈ 9, we guess 9.

How many 307s are in 1132? Since 11 ÷ 3 ≈ 3, we guess 3.

Since 211 < 307, 93 is the quotient and 211 is the remainder.

Check: $28{,}762 = 307 \cdot 93 + 211$.

The Short-Division Algorithm An algorithm for division that is quite useful, even in this calculator age, is the **short-division** algorithm. This is a much simplified version of the long-division algorithm and is quite useful and quick when the divisor is a single digit. It can be developed from the scaffold method as we did a moment ago. Also, it follows directly from the long-division algorithm if that is already known.

EXAMPLE 3.14

Using the Short-Division Algorithm

Divide 2834 by 3 and check your answer.

Solution

Consider the following divisions, which show how the short-division algorithm comes from the long-division algorithm:

```
      944
   3)2834
     27
      13
      12
       14
       12
        2
```

```
       9 4 4  R2
   3)2 8¹3¹4
```

```
     9          4          4
   3)28       3)13       3)14
     27         12         12
      1          1          2
```

3.3 Problem Set

Understanding Concepts

1. (a) Make a suitable drawing of units and strips to illustrate the product $4 \cdot 8 = 32$.

   (b) Make a suitable sketch of place-value cards to illustrate the product $4 \cdot 8 = 32$.

2. Make a suitable sketch of place-value cards to illustrate the product $3 \cdot 254 = 762$.

3. (a) In the product shown, what does the red 2 actually represent?

```
     2
    274
  ×  34
   1096
   8220
   9316
```

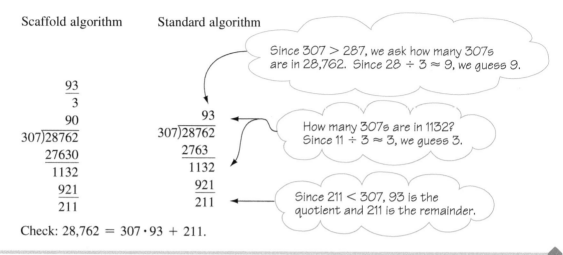

(b) In the product shown in part (a), when multiplying $4 \cdot 7$, one "exchanges" a 2. What is actually being exchanged?

4. The accompanying diagram illustrates the product $27 \cdot 32$. Discuss how this product is related to finding the product by the instructional algorithm for multiplication.

5. Justify each step in this calculation by stating a property of the whole numbers:

$17 \cdot 4 = (10 + 7) \cdot 4$	expanded notation
$= 10 \cdot 4 + 7 \cdot 4$	**(a)** ________________
$= 10 \cdot 4 + 28$	one-digit multiplication fact
$= 10 \cdot 4 + (2 \cdot 10 + 8)$	expanded notation
$= 4 \cdot 10 + (2 \cdot 10 + 8)$	**(b)** ________________
$= (4 \cdot 10 + 2 \cdot 10) + 8$	**(c)** ________________
$= (4 + 2) \cdot 10 + 8$	**(d)** ________________
$= 6 \cdot 10 + 8$	one-digit addition fact
$= 68$	expanded notation

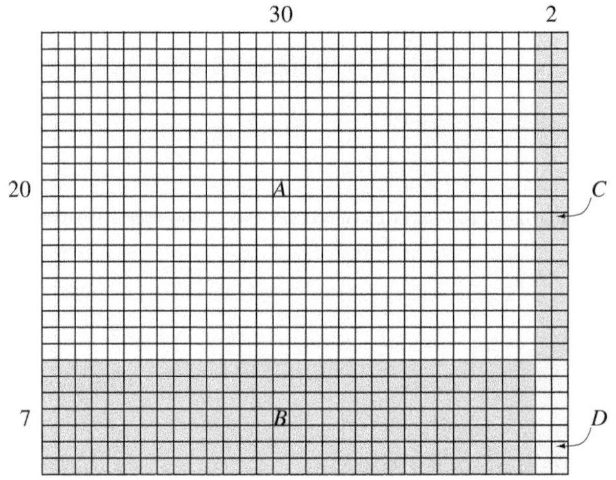

6. Draw a sequence of sketches of units, strips, and mats to illustrate dividing 429 by 3.

7. What calculation does this sequence of sketches illustrate? Explain briefly.

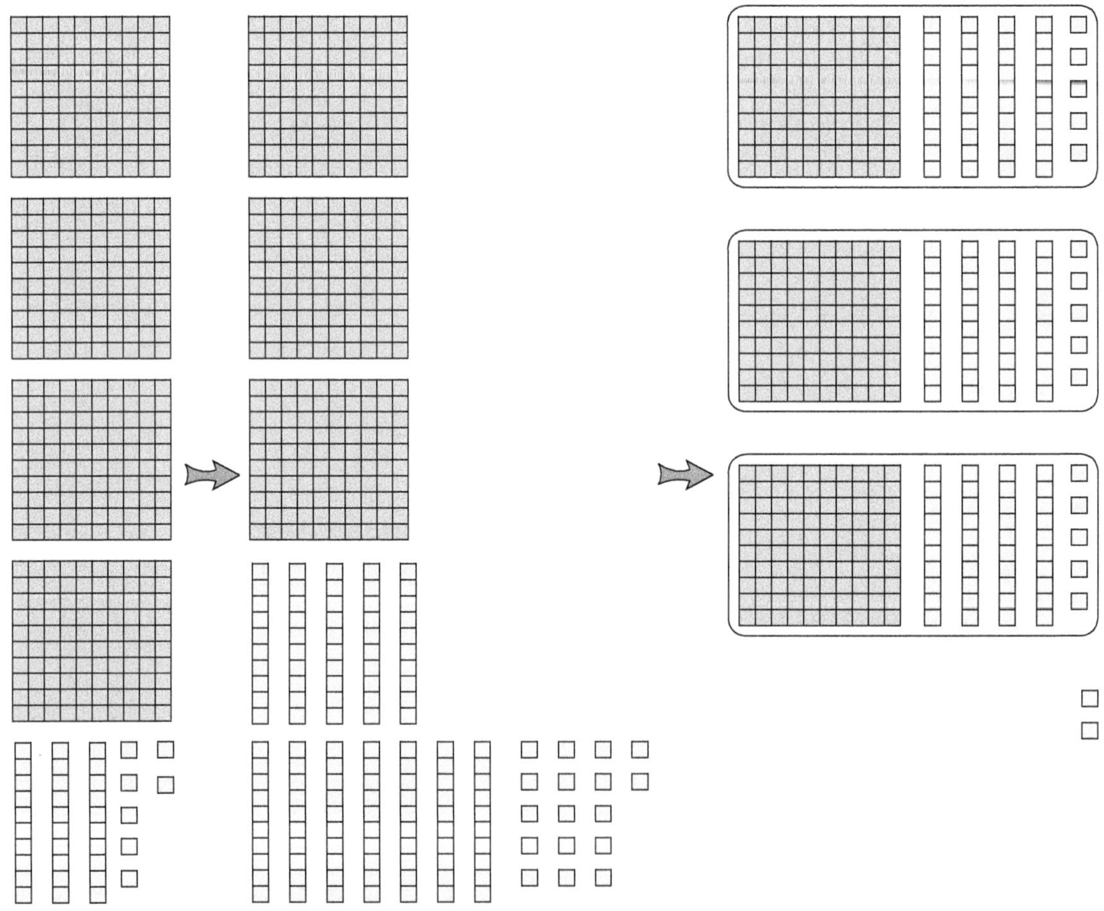

8. Multiply 352 by 27, using the instructional algorithm for multiplication.

9. Multiply 241 by 35 using the algorithm for multiplication.

10. Find the quotient q and a remainder r when a is divided by b, and write the result in the form $a = bq + r$ of the division algorithm for each of these choices of a and b:

(a) $a = 27, b = 4$

(b) $a = 354, b = 29$

11. Find the quotient q and the remainder r if $a = bq + r$ where

(a) $a = 871$ and $b = 17$

(b) $a = 723$ and $b = 21$

12. Perform each of the given divisions by the scaffold method. In each case, check your results by using the equation of the division algorithm.

(a) $351\overline{)7425}$ (b) $23\overline{)6814}$

13. Perform each of the given divisions by the scaffold method. In each case, check your results by using the equation of the division algorithm.

(a) $213\overline{)3175}$ (b) $43\overline{)8250}$

14. Use short division to find the quotient and remainder for each division given. Check each result.

(a) $5\overline{)873}$ (b) $7\overline{)2432}$

15. Use short division to find the quotient and remainder for each division problem given. Check each result.

(a) $8\overline{)10095}$ (b) $6\overline{)8432}$

Into the Classroom

16. Problem 4 of this problem set shows a diagram to represent the product $27 \cdot 32$. Working in groups,

(a) Draw a similar diagram to represent the product $17 \cdot 23$.

(b) In expanded notation, the product appears as follows:

$$\begin{aligned}(10 + 7)\cdot(20 + 3) &= (10 + 7)\cdot 20 + (10 + 7)\cdot 3\\ &= 10\cdot 20 + 7\cdot 20 + 10\cdot 3 + 7\cdot 3\\ &= 200 + 140 + 30 + 21\end{aligned}$$

Identify each of 200, 140, 30, and 21 with the regions in your diagram for part (a).

(c) Using the instructional algorithm for multiplication, we find the product as follows:

$$\begin{array}{r} 23\\ \underline{17}\\ 21\\ 140\\ 30\\ \underline{200}\\ 391\end{array}$$

This is clearly an alternative representation of the product, using expanded notation. In the final algorithm, the product appears as follows:

$$\begin{array}{r} \overset{2}{2}3\\ \underline{17}\\ 161\\ \underline{23}\\ 391\end{array}$$

Explain the 2 that is exchanged in terms of the instructional algorithm, and show how the numbers 161 and 23 derive from the instructional algorithm.

17. Another multiplication algorithm is the **lattice algorithm.** Suppose, for example, you want to multiply 324 by 73. Form a two-by-three rectangular array of boxes with the 3, 2, and 4 across the top and the 7 and 3 down the right side as shown in the accompanying figure. Now compute the products $3\cdot 7 = 21$, $2\cdot 7 = 14$, $4\cdot 7 = 28$, $3\cdot 3 = 9$, $2\cdot 3 = 6$, and $4\cdot 3 = 12$. Place the products in the appropriate boxes (e.g., $3\cdot 7$ is in the 3 column and the 7 row, for example), with the units digit of the product below the diagonal in each box and the tens digit (if there is one) above the diagonal. Now add down the diagonals, and add any "exchanges" to the sum above the next diagonal. The result of 23,652 is the desired product.

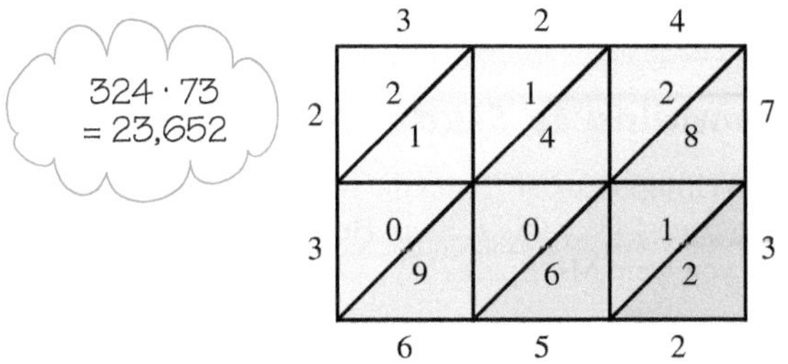

(a) Multiply 374 by 215, using the lattice algorithm.

(b) Write a short paragraph comparing the lattice algorithm with the standard pencil-and-paper algorithm.

(c) Would presenting this algorithm to students in a class you might teach help them gain a better understanding of positional notation? Explain.

18. Consider the following computation:

$$\begin{array}{r} 374\\ \underline{\times\ \ 23}\\ 748\\ \underline{1122}\\ 8602\end{array}$$

(a) Is the algorithm correct? Explain briefly.

(b) Multiply 285 by 362 with this method.

(c) What would you say to one of your students who multiplied multidigit numbers in the manner shown here? Would you insist that there is really just one correct way to perform the pencil-and-paper calculation? Explain.

Divide into small groups to work each of the next two problems. In each case, discuss possible strategies among the members of your group and develop an answer agreed upon by the entire group.

19. Use each of 1, 3, 5, 7, and 9 once, and only once, in the boxes to obtain the largest possible product in each case:

(a)

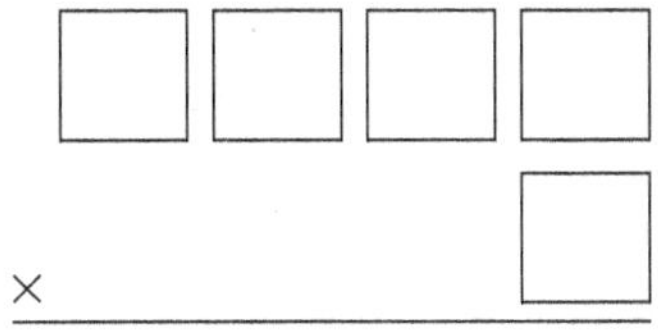

(b)

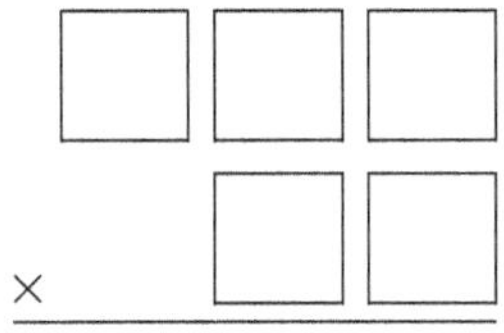

20. Use each of 1, 3, 5, 7, and 9 once, and only once, in the boxes to obtain the smallest possible product in each case:

(a)

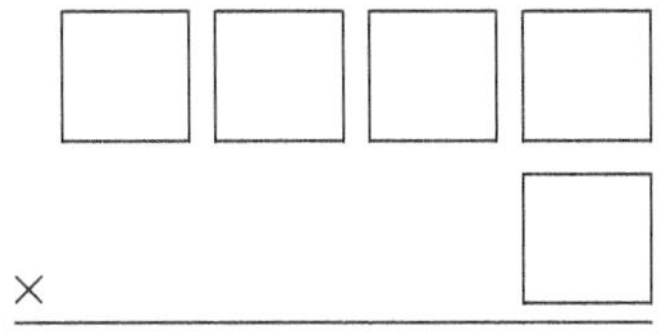

(b)

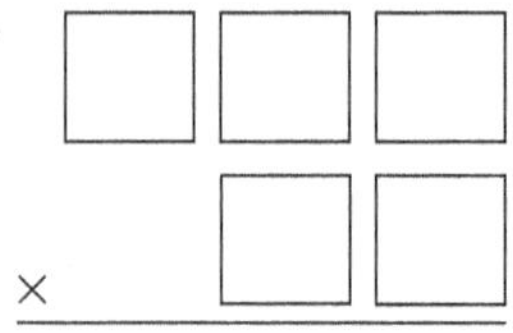

Responding to Students

21. (**Writing**) Look at "From the NCTM Principles and Standards" on page 148 of this section. How would you respond to Henry if you were Ms. Sparks?

22. Hector, a fourth-grade student, has not had previous instruction in two-digit–by–two-digit multiplication and attempted a two-digit multiplication problem as follows:

$$\begin{array}{r} 39 \\ \times\ 42 \\ \hline 15{,}276 \end{array}$$

What algorithm was Hector trying to use but did so incorrectly?

23. What division errors by the fourth graders do you see in each problem?

(a)
$$\begin{array}{r} 220 \\ 3\overline{)862} \end{array}$$

(b)
$$\begin{array}{r} 2 \\ 3\overline{)862} \\ -6\downarrow\downarrow \\ \hline 262 \end{array}$$

(c)
$$\begin{array}{r} 28R2 \\ 3\overline{)806} \\ -6\ \downarrow \\ \hline 26 \\ -24 \\ \hline 2 \end{array}$$

24. Why might a third- or fourth-grade student struggle with the following division problem? Find the number for the blank if $\_\_\_ \div 3 = 8$.

25. Marsha multiplied 53 by 45 and wrote

$$\begin{array}{r} 53 \\ \times\ 45 \\ \hline 2120 \\ 265 \\ \hline 2385 \end{array}$$

Is Marsha's algorithm correct? Explain briefly.

Thinking Critically

26. The ***Egyptian algorithm*** for multiplication was one of the interesting subjects explained in the Rhind papyrus mentioned in Chapters 1 and 2. We will explain the algorithm with an example. Suppose we want to compute 19 times 35. Successively doubling 35, we obtain this list:

$$\begin{array}{rl} \rightarrow & 1 \cdot 35 = 35 \\ \rightarrow & 2 \cdot 35 = 70 \\ & 4 \cdot 35 = 140 \\ & 8 \cdot 35 = 280 \\ \rightarrow & 16 \cdot 35 = 560 \end{array}$$

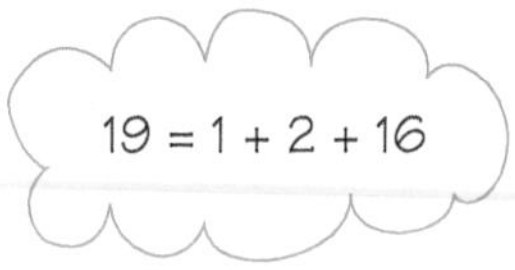

Adding the results in the indicated rows gives us 665 as the desired product.

(a) After carefully considering the preceding computation, write a short paragraph explaining how and why the process always works.

(b) This scheme is also known as the *duplation algorithm*. Use duplation to find the product of 24 and 71.

27. The ***Russian peasant algorithm*** for multiplication is similar to the duplation algorithm described in problem 26. To find the product of 34 and 54, for example, successively divide the 34 by 2 (ignoring remainders if they occur) and successively multiply 54 by 2. This gives the following lists:

~~34~~	~~54~~
17	108
~~8~~	~~216~~
~~4~~	~~432~~
~~2~~	~~864~~
1	1728
	1836

Now cross out the even numbers in the left-hand column and the companion numbers in the right-hand column. Add the remaining numbers in the right-hand column to obtain the desired product. To see why the process works, consider the products $34 \cdot 54 = 1836$ and $17 \cdot 108 = 1836$. Also, consider $8 \cdot 216, 4 \cdot 432$, and $2 \cdot 864$.

(a) Why are $34 \cdot 54$ and $17 \cdot 108$ the same?

(b) Why are $17 \cdot 108$ and $8 \cdot 216$ different? How much do they differ?

(c) Why are $8 \cdot 216, 4 \cdot 432, 2 \cdot 864$, and $1 \cdot 1728$ all the same?

(d) Write a short paragraph explaining why the Russian peasant algorithm works.

(e) Use the Russian peasant algorithm to compute $29 \cdot 81$ and $11 \cdot 243$.

Making Connections

28. Alicia, Arturo, and Adam took part in a walk for a charity. Each had sponsors who agreed to contribute 5 dollars for each mile their student walked. Alicia walked 4 miles and had 6 sponsors. Arturo walked 7 miles and had 11 sponsors. Adam walked 8 miles and had 7 sponsors. In all, how much money was contributed to the charity on the students' behalf?

29. On Monday, Melody practiced her piano lesson for 48 minutes. On each of Tuesday and Wednesday, she practiced for a full hour. On Thursday, she practiced for only 35 minutes.

(a) All told, how many minutes had she practiced before her piano lesson on Friday?

(b) Express the time Melody practiced in hours and minutes, using as few minutes as possible.

30. The students in the third-grade class at Franklin Grade School wanted to earn $165 to buy a wheelchair for one of their classmates. The students planned to sell boxes of cookies at $2.25 per box. How many boxes would they need to sell? (*Suggestion:* Express the amounts of money in pennies.)

State Assessments

31. (Washington State, Grade 4)
When Lori tries multiplying with her calculator, she gets the following results:

$8 \times 3 = 34$ $\quad 4 \times 2 = 18$

$9 \times 5 = 55$ $\quad 8 \times 4 = 42$

Lori knows her multiplication facts and knows that something is wrong with her calculator. She wants to multiply 9×6. Explain what is wrong with Lori's calculator and tell what she needs to do to get the correct answer. Use words, numbers, or pictures.

32. (Grade 4)
Which number fact goes with this smiley picture?

☺	☺	☺
☺	☺	☺
☺	☺	☺
☺	☺	☺

A. $4 + 3 =$
B. $4 \div 3 =$
C. $4 \times 3 =$
D. $4 - 3 =$

33. (Grade 4)
Which number sentence should **NOT** have a 7 in the box?

A. $21 \div \square = 3$
B. $32 \div 4 = \square$
C. $35 \div \square = 5$
D. $42 \div 6 = \square$

34. (Grade 5)
Grammy gave her collection of 438 pennies to her 6 grandchildren. Each grandchild received the same number of pennies. Which computation can be used to solve the question "how many pennies did each child get?"

A. 438×6
B. $438 \div 6$
C. $438 + 6$
D. $438 - 6$

3.4 Mental Arithmetic and Estimation

We start with mental arithmetic by looking at techniques such as easy combinations, making adjustments, and working from left to right.

The One-Digit Facts

It is essential that the basic addition and multiplication facts be memorized, since all other numerical calculations and estimations depend on that foundation. At the same time, this should not be rote memorization of symbols. Using a variety of concrete objects, students should actually *experience* the fact that $8 + 7 = 15$, that $9 \cdot 7 = 63$, and so on. Moreover, rather than having children simply memorize the addition and multiplication tables, the tables should be learned by the frequent and long-term use of manipulatives, games, puzzles, oral activities, and appropriate problem-solving activities. In the same way, children learn the basic properties of the whole numbers, which, in turn, can be used to recall some momentarily forgotten arithmetic fact. For example, $7 + 8$ can be recalled as $7 + 7 + 1$, $6 \cdot 9$ can be recalled as $5 \cdot 9 + 9$ or as $6 \cdot 10 - 6$, and so on. In the same way, the properties of whole numbers, along with the **one-digit facts,** form the basis for mental calculation.

Easy Combinations

There are several strategies for mental calculation. One is to always look for **easy combinations,** especially regrouping to find multiples of 10. The next example shows how this works.

EXAMPLE 3.15 **Using Easy Combinations**

Use mental processes to perform these calculations:

(a) $35 + 7 + 15$ **(b)** $8 + 3 + 4 + 6 + 7 + 12 + 4 + 3 + 6 + 3$
(c) $25 \cdot 8$ **(d)** $4 \cdot 99$ **(e)** $57 - 25$ **(f)** $47 \cdot 5$

Solution

(a) Using the commutative and associative properties, we have

$$35 + 7 + 15 = 35 + 5 + 10 + 7 = 40 + 10 + 7 = 50 + 7 = 57.$$

Think: 35, 40, 50, 57 The answer is 57.

(b) Note numbers that add to 10 or multiples of 10:

$$8 + 3 + 4 + 6 + 7 + 12 + 4 + 3 + 6 + 3 = 56$$

Think: 20, 30, 40, 50, 53, 56 The answer is 56.

(c) $25 \cdot 8 = 25 \cdot 4 \cdot 2 = 100 \cdot 2 = 200$

Think: 25, 100, 200 The answer is 200.

(d) $4 \cdot 99 = 4(100 - 1) = 400 - 4 = 396$

Think: 400 − 4 = 396

(e) $57 - 25 = 50 - 25 + 7 = 25 + 7 = 32$

Think: 50 = 2 · 25, so 50 − 25 plus 7 gives 32.

Think: Two quarters are worth 50¢.

(f) $47 \cdot 5 = 47 \cdot 10 \div 2 = 470 \div 2 = 235$

Think: 47, 470, 235

Adjustment

In parts (d) and (e) of Example 3.15, we made use of the fact that 99 and 57 are close to 100 and 50, respectively. This is an example of adjustment. **Adjustment** simply means that, at the beginning of a calculation, we modify numbers to minimize the mental effort required by adding zero. In Example 3.16a, the key is that $3 - 3 = 0$.

EXAMPLE 3.16

Using Adjustment in Mental Calculation

Use mental processes to perform these calculations:

(a) $57 + 84$ **(b)** $83 - 48$ **(c)** $286 + 347$
(d) $493 \cdot 7$ **(e)** $2646 \div 9$ **(f)** $639 \div 7$

Solution

(a) $57 + 84 = (57 + 3) + (84 - 3)$
$= 60 + 81 = 60 + 80 + 1$
$= 140 + 1 = 141$

Think: 57 + 84, 60 + 81, 140, 141

(b) $83 - 48 = (83 + 2) - (48 + 2) = 85 - 50 = 35$

Think: 83 − 48, 85 − 50, 35

(c) $286 + 347 = (286 + 14) + (347 - 14)$
$= 300 + 300 + 47 - 14$
$= 600 + 33 = 633$

Think: 300, 647 − 14, 633

(d) $493 \cdot 7 = (500 - 7) \cdot 7 = 3500 - 49 = 3451$

Think: (500 − 7) · 7, 3500 − 49, 3451

(e) $2646 \div 9 = (2700 - 54) \div 9 = 300 - 6 = 294$

Think: ÷ 9, 2700 − 54, 300 − 6, 294

(f) $639 \div 7 = (630 + 7 + 2) \div 7 = 90 + 1\text{ R }2 = 91\text{ R }2$

Think: ÷ 7, 630 + 7 + 2, 90 + 1 R 2, 91 R 2

Working from Left to Right

Because it tends to reduce the amount one has to remember, many expert mental calculators **work from left to right,** rather than the other way around as in most of our standard algorithms.

EXAMPLE 3.17

Working from Left to Right

Use mental processes to perform these calculations:

(a) $352 + 647$ **(b)** $739 - 224$ **(c)** $4 \cdot 235$

Solution

(a)
$$352 + 647 = (300 + 50 + 2) + (600 + 40 + 7)$$
$$= (300 + 600) + (50 + 40) + (2 + 7)$$
$$= 900 + 90 + 9 = 999$$

Think: 900, 990, 999

(b)
$$739 - 224 = (700 + 30 + 9) - (200 + 20 + 4)$$
$$= (700 - 200) + (30 - 20) + (9 - 4)$$
$$= 500 + 10 + 5 = 515$$

Think: 500, 510, 515

(c)
$$4 \cdot 235 = 4(200 + 30 + 5)$$
$$= 800 + 120 + 20$$
$$= 920 + 20$$
$$= 940$$

Think: 800, 120, 920, 940

Left-to-right methods often combine nicely with an understanding of positional notation to simplify mental calculation. For example, since $4200 = 42 \cdot 100$, we might compute the sum

$$\begin{array}{r} 3700 \\ 900 \\ 2800 \\ +\ 5600 \\ \hline \end{array}$$

by thinking of

$$\begin{array}{r} 37 \\ 9 \\ 28 \\ +\ 56 \\ \hline \end{array}$$

Then, working from left to right, we think

30, 50, 100, 107, 116, 124, 130
times 100. The answer is 13,000.

Estimation

The ever-increasing use of calculators and computers makes it essential that students develop skill at **computational estimation** (or **estimation,** for short.) On the one hand, if you are asked to multiply .983 by 1.17, you will need a calculator, of course. On the other hand, since both numbers are about 1, you should expect the answer to be close to 1. If the number on the calculator comes out to be 234.5623 for the product, then, clearly, you have entered at least one of the two numbers incorrectly. Your skill at estimation enables you to recognize, without much work, that you couldn't possibly have the correct answer.

The goal of estimation is to be able to see, without doing much computation yourself, how large an answer should be or what it should be close to. In order to have that ability, comfort with doing simple arithmetic should be of second nature. For example, the *NCTM Principles and Standards* says that students should "compute fluently and makes reasonable estimates." (For more details, see the NCTM's view of estimation that follows.)

FROM THE NCTM PRINCIPLES AND STANDARDS

Estimation

Estimation serves as an important companion to computation. It provides a tool for judging the reasonableness of calculator, mental, and paper-and-pencil computations. However, being able to compute exact answers does not automatically lead to an ability to estimate or judge the reasonableness of answers, as Reys and Yang (1998) found in their work with sixth and eighth graders. Students in grades 3–5 will need to be encouraged to routinely reflect on the size of an anticipated solution. Will 7×18 be smaller or larger than 100? If 3/8 of a cup of sugar is needed for a recipe and the recipe is doubled, will more or less than one cup of sugar be needed? Instructional attention and frequent modeling by the teacher can help students develop a range of computational estimation strategies including flexible rounding, the use of benchmarks, and front-end strategies. Students should be encouraged to frequently explain their thinking as they estimate. As with exact computation, sharing estimation strategies allows students to access others' thinking and provides many opportunities for rich class discussions.

SOURCE: Principles and Standards for School Mathematics by NCTM, *pp. 155–156. Copyright © 2000 by the National Council of Teachers of Mathematics. Reproduced with permission of the National Council of Teachers of Mathematics via Copyright Clearance Center. NCTM does not endorse the content or validity of these alignments.*

COOPERATIVE INVESTIGATION

Estimating the Number of Cars in a Traffic Jam

There is a 2-mile-long traffic jam on the highway. Working together in pairs, how would you decide how many cars are in the traffic jam?

Discussions

This problem is purposely vague and open-ended. In order to think logically about what might be a description of the traffic jam, you and your partner will need to set up a model (or description) of what you think a traffic jam means. Then you will need to do some mathematics to get an estimate of the number of cars. Of course, you must write out all of your assumptions in constructing your model.

The notions of estimation and of modeling a situation are intimately connected. In fact, the Standard for Mathematical Practice 4 (SMP 4) plays a strong role in considering modeling and estimation. Both future teachers and K–8 students find these approaches interesting. This problem the Traffic Jam Problem has been used successfully with fourth-grade students in Germany.* Here is a brief view of SMP 4. Problem 20 in this problem set asks how you might find results to this problem.

*Peter-Koop, A., "Teaching and Understanding Mathematical Modelling through Fermi-Problems" in B. Clarke, B. Grevholm, and R. Millman, *Tasks in Primary Mathematics Teacher Education*, Springer, New York, 2009.

SMP 4 Model with mathematics.

"Mathematically proficient students can apply the mathematics they know to solve problems arising in everyday life, society, and the workplace . . . [They] who apply what they know are comfortable making assumptions and approximations to simplify a complicated situation, realizing that these may need revision later. . . . They routinely interpret their mathematical results in the context of the situation and reflect on whether the results make sense, possibly improving the model if it has not served its purpose."

On the other hand, there is more to estimation than meets the eye and that requires using appropriate tools strategically. We must be able to find what data apply to a problem. For the Traffic Jam Problem, how would you (or your students) find out the size of the cars? Would it depend on where the highway is? (After all, the economic status of the area might affect what automobiles would be in the jam.) With that in mind, the modeling of SMP 4 is very important for SMP 5.

SMP 5 Use appropriate tools strategically.

"Mathematically proficient students consider the available tools when solving a mathematical problem . . . [They] at various levels are able to identify relevant external mathematical resources, such as digital content . . . and use them to pose or solve problems."

Front-End Method

Another method, which is even easier than the previous one, also starts on the left but then almost ignores the other digits. This approach, called **front-end estimation,** adds the numbers under the assumption that all their digits except the first on the left are zero and then makes some changes on the basis of the second digit. Here's an example:

EXAMPLE 3.18 Front-End Addition and Subtraction

Use front-end estimation to give an approximate answer to each of the following problems:

(a) $352 + 647$ **(b)** $739 - 224$

Solution

(a) By ignoring the second and third digits, we first think of $300 + 600 = 900$. However (and this is the "make some changes" part), considering that there are 5 tens in the first number and 4 tens in the second, we estimate that we need to adjust our original estimation of 900 by adding 9 tens, or about 100. Thus, we have the better estimate of 1000 for an answer.

(b) Here we think of the answer as being close to $700 - 200 = 500$. When we look at the second digits, there are 3 tens from which we could subtract 2 tens, so no adjustment is needed.

Note that, on the one hand, the answers to Examples 3.17 and 3.18 are close but not the same. That shouldn't be surprising, since we are only estimating, not trying to find the exact answer. On the other hand, the two examples should end up with numbers that are close.

Rounding

Often, we are *not* interested in exact values. This is certainly true when we are *estimating* the results of numerical calculations, and it is frequently the case that exact values are actually unobtainable. What does it mean, for example, to say that the population of California in 2010 was 37,253,956, using U.S. census data? Even if this is supposed to be the actual count on a given day, it is almost surely in error because of the sheer difficulty in conducting a census.

What does "population" mean? At first, it seems obvious, but does the term include transients, those who have a home in California and in other states, or those who are homeless? How does one keep track of births and deaths during the year? Just defining the term *population* is difficult.

We will now show what rounding is by first deciding at what level (place value) to estimate the population of California in 2010. We decide to approximate the population in terms of millions or, said more carefully, is 37,253,956 closer to 37,000,000 or 38,000,000? Since the actual census figure is closer to 37 million (not 38 million), the answer is (rounding to millions position) 37 million.

Note that there's a difference between rounding to the nearest million versus rounding to the nearest hundred thousand. Your answers will be different but both correct—37 million (rounding to the nearest million) versus 37,300,000 (rounding to the nearest hundred thousand).

To see which algorithm to use to round off, first we decide on the position. We will now **round** the California population of 37,253,956 to the nearest million. This is accomplished by considering the digit in the hundred thousands position. If this digit is 5 or more, we increase the digit in the millions position by 1 and replace all the digits to the right of this position by 0s. If the hundred thousands digit is 4 or less, we leave the millions digit unchanged and replace all the digits to its right by 0s.

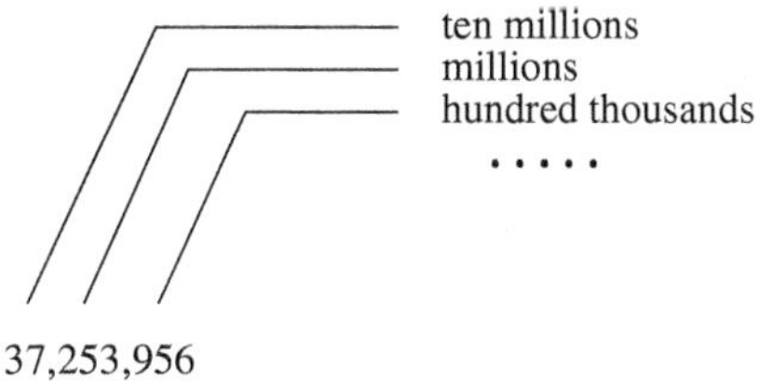

Rounding by Using the 5-Up Rule

(i) Determine to which position you are rounding.
(ii) If the digit to the right of this position is 5 or more, add 1 to the digit in the position to which you are rounding. Otherwise, leave the digit unchanged.
(iii) Replace with 0s all digits to the right of the position to which you are rounding.

EXAMPLE 3.19

Using the 5-Up Rule to Round Whole Numbers

Round 27,250 to the position indicated.

(a) The nearest ten thousand **(b)** The nearest thousand
(c) The nearest hundred **(d)** The nearest ten

Solution

(a) The digit in the ten thousands position is 2. Since the digit to its right is 7 and $7 > 5$, we add 1 to 2 and replace all digits to the right of the rounded digit with 0s. This gives 30,000.
(b) The time 7 is the critical digit, and 2 is the digit to its right. Thus, to the nearest thousand, 27,250 is rounded to 27,000.
(c) Here, 2 is the digit in the hundreds position, and 5 is to its right. Thus, 27,250 is rounded to 27,300 to the nearest hundred.
(d) This time 5 is the critical digit, and 0 is the digit on its right. Thus, no change need be made because 27,250 is already rounded to the nearest ten.

Approximating by Rounding

Rounding is often used in finding estimates. The advantage of **approximating by rounding** is that it gives a single estimate that is reasonably close to the desired answer. The idea is to round the numbers involved in a calculation to the position of the leftmost one or two digits and to use these rounded numbers in making the estimate. For example, consider the sum 467 + 221. Rounding to the nearest hundred, we have

$$467 \approx 500 \quad \text{and} \quad 221 \approx 200,$$

where we use the symbol $\approx$ to mean "is approximated by." Thus, we obtain the approximation

$$467 + 221 \approx 500 + 200 = 700.$$

The actual answer is 688, so the approximation is reasonably good. Rounding to the nearest ten usually gives an even closer approximation if it is needed. Thus,

$$467 \approx 470, \qquad 221 \approx 220,$$

and $470 + 220$ give the very close approximation 690.

EXAMPLE 3.20 Approximating by Rounding

Round to the leftmost digit to find approximate answers to each of these calculations. Also, compute the exact answer in each case.

(a) $681 + 241$ **(b)** $681 - 241$
(c) $681 \cdot 241$ **(d)** $57{,}801 \div 336$

Solution

To the nearest hundred, $681 \approx 700$ and $241 \approx 200$. Also, $57{,}801 \approx 60{,}000$ and $336 \approx 300$. Using these rounded numbers, we obtain the approximations shown.

Approximation	Exact Answer
(a) $681 + 241 \approx 700 + 200 = 900$	$681 + 241 = 922$
(b) $681 - 241 \approx 700 - 200 = 500$	$681 - 241 = 440$
(c) $681 \cdot 241 \approx 700 \cdot 200 = 140{,}000$	$681 \cdot 241 = 164{,}121$
(d) $57{,}801 \div 336 \approx 60{,}000 \div 300 = 200$	$57{,}801 \div 336 = 172 \text{ R } 9$

Here, the quotient is approximately 200.

Note that we round first and then perform the computation to obtain an estimate. Students will mistakenly add first and then round off, defeating the purpose of the method.

3.4 Problem Set

Understanding Concepts

1. Calculate the given expression mentally, using easy combinations. Write a sequence of numbers indicating intermediate steps in your thought process. The first one is done for you.

(a) $7 + 11 + 5 + 3 + 9 + 16 + 4 + 3$

Think: 10, 30, 50, 55, 58

(b) $6 + 9 + 17 + 5 + 8 + 12 + 3 + 6$

(c) $27 + 42 + 23$

2. Calculate the given expression mentally, using easy combinations. Write a sequence of numbers indicating intermediate steps in your thought process.

(a) $47 - 23$ **(b)** $48 \cdot 5$ **(c)** $21{,}600 \div 50$

3. Calculate mentally, using adjustment. Write down a sequence of numbers indicating intermediate steps in your thought process. The first one is done for you.

(a) $78 + 64$

Think: 80 + 62, 140, 142

(b) $294 + 177$ **(c)** $306 - 168$ **(d)** $294 - 102$

4. Calculate mentally, using adjustment. Write down a sequence of numbers indicating intermediate steps in your thought process.

(a) $479 + 97$ **(b)** $3493 \div 7$ **(c)** $412 \cdot 7$

5. Perform the given calculations mentally from left to right. Write down a sequence of numbers indicating intermediate steps in your thought process.
   (a) $425 + 362$ (b) $363 + 274$ (c) $572 - 251$

6. Perform the given calculations mentally from left to right. Write down a sequence of numbers indicating intermediate steps in your thought process.
   (a) $764 - 282$
   (b) $3 \cdot 342$
   (c) $47 + 32 + 71 + 9 + 26 + 32$

7. Using the technique of front-end estimation, find an approximate value for each of the following:
   (a) $425 + 362$ (b) $363 + 274$
   (c) $572 - 251$ (d) $764 - 282$
   (e) Are you surprised that the answers to this question are not the same as those of number 5?

8. Round 631,575 to the
   (a) nearest ten thousand.
   (b) nearest thousand.
   (c) nearest hundred.

9. Round each of these numbers to the position indicated:
   (a) 947 to the nearest hundred.
   (b) 850 to the nearest hundred.
   (c) 27,462,312 to the nearest million.
   (d) 2461 to the nearest thousand.

10. Find the rounded 2010 population of each of the following.
    (a) California, rounding to the nearest hundred thousand. (See the paragraph on California on pp. 162–163.)
    (b) Georgia, whose population was 9,687,653, rounding to the nearest million.

11. Find the rounded 2010 population of each of the following.
    (a) Kentucky, whose population was 4,339,367, rounding to the nearest ten thousand.
    (b) Texas, whose population was 25,145,561, rounding to the nearest million.

12. Rounding to the leftmost digit, calculate approximate values for each of these sums and differences:
    (a) $478 + 631$ (b) $782 + 346$
    (c) $678 - 431$ (d) $257 \cdot 364$
    (e) $7403 \cdot 28$ (f) $28{,}329 \div 43$
    (g) $71{,}908 \div 824$

13. Rounding to the nearest thousand and using mental arithmetic, estimate each of these sums and differences:

    (a) $\begin{array}{r} 17{,}281 \\ 6\,564 \\ 12{,}147 \\ 2\,481 \\ +\ 13{,}671 \\ \hline \end{array}$ (b) $\begin{array}{r} 2734 \\ 3541 \\ 2284 \\ 3478 \\ +\ 7124 \\ \hline \end{array}$

    (c) $\begin{array}{r} 4270 \\ -\ 1324 \\ \hline \end{array}$

14. Rounding to the nearest thousand and using mental arithmetic, estimate each of these sums and differences:

    (a) $\begin{array}{r} 28{,}341 \\ 942 \\ 2\,431 \\ 4\,716 \\ +\ 12{,}824 \\ \hline \end{array}$ (b) $\begin{array}{r} 21{,}243 \\ -\ 7\,824 \\ \hline \end{array}$

    (c) $\begin{array}{r} 37{,}481 \\ -\ 16{,}249 \\ \hline \end{array}$

15. Using rounding to the leftmost digit, estimate these products:
    (a) $2748 \cdot 31$ (b) $4781 \cdot 342$
    (c) $23{,}247 \cdot 357$
    (d) Use your calculator to determine the exact values of the products in parts (a) through (c).

16. Use rounding to the leftmost digit to estimate the quotient in each of the following divisions:
    (a) $29{,}342 \div 42$ (b) $7431 \div 37$ (c) $79{,}287 \div 429$
    (d) Use your calculator to determine the exact (in the case of nonrepeating decimals) values of the quotients in parts (a) through (c).

Into the Classroom

17. (a) Would rounding to the leftmost digit give a very good estimate of this sum? Why or why not?

    $\begin{array}{r} 1478 \\ 2395 \\ 1492 \\ +\ 5481 \\ \hline \end{array}$

    (b) What would you suggest that your students do to obtain a more accurate estimate?
    (c) Compute the accurate answer to the addition in part (a).

18. How would you respond to one of your students who rounded 27,445 to the nearest thousand as follows: $27{,}445 \rightarrow 27{,}450 \rightarrow 27{,}500 \rightarrow 28{,}000$?

19. (**Writing**) Read the Number and Operations Standard for Grades 3–5, pages 148–156, from NCTM's *Principles and Standards for School Mathematics,* and write a short critique of the standard discussed there, emphasizing your own reaction. How do these recommendations compare with your own school experience?

20. (**Writing**) Working with another future teacher, find a solution for estimating the number of cars in a traffic jam in the Cooperative Investigation on page 161. Make sure that you write out carefully what assumptions you make and what answer the two of you will give to your classmates. Do you think that fourth-grade students would be able to do this problem?

Responding to Students

21. Diley is solving her problems by using rounding. Here is how she solved two problems:

    (a) $\begin{array}{r} 449 \\ +\ 333 \\ \hline 782 \end{array}$ She says that it rounds to 800.

    (b) $\begin{array}{r} 881 \\ +\ 449 \\ \hline 1330 \end{array}$ She says that it rounds to 1300.

Is she following rounding procedures for addition? Does rounding work for her? What would you suggest?

22. Richard asks his eighth-grade teacher, Ms. Rocchio, why he got an answer different from hers when it came to rounding 845. He rounded 845 to 850 and then rounded 850 to 900. Her answer was 800. Was Richard or Ms. R. correct and why?

23. When asked to round 1268 to the nearest ten, Raphael replied that it was 1278. What was the fifth grader thinking?

24. When asked to round 8452 to the nearest one hundred, Sally says that the answer is 8000. How would you respond?

25. Elementary school students will be asked to round 217 to the nearest ten. Some will write

200 300
↖ ↗
217

and try to decide whether the answer should be 200 or 300. How would you help them move to a correct approach?

26. When asked to take the number 29,457,300 and add 1 million more, fourth graders Drew and Alonzo gave the following responses:

(a) 210,457,300 (Drew) **(b)** 39,457,300 (Alonzo)

What error did each of the boys make, and why do you think he made it? How would you help him correct his work?

Thinking Critically

27. **(Writing)** Theresa and Fontaine each used their calculators to compute 357 + 492. Fontaine's answer was 749 and Theresa's was 849. Who was more likely correct? In two brief sentences, tell how estimation can help you decide whose answer was probably correct.

28. **(a)** Use rounding to estimate the results of each of the following calculations:

(i) $\dfrac{452 + 371}{281}$

(ii) $\dfrac{3 \cdot 271 + 465}{74 + 9}$

(iii) $\dfrac{845 \cdot 215}{416}$

(b) Use your calculator to determine the exact answer to each of parts (i) through (iii).

29. Sometimes the last digits of numbers can help you decide whether calculator computations are correct.

(a) Given that one of 27,453; 27,587; and 27,451 is the correct result of multiplying 283 by 97, which answer is correct?

(b) **(Writing)** In two brief sentences, tell how a consideration of last digits helped you answer part (a).

30. Since 25,781; 24,323; 26,012; and 25,243 are all about the same size, about how large is their sum? Explain briefly.

Making Connections

31. Utah, Colorado, New Mexico, and Arizona are the only four states in the United States that meet at a common corner. The areas of the four states are, respectively, 82,168 square miles, 103,730 square miles, 121,365 square miles, and 113,642 square miles.

(a) Mentally estimate the combined area of the four states, writing down a sequence of numerals to indicate your thought process.

(b) Compute the actual sum of the areas in part (a).

32. While grocery shopping with $40, you buy the following items at the prices listed:

2 gallons of milk	$2.29 a gallon
1 dozen eggs	$1.63 per dozen
2 rolls of paper towels	$1.21 per roll
1 5-pound pork roast	$1.98 per pound
2 boxes of breakfast cereal	$3.19 each
1 azalea	$9.95 each

(a) About how much will all of this cost?

(b) If you don't buy the azalea, about how much change should you receive?

State Assessments

33. (Grade 4)
We would like the class to subtract using estimation, rather than actually doing the subtraction. What would be the best way to <u>estimate</u> the difference between 599 and 218?

A. 500 − 200 B. 500 − 300
C. 600 − 200 D. 600 − 300

34. (California, Grade 4)
What is 67,834,519 rounded to the nearest hundred thousand?

A. 67,000,000 B. 67,800,000
C. 67,830,000 D. 67,900,000

35. (Washington State, Grade 4)
Estimate the answer. Show how you found your estimate.

$$9\overline{)820}$$

36. (Grade 4)
Using mental arithmetic, what strategy would you best use to find 6 × 49?

A. 6 × 7 × 7 B. 6 × 40 + 6 × 9
C. 6 × 40 + 9 D. 2 × 3 × 49

37. (Grade 4)
There are four different multiplications. Which product is between 400 and 450?

A. 42 × 11 B. 38 × 11
C. 35 × 11 D. 46 × 11

38. (California, Grade 4)
What is 583,607 rounded to the nearest hundred?

A. 583,000 B. 583,600
C. 583,700 D. 84,000

3.5 Nondecimal Positional Systems

In this chapter so far, we have discussed the numeration system in common use now, the decimal system, including various ways to do arithmetic operations and the use of manipulatives. Base ten probably came about because we have ten fingers and children often count on their fingers. Although bases other than base ten were once part of the elementary school curriculum, they are not present now. However, nondecimal positional systems are a part of the required "math methods" course to deepen understanding of place value.

The decimal system (i.e., base ten) is a most important aspect of elementary school mathematics. We will also enrich our understanding of place value by looking at what happens with bases other than ten. In this section, we will look through aspects of base five and base six, show how they can be converted into base-ten numbers, and illustrate how to perform operations such as addition, subtraction, and multiplication in bases other than ten. Division is not covered in this text due to length. However, the approach to division would follow the same operation for division in base ten as it would in other bases.

Base-Five and Base-Six Place Value Including Conversion to the Decimal System and Operations

Base-Five and Base-Six Conversion to the Decimal System For our purposes, perhaps the quickest route to understanding **base-five notation** is to consider again the abacus of Figure 3.1. This time, however, we allow only 5 beads to be moved forward on each wire. As before, we start with the wire to the students' right (our left) and move 1 bead forward each time as we count 1, 2, 3, and so on. When we reach 5, we have counted the first wire once, and we record this on the abacus by moving 1 bead forward on the second wire while moving all 5 beads on the first wire to the back. We continue to count, and when the count reaches 10, we will have counted all the beads on the first wire a second time. We move a second bead forward on the second wire and again move all the beads on the first wire to the back. Thus, each bead on the second wire counts for 5; that is, the second wire is the "fives" wire. When the count reaches 19, the abacus will appear as in Figure 3.9, and

$$19 = 3 \cdot 5 + 4.$$

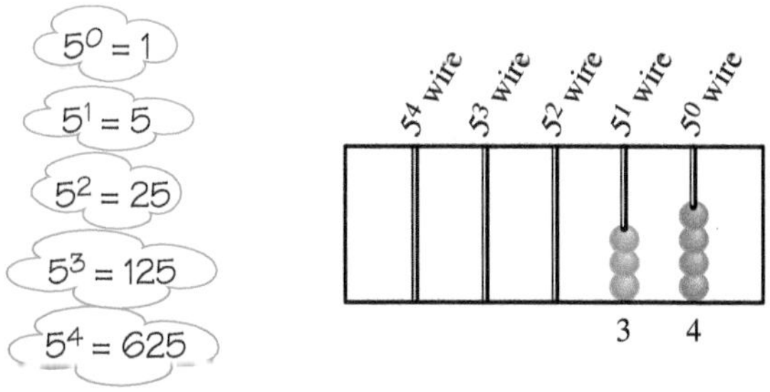

Figure 3.9
A count of 19 on a five-bead abacus

If we continue the count to 25, all the beads on the first wire will have been counted five times. This means that we have moved all 5 beads to the front on the *second* wire. We record that fact on the abacus by returning these beads to the back of the abacus and moving 1 bead forward on the *third* wire. Thus, the third wire becomes the $25 = 5^2$ wire. The count can be continued in this way, and it is apparent that any whole number will eventually be counted and can be recorded on the abacus with only 0, 1, 2, 3, or 4 beads per wire showing at the front; that is, we need only the digits 0, 1, 2, 3, and 4 in base five. For example, if we continue the counting process up to 113, the abacus will have 3 beads on the units wire, 2 on the fives wire, and 4 on the twenty-fives wire—that is,

$$113 = 4 \cdot 5^2 + 2 \cdot 5 + 3.$$

Just as we shorten $4 \cdot 10^2 + 2 \cdot 10 + 3$ to 423, someone working in base five also shortens $4 \cdot 5^2 + 2 \cdot 5 + 3$ to 423. Thus, the three-digit sequence 423 can represent many different numbers, depending on which base is chosen. To avoid confusion in talking about base-five numeration, we agree that we will *not* say "four hundred twenty-three" when we read 423 as a base-five numeral.

Instead, we will say "four two three, base five" and will write 423_{five}, where the subscript "five" indicates the base.

We now turn to base six and define place value. Note first the analogy of the numerals, 423, and see how the number itself (one in base five and one in base six) changes with the base. The digits of base six are in the set {1, 2, 3, 4, 5} and you should remember that there is a 5 as one of the digits, although there is not a 5 in base five.

In base six, by analogy, the place value of 423_{six} is based on powers of 6. Thus, we can compute 423_{six} by comparing this case where the coefficients are the products of 6^0, 6^1, 6^2,

$$423_{\text{six}} = 4 \cdot 6^2 + 2 \cdot 6 + 3,$$

and we read the numeral as "four two three, base six." Doing the arithmetic on the right side makes it clear that $423_{\text{six}} = 159_{\text{ten}}$. This is called a **conversion** from base six to base ten.

Notice the difference—when we change the base (from five to six), the number changes. More precisely, $423_{\text{five}} = 113_{\text{ten}}$, whereas $423_{\text{six}} = 159_{\text{ten}}$—they are different numbers! Unless expressly stated otherwise, a numeral written *without* a subscript should be read as a base-ten numeral.

Observe that, in base five, we need the digits 0, 1, 2, 3, and 4, and we need to know the values of the positions in base five. In base six, the digits are 0, 1, 2, 3, 4, and 5, and we need to know the positional values in base six. For these two bases, the positional values are given in Table 3.7.

TABLE 3.7 DIGITS AND POSITIONAL VALUES IN BASES FIVE AND SIX

			DECIMAL VALUES OF POSITIONS				
Base b	**Digits Used**	...	b^4	b^3	b^2	b^1	**Units ($b^0 = 1$)**
Five	0, 1, 2, 3, 4	...	$5^4 = 625$	$5^3 = 125$	$5^2 = 25$	$5^1 = 5$	$5^0 = 1$
Six	0, 1, 2, 3, 4, 5	...	$6^4 = 1296$	$6^3 = 216$	$6^2 = 36$	$6^1 = 6$	$6^0 = 1$

EXAMPLE 3.21

Converting from Base-Five to Base-Ten Notation

Write the base-ten representation of 3214_{five}.

Solution

We use the place values of Table 3.7 along with the digit values. Thus,

$$\begin{aligned} 3214_{\text{five}} &= 3 \cdot 5^3 + 2 \cdot 5^2 + 1 \cdot 5^1 + 4 \cdot 5^0 \\ &= 3 \cdot 125 + 2 \cdot 25 + 1 \cdot 5 + 4 \cdot 1 \\ &= 375 + 50 + 5 + 4 \\ &= 434. \end{aligned}$$

Therefore, "three two one four, base five" is four hundred thirty-four.

EXAMPLE 3.22

Converting from Base-Ten to Base-Five Notation

Write the base-five representation of 97.

Solution

Recall that 97 without a subscript has its usual meaning as a base-ten numeral.

We begin with the notion of grouping. Suppose that we have 97 beans and want to put them into groups of single beans (units), groups of 5 beans (fives), groups of five groups of 5 beans each (twenty-fives), and so on. The problem is to complete the grouping by using the least possible number of groups. This means that we must use as many of the larger groups as possible. Diagrammatically, we have

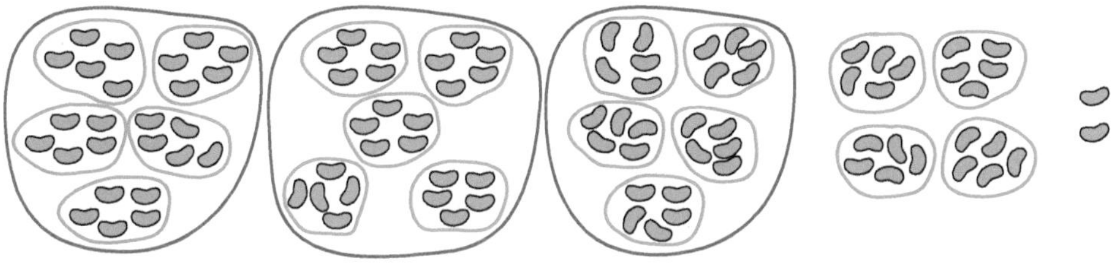

so $97_{ten} = 342_{five}$. Arithmetically, this corresponds to determining how many of each position value in base five are required to represent 97. The answer can be determined by successive divisions. Referring to Table 3.7 for position values, we see that 125 is too big. Thus, we begin with 25 and divide successive remainders by successively lower position values:

$$\begin{array}{ccc} 3 & 4 & 2 \\ 25\overline{)97} & 5\overline{)22} & 1\overline{)2} \\ \underline{75} & \underline{20} & \underline{2} \\ 22 \nearrow & 2 \nearrow & 0 \end{array}$$

These divisions reveal that we need three 25s, four 5s, and two 1s, or, in tabular form,

25s	5s	1s
3	4	2

Thus,

$$97_{ten} = 342_{five}.$$

As a check, we note that

$$\begin{aligned} 342_{five} &= 3 \cdot 25 + 4 \cdot 5 + 2 \\ &= 75 + 20 + 2 \\ &= 97_{ten}. \end{aligned}$$

INTO THE CLASSROOM
Tara Morey Comments on Base Five

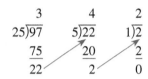

To engage students in exploring base five, I use a scenario in which students compete as members of a one-handed tribe. This means that the students can use only one hand. Begin with students counting a certain number of dots in base five, allowing students to work in groups and begin conversations about vocabulary and methodology. This is followed by a group discussion about how they, the "tribe," will discuss base-five numbers. Small groups are then given cards with numbers, dots, expanded form notation, and a problem with a missing value. Students are asked to match the number, dot, and expanded form cards and glue them on poster paper.

As a large group, students present and discuss their work on adding and subtracting numbers in base five. Students are asked to critique their work and pick one method for their own use in solving for the missing value on the addition and subtraction problem cards. Students work to match the final card with the number.

If students struggle with the symbolic representation of base-five numbers, provide them with base-five manipulatives. For more information on creating formative assessment lessons, visit https://sites.google.com/site/moreygoodman/.

Adding in Base Five The next example shows how the addition algorithm works in base five. Let's see how the process works with place-value cards, which are examples of a manipulative.

EXAMPLE ◆ 3.23 **Adding in Base Five**

Compute the sum of 143_{five} and 234_{five} in base-five notation.

Solution

With Place-Value Cards

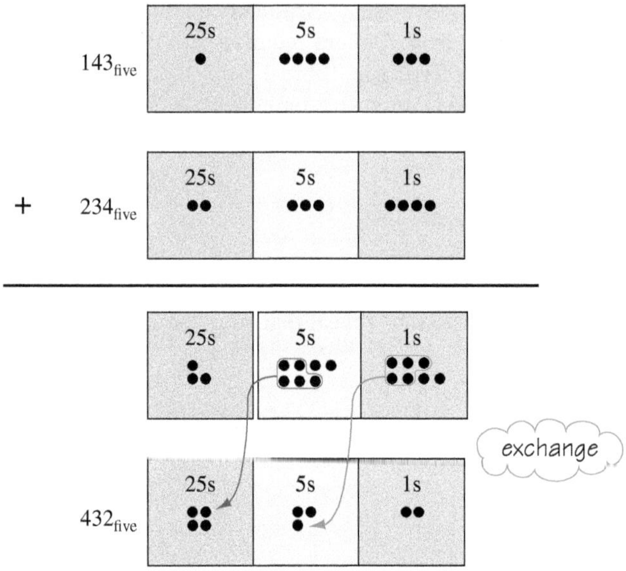

With Place-Value Diagrams and Instructional Algorithms

Place-Value Diagram	Instructional Algorithm		Final Algorithm
25s 5s 1s	143	$3_{five} + 4_{five} = 12_{five}$	1 1
1 4 3	+ 234		1 4 3
+ 2 3 4	12		one 25 + 2 3 4 one 5
3 12 12	120	$1_{five} + 2_{five} = 3_{five}$	4 3 2
4 3 2	exchange 300		
	432		Compress the result to one line.
		$3_{five} + 1_{five} = 4_{five}$	

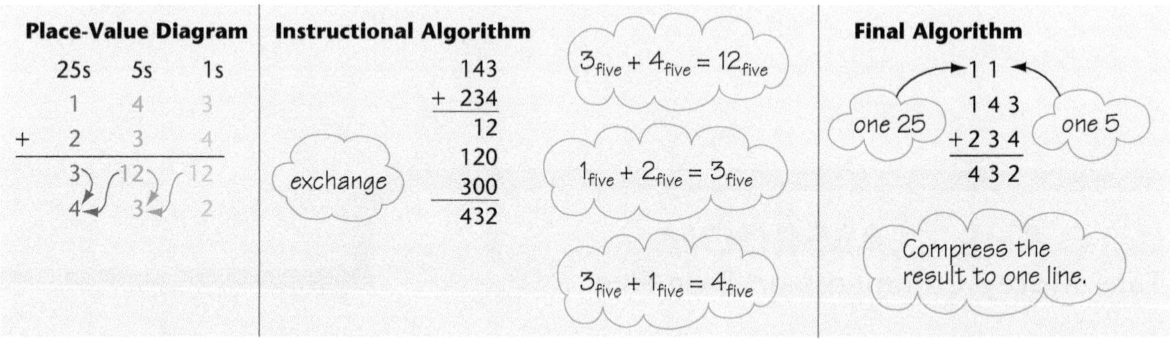

So the sum is 432_{five}.

To check the addition, we convert everything to base ten, in which we feel comfortable:

$$143_{five} = 1 \cdot 25 + 4 \cdot 5 + 3 \cdot 1 = 48_{ten}$$
$$234_{five} = 2 \cdot 25 + 3 \cdot 5 + 4 \cdot 1 = 69_{ten}$$
$$432_{five} = 4 \cdot 25 + 3 \cdot 5 + 2 \cdot 1 = 117_{ten}$$

The result is confirmed, since $48_{ten} + 69_{ten} = 117_{ten}$.

The difficulty in doing base-five arithmetic is unfamiliarity with the meaning of the symbols—we do not recognize at a glance, for example, that $8_{ten} = 13_{five}$, since 13_{five} is 1 five and 3 ones. By contrast, thinking of units, strips with 5 units per strip, and so on, we do not find it hard to think of 1 strip and 3 units as 8. In fact, doing arithmetic in another base gives good practice in mental arithmetic, which is a desirable end in itself. Of course, elementary school children memorize the addition and multiplication tables in base ten so that the needed symbols are readily recalled, and we could do the same thing here. The addition table in base five, for example, is as follows:

+	0	1	2	3	4
0	0	1	2	3	4
1	1	2	3	4	10
2	2	3	4	10	11
3	3	4	10	11	12
4	4	10	11	12	13

These numerals are in base five.

This table could be used to make addition in base five easier and more immediate. For example, from the table, we see that $3 + 4 = 12_{\text{five}}$. So we write down the 2 and exchange the five 1s for a 5 in the 5s column, as shown in the final algorithm in Example 3.23. Next, $4 + 3 = 12_{\text{five}}$, and the 1 from the exchange gives 13_{five}. Thus, we write down the 3 and exchange the five 5s for one 25 in the next column. Finally, $1 + 2 = 3$, and 1 from the exchange makes 4, so we obtain 432_{five} as before. Still, we suggest that you not memorize the addition table in base five; it is far better just to think carefully about what is going on.

Subtracting and Multiplying in Base Five

EXAMPLE 3.24

Subtracting in Base Five

Subtract 143_{five} from 234_{five} in base-five notation.

Solution

With Place-Value Cards

There is no problem in taking away 3 markers from the 1s square, but we cannot remove 4 markers from the 5s square without exchanging 1 marker on the 25s square for 5 markers on the 5s square. Taking away the desired markers, we are left with zero 25s, four 5s, and one 1, for 41_{five}:

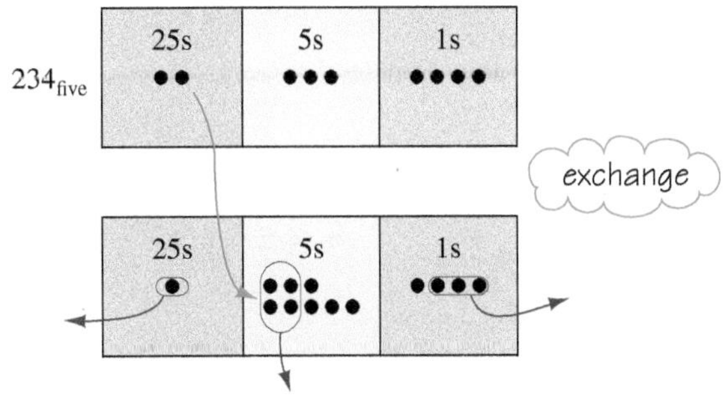

$234_{\text{five}} - 143_{\text{five}} = 41_{\text{five}}$

With Place-Value Diagrams and Instructional Algorithms in Base Five

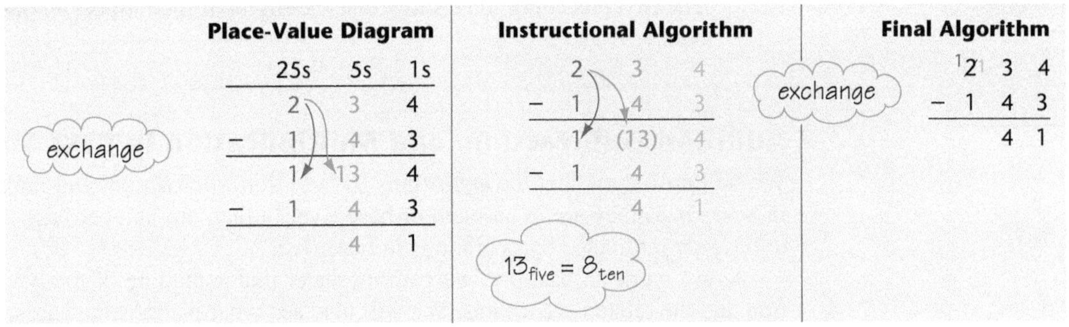

So the answer is 41_{five}.

As a check, we already know that $234_{\text{five}} = 69_{\text{ten}}$ and $143_{\text{five}} = 48_{\text{ten}}$. Since $69 - 48 = 21$ and $41_{\text{five}} = 4 \cdot 5 + 1 \cdot 1 = 21$, the calculation checks.

EXAMPLE 3.25

Multiplying in Base Five

Compute the product of each of two numbers in base five.

(a) What is $3_{\text{five}} \cdot 4_{\text{five}}$ in base five?
(b) What is $21_{\text{five}} \cdot 3_{\text{five}}$ using expanded notation?

Solution

(a) $3_{\text{five}} \cdot 4_{\text{five}}$ is $4 + 4 + 4$

adding 4 to itself 3 times

$$4 + 4 + 4 = 4 + 3 + 5 = 2 + 2 + 3 + 5 = 2 + 5 + 5$$

breaking into 5s and units *breaking into 5s and units*

$$= 2.5 + 2 = 22_{\text{five}}$$

Write in positional values from Table 3.7. Thus, 22_{five} is the result. Note that in base ten, $3_{\text{five}} = 3$, and $4_{\text{five}} = 4$, so in base ten, their product is 12. On the other hand,

$$22_{\text{five}} \text{ is } 2 \cdot 5 + 2 = 12_{\text{ten}}$$

The answers are the same, of course.

(b) With more than one digit, using expanded notation is useful (as are the instructional and final algorithms).

$$(21)_{\text{five}} = 2 \cdot 5 + 1 \text{ and } 3_{\text{five}} = 3$$

thus

$$\begin{aligned}(21)_{\text{five}} \cdot 3_{\text{five}} &= (2 \cdot 5 + 1)3 = 6 \cdot 5 + 3 \\ &= (5 + 1) \cdot 5 + 3 \\ &= 5^2 + 5 + 3 = 1 \cdot 5^2 + 1 \cdot 5 + 3 \\ &= 113_{\text{five}}.\end{aligned}$$

distributive property

expanded notation

place value

The two preceding lines show that the digits of the number in base five are the purple numbers 1, 1, and 3.

Addition, Subtraction, and Multiplication in Base Six

We will now show that the algorithms for addition, subtraction, and multiplication are just as valid in base six as they are in base ten or base five. In fact, any integer base can be used. We now move to base six motivated by operations in base five.

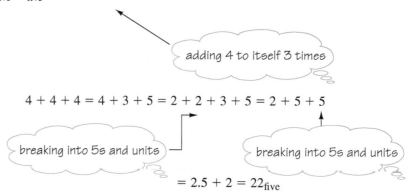

As we work on base six, we gain a greater understanding of the whole idea of positional notation and the related algorithms. We will also see what place-value cards, place-value diagrams, and instructional and final algorithms look like after an arithmetic operation is done in base six. In base five, for example, the place-value cards would have a units or 1s square; a 5s square; a 5^2, or 25s, square; and so on. With sticks, we could use loose sticks, bundles of 5 sticks, bundles of 5 bundles of sticks, and so on. With units, strips, and mats, we would have units, strips, with 5 units each, and mats with 5 strips per mat.

Adding in Base Six The next example shows how the addition algorithm works in base six. The place-value cards in base six are shown in Example 3.26.

EXAMPLE 3.26

Adding in Base Six

Compute the sum of 143_{six} and 234_{six} in base-six notation.

Solution

With Place-Value Cards

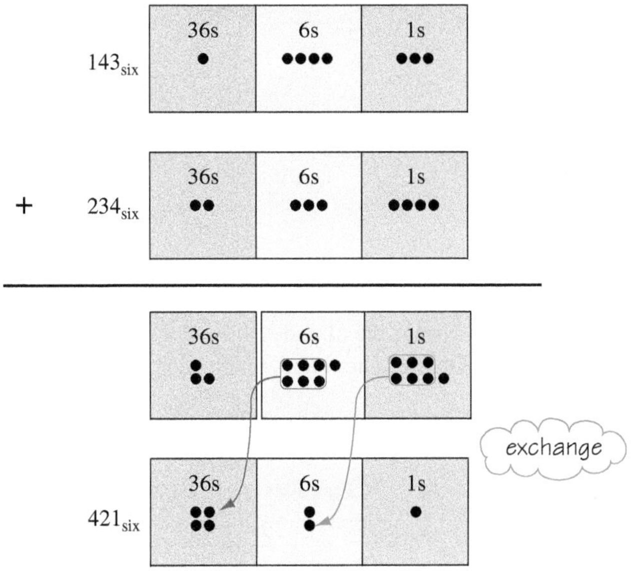

With Place-Value Diagrams and Instructional Algorithms

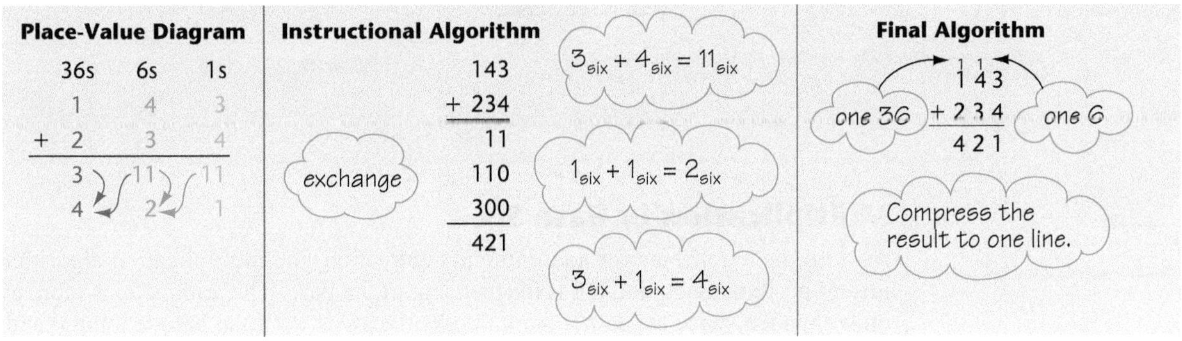

So the sum is 421_{six}.

To check the addition, we convert everything to base ten, in which we feel comfortable:

$$\begin{aligned} 143_{six} &= 1\cdot 36 + 4\cdot 6 + 3\cdot 1 = 63_{ten} \\ 234_{six} &= 2\cdot 36 + 3\cdot 6 + 4\cdot 1 = 94_{ten} \\ \hline 421_{six} &= 4\cdot 36 + 2\cdot 6 + 1\cdot 1 = 157_{ten} \end{aligned}$$

The result is confirmed, since $63_{ten} + 94_{ten} = 157_{ten}$.

Subtracting in Base Six

EXAMPLE 3.27

Subtracting in Base Six

Subtract 143_{six} from 234_{six} in base-six notation.

Solution

With Place-Value Cards

There is no problem in taking away 3 markers from the 1s square of 143_{six}, but we cannot remove 4 markers from the 6s square without exchanging 1 marker on the 36s square for 6 markers on the 6s square. Taking away the desired markers, we are left with zero 36s, four 6s, and one 1, for 51_{six}:

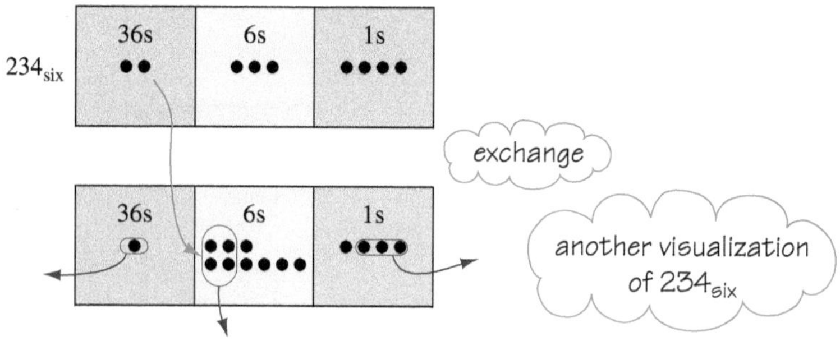

We can remove marks to find that $234_{six} - 143_{six} = 51_{six}$.

So the answer is 51_{six}.

As a check, we already know that $234_{six} = 94_{ten}$ and $143_{six} = 63_{ten}$. So $94 - 63 = 31$ gives us the answer. On the other hand, using base six shows $51_{six} = 5 \cdot 6 + 1 \cdot 1 = 31_{ten}$ so the calculation checks.

With Place-Value Diagrams and Instructional Algorithms

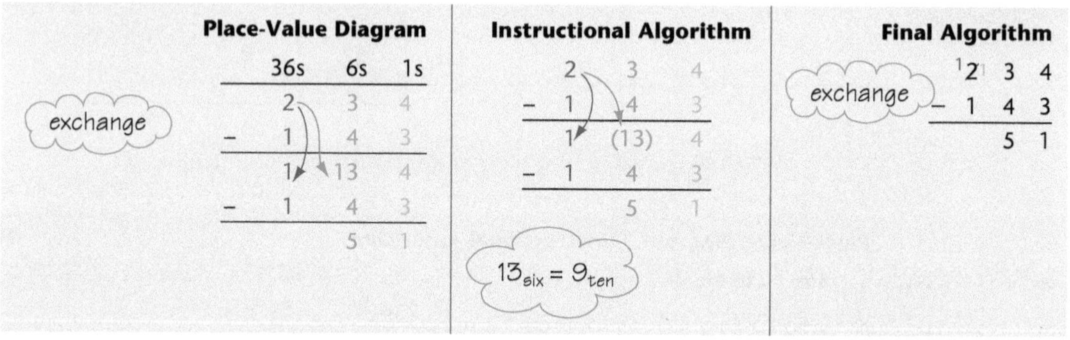

Multiplication in Base Six

As with the algorithms for addition and subtraction, the multiplication algorithm depends on the idea of positional notation but is independent of the base. The arithmetic is more awkward for bases other than ten, since we do not think in other bases as we do in base ten, but the ideas are the same. However, working in other bases not only enhances our understanding of base ten but also provides an interesting activity that helps to improve mental arithmetic skills. Let's consider an example.

EXAMPLE 3.28

Multiplying in Base Six

Compute the product $324_{six} \cdot 15_{six}$ in base six.

Solution

We use the base-six positional values from Table 3.7 and the instructional algorithm of Example 3.12. To simplify notation, we will write all *numerals* in base-six notation so that the subscripts will be omitted. Finally, the numeral *words* will have their usual base-ten meaning. The work proceeds as follows, with descriptive comments in the think cloud:

$$\begin{array}{r} 324 \\ \times\ 15 \\ \hline 32 \\ 140 \\ 2300 \\ 40 \\ 200 \\ 3000 \\ \hline 10152 \end{array}$$

Think

$5 \times 4 = \text{twenty} = 3 \text{ sixes} + 2 \text{ units} = 3 \cdot 6^1 + 2 \cdot 6^0 = 32_{six}$

$5 \times 20 = 5 \times 2 \text{ sixes} = 10 \text{ sixes} = (1 \cdot 6 + 4) \cdot 6^1 = 1 \cdot 6^2 + 4 \cdot 6^1 + 0 \cdot 6^0 = 140_{six}$

$5 \times 300 = 5 \times 3 \text{ thirty-sixes} = 15 \text{ thirty-sixes} = (2 \cdot 6 + 3) \cdot 6^2 = 2 \cdot 6^3 + 3 \cdot 6^2 + 0 \cdot 6^1 + 0 \cdot 6^0 = 2300_{six}$

$10 \times 4 = 1 \text{ six} \times 4 = 4 \cdot 6^1 + 0 \cdot 6^0 = 40_{six}$

$10 \times 20 = 1 \text{ six} \times 2 \text{ sixes} = 2 \cdot 6^2 + 0 \cdot 6^1 + 0 \cdot 6^0 = 200_{six}$

$10 \times 300 = 1 \text{ six} \times 3 \text{ thirty-sixes} = 3 \cdot 6^3 + 0 \cdot 6^2 + 0 \cdot 6^1 + 0 \cdot 6^0 = 3000_{six}$

To check, we convert all numerals to base ten:

$$\begin{aligned} 324_{six} &= 3 \cdot 6^2 + 2 \cdot 6^1 + 4 \cdot 6^0 \\ &= 3 \cdot 36 + 2 \cdot 6 + 4 \cdot 1 \\ &= 108 + 12 + 4 = 124_{ten} \\ 15_{six} &= 1 \cdot 6^1 + 5 \cdot 6^0 \\ &= 1 \cdot 6 + 5 \cdot 1 \\ &= 6 + 5 = 11_{ten} \\ 10{,}152_{six} &= 1 \cdot 6^4 + 0 \cdot 6^3 + 1 \cdot 6^2 + 5 \cdot 6^1 + 2 \cdot 6^0 \\ &= 1 \cdot 1296 + 0 \cdot 216 + 1 \cdot 36 + 5 \cdot 6 + 2 \cdot 1 \\ &= 1296 + 36 + 30 + 2 = 1364_{ten} \end{aligned}$$

Since $11_{ten} \cdot 124_{ten} = 1364_{ten}$, the check is complete.

3.5 Problem Set

Understanding Concepts

1. In a long column, write the base-five numerals for the numbers from 0 through 25.

2. Briefly describe the pattern or patterns you observe in the list of numerals in problem 1.

3. Here are the base-six representations of the numbers from 0 through 35, arranged in a rectangular array:

0	1	2	3	4	5
10	11	12	13	14	15
20	21	22	23	24	25
30	31	32	33	34	35
40	41	42	43	44	45
50	51	52	53	54	55

Briefly describe any patterns you observe in this array.

4. What would be the entries in the next two rows of the table in problem 3?

5. Write the base-ten representations of each of the following:
   (a) 413_{five} (b) 2004_{five} (c) 10_{five}

6. Write the base-ten representations of each of the following:
   (a) 100_{five} (b) 1000_{five} (c) 2134_{five}

7. The abacus of Figure 3.9 has "five" playing the role of "ten." Thus, 42 would be represented on the abacus as shown here. Moreover, it would be recorded as 132 and read as "one three two" and *not* as "one hundred thirty-two." What numbers would be represented if the beads on the abacus described are as shown in the following diagrams?

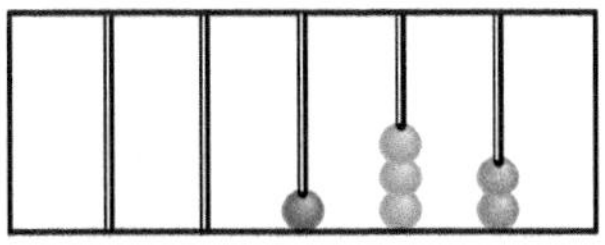

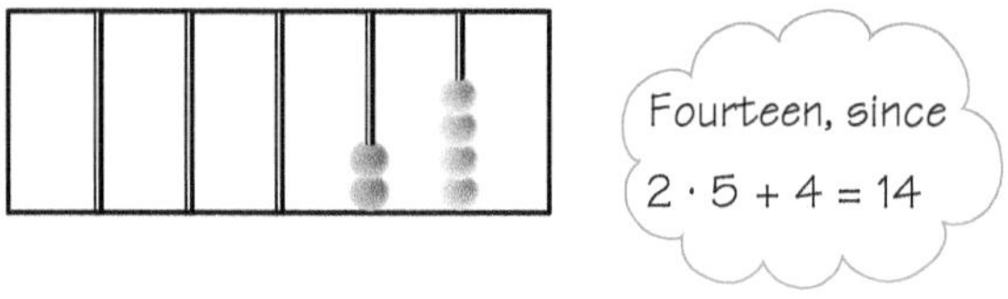

Fourteen, since
2 · 5 + 4 = 14

8. Read problem 7 and find out what numbers would be represented by the beads on the two abacuses described here. Remember, this is a base-five abacus.

(a)

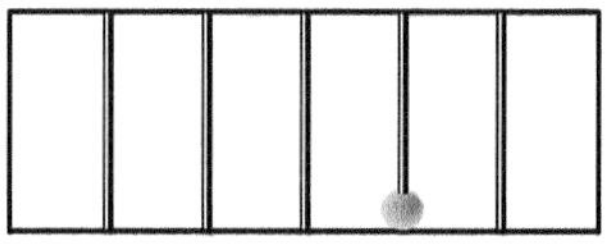

(b)

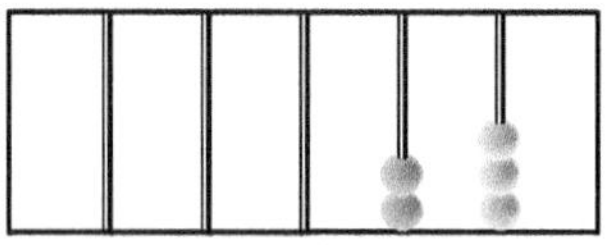

9. Suppose the abacus of problem 7 is configured as shown here:

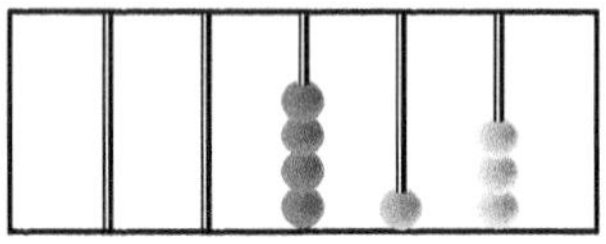

(a) What number does this arrangement represent?

(b) How would this number naturally be recorded?

(c) Suppose the number shown in part (a) is increased by 7. Draw a diagram to show how the abacus will then be configured.

10. Write the base-ten representations of each of the following numbers that are in base six:

(a) 413_{six} (b) 2004_{six} (c) 10_{six}
(d) 100_{six} (e) 1000_{six} (f) 2134_{six}

11. Determine the base-five representation for each of the given numbers. Remember that a numeral with no subscript is understood to be in base ten.

(a) 362 (b) 27
(c) 5 (d) 25

12. Determine the base-six representation for each of the following numbers which are in base ten:

(a) 342 (b) 21
(c) 6 (d) 216

13. Carry out the addition shown using base-five notation and show that your answer is correct in decimal notation.

(a) $41_{five} + 14_{five}$ (b) $213_{five} + 432_{five}$

14. Carry out the addition shown using base-six notation and show that your answer is correct in decimal notation.

(a) $41_{six} + 53_{six}$ (b) $123_{six} + 44_{six}$

15. Carry out the addition shown using base-six notation and show that your answer is correct in decimal notation.

(a) $23_{six} + 35_{six}$ (b) $423_{six} + 43_{six}$

16. Carry out the addition shown using base-five notation and show that your answer is correct in decimal notation.

(a) $44_{five} + 14_{five}$ (b) $314_{five} + 132_{five}$

17. Carry out the subtraction shown using base-five notation and show that your answer is correct in decimal notation.

(a) $41_{five} - 14_{five}$ (b) $213_{five} - 32_{five}$

18. Carry out the subtraction shown using base-five notation and show that your answer is correct in decimal notation.

(a) $43_{five} - 34_{five}$ (b) $213_{five} - 132_{five}$

19. Carry out the subtraction shown using base-six notation and show that your answer is correct in decimal notation.

(a) $505_{six} - 35_{six}$ (b) $423_{six} - 43_{six}$

20. What is being illustrated by the given sequence of sketches of place-value cards? Explain briefly.

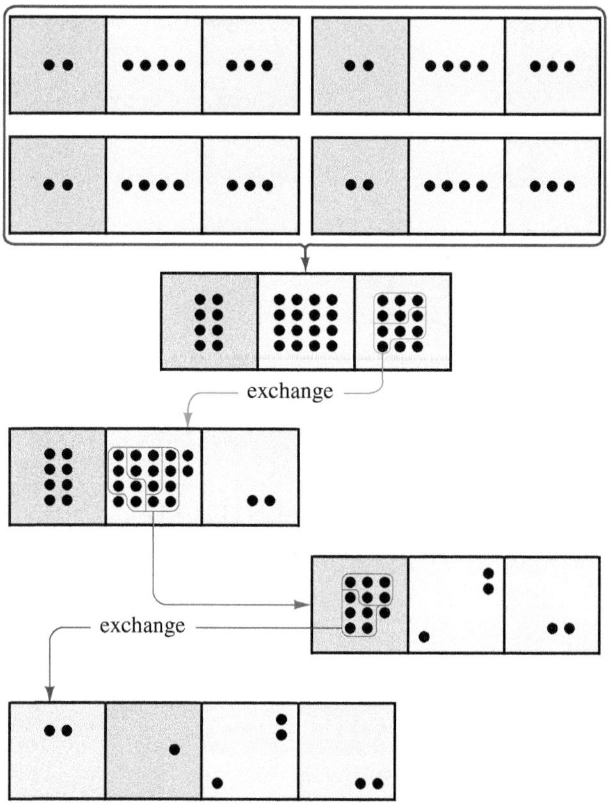

exchange
exchange

21. Carry out the multiplications shown, using base-five notation. All the numerals are already written in base five, so no subscript is needed.

(a) $\begin{array}{r} 23 \\ \times\ 3 \\ \hline \end{array}$ (b) $\begin{array}{r} 432 \\ \times\ 41 \\ \hline \end{array}$ (c) $\begin{array}{r} 2013 \\ \times\ 23 \\ \hline \end{array}$

(d) Convert the numerals in parts (a), (b), and (c) to base ten, and check the results of your base-five computations.

Into the Classroom

22. Explain how you would respond to one of your students who claims that the base-five representation of 188_{ten} is 723_{five}. Note that $7 \cdot 5^2 + 2 \cdot 5 + 3 = 188_{ten}$.

23. (a) In trying to help your students to better understand positional notation, you might ask them what sort of number system people who counted on both their fingers and toes might have invented.

(b) If your students had trouble answering the question in part (a), what might you do to help?

The Chapter in Relation to Future Teachers

Numerals, or symbols for writing numbers, and *algorithms* (methods for calculating), together called a *numeration system*, are the basis of Chapter 3. Success in learning math requires, first, that your students have a deep understanding of the concept of positional notation for numbers. A deep understanding of positional notation means realizing that 423 (base ten) is really 4 times $10^2 + 2$ times $10 + 3$ ones, and that is the foundational idea that allows children to learn the basic arithmetic operations. Being able to introduce, understand, and use algorithms is what teachers can bring to their students to make mathematics easier and more useful.

Knowledge of the historical background of mathematics is quite helpful in understanding the concepts discussed in any mathematics course. For numeration, a historical approach is especially popular, and we have introduced the reader to the use of the Egyptian, Roman, Babylonian, Mayan, and Indo-Arabic (decimal) systems as well as nondecimal systems in Section 3.5.

Chapter 3 Summary

Section 3.1 Numeration Systems Past and Present	Page Reference
CONCEPTS	
• **Egyptian system:** An additive system using hieroglyphic symbols that was developed for recording numbers on stone tablets as early as 3400 B.C.	127
• **Roman system:** An additive system using letters or Roman numerals to represent numbers. They are somewhat familiar from their current usage on the faces of analog watches and clocks.	128
• **Babylonian system:** A base-sixty positional system with just two symbols combined additively to represent the digits 1 through 59. Although a symbol to represent zero was not present originally, the system eventually acquired such a symbol as a placeholder but not as a number. It introduced the notion of place value.	129
• **Mayan system:** A positional system to base twenty for the first two positions, while, from the third position on, the positions are $18 \cdot 20, 18 \cdot 20^2, \ldots$. The digits are a rather simple set of well-chosen symbols.	130
• **Indo-Arabic, or decimal, system:** Also known as the Hindu-Arabic system, this is our present positional base-ten system.	132
• **Digits and expanded notation:** For example, in base ten, the 10 digits are $0, 1, 2, 3, \ldots, 9$, and $4073 = 4 \cdot 10^3 + 0 \cdot 10^2 + 7 \cdot 10^1 + 3 \cdot 10^0$ is in expanded form.	132
• **Physical models for positional systems:** Abacus; sticks in a bundle; Unifix™ cubes; units, strips, mats; and base-ten blocks.	133
DEFINITIONS	
• **Numerals** are the symbols for writing numbers.	126
• **Algorithms** are methods for calculating.	126
• A **numeration system** is any particular system of numerals and algorithms.	126
• **Place,** or **positional, value** is the notion that the position of a symbol in a numeral determines the value of the numeral.	129
FORMULA	
• The **decimal (base-ten) system:** The Indo-Arabic system is a positional numeration system based on the number 10 and is the modern system of numeration.	132

• **Roman system of numeration:** The Roman system consisted of a few letters and made use of a subtractive principle but calculations were cumbersome.	128
• **Mayan system of numeration:** The Mayan system consisted of two positional systems, one that used involved hieroglyphics and was based on the number 20, and another that was devised to facilitate computations related to the Mayan calendar.	130
NOTATION	
• **Expanded notation:** The amount each digit contributes to the number is the value of the digit times the value of the position the digit occupies in the representation. For example, $2572 = 2 \cdot 1000 + 5 \cdot 100 + 7 \cdot 10 + 2 \cdot 1$.	133

Section 3.2 Algorithms for Adding and Subtracting Whole Numbers	Page Reference
CONCEPTS	
• **Use of manipulative devices:** Units, strips, and mats; place-value cards; place-value diagrams as models.	138
• **Understanding of positional notation and place value:** Knowing that a number is given in terms of the sum of each of its decimal digits times a power of its base.	140
• **Instructional algorithms:** Greatly expanded algorithms that enhance student conceptual understanding of standard algorithms.	140
• **Exchanging:** Ten units can be exchanged for 1 ten, 10 tens can be exchanged for 1 hundred, and so on.	141
FORMULA	
• **Exchange** is used to simplify the tasks of addition, subtraction, multiplication, and division. It clarifies the idea of "carrying a 1." For example, in the addition algorithm and using base ten, one can exchange 10 units for 1 strip or 10 strips for 1 mat.	141
NOTATION	
• **Positional notation** is the notion that the position of a symbol in a numeral determines the value of the numeral. This is the foundation on which the operations of arithmetic are built.	138
PROCEDURES	
• **Algorithms for addition and subtraction** involve column-by-column addition or subtraction of integers in order to simplify the tasks of a complex addition or subtraction problem. There are many different correct procedures. Exchanging is used in both addition and subtraction.	139
• **Partial sum or instructional algorithm** uses simple, one-digit subtraction of numbers with corresponding place values to help complete subtractions of numbers with two or more digits.	140
• **Traditional, or final, algorithm** shows addition or subtraction in the usual, familiar column-by-column algorithm.	140
• **Partial-difference algorithm** uses the idea of "taking away" to help clarify the subtraction operation.	142

Section 3.3 Algorithms for Multiplication and Division of Whole Numbers	Page Reference
CONCEPTS	
• **Multiplication:** Various algorithms for multiplying in base-ten notation.	148
• **Multiplication in nondecimal bases such as base five or base six:** Similar to base-ten multiplication algorithms.	172

• **Division:** Various algorithms, including long division, for dividing in base-ten notation.	151
• **Short-division algorithm:** A neat algorithm for dividing by a single digit.	153
DEFINITION AND THEOREM	
• **The division algorithm:** If a and b are whole numbers with $b \neq 0$, there exists exactly one pair of whole numbers q and r with $0 \leq r < b$ such that $a = bq + r$. q is the **quotient** and r is the **remainder,** of a divided by b.	151
PROCEDURES	
• **Expanded notation:** The amount each digit contributes to the number is the value of the digit times the value of the position the digit occupies in the representation. For example, in base ten, $2572 = 2 \cdot 1000 + 5 \cdot 100 + 7 \cdot 10 + 2 \cdot 1$.	149
• **Partial-products, or instructional, algorithm** uses a step-by-step process of the multiplication of one-digit numbers in order to solve the multiplication of numbers with two or more digits through the use of place value.	149
• **Multiply with exchanging** uses the idea of trading 10 items of a certain place value with one in the next highest place value in order to complete the multiplication.	150
• **Scaffold or pyramid algorithm** is another version of the long-division algorithm where the quotient is placed above the division symbol.	152
• **Short-division algorithm** is a compressed version of the long-division algorithm, eliminating the process of writing down subtraction operations.	153
• **Lattice algorithm** is another method for solving complex multiplication problems using a rectangular array of boxes to organize simple multiplication operations.	155
• **Egyptian algorithm** is another method for solving complex multiplication problems involving the breakdown of one of the factors and the doubling of the other. It is also known as the duplation algorithm.	156
• **Russian peasant algorithm** is another method for solving complex multiplication problems, which involves successively dividing one of the factors by two, then crossing out even quotients and adding odd quotients.	156

Section 3.4 Mental Arithmetic and Estimation	Page Reference
CONCEPTS	
• **Easy combinations:** In doing a calculation, combine numbers that add to 10 or to a multiple of 10.	157
• **Adjustment:** Change numbers to 10 or multiples of 10, with subsequent adjustment to allow for those changes by adding zero in a clever way.	158
• **Work from left to right:** This approach is often easier than working from right to left.	159
• **Estimation:** If an exact answer is not required, do mental arithmetic with rounded numbers.	160
• **Rounding:** The 5-up rule, rounding to the leftmost digit.	162
PROCEDURES	
• **Adjustment:** Modify numbers by adding zero at the beginning of a calculation to minimize the mental effort required.	158
• **Working from left to right:** Work from left to right to reduce the amount one has to remember.	159

- **Front-end method of estimation:** Add the numbers under the assumption that all their digits except the first on the left are zero and then make some changes on the basis of the second digit. 162
- **Obtaining approximate answers by rounding:** Consider the place you want to round. If the digit to its right is 5 or more, then increase the original digit by 1 and replace all the digits to its right with zeros. If it is 4 or less, the digit remains unchanged and replace the digits to its right with zeros. You then carry out your operation with the rounded numbers. 163

STRATEGIES

- **Easy combinations:** Look for easy combinations, especially regrouping to find multiples of 10, when doing mental calculation. 157
- **Estimation, or computational estimation:** Estimate what you think the answer of the problem is likely to be to determine if your computation has had any errors. 160

BASIC

- **One-digit facts** are those basic addition and multiplication facts upon which all other numerical calculations and estimations depend. 157

Section 3.5 Nondecimal Positional Systems

CONCEPTS

- **Positional systems:** Notational systems like our present Indo-Arabic system but with bases other than just ten. 167
- **Converting from base ten to another base:** The method for converting from base-ten notation to notation in another base. 167
- **Converting from other bases to base ten:** The method for converting from notation in some other base to base-ten notation. 170
- Adding, subtracting and multiplying in base five or base six. 170

NOTATIONS

- **Base-five notation** is a positional numeration system whose base is the number 5. 167
- **Base-six notation** is a positional numeration system whose base is the number 6. 167

Chapter Review Exercises

Section 3.1

1. Write the Indo-Arabic equivalent of each of these numbers:

(a) 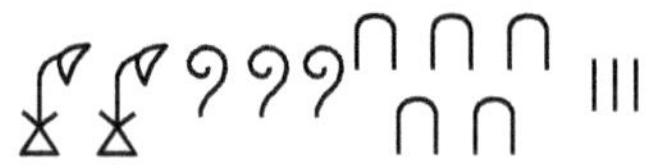

(b) 

(c) MCMXCVIII

2. Write 234,572 in Mayan notation.

3. Suppose you have 5 mats, 27 strips, and 32 units, for a total count of 802. Briefly describe the exchanges that must be made to represent this number with the smallest possible number of manipulative pieces. How many mats, strips, and units result?

Section 3.2

4. Sketch the solution to 47 + 25, using mats, strips, and units. Draw a square for each mat, a vertical line segment for each strip, and a dot for each unit.

5. Use the instructional algorithm for addition to perform the following additions:

(a) 42 + 54 (b) 47 + 35 (c) 59 + 63

6. Use sketches of place-value cards to illustrate each of these subtractions:

   (a) 487 − 275 (b) 547 − 152

Section 3.3

7. Perform these multiplications, using the instructional algorithm for multiplication:

   (a) 4 × 357 (b) 27 × 642

8. Use the scaffold method to perform each of these divisions:

   (a) $7\overline{)895}$ (b) $347\overline{)27{,}483}$

9. Use the short-division algorithm to perform each of these divisions:

   (a) $5\overline{)27{,}436}$ (b) $8\overline{)39{,}584}$

Section 3.4

10. Round 274,535

    (a) to the nearest one hundred thousand.

    (b) to the nearest ten thousand.

    (c) to the nearest thousand.

11. Rounding to the leftmost digit, compute approximations to the answers to each of these expressions:

    (a) 657 + 439 (b) 657 − 439

    (c) 657 · 439 (d) 1657 ÷ 23

Section 3.5

12. Perform the given calculations in base-five notation. Assume that the numerals are already written in base five.

    (a) $\begin{array}{r} 2433 \\ +\ 141 \\ \hline \end{array}$ (b) $\begin{array}{r} 2433 \\ -\ 141 \\ \hline \end{array}$ (c) $\begin{array}{r} 243 \\ \times\ 42 \\ \hline \end{array}$

13. Carry out each of the given multiplications in base five. Assume that the numerals are already written in base five.

    (a) 23 · 42 (b) 2413 · 332

4 Number Theory

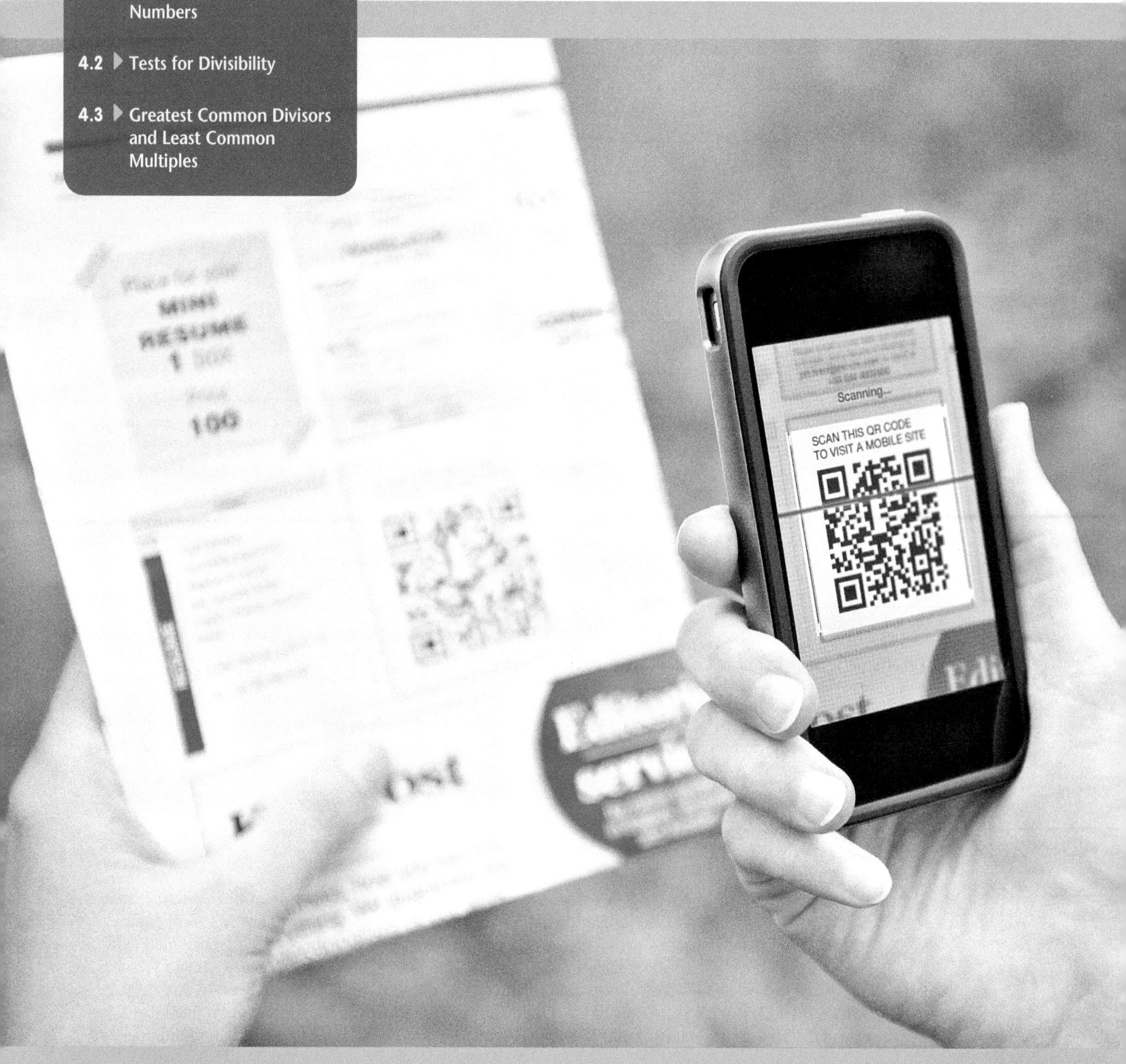

COOPERATIVE INVESTIGATION

Primes and Composites via Rectangular Arrays

Materials Needed

1. Twenty-five small cubes or number tiles for each student or small group of students.
2. One record sheet like this for each student:

Values of n	Dimensions of Rectangles	Number of Rectangles	Factors of n
1			
2			
3			
*4	$1 \times 4,\ 2 \times 2,\ 4 \times 1$	3	1, 2, 4
5			
6			
7			
8			
9			
10			
11			
12			
19			
20			
21			
22			
23			
24			
25			

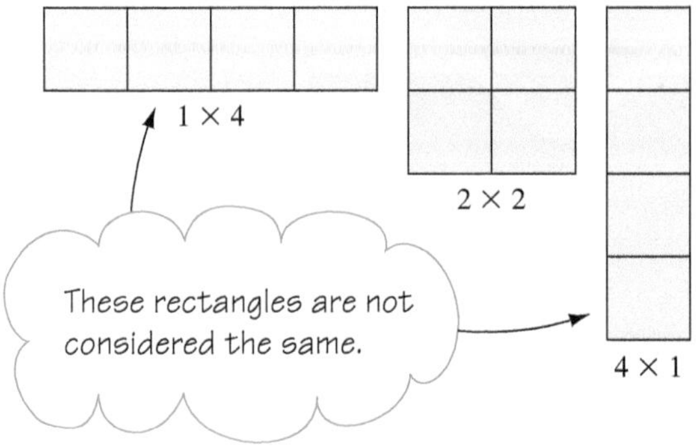

Directions

1. For each value of n, make up all possible rectangular arrays of n tiles. Then, on your record sheet, record the dimensions of each rectangle and the number of rectangles. For $n = 4$, we have the rectangles shown here, and we fill in the fourth row of the record sheet as shown.
2. In Chapter 2, we used diagrams like this to illustrate the product $3 \cdot 5 = 15$:

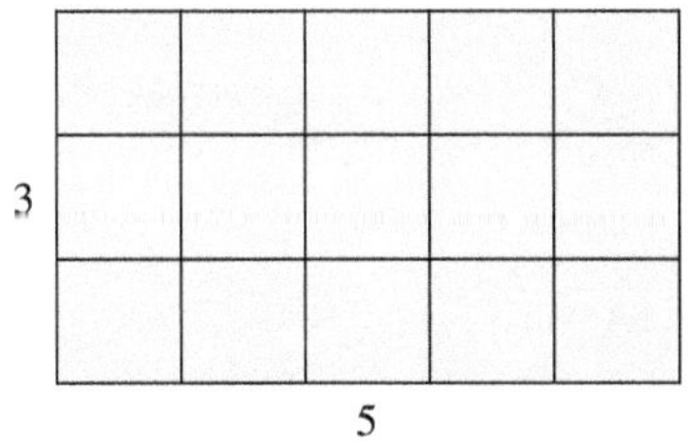

We call 3 and 5 *factors* of 15. Label the last column of your record sheet "Factors of n," and list the factors in increasing order for each value of n. The row for $n = 4$ is done for you.

3. Natural numbers that have exactly two factors are called *prime numbers.* Place a P along the left side of your record sheet next to each prime number.
4. Numbers with more than two factors are called *composite numbers.* Place a C along the left-hand side of your record sheet next to each composite number.
5. Is 1 a prime or composite number or neither? Why?
6. Put an asterisk just to the left of each n that has an odd number of factors. Do these numbers seem to share some other property? State a guess (conjecture) about natural numbers with an odd number of factors.
7. Carefully considering all the data on your record sheet, see if you can guess a number that has just seven factors. Check to see that your guess is correct.

CHAPTER PREVIEW

The whole numbers $W = \{0, 1, 2, \ldots\}$ were introduced in Chapter 2, where we examined the basic operations of addition and multiplication together with their respective inverse operations of subtraction and division. Then, in Chapter 3, we examined how a whole number can be represented symbolically within a numeration system, and we used the place-value numeration system to devise algorithms to calculate sums, differences, products, and quotients of whole numbers. In this chapter, we examine additional properties of the whole numbers that revolve around two interrelated themes: the divisibility of one natural number by another and the representation of a given natural number as a product of other natural numbers. In particular, you will discover how to answer the two questions in the accompanying excerpt from the NCTM *Principles and Standards* shown above.

The study of the properties of whole numbers is known as **number theory.**

KEY IDEAS

- Divisors, factors, and multiples of whole numbers
- The classification of a natural number $n \geq 2$ as a prime or composite number
- The representation of a composite number as a product of prime numbers
- How to test a natural number for primality and divisibility
- Given two (or more) natural numbers, how to find the largest natural number that divides each of the given numbers, a number called the **greatest common divisor.** Since a divisor of a number is also a factor of the number, the greatest common divisor is also known as the **greatest common factor.**
- Given two (or more) natural numbers, how to find the smallest natural number that can be divided by each of the given numbers, a number called the **least common multiple.**

FROM THE NCTM PRINCIPLES AND STANDARDS

Throughout their study of numbers, students in grades 3–5 should identify classes of numbers and examine their properties. For example, integers that are divisible by 2 are called *even numbers* and numbers that are produced by multiplying a number by itself are called *square numbers.* Students should recognize that different types of numbers have particular characteristics; for example, square numbers have an odd number of factors and prime numbers have only two factors.

Students can also work with whole numbers in their study of number theory. Tasks, such as the following, involving factors, multiples, prime numbers, and divisibility, can afford opportunities for problem solving and reasoning.

4.1 Divisibility of Natural Numbers

Divides, Divisors, Factors, Multiples

In Chapter 3, we considered the division algorithm. If a and b are whole numbers with b not zero, then, when we divide a by b, we obtain a unique quotient q and remainder r such that $a = bq + r$ and $0 \leq r < b$. Thus, the division

$$\begin{array}{r} 4 \text{ R } 2 \\ 3\overline{)14} \end{array}$$

is equivalent to the equation

$$14 = 3 \cdot 4 + 2.$$

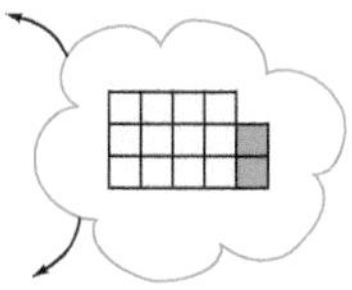

Of special interest in this chapter is the case when the remainder r is zero. Then $a = bq$, and we say that ***b* divides *a* evenly** or, more simply, ***b* divides *a*.** This relationship is expressed in other terminology as indicated here:

The words **factor** and **divisor** have the same meaning.

DEFINITION Divides, Factor, Divisor, Multiple

If a and b are whole numbers with $b \neq 0$ and there is a whole number q such that $a = bq$, we say that b **divides** a. We also say that b is a **factor** of a or a **divisor** of a and that a is a **multiple** of b. If b divides a and b is less than a, it is called a **proper divisor** of a.

A useful model for the ideas "b divides a" and "a is a multiple of b" has already been provided by the array models for multiplication of natural numbers in Chapter 2. Thus, the 5-by-7 rectangular array in Figure 4.1 illustrates the fact that 5 and 7 are both divisors of 35 and that 35 is a multiple of 5 and also of 7. We also say that 5 and 7 are factors of 35.

Figure 4.1
Array model showing that 5 and 7 are factors of 35 and that 35 is a multiple of both 5 and 7

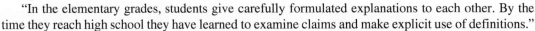

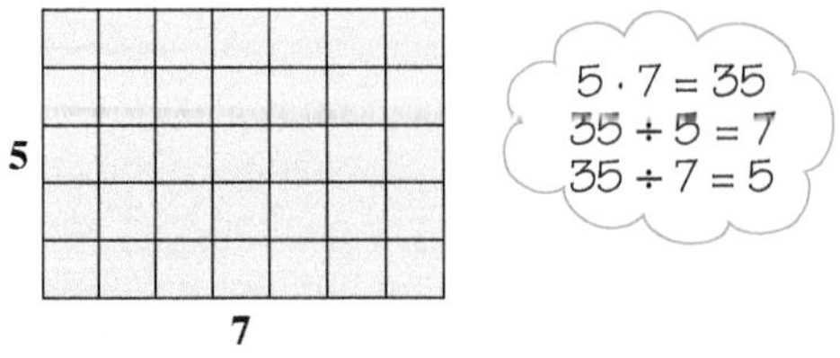

The whole numbers that are divisible by 2 or, equivalently, that are multiples of 2 are known as the **even** whole numbers. That is, a number is even when 2 is one of its factors. Divisibility allows us to make the following precise definition:

DEFINITION Even and Odd Whole Numbers
A whole number a is **even** precisely when it is divisible by 2. That is, a is even if and only if 2 is one of its factors. A whole number that is not even is called an **odd** whole number.

The sixth Standard for Mathematical Practice (SMP 6) from the **Common Core** calls for an attention to precision. This not only includes the precision of a numerical answer but far more as well. For example, the standard states that

SMP 6 Attend to precision.

"In the elementary grades, students give carefully formulated explanations to each other. By the time they reach high school they have learned to examine claims and make explicit use of definitions."

It is therefore important that the teacher understand and be able to convey the exact meaning of a mathematical term. Imagine how you might respond to a student who claims that 6 is both even—since it is divisible by 2—and odd, since it is divisible by the odd number 3.

Often the meaning of a definition becomes clearer when some of the consequences of a definition are examined. The example that follows uses the definition of even and odd to derive a very useful algebraic characterization of even and odd whole numbers.

EXAMPLE 4.1 Representing Even and Odd Whole Numbers

(a) Explain why a whole number a is even if and only if there is a whole number k such that

$$a = 2k.$$

(b) Explain why a whole number b is odd if and only if there is a whole number j such that

$$b = 2j + 1.$$

(c) Let c be a whole number for which c^2 is even. Explain why c is even.

Solution

(a) By definition, a is even if and only if a is divisible by 2. By the definition of divisibility by 2, this is the case precisely when there is a whole number k for which $a = 2k$. Thus, a is even when it can be written as the product of 2 and another factor k.

(b) The whole number b is odd precisely when 2 is *not* a divisor of b. Therefore, b is odd if and only if it leaves a nonzero remainder r when divided by 2. That is, $b = 2j + r$, where j is some whole number and $0 < r < 2$. Evidently, $r = 1$ and we see that b is odd if and only if there is a representation of the form $b = 2j + 1$.

(c) Suppose, to the contrary, that c is odd. By part (b), there is a whole number j for which $c = 2j + 1$. But then $c^2 = 4j^2 + 4j + 1 = 2(2j^2 + 2j) + 1$, so $c^2 = 2i + 1$ for the whole number $i = 2j^2 + 2j$. Again by part (b), this means that c^2 is odd. This is a contradiction, so c must be even.

Example 4.1 can be used in two ways. To prove that a given integer is even or odd, investigate whether it can be written in the form $2k$ for some whole number k, or if it can be written as $2j + 1$ for some whole number j. In the opposite direction, if you are investigating a property involving a general even or odd number, it is often helpful to know that it can be written as either $2k$ or $2j + 1$.

EXAMPLE 4.2

The Divisors, or Factors, of 6

List all the factors of 6.

Solution

We must find all the whole numbers b for which there is another whole number q for which $6 = bq$. This is equivalent to finding all rectangular arrays with area 6 and sides of length b and q.

We find that there are only four such rectangles, as shown below, and accordingly, the desired factors of 6 are 1, 2, 3, and 6. We say that 1, 2, 3, and 6 are factors, or divisors, of 6 and that 6 is a multiple of each of 1, 2, 3, and 6. It is also correct to say that 1 divides 6, 2 divides 6, 3 divides 6, and 6 divides 6. The proper divisors of 6 are 1, 2, and 3.

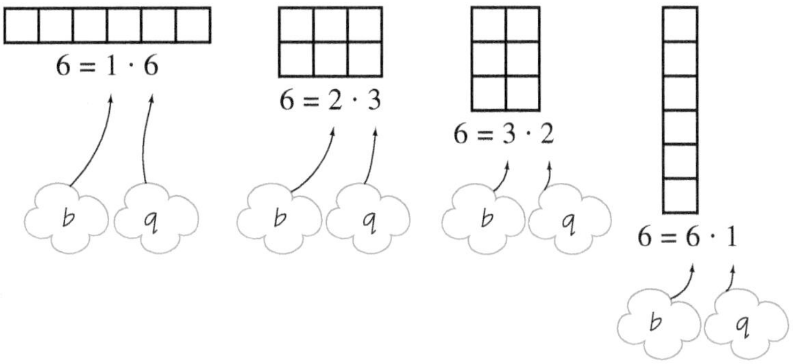

Prime and Composite Numbers

Since $1 \cdot a = a$, 1 and a are *always* factors of a for every natural number a. For this reason, 1 and a are often called trivial factors of a, and $1 \cdot a$ and $a \cdot 1$ are trivial factorings. Some numbers, such as 2, 3, 5, and 7, have only trivial factorings. Other numbers, such as 6, have nontrivial factorings. The number 1 stands alone, since it has only one factor: 1 itself. All this is summarized in the following definition:

DEFINITION Primes, Composite Numbers, and Units

A natural number that possesses exactly two different factors—itself and 1—is called a **prime number.** A natural number that possesses more than two different factors is called a **composite number.** The number 1 is called a **unit;** it is neither prime nor composite.

The primes are sometimes called the building blocks of the natural numbers, since every natural number other than 1 is either a prime or a product of primes. For example, consider the number 180. This number is composite, since, for example, we can write $180 = 10 \cdot 18$. Moreover, 10 and 18 are both composite, since $10 = 2 \cdot 5$ and $18 = 2 \cdot 9$. Now, 2 and 5 are both primes and cannot be factored further. But $9 = 3 \cdot 3$, and 3 is a prime. We simply continue to factor a composite number into smaller and smaller factors and stop when we can proceed no further—that is, when the factors are all primes. In the case of the number 180, we see that

$$\begin{aligned} 180 &= 10 \cdot 18 \\ &= 2 \cdot 5 \cdot 2 \cdot 9 \\ &= 2 \cdot 5 \cdot 2 \cdot 3 \cdot 3, \end{aligned}$$

and this is a product of primes.

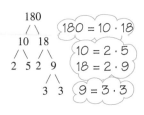

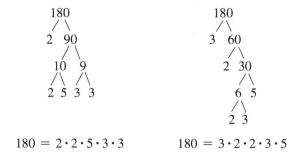

Highlight from History: Sophie Germain (1776–1831)

Sophie Germain grew up in a time of social, political, and economic upheaval in France. To shield Germain from the violence in the streets of Paris during the time of the fall of the Bastille, her wealthy parents confined their 13-year-old daughter to the family's library. There, she chanced upon J. E. Montucla's *History of Mathematics,* which recounts the legend of Archimedes' death. The story tells how a Carthaginian soldier, heedless of orders to spare the renowned mathematician, killed the unsuspecting Archimedes, who remained absorbed in a geometry problem. Germain wished to explore for herself a subject of such compelling interest.

Germain's family initially resisted her determination to study mathematics, but eventually they gave her the freedom to follow her intellectual instincts. Since women were not permitted to enroll in the École Polytechnique, which opened in Paris in 1794, she resorted to collecting lecture notes from various professors at the university. Her lack of a formal mathematical education was compensated for by her courage to overcome strenuous challenges.

Germain's early research was in number theory. She corresponded regularly with the great Carl Friedrich Gauss, who gave her work high praise. At the turn of the century, her attention turned increasingly to the mathematical theory of vibrating elastic surfaces. Her prizewinning paper on vibrating elastic plates in 1816 placed her in the ranks of the most celebrated mathematicians of the time. Gauss recommended that she be awarded an honorary doctorate from the University of Göttingen, but unfortunately Sophie Germain's death came before the awarding of the degree.

```
   180
   / \      (180 = 10 · 18)
  10  18
 / \  / \   (10 = 2 · 5)
2  5 2  9   (18 = 2 · 9)
       / \
      3   3 (9 = 3 · 3)
```

180 = 2 · 5 · 2 · 3 · 3

Figure 4.2
A factor tree for 180

A convenient way of organizing this work is to develop a **factor tree,** as shown in Figure 4.2, to keep track of each step in the process. But there are other ways to factor 180, as these factor trees show:

```
     180                180
     / \                / \
    2   90             3   60
        / \                / \
      10    9             2   30
      / \  / \                / \
     2  5  3  3              6    5
                            / \
                           2   3
```

$180 = 2 \cdot 2 \cdot 5 \cdot 3 \cdot 3$ $\quad$ $180 = 3 \cdot 2 \cdot 2 \cdot 3 \cdot 5$

Another method for finding the prime factors of a number is to use prime divisors and short division until you arrive at a quotient that is a prime. For 180, we have these three sequences of divisions, each read from the bottom upward:

```
    5          3          5
3)15        3)9       2)10
3)45       2)18       2)20
2)90       2)36       3)60
2)180      5)180      3)180
```

$180 = 2 \cdot 2 \cdot 3 \cdot 3 \cdot 5$ $\quad$ $180 = 5 \cdot 2 \cdot 2 \cdot 3 \cdot 3$ $\quad$ $180 = 3 \cdot 3 \cdot 2 \cdot 2 \cdot 5$

This procedure is sometimes called **stacked short division.**

Both the factor tree and the stacked short-division method leave a written record of the calculations, which can often be done by mental arithmetic.

The most important thing about all the factorings shown for the number 180 is that, no matter what method is used and no matter how it is carried out, the *same prime factors always result.* That this is always the case is stated here without proof:

THEOREM The Fundamental Theorem of Arithmetic
Every natural number greater than 1 is a prime or can be expressed as a product of primes in one, and only one, way, apart from the order of the prime factors.

The fundamental theorem of arithmetic tells us that every natural number larger than 1 can be written as a product of prime numbers. This product is known as the **prime factorization** of the natural number. It makes no difference in which order the prime factors appear, but it is generally best to list the primes from smallest to largest. Thus, the prime factorization of 180 would be written $180 = 2 \cdot 2 \cdot 3 \cdot 3 \cdot 5$. Since 1 is not a prime number, no prime factorization includes the factor 1. The prime factorization of a prime number is just the prime itself.

The preceding theorem is the reason that we do not think of 1 as either a prime or a composite number. If 1 were considered a prime, the theorem would not be true. For example, we could multiply 180 by any number of factors of 1, and the prime factorization of 180 would not be unique.

$$\begin{aligned} 180 &= 2 \cdot 2 \cdot 3 \cdot 3 \cdot 5 \\ &= 1 \cdot 2 \cdot 2 \cdot 3 \cdot 3 \cdot 5 \\ &= 1 \cdot 1 \cdot 2 \cdot 2 \cdot 3 \cdot 3 \cdot 5 \\ &= \cdots \end{aligned}$$

EXAMPLE 4.3

The Prime Factorization of 600

Determine the prime factorization of 600.

Solution

Using a factor tree:

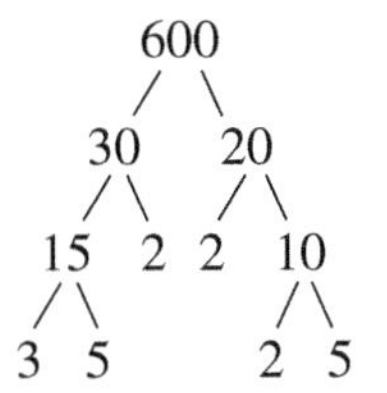

Using stacked short division:

```
     5
  5)25
  2)50
  2)100
  3)300
  2)600
```

The prime factorization is $600 = 2 \cdot 2 \cdot 2 \cdot 3 \cdot 5 \cdot 5$.

In Example 4.3, we can also write 600 as a product of prime factors by collecting like primes together and writing their products as powers; that is, we write $600 = 2^3 \cdot 3^1 \cdot 5^2$. Since this could be done for any natural number, it is useful to give an alternative version of the **fundamental theorem of arithmetic.**

THEOREM Prime-Power Form of the Fundamental Theorem of Arithmetic

Every natural number n greater than 1 is a power of a prime or can be expressed as a product of powers of different primes in one, and only one, way, apart from order. This representation is called the **prime-power factorization of *n*.**

EXAMPLE 4.4

The Prime-Power Factorization of 675

Determine the prime-power factorization of 675.

Solution

```
     3
  3)9
  3)27
  5)135
  5)675
```

Using the stacked short-division method, we see that $675 = 3^3 \cdot 5^2$.

INTO THE CLASSROOM

Marianne Strayton Discusses Teaching Prime and Composite Numbers with *Factor Captor*

Factor Captor helps my students recognize prime and composite numbers through the context of a competitive or collaborative game. The object is to capture numbers that will have the greatest sum at the end of the game. On a board numbered 1 to 24 (or greater), one player takes a number and the other player can take all the numbers that are factors of player 1's chosen number. For example, if a player takes 24, the opponent can take 1, 2, 3, 4, 6, 8, and 12. If a factor is overlooked, the first player can "steal" it. As players take turns, numbers are taken off the board, and players continue to add up their points. Students quickly realize that taking prime numbers is best, since their opponent can get at most 1 point. They start to recognize they should avoid composite numbers that have many factors, like 24, and instead choose numbers that have only a few factors left on the board.

The Divisors of a Natural Number

The prime-power factorization of a number makes it easy to determine the divisors of the number. As an example, from Example 4.4 we know that $675 = 3^3 \cdot 5^2$. If r is a divisor of say, 675, with a quotient s, then

$$rs = 675 = 3^3 \cdot 5^2.$$

Since $3^0 = 1, 3^1 = 3, 3^2 = 9$, and $3^3 = 27$ are the only powers of 3 that divide 675, they are also the only powers of 3 that divide r. Similarly, $5^0 = 1, 5^1 = 5$, and $5^2 = 25$ are the only powers of 5 that divide r. Altogether, we conclude that any divisor r of 675 is the product of one of the four numbers $1, 3^3, 3^2$, and 3^3 multiplied by one of the three numbers $5^0, 5^1$, and 5^2. Moreover, any of these products is a divisor of 675. For example, the product $3^2 \cdot 5^1 = 45 = d$ is a divisor of 675, with the quotient $3^1 \cdot 5^1 = 15 = q$. All of the divisors of 675 can be written down in the following list of $4 \cdot 3 = 12$ numbers:

(3 + 1) (2 + 1) = 12

$1 = 3^0 \cdot 5^0$	$5 = 3^0 \cdot 5^1$	$25 = 3^0 \cdot 5^2$
$3 = 3^1 \cdot 5^0$	$15 = 3^1 \cdot 5^1$	$75 = 3^1 \cdot 5^2$
$9 = 3^2 \cdot 5^0$	$45 = 3^2 \cdot 5^1$	$225 = 3^2 \cdot 5^2$
$27 = 3^3 \cdot 5^0$	$135 = 3^3 \cdot 5^1$	$675 = 3^3 \cdot 5^2$

The argument just given extends to the case where there are more than two prime factors. For example, since $140 = 2^2 \cdot 5^1 \cdot 7^1$, it follows that the divisors of 140 are all numbers that can be written in the form $2^a \cdot 5^b \cdot 7^c$, where $0 \le a \le 2$, $0 \le b \le 1$, and $0 \le c \le 1$. Altogether, there are $3 \cdot 2 \cdot 2 = 12$ different divisors of 140.

The general case is described in this theorem:

THEOREM The Divisors of a Natural Number

Let n be a natural number whose prime-power factorization has the form $n = p^a q^b \cdots v^c$, where the primes $p, q, \ldots, v$ are raised to the respective powers $a, b, \ldots, c$. Then a number r is a divisor of n if and only if r has a prime factorization of the form $r = p^j q^k \cdots v^l$, where $0 \le j \le a$, $0 \le k \le b, \ldots, 0 \le l \le c$. The number of divisors of n is given by $N = (a + 1)(b + 1) \cdots (c + 1)$.

EXAMPLE 4.5

The Divisors of 600

Count the number of divisors of 600, and show them in a list.

Solution

Since $600 = 2^3 \cdot 3^1 \cdot 5^2$, the divisors must be all the numbers of the form $2^r \cdot 3^s \cdot 5^t$ with $0 \le r \le 3$, $0 \le s \le 1$, and $0 \le t \le 2$. There are $3 + 1$ choices for r, $1 + 1$ choices for s, and $2 + 1$ choices for t to show that there are $4 \cdot 2 \cdot 3 = 24$ divisors of 600. We make a systematic list of the divisors as follows:

$1 = 2^0 \cdot 3^0 \cdot 5^0$	$5 = 2^0 \cdot 3^0 \cdot 5^1$	$25 = 2^0 \cdot 3^0 \cdot 5^2$
$2 = 2^1 \cdot 3^0 \cdot 5^0$	$10 = 2^1 \cdot 3^0 \cdot 5^1$	$50 = 2^1 \cdot 3^0 \cdot 5^2$
$4 = 2^2 \cdot 3^0 \cdot 5^0$	$20 = 2^2 \cdot 3^0 \cdot 5^1$	$100 = 2^2 \cdot 3^0 \cdot 5^2$
$8 = 2^3 \cdot 3^0 \cdot 5^0$	$40 = 2^3 \cdot 3^0 \cdot 5^1$	$200 = 2^3 \cdot 3^0 \cdot 5^2$
$3 = 2^0 \cdot 3^1 \cdot 5^0$	$15 = 2^0 \cdot 3^1 \cdot 5^1$	$75 = 2^0 \cdot 3^1 \cdot 5^2$
$6 = 2^1 \cdot 3^1 \cdot 5^0$	$30 = 2^1 \cdot 3^1 \cdot 5^1$	$150 = 2^1 \cdot 3^1 \cdot 5^2$
$12 = 2^2 \cdot 3^1 \cdot 5^0$	$60 = 2^2 \cdot 3^1 \cdot 5^1$	$300 = 2^2 \cdot 3^1 \cdot 5^2$
$24 = 2^3 \cdot 3^1 \cdot 5^0$	$120 = 2^3 \cdot 3^1 \cdot 5^1$	$600 = 2^3 \cdot 3^1 \cdot 5^2$

Two Questions about Primes

Since we have just seen how important the primes are as "building blocks" for the natural numbers, it is reasonable to ask the following two questions:

- Is there a largest prime number, or is the set of primes an infinite set?
- How does one determine whether a given natural number is a prime?

We answer these questions in the order asked.

There Are Infinitely Many Primes

The answer to the first question is that there are infinitely many primes, a result shown by the Greek mathematician Euclid in ancient times. To see this, we describe a step-by-step process that determines at least one new prime at each step. Since the process can be continued without end, it follows that the set of primes is infinite.

THEOREM The Number of Primes
There are infinitely many primes.

PROOF (Euclid's proof by contradiction). Suppose that there is a finite number of primes, say, $2, 3, 5, 7, 11, 13, \ldots, p$. Here, p is the largest prime, so any natural number larger than p must be a composite number. In particular, the number $n = (2 \cdot 3 \cdot 5 \cdot 7 \cdot 11 \cdot 13 \cdot \cdots \cdot p) + 1$ is larger than p, so n must be a composite number. Now consider any prime number P that is included in the prime factorization of n. Then $P \ne 2$, since $n = 2 \cdot (3 \cdot 5 \cdot 7 \cdot 11 \cdot 13 \cdot \cdots \cdot p) + 1$, which shows n divided by 2 leaves a remainder of 1. In the same way, $P \ne 3$, since $n = 3 \cdot (2 \cdot 5 \cdot 7 \cdot 11 \cdot 13 \cdot \cdots \cdot p) + 1$, which shows n divided by 3 also leaves a remainder of 1. Indeed, if we divide n by any of the primes $2, 3, 5, 7, \ldots, p$, there is always a remainder of 1. This means the prime P is not in what we assumed was the complete list of all primes, and we have arrived at a contradiction.

Euclid's theorem shows that arbitrarily large prime numbers exist, but it is not too helpful to construct an explicitly given large number that is known to be prime. There is an ongoing search for very large prime numbers, with the current record being $p = 2^{57{,}885{,}161} - 1$. When p is written as a decimal numeral, it contains 17,425,170 digits.

Determining Whether a Given Natural Number Is Prime

We now answer the second question posed about primes: How can one determine whether a given natural number n is prime? Assume that n is a composite natural number, so that it has at least two prime factors. If p and q are the two smallest prime factors of n, with $p \le q$, then $p^2 \le p \cdot q \le n$. That is, the smallest prime factor p of n satisfies $p^2 \le n$. Taking square roots gives us this result.

THEOREM Prime Divisors of n

If n is composite, then n has a prime factor p for which $p \le \sqrt{n}$ (i.e., $p^2 \le n$).

Here is how the theorem can be interpreted to form a test for primality.

TEST FOR PRIMALITY

Let n be any natural number, with $n > 1$. Compute $\sqrt{n}$ and make a list of the prime numbers $2, 3, 5, \ldots, p$ for which $p \le \sqrt{n}$. If none of the primes in the list divide n, then n is a prime number.

Let's now use this test to make a list of prime numbers no larger than 100. Start by imagining a list of the numbers 2 through 100 as shown below but with no numbers yet circled or crossed off. We know that 2 is a prime number, so it is circled and its multiples 4, 6, 8, . . . are composite numbers that are crossed off the list. The next number not yet crossed off is 3. It is not divisible by 2 so 3 is a prime number that is circled and its multiples are the composite numbers 6, 9, 12, . . . that are crossed off the list, perhaps for the second time. The next number not yet crossed off is 5. Since neither 2 nor 3 is a factor of 5, it is a prime number that is circled and its multiples 10, 15, 20, . . . are composite numbers that are crossed off. In the same way, the next number not crossed off is 7. None of 2, 3, or 5 is a factor of 7, so 7 is a prime number. It is circled, and its multiples 14, 21, 28, . . . , 91, and 98 are crossed off the list. Since $\sqrt{100} = 10$, any composite number no larger than 100 has a prime factor no larger than 10. But we have already found all of these primes—namely, 2, 3, 5, and 7. This means all of the composite numbers through 100 have already been crossed off, and those that remain can be circled since they must be prime numbers.

This method of determining a list of primes was devised by the ancient Greek mathematician Eratosthenes (276–195 B.C.). The method is known as the **sieve of Eratosthenes**, as shown in Figure 4.3.

Figure 4.3 The sieve of Eratosthenes for $n = 100$

	(2)	(3)	~~4~~	(5)	~~6~~	(7)	~~8~~	~~9~~	~~10~~
(11)	~~12~~	(13)	~~14~~	~~15~~	~~16~~	(17)	~~18~~	(19)	~~20~~
~~21~~	~~22~~	(23)	~~24~~	~~25~~	~~26~~	~~27~~	~~28~~	(29)	~~30~~
(31)	~~32~~	~~33~~	~~34~~	~~35~~	~~36~~	(37)	~~38~~	~~39~~	~~40~~
(41)	~~42~~	(43)	~~44~~	~~45~~	~~46~~	(47)	~~48~~	~~49~~	~~50~~
~~51~~	~~52~~	(53)	~~54~~	~~55~~	~~56~~	~~57~~	~~58~~	(59)	~~60~~
(61)	~~62~~	~~63~~	~~64~~	~~65~~	~~66~~	(67)	~~68~~	~~69~~	~~70~~
(71)	~~72~~	(73)	~~74~~	~~75~~	~~76~~	~~77~~	~~78~~	(79)	~~80~~
~~81~~	~~82~~	(83)	~~84~~	~~85~~	~~86~~	~~87~~	~~88~~	(89)	~~90~~
~~91~~	~~92~~	~~93~~	~~94~~	~~95~~	~~96~~	(97)	~~98~~	~~99~~	~~100~~

EXAMPLE 4.6 Determining the Primality of 439

Show that 439 is a prime.

Solution

Understand the Problem

We must show that 439 has no factors other than itself and 1.

Devise a Plan

By the test for primality, 439 is prime if it is not divisible by any prime.

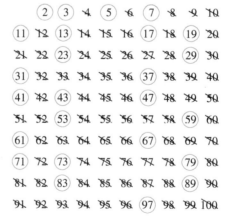

Carry Out the Plan

First, compute the square root: $\sqrt{439} \doteq 21$. From the sieve of Eratosthenes, we see that 2, 3, 5, 7, 11, 13, 17, and 19 are all of the prime numbers through 21. Putting 439 in the memory of our calculator and using the memory recall button [MR] repeatedly, we easily complete the desired divisions and determine that none of these primes divides 439. Thus, 439 is a prime by the test for primality.

Look Back

There was no need to check for divisibility by all of the prime numbers through 439. We only had to check that the eight prime numbers 2, 3, 5, . . . , 19 were not divisors of 439.

COOPERATIVE INVESTIGATION

Representing Integers as Sums

Materials Needed

One copy of the sheet outlined next for each student.

Directions

Divide the class into groups of three or four students each and have them carry out the directions on the handout.

Investigation 1. Representing Natural Numbers as a Sum of Consecutive Integers

Some numbers can be written as a sum of two or more consecutive positive integers. For example, $3 = 1 + 2$ and $15 = 4 + 5 + 6$ (alternatively, $15 = 1 + 2 + 3 + 4 + 5$). By contrast, 2 cannot be written as the sum of two or more consecutive integers. Fill in the rest of this table, and then conjecture what pattern is revealed in your table.

n		*n*		*n*		*n*	
1	X	11		21		31	
2	X	12		22		32	
3	1 + 2	13		23		33	
4		14		24		34	
5		15		25		35	
6		16		26		36	
7		17		27		37	
8		18		28		38	
9		19		29		39	
10		20		30		40	

Investigation 2. Representing Numbers with Sums of Consecutive Evens or Odds

Some numbers can be written as a sum of two or more consecutive even or consecutive odd positive integers. For example, $6 = 2 + 4$, and $45 = 5 + 7 + 9 + 11 + 13$ (alternatively, $45 = 13 + 15 + 17$). On the other hand, 2 cannot be written as the sum of two or more consecutive integers that are all even or all odd. Fill in this table and then conjecture what pattern is revealed in your table:

n		*n*		*n*		*n*	
1	X	11		21		31	
2	X	12		22		32	
3		13		23		33	
4		14		24		34	
5		15		25		35	
6	2 + 4	16		26		36	
7		17		27		37	
8		18		28		38	
9		19		29		39	
10		20		30		40	

4.1 Problem Set

Understanding Concepts

1. Draw array diagrams to show that
   (a) 4 is a factor of 36. (b) 6 is a factor of 36.

2. Draw array diagrams to illustrate all the ways that 35 can be written as a product of two factors, taking order into account; that is, think of $1 \cdot 35$ as different from $35 \cdot 1$.

3. (a) List the first 10 positive multiples of 8, starting with $1 \cdot 8 = 8$.
   (b) List the first 10 positive multiples of 6, starting with $1 \cdot 6 = 6$.
   (c) Use parts (a) and (b) to determine the least natural number that is a multiple of both 8 and 6.

4. (a) Make lists of the first eight multiples of 15 and 18.
   (b) What is the least multiple of 15 and 18 that is common to both lists?

5. Suppose that the product of two natural numbers is odd. Use the representation given in Example 4.1 to explain why both factors must be odd.

6. Use the representation of an even number given in Example 4.1 to prove that the square of any even natural number is divisible by 4.

7. Use the representation of an odd number given in Example 4.1 to prove that the square of any odd natural number leaves a remainder of 1 when divided by 4.

8. Use the representations shown in Example 4.1 to show why the sum of the squares of two consecutive whole numbers is 1 larger than a multiple of 4.

9. Complete this table of all factors of 18 and their corresponding quotients:

Factors of 18	1	2				
Corresponding quotients	18	9				

10. Construct factor trees for each of these numbers:
    (a) 72 (b) 126 (c) 264 (d) 550

11. Use stacked short division to find all the prime factors of each of these numbers:
    (a) 700 (b) 198 (c) 450 (d) 528

12. (a) List all the divisors (factors) of 48.
    (b) List all the divisors (factors) of 54.
    (c) Use parts (a) and (b) to find the largest common divisor of 48 and 54.

13. (a) Determine the prime-power factorizations of both 136 and 102.
    (b) Determine the set of all divisors (factors) of 136.
    (c) Determine the set of all divisors of 102.
    (d) Determine the greatest divisor of both 136 and 102.

14. Determine the prime-power factorization of each of these numbers:
    (a) 48 (b) 108 (c) 2250 (d) 24,750

15. Give the prime-power factorization of each of these natural numbers.
    (a) 91 (b) 6125 (c) 23,000

16. Give the prime-power factorization of each of these natural numbers.
    (a) 1500 (b) 4840 (c) 9999

17. Let $a = 2^3 \cdot 3^1 \cdot 7^2$.
    (a) Is $2^2 \cdot 7^1 = 28$ a factor of a? Why or why not?
    (b) Is $2^1 \cdot 3^2 \cdot 7^1 = 126$ a factor of a? Why or why not?
    (c) One factor of a is $b = 2^2 \cdot 3^1$. What is the quotient when a is divided by b?
    (d) How many different factors does a possess?
    (e) Make an orderly list of all of the factors of a.

18. To determine whether 599 is a prime, which primes must you check as possible divisors?

19. Use the test for primality to determine which of the following natural numbers are prime or composite. Be sure to show all of the steps you have taken to apply the test.
    (a) 271 (b) 319 (c) 731 (d) 1801

20. Classify each of these natural numbers as prime or composite. Show how you have applied the test for primality.
    (a) 361 (b) 1013 (c) 4049 (d) 6083

Into the Classroom

21. **(Writing)** Carefully explain how you would convince your class that it is sometimes the case that one or more prime factors of a composite number n are greater than $\sqrt{n}$ even though it is always the case that at least one prime factor of n must be less than $\sqrt{n}$.

22. **(Writing)** Students often think that if a product ab is divisible by a natural number n, then n must be a factor of either a or b. Carefully explain, as if to a student, why this may not be so.

23. **(Writing)** Give a brief but careful explanation of this fact: If a prime p divides a product ab of two natural numbers a and b, then p must be a factor of at least one of a or b.

24. **(Writing)** Give brief but careful explanations, as if to a class, why each of these statements is true or false.
    (a) n divides 0 for every natural number n.
    (b) 0 divides n for every natural number n.

(c) 1 divides n for every natural number n.

(d) n divides n for every natural number n.

(e) 0 divides 0.

Responding to Students

25. Emma has checked that 90 is divisible by the primes 2, 3, and 5 and therefore thinks that the prime factorization of 90 is $2 \cdot 3 \cdot 5$. What is Emma forgetting?

26. Respond to the child who claims that 6 is both even and odd since it is divisible by both 2 and the odd number 3.

27. A fourth grader claims that "Zero is nothing, so zero is neither even nor odd." What's your response?

Thinking Critically

28. (a) Any whole number n divisible by 2 has the representation $n = 2k$ for some whole number k. How can a number n divisible by 3 be represented?

(b) Use part (a) to prove that the sum of any three consecutive whole numbers is divisible by 3.

(c) Is the sum of the squares of any three consecutive whole numbers ever divisible by 3?

29. (a) List all the distinct factors of $9 = 3^2$.

(b) Find three natural numbers (other than 9) that have precisely three distinct factors.

(c) Find three natural numbers that have precisely four distinct factors.

30. A prime number p such that $2p + 1$ is also prime is called a **Germain prime,** after the eminent 19th-century German mathematician Sophie Germain (see the Highlight from History in this section). Which of 11, 13, 97, 241, and 359 are Germain primes?

31. Assume that p and q are different prime numbers and that n is a natural number. Argue briefly that if p divides n and q divides n, then pq divides n.

32. Draw a square measuring 10 centimeters on a side. Draw vertical and horizontal line segments dividing the square into rectangles of areas 12, 18, 28, and 42 square centimeters, respectively. Where should A, B, C, and D be located?

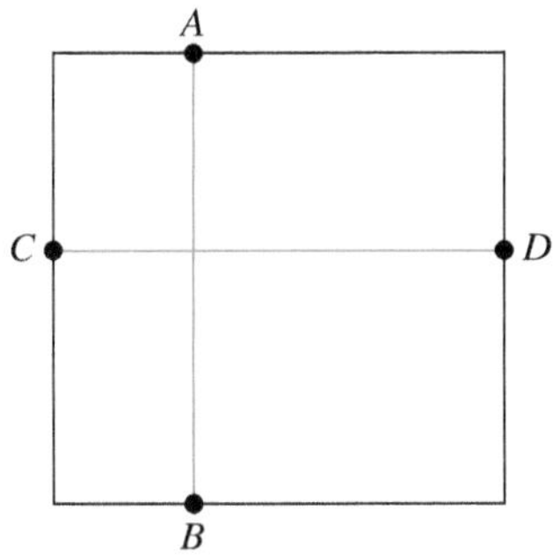

33. (a) The following rectangle has been partially covered with 1-by-2 dominoes:

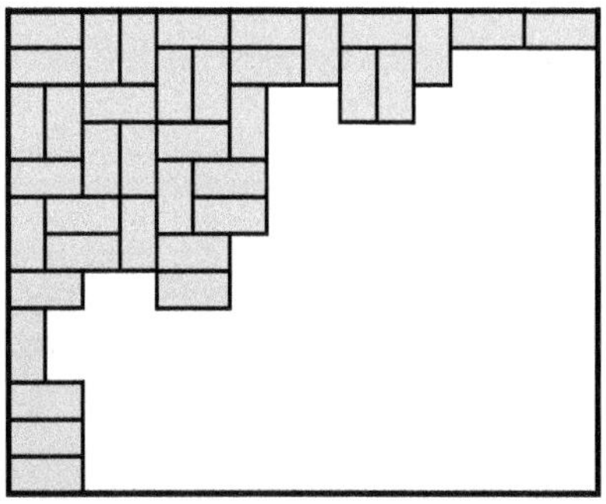

Without actually completing the tiling, describe how to determine the total number of dominoes that are needed to tile the rectangle.

(b) The two types of triominoes, each made from three unit squares, are shown here:

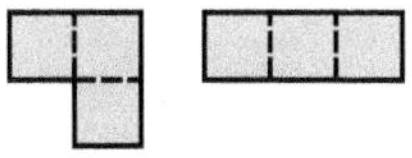

Can triominoes, of either or both types, be used to tile the square shown in part (a)? Explain carefully.

34. What is the smallest natural number whose factors include 1, 2, 3, . . . , 10?

35. Recall that the Fibonacci numbers are $F_1 = 1, F_2 = 1, F_3 = 2, F_4 = 3, \ldots$, where each new term in the sequence is the sum of its two immediate predecessors.

(a) What is the parity (even or odd) of F_{999}? Of F_{1000}?

(b) What is an easy way to know if F_n is even or odd?

36. **(Writing)** The first 12 Fibonacci numbers are shown in the following table:

n	1	2	3	4	5	6	7	8	9	10	11	12
F_n	1	1	2	3	5	8	13	21	34	55	89	144

The divisors of $n = 4$ are 1, 2, and 4, and we see that the Fibonacci number $F_4 = 3$ is divisible by F_1, F_2, and F_4. Write a report that investigates this question: *If d is a factor of n, then F_d is a factor of F_n.* If this seems to be true, formulate a conjecture, but if it is false, provide a counter example.

37. The Perrin sequence is defined similarly to the Fibonacci numbers, except that consecutive pairs of numbers are added to form the term that *follows* the next term in the sequence. Starting with $P_0 = 3$, $P_1 = 0$, and $P_2 = 2$, the sequence begins as shown in the following table, where $P_3 = 3 + 0 = 3$, $P_4 = 0 + 2 = 2$, $P_5 = 2 + 3 = 5$, and so on:

n	0	1	2	3	4	5	6	7	8	9	10	11	12
P_n	3	0	2	3	2	5	5	7	10	12	17	22	29

(a) Use a calculator to extend the table to show the Perrin numbers P_n for $n \leq 32$.

(b) Formulate a conjecture about when n divides P_n.

38. Determine the prime-power representation of each of these integers:

(a) 894,348 **(b)** 245,025 **(c)** 1,265,625

(d) Which of the numbers in parts (a), (b), and (c) are squares?

(e) What can you say about the prime-power representation of a square? Explain briefly.

(f) Guess how you might know from a glance at its prime-power representation that 93,576,664 is the cube of a natural number. Explain.

Making Connections

39. Suppose that the digit 1 corresponds to a black square and a 0 to a white square. Then the 15-symbol string 101011110110101 corresponds to this message:

(a) What string of symbols represents this picture?

(b) What message is hidden in the following? 11101011010101110011110100101110010

(c) Why would a string of 96 digits be ambiguous?

40. In 1974, as part of the Search for Extraterrestrial Intelligence (SETI), the Arecibo message shown here was transmitted (without the colors) into space in the form of a string of 1679 binary digits—zeros and ones. If intelligent life were to receive the message, it was assumed it could form the diagram on the right that contains the numerals 1 through 10, the atomic numbers of the elements found in DNA, and other information.

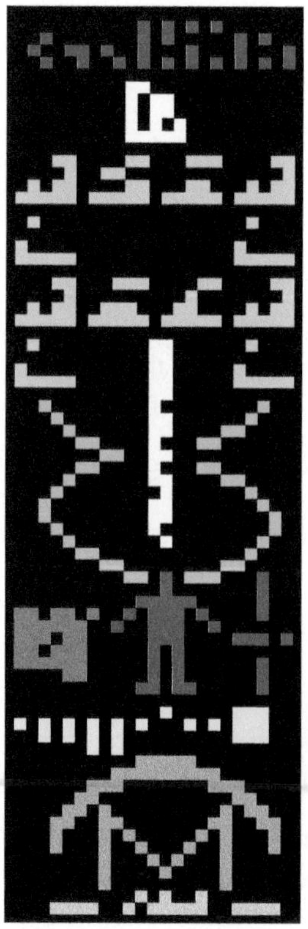

(a) Determine the prime factorization of 1679.

(b) Why do you think 1679 was chosen?

(c) How are the numbers 1 through 10 being described by the pattern in the first four rows of the message?

State Assessments

41. (Grade 6)

What is the prime factorization of 440?

A. $2 \cdot 5 \cdot 11$ B. $2^3 \cdot 5 \cdot 11$
C. $8 \cdot 55$ D. $2^2 \cdot 5^2 \cdot 11$

42. (Massachussetts, Grade 5)

What is the total number of factors of 12?

A. 4 B. 6 C. 8 D. 12

43. (Massachussetts, Grade 8)

Which of the following is the prime factorization of 72?

A. $2^3 \cdot 3^2$ B. $2^4 \cdot 3^3$ C. $8 \cdot 3^2$ D. $2^3 \cdot 9$

44. (Grade 5)

The factors of a composite number can be found by making a *paired factor diagram.* For example, the paired factor diagram of 75 starts with $1 \times 75 = 75$, and then shows that $3 \times 25 = 75$ and $5 \times 15 = 75$.

The paired factor diagram for 75 is below.

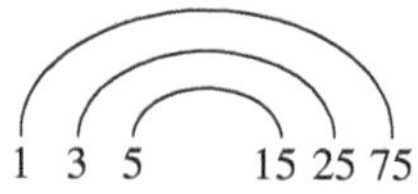

Similarly, the paired factor diagram for 80 below shows that $1 \times 80 = 80$, $2 \times 40 = 80$, $4 \times 20 = 80$, $5 \times 16 = 80$, and $8 \times 10 = 80$.

The paired factor diagram for 80 is below.

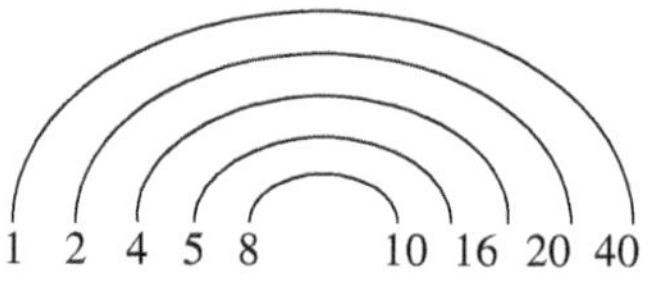

All of the prime factors of a number can be found by creating a *factor tree*, where the first branch shows any pair of factors of the number shown in the paired factor diagram. For example, choosing 3×25 from the paired factor diagram gives this prime factor tree.

The prime factor tree for 75 shows that 3 and 5 are its prime factors, and $3 \times 5 \times 5 = 75$.

Prime factor Tree for 75

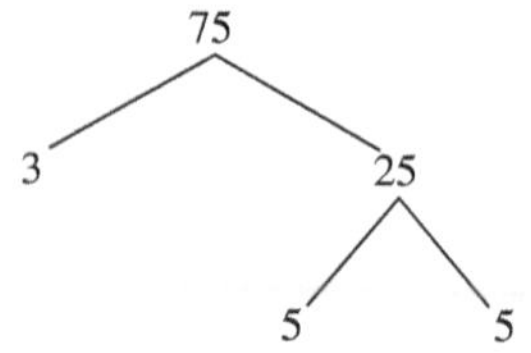

Similarly, the prime factor tree for 80 shows that 2 and 5 are its prime factors, and $2 \times 2 \times 2 \times 2 \times 5 = 80$.

Prime factor Tree for 80

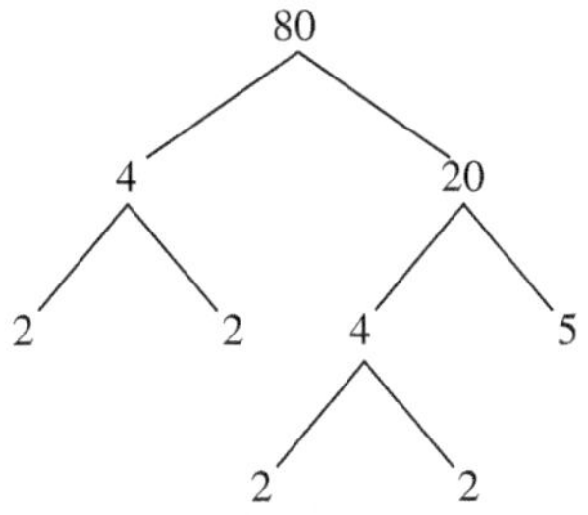

(a) Choose a composite number between 61 and 99 and create the paired factor diagram for this number.

(b) Choose any pair of factors from the factor diagram of part (a), and use these factors to draw a prime factor tree for your composite number.

45. (California, Grade 5)

Determine the prime factors of all numbers through 50, and write the numbers as the product of their prime factors by using exponents to show multiples of a factor (e.g., $24 = 2 \times 2 \times 2 \times 3 = 2^3 \times 3$)

4.2 Tests for Divisibility

It is usually difficult to determine the divisors, or factors, of a given natural number. However, the divisibility tests described in this section can often identify at least some of the smaller divisors. Children typically enjoy using divisibility tests, since they are carried out quickly and usually involve only mental arithmetic. The tests are also a topic that encourages students to think mathematically as they answer the question "Why does this test work?" The rationale behind nearly every test presented in this section can be explained by understanding how divisibility relates to sums and differences.

Divisibility of Sums and Differences

It is clear that 3 divides both 66 and 18. Indeed, $66 = 3 \cdot 22$ and $18 = 3 \cdot 6$. Since multiplication distributes over addition and subtraction, we have

$$84 = 66 + 18 = 3 \cdot 22 + 3 \cdot 6 = 3(22 + 6) = 3 \cdot 28$$

and

$$48 = 66 - 18 = 3 \cdot 22 - 3 \cdot 6 = 3(22 - 6) = 3 \cdot 16.$$

That is, since 3 divides both 66 and 18, then it also divides their sum and difference.

More generally, we have the following useful theorem:

THEOREM Divisibility of Sums and Differences
Let d, a, and b be natural numbers. Then if d divides both a and b, it also divides both their sum, $a + b$, and their difference, $a - b$.

PROOF Since d divides a and b, it follows that $a = dj$ and $b = dk$ for some natural numbers j and k. By the distributive property of multiplication over addition, we then have $a + b = dj + dk = d(j + k)$. That is, $a + b = dq$, where $q = j + k$. Therefore, d divides the sum $a + b$. Similarly, using the distributive property of multiplication over subtraction gives $a - b = dj - dk = d(j - k) = dr$, where $r = j - k$. This shows that if d divides a and b, then d divides their difference, $a - b$.

Divisibility by 2, 5, and 10

Is 9276 divisible by 2? You may already know that it is, because the units digit 6 is an even number. But why does this simple test work, and how would you explain it to a youngster? The key idea is to write $9276 = 9270 + 6$ in the form

$$9276 = 927 \cdot 10 + 6.$$

Since 2 divides 10 and 6, 2 also divides $927 \cdot 10$ and 6. Then, by the divisibility-of-sums-and-differences theorem, it follows that 2 divides 9276.

The analysis can be generalized to any natural number n. Dividing n by 10, we see that

$$n = q \cdot 10 + r,$$

where r is the units digit of n. Using the divisibility-of-sums-and-differences theorem, together with the fact that 10 is divisible by 2, we see that n is divisible by 2 precisely when r is divisible by 2—that is, when the units digit r is 0, 2, 4, 6, or 8.

Similarly, since 10 is divisible by 5, it follows that n is divisible by 5 exactly when r is also divisible by 5—that is, when the units digit is 0 or 5. Finally, n is divisible by 10 only when the units digit is 0. Thus, we have proved the following theorem:

THEOREM Tests for Divisibility by 2, 5, and 10

Let n be a natural number. Then n is

- divisible by 2 exactly when its base-ten units digit is 0, 2, 4, 6, or 8;
- divisible by 5 exactly when its base-ten units digit is 0 or 5;
- divisible by 10 exactly when its base-ten units digit is 0.

Divisibility by 4, 8, and Other Powers of 2

Our successful justification of the divisibility tests for 2, 5, and 10 suggests that we reexamine our reasoning to see if it can be modified to derive other divisibility tests. Since our initial idea was to divide a given natural number n by 10, let's investigate what divisibility tests we can obtain by dividing n by 100.

Dividing a natural number n by 100, we can write n in the form

$$n = q \cdot 100 + r, \quad 0 \le r \le 99,$$

where the remainder r is the number formed from the last two digits of n. Of course, 2, 5, and 10 are divisors of 100, but there are several new divisors as well, such as 4, 20, 25, and 50. Arguing just as before, we conclude that n is divisible by 4, or 20, or 25, or 50 precisely when the number r formed from the last two digits of n is divisible, respectively, by 4, or 20, or 25, or 50. For example, the number 9276 is divisible by 4, since $76 = 4 \cdot 19$. However, 9276 is not divisible by 20, 25, or 50, since 76 is not divisible by any of these numbers.

If n is divided by 1000, we obtain

$$n = q \cdot 1000 + r, \quad 0 \le r \le 999,$$

where r is the number formed by the last three digits of n. As before, if d is any divisor of 1000, then it is also a divisor of n exactly when d is a divisor of r. Since $1000 = 2 \cdot 2 \cdot 2 \cdot 5 \cdot 5 \cdot 5$, among the choices for d are 8, 125, and 500. For example, $n = 4{,}357{,}832$ is divisible by 8 (since $832 = 8 \cdot 104$), but it is not divisible by either 125 or 500, since neither 125 nor 500 divides $r = 832$.

Divisibility tests for any power 2^m can be obtained by division by 10^m, showing us that a natural number n is divisible by 2^m if and only if the number r formed by the last m digits of the base-ten representation of n is divisible by 2^m. Since divisibility by 4 and 8 are the most useful cases, we state the following theorem:

THEOREM Tests for Divisibility by 4 and 8

Let n be a natural number. Then n is divisible by 4 exactly when the number represented by its last two base-ten digits is divisible by 4. Similarly, n is divisible by 8 exactly when the number represented by its last three base-ten digits is divisible by 8.

EXAMPLE 4.7

Using the Divisibility Tests for 2, 4, 5, 8, and 10

Decide whether each of the following statements is true or false, giving a reason for each answer:

(a) 2 divides 54,628.
(b) 4 divides 54,628.

(c) Since both 2 and 4 are divisors of 54,628, 8 is also a divisor, because $2 \cdot 4 = 8$.
(d) 5 divides $2439 + 8206$.

Solution

(a) True, since 2 divides the units digits 8.
(b) True, since 4 divides 28.
(c) False, since $628 = 8 \cdot 78 + 4$ shows that 8 does not divide 628 and therefore 8 does not divide 54,628. Note that if a number is divisible by 4, then $4 = 2 \cdot 2$ appears in the prime factorization of the number and therefore the number is also divisible by 2. However, we have no reason to know if $2 \cdot 2 \cdot 2$ appears in the prime factorization of 54,628, so we do not know if that number is divisible by 8.
(d) True. Since $9 + 6 = 15$, the last digit of the sum is a 5. Note that a sum may be divisible by 5 even though neither addend is divisible by 5.

Divisibility by 3 and 9

The tests for **divisibility by 3 and 9** depend on the simple observation that every power of 10 is 1 larger than an obvious multiple of 3 and 9:

$$10 = 9 + 1,\ 100 = 99 + 1,\ 1000 = 999 + 1,\ 10{,}000 = 9999 + 1, \ldots$$

Let's use this observation to test whether $n = 27{,}435$ is divisible by 3 or 9. Using expanded notation, we have

$$\begin{aligned} n = 27{,}435 &= 2 \cdot 10{,}000 + 7 \cdot 1{,}000 + 4 \cdot 100 + 3 \cdot 10 + 5 \\ &= 2 \cdot (9999 + 1) + 7 \cdot (999 + 1) + 4 \cdot (99 + 1) + 3 \cdot (9 + 1) + 5 \\ &= \underbrace{(2 \cdot 9999 + 7 \cdot 999 + 4 \cdot 99 + 3 \cdot 9)}_{\text{a number divisible by 3 and by 9}} + \underbrace{(2 + 7 + 4 + 3 + 5)}_{\text{the sum of the digits of } n} \end{aligned}$$

By the divisibility-of-sums-and-differences theorem, the preceding equation shows that n is divisible by 3 or 9 precisely when the sum s of its digits is also divisible by 3 or 9. For example, the sum of the digits of $n = 832{,}452$ is $8 + 3 + 2 + 4 + 5 + 2 = 24$. Since 24 is divisible by 3 but is not divisible by 9, it follows that 832,452 is divisible by 3 but is not divisible by 9.

The same procedure can be applied to any natural number n; therefore, we have the following theorem:

THEOREM Tests for Divisibility by 3 and 9
A natural number n is divisible by 3 if and only if the sum of its digits is divisible by 3. Similarly, n is divisible by 9 if and only if the sum of its digits is divisible by 9.

Combining Divisibility Tests

Consider the natural number $n = 6735$. Since the last digit is 5, it follows that n is divisible by 5. Moreover, the sum of the digits of 6735, $6 + 7 + 3 + 5 = 21$, is divisible by 3, so n is also divisible by 3. Thus, the prime factorization of n includes at least one factor of 3 and one factor of 5, so their product $3 \cdot 5 = 15$ is also a divisor of n.

It may seem at first that if we know that a number n has two divisors a and b, then their product $a \cdot b$ must also be a divisor of n. However, this may not be the case if a and b share a common prime divisor. As an example of this situation, consider $m = 5396$. The last digit of m is even, so m is divisible by 2. Also, 4 divides the number 96 formed from the last two digits, so m is also divisible by 4. It is tempting to claim that m is therefore divisible by $2 \cdot 4 = 8$, but this is not so, since $5396 = 8 \cdot 674 + 4$. We can be assured that the product of two divisors a and b is also a divisor only when a and b have no common prime divisor. In the example, the two divisors 2 and 4 of m have the common prime divisor 2, and we know for sure only that there are two occurrences of the prime number 2 in the prime factorization of m.

The following theorem summarizes what can be said about the possibility of multiplying two divisors to get a third divisor:

THEOREM Divisibility by Products
Let a and b be divisors of a natural number n. If a and b have no common divisor other than 1, then their product ab is also a divisor of n.

A simple case of this theorem occurs when $a = 2$ and $b = 3$. Since only 1 divides both 2 and 3, we know that a natural number n is divisible by $2 \cdot 3 = 6$ if and only if n is divisible by 3—that is, its digits sum to a multiple of 3—and n is even—that is, the units digit is even. This gives us a test for divisibility by 6.

COROLLARY A Divisibility Test for 6
A natural number n is divisible by 6 if and only if it is divisible by both 2 and 3. That is, the sum of the digits of n is a multiple of 3 and the units digit is 0, 2, 4, 6, or 8.

EXAMPLE 4.8

Combining Divisibility Tests

Decide whether each statement is true or false without actually dividing. Give a reason for each answer.

(a) 45 divides 43,695.
(b) 36 divides 9486.
(c) 6 divides 876,324.
(d) 4 divides 876,324.
(e) 24 divides 876,324.
(f) 60 divides 395,220.

Solution

(a) True. 5 is a divisor, since the last digit is 5, and 9 is a divisor, since the sum of the digits of 43,695, $4 + 3 + 6 + 9 + 5 = 27$, is divisible by 9. Since 5 and 9 have no common factor other than 1, $5 \cdot 9 = 45$ divides 43,695.
(b) False. If 36 were a divisor, then 4 would also be a divisor. However, 86 is not divisible by 4.
(c) True. 2 is a divisor since the last digit 4 is even, and 3 is also a divisor, since the sum of the digits of 876,324, $8 + 7 + 6 + 3 + 2 + 4 = 30$, is divisible by 3. Therefore, 6 divides 876,324 by the divisibility by 6 test.
(d) True. 4 divides the number 24 formed by the last two digits.
(e) False (even though 4 and 6 are divisors). If 24 were a divisor, then 8 would also be a divisor. However, the number 324 formed from the last three digits of 876,324 is not divisible by 8.
(f) True. Let $n = 395{,}220$. Since $3 + 9 + 5 + 2 + 2 + 0 = 21$ is divisible by 3, n is also divisible by 3. Also, 20 is divisible by 4, so n is also divisible by 4. Finally, since the last digit of n is 0, n is divisible by 5. But 3, 4, and 5 have no common factor other than 1, so n is divisible by the product $3 \cdot 4 \cdot 5 = 60$.

Highlight from History: Srinivasa Ramanujan (1887–1920)

Perhaps the most exotic and mysterious of all mathematicians was Srinivasa Ramanujan, born to a high-caste family of modest means in Kumbakonam in southern India in 1887. Largely self-taught, Ramanujan wrote a letter in 1913 to the eminent English mathematician G. H. Hardy at Cambridge University in which he included a list of formulas he had discovered. Of the formulas, Hardy wrote, "[such formulas] defeated me completely . . . a single look at them is enough to show that they were written by a mathematician of the highest class. They must be true because, if they were not true, no one would have had the imagination to invent them. Finally (you must remember that I knew nothing about Ramanujan and had to think of every possibility), the writer must be completely honest, because great mathematicians are commoner than thieves and humbugs of such incredible skill."

In any event, Hardy arranged for Ramanujan to come to England in 1914, and the two collaborated intensively for the next three years, with Ramanujan using his unorthodox methods and fantastic intuition to come up with deep and totally unexpected results that Hardy, with his considerable intellectual power and formal training, then proved. Ramanujan fell ill in 1917 and returned to India, where he died in 1920.

The English mathematician J. E. Littlewood once remarked that every positive integer was one of Ramanujan's friends. Once, during Ramanujan's illness, Hardy visited him in the hospital in Putney. Trying to find a way to begin the conversation, Hardy remarked that he had come to the hospital in cab number 1729 and that he could not imagine a more uninteresting number, to which Ramanujan replied, "No, it is a very interesting number; it is the smallest number that can be written as the sum of two cubes in two different ways!"*

*$1729 = 9^3 + 10^3 = 1^3 + 12^3$

SOURCE: Quote by G.H. Hardy, public domain.

Summary of Useful Divisibility Tests

Divisibility tests can be derived for many other natural numbers, but the tests of most value are collected in Table 4.1.

TABLE 4.1 TESTS FOR DIVISIBILITY BY 2, 3, 4, 5, 6, 8, 9, 10, AND 11

Divisor	Test
2	Last digit must be 0, 2, 4, 6, or 8
3	Sum of digits must be divisible by 3
4	Last two digits must form a number divisible by 4
5	Last digit must be 0 or 5
6	Number must be divisible by 2 and 3
8	Last three digits must form a number divisible by 8
9	Sum of digits must be divisible by 9
10	Last digit must be 0

Problem 17 describes a test for 7, with its derivation outlined in problem 18.

Applications of Divisibility

Most people are unaware of the role that divisibility plays in the modern world. Here are some of the more common ways that divisibility is encountered:

- If you purchase an item at the store, it is scanned to read the Uniform Product Code (UPC). The scan is checked for errors by making sure it passes a divisibility test.
- If you enter your credit card number to make a purchase or donation at an Internet site, the number on your card must pass a divisibility test to be accepted.
- When the bar code at the bottom of a letter is read by a scanner at the post office, the validity of the zip code is checked by seeing if the sum of the digits is a multiple of 10.
- The last digit of the International Standard Book Number (ISBN) that identifies nearly every book published throughout the world is chosen so a special sum of the 13 digits is a multiple of 10.

The ISBN divisibility check is described in more detail in the following example.

EXAMPLE 4.9

Working with the ISBN-13 Code

Since 2007, an ISBN is a 13-digit code separated into five parts. In particular, the first 12 digits are divided into four groups that identify the type of industry (usually 978, indicating a publisher), a language group, the publisher, and the specific book. The last symbol is the *check digit*, a 0, 1, 2, . . . , 9. The check digit is chosen so that when the 13 digits are multiplied by the alternating factors 1, 3, 1, . . . , 3, 1, the sum of these 13 products is a multiple of 10. Here is an example of how the ISBN-13 code 978-0-321-69312-9 is checked for accuracy:

ISBN-13	9	7	8	0	3	2	1	6	9	3	1	2	9	
Multiplication factors	1	3	1	3	1	3	1	3	1	3	1	3	1	
Products	9	21	8	0	3	6	1	18	9	9	1	6	9	sum of products = 100

Now verify which of these codes are correct or supply the correct check digit if it is omitted.

(a) ISBN-13: 978-1-908-47874-0
(b) ISBN-13: 978-0-321-52862-4
(c) ISBN-13: 978-0-10006-253-?

Solution

(a) $9 \times 1 + 7 \times 3 + 8 \times 1 + 1 \times 3 + 9 \times 1 + 0 \times 3 + 8 \times 1 + 4 \times 3 + 7 \times 1 + 8 \times 3 + 7 \times 1 + 4 \times 3 + 0 \times 1 = 120$, so correct
(b) $9 \times 1 + 7 \times 3 + 8 \times 1 + 0 \times 3 + 3 \times 1 + 2 \times 3 + 1 \times 1 + 5 \times 3 + 2 \times 1 + 8 \times 3 + 6 \times 1 + 2 \times 3 + 4 \times 1 = 105$, so incorrect
(c) $9 \times 1 + 7 \times 3 + 8 \times 1 + 0 \times 3 + 1 \times 1 + 0 \times 3 + 0 \times 1 + 0 \times 3 + 6 \times 1 + 2 \times 3 + 5 \times 1 + 3 \times 3 + ? \times 1 = 65 + ?$, so $? = 5$

Illustrating Factors and Divisibility with a Manipulative

Manipulatives can help children understand the concepts of primes, factors, and divisibility. For example, ask children to take a handful of beans (say, 15 to 25 or so) with the instruction that they do not count the number of beans in their hand. Now have them make piles of, say, 5 beans per pile. Finally, have them determine the number of beans they started with by seeing the number of complete piles they formed and the number of beans left that weren't enough to make a complete pile of 5. Point out that if no beans were left over, then 5 is a factor of the number of beans taken originally. If there is a remainder, start again but this time put a different number of beans, say 6, in each pile.

There are several questions to accompany this activity:

- If you made 3 piles of 5 beans, what other way can you form piles with none left over? (Answer: 1 pile of 15, 5 piles of 3, or 15 piles with 1 bean each)
- How many ways can 11 beans be made into piles of equal size with none left over? (Answer: two ways, with 1 pile of 11 or 11 piles with 1 each)
- What number can be piled in just one way? (Answer: just the number 1)

Number rods, whose length in centimeters represents a natural number 1, 2, 3, . . . , also effectively convey the concepts of factors and divisibility. We say that a length of n centimeters is *measured* by rods of length d if a train of length n can be made with rods of length d. In Figure 4.4, we see that a length of $n = 12$ can be measured with rods of lengths 3 and 4 but not with rods of length 5. Thus, 12 is divisible by 3 and 4 but not 5. Or, equivalently, 12 is a multiple of 3 or 4 but not 5.

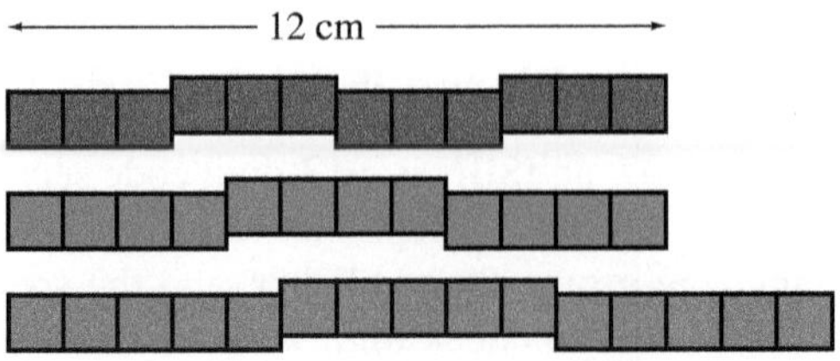

Figure 4.4 Showing that a length of 12 cm can be measured with 3-rods and 4-rods but not 5-rods

4.2 Problem Set

Understanding Concepts

1. Use the divisibility tests of Table 4.1 to decide which of the numbers 2, 3, 4, 5, 6, 8, 9, 10 is a divisor of the numbers in the left column of the following table. Write ✓ if divisible and ✗ if not. The number 684 is done for you.

	2	3	4	5	6	8	9	10
684	✓	✓	✓	✗	✓	✗	✓	✗
(a) 1950								
(b) 2014								
(c) 2015								
(d) 51,120								

2. Construct a divisibility table as shown in problem 1 for the following natural numbers:

   (a) 27,840 **(b)** 536,301
   (c) 2,783,428 **(d)** 9,876,432

3. Let the natural number n have the decimal numeral 123,456,78d, where d is the units digit. Use divisibility tests to give all of the choices of d by which n is divisible.

   (a) 2 **(b)** 3 **(c)** 4
   (d) 5 **(e)** 6 **(f)** 8
   (g) 9 **(h)** 10

4. Let d be the last digit of the base-ten numeral m = 73,486,30d. Use divisibility tests to give all of choices of d by which m is divisible.

   (a) 2 **(b)** 3 **(c)** 4
   (d) 5 **(e)** 6 **(f)** 8
   (g) 9 **(h)** 10

5. **(a)** Why, if given any seven different natural numbers, must there be at least two of them whose sum or difference is divisible by 10?

   (b) Make a list of six different numbers for which the sum or difference of any pair is not divisible by 10.

6. **(a)** Why, if given any six different natural numbers, must there be at least two of them whose difference is divisible by 5?

   (b) Make a list of five different numbers so none of the differences between any two of them is divisible by 5.

7. Let n = 34,d40,318. For what value(s) of the digit d is n divisible by 3 but not 9?

8. Let m = 76,1d8,071. For what value(s) of the digit d is m divisible by 3 but not 9?

9. Let n = 473,246,8de be a base-ten numeral with d and e its last two digits. Give all of the choices of the two-digit numbers de for which n is divisible by 12.

10. Let m = 57,609,9de. Give all of the choices of the two-digit numbers de for which n is divisible by 36.

Into the Classroom

11. Number rods are an effective manipulative to illustrate divisibility properties. For example, the following number-rod diagram illustrates why 3 is a divisor of 9:

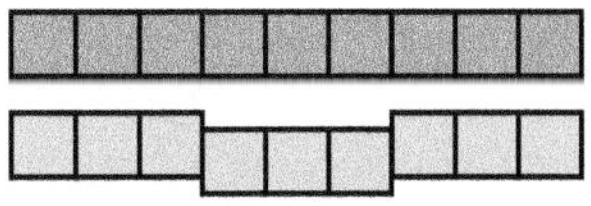

   (a) Use a number-rod diagram to show that 6 is divisible by 3.

   (b) Use a number-rod diagram to show that, since 3 divides 9 and 6, 3 must also divide the sum 9 + 6.

   (c) Use number-rod diagrams to show that 2 is a divisor of 6 and 10.

   (d) Use the diagrams in part (c) to explain why 2 is a divisor of 16 = 10 + 6 and 4 = 10 − 6.

Responding to Students

12. A student claims that 157,163 is divisible by 3 since the last digit in the number is 3. Explain how you would correct the student's thinking.

13. A student claims that, to multiply a two-digit number by 11, all you have to do is write a three-digit number whose first and last digits are the first and last digits of the two-digit number and whose middle digit is the sum of the digits of the two-digit number (e.g., $32 \cdot 11 = 352$).

   (a) Is this ever true?

   (b) Is this always true?

   (c) If a three-digit number is such that the middle digit is the sum of its first and last digits, is the number divisible by 11?

   (d) How would you discuss (a), (b), and (c) with a class of fourth graders?

14. Tia claims that 24 divides 420 since 6 and 4 both divide 420. Explain how you would correct Tia's thinking.

15. Max claims that 15 does not divide 177 + 48 since 15 does not divide 177 and 15 does not divide 48.

   (a) Is Max correct?

   (b) How would you respond to Max?

Thinking Critically

16. Prime numbers do not appear with any obvious regularity. For example, it seems possible to find quite long strings of consecutive composite numbers, such as the string of seven consecutive composite numbers 90, 91, 92, 93, 94, 95, and 96. It is important to be alert to asking such questions as "Is it always the case that . . . ?" "What if . . . ?" "Is there a . . . ?" and so on. Looking over the list of prime numbers suggests that we ask this question:

Is there a limit on the length of a string of consecutive composite numbers? To help you answer this question, it is useful to look at products of successive natural numbers known as *factorials,* where $n!$ (read as "n factorial") is defined by the product of the first n natural numbers. That is, $n! = n \cdot (n-1) \cdot (n-2) \cdot \cdots \cdot 3 \cdot 2 \cdot 1$. As an example, $11! = 11 \cdot 10 \cdot 9 \cdot 8 \cdot 7 \cdot 6 \cdot 5 \cdot 4 \cdot 3 \cdot 2 \cdot 1$.

(a) It is clear that 11! is divisible by the ten divisors 2, 3, . . . , 11. Why are these ten numbers also divisors of the ten consecutive natural numbers $11! + 2, 11! + 3, \ldots, 11! + 11$?

(b) Part (a) has shown that $11! + 2, 11! + 3, \ldots, 11! + 11$ are ten consecutive composite natural numbers. Describe how to create a list of 100 consecutive composite natural numbers.

(c) Explain why there are arbitrarily long strings of consecutive composite natural numbers.

17. Here is a test for divisibility of a number n by 7:

> **A Divisibility-by-7 Test,** Let r be the last digit of n, so n has the form $n = 10q + r$. Then n is divisible by 7 if and only if the number $m = q - 2r$ is divisible by 7.

For example, $n = 40{,}061$ is divisible by 7 if and only if $4006 - 2 = 4004$ is divisible by 7; 4004 is divisible by 7 if and only if $400 - 8 = 392$ is divisible by 7; and finally, 392 is divisible by 7 if and only if $39 - 4 = 35$ is divisible by 7. Since $35 = 7 \cdot 5$, it follows that 40,061 is divisible by 7.

Use the test to check each of the following numbers for divisibility by 7, without actually doing the division:

(a) 686 **(b)** 2951 **(c)** 18,487

18. The divisibility test for 7 given in problem 17 makes the following claim: $n = 10q + r$ is divisible by 7 if and only if $m = q - 2r$ is divisible by 7.
Use *algebraic reasoning* to prove this claim, following these two steps:

(a) Assume that $10q + r$ is divisible by 7, so that $10q + r = 7k$ for some natural number k. Now show that $10m$ is divisible by 7, where $m = q - 2r$. Why do you then know that m is divisible by 7?

(b) Assume that $m = q - 2r$ is divisible by 7. Then show by algebraic reasoning that $n = 10q + r$ is divisible by 7.

19. A Divisibility by 11 Test The powers of 10 alternate between being one more and one less than a multiple of 11. That is, $10 = 1 \times 11 - 1$, $100 = 9 \times 11 + 1$, $1000 = 91 \times 11 - 1$, and so on. This means the alternating sum of the digits of a number must be divisible by 11 if the number itself is divisible by 11. For example, $627 = 6(100) + 2(10) + 7 = 6(9 \times 11 + 1) + 2(11 - 1) + 7 = (6 \times 9 + 2)11 + 6 - 2 + 7$, which shows that 627 is a multiple of 11 if and only if the alternating sum $6 - 2 + 7 = 11$ is divisible by 11. Since it is, 627 is divisible by 11. On the other hand, the number 7859 is not divisible by 11 since its alternating sum of digits $-7 + 8 - 5 + 9 = 5$ is not divisible by 11. Use the alternating-sum-of-digits test to decide which of the following numbers is divisible by 11.

(a) 2,262,942 **(b)** 48,360,219
(c) 7,654,592 **(d)** 8,352,607
(e) 718,161,906

20. The number $n = 6{,}34d{,}217$ is divisible by 11. Since its alternating sum of digits is $6 - 3 + 4 - d + 2 - 1 + 7 = (6 + 4 + 2 + 7) - (3 + d + 1) = 15 - d$, the divisibility by 11 test in the previous problem means that $d = 4$. Determine the unknown digit a, b, or c, in each of these numbers, knowing that each is divisible by 11.

(a) 897,650,243,28a **(b)** 56,b39,975
(c) 678,2c6,322,904

21. A palindrome is a number that reads the same forward and backward, such as 2,743,472.

(a) Give a clear but brief argument showing that every palindrome with an even number of digits is divisible by 11.

(b) Is it possible for a palindrome with an odd number of digits to be divisible by 11? Explain.

22. (Writing) The digits of $n = 354{,}278$ sum to $3 + 5 + 4 + 2 + 7 + 8 = 29$, which leaves a remainder of 2 when divided by 9. Does n also leave a remainder of 2 when divided by 9? Write a report answering, the following question: *Can remainders also be determined from a divisibility test?*

23. (Writing) The Fibonacci numbers $F_1 = 1, F_2 = 1, F_3 = 2, \ldots, F_{n+2} = F_{n+1} + F_n \ldots$ are shown in the following table.

n	1	2	3	4	5	6	7	8
F_n	1	1	2	3	5	8	13	21

n	9	10	11	12	13	14	15
F_n	34	55	89	144	233	377	610

The highlighted columns show that $F_5 = 5$ divides the Fibonacci numbers $F_5 = 5$, $F_{10} = 55$, and $F_{15} = 610$.

(a) Conjecture which Fibonacci numbers in general are divisible by $F_5 = 5$.

(b) Which Fibonacci numbers seem to be divisible by $F_3 = 2$? By $F_4 = 3$?

(c) Make and discuss a general conjecture about which Fibonacci numbers divide a Fibonacci number F_n.

Making Connections

24. A common error in banking is to interchange, or transpose, some of the digits in a number involved in a transaction. For example, a teller may pay out \$43.34 on a check actually written for \$34.43 and hence be short by \$8.91 at the end of the day. Show that such a mistake always causes the teller's balance sheet to show an error, in terms of pennies—here 891—that is divisible by 9. (*Hint:* Recall that, in the proof of the divisibility test for 9, we showed that every number differs from the sum of its digits by a multiple of 9.)

25. Casting Out Nines Before the development of mechanical and electronic calculators, arithmetic calculations were very vulnerable to errors. As early as the 3rd century A.D. computations were often checked by a method called "casting out nines." The idea is simple: Replace every number in the calculation by its remainder when divided by

9 and repeat the calculation using the remainders. The remainders are easy to find by just adding the sum of the digits, perhaps repeatedly. Since the remainders are single-digit numbers, the calculation with remainders is almost surely accurate. If the remainders do not agree, there is a mistake in the calculation (though there may be a mistake even if the remainders happen to agree). Here are two examples of calculations checked by casting out nines, one correct and one incorrect.

6735	↔	3	6975	↔	0
245	↔	2	265	↔	1
+681	↔	+6	+834	↔	+6
7661		11	8064		10
↕		↕	↕	does not	↕
2	checks	2	0	check	1

Use casting out nines to see if these calculations may have an error. If incorrect, give the correct sum, difference, or product.

(a) $35{,}874 + 7531 + 69{,}450 = 113{,}855$

(b) $8514 + 6854 + 2578 + 6014 = 23{,}860$

(c) $78{,}962 - 3621 = 75{,}331$

(d) $358 \times 592 = 221{,}936$

State Assessments

26. (Massachussetts, Grade 6)

The clues below describe a three-digit number.

> The hundreds digit is 4. The ones digit is 3. The three-digit number is divisible by 3.

Which of the following could be the tens digit of the number?

A. 2 B. 3 C. 6 D. 9

27. (Massachussetts, Grade 4)

Classes that visit the Life Science Museum are divided into groups of 4 students for each tour guide. Which of the following classes would **not** be able to form groups of 4 students with none left over?

A. A class of 36 students B. A class of 40 students

C. A class of 46 students D. A class of 52 students

28. (Grade 7)

The 9th grade class at Oak Grove School can be divided into equal size groups of 4, 8, and 13 students. What is the smallest possible size of the class?

A. 52 B. 64 C. 104 D. 416

4.3 Greatest Common Divisors and Least Common Multiples

An architect is designing an elegant display room for an art museum. One wall is to be covered with large square marble tiles. To obtain the desired visual effect, the architect wants to use the largest tiles possible. If the wall is 12 feet high and 42 feet long, how large can the tiles be?

The length of the side of a tile must be a divisor of both the height and length of the wall (that is, a common factor of both 12 and 42). Since the sets of divisors of 12 and 42 are $D_{12} = \{1, 2, 3, 4, 6, 12\}$ and $D_{42} = \{1, 2, 3, 6, 7, 14, 21, 42\}$, respectively, the tile size must be chosen from the set $D_{12} \cap D_{42} = \{1, 2, 3, 6\}$, the set of common divisors of both 12 and 42. Thus, if the tiles are to be as large as possible, they must measure 6 feet on a side, since 6 is the largest of the common divisors, or factors, of 12 and 42.

Considerations like those in the previous paragraph lead to the notion of the greatest common divisor of two natural numbers, defined as follows:

Elementary school textbooks most often use **GCF**. Calculators and spreadsheets most often use **GCD** (or gcd).

DEFINITION Greatest Common Divisor and Greatest Common Factor

Let a and b be whole numbers not both 0. The greatest natural number d that divides both a and b is called their **greatest common divisor,** and we write $d = \text{GCD}(a, b)$. Alternatively d is the **greatest common factor** of a and b, and we write $d = \text{GCF}(a, b)$.

For example, $\text{GCD}(12, 42) = 6$, which can also be written as $\text{GCF}(12, 42) = 6$.

The second of the Standards for Mathematical Practice (SMP 2) from the **Common Core** states that mathematically proficient students should have the ability

". . . to abstract a given situation and represent it symbolically and manipulate the representing symbols as if they have a life of their own."

SMP 2 Reason abstractly and quantitatively.

For example, with a clear understanding of the meaning of GCD and GCF, it should be clear that $\text{GCF}(a, a) = a$ and $\text{GCF}(a, 0) = a$ for any natural number a. It should also be evident why "GCD(0, 0)" is not defined: *All* natural numbers are divisors of 0 and there is no largest divisor. If $a \leq b$, we see that

$$1 \leq \text{GCD}(a, b) \leq a.$$

This is because 1 is a common divisor of every pair of natural numbers, and no natural number larger than a can be a divisor, or factor, of a.

The concept of the greatest common divisor has many applications, but one of the most important is its connection to expressing fractions in simplest form. For example, the fraction $\frac{12}{42}$ can be simplified by factoring a 6 from its numerator and denominator, the GCD (or GCF) of 12 and 42, to obtain the simplest form $\frac{12}{42} = \frac{6 \cdot 2}{6 \cdot 7} = \frac{2}{7}$. It is therefore important to be able to compute the greatest common divisor; fortunately, there are several useful methods that can be employed.

GCD Method 1: Greatest Common Divisors by Intersection of Sets

This method works well when the numbers involved are small and all of the divisors of both numbers are easily written down. By way of explanation, it is probably best to present an example.

EXAMPLE 4.10 Finding the Greatest Common Divisor by Intersection of Sets

Find the greatest common divisor of 18 and 45.

Solution

Let D_{18} and D_{45} denote the sets of divisors of 18 and 45, respectively. Since $D_{18} = \{1, 2, 3, 6, 9, 18\}$ and $D_{45} = \{1, 3, 5, 9, 15, 45\}$, the set of common divisors of 18 and 45 is $D_{18} \cap D_{45} = \{1, 3, 9\}$. Thus, GCD(18, 45) = 9 or, equivalently, GCF(18, 45) = 9.

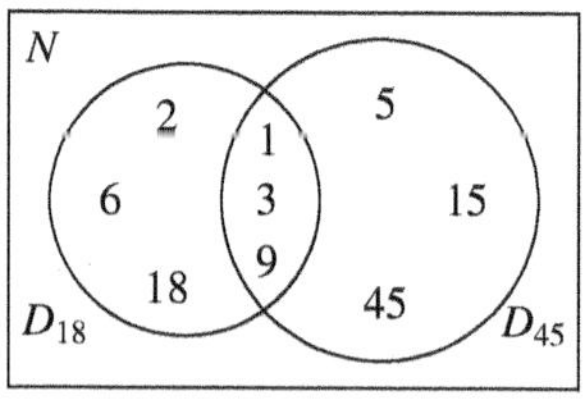

The sets of divisors can be visualized with a Venn diagram with loops representing the two sets of divisors D_{18} and D_{45} in the universe N of the natural numbers. The common divisors are those natural numbers in the overlapped region representing the intersection of sets, and the largest number in the intersection is the greatest common divisor.

GCD Method 2: Greatest Common Divisor from Prime Factorizations

The greatest common divisor can also be found by using prime factorizations. For example, to find GCD(270, 630), first determine the prime factorizations $270 = 2 \cdot 3 \cdot 3 \cdot 3 \cdot 5$ and $630 = 2 \cdot 3 \cdot 3 \cdot 5 \cdot 7$, and then write the factorizations, aligning each repeated factor vertically. The GCD is then the product of the prime factors that appear in both factorizations. That is,

$$\begin{aligned} 270 &= 2 \cdot 3 \cdot 3 \cdot 3 \cdot 5 \\ 630 &= 2 \cdot 3 \cdot 3 \quad\;\; \cdot 5 \cdot 7 \\ \hline \text{GCD}(270, 630) &= 2 \cdot 3 \cdot 3 \quad\;\; \cdot 5 \end{aligned}$$

so $\text{GCD}(270, 630) = 2 \cdot 3 \cdot 3 \cdot 5 = 90$.

It is even easier to use prime-power factorizations, where the GCD is the product of primes that appear in each factorization raised to the smaller exponent. It is helpful to allow 0 as an exponent so that both factorizations have the same sequence of primes. Therefore,

$$\begin{aligned} 168 &= 2^3 \cdot 3^1 \cdot 5^0 \cdot 7^1 \\ 90 &= 2^1 \cdot 3^2 \cdot 5^1 \cdot 7^0 \end{aligned}$$

so choosing the smaller exponent in the factorizations, we find that $\text{GCD}(168, 90) = \text{GCF}(168, 90) = 2^1 \cdot 3^1 \cdot 5^0 \cdot 7^0 = 6$.

The method works in general and is described by the following theorem:

THEOREM Greatest Common Divisor from Prime Factorizations

Let a and b be natural numbers. Then the GCD(a, b) is the product of the prime powers in the prime-power factorizations of a and b that have the smaller exponents (including zero).

EXAMPLE 4.11

Finding the Greatest Common Divisor by Using Prime-Power Representations

Compute the greatest common divisor of 504 and 3675.

Solution

We first find the prime-power representation of each number:

$$\begin{array}{r} 3\\ 3\overline{)9}\\ 7\overline{)63}\\ 2\overline{)126}\\ 2\overline{)252}\\ 2\overline{)504} \end{array} \qquad\qquad \begin{array}{r} 3\\ 7\overline{)21}\\ 7\overline{)147}\\ 5\overline{)735}\\ 5\overline{)3675} \end{array}$$

$$504 = 2^3 \cdot 3^2 \cdot 5^0 \cdot 7^1 \qquad\qquad 3675 = 2^0 \cdot 3^1 \cdot 5^2 \cdot 7^2$$

Choosing the smaller of the exponents on each prime that appears in both factorizations, we see that

$$\text{GCD}(504, 3675) = 2^0 \cdot 3^1 \cdot 5^0 \cdot 7^1 = 21.$$

As a check, we note that $504 \div 21 = 24$, that $3675 \div 21 = 175$, and that 24 and 175 have no common factor other than 1.

GCD Method 3: Greatest Common Divisor from the Euclidean Algorithm

Both of the preceding methods for finding a GCD (or GCF)—sets of divisors and prime factorizations—require extensive and sometimes difficult calculations. Fortunately, there is a simpler method that depends only on repeated subtraction; that is, on division with remainders. The method is called the **Euclidean algorithm**, since it first appeared in Book IV of Euclid's *Elements,* written about 300 B.C.

Suppose, as in Example 4.11, that we wish to calculate $d = \text{GCD}(3675, 504)$. By the divisibility-of-differences theorem, since d divides both 3675 and 504, d also divides $3675 - 504$. This means it also divides $3675 - 2 \times 504$, and divides $3675 - 3 \times 504$, and so on. Since, by division with remainders, we know that $3675 - 7 \times 504 = 147$, we see that d divides 147. By the same reasoning, because d divides both 504 and 147, we also know that d is a divisor of $504 - 147$, and $504 - 2 \times 147$, and so on. The division with remainders gives us $504 - 3 \times 147 = 63$, and we see that d divides both 63 and 147. Now continue this procedure for two more steps:

Since	$147 = 2 \times 63 + 21,$	d divides both 63 and 21,
Since	$63 = 3 \times 21 + 0,$	d divides 21 and 0.

Therefore, $d = 21$.

The procedure we have followed to show that $\text{GCD}(3675, 504) = 21$ works equally well to calculate any greatest common divisor $\text{GCD}(a, b)$ or greatest common factor $\text{GCF}(a, b)$. Such a step-by-step method is called an **algorithm**, and we have the following result.

THEOREM The Euclidean Algorithm

Let a and b be any two natural numbers, with $a \geq b$. To calculate $d = \text{GCD}(a, b)$ (or GCF),

Step 1. Divide a by b to get the remainder r, where $0 \leq r < b$. Then $d = \text{GCD}(b, r)$.
Step 2. If $r > 0$, repeat step 1, but now divide b by r; if $r = 0$; then $d = b$.

Since the remainders are always nonnegative and become smaller at each division, we can be sure there is some point at which the remainder is zero. The previous positive remainder is the GCD.

The following calculations show how repeated division with remainders determines the greatest common divisor of of 3144 and 1539 in three steps.

Division with Remainder	GCD Equality Shown
$1539\overline{)3144}$ 2 R 66	GCD(3144, 1539) = GCD(1539, 66)
$66\overline{)1539}$ 23 R 21	GCD(1539, 66) = GCD(66, 21)
$21\overline{)66}$ 3 R 3	GCD(66, 21) = GCD(21, 3)
$3\overline{)21}$ 7 R 0	GCD(21, 3) = GCD(3, 0) = 3

Since the last nonzero remainder is 3, it follows that GCD(3144, 1539) = 3.

EXAMPLE 4.12

Using the Euclidean Algorithm

Compute GCF(18,411, 1649), using the Euclidean algorithm.

Solution

Using the Euclidean algorithm, we have

$$\begin{array}{r}11 \text{ R } 272\\ 1649\overline{)18{,}411}\end{array} \qquad \begin{array}{r}6 \text{ R } 17\\ 272\overline{)1649}\end{array} \qquad \begin{array}{r}16 \text{ R } 0\\ 17\overline{)272}\end{array}$$

so that 17 is the last positive remainder. Therefore, GCF(18,411, 1649) = 17.

An Application of the Greatest Common Factor

EXAMPLE 4.13

Making Party Favors

Molly is inviting friends to her party and hopes to give each guest a gift bag containing some stickers, fun-size candy bars, and tangerines. She has 14 stickers, 21 candy bars, and 35 tangerines. What is the largest number of gift bags she can make if each bag is filled in the same way and all the stickers, candy bars, and tangerines are used?

Solution

Since there are equally many stickers in each bag, the number of bags is a factor of 14. By the same reasoning, the number of bags is a factor of 21 and 35 as well. Therefore, the largest number of bags is the largest common factor of 14, 21, and 35. That is, the number of bags is GCF(14, 21, 35). There are now three natural numbers to consider, but the same principles apply. In particular, the GCF is unchanged if from any number we subtract from it a multiple of any smaller number. Thus, GCF(14, 21, 35) = GCF(14, 7, 7) if 35 is replaced with $35 - 2 \times 14 = 7$. There is no need to list 7 twice, and GCF(14, 7) = GCF(0, 7) since we can replace 14 with $14 - 2 \times 7 = 0$. Since any natural number is a factor of 0, we see that GCF(0, 7) = 7. Therefore, 7 gift bags can be filled, each containing 2 stickers, 3 candy bars, and 5 tangerines. Hopefully, Molly is not inviting more than 7 to her party.

COOPERATIVE INVESTIGATION

Euclid's Game

Euclid's Game is a two-player game that starts with two natural numbers written on the game board. Taking alternate turns, each player chooses any two different numbers on the board whose difference does not yet appear. The result of the subtraction is written on the board. The first player unable to make a move loses the game. For example, if the two starting numbers are 76 and 52 (shown circled), then, after five plays the board may look as shown. Play Euclid's Game several times with your partner. You may choose the starting pair of numbers from this list or make up your own:

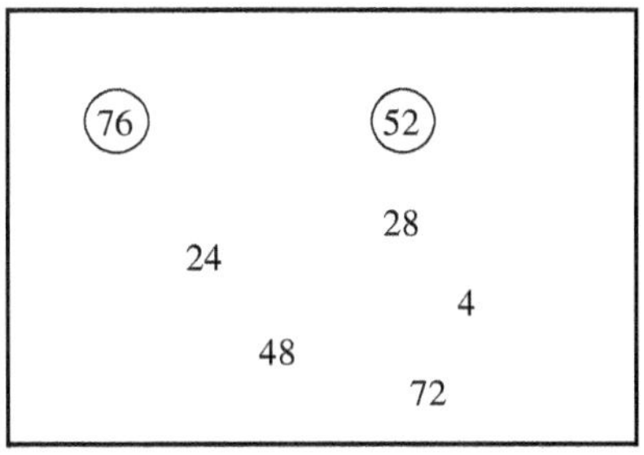

17 and 51, 28 and 24, 15 and 8, 75 and 39, 21 and 18, 70 and 21.

Now respond to these questions, working with your game partner:

(a) Is it possible to determine all of the numbers that will appear on the board ahead of time? If so, describe how this can be done.
(b) Given the starting numbers 52 and 76, should you choose to play first or second?
(c) Given the two starting numbers m and n, how should you determine to play first or second?

[*Note*: Euclid's Game can be played online by going to the site http://www.cut-the-knot.org.]

The Least Common Multiple

The multiples of a positive integer can be nicely illustrated using number rods as the manipulative. The idea is to form trains using cars of just one length. For example, with 8-rods (the brown rods of length 8 cm), the multiples 8, 16, 24, 32, 40, 48, 56, . . . can be represented by trains with all brown rods. Similarly, trains made up of 6-rods can be used to form the multiples 6, 12, 18, 24, 30, 36, 42, 48, 54, ... Some multiples, such as 24, 48, 72, . . . are common to both lists. Trains with these lengths are common multiples of 6 and 8. Of special interest is the train of shortest length for which this occurs. This length is called the **least common multiple** of 6 and 8. An example is shown in Figure 4.5, where we see that 24-cm is the shortest train that can be formed from both 8-rods and 6-rods.

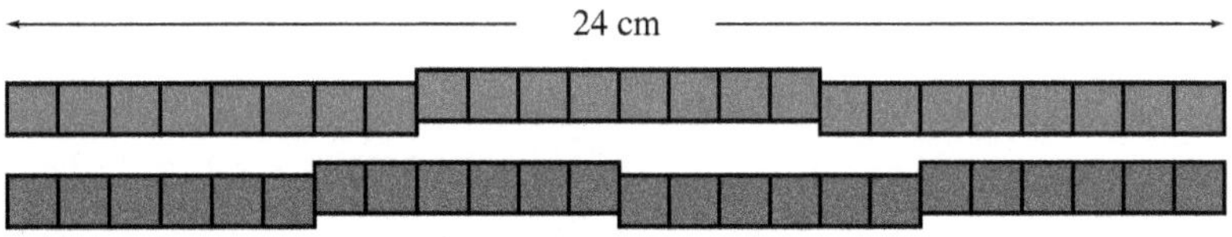

Figure 4.5
The shortest distance that can be measured by both the 8-rod and the 6-rod is 24 cm. Thus, 24 is the *least common multiple* of 8 and 6:
LCM(8, 6) = 24

In general, we have the following definition:

DEFINITION Least Common Multiple
Let a and b be natural numbers. The least natural number m that is a multiple of both a and b is called their **least common multiple,** and we write $m = \text{LCM}(a, b)$.

The least common multiple is especially important for computing with fractions, since the least common multiple is used to find the least common denominator.

For example, in Chapter 6 we will discover that

$$\frac{1}{6} + \frac{5}{8} = \frac{4}{24} + \frac{15}{24} = \frac{19}{24}$$

There are several methods available to determine an LCM.

LCM Method 1: Least Common Multiples by Intersection of Sets

The following method is simple and works particularly well if the two numbers are not large: Let a and b be any two natural numbers. The sets of natural-number multiples of a and b are

$$M_a = \{a, 2a, 3a, \ldots\} \qquad \text{and} \qquad M_b = \{b, 2b, 3b, \ldots\}.$$

Therefore, $M_a \cap M_b$ is the set of all natural-number common multiples of a and b, and the least number in this set is the least common multiple of a and b. Since ab is clearly a common multiple of both a and b, one need not extend the sets beyond this product. Indeed, it is always the case that $\text{LCM}(a, b) \leq ab$.

EXAMPLE 4.14

Finding a Least Common Multiple by Set Intersection

Find the least common multiple of 9 and 15.

Solution

Since $9 \cdot 15 = 135$, we consider the sets

$$M_9 = \{9, 18, 27, 36, 45, 54, 63, 72, 81, 90, 99, 108, 117, 126, 135, \ldots\};$$
$$M_{15} = \{15, 30, 45, 60, 75, 90, 105, 120, 135, \ldots\};$$
$$M_9 \cap M_{15} = \{45, 90, 135, \ldots\}.$$

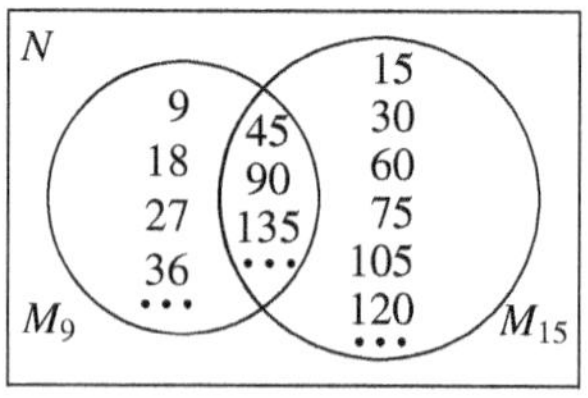

Since $M_9 \cap M_{15}$ is the set of all natural-number common multiples of 9 and 15, the least element of this set is LCM(9, 15). Therefore, LCM(9, 15) = 45. Notice that GCD(9, 15) = 3, so GCD(9, 15) · LCM(9, 15) $= 3 \cdot 45 = 135 = 9 \cdot 15$.

The set intersection method of finding the LCM can also be illustrated by a Venn diagram, as shown at the left. The numbers in the overlapped loops are the common multiples, and the smallest of these is the least common multiple.

LCM Method 2: Least Common Multiples from Prime Factorizations

The least common multiple can be found by using prime factorizations, a method similar to the procedure used earlier to find the GCD. For example, to find LCM(270, 630), write the factorizations aligned vertically to show the prime factors that are repeated. The LCM is then the product of the factors that appear in at least one of the factorizations. That is,

$$\begin{aligned} 270 &= 2 \cdot 3 \cdot 3 \cdot 3 \cdot 5 \\ 630 &= 2 \cdot 3 \cdot 3 \cdot \, 5 \cdot 7 \\ \hline \text{LCM}(270, 630) &= 2 \cdot 3 \cdot 3 \cdot 3 \cdot 5 \cdot 7, \end{aligned}$$

so $2 \cdot 3 \cdot 3 \cdot 3 \cdot 5 \cdot 7 = 1890$ is the least common multiple.

Prime-power expansions can also be used, but this time the prime powers with the larger exponents are multiplied to find the LCM. Again, it is helpful to allow the exponent zero so that the same sequence of primes appears in both factorizations. For example

$$\begin{aligned} 168 &= 2^3 \cdot 3^1 \cdot 5^0 \cdot 7^1 \\ 90 &= 2^1 \cdot 3^2 \cdot 5^1 \cdot 7^0 \\ \hline \text{LCM}(168, 90) &= 2^3 \cdot 3^2 \cdot 5^1 \cdot 7^1 \end{aligned}$$

so $2^3 \cdot 3^2 \cdot 5^1 \cdot 7^1 = 2520$ is the least common multiple.

> **THEOREM Least Common Multiple from Prime-Power Factorizations**
> Let m and n be natural numbers. Then the LCM(m, n) is the product of the prime powers in the prime-power factorizations of m and n that have the larger exponents.

EXAMPLE 4.15

Finding LCMs and GCDs by Using Prime-Power Representations

Compute the least common multiple and greatest common divisor of $m = 2268 = 2^2 \cdot 3^4 \cdot 7^1$ and $n = 77{,}175 = 3^2 \cdot 5^2 \cdot 7^3$.

Solution

$5^0 = 1$
$2^0 = 1$

Both the least common multiple and the greatest common divisor will be of the form $2^a \cdot 3^b \cdot 5^c \cdot 7^d$. For the greatest common divisor, we choose the *smaller* of the two exponents with which each prime appears in m and n, and for the least common multiple, we choose the *larger* of each pair of exponents. Thus, since

$$m = 2^2 \cdot 3^4 \cdot 5^0 \cdot 7^1 \quad \text{and} \quad n = 2^0 \cdot 3^2 \cdot 5^2 \cdot 7^3,$$

we have

$$\text{GCD}(m, n) = 2^0 \cdot 3^2 \cdot 5^0 \cdot 7^1 = 63$$

and

$$\text{LCM}(m, n) = 2^2 \cdot 3^4 \cdot 5^2 \cdot 7^3 = 2{,}778{,}300.$$

Since both the smaller and larger exponents on each prime were used in finding GCD(m, n) and LCM(m, n), it follows that

$$mn = \text{GCD}(m, n) \cdot \text{LCM}(m, n).$$

Thus,

$$mn = 2268 \cdot 77{,}175 = 175{,}032{,}900$$

and

$$\text{GCD}(m, n) \cdot \text{LCM}(m, n) = 63 \cdot 2{,}778{,}300 = 175{,}032{,}900.$$

The last part of the solution in Example 4.15, showing that $mn = \text{GCD}(m, n) \cdot \text{LCM}(m, n)$, can be repeated in general, so we have the following important theorem:

THEOREM $mn = \text{GCD}(m, n) \cdot \text{LCM}(m, n)$

If m and n are any two natural numbers, then $mn = \text{GCD}(m, n) \cdot \text{LCM}(m, n)$.

LCM Method 3: Least Common Multiples by Using the Euclidean Algorithm

As noted in the previous theorem, if m and n are natural numbers, then $\text{GCD}(m, n) \cdot \text{LCM}(m, n) = mn$. Thus,

$$\text{LCM}(m, n) = \frac{mn}{\text{GCD}(m, n)},$$

and we have already learned that GCD(m, n) can always be found by means of the Euclidean algorithm.

EXAMPLE 4.16

Finding a Least Common Multiple by the Euclidean Algorithm

Find the least common multiple of 2268 and 77,175 by using the Euclidean algorithm.

Solution

These are the same two numbers treated in Example 4.15. However, here our procedure is totally different, since we no longer need to know the prime factorizations of the given numbers. Using the Euclidean algorithm, we have these divisions:

$$2268\overline{)77{,}175}\quad 34 \text{ R } 63 \qquad\qquad 63\overline{)2268}\quad 36 \text{ R } 0$$

Since the last nonzero remainder is 63, GCD(2268, 77,175) = 63 and

$$\begin{aligned}\text{LCM}(2268, 77{,}175) &= \frac{2268 \cdot 77{,}175}{63}\\ &= 2{,}778{,}300,\end{aligned}$$

as before.

An Application of the LCM

EXAMPLE 4.17

Using the LCM to Investigate the "Calendar Round"

The ancient Mayan culture* of Central America adopted and refined a calendar first devised by the Zapotecs, an ancient indigenous culture centered at Monte Albán in southern Mexico. The Calendar Round, represented by the wheel calendar shown in the accompanying photo, is based on a combination of two cycles: the Tzlok'in ("count of days") 260-day cycle and the Haab', a cycle consisting of 18 months of 20 days each plus a period of 5 "nameless days." The Calendar Round was the period required for the Tzlok'in and Haab' cycles to end together. What is the length of the Calendar Round period, as measured in both days and Haab' "years"?

Solution

The Haab' is a period of $18 \times 20 + 5 = 365$ days. Thus, the Calendar Round has a duration of LCM(260, 365). Using the prime factorizations $260 = 2^2 \cdot 5 \cdot 13$ and $365 = 5^1 \cdot 73^1$, we see that the Calendar Round is $2^2 \cdot 5^1 \cdot 13^1 \cdot 73^1 = 18{,}980$ days long. Since $365 = 5 \cdot 73$ is one Haab', the Calendar Round is $2^2 \cdot 13^1 = 52$ Haab's, or about 52 years in duration. The Mayans were well aware that a true solar year was about a quarter of a day longer than a Haab'.

*The Mayan calendric systems are very interesting mathematically, and more information about them is easily available on the Internet. Of considerable interest is the Long Count Calendar, which began (in its Gregorian calendar equivalent) on September 6, 3114 B.C., and ended its first complete cycle on December 21, 2012. It's amazing that the Mayans could pinpoint a winter solstice so far into the future.

4.3 Problem Set

Understanding Concepts

1. Find the greatest common divisor of each of these pairs of numbers by the method of intersection of sets of divisors.
   (a) 24 and 27
   (b) 14 and 22
   (c) 48 and 72

2. Find the greatest common factors of these pairs of natural numbers by the method of intersection of sets.
   (a) 18 and 45
   (b) 36 and 54
   (c) 20 and 75

3. Find the least common multiple of each of these pairs of numbers by the method of intersection of sets of multiples.
   (a) 24 and 27
   (b) 14 and 22
   (c) 48 and 72

4. Find the least common multiples of pairs of numbers by the method of intersection of sets of multiples.
   (a) 18 and 45
   (b) 36 and 54
   (c) 20 and 75

5. Use the methods based on prime factorizations to find the greatest common divisor and least common multiple of each of these pairs of numbers.
(a) $r = 2^2 \cdot 3^1 \cdot 5^3$ and $s = 2^1 \cdot 3^3 \cdot 5^2$
(b) $u = 5^1 \cdot 7^2 \cdot 11^1$ and $v = 2^2 \cdot 5^3 \cdot 7^1$
(c) $w = 2^2 \cdot 3^3 \cdot 5^2$ and $x = 2^1 \cdot 5^3 \cdot 7^2$

6. Use the method of prime factorization to find the GCF and LCM of each of these pairs of numbers.
(a) $i = 3^2 \cdot 7^1 \cdot 11^1 \cdot 13^3$ and $j = 3^1 \cdot 5^2 \cdot 13^2$
(b) $m = 2^4 \cdot 7^2 \cdot 13^2$ and $n = 2^2 \cdot 3^1 \cdot 5^2 \cdot 13^2$
(c) $p = 2^3 \cdot 5^1 \cdot 19^2 \cdot 29^1$ and $q = 2^3 \cdot 3^1 \cdot 7^2 \cdot 11^2$

7. Use the Euclidean algorithm to find each of the following:
(a) GCD(3500, 550) and LCM(3500, 550)
(b) GCD(3915, 825) and LCM(3915, 825)
(c) GCD(624, 1044) and LCM(624, 1044)

8. Use the Euclidean algorithm to find each of the following:
(a) GCF(9025, 425) and LCM(9025, 425)
(b) GCF(4267, 4403) and LCM(4267, 4403)
(c) GCF(18,513, 16,170) and LCM(18,513, 16,170)

9. Use the definition of GCD, not a calculation, to find
(a) GCD(40, 40).
(b) GCD(19, 190).
(c) GCD(59, 0).

10. Use the definition of LCM, not a calculation, to find
(a) LCM(25, 25).
(b) LCM(13, 130).
(c) LCM(47, 1).

11. **(Writing)** Write a statement that generalizes each part of problems 9 and 10. As an example, one statement would be "If n is any natural number, then GCD$(n, n) = n$."

Into the Classroom

12. Cuisenaire® rods are a versatile manipulative for the classroom. They consist of colored rods 1 centimeter square and of lengths 1 cm, 2 cm, 3 cm, . . . , 10 cm, as shown here:

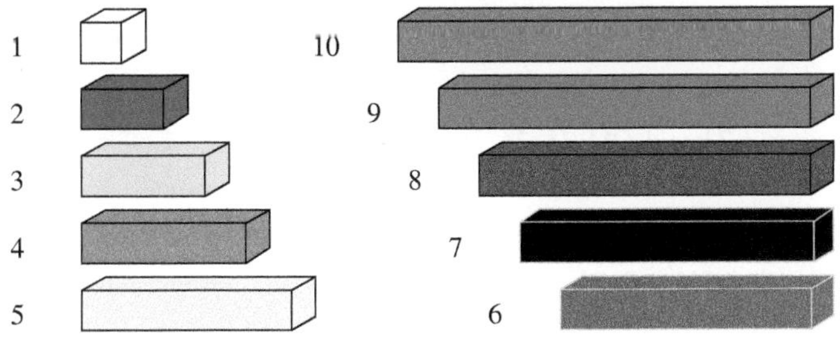

Recall that a smaller rod *measures* a longer rod if a number of the smaller rods placed end to end are the same length as the given rod.
(a) Which rods will measure the 9-rod?
(b) Which rods will measure the 6-rod?
(c) Which is the greatest common divisor of 9 and 6?
(d) Place the 10-rod and the 8-rod end to end to form an "18-train" as in the "Counting Cars and Trains" Cooperative Investigation activity of Chapter 2. Which rods or trains will measure the 18-train?
(e) Place the 7-rod and the 5-rod end to end to form a 12-train. Which rods or trains will measure the 12-train?
(f) What is the greatest common divisor of 12 and 18?
(g) Describe how you might use Cuisenaire® rods to demonstrate the notion of greatest common divisor to children.

13. Cuisenaire® rods are described in the preceding problem.
(a) What is the shortest train or length that can be measured by both 4-rods and 6-rods?
(b) What is the least common multiple of 4 and 6?
(c) What is the shortest train or length that can be measured by both 6-rods and-9 rods?
(d) What is the least common multiple of 6 and 9?
(e) Briefly describe how you could use Cuisenaire® rods to demonstrate the notion of the least common multiple to children.

14. **(Writing)**
(a) Indicate how you could use a number line to illustrate the notion of the greatest common divisor to children.

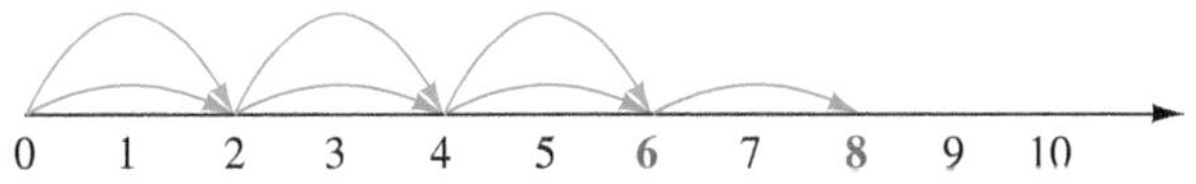

(b) Indicate how you could use a number line to illustrate the notion of the least common multiple to children.

Responding to Students

15. How would you respond to a student who claims that GCD$(0, n) = n$ for every whole number n?

16. Since n divides 0 for every natural number n, how would you respond to a student who claims that LCM$(0, n) = 0$ for every natural number n?

17. How would you respond to Tawana, who claims that mn is the least common multiple of m and n in every case?

18. Paulita says, "Why do we have to learn about greatest common divisors, anyway? It seems like a waste of time to me." How would you respond to Paulita?

Thinking Critically

19. The following figure shows that a 5″ by 5″ square tile can be used to cover a floor that measures 60″ by 105″:

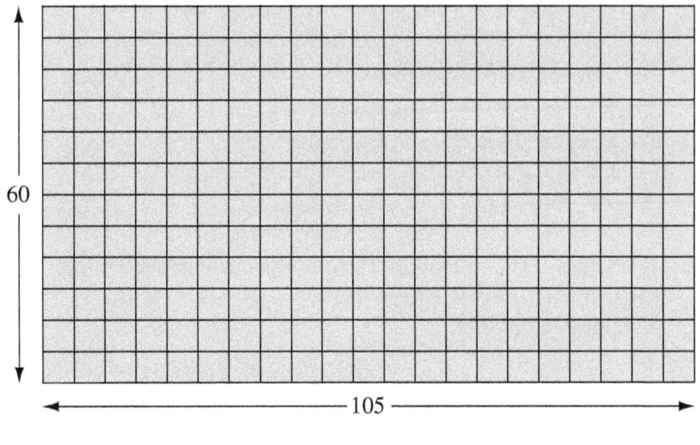

(a) List all of the sizes of squares tiles that can be used to tile the floor. What is the largest size of square tile that can be used?

(b) How many square tiles of the largest size are used to cover the floor?

20. **(a)** Suppose you wish to tile a hallway measuring 48″ wide by 216″ long. What is the largest size of square tile that can be used, and how many tiles of this size are required?

(b) Generalize your results: What is the largest size of square tile that can be used to cover an $m \times n$ rectangle, and how many tiles of that size are required?

21. The following figure shows that an 8″ by 12″ rectangular brick can tile a 48″ by 48″ square.

(a) Is there a smaller size square that can be tiled with 8″ by 12″ bricks, all placed horizontally? Explain why or why not.

(b) What is the smallest square that can be tiled with horizontal 9″ by 12″ bricks? Explain your reasoning.

22. **(a)** More generally, what is the smallest square that can be tiled with m-by-n rectangular bricks, with all bricks having the same orientation?

(b) How many bricks are required to tile the smallest square with bricks of size m by n? Obtain a simple formula involving both the GCD and LCM of m and n, and check your answer against your answers to parts (a) and (b).

23. The notions of the greatest common divisor and the least common multiple extend naturally to more than two numbers. Moreover, the prime-factorization method extends naturally to finding GCD(a, b, c) and LCM(a, b, c).

(a) If $a = 2^2 \cdot 3^1 \cdot 5^2$, $b = 2^1 \cdot 3^3 \cdot 5^1$, and $c = 3^2 \cdot 5^3 \cdot 7^1$, compute GCD($a, b, c$) and LCM($a, b, c$).

(b) Is it necessarily true that GCD(a, b, c) · LCM(a, b, c) = abc?

(c) Find numbers r, s, and t such that GCD(r, s, t) · LCM(r, s, t) = rst.

24. Use the method of intersection of sets to compute the following:

(a) GCD(18, 24, 12) and LCM(18, 24, 12)

(b) GCD(8, 20, 14) and LCM(8, 20, 14)

(c) Is it true that GCD(8, 20, 14) · LCM(8, 20, 14) = $8 \cdot 20 \cdot 14$?

25. **(a)** Compute GCD(6, 35, 143) and LCM(6, 35, 143).

(b) Is it true that GCD(6, 35, 143) · LCM(6, 35, 143) = $6 \cdot 35 \cdot 143$?

(c) Guess under what conditions GCD(a, b, c) · LCM(a, b, c) = abc.

26. Example 4.13 computed the GCD (or GCF) of three natural numbers by replacing any number with the difference it makes with any smaller number. If two (or more) numbers are the same, all but one can be discarded. The last positive number remaining is the GCF, since GCF(a) = a for any natural number a. For example, the GCD(18, 24, 42) can be found by the following steps.

GCD(18, 24, 42) = GCD(18, 24, 18) = GCD(18, 24) = GCD(18, 6) = GCD(12, 6) = GCD(6, 6) = GCD(6) = 6.

Use the pairwise difference method to compute these GCDs and GCFs.

(a) GCF(30, 75, 225)

(b) GCD(42, 63, 105)

(c) GCD(34, 51, 85, 136)

(d) GCF(54, 26, 33, 44)

27. The LCM method demonstrated in the table that follows is popular in Asian countries. The idea is quite simple: Factor out any prime divisor of one or more terms, and replace each term containing the prime factor by its quotient. Continue until the last row contains only distinct prime numbers and 1s. Then multiply the extracted primes and the primes in the last row to obtain the prime-power form of the LCM. The method works equally well with any number of terms, so we compute LCM(75, 24, 30, 42).

2	75	24	30	42
3	75	12	15	21
2	25	4	5	7
5	25	2	5	7
	5	2	1	7

Thus, LCM(75, 24, 30, 42) = $2 \cdot 3 \cdot 2 \cdot 5 \cdot 5 \cdot 2 \cdot 7 = 2^3 \cdot 3^1 \cdot 5^2 \cdot 7^1$.

Use this method to find these LCMs in prime-power form:

(a) LCM (48, 25, 35)

(b) LCM (40, 28, 12, 63)

(c) LCM (250, 28, 44, 110)

28. **(Writing)** The following table of Fibonacci numbers shows that $F_{10} = 55$ and $F_{15} = 610$. Their greatest common divisor is GCD(55, 610) = GCD(5 × 11, 2 × 5 × 61) = 5. But $F_5 = 5$, so we see that GCD(F_{10}, F_{15}) = $F_{\text{GCD}(10,\,15)}$. Similarly, using the Euclidean algorithm,

$$\text{GCD}(F_{21}, F_{28}) = \text{GCD}(10{,}946, 317{,}811) = \text{GCD}(10{,}946, 377) = \text{GCD}(13, 377) = 13 = F_7 = F_{\text{GCD}(21,\,28)}.$$

Explore the possibility that these are two examples of this general result:

For all natural numbers m and n, $\text{GCD}(F_m, F_n) = F_{\text{GCD}(m,n)}$.

Write a careful description of your investigation. Be sure to compute several additional examples.

1	2	3	4	5	6	7	8	9	10
1	1	2	3	5	8	13	21	34	55
11	12	13	14	15	16	17	18	19	20
89	144	233	377	610	987	1597	2584	4181	6765
21	22	23	24	25	26	27	28	29	30
10,946	17,711	28,657	46,368	75,025	121,393	196,418	317,811	514,229	832,040

Making Connections

29. Fractions will not be discussed in this text until Chapter 6. However, in elementary school you should have learned that a fraction $\frac{a}{b}$ can be written in simplest terms (i.e., as $\frac{c}{d}$, where c and d have no common factor other than 1) by dividing both a and b by GCD(a, b). For example, $\frac{16}{20} = \frac{4}{5}$, since GCD(16, 20) = 4, 16 ÷ 4 = 4, and 20 ÷ 4 = 5. Compute the following:

(a) GCD(6, 8) **(b)** GCD(18, 24)

(c) GCD(132, 209) **(d)** GCD(315, 375)

Find a fraction in simplest terms that is equivalent to each of these:

(e) $\frac{6}{8}$ **(f)** $\frac{18}{24}$

(g) $\frac{132}{209}$ **(h)** $\frac{315}{375}$

30. John liked his sister Molly's idea of gift bags (see Example 4.13), and he plans to do the same at his birthday party. He has 91 marbles, 65 sticks of gum, 78 walnuts, and 52 baseball cards. What is the largest number of bags he can make, where each bag has the same number of marbles, the same number of sticks of gum, and so on? All of the marbles, gum, walnuts, and cards must be given away.

31. Fractions will be explored in depth in Chapter 6, but you should already know that adding and subtracting fractions depends on finding a common denominator. The least common denominator is the least common multiple of the denominators that occur in the computation. For example, since LCM(12, 8) = 24, we see that $\frac{5}{12} + \frac{3}{8} = \frac{10}{24} + \frac{9}{24} = \frac{19}{24}$. Similarly, $\frac{5}{12} - \frac{3}{8} = \frac{10}{24} - \frac{9}{24} = \frac{1}{24}$. Find least common denominators and compute the following additions and subtractions.

(a) $\frac{3}{4} + \frac{1}{6}$ **(b)** $\frac{7}{10} + \frac{2}{15}$

(c) $\frac{8}{9} - \frac{4}{15}$ **(d)** $\frac{11}{25} - \frac{3}{10}$

32. The front wheel of a tricycle has a circumference of 54 inches, and the back wheels have a circumference of 36 inches.

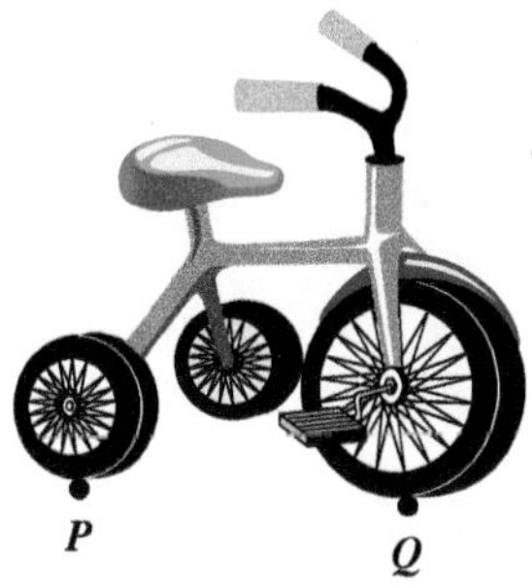

If points P and Q are both touching the sidewalk when Marja starts to ride, how far will she have ridden when P and Q first touch the sidewalk at the same time again?

33. In a machine, a gear with 45 teeth is engaged with a gear with 96 teeth, with teeth A and B on the small and large gears, respectively, in contact as shown. How many more revolutions will the small gear have to make before A and B are again in the same position?

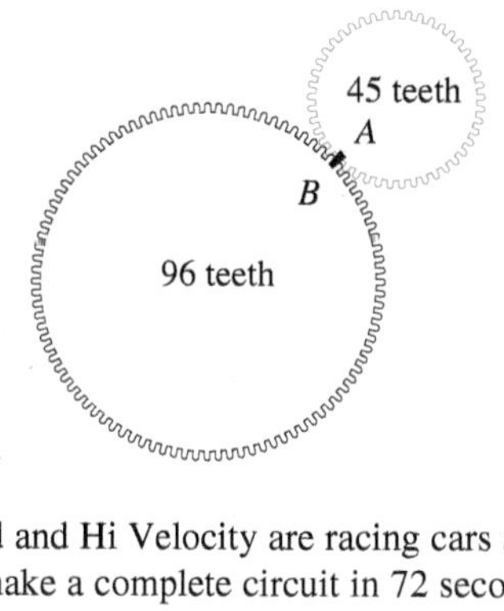

34. Sarah Speed and Hi Velocity are racing cars around a track. If Sarah can make a complete circuit in 72 seconds and Hi completes a circuit in 68 seconds,

(a) how many seconds will it take for Hi to first pass Sarah at the starting line?

(b) how many laps will Sarah have made when Hi has completed one more lap than she has?

35. Cicadas and Prime Numbers There are two common types of periodical cicadas. The most prevalent is the 17-year cicada (genus *Magicicada*), for which a brood of nymphs emerges from the ground to breed and lay eggs after a 17-year underground wait. The 13-year cicada (also of genus *Magicicada*) has a similar life cycle but with a 13-year interval between one emergence and the next. Studies show that having a prime-number interval, such as 13 or 17, offers evolutionary advantages to each species. In particular, there is less interference between the two species when they live in the same region.

(a) If broods of 13- and 17-year species happened to emerge in the same year, how many years will pass before this interference is repeated?

(b) Suppose, hypothetically, that there exist 12- and 18-year species of cicadas. If both species happened to emerge the same year, in how many years will they interfere with one another again?

(c) Answer part (b) again but for hypothetical 9- and 15-year species.

State Assessments

36. (Grade 4)
What is the least common multiple of 12 and 30?

A. 42 B. 48 C. 60 D. 120

37. (Grade 7)
What is the greatest common factor (GCF) of these numbers?
12, 16, 24, 40

A. 3 B. 4 C. 8 D. 12

38. (Grade 6)
What is the least common multiple (LCM) of 6 and 15?

A. 15 B. 30 C. 60 D. 90

39. (Massachussetts, Grade 6)
This is an open-response question.

- BE SURE TO ANSWER AND LABEL ALL PARTS OF THE QUESTION.
- Show all your work (diagrams, tables, or computations) in your Student Answer Booklet.
- If you do the work in your head, explain in writing how you did the work.

A local bakery celebrated its one-year anniversary on Saturday. On that day, every 4th customer received a free cookie. Every 6th customer received a free muffin.

(a) Did the 30th customer receive a free cookie, a free muffin, both, or neither? Show or explain how you got your answer.

(b) Casey was the first customer to receive both a free cookie and a free muffin. What number customer was Casey? Show or explain how you got your answer.

(c) Tom entered the bakery after Casey. He received a free cookie only. What number customer could Tom have been? Show or explain how you got your answer.

(d) On that day, the bakery gave away a total of 29 free cookies. What was the total number of free muffins the bakery gave away on that day? Show or explain how you got answer.

40. (Grade 6)
There are 24 pairs of dice, 36 rulers, and 60 marker pens in the supply cabinet. What is the largest number of envelopes needed to hold all of the items, so that each envelope contains the same number of items of each kind?

A. 6 B. 12 C. 60 D. 120

41. (Grade 7)
The following puzzle was put on the board.

> We are two natural numbers. Our difference is 30, and our greatest common factor is 15, and our sum is 120. Who are we?

A. 15 and 30, since our greatest common divisor is 15.

B. 15 and 45, since our difference is 30 and our greatest common factor is 15.

C. 15 and 105, since our sum is 120 and our greatest common factor is 15.

D. 45 and 75, since our difference is 30 and our greatest common factor is 15.

The Chapter in Relation to Future Teachers

In this chapter, we have introduced the basic number-theoretic notions of divisibility, factoring, factors and multiples, primes and composite numbers, least common multiples, greatest common factors, and greatest common divisors. These ideas not only are useful in other parts of mathematics and in disciplines like computer science but also provide interesting and stimulating motivational material for the elementary mathematics classroom. Increasingly, number-theoretic notions are appearing in elementary school texts, and these ideas must be thoroughly understood by teachers.

Understandably, most current research in mathematics is far too technical to describe to younger students, but number theory is an exception, since many of its results and unsolved problems are quite easily explained to students in the upper elementary or middle school years. For example, it is not too difficult to find triples of natural numbers (a, b, c) for which $a^2 + b^2 = c^2$. For example, $(3, 4, 5)$ is such a triple, corresponding to a right triangle with sides of length 3, 4, and 5. Indeed, even the ancient Greek mathematicians knew that infinitely many such so-called **Pythagorean triples** exist. But what if the Pythagorean relation is modified just a little, so that the goal is to find triples of natural numbers for which $a^n + b^n = c^n$ for other values of n? It wasn't until 1993, after hundreds of years of effort, that the mathematician Andrew Wiles proved that no triples exist for $n > 2$.

Here are some other questions that are easy to ask but have yet to be answered:

- **Twin-Prime Conjecture** If both p and $p + 2$ are prime, they are called twin primes. *Are there infinitely many twin primes?*
- **Germain Primes** If both p and $2p + 1$ are prime, then p is a *Germain prime*, named for the mathematician Sophie Germain. (See problem 30 from Problem Set 4.1.) *Are there infinitely many Germain primes?* The largest known Germain prime is currently $p = 183{,}027 \cdot 2^{265440} - 1$, discovered in 2010. This Germain prime has 79,911 digits.
- **Goldbach Conjecture** *Is every even number that is larger than 4 the sum of two prime numbers?* It was shown in 2009 that every even natural number up to 16×10^{17} is the sum of two primes!
- **Odd Perfect Numbers** The ancient Greeks called a number *perfect* if it equals the sum of its proper divisors. For example, $6 = 1 + 2 + 3$, $28 = 1 + 2 + 4 + 7 + 14$, and $496 = 1 + 2 + 4 + 8 + 16 + 31 + 62 + 124 + 248$ are the first three perfect numbers. Currently, there are 47 known perfect numbers, and all are even. *Are there any odd perfect numbers?* (If one exists, then it is larger than 10^{300}.)

Chapter 4 Summary

In what follows, unless expressly stated to the contrary, all symbols represent whole numbers.

Section 4.1 Divisibility of Natural Numbers	Page Reference
CONCEPTS	
• **Divisibility and factorization:** Given a natural number, n, the determination of the divisors of n and its representation as a product of natural numbers.	190
• **Determination of primality:** A test to classify a given natural number as a prime number, a composite number, or the unit number 1.	192
DEFINITIONS	
• b **divides** a when they are two whole numbers and $b \neq 0$, and there is a whole number q such that $a = bq$.	185
• When $a = bq$, b is a **factor,** or **divisor,** of a.	185
• A **proper divisor** is when b divides a and b is less than a when $a = bq$.	185
• When $a = bq$, a is a **multiple** of b.	185
• Whole numbers are **even** when they are divisible by 2. A whole number that is not even is called an **odd** whole number.	186
• A **prime number** is a natural number that possesses exactly two different factors, itself and 1.	187
• A **composite number** is a natural number that possesses more than two different factors.	187
• The number 1 is a **unit;** it is neither prime nor composite.	187
THEOREMS	
• **Fundamental theorem of arithmetic:** Every natural number greater than 1 is a prime or can be expressed as a product of primes in one, and only one, way, apart from the order of the prime factors.	188
• **Prime-power factorization of n:** Every natural number greater than 1 is a power of a prime or can be expressed as a product of powers of different primes in one, and only one, way, apart from order.	189

• **Prime-power factorization of divisors of n:** Any divisor of n is a product of prime powers of the form p^j, $0 \leq j < k$, where p^k is a term in the prime-power representation of n.	189
• **Infinitude of primes:** There are infinitely many prime numbers.	191
• **Primality test:** If a natural number $n \geq 2$ has no prime divisor p for $2 \leq p \leq \sqrt{n}$, then n is a prime number.	192
PROCEDURE	
• **Factor tree:** A method of factoring composite numbers into smaller and smaller ones until only primes remain.	188
• **Stacked short division:** A method of factoring composite numbers into prime numbers using prime divisors and short division.	188
• **Sieve of Eratosthenes:** A systematic counting method for determining a table of all of the primes up to a given limit derived by the Greek mathematician Eratosthenes.	192

Section 4.2 Tests for Divisibility	Page Reference
CONCEPTS	
• **Divisibility test:** A method to quickly determine, often with mental arithmetic, whether a natural number is divisible by a specified number.	197
THEOREMS	
• **Divisibility by 2, 5, and 10:** Let n be a natural number. Then n is	
• divisible by 2 exactly when its base-ten units digit is 0, 2, 4, 6, or 8.	197
• divisible by 5 exactly when its base-ten units digit is 0 or 5.	197
• divisible by 10 exactly when its base-ten units digit is 0.	197
• **Divisibility of sums and differences:** If n divides both a and b, then n also divides their sum $a + b$ and their difference $a - b$.	197
• **Divisibility by 4, 8, and other powers of 2:** Let n be a natural number. Then n is divisible by 4 exactly when the number represented by its last two base-ten digits is divisible by 4. Similarly, n is divisible by 8 exactly when the number represented by its last three base-ten digits is divisible by 8.	198
• **Divisibility by 3 and 9:** A natural number n is divisible by 3 if and only if the sum of its digits is divisible by 3. Similarly, n is divisible by 9 if and only if the sum of its digits is divisible by 9.	199
• **Divisibility by 6:** A natural number n is divisible by 6 exactly when it is even and divisible by 3 (its sum of digits is a multiple of 3).	200
• **Divisibility by products:** Let a and b be divisors of a natural number n. If a and b have no common divisor other than 1, then their product ab is also a divisor of n.	200

Section 4.3 Greatest Common Divisors and Least Common Multiples	Page Reference
DEFINITIONS	
• The **greatest common divisor, GCD(a, b),** is the greatest natural number d that divides both a and b.	205
• The **greatest common factor, GCF(a, b),** is an alternative description of the greatest common divisor, with GCD(a, b) = GCF(a, b).	205
• The **least common multiple, LCM(a, b),** is the least natural number m that is a multiple of both a and b.	209

THEOREMS

- **The Euclidean algorithm:** Let a and b be any two natural numbers, with $a \geq b$. Divide a by b to get a remainder r. If $r = 0$, the $b = \text{GCD}(a, b)$, but if $r > 0$, divide b by r to get a remainder s. If $s = 0$, then $r = \text{GCD}(a, b)$, but if $s > 0$, divide r by s to get a remainder t. Continue the division with remainder process until a remainder of 0 results. Then the last nonzero remainder is GCD(a, b). 207
- **Product of the GCD and LCM:** For any two natural numbers, $\text{GCD}(a, b) \cdot \text{LCM}(a, b) = ab$. 211

PROCEDURES

- **GCD (or GCF) by set intersection:** GCD(a, b) is the largest member of the intersection of the sets of divisors of a and b. 206
- **GCD from prime-power factorizations:** GCD(a, b) is the product of the prime powers of the form p^s, $0 \leq s$, where s is the smaller power of the prime p in the prime-power representations of a and b. 206
- **GCD by the Euclidean algorithm:** Given a and b, repeatedly use division with remainder, discarding quotients, to replace the larger number by the remainder. Continue this process until a remainder is zero. Then the last nonzero remainder is the greatest common divisor. 207
- **LCM by set intersection:** LCM(a, b) is the least member of the intersection of the sets of multiples of a and b. 209
- **LCM from prime-power representations:** LCM(a, b) is the product of the prime powers of the form p^t, where t is the larger power of the prime p in the prime-power representations of a and b. 210
- **LCM by the Euclidean algorithm:** First use the Euclidean algorithm to find the greatest common divisor GCD(a, b) of a and b, and then calculate the least common multiple using the formula 211

$$\text{LCM}(a, b) = \frac{ab}{\text{GCD}(a, b)}.$$

Chapter Review Exercises

Section 4.1

1. Draw rectangular diagrams to illustrate the factorings of 15, taking order into account; that is, think, for example, of $1 \cdot 15$ as different from $15 \cdot 1$.

2. Construct a factor tree for 96.

3. (a) Determine the set D_{60} of all divisors of 60.
   (b) Determine the set D_{72} of all divisors of 72.

4. (a) Determine the prime-power factorization of 1200.
   (b) Determine the prime-power factorization of 2940.
   (c) How many natural numbers divide 1200?
   (d) How many natural numbers divide 2940?

5. How many different factors does each of these numbers have?
   (a) $2310 = 2^1 \cdot 3^1 \cdot 5^1 \cdot 7^1 \cdot 11^1$
   (b) $5{,}336{,}100 = 2^2 \cdot 3^2 \cdot 5^2 \cdot 7^2 \cdot 11^2$

6. Use information from the sieve of Eratosthenes in Figure 4.3 to determine whether 847 is prime or composite.

7. (a) Determine a composite natural number n with a prime factor greater than $\sqrt{n}$.
   (b) Does the n in part (a) have a prime divisor less than $\sqrt{n}$? If so, what is it?

8. Determine natural numbers r, s, and m such that r divides m and s divides m, but rs does not divide m.

9. Let $m = 3^4 \cdot 7^2$.

(a) How many divisors does m have?

(b) List all the divisors of m.

Section 4.2

10. Use mental arithmetic to test each of these numbers for divisibility by 2, 3, 4, 5, 6, 8, 9, and 10.

(a) 62,418

(b) 222,789

(c) 726,840

(d) 874,273.

11. Use mental arithmetic and combined divisibility tests to decide if these statements are true or false.

(a) 67,275 is divisible by 15.

(b) 578,940 is divisible by 12.

(c) 42,720 has 24 as a factor.

(d) 62,730 has 36 as a factor.

12. The number 12,3de is divisible by 90. What are its last two digits, d and e?

13. Draw a number rod diagram that shows that 15 is divisible by 3 but is not divisible by 4.

Section 4.3

14. Let D_n denote the set of divisors of n and let M_n denote the set of positive multiples of n. Explain why

(a) $D_a \cap D_b = D_{\text{GCD}(a,\, b)}$,

(b) $M_a \cap M_b = M_{\text{LCM}(a,\, b)}$.

15. **(a)** Find the greatest common divisor of 63 and 91 by the method of intersection of sets of divisors.

(b) Find the least common multiple of 63 and 91 by the method of intersection of sets of multiples.

(c) Demonstrate that $\text{GCD}(63, 91) \cdot \text{LCM}(63, 91) = 63 \cdot 91$.

16. If $r = 2^1 \cdot 3^2 \cdot 5^1 \cdot 11^3$, $s = 2^2 \cdot 5^2 \cdot 11^2$, and $t = 2^3 \cdot 3^1 \cdot 7^1 \cdot 11^3$, determine each of the following:

(a) $\text{GCF}(r, s, t)$

(b) $\text{LCM}(r, s, t)$

17. Determine each of the following, using the Euclidean algorithm:

(a) GCD(119,790, 12,100)

(b) LCM(119,790, 12,100)

18. Seventeen-year locusts and 13-year locusts both emerged in 1971. When will these insects' descendants next emerge in the same year?

5 Integers

COOPERATIVE INVESTIGATION
The Debit/Credit Game

Materials Needed

1. Red and black markers and a pair of dice.

2. Scoring sheets as shown in the following sample, three or four for each student:

Debit (red)	Credit (black)
IIII	JHT II
JHT IIII	III

Net
JHT IIII IIII = III JHT II III

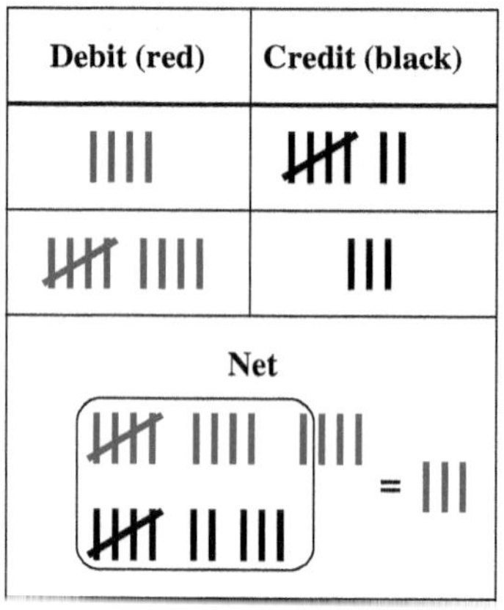

Directions

Each student chooses an opponent. Play several games with the following rules:

1. Take turns rolling the dice, entering your score with tally marks as shown in the sample. You may enter your score in any of the four boxes. If it is recorded as a credit, use the black marker; if it is recorded as a debit, use the red marker.

2. The game ends when each player has made four rolls and has recorded two credits and two debits on the score sheet. Use a loop as shown in the sample to cancel a red debit tally and a black credit tally in pairs, leaving the net value shown by the tally marks that have not been canceled.

3. The winner is the player with the highest net worth.

Question

What strategy will help you win the game?

The **integers** are defined and investigated in this chapter. This system of numbers, denoted by I, includes the whole numbers $W = \{0, 1, 2, 3, \ldots\}$ and the **negative integers,** denoted by $-1, -2, -3, \ldots$. Altogether, the set of integers is written

$$I = \{\ldots, -3, -2, -1, 0, 1, 2, 3, \ldots\}.$$

Note that $N \subset W \subset I$:

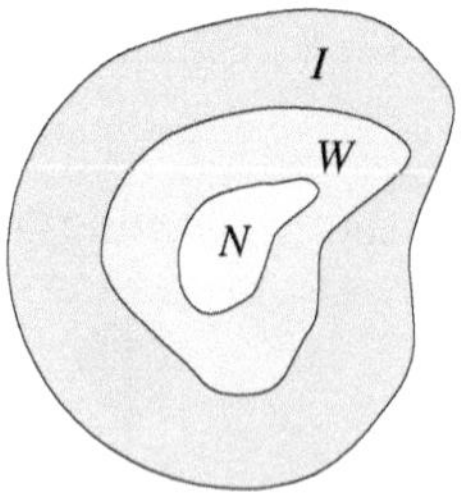

The set of natural numbers $N = \{1, 2, 3, \ldots\}$ is known as the set of **positive integers.** Each positive integer n is associated with a corresponding negative integer, denoted by $-n$, that is called its **opposite.** The association can also be reversed, so that, for example, the opposite of -4, denoted by $-(-4)$, is the positive integer 4. The most important property of the integers is that the sum of an integer n and its opposite $-n$ is 0; that is, $n + (-n) = 0$. Another way to describe this property is that every integer n has an **additive inverse** $-n$. Since $0 + 0 = 0$, the additive inverse of 0 is itself; that is, $-0 = 0$.

The following excerpt from the NCTM *Principles and Standards* gives a brief description of what is expected of elementary and middle school students with regard to the integers:

FROM THE NCTM PRINCIPLES AND STANDARDS

Middle-grades students should also work with integers. In lower grades, students may have connected negative integers in appropriate ways to informal knowledge derived from everyday experiences, such as below-zero winter temperatures or lost yards of football plays. In the middle grades, students should extend these initial understandings of integers. Positive and negative integers should be seen as useful for noting relative changes or values. Students can also appreciate the utility of negative integers when they work with equations whose solution requires them, such as $2x + 7 = 1$.

SOURCE: Principles and Standards for School Mathematics by NCTM, *pp. 217–218. Copyright © 2000 by the National Council of Teachers of Mathematics. Reproduced with permission of the National Council of Teachers of Mathematics via Copyright Clearance Center. NCTM does not endorse the content or validity of these alignments.*

KEY IDEAS

- Additive inverses (opposites)
- Absolute value
- Representations of the integers with colored counters, mail-time situations, number line
- Using manipulatives, pictorial representations, and conceptual models to understand addition, subtraction, multiplication, and division in the system of integers
- Why subtracting is equivalent to adding the opposite
- Why the multiplication of two negative integers is a positive integer
- Applications of the integers to real-life problems

5.1 Representations of Integers

The integers are an extension of the system of whole numbers, obtained by starting with the whole numbers $W = \{0, 1, 2, 3, \ldots\}$ and then creating additional numbers denoted by $-1, -2, -3, \ldots$ to form the set of integers $I = \{\ldots, -3, -2, -1, 0, 1, 2, 3, \ldots\}$.* The integers I by definition, have the additive-inverse property, a property that does not hold for the whole numbers:

DEFINITION The Integers

The integers $I = \{\ldots, -3, -2, -1, 0, 1, 2, 3, \ldots\}$ consist of

- the natural numbers $\{1, 2, 3, \ldots\}$, now called the **positive integers,**
- the **negative integers** $\{\ldots, -3, -2, -1\}$, and
- the number 0, which is neither negative nor positive.

Moreover,

- 0 is the **additive identity** in I, with $0 + n = n + 0 = n$ for all integers n, and
- $-n$ is the **additive inverse** of the positive integer n, having the property that

$$n + (-n) = (-n) + n = 0.$$

If we start with a negative integer, $-n$, then the formula $n + (-n) = (-n) + n = 0$ shows that n is the additive inverse of $-n$. For example, the additive inverse of -3 is 3, so $-(-3) = 3$. More generally, *every* integer, whether positive, negative, or zero, can be paired with its additive inverse, $-n$. The additive inverse, $-n$, is also called the **negative** of n or the **opposite** of n. It is important to observe how parentheses are used, and expressions such as "$--3$" and "$3 + -3 = 0$" are incorrect. Instead, we write $-(-3) = 3$ and $3 + (-3) = 0$. Even 0 has an opposite, namely, itself, since $0 + 0 = 0$; that is, $-0 = 0$.

*Sometimes the negative integers are denoted as ^{-}n, but this is replaced with $-n$ by the time students enter the middle grades.

Integers have the following property:

PROPERTY The Additive-Inverse Property of a Number System
A number system has the **additive-inverse property** if and only if for every number n in the system, there is a number m in the system for which $n + m = m + n = 0$.

The whole numbers do not have the additive-inverse property, since for example, adding any whole number to 1 never results in 0. In the following chapters, we will see that the rational and real numbers do have the additive-inverse property.

Absolute Value of an Integer

Given an integer n, either n or its opposite, $-n$, is a nonnegative integer known as the absolute value of n, written as $|n|$. The absolute value is also called the magnitude of an integer.

$|-2| = -(-2) = 2$

DEFINITION Absolute Value of an Integer
If n is an integer, then its **absolute value,** or **magnitude,** is the nonnegative integer $|n|$ given by

$$|n| = \begin{cases} n \text{ if } n \text{ is positive or } 0 \\ -n \text{ if } n \text{ is negative.} \end{cases}$$

EXAMPLE 5.1 Determining Absolute Values

Determine the absolute values of these integers.

(a) -71 **(b)** 29 **(c)** 0 **(d)** -852

Solution

(a) Since -71 is negative, $|-71| = 71$.
(b) Since 29 is positive, $|29| = 29$.
(c) $|0| = 0$.
(d) Since -852 is negative, $|-852| = 852$.

Criteria for the Representation of the Integers

Any representation of the integers $I = \{\ldots, -2, -1, 0, 1, 2, 3, \ldots\}$, whether by means of a manipulative, a visualization, or a conceptual model, must make the following fundamental ideas clear:

- how each integer is represented, including the possibility that an integer can be represented in multiple ways;
- how the representation of any nonzero number describes both the magnitude (or *absolute value*) and the *sign*—positive or negative—of the number;
- how 0, the unique integer with a zero magnitude, is represented;
- why the addition-by-zero property holds, so that $n + 0 = 0 + n = n$ for all integers; and finally,
- how every integer n is paired with its *opposite* (also known as its *negative* or its *additive inverse*), $-n$, so that $n + (-n) = 0$.

Three models for the integers will now be introduced: colored counters, mail-time stories, and the number line. Yet a fourth model is described by the balloon model found in the Cooperative Investigations of Sections 5.2 and 5.3. Throughout this chapter, we will rely on these models to understand the fundamental properties of the integers and their arithmetic.

The second of the Standards for Mathematical Practice (SMP 2) from the Common Core emphasizes the importance of representations, not simply as a means of calculation but also as an adjunct to give deeper meaning. We read that

SMP 2 Reason abstractly and quantitatively.

"Quantitative reasoning entails habits of creating a coherent representation of the problem at hand; considering the units involved; attending to the meaning of quantities, not just how to compute them; and knowing and flexibly using different properties of operations and objects."

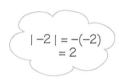

Representing Integers with Colored Counters

In the colored-counter model, an integer is represented by a set of colored counters—red and black discs, or sometimes red and yellow, are often used. Some examples are shown in Figure 5.1.

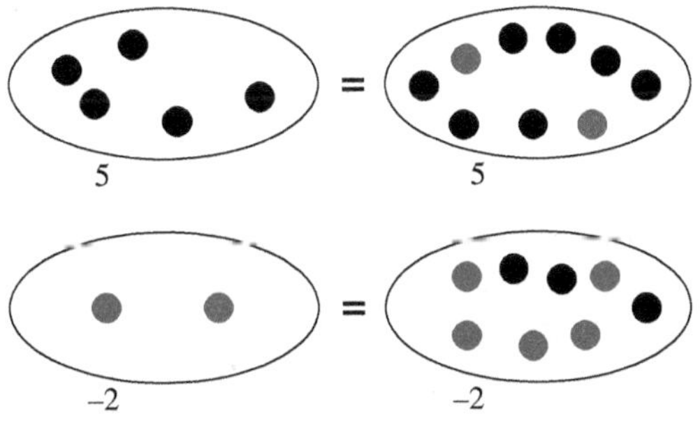

Figure 5.1 Colored-counter representations of positive 5 and negative 2

It is important to understand that a pair of oppositely colored counters, one red and one black, neutralize one another. For this reason, the colored-counter model is sometimes called a "charge" model, in which a black counter is a "unit positive charge" that is annihilated by a "unit negative charge" red counter. Since oppositely colored counters neutralize one another, any colored-counter representation of an integer can be altered either by deleting or by adding pairs of oppositely colored counters. For example, starting with the 5B loop with five black counters that represents 5, we can add two oppositely colored BR pairs to the loop to give the equivalent 7B2R representation of 5 shown in Figure 5.1. In the other direction, starting with a 3B5R loop as shown at the bottom right of Figure 5.1, we can remove three BR pairs of counters from the loop, to leave the equivalent, but simpler, representation 2R. Thus, both of the bottom loops represent negative 2, which is written -2.

A loop of colored counters *represents a positive integer* if there are more black counters than red, and the number of unpaired black counters is the absolute value of the positive integer. The positive integers are identified as the natural numbers. Similarly, if more red than black counters are in the loop, the number represented is a negative integer whose absolute value is the excess number of red counters.

Of special importance are loops that have an equal number of red and black counters, since these each represent 0. Three equivalent representations of 0 are shown in Figure 5.2.

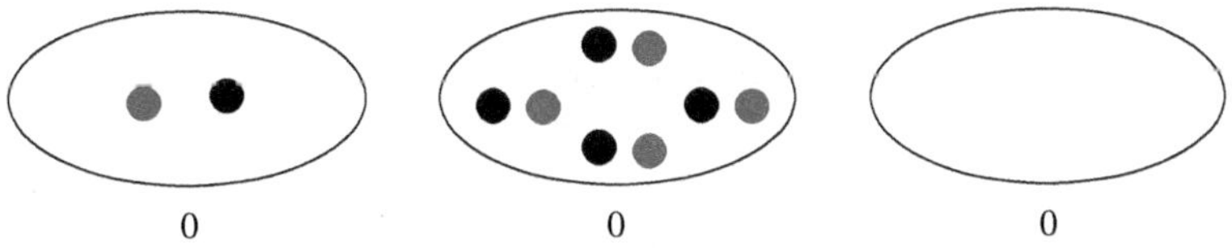

Figure 5.2 Three colored-counter representations of 0

Since there is neither an excess of black nor an excess of red counters, 0 is neither positive nor negative, and it has an absolute value of 0.

EXAMPLE 5.2

Interpreting Sets of Colored Counters as Integers

Determine the integer represented by each of these loops of colored counters:

(a) 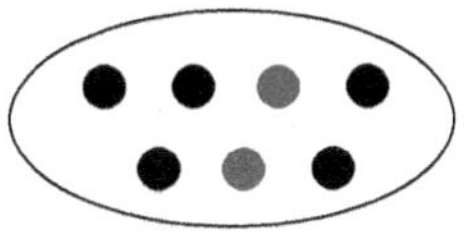
(b) 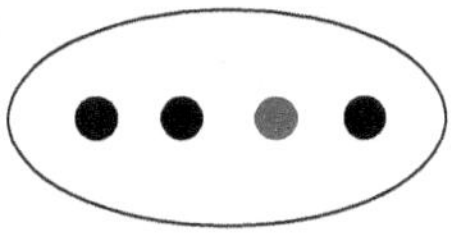

(c) 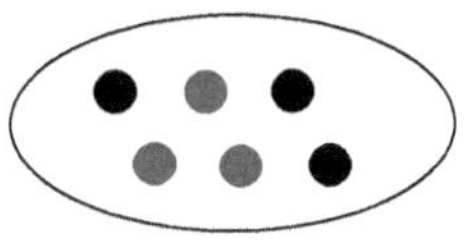
(d) 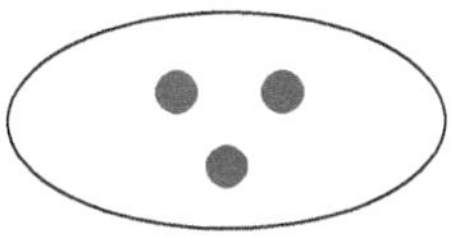

Solution

(a) Matching the two red counters with two of the black counters, we get 3B, representing the integer 3.

(b) Matching the red counter with a black counter, we get 2B, representing the integer 2.

(c) Matching the three red with the three black counters, we get 0; we have no red or black counters unmatched.

(d) Since there are no black counters, 3R represents the negative integer -3.

EXAMPLE 5.3

Representing Integers with Sets of Colored Counters

Illustrate two different colored-counter representations for each of these integers:

(a) 2 **(b)** -3 **(c)** 0 **(d)** 5

Solution

(a) Here, the loops must have two more black counters than red counters. Two possibilities are

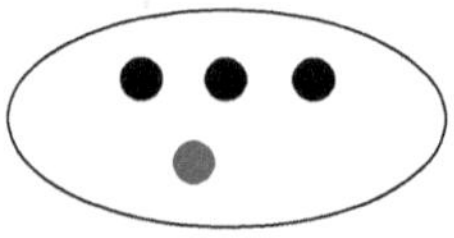 and 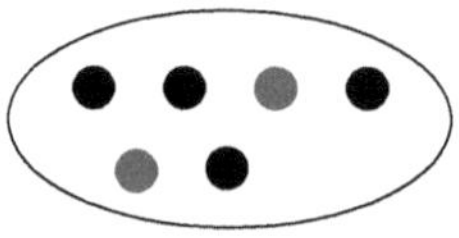

2 2

(b) Here, the loop must have three fewer black counters than red counters. Possibilities include

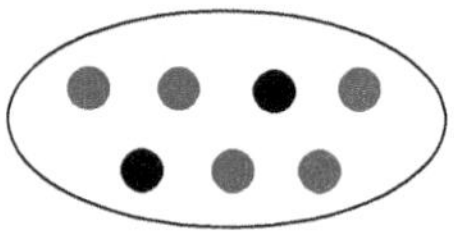 and 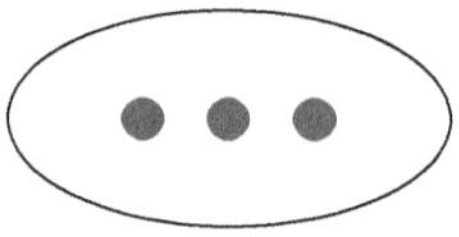

-3 -3

(c) Here, the loop must contain the same number of red and black counters or no counters at all. Possibilities include

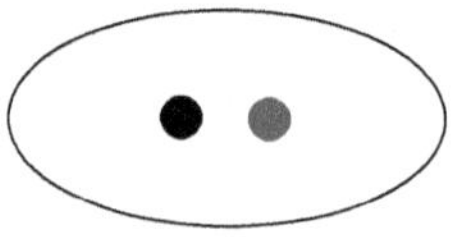 and 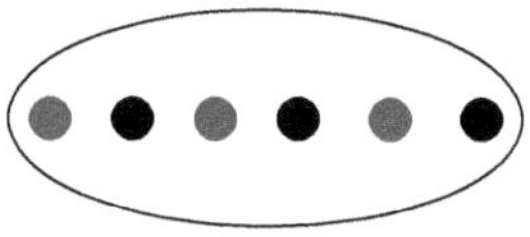

0 0

(d) Here, the loop must have five more black than red counters. Possibilities include

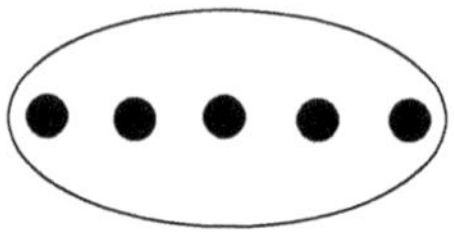 and 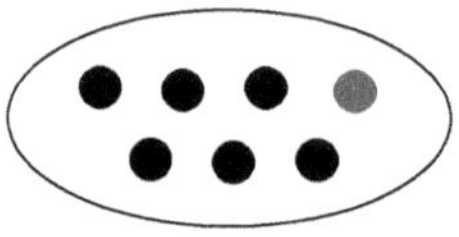

5 5

The Addition-by-0 Property with Colored Counters

Viewing addition as "putting together," the colored-counter model interprets addition straightforwardly: Combine the counters in the loops to get a loop that represents the sum. This idea will be explored in detail in the next section, but the case of addition by 0 deserves special attention. Two examples are shown in Figure 5.3, where we see that $3 + 0 = 3$ and $0 + (-2) = -2$.

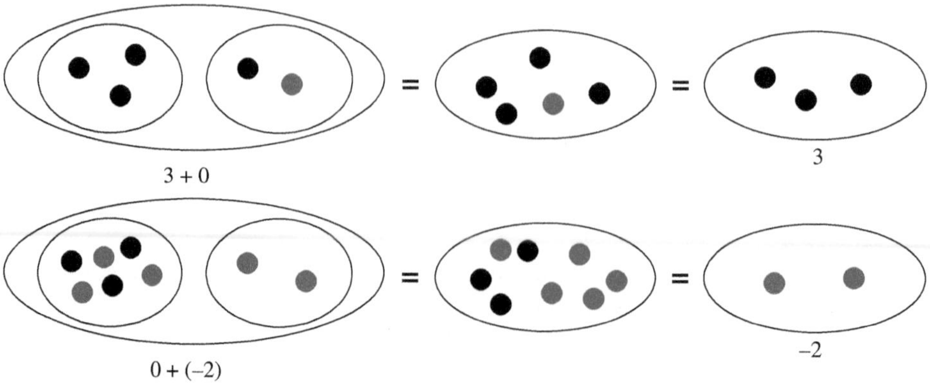

Figure 5.3 Illustrating the addition-by-0 property with colored counters

More generally, we see that the addition-by-0 property holds for the integers. That is, for every integer n—positive, zero, or negative—we have

$$n + 0 = 0 + n = n.$$

Taking Opposites with Colored Counters

Let the integer n be represented by a loop of colored counters. Then the integer represented by the loop with reversed colors represents the negative, or opposite, $-n$. For example, if 5 is represented by 5B, then its opposite is the integer -5 represented by 5R. Similarly, if -3 is represented by 3R, then its opposite is the integer 3 represented by 3B. Since reversing colors twice returns each counter to its original color, we have this result:

THEOREM The Opposite of the Opposite of an Integer

For every integer n, $-(-n) = n$.

EXAMPLE 5.4

Determining Negatives

Determine the negative, or opposite, of each of these integers:

(a) 4 **(b)** -2 **(c)** 0 **(d)** -320

Solution

(a) The negative of 4 is -4.
(b) The negative of -2 is $-(-2) = 2$.
(c) The negative of 0 is $-0 = 0$, since 0 is neither positive nor negative.
(d) The negative of -320 is $-(-320) = 320$.

Note that the negative of a negative integer is positive.

Because addition is modeled by combining colored counters, it becomes clear that, when an integer n is added to its opposite, $-n$, every counter of one color is matched to a counter of the opposite color. That is, the result is 0. The case $5 + (-5) = 0$ is shown in Figure 5.4.

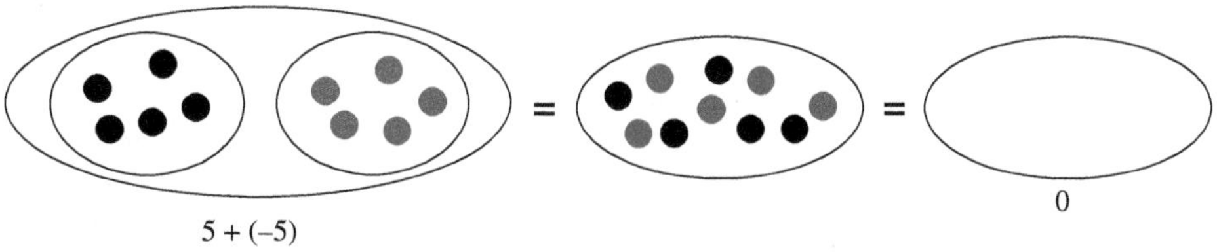

Figure 5.4 Illustrating that $5 + (-5) = 0$ with colored counters

It is now evident how the colored-counter model demonstrates that every integer has an additive inverse.

Mail-Time Representations of Integers

Integers can also be represented in other real-life situations. For example, a golf score is above or below par and the result of a football play may be a gain or loss of yardage. In personal finance, we deal with assets and debts. The following examples illustrate how the mail delivery of a check or a bill affects your overall net worth or the value of your assets at any given time:

At mail time, suppose that you are delivered a check for \$20. What happens to your net worth? Answer: It goes up by \$20.

At mail time, you are delivered a bill for \$35. What happens to your net worth? Answer: It goes down by \$35.

At mail time, you received a check for \$10 and a bill for \$10. What happens to your net worth? Answer: It stays the same.

These examples can be easily generalized. A positive integer n is represented by the delivery of a check for n dollars, a negative integer, $-m$, is represented by receiving a bill for m dollars, and 0 is represented by either no checks or bills or by a check and a bill, each for the same amount. The absolute value is the number of dollars shown on the amount payable or amount due.

EXAMPLE 5.5 Interpreting Mail-Time Situations

(a) At mail time, you are delivered a check for \$27. What happens to your net worth?
(b) At mail time, you are delivered a bill for \$36. Are you richer or poorer? By how much?

Solution

(a) Your net worth goes up \$27.
(b) Poorer. Your net worth goes down by \$36.

EXAMPLE 5.6 Describing Mail-Time Situations for Given Integers

Describe a mail-time question corresponding to each of these integers:

(a) -42 **(b)** 75 **(c)** 0

Solution

(a) At mail time, the letter carrier brought you a bill for \$42. Are you richer or poorer, and by how much?
(b) At mail time, you were delivered a check for \$75. What happens to your net worth?
(c) Quite to your surprise, at mail time the mail carrier skipped your house, so you received no checks and no bills. Are you richer or poorer, and by how much?

Number-Line Representations of Integers

Figure 5.5 shows how the number line for the whole numbers is extended to include all the integers. The points on the line representing n and its opposite $-n$ are the same distance from but on opposite sides of 0. The arrow pointing to the right on the line indicates the ordering of the integers, so $m < n$ when the point corresponding to m is to the left of the point corresponding to n. For example, $-4 < -2$, since -4 is to the left of -2. In some elementary texts, arrows are placed at both ends of the number line, though the left arrow is later dropped when the number line becomes identified as the x-axis.

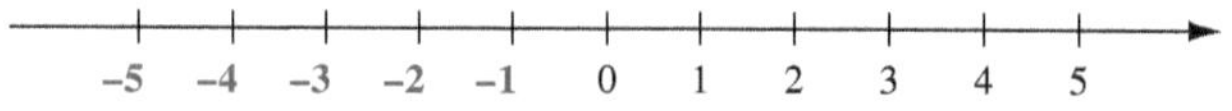

Figure 5.5
Representing integers on a number line

The number-line representation of the integers corresponds nicely to marking thermometers with degrees above 0 and degrees below 0, and with the practice in most parts of the world (with the notable exceptions of North America and Russia) of numbering floors above ground and below ground in a skyscraper. It also corresponds to the countdown of the seconds to liftoff and beyond in a space shuttle launch. The count 9, 8, 7, 6, 5, 4, 3, 2, 1, liftoff!, 1, 2, 3, . . . is not really counting backward and then forward, but forward all the time. The count is actually

9 seconds before liftoff	-9
8 seconds before liftoff	-8
⋮	⋮
1 second before liftoff	-1
Liftoff!	0
1 second after liftoff	1
2 seconds after liftoff	2

and so on, where -9 is announced as "tee minus nine" in the jargon of NASA. This example is familiar to children and helps to make positive and negative numbers real and understandable.

Integers are also represented by arrows. For example, an arrow from any point to a point 5 units *to the right* represents 5, and an arrow from any point to a point 5 units *to the left* represents -5, as illustrated in Figure 5.6. The figure also shows that 0 is the arrow whose head and tail coincide.

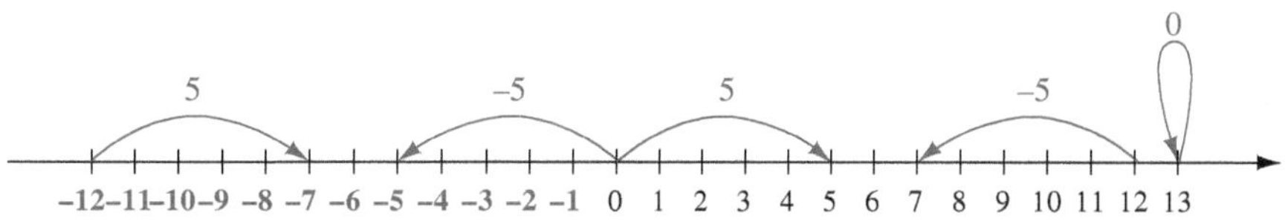

Figure 5.6
Using arrows to represent integers

It is important to understand that each integer is represented in two ways: either as a *position* along the number line or as an *arrow* that describes a "jump" along the number line. A point on the line that is n units to the right of 0 represents the positive integer n, and its opposite, $-n$, is the point on the opposite side of 0 that is equally far from 0. Alternatively, the arrow representing a positive integer n is a right-pointing arrow n units long, and its opposite, $-n$, is the left-pointing arrow of the same length. The 0 arrow is like a jump straight up, which makes no net forward or backward motion along the number line.

Unlike points on the number line, which remain in a fixed position, arrows are free to translate from side to side. It is this freedom to translate arrows that allows them to be combined to define new arrows. Therefore, arrows are used to illustrate how integers are added and subtracted, as will be shown in detail in the next section. For now, it is enough to notice that combining an arrow representing n with the 0 arrow does not change the length or direction of the n arrow. That is, $n + 0 = 0 + n = n$. Also, if jumps of equal length, but opposite direction, are combined, there is no net motion along the number line, so $n + (-n) = 0$. These fundamental ideas are shown in Figure 5.7.

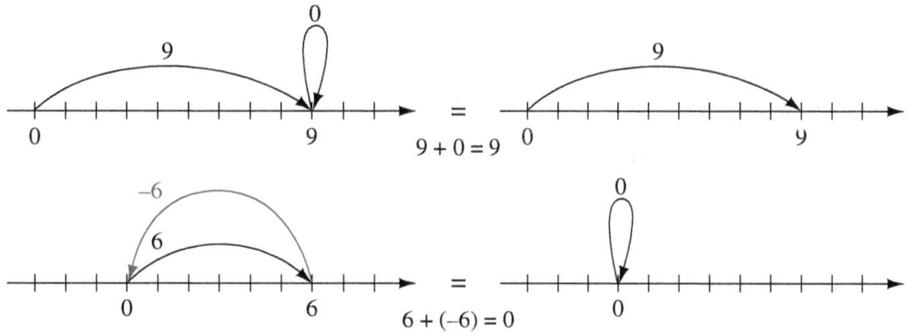

Figure 5.7
Using the number-line model to show that $9 + 0 = 9$ and $6 + (-6) = 0$

On the number line, the absolute value is the length of the jump given by the arrow that represents the integer in question. Or, if integers are viewed as points along the number line, then the absolute value is the distance from the point in question to 0, as shown in Figure 5.8.

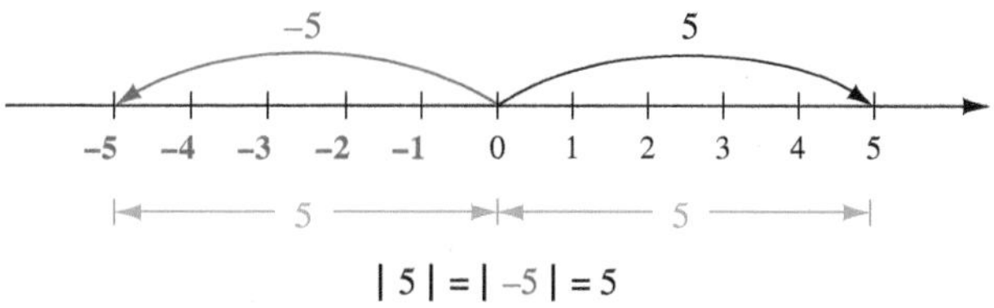

Figure 5.8
Showing the absolute value of both 5 and -5 is 5

EXAMPLE 5.7

Determining Absolute Values with the Number-Line Model

Determine the absolute values of these integers:

(a) -11 **(b)** 13 **(c)** 0 **(d)** -9

Solution

We plot the numbers on the number line and determine the distance of the points from 0.

(a) $|-11| = 11$ since -11 is 11 units from 0.
(b) $|13| = 13$ since 13 is 13 units from 0.
(c) $|0| = 0$ since 0 is 0 units from 0.
(d) $|-9| = 9$ since -9 is 9 units from 0.

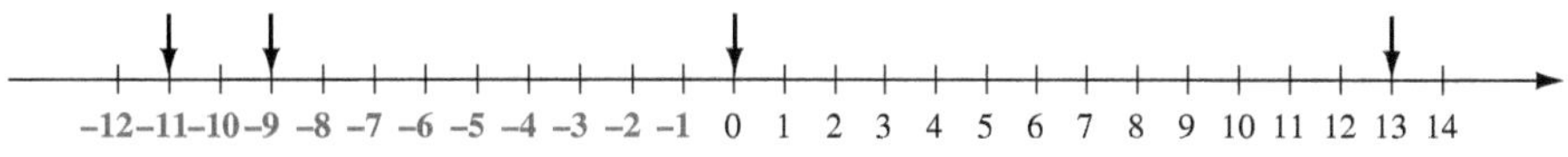

5.1 Problem Set

Understanding Concepts

1. Draw two colored-counter diagrams to represent each of these integers:
   (a) 5 (b) -2

2. Draw two colored-counter diagrams to represent each of these integers:
   (a) 0 (b) -3 (c) 3

3. Draw a colored-counter diagram with the least number of counters to represent each of the following:
   (a) 3 (b) -4

4. Draw a colored-counter diagram with the least number of counters to represent each of the following:
   (a) 0 (b) 2

5. (a) Describe a mail-time situation that illustrates 14.
   (b) Describe a mail-time situation that illustrates -27.

6. (a) At mail time, you receive a check for \$25 and an envelope with neither a bill nor a check. What addition-by-0 fact has been illustrated?
   (b) At mail time, you receive a postcard from your friend and a bill for \$32. What addition fact is illustrated?
   (c) What is illustrated if your mailbox has a bill for \$10 and a check for \$10?

7. At mail time, you are delivered a check for \$48 and a bill for \$31.
   (a) Are you richer or poorer, and by how much?
   (b) What integer addition does this situation illustrate?

8. (a) At mail time, you are delivered a check for \$27 and a bill for \$42. What integer does this situation illustrate?
   (b) Describe a different mail-time situation that illustrates the same integer as in part (a).

9. What integers are represented by the curved arrow on each of these number-line diagrams?

   (a)
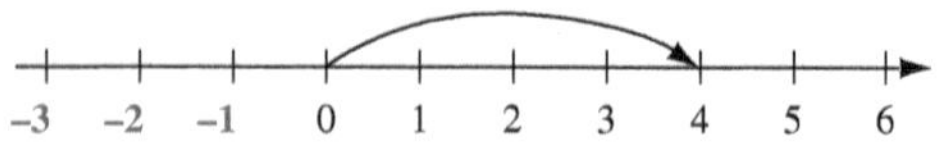

   (b)
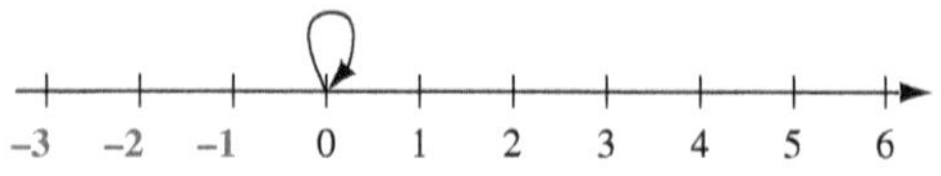

   (c)
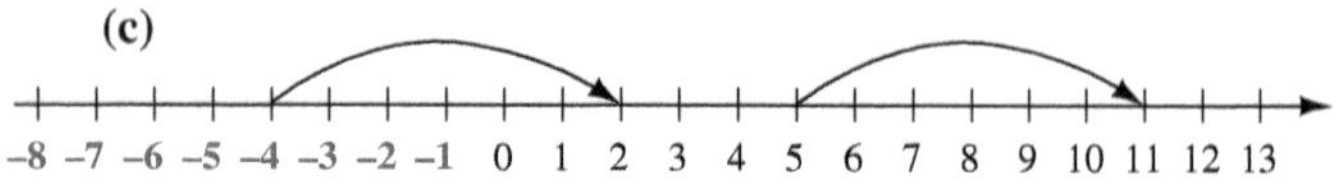

   (d)
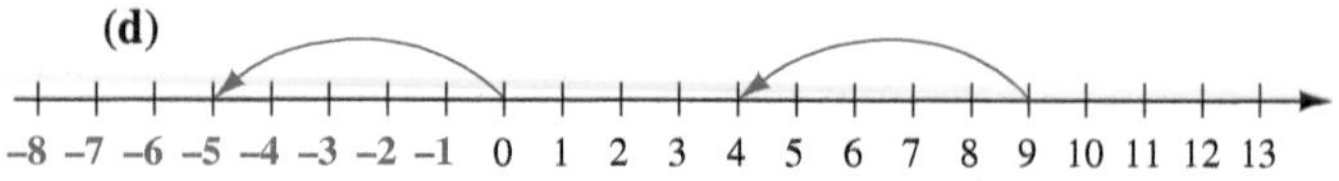

10. Draw a number line and plot the points representing these integers: 0, 4, -4, 8

11. Draw number-line arrow diagrams to represent each of these integers:
    (a) 7 (b) 0
    (c) -9 (d) 9

12. Find the absolute values of these quantities:
    (a) 34 (b) 0
    (c) -76 (d) -12

13. For what values of x are these equations true?
    (a) $|x| = 13$
    (b) $|x| + 1 = 2$
    (c) $|x| + 5 = 0$

14. Determine all pairs (x, y) of integer values of x and y for which $|x| + |y| = 2$.

Into the Classroom

15. (**Writing**)
    (a) How would you use colored counters to help students understand that $-(-4) = 4$?
    (b) How would you use colored counters to help students understand that $-(-n) = n$ for every integer n (positive, negative, or 0)?

16. (**Writing**) The definition of absolute value is often confusing to students. On the one hand, they understand that the absolute value of a number is always positive. On the other hand, the definition states that $|n| = -n$ sometimes. How would you explain this seeming contradiction?

Responding to Students

17. Althea claims that $-0 = 0$. How would you respond to Althea? Is she correct?

18. (**Writing**) Lili doesn't think she has any need for negative integers and she can get along fine without them. What questions might you ask to get her to rethink her opinion?

Thinking Critically

19. (a) What colored counters would have to be added to this set in order to represent -3?

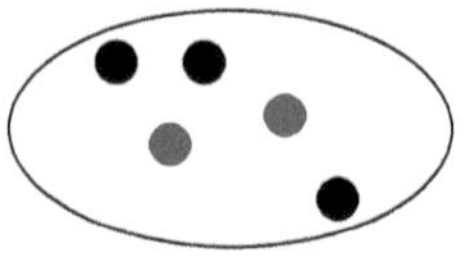

    (b) Could the question in part (a) be answered in more than one way?

20. How many different representations of -3 can be made with 20 or fewer counters?

21. If all the counters are used each time, list all the integers that can be represented with

(a) 12 counters. **(b)** 11 counters.

22. **(a)** If all the counters are used each time, describe the set of integers that can be represented with n counters.

(b) How many different integers are representable by using n counters as in part (a)?

23. You have a set of 10 red and 10 black counters. What integers are represented by the subsets of these 20 counters?

24. If some or all of the counters are used each time, describe the set of integers that can be represented

(a) with 12 counters.

(b) with 11 counters.

(c) with n counters.

25. **(a)** What integers are represented by these arrays of counters?

(b) Considering the pattern of answers to part (a), what integer would be represented by a similar array with n rows and n columns?

26. **(a)** How many black and how many red counters are there in a triangular array like this, but with 20 rows?

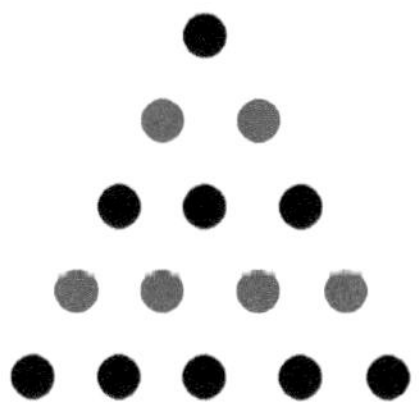

You should not actually need to make a diagram with 20 rows in order to answer this question.

(b) **(Writing)** Explain how you solved part (a).

(c) Repeat parts (a) and (b) but with a triangular array with 21 rows.

(d) What integers are represented by the triangular arrays in parts (a) and (c)?

(e) Make a table of integers represented by triangular arrays like those in parts (a) and (c), but with n rows for $n = 1$, 2, 3, 4, 5, 6, 7, and 8.

(f) Carefully considering the table of part (e), conjecture what integer is represented in a triangular array like those in parts (a) and (c), but with n rows, where n is any natural number. (*Suggestion:* Consider n odd and n even separately.)

Making Connections

27. A delivery truck leaves the warehouse on Xavier at East 2nd Street and drives east on Xavier to pick up a package at East 9th Street. The truck then heads west on Xavier to deliver the package at West 5th Street, first crossing East 1st, then Main Street, and then West 1st Street.

(a) Draw a number-line–and–arrow diagram that shows the truck's journey.

(b) Use absolute values to compute the total distance traveled from the warehouse to West 5th Street.

(c) Suppose the truck now drives east three blocks to a gas station on Xavier. What is the cross street of the gas station, and how many blocks is it from the warehouse?

28. The highest point on the North American continent is the summit of Mount McKinley, at 20,320 feet above sea level. The lowest point is Badwater Basin in Death Valley at 282 feet below sea level.

(a) What is the difference in elevation between these two extreme elevations in North America?

(b) The diameter of Earth is about 8000 miles and a mile is 5280 feet. What percentage of Earth's diameter is your answer to part (a)?

29. In Europe, the floor of a building at ground level is called the ground floor. What is called the second floor in America is called the first floor in Europe, and so on.

(a) If an elevator in a tall building in Paris, France, starts on the fifth basement level below ground, B5, and goes up 27 floors, on which numbered floor does it stop?

(b) What would the answer to part (a) be if the building were located in New York?

30. If an elevator starts on basement level B3 and goes down to B6, how far down has it gone?

31. The Wildcats made a first down on their own 33-yard line. On the next three plays, they lost 9 yards on a fumble, lost 6 yards when their quarterback was sacked, and completed a 29-yard pass play. Where was the line of scrimmage for the next play?

State Assessments

32. (California, Grade 7)
Write the absolute value of each number:

(a) $|-15| =$ _____

(b) $|+8| =$ _____

(c) $|12| =$ _____

(d) $|-20| =$ _____

(e) Which of the four numbers above has the largest absolute value?

(f) Show the absolute value of $(+4)$ as a distance on the number line below.

(g) Trace a line segment on the number line below to show the absolute value of (-7) as a distance.

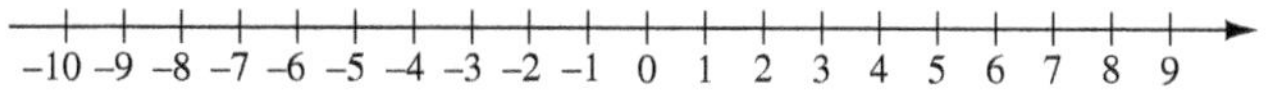

33. (Grade 5)
The highest point in the world is the top of Mount Everest in Nepal, whose summit is 29,028 feet of

elevation above sea level. The lowest point in the world is the Dead Sea in Israel, which is 1384 feet below sea level. In 2012, the countries of Nepal and Israel jointly issued a stamp contrasting these geographic wonders. If an adventurer travels from the Dead Sea to the top of Mount Everest, how much elevation is gained? Show how you get your answer with a simple diagram.

34. (Massachusetts, Grade 8. Note that students were *not* allowed to use calculators to solve this problem.)

Use the number line below to answer the question.

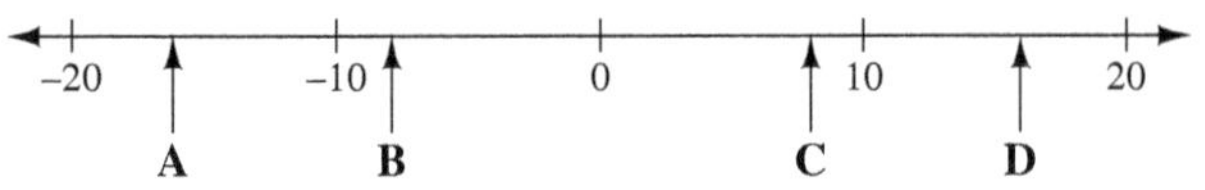

Which point represents the number $(-2)^4$?

5.2 Addition and Subtraction of Integers

Addition of Integers

In the previous section, we introduced the integers together with representations such as colored counters, mail-time stories, and number lines. In this section, we consider the addition and subtraction of integers, using the same representations to give meaning to these operations.

Addition of Integers by Using Sets of Colored Counters

Addition in the integers can be viewed as the union of sets of colored counters. For example, to add 8 and -3, we combine a loop of eight black counters with a loop of three red counters, as shown in Figure 5.9. The combined loop, 8B3R, contains five more black than red counters, so we see that $8 + (-3) = 5$. Moreover, the order of the two addends makes no difference, so the same diagram shows that $(-3) + 8 = 5$.

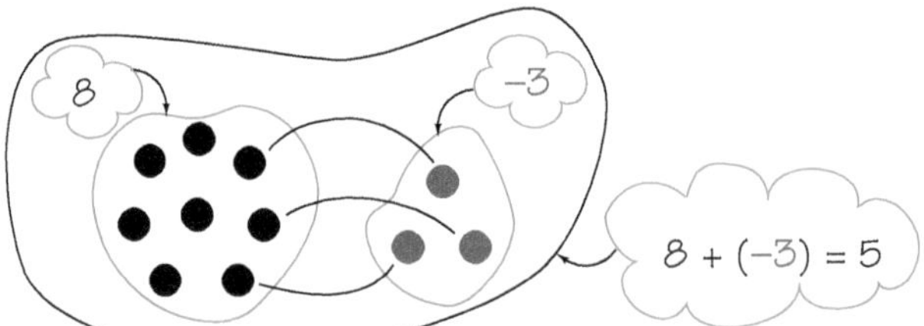

Figure 5.9
Diagram of colored counters illustrating $8 + (-3) = 5$

It would not have made any difference in the outcome if we had represented 8 and -3 differently. In Figure 5.10, 8 is represented by the loop 10B2R with ten black and two red counters, and -3 is represented by the loop 3B6R with three black and six red counters. The combined loop 13B8R, with thirteen black and eight red counters, represents the integer 5.

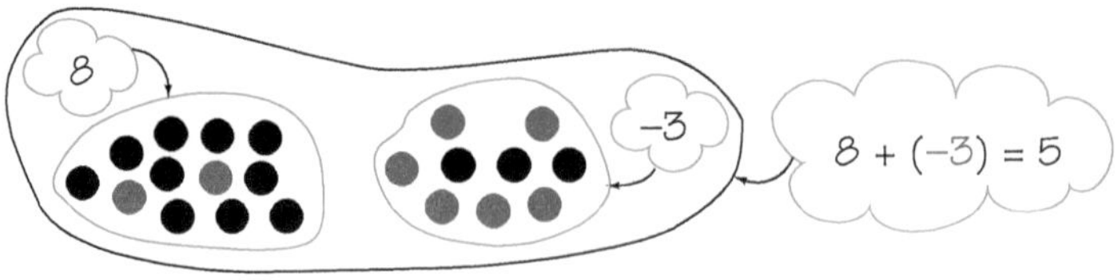

Figure 5.10
Another representation of $8 + (-3) = 5$, using sets of colored counters

EXAMPLE 5.8 **Representing Sums of Integers by Using Colored-Counter Diagrams**

Draw appropriate diagrams of colored counters to illustrate each of these sums:

(a) $(-2) + (-4)$ **(b)** $5 + (-3)$
(c) $5 + (-7)$ **(d)** $4 + (-4)$

Solution

(a) This sum can be represented as follows:

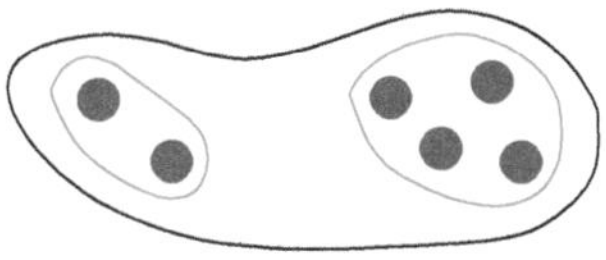

Since the combined set has a score of 6R, the diagram illustrates the sum

$$(-2) + (-4) = -6.$$

(b) Using the simplest representations of 5 and -3, we draw this diagram:

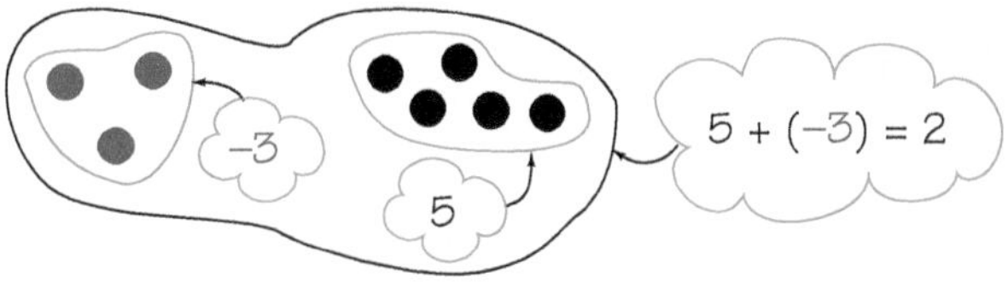

Since the combined set has a score of 2B, this diagram represents 2. Thus, $(-3) + 5 = 2$.

(c) $5 + (-7) = -2$.

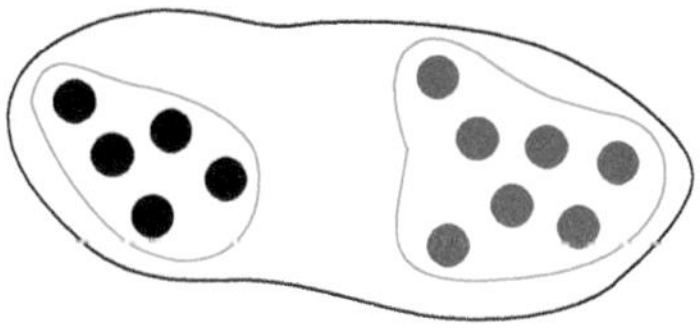

(d) $4 + (-4) = 0$.

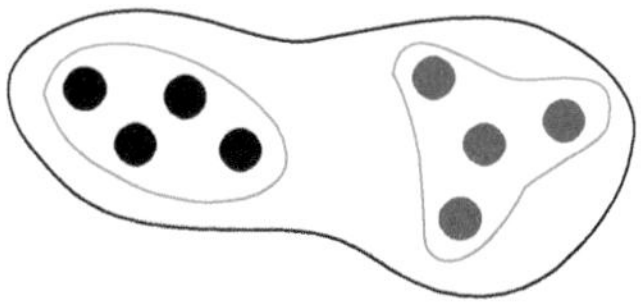

Since loops of colored counters can be combined in either order, we see that addition is a commutative operation. For example, the preceding example shows that

$$(-2) + (-4) = (-4) + (-2) \quad \text{and} \quad 5 + (-7) = (-7) + 5.$$

Example 5.8 also shows that the outcome of an addition is determined by comparing the number of black counters and the number of red counters in the combined loop. We see that

(a) combining two red and four red counters gives $2 + 4$ red counters, so

$$(-2) + (-4) = -(2 + 4).$$

(b) since $5 > 3$, combining five black and three red counters gives a loop with $5 - 3$ more black counters than red; that is,

$$5 + (-3) = 5 - 3.$$

(c) since $5 < 7$, combining five black and seven red counters gives a loop with $7 - 5$ more red counters than black; that is,

$$5 + (-7) = -(7 - 5).$$

(d) combining loops with four black and four red counters gives a loop that represents 0; that is,

$$4 + (-4) = 0.$$

The following definition is motivated by viewing the addition of integers as combining loops of colored counters.

DEFINITION Addition of Integers

Let m and n be positive integers, where $m \geq 0$ and $n \geq 0$. Then the following are true:

- $m + n = n + m$.
- $(-m) + (-n) = -(m + n)$.
- If $m > n$, then $m + (-n) = m - n$.
- If $m < n$, then $m + (-n) = -(n - m)$.
- $n + (-n) = (-n) + n = 0$.

EXAMPLE 5.9

Adding Integers

Compute these sums:

(a) $7 + 11$	**(b)** $(-6) + (-5)$	**(c)** $7 + (-3)$
(d) $4 + (-9)$	**(e)** $6 + (-6)$	**(f)** $(-8) + 3$

Solution

(a) $7 + 11 = 18$
(b) $(-6) + (-5) = -(6 + 5) = -11$
(c) Since $7 > 3$, $7 + (-3) = 7 - 3 = 4$.
(d) Since $4 < 9$, $4 + (-9) = -(9 - 4) = -5$.
(e) $6 + (-6) = 0$
(f) $(-8) + 3 = 3 + (-8)$, and since $8 > 3$, $3 + (-8) = -(8 - 3) = -5$. Therefore, $(-8) + 3 = -5$.

The first four properties listed in the theorem that follows are the same as for the whole numbers. However, the "existence of negatives" property is true of the integers but not of the whole numbers.

THEOREM Properties of the Addition of Integers

Let m, n, and r be integers. Then the following properties hold:

Closure Property	$m + n$ is an integer.
Commutative Property	$m + n = n + m$.
Associative Property	$m + (n + r) = (m + n) + r$.
Additive-Identity Property of 0	$0 + m = m + 0 = m$.
Existence of Negatives	$(-m) + m = m + (-m) = 0$.

PROOF Since we have defined integers in terms of unions of sets of colored counters, the first four properties in the theorem follow from the fact that, for any sets M, N, and R,

$$M \cup N \text{ is a set,}$$
$$M \cup N = N \cup M,$$
$$M \cup (N \cup R) = (M \cup N) \cup R, \text{ and}$$
$$\varnothing \cup M = M \cup \varnothing = M.$$

The existence of the negative, or additive inverse, of m for every integer m follows by observing that the union of a set with m red counters with a set with m black counters is a set that represents 0.

Bringing something *to* you is adding.

Addition of Integers by Using Mail-Time Stories

A second useful approach to the addition of integers is by means of mail-time stories.

At mail time, suppose you receive a check for \$13 and another check for \$6. Are you richer or poorer, and by how much? Answer: Richer by \$19. This story illustrates that $13 + 6 = 19$.

EXAMPLE 5.10

Adding Integers by Using Mail-Time Stories

Write the addition equation illustrated by each of these stories:

(a) At mail time, you receive a bill for \$2 and another bill for \$4. Are you richer or poorer, and by how much?
(b) At mail time, you receive a bill for \$3 and a check for \$5. Are you richer or poorer, and by how much?
(c) At mail time, you receive a check for \$5 and a bill for \$7. Are you richer or poorer, and by how much?
(d) At mail time, you receive a check for \$4 and a bill for \$4. Are you richer or poorer, and by how much?

Solution

(a) Receiving a bill for \$2 and another bill for \$4 makes you \$6 poorer. The story illustrates that $(-2) + (-4) = -6$.
(b) Receiving a bill for \$3 and a check for \$5 makes you \$2 richer. The story illustrates that $(-3) + 5 = 2$.
(c) Receiving a check for \$5 makes you richer by \$5, but receiving a bill for \$7 makes you \$7 poorer. The net effect is that you are \$2 poorer. The story illustrates that $5 + (-7) = -2$.
(d) Receiving a \$4 check and a \$4 bill exactly balances out, and you are neither richer nor poorer. The story illustrates that $4 + (-4) = 0$.

Note that these results are exactly the results of Example 5.8. Moreover, the arguments hold in general, and we are again led to the same definition of addition of integers.

INTO THE CLASSROOM

Kristin Hanley Using Contextual Narratives to Build Understanding of Integers

Similar to mail-time stories, I use contextual narratives to build understanding in my classroom. Let's say I had \$3 but wanted to buy a book at the book fair that cost \$5. My friend loaned me exactly how much I needed in order to buy the book. How much would I owe my friend? Most students can quickly determine I would owe \$2. Then I applaud them for their knowledge of negative numbers! They always look surprised. We introduce a way to write the equation they solved $3 + (-5) = -2$. I use the colored counters to represent the amounts of money spent and money owed in the story. Then the students are given sets of counters to complete new stories. They love to create their own stories where they gain and lose points while playing video games. These are similar to the mail-time stories where points are gained and lost continuously and present a high level of engagement for my students.

Addition of Integers by Using a Number Line

Suppose we want to illustrate 5 + 4 on a number line. This addition can be thought of as starting at 0 and then combining a jump of 5 with a jump of 4. Figure 5.11 shows that we arrive at the point 9 on the number-line. Thus, 5 + 4 = 9.

Figure 5.11
Illustrating 5 + 4 = 9 on a number line

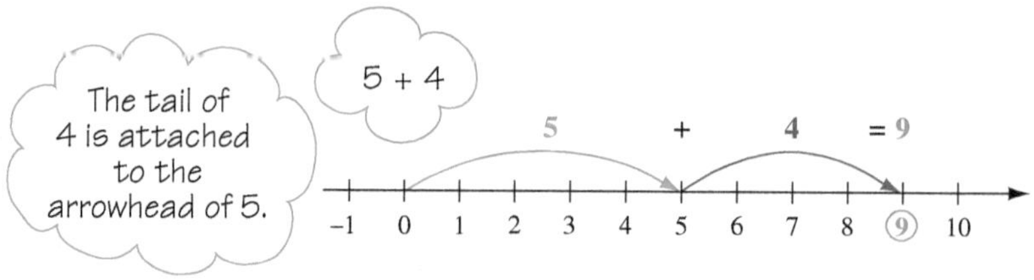

The integer −4 is represented by a left-pointing arrow of length 4. Thus, the addition 5 + (−4) is depicted on the number line as in Figure 5.12, where the right-pointing 5 arrow is combined with the left-pointing −4 arrow to show that 5 + (−4) = 1.

Figure 5.12
Illustrating 5 + (−4) = 1 on a number line

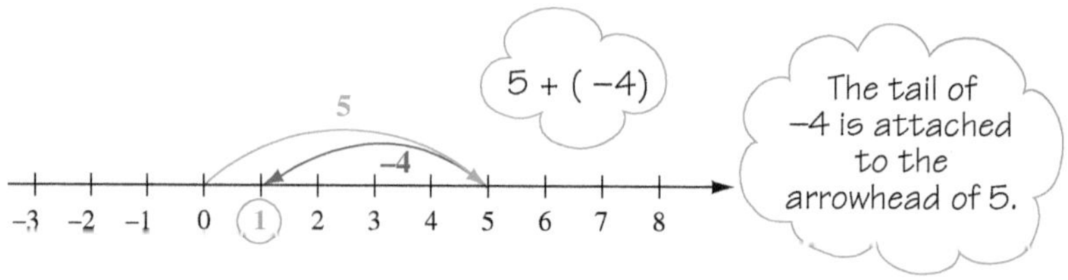

If we had first made a jump −4 to the left and then a jump 5 to the right, we would still arrive at 1. That is, (−4) + 5 = 1.

EXAMPLE 5.11

Adding Integers on a Number Line

What addition fact is illustrated by each of these diagrams?

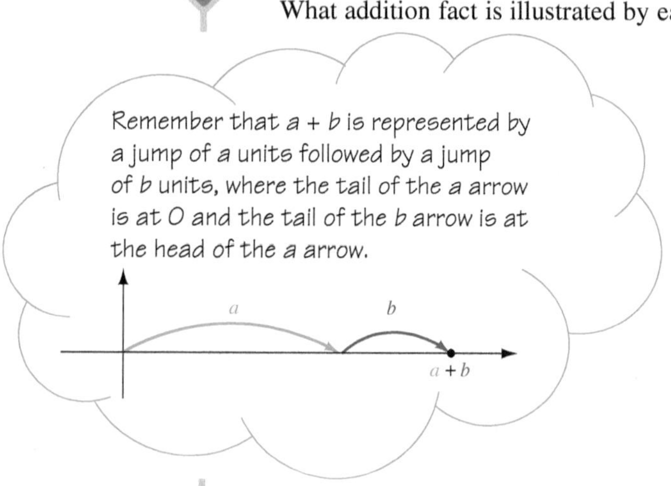

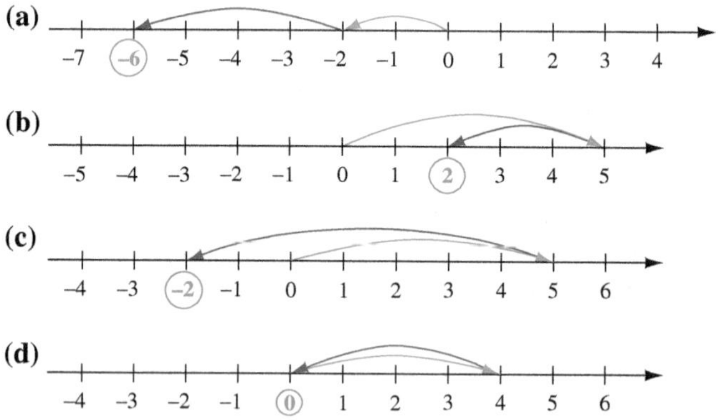

Solution

Since counting in the direction indicated by an arrow means adding, these diagrams represent the following sums:

(a) (−2) + (−4) = −6
(b) 5 + (−3) = 2
(c) 5 + (−7) = −2
(d) 4 + (−4) = 0

The results are the same as in Example 5.8.

EXAMPLE 5.12

Drawing Number-Line Addition Diagrams

Draw diagrams to illustrate the following integer additions:

(a) 2 + (−5) **(b)** (−5) + 2

Solution

(a) A jump of 2 units to the right is followed by a jump to the left of 5 units, arriving at the point −3 on the number line. Therefore, 2 + (−5) = −3.

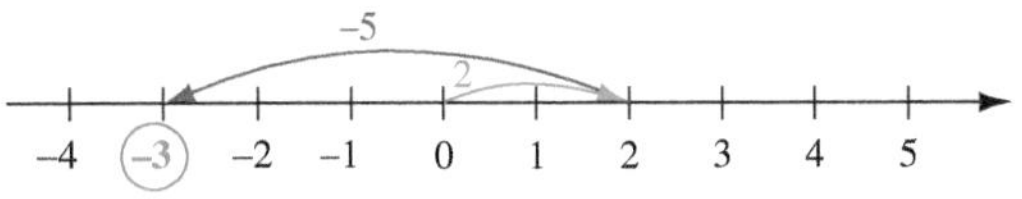

(b) The arrows are now drawn in the reverse order, with a jump to the left of 5 units followed by a jump to the right by 2 units. This shows that (−5) + 2 = −3.

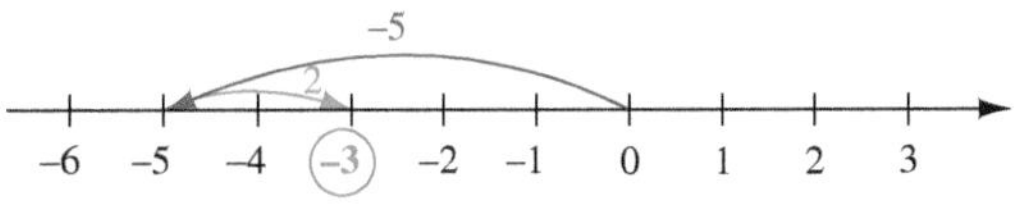

The diagrams confirm an instance of the commutative property of addition:

$$2 + (-5) = (-5) + 2.$$

Subtraction of Integers

Our conceptual models of subtraction—take away, missing addend, number line—will continue to be our guide as we extend subtraction to the integers. As with addition, we make subtraction meaningful by interpreting the operation with colored counters, mail-time stories, and jumps on the number line.

Subtraction of Integers with Colored Counters

Consider the subtraction

$$7 - (-3).$$

Modeling 7 with counters, we must "take away" a representation of −3. Since any representation of −3 must have at least three red counters, we must use a representation of 7 with at least three red counters. The simplest representation of this subtraction is shown in Figure 5.13.

Figure 5.13
Colored-counter representation of 7 − (−3) = 10

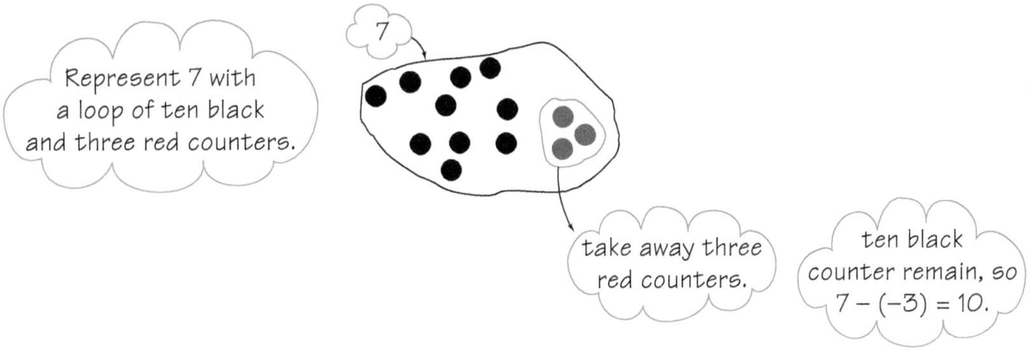

Thus, 7 − (−3) = 10. The result does not change if we use another representation of 7 with at least three red counters. A second representation of 7 − (−3) = 10 is shown in Figure 5.14, where 7 is represented with a loop containing 14 black and 7 red counters.

Figure 5.14
A second colored-counter representation of $7 - (-3) = 10$

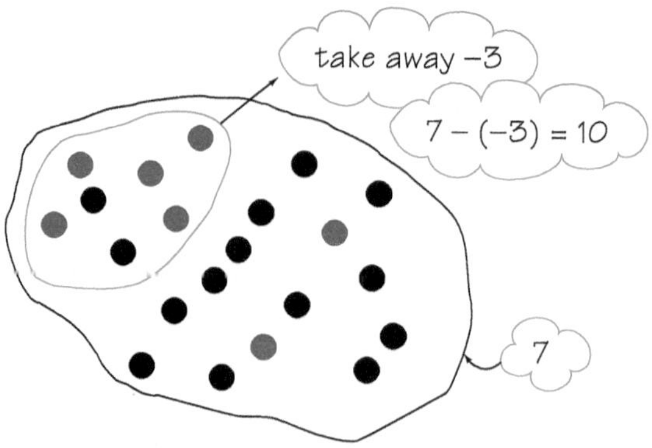

EXAMPLE 5.13

Subtracting by Using Colored Counters

Write out the subtraction equations illustrated by these diagrams:

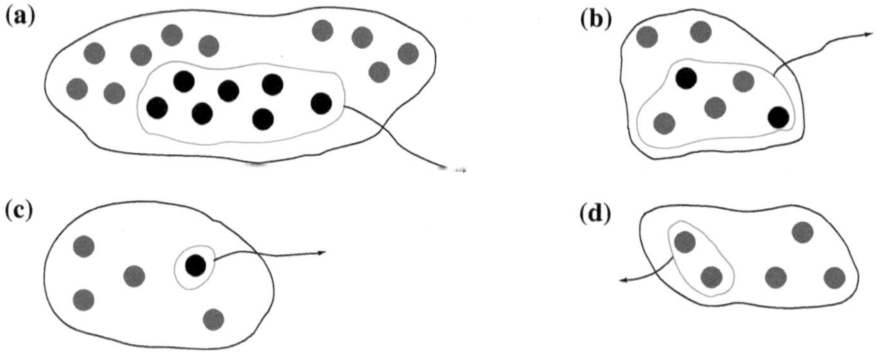

Solution

(a) Since the large loop contains ten red counters and seven black counters, it represents -3. The small loop contains seven black counters and thus represents 7. If we remove the small loop of counters as indicated, we are left with ten red counters, representing -10. Thus, this diagram illustrates the subtraction $(-3) - 7 = -10$.
(b) This diagram represents the subtraction $(-3) - (-1) = -2$.
(c) This diagram represents the subtraction $(-3) - 1 = -4$.
(d) This diagram represents the subtraction $(-5) - (-2) = -3$.

EXAMPLE 5.14

Drawing Diagrams for Given Subtraction Problems

Draw a diagram of colored counters to illustrate each of these subtractions, and determine the result in each case:

(a) $7 - 3$ **(b)** $(-7) - (-3)$ **(c)** $7 - (-3)$ **(d)** $(-7) - 3$

Solution

(a) Many different diagrams could be drawn, but the simplest is shown here:

(**b**) As in part (a), we can use a diagram with counters of only one color:

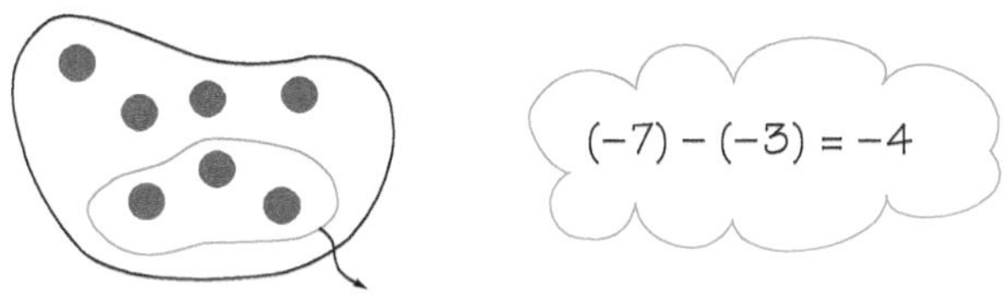

(**c**) This subtraction is illustrated in Figures 5.13 and 5.14.

(**d**) Here, in order to remove three black counters (that is, subtract 3), the representation for -7 must have at least three black counters. The simplest diagram is the following:

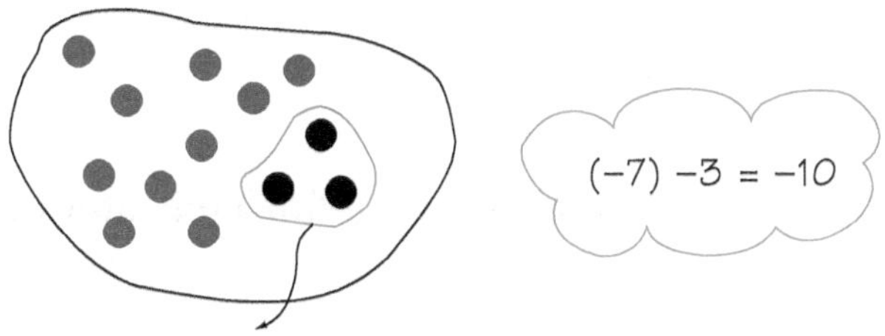

The colored-counter model illustrates why subtraction is the inverse of addition. For example, Figure 5.15a shows that $(-2) - 4 = -6$, since removing four black counters from the loop 4B6R that represents -2 leaves a loop with six red counters. However, if the procedure is done in reverse order, as in Figure 5.15b, we see that a loop representing -2 is obtained by adding four black counters to a loop of six red counters. Thus, $-2 = (-6) + 4$.

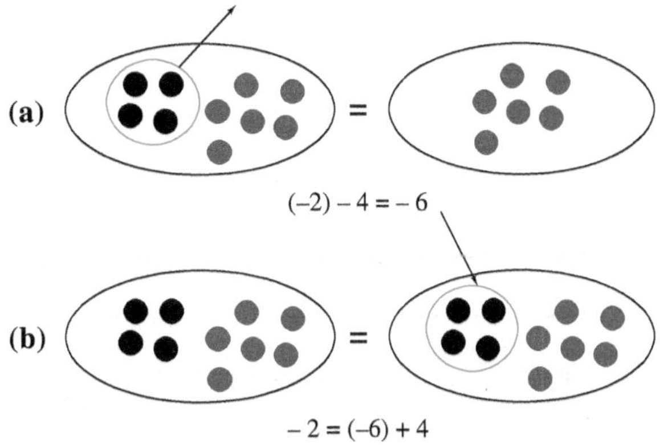

Figure 5.15
Showing the subtraction $(-2) - 4 = -6$ is equivalent to the addition fact $-2 = (-6) + 4$

Since the addition of integers has already been defined, subtraction is defined as the inverse of addition.

DEFINITION Subtraction of Integers

If a, b, and c are integers, then

$$a - b = c \text{ if and only if } a = c + b.$$

By the commutative property of addition, $b + c = c + b$, we obtain the fact family

$$a - b = c, a = c + b, a = b + c, a - c = b.$$

If one of these equations is true, then all the equations are true.

The Equivalence of Subtraction with Addition of the Opposite

We know that $5 - 2 = 3$, since removing two black counters from a loop of five black counters leaves three black counters in the loop. However, there is a second way to remove two black counters from the loop: Annihilate them by adding two red counters. This still leaves two black counters, so the subtraction $5 - 2$ can be replaced by the equivalent addition $5 + (-2)$. The addition and subtraction diagrams in Figure 5.16 show that $5 - 2 = 5 + (-2) = 3$.

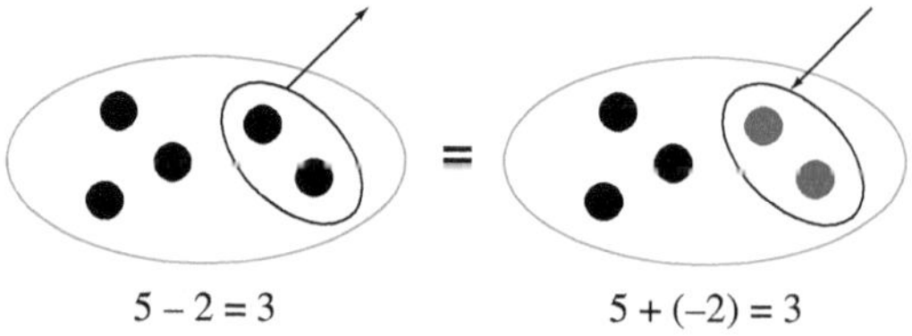

Figure 5.16
Showing subtraction by 2 is equivalent to addition by −2

Since removing counters of one color is equivalent to adding counters of the opposite color, we obtain the following theorem.

To subtract, add the negative.

THEOREM Subtracting by Adding the Opposite
Let a and b be any integers. Then

$$a - b = a + (-b).$$

Since the set of integers is closed under addition, a consequence of this theorem is that *the set of integers is closed under subtraction.*

THEOREM Closure Property for the Subtraction of Integers
The set of integers is closed under subtraction.

For example, in the set of whole numbers, there is no answer to $3 - 5$. But in the set of integers, $3 - 5 = -2$.

EXAMPLE 5.15

Subtracting by Adding the Opposite

Perform each of these subtractions as additions:

(a) $7 - 3$ **(b)** $(-7) - (-3)$ **(c)** $7 - (-3)$ **(d)** $(-7) - 3$

Solution

We make use of the theorem stating that $a - b = a + (-b)$ for any integers a and b.

(a) $7 - 3 = 7 + (-3) = 4$
(b) $(-7) - (-3) = (-7) + 3 = -4$
(c) $7 - (-3) = 7 + 3 = 10$
(d) $(-7) - 3 = -7 + (-3) = -10$

Subtraction of Integers by Using Mail-Time Stories

For this model of subtraction to work, we must imagine a situation where checks and bills are immediately credited or debited to your account as soon as they are delivered, whether they are really intended for you or not. If an error has been made by the mail carrier, he or she must return and reclaim delivered mail and take it to the intended recipient. Thus,

bringing a check adds a positive number,
bringing a bill adds a negative number,
taking away a check subtracts a positive number, and
taking away a bill subtracts a negative number.

EXAMPLE 5.16

Subtraction Facts from Mail-Time Stories

Indicate the subtraction facts that are illustrated by each of the following mail-time stories:

(a) The mail carrier brings you a check for \$7 and takes away a check for \$3. Are you richer or poorer, and by how much?
(b) The mail carrier brings you a bill for \$7 and takes away a bill for \$3. Are you richer or poorer, and by how much?
(c) The mail carrier brings you a check for \$7 and takes away a bill for \$3. Are you richer or poorer, and by how much?
(d) The mail carrier brings you a bill for \$7 and takes away a check for \$3. Are you richer or poorer, and by how much?

Solution

(a) You are \$4 richer. This illustrates the subtraction fact $7 - 3 = 4$.
(b) You are \$4 poorer. This illustrates the subtraction fact $(-7) - (-3) = -4$.
(c) You are \$10 richer. This illustrates the subtraction fact $7 - (-3) = 10$.
(d) You are \$10 poorer. This illustrates the subtraction fact $(-7) - 3 = -10$.

These are the same subtractions that were illustrated in Example 5.14 with diagrams of colored counters and in Example 5.15 by adding negatives.

Subtraction of Integers by Using the Number Line

The addition $5 + 3 = 8$ is illustrated on a number line in Figure 5.17. We see that combining rightward jumps of 5 and 3 is the same as a jump of 8 to the right.

Figure 5.17
The sum $5 + 3 = 8$ on the number line

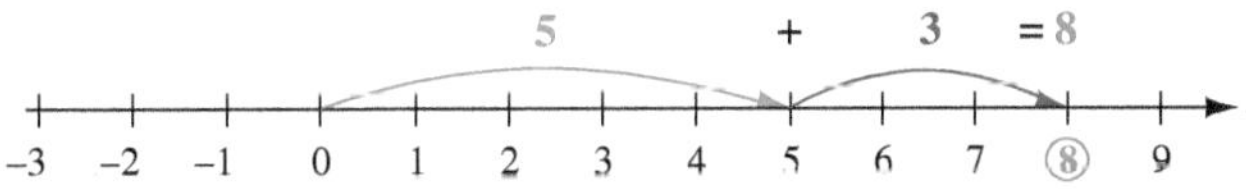

Now consider the subtraction

$$5 - 3.$$

In this instance, we start at 0 and jump five units to the right as before, and next we *remove* three units, as represented by the right-pointing arrow of length 3. Figure 5.18 shows that the result is then 2, as we already know from the subtraction of whole numbers. In particular, this diagram models the missing-addend approach to subtraction. For example, since $5 - 3$ is the same as asking "What must be added to a jump of 3 to get to 5?" we see it is a rightward jump of 2. That is, $5 - 3 = 2$ since $2 + 3 = 5$.

Figure 5.18
The subtraction $5 - 3 = 2$ on the number line

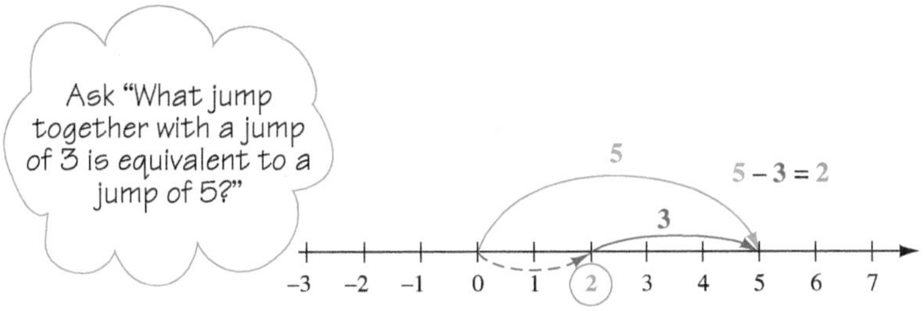

EXAMPLE 5.17

Drawing Diagrams to Illustrate Subtraction on the Number Line

Illustrate each of these subtractions on a number line, and give the result in each case:

(a) $7 - 3$ **(b)** $(-7) - (-3)$
(c) $7 - (-3)$ **(d)** $(-7) - 3$

Solution

(a)

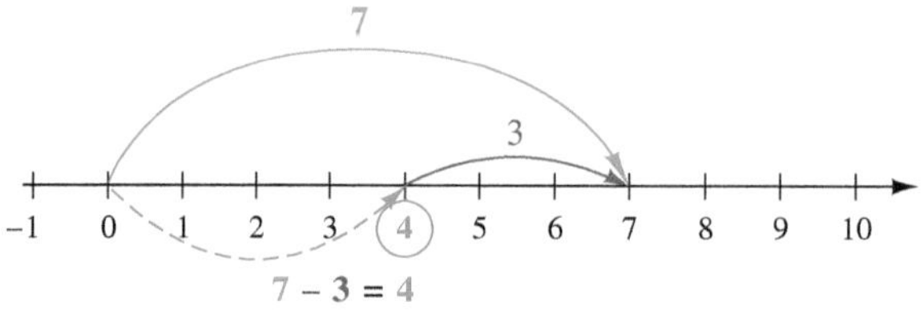

(b)

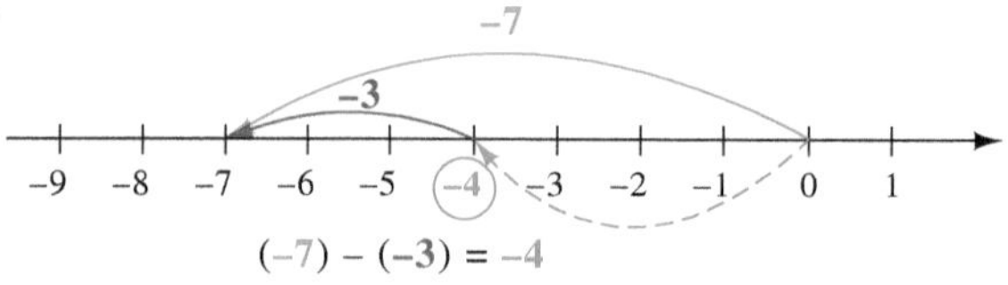

(c)

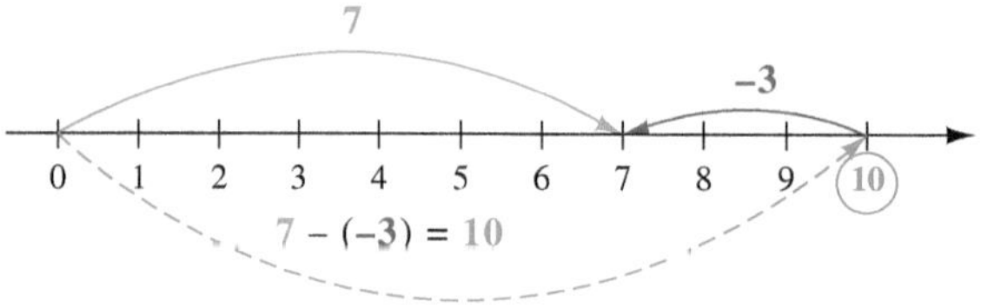

(d)

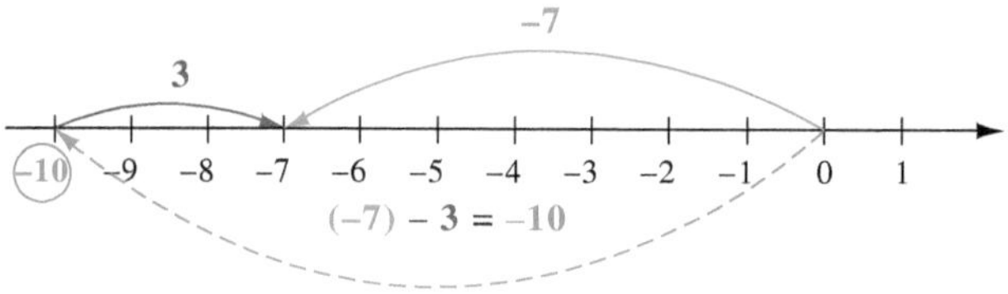

Number-line diagrams for addition and subtraction are not the same. Figure 5.19a shows the addition diagram for $4 + (-7) = -3$: A rightward jump of 4 is followed by a leftward jump of -7 to arrive at the point -3. Figure 5.19b is the subtraction diagram showing that $4 - 7 = -3$: The rightward jump of 4 is followed by the removal of a rightward jump of 7.

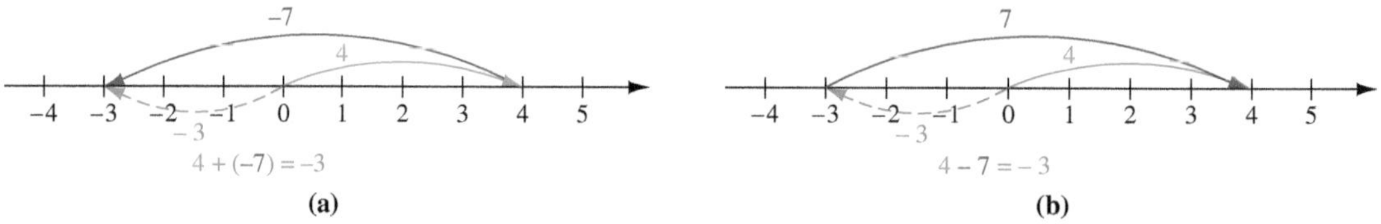

Figure 5.19
Comparing diagrams for addition and subtraction

Figure 5.19 also shows that subtraction is equivalent to adding the opposite, since $4 - 7 = 4 + (-7)$.
Number-line diagrams obey these rules:

- Positive integers are always shown with right-pointing arrows.
- Negative integers are always shown with left-pointing arrows.
- For the addition $a + b$, first draw the arrow for a and then place the tail of the b arrow at the head of the a arrow, so that $a + b$ is the combined jump.
- For the subtraction $a - b$, first draw the arrow for a, and then draw the arrow for b with its head at the head of the arrow for a, so that removing the b arrow (following it backward) arrives at the point given by $a - b$.

With these rules, it is easy to distinguish an addition diagram from a subtraction diagram.

COOPERATIVE INVESTIGATION

Extending the Fibonacci Sequence with Integer Subtraction

The goal of this activity is to extend the Fibonacci numbers to become a doubly infinite sequence and then investigate patterns in the extended Fibonacci numbers. This requires addition and subtraction of integers.

Materials

Each group of two to four students should have several strips of paper divided into uniform squares, about 20 to 25 squares per strip. Each group should also have some marker pens to write numbers to the squares.

Directions

Working as a group to check the arithmetic, write the Fibonacci numbers $F_1 = 1$ and $F_2 = 1$ in two adjacent squares about halfway along a strip and then use addition to fill in more of the Fibonacci sequence in the squares to the right. The strip will therefore begin this way.

									1	1	2	3	5	8					

Activities and Questions

(a) Describe how the squares on the left portion of the strip can be filled using subtraction, and fill in these squares once a decision is made. *Hint:* The first square to the left of the 1s should be $1 - 1 = 0$.

(b) The sixth and seventh squares to the right of the 0 square should contain the numbers $F_6 = 8$ and $F_7 = 13$. What numbers, F_{-6} and F_{-7}, did you place in the sixth and seventh squares to the left of the 0 square? In general, how do you think the numbers F_{-n} and F_n on the opposite sides of the 0 square are related?

(c) On another paper strip, placed just below the first strip, write in the sums of the Fibonacci numbers beginning with 0. That is, the second strip has the sum $S_n = F_0 + F_1 + \cdots + F_n$ in the nth square to the right and $S_{-n} = F_0 + F_{-1} + \cdots + F_{-n}$ in the nth square to the left. The strips should include the numbers shown here, as well as a number in each of the other squares. Your strips should show that $S_8 = F_0 + F_1 + \cdots + F_8 = 54$ and $S_{-5} = F_0 + F_{-1} + \cdots + F_{-5} = 4$.

				−3				0	1	1	2	3	5					55	
			4					0	1	2	4	7				54			

On the basis of the pattern seen in the strips, conjecture a formula of S_n and S_{-n} that is related to the Fibonacci numbers in the upper strip.

For Further Exploration

Make a similar exploration in which the Fibonacci sequence is replaced with the Lucas sequence $L_1 = 1$, $L_2 = 3$, $L_3 = 4$, $L_4 = 7, \ldots$, where $L_{n+2} = L_{n+1} + L_n$ for all integers. Thus, $L_0 = 3 - 1 = 2$.

Note: Save your strip of Fibonacci numbers (and the strip of Lucas numbers if it was created) for the Cooperative Investigation in the next section.

Ordering the Set of Integers

The integers are ordered in the same way the whole numbers are ordered. For example, we say that $-3 < 2$ since positive number 5 added to negative number 3 is 2. Similarly, $-8 < 1$ since $7 + (-8) = -1$, and 7 is positive. In general, we have this definition.

DEFINITION Less Than and Greater Than for the Set of Integers

Let a and b be integers. Then ***a* is less than *b***, written $a < b$, if and only if there is a positive integer c for which $c + a = b$. Similarly, ***a* is greater than *b***, written $a > b$ if and only if $b < a$.

Since the value of c is given by $c = b - a$, and subtraction is always defined in the integers, we see that

$$a < b \textit{ if and only if } b - a > 0.$$

The order relation is also easy to visualize on the number line by noting that $a < b$ precisely when the point b is reached by a rightward jump from a. Thus,

$a < b$ if and only if the point a is to the left of point b on the number line.

If a and b are any two integers, they are equal, or a is to the left of b, or b is to the left of a. This observation is known as the *law of trichotomy.*

THEOREM The Law of Trichotomy

If a and b are any two integers, then precisely one of the following relations holds:

$$a < b, \text{ or } a = b, \text{ or } a > b.$$

EXAMPLE 5.18

Ordering a List of Integers

Locate the integers $A, B, C, D, E,$ and F that follow on the number line and then write them in nondecreasing order using the symbols $<$ and $=$ as needed.

$$A = 5 + (-5), B = 2 - 7, C = 7 - 2, D = |3 - 6|, E = (-2) + 5, F = (-4) - 2$$

Solution

Since

$$A = 0, B = -5, C = 5, D = 3, E = 3, F = -6,$$

we see that $F < B < A < D = E < C$.

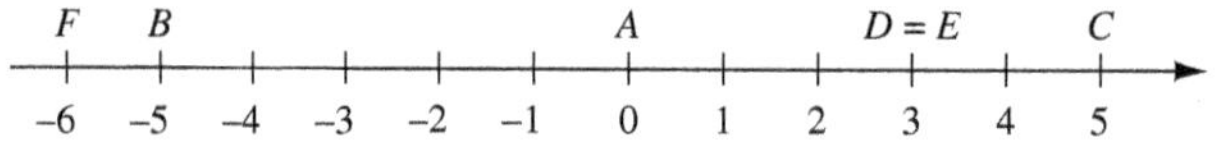

COOPERATIVE INVESTIGATION

Balloon Rides* I: Addition and Subtraction

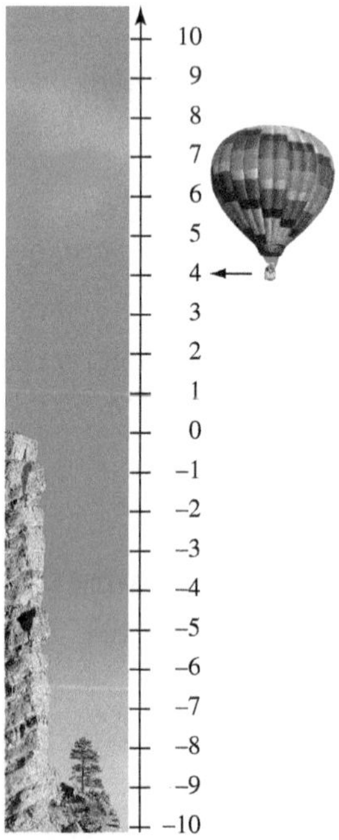

Let's go on a balloon ride over the Grand Canyon! The rim of the canyon is considered level 0, and we use a vertical number line to show our altitude above the rim (by a positive integer) or below the rim (by a negative integer). Our altitude is adjusted by adding or removing gasbags or sandbags. Indeed, let's suppose that

- **adding** 1 gasbag increases our height by 1;
- **adding** 1 sandbag decreases our height by 1;
- **subtracting** 1 gasbag decreases our height by 1;
- **subtracting** 1 sandbag increases our height by 1.

Our balloon ride can now be described with the addition and subtraction of integers, where a positive integer represents the number of gasbags moved and a negative integer represents the number of sandbags moved. For example, starting at level 4, adding 3 gasbags increases our height to level 7. That is, $4 + 3 = 7$. Similarly, $7 - 10 = -3$, since, in "balloon language," if we start at level 7 and remove 10 gasbags, we lower our altitude by 10 units to end up at level -3, 3 units below the canyon rim.

Problems and Questions for Small Groups

1. Translate $4 + (-9) = -5$ into balloon language.
2. Translate the balloon language statement, "Starting at -3, we added 7 sandbags to reach level -10."
3. Translate $8 - 3 = 5$ into balloon language.
4. Translate "Starting at -5, we removed 8 sandbags to get to level 3" into an integer equation.
5. Give two ways, each described in both balloon language and an integer equation, to start at level 3 and go to level 11. You are allowed to move bags of only one type.
6. What theorem about integers is illustrated in question 5?
7. Your balloon ride is over, so you must return to level 0 to land safely on the canyon rim. If your current height is 5, what bags should you add? If your current height is -7, what bags should you add?
8. What property of integers is illustrated by your response to question 7?

*This activity is adapted from a presentation by Bill Kring.

5.2 Problem Set

Understanding Concepts

1. Draw diagrams of colored counters to illustrate these computations, and state the answer in each case:

(a) $8 + (-3)$ (b) $(-8) + 3$
(c) $(-8) - (-3)$ (d) $8 - (-3)$

2. Draw diagrams of colored counters to illustrate these computations, and state the answer in each case:

(a) $9 + 4$ (b) $9 + (-4)$
(c) $(-9) + 4$ (d) $(-9) - (-4)$

3. Describe mail-time situations that illustrate each computation, and state the answer in each case:

(a) $(-27) + (-13)$ (b) $(-27) - 13$
(c) $27 + 13$ (d) $27 - 13$

4. Describe mail-time situations that illustrate each computation, and state the answer in each case:

(a) $(-41) + 13$ (b) $(-41) - 13$
(c) $(-13) + 41$ (d) $13 - 41$

5. Draw number-line diagrams that illustrate each computation, and state the answer in each case:

(a) $8 + (-3)$ (b) $8 - (-3)$
(c) $(-8) + 3$ (d) $(-8) - (-3)$

6. Draw number-line diagrams that illustrate each computation, and state the answer in each case:

(a) $4 + (-7)$ (b) $4 - (-7)$
(c) $(-4) + 7$ (d) $(-4) - (-7)$

7. Write each of these subtractions as an addition:

(a) $13 - 7$ (b) $13 - (-7)$
(c) $(-13) - 7$ (d) $(-13) - (-7)$

8. Write each of these subtractions as an addition:

(a) $3 - 8$ (b) $8 - (-3)$
(c) $(-8) - 13$ (d) $(-8) - (-13)$

9. Perform each of these computations:

(a) $27 - (-13)$ (b) $12 + (-24)$
(c) $(-13) - 14$ (d) $-81 + 54$

10. Perform each of these computations:

(a) $(-81) - 54$ (b) $(-81) - (-54)$
(c) $(-81) + (-54)$ (d) $27 + (-13)$
(e) $(-27) - 13$

11. Use mental arithmetic to estimate the given sums and differences. Indicate the reasoning you have used to make your estimates.

(a) $(-356) + 148$ (b) $728 + (-273)$
(c) $298 - (-454)$ (d) $-827 - 370$

12. By 2 P.M., the temperature in Cutbank, Montana, had risen 31° from a nighttime low of 41° below 0.

(a) What was the temperature at 2 P.M.?
(b) What computation does part (a) illustrate?

13. (a) If the high temperature on a given day was 2° above 0 and the morning's low was 27° below zero, how much did the temperature rise during the day?
(b) What computation does part (a) illustrate?

14. (a) If the high temperature for a certain day was 8° above 0 and that night's low temperature was 27° below 0, how much did the temperature fall?
(b) What computation does part (a) illustrate?

15. During the day, Sam's Soda Shop took in \$314. That same day, Sam paid a total of \$208 in bills.

(a) Was Sam's net worth more or less at the end of the day? By how much?
(b) What computation does part (a) illustrate?

16. During the day, Sam's Soda Shop took in \$284. Also, Sam received a check in the mail for \$191 as a refund for several bills that he had inadvertently paid twice.

(a) Was Sam's net worth more or less at the end of the day? By how much?
(b) What computation does part (a) illustrate?
(c) If you think of the \$191 check as removing or taking away the bills previously paid, what computation does this represent? Explain.

17. Order these pairs of integers:

(a) $-117, -24$ (b) $0, -4$ (c) $18, 12$

18. Order these pairs of integers:

(a) $18, -12$ (b) $-5, 1$ (c) $-5, -9$

19. List these numbers in increasing order from least to greatest: $-5, 27, 5, -2, 0, 3, -17$.

Into the Classroom

20. (**Writing**) Maximus Planudes (c. A.D. 1255–1305) helped introduce Indo-Arabic numerals and arithmetic calculation into Europe. In *The Great Calculation According to the Indians,* he states, "It is not possible to take a greater from a lesser number, for it is not possible to take away what is not there." The question still arises even today. Design a poster for the upper elementary school classroom that shows how integers—both positive and negative—are used today.

21. (**Writing**) Students are often puzzled by the similarly labeled buttons [(−)] and [−] on a calculator. Create a classroom lesson that distinguishes between negation ("take the opposite") and subtraction ("take the difference").

Responding to Students

22. Keyshawn drew the following diagram to illustrate the subtraction of -3 from 7, arguing that you have to count from 7 on the number line 3 steps in the opposite direction from that indicated by -3:

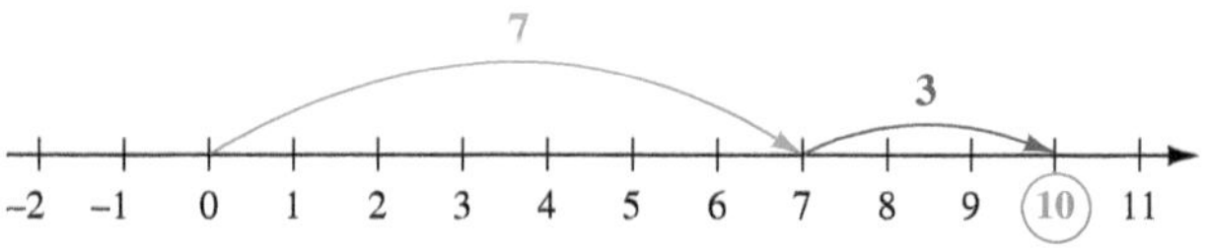

How would you respond to Keyshawn? Compare Keyshawn's response with the solutions to part (c) of both Example 5.16 and Example 5.17.

23. Melanie says that adding the opposite, $-n$, of a number n to $m + n$ just exactly nullifies the effect of adding n to m. Is Melanie right? How would you respond to her statement?

24. Julliene answered "No" to the question "Is the inequality $10 \geq 7$ a correct assertion?" She maintained that the inequality should be $10 > 7$. Is Julliene correct? How would you respond to her answer?

25. Armand said that to represent subtracting -3 from 7 by using colored counters, you could add three black counters to seven black counters.

(a) Is Armand correct?
(b) Determine the difference $7 - (-3)$.
(c) How would you respond to Armand?

Thinking Critically

26. Which of the following are *true*?

(a) $3 < 12$ **(b)** $-3 < -12$
(c) $-3 < 12$ **(d)** $3 < -12$

27. **(a)** If $a < b$, is $a \leq b$ true? Explain.
(b) If $a \geq b$, is $a > b$ true? Explain.
(c) Is $2 \geq 2$ true? Explain.

28. **(a)** For which integers x is it true that $|x| < 7$?
(b) For which integers x is it true that $|x| > 99$?

29. Compute each of these absolute values:

(a) $|5 - 11|$
(b) $|(-4) - (-10)|$
(c) $|8 - (-7)|$
(d) $|(-9) - 2|$

30. Draw a number line, and determine the distance between the points on the number line for each of these pairs of integers.

(a) 5 and 11 **(b)** -4 and -10
(c) 8 and -7 **(d)** -9 and 2

31. Given two points a and b on the number line, what is the meaning of $|a - b|$?

32. Make a magic square using the integers $-4, -3, -2, -1, 0, 1, 2, 3,$ and 4.

33. Complete the magic square using the integers $-7, -6, \ldots, 0, 1, \ldots, 8$. What is the magic sum?

	-6	8	1
4			-2
-7		2	
	3		

34. In each of the following diagrams, determine numbers to place in the red circles so that the sum of the numbers in each pair of adjacent red circles is the number in the corresponding large blue circle:

(a)

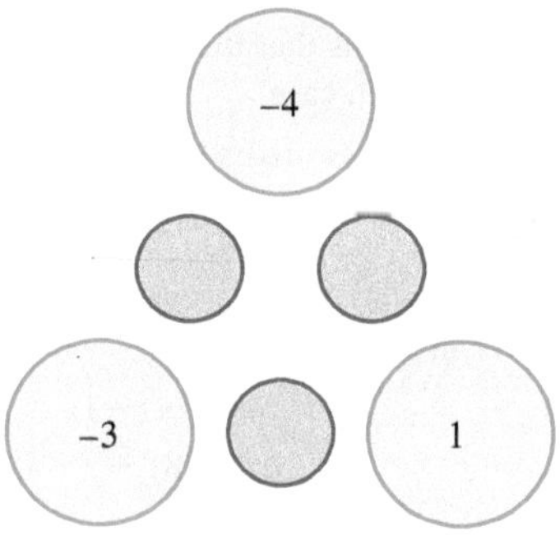

(b)

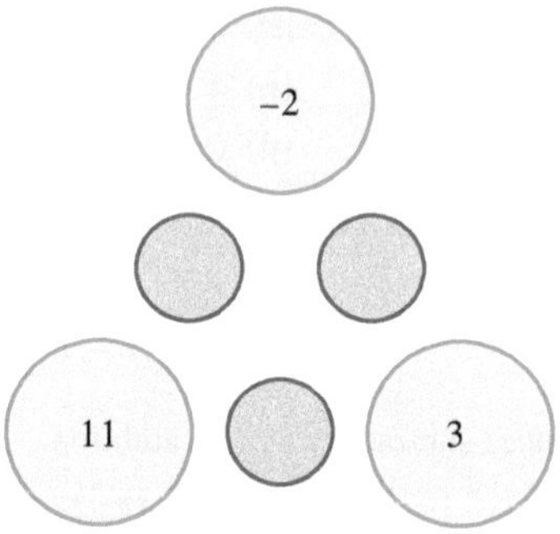

35. **(a)** Perform these pairs of computations:
(i) $7 - (-3)$ and $(-3) - 7$
(ii) $(-2) - (-5)$ and $(-5) - (-2)$
(b) Does the commutative law for subtraction hold for the set of integers? Explain briefly.

36. Does the associative law for subtraction hold for the set of integers (i.e., is $a - (b - c) = (a - b) - c$ true for all integers a, b, and c)? If so, explain why. If not, give a counterexample.

37. Let a, b, and n be integers. Prove that $a < b$ if and only if $a + n < b + n$.

38. Let a and b be integers. Prove that $a < b$ if and only if $-b < -a$.

Making Connections

39. One day Anne had the flu. At 8 A.M. her temperature was 101°. By noon her temperature had increased by 3°, and then it fell 5° by 6 in the evening.

(a) Write a single addition equation to determine Anne's temperature at noon.

(b) Write a single equation using both addition and subtraction to determine Anne's temperature at 6 P.M.

40. Vicky was 12 years old on her birthday today.

(a) How old was Vicky on her birthday 7 years ago?

(b) How old will Vicky be on her birthday 7 years from now?

(c) Write addition equations that answer both parts (a) and (b) of this problem.

41. Greg's bank balance was \$4500. During the month, he wrote checks for \$510, \$87, \$212, and \$725. He also made deposits of \$600 and \$350. What was his balance at the end of the month?

42. A ball is thrown upward from the top of a building 144 feet high. Let h denote the height of the ball above the top of the building t seconds after it was thrown. It can be shown that $h = -16t^2 + 96t$ feet.

(a) Complete this table of values of h:

t	0	1	2	3	4	5	6	7
h	0	80						

(b) Give a carefully worded plausibility argument (not a proof) that the greatest value of h in the table is the greatest height the ball reaches.

(c) Carefully interpret (explain) the meaning of the value of h when $t = 7$.

43. The velocity, in feet per second, of the ball in problem 42 is given by the equation $v = -32t + 96$.

(a) Complete the table of values of v in the table shown.

(b) Carefully interpret (explain) the value of v when $t = 0$.

(c) Interpret the value of v when $t = 3$.

t	0	1	2	3	4	5	6	7
r	96	64						

(d) Carefully interpret the meaning of the values of v for $t = 4, 5, 6,$ and 7.

(e) Compare the values of v for $t = 0$ and 6, 1 and 5, and 2 and 4. What do these values tell you about the motion of the ball?

State Assessments

44. (California, Grade 4)
True or false?

1. $-9 > -10$ **2.** $-31 < -29$

45. (Massachusetts, Grade 6)
Corazón used the number-line model shown below to help her write a true number sentence.

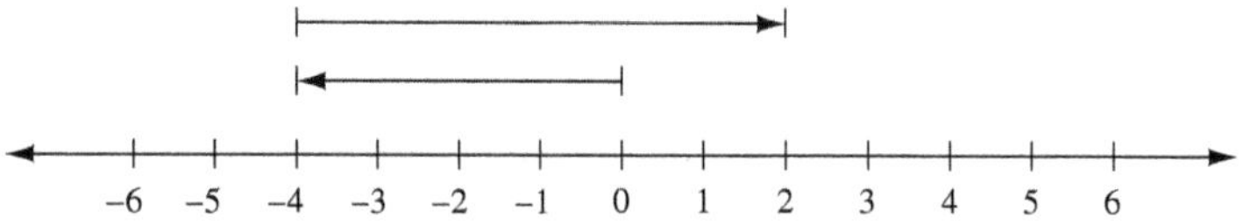

Which of the following could be Corazón's number sentence?

A. $-4 + 2 = 6$ B. $-4 + 6 = 2$
C. $2 + 6 = -4$ D. $2 + -4 = 6$

46. (Grade 7)
Which expression below is equivalent to $(-4) - (-9)$?

A. $(-4) - 9$ B. $4 - 9$
C. $(-4) + 9$ D. $(-9) - (-4)$

5.3 Multiplication and Division of Integers

Multiplication of Integers

Multiplication and division in the set of integers are direct extensions of these operations for whole numbers. In the whole numbers, $4 \cdot 3$ is defined by the repeated addition

$$4 \cdot 3 = 3 + 3 + 3 + 3 = 12.$$

This definition is extended to the integers by defining $4 \cdot (-3)$ as the repeated addition

$$4 \cdot (-3) = (-3) + (-3) + (-3) + (-3) = -12.$$

So far so good, but what meaning can be given to multiplication by a negative integer? For example, what is meant by $(-4) \cdot 3$ and by $(-4) \cdot (-3)$? Let's see what is suggested by representing the integers with colored counters.

Multiplication of Integers by Using Loops of Colored Counters

We can start with an empty set of counters and then add four sets of either three black or three red counters to see that $4 \cdot 3 = 12$ and $4 \cdot (-3) = -12$, as in Figure 5.20.

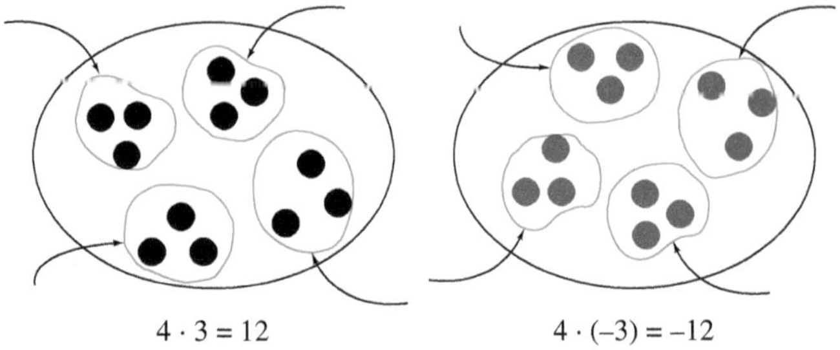
4 · 3 = 12
4 · (–3) = –12

Figure 5.20
Using colored counters to show that $4 \cdot 3 = 12$ and $4 \cdot (-3) = -12$

More generally, we see that

$$m \cdot n = mn \quad \text{and} \quad m \cdot (-n) = -mn.$$

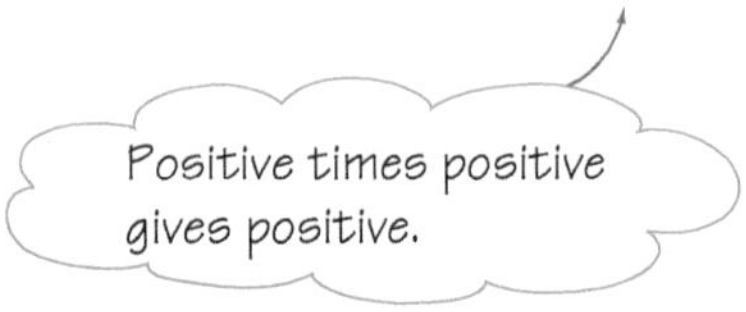
Positive times positive
gives positive.

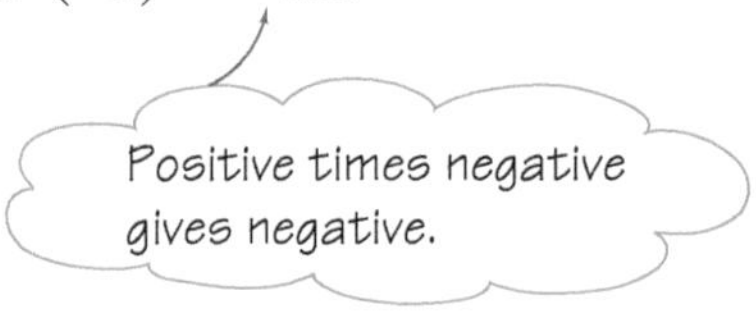
Positive times negative
gives negative.

Since multiplication by a positive integer is repeated addition, the most natural interpretation of multiplication by a negative integer is repeated subtraction. For example, let's start with zero represented by a loop with 12 black and 12 red counters, as shown in Figure 5.21.

Figure 5.21
Using colored counters to show that $(-4) \cdot 3 = -12$ and $(-4) \cdot (-3) = 12$

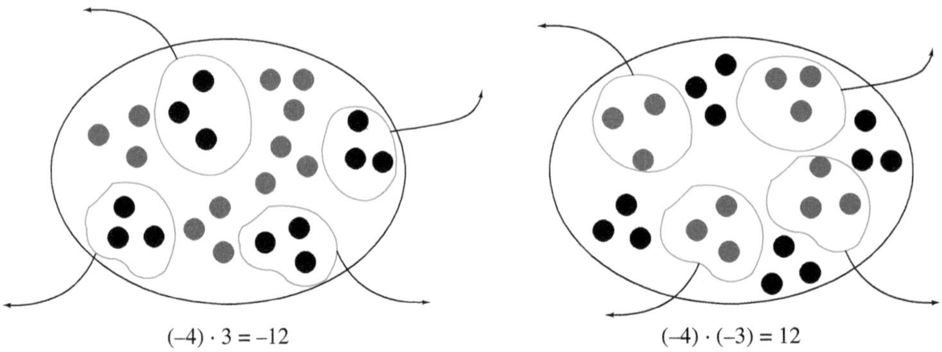
(–4) · 3 = –12
(–4) · (–3) = 12

If four loops each with three black counters are removed, we are left with 12 red counters; that is, $(-4) \cdot 3 = -12$. Similarly, if four loops each with three red counters are removed, then we are left with 12 black counters. Thus, $(-4) \cdot (-3) = 12$.

More generally, we see that

$$(-m) \cdot n = -m \quad \text{and} \quad (-m) \cdot (-n) = mn.$$

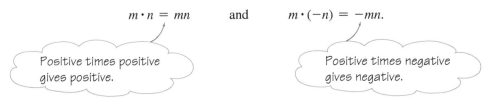

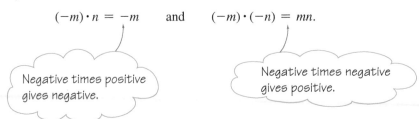

Negative times positive
gives negative.

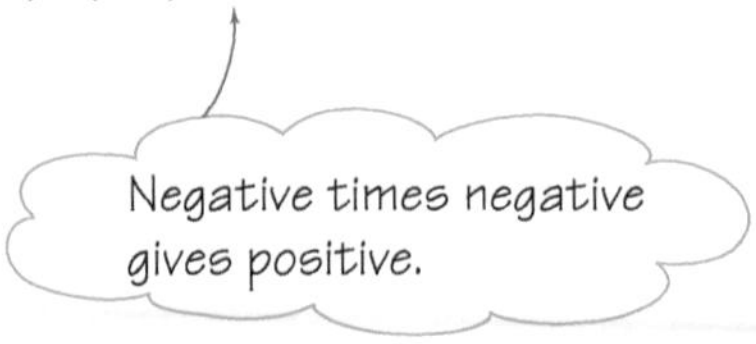
Negative times negative
gives positive.

The colored-counter model motivates the definition of multiplication in the integers.

DEFINITION The Rule of Signs for Multiplication of Integers

Let m and n be positive integers so that $-m$ and $-n$ are negative integers. Then the following **rules of signs** are true:

- $m \cdot (-n) = -mn$;
- $(-m) \cdot n = -mn$;
- $(-m) \cdot (-n) = mn$.

EXAMPLE 5.19

Multiplying Integers

Compute these products:

(a) $(-7) \cdot (-8)$ **(b)** $(-8) \cdot (-7)$
(c) $3 \cdot (-7)$ **(d)** $(-7) \cdot 3$
(e) $[8 \cdot (-5)] \cdot 6$ **(f)** $8 \cdot [(-5) \cdot 6]$
(g) $(-4) \cdot [(-5) + 7]$ **(h)** $(-4) \cdot (-5) + (-4) \cdot 7$

Solution

(a) $(-7) \cdot (-8) = 7 \cdot 8 = 56$
(b) $(-8) \cdot (-7) = 8 \cdot 7 = 56$
(c) $3 \cdot (-7) = -(3 \cdot 7) = -21$
(d) $(-7) \cdot 3 = -(7 \cdot 3) = -21$
(e) $[8 \cdot (-5)] \cdot 6 = [-(8 \cdot 5)] \cdot 6 = (-40) \cdot 6 = -(40 \cdot 6) = -240$
(f) $8 \cdot [(-5) \cdot 6] = 8 \cdot [-(5 \cdot 6)] = 8 \cdot (-30) = -(8 \cdot 30) = -240$
(g) $(-4) \cdot [(-5) + 7] = (-4) \cdot 2 = -(4 \cdot 2) = -8$
(h) $(-4) \cdot (-5) + (-4) \cdot 7 = 4 \cdot 5 + [-(4 \cdot 7)] = 20 + (-28) = -8$

The results shown in the preceding example illustrate that the closure, commutative, associative, and distributive properties of multiplication hold for the set of integers. We have also seen that $1 \cdot r = r \cdot 1 = r$ and that $0 \cdot r = r \cdot 0 = 0$, for every integer r. These properties are collected in the following theorem.

THEOREM Multiplication Properties of Integers

Let r, s, and t be any integers. Then

Closure Property	rs is an integer.
Commutative Property	$rs = sr$
Associative Property	$r(st) = (rs)t$.
Distributive Property	$r(s + t) = rs + rt$.
Multiplicative-Identity Property of 1	$1 \cdot r = r \cdot 1 = r$.
Multiplicative Property of 0	$0 \cdot r = r \cdot 0 = 0$.

Multiplication of Integers by Using Mail-Time Stories

Suppose the letter carrier brings five bills for \$11 each. Are you richer or poorer, and by how much? Answer: Poorer by \$55. This illustrates the product

$$5 \cdot (-11) = -55.$$

Suppose the mail carrier takes away four bills for \$13 each. Are you richer or poorer, and by how much? Answer: Richer by \$52. This illustrates the product

$$(-4) \cdot (-13) = 52,$$

since four bills for the same amount are *taken away*.

EXAMPLE 5.20

Writing Mail-Time Stories for Multiplication of Integers

Write a mail-time story to illustrate each of these products:

(a) $(-4) \cdot 16$ **(b)** $(-4) \cdot (-16)$
(c) $4 \cdot (-16)$ **(d)** $4 \cdot 16$

Solution

(a) The letter carrier takes away four checks for \$16 each. Are you richer or poorer, and by how much? Answer: \$64 poorer; $(-4) \cdot 16 = -64$.
(b) The letter carrier takes away four bills for \$16 each. Are you richer or poorer, and by how much? Answer: \$64 richer; $(-4) \cdot (-16) = 64$.
(c) The letter carrier brings four bills for \$16 each. Are you richer or poorer, and by how much? Answer: Poorer by \$64; $4 \cdot (-16) = -64$.
(d) The letter carrier brings you four checks for \$16 each. Are you richer or poorer, and by how much? Answer: Richer by \$64; $4 \cdot 16 = 64$.

Mail-time stories are an especially effective way to interpret multiplication by a negative integer simply as repeated subtraction. For example, $(-3)5 = -15$, since subtracting 5 three times gives -15. Similarly, $(-3)(-5) = 15$, since subtracting -5 three times gives the positive integer 15.

Multiplication of Integers by Using a Number Line

Think of

$3 \cdot 4$ as $4 + 4 + 4$,

$3 \cdot (-4)$ as $(-4) + (-4) + (-4)$,

$(-3) \cdot (4)$ as $-4 - 4 - 4$, and

$(-3) \cdot (-4)$ as $-(-4) - (-4) - (-4)$.

These multiplications can be illustrated on the number line as shown in Figure 5.22. Remember that jumps are added by tracing the jump arrow forward from tail to head and jumps are subtracted by tracing the jump arrow in reverse from head to tail.

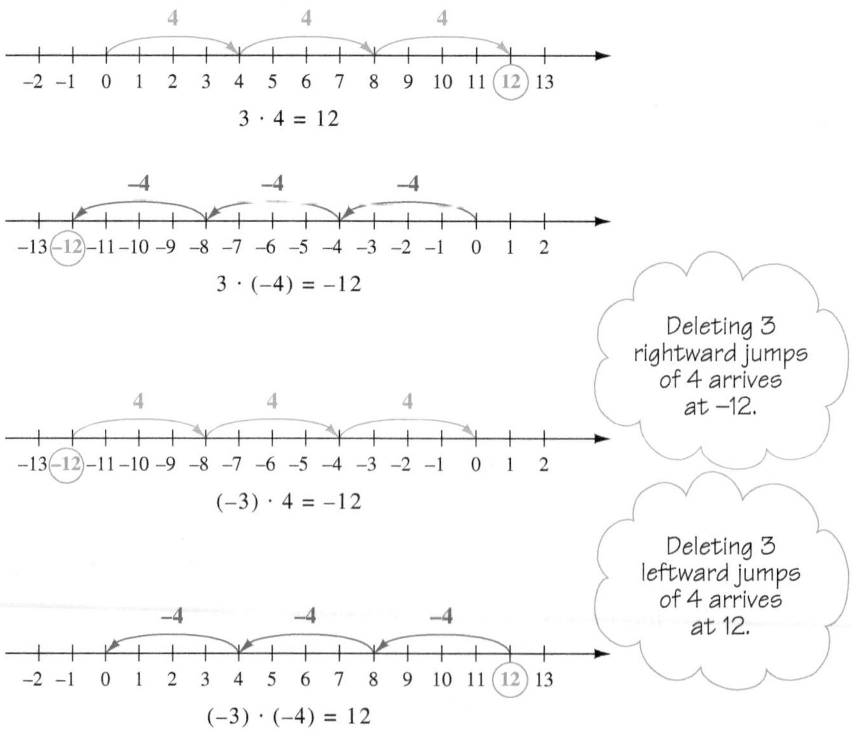

Figure 5.22 Multiplication on the number line

EXAMPLE 5.21

Multiplying Integers by Using a Number Line

What products do these diagrams illustrate?

(a)

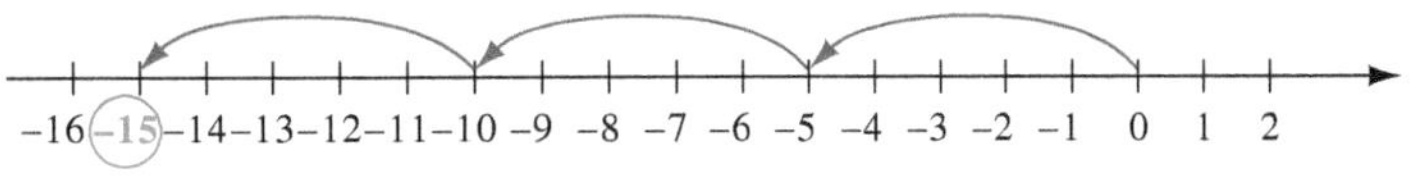

(b)

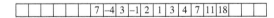

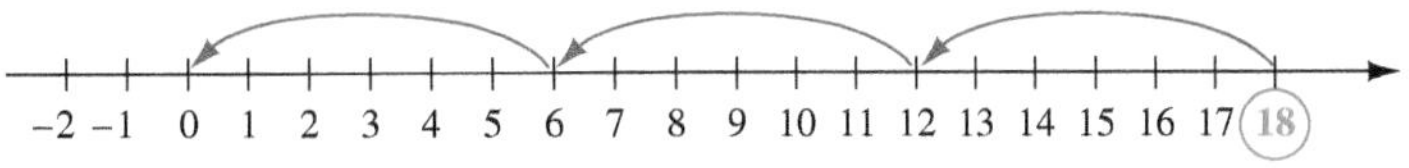

Solution

(a) Since 3 leftward jumps of distance 5 are combined, we are *adding* -5 to -5 to -5, and the result is $3 \cdot (-5) = -15$.

(b) Three leftward jumps of length 6 are removed to arrive at 18. Thus, $(-3) \cdot (-6) = 18$.

COOPERATIVE INVESTIGATION

Exploring the Extended Fibonacci Sequence with Integer Arithmetic

This activity is a continuation of the Cooperative Investigation in the previous section, in which the extended Fibonacci numbers were entered on a strip of paper divided into squares as shown below. The goal here is to discover more patterns of this sequence using the arithmetic of the integers.

Materials

Paper strips and marker pens for each group of two to four students.

Directions

The paper strips shown below contain rows 2, 3, and 4 of Pascal's triangle (see Chapter 1.3). Use marker pens to fill in the entries for rows 2, 3, . . . , 7 of Pascal's triangle.

1	2	1		
1	3	3	1	
1	4	6	4	1

Activities and Questions

Place the row 2 strip anywhere under the strip of extended Fibonacci numbers. Now add the products of the vertically aligned numbers. Two examples are shown here.

–21	13	–8	5	–3	2	–1	1	0	1	1	2	3	5	8	13	21	34	55	89
				1	2	1					1	2	1						

$1 \cdot (-3) + 2 \cdot 2 + \cdot 1 \cdot (-1) = 0$ $\quad$ $1 \cdot 2 + 2 \cdot 3 + 1 \cdot 5 = 13$

(a) Place the row 2 strip at other positions and calculate the sum of products. What pattern do you see?

(b) Place the strip with row 3 of Pascal's triangle at several places along the extended Fibonacci numbers and compute the sum of products. Describe the pattern you observe.

(c) Use strips with other rows of Pascal's triangle to make a conjecture about the sum of the products for row *n* of Pascal's triangle.

For Further Exploration

Explore what happens if you use the sequence of extended Lucas numbers, as shown in this partially completed strip.

						7	–4	3	–1	2	1	3	4	7	11	18			

Division of Integers

Division in the whole numbers is defined this way:

If *a*, *b*, and *c* are whole numbers, with $b \neq 0$, then $a \div b = c$ if and only if $a = bc$.

For example, since $12 = 3 \cdot 4$ and $12 = 4 \cdot 3$, we see that the whole numbers 3, 4, and 12 are related by the fact family

$$3 \cdot 4 = 12, \qquad 12 \div 3 = 4, \qquad 4 \cdot 3 = 12, \qquad 12 \div 4 = 3.$$

SMP 8 Look for and express regularity in repeated reasoning

Since we want the integers to be an extension of the whole numbers, we should define division in the integers in the same way. This type of mathematical thinking is emphasized in the eighth Standard for Mathematical Practice (SMP 8) from the Common Core:

"Mathematically proficient students notice if calculations are repeated, and look . . . for general methods. . . . "

DEFINITION Division of Integers

If a, b, and c are integers, with $b \neq 0$, then $a \div b = c$ if and only if $a = bc$.

Since we know that $3 \cdot (-4) = -12$ and $(-4) \cdot 3 = -12$, we then have the fact family

$$3 \cdot (-4) = -12, \quad -12 \div 3 = -4, \quad (-4) \cdot 3 = -12, \quad -12 \div (-4) = 3.$$

Similarly, we know that $(-3) \cdot (-4) = 12$ and $(-4) \cdot (-3) = 12$, giving us the fact family

$$(-3) \cdot (-4) = 12, \quad 12 \div (-3) = -4, \quad (-4) \cdot (-3) = 12, \quad 12 \div (-4) = -3.$$

It continues to be important never to divide any integer by 0. For example, suppose you want to find *all* integers n for which $n^2 = -4n$. It is tempting to divide each side of the equation by n and say that $n = -4$. But this assumes you are looking for a nonzero n, and the second solution, $n = 0$, will be overlooked.

EXAMPLE 5.22 Dividing Integers

Perform the following divisions:

(a) $28 \div 4$ **(b)** $28 \div (-4)$ **(c)** $(-28) \div 4$ **(d)** $(-28) \div (-4)$

Solution

(a) $28 \div 4 = 7$, since $28 = 4 \cdot 7$.
(b) $28 \div (-4) = -7$, since $28 = (-4) \cdot (-7)$.
(c) $(-28) \div 4 = -7$, since $-28 = 4 \cdot (-7)$.
(d) $(-28) \div (-4) = 7$, since $-28 = (-4) \cdot 7$.

Since these results are entirely typical, we state here the rule of signs for division of integers:

THEOREM Rule of Signs for Division of Integers

Let m and n be positive integers so that $-m$ and $-n$ are negative integers, and suppose that n divides m. Then the following are true:

- $m \div (-n) = -(m \div n)$;
- $(-m) \div n = -(m \div n)$;
- $(-m) \div (-n) = m \div n$.

Thus, *given that n divides m,* it follows that

- a positive integer divided by a negative integer is a negative integer,
- a negative integer divided by a positive integer is a negative integer,
- a negative integer divided by a negative integer is a positive integer, and
- a positive integer divided by a positive integer is a positive integer.

EXAMPLE 5.23

Performing Division of Integers

Compute each quotient if possible:

(a) $(-24) \div (-8)$ **(b)** $24 \div (-8)$
(c) $48 \div 12$ **(d)** $(-48) \div 12$
(e) $(-57) \div 19$ **(f)** $(-12) \div 0$
(g) $(-51) \div (-17)$ **(h)** $28 \div (9 - 5)$
(i) $(27 + 9) \div (-4)$

Solution

We use the preceding theorem and the definition of division of integers.

(a) $(-24) \div (-8) = 3$. Check: $-24 = (-8) \cdot 3$.
(b) $24 \div (-8) = -3$. Check: $24 = (-8) \cdot (-3)$.
(c) $48 \div 12 = 4$. Check: $48 = 12 \cdot 4$.
(d) $(-48) \div 12 = -4$. Check: $-48 = 12 \cdot (-4)$.
(e) $(-57) \div 19 = -3$. Check: $-57 = 19 \cdot (-3)$.
(f) $(-12) \div 0$ is not defined, since there is no number c such that $-12 = 0 \cdot c$.
(g) $(-51) \div (-17) = 3$. Check: $-51 = (-17) \cdot 3$.
(h) $28 \div (9 - 5) = 28 \div 4 = 7$. Check: $28 = 4 \cdot 7$.
(i) $(27 + 9) \div (-4) = 36 \div (-4) = -9$. Check: $36 = (-4) \cdot (-9)$.

Multiplication and Division with Colored-Counter Arrays

Colored counters are an effective classroom manipulative to illustrate that multiplication and division are inverse operations in the integers. This can be shown with loops of colored counters (see problems 16 and 17). An alternative method is to form rectangular arrays of colored counters. To get started, we can illustrate the four products

$$1 \cdot 1 = 1, \quad 1 \cdot (-1) = -1, \quad (-1) \cdot 1 = -1, \quad \text{and } (-1) \cdot (-1) = 1$$

this way.

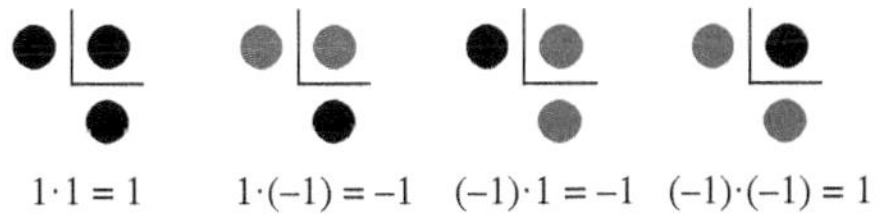

More general products of integers $a \cdot b = c$ are illustrated with a columns and b rows, and each array also illustrates the corresponding division $c \div a = b$. The fact family for the integers −12, 3, and −4 is shown in Figure 5.23.

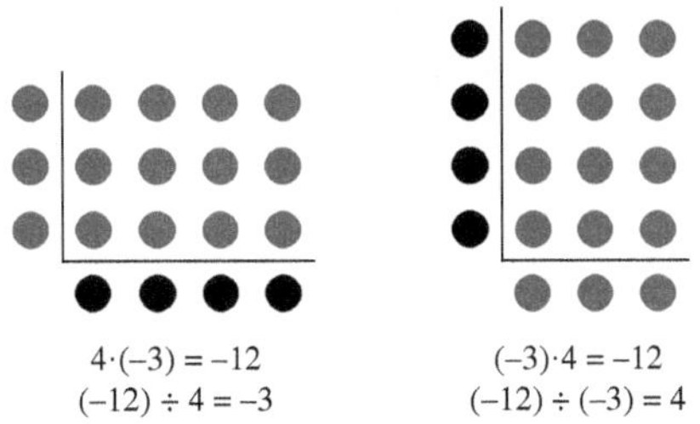

Figure 5.23
Illustrating a fact family of integers with colored-counter arrays

EXAMPLE 5.24

Illustrating Multiplication and Division with Colored-Counter Arrays

Form colored-counter arrays showing the fact families of the integers 15, −3, and −5.

Solution

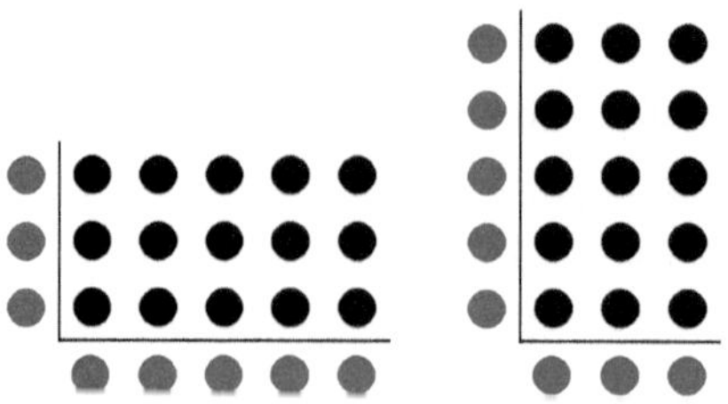

$(-5)\cdot(-3) = 15$
$15 \div (-5) = -3$

$(-3)\cdot(-5) = 15$
$15 \div (-3) = -5$

COOPERATIVE INVESTIGATION

Balloon Rides* II: Multiplication and Division

Balloon Rides I of Section 5.2 explored the balloon model of integer addition and subtraction. In this activity, that model is extended to multiplication and division. Once again, the altitude of our balloon is controlled by adding or removing either gasbags or sandbags to or from the balloon. However, there is one important difference: We now assume that *we always start at level 0*, the height of the canyon rim. Given next are some integer multiplication and division examples to help you get started. You want to translate back and forth between "balloon language" and the corresponding equations in integers.

Example 1

In balloon language, suppose that "Three times, you add a pair of gasbags to your balloon, so, starting at level 0, you rise to level 6." This statement translates to the integer multiplication formula $3 \times 2 = 6$.

Example 2

The equation $(-4) \times 3 = -12$ translates into balloon language as "Four times, you remove a group of three gasbags, so, starting at level 0, the balloon sinks to level -12."

Example 3

"To reach an altitude of -15 starting at level 0, what can we do with groups of gasbags, where there are 5 gasbags per group?"

Since the removal of each group of 5 gasbags lowers the height by 5, we need to remove three groups. Therefore, $-15 \div 5 = -3$, or equivalently, $-15 = (-3) \times 5$.

Example 4

Starting at level 0, how can we reach an altitude of -15 by using groups of sandbags, with 5 sandbags per group? This time, we must add three groups of sandbags. Therefore, $-15 \div (-5) = 3$, or equivalently, $-15 = 3 \times (-5)$.

Problems and Questions for Small Groups

1. Evaluate $4 \times (-2)$, using both balloon language and integer equations.
2. Starting at level 0, suppose 6 pairs of sandbags are added. What is your new altitude? Give your answer both in balloon language and an integer equation.
3. There are two balloons, both starting at level 0. Balloon A adds 3 pairs of gasbags, whereas balloon B removes 3 pairs of sandbags. What are the new heights of balloons A and B?
4. Discuss what general property of integer multiplication is illustrated in question 3.
5. Suppose both gasbags and sandbags come in groups of 4 bags. Starting at level 0, how can you reach a height of -20? Give more than one answer.
6. Suppose that both gasbags and sandbags come either individually or bundled in groups of a single size. Starting at level -8, you can reach level 7 by removing three groups of the same type of bag. What type of bag has been removed, and how many bags have been bundled into each group?
7. Using the balloon model, invent and solve your own algebra problem.

*This activity is adapted from a presentation by Bill Kring.

5.3 Problem Set

Understanding Concepts

1. Perform these multiplications:

(a) $7 \cdot 11$ (b) $7 \cdot (-11)$
(c) $(-7) \cdot 11$ (d) $(-7) \cdot (-11)$

2. Perform these multiplications:

(a) $12 \cdot 9$ (b) $12 \cdot (-9)$
(c) $(-12) \cdot 9$ (d) $(-12) \cdot (-9)$
(e) $(-12) \cdot 0$

3. Perform these divisions:

(a) $36 \div 9$ (b) $(-36) \div 9$
(c) $36 \div (-9)$ (d) $(-36) \div (-9)$

4. Perform these divisions:

(a) $(-143) \div 11$ (b) $165 \div (-11)$
(c) $(-144) \div (-9)$ (d) $275 \div 11$
(e) $72 \div (21 - 19)$

5. Write another multiplication equation and two division equations that are equivalent to $(-11) \cdot (-25{,}753) = 283{,}283$.

6. Write two multiplication equations and another division equation that are equivalent to $(-1001) \div 11 = -91$.

7. What computations do these mail-time stories illustrate?

(a) The mail carrier brings you six checks for \$13 each. Are you richer or poorer? By how much?

(b) The mail carrier brings you four bills for \$23 each. Are you richer or poorer? By how much?

8. What computations do these mail-time stories illustrate?

(a) The mail carrier takes away three bills for \$17 each. Are you richer or poorer? By how much?

(b) The mail carrier takes away five checks for \$20 each. Are you richer or poorer? By how much?

9. What computations do these number-line diagrams represent?

(a)
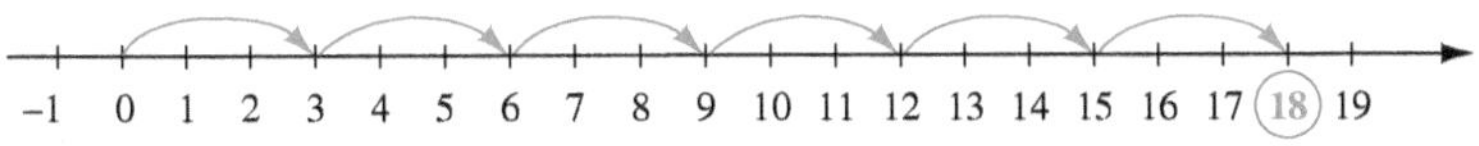

(b)
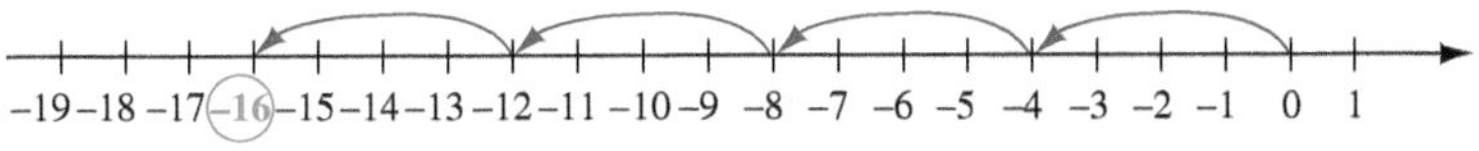

(c)
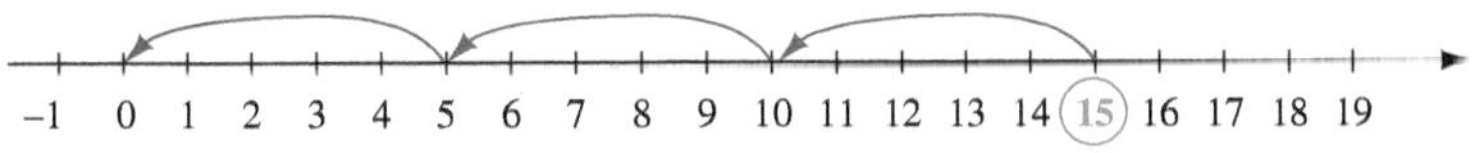

10. Draw a number-line diagram to illustrate $(-4) \cdot 3 = -12$.

11. Illustrate the following operations with loops of colored counters:

(a) $2 \times (-3)$
(b) $(-2) \times 3$
(c) $(-2) \times (-3)$

12. What fact family is shown in these arrays of colored counters?

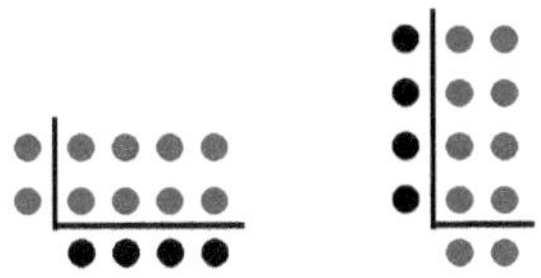

13. Draw arrays of colored counters that illustrate the fact family

$$(-3) \cdot (-2) = 6, \quad 6 \div (-3) = -2,$$
$$(-2) \cdot (-3) = 6, \quad 6 \div (-2) = -3.$$

14. Use mental arithmetic to give estimates of the given products and quotients. Indicate how you obtained each of your estimates.

(a) -21×18 (b) $(-32) \times (-28) \times 31$
(c) $254 \div (-49)$ (d) $-6642 \div -223$

Into the Classroom

15. (**Writing**) Make a report describing "bank" arithmetic, for which a positive integer is the amount of a deposit and a negative is the amount of a loan. For example, the expression "$30 + (-100)$" represents the change in your net assets in the bank when you make a \$30 deposit and take out a loan for \$100, and the expression "$(-3) \cdot (-50)$" is the change made by making three payments of \$50 on your loans. Create a collection of expressions and problems involving all four operations: $+, \times, -, \div$. Be sure to include examples that show why "a negative of a negative is a positive" and "a negative times a negative is a positive."

16. **(Writing)** Division by a positive integer is easily shown with loops of colored counters, where division is described by the partitive (sharing) model. For example, $-12 \div 4 = -3$ is shown this way: Partition a loop of 12 red counters into four groups of equal size, showing that there are three red counters in each group.

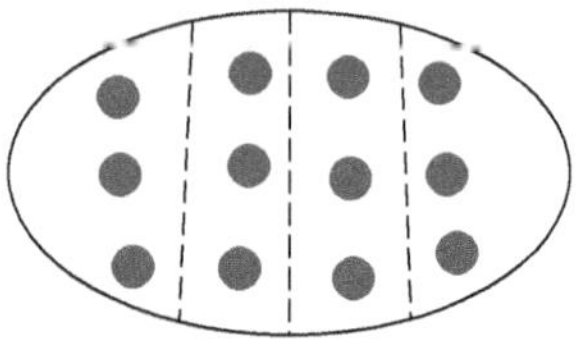

Write a lesson on division by a positive integer using this model.

17. **(Writing)** Division by a negative integer can be illustrated with colored counters if division is described by repeated subtraction. For example, $12 \div (-4) = 3$ is shown this way: Start with a loop of 12 red counters and show that three groups, each with four red counters, can be removed.

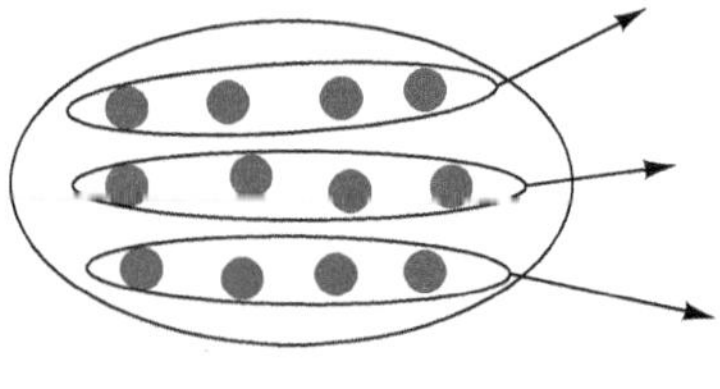

Write a lesson on division by a negative integer using this model.

Responding to Students

18. Marni claims that $\sqrt{4} = -2$ because $(-2) \cdot (-2) = 4$. Daniel claims that Marni is wrong, since everyone knows that $\sqrt{4} = 2$. How would you settle the disagreement between Marni and Daniel?

19. Kris obtained the equation $n(n - 3) = n$ and claims that, after dividing each side by n, she gets the equation $n - 3 = 1$, which is solved by $n = 4$. Is Kris correct, or has she overlooked a possibility?

20. Vijay obtained the inequality $n^2 \leq 9$, where n must be an integer. He claims that there are four possible values of n, namely, 0, 1, 2, and 3. Is Vijay correct, or has he overlooked other solutions?

Thinking Critically

21. Let n be any integer. What, if anything, can be said about the sign of each of the expressions that follow? Your answers may be "always positive," "always nonnegative," "can't tell," and so on.

(a) $n + 5$ **(b)** n^2 **(c)** $n^2 + 1$
(d) n^3 **(e)** $-n^4$

22. Let a and b be positive integers, with $a < b$. Prove that if c is a negative integer, then $ac > bc$. (*Suggestion:* Try using specific numbers first.)

23. The equation $a(b - c) = ab - ac$ expresses the distributive law for multiplication over subtraction. Is it true for all integers a, b, and c? If so, explain why. If not, give a counterexample.

24. Decide whether the properties that follow are true or false in the integers, where a, b, and c can be arbitrarily chosen integers. If the statement is true, give a proof: If the statement is false, give a counterexample by choosing explicit values for a, b, and c.

(a) If $a < b$, then $a^2 < b^2$.
(b) If $a^2 < b^2$, then $a < b$.
(c) If $ac \leq bc$, then $a \leq b$.
(d) $(a - b)(a + b) = a^2 - b^2$.

25. Another way to discover the rule of signs for integer multiplication is to look at patterns. Here is an example: Starting with $4 \cdot 3 = 12, 4 \cdot 2 = 8, 4 \cdot 1 = 4$, and $4 \cdot 0 = 0$, and continuing to decrease each product by four, we see that the next terms in the pattern are $4 \cdot (-1) = -4, 4 \cdot (-2) = -8, 4 \cdot (-3) = -12$. That is, a positive integer m times a negative integer, $-n$, is the negative integer given by $m \cdot (-n) = -mn$. Continue each of the following patterns, and then name the corresponding rule of signs that emerges from the pattern:

(a) $4 \cdot (-3) = -12, 3 \cdot (-3) = -9, 2 \cdot (-3) = -6,$
$1 \cdot (-3) = -3$
(b) $(-3) \cdot (-4) = 12, (-3) \cdot (-3) = 9, (-3) \cdot (-2) = 6,$
$(-3) \cdot (-1) = 3$

26. Complete each of the number patterns that follow so that the number in each blue circle is the sum of the two numbers in the adjacent red circles. Indicate whether the solution to each pattern is unique.

(a)

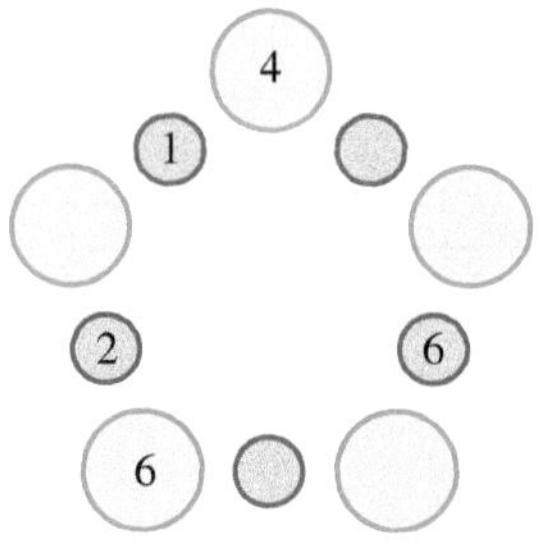

(b)

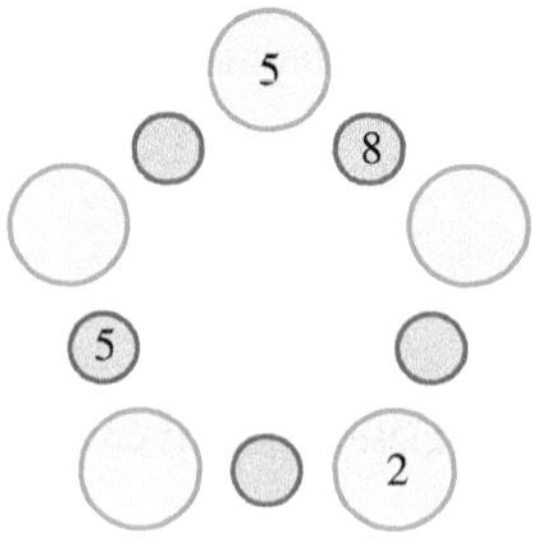

(c)

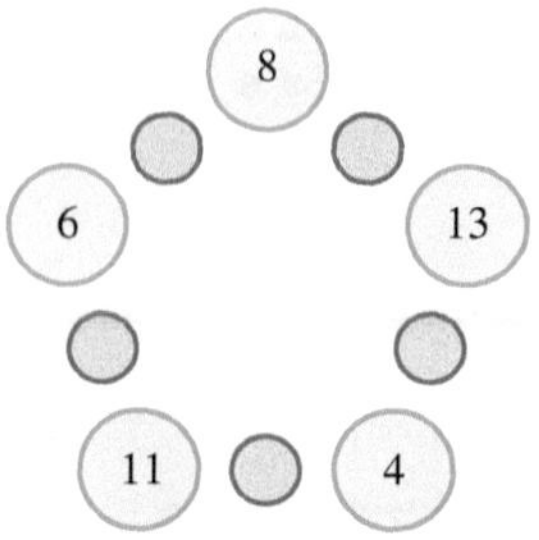

27. Use the numbers −8, −6, −4, −2, 0, 2, 4, 6, and 8 to make a magic square. What should the sum in each row, column, and diagonal be? What should the middle number be?

28. Use the numbers −7, −6, −5, −1, 0, 1, 5, 6, and 7 to make a magic square.

29. Exploring Clock Arithmetic. In 12-hour clock arithmetic, operations are first evaluated with integer arithmetic, but then the resulting sum, difference, product, or quotient is replaced with its remainder when the result is divided by 12. For example, in clock arithmetic, since $8 + 11 = 19$ and $19 \div 12 = 1 \text{ R } 7$, we write $8 +_{12} 11 = 7$. Notice that starting at 0 (i.e., at 12 o'clock) and moving forward first by 8 hours and then by 11 hours, you arrive at 7 o'clock. Similarly, $8 \times 11 = 88$ and $88 \div 12 = 7 \text{ R } 4$, so $8 \times_{12} 11 = 4$, meaning that starting at 12 o'clock and moving clockwise 8 jumps of 11 hours each, the clock will then show 4 o'clock. Thus, in 12-hour clock arithmetic, there are just 12 numbers, namely, those in the set

$$I_{12} = \{0, 1, 2, 3, 4, 5, 6, 7, 8, 9, 10, 11\}.$$

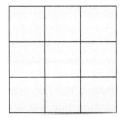

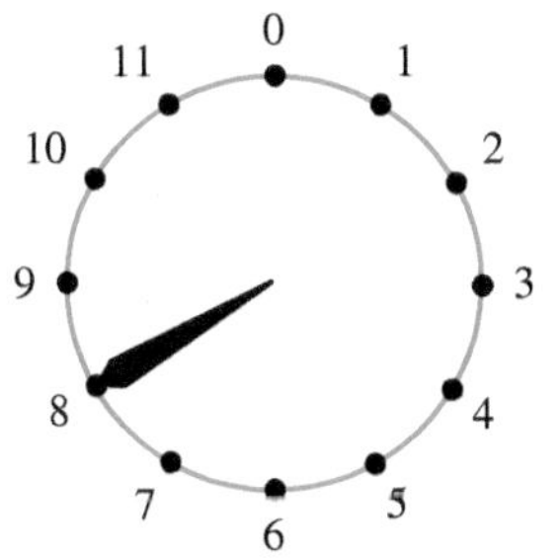

(a) Create an addition $+_{12}$ table and a multiplication $\times_{12}$ table for 12-hour clock arithmetic.

(b) What properties hold for clock addition, $+_{12}$, in the set I_{12}?

(c) Does the number system I_{12} have the additive-inverse property?

(d) What properties hold for clock multiplication, $\times_{12}$, in the set I_{12} ?

(e) If $m, n \in I_{12}$ and $m \times_{12} n = 1$, does it necessarily follow that $m = n = 1$?

(f) If $m, n \in I_{12}$ and $m \times_{12} n = 0$, does it necessarily follow that either $m = 0$ or $n = 1$?

Making Connections

30. The members of the Pep Club tried to raise money by raffling off a pig, agreeing that if they actually lost money on the enterprise, they would share the loss equally.

(a) If they lost \$105 and there are 15 members in the club, how much did each club member have to pay?

(b) What arithmetic might be used to illustrate this story? Explain.

31. It cost \$39 each to buy sweatshirts for members of the Pep Club.

(a) If there are 15 members in the Pep Club, what was the total cost?

(b) What arithmetic might be used to illustrate this story? Explain.

32. It is common to report a golfer's score on a hole in relation to par. For example, one stroke over par is a bogey and two strokes over par is a double bogey. Scores one, two, or three strokes below par are respectively a birdie, an eagle, or the extremely rare albatross.

(a) Erratic Ernie scored 1 albatross, 2 eagles, 5 birdies, 7 pars, 1 bogey, and 2 double bogeys on a par-72 eighteen-hole course. What was Ernie's score for the round?

(b) In July 2009, PGA golfer Mark Calcavecchia started with 2 pars but then shot a record 9 straight birdies on holes 3 through 11. He followed this feat with bogeys on the next two holes and finished the round with pars. What was his final score on the par-72 course?

33. The Diffy process described in the Cooperative Investigation in Section 2.3 employed whole numbers. Explain why the process is essentially unchanged if we start with all negative or a mixture of positive and negative integers.

State Assessments

34. (Grade 8)
What is the value of the expression $(2 \cdot 3)^2 - (12 - 7)^2 + (4 + 5)$?

A. 20   B. 70   C. 40   D. 100

35. (Massachusetts, Grade 8)
(Calculators are not allowed here because one can obtain the correct answer without understanding.)
Compute: $(-2)(-5)(-1) =$

The Chapter in Relation to Future Teachers

History shows that the concept of a negative number is not a simple one. For example, the ancient Greek mathematician Diophantus recognized negative numbers as early as A.D. 275, but rejected them as absurd. Thus, he refused to accept $x = -4$ as a meaningful solution of the equation $4x + 20 = 4$.

Therefore, the classroom teacher is faced with a nontrivial task: How can negative integers be made meaningful and useful to young children? As shown throughout this chapter, the integers can be effectively represented by manipulatives and pictorial models, including colored counters and mail-time stories. In addition, the integers can be visualized as positions and movements on a number line, an approach that builds naturally on the whole-number line that is already familiar to students. Within the framework of each of these representations, the arithmetic operations are illustrated in a natural setting that will seem sensible to a child. Especially in our contemporary world, it is easy to find applications of the integers. In particular, the teacher will want to relate how integers are used to record yards lost and gained in football, a golf score relative to par, a temperature relative to zero, profits and losses, amounts saved and owed, and so on.

Finally, it is important to understand that the system of integers has important new properties not found in the whole-number system. The most important is that subtraction is now a closed operation, so an equation such as $x + 5 = 3$ is now solved by the negative integer -2. However, an equation such as $5x = 3$ still has no solution x, even in the integers. In Chapter 6, the integers are extended to the still larger system of rational numbers, where there is a solution of $5x = 3$, namely, the fraction $x = \frac{3}{5}$.

Chapter 5 Summary

In what follows, all symbols represent integers.

Section 5.1 Representations of Integers	Page Reference
CONCEPTS	
• **Integers:** An extension of the whole-number system to include both the negative integers and the whole numbers.	223
• **Colored counters:** Each integer is represented by a loop containing counters of two colors, often red and black or yellow and black. If n is positive, the loop contains n more black than red counters. A negative integer, $-m$, is a loop with m more red than black counters. A loop with equally many red and black counters represents 0.	225
• **Mail-time stories:** A positive integer n is represented by a check in the amount of n dollars, and a negative integer $-m$ is represented by a bill in the amount of m dollars.	227
• **Number line:** A positive integer n is represented by a point n units to the right of the point that represents 0, or alternatively as a right-pointing arrow of length n. A negative integer $-m$ is represented by a point m units to the left of 0, or alternatively as a left-pointing arrow of length m.	228
DEFINITIONS	
• **Integers:** The set of numbers $I = \{\ldots, -3, -2, -1, 0, 1, 2, 3, \ldots\}$.	223
• **Positive integers:** The term given to the set of natural numbers $\{1, 2, 3, \ldots\}$ when considered as a subset of the integers.	223
• **Negative integers,** or **opposites:** The numbers $-1, -2, -3, \ldots$ for which $n + (-n) = (-n) + n = 0$.	223
• **Additive-identity property of 0:** For all integers n, $n + 0 = 0 + n = n$.	223
• **Additive-inverse property of a number system:** A system of numbers for which every element n has an additive inverse m, so that $n + m = m + n = 0$. The number m is called the **additive inverse** of n, written as $m = -n$.	224
• **Absolute value,** or **magnitude,** of an integer n is the nonnegative integer $\lvert n \rvert = n$ if n is nonnegative and $\lvert n \rvert = -n$ if n is negative.	224
THEOREM	
• **Opposite of the opposite:** For all integers n, $-(-n) = n$.	223

Section 5.2 Addition and Subtraction of Integers	Page Reference
CONCEPTS	
• **Addition of integers:** The binary operation modeled by combining loops in the colored-counter representation, placing several checks or bills in the mailbox, or combining jumps on the number line.	232
• **Subtraction of integers:** The binary operation modeled by removing one loop from another in the colored-counter representation, removing checks or bills from a mailbox, or removing one jump from another on the number line.	237
DEFINITIONS	
• **Addition of integers:** If m and n are positive integers, then $(-m) + (-n) = -(m + n)$, $m + (-n) = m - n$ if $m > n$, $m + (-n) = -(n - m)$ if $m < n$, and $m + (-m) = 0$.	234
• **Subtraction of integers:** For any integers a, b, and c, $a - b = c$ if and only if $a = b + c$.	239
• The integer a is **less than** b, $a < b$, if $a + c = b$ for some positive integer c, and a is **greater than** b, $a > b$, if $b < a$.	243
THEOREMS	
• **Properties of integer addition:** Given any integers m, n, and r, addition is closed ($m + n$ is an integer), commutative ($m + n = n + n$), associative ($m + (n + r) = (m + n) + r$), with 0 the additive identity ($m + 0 = 0 + m = m$), and $-m$ the additive inverse ($m + (-m) = (-m) + m = 0$).	234
• **Subtraction by adding the opposite:** Let a and b be any integers. Then $a - b = a + (-b)$.	240
• **Closure of the integers under subtraction:** Let a and b be any integers. Then their difference $a - b$ is an integer.	240
• **Law of trichotomy:** Given any integers, a and b, precisely one of the three possibilities $a < b$, $b > a$, or $a = b$ holds.	244

Section 5.3 Multiplication and Division of Integers	Page Reference
CONCEPTS	
• **Multiplication** by a positive integer is repeated addition, and multiplication by a negative integer is repeated subtraction.	247
• **Division** is the inverse of multiplication, so that $a \div b$ is the unique integer c for which $a = b \cdot c$.	252
DEFINITIONS	
• **Multiplication of integers:** If m and n are integers, with m positive, then $m \cdot n = n + n + \cdots + n$ and $(-m) \cdot n = -(n + n + \cdots + n)$.	247
• **The rule of signs for multiplication:** If m and n are positive integers, then $m \cdot (-n) = -mn$, $(-m) \cdot n = -mn$, and $(-m) \cdot (-n) = mn$.	249
• **Division of integers:** If a, b, and c are integers, with $b \neq 0$, then $a \div b = c$ if, and only if, $a = b \cdot c$.	252

THEOREMS

- **Properties of integer multiplication:** Given any integers m, n, and r, multiplication is closed ($m \cdot n$ is an integer), commutative ($m \cdot n = n \cdot m$), associative ($m \cdot (n \cdot r) = (m \cdot n) \cdot r$), distributes over addition ($m \cdot (n + r) = m \cdot n + m \cdot r$), has the multiplicative identity 1 ($m \cdot 1 = 1 \cdot m = m$), and has the multiplication by 0 property ($m \cdot 0 = 0 \cdot m = 0$). 249
- **The rule of signs for division:** If m and n are positive integers, then $m \div (-n) = -(m \div n)$, $(-m) \div n = -(m \div n)$, and $(-m) \div (-n) = m \div n$. 252

Chapter Review Exercises

Section 5.1

1. You have 15 counters colored black on one side and red on the other.
   (a) If you drop them on your desktop and 7 come up black and 8 come up red, what integer is represented?
   (b) If you drop them on your desktop and twice as many come up black as red, what number is being represented?
   (c) What numbers are represented by all possible drops of the 15 counters?

2. (a) If the mail carrier brings you a check for \$12, are you richer or poorer, and by how much? What integer does this situation illustrate?
   (b) If the mail carrier brings you a bill for \$37, are you richer or poorer, and by how much? What integer does this situation illustrate?

3. (a) 12° above 0 illustrates what integer?
   (b) 24° below 0 illustrates what integer?

4. (a) List five different loops of colored counters that represent the integer -5.
   (b) List five different loops of colored counters that represent the integer 6.

5. (a) Give a mail-time story that illustrates -85.
   (b) Give a mail-time story that illustrates 47.

6. (a) What number must you add to 44 to obtain 0?
   (b) What number must you add to -61 to obtain 0?

Section 5.2

7. What addition is represented by this diagram?

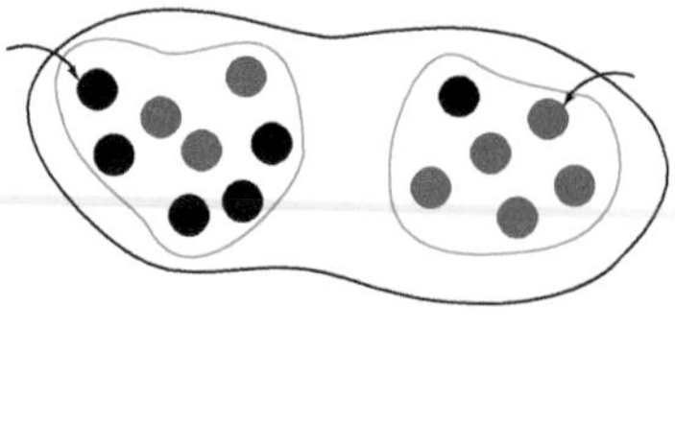

8. What subtraction is represented by this diagram?

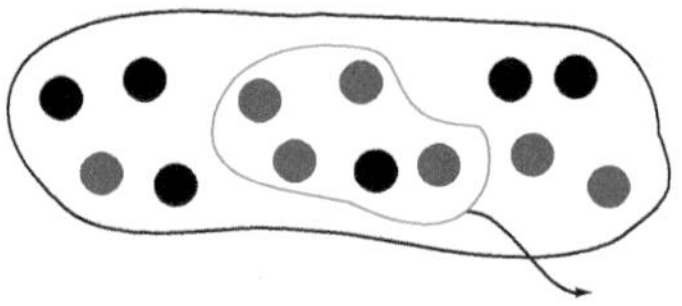

9. What additions and subtractions are represented by these mail-time stories?
   (a) At mail time, the letter carrier brings you a check for \$45 and a bill for \$68. Are you richer or poorer, and by how much?
   (b) At mail time, the letter carrier brings you a check for \$45 and takes away a bill for \$68 left previously. Are you richer or poorer, and by how much?

10. What additions and/or subtractions do these number-line diagrams represent?

(a)

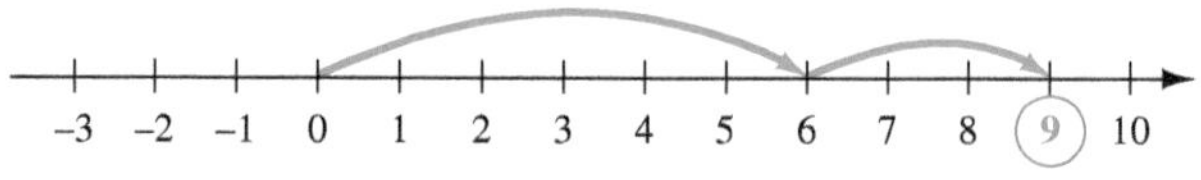

(b)

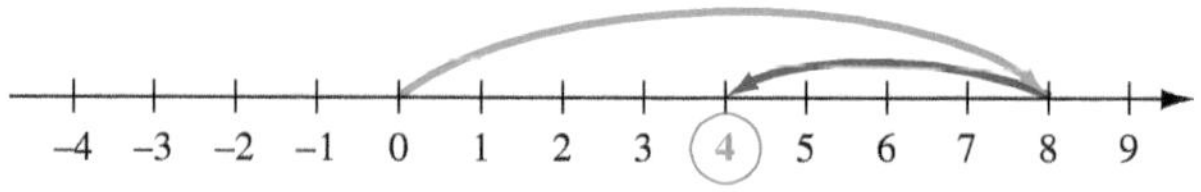

(c)

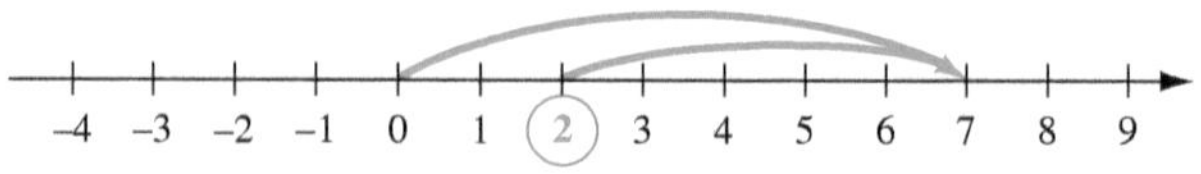

(d)

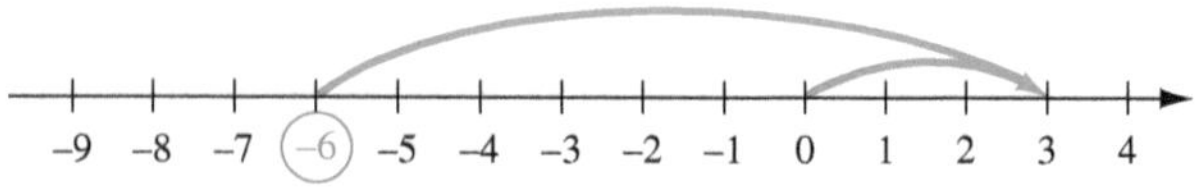

(e)

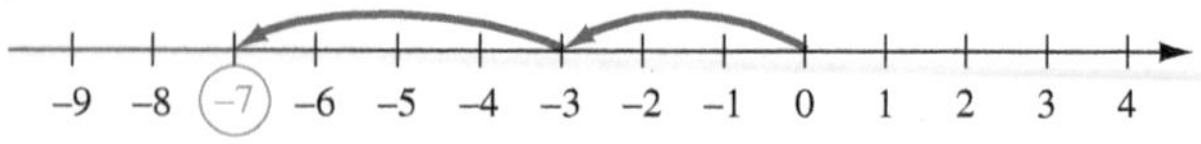

(f)

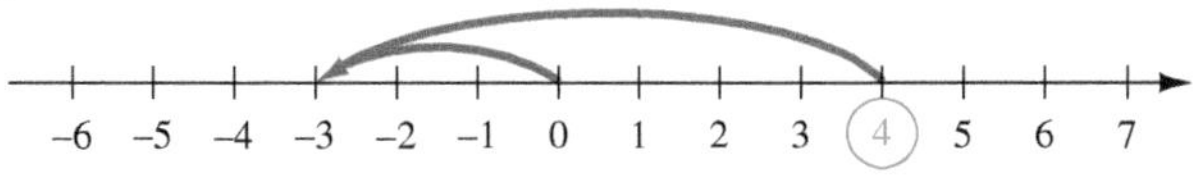

(g)

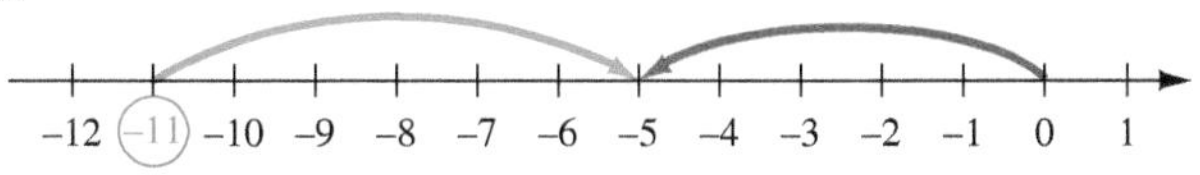

11. Perform these additions and subtractions:

(a) $5 + (-7)$ **(b)** $(-27) - (-5)$

(c) $(-27) + (-5)$ **(d)** $5 - (-7)$

(e) $8 - (-12)$ **(f)** $8 - 12$

12. **(a)** If it is 15° below 0 and the temperature falls 12°, what temperature is it?

(b) What arithmetic does this situation illustrate?

13. **(a)** Dina's bank account was overdrawn by \$12. What was her balance after she deposited the \$37 she earned working at a local pizza parlor?

(b) What arithmetic does this situation illustrate?

14. **(a)** Plot these numbers on a number line: -2, 7, 0, -5, -9, and 2.

(b) List the numbers in part (a) in increasing order.

(c) Determine what integer must be added to each number in your list from part (b) to obtain the next.

Section 5.3

15. What products do these number-line diagrams represent?

(a)

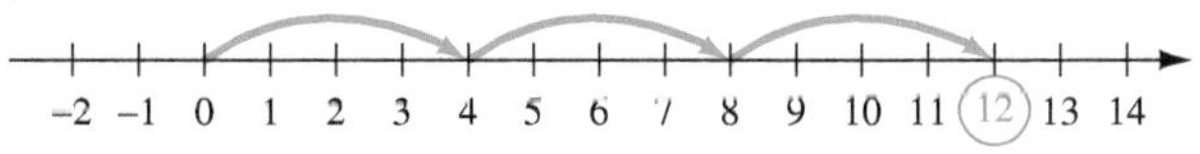

(b)

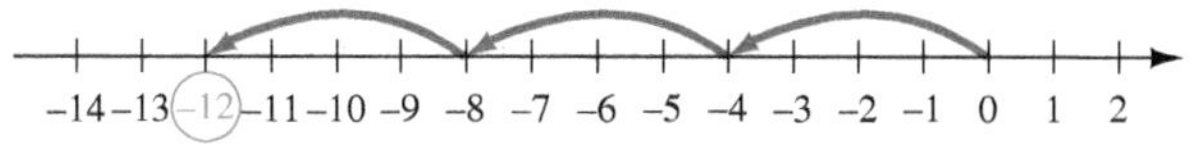

(c)

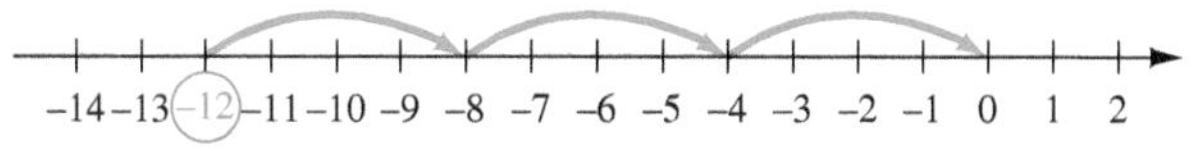

(d)

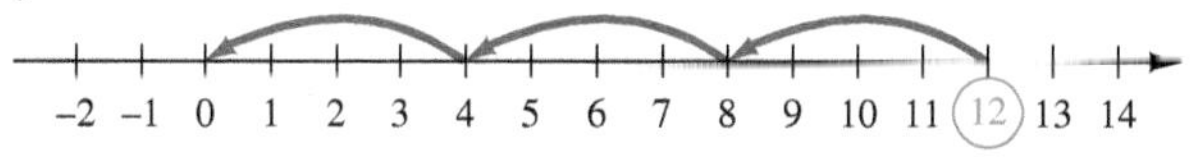

16. At mail time, if the mail carrier took away five bills for \$27 each, are you richer or poorer, and by how much? What calculation does this illustrate?

17. Perform each of these computations:

(a) $(-8)\cdot(-7)$ **(b)** $8\cdot(-7)$

(c) $(-8)\cdot 7$ **(d)** $84 \div (-12)$

(e) $(-84) \div 7$ **(f)** $(-84) \div (-7)$

18. Write a mail-time story to illustrate each of these products:

(a) $7\cdot(12)$ **(b)** $(-7)\cdot(13)$ **(c)** $(-7)\cdot(-13)$

19. Prove that if d divides n, then d divides $-n$, $-d$ divides n, and $-d$ divides $-n$.

20. What multiplication/division fact family is illustrated by these arrays of colored counters?

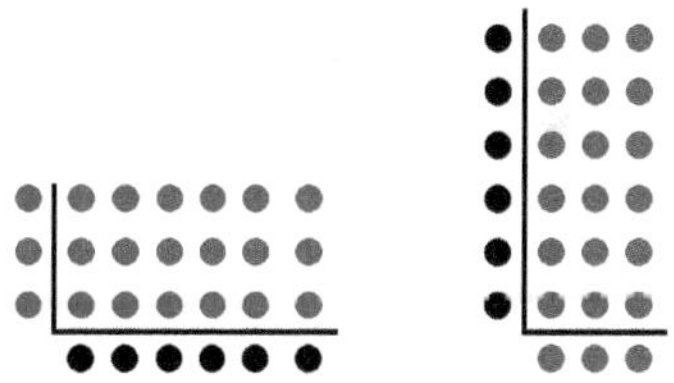

6 Fractions and Rational Numbers

COOPERATIVE INVESTIGATION
Folded Fractions

Materials Needed

Each student (or cooperating pair of students) should have about six paper squares (4- to 5-inch side length) and several colored pencils. Patty Paper (waxed paper that serves as meat-patty separators) works very well.

Example: Folding Quarters

A *unit of area* is defined as the area of a square.

 $= 1$

Fold the unit square in half twice vertically and then unfold. Since the four rectangles are congruent (i.e., have the same shape and size), each rectangle represents the fraction $\frac{1}{4}$.

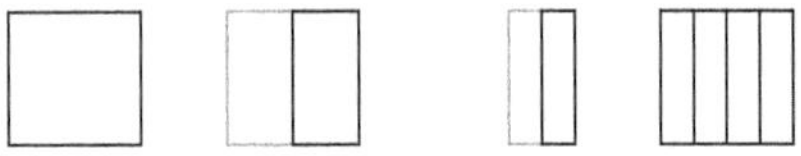

Coloring individual rectangles gives representations of the fractions $\frac{0}{4}, \frac{1}{4}, \frac{2}{4}, \frac{3}{4}$, and $\frac{4}{4}$. (Notice that we count regions, not folds or fold lines.)

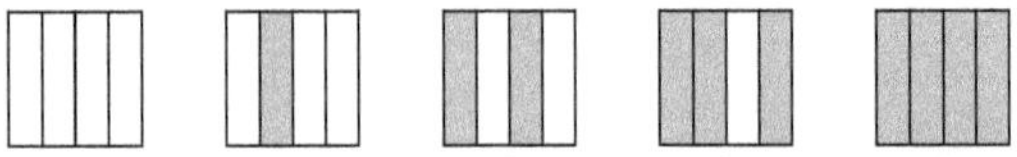

$\frac{0}{4}$ $\frac{1}{4}$ $\frac{2}{4}$ $\frac{3}{4}$ $\frac{4}{4}$

Quarters can also be obtained by other folding procedures. For example, we can make a vertical half fold followed by a horizontal half fold to create four small squares within the unit square. This technique illustrates why $\frac{1}{2} \times \frac{1}{2} = \frac{1}{4}$.

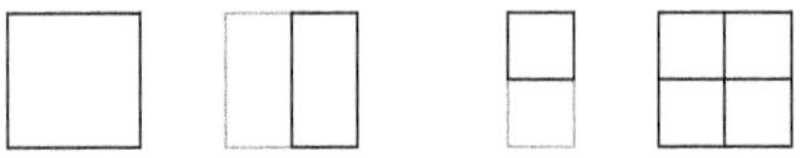

If one colored fraction pattern can be rearranged and regrouped to cover the same region of the unit as another fraction pattern, we say that the two fractions are equivalent. For example, the following rearrangement and regrouping show that $\frac{2}{4}$ is equivalent to $\frac{1}{2}$:

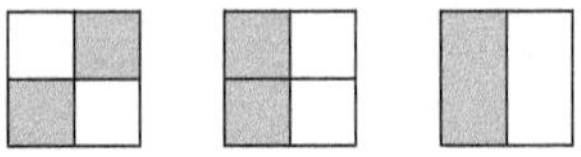

Therefore, we write $\frac{2}{4} = \frac{1}{2}$.

Fraction Folding Activities and Problems

1. **Eighths.** Fold a square into quarters as in the example, and also fold along the two diagonals.

   (a) Identify these fractions:

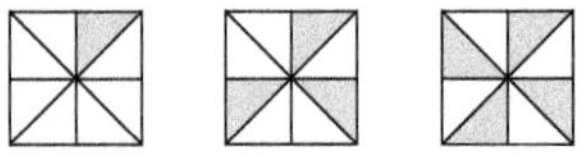

   (b) How many *different* ways can two of the eight regions be colored to give a representation of $\frac{2}{8}$?

   Two colorings are considered identical if one pattern can be rotated to become identical to the second pattern.

2. **Sixths.** "Roll" the paper square into thirds, flatten, and then fold in half in the opposite direction to divide the square into six congruent rectangles.

   This procedure illustrates that $\frac{1}{2} \times \frac{1}{3} = \frac{1}{6}$.

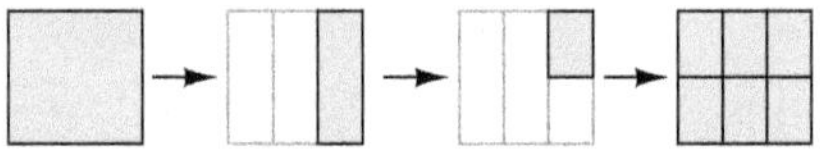

   How many *different* ways can you represent the fraction $\frac{3}{6}$?

   Remember that two patterns are different only if one cannot be rotated to become identical to another.

3. **Identifying fractions.** The leftmost shaded region shown in the figure is bounded by segments joining successive midpoints of the sides of the square. What is the area of the shaded region? The region can be divided into smaller regions that are easy to rearrange into a pattern that shows that $\frac{1}{2}$ of the square is shaded.

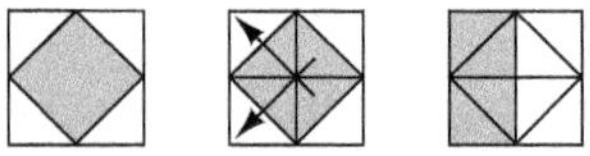

   Use similar multiple representations of the same amount of shaded region to identify the fractions represented by the following colored regions:

   (a) The shaded region obtained by joining the midpoints of the vertical edges of the square with the $\frac{1}{3}$ and $\frac{2}{3}$ points along the horizontal edges:

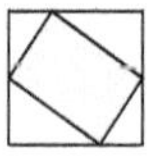

   (b) The shaded region obtained by joining the $\frac{1}{3}$ and $\frac{2}{3}$ points along the vertical and horizontal edges:

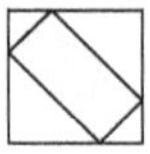

   (c) The shaded square whose corners are the intersections of the segments joining the corners of the unit square to the midpoint of an opposite side of the unit square. (*Suggestion:* Rearrange the unshaded regions to create squares, each congruent to the shaded square.)

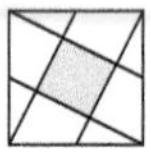

CHAPTER PREVIEW

In this chapter, we focus on the properties, computational procedures, and applications of fractions and rational numbers. We'll see that rational numbers help us solve problems that could not be answered in integers.

KEY IDEAS

- Fraction basics: unit, denominator, numerator
- Fraction models: colored regions (area model), fraction strips, fraction circles, set model, number-line model
- Equivalence of fractions and the fundamental law of fractions
- Simplest form
- Common denominator, least common denominator
- Order relation on the fraction
- Arithmetic operations on fractions and their illustration with fraction models
- Proper fractions and mixed numbers
- Reciprocals
- Algorithms for division, including the "invert-and-multiply" rule
- The meaning of a fraction as an operator
- Rational number system
- Properties of the rational number system
- Existence of multiplicative inverses of nonzero rational numbers
- Density property of the rational numbers
- Computations (exact, approximate, mental) with rational numbers
- Applications of rational numbers

6.1 The Basic Concepts of Fractions and Rational Numbers

Fractions first arise in measurement problems, where they can express a quantity that is less than a whole unit. The word *fraction* comes from the Latin word *fractio,* meaning "the act of breaking into pieces."

Fractions indicate capacity, length, weight, area, or any quantity formed by dividing a unit of measure into b equal-size parts and then collecting a of those parts. To interpret the meaning of the fraction $\frac{a}{b}$, we must

- agree on the **unit** (such as a cup, an inch, the area of the hexagon in a set of pattern blocks, or a whole pizza);
- understand that the unit is **subdivided into b parts of equal size;** and
- understand that we are considering **a of the parts of the unit.**

For example, suppose the unit is one pizza, as shown in Figure 6.1. The pizza has been divided into 8 parts, and 3 parts have been consumed. Using fractions, we can say that $\frac{3}{8}$ of the pizza was eaten and $\frac{5}{8}$ of the pizza remains.

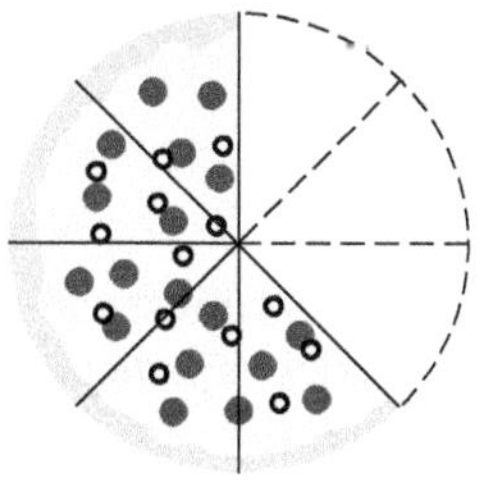

Figure 6.1
Choosing a whole pizza as the unit and dividing it into 8 equal parts shows that $\frac{3}{8}$ of the pizza has been eaten and $\frac{5}{8}$ remains

DEFINITION Fractions

A **fraction** is an ordered pair of integers a and b, $b \neq 0$, written $\frac{a}{b}$ or a/b. The integer a is called the **numerator** of the fraction, and the integer b is called the **denominator** of the fraction.

The word *numerator* is derived from the Latin *numeros*, meaning "number." *Denominator* is from the Latin *denominare,* meaning "namer."

This definition permits the numerator or denominator to be a negative integer and the numerator can be zero. For example, $\frac{-3}{-8}, \frac{-4}{12}, \frac{31}{-4}$, and $\frac{0}{1}$ are all fractions according to the definition just given.

Fraction Models

Some popular representations of fraction concepts are colored regions, fraction strips, fraction circles, the set model, and the number line. A representation of a fraction must clearly answer these three questions:

- What is the unit? (What is the "whole"?)
- Into how many equal parts (the denominator) has the unit been subdivided?
- How many of these parts (the numerator) are under consideration?

Errors and misconceptions about fractions suggest that at least one of these questions has not been properly answered or clearly considered. A common error is the failure to identify the fraction's unit, since the notation $\frac{a}{b}$ makes the numerator and denominator explicit but leaves the unit unidentified.

Colored Regions Choose a shape to represent the unit and then divide that shape into subregions of equal size. Represent a fraction by coloring some of the subregions. See Figure 6.2. Colored-region models are sometimes called *area* models.

Figure 6.2 Some colored-region models for fractions

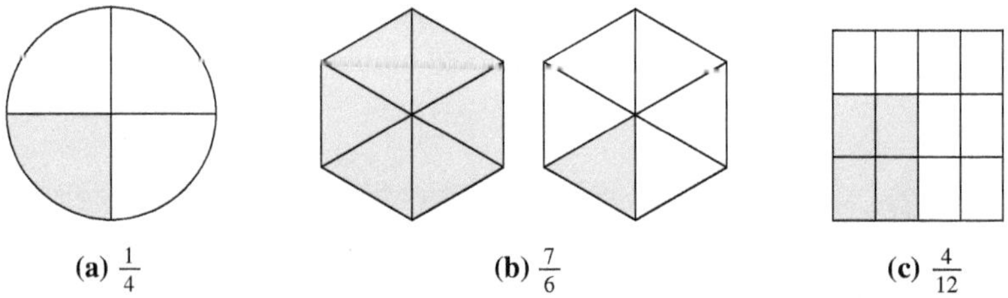

Fraction Strips Define the unit by a rectangular strip. Model a fraction such as $\frac{3}{6}$ by shading 3 of 6 equally sized subrectangles of the rectangle. See Figure 6.3. A set of fraction strips typically contains strips for the denominators 1, 2, 3, 4, 6, 8, and 12.

Figure 6.3 Fractions modeled by fraction strips

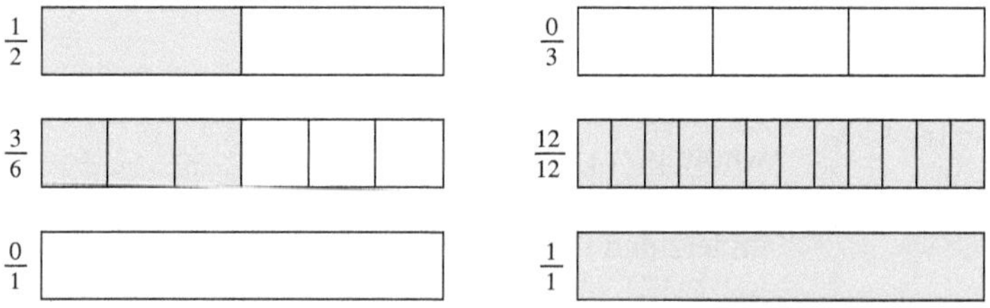

Fraction Circles Divide a circular region into uniformly spaced radial segments corresponding to the denominator. Shade the number of sectors in the numerator. This visual model is particularly effective, since children can easily draw and interpret it. See Figure 6.4.

Figure 6.4 Fractions illustrated by fraction circles

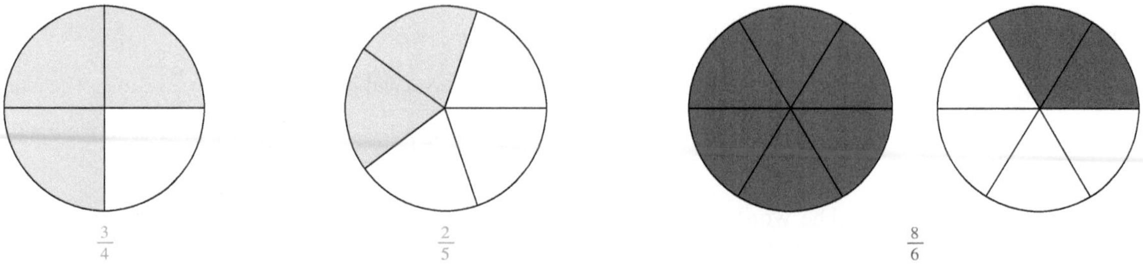

The Set Model Let the unit be a finite set U of objects. Recall that $n(U)$ denotes the number of elements in U. A subset A of U represents the fraction $\frac{n(A)}{n(U)}$. For example, the set of 10 apples shown in Figure 6.5 contains a subset of 3 that are wormy. Therefore, we would say that $\frac{3}{10}$ of the apples are wormy. In Chapter 14, we will see that the set model of fractions is particularly useful in probability. An apple drawn at random from the 10 apples has a $\frac{3}{10}$ probability of being wormy. Sometimes the set model is called the *discrete model.*

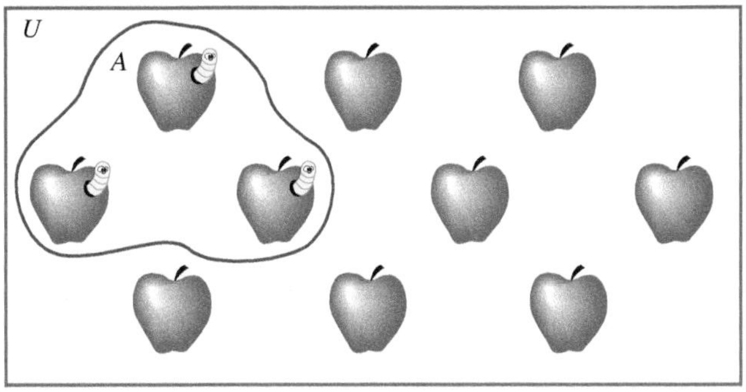

Figure 6.5
The set model showing that $\frac{3}{10}$ of the apples are wormy

The Number-Line Model Assign the points 0 and 1 on a number line. Doing so determines all of the points corresponding to the integers. The unit is the length of the line segment from 0 to 1; it is also the distance between successive integer points. Assign a fraction such as $\frac{5}{2}$ to a point along the number line by subdividing the unit interval into two equal parts and then counting off five of these lengths to the right of 0. See Figure 6.6. Notice that you can name the same point on the fraction number line by different fractions. For example, $\frac{1}{2}$ and $\frac{2}{4}$ correspond to the same distance. The number-line model also illustrates negative fractions such as $-\frac{3}{4}$.

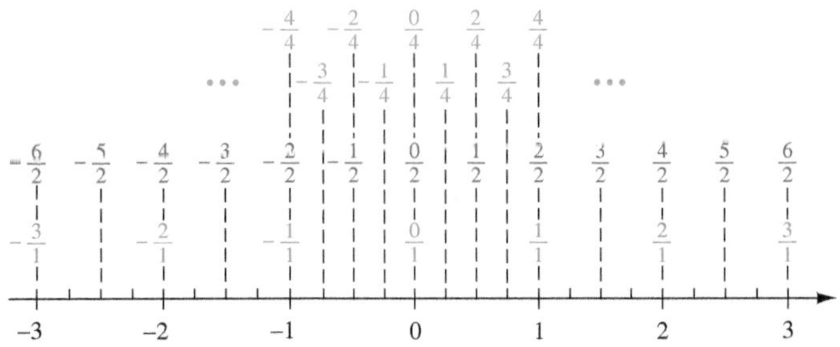

Figure 6.6
Fractions as points on the number-line model

As we did with the number-line model for the whole numbers and the integers, we can represent fractions as "jumps" along the number line, using right-pointing or left-pointing arrows. See Figure 6.7.

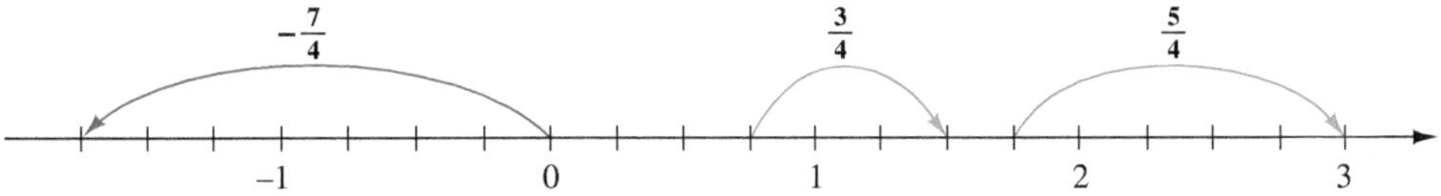

Figure 6.7
Fractions as "jumps" along the number line

Equivalent Fractions

We know that the same point on the number line corresponds to infinitely many different fractions. For example, the point one-half unit to the right of 0 corresponds to $\frac{1}{2}, \frac{2}{4}, \frac{3}{6}$, and so on. Similarly, in Figure 6.8, the fraction strip representing $\frac{2}{3}$ is subdivided by vertical dashed lines to show that $\frac{4}{6}, \frac{6}{9}$, and $\frac{8}{12}$ express the *same* shaded portion of a whole strip.

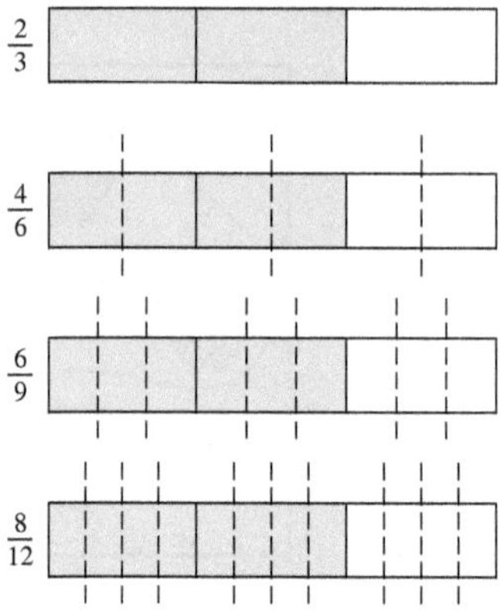

Figure 6.8
The fraction-strip model showing that $\frac{2}{3}, \frac{4}{6}, \frac{6}{9}$, and $\frac{8}{12}$ are equivalent fractions

Fractions that express the same quantity are **equivalent fractions.** The equality symbol, =, signifies that fractions are equivalent, so we write

$$\frac{2}{3} = \frac{4}{6} = \frac{6}{9} = \frac{8}{12}.$$

We can use any number of additional dashed lines between each vertical pair of solid lines in Figure 6.8; therefore,

$$\frac{2}{3} = \frac{2 \cdot n}{3 \cdot n}$$

for natural number n. So given a fraction $\frac{a}{b}$, we obtain an infinite list, $\frac{2a}{2b}, \frac{3a}{3b}, \frac{4a}{4b}, \ldots$, of equivalent fractions.

This idea leads to the following important property.

PROPERTY The Fundamental Law of Fractions

Let $\frac{a}{b}$ be a fraction. Then

$$\frac{a}{b} = \frac{an}{bn}, \quad \text{for any integer } n \neq 0.$$

Equivalent fractions can also be found by dividing both the numerator and denominator by a common factor. For example, the numerator and denominator of $\frac{35}{21}$ are each divisible by 7, so, by the fundamental law,

$$\frac{35}{21} = \frac{5 \cdot 7}{3 \cdot 7} = \frac{5}{3}.$$

Now suppose we are given two fractions, say, $\frac{3}{12}$ and $\frac{2}{8}$, and we wish to know if they are equivalent. From the fundamental law of fractions, we see that

$$\frac{3}{12} = \frac{3 \cdot 8}{12 \cdot 8} \quad \text{and} \quad \frac{2}{8} = \frac{2 \cdot 12}{8 \cdot 12}.$$

Since $\frac{3 \cdot 8}{12 \cdot 8}$ and $\frac{2 \cdot 12}{8 \cdot 12}$ have the same denominator, they are equivalent because they also have the same numerator: $24 = 3 \cdot 8 = 2 \cdot 12$.

Thus, $\frac{3}{12}$ and $\frac{2}{8}$ are equivalent fractions. More generally, the fractions $\frac{a}{b}$ and $\frac{c}{d}$ are equivalent to $\frac{ad}{bd}$ and $\frac{bc}{bd}$. Since the denominators are the same, the fractions are equivalent if and only if the numerators ad and bc are equal. This result, known as the **cross-product property,** is stated in the following theorem.

THEOREM The Cross-Product Property of Equivalent Fractions

The fractions $\frac{a}{b}$ and $\frac{c}{d}$ are **equivalent** if and only if $ad = bc$. That is,

$$\frac{a}{b} = \frac{c}{d} \quad \text{if and only if} \quad ad = bc$$

EXAMPLE 6.1

Examining Fraction Equivalence with Pizza Diagrams

Enrique and Ana each ordered a large pizza. Enrique likes large pieces, so he requested that his pizza be cut into 8 pieces. Ana, however, asked that her pizza be cut into 12 pieces. Enrique ate 6 pieces of his pizza, and Ana ate 9 pieces. Enrique claims that he ate less than Ana, since he ate 3 fewer pieces. Use words and diagrams (similar to Figure 6.1) to see if Enrique is correct.

Solution

The following pizza diagrams show that both Enrique and Ana have eaten $\frac{3}{4}$ of their pizzas and both have $\frac{1}{4}$ of their pizza remaining:

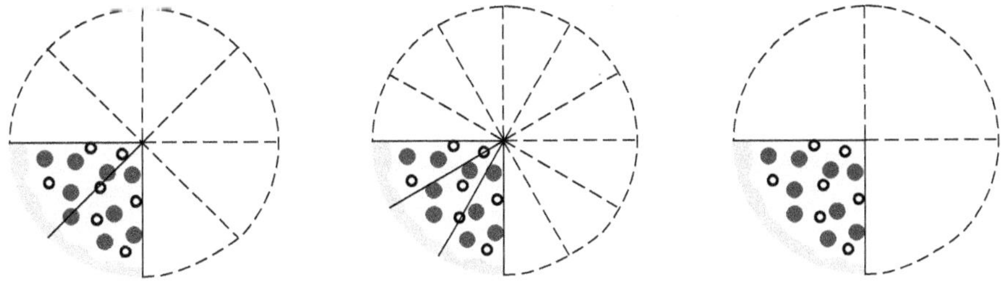

Enrique's pizza   Ana's pizza   A pizza that is $\frac{3}{4}$ eaten, with $\frac{1}{4}$ remaining

The fundamental law of fractions gives the same result: $\frac{6}{8} = \frac{2 \cdot 3}{2 \cdot 4} = \frac{3}{4}$ and $\frac{9}{12} = \frac{3 \cdot 3}{3 \cdot 4} = \frac{3}{4}$.

Alternatively, we can check that $\frac{6}{8} = \frac{9}{12}$ using the cross-product property: $6 \cdot 12 = 8 \cdot 9$.

Fractions in Simplest Form

Often, it is preferable to use the simplest equivalent form of a fraction, which is the one with the smallest positive denominator. For example, it may be best to use $\frac{2}{3}$ instead of $\frac{400}{600}$ and to use $\frac{-3}{4}$ instead of $\frac{75}{-100}$.

DEFINITION Fractions in Simplest Form

A fraction $\frac{a}{b}$ is in **simplest form** if a and b have no common divisor larger than 1 and b is positive.

A fraction in simplest form is also said to be in **lowest terms**. (We avoid using the term *reduced form* because it mistakenly suggests that the fraction is smaller than an equivalent one that is not in simplest form.)

There are several ways to determine the simplest form of a fraction $\frac{a}{b}$. Here are three of the most popular.

Method 1. Divide successively by common factors. Suppose we want to write $\frac{560}{960}$ in simplest form. Using the fundamental law of fractions, we repeatedly divide both numerator and denominator by common factors. Since 560 and 960 are both divisible by 10, it follows that

Both 560 and 960 are divisible by: $\frac{560}{960} = \frac{56 \cdot 10}{96 \cdot 10} = \frac{56}{96}$.

But both 56 and 96 are even (that is, divisible by 2): $\frac{56}{96} = \frac{28 \cdot 2}{48 \cdot 2} = \frac{28}{48}$.

Both 28 and 48 are divisible by 4: $\frac{28}{48} = \frac{7 \cdot 4}{12 \cdot 4} = \frac{7}{12}$.

Since 7 and 12 have no common factor other than 1, $\frac{7}{12}$ is the simplest form of $\frac{560}{960}$. Indeed, one might quickly and efficiently carry out this simplification as follows to obtain the desired result.

$$\frac{\overset{7}{\cancel{\overset{28}{\cancel{560}}}}}{\underset{12}{\cancel{\underset{48}{\cancel{960}}}}} \quad \text{or} \quad \frac{560}{960} = \frac{56}{96} = \frac{28}{48} = \frac{7}{12}.$$

Method 2. Divide by the common factors in the prime-power factorizations of *a* and *b*. Using this method, we have

$$\frac{560}{960} = \frac{2^4 \cdot 5 \cdot 7}{2^6 \cdot 3 \cdot 5} = \frac{7}{2^2 \cdot 3} = \frac{7}{12}.$$

Method 3. Divide *a* and *b* by GCD (*a*, *b*). Using the ideas of Section 4.3, calculate the greatest common divisor GCD(560, 960) $= 80$. Therefore, $\frac{560}{960} = \frac{560 \div 80}{960 \div 80} = \frac{7}{12}$.

EXAMPLE 6.2

Simplifying Fractions

Find the simplest form of each fraction.

(a) $\frac{240}{72}$ **(b)** $\frac{294}{-84}$ **(c)** $\frac{48}{64}$

Solution

(a) By Method 1, $\frac{240}{72} = \frac{120}{36} = \frac{60}{18} = \frac{10}{3}$, where the successive common factors 2, 2, and 6 were divided into the numerator and denominator.

(b) Using Method 2 this time, we find that

$$\frac{294}{-84} = \frac{2 \cdot 3 \cdot 7^2}{(-2) \cdot 2 \cdot 3 \cdot 7} = \frac{7}{-2} = \frac{-7}{2}.$$

(c) By Method 3, GCD(48, 64) $= 16$, $\frac{48}{64} = \frac{48 \div 16}{64 \div 16} = \frac{3}{4}$.

Common Denominators

Fractions with the same denominator are said to have a **common denominator.** In working with fractions, it is often helpful to rewrite them with equivalent fractions that share a common denominator. For example, $\frac{5}{8}$ and $\frac{7}{10}$ can each be rewritten by equivalent fractions with the common denominator $8 \cdot 10 = 80$, so

$$\frac{5}{8} = \frac{5 \cdot 10}{8 \cdot 10} = \frac{50}{80} \quad \text{and} \quad \frac{7}{10} = \frac{7 \cdot 8}{10 \cdot 8} = \frac{56}{80}.$$

In the same way, any two fractions $\frac{a}{b}$ and $\frac{c}{d}$ can be rewritten with the common denominator $b \cdot d$, since $\frac{a}{b} = \frac{a \cdot d}{b \cdot d}$ and $\frac{c}{d} = \frac{c \cdot b}{d \cdot b}$.

Sometimes we need to find the common positive denominator that is as small as possible. Assuming that $\frac{a}{b}$ and $\frac{c}{d}$ are in simplest form, we require a common denominator that is the least common multiple of both b and d. This denominator is called the **least common denominator.** For the example $\frac{5}{8}$ and $\frac{7}{10}$, we would calculate LCM(8, 10) $= 40$, so 40 is the least common denominator. Therefore,

$$\frac{5}{8} = \frac{5 \cdot 5}{8 \cdot 5} = \frac{25}{40} \quad \text{and} \quad \frac{7}{10} = \frac{7 \cdot 4}{10 \cdot 4} = \frac{28}{40}.$$

We can often find the least common denominator using mental arithmetic. Consider, $\frac{5}{6}$ and $\frac{3}{8}$, and notice that $4 \cdot 6 = 24$ and $3 \cdot 8 = 24$ is the least common denominator. Thus, $\frac{5}{6} = \frac{20}{24}$ and $\frac{3}{8} = \frac{9}{24}$ when written with the least common denominator.

EXAMPLE 6.3

Finding Common Denominators

Find equivalent fractions with a common denominator.

(a) $\frac{5}{6}$ and $\frac{1}{4}$

(b) $\frac{9}{8}$ and $\frac{-12}{7}$

(c) $\frac{3}{4}$, $\frac{5}{8}$, and $\frac{2}{3}$

Solution

(a) By mental arithmetic, the least common denominator of $\frac{5}{6}$ and $\frac{1}{4}$ is 12. Therefore, $\frac{5}{6} = \frac{10}{12}$ and $\frac{1}{4} = \frac{3}{12}$. This can be shown with fraction strips:

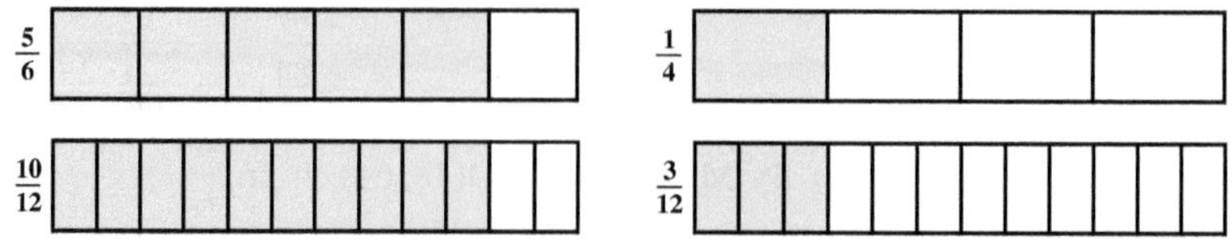

(b) Since 8 and 7 have no common factors other than 1, $8 \cdot 7 = 56$ is the least common denominator:

$$\frac{9}{8} = \frac{9 \cdot 7}{8 \cdot 7} = \frac{63}{56} \quad \text{and} \quad \frac{-12}{7} = \frac{-12 \cdot 8}{7 \cdot 8} = \frac{-96}{56}.$$

(c) With three fractions, it is still possible to use the product of all of the denominators as a common denominator. Since $4 \cdot 8 \cdot 3 = 96$, we have

$$\frac{3}{4} = \frac{3 \cdot 8 \cdot 3}{4 \cdot 8 \cdot 3} = \frac{72}{96}, \quad \frac{5}{8} = \frac{5 \cdot 4 \cdot 3}{8 \cdot 4 \cdot 3} = \frac{60}{96}, \quad \text{and} \quad \frac{2}{3} = \frac{2 \cdot 4 \cdot 8}{3 \cdot 4 \cdot 8} = \frac{64}{96}.$$

Alternatively, LCM(4, 8, 3) = 24, so

$$\frac{3}{4} = \frac{3 \cdot 6}{4 \cdot 6} = \frac{18}{24}, \quad \frac{5}{8} = \frac{5 \cdot 3}{8 \cdot 3} = \frac{15}{24}, \quad \text{and} \quad \frac{2}{3} = \frac{2 \cdot 8}{3 \cdot 8} = \frac{16}{24}.$$

These equations express the three fractions as equivalent fractions with the least common denominator.

Rational Numbers

We have seen that different fractions can express the same number amount. For example, the fraction strips in Figure 6.8 show that $\frac{2}{3}, \frac{4}{6}, \frac{6}{9}$, and $\frac{8}{12}$ represent the same number. In Figure 6.6, one point on the number line can be expressed by infinitely many fractions.

Any number that can be represented by a fraction *or* any of its equivalent fractions is called a **rational number.** Any one of these equivalent fractions can be used to represent that rational number. For example, $\frac{2}{3}$ and $\frac{10}{15}$ are different fractions, since their numerators and denominators are different. However, since $\frac{2}{3}$ and $\frac{10}{15}$ are equivalent fractions, they both represent the *same rational number*. When we say "the rational number $\frac{2}{3}$," we really mean the rational number represented by the fraction $\frac{2}{3}$ or any fraction that is equivalent to $\frac{2}{3}$.

DEFINITION Rational Numbers

A **rational number** is a number that can be represented by a fraction $\frac{a}{b}$, where a and b are integers, $b \neq 0$.

Two rational numbers are **equal** if and only if they can be represented by equivalent fractions, and equivalent fractions represent the same rational number.

Figure 6.9
The Venn diagram of the number systems N, W, I, and Q

The **set of rational numbers** is denoted by Q. The letter Q reminds us that a fraction is represented as a quotient. A rational number such as $\frac{3}{4}$ can also be represented by $\frac{6}{8}, \frac{30}{40}$, or any other fraction equivalent to $\frac{3}{4}$. An integer $n \in I$ is identified with the rational number $\frac{n}{1}$, so the integers I are a proper subset of the rational numbers Q. Figure 6.9 shows the Venn diagram of the sets of natural numbers N, whole numbers W, integers I, and rational numbers Q.

EXAMPLE 6.4

Representing Rational Numbers

How many different rational numbers are given in this list of five fractions?

$$\frac{2}{5}, \quad 3, \quad \frac{-4}{-10}, \quad \frac{39}{13}, \quad \text{and} \quad \frac{7}{4}$$

Solution

Since $\frac{2}{5} = \frac{-4}{-10}$ and $\frac{3}{1} = \frac{39}{13}$, there are three different rational numbers: $\frac{2}{5}$, 3, and $\frac{7}{4}$.

Ordering Fractions and Rational Numbers

Two fractions with the same positive denominator are ordered by comparing their numerators. For example, $\frac{3}{7} < \frac{4}{7}$ and $\frac{-4}{5} < \frac{3}{5}$. To order two fractions with different denominators, we can rewrite them with a common positive denominator and then compare their numerators. For example, by rewriting $\frac{3}{4}$ and $\frac{5}{6}$ as $\frac{9}{12}$ and $\frac{10}{12}$, we see that $\frac{3}{4} < \frac{5}{6}$.

Sometimes it is helpful to compare two fractions with a third one. For example, to order $\frac{3}{7}$ and $\frac{5}{9}$, we see that $\frac{3}{7} < \frac{1}{2}$ and $\frac{1}{2} < \frac{5}{9}$. Therefore, $\frac{3}{7} < \frac{1}{2} < \frac{5}{9}$, and we conclude that $\frac{3}{7} < \frac{5}{9}$.

Manipulatives and pictorial representations of fractions also reveal the order relation. The following excerpt shows how parallel number lines help visualize the order relation:

FROM THE NCTM PRINCIPLES AND STANDARDS

During grades 3–5, students should build their understanding of fractions as parts of a whole and as division. They will need to see and explore a variety of models of fractions, focusing primarily on familiar fractions such as halves, thirds, fourths, fifths, sixths, eighths, and tenths. By using an area model in which part of a region is shaded, students can see how fractions are related to a unit whole, compare fractional parts of a whole, and find equivalent fractions. They should develop strategies for ordering and comparing fractions, often using benchmarks such as 1/2 and 1. For example, fifth graders can compare fractions such as 2/5 and 5/8 by comparing each with 1/2—one is a little less than 1/2, and the other is a little more. By using parallel number lines, each showing a unit fraction and its multiples, students can see fractions as numbers, note their relationship to 1, and see relationships among fractions, including equivalence.

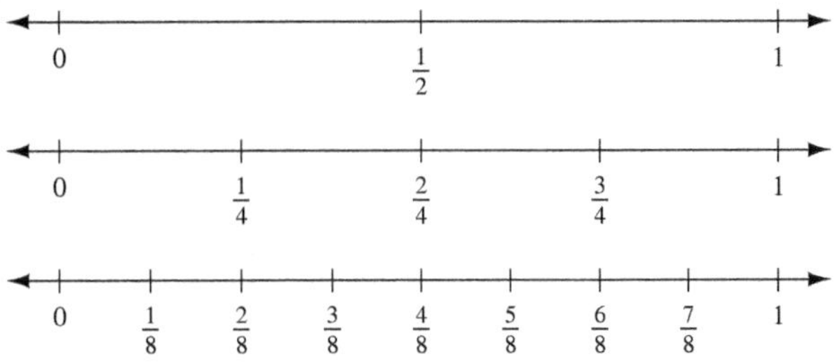

SOURCE: Principles and Standards for School Mathematics by NCTM, p. 150. *Copyright © 2000 by the National Council of Teachers of Mathematics. Reproduced with permission of the National Council of Teachers of Mathematics via Copyright Clearance Center. NCTM does not endorse the content or validity of these alignments.*

We can also use fraction strips to order fractions. For example, by aligning the fraction strips for $\frac{3}{4}$ and $\frac{5}{6}$, as shown in Figure 6.10, we see that $\frac{3}{4}$ represents a smaller shaded portion of a whole strip than $\frac{5}{6}$ does.

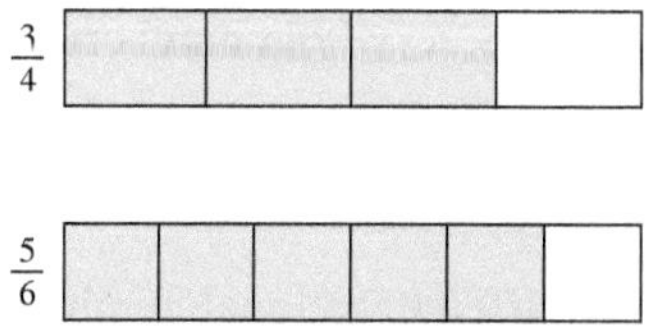

Figure 6.10
Showing that $\frac{3}{4} < \frac{5}{6}$ with fraction strips

The ordering just described for fractions is used to define an ordering of rational numbers. For example, suppose we were told to make a right turn in three-quarters of a mile, and the odometer on our car shows we have gone six-tenths of a mile. Did we miss our turn, or do we need to drive further? To decide, we need to compare the rational number "six-tenths" with the rational number "three-quarters." We can represent these rational numbers by the fractions $\frac{6}{10}$ and $\frac{3}{4}$. To compare these two fractions, rewrite them with a common denominator, $\frac{6}{10} = \frac{6 \cdot 4}{10 \cdot 4}$ and $\frac{3}{4} = \frac{10 \cdot 3}{10 \cdot 4}$. Since $6 \cdot 4 < 10 \cdot 3$, we see that $\frac{6}{10} < \frac{3}{4}$.

To compare any two rational numbers, say, r and s, first write r and s as fractions $r = \frac{a}{b}$ and $s = \frac{c}{d}$, both with positive denominators b and d. Since $r = \frac{a}{b} = \frac{ad}{bd}$ and $s = \frac{c}{d} = \frac{bc}{bd}$, we see that $\frac{a}{b} < \frac{c}{d}$ if and only the "cross products" ad and bc satisfy $ad < bc$. Earlier, we had seen that $\frac{a}{b} = \frac{c}{d}$ if and only if $ad = bc$.

DEFINITION Ordering Rational Numbers

Let two rational numbers be represented by the fractions $\frac{a}{b}$ and $\frac{c}{d}$, with b and d positive. Then $\frac{a}{b}$ is **less than** $\frac{c}{d}$, written $\frac{a}{b} < \frac{c}{d}$, if and only if $ad < bc$.

The corresponding relations "less than or equal to," $\leq$; "greater than," $>$; and "greater than or equal to," $\geq$, are defined similarly.

It makes no difference which equivalent fractions are used to represent the rational numbers in the definition; the ordering is always the same. For example, if the rational numbers "six-tenths" and "three-quarters" are represented by $\frac{3}{5}$ and $\frac{9}{12}$, their cross products are $3 \cdot 12 = 36$ and $5 \cdot 9 = 45$. Since $36 < 45$, we still see that "six-tenths" is less than "three-quarters."

EXAMPLE 6.5

Comparing Rational Numbers

Are the following relations correct? If not, what is the correct relation?

(a) $\frac{3}{4} > \frac{2}{5}$ **(b)** $\frac{15}{29} = \frac{6}{11}$ **(c)** $\frac{2106}{7047} = \frac{234}{783}$ **(d)** $\frac{-10}{13} < \frac{-22}{29}$

Solution

(a) Correct. Since $3 \cdot 5 = 15 > 8 = 2 \cdot 4$, we have $\frac{3}{4} > \frac{2}{5}$. Alternatively, $\frac{1}{2}$ can be used as a benchmark, with $\frac{3}{4} > \frac{1}{2}$ and $\frac{2}{5} < \frac{1}{2}$. Thus, $\frac{3}{4} > \frac{2}{5}$ by the transitive strategy.

(b) Incorrect. Since $15 \cdot 11 = 165 < 174 = 6 \cdot 29$, we have $\frac{15}{29} < \frac{6}{11}$.

(c) Using a calculator, we find that $2106 \cdot 783 = 1{,}648{,}998 = 7047 \cdot 234$. Thus, the two rational numbers are equal: $\frac{2106}{7047} = \frac{234}{783}$.

(d) Since $-10 \cdot 29 = -290 < -286 = -22 \cdot 13$, we conclude that $\frac{-10}{13} < \frac{-22}{29}$.

COOPERATIVE INVESTIGATION

Exploring Fraction Concepts with Fraction Tiles

The basic concepts of fractions are best taught with a variety of manipulative materials that can be purchased from education supply houses or, alternatively, can be homemade from patterns printed and cut from card stock. The commercially made manipulatives are usually colored plastic, with the unit a square, a circle, a rectangle, or some other shape that can be subdivided in various ways to give denominators such as 2, 3, 4, 6, 8, and 12. Usually, the shapes are also available in translucent plastic for demonstrations on an overhead projector. In the following activity, you will make and then work with a set of *fraction tiles*:

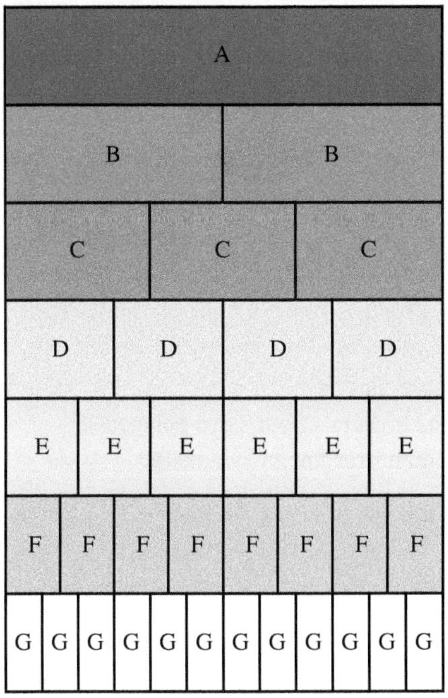

Materials

Each group of three or four students needs a set of fraction tiles, either a plastic set or a set cut from card stock using a pattern such as the one shown (at reduced scale) to the left. The pieces at the right are lettered A through G but can also be described by their color.

Directions

If a B tile is chosen as the unit, then the fraction $\frac{1}{2}$ is represented by a D tile, which can be verified by comparing two D tiles and one B tile. Similarly, an A tile represents 2, and $\frac{2}{3}$ is represented by two E tiles or by a C tile. Now use your fraction tile set to answer and discuss the questions that follow. Compare and discuss your answers within your group.

Questions

1. If an A tile is the unit, find (one or more) tiles to represent $\frac{1}{3}, \frac{2}{3}, \frac{3}{3}, \frac{4}{3}, \frac{2}{6}$, and $\frac{5}{12}$.
2. If a C tile is $\frac{1}{6}$, what fractions are represented by each of the other tiles?
3. If a B tile represents $\frac{3}{2}$, represent these fractions with tiles: $\frac{1}{2}, \frac{1}{1}, \frac{1}{4}, \frac{3}{1}$, and $\frac{3}{4}$.

6.1 Problem Set

Understanding Concepts

1. What fraction is represented by the shaded portion of the following figures?

(a) 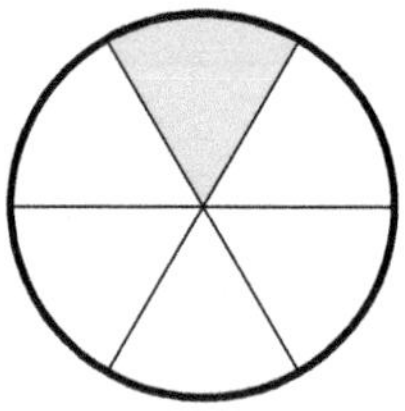

(b) 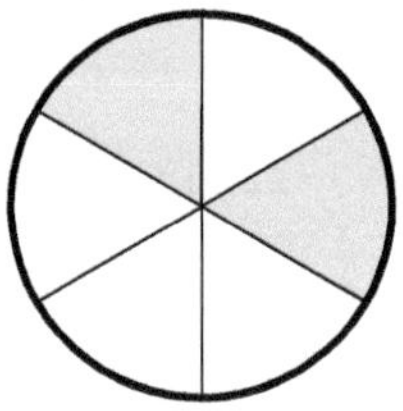

(c) 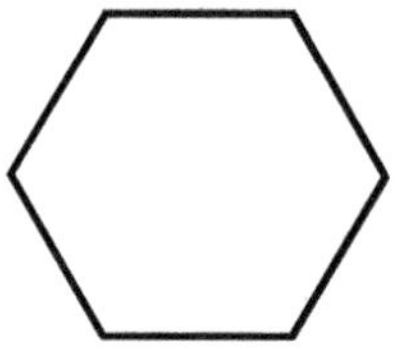

2. What fraction is represented by the shaded portion of the following figures?

(a) 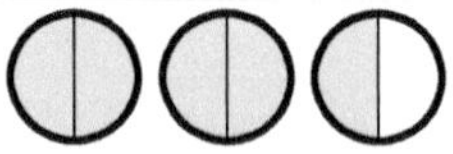

Unit = one disc

(b) 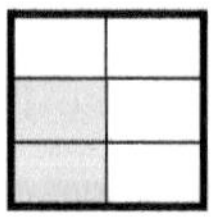

(c) 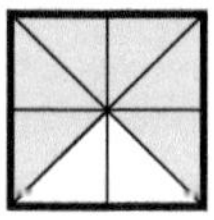

3. Subdivide and shade the unit octagons shown to represent the given fraction.

(a) 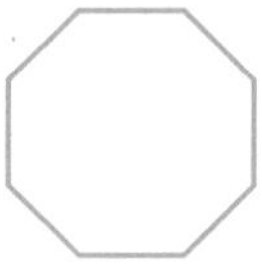

$\frac{1}{8}$

(b) 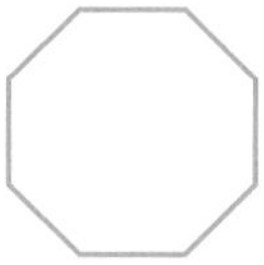

$\frac{6}{8}$

4. Subdivide and shade the unit octagons shown to represent the given fraction.

(a) 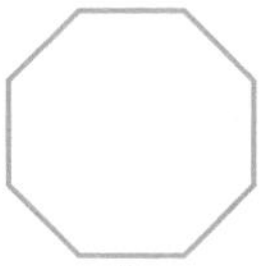

$\frac{3}{4}$

(b) 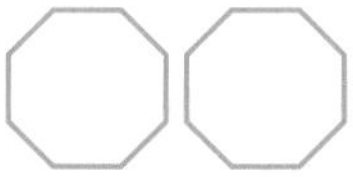

$\frac{11}{8}$

5. What fraction is illustrated by the shaded region of these rectangles and circles, each having one unit of area?

(a) (b) (c)

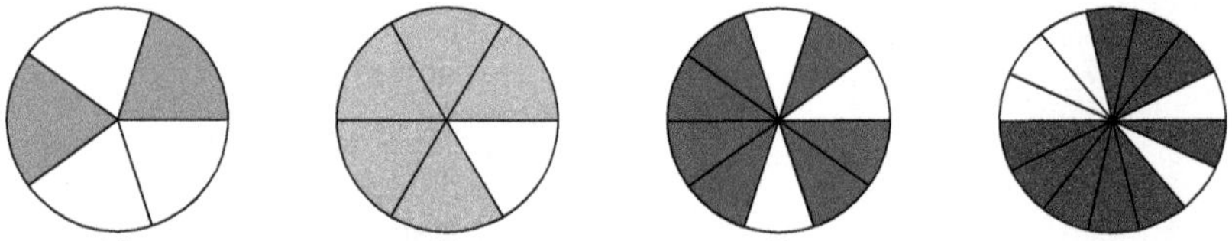

(d) (e) (f) (g)

6. Draw fraction strips and fraction circle diagrams to represent these fractions:

(a) $\frac{2}{3}$ (b) $\frac{1}{4}$ (c) $\frac{2}{8}$ (d) $\frac{7}{12}$

7. Consider the following set of objects:

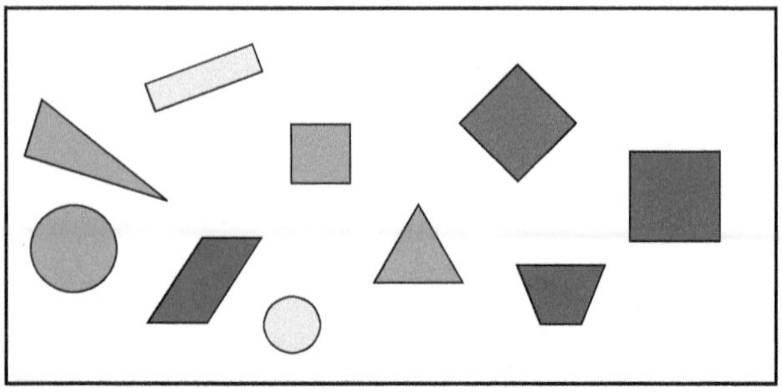

What is the fraction of the objects in the set that

(a) are green?
(b) are squares?
(c) are rectangles?
(d) are quadrilaterals (four-sided polygons)?
(e) have no mirror line of symmetry?

8. For each lettered point on the number lines shown, express its position by a corresponding fraction. Remember, it is the number of subintervals in a unit interval that determines the denominator, not the number of tick marks.

(a) 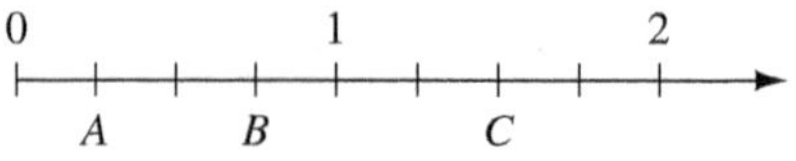

(b)

(d)

(c)

(e)

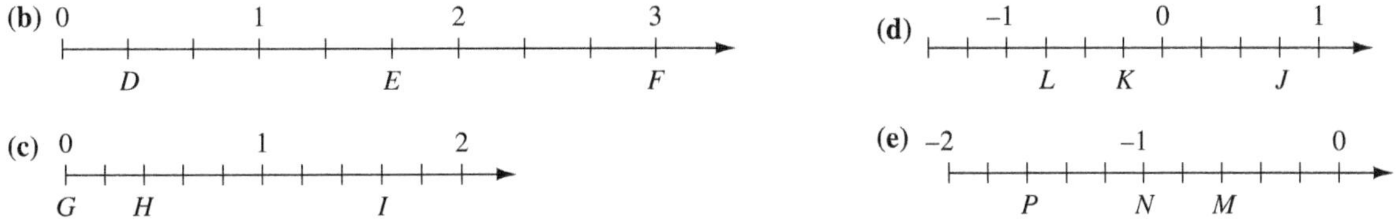

9. For each number-line diagram, list the fractions represented by the arrows:

(a)

(b)

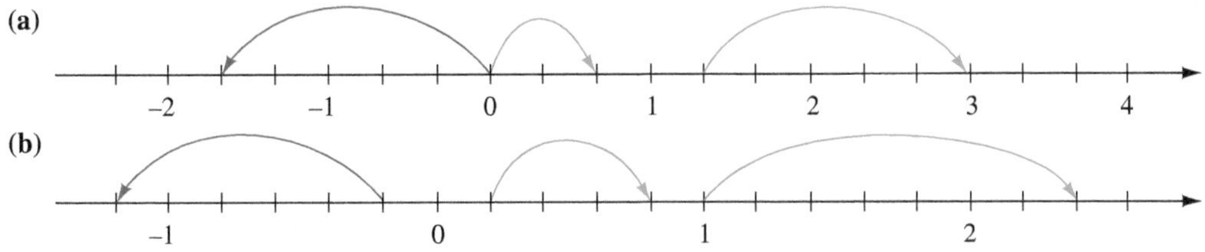

10. Choose a fraction from the set $\left\{\dfrac{1}{6}, \dfrac{1}{3}, \dfrac{1}{2}, \dfrac{3}{4}, \dfrac{6}{13}, \dfrac{17}{20}, \dfrac{17}{30}\right\}$ that best represents the shaded area of each of these unit rectangles.

(a)

(b)

(c)

(d)

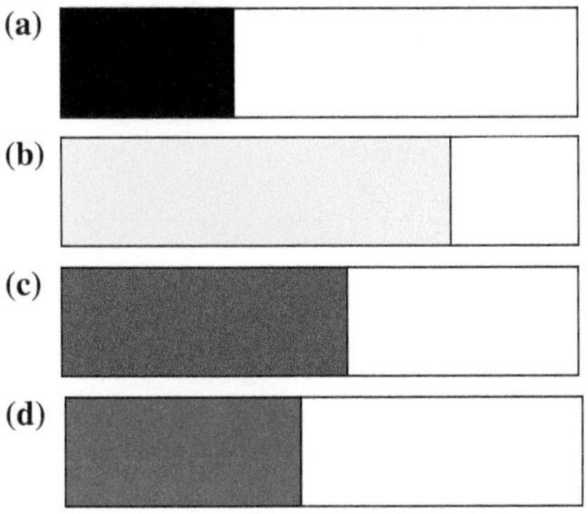

11. Depict the fraction $\dfrac{4}{6}$ with the following models:

(a) Fraction-circle model　　**(b)** Set model
(c) Fraction-strip model　　**(d)** Number-line model

12. In Figure 6.8, fraction strips show that $\dfrac{2}{3}, \dfrac{4}{6}, \dfrac{6}{9}$, and $\dfrac{8}{12}$ are equivalent fractions. Use a similar drawing of fraction strips to show that $\dfrac{3}{4}, \dfrac{6}{8}$, and $\dfrac{9}{12}$ are equivalent fractions.

13. What equivalence of fractions is shown in these pairs of colored-region models?

(a)　　　　　**(b)**　　　　　**(c)**

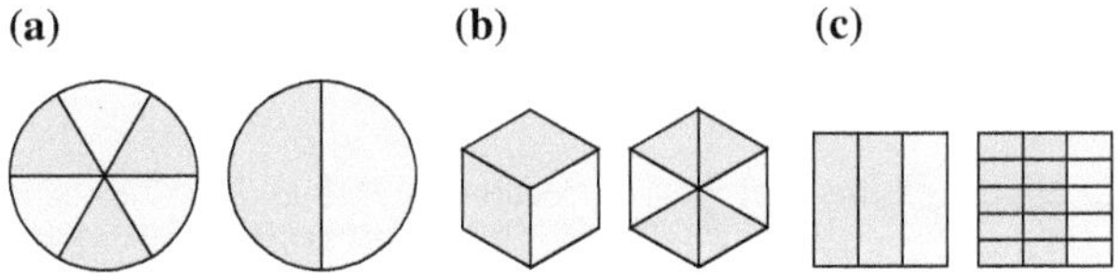

14. Find four different fractions equivalent to $\dfrac{4}{9}$.

15. Subdivide and shade the unit square on the right to illustrate that the given fractions are equivalent.

(a)　　　　　**(b)**　　　　　**(c)**

$\dfrac{2}{4} = \dfrac{4}{8}$　　$\dfrac{1}{3} = \dfrac{3}{9}$　　$\dfrac{1}{4} = \dfrac{2}{8}$

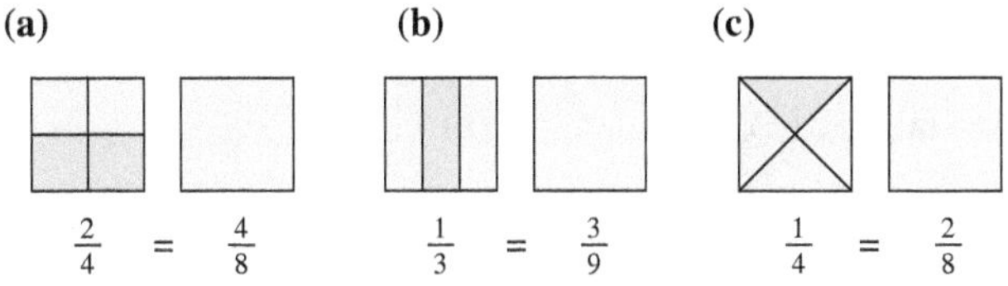

16. Fill in the missing integer to make the fractions equivalent.

(a) $\dfrac{4}{5} = \dfrac{}{30}$ 　　(b) $\dfrac{6}{9} = \dfrac{2}{}$

(c) $\dfrac{-7}{25} = \dfrac{}{500}$ 　　(d) $\dfrac{18}{3} = \dfrac{-6}{}$

17. Determine whether each set of two fractions is equivalent by calculating equivalent fractions with a common denominator.

(a) $\dfrac{18}{42}$ and $\dfrac{3}{7}$ 　　(b) $\dfrac{18}{49}$ and $\dfrac{5}{14}$

(c) $\dfrac{9}{25}$ and $\dfrac{140}{500}$ 　　(d) $\dfrac{24}{144}$ and $\dfrac{32}{96}$

18. Determine which of these pairs of fractions are equivalent:

(a) $\dfrac{78}{24}$ and $\dfrac{546}{168}$ 　　(b) $\dfrac{243}{317}$ and $\dfrac{2673}{3487}$

(c) $\dfrac{412}{-864}$ and $\dfrac{-308}{616}$

19. (a) Is it true that $\dfrac{4 \cdot 3}{9 \cdot 3} = \dfrac{12}{27}$?

(b) Is it true that $\dfrac{4 \cdot 3}{9 \cdot 3} = \dfrac{4}{9}$?

(c) Is it true that $\dfrac{4 + 3}{9 + 3} = \dfrac{7}{12}$?

(d) Is it true that $\dfrac{4 + 3}{9 + 3} = \dfrac{4}{9}$?

20. Rewrite the following fractions in simplest form:

(a) $\dfrac{84}{144}$ 　　(b) $\dfrac{208}{272}$

(c) $\dfrac{-930}{1290}$ 　　(d) $\dfrac{325}{231}$

21. Find the prime factorizations of the numerators and denominators of these fractions, and use them to express the fractions in simplest form:

(a) $\frac{96}{288}$ (b) $\frac{247}{-75}$ (c) $\frac{2520}{378}$

22. For each of these sets of fractions, determine equivalent fractions with a common denominator:

(a) $\frac{3}{11}$ and $\frac{2}{5}$ (b) $\frac{5}{12}$ and $\frac{2}{3}$

(c) $\frac{4}{3}, \frac{5}{8}$, and $\frac{1}{6}$ (d) $\frac{1}{125}$ and $\frac{-3}{500}$

23. For each of these sets of fractions, determine equivalent fractions with the least common denominator.

(a) $\frac{3}{8}$ and $\frac{5}{6}$ (b) $\frac{1}{7}, \frac{4}{5}$, and $\frac{2}{3}$

(c) $\frac{17}{12}$ and $\frac{7}{32}$ (d) $\frac{17}{51}$ and $\frac{56}{42}$

24. Order each pair of fractions.

(a) $\frac{11}{7}, \frac{11}{9}$ (b) $\frac{3}{13}, \frac{4}{13}$

(c) $\frac{13}{18}, \frac{7}{12}$ (d) $\frac{9}{11}, \frac{8}{10}$

25. Order the rational numbers from least to greatest in each part.

(a) $\frac{2}{3}, \frac{7}{12}$ (b) $\frac{2}{3}, \frac{5}{6}$ (c) $\frac{5}{6}, \frac{29}{36}$

(d) $\frac{-5}{6}, \frac{-8}{9}$ (e) $\frac{2}{3}, \frac{5}{6}, \frac{29}{36}, \frac{8}{9}$

26. Find the numerator or denominator of these fractions to create a fraction that is close to $\frac{1}{2}$ but is slightly larger:

(a) $\frac{\quad}{20}$ (b) $\frac{7}{\quad}$ (c) $\frac{\quad}{17}$ (d) $\frac{30}{\quad}$

27. How many different rational numbers are in this list?

$$\frac{27}{36}, 4, \frac{21}{28}, \frac{24}{6}, \frac{3}{4}, \frac{-8}{-2}$$

Into the Classroom

28. Number rods (or Cuisenaire® rods) can be used to illustrate the basic concepts of fractions. For example, suppose that the brown rod is the whole (that is, the unit). Then, since a train of two purple rods has the length of the brown rod, we see that the purple rod represents the fraction $\frac{1}{2}$:

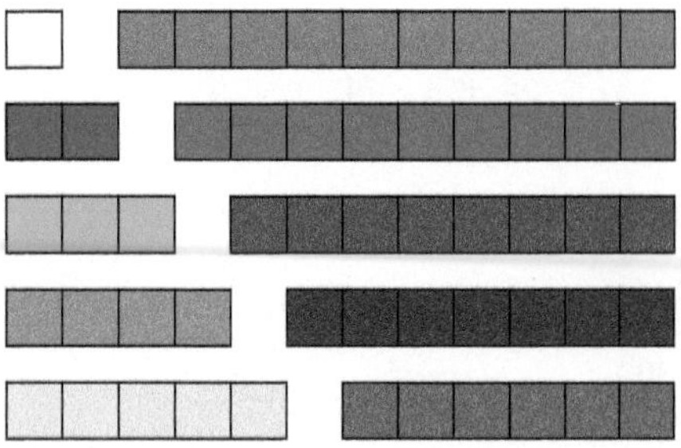

(a) Which rod represents $\frac{1}{4}$? $\frac{3}{4}$?

(b) What fraction does a train made from the red and orange rods laid end to end represent?

(c) Using the discussion and questions in this problem as a model, write several more activities and accompanying questions that you feel would be effective in the elementary school classroom to teach basic fraction concepts with the use of number rods as a manipulative.

29. Pattern blocks are a popular and versatile manipulative that is used successfully in many elementary school classrooms. The following four shapes are included in any set.

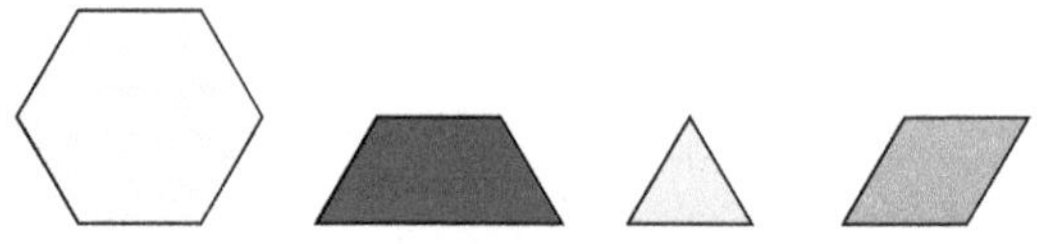

Use your pattern blocks to answer these questions:

(a) Choose the hexagon as the unit. What fraction is each of the other three pattern blocks?

(b) Choose the trapezoid as the unit. What fraction is each of the other three pattern blocks?

(c) Choose the rhombus as the unit. What fraction is each of the other three pattern blocks?

Responding to Students

30. When asked to illustrate the concept of $\frac{2}{3}$ with a colored-region diagram, Shanti drew the figure shown. How would you respond to Shanti?

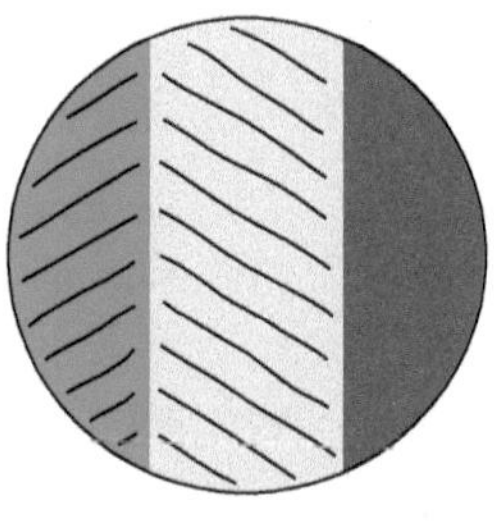

31. (**Writing**) Like most fifth graders, Dana likes pizza. When given the choice of $\frac{1}{4}$ or $\frac{1}{6}$ of a pizza, Dana says, "Since 6 is bigger than 4 and I'm really hungry, I'd rather have $\frac{1}{6}$ of the pizza." Write a dialogue, including useful diagrams, to clear up Dana's misconception about fractions.

32. Nicole claims that $\frac{3}{4} < \frac{5}{8}$ because $3 < 5$ and $4 < 8$. Is Nicole correct, or how can you help her understanding?

33. Miley has been asked to compare $\frac{3}{8}$ and $\frac{4}{13}$. She says that if she had three $\frac{3}{8}$, she would have $\frac{9}{8}$, a bit more than 1, but if she had three $\frac{4}{13}$, she would have $\frac{12}{13}$, which is less than 1. Therefore, Miley claims that $\frac{3}{8} > \frac{4}{13}$. Is Miley's reasoning valid?

Thinking Critically

34. **(a)** The rectangle shown is $\frac{2}{3}$ of a unit. What is the unit?

(*Suggestion:* Divide the rectangle into two identical rectangles.)

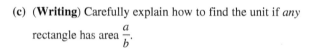

(b) The rectangle shown is $\frac{5}{2}$ of a whole. What is the whole?

(*Suggestion:* Use the idea from part (a), but first decide the number of parts into which to subdivide the rectangle.)

(c) (**Writing**) Carefully explain how to find the unit if *any* rectangle has area $\frac{a}{b}$.

35. **(a)** The set shown is $\frac{3}{4}$ of a unit. What is the unit?

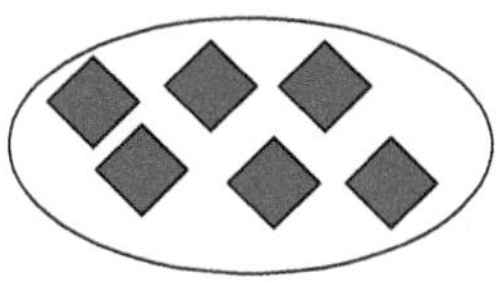

(b) The set shown is $\frac{5}{3}$ of a unit. What is the unit?

36. Decide whether each statement is *true* or *false*. Explain your reasoning in a brief paragraph.

(a) There are infinitely many ways to replace two fractions with two equivalent fractions that have a common denominator.

(b) There is a unique least common denominator for a given pair of fractions.

(c) There is a least positive fraction.

(d) There are infinitely many fractions between 0 and 1.

37. What fraction represents the part of the whole region that has been shaded? Draw additional lines to make your answer visually clear. For example, $\frac{2}{6}$ of the regular hexagon on the left is shaded, since the entire hexagon can be subdivided into six congruent regions, as shown on the right.

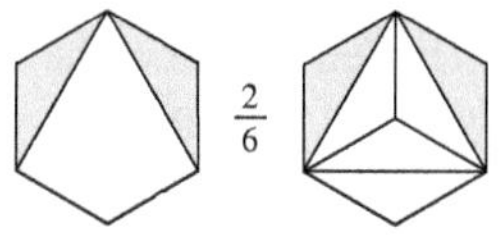

(a)

(b)

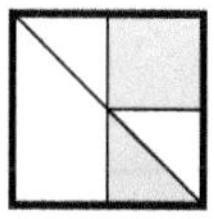

(c)

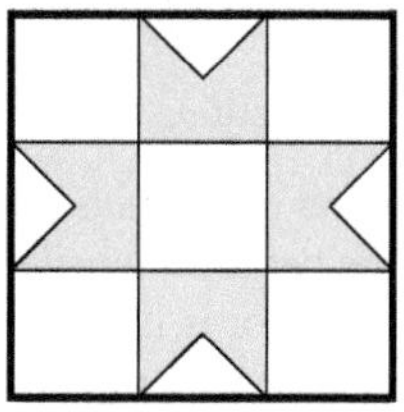

(d)

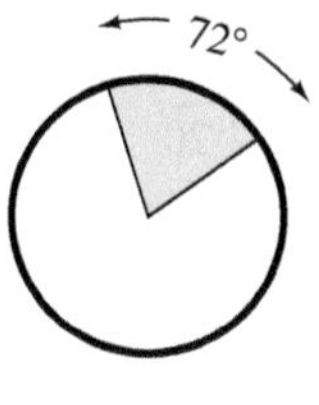

(e)

(f)

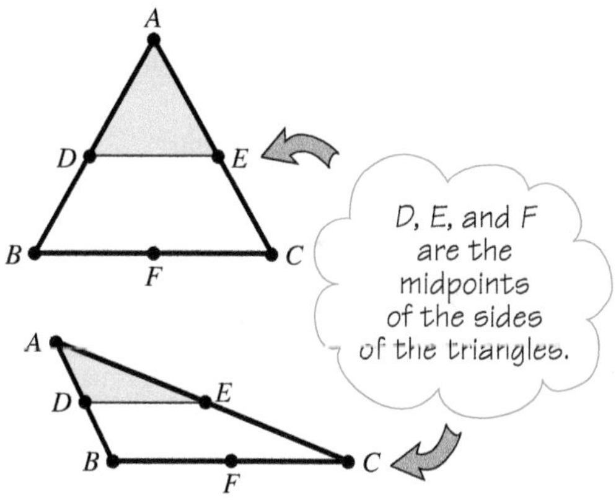

(g)

38. Solve this fraction problem from ancient India: "In what time will four fountains, being let loose together, fill a cistern, which they would severally [i.e., individually] fill in a day, in half a day, in a quarter and in a fifth part of a day?" (Reference: *History of Hindu Mathematics,* by B. Datta and A. N. Singh. Bombay, Calcutta, New Delhi, Madras, London, New York: Asia Publishing House, 1962, p. 234.)

39. Andrei bought a length of rope at the hardware store. He used half of it to make a bow painter (a rope located at the front of the canoe) for his canoe and then used a third of the remaining piece to tie up a roll of carpet. He now has 20 feet of rope left. What was the length of rope Andrei purchased? Since it is often helpful to visualize a problem, obtain your answer pictorially by adding additional marks and labels to the following drawing:

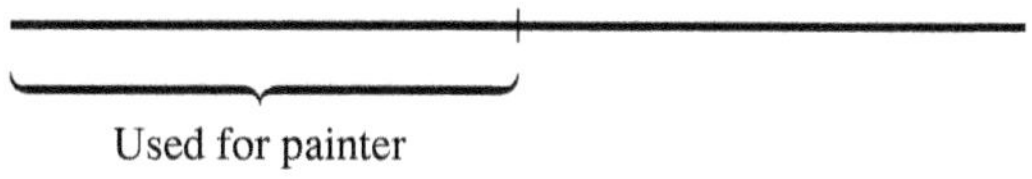

40. Use a diagram similar to the previous problem to answer this question. Andrei used one-third of the rope to make a leash for his dog and one-fourth of the remaining rope to tie up a bundle of stakes. This left 15 feet of the rope unused. What was the original length of the rope?

41. **(Writing)** The following row of Pascal's triangle (see Chapter 1) has been separated by a vertical bar drawn between the 15 and the 20, with 3 entries of the row to the left of the bar and 4 entries to the right of the bar:

$$1 \quad 6 \quad 15 \;\big|\; 20 \quad 15 \quad 6 \quad 1$$

Notice that $\frac{15}{20} = \frac{3}{4}$. That is, the two fractions formed by the entries adjacent to the dividing bar and by the number of entries to the left and the right of the bar are equivalent fractions. Was this equivalence an accident, or is it a general property of Pascal's triangle? Investigate the property further by placing the dividing bar in new locations and examining other rows of Pascal's triangle. Write a report summarizing your calculations.

Making Connections

42. Express the following quantities by a fraction placed in the blank space:

(a) 20 minutes is __________ of an hour.
(b) 30 seconds is __________ of a minute.
(c) 5 days is __________ of a week.
(d) 25 years is __________ of a century.
(e) A quarter is __________ of a dollar.
(f) 3 eggs is __________ of a dozen.
(g) 2 feet is __________ of a yard.
(h) 3 cups is __________ of a quart.

43. Francisco's pickup truck has a 24-gallon gas tank and an accurate fuel gauge. Estimate the number of gallons in the tank at these readings:

(a) (b) (c)

44. If 153 of the 307 graduating seniors go on to college, it is likely that a principal would claim that $\frac{1}{2}$ of the class is college bound. Give simpler convenient fractions that approximately express the data in these situations:

(a) Estebán is on page 310 of a 498-page novel. He has read __________ of the book.

(b) Myra has saved \$73 toward the purchase of a \$215 plane ticket. She has saved __________ of the amount she needs.

(c) Nine students in Ms. Evaldo's class of 35 students did perfect work on the quiz. __________ of the class scored 100% on the quiz.

(d) The Math Club has sold 1623 of the 2400 raffle tickets. It has sold __________ of the available tickets.

45. **Fractions in Probability.** If a card is picked at random from an ordinary deck of 52 playing cards, there are 4 ways it can be an ace, since it could be the ace of hearts, diamonds, clubs, or spades. To measure the chances of drawing an ace, it is common to give the probability as the rational number $\frac{4}{52}$.

In general, if n equally likely outcomes are possible and m of these outcomes are successful for an event to occur, then the probability of the event is $\frac{m}{n}$. As another example, $\frac{5}{6}$ is the probability of rolling a single die and having more than one spot appear. Give fractions that express the probability of the following events:

(a) Getting a head in the flip of a fair coin
(b) Drawing a face card from a deck of cards
(c) Rolling an even number on a single die
(d) Drawing a green marble from a bag that contains 20 red, 30 blue, and 25 green marbles
(e) Drawing either a red or a blue marble from the bag of marbles described in part (d)

46. What fraction represents the probability that the spinner shown comes up (a) yellow? (b) red? (c) blue? (d) not blue?

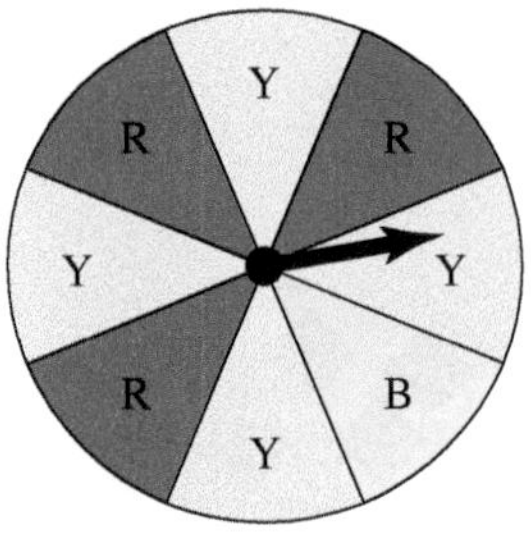

State Assessments

47. (Grade 5)

Which figure has $\frac{2}{3}$ of the area of the diamond shape shaded?

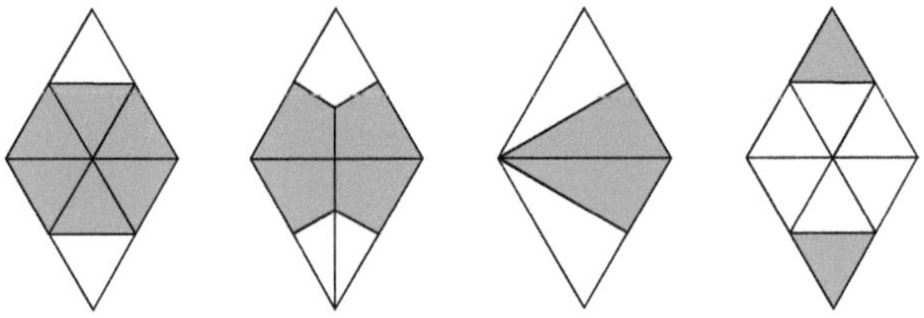

(a) (b) (c) (d)

48. **(Writing)** (Grade 5)

LuAnn thinks that $\frac{3}{5}$ of the squares are shaded and Jim thinks that $\frac{9}{15}$ of the squares are shaded.

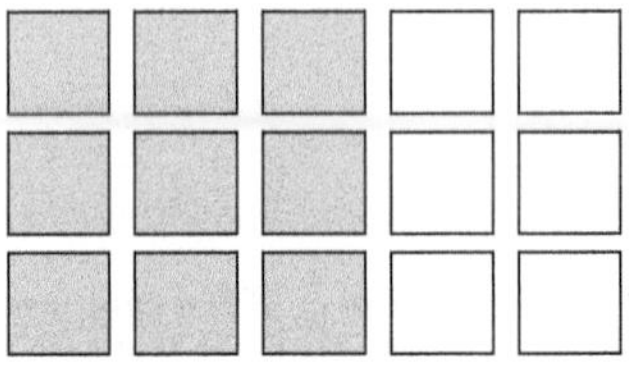

(a) Who is correct—LuAnn, Jim, or both?

(b) In words and diagrams, explain how you answered part (a).

49. (Grade 4)
Two pizzas were partially eaten at the party, as shown below.

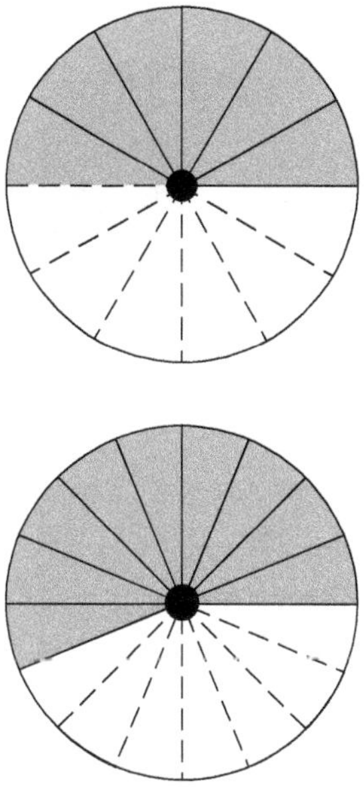

Which of these fraction inequalities compares the portion of pizza that was not eaten?

A. $\frac{5}{12} < \frac{9}{16}$ B. $\frac{9}{12} < \frac{5}{16}$

C. $\frac{5}{12} < \frac{9}{12}$ D. $\frac{5}{16} < \frac{9}{12}$

50. (Grade 5)
A dart is thrown repeatedly at this rectangle. How likely is it that it lands in a shaded region?

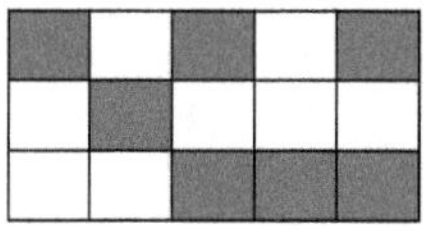

A. Rarely
B. Most of the time
C. Not quite half the time
D. Over half the time

51. (Grade 5)
Which of the following lists the mixed numbers in order from **least to greatest**?

A. $7\frac{1}{5}$ $4\frac{2}{5}$ $2\frac{3}{5}$ $1\frac{4}{5}$ B. $5\frac{2}{5}$ $6\frac{3}{5}$ $6\frac{4}{5}$ $7\frac{1}{5}$

C. $3\frac{3}{5}$ $2\frac{3}{5}$ $3\frac{4}{5}$ $4\frac{1}{5}$ D. $6\frac{4}{5}$ $4\frac{2}{5}$ $3\frac{2}{5}$ $1\frac{1}{5}$

---

52. (Massachusetts, Grade 6)

Which point on the number line shown below appears to be located at $1\frac{3}{8}$?

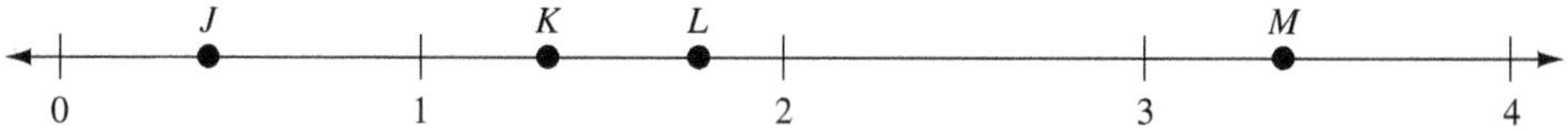

6.2 Addition and Subtraction of Fractions

The geometric and physical models of fractions motivate the definitions of the addition and subtraction of fractions. Since rational numbers are represented with fractions, we are simultaneously defining sums and differences of rational numbers.

Addition of Fractions

The sum of $\frac{3}{8}$ and $\frac{2}{8}$ is illustrated in two ways in Figure 6.11. The fraction-circle and number-line models both show that $\frac{3}{8} + \frac{2}{8} = \frac{5}{8}$. The models suggest that the sum of two fractions with a common denominator is found by adding the two numerators. This motivates the following definition:

DEFINITION Addition of Fractions with a Common Denominator

Let two fractions $\frac{a}{b}$ and $\frac{c}{b}$ have a common denominator. Then their **sum** is the fraction

$$\frac{a}{b} + \frac{c}{b} = \frac{a + c}{b}.$$

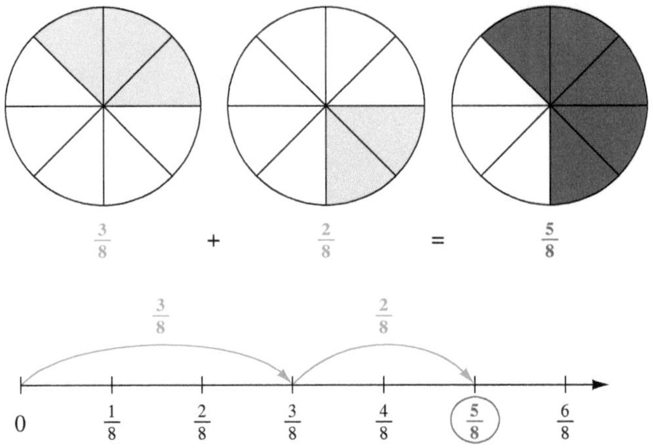

Figure 6.11 Showing that $\frac{3}{8} + \frac{2}{8} = \frac{5}{8}$ with the fraction-circle and number-line models

To add fractions with unlike denominators, we first rewrite the fractions with a common denominator. For example, to add $\frac{1}{4}$ and $\frac{2}{3}$, rewrite them with 12 as a common denominator:

$$\frac{1}{4} = \frac{1 \cdot 3}{4 \cdot 3} = \frac{3}{12} \quad \text{and} \quad \frac{2}{3} = \frac{2 \cdot 4}{3 \cdot 4} = \frac{8}{12}.$$

According to the preceding definition, we then have

$$\frac{1}{4} + \frac{2}{3} = \frac{1 \cdot 3}{4 \cdot 3} + \frac{2 \cdot 4}{3 \cdot 4} = \frac{1 \cdot 3 + 2 \cdot 4}{4 \cdot 3} = \frac{3 + 8}{12} = \frac{11}{12}.$$

The procedure just followed can be modeled with fraction strips, as shown in Figure 6.12. It is important to see how the fraction strips are aligned.

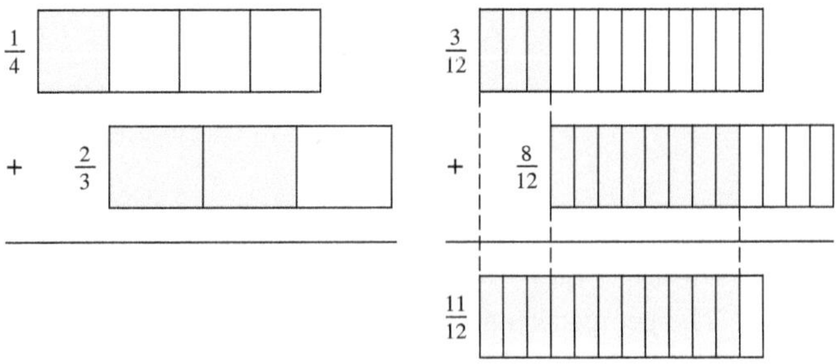

Figure 6.12 The fraction-strip model showing that $\frac{1}{4} + \frac{2}{3} = \frac{3}{12} + \frac{8}{12} = \frac{11}{12}$

The same procedure can be followed to add any two fractions $\frac{a}{b}$ and $\frac{c}{d}$. We have

$$\frac{a}{b} + \frac{c}{d} = \frac{a \cdot d}{b \cdot d} + \frac{c \cdot b}{d \cdot b} = \frac{ad + bc}{bd}.$$

Rewrite the fractions with a common denominator.

Add the numerators and retain the common denominator.

FORMULA Addition of Fractions with Arbitrary Denominators

Let $\frac{a}{b}$ and $\frac{c}{d}$ be fractions. Then their sum is the fraction

$$\frac{a}{b} + \frac{c}{d} = \frac{ad + bc}{bd}.$$

There are two important observations about this formula:

- The formula also applies to the addition of fractions with common denominators, since

$$\frac{a}{b} + \frac{c}{b} = \frac{ab + bc}{bb} = \frac{(a + c)b}{bb} = \frac{a + c}{b}.$$

- The formula is a useful computational rule, but it obscures the conceptual meaning of the addition of two numbers. Therefore, it is often better to continue to use the common denominator approach to the addition of fractions and to model fraction addition with manipulatives and pictorial representations.

EXAMPLE 6.6 Adding Fractions

Show how to compute each of the sums of fractions that follow. Imagine that you are giving a careful explanation to a fifth grader, using an appropriate representation of fraction addition to illustrate how you obtain the answer.

(a) $\frac{1}{2} + \frac{3}{8}$ **(b)** $\frac{3}{8} + \frac{-7}{12}$ **(c)** $\left(\frac{3}{4} + \frac{5}{6}\right) + \frac{2}{3}$ **(d)** $\frac{3}{4} + \left(\frac{5}{6} + \frac{2}{3}\right)$

Solution

(a) Since the two fractions have different denominators, we first look for a common denominator. One choice is $2 \cdot 8 = 16$, but it's even simpler to use the least common denominator, 8. Then $\frac{1}{2} + \frac{3}{8} = \frac{4}{8} + \frac{3}{8} = \frac{7}{8}$. This relationship can also be shown with fraction circles:

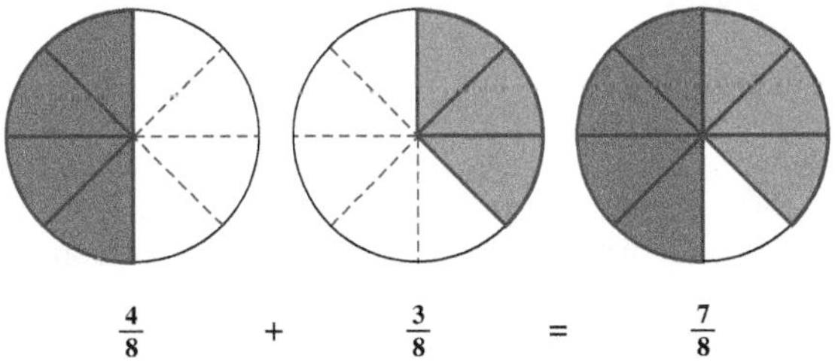

(b) We could use the common denominator $8 \cdot 12 = 96$, but since 4 is a common divisor of 8 and 12, we can also use the common denominator $96 \div 4 = 24$. This is the least common denominator, and we get $\frac{3}{8} + \frac{-7}{12} = \frac{9}{24} + \frac{-14}{24} = \frac{9 + (-14)}{24} = \frac{-5}{24}$. The following number-line diagram shows the same result:

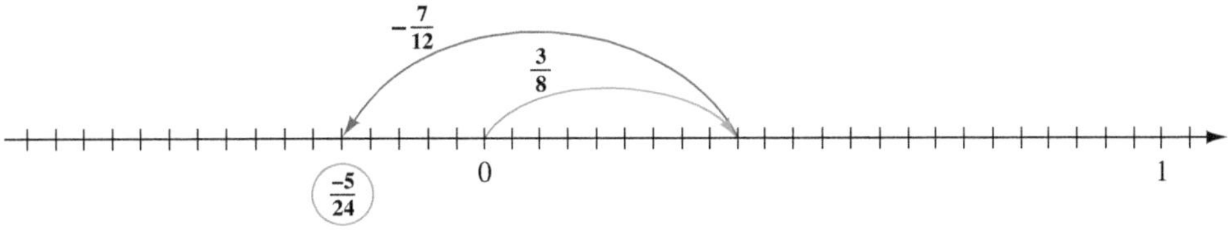

(c) The parentheses tell us first to compute

$$\frac{3}{4} + \frac{5}{6} = \frac{9}{12} + \frac{10}{12} = \frac{19}{12}.$$

Then we compute

$$\left(\frac{3}{4}+\frac{5}{6}\right)+\frac{2}{3}=\frac{19}{12}+\frac{2}{3}=\frac{19}{12}+\frac{8}{12}=\frac{19+8}{12}=\frac{27}{12},$$

which simplifies to $\frac{9}{4}$.

(d) The sum in parentheses is

$$\frac{5}{6}+\frac{2}{3}=\frac{5}{6}+\frac{4}{6}=\frac{9}{6}=\frac{3}{2}.$$

Then we have

$$\frac{3}{4}+\left(\frac{5}{6}+\frac{2}{3}\right)=\frac{3}{4}+\frac{3}{2}=\frac{3}{4}+\frac{6}{4}=\frac{9}{4}.$$

Parts (c) and (d) of Example 6.6 show that

$$\left(\frac{3}{4}+\frac{5}{6}\right)+\frac{2}{3}=\frac{3}{4}+\left(\frac{5}{6}+\frac{2}{3}\right),$$

since each side is $\frac{9}{4}$. This result is a particular example of the associative property for the addition of fractions. The properties of addition and subtraction of fractions and, therefore, of rational numbers will be explored in Section 6.4.

INTO THE CLASSROOM

Ann Hlabangana-Clay Discusses the Addition of Fractions

I use red shoelace licorice to introduce adding fractions. It is flexible and tangible for small fingers to demonstrate whole to part. To start the lesson, I give each student one whole red shoelace licorice. I ask them to spread it out from end to end and use it to measure a starting line and an ending line. To find $\frac{1}{2}+\frac{3}{4}$, I give each student a $\frac{1}{2}$ length shoelace and have them measure it against the whole. Each student also gets a $\frac{3}{4}$ length shoelace to measure against the whole and to compare to the $\frac{1}{2}$. After comparing the $\frac{1}{2}$ and the $\frac{3}{4}$, I have the students connect the two shoelaces together and share their findings with their partner.

Proper Fractions and Mixed Numbers

The sum of a natural number and a positive fraction is most often written as a **mixed number.** For example, $2+\frac{3}{4}$ is written $2\frac{3}{4}$ and is read "two and three-quarters." It is important to realize that it is the addition symbol, +, that is suppressed, since the common notation xy for multiplication might suggest, incorrectly, that $2\frac{3}{4}$ is $2\cdot\frac{3}{4}$. Thus, $2\frac{3}{4}=2+\frac{3}{4}$, not $\frac{6}{4}$.

Mixed numbers have a hidden + sign: $A\frac{b}{c}=A+\frac{b}{c}$

A mixed number can always be rewritten in the standard form $\frac{a}{b}$ of a fraction. For example,

$$2\frac{3}{4} = 2 + \frac{3}{4} = \frac{8}{4} + \frac{3}{4} = \frac{11}{4}.$$

Thus, $-2\frac{3}{4} = -\left(2\frac{3}{4}\right) = \frac{-11}{4}$.

Mixed numbers and their equivalent forms as a fraction can be visualized this way:

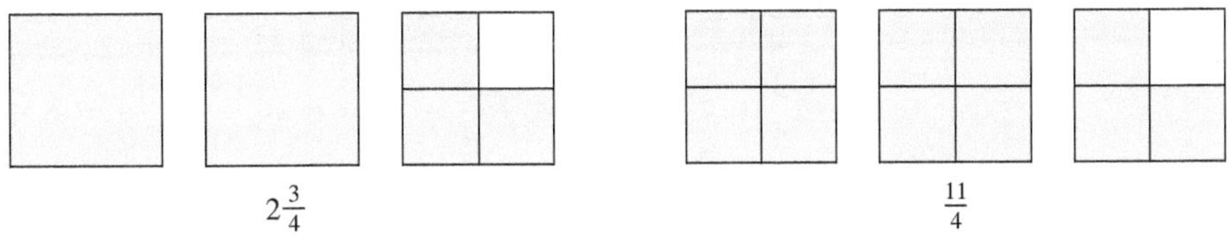

Since $A\frac{b}{c} = \frac{Ac}{c} + \frac{b}{c} = \frac{Ac + b}{c}$, it is an easy calculation to write a mixed number as a fraction: $A\frac{b}{c} = \frac{Ac + b}{c}$.

A fraction $\frac{a}{b}$ for which $0 \leq |a| < b$ is called a **proper fraction.** For example, $\frac{2}{3}$ is a proper fraction, but $\frac{3}{2}, \frac{-8}{5}$, and $\frac{6}{6}$ are not proper fractions. It is common, though not necessary, to rewrite fractions that are not proper as mixed numbers. For example, to express $\frac{439}{19}$ as a mixed number, we first use division with a remainder to find that $439 = 23 \cdot 19 + 2$. Then we have

$$\frac{439}{19} = \frac{23 \cdot 19 + 2}{19} = \frac{23}{1} + \frac{2}{19} = 23 + \frac{2}{19} = 23\frac{2}{19}.$$

In mixed-number form, it is obvious that $23\frac{2}{19}$ is just slightly larger than 23; this fact was not evident in the original fraction form $\frac{439}{19}$. Nevertheless, it is perfectly acceptable to express rational numbers as "improper" fractions. In general, the fractional form $\frac{a}{b}$ is the more convenient form for arithmetic and algebra, and the mixed-number form is easiest to understand for practical applications. For example, it would be more common to buy $2\frac{1}{4}$ yards of material than to request $\frac{9}{4}$ yards.

EXAMPLE 6.7

Working with Mixed Numbers

(a) Give an improper fraction for $3\frac{17}{120}$.

(b) Give a mixed number for $\frac{355}{113}$.

(c) Give a mixed number for $\frac{-15}{4}$.

(d) Compute $2\frac{3}{4} + 4\frac{2}{5}$.

Solution

(a) $3\frac{17}{120} = \frac{3}{1} + \frac{17}{120} = \frac{3 \cdot 120 + 1 \cdot 17}{120} = \frac{360 + 17}{120} = \frac{377}{120}$.

This rational number was given by Claudius Ptolemy around A.D. 150 to approximate π, the ratio of the circumference of a circle to its diameter.

It has better accuracy than $3\frac{1}{7}$, the value proposed by Archimedes in about 240 B.C.

(b) Using the division algorithm, we calculate that $355 = 3 \cdot 113 + 16$. Therefore,

$$\frac{355}{113} = \frac{3 \cdot 113 + 16}{113} = \frac{3}{1} + \frac{16}{113} = 3\frac{16}{113},$$

which corresponds to a point somewhat to the right of 3 on the number line. The value $\frac{355}{113}$ was used around A.D. 480 in China to approximate π; as a decimal number, it is correct to six places!

(c) $\frac{-15}{4} = \frac{-(3 \cdot 4 + 3)}{4} = -\left(3 + \frac{3}{4}\right) = -3\frac{3}{4}.$

(d) $2\frac{3}{4} + 4\frac{2}{5} = 2 + 4 + \frac{3}{4} + \frac{2}{5} = 6 + \frac{15}{20} + \frac{8}{20} = 6 + \frac{23}{20} = 7\frac{3}{20}.$

Subtraction of Fractions

Figure 6.13 shows how the take-away, measurement, and missing-addend conceptual models of the subtraction operation can be illustrated with colored regions, the number line, and fraction strips. In each case, we see that $\frac{7}{6} - \frac{3}{6} = \frac{4}{6}$.

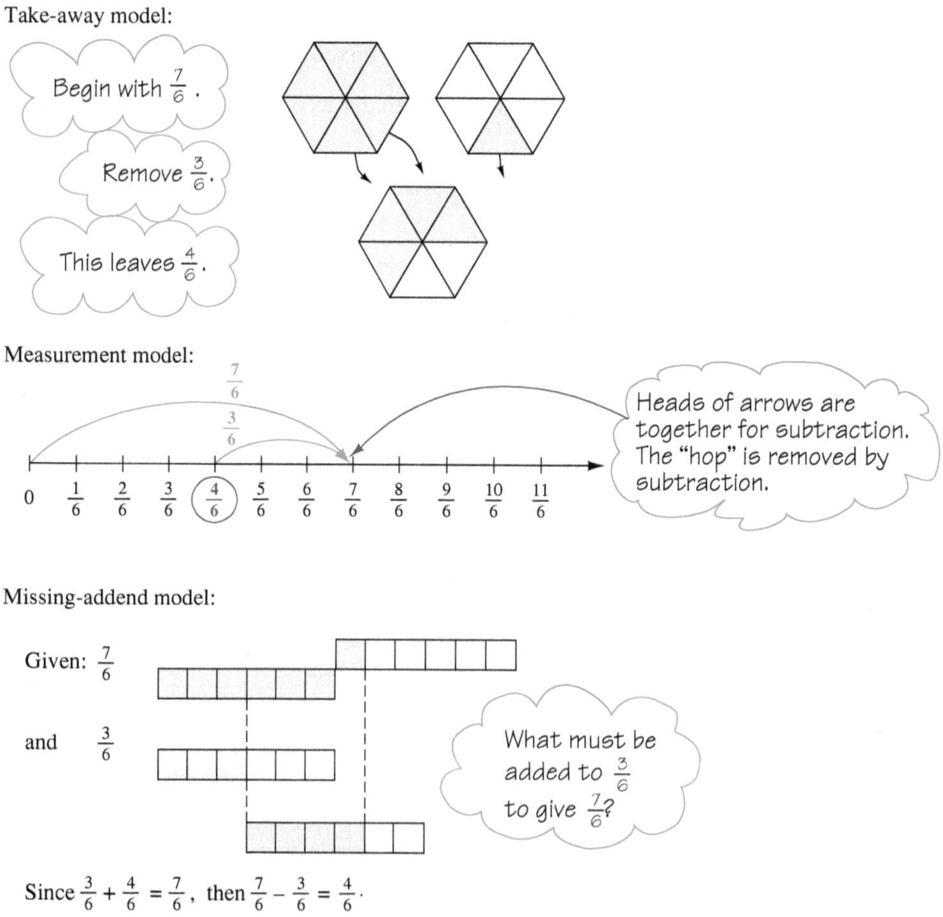
Take-away model:
Begin with $\frac{7}{6}$.
Remove $\frac{3}{6}$.
This leaves $\frac{4}{6}$.
Measurement model:
$\frac{7}{6}$
$\frac{3}{6}$
0 $\frac{1}{6}$ $\frac{2}{6}$ $\frac{3}{6}$ $\frac{4}{6}$ $\frac{5}{6}$ $\frac{6}{6}$ $\frac{7}{6}$ $\frac{8}{6}$ $\frac{9}{6}$ $\frac{10}{6}$ $\frac{11}{6}$
Heads of arrows are together for subtraction. The "hop" is removed by subtraction.
Missing-addend model:
Given: $\frac{7}{6}$
and $\frac{3}{6}$
What must be added to $\frac{3}{6}$ to give $\frac{7}{6}$?
Since $\frac{3}{6} + \frac{4}{6} = \frac{7}{6}$, then $\frac{7}{6} - \frac{3}{6} = \frac{4}{6}$.

Figure 6.13
Models showing that $\frac{7}{6} - \frac{3}{6} = \frac{4}{6}$

Subtraction of whole numbers and integers was defined on the basis of the missing-addend approach, which emphasizes that subtraction is the inverse operation to addition. Subtraction of fractions is defined in the same way:

DEFINITION Subtraction of Fractions

Let $\frac{a}{b}$ and $\frac{c}{d}$ be fractions. Then $\frac{a}{b} - \frac{c}{d} = \frac{e}{f}$ if and only if $\frac{a}{b} = \frac{c}{d} + \frac{e}{f}$.

For two fractions $\frac{a}{b}$ and $\frac{c}{b}$ with the same denominator, we see that $\frac{c}{b} + \frac{a - c}{b} = \frac{c + (a - c)}{b} = \frac{a}{b}$ and, therefore, from the definition, $\frac{a}{b} - \frac{c}{b} = \frac{a - c}{b}$. To subtract fractions $\frac{a}{b}$ and $\frac{c}{d}$ with arbitrary denominators, we can rewrite them with common denominators as $\frac{ad}{bd}$ and $\frac{bc}{bd}$ to obtain the formula $\frac{a}{b} - \frac{c}{d} = \frac{ad - bc}{bd}$. However, just as with the addition, the formula obscures the meaning of the subtraction of fractions. Therefore, it important to explain subtraction using common denominators, manipulatives, and pictorial representations. These approaches emphasize the meaning of the subtraction operation.

EXAMPLE 6.8

Subtracting Fractions

Show, as you might to a student, the steps to compute these differences:

(a) $\frac{4}{5} - \frac{2}{3}$ **(b)** $\frac{103}{24} - \frac{-35}{16}$ **(c)** $4\frac{1}{4} - 2\frac{2}{3}$

Solution

(a) $\frac{4}{5} - \frac{2}{3} = \frac{4 \cdot 3}{5 \cdot 3} - \frac{5 \cdot 2}{5 \cdot 3} = \frac{12}{15} - \frac{10}{15} = \frac{12 - 10}{15} = \frac{2}{15}$.

The subtraction can also be shown with a number-line diagram. Since both fractions are positive, both arrows point to the right, and since this is a subtraction, the arrowheads are together. The diagram shows that removing $\frac{10}{15}$ from $\frac{12}{15}$ results in $\frac{2}{15}$.

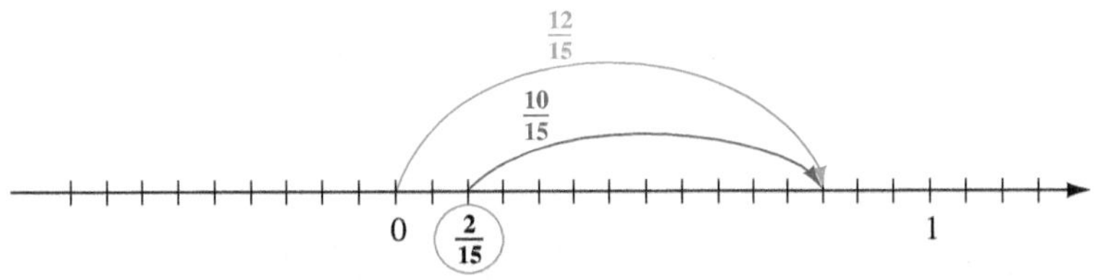

(b) Since LCM(24, 16) = 48, the least common denominator, 48, can be used to give

$$\frac{103}{24} - \frac{-35}{16} = \frac{206}{48} - \frac{-105}{48} = \frac{206 - (-105)}{48} = \frac{311}{48}.$$

(c) The mixed forms can be converted to standard form, showing that

$$4\frac{1}{4} - 2\frac{2}{3} = \frac{17}{4} - \frac{8}{3} = \frac{17 \cdot 3}{4 \cdot 3} - \frac{4 \cdot 8}{4 \cdot 3} = \frac{51}{12} - \frac{32}{12} = \frac{51 - 32}{12} = \frac{19}{12} = 1\frac{7}{12}.$$

Alternatively, subtraction of mixed numbers can follow the familiar regrouping algorithm:

$$\begin{array}{r} 4\frac{1}{4} \\ -2\frac{2}{3} \\ \hline \end{array} \Rightarrow \begin{array}{r} 4\frac{3}{12} \\ -2\frac{8}{12} \\ \hline \end{array} \Rightarrow \begin{array}{r} 3\frac{15}{12} \\ -2\frac{8}{12} \\ \hline 1\frac{7}{12} \end{array}$$

$4\frac{3}{12} = 3 + \frac{12}{12} + \frac{3}{12} = 3\frac{15}{12}$

6.2 Problem Set

Understanding Concepts

1. What addition fact is illustrated by the following fraction-strip models?

(a)

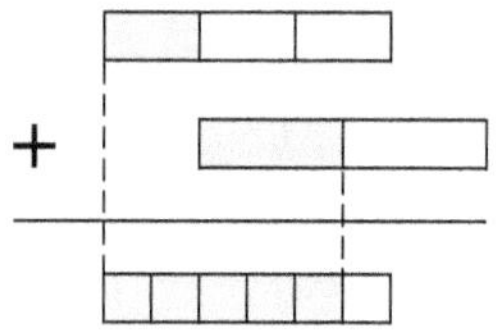

(b)

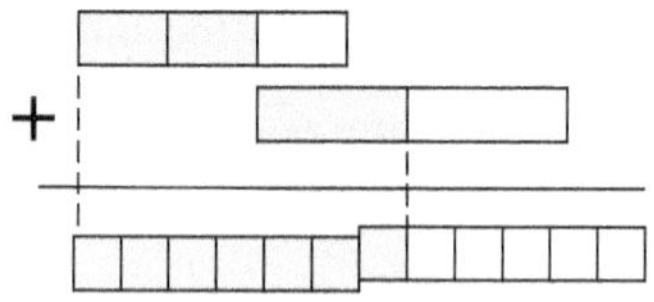

2. (a) Illustrate $\frac{1}{6} + \frac{1}{4}$ with the fraction-strip model.

(b) Illustrate $\frac{2}{3} + \frac{3}{4}$ with the fraction-strip model. (The sum will require two strips.)

3. Use fraction circles to illustrate these sums:

(a) $\frac{2}{5} + \frac{6}{5}$ (b) $\frac{1}{4} + \frac{1}{2}$ (c) $\frac{2}{3} + \frac{1}{4}$

4. The points A, B, C, . . . , G, and H are equally spaced along the number line:

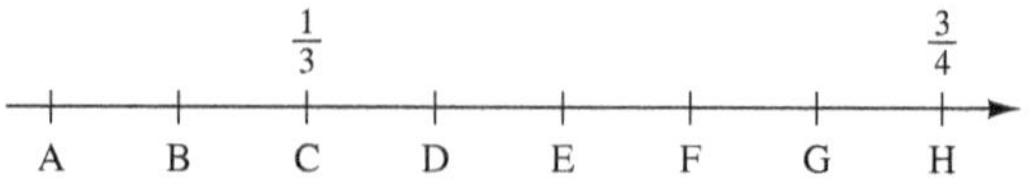

(a) What rational number corresponds to point G?

(b) What point corresponds to $\frac{1}{2}$?

(c) Are there lettered points that correspond to $\frac{1}{4}$? to $\frac{1}{6}$?

(d) Which lettered points are nearest to either side of $\frac{4}{7}$?

5. Represent each of these sums with a number-line diagram:

(a) $\frac{1}{8} + \frac{3}{8}$ (b) $\frac{1}{4} + \frac{5}{4}$ (c) $\frac{3}{4} + \frac{-2}{4}$

6. Use the number-line model to illustrate the given sums. Recall that negative fractions are represented by arrows that point to the left.

(a) $\frac{2}{3} + \frac{1}{2}$ (b) $\frac{-3}{4} + \frac{2}{4}$ (c) $\frac{-3}{4} + \frac{-1}{4}$

7. Perform the given additions. Express each answer in simplest form.

(a) $\frac{2}{7} + \frac{3}{7}$ (b) $\frac{6}{5} + \frac{4}{5}$

(c) $\frac{3}{8} + \frac{11}{24}$ (d) $\frac{6}{13} + \frac{2}{5}$

8. Perform the given additions. Express each answer in simplest form.

(a) $\frac{5}{12} + \frac{17}{20}$ (b) $\frac{6}{8} + \frac{-25}{100}$

(c) $\frac{-57}{100} + \frac{13}{10}$ (d) $\frac{213}{450} + \frac{12}{50}$

9. Express these fractions as mixed numbers:

(a) $\frac{9}{4}$ (b) $\frac{17}{3}$ (c) $\frac{111}{23}$ (d) $\frac{3571}{-100}$

10. Express these mixed numbers as fractions:

(a) $2\frac{3}{8}$ (b) $15\frac{2}{3}$ (c) $111\frac{2}{5}$ (d) $-10\frac{7}{9}$

11. (a) What subtraction fact is illustrated by this fraction-strip model?

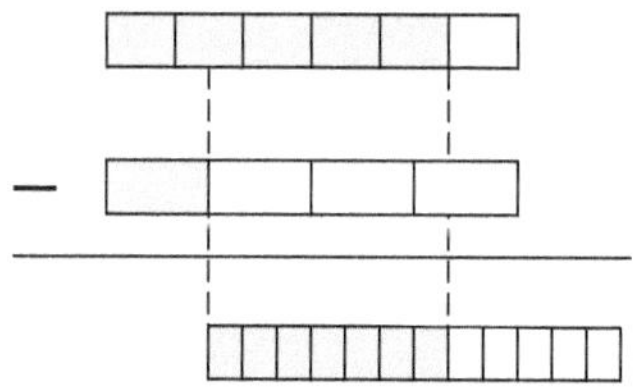

(b) Use the fraction-strip model to illustrate $\frac{2}{3} - \frac{1}{4}$.

12. (a) What subtraction fact is illustrated by this colored-region model?

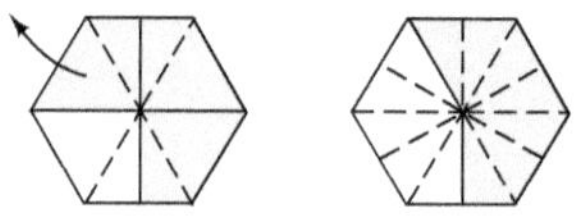

(b) Use a fraction-circle model to illustrate $\frac{2}{3} - \frac{1}{4}$.

13. (a) What subtraction fact is illustrated by this number-line model?

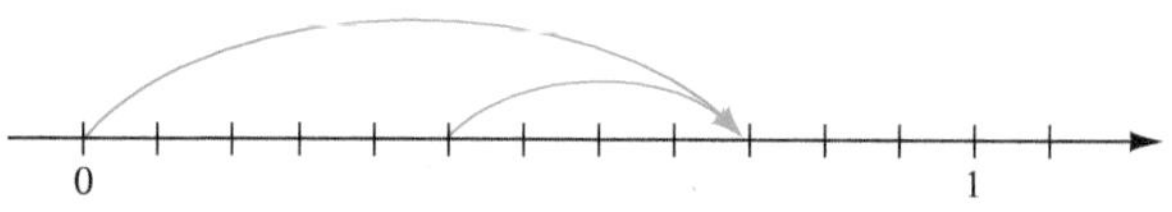

(b) Illustrate $\frac{2}{3} - \frac{1}{4}$ with the number-line model.

14. Compute these differences, expressing each answer in simplest form:

(a) $\frac{5}{8} - \frac{2}{8}$ (b) $\frac{3}{5} - \frac{2}{4}$

(c) $2\frac{2}{3} - 1\frac{1}{3}$ (d) $4\frac{1}{4} - 3\frac{1}{3}$

15. Compute these differences, expressing each answer in simplest form.

(a) $\frac{6}{8} - \frac{5}{12}$ (b) $\frac{1}{4} - \frac{14}{56}$

(c) $\frac{137}{214} - \frac{-1}{3}$ (d) $\frac{-23}{100} - \frac{198}{1000}$

16. Use mental arithmetic to estimate the following sums and differences to the nearest integer:

(a) $\frac{19}{22} + \frac{31}{15}$ **(b)** $1\frac{5}{11} + 4\frac{1}{2}$

(c) $7\frac{53}{97} - 2\frac{5}{9}$ **(d)** $6\frac{2}{3} + 8\frac{11}{32} - 2\frac{1}{29}$

17. An alternative but equivalent definition of rational number inequality is the following:

$$\frac{a}{b} < \frac{c}{d} \quad \text{if and only if} \quad \frac{c}{d} - \frac{a}{b} > 0.$$

Use the alternative definition to verify these inequalities.

(a) $\frac{2}{3} < \frac{3}{4}$ **(b)** $\frac{4}{5} < \frac{14}{17}$ **(c)** $\frac{19}{10} < \frac{99}{50}$

Into the Classroom

18. In the number-rod diagram shown, the brown rod has been adopted as the unit. Therefore, the red rod is $\frac{1}{4}$, the purple rod is $\frac{1}{2}$, and the dark-green rod is $\frac{3}{4}$. Since the train formed with the purple and dark-green rods has the length of the brown-plus-red train, we have $\frac{1}{2} + \frac{3}{4} = 1\frac{1}{4}$.

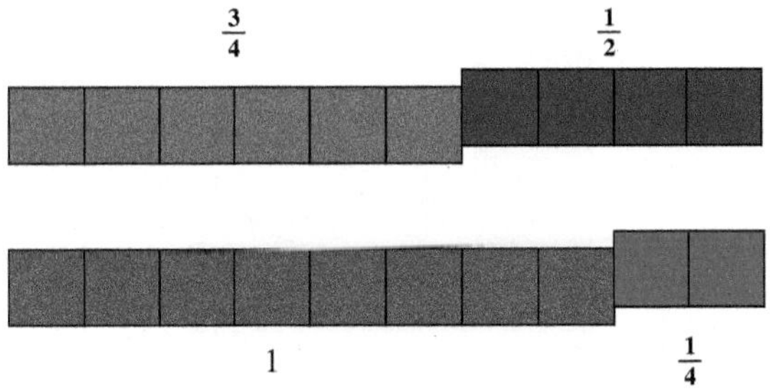

(a) Use number rods to create another illustration of the addition of fractions. Use both words and diagrams to describe your example clearly.

(b) Use number rods to create an illustration of the subtraction of fractions.

19. **(Writing)** Create instructions for an elementary school class activity in which students use strips of paper 1″ wide that are cut into smaller lengths and then measured to the nearest $\frac{1}{2}$″, $\frac{1}{4}$″, or $\frac{1}{8}$″, and the length is written on the strip. For addition, two strips are chosen and their lengths added. Then the strips are laid end-to-end and measured to see if the total length agrees with the result of the addition calculation. A similar procedure for subtraction should be described in both words and diagrams.

Responding to Students

The "pizza stories" discussed in problems 20–22 suggest common misconceptions students have about fractions and their arithmetic. Use words and diagrams to formulate helpful responses for each situation described.

20. The Estevez family ordered a large pepperoni pizza and a medium tomato-and-green-pepper pizza. Since $\frac{1}{3}$ of the pepperoni pizza was not eaten and $\frac{1}{2}$ of the other pizza remains, Paco claims that $\frac{1}{3} + \frac{1}{2} = \frac{5}{6}$ of a pizza remains. Is Paco correct?

21. After the pizza party, it was discovered that just 2 pieces of one of the pizzas were consumed, leaving 6 more slices. Dannea was told that she could have half of the leftover pizza but to leave the rest for her brother. Dannea reasoned that the whole pizza was 8 slices, so that half of the pizza was 4 slices. Does her brother have a reason to complain when he discovered that Dannea left him just 2 slices? In particular, what principle of fractions has Dannea violated?

22. Katrina claims that since $\frac{2}{6}$ of one pizza and $\frac{3}{4}$ of a second pizza were not eaten, "$\frac{2}{6} + \frac{3}{4} = \frac{5}{10} = \frac{1}{2}$" of a pizza is left over for tomorrow's lunch. Respond to Katrina in a helpful way.

23. Melanie got 1 hit in 3 times at bat in the first baseball game, giving her a hitting average of $\frac{1}{3}$. In the second game, her average was $\frac{2}{5}$, since she had 2 hits in 5 times at bat. She then claims that her two-game average is $\frac{1}{3} + \frac{2}{5} = \frac{3}{8}$, because she had 3 hits in her 8 times at bat. Melanie's brother John objects, since he knows that $\frac{1}{3} + \frac{2}{5} = \frac{5}{15} + \frac{6}{15} = \frac{11}{15}$ and $\frac{11}{15} \neq \frac{3}{8}$. Who is right, Melanie or John?

Thinking Critically

24. **(Writing)** A worm is on page 1 of Volume 1 of a set of encyclopedias neatly arranged on a shelf. He (or she—how do you tell?) eats straight through to the last page of Volume 2. If the covers of each volume are $\frac{1}{8}$″ thick and the pages are a total of $\frac{3}{4}$″ thick in each of the volumes, how far does the worm travel?

25. **(Writing)** When the sultan died, he left a stable with 17 horses. His will stipulated that half of his horses should go to his oldest son, a third to his middle son, and a ninth to his youngest son. The executor of the estate decided to make the distribution easier by contributing his own horse to the estate. With 18 horses now available, he gave 9 horses (a half) to the oldest son, 6 horses (a third) to the middle son, and 2 horses (a ninth) to the youngest son. This satisfied the terms of the will, and since only $9 + 6 + 2 = 17$ horses were given to the sons, the executor was able to retain his own horse. Explain what has happened.

26. Find the missing fractions in the following Magic Fraction Squares so that the entries in every row, column, and diagonal add to 1:

(a)

$\frac{1}{2}$	$\frac{1}{12}$	
$\frac{1}{4}$		

(b)

	$\frac{1}{3}$	$\frac{3}{5}$
		$\frac{2}{15}$

27. Start with any triangle, and let its area be the unit. Next, trisect each side of the triangle and join these points to the opposite vertices. This creates a dissection of the unit triangle, as illustrated in the accompanying figure. Amazingly, the areas of the regions shown are precisely the fractions shown in the figure. For example, the inner lavender hexagon has area $\frac{1}{10}$, and the area of each green triangle is $\frac{1}{21}$.

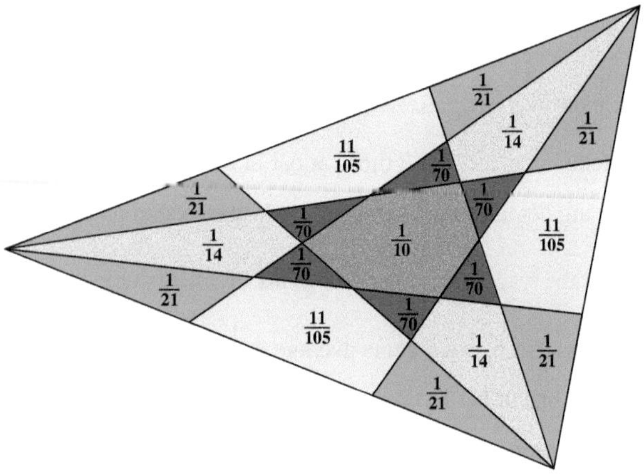

(a) Verify that the large triangle is a unit triangle by summing the areas of the regions within the triangle.

(b) What is the area of the three-pointed star composed of the hexagon, the six red triangles, and the three yellow quadrilaterals?

(c) What is the area of the triangle composed of the hexagon and the three red triangles on every other side of the hexagon?

28. With the exception of $\frac{2}{3}$, which was given the hieroglyph , the ancient Egyptians attempted to express all fractions as a sum of different fractions, each with 1 as the numerator.

(a) Verify that $\frac{23}{25} = \frac{1}{2} + \frac{1}{3} + \frac{1}{15} + \frac{1}{50}$.

(b) Verify that $\frac{7}{29} = \frac{1}{6} + \frac{1}{24} + \frac{1}{58} + \frac{1}{87} + \frac{1}{232}$.

(c) Verify that $\frac{7}{29} = \frac{1}{5} + \frac{1}{29} + \frac{1}{145}$.

(d) If represents $1\frac{1}{2}$, and  is interpreted as "one over," does the symbol  seem reasonable for $\frac{2}{3}$?

29. **Diffy with Fractions.** We previously played the Diffy game, described in Section 2.3, with whole numbers. Shown here are the beginning lines of Diffy when the entries are fractions. A new line is formed by subtracting the smaller fraction from the larger.

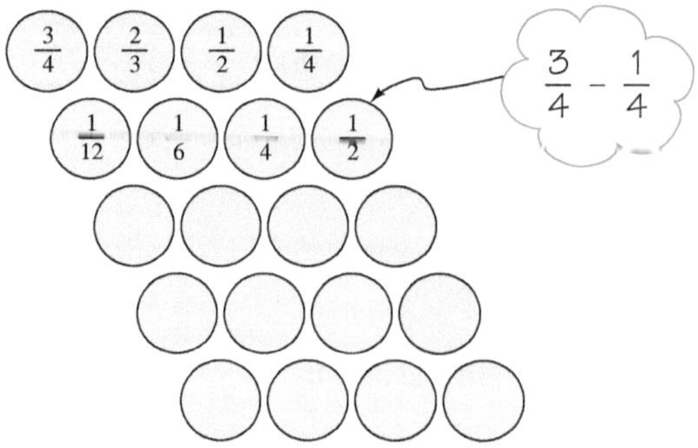

(a) Fill in additional lines of the Diffy array. Does it terminate?

(b) Try fraction Diffy with these fractions in your first row: $\frac{2}{7}, \frac{4}{5}, \frac{3}{2}$, and $\frac{5}{6}$.

(c) (**Writing**) Suppose you know that Diffy with whole-number entries always terminates with 0, 0, 0, and 0. Does it necessarily follow that Diffy with fractions must terminate? Explain your reasoning carefully.

Making Connections

30. A "2 by 4" piece of lumber is planed from a rough board to a final size of $1\frac{1}{2}''$ by $3\frac{1}{2}''$. Find the dimensions x and y of the shape created with two such boards.

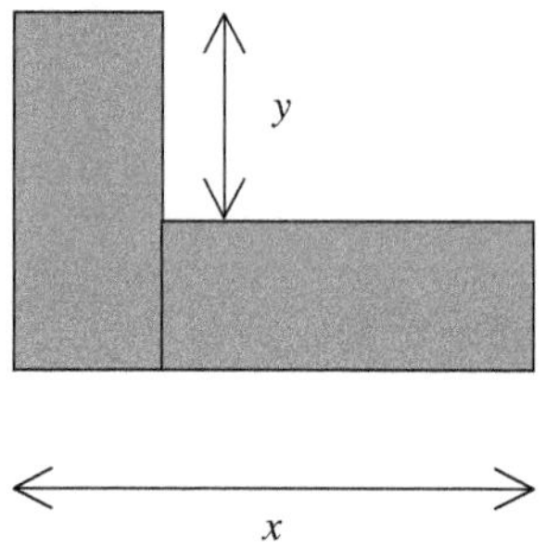

31. A wall in a house is made with sheets of $\frac{5}{8}$-inch drywall screwed to the opposite short edges of the vertical 2-by-4 studs within the wall. (See problem 30). How thick is the resulting wall?

32. A $1\frac{1}{4}''$ drywall screw is countersunk (i.e., screwed until it is below the surface of the drywall) by $\frac{1}{16}''$ to fasten a $\frac{5}{8}''$ sheet of drywall to the ceiling joist.

(a) How much of the screw extends into the joist?

(b) How much of the screw is within the drywall?

33. A board $10\frac{1}{2}$ inches long is sawn off a board that is 2 feet long. If the width of the saw cut is $\frac{1}{16}''$, what is the length of the remaining piece?

34. A picture is printed on an $8\frac{1}{2}''$-by-$11''$ piece of photographic paper and placed symmetrically in a frame whose opening is $8''$ by $10''$. What are the dimensions of the strips on the sides of the picture that are hidden by the frame?

State Assessments

35. (Massachusetts, Grade 4)
What is the solution to the problem shown below?

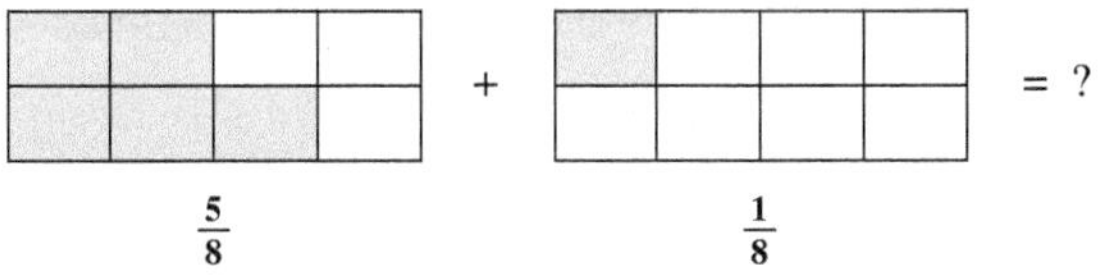

A. $\frac{6}{8}$ B. $\frac{6}{10}$ C. $\frac{6}{16}$ D. $\frac{5}{64}$

36. (Massachusetts, Grade 6)

Henry had a piece of rope that was $23\frac{1}{2}$ inches long. He cut the rope into two pieces so that one piece was $8\frac{1}{4}$ inches long. What was the length of the other piece of rope?

A. $15\frac{1}{4}$ B. $15\frac{1}{2}$ C. $31\frac{1}{3}$ D. $31\frac{3}{4}$

37. (Grade 5)

Abe, Bert, and Carol shared a pizza. Abe ate $\frac{3}{8}$ of the pizza, Bert ate $\frac{1}{4}$ of the pizza, and Carol ate the rest. How much of the pizza did Carol eat?

A. $\frac{1}{2}$ B. $\frac{3}{4}$ C. $\frac{5}{8}$ D. $\frac{3}{8}$

6.3 Multiplication and Division of Fractions

As a pleasant surprise, multiplication and division of fractions are computationally easier than addition and subtraction. For example, we will discover that the product of $\frac{3}{4}$ and $\frac{2}{5}$ is computed simply by multiplying the respective numerators and denominators to get $\frac{6}{20}$. Division is only slightly trickier: To compute $\frac{2}{5} \div \frac{3}{4}$, we can use the "invert and multiply rule" to get the answer, $\frac{2}{5} \times \frac{4}{3} = \frac{8}{15}$. However, even though multiplication and division of fractions are easy computationally, these operations are quite difficult to understand conceptually. To successfully impart a deep understanding of the multiplication and division of fractions, teachers need to be able to answer these questions:

- *How* can multiplication be illustrated with manipulatives? with pictorial representations?
- *Why* is multiplication defined with such a simple formula?
- *How* is a fraction interpreted as a multiplicative operator?
- *In what way* can division still be viewed as repeated subtraction or as a missing factor?
- *What* justifies the "invert and multiply rule"?
- *What* real-life problems require the multiplication or division of fractions?

It is important to understand that the multiplication of fractions is an extension of the multiplication operation on the integers. It is helpful to take a stepwise approach in which first a fraction is multiplied by an integer and then an integer is multiplied by a fraction. These two preliminary steps prepare the student for the general definition of how two fractions are multiplied.

Similarly, division of fractions is made meaningful by continuing to view division as repeated subtraction (grouping), as a partition (sharing), and as a missing factor.

Multiplication of a Fraction by an Integer

Three family-size pizzas were ordered for a party, and a quarter of each pizza was not consumed by the partygoers. Viewing multiplication by a positive integer as repeated addition, we see from the fraction-circle diagram in Figure 6.14 that $3 \cdot \frac{1}{4} = \frac{1}{4} + \frac{1}{4} + \frac{1}{4} = \frac{3}{4}$ of a pizza is left over.

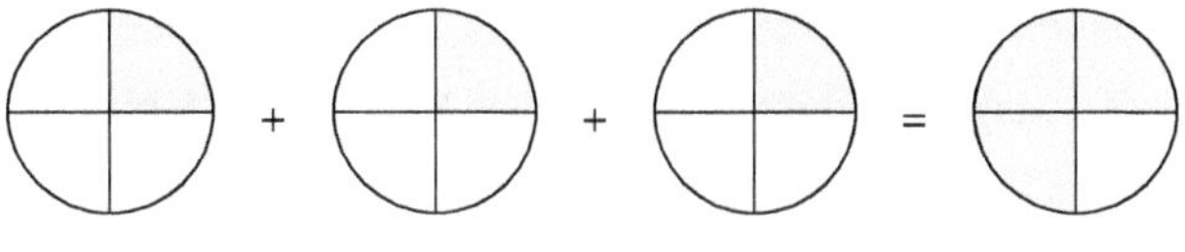

Figure 6.14
A fraction-circle diagram showing that $3 \cdot \frac{1}{4} = \frac{3}{4}$

EXAMPLE 6.9

Multiplying a Fraction by an Integer

(a) Use fraction strips to show that $3 \cdot \frac{3}{8} = \frac{9}{8}$.

(b) Use a number line to show that $4 \cdot \frac{2}{3} = \frac{8}{3}$.

(c) Use a number line to show that $-3 \cdot \frac{5}{6} = \frac{-15}{6}$.

Solution

(a) Three fraction strips, each representing $\frac{3}{8}$, are aligned to show that their combined shaded region represents $\frac{9}{8}$:

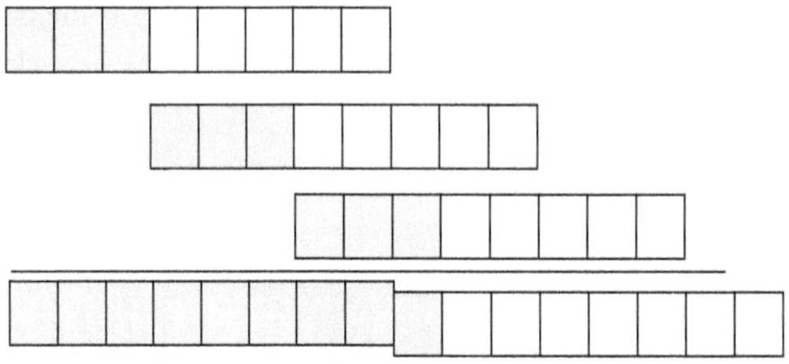

(b) Four jumps of length $\frac{2}{3}$ arrive at the point $\frac{8}{3}$:

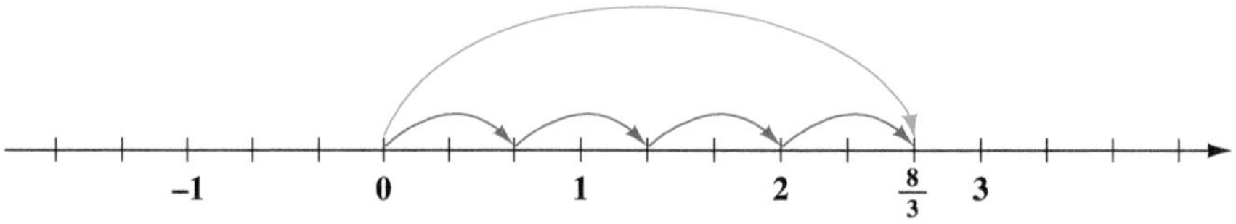

(c) The repeated subtraction of three jumps, each of length $\frac{5}{6}$, takes us to the point $\frac{-15}{6}$ on the number line:

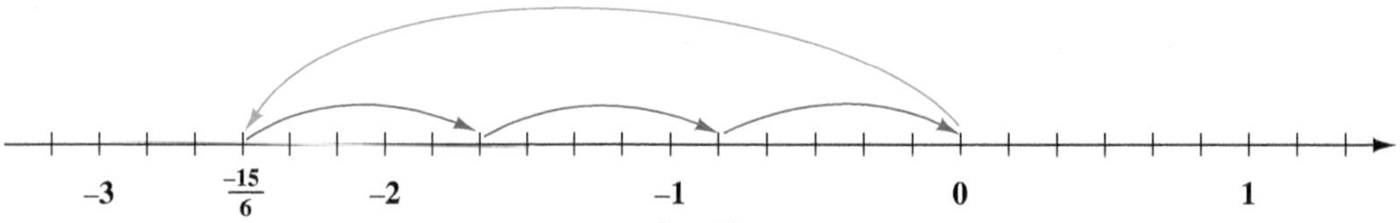

Since $\frac{5}{6}$ is positive, it is represented by a right-pointing arrow. Removing three of these arrows takes us to the point $\frac{-15}{6}$; this is equivalent to a leftward jump to $\frac{-15}{6}$, which is shown by a left-pointing arrow.

Multiplication of an Integer by a Fraction

Imagine that we are taking a 5-mile hike and we meet someone on the trail who estimates that we have come two-thirds of the way to our destination. How many miles have we covered? This is an easy question to answer: One-third of the 5-mile trail is $\frac{5}{3}$ miles long, and we have covered twice this distance, so we have hiked $\frac{10}{3}$ miles so far.

We see that "two-thirds of" gives meaning to $\frac{2}{3}$ as an **operator:** To take two-thirds of 5, first divide 5 into 3 equal parts and then take 2 of those parts. That is, $\frac{2}{3} \cdot 5 = 2 \cdot \frac{5}{3} = \frac{10}{3}$. Here are some other examples related to the hike:

- If the entire hike will take 6 hours, we had been hiking $\frac{2}{3} \cdot 6 = \frac{12}{3} = 4$ hours when we met the other hiker.
- If we brought 2 liters of water for the hike, we might estimate that we have used $\frac{2}{3} \cdot 2 = \frac{4}{3}$ liters of our water.

Additional examples of how a fraction is used as a multiplicative operator are shown in Figure 6.15. In each case, we multiply by a fraction $\frac{a}{b}$ by first making a partition into b equal parts and then taking a of those parts. That is, $\frac{a}{b} \cdot n = a \cdot \frac{n}{b} = \frac{an}{b}$.

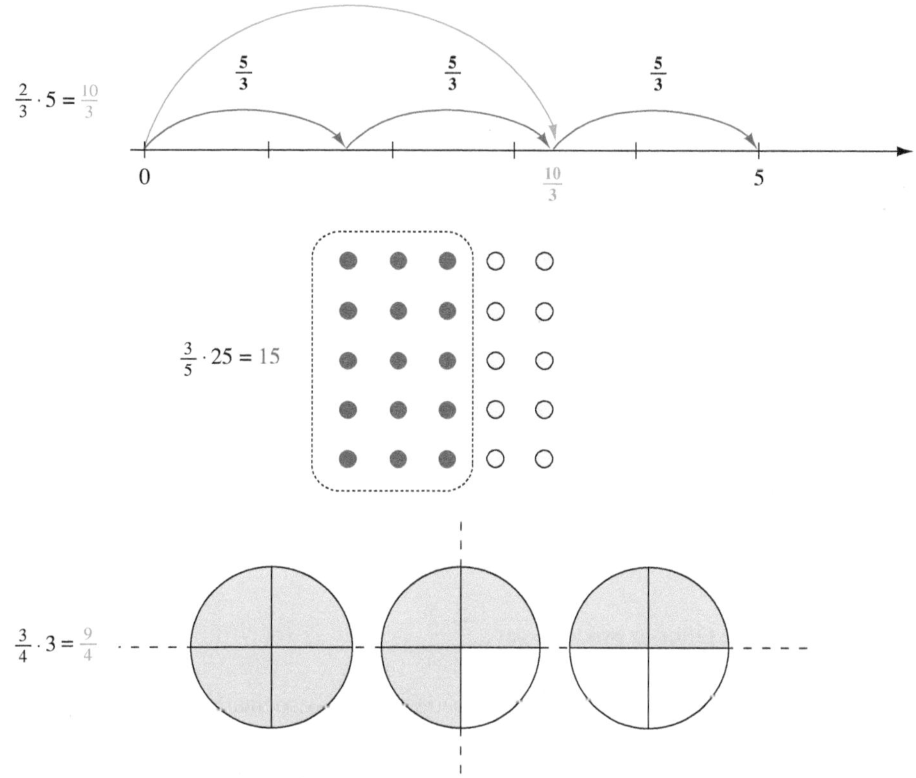

Figure 6.15
Depicting the product of a fraction and an integer

Multiplication of a Fraction by a Fraction

In earlier chapters, the rectangular area model provided a useful visualization of multiplication for the whole numbers and integers. This model works equally well to motivate the general definition of the multiplication of two fractions.

In Figure 6.16a, there is a reminder of how the rectangular area model is used with the whole numbers: We simply create a shaded rectangle that is 2 units high and 3 units long and notice that 6 unit squares fill the rectangle. That is, $2 \cdot 3 = 6$. In Figure 6.16b, the rectangular area model reconfirms the result $2 \cdot \frac{3}{2} = \frac{2 \cdot 3}{2} = \frac{6}{2}$ for how to multiply a fraction by a whole number. Similarly, Figure 6.16c shows us that the product of the whole number 3 and the fraction $\frac{7}{4}$ is given by $\frac{7}{4} \cdot 3 = \frac{7 \cdot 3}{4} = \frac{21}{4}$.

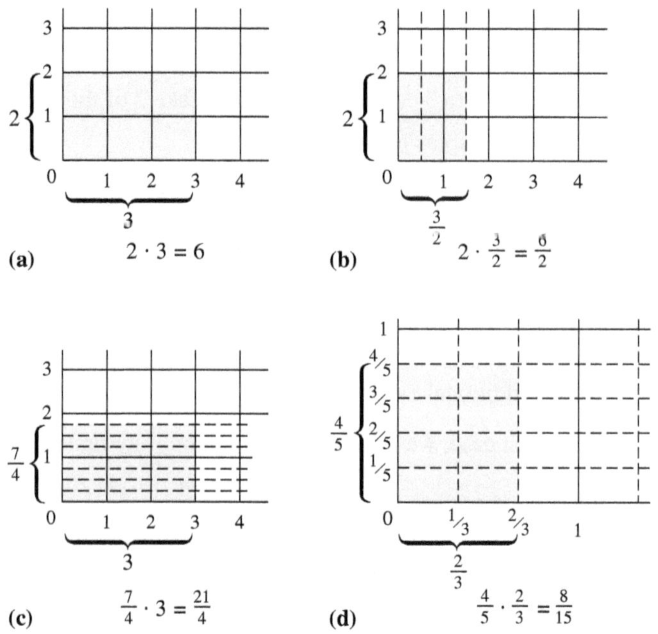

Figure 6.16 Extending the area model of multiplication to fractions

Finally, in Figure 6.16d, we see that the product of the fractions $\frac{4}{5}$ and $\frac{2}{3}$ is given by $\frac{4}{5} \cdot \frac{2}{3} = \frac{8}{15}$, since 8 rectangles are shaded and there are 15 congruent rectangles that fill a 1-by-1 unit square. The product of the numerators, $4 \cdot 2 = 8$, gives us the number of small shaded rectangles and forms the numerator of the answer. The denominator of the answer is the product $5 \cdot 3 = 15$ of the denominators and counts the number of small rectangles in a unit square.

DEFINITION Multiplication of Fractions

Let $\frac{a}{b}$ and $\frac{c}{d}$ be fractions. Then their **product** is given by

$$\frac{a}{b} \cdot \frac{c}{d} = \frac{ac}{bd}.$$

Often a product, say, $\frac{4}{5} \cdot \frac{2}{3}$, is read as four-fifths "of" two-thirds, which emphasizes how $\frac{4}{5}$ is an operator applied to the fraction $\frac{2}{3}$. The association between "of" and "times" is natural for multiplication by whole numbers and extends to multiplication of fractions. For example, "I'll buy three *of* the half-gallon-size bottles" is equivalent to buying $3 \cdot \frac{1}{2} = \frac{3}{2} = 1\frac{1}{2}$ gallons.

EXAMPLE 6.10 Calculating Products of Fractions

Illustrate these two products with the area model:

(a) $\frac{5}{8} \cdot \frac{2}{3}$ **(b)** $3\frac{1}{7} \cdot 5\frac{1}{4}$

Solution

(a) Let a rectangle denote a whole unit of area. Next, use vertical lines to divide the unit into 3 equal parts, so that shading the 2 leftmost rectangular columns represents $\frac{2}{3}$. Similarly, use horizontal lines to divide the unit rectangle into 8 equal parts, and shade the lower 5 rectangular rows to represent $\frac{5}{8}$ of the whole rectangle. The doubly shaded part of the whole represents $\frac{5}{8}$ of $\frac{2}{3}$ of the whole. Observing that the whole rectangle has been divided into 24 small rectangles, and 10 of these are within the overlapped doubly shaded region, we see that $\frac{5}{8} \cdot \frac{2}{3} = \frac{10}{24}$. We also see that the product of the numerators, $5 \cdot 2 = 10$, gives the number of small rectangles and is the numerator of our answer. Also, the product of the denominators, $8 \cdot 3 = 24$, gives the number of small rectangles in a unit rectangle and is the denominator of the fraction that represents the product.

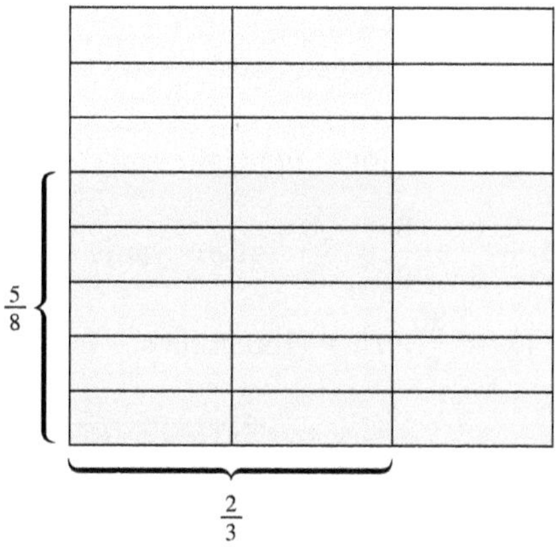

(b) Equivalently, we want to model the product $\frac{22}{7} \cdot \frac{21}{4}$ by a rectangular area diagram. The doubly shaded (green) region contains $22 \cdot 21$ small rectangles, where each unit square contains $4 \cdot 7 = 28$ small rectangles. Thus, $3\frac{1}{7} \cdot 5\frac{1}{4} = \frac{22 \cdot 21}{7 \cdot 4}$, which simplifies to $\frac{22 \cdot 21}{4 \cdot 7} = \frac{11 \cdot 3}{2 \cdot 1} = \frac{33}{2} = 16\frac{1}{2}$. This answer can also be seen in the figure by noticing that the green region contains 15 unit squares and 42 small rectangles. Since $42 = 28 + 14$, the small rectangles cover $1\frac{1}{2}$ unit squares. Altogether, we again arrive at the final answer of $16\frac{1}{2}$.

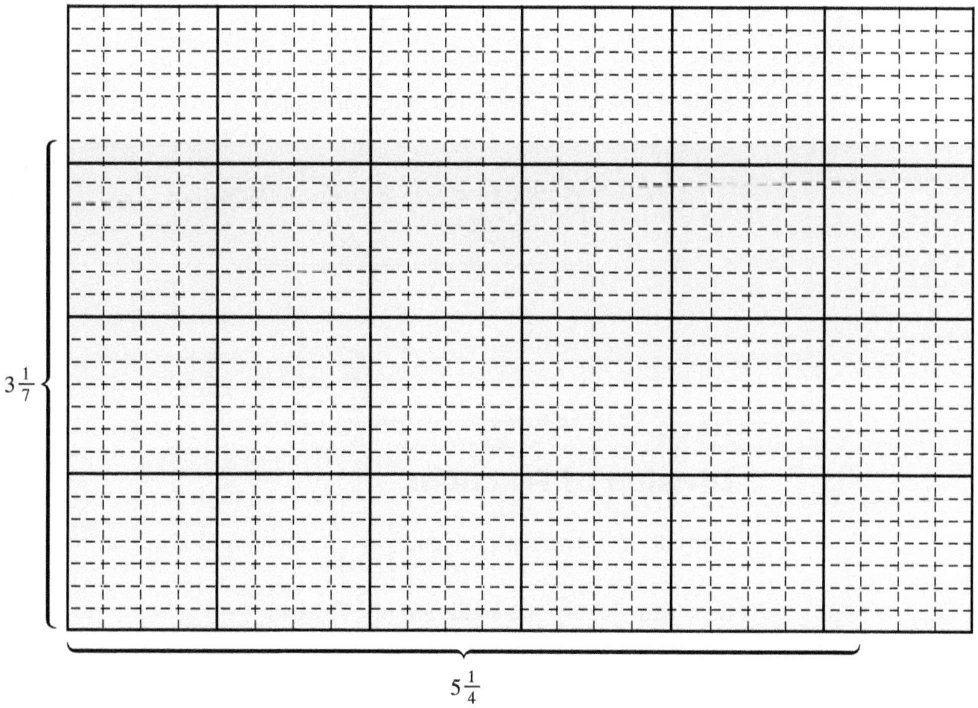

EXAMPLE 6.11

Computing the Area and Cost of a Carpet

The hallway in the Bateks' house is a rectangle 4 feet wide and 20 feet long; that is, it measures $\frac{4}{3}$ yards by $\frac{20}{3}$ yards. What is the area of the hallway in square yards? Mrs. Batek wants to know how much she will pay to buy carpet priced at $18 per square yard to carpet the hall.

Solution

Since $\frac{4}{3} \cdot \frac{20}{3} = \frac{80}{9} = 8\frac{8}{9}$, the area is $8\frac{8}{9}$ square yards, or nearly 9 square yards. This can be seen in the accompanying diagram, which shows the hallway divided into six full square yards, six $\frac{1}{3}$-square-yard rectangular regions, and eight square regions that are each $\frac{1}{9}$ of a square yard. Thus, the total area is $6 + \frac{6}{3} + \frac{8}{9} = 8\frac{8}{9}$ square yards. At \$18 per square yard, the cost of the carpet will be $8\frac{8}{9} \cdot 18 = \frac{80}{9} \cdot 18 = 160$ dollars.

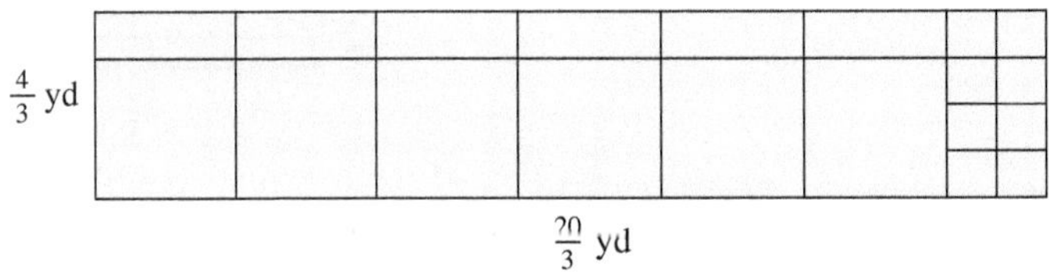

Multiplication of fractions can also be illustrated on the number line, as shown in the next example.

EXAMPLE 6.12

Multiplying Fractions on the Number Line

Illustrate why $\frac{2}{3} \cdot \frac{4}{5} = \frac{8}{15}$ with a number-line diagram.

Solution

It's helpful to view $\frac{2}{3}$ as a multiplicative operator. That is, first partition $\frac{4}{5}$ into three equal intervals, each of length $\frac{4}{15}$. Since two jumps of length $\frac{4}{15}$ arrive at $\frac{8}{15}$, the diagram confirms that $\frac{2}{3} \cdot \frac{4}{5} = \frac{8}{15}$.

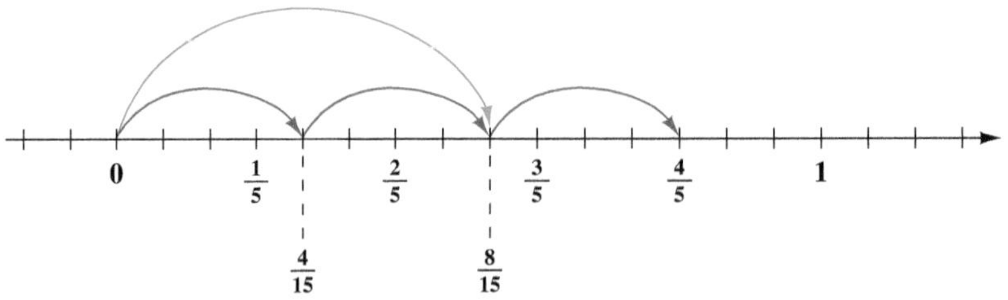

Division of Fractions

Consider the division problem $\frac{4}{3} \div \frac{1}{6}$. Viewing division as repeated subtraction, we want to know this: "How many one-sixths are in $\frac{4}{3}$?" Since $\frac{4}{3} = \frac{8}{6} = 8 \cdot \frac{1}{6}$, there are eight one-sixths in $\frac{4}{3}$. That is, $\frac{4}{3} \div \frac{1}{6} = 8$, since $\frac{4}{3} = 8 \cdot \frac{1}{6}$.

Let's try another example: Consider the division problem $\frac{7}{12} \div \frac{1}{6}$. We want to determine how many one-sixths are in $\frac{7}{12}$. Since $\frac{7}{12} = \frac{7}{2 \cdot 6} = \frac{7}{2} \cdot \frac{1}{6}$, there are $\frac{7}{2}$, or $3\frac{1}{2}$, one-sixths in $\frac{7}{12}$. That is, $\frac{7}{12} \div \frac{1}{6} = \frac{7}{2}$, since $\frac{7}{12} = \frac{7}{2} \cdot \frac{1}{6}$.

In general, to find the fraction $\frac{m}{n}$ that solves the division $\frac{a}{b} \div \frac{c}{d} = \frac{m}{n}$, we must find the missing factor $\frac{m}{n}$ in the multiplication formula $\frac{a}{b} = \frac{m}{n} \cdot \frac{c}{d}$. The missing-factor interpretation is used to define the division of fractions.

DEFINITION Division of Fractions

Let $\frac{a}{b}$ and $\frac{c}{d}$ be fractions, where $\frac{c}{d}$ is not zero. Then $\frac{a}{b} \div \frac{c}{d} = \frac{m}{n}$ if and only if

$$\frac{a}{b} = \frac{m}{n} \cdot \frac{c}{d}.$$

This definition stresses that division is the inverse operation of multiplication. However, other conceptual models of division—as a measurement (that is, as a repeated subtraction or grouping) and as a partition (or sharing)—continue to be important. These models can be represented with manipulatives and diagrams, enabling the teacher to convey a deep understanding of what division of fractions really means and how it is used to solve problems.

This is reflected in the first of the Standards of Mathematical Practice (SMP 1) from the **Common Core:**

SMP 1 Make sense of problems and persevere in solving them.

Younger students might rely on using concrete objects or pictures to help conceptualize and solve a problem. Mathematically proficient students check their answers to problems using a different method, and they continually ask themselves, "Does this make sense?"

The next two problems are solved by reasoning directly from the concept of division either as a grouping or as a sharing. No formula or computational rule is required to obtain the answer. It is very important to keep the conceptual notion of division in mind and not rely simply on rules and formulas that have no intrinsic meaning.

EXAMPLE 6.13

Seeding a Lawn: Division of Fractions by Grouping

The new city park will have a $2\frac{1}{2}$-acre grass playfield. Grass seed can be purchased in large bags, each sufficient to seed $\frac{3}{4}$ of an acre. How many bags are needed? Will there be some grass seed left over to keep on hand for reseeding worn spots in the field?

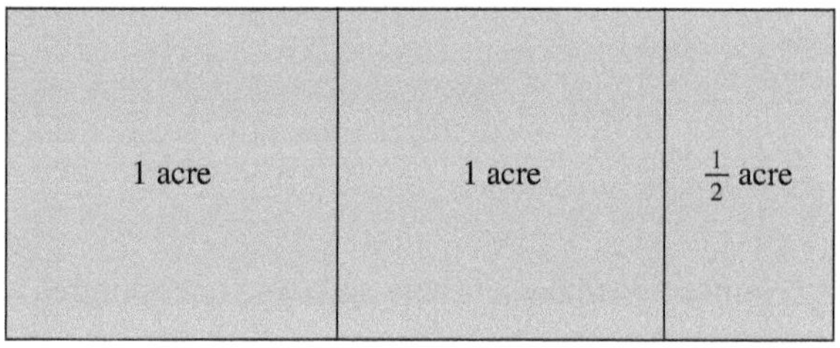

Solution

We need to determine the number of $\frac{3}{4}$-acre regions in the $2\frac{1}{2}$-acre field. That is, we must compute $2\frac{1}{2} \div \frac{3}{4}$. In the diagram that follows, the field is grouped into four regions. Three of these regions each cover $\frac{3}{4}$ of an acre and, therefore, require a whole bag of seed. There is also a $\frac{1}{4}$-acre region that will use $\frac{1}{3}$ of a bag. Thus, a total of $2\frac{1}{2} \div \frac{3}{4} = 3\frac{1}{3}$ bags of seed are needed for the playfield. We conclude that four bags of seed should be ordered, which is enough for the initial seeding and leaves $\frac{2}{3}$ of a bag on hand for reseeding.

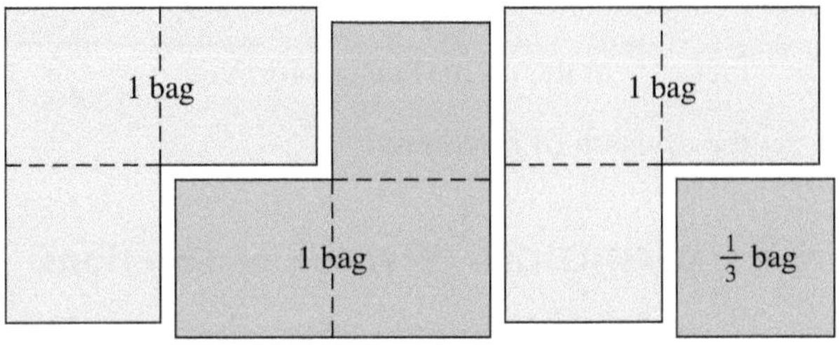

EXAMPLE 6.14

Making Cookies: Division of Fractions by Sharing

Jesse has enough flour to make $2\frac{1}{2}$ recipes of chocolate chip cookies and $1\frac{1}{3}$ cups of chocolate chips. How many cups of chocolate chips will be in each recipe?

Solution

The chocolate chips and recipes are shown on the following two fraction strips, where each small brown rectangle represents a third of a cup of chocolate chips and each small yellow rectangle represents a half recipe of cookies:

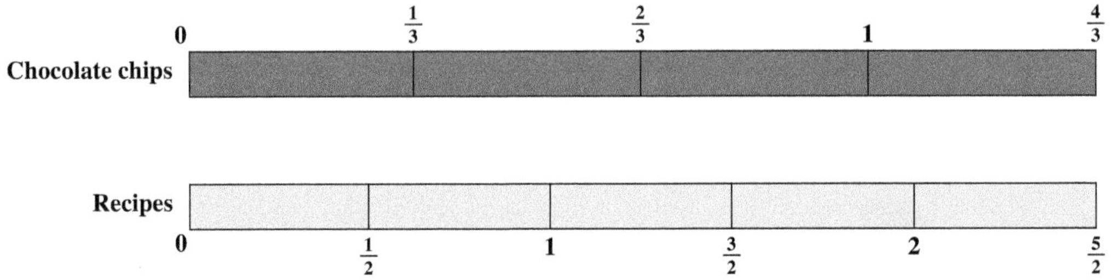

If we subdivide each small brown rectangle into 5 parts and each small yellow rectangle into 4 parts, then both the amounts of chocolate chips and recipes are divided into 20 parts.

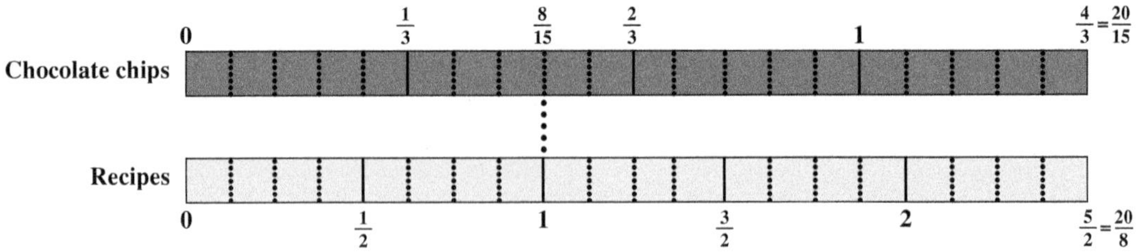

Since 15 smallest brown squares correspond to a cup of chocolate chips, each of these represents $\frac{1}{15}$ of a cup of chips. Therefore, $\frac{8}{15}$ of a cup of chips corresponds to one recipe, and we conclude that $\frac{4}{3} \div \frac{5}{2} = \frac{8}{15}$.

Algorithms for Calculating the Division of Fractions

There are three ways to simplify the calculation of a quotient of fractions:

- **Division by finding common numerators**

  In Example 6.14, it was shown that $\frac{4}{3} \div \frac{5}{2} = \frac{20}{15} \div \frac{20}{8} = \frac{8}{15}$.

  To verify the formula $\frac{a}{b} \div \frac{a}{c} = \frac{c}{b}$, notice that $\frac{a}{b} = \frac{a}{c} \cdot \frac{c}{b}$.

- **Division by finding common denominators**

  We showed earlier that $\frac{7}{12} \div \frac{1}{6} = \frac{7}{12} \div \frac{2}{12} = \frac{7}{2}$.

  To verify the formula $\frac{a}{b} \div \frac{c}{b} = \frac{a}{c}$, we simply observe that $\frac{a}{b} = \frac{c}{b} \cdot \frac{a}{c}$.

- **Division by the "invert and multiply" rule**

  In Example 6.13, we showed that $\frac{5}{2} \div \frac{3}{4} = \frac{20}{6}$, since $\frac{20}{6} = \frac{10}{3} = 3\frac{1}{3}$.

  This calculation suggests that $\frac{a}{b} \div \frac{c}{d} = \frac{a}{b} \cdot \frac{d}{c}$, where the divisor $\frac{c}{d}$ is "inverted" and becomes a factor $\frac{d}{c}$ that is multiplied by $\frac{a}{b}$. This "invert and multiply" formula may be verified by observing that $\left(\frac{a}{b} \cdot \frac{d}{c}\right) \cdot \frac{c}{d} = \frac{adc}{bcd} = \frac{a}{b}$.

  The following theorem summarizes these useful formulas for the division of fractions:

THEOREM Formulas for the Division of Fractions

Common-numerator rule	$\frac{a}{b} \div \frac{a}{c} = \frac{c}{b}$
Common-denominator rule	$\frac{a}{b} \div \frac{c}{b} = \frac{a}{c}$
Invert-and-multiply rule	$\frac{a}{b} \div \frac{c}{d} = \frac{a}{b} \cdot \frac{d}{c}$

EXAMPLE 6.15

Dividing Fractions

Compute these division problems:

(a) $\frac{3}{4} \div \frac{1}{8}$ **(b)** $\frac{2}{5} \div \frac{2}{3}$ **(c)** $3 \div \frac{4}{3}$

(d) $39 \div 13$ **(e)** $13 \div 39$ **(f)** $4\frac{1}{6} \div 2\frac{1}{3}$

Solution

(a) $\frac{3}{4} \div \frac{1}{8} = \frac{6}{8} \div \frac{1}{8} = 6$, or $\frac{3}{4} \div \frac{1}{8} = \frac{3}{4} \cdot \frac{8}{1} = \frac{24}{4} = 6$.

(b) $\frac{2}{5} \div \frac{2}{3} = \frac{3}{5}$, or $\frac{2}{5} \div \frac{2}{3} = \frac{2}{5} \cdot \frac{3}{2} = \frac{2 \cdot 3}{5 \cdot 2} = \frac{3}{5}$.

(c) $3 \div \frac{4}{3} = \frac{3}{1} \cdot \frac{3}{4} = \frac{9}{4}$.

(d) $39 \div 13 = \frac{39}{1} \cdot \frac{1}{13} = \frac{39}{13} = 3$.

(e) $13 \div 39 = \frac{13}{1} \cdot \frac{1}{39} = \frac{13}{39} = \frac{1}{3}$.

(f) $4\frac{1}{6} \div 2\frac{1}{3} = \frac{25}{6} \div \frac{7}{3} = \frac{25}{6} \cdot \frac{3}{7} = \frac{25 \cdot 3}{6 \cdot 7} = \frac{25 \cdot 1}{2 \cdot 7} = \frac{25}{14}$.

Reciprocals as Multiplicative Inverses in the Rational Numbers

The "invert and multiply" rule, though not the definition of division, has an important consequence: The division of any rational by a nonzero rational number has a unique quotient that is also a rational number. That is, the nonzero rational numbers are closed under division. This was not true of the system of integers. For example, $3 \div 8$ cannot be written as any integer. But when 3 and 8 are interpreted as the fractions $\frac{3}{1}$ and $\frac{8}{1}$, the we see that $3 \div 8 = \frac{3}{1} \div \frac{8}{1} = \frac{3}{8}$.

More precisely stated, the nonzero fraction $\frac{c}{d}$ is "inverted" by forming its **reciprocal.**

DEFINITION Reciprocal of a Fraction

The **reciprocal** of a nonzero fraction $\frac{c}{d}$ is the fraction $\frac{d}{c}$.

Since $\frac{c}{d} \cdot \frac{d}{c} = \frac{d}{c} \cdot \frac{c}{d} = 1$, and 1 is the multiplicative identity, we say that the reciprocal of a nonzero fraction is its **multiplicative inverse.** Multiplicative inverses are very useful for solving equations, as shown in the next two examples.

EXAMPLE 6.16 Using the Reciprocal

One morning at the preschool for 3- and 4-year-olds, $\frac{2}{3}$ of the 3-year-old children and $\frac{5}{6}$ of the 4-year-old children went on a walk. For safety, the children were paired so that each 3-year-old child held the hand of a 4-year-old. Among the children who did not go on the walk, there were 9 more 3-year-olds than 4-year-olds. What is the enrollment of the school?

Solution

Understand the Problem

Information is given about the number of 3- and 4-year-olds. Our goal is to determine the number of children in the school.

Devise a Plan

Although we are asked only for the total number of children, it is probably helpful to determine both the number of 3-year-olds and the number of 4-year-olds. Summing these two numbers will give us our answer. Thus, we can introduce variables, use the given information to form equations, and finally solve these equations to find our answer. Since the given information involves fractions, we expect that we need to be skillful working with the arithmetic of fractions.

Carry Out the Plan

Let the variables T and F represent the respective number of 3- and 4-year-old children. The information given is equivalent to the equations $\frac{2}{3}T = \frac{5}{6}F$ and $T = F + 9$. The first equation can be solved for T by multiplying both sides by the reciprocal of $\frac{2}{3}$, namely, $\frac{3}{2}$, so

$$T = \left(\frac{3}{2} \cdot \frac{2}{3}\right) \cdot T = \frac{3}{2} \cdot \left(\frac{2}{3} \cdot T\right) = \frac{3}{2} \cdot \left(\frac{5}{6} \cdot F\right) = \left(\frac{3}{2} \cdot \frac{5}{6}\right) \cdot F = \frac{5}{4} \cdot F.$$

Comparing this equation with $T = F + 9$, we obtain the new relation $\frac{5}{4}F = F + 9$. Subtracting F from both sides gives $\frac{1}{4} \cdot F = 9$. To solve for F, we multiply this equation by the reciprocal of $\frac{1}{4}$, which is 4, to get $F = 4 \cdot 9 = 36$. Therefore, $T = 36 + 9 = 45$. The school has $36 + 45 = 81$ children in all.

Look Back

It is always a good idea to ask if a problem can be solved differently. The figure that follows shows clearly all of the information given in the problem: $\frac{2}{3}$ of the 3-year-olds walk with $\frac{5}{6}$ of the 4-year-olds, and there are 9 more older children. The dotted lines have been added to divide each region into rectangles of equal size.

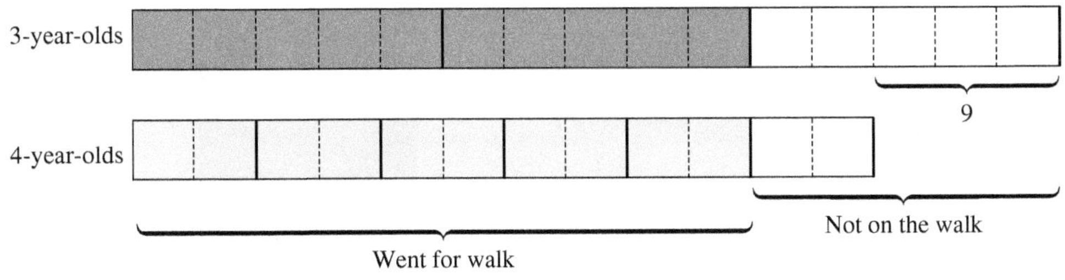

Evidently, 3 small rectangles represent 9 children, so each small rectangle represents 3 children. The diagram now makes it clear there are $3 \cdot 3 \cdot 5 = 45$ children of age 3, $3 \cdot 6 \cdot 2 = 36$ of age 4, and 81 children in all.

The missing-factor model of division is used to solve the real-life problem in the next example. The reciprocal, being a multiplicative inverse, allows us to solve for the missing factor.

EXAMPLE 6.17 Bottling Root Beer: Division of Fractions with the Missing-Factor Model

Ari is making homemade root beer. The recipe he followed nearly fills a 5-gallon glass jug, and he estimates that it contains $4\frac{3}{4}$ gallons of root beer. He is now ready to bottle his root beer. How many $\frac{1}{2}$-gallon bottles can he fill?

Solution

Let x denote the number of half-gallon bottles required, where x will be allowed to be a fraction, since we expect that some bottle may be only partially filled. We must then solve the equation

$$x \cdot \frac{1}{2} = 4\frac{3}{4},$$

which is the missing-factor problem that is equivalent to the division problem $4\frac{3}{4} \div \frac{1}{2}$.

Since

$$4\frac{3}{4} = \frac{19}{4} = \frac{19}{2} \cdot \frac{1}{2},$$

it follows that the missing factor is $x = \frac{19}{2} = 9\frac{1}{2}$.

The situation is shown in the accompanying figure. Ari will need 9 half-gallon bottles, and he will probably see if he can also find a quart bottle to use.

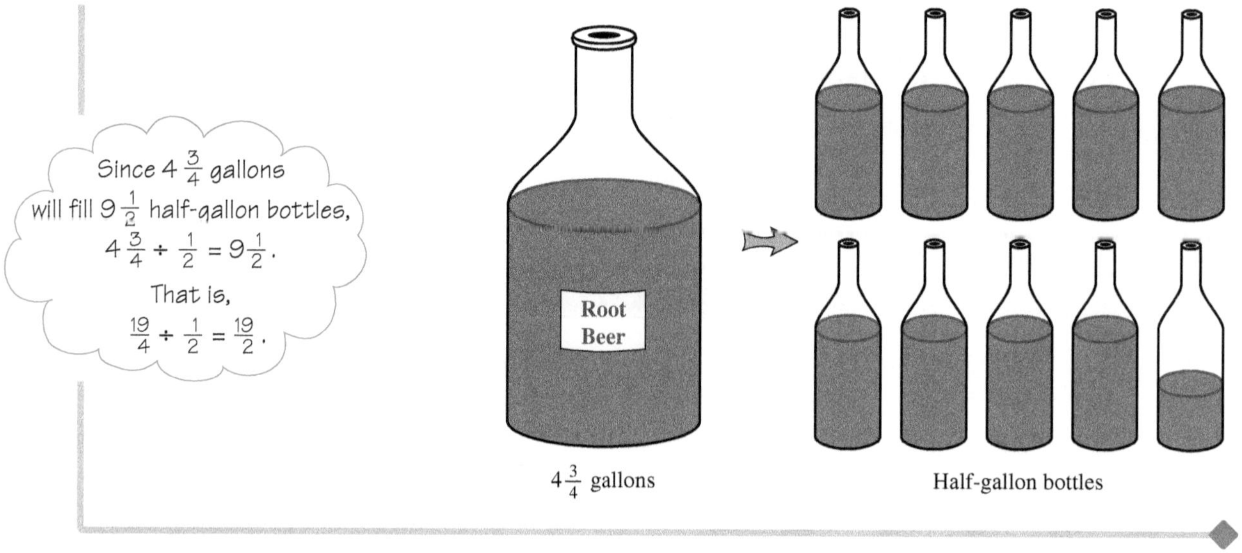

Since $4\frac{3}{4}$ gallons will fill $9\frac{1}{2}$ half-gallon bottles, $4\frac{3}{4} \div \frac{1}{2} = 9\frac{1}{2}$.
That is,
$\frac{19}{4} \div \frac{1}{2} = \frac{19}{2}$.

Root Beer

$4\frac{3}{4}$ gallons

Half-gallon bottles

INTO THE CLASSROOM

Teachers' Understandings of the Division of Fractions in the United States and China

As part of her comparative study of the differences in mathematical knowledge between teachers in the United States and China, Liping Ma asked teachers in both countries to calculate the division $1\frac{3}{4} \div \frac{1}{2}$. She reported that just 43% of the U.S. teachers were successful, and none showed an understanding of the rationale of the algorithm that was used. This was in sharp contrast to the teachers in China, all of whom computed the division correctly. The Chinese teachers were not satisfied just to obtain the answer but also enjoyed presenting and commenting upon various ways to do the division. They showed they had a deep conceptual understanding of all of the operations on fractions. Liping Ma's book* makes the case that elementary school teachers must acquire a "profound understanding of fundamental mathematics" to support the ways they present mathematics to their students, particularly the ways demanded by the standards set forth by the National Council of Teachers of Mathematics and by the Common Core State Standards in Mathematics.

*From *Knowing and Teaching Elementary Mathematics: Teacher's Understanding of Fundamental Mathematics in China and in the United States*, © 1999, by Liping Ma.

6.3 Problem Set

Understanding Concepts

1. What fraction multiplication is illustrated with this fraction-circle diagram?

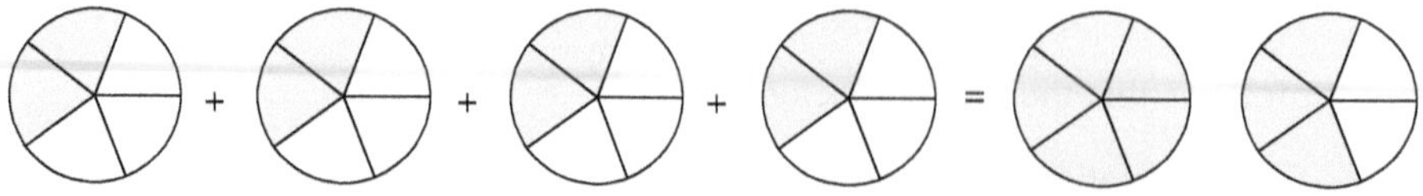

2. Draw a fraction-circle diagram that illustrates $5 \cdot \frac{3}{4}$.

3. Draw number-line diagrams that illustrate these products:

(a) $4 \cdot \frac{3}{8}$ **(b)** $-4 \cdot \frac{3}{8}$

4. Draw number-line diagrams that illustrate these products.

(a) $4 \cdot \left(\frac{-3}{8}\right)$

(b) $-4 \cdot \left(\frac{-3}{8}\right)$

---

5. What fraction multiplication is illustrated by this number-line diagram?

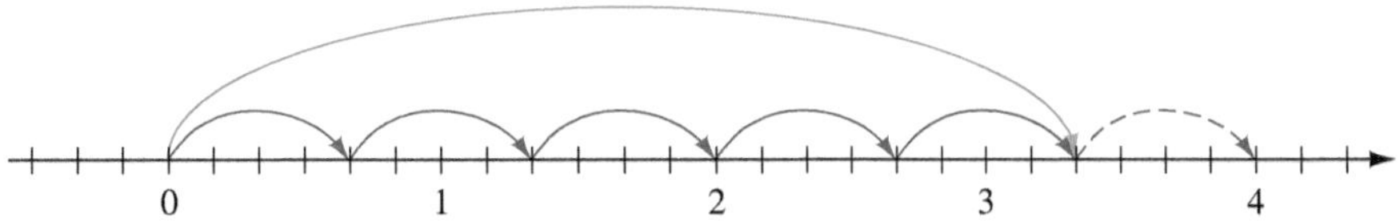

6. Draw a number-line diagram to illustrate that $\frac{3}{4} \times 5 = \frac{15}{4} = 3\frac{3}{4}$.

---

7. What multiplication facts are illustrated by these rectangular area models?

(a)

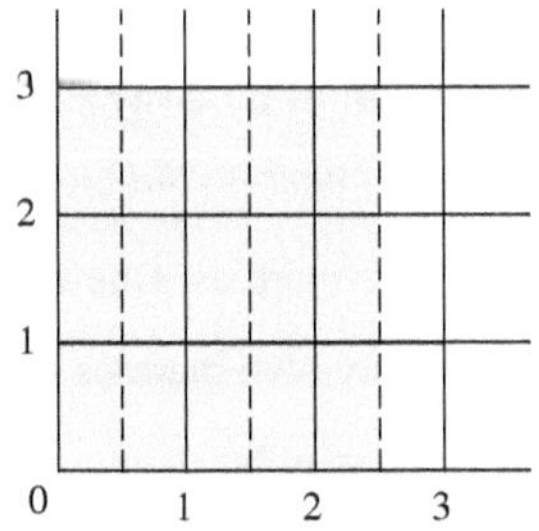

(b)

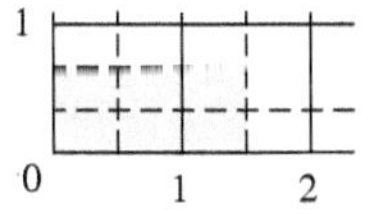

(c)

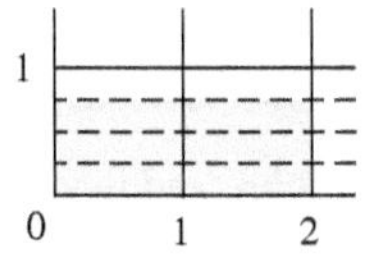

8. Illustrate these multiplications with the rectangular area model.

(a) $2 \times \frac{3}{5}$ **(b)** $\frac{3}{2} \times \frac{3}{4}$ **(c)** $1\frac{2}{3} \times 2\frac{1}{4}$

9. A rectangular plot of land is $2\frac{1}{4}$ miles wide and $3\frac{1}{2}$ miles long. What is the area of the plot in square miles? Draw a sketch that verifies your answer.

10. Find the reciprocals of the following fractions:

(a) $\frac{3}{8}$ **(b)** $\frac{4}{3}$ **(c)** $2\frac{1}{4}$

11. Find the reciprocals of the following fractions:

(a) $\frac{1}{8}$ **(b)** 5 **(c)** 1

12. Compute each of these products, expressing each answer in lowest terms:

(a) $\frac{5}{4} \times \frac{12}{25}$ **(b)** $\frac{7}{8} \times \frac{2}{21}$

(c) $\frac{-3}{5} \times \frac{10}{21}$ **(d)** $\frac{4}{7} \times \frac{21}{16}$

13. Use the common-numerator and common-denominator rules to compute these divisions, expressing each answer in lowest terms:

(a) $\frac{5}{4} \div \frac{3}{4}$ **(b)** $\frac{7}{8} \div \frac{7}{11}$

(c) $\frac{2}{5} \div \frac{7}{5}$ **(d)** $\frac{4}{7} \div \frac{4}{21}$

14. Compute these divisions, expressing each answer in simplest form:

(a) $\frac{2}{5} \div \frac{3}{4}$ **(b)** $\frac{6}{11} \div \frac{4}{3}$

(c) $\frac{100}{33} \div \frac{10}{3}$ **(d)** $2\frac{3}{8} \div 5$

(e) $3 \div 5\frac{1}{4}$ **(f)** $\frac{21}{25} \div \frac{7}{25}$

15. Compute the fraction with the simplest form that is equivalent to the given expression.

(a) $\frac{2}{3} \cdot \left(\frac{3}{4} + \frac{9}{12}\right)$ **(b)** $\left(\frac{3}{5} - \frac{3}{10}\right) \div \frac{6}{5}$

(c) $\left(\frac{2}{5} \div \frac{4}{15}\right) \cdot \frac{2}{3}$

16. Set up and evaluate expressions to solve these map problems:

(a) Each inch on a map represents an actual distance of $2\frac{1}{2}$ miles. If the map shows Helmer as being $3\frac{3}{4}$ inches due east of Deary, how far apart are the two towns?

(b) A map shows that Spokane is 60 miles north of Colfax. A ruler shows that the towns are $3\frac{3}{4}$ inches apart on the map. How many miles are represented by each inch on the map?

17. Carefully describe each step in the arithmetic of fractions that is required to solve these equations:

(a) $\frac{2}{5}x - \frac{3}{4} = \frac{1}{2}$ **(b)** $\frac{2}{3}x + \frac{1}{4} = \frac{3}{2}x$

18. I am a fraction whose product with 7 is the same as my sum with 7. Who am I?

Into the Classroom

The division of fractions is one of the most difficult concepts to teach in the elementary school curriculum. In each problem that follows, provide pictures and word descriptions that you believe would effectively convey the central ideas to a youngster. Some of the examples in the section can be used as models.

19. (**Writing**) Sean has a job mowing grass for the city, using a riding mower. He can mow $\frac{5}{6}$ of an acre per hour. How long will he need to mow the 3-acre city park?

20. (**Writing**) Angie, Bree, Corrine, Dot, and Elaine together picked $3\frac{3}{4}$ crates of strawberries. How many crates should each be allotted in order to distribute the berries evenly?

21. (**Writing**) Gerry is making a pathway out of concrete stepping-stones. The path is 25 feet long, and each stone extends $1\frac{2}{3}$ of a foot along the path. By letting x denote the number of stones, Gerry knows that he needs to solve the equation $x \cdot 1\frac{2}{3} = 25$, but he isn't sure how to solve for x. Provide a careful explanation.

Responding to Students

22. (**Writing**) Respond to Eva's statement:

> I know that multiplication by $\frac{3}{4}$ means splitting into 4 equal parts and then taking 3 of the parts. I also know that division is the inverse of multiplication, so division by $\frac{3}{4}$ means splitting into 3 equal parts and then taking 4 of the parts. So, instead of dividing by $\frac{3}{4}$, I might just as well multiply by $\frac{4}{3}$.

Children (and adults!) often have difficulties relating fractions and their arithmetic to practical situations. In problems 23–25, respond to the children who have been asked to pose a multiplication problem that requires the answer to $\frac{1}{4}$ of $\frac{2}{3}$.

23. There is $\frac{2}{3}$ of a pizza left. Suzanne then ate a quarter of a pizza. How much pizza is left?

24. There is $\frac{2}{3}$ of a pizza left. Jamie ate $\frac{1}{4}$ of the remaining pieces. How many pieces of pizza now remain?

25. There is $\frac{2}{3}$ of a cake left. One-fourth of Miguel's classmates want some of the cake. How much cake is each child given?

In problems 26 and 27, respond to the children who have been asked to pose a problem that corresponds to the division of $1\frac{3}{4}$ by $\frac{1}{2}$.

26. There is $1\frac{3}{4}$ of a pizza left, and Kelly needs to share it equally with her brother. How much pizza does each child get?

27. There are $1\frac{3}{4}$ cups of chocolate chips in a bag. If a cookie recipe calls for a half-cup of chocolate chips, how many recipes can be made?

Thinking Critically

28. Jason and Kathy are driving in separate cars from home to Grandma's house, each driving the same route that passes through Midville. They both travel at 50 miles per hour, and Kathy left 4 hours ahead of Jason. They have the following conversation on their cell phones:

Jason: "Home is half as far away as Midville."
Kathy: "Grandma's house is half as far away as Midville."

How far is Grandma's from home? (*Suggestion:* Draw a picture.)

29. A bag contains red, green, and blue marbles. One-fifth of the marbles are red, there are $\frac{1}{3}$ as many blue marbles as green marbles, and there are 10 fewer red marbles than green marbles. Determine the number of marbles in the bag in these two approaches.

(a) Use algebraic reasoning (i.e., introduce variables, form equations, and then solve the equations).

(b) Use the set model of fractions, labeling the following Venn diagram:

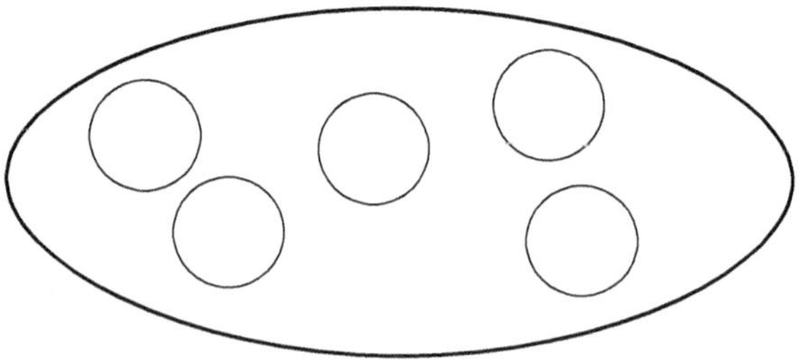

30. The positive rational numbers 3 and $1\frac{1}{2}$ are an interesting pair because their sum is equal to their product: $3 + \frac{3}{2} = \frac{6}{2} + \frac{3}{2} = \frac{9}{2}$ and $3 \cdot \frac{3}{2} = \frac{9}{2}$.

(a) Show that $3\frac{1}{2}$ and $1\frac{2}{5}$ have the same sum and product.

(b) Show that $2\frac{3}{5}$ and $1\frac{5}{8}$ have the same sum and product.

(c) If two positive rational numbers $\frac{a}{b}$ and $\frac{c}{d}$ have the same sum and product, what must be true of the sum of their reciprocals?

31. Three children had just cut their rectangular cake into three equal parts, as shown, to share, when a fourth friend joined them. Describe how to make one additional straight cut through the cake so that all four can share the cake equally.

32. **(a)** Verify that $\frac{21}{8} \div \frac{7}{4} = \frac{21 \div 7}{8 \div 4} = \frac{3}{2}$.

(b) Show that if $a \div c = m$ and $b \div d = n$, then $\frac{a}{b} \div \frac{c}{d} = \frac{a \div c}{b \div d} = \frac{m}{n}$.

33. **Illustrating Fraction Division with the Rectangular Area Model.** The following sequence of diagrams illustrates why $\frac{3}{8} \div \frac{2}{5} = \frac{15}{16}$, since it shows that the number of two-fifths in three-eighths is $\frac{15}{16}$.

A. Shade $\frac{3}{8}$ of a unit rectangle, using vertical lines.

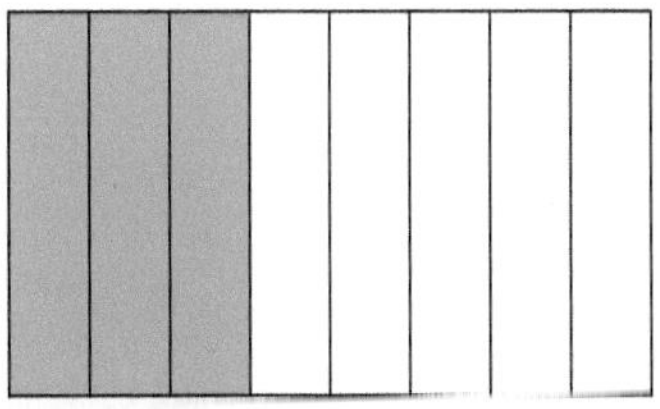

B. Use horizontal lines to create a rectangle of area $\frac{2}{5}$.

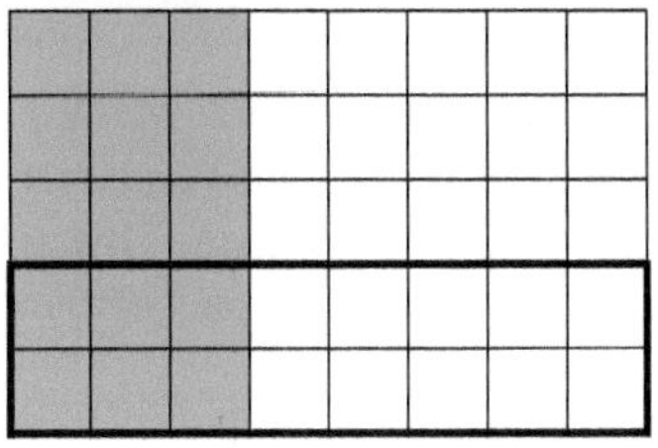

C. Move the shaded boxes into the $\frac{2}{5}$ rectangle.

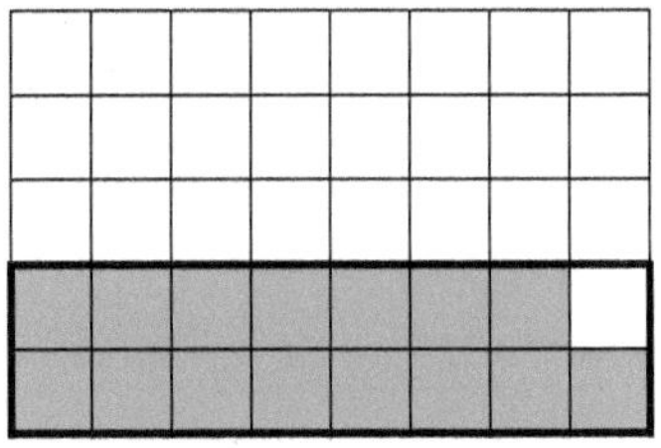

Make similar diagrams that illustrate each of these division problems:

(a) $\frac{2}{5} \div \frac{3}{4}$ **(b)** $\frac{3}{5} \div \frac{5}{6}$

(c) $\frac{7}{8} \div \frac{1}{3}$ [*Suggestion*: Make a row of rectangles, each of area $\frac{1}{3}$, to wholly or partially cover the shaded boxes.]

34. Solve this problem found in the Rhind papyrus: "A quantity and its $\frac{1}{7}$th added together become 19. What is the quantity?"

35. **Divvy.** The process called Divvy is like Diffy, except that the larger fraction is *divided* by the smaller. The first few rows of a sample Divvy are shown, where all of the fractions are positive.

(a) Continue to fill in additional rows, using a calculator if you like.

(b) Try Divvy with $\frac{2}{7}, \frac{4}{5}, \frac{3}{2}$, and $\frac{5}{6}$ in the first row.

Don't let complicated fractions put you off! Things should get better if you persist.

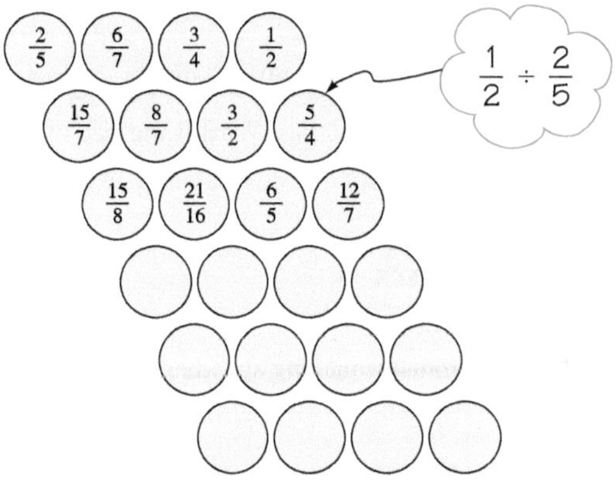

36. **(Writing)** For each fraction operation that follows, make up a realistic word problem whose solution requires the computation shown. Try to create an interesting and original situation.

(a) $\frac{4}{5} \times \frac{7}{8}$ **(b)** $\frac{9}{10} \div \frac{3}{5}$

Making Connections

37. At a certain university, a student's senior thesis is acceptable if at least $\frac{3}{4}$ of the student's committee votes in its favor. What is the smallest number of favorable votes needed to accept a thesis if the committee has 3 members? 4 members? 5 members? 6 members? 7 members? 8 members?

38. A sign on a rolled-up canvas says that the canvas contains 42 square yards. The width of the canvas, which is easily measured without unrolling, is 14 feet (or $4\frac{2}{3}$ yards). What is the length of the piece of canvas, in yards?

39. Tongue-and-groove decking boards are each $2\frac{1}{4}$ inches wide. How many boards must be placed side by side to build a deck 14 feet in width?

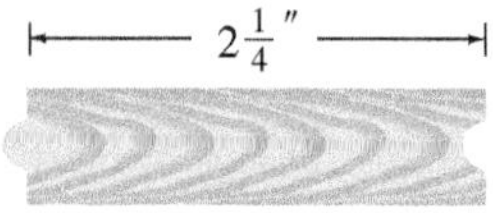

40. Six bows can be made from $1\frac{1}{2}$ yards of ribbon. How many bows can be made from $5\frac{3}{4}$ yards of ribbon?

41. Andre has 35 yards of material available to make aprons. Each apron requires $\frac{3}{4}$ yard. How many aprons can Andre make?

42. Gisela paid \$28 for a skirt that was "$\frac{1}{3}$ off." What was the original price of the skirt?

43. A soup recipe calls for $2\frac{3}{4}$ cups of chicken broth and will make enough to serve 8 people. How much broth is required if the recipe is modified to serve 6 people?

44. The label on a large 4-liter bottle claims to be the equivalent of $5\frac{1}{3}$ small bottles. What is the size of a small bottle?

State Assessments

45. (Grade 4)
Jennie spent 15 minutes picking strawberries from the plants that are within the loop shown below. She has asked her older brother Kyle how much time she can anticipate it will take to pick all of the strawberries in the plot.

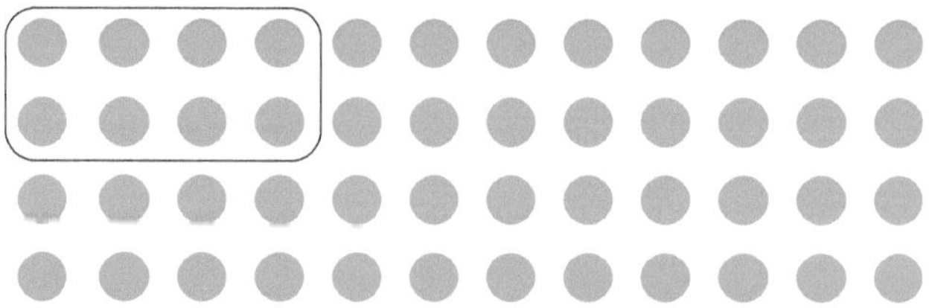

(a) What should Kyle tell Jennie about the total time to expect?

(b) What should Kyle say to Jennie to explain how he found his answer.

46. Writing (Grade 5)
Four children want to share seven large peanut butter cookies. How many cookies does each child get if the cookies are shared equally? Explain in words and pictures how to get your answer.

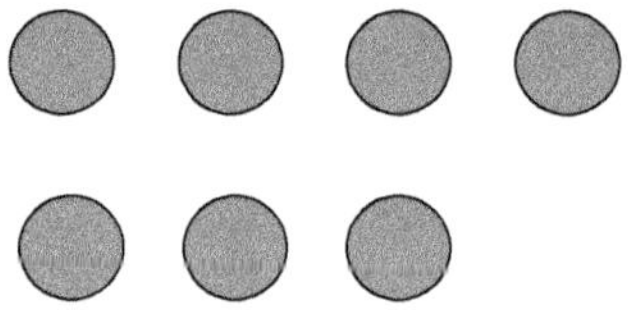

47. (Massachusetts, Grade 7)
It took a ball 1 minute to roll 90 feet. What was this ball's average rate of speed, in feet per second?

A. $\frac{2}{3}$ feet per second

B. $1\frac{1}{2}$ feet per second

C. 2 feet per second

D. 3 feet per second

6.4 The Rational Number System

This section explores the properties of the rational numbers. Many properties will be familiar, since the integers have the same properties. However, we will also discover some important new properties of rational numbers that have no counterpart in the integers. This section also gives techniques for estimation and computation and presents additional examples of the application of rational numbers to the solution of practical problems.

Properties of Addition and Subtraction

To add two rational numbers, we first represent each rational number by a fraction and then add the two fractions. For example, to add the rational numbers represented by $\frac{5}{6}$ and $\frac{3}{10}$, we could use the common denominator 30; thus,

$$\frac{5}{6} + \frac{3}{10} = \frac{25}{30} + \frac{9}{30} = \frac{34}{30}.$$

We could also have used the addition formula to find that

$$\frac{5}{6} + \frac{3}{10} = \frac{5 \cdot 10 + 6 \cdot 3}{6 \cdot 10} = \frac{50 + 18}{60} = \frac{68}{60}.$$

The two answers, namely, $\frac{34}{30}$ and $\frac{68}{60}$, are different fractions. However, they are equivalent fractions, and both represent the *same* rational number, $\frac{17}{15}$, when expressed by a fraction in simplest form. More generally, *any* two rational numbers have a unique rational number that is their sum. That is, the rational numbers are closed under addition.

It is also straightforward to check that addition in the rational numbers is commutative and associative. For example, $\frac{5}{6} + \frac{3}{10} = \frac{3}{10} + \frac{5}{6}$ and $\frac{3}{4} + \left(\frac{-1}{3} + \frac{2}{5}\right) = \left(\frac{3}{4} + \frac{-1}{3}\right) + \frac{2}{5}$. Similarly, 0 is the additive identity. For example, $\frac{7}{9} + 0 = \frac{7}{9}$, since $0 = \frac{0}{9}$.

The rational numbers share one more property with the integers: the existence of **negatives,** or **additive inverses.** We have the following definition:

DEFINITION Negative or Additive Inverse

Let r be a rational number represented by the fraction $\frac{a}{b}$. Its **negative,** or **additive inverse,** written $-r$, is the rational number represented by the fraction $\frac{-a}{b}$.

For example, $-\frac{4}{7} = \frac{-4}{7}$. This is the additive inverse of $\frac{4}{7}$, since

$$\frac{4}{7} + \left(-\frac{4}{7}\right) = \frac{4}{7} + \frac{-4}{7} = \frac{4 + (-4)}{7} = \frac{0}{7} = 0.$$

As another example, $-\left(-\frac{3}{4}\right) = -\left(\frac{-3}{4}\right) = \frac{-(-3)}{4} = \frac{3}{4}$, which illustrates the general property $-\left(-\frac{a}{b}\right) = \frac{a}{b}$.

A negative is also called an **opposite.** This term describes how a rational number and its negative are positioned on the number line: $-r$ is on the opposite side of 0 from r, at the same distance from 0.

The properties of addition on the rational numbers are listed in the following theorem:

THEOREM Properties of Addition of Rational Numbers

Let r, s, and t be rational numbers that are represented by the fractions $\frac{a}{b}$, $\frac{c}{d}$, and $\frac{e}{f}$. Then following properties hold:

Closure Property $r + s$ is a rational number; that is, $\frac{a}{b} + \frac{c}{d}$ is a fraction.

Commutative Property $r + s = s + r$; that is, $\frac{a}{b} + \frac{c}{d} = \frac{c}{d} + \frac{a}{b}$

Associative Property $(r + s) + t = r + (s + t)$; that is, $\left(\frac{a}{b} + \frac{c}{d}\right) + \frac{e}{f} = \frac{a}{b} + \left(\frac{c}{d} + \frac{e}{f}\right)$

Zero Is an Additive Identity $r + 0 = r$; that is, $\frac{a}{b} + 0 = \frac{a}{b}$

Existence of Additive Inverses $r + (-r) = 0$; that is, $\frac{a}{b} + \left(-\frac{a}{b}\right) = 0$, where $-\frac{a}{b} = \frac{-a}{b}$

In the integers, we discovered that subtraction was equivalent to addition of the negative. The same result holds for the rational numbers. Since the rational numbers are closed under addition and every rational number has a negative, the rational numbers are closed under subtraction.

THEOREM Subtraction Is Equivalent to Addition of the Negative

Let r and s be rational numbers represented by the fractions $\frac{a}{b}$ and $\frac{c}{d}$. Then $r - s = r + (-s)$; that is,

$$\frac{a}{b} - \frac{c}{d} = \frac{a}{b} + \left(-\frac{c}{d}\right) = \frac{ad - bc}{bd}.$$

Subtraction is neither commutative nor associative, as examples such as $\frac{1}{2} - \frac{1}{4} \neq \frac{1}{4} - \frac{1}{2}$ and $1 - \left(\frac{1}{2} - \frac{1}{4}\right) \neq \left(1 - \frac{1}{2}\right) - \frac{1}{4}$ show. This means that careful attention must be given to the order of the terms and the placement of parentheses.

EXAMPLE 6.18 Subtracting Rational Numbers

Compute the following differences:

(a) $\frac{3}{4} - \frac{7}{6}$ **(b)** $\frac{2}{3} - \frac{-9}{8}$ **(c)** $\left(-2\frac{1}{4}\right) - \left(4\frac{2}{3}\right)$

Solution

(a) $\frac{3}{4} - \frac{7}{6} = \frac{3 \cdot 6}{4 \cdot 6} - \frac{4 \cdot 7}{4 \cdot 6} = \frac{18 - 28}{24} = \frac{-10}{24} = \frac{-5}{12} = -\frac{5}{12}.$

(b) $\frac{2}{3} - \frac{-9}{8} = \frac{2}{3} + \frac{9}{8} = \frac{2 \cdot 8}{3 \cdot 8} + \frac{3 \cdot 9}{3 \cdot 8} = \frac{16}{24} + \frac{27}{24} = \frac{43}{24}.$

(c) $\left(-2\frac{1}{4}\right) - \left(4\frac{2}{3}\right) = (-2 - 4) - \left(\frac{1}{4} + \frac{2}{3}\right) = -6 - \left(\frac{3}{12} + \frac{8}{12}\right) = -6\frac{11}{12}.$

Properties of Multiplication and Division

Multiplication of rational numbers includes all of the properties of multiplication on the integers. For example, let's investigate the distributive property of multiplication over addition by considering a specific case:

$$\frac{2}{5} \cdot \left(\frac{3}{4} + \frac{7}{8}\right) = \frac{2}{5} \cdot \left(\frac{6}{8} + \frac{7}{8}\right) = \frac{2}{5} \cdot \frac{13}{8} = \frac{26}{40}.$$

Add first; then multiply.

$$\frac{2}{5} \cdot \frac{3}{4} + \frac{2}{5} \cdot \frac{7}{8} = \frac{6}{20} + \frac{14}{40} = \frac{12}{40} + \frac{14}{40} = \frac{26}{40}.$$

Multiply first; then add.

These computations show that multiplication by $\frac{2}{5}$ distributes over the sum $\frac{3}{4} + \frac{7}{8}$. A similar calculation proves the general distributive property $r \cdot (s + t) = r \cdot s + r \cdot t$; that is, $\frac{a}{b} \cdot \left(\frac{c}{d} + \frac{e}{f}\right) = \frac{a}{b} \cdot \frac{c}{d} + \frac{a}{b} \cdot \frac{e}{f}$.

An important new property of multiplication of rational numbers that is *not* true for the integers is the existence of multiplicative inverses. For example, the nonzero rational number $\frac{5}{8}$ has the multiplicative inverse $\frac{8}{5}$, since

$$\frac{5}{8} \cdot \frac{8}{5} = 1.$$

This property does not hold in the integers. For example, since there is no integer m for which $2 \cdot m = 1$, the integer 2 does not have a multiplicative inverse in the set of integers.

THEOREM Properties of Multiplication of Rational Numbers

Let r, s, and t be rational numbers represented by the fractions $\frac{a}{b}, \frac{c}{d}$, and $\frac{e}{f}$. Then the following properties hold:

Closure Property rs is a rational number; that is, $\frac{a}{b} \cdot \frac{c}{d}$ is a fraction.

Commutative Property $rs = sr$; that is, $\frac{a}{b} \cdot \frac{c}{d} = \frac{c}{d} \cdot \frac{a}{b}$

Associative Property $(rs)t = r(st)$; that is, $\left(\frac{a}{b} \cdot \frac{c}{d}\right) \cdot \frac{e}{f} = \frac{a}{b} \cdot \left(\frac{c}{d} \cdot \frac{e}{f}\right)$

Distributive Property of Multiplication over Addition and Subtraction

$r(s + t) = rs + rt$ and $r(s - t) = rs - rt$; that is,

$$\frac{a}{b} \cdot \left(\frac{c}{d} + \frac{e}{f}\right) = \frac{a}{b} \cdot \frac{c}{d} + \frac{a}{b} \cdot \frac{e}{f} \quad \text{and} \quad \frac{a}{b} \cdot \left(\frac{c}{d} - \frac{e}{f}\right) = \frac{a}{b} \cdot \frac{c}{d} - \frac{a}{b} \cdot \frac{e}{f}$$

Multiplication by 0 $0 \cdot r = 0$; that is, $0 \cdot \frac{a}{b} = 0$

One Is a Multiplicative Identity $1 \cdot r = r$; that is, $1 \cdot \frac{a}{b} = \frac{a}{b}$

Existence of Multiplicative Inverse If $r = \frac{a}{b} \neq 0$, then there is a unique rational number, namely, $r^{-1} = \frac{b}{a}$, for which $r \cdot r^{-1} = \frac{a}{b} \cdot \frac{b}{a} = 1$.

EXAMPLE 6.19

Solving an Equation with the Multiplicative Inverse

Consuela paid \$36 for a pair of shoes at a "one-fourth off" sale. What was the price of the shoes before the sale?

Solution

Let x be the original price of the shoes, in dollars. Consuela paid $\frac{3}{4}$ of this price, so $\frac{3}{4} \cdot x = 36$. To solve this equation for the unknown x, multiply both sides by the multiplicative inverse $\frac{4}{3}$ to get

$$\frac{4}{3} \cdot \frac{3}{4} \cdot x = \frac{4}{3} \cdot 36.$$

But $\frac{4}{3} \cdot \frac{3}{4} = 1$ by the multiplicative inverse property, so

$$1 \cdot x = \frac{4 \cdot 36}{3} = 48.$$

Since 1 is a multiplicative identity, it follows that $1 \cdot x = x = 48$. Therefore, the original price of the pair of shoes was \$48.

Properties of the Order Relation

The order relation has many useful properties that are not difficult to prove. The most useful are the **transitive property,** the **addition property,** the **multiplication property,** and the **trichotomy property.**

THEOREM Properties of the Order Relation on the Rational Numbers

Let r, s, and t be rational numbers represented by the fractions $\frac{a}{b}$, $\frac{c}{d}$, and $\frac{e}{f}$.

Transitive Property If $r < s$ and $s < t$, then $r < t$; that is,

$$\frac{a}{b} < \frac{c}{d} \text{ and } \frac{c}{d} < \frac{e}{f}, \text{ then } \frac{a}{b} < \frac{e}{f}.$$

Addition Property If $r < s$, then $r + t < s + t$; that is,

$$\frac{a}{b} < \frac{c}{d}, \text{ then } \frac{a}{b} + \frac{e}{f} < \frac{c}{d} + \frac{e}{f}.$$

Multiplication Property If $r < s$ and $t > 0$, then $rt < st$; that is,

$$\frac{a}{b} < \frac{c}{d} \text{ and } \frac{e}{f} > 0, \text{ then } \frac{a}{b} \cdot \frac{e}{f} < \frac{c}{d} \cdot \frac{e}{f}.$$

If $r < s$ and $t < 0$, then $rt > st$; that is,

$$\frac{a}{b} < \frac{c}{d} \text{ and } \frac{e}{f} < 0, \text{ then } \frac{a}{b} \cdot \frac{e}{f} > \frac{c}{d} \cdot \frac{e}{f}.$$

Trichotomy Property Exactly one of the following holds: $r < s$, $r = s$, or $r > s$, that is,

exactly one of $\frac{a}{b} < \frac{c}{d}$, $\frac{a}{b} = \frac{c}{d}$, or $\frac{a}{b} > \frac{c}{d}$ holds.

The Density Property of Rational Numbers

Since 1 and 2 are successive integers with $1 < 2$, there is no integer between 1 and 2. But consider the rational numbers $\frac{1}{2}$ and $\frac{2}{3}$, for which,

$$\frac{1}{2} < \frac{2}{3}.$$

Alternatively, using 6 as a common denominator, we have

$$\frac{3}{6} < \frac{4}{6},$$

or with 12 as a common denominator, we have

$$\frac{6}{12} < \frac{8}{12}.$$

In the last form, we see that $\frac{7}{12}$ is a rational number that is between $\frac{6}{12}$ and $\frac{8}{12}$. That is,

$$\frac{1}{2} < \frac{7}{12} < \frac{2}{3},$$

as shown in Figure 6.17.

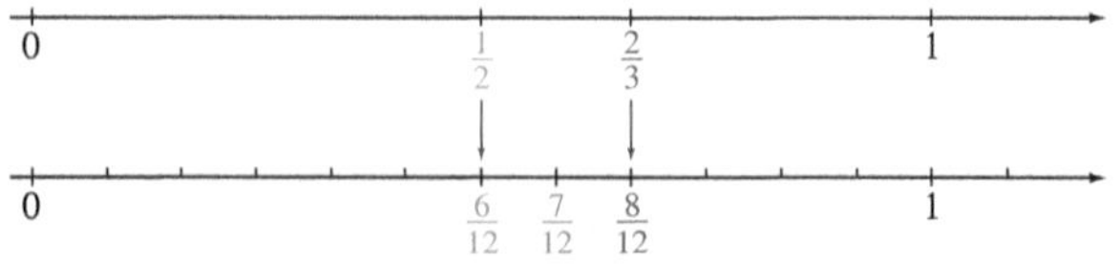

Figure 6.17
The rational number $\frac{7}{12}$ is between $\frac{1}{2}$ and $\frac{2}{3}$

The idea used to find a rational number that is between $\frac{1}{2}$ and $\frac{2}{3}$ can be extended to show that between *any* two rational numbers there is some other rational number. This interesting fact is called the **density property** of the rational numbers.

THEOREM The Density Property of Rational Numbers

Let r and s be any two rational numbers, with $r < s$. Then there is a rational number t between r and s; that is, $r < t < s$. Equivalently, if $\frac{a}{b} < \frac{c}{d}$, then there is a fraction $\frac{e}{f}$ for which $\frac{a}{b} < \frac{e}{f} < \frac{c}{d}$.

EXAMPLE 6.20

Finding Rational Numbers between Two Rational Numbers

Find a rational number between the two given rational numbers.

(a) $\frac{2}{3}$ and $\frac{3}{4}$ **(b)** $\frac{5}{12}$ and $\frac{3}{8}$

Solution

(a) Using $2 \cdot 3 \cdot 4 = 24$ as a common denominator, we have $\frac{2}{3} = \frac{16}{24}$ and $\frac{3}{4} = \frac{18}{24}$. Since $\frac{16}{24} < \frac{17}{24} < \frac{18}{24}$, it follows that $\frac{17}{24}$ is one answer.

(b) If we use 48 as a common denominator, we have $\frac{5}{12} = \frac{20}{48}$ and $\frac{3}{8} = \frac{18}{48}$. Thus, $\frac{19}{48}$ is one answer. Alternatively, we could use 480 as a common denominator, writing $\frac{5}{12} = \frac{200}{480}$ and $\frac{3}{8} = \frac{180}{480}$. This makes it clear that $\frac{181}{480}, \frac{182}{480}, \ldots, \frac{199}{480}$ are all rational numbers between $\frac{3}{8}$ and $\frac{5}{12}$. Using a larger common denominator allows us to identify even more rational numbers between the two given rationals.

Computations with Rational Numbers

To work confidently with rational numbers, it is important to develop skills in estimation, rounding, mental arithmetic, and efficient pencil-and-paper computation.

Estimations

In many applications, the exact fractional value can be rounded off to the nearest integer value; if more precision is required, values can be rounded to the nearest half, third, or quarter.

EXAMPLE 6.21

Using Rounding of Fractions to Convert a Brownie Recipe

Krishna's recipe, shown in the box, makes two dozen brownies. He'll need five dozen for the Math Day picnic, so he wants to adjust the quantities of his recipe. How should this be done?

Solution

Brownie Recipe (2 dozen)

- 4 squares chocolate
- $\frac{3}{4}$ cup butter
- 2 cups sugar
- 3 eggs
- 1 teaspoon vanilla
- 1 cup flour
- $1\frac{1}{4}$ cups chopped walnuts

Since Krishna needs $2\frac{1}{2}$ times the number of brownies given by his recipe, he will multiply the quantities by $\frac{5}{2}$. For example, $\frac{5}{2} \times 4 = 10$, so he will use 10 squares of chocolate. Similarly, he will use $\frac{5}{2} \times 2 = 5$ cups of sugar, $2\frac{1}{2}$ teaspoons of vanilla, and $2\frac{1}{2}$ cups of flour. However, $\frac{5}{2} \times \frac{3}{4} = \frac{15}{8} = 1\frac{7}{8}$, so Krishna will use "just short" of 2 cups of butter. Similarly, $\frac{5}{2} \times \frac{5}{4} = \frac{25}{8} = 3\frac{1}{8}$, so Krishna will use about 3 cups of chopped walnuts. Finally, $\frac{5}{2} \times 3 = \frac{15}{2} = 7\frac{1}{2}$, so he will use either 7 or 8 eggs.

Mental Arithmetic

By taking advantage of the properties, formulas, and algorithms associated with the various operations, it is often possible to simplify the computational process. Some useful strategies are demonstrated in the next example.

EXAMPLE 6.22

Computational Strategies for Rational Number Arithmetic

Perform these computations mentally:

(a) $53 - 29\frac{3}{5}$ **(b)** $\left(2\frac{1}{8} - 4\frac{2}{3}\right) + 7\frac{7}{8}$

(c) $\frac{7}{15} \times 90$ **(d)** $\frac{3}{8} \times 14 + \frac{3}{4} \times 25$

(e) $4\frac{1}{6} \times 18$ **(f)** $\frac{5}{8} \times \left(\frac{7}{10} \times \frac{24}{49}\right)$

Solution

(a) Adding $\frac{2}{5}$ to each term gives $53 - 29\frac{3}{5} = 53\frac{2}{5} - 30 = 23\frac{2}{5}$.

(b) $\left(2\frac{1}{8} - 4\frac{2}{3}\right) + 7\frac{7}{8} = \left(2 + 7 + \frac{1}{8} + \frac{7}{8}\right) - 4\frac{2}{3}$

$= 10 - 4\frac{2}{3} = 10\frac{1}{3} - 5 = 5\frac{1}{3}$.

(c) $\frac{7}{15} \times 90 = 7 \times \frac{90}{15} = 7 \times 6 = 42$.

(d) $\frac{3}{8} \times 14 + \frac{3}{4} \times 25 = \frac{3}{8} \times 14 + \frac{3}{8} \times 50$

$= \frac{3}{8} \times (14 + 50) = \frac{3}{8} \times 64$

$= 3 \times \frac{64}{8} = 3 \times 8 = 24$.

(e) $4\frac{1}{6} \times 18 = \left(4 + \frac{1}{6}\right) \times 18 = 4 \times 18 + \frac{1}{6} \times 18 = 72 + 3 = 75$.

(f) $\frac{5}{8} \times \left(\frac{7}{10} \times \frac{24}{49}\right) = \frac{5}{10} \times \frac{7}{49} \times \frac{24}{8} = \frac{1}{2} \times \frac{1}{7} \times 3 = \frac{3}{14}$.

EXAMPLE 6.23

Designing Wooden Stairs

A deck is 4′ 2″ above the surface of a patio. How many steps and risers should there be in a stairway that connects the deck to the patio? The decking is $1\frac{1}{2}''$ thick, the stair treads are $1\frac{1}{8}''$ thick, and the steps should each rise the same distance, from one to the next. Calculate the vertical dimension of each riser.

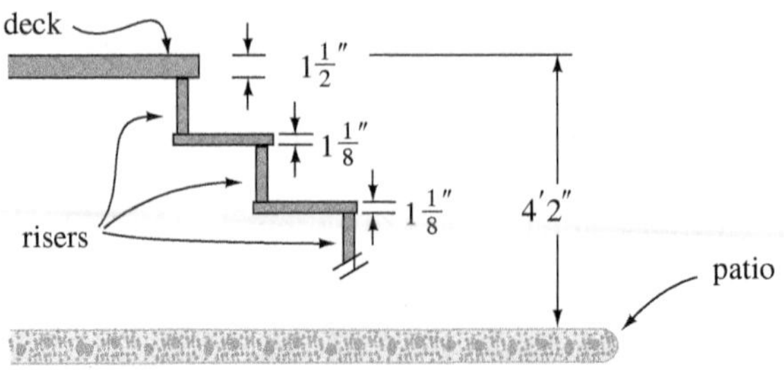

Solution

Understand the Problem

We have been given some important dimensions, including the thickness of the treads and the deck. We have *not* been told what dimension to cut the risers. This, we see from the figure, depends on the number of steps we choose, with a lower rise corresponding to a greater number of steps. We must choose a number of steps that feels natural to walk on. We must also be sure that each step rises the same amount.

Devise a Plan

If we know what vertical rise from step to step is customary, we can first get a reasonable estimate of the number of steps to use. Once it is agreed what number of steps to incorporate into the design, we can calculate the height of each riser. The riser meeting the deck must be adjusted to account for the decking being thicker than the stair tread.

Carry Out the Plan

A brief survey of existing stairways shows that most steps rise 5″ to 7″ from one to the next. Since $4'\,2'' = 50''$, $\frac{50}{5} = 10$, and $\frac{50}{7} = 7\frac{1}{7}$, then 8, or perhaps 9, steps will work well, including the step onto the deck itself. Let's choose 8. Then $\frac{50''}{8} = 6\frac{2''}{8} = 6\frac{1''}{4}$; that is, each combination of riser plus tread is to be $6\frac{1''}{4}$. Since the treads are $1\frac{1''}{8}$ thick, the first seven risers require the boards to be cut $6\frac{1''}{4} - 1\frac{1''}{8} = 5\frac{1''}{8}$ wide. Since the decking is $1\frac{1''}{2}$ thick, the uppermost riser is $6\frac{1''}{4} - 1\frac{1''}{2} = 4\frac{3''}{4}$ wide.

Look Back

If we are concerned that the steps will take up too much room on the patio, we might use a steeper, 7-step design. Each riser plus tread would then rise $7\frac{1''}{7}$. Experienced carpenters would think of this value as a "hair" more than $7\frac{1''}{8}$ and cut the lower six risers from a board 8″ wide, for a total rise of $7\frac{1''}{8}$. The last riser is cut to make any final small adjustment.

6.4 Problem Set

Understanding Concepts

1. Explain what properties of addition of rational numbers can be used to make this sum very easy to compute:

$$\left(3\frac{1}{5} + 2\frac{2}{5}\right) + 8\frac{1}{5}$$

2. What properties can you use to make these computations easy?

(a) $\frac{2}{5} + \left(\frac{3}{5} + \frac{2}{3}\right)$

(b) $\frac{1}{4} + \left(\frac{2}{5} + \frac{3}{4}\right)$

(c) $\frac{2}{3}\cdot\frac{1}{8} + \frac{2}{3}\cdot\frac{7}{8}$

(d) $\frac{3}{4}\cdot\left(\frac{5}{9}\cdot\frac{4}{3}\right)$

3. Find the negatives (i.e., the additive inverses) of the given rational numbers. Show each number and its negative on the number line.

(a) $\frac{4}{5}$

(b) $\frac{-3}{2}$

(c) $\frac{8}{-3}$

(d) $\frac{4}{2}$

4. Compute the given sums of rational numbers. Explain what properties of rational numbers you find useful. Express your answers in simplest form.

(a) $\frac{1}{6} + \frac{2}{-3}$ (b) $\frac{-4}{5} + \frac{3}{2}$

(c) $\frac{9}{4} + \frac{-7}{8}$ (d) $\frac{3}{4} + \frac{-5}{8} + \frac{7}{-12}$

5. Compute the given differences of rational numbers. Explain what properties you find useful. Express your answers in simplest form.

(a) $\frac{2}{5} - \frac{3}{4}$ (b) $\frac{-6}{7} - \frac{4}{7}$ (c) $\frac{3}{8} - \frac{1}{12}$

6. Compute the given differences of rational numbers. Explain properties you find useful. Express your answers in simplest form.

(a) $3\frac{2}{5} - \frac{7}{10}$ (b) $2\frac{1}{3} - 5\frac{3}{4}$ (c) $-4\frac{2}{3} - \frac{-19}{6}$

7. Calculate the given products of rational numbers. Explain what properties of rational numbers you find useful. Express your answers in simplest form.

(a) $\frac{3}{5} \cdot \frac{7}{8} \cdot \left(\frac{5}{3}\right)$ (b) $\frac{-2}{7} \cdot \frac{3}{4}$ (c) $\frac{-4}{3} \cdot \frac{6}{-16}$

8. Calculate the given products of rational numbers. Explain what properties of rational numbers you find useful. Express your answers in simplest form.

(a) $3\frac{1}{8} \cdot 2\frac{1}{5} \cdot (40)$

(b) $\frac{14}{15} \cdot \frac{60}{7}$

(c) $\left(\frac{4}{11} \cdot \frac{22}{7}\right) \cdot \left(\frac{-3}{8}\right)$

9. Find the reciprocals (i.e., the multiplicative inverses) of the given rational numbers. Show each number and its reciprocal on the number line.

(a) $\frac{3}{2}$ (b) $\frac{4}{-9}$ (c) $\frac{-4}{-11}$

10. Find the reciprocals (i.e., the multiplicative inverses) of the given rational numbers. Show each number and its reciprocal on the number line.

(a) 5 (b) −2 (c) $2\frac{1}{2}$

11. Use the properties of the operations of rational number arithmetic to perform these calculations. Express your answers in simplest form.

(a) $\frac{2}{3} \cdot \frac{4}{7} + \frac{2}{3} \cdot \frac{3}{7}$ (b) $\frac{4}{5} \cdot \frac{2}{3} - \frac{3}{10} \cdot \frac{2}{3}$

12. Use the properties of the operations of rational number arithmetic to perform these calculations. Express your answer in simplest form.

(a) $\frac{4}{7} \cdot \frac{3}{2} - \frac{4}{7} \cdot \frac{6}{4}$ (b) $\left(\frac{4}{7} \cdot \frac{2}{5}\right) \div \frac{2}{7}$

13. Justify each step in the proof of the distributive property of multiplication over addition that follows. In this case, the two addends have a common denominator, but the operation is always possible for any rational addends by using a common denominator for the two fractions.

$$\frac{a}{b} \cdot \left(\frac{c}{d} + \frac{e}{d}\right) = \frac{a}{b} \cdot \frac{c + e}{d} \quad \text{(a) Why?}$$

$$= \frac{a \cdot (c + e)}{b \cdot d} \quad \text{(b) Why?}$$

$$= \frac{a \cdot c + a \cdot e}{b \cdot d} \quad \text{(c) Why?}$$

$$= \frac{a \cdot c}{b \cdot d} + \frac{a \cdot e}{b \cdot d} \quad \text{(d) Why?}$$

$$= \frac{a}{b} \cdot \frac{c}{d} + \frac{a}{b} \cdot \frac{e}{d} \quad \text{(e) Why?}$$

14. If $\frac{a}{b} \cdot \frac{4}{7} = \frac{2}{3}$, what is $\frac{a}{b}$? Carefully explain how you obtain your answer. What properties do you use?

15. Solve each equation for the rational number x. Show your steps and explain what property justifies each step.

(a) $4x + 3 = 0$ (b) $x + \frac{3}{4} = \frac{7}{8}$

(c) $\frac{2}{3}x + \frac{4}{5} = 0$ (d) $3\left(x + \frac{1}{8}\right) = -\frac{2}{3}$

16. Arrange each group of rational numbers in increasing order. Show, at least approximately, the numbers on the number line.

(a) $\frac{4}{5}, -\frac{1}{5}, \frac{2}{5}$ (b) $\frac{-3}{7}, \frac{4}{7}, \frac{-5}{7}$

(c) $\frac{3}{8}, \frac{1}{2}, \frac{3}{4}$ (d) $\frac{-7}{12}, \frac{-2}{3}, \frac{3}{-4}$

17. Verify these inequalities:

(a) $\frac{-4}{5} < \frac{-3}{4}$

(b) $\frac{1}{10} > -\frac{1}{4}$

(c) $\frac{-19}{60} > \frac{-1}{3}$

18. Solve the inequalities that follow. Show all your steps.

(a) $x + \frac{2}{3} > -\frac{1}{3}$ (b) $x - \left(-\frac{3}{4}\right) < \frac{1}{4}$

(c) $\frac{3}{4}x < -\frac{1}{2}$ (d) $-\frac{2}{5}x + \frac{1}{5} > -1$

19. Find a rational number that is between the two given rational numbers.

(a) $\frac{4}{9}$ and $\frac{6}{11}$ (b) $\frac{1}{9}$ and $\frac{1}{10}$

(c) $\frac{14}{23}$ and $\frac{7}{12}$ (d) $\frac{141}{568}$ and $\frac{183}{737}$

20. Find three rational numbers between $\frac{1}{4}$ and $\frac{2}{5}$.

21. For each given rational number, choose the best estimate from the list provided.

(a) $\frac{104}{391}$ is approximately $\frac{1}{3}, \frac{1}{4}$, or $\frac{1}{2}$.

(b) $\frac{217}{340}$ is approximately $\frac{1}{3}, \frac{1}{2}$, or $\frac{2}{3}$.

(c) $\frac{-193}{211}$ is approximately $-\frac{1}{2}, -1, 1$, or $\frac{1}{2}$.

(d) $\frac{453}{307}$ is approximately $\frac{3}{4}, 1, 1\frac{1}{3}$, or $1\frac{1}{2}$.

22. Use estimations to choose the best approximation of the following expressions:

(a) $3\frac{19}{40} + 5\frac{11}{19}$ is approximately $8, 8\frac{1}{2}, 9$, or $9\frac{1}{2}$.

(b) $2\frac{6}{19} + 5\frac{1}{3} - 4\frac{7}{20}$ is approximately $3, 3\frac{1}{3}, 3\frac{1}{4}$, or 4.

(c) $17\frac{8}{9} \div 5\frac{10}{11}$ is approximately $2, 3, 3\frac{1}{2}$, or 4.

23. **(Writing)** First do the given calculations mentally. Then use words and equations to explain your strategy.

(a) $\frac{1}{2} + \frac{1}{4} + \frac{3}{4}$

(b) $\frac{5}{2} \cdot \left(\frac{2}{5} - \frac{2}{10}\right)$

(c) $\frac{3}{4} \cdot \frac{12}{15}$

(d) $\frac{2}{9} \div \frac{1}{3}$

24. **(Writing)** First do the given calculations mentally. Then use words and equations to explain your strategy.

(a) $2\frac{2}{3} \times 15$

(b) $3\frac{1}{5} - 1\frac{1}{4} + 7\frac{4}{5}$

(c) $6\frac{1}{8} - 8\frac{1}{4}$

(d) $\frac{2}{3} \cdot \frac{7}{4} - \frac{2}{3} \cdot \frac{1}{4}$

25. (a) Hal owns $3\frac{1}{2}$ acres and just purchased an adjacent plot of $\frac{3}{4}$ acres. The answer is $4\frac{1}{4}$. What is the question?

(b) Janet lives $1\frac{3}{4}$ miles from school. On the way to school, she stopped to walk the rest of the way with Brian, who lives $\frac{1}{2}$ mile from school. The answer is $1\frac{1}{4}$. What is the question?

(c) A family room is $5\frac{1}{2}$ yards wide and 6 yards long. The answer is 33 square yards. What is the question?

(d) Clea made $3\frac{1}{2}$ gallons of ginger ale, which she intends to bottle in "fifths" (i.e., bottles that contain a fifth of a gallon). The answer is 17 full bottles and $\frac{1}{2}$ of another bottle. What is the question?

26. Invent interesting word problems that lead you to the given expression.

(a) $\frac{3}{4} + \frac{1}{2}$

(b) $4\frac{1}{2} \div \frac{3}{4}$

(c) $7\frac{2}{3} \times \frac{1}{4}$

Into the Classroom

The area model of fractions is a useful visualization of the meaning of fractions. The unit of area is a simple shape that is partitioned into b congruent pieces and then a of the pieces are shaded to give a pictorial representation of the fraction $\frac{a}{b}$. A challenging problem is to reverse the model: Shade part of a figure of unit area and determine the fraction that describes this area. The following two problems can initiate a lively class discussion.

27. A unit square is partitioned into the seven tangram pieces A, B, C, D, E, F, and G. What fraction gives the area of each piece?

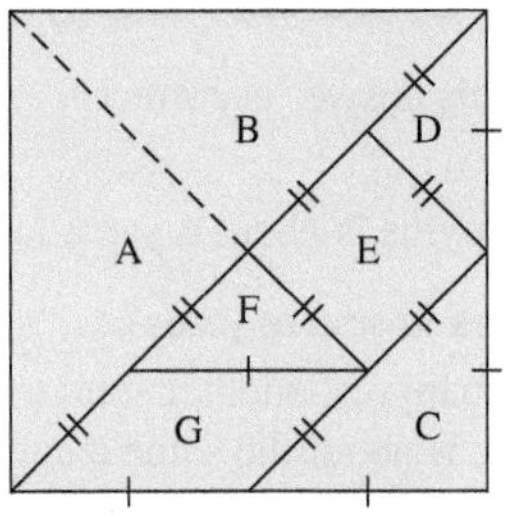

28. Start with a square piece of paper and join each corner to the midpoint of the opposite side, as shown in the left-hand figure. Shade the slanted square that is formed inside the original square. What fraction of the large square have you shaded? It will help to cut the triangular pieces as shown on the right and reattach them to form some new squares.

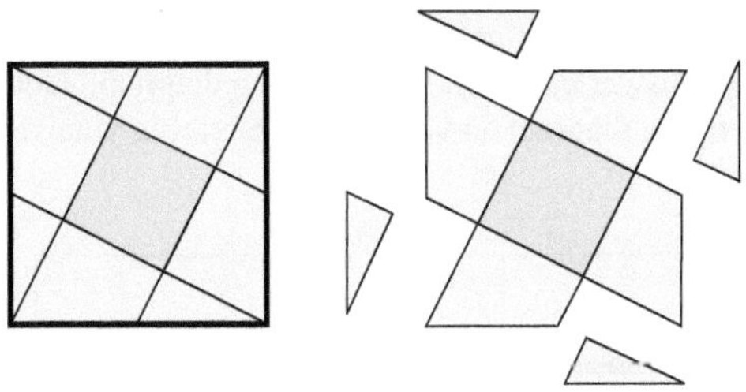

Responding to Students

29. When asked to evaluate the sum $\frac{1}{3} + \frac{3}{5}$, a student claimed that the answer, when simplified, is $\frac{1}{2}$. How do you suspect the student arrived at this answer? Describe what you might do to help this student's understanding.

30. When asked to evaluate the difference $\frac{9}{11} - \frac{3}{22}$, a student gave the answer $\frac{165}{242}$. What would you suggest to the student?

Thinking Critically

In 1858, the Scotsman H. A. Rhind purchased an Egyptian papyrus copied by the scribe A'h-mose (or Ahmes) in 1650 B.C. from an earlier document written in about 1850 B.C. Problems 31–34 are adapted from the Rhind papyrus.

31. "A quantity, its $\frac{2}{3}$, its $\frac{1}{2}$, its $\frac{1}{7}$, its whole, amount to 33." That is, in modern notation, $\frac{2}{3}x + \frac{1}{2}x + \frac{1}{7}x + x = 33$. What is the quantity?

32. "Divide 100 loaves among five men in such a way that the share received shall be in arithmetic progression and that one-seventh of the sum of the largest three shares shall be equal to the sum of the smallest two." (*Suggestion:* Denote the shares as $s, s + d, s + 2d, s + 3d$, and $s + 4d$.)

33. The ancient Egyptians measured the steepness of a slope by the fraction $\frac{x}{y}$, where x is the number of hands of horizontal "run" and y is the number of cubits of vertical "rise." Seven hands form a cubit. Problem 56 of the Rhind papyrus asks for the steepness of the face of a pyramid 250 cubits high and having a square base 360 cubits on a side. The papyrus gives the answer $5\frac{1}{25}$. Show why this answer is correct.

34. **(a)** The Rhind papyrus is about 6 yards long and $\frac{1}{3}$ yard wide. What is its area in square yards?

(b) The Moscow papyrus, another source of mathematics of ancient Egypt, is about the same length as the Rhind papyrus, but has $\frac{1}{4}$ the area. What is the width of the Moscow papyrus?

35. (**Writing**) Three quarters of the pigeons occupy two-thirds of the pigeonholes in a pigeon house (one pigeon per hole), and the other one-quarter of the pigeons are flying around. If all the pigeons were in the pigeonholes, just five holes would be empty. How many pigeons and how many holes are there? Give an answer based on the following diagram, adding labels to the diagram and carefully describing your reasoning in words:

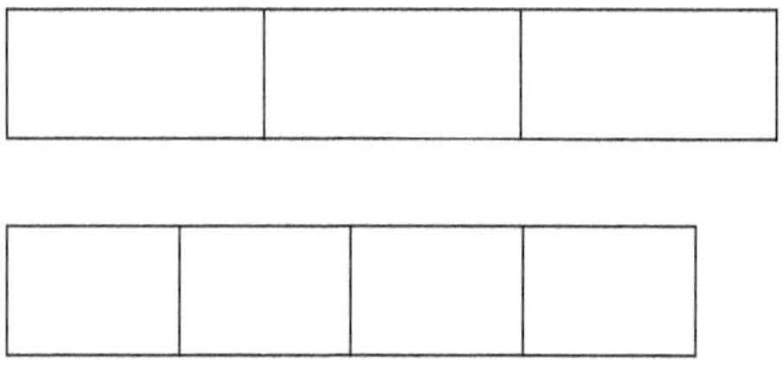

36. (**Writing**) A bag contains red and blue marbles. One-sixth of the marbles are blue, and there are 20 more red marbles than blue marbles. How many red marbles and how many blue marbles are there?

(a) Give an answer based on the following set diagram, adding labels and describing how to use the diagram to obtain an answer:

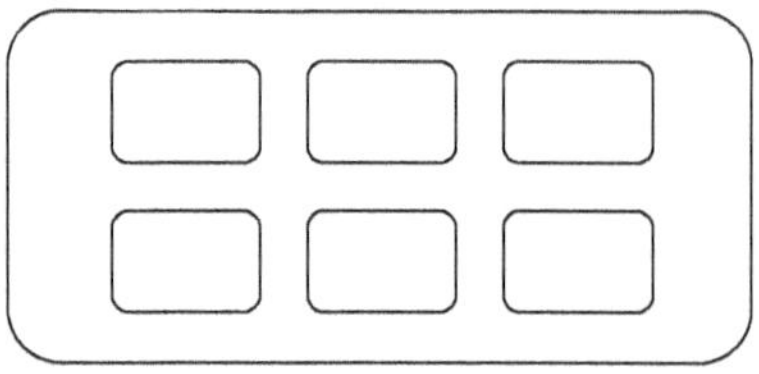

(b) Give an answer based on finding and solving equations.

37. **(a)** Let x, y, and z be arbitrary rational numbers. Show that the sums of the three entries in every row, column, and diagonal in the Magic Square are the same.

$x - z$	$x - y + z$	$x + y$
$x + y + z$	x	$x - y - z$
$x - y$	$x + y - z$	$x + z$

(b) Let $x = \frac{1}{2}$, $y = \frac{1}{3}$, and $z = \frac{1}{4}$. Find the corresponding Magic Square.

(c) Here is a partial Magic Square that corresponds to the form shown in part (a). Find x, y, and z, and then fill in the remaining entries of the square.

$-\frac{5}{12}$		1
$-\frac{1}{3}$		

(*Hint:* $2x = (x + y) + (x - y)$.)

Making Connections

38. The Fahrenheit and Celsius temperature scales are related by the formula $F = \frac{9}{5}C + 32$. For example, a temperature of 20° Celsius corresponds to 68° Fahrenheit, since

$$\frac{9}{5} \cdot 20 + 32 = 36 + 32 = 68.$$

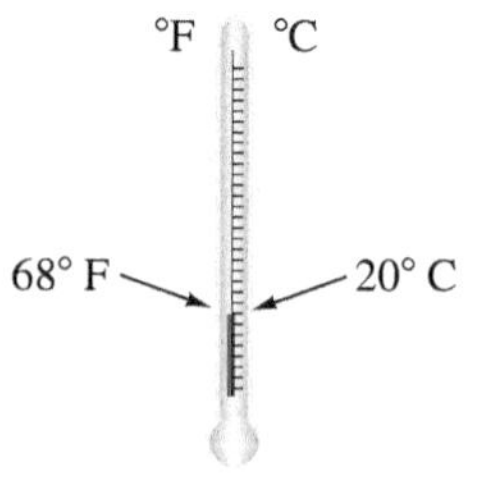

(a) Derive a formula for C as it depends on F. Deduce the equivalent formula $C = \frac{5}{9}(F - 32)$ relating the two temperature scales.

(b) Fill in the missing entries in this table:

°C	−40°		0°	10°	20°		
°F		−13°			68°	104°	212°

(c) Electronic signboards frequently give the temperature in both Fahrenheit and Celsius degrees. What is the temperature if both readings are the same (i.e., F = C)? The negative of one another (i.e., F = −C)? Answer the latter question with an exact rational number and the approximate integer that would be seen on the signboard.

39. Earth revolves around the sun once in 365 days, 5 hours, 48 minutes, and 46 seconds. Since this *solar* year is more than 365 days long, it is important to have 366-day-long *leap* years to keep the seasons in the same months of the calendar year.

(a) The solar year is very nearly 365 days and 6 hours, or $365\frac{6}{24} = 365\frac{1}{4}$ days, long. Explain why having the years divisible by 4 as leap years is a good rule to decide when leap years should occur.

(b) Show that a solar year is $365\frac{20,926}{24 \cdot 60 \cdot 60}$ days long.

(c) Estimate the value given in part (b) with $365\frac{20,952}{24 \cdot 60 \cdot 60}$, and then show that this number can also be written as $365\frac{1}{4} - \frac{1}{100} + \frac{1}{400}$.

(d) What rule for choosing leap years is suggested by the approximation $365\frac{1}{4} - \frac{1}{100} + \frac{1}{400}$ to the solar year? (*Hint:* 1900 was not a leap year, but 2000 was a leap year.)

40. Fractions have a prominent place in music, where the time value of a note is given as a fraction of a whole note. The note values and their corresponding fractions are shown in this table:

Name	Whole	Half	Quarter	Eighth	Sixteenth
Note					
Time Value	1	$\frac{1}{2}$	$\frac{1}{4}$	$\frac{1}{8}$	$\frac{1}{16}$

The time signature of the music can be interpreted as a fraction that gives the duration of the measure. Here are two examples:

$\frac{3}{4} = \frac{1}{4} + \frac{1}{2}$ $\qquad$ $\frac{6}{8} = \frac{1}{4} + \frac{1}{8} + \frac{2}{16} + \frac{1}{4}$

In each measure shown, fill in the upper number of the time signature by adding the note values shown in the measure.

(a) 

(b) 

(c) 

41. (**Writing**) Design a stairway that connects the patio to the deck. (See Example 6.23.)

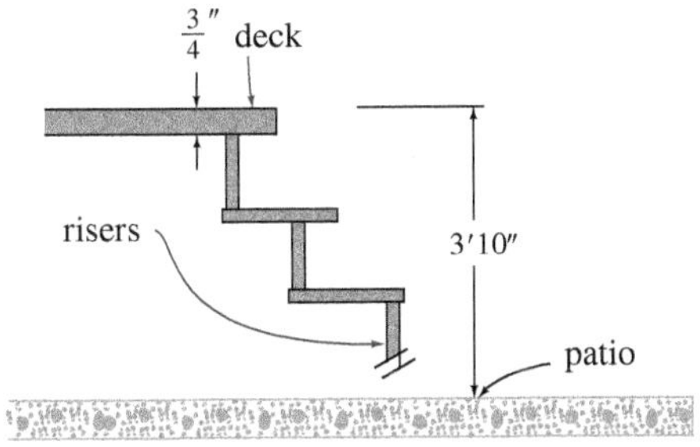

Assume that the tread of each stair is $\frac{1}{2}$″ thick.

42. This recipe makes six dozen cookies. Adjust the quantities to have a recipe for four dozen cookies.

1 cup shortening	$1\frac{1}{2}$ cup sugar
1 tsp baking soda	3 eggs
3 cups unsifted flour	$\frac{1}{2}$ tsp salt
9 oz mincemeat	

43. Anja has a box of photographic print paper. Each sheet measures 8 by 10 inches. She could easily cut a sheet into four 4-by-5-inch rectangles. However, to save money, she wants to get six prints, all the same size, from each 8-by-10-inch sheet. Describe how she can cut the paper.

44. Krystoff has an 8′ piece of picture frame molding, shown in cross section as follows:

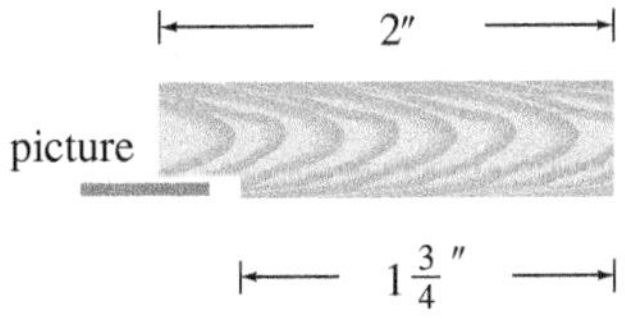

Is this 8′ piece sufficient to frame a 16″-by-20″ picture? Allow for saw cuts and some extra space to ensure that the picture fits easily into the frame.

State Assessments

45. (Grade 5)

Which figure shows $\frac{2}{5}$ has been shaded?

A. 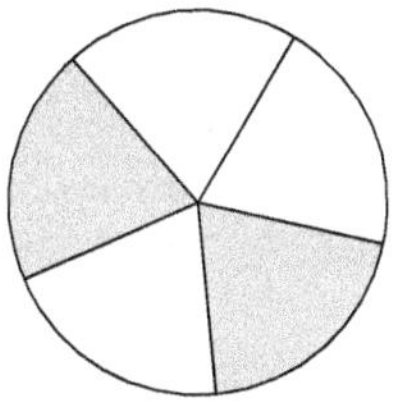

B. 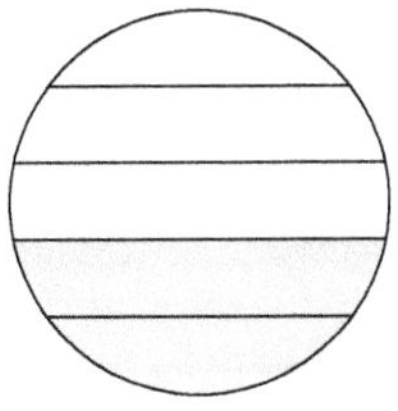

C. 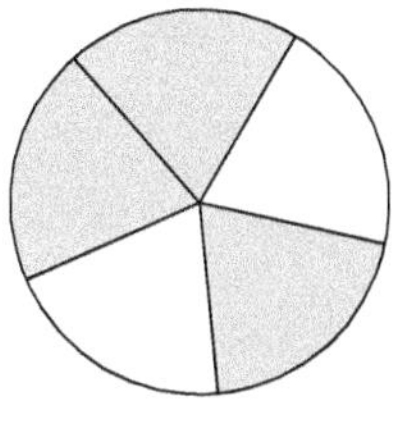

D. 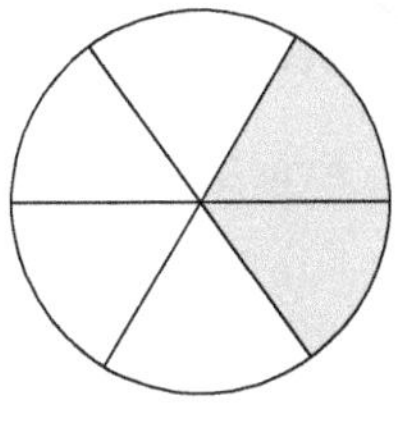

46. **(Writing)** (Grade 5)
Becky cut a rectangular dish of brownies into four pieces as shown below. Not all of the pieces are the same shape, but she knows that the pieces are the same size.

Explain how Becky knows this.

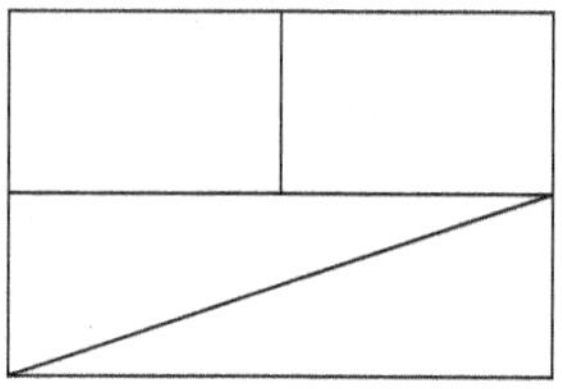

47. (Grade 7)

In Hank's garden, $\frac{3}{4}$ of his tomato plants have blossoms, and of these $\frac{2}{3}$ have green tomatoes. How many of Hank's tomato plants have green tomatoes?

A. More than $\frac{2}{3}$

B. Between $\frac{2}{3}$ and $\frac{3}{4}$

C. Less than $\frac{1}{2}$

D. Exactly $\frac{1}{2}$

The Chapter in Relation to Future Teachers

If elementary school teachers were asked to list topics in the mathematics curriculum that they felt were the most challenging, both for themselves and their students, fractions will often head their lists. Unlike the integers, which are usually associated with the familiar idea of counting, fractions play a much wider range of roles. Here is a list of three roles highlighted in this chapter:

- *Fractions are an extension of measurement.* The unit is subdivided, and the measurement of a quantity (distance, area, time, etc.) is given by counting both the number of whole units and the number of subdivisions of a unit.
- *Fractions represent the rational numbers.* Indeed, any given rational number can be represented by any of infinitely many equivalent fractions. Moreover, the rational numbers *are* a system of numbers. That is, like the integers, the rational numbers can be identified with a point or distance on the number line, their arithmetic operations obey certain algebraic properties, and the result of an arithmetic operation $+$, $\times$, $-$, or $\div$ is computed by an accompanying formula.
- *Fractions are operators.* That is, a fraction is an action as opposed to an object, much in the way a verb is opposed to a noun. For example, at a storewide $\frac{1}{4}$-off sale, we would determine the price of any item by multiplying its current price by $\frac{3}{4}$. Now $\frac{3}{4}$ has become an operator.

Connecting the Standards for Mathematical Practice and the Standards for Mathematical Content.

These meanings of fractions have been illustrated throughout the chapter by physical models and visualizations: paper folding, colored regions, fraction strips, fraction circles, rectangular arrays, and the like. It is important to understand how the representations and models are of value as instructional aids.

The **Common Core** makes it clear why instruction must focus on conveying understanding: Students who lack understanding of a topic may rely on procedures too heavily. Without a flexible base from which to work, they may be less likely to consider analogous problems, represent problems coherently, justify conclusions, apply the mathematics to practical situations, use technology mindfully to work with the mathematics, explain the mathematics accurately to other students, step back for an overview, or deviate from a known procedure to find a shortcut.

Our work with fractions and rational numbers is not yet complete. In Chapter 7, we will explore which decimal numbers represent rational numbers, see how fractions are related to ratios and proportional reasoning, and learn that a percent is a fraction with 100 as its denominator.

Chapter 6 Summary

Section 6.1 The Basic Concepts of Fractions and Rational Numbers	Page Reference
CONCEPTS	
• **Fraction:** A number of the form $\frac{a}{b}$ that conveys a quantity obtained by splitting a unit into b equal parts, and then taking a of those parts.	265

- **Colored regions, or area model:** The unit is a geometric figure such as a polygon or disc, sectioned into b parts of equal area, of which a parts are colored to represent the fraction $\frac{a}{b}$. 266
- **Fraction strip:** The unit is a rectangle partitioned by vertical lines that partition it into b identical subrectangles, of which a are shaded to represent $\frac{a}{b}$. 266
- **Fraction circle:** The unit is a circle partitioned into b congruent sectors, so that shading a of the sectors gives a representation of the fraction $\frac{a}{b}$. 266
- **Set model:** The unit is a set of b elements, so that a subset of a elements represents the fraction $\frac{a}{b}$. 267
- **Number line:** Each unit interval of the integer number line is partitioned into b segments of the same length, so that the fraction $\frac{a}{b}$ is the length of a jump by a of these segments. Moreover, the position of the point at the distance of this jump from 0 is also a representation of $\frac{a}{b}$. 267
- **Strategies for ordering fractions:** Look for a common denominator. 273

DEFINITIONS

- A **fraction** is a pair of integers a and b, $b \neq 0$, written a/b or $\frac{a}{b}$. 265
- In the expression $\frac{a}{b}$, a is **the numerator** and b is the **denominator.** 265
- **Equivalent fractions** are fractions that express the same quantity. 268
- A fraction $\frac{a}{b}$ is in **simplest form** if a and b have no common divisor larger than 1 and b is positive. 270
- Fractions with the same denominator are said to have a **common denominator.** 271
- The **least common denominator** is the least common multiple of the denominators of two (or more) fractions. 271
- A **rational number** is any number that can be represented by a fraction or any of its equivalent fractions. 272

THEOREM

- **Cross-product property:** The fractions $\frac{a}{b}$ and $\frac{c}{d}$ are equivalent if and only if $ad = bc$. 269

PROPERTY

- **Fundamental law of fractions:** Let $\frac{a}{b}$ be a fraction. Then $\frac{a}{b} = \frac{an}{bn}$, for any integer $n \neq 0$. 268

NOTATION

- **Set of rational numbers, Q:** The set of all numbers that can be expressed as the fraction a/b where a and b are integers and b is not zero. 272

Section 6.2 Addition and Subtraction of Fractions — Page Reference

DEFINITIONS

- **Addition of fractions:** For fractions $\frac{a}{b}$ and $\frac{c}{d}$ with the common denominator b, let $\frac{a}{b} + \frac{c}{b} = \frac{a+c}{b}$. For fractions $\frac{a}{b}$ and $\frac{c}{d}$ with unlike denominators b and d, use a common denominator such as bd so that $\frac{a}{b} + \frac{c}{d} = \frac{ad}{bd} + \frac{bc}{bd} = \frac{ad+bc}{bd}$. — 282
- **Subtraction of fractions, $\frac{a}{b} - \frac{c}{d}$:** Let $\frac{a}{b}$ and $\frac{c}{d}$ be fractions. Then $\frac{a}{b} - \frac{c}{d} = \frac{e}{f}$ if and only if $\frac{a}{b} = \frac{c}{d} + \frac{e}{f}$. This implies the formula $\frac{a}{b} - \frac{c}{d} = \frac{ad-bc}{bd}$. — 287
- A **mixed number** is the sum of a natural number and a positive fraction. For example, $2\frac{3}{4}$ is a mixed number. — 284
- A **proper fraction** is a fraction a/b for which $0 \le |a| < b$. — 285

Section 6.3 Multiplication and Division of Fractions — Page Reference

CONCEPTS

- **Multiplication:** Extending the area model of multiplication motivates the definition of multiplication of fractions: $\frac{a}{b} \cdot \frac{c}{d} = \frac{ac}{bd}$. — 291
- **Division:** Given fractions $\frac{a}{b}$ and $\frac{c}{d} \neq 0$, the division $\frac{a}{b} \div \frac{c}{d}$ represents: — 296
  - **(a)** the fractional number of times $\frac{c}{d}$ can be removed from $\frac{a}{b}$ (repeated-subtraction, or grouping, model)
  - **(b)** the fractional amount per unit when the quantity $\frac{a}{b}$ is partitioned into $\frac{c}{d}$ parts (sharing, or partitive, model)
  - **(c)** the missing fractional factor $x = \frac{m}{n}$ for which $\frac{a}{b} = x \cdot \frac{c}{d}$ (missing-factor model)

DEFINITIONS

- **Multiplication of fractions, $\frac{a}{b} \times \frac{c}{d}$:** Let $\frac{a}{b}$ and $\frac{c}{d}$ be fractions. Then their product is given by $\frac{a}{b} \cdot \frac{c}{d} = \frac{ac}{bd}$. — 294
- **Division of fractions, $\frac{a}{b} \div \frac{c}{d}$:** Let $\frac{a}{b}$ and $\frac{c}{d}$ be fractions, where $\frac{c}{d}$ is not zero. Then $\frac{a}{b} \div \frac{c}{d} = \frac{m}{n}$ if and only if $\frac{a}{b} = \frac{m}{n} \cdot \frac{c}{d}$. — 297
- The **reciprocal,** or **multiplication inverse,** of a nonzero fraction $\frac{c}{d}$ is the fraction $\frac{d}{c}$, so that $\frac{c}{d} \cdot \frac{d}{c} = 1$. — 300

FORMULA

- **Invert-and-multiply algorithm for division:** The division $\frac{a}{b} \div \frac{c}{d}$ is equivalent to the product $\frac{a}{b} \cdot \frac{d}{c}$ obtained by multiplying $\frac{a}{b}$ by the reciprocal of the divisor $\frac{c}{d}$. That is, $\frac{a}{b} \div \frac{c}{d} = \frac{a}{b} \cdot \frac{d}{c}$. 299

Section 6.4 The Rational Number System	Page Reference

CONCEPT

- **Rational numbers:** The system of numbers represented by fractions, so that the arithmetic operations (addition, subtraction, multiplication, division), properties, and order relations (less than and greater than) correspond to the same operations, properties, and order relations for fractions. A rational number is represented by any choice of the equivalent fractions corresponding to the rational numbers. 306

DEFINITION

- The **negative,** or **additive inverse,** of rational number $\frac{a}{b}$ is written $-\frac{a}{b}$ and is the rational number corresponding to the fraction $\frac{(-a)}{b}$. 307

THEOREMS

- **Properties of addition of rational numbers:** Given three rational numbers r, s, and t, addition is closed ($r + s$ is a rational number), commutative ($rs = sr$), associative ($r + (s + t) = (r + s) + t$), 0 is the additive identity ($r + 0 = 0 + r = r$), and $-r$ is the additive inverse $\left(r + (-r) = (-r) + r = 0, \text{ where } -r = \frac{-a}{b} \text{ if } r = \frac{a}{b}\right)$. 307
- **Subtraction is equivalent to the addition of the negative:** $\frac{a}{b} - \frac{c}{d} = \frac{a}{b} + \left(-\frac{c}{d}\right) = \frac{ad - bc}{bd}$. 308
- **Properties of multiplication of rational numbers:** Given three rational numbers r, s, and t, multiplication is closed ($r \cdot s$ is a rational number), commutative ($r \cdot s = s \cdot r$), associative ($r \cdot (s \cdot t) = (r \cdot s) \cdot t$), and distributive over addition ($r \cdot (s + t) = r \cdot s + r \cdot t$), has the multiplicative identity 1 ($r \cdot 1 = 1 \cdot r = r$), and has the multiplication by 0 property ($r \cdot 0 = 0 \cdot r = 0$). Moreover, each nonzero rational number $r = \frac{a}{b} \neq 0$ has the multiplicative inverse r^{-1} given by its reciprocal $\frac{b}{a}$, so that $r \cdot r^{-1} = \frac{a}{b} \cdot \frac{b}{a} = 1$. 309
- **Properties of the order relation on rational numbers:** Given three rational numbers r, s, and t, the order relation $<$ is transitive (if $r < s$ and $s < t$, then $r < t$), has the addition property (if $r < s$, then $r + t < s + t$), has the multiplication by a positive property (if $r < s$ and $t > 0$, then $r \cdot t < s \cdot t$), and satisfies the law of trichotomy (exactly one of the relations $r < s$, $r = s$, $r > s$ holds). 310
- **Density property:** Given two different rational numbers r, s, with $r < s$, there is a rational number t between r and s. Indeed, there are infinitely many rational numbers t for which $r < t < r$. 311

Chapter Review Exercises

Section 6.1

1. What fraction is represented by the darker blue shading in each of the colored-region models shown? In (a) and (d), the unit is the region inside one circle. In (b) and (c), the units are the regions inside the hexagon and square, respectively.

(a)
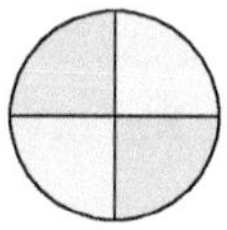

(b)
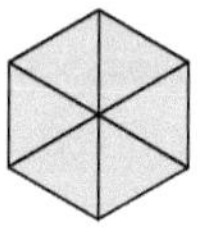

(c)
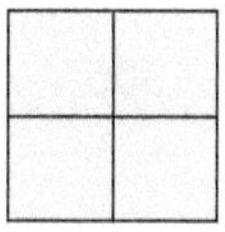

(d)
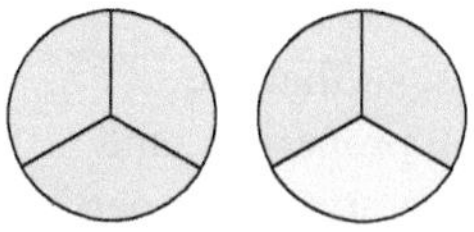

2. Illustrate $\frac{2}{3}$

(a) on the number line,

(b) with a fraction strip,

(c) with a colored-region model, and

(d) with the set model.

3. Give three different fractions, each equivalent to $-\frac{3}{4}$.

4. Label the points on the number line that correspond to these rational numbers:

(a) $\frac{3}{4}$ (b) $\frac{12}{8}$

(c) 1 (d) $2\frac{3}{8}$

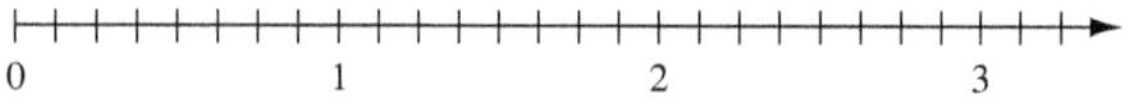

5. Express each rational number by a fraction in simplest form.

(a) $\frac{27}{81}$ (b) $\frac{100}{825}$

(c) $\frac{378}{72}$ (d) $\frac{3^5 \cdot 7^2 \cdot 11^3}{3^2 \cdot 7^3 \cdot 11^2}$

6. Order these fractions from smallest to largest:

$$\frac{1}{2}, \frac{13}{27}, \frac{25}{49}, \frac{13}{30}, \frac{26}{49}$$

7. Find the least common denominator of each set of fractions.

(a) $\frac{4}{9}, \frac{5}{12}$ (b) $\frac{7}{18}, \frac{5}{6}, \frac{1}{3}$

Section 6.2

8. Illustrate $\frac{3}{4} + \frac{7}{8}$ on the number line.

9. Illustrate $\frac{3}{4} - \frac{1}{3}$ with fraction strips.

10. Compute these sums and differences:

(a) $\frac{3}{8} + \frac{1}{4}$ (b) $\frac{2}{9} + \frac{-5}{12}$

(c) $\frac{4}{5} - \frac{2}{3}$ (d) $5\frac{1}{4} - 1\frac{5}{6}$

11. Perform these calculations:

(a) $\frac{1}{3} + \frac{5}{8} - \frac{5}{6}$ (b) $\left(\frac{2}{3} - \frac{5}{4}\right) \div \frac{3}{4}$

(c) $\frac{4}{7} \cdot \left(\frac{35}{4} + \frac{-42}{12}\right)$ (d) $\frac{123}{369} \div \frac{1}{3}$

Section 6.3

12. Illustrate these products by labeling and coloring appropriate rectangular regions:

(a) $3 \times \frac{1}{3}$

(b) $\frac{2}{3} \times 4$

(c) $\frac{5}{6} \times \frac{3}{2}$

13. Gina, Hank, and Igor want to share $2\frac{1}{4}$ pizzas equally. Draw an appropriate diagram to show how much pizza each should be given. What conceptual model of division should you use?

14. Each quart of soup calls for $\frac{2}{3}$ of a cup of pinto beans. How many quarts of soup can be made with 3 cups of beans? Draw an appropriate diagram to find your answer. What conceptual model of division are you using?

15. On a map, it is $7\frac{1}{8}''$ from Arlington to Banks. If the scale of the map is $1\frac{1}{4}''$ per mile, how far is it between the two towns?

Section 6.4

16. Perform these calculations, expressing your answers in simplest form:

(a) $\frac{-3}{4} + \frac{5}{8}$ (b) $\frac{4}{5} - \frac{-7}{10}$

(c) $\left(\frac{3}{8} \cdot \frac{-4}{27}\right) \div \frac{1}{9}$ (d) $\frac{2}{5} \cdot \left(\frac{3}{4} - \frac{5}{2}\right)$

17. Solve each equation. Give the rational number x as a fraction in simplest form.

(a) $3x + 5 = 11$ (b) $x + \frac{2}{3} = \frac{1}{2}$

(c) $\frac{3}{5}x + \frac{1}{2} = \frac{2}{3}$ (d) $-\frac{4}{3}x + 1 = \frac{1}{4}$

18. Solve the given equations and inequalities for all possible rational numbers x. Show all of your steps.

(a) $2x + 3 > 0$ (b) $\frac{5}{4}x > -\frac{1}{3}$

(c) $\frac{1}{2} < 4x + \frac{5}{6}$

19. Find two rational numbers between $\frac{5}{6}$ and $\frac{10}{11}$.

7 Decimals, Real Numbers, and Proportional Reasoning

COOPERATIVE INVESTIGATION
Squares, Areas, and Side Lengths

Materials Needed

Squared paper with 1″ × 1″ squares, scissors, ruler, marker pen, and calculator.

Directions

Recall that if a square has sides of length s, then the area of the square is s^2.

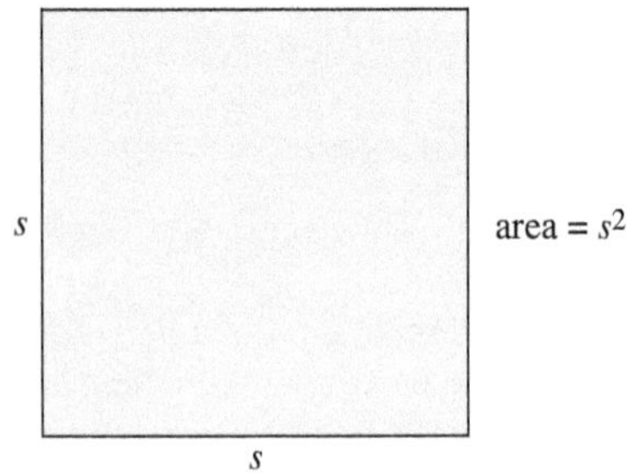

Investigation 1

Cut a 2″ × 2″ square from the squared paper and draw the tilted square that has its corners at the midpoints of the sides of the larger square. Label each side of the smaller square with its length c. Now show that $c^2 = 2$ by finding the area of the smaller square using this idea: Fold the four corners of the large square to the center of the square.

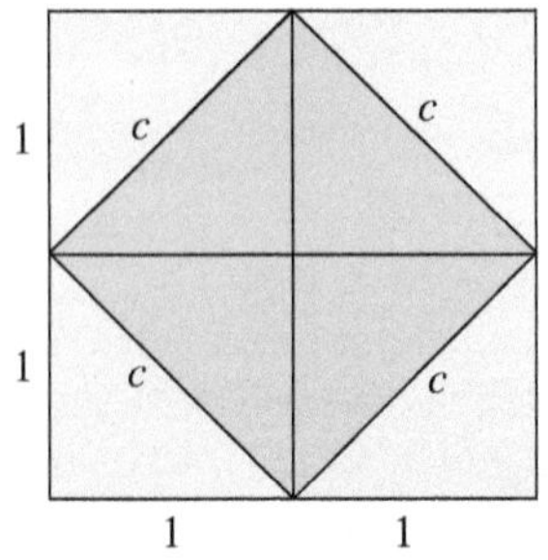

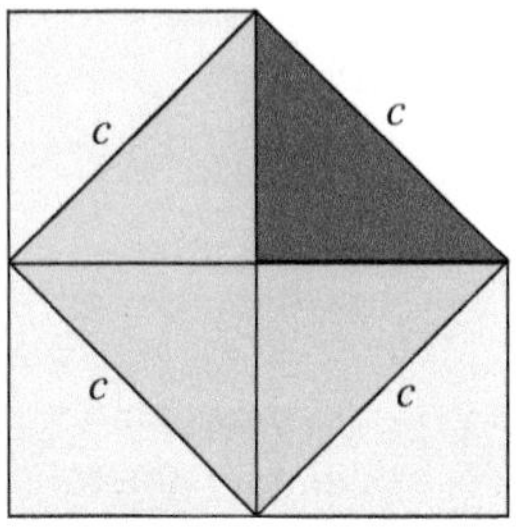

after first fold

Investigation 2

Use your ruler to show that c is about 1.4″. Since $1.4^2 = 1.96$, c must be a little larger than 1.4. Use your calculator to see if c is either larger or smaller than 1.45. Continue to guess and check and see if you can find a decimal number that, when squared, is exactly 2.

Investigation 3

Check that the square root function on your calculator will claim (depending on the calculator) that c is 1.4142136. Enter the number given in the display and square it. Is c^2 exactly 2? (If so, this is due to a round-off error, since we will show in this chapter that there is no decimal with a finite number of digits that gives $c = \sqrt{2}$ exactly. More precisely, it is impossible to represent $\sqrt{2}$ as a fraction, so it is a real number known as an irrational number.)

Further Work

Repeat the three preceding investigations but this time start with a 3″ × 3″ square, and let d be the length of the sides of a smaller square whose corners are at the points 1″ from the corners of the larger square as shown here. (You should discover that $d^2 = 5$, where it can be proved that $\sqrt{5}$ is an irrational real number.)

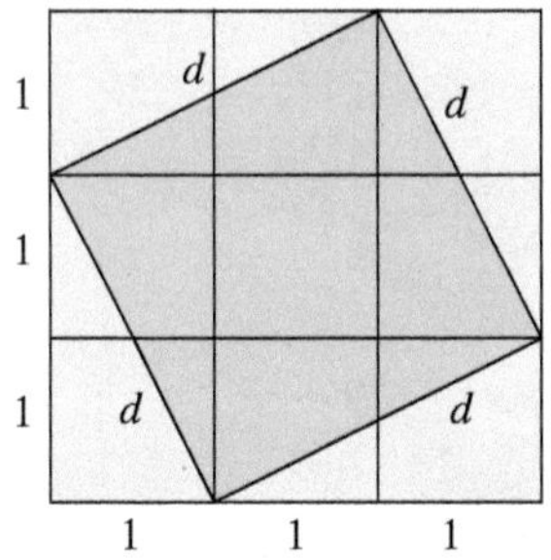

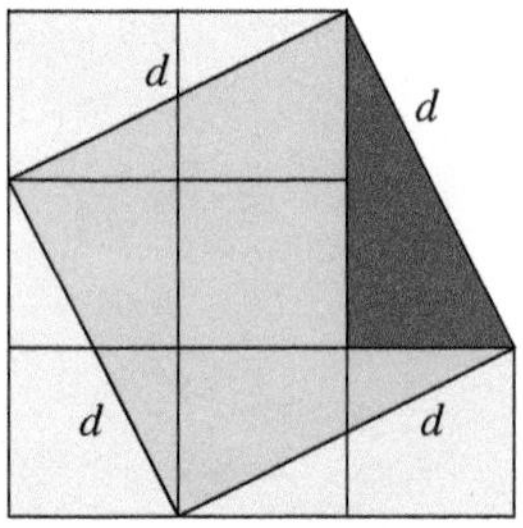

after first fold

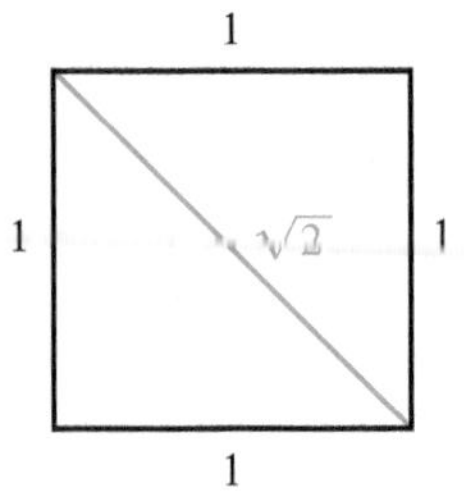

Previous chapters have introduced the natural numbers N, the whole numbers W, the integers I, and the rational numbers Q. This chapter introduces yet one more number system: the real numbers R. The set of real numbers is an extension of the previous number systems, so $N \subset W \subset I \subset Q \subset R$.

In Section 1 of this chapter, the real numbers are defined as those numbers that can be represented as decimal numerals. Such a definition gives a new representation of a rational number not only as a fraction but also as a decimal numeral that either terminates or has an infinitely repeated string of decimal digits. What is new are those real numbers represented by a decimal with infinitely many digits that do not have a repeating pattern. For example, $\sqrt{2} = 1.41421356\ldots$ is the irrational real number that gives the length of the diagonal of a square with sides of length 1; the list of decimal digits following the decimal point is never ending and has no repeating pattern. In Section 2, we see that we can adapt previously learned computational algorithms to perform arithmetic operations with decimal numbers. Section 3 explores proportional reasoning, which is applied in the concluding Section 4 to develop the important concept of percent.

KEY IDEAS

- Decimal numbers and their representations with manipulatives such as base-ten blocks and dollars–dimes–pennies
- Decimals written in expanded form
- Terminating and periodic decimals and their conversion to a representation by a fraction
- The irrationality of $\sqrt{2}$; that is, if $c^2 = 2$, then c cannot be represented as a fraction $\frac{a}{b}$ for any natural numbers a and b
- The order relation on decimals
- Computing with decimals: rounding, addition, subtraction, multiplication, and division
- Ratios, proportions, and applications of proportional reasoning
- Percent and its applications

FROM THE NCTM PRINCIPLES AND STANDARDS

When students leave grade 5, they should be able to solve problems involving whole-number computation and should recognize that each operation will help them solve many different types of problems. They should be able to solve many problems mentally, to estimate a reasonable result for a problem, to efficiently recall or derive the basic number combinations for each operation, and to compute fluently with multi-digit whole numbers. They should understand the equivalence of fractions, decimals, and percents and the information each type of representation conveys. With these understandings and skills, they should be able to develop strategies for computing with familiar fractions and decimals.

SOURCE: Principles and Standards for School Mathematics by NCTM, *p. 149. Copyright © 2000 by the National Council of Teachers of Mathematics. Reproduced with permission of the National Council of Teachers of Mathematics via Copyright Clearance Center. NCTM does not endorse the content or validity of these alignments.*

7.1 Decimals and Real Numbers

The term *decimal system* (from the Latin *decimus,* meaning "tenth") refers generally to the Indo-Arabic system of numeration in common use today. For example, we may see that an ice cream cone is priced at \$1.79, or we can buy 12.83 gallons of gas paying \$3.899 per gallon. Even at a restaurant, the sandwich listed on the menu for \$9 is a decimal numeral that could be written as 9.00. Here is the general definition of a decimal numeral.

DEFINITION Decimal

A **decimal** is a base-ten positional numeral, either positive or negative, in which there are finitely many digits to the left of a point called the **decimal point** that represent units, tens, hundreds, and so on, and a finite or infinite sequence of digits extending to the right of the decimal point that represent tenths, hundredths, thousandths, and so on.

By definition, any decimal numeral has an **expanded form** as a sum of units, tens, hundreds, . . . and a sum of tenths, hundredths, thousandths, ... For example, the decimal 23.47 represents the number with the expanded form

$$\begin{aligned}23.47 &= 2\cdot 10^1 + 3\cdot 10^0 + 4\cdot 10^{-1} + 7\cdot 10^{-2}\\ &= 20 + 3 + \frac{4}{10} + \frac{7}{100}.\end{aligned}$$

We see that the decimal point separates the numeral 23.47 into an integer part 23 to the left of the decimal point and a fractional part $0.47 = \frac{47}{100}$ to the right of the decimal point, so 23.47 is read as "twenty-three and forty-seven hundredths."

Representations of Decimals

For beginning elementary school students, it is helpful to introduce the study of decimals by considering concrete physical manipulative devices and pictorial representations. We discuss three such representations here.

Unit Squares, Strips, and Small Squares In this model, a square represents one unit, or 1, a strip represents one-tenth, or $\frac{1}{10}$, and a small square represents one-hundredth, or $\frac{1}{100}$. See Figure 7.1.

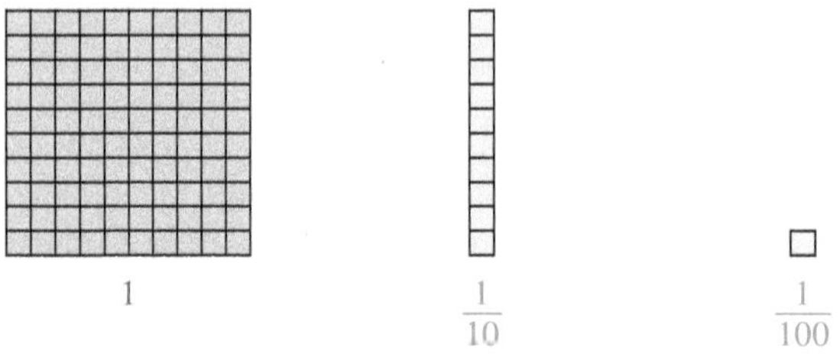

Figure 7.1
Representation of 1, $\frac{1}{10}$, and $\frac{1}{100}$ by using a unit square, a strip, and a small square

Thus, a display of 3 unit squares, 2 strips, and 5 small squares, as shown in Figure 7.2, represents $3 \cdot 1 + 2 \cdot \frac{1}{10} + 5 \cdot \frac{1}{100} = 3 + \frac{2}{10} + \frac{5}{100}$, which we write as the decimal numeral 3.25.

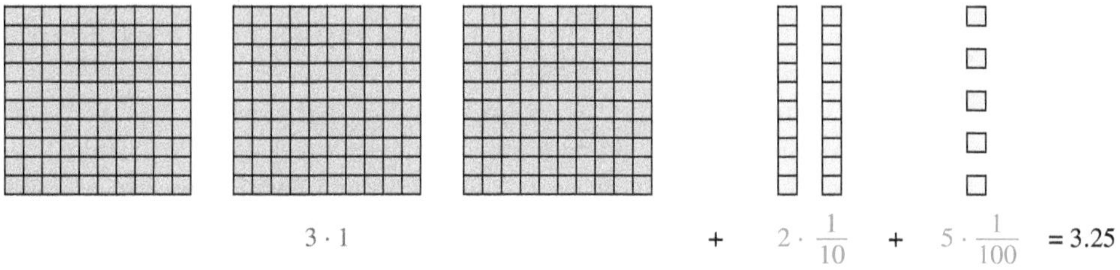

Figure 7.2
Representation of 3.25 by using unit squares, strips, and small squares

Base-Ten Blocks In this model, the block is chosen as the unit. Then, since 10 flats are equivalent to a block, a flat represents $\frac{1}{10}$. See Figure 7.3.

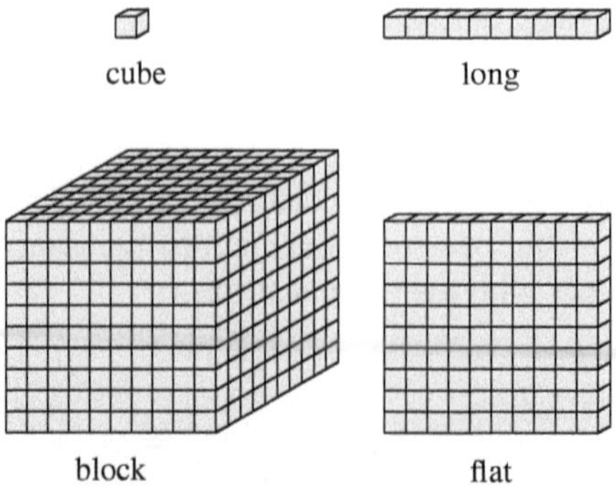

Figure 7.3
Base-ten blocks

Similarly, 10 longs are equivalent to a flat, and $10 \cdot 10 = 100$ longs are equivalent to a block. Thus, a long represents $\frac{1}{100}$. Finally, since there are $10 \cdot 10 \cdot 10 = 1000$ cubes in a unit block, a cube represents $\frac{1}{1000}$. With all this in mind, a display of 2 blocks, 1 flat, 3 longs, and 2 cubes represents $2 \cdot 1 + 1 \cdot \frac{1}{10} + 3 \cdot \frac{1}{100} + 2 \cdot \frac{1}{1000} = 2 + \frac{1}{10} + \frac{3}{100} + \frac{2}{1000} = \frac{2132}{1000}$, which we write as the decimal 2.132. See Figure 7.4.

Figure 7.4
Representing 2.132 using base-ten blocks

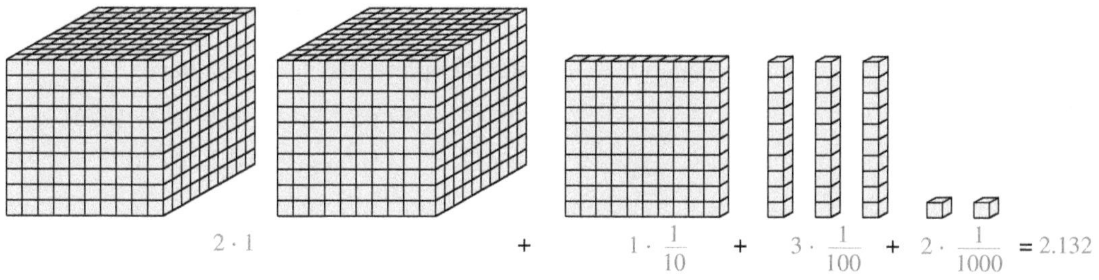

Dollars, Dimes, and Pennies A particularly apt manipulative, already familiar to children, is money. Since 10 dimes are worth 1 dollar and 100 pennies are worth 1 dollar, a dollar represents 1, a dime represents $\frac{1}{10}$, and a penny represents $\frac{1}{100}$. Thus, 3.25 would be represented as in Figure 7.5. Such a representation is particularly useful because students are already familiar with the fact that this set of coins is worth 3 dollars and 25 cents, as well as the fact that the value of the coins is written \$3.25, thus making the understanding of the expanded notation for 3.25 quite natural.

Figure 7.5
Representing 3.25 by using dollars, dimes, and pennies

EXAMPLE 7.1

Using Money to Represent Decimals

How would you use money to explain the decimal 23.75 to elementary school students?

Solution

Use 2 ten-dollar bills, 3 one-dollar bills or coins, 7 dimes, and 5 pennies. The students will readily see that this collection of bills and coins is worth \$23.75, making it easy to explain the expanded notation

$$23.75 = 2 \cdot 10 + 3 \cdot 1 + 7 \cdot \frac{1}{10} + 5 \cdot \frac{1}{100},$$

since a dime is one-tenth of a dollar and a penny is one-hundredth of a dollar.

By using the representations discussed, students can be led to understand that the shorthand notation for $\frac{1}{10}$ is 0.1, for $\frac{1}{100}$ is 0.01, for $\frac{1}{1000}$ is 0.001, for $\frac{2}{10}$ is 0.2, for $\frac{3}{100}$ is 0.03, and so on. The names of the positional values are given in Table 7.1, which also gives their decimal, fractional, and power of 10 forms.

TABLE 7.1 POSITIONAL VALUES IN THE DECIMAL SYSTEM

	Position Names								
Form	. . .	**Hundreds**	**Tens**	**Units**	**Tenths**	**Hundredths**	**Thousandths**	**Ten-Thousandths**	. . .
Decimal Form	. . .	100	10	1	0.1	0.01	0.001	0.0001	. . .
Fractional Form	. . .	100	10	1	$\frac{1}{10}$	$\frac{1}{100}$	$\frac{1}{1000}$	$\frac{1}{10{,}000}$	. . .
Power of 10	. . .	10^2	10^1	10^0	10^{-1}	10^{-2}	10^{-3}	10^{-4}	. . .

For example, in **expanded exponential form,** we have:

$$0.235 = 2 \cdot 10^{-1} + 3 \cdot 10^{-2} + 5 \cdot 10^{-3};$$
$$23.47 = 2 \cdot 10^{1} + 3 \cdot 10^{0} + 4 \cdot 10^{-1} + 7 \cdot 10^{-2}.$$

In this form, the pattern of decreasing exponents is both clear and neat.

EXAMPLE 7.2

Writing Decimals in Expanded Exponential Form

Write each of these decimals in expanded exponential form:

(a) 234.72 **(b)** 30.0012

Solution

(a) $234.72 = 2 \cdot 10^{2} + 3 \cdot 10^{1} + 4 \cdot 10^{0} + 7 \cdot 10^{-1} + 2 \cdot 10^{-2}$

(b) $30.0012 = 3 \cdot 10^{1} + 0 \cdot 10^{0} + 0 \cdot 10^{-1} + 0 \cdot 10^{-2} + 1 \cdot 10^{-3} + 2 \cdot 10^{-4}$

Multiplying and Dividing Decimals by Powers of 10

Consider the decimal

$$25.723 = 2 \cdot 10^{1} + 5 \cdot 10^{0} + 7 \cdot 10^{-1} + 2 \cdot 10^{-2} + 3 \cdot 10^{-3}.$$

If we multiply by 10^2 then, using the distributive property, we obtain

$$\begin{aligned}(10^2)(25.723) &= (10^2)(2 \cdot 10^{1} + 5 \cdot 10^{0} + 7 \cdot 10^{-1} + 2 \cdot 10^{-2} + 3 \cdot 10^{-3}) \\ &= 2 \cdot 10^{2+1} + 5 \cdot 10^{2+0} + 7 \cdot 10^{2+(-1)} + 2 \cdot 10^{2+(-2)} + 3 \cdot 10^{2+(-3)} \\ &= 2 \cdot 10^{3} + 5 \cdot 10^{2} + 7 \cdot 10^{1} + 2 \cdot 10^{0} + 3 \cdot 10^{-1} \\ &= 2572.3,\end{aligned}$$

and the notational effect is to move the decimal point two places to the right. Note that 2 is both the exponent of the power of 10 by which we are multiplying and the number of zeros in 100.

$10^2 = 100$

The result is analogous if we *divide* 25.723 by 10^2, except that the notational effect is to move the decimal point two places *to the left.* To see this, recall that we can divide by multiplying by the multiplicative inverse. Thus,

$10^{-2} = \frac{1}{10^2}$

$$\begin{aligned}(25.723) \div 10^2 &= (25.723) \cdot (1/10^2) \\ &= (25.723) \cdot 10^{-2} \\ &= (2 \cdot 10^{1} + 5 \cdot 10^{0} + 7 \cdot 10^{-1} + 2 \cdot 10^{-2} + 3 \cdot 10^{-3}) \cdot 10^{-2} \\ &= 2 \cdot 10^{1+(-2)} + 5 \cdot 10^{0+(-2)} + 7 \cdot 10^{(-1)+(-2)} + 2 \cdot 10^{(-2)+(-2)} \\ &\quad + 3 \cdot 10^{(-3)+(-2)} \\ &= 2 \cdot 10^{-1} + 5 \cdot 10^{-2} + 7 \cdot 10^{-3} + 2 \cdot 10^{-4} + 3 \cdot 10^{-5} \\ &= 0.25723.\end{aligned}$$

These results are typical of the general case, which we state here as a theorem:

THEOREM Multiplying and Dividing Decimals by Powers of 10
If r is a positive integer, the notational effect of multiplying a decimal by 10^r is to move the decimal point r places to the right. The notational effect of dividing a decimal by 10^r (that is, multiplying by 10^{-r}) is to move the decimal point r places to the left.

It is important that elementary school children understand *why* it is so easy to multiply or divide by a power of 10.

EXAMPLE 7.3

Multiplying and Dividing Decimals by Powers of 10

Compute each of the following:

(a) $(10^3)(253.26)$ **(b)** $(253.26) \div 10^3$
(c) $(100)(34.764)$ **(d)** $(34.764) \div 10{,}000$

Solution

The preceding theorem gives the desired results:

(a) $(10^3)(253.26) = 253{,}260$
(b) $(253.26) \div 10^3 = 0.25326$
(c) $(100)(34.764) = (10^2)(34.764) = 3476.4$
(d) $(34.764) \div 10{,}000 = 34.764 \div (10^4) = 0.0034764$

Terminating Decimals as Fractions

A decimal such as 24.357, which has a finite number of digits, is called a **terminating decimal.** Every terminating decimal represents a rational number; that is, any terminating decimal can be represented by a fraction with integers in the numerator and denominator. For example, using expanded notation, we write

$$\begin{aligned}24.357 &= 20 + 4 + \frac{3}{10} + \frac{5}{100} + \frac{7}{1000}\\ &= \frac{20{,}000}{1000} + \frac{4000}{1000} + \frac{300}{1000} + \frac{50}{1000} + \frac{7}{1000}\\ &= \frac{24{,}357}{1000}.\end{aligned}$$

Alternatively,

$$\begin{aligned}24.357 &= (24.357) \cdot \frac{1000}{1000}\\ &= \frac{24{,}357}{1000},\end{aligned}$$

$a \cdot \frac{b}{c} = \frac{ab}{c}$

as before. In each case, the denominator is determined by the position of the rightmost digit. (In this case, the 7 is in the thousandths position.)

Any fraction whose denominator is a power of 10 may be represented by a terminating decimal. This is also true for a fraction whose denominator has a prime factorization involving only powers of 2, powers of 5, or some combination of both 2s and 5s. As an example, consider

$$\frac{17}{40} = \frac{17}{2^3 \cdot 5^1}.$$

This fraction can be written so that the denominator is a power of 10 by multiplying the numerator and denominator by 5^2. Thus,

Multiply numerator and denominator by 5^2.

$$\frac{17}{40} = \frac{17}{2^3 \cdot 5^1} = \frac{17 \cdot 5^2}{2^3 \cdot 5^3} = \frac{17 \cdot 25}{10^3} = \frac{425}{1000} = 0.425.$$

Since the numbers presented are typical of the general case, the results can be summarized as a theorem:

THEOREM Rational Numbers Represented by Terminating Decimals

If a and b are integers with $b \neq 0$, if $\frac{a}{b}$ is in simplest form, and if no prime other than 2 and/or 5 divides b, then $\frac{a}{b}$ can be represented as a terminating decimal, and conversely.

EXAMPLE 7.4

Writing a Terminating Decimal as a Fraction

Express each of these decimals in the form $\frac{a}{b}$, where the fraction is in simplest form:

(a) 31.75 **(b)** 4.112 **(c)** -0.035

Solution

(a) $31.75 = 31.75 \cdot \frac{100}{100} = \frac{3175}{100} = \frac{127}{4}$. Note that GCD(3175, 100) = 25.

(b) $4.112 = 4.112 \cdot \frac{1000}{1000} = \frac{4112}{1000} = \frac{514}{125}$. Note that GCD(4112, 1000) = 8.

(c) $-0.035 = -0.035 \cdot \frac{1000}{1000} = \frac{-35}{1000} = \frac{-7}{200}$. Note that GCD(35, 1000) = 5.

The decimal equivalent of $\frac{17}{40}$ can also be found by division. The standard long-division algorithm can be followed, with the difference that the digits are arranged vertically into columns that represent tens, units, tenths, hundredths, and thousandths. We have

$$\begin{array}{rl}
0.425 & \\
40\overline{)17.000} & \\
-16.0 & \text{Remove } 0.4 \cdot 40 = 16.0 \\
1.00 & \\
-0.80 & \text{Remove } 0.02 \cdot 40 = 0.80 \\
0.200 & \\
-0.200 & \text{Remove } 0.005 \cdot 40 = 0.200 \\
0 &
\end{array}$$

EXAMPLE 7.5

Converting Certain Fractions to Decimals

Convert each of the given fractions to decimals by writing each as an equivalent fraction whose denominator is a power of 10. Check by long division.

(a) $\frac{37}{40}$ **(b)** $\frac{29}{200}$

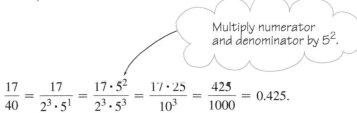

Solution

(a) $\dfrac{37}{40} = \dfrac{37}{2^3 \cdot 5^1} = \dfrac{37 \cdot 5^2}{2^3 \cdot 5^3}$
$= \dfrac{37 \cdot 25}{10^3}$
$= \dfrac{925}{1000}$
$= 0.925$

$$\begin{array}{r} 0.925 \\ 40\overline{)37.0} \\ -36.0 \\ \hline 1.00 \\ -0.80 \\ \hline 0.200 \\ -0.200 \\ \hline 0 \end{array}$$

(b) $\dfrac{29}{200} = \dfrac{29}{2^3 \cdot 5^2} = \dfrac{29 \cdot 5}{2^3 \cdot 5^3}$
$= \dfrac{145}{10^3}$
$= \dfrac{145}{1000}$
$= 0.145$

$$\begin{array}{r} 0.145 \\ 200\overline{)29.000} \\ -20.0 \\ \hline 9.00 \\ -8.00 \\ \hline 1.00 \\ -1.00 \\ \hline 0 \end{array}$$

Repeating Decimals and Rational Numbers

Not all rational numbers have decimal expansions that terminate. For example,

$$\frac{1}{3} = 0.333\ldots = 0.\overline{3},$$

$$\begin{array}{r} 0.333\ldots \\ 3\overline{)1.000\ldots} \\ -0.9 \\ \hline 0.10 \\ -0.09 \\ \hline 0.010 \\ -0.009 \\ \hline 0.0010 \\ \ldots \end{array}$$

where the three dots indicate that the decimal continues *ad infinitum* and the bar over the 3 indicates the digit or group of digits that repeats. We can also use algebraic reasoning by setting

$$x = 0.333\ldots.^*$$

Then, multiplying by 10, we obtain

$$10x - 3.333\ldots,$$

so that

$$10x - x = 9x = 3.$$

$$\begin{array}{r} 3.333\ldots \\ -0.333\ldots \\ \hline 3.000\ldots \end{array}$$

But this equation implies that

$$x = \frac{3}{9} = \frac{1}{3}.$$

The decimal expansion of $\frac{1}{3}$ is a nonterminating, but *repeating,* decimal; that is, the stream of 3s repeats without end. In general, we have the following definition:

DEFINITION A Repeating Decimal
A nonterminating decimal with the property that a digit or string of adjacent digits repeats *ad infinitum* from some point on is called a **periodic,** or **repeating, decimal.** The number of digits in the repeating group is called the **length of the period.**

---

*Actually, there is a touchy point here that we gloss over. The question is whether $0.333\ldots = \frac{3}{10} + \frac{3}{100} + \frac{3}{1000} + \cdots$ means anything at all, since it is the sum of an *infinite* number of numbers. That the answer is "yes" really depends on ideas from calculus!

For example, $0.\overline{24}$ is a repeating decimal for which the repeating part, 24, has length 2. That is, $0.\overline{24} = 0.24242424\ldots$ Similarly, $3.14\overline{5} = 3.14555555\ldots$ is a repeating decimal whose repeating part, 5, has length 1. The next example uses algebraic reasoning to rewrite repeating decimals as fractions. That is, the repeating decimals represent rational numbers. The same reasoning shows that *every* repeating decimal represents a rational number.

EXAMPLE 7.6

Repeating Decimals as Rational Numbers

Write each of the given repeating decimals in the form $\dfrac{a}{b}$, where a and b are integers and the fraction is in simplest form. Check by dividing a by b with your calculator.

(a) $0.242424\ldots = 0.\overline{24}$ (b) $3.14555\ldots = 3.14\overline{5}$ (c) $0.9999\ldots = 0.\overline{9}$

Solution

(a) Let $x = 0.242424\ldots$ *a decimal of period 2*
Then

Multiply by $10^2 = 100$ to move the decimal point to the right of the repeated part.

$$100x = 24.242424\ldots$$

and

$$100x - x = 99x = 24.$$

24.242424...
− 0.242424...
24.000000...

But then

$$x = \frac{24}{99} = \frac{8}{33}.$$

Also, by calculator, $8 \div 33 \doteq 0.2424242$.

Here, the calculator answer is approximate, since only finitely many digits can appear in the calculator display.

(b) Let $x = 3.14555\ldots$
Then

Multiply by 10^2 to move the decimal point to the start of the repeated part.

$$100x = 314.555\ldots$$

and

a decimal of period 1

$$1000x = 3145.555\ldots$$

10 · 100 = 1000

Multiply by 10 again to move the decimal point to the start of the second repeated part.

Then

$$1000x - 100x = 900x = 2831$$

3145.555...
− 314.555...
2831.000...

and

$$x = \frac{2831}{900}.$$

Also, by calculator, $2831 \div 900 \doteq 3.1455556$. Do you see why the last digit in the calculator display is a 6?

(c) Let $x = 0.9999\ldots$.

Then $10x = 9.9999\ldots$ and $10x - x = 9.9999\ldots - 0.9999\ldots = 9$.

This shows that $9x = 9$ and, therefore, $x = 1$. In general, any decimal for which 9 is the repeating part is better represented by the terminating decimal found by increasing the digit just to the left of the repeating part by one. Two examples are $7.203\overline{9} = 7.204$ and $5.678\overline{9} = 5.679$.

The procedure followed in Example 7.6 can be adjusted to show that *every* repeating decimal numeral can also be represented by a fraction. We also know that any fraction with only 2s and 5s in its denominator can be written as a terminating decimal numeral. The following example shows that if the denominator of a fraction in simplest form has a prime factor other than 2 or 5, then it can be represented with a repeating decimal.

EXAMPLE 7.7 **The Decimal Expansions of $\dfrac{3}{7}$ and $\dfrac{23}{54}$**

Use long division to find the decimal expansions of **(a)** $\dfrac{3}{7}$ and **(b)** $\dfrac{23}{54}$.

Solution

(a) Since $\dfrac{3}{7} = 3 \div 7$, we obtain the desired decimal expansion by dividing 3 by 7. We have

$$
\begin{array}{r}
.428571\ldots \\
7\overline{)3.000000} \\
2\,8 \\
\hline
20 \\
14 \\
\hline
60 \\
56 \\
\hline
40 \\
35 \\
\hline
50 \\
49 \\
\hline
10 \\
7 \\
\hline
3
\end{array}
$$

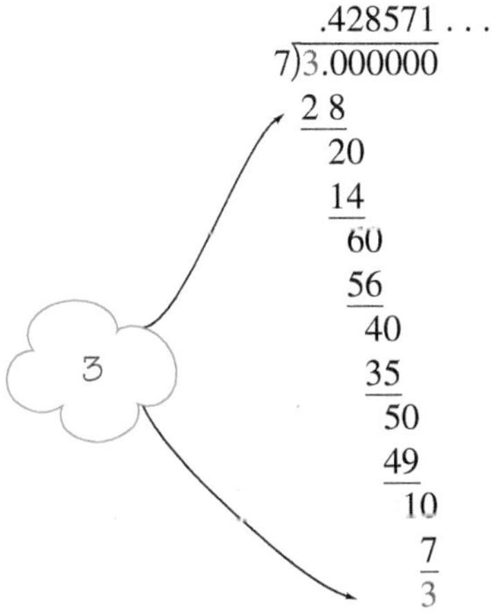

and, since the remainder at this stage is 3, the number with which we began, the division will continue *ad infinitum* with that repeating pattern. Thus,

$$
\begin{array}{r}
.428571428571428571\ldots \\
3\overline{)3,000000000000000\ldots}
\end{array}
$$

so

$$
\frac{3}{7} = 0.428571428571428571\ldots = 0.\overline{428571}.
$$

Since we are dividing by 7, and we know the decimal does not terminate, the only possible remainders at each step are 1, 2, 3, 4, 5, and 6. Thus, the length of the period can be no longer than 6.

(b) The decimal long division that follows shows that the first repeated remainder is 14, so the quotient repeats the three-digit pattern 259 without end. That is, $\frac{23}{54} = 0.4259259259259\ldots = 0.4\overline{259}$. We know in advance that the only possible remainders are between 1 and 53, so there cannot be more than 53 digits in the repeated pattern.

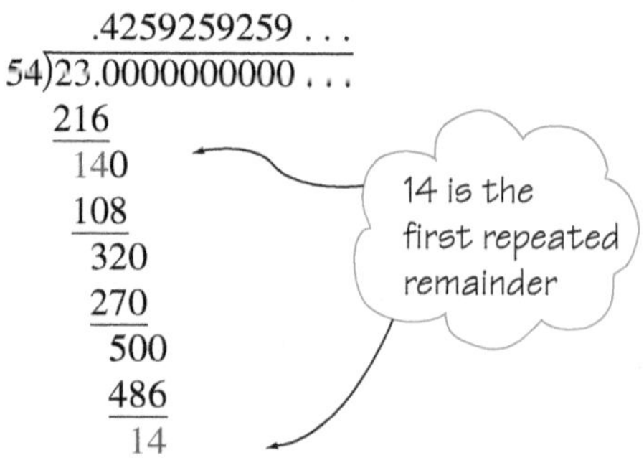

The following theorem describes which rational numbers have a repeating decimal representation:

THEOREM Rational Numbers and Periodic Decimals

Every repeating decimal represents a rational number $\frac{a}{b}$. If $\frac{a}{b}$ is in simplest form, and $b > 1$, then b must contain a prime factor other than 2 or 5. Conversely, if $\frac{a}{b}$ is such a rational number, its decimal representation must be repeating, with a period no longer than b.

The last two theorems can be combined to give a decimal description of the rational numbers:

THEOREM Characterizing Rational Numbers as Decimals

Every rational number can be written as either a terminating or a repeating decimal. Conversely, every terminating or repeating decimal represents a rational number.

The Set of Real Numbers

Earlier, we showed that the decimal expansion of a rational number either terminates or is nonterminating and repeating. This property raises an interesting question: What kind of numbers are represented by **nonterminating, nonrepeating decimals** such as 0.101001000100001 . . . ? These numbers cannot be represented as ratios of integers and so are called **irrational numbers.** This property distinguishes them from the rational numbers considered in Chapter 6. The set consisting of all rational and irrational numbers is called the set of **real numbers.**

DEFINITION Irrational and Real Numbers

Numbers represented by **nonterminating, nonperiodic decimals** are called **irrational numbers.** The set R consisting of all rational numbers and all irrational numbers is called the set of **real numbers.**

Thus, the set of real numbers is another extension of the number system. The set of real numbers contains all the earlier systems, as illustrated by the Venn diagram in Figure 7.6.

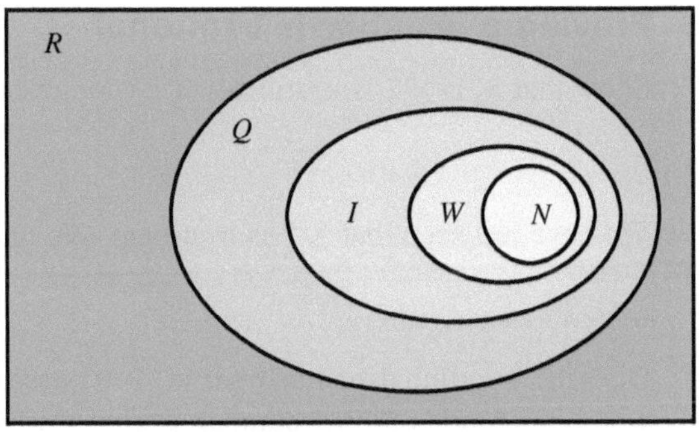

Figure 7.6
Venn diagram for the natural numbers N, the whole numbers W, the integers I, the rational numbers Q, and the real numbers R, showing that $N \subset W \subset I \subset Q \subset R$

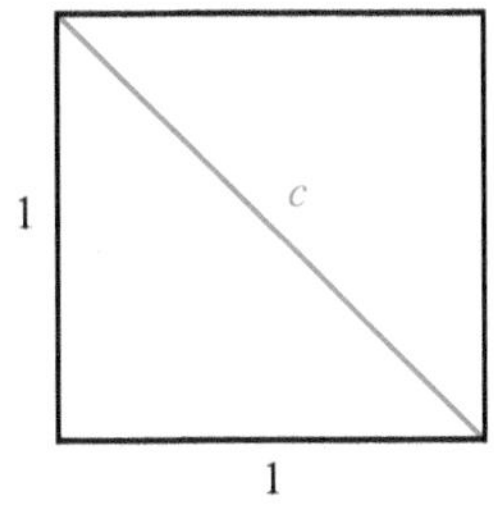

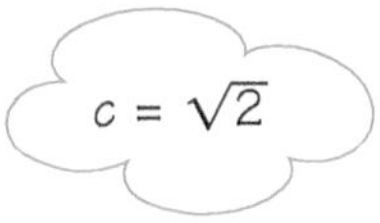

Irrationality of $\sqrt{2}$

As discovered in the Cooperative Investigation at the beginning of this chapter, if c is the length of the diagonal of a square 1 unit on a side, then

$$c^2 = 2,$$

Therefore, $c = \sqrt{2}$. We now show that $\sqrt{2}$ is not a rational number.

THEOREM Irrationality of $\sqrt{2}$
If $c^2 = 2$, then c is not a rational number.

PROOF (BY CONTRADICTION) Suppose, to the contrary, that there is a rational number c for which $c^2 = 2$. Since any rational number can be expressed as a fraction in simplest form, let

$$c = \frac{u}{v},$$

where u and v have no common factor other than 1. But $c^2 = 2$, so

$$2 = \frac{u^2}{v^2}.$$

Thus,

$$2v^2 = u^2.$$

This equation says that 2 is a prime factor of u^2. Hence, by the fundamental theorem of arithmetic, 2 is a prime factor of u. But then $u = 2k$ for some integer k, and

$$2v^2 = u^2 = (2k)^2 = 4k^2.$$

This relationship implies that

$$v^2 = 2k^2,$$

so 2 is also a factor of v^2 and hence of v. But then u and v have a factor of 2 in common, and $\frac{u}{v}$ is not in simplest form. This contradicts our assumption that $\frac{u}{v}$ is in simplest form. In view of this contradiction, the assumption that c is rational, which started this chain of reasoning, must be false. Therefore, c is irrational, as was to be proved.

EXAMPLE 7.8

Proving a Number Is Irrational

Show that $3 + \sqrt{2}$ is irrational.

Solution

Understand the Problem

We have just seen that $\sqrt{2}$ is irrational. We must show that $3 + \sqrt{2}$ is irrational.

Devise a Plan

Does the assertion even make sense? Is it possible that $3 + \sqrt{2}$ is rational? If so, then $3 + \sqrt{2} = s$, where s is some rational number. Perhaps we can use this equation, together with the fact that we already know that $\sqrt{2}$ is irrational, to arrive at the desired conclusion.

Carry Out the Plan

Since $3 + \sqrt{2} = s$, it follows that $\sqrt{2} = s - 3$. But 3 and s are rational, and we know that the rational numbers are closed under subtraction. This implies that $\sqrt{2}$ is rational, and we know that that is not so. Therefore, the assumption that $3 + \sqrt{2} = s$, where s is rational, must be false. Hence, $3 + \sqrt{2}$ is irrational, as was to be shown.

Look Back

Initially, it was not clear that the assertion stated in the problem had to be true, so we assumed briefly that it was not true. But this assumption led directly to the contradiction that $\sqrt{2}$ was rational, so the assumption that $3 + \sqrt{2}$ was rational had to be false; that is, $3 + \sqrt{2}$ had to be irrational, as we were to prove.

The preceding argument could be repeated exactly with 3 replaced by any rational number r. Thus, since there are infinitely many rational numbers, the preceding example shows that there are infinitely many irrational numbers. Moreover, in a very precise sense, which we will not discuss here, there are many more irrational numbers than rational numbers. Some of these irrationals may be familiar to you. For example, the number

$$\pi = 3.14159265\ldots$$

occurring in the formula $A = \pi r^2$ for the area of a circle of radius r and in the formula $C = 2\pi r$ for the circumference of a circle has been shown to be irrational.

Real Numbers and the Number Line

Each real number corresponds to a point on the number line. Figure 7.7 illustrates how to determine $\sqrt{2}$ on the number line.

Figure 7.7
$\sqrt{2}$ on the number line

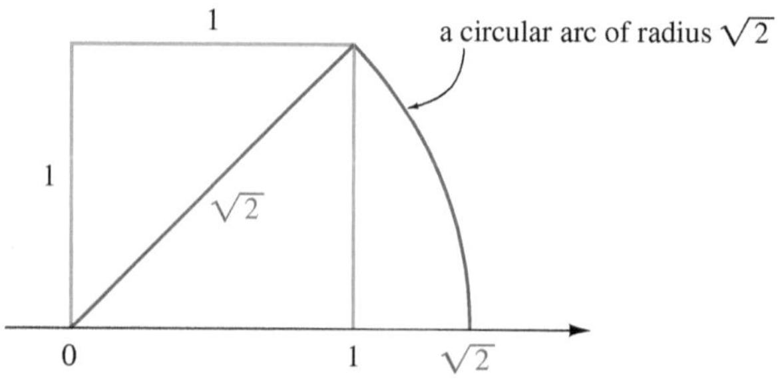

Also, the decimal expansion of $\sqrt{2}$ can be found step-by-step as follows: Using a calculator and trial and error, we find that

$$1^2 = 1 < 2 < 4 = 2^2,$$
$$1.4^2 = 1.96 < 2 < 2.25 = 1.5^2,$$
$$1.41^2 = 1.9881 < 2 < 2.0164 = 1.42^2,$$
$$1.414^2 = 1.999396 < 2 < 2.002225 = 1.415^2,$$

and so on. Thus, the decimal expansion of $\sqrt{2}$ is

$$\sqrt{2} = 1.414\ldots$$

Note that this value is determined by the lower estimates shown.

where successive digits are found by catching $\sqrt{2}$ between successive units, between successive tenths, between successive hundredths, and so on. To show that this can be done for any real number is somewhat tricky, but the following theorem can be proved:

THEOREM Real Numbers and the Number Line

There is a one-to-one correspondence between the set of real numbers and the set of points on a number line. The absolute value of the number associated with any point on the line gives the point's distance from 0.

7.1 Problem Set

Understanding Concepts

1. Write the following decimals in expanded form and in expanded exponential form:

   (a) 273.412 (b) 0.000723

2. Write the following decimals in expanded form and in expanded exponential form:

   (a) 42.307 (b) 0.20305

3. Write the decimal of the four points shown on the number line below.

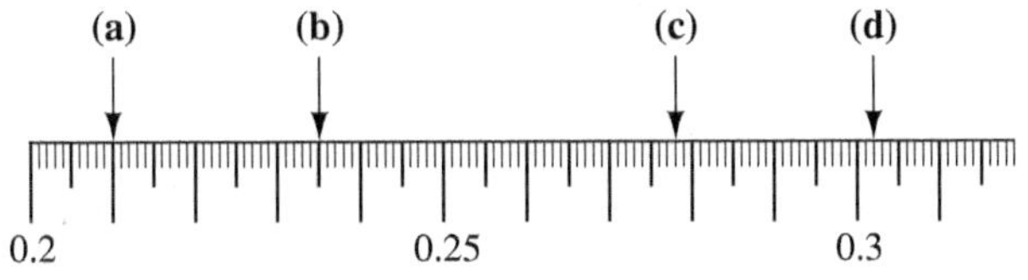

4. Match the given fractions to the nearest decimal.

   (a) $3\frac{3}{8}$ (b) $3\frac{1}{8}$ (c) $3\frac{1}{3}$ (d) $\frac{25}{7}$

   A. 3.1 B. 3.57 C. 3.4 D. 3.3

5. What decimal numeral is represented by this diagram?

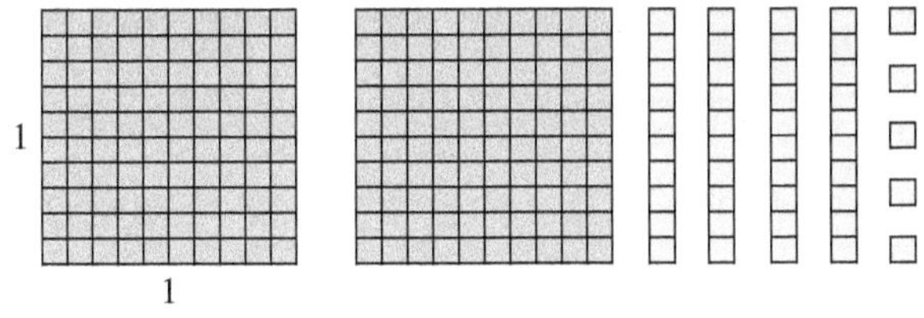

6. (a) Represent 2.14 with a coin diagram.

   (b) Using the fewest dollar coins, dimes, and pennies possible, what coins are needed to represent 5.72?

   (c) Using the fewest bills (hundreds, tens, and ones) and coins (dimes and pennies), how can 206.85 be represented?

7. Write these decimals as fractions in lowest terms, and determine the prime factorization of the denominator in each case:

   (a) 0.324 (b) 0.028

8. Answer the question in problem 7 for these decimals.

   (a) 4.25 (b) 0.95

9. Write these fractions as terminating decimals:

   (a) $\frac{7}{20}$ (b) $\frac{7}{16}$

10. Write these fractions as terminating decimals:

    (a) $\frac{3}{75}$ (b) $\frac{18}{2^2 \cdot 5^4}$

11. Represent each of these fractions as a decimal by using the long-division algorithm.

    (a) $\frac{7}{8}$ (b) $\frac{53}{40}$

12. Use long division to write these fractions as decimals.

    (a) $\frac{15}{48}$ (b) $\frac{71}{125}$

13. Represent each of these fractions as a repeating decimal by using the long-division algorithm:

    (a) $\frac{5}{6}$ (b) $\frac{7}{12}$

14. Write these fractions as decimals by using division.

(a) $\frac{3}{11}$ (b) $\frac{2}{27}$ (c) $\frac{4}{7}$

15. Determine the fraction in lowest terms represented by each of these periodic decimals:

(a) $0.321321\ldots = 0.\overline{321}$

(b) $0.12414141\ldots = 0.12\overline{41}$

(c) $3.262626\ldots = 3.\overline{26}$

16. In lowest terms, give the fraction represented by each of these repeating decimals.

(a) $0.666\ldots = 0.\overline{6}$

(b) $0.142857142857142857\ldots = 0.\overline{142857}$

(c) $0.153846153846153846\ldots = 0.\overline{153846}$

17. For each of the following, determine the fraction in lowest terms represented by the repeated decimal:

(a) $0.3\overline{54}$ (b) $5.21\overline{6}$

18. In lowest terms, write the following decimals as fractions:

(a) $2.3\overline{402}$ (b) $7.24\overline{86}$

19. Give an example of a fraction whose decimal expansion is repeating and has a period of length 3.

20. Find a fraction whose decimal repeats with length 4.

21. Show that $\sqrt{3}$ is irrational. (*Hint:* Have you seen a similar result proved?)

22. (a) Show that $3 - \sqrt{2}$ is irrational.

(b) Show that $2\sqrt{2}$ is irrational.

Into the Classroom

23. (**Writing**) Children often think that a decimal is more like a number than a fraction. For example, 0.75 is easy to find on a ruler, but $\frac{3}{4}$ is the amount *of* something. How would you show youngsters that fractions and decimals are both numbers?

24. (**Writing**) In Chapter 3, whole numbers were represented with counters and place-value cards. Modify this idea to represent decimal numbers. Is there a way to adjust the position of the decimals? Devise several problems and activities that teach decimal concepts with place-value cards.

Responding to Students

25. Janeshia was asked to say the number 56.127. Her response was "fifty-six tenths, one hundred twenty-seven thousandths." What error did Janeshia make with her answer?

Thinking Critically

26. Compute the decimal expansions of each of the following numbers:

(a) $\frac{1}{11}$ (b) $\frac{1}{111}$ (c) $\frac{1}{1111}$

(d) Guess the decimal expansion of $1/11111$.

(e) Check your guess in part (d) by converting the decimal of your guess back into a ratio of two integers.

(f) Carefully describe the decimal expansion of $1/N_n$, where $N_n = 111\ldots1$ with n 1s in its representation.

27. For each of the given periodic decimals, what rational number corresponds to that decimal? (Refer to Example 7.6 if you need help.)

(a) $0.747474\ldots = 0.\overline{74}$

(b) $0.777\ldots = 0.\overline{7}$

(c) $0.235235235\ldots = 0.\overline{235}$

(d) If a, b, and c are digits, what rational numbers are represented by the given periodic decimals? You should be able to guess the answers on the basis of patterns observed in parts (a), (b), and (c).

(i) $0.aaa\ldots = 0.\overline{a}$

(ii) $0.ababab\ldots = 0.\overline{ab}$

(iii) $0.abcabcabc\ldots = 0.\overline{abc}$

28. Write the decimals representing these rational numbers without doing any calculation, or at most doing only mental calculation:

(a) $\frac{5}{9}$ (b) $\frac{22}{99}$ (c) $\frac{317}{999}$

(d) $\frac{17}{33}$ (e) $\frac{14}{11}$

(f) Check the answers in parts (a) through (e), using your calculator. Is this check foolproof? Explain.

29. (a) Give an example showing that the sum of two irrational numbers is sometimes rational.

(b) Give an example showing that the sum of two irrational numbers is sometimes irrational.

30. Give an example showing that the product of two irrational numbers is sometimes rational.

31. Give an example showing that the quotient of two irrational numbers is sometimes rational.

32. (**Writing**)

(a) If a is an integer, what are the possibilities for the last digit of the decimal representation of a^2?

Last digit of a	0	1	2	3	4	5	6	7	8	9
Last digit of a^2										

(b) If b is an integer, what are the possibilities for the last digit of the decimal representation of $2b^2$?

Last digit of b	0	1	2	3	4	5	6	7	8	9
Last digit of $2b^2$										

(c) Using the results of parts (a) and (b), make a careful argument that $\sqrt{2} = \frac{a}{b}$, with a and b integers, is impossible.

33. (**Writing**) If there is a fraction $\frac{a}{b}$ in lowest terms for which $\sqrt{5} = \frac{a}{b}$, then $a^2 = 5b^2$.

(a) Show that no such fraction exists by comparing the last digits of $5b^2$ and a^2.

(b) What kind of a number is $\sqrt{5}$?

34. **(Writing)** In Chapter 4, you learned that if two natural numbers a and b have no common divisor other than 1, and if b divides a^2, then b also divides a. Use this fact to give a very short proof that $\sqrt{2}$ is an irrational number. [*Suggestion*: Assume that there is a fraction $\frac{a}{b}$ in simplest form for which $\sqrt{2} = \frac{a}{b}$. Next, square both sides, multiply by b, and carefully examine the equation you obtain.]

Making Connections

35. **(Writing)** Consider the unending sequence of shaded unit squares shown here.

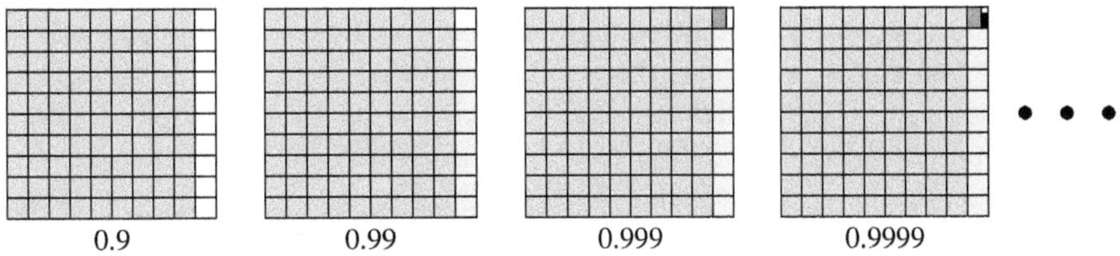

Explain why the infinite sequence of diagrams shows that

$$\frac{9}{10} + \frac{9}{100} + \frac{9}{1000} + \frac{9}{10000} + \cdots = 1.$$

36. Modify the sequence of shaded unit squares in problem 35 to explain why it is reasonable to write

$$\frac{1}{2} + \frac{1}{4} + \frac{1}{8} + \frac{1}{16} + \cdots = 1.$$

37. Use a calculator to find the decimal representations of these fractions:

(a) $\frac{1}{11}$ **(b)** $\frac{15}{22}$ **(c)** $\frac{5}{21}$ **(d)** $\frac{2}{13}$

State Assessments

38. (Grade 6)

(a) What decimal number is represented by the shaded portion of this diagram?

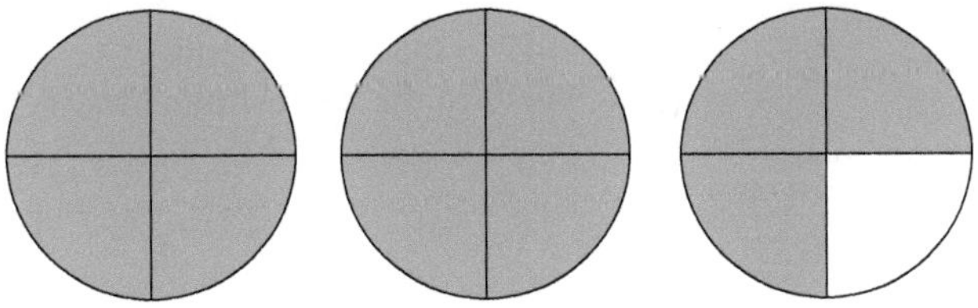

(b) Which diagram of unit squares is shaded to represent 1.47?

A.

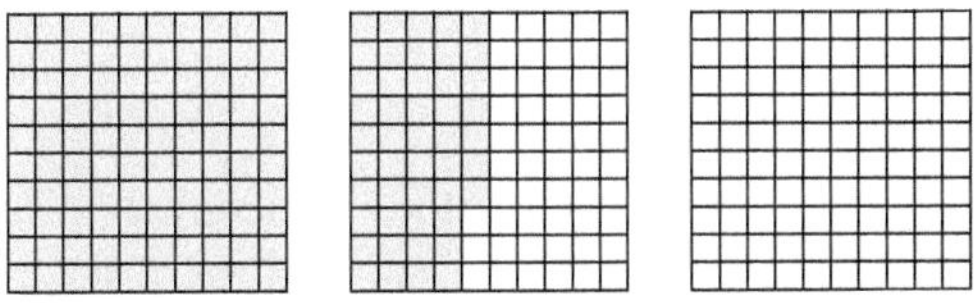

B.

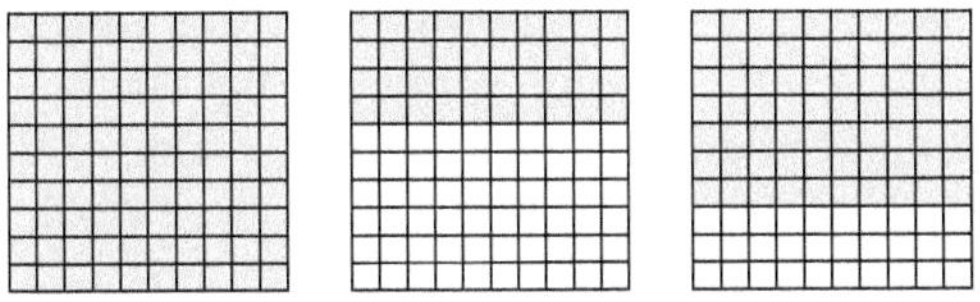

C.

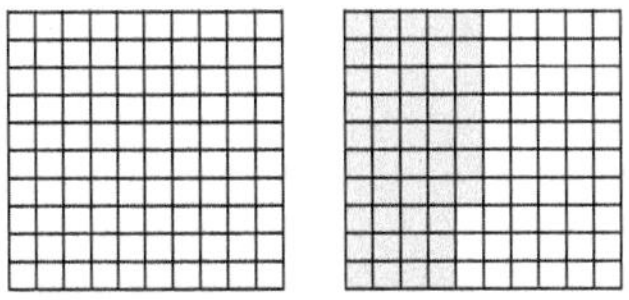

D.

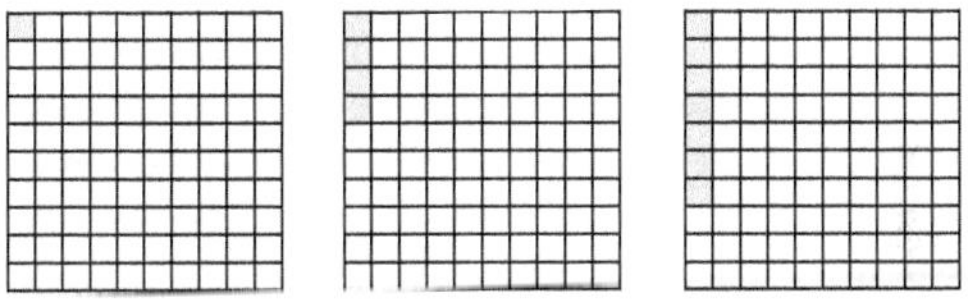

39. (Grade 5)
Use the picture below to answer the question.
Radio station **KXYZ** can be found at 94.3 on your radio dial. Which letter shown represents where you would find **KXYZ**?

A. letter A B. letter B C. letter C D. letter D

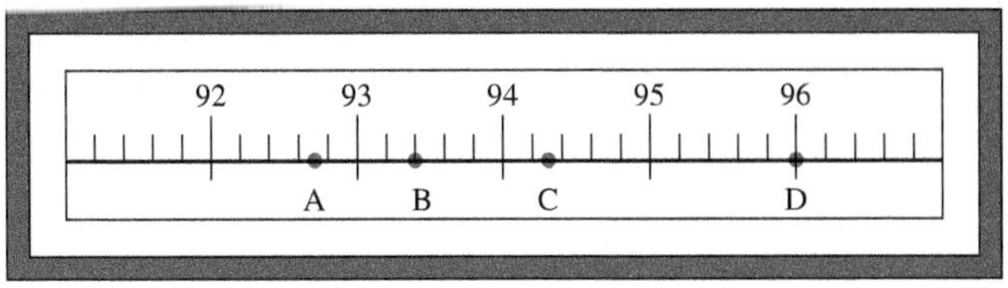

40. (Grade 7)
Between which two consecutive integers is the value of this irrational number? $\sqrt{201}$

A. 10 and 11 B. 12 and 13
C. 14 and 15 D. 20 and 21

41. (Grade 8)
Which number is irrational?

A. $(2.5)^2$ B. $\sqrt{26}$
C. 3.1 D. $\sqrt{36}$

7.2 Computations with Decimals

Rounding Decimals

The real numbers satisfy the same properties of arithmetic that hold for the rational numbers. For example, both addition and multiplication are closed, commutative, and associative operations, and every real number has an additive inverse—its negative—and every nonzero real number has a multiplicative inverse—its reciprocal. Fortunately, too, the familiar computational algorithms for the whole numbers continue to apply to decimal numerals, with the only change being that the decimal point for the outcome of an operation must be placed correctly. Often, errors can be detected by first estimating a reasonable outcome and checking that the computed answer agrees with the estimate. The estimates can be calculated by mental arithmetic using rounded values of the decimals.

Consider this example: A dressmaker wants to buy material to make into a dress for an upcoming fashion show. The material comes in 40-inch widths and the pattern requires a piece 3.75 yards long. If the material costs $15.37 per yard, how much will the dressmaker be charged? Using a calculator, we find that the cost would be

$$(3.75)(\$15.37) = \$57.6375,$$

and the dressmaker would be charged $57.64, the cost rounded to the nearest cent. The process of rounding here is precisely the same as it was for integers. Thus, 57.6375 rounded

- to the nearest integer is 58,
- to the nearest tenth is 57.6,
- to the nearest hundredth is 57.64,
- to the nearest thousandth is 57.638,

and so on. As before, we are using the 5-up rule.

RULE The 5-Up Rule for Rounding Decimals

To round a decimal to a given place, consider the digit in the next place to the right. If it is smaller than 5, replace it and all of the digits to its right with 0. If it is 5 or larger, replace it and all digits to the right by 0 and increase the digit in the given place by one. Replaced digits to the right of the decimal are then dropped to give the rounded decimal.

EXAMPLE 7.9

Rounding Decimals

Round each of these decimals to the indicated position.

(a) 23.2047 to the nearest integer
(b) 3.6147 to the nearest tenth
(c) 0.015 to the nearest hundredth
(d) 8.53972 to the nearest thousandth
(e) 3482.3 to the nearest hundred

Solution

(a) 23 **(b)** 3.6 **(c)** 0.02
(d) 8.540 **(e)** 3500

Adding and Subtracting Decimals

Addition and subtraction can be illustrated with any of the representations of decimals discussed in the previous section. For example, an addition with ten-dollar bills, one-dollar coins, dimes, and pennies is shown in the following figure. Here an intermediate step is shown, where regrouping the pennies and dollars is needed to obtain the final answer. It is useful to include the decimal point to reinforce the need to align the numerals into columns that correspond to the place values of the digits. The answer can be checked by rounding to the nearest dollar: We expect the answer should be close to $18 + $6 = $24, and this is true, since the exact answer of $23.71 rounds off to $24.

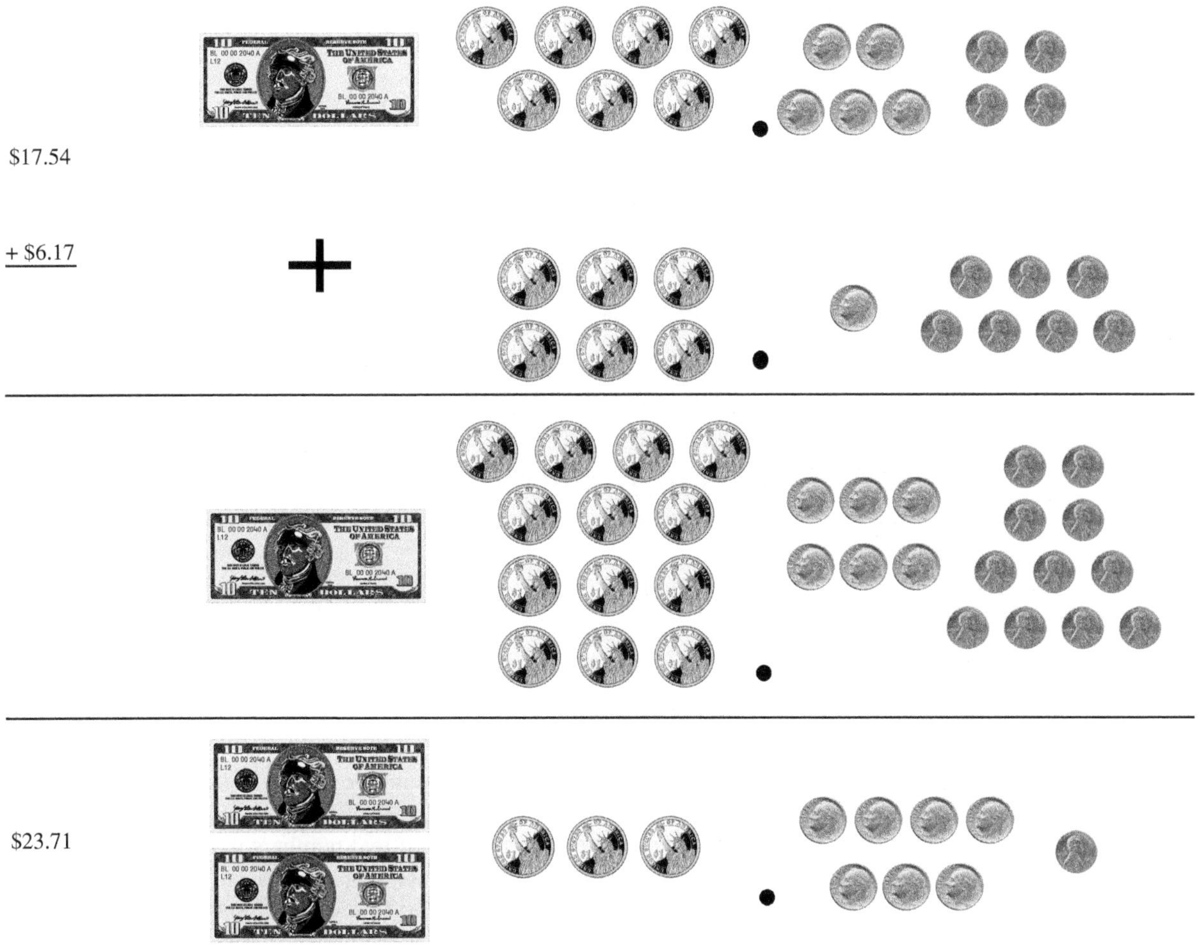

Figure 7.8
The sum 17.54 + 6.17 = 23.71 illustrated in U.S. currency, with the final answer requiring regrouping

The subtraction of decimals can also be computed with the usual algorithms of subtraction, again with attention given to careful vertical alignment of the digits columns to both the left and right of the decimal point. As an example, suppose that we want to subtract 2.71 from 37.762. The calculation is much the same as before, and we have

$$\begin{array}{r} 37.762 \\ -\ \ 2.710 \\ \hline 35.052 \end{array}$$

Notice how we write the problem in vertical style, lining up the decimal points, and then subtract essentially as we subtract integers.

We should estimate and perform mental calculation as well. Thus, we think

approximately 38 minus approximately 3 gives approximately 35

and thereby avoid gross errors.

EXAMPLE 7.10

Adding and Subtracting Decimals

Compute each of these by estimating and by hand:

(a) 23.47 + 7.81 **(b)** 351.42 − 417.815

Solution

(**a**) By estimating: Approximately 23 plus approximately 8 gives approximately 31.

By hand:

$$\begin{array}{r} \overset{1}{2}\overset{1}{3}.47 \\ +\ 7.81 \\ \hline 31.28 \end{array}$$

(**b**) By estimating: Approximately 350 minus approximately 400 gives approximately −50.

By hand:

$$\begin{array}{r} \overset{3}{\cancel{4}}{}^{1}17.\overset{7}{\cancel{8}}{}^{1}15 \\ -3\,51.4\;2 \\ \hline 66.395 \end{array} \qquad \text{or} \qquad \begin{array}{r} \overset{3}{\cancel{4}}\overset{11}{\cancel{1}}7.\overset{7}{\cancel{8}}\overset{11}{\cancel{1}}5 \\ -\ 3\,51.42 \\ \hline 66.395 \end{array}$$

Since $417.815 > 351.42$, the desired answer is -66.395.

Ordering Decimals and the Real Numbers

The real numbers are ordered in the same way as the rational numbers.

DEFINITION Order of the Real Numbers

Given two real numbers a and b, we say that ***a* is less than *b***, and write $a < b$, if and only if there is a positive real number c for which $a + c = b$.

Since $c = b - a$, we can also say that $a < b$ if and only if the difference $b - a$ is positive. When a and b are identified as points on the number line, $a < b$ if and only if point a is to the left of point b.

It is also important to determine the order of real numbers that is given by their decimal numerals. Certainly, every negative real number is less than every positive real number. Now suppose that a and b are both positive. If the integer part of a (the whole number to the left of the decimal point) is less than the integer part of b, then $a < b$. If the integer parts of the positive real numbers a and b are the same, we then examine the digits of a and b to the right of the decimal point. It the first disagreement has a larger digit for b than for a, then $a < b$. If a and b are not both positive, then $a < b$ if and only $-b < -a$.

Here are four examples of ordering pairs of decimals:

- $-7.89 < 2.145$ since −7.89 is negative and 2.145 is positive
- $4.98 < 6.21$ since both numbers are positive and the integer part 4 is less than 6
- $83.5677 < 83.5683$ since the integer parts 83 are the same and the first unequal digits to the right of the decimal point are 7 and 8 and $7 < 8$
- $-6.3 < -4.7$ since $4.7 < 6.3$

EXAMPLE 7.11

Ordering Decimals

In each case, decide which of the decimals represents the lesser number.

(**a**) 2.35714 and 2.3570946 (**b**) 23.45 and $23.4\overline{5}$ (**c**) 4.98 and 12.3

Solution

(**a**) Here, the first digits from the left that differ are 1 and 0. Since $0 < 1$, it follows that $2.3570946 < 2.35714$.

(**b**) Since $23.4\overline{5} = 23.4555\ldots$ and $23.45 = 23.45\overline{0}$, the first digits from the left that differ are 0 and 5. Since $0 < 5$, it follows that $23.45 < 23.4\overline{5}$.

(**c**) The integer part 4 is less than the integer part 12. Thus, $4.98 < 12.3$.

Multiplying Decimals

Tom Swift tried out his new Ferrari on a racetrack, and drove at 91.7 miles per hour for 15 minutes. How far did he go? Since 15 minutes equals 0.25 hour and distance traveled equals rate times elapsed time, Tom traveled (91.7) · (0.25) miles. For the exact answer, we need to be able to multiply decimals. We see that

$$\begin{aligned} 91.7 \times 0.25 &= 917 \times 10^{-1} \times 25 \times 10^{-2} \\ &= 917 \times 25 \times 10^{-1} \times 10^{-2} \\ &= 22{,}925 \times 10^{-3} \\ &= 22.925. \end{aligned}$$

The product $917 \times 25 = 22{,}925$ is found by the multiplication algorithm for whole numbers, and the product $10^{-1} \times 10^{-2} = 10^{-3}$ determines the placement of the decimal point. It is common to write the multiplication in this way:

$$\begin{array}{r} 91.7 \\ \times 0.25 \\ \hline 4585 \\ 1834 \\ \hline 22.925 \end{array}$$

One digit is to the right of the decimal point.

Two digits are to the right of the decimal point.

Three digits are to the right of the decimal point, since 1 + 2 = 3.

Since this type of calculation is typical of the general case, we summarize it as a computational algorithm.

ALGORITHM Multiplying Decimals

To multiply two decimals, do the following:

1. Multiply as with integers.
2. Count the number of digits to the right of the decimal point in each factor in the product, add these numbers, and call their sum t.
3. Place the decimal point in the product that is obtained so that there are t digits to the right of the decimal point.

Since Tom traveled at almost 100 miles per hour for a quarter of an hour, we should expect he traveled a bit under 25 miles.

EXAMPLE 7.12 Multiplying Decimals

Compute these products by estimating and by hand:

(a) (471.2) · (2.3) **(b)** (36.34) · (1.02)

Solution

(a) By estimating: Approximately 500 times approximately 2 gives approximately 1000.

By hand:

$$\begin{array}{r} 471.2 \\ \times \quad 2.3 \\ \hline 14136 \\ 9424 \\ \hline 1083.76 \end{array}$$

Two digits are to the right of the decimal point.

1 + 1 = 2

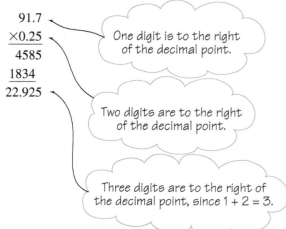

(b) By estimating: Approximately 36 times approximately 1 gives approximately 36.

By hand:
```
   36.34
 × 1.02
    7268
  36340
 37.0668
```

Dividing Decimals

Tom Swift also has his own airplane. If Tom traveled 537.6 miles in 2.56 hours in his airplane, how fast did he travel? The rate equals distance traveled divided by elapsed time. Therefore, we need to compute the quotient 537.6 ÷ 2.56.

$$\frac{537.6}{2.56} = \frac{53{,}760 \times 10^{-2}}{256 \times 10^{-2}} = \frac{53{,}760}{256} = 210.$$

Thus, the problem is reduced to that of dividing 53,760 by 256—that is, to dividing integers. Recall that when confronted by a division such as

$$2.56\overline{)537.6}$$

students are often told to move the decimal point in both the divisor and the dividend two places to the right so that the divisor becomes an integer. The preceding calculation with fractions justifies this rule, and, by hand, we have

Moving the decimal points this way amounts to multiplying both dividend and divisor by 100.

```
            210
2.56.)537.60.
      512
       25 6
       25 6
          0
```

Check:
```
  2.56
 × 210
  2560
  512
537.60
```

7.2 Problem Set

Understanding Concepts

1. Perform these additions and subtractions by hand:

   (a) 32.174 + 371.5 **(b)** 371.5 − 32.174

2. Perform these additions and subtractions by hand:

   (a) 0.057 + 1.08 **(b)** 0.057 − 1.08

3. Perform these multiplications and divisions by hand:

   (a) (37.1) · (4.7) **(b)** (3.71) · (0.47)

4. Perform these multiplications and divisions by hand:

   (a) 138.33 ÷ 5.3 **(b)** 1.3833 ÷ 0.53

5. Estimate the result of each of these computations mentally, and then perform the calculations with a calculator.

   (a) 4.112 + 31.3 **(b)** 31.3 − 4.112

6. Estimate the result of each of these computations mentally, and then perform the calculations with a calculator.

   (a) (4.112) · (31.3) **(b)** 31.3 ÷ 4.112

7. Use mental arithmetic to estimate the value of each of the following decimal expressions, and then calculate the exact value by hand and by calculator:

   (a) 23.07 + 4.8 + 0.971 **(b)** 41.5 − 6.48 + 13.013

8. Use mental arithmetic to estimate the value of each of the following decimal expressions, and then calculate the exact value by hand and by calculator:

   (a) (13.12 × 21.3) ÷ 4.1 **(b)** $17.8 + \left(\frac{3.27}{1.5}\right)$

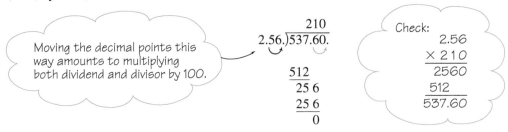

9. When the gasoline pump shut off automatically, 13.6 gallons at $\$3.19^{9/10}$ per gallon had been delivered. What is the amount that will show on the cost dial?

10. How many gallons of gas at $\$3.39^{9/10}$ per gallon can be purchased for $30?

11. Evaluate each of the following multiplications and divisions by correctly positioning the decimal point:

(a) 34.796×10^3 **(b)** 34.796×10^{-3}

12. Evaluate each of the following multiplications and divisions by correctly positioning the decimal point:

(a) $34.796 \div 10^2$ **(b)** $34.796 \div 10^{-2}$

13. The number 456,123,789 can be written in **scientific notation** as the product of 4.56123789 and 10^8. That is, $456{,}123{,}789 = 4.56123789 \times 10^8$. The decimal point always immediately follows the first nonzero digit. Determine the power of 10 needed to write these decimals in scientific notation:

(a) $34{,}762 = 3.4762 \times 10^a$
(b) $4{,}256{,}000 = 4.256 \times 10^b$
(c) $0.009031 = 9.031 \times 10^c$
(d) $0.000004320017 = 4.320017 \times 10^d$

14. If a number written in scientific notation (see problem 13) is rounded, then the digits that remain are called the **significant digits.** For example, 4.0285×10^6 becomes 4.03×10^6 when rounded to three significant digits. Write each of these numbers in scientific notation with the indicated number of significant digits:

(a) 276,543,421 to three significant digits.
(b) 0.000005341 to two significant digits.
(c) 376,712.543248 to two significant digits.

15. Calculate these multiplications and divisions with a suitable calculator, and write the answer in scientific notation to three significant digits:

(a) $0.0000127 \times 0.000008235$
(b) $98{,}613{,}428 \times 5{,}746{,}312$
(c) $0.0000127 \div 98{,}613{,}428$
(d) $98{,}613{,}428 \div 0.000008234$

16. For each of these multiplications and divisions, estimate the answer, calculate the result to three significant digits on a suitable calculator, and then write the answer in scientific notation:

(a) $(7.123 \times 10^5) \cdot (2.142 \times 10^4)$
(b) $(7.123 \times 10^5) \div (2.142 \times 10^4)$
(c) $(7.123 \times 10^5) \cdot (2.142 \times 10^{-9})$
(d) $(7.123 \times 10^{-2}) \div (2.142 \times 10^8)$

17. For each of these sequences of numbers, write the numbers in order of increasing size (from least to greatest):

(a) $0.017, 0.007, 0.01\overline{7}, 0.027$
(b) $25.412, 25.312, 24.999, 25.41\overline{2}$

18. For each of these sequences of numbers, write the numbers in order of increasing size (from least to greatest):

(a) $\frac{9}{25}, 0.35, 0.36, \frac{10}{25}, 0.\overline{35}$ **(b)** $\frac{1}{4}, \frac{5}{24}, \frac{1}{3}, \frac{1}{6}$

Into the Classroom

19. **(Writing)** Create a lesson plan that teaches the addition and subtraction of decimals using place-value cards, where a moveable marker (small pebble, for example) indicates the position of the decimal point. The markers on this place-value card represent the decimal numeral 40.723.

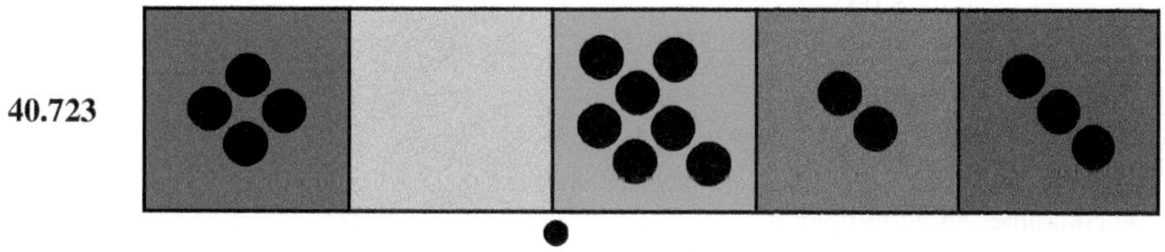

20. One can use the rectangular area model of multiplication to help students understand multiplication of decimals, just as was done to promote their understanding of multiplication of natural numbers and rational numbers. The diagram shown illustrates the product 2.3×3.2.

(a) Identity the colored regions in the diagram with the numbers shown in the hand calculation of the product.

(b) How would you use the diagram to justify the exchange shown in the hand calculation?

(c) Discuss how the diagram helps to illustrate the rule for the proper placement of the decimal point in the final answer.

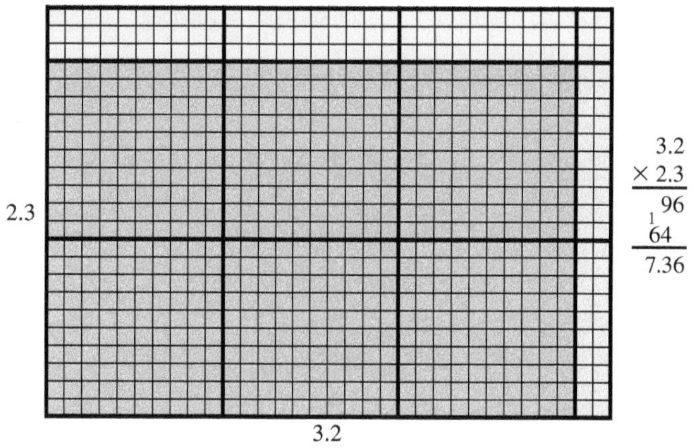

Responding to Students

21. When asked to round 7.2447 to the nearest tenth, Toni proceeded as follows: 7.245, 7.25, 7.3. Toni seems to need help. How would you help her?

22. When asked to perform the decimal computations 47 + 3.2, 6.9 − 5, 3.6 × 0.2, and 8.36 ÷ 0.4, some students submitted the answers shown here. Respond to each student in helpful ways that provide an understanding of the error made and how it can be avoided in the future.

(a)	(b)	(c)	(d)
47	6.9	3.6	2.9
+3.2	−5	×0.2	0.4)8.36
7.9	6.4	7.2	

23. (**Writing**) A student makes the following error pattern in several problems:

4.8 + 32 + 0.79 + 7.8 = 23.7

46.325 + 234.56 + 13.567 + 2.7964 = 111.312

Describe the error the student is making, and state what steps you would use to assist the student in learning how to do the problem correctly.

24. Ordering decimal numbers is sometimes difficult for children. Analyze what you believe might lead to the following mistakes and how you might correct them:

(a) Jorge thinks that 0.5 is smaller than 0.06.

(b) Jenna thinks that 0.34 is smaller than 0.334.

(c) Zach thinks that 0.4567 is smaller than 0.45.

Thinking Critically

25. The sum of the numbers in any two adjacent blanks is the number immediately below and between the two numbers. Complete each problem so that the same pattern holds. The first one has been completed for you.

(a)
2.107 1.3 4.26
3.407 5.56
8.967

(b)
21.06 3.21 ____
____ 5.00
____

(c)
____ 0.041 ____
2.415 ____
7.723

(d)
____ 1.414 ____
____ ____
3.142

(e) Can any of the problems be completed in more than one way? Explain.

26. Fill in the blanks so that each of these sequences is an arithmetic progression:

(a) 3.4, 4.3, 5.2, ____, ____, ____

(b) −31.56, ____, −21.10, ____, ____, ____

(c) 0.0114, ____, ____, 0.3204, ____, ____

(d) 1.07, ____, ____, ____, −9.21, ____

27. Fill in the blanks so that each of these sequences is a geometric progression.

(a) 2.11, 2.321, ____, ____, ____

(b) 35.1, ____, 1.404, −0.2808, ____

(c) 6.01, ____, ____, 0.75125, ____

28. Recall that the Fibonacci numbers are the numbers 1, 1, 2, 3, 5, 8, 13, 21, 34, 55, 89, . . .

(a) Compute the decimal expansion of $\frac{1}{89}$ correct to 10 decimal places. Note that 89 is the 11th Fibonacci number.

(b) Are you surprised at the decimal expansion of $\frac{1}{89}$, particularly in the 7th, 8th, 9th, and 10th places?

(c) Compute this sum by hand:

0.0
0.01
0.001
0.0002
0.00003
0.000005
0.0000008
0.00000013
0.000000021
0.0000000034
0.00000000055
0.000000000089
0.0000000000144
0.00000000000233
+0.000000000000377

the Fibonacci numbers divided by powers of 10

(d) Make a conjecture on the basis of the result in part (c). Recall that many decimals never end and consequently are actually sums of infinitely many terms.

(e) Does it make sense to set $F_0 = 0$?

29. Place numbers in the circles in these diagrams so that the numbers in the large circles are the sums of the numbers in the two adjacent smaller circles:

(a)
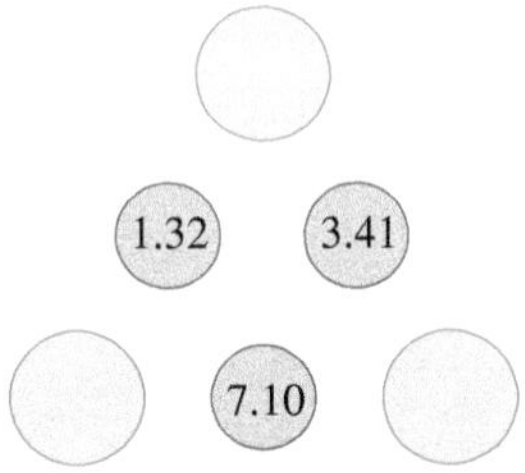

(b)
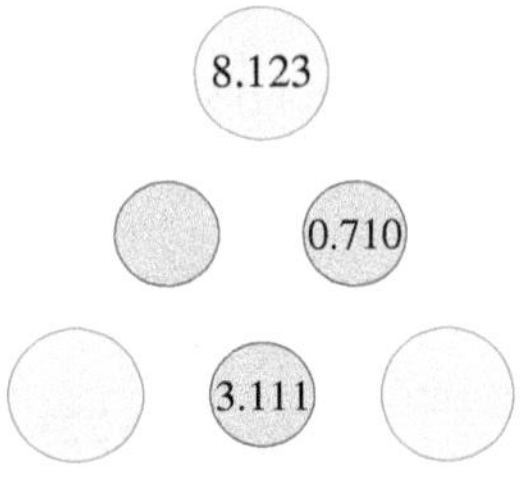

(c)
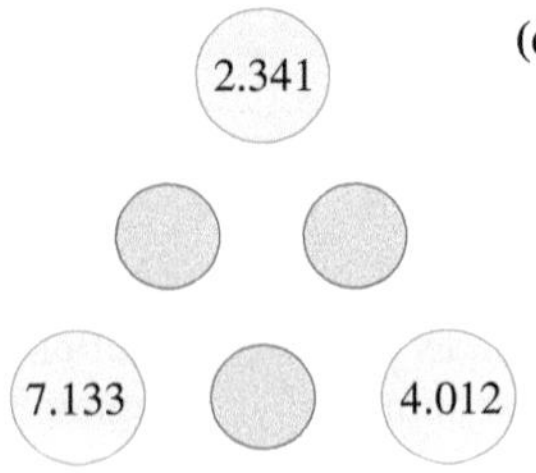

(d)
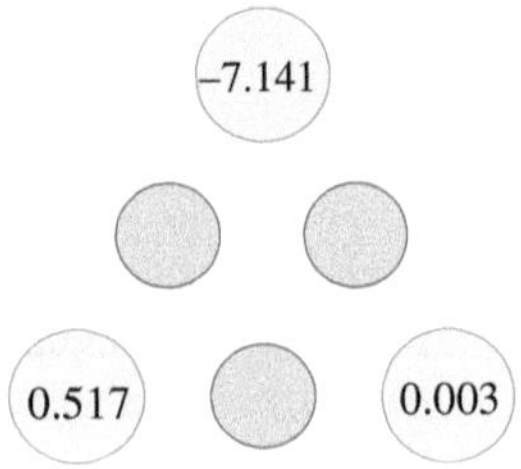

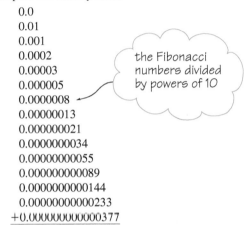

Making Connections

30. Kristina bought pairs of gloves as Christmas presents for three of her best friends. If the gloves cost $9.72 a pair, how much did she spend for these presents?

31. Yolanda also bought identical pairs of gloves for each of her four best friends. If her total bill was $44.92, how much did each pair of gloves cost?

32. Dante cashed a check for \$74.29 and then bought an MP3 player for \$42.91 and a special pair of earphones for \$17.02. After paying for his purchases, how much did he have left?

33. A picture frame 2.25 inches wide surrounds a picture 17.5 inches wide by 24.75 inches high.

(a) What is the area of the picture frame?

(b) What is the area of the picture?

34. When very large or very small numbers occur, it is common to write them in scientific notation, as described in problem 13. That is, the number is written as the product of a decimal with a single nonzero digit to the left of the decimal point and a power of 10. For example, instead of expressing Earth's mass as about 5,973,600,000,000,000,000,000,000 kilograms, it is more appropriate to write the mass as 5.9736×10^{24}, which indicates that the estimate is accurate to five **significant digits.** Similarly, it is better to write very small numbers in scientific notation. For example, the mass of an electron is $9.1093826 \times 10^{-31}$ kilograms in scientific notation with eight significant digits—a number that is simpler to read than the decimal 0.00000000000000000000000000000091093826 kilogram. Convert each of the following numbers either from or to scientific notation, and determine the number of significant digits:

(a) 200,000,000,000 (number of stars in the Andromeda galaxy)

(b) 0.000000000753 kilogram (mass of a dust particle)

(c) 10,300,000,000,000,000,000,000 (number of carbon atoms in a 1-carat diamond)

(d) 2×10^{-23} (mass of a carbon atom in grams)

(e) 7.1×10^{9} (current world population)

(f) 6.022×10^{23} (Avogadro's number)

State Assessments

35. (Grade 5)
Which number is between 4.8 and 4.9?

A. 4.91 B. 3.85 C. 4.86 D. 4.79

36. (Grade 4)
Mr. Vesco's class has a rain gauge that measures the amount of rain each day and is then emptied. During the last five days it has rained 0.231, 0.782, 0.481, 0.093, and 0.898 in inches. Which is the best estimate of the total rainfall over these five days?

A. 1.5 in. B. 2 in. C. 2.5 in. D. 3 in.

37. (Grade 5)
You have a summer job selling tickets to the local car wash. Each ticket is \$8.50 and you are paid by multiplying the amount of money taken in by 0.1. How much money have you earned if you have sold 112 tickets?

A. \$9.52 B. \$95.20 C. \$952 D. \$9520

38. (Grade 4)
Jason worked 4.6 hours at his after school this week, and 5.9 hours the week before. How many more hours did Jason work the week before?

A. 1.3 B. 0.3 C. 0.7 D. 1.7

39. (Grade 5)
It cost \$43.23 for Kevin to fill the gas tank on his car, for which he paid \$3.93 per gallon. How many gallons of gas did Kevin buy?

A. 10 B. 11 C. 12 D. 13

40. (Grade 4)
Find the sum 1.54 + 0.66

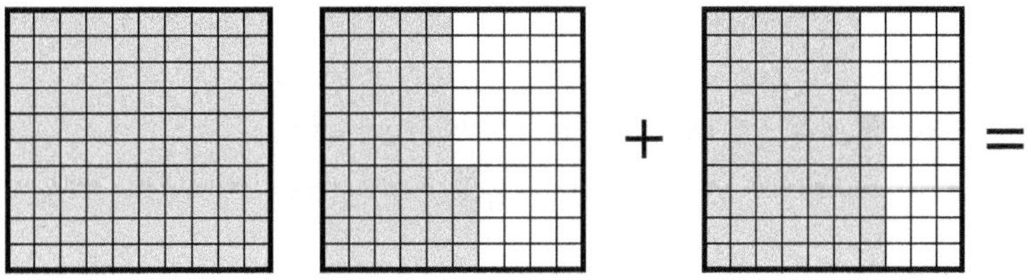

A. 1.0 B. 1.20 C. 2.00 D. 2.20

41. (Grade 4)
Find the difference 2.50 − 0.95

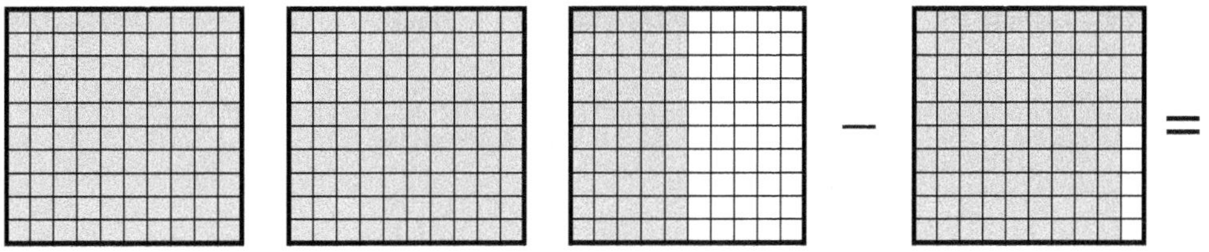

A. 0.45 B. 1.45 C. 1.55 D. 2.05

42. (Grade 6)
Amy bought four ice cream sundaes for \$3.49 each. What was the total amount of her purchase?

A. \$3.53 B. \$12.49 C. \$12.98 D. \$13.96

7.3 Proportional Reasoning

In grades K–4, a main focus is the development of the additive principles of arithmetic. For example, if 6 new students join a history class of 18 students, there are now 24 students in the classroom, since $18 + 6 = 24$. Even multiplication, since it is viewed as repeated addition, is initially considered within an additive conceptual framework. However, in the upper elementary and middle grades 5–8, a new goal emerges: Students should now see that multiplicative relations are essential in understanding how *relative* quantities can be compared and how changes in quantities can be measured by a *rate*.

The fourth of the Standards for Mathematical Practice (SMP 4) from the **Common Core** states that

In middle grades, a student might apply proportional reasoning to plan a school event or analyze a problem in the community.

SMP 4 Model with mathematics.

For example, suppose 6 new students join a PE class of 24 students, resulting in 30 students. In additive reasoning, both the history and PE class underwent the same change in number, since both classes have 6 additional students. Proportional reasoning, however, gives a different answer: Because $\frac{6}{24} = \frac{1}{4}$, a "quarter" of the students are new to history, since $\frac{6}{30} = \frac{1}{5}$, a "fifth" of the students are new to the PE class. Therefore, there is a larger *rate* of change in the history class. If 8 new students were added to the PE class, then, since $\frac{6}{24} = \frac{8}{32}$, proportional reasoning allows us to say that the history and PE classes have increased by the same *proportional* amount.

When a part is compared to another part or to a whole, the relative sizes of the two parts can be represented by a fraction known as a *ratio.* What is new is that the individual sizes of the parts and the whole no longer matter in the ratio. For example, if 8 new students join the PE class of 24, then the history and PE classes have the same ratio, $\frac{1}{4}$, of new students to total students, even though the classes are not the same size. In terms of the definition that will be given shortly, the two classes will be said to have grown by the same *proportion.*

In this section, we begin with ratios and proportions and then apply these concepts to solve problems by proportional reasoning. It is important to understand when and how proportional reasoning can be used as a problem-solving strategy. It is equally important to know when proportional reasoning is inappropriate. For example, it may be tempting to say that a "family-size" 20-inch-diameter pizza will feed twice as many people as the "small" 10-inch pizza. But we see from the formula πr^2 for the area of a circle that the respective areas of the pizzas are $\pi(10)^2 = 100\pi$ and $\pi(5)^2 = 25\pi$, so the larger pizza will feed four times the number of people as the smaller one.

Ratio

At basketball practice, Caralee missed 18 free throws out of 45 attempts. Since she made 27 free throws, we say that the **ratio** of the number missed to the number made was 18 to 27. This ratio can be expressed by the fraction $\frac{18}{27}$ or by the notation 18 : 27. The notation 18 : 27 is read "18 to 27," as in the statement "The ratio of the number of free throws Caralee missed to the number she made was 18 : 27." We will always use the fraction notation in what follows.

Other ratios from Caralee's basketball practice are

- the ratio of the number of shots made to the number attempted, $\frac{27}{45}$;
- the ratio of the number of shots missed to the number attempted, $\frac{18}{45}$; and
- the ratio of the number of shots made to the number missed, $\frac{27}{18}$.

DEFINITION Ratio

If a and b are real numbers with $b \neq 0$, then the **ratio of a to b** is the quotient $\frac{a}{b}$.

Ratios occur with great frequency in everyday life. If you use 10.4 gallons of gasoline in driving 400.4 miles, the efficiency of your car is measured in miles per gallon given by the ratio $\frac{400.4}{10.4}$, or 38.5 miles per gallon. If Lincoln Grade School has 405 students and 15 teachers, the student–teacher ratio is the quotient $\frac{405}{15}$. If José Varga got 56 hits in 181 times at bat, his batting average is the ratio $\frac{56}{181}$. The number of examples that could be cited is almost endless.

The preceding examples show that there are at least three applications of a ratio:

1. **A ratio measures the relative size of different parts.**
   Here is an example from geometry:
   - Given a circle, the ratio of its circumference C (distance around) to its diameter D (distance across) is the same for all circles, namely, the real number universally denoted by π. That is, $\frac{C}{D} = \pi$.

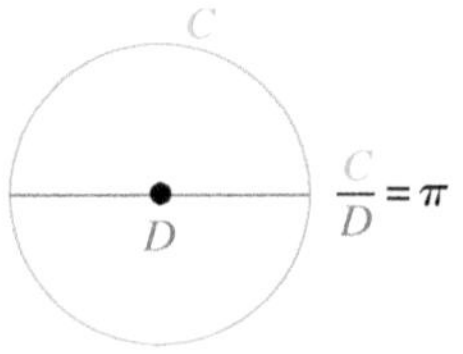

2. **A ratio measures the relative size of a part to a whole.**
   Here is an example that is well known to baseball fans:
   - A player's batting average is the ratio of the number of times the player made a hit to the total number of times at bat.
3. **A ratio measures the rate of the change in one quantity with respect to a corresponding change in a second quantity.**
   For example, commodities are often priced as a rate, so we may pay $3.85[9] for every gallon of gas or we may pay $8.59 for every pound of coffee. A common rate from plane coordinate geometry is the slope of a line:
   - Given a line in the coordinate x,y-plane, its slope (steepness) is the ratio of the change made in the vertical (y-axis) direction (rise) to the change made in the horizontal (x-axis) direction (run).

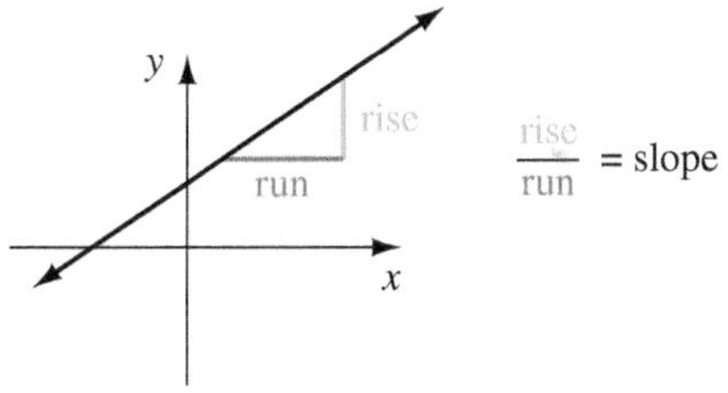

When the amount of change of a quantity is compared to a change by 1 unit of the second quantity, the rate is called a **unit rate.** For example, if a car traveled 408 miles and used 16 gallons of gasoline, then its unit rate of fuel consumption is 25.5 miles per gallon (mpg), as shown with the following calculation:

$$\frac{408 \text{ miles}}{16 \text{ gallons}} = \frac{408 \text{ miles} \div 16}{16 \text{ gallons} \div 16} = \frac{25.5 \text{ miles}}{1 \text{ gallon}} = 25.5 \text{ mpg}.$$

EXAMPLE 7.13

Determining Ratios

Determine these ratios:

(a) The ratio of the number of boys to the number of girls in Martin Luther King High School if there are 285 boys and 228 girls
(b) The ratio of the number of boys to the number of students in part (a)

Solution

(a) The desired ratio is $\frac{285}{228}$.

(b) Since the total number of students is $285 + 228 = 513$, the desired ratio is $\frac{285}{513}$.

The ratio of the number of boys to the number of students in Martin Luther King High School was shown to be $\frac{285}{513}$, or 285 to 513. This is certainly correct, but it is not nearly as informative as it would be if the ratio were written in simplest form. Thus,

$$\frac{285}{513} = \frac{5}{9},$$

$285 \div 57 = 5$
$513 \div 57 = 9$

and this ratio says that $\frac{5}{9}$, or a little more than $\frac{1}{2}$, of the students in Martin Luther King High School are boys. Expressing a ratio by a fraction in simplest form, or even an approximate fraction, is often useful and informative.

EXAMPLE 7.14

Expressing Ratios in Simplest Form

Express these ratios in simplest form:

(a) The ratio of 385 to 440

(b) The ratio $\frac{432}{504}$

Solution

(a) The ratio of 385 to 440 is the quotient $\frac{385}{440}$. Expressing this quotient in simplest form, we have

$$\frac{385}{440} = \frac{7}{8}.$$

Thus, in simplest form, the ratio is 7 to 8.

(b) Expressing the quotient in simplest form, we have

$$\frac{432}{504} = \frac{6}{7}.$$

Thus, in simplest form, the ratio is $\frac{6}{7}$, or 6 to 7.

EXAMPLE 7.15

Determining a Less Obvious Ratio

If one-seventh of the students at Garfield High are nonswimmers, what is the ratio of nonswimmers to swimmers?

Solution

Understand the Problem

The desired ratio is the number of nonswimmers attending Garfield High School divided by the number of swimmers.

Devise a Plan

The needed numbers are not given in the problem, and we don't even know how many students attended the school. If, for example, there were 700 students attending the school, then $\frac{1}{7} \cdot 700 = 100$

Use algebraic reasoning by introducing a variable.

would be nonswimmers and the remaining $600 = \frac{6}{7} \cdot 700$ would be swimmers. But suppose there are n students attending Garfield. Then $\frac{1}{7} \cdot n = \frac{n}{7}$ are nonswimmers and $\frac{6}{7}n = \frac{6n}{7}$ are swimmers. Perhaps we can use these expressions to determine the desired ratio.

Carry Out the Plan

Letting n denote the number of students attending Garfield High, we just saw that $\frac{n}{7}$ students are nonswimmers and $\frac{6n}{7}$ are swimmers. Therefore, the desired ratio is

$$\frac{\frac{n}{7}}{\frac{6n}{7}} = \frac{n}{7} \cdot \frac{7}{6n} = \frac{1}{6}.$$

Look Back

Since the ratio we need to determine is independent of the number of students, we could just assume that 700 students (or even just 7!) attend the school. There would be 100 nonswimmers and 600 swimmers, and the desired ratio is $\frac{100}{600} = \frac{1}{6}$. We can also solve the problem with a pictorial model. In the figure below, we group the students into seven rectangles, each containing the same number of students. The brown "beach" rectangle contains the nonswimmers, and the remaining six blue "pools" contain the swimmers. The figure makes it clear that the ratio of nonswimmers to swimmers is 1 to 6, or $\frac{1}{6}$.

Proportion

Ratios are especially useful when we make comparisons. For example, at basketball practice, Caralee made 27 of 45 free throws attempted and Sonja made 24 of 40 attempts. Which player appears to be the better foul-shot shooter? Since

$$\frac{27}{45} = \frac{3}{5},$$

$\frac{27}{45}$ and $\frac{3}{5}$ are equivalent fractions.

we see that Caralee made $\frac{3}{5}$ of her shots. Similarly, for Sonja, the ratio of shots made to shots attempted is

$$\frac{24}{40} = \frac{3}{5},$$

$\frac{27}{45}$, $\frac{24}{40}$, and $\frac{3}{5}$ are all equivalent fractions.

and these ratios suggest that the two girls are equally skilled at shooting foul shots. Because of its importance in such comparisons, the equality of two ratios is called a **proportion.**

DEFINITION Proportion

If $\frac{a}{b}$ and $\frac{c}{d}$ are two ratios and

$$\frac{a}{b} = \frac{c}{d},$$

this equality is called a **proportion.**

From Chapter 6, we know that

$$\frac{a}{b} = \frac{c}{d}$$

for integers a, b, c, and d if and only if $ad = bc$. But the same argument holds if a, b, c, and d are real numbers. This fact leads to the next theorem.

THEOREM Conditions for a Proportion

The equality

$$\frac{a}{b} = \frac{c}{d}$$

is a proportion if and only if $ad = bc$.

EXAMPLE 7.16

Determining Proportions

In each of these equations, determine x so that the equality is a proportion:

(a) $\frac{28}{49} = \frac{x}{21}$ **(b)** $\frac{2.11}{3.49} = \frac{1.7}{x}$

Solution

We use the preceding theorem, which amounts to multiplying both sides of the equality by the product of the denominators, or "cross multiplying," as we often say.

(a)

$$\frac{28}{49} = \frac{x}{21}$$
$$28 \cdot 21 = 49x$$
$$\frac{28 \cdot 21}{49} = x$$
$$12 = x$$

(b)

$$\frac{2.11}{3.49} = \frac{1.7}{x}$$
$$2.11x = (1.7)(3.49)$$
$$x = \frac{(1.7)(3.49)}{2.11}$$
$$x \doteq 2.81$$

EXAMPLE 7.17

Proving a Property of Proportions

Prove that if

$$\frac{a}{b} = \frac{c}{d},$$

then

$$\frac{a+b}{b} = \frac{c+d}{d}.$$

Solution

Understand the Problem

We are given that $\frac{a}{b} = \frac{c}{d}$ is a proportion and are asked to show that $\frac{a+b}{b} = \frac{c+d}{d}$ is also a proportion.

Devise a Plan

Since it is not immediately clear what to do, we ask what it means to say that $\frac{a}{b} = \frac{c}{d}$ and $\frac{a+b}{b} = \frac{c+d}{d}$ are proportions. Perhaps answering this question will put the problem in a form that is easier to understand and to work on. By the preceding theorem,

$$\frac{a}{b} = \frac{c}{d} \text{ if and only if } ad = bc$$

and

$$\frac{a+b}{b} = \frac{c+d}{d}$$

if and only if $(a + b)d = b(c + d)$. Perhaps we can use the first of these equations to prove the second.

Carry Out the Plan

We want to show that $(a + b)d = b(c + d)$; that is, using the distributive property, we get

$$ad + bd = bc + bd.$$

Say it in a different way.

But we know that

$$ad = bc,$$

and adding bd to both sides of this equation gives

$$ad + bd = bc + bd.$$

Hence,

$$\frac{a+b}{b} = \frac{c+d}{d},$$

as was to be shown.

Look Back

Here, our principal strategy was simply to ask, "What does it mean to say that $\frac{a}{b} = \frac{c}{d}$ and $\frac{a+b}{b} = \frac{c+d}{d}$ are proportions?" Answering this question allowed us to "say it in a different way"—that is, to state an equivalent problem that proved to be quite easy to solve. The strategy **say it in a different way** is often very useful.

Applications of Proportional Reasoning

Suppose that a car is traveling at a constant unit rate of speed of 55 miles per hour. Table 7.2 gives the distances the car will travel in different periods.

TABLE 7.2 DISTANCE TRAVELED IN *t* HOURS AT 55 MILES PER HOUR

Time *t*	1	2	3	4	5	6	7	8
Distance *d*	55	110	165	220	275	330	385	440

The ratios $\frac{d}{t}$ are all equal for the periods shown; that is,

$$\frac{55}{1} = \frac{110}{2} = \frac{165}{3} = \frac{220}{4} = \frac{275}{5} = \frac{330}{6},$$

and so on. Thus, each pair of ratios from the list forms a proportion. Indeed, $\frac{d}{t} = 55$ for every pair d and t. This relationship is also expressed by saying that the distance traveled at a constant rate is proportional to the elapsed time. In the current example,

$$d = 55t$$

for every pair d and t. The number 55 is called the **constant of proportionality.**

DEFINITION ***y* is Proportional to *x***

If the variables x and y are related by the equation

$$\frac{y}{x} = k$$

$$y = kx,$$

then ***y* is proportional to *x*** and k is called the **constant of proportionality.**

This situation is extremely common in everyday life. Gasoline consumed by your car is proportional to the number of miles traveled, the cost of pencils purchased is proportional to the number of pencils purchased, income from the school raffle is proportional to the number of tickets sold, and so on.

Sometimes it is said that "y is *directly* proportional to x" when $y = kx$, to distinguish that situation from "y is *inversely* proportional to x," when the variables are related by the equation $y = \frac{k}{x}$ for some constant k. For example, if you always commute 12 miles to work, then your travel time t is inversely proportional to your rate of travel r, since we have the equation $t = \frac{12}{r}$. A famous example is Newton's "inverse-square law" of gravity, which states that the force F of gravity on an object is inversely proportional to the square of the distance r to Earth's center. That is, there is a constant k such that $F = \frac{k}{r^2}$.

EXAMPLE 7.18

Comparing Distances

Khalid and his brother Ahmed both left the same city at 1 P.M., but Khalid averaged 63 miles per hour in his new car and Ahmed averaged just 51 miles per hour in his older car. How much farther has Khalid traveled than Ahmed at 4:30 P.M. that afternoon?

Solution

One way to find the answer is to compute how far each brother traveled and then take the difference. Thus, Khalid traveled $3.5 \times 63 = 220.5$ miles and Ahmed traveled $3.5 \times 51 = 178.5$ miles, so Khalid traveled $220.5 - 178.5 = 42$ miles farther than his brother.

A much better method is to apply proportional reasoning directly, since the distance of separation is proportional to the time of travel. Khalid travels $k = 12$ miles per hour faster than his brother, so in $t = 3.5$ hours he is $kt = 3.5 \times 12 = 42$ miles ahead of Ahmed. This calculation is so simple that it can be done with mental arithmetic.

EXAMPLE 7.19

Going Nutty with Proportional Reasoning

Marci liked the mixed nuts she made by combining 6 cups of cashews with 2 cups of pecans. She hopes to make gift boxes with the same mixture of nuts.

(a) If Marci has 5 cups of pecans, how many cups of cashews does she need?
(b) Marci has several tin gift boxes on hand of various sizes, holding 4 cups, 8 cups, 12 cups, 16 cups, and 20 cups. How many cups of each type of nut should be added to these boxes so that each box contains the same mixture of nuts?

Solution

(a) Since 6 cups of cashews and 2 cups of pecans make 8 cups of mixed nuts, a "half" recipe of 3 cups of cashews and 1 cup of pecans makes 4 cups of mixed nuts. Since she wants 3 times as many cashews as pecans, Marci needs to add $3 \times 5 = 15$ cups of cashews to her 5 cups of pecans. This will create 20 cups of mixed nuts.
(b) The examples in part (a) can be organized into a table in which each value in the second row is 3 times the value in the top row and each value in the bottom row is 4 times the value in row 1:

Pecans	1	2	3	4	5
Cashews	3	6	9	12	15
Mixture	4	8	12	16	20

Alternatively, the values in the first and second row are, respectively, $\frac{1}{4}$ and $\frac{3}{4}$ of the value in the third row.

A table with the property that the values in one row are a constant multiple of the values in another row is called a **ratio table.**

EXAMPLE 7.20

Proportional Reasoning with Diagrams

Hannah and her brother Ethan went to the mall with an equal amount of money to buy their mother a birthday present. Hannah purchased a \$26 scarf and Ethan bought his mother a jewelry box for \$34. After their purchases, Ethan has $\frac{2}{3}$ of the money that Hannah has remaining. How much money do Hannah and Ethan now have, and how much did they start with before making their purchases?

Solution

The diagram that follows shows that, after spending \$34, Ethan has $\frac{2}{3}$ of the money that Hannah has remaining after her \$26 purchase. Since Hannah has $\$8 = \$34 - \$26$ more than Ethan, we see that Hannah has \$24, Ethan has \$16, and each went to the mall with \$50.

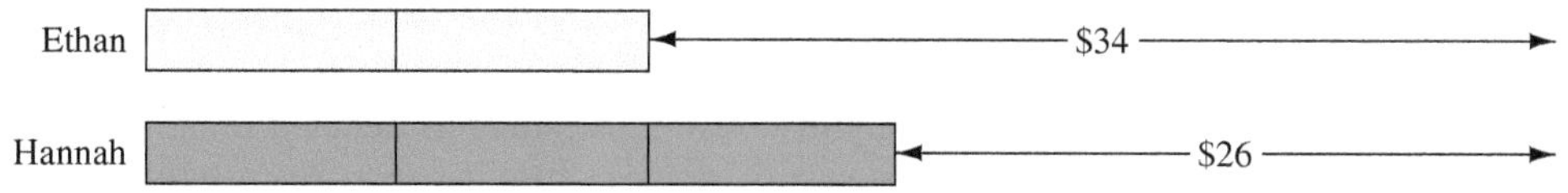

EXAMPLE 7.21

Springs and Weights

The students in Mrs. Stratton's class have created the data plot shown next. The horizontal axis shows the weight they have suspended from a spring, and the vertical axis shows how many inches the spring is stretched. They are about to measure the amount of stretch of a 5-pound weight, but Michelle thinks that she already knows how much stretch to expect. What do you think Michelle's guess is, and what reasoning did she employ to make her guess?

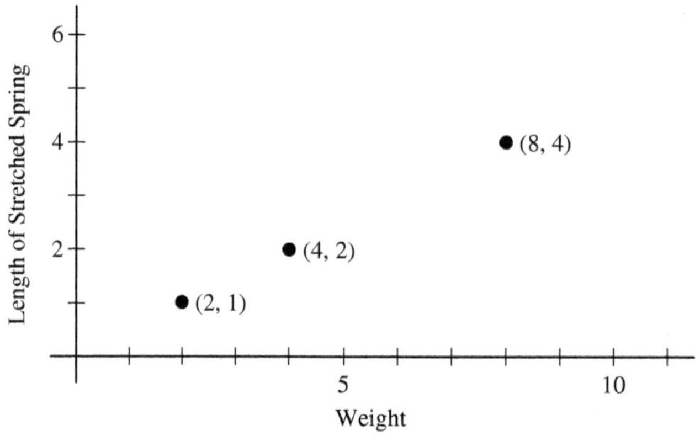

Solution

Michelle has noticed that the amount of stretch in inches is one-half of the weight in pounds. Therefore, a 5-pound weight can be expected to stretch the spring $\frac{5}{2}$, or $2\frac{1}{2}$, inches. Michelle has also noticed that every data point will fall on a line, as shown on the graph that follows. For example, a 10-pound weight will stretch the spring 5 inches, and the point (10, 5) will be on the line.

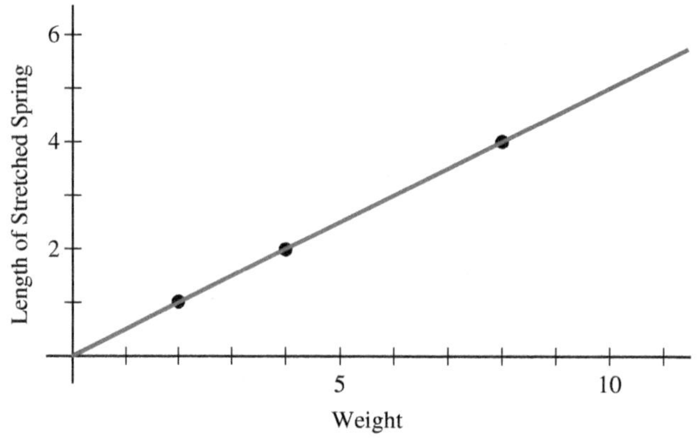

The graphs in Example 7.21 show that the y-coordinate is proportional to the x-coordinate. That is, we have a line with an equation of the form $y = kx$ that passes through the origin. The value of k is the constant of proportionality and gives the ratio of the amount of change in y that corresponds to a constant change in x. That is, k is the *slope* of the line with equation $y = kx$. Lines and their equations will be explored in more detail in Chapter 8.

For the next example, recall that in geometry two figures are said to be similar if they are the same shape but not necessarily the same size—that is, one is a magnification of the other.

EXAMPLE 7.22

Computing the Height of a Tree

Ms. Gulley-Pavey's fifth-grade class had been studying the concepts of ratio and proportion. One afternoon, she took her students outside and challenged them to find the height of a tree in the school yard. After a lively discussion, the students decided to measure the length of the shadow cast by a yardstick and that cast by the tree, arguing that these should be proportional. To help convince the class that this was so, Omari drew the picture shown. If the lengths of the shadows are 4′7″ and 18′9″, respectively, complete the calculation to determine the height of the tree.

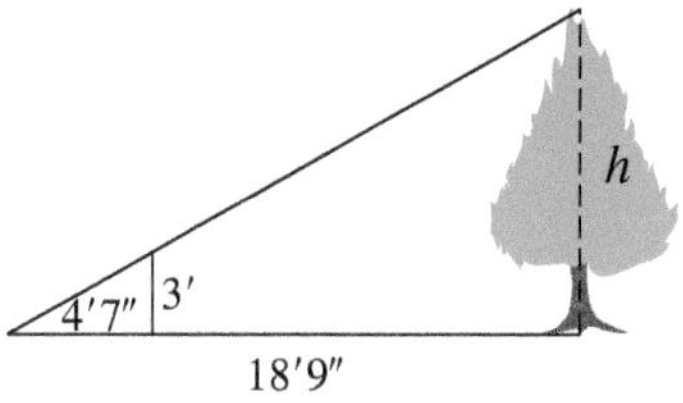

Solution

Since the two triangles shown in the diagram are similar, the lengths of the sides are proportional. Also, $7'' = \frac{7}{12}$ feet and $9'' = \frac{9}{12}$ feet. Therefore, we have the proportion

$$\frac{h}{3} = \frac{18\frac{9}{12}}{4\frac{7}{12}}$$

The units of both ratios must be the same.

$$= \frac{\frac{225}{12}}{\frac{55}{12}} = \frac{225}{12} \cdot \frac{12}{55} = \frac{225}{55} = \frac{45}{11}$$

and it follows that

$$h = \frac{3 \cdot 45'}{11} = \frac{135'}{11} = 12\frac{3'}{11} \approx 12\frac{3'}{12} = 12'3''.$$

Thus, the tree is approximately 12′3″ tall.

COOPERATIVE INVESTIGATION

Discovering a Proportional Relation for Pendulums

Materials Needed

String, heavy washers (or other weights).

Directions

Working with your group, make a pendulum from a long piece of string (say, 150 centimeters long) by tying three or four heavy washers onto the end of the string. Mark the string at 10-centimeter intervals starting from the center of the washers. Hold the pendulum from a suitable support and measure the time T in seconds for the pendulum to make 10 complete swings for various lengths L of string from the support to the center of the washers.

Complete the following table, computing L to the nearest hundredth.

L	T	$\sqrt{L}$
10		
20		
30		
40		
50		

Questions

1. What proportional relation is seen in your table?
2. How long it will take for 10 swings if the pendulum is 100 centimeters long? Check your guess by actually timing 10 swings. Is the time obtained about what you expected?

7.3 Problem Set

Understanding Concepts

1. There are 10 girls and 14 boys in Mr. Tilden's fifth-grade class. What is the ratio of
   (a) boys to girls?
   (b) girls to students?
   (c) boys to students?
   (d) girls to boys?
   (e) students to girls?
   (f) students to boys?

2. There are 32 red, 24 blue, and 16 green candies in a bag. Express each of the following ratios in simplest form.
   (a) red to green candies
   (b) green to blue candies
   (c) blue to red candies
   (d) red candies to candies of any color

3. Determine which of these equations are proportions:
   (a) $\frac{2}{3} = \frac{8}{12}$
   (b) $\frac{21}{28} = \frac{27}{36}$
   (c) $\frac{7}{28} = \frac{8}{31}$
   (d) $\frac{51}{85} = \frac{57}{95}$

4. Determine which of these are proportions:
   (a) $\frac{14}{49} = \frac{18}{60}$
   (b) $\frac{20}{35} = \frac{28}{48}$
   (c) $\frac{1.5}{2.1} = \frac{11.5}{16.1}$
   (d) $\frac{17.1}{6.2} = \frac{31.2}{9.7}$
   (e) $\frac{0.84}{0.96} = \frac{91.7}{104.8}$

5. Determine values of r and s so that each of these equations is a proportion.
   (a) $\frac{6}{14} = \frac{r}{21}$
   (b) $\frac{8}{12} = \frac{10}{r}$
   (c) $\frac{47}{3.2} = \frac{s}{7.8}$

6. Express each of these ratios as fractions in simplest form.
   (a) A ratio of 24 to 16
   (b) A ratio of 296 to 111
   (c) A ratio of 248 to 372
   (d) A ratio of 209 to 341
   (e) A ratio of 3.6 to 4.8
   (f) A ratio of 2.09 to 3.41
   (g) A ratio of 6.264 to 9.396

7. Collene had an after-school job at Taco Time at $9.50 per hour.
   (a) How much did she earn on Monday if she worked $3\frac{1}{2}$ hours?
   (b) On Tuesday she earned $47.50. How long did she work?

8. David bought four sweatshirts for $119.92. How much would it cost him to buy nine sweatshirts at the same price per sweatshirt?

9. Brand A is 43¢ per ounce and Brand B is $7.19 per pound. Compare the two unit rates to determine which brand is more expensive.

10. A 14-ounce box of cornflakes is $2.66, and the 1-pound, 2-ounce box is $3.42. Which size box is the better deal?

11. If s is proportional to t and $s = 62.5$ when $t = 7$, what is s when $t = 10$?

12. The flagpole at Sunnyside Elementary School casts a shadow 9′8″ long at the same time that Mr. Schaal's shadow is 3′2″ long. If Mr. Schaal is 6′3″ tall, how tall is the flagpole, to the nearest foot?

13. A kilometer is a bit more than six-tenths of a mile. If the speed limit along a stretch of highway in Canada is 90 kilometers per hour, about how fast can you travel in miles per hour and still not break the speed limit?

14. The "squareness" of a rectangle can be measured by the ratio of the lengths of the short to the long side. Consider these rectangles: 5 by 7, 8 by 10, 12 by 15, 15 by 21, and 16 by 20. A square has squareness one.
   (a) Which rectangle (or rectangles) is the most squarelike? the least squarelike?
   (b) Are there any rectangles with the same squareness?

15. The Green Cab Company charges $2.50 plus 25¢ per quarter mile traveled. The Red Taxi Company charges $1.80 plus 30¢ per quarter mile. Apparently, you should take a Red Taxi for short trips and a Green Cab for longer trips. Use proportional reasoning to determine the break-even distance.

16. For the given x and y variables, decide whether y is or is not proportional to x.
   (a) y is the cost of x gallons of gasoline.
   (b) y is the perimeter of a square whose sides have length x.
   (c) y is the area of a square whose sides have length x.
   (d) y is the sales tax in Idaho of an item that costs x dollars.
   (e) y is the taxi fare for a ride of x miles.

Into the Classroom

17. How could you use measurement and a diagram like the following to help your students understand proportions and the phrase "is proportional to"?

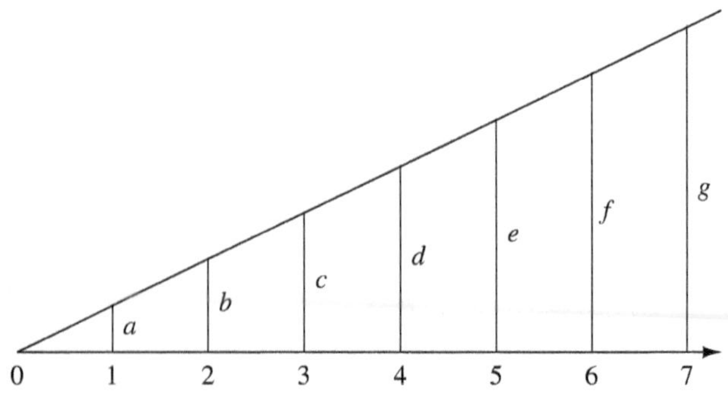

18. Some teachers use diagrams like the following, depicting $a : b = c : d$, or $a/b = c/d$, to help their students visualize proportions:

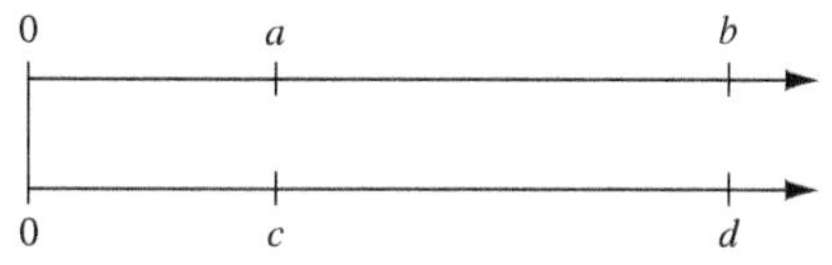

(a) If $a = 6$ when $b = 24$, determine c when $d = 72$.

(b) If $b = 2400$ when $a = 128$, determine d when $c = 512$.

(c) If $b = 24.7$ when $a = 1.4$, determine d when $c = 4.5$.

Responding to Students

19. Jo was informed that data from the American Water Works Association show that the average person uses 20 gallons of water to take a shower. Jo was then asked to find the amount of water used by a person in a week, a month, and a year. She made the following table:

Gallons	20	140	4200	51,100
Days	1	7 (1 week)	30 (1 month)	365 (1 year)

Give helpful advice to Jo.

20. Linda helped with her mother's 20-foot–by–20-foot garden plot. Since Linda thought it would be nice to get twice the amount of produce, she wanted a 40-foot–by–40-foot garden next year. How would you use diagrams to help Linda?

21. (**Writing**) Allison has a 4-inch by 6-inch digital photo. To make an enlargement, she has dragged the size handle on edges of the photo to create an 8-inch by 10-inch photo. She thinks that the picture now looks distorted and has asked you to help with her project. Write a paragraph explaining to Allison what went wrong and how the photo can be enlarged without distortion.

Thinking Critically

22. Suppose a is to b as c is to d; that is,

$$\frac{a}{b} = \frac{c}{d}.$$

(a) Show that b is to a as d is to c.

(b) Show that $a - b$ is to b as $c - d$ is to d.

(c) Show that a is to $a + b$ as c is to $c + d$.

(d) Show that $a + b$ is to $c + d$ as b is to d.

23. Suppose that the variables x and y are related by one of the formulas (a) $y = \frac{x}{2}$, (b) $y = -\frac{x}{2}$, (c) $y = 3 + x$, and (d) $y = x^2$. Match each formula to its plot, shown here:

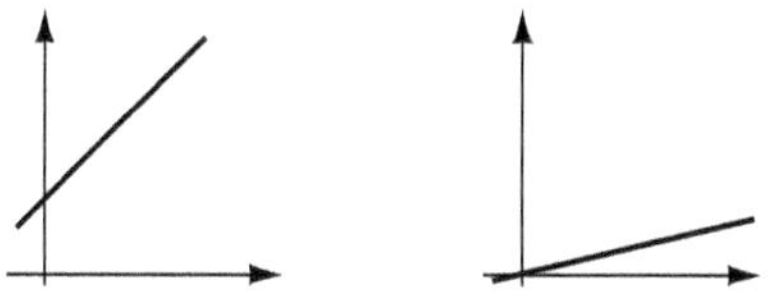

A. B.

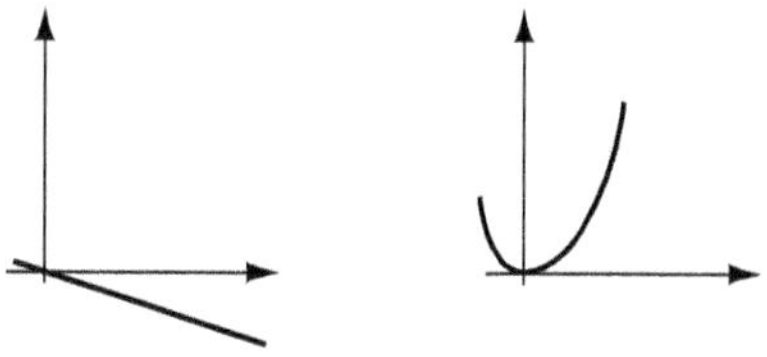

C. D.

24. Pose a problem (inspired by Example 7.20) whose solution becomes evident from the figure shown. Describe how the diagram is used to answer your problem.

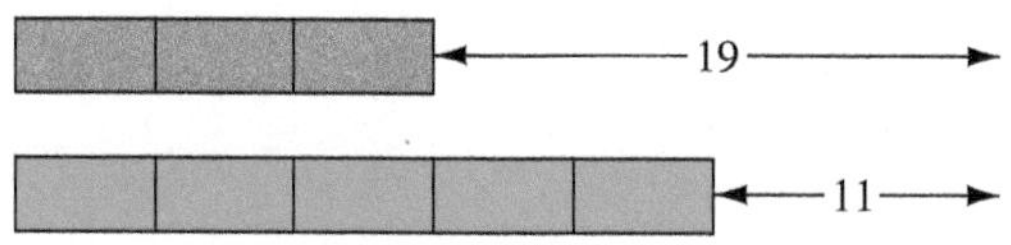

25. **(a)** If y is proportional to x^2 and $y = 27$ when $x = 6$, determine y when $x = 12$.

(b) Determine the ratio of the y-values in part (a).

(c) If y and x are related as in part (a), what happens to the value of y if the value of x is doubled? Explain.

26. Consider the sequence consisting of the odd natural numbers $1, 3, 5, 7, 9, 11, 13, 15, \ldots$.

(a) What is interesting about the ratio of the sum of the first k odds compared with the sum of the next k odds? For example, if $k = 2$, then the sum of the first two odds is $1 + 3$ and the sum of the next two odds is $5 + 7$, so the ratio of sums is $\frac{1 + 3}{5 + 7} = \frac{4}{12} = \frac{1}{3}$. Try other values of k and make a conjecture about the ratios.

(b) Repeat part (a), but this time form ratios of the sum of the first k odds to the sum of the next $2k$ odds.

(c) Investigate the ratio of the sum of the first k odds to the sum of the next nk odds for any given natural number n. Make a conjecture about these ratios.

Making Connections

27. Celeste won a 100-meter race with a time of 11.6 seconds, and Michelle came in second with a time of 11.8 seconds. Given that each girl ran at a constant rate throughout the race, determine the ratio of Celeste's speed to Michelle's in simplest terms.

28. On a trip of 320 miles, Sunao's truck averaged 9.2 miles per gallon. At the same rate, how much gasoline would his truck use on a trip of 440 miles?

29. Which is the better buy in each case?

(a) 32 ounces of cheese for 90¢ or 40 ounces of cheese for \$1.20

(b) A gallon of milk for \$2.21 or two half-gallons at \$1.11 per half-gallon

(c) A 16-ounce box of bran flakes at \$3.85 per box or a 12-ounce box at \$2.94 per box

30. The ratio of Dexter's salary to Claudine's is 4 to 5. If Claudine earns \$3200 per month, how much of a raise will Dexter have to receive to make the ratio of his salary to Claudine's 5 to 6?

31. The ratio of boys to girls in Ms. Zombo's class is 3 to 2. In Mr. Stolarski's class it is 4 to 3. If there are 30 students in Ms. Zombo's class and 28 students in Mr. Stolarski's class, what is the ratio of boys to girls in the combined classes?

32. The accompanying table shows U.S. hat sizes from XS to XL, together with the circumference of the hat.

(a) Is the circumference proportional to the hat size?

(b) Is there a simple relationship between hat size and circumference?

(*Suggestion*: Consider subtracting 1 inch from the circumference.)

Hat Size	$6\frac{5}{8}$	$6\frac{3}{4}$	$6\frac{7}{8}$	7	$7\frac{1}{8}$	$7\frac{1}{4}$	$7\frac{3}{8}$	$7\frac{1}{2}$
Circumference (in Inches)	$20\frac{7}{8}$	$21\frac{1}{4}$	$21\frac{5}{8}$	22	$22\frac{3}{8}$	$22\frac{3}{4}$	$23\frac{1}{8}$	$23\frac{1}{2}$

33. In the diagram that follows, use a metric ruler to *very carefully* measure the lengths *AC*, *BC*, *CD*, *BE*, and *DE* (even trying to estimate to the nearest tenth of a millimeter—that is, to the nearest hundredth in the decimal representation of each length). Then compute the ratios

$$\frac{AC}{BC}, \frac{BC}{CD}, \frac{CD}{DE}, \frac{AD}{DC}, \text{ and } \frac{BE}{ED}.$$

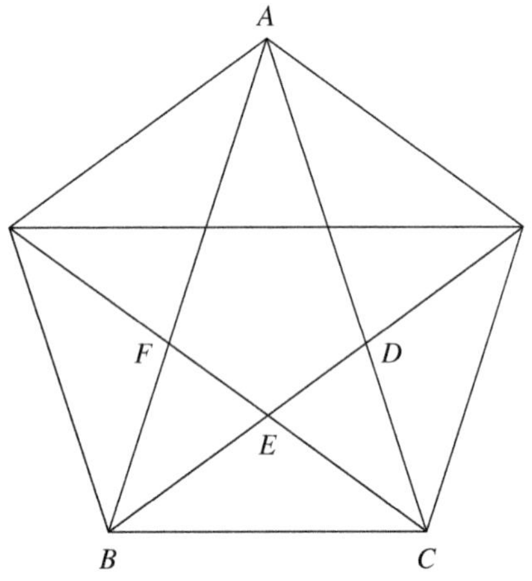

Given the difficulty in obtaining really accurate measurements, what seems to be true of all of these ratios?

34. If a 12″ pepperoni pizza from Ricco's costs $9.56, what should a 14″ pizza from Ricco's cost?

35. Suppose your car uses 8.7 gallons of gas traveling 192 miles. Determine approximately how many gallons it would use traveling a distance of 305 miles.

36. If it takes $1\frac{1}{3}$ cups of sugar to make a batch of cookies, how much sugar would be required to make four batches?

37. If three equally priced shirts cost a total of $59.97, how much would seven shirts cost at the same price per shirt?

38. If 12 erasers cost $8.04 and Mrs. Orton bought $14.07 worth of erasers, how many erasers did she buy?

39. One day Kenji took his class of sixth graders outside and challenged them to find the distance between two rocks that could easily be seen, one above the other, on the vertical face of a bluff near the school. After some discussion, the children decided to hold a rod in a vertical position at a point 100′ from the base of the cliff, as shown in the following diagram:

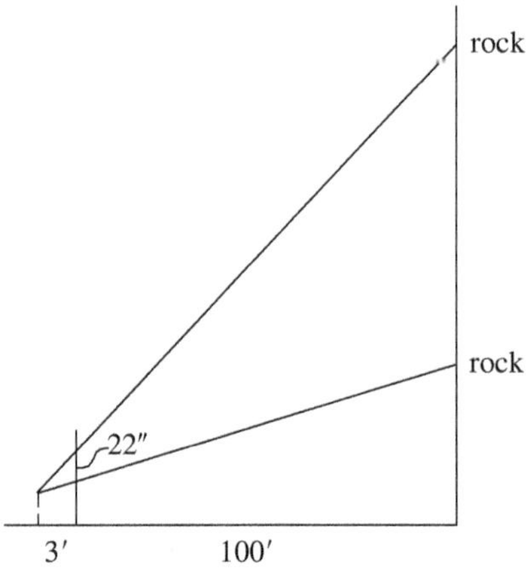

They also decided to mark the points on the rod where the lines of sight of a student standing 3′ farther from the cliff and looking at the rocks cut the rod. If the marks on the rod were 22″ apart, what was the distance between the two rocks? (*Caution:* Convert all measurements to feet.)

40. *The New York Times* crossword puzzle is constructed on a 21 × 21 grid on Sundays and on a 15 × 15 grid the other days of the week. "About" a sixth of the squares must be black and must form a pattern with 180° rotational symmetry.

(a) Find a newspaper that publishes *The New York Times* crossword puzzle, and report on how closely the "one-sixth" rule is observed.

(b) Why can't the "one-sixth" rule ever be obeyed exactly? What is the number of black squares that comes closest to following the rule on a 15 × 15 grid? On a 21 × 21 grid?

41. Bike Speeds and Gear Ratios. A modern touring bike often has a system of three "chainrings" turned by the pedal on the front and a set of seven "cogs" on the rear wheel hub. A common setup is to have a choice of 24, 35, or 51 teeth on the chainrings, and 34, 28, 23, 19, 16, 13, or 11 teeth on the cogs. The "speed" of the bike is then proportional to the ratio of the number of teeth of the chainring engaged to that of the cog. For example, the fastest speed is expressed by the ratio $\frac{51}{11} \doteq 4.63$, for which each complete revolution of the pedals rotates the rear wheel 4.63 revolutions. For climbing hills, a rider may want to use the smallest chainring and the largest cog, giving a ratio $\frac{24}{34} \doteq 0.71$ that turns the rear wheel slowly, but with more force.

(a) Create a table showing the ratios of all 21 speeds.

(b) Does a "21-speed" bicycle truly have 21 speeds? Is there a more accurate number of speeds? Write a brief summary of your findings.

(c) Count the number of teeth on the chainrings and the number on cogs of your bike or that of a friend. Use your data to create a table of ratios for the bike, and report on the number of "speeds" of the bike.

42. The ancient Mayan astronomers were elated when they discovered how the 584-day "year" of Venus compared with the 365-day solar year. What proportion did their observations reveal?

State Assessments

43. **Writing** (Grade 4)
Kerri usually buys the 20 ounce box of oatmeal, which gives her enough for breakfast for 8 days. However, she thinks the 60 ounce box is a less expensive choice, but wonders how long it will last.

A. How many breakfasts can Kerri expect to get from the large box?

B. Explain to Kerri how you get your answer.

44. (Grade 5)
A ruler was laid on a map with town A at the 3″ mark and town B at the 6″ mark. The map has a scale of 1 inch = 5 miles.

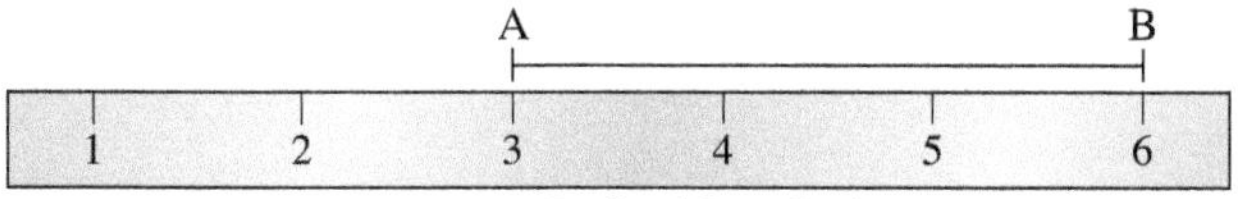

Scale: 1 in = 5 mi

What is the actual distance between towns A and B?

A. 3 miles B. 5 miles
C. 10 miles D. 15 miles

45. (Grade 6)
At Freeport Middle School, 25 of every 100 students participate in sports. Which of the following ratios describes this rate of participation?

A. $\frac{4}{25}$ of the students participate in sports.

B. $\frac{1}{4}$ of the students participate in sports

C. $\frac{1}{5}$ of the students participate in sports

D. $\frac{25}{75}$ of the students participate in sports

46. (Grade 5)
Jenny chose 360 photos to be printed in a book of photos of her African safari. She estimates she used about 2/5 of all the photos she took in her book. How many pictures did she take on her trip?

A. 144 B. 540 C. 720 D. 900

47. (Grade 8)
A tree casts a shadow 45 feet long, compared to a yard stick that casts a shadow 2 feet long. Which of these equations can be used to determine the height h of the tree?

A. $\frac{h}{2} = \frac{45}{3}$ B. $\frac{45}{h} = \frac{2}{3}$
C. $\frac{h}{45} = \frac{2}{3}$ D. $\frac{3}{h} = \frac{45}{2}$

7.4 Percent

Percent

It is essentially impossible to live in today's society and not be conversant with the notion of **percent.**

- How much will a coat regularly priced at \$79.99 be when it is offered at a 30%-off sale?
- What is your new salary when you receive a 3.8% raise?
- What percent interest do you pay on the outstanding balance on your credit card?
- If you are required to make a down payment of 11% of the purchase price of \$175,000 when buying a home, how much do you have to put down?
- How rapidly will your retirement account grow if the account pays 5% interest compounded quarterly?

Percent (from the Latin *per centum*, meaning "per hundred") is one of the most important ratios in school mathematics. Thus, 50% is the ratio $\frac{50}{100}$, and this is quickly simplified to the fraction $\frac{1}{2}$ or written as the decimal 0.50. As another example, if I have \$98 and give you 50% of what I have, I give you

$$\frac{1}{2} \cdot \$98 = \$49, \quad \text{or} \quad 0.5 \times \$98 = \$49.$$

The "of" in the preceding sentence translates into "times." Thus,

$$50\% \text{ of} \quad \text{means} \quad 50\% \times,$$

$$\frac{1}{2} \text{ of} \quad \text{means} \quad \frac{1}{2} \times,$$

and

$$0.5 \text{ of} \quad \text{means} \quad 0.5 \times.$$

$r\% = \frac{r}{100}$

DEFINITION Percent

If r is any nonnegative real number, then r **percent**, written $r\%$, is the ratio

$$\frac{r}{100}.$$

Since $r\%$ is defined as the ratio $\frac{r}{100}$ and the notational effect of dividing a decimal by 100 is to move the decimal point two places to the left, it is easy to write a given percent as a decimal. For example, $12\% = 0.12$, $25\% = 0.25$, $130\% = 1.3$, and so on. Conversely, $0.125 = 12.5\%$, or $12\frac{1}{2}\%$; $0.10 = 10\%$; $1.50 = 150\%$; and so on.

EXAMPLE 7.23

Expressing Decimals as Percents

Express these decimals as percents:

(a) 0.25 **(b)** 0.333 . . . **(c)** 1.255 **(d)** 0.0035 **(e)** 7

Solution

(a) $0.25 = 25\%$
(b) $0.33\ldots = 33.333\ldots\%$ $= 33\frac{1}{3}\%$
(c) $1.255 = 125.5\%$
(d) 0.35%
(e) 700%

EXAMPLE 7.24

Expressing Percents as Decimals

Express these percents as decimals:

(a) 40% **(b)** 12% **(c)** 127% **(d)** 0.5%

Solution

(a) $40\% = 0.40$
(b) $12\% = 0.12$
(c) $127\% = 1.27$
(d) 0.005

$40\% = \frac{40}{100} = 0.40$

EXAMPLE 7.25

Expressing Percents as Fractions

Express these percents as fractions in lowest terms:

(a) 60% **(b)** $66\frac{2}{3}\%$ **(c)** 125%

Solution

(a) By definition, 60% means $\frac{60}{100}$. Therefore, $60\% = \frac{60}{100} = \frac{3}{5}$.

(b) $66\frac{2}{3}\% = \frac{66\frac{2}{3}}{100} = \frac{\frac{200}{3}}{100} = \frac{2}{3}$.

(c) $125\% = \frac{125}{100} = \frac{5}{4}$, or $1\frac{1}{4}$.

EXAMPLE 7.26

Expressing Fractions as Percents

Express these fractions as percents:

(a) $\frac{1}{8}$ **(b)** $\frac{1}{3}$ **(c)** $\frac{16}{5}$

Solution 1

Using Proportions

Since percents are ratios, we can use variables to determine the desired percents.

(a) Suppose $\frac{1}{8} = r\% = \frac{r}{100}$. Then

$$r = 100 \cdot \frac{1}{8} = 12.5$$

and

$$\frac{1}{8} = 12.5\%.$$

(b) If $\frac{1}{3} = s\% = \frac{s}{100}$, then

$$s = \frac{100}{3} = 33\frac{1}{3}$$

$= 33\frac{1}{3}\%$

and

$$\frac{1}{3} = 33\frac{1}{3}\%.$$

(c) If $\frac{16}{5} = u\% = \frac{u}{100}$, then

$$u = \frac{16}{5} \cdot 100 = 320$$

and

$$\frac{16}{5} = 320\%.$$

Solution 2

Using Decimals

Here, we write the fractions as decimals and then as percents.

(a) By division,

$$\frac{1}{8} = 0.125 = 12.5\%.$$

(b) Here,

$$\frac{1}{3} = 0.333\ldots = 33.\bar{3}\% = 33\frac{1}{3}\%.$$

(c) $\frac{16}{5} = 3.2 = 320\%.$

Solving the Three Basic Types of Percent Problems

A percent p is a way to express a ratio of a to b, with $\frac{p}{100} = \frac{a}{b}$. Thus, if any two of the three values a, b, or p are known, then the remaining value can be determined. Each of these three cases arises naturally in realistic problems, as the next three examples illustrate. In each case, the key idea used to obtain the solution is to identify the whole that is represented by 100%. Most errors made with percent occur when the whole has not been correctly defined. The most important question to ask when working with percent is "What is the amount that corresponds to 100%?" In the three examples that follow, a useful diagram helps make it clear how the whole corresponds to 100% and how a part of the whole corresponds to p percent of the whole.

EXAMPLE 7.27

Calculating a Percentage of a Number

The Smetanas bought a house for \$175,000. If a 15% down payment was required, how much was the down payment?

Solution 1 **Using an Equation**

The down payment is 15% of the cost of the house. Thus, if d is the down payment,

$$\begin{aligned} d &= 15\% \times \$175{,}000 \\ &= 0.15 \times \$175{,}000 \\ &= \$26{,}250. \end{aligned}$$

The "of" translates into "times."

Solution 2 **Using a Ratio and a Proportion**

The size of the down payment is proportional to the cost of the house; that is, $\frac{d}{175{,}000}$ is a ratio that is equivalent to the ratio 15%. The following diagram makes the desired proportion clear:

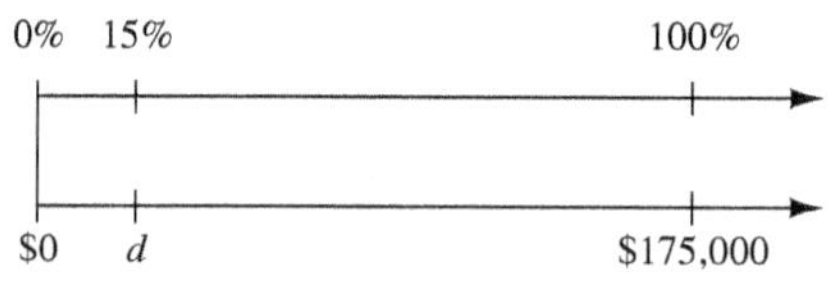

Thus,

$$\frac{d}{\$175{,}000} = \frac{15}{100} = 0.15$$

and

$$\begin{aligned} d &= 0.15 \times \$175{,}000 \\ &= \$26{,}250. \end{aligned}$$

EXAMPLE 7.28

Calculating a Number of Which a Given Number Is a Given Percentage

Soo Ling scored 92% on her last test. If she answered 23 questions correctly, how many problems were on the test?

Solution 1

Using a Variable

Let n denote the number of questions on the test. Since Soo Ling answered 23 questions correctly for a score of 92%, we know that 23 is 92% of n; that is,

$$23 = 92\% \times n = 0.92n.$$

Hence,

$$n = \frac{23}{0.92} = 25.$$

Solution 2

Using a Ratio and a Proportion

As seen in the diagram, the ratio of 23 to the number of questions on the test must be the same as the ratio 92%:

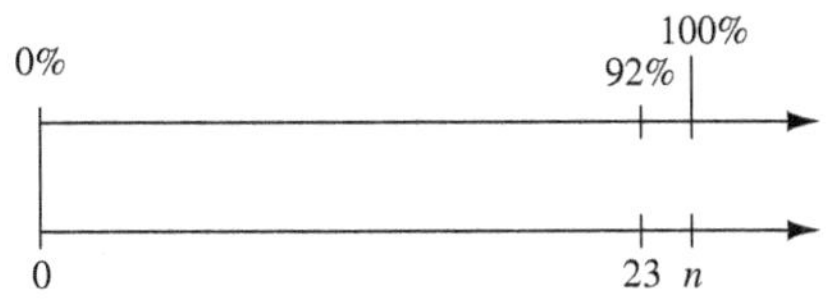

Thus,

$$\frac{n}{100} = \frac{23}{92}.$$

Therefore,

$$n = \frac{23}{92} \times 100 = \frac{2300}{92} = 25.$$

EXAMPLE 7.29 Calculating What Percentage One Number Is of Another

Tara received 28 out of 35 possible points on her last math test. What percentage score did the teacher record in her grade book for Tara?

Solution 1

Using the Definition

Tara received 28 out of 35 points on the test. Since

$$\frac{28}{35} = 0.80 = 80\%,$$

the teacher recorded 80% in her grade book.

Solution 2

Using a Ratio and a Proportion

Let x be the desired percentage. Then, as in Examples 7.27 and 7.28, the diagram

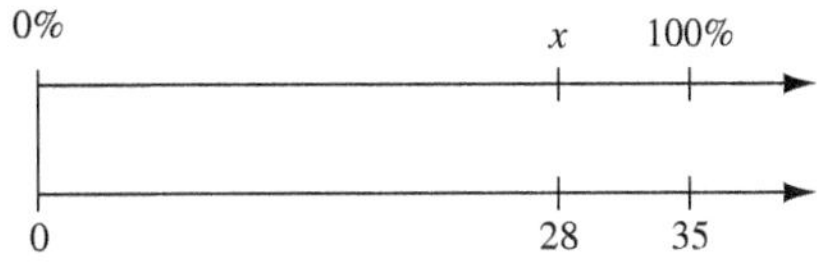

helps one visualize the proportion

$$\frac{x}{28} = \frac{100\%}{35}.$$

Thus,

$$x = \frac{28 \cdot 100\%}{35} = 80\%.$$

Percentage Increase and Decrease

Relative gain or loss is often measured as a percent change. For example, a company may grant a 3.5% raise in salary, or a community may have experienced a loss of 15% of the value of homes in a certain year. The next example illustrates how to answer questions associated with percentage increases and losses.

EXAMPLE 7.30

Computing Percentage Increases and Decreases

Chad is a third-grade teacher currently earning \$53,000 a year. However, because he was recently granted National Board Certification, he will receive a 7% increase in salary. This situation is fortunate, since the value of his home recently dropped by 15% and is now valued at \$255,000.

(a) What will Chad's salary be next year?
(b) What was the value of Chad's house before the downturn in price?

Solution

(a) Since 7% of \$53,000 is $0.07 \times \$53{,}000 = \$3710$, Chad's salary will be \$3710 higher than it is for the current year, making his total salary $\$53{,}000 + \$3710 = \$56{,}710$. Notice that his new salary can be computed directly by the calculation $(1 + 0.07) \times \$53{,}000 = 1.07 \times \$53{,}000 = \$56{,}710$, showing how the percentage increase is added to the current salary figure.
(b) The current value of \$255,000 for Chad's house is $1 - 15\% = 85\%$ of its former valuation. Therefore, Chad's house was valued at $\$255{,}000 \div 0.85 = \$300{,}000$.

It's always important to keep in mind when working with percentage increases and decreases that the "whole" to which the percentage change is applied is clearly understood. For example, suppose you have a retirement portfolio of \$100,000 that dropped 50% one year and rebounded by 50% the following year. Have you completely recovered the value of your investment? The answer is no, since a 50% drop applied to \$100,000 reduces your portfolio to \$50,000, but a 50% gain made on \$50,000 brings you back to only \$75,000.

Compound Interest

If you keep money in a savings account at a bank, the bank pays you interest at a fixed rate (percentage) for the privilege of using your money. For example, suppose you invest \$5000 for a year at 7% interest. How much is your investment worth at the end of the year? Since the interest earned is 7% of \$5000, the interest earned is

$$7\% \times \$5000 = 0.07 \times \$5000 = \$350,$$

and the value of your investment at the end of the year is

$$\begin{aligned}\$5000 + \$350 &= \$5000 + 0.07 \times \$5000\\ &= \$5000 \cdot (1.07)\\ &= \$5350.\end{aligned}$$

If you leave the total investment in the bank, its value at the end of the second year is

$$\begin{aligned}\$5350 + 0.07 \times \$5350 &= \$5350 \cdot (1.07)\\ &= \$5000 \cdot (1.07)(1.07)\\ &= \$5000 \cdot (1.07)^2\\ &= \$5724.50.\end{aligned}$$

Similarly, at the end of the third year, your investment would be worth

$$\$5000 \cdot (1.07)^3 \doteq \$6125.22$$

when rounded to the nearest penny. In general, it would be worth

$$\$5000 \cdot (1.07)^n$$

at the end of n years. This is an example of **annual compound interest,** where the term *compound* implies that each year you earn interest on all the interest earned in preceding years as well as on the original amount invested (the **principal**).

Usually, interest is compounded more than once a year. Suppose the \$5000 investment just discussed was made in a bank that pays interest at the rate of 7% compounded semiannually—that is, twice a year. Since the rate for a year is 7%, the rate for half a year is 3.5%. Thus, the value of the investment at the end of the year (that is, at the end of *two* interest periods) is

$$\$5000(1.035)^2 \doteq \$5356.13,$$

and the values at the end of two years and three years, respectively, are

$$\$5000(1.035)^4 \doteq \$5737.62$$

and

$$\$5000(1.035)^6 \doteq \$6146.28.$$

Compounding more and more frequently is to your advantage, and most banks compound monthly or even daily.

The preceding calculations are typical and are summarized in this theorem:

$\frac{r}{n}$ is the rate per interest period and nt is the number of interest periods.

THEOREM Calculating Compound Interest

The value of an investment of P dollars at the end of t years, if interest is paid at the annual rate of r% compounded n times a year, is

$$P\left(1 + \frac{r}{100n}\right)^{nt}.$$

INTO THE CLASSROOM

Debbie Goodman Applies Percent to the Calculation of Tips

Giving students an application for percent that their parents use is very important to strengthen students' understanding. I like to use the real-life application of the use of percent in the food-service industry. Students, working in small groups, are given several actual restaurant receipts. Each group is asked to decide an appropriate tip, usually 12% to 20% of the total bill. Next, students determine the "tip-out," the dollar amount of the tip that is passed on to the host, the kitchen staff, and the beverage staff. For example, the policy at a restaurant may be that 5% of the food cost goes to the kitchen staff, 5% of the beverage cost goes to the beverage staff, and 1% of the total bill goes to the restaurant host. The remaining part of the total tip is shared equally by all of the servers.

An example is shown below, where the students have decided to tip \$6.00.

Mama's Restaurant	
February 23, 2015	2:05 P.M.
Server: TED	
Turkey Melt	\$10.00
Burger	\$11.00
Soda	\$2.00
Soda	\$2.00
Dessert	\$9.00
Sub-Total	\$34.00
Tip	______
TOTAL	______

Calculations to be completed by each group:

Total Tip	\$6.00
Food Cost	\$30.00
Beverage Cost	\$4.00
Total Bill	\$34.00
5% Tip-Out to Kitchen Staff	
(30.00 × 0.05 = 1.50)	\$1.50
5% Tip-Out to Beverage Staff	
(4.00 × 0.05 = 0.20)	\$0.20
1% Tip-Out to Host	
(34.00 × 0.01 = 0.20)	\$0.34
Tip-Shared by Servers	
(6.00 − 1.50 − 0.20 − 0.34 = 3.96)	\$3.96

EXAMPLE 7.31

Computing the Cost of Debt

Many credit card companies charge 18% interest compounded monthly on unpaid balances. Suppose your card was "maxed out" at your credit limit of $10,000 and that you were unable to make any payments for two years. Aside from penalties, how much debt would you owe, on the basis of compound interest alone?

Solution

Since the interest is computed at 18% compounded monthly, the effective rate per month is $\frac{18\%}{12} = 1.5\% = 0.015$ and the number of interest periods in two years is $12 \cdot 2 = 24$. Thus, your debt, to the nearest penny, would be

$$\$10{,}000(1.015)^{24} \doteq \$14{,}295.03.$$

If the debt went unpaid for six years, you would owe

$$\$10{,}000(1.015)^{72} \doteq \$29{,}211.58.$$

This is almost triple what you originally owed!

The Mathematics of Growth

Population growth occurs in exactly the same way that an investment earning compound interest grows. Suppose, for example, that the population in the Puget Sound region in northwest Washington is approximately 2.2 million and that it is growing at the rate of 5% per year. In one year, the population will be approximately

$$2.2(1.05) \doteq 2.3$$

million. If the growth continues unabated, in 14 years it will be approximately

$$2.2(1.05)^{14} \doteq 4.4$$

million, twice what it is today. Given the fact that the area has already experienced several years of water shortages, these figures are cause for concern among officials.

Like population growth, prices of commodities also rise with inflation as an investment grows by drawing compound interest.

EXAMPLE 7.32

Pricing a Car

If the economy were to experience a steady inflation rate of 2.5% per year, what would be the price of a new car in five years if the same-quality car sells today for $18,400?

Solution

Using the same formula as in computing compound interest, we find that the price of the car five years from now would be approximately

$$\$18{,}400(1.025)^5 \doteq \$20{,}818.$$

7.4 Problem Set

Understanding Concepts

1. Write each of these ratios as a percent:

(a) $\frac{3}{16}$ (b) $\frac{7}{25}$

(c) $\frac{37}{40}$ (d) $\frac{5}{6}$

2. Write each of these ratios as a percent:

(a) $\frac{3.24}{8.91}$ (b) $\frac{7.801}{23.015}$

(c) $\frac{1.6}{7}$ (d) $\frac{\sqrt{2}}{\sqrt{6}}$

3. Write each of these decimal numbers as a percent:
   (a) 0.19 (b) 0.015 (c) 2.15 (d) 3

4. Write each of these percents as a fraction in simplest form:
   (a) 10% (b) 25% (c) 62.5% (d) 137.5%

5. Compute each of the following percents:
   (a) 70% of 280 (b) 120% of 84
   (c) 38% of 751 (d) $7\frac{1}{2}$% of $20,000
   (e) 0.02% of 27,481 (f) 1.05% of 845

6. Compute each of these percents mentally:
   (a) 50% of 840 (b) 10% of 2480
   (c) 12.5% of 48 (d) 125% of 24
   (e) 200% of 56 (f) 110% of 180

7. (a) Michelle's salary jumped to $50,400 after her 5% raise. What was her former salary?
   (b) Jason's investment of $12,000 is now worth $11,160. What was his percentage loss?

8. Jerry and Cynthia both started with the same salary. Then Jerry received a 5% raise one year followed by a 3% raise the next year. Cynthia received a 3% raise followed by a 5% raise.
   (a) Who has the higher salary following the two raises?
   (b) Would you rather be Jerry or Cynthia? Explain carefully.

9. First guess the percent of each square board that is colored blue, and then determine the exact percent:

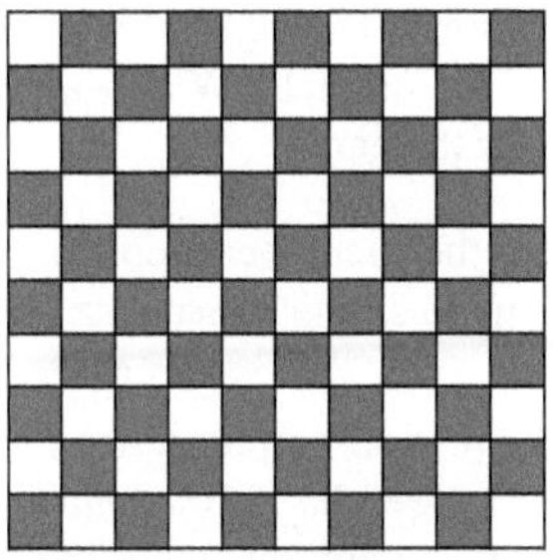
(a)

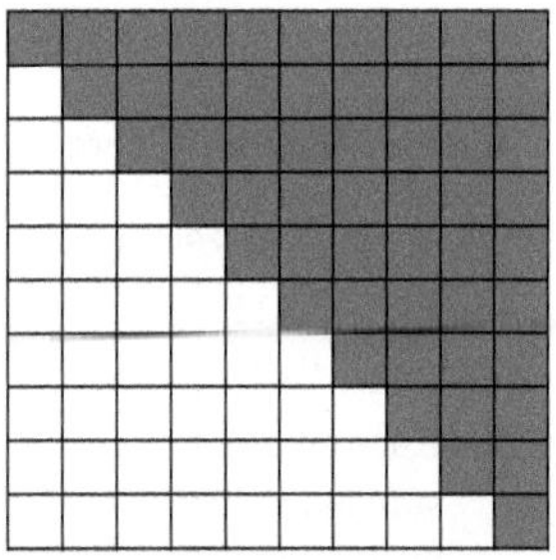
(b)

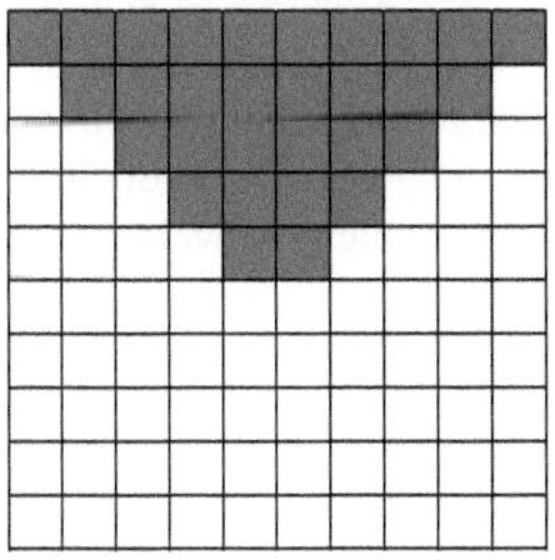
(c)

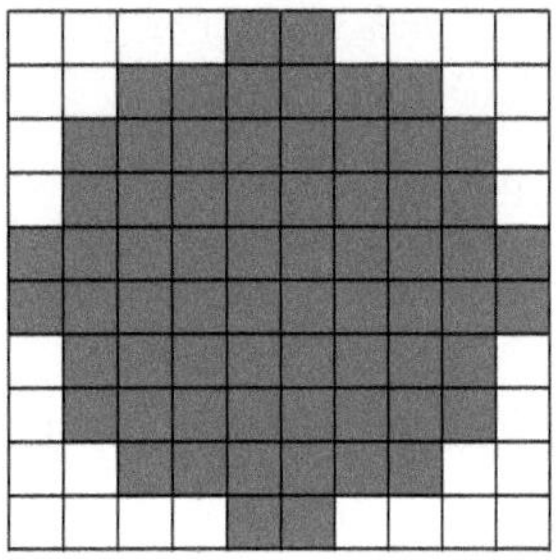
(d)

10. If the following 20-mm–by–20-mm square represents 100%, draw rectangles 20 mm wide that represent each of the given percentages.

   (a) 75% (b) 125% (c) 200% (d) 20%

11. What percentage of each of these figures is red?

(a)
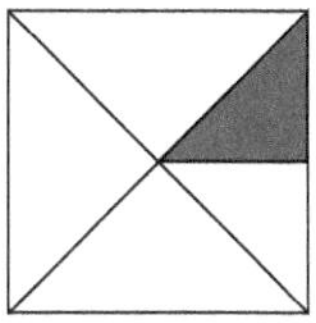

(b)
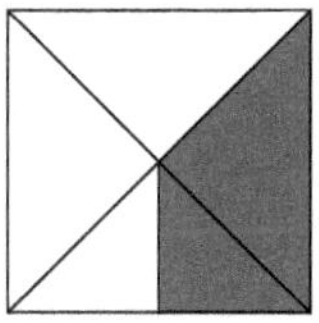

(c)
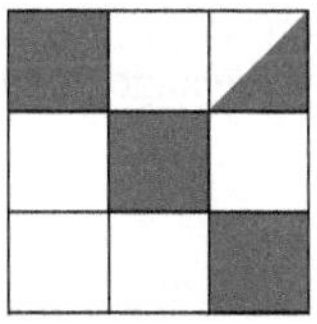

(d)

12. How many small squares must be shaded to represent the given percentages of the large square shown?
   (a) 100% (b) 0% (c) 25% (d) 87%

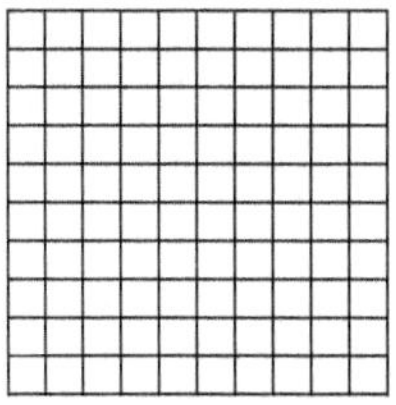

13. How many of the small rectangles must be shaded to represent the given percentages of the large rectangle shown?
   (a) 50% (b) 25% (c) 20% (d) 37.5%

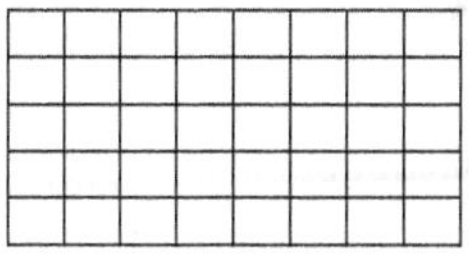

14. If the area of the yellow hexagon is the whole, then the remaining pattern blocks have areas that are approximately 16.67%, 19.2%, 33.3%, 38.5%, and 50%. Match the pattern block that corresponds to the given percent.

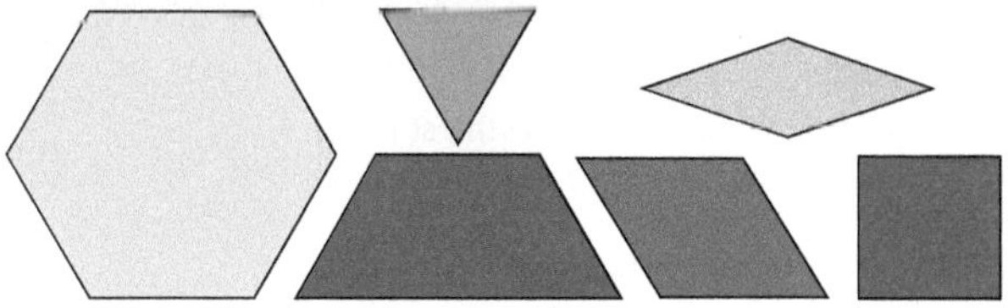

15. Mentally convert each of these fractions to a percent:
   (a) $\frac{7}{28}$ (b) $\frac{11}{33}$
   (c) $\frac{72}{144}$ (d) $\frac{44}{66}$

16. Mentally estimate the numbers that should go in the blanks to make these statements *true*.
   (a) 27% of ____ equals 16.
   (b) 4 is ____% of 7.5.
   (c) 41% of 120 = ____.

Into the Classroom

17. (**Writing**) There are three common types of percentage problems:

    (i) Find the percent: Given a and b, what percent p of b is a?

    (ii) Find the part: Given a percent p and the whole b, what is a if it is p percent of b?

    (iii) Find the whole: Given a part a that is p percent of the whole b, what is the whole b?

    Design a classroom poster that gives a realistic problem and its solution for each of these types of percentage problems.

Responding to Students

18. Tony and Maria want to buy a CD to share. Tony favors store A, which advertises that it always marks down CDs by 15% and now has a storewide 20%-off sale. Maria thinks that store B offers a better deal with its 35%-off sale. Their friend Betty says it doesn't matter. How would you respond to the three students who have come to you with their dispute?

19. Patrick and Heather were working on their science fair project, and they were trying to decide which plant grew faster. They recorded the growth of two plants. Plant A grew from 9 centimeters to 12 centimeters. Plant B grew from 13 centimeters to 16 centimeters. Patrick said that they both grew 3 centimeters, so there was no difference in the variable they were testing. Heather wasn't sure. How can you use the idea of percent to help these students?

20. Kelsey got all 20 words right on her spelling test and attempted 2 bonus words, which were also correct. Her friend Elizabeth graded Kelsey's test and wrote "Good Job, Kelsey, 102%!"

    (a) What would actually be the correct percentage for 22 out of 20?

    (b) How is Elizabeth improperly applying percents to Kelsey's grade?

21. Students in Mrs. Giles's class were asked which was the better deal: (1) 35% off the original price of \$10 or (2) pay 65% of the original price of \$10. Louis and Arnold argued that 35% off was better because the price was going "down."

    (a) What is confusing the students (and often many shoppers)?

    (b) How would you show the students that the answers are actually the same?

22. Using a regular household calculator to complete his mathematics homework on percents, Anthony had the following problem: "What percent is 20 out of 25?" He entered 20 and divided by 25, so 0.8 showed on his calculator display. He then wrote 8% as his answer.

    (a) What is Anthony doing incorrectly?

    (b) What would be your next step as a teacher?

23. Yoshi was paid \$3 a day to water his neighbor's plants last year during their vacation. This year he will also look after their new dog, so he will be paid \$6 a day. Since he is paid twice as much, he claims that his pay has increased by 200%, since 200% = 2. How would you respond to Yoshi?

Thinking Critically

24. In a given population of men and women, 40% of the men are married and 30% of the women are married. What percentage of the adult population is married?

25. A garage advertised car repairs at 10% off the usual price—5% on parts and 5% on labor. The Consumer Protection Agency (CPA) chided the garage for false advertising. Was the CPA correct? Explain.

26. A sign at the cookie store in the airport read "Increase your ecstasy by 25%—4 cookies for the price of 3." Is there anything amiss with this claim?

27. A customer made a purchase in a store in Washington state, where the sales tax is 8%. The customer then noticed that her 20% coupon hadn't been taken into account, so the cashier took 20% off the bill that included the sales tax.

    (a) Should the customer insist that the 20% off coupon be applied *before* the sales tax is added?

    (b) Should the cashier redo the calculation, first applying the 20%-off coupon and then adding the sales tax?

28. During the first half of a basketball game, the basketball team at Red Cloud High School made 60% of its 40 field goal attempts. During the second half, the team scored on only 25% of 44 attempts from the field. To the nearest 1%, what was Red Cloud High's field goal shooting percentage for the entire game?

29. During the first half of a basketball game, Skeeter Thoreson missed all 5 of her field goal attempts. During the second half, she hit 75% of her 16 attempts from the field. What was her field goal shooting percentage for the game?

30. **The Cucumber Problem.** Solve the cucumber problem, posed by the Hungarian-born American mathematician Paul Halmos (1916–2006):

    > Cucumbers are assumed, for present purposes, to be a substance that is 99% water by weight. If 500 pounds of cucumbers are allowed to stand overnight, and if the partially evaporated substance that remains in the morning is 98% water, how much is the morning weight?

    [*Be careful!* The answer, which is not 495 pounds, will surprise you. Try using variables such as W for the weight of the water in the evening and S for the unchanging weight of the cucumbers that is not water.]

Making Connections

31. Tabata's Furniture calculates the retail price of furniture that the company sells by marking up its cost at wholesale by a full 100%.

    (a) If Tabata's had a sale with all items marked down 20% from the retail price, what percentage profit did the company actually make on each item sold during the sale?

    (b) If Tabata's had a sale with all items marked down 50% from the retail price, what percentage profit did the company make on each item sold during the sale?

32. A merchant obtains the retail price of an item by adding 20% to the wholesale price. Later, he has a sale and marks every item down 20% from the retail price. Is the sale price the same as the wholesale price? Explain briefly. (*Hint:* Consider a specific item whose wholesale price is \$100.)

33. Acme Electric in Boise, Idaho, purchases hot-water heaters at wholesale for \$185 each and sells them after marking up the wholesale price by 45%. Since Idaho has a 5% sales tax, how much would you have to pay Acme for one of its water heaters?

34. There is \$100 in the cash register at the start of a cashier's shift at a supermarket. At the end of the shift, the register contains \$800. If the state has a 5% sales tax on all sales, how much of the \$800 is for tax?

35. The mortgage company requires an 11% down payment on houses it finances. If the Sumis bought a house for \$158,000, how much down payment did they have to make?

36. Mr. Swierkos invested \$25,000 in a mutual fund. If his broker deducted a 6% commission before turning the rest of the money over to the mutual fund, and the value of each share increased by 18% during the year, what percentage return on his investment did Mr. Swierkos realize at the end of the first year?

37. Find the value of each of these investments at the end of the period specified:

(a) \$2500 invested at $5\frac{1}{4}\%$ interest compounded annually for seven years

(b) \$8000 invested at 7% interest compounded semiannually for ten years

38. The *progress bar* was invented in 1896, though it remained obscure until its rediscovery about 15 years later. Now commonplace in computing, progress bars visually show how much information has been transmitted, together with the percent copied, the transfer rate, and the size of files. In each of the following examples, first estimate x, the percent copied so far, to the nearest 10%, and then determine the values of y and z.

(a) 

Percent copied: x%
Estimated time left: z sec (5.0 MB of y MB copied)
Transfer Rate: 2.50 MB/sec

(b) 

Percent copied: x%
Estimated time left: z sec (36.0 MB of y MB copied)
Transfer Rate: 2.00 MB/sec

(c) 

Percent copied: x%
Estimated time left: 3 sec (y MB of 30 MB copied)
Transfer Rate: z MB/sec

39. How much would you have to invest at 6% interest compounded annually in order to have \$16,000 at the end of five years? (*Note:* If P is the amount invested, then $16{,}000 = P \cdot (1.06)^5$.)

40. Suppose you place \$1000 in a bank account on January 1 of each year for nine years. If the bank pays interest on such accounts at the rate of 5.3% interest compounded annually, how much will your investment be worth on January 1 of the tenth year?

41. The value of an investment at r% interest compounded annually for n years is $P(1 + \frac{r}{100})^n$. Determine the least number of years for the value of an investment to double at each of these rates:

(a) 5% **(b)** 7% **(c)** 14% **(d)** 20%

(e) The "rule of 72" states that you divide 72 by the interest rate to obtain the doubling time for an initial investment. This rule is used by many bankers to give a crude approximation of the time for the value of an investment to double at a given rate. Does it seem like a reasonable rule? Explain briefly.

State Assessments

42. (Grade 8)
Megan scored of 75% on her first test and wants to increase her scored to 90% on the next test. What is the percentage change in the number of correct answers she needs to go from a 75% score to a 90% score?

A. 10% B. 15% C. 20% D. 25%

43. (Grade 5)
Elaine correctly answered 11 of the 16 questions on her quiz. What is the approximate percentage grade she received?

A. 55% B. 60% C. 70% D. 75%

44. (Massachusetts, Grade 8)
Use the advertisement below to answer the question.

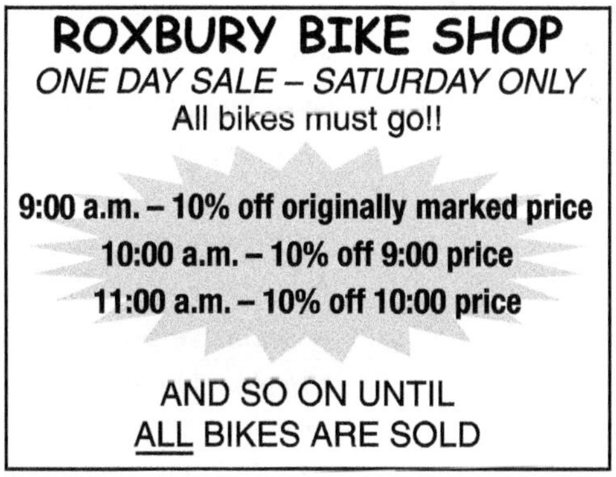
ROXBURY BIKE SHOP
ONE DAY SALE – SATURDAY ONLY
All bikes must go!!
9:00 a.m. – 10% off originally marked price
10:00 a.m. – 10% off 9:00 price
11:00 a.m. – 10% off 10:00 price
AND SO ON UNTIL
ALL BIKES ARE SOLD

Mr. Howard bought a bike with an originally marked price of \$400. What was the price of the bike at 12:15 P.M.?

A. \$262.44 B. \$240.00 C. \$291.60 D. \$280.00

45. (Grade 7)
A pair of hiking shoes are on sale at 30% off. The sale price is \$48.99 What is the approximate regular price of the shoes when they are not on sale?

A. \$52 B. \$70 C. \$118 D. \$163

46. (Grade 7)
The bill for dinner was \$66. In addition to the bill, how much should be left for a tip of about 15%?

A. \$10 B. \$15 C. \$20 D. \$81

The Chapter in Relation to Future Teachers

This chapter has addressed topics of considerable importance in the development of an elementary or middle school student's number sense and quantitative reasoning skills. In particular, the idea of the decimal representation of numbers leads naturally to the real number system, which is the last number system introduced in the K–8 mathematics[*] curriculum. Computations with decimals highlight the importance of acquiring a good understanding of the corresponding algorithms in integer arithmetic; place value is of paramount importance. As has been shown with many real-life examples, proportional reasoning is a skill that will often play an effective role in solving problems, understanding relative comparisons of quantities, and expressing a rate of change. Percent is a specialization of proportional reasoning in which a comparison is made to the one hundred parts that make up the whole.

Table 7.3 presents a review of all the number systems introduced in this and the foregoing chapters. As you examine the table, think about how you would answer questions such as the following:

- How can the numbers within a system be modeled with manipulatives and visualizations?
- How can the properties of a number system be illustrated with appropriate models?
- Why do some properties hold for one number system but not another?

TABLE 7.3 NUMBER SYSTEMS AND THEIR PROPERTIES

Properties	Natural Numbers	Whole Numbers	Integers	Rational Numbers	Real Numbers
Closure property for addition	•	•	•	•	•
Closure property for multiplication	•	•	•	•	•
Closure property for subtraction			•	•	•
Closure property for division, except for division by zero				•	•
Commutative property for addition	•	•	•	•	•
Commutative property for multiplication	•	•	•	•	•
Commutative property for subtraction					
Commutative property for division					
Associative property for addition	•	•	•	•	•
Associative property for multiplication	•	•	•	•	•
Associative property for subtraction					
Associative property for division					
Distributive property for multiplication over addition	•	•	•	•	•
Distributive property for multiplication over subtraction	•	•	•	•	•
Contains the additive identity, 0		•	•	•	•
Contains the multiplicative identity, 1	•	•	•	•	•
The multiplication property by 0 holds		•	•	•	•
Each element possesses an additive inverse			•	•	•
Each element except 0 possesses a multiplicative inverse				•	•

[*]Other systems are of value in higher mathematics. Of particular importance in advanced mathematics are the *complex numbers*. One reason for this is that every polynomial equation has a solution in the complex numbers.

Chapter 7 Summary

Section 7.1 Decimals and Real Numbers	Page Reference
CONCEPTS	
• **Decimal:** In general, the Indo-Arabic system of positional notation for numerals. Colloquially, nonintegers represented in positional notation to base ten.	325
• **Manipulatives to represent decimals:** Decimals can be represented with mats, strips, and units; base-ten blocks; dollars, dimes, and pennies; place-value cards; and so on.	326
• **Real numbers:** The set of all numbers that can be represented as decimals. Alternatively, the set of all rational and irrational numbers.	334
DEFINITIONS	
• A **decimal** is a base-ten position numeral, either positive or negative, in which there are finitely many digits extending to the left of the decimal point and either finitely or infinitely many digits extending to the right of the decimal point.	325
• **Expanded form** for integers is the written notation for the positional values such as $29.437 = 2 \cdot 10 + 9 + \frac{4}{10} + \frac{3}{100} + \frac{7}{1000}$.	325
• **Expanded exponential form** is the written notation for the positional values of decimals using both positive and negative powers of 10. For example, $2.437 = 2 \cdot 10^0 + 4 \cdot 10^{-1} + 3 \cdot 10^{-2} + 7 \cdot 10^{-3}$.	328
• A **terminating decimal** is a decimal that has a finite number of digits.	329
• A **periodic decimal,** or **repeating decimal,** is a nonterminating decimal with infinitely many nonzero digits, where one digit or string of digits repeats *ad infinitum.*	331
• **Repeating decimal:** A nonterminating decimal for which a digit or a string of adjacent digits repeats ad *infinitum.*	331
• A **nonterminating, nonrepeating decimal** is a decimal with infinitely many digits, no pattern of which repeats *ad infinitum.*	334
• **Irrational numbers** are the numbers that are represented by nonterminating, nonrepeating decimals.	334
• The set of all **real numbers, *R*,** is the set containing all rational and irrational numbers.	334
THEOREMS	
• **Multiplying decimals by powers of 10:** If n is a natural number, multiplying a number by 10^n notationally moves the decimal point n places to the right. Multiplying by 10^{-n} moves the decimal point n places to the left.	329
• **Rational numbers represented by terminating decimals:** A rational number represented by the fraction $\frac{a}{b}$ in simplest form can be written as a terminating decimal if and only if the denominator b has no prime divisors other than 2 and 5.	330
• **Rational numbers represented by repeating decimals:** A rational number represented by the fraction $\frac{a}{b}$ in simplest form can be written as a periodic (or repeating) decimal if and only if the denominator b has a prime divisor other than 2 and 5.	331
• **$\sqrt{2}$ is irrational:** $\sqrt{2}$ and infinitely many other numbers are irrational.	335

• **Real numbers as points on the number line:** Every real number corresponds to a unique point on the number line, and conversely.	337

Section 7.2 Computations with Decimals	Page Reference
PROCEDURES	
• **Adding and subtracting decimals:** Align decimal points and then add or subtract the decimals as is done with integers.	340
• **Multiplying decimals:** Multiply as integers. Let r denote the total number of digits to the right of the decimal points in the two decimals. Place the decimal point r places from the right in the answer.	342
• **Dividing decimals:** Multiply both dividend and divisor by the appropriate power of 10 to make the divisor an integer. Using the long-division algorithm, divide as integers, placing the decimal point above the decimal point in the dividend.	344
• **The 5-Up Rule:** To round a decimal to a given place consider the digit in the next place to the right. If it is smaller than 5, replace it and all of the digits to its right with 0. If it is 5 or larger, replace it and all digits to the right by 0 and increase the digit in the given place by one. Replaced digits to the right of the decimal are then dropped to give the rounded decimal.	340
DEFINITION	
• **Order relation:** If a and b are real numbers, then $a < b$ if and only if $b - a$ is positive.	342

Section 7.3 Proportional Reasoning	Page Reference
CONCEPTS	
• **Ratio:** A ratio is a relative comparison of two quantities.	348
• **Applications of ratios:** A ratio measures the relative size of different parts, the relative size of a part to a whole, and a rate of change of one quantity compared to the amount of change of a second quantity.	349
DEFINITIONS	
• A **ratio** is a quotient written as a fraction $\frac{a}{b}$ where a and b, $b \neq 0$, are real numbers.	348
• The **rate of change** is a ratio that measures the change in one quantity with respect to a corresponding change in a second quantity.	349
• A **unit rate** is a rate given by the amount of change of a quantity compared to a change by 1 unit of the second quantity.	349
• A **proportion** is the equality of two ratios: $\frac{a}{b} = \frac{c}{d}$.	352
• If $y = kx$ for some constant k, then **y is proportional to x.**	352
• If $y = kx$ for some constant k, then k is the **constant of proportionality.**	354
THEOREM	
• **Conditions for a proportion.** $\frac{a}{b} = \frac{c}{d}$ if and only if $ad = bc$.	352

Section 7.4 Percent	Page Reference
CONCEPTS	
• **Three basic percent problems:** Given a percent p, part a, or whole b, where $\frac{p}{100} = \frac{a}{b}$, any two of p, a, and b determine the third.	364
• **Interest:** The percentage rate paid to borrow money or earned on an investment.	366
• **Percentage increase or decrease:** The relative change of a gain or loss, expressed as a percent.	366
• **Compound interest:** Interest calculated so that the interest paid in any given year is paid both on the original amount and on the interest earned in previous years.	366
DEFINITIONS	
• A **percent** for any real number r, $r\% = \frac{r}{100}$.	362
• The **percentage increase/decrease** is the relative change of a gain or loss, expressed as a percent.	366
• **Interest** is the percentage rate paid to borrow money or earned on an investment.	366
• **Compound interest** is the interest calculated so that the interest paid in any given year is paid both on the original amount and on the interest earned in previous years.	366
THEOREM	
• **Calculating compound interest:** The value of an investment of P dollars at the end of t years, if interest is paid at the annual rate of $r\%$ and compounded n times per year, is $P\left(1 + \frac{r}{100n}\right)^{nt}$.	367

Chapter Review Exercises

Section 7.1

1. Write these decimals in expanded exponential form:

(a) 273.425 **(b)** 0.000354

2. Write these fractions in decimal form:

(a) $\frac{7}{125}$ **(b)** $\frac{6}{75}$ **(c)** $\frac{11}{80}$

3. Write these fractions in decimal form:

(a) $\frac{84}{175}$ **(b)** $\frac{24}{99}$ **(c)** $\frac{7}{11}$

4. Write these decimals as fractions in simplest form:

(a) 0.315 **(b)** 1.206 **(c)** 0.2001

5. Write these numbers as fractions in simplest form:

(a) $10.\overline{363}$ **(b)** $2.1\overline{42}$

6. Suppose $a = 0.202002000200002000002\ldots$, continuing in this way with one more 0 between each successive pair of 2s. Is this number rational or irrational? Explain briefly.

7. Using only mental arithmetic, determine the numbers represented by these base-ten numerals as fractions in simplest form:

(a) $0.222\ldots = 0.\overline{2}$ **(b)** $0.363636\ldots = 0.\overline{36}$

Section 7.2

8. Perform these computations by hand:

(a) $21.734 + 3.2145 + 71.24$

(b) $23.471 - 2.89$

(c) 35.4×2.37

(d) $24.15 \div 3.45$

9. Compute the following, using a calculator:

(a) $31.47 + 3.471 + 0.0027$

(b) $31.47 - 3.471$

(c) 31.47×3.471

(d) $138.87 \div 23.145$

10. Write estimates of the results of these calculations, and then do the computing within the accuracy of your calculator:

(a) $47.25 + 13.134$

(b) $52.914 - 13.101$

(c) 47.25×13.134

(d) $47.25 \div 13.134$

11. Arrange these numbers in order from least to greatest:

$$\frac{4}{12}, 0.33, 0.3334, \frac{5}{13}, \frac{2}{66}.$$

12. Show that $3 - \sqrt{2}$ is irrational.

13. Show that the sum of two irrational numbers can be rational. (*Hint:* Consider problem 12.)

14. What can you say about the decimal expansion of an irrational number?

15. (a) A wall measures 8.25 feet by 112.5 feet. What is the area of the wall?

(b) If it takes 1 quart of paint to cover 110 square feet, how many quarts of paint must be purchased to paint the wall in part (a)?

16. Give an example of a fraction whose decimal is repeating and has a period of length 4.

Section 7.3

17. Maria made 11 out of 20 free-throw attempts during a basketball game. What was the ratio of her successes to failures on free throws during the game?

18. Determine which of these equations are proportions:

(a) $\frac{775}{125} = \frac{155}{25}$ (b) $\frac{31}{64} = \frac{15}{32}$ (c) $\frac{9}{24} = \frac{12}{32}$

19. If Che bought 2 pounds of candy for \$3.15, how much would it cost him to buy 5 pounds of candy at the same price per pound?

20. It took Donnell 7.5 gallons of gas to drive 173 miles. Assuming that he gets the same mileage per gallon, how much gasoline will he need to travel 300 miles?

21. If y is proportional to x and $y = 7$ when $x = 3$, determine y when $x = 5$.

22. If a flagpole casts a shadow 12′ long when a yardstick casts a shadow 10″ long, how tall is the flagpole?

Section 7.4

23. Convert each of these numbers to percents:

(a) $\frac{5}{8}$ (b) 2.115 (c) 0.015

24. Convert each of these percents to decimals:

(a) 28% (b) 1.05% (c) $33\frac{1}{3}\%$

25. If the sales tax is calculated at 7.2%, how much tax is due on a \$49 purchase?

26. If a tax of \$6.75 is charged on an \$84.37 purchase, what is the sales tax rate?

27. Refer to problem 17. What percent of free throws attempted did Maria make during the game?

28. (a) Alex made \$48,000 last year but has received a letter saying that his new salary is \$51,000. What percent raise did Alex get?

(b) After undergoing a drop in the market by 10%, Monique's stock is now valued at \$8100. What was the value of her stock before the downturn?

29. If you invest \$3000 at 8% interest compounded every three months (quarterly), how much is your investment worth at the end of two years?

8 Algebraic Reasoning, Graphing, and Connections with Geometry

COOPERATIVE INVESTIGATION

The Equation Balance Scale

Materials Needed

Each student needs an enlarged copy of the materials sheet shown here. Cut out the "unknowns" marked "x," the unit squares, and the trapezoid depicting the equation balance scale.

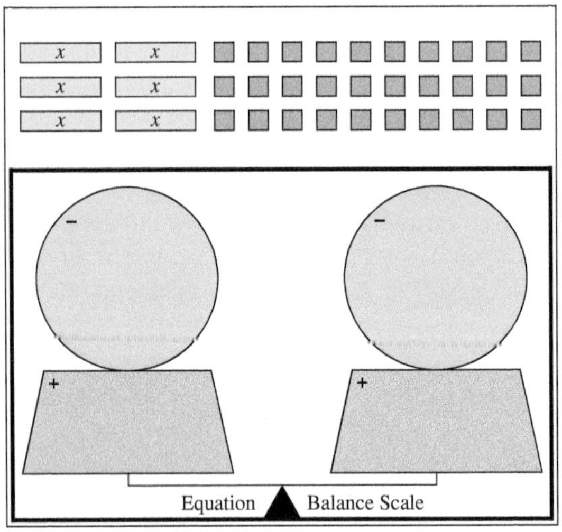

Using the Equation Balance Scale

An algebraic equation such as $3x - 5 = x + 7$ is represented on the Equation Balance Scale by placing three "x" shapes in the + weight area and five unit squares in the negative "balloon" on the left side of the scale. On the right side, an "x" and seven unit squares are placed in the + weight area. Thinking of the scale as balanced gives the following initial representation of the equation:

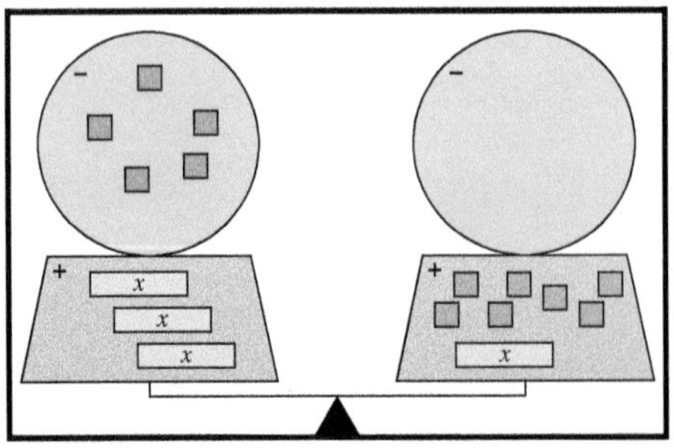

$3x - 5 = x + 7$

To solve the equation, we manipulate the pieces to maintain the balance and we modify the equation to reflect the manipulations. We can begin by removing an "x" from each side of the balance:

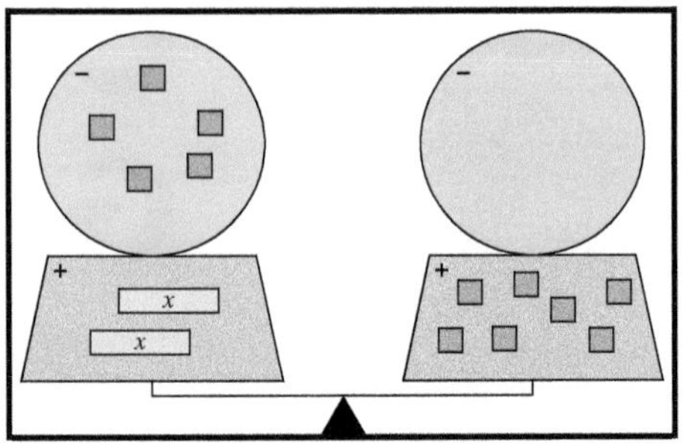

$2x - 5 = 7$

Next, add five unit squares to both sides:

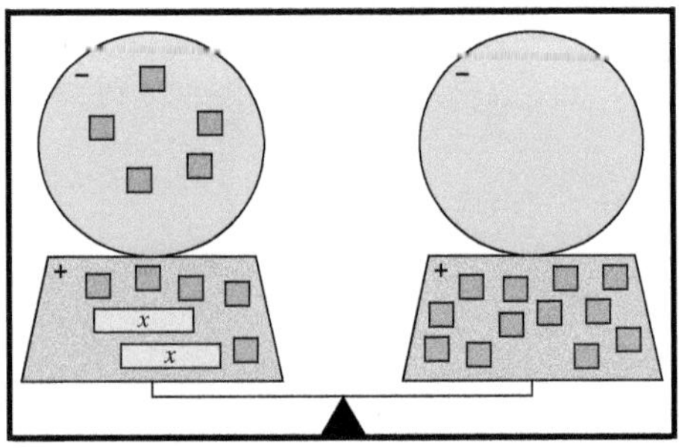

$(2x - 5) + 5 = 7 + 5$

On the left side of the balance, the five + weight units are matched by the five − lifting units in the balloon. Each weight-unit–lifting-unit pair with the same units is usually referred to as a "zero pair." We get the following diagram and equation:

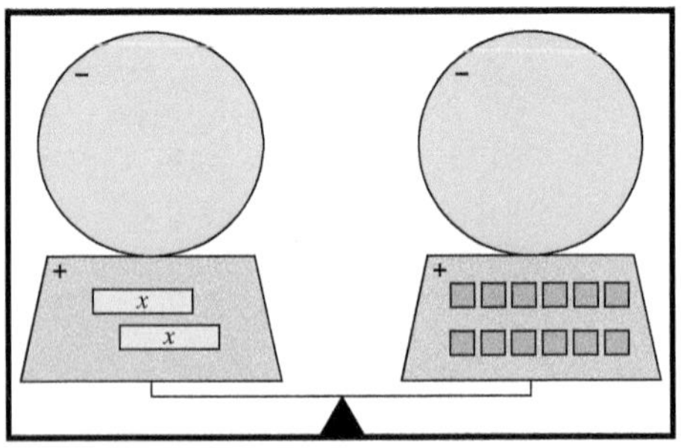

$2x = 12$

An equivalent manipulation would be to move the five negative units in the balloon on the left to the positive weight on the right. That is, $2x - 5 = 7$ becomes $2x = 7 + 5$, so we have $2x = 12$ as before.

Finally, we see that there are two rows of identical arrangements of an "x" on the left and two rows of six unit squares on the right. Therefore, we divide the number of pieces on each side of the balance by 2, and we have the following solution:

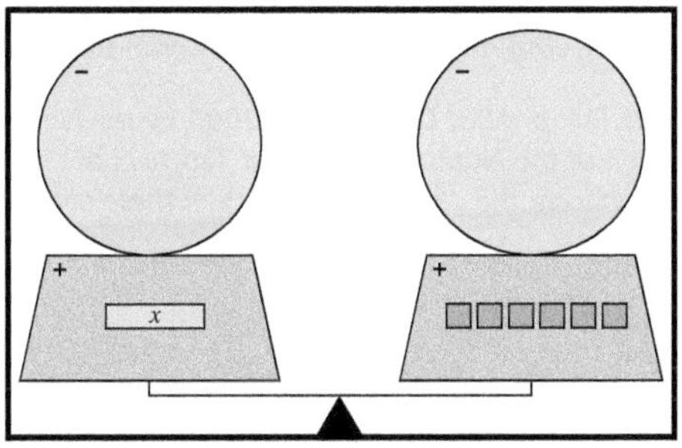

$x = 6$

Equations to Solve with the Equation Balance Scale

Work in pairs or small groups to solve the given equations, using your Equation Balance Scale. Carefully describe each step of your solution, and show the new equation that expresses the step just completed.

1. $x + 4 = 9$
2. $x + 6 = 1$
3. $3x + 5 = 14$
4. $4x - 3 = 2x + 5$
5. $2x - 1 = 5 - x$
6. $2 - x = 10 - 3x$

CHAPTER PREVIEW

Most people are surprised that the beginnings of algebra are introduced in elementary school nowadays. The NCTM algebra standards for grades 3–5 that follow describe how algebra enters the life of a third grader. In this chapter, we review topics from algebra with special attention to the concepts that are a part of grades K–8. However, this chapter is not meant to be a comprehensive discussion of coordinate geometry or algebra, as that would lead us away from the mathematics of elementary school. Instead, building on Section 1.4, we review topics such as straight lines and slopes, circles, and some nonlinear functions (including quadratic equations and exponentials) that can appear in elementary and middle school mathematics. We also examine the productive way in which algebra and geometry can work together. We use algebra freely in the rest of the book.

KEY IDEAS

- Algebra explains patterns and solves problems (Section 1.4).
- Algebra describes generality, conditions that lead to equations, and the relationship between quantities (formulas).
- Algebra and the geometry of the Cartesian plane combine to form a powerful problem-solving strategy that gives surprising results from geometry.

FROM THE NCTM PRINCIPLES AND STANDARDS

Algebra Standard for Grades 3–5

Instructional programs from prekindergarten through grade 12 should enable all students to—	**In grades 3–5 all students should—**
Understand patterns, relations, and functions	• *describe, extend, and make generalizations about geometric and numeric patterns;* • *represent and analyze patterns and functions, using words, tables, and graphs.*
Represent and analyze mathematical situations and structures using algebraic symbols	• *identify such properties as commutativity, associativity, and distributivity and use them to compute with whole numbers;* • *represent the idea of a variable as an unknown quantity using a letter or a symbol;* • *express mathematical relationships using equations.*
Use mathematical models to represent and understand quantitative relationships	• *model problem situations with objects and use representations such as graphs, tables, and equations to draw conclusions.*

Analyze change in various contexts	• *investigate how a change in one variable relates to a change in a second variable;* • *identify and describe situations with constant or varying rates of change and compare them.*

SOURCE: *Principles and Standards for School Mathematics by NCTM, p. 158. Copyright © 2000 by the National Council of Teachers of Mathematics. Reproduced with permission of the National Council of Teachers of Mathematics via Copyright Clearance Center. NCTM does not endorse the content or validity of these alignments.*

8.1 Variables, Algebraic Expressions, and Functions

In this section, we review algebraic reasoning with an emphasis on the mathematics in grades K–8. Topics include

- the meaning and uses of variables,
- how to form algebraic expressions involving variables,
- the definition and visualization of functions.

We now begin with the use of variables in constructing algebraic expressions.

Variables

Mathematical expressions in elementary arithmetic involve fixed values called **constants.** For example, there are 12 in a dozen, there are four sides of a rectangle, and $2 + 5 = 7$. Constants describe static situations in which change, generality, and variability are not under consideration. Mathematical expressions in algebra still include constants but also include quantities that are unknown, that can change, or whose value depends on how other, related values are determined or selected. Changeable quantities are referred to as **variables** and are denoted by a symbol. Usually the symbols are letters, such as the x in the equation $x - 3 = 5$, but sometimes other symbols are used. For example, in elementary school, students are asked what number must be put in the placeholder indicated by the box in the equation $\square - 3 = 5$.

Introducing quantities can give a dynamic aspect to algebraic reasoning that goes far beyond the reasoning of simple computation. Variables are used in at least four different ways in algebra.

Variables Describe Generalized Properties For example, the distributive property of multiplication over addition in the real numbers is succinctly stated as

$$a(b + c) = ab + ac, \quad \text{for all real numbers } a, b, \text{ and } c$$

Here, the symbols a, b, and c are *generalized variables,* used to describe a general property or pattern. A generalized variable represents an arbitrary member of the set of elements for which the property or pattern holds.

How would you explain the distributive property of multiplication over addition to your students in late elementary or middle school? Of course, there are many ways to do this. One that is appealing is to lead them through the process as was sketched out in Figure 2.25 of Section 2.4. Another way is to ask them to compute the area of a swimming pool of width 4 feet and length $(6 + 3)$ feet. Follow that question with a few more that bring them to finding the area of a pool of width a and length $(b + c)$, computed in two ways. Their result will be the distributive property. This mathematical habit of the mind shows how thinking with examples generates generalities.

Variables Express Relationships Jolie was born on her three-year-old sister Kendra's birthday. How are their ages related? Letting J and K denote their respective ages, we have several choices to express the age relationship:

$J = K - 3$ ("Jolie is three years younger than Kendra.")
$K = J + 3$ ("Kendra is three years older than Jolie.")
$K - J = 3$ ("The difference in age between Kendra and her younger sister Jolie is three years.")

Variables Serve as Unknowns in Equations For example, in the early grades students can be asked to find the number to place in the box that makes the sentence $\square + 5 = 9$ true. In the middle grades a student can be asked to find all values of x for which $(2x - 6)(x - 1) = 0$. The second example shows why it is appropriate to think of x as a variable, since it can be replaced either with 3 or with 1 to make the equation true.

Variables Express Formulas Variables are used in formulas, as you have seen in your earlier math courses. To give examples, several useful formulas are collected in Table 8.1.

TABLE 8.1 SOME COMMONLY USED FORMULAS

Topic	Variables	Formula
Distance	d = distance traveled r = rate of travel (speed) t = time of travel	$d = rt$
Triangular numbers (Gauss' insight)	$t_n = 1 + 2 + \cdots + n$ (sum of the first n integers) n = number of terms in the sum	$t_n = \frac{1}{2}n(n + 1)$
Perimeter of a rectangle	P = perimeter L = length W = width	$P = 2L + 2W$
Area of a rectangle	A = area	$A = LW$
Circumference of a circle	C = circumference r = radius	$C = 2\pi r$
Area of a circle	A = area r = radius	$A = \pi r^2$

The next example gives you an opportunity to check your understanding of the meaning and uses of variables. It would also be useful for you to review Figure 1.6 (Steps in Algebraic Reasoning) from Section 1.4.

EXAMPLE 8.1

Identifying the Role of Variables

In each part, identify the variable and the role it plays.

(a) The length of a rectangle is twice its width w, and the area of the rectangle is 18 square feet. What is the width w of the rectangle?

(b) The associative law of addition in the set of integers states that

$$a + (b + c) = (a + b) + c, \quad \text{for all integers } a, b, \text{ and } c$$

(c) A single taxpayer has an adjusted gross income of X dollars in 2003, where X is between \$28,400 and \$68,800. The taxpayer owes the Internal Revenue Service \$3910 + (0.25)(X − \$28,400).

(d) Milos is twice as tall as his little sister, Anke, so $m = 2a$, where m and a denote the respective heights of the siblings.

Solution

(a) w has the role of an unknown that satisfies the equation $w(2w) = 18$, or $w^2 = 9$. Since w is positive, the value of the unknown, namely, $w = 3$ feet, can be determined from the equation.

(b) a, b, and c play the role of generalized variables in the set of integers.

(c) X has the role of a variable in the formula used to compute the federal income tax owed.

(d) m and a are variables in the equation and express a relationship between the heights of the children.

Algebraic Expressions

A **numerical expression** is any representation of a number that involves numbers and operation symbols. For example, $3 + 8$ and $22 \div 2$ are both numerical expressions for 11. An **algebraic expression** is a representation that involves variables, numbers, and operation symbols. For example, $4x + 8y$ is an algebraic expression. It might represent the cost of taking x children and y adults to a play, where children's tickets are \$4 each and adults' tickets are \$8 each.

DESCRIPTION Algebraic Expression

An **algebraic expression** is a mathematical expression involving variables, numbers, or operation symbols.

In the examples that follow, you are challenged to create algebraic expressions.

EXAMPLE 8.2

Forming Algebraic Expressions

For each situation, form an algebraic expression that represents the requested values.

(a) The cost of every item in a store is increased by 15 cents. What is the cost of an item that used to cost c dollars? What is the old cost of an item that now costs d dollars?

(b) There was 3% inflation each of the last two years. What is the current price of an item that cost p dollars last year? If an item is q dollars today, what was its price two years ago? Assume that costs exactly followed the inflation rate.

(c) The electric power company charges \$5 a month plus 7¢ per kilowatt-hour of electricity used. What is the monthly cost to use K kilowatt-hours?

(d) Joan earned a credit hours of A work, b credit hours of B work, and so on. What is her grade point average (GPA)? The scale of the grading system is that each A contributes four points, each B three points, and so on.

Solution

(a) $c + 0.15, d - 0.15$ (in dollars)

(b) $1.03p, q/(1.03)^2$

(c) $5 + .07K$ (in dollars)

(d) The number of grade points earned is $4a + 3b + 2c + d$. The number of credit hours is $a + b + c + d + f$. Therefore, Joan's GPA is given by the quotient

$$\frac{4a + 3b + 2c + d}{a + b + c + d + f}.$$

The **domain of a variable** is the set of values for which the expression is meaningful. For example, if n denotes the number of students in a room, the domain is the set of whole numbers. If x is the width of a rectangle, the domain is the set of positive real numbers. An algebraic expression is **evaluated** by replacing each of its variables with particular values from the domain of the variables.

For example, consider again the expression $4x + 8y$ that gives the cost of x children and y adults attending the theater. Here the domain of each of the variables x and y is the set of whole numbers. When evaluated at $x = 5$ and $y = 3$, the expression has the value $4 \cdot 5 + 8 \cdot 3 = 44$.

One of the most popular ways to introduce the concept of algebraic expressions to elementary school children is to play the game *Guess My Rule.*

EXAMPLE 8.3

Guessing Erica's Rule

Erica is "it" in a game of *Guess My Rule*. As the children pick an input number, Erica tells what number her rule gives back, as shown in this table:

Children's Choice, x	Result of Erica's Rule, y
2	−1
5	8
6	11
0	−7
1	−4

Can you guess Erica's rule? What is the result of Erica's rule if a child picks $x = 12$?

Solution

Understand the Problem

Erica has chosen a rule that assigns to each choice (input) an answer (output). Given a value of the input variable x, she uses her function to determine the value of the output variable y, after which she reveals that to the class. We must guess her function. We wish to express the formula for y in terms of a variable x or n.

Devise a Plan

The children's choices are somewhat random. Perhaps a pattern will become more apparent if we arrange their values in order of increasing size. We anticipate that Erica's function is given by a formula.

Carry Out the Plan

Rearranging the children's choices in order of increasing size, we have the following table:

Children's Choice, x	Result of Erica's Rule, y
0	−7
1	−4
2	−1
5	8
6	11

When $x = 0$ is the input, the formula reads $y = -7$. This suggests that -7 is a separate term in the formula; when $x = 0$, all the other terms are 0. Also, we observe that when x increases by 1 from 0 to 1, from 1 to 2, and from 5 to 6, the output number increases by 3. This suggests that the formula also contains the term $3x$, since this quantity increases by 3 each time x increases by 1. Combining these observations, we guess that Erica's function (rule), E, is given by the formula

$$E(x) = 3x - 7.$$

Checking, we see that $E(0) = 3 \cdot 0 - 7 = -7$, $E(1) = 3 \cdot 1 - 7 = -4$, $E(2) = -1$, $E(5) = 8$, and $E(6) = 11$, as in Erica's table. When challenged, Erica reveals that we have guessed correctly.

Look Back

Erica's rule is really a function: Given a value of x, her rule returns a single value for y. We guessed the rule by arranging the data in a more orderly way, noting that her rule associated -7 with 0 and that the output number increased by 3 each time the input number increased by 1. Thus, we correctly guessed Erica's function to be $E(x) = 3x - 7$. To answer the second question, since $x = 12$, $E(x) = 3(12) - 7$. Thus, $y = E(12) = 36 - 7 = 29$.

Solving Equations

Two algebraic expressions with the same value form an **equation**, symbolized with the equal sign, =, placed between the expressions.

DEFINITION Equation

An **equation** is a mathematical expression stating that two algebraic expressions have the same value. The equal sign, =, indicates that the expression on the left side of the symbol has the same value as the expression on the right side.

Every equation is one of two types, either an identity or a **conditional equation.** On the one hand, an **identity** is an equation that is true for all evaluations of the variables from their domains. For example, $(x + y)^2 = x^2 + 2xy + y^2$ is an identity in the real numbers, since the expressions on both sides of the equality sign have the same value for all real numbers x and y.

On the other hand, if only certain values of the variables give equality, then the equation is **conditional.** In this case, x is referred to as the *unknown* and one hopes that all of the values of x can be determined from the equation. For example, $x^2 = 9$ is a conditional equation in the unknown x with the possible two solutions 3 and -3. If the domain of x is more restricted—say, to the positive numbers—there would be just one solution, $x = 3$. Determining the set of values that makes an equation true is an important step in algebraic reasoning and the notion of a solution set is necessary in high school.

DEFINITION Solution Set of an Equation, Equivalent Equations

The **solution set** of an equation is the set of *all* values in the domain of the variables that satisfy the given equation. Two equations are **equivalent** if they have the same solution set.

Some equations, such as $x^2 + 5 = 0$ in the domain of all real numbers, do not have a solution. In this case, their solution set is the empty set. We have included two examples: $x^2 + 5 = 0$ and Example 8.3.

Next, we discuss how the value of a variable can depend on the value of another variable or other variables. The rule that connects these variables is called a **function.** The concept of a function adds an important dynamic aspect to algebraic reasoning, providing a way to show relationships between quantities, describe change, and make predictions. Coming right after the ideas of number and operations, the concept of a function is a fundamental building block of elementary school mathematics. Functions appear frequently in state assessment exams in such forms as guessing patterns or describing a rule. More importantly, research shows that the basis for a thorough understanding of functions needs to start in elementary school in order to provide a solid foundation for the remainder of a student's mathematical education.

Highlight from History: Emmy Noether (1882–1935)

Emmy Noether was born in Erlangen, Germany, to a family noted for mathematical talent. Much of her life was spent at the University of Göttingen, exploring, teaching, and writing about algebra. This university—where Carl Gauss had taught a century earlier—was the first in Germany to grant a doctoral degree to a woman. Yet Noether met with frustrating discrimination there. For many years she was denied [an] appointment to the faculty; finally she was given an impressive title, "extraordinary professor"—but with no salary. Her abilities overcame the obstacles that daunted many other women in mathematics. Her work in the 1920s brought invitations to lecture throughout Europe and in Moscow. In 1933, as the Nazi party came to power in Germany, Noether met with persecution not only as a woman but also as an intellectual, a Jew, a pacifist, and a political liberal. She fled to the United States, where she taught and lectured at Bryn Mawr College and Princeton until her death in 1935.

> *"In the judgement of the most competent living mathematicians, Fräulein Noether was the most significant creative mathematical genius thus far produced since the higher education of women began. In the realm of algebra . . . she discovered methods which have proved of enormous importance. . . ."*
>
> —ALBERT EINSTEIN, 1935

SOURCE: *Biographical information can be found in Lynn Osen,* Women in Mathematics *(MIT Press, 1974). Quotations are cited in that source, original references include Einstein,* New York Times, *May 4, 1935.*

Quote by Albert Einstein, public domain.

Defining and Visualizing Functions

Suppose you are at the service station, filling your tank with gasoline priced at \$3.79 per gallon. The general principle is that your bill that day at the station is a *function* of the amount of gasoline you purchase. If we let the variable x (the input) represent the number of gallons of gasoline purchased and let the variable y (the output) represent the final bill in dollars, then y is defined by the simple rule $y = 3.79x$. Nicely enough, the pump carries out this calculation right before our eyes, and as the values of x whirl by on one display of the pump, we can simultaneously watch the corresponding values of y on another display.

It is easy to make a list of functions used in our lives, and you've seen many in algebraic courses. Here are three examples:

- The amount of postage on a first-class letter is a function of the weight of the envelope, rounded upward to the nearest ounce.
- The recommended amount of lawn fertilizer to be applied is a function of the number of square feet of lawn area.
- The amount of federal income tax you will owe is a function of your taxable income the previous year.

These examples should help you understand the general definition of a function.

DEFINITION Function
A **function** on a set D is a rule that associates, with each element $x \in D$, precisely one value y. The set D is called the **domain** of the function.

The definition of *function* requires that a single value y be assigned to each x-value in the domain. In the preceding example of the gasoline purchase, x is the number of gallons and the cost of gasoline is $y = 3.79$ (in dollars). The reason that there is precisely one value of y for each x is important and practical. If we use the gasoline example, there is only one answer to the question, "How much does 10 gallons cost?" and that answer is \$37.90. There can be no other price for that amount.

Another method of looking at a rule to see if the assignment that it makes is a function is called the **vertical-line test.** If you draw a graph, see whether a vertical line intersects the graph. If, for any vertical line, the line doesn't intersect the graph in two or more points, then the graph is the graph of a function.

A function can also be viewed as a set of ordered pairs with a certain condition: $\{(x, y) \mid x \in D$ and $y = f(x)\}$. Since just one y value is associated with each x, a set of ordered pairs represents a function when, and only when, there are no two ordered pairs with the same x-value but different y-values. This is a restatement of the "vertical-line criterion" previously mentioned.

If a function assigns the value y to an element x in the domain, then y is called the **image**, or **value**, of f at x. For example, if f denotes the function that calculates our bill at the gasoline station and if gas costs \$3.79 a gallon and we have purchased $x = 10$ gallons of gas, then $f(10) = 37.90$ is the value of the function at $x = 10$. We can also write $f(x) = 3.79 \cdot x$ and $D = \{x \mid x \geq 0\}$ or, more simply, D is all nonnegative numbers, which describes the gasoline-buying function by an algebraic equation and the domain of permissible values of x.

The set of all image values is called the **range** of the function.

DEFINITION Range of a Function
The **range** of a function f on a set D is the set of images of f. That is,

$$\text{range} f = \{y \mid y = f(x) \text{ for some } x \in D\}.$$

For example, a first-class letter mailed in the United States costs \$0.46 for the first ounce and \$0.20 for each additional ounce. Therefore, the range of the function that gives the cost of mailing a first-class letter is {\$0.46, \$0.66, \$0.86, . . .}.

Highlight from History: Two Women from Early Mathematics: Theano (ca. sixth century B.C.) and Hypatia (A.D. 370–415)

The early history of mathematics mentions few women's names. Indirect evidence, however, suggests that at least some women of ancient times had access to mathematical knowledge and likely made contributions to it. For example, the misnamed "brotherhood" of Pythagoreans, at the insistence of Pythagoras himself, included women in the order, both as teachers and as scholars. Indeed, Theano, the wife and former student of Pythagoras, assumed leadership of the school at the death of her husband. Theano wrote several treatises on mathematics, physics, medicine, and child psychology.

Hypatia was the first woman to attain lasting prominence in mathematics history. Her father, Theon, was a professor of mathematics at the Alexandrian Museum. Theon took extraordinary interest in his daughter and saw to it that she was thoroughly educated in arts, literature, science, philosophy, and, of course, mathematics. Hypatia's fame as a mathematician was secured in Athens, where she studied with Plutarch the Younger and his daughter Asclepigenia. Later she returned to the university at Alexandria, where she lectured on Diophantus' *Arithmetica* and Apollonius' *Conic Sections* and wrote several treatises of her own. The account of Hypatia's life of accomplishments in mathematics, astronomy, and teaching ends on a tragic note, for in March 415 she was seized by a mob of religious fanatics and brutally murdered.

EXAMPLE 8.4

Finding the Range of a Function

(a) Let f be the function defined by the formula $f(x) = x(10 - x)$ on the domain $D = \{1, 2, 3, 4, 5, 6, 7, 8, 9, 10.\}$ Find the range of f.

(b) The domain of $f(x) = x^2$ is the set of all real numbers between -2 and 5. What is the range of f?

Solution

(a) The image of f at $x = 1$ is $f(1) = 1 \cdot (10 - 1) = 9$. Similarly, $f(2) = 2 \cdot 8 = 16$, $f(3) = 3 \cdot 7 = 21$, $f(4) = 4 \cdot 6 = 24$, $f(5) = 5 \cdot 5 = 25$, $f(6) = 6 \cdot 4 = 24$, $f(7) = 7 \cdot 3 = 21$, $f(8) = 8 \cdot 2 = 16$, $f(9) = 9 \cdot 1 = 9$, and $f(10) = 10 \cdot 0 = 0$. Thus, the range of f is $\{0, 9, 16, 21, 24, 25\}$. Some image values, such as 9, 16, and 21, correspond to more than one x in the domain, but any x yields a single y-value in the range.

(b) The function f squares each number. For example, at $x = -2$, f gives 4, a positive number. $f(0)$ is zero and all other values are positive, with the largest value being $f(5) = 25$. Checking out a few more values, we easily see that the range is all numbers between 0 and 25 inclusive. Written a bit more formally in set notation, range $(f) = \{y \in R$ such that $0 \le y \le 25\}$.

Describing and Visualizing Functions

There are several useful ways to describe and visualize functions. We've already seen one, Erica's rule, and now will list functions as tables, arrow diagrams, machines, and graphs.

- **Functions as formulas.** Let's start off with some formulas that involve geometry. Consider a circle of radius r, where r is any positive number. The function $A = \pi r^2$ is the formula for the area of a circle, where r is the input. If we have a rectangle of length l and width w, then the function P written as the perimeter of a rectangle is $P = 2(l + w)$.

  Functions that are defined on the domain N of the natural numbers are called **sequences** and it is common to use the notation f_n instead of $f(n)$. How would we define sequences such as all the odd numbers? After trying a couple of values for n, we can see that the nth odd number is given by the formula $h(n) = 2n - 1$, so the 50th odd number is $h(50) = 2(50) - 1 = 99$. Another example is the sum $1 + 2 + 3 + \cdots + n$ of the first n natural numbers (i.e., the nth triangular number), which is given by Gauss' insight as $t(n) = n(n + 1)/2$. Thus, the sum of the first 100 natural numbers is $1 + 2 + 3 + \cdots + 99 + 100 = t(100) = 100(100 + 1)/2 = 5050$.

- **Functions as tables.** The following table gives the grades of three students on an essay question:

Student	**Grade**
Raygene	8
JaiWoo	7
Leticia	10

No algebraic formula connects the student to the grade. The table assigns a unique grade to each student; it is a function. The domain of this function is $D = \{\text{Raygene, JaiWoo, Leticia}\}$ and its range is $\{7, 8, 10\}$.

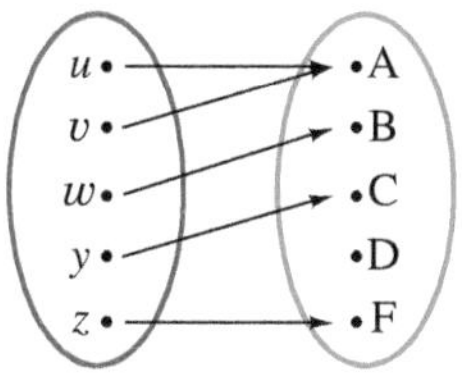

Figure 8.1
A function represented with an arrow diagram

- **Functions as arrow diagrams.** In an arrow diagram, one oval represents the domain and the second represents a set that contains the values of the function and possibly other points as well. Every point in the domain must have exactly one arrow that extends from that point to one of the points in the second oval. The arrow diagram in Figure 8.1 shows that the students Ursula, Vincent, Whitney, Yolanda, and Zach have been assigned grades on their class project. We see that Ursula and Vincent both received As and that no student received a D grade.
- **Functions as machines.** Viewing a function as a machine gives students an attractive dynamic visual model. The machine has an input hopper that accepts any domain element x and an output chute that gives the image $y = f(x)$. A few function machines are shown in Figure 8.2.

Figure 8.2
Three function machines

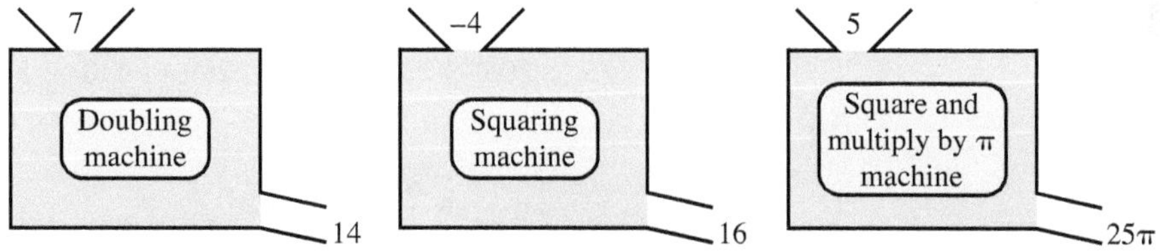

- **Functions as graphs.** A function whose domain and range are sets of numbers can be graphed on a set of x- and y-axes. Here are two graphs similar to what you see in middle school. We go into more depth about (x, y) coordinates in the next section. If $f(x) = y$, plot the points (x, y) for all x in the domain. The Fibonacci sequence (from Section 1.3) $F_1 = 1$, $F_2 = 1$, $F_3 = 2$, $F_4 = 3, \ldots$ and the doubling function $y = 2x$ are both plotted in Figure 8.3. Note that these two functions have different domains.

Figure 8.3
Graphs of (a) the Fibonacci sequence and (b) the doubling function

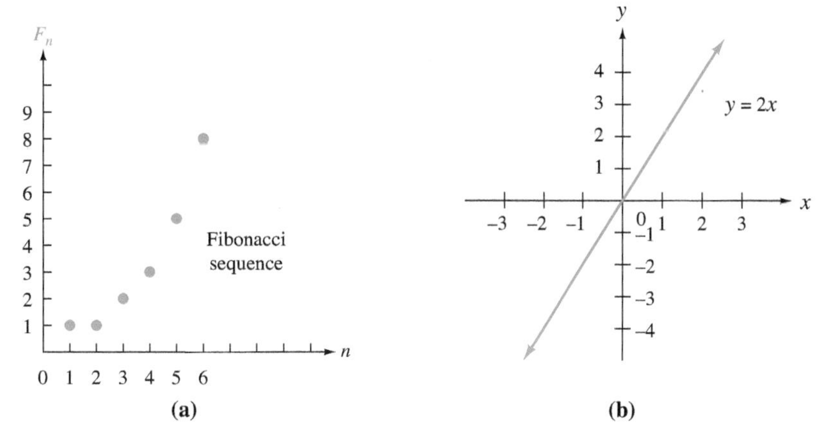

Computers and graphing calculators are useful tools for creating graphs and investigating the properties of a function represented by a graph. However, it is still instructive to make graphs by hand using graph paper. Begin by making a table of values of x from the domain and calculating the corresponding function values y. The table will help you choose an appropriate range of x- and y-values on the axes. Next, plot the points (x, y) from your table onto your graph. Finally, fill in the rest of the graph by smoothly connecting the points, as dictated by the domain.

We will now graph the points of the function $f(x) = 0.31x + 175$ in Example 8.5 that lie along a line. More generally, the graph of any function of the form $f(x) = mx + b$ is a line, where m and b are constants. For this reason, such functions are called **linear functions.** In the next section, additional attention is given to lines and their equations as well as two kinds of functions that appear in middle school: quadratic equations and exponentials, neither of which are linear.

EXAMPLE 8.5

Calculating the Cost of Owning and Driving a Car

A survey of car owners shows that the monthly cost (in dollars) to own and drive an automobile is given by the function $f(x) = 0.31x + 175$. Here, x represents the number of miles driven throughout the month and 175 represents the monthly cost of ownership that is independent of the miles driven (insurance, vehicle license fees, and so on).

(a) Make a table that shows the cost of having a car that is driven 0, 100, 200, . . . , 1000 miles per month.

(b) Use the table of values in part (a) to draw a graph that shows the cost of driving a car for up to 1000 miles in a month.

(c) Use your graph to estimate the corresponding limit on the number of miles driven in a month if your budget limits your car expenditures to \$350.

(d) What is the exact number of miles driven throughout the month if the expenditures are \$350?

Solution

(a)

Miles, x	0	100	200	300	400	500	600	700	800	900	1000
Cost, y	175	206	237	268	299	330	361	392	423	454	485

(b)

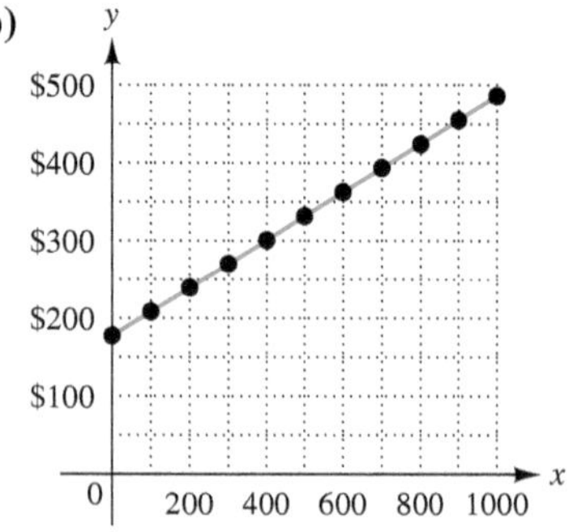

(c) The graph indicates that about 570 miles can be driven.

(d) This part requires the use of algebra. The exact value can be obtained from the equation $350 = 0.31x + 175$. Subtracting 175 from both sides and then dividing by 0.31 shows that $x = \dfrac{350 - 175}{0.31} = \dfrac{175}{0.31}$ miles (about 565 miles).

8.1 Problem Set

Understanding Concepts

1. Classify the quantities in each part as a constant or a variable.
   (a) The number of feet in a mile
   (b) The number of hours of daylight in a day

2. Classify the quantities in each part as a constant or a variable.
   (a) The price of a gallon of gasoline
   (b) The speed of light in empty space
   (c) The distance from Earth's center to the moon

3. In each of the given situations, classify the role of the variables as one of the following types: generalized, expressing a relationship, expressing a formula, an unknown.
   (a) For all real numbers x, y, and z, $x(y + z) = xy + xz$.
   (b) Elena is 4 inches shorter than her husband, Joe, so $E = J - 4$.

4. In each of the given situations, classify the role of the variables as one of the following types: generalized, expressing a relationship, expressing a formula, an unknown.

(a) To construct a circular flower bed covering 400 square feet of ground, the radius r must satisfy $\pi r^2 = 400$.

(b) Tickets to the play are \$5 for adults and \$3 for children. If A adults and C children attend the play, the total proceeds are $5A + 3C$.

5. Alicia, Boris, Carlos, Dan, and Xavier all collect cards. If x denotes the number of cards in Xavier's collection, form expressions for the number of cards in the other children's collections, using the information provided. For example, Alicia has six more cards than Xavier, so Alicia has $x + 6$ cards.

(a) Boris is three cards short of having twice the number of cards as Xavier.

(b) Carlos has two more than half the number of cards owned by Xavier.

(c) Dan has the same number of cards as Alicia and Xavier combined.

6. Penny is p years old. Form algebraic expressions with the variable p that represent the ages requested.

(a) Penny's age in five years

(b) Penny's age eight years ago

(c) The *current* age of Penny's little brother, who will be half of Penny's age in two more years

(d) The *current* age of Penny's mother, who was four times Penny's age three years ago

7. Richie weighs q pounds today, which is the last day of February. Form algebraic expressions with the variable that represent the weight requested.

(a) Richie's father is 10 pounds less than twice Richie's weight.

(b) Richie would like to lose .25 pound per day for the next week to make wrestling weight. What does he weigh at the end of the week?

(c) Richie would like to gain an eighth of a pound a day until the end of March. What will he weigh on April 1 if he started on March 1?

8. Five children, say, A, B, C, D, and E, are in a line. A writes a number on a slip of paper and hands it to B. B squares the number received and writes that value on a new slip of paper that is handed to C. In a similar way, C adds 5 to the number obtained from B and passes that number on to D. D multiplies the number by 7 and writes it on a slip of paper that is handed to E.

(a) If A writes a 3 on a slip of paper, what number will be handed to E?

(b) If child A puts x on the first slip of paper, what algebraic expression is handed to child E?

(c) If E is given the number 35, what number did A write on the initial slip of paper?

(d) Suppose the children line up in the order A, D, B, C, and E. If A writes y on a slip of paper, what expression is eventually handed to E?

9. As with problem 8, children are in a line and each is assigned one operation to perform. If the first child writes x on a slip of paper and the last child is handed a slip on which the expression $(3 - 5x)^3 + 4$ appears, describe how many children are in the line and what single operation each performs.

10. Bernie weighs 90 pounds plus half his own weight. What does Bernie weigh?

11. Consider the pair of equations $y - 8x = 2$ and $y + 4 = 3x$.

(a) Solve each equation for y as an expression in x.

(b) Equate the expressions found in part (a) to obtain an equation in the variable x.

(c) Solve the equation in part (b) for the unknown x.

(d) Solve for y.

12. Little Red Riding Hood rode her bike 15 miles to Grandma's house, arriving in 3 hours. When she discovered the Big Bad Wolf, she turned around and scampered home at 15 miles per hour. The *distance* = *rate* × *time* formula will be helpful to answer these questions.

(a) What was Little Red Riding Hood's average speed to Grandma's house?

(b) How long did it take to get home?

(c) How long was the round-trip?

(d) How many hours did the round-trip take?

(e) What is the average speed for the round-trip?

(f) Is the average speed for the round-trip equal to the average of the speeds to and from Grandma's house? Why or why not?

13. Decide which of these arrow diagrams represent(s) a function with domain A. If the diagram does not define a function, explain why. If a function is defined, give the range of the function.

(a)

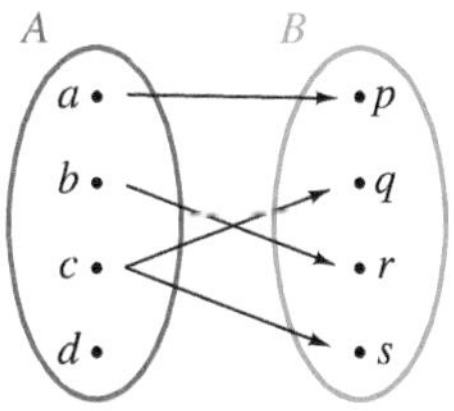

(b)

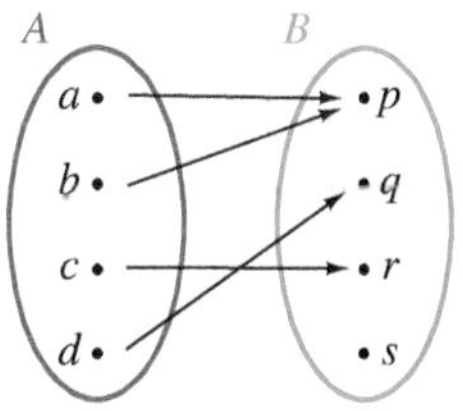

14. Decide which of these arrow diagrams represent(s) a function with domain A. If the diagram does not define a function, explain why. If a function is defined, give the range of the function.

(a)

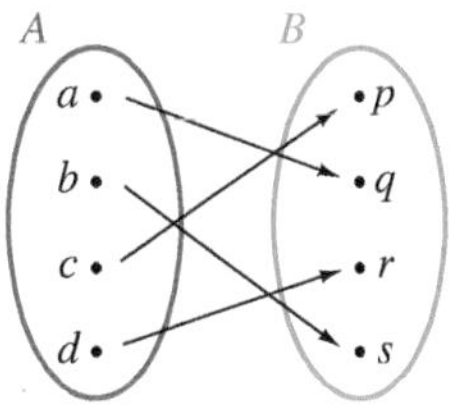

(b)

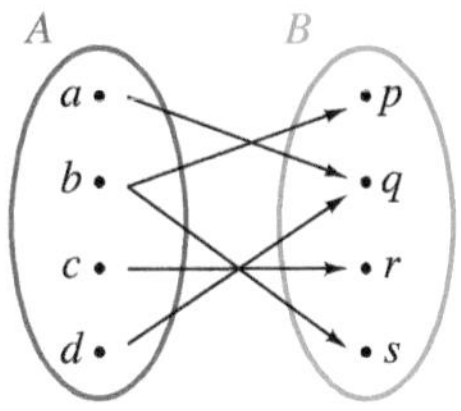

15. Draw an arrow diagram for each of the given sets of ordered pairs in cases (a), (b), (c), where, in each diagram, the left oval represents the set $A = \{1, 2, 3, 4, 5\}$ and the right oval represents the set $B = \{1, 2, 3, 4\}$. For each set, explain why the set of ordered pairs does or does not represent a function with domain A.

(a) $\{(1, 3), (2, 3), (1, 2), (3, 2), (4, 3), (5, 4)\}$

(b) $\{(1, 2), (2, 1), (3, 3), (4, 2), (5, 2)\}$

(c) $\{(1, 4), (2, 3), (4, 1), (5, 2)\}$

16. Let f be the function given by the formula $f(x) = 8 - 2x$ on the domain $D = \{x | x = 2, 3, 4, 5\}$.

(a) Make a table of the x- and y-values.

(b) What is the range of f?

(c) Sketch the graph of f.

17. Let g be the function given by the formula $g(x) = 2x - 2$, defined for all the real (decimal) numbers x in the interval $-2 \le x \le 3$.

(a) Make a table of values of $g(x)$ for $x = -2, -1, 0, 1, 2$, and 3.

(b) Plot the points in your table from part (a) and then sketch the entire graph.

(c) What is the range of g?

18. Let g be the function from $S = \{0, 1, 2, 3, 4\}$ to the whole numbers W given by the formula $g(x) = 5 - 2x + x^2$.

(a) Find $g(0), g(1), g(2), g(3), g(4)$.

(b) What is the range of g?

19. Let h be the function defined by $h(x) = x^2 - 1$, where the domain is the set of real numbers.

(a) Find $h(2)$.

(b) Find $h(-2)$.

(c) If $h(t) = 15$, what are the possible values of t?

(d) Find $h(7.32)$.

20. Karalee made a trip to the store one afternoon. Her trip is shown in the graph that follows, where her distance (in miles) from home is plotted as a function of the hours past noon. Refer to the graph to answer these questions about her trip.

(a) What time did she leave home?

(b) When did she realize that she forgot her checkbook?

(c) What did she do about the forgotten checkbook?

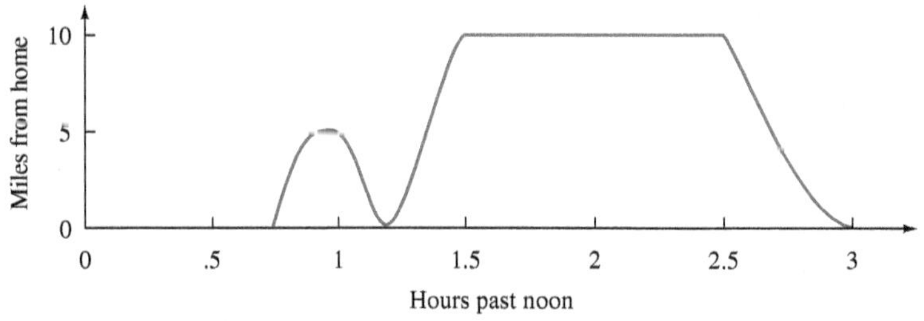

21. Look at Karalee's trip to the store one afternoon. Her trip is shown in the graph of problem 20. Her distance (in miles) from home is plotted as a function of the hours past noon. Refer to the graph to answer these questions about her trip.

(a) When did Karalee park the car at the store?

(b) How long did she shop?

(c) How far away was the store?

(d) Did Karalee encounter slower traffic going to or coming from the store? Explain how you can determine this from the graph.

22. Four graphs are shown, labeled G1, G2, G3, and G4. Which of these graphs matches the function described?

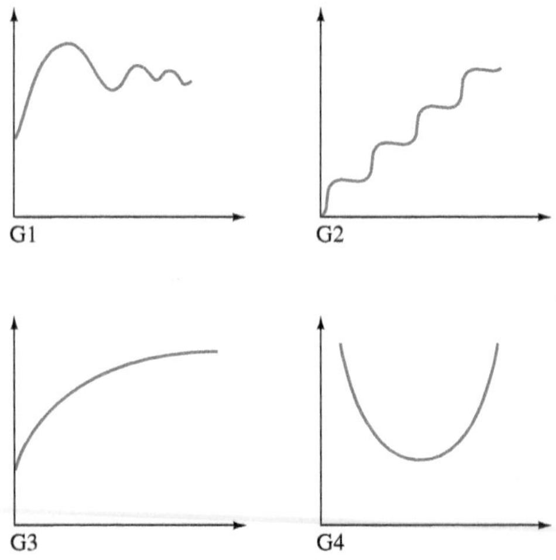

(a) Temperature of a pan of water placed on a hot burner

(b) Height of a flag being run up the flagpole

(c) Height of a weight hung from a Slinky®

(d) Length of the shadow of a flagpole throughout a sunny day in Dallas

23. Make simple approximate sketches of graphs (such as those in problem 22) that correspond to these functions of time:

(a) Temperature of a forgotten cup of hot tea

(b) Perceived pitch of a train whistle as the train passes

(c) Height of the water in a bathtub during the time someone takes a bath

(d) Hours of daylight in Chicago during a calendar year; that is, domain $= \{1, 2, \ldots, 365\}$

24. Consider the "double-square" rectangle of height x and width $2x$, as shown.

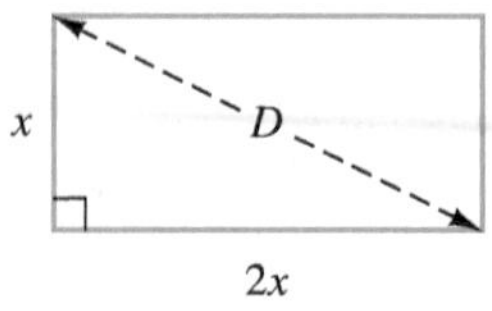

(a) Find the function $A(x)$ that gives the area of a double square of width x.

(b) Find the function $P(x)$ that gives the perimeter of a double square of width x.

(c) Use the Pythagorean theorem to obtain the function $D(x)$ for the length of the diagonal of a double square of width x.

25. Solve each of these *Guess My Rule* games:

(a)

Guess	4	7	2	0	10	1
Response	9	12	7	5	15	6

(b)

Guess	4	2	0	7	3	9
Response	10	6	2	16	8	20

(c)

Guess	4	5	7	3	0	10
Response	17	26	50	10	1	101

26. Che is "It" in a game of *Guess My Rule*. The students' inputs and Che's outputs are shown in the following table:

Input Given to Che	5	2	4	0	−2	−5
Output Reported by Che	24	3	15	−1	3	24

(a) Guess Che's rule (function), $y = C(x)$.

(b) Antonio says that Che's rule is $C(x) = x^2 - 1$, but Claudette claims that it is $C(x) = (x + 1)(x - 1)$. Who is correct, Antonio or Claudette? Explain.

27. Suppose two function machines are hooked up in sequence so that the output chute of machine g empties into the input hopper of machine f. Such a coupling of machines, which is defined if the range of g is a subset of the domain of f, is called the *composition* of f and g and can be written $F(x) = f(g(x))$ and depicted as follows:

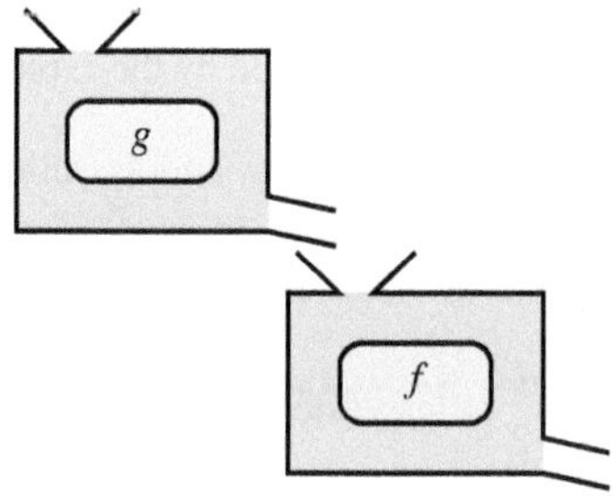

Suppose f is the doubling function $f(x) = 2x$ and g is the "add 3" function $g(t) = t + 3$. Then $F(4) = f(g(4)) = f(4 + 3) = f(7) = 2 \times 7 = 14$. Evaluate $f(g(x))$ for $x = 0, 1, 2$, and 3.

28. Using the two function machines of the previous problem, suppose the doubling and "add 3" function machines are coupled in reverse order to define the composition of g and f, given by $G(x) = g(f(x))$. Then $G(4) = g(f(4)) = g(2 \times 4) = g(8) = 8 + 3 = 11$. Evaluate $g(f(x))$ for $x = 0, 1, 2$, and 3.

Into the Classroom

29. A student thinks that the expressions $(a + b)^2$ and $a^2 + b^2$ are equal to one another. How would you explain to the student why this is not so?

30. Function concepts can be taught to young children by incorporating appropriate manipulatives and activities. For example, suppose a student plays the role of a function machine whose inputs are lengths of Unifix™ cubes. When handed a length of five cubes, the student returns a length of eight cubes. When given a length of two cubes, the student returns a length of five cubes.

(a) What function is the student apparently evaluating? How could it be tested? How many cubes should be put into the machine so that the output is ten cubes snapped together?

(b) Describe a function machine, again employing Unifix™ cubes, that illustrates the addition function $f(a, b) = a + b$ on the domain of pairs of natural numbers.

31. Create an activity, using commonly available manipulatives, that will illustrate a function concept. Use the Unifix™ cube activity described in problem 30 as an example, but be original.

Responding to Students

32. Many fourth-grade students have an extremely difficult time solving the equation $7 = \frac{x}{4}$ for x. Often many children won't generate an answer for x at all. However, if students are given $\frac{x}{4} = 7$, more can solve the problem.

(a) What is confusing the students?

(b) What is your next step as a teacher to help students realize that this is the same problem written two different ways?

33. Josiah was asked to identify which of the following statements shows that a is 9 less than b.

A. $9a = b$

B. $a + 9 = b$

C. $9 - a = b$

D. $\frac{a}{9} = b$

He answered C, $9 - a = b$. What is Josiah doing incorrectly?

34. AnnElise was asked to analyze population data for a nearby town that was growing each year. She was given the following table:

Year	2007	2008	2009	2010
Number of People	267	305	343	381

AnnElise was then asked to tell how many people would be living in the town in the year 2012 if the population increased at a constant rate as shown in the table. AnnElise decided that the population would be 419.

(a) What common error did she make when computing her answer?

(b) How would you guide her to use the table to find the correct answer?

35. Students created the following table:

x	1	2	3	4	5	10
y	3	5	7	9	11	?

They decided that the rule was to add 2 more each time, and they gave the answer as 13.

(a) What pattern do most elementary school students see first?

(b) How can you help them to see the function rule for any x?

36. Tara is working to simplify the expression $4x + 2$ and gives the answer 8. She explains that 4 times 2 is 8. What is confusing her? What are some other ways to write 4 times x?

Thinking Critically

37. Huong spent 90 cents at the store, buying some pencils at 15 cents each and some erasers at 6 cents each. He purchased several more erasers than pencils. How many pencils and erasers did he buy? Use algebraic reasoning to find your answer.

38. A commuter travels to work at an average speed u and returns home at an average speed v. Show that the average speed for the round-trip is $\frac{2uv}{u + v}$. Are you surprised that it is not $\frac{u + v}{2}$?

39. Suppose that n^3 sugar cubes are stacked to form an $n \times n \times n$ cube, as shown. The large cube is now painted yellow. Depending on where a sugar cube is positioned in the large cube, it may have one or several of its six faces painted yellow. For example, if $n = 1$, all six faces are painted. For larger n, there are some sugar cubes with unpainted faces.

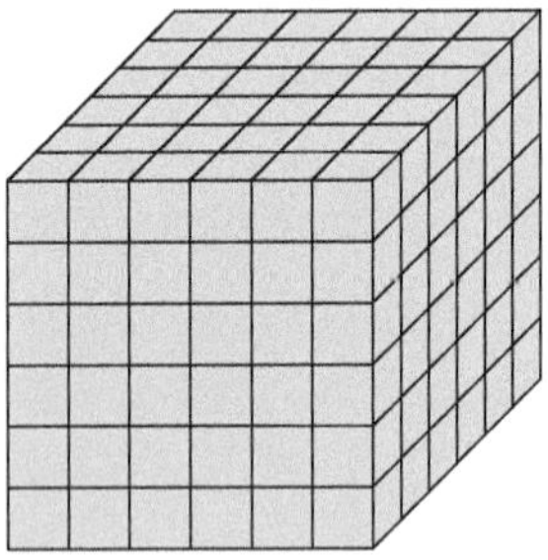

(a) Describe how the faces of the sugar cubes in a $2 \times 2 \times 2$ cube are painted.

(b) Count the number of sugar cubes in a $3 \times 3 \times 3$ cube with 0, 1, 2, or 3 painted faces. Do you account for all of the sugar cubes that form the large cube?

(c) Give expressions in the variable n, $n \geq 3$, for the number of sugar cubes in the $n \times n \times n$ cube with 0, 1, 2, or 3 painted faces, respectively.

(d) The four expressions obtained in part (c), when added together, should give n^3 for all n larger than or equal to 3. Why? Use algebraic simplification to check your identity in the variable n.

40. Find the solution set in the real numbers of each of these equations.

(a) $2(3x - 6x + 1) = 3(1 - 2x) - 1$

(b) $x(x - 2) = 2(x^2 - x + 1)$

41. The bank robbers are 9 miles north of Dodge City and are continuing to head north at 8 miles per hour on their horses. If the posse begins the chase from Dodge City on fresher horses at 10 miles per hour, how many hours will it take for the posse to catch the thieves? Show your answer with a graph that plots both the "robber" function $9 + 8t$ and the "posse" function $10t$, each of which gives the distance north of Dodge after t hours of the chase.

Making Connections

42. Rug remnants can be purchased for \$3.60 per square yard, and the edges can be finished at a cost of 12¢ per foot. Let L and W respectively denote the length and width of a rectangular remnant, given in feet. Give an expression for the cost, in dollars, to buy and finish an L-by-W remnant.

43. Zal is paid \$18 an hour for a 40-hour workweek and is paid time and a half for overtime work. Give expressions for Zal's pay if he works t hours. You'll need to distinguish between the cases where t is less than or equal to 40 and t is greater than 40.

State Assessments

44. (California, Grade 5)
Sophie caught twice as many fish as her dad. If her dad caught F fish, how many fish did Sophie catch?

A. $F + 2$ B. $F - 2$ C. $F \times 2$ D. $F \div 2$

45. (California, Grade 5)

$$c + 2\frac{1}{2}$$

Which situation could be described by the expression above?

A. Lia jogged c miles yesterday, and $2\frac{1}{2}$ miles farther today.

B. Lia jogged c miles yesterday, and $2\frac{1}{2}$ miles fewer today.

C. Lia jogged $2\frac{1}{2}$ miles yesterday, and c miles fewer today.

D. Lia jogged $2\frac{1}{2}$ miles yesterday, and c times as far today.

46. (Grade 7)
Which phrase below *best* represents the following expression?

$$(y + 4)(y - 3)$$

A. 3 less than 4 more than a number

B. 4 more than the product of a number and 3

C. The product of a number and 3 less than the number

D. The product of 4 more than a number and 3 less than the number

47. (Massachusetts, Grade 8)
Four friends earned money by painting a house. After they divided the money equally, they each received \$315. Which of

the following equations could be used to determine x, the total amount, in dollars, that the four friends earned by painting the house?

A. $\frac{x}{4} = 315$ B. $4x = 315$

C. $x - 4 = 315$ D. $x + 4 = 315$

48. (Grade 4)
The table below gives the number of milligrams of sodium in each of the three sizes of a breakfast cereal. A bowl is 28 grams.

Sodium Amounts in Cereal Bowl Sizes

Cereal Size (Grams)	Sodium Amount (Milligrams)
7	42
14	84
21	126

Based on the table, what is the total number of milligrams of sodium in a 28 gm cup of cereal?

A. 84 mg
B. 168 mg
C. 63 mg
D. 252 mg

49. (California, Grade 5)
What value for z makes this equation true?

$$8 \times 37 = (8 \times 30) + (8 \times z)$$

A. 7
B. 8
C. 30
D. 37

8.2 Graphing Points, Lines, and Elementary Functions

In this section, we explore the graphs of equations of functions from a geometric point of view. In the next section, we add algebra to the mix through the use of the Cartesian plane. Much of the current section is a review of your previous coursework but is included to remind you of the material to provide a platform for the remaining chapters and to introduce a new problem-solving strategy: using Cartesian coordinates to do geometric problems. Although your first introduction to these topics might have been in middle school (or even high school), their beginnings are now included in the curriculum and standardized tests of elementary school.

We start with the introduction of the **Cartesian plane,** also called the **coordinate** or **Cartesian coordinate plane,** in honor of the philosopher and mathematician René Descartes. (See the Highlight from History box, p. 395.)

The Cartesian Coordinate Plane

Consider the two perpendicular number lines illustrated in Figure 8.4. Any point in the plane can be uniquely located by giving its distance to the right or left of the vertical number line and its distance above or below the horizontal number line. In the figure, the point P is 5 units to the right of the vertical number line and 3 units above the horizontal number line; there is only one such point. Thus, P is identified by the **ordered pair** (5, 3). We may also write (5, 3) as $P(5, 3)$, as shown. Other times we will just write (5, 3).

Figure 8.4
The Cartesian coordinate plane

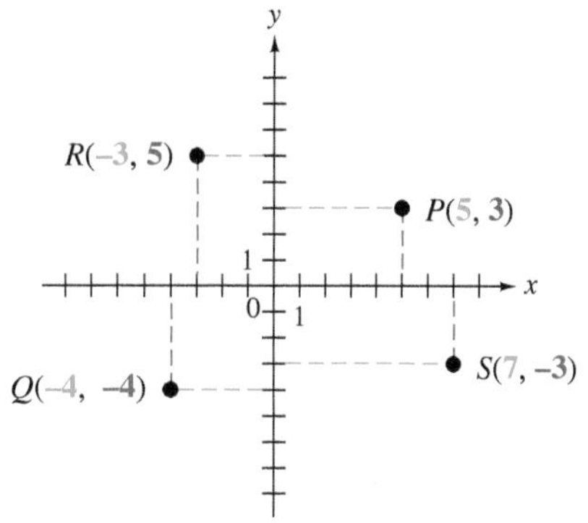

The two number lines are called **coordinate axes** or just **axes.** Typically, the horizontal number line is called the ***x*-axis** and the vertical number line is called the ***y*-axis.** The two numbers in the number pair (a, b) that locate a point are called the **Cartesian coordinates** of the point. The first number in the pair is called the ***x*-coordinate** and gives the distance of the point to the right or left of the vertical axis—that is, in the direction of the x-axis. The second number in the ordered pair is called the ***y*-coordinate** and gives the distance of the point above or below the horizontal axis—that is, in the direction of the y-axis. The axes divide the plane into four regions, or **quadrants,** numbered I, II, III, and IV counterclockwise from the upper right-hand quadrant. A point lies in

- quadrant I if both its coordinates are positive,
- quadrant II if the first coordinate is negative and the second coordinate is positive,
- quadrant III if both coordinates are negative, and
- quadrant IV if the first coordinate is positive and the second coordinate is negative.

All of the notions are summarized in Figure 8.5, especially the point (0, 0) where the axes intersect and is called the **origin** of the coordinate system. We can also write the Cartesian plane in set notation as $R^2 = \{(x, y) \mid x \in R \text{ and } y \in R\}$.

Figure 8.5
Key features of the Cartesian coordinate system

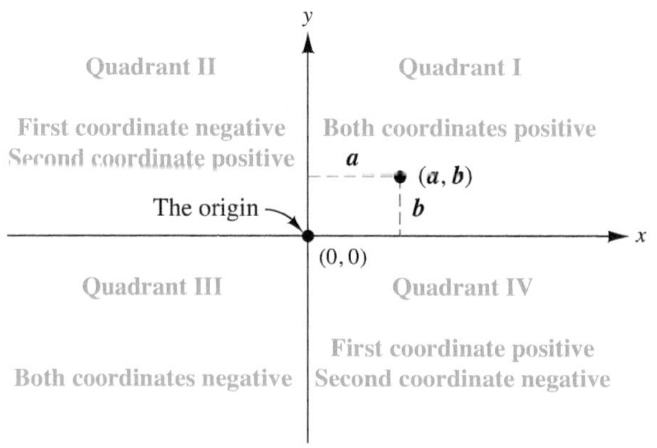

EXAMPLE 8.6

Plotting Points

The following diagram shows the partial outline of a house:

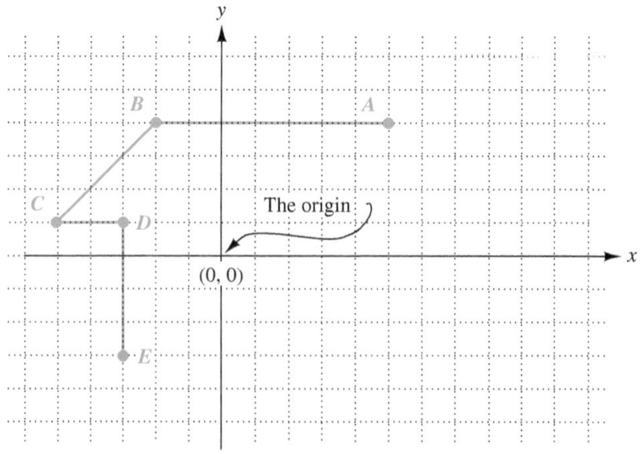

(a) Give the coordinates (i.e., the ordered pair) naming each of A, B, C, D, and E.
(b) Plot the points $F(8, -3)$, $G(8, 1)$, $H(10, 1)$, $I(7, 4)$, $J(6, 4)$, $K(6, 5)$, and $L(5, 5)$.
(c) Draw the segments $\overline{EF}$, $\overline{FG}$, $\overline{GH}$, $\overline{HI}$, $\overline{IJ}$, $\overline{JK}$, $\overline{KL}$, and $\overline{LA}$.

Solution

(a) The point A is 5 units to the right of (0, 0) (that is, in the x-direction from the origin) and 4 units above (0, 0) in the y-direction so A is the point (5, 4). The x-coordinate of A is 5 and the y-coordinate of A is 4. Similarly, we determine that the coordinates of the other points are $B(-2, 4)$, $C(-5, 1)$, $D(-3, 1)$, and $E(-3, -3)$.

(b) The point $F(8, -3)$ is 8 units to the right of (0, 0) and 3 units *below* (0, 0). Similarly, the other points are located as shown:

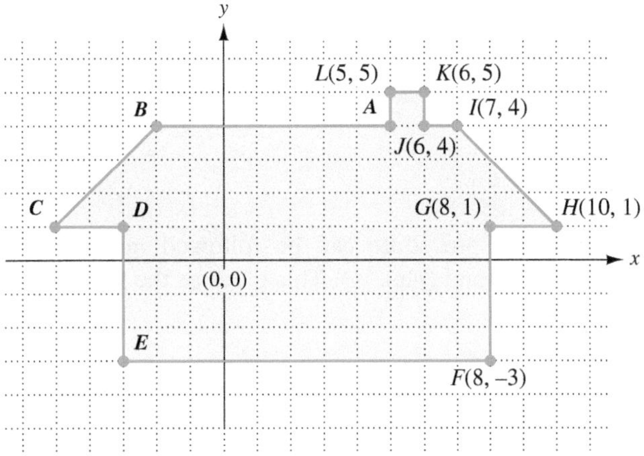

(c) The line segments form the outline of a house.

Highlight from History: René Descartes and Coordinate Geometry

It is difficult to ascribe the development of any particular body of mathematics to its originator. The invention of coordinate geometry is usually ascribed to René Descartes (1596–1650), a leading 17th-century mathematician and philosopher. He made great strides in combining the ideas of geometry and algebra, as explained in a book titled *La géométrie*. However, Descartes never thought of an ordered pair (a, b) as the coordinates of a point in the plane. Ascribing the Cartesian coordinate system to Descartes is largely misplaced. The idea of coordinates goes back at least as far as the third century B.C. The idea was also known to the amateur, but inspired, Pierre de Fermat, a contemporary of Descartes, and was popularized by the Dutch mathematician Frans van Schooten in his *Geometria a Renato Des Cartes* (*Geometry by René Descartes*) in 1649. It is probably fair to say that our modern ideas of coordinate geometry were inspired by Descartes, but organized and popularized by Schooten.

The Distance Formula

Consider the points $P(2, 5)$ and $Q(7, 8)$ shown in Figure 8.6.

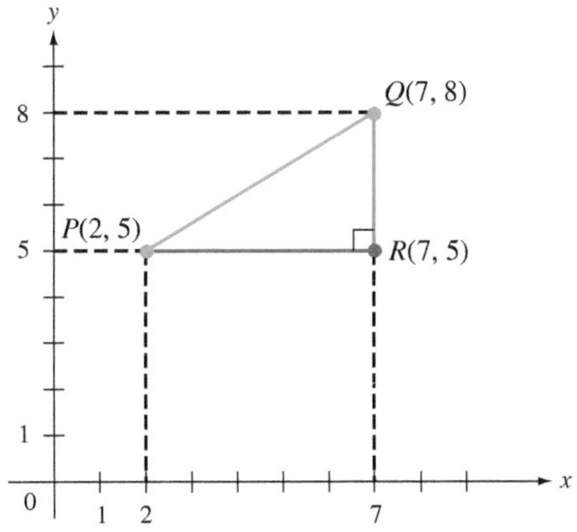

Figure 8.6
The distance between points P and Q is found by using the Pythagorean theorem for the right triangle PQR

The vertical line through the upper point and the horizontal line through the lower point intersect at the point $R(7, 5)$. The two points $P(2, 5)$ and $R(7, 5)$ lie on the horizontal line of points whose y-coordinates are $y = 5$. The distance between P and R, which is denoted by PR, is then just the absolute value of the difference in the x-coordinates of the points. That is, $PR = |7 - 2| = 5$. Similarly, the points $Q(7, 8)$ and $R(7, 5)$ both are on the vertical line for which $x = 7$, so the distance between Q and R is $QR = |8 - 5| = 3$.

To find the distance between P and Q, we use the fact that PQR is a right triangle with hypotenuse PQ and legs PR and RQ. The Pythagorean theorem, mentioned in Section 7.1 and discussed in depth in Section 10.3, then tells us that $PQ^2 = PR^2 + RQ^2$. That is,

$$\begin{aligned} PQ &= \sqrt{PR^2 + RQ^2} \\ &= \sqrt{(7-2)^2 + (8-5)^2} \\ &= \sqrt{25 + 9} = \sqrt{34} \doteq 5.8. \end{aligned}$$

This procedure can be followed in exactly the same way beginning with any two points $P(x_1, y_1)$ and $Q(x_2, y_2)$. This gives us the following theorem:

THEOREM The Distance Formula

Let P and Q be the points (x_1, y_1) and (x_2, y_2). Then the distance between P and Q is

$$PQ = \sqrt{(x_2 - x_1)^2 + (y_2 - y_1)^2}.$$

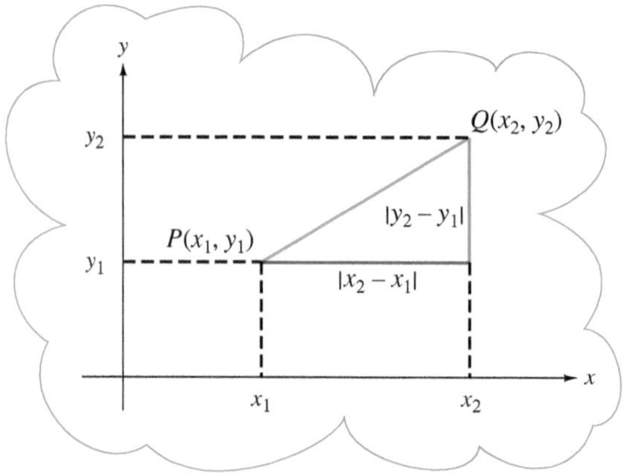

In the distance formula, it does not make any difference which point is chosen for P and which is chosen for Q. For example, the distance between (4, 7) and (1, 3) is given by both

$\overleftrightarrow{PQ}$ is a line, $\overline{PQ}$ is a segment, and PQ is the length of the line $\overleftrightarrow{PQ}$.

$$\begin{aligned} \sqrt{(4-1)^2 + (7-3)^2} &= \sqrt{3^2 + 4^2} \\ &= \sqrt{9 + 16} = \sqrt{25} = 5 \end{aligned}$$

and

$$\begin{aligned} \sqrt{(1-4)^2 + (3-7)^2} &= \sqrt{(-3)^2 + (-4)^2} \\ &= \sqrt{9 + 16} = \sqrt{25} = 5. \end{aligned}$$

It is important to notice that the notation makes a distinction between the geometric object $\overline{PQ}$ (a set of points in the plane) and its length PQ (a nonnegative real number), as you can see via the think cloud and on the next page.

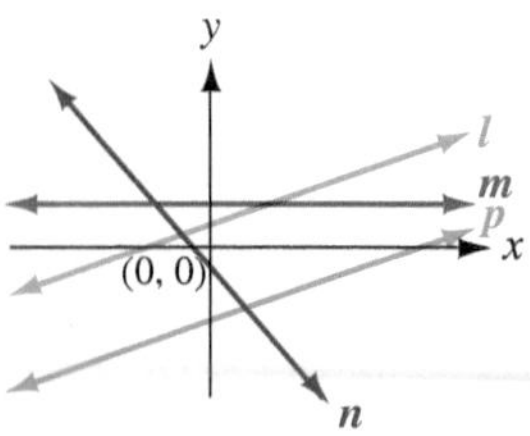

Figure 8.7
Lines in the plane

Slope

Consider the lines l, m, n, and p in Figure 8.7. The properties that distinguish lines from one another are their location on the coordinate system and their direction, or steepness. The direction, or steepness, of a line leads to the notion of *slope*, which will be defined carefully.

Now, we define *line, line segment*, and *length of a line segment* and then build on the concept of slope.

DEFINITION Line, Line Segment, and Length of Line Segment

If P and Q are distinct points, then the **line** is denoted by $\overleftrightarrow{PQ}$. The points on the line $\overleftrightarrow{PQ}$ that are between P and Q form a **line segment** that includes the **endpoints** P and Q. The **length,** PQ, of the line segment is the **distance between its endpoints.**

Of course, we can compute the length PQ by the use of the distance formula.

As an application, carpenters make good use of the ratio

$$s = \frac{h}{r},$$

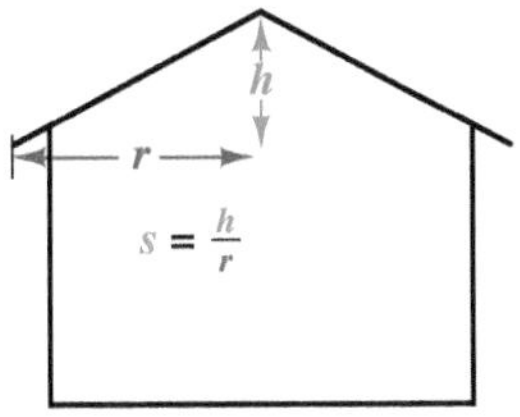

Figure 8.8
Slope of a roof

where h is the vertical distance a roof rises and r is the horizontal distance over which the rise takes place, to compute the slope of a roof, as shown in Figure 8.8. This relationship is sometimes expressed by saying that the slope of a roof is "the rise over the run." Surveyors use the same idea to calculate the slope of a road. If a road rises 5 feet while moving forward horizontally 100 feet, the road has a slope of 0.05. In surveying, slopes are usually expressed as percents. Thus, a grade with a slope of 0.05 is said to be a 5% grade.

We also use the idea *rise over run* to find the slope of a line segment. Consider the points $P(3, 5)$ and $Q(9, 7)$ shown in Figure 8.9. We move up 2 units while moving to the right 6 units. The rise over the run gives a slope of $\frac{1}{3}$, indicated by the letter m:

$$m = \frac{7 - 5}{9 - 3} = \frac{2}{6} = \frac{1}{3}.$$

Figure 8.9
Slope of a segment

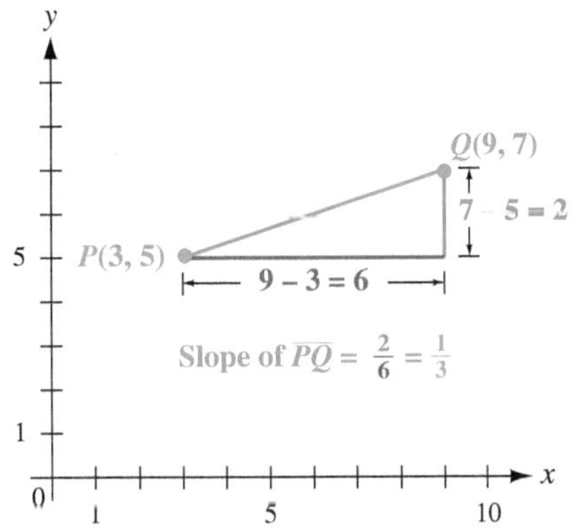

In general, the reasoning is similar and leads to this definition of slope:

DEFINITION Slope of a Line Segment or Line

Let $P(x_1, y_1)$ and $Q(x_2, y_2)$, with $x_1 \neq x_2$, be two points. Then the **slope of the line segment** $\overline{PQ}$, or the line $\overleftrightarrow{PQ}$, is given by

$$m = \frac{y_2 - y_1}{x_2 - x_1}.$$

If $x_1 = x_2$ in the preceding definition, then $x_2 - x_1 = 0$ and $\overline{PQ}$ is vertical. Since division by 0 is undefined, we must say that *the slope of a vertical segment (or line) is undefined.* Another common usage is that a vertical segment has "no slope." Be sure not to confuse "no slope" with having "zero slope" (which describes a horizontal, rather than a vertical, segment).

The Standard for Mathematical Practice 6 (SMP 6) shows the importance of a precise definition. For example, a subtle concept is "slope of a straight line." Notice that the formal definition of *slope* was applied to every pair of points that had different x-coordinates and included a separate statement for those pairs that had the same x-coordinate: Their slope was undefined (they determine the vertical lines). Furthermore, there are efficient and accurate ways of computing the slope of a line: point–slope form, slope–intercept form, and two-point form of the equation of a line. This section shows three ways of determining the slope of a line that all three will end up with the same answer.

SMP 6 Attend to precision.

"Mathematically proficient students try to communicate precisely to others. They try to use clear definitions in discussion with others and in their own reasoning. . . . They calculate accurately and efficiently, express numerical answers with a degree of precision appropriate for the problem context."

If $y_1 = y_2$ in the preceding definition, then $\overline{PQ}$ is horizontal and $m = 0$. Thus, saying that a line segment is horizontal is the same as saying that it has zero slope.

$\frac{5-7}{3-9} = \frac{7-5}{9-3}$

As we said, in computing the slope of a segment, it makes no difference which point is chosen as P and which is chosen as Q. However, once the choice is made, we must stick with it and always subtract *in the same direction* in both numerator and denominator. For example, in computing the slope of the segment in Figure 8.9, we identified P and Q as (3, 5) and (9, 7), respectively. But this could have been reversed to obtain

$$m = \frac{5-7}{3-9} = \frac{-2}{-6} = \frac{1}{3}$$

as before.

Notice in Figure 8.9 the slope is positive and the segment slopes upward to the right. This is always the case for segments with positive slopes. If $y_1 = y_2$ in the definition of slope, then the slope is 0 and the segment $\overline{PQ}$ is horizontal. If the slope of a segment is negative, the segment slopes *downward to the right.* In the next example, $\overline{PS}$ has negative slope.

EXAMPLE 8.7 Finding the Slopes of Line Segments

The points $P(-4, 3)$, $Q(5, 6)$, $R(5, -1)$, and $S(-1, -3)$ are the vertices of the four-sided polygon $PQRS$ shown in the given figure. Find the slope of each side of the polygon.

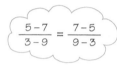

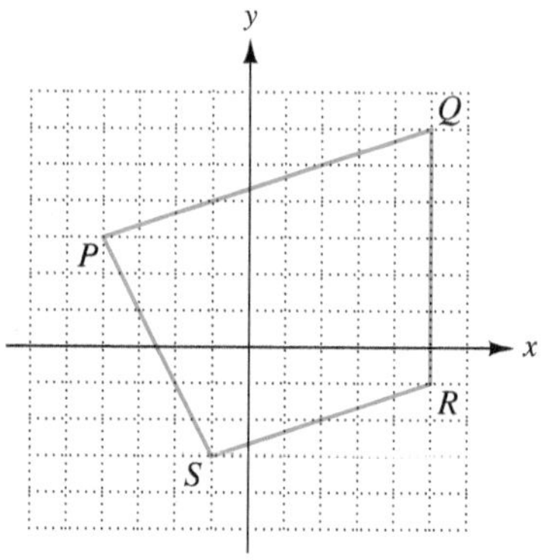

Solution

Using the slope formula $m = \frac{y_2 - y_1}{x_2 - x_1}$, we find that

$$m(\overline{PQ}) = \frac{6-3}{5-(-4)} = \frac{3}{9} = \frac{1}{3},$$

$$m(\overline{RS}) = \frac{(-3)-(-1)}{(-1)-(5)} = \frac{-2}{-6} = \frac{1}{3}, \text{ and}$$

$$m(\overline{PS}) = \frac{(-3)-(3)}{(-1)-(-4)} = \frac{-6}{3} = -2.$$

The side $\overline{QR}$ is vertical, since both R and Q have the same x-coordinate, 5. Therefore, $\overline{QR}$ has undefined slope.

Equations of Lines

With the tools of coordinate geometry, it is now possible to write equations whose graphs are lines. We begin by considering a particular example of finding the equation of a line, which requires a point and a slope, then the slope–intercept form, followed by the two-point form.

EXAMPLE 8.8

Determining the Equation of a Line Through (2, 3) with Slope $\frac{4}{3}$

Derive an equation of the line through point $P(2, 3)$ and having slope $\frac{4}{3}$.

Solution

Understand the Problem

There is exactly one line through the point (2, 3) and we know the slope is $\frac{4}{3}$. We can draw the line by plotting the point $P(2, 3)$ and then plotting the point Q 3 units to the right and 4 units above P. The line segment through these points must have slope $\frac{4}{3}$, so the line through these points must be the desired line. Suppose $R(x, y)$ is *any* point on the line. We must find an equation that is satisfied by the points R on the line.

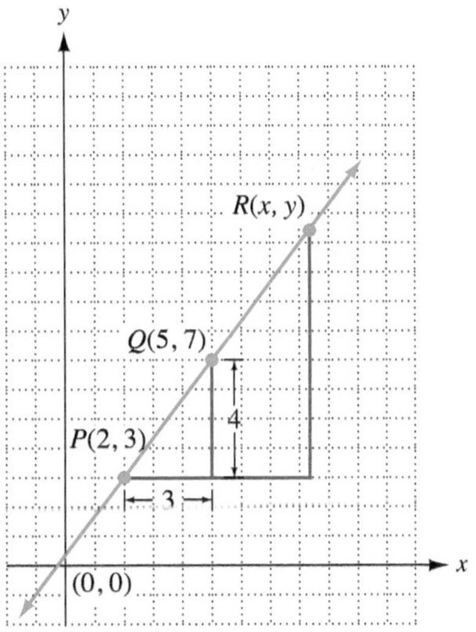

Devise a Plan

Since the slope of a line can be determined by *any* two points on the line, $R(x, y)$ is on the line precisely when the

$$\text{slope } \overline{PR} = \frac{4}{3}.$$

Perhaps we can use this fact to derive the desired equation.

Carry Out the Plan

We can compute the slope of $\overline{PR}$ since $R = (x, y)$ so that

$$\text{slope } \overline{PR} = \frac{y - 3}{x - 2}.$$

It follows that R is on the line in question if and only if

$$\frac{y - 3}{x - 2} = \frac{4}{3},$$

or, alternatively, multiplying both sides by $(x - 2)$,

$$y - 3 = \frac{4}{3}(x - 2).$$

Hence, we have found the desired equation.

Look Back

Since there is one, and only one, line through a given point and having a given slope, $R(x, y)$ is on the line if and only if the slope $\overline{PR} = \frac{4}{3}$. By expressing slope $\overline{PR}$ in terms of x and y, we obtained the desired equation.

THEOREM Point–Slope Form of the Equation of a Line

The equation of the line through $P(x_1, y_1)$ and having slope m is

$$y - y_1 = m(x - x_1).$$

This is called the **point–slope form** of the equation of a line and depends exactly on the slope m and the point, P.

As this theorem suggests, there are several forms of the equation of a line. Another particularly useful form is stated in the next theorem. First we note that if a line crosses the y-axis at the point $(0, b)$, b is called the ***y*-intercept** of the line.

THEOREM Slope–Intercept Form of the Equation of a Line

The **slope–intercept form** of the equation of a line is

$$y = mx + b,$$

where m is the slope and b is the y-intercept.

PROOF Since b is the y-intercept, the line passes through the point $(0, b)$. Also, the line has slope m. Therefore, by the point–slope form of the equation of a line, the desired equation is

$$y - b = m(x - 0),$$

or, equivalently,

$$y = mx + b.$$

In the theorem just proved, we started with a line in the plane, defined by its slope m and its y-intercept, and deduced that its y-values are given by the function $y = mx + b$. On the other hand, we see that the graph of any function of this form is a line of slope m that intersects the y-axis at $y = b$. For example, the graph of the function $f(x) = -\frac{x}{2} + 3$ has slope $-\frac{1}{2}$ and intersects the y-axis at $y = 3$. It has now become clear why functions of the form $f(x) = mx + b$ are called linear functions: Their graphs are straight lines.

EXAMPLE 8.9

Using the Slope–Intercept Form of the Equation of a Line

Write the equations of the given lines with slope and y-intercept as indicated. Also, draw each line on a coordinate system.

(a) $m = -3, b = 5$
(b) $m = 0, b = -4$

Solution

(a) Using the preceding theorem of the slope–intercept form of the equation of a line, we obtain the equation $y = -3x + 5$. To draw the line, we plot the point (0, 5), which is the y-intercept, and the point (1, 2), which is 3 units *below* and 1 unit to the right of (0, 5). Then we draw the line through these two points:

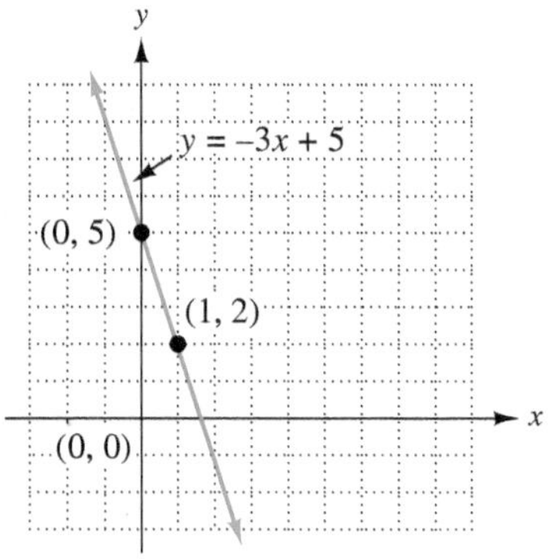

(b) This line goes through the point (0, −4) and has slope 0. Therefore, using the slope–intercept form, we obtain the equation

$$y = 0x + (-4),$$

or just

$$y = -4.$$

This equation says that the line is horizontal and that a point on this line always has its y-coordinate as −4. These points can have any x-coordinates and the line is as shown here:

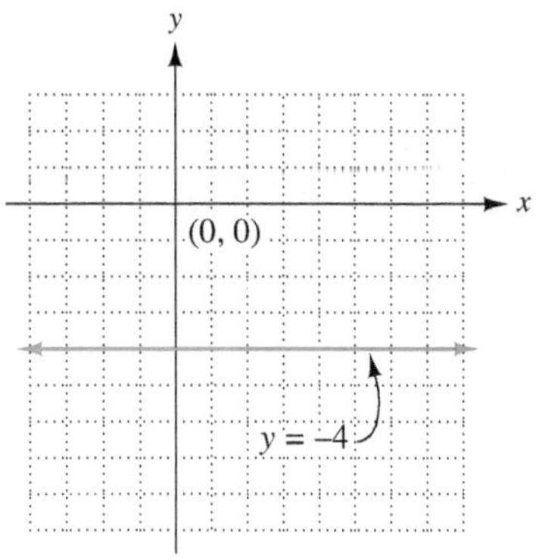

The preceding example suggests that *every* horizontal line has an equation of the form $y = b$; the slope is 0 and the x-values of points on the lines can be any number. By analogy, *every* vertical line has an equation of the form $x = a$; the slope is undefined and the y-values of points on the lines are left unrestricted.

EXAMPLE 8.10

Determining the Equation of a Line through Two Points

Determine the equation of the line through $P(-2, 3)$ and $Q(6, 7)$.

Solution

First, we will find the slope of $\overline{PQ}$:

$$\text{slope } \overline{PQ} = \frac{7 - 3}{6 - (-2)} = \frac{4}{8} = \frac{1}{2}.$$

We now employ the point–slope form of the equation of the line, using either point. Trying $P(-2, 3)$ gives

$$y - 3 = \frac{1}{2}(x - (-2)), \quad \text{or, equivalently,} \quad y - 3 = \frac{1}{2}(x + 2).$$

This equation can be rewritten in the form $y = \frac{1}{2}x + 1 + 3$, or $y = \frac{1}{2}x + 4$, which, from the slope–intercept form, tells us that the line intersects the y-axis at $y = 4$. Using the point $Q(6, 7)$ instead of P, we have the same equation,

$$y - 7 = \frac{1}{2}(x - 6).$$

The equation of the line through *any* two points $P(x_1, y_1)$ and $Q(x_2, y_2)$ can be found in the same way as shown in Example 8.10. Assuming that $\overline{PQ}$ is not a vertical segment, we see that it has the slope given by $m = \frac{y_2 - y_1}{x_2 - x_1}$. Then, using point $P(x_1, y_1)$, we find that the point–slope equation of the line is $y - y_1 = m(x - x_1)$, so we have the following theorem:

THEOREM Two-Point Form of the Equation of a Line

The equation of the line through $P(x_1, y_1)$ and $Q(x_2, y_2)$, where $x_1 \neq x_2$, is

$$y - y_1 = m(x - x_1), \quad \text{where} \quad m = \frac{y_2 - y_1}{x_2 - x_1}.$$

This is called the **two-point form** of the equation of a line with slope m.

Nonlinear Functions

We have seen that it is unexpectedly easy to graph a straight line. For example, the function $f(x) = -2x + 7$ can be graphed by observing that the two points $P(0, 7)$ and $Q(1, 5)$ are on the graph of the function. The rest of the graph is obtained by drawing the line through P and Q.

DEFINITION Linear Function

A **linear function** is a function that is a straight-line graph.

Another way to give a definition of a linear function is say that the relationship between y and x is $y = mx + b$. We have seen this equation many times in this text and in grades K–12; it is, of course, the graph of a straight line.

DEFINITION Nonlinear Function

A **nonlinear function** is a function that is not a linear function.

Some examples of nonlinear functions are $g(x) = x^2 - 3x + 5$, $h(x) = \dfrac{x}{x^2 + 2}$, and $k(x) = 2^x$. The graphs of these functions, each created with a graphing calculator, are shown in Figure 8.10.

Figure 8.10
The graphs of three nonlinear functions

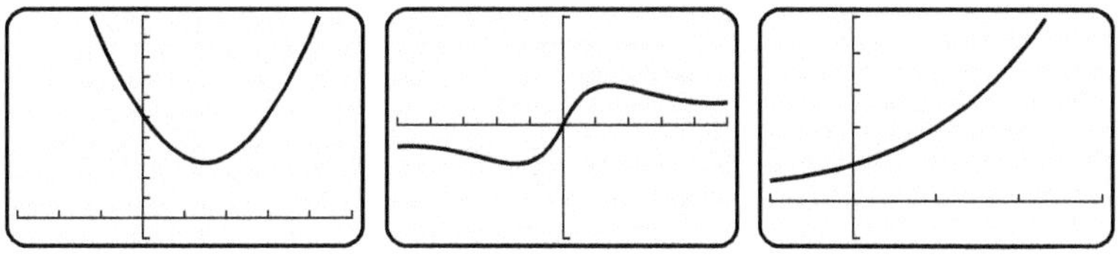

$g(x) = x^2 - 3x + 5$ $\qquad h(x) = \dfrac{x}{x^2 + 2}$ $\qquad k(x) = 2^x$

As expected, none of the graphs is a straight line.

You have already studied some nonlinear functions in high school or college. However, since our emphasis is on the mathematics of grades K–8, we will concentrate exclusively on **quadratic functions,** and **exponential functions,** which are both a part of the middle school curriculum.

DEFINITION Quadratic Function

A **quadratic function** is defined as $f(x) = ax^2 + bx + c$ for any number a, b, c if $a \neq 0$.

We will now do an "applied problem" that uses a quadratic function.

To graph nonlinear functions by hand, it is necessary to make a table of *many* points—not just two!—that are on the graph. It is now easy to do this with a graphing calculator. Plotting these points will then suggest the shape of the graph that can be approximated by smoothly connecting the points with a curve. This procedure is used in Example 8.11 for a quadratic function.

EXAMPLE 8.11 Making the Biggest Animal Pens (Quadratic)

The third-grade class wants to make pens for its pet rabbits and guinea pigs. A parent has donated 24 feet of chain-link fencing material that will be used to make two side-by-side rectangular pens along a wall, as shown in the figure that follows. What dimensions of x and w will give the pens their largest total area?

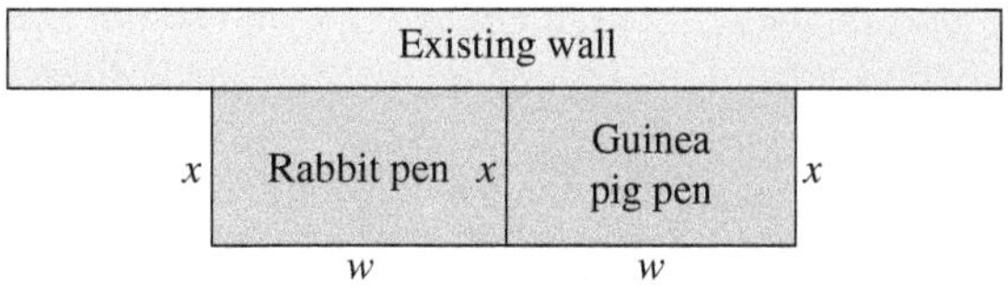

Solution

As seen in the figure, the pens form a rectangle x feet wide and $2w$ feet long. Therefore, the total area of the pens is given by $A = 2wx$. The pens will require $3x + 2w$ feet of fencing. Since 24 feet of fencing is available, this gives the equation $3x + 2w = 24$, which can be rewritten as $2w = 24 - 3x$. Thus, the area A of the pens is given by the following equation in x:

$$A = 2wx = (24 - 3x)x = 24x - 3x^2.$$

We can now see how the total area A of the pens is given as the value of the function $f(x) = 24x - 3x^2$. Since the variable x is a length of a side of the pen, we must have $x \geq 0$. Also, the three sides perpendicular to the wall, each of length x, must use less than the 24 feet of available fence, so $3x \leq 24$, or $x \leq 8$. Therefore, the domain of the function is $D = \{x | 0 \leq x \leq 8\}$.

The following table of values of the function indicates that the largest area occurs close to when $x = 4$:

x	0	1	2	3	4	5	6	7	8
Area A	0	21	36	45	48	45	36	21	0

The total area of the two pens is 48 square feet and $2w = 24 - 3 \cdot 4 = 12$. Therefore, $w = 6$, so each pen is 4 feet by 6 feet.

The following graph of the function $y = 24x - 3x^2$ shows that the maximum area actually occurs when $x = 4$:

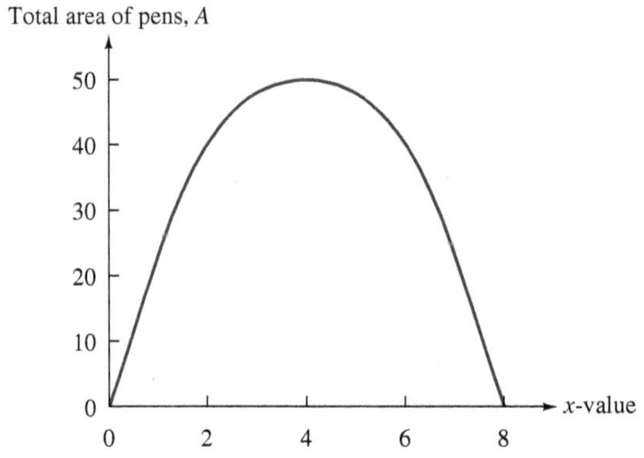

Looking back, we see that this function is a quadratic function whose maximum is the answer to our question. The maximum is $x = 4$.

Exponentials are functions in which a fixed positive real number is raised to a variable power. Exponentials exhibit rapid growth (e.g., rabbit populations) or decay (radioactivity) and are topics that you have worked with already. In fact, $k(x)$ of Figure 8.10 is an exponential function. We will not go into detail about these two types of nonlinear functions but will include a worked example of each. They will be in the problem set at the end of this section.

First, we give an example of an exponential function and then give its formal definition.

EXAMPLE 8.12

A Choice on a Checkerboard (Exponential)

My father put a dollar on the first square of a checkerboard. On the second square, Dad doubled to \$2, on the third \$4, the fourth \$8, and so on. Dad gives me a choice: Receive either \$1 million or the money from the last square on the checkerboard. What should I do?

Solution

We are using the usual 8×8 checkerboard, and we have a pattern for the amount A_n on the nth square. The last square is the 64th. The following table shows the amount of money on the first few squares:

n	1	2	3	4	5	6
A_n	\$1	\$2	\$4	\$8	\$16	\$32

The pattern from the table gives $A_n = 2^{n-1}$, where $n \geq 1$. We answer the question posed in this example by deciding which is larger, \$1 million or A_{64}. Using a calculator, we find that $A_{64} = 2^{63}$, which is close to 9.2234×10^{18}, or about 9 with 18 zeros after it! Certainly, as generous as Dad is, the checkerboard option is the best!

Motivated by this example, we define an exponential function, which is really about raising a fixed number, called the base to a power:

DEFINITION Exponential Function

Let a be a positive real number called the **base.** A function $f(k)$ is an **exponential function** if it is of the form $f(k) = Ba^k$, where B is a real number.

Since $a^0 = 1$ if $a \neq 0$, an exponential function is a multiple of a power of its base and the value of the function "when time starts," which is B; that is, $B = f(0)$, or B is the value when $k = 0$. The domain of an exponential function is usually the integers, the positive real numbers, or all real numbers. Note that if the base is less than one, the function gets smaller with time, not larger.

COOPERATIVE INVESTIGATION

The Open-Top Box Problem

Materials

Each group of four to seven students needs eight sheets of centimeter-squared grid paper, scissors, tape, and two sheets of graph paper.

Directions for Making an Open-Top Box from a Rectangle

Each group will make open-top boxes by cutting x- by x-sized squares from the corners of a 16- by 21-centimeter rectangle cut from the centimeter-squared grid paper. Fold the rectangular sides of the box upward and tape the corners to create the box. The group should make a set of boxes corresponding to $x = 1, 2, 3, 4, 5, 6$, and 7 centimeters.

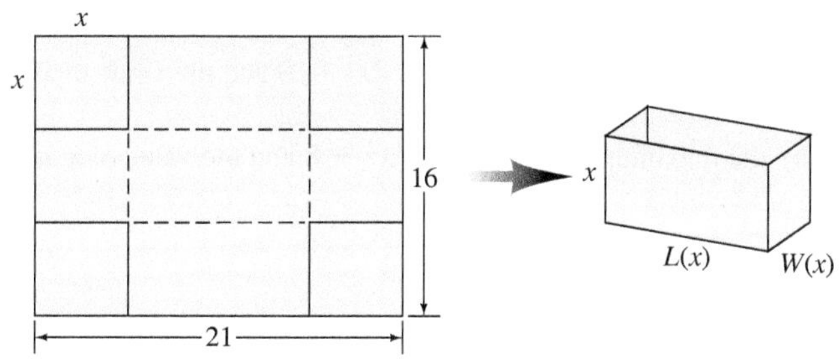

Activities and Questions

1. Guess which box has the largest volume. Which box apparently has the smallest volume?

2. Fill in the following table, and in the last two rows compute the area $B(x)$ of the bottom of your box and the total area $S(x)$ of the sides of your box:

x	1	2	3	4	5	6	7
L							
W							
V							
B							
S							

3. Divide your group into two subgroups. One subgroup graphs the volume of the box as a function of x for $1 \leq x \leq 7$ and, on the basis of its graph, decides which value of x results in the open-top box of largest volume. The other subgroup graphs the two functions $B(x)$ and $S(x)$ (on the same set of axes) and reports on the value of x for which $B(x) = S(x)$. As a group, report a connection on the two reported values of x from the subgroups.

4. If graphing calculators are available, express each of $V(x)$, $B(x)$, and $S(x)$ as functions of x. Then enter these functions on a graphing calculator to graph the functions $V(x)$, $B(x)$, and $S(x)$.

8.2 Problem Set

Understanding Concepts

1. Plot and label the following points on a Cartesian coordinate system drawn on a sheet of graph paper:
   (a) $P(5, 7)$ (b) $Q(5, -7)$
   (c) $R(-5, 7)$ (d) $S(-5, -7)$

2. Plot and label the following points on a Cartesian coordinate system drawn on a sheet of graph paper:
   (a) $T(0, 5)$ (b) $U(7, 1)$ (c) $V(0, -5.2)$
   (d) $W(-7, 1)$ (e) $X(0, 0)$

3. Plot these points and connect them in order with line segments: (1, 1), (1, 11), (4, 13), (5, 15), (6, 13), (7, 12), (10, 11), (9, 10), (6, 9), (4, 7), and (1, 1).

4. (a) Plot the points (5, 0), (4, 3), (3, 4), (0, 5), (−3, 4), (−4, 3), (−5, 0), (−4, −3), (−3, −4), (0, −5), (3, −4), and (4, −3).
   (b) What do you observe about the points in part (a)?

5. Compute the distance between these pairs of points:
   (a) (−2, 5) and (4, 13) (b) (3, −4) and (8, 8)

6. Compute the distance between these pairs of points:
   (a) (0, 7) and (8, −8) (b) (3, 5) and (2, −4.3)

7. Compute the slopes of the line segments determined by the given pairs of points. In each case, tell whether the segment is vertical, is horizontal, slopes upward to the right, or slopes downward to the right.
   (a) $P(1, 4), Q(3, 8)$ (b) $R(-2, 5), S(-2, -6)$
   (c) $U(-2, -3), V(-4, -7)$

8. Compute the slopes of the line segments determined by the given pairs of points. In each case, tell whether the segment is vertical, is horizontal, slopes upward to the right, or slopes downward to the right.
   (a) $C(3, 5), D(-3, 5)$ (b) $E(1, -2), F(-2, -5)$
   (c) $G(-2, -2), H(4, -5)$

9. Determine b so that the slope of $\overline{PQ}$ is 2, where P and Q are the points $(b, 3)$ and $(4, 7)$, respectively.

10. Determine d so that the slope of $\overline{CD}$ is undefined if C and D are the points $(d, 3)$ and $(-5, 5)$, respectively.

11. What is the value of a if the point $(a, 3)$ is on the line $2x + 3y = 18$?

12. (a) Find two different points on the line $3x + 5y + 15 = 0$.
    (b) Draw the graph of the line in part (a) in a coordinate system.

13. Graph each of these lines in a single coordinate system and label each line:
    (a) $3x + 5y = 12$ (b) $6x = -10y + 12$
    (c) $5y - 3x = 15$ (d) $6x + 10y = 24$
    (e) What do you conclude about the lines of parts (a) and (d)?

14. (a) Draw the graph of $3x + 2y + 6 = 0$.
    (b) Does the equation of part (a) define y as a function of x? If so, identify the function.

15. (a) Draw the graph of the equation $5x - 3y - 15 = 0$.
    (b) Does the equation of part (a) define y as a function of x? If so, identify the function.
    (c) Does the equation of part (a) define x as a function of y? If so, define the function.

16. (a) On a single set of axes, draw the graphs of $y = 2x + 3$, $y = 2(x - 3) + 3$ and $y = 2(x + 4) + 3$.
    (b) Compare the graphs of part (a).

17. Graph each of these functions:
    (a) $y = x^2$ (b) $y = (x - 2)^2$ (c) $y = (x + 3)^2$
    (d) Discuss the relationship between the graphs of parts (a), (b), and (c).

18. Graph each of these functions:
    (a) $y = x^2 - 4x + 4$
    (b) $y = x^2 + 6x + 9$
    (c) $y = x^2 + 4x + 4$
    (d) Could you write each function in a different, more concise form? (*Hint:* Consider problem 17.)

19. (a) On a single set of axes, graph the equations $y = x^2$, $y = x^2 + 4$, and $y = x^2 - 3$.
    (b) Compare the graphs of part (a).

20. By sketching a graph, find
    (a) the minimum value of the quantity $x^2 + 10x$ and the value of x at which it occurs.
    (b) the maximum value of $8x - 2x^2$ and the value of x at which it occurs.

21. By sketching the graph of the function, find
    (a) the minimum value of $2x^2 - 4x + 10$ and the value of x at which it occurs.
    (b) the maximum value of $2x - x^2 + 8$ and the value of x at which it occurs.

Into the Classroom

22. Kristi has a large detailed map showing the winding 6-mile-long trail up to the top of Pyramid Peak. She knows that she started up the trail at 8 A.M., hiked at a steady rate, and reached the summit at noon. Kristi would like to make a graph of her distance up the trail during the hike so that she can see on the map where she was at each hour and half hour along the trail. There are mileage markers on the map, but she is

confused about how the graph will look because the trail was very crooked. Carefully describe how you would work with Kristi to help her accomplish her goals.

23. Children often think that all graphs are linear. Devise an activity in which measurements are made and entered into a table and graphs are made that clearly show that some data vary linearly and other related data vary nonlinearly.

Thinking Critically

24. Find an equation for each of the lines **(a)**, **(b)**, **(c)**, and **(d)** shown in the following coordinate system:

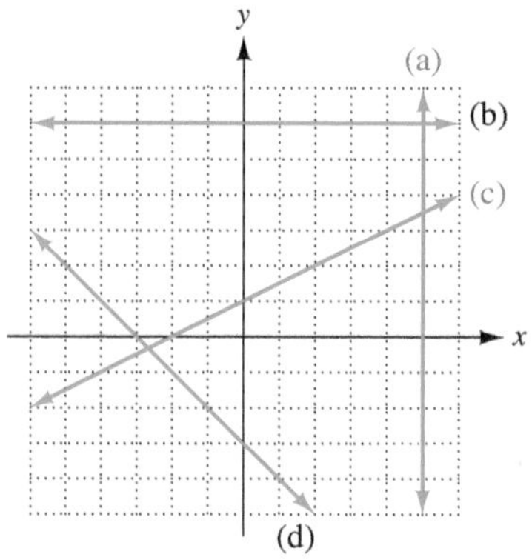

25. If $A(0, 0)$, $B(3, 5)$, $C(r, s)$, and $D(7, 0)$ are the vertices of a parallelogram, determine r and s.

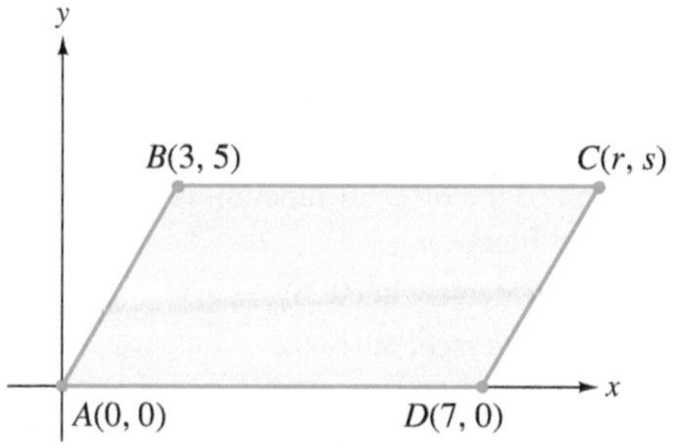

26. A square $ABCD$ in the coordinate plane has vertices at $A(2, 3)$, $B(6, 2)$, $C(r, s)$, and $D(u, v)$. Find all possible choices of r, s, u, and v.

27. Suppose the third graders in Example 8.11 want to make a third pen for baby chicks, still using the 24 feet of fence. That is, there are now to be three side-by-side pens along the wall, each measuring x feet by w feet. Draw a figure showing the pens along the wall, with appropriate labels, and plot a graph to show the dimensions that give the pens the largest possible area.

28. In the checkerboard example (Example 8.12), at what square would the value be more than $1 million for the first time?

29. If Dad were to put a quarter on the first square and then follow the process of Example 8.12, would you choose the checkerboard or the $1 million?

30. If Dad followed a checkerboard process as in Example 8.12, but changed the rule so that he squared the amount and put that in the next square, would you pick the checkerboard rather than the $1 million

(a) if the first square had $2 on it?
(b) if the first square had $1 on it?
(c) if the first square had a quarter on it?

Making Connections

31. (a) **ADA-Approved Ramps.** The Americans with Disabilities Act states, "The maximum slope of a ramp in new construction shall be 1:12. The maximum rise for any run shall be 30 in." What is the minimum amount of run for a rise of 30 inches?

(b) **Highway Design.** Highway 195 into Lewiston, Idaho, undergoes a difference in elevation of 1800 feet in 7 miles. What is the average percent slope of the Lewiston grade?

32. **World Population.** The following table gives the world population (in billions) every 20 years since 1900 and an estimate of the population for 2020:

Year	1900	1920	1940	1960	1980	2000	2020
Pop.	1.6	1.9	2.3	3.0	3.7	6.0	7.6

(a) Make a graph of the world population as given in the table.
(b) Letting x denote the number of decades (10-year periods) since 1900, plot the graph of the function $y = (1.45)1.14^x$ on the graph drawn in part (a). Does the formula mimic the population data well in your opinion?
(c) If a graphing calculator is available, use it to plot the table and the function given in part (b) on the same screen and then compare the two graphs.

Making Connections (with Linear Functions)

33. **Hooke's Law.** If a (small) weight w is suspended from a spring, the length L of the stretched spring is a linear function of w. Find m and b in the formula $L = mw + b$ if the unstretched spring has length 10″ and a weight of 2 pounds stretches it to 14″.

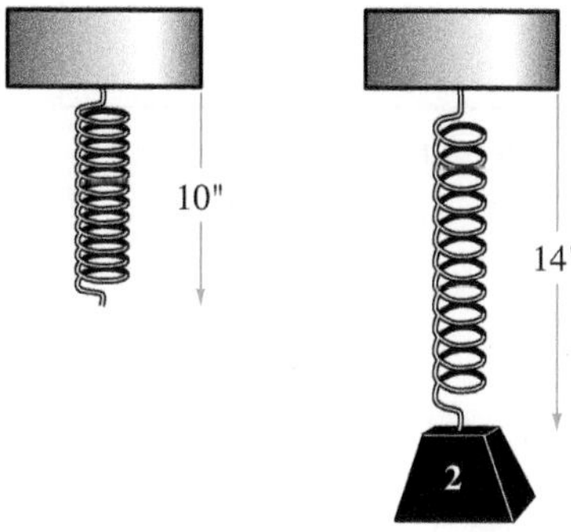

34. **Temperature Conversion.** The temperature at which water freezes is 32° Fahrenheit (0° Celsius), and the temperature at which water boils is 212° Fahrenheit (100° Celsius). Find the constants m and b in the formula $F = mC + b$ that expresses the Fahrenheit temperature F as a function of the Celsius temperature C.

35. **The Lightning Distance Function.** The speed of sound is about 760 miles per hour. Assuming that a lightning flash takes no appreciable time to be seen and the corresponding peal of

thunder is heard after t seconds, show that $d = \frac{t}{5}$ gives the approximate distance d (in miles) to the lightning strike.

State Assessments

36. (Grade 4)
Here is a game which is played on a board.

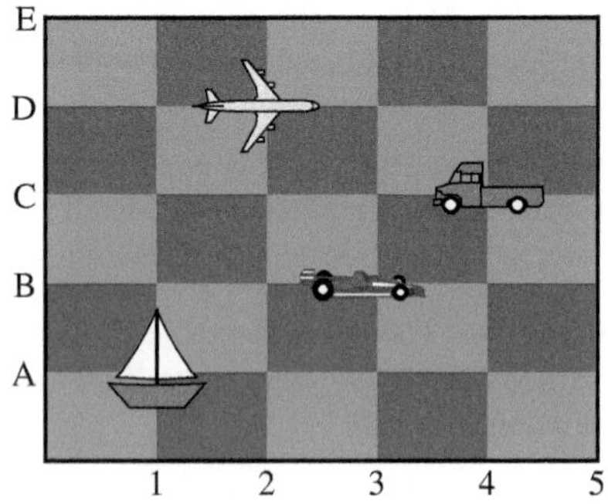

Which object is located at (2, D)?

A. The plane    B. The truck 

C. The race car    D. The boat 

37. (Grade 8)
To the nearest whole number, what is the distance between points C and D?

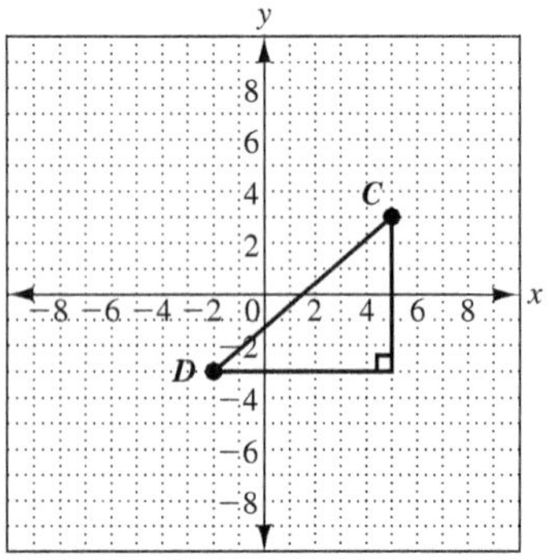

A. −9   B. 11   C. 9   D. 10

38. (Grade 7)
Mark drew a triangle on the coordinate plane shown below.

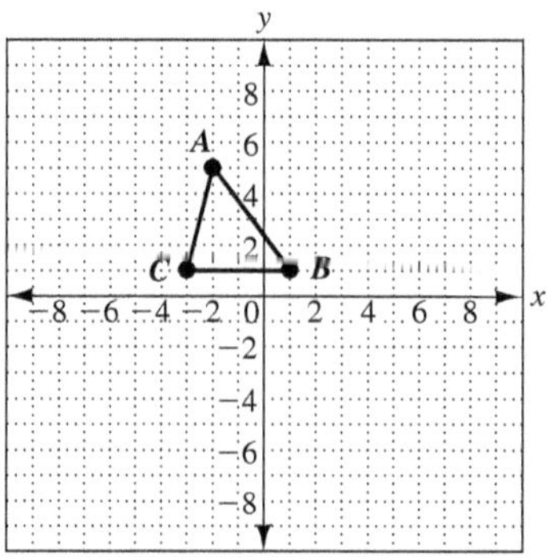

Which of the following best represents the coordinates of the vertices of $\triangle ABC$?

A. $(-2, 5), (1, 1), (-3, 1)$
B. $(-2, 5), (1, 1), (1, -3)$
C. $(5, -2), (1, 1), (1, -3)$

39. (Grade 8)
Which of the following ordered pairs represents a point in the third quadrant of a coordinate plane?

F. $(-3, 14)$ G. $(6, 7)$ H. $(4, -1)$ J. $(-2, -3)$

40. (Grade 9)
The original function $3x + 6$ is graphed on the same grid as the new function $\frac{x}{3} + 6$. Which of the following statements about these graphs is true?

F. The graph of the original function is steeper than the graph of the new function.
G. The graph of the original function is parallel to the graph of the new function.
H. The graphs intersect at (0, 6).
J. The graphs intersect at (6, 0).

8.3 Connections between Algebra and Geometry

There is a tendency for students to think of mathematics as compartmentalized into algebra, geometry, probability, statistics, and other areas. In fact, in nearly all of the applications of mathematics, one uses many different subareas and, in fact, integrates them.

This section is devoted to one approach to combining two or more areas to get results: the use of algebra and Cartesian coordinates in geometric problems.

As usual, we need to start with definitions even though we already "know" what they are. Informally, a triangle, $\triangle ABC$, is the union of the three line segments defined by the points A, B, and C (provided that the three points do not lie on the same straight line). More formally,

DEFINITION Triangle, Side

Let A, B, and C be three points in the Cartesian plane that do not lie on the same straight line. The **triangle,** $\triangle ABC$, is the union of the three line segments determined by A, B, and C. Each of the segments is called a **side** of the triangle. Using set notation,

$$\triangle ABC = \overline{AB} \cup \overline{BC} \cup \overline{CA}.$$

DEFINITION Isosceles Triangle

An **isosceles triangle** is a triangle in which at least two sides have the same length.

Our first example is to show that a certain triangle, $\triangle RST$, is isosceles by the use of Cartesian coordinates.

EXAMPLE 8.13

Using Cartesian Coordinates to Prove That the Triangle $\triangle RST$ Is Isosceles

Show that the triangle with vertices $R(1, 4)$, $S(5, 0)$, and $T(7, 6)$ is isosceles.

Solution

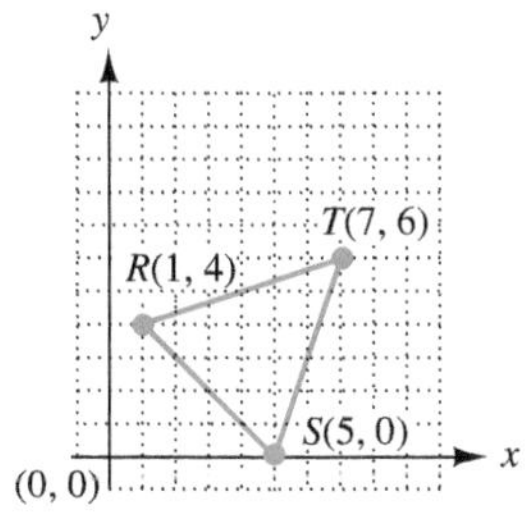

We compute the length of the three sides:

$$RS = \sqrt{(1-5)^2 + (4-0)^2} = \sqrt{16+16} = \sqrt{32};$$
$$RT = \sqrt{(1-7)^2 + (4-6)^2} = \sqrt{36+4} = \sqrt{40};$$
$$ST = \sqrt{(5-7)^2 + (0-6)^2} = \sqrt{4+36} = \sqrt{40}.$$

Since RT and ST are both $\sqrt{40}$, these two segments are the same length. It follows that $\triangle RST$ is isosceles.

PROBLEM-SOLVING STRATEGY Use Cartesian Coordinates to Do Geometric Problems

In solving geometric problems, it is sometimes easier and faster to place the figures in a Cartesian plane and use algebra.

We will use the preceding problem-solving strategy to show that the slope of two lines determines if the lines are parallel, perpendicular, or neither. This habit is a way to think mathematically: understanding a geometric condition ("being parallel") to be the same as an algebraic condition ("slopes are the same").

Parallel and Perpendicular Lines

It seems obvious that two different but equally steep lines never meet; that is, they are parallel. As noted earlier, the slope of a line is a measure of its steepness. Therefore, it seems obvious that two lines with equal slope are parallel.

DEFINITION Parallel Lines

The lines l and m in the Cartesian plane are **parallel** if they have no points in common or if they are equal. We write $l \parallel m$ if l and m are parallel lines and $l \nparallel m$ if l and m are not parallel.

There are only three alternatives for the intersection of two lines l and m:

- l and m are the same line,
- l and m are distinct lines and $l \parallel m$, so l and m have no points in common, or
- $l \nparallel m$ so that l and m intersect in a single point (one point in common).

Given the importance of definitions, some care must be taken when one looks at the definition of parallel lines as it appears in different texts. The case in which $l = m$ is sometimes not included in the term *parallel*, but we do so in this text. In addition, note that the lines that we discuss are always in the same plane. Although the proof of the next theorem is optional, we suggest that future middle school teachers work through it, as it is a key conceptual link and gives a good example of the relationship between algebra and geometry.

THEOREM Condition for Parallelism

Two lines in the plane p and q, are parallel if and only if they both have the same slope or both are vertical lines.

PROOF (OPTIONAL) Two vertical lines are certainly parallel. Furthermore, if $p \parallel q$ and p is vertical, then q must also be vertical.

Let's now explore the case where p and q are lines and neither is vertical. If p and q are equal, then their slopes are, of course, the same. We now assume that the slope of p and the slope of q are equal and $p \neq q$. We let m_1 be the slope of p (and b_1 its y-intercept) and m_2 the slope of q (with b_2 its y-intercept). The condition that $p \parallel q$ means that there is no point $P \in p \cap q$. Assume that such a point exists and let (c, d) be the coordinates of P. The next steps will show what this assumption means about the relationship between m_1 and m_2.

The critical condition now is that $(c, d) = P \in p \cap q$ if and only if (c, d) satisfies the slope–intercept form of a line for *both* the line p and the line q. The last phrase means that the real numbers c and d must satisfy the two equations

$$d = m_1c + b_1 \quad \text{and} \quad d = m_2c + b_2.$$

Subtracting the second equation from the first yields

$$0 = m_1c - m_2c + b_1 - b_2 \quad \text{or} \quad (m_1 - m_2)c = b_2 - b_1.$$

If $m_1 \neq m_2$, then there is a solution for c, which is $c = \dfrac{(b_2 - b_1)}{m_1 - m_2}$, and also one for d. Thus, p and q intersect at $P(c, d)$ and the lines are not parallel. If $m_1 = m_2$, then the only way the two equations can both be solved is if $b_1 = b_2$. The lines $y = m_1x + b_1$ and $y = m_2x + b_2$ are then the same, since $m_1 = m_2$ and $b_1 = b_2$. Thus, the only way the nonvertical lines p and q can both have the same slope and y-intercept is if $p = q$, but p and q are different lines, so the result is proven.

We will now use algebra and the previous theorem to prove a geometric result. We will show that the line through the midpoints of two sides of a triangle is parallel to the third side. Recall from previous courses that the **midpoint** of the line segment $\overline{AB}$ is the point in $\overline{AB}$ that is the same distance from A as it is from B.

The technique of rigidly moving the geometric shapes is quite useful. Since rigid motions don't change geometric properties, we have come up with a new strategy. Ideas associated with rigid motions of the plane will be discussed fully in Chapter 11. Example 8.14 shows the valuable nature of "Use Rigid Motions."

PROBLEM-SOLVING STRATEGY **Use Rigid Motions**

To solve a geometric problem, it may be helpful to use a rigid motion (translation, rotation, or a combination of the two) to move the geometric figure to another position in which the solution becomes easier to see.

EXAMPLE 8.14 Line Joining the Midpoints of Two Sides of a Triangle

Using algebra, show that the line joining the midpoint of any two sides of a triangle is parallel to the third side.

Solution

Understand the Problem

Let $\triangle ABC$ be given, and let M be the midpoint of $\overline{AB}$ and N be the midpoint of $\overline{AC}$. We must show that the lines $\overleftrightarrow{MN}$ and $\overleftrightarrow{BC}$ are parallel. The first step is to draw Figure 8.11a, which shows the problem.

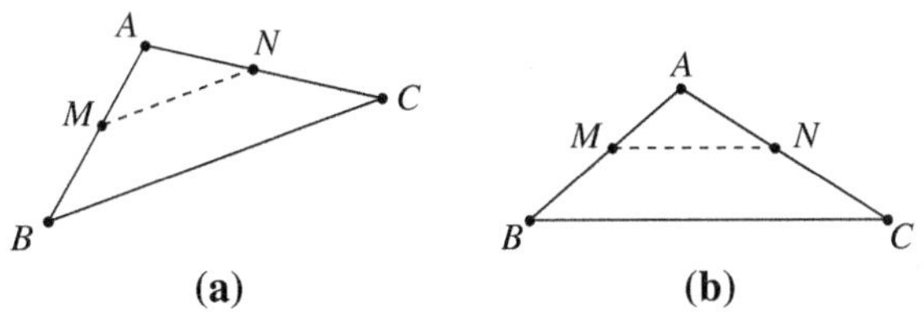

Figure 8.11
Lines joining the midpoints of two sides of a triangle

Devise a Plan

Since rotating or moving the entire drawing up, down, or sideways (translating) would not affect the relationship between $\overleftrightarrow{MN}$ and $\overleftrightarrow{BC}$, we may place $\triangle ABC$ as in Figure 8.11b as we've used rigid motions. If we put an (x, y)-coordinate system on the figure, we will be able to use algebra to show that $\overleftrightarrow{MN}$ is parallel to $\overleftrightarrow{BC}$. Since $\overleftrightarrow{BC}$ is not vertical, we must show that the lines have the same slope.

Carry Out the Plan

We have placed the origin of the (x, y)-coordinate system at the point B and located $\overleftrightarrow{BC}$ on the x-axis, as in Figure 8.12. Certainly, $\overleftrightarrow{BC}$ has slope 0 (because the line is horizontal), so we must compute the slope of $\overleftrightarrow{MN}$ and show that it is 0 also. Since non-vertical lines are parallel when they have the same slope, we will be done.

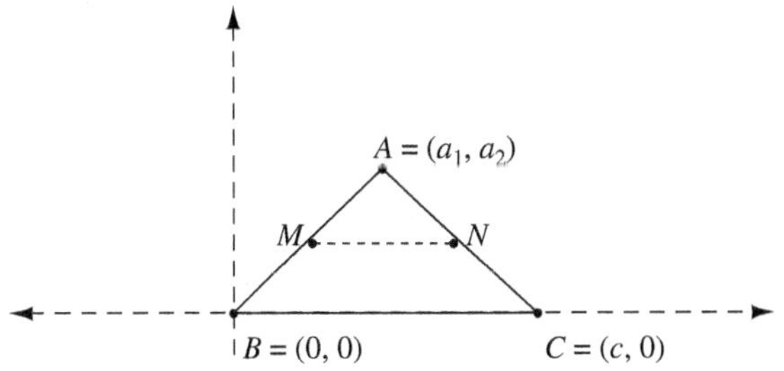

Figure 8.12
Lines joining the midpoints of two sides of a triangle placed in a Cartesian coordinate system

Let's name the points so that we can first compute the two midpoints. Since B is the origin and C is on the x-axis, we may write $B = (0, 0)$, $C = (c, 0)$, and $A = (a_1, a_2)$, where c, a_1, and a_2 are real numbers. (For simplicity, we have drawn a picture with a_1, a_2, and c positive.)

What are the coordinates of M in Figure 8.12? The x-coordinate of the midpoint must be the average of the x-coordinates of A and B, and the y-coordinate is the average of the two y-coordinates. Thus,

$$M = \left(\frac{a_1 + 0}{2}, \frac{a_2 + 0}{2}\right) = \left(\frac{a_1}{2}, \frac{a_2}{2}\right) \quad \text{and} \quad N = \left(\frac{c + a_1}{2}, \frac{0 + a_2}{2}\right) = \left(\frac{c + a_1}{2}, \frac{a_2}{2}\right).$$

The slope of the line $\overleftrightarrow{MN}$ is 0, since the points M and N have the same y-coordinate (so that the rise is 0). Thus, both $\overleftrightarrow{MN}$ and $\overleftrightarrow{BC}$ have the same slope, 0, and so they are parallel.

Look Back

A little thought led us to use a coordinate system and apply algebra. We noticed that rotation and translation allowed us to move the triangle from the general one of Figure 8.11a to one in a simpler position with a coordinate system available (Figure 8.12). This process did not change the problem but allowed for a much easier computation of slopes. We found that the two slopes were the same (both are 0) so that the lines $\overleftrightarrow{MN}$ and $\overleftrightarrow{BC}$ are parallel.

There is a technical point about Example 8.14 that needs to be made. In both parts of Figure 8.11, we used the same names for the vertices of the triangle. To be rigorous, in the second triangle we should have used different names, such as A', B', and C', to reflect the fact that the triangle moved (and so has different coordinates for its vertices). In the interests of clarity and because there is little ambiguity, we kept the same names for the vertices.

We now move from parallel lines to perpendicular ones. In earlier coursework, you have been taught that, provided that neither line is vertical, "Two lines are perpendicular if and only if their slopes are negative reciprocals." We shall now give a proof of that statement.

DEFINITION Perpendicular Lines

The lines l_1 and l_2 are **perpendicular** if they intersect at a 90° angle.

THEOREM Condition for Perpendicular Lines

Assume that l_1 and l_2 are lines that are not parallel to the x-axis or the y-axis. If m_1 is the slope of l_1 and m_2 is the slope of l_2 and the two lines intersect, then l_1 is perpendicular to l_2 if and only if $m_2 = -\dfrac{1}{m_1}$. (Said another way, m_1 and m_2 are negative reciprocals.)

PROOF (OPTIONAL) Assume that l_1 and l_2 meet at a 90° angle, as shown in the accompanying figure. Then the triangle along l_1 with a horizontal leg of length a and a vertical leg of length b shows that the line l_1 has slope $m_1 = \dfrac{b}{a}$. Line l_2 is perpendicular to l_1 and can be viewed as the line obtained by rotating line l_1 about the intersection point P through 90° and so rotates the triangles as pictured. What, then, is the slope of l_2? From the picture, the change in the y-coordinates is $-a$ (since the values along l_2 are getting smaller as x gets larger) and the change in the x-coordinates is b. Thus, the slope of l_2 is $m_2 = \dfrac{-a}{b}$.

Since the slope of l_1 is $m_1 = \dfrac{b}{a}$, it follows that

$$m_2 = -\frac{a}{b} = -\frac{1}{\frac{b}{a}} = -\frac{1}{m_1},$$

and the statement before the previous definition is proved.

(It can be shown that if $m_2 = -\dfrac{1}{m_1}$, then l_1 and l_2 are perpendicular, but we omit the proof here.)

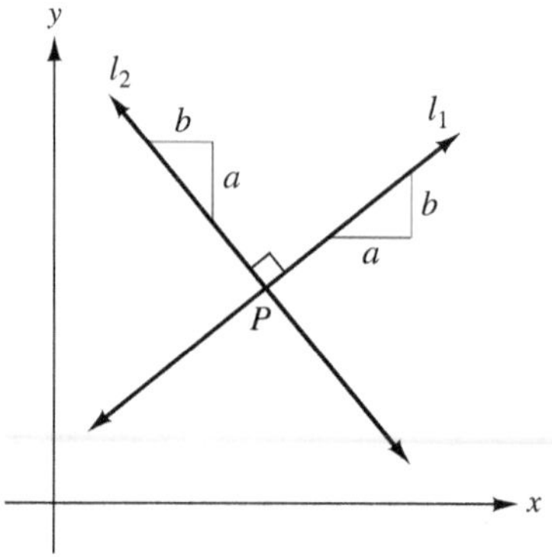

We use the condition for perpendicularity, which gives a quite surprising geometric result. First, we need two definitions.

DEFINITION Altitude of a Triangle
An **altitude of a triangle** is a line through a vertex of a triangle that is perpendicular to the line containing the opposite side of the triangle.

Figure 8.13
Altitude of triangles

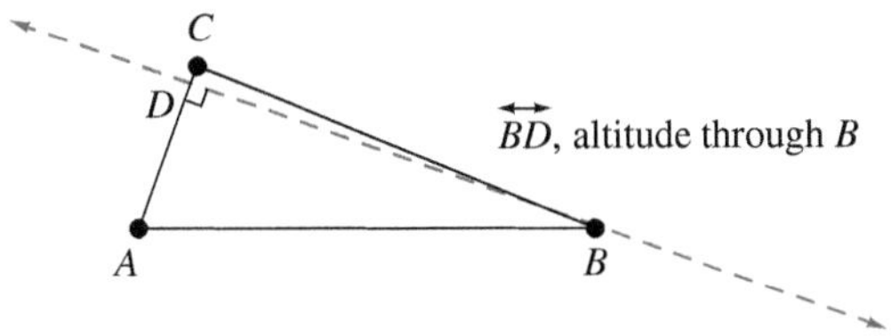

(a) The line $\overleftrightarrow{BD}$ is the altitude of $\triangle ABC$ (acute triangle) through B.

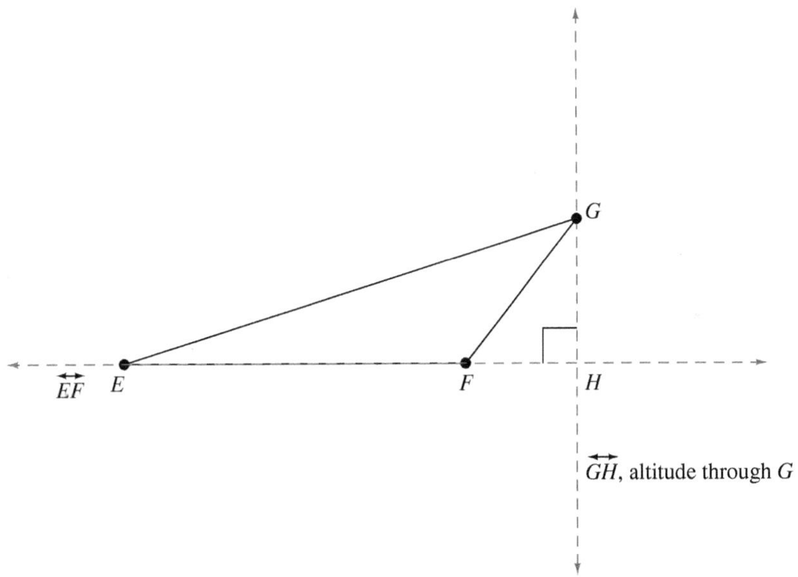

(b) The line $\overleftrightarrow{GH}$ is the altitude of $\triangle EFG$ (obtuse triangle) through G.

DEFINITION Concurrent Set of Lines
A collection of lines is **concurrent** if the same point is on each of the lines.

The next example has two deep aspects: First, there are two problem-solving strategies, and second, we use algebra (via the equation of straight lines) to show that certain lines are concurrent. One way involves using algebra to show that the altitudes are concurrent, although this is more like a high school problem and is optional.

On the other hand, it is fun to have middle school students experiment by drawing the three altitudes of a given triangle. They will see that because the three altitudes have a common point, the altitudes are concurrent.

EXAMPLE 8.15 Showing That the Altitudes of a Triangle Are Concurrent

Solution

We will use the problem-solving strategies "Use Cartesian Coordinates to Do Geometric Problems" and "Use a Rigid Motion." Our plan is to use the point–slope form of a line to find the equations for each of the three altitudes. We then show that the three have a point in common because there is a point whose coordinates lie on each of the altitudes.

Using a rigid motion, we first orient the triangle on a coordinate system with one vertex at the origin, one vertex on the positive x-axis, and one vertex in the upper half plane. Let $A(0, 0)$, $B(b, d)$, and $C(c, 0)$, with $c \neq b$, be these vertices. (In the $c = b$ case, the triangle is a right triangle and can be handled separately.)

We will also use the notion of slope and the point–slope form of the equation of a line. Suppose that $\overleftrightarrow{AE}$, $\overleftrightarrow{BF}$, and $\overleftrightarrow{CD}$ are the altitudes of the triangle.

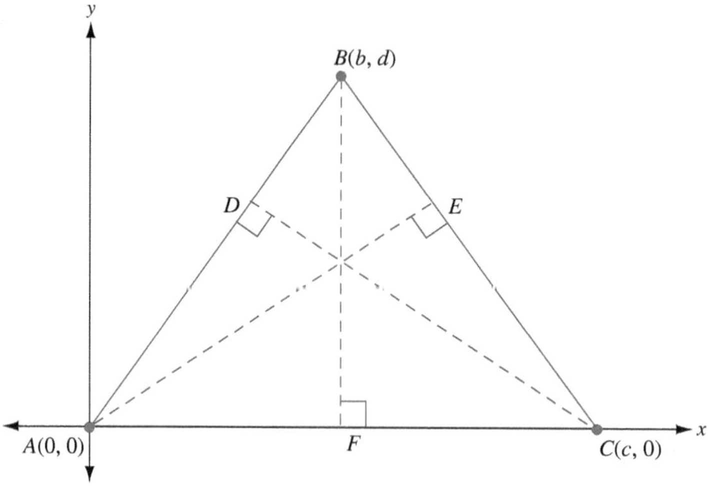

Then, because the slope of the perpendicular is the negative reciprocal of the slope of the line segment,

$$\text{slope of } \overline{BC} = \frac{d}{b - c}, \text{ so the slope of } \overline{AE} = -\frac{(b - c)}{d}.$$

We write the point–slope form of the equation of $\overleftrightarrow{AE}$ and simplify it,

$$y - 0 = -\frac{b - c}{d}(x - 0), \text{ so that } dy = cx - bx.$$

Similarly, because

$$\text{slope } \overline{AB} = \frac{d}{b}, \text{ so the slope } \overline{DC} = -\frac{b}{d}.$$

Now we can write the equation of $\overleftrightarrow{DC}$ in point–slope form and then simplify it:

$$y - 0 = -\frac{b}{d}(x - c) \text{ or } dy = -bx + bc.$$

To find where $\overleftrightarrow{AE}$ and $\overleftrightarrow{DC}$ intersect, we determine the simultaneous solution of the equations of these two lines. Subtracting the equation for $\overleftrightarrow{DC}$ $(dy = -bx + c)$ from the equation for $\overleftrightarrow{AE}$ $(dy = cx - bx)$, we obtain

$$0 = (cx - bx) - (-bx + bc) = cx - bc.$$

Therefore, the x-coordinate of the point of intersection of $\overleftrightarrow{AE}$ and $\overleftrightarrow{DC}$ is

$$x = \frac{bc}{c} = b.$$

Hence, without even determining the y-coordinate of the point of intersection of $\overleftrightarrow{AE}$ and $\overleftrightarrow{DE}$, it follows that the point of intersection lies on $\overleftrightarrow{BF}$, since $\overleftrightarrow{BF}$ is vertical, passes through the point $B(b, d)$, and so has equation $x = b$. Thus, the three altitudes are concurrent, as was to be shown. (The point that the altitudes have in common is called the **orthocenter** of the triangle.)

Circles

We now turn to the notion of circles, a concept that permeates the mathematics of elementary school. First informally and then using algebra, we will give the definition of *circle, diameter,* and *radius.* These three definitions, as well as those of the *area* and *circumference* of a circle (Chapter 10.2), will play a major role in the remainder of the text.

For children who have played with circles, a circle starts with a center C and a radius r (say, 1). A child looks to various points to see if the point that she is working with is distance 1 from C. We thus see that "distance away from the center" is the important concept that defines a circle.

More formally, by definition, a **circle** is the set of all points in the plane that are a fixed distance from a given point. The given point is the **center** of the circle and the fixed distance is the **radius.** The definitions follow, using first a figure and then, more rigorously, set notation and algebra. Note that the word *radius* is used in two ways: It is both a segment from the center to a point on the circle (segment $\overline{QC}$, where C is the center of the circle in Figure 8.14) and the length of such a segment. Likewise, a *diameter* is both a segment and a length. The interior of a circle is a **disc.**

Figure 8.14
The parts of a circle and disc

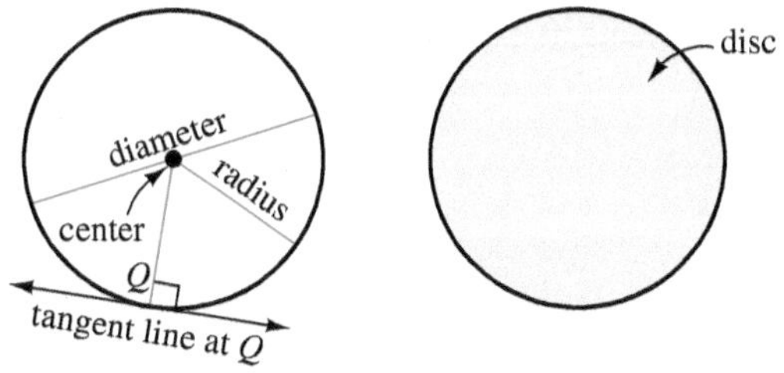

DEFINITION Circle, Center, Radius, and Diameter of a Circle

Let $C = (a, b)$ be a fixed point (called the **center**) and r a positive number (called the **radius**). A point $P = (x, y)$ is a point on the **circle** $\mathcal{C}$ if and only if

$$(x - a)^2 + (y - b)^2 = r^2.$$

The **diameter** of $\mathcal{C}$ is twice its radius. Note that the preceding equation uses exactly the distance formula from Section 8.2 so that $\mathcal{C}$ is precisely the points that satisfy the preceding equation.

The **tangent line to the circle** $\mathcal{C}$ at the point Q on $\mathcal{C}$ is the line that goes through the point Q and is perpendicular to the radius.

The center, radius, diameter, and tangent line are pictured in Figure 8.14 to remind the reader of these terms. The tangent line is useful if you want to know how far you can see from a mountaintop, as we will see Example 10.18.

EXAMPLE 8.16

Equation of a Tangent Line to a Circle at a Point

If $\mathcal{C}$ is a circle with center $C = (a, b)$, and if $Q = (c, d) \in \mathcal{C}$, then the equation of the tangent line to $\mathcal{C}$ at Q is

$$y - d = \left(\frac{a - c}{d - b}\right)(x - c).$$

Solution In order to find out the equation of the tangent line, we must know its slope and a point on it. However, from the definition of *tangent line* and the condition of perpendicularity that we just proved, the slope of the tangent line is the negative reciprocal of that of the radius $\overline{QC}$.

We first find the slope of the tangent line. Because $C = (a, b)$ and $Q = (c, d)$ are on the radius the slope of $\overline{QC}$ is $\frac{d - b}{c - a}$. Therefore,

$$\text{slope of tangent line} = \frac{-1}{\text{slope of } \overline{QC}} = \frac{a - c}{d - b}.$$

We have the slope all ready for insertion into the point–slope form of a line (p. 400). We'll now use the only point we know, $Q = (c, d)$, that is on the tangent line to $\mathcal{C}$ at Q. The equation is then $y - d = \left(\frac{a - c}{d - b}\right)(x - c)$, as we were to show.

SMP 3 Construct viable arguments and critique the reasoning of others.

The Standard for Mathematical Practice 3 (SMP 3) shows in Example 8.16 that there is a formula which establishes an equation of a tangent line to a circle. We can depend on its result because of the definition of *tangent line*, the condition for perpendicular lines theorem (p. 412), and the formula for the point–slope form of a line (p. 400).

"Mathematically proficient students understand and use stated assumptions, definitions, and previously established results in constructing arguments. They . . . build a logical progression of statements to explore the truth of their conjectures."

8.3 Problem Set

Understanding Concepts

1. (a) Find two points on the line $4x + 2y = 6$.
   (b) Use the points determined in part (a) to compute the slope of the line.
   (c) Solve the equation in part (a) for y in terms of x and thereby again determine the slope of the line, as well as the y-intercept. (*Hint:* Solving for y in terms of x gives the slope–intercept form of the equation of a line.)
   (d) Show your answer to parts (a) and (b) to your neighbor. Although there may be a difference in the points that you two chose, what does the slope turn out to be?

2. (a) Find two points on the line $-2x + y = 6$.
   (b) Use the points determined in part (a) to compute the slope of the line.
   (c) Solve the equation in part (a) for y in terms of x and thereby again determine the slope of the line, as well as the y-intercept. (*Hint:* Solving for y in terms of x gives the slope–intercept form of the equation of a line.)
   (d) Show your answer to parts (a) and (b) to your neighbor. Although there may be a difference in the points that you two chose, what does the slope turn out to be?

3. In each case, determine k so that the line is parallel to the line $3x - 5y + 45 = 0$.
   (a) $7x + ky = 21$ (b) $kx - 8y - 24 = 0$

4. In each case determine k so that the line is parallel to the line $3x - 5y + 45 = 0$.
   (a) $y = kx + 5$ (b) $x = ky + 5$

5. Draw the graphs of each of these linear functions.
   (a) $y = 2x - 3$ (b) $y = 0.5x + 2$ (c) $y = -3x$

6. Draw the graphs of each of these linear functions.
   (a) $y = 3x + 1$ (b) $2y + 5x = 1$ (c) $2x = y + 4$

7. (a) On the same coordinate system, draw the graphs of these three linear functions: $y = 4x$, $y = 4x + 5$, and $y = 4x - 3$.
   (b) Briefly discuss the graphs in part (a) and find their slope.

8. (a) On the same coordinate system, draw the graphs of these three linear functions: $y = -2x$, $y = -2x + 5$, and $y = -2x - 3$.
   (b) Briefly discuss the graphs in part (a) and find their slope.

9. (a) Prove that $R(1, 2)$, $S(7, 10)$, and $T(5, -1)$ are the vertices of a right triangle. (*Hint:* Show that the square of one side is the sum of the squares of the other two sides.)
   (b) Draw the triangle RST of part (a) on graph paper.

10. In each part that follows, you are given an equation of a line and a point. Find the equation of the line through the given point that is perpendicular to the given line. (The slope of the perpendicular line is the negative reciprocal of the slope of the given line if the given line is neither vertical nor horizontal, as described in the theorem on p. 412.)

(a) $y = 2x$, $P(0, 0)$

(b) $y = 3x + 5$, $Q(1, 2)$

11. In each part that follows, you are given an equation of a line and a point. Find the equation of the line through the given point that is perpendicular to the given line.

(a) $y = -\frac{2}{3}x + 7$, $R(4, -1)$

(b) $2y - 6x - 5 = 0$, $S(0, 3)$

12. A square in the coordinate plane is shown below, with its vertices at $A(1, 0)$, $B(1, 1)$, $C(0, 1)$, and $D(0, 0)$. Since the squared distance between points $P(x, y)$ and $Q(a, b)$ is given by the formula $PQ^2 = (x - a)^2 + (y - b)^2$, we see that $PA^2 = (x - 1)^2 + (y - 0)^2 = x^2 - 2x + 1 + y^2$ is the squared distance between P and vertex A of the square.

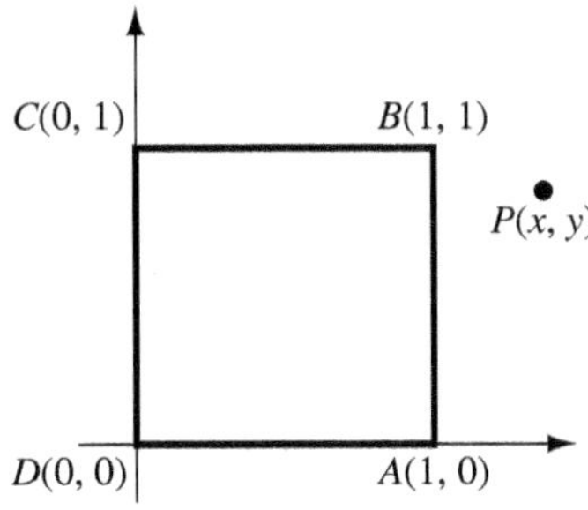

(a) Find PB^2, PC^2, and PD^2.

(b) Describe the set of points P in the coordinate plane for which $PA^2 + PC^2 = PB^2 + PD^2$. (Yes, the answer is a surprise!)

13. Is the triangle of Example 8.13 equilateral?

14. What is the altitude through the point S in the triangle of Example 8.13?

15. Let $A = (1, -2)$ and $B = (2, 4)$. What is the equation of the line through the midpoint of $\overline{AB}$ that is perpendicular to $\overline{AB}$? This line is called the **perpendicular bisector** of $\overline{AB}$.

16. (a) What is the perpendicular bisector of the segment between $C = (5, -1)$ and $D = (5, 7)$?

(b) Did you do part (a) algebraically or geometrically?

17. (a) What is the equation of the line that is tangent to the circle of radius 3 at $(0, -3)$ and whose center is at the origin?

(b) What is the equation of the line that is tangent to the circle of radius 3 at $(0, 3)$ and whose center is at the origin?

(c) Is the line in part (b) the same line as that in part (a)?

18. (a) Show that the point $Q = (2, -1 + 2\sqrt{2})$ is on the circle of diameter 6 and whose center is $(1, -1)$.

(b) What is the equation of the line that is tangent to the circle of diameter 6 at $(2, -1 + 2\sqrt{2})$ and whose center is $(1, -1)$?

19. Draw three lines l, m, and n in the plane in such a way that l is perpendicular to m and m is perpendicular to n. For the set of lines that you picked, what is the relationship between l and n? Try the experiment again and see whether your answer is the same.

20. Suppose that l, m, and n are lines in the plane. Show that if l is perpendicular to m and m is perpendicular to n, then l is parallel to n. (*Hint:* First draw a picture, and then use the theorem about the condition for parallelism and perpendicularity in this section.)

21. Which of the following statements are true and which are false? If the statement is true, draw a picture.

(a) Two circles can intersect at exactly one point.

(b) Two circles can intersect at exactly two points.

(c) Two circles can intersect in no points.

(d) Two circles can intersect in exactly three points.

(e) Two circles can intersect in four or more points.

22. Draw a circle of radius 2 whose center is at the point $(2, -1)$. What does a circle of radius 3 but the same center look like when drawn on the same piece of graph paper?

Into the Classroom

23. (**Writing**) Constructing definitions from your intuition and then making sure the definition you've given reflects your intuition is an excellent example of a mathematical habit of the mind. Let's try that approach now. Without looking at another text, give a formal definition, based on your intuition, of the following parts of a circle:

(a) Diameter

(b) Tangent line

(c) Go to the library at your institution and compare your definitions to those of a high school, middle school, or elementary school text. Discuss the differences and similarities.

24. Follow the directions given in the previous section's Cooperative Investigation activity, the Open-Top Box Problem (p. 405), but this time start with a 12- by 12-centimeter square cut from centimeter-squared grid paper. Work in small groups to cut small x by x squares from the corners of the starting square to create an open-top box x centimeters high.

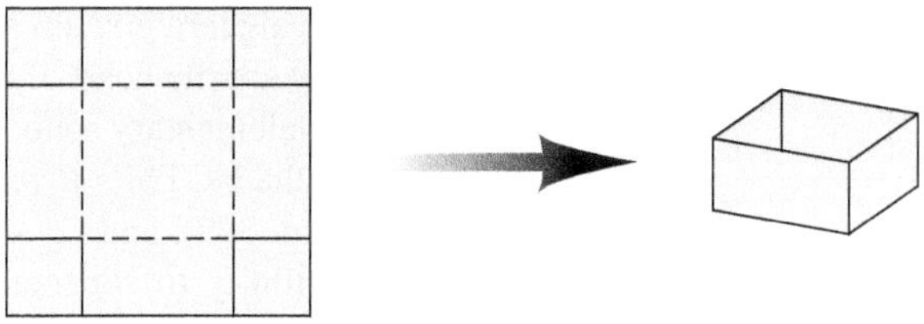

(a) Within your group, make boxes corresponding to $x = 1, 2, 3, 4$, and 5 centimeters. Guess which box has the largest volume.

(b) Derive a formula for the volume, $V(x)$, of the box as a function of x and make a graph of the volume function. On the basis of your graph, what value of x will give the box the largest volume?

(c) Derive a formula for the area, $B(x)$, of the bottom of the box and the total area, $S(x)$, of the sides of the box. Plot the graphs of $B(x)$ and $S(x)$ on the same axes, over the domain $0 \le x \le 6$. At what value of x is $B(x) = S(x)$?

(d) Do you see a connection between your answers to parts (b) and (c)? Discuss within your group and form a careful description of your observations.

Using Algebra in Geometry

25. Any triangle can be placed on a coordinate system so that one point is at the origin, one point is on the positive x-axis, and one point is in the first quadrant. Thus, without loss of generality, we can take $A = A(0, 0)$, $B = B(2a, 2b)$, and $C = C(2c, 0)$. Show that the three medians of the triangle meet at the same point whose coordinates are

$$G\left(\frac{2a + 2c}{3}, \frac{2b}{3}\right).$$

What is surprising is that all three lines go through the same point, G, called the **centroid** of the triangle.

26. Consider the triangle SPQ inscribed in a semicircle as shown. Use coordinate methods to prove Thales's theorem; that is, show that $\overline{PQ}$ is perpendicular to $\overline{PS}$. (*Hint:* Recall that $x^2 + y^2 = r^2$ and $(x + r)(x - r) = x^2 - r^2$.)

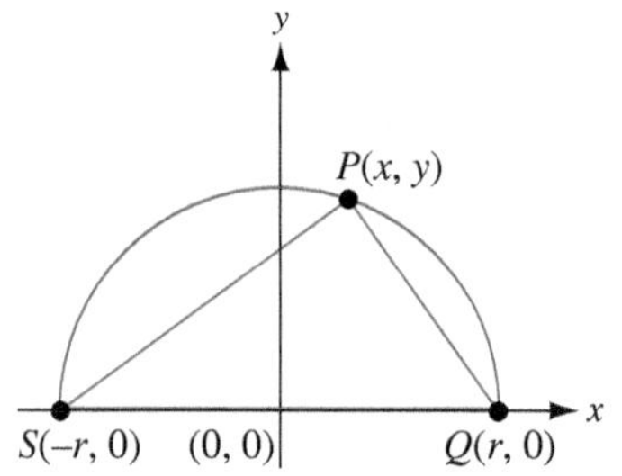

27. In "taxicab geometry," the points are the corners of a square grid of "city blocks" in a plane. In the figure shown, the shortest trip from A to B must cover five blocks, and so the **taxi distance** from A to B is 5. A "taxi segment" is the set of points on a path of shortest taxi distance from one point to another, and so $\{A, W, X, Y, Z, B\}$ is a taxi segment from A to B.

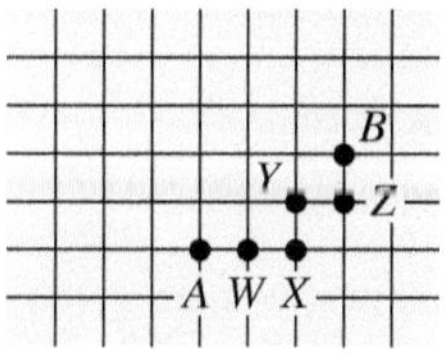

(a) How many taxi segments join A and B?

(b) Find all points that are at a taxi distance of 5 from A. Does your "taxi circle" look like a circle drawn with a compass?

(c) Use pencils of different colors to draw the concentric taxi circles of taxi radius 1, 2, 3, 4, 5, and 6. Describe the pattern you see.

State Assessments

28. (Grade 6)
Which of the following statements about angle measures is true?

A. An angle that measures 90° is a straight angle.
B. An angle that measures 35° is an obtuse angle.
C. An angle that measures 85° is an acute angle.
D. An angle that measures 180° is a right angle.

29. (Washington State, Grade 4)
Raul is going to a friend's house. Raul remembers that his friend's house is on a street parallel to Southport. On which street does Raul's friend live?

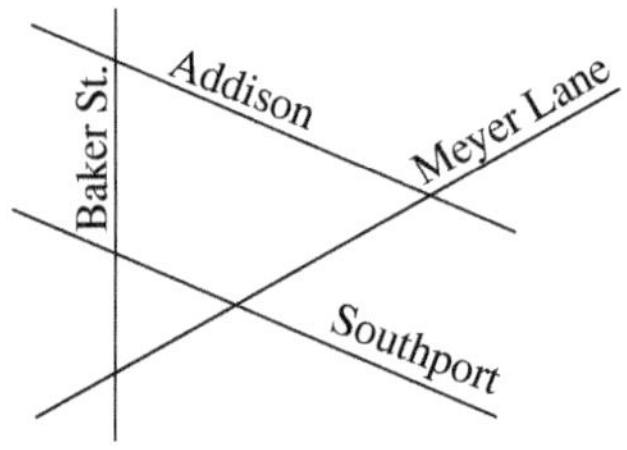

A. Addison B. Baker Street C. Meyer Lane

The Chapter in Relation to Future Teachers

This chapter focuses on the mathematics of elementary and middle school and is not meant to be a review of the algebra you already have seen. The power of algebra and the interplay between algebra and geometry are the heart of the chapter. We have two points to be made here: (1) The introduction of algebra in elementary school ("early algebra") is important for children's future learning of mathematics (see the NCTM's *Principles and Standards for School Mathematics* for grades 3–5 that starts this chapter, p. 379), and (2) **social justice** demands that we make sure that *all* of our students have every opportunity to succeed educationally. Knowing algebra opens careers for your students, whereas not understanding or not having opportunities for competence and beyond algebra severely limits their career choices. One way that governments have denied individuals or groups professional opportunities is by simply excluding them from a good education. That approach frequently translates into "no need for algebra or higher mathematics for the group" as a matter of repressive policy.

In addition, unfortunately, some students self-select coursework that does not lead to higher mathematics and thereby limits their opportunities. We urge you to provide and encourage opportunities for your students to learn algebra that will then open many paths to their future success.

Chapter 8 Summary

Section 8.1 Variables, Algebraic Expressions, and Functions	Page Reference
CONCEPTS	
• **Algebraic reasoning:** Algebraic reasoning is used to solve problems and understand patterns by following these steps: Introduce variables, derive algebraic expressions, form equations, solve equations, and interpret the solution of the equations in the context of the original problem or pattern.	380
• **Variables:** Variables represent quantities that are unknown, that can change, or that depend on varying choices of related quantities. In particular, variables describe generalized properties, express relationships, express formulas, or serve as unknowns.	380
• **Algebraic expressions:** An algebraic expression is a mathematical expression involving variables, numbers, and operation symbols.	382
• **Equation:** Setting two algebraic expressions that represent the same quantity equal to one another creates an equation.	383
• **Solution of equations:** The values of the variables for which the equations are true are the solutions of the equations and form a set called the solution set of the equations.	383
• **Function:** A function is a rule that assigns exactly one value to each element x in a set D. The set D is the domain of the function.	384
• **Describing and visualizing functions:** Functions can be represented by the following means: a formula, a table of values, an arrow diagram, a machine into which x is input and $f(x)$ is output, and a graph.	386
• **Linear function:** A linear function is a function whose graph is a straight line. See the formal definition that follows.	388
DEFINITIONS	
• **Constants** are fixed values in mathematical expressions.	380
• **Variables** are quantities that vary. They are denoted by a symbol.	380
• A **numerical expression** is any representation of a number that involves numbers and operation symbols. For example, $44 \div 4$ is a numerical expression for 11.	382
• An **algebraic expression** is a mathematical expression involving variables, numbers, and operation symbols.	382
• The **domain of a variable** is the set of values for which the expression is defined.	382
• An **equation** is a mathematical expression stating that two algebraic expressions have the same value.	384
• A **conditional equation** is when only certain values of the variables give equality.	384
• The **solution set** of an equation is the set of all values in the domain of the variables that satisfy the given equation.	384
• Two equations are **equivalent** if they have the same solution set.	384

• A **function** on a set D is a rule that assigns exactly one value to each element x in D, precisely one value y. If the function is denoted by f, and if x is an element of the domain D, then the value assigned to x is denoted by $f(x)$.	385
• The **domain of a function** is the set of a function D.	385
• The **image, or value,** of f at x is the value y when a function assigns the y to an element x in the domain.	385
• The **range of a function** is the set of all values assigned by the function.	385
• A **linear function** is a function of the form $f(x) = mx + b$, where m and b denote constants not both zero.	388

PROCEDURES

• **Evaluation of an algebraic expression:** Replace each of the variables with particular values from the domain of the variables to evaluate the expression.	382
• **Vertical-line test:** To see if the equation is a function, draw a graph and see how often a vertical line intersects the graph. If no line intersects the graph in two or more points, the graph is a function.	385

Section 8.2 Graphing Points, Lines, and Elementary Functions	Page Reference

CONCEPTS

• **Cartesian coordinates:** The Cartesian coordinate system consists of two perpendicular axes, with the horizontal axis typically called the x-axis and the vertical axis typically called the y-axis. Any point $P(x, y)$ in the plane is uniquely described by its x- and y-coordinates.	393
• **Distance formula:** The distance between $P(x_1, y_1)$ and $Q(x_2, y_2)$ is given by $PQ = \sqrt{(x_2 - x_1)^2 + (y_2 - y_1)^2}$.	395
• **Line segment:** $\overline{PQ}$ denotes the line segment with endpoints P and Q.	396
• **Slope:** The slope of the line or line segment through the points $P(x_1, y_1)$ and $Q(x_2, y_2)$ is the ratio of the line's or line segment's "rise over run"; that is, slope $(\overline{PQ}) = \dfrac{y_2 - y_1}{x_2 - x_1}$ when $x_1 \neq x_2$. Slopes are not defined for vertical segments or lines. Lines or segments are parallel precisely when they have the same slope or are both vertical.	396
• **Equation of a line:** The equation of a nonvertical line can be given in point–slope, slope–intercept, and two-point forms.	397
• **Nonlinear function:** Nonlinear functions are graphed by making a table of values. The points listed in a table are plotted and then connected to approximate the graph of the function. Quadratic and exponential equations are given emphasis as they are a part of K–8 mathematics.	402

DEFINITIONS

• The horizontal axis is typically called the ***x*-axis.**	394
• The vertical axis is typically called the ***y*-axis.**	394
• Any point $P(x, y)$, known as a **Cartesian coordinate** in the plane, is uniquely described by its ***x*-** and ***y*-coordinates.** The collection of the points defined by Cartesian coordinates is called the **Cartesian plane** or **coordinate plane.**	393
• The **origin** is the point (0, 0).	394

• **Quadrants** are the four regions that the axes divide the plane into, as in Figure 8.5.	394
• The set of points between and including the two distinct endpoints, P and Q, is the **line segment** $\overline{PQ}$.	397
• The **slope,** m, of the line or line segment through the points $P(x_1, y_1)$ and $Q(x_2, y_2)$ is the ratio of the line or line segment's "rise over run," given by $m = (y_2 - y_1)/(x_2 - x_1)$ if $x_1 \neq x_2$. If $x_1 = x_2$, then the line is vertical and its slope is undefined.	397
• **Nonlinear functions** are functions not of the form $f(x) = mx + b$.	403
• **Quadratic functions** are polynomials of degree 2.	403
• **Exponential functions** are functions that are a multiple of a constant by a fixed positive real number raised to a variable power.	403

FORMULAS

• **Distance formula:** If $P(x_1, y_1)$, $Q(x_2, y_2)$ are points, then the distance between them is $$PQ = \sqrt{(x_2 - x_1)^2 + (y_2 - y_1)^2}.$$	396
• **Equation of lines:** If $P(x_1, y_1)$ and $Q(x_2, y_2)$ are on a line then • Two-point form $$y - y_1 = m(x - x_1), \text{ where } m = \frac{y_2 - y_1}{x_2 - x_1} \text{ if } x_1 \neq x_2.$$	400
• Point–slope form $y - y_1 = m(x - x_1)$.	
• Slope–intercept form $y = mx + b$ where b is called the y-intercept.	400

Section 8.3 Connections between Algebra and Geometry	Page Reference
CONCEPTS	
• **Use Cartesian coordinates to solve geometric problems:** Geometric relationships can be solved by algebraic processes.	409
• **Parallel and perpendicular lines and their slopes:** The relationship between parallel lines and their slopes can be translated into statements about their slopes. The same goes for perpendicular lines and their slopes.	409
• **Rigid motions:** The use of rigid motions to solve a geometric problem is based on moving a figure to another position without changing its geometric properties but in which the solution becomes easier to see.	491
• **Circles:** Circles are the set of all points in the plane that are at a fixed distance from a given point and play a strong role in K–8 geometry.	415
DEFINITIONS	
• **Parallel lines** are lines on a Cartesian plane that have either no points in common or are equal.	409
• The **midpoint** of a line segment is the point on the segment in which the distance from the point to each endpoint is equal.	410
• **Perpendicular lines** are lines that intersect at a 90° angle.	412
• The **altitude of a triangle** is a line through a vertex of a triangle that is perpendicular to the line containing the opposite side of the triangle.	413
• A collection of lines is **concurrent** if each of the lines contains the same point.	413
• The point at which the three altitudes of a triangle are concurrent is called the **orthocenter.**	415

- A **circle** is the set of all points in the plane that are a fixed distance from a given point. The given point is the **center** of the circle. The fixed distance is the **radius.** — 415
- The **diameter** of a circle is twice its radius. — 415
- A **tangent line to the circle** at a point Q is the line that goes through the point Q and is perpendicular to the radius of the circle. — 415
- **Social justice** is the concept that all students should have every opportunity for success in the worlds of career and education. — 418

THEOREM

- **Condition for parallelism:** Two lines in the plane are parallel if they both have the same slope or both are vertical lines. — 410
- **Condition for perpendicularity:** Two lines in the plane are perpendicular if their slopes are negative reciprocals of one another or if one is parallel to the z-axis and the other to the y-axis. — 412

STRATEGIES

- **Cartesian coordinates to solve geometric problems:** Place the figures in a Cartesian plane and use algebra. — 409
- **Rigid motions:** Use translation, rotation, or a combination of the two to move the geometric figure to another position in which the solution becomes easier to see. — 411

Chapter Review Exercises

Section 8.1

1. Let a, b, and c denote the current ages of Alicia, Ben, and Cory, respectively. Write expressions in the variables a, b, and c that express the given quantity.

(a) Alicia's age in five years

(b) The fact that Ben is younger than Cory

(c) The difference in age between Ben and Cory

(d) The average age of Alicia, Ben, and Cory

2. Let a, b, and c denote the current ages of Alicia, Ben, and Cory as in problem 1. Write equations in the variables a, b, and c that express the given relationship.

(a) Alicia will be 11 years old in two more years.

(b) Three years ago, Ben's age was half of Cory's age last year.

(c) Alicia's age is the average of Ben's and Cory's ages.

(d) The average age of Alicia, Ben, and Cory is 10.

3. Solve these two *Guess My Rule* games, carefully describing a function that agrees with the table of values:

(a)

x	4	2	0	5	1
y	14	8	2	17	5

(b)

x	4	2	5	0	1
y	20	6	30	0	2

4. Let f be a function defined by the formula $f(x) = 2x(x - 3)$.

(a) Find $f(3), f(0.5)$, and $f(-2)$.

(b) If $f(x) = 0$, what are the possible values of x?

Sections 8.2 and 8.3

5. **(a)** Plot the seven-sided polygon $ABCDEFG$ on graph paper, where the coordinates of the vertices (corners) of the polygon are $A(4, 2)$, $B(3, 3)$, $C(0, 3)$, $D(-2, 2)$, $E(-3, -2)$, $F(0, -2)$, and $G(4, 0)$.

(b) Point A is in the first quadrant, and point G is on the positive x-axis. Give similar descriptions for the other vertices of the polygon.

(c) The slope of side $\overline{AB}$ is -1, since the "rise over run" ratio is $\frac{-1}{1}$. Find the slopes of the other six sides of the polygon when the slope is defined. Are there any parallel sides?

6. Plot the hexagon $ABCDEF$ on graph paper, where the vertices of the hexagon are $A(0, 0)$, $B(7, 0)$, $C(16, 12)$, $D(7, 24)$, $E(0, 24)$, and $F(-9, 12)$.

(a) Find the lengths AB, AC, AD, AE, AF, and CF.

(b) Draw all the diagonals of the hexagon, using the symmetry of the figure, and label the lengths of all the sides and diagonals. Why do you think this figure is called an "integer hexagon"?

7. Find an equation of each of the lines described.
 (a) The line through $R(3, 4)$ with slope 2
 (b) The line through $T(6, -1)$ and $U(-2, 5)$
 (c) The line of slope 3 that intersects the y-axis at $y = -4$

8. Find a line perpendicular to $x + 7 = 2y$ that goes through $(7, 4)$.

9. What is the altitude through the point R in the triangle of Example 8.13?

10. Suppose that C_1 and C_2 are circles. How many members can the set $C_1 \cap C_2$ have? Draw an example of each.

11. Consider the exponential function $y = \left(\frac{1}{4}\right)^x$.
 (a) Find the values of y when x is 0, 1, 2, and 3, and sketch the graph of y.
 (b) What is the relationship between (a) and problem 30(c) of Problem Set 8.2?

12. Consider the function $f(x) = 2x^2 + 4x$.
 (a) Graph the function.
 (b) Determine the smallest value of the function, and find the corresponding value of x at which this minimum value occurs.

9 Geometric Figures

COOPERATIVE INVESTIGATION

Investigating Triangles via Paper Folding

Materials Needed

Paper (thin and colorful paper similar to origami paper works well), scissors, rulers, protractors, pencils, and tape.

Directions

Many important geometric concepts, figures, and relationships can be investigated with paper folding. In this activity, you will first practice some basic constructions using paper folding—creating perpendicular lines, angle bisectors, and perpendicular bisectors. You will then use these basic constructions to discover some properties of triangles and quadrilaterals. (A quadrilateral is a polygon with four sides, as we'll see here formally.)

The Basic Folds

Use your measuring tools to check that these constructions work as claimed:

1. **Perpendicular line.** Draw a line *l* and place a point *P* on a sheet of paper. Crease the paper so that the line is folded onto itself and the crease passes through the point *P*. Unfold your paper, and then use your protractor to check that the crease makes a 90° angle with the line *l*.

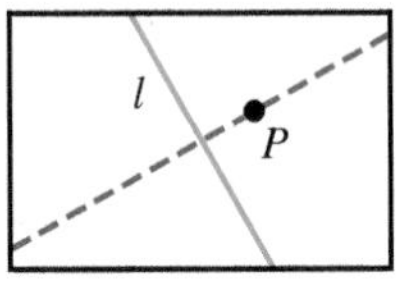

Perpendicular line

2. **Angle bisector.** Draw two rays from a point *A*, and make a crease through *A* so that one ray is folded onto the other ray. Unfold your paper, and use your protractor to show that the crease makes angles of equal size with the two rays.

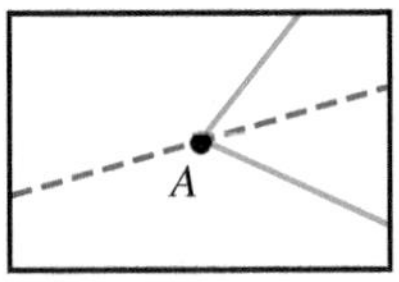

Angle bisector

3. **Perpendicular bisector.** Draw a line segment joining two points *C* and *D*. Crease your paper so that point *C* is folded on top of point *D*. Unfold your paper, and label the point where the crease crosses the segment $\overline{CD}$ as point *M*. Use your measuring tools to check that *M* is the same distance from both *C* and *D* and that the crease is perpendicular to the segment.

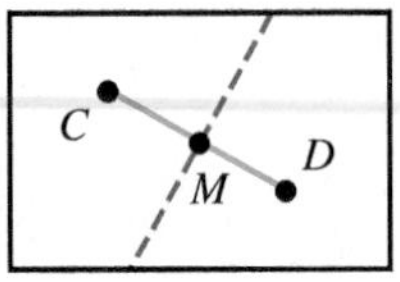

Perpendicular bisector

Exploring Properties of Triangles

The basic folds make it simple to construct some interesting lines and line segments associated with a triangle, as shown in the following diagram:

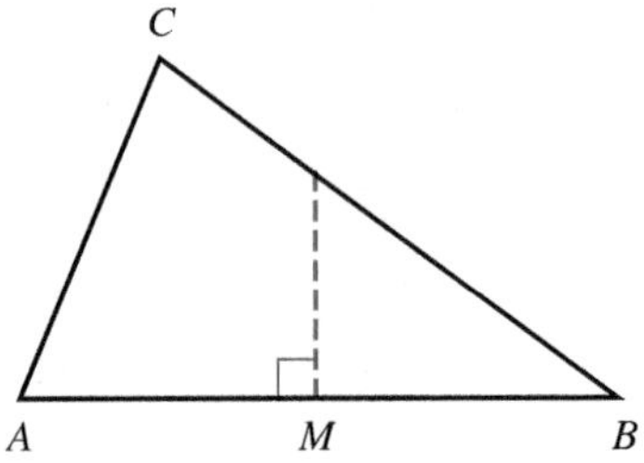

Midpoint *M* and perpendicular bisector of side $\overline{AB}$

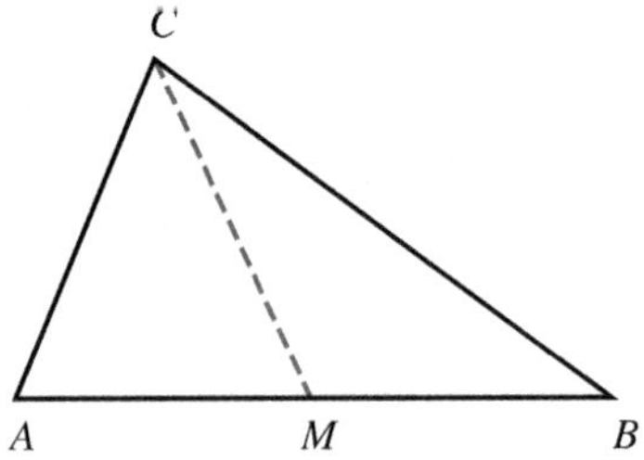

Median through vertex *C* to midpoint of side $\overline{AB}$

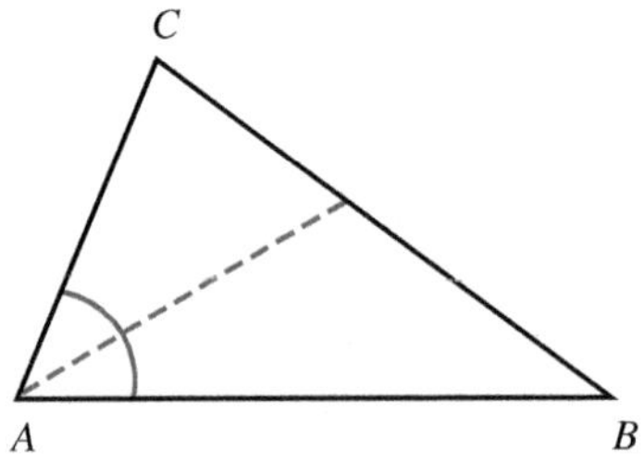

Angle bisector at vertex *A*

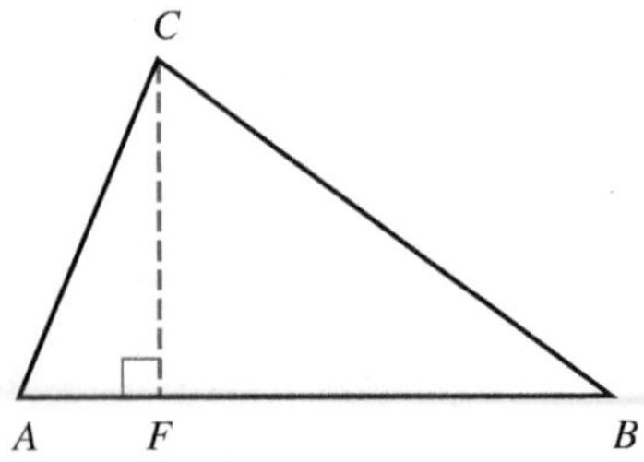

Altitude from vertex *C* perpendicular to side $\overline{AB}$

1. **The Midpoint of a Hypotenuse of a Right Triangle.** Cut off the corner of a rectangular sheet of paper to create a right triangle ABC, where the 90° angle is at vertex C. Use folding to construct the midpoint M of the hypotenuse $\overline{AB}$. Now find the perpendicular bisectors of legs $\overline{AC}$ and $\overline{BC}$. What do you find interesting? How do the respective distances from M to A, B, and C compare with one another?
2. **The Intersection of the Perpendicular Bisectors of the Sides of a Triangle.** Cut out a triangle from a sheet of paper. It may have any shape at all, but begin with a triangle with no angle larger than 90°. Next, construct the perpendicular bisectors of each of the sides of the triangle. What is interesting about how your lines intersect? Repeat your investigation with a triangle with one angle larger than 90°. Tape your unfolded triangle onto a large sheet of paper, and use a ruler to extend the crease lines. Describe what property you observe.
3. **The Intersection of the Medians of a Triangle.** Use short creases to find the midpoints of all three sides of a paper triangle, and then make creases to construct all three medians of your triangle. What special property do you discover?
4. **The Intersection of the Angle Bisectors of a Triangle.** Use paper folding to construct the angle bisectors of all three angles of a paper triangle. What special property do you observe?
5. **The Intersection of the Altitudes of a Triangle.** Use paper folding to construct the altitudes through each of the vertices of a triangle. What seems special about your three lines? If your triangle has an angle larger than 90°, you will want to tape your unfolded triangle to a large sheet of paper and use a ruler to extend your crease lines.
6. **The Angle Sum and Area of a Triangle.** Use folding to construct the altitude $\overline{CF}$ of a paper triangle ABC. Measure the height $h = CF$ and the length of the base $b = AB$ of your triangle. Next, fold all three vertices of your triangle to F, as shown in the right-hand diagram:

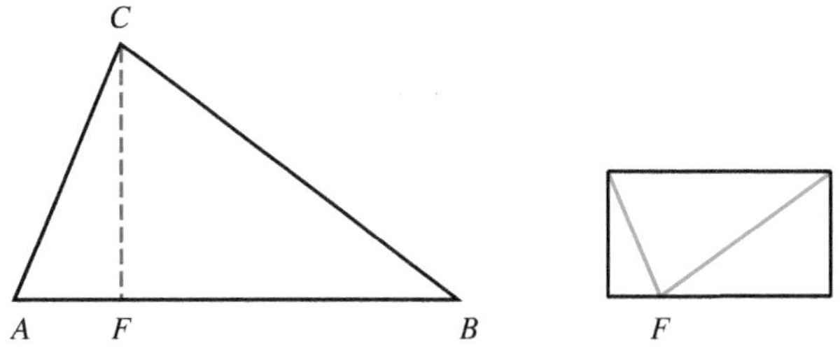

Why is the new figure a rectangle? (A rectangle is a four-sided polygon with a 90° angle at each vertex.) How do the lengths of the sides of your rectangle compare with the height and base of your triangle? How is the area of your rectangle related to the area of your triangle? What has your folding revealed about the sum of the measures of the angles of your triangle?

CHAPTER PREVIEW

This chapter is the first of four dealing with topics in geometry that are very much a part of the life of an elementary or middle school teacher or student. Our philosophy, then, in these four chapters, is not only to study shapes in geometry in an informal manner but also to be careful mathematically with the material.

Before 600 B.C., geometry was both informal and practical. In the period from 600 B.C. to 300 B.C., Pythagoras, Euclid, and others organized the knowledge and experience that had been accumulated and transformed geometry into a theoretical science. Pragmatic approaches gave way to abstraction and general methods. With Euclid's *Elements,* geometry became a formal system in which geometric theorems were deduced logically from a list of statements called *axioms* that were accepted without proof. Many people believe that geometry is restricted to a Euclidean formalism in which exacting standards of proof and logical development must be met, but that is not true.

In this text, however, because our goal is to prepare K–8 teachers, we return to learning by trusting our intuition and experience. First, undefined terms are identified and careful definitions of other concepts are made from those terms. Geometric facts are then discovered by explorations of pictorial representations and physical models, with little attention given to the overall formal logical structure. This approach models the levels of learning geometry that were identified by the van Hieles' research. (See Into the Classroom on the next page.) However, there will be many opportunities to verify patterns and conjectures by examining the consequences of properties and facts that have already been accepted.

As we discussed in the previous two paragraphs, there is both a formal approach to geometry and an informal one. How, then, does one reason or give plausible arguments to show that results are true without using the rigorous formalism of Euclidean geometry? The Reasoning and Proof Standard and Representative Standard of NCTM provide some guidance, as the excerpt "Geometry in Grades Pre-K–2" on page 428 shows.

Our goals are to help young children recognize differences and similarities among shapes; to analyze the properties of a shape or class of shapes; and to model, construct, and draw shapes in a variety of ways. These goals are inseparably intertwined, but it will be seen that the discussion follows three threads of development: *classification, analysis,* and *representation* of the plane and space curves and surfaces that are presented in elementary school.

KEY IDEAS

- Understand definitions of geometric objects in the plane and what they mean.
- Recognize the similarities and differences between plane geometric objects; that is, classify them.
- Analyze the various relationships between angles in plane figures.
- Understand the differences among geometric objects in space, such as spheres, pyramids, prisms, cylinders, and especially polyhedra.
- The nature and difference of geometric shapes are very interesting but, as Marjorie Senechal* points out, *shape* is an indefinable term.

How do children look at various kinds of geometric objects? These thoughts will be explored in the three problem sets in this chapter. For example, how does a child learn to think about the differences among a triangle, quadrilateral, pentagon, and other shapes as recognized by the seventh of the Standards for Mathematical Practice? The "significance of an existing line in a geometric setting" works extremely well for young children in the preceding Cooperative Investigation, "Investigating Triangles via Paper Folding." In fact, older children can use "existing line" constructions to show how either paper folding or a formal theorem applies, which can be enormously important. Look at Figure 9.12 on page 438 for the theorem that shows the sum of the measures of the angles of a triangle is 180°, or do the paper folding and see how conceptually similar these two approaches are!

SMP 7 Look for and make use of structure.

"Mathematically proficient students look closely to discern a pattern or structure. Young students, for example, ... may sort a collection of shapes according to how many sides the shapes have . . . [Older students] . . . recognize the significance of an existing line in a geometric figure and can use the strategy of drawing an auxiliary line for solving problems."

*See L. A. Steen, ed., "Shape," in *On the Shoulders of Giants: New Approaches to Numeracy* (Washington, DC: National Academy Press, 1990), pp. 139–182.

INTO THE CLASSROOM
Activity-Based Learning and the van Hiele Levels

From kindergarten onward, geometry is learned best through hands-on activities. A successful teacher will take advantage of the enjoyment children experience when working with colored paper, straws, string, crayons, toothpicks, and other tangible materials. Children learn geometry by doing geometry as they construct two- and three-dimensional shapes, combine their shapes to create attractive patterns, and build interesting space figures out of plane shapes. By its nature, informal geometry provides unlimited opportunities to construct shapes, designs, and structures that all work to capture a child's interest.

According to pioneering research of the van Hieles in the late 1950s, the knowledge children construct for themselves through hands-on activities is essential to learning geometry. Dr. Pierre van Hiele and Dr. Dina van Hiele-Geldof, both former mathematics teachers in the Netherlands, theorized that learning geometry progresses through five levels, which can be described briefly as follows:

Level 0—Recognition of shape

Children recognize shapes holistically. Only the overall appearance of a figure is observed, with no attention given to the parts of the figure. For example, a figure with three curved sides would likely be identified as a triangle by a child at Level 0. Similarly, a square tilted point downward may not be recognized as a square.

Level 1—Analysis of single shapes

Children at Level 1 are cognizant of the parts of certain figures. For example, a rectangle has four straight sides that meet at "square" corners. However, at Level 1, the interrelationships between figures and properties are not understood.

Level 2—Relationships among shapes

At Level 2, children understand how common properties create abstract relationships among figures. For example, a square is both a rhombus and a rectangle. Also, children can make simple deductions about figures, using the analytic abilities acquired at Level 1.

Level 3—Deductive reasoning
The student at Level 3 views geometry as a formal mathematical system and can write deductive proofs.

Level 4—Geometry as an axiomatic system
This is the abstract level, reached only in high-level university courses. The focus is on the axiomatic foundations of a geometry, and no dependence is placed on concrete or pictorial models.

Ongoing research supports the thesis that students learn geometry by progressing through the van Hiele levels and need to "play" or think about geometric shapes. This text—by means of hands-on activities and examples and problems that require constructions and drawings—promotes the spirit of the van Hiele approach. However, it is the elementary school classroom teacher who must bring geometry to life for his or her students by creating interesting activities that support each child's progression through the first three van Hiele levels.

Materials for Explorations

Many examples will be presented in the form of an *exploration.* First, you will represent a shape, perhaps with a drawing or physical model, that satisfies the stated conditions. Next, you will be asked to discover, analyze, and describe the properties of the shape. Often, you will not want to read further until you have followed the directions and made some discoveries for yourself. Only then should you read on to see if the patterns and relationships you have uncovered agree with those discussed in the text.

The following tools and materials will be useful in drawing, constructing, or creating the shapes you will explore:

- colored pencils
- ruler (best if marked in both inches and millimeters)
- compass (be sure it is of good quality)
- tape
- glue
- protractor
- drafting triangles (30°–60°–90° and 45°–45°–90°)
- scissors
- unlined paper
- graph paper
- dot paper in both square and triangular patterns
- patty paper (waxed meat-patty separating sheets)

Exploring shapes, figures, and their properties should start even before kindergarten. See the following excerpt from NCTM Principles and Standards.

FROM THE NCTM PRINCIPLES AND STANDARDS

Geometry in Grades Pre-K–2

Children begin forming concepts of shape long before formal schooling. The primary grades are an ideal time to help them refine and extend their understandings. Students first learn to recognize a shape by its appearance as a whole or through qualities such as "pointiness." They may believe that a given figure is a rectangle because "it looks like a door."

Pre-K–2 geometry begins with describing and naming shapes. Young students begin by using their own vocabulary to describe objects, talking about how they are alike and how they are different. Teachers must help students gradually incorporate conventional terminology into their descriptions of two- and three-dimensional shapes. However, terminology itself should not be the focus of the pre-K–2 geometry program. The goal is that early experiences with geometry lay the foundation for more-formal geometry in later grades. Using terminology to focus attention and to clarify ideas during discussions can help students build that foundation.

Teachers must provide materials and structure the environment appropriately to encourage students to explore shapes and their attributes. For example, young students can compare and sort building blocks as they put them away on shelves, identifying their similarities and differences. They can use commonly available materials such as cereal boxes to explore attributes or shapes of folded paper to investigate symmetry and congruence . . . . Technology can help all students understand mathematics, and interactive computer programs may give students with special instructional needs access to mathematics they might not otherwise experience.

SOURCE: *Principles and Standards for School Mathematics by NCTM, pp. 97–98. Copyright © 2000 by the National Council of Teachers of Mathematics. Reproduced with permission of the National Council of Teachers of Mathematics via Copyright Clearance Center. NCTM does not endorse the content or validity of these alignments.*

A Picture Gallery of Plane Figures

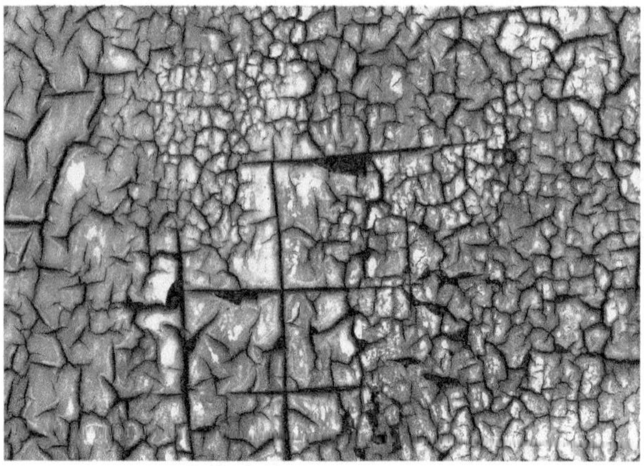

Fissures in a gelatinous preparation of tin oil

Butterfly wings

Pottery: a Talavera (Mexican) plate

A *fractal*, an example of a complex, beautiful image created with a computer

A snow crystal

The pattern of seeds in the head of a sunflower

A variety of manipulatives are available from commercial suppliers and are of great value in the study of geometry. Having access to such items as the following is desirable:

- geoboards
- tangrams
- pattern blocks
- geometric solids (wood or plastic)
- pentominoes
- reflective drawing tools such as a Mira®

9.1 Figures in the Plane

The shapes in the picture gallery of plane figures (page 430) and the picture gallery of space figures (page 466) are each quite different when viewed as a whole, but underlying this complexity is an orderly arrangement of simpler parts. In this section, we consider the most basic shapes of geometry: points, lines, segments, rays, and angles. We will also introduce a large number of notations and terms that are essential for the communication of geometric concepts and relationships.

Some terms have no formal definition (*line* or *plane,* for example). Other concepts, which you have seen many times (such as *line segment*, *ray*, and *vertical angles*), require careful definitions. We will define common terms in the body of a paragraph and save the displayed definitions for those concepts with which you may be less familiar.

Points and Lines

At first, we think that **points** are just locations. We are now going to think about a **plane** as a two-dimensional flat object or a piece of paper that is infinite in all directions. In our informal way of thinking about geometry, we consider a Cartesian coordinate plane on which a figure is placed. When we put a figure on a plane, we have a geometric object that is the set of all points that are a part of the figure. The figure could be an object as simple as a triangle, circle, or rectangle, or as complex as the preceding examples in the picture gallery of plane figures.

When we think of a **line** informally, we look at two different points A and B, and we see in Figure 9.1 that a line is all points between A and B (red), all points from A to B and beyond (green), which is a ray, and all points from B to A and beyond (blue), which would also be a ray. Rays will be discussed in more depth later in the chapter. Figure 9.1 shows that the line is denoted by $\overleftrightarrow{AB}$ and consists of three segments. Lines will often be denoted by lowercase letters such as l, for example. In Figure 9.1, these representations are shown.

Figure 9.1
Points, lines, and rays

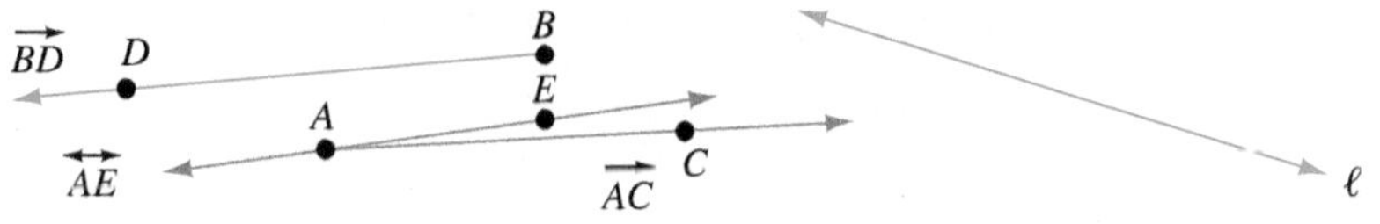

The arrows in the drawings and in the notation $\overleftrightarrow{AB}$ indicate that lines extend infinitely far in two directions. On paper, lines can be drawn with either a ruler or a **straightedge.** A straightedge is like a ruler but without any marks on it.

If three or more points lie on the same line, then we say the points are **collinear,** as shown in Figure 9.2. If three points are not on a line, they are called **noncollinear points.**

Figure 9.2
Three points determine either three lines or one line

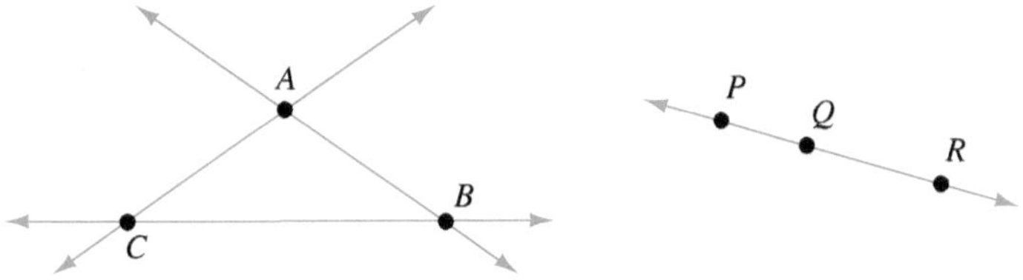

A, *B*, and *C* are noncollinear points.

P, *Q*, and *R* are collinear points.

Recall from Section 8.3 that two lines in the same plane are parallel if and only if either they are the same line or they have no points in common. Two distinct lines p and q in a plane that are not parallel must have a single point in common, called their **point of intersection.** The terms that we define next show some of the different ways in which two or three lines can be arranged in the plane (Figure 9.3).

DEFINITION Concurrent Lines

If there is a point B that is on each of the lines i, j, and k, then the three lines are said to be **concurrent.**

Figure 9.3
The possible arrangements of two and three lines in a plane

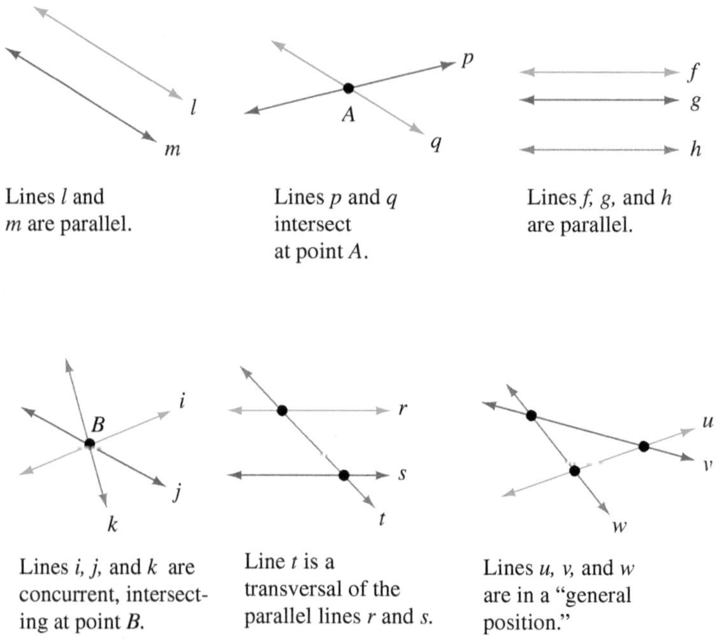

DEFINITION Transversal

If r and s are distinct lines and t is a line that intersects each of them, but not at the same point, then t is called a **transversal** to r and s.

EXAMPLE 9.1 Exploring Collinearity and Concurrency

(a) Place three circular objects (e.g., coins and cups) of different size on a sheet of paper, and trace them to create three circles, labeled C_1, C_2, and C_3. Next, place a ruler tightly against one side of the objects you used to trace C_1 and C_2, and draw the line l that just touches both circles. Move the ruler to the other side of your objects to draw the line m, as shown in the accompanying figure. The two lines you've drawn are called the external tangents to the circles C_1 and C_2. Label their point of intersection as P.

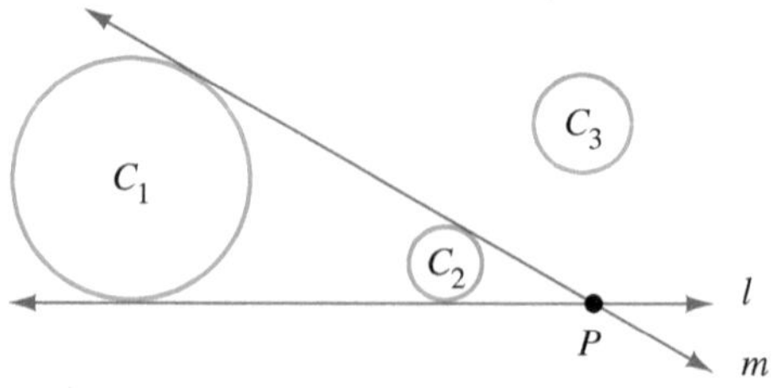

In the same way, use your ruler to draw the two lines externally tangent to C_2 and C_3, and let Q be their point of intersection. Finally, draw the external tangents of C_1 and C_3, and let R be their point of intersection. What conjecture do you have concerning P, Q, and R?

(b) Trace around a cup bottom (or use a compass) to draw an accurate circle. Then use a ruler to draw any three lines that are tangent to the circle at points labeled X, Y, and Z and that intersect in pairs at the points labeled A, B, and C. Finally, draw the lines $\overleftrightarrow{AX}$, $\overleftrightarrow{BY}$, and $\overleftrightarrow{CZ}$ (in the accompanying figure). What conjecture can you make about $\overleftrightarrow{AX}$, $\overleftrightarrow{BY}$, and $\overleftrightarrow{CZ}$?

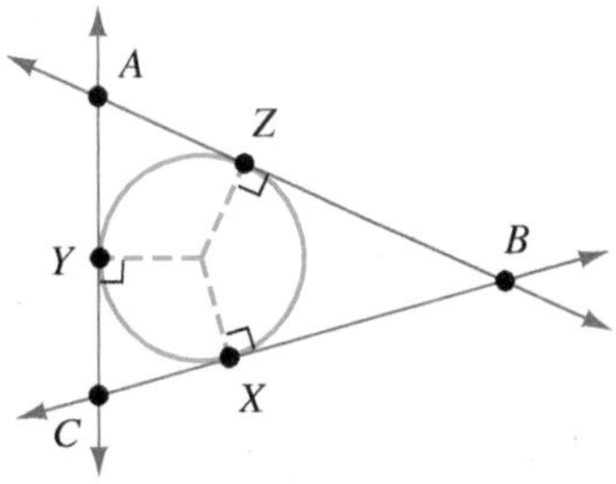

Solution

(a) P, Q, and R are collinear.
(b) $\overleftrightarrow{AX}$, $\overleftrightarrow{BY}$, and $\overleftrightarrow{CZ}$ are concurrent.

Line Segments and the Distance between Points

Let A and B be any two points. The line $\overleftrightarrow{AB}$ can be viewed as a copy of the number line. That is, every point on $\overleftrightarrow{AB}$ corresponds to a unique real number, and every real number corresponds to a unique point on $\overleftrightarrow{AB}$. If A and B correspond to the real numbers x and y, respectively, then the absolute value, $|x - y|$, gives the **distance** between A and B. We denote this distance by AB:

AB with no overbar denotes the length of segment $\overline{AB}$.

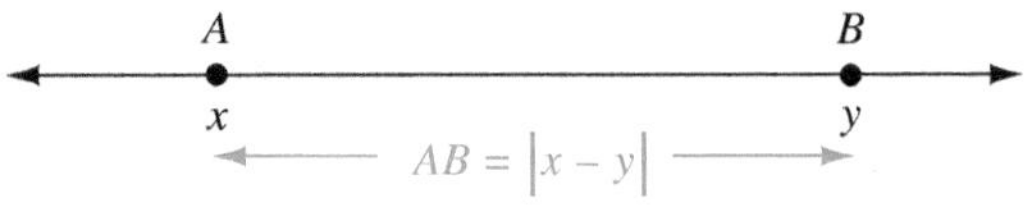

The points on the line $\overleftrightarrow{AB}$ that are between A and B, together with A and B themselves, form the **line segment** $\overline{AB}$. Points A and B are called the **endpoints** of $\overline{AB}$, and the distance AB is the **length** of $\overline{AB}$. It is important to see that the overbar used in the notation distinguishes the real number AB from the line segment $\overline{AB}$. All of these are good to review by looking in Section 8.2, p. 393. Note that if the *line* is not horizontal, we can compute the length of $\overline{AB}$ by using the distance function between A and B.

DEFINITION Congruent, Midpoint

Two segments $\overline{AB}$ and $\overline{CD}$ are said to be **congruent** if they have the same length. This relationship is symbolized by writing $\overline{AB} \cong \overline{CD}$. Thus, $\overline{AB} \cong \overline{CD}$ if and only if $AB = CD$. The point M in $\overline{AB}$ that is the same distance from A and B is called the **midpoint** of $\overline{AB}$. This information is summarized in Figure 9.4.

Figure 9.4
A segment $\overline{AB}$, its length AB, congruent segments, and the midpoint M of $\overline{AB}$

Tick marks are used to indicate congruent line segments.

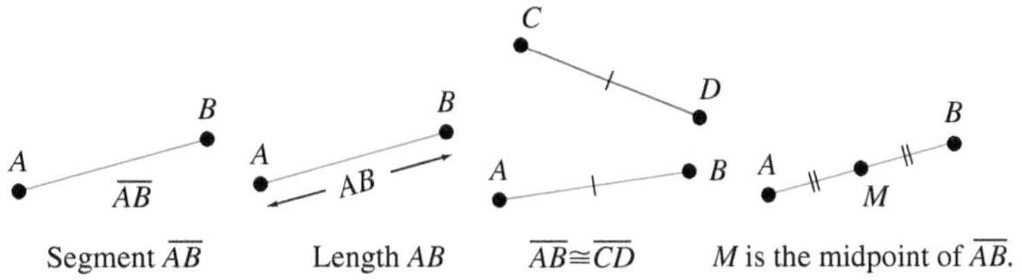

Rays, Angles, and Angle Measure

A **ray** is a subset of a line that contains a point P, called the **endpoint** of the ray, and all points on the line lying to one side of P. If Q is any point on the ray other than P, then $\overrightarrow{PQ}$ denotes the ray. The union of two rays with a common endpoint is an **angle.** If the rays are $\overrightarrow{AB}$ and $\overrightarrow{AC}$, then the angle is denoted by $\angle BAC$. The common endpoint of the two rays is called the **vertex** of the angle and is the middle letter in the symbol for the angle (for example, A in $\angle BAC$). The points B and C not at the vertex can be written in either order, so that $\angle CAB$ denotes the same angle as $\angle BAC$. The rays $\overrightarrow{AB}$ and $\overrightarrow{AC}$ are called the **sides** of the angle. (See Figure 9.5.)

Figure 9.5
A ray $\overrightarrow{PQ}$ and an angle $\angle BAC$

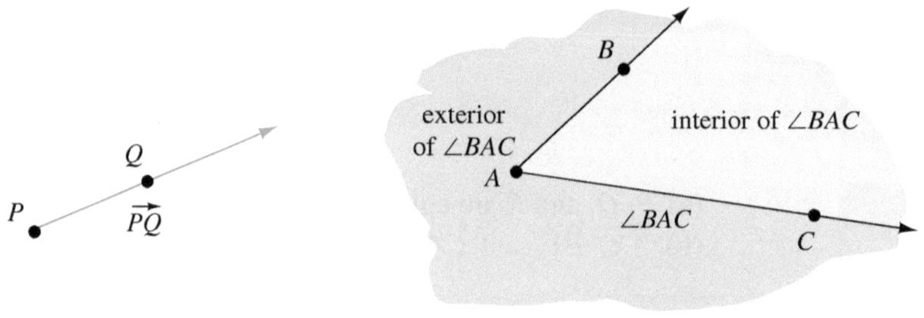

A ray with endpoint P that contains point Q

An angle with vertex A and sides $\overrightarrow{AB}$ and $\overrightarrow{AC}$

An angle whose sides are not on the same line partitions the remaining points of the plane into two parts: the **interior** and the **exterior** of the angle. The points along a line segment that joins an endpoint on side $\overrightarrow{AB}$ to an endpoint on $\overrightarrow{AC}$ are all interior points of $\angle BAC$.

If $\angle BAC$ is the only angle with its vertex at A, it is common to write $\angle A$ in place of $\angle BAC$. When more than one angle has a vertex at A, it is essential to use the full three-letter symbol. Sometimes it is useful to number the angles that appear in a drawing and refer to $\angle 1$, $\angle 2$, $\angle 3$, and so on.

m(∠A) denotes the measure of the angle A with vertex at point A.

The size of an angle is measured by the amount of rotation required to turn one side of the angle to the other by pivoting about the vertex. The **measure of an angle** is generally given in **degrees,** where there are 360° in a full revolution. The measure of $\angle A$ is denoted by $m(\angle A)$. If the rotation is imagined to pass through the interior of the angle, the measure is a number between 0° and 180°. Unless stated otherwise, $m(\angle A)$ is the measure of $\angle A$ not larger than 180°.

An angle of measure 180° is a **straight angle,** an angle of measure 90° is a **right angle,** and an angle of measure 0° is a **zero angle.** Angles measuring between 0° and 90° are **acute,** and angles measuring between 90° and 180° are **obtuse.**

In some applications, the measure of interest corresponds to the rotation through the exterior of the angle and is therefore a number between 180° and 360°. An angle with measure greater than 180°, but less than 360°, is called a **reflex angle.**

The classification of angles according to their measure is summarized in Figure 9.6.

Figure 9.6
The classification of angles by their measure

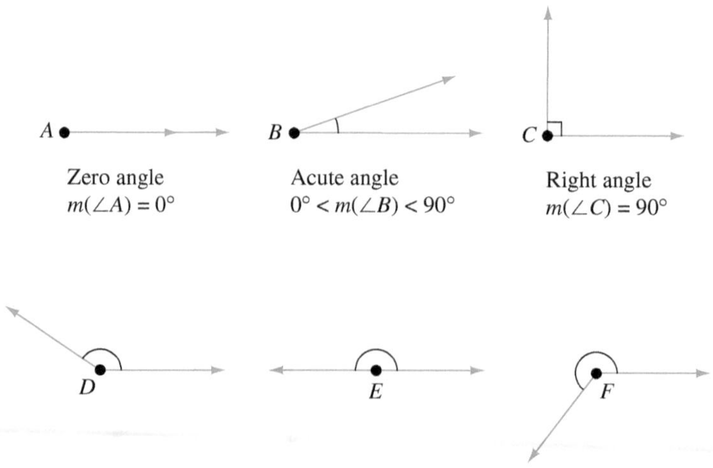

Zero angle $m(\angle A) = 0°$

Acute angle $0° < m(\angle B) < 90°$

Right angle $m(\angle C) = 90°$

Obtuse angle $90° < m(\angle D) < 180°$

Straight angle $m(\angle E) = 180°$

Reflex angle $180° < m(\angle F) < 360°$

Right angles in drawings are indicated by a small square placed at the vertex. A circular arc is required to indicate reflex angles.

Two lines l and m that intersect at right angles are called **perpendicular lines.** This relationship is indicated in writing by $l \perp m$. Similarly, two rays, two segments, or a segment and a ray are perpendicular if they are contained in perpendicular lines.

DEFINITION Congruent Angles

Two angles are **congruent** if and only if they have the same measure.

It is important to remember that there is a significant difference between two angles being equal (meaning that the two rays defining them are the same) and being congruent (their measures being equal). There are, for example, many different angles whose measure is 30° (so they are congruent). Of course, if two angles are equal, then they are also congruent. The symbol $\cong$ is used to denote the congruence of angles. Therefore,

$$\angle P \cong \angle Q \quad \text{if and only if} \quad m(\angle P) = m(\angle Q).$$

We can use a **protractor** to measure angles and to draw angles that have a given measure. A protractor and other traditional drawing and measuring tools are shown in Figure 9.7. Increasingly, these tools are being supplemented and replaced by geometry software.

Figure 9.7
Some useful tools for drawing and measuring geometric figures

Circle master compass

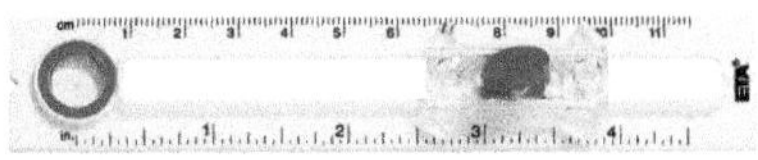

ETA/Cuisenaire® Safe Drawing Compass

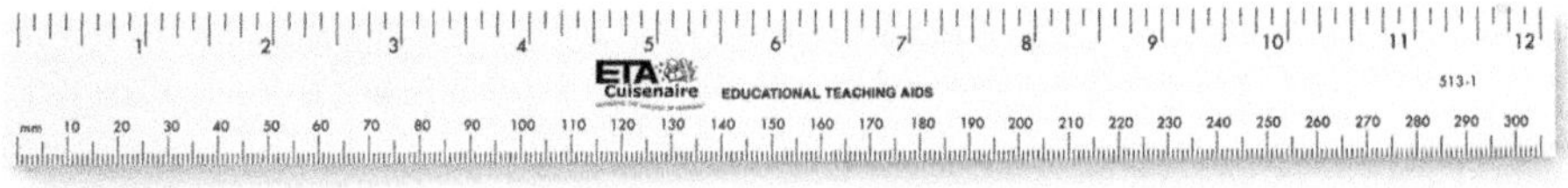

Ruler

Drafting triangles

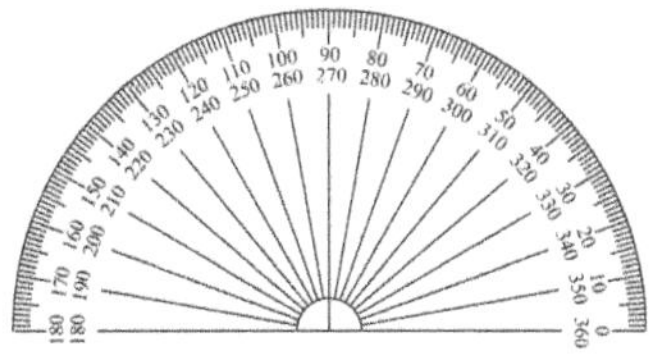

Protractor

EXAMPLE 9.2

Exploring Angles and Distances in a Circle

(a) Draw a large circle and choose any point P, other than the center, inside the circle. Any line l through P intersects the circle in two points, say, A and B. Measure the distances AP and PB (to the nearest millimeter) and compute the product $AP \cdot PB$. Draw several other lines through P and measure the distances of the two segments. Which line through P makes the product of distances $AP \cdot PB$ as large as possible?

(b) Draw a circle with center C. Draw a line through C, and let A and B denote its intersections with the circle. Choose any three points P, Q, and R on the circle other than A or B. Use a protractor to measure $\angle APB$, $\angle AQB$, and $\angle ARB$. Compare with other choices of points. What general result does this activity suggest?

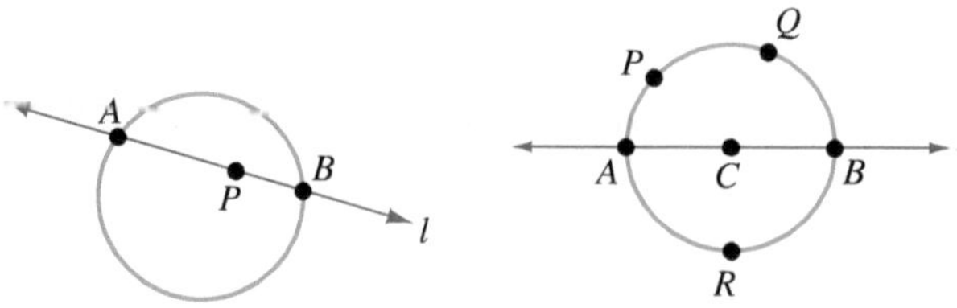

Solution

(a) For every choice of l, the product $AP \cdot PB$ is the same. Therefore, no line through P gives a larger product than any other line.

(b) Each angle is a right angle. This is one of geometry's earliest theorems, attributed to Thales of Miletus (ca. 600 B.C.).

Pairs of Angles and the Corresponding-Angles Theorem

As shown in Figure 9.8, two angles are **complementary** if the sum of their measures is 90°. Similarly, two angles are **supplementary** if their measures sum to 180°.

Figure 9.8
Examples of complementary and supplementary angles

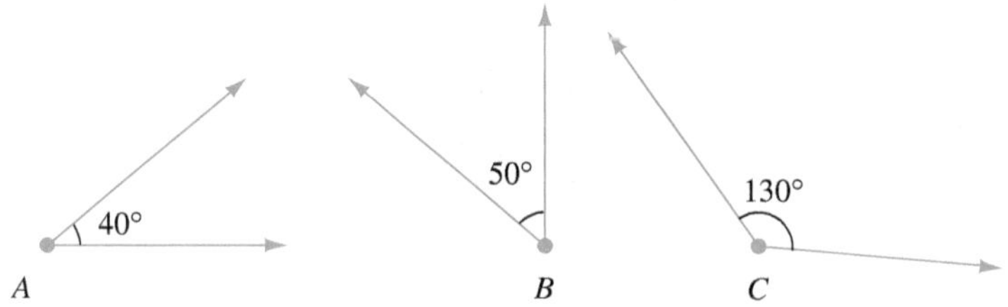

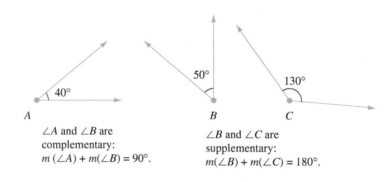

$\angle A$ and $\angle B$ are complementary: $m(\angle A) + m(\angle B) = 90°$.

$\angle B$ and $\angle C$ are supplementary: $m(\angle B) + m(\angle C) = 180°$.

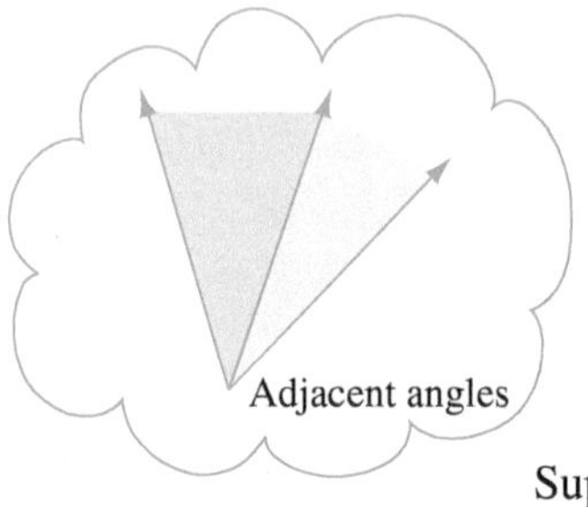

Two angles that have a common side and nonoverlapping interiors are called **adjacent angles.** Supplementary and complementary angles frequently occur as adjacent angles. See Figure 9.9.

Figure 9.9
Adjacent supplementary and complementary angles

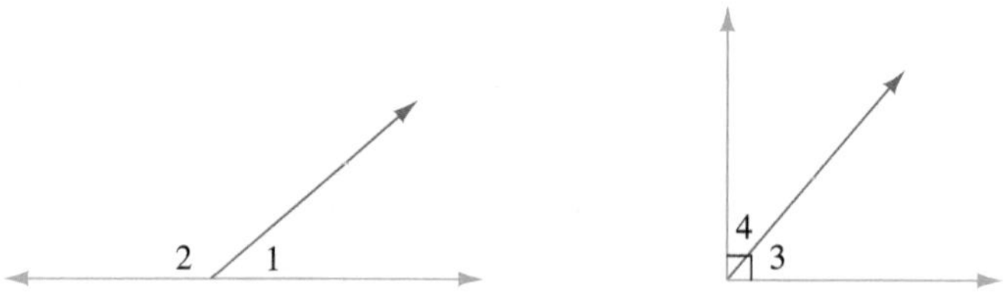

$\angle 1$ and $\angle 2$ are adjacent supplementary angles.

$\angle 3$ and $\angle 4$ are adjacent complementary angles.

Two nonadjacent angles formed by two intersecting lines are called **vertical angles,** as shown in Figure 9.10. Since $\angle 1$ and $\angle 2$ are supplementary, we know that $m(\angle 1) + m(\angle 2) = 180°$. Likewise, $\angle 2$ and $\angle 3$ are supplementary, so we also have $m(\angle 2) + m(\angle 3) = 180°$. Comparing these two equations shows that $m(\angle 1) = m(\angle 3)$. This proves another theorem of Thales.

Figure 9.10
Intersecting lines form two pairs of vertical angles

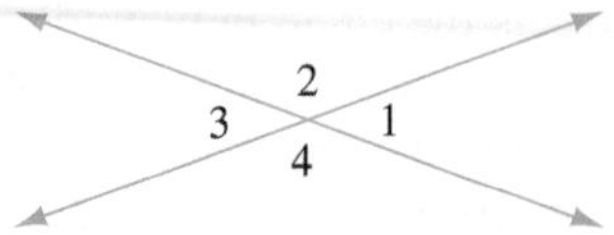

$\angle 1$ and $\angle 3$ are vertical angles.
$\angle 2$ and $\angle 4$ are vertical angles.

THEOREM Vertical-Angles Theorem
Vertical angles have the same measure.

Now suppose two lines l and m intersect a transversal at two points. There are eight angles formed, in four pairs of **corresponding angles.**

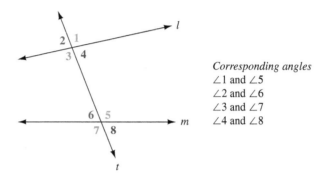

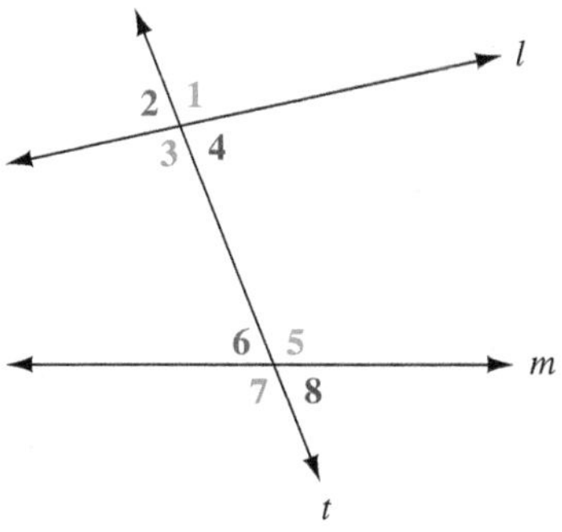

Corresponding angles
$\angle 1$ and $\angle 5$
$\angle 2$ and $\angle 6$
$\angle 3$ and $\angle 7$
$\angle 4$ and $\angle 8$

A case of special importance occurs when l and m are parallel lines, as shown in Figure 9.11. It would appear that each pair of corresponding angles is a pair of congruent angles. Conversely, if any one pair of corresponding angles is a congruent pair of angles, then the lines l and m appear to be parallel. We will accept the truth of these observations, giving us the **corresponding-angles property.** Many formal treatments of Euclidean geometry introduce the corresponding-angles property as an axiom.

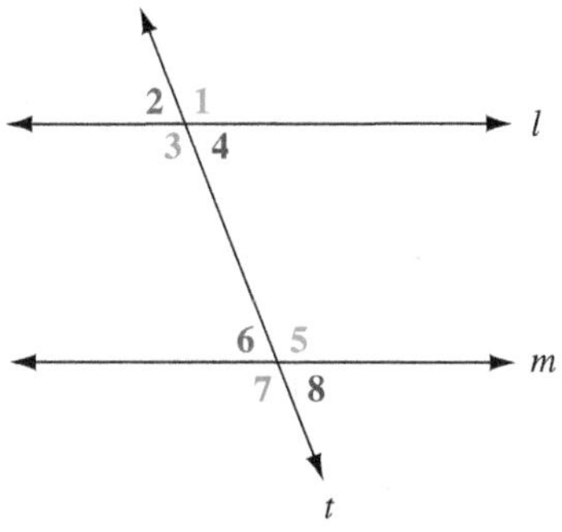

Figure 9.11
Lines l and m are parallel if and only if the angles in some corresponding pair have the same measure

PROPERTY Corresponding-Angles Property

- If two parallel lines are cut by a transversal, then their corresponding angles have the same measure.
- If two lines in the plane are cut by a transversal and some pair of their corresponding angles has the same measure, then the lines are parallel.

EXAMPLE 9.3

Using the Corresponding-Angles Property

(a) Lines l and m are parallel and $m(\angle 6) = 35°$. Find the measures of the remaining seven angles.

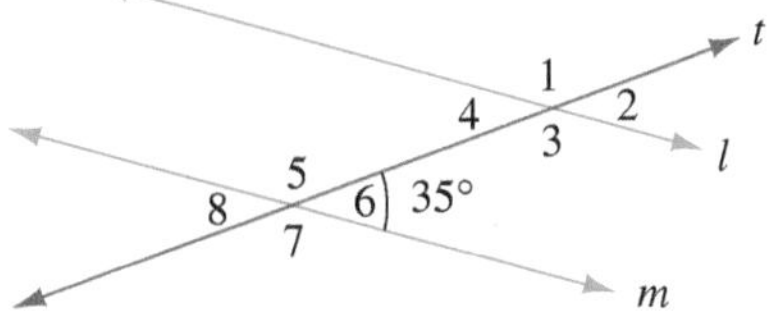

(b) Lines t and j intersect at P and form an angle measuring 122°. Describe how to use a protractor and straightedge to draw a line k through Q that is parallel to line j.

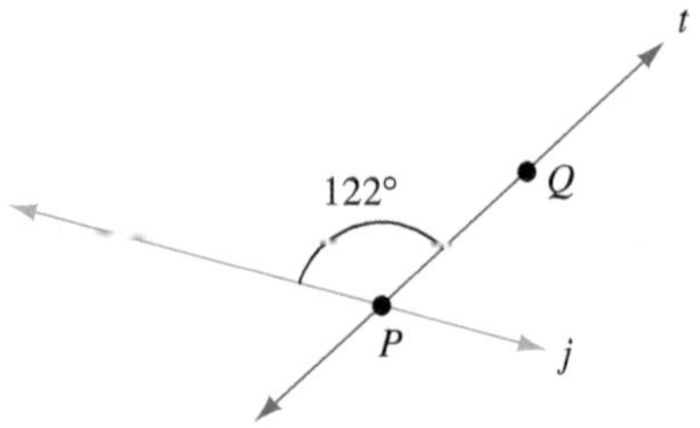

Solution

(a) Since $\angle 8$ and $\angle 6$ are vertical angles, $m(\angle 8) = 35°$. Also, $\angle 5$ and $\angle 7$ are supplements of $\angle 6$, so $m(\angle 5) = m(\angle 7) = 180° - 35° = 145°$. By the corresponding-angles property, $m(\angle 1) = m(\angle 5) = 145°$, $m(\angle 2) = m(\angle 6) = 35°$, $m(\angle 3) = m(\angle 7) = 145°$, and $m(\angle 4) = m(\angle 8) = 35°$.

(b) Use the protractor to form the corresponding angle measuring 122° at point Q.

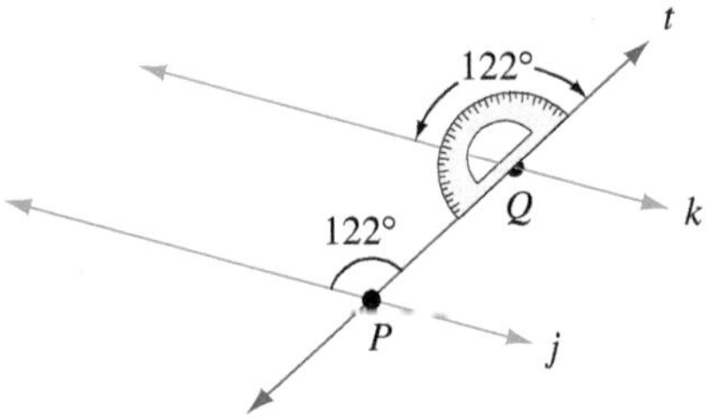

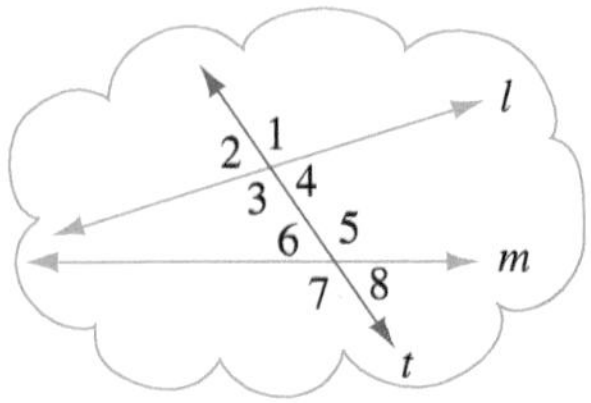

The pair of angles $\angle 4$ and $\angle 6$ between l and m, but on opposite sides of the transversal t, is called a pair of **alternate interior angles.** Since $\angle 2$ and $\angle 4$ are vertical angles, they are congruent by the vertical-angles theorem. Thus, the corresponding angles $\angle 2$ and $\angle 6$ are congruent if and only if the alternate interior angles $\angle 4$ and $\angle 6$ are congruent. This statement gives the following consequence of the corresponding-angles property:

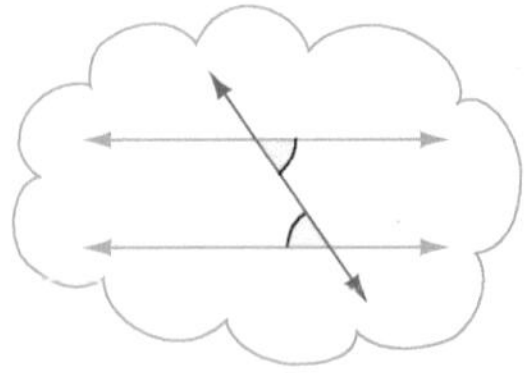

THEOREM Alternate-Interior-Angles Theorem

Two lines cut by a transversal are parallel if and only if a pair of alternate interior angles is congruent.

The Measure of Angles in Triangles

We shall now show why certain geometric facts (such as the sum of the angle measures of a triangle being 180°) become very clear when we use simple constructions together in a classroom. This approach gives rise to a major result in geometry, even at an elementary school level. The Cooperative Investigation that starts this chapter (p. 426) is full of such ideas, and we hope that you will fold paper many times to explore the marvelous insights of this chapter.

If a triangle ABC is cut from paper, and its three corners are torn off, it is soon discovered that the three pieces will form a straight angle along a line l. (See Figure 9.12.) Thus, $m(\angle 1) + m(\angle 2) + m(\angle 3) = 180°$, and we have a physical demonstration that the sum of the measures of the angles of a triangle is 180°. Of course, an "angle of a triangle" means an *interior* angle of that triangle, as in Figure 9.12.

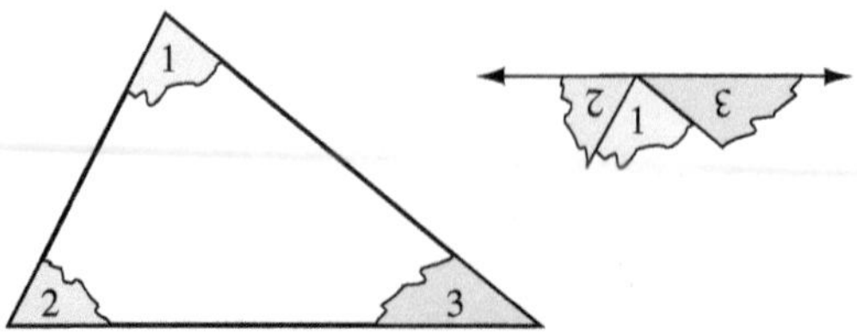

Figure 9.12
The torn corners of a triangle cut from paper can be placed along a line to show that $m(\angle 1) + m(\angle 2) + m(\angle 3) = 180°$

We will now derive the same result by using the alternate-interior-angles theorem.

THEOREM Sum of Angle Measures in a Triangle
The sum of the measures of the angles in a triangle is 180°.

PROOF Consider the line l through point A that is parallel to the line $m = \overleftrightarrow{BC}$:

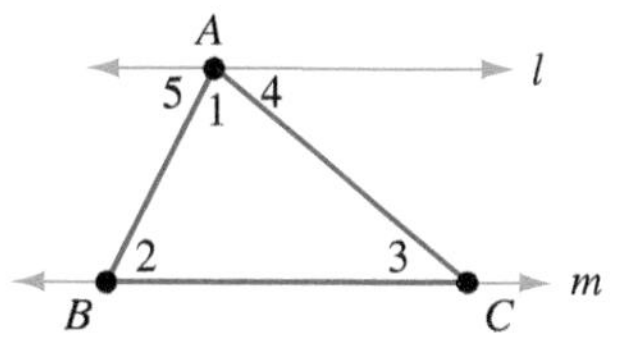

Line $\overleftrightarrow{AB}$ is a transversal to l and m for which $\angle 5$ and $\angle 2$ are alternate interior angles. Thus, $m(\angle 5) = m(\angle 2)$ by the alternate-interior-angles theorem. Similarly, $\angle 4$ and $\angle 3$ are alternate interior angles for the transversal $\overleftrightarrow{AC}$, so $m(\angle 4) = m(\angle 3)$. Since $\angle 5$, $\angle 1$, and $\angle 4$ form a straight angle at vertex A, we know that $m(\angle 5) + m(\angle 1) + m(\angle 4) = 180°$. Thus, by substitution, $m(\angle 2) + m(\angle 1) + m(\angle 3) = 180°$.

EXAMPLE 9.4

Measuring an Opposite Exterior Angle of a Triangle

In the accompanying figure, $\angle 4$ is called an **exterior angle** of triangle PQR and $\angle 1$ and $\angle 2$ are its **opposite interior angles.** Show that the measure of the exterior angle is equal to the sum of the measures of the opposite interior angles; that is, prove that $m(\angle 4) = m(\angle 1) + m(\angle 2)$.

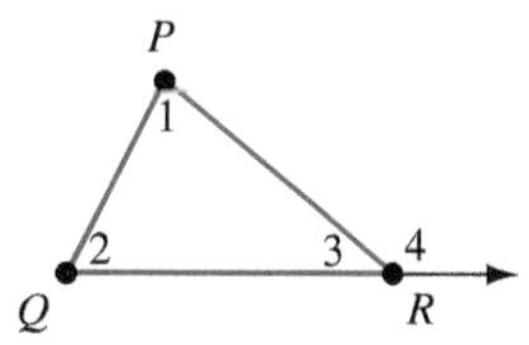

Solution

By the preceding "sum of the angle measures" theorem, we have $m(\angle 1) + m(\angle 2) + m(\angle 3) = 180°$. Also, $\angle 3$ and $\angle 4$ are supplementary, so $m(\angle 3) + m(\angle 4) = 180°$. Therefore, $m(\angle 1) + m(\angle 2) + m(\angle 3) = m(\angle 3) + m(\angle 4)$. Subtracting $m(\angle 3)$ from both sides of this equation gives $m(\angle 1) + m(\angle 2) = m(\angle 4)$.

Directed Angles

Until now, we have measured angles without considering a *direction*—clockwise or counterclockwise. One side rotates until it coincides with the second side. Often, it is useful to specify one side as the **initial side** and the other side as the **terminal side.** Angles are then measured by specifying the number of degrees to rotate the initial side to the terminal side.

Mathematicians usually assign a positive number to counterclockwise turns, and a negative number to clockwise turns. Angles that specify an initial and final side and a direction of turn are called **directed angles.** Some examples are shown in Figure 9.13, where the arrows on the circular arcs indicate the direction of turn. Notice that an angle measure of $-90°$ could also be assigned the measure $+270°$.

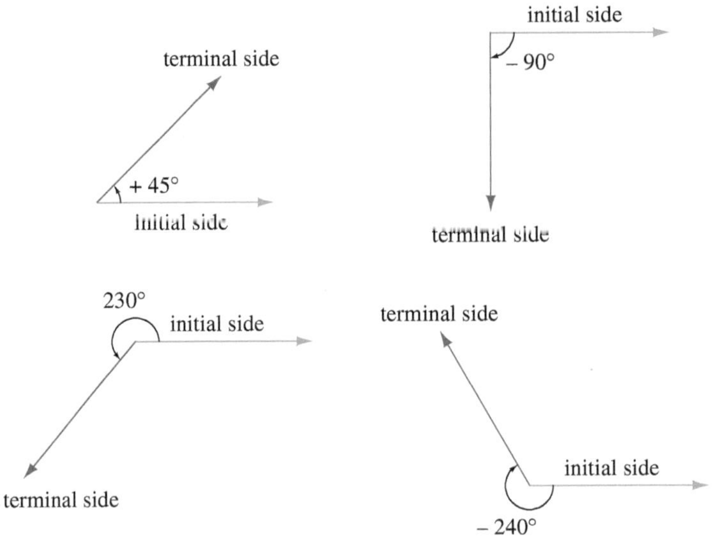

Figure 9.13
Directed angles, measured positively for counterclockwise turns and negatively for clockwise turns

EXAMPLE 9.5

Measuring Directed Angles

Paul Pathfinder's trip through the woods to Grandmother's house started and ended in an easterly direction but zigzagged through Wolf Woods to avoid trouble. The first two angles Paul turned through are 45° and −60°, as shown. Use a protractor to measure the three remaining turns. What is the sum of all five directed angles? Explain your surprise—or lack of surprise.

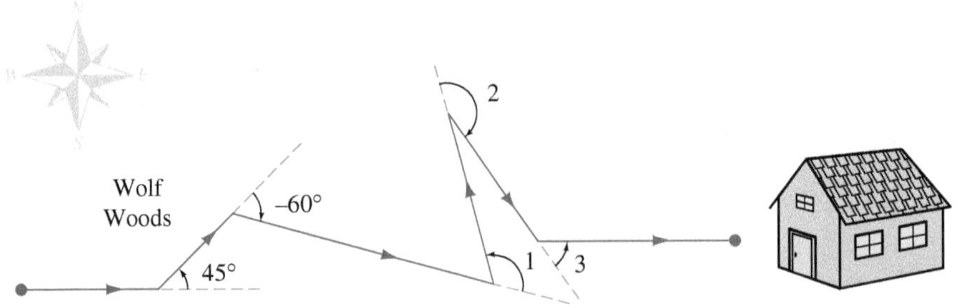

Solution

$m(\angle 1) = 120°$, $m(\angle 2) = -155°$, and $m(\angle 3) = 50°$. The sum of all five directed angles is $45° - 60° + 120° - 155° + 50° = 0°$. This is not surprising, since Paul's path started and stopped in the same direction and his path didn't make any loops.

9.1 Problem Set

Understanding Concepts

1. Use symbols to name each of the figures shown. If more than one symbol is possible, give all possible names.

   (a)

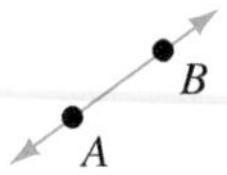

   (b)

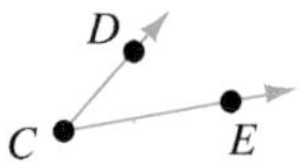

2. (a)

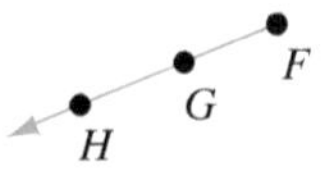

   (b)

3. The points E, U, C, L, I, and D are as shown:

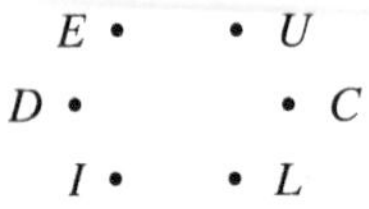

Draw the following figures:

(a) $\overleftrightarrow{EU}$ (b) $\overrightarrow{CL}$ (c) $\overline{ID}$

4. Using the points of problem 3, draw the following figures:

(a) $\overleftrightarrow{EL}$ (b) $\overrightarrow{LC}$ (c) $\overline{LD}$ (d) $\overline{DL}$

5. Trace the 5-by-5 square lattice shown, and draw the line segment $\overline{AB}$:

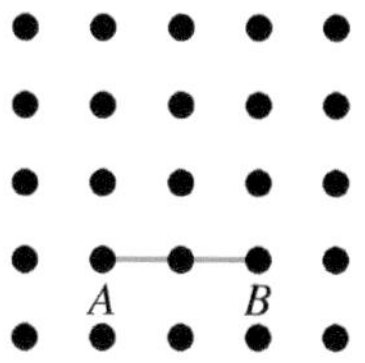

Use colored pencils to circle all of the points C of the lattice that make $\angle BAC$

(a) a right angle,
(b) an acute angle,
(c) an obtuse angle,
(d) a straight angle, and
(e) a zero angle.

6. In the figure shown, $\angle BXD$ is a right angle and $\angle AXE$ is a straight angle:

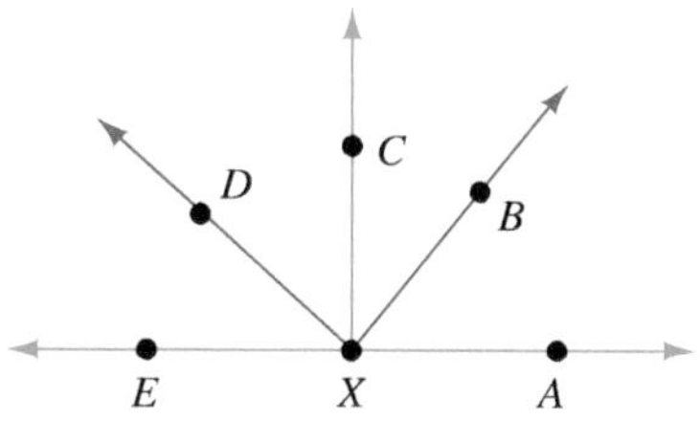

If $m(\angle BXC) = 45°$ and $m(\angle BXE) = 140°$, explain how you can determine the measures of $\angle AXB$, $\angle CXD$, and $\angle DXE$ without using a protractor.

7. The point P shown in the diagram that follows is the intersection of the two ***external*** **tangent** lines to a pair of circles. (See Example 9.1.) The point Q is the intersection of the two ***internal*** **tangent** lines to two circles.

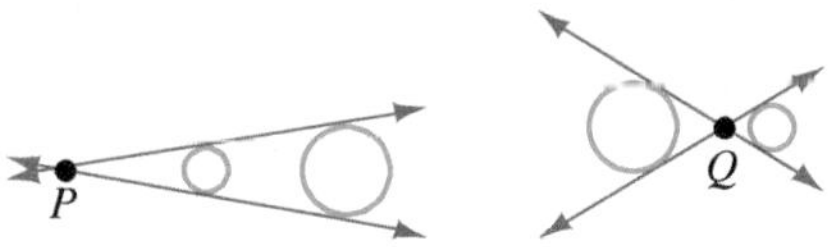

Draw three circles C_1, C_2, and C_3, all of different radii and with no circle containing or intersecting either of the other circles. Let P be the intersection of the external tangent lines of C_1 and C_2, and let Q and R be the respective intersections of the internal tangent lines of the pairs of circles C_2, C_3 and C_1, C_3. What conclusion is suggested by your drawing? Compare with others.

8. Two intersecting circles determine a line, as shown as on the next column in the accompanying diagram. Draw three circles, where each circle intersects the other two circles, as in the example shown on the right. Next, draw the three lines determined by each pair of intersecting circles. What conclusion is suggested by your drawing?

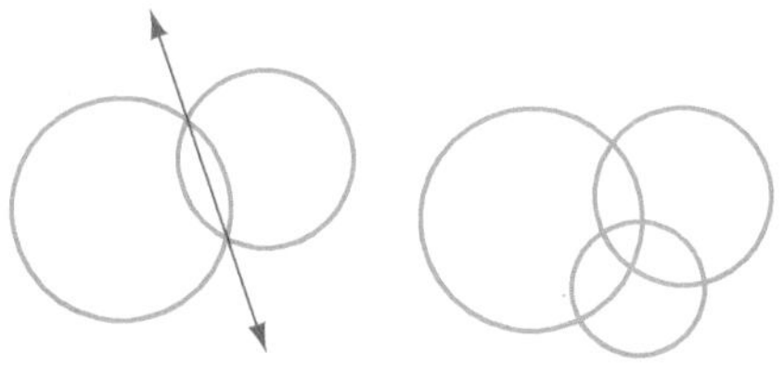

9. Use a compass (or carefully trace around a cup bottom or some other circular object) to draw three circles of the same size through point A. Let B, K, and L be the other points of intersection of pairs of circles. Now draw a fourth circle of the same size as the other three that passes through B and creates points of intersection M and N, as shown:

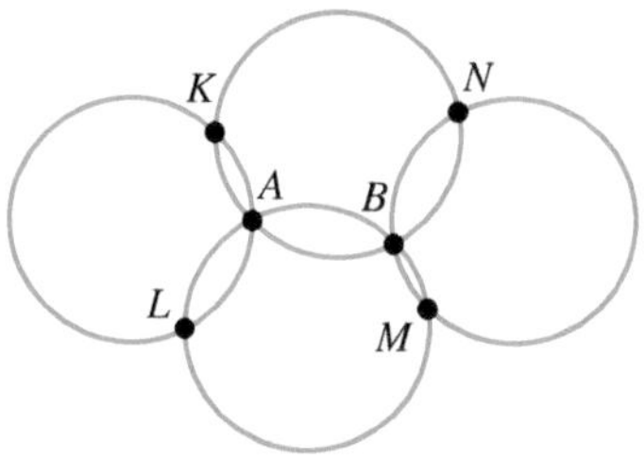

(a) Draw the lines $\overleftrightarrow{KL}$ and $\overleftrightarrow{MN}$. What can you say about these two lines?

(b) Draw the lines $\overleftrightarrow{LM}$ and $\overleftrightarrow{KN}$. What can you say about these two lines?

(c) Draw the line $\overleftrightarrow{AB}$. What connection does it seem to have to any of the lines drawn earlier?

10. Draw a circle with center at point A and a second circle of the same radius with center at point B, where the radius is large enough to cause the circles to intersect at two points C and D. Let M be the point of intersection of $\overline{CD}$ and $\overline{AB}$.

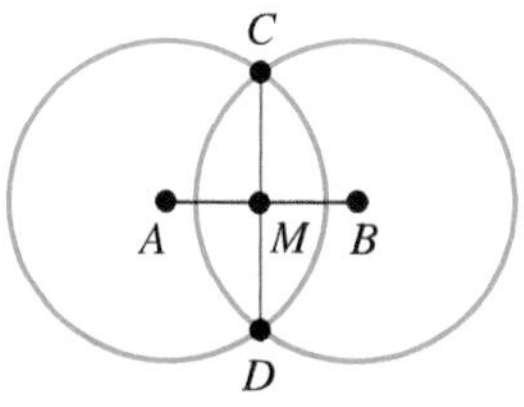

(a) Use a protractor to measure the angles at M. What can you say about how $\overline{CD}$ and $\overline{AB}$ intersect?

(b) Use a ruler to measure $\overline{MA}$ and $\overline{MB}$. What can you say about point M?

11. Draw two circles, and locate four points on each circle. Draw the segments as shown:

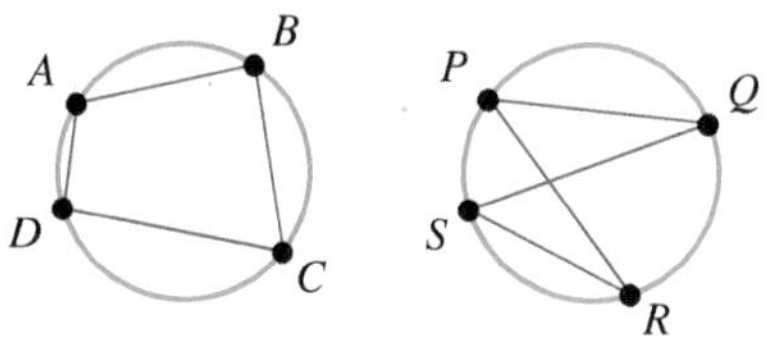

(a) Carefully measure the angles with vertices at A, B, C, and D with a protractor. What relationships do you see on the basis of your measurements?

(b) Measure the angles with vertices at P, Q, R, and S, and discuss what relationships hold among the angles in that figure.

12. The hour and minute hands of a clock form a zero angle at noon and midnight. Between noon and midnight, how many times do the hands again form a zero angle?

13. How many degrees does the minute hand of a clock turn through

(a) in 60 minutes?

(b) in 10 minutes?

(c) in 2 minutes?

How many degrees does the hour hand of a clock turn through

(d) in 120 minutes?

(e) in 5 minutes?

14. Find the angle formed by the minute and hour hands of a clock at these times:

(a) four o'clock **(b)** seven o'clock

(c) 4:30 **(d)** 10:20

15. The lines l and m are parallel. Find the measures of the numbered angles shown.

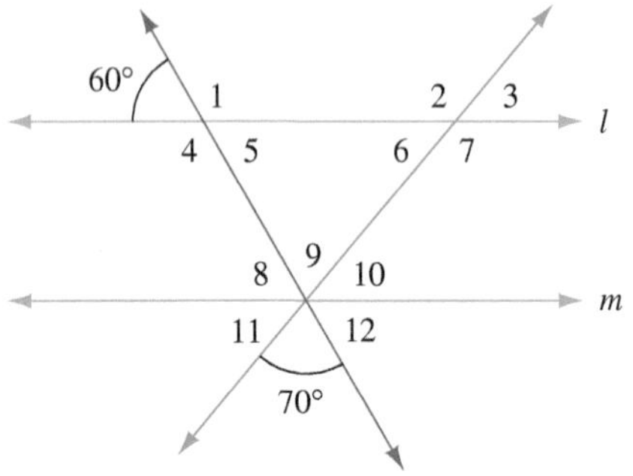

16. Determine the measure of $\angle P$ if $\overrightarrow{AB}$ and $\overrightarrow{CD}$ are parallel.

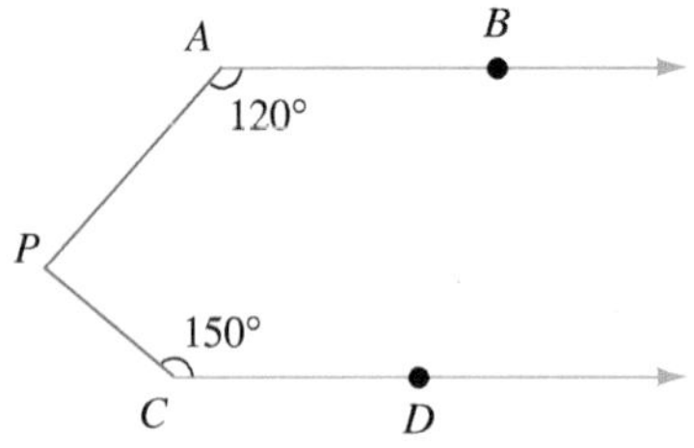

17. Find the measures of the numbered angles in the triangles shown.

(a)

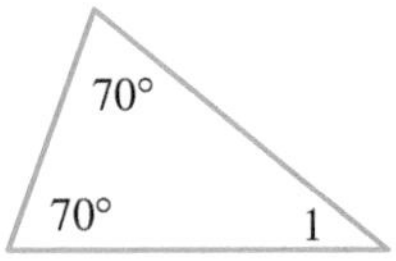

(b)

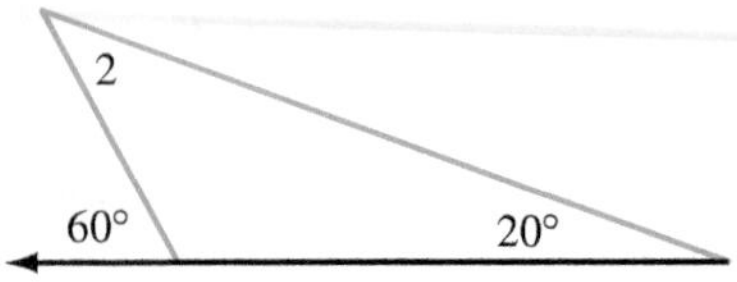

18. Find the measure of the numbered angles in the triangles shown.

(a)

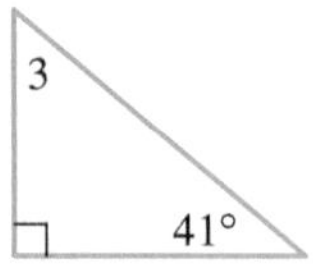

(b)

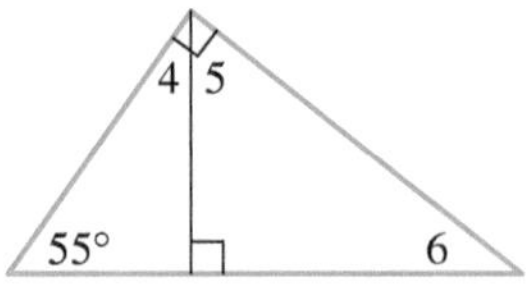

19. Find the measures of the interior angles of the following triangles:

(a)

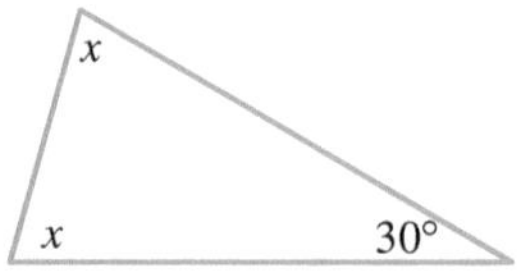

(b)

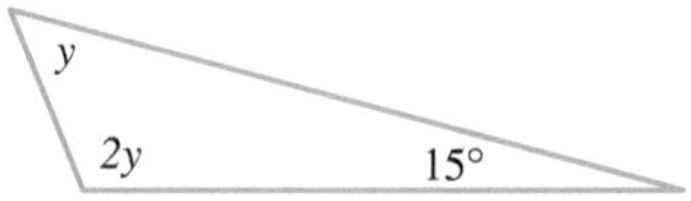

20. Find the measures of the interior angles:

(a)

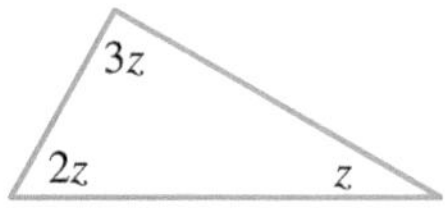

(b) A triangle has the property that the measure of $\angle A$ is four times the measure of $\angle B$ and half of the measure of $\angle C$. What is the measure of each angle?

21. **(a)** Can a triangle have two obtuse angles? Explain your answer.

(b) Can a triangle have two right angles? Explain your answer.

(c) Suppose no angle of a triangle measures more than 60°. What do you know about the triangle?

22. A hiker started heading due north, then turned to the right 38°, then turned to the left 57°, and next turned to the right 9°. To resume heading due north, what turn must the hiker make?

23. In the accompanying figure, $\angle APC$ and $\angle BPD$ are right angles. Show that $\angle 1 \cong \angle 3$.

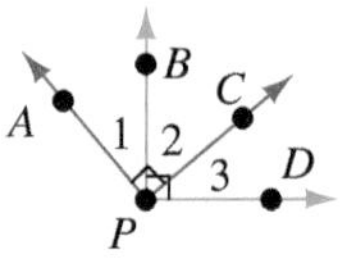

24. Let $\overline{PQ} \parallel \overline{AB}$ and $\overline{RQ} \parallel \overline{AC}$. Find the measures of $\angle 1, \angle 2, \ldots, \angle 8$. Explain how you found your answers.

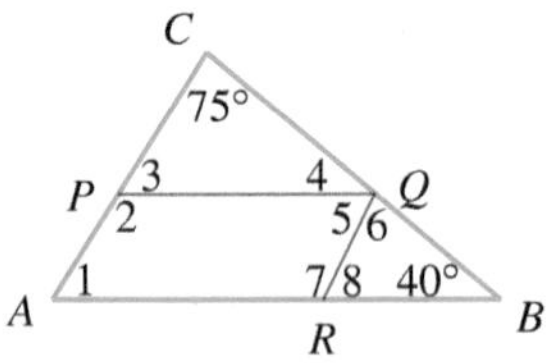

Into the Classroom

25. Youngsters (and college students, too!) often learn geometric concepts best when working with common three-dimensional objects—say, a cardboard shoebox. The corners of a box model points, and each edge of the box represents a line segment. Additional lines can be drawn on the box with a ruler. Create a lesson that investigates several of the concepts of this section, using a shoebox as a manipulative. Do you see some congruent segments? some right angles? and so on.

26. **Treasure Hunt.** Create a treasure hunt game in which groups of children are given starting points and directions in the classroom. Each group uses protractors and rulers to follow directions that, when accurately carried out, will lead them to "treasures" hidden throughout the room. For example, the directions might start, "Begin at point A facing the front of the room. Turn 40° and go 8 feet to arrive at point B. Turn −120° and go 20 feet to . . . ." In particular, show a diagram of a classroom, a starting point, and a route leading to the location T of the hidden treasure. Give the corresponding directions for the route.

27. Children frequently have difficulty understanding the difference between two segments being equal and two segments being congruent. What problem would you give your class to show the students the difference between the two concepts?

28. Children have difficulty understanding the difference between two angles being equal and two angles being congruent. What problem would you give your class to show the students the difference between the two concepts?

Responding to Students

29. Larisa, a second grader, is asked whether the following segments drawn on a page are parallel:

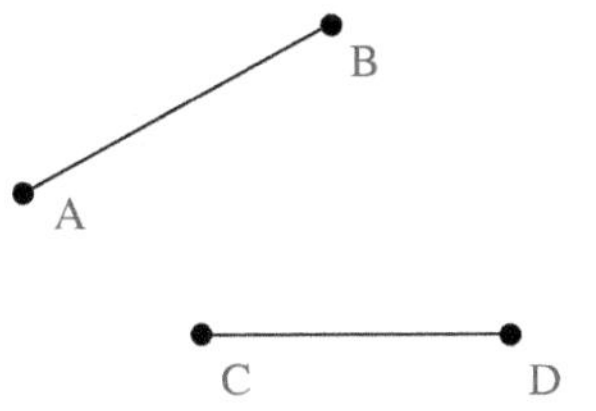

After a brief pause, she says, "Yes, they are because they don't meet." How would you respond to Larisa?

30. Aja's teacher asked her to name the line shown. She answered the question by naming it line ABCDEF.

(a) What does Aja not seem to understand about naming lines (and probably rays and line segments)?

(b) How would you guide her in correctly naming lines, rays, and line segments?

31. Sebastian was given the following problem on his math homework last night: "Which of the following shapes does not have a right angle? Explain your answer."

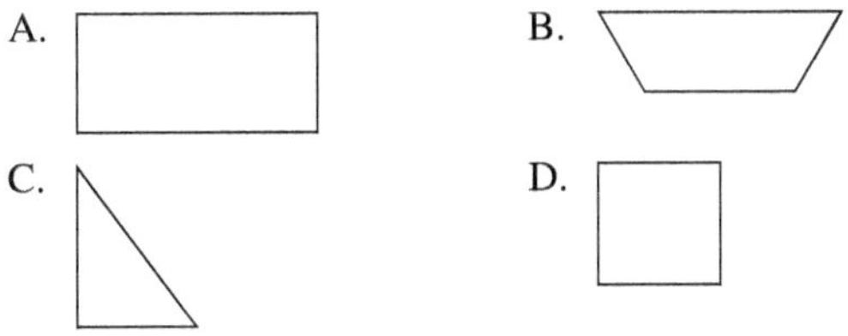

His answer was "(C) because it is not a quadrilateral."

(a) What incorrect statement did Sebastian make about quadrilaterals?

(b) How would you help guide him to the correct answer?

Thinking Critically

32. Five lines are drawn in the plane.

(a) What is the smallest number of points of intersection of the five lines?

(b) What is the largest number of points of intersection?

(c) If m is an integer between the largest and smallest number of intersection points, can you arrange the lines to have m points of intersection? You are being asked to decide which values of m have such an arrangement and which don't.

33. Three noncollinear points determine three lines, as was shown in Figure 9.2.

(a) How many lines are determined by four points, no three of which are collinear?

(b) How many lines are determined by five points, no three of which are collinear?

(c) How many lines are determined by n points? Assume that no three points are collinear.

34. The arrangement of five points shown here has two lines that each pass through three of the points:

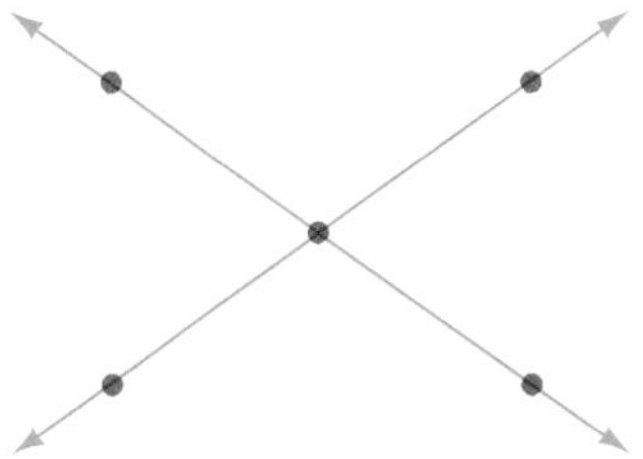

(a) Find an arrangement of six points that has four lines that each pass through three of your six points.

(b) Find an arrangement of seven points that has six lines that each pass through three of your seven points.

35. Suppose that a large triangle is drawn and a pencil is placed along an edge. What property of triangles is illustrated by the sequence A through H of slides and turns shown below? Explain in a carefully worded paragraph.

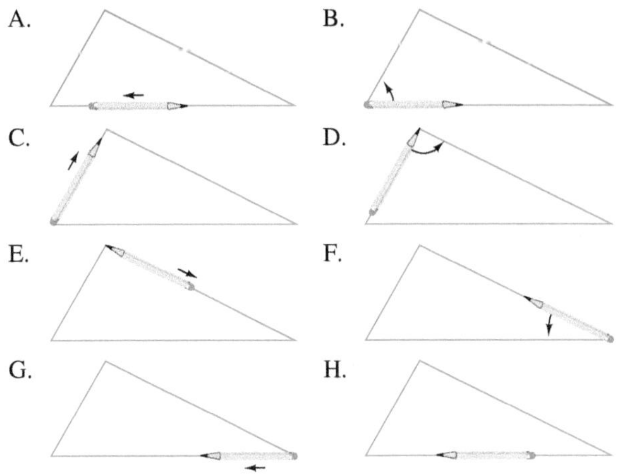

36. Use a ruler to draw a large four-sided polygon (a *quadrilateral*) such as the one shown here. Starting with a pencil laid along one side of the quadrilateral, slide the pencil to a corner,

rotate it to the next side, slide the pencil to the next corner, and so on. (See problem 35.)

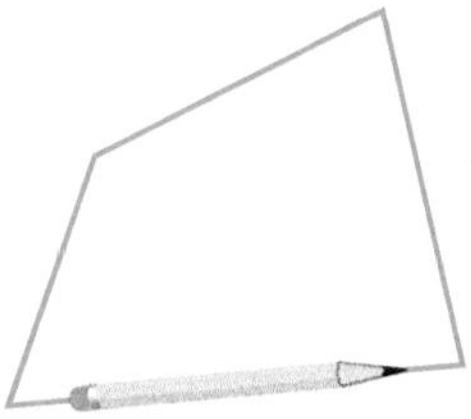

(a) What direction will the pencil point to when it returns to the initial side? What does this activity say about the sum of the measures of the angles at the four corners of a quadrilateral?

(b) Draw a diagonal across the quadrilateral to form two triangles. What does the sum of angle measures in the triangles tell you about the sum of angle measures in the quadrilateral?

(c) Cut the quadrilateral out with scissors and rip off the four corners. What angle can be covered with the four pieces?

37. The incident (incoming) and reflected (outgoing) rays of a light beam make congruent angles with a flat mirror:

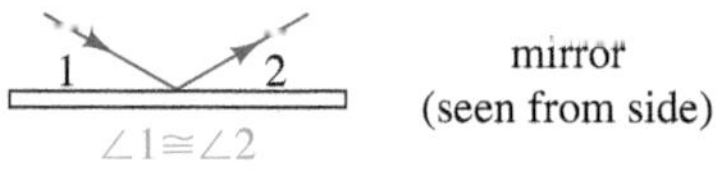

Suppose two mirrors are perpendicular to one another. Show that, after the second reflection, the outgoing ray is parallel to the incoming ray. (*Hint:* Show that ∠5 and ∠6 are supplementary. Why does this imply that the rays are parallel?)

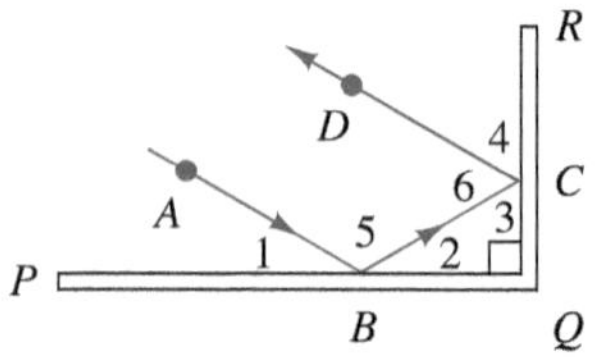

Making Connections

38. An explorer made the following trip from base camp:

First leg: north 2 miles.

Second leg: southeast 5 miles.

Third leg: west 6 miles.

Fourth leg: south 1 mile.

(a) Make a scale drawing of the trip.

(b) Show each angle the explorer turned through to go from one leg of the journey to the next.

(c) Estimate the compass heading and approximate distance the explorer needs to follow to return most directly to base camp.

39. A plumb bob (a small weight on a string) suspended from the center of a protractor can be used to measure the angle of elevation of a treetop. If the string crosses the protractor's scale at the angle marked *P*, what is the measure of the angle of elevation?

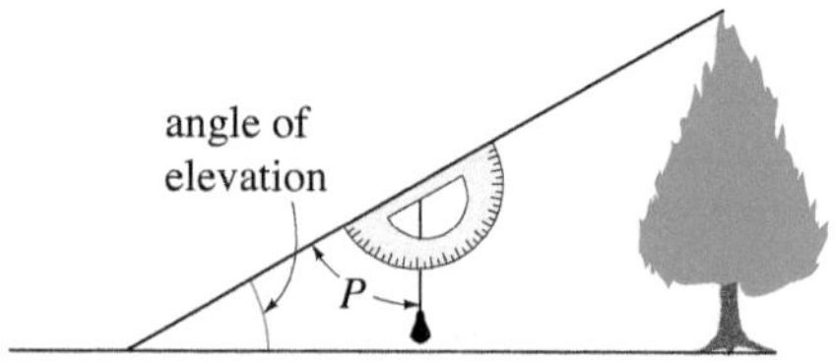

40. **(a)** How many degrees does Earth turn in one hour?

(b) How many degrees does Earth turn in one minute?

(c) On a clear night with a full moon, it can be observed that Earth's rotation makes it appear that the moon moves a distance equal to its own diameter in two minutes of time. What angle does a diameter of the moon make as seen from Earth?

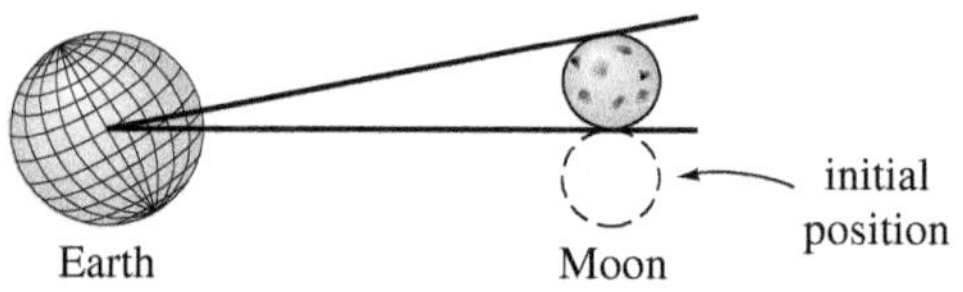

State Assessments

41. (Massachusetts, Grade 4)
To answer parts (a) through (d), connect dots on the dot patterns shown here. Always connect the dots to draw closed shapes with STRAIGHT SIDES.

(a) On the dot pattern labeled Part *a*, draw a shape that has EXACTLY ONE right angle. DRAW A RING around the RIGHT ANGLE.

(b) On the dot pattern labeled Part *b*, draw a shape that has NO right angles.

(c) On the dot pattern labeled Part *c*, draw a shape that has at least ONE acute angle. DRAW A RING around the ACUTE ANGLE.

(d) On the dot pattern labeled Part *d*, draw a shape that has EXACTLY TWO right angles. DRAW A RING around each RIGHT ANGLE.

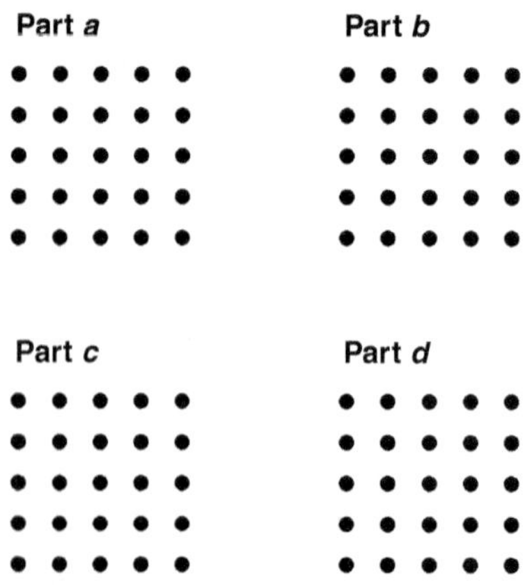

42. (Grade 7) (See Cooperative Investigation, p. 426)
Determine which point lies on the perpendicular bisector of the following line segment:

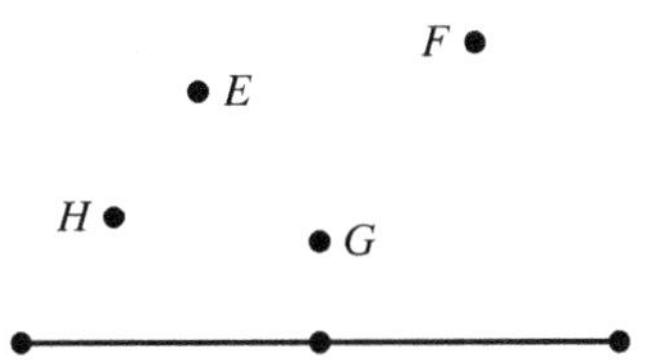

A. Point E B. Point F

C. Point G D. Point H

43. (Grade 8)
Parallel lines p and q are cut by transversal t. In addition, $m\angle 4 = m\angle 5$ and $m\angle 6 = m\angle 7$.

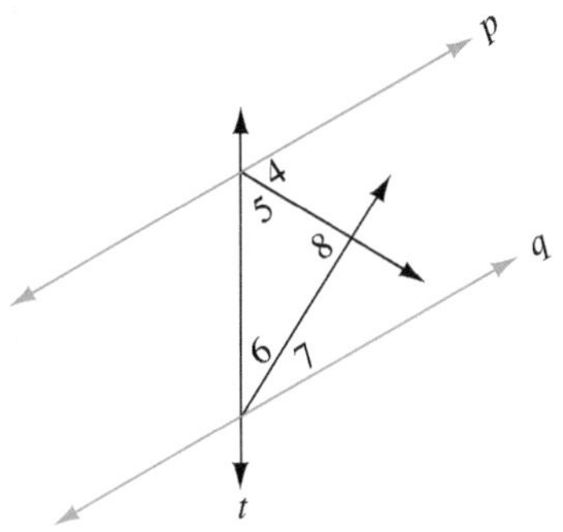

What is the measure of $\angle 8$?

A. 120° B. 90°

C. 35° D. 60°

44. (Grade 6)
For triangle GXU, what is the value of the following expression?

$$m\angle C + m\angle B + m\angle A$$

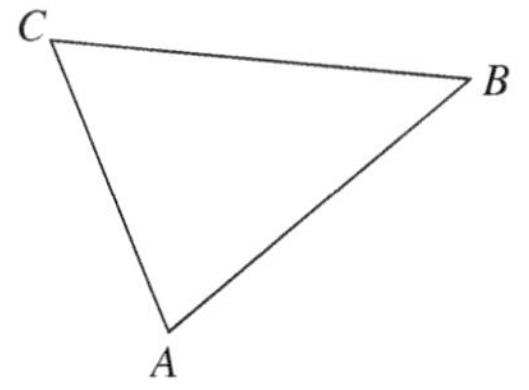

D. 90° E. 180°

F. 100° G. 360°

45. (Grade 6)
What is the measure of $\angle X$ in the right isosceles triangle shown below?

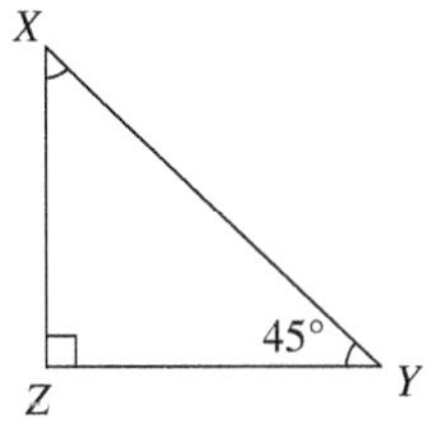

A. 90° B. 45°

C. 55° D. 180°

9.2 Curves and Polygons in the Plane

Curves and Regions

A **curve** in the plane can be described informally as a set of points that a pencil can trace without lifting until all points in the set are covered. A more precise definition is required for advanced mathematics, but this intuitive idea of curve meets our present needs. If the pencil never touches a point more than once, then the curve is **simple.** If the pencil is lifted at the same point at which it started tracing the curve, then the curve is **closed.** If the common initial and final point of a closed curve is the only point touched more than once in tracing the curve, then the curve is a **simple closed curve.** We require that a curve have both an initial and a final point, so lines, rays, and angles are *not* curves for us.

Several examples of curves are shown in Figure 9.14.

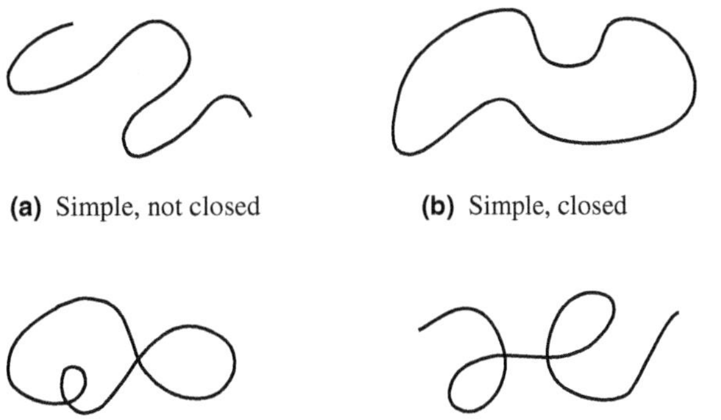

Figure 9.14
The classification of curves

(a) Simple, not closed

(b) Simple, closed

(c) Closed, not simple

(d) Not simple, not closed

Any simple closed curve partitions the points of the plane into three disjoint pieces: the curve itself, the **interior,** and the **exterior,** as shown in Figure 9.15. This property of a simple closed curve may seem obvious, but in fact it is an important theorem of mathematics.

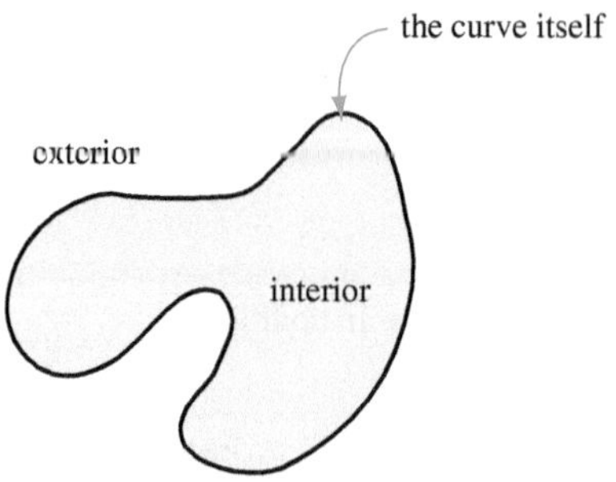

Figure 9.15
A simple closed curve and its interior and exterior

THEOREM Jordan Curve Theorem
A simple closed plane curve partitions the plane into three disjoint subsets: the curve itself, the interior of the curve, and the exterior of the curve.

The French mathematician Camille Jordan (1838–1922) was the first to recognize that such an "obvious" result needed proof. To see why the theorem is difficult to prove (even Jordan's own proof was incorrect!), try to determine whether the points G and H are inside or outside the very crinkly, but still simple, closed curve shown in Figure 9.16.

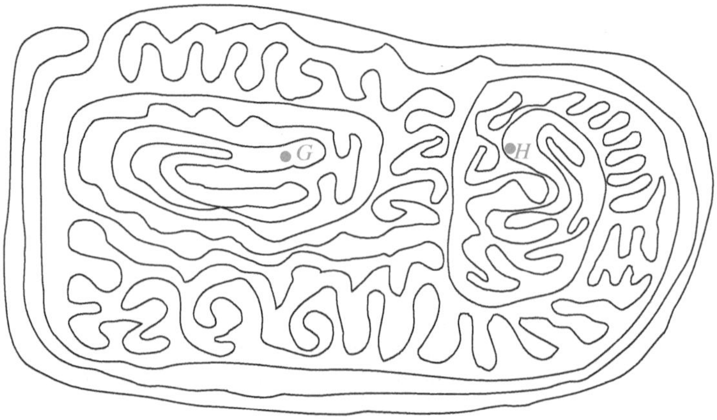

Figure 9.16
Is G in the interior or exterior of this simple closed curve? What about H?

EXAMPLE 9.6 Determining the Interior Points of a Simple Closed Curve in the Plane

Devise a method to determine whether a given point is in the interior or the exterior of a given simple closed curve.

Solution

Think of the curve as a fence. If we jump over the fence, we go from the interior to the exterior of the curve or vice versa. Now draw any ray from the given point.

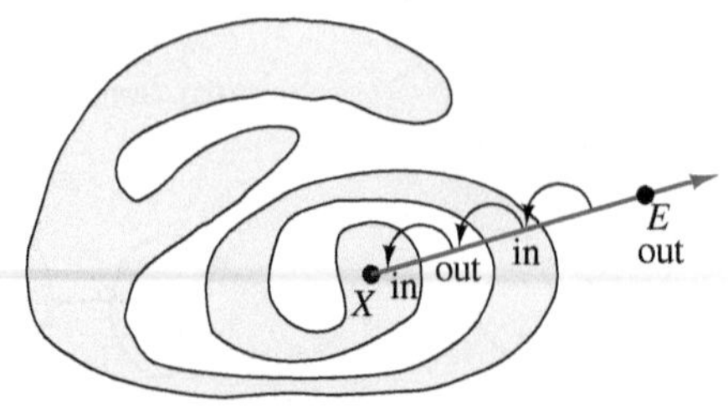

X is an interior point

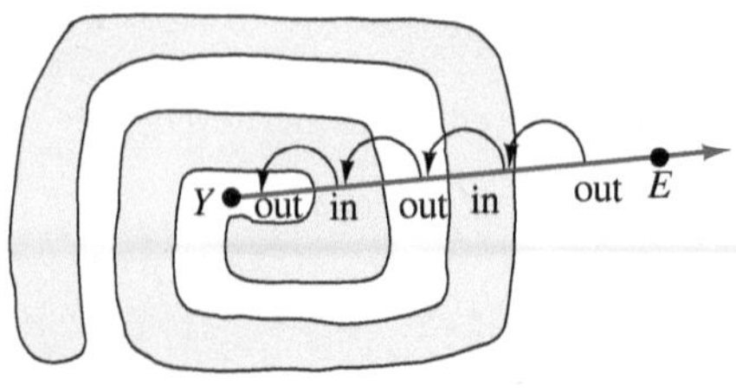

Y is an exterior point

Start at an exterior point E on the ray, and see how many times the fence is crossed as you move to the endpoint of the ray. If the curve is crossed an odd number of times to reach the point X from the exterior point E, then the given point X is an interior point. An even number of crossings from E to reach a point Y means that Y is an exterior point. It should now be simple to verify that G is exterior and H is interior to the curve in Figure 9.16.

We defined informally the "interior" and "exterior" of a simple closed curve. (See Figure 9.15.) We now use those terms to define the idea of **region,** as it will appear often in the chapters that follow.

DEFINITION Regions Defined by a Simple Closed Curve
The interior and exterior of a simple closed curve are called the **regions defined by the curve.**

Continuing informally, the complement of some system of lines, rays, and curves will be composed of one or more regions. For example, a line partitions the plane into two regions called **half planes.** An angle, if not zero or straight, partitions the plane into two regions called the **interior and exterior of the angle,** as in Figure 9.5 of the previous section. Since we are taking an informal approach to geometry, we will use the word **region** to mean one of the areas defined by a curve and for the area occurring in the examples of the sentences of this paragraph. While it is intuitively clear to children and adults what regions are, to define a region rigorously requires mathematics well beyond the level of this text.

EXAMPLE 9.7

Counting Regions in the Plane

Count the number of regions into which the plane is partitioned by the following shapes:

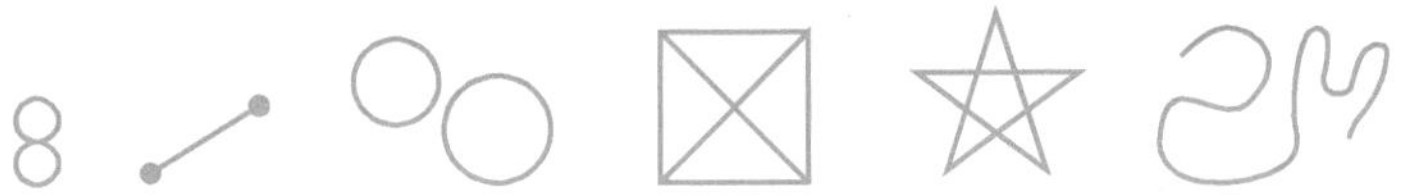

(a) a figure 8 (b) a segment
(c) two nonintersecting circles (d) a square and its two diagonals
(e) a pentagram (f) any simple nonclosed curve

Solution

(a) 3 (b) 1 (c) 3
(d) 5 (e) 7 (f) 1

COOPERATIVE INVESTIGATION

Regions Interior to Two Simple Curves and External to Both of the Curves

Work in pairs, with one partner using a black pencil and the other partner a red pencil. Each partner draws a *simple* closed curve on his or her own clean sheet of paper. The papers are exchanged, and a second simple closed curve is drawn. The newly drawn curve should cross the previously drawn curve several times. At each intersection point, the newly drawn curve must cross from the inside to the outside of the previously drawn curve or vice versa. (It cannot just touch and turn away.) Each partner then classifies and marks each region of the plane by its type:

✓: Interior to both curves
×: Exterior to both curves
□: Interior to black curve and exterior to red curve
■: Interior to red curve and exterior to black curve

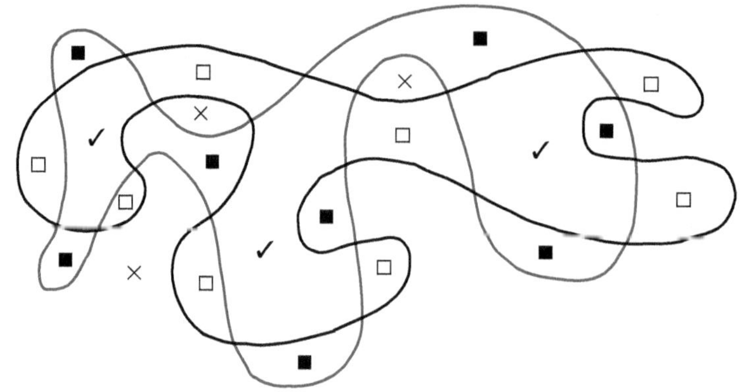

Count the number of regions of each of the four types. Compare your counts with the values obtained by other partners, and draw several fresh examples to provide additional data. What connections do you see among the numbers of regions of each type in any crossing pattern created by two simple closed curves? Make a conjecture, and create additional examples to investigate your conjecture.

Convex Curves and Figures

The interior of an angle has the property that the segment between any two interior points does not leave the interior of the angle. This means that the interior of an angle is a convex figure according to the following definition:

DEFINITION Convex and Concave Figures
A figure is **convex** if and only if it contains the segment $\overline{PQ}$ for each pair of points P and Q contained in the figure. A figure that is not convex is called **concave** or, occasionally, **nonconvex.**

Several convex and concave shapes are shown in Figure 9.17. To show that a figure is concave, it is enough to find two points P and Q within the figure whose corresponding line segment $\overline{PQ}$ contains at least one point not in the figure.

Figure 9.17
Convex and concave plane figures

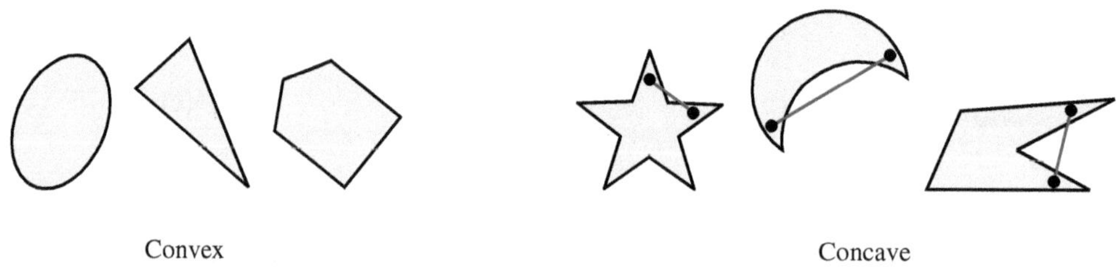

Polygonal Curves and Polygons

A curve that consists of a union of finitely many line segments is called a **polygonal curve.** The endpoints of the segments are called **vertices,** and the segments themselves are the **sides** or **edges** of the polygonal curve.

DEFINITION Polygon, Polygonal Region, Convex Polygon
A **polygon** is a simple closed polygonal curve. The interior of a polygon is called a **polygonal region.** A **convex polygon** is a polygon whose interior is convex. Figure 9.18 illustrates examples of these terms.

Figure 9.18
The classification of polygonal curves

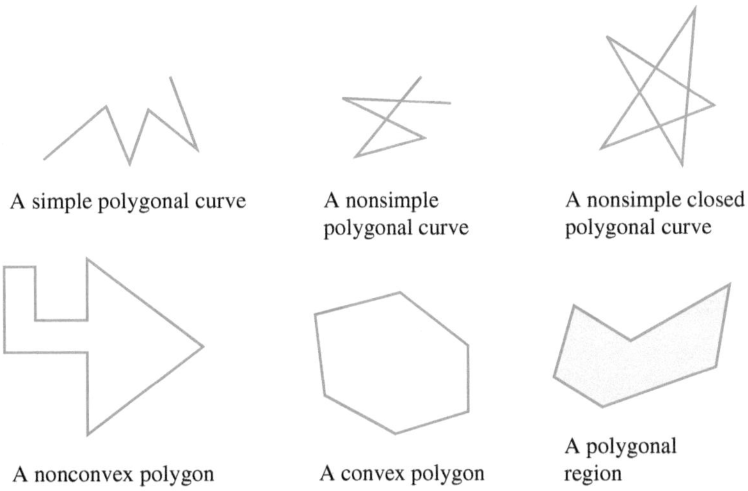

Polygons are named according to the number of sides or vertices they have. For example, a polygon of 17 sides is sometimes called a *heptadecagon* (*hepta* = seven, *deca* = 10). With more directness, it can also be called a 17-gon. The common names of polygons are shown in Table 9.1.

TABLE 9.1 NAMES OF POLYGONS

Polygon	Number of Sides	Example
Triangle	3	
Quadrilateral	4	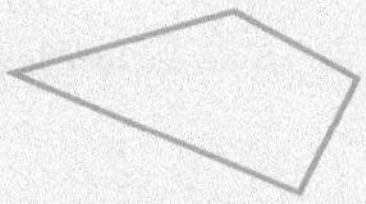
Pentagon	5	
Hexagon	6	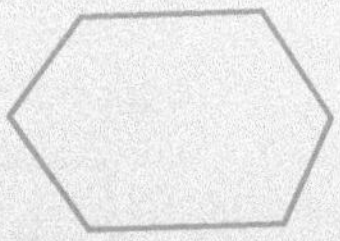
Heptagon	7	
Octagon	8	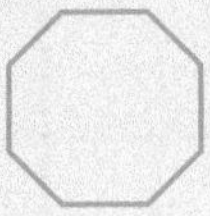
Nonagon (or enneagon)	9	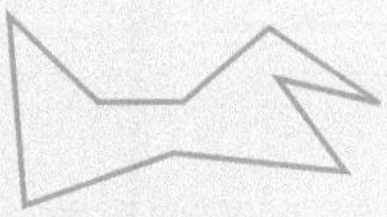
Decagon	10	
n-gon	*n*	

The rays along two sides with a common vertex determine an **angle of the polygon.** For a convex polygon, the interiors of these angles include the interior of the polygon. The angles are also called **interior angles,** as shown in Figure 9.19. An angle formed by replacing one of these rays with its opposite ray is an **exterior angle** of the polygon. The two exterior angles at a vertex are congruent by the vertical-angles theorem. The interior angle and either of its adjacent exterior angles are supplementary.

Figure 9.19
Angles in a convex polygon

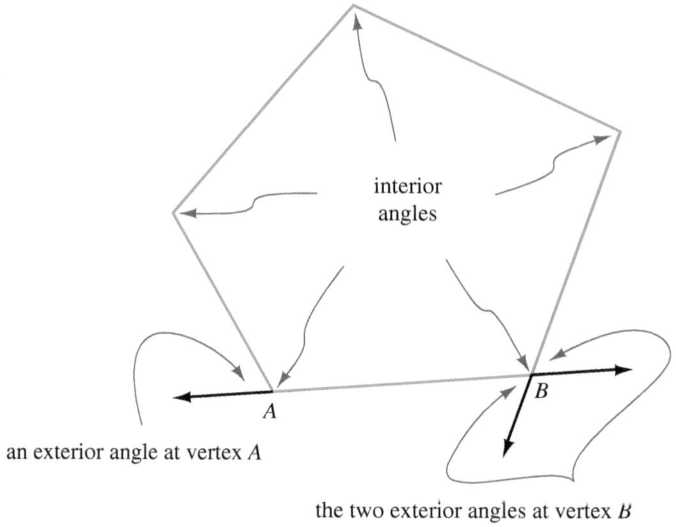

In the following theorem, we consider the interior angle and *one* of its supplementary exterior angles at each vertex of a convex polygon:

THEOREM Sums of the Angle Measures in a Convex Polygon

(a) The sum of the measures of the exterior angles of a convex polygon is 360°.

(b) The sum of the measures of the interior angles of a convex n-gon is $(n - 2)180°$.

PROOF

(a) We will picture the case for $n = 5$. Imagine a walk completely around a polygon. At each vertex, we must turn through an exterior angle. At the conclusion of the walk we are heading in the same direction as we began, so our total turn is through 360°. If $\angle 1', \angle 2', \ldots, \angle n'$ denote the exterior angles, then our walk around the polygon shows that $m(\angle 1') + m(\angle 2') + \cdots + m(\angle n') = 360°$.

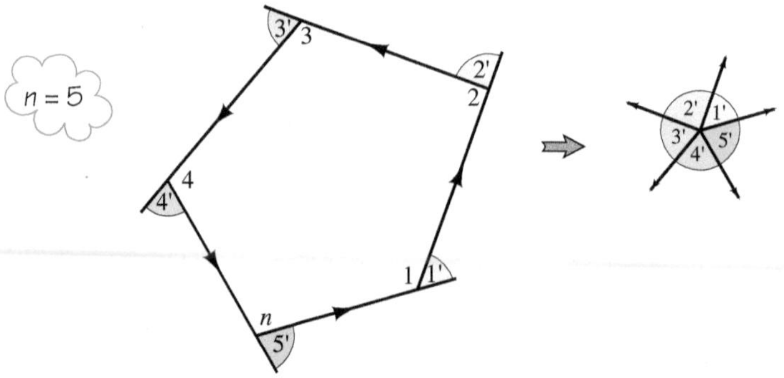

(b) Since an interior angle and an exterior angle at a vertex are supplementary, we have the equations $m(\angle 1) = 180° - m(\angle 1')$, $m(\angle 2) = 180° - m(\angle 2')$, . . . , $m(\angle n) = 180° - m(\angle n')$. Adding these n equations gives us, by part (a),

$$\begin{aligned} m(\angle 1) + m(\angle 2) + \cdots + m(\angle n) &= n \cdot 180° - [m(\angle 1') + m(\angle 2') + \cdots + m(\angle n')] \\ &= n \cdot 180° - 360° = (n-2) \cdot 180°, \end{aligned}$$

where we need the result of part (a) in the second equality.

When we look at the case in which $n = 3$ in part (b) of the theorem, its statement becomes that the sum of the interior angles of a 3-gon is $(n - 2)180 = 180°$. Since a 3-gon is a triangle, the theorem generalizes the "angle sum" theorem from triangles to all convex polygons.

INTO THE CLASSROOM

Teri M. Rodriguez Discusses Geometry

I like to take my third graders for a walk around the school to look for examples of polygons. The students really enjoy being math explorers in a quest to look for shapes. They not only enjoy leaving the classroom environment, but they also look forward to exploring math in a tangible way outdoors. They are especially thrilled to find so many shapes on the playground and in many places that they see every day. Now they see the world through their "math eyes." I particularly enjoy seeing the faces of the English language learners as they connect new mathematical vocabulary like *cylinder, sphere, rectangular prism, cone,* and *cube* to the shapes found in the school environment.

Teri M. Rodriguez, Third Grade Teacher, Fairview Elementary

SOURCE: Reprinted with permission.

EXAMPLE 9.8

Finding the Angles in a Pentagonal Arch

Find the measures $3x$, $8x$, y, and z of the interior and exterior angles of the pentagon below.

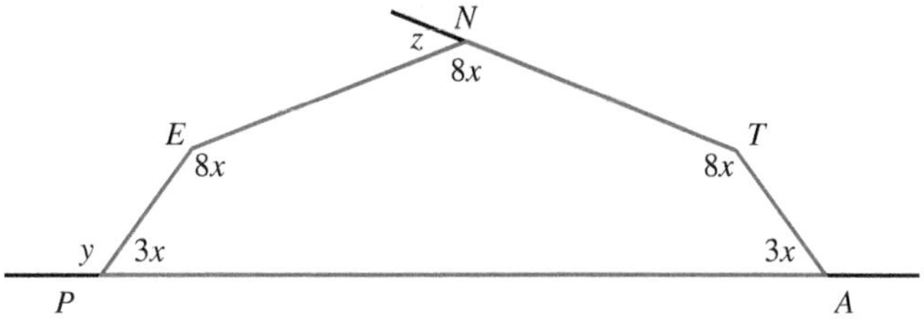

Solution

By the theorem just proved, we know that the sum of the measures of the five interior angles is

$$3x + 8x + 8x + 8x + 3x = (5 - 2)180°.$$

That is, $30x = 3 \cdot 180°$, so that $x = (3 \cdot 180°/30) = 18°$.

The interior angles at P and A therefore measure $3 \cdot 18° = 54°$ and at E, N, and T measure $8 \cdot 18° = 144°$. The measures of the exterior angles are $y = 180° - 54° = 126°$ and $z = 180° - 144° = 36°$.

In a concave (that is, not convex) polygon, some of the interior angles are reflex angles with measures larger than 180°. Nevertheless, it can be proven (see problems 37 and 38 in Problem Set 9.2) that the sum of the measures of the n interior angles is still given by $(n - 2)180°$.

THEOREM Sum of Interior Angle Measures of a General Polygon
The sum of the measures of the interior angles of any n-gon is $(n - 2)180°$.

An example of a concave heptagon is shown in Figure 9.20.

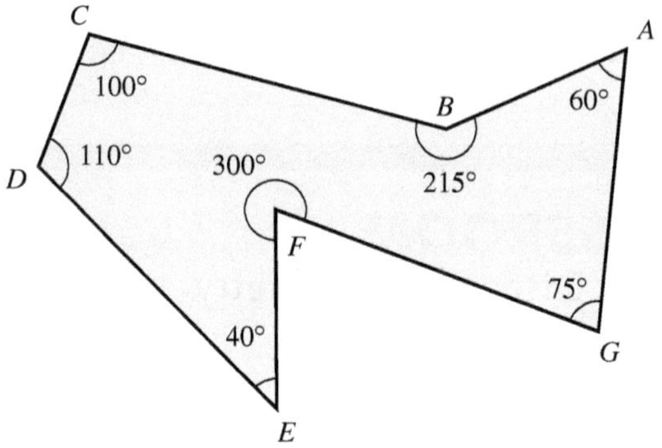

Figure 9.20
The sum of the interior angle measures of the 7-gon is $60° + 215° + 100° + 110° + 40° + 300° + 75° = 900° = (7 - 2) \cdot 180°$

EXAMPLE 9.9 Measuring the Angles in a Five-Pointed Star

The reflex angles at each of the five "inward" points of the star shown have three times the measure of the angles of the "outward" points. What is the measure of the angle at each point of the stars pictured here?

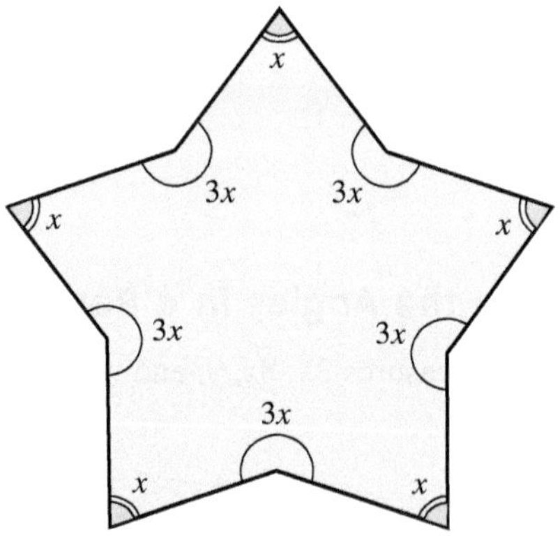

Solution

The sum of the measures of all 10 interior angles of the star is $5x + 5 \cdot (3x)$, or $20x$. Since the star is a decagon (or 10-gon), the sum must equal $(10 - 2) \cdot 180° = 1440°$. This gives us the equation $20x = 1440°$. Solving for x shows that $x = 1440° \div 20 = 72°$. Thus, each acute interior angle measures 72°, and each of the reflex angles at the inward points measures $3x = 216°$.

A walk around any closed curve, simple or nonsimple, that returns to the starting point and to the same orientation as the walk began must have turned through some integer multiple of 360°. It is customary to measure turns to the left (counterclockwise) as positive and turns to the right (clockwise) as negative.

THEOREM The Total-Turn Theorem

The total turn around any closed curve is an integral multiple of 360°.

To help determine the total turn, draw a point S at an arbitrary point along the curve and lay a pencil at that point, with the point of the pencil oriented in the direction of travel. Now trace the curve with the pencil and count the net number of rotations the pencil has made when it returns to its initial position at point S. For a polygonal curve, the pencil turns only when it reaches a vertex of the curve.

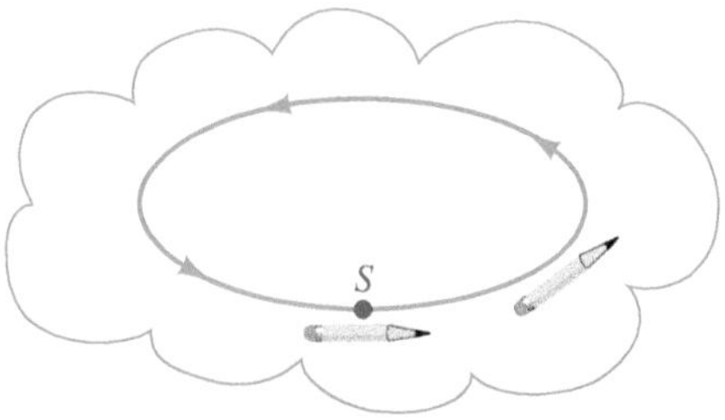

EXAMPLE 9.10 Finding Total Turns

Find the total turn for each of the following closed curves:

(a)

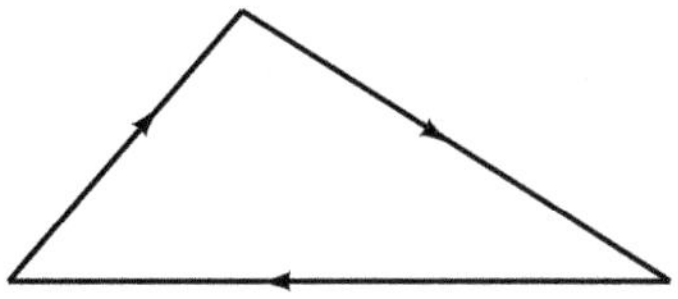

(b)

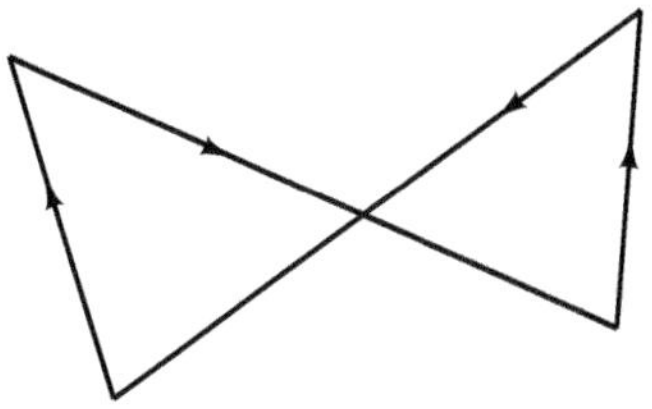

(c)

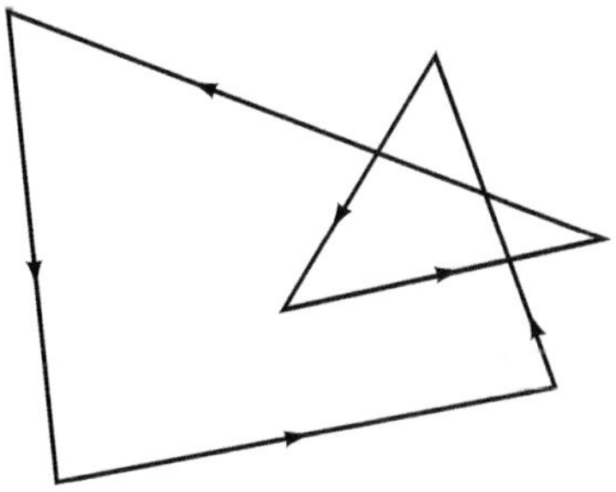

(d)

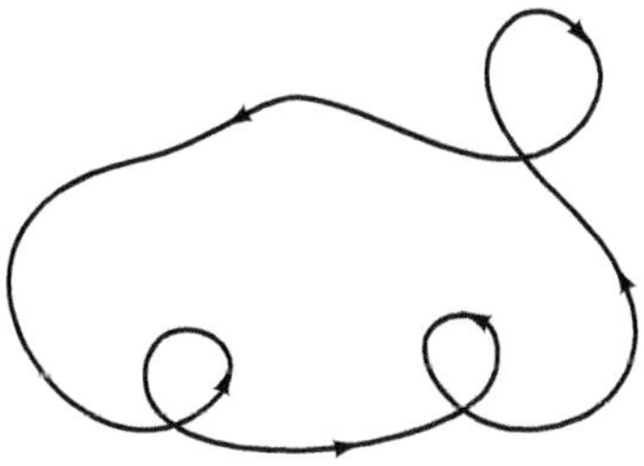

Solution

(a) −360°

(b) 0°

(c) 720°

(d) 720°

Triangles

Triangles are classified by the measures of their angles or sides, as shown in Table 9.2. The classifications of both triangles and quadrilaterals are very important, as they are included in statewide exams.

TABLE 9.2 THE CLASSIFICATION OF TRIANGLES

Classification by Angle Measure A triangle is	
acute if all three interior angles are acute;	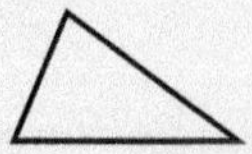
right if one angle is a right angle;	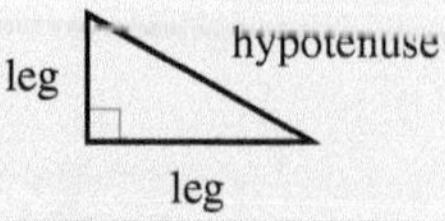
obtuse if an interior angle is obtuse.	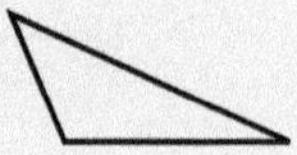
Classification by Side Lengths A triangle is	
scalene if no two sides have the same length;	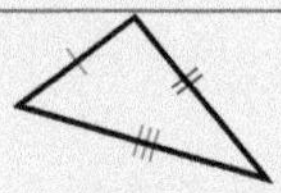
isosceles if at least two sides have the same length (*Note:* Some texts define an isosceles triangle to have *exactly* two, but not three, sides of the same length);	
equilateral if all three sides have the same length.	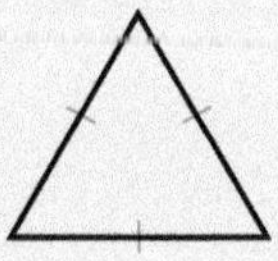

EXAMPLE 9.11

Classifying Triangles

In the figure shown, there are a number of triangles with vertices at A, B, C, D, E, and F. Classify the triangles according to Table 9.2. Use the corner of an index card to check for right angles, and use a ruler or mark on the edge of an index card to check for congruent sides.

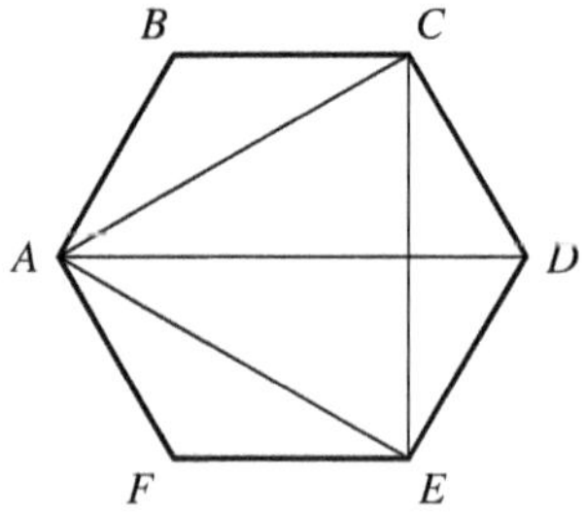

Solution

Acute: ΔACE; Right: ΔACD and ΔAED; Obtuse: ΔABC, ΔCDE, and ΔAFE; Scalene: ΔACD and ΔAED; Isosceles: ΔABC, ΔCDE, ΔAFE, and ΔACE; Equilateral: ΔACE.

Quadrilaterals

The four-sided polygons are classified as shown in Table 9.3. This classification allows a parallelogram to be described as a trapezoid; similarly, a square is a rectangle and a rectangle is a parallelogram.

The classification hierarchy is summarized by a Venn diagram. Notice that the squares are the intersection of the rhombus and rectangle loops. Arranging figures in classes that are subsets of one

TABLE 9.3 THE CLASSIFICATION OF QUADRILATERALS

a. A **kite** is a quadrilateral with two distinct pairs of congruent adjacent sides. A kite can be either convex or concave.	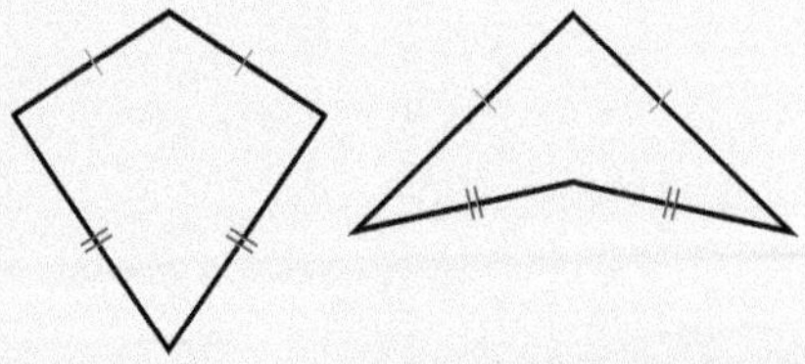
b. A **trapezoid** is a quadrilateral with at least one pair of parallel sides. (*Note:* Some dictionaries and texts require a trapezoid to have *exactly* one pair of parallel sides.)	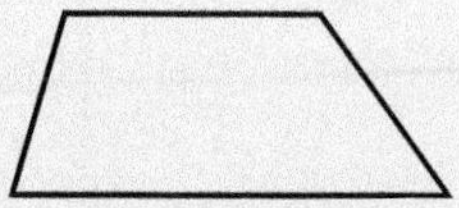
c. An **isosceles trapezoid** is a trapezoid with a pair of congruent angles along one of the parallel sides.	
d. A **parallelogram** is a quadrilateral in which each pair of opposite sides is parallel.	
e. A **rhombus** is a parallelogram with all of its sides the same length.	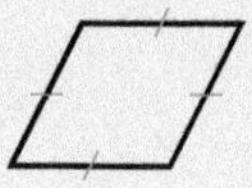
f. A **rectangle** is a parallelogram with a right angle.	
g. A **square** is a rectangle with all sides of equal length.	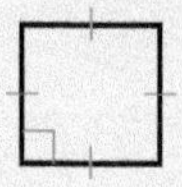

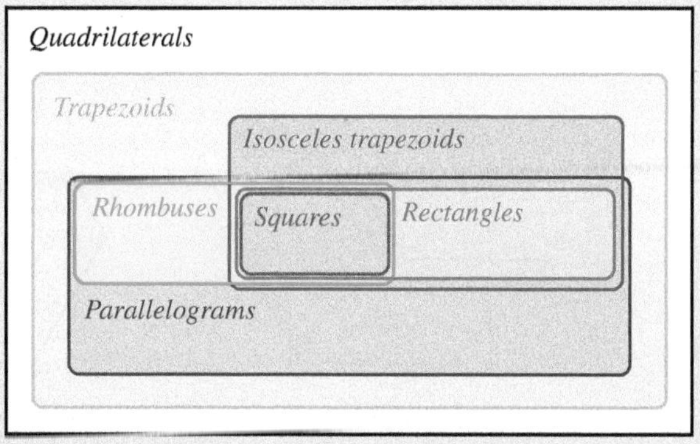

another can be very useful. For example, suppose we wish to show that the points A, B, C, and D are the vertices of a square. Step 1 may be to show that one pair of sides is parallel, telling us that $ABCD$ is a trapezoid. Step 2 may show that the remaining pair of opposite sides is parallel, and now we know that $ABCD$ is a parallelogram. If step 3 shows that $ABCD$ is a kite and step 4 shows that $\angle A$ is a right angle, then we correctly deduce that $ABCD$ is a square.

EXAMPLE 9.12 Exploring Quadrilaterals

Draw a convex quadrilateral $ABCD$. Use a compass to erect equilateral triangles on each side, alternately pointing to the interior and exterior of the quadrilateral. Two such triangles are shown here, determining points P and Q. The equilateral triangles on the remaining sides determine points R and S.

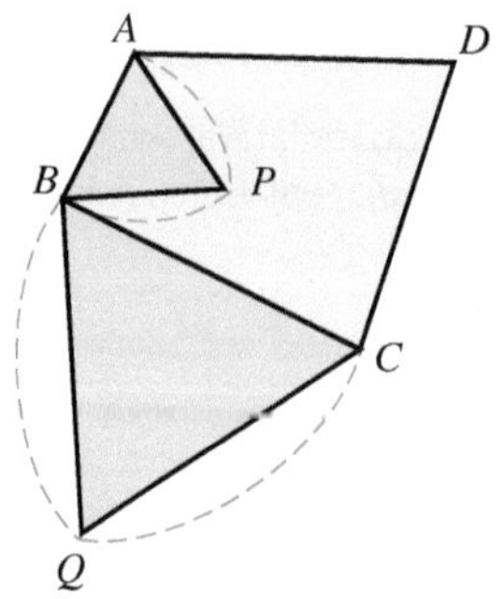

What can you say about quadrilateral $PQRS$? Support your conjecture by drawing another quadrilateral and its system of equilateral triangles. Use a ruler and protractor to measure the lengths of sides and the measure of the angles of $PQRS$.

Solution You should discover that $PQRS$ is a parallelogram; that is, each pair of opposite sides is parallel.

Regular Polygons

Polygons that are regular or that exhibit some degrees of symmetry are pleasing to the eye and have special names.

> **DEFINITION Equilateral, Equiangular, and Regular Polygons**
> A polygon with all of its sides congruent to one another is **equilateral** (that is, "equal sided"). A convex polygon whose interior angles are all congruent is **equiangular** (that is, "equal angled"). A **regular** polygon is a convex polygon that is both equilateral and equiangular.

To construct an equilateral triangle with side $\overline{AB}$, draw a circular arc centered at A through B and another arc centered at B through A. The arcs intersect at the point P needed to complete the equilateral triangle ABP.

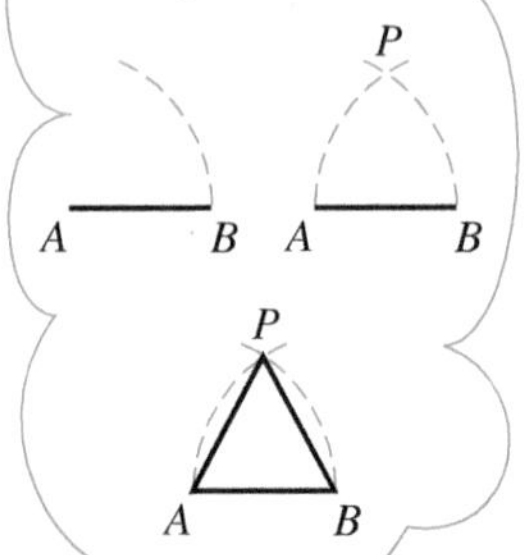

Some hexagonal examples are shown in Figure 9.21.

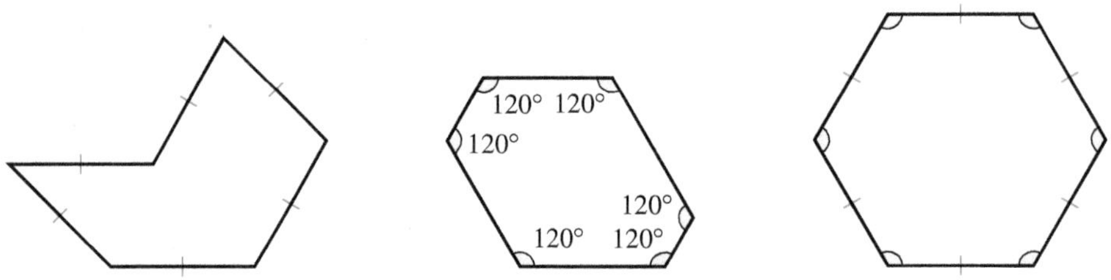

(a) An equilateral hexagon **(b)** An equiangular hexagon **(c)** A regular hexagon

Figure 9.21
Hexagons that are equilateral, equiangular, and regular

Since the six congruent interior angles in an equiangular hexagon have measures that add up to $(6 - 2) \cdot 180° = 720°$, each interior angle measures $720°/6 = 120°$. Similarly, the measures of the interior angles of an equiangular n-gon add up to $(n - 2) \cdot 180°$, so each of the n congruent interior angles measures $(n - 2) \cdot 180°/n$.

In a regular n-gon, any angle with a vertex at the center of a regular polygon and sides containing adjacent vertices of the polygon is called a **central angle** of the polygon (see Figure 9.22):

Figure 9.22
A regular n-gon where $n = 6$

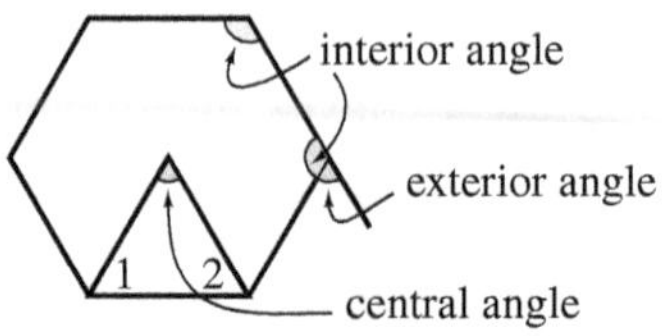

The following formulas give the measures of the exterior, interior, and central angles of a regular polygon:

THEOREM Angle Measure in a Regular *n*-gon

In a regular n-gon,

(a) each interior angle has measure $(n - 2) \cdot 180°/n$;
(b) each exterior angle has measure $360°/n$;
(c) each central angle has measure $360°/n$.

PROOF

(a) The sum of the measures of the interior angles of an n-gon is $(n - 2)(180)$ and each interior angle is the same (since the regular polygon is equiangular). Thus, the measure of each of the interior angles is $\frac{(n-2)(180)}{n}$.

(b) The measures of the external angle and internal angle must add up to 180° since they are supplementary. We first notice that, thanks to part (a), the measure of the internal angle must be

$$\frac{(n-2)(180)}{n} = \frac{180n - 360}{n} = 180 - \frac{360}{n}.$$

But then (the measure of the external angle) $+ \left(180 - \frac{360}{n}\right) = 180$. Thus, with simple algebra, the measure of the external angle is $\frac{360}{n}$, as promised.

(c) Finding the formula for the measure of the central angle of a regular polygon follows the same logic as part (b).

It is useful to observe that an interior angle and an exterior angle are supplementary. Thus, the measure of an interior angle is also given by $180° - 360°/n$.

EXAMPLE 9.13

Working with Angles in Regular *n*-Gons

(a) The Baha'i House of Worship in Wilmette, Illinois, has the unusual floor plan shown in the accompanying diagram. What are the measures of $\angle ABC$ and $\angle AOB$?

(b) Suppose an archeologist found a broken piece of pottery such as that shown on the right. If the angle measures 160° and it is assumed that the plate had the form of a regular polygon, how many sides would the complete plate have had?

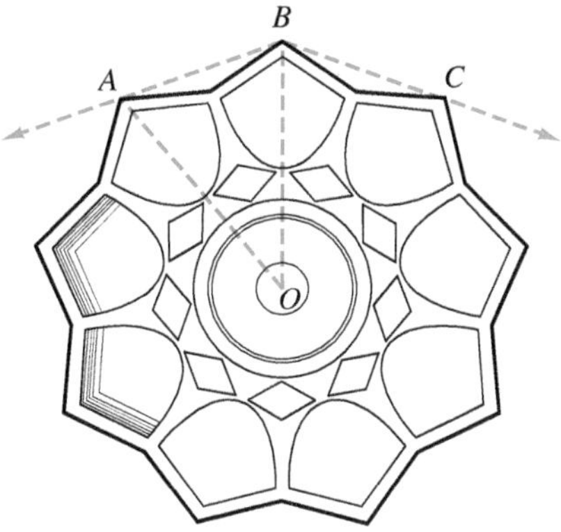

Solution

(a) $\angle ABC$ is the interior angle of a regular 9-gon and it has measure $(9 - 2) \cdot 180°/9 = 140°$. $\angle AOB$ is a central angle of a regular 9-gon, so its measure is $360°/9 = 40°$.

(b) The corresponding exterior angle measures 20°. Since $20° = 360°/n$, it follows that $n = 360°/20° = 18$. Under the assumptions stated, the plate would have had 18 sides.

COOPERATIVE INVESTIGATION

From Paper Discs to Polygons

Materials Needed

1. Two paper discs per student, each 7 to 8 inches (or 18 to 20 centimeters) in diameter. All discs used in a group should be the same size.
2. Drawing and measuring tools (pencils, protractors, rulers).

Directions

Work in groups of four. There are two sets of explorations, each using one paper disc per student.

Explorations with the First Paper Disc

1. Make a light pencil mark on the disc that you think estimates the center of the disc. To check how close you are, lightly (do not make a heavy crease) fold the disc in half. Undo your fold and lay the disc out flat. Is your mark along the diameter you have folded? Now fold the disc in half once more in a new direction and then unfold. Is your mark at the intersection of the two diameters? Clearly mark the true center of your disc and label it O.

2. Fold across a chord of the disc so that the folded arc of the circle passes through the center O, as shown. Without unfolding the first fold, make a second fold across a chord that has the same endpoint as the first chord and again with the circular arc passing through point O. Finally, fold the remaining arc of the circle. Does it also pass through point O? What kind of a triangle seems to have been created? Check out your guess by measuring the three sides and the three angles of the triangle. Compare with others in your group.

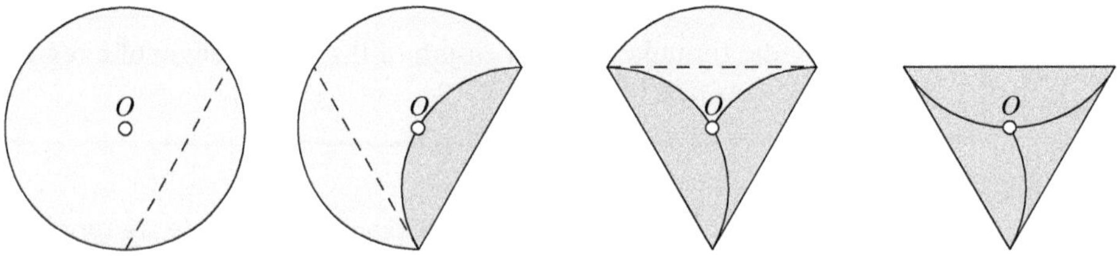

3. Find the midpoints of the sides of your triangle (how can this be done with folding?), and mark them with a pencil. Fold a vertex of your triangle to the midpoint on the opposite side. What polygon have you created? Without unfolding, fold another vertex of a triangle to the marked midpoint on the opposite side. What is the polygon now? Fold the third vertex to the midpoint marked on the opposite side. What is the name of this polygon?

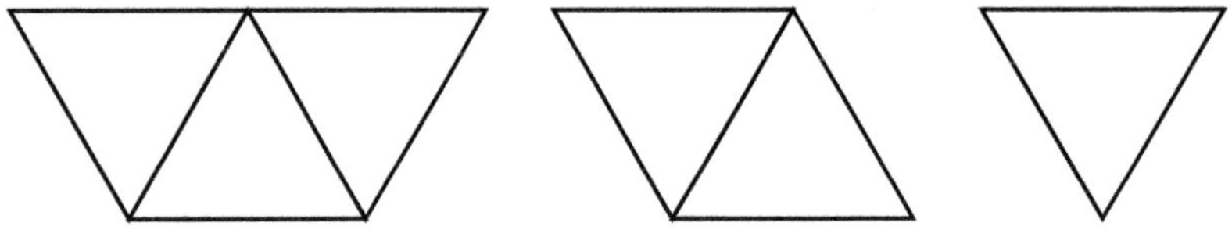

4. Unfold your disc to return to the large equilateral triangle made in Exploration step 2. Now fold each vertex to the center point O. Describe the polygon you have created.

Explorations with the Second Paper Disc

5. Mark a point well away from the center of the paper disc and label it H. Fold two arcs, sharing a common endpoint, so that both arcs pass through point H. Next, fold the remaining arc of the disc to form a triangle. Did the third arc you folded also pass through point H? Compare with others in your group.

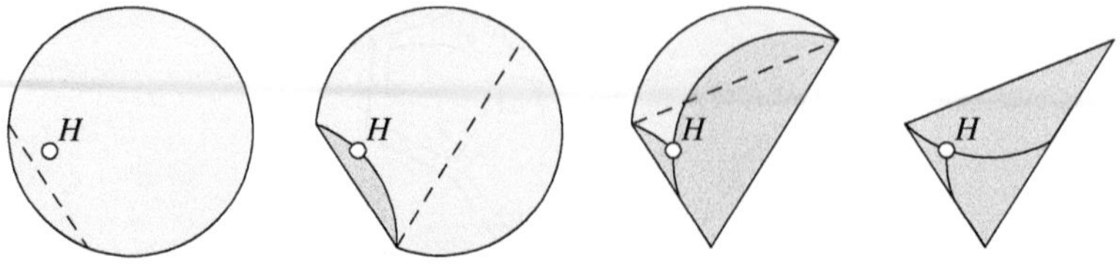

6. Unfold your triangle, and use a ruler to draw the chord that begins at a vertex of the triangle and passes through point *H*. Similarly, draw the chords through *H* from the other two vertices of the triangle. At what angle does each chord intersect the opposite side of the triangle? Use a protractor to measure the angle, and compare your answer with those of others in your group.

9.2 Problem Set

Understanding Concepts

1. If the figure shown has the property listed, place a check in the table that follows.

(a)
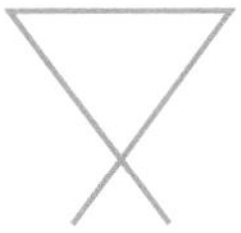
(b)
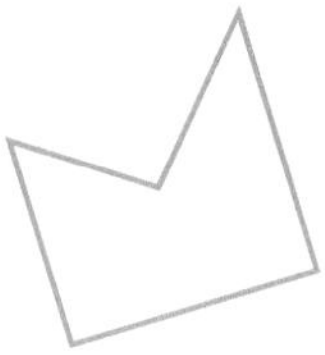
(c)

(d)

(e)
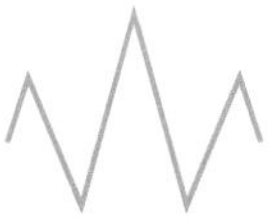
(f)
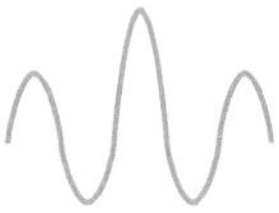

	(a)	(b)	(c)	(d)	(e)	(f)
Simple Curve						
Closed Curve						
Polygonal Curve						
Polygon						

2. If the figure shown has the property listed, place a check in the table that follows.

(a)

(b)
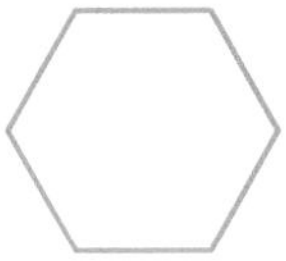
(c)
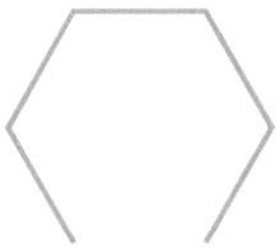
(d)

(e)

(f)
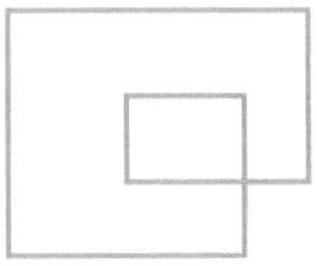

	(a)	(b)	(c)	(d)	(e)	(f)
Simple Curve						
Closed Curve						
Polygonal Curve						
Polygon						

3. Draw figures that satisfy the given description.
   (a) A nonsimple closed four-sided polygonal curve
   (b) A concave pentagon
   (c) An equiangular quadrilateral
   (d) A convex octagon

4. Classify each region as convex or concave.

(a)
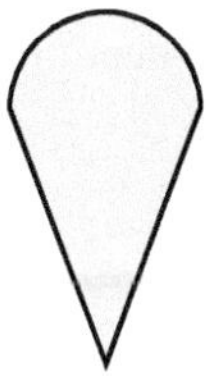
(b)

(c)
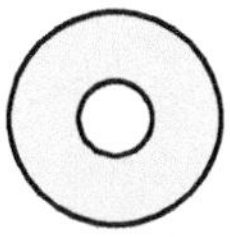
(d)
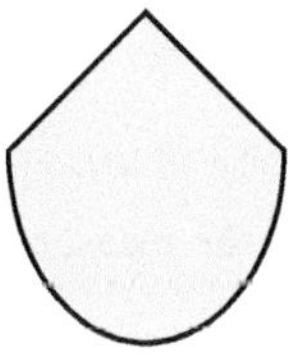
(e)
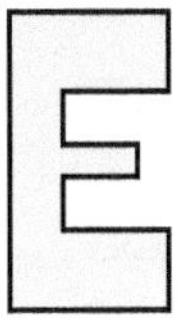

5. Imagine stretching a rubber band tightly around each figure shown. Shade the region within the band with a colored pencil. Is the shaded region always convex?

(a)
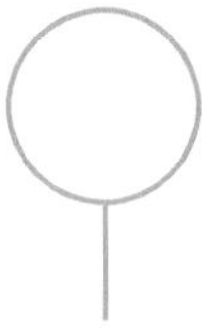
(b)

6. Follow problem 5. Is the shaded region always convex?

(a) 

(b) 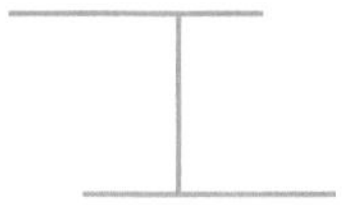

7. How many different regions in the plane are determined by these figures?

(a) 

(b) 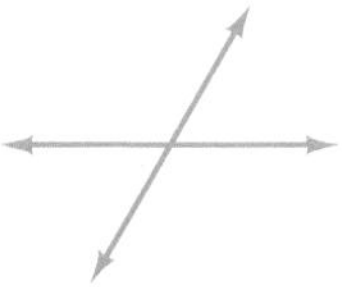

8. How many different regions in the plane are determined by these figures?

(a) 

(b) 

9. Determine the measures of the interior angles of this polygon.

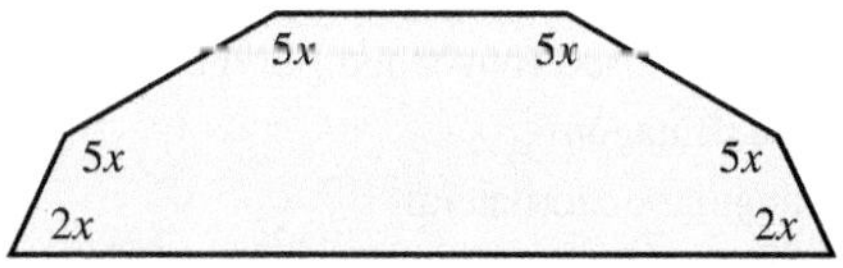

10. The following equilateral nonagon (9-gon) can form spiral tiling similar to the one you will see in Figure 11.26 on page 599. Find the measures of all of the interior angles.

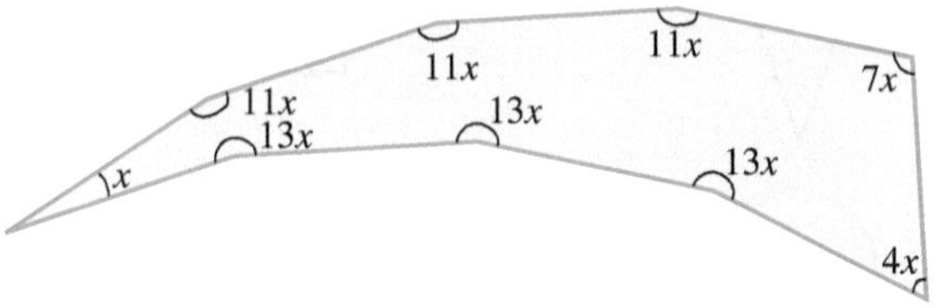

11. Calculate the measures of the angles in this concave polygon:

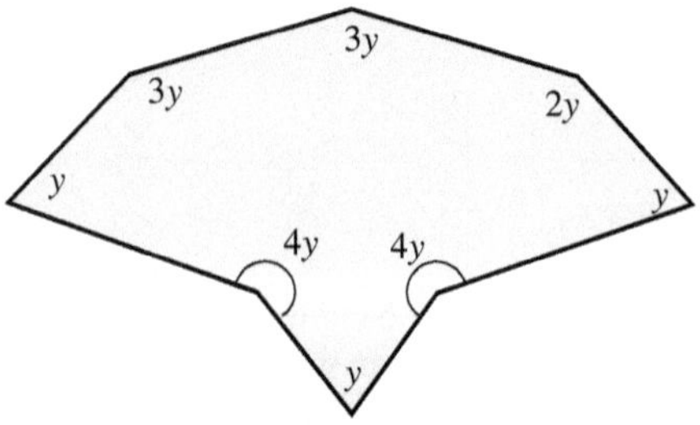

12. A **lattice polygon** is formed by a rubber band stretched over the nails of a geoboard. Find the sum of the measures of the interior angles of the following lattice polygons:

(a) 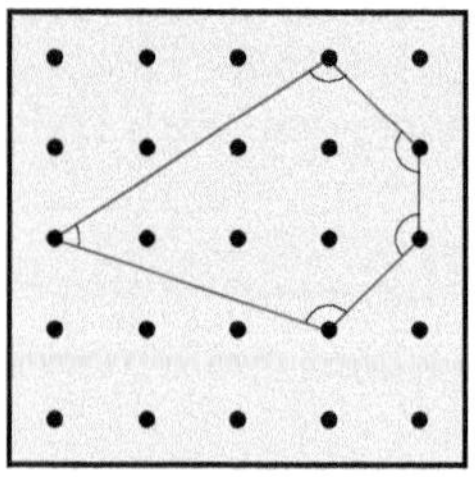

(b) 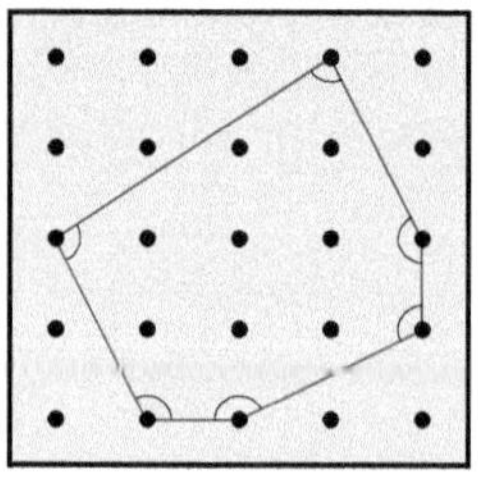

13. A lattice polygon is formed by a rubber band stretched over the nails of a geoboard. Find the sum of the measures of the interior angles of the following lattice polygons:

(a) 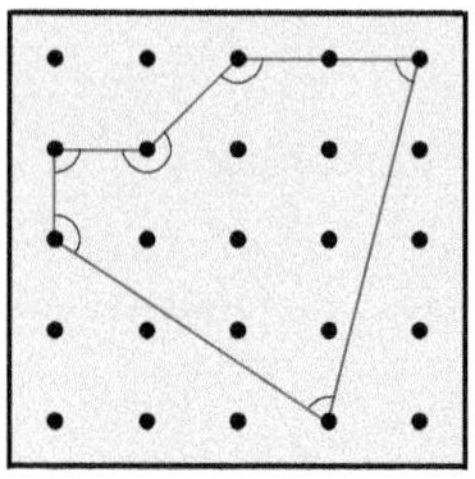

(b) 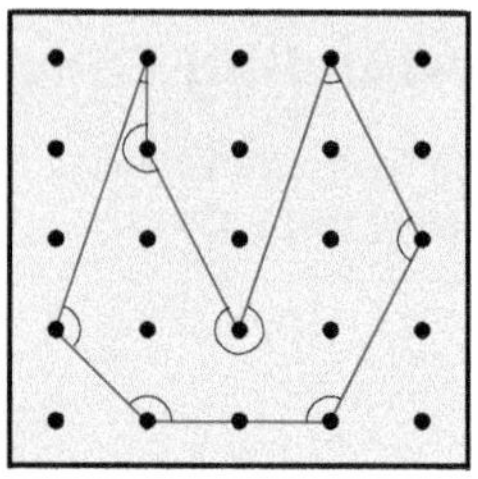

14. On squared dot paper, draw lattice polygons (see previous problem) whose interior angles have the given sum of their measures.

(a) 180° (b) 1080°

(c) 1440° (d) 1800°

15. The interior angles of an n-gon have an average measure of 175°.

(a) What is n?

(b) Suppose the polygon has flexible joints at the vertices. As the polygon is flexed to take on new shapes, what happens to the average measure of the interior angles? Explain your reasoning.

16. The interior angles of an n-gon have an average measure of 155°. How many sides can the n-gon have?

17. Suppose you walk north 10 paces, turn left through 24°, walk 10 paces, turn left through 24°, walk 10 paces, and so on.

(a) Will you return to your starting point?

(b) What is the shape of the path?

18. Fill in the missing vertices to give the type of triangle required, choosing vertices from A, B, C, D, E, F, G, and H. There may be more than one way to answer. Use a ruler and protractor to measure lengths and angles.

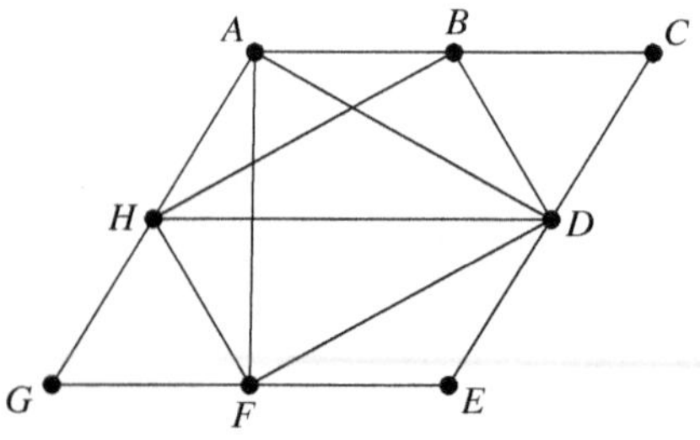

(a) Equilateral triangle D ___ ___
(b) Right triangle F ___ ___
(c) Obtuse triangle F ___ ___
(d) Isosceles triangle E ___ ___
(e) Acute triangle H ___ ___

19. Refer to the figure shown in the previous problem to give the type of quadrilateral required, filling in vertices from $A, B, C, D, E, F, G,$ and H.

(a) Rhombus A ___ ___ ___
(b) Rectangle B ___ ___ ___
(c) Isosceles trapezoid A ___ ___ ___
(d) Nonisosceles trapezoid G ___ ___ ___
(e) Kite E ___ ___ ___

20. For each regular n-gon shown, give the measures of the interior, exterior, and central angles.

(a)
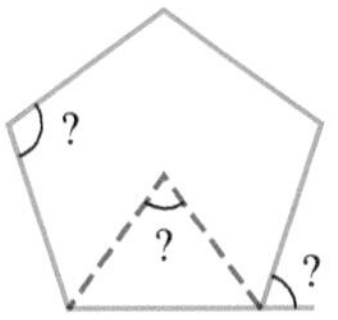

(b)
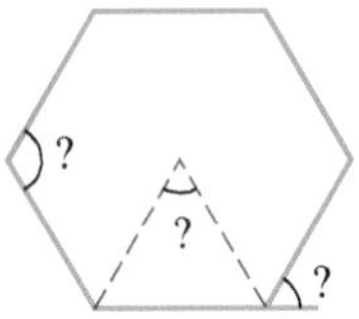

21. For each regular n-gon shown, give the measures of the interior, exterior, and central angles.

(a)
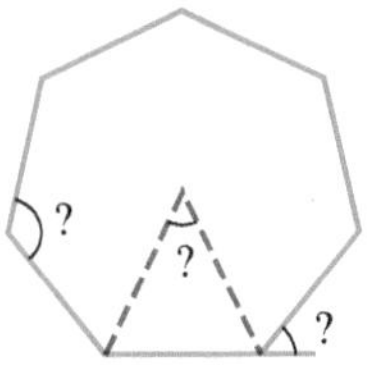

(b)
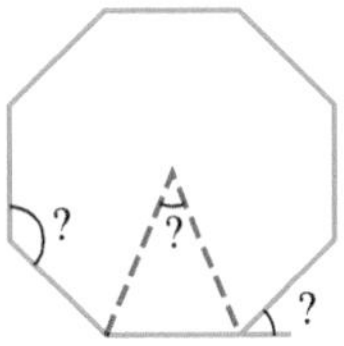

22. (a) A regular n-gon has exterior angles of measure 15°. What is n?

(b) A regular n-gon has interior angles each measuring $172\frac{1}{2}°$. What is n?

Using Algebra in Geometry

23. Show that the quadrilateral whose vertices are $A = (0, 0)$, $B = (1, \sqrt{3})$, $C = (3, \sqrt{3})$, and $D = (2, 0)$ is a rhombus, and graph $ABCD$.

24. Show that the diagonals of a rhombus are perpendicular. (*Hint:* A plan includes the problem-solving strategies from Section 8.3 called "use rigid motions" and "use Cartesian coordinates to do geometric problems.") Start by placing the rhombus $ABCD$ in such a way that $A = (0, 0)$, $B = (a, b)$, $C = (r, s)$, and $D = (c, 0)$.

(a) Graph the rhombus.
(b) Find the slopes of $\overline{AB}$ and $\overline{BD}$.
(c) What is the relationship between the two answers of part (b)?

25. Use coordinate methods to show that the sum of the squares of the lengths of the diagonals of a parallelogram equals the sum of the squares of the lengths of the sides. (*Hint*: Any parallelogram can be placed in a coordinate system as shown. Let P be the point (a, b) and R the point $(c, 0)$.)

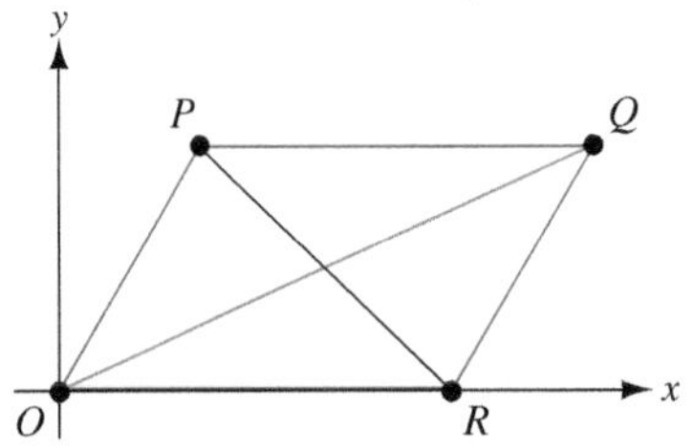

Into the Classroom

26. Pattern-Block Shapes. Pattern blocks are an especially effective manipulative to explore polygons and angles. The six shapes of pattern blocks are as follows:

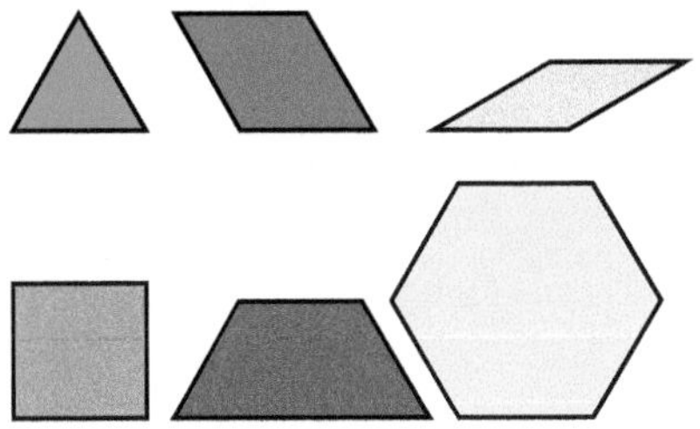

Which shapes

(a) are quadrilaterals?
(b) are parallelograms?
(c) are rhombuses?
(d) contain an obtuse angle?
(e) are regular polygons?

27. Pattern-Block Angles. The interior angles of pattern blocks (see problem 26) can be investigated by creating designs such as the ones shown here:

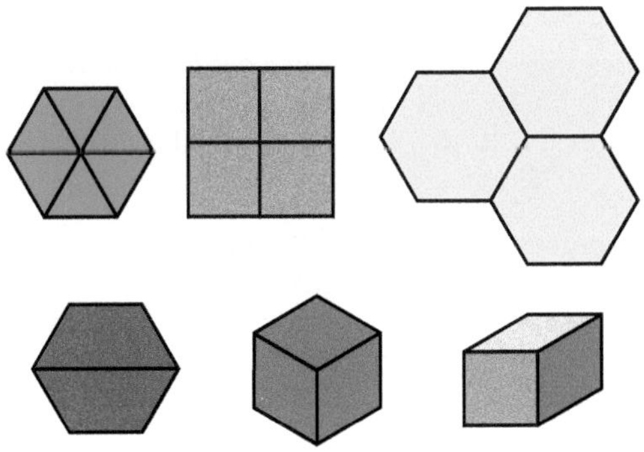

Explain how to use these designs to determine the measures of the interior angles

(a) of the equilateral triangle, square, and regular hexagon.
(b) of the red trapezoid.
(c) of the blue rhombus.
(d) of the small rhombus.

Responding to Students

28. Omar has made the following statements about rectangles, squares, rhombuses, and trapezoids:

Rectangles are *always* squares.

Rhombuses are *never* squares.

Trapezoids are *sometimes* squares.

(a) What incorrect statements did Omar make about squares?

(b) How would you explain to Omar the relationships among rectangles, squares, rhombuses, and trapezoids?

29. JaVonte was drawing some shapes during math class. He showed his drawing to his teacher, Mr. Hernandez, and Mr. Hernandez asked him to identify the shapes. JaVonte told him that triangles (a) and (b) were right triangles and that triangle (c) was an obtuse triangle. When Mr. Hernandez asked how he knew that (a) and (b) were right triangles, JaVonte explained that both triangles had a right angle in them, making them right triangles. Then he said that he knew (c) was an obtuse triangle because it was bigger than the other two right triangles, so that made it obtuse.

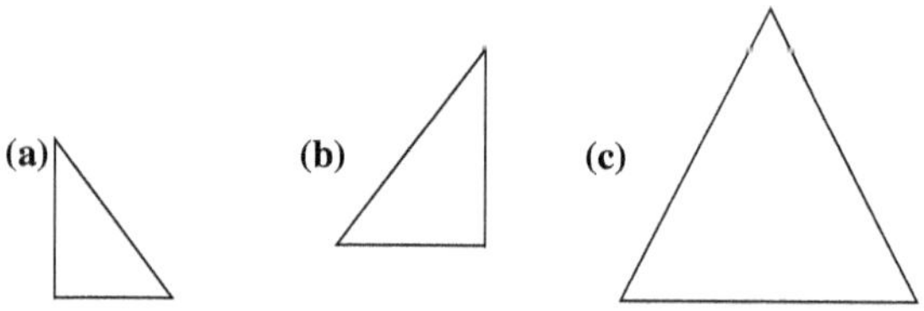

What should Mr. Hernandez say to JaVonte to correct his misunderstanding of obtuse triangles?

30. JaVonte's classmate, Ralph, from Mr. Hernandez's class in the previous problem, says that JaVonte is wrong about both (a) and (b) being right triangles. Ralph says that only (a) is a right triangle and that (b) isn't because it goes in the wrong direction. What should Mr. Hernandez say to Ralph to correct his misunderstanding?

31. Samuel knows that each side of a quadrilateral is 1 mm in length and it has four right angles. He tells you that the figure is a rectangle. Is he right or wrong, and how would you explain to him what the figure is?

32. Emile, a fourth-grade student, is asked to draw a picture of a figure with four congruent sides. His picture is a square. When asked for other pictures, he changes the length of the side, but the figure is always a square. What discussion would you have with Emile?

33. Martine believes that if two rectangles have diagonals that are the same length, then the two rectangles are the same size (that is, have the same width and length). Is Martine right, and how would you explain your answer to her?

Thinking Critically

34. A goat is tethered at the corner of an 80-foot by 30-foot rectangular barn. The rope is 50 feet long. Describe the region the goat can reach.

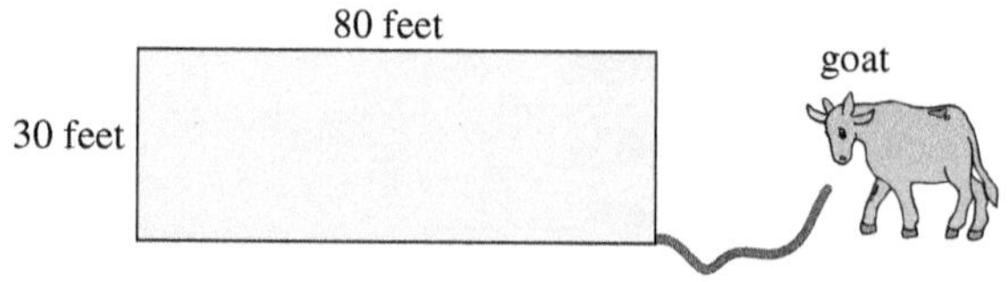

35. **(a)** A boat B is anchored at point A. Describe, in words and a sketch, the region where the boat can drift due to wind and currents.

(b) Suppose a second anchor at point C has been set, as shown in the next diagram. Describe, in words and a sketch, the region to which the boat is now confined.

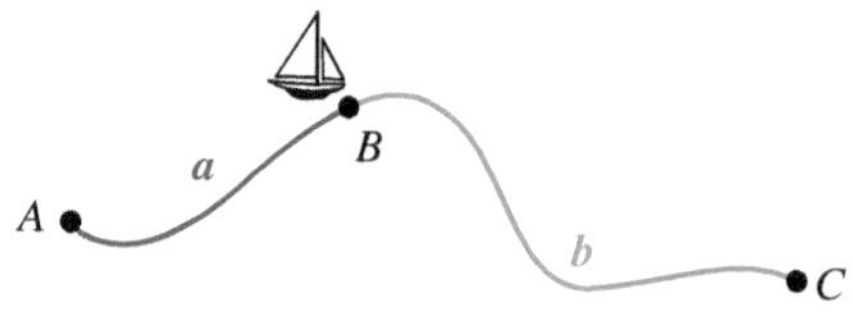

36. The following segment $\overline{AB}$ is to be completed to become a side of a triangle ABC:

Describe, in words and sketches, the set of points C so that

(a) $\triangle ABC$ is a right triangle and $\overline{AB}$ is a leg.

(b) $\triangle ABC$ is a right triangle and $\overline{AB}$ is a hypotenuse. (*Hint:* For (b), proceed experimentally, using the corner of a sheet of paper as a right angle.)

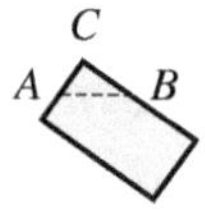

(c) $\triangle ABC$ is an acute triangle.

(d) $\triangle ABC$ is an obtuse triangle.

37. The heptagonal region shown on the left has been broken into five triangular regions by drawing four nonintersecting diagonals across the interior of the polygon, as shown on the right. In this way, we say that the polygon is triangulated by diagonals.

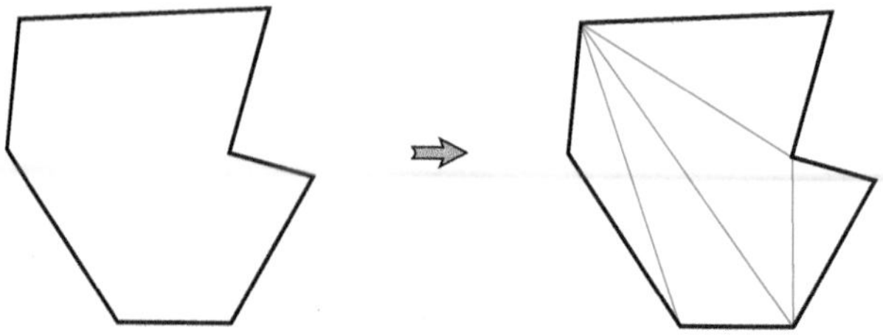

(a) Investigate how many diagonals are required to triangulate any n-gon.

(b) How many triangles are in any triangulation of an n-gon by diagonals?

(c) Explain how a triangulation by diagonals can give a new derivation of the formula $(n - 2) \cdot 180°$ for the sum of the measures of the interior angles of any n-gon.

38. The polygon shown contains a point S in its interior that can be joined to any vertex by a segment that remains inside the polygon. Drawing all such segments produces a triangulation of the interior of the polygon. (Compare with problem 37.)

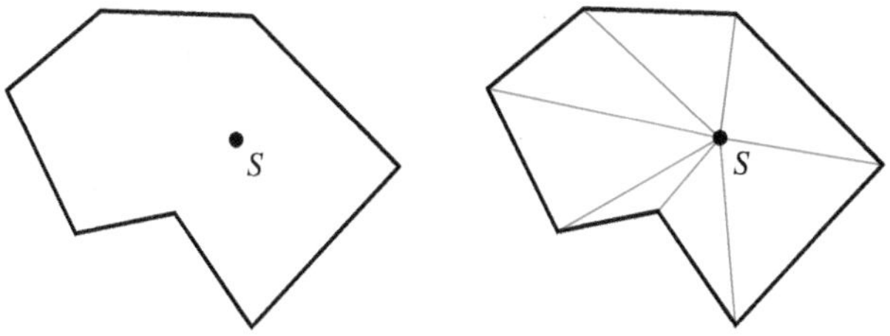

If an n-gon contains such a point S, explain how the corresponding triangulation can be used to derive the $(n - 2) \cdot 180°$ formula for the sum of interior angle measures.

39. (a) Find the sum of the angle measures in the following five-pointed star shown on the left. Explain how the total-turn theorem can be used to obtain your answer.

(b) What is the measure of the angle at each point of the pentagram shown on the right?

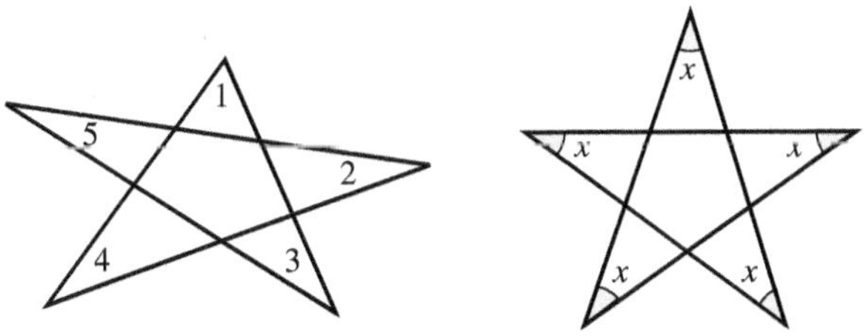

40. The wall mosaic shown here contains two nested stars, each with 16 outward points:

(a) What is the measure of each angle at the points of the tan outer star?

(b) What is the measure of each angle at three points of the blue inner star? (*Hint:* The inner star is really two 8-pointed stars.)

41. What is the largest number of regions you can form with a system of 10 circles?

42. Regions can be formed in a circle by drawing chords, no three of which are concurrent. If C chords are drawn and they intersect in I points, determine a formula for the number of pieces P (that is, the number of regions) that are formed inside the circle.

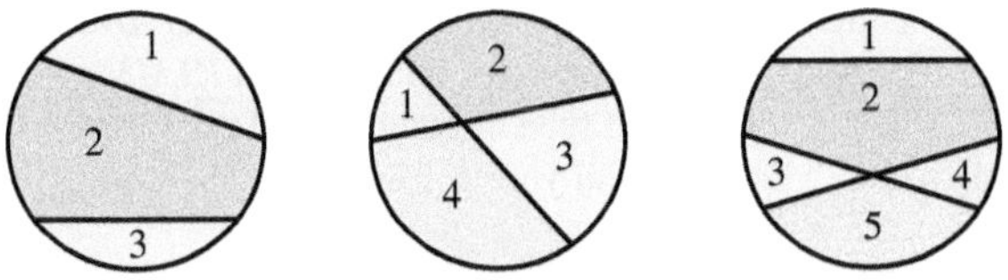

P	*C*	*I*
3	2	0
4	2	1
5	3	1

43. Let $ABCD$ be a parallelogram.

(a) Show that $\angle A$ and $\angle B$ are supplementary.

(b) Show that $\angle A \cong \angle C$ and $\angle B \cong \angle D$.

44. Let $PQRS$ be a convex quadrilateral for which $\angle P \cong \angle R$ and $\angle Q \cong \angle S$.

(a) Show that $\angle P$ and $\angle Q$ are supplementary.

(b) Show that $PQRS$ is a parallelogram.

Making Connections

45. Access to underground utility cables, water pipes, and storm drains is usually provided by circular holes covered by heavy metal circular covers. What unsafe condition would be present if a square shape were used instead of circular one?

46. The valve stems on fire hydrants are usually triangular or pentagonal. Fire trucks carry a special wrench with a triangular or pentagonal hole that fits over the valve stem.

(a) Why are squares and regular hexagons not used? (*Hint:* What is the shape of the jaws of ordinary adjustable wrenches?)

(b) Why are squares and regular hexagons the standard shape found in bolt heads and nuts?

47. Construct a convex quadrilateral $ABCD$ and the midpoints K, L, M, and N of its sides. The quadrilateral $KLMN$ is called the **medial quadrilateral** of $ABCD$. This problem can be done by making a drawing yourself or you can use geometric software.

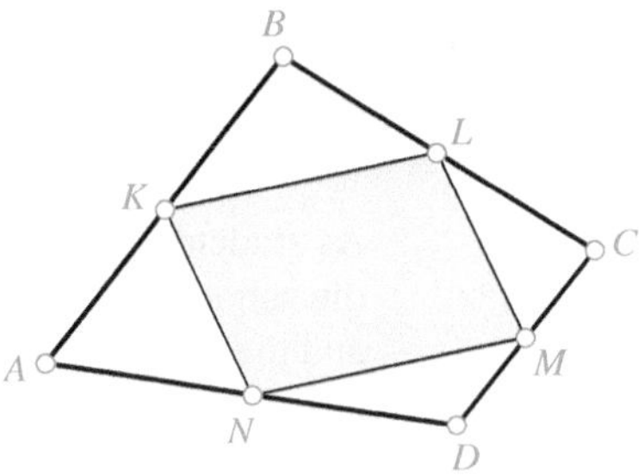

(a) What type of quadrilateral does *KLMN* appear to be? Explore with your geometry software.

(b) Construct two lines perpendicular to one another. Next, construct a quadrilateral *ABCD* with vertices *A* and *C* on one of your lines and vertices *B* and *D* on the other line. Finally, construct the medial quadrilateral *KLMN* of *ABCD*. What type of quadrilateral does the medial quadrilateral now appear to be? Explore with your software.

(c) Drag one of the vertices of the quadrilateral *ABCD* drawn in part (b) until the medial quadrilateral *KLMN* appears to be a square. What type of quadrilateral does *ABCD* appear to be? Explore with your software.

State Assessments

48. (Massachusetts, Grade 4)
These shapes are quadrilaterals.

These shapes are not quadrilaterals.

(a) Write a complete definition for a quadrilateral. Be sure to tell all the important ideas about what makes a quadrilateral.

(b) This is a rhombus:

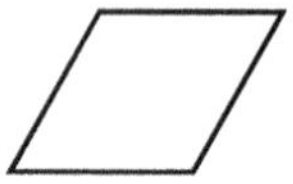

A rhombus is a special kind of quadrilateral. Explain what makes it different from other quadrilaterals.

49. (Grade 8)
What is the best name of the geometric figure of this problem?

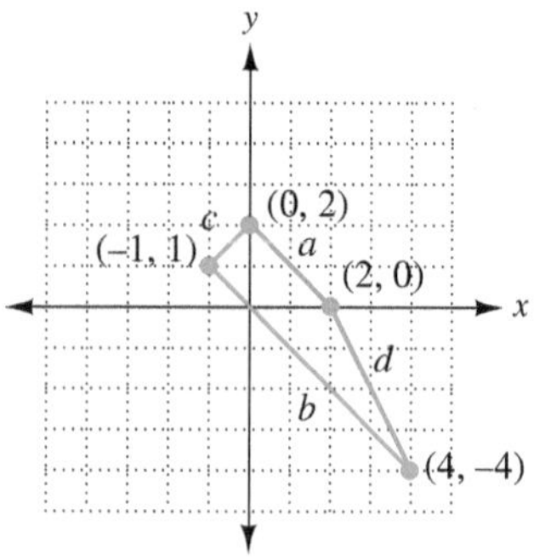

A. Rhombus
B. Parallelogram
C. Trapezoid
D. Quadrilateral

50. (Grade 7)
Which term does *not* apply to the figure below?

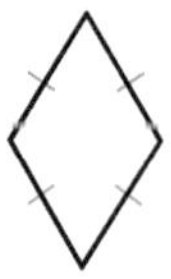

A. Rhombus
B. Trapezoid
C. Rectangle
D. Parallelogram

51. (Grade 7)
The quadrilateral *ABCD* has $\angle A$ congruent to $\angle C$ and $\angle B$ congruent to $\angle D$. The other quadrilateral, *EFGH*, has $\angle E$ congruent to $\angle F$ and $\angle H$ congruent to $\angle G$. Which statement about the quadrilaterals is correct?

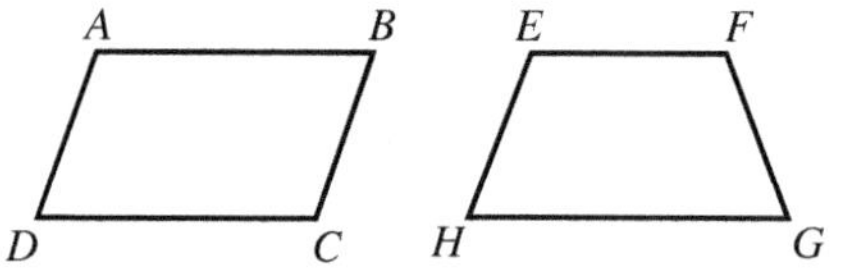

A. Both quadrilaterals have 2 sets of opposite angles that are congruent.
B. Each quadrilateral has a right angle.
C. Both quadrilaterals have 2 obtuse angles.
D. Both quadrilaterals have opposite angles congruent.

52. (Grade 7)
In this problem you are to use the two-dimensional figures below to explain the geometric relationships of the figures.

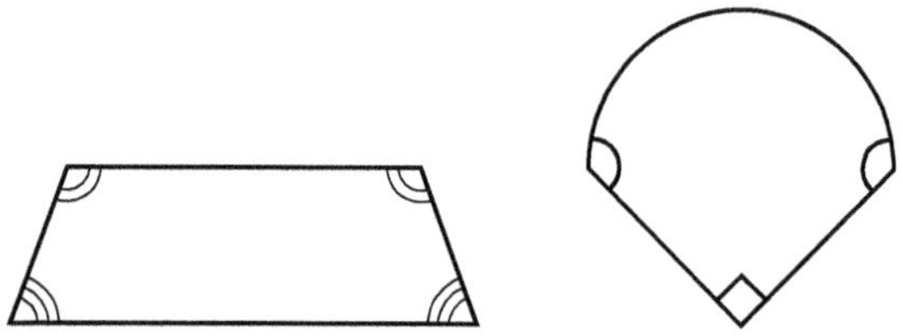

(a) Explain two ways the figures shown about are different.

(b) Explain one way they are the same.

Show all your work or explain your reasoning *for each part.* You can use sentences, numbers, or drawings (sketches) in both (a) and (b). Your answer should be clear enough so that others (whether they be sketches or not) could understand your reasoning.

9.3 Figures in Space

As students, you may not have encountered three-dimensional geometry in your elementary or middle school days but your elementary school students certainly will. The inclusion of solid geometry and the notions of volume and surface area in K–5 education, for example, is relatively recent and so is now a part of content math courses for future teachers.

The picture gallery of space figures that follows shows several interesting examples of shapes whose points do not belong to a single plane. Intuitively, we think of space as having three dimensions. For example, the shape of a shoebox requires us to know not just width and length but height as well. In this section, we classify, analyze, and represent some of the basic figures in space. Look at the State Assessment questions in the exercises to show that these concepts really are presented in elementary school!

Planes and Lines in Space

There are infinitely many planes in space. Each plane separates the points of space into three disjoint sets: the plane itself and two regions called **half-spaces.** Two planes either are **parallel** or intersect in a line, as shown in Figure 9.23.

Figure 9.23
Parallel and intersecting planes

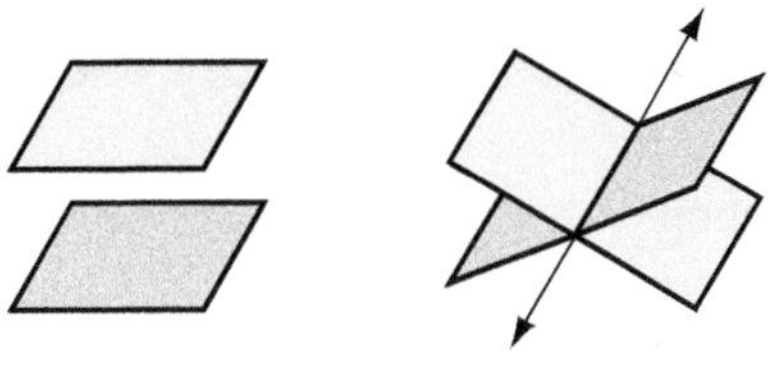

Parallel planes Intersecting planes

Two distinct, nonintersecting lines in space are **parallel lines** if they belong to a common plane. Two nonintersecting lines that do not belong to a common plane are called **skew lines.** A line l that does not intersect a plane P is said to be **parallel to the plane** P. A line m is **perpendicular to a plane** Q at a point A if every line in the plane through A intersects m at a right angle. Diagrams illustrating these terms are shown in Figure 9.24.

Figure 9.24
Lines and planes in space

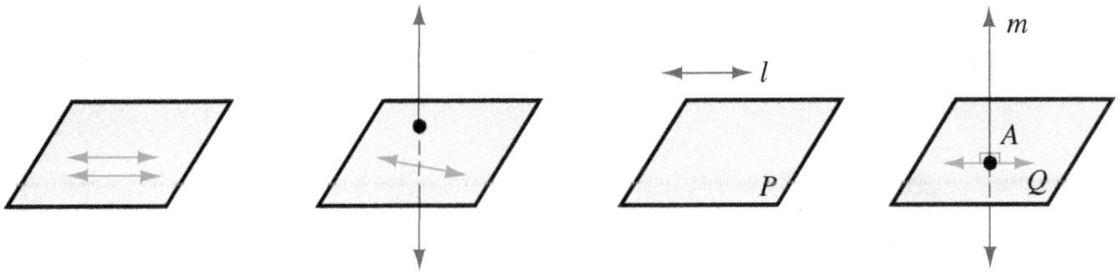

Parallel lines in space Skew lines Line l is parallel to plane P. Line m is perpendicular to plane Q.

Curves, Surfaces, and Solids

The intuitive concept of a curve can be extended from the plane to space by imagining figures drawn with a "magic" pencil whose point leaves a visible trace in the air. Two examples are shown in Figure 9.25: a helix (corkscrew) and a space octagon whose sides are 8 of the 12 edges of a cube. Many of the definitions in the remainder of this section should remind you of similar figures in the plane.

Figure 9.25
Two curves in space

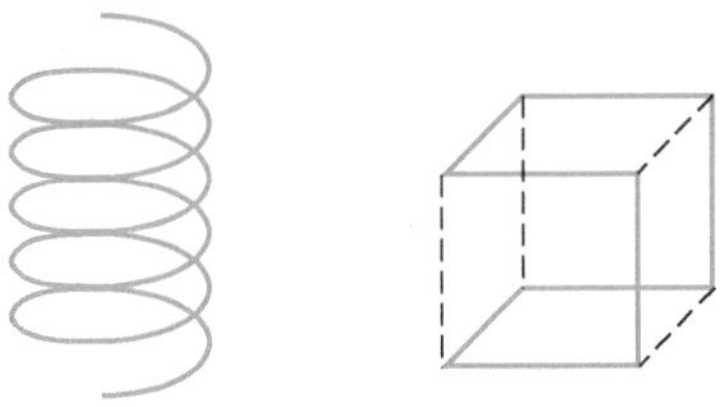

Helix Space octagon (blue)

A Picture Gallery of Space Figures

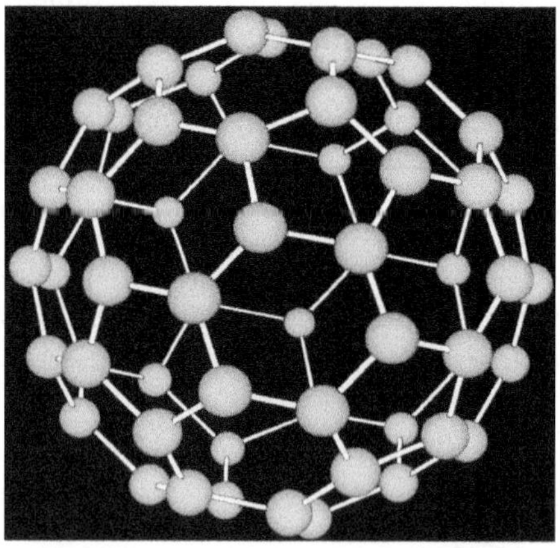

A buckyball (named for Buckminster Fuller), the third form of pure carbon

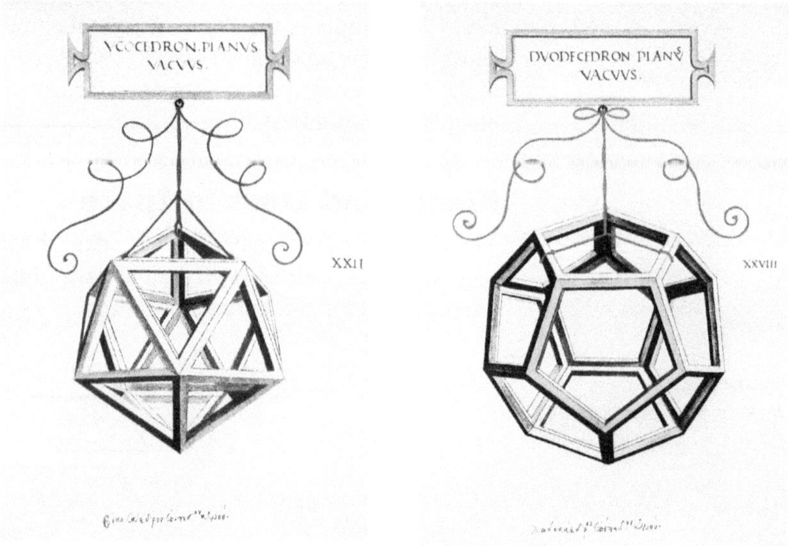

Leonardo da Vinci's drawings of an icosahedron and a dodecahedron for Fra Luca Pacioli's *Di Divina Proportione*

A seashell

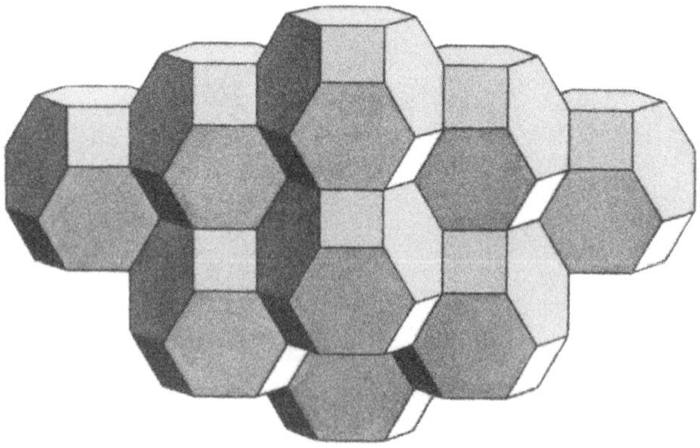

A partial filling of space by truncated octahedra

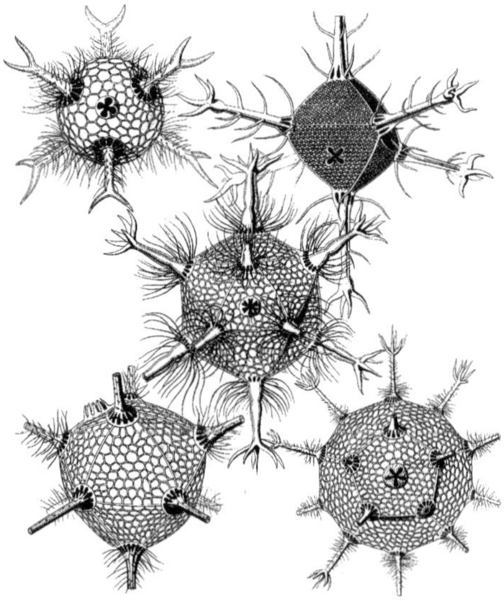

Skeletons of microscopic radiolaria

A **sphere** is the set of points in space at a constant distance from a single point called the **center.** Informally, the surface of a basketball is a sphere. A sphere partitions the remaining points of space into two disjoint regions, namely, the points inside the sphere and the points outside the sphere. Any surface without holes and that encloses a hollow region—its interior—is called a **simple closed surface.** An easy check to see if a figure is a simple closed surface is to imagine what shape it would take if it were made of stretchy rubber. If it can be "blown up" into a sphere, then it is a simple closed surface.

The union of all points on a simple closed surface and all points in its interior forms a space figure called a **solid.** For example, the shell of a hardboiled egg can be viewed, overlooking its thickness, as a simple closed surface; the shell, together with the white and yolk of the egg, forms a solid.

A simple closed surface is **convex** if the line segment that joins any two of its points contains no point that is in the region exterior to the surface. The definition of a convex shape in space is quite similar to a convex region in the plane. One of them lies in three dimensions and the other in two dimensions (a plane). A solid bounded by the surface is a convex shape set in space.

The sphere, soup can, and box shown in Figure 9.26 are all convex. The potato skin surface shown in the figure is not convex, however, since it is possible to find two points on this surface for which the line segment connecting them contains points in the exterior region.

Figure 9.26
(a), (b), (c), and (d) are simple closed surfaces; (e) is a nonclosed surface; and (f) is closed but not simple

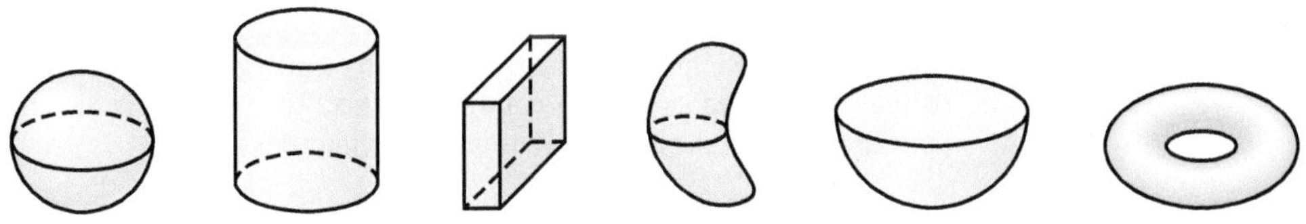

(a) Sphere **(b)** Soup can (circular cylinder) **(c)** Box (rectangular prism) **(d)** Skin of potato **(e)** Bowl (hemisphere) **(f)** Torus (doughnut)

Polyhedra

Joining plane polygonal regions from edge to edge forms a simple closed surface called a **polyhedron.** The name comes from *poly*, meaning "many," and *hedron* meaning "face." Polyhedra play the same role in three dimensions that polygons play in two dimensions. Look at the Into the Classroom feature on p. 428. There is also a fine math book for elementary school children about polyhedra. (See problem 21 of this section.)

DEFINITION Polyhedron

A **polyhedron** is a simple closed surface formed from planar polygonal regions. Each polygonal region is called a **face** of the polyhedron. The vertices and edges of the polygonal regions are called the **vertices** and **edges,** respectively, of the polyhedron.

Polyhedra (*polyhedra* is the plural of *polyhedron*) are named according to the number of faces. For example, a **tetrahedron** has four faces, a **pentahedron** has five faces, a **hexahedron** has six faces, and so on. Several polyhedra are pictured in Figure 9.27.

Figure 9.27
Examples of polyhedra

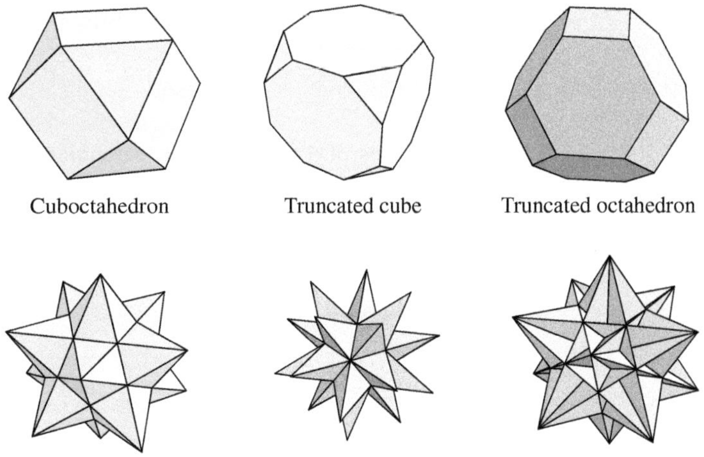

Cuboctahedron — Truncated cube — Truncated octahedron

Small stellated dodecahedron — Great stellated dodecahedron — Great icosahedron

The word *truncated* in the name of a polyhedron means that the polyhedron is formed by removing the corners of another polyhedron. For example, removing the eight corners of a cube creates a truncated cube with six hexagonal faces and eight triangular faces. The word *stellated* in the name of a polyhedron means that pyramids have been erected on the faces of another polyhedron.

The most spectacular polyhedral shapes on earth are the Egyptian and Mayan pyramids. Egyptian pyramids have a square base and four congruent triangular faces that meet at a common vertex. Mayan pyramids have a stepped form. In geometry, a **pyramid** can have any polygonal region as a base.

DEFINITION Pyramid

A **pyramid** is a simple closed surface given by a polygon (called its **base**) and a point not in the plane of the polygon, called its **apex** or **common vertex.** The pyramid is the union of the base with all of the triangular faces that rise from the base edges to the apex.

Examples of pyramids and their names are shown in Figure 9.28.

Figure 9.28
Pyramids and their names

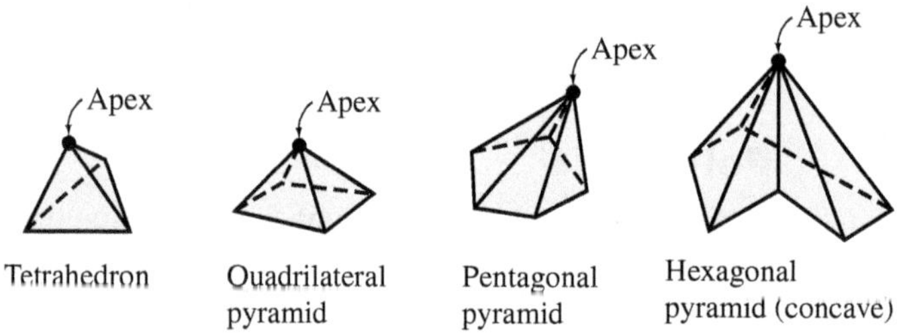

Another commonly occurring polyhedral shape is a prism, which can be imagined as a box with polygons as bases and is quite different than a pyramid. More precisely, we have the following definition:

DEFINITION Prism

A **prism** is a simple closed surface that consists of two congruent polygons that are in parallel planes (called the **bases**) and the lateral faces joining the bases, which are parallelograms.

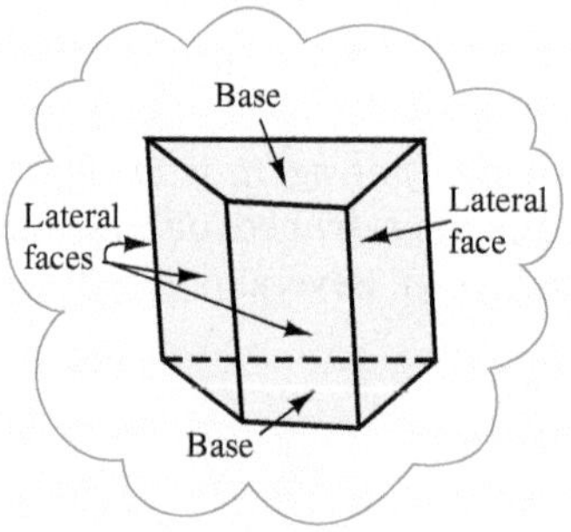

If the lateral faces of a prism are all rectangles, it is a **right prism;** otherwise, it is an **oblique prism** and the edges between the bases are not perpendicular to the plane of the base. Examples of prisms and their names are shown in Figure 9.29.

Figure 9.29
Right and oblique prisms

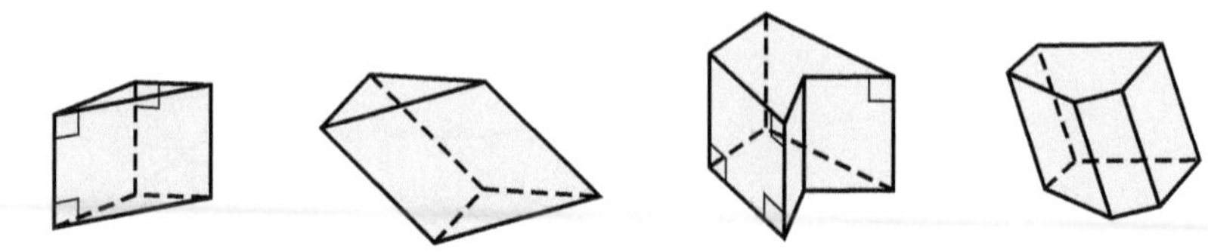

EXAMPLE 9.14

Determining Angles and Planes in a Hexagonal Prism

The bases of the right prism shown are regular hexagonal regions. How many pairs of parallel planes contain the faces of this prism?

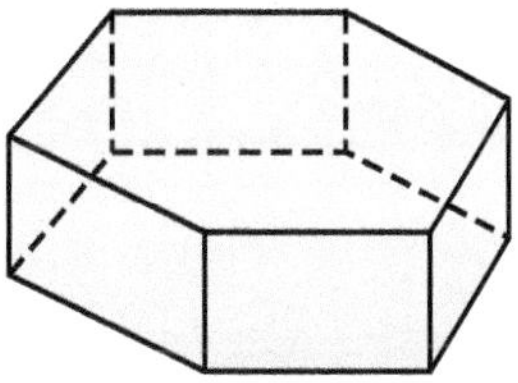

Solution The opposite sides of a regular hexagon are parallel, so there are three pairs of parallel planes containing the opposite lateral faces of the prism. A fourth pair of parallel planes contains the hexagonal bases of the prism. The answer is 4.

Regular Polyhedra

A regular polyhedron is quite special because each face "looks the same" and the same number of faces meet at each vertex of this convex surface. The five regular polyhedra were known to the ancient Greeks. These polyhedra are described in Plato's book *The Republic,* so the shapes are often referred to as the **Platonic solids.** Theaetetas (ca. 415–369 B.C.), a member of the Platonic school, is credited with the first proof that there are no regular polyhedra other than the five known to Plato.

DEFINITION Regular Polyhedron

A **regular polyhedron** is a polyhedron with these properties:

- The surface is convex.
- The faces are congruent regular polygonal regions.
- The same number of faces meet at each vertex of the polyhedron.

The most commonly seen regular polyhedron is the cube: The six faces are congruent squares, and three squares meet at each of the eight vertices. The cube is the only regular polyhedron with square faces, since, if we were to attempt to put four squares about a single vertex, their interior angle measures would add up to 360°. That is, four edge-to-edge squares with a common vertex lie in a common plane and therefore cannot form a "corner" figure of a regular polyhedron.

Similar reasoning with equilateral triangles shows that corner figures in space can be formed with either three, four, or five congruent copies of the triangle. However, a convex corner cannot be formed with six or more equilateral triangles. Likewise, there is just one way to form a corner with congruent regular pentagons, and it is impossible to form a corner figure from regular n-gons for any $n \geq 6$. The possible corner figures of regular polyhedra are shown in Figure 9.30.

Figure 9.30
The five ways to form corner figures with congruent regular polygons

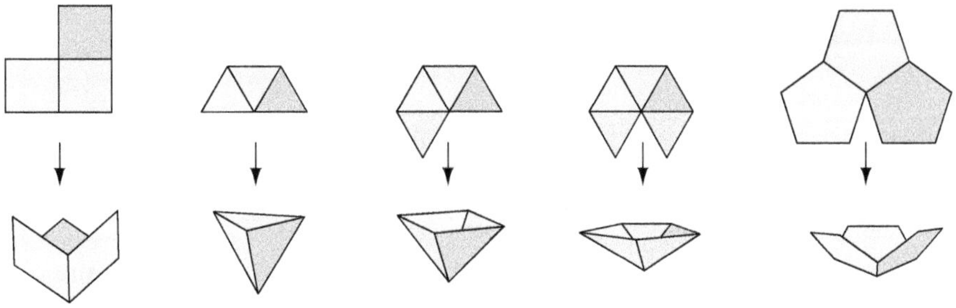

Each of the five corner figures depicted in Figure 9.30 can be completed to form a regular polyhedron. These are shown in Table 9.4, which also shows patterns called **nets** used in elementary school. Models of the polyhedra can be made by cutting the net from heavy paper, folding, and gluing. It helps to include flaps on every other outside edge of the net; these are then coated with glue and tucked under the adjoining face, forming a sturdy model.

TABLE 9.4 THE FIVE REGULAR POLYHEDRA

Polyhedron Name	Face Polygons	Net	Model
Cube	6 squares	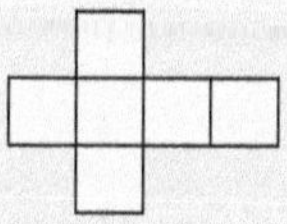	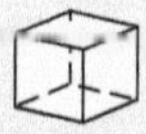
Tetrahedron	4 equilateral triangles	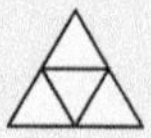	
Octahedron	8 equilateral triangles	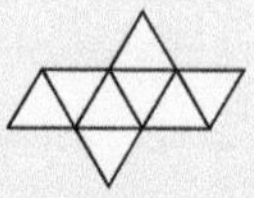	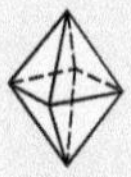
Icosahedron	20 equilateral triangles		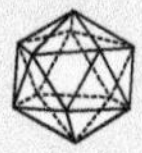
Dodecahedron	12 regular pentagons	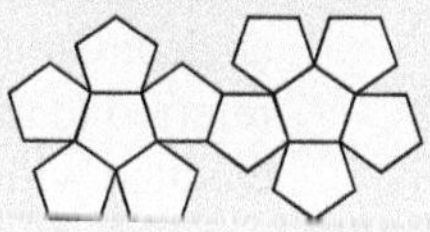	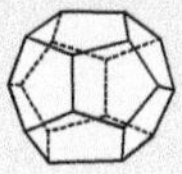

Euler's Formula for Polyhedra

The name given to a polyhedron usually indicates the number of its faces. For example, an octahedron has eight faces. A more complete description of a polyhedron includes the number of its vertices and edges. The following notation will be useful:

$$F = \text{the number of faces of a polyhedron;}$$
$$V = \text{the number of vertices of a polyhedron;}$$
$$E = \text{the number of edges of a polyhedron.}$$

For a regular octahedron, we have $F = 8$, $V = 6$, and $E = 12$.

In 1752, the great Swiss mathematician Leonhard Euler discovered that the number of faces F, the number of vertices V, and the number of edges E are related to one another. Euler was unaware that he had rediscovered a formula found about 1635 by the French mathematician–philosopher René Descartes. Euler (1707–1783) was extraordinarily prolific, publishing new ideas in 886 papers and books. Although he went blind 17 years before his death, he was able to have friends write down complicated formulas that were totally in his head. As an example of his insights, he earlier outlined the calculation of the orbit of the newly discovered planet Uranus.

COOPERATIVE INVESTIGATION

The Envelope Tetrahedron Model

Diagrams and photos of polyhedra are certainly useful, but physical models that can be seen and touched are much better. The construction of models of polyhedra is a worthwhile classroom activity; useful geometric principles are learned as students create beautiful and interesting shapes. Skeletal models are formed easily from drinking straws joined by thin string run through the straws and tied at the vertices. Paper models, in which a carefully drawn net of the polyhedron is cut, folded, and glued, can be colored in interesting ways.

Here is a quick way to construct a regular tetrahedron from an ordinary envelope:

1. Glue the flap of the envelope down.
2. Fold the envelope in half lengthwise, forming a crease $\overline{AB}$ along the centerline.
3. Fold a corner point C upward from corner D, so that C determines point E on the centerline. Once E is found, flatten out the fold.
4. Fold the envelope straight across at E and then cut the envelope off at the height of E.
5. Make sharp folds along $\overline{DE}$ and $\overline{CE}$.
6. Open up the envelope by pulling the two sides of the envelope at E apart; a regular tetrahedron should appear!

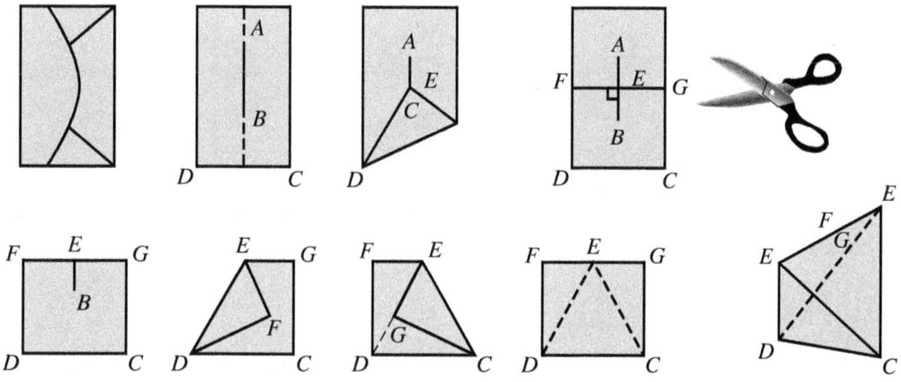

After completing your model, justify the construction procedure.

EXAMPLE 9.15

Discovering Euler's Formula

Let V, F, and E denote the respective number of vertices, faces, and edges of a polyhedron. What relationship holds among V, F, and E?

Solution

Understand the Problem

The numbers V, F, and E are not independent of one another. The goal is to uncover a formula that relates the three numbers corresponding to *any* polyhedron.

Devise a Plan

Formulas are often revealed by seeing a pattern in specific cases. By making a table of values of V, F, and E, we have a better chance to see what this pattern may be. To be confident that the pattern holds for all polyhedra, we need to examine polyhedra of varied kinds.

Carry Out the Plan

A pentagonal pyramid, a hexagonal prism, a "house," and a truncated icosahedron are shown. The truncated icosahedron, formed by slicing off the corners of an icosahedron to form pentagons, may look familiar; it is a common pattern on soccer balls. It is also the pattern of a buckyball, as shown in the picture gallery of space figures at the beginning of this section.

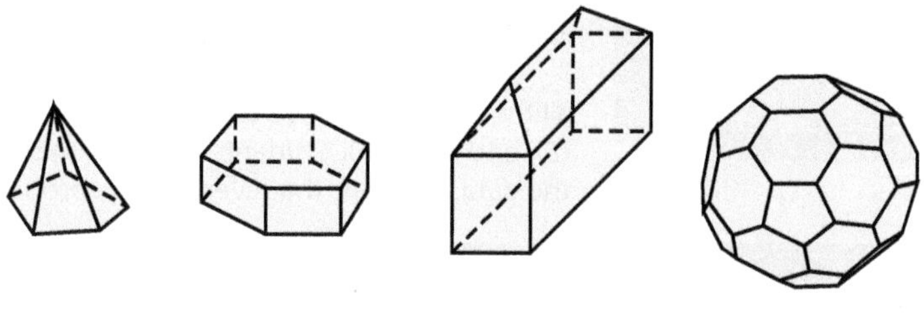

The following table lists the number of vertices, faces, and edges for these polyhedra, as well as for some of the regular polyhedra depicted in Table 9.4:

Polyhedron	V	F	E
Pentagonal pyramid	6	6	10
Hexagonal prism	12	8	18
"House"	10	9	17
Cube	8	6	12
Tetrahedron	4	4	6
Octahedron	6	8	12
Truncated icosahedron	60	32	90

The table reveals that the sum of the number of vertices and faces is 2 more than the number of edges. That is, $V + F = E + 2$.

Look Back

If any two values of V, F, and E are known, the remaining value can be found with the use of Euler's formula $V + F = E + 2$. For example, the dodecahedron has $F = 12$ pentagonal faces. The product $5 \cdot 12$ counts *twice* the number of edges, since each edge borders two of the pentagonal faces. Thus, $E = 5 \cdot 12/2 = 30$ for the dodecahedron. Euler's formula can now be used to compute the number of vertices. Solving for V, we get $V = E + 2 - F = 30 + 2 - 12 = 20$, so a dodecahedron has 20 vertices.

The evidence gathered in Example 9.15 supports the next theorem although we won't prove it. Look at the Into the Classroom box about Euler's formula for more insight.

THEOREM Euler's Formula for Polyhedra

Let V, F, and E denote the respective number of vertices, faces, and edges of a polyhedron. Then

$$V + F = E + 2.$$

INTO THE CLASSROOM

Ralph Pantozzi Discusses Euler's Formula

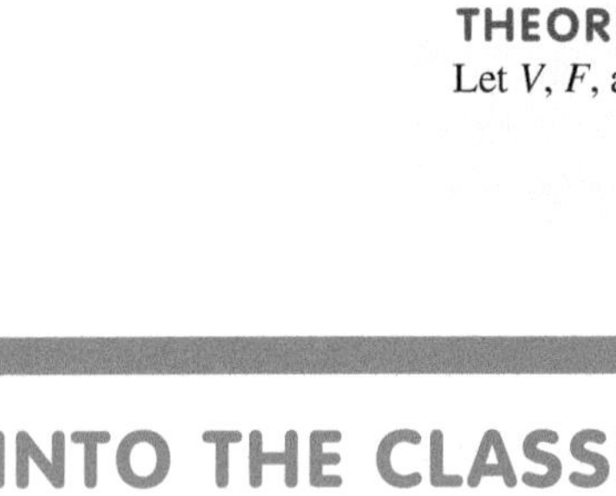

Polyhedra are all around us—from boxes of tissues, to the Washington Monument, to diamonds. When students go out into their own world and find polyhedra to examine, a student debate about "what counts" as being a polyhedron ensues. Students make additional examples using straws (edges) and minimarshmallows (vertices) to see what properties the polyhedron has. The whole class also contributes to the construction of a 3-foot-diameter geodesic dome. For additional fun, we slice the corners off a cube of cheese or clay to form a truncated cube.

Students examine their collection of shapes and make observations about them. In all the examples found or constructed, students count the number of vertices (V), edges (E), and faces (F). Students organize their data into a table and, given time to examine the data, discover that every polyhedron obeys Euler's formula ($V + E - F = 2$).

EXAMPLE 9.16

Searching for Polyhedra with Hexagonal Faces

The Epcot Center in Florida is the site of one of the world's largest geodesic domes. The surface of the 165-foot-diameter structure is covered with both hexagons and pentagons. Similarly, as an example from biology, the microscopic frame of the radiolarian (a unicellular planktonic organism) is covered by pentagons and hexagons. These shapes suggest the following question: *Can all the faces of a polyhedron be hexagonal?* Show that this is not possible, even if the hexagons need not all be congruent and are permitted to be irregular.

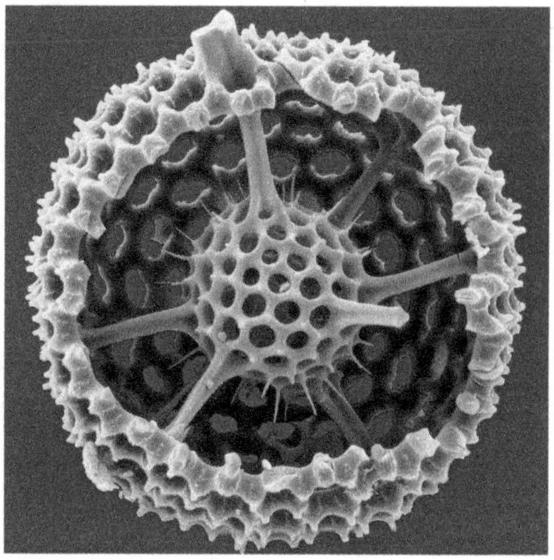

Solution

Suppose, to the contrary, that hexagonal faces can form a polyhedron. As usual, let F, E, and V denote the number of faces, edges, and vertices. Since each face is bordered by six edges and each edge touches two faces, we obtain the formula $6F = 2E$. Thus, we have $F = \frac{2E}{6} = \frac{E}{3}$. Moreover, each of the E edges has two ends, so there are $2E$ ends of edges altogether. Since at least three ends of edges meet at each of the V vertices, it follows that $3V \leq 2E$, or, equivalently, $V \leq \frac{2E}{3}$.

Adding $V \leq \frac{2E}{3}$ to $F = \frac{E}{3}$, we get the inequality $V + F \leq \frac{2E}{3} + \frac{E}{3} = E$. But this inequality contradicts Euler's formula, which tells us that $V + F = E + 2 > E$. We conclude that there is no polyhedron whose faces are all hexagons.

The Standard for Mathematical Practice 3 (SMP 3) allows some careful types of reasoning in middle grades and beyond. One form of reasoning called *proof by contradiction* or *indirect proof* was introduced in Section 1.6. Notice that at the beginning of the solution to Example 9.16, we write "Suppose, to the contrary. . . ." Then the inequality at the end of the proof contradicts Euler's formula, which means that the hexagonal faces do NOT form a polyhedron. The technique of proof by contradiction forms an effective but subtle argument.

SMP 3 Construct viable arguments and critique the reasoning of others.

"Mathematically proficient students are also able to compare the effectiveness of two plausible arguments, distinguish correct logic or reasoning from that which is flawed, ... Later, students learn to determine domains to which an argument applies."

Cones and Cylinders

Note that a pyramid or prism has a base (or bases) that is a polygon. If a surface has a base that is a region in the plane, rather than being restricted to being a polygon, it is called a **cone.** More formally, we have the following definition:

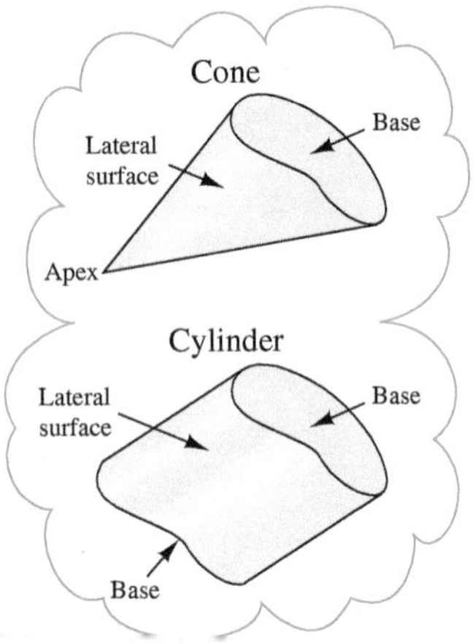

DEFINITION Cone

A **cone** is given by a simple closed curve in the plane (called its **base**) and a point, called its **apex,** which is not in the plane of the curve. A cone is a simple closed surface that is the union of its base and all line segments from the base to its apex. The **lateral surface** of the cone is generated by the line segments from the base to the apex.

A **right circular cone,** an **oblique circular cone,** and a **general cone** are shown in Figure 9.31. The line segment $\overline{AB}$ through the apex A of a cone that intersects the plane of the base perpendicularly at B is called the **altitude** of the cone.

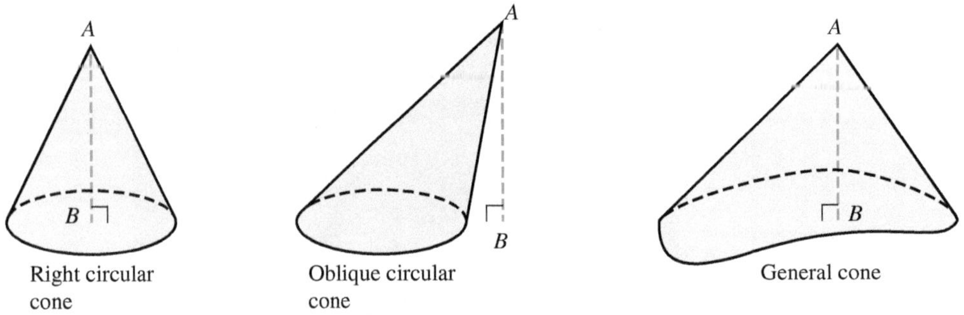

Figure 9.31
A right circular cone, an oblique circular cone, and a general cone

It will be helpful for you to look at Figure 9.32 for examples of the next concept, **cylinders,** before reading the definition. Cylinders look like cans (tilted or straight).

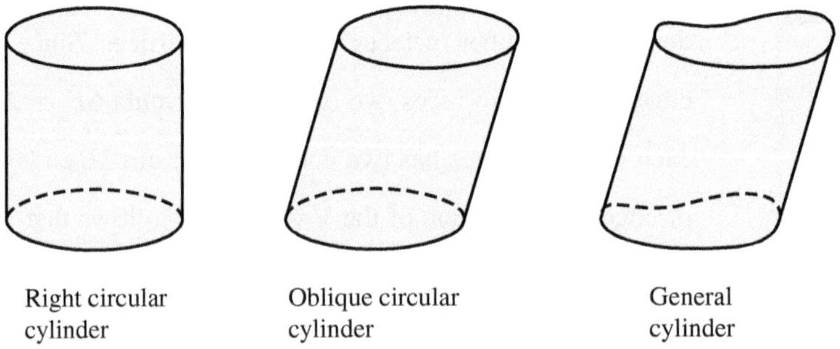

Figure 9.32
A right circular cylinder, an oblique circular cylinder, and a general cylinder

DEFINITION Cylinder

A **cylinder** is a simple closed surface generated by translating the points of a simple closed region in one plane to a parallel plane.

The points joining corresponding points on the curves bounding the bases form the **lateral surface** of the cylinder. If the line segments joining corresponding points in the two bases are perpendicular to the planes of the bases, the cylinder is a **right cylinder.** Cylinders that are not right cylinders are **oblique cylinders.**

9.3 Problem Set

Understanding Concepts

1. Which of the following figures are polyhedra?

(a) 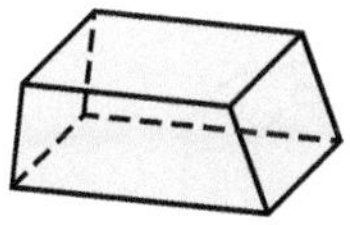 (b) 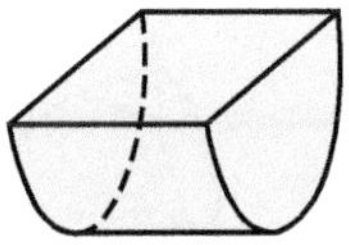 (c) 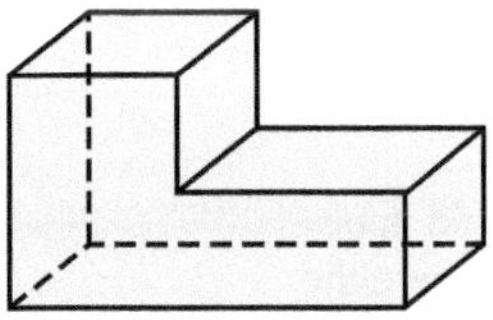

2. Which of the following figures are polyhedra?

(a) 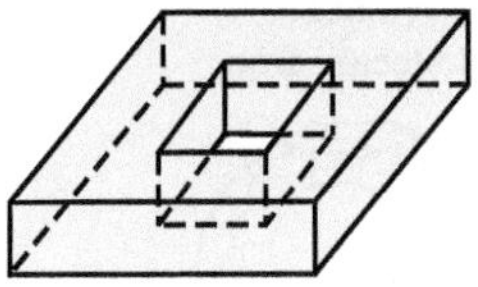 (b) 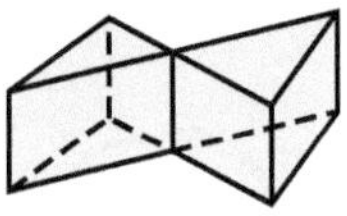 (c) 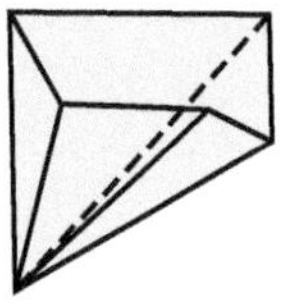

3. Name each of these surfaces:

(a) 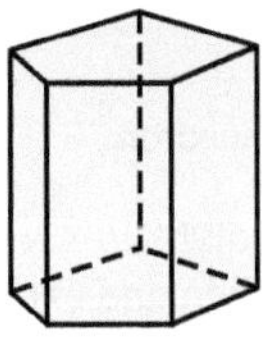 (b) 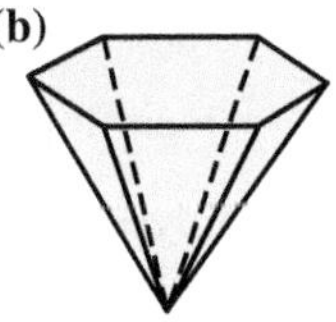 (c) 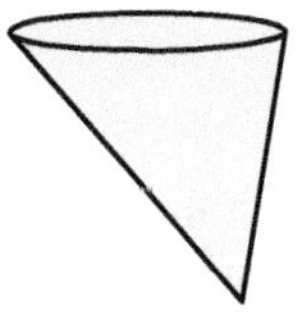 (d) 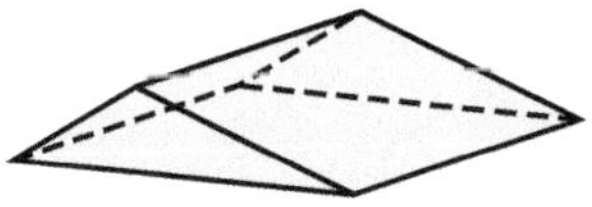

4. Name each surface:

(a) 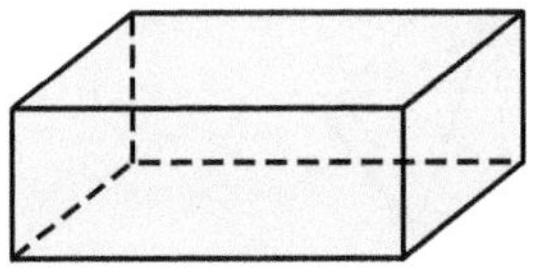 (b) 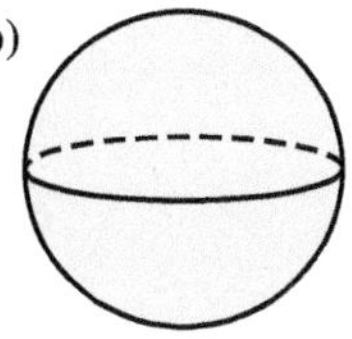 (c)  (d) 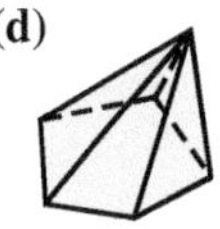

5. A tetrahedron is shown below.
   (a) How many planes contain its faces?
   (b) Name all of the edges.
   (c) Name all of the vertices.
   (d) Name all of the faces.

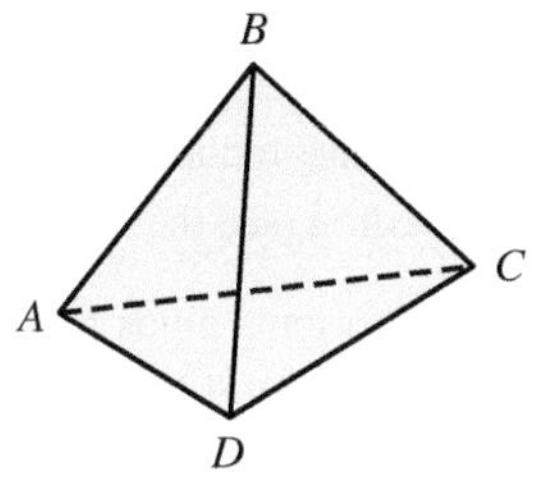

6. Draw freehand pictures of the figures that follow, using dashed lines to indicate hidden edges. Don't just copy—transfer the image in your mind to paper.
   (a) Cube (b) Tetrahedron (c) Square pyramid

7. Follow the same instructions as problem 6.
   (a) Pentagonal right prism
   (b) Oblique hexagonal prism
   (c) Octahedron
   (d) Right circular cone

8. A cube with vertices *A*, *B*, *C*, *D*, *E*, *F*, *G*, and *H* is shown in the accompanying figure. Vertices *D*, *E*, *G*, and *H* are the vertices of a tetrahedron inscribed in the cube.

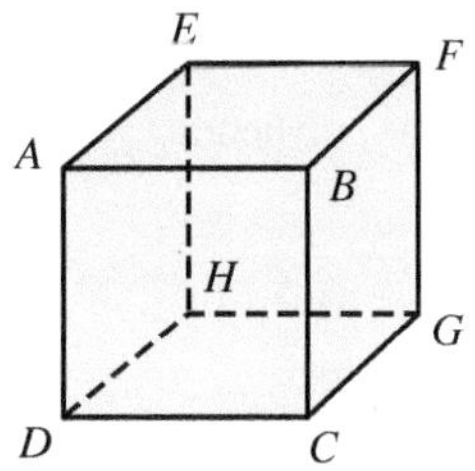

   (a) Trace the cube in one color and then draw the tetrahedron *DEGH* in a different color.
   (b) Three of the faces of the tetrahedron *DEGH* are subsets of the faces of the cube. Find a tetrahedron inscribed in the cube that has none of its faces in the planes of the faces of the cube. Sketch your tetrahedron and the surrounding cube.

9. The pattern shown on the left folds up to form the cube on the right:

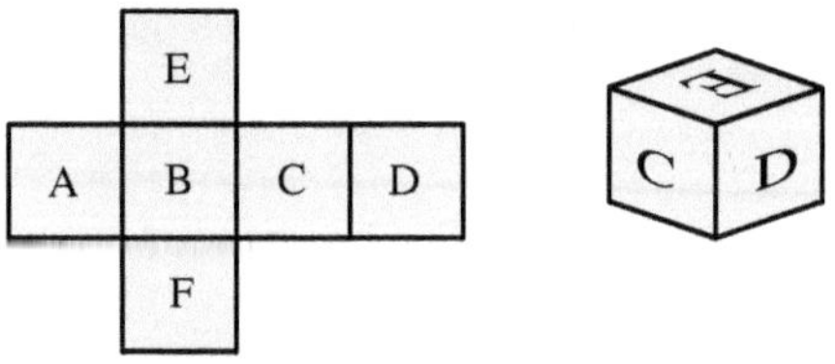

Sketch the letter, *in its correct orientation,* that should appear on each blank face of each of the following cubes, where the same pattern is used:

(a) **(b)** **(c)**

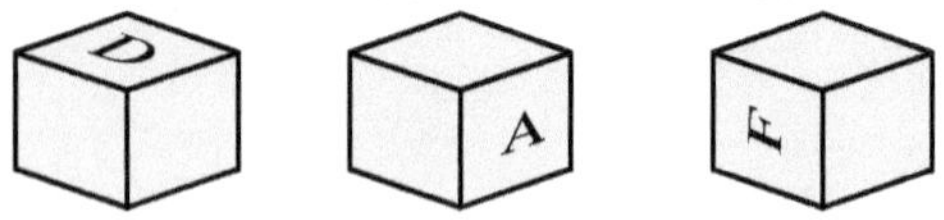

10. The numbers in the 2-by-3 grid of squares correspond to the pattern of stacked cubes shown in the **isometric drawing** on the right.

3	2	1
1	1	0

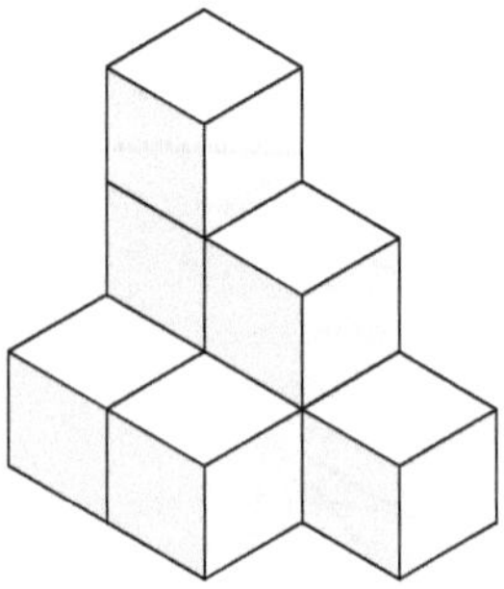

Make an isometric drawing of these patterns:

(a)

2	3	2
2	2	1

(b)

4	2
3	0
1	1

11. Sketch a cube. Color the eight vertices either red or blue (or mark *R* or *B* at the vertices) so that no plane through any four vertices of the cube has the same color at all four vertices.

12. Are any of the regular polyhedra

(a) prisms?

(b) pyramids?

13. Verify Euler's formula for

(a) a pyramid with a hexagonal base.

(b) a prism with octagonal bases.

(c) a regular icosahedron. (*Suggestion:* Modify the counting method used for the regular dodecahedron in the Look Back step of Example 9.15.)

14. Use Euler's formula to complete the accompanying table. These four polyhedra are representative of a class of 13 polyhedra discovered by Archimedes.

	Number of Vertices, Faces, and Edges		
Polyhedron	***V***	***F***	***E***
Truncated tetrahedron	12	8	—
Truncated dodecahedron	—	32	90
Snub cube	24	—	60
Great rhombicosidodecahedron	120	62	—

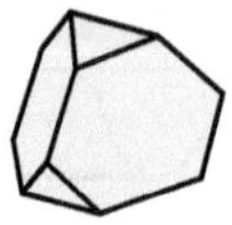

Truncated tetrahedron

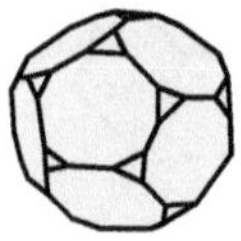

Truncated dodecahedron

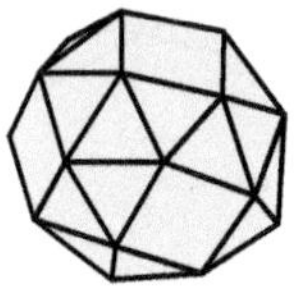

Snub cube

Great rhombicosidodecahedron

15. **A double pyramid** (or **dipyramid**) is a polyhedron with triangular faces arranged about a plane polygon and extending both upward to an upper apex and downward to a lower apex:

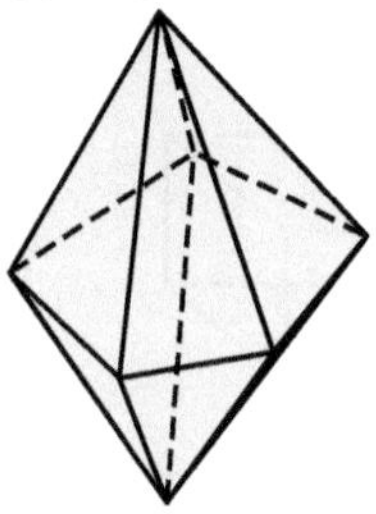

(a) Find the number of vertices, edges, and faces for the pentagonal double pyramid shown. Then verify that Euler's formula is satisfied.

(b) Repeat part (a) for the double pyramid built from a polygon with 20 sides.

16. Draw nets corresponding to the following polyhedra (see Table 9.4):

(a) A square pyramid with equilateral triangular faces

(b) A truncated tetrahedron (see the figure in problem 14.)

17. Check whether Euler's formula holds for the figures that follow. If not, explain what assumption required for Euler's formula is not met.

(a)

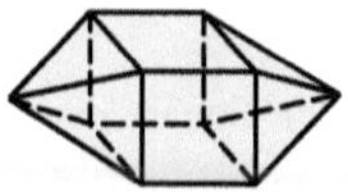

(b)

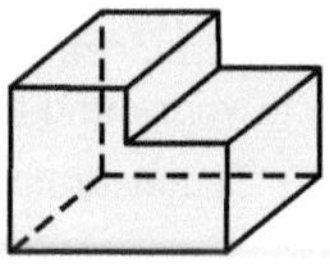

(c)

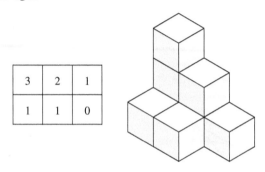

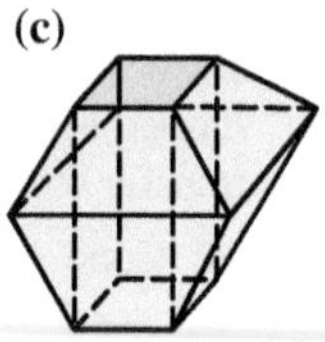

(An octahedron with a square prism hole)

Into the Classroom

18. **Shape Search.** Go on a geometry walk; search for interesting three-dimensional shapes around campus, at the grocery store, in local sculptures, in nature, . . . , wherever! Photograph or make accurate drawings of three or four shapes that you find especially interesting, and then write careful descriptions that discuss the properties of your shapes. For example, if your shape is a polyhedron, count its vertices, faces, and edges and verify Euler's formula. If possible, include some shapes that can be contributed to a class collection. For example, many products at the store are packaged in ingeniously shaped boxes.

19. **Make a Shape.** Children (and college students, too) can learn important principles of geometry by constructing their own three-dimensional shapes. Often, just ordinary materials, such as toothpicks and minimarshmallows, suffice, say, to make one of the regular polyhedra or even a figure as complex as a buckyball (the truncated icosahedron). Instructions can be found readily from several interesting sites on the Internet. A well-equipped math lab might also include some commercially manufactured kits that can be used to create fun shapes, such as Polydrons™ or the Zome System. Your challenge: make an interesting shape and write a brief report about its properties and why you find it interesting.

20. **Thinking about Definitions.** Trying to define mathematical ideas by working in groups is very helpful in getting an intuitive feel for what the terms mean. A number of studies show that discussing, arguing, disagreeing, and agreeing about what the definition should mean before you have actually seen it formally helps the learner understand conceptually what the topic is. With that in mind, after looking at some polyhedrals that you bring with you,

(a) Ask a group of adults to try to define *vertex, edge, face, prism,* and *pyramid.* Write a short description of both what definitions were arrived at and on what the discussion centered.

(b) Ask a group of elementary or middle school children the same questions and write about that experience.

21. **(Writing, Children's Literature)** Read a work that is accessible to elementary school children and that involves polyhedra. In a short paper, discuss what you read from both its content and its use in the classroom. One such work is *Sir Cumference and the Sword in the Cone* by Cindy Newschwander.

---

22. Five congruent squares can be joined along their edges to form 12 distinct shapes known as **pentominoes.** Pentominoes were invented by mathematician and electrical engineer Solomon Golomb in 1953 in a talk to the Harvard Mathematics Club. They are named by the letters they somewhat resemble. Have your class work on this problem and the next one in small groups.

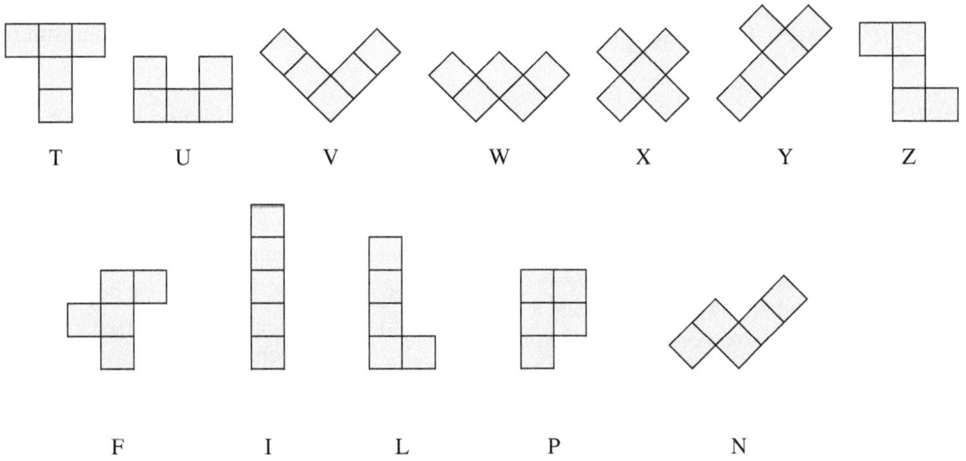

SOURCE: Golomb, Solomon; Polyominoes. © 1994 Princeton University Press. Reprinted by permission of Princeton University Press.

(a) (Individually) Which of the 12 pentominoes shown can be folded to form a cube-shaped box with an open top?

(b) (In small groups) Copy the *T* pentomino onto paper, cut the pattern with scissors, and construct an open-topped cubical box by taping edges together. Now choose a target pentomino (not the *T*, since it's too easy!) and sketch it on the bottom of the box. Then use scissors to cut the box apart to form the targeted pentomino shape. (*Note:* This activity was created by Marion Walter of the University of Oregon, who makes open cubical boxes from the bottoms of clean milk cartons, discarding the tops.)

---

23. (Read problem 22 first.) A **hexomino** is formed by joining six congruent squares along their edges. There are 35 different hexominoes in all, including the 5 shown. Do the following exercise in small groups.

(a) Hexomino (i) is a net for a cube. Which of the other 4 hexominoes shown can be folded to form a cube?

(b) There are 11 hexominoes that can be folded to form a cube. Try to find all 11 shapes. Be careful not to repeat a shape; two congruent shapes may at first appear to be different when one is rotated or flipped over.

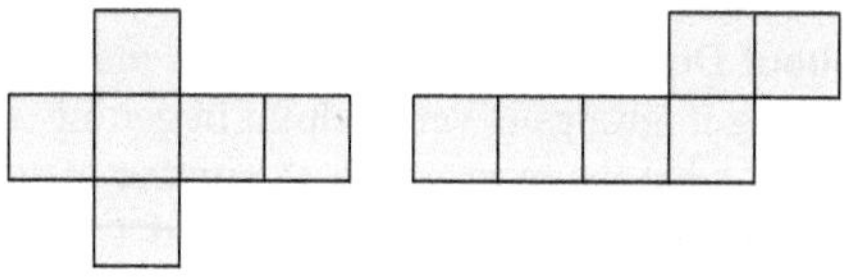

(i) (ii)

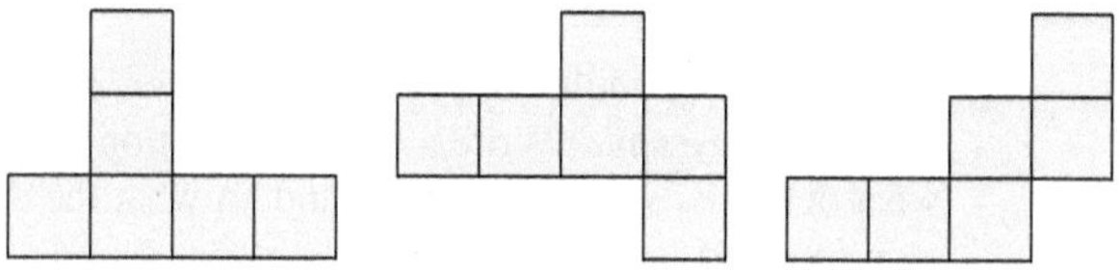

(iii) (iv) (v)

Responding to Students

24. Frank was asked to read the following list of geometric solids:

Cylinder, cone, and square pyramid.

His teacher then asked him to come up with at least two ways to show that the square pyramid is different from the other two solids. His answer was the following:

- The square pyramid has no curved surfaces.
- The square pyramid has a point or vertex.

The teacher then redirected him about his second idea and asked Frank to rethink whether a cone has a vertex. Frank answered that it does not because it doesn't have flat faces on the side.

(a) What would you do next as Frank's teacher?

(b) What is Frank misunderstanding about solid geometric shapes?

25. Polina is asked to solve the following riddle:

I have the same number of faces as vertices. What am I?

Polina said that a triangular pyramid was the correct answer. The teacher agreed and then asked Polina to come up with a few more answers to the riddle. She told Polina to consider a square pyramid. Polina thought and then said that a square pyramid couldn't answer the riddle because it was made of four triangles and one square and, since the shapes weren't the same, it couldn't have the same number of faces as vertices.

(a) What is Polina misunderstanding about the definition of faces and possibly vertices in geometric shapes?

(b) How could you guide her to see that a square pyramid, pentagonal pyramid, hexagonal pyramid, and so on all are correct answers to the riddle?

26. John is shown a picture of a cube. He is asked to tell how many vertices, faces, and edges it has. His answer follows:

Faces: 3

Vertices: 7

Edges: 9

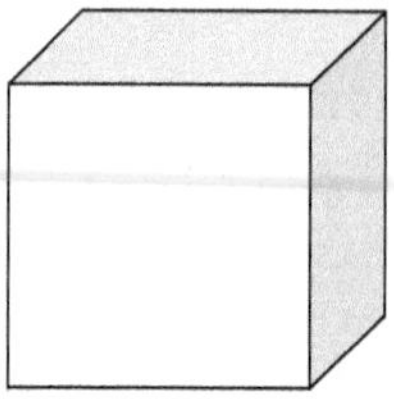

(a) Explain what John knows about geometric definitions and is identifying correctly.

(b) Identify what John is doing incorrectly.

(c) How would you guide him in approaching this problem in the future?

27. Tianna is given a set of geometric solids. She correctly identifies the various pyramids but calls all of the prisms in the set rectangular prisms. The set Tianna was given actually contains a triangular prism, a rectangular prism, a pentagonal prism, and a hexagonal prism.

(a) What is Tianna thinking when she improperly names the prisms?

(b) How would you guide her so that in the future she is able to properly identify geometric solids?

Thinking Critically

28. Let V, E, and F denote the number of vertices, edges, and faces, respectively, of a polyhedron.

(a) Explain why $2E \geq 3F$. (*Hint:* Every face has at least three sides, and every edge borders two faces.)

(b) Explain why $2E \geq 3V$. (*Hint:* Every vertex is the endpoint of at least three edges.)

(c) Show that every polyhedron has at least six edges. (*Hint:* Add the inequalities of parts (a) and (b), and use Euler's formula.)

(d) Use (a) and (b) to prove that no polyhedron can have seven edges. (*Hint:* Use Euler's formula.)

(e) Show that there are polyhedra with 6, 8, 9, 10, . . . edges.

29. A convex polyhedron with five faces is called a **pentahedron.** Find and sketch two different types of pentahedra. (*Hint:* One is "easy as pie.")

30. A polyhedron with six faces is a **hexahedron.** For example, a cube is a hexahedron.

(a) Draw a pyramid that is a hexahedron.

(b) Draw a double pyramid (see problem 15) that is a hexahedron.

31. Suppose a skeletal model of a convex polyhedron is made, where just the edges of the polyhedron are outlined. If the model is viewed in perspective from a position just outside the center of a face, the edges of that face form a bounding polygon inside which the remaining edges are seen. The resulting pattern of edges is called a **Schlegel diagram,** named for the German mathematician Viktor Schlegel, who invented the diagram in 1883. Schlegel diagrams for the cube and dodecahedron are as follows:

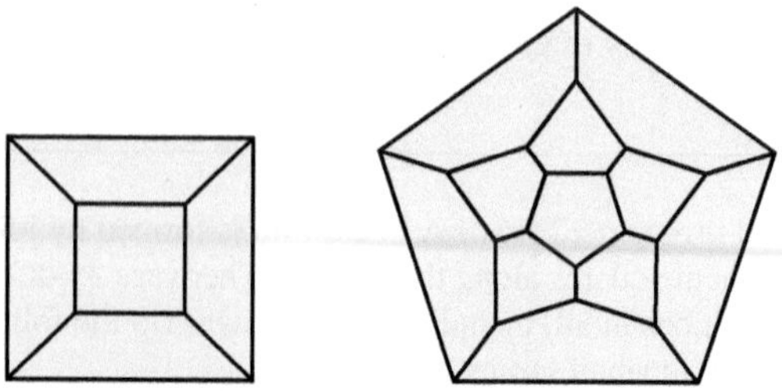

Draw Schlegel diagrams for

(a) the tetrahedron

(b) the octahedron

(c) the icosahedron (*Hint:* Start by drawing a large equilateral triangle, and keep in mind that each vertex must touch five edges.)

Making Connections

32. The round door shown has a problem. What facts of space geometry create a difficulty?

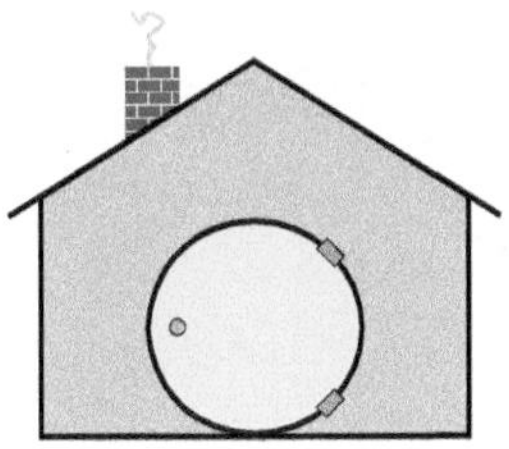

What is important about the placement of door hinges?

33. The ancient Greeks divided physical space into five parts: the universe, earth, air, fire, and water. Each of these was associated with one of the five regular polyhedra. Using the Internet, investigate what correspondence was made.

34. Biologists and physical scientists frequently become involved with the analysis of shape and form. For example, many viruses have an icosahedral structure, and the crystalline structures of minerals are often polyhedral forms of considerable beauty. An example of a pyrite crystal is shown here:

Browse through your school library, and see what three-dimensional shapes are receiving interest and attention. Report on your findings.

State Assessments

35. (Washington State, Grade 4)
Look at the cube below.

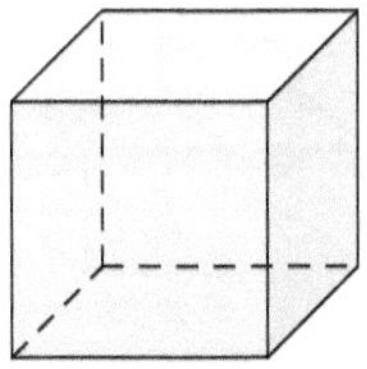

Now look at figures A, B, and C.

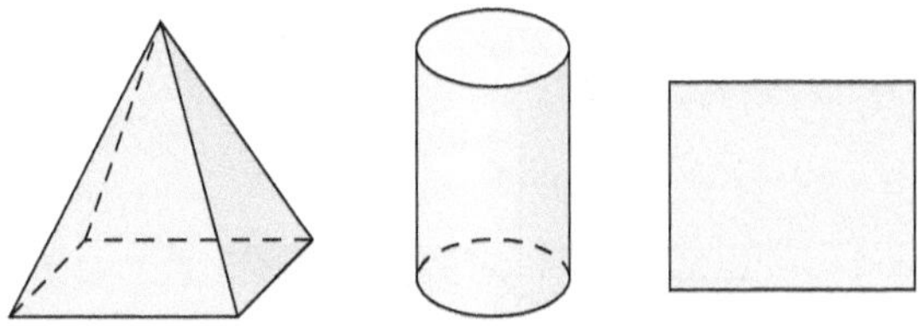

A. B. C.

Choose one figure. Tell what that figure has in common with the cube. Explain your answer using words, numbers, or pictures. Now choose a different figure. Tell something it has in common with the cube. Explain your answer using words, numbers, or pictures.

36. (Grade 5)
Marilyn likes to use paper, especially to build cylinders. What paper shapes which are cylinders could she have? *Hint*: There may be more than one correct answer.

A. 2 circles and 1 rectangle
B. 4 triangles and 1 square
C. 2 triangles and 1 rectangle
D. 2 hexagons and 2 rectangles
E. 3 circles and 3 triangles

37. (Massachusetts, Grade 4)
Which shape CAN be folded to form an OPEN box?

A. 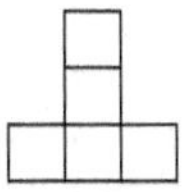

B. 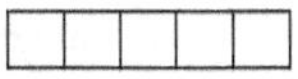

C. 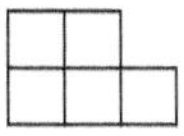

D. 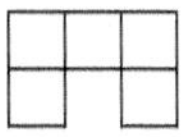

38. (Grade 7)
Using identical cubes, the drawings below show the top, front, and right-side views of a 3-dimensional figure built.

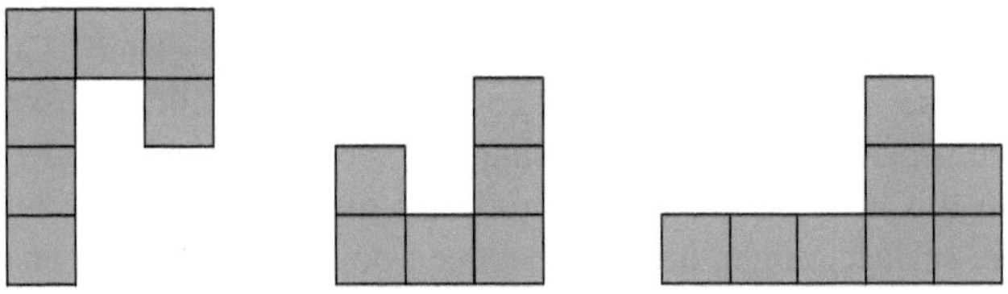

Top view　　Front view　　Right-side view

Which 3-dimensional figure do these views best represent?

F.

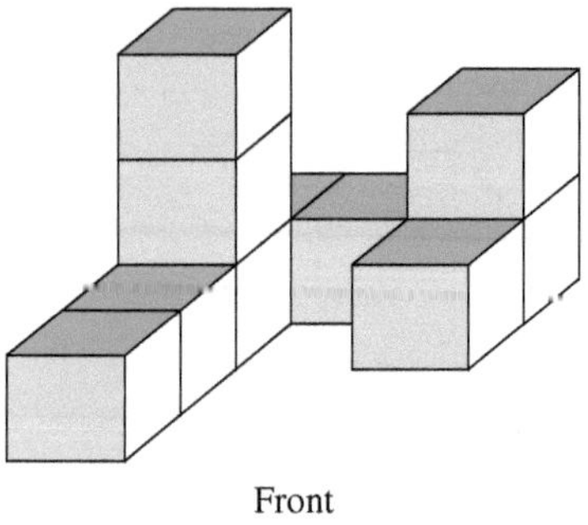

G.

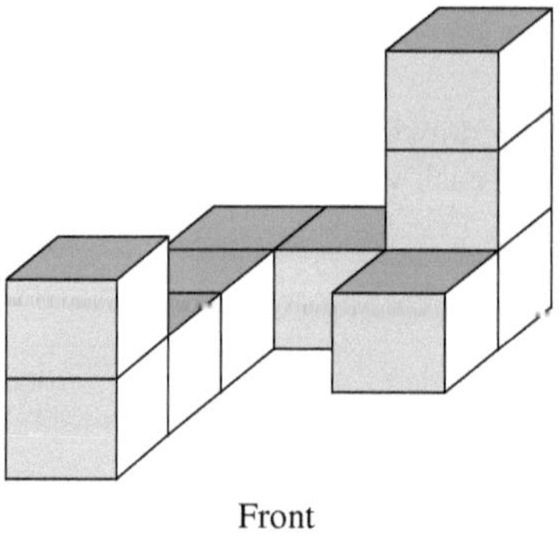

H.

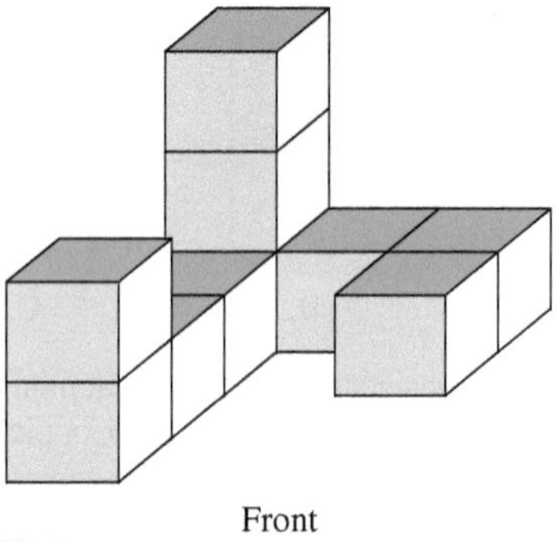

J.

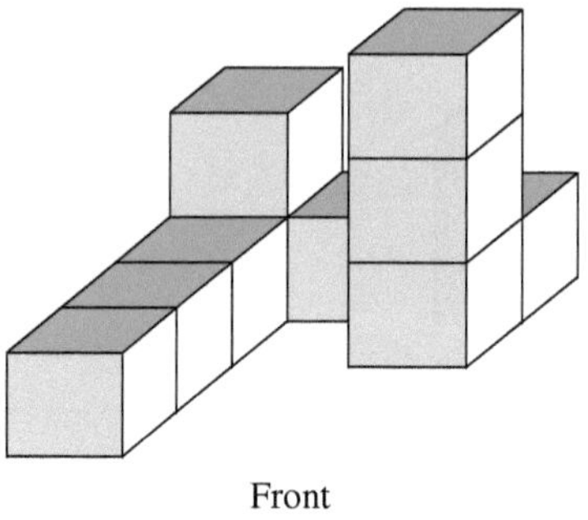

39. (Grade 7)
Look carefully at the three-dimensional figures shown below to explain in writing the geometric relationships of the figures. Another student must be able to read and understand what you have written.

This problem requires you to explain all your reasoning in written form. You may use drawings, words, or numbers in your answer.

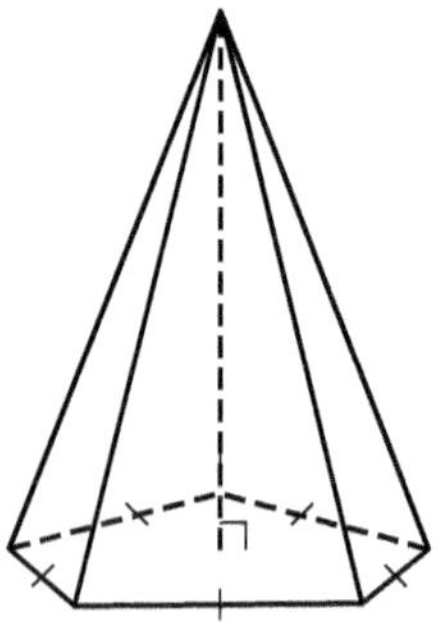

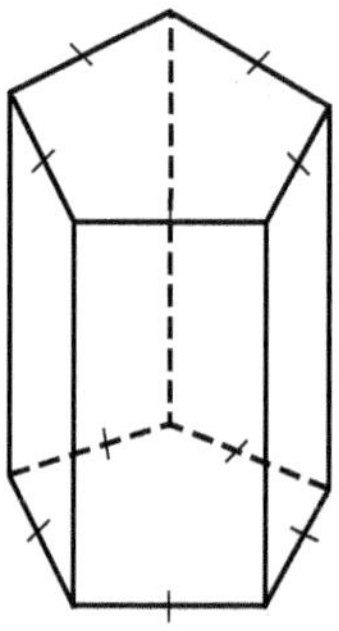

(a) Explain, using geometric properties, one way the figures shown above are the same.

(b) Explain two ways they are different.

Explain your reasoning *for each part.*

40. (Grade 7)
A hexagonal prism is shown below.

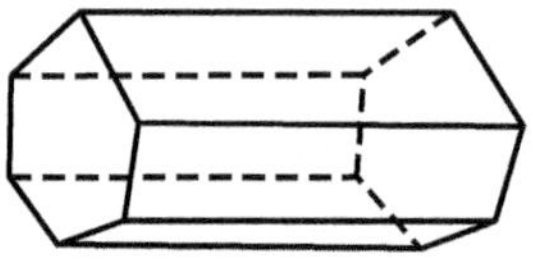

Which is a property of a hexagonal prism?

A. It has two pentagons as bases.
B. It has five hexagons as faces.
C. It has exactly nine edges.
D. It has twelve vertices.

The Chapter in Relation to Future Teachers

This chapter deals with the beginning concepts of geometry in elementary and middle school. We have taken an informal approach because children learn geometry by looking at and experimenting with figures on the plane and in space. Their teachers therefore need to understand and use definitions (which abound in this chapter) so that they help their students comprehend what the terms mean and how they are related to each other (when they are so related). Geometry will be covered in your "methods course," but an example of a good way to start children "experimenting with math" is to have them search for and make geometric shapes. (See problems 18 and 19 of Section 9.3, for example.)

In this chapter, we have introduced many of the basic notions of geometry: point, line, plane, curve, surface, angle, distance between points, measure of an angle, region, and space. In the chapters that follow, these ideas are developed in more depth as we encounter the ideas of measurement, transformations, tilings, symmetries, congruence, constructions, and similarity.

Chapter 9 Summary

Section 9.1 Figures in the Plane	Page Reference
CONCEPTS	
• **Line:** Two distinct points A and B uniquely determine a line $\overleftrightarrow{AB}$.	431
• **Length of a segment:** The length of the line segment $\overline{AB}$ is the distance between the points A and B and is written AB.	433
• **Corresponding-angle criterion for parallel lines:** There is an equivalence between two lines being parallel and two corresponding angles being congruent.	437
• **Angle sum in a triangle:** The sum of the measures of the three interior angles of a triangle is 180°.	439
DEFINITIONS	
• **Collinear** points are three or more points that lie on the same line.	431
• **Plane figures** are subsets of a plane.	431
• The **point of intersection** is the single point in a plane at which nonparallel lines p and q meet.	432
• A **transversal** is a line that intersects two distinct lines but not at the same point.	432
• **Congruent** line segments, $\overline{AB}$ and $\overline{CD}$, are line segments that have the same length.	433
• A **ray,** $\overrightarrow{PQ}$**,** is a subset of the line $\overleftrightarrow{PQ}$ that contains the point P and all points on the line to one side of P.	434
• An **angle,** $\angle BAC$, is the points of $\overrightarrow{AB} \cup \overrightarrow{AC}$.	434
• The **vertex of an angle** is the common endpoint of two rays.	434
• The **measure of an angle,** $m(\angle BAC)$, is the number of degrees of turn to rotate about the vertex A side $\overrightarrow{AC}$ onto side $\overrightarrow{AB}$.	434
• A **straight angle** has a measure of 180°. A **right angle** has an angle measure of 90°. A **zero angle** has a measure of 0°.	434
• An **acute angle** has a measure between 0° and 90°. An **obtuse angle** has a measure between 90° and 180°. A **reflex angle** has a measure greater than 180° but less than 360°.	434
• **Perpendicular lines** are two lines that intersect at right angles.	435
• **Congruent angles** are two angles that have the same measure.	435
• Two angles are **complementary** if the sum of their measures is 90°.	436
• Two angles are **supplementary** if the sum of their measures is 180°.	436
• **Adjacent angles** are two angles that share a common side and nonoverlapping interiors.	436
• **Vertical angles** are two nonadjacent angles formed by two intersecting lines.	436
• **Interior** and **exterior angles of a triangle:** In triangle BAC, the angles formed by two intersecting rays (for example, $\overrightarrow{AB}$ and $\overrightarrow{AC}$) are **interior angles.** An **exterior angle** of triangle BAC is a supplementary adjacent angle.	438
• **Directed angles** are angles that specify an initial side and final side and a direction of turn.	439
• The **initial side** of a directed angle is the side that is being directed to turn, and the **terminal side** is the side that indicates where the direction of turn stops.	439

THEOREMS	
• **Vertical-angles theorem:** Vertical angles have the same measure (or are congruent).	437
• **Corresponding-angles property:** If two parallel lines are cut by a transversal, then their corresponding angles have the same measure. If two lines in the plane are cut by a transversal and some pair of their corresponding angles is congruent, then the lines are parallel.	437
• **Alternate-interior-angles theorem:** Two lines cut by a transversal are parallel if, and only if, a pair of alternate interior angles is congruent.	438

Section 9.2 Curves and Polygons in the Plane	**Page Reference**
CONCEPTS	
• **A partial classification of curves:** Curves can be simple (non–self-intersecting), closed (no endpoints), or both simple and closed.	445
• **Polygonal curves:** A polygonal curve is a curve formed by sequentially joining points called **vertices** with line segments called **sides** of the polygonal curve.	448
• **Polygon:** A polygon is a simple closed polygonal curve. A polygon with 3, 4, 5, 6, 7, 8, 9, 10, . . . , n sides is called a triangle, quadrilateral, pentagon, hexagon, heptagon, septagon, octagon, nonagon (or enneagon), decagon, . . . , n-gon, respectively.	448
• **Sum of angle measures in a polygon:** The sum of the measures of the interior angles of an n-gon is $(n - 2)180°$. This formula is the generalization of the statement, for $n = 3$, that "the angle sum of a triangle is 180."	450
• **Regular polygon:** A polygon is regular if its sides are congruent and its interior angles are congruent. In a regular n-gon, the exterior and central angles each measure $360°/n$ and each interior angle measures $(n - 2)180°/n$.	452
DEFINITIONS	
• **Curves** are defined as **simple** (non–self-intersecting), **closed** (no endpoints), or **simple closed** (simple and closed).	445
• Regions defined by a simple closed curve are the **interior and exterior regions.** A line partitions the plane into two regions called **half planes.** An angle partitions the plane into two regions called the **interior** and **exterior of the angle.**	446
• A **region** is one of the areas defined by a curve.	447
• A **convex figure** is a figure that contains the segment $\overline{PQ}$ for each pair of points P and Q contained in the figure.	448
• A **concave figure** (or **nonconvex figure)** is a figure that is not convex.	448
• A **polygonal curve** is a curve that consists of a union of finitely many line segments joined at the vertices.	448
• **Vertices** are the endpoints of the segments in a polygonal curve.	448
• **Sides,** or **edges,** are the segments themselves in a polygonal curve.	448
• A **polygon** is a simple closed polygonal curve. A polygon with 3, 4, 5, 6, 7, 8, 9, 10, . . . , n sides is called a **triangle, quadrilateral, pentagon, hexagon, heptagon** (or septagon), **octagon, nonagon** (or enneagon), **decagon,** . . . , ***n*-gon,** respectively.	448
• The **polygonal region** is the interior of a polygon.	448
• A **convex polygon** is a polygon that is convex.	448

• An **interior angle** of a polygon is the angle at a vertex formed by the two sides of the polygon that intersect at the vertex and is in the polygonial region. An **exterior angle** is a supplementary adjacent angle to an interior angle.	450
• A triangle is **acute** if all three interior angles are acute, **right** if one angle is a right angle, or **obtuse** if an interior angle is obtuse.	454
• A triangle is **scalene** if no two sides have the same length, **isosceles** if at least two sides have the same length, or **equilateral** if all three sides have the same length.	454
• A **quadrilateral** can be a **kite** (two pairs of adjacent congruent sides), a **trapezoid** (two sides parallel), an **isosceles trapezoid** (two sides parallel and two congruent angles), a **parallelogram** (two pairs of opposite parallel sides), a **rhombus** (parallelogram with all sides of the same length), a **rectangle** (parallelogram with all right angles), or a **square** (all sides of the same length and all right angles).	455
• An **equilateral polygon** is a polygon with all of its sides congruent.	456
• An **equiangular polygon** is a convex polygon whose interior angles are all congruent.	456
• A **regular polygon** is a convex polygon that is both equilateral and equiangular.	456
• A **central angle** is an angle in a regular n-gon with a vertex at the center of a regular polygon and sides containing adjacent vertices of the polygon.	456
THEOREMS	
• **Jordan curve theorem:** A simple closed plane curve partitions the plane into three disjoint subsets: the curve itself, the interior of the curve, and the exterior of the curve.	446
• **Sums of the angle measures in a convex polygon:** The sum of the measures of the exterior angles of a convex polygon is 360°. The sum of the measures of the interior angles of a convex n-gon is $(n - 2)180°$.	450
• **Total-turn theorem:** The total turn made when traversing any closed curve is an integer multiple of 360°.	453

Section 9.3 Figures in Space	Page Reference
CONCEPTS	
• A **plane** is a flat surface that is infinite in all directions and "looks like" the Cartesian plane but is not three dimensional.	465
• A **simple closed surface** is a surface without holes or boundary edges that encloses a region called its **interior.**	466
• There are only five regular polyhedra in the plane and they were known in the time of Plato.	469
• **Nets** are models of polyhedrons that depict the patterns of figures that make up the polyhedron.	469
• Example of **curved surfaces** are Spheres, hemispheres, cones, prisms, and so on.	474
DEFINITIONS	
• **Half-spaces** are the two regions that are separated by a plane.	465
• **Skew lines** are two lines that do not belong to a common plane.	465
• A **sphere** is the set of points in space at a constant distance from a single point called its **center.**	466
• A **solid** is a space figure that is the union of all points on a simple closed surface and all points in its interior.	466

• A **polyhedron** (*pl.* polyhedra) is a simple closed surface formed by planar polygonal regions. These polygons are called **faces,** and the sides and vertices of the faces are called the **edges** and **vertices,** respectively, of the polyhedron. There is a wide variety of polyhedra including prisms, pyramids, and the five regular polyhedra (tetrahedron, cube, octahedron, dodecahedron, and icosahedron).	467
• A **pyramid** is a simple closed surface given by a polygon and a point not in the plane of the polygon, called its apex.	468
• A **prism** is a simple closed surface that consists of two congruent polygons in parallel planes together with the lateral faces joining the bases, which are parallelograms.	468
• A **right prism** is a prism whose lateral faces are all rectangles.	468
• An **oblique prism** is a prism whose lateral faces are not perpendicular to the plane of the base.	468
• A **regular polyhedron** is a polyhedron that has a convex surface, faces that are congruent regular polygonal regions, and the same number of faces meeting at each vertex of the polyhedron.	469
• **Tetrahedron** is a three-dimensional shape with four faces.	467
• **Pentahedron** is a three-dimensional shape with five faces.	467
• **Hexadedron** is a three-dimensional shape with six faces.	467
• A **cone** is a simple closed planar curve, called the **base,** and a point, called the **apex** or the **vertex,** which is not in the plane of the curve.	474
• A **cylinder** is a simple closed surface generated by translating the points of a simple closed region in one plane to a parallel plane. If the line segments joining corresponding points in the two bases are perpendicular to the planes of the bases, it is a **right cylinder.** If they are not perpendicular, the cylinder is an **oblique cylinder.**	474
THEOREM	
• **Euler's formula:** The numbers V of vertices, F of faces, and E of edges of a polyhedron are related by the formula $V + F = E + 2$.	472

Chapter Review Exercises

Section 9.1

1. Let $ABCD$ be the following quadrilateral:

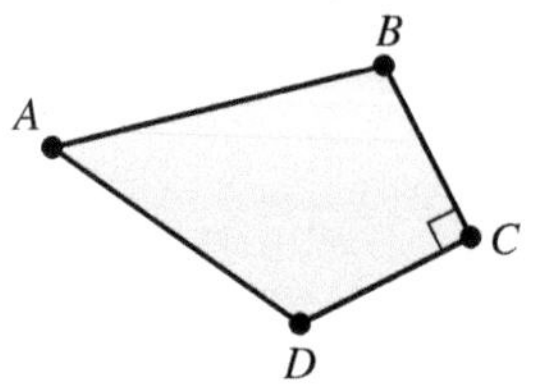

Give symbols for the following:

(a) The line containing the diagonal through C

(b) The diagonal containing B

(c) The length of the side containing A and D

(d) The angle *not* containing D

(e) The measure of the interior angle at C

(f) The ray that has vertex D and is perpendicular to a side of the quadrilateral

2. For the quadrilateral shown in problem 1, which angle(s) appear to be

(a) acute?

(b) right?

(c) obtuse?

3. An angle measures 37°. What is the measure of

(a) its supplementary angle?

(b) its complementary angle?

4. Lines l and m are parallel. Find the measures p, q, r, and s of the angles shown.

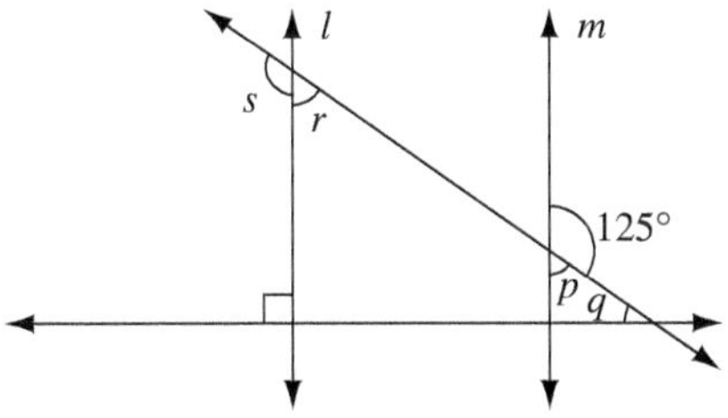

5. Find the measures x, y, and z of the angles in the following figure:

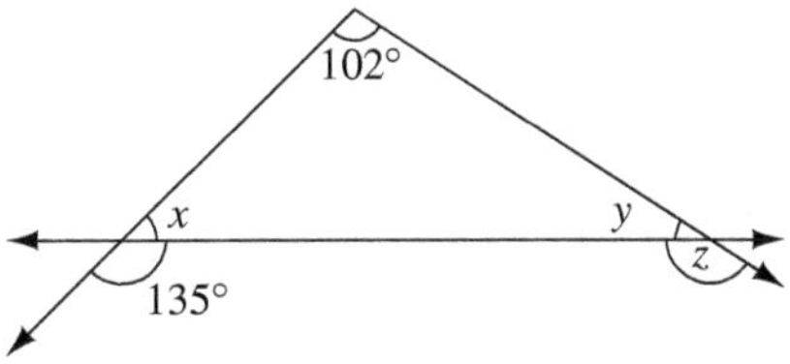

Section 9.2

6. Match each curve to one of the descriptions that follow:

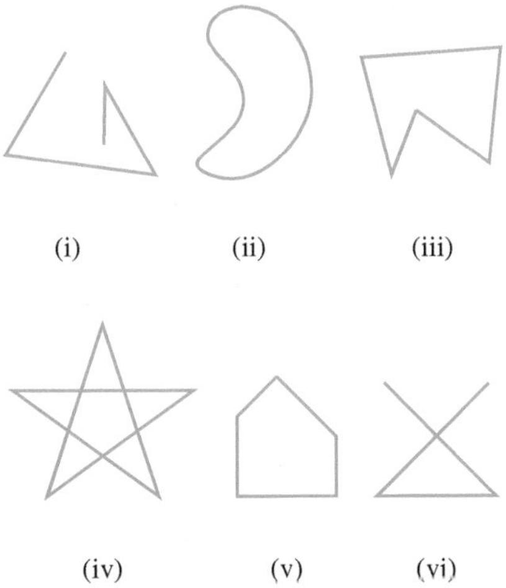

(a) Nonconvex nonsimple polygonal curve
(b) Nonclosed simple curve
(c) Nonsimple nonclosed polygonal curve
(d) Convex polygon
(e) Simple closed nonconvex nonpolygonal curve
(f) Nonconvex polygon

7. For each part, carefully explain your reasoning:
(a) Can a triangle have two obtuse angles?
(b) Can a convex quadrilateral have three obtuse interior angles?
(c) Is there an "acute" quadrilateral (that is, a quadrilateral whose angles are all acute)?

8. Find the measures of the interior angles of this polygon:

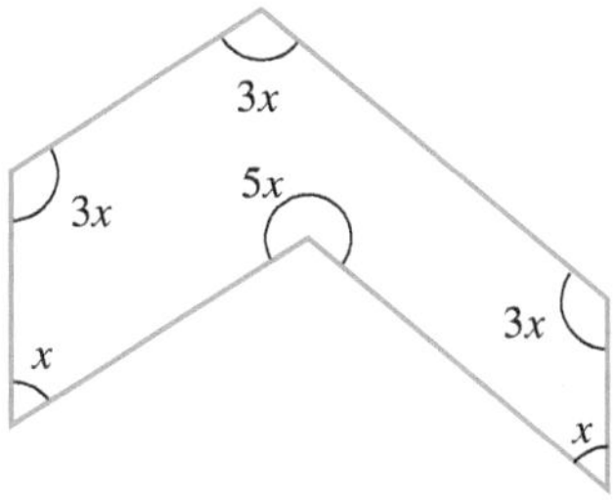

9. A turtle walks along the path $ABCDEFGA$ in the direction of the arrows, returning to the starting point and initial direction. What total angle does the turtle turn through?

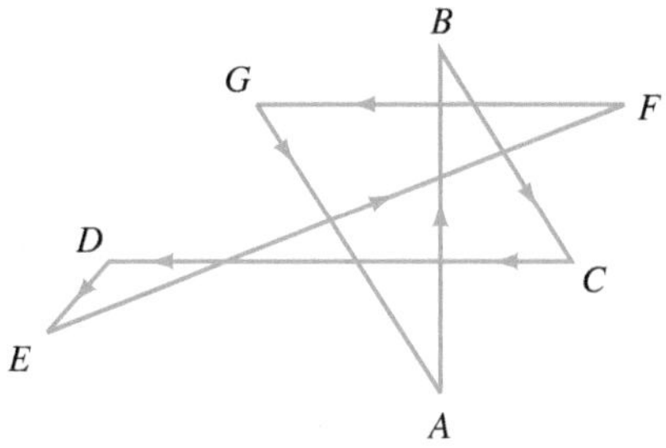

Section 9.3

10. Let $ABCDEFGH$ be the vertices of a cube, as follows:

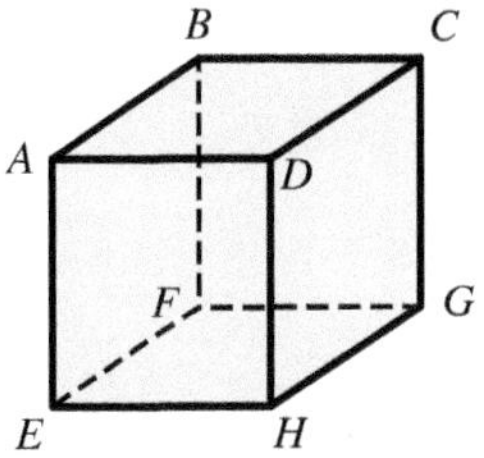

(a) How many planes are determined by the faces of the cube?
(b) Which edges of the cube are parallel to edge $\overline{AB}$?
(c) Which edges of the cube are contained in lines that are skew to the line $\overleftrightarrow{AB}$?
(d) What is the measure of the dihedral angle between the plane containing $ABCD$ and the plane containing $ABGH$?

11. Name the following surfaces in space:

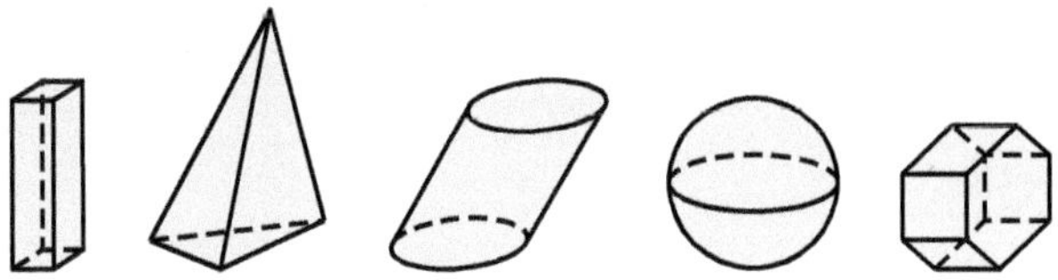

12. Draw the following shapes:
(a) A right circular cone
(b) A pentagonal prism
(c) A nonconvex quadrilateral pyramid

13. (a) Draw a regular octahedron.
(b) Using your drawing in part (a), count the number of vertices, faces, and edges of the octahedron and then verify that Euler's formula holds for the octahedron.

14. A polyhedron has 14 faces and 24 edges. How many vertices does it have?

10 Measurement: Length, Area, and Volume

COOPERATIVE INVESTIGATION

The Metric Measurement and Estimation Tournament

Directions

Set up the six event stations with the materials described in the forthcoming table. For each event, put a copy of the directions for that event at the event's station. Divide the class into six teams, with each team assigned to one of the event stations, and give the team the score sheet that follows the table. Each team then competes in its assigned event, records its scores on the score sheet, and rotates to its next event station. When all teams have completed all of the events, the true distances between cities are made known for the Distance Flier event and the true areas of the foot patterns are revealed for the Big Foot event. After the teams have completed filling out their score sheets, they are ranked in each event according to the lowest average error made by the team. For each event, the most accurate team is awarded 10 points, with the first and second runner-up teams given 7 and 3 points, respectively. The team with the most total points for all six events wins the tournament.

Materials and Directions for the Events

Event	Materials	Directions
String Cut 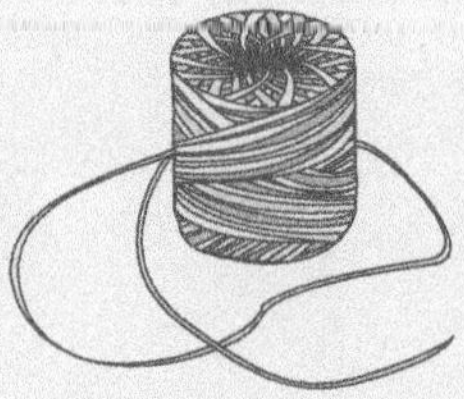	Ball of string, scissors, die, centimeter ruler	In turn, each team member rolls the die and cuts a piece of string whose length in centimeters is estimated to be 10 times the number on the die. Measure the actual length of the string and record the error for each team member. Then compute and record the average team error.
Shot Put 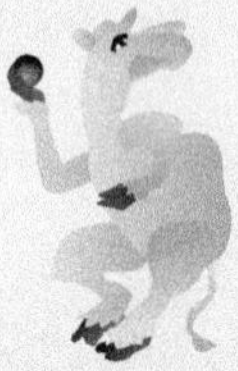	Foam ball (or wad of newspaper), 10-meter-long tape measure (if a metric tape measure is unavailable, mark meters on a tape measure that shows units in feet)	Each team member throws the ball like a shot put and estimates the distance of the throw in meters. The true distance is measured, and the error made in the estimates is recorded for each team member. Compute and record the average team error.
Weight Lift 	Scale (reading in grams), collection of objects of various weights (rocks, canned goods, books, etc.)	Each team member chooses an object and estimates its weight in grams. The objects are then weighed, the errors are recorded, and the average team error is computed.
Pour It On 	Bag of rice (or small dry beans), die, bowl, measuring cup marked in millilitres	In turn, each team member rolls the die to get a value $d = 1, 2, 3, 4, 5$, or 6. The team member then pours rice into the bowl, attempting to fill it with $d \times 100$ milliliters of rice. Finally, pour the rice in the bowl into the measuring cup to measure the actual volume of rice. Determine the individual error made and the team's average error.

Distance Flier 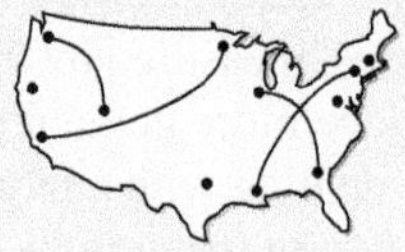	Map of the United States (or a state), paper bag containing cards showing the names of about a dozen cities on the map (whose true distances apart are on an overhead transparency shown at the end of the tournament to determine the teams' errors and rankings in the event)	Each student draws two cards from the bag and estimates the distances between the cities in kilometers. The individual errors and average team error are determined when the instructor reveals the true distances between the cities.
Big Foot Area 	Centimeter-squared paper, die, collection of six foot shapes whose areas are known only to the class instructor (who reveals the areas at the end of the activity to determine the teams' errors and rankings in the event)	A die is rolled to select the team's foot pattern. The pattern is traced onto centimeter-squared paper, and the entire team estimates its area in square centimeters. The team's error is determined when the instructor reveals the true areas of the foot shapes.

Team Score Sheet

Event		Team Members					Total of errors	Average errors	Tournament points
String Cut	Estimate								
	Actual								
	Error								
Shot Put	Estimate								
	Actual								
	Error								
Weight Lift	Estimate								
	Actual								
	Error								
Pour It On	Estimate								
	Actual								
	Error								
Distance Flier	City A								
	City B								
	Est. dist.								
	True dist.								
Big Foot	# of foot measured is ______	Estimated area of foot = cm^2					Error of measurement =		
		Actual area of foot = cm^2							

CHAPTER PREVIEW

The Principles and Processes of Measurement

Measurement played a limited, but important, role in Chapter 9. Only two elements were measured: line segments, measured by the distance between their endpoints; and angles, measured by the degrees of rotation needed to turn one side to the other. In this chapter, we introduce more general notions of measurement of geometric figures. The Pythagorean theorem is a key to dealing with a large number of problems. Plane regions will be measured by area and perimeter. Space figures will be measured by surface area and volume.

The process starts by asking what **attribute** of the geometric figure we want to measure. Next, we begin to discuss the general process of measurement and the concept of a unit of measurement. The two principal systems of measurement are then described: the U.S. Customary (English) System, used in the United States, but in almost no other country; and the International System (metric), used by all countries worldwide, including the United States.

10.1 The Measurement Process

First, here's a bit of the history of measurement. The geometry of the Babylonians and ancient Egyptians always had a practical purpose, and often this purpose was dependent on having knowledge of size and capacity. Then—as now—it was important to know the areas of fields, the volumes of granaries, and so on. Many engineering projects gave rise to geometric problems concerned with magnitudes.

Determining size of an item requires that a comparison be made between what you have (say, a large field) and with a special size—such as an acre of land. The original definition of an *acre* is the area of land that could be plowed in one day with one team of oxen. In this case, the unit is called an *acre*.

In early times, **units** of measurement were defined more for convenience than accuracy. For example, many units of length correspond to parts of the human body, some of which are shown in Figure 10.1. The hand, span, foot, and cubit all appear in early records of Babylonia and Egypt.

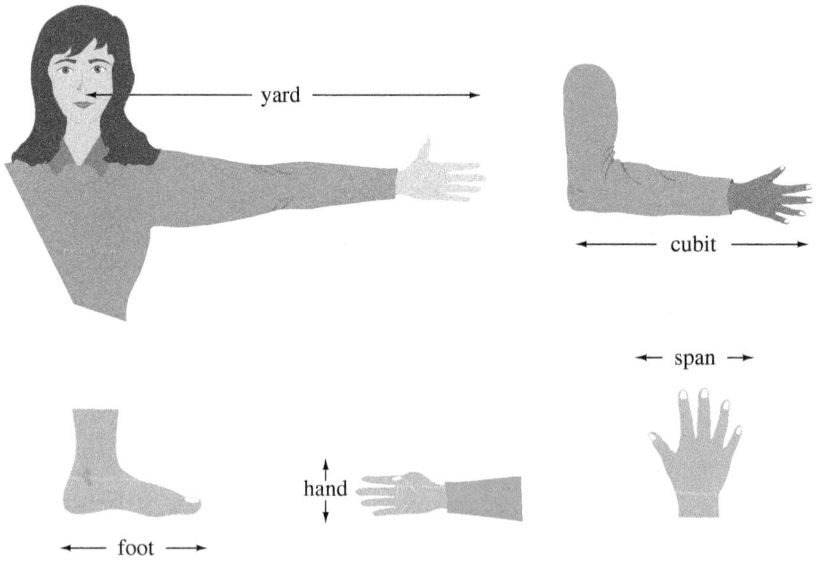

Figure 10.1
Examples of traditional units of length based on the human body

Many of these units later became **standardized** and persist today. For example, horses are still measured in hands, where a hand is now 4 inches. Originally, an inch was the length of three barleycorns (a major cereal grain) placed end to end.

Learning the **measurement process** can be viewed as a sequence of steps—steps that start even before a child goes to school. Many adults think only of concepts of measurement such as length, area, and volume as topics for the later grades. However, the NCTM takes the strong position that the study of measurement should start even before schooling does, and there is much research to support that assertion! (See the NCTM's *Measurable Attributes and the Processes of Measurement in Grades Pre-K–2* that follows.)

The Measurement Process

(i) Choose the property, or attribute (such as length, area, volume, capacity, temperature, time, or weight), of an object or event that is to be measured.
(ii) Select an appropriate unit of measurement.
(iii) Use a measurement device to "cover," "fill," "time," or otherwise provide a comparison of the object with the unit.
(iv) Express the measurement as the number of units used.

FROM THE NCTM PRINCIPLES AND STANDARDS

Measurable Attributes and the Processes of Measurement in Grades Pre-K–2

Children should begin to develop an understanding of attributes by looking at, touching, or directly comparing objects. They can determine who has more by looking at the size of piles of objects or identifying which of two objects is heavier by picking them up. They can compare shoes, placing them side by side, to check which is longer. Adults should help young children recognize attributes through their conversations. "That is a *deep* hole." "Let's put the toys in the *large* box." "That is a *long* piece of rope." In school, students continue to learn about attributes as they describe objects, compare them, and order them by different attributes. Seeing order relationships, such as that the soccer ball is bigger than the baseball but smaller than the beach ball, is important in developing measurement concepts.

Teachers should guide students' experiences by making the resources for measuring available, planning opportunities to measure, and encouraging students to explain the results of their actions. Discourse builds students' conceptual and procedural knowledge of measurement and gives teachers valuable information for reporting progress and planning next steps. The same conversations and questions that help students build vocabulary help teachers learn about students' understandings and misconceptions. For example, when students measure the length of a desk with rods, the teacher might ask what would happen if they used rods that were half as long. Would they need more rods or fewer rods? If students are investigating the height of a table, the teacher might ask what measuring tools would be appropriate and why.

SOURCE: Principles and Standards for School Mathematics by NCTM, *page 103. Copyright © 2000 by the National Council of Teachers of Mathematics. Reproduced with permission of the National Council of Teachers of Mathematics via Copyright Clearance Center. NCTM does not endorse the content or validity of these alignments.*

In Example 10.1, a **tangram** piece is chosen to be the unit of area measurement. Tangrams originated in ancient China and continue to be a versatile manipulative that is used in the present-day classroom. The instructions that follow show how to make a set of tangram pieces by folding and cutting a square sheet of paper. A more sturdy set of tangrams can be cut from cardboard or vinyl tile, using the paper shapes as templates.

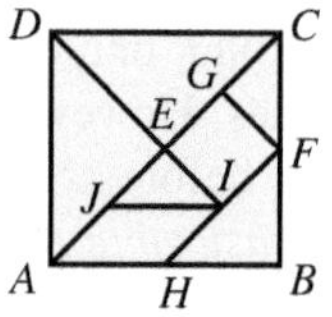

The seven tangram pieces.

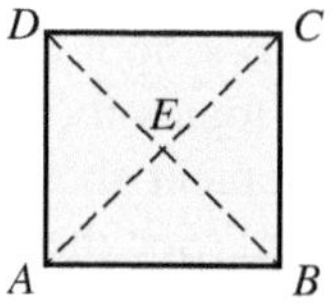

Step 1: Fold the two diagonals of the square $ABCD$. Let E denote its center.

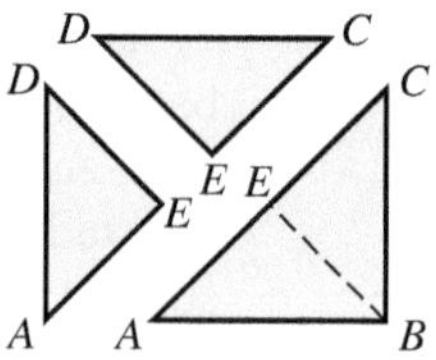

Step 2: Cut out the tangram pieces CDE and ADE, leaving triangle ABC.

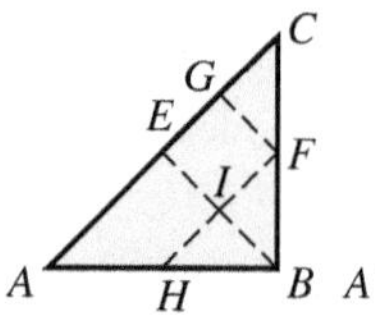

Step 3: Fold B and C to center point E to create folds FG and FH.

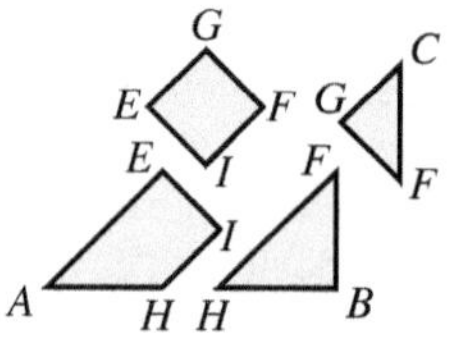

Step 4: Cut out tangram pieces BFH, CFG, and $FGEI$, leaving trapezoid $AHIE$.

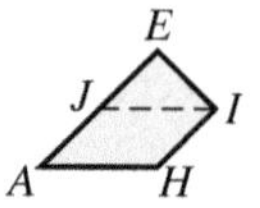

Step 5: Fold E to H to create fold IJ.

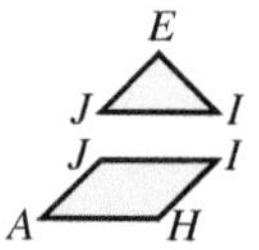

Step 6: Cut out tangram pieces $AHIJ$ and EIJ.

EXAMPLE 10.1

Investigating Tangram Measurements

Label the tangram pieces I, II, . . . , VII, as shown. Use shape I, the small isosceles right triangle, as the unit of "one tangram area" (abbreviated 1 tga) to measure the following:

(a) the area of each of the tangram pieces,
(b) the area of the "fish," and
(c) the area of the circle that circumscribes the square.

Are your measurements exact or only approximate?

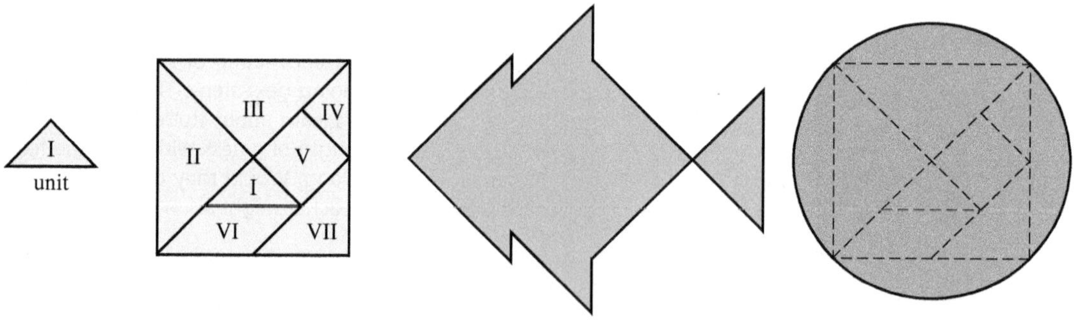

Solution

(a) Each tangram piece can be covered by copies of the unit shape I, giving the exact measurements in the table.

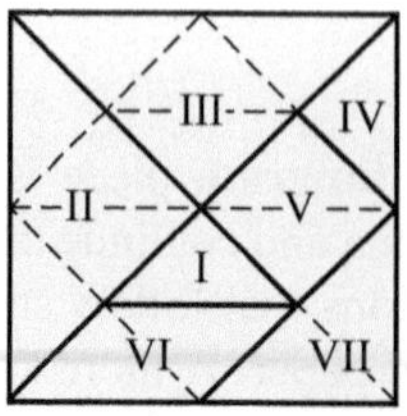

Piece	Area
I, IV	1 tga
II, III	4 tga
V, VI, VII	2 tga

(b) The fish is covered by the seven tangram pieces, so its area is exactly $(4 + 2 + 1 + 2 + 1 + 4 + 2)$ tga $= 16$ tga.

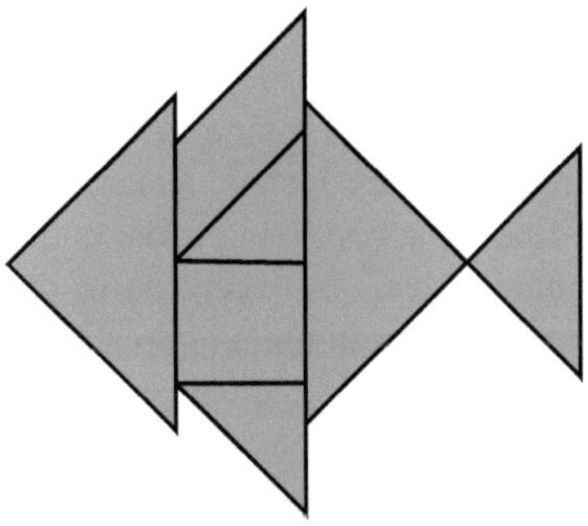

(c) The circle can be covered by the seven tangram pieces, together with 12 additional copies of the unit shape. This shows that the circle's area is between 16 tga and 28 tga. Thus, we might estimate the area at about 25 tga.

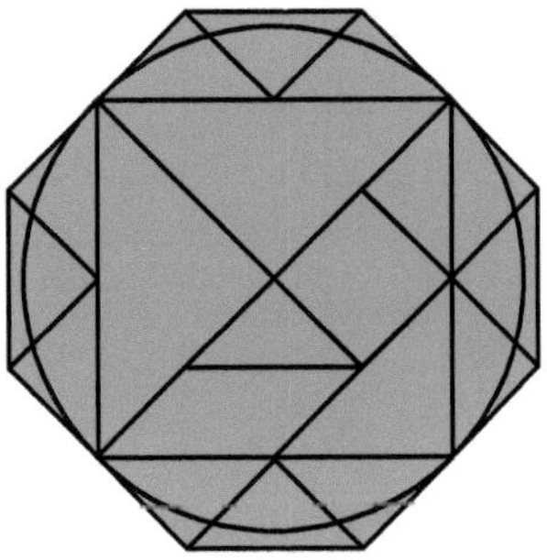

An important practical purpose of measurement is communication. By agreeing on common units of measurement, people are able to express and interpret information about size, quantity, capacity, and so on. Historically, as commerce developed and goods were traded over increasingly large distances, the need for a standard system of units became more and more apparent. In the seventeenth and eighteenth centuries, the rise of science and the beginning of the Industrial Revolution gave further impetus to the development of universal systems of measurement.

The U.S. Customary, or "English," System of Measures

The English system arose from a hodgepodge of traditional informal units of measurement. Table 10.1 lists some of the units of **length** with this system. The ratios comparing one unit of length with another are clearly the result of accident, not planning. Learning the customary system requires extensive memorization, and using the system involves computations with cumbersome numerical factors.

TABLE 10.1 **UNITS OF LENGTH IN THE CUSTOMARY SYSTEM**

Unit	Abbreviation	Equivalent Measurement in Feet
Inch	in.	$\frac{1}{12}$ ft
Foot	ft	1 ft
Yard	yd	3 ft
Rod	rd	$16\frac{1}{2}$ ft
Furlong*	fur	660 ft
Mile	mi	5280 ft

*The *furlong* is a shortened form of "furrow long," revealing its origin in agriculture.

Area is a measure of the region bounded by a closed plane curve. Any shape could be chosen as a unit, but the square is the most common. The size of the square is arbitrary, but it is natural to choose the length of a side to correspond to a unit measure of length. Areas are therefore usually measured in square inches, square feet, and so on. A moderate-sized house may have 1800 square feet of floor space, a living room carpet may cover 38 square yards, and a national forest may cover 642 square miles. An exception to this pattern is the acre: 640 acres have a total area of 1 square mile. Some common units of area are listed in Table 10.2. The superscript-2 notation shows a square unit; for example, ft^2 indicates square feet.

TABLE 10.2 UNITS OF AREA IN THE CUSTOMARY SYSTEM

Unit	Abbreviation	Equivalent Measure in Other Units
Square inch	$in.^2$	$\frac{1}{144}\ ft^2$
Square foot	ft^2	$144\ in.^2$, or $\frac{1}{9}\ yd^2$
Square yard	yd^2	$9\ ft^2$
Acre	acre	$\frac{1}{640}\ mi^2$, or $43{,}560\ ft^2$
Square mile	mi^2	640 acres, or $27{,}878{,}400\ ft^2$

The ratios comparing one unit of area with another can be visualized, as shown in Figure 10.2. We see that the area of a 3-ft by 3-ft square is obtained by the multiplication $3\ ft \times 3\ ft = 3 \times 3 \times ft \times ft = 9\ ft^2$. *When computing with dimensioned quantities, it is essential to retain the units in all equations and expressions.* For example, it is correct to write 12 in. = 1 ft; without the dimensions, this equation would be incorrect, since $12 \neq 1$. Omitting the units in expressions is a common source of errors.

Figure 10.2
Comparing units of area measure

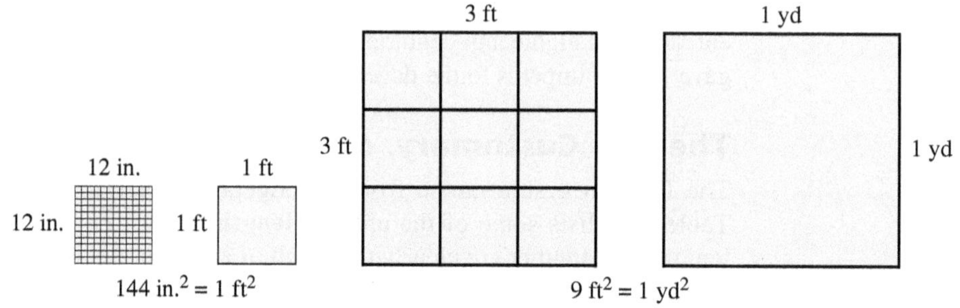

Volume is the measure of space taken up by a solid in three-dimensional space. The unit, as shown in Table 10.3, is the volume of a cube whose side length is one of the standard units of length.

TABLE 10.3 UNITS OF VOLUME IN THE CUSTOMARY SYSTEM

Unit	Abbreviation	Equivalent Measure in Other Units
Cubic inch	$in.^3$	$\frac{1}{1728}\ ft^3$
Cubic foot	ft^3	$1728\ in.^3$, or $\frac{1}{27}\ yd^3$
Cubic yard	yd^3	$27\ ft^3$

The ratios comparing units of volume are illustrated in Figure 10.3.

Figure 10.3
Comparing units of volume measure

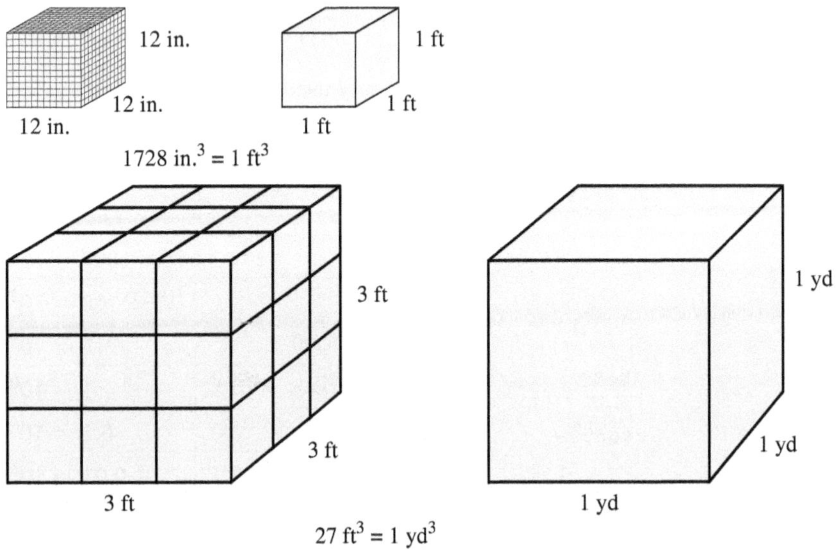

Have you ever wondered why x^2 is called "x squared" and x^3 is called "x cubed"? The right answer comes from thinking about the questions "What is the area of a square of side x?" and "What is the volume of a cube of side x?"

Capacity is the volume that can be held in a container such as a bottle, pan, basket, or tank. Capacity is often expressed in in.^{3}, ft^3, or yd^3, but other units are also in common use. Some common examples are shown in Table 10.4.

TABLE 10.4 UNITS OF CAPACITY IN THE U.S. CUSTOMARY SYSTEM

Unit	Abbreviation	Equivalent Measure in Other Units
Teaspoon	tsp	$\frac{1}{3}$ tablespoon
Tablespoon	T or tbl	Ø.5 fl. oz.
Fluid ounce	fl. oz.	$\frac{1}{8}$ cup
Cup	C	8 fl. oz., or $\frac{1}{4}$ quart
Quart	qt	4 cups, or $\frac{1}{4}$ gallon
Gallon	gal	4 quarts, or 231 cubic inches

Metric Units: The International System

The metric system of measurement originated in France shortly after the revolution of 1789. The definitions of the units have been modified over succeeding years, taking advantage of scientific and technological advances. The system was codified in 1981 by the International Standardization Organization. The International System of Units, also called the **SI system** has now achieved worldwide acceptance. In the 1970s, a movement in the United States to replace customary units with metric units was unsuccessful. About the same time, most other English-speaking countries, including Great Britain, Canada, Australia, and New Zealand, did change to metric units. Even day-to-day measurements in those countries—speed limits, distances between cities, and amounts in recipes—were replaced with metric units.

The principal advantage of the metric system—other than its universality—is the ease of comparison of units. The ratio of one unit to another is always a power of 10, which ties the metric system conveniently to the base-ten numeration system. This relationship makes it quite simple to convert a measurement in one metric unit to the equivalent measurement in another metric unit.

Each power of 10 is given a prefix that modifies the fundamental unit. For example, the factor 1000 (that is, 10^3) is expressed by the prefix *kilo*. Thus, a kilometer is 1000 meters. Similarly, the factor $\frac{1}{100}$ (that is, 10^{-2}, or 0.01) is expressed by the prefix *centi.* Therefore, a centimeter is $\frac{1}{100}$ of a meter. The more commonly used prefixes and their symbols are listed in Table 10.5.

TABLE 10.5 THE SI DECIMAL PREFIXES

Prefix	Factor	Symbol
kilo	$1000 = 10^3$	k
hecto	$100 = 10^2$	h
deka (or deca)	$10 = 10^1$	da
(none for basic unit)	$1 = 10^0$	(none)
deci	$0.1 = 10^{-1}$	d
centi	$0.01 = 10^{-2}$	c
milli	$0.001 = 10^{-3}$	m
micro	$0.000001 = 10^{-6}$	μ (Greek *mu*)

Length

The fundamental unit of length in the SI system is the **meter,** abbreviated by the symbol m. The unit symbol is always written last in SI, so there can be no confusion with the prefix *milli,* which is also given the symbol m. For example, one-thousandth of a meter is a millimeter, written as 1 mm. There is no space between the first and second m, and there are no periods between or after the symbols. When typed, the symbols are always in roman font, not italic. The most commonly used metric units of length are listed in Table 10.6.

TABLE 10.6 METRIC UNITS OF LENGTH

Unit	Abbreviation	Multiple or Fraction of 1 Meter
1 kilometer	1 km	1000 m
1 hectometer	1 hm	100 m
1 dekameter	1 dam	10 m
1 meter	1 m	1 m
1 decimeter	1 dm	0.1 m
1 centimeter	1 cm	0.01 m
1 millimeter	1 mm	0.001 m
1 micrometer (or micron)	1 μm	0.000001 m
1 nanometer	1 nm	10^{-9} m

The prefix (for example, the "c" in the notation "cm") in a metric measurement can be replaced with its corresponding numerical factor. For example,

$$251 \text{ cm} = 251 \times 10^{-2} \text{ m} = 2.51 \text{ m}.$$

Similarly, a power of 10 can be replaced with the corresponding prefix, as in

$$0.179 \text{ m} = 179 \times 10^{-3} \text{ m} = 179 \text{ mm}.$$

Some metric measurements are shown in Figure 10.4. Since the items shown differ so dramatically in size, the scale for each image (except for the nickel) is greatly reduced or enlarged. The length shown as representing one micrometer is intended to make it clear that waves of red and violet light, the colon bacillus, and the smallpox virus are so tiny as to be determinable only by using sophisticated scientific equipment.

Your students and you will need to estimate lengths, areas, and volumes by taking measurements in the metric system. Another important practical application in understanding what sizes look like when metric measurements are used is their appearance in Statewide Assessment exams. (See the Student State Assessment problems at the end of this section.)

The nanometer (see Table 10.6) is often associated with the field of nanotechnology. Since the late 1980s, nanotechnology has been used to describe devices such as cell phones and computer chips as well as many aspects of modern manufacturing and semiconductor industries. The key is the "small scale," as 1 nm is one-billionth of a meter. The diameter of a helium atom is about one-tenth of a nanometer, and that is truly small!

Figure 10.4
Examples of metric measurements of length

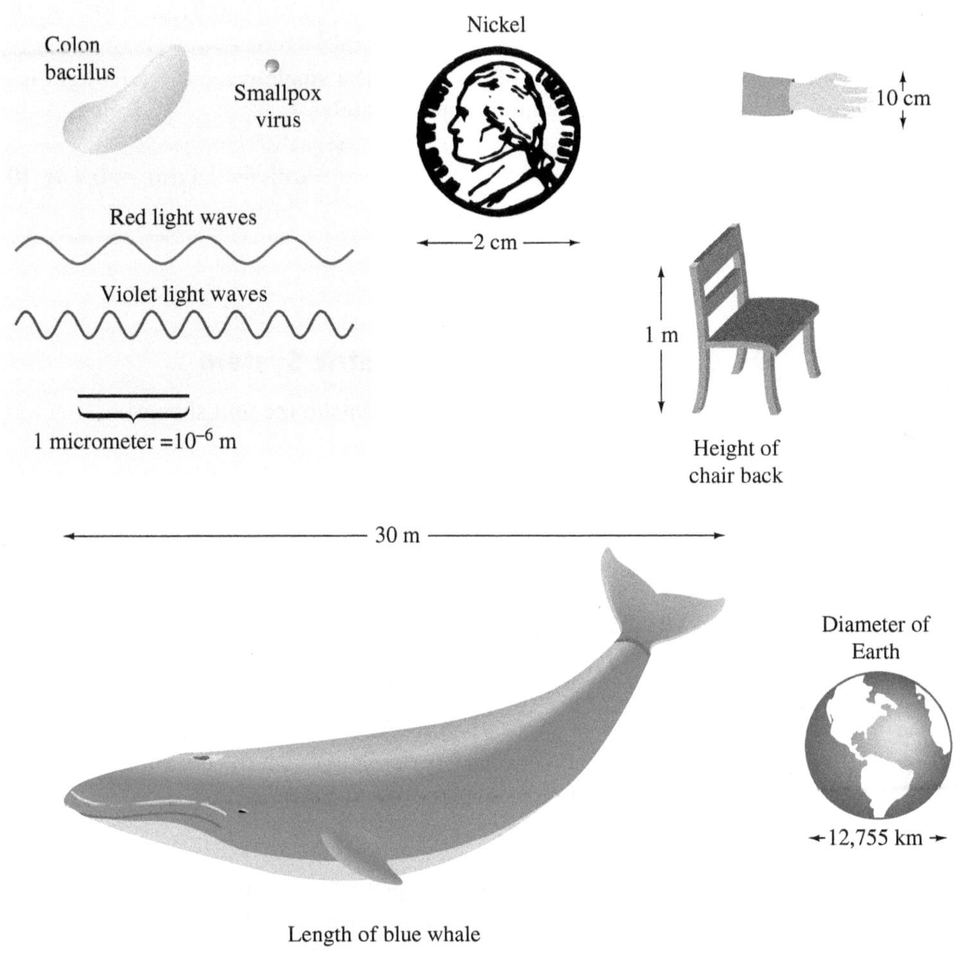

EXAMPLE 10.2

Determining Tiny Lengths in Figure 10.4

Using a metric ruler, measure the following elements from Figure 10.4 as carefully as possible and give the measurements in micrometers and meters:

(a) The length of the colon bacillus
(b) The diameter of the smallpox virus
(c) The length of one wave of red light (i.e., the distance from one peak to the next)
(d) The length of one wave of violet light

Solution

Recall that it is common to use both $\doteq$ and $\approx$ as symbols for "is approximately equal to."

(a) The length of the image of the colon bacillus is approximately 19 mm. Since the length of the unit representing one micrometer is approximately 14 mm, the length of the colon bacillus is approximately

$$\frac{19}{14}\mu\text{m} \doteq 1.4\mu\text{m} = 1.4 \times 10^{-6}\text{ m}.$$

(b) Estimating as closely as possible, it appears that the diameter of the image of the smallpox virus is 1.5 mm. Thus, the actual length of the diameter is approximately

$$\frac{1.5}{14}\,\mu\text{m} \doteq 0.1\,\mu\text{m} = 10^{-7}\text{ m}.$$

(c) The length of the image of a single wave of red light is approximately 11 mm. Hence, the actual length is approximately

$$\frac{11}{14}\,\mu\text{m} \doteq 0.8\,\mu\text{m} = 0.8 \times 10^{-6}\text{ m}.$$

(d) The length of the image of a single wave of violet light is approximately 5.5 mm. Thus, the actual length is approximately

$$\frac{5.5}{14}\,\mu\text{m} \approx 0.4\,\mu\text{m} = 0.4 \times 10^{-6}\text{ m}.$$

EXAMPLE 10.3

Changing Units in the Metric System

Convert each of these measurements to the unit shown:

(a) 1495 mm = _______ m
(b) 29.4 cm = _______ mm
(c) 38,741 m = _______ km

Solution

(a) $1495 \text{ mm} = 1495 \times 10^{-3}\text{ m} = 1.495\text{ m}$
(b) $29.4 \text{ cm} = (294 \times 10^{-1}) \times 10^{-2}\text{ m} = 294 \times 10^{-3}\text{ m} = 294\text{ mm}$
(c) $38{,}741 \text{ m} = 38.741 \times 10^{3}\text{ m} = 38.741\text{ km}$

For conversion between metric system and the customary system, 1 m is about 3.28 ft (a bit larger than a yard); 1 ft is about 0.3048 m.

Area

Area is usually expressed in square meters (m^2) or square kilometers (km^2). Another common unit is the hectare. A **hectare** (ha) is the area of a 100-m square; that is, 1 ha = 10,000 m^2 (see Table 10.7).

The floor space of a classroom might typically be about one **are** (pronounced "air"). A hectare is about 2.5 acres, so the area of farmland is measured in hectares in metric countries. The name *hectare* comes about because "hecto" is the prefix for 100 in SI (see Table 10.5) and one hectare is 100 ares.

TABLE 10.7 METRIC UNITS OF AREA

Unit	Abbreviation	Multiple or Fraction of 1 Square Meter
1 square centimetre	1 cm^2	0.0001 m^2
1 square meter	1 m^2	1 m^2
1 are (1 square dekameter)	1 a	100 m^2
1 hectare (1 square hectometer)	1 ha	10,000 m^2
1 square kilometre	1 km^2	1,000,000 m^2

Volume and Capacity

Small volumes are typically measured in cubic centimeters (abbreviated cm^3). Large volumes are often measured in cubic meters (m^3). **Capacity** refers to the ability to hold a fluid, such as the capacity of a milk bottle. A convenient unit of capacity is the **liter,** which is used worldwide.

DEFINITION Liter

A **liter** is the volume of a cube, each of whose sides is 10 centimeters or 1 liter is 1000 cm^3.

Thus, a liter holds as much liquid as a cube that is 10 cm (about the length of five fingers) on each side. (See Figure 10.5.) The liter abbreviation is written as either the letter ell, l, or L, with capital L preferred in the United States to avoid confusion with the numeral 1. Since 10 cm is a decimeter, a liter can also be defined as a cubic decimeter.

Since $10\text{ cm} \times 10\text{ cm} \times 10\text{ cm} = 1000\text{ cm}^3$, a liter is also 1000 cubic centimeters. Recalling that *milli* is the prefix for $\frac{1}{1000}$, we see that a milliliter (mL or ml) is the same as one cubic centimeter:

$$1\text{ L} = 1\text{ liter} = 1000\text{ cm}^3;$$
$$1\text{ mL} = 1\text{ milliliter} = 1\text{ cm}^3.$$

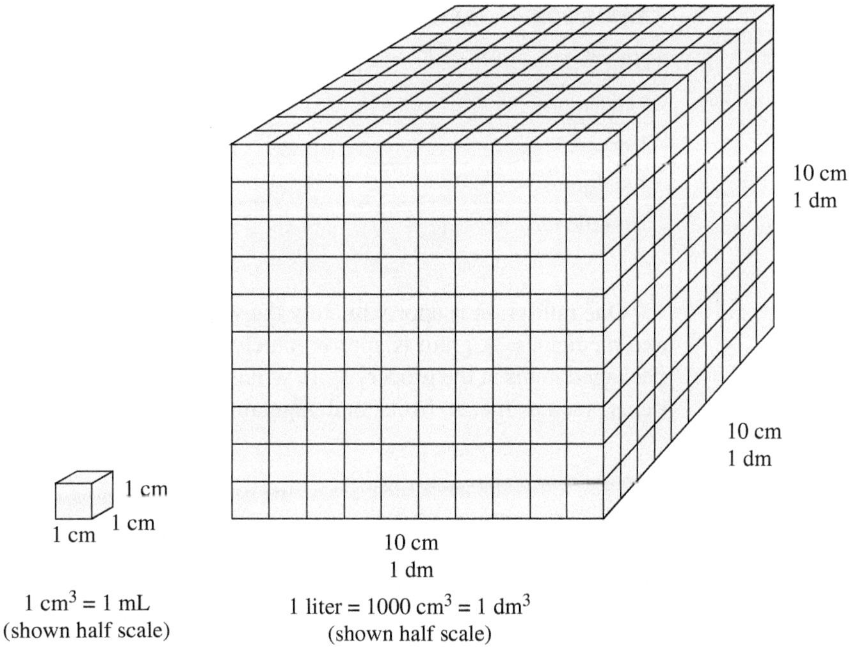

Figure 10.5
A liter is a cubic decimeter, or, equivalently, 1000 cubic centimeters

Large plastic bottles of soda usually contain 2 liters, while a typical soft-drink can contains about 354 milliliters. A child's dose of cough medicine may be 3 mL. A recipe may call for 0.5 liter of water. In metric countries, gasoline is priced by the liter, and to fill a car's gas tank takes about 40 to 60 liters.

Weight and Mass

The **weight** of an object is the force exerted on the object by gravity. Because gravity changes from place to place, a brick on the surface of Earth may weigh 6 pounds, but on the surface of the moon it would weigh only about 1 pound. During the journey from Earth to moon, the brick would weigh nearly nothing at all. Nevertheless, an astronaut would not care to be hit by a weightless brick, since the brick never loses its mass.

In science, the distinction between mass and weight is very important: **Mass** is the amount of matter of an object, and weight is the force of gravity on the object. But on the surface of Earth and in everyday-life situations, the weight of an object is proportional to its mass. That is, the mass of an object is accurately estimated by weighing it.

The U.S. customary unit of weight is the familiar pound. Lighter weights are often given in ounces, and very heavy weights are given in tons:

$$16 \text{ ounces (oz)} = 1 \text{ pound (lb)};$$
$$2000 \text{ pounds} = 1 \text{ ton}.$$

A base unit of weight in the metric system is the **kilogram,** which is about 2.2 pounds.

DEFINITION Kilogram
A **kilogram** is the weight of one liter of water.

Table 10.8 lists some metric units of weight. Since there are 1000 cubic centimeters in one liter and a liter of water weighs 1 kilogram, one cubic centimeter of water weighs 0.001 of a kilogram or, equivalently, 1 gram.

TABLE 10.8 METRIC UNITS OF WEIGHT

Unit	Abbreviation	Multiples of Other Metric Units
1 milligram	1 mg	0.001 g
1 gram	1 g	0.001 kg
1 kilogram	1 kg	1000 g
1 metric ton	1 t	1000 kg

One milligram is approximately the weight of a grain of salt. It is a common measure of vitamins and medicines. A gram is approximately the weight of half a cube of sugar. Canned goods and dry packaged items at the grocery store will usually be weighed in grams. In metric countries, larger food items, such as meats, fruits, and vegetables, are priced by the kilogram (about 2.2 pounds).

EXAMPLE 10.4

Estimating Weights in the Metric System

Match each item to the approximate weight of the item taken from the list that follows:

(a) Nickel
(b) Compact automobile
(c) Two-liter bottle of soda
(d) Recommended daily allowance of vitamin B-6
(e) Size D battery
(f) Large watermelon

List of weights: 2 mg, 2 kg, 100 g, 1200 kg, 9 kg, 5 g

Solution

(a) 5 g **(b)** 1200 kg **(c)** 2 kg **(d)** 2 mg **(e)** 100 g **(f)** 9 kg

Temperature

There are two commonly used scales to measure temperature. According to the **Fahrenheit scale,** 32°F represents the freezing point of water and 212°F the boiling point of water. Thus, the Fahrenheit scale introduces 180 degrees of division between the freezing and boiling temperatures. The **Celsius scale** divides this temperature range into 100 degrees: The freezing point is 0° Celsius, and the boiling point is 100° Celsius.

EXAMPLE 10.5 Translating Temperature in Degrees between Celsius and Fahrenheit

Solution

The thermometers shown side by side in Figure 10.6 can be used to derive a formula that relates C degrees Celsius to the equivalent of F degrees Fahrenheit because it is a linear relationship. We will find the straight line that translates Celsius to Fahrenheit. The formula will come from the use of the two-point form of the equation of a line as was done in Section 8.2 (p. 000) where $P(x_1, y_1)$ and $Q(x_2, y_2)$ are the points. In that case, the equation of the line is

$$y - y_1 = m(x - x_1) \text{ where } m \text{ is the slope of the line.}$$

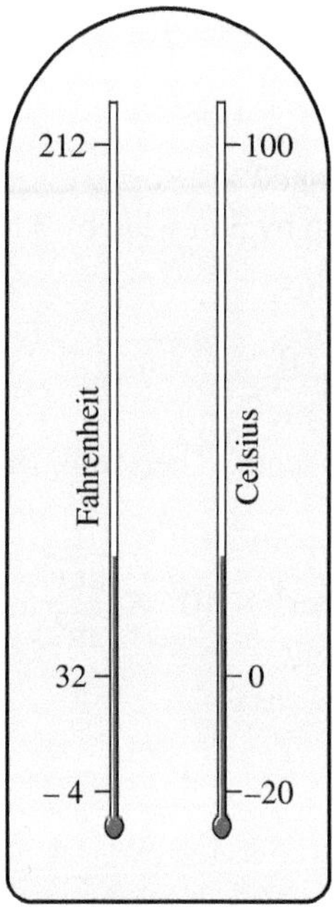

Figure 10.6
Thermometers showing Fahrenheit and Celsius scales

We will now translate the variables, x and y, to C instead of x and F instead of y. Thus, we are looking for the formula that gives the dependent variable F in terms of the independent variable C.

The key is that we already have two points. When $C = 0$ at freezing, F is 32, which is written as (0, 32). At boiling point, $C = 100$ and $F = 212$, so (100, 212) is another point. Thus, we have $P = (0, 32)$ and $Q = (100, 212)$. These two points have their Celsius coordinates listed first and Fahrenheit second.

We first show that the slope m is 1.8. We use the point $P = (0, 32)$ in the two-point form to start the formula

$$F - 32 = m(C - 0).$$

We now use the other point $Q = (100, 212)$ to finish finding the slope:

$$m = \frac{212 - 32}{100 - 0} = 1.8.$$

The two-point form of the equation of a line translating C to F is then $F - 32 = 1.8C$, or

$$F = 1.8\,C + 32.$$

It is not hard to solve the foregoing equation for Celsius in terms of Fahrenheit, which gives the other translation

$$C = \frac{5}{9}(F - 32).$$

Unit Analysis

It is often of interest to express a measurement given in one unit by the equivalent measurement in a new unit. A procedure known as **unit analysis** (or **dimensional analysis**) can help arrange the calculation to make it clear if the factors comparing units are used as multipliers or divisors. The idea of unit analysis can be explained by an example. Suppose a distance has been given as 3.75 miles, and you would like the distance in yards. You recall that 1 mi = 5280 ft and 3 ft = 1 yd. These equations can also be written as $1 = \frac{5280 \text{ ft}}{1 \text{ mi}}$ and $1 = \frac{1 \text{ yd}}{3 \text{ ft}}$. Therefore,

$$3.75 \text{ mi} = 3.75 \cancel{\text{mi}} \times \frac{5280 \cancel{\text{ft}}}{1 \cancel{\text{mi}}} \times \frac{1 \text{ yd}}{3 \cancel{\text{ft}}} = \frac{3.75 \times 5280}{3} \text{yd},$$

Since $\frac{3.75 \times 5280}{3} = 6600$, it follows that 3.75 miles = 6600 yards.

The Standard for Mathematical Practice 2 (SMP 2) cautions students to be careful about specific quantities and their operation including differing units, such as cm, cm^2, cm^3, Fahrenheit versus Celsius temperature, and so on. Computations with different units can cause a real change in a problem. Unfortunately, you will see an example of a disaster in the paragraph immediately after SMP 2.

SMP 2 Reason abstractly and quantitatively.

"Mathematically proficient students make sense of quantities and their relationships in problem situations . . . Quantitative reasoning entails habits of creating a coherent representation of the problem at hand; considering the units involved; attending to the meaning of quantities, not just how to compute them. . . ."

Unit conversion, even using unit analysis, should always be accompanied by careful reasoning. Always ask, "Is this answer reasonable? Does this answer agree with an approximate mental estimation?" Mistakes with unit conversions have led to some unfortunate disasters. For example, in 1999 engineers mistook a measurement of force given in pounds to be given in the metric unit of newtons of force. As a result, over four times the correct amount of thrust was applied to the *Mars Climate Orbiter,* sending the $125 million space probe to an early demise in the atmosphere of the Red Planet.

EXAMPLE 10.6 Computing Speed and Capacity with Unit Analysis

(a) A cheetah can run 60 miles per hour. What is the speed in feet per second?

(b) A fish tank at the aquarium has the shape of a rectangular prism 2 m deep by 3 m wide by 3 m high. What is its capacity in liters?

Solution

(a) $60\dfrac{\text{mi}}{\text{hr}} = 60\dfrac{\cancel{\text{mi}}}{\cancel{\text{hr}}} \times \dfrac{5280 \text{ ft}}{1 \cancel{\text{mi}}} \times \dfrac{1 \cancel{\text{hr}}}{60 \cancel{\text{min}}} \times \dfrac{1 \cancel{\text{min}}}{60 \text{ sec}} = \dfrac{60 \times 5280 \text{ ft}}{60 \times 60 \text{ sec}} = 88\dfrac{\text{ft}}{\text{sec}}.$

(b) Recall that a liter is a cubic decimeter, and *deci* is the prefix for one-tenth. Therefore, the volume of the tank is

$$(2 \text{ m}) \times (3 \text{ m}) \times (3 \text{ m}) = 18 \text{ m}^3 = 18 \text{ m}^3 \times \left(\frac{10 \text{ dm}}{1 \text{ m}}\right)^3 = 18 \times \cancel{\text{m}^3} \times 10^3 \times \frac{\text{dm}^3}{\cancel{\text{m}^3}}$$

$$= 18{,}000 \cancel{\text{dm}^3} \times \frac{1 \text{ liter}}{\cancel{\text{dm}^3}} = 18{,}000 \text{ liters.}$$

10.1 Problem Set

Understanding Concepts

1. For each of the following objects, make a list of some of the measurable properties:

(a) A bulletin board **(b)** An extension cord

(c) A file box **(d)** A table

2. Suppose you are designing a house. Give examples of measurements you believe are important to consider. For example, the height of the house may be needed to satisfy a zoning regulation. Discuss examples of measurements of (a) length, (b) area, and (c) volume and capacity. What units are appropriate?

3. Find the area of each tangram figure shown, where a unit is the area of the small isosceles right triangle:

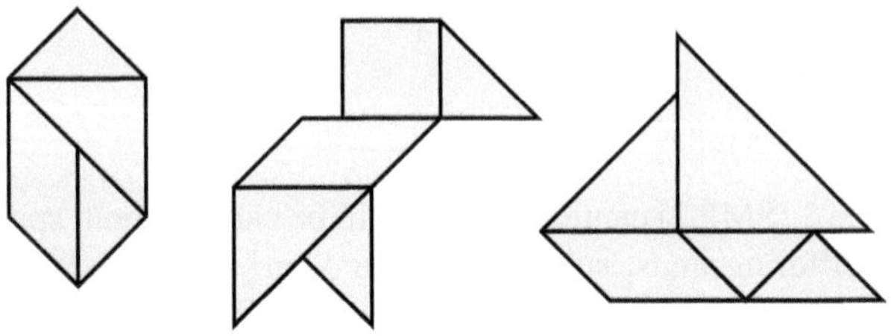

4. Let's use the term *pen* to refer to the area of a penny.

(a) Estimate the area of a 4″ × 6″ card in pens.

(b) Discuss why pens are a difficult unit of area to use.

5. Arrange the solids shown into a list according to volume from smallest to largest. Are there any ties?

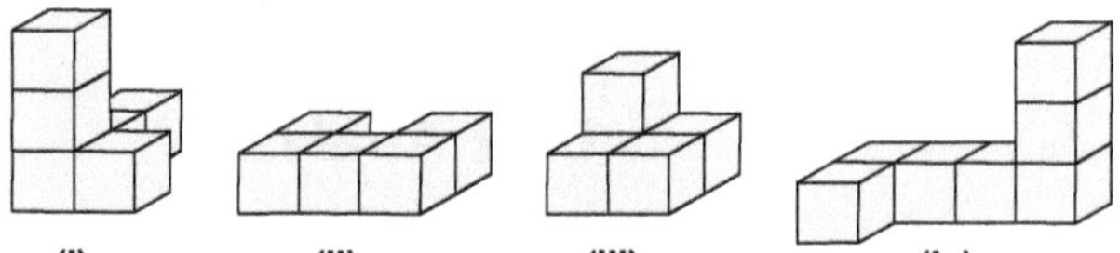

(i) (ii) (iii) (iv)

6. (a) Verify that an acre contains 43,560 square feet. Show your computation.

(b) A square lot contains 1 acre. What is the length of each side, to the nearest foot?

7. (a) A football field is 120 yards long (including the end zones) and 160 feet wide. What is the area of a football field in acres?

(b) A soccer field measures 110 meters by 70 meters. What is its area in ares? in hectares? (See Table 10.7.)

8. A small bottle of sparkling mineral water contains 33 cL.

(a) What is the volume in milliliters?

(b) Will three small bottles fill a 1-liter bottle?

9. Fill in the blanks:

(a) 58,728 g = ________ kg

(b) 632 mg = ________ g

10. Fill in the blanks:

(a) 0.23 kg = ________ g

(b) A cubic meter of water weighs = ________ kg.

11. In each part, give the most reasonable answer listed.

(a) A newborn baby weighs about: 8.3 kg, 3.5 kg, 750 g, 1625 mg.

(b) A compact car weighs about: 5000 kg, 2000 g, 1200 kg, 50 kg.

12. In each part, give the most reasonable answer listed.

(a) The recommended daily allowance of vitamin C is: 250 g, 60 mg, 0.3 kg, 0.002 mg.

(b) A grain of salt is about: 1 kg, 1 mg, 750 g, 1 kg.

13. In each of the following, select the most reasonable metric measurement:

(a) The height of a typical center in the National Basketball Association is: 6.11 m, 3 m, 95 cm, 212 cm.

(b) The diameter of a coffee cup is about: 50 m, 50 mm, 500 mm, 5 km.

14. (a) A coffee cup has a capacity of about: 8 L, 8 mL, 240 mL, 500 mL.

(b) Is the capacity of a liter of milk about the same as the capacity of a liter of water?

15. Use a metric ruler to measure these items:

(a) The size of a sheet of standard notebook paper

(b) The length and width of the cover of this textbook

(c) The diameter of a nickel

(d) The perimeter of (distance around) your wrist

16. The dimensions of Noah's Ark are given in the Bible as 300 cubits long, 50 cubits wide, and 30 cubits high. Give the dimensions in (a) feet and (b) meters. Use a meterstick and ruler to measure your own cubit, as shown in Figure 10.1.

17. (a) In Atlanta, if the temperature is 68 degrees Fahrenheit, what is the temperature in Celsius?

(b) If the temperature in San Francisco is 13 degrees Celsius, what is the temperature in Fahrenheit?

(c) Would you let your students go outside if the temperature were zero? (This is a trick question.)

18. (a) If the temperature in Louisville is 30 degrees Fahrenheit, what is the temperature in Celsius?

(b) If the temperature in Auburn, Alabama, is 3 degrees Celsius, what is the temperature in Fahrenheit?

19. Think of two points: (0, 32) and (100, 212). Draw a straight line through these two points.

(a) What is the equation of this straight line using x- and y-coordinates?

(b) Instead of naming the variables x- and y-coordinates as in part (a), use F and C for the variables. What is the relationship between your answer to part (a) and the relationship between Fahrenheit and Celsius?

Into the Classroom

20. Children often enjoy word games, and this exercise can provide a fun way to learn some of the metric prefixes. For example, the prefix for 10^{-12} in the metric system is *pico*, so what are 10^{-12} boos? One picoboo (peekaboo), of course. Now try these:

(a) 10 millipedes

(b) 10^{-6} phones

(c) 2000 mockingbirds

(d) 10 cards

(e) 10^{-9} goat

21. Most of us can visualize the length of an inch, a foot, a yard, and even a football field because we have repeatedly experienced these measurements. To develop the same feeling for metric measurements of length, have children use a metric ruler and meterstick or metric tape measure to determine each of the following in metric units:

(a) The length of a new pencil

(b) The length of your shoe

(c) Your own height

(d) The height and width of a door

(e) The length of a yardstick

(f) The length of a football field

22. A quart and a liter of the same brand of sparkling water are on a shelf. If they are the same price, which one should you pick?

Responding to Students

23. Majandra was asked to decide which of the following choices would be the most reasonable length for a dining room table:

(a) 6 feet

(b) 6 miles

(c) 6 yards

(d) 6 inches

Majandra decided that the answer must be 6 yards. When her teacher asked her to explain, Majandra said that 6 inches is very small, 6 miles is very long, and 6 feet is how we would measure a person, not a table.

(a) What is Majandra doing incorrectly when applying measurements to objects?

(b) How would you guide her to understand the length of these different amounts and help her approach such problems in the future?

24. Brent was solving metric conversion problems in class. He was successful when converting whole-number large metric measurements such as meters and rewriting them in smaller units such as centimeters. The teacher then gave Brent the conversion

$$5 \text{ cm} = \_\_\_\_ \text{ km}$$

and asked him to complete the problem. Brent answered that you cannot do this problem because you will end up with a decimal and you can't have a decimal amount when measuring.

(a) What is the correct answer to this metric problem?

(b) It is common for students who have been practicing metric conversions going from "big to small" to suddenly become very confused when switching and converting from "small to big." How can you guide Brent toward a better understanding of this conversion?

25. Corina thought she correctly solved the following metric problems:

18 cm = **180 mm**

4 m = **400 cm**

40 mm = **400 cm**

3000 mm = **3,000,000 m**

(a) Which two conversion problems did Corina solve?

(b) Which two conversion problems did Corina fail to solve?

(c) Why is Corina's mistake common and how can you guide her in the future?

26. Sandy, a sixth grader, is asked to compute the degrees in Celsius if the temperature is 20 degrees Fahrenheit. She immediately says that the answer is about 72 degrees Fahrenheit. When the teacher, Mr. Monge, says that the answer is wrong, Sandy says her answers are approximations and she can do them in her head.

(a) Mr. Monge asks her what 10 degrees Celsius is in Fahrenheit and she immediately says 52 degrees Fahrenheit. How close is Sandy to the exact answer?

(b) Sandy has looked at the equation relating Celsius to Fahrenheit and says that she uses 2.0 instead of 1.8 in the equation on page 501. Is that a good approximation? It is certainly a very clever way for finding an estimate to convert Celsius to Fahrenheit!

Thinking Critically

27. Pints, quarts, and gallons are part of a larger "doubling" system of capacity measure:

1 jigger = 2 mouthfuls	1 pint = 2 cups
1 jack = 2 jiggers	1 quart = 2 pints
1 jill = 2 jacks	1 bottle = 2 quarts
1 cup = 2 jills	1 gallon = 2 bottles
	1 pail = 2 gallons

(a) How many mouthfuls are in a jill? a cup? a pint?

(b) Suppose one mouthful, one jigger, one jack, . . . , and one gallon are poured into an empty pail. Does the pail overflow, is it exactly filled, or is there room for more? (*Hint:* Draw an empty pail. Put one gallon in, then one bottle, and so on.)

28. The weight of diamonds and other precious gemstones is given in *carats,* where 1 carat = 200 mg. The largest diamond discovered thus far is the Cullinan, found in 1906 at the Premier mine in South Africa. It weighed 3106 carats. Using the conversion 1 kg = 2.2 lb, estimate the weight of the Cullinan diamond in pounds.

29. Can you cut the board shown into just two pieces to exactly cover the 60-cm by 12-cm hole?

Board

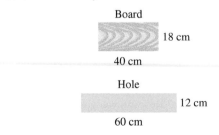

Hole

30. Nearly 4700 years ago, the Great Pyramid of Khufu was built to astonishing accuracy with the use of a measuring unit called the cubit. (See Figure 10.1.) For further accuracy, the cubit was divided into seven *palms,* and each palm was further subdivided into four *digits* (finger widths). Longer distances were measured by the *hayt,* equal to 100 cubits.

(a) Use a meterstick to measure the cubit (elbow-to-fingertip distance, to the nearest centimeter) and palm (distance across four fingers, to the nearest millimeter) of 10 classmates, and make a histogram of your data. Compute the average and standard deviation of the cubit and palm measurements. Does it seem accurate that seven palms are in a cubit?

(b) Give reasonable ancient Egyptian measurements for the height of the ceiling in your classroom, the length of a piece of notebook paper, and the length of a football field.

Making Connections

31. Metric countries rate the fuel efficiency of a car by the number of liters of gasoline required to drive 100 kilometers. If a car takes 9 liters per 100 kilometers, what is its efficiency in miles per gallon? Use the conversions 1 gal, which is about 3.7854 L, and 1 mile, which is close to 1.6 km.

32. A light-year is the distance light travels in empty space in one year.

(a) Light travels at a speed of 186,000 miles per second. The star nearest the sun is Proxima Centauri, in the constellation Centaurus, whose distance from Earth is 4 light years. What is the distance to Proxima Centauri in miles?

(b) In metric measurements, the speed of light is 3.00×10^8 meters per second. Verify that a light-year is about 10^{16} meters.

33. An herbicide is bottled in concentrated form. A working solution is mixed by adding 1 part concentrate to 80 parts water.

(a) How many liquid ounces of concentrate should be added to 5 gallons of water?

(b) How many liters of water should be added to 65 milliliters of concentrate?

34. Lumber is measured in board feet, where a board foot is the volume of a piece of lumber one foot square and one inch thick.

(a) How many board feet are in a two-by-four (2″ by 4″) that is 10 feet long? (The volume of a rectangular solid is length times width times height.)

(b) Lumber is priced in dollars per thousand board feet. Suppose two-by-fours 10 feet long are $690 per thousand board feet. What is the cost of 144 two-by-fours, each 10 feet long?

35. On July 23, 1983, Air Canada Flight 143 was on the ground in Montreal. The pilot had 7682 liters of fuel on board but knew the Boeing 767 would need 22,300 kilograms to reach its destination, Edmonton, Alberta. Since airliners measure fuel by weight, not volume, the pilot asked for the weight of a liter of fuel and was told it was 1.77 kilograms.

(a) Calculate how many liters of fuel were added to the plane's tanks.

(b) Actually, 1.77 is the number of pounds, not kilograms, of fuel per liter. There is really just 0.803 kilogram per liter. How much fuel should have been added to the plane's tanks?

(Yes, the plane ran out of fuel, but the pilots managed to glide 22 miles to a safe landing in Winnipeg.)

State Assessments

36. (Grade 3)
Which is the best estimate for the capacity of a bottle of fruit juice?

A. 1 milliliter
B. 1 cup
C. 1 liter
D. 4 gallons

37. (Washington State, Grade 4)
Your class project is to build a bird feeder.

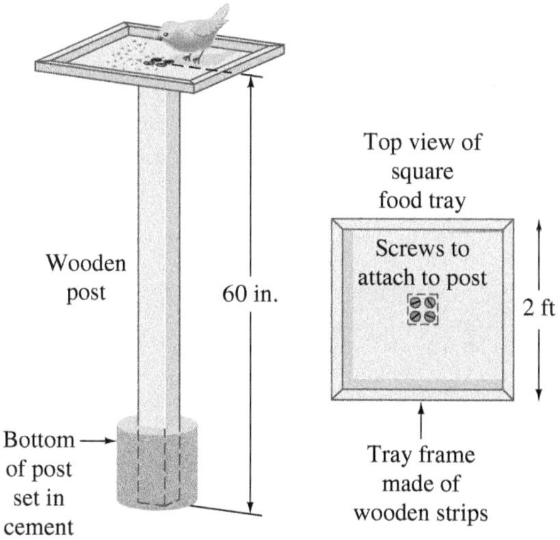

Item	Cost per Unit
wooden post	$2.50 per foot
wooden strips for tray frame	$1.00 per foot
tray bottom	$6.00
tools, screws, nails, wood glue, and cement mix	loaned or donated by parents

Explain how you could use the information given to find the total cost of materials. Use words, numbers, or pictures.

38. (Grade 6)
Autumn Cortney, a baby, is 83.84 centimeters tall. Which of the following is equivalent to 83.84 centimeters?

A. .08384 m
B. 838.4 m
C. .8384 m
D. 8.384 m

39. (Grade 6)
A bag of Georgia peanuts has total weight of 27 ounces. Which of the following is equivalent to 27 ounces?

A. 3.375 pounds
B. 1.6875 pounds
C. 1.125 pounds
D. 0.5625 pounds

40. (Grade 6)
Samantha made 6 liters of lemonade. Which of the following is equivalent to 6 liters?

A. .6 milliliters
B. 6 milliliters
C. 60 milliliters
D. 6,000 milliliters

10.2 Area and Perimeter

Measurements in Nonstandard Units

We now turn our attention to the notions of the **area** and **perimeter** of a region in the plane and follow the steps of the measurement process as described in Section 10.1, page 490. In order to discuss what the area of a region is, we must first specify the unit of measurement. Usually, squares are chosen to define a **unit of area,** so units such as cm^2, in^2, ft^2, and so on, are common, but any shape that tiles the plane (that is, covers the plane without gaps or overlaps) can serve equally well. In Example 10.7, we will work with a nonstandard unit rather than squares. This practice allows students to discover important general principles of the measurement process.

DEFINITION Area of a Region in the Plane

Let R be a region and assume that a unit of area is chosen. The number of units required to cover a region in the plane without overlap is the **area** of the region R.

EXAMPLE 10.7

Making Measurements in Nonstandard Units

Find the area of each figure A, B, C, and D in terms of the unit of area shown at the right.

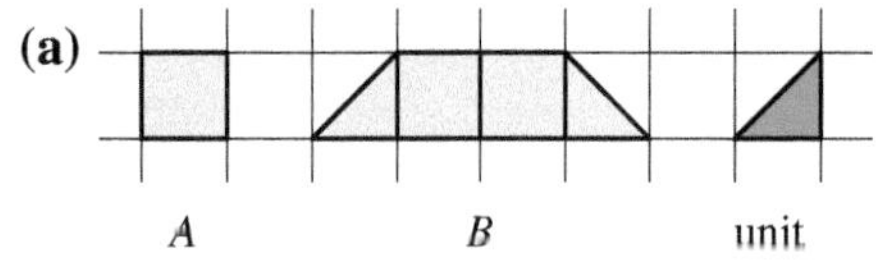

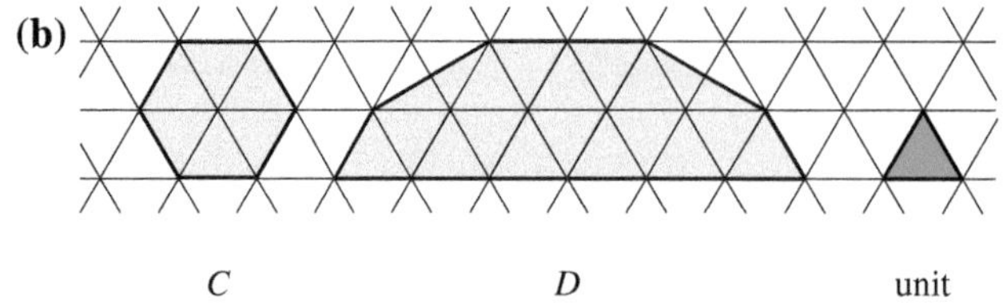

Solution

(a) The full square A can be covered by 2 of the unit shapes, so area(A) = 2 units. Region B is covered by 6 units, so area(B) = 6 units.

(b) The hexagon C is covered by 6 of the triangular units, so area(C) = 6 units. Region D cannot be covered directly by the triangular units, although it is evident that the area of D is between 16 and 20 units. To find the exact area, remove and then rejoin a triangular piece, as shown in the diagram that follows, to form a new shape D' of the same area as D. That is, area(D') = area(D). Since D' can be covered by 18 triangular units, it follows that area(D) = 18 units.

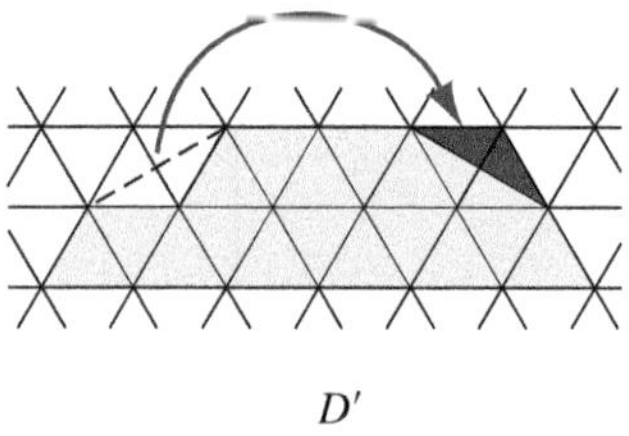

The Congruence and Addition Properties of Area

The solution given in Example 10.7 shows two important principles for the calculation of areas, the first of which is **congruence.** The notion of two regions being congruent will be defined carefully in the next chapter by saying that they are congruent if, using rigid motions like rotation and translation, we can superimpose either region on the other. For now, we'll use more informal language.

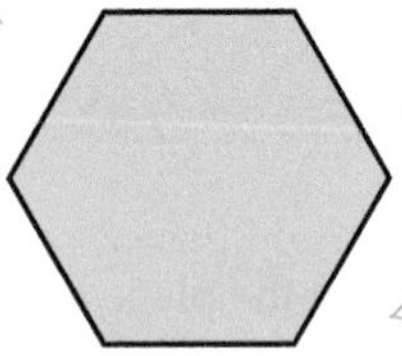

A hexagonal region

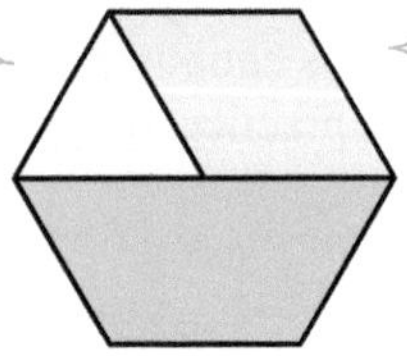

A hexagonal region dissected into three subregions

DEFINITION Congruence of Two Regions in the Plane

If R and S are regions in the plane that have the same size and shape, then they are **congruent** and we use the symbol $\cong$ to write $R \cong S$.

The second principle of Example 10.7 comes from a figure being **dissected** or **partitioned**—that is, the figure is cut into a collection of nonoverlapping regions that covers the larger figure. The **congruence** and **addition properties of area** that follow will be used repeatedly throughout the rest of this chapter.

PROPERTIES The Congruence and Addition Properties of Area

Congruence property

If region R is congruent to region S, then the two regions have the same area:

$$\text{area}(R) = \text{area}(S).$$

Addition property
If a region R is dissected into nonoverlapping subregions $A, B, \ldots, F$, then the area of R is the sum of the areas of the subregions:

$$\text{area}(R) = \text{area}(A) + \text{area}(B) + \cdots + \text{area}(F).$$

These properties are illustrated in Figure 10.7.

Figure 10.7
The congruence and addition properties of area

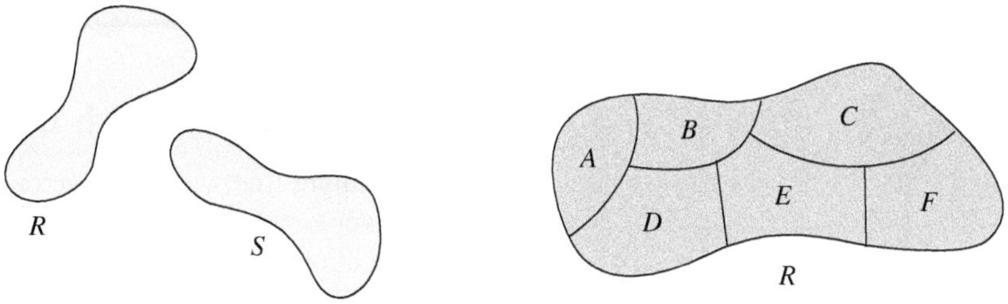

(a) If $R \cong S$, then area $(R) = \text{area}(S)$

(b) $\text{area}(R) = \text{area}(A) + \text{area}(B) + \cdots + \text{area}(F)$

The congruence and addition properties show that rearranging the pieces of a figure forms a new figure with the same area as the original figure.

EXAMPLE 10.8

Solving Leonardo's Problems

Leonardo da Vinci (1452–1519) once became absorbed in showing how the areas of certain curvilinear (curved-sided) regions could be determined and compared among themselves and with rectangular regions. The pendulum and the ax are two of the examples he included in notes for his book *De Ludo Geometrico* (roughly meaning "Fun with Geometry"), which he never completed. The dots show the centers of the circular arcs that form the boundaries of the regions.

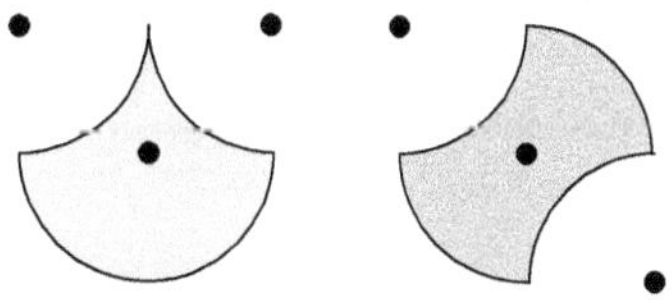

If the arcs forming the pendulum and the ax have radius 1, show that the areas of both figures are equal to that of a 1-by-2 rectangle.

Solution

After inscribing the figures in a square, we use the congruence and addition properties of area to rearrange the subregions to form a 1-by-2 rectangle:

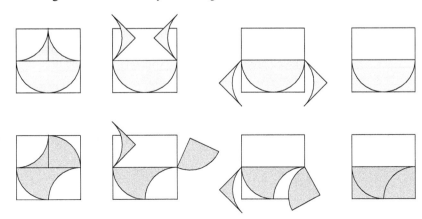

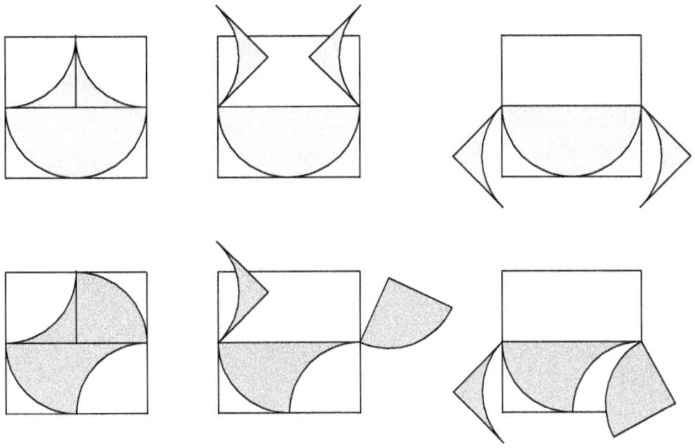

It is surprising to learn that the pendulum and ax both have an area of 2 square units. It is also a tribute to da Vinci's genius that he could see this in the fifteenth century!

SOURCE: These examples and others are described in a booklet by Herbert Wills III published by the National Council of Teachers of Mathematics.

Unlike Examples 10.7 and 10.8, most area measurement problems are answered by giving a reasonable *estimate* of the area. Units of square shape are easy to subdivide into smaller squares to give a more precise estimation.

EXAMPLE 10.9 Investigating the Area of a Cycloidal Arch of Galileo

Imagine rolling a wheel along a straight line, with a reflector at point P on the rim. Point P traces an arch-shaped curve. In the early seventeenth century, Galileo investigated this curve and named it the **cycloid**. Discover for yourself the conjecture Galileo made about how the area of the cycloidal arch compares to the area of the circle used to generate the arch.

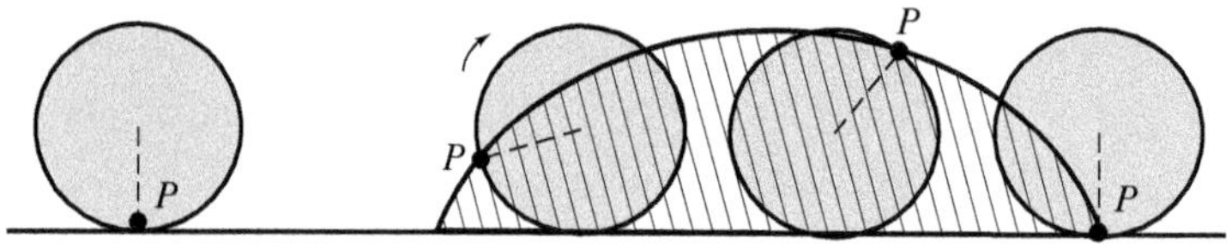

Solution

Understand the Problem

It is visually clear that the cycloidal arch has an area that is much larger than that of the circle. Our goal is to guess the ratio area(cycloid)/area(circle) that gives the comparison between the areas.

Devise a Plan

The areas of the arch and circle must both be measured in some unit of area. For example, we can use squares of size U, where the diameter of the circle is equal to the sum of four side lengths of U. For better accuracy, we can also use small square units of size u, where the side length of u is half that of U.

Carry Out the Plan

The circle and arch are overlaid by a square grid, with squares of unit area U.

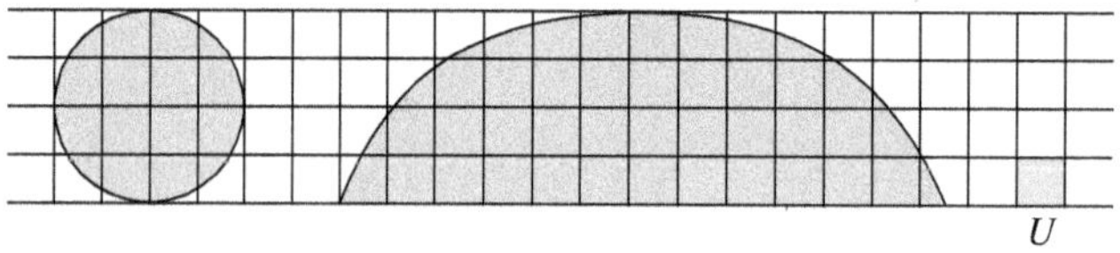

The circle is entirely within 16 squares of size U but does not entirely cover about one unit of area in each of the four corners. Thus, we estimate that area (circle) is close to 12 U. Similarly, we see that area (cycloid) is of about 37 U is a reasonable estimate.

Better accuracy is given by the grid of squares of unit area u. The following diagram leads us to the estimated area (circle), approximately 50 u, and area (cycloid) is very close to 149 u:

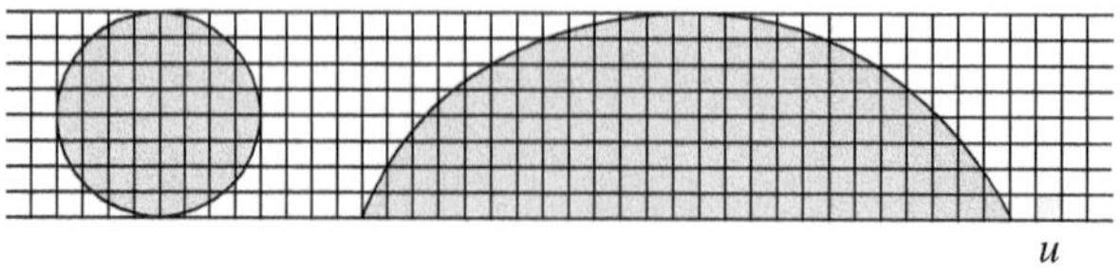

After cancellation, both $\frac{37\,U}{12\,U}$ and $\frac{149\,u}{50\,u}$ are nearly 3, which in fact was Galileo's conjecture. The correctness of Galileo's idea was proved in 1634 by Gilles Persone de Roberval.

Look Back

The finer grid of squares gave us additional precision in our measurements, but this required considerably more time and effort to obtain. The measurement process nearly always requires us to make a judgment about how to balance the conflicting needs of precision versus cost.

Areas of Polygons: A Conceptual Understanding

In this part of Section 10.2, we'll derive formulas for the areas of rectangles, parallelograms, triangles, and trapezoids based on the square as a unit. All of the formulas are very familiar to you—you will now see conceptually why they are true. This approach will help you give your students a conceptual understanding of the formulas for area.

Rectangles A 3-cm by 5-cm rectangle can be covered by 15 unit squares when the unit square is 1 cm^2, as shown in Figure 10.8. Similarly a 2.5-cm by 3.5-cm rectangle can be covered by six whole units, five half-unit squares, and one quarter-unit square, giving a total area of 8.75 square centimeters. This is also the product of the width and the length, since 2.5 cm $\times$ 3.5 cm $=$ 8.75 cm^2.

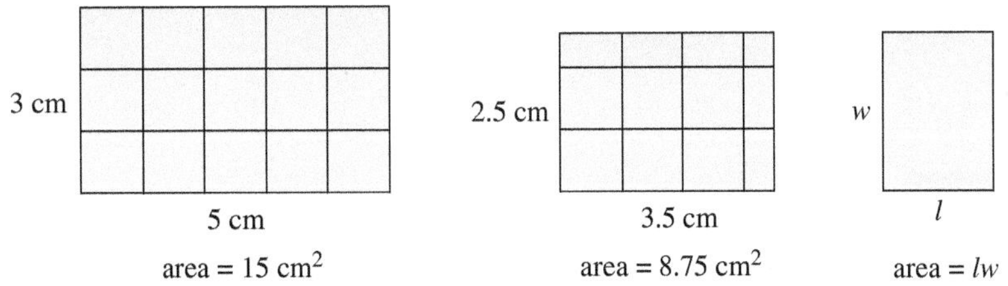

Figure 10.8
The area of a rectangle is the product of its length and width

For any rectangle, the formula for the area A is as follows:

FORMULA Area of a Rectangle
A rectangle of length l and width w has area A given by the formula $A = lw$.

Parallelograms We now move from a rectangle to a more general quadrilateral, a parallelogram. Suppose a parallelogram has a pair of opposite sides b units long and these sides are h units apart; an example is shown in Figure 10.9. We say that b is the **base** of the parallelogram and h is the **altitude**, or **height**. (Unless the parallelogram is a rectangle, the altitude is *not* the same as the length of the other two sides of the parallelogram.) Removing and replacing a right triangle T forms a rectangle of the same area as the parallelogram. The rectangle has length b and width h, so its area is bh. Therefore, the area of the parallelogram in Figure 10.9 is also bh.

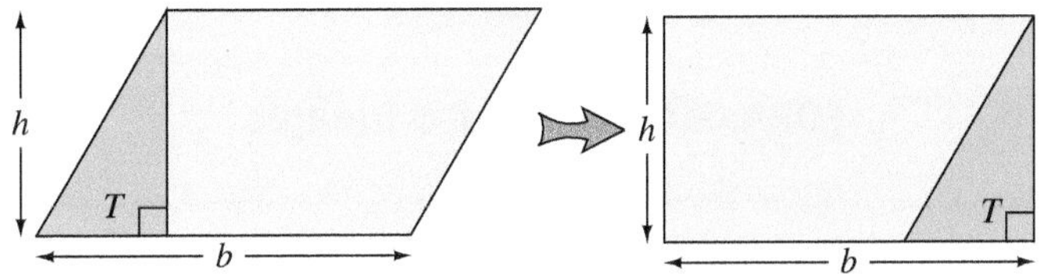

Figure 10.9
A parallelogram of base b and altitude h has the same area as a b-by-h rectangle. Therefore, the area of the parallelogram is bh

Any parallelogram with base b and altitude h can be dissected and rearranged to form a rectangle of length b and width h in a way similar to that shown in Figure 10.9. (See problem 32 in Problem Set 10.2 for a more general case.) This discussion gives the following formula:

FORMULA Area of a Parallelogram

A parallelogram of base b and altitude h has area A given by $A = bh$.

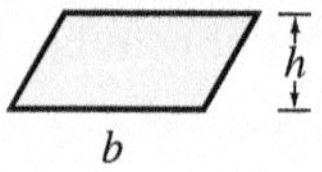

Thus, the parallelogram shown has the same area as a rectangle of width h and length b.

EXAMPLE 10.10

Using the Parallelogram Area Formula

Find the area of each parallelogram and then compute the lengths x and y.

(a)

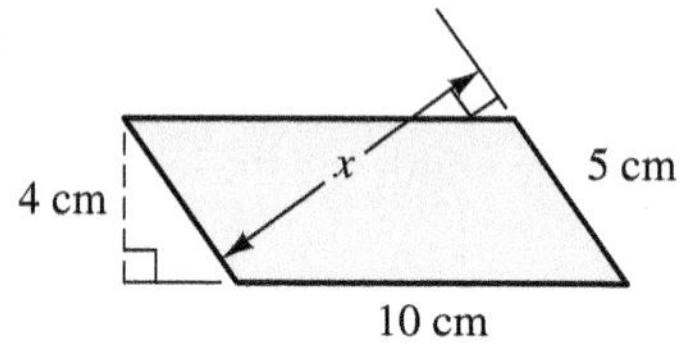

(b)

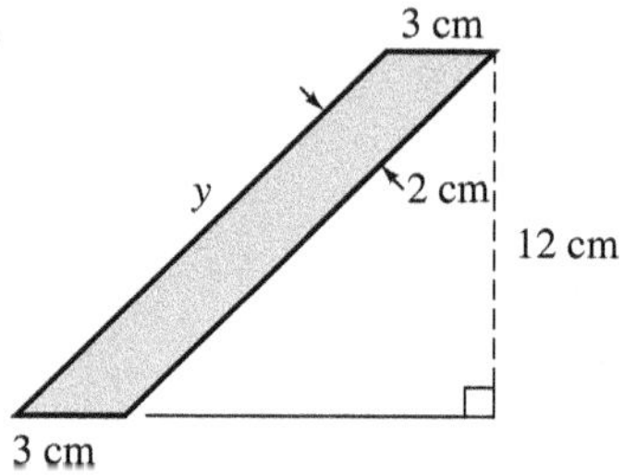

Solution

(a) The parallelogram has base 10 cm and height 4 cm, so its area is $A = (10 \text{ cm})(4 \text{ cm}) = 40 \text{ cm}^2$. If the side of length 5 cm is considered the base, then x is the corresponding height and $A = (5 \text{ cm})x$. Since $A = 40 \text{ cm}^2$, we find that $x = 40 \text{ cm}^2/5 \text{ cm} = 8 \text{ cm}$.

(b) The procedure for (a) is followed. The area is $A = (3 \text{ cm})(12 \text{ cm}) = 36 \text{ cm}^2$. Viewing the side of length y as the base with corresponding altitude 2 cm, we have $36 \text{ cm}^2 = y(2 \text{ cm})$. Therefore, $y = 36 \text{ cm}^2/2 \text{ cm} = 18 \text{ cm}$.

Triangles We now look for the area of a triangle. Figure 10.10 shows that a triangle of base b and altitude h can be dissected and rearranged to form a parallelogram of base $\frac{b}{2}$ and altitude h. The formula $\frac{1}{2}bh$ for the area of the triangle then follows from the area formula already derived for the parallelogram.

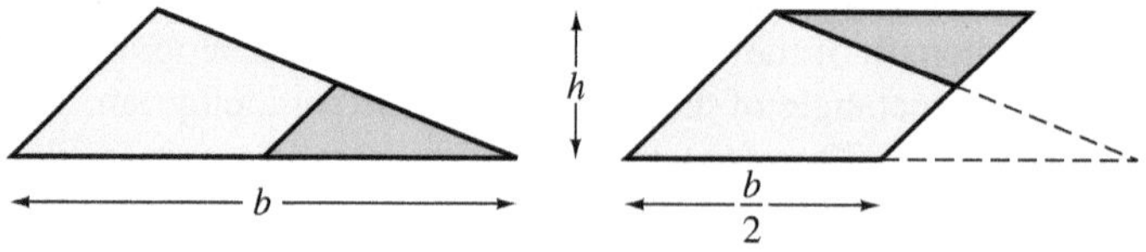

Figure 10.10
A triangle of base b and altitude h can be dissected and rearranged to form a parallelogram of base $\frac{b}{2}$ and altitude h

FORMULA Area of a Triangle

A triangle of base b and altitude h has area $A = \frac{1}{2}bh$.

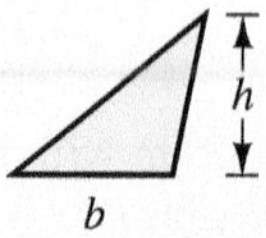

As an aside, any side of a triangle can be considered as the base, so there are three pairs of bases and altitudes.

EXAMPLE 10.11

Using the Triangle Area Formula

Find the area of each triangle and the distances v and w.

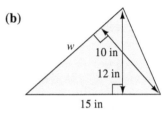

Solution

(a) The formula $A = \frac{1}{2}bh$ shows that the area of the triangle is $A = \frac{1}{2}(10\text{ cm})\cdot(7\text{ cm}) = 35\text{ cm}^2$. If the side of length 14 cm is considered the base, then the corresponding altitude is v. Since $A = \frac{1}{2}(14\text{ cm})\cdot(v)$, we have $v = A/(7\text{ cm}) = (35\text{ cm}^2)/(7\text{ cm}) = 5\text{ cm}$.

(b) The area of the triangle is $A = \frac{1}{2}(15\text{ in})\cdot(12\text{ in}) = 90\text{ in}^2$. Taking the side of length w as the base, we find that the corresponding altitude is 10 in and $A = \frac{1}{2}w(10\text{ in})$. Therefore, $w = (90\text{ in}^2)/(5\text{ in}) = 18\text{ in}$.

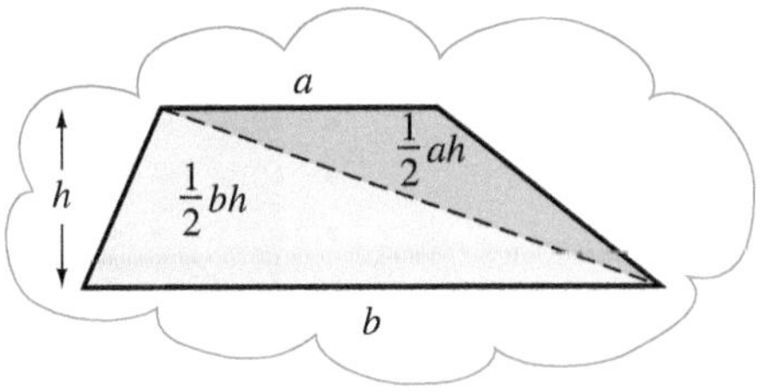

Trapezoids There are several ways to derive the formula for the area of a trapezoid with altitude h and bases a and b. For example, a diagonal drawn through the trapezoid dissects it into two triangles. The altitudes of both triangles are h, and their bases are a and b, so the areas of the triangles are $\frac{1}{2}ah$ and $\frac{1}{2}bh$. Adding the areas gives the following formula, as is also shown in Figure 10.11:

FORMULA Area of a Trapezoid

A trapezoid with bases of length a and b and altitude h has area $A = \frac{1}{2}(a + b)h$.

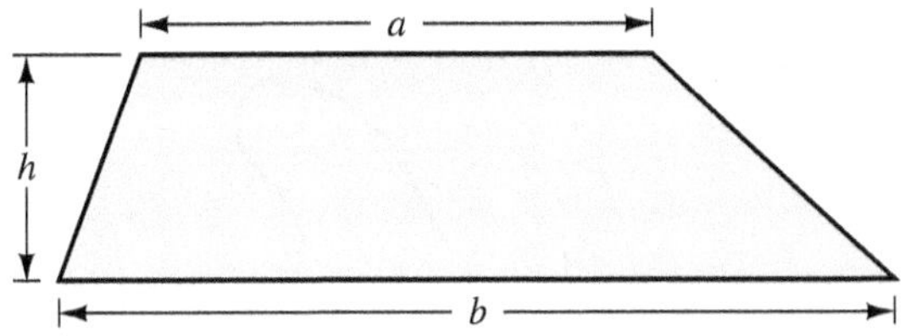

Figure 10.11
The area of a trapezoid is $\frac{1}{2}(a + b)h$

This result says that the area of a trapezoid of altitude h is the same as the area of a rectangle of width h and whose length l is the *average* of the lengths of the trapezoid; in other words, $l = \frac{1}{2}(a + b)h$.

EXAMPLE 10.12

Finding the Areas of Lattice Polygons

A polygon formed by joining points of a square array is called a **lattice polygon**. Lattice polygons are easy to draw on dot paper, or they can be formed with rubber bands on a geoboard. Find the areas of the following lattice polygons, where the unit of area is the area of a small square of the array:

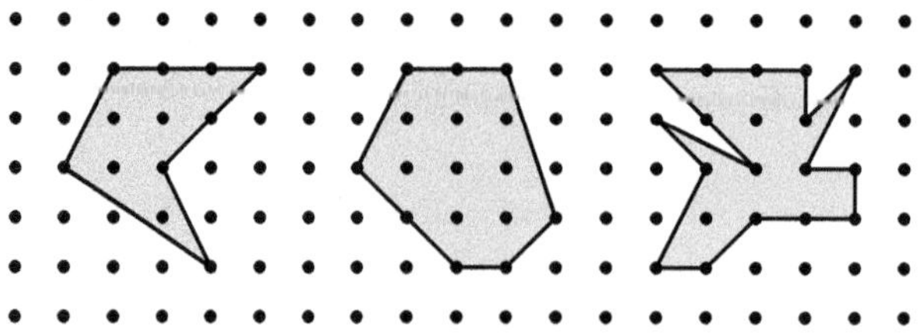

(a) (b) (c)

Solution

To find the area of each of these lattice polygons, we break each one up (partition) and then add up the areas of the pieces. Note the pictures on the side.

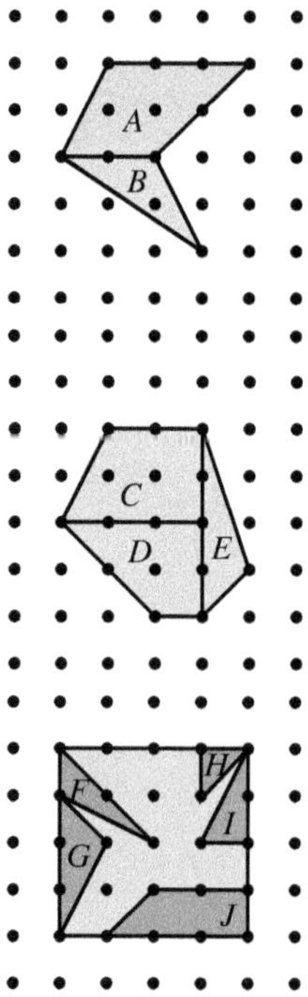

(a) A horizontal line dissects the polygon into a trapezoid A of area $\frac{1}{2}(3 + 2) \cdot (2) = 5$ and a triangle B of area $\frac{1}{2}(2)(2) = 2$. The area of the polygon is therefore 7.

(b) The lattice hexagon can be dissected into trapezoids C and D and triangle E. The total area of the hexagon is therefore $\frac{1}{2}(2 + 3) \cdot (2) + \frac{1}{2}(3 + 1) \cdot (2) + \frac{1}{2}(4) \cdot (1) = 11$. Other dissections of the hexagon can be used, but the total area will always be the same.

(c) We could solve the problem in the same way as in parts (a) and (b), but there is another useful technique: Construct a square about the polygon, and then subtract the areas of the regions F, G, H, I, and J. Therefore, the area of the polygon is $16 - \left(1 + 1\frac{1}{2} + \frac{1}{2} + 1 + 2\frac{1}{2}\right) = 9\frac{1}{2}$.

Length of a Curve

The **length of a polygonal curve** is obtained by summing the lengths of its sides. The key to finding (and, in fact, the definition of) the **length of a nonpolygonal curve** is to measure, or at least estimate, by calculating the length of an approximating polygonal curve with vertices on the given curve. The accuracy of the estimation is improved by using an approximating polygonal curve with ever more vertices, as shown in Figure 10.12.

(a) (b) (c)

Figure 10.12 The length of the curve in (a) is estimated by measuring the length of a polygonal approximation, as in (b). Increasing the number of vertices gives an improved estimate carefully, as in (c)

The length of a curve can also be measured by first laying a string along the curve and then straightening the string along a ruler. This is the principle that makes the flexible tape measure used for sewing so useful.

EXAMPLE 10.13 Determining the Length of a Cycloid

A circle and the cycloid it generates (see Example 10.9) are shown next. Use a marker pen and a piece of string (or thin strip of paper) to make a tape measure, where the unit of length is the diameter d of the circle.

(a) According to your tape measure, what is the length from point A to point B along the cycloid?
(b) What is the approximate length of the line segment $\overline{AB}$?

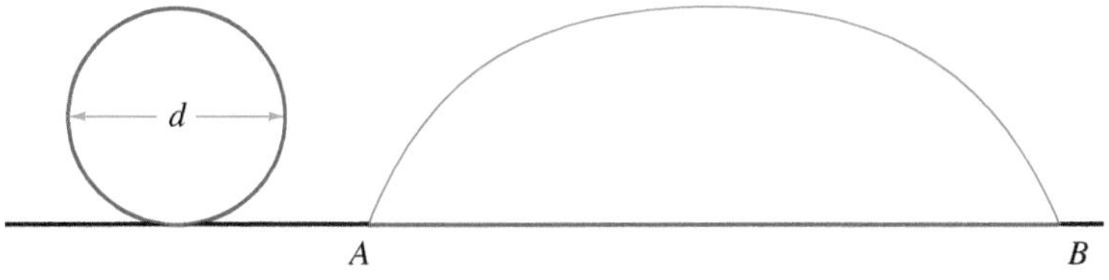

Solution

(a) The tape measure shows that the length of the cycloid is very nearly four diameters of the circle. In 1658, Christopher Wren (1632–1723) proved that the length of a cycloid is *exactly* four diameters. Wren is perhaps best known as the architect of St. Paul's Cathedral in London.
(b) The segment $\overline{AB}$ is a bit over three diameters. Because $\overline{AB}$ is covered by rolling the circle once around, AB is the length around the circle; that is, AB is the circumference, and "a bit over three" is the famous value now written as π.

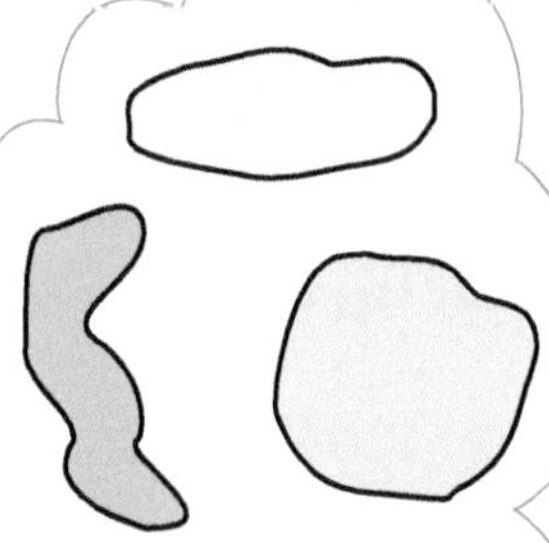

A loop of string forms regions with the same perimeter, but different enclosed areas.

Perimeter The length of a simple closed plane curve is called the curve's **perimeter.** It is quite hard to formally define the perimeter of a region. The difficulties are in the definition of the terms *region*, *length*, and *boundary*. Because these ideas are intuitive and you will be using them in elementary or middle school, we write the following more informal definition of perimeter:

DEFINITION Perimeter of a Region
If a region is bounded by a simple closed curve, then the **perimeter** of the region is the length of the curve. More generally, the **perimeter** of a region is the length of its boundary.

The perimeter is a *length* measurement and is given in centimeters, inches, feet, meters, and so on. It is important not to confuse the *area* of the region enclosed by a simple closed curve with the perimeter of the figure. Area is given in cm^2, in^2, ft^2, m^2, and so on. In summary, the perimeter is the measure of the distance around a region, and the area is the measure of the size of the region within a boundary.

EXAMPLE 10.14 Finding Perimeters and Areas

The figures that follow have been drawn on a square grid, where each square is 1 cm on a side. Give the perimeter and area of each figure.

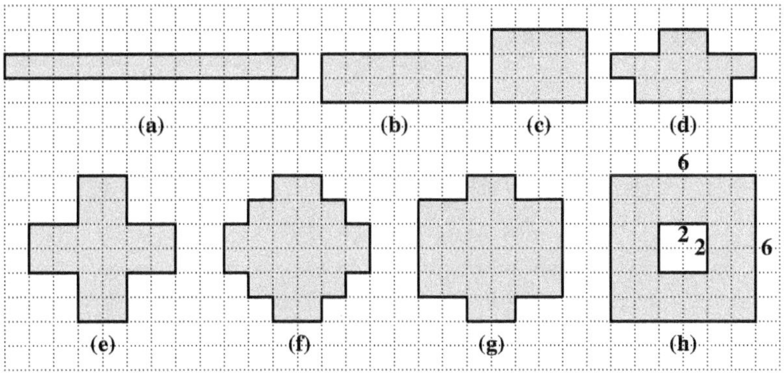

Solution

Figure	(a)	(b)	(c)	(d)	(e)	(f)	(g)	(h)	
Perimeter	26	16	14	18	24	24	24	32	centimeters
Area	12	12	12	12	20	24	28	32	square centimeters

Figures (a), (b), (c), and (d) have the same area but different perimeters. Figures (e), (f), and (g) have the same perimeter but different areas. Note that Figure (h) is not bounded by a simple closed curve.

The Circumference of a Circle The perimeter of a circle is called the circle's **circumference.** By using a piece of string or a tape measure, or by rolling a disc along a line (as in Example 10.13(b), it is easy to rediscover a fact known even in ancient times: *The ratio of the circumference of a circle to the circle's diameter is the same for all circles.* Two examples are shown in Figure 10.13.

This ratio, which is somewhat larger than 3, is given by the symbol π, the lowercase Greek letter *pi*.

DEFINITION π

The ratio of the circumference C to the diameter d of a circle is π. Therefore,

$$\frac{C}{d} = \pi \quad \text{and} \quad C = \pi d.$$

Figure 10.13
The ratio of the circumference C to the diameter d is the same for all circles: $c/d = \pi$, or $C = \pi d$

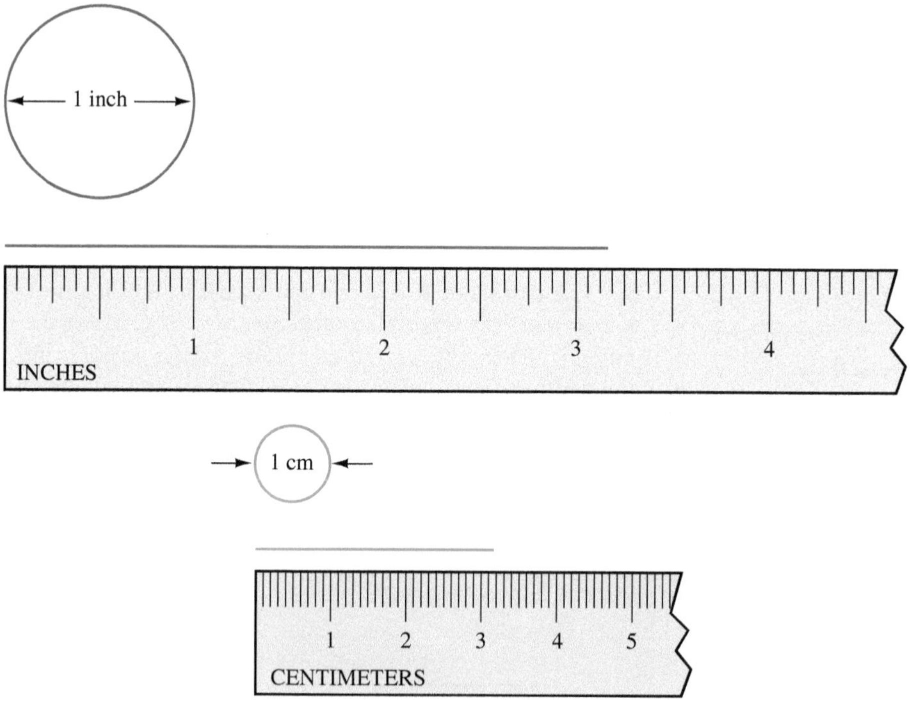

Since the diameter d is twice the radius r of the circle, we also have the formula $C = 2\pi r$.

The values $3\frac{1}{7}$ and 3.14 are useful approximate values, but precision measurements require the use of more decimal places in the unending decimal expansion $\pi = 3.1415926\ldots$. A circle 100 feet in diameter has an *approximate* circumference of 314 feet, but the exact circumference is 100π feet. It is acceptable to use the symbol π to express results, since this gives exact values. When an approximate numerical value is needed, an appropriate estimate of π, such as 3.1416, can be used in the calculations. In 1761, John Lambert proved that π is an irrational number, so it is impossible to express π exactly by a fraction or as a terminating or repeating decimal.

EXAMPLE 10.15

Calculating the Equatorial Circumference of Earth

The equatorial diameter of Earth is 7926 miles. Calculate the distance around Earth at the equator, using the following approximations for π: **(a)** 3.14 **(b)** 3.1416.

Solution

(a) (3.14) (7926 miles) = 24,887.64 miles
(b) (3.1416) (7926 miles) = 24,900.322 miles

The two different approximations of π show a difference of about 12.7 miles.

Highlight From History: A Brief History of π

In the third century B.C., Archimedes showed that π is approximately $3\frac{1}{7}$. To estimate π, Archimedes inscribed a regular polygon in a circle and then calculated the ratio of the polygon's perimeter to the diameter of the circle. An inscribed hexagon shows that π is about 3, but by using a 96-gon, Archimedes proved that π is between $3\frac{10}{71}$ and $3\frac{10}{70}$. In the eighteenth century, the English mathematician John Machin took advantage of the invention of calculus to calculate π to 100 decimal places, and this method was used well into the twentieth century. π has now been calculated to more than a trillion decimal places.

The Area of a Circle

The area of a circle of radius r is given by the formula πr^2 and was first proved rigorously by Archimedes.

FORMULA Area of a Circle
The area A enclosed by a circle of radius r is $A = \pi r^2$.

Since π is defined as a ratio of lengths, it seems surprising to find that π also occurs in the formula for the area of a circle. A convincing, but informal, derivation of the formula $A = \pi r^2$ is shown in Figure 10.14. The circle of radius r and circumference $C = 2\pi r$ is dissected into congruent sectors that are rearranged to form a "parallelogram" of base $\frac{1}{2}C = \pi r$ and altitude r. By the formula for the area of a parallelogram, the wavy-based "parallelogram" has area $\pi r \times r = \pi r^2$. If the number of sections is made larger and larger, the sum of the areas of the thin sectors forms an increasingly exact approximation to a true parallelogram of area πr^2.

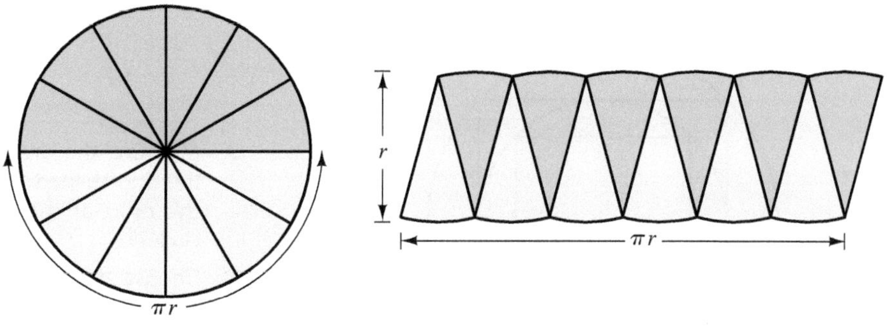

Figure 10.14 The sectors of a circle can be rearranged to approximate a parallelogram of area πr^2

EXAMPLE 10.16

Determining the Size of a Pizza π

A 14″ pizza has the same thickness as a 10″ pizza. How many times more ingredients are there on the larger pizza? Said more mathematically, what is the ratio of the area of the 14″ pizza to the 10″ pizza?

Solution Pizzas are measured by their diameters, so the radii of the two pizzas are 7″ and 5″, respectively. Since the thicknesses are the same, the amount of ingredients used is proportional to the areas of the pizzas. The larger pizza has area $\pi(7\text{ in})^2 = 49\pi\text{ in}^2$, and the smaller pizza has area $\pi(5\text{ in})^2 = 25\pi\text{ in}^2$. The ratio of areas is $49\pi\text{ in}^2/25\pi\text{ in}^2 = 1.96$, showing that the 14″ pizza has about twice the ingredients of the 10″ one.

COOPERATIVE INVESTIGATION

Measurements in Beanland

Materials Needed

Dry beans (small red kidney or white navy beans); enlarged copies of the figures shown in the two activities that follow.

Directions

In Beanland, the lengths of curves are measured in *beanlengths,* abbreviated "bl." Similarly, the areas of regions are measured in *beanareas,* abbreviated "ba." The following diagram shows that the length of the curve is about 15 bl.

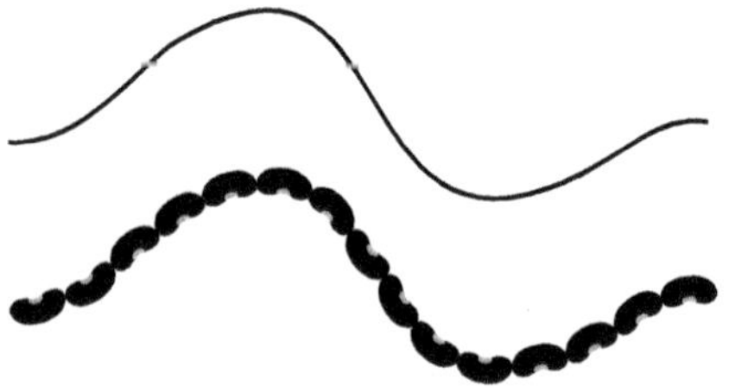

The next diagram shows a region bounded by a simple closed curve. By counting the beans, we see that the region has an area of about 55 ba.

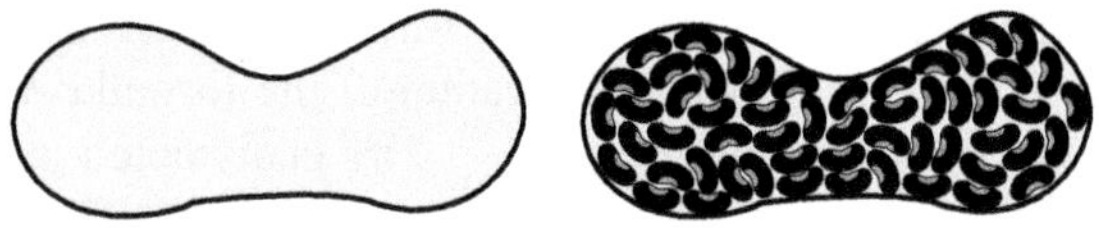

Carry out the following activities in pairs:

Activities

1. Consider the following system of squares and circles (use an enlarged copy of this figure, with the larger circle about 14 to 16 beanlengths in diameter):

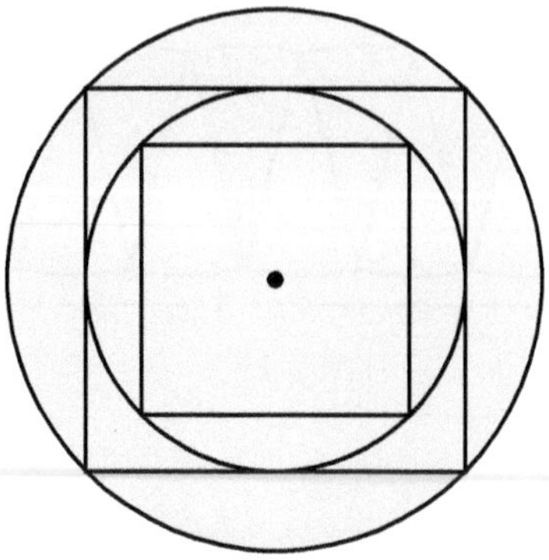

(a) Measure the area of the ring-shaped region between the two circles, using your beans.

(b) Measure the area of the smaller circle, and then compare this area with that of the ring.

(c) Measure the area of the small square and then the area of the region between the two squares. How do these areas compare?

(d) Measure the perimeter of (the distance around) the small square in beanlengths. Next measure the length of the diagonal of the large square. How do these lengths compare?

(e) Measure the circumference and the diameter of the large circle. What is the ratio of the circumference to the diameter? Compare with other groups and determine the average ratio.

2. Consider an equilateral triangle *ABC* and its circumscribed and inscribed circles (again, use an enlargement so that the larger circle has a diameter of about 14 to 16 beanlengths):

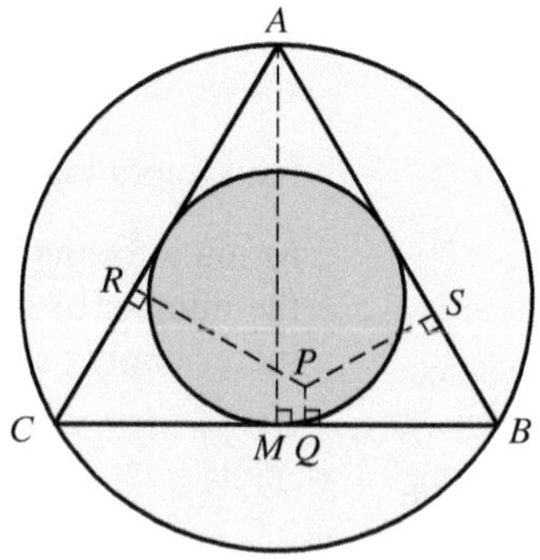

3. Measure the area of the small circle and the area of the ring-shaped region between the two circles. What is the ratio of the area of the ring to that of the small circle?

4. Choose an arbitrary point *P* inside the triangle, and measure the three distances *PQ, PR,* and *PS* to the sides of the triangle. How does the sum $PQ + PR + PS$ of these three distances compare with the length *AM* of the altitude of the triangle?

10.2 Problem Set

Understanding Concepts

1. Botanists often need to measure the rate at which water is lost by transpiration through the leaves of a plant. For this measurement, it is necessary to know the leaf area of the plant. Estimate the area of the leaf shown. It has been overlaid with a grid of squares 1 cm on a side, shown at reduced scale.

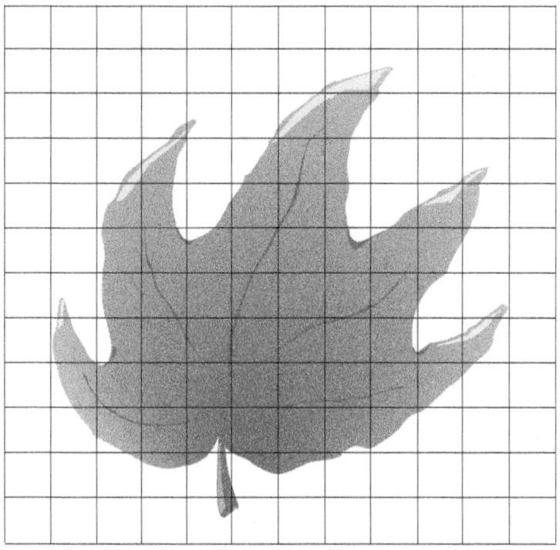

2. Cut a convex quadrilateral from card stock, locate the midpoints of its sides, and then cut along the segments joining successive midpoints to give four triangles T_1, T_2, T_3, and T_4 and a parallelogram P.

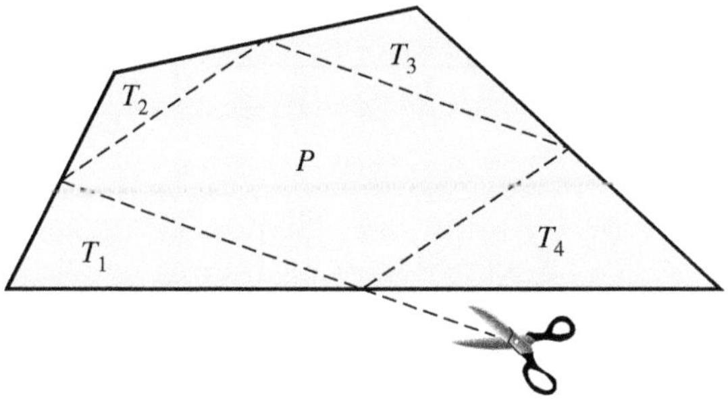

   (a) Show that the four triangles can be arranged to cover the parallelogram.

   (b) How does the area of the parallelogram compare with the area of the original quadrilateral?

3. The regular dodecagon shown next is dissected into six subregions. Carefully trace the pattern, cut out the subregions, and show how the pieces can be reassembled into a square. How are the areas of the dodecagon and square related?

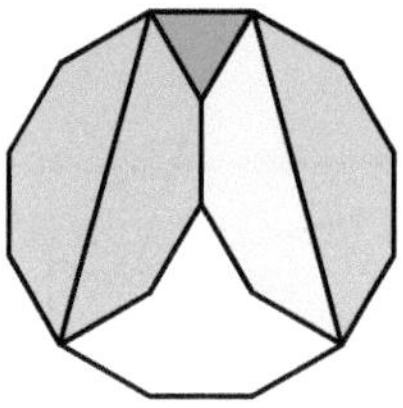

4. The goblet shown is drawn with circular 90° arcs centered at the black dots in a grid of squares of unit area. Redraw the vase on squared paper, and then use a dissection argument similar to that in Example 10.8 to find the area of the goblet.

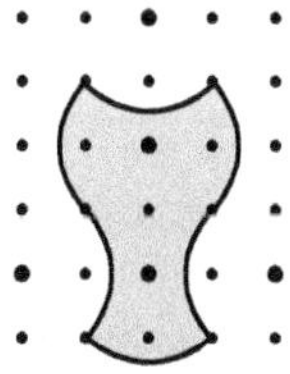

5. Measure the left-hand figure F shown in each of the three nonstandard units of area **(a)**, **(b)**, and **(c)** shown to the right of F. Do so by tracing F and then dissecting the region into subregions with the unit area shape.

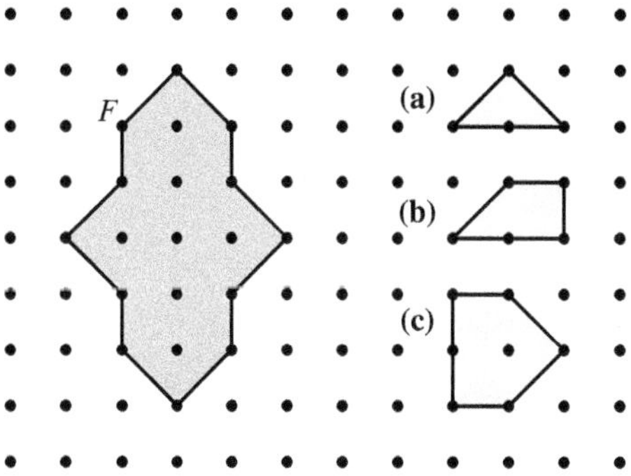

6. Find the area of each of these figures:

   (a)

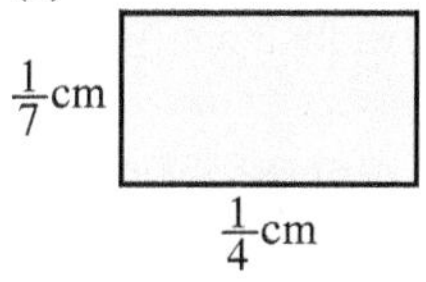

Rectangle

   (b)

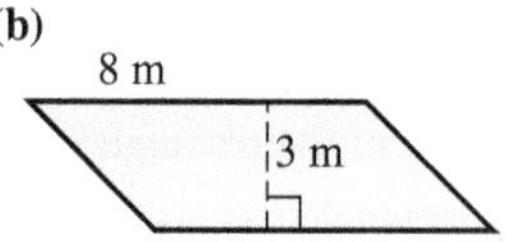

Parallelogram

7. Find the area of each of these figures:

   (a)

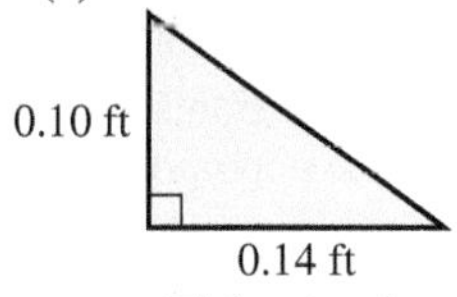

Right triangle

   (b)

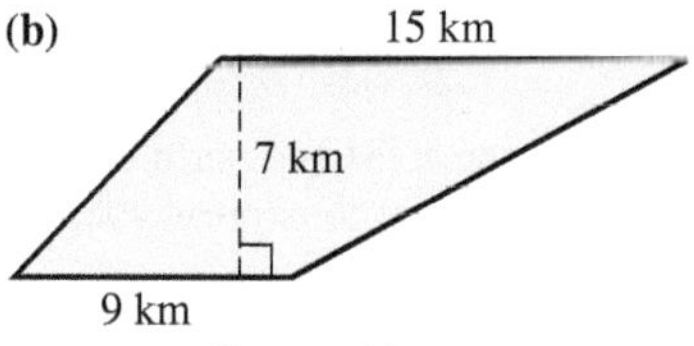

Trapezoid

8. Find the area of each of these figures:

   (a)

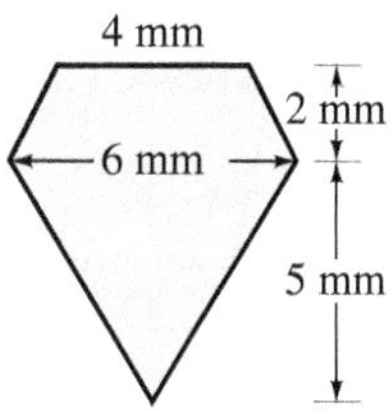

   (b)

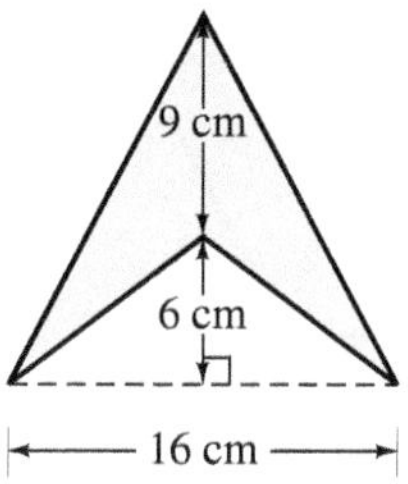

9. Find the area of the following figure:

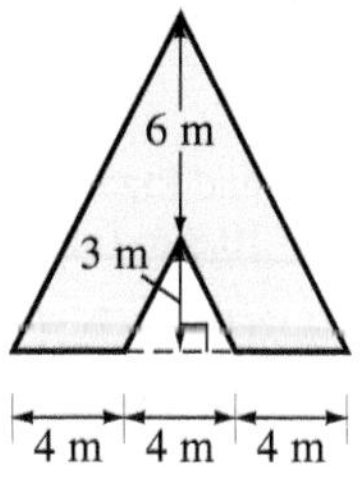

10. A rectangle has a length that is twice its width.

(a) If its perimeter is 14 in, what are the dimensions of the rectangle?

(b) If its area is 32 in^2, what are the dimensions of the rectangle?

11. A rectangle has a length 1 cm larger than its width.

(a) If the perimeter of the rectangle is 20 cm, what are the dimensions of the rectangle?

(b) If the area of the rectangle is 6 cm^2, what are the dimensions of the rectangle?

12. Fill in the blanks.

(a) 3.45 m^2 = ——— cm^2

(b) 56,000 mm^2 = ——— cm^2

(c) 56,700 ft^2 = ——— yd^2

13. Fill in the blanks.

(a) 0.085 mi^2 = ——— ft^2

(b) 47,000 a = ——— ha = ——— m^2

(c) 5,800,000 m^2 = ——— ha = ——— km^2

14. **(a)** A rectangle has area 36 cm^2 and width 3 cm. What is the length of the rectangle?

(b) A rectangle has area 60 cm^2 and perimeter 38 cm. Use the guess and check method to find the length and width of the rectangle.

15. Twenty-four 1-cm by 1-cm squares are used to tile a rectangle.

(a) Find the dimensions of all possible rectangles.

(b) Which rectangle has the smallest perimeter?

(c) Which rectangle has the largest perimeter?

16. Find the areas and perimeters of the parallelograms that follow. Be sure to express your answer in the appropriate units of measurement.

(a)

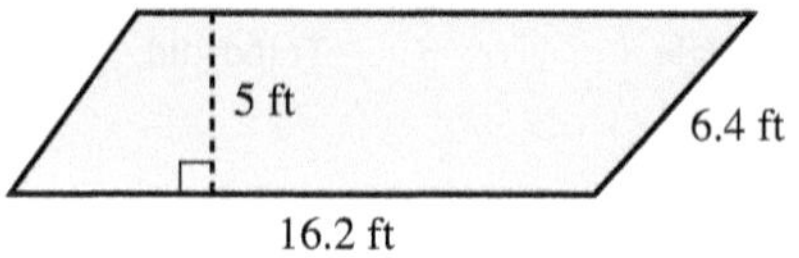

(b)

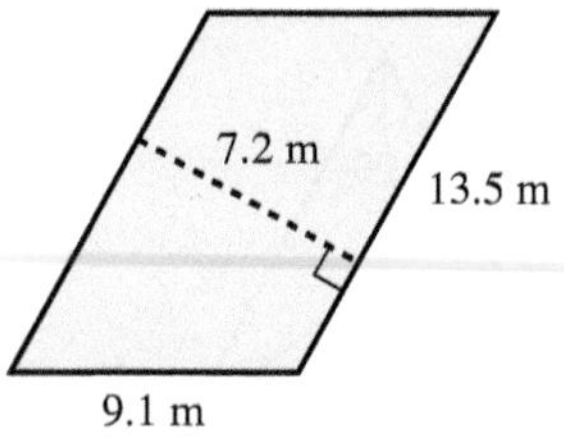

(c)

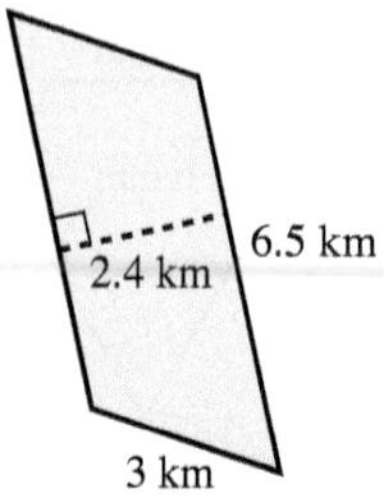

17. Find the areas and perimeters of these triangles:

(a)

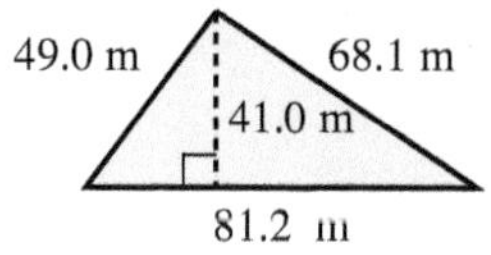

(b)

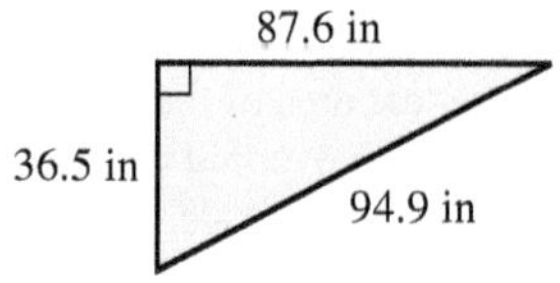

(c)

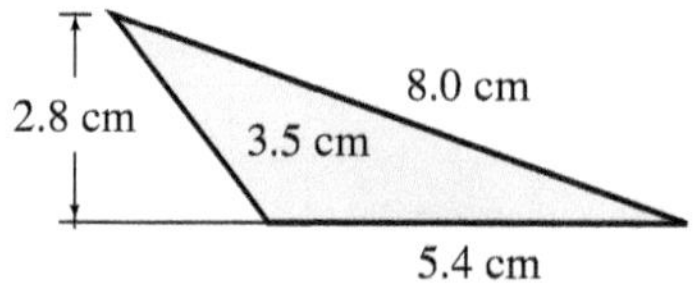

18. Find the areas of the figures shown. Express each area in square units.

(a)

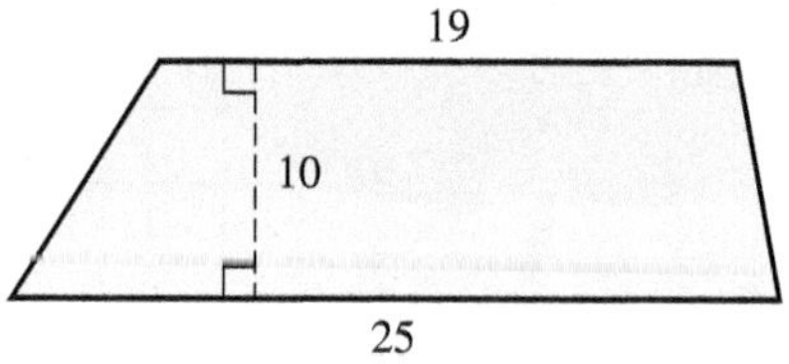

(b)

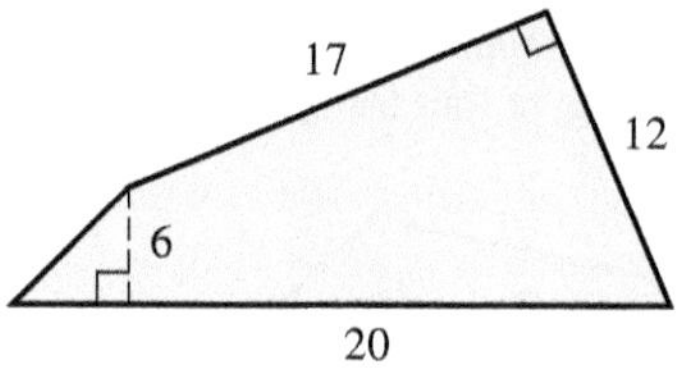

(c)

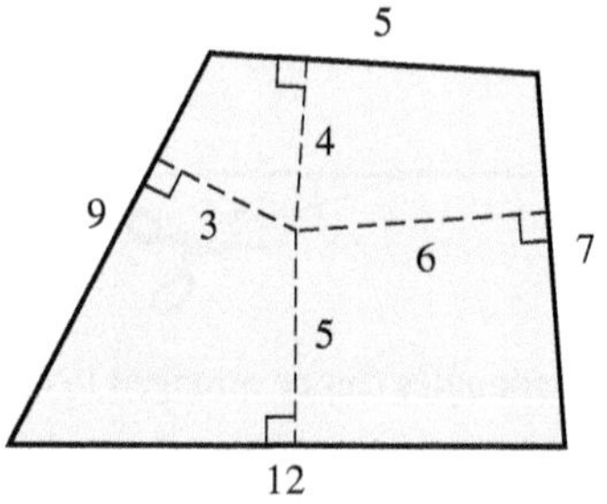

(d)

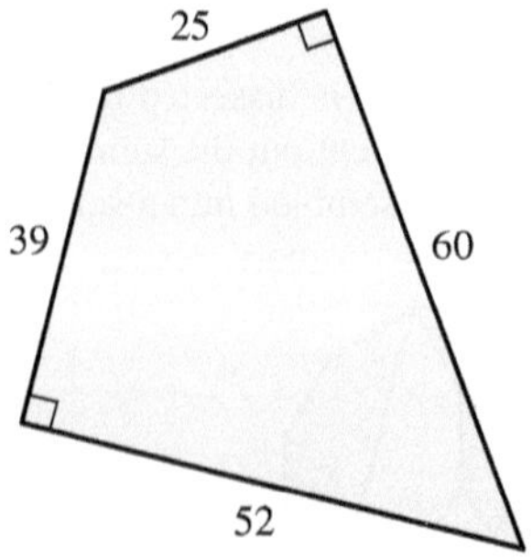

19. **(a)** Find the area and the perimeter of each of the four right triangles.

(i) 2 cm by 2 cm

(ii) 1 cm by 4 cm

(iii) .5 cm by 8 cm
(iv) .25 cm by 16 cm

(b) What is happening to the area and perimeter as you proceed from (i) to (iv), and what do the triangles look like?

20. Lines k, l, and m are parallel to the line containing the side $\overline{AB}$ of the triangles shown here:

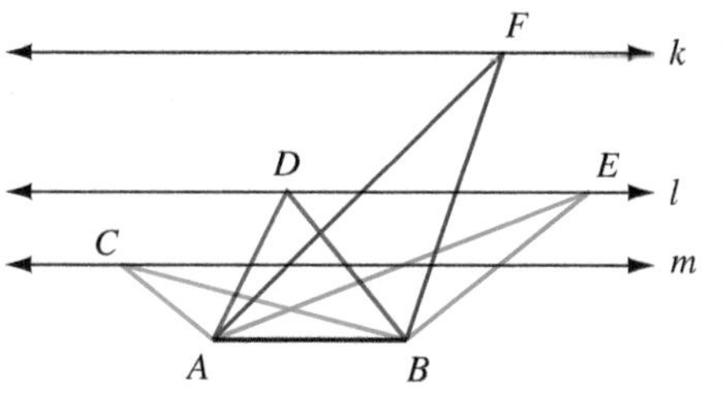

(a) What triangle has the smallest area? Why?
(b) What triangle has the largest area? Why?
(c) Which two triangles have the same area? Why?

21. Lines k, l, and m are equally spaced parallel lines. Let $ABCD$ be a parallelogram of area 12 square units.

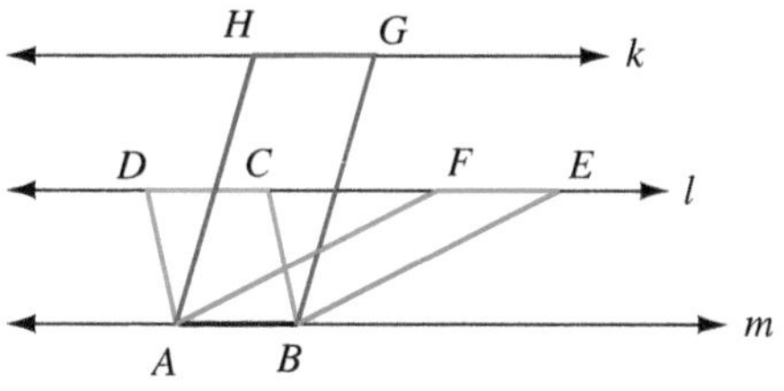

(a) What is the area of the parallelogram $ABEF$?
(b) What is the area of the parallelogram $ABGH$?
(c) If $AB = 3$ units of length, what is the distance between the parallel lines?

22. Find the area of each lattice polygon shown:

(a)

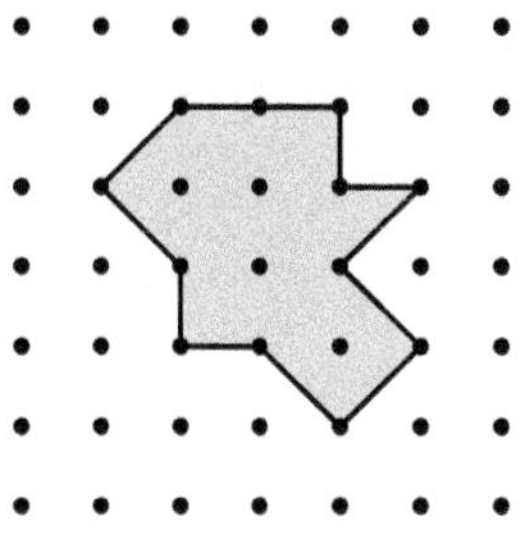

(b)

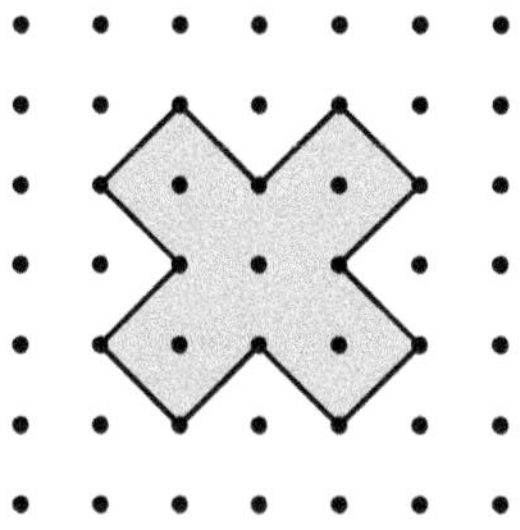

23. Find the area of this lattice polygon shown:

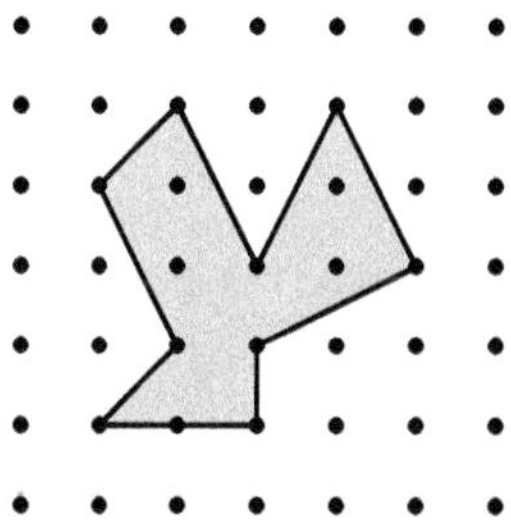

24. An oval track is made by erecting semicircles on each end of a 50-m–by–100-m rectangle.

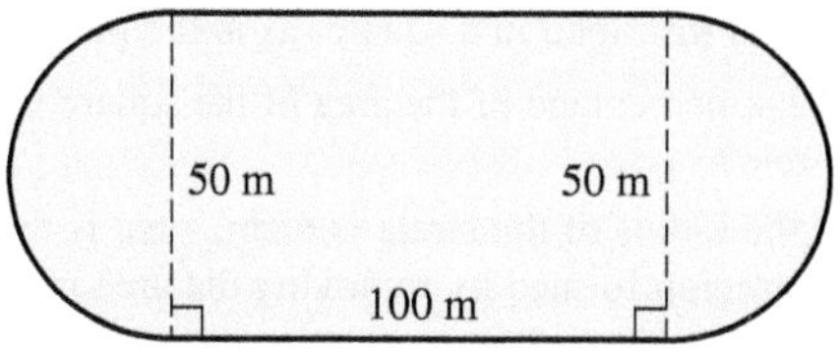

(a) What is the length of the track?
(b) What is the area of the region enclosed by the track?

25. A track has lanes 1 meter wide. The turn radius of the inner lane is 25 meters. To make a fair race, the starting lines in each lane must be staggered so that each competitor runs the same distance to the finish line. Find the distance between the starting line in one lane to the starting line in the next lane. Is the same distance used between the first and second lanes and between the second and third lanes?

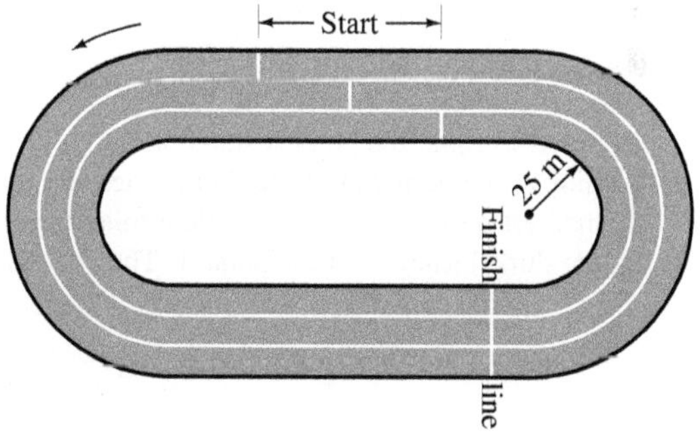

26. An **annulus** is the region bounded by two concentric circles.

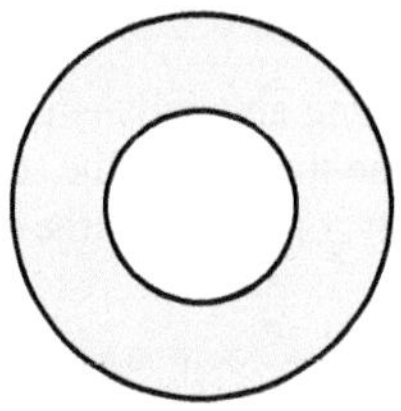

(a) If the radius of the small circle is 1 and the radius of the larger circle is 2, what is the area of the annulus?
(b) What is the perimeter of the annulus?
(c) A dartboard has four annular rings surrounding a bull's-eye.

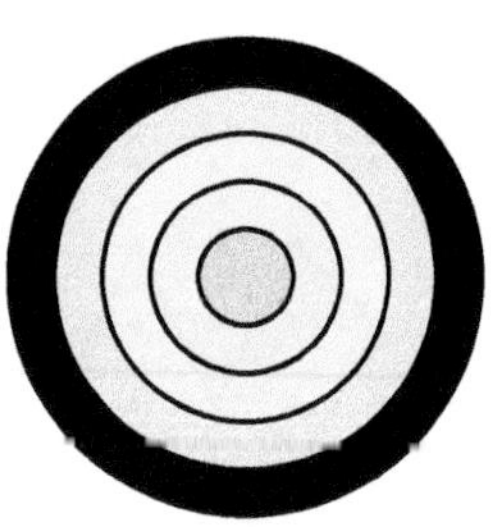

The circles have radii 1, 2, 3, 4, and 5. Suppose a dart is equally likely to hit any point of the board. Is the dart more likely to hit in the outermost ring (shown black) or inside the region consisting of the bull's-eye and the two innermost rings?

27. A circle is inscribed in a square, as in the picture shown.

(a) What percentage of the area of the square is inside the circle?

(b) If the radius of the circle is 6 cm, what is the perimeter of the region formed by removing the area inside the circle from the square?

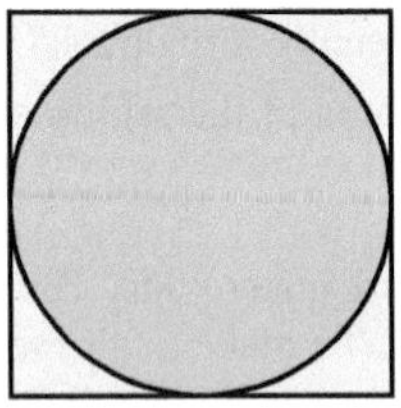

28. If a rectangle with length l and width w has a perimeter of 20 cm, what are the dimensions of the rectangle with the largest area and what is that area? (*Hint*: See Example 8.11 on page 403.)

29. This problem looks similar to the preceding one. Is it true that there is a maximum value for the perimeter of a rectangle whose area is 4 m^2? More precisely, if a rectangle with length l and width w has area 4 m^2, then

(a) Write an equation for the perimeter of the rectangle in terms of only one of the variables.

(b) Is your answer for part (a) a function that can become infinitely large?

30. Two semicircular arcs, of radius 3 m and 5 m, are centered on the diameter $\overline{AB}$ of a large semicircle as shown. Which route from A to B is shorter: along the large semicircle or along the two smaller semicircles that touch at C?

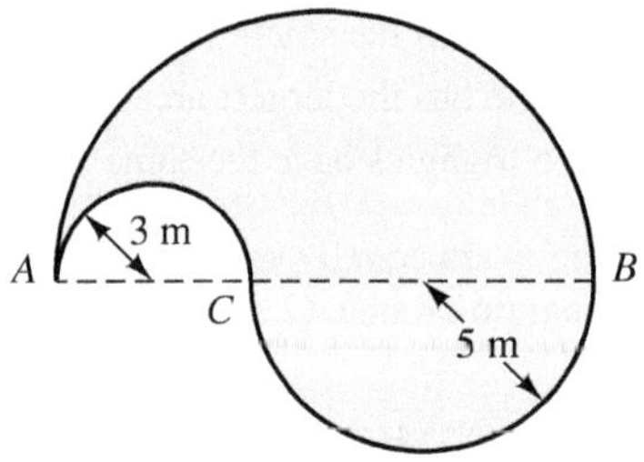

Into the Classroom

31. In teaching any topic, it is extremely helpful if surprising results of interest can be included. The result of Example 10.9 is such a case, as are the results of the following problems:

(a) Instead of rolling a wheel along a straight line, imagine "rolling" an equilateral triangle, as shown below. As the triangle "rolls" from left to right, the point A starts on the line, moves to the top in the middle position of the triangle, and again comes to rest on the line at the right end of the figure. Measure the figure shown *very carefully* with a metric ruler (estimate your measurements to the nearest millimeter), and then determine the area in square centimeters of the equilateral triangle and of the triangle AAA formed by the three locations of the point A. Then calculate the ratio of the area of triangle AAA to the area of the equilateral triangle.

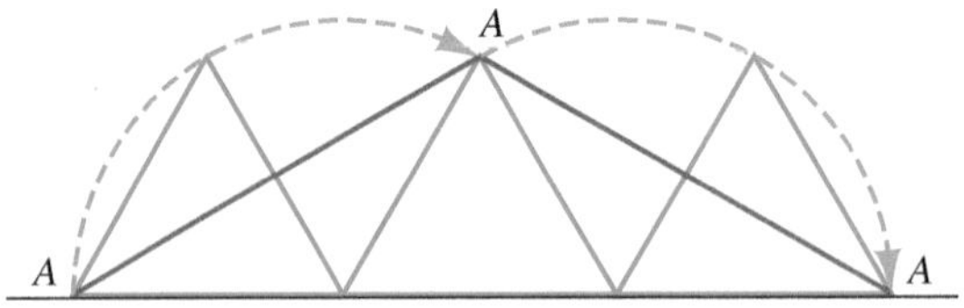

(b) Instead of rolling an equilateral triangle, this time "roll" a square, as shown. Again, measure this figure with a metric ruler, and then determine the area of the square and the area under the polygonal arch $AAAA$. As before, also determine the ratio of the area under the polygonal arch to the area of the square. (See Example 10.7, part (a).)

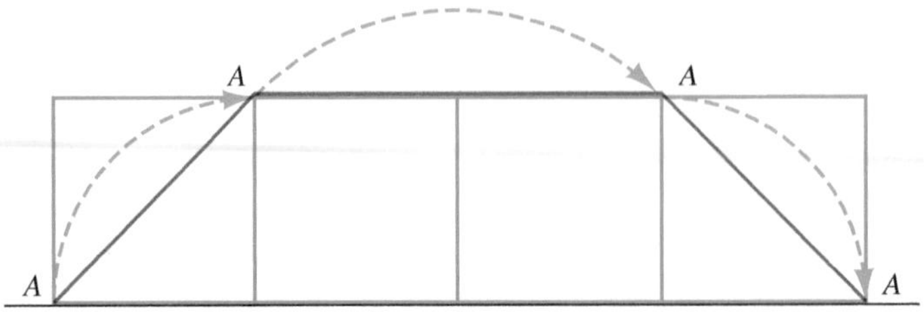

(c) Repeat parts (a) and (b), but this time roll a regular pentagon, as shown.

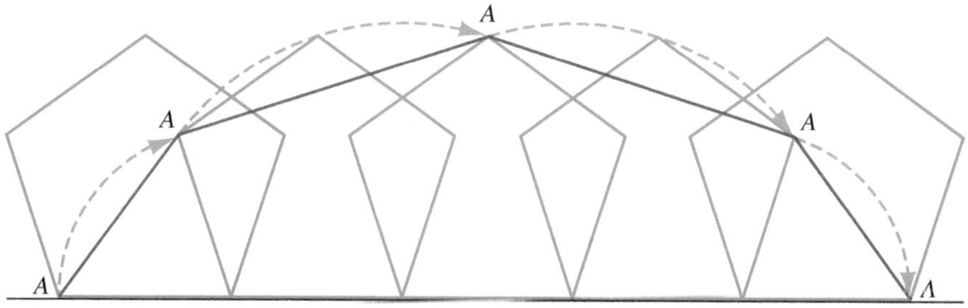

(d) Repeat parts (a), (b), and (c), but for a regular hexagon. (See Example 10.7, part (b).)

(e) Make a conjecture based on parts (a) through (d).

For the next four problems and activities, you will need several sheets of paper (including a sheet of ruled notebook paper), scissors, and a ruler.

32. The derivation of the formula for the area of a parallelogram depicted in Figure 10.9 does not apply to a tall, slanted parallelogram, since more than two pieces are required to form a rectangle. Draw a parallelogram something like the one shown, where the base is, say, three vertical ruled lines long. Cut out the parallelogram and make vertical cuts along every third ruled line. Show that the pieces you obtain can be reassembled into a rectangle.

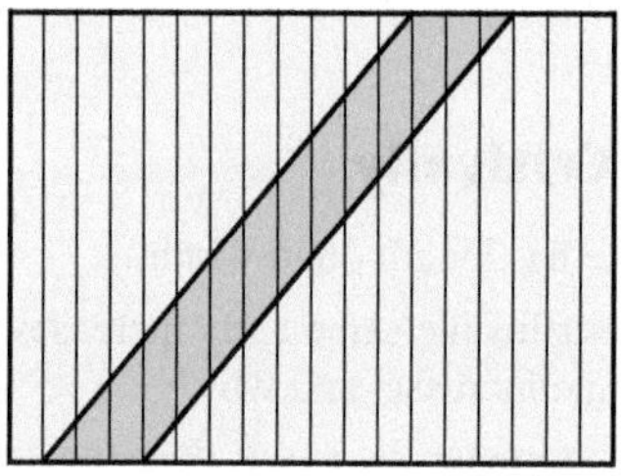

33. Fold a sheet of paper in half and cut out a pair of congruent triangles.

(a) Show that the two triangles can be arranged to form a parallelogram.

(b) Use the construction in part (a) to obtain a new explanation of how the formula for the area of a triangle follows from the formula for the area of a parallelogram.

34. Cut several triangles, as illustrated, from paper.

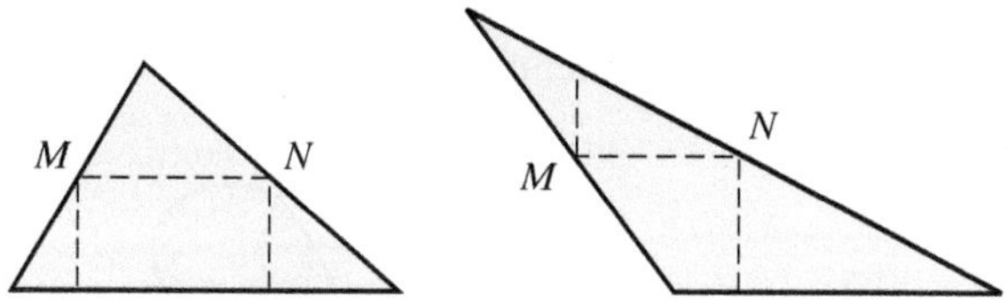

Find the midpoints M and N of the slanted sides (that is, the sides, not the base) by folding.

(a) Fold on the horizontal and vertical lines (shown dashed) to form a doubly covered (two layers of paper) rectangle.

(b) If the triangle has base b and altitude h, what are the lengths of the sides of the rectangle you formed by folding? Obtain the triangle area formula $A = bh/2$.

35. The formula for the area of a trapezoid can be obtained in several ways with paper folding and cutting. Discuss how to obtain the formula by each of these methods:

(a) Fold a sheet of paper in half and cut out, simultaneously, a pair of congruent trapezoids. Show how to arrange the two trapezoids into a parallelogram, and then explain how the area formula for a trapezoid can be derived from the area formula for a parallelogram.

(b) Fold one of the bases of a paper trapezoid onto the other, and crease along the midline between the two bases. Now cut along the crease to create two trapezoids. Show how to arrange them into a parallelogram, and then derive the area formula for the original trapezoid from that of the parallelogram you have formed.

36. In groups of three people, explore the relationship between the area and perimeter of a rectangle. This question is far too imprecise in this form so we will be more specific:

(a) Start with a length of string that is 4 meters long, place it on the floor in the shape of a square, and calculate its area.

(b) Change the dimensions of the rectangle by using the string of 4 m, and still maintaining its character as a rectangle so that its area (in square meters) is less than the area (in square meters) of the rectangle of part (a).

(c) Now change the dimensions of the rectangle by using the string of 4 m, still maintaining its character as a rectangle so that its area (in square meters) is more than the area (in square meters) of the rectangle of part (a).

(d) Is it possible to shape the string into a rectangle so that its area (in square meters) is .19 m^2?

(e) Suppose that a is any positive real number. Is it possible to shape the string into a rectangle so that its area (in square meters) is a m^2? (This assertion is saying that, for a fixed perimeter, you can construct a rectangle of any given area. In other words, there is no relationship between area and perimeter for rectangles!) It is important to point out that even the units don't make sense: Area is in square meters, for example, whereas perimeter is in meters. That is why we phrased things carefully.

37. In groups of three people, explore the relationship between the area and perimeter of a rectangle. This question is far too imprecise so we will be specific:

(a) Start with a 1-m–by–1-m (paper) tile, place it on the floor, and calculate its perimeter.

(b) Change the dimensions of the paper tile by cutting and pasting, and still maintain its character as a rectangle of area 1 m^2 so that its perimeter (in meters) is less than the perimeter (in meters) of the rectangle of part (a).

(c) Change dimensions of the paper tile by cutting and pasting, still maintaining the character of the rectangle as a rectangle of 1 m^2 so that its perimeter (in meters) is more than the perimeter (in meters) of the rectangle of part (a).

(d) Is it possible to shape the string into a rectangle so that its perimeter (in meters) is 0.1 m? A "yes" answer implies that, for a fixed area, you can construct a rectangle of any given perimeter. In other words, there is no relationship between area and perimeter for rectangles.

Responding to Students

38. Andrew was asked to find the area of the following figure:

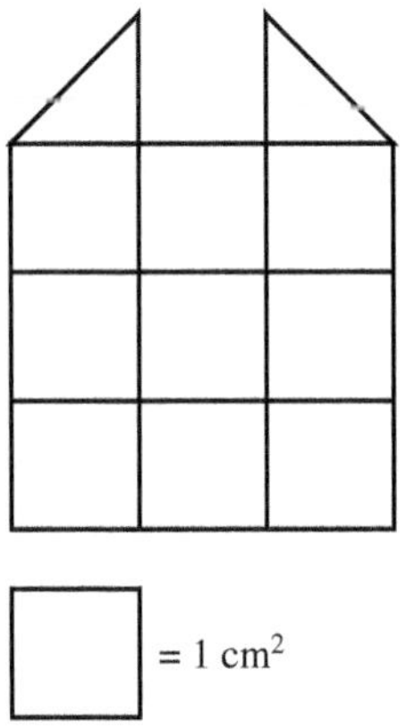

Andrew answered 11 cm^2.

(a) What does Andrew correctly understand about figuring area?

(b) What is Andrew doing incorrectly when figuring area, and how would you guide him in the future?

39. A packing box is 11 inches long by 7 inches wide by 6 inches deep. Taneisha is asked to find the perimeter of the bottom of the box. Taneisha gets out her paper, does some addition, and then writes 24 inches.

(a) What is the correct perimeter of the bottom of the box?

(b) Where did Taneisha's answer of 24 inches come from?

(c) How could you guide Taneisha in the future when approaching such problems?

40. Estelle had been practicing area and perimeter problems in math for several days. The class first learned all about perimeters and Estelle was successful. The class later learned about areas and again Estelle was successful when practicing. The teacher gave her class the following problem in which they had to calculate both the area and the perimeter for the same figure:

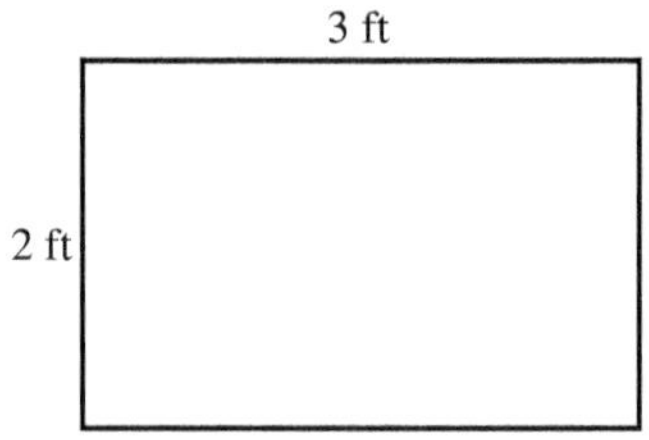

Here are Estelle's answers:

Perimeter: 5 feet

Area: 5 feet

(a) What is Estelle doing incorrectly when applying the formulas she learned for area and perimeter?

(b) Some students are often successful when solving area and perimeter problems in isolation but can make many mistakes when asked to complete both in the same problem. Why do you suppose this happens? What could a teacher do to remedy this confusion?

41. Mr. Richard, who, unfortunately, is known for his math jokes, told his eighth-grade class that March 14 at 1:59 A.M. is his favorite time every year. What responses do you think that he might have had in mind? (*Hint:* Many math departments will serve something that is approximate to π on that day.)

Thinking Critically

42. A rectangle has length l and width w.

(a) If w remains the same and l increases, must the area of the rectangle increase, and why?

(b) If the perimeter increases, does the area increase also? In this case, the conditions are that the length l and width w of the rectangle can vary, but if the perimeter increases, then does the area of the new rectangle always increase?

43. Find formulas for the perimeter and area of a regular hexagon with sides of length b.

44. Two regions A and B are cut from paper. Suppose the area of region A is 20 cm^2 larger than that of region B. If the regions are overlapped, by how much does the area of the nonoverlapped part of region A exceed the nonoverlapped part of region B? Explain carefully.

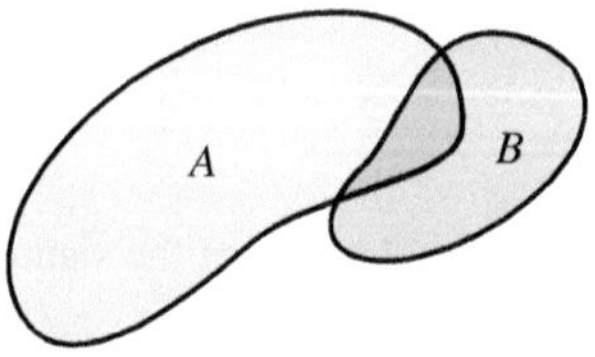

45. Four mutually tangent circles of diameter 10 cm with their centers at the vertices of a square are used to draw a vase, as follows:

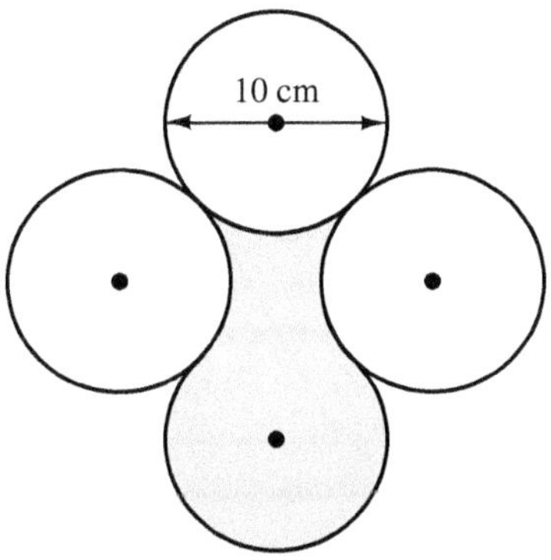

(a) Use dissection and rearrangement to form a square of the same area as the vase. (Show a sequence of steps similar to those in the solution to Example 10.8.)

(b) Show that the vase has area 100 cm^2.

(c) As an extra challenge, see if you can do part (a) by cutting the vase into just three pieces.

46. A square cake measures 8″ by 8″. A wedge-shaped piece is cut by two slices meeting at 90° at the cake's center. What is the area of the top of the piece? Explain your reasoning carefully.

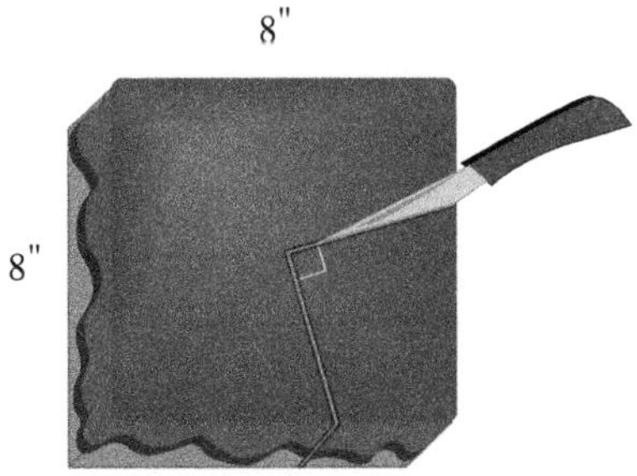

47. For reasons lost in history, the two cornfields R and S were divided by two line segments $\overline{AB}$ and $\overline{BC}$. The friendly owners of the fields would like to divide their adjoining fields by a single straight boundary line. Carefully describe how to divide the quadrilateral into two fields R' and S' with a single segment so that the area of each new cornfield is the same as before.

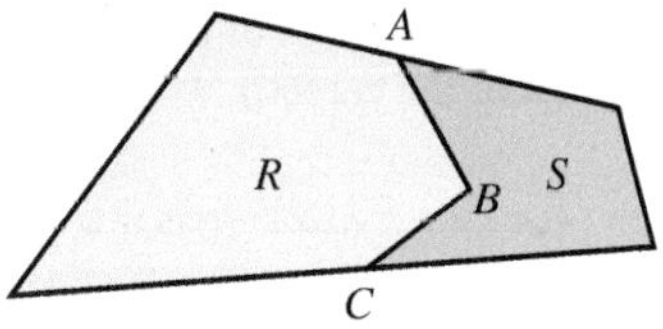

48. **(a)** The colored region shown is formed by circular arcs drawn from two opposite corners of a 1-by-1 square. What is the area of the region?

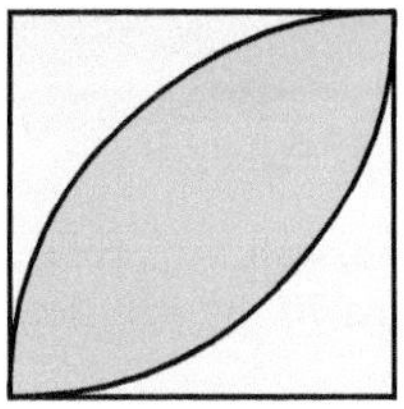

(b) Four semicircles are drawn with centers at the midpoints of the sides of a 1-by-1 square. What is the area of the shaded region?

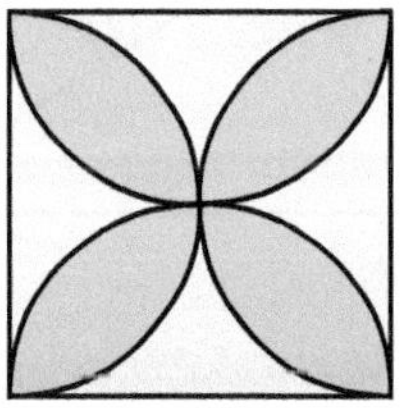

49. A sidewalk 8 feet wide surrounds the polygonally shaped garden of perimeter 300 feet, as shown. The sidewalk makes circular sectors of 8-foot radius at the vertices of the polygon. Explain why the area covered by the walk is $2400 + 64\pi$ square feet.

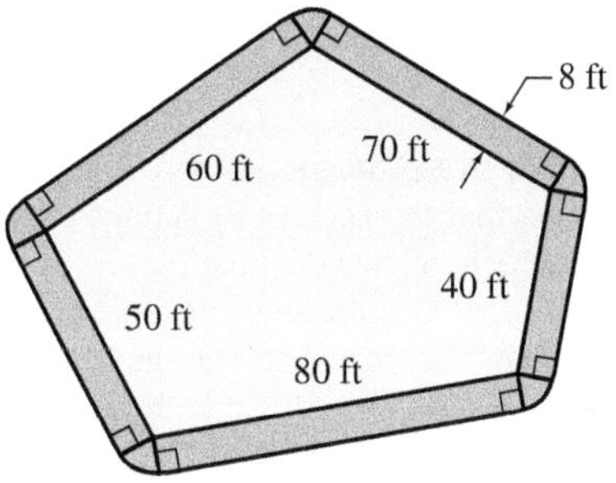

50. Erin walks her dog Nerd with a leash of length L. Nerd is very obedient and always walks directly to Erin's right at the end of his leash. Erin follows several different routes and wants to compare the length of her walk with that of Nerd. For the following routes, how much farther does Nerd walk than Erin?

(a) Around a circle of radius R

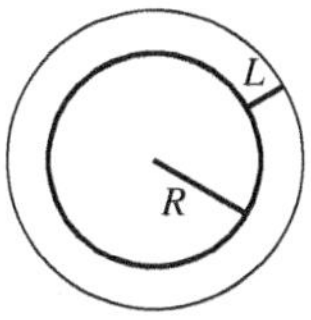

(b) Around a square with sides of length S

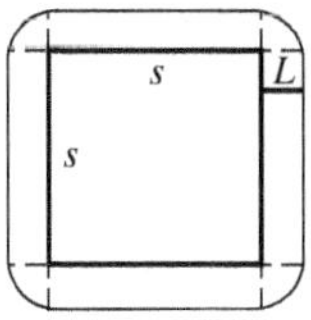

(c) Around a track shaped like a rectangle of length A with half circles of radius R on each end of the rectangle

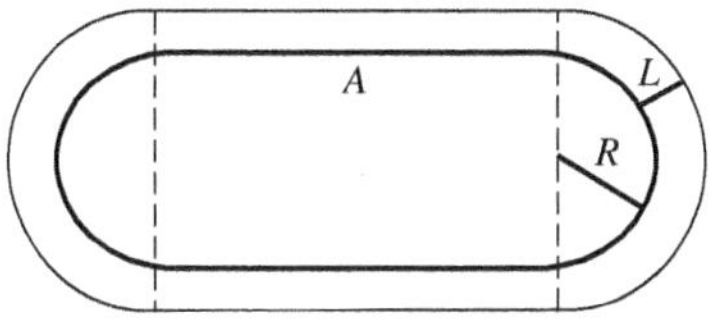

(d) Around a triangle with sides of length A, B, and C

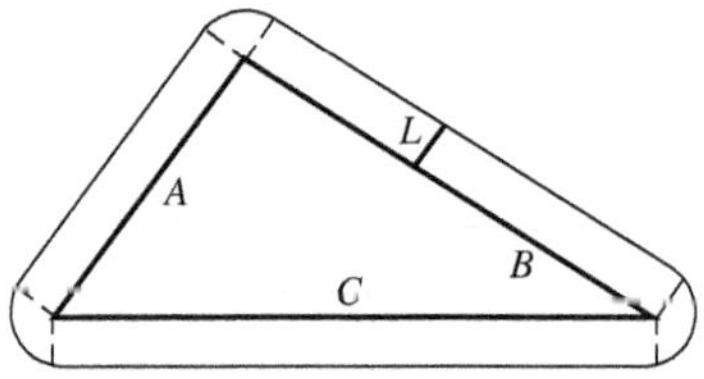

(e) Around any simple closed curve. (*Hint:* Approximate the curve by a polygon and review Section 9.2.)

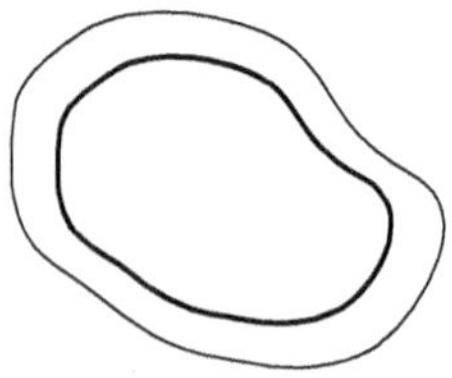

51. We will now see that problems 48 and 50 of the Rhind papyrus suggest that the ancient Egyptians approximated π with $(16/9)^2$, which, to two decimal places, is the quite accurate value 3.16. Use the sequence of figures that follows to explain the reasoning that may have been used to derive this estimation. Notice that the area of the circle with diameter 9 is approximated by an octagon and that the octagon's area is then approximated by a square of side length 8.

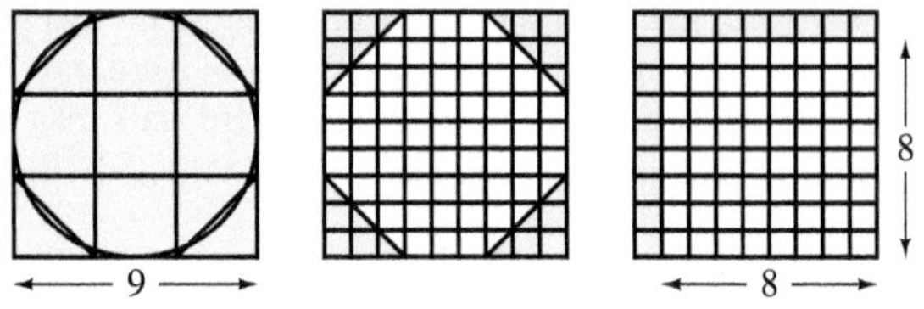

52. The commentaries of the Talmud (Tosfos Pesachim 109a, Tosfos Succah 8a, Marsha Babba Bathra 27a) present a nice approach to the formula $A = \pi r^2$ for the area of a circle. Imagine that the interior of a circle is covered by concentric circles of yarn. The yarn circles are clipped along a vertical radius, and each strand is straightened to cover an isosceles triangle, as shown in the accompanying figure. Find the area of the triangle and then explain how the area formula for a circle follows.

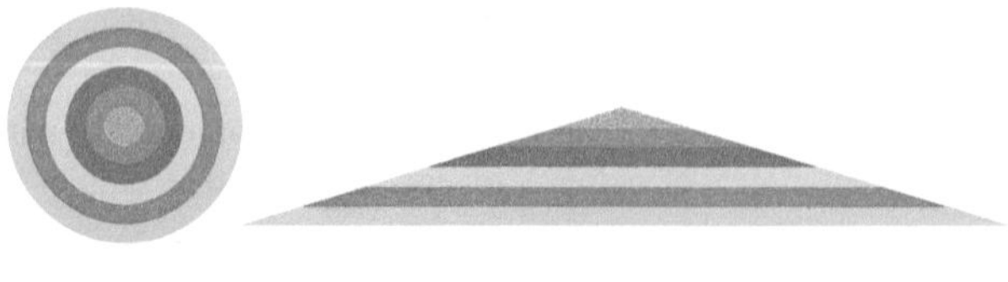

53. Let P be an arbitrary point in an equilateral triangle ABC of altitude h and side s, as shown. What is the sum $x + y + z$ of the distances to the sides of the triangle? (*Hint:* The areas of $\triangle ABP$, $\triangle BCP$, and $\triangle ACP$ add up to the area of $\triangle ABC$.)

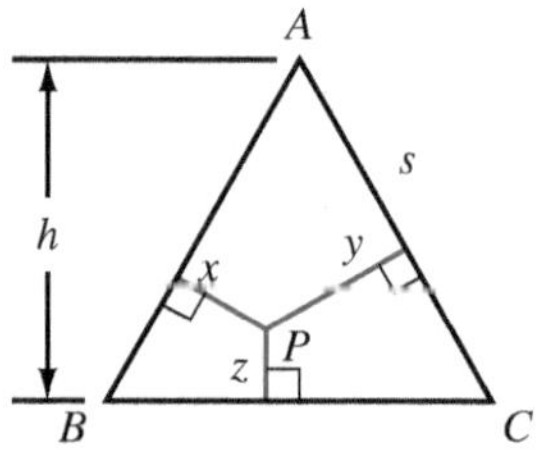

54. Joining each vertex of the triangle shown here to the midpoint of the opposite side divides the triangle into six small triangles:

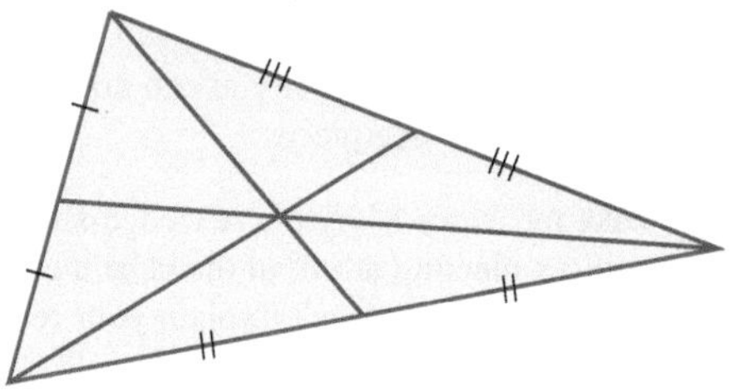

Show that all six triangles have the same area. (*Hint:* Look for pairs of triangles with the same base and height.)

Making Connections

55. Kelly has been hired to mow a large rectangular lawn measuring 75 feet by 125 feet. The lawn mower cuts a path 21 inches wide. Estimate how far (in feet) Kelly must walk to complete the mowing job.

56. Roll ends of carpet are on sale for six dollars per square yard. To finish the rough-cut edges, edging material costing 10 cents per foot is glued in place. Compute the total cost of a roll end measuring 8 feet by 10 feet.

57. A carpet is made by sewing a 1-inch-wide braid around and around until the final shape is an oval with semicircular ends, as shown. Estimate the length of braid required. (*Hint:* Estimate the area of the carpet.)

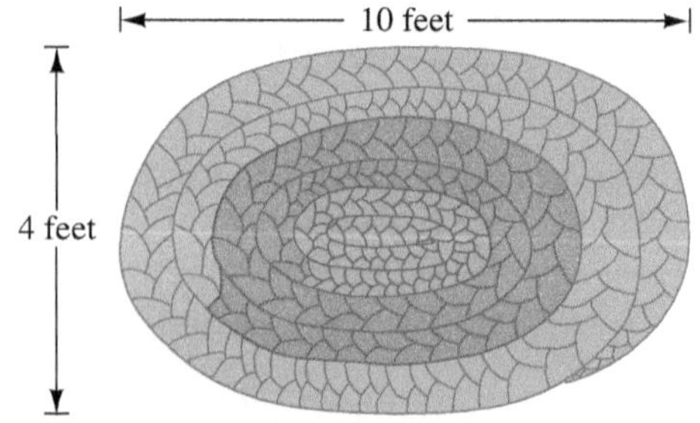

58. An L-shaped house, walkway, garage, and driveway are shown situated on a 70′ by 120′ lot. How many bags of fertilizer are needed for the lawn? Assume that the bags are each 20 pounds and 1 pound of fertilizer will treat 200 square feet of lawn.

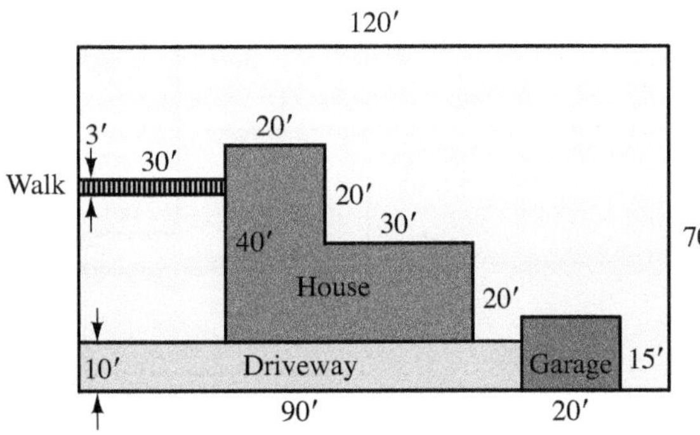

59. A 10′ by 12′ kitchen floor is to be tiled with 8″ square tiles. Estimate the number of tiles this will require.

60. **(a)** The normal-size tires on a truck have a 14-inch radius. If oversized tires of 15-inch radius are used, how much farther does the truck travel per revolution of the wheel?

(b) If the speedometer indicates that a truck is traveling at 56 miles per hour, what is the true speed when the truck is running on the oversize tires?

61. Sunaina has 600 feet of fencing. She wishes to build a corral along an existing high, straight wall. She has already decided to make the corral in the shape of an isosceles triangle, with two sides each 300 feet long. What is the measure of $\angle A$ that will give Sunaina the corral of most area? (*Hint:* Consider one of the sides of length 300 feet as a base of the triangle.)

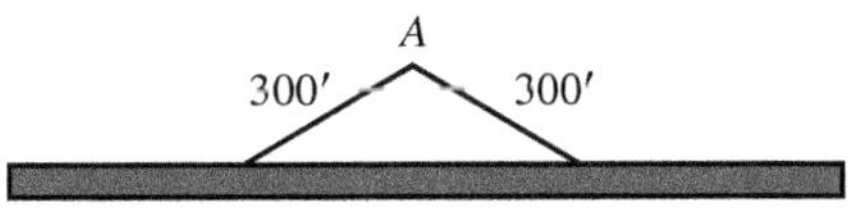

State Assessments

62. (Washington State, Grade 4)
Casey is making a quilt. Quilts are made up of quilt blocks. Each quilt block will look like the one below.

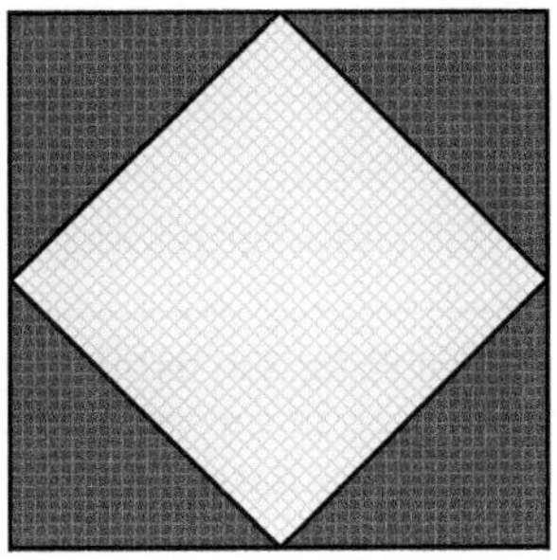

Her quilt will have 25 blocks. Casey knows how much fabric she needs to make the patterned inside squares. Tell the steps she could take to figure out how much fabric she will need to make all of the shaded corner pieces. Explain your thinking using words, numbers, or pictures.

63. (Grade 5)
The length of a rectangular swimming pool is 10 feet. If the perimeter of the swimming pool is 32 feet, which of the following answers is the width of the swimming pool?

A. 16 feet
B. 10 feet
C. 6 feet
D. 132 feet
E. 64 feet

64. (Grade 8)
There is a swimming pool in a large park. The surface of the swimming pool has a fence around it. The first of three pieces of the pool is a square wading pool of 30 ft. The second piece is a right triangle with legs of 30 ft and 40 ft for older children. The third piece is a very deep part of the pool which is surrounded by a semi-circle. The pool is sketched below.

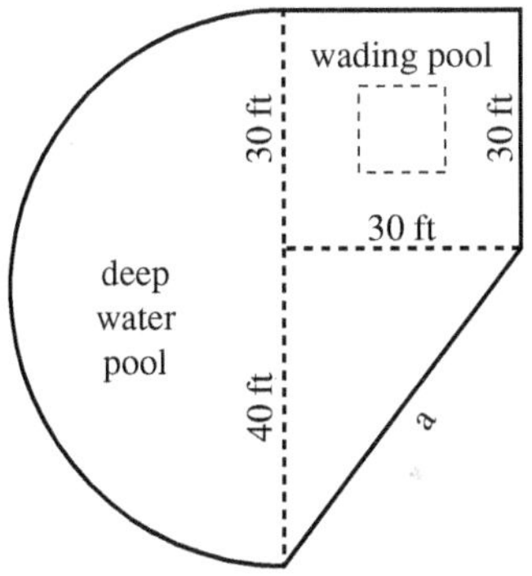

The solid line shows the fence. The dotted line gives one side of the wading pool and the 40ft leg of the right triangle.

(a) One side of the fence doesn't have a given length. However, it is designated with 'a' in the picture. What is the length of that part of the fence?

(b) What is the length of the fence around the pool?

65. (Washington State, Grade 8)
Omari is making a string design for art class. Look at the design.

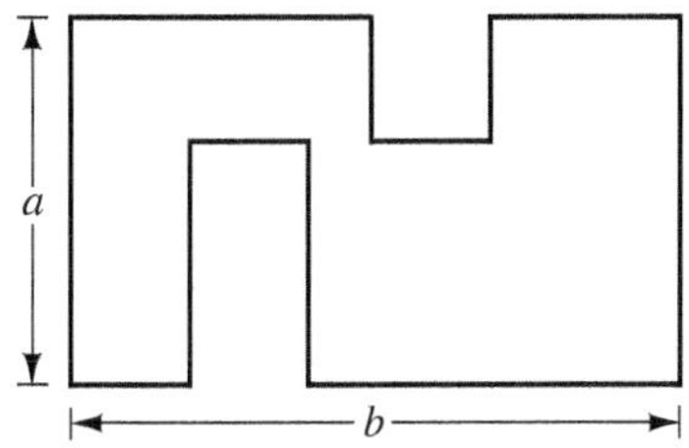

Which expression can he use to figure how much string he needs to create the design?

A. $a + b$
B. $2a + 2b$
C. $3a + 2b$
D. $4a + 2b$

66. (Washington, Grade 6)
Lacey's family is buying a new home. It will be her job to mow the lawn. She wants to know the size (area) of the lawn.

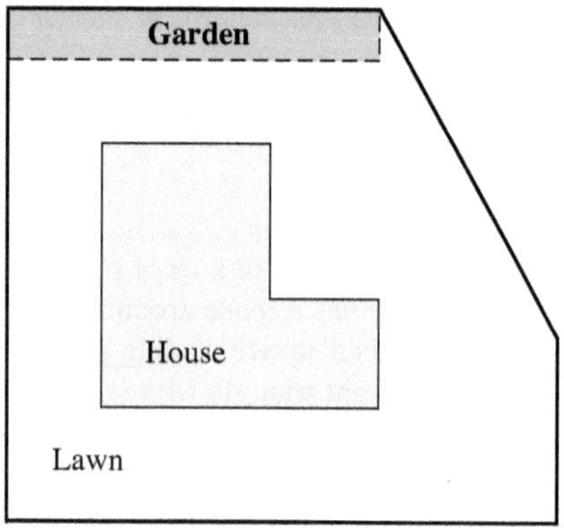

Describe at least five steps Lacey would need to take to determine the size (area) of the lawn.

67. (Grade 6)
What is the perimeter, in centimeters, of the figure shown below?

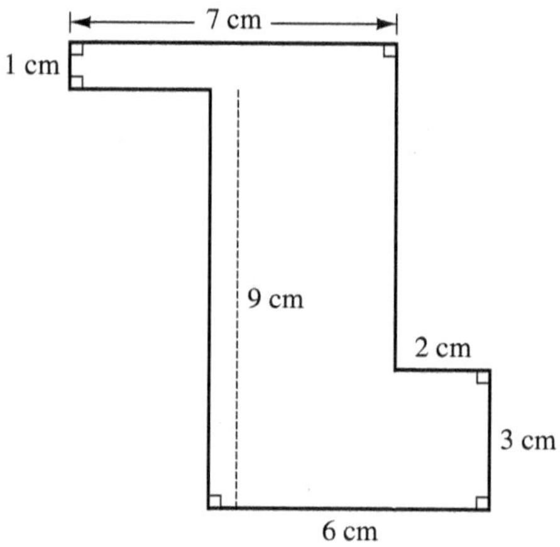

68. (Grade 6)
My house has a square foot print and the lawn around it is rectangular.

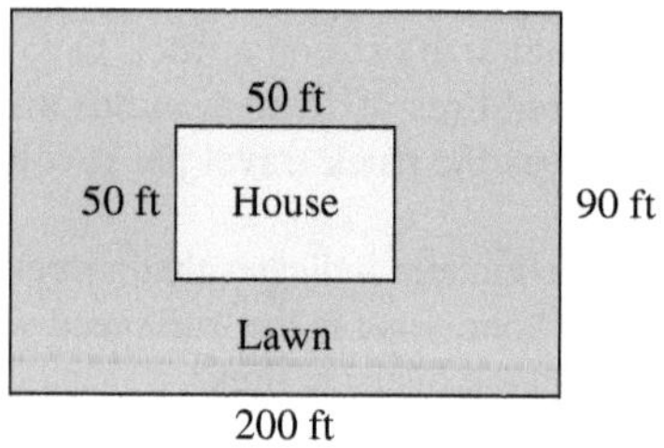

What is the area, in square feet, of the lawn?

COOPERATIVE INVESTIGATION
Discovering Pick's Formula

In 1899, the German mathematician Georg Pick discovered a remarkable formula for the area of a polygon drawn on square dot paper. Polygons of this special type are known as lattice polygons. Stretching a rubber band onto a geoboard is an easy way to form lattice polygons.

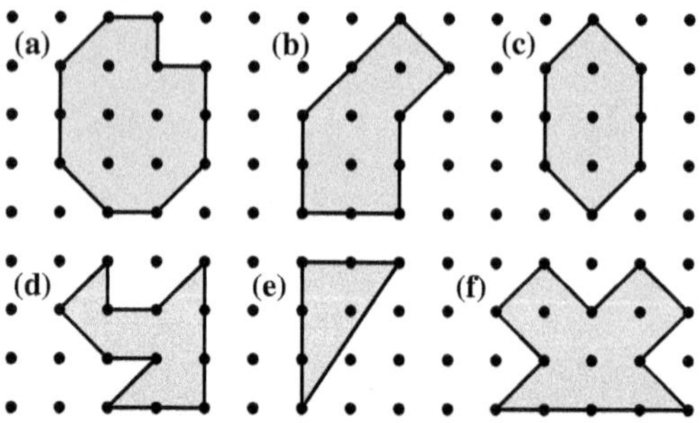

1. Complete the table of values for each polygon, where

b = number of dots on the boundary of the polygon,
i = number of dots in the interior of the polygon, and
A = area of the polygon:

Polygon	b	i	A
(a)	11	5	$9\frac{1}{2}$
(b)			
(c)			
(d)			
(e)			
(f)			

The values of b, i, and A for polygon (a) are given as an example.

2. Try to guess a formula for A in terms of b and i. (If you have trouble, add a column of the values of $b/2$. You may also want to obtain more data by drawing other lattice polygons.)

10.3 The Pythagorean Theorem

The Pythagorean theorem is a most remarkable result in geometry. We will see that the theorem is very useful in solving practical problems and is also aesthetically pleasing. A tremendous number of proofs have been devised since the time of Pythagoras (ca. 572–501 B.C.), including one by a president of the United States (James Garfield.) In the second edition of *The Pythagorean Proposition*, E. S. Loomis catalogs 370 different proofs.

THEOREM The Pythagorean Theorem

If a right triangle has legs of length a and b and its hypotenuse has length c, then

$$a^2 + b^2 = c^2.$$

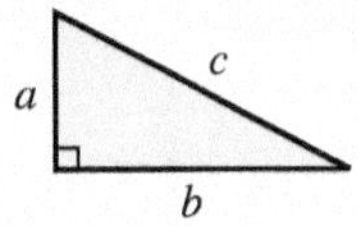

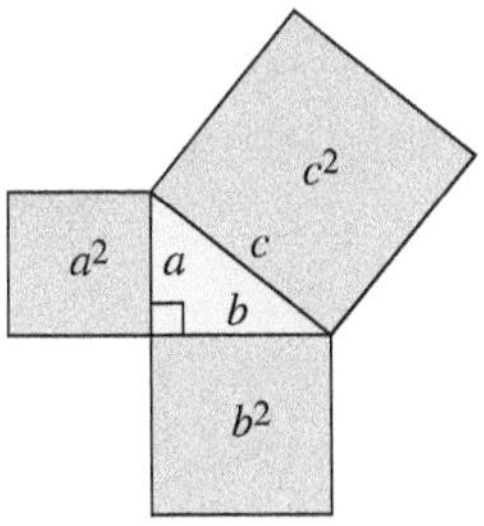

Figure 10.15
The sum of the areas of the squares on the legs of a right triangle equals the area of the square on the hypotenuse

Proving the Pythagorean Theorem

By erecting squares on the sides of a right triangle, the Pythagorean relation $a^2 + b^2 = c^2$ can be interpreted as a result about areas: *The sum of the areas of the squares on the legs of a right triangle is equal to the area of the square on the hypotenuse.* The area interpretation of the Pythagorean theorem is shown in Figure 10.15 and should be looked at carefully by you and, ultimetely, your students.

The area interpretation probably led to the discovery of the theorem, at least in special cases. For example, it is apparent by looking at Figure 10.16 that the area of the square on the hypotenuse equals the area of the two squares on the legs for an isosceles right triangle.

Figure 10.16
A special case of the Pythagorean theorem

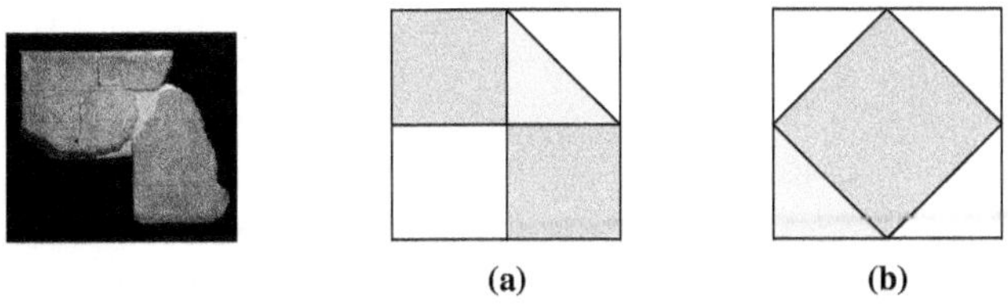

The demonstration depicted in Figure 10.16 is not a general proof of the Pythagorean theorem because the right triangle is isosceles. However, a similar idea can be followed for arbitrary right triangles, using the *dissection method.* In Figure 10.17a, we begin with any right triangle, letting a and b denote the respective lengths of the legs and c the length of the hypotenuse. Next, consider two squares with sides $a + b$, as in Figures 10.17b and 10.17c. Four congruent copies of the right triangle are placed inside the squares in two different ways.

In Figure 10.17b, the four triangles leave two squares uncovered, with respective areas a^2 and b^2. In Figure 10.17c, the four triangles leave one square of area c^2 uncovered. Since the four triangles must leave the same area uncovered in both arrangements, we conclude that $a^2 + b^2 = c^2$ and the theorem is proved using areas.

Figure 10.17
A dissection proof of the Pythagorean theorem

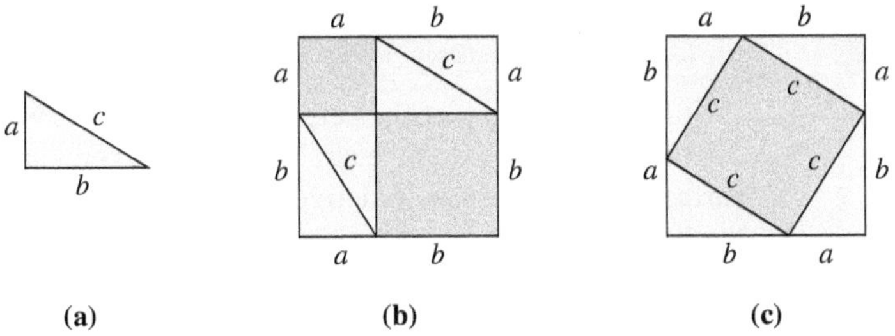

SMP

SMP 3 Construct viable arguments and critique the reasoning of others.

The Cooperative Investigation exercise at the beginning of Chapter 7 showed another proof of the Pythagorean theorem, one that uses algebra rather than dissection.

The Standard for Mathematical Practice 3(SMP 3) involves the ability to make a plausible argument of a theorem involving areas. Paraphrasing SMP 3, we will construct a viable argument that involves objects and drawings (the Pythagorean theorem and the use of area).

"Mathematically proficient students understand and use stated assumptions, [and] definitions . . . Elementary students can construct arguments using concrete referents such as objects, drawings, diagrams, and actions. Such arguments can make sense and be correct, even though they are not generalized or made formal until later grades. Later, students learn to determine domains to which an argument applies. Students at all grades can listen or read the arguments of others, decide whether they make sense, . . ."

No records have survived to indicate what proof, if any, Pythagoras may have offered. The dissection proof requires showing that the inner quadrilateral of Figure 10.17c is actually a square (why is it?), and Pythagoras's knowledge of angles in a right triangle was sufficient to do this.

Applications of the Pythagorean Theorem

EXAMPLE 10.17

Using the Pythagorean Theorem

Find the lengths x and y in the following figures:

(a)

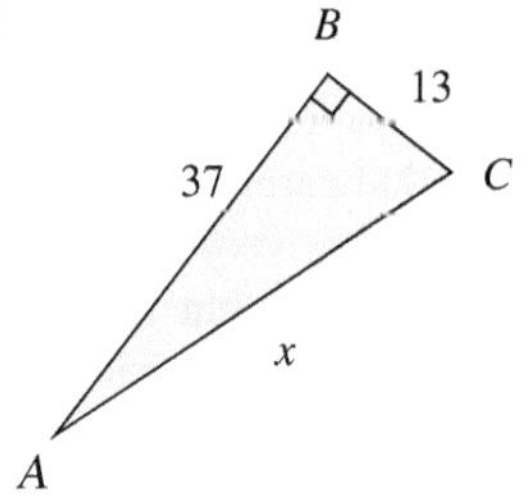

(b)

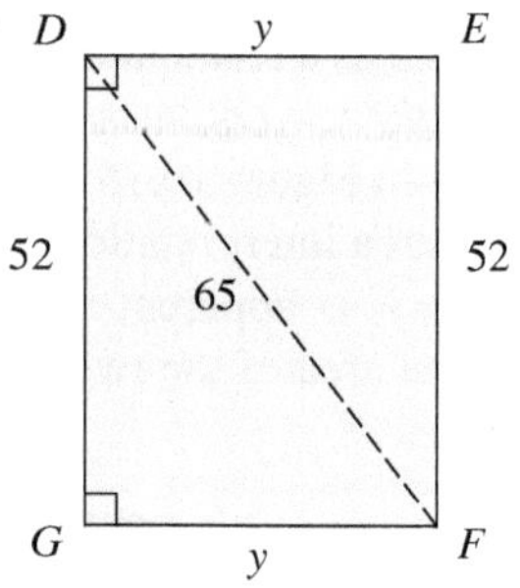

Solution

(a) By the Pythagorean theorem, $x^2 = 13^2 + 37^2 = 169 + 1369 = 1538$. Therefore, $x = \sqrt{1538} \doteq 39.2$.

(b) The diagonal $\overline{DF}$ of rectangle $DEFG$ is the hypotenuse of the right triangle DEF. The Pythagorean theorem, applied to $\triangle DEF$, gives $y^2 + 52^2 = 65^2$. Therefore, $y^2 = 65^2 - 52^2 = 4225 - 2704 = 1521$, and $y = \sqrt{1521} = 39$.

For many applications, it is necessary to write and solve equations based on the Pythagorean relation. Here is an example, and we'll see many more in the next section.

EXAMPLE 10.18

Determining How Far You Can See

Imagine yourself on top of a mountain, or perhaps in an airplane, at a known altitude given in feet. Approximately how far away, in miles, is the horizon?

Solution

Understand the Problem

Altitude is a measure of the perpendicular distance above the surface of Earth (which is very nearly a sphere). The horizon is the circle of points where our line of sight is tangent to the sphere of Earth's surface. The problem is to derive a formula that expresses, or at least approximates, the distance to the horizon in relation to the altitude of the observer. Since the altitude is given in feet, while the distance to the horizon is to be given in miles, special care must be taken to handle the units of measure properly.

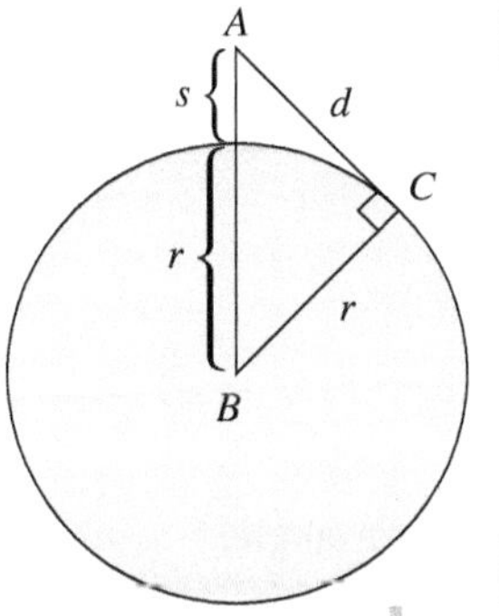

Devise a Plan

A line of sight to the horizon forms a leg of a right triangle, as the diagram shows. Since the radius r of Earth is about 4000 miles and the altitude s is known in miles, the Pythagorean theorem can be used to solve for the distance to the horizon. In the diagram, all distances, including s, are expressed in miles; if h is the altitude in feet, we can use the conversion formula $h = 5280s$. (Recall that 1 mile $=$ 5280 feet.) Note that $\overleftrightarrow{AC}$ is the tangent line at C to the circle whose center, B, is the center of Earth. See Example 8.16 on page 415.

Carry Out the Plan

Applying the Pythagorean theorem to the right triangle ABC gives $d^2 + r^2 = (s + r)^2$. The squared term on the right can be written $s^2 + 2sr + r^2$, so $d^2 + r^2 = s^2 + 2sr + r^2$. Subtracting r^2 from both sides shows that $d^2 = s^2 + 2sr$. Therefore, the exact distance d, in miles, is given by the formula

$$d = \sqrt{s^2 + 2rs}.$$

Since $r = 4000$ miles, the formula can also be written

$$d = \sqrt{s^2 + 8000s} \quad \text{or} \quad d = \sqrt{s(s + 8000)}.$$

From the top of a mountain (say, $s = 2$ miles), or from an airplane (say, $s = 7$ miles), or even from the International Space Station ($s = 200$ miles), it is evident that the altitude s is much smaller than 8000. Therefore, little accuracy is lost if the term $(s + 8000)$ in the exact formula is replaced with simply 8000. This gives us the approximate equation

$$d \doteq \sqrt{8000s}.$$

Using the equation $s = \dfrac{h}{5280}$ gives $d \doteq \sqrt{\dfrac{8000}{5280}h}$. Finally, since $\sqrt{\dfrac{8000}{5280}}$ is about 1.2, we obtain a simple approximatel formula for the number of miles, d, to the horizon as seen from an altitude of h feet:

$$d \doteq 1.2\sqrt{h}.$$

For example, the distance to the horizon as seen from an airplane flying at 40,000 feet is about $1.2\sqrt{40{,}000} = (1.2)(200) = 240$ miles.

Look Back

This problem involved several steps that are typical of the way the Pythagorean theorem is used:

- Draw a figure and label all the distances.
- Identify all the right triangles in the drawing.
- Write the Pythagorean relationships for all of the right triangles.
- Solve the Pythagorean formulas to determine unknown values needed for the solution of the problem.
- Use an estimation technique while carrying out the plan.

The Converse of the Pythagorean Theorem

The numbers 5, 12, and 13 satisfy $5^2 + 12^2 = 13^2$. Is the triangle with sides of length 5, 12, and 13 a right triangle? The answer is yes, since the Pythagorean relation $a^2 + b^2 = c^2$ holds if *and only if* a, b, and c are the side lengths of a right triangle. That is, the converse of the Pythagorean theorem is true and is stated without proof in the following theorem. The proof of this converse is beyond this course, but we will now show an example.

THEOREM Converse of the Pythagorean Theorem

Let a triangle have sides of length a, b, and c. If $a^2 + b^2 = c^2$, then the triangle is a right triangle and the angle opposite the side of length c is its right angle.

EXAMPLE 10.19 Checking for Right Triangles

Determine whether the three lengths given are the lengths of the sides of a right triangle.

(a) 15, 17, 8 **(b)** 10, 5, $5\sqrt{3}$ **(c)** 231, 520, 568

Solution

(a) $8^2 + 15^2 = 64 + 225 = 289 = 17^2$, so 8, 15, and 17 are the lengths of the sides of a right triangle.

(b) $5^2 + (5\sqrt{3})^2 = 25 + 25 \cdot 3 = 25 + 75 = 100 = 10^2$, so 5, $5\sqrt{3}$, and 10 are the lengths of the sides of a right triangle.

(c) Using a calculator, we can show that $231^2 + 520^2 = 53{,}361 + 270{,}400 = 323{,}761 \neq 322{,}624 = 568^2$, so 231, 520, and 568 are not the lengths of sides of a right triangle. This would be difficult to see by measuring angles with a protractor, since this triangle closely resembles the right triangle with sides of length 231, 520, and 569. Note that $569^2 = 323{,}761$.

INTO THE CLASSROOM
Simone Wells-Heard on Connecting All the Dots

The Pythagorean theorem provides an opportunity for middle school students to discover geometric principles that are foundational for higher mathematics. Given a rectangle, students can split it into two right triangles and discover that the diagonals of a rectangle are congruent long before they are asked in high school math class to write a proof. I ask my students to plot the points (10, 5), (10, −1), and (2, −1) on the coordinate plane and then find the distance between (10, 5) and (2, −1). Next I ask them to pick any two points that do not form a vertical or horizontal line and calculate the distance between them. The goal is that my students will be able to apply the Pythagorean theorem to find the distance between any two points years before they are presented with the distance formula. As another example, inscribing a right triangle in a circle where the legs are the radii of that circle gives students experience in calculating the length of a chord. The possibilities and connections are endless, providing a great way to break the repetition of the traditional Pythagorean theorem problem set.

10.3 Problem Set

Understanding Concepts

1. Find the distance x in each figure.

(a)

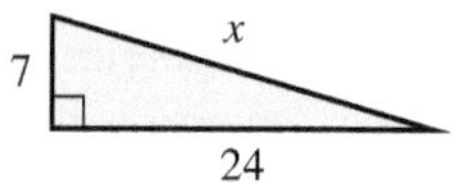

(b)

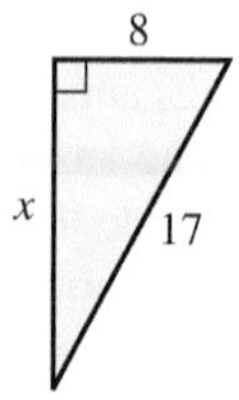

(c)

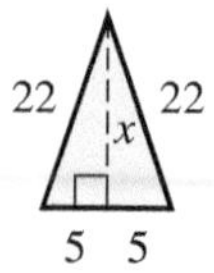

2. Find the distance x in each figure.

(a)

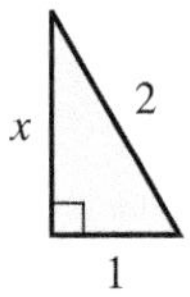

(b)

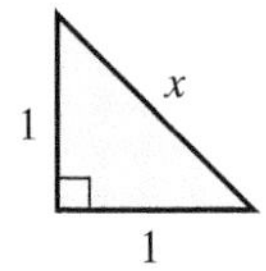

(c)

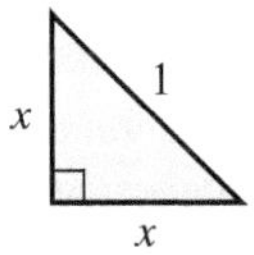

3. Find the distance x in each figure.

(a)

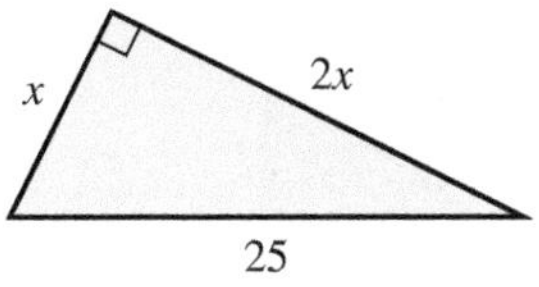

(b)

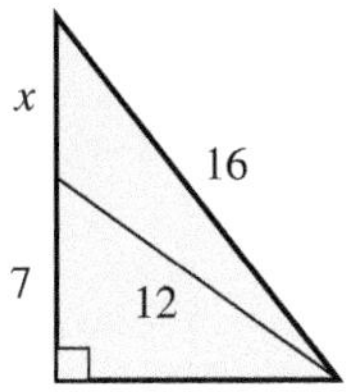

(c)

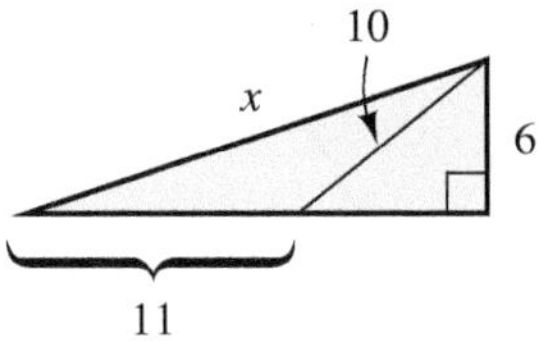

4. Find the area of a right triangle if

(a) Its hypotenuse is 8 in and one of its legs is 3 in.

(b) One leg is three times the other and the hypotenuse is 10 cm.

5. Find the distances x and y in the rectangular prism and the cube.

(a)

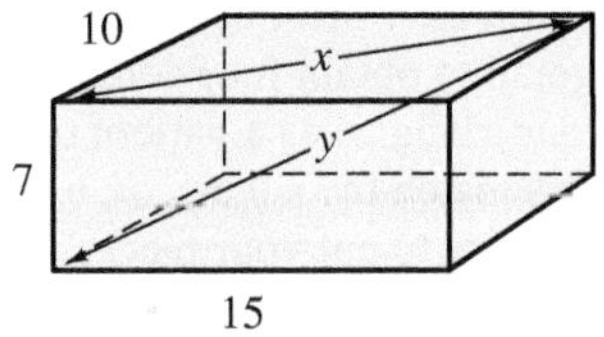

(b)

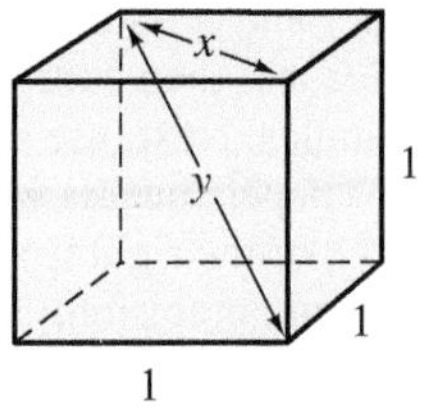

6. Find the distance x in these space figures.

(a)

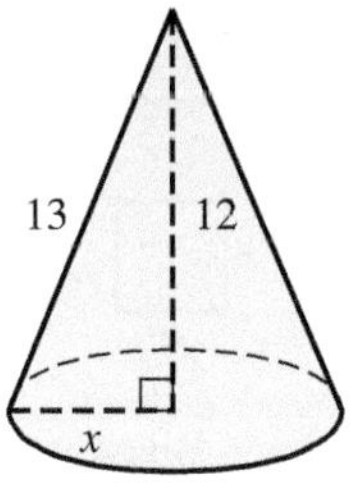

Cone

(b)

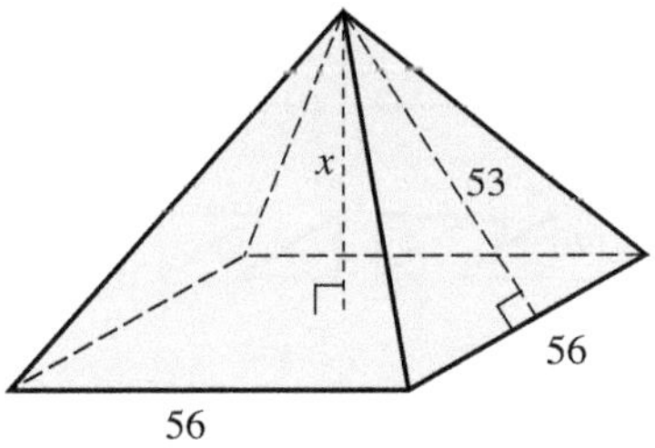

Square-based right regular pyramid, with sides 56 and slant height 53

(c)

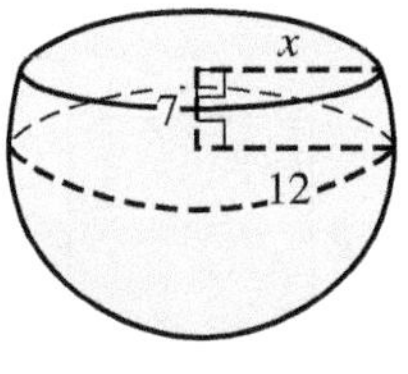

Sphere cut by plane

7. Find the areas of these figures:

(a)

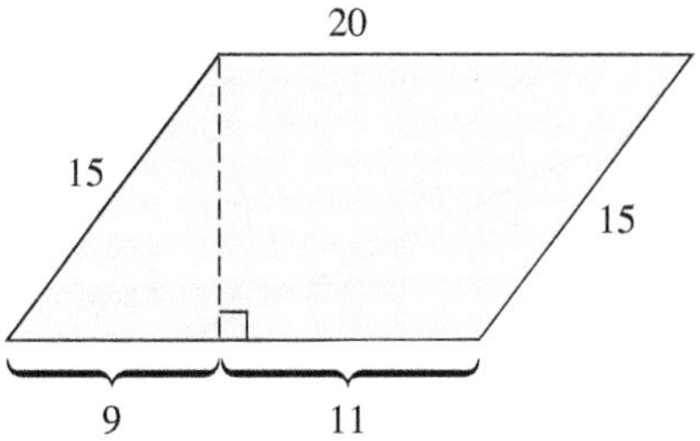

(b)

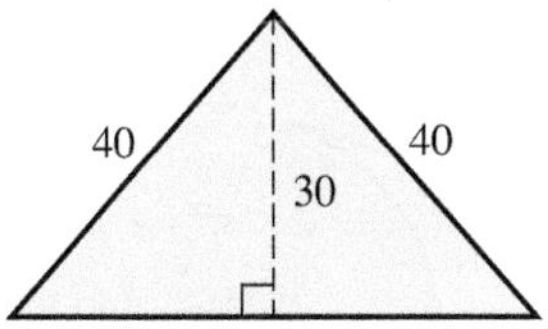

8. Find the area of the triangle:

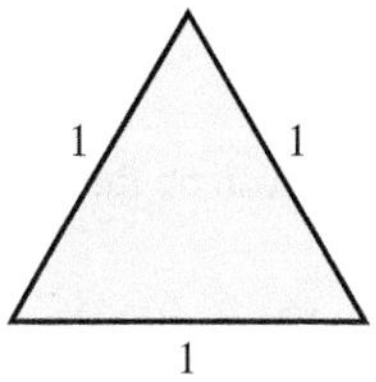

9. Françoise and Maurice cut diagonally across a 50-foot by 100-foot vacant lot on their way to school. How much distance do they save by not staying on the sidewalk?

10. A square with sides of length 2 is inscribed in a circle and circumscribed around another circle. Which is larger, the area of the region between the circles or the area inside the smaller circle?

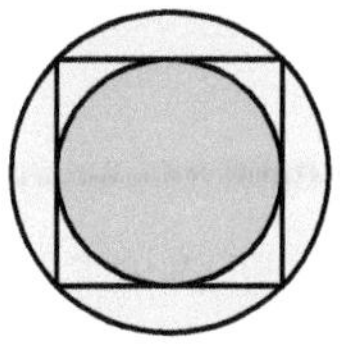

11. At noon, car A left town heading due east at 50 miles per hour. At 1 P.M., car B left the same town heading due north at 40 miles per hour. How far apart were the two cars at

(a) 2 P.M.? **(b)** 3:30 P.M.?

12. Find AG in this spiral of right triangles.

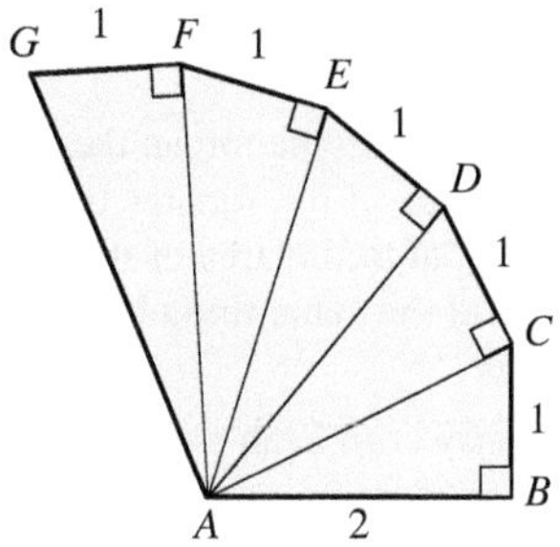

13. **(a)** What is the length of the side of a square inscribed in a circle of radius 1?

(b) What is the length of the side of a cube inscribed in a sphere of radius 1?

14. What is the distance between the centers of these circles?

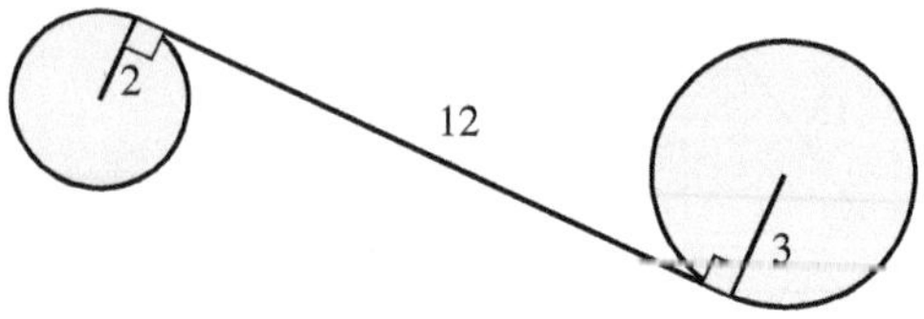

15. What is the radius of the following circle?

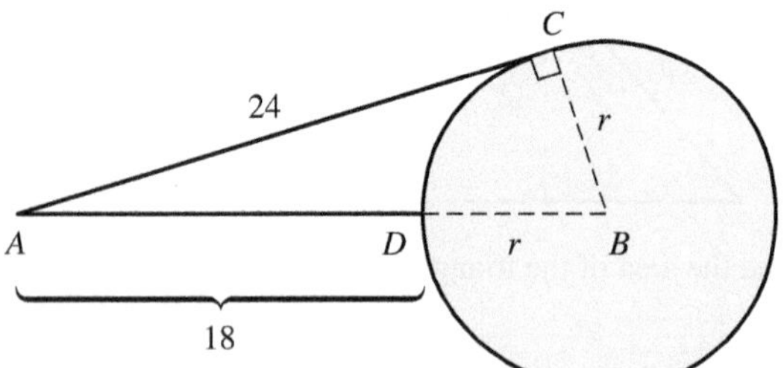

16. Which of the following can be the lengths of the sides of a right triangle?

(a) 21, 28, 35 **(b)** 9, 40, 41 **(c)** 12, 35, 37

17. Which of the following can be the lengths of the sides of a right triangle?

(a) 14, 27, $\sqrt{533}$

(b) $7\sqrt{2}, 4\sqrt{7}, 2\sqrt{77}$

(c) 9.5, 16.8, 19.3

18. If a, b, and c are the lengths of the sides of a right triangle, explain why $10a$, $10b$, and $10c$ are also the lengths of the sides of a right triangle.

19. **(a)** Show that the altitude h of an equilateral triangle with sides of length s is given by $h = \frac{\sqrt{3}}{2}s$.

(b) Find a formula for the area of an equilateral triangle of side length s.

(c) Find a formula for the area of a regular hexagon of side length s.

(d) Show that the area of the inscribed circle of a regular hexagon is $\frac{3}{4}$ the area of the circumscribed circle.

Into the Classroom

20. During a lesson on the Pythagorean theorem, Sean said, "The sum of the areas of the squares on the sides of a right triangle is equal to the area of the square on the hypotenuse. Would the same thing be true of semicircles instead of squares?"

(a) Is the short answer to Sean's question "yes" or "no"? Justify your answer.

(b) How would you respond to Sean? Would you simply tell him the answer, or would you suggest that he see if he can discover the answer for himself? Or might you perhaps challenge the entire class to try to discover the answer to Sean's question?

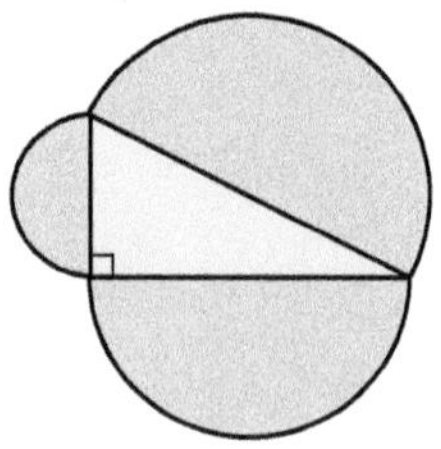

21. Repeat problem 20 (*a*) and (*b*), but for equilateral triangles drawn on the sides of the right triangle. (*Suggestion:* Use the assertion that the altitude h of an equilateral triangle of side s is $h = \frac{\sqrt{3}}{2}s$; see part (a) of problem 19.)

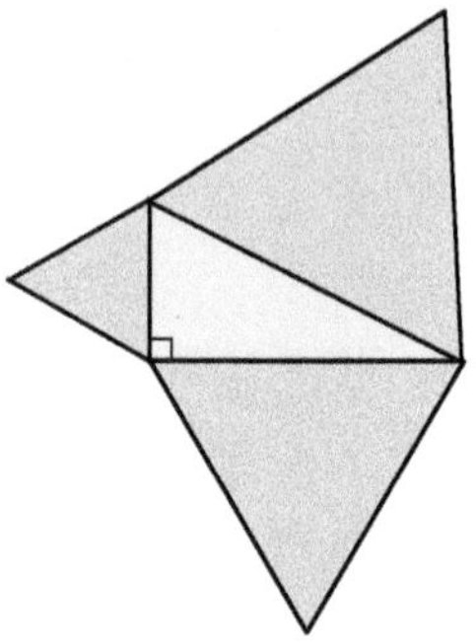

22. Do problem 16, but for equilateral triangles drawn on the sides of the original triangle. (*Suggestion:* Again use the assertion of problem 19(a).)

Thinking Cooperatively

For problems 23 through 25, you will need several sheets of paper and scissors. Begin by folding a sheet of paper in half twice and then cutting a diagonal to obtain four congruent right triangles. Next, use one of your triangles as a pattern to cut four paper squares whose sides match the sides a, b, $b - a$, and c of the right triangles. Be sure to cut your pieces with care and precision.

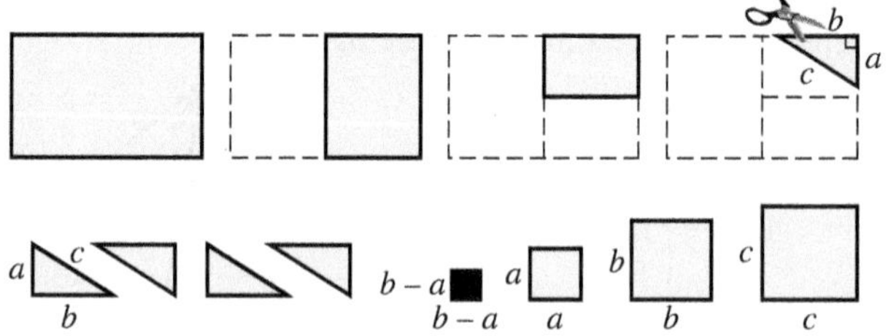

23. The twelfth-century Hindu mathematician Bhaskara arranged four copies of a right triangle of side lengths a, b, and c into a c-by-c square, filling in the center with the $b - a$–sided square.

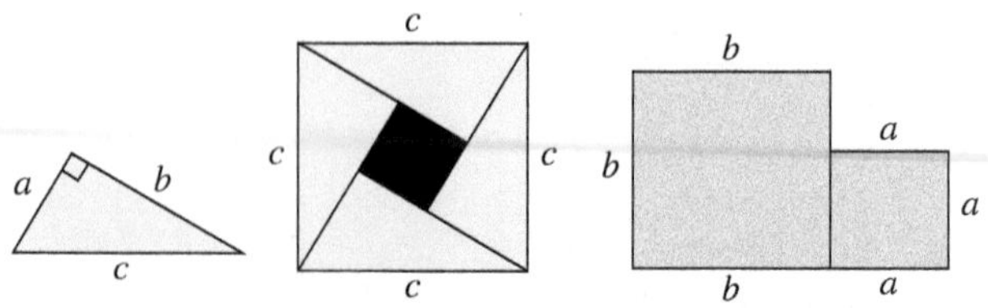

(a) Show how the five pieces in the c-by-c square can be arranged to fill the "double" square region on the right.

(b) Explain how the Pythagorean theorem follows from part (a).

24. Tile the pentagon shown here in two ways:

(a) with two triangles and the squares of sides a and b;

(b) with two triangles and the square of side c.

(c) Explain why the two tilings in parts (a) and (b) prove the Pythagorean theorem.

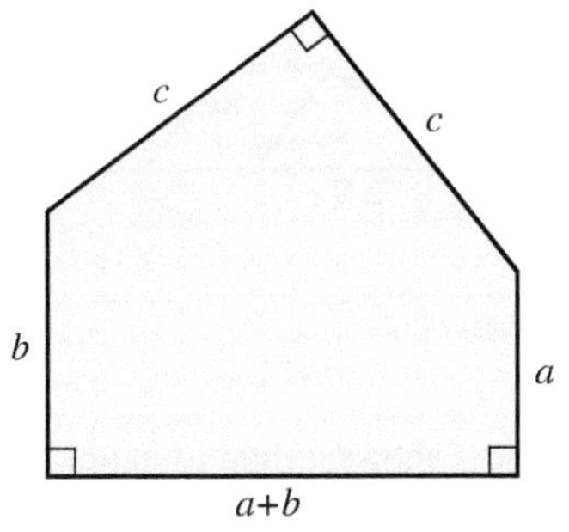

25. In the nineteenth century, Henry Perigal, a London stockbroker and amateur astronomer, discovered a beautiful scissors-and-paper demonstration of the Pythagorean theorem. Follow these steps to complete your own demonstration: Through the center of the larger square on the leg of the right triangle, draw one line perpendicular to the hypotenuse and a second line parallel to the hypotenuse. Cut along these two lines to divide the square into four congruent pieces and then show how to arrange the four pieces, together with the square on the shorter leg, to form a square on the hypotenuse of the right triangle.

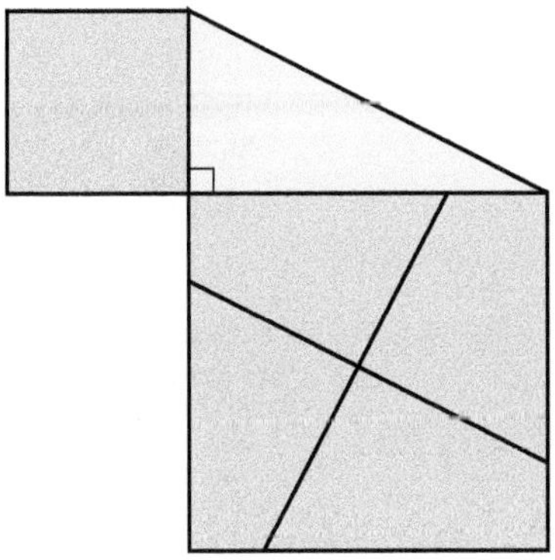

Responding to Students

26. Tia says, "Suppose we draw semicircles on the sides of a triangle and compute their areas. If the sum of the areas of two of the semicircles equals the area of the third, is the triangle a right triangle?"

(a) Is the short answer to Tia's question "yes" or "no"? Justify your answer.

(b) How would you respond to Tia?

27. Sammy looks at the right triangle pictured here. He says that the area of the triangle is 10 because it is 4×5 divided by 2. How would you respond to Sammy?

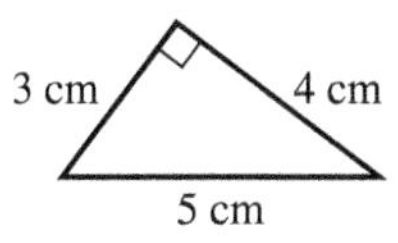

28. Larisa looks at the right triangle pictured here. She writes its area as $\frac{1}{2}(3 \times 5)\text{ cm}^2$. The substitute teacher says that the correct answer is $\frac{1}{2}(5 \times 3)\text{ cm}^2$ and marks her solution as incorrect. How would you, as her teacher, respond to Larisa when she asks why it was graded as wrong?

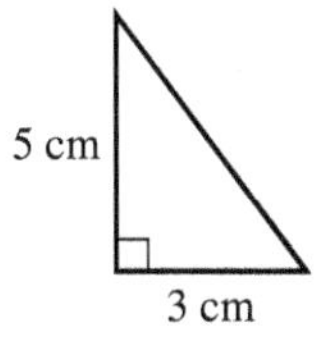

Thinking Critically

29. Find the length of the diagonals of this isosceles trapezoid:

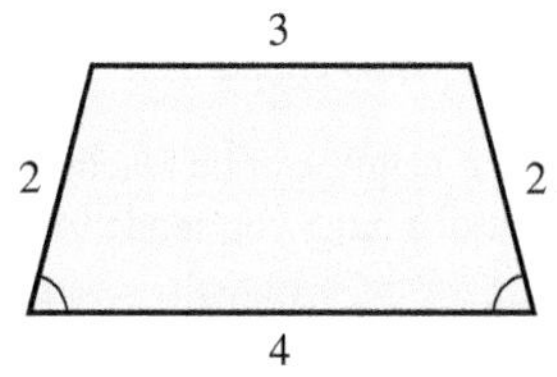

30. An ant is at corner A of a shoebox that is 9 inches long, 5 inches wide, and 3 inches high. What route should the ant follow over the surface of the box to reach the opposite corner C in the shortest distance? (*Suggestion:* It will help to tear along the vertical edges of the box and flatten the top.)

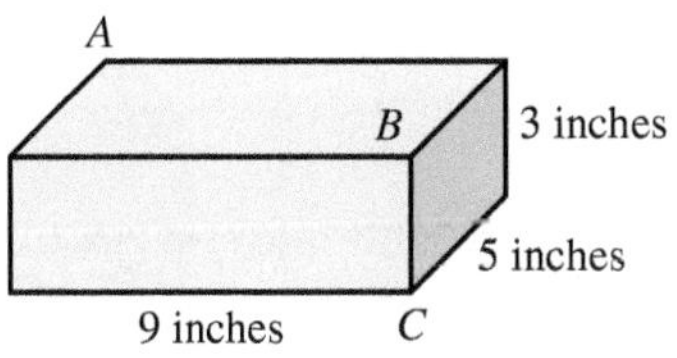

31. A chord of the large circle is tangent to the inner concentric circle. If the chord is 20 cm long, what is the area of the annulus (the region between the two circles)?

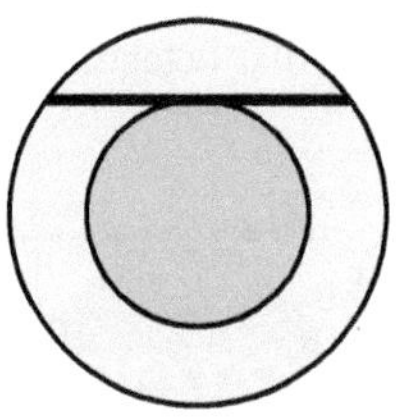

32. Justify why the light-brown regions (which are all parallelograms) in the sequence of diagrams shown have the same area. Since the same reasoning shows that the dark-brown regions also have the same area, this approach provides a striking dynamic proof of the Pythagorean theorem.

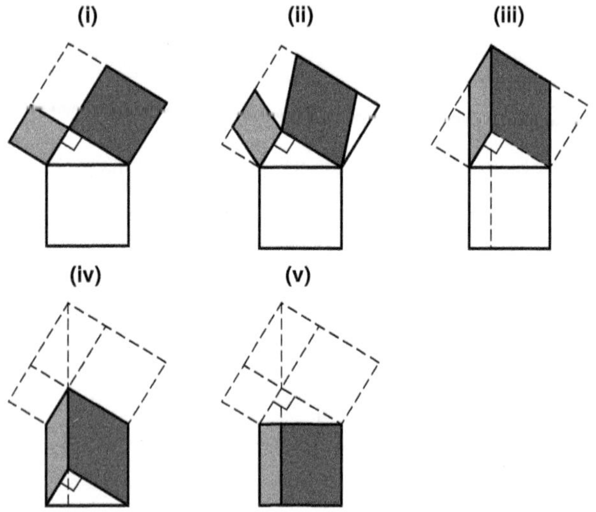

Making Connections

33. A baseball diamond is actually a square 90 feet on a side. What distance must a catcher throw the ball to pick off a runner attempting to steal second base?

34. Approximately what height can be reached from a 24-foot ladder? What assumptions have you made to arrive at your answer?

35. The ancient Egyptians squared off fields with a rope 12 units long, with knots tied to indicate each unit. Explain how such a rope could be used to form a right angle. What theorem justifies their procedure?

36. A water lily floating in a murky pond is rooted on the bottom of the pond by a stem of unknown length.

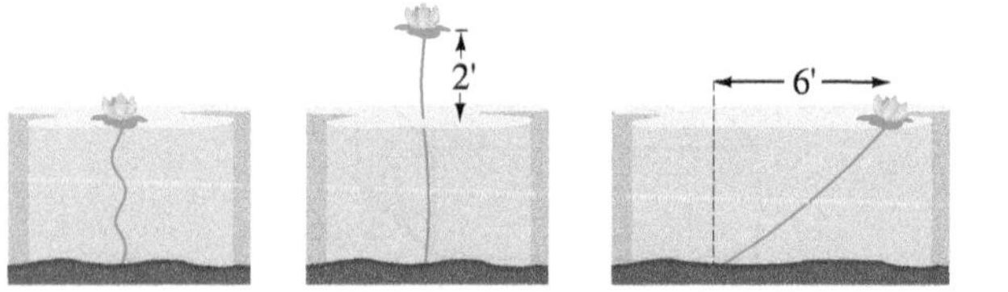

The lily can be lifted 2 feet over the water and moved 6 feet to the side. What is the depth of the pond?

37. A stop sign is to be made by cutting off triangles from the corners of a square sheet of metal 32 inches on a side. What length x will leave a regular octagon? Give your answer to the nearest eighth inch.

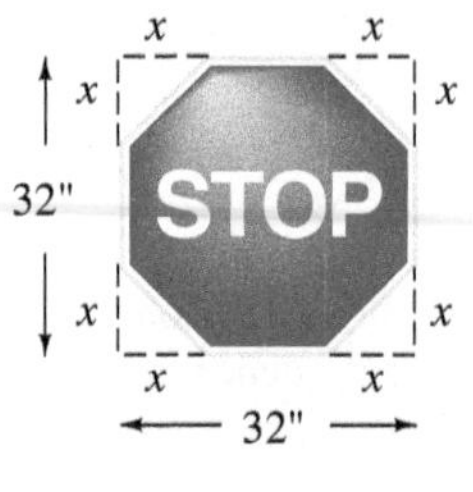

38. A 12-foot-wide crosswalk diagonally crosses a street whose curbs are 40 feet apart, intersecting the opposite curb with a 30-foot displacement in the direction of the street. Reflective striping tape will be applied to all four sides of the crosswalk. How many feet of striping should be ordered? (*Suggestion:* First find the length x of the crosswalk. Next, find the area of the crosswalk in order to help you calculate the curb length y.)

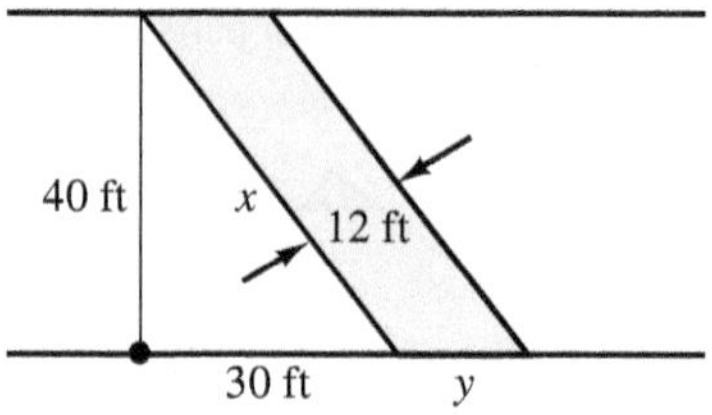

State Assessments

39. (Grade 9)
The drawing below shows a right triangle, $\triangle ABC$. There are three squares each of which is attached to the side of the triangle. What is area of the shaded square?

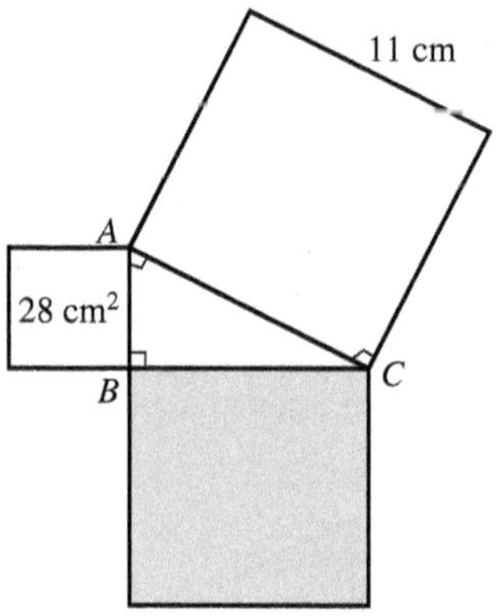

D. 135 cm² E. 28 cm²
F. 121 cm² G. 93 cm²

40. (Grade 9)
Ms. Russell built very well a wooden gate which is diagrammed below. She used a braced wooden piece along the diagonal.

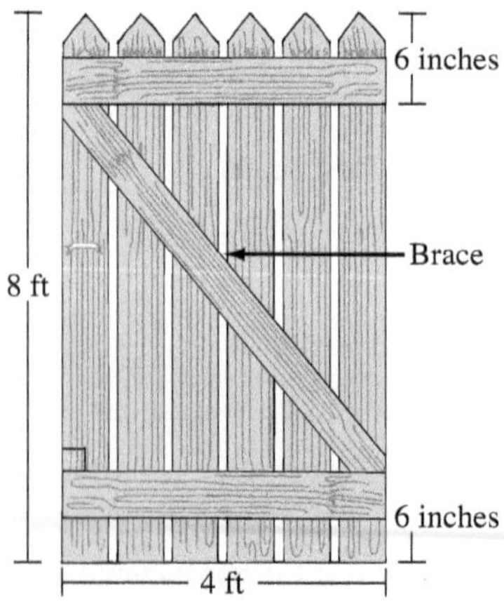

What is the length of the diagonal board (in feet)?

41. (Grade 8)
What is the value of y in the diagram below?

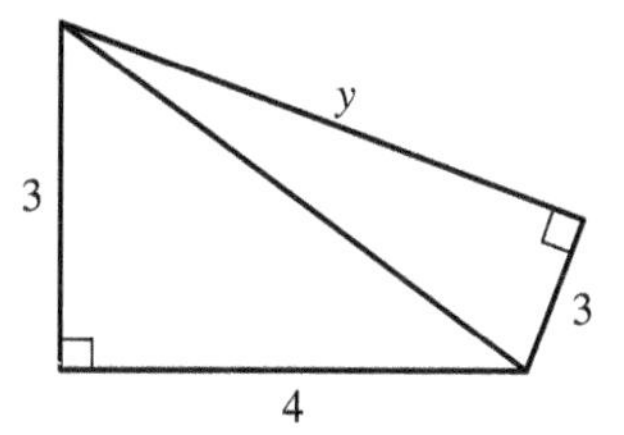

A. 3 B. 4 C. 9 D. 11

42. (Grade 8)
Trapezoid $ABCD$ is shown.

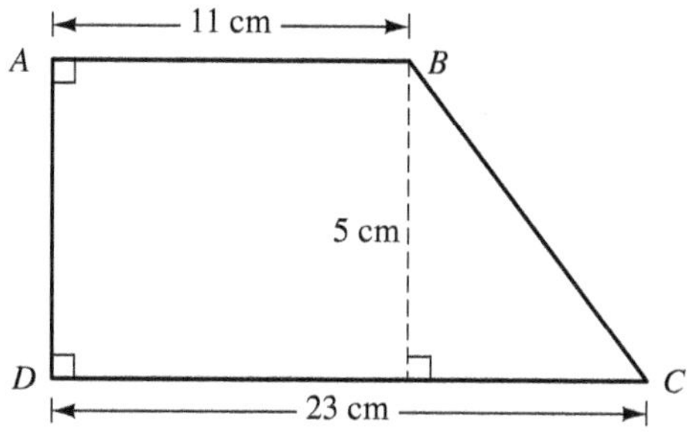

What is the length, in centimeters, of $\overline{BC}$?

A. 16 cm B. 10 cm
C. 11 cm D. 13 cm

10.4 Volume

It would be useful to review Section 9.3 before starting this section.

The two most important measures of a figure in space are its surface area (Section 10.5) and volume (Section 10.4). These are the counterparts of the perimeter and area, respectively, of a figure in the plane. The **surface area,** often denoted by *SA*, or just *S*, measures the boundary of the space figure. The **volume,** often denoted by *V*, measures the amount of space enclosed within the boundary.

It is important to understand how surface area and volume are different. For example, the amount of aluminum in a soda can is closely related to the can's surface area, whereas the amount of fluid within the can is given by the volume. Even the units are different. Surface area is given in in^2, cm^2, and so on. Volume is given in^3, cm^3, and the like, or possibly in a unit of capacity, such as fluid ounces, gallons, milliliters, and so on.

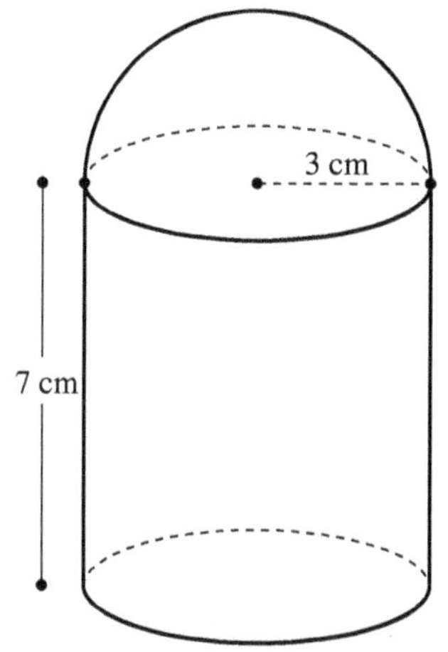

Figure 10.18
Right circular cylinder surmounted by a hemisphere

When we teach this material, we find that many students keep asking, as your own students will, "What is the right formula to use?" This is exactly the wrong question for someone who wants a conceptual understanding of surface or volume (or mathematics in general). Remember to apply Pólya's principles and first understand the problem and devise a plan (usually, to break down the given solid into pieces, for each of which we can easily find the volume). Of course, at that point, you will need to use some formulas (which will be derived in this section) to finish your plan. However, the formulas obtained next, and indeed all of the formulas in this section, are of far less importance than the ideas used to derive them.

For example, if you were asked to find the volume of the solid in Figure 10.18, how should you go about it? The problem asks for the volume of a solid that is made of two basic solids: half of a sphere of radius 3 cm and a right circular cylinder of radius 3 cm and height 7 cm. Thus, our plan would be to partition the given solid into two: a hemisphere and a right circular cylinder.

PROBLEM-SOLVING STRATEGY **Decompose one complex problem into a number of simpler ones (with applications to volume).**

In this case, to find the volume of a solid, first break up the solid appropriately into a collection of solids whose volume is easy to compute, and then add those volumes.

In this section, we will derive formulas for the volume of some basic figures in space, such as right and oblique prisms and cylinders, pyramids, right circular cones, and spheres. These formulas allow us to finish problems such as the one mentioned in the previous paragraph (see problem 6 in Problem Set 10.4) by using the preceding problem-solving strategy.

Volumes of Right Prisms and Right Cylinders

The rectangular box shown in Figure 10.19 is given by its three dimensions—length, width, and height—which are in order l, w, and h. Note that h is the height of the box and is perpendicular to its base just as the right prisms and cylinders are. See Figures 10.19 and 10.20.

FORMULA Volume of a Rectangular Box

The volume V of a rectangular box is the product of its length, width, and height, or $V = lwh$. The volume can also be written as $V = Bh$, since $B = lw$ is the area of the base.

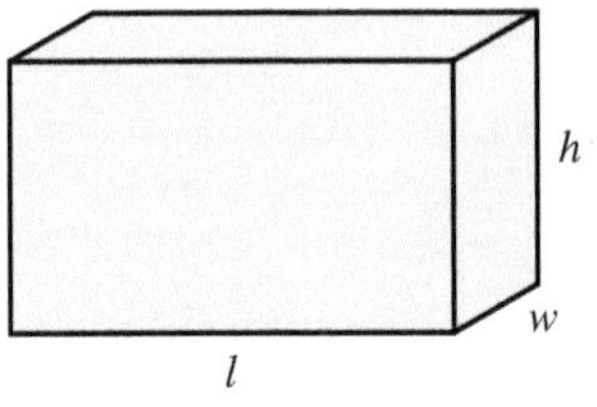

Figure 10.19
A rectangular box has volume $V = lwh$. Equivalently, $V = Bh$, where B is the area of the base and h is the height

Figure 10.20a shows a solid composed of many (say, n) small right rectangular prisms, all of height h. If $B_1, B_2, \ldots, B_n$ are the areas of the bases, the total volume V of the prisms is $B_1h + B_2h + \cdots + B_nh = (B_1 + B_2 + \cdots + B_n)h$. That is, the volume V is given by $V = Bh$, where $B = B_1 + B_2 + \cdots + B_n$ is the total area of the base. The right cylinder depicted in Figure 10.20b can be approximated to arbitrary accuracy by prisms of height h as shown.

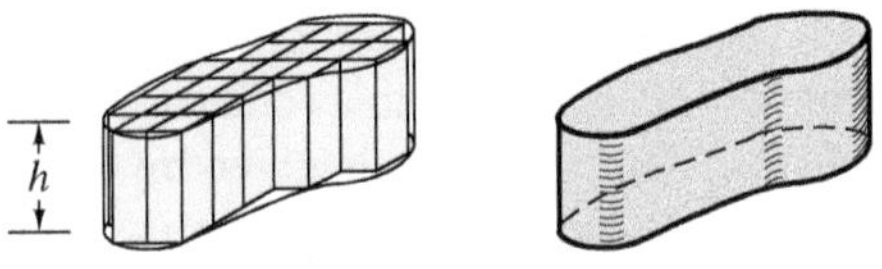

Figure 10.20
Right rectangular prisms can approximate a right prism or a right cylinder

FORMULA Volume of a Right Prism or a Right Cylinder

Suppose a right prism or right cylinder has height h and a base of area B. Then its volume V is given by

$$V = Bh.$$

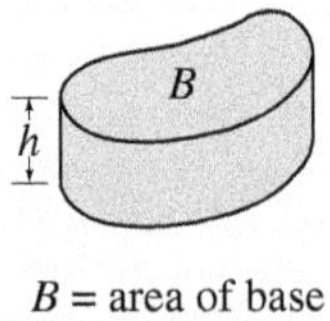

B = area of base

EXAMPLE 10.20 Computing the Volume of a Right Prism and a Right Cylinder

Find the volume of the gift box and the juice can.

(a)

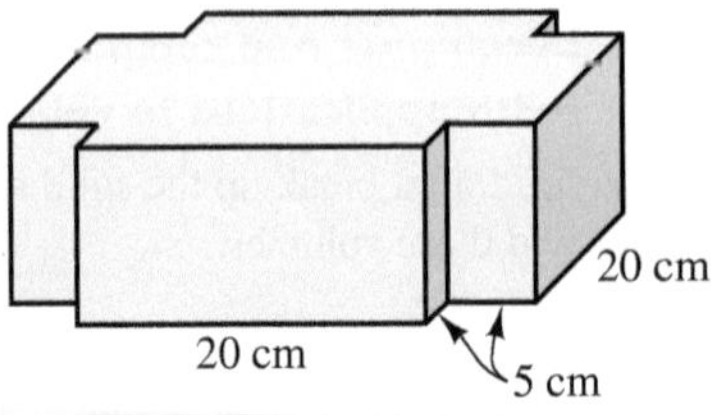

(b)

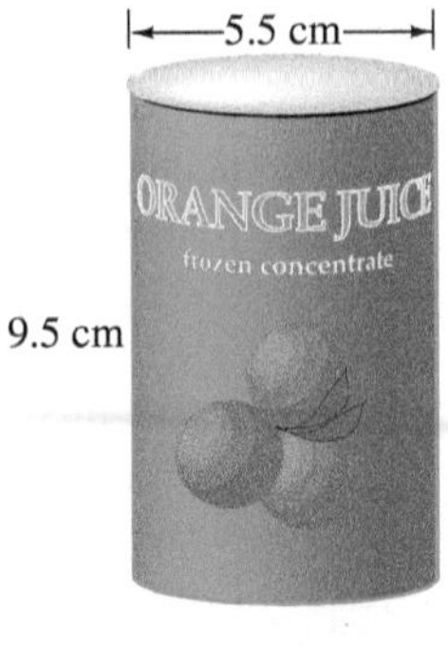

Solution

(a) The base area B consists of a square of length 20 cm and four 5-cm by 20-cm rectangles, so $B = 800\ \text{cm}^2$. The height is $h = 10\ \text{cm}$, so the volume is $V = Bh = 8000\ \text{cm}^3$, which can also be expressed as 8 liters.

(b) The area of the circular base of the juice can is $\pi(2.75\ \text{cm})^2 = 7.5625\pi\ \text{cm}^2$. The height is $h = 9.5\ \text{cm}$, so the volume $V = Bh = 71.84375\pi\ \text{cm}^3$, or about $226\ \text{cm}^3$.

Volumes of Oblique Prisms and Cylinders

A deck of neatly stacked playing cards forms a right rectangular prism as shown in Figure 10.21. The total volume of the deck is the sum of the volumes of each card. If the cards slide easily on one another, it is easy to tilt the deck to form an oblique prism of the same base area B and the same height h. The solid is still made up of the same cards, so the oblique prism still has volume Bh.

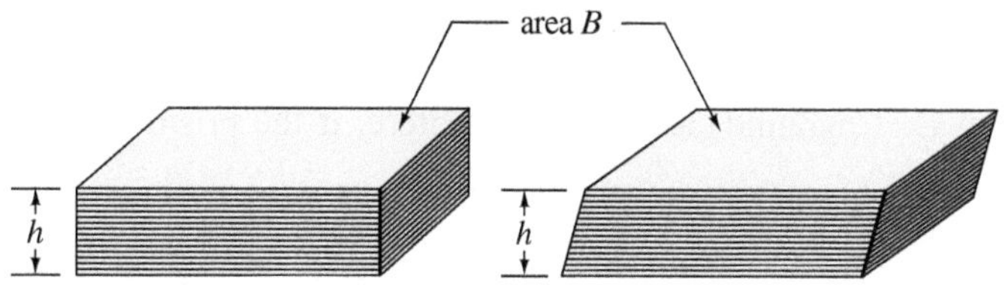

Figure 10.21
An oblique prism of base area B and height h has the same volume $V = Bh$ as the corresponding right prism

Any oblique prism or cylinder can be imagined as a stack of very thin cards, all shaped like the base of the solid. With no change of volume, the oblique stack can be straightened to form a right prism or right cylinder of the same height h and base area B. Both the right and oblique shapes therefore have the same volume, namely, $V = Bh$. This relationship is illustrated in Figure 10.22.

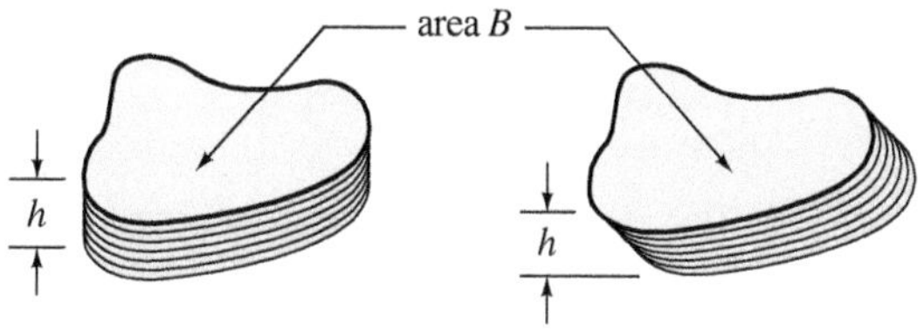

Figure 10.22
Any prism or cylinder, either right or oblique, has volume $V = Bh$, where B is the base area and h is the height

FORMULA Volume of a General Prism or Cylinder

A prism or cylinder of height h and base area B has volume $V = Bh$.

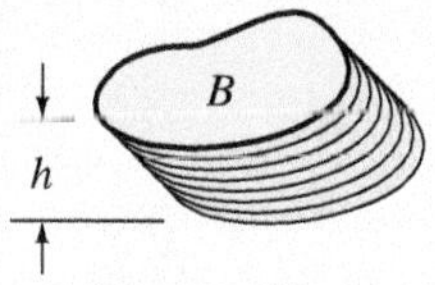

Volumes of Pyramids and Cones

In two-dimensional space (that is, in the plane), a diagonal dissects a square into two congruent right triangles. Thus, the area of each triangle is one-half that of the corresponding square.

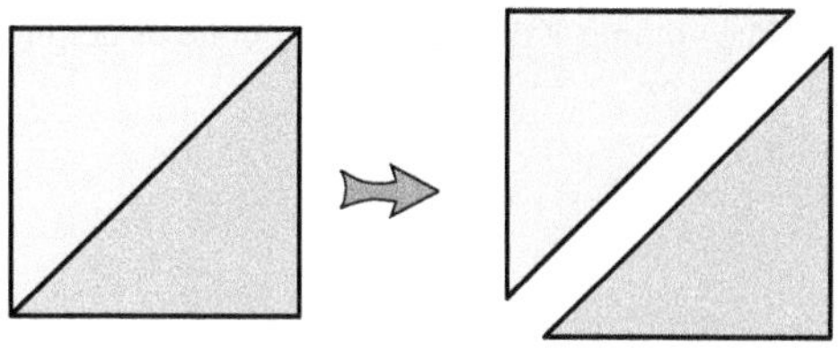

In three-dimensional space, the diagonals from one corner of a cube form the edges of three congruent pyramids that fill the cube. Therefore, each pyramid has one-third the volume of the corresponding cube, as shown in Figure 10.23.

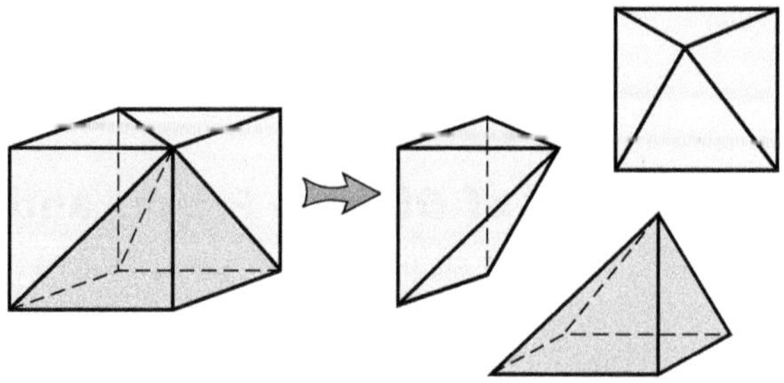

Figure 10.23
A cube can be dissected into three congruent pyramids

Instead of a cube, consider a rectangular solid and use the diagonals from one corner to decompose the solid into three pyramids. An example is shown in Figure 10.24. In general, the three pyramids are not congruent to one another. However, it can be shown that the volumes of the three pyramids are equal. Therefore, if the prism has base area B and height h, we conclude that each pyramid has volume $\frac{1}{3}Bh$.

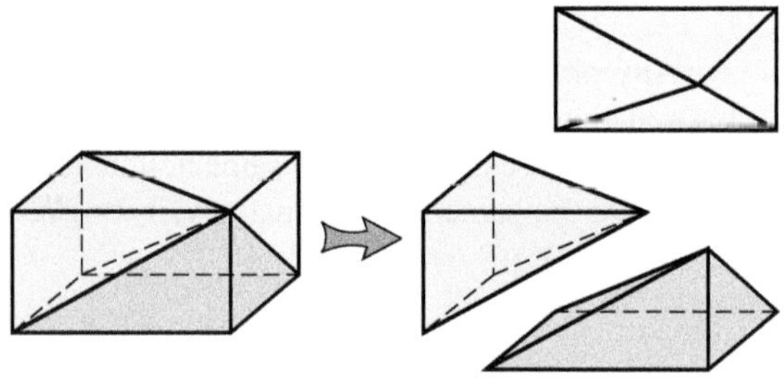

Figure 10.24
A rectangular prism can be dissected into three pyramids of equal volume

Similar reasoning shows that *all* pyramids of base B and height h have volume $\frac{1}{3}Bh$. The base can be any polygon, and the apex can be any point at distance h to the plane of the base, as shown in Figure 10.25.

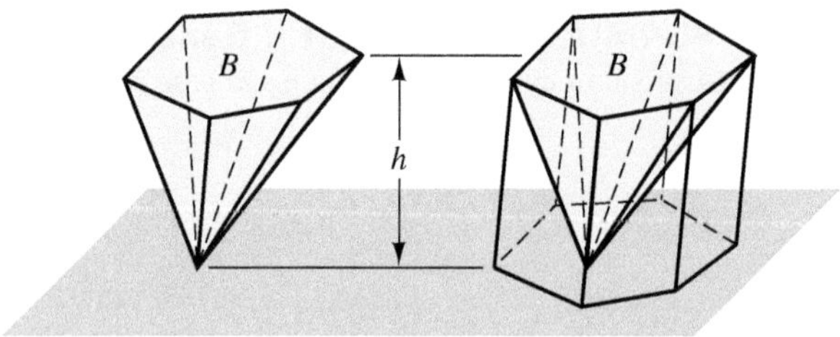

Figure 10.25
A pyramid of height h and base of area B has one-third the volume of a corresponding prism of base area B and height h. Therefore, the pyramid has volume $\frac{1}{3}Bh$

The base of a cone can be approximated to arbitrary accuracy by a polygon with sufficiently many sides, so the volume of a cone of base area B and height h is also given by the formula $\frac{1}{3}Bh$.

FORMULA Volume of a Pyramid or Cone
The volume V of a pyramid or cone of height h and base area B is given by

$$V = \frac{1}{3}Bh.$$

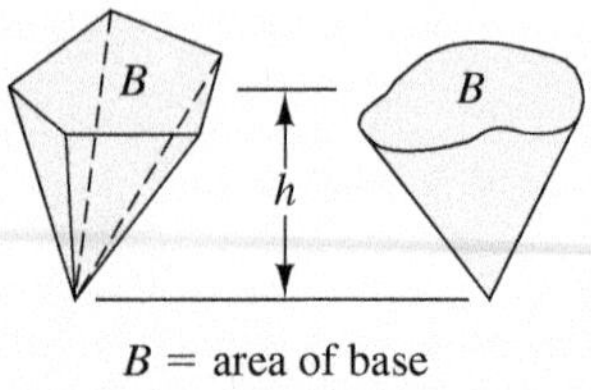

EXAMPLE 10.21

Determining the Volume of an Egyptian Pyramid

The pyramid of Khufu is 147 m high, and its square base is 231 m on each side. What is the volume of the pyramid?

Solution

The area of the base is $(231 \text{ m})^2 = 53{,}361 \text{ m}^2$. Therefore, the volume is

$$\frac{1}{3}(53{,}361 \text{ m}^2)\cdot(147 \text{ m}) = 2{,}614{,}689 \text{ m}^3.$$

If the stones were stacked on a football field, a rectangular prism nearly 2000 feet high would result.

Volume of a Sphere

Suppose that a solid sphere of radius r is placed in a right circular cylinder of height $2r$ that just contains it. Filling the remaining space in the cylinder with water, we find that removing the sphere leaves the cylinder one-third full, as illustrated in Figure 10.26. This means that the sphere takes up two-thirds of the volume of the cylinder. Since the volume of the cylinder is $Bh = (\pi r^2)(2r) = 2\pi r^3$, the experiment suggests that the volume of a sphere of radius r is given by $\frac{2}{3}(2\pi r^3) = \frac{4}{3}\pi r^3$. The first rigorous proof of this remarkable formula was given by Archimedes.

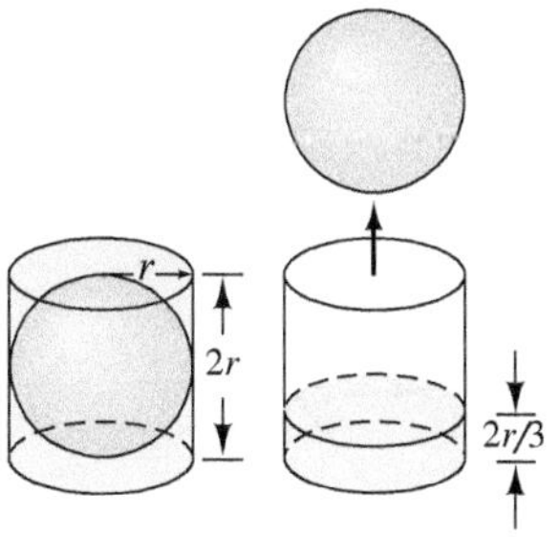

Figure 10.26
A sphere fills two-thirds of the circular cylinder containing the sphere

FORMULA Volume of a Sphere

The volume V of a sphere of radius r is given by the formula

$$V = \frac{4}{3}\pi r^3.$$

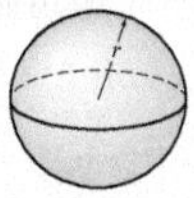

EXAMPLE 10.22

Using the Sphere Volume Formula

An ice cream cone is 5 inches high and has an opening 3 inches in diameter. If filled with ice cream and given a hemispherical top, how much ice cream is there?

Solution

The hemisphere has radius 1.5 inches, so its volume is one-half that of a sphere of radius 1.5 inches, or $\frac{2}{3}\pi(1.5 \text{ in})^3 = 2.25\pi \text{ in}^3$. The cone has volume $\frac{1}{3}Bh = \frac{1}{3}\pi(1.5 \text{ in})^2\cdot(5 \text{ in}) = 3.75\pi \text{ in}^3$. Thus, the total volume is $2.25\pi \text{ in}^3 + 3.75\pi \text{ in}^3 = 6\pi \text{ in}^3$, or about 19 in^3. Since a gallon is 231 in^3, we see that the cone holds very close to a third of a quart of ice cream.

10.4 Problem Set

Understanding Concepts

1. Identify which measure—surface area or volume—is most important to know
   (a) about a room, to buy the correct amount of paint;
   (b) about a chair, to determine the amount of stuffing required to reupholster the chair;
   (c) about a swimming pool, to add the correct amount of chlorine;
   (d) about a lawn, to apply the correct amount of weed killer.

2. Find the volume of each of these prisms and cylinders.

(a)

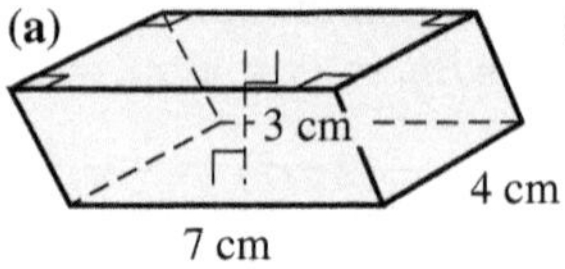

Oblique prism

(b)

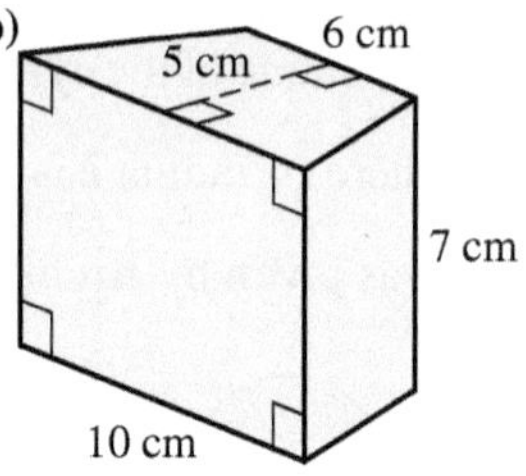

Trapezoidal right prism

3. Find the volume of each of these prisms and cylinders:

(a)

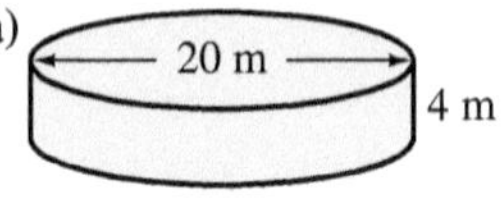

Right circular cylinder

(b)

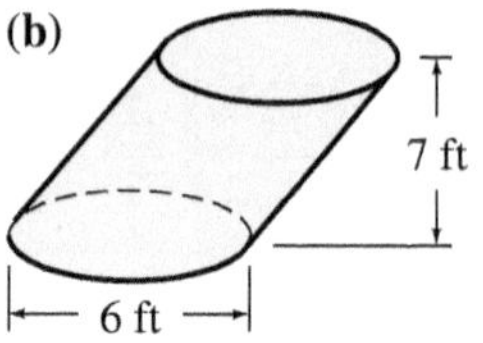

Oblique circular cylinder

4. Find the volume of each of these pyramids and cones:

(a)

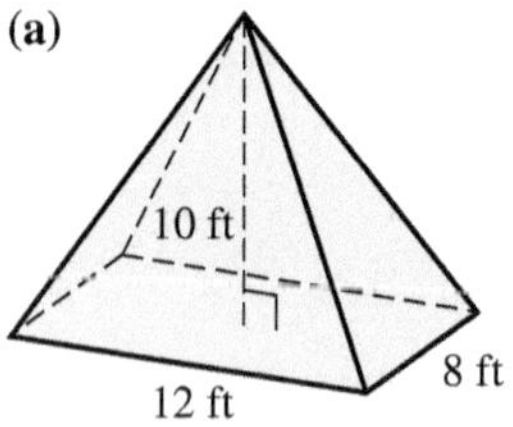

Rectangular pyramid

(b)

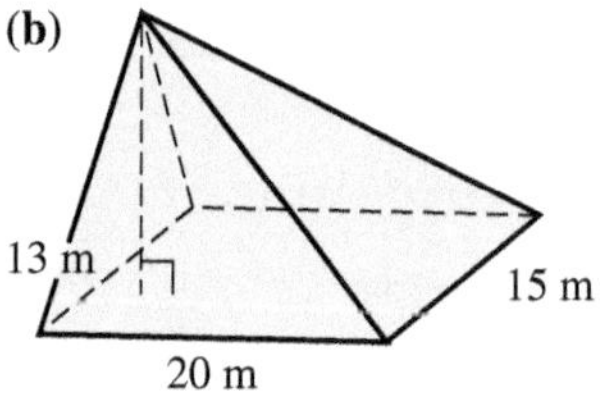

Rectangular pyramid

5. Find the volume of each of these pyramids and cones:

(a)

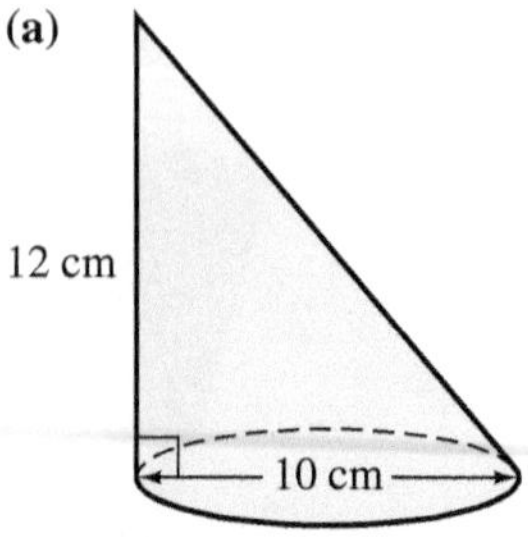

Oblique circular cone

(b)

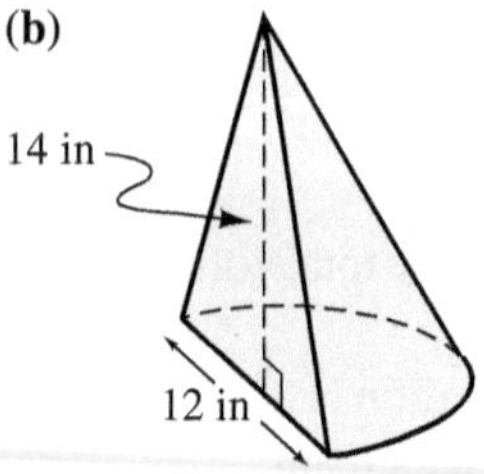

Semicircular cone

6. What is the volume of the solid of Figure 10.18?

7. A circular bowl is built to hold 2 cups of mint chocolate chip ice cream. How many inches should the radius of the bowl be? (*Hint:* Use Table 10.4 of Section 10.1 to find the number of cubic inches in 2 cups.)

Into the Classroom

8. Eno is having trouble understanding that the volume of a solid nonrectangular shape (say, a pyramid, cone, or sphere) can be given in terms of a unit cube. How would you help Eno understand that this can be so?

9. Ruth Ann doesn't believe that the volume of *any* pyramid is given by the formula $V = \frac{1}{3}Bh$. What might you do to help Ruth Ann?

In problems 10 and 11, you will need paper, tape, scissors, and drawing tools (ruler, compass, and pencils). Work in pairs, measuring your constructed models and verifying your observations with calculations.

10. A sheet of $8\frac{1}{2}''$ by $11''$ notebook paper can be rolled into a cylinder in either of two ways:

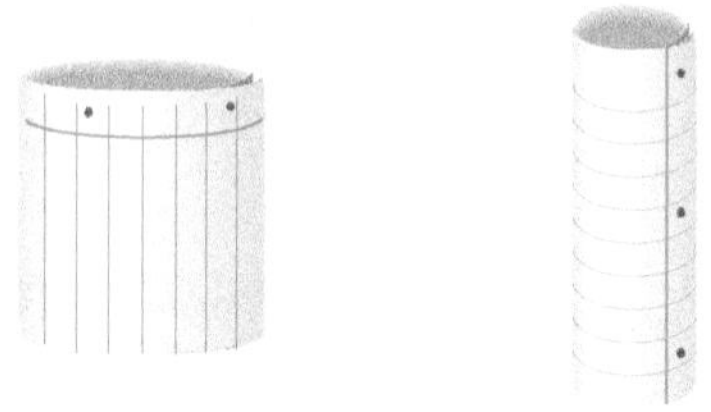

Which way encloses the larger volume? Make a prediction and then check it.

11. Cut out the circular sector shown in the accompanying figure and roll it into a right circular cone in which the two 4-inch radial segments are joined.

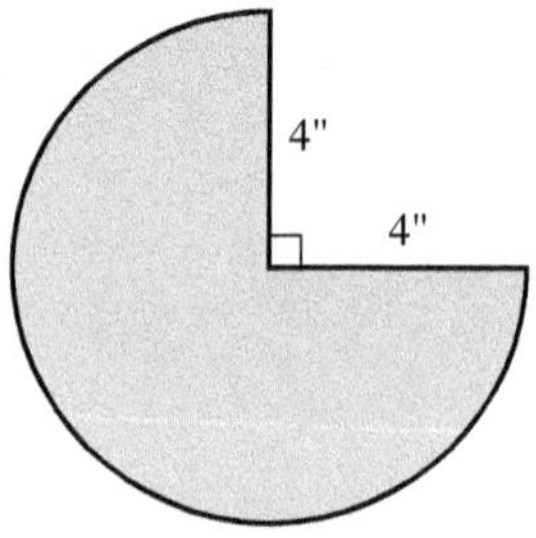

Find the following measurements, both by exact computation and by measuring your paper model:
   (a) The radius of the base of the cone.
   (b) The height of the cone.
   (c) Use your answers to parts (a) and (b) to compute the volume of the cone.

Responding to Students

12. Urmi says that, since there's a third power in the volume formula and a square in the surface area formula, the volume of a sphere of radius r is always greater than the surface area of the sphere.

(a) Is Urmi's statement true?

(b) How would you explain your answer to part (a) to Urmi, who is in fifth grade?

13. While experimenting with volume problems, Samuel, a fifth grader, computes the volume of a right circular cylinder of height 6 cm four times, using radii of 2 cm, 4 cm, 6 cm, and 8 cm, and fills in the following table:

Radius in cm	2	4	6	8
Volume in cm^3				

From these data, he concludes that the volume of any cylinder increases as its radius does.

(a) Fill in Samuel's table.

(b) Is Samuel correct?

(c) How would you respond to him?

14. Vera was asked to calculate the volume of the following rectangular prism, which is 4 by 5 by 8 meters:

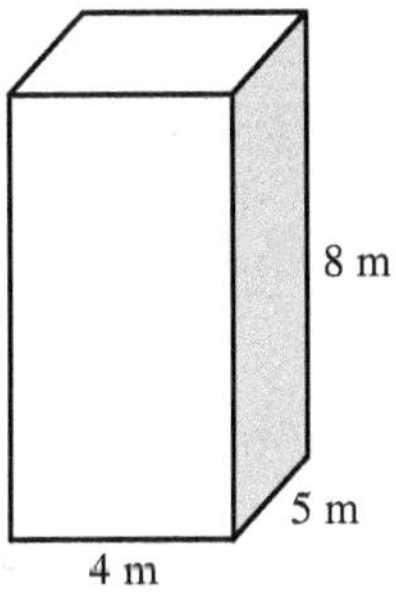

Vera's answer was 17 m^3.

(a) What is the correct volume for the rectangular prism?

(b) What did Vera do when figuring her answer?

(c) How might you explain to Vera what she needs to do in a future problem?

15. Maurice was given the following picture of a prism, whose base has area 6 ft^2, and asked to calculate its volume:

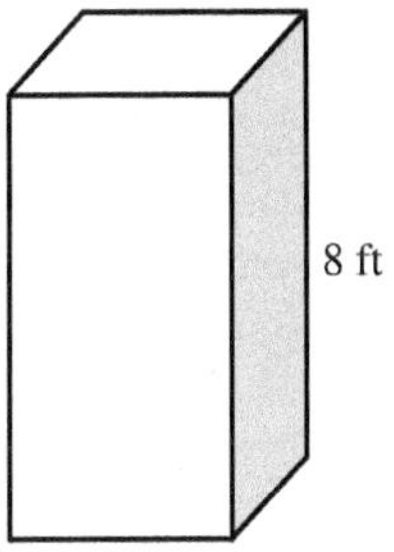

Maurice's answer was 48 ft^2. His teacher looked over his answer, praised it for being partly correct, and then asked him to relabel it. Maurice said that he didn't understand why he needed to change the label, because he multiplied two things and so put a little 2 behind his answer. When he multiplies three things to find out the volume, then he puts a little 3.

(a) What did Maurice do correctly?

(b) What did Maurice not understand about the figure he was given?

(c) What can his teacher do to help him calculate this problem so that Maurice understands why he is using "little numbers" behind his answer?

Thinking Critically

16. A right circular cone has height r and a circular base of radius $2r$. Compare the volume of the cone with that of a sphere of radius r. Sketch both solids, using the same scale.

17. The right circular cylinder and cone shown in the accompanying figure both have a base of radius r and height $2r$, and the sphere has radius r. Show that the ratio of the volumes of the cone to the sphere to the cylinder is 1 to 2 to 3.

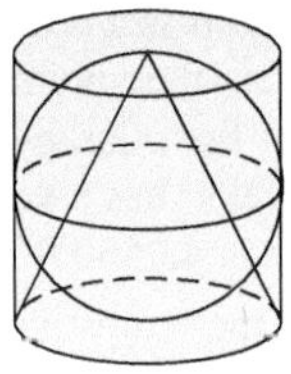

18. A right circular cylinder of height 7 inches has a hemisphere surmounted on top of it similar to that shown in Figure 10.18.

(a) Show that, if we want the volume of the solid to be $\frac{23\pi}{3}$ in^3, then we can use a radius of 1 inch.

(b) Are there other possible dimensions for the solid of this problem with the same height and volume? This more difficult question asks if there are other values for the radius that give a volume of $\frac{23\pi}{3}$. If so, what are they? If not, why not? (*Hint:* Use algebra, including the quadratic formula and part (a).)

Making Connections

In these problems, you will need to use a calculator

19. A napkin ring is being made of cast silver. It has the shape of a cylinder 1.25 inches high, with a cylindrical hole 1 inch in diameter and a thickness of $\frac{1}{16}$ inch. How many ounces of silver are required? It will help to know that silver weighs about 6 ounces per cubic inch.

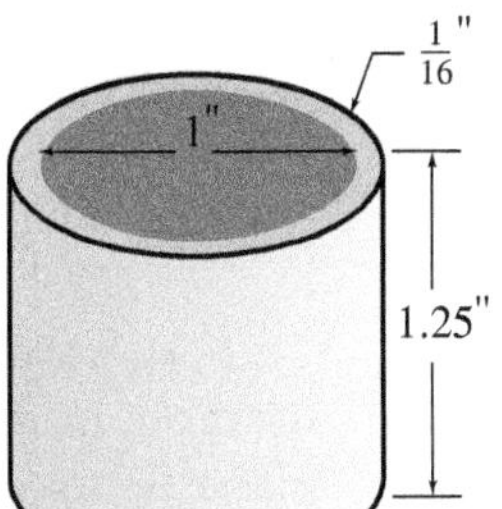

20. Give the dimensions of a rectangular aquarium 40 cm high that holds 48 liters of water.

21. A theater sells 4″ by 5″ by 8″ boxes of popcorn for \$1.75. It also sells cylindrical "tubs" of popcorn for \$3.50, where the tub is 10″ high and has a diameter of 6″. Is it better to buy one tub or two boxes of popcorn?

22. Small grapefruits of diameter 3 inches are on sale at five for a dollar. The large, 4-inch-diameter grapefruits are three for a dollar. If you are buying \$5 worth of grapefruits, should you choose small ones or large ones in order to get a better deal?

23. **World Records.** *The Guinness Book of World Records,* published annually by Facts on File, New York, contains a fascinating collection of measurements.

(a) The world's largest flawless crystal ball weighs 106.75 pounds and is 13 inches in diameter. What is the weight of a crystal ball 5 inches in diameter?

(b) The largest pyramid is the Quetzalcoatl, 63 miles southeast of Mexico City. It is 177 feet tall and covers an area of 45 acres. Estimate the volume of the pyramid. By comparison, the largest Egyptian pyramid of Khufu (called Cheops by the Greeks) has a volume of 88.2 million ft^3. Recall that an acre is 43,560 ft^2.

(c) The building with the largest volume in the world is the Boeing Company's main assembly plant in Everett, Washington. The building encloses 472 million ft^3 and covers 98.3 acres. What is the size of a cube of equal volume?

24. A water pipe with an inside diameter of $\frac{3}{4}$ inches is 50 feet long in its run from the hot-water tank to the faucet. How much hot water is wasted when the water inside the pipe cools down? Give your answer in gallons, where 1 gallon = 231 in^3.

State Assessments

25. (Grade 8)
Barb is moving and has a number of rectangular storage boxes with a height of 2 feet and a square base of 10 inches for each side. When she cuts a 4-inch strip around the top of the box, what will be the volume of the new box (in cubic inches)?

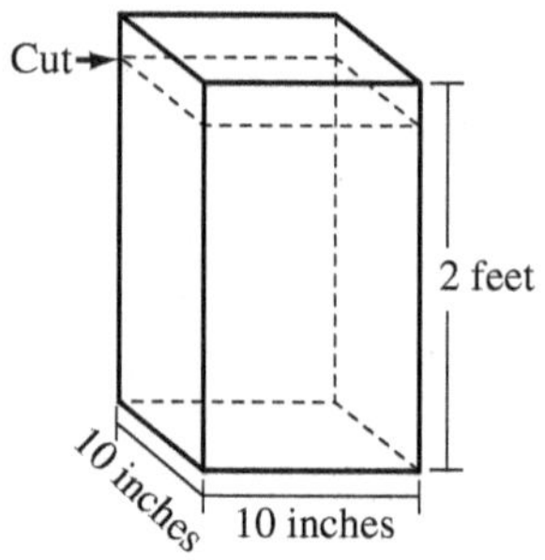

A. 1600 in^3 B. 200 in^3
C. 2400 in^3 D. 2000 in^3

26. (Grade 8)
Isaac purchased a small rectangular box that contains 12 tightly packaged pencils which look like rectangular prisms as we see below.

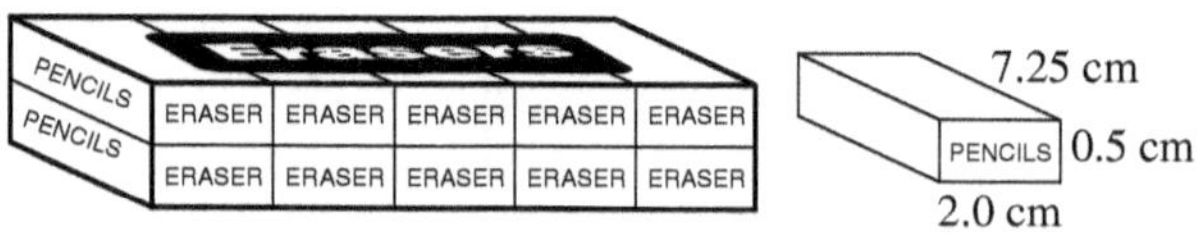

In cubic centimeters, what is the volume of the 10 boxes of pencils together?

10.5 Surface Area

In this section, we will derive formulas for the surface area of some basic figures in space, such as polyhedra, right and oblique prisms and cylinders, pyramids, right circular cones, and spheres. These formulas allow us to solve surface area problems in the same way that we did volume problems—that is, by appropriately breaking the solid into pieces.

PROBLEM-SOLVING STRATEGY **Decompose one complex problem into a number of simpler ones (with applications to surface area).**

In this case, to find the surface area of a solid, first break up the solid appropriately into a collection of solids whose surface areas are easy to compute and then add those surface areas.

For a polyhedron, the surface area is simply the sum of the areas of each of its faces. It's often useful to imagine cutting a polyhedron apart and placing its faces in a plane. In many cases, a plane figure or set of figures is formed in this manner, with areas that are easy to determine.

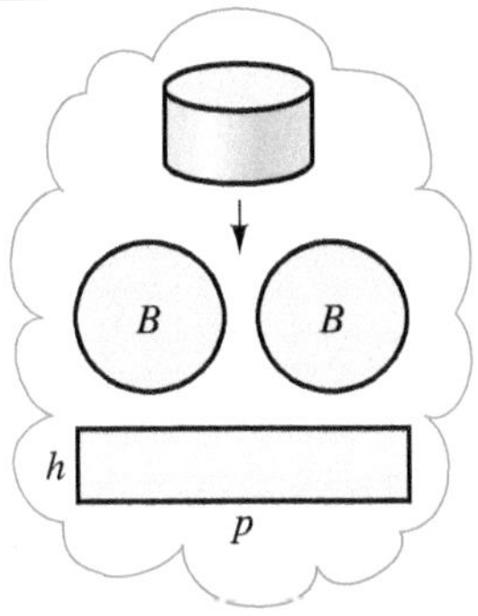

For space figures with curved boundaries, it is still sometimes possible to imagine cutting and unrolling the surface to create plane figures whose areas can be easily found. For example, to find the surface area of a tin can, we can imagine cutting off both ends and down the seam of the can. This set of steps creates two discs and, when the can is unrolled, a rectangle, all of whose areas are easy to calculate.

For a sphere and other more general solids in space, it becomes necessary to approximate the figure with a sequence of a polyhedra whose surface areas approximate that of the space figure. The surface area of the space figure is the limiting value of the surface areas of the approximating polyhedra. We will use this approach by finding the surface area of a right circular cone and that of a sphere. This technique for finding surface area is the same type of computation used to find volume in the previous section.

Surface Area of Right Prisms and Cylinders

Figure 10.27 shows how the surface of a right prism is cut into two congruent bases, with the lateral surfaces of the prism unfolded to form a rectangle. If the right prism has height h and the perimeter of the base is p, then the rectangle has area hp. The same reasoning applies to any right cylinder, where the lateral surface (surface on the side) is imagined to be unrolled onto the plane to form a rectangle.

Figure 10.27
The surface of a right prism can be cut and unfolded to form two congruent bases and a rectangle

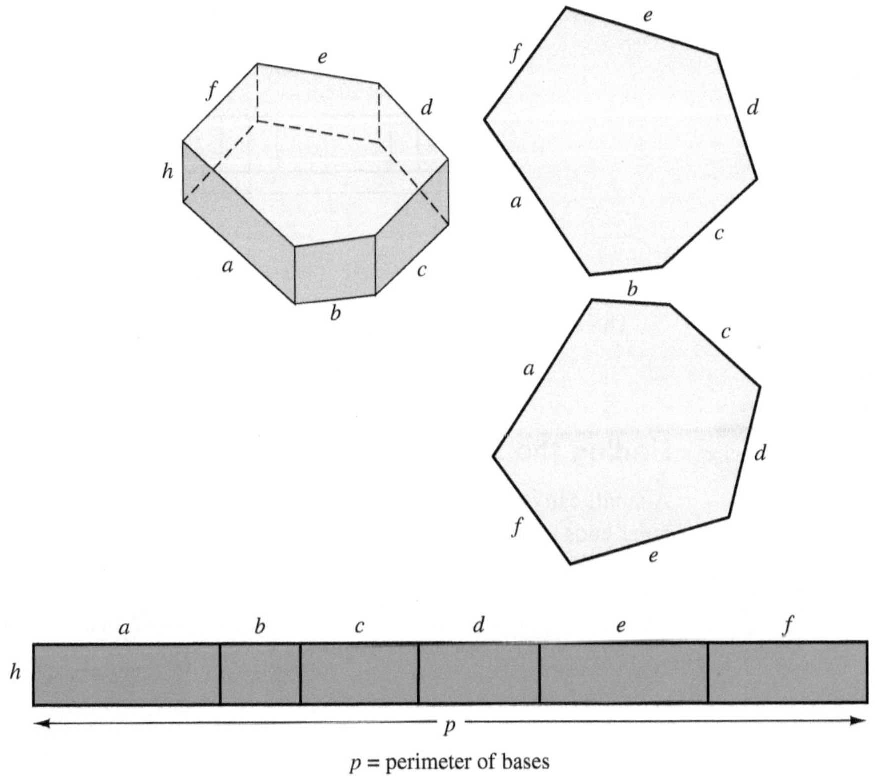

It is helpful to understand that the total surface area of a three-dimensional solid is often the sum of the **base area** and the **lateral surface area** of the solid (recall that *lateral* means "side").

FORMULA Surface Area of a Right Prism or Right Cylinder

Let a right prism or cylinder have height h and bases of area B, and let p be the perimeter of each base. Then the surface area SA is given by

$$SA = 2B + ph.$$

EXAMPLE 10.23

Finding the Surface Area of a Prism-Shaped Gift Box

A gift box has the shape shown. The height is 10 cm, the longer edges are 20 cm long, and the short edges of the square corner cutouts are each 5 cm long. What is the surface area of the box?

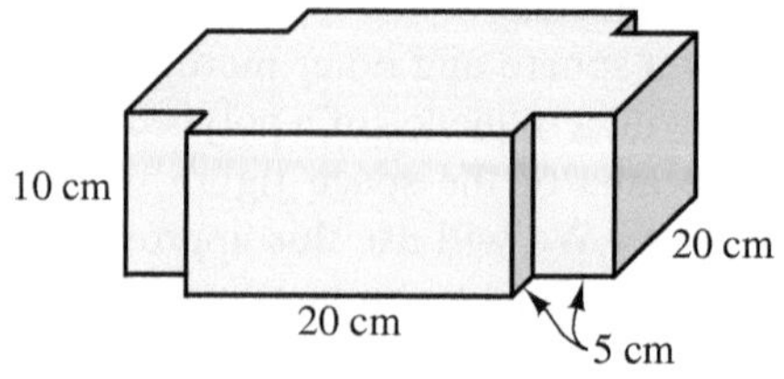

Solution

Each base has the area $B = (20)(20 + 5 + 5) + 20(5) + 20(5) = 800\text{ cm}^2$ as can be seen in the picture that follows. The lateral surface area is that of a rectangle 10 cm high and 120 cm long. That is, the lateral surface area is 1200 cm^2. Altogether, the surface area of the box is $SA = 2 \times 800\text{ cm}^2 + 1200\text{ cm}^2 = 2800\text{ cm}^2$.

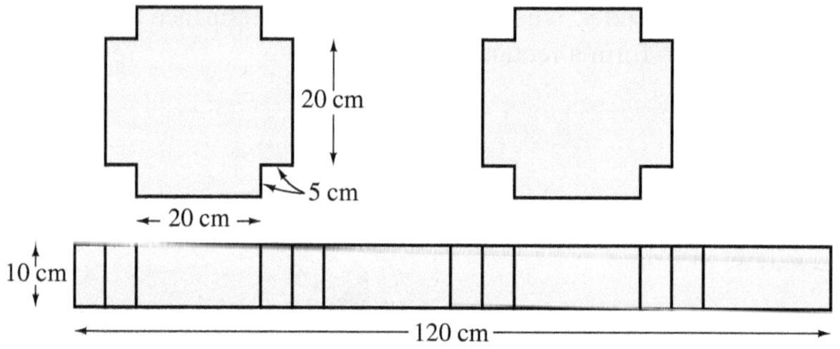

The next problem uses the problem-solving strategy discussed at the beginning of this section.

EXAMPLE 10.24

Finding the Surface Area of a Cylindrical Juice Can

A small can of frozen orange juice is about 9.5 cm tall and has a diameter of about 5.5 cm. The circular ends are metal, and the rest of the can is cardboard. How much metal and how much cardboard are needed to make a juice can?

Solution

The rectangle that follows is 9.5 cm wide and 5.5π cm long, so the area of cardboard is $(9.5) \cdot (5.5)\pi\text{ cm}^2 = 52.25\pi\text{ cm}^2$, or about 164 cm^2. The circles each have a radius of 2.75 cm, so each circle has area $\pi(2.75\text{ cm})^2$. Twice this is $15.125\pi\text{ cm}^2$, so the area of the two metal ends is about 47.5 cm^2.

Surface Area of Pyramids

The surface area of a pyramid is computed by adding the area of the base to the sum of the areas of the triangles forming the lateral surface of the pyramid. Of special importance is the **right regular pyramid**, for which the base is a regular polygon and the lateral surface is formed by congruent isosceles triangles. The altitude of the triangles is called the **slant height** of the pyramid. The formula for the surface area of a right regular pyramid can be obtained from Figure 10.28. The triangles each have altitude s, and the sum of the lengths of the bases is the perimeter p of the base polygon. The total area of the triangles is therefore $\frac{1}{2}ps$. If the base of the pyramid has area B, then the total surface area is $SA = B + \frac{1}{2}ps$.

Figure 10.28
A right regular pyramid has total surface area $SA = B + \frac{1}{2}ps$

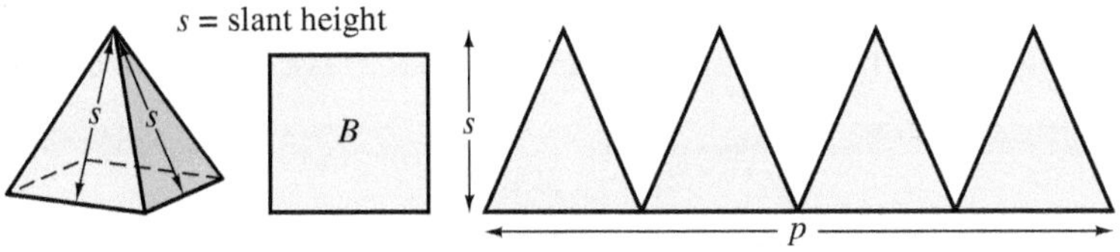

FORMULA Surface Area of Right Regular Pyramid

Let a right regular pyramid have slant height s and a base of area B and perimeter p. Then the surface area SA of the pyramid is given by the formula

$$SA = B + \frac{1}{2}ps.$$

Once again, the reasoning that leads to the formula is much more important than the formula itself.

EXAMPLE 10.25

Finding the Surface Area of a Right Regular Pyramid

A pyramid has a square base that is 10 cm on a side. The edges that meet at the apex have length 13 cm. Find the slant height of the pyramid, and then calculate the total surface area (including the base) of the pyramid.

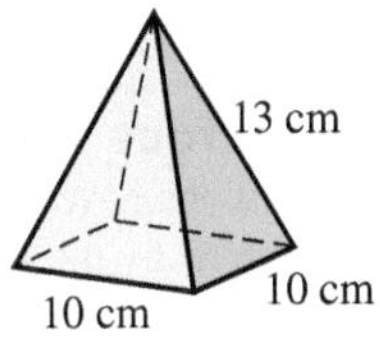

Solution

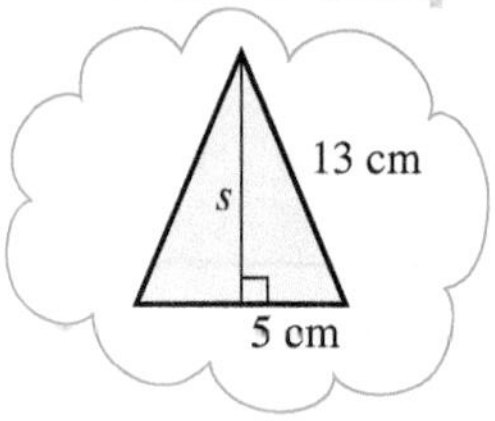

It is clear that the base of the pyramid has area $B = 100\text{ cm}^2$ and perimeter $p = 40\text{ cm}$. The slant height can be calculated with the Pythagorean theorem, which shows that

$$s = \sqrt{13^2 - 5^2} = \sqrt{169 - 25} = \sqrt{144} = 12\text{ cm}.$$

Thus, the surface area of the pyramid is $SA = 100\text{ cm}^2 + \frac{1}{2} \cdot 40 \cdot 12\text{ cm}^2 = 340\text{ cm}^2$.

INTO THE CLASSROOM
Problem Solving with Measurement

Each pair of students has an orange and a sheet of centimeter-squared graph paper. The following challenge is then given: *Find the area of the peel of the orange.* Students well versed in the basic principles of measurement may solve the problem in a direct, yet appropriate, way: The orange is peeled, and then the peeling is cut or torn into small pieces to tile a region of the graph paper; the region's boundary is traced; and then its area, which equals that of the orange peel, is estimated by counting the number of square centimeters covered.

Problem solving with measurement reinforces both the principles and the processes of measurement. By contrast, overemphasis on exercises that require only a routine application of a formula reduces measurement to a mechanistic level. Here are two examples illustrating the difference between a routine exercise and a problem:

Exercise: A right triangle has legs of length 6″ and 10″. What is the area of the triangle?
Problem: Two straws, of lengths 6″ and 10″, are joined with paper clips to form a flexible hinge. At what angle should the straws meet to form the sides of a triangle of largest possible area?

Exercise: A rectangular solid has length 6 cm, width 2 cm, and height 2 cm. What is the surface area of the solid?
Problem: The Math Manipulative Supply House sells wooden centimeter cubes in sets of 24 cubes. What shape of a rectangular box that holds one set of cubes requires the least amount of cardboard?

Surface Area of Right Circular Cones

Consider a right circular cone of slant height s and with a circular base of radius r. The formula for the surface area of the cone can be derived from the surface area formula for a pyramid. To see how, imagine that the cone is closely approximated by a right regular pyramid. An example is shown in Figure 10.29, where a cone is approximated by a pyramid with a dodecagon (12-gon) as its base.

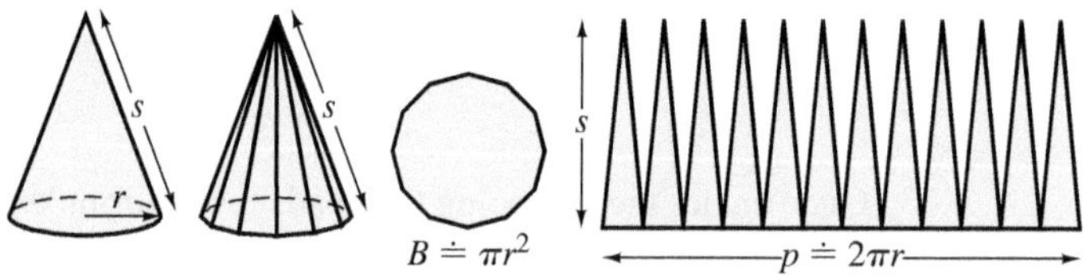

Figure 10.29
A right circular cone has total surface area $SA = \pi r^2 + \pi rs$

The area of the circular base of the pyramid is B, which is approximately πr^2, and the perimeter of the base p is about $2\pi r$. Therefore, the surface area of the pyramid $SA = B + \frac{1}{2}ps$ can be estimated by $\pi r^2 + \pi rs$. The pyramid's approximation to the cone becomes increasingly exact as the number of sides in the base polygon is increased, giving us the following formula:

FORMULA Surface Area of a Right Circular Cone

Let a right circular cone have slant height s and a base of radius r. Then the surface area SA of the cone is given by the formula

$$SA = \pi r^2 + \pi rs.$$

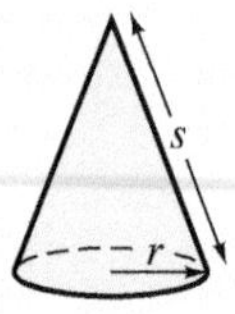

The Surface Area of a Sphere

We will now derive a formula for the surface area of a sphere of radius r by making use of the formula that we already know for the volume of the sphere. This derivation is done through reasoning, rather than by formal proof, as is our approach of this chapter. The key concept will be to relate the volume of the sphere to the volume of pyramid-like solids whose base is a small piece of surface area. We first divide the sphere's surface into many (say, n) tiny regions of area $B_1, B_2, B_3, \ldots, B_n$. The sum $B_1 + B_2 + B_3 + \cdots + B_n$ is the surface area S of the sphere. Each region can also be viewed as the base of a pyramid-like solid whose apex is the center of the sphere. Each of the pyramids has height r, so the pyramids have volumes $\frac{1}{3}B_1r, \frac{1}{3}B_2r, \frac{1}{3}B_3r$, and so on. An example is shown in Figure 10.30.

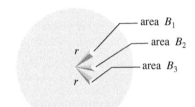

Figure 10.30
A solid sphere can be viewed as made up of pyramid-like pieces

We have the following relationships:

$$S = B_1 + B_2 + B_3 + \cdots + B_n \qquad \text{(surface area of the sphere)}$$

and

$$V = \frac{1}{3}B_1r + \frac{1}{3}B_2r + \frac{1}{3}B_3r + \cdots + \frac{1}{3}B_nr. \qquad \text{(volume of the sphere)}$$

Therefore, since $\frac{r}{3}$ is a common factor,

$$V = \frac{r}{3}(B_1 + B_2 + B_3 + \cdots + B_n) = \frac{r}{3}S.$$

Since $V = \frac{4}{3}\pi r^3$, the preceding formula gives us the equation

$$\frac{4}{3}\pi r^3 = \frac{r}{3}S.$$

Multiplying both sides by 3 and dividing both sides by r, we can solve for the surface area S.

FORMULA Surface Area of a Sphere

The surface area S of a sphere of radius r is given by the formula

$$S = 4\pi r^2.$$

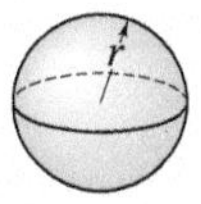

EXAMPLE 10.26

Comparing Earth with Jupiter

The diameter of Jupiter is about 11 times the diameter of our planet Earth. How many times greater is **(a)** the surface area of Jupiter? **(b)** the volume of Jupiter?

Solution

(a) Let r denote the radius of Earth and R the radius of Jupiter. Therefore, $R = 11r$. Using the formula for the surface area of a sphere, we find that the ratio of the surface area of Jupiter to that of Earth is

$$\frac{4\pi R^2}{4\pi r^2} = \frac{R^2}{r^2} = \left(\frac{R}{r}\right)^2 = (11)^2.$$

That is, the surface area of Jupiter is 11^2, or 121, times the surface area of Earth.

(b) Using the sphere volume formula, we find that the ratio of volumes is

$$\frac{\frac{4}{3}\pi R^3}{\frac{4}{3}\pi r^3} = \frac{R^3}{r^3} = \left(\frac{R}{r}\right)^3 = (11)^3.$$

Therefore, the volume of Jupiter is about 11^3, or 1331, times the volume of Earth. Precise measurements of the two not-quite-spherical planets show that the volume ratio is 1323.3, which is within 1% of our estimate of 1331.

Comparing Measurements of Similar Figures

Two figures are similar if they have the same shape but may be different in size. The ratio of all pairs of corresponding lengths in the two figures is a constant value called the *scale factor,* which we will denote by the letter k. The ratio of *any linear measurement* of the two figures—perimeter, height, diameter, slant height, and so on—is also the scale factor k. For example, since the diameter of Jupiter is 11 times the diameter of Earth, the scale factor is $k = 11$. We then also know that the equator of Jupiter is 11 times as long as Earth's equator.

The ratio of *areas* of similar figures is given by the *square,* k^2, of the scale factor k. The ratio of volumes is given by the *cube,* k^3, of the scale factor. This basic fact is evident for the cubes shown in Figure 10.31.

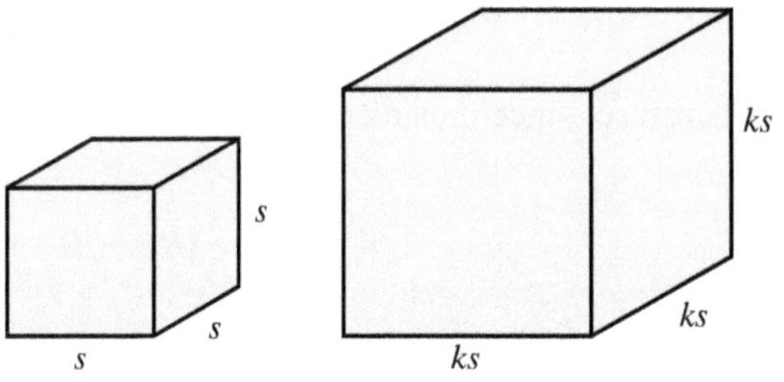

Scale factor = k

	Cube I	Cube II
Length of edge	s	ks
Area of each face	s^2	k^2s^2
Volume	s^3	k^3s^3

Figure 10.31
Area varies by the square, k^2, of the scale factor k, and volume varies by the cube, k^3, of the scale factor

The same comparison of areas and volumes holds for any pair of similar figures, not just the cube shown in Figure 10.31. The following theorem is an important principle for the comparison of the measurements of similar figures:

THEOREM The Similarity Principle of Measurement

Let Figures I and II be similar. Suppose some length dimension of Figure II is k times the corresponding dimension of Figure I; that is, k is the scale factor. Then

(i) *any* length measurement—perimeter, diameter, height, slant height, and so on—of Figure II is k times that of the corresponding length measurement of Figure I;

(ii) *any* area measurement—surface area, area of a base, lateral surface area, and so on—of Figure II is k^2 times that of the corresponding area measurement of Figure I; and

(iii) *any* volume measurement—total volume, capacity, half-fullness, and so on—of Figure II is k^3 times the corresponding volume measurement of Figure I.

EXAMPLE 10.27 Using the Similarity Principle

(a) Television sets are measured by the length of the diagonal of the rectangular screen. How many times larger is the screen area of a 40-inch model than that of a 13-inch table model?

(b) A 2″ by 4″ by 8″ rectangular brick of gold weighs about 44 pounds. What are the dimensions of a similarly shaped brick that weighs 10 pounds?

Solution

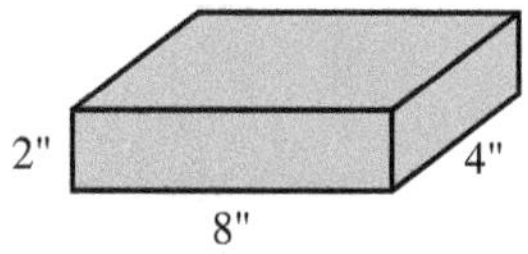

(a) The scale factor k is $40/13 \doteq 3.08$. Since area varies by $k^2 \doteq (3.08)^2 \doteq 9.5$, the large screen has about 9.5 times the area of the similarly shaped small screen.

(b) The weight of a gold brick is proportional to its volume, and the volume of similarly shaped bricks varies by the factor k^3, the cube of the scale factor k. Therefore, $10 = k^3 44$, so $k^3 = 10/44$ and $k = (10/44)^{1/3} \doteq 0.6$. Multiplying the dimensions of the 44-pound brick by 0.6 gives the approximate dimensions of a similarly shaped 10-pound brick of gold, namely, 1.2″ by 2.4″ by 4.8″.

10.5 Problem Set

The Problem Set for Section 10.4 includes only volume problems. However, Problem Set 10.5 purposely has both volume and surface area problems in it because people (including the children you will be teaching) do confuse the two concepts. Reread each problem carefully before you decide whether it is a volume or a surface area problem.

Understanding Concepts

1. Find the surface areas of these prisms:

(a)

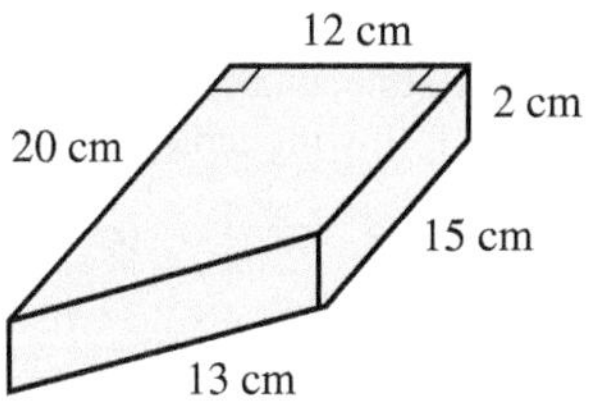

Right trapezoidal prism

(b)

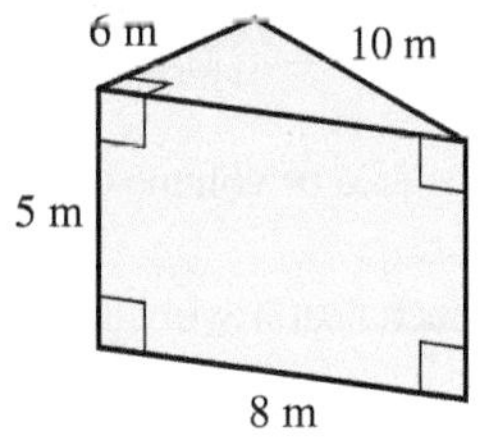

Right triangular prism

2. Find the surface areas of these cylinders and the prisms:

(a)

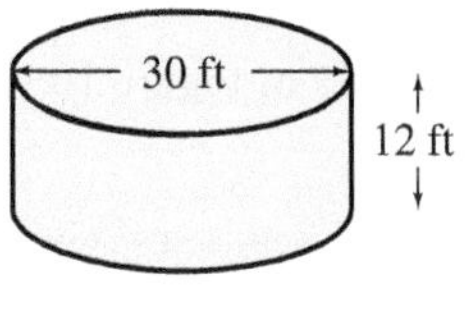

Right circular cylinder

(b)

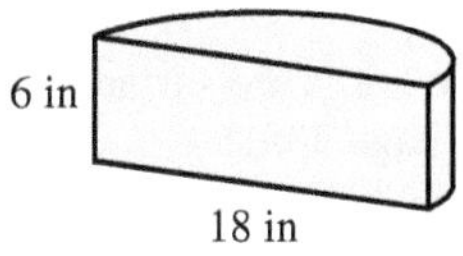

Semicircular right prism

3. Find the surface areas of these right regular pyramids:

(a)

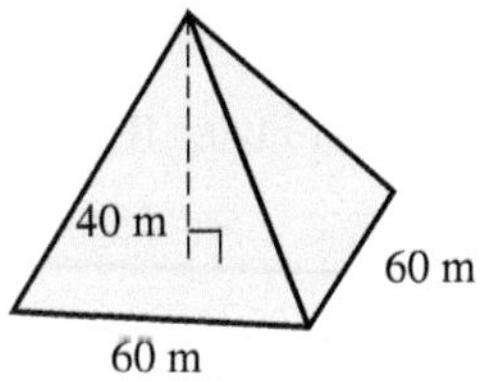

Right square pyramid

(b)

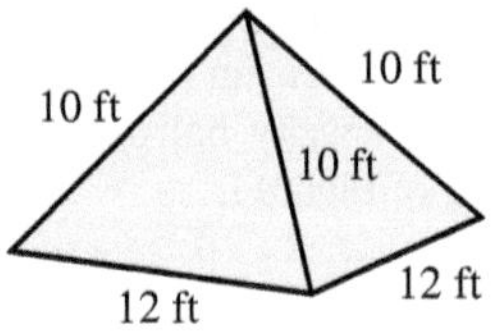

Right square pyramid

4. Find the surface areas of these right circular cones:

(a)

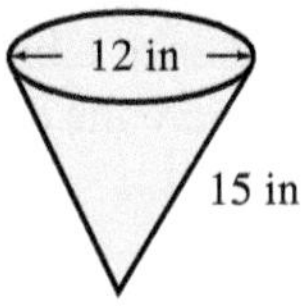

Right circular cone

(b)

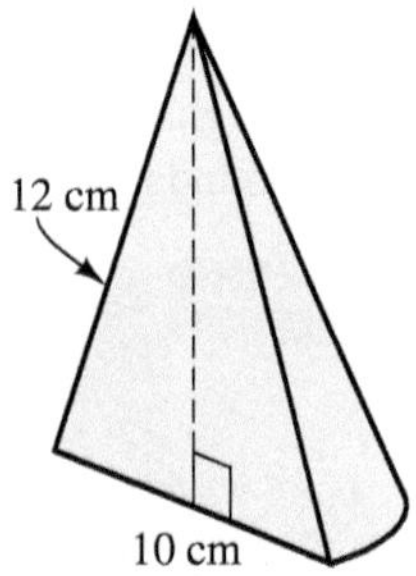

Semicircular right cone

5. Find the surface areas and volumes of these solids:

(a)

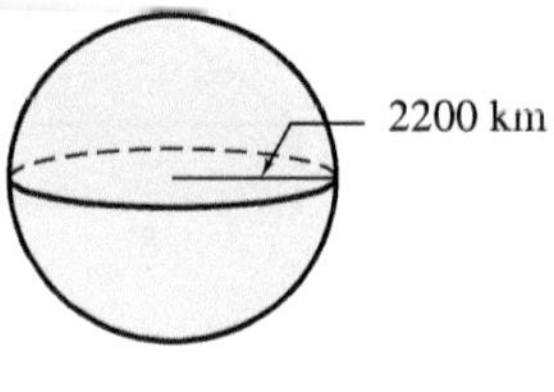

Sphere

(b)

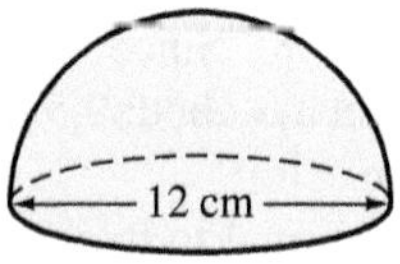

Hemisphere

6. Find the surface areas and volumes of these solids:

(a)

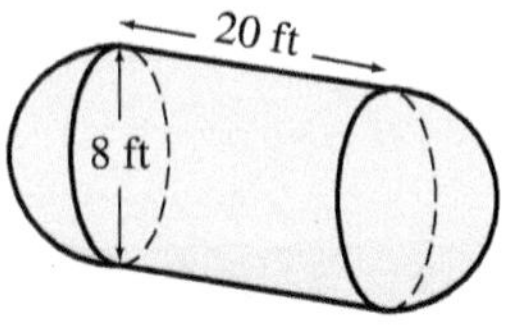

Cylindrical storage tank with hemispherical ends

(b)

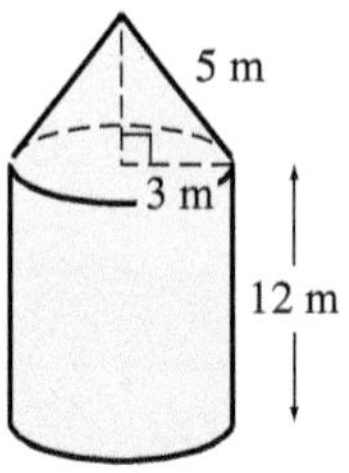

Cylindrical grain silo with conical top

7. Identify which measure—surface area or volume—is most important to know

(a) about a room, to see how much heat it will take to heat the room.

(b) about a child's favorite stuffed animal, to determine the amount of stuffing required to mend the animal.

(c) about a swimming pool, to purchase a pool cover.

(d) about a lawn, to see how long it will take to mow it.

8. What is the surface area of the solid of Figure 10.18 (page 535)?

9. An aluminum soda pop can has a diameter of 6.5 cm and a height of 11 cm. If there are 30 milliliters in a fluid ounce, verify that the capacity of the can is 12 fluid ounces, as printed on the can's label.

10. If it takes a quart of paint to cover the base of a hemisphere, how many quarts does it take to paint the spherical part of the same hemisphere?

11. Archimedes showed that the volume of a sphere is two-thirds the volume of the right circular cylinder just containing the sphere. Show that the area of the sphere is also two-thirds the surface area of the cylinder. (Archimedes was so pleased with these discoveries that he requested that the figure shown be placed on his tombstone.)

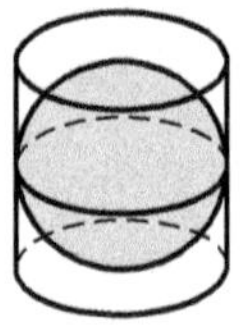

12. A square right regular pyramid is formed by cutting, folding, and gluing the following pattern:

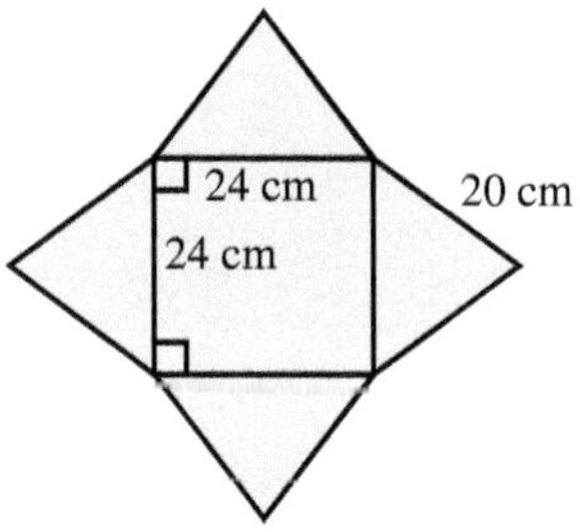

(a) What is the slant height of the pyramid?

(b) What is the lateral surface area of the pyramid?

(c) Use the Pythagorean theorem to find the height of the pyramid.

(d) What is the volume of the pyramid?

Use the similarity principle to answer problems 13 through 17. Explain carefully how the principle is used.

13. (a) An 8″ (diameter) pizza will feed one person. How many people will a 16″ pizza feed?

(b) Is it better to buy one 14″ pizza at \$10 or two 10″ pizzas at \$6 each? (*Hint:* 1.4^2 is about 2.0.)

14. What fraction of the area of the large circle is shaded in each figure?

(a)

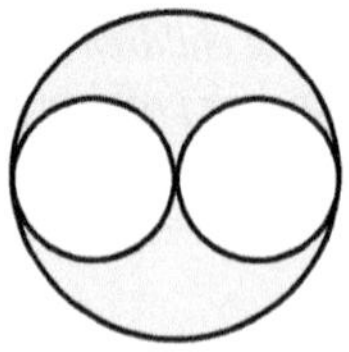

(b)

(*Hint:* First compare each unshaded circle with the large circle.)

15. Cones I, II, and III are similar to one another. Fill in the measurements left blank in the following table:

	I	II	III
Height	6	18	cm
Perimeter of Base		30	15 cm
Lateral Surface Area	40		cm^2
Volume			10 cm^3

16. A cylindrical can holds 100 milliliters.

(a) If the radius of the base is doubled and the height halved, what is the new volume of the can?

(b) If the radius of the base is halved and the height doubled, what is the new volume of the can?

17. A cube 10 cm on a side holds 1 liter.

(a) How many liters does a cube 20 cm on a side hold?

(b) What is the length of each side of a cube that holds 2 liters?

18. A right circular cone has a surface area of 16 cm^2. Its radius is the same as its slant height. What is the radius of the cone?

19. If the surface area of a right circular cylinder is 392 cm^2 and its height is three times its radius, what are the dimensions of the cylinder?

20. (a) Is there a sphere whose volume (in in^3) is the same number as its surface area (in in^2)?

(b) Is there more than one solution to this problem? That is, are there spheres of different radii that have the property of part (a)?

Into the Classroom

In these problems, you will need paper, tape, scissors, and drawing tools (ruler, compass, and pencils). Work in pairs, measuring your constructed models and verifying your observations with calculations.

21. A pyramid is formed by joining a vertex of a cube of side length 8 cm to the four vertices of an opposite face.

(a) Use your drawing tools to accurately make a pattern that, when cut and folded, will form the pyramid. What are the lengths of each edge in your pattern? Use the Pythagorean theorem to find out. Give your answer both exactly, using square roots, and as a decimal approximation.

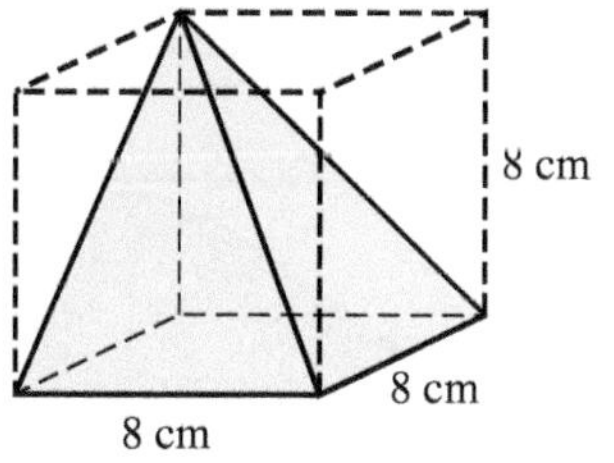

(b) Use the lengths of the edges found in part (a) to obtain the surface area of the pyramid, giving both an exact answer, using square roots, and a decimal approximation.

(c) Trace your pattern onto heavy paper and cut, fold, and tape three paper models of the pyramid. Show that the three congruent copies can be arranged to form a cube. What is the volume of the cube? What is the volume of each pyramid?

22. Cut out a semicircular sector, roll it up, and join the two radial segments to form a cone. Show that the diameter of the cone is equal to the slant height of the cone, both by measuring your paper model and by making a calculation.

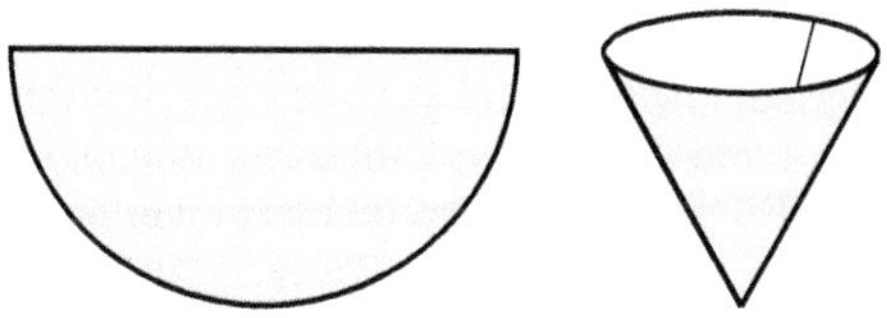

Responding to Students

23. A can of diameter 4 cm and height 6 cm is sitting on a table. Jerry is asked what the surface area of the can is and, after looking at the table and thinking, responds "28π cm^2." How do you think that Jerry got that answer, and how would you explain to Jerry that his answer is incorrect?

24. A cube whose sides are 8 cm is sitting on a table. If Annie says that the surface area of the cube is 256 cm^2, how would you explain her error to her?

25. Inez knows that the volume of a sphere involves its radius to the third power, whereas its surface area is the square of its radius. She therefore concludes that the volume is always larger than the surface area of a cube. How would you respond to her?

26. Wade is working on a right circular cone and measures the various sides and radius. He now knows that the surface area has a base radius of 1 cm and slant height of 4 cm. He concludes that the surface area of the cone is π cm + 4π cm^2. Why do you think Wade's response is wrong?

Thinking Critically

27. The right circular cylinder and cone shown in the accompanying figure both have a base of radius r and height $2r$, and the sphere has radius r. Show that the ratio of the surface areas of the cone to the sphere to the cylinder is τ to 2 to 3, where $\tau = (1 + \sqrt{5})/2$ is the famous golden ratio.

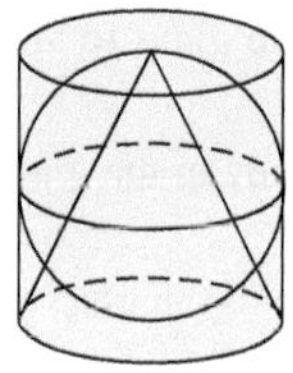

28. A birthday cake has been baked in a 7″ by 7″ by 2″ pan. Frosting covers the top and sides of the cake.

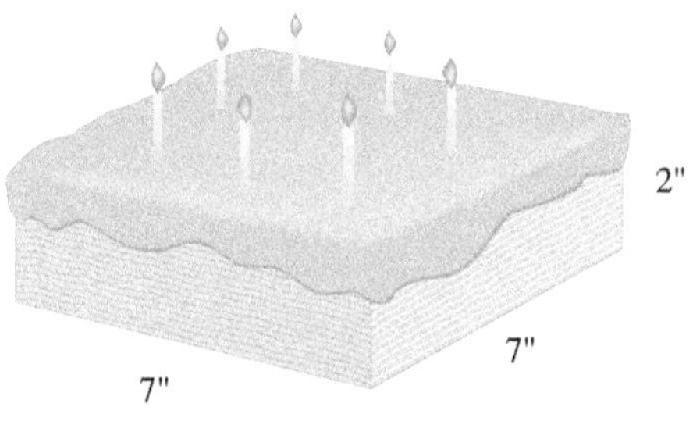

(a) What is the volume of the cake?

(b) What is the area covered by the frosting?

(c) Describe how to cut the cake into eight pieces so that each piece is the same size (measured by volume) *and* has the same amount of frosting (measured by area covered with frosting).

(d) Describe how to cut the cake into seven pieces, each of the same size and with the same amount of frosting. (*Hint:* Consider a slice made by two vertical cuts from the center that also intercepts 4 inches of the perimeter.)

29. The cube $ABCDEFGH$ with edges of length s contains the tetrahedron $ACEG$.

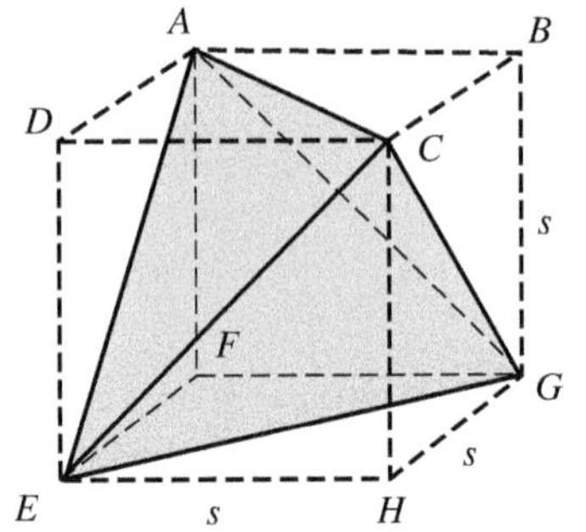

(a) Explain why $ACEG$ is a regular tetrahedron with edges of length $\sqrt{2}\,s$.

(b) Show that the volume of tetrahedron $ACDE$ is $\frac{1}{6}s^3$.

(c) Show that the volume of $ACEG$ is $\frac{1}{3}s^3$.

(d) Use the similarity principle to explain why a regular tetrahedron with edges of length b has volume $\frac{\sqrt{2}b^3}{12}$.

30. An ice cream soda glass is shaped like a cone of height 6 inches and has a capacity of 16 fluid ounces when filled to the rim. Use the similarity principle and the fact that the cone of liquid is similar to the cone of the entire region inside the glass to answer the following questions:

(a) How high is the soda in the glass when it contains 2 fluid ounces?

(b) How much soda is in the glass when it is filled to a level 1 inch below the rim?

State Assessments

31. (Grade 5)
The drawing for a new physics lab is made to scale so that 1 cm = 1 meter. If the picture is drawn of the lab with an area of 100 square cm, what will the actual area of the lab be?

A. 1 sq. meter
B. 100 meters
C. 10 meters
D. 10 sq. meters
E. 100 sq. meters

32. (Grade 8)
Cameron bought a fish tank in the shape of a right rectangular prism like the one pictured below.

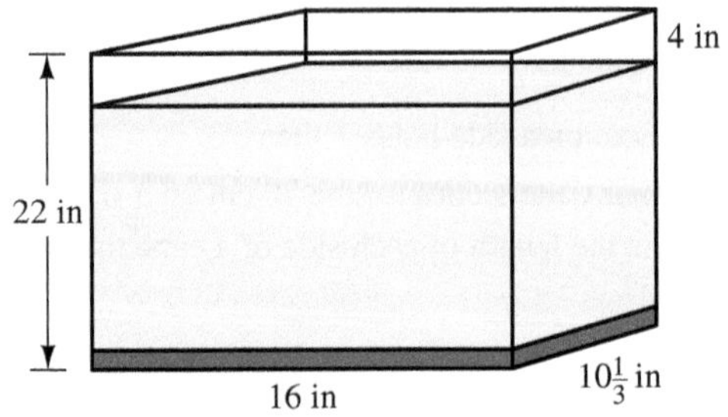

(a) Cameron wants to put 3 inches of sand evenly in the bottom of the tank. What is the volume of sand that Cameron will put in his fish tank?

(b) After Cameron puts 3 inches of sand in the bottom of the fish tank, he will put water in the tank, leaving 4 inches of space at the top. What is the volume of water Cameron will put into his fish tank?

33. (Grade 8)
Ace Tennis Cans are in the shape of right circular cylinders. Their tennis cans are shown as a net which is described below. From the information of the net, what is the total surface area of an Ace Tennis Can?

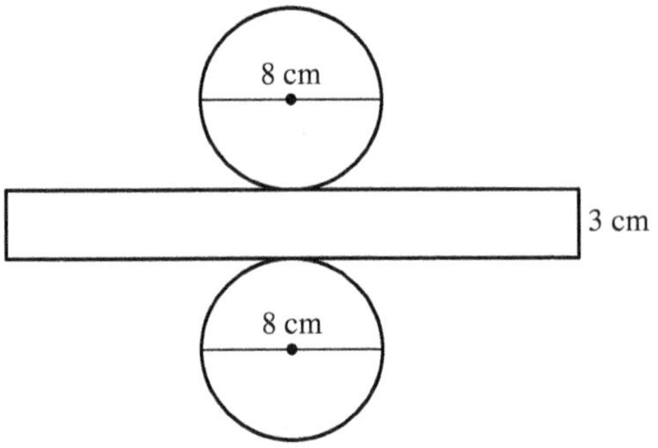

34. (Washington, Grade 8)
Bella Restaurant is building a curved awning for the entrance to their restaurant. They need material for only the top and the front of the awning.

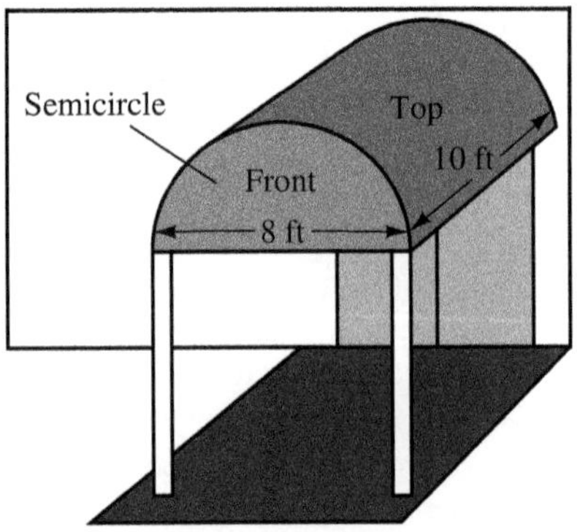

Area of a circle = πr^2
Circumference of a circle = πd

Find the surface area of the awning to determine the total amount of canvas necessary to make the awning.

Show your work using words, numbers, and or pictures. Be sure to label your answer.

How much canvas is necessary to make the awning? _____

35. (Washington, Grade 6)
Gavin is building a storage shed attached to the side of his house. He needs to calculate the total surface area of the three walls of the shed to help determine how much paint he needs to purchase.

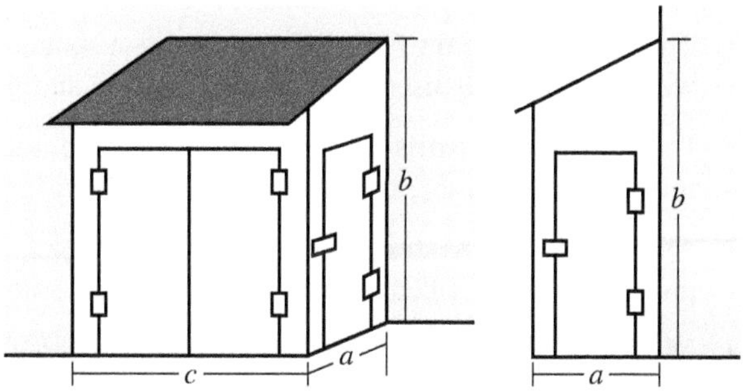

Which expression will provide a good estimate of the total surface area of all three shed walls?

A. $(c \times b) + 2(a \times b)$ B. $2(c \times b) + (a \times b)$
C. $2a \times 2b + c$ D. $a \times b \times c$

The Chapter in Relation to Future Teachers

This chapter has introduced the basic notions of measurement, especially for length, area, and volume. Measurement is concerned with how size is determined and communicated. For some ideal shapes—triangles, prisms, pyramids, circles, and spheres, to name a few—the measurement process can be done through the use of a formula. However, these formulas must be carefully derived before they are used; they are not simply a collection of formulas. We will often need to return to the basic principles of the measurement process and apply them to complicated solids by breaking down the solids into pieces that are either easily measured or for which there is an established formula.

Chapter 10 Summary

Section 10.1 The Measurement Process	Page Reference
CONCEPTS	
• **Unit** is an established amount used to determine size to allow comparison.	490
• **Standardized system:** A measurement system with well-defined units used to communicate size and magnitude. The United States uses the customary system, and the entire world, including the United States, uses the metric system.	490
• **The metric (SI) system:** A system with units related by powers of 10 described by a prefix system; for example, *milli* $= \frac{1}{1000}$, *centi* $= \frac{1}{100}$, and *kilo* $= 1000$.	495
• **Metric units:** The basic unit of length is the meter, with area given in square meters or hectares (1 hectare $= 10{,}000 \text{ m}^2$) and volume in cubic meters. Capacity is also given in liters (1 liter $= 1000 \text{ cm}^3$). Mass is given in grams.	495
• The **Fahrenheit temperature scale** is a scale used to measure temperature where 32°F represents the freezing point of water and 212°F represents the boiling point of water.	500
• The **Celsius temperature scale** is a scale used to measure temperature with 0°C as the freezing point and 100°C as the boiling point of water.	500
DEFINITIONS	
• A **measurable attribute** is a property of an object, such as length, area, volume, capacity, temperature, time or weight, which is to be measured.	490
• A **tangram** piece is a certain unit of measurement that originated in ancient China.	491

- The **U.S. customary (English) system** is a standardized measurement system with well-defined units used to communicate size and magnitude. 495
- **U.S. customary units** are inch, foot, yard, and mile for length; in^2, ft^2, yd^2, acre, and mi^2 for area; and in^3, ft^3, yd^3, quart, and gallon for volume and capacity. 493
- The **metric (SI) system** is a measurement system used worldwide with units related by powers of 10 described by a prefix system. For example, *milli* = 1/1000 and *kilo* = 1000. 495
- **Metric units** are meter, centimeter, and kilometer for length; m^2, cm^2, km^2, and hectare for area; and m^3, cm^3, km^3, and liter for volume and capacity. 495
- A **liter** is the volume of a cube, each of whose sides is 10 centimeters or, equivalently, 1 liter is 1000 cm^3. 499
- The **weight** of an object is the force exerted on the object by gravity. 499
- **Mass** is the amount of matter an object has. 499
- A **kilogram** is the weight of 1 liter of water. 500

PROCEDURES

- **The measurement process:** Step 1: Choose the property (length, area, volume, and so on) to be measured. Step 2: Select a unit of measurement. Step 3: Compare the size of the object being measured with the size of the unit by covering, filling, and so on. Step 4. Express the measurement as the number of units used. 491
- **Converting between customary system and metric system:** 1 m is about 3.28 ft and 1 ft is about 0.3048 m. 500
- **Unit,** or **dimensional, analysis:** A process to convert a measurement given in one unit to the equivalent measurement in a new unit. 501

Section 10.2 Area and Perimeter — Page Reference

CONCEPTS

- **Area:** Area is the amount of the plane covered by a region in the plane. The unit of area is arbitrary, but usually a square one unit of length on a side is chosen. To compare the area of one region with that of another, the congruence and addition properties are often useful. 505
- The **units of area** are usually given as cm^2, in^2, ft^2, and so on. 505
- **Area formulas for common polygons:**
  - Rectangle of length l and width w: $A = lw$. 509
  - Parallelogram of base b and height h: $A = bh$. 510
  - Triangle of base b and altitude h: $A = \frac{1}{2}bh$. 510
  - Trapezoid with bases a and b and altitude h: $A = \frac{1}{2}(a + b)h$. 511
- **Length:** Length is the distance along a curve. 512
- **Perimeter:** Perimeter is the length of a simple closed curve or, more generally, the length of the boundary of a region. 513
- **Perimeter, circumference, and area of a circle:** The perimeter of a circle is called the circumference of the circle, given by $2\pi r$, where r is the radius of the circle and, by definition, π (pi) is the ratio of the circumference to the diameter of a circle. The area of a circle is πr^2. 514

DEFINITIONS

- Two regions in the plane are **congruent** if they have the same size and shape. 506

- A **cycloid** is an arch-shaped curve that is created by the path a point on a circle travels when it is rolled along a flat surface. 508
- An **altitude of a parallelogram** is the distance that separates a pair of parallel sides. 509
- A **lattice polygon** is a polygon formed by joining points of a square array. 512
- The **length of a polygonal curve** is obtained by summing the lengths of its sides. The **length of a nonpolygonal curve** is measured or estimated by calculating the length of an approximating polygonal curve with vertices on the original curve. 512
- The **perimeter** of a region is the length of its boundary. 513

PROPERTIES

- **Congruence property of area:** If two regions are congruent, then they have the same area. 506
- **Addition property of area:** If a region R is divided into nonoverlapping subregions, then the area of R is the sum of the area of the subregions. 507

PROCEDURE

- **Dissection of a plane region into subregions:** Cut the figure into a collection of nonoverlapping regions that covers the larger figure. 506

Section 10.3 The Pythagorean Theorem — Page Reference

CONCEPTS

- **Pythagorean theorem:** The lengths a, b, and c of the sides of a right triangle with hypotenuse of length c satisfies $a^2 + b^2 = c^2$. Therefore, the sum of the areas of the squares on the legs of a right triangle equals the area of the square on the hypotenuse. 527
- **The converse of the Pythagorean theorem:** This statement is also true: If $a^2 + b^2 = c^2$, then a triangle with sides of length a, b, and c is a right triangle. 530

Section 10.4 Volume — Page Reference

CONCEPTS

- **Decomposition of solids:** Use the problem-solving strategy of breaking complicated solids into a collection of more manageable solids (decomposing one complex problem into a number of simpler ones). 535
- **Volume formulas:**
  Prism or cylinder of base area B and height h: $V = Bh$. 536
  Pyramid or cone of base area B and height h: $V = \frac{1}{3}Bh$. 538
  Sphere of radius r: $V = \frac{4}{3}\pi r^3$. 539

DEFINITION

- The **volume, V,** of a solid measures the amount of space enclosed within its boundary. 535

Section 10.5 Surface Area — Page Reference

CONCEPTS

- **Surface area of a polyhedron:** The surface area of a polyhedron is the sum of the areas of its plane faces. For some polyhedra, such as right prisms and right regular pyramids, it is useful to imagine that the surface is cut and unfolded into the plane. 543

- **Surface area of a general space figure:** This quantity is found as the limiting value of the surface areas of approximating polyhedra. 543
- **Surface area formulas:**
  Right prism or right cylinder of height h, bases of area B, and perimeter p: $SA = 2B + ph$. 543
  Right regular pyramid of slant height s, base area B, and perimeter p: $SA = B + \frac{1}{2}ps$. 545
  Right circular cone of slant height s, and base radius r: $SA = \pi r^2 + \pi rs$. 546
  Surface area of a sphere of radius r: $S = 4\pi r^2$. 547
- **Similarity principle of measurement:** If two space figures are similar with a scale factor k, then all corresponding linear measurements vary by the factor k, all area measurements vary by k^2, and all volume measurements vary by k^3. 548

DEFINITIONS

- The **surface area of a surface in space** is the measure of its boundary. 543
- **Lateral surface area** of a prism or cylinder is the surface area of the sides, not the base, of the solid. 543
- A **right regular pyramid** is a pyramid whose base is a regular polygon and the lateral surface area is formed by congruent isosceles triangles. 545
- The **slant height** of a right, regular pyramid is the distance measured along a lateral face. 545

Chapter Review Exercises

Section 10.1

1. Select an appropriate metric unit of measurement for each of the following:
   (a) The length of a sheet of notebook paper
   (b) The diameter of a camera lens
   (c) The distance from Los Angeles to Mexico City
   (d) The height of the Washington Monument
   (e) The area of Central Park
   (f) The area of the state of Kentucky
   (g) The volume of a raindrop
   (h) The capacity of a punch bowl

2. Give the most likely answer.
   (a) A bottle of cider contains 30 mL, 4 L, 15 L.
   (b) Cross-country skis have length 190 cm, 190 km, 190 m.
   (c) The living area of a house is 2000 cm^2, 1.2 ha, 200 m^2.

3. An aquarium is a rectangular prism 60 cm long, 40 cm wide, and 35 cm deep. What is the capacity of the aquarium in liters?

4. A sailfish off the coast of Florida took out 300 feet of line in 3 seconds. Estimate the speed of the fish in miles per hour.

Section 10.2

5. Let M be the midpoint of side $\overline{AD}$ of the trapezoid $ABCD$. What is the ratio of the area of triangle MBC to the area of the trapezoid? (*Hint:* Dissect the trapezoid by a horizontal line through M and rearrange the two pieces to form a parallelogram.)

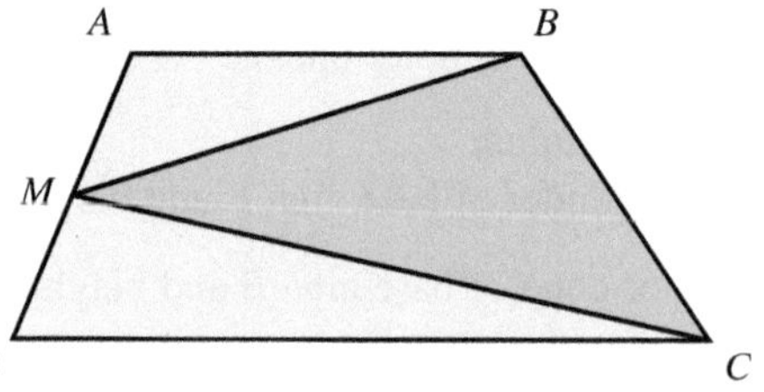

6. A square cake 20 inches on a side is being shared evenly among five people. Two cuts have been made to the center as shown.

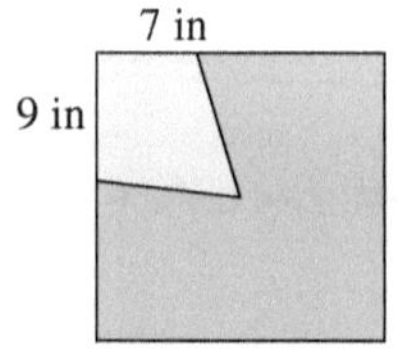

   (a) Is the piece of cake shown a fair piece?
   (b) Draw a figure showing where three more cuts from the edge to the center create pieces of the same size.

7. Find the areas of these figures:

(a)

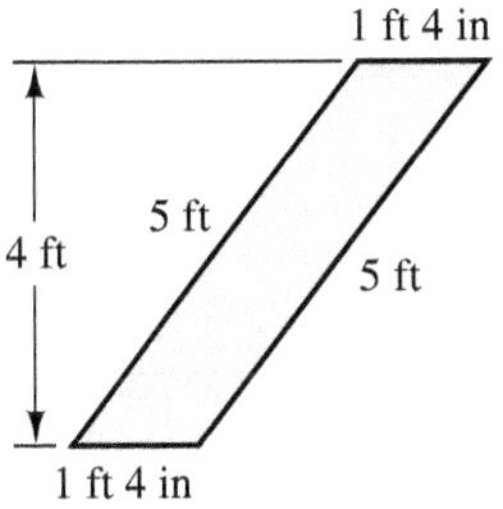

(b)

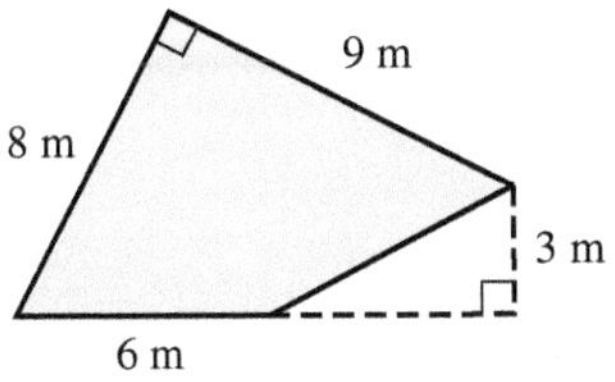

(c)

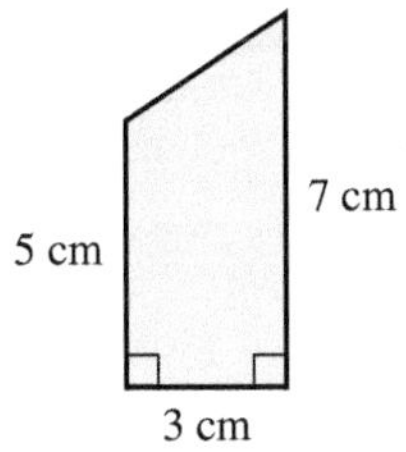

8. Find the areas of these lattice polygons:

(a)

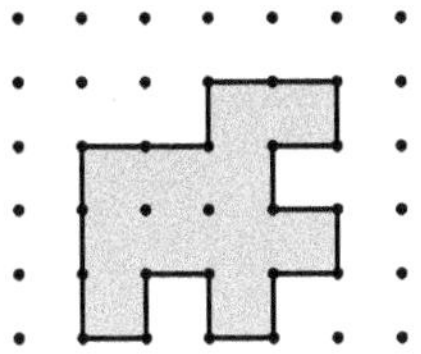

(b)

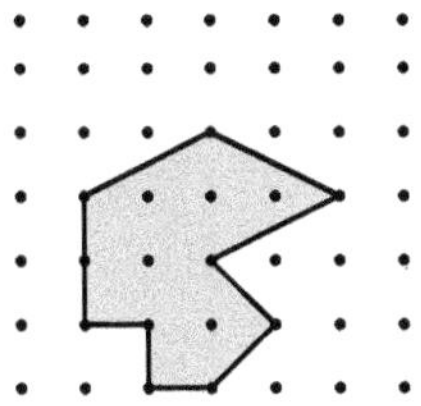

(c)

9. Find the areas and perimeters of these figures:

(a)

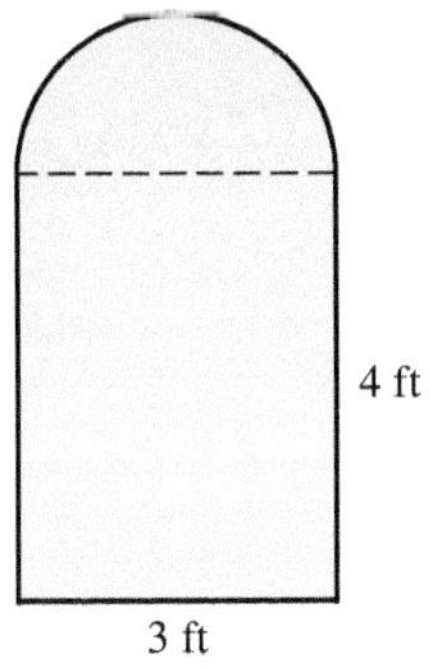

(b)

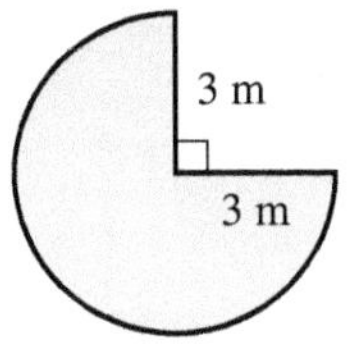

Section 10.3

10. Solve for x and y in the figure.

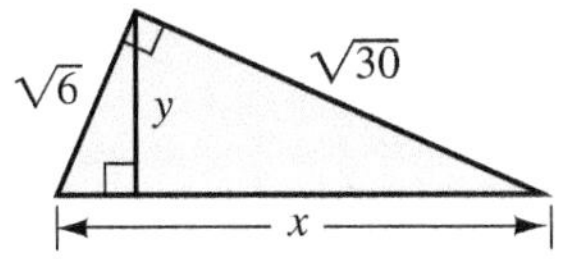

11. A right circular cone has slant height 35 cm and a base of diameter 20 cm. What is the height of the cone?

12. A rectangular box has sides of length 4 inches, 10 inches, and 12 inches. What are the lengths of each of the four diagonals of the box?

13. Find the perimeter of the following lattice polygon:

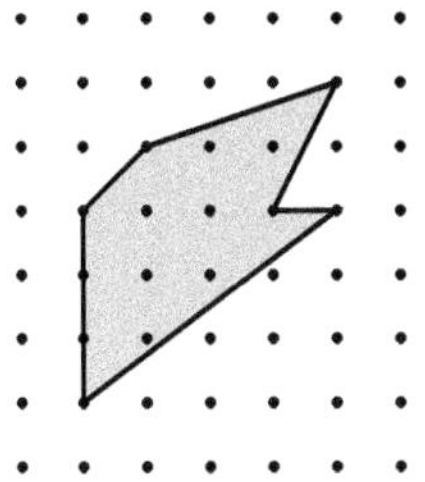

Section 10.4 and Section 10.5

14. Find the volumes and surface areas of these figures:

(a)

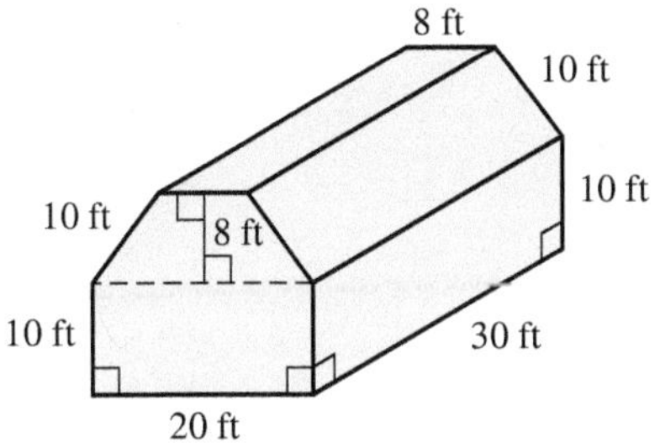

(b)

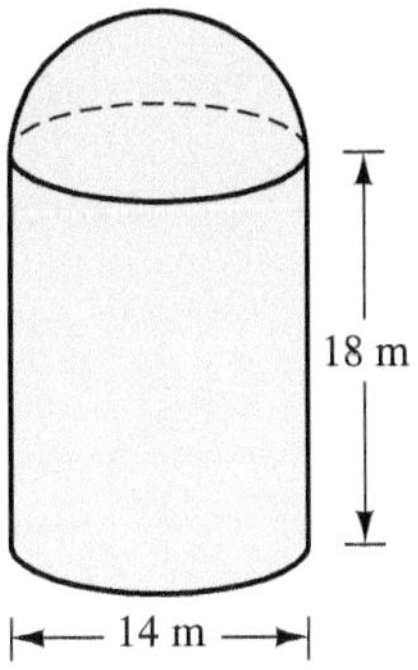

(c)

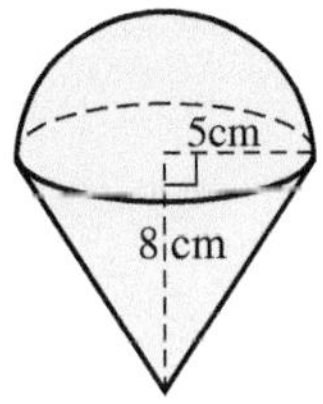

15. Which has the larger volume, a sphere of radius 10 meters or four cubes with sides of length 10 meters?

11

Transformations, Symmetries, and Tilings

11.1 ▶ Rigid Motions and Similarity Transformations

11.2 ▶ Patterns and Symmetries

11.3 ▶ Tilings and Escher-like Designs

COOPERATIVE INVESTIGATION

Exploring Reflection and Rotation Symmetry

Materials Needed

Clear acetate sheets, overhead transparency pens, tissues to clean acetate sheets for reuse, and Mira® (if available; otherwise, small plastic or metal rectangular mirror).

How to Check for Reflection and Rotation Symmetry

A figure has **reflection** (or **line**) **symmetry** if there is a mirror line that reflects the figure onto itself. For example, the parafoil kite below has a vertical line of symmetry. The wheel cover at the right does not have reflection symmetry, but it does have **rotation symmetry,** since the figure turns onto itself when rotated through 72° about the center point.

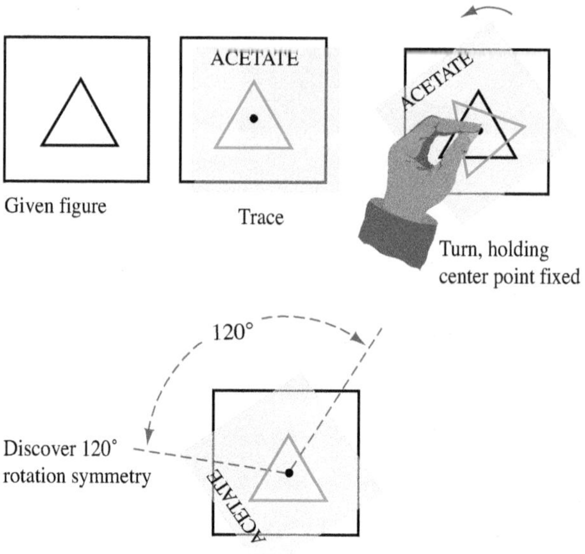

Reflection symmetry can be verified by the "trace-and-flip" test. The figure and its line of symmetry are traced onto an acetate sheet. The acetate is then flipped over across the proposed symmetry line to check that the points of the traced figure coincide with those of the original figure. Reflection symmetry can also be investigated by placing the drawing line of a Mira on a proposed line of symmetry of a figure. Alternatively, looking in the mirror placed on the line of symmetry should make the whole figure appear.

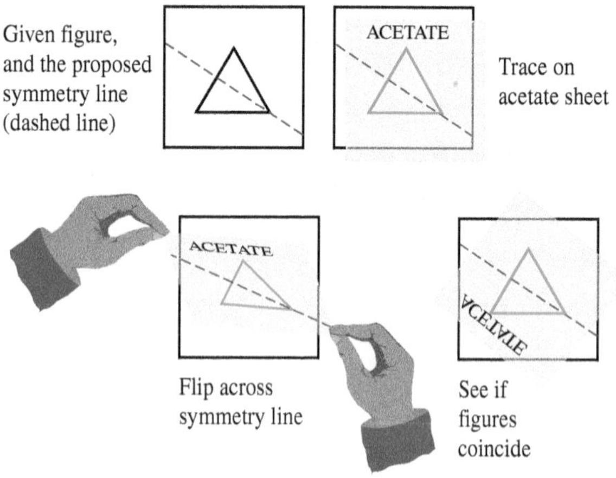

Rotation symmetry can be investigated by the "trace-and-turn" test, illustrated next. The point held fixed is the **center of rotation.** Since the tracing coincides with the original figure after a 120° turn, the trace-and-turn test shows that an equilateral triangle has 120° rotation symmetry.

Activities

1. Use either a Mira, a mirror, or the trace-and-flip test to find all lines of symmetry of the given figures. Use dashed lines to draw the symmetry lines.

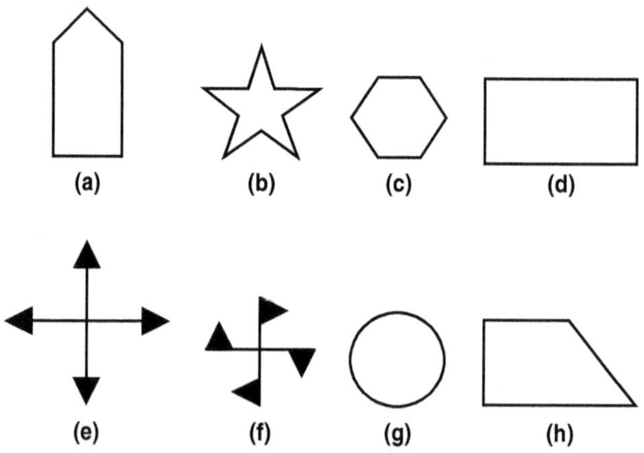

2. Use the trace-and-turn test to describe the rotation symmetries of the figures in Activity 1. Indicate the center of rotation and the angle measure of the rotation.
3. Sketch all lines of symmetry and describe all rotation symmetries for the following polygons:
   (a) Triangles: equilateral, isosceles, scalene
   (b) Quadrilaterals: square, rhombus, rectangle, parallelogram, trapezoid, isosceles trapezoid, kite

This chapter investigates **transformational geometry,** a dynamic approach to geometry in which every point P is moved from a starting position to a final position P'. Of special importance is a **rigid motion,** in which every pair of points P and Q moves to P' and Q' in a way that leaves the distance unchanged; that is, $PQ = P'Q'$. Any rigid motion will be shown to be one of just four basic types: a translation, a rotation, a reflection, or a glide–reflection. In elementary school terminology, the four basic motions are more informally called, respectively, a slide, a turn, a flip, and a glide. The concept of a rigid motion then allows us, in Section 2, to investigate and classify symmetries and patterns in a precise way. In Section 3, interesting and useful patterns are created by repeatedly transforming a figure to new positions by means of rigid motions. With practice, we will even create artistic tilings that are suggestive of the graphic work pioneered by M. C. Escher.

KEY IDEAS

- Transformation of the plane
- Rigid motion
- The four basic rigid motions: translation (slide), rotation (turn), reflection (flip), glide–reflection (glide)
- Classification theorem: Any rigid motion is one of the four basic types
- Dilations and similarity transformations
- Congruent and similar figures, as described by rigid motions and similarity transformations
- Symmetric figures
- Types of symmetry: reflection, rotation, point (half-turn)
- Border patterns and their classification
- Tilings (tessellations) of the plane, including regular and semiregular tilings
- Escher-like tilings

11.1 Rigid Motions and Similarity Transformations

Imagine that each point P of the plane is "moved" to a new position P' in the same plane. Call P' the **image** of P, and call P the **preimage** of P'. If distinct points P and Q have distinct images P' and Q', and if every point of the plane has a unique preimage point, then the association $P \leftrightarrow P'$ defines a one-to-one correspondence of the plane onto itself. Such a correspondence is called a **transformation of the plane.**

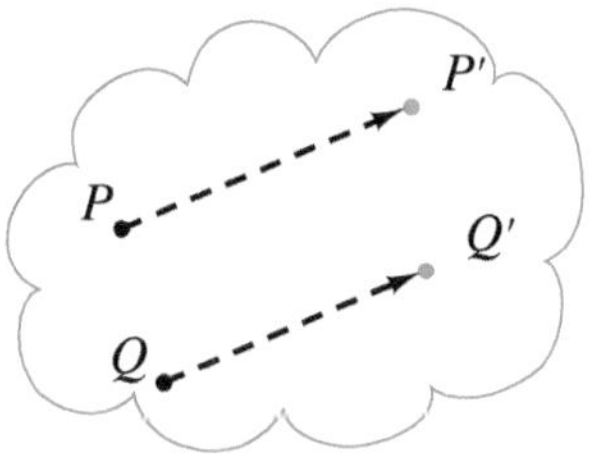

DEFINITION Transformation of the Plane

A one-to-one correspondence of the set of points in the plane onto itself is a **transformation of the plane.** If point P corresponds to point P', then P' is called the **image** of P under the transformation. Point P is called the **preimage** of P'.

A type of transformation called a **rigid motion** is of special importance. As the name implies, a rigid motion does not allow stretching or shrinking of distances.

DEFINITION Rigid Motion of the Plane

A transformation of the plane is a **rigid motion** if and only if the distance between any two points P and Q equals the distance between their image points P' and Q'. That is, $PQ = P'Q'$ for all points P and Q.

A rigid motion is also called an **isometry,** meaning "same measure" (*iso* = same, *metry* = measure).

A useful physical model of a rigid motion of the plane can be realized with a sheet of clear acetate and a sheet of paper containing figures with certain points labeled A, B, C, ... The figures are traced onto the transparency, and a rigid motion is modeled by moving the transparency to a new position in the plane of the paper. An example is shown in Figure 11.1, where primed letters $A', B', C', \ldots$ indicate the points in the image figure that correspond to the respective points A, B, $C, \ldots$ in the original figure. A rigid motion actually maps *all* of the points of the plane, but usually it is enough to show how a simple figure such as a triangle is moved to describe the motion. It is allowable to turn the transparency upside down before it is returned to the plane of the paper, since the definition of rigid motion is still satisfied.

Figure 11.1
Illustrating a rigid motion of the plane

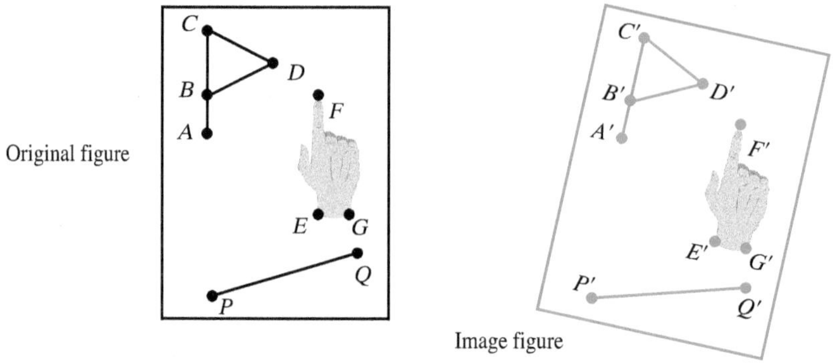

Perhaps the simplest "motion" of all is to leave the acetate sheet in place, so that $P = P'$ for all points of the plane. This is the **identity** transformation.

Note that only the initial and final positions of a transformation are of interest. The transparency could have been taken for a roller-coaster ride before reaching its final position. When the net outcomes of two motions are the same, the transformations are said to be **equivalent.**

The Four Basic Rigid Motions

Four transformations of the plane have special importance. They are the four **basic rigid motions of the plane:** translations, rotations, reflections, and glide–reflections.

Translations

A **translation,** also known as a **slide,** is the rigid motion in which all points of the plane are moved the same distance in the same direction. An arrow drawn from a point P to its image point P' completely specifies the two pieces of information required to define a translation: The direction of the slide is the direction of the arrow, and the distance moved is the length of the arrow. The arrow is called the **slide arrow** or **translation vector.**

A translation is illustrated by the "trace-and-slide" model in Figure 11.2.

Figure 11.2
A slide, or translation, moves each point of the plane through the same distance in the same direction

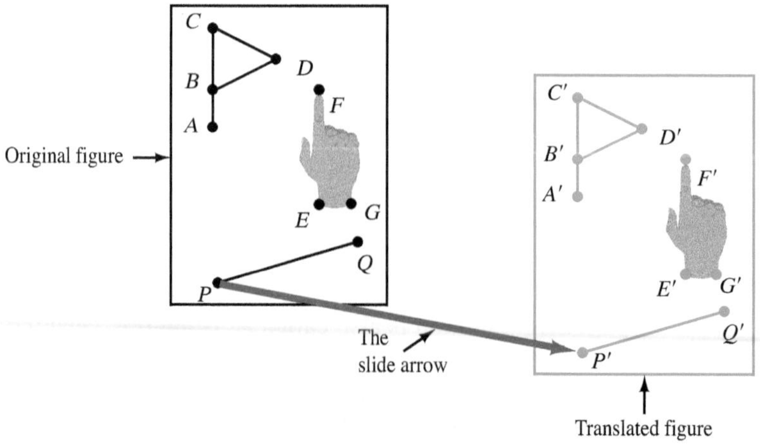

EXAMPLE 11.1

Finding the Image Under a Translation

Find the image of the pentagon $ABCDE$ under the slide that takes the point C to C'.

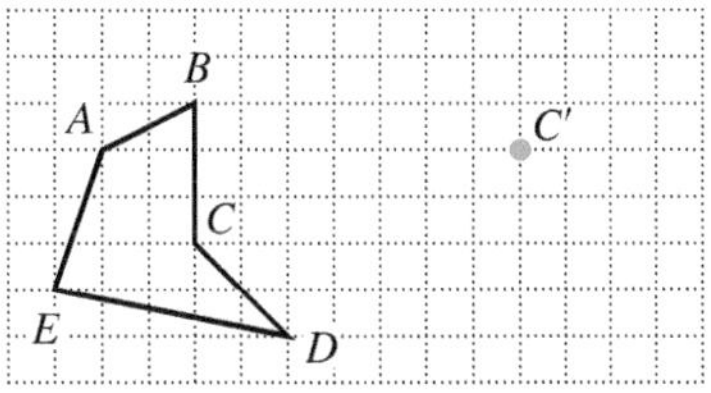

Solution

The slide arrow from C to C' is seven units to the right and two units up. Therefore, A' is found by moving seven units to the right of A and then two units up. The remaining points are found in the same way, always with the same slide arrow. Here is the image under the given translation:

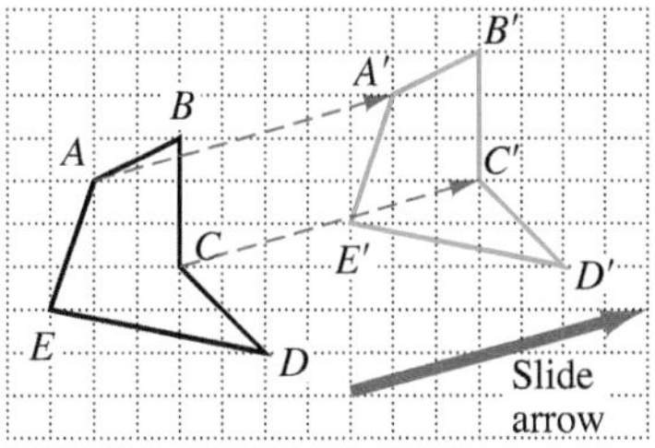

Rotations

A **rotation,** also called a **turn,** is another basic rigid motion. One point of the plane—called the **turn center** or the **center of rotation**—is held fixed, and the remaining points are turned about the center of rotation through the same number of degrees—the **turn angle** or **angle of rotation.** A counterclockwise turn about point O through 120° is shown in Figure 11.3. Note that the right-hand EFG is taken to the right-hand $E'F'G'$.

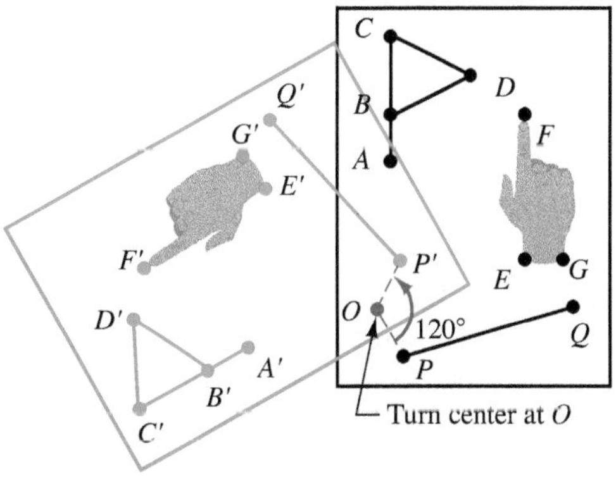

Figure 11.3
A turn, or rotation, rotates each point of the plane about a fixed point O—the turn center—through the same number of degrees and in the same direction of rotation

A rotation is determined by giving the turn center and the directed angle corresponding to the turn angle. This information can be pictured by a **turn arrow,** as shown in Figure 11.4. The turn angle is determined by joining a point and its image to the turn center, so $x°$ is the counterclockwise angle that rotates $\overline{OA}$ onto $\overline{OA'}$.

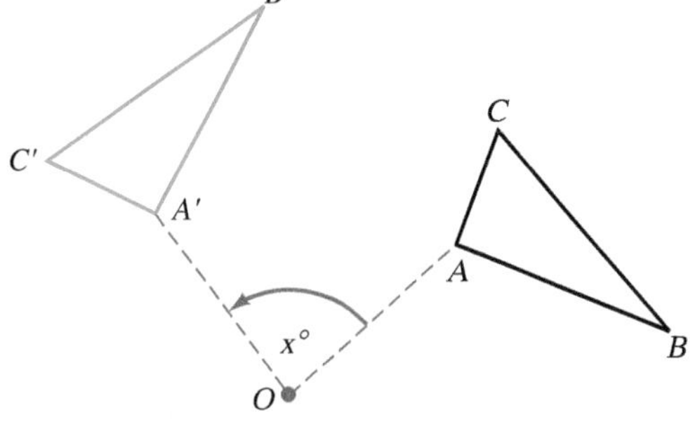

Figure 11.4
The rotation about center O by $x°$ can be indicated by a turn arrow

Counterclockwise turn angles are assigned positive degree measures, whereas negative measures indicate that the rotation is clockwise. In this way, a $-120°$ turn is equivalent to a $240°$ turn about the same center.

EXAMPLE 11.2

Finding Images Under Rotations

Find the image of each figure under the indicated turn.

(a) $90°$ rotation about P **(b)** $180°$ rotation about Q **(c)** $-90°$ rotation about R

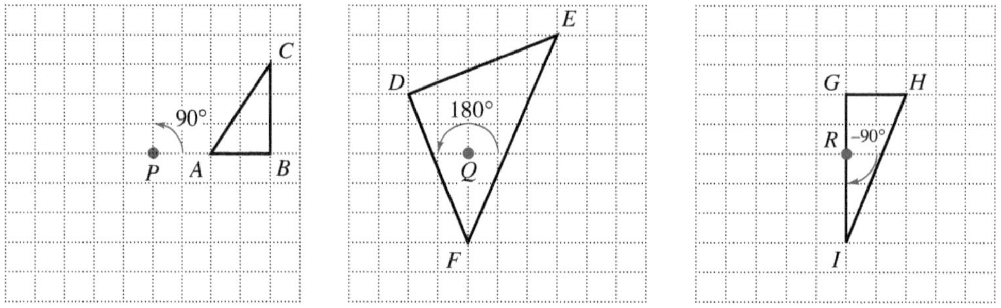

Solution

(a) **(b)** **(c)**

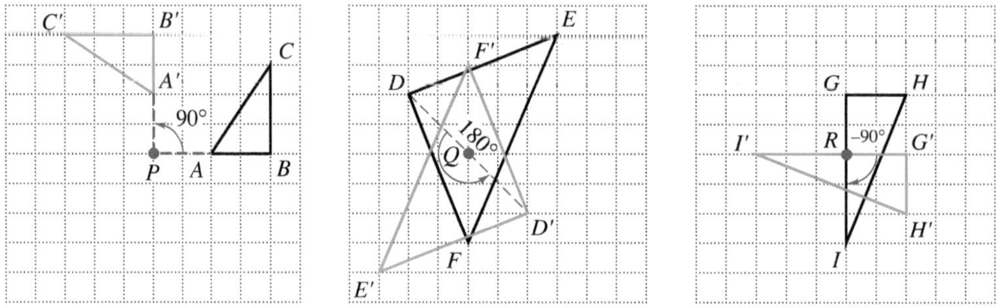

Reflections

The third basic rigid motion is a **reflection,** which is also called a **flip** or a **mirror reflection.** A reflection is determined by a line in the plane called the **line of reflection** or the **mirror line**. Each point P of the plane is transformed to the point P' on the opposite side of the mirror line m and at the same distance from m, as shown in Figure 11.5. Note that P' is located so that m is the perpendicular bisector of $\overline{PP'}$. Every point Q on m is transformed into itself; that is, $Q' = Q$ if Q is any point on m. Observe that a right hand is reflected to a left hand, and vice versa.

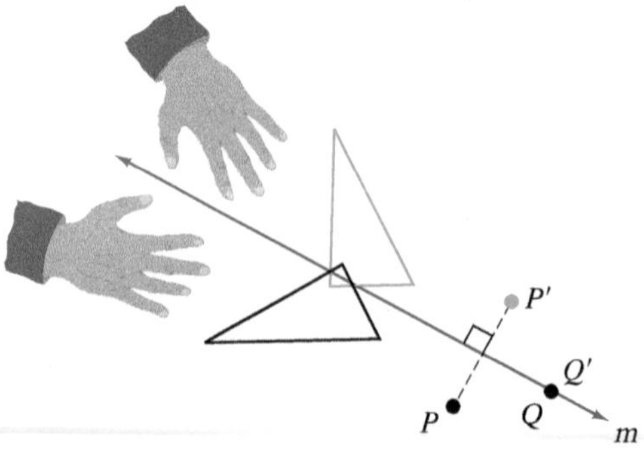

Figure 11.5
A flip, or reflection, about a line m transforms each point of the plane to its mirror image on the opposite side of m

Reflections can be performed with a trace-and-flip procedure by using an acetate transparency. First, the original figure is traced, including the line of reflection and a reference point (such as Q in Figure 11.5). The transparency is then turned over to perform the flip, and the points along the line of reflection are placed over their original position. The alignment of the reference point ensures that no sliding along the reflection line occurs. Each point along the mirror line is mapped to itself.

Reflections can also be carried out through paper folding, where the fold line serves as the mirror line. For example, draw a figure with ink (or drop a splotch of ink or poster paint) on one-half of a piece of paper, and then, while the ink or paint is still wet, fold the other half over to cover the figure. If tracing paper (or patty paper) is available, first draw an image on the outside of one-half of a folded sheet. Then turn the folded sheet upside down and trace the figure as seen through the two folded layers.

The Mira is ideally suited to draw reflections. As shown in Figure 11.6, the plastic surface of a Mira both reflects a figure in front and still allows points behind the surface to be seen. This makes it simple to draw the reflected image of a given figure, with the bottom edge of the Mira acting as the line of reflection.

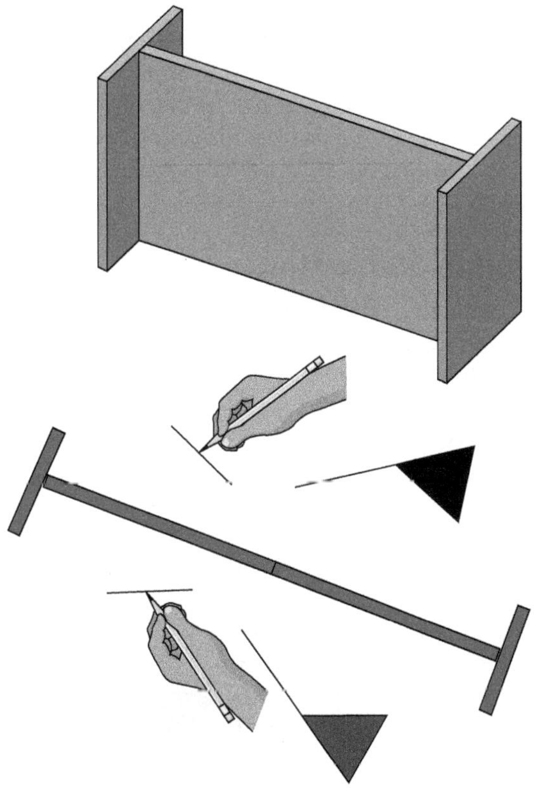

Figure 11.6
The Mira on the top can be used to draw reflections, as shown on the bottom

EXAMPLE 11.3

Finding Images Under Reflections

Sketch the image of the flag under a flip across line *m*.

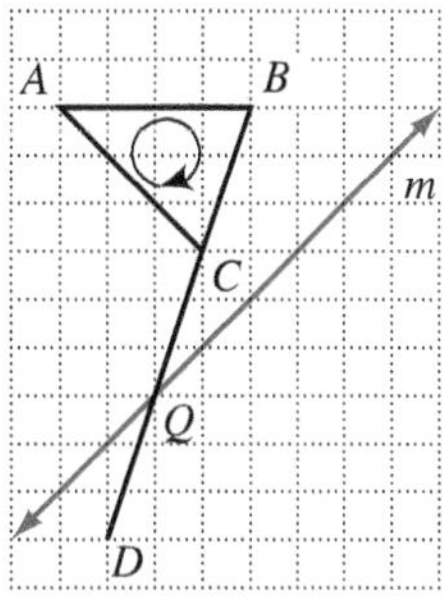

Solution

We can follow the trace-and-flip method if an acetate sheet or tracing paper is available, or we can use a Mira or paper folding. Alternatively, the point A' that is the mirror point of A across line m can be plotted on the square grid. Similarly, B', C', and so on can be plotted, until the entire image can be sketched accurately.

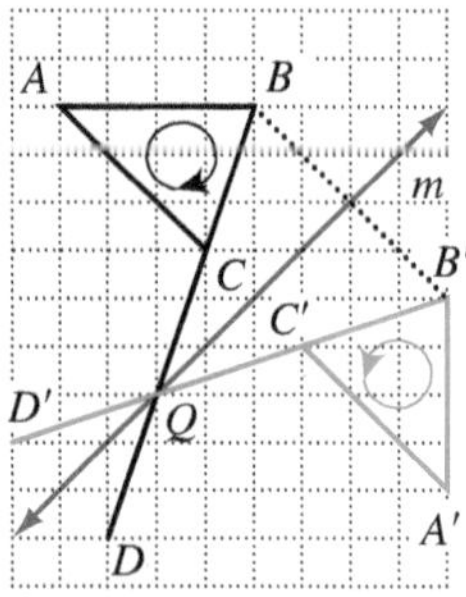

It's important to notice that a reflection reverses left-handed and right-handed orientations. For example, the left-pointing flag in Example 11.3 is transformed into a right-pointing flag, and the clockwise-pointing arrow on the circle that follows A, B, C in that order becomes a counterclockwise-pointing arrow in the image as it follows A', B', C' in order. A rigid motion that interchanges "handedness" is called **orientation reversing.** Thus, a reflection is orientation reversing. Translations and rotations, since they do not reverse handedness, are **orientation-preserving** transformations.

Glide–Reflections

The fourth, and last, basic rigid motion is the **glide–reflection.** As the name suggests, a glide–reflection combines both a slide and a reflection. The example most easily recalled is the motion that carries a left footprint into a right footprint, as depicted in Figure 11.7. It is required that the line of reflection, called the **glide mirror,** be parallel to the direction of the slide. The slide is usually called a **glide,** and its vector is called the **glide arrow** or **glide vector.** In Figure 11.7, the slide came before the reflection, but if the reflection had preceded the slide, the net outcome would have been the same. Note that the image of the left footprint under the glide–reflection is the blue right footprint. The green footprint is only an intermediate step used in completing the transformation.

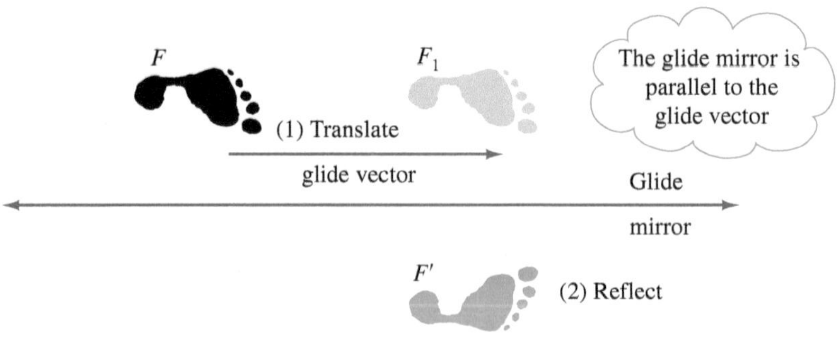

Figure 11.7
A glide–reflection combines (1) a slide and (2) a reflection, where the line of reflection is parallel to the direction of the slide

A glide–reflection changes handedness, so it is an orientation-reversing rigid motion. This property is due to the reflection part of the motion.

To determine a glide–reflection, it is useful to observe that the midpoint M of the segment $\overline{PP'}$ lies on the glide mirror. This is seen by noticing that triangle PP_1P' has the same shape but twice the size of triangle PNM. (See Figure 11.8.) This information is the key to solving the problem in the next example.

Figure 11.8
If points P and P' correspond under a glide–reflection, then the midpoint M of PP' lies on the glide mirror

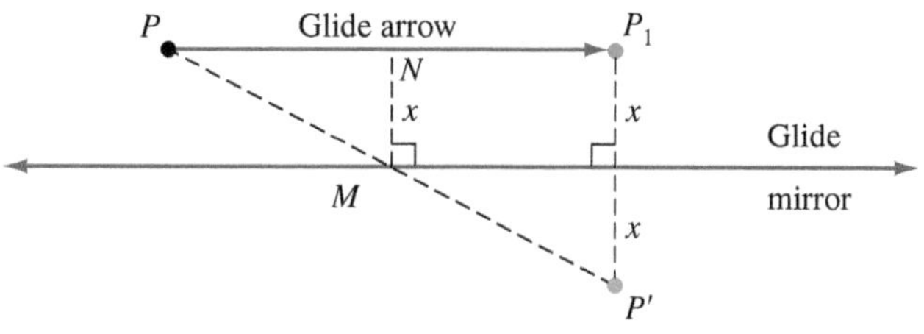

EXAMPLE 11.4

Determining a Glide–Reflection

A glide–reflection has taken points A and B of triangle ABC to the points A' and B', as shown in the accompanying grid. Find the glide mirror and the glide arrow of the glide–reflection, and then sketch the image triangle $A'B'C'$ under the glide–reflection.

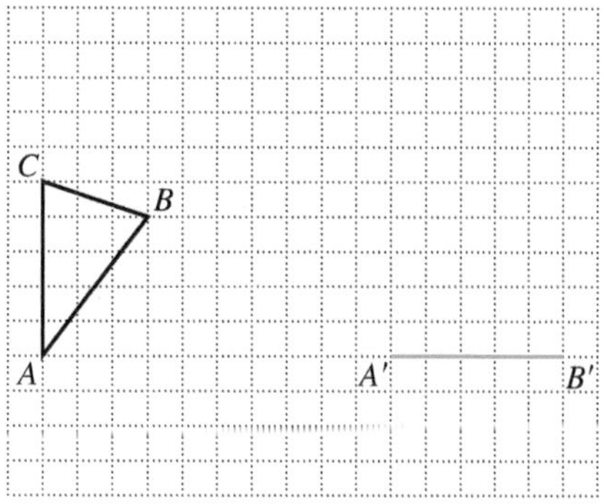

Solution

The square grid makes it easy to draw the respective midpoints M and N of the line segments $\overline{AA'}$ and $\overline{BB'}$. Since both M and N lie on the mirror line, $m = \overleftrightarrow{MN}$ is the glide mirror. Reflecting A' across m determines the point A_1, and the glide arrow is drawn by connecting A to A_1. The glide arrow has components eight units to the right and four units up, which allows us to find C_1. Reflecting C_1 across the glide mirror locates C'; therefore, $\triangle A'B'C'$ can be completed.

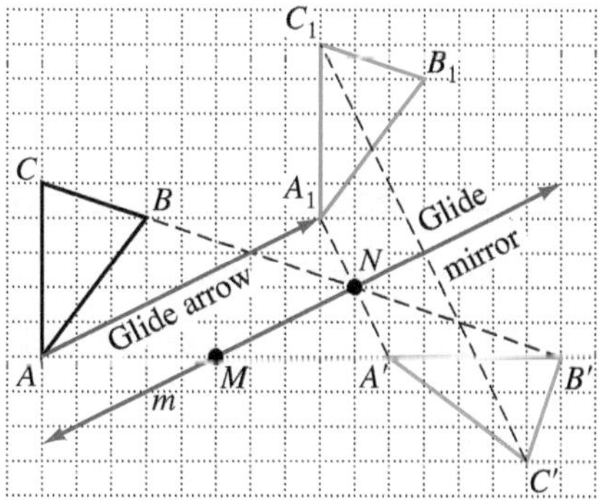

Table 11.1 on the next page summarizes useful information about the four basic rigid motions.

The Net Outcome of Two Successive Reflections

Recall that any two rigid motions that have the same net outcome are called **equivalent.** For example, rotations of $-120°$ and $+240°$ about the same center O are equivalent. Similarly, the motion consisting of two consecutive $180°$ rotations about a point O is equivalent to the **identity transformation,** which is the rigid "motion" that leaves all points of the plane fixed.

Two consecutive reflections across the same line of reflection also bring each point back to its original position, so such a "double flip" is also equivalent to the identity transformation. Suppose, however, that two flips are taken in succession across two *different* lines of reflection—say, first over m_1 and next over m_2. There are two cases to consider: m_1 and m_2 are parallel, and m_1 and m_2 intersect.

TABLE 11.1 THE FOUR BASIC RIGID MOTIONS

Name (alternate name) and Sketch	Information Needed	Description	Orientation Property
Translation (slide) 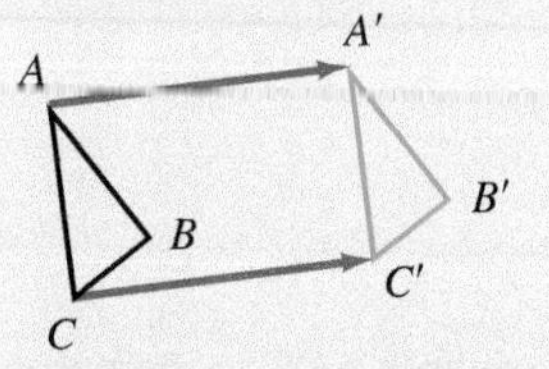	Slide arrow, indicating distance and direction	Every point of the plane is moved the same distance in the same direction.	Orientation is preserved.
Rotation (turn) 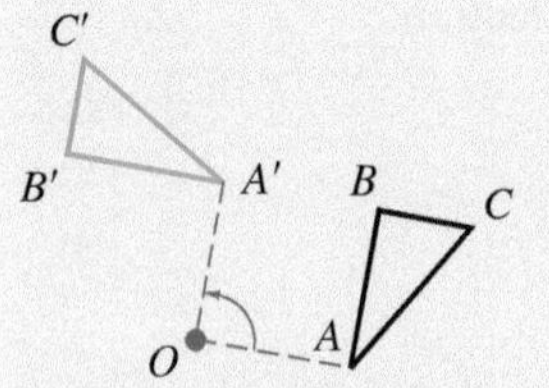	Turn center and turn angle	Every point of the plane is rotated through the same directed angle about the turn center.	Orientation is preserved.
Reflection (flip) 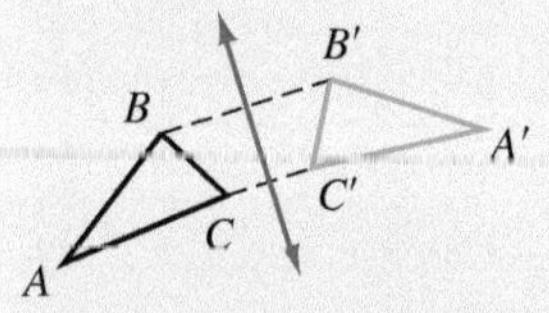	Line of reflection (mirror line)	Every point of the plane not on the mirror line is moved to its mirror image on the opposite side of the line of reflection. Points on the mirror line are fixed.	Orientation is reversed.
Glide–reflection (glide) 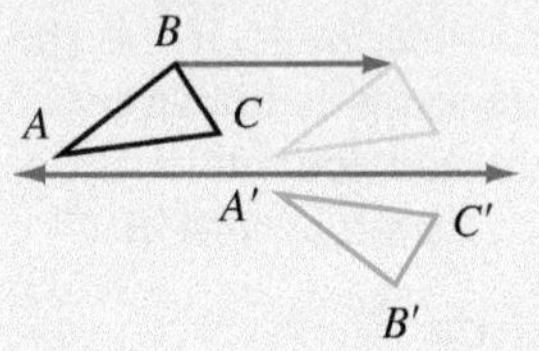	Glide arrow and a glide mirror parallel to the glide direction	Every point of the plane is moved by the same translation (glide) and reflected across the same line (glide mirror) parallel to the glide direction.	Orientation is reversed.

EXAMPLE 11.5

Exploring Consecutive Reflections Across Parallel Lines

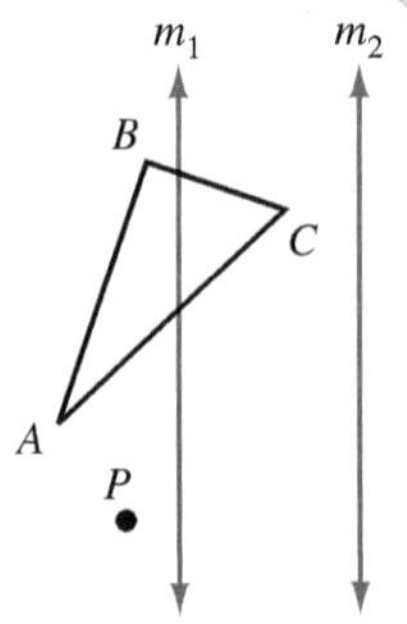

Let m_1 and m_2 be parallel lines of reflection.

(a) Sketch the images of $\triangle ABC$ and point P under the reflection across m_1; let the images be labeled $\triangle A_1B_1C_1$ and P_1, respectively.
(b) Sketch the images of $\triangle A_1B_1C_1$ and P_1 under reflection across line m_2; let the images be labeled $\triangle A'B'C'$ and P', respectively.
(c) Describe the net outcome of the rigid motion consisting of the two successive reflections, first across m_1 and next across m_2.

Solution

(a) and **(b)** Each reflection can be drawn with a Mira, by the trace-and-flip or paper-folding method, or, easiest of all, by using geometry software. Whatever method is used will result in the images that follow.

(c) If d is the directed distance from line m_1 to line m_2, then point P is moved a distance $2d$ in the direction perpendicular to m_1 and m_2 and pointing from m_1 toward m_2. In fact, *all* points of the plane are moved in this direction through the same distance $2d$, so the net outcome of two successive reflections across the pair of parallel lines m_1 and m_2 is equivalent to the translation shown on the right side of the figure that follows.

(a)
Two reflections across parallel lines

(b)
The equivalent translation

A similar investigation can be carried out for the motion consisting of two successive reflections across lines m_1 and m_2 that intersect at a point O. Most people find the result very surprising: *The net outcome of the two reflections across intersecting lines is equivalent to a rotation about the point O of intersection of m_1 and m_2. The angle of rotation has twice the measure of the directed angle that turns line m_1 onto line m_2.* This motion is illustrated in Figure 11.9.

Figure 11.9
Two reflections across intersecting lines are equivalent to a rotation about the point of intersection of the two lines

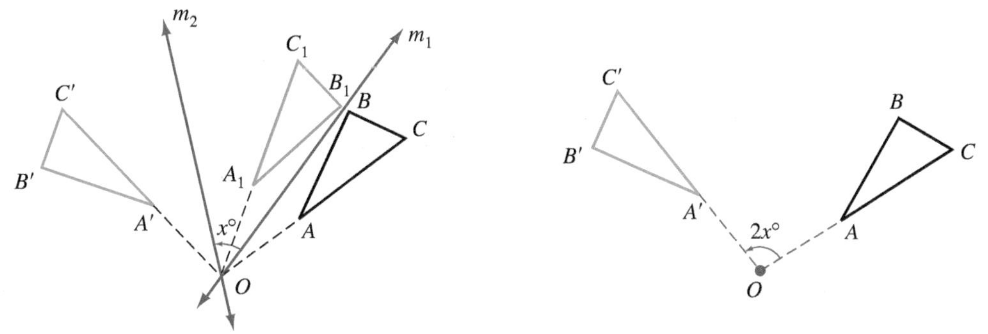

(a)

(b)

When two reflections are performed, the orientation is reversed twice, so the final image has the *same* orientation as the initial image. The following theorem summarizes the two possible net outcomes of a pair of successive reflections:

THEOREM The Net Outcome of Two Reflections in Distinct Lines

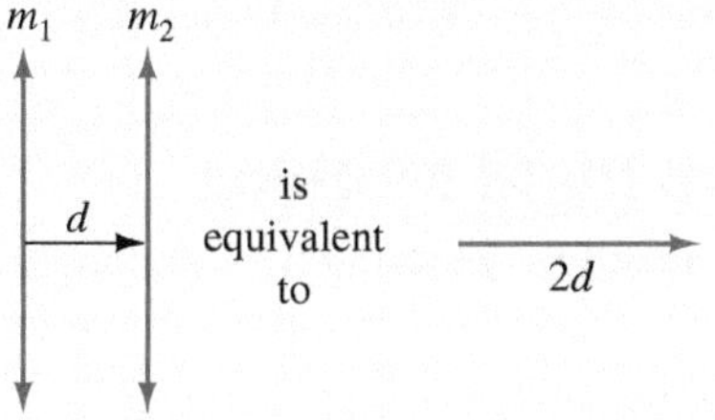

- The net outcome of reflections across two parallel lines is equivalent to a translation perpendicular to the lines and twice the directed distance from the first line to the second line.

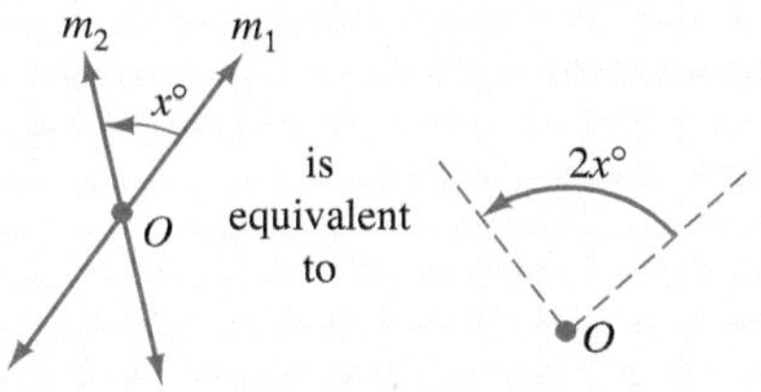

- The net outcome of reflections across two intersecting lines m_1 and m_2 is equivalent to a rotation about their point of intersection through an angle twice the directed angle from the first line to the second.

The Net Outcome of Three Successive Reflections

Three lines can be arranged in several ways in the plane. For example, the lines m_1, m_2, and m_3 in Figure 11.10a are parallel to one another. The successive images of $\triangle ABC$ across the lines are shown, with $\triangle A'B'C'$ the image at the completion of all three reflections. In Figure 11.10b, we see that $\triangle ABC$ can be taken to $\triangle A'B'C'$ by a *single* reflection over the line l. Line l is the image of line m_1 under the translation that takes m_2 to m_3. Thus, three successive reflections across parallel lines are equivalent to a single reflection.

Figure 11.10
Three reflections across parallel lines m_1, m_2, and m_3, are equivalent to a single reflection across line l

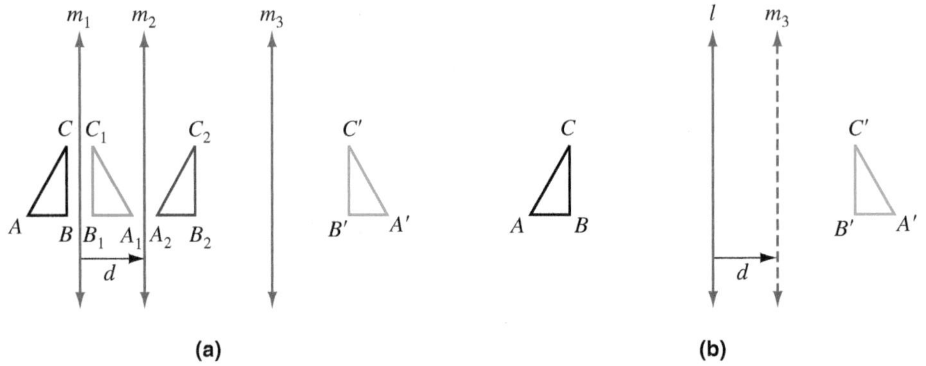

Three reflections across concurrent lines m_1, m_2, and m_3 can also be discovered to be equivalent to a single reflection across a certain line l that passes through the point O of concurrence. (See problem 31 in Problem Set 11.1.) In all other cases, where the three lines are neither parallel nor concurrent, it can be shown that the net outcome of three successive reflections is equivalent to a glide–reflection. (See problem 32 in Problem Set 11.1.)

In sum, we have the following theorem:

THEOREM The Net Outcome of Three Reflections

The net outcome of three successive reflections across lines m_1, m_2, and m_3 is equivalent to either

- a reflection, if m_1, m_2, and m_3 are parallel or concurrent;

or

- a glide–reflection, if m_1, m_2, and m_3 are neither parallel nor concurrent.

Classification of General Rigid Motions

Any rigid motion is uniquely determined by knowing how a triangle ABC is mapped onto its image triangle $A'B'C'$. Figure 11.11 shows this transformation can be accomplished with no more than three reflections. The first reflection is taken across the perpendicular bisector m_1 of $\overline{AA'}$, so point A is mapped to point A'. If this reflection happens to also take B to B' and C to C', our rigid motion is a single reflection. Otherwise, suppose that B is taken to the point $B_1 \neq B'$; then a second reflection across the perpendicular bisector m_2 of $\overline{B_1B'}$ maps B_1 onto B'. Since $A'B_1 = A'B'$, we see that A' is a point along m_2, so A' remains fixed by reflection across m_2. If it happens that this second reflection maps C_1 to C', then our rigid motion is equivalent to two successive reflections, first across m_1 and then across m_2. But if instead C_1 is mapped to a point $C_2 \neq C'$, then a third reflection across the perpendicular bisector m_3 of $\overline{C_2C'}$ is required. Both A' and B' are points of m_3, so this reflection leaves each of these points fixed.

Figure 11.11
Any rigid motion is equivalent to a sequence of at most three reflections

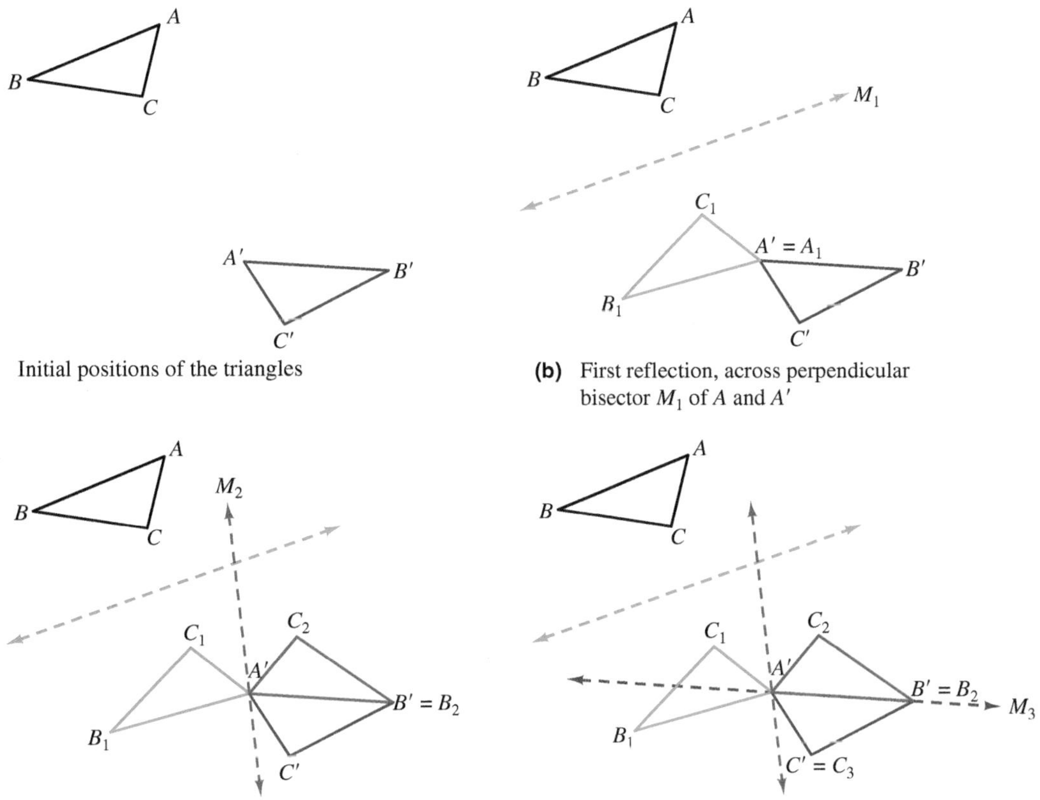

(a) Initial positions of the triangles

(b) First reflection, across perpendicular bisector M_1 of A and A'

(c) Second reflection, across perpendicular bisector M_2 of B_1 and B'

(d) Third reflection, across perpendicular bisector M_3 of C_2 and C'

Since two reflections are equivalent to a translation or rotation, and three reflections are equivalent to a single reflection or a glide–reflection, we have proved the following remarkable theorem:

THEOREM Classification of General Rigid Motions
Any rigid motion of the plane is equivalent to one of the four basic rigid motions: a translation, a rotation, a reflection, or a glide–reflection.

If a rigid motion is described or shown to us, a natural question to ask is, "What is the basic type of the transformation?" A good initial step in such a classification is to first observe whether the motion preserves or reverses orientation, since this characteristic is often easily determined:

- If orientation is preserved, then the transformation must be either a translation or a rotation. A translation is easy to detect, since the image figure will be parallel to and face in the same direction as its preimage. A rotation will change the direction of the figure. An image figure that is parallel to its preimage, but "upside down," is related to the original by a 180° rotation.

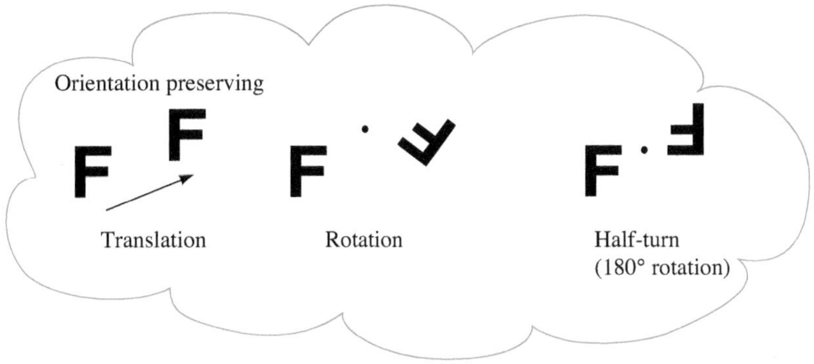

- If orientation is reversed, the rigid transformation is a reflection or a glide–reflection. Reflections are quickly identified since the image figure will be a mirror reflection of its preimage and the mirror line will be halfway between the image and its preimage. If the orientation-reversing transformation has no mirror line, it is necessarily a glide–reflection.

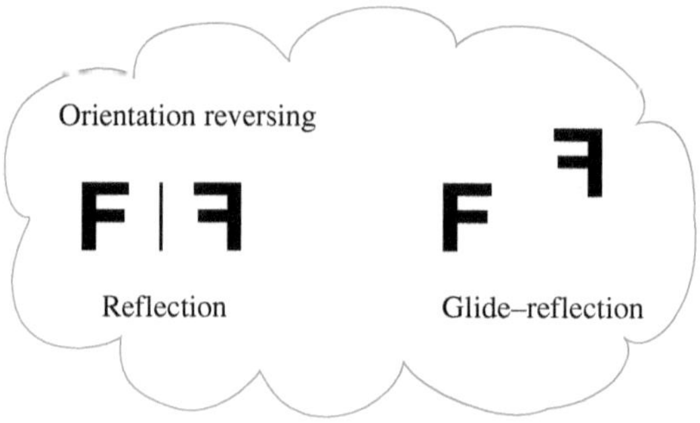

Rigid motions have many applications in geometry. For example, the informal definition of congruence to mean "same size and shape" can now be made precise:

DEFINITION Congruent Figures

Two figures are **congruent** if and only if one figure is the image of the other under a rigid motion.

EXAMPLE 11.6

Classifying Rigid Motions

Describe the type of rigid motion, if any, that maps the dog X to the dogs $A, B, \ldots, F$. Give a reason for each choice of a rigid motion or why no rigid motion exists that maps dog X to the other dogs.

Solution

(a) Translation (orientation preserved and not a rotation)
(b) Reflection (over vertical line)
(c) Glide reflection (orientation reversed; dog C is a translation of dog B)
(d) Not a rigid motion (dog D is smaller than dog X)
(e) Rotation (since orientation is preserved and directions are changed)
(f) Not a rigid motion (dog F is larger than dog X)

Dilations and Similarity Motions

Any figure mapped by a rigid motion is unchanged in both size and shape. Suppose, however, we wish to find transformations of the plane that preserve shape but change the size of figures. For example, now can we describe a transformation that maps dog X of Example 11.16 to the smaller dog D or to the larger dog F? Transformations that change size but not shape are quite common: Consider making an enlargement or reduction on a photocopy machine or using the Zoom command from the View menu of a computer program.

The simplest transformation to change the size of a figure, but preserve its shape and orientation, is a **dilation** (also called a **size transformation**). A point O is chosen as the center, and the points of the plane are all moved toward ($k < 1$) or away from ($k > 1$) the center O by the same proportional factor k. More precisely, we have the following definition:

DEFINITION Dilation, or Size Transformation

Let O be a point in the plane and k a positive real number. A **dilation,** or **size transformation,** with **center O** and **scale factor k** is the transformation that takes each point $P \neq O$ of the plane to the point P' on the ray $\overrightarrow{OP}$ for which $OP' = k \cdot OP$ and takes the point O to itself.

Two examples of dilations are shown in Figure 11.12.

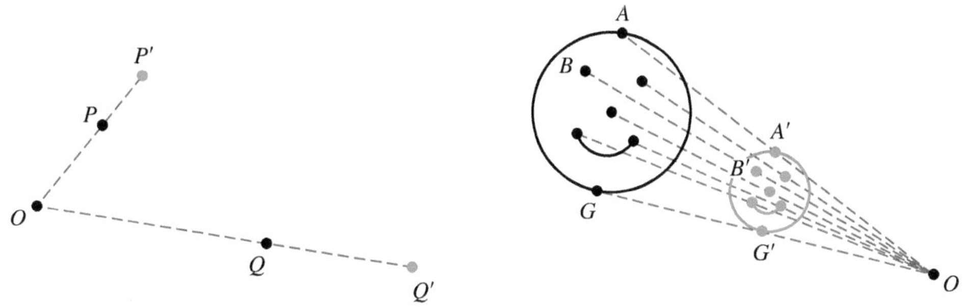

(a) $\frac{OP'}{OP} = \frac{OQ'}{OQ} = k, k > 1$

(b) $\frac{OA'}{OA} = \frac{OB'}{OB} = \cdots = \frac{OG'}{OG} = k, k < 1$

Figure 11.12
Two dilations, or size transformations

When the scale factor k is larger than 1, the image of a figure is larger than the original and the dilation is an **expansion.** If $k < 1$, the dilation is a **contraction.** If $k = 1$, then all points are left unmoved—that is, $P = P'$ for all P—and the dilation is the identity transformation.

The most important fact about dilations is contained in the next theorem. Recall that if a point P is moved to the image point P', we say that P is the preimage of P'.

THEOREM Distance Change Under a Dilation

Under a dilation with scale factor k, the distance between any two image points is k times the distance between their preimages. That is, for all points P and Q, $P'Q' = k \cdot PQ$.

Dilations change size but, like translations, produce images in which corresponding line segments are parallel. This property allows you to determine both the center and the scale factor of the dilation. Simply draw some lines through pairs of corresponding points in the image and preimage figures; the lines will intersect at the center O of the dilation. The scale factor k is then given by the ratio of distances OP' to OP to a pair of corresponding points P' and P. That is, $k = \frac{OP'}{OP}$.

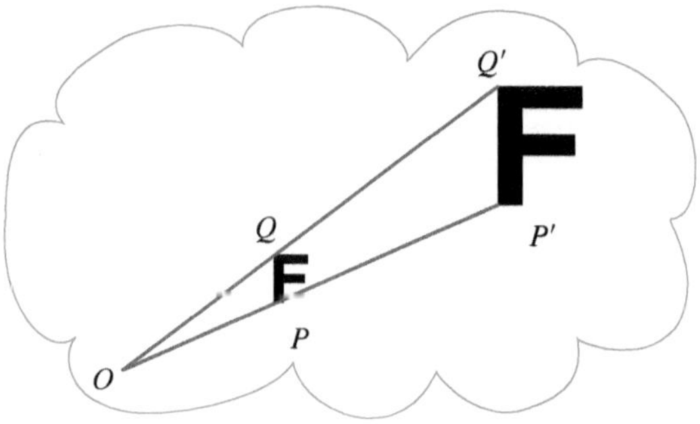

A sequence of dilations and rigid motions performed in succession is called a **similarity transformation.**

DEFINITION Similarity Transformation
A transformation is a **similarity transformation** if and only if it is a sequence of dilations and rigid motions.

We can now give a precise definition of similarity of figures.

DEFINITION Similar Figures
Two figures F and G are **similar,** written $F \sim G$, if and only if there is a similarity transformation that takes one figure onto the other figure.

The fish F and G in Figure 11.13 are similar to one another. A dilation centered at point O maps F to F', and a reflection maps F' to G.

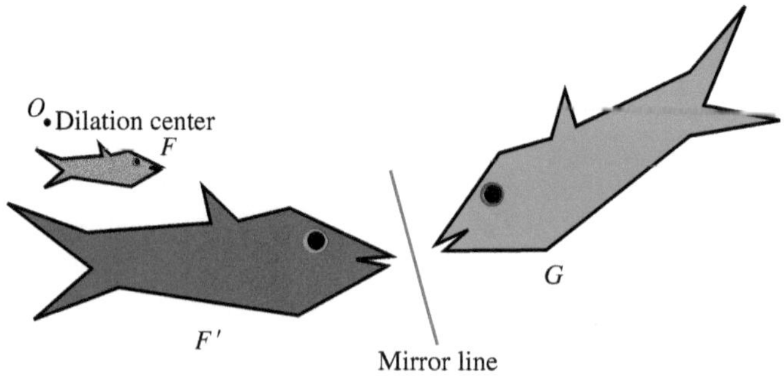

Figure 11.13
A dilation centered at O and followed by a reflection defines a similarity transformation taking figure F onto figure G. Therefore, figures F and G are similar

EXAMPLE 11.7

Verifying Similarity

Show that the small letter F and the larger letter F are similar by describing a similarity transformation that takes the smaller figure onto the larger one.

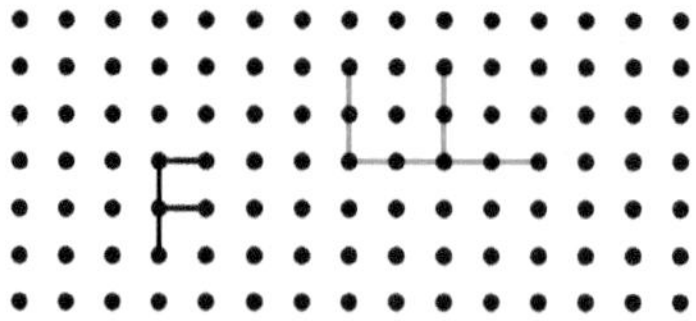

Solution

We need a sequence of dilations and rigid motions that rotates, stretches, and positions the smaller letter onto the larger. This can be done in three steps: (a) rotate 90°, (b) dilate with scale factor $k = 2$, and (c) translate.

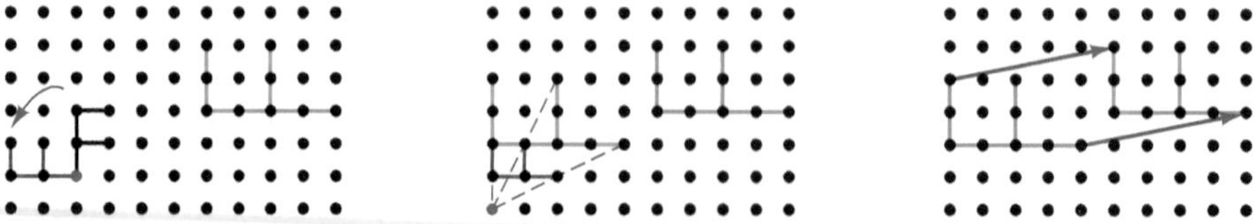

(a) Rotate 90° **(b)** Dilate **(c)** Translate

These three steps are not unique. For example, a translation could have been taken first, then a 90° rotation, and finally a dilation with a properly chosen center. Showing that *some* similarity transformation takes one figure onto the other is all that is required.

11.1 Problem Set

Understanding Concepts

1. Which of the following "transformations" correspond to a rigid motion? Explain the reasoning you have used to give your answer.

(a) A deck of cards is shuffled.

(b) A completed jigsaw puzzle is taken apart and then put back together.

(c) A jigsaw puzzle is taken from the box, assembled, and then placed back in its box.

(d) A painting is moved to a new position on the same wall.

(e) Bread dough is allowed to rise.

2. Which of the following motions are rigid?

(a) A book is moved to a new shelf in a bookcase.

(b) A book is opened to page 34.

(c) A balloon is inflated.

(d) A bowling ball is rolled down the alley.

(e) A length of yarn is wound into a ball.

3. Find the image of the polygon shown under the translation that takes P onto P'.

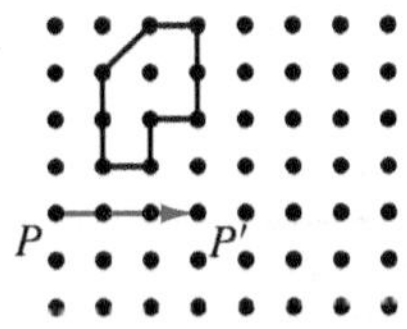

4. Find the image of the polygon shown under the translation that takes P onto P'.

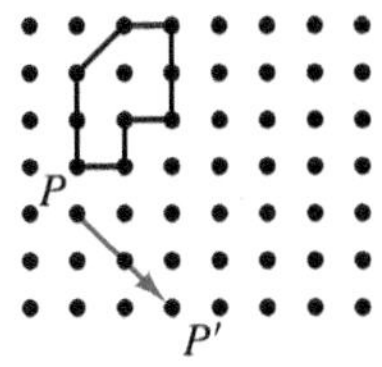

5. The translation that takes point P onto P' transforms triangle ABC (not shown) onto its image $A'B'C'$ (shown in the accompanying diagram).

(a) Draw triangle ABC.

(b) Describe the rigid motion that transforms $\triangle A'B'C'$ onto $\triangle ABC$, and compare the motion with the translation that takes $\triangle ABC$ onto $\triangle A'B'C'$.

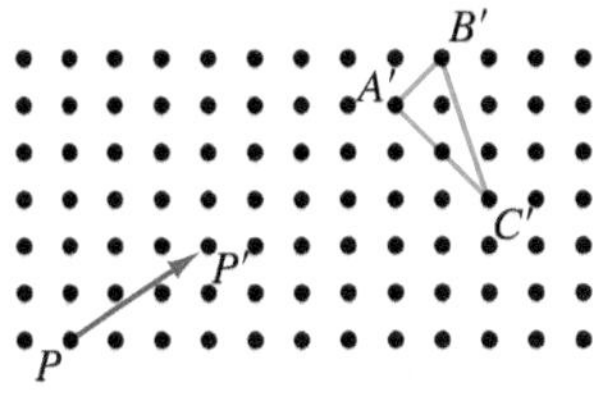

6. In each of the following statements, give an answer between 0° and 360°:

(a) A clockwise rotation of 60° is equivalent to a counterclockwise rotation of ___________.

(b) A clockwise rotation of 433° is equivalent to a clockwise rotation of ___________.

(c) A clockwise rotation of 3643° is equivalent to a clockwise rotation of ___________.

(d) A sequence of two consecutive clockwise rotations, first of 280° and next of 120°, about the same center is equivalent to a single clockwise rotation of ___________.

(e) A rotation of −260° is equivalent to a rotation of ___________.

7. Sketch the image of $\triangle ABC$ under a 90° counterclockwise rotation about O.

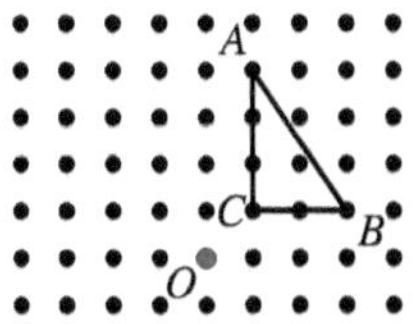

8. Sketch the image of $\triangle ABC$ under a rotation of 180° about P.

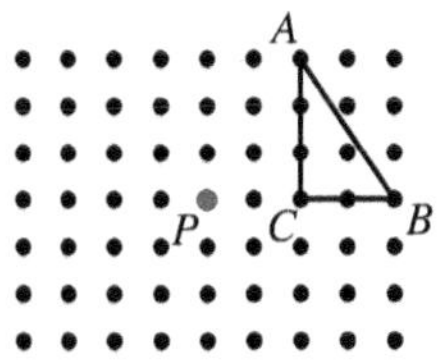

9. A rotation sends A to A' and B to B', as follows:

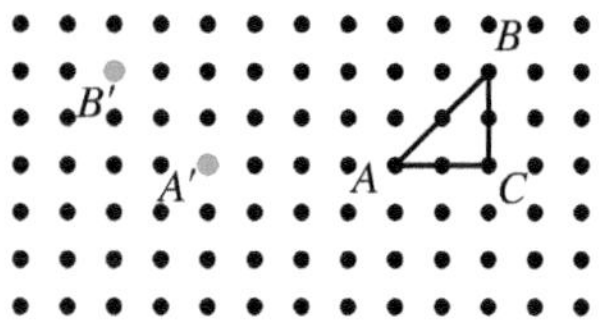

(a) Find the center of the rotation.

(b) Find the turn angle.

(c) Sketch the image triangle $A'B'C'$.

10. The equilateral triangle shown has center O and is pointing upward from its horizontal base. Describe *all* nonequivalent rigid motions that move the triangle to point downward and leave point O fixed.

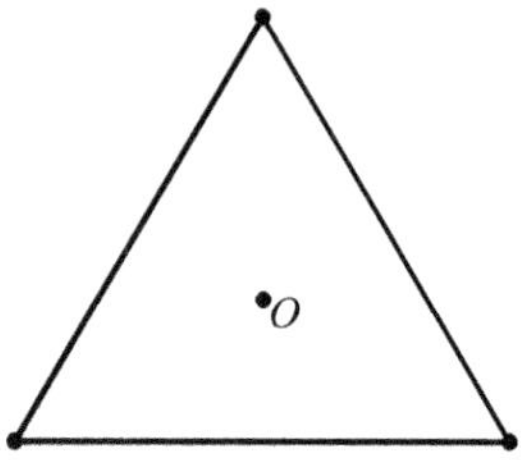

11. A point P that is mapped onto itself by a transformation such that $P = P'$ is called a **fixed point.** What can be said about the type of rigid motion that

(a) has no fixed points?

(b) has exactly one fixed point?

(c) has at least two fixed points and a nonfixed point?

(d) has only fixed points.

12. Trace the given figure, which shows $\triangle ABC$ and its image under a rotation. Use any drawing tools you wish (Mira, compass, or straightedge) to construct the center of rotation. (*Hint:* Why is the center of rotation on the perpendicular bisector of $\overline{AA'}$?)

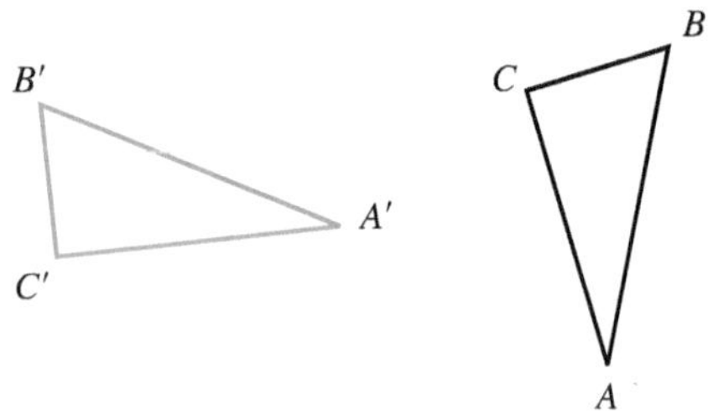

13. Redraw the given figure on graph paper. Then sketch the reflection of $\triangle ABC$ across the mirror line m.

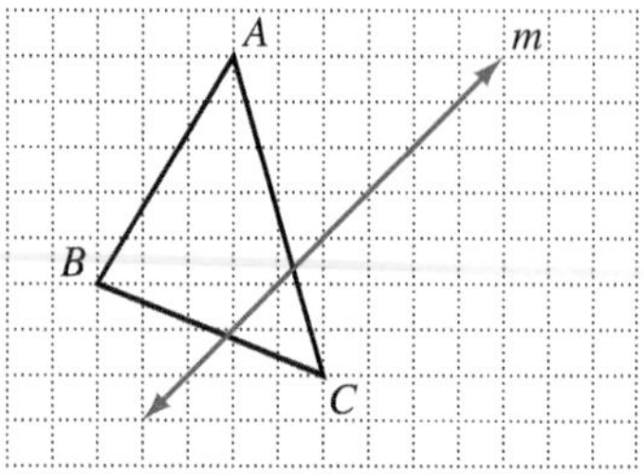

14. In the image that follows, a reflection sends P to P'.

(a) Find the line of reflection.

(b) Find the image of the polygon $PQRST$ under the reflection that takes P to P'.

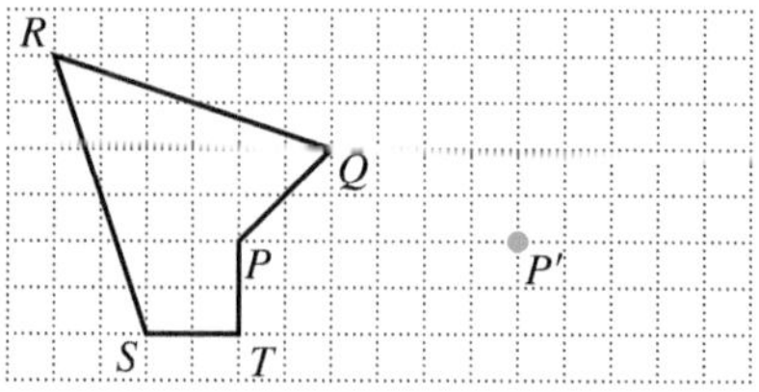

15. (a) A reflection across line m leaves point A fixed, so that $A' = A$. What can be said about A's relationship to m?

(b) A reflection across line m leaves two points A and B fixed, so that $A' = A$ and $B' = B$. What can you say about A, B, and m?

(c) A reflection takes point C to point D. Where does the reflection take point D?

16. A glide–reflection is defined by the glide arrow and glide mirror m shown. Draw the following images of the polygon $ABCDE$:

(a) The image $A_1B_1C_1D_1E_1$ under the slide

(b) The image $A'B'C'D'E'$ under the glide–reflection

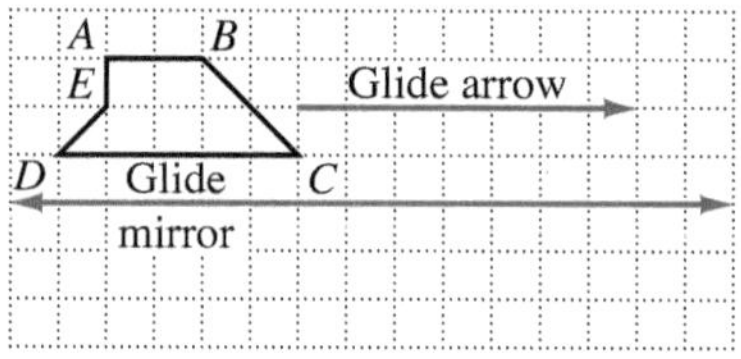

17. A glide–reflection has taken B to B' and E to E'. Find

(a) the glide mirror.

(b) the glide arrow.

(c) the image $A'B'C'D'E'$ of the polygon $ABCDE$.

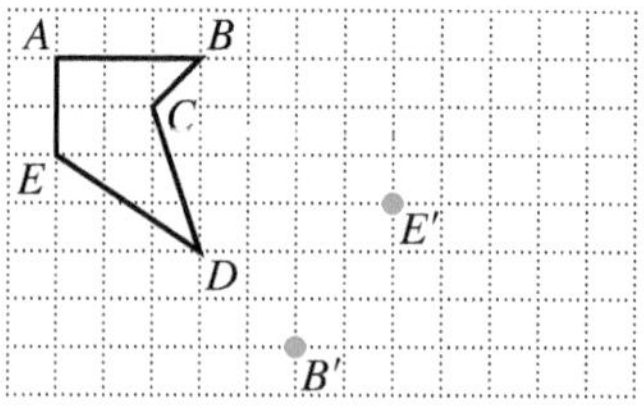

18. In each part that follows, draw a line m_2 so that the net outcome of successive reflections about m_1 and then m_2 is equivalent to the translation specified by the slide arrow shown.

(a)

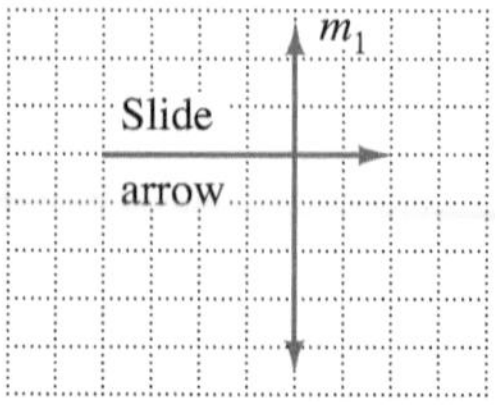

(b)

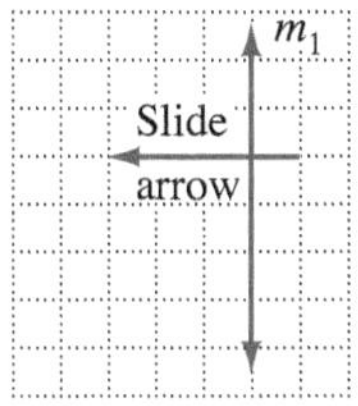

(c)

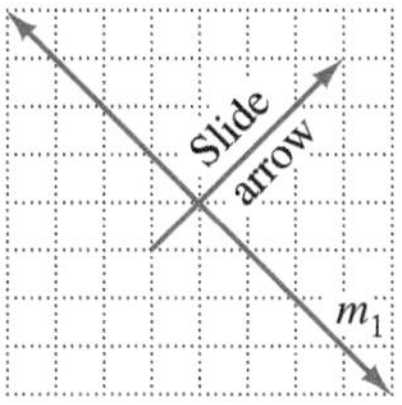

19. A reflection across line j followed by a reflection across line k as shown here is equivalent to the translation by four units to the right, as shown in the table that follows:

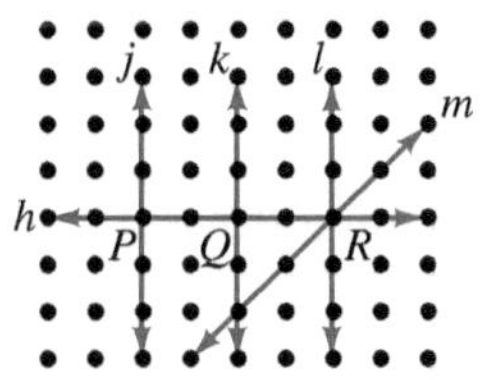

Fill in the missing entries (a) through (f) in the following table.

Reflection Lines

First	Second	Equivalent Transformation
j	k	Translate right four units
l	m	Rotate clockwise 90° around point R
j	l	(a)
k	(b)	Translate left four units
(c)	k	Translate left four units
h	m	(d)
m	(e)	Rotate 180° about point P
k	(f)	Identity transformation (all points fixed)

20. Use dot or graph paper to copy $\triangle ABC$ and points O and P. Then draw the image of $\triangle ABC$ for

(a) the dilation with center O and scale factor 2.

(b) the dilation with center P and scale factor $1/2$.

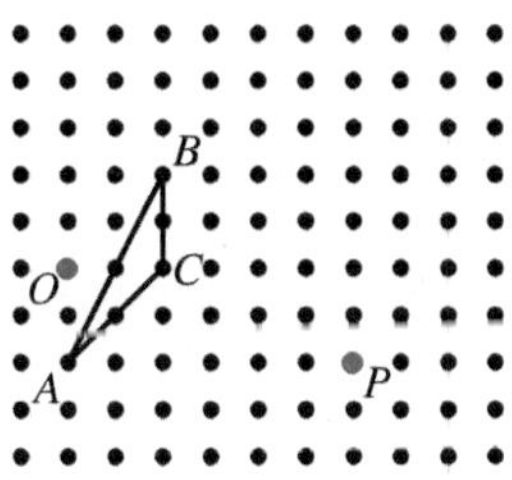

21. In the figure shown, $\triangle DEF$ is the image of $\triangle ABC$ under the dilation with center O and scale factor 2.

(a) Describe the size transformation that takes $\triangle DEF$ onto $\triangle GHI$. Show the center of the transformation on a sketch and give the scale factor.

(b) Describe the dilation that takes $\triangle ABC$ onto $\triangle GHI$ by locating the center and giving the scale factor.

(c) The Pythagorean theorem shows that $AB = \sqrt{5}$, so the perimeter of $\triangle ABC$ is $3 + \sqrt{5}$. Explain how to use the scale factors determined in parts (a) and (b) to obtain the perimeters of $\triangle DEF$ and $\triangle GHI$.

(d) A dilation with scale factor 4 takes $\triangle ABC$ onto $\triangle JKL$. What is the perimeter of $\triangle JKL$?

(e) Find the areas of the three triangles shown. Explain how the areas are related to the scale factors.

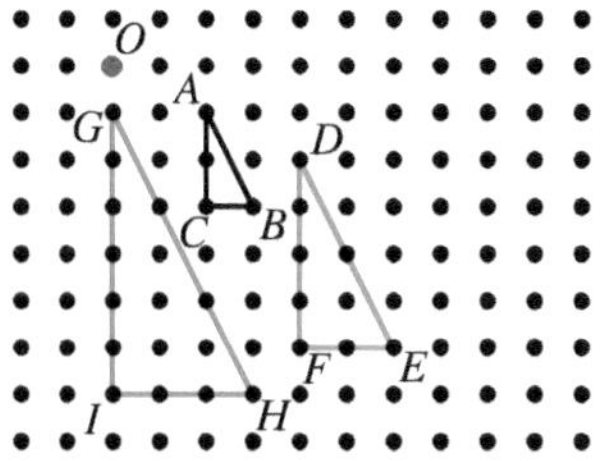

22. Sketch the image of $\triangle JKL$ under the similarity transformation composed of a dilation centered at point P with scale factor $2/3$ followed by a reflection across line m.

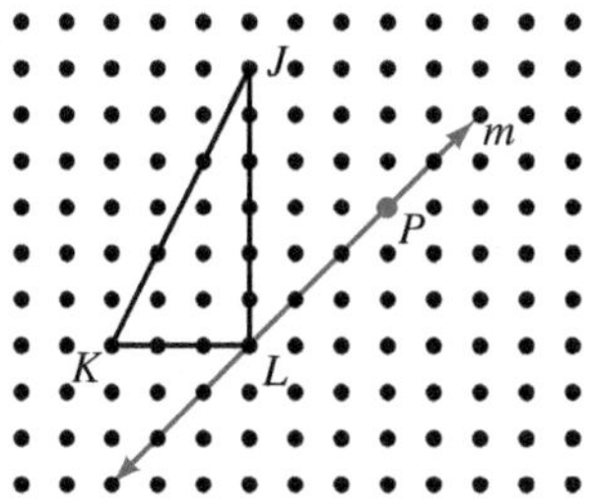

23. Describe a similarity transformation that takes quadrilateral $ABCD$ onto quadrilateral $A'B'C'D'$ as shown. Sketch the intermediate images of the dilation and rigid motions that compose the similarity transformation.

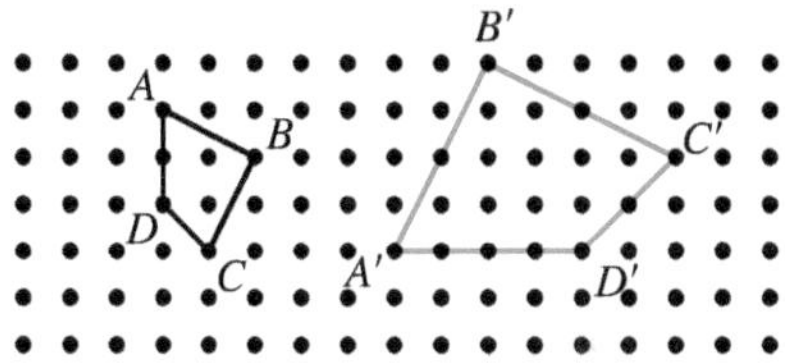

Into the Classroom

24. (**Writing**) Create an activity that explores rigid motions by expanding on the following idea adapted from John A. Van de Walle: Each student is given, in Van de Walle's terminology, a "Motion Man," as shown at the left in the accompanying drawing. The figure, on 10 × 10 centimeter dot paper, can be copied onto overhead transparency sheets (several at a time, then cut). Alternatively, the figure can be copied onto both sides of a sheet of paper, so that the figures on the opposite side and front sides will match when held up to the light. Students can now be asked to carry out transformations—for example, half- and quarter-turns, and flips over the lines—and observe how the transformed figure compares against its initial upright position with the flag in the right hand. The right-hand figure in the drawing shows a flip over the lower left to upper right diagonal. To include translations and glide–reflections, have students move Motion Man around a 20 × 20 sheet of centimeter dot paper.

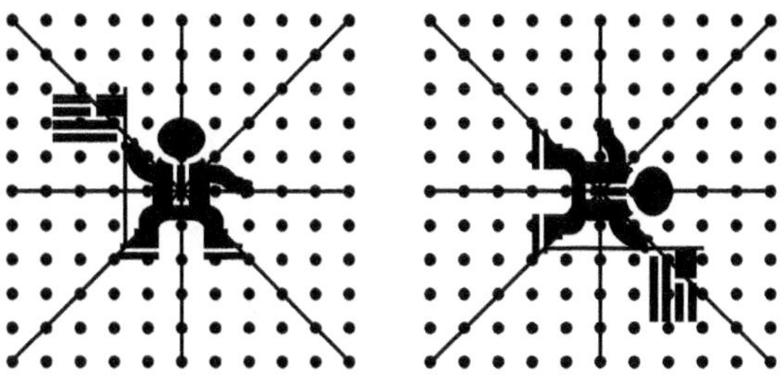

Responding to Students

25. Ephraim and DeVonte were asked to reflect the pattern-block figure shown on the left over the horizontal line through the center of the square. They rearranged the blocks to form the figure on the right.

(a) What type of rigid motion did Ephraim and DeVonte use?

(b) How would you help clear up their misunderstanding?

26. Lisa was asked to identify the type of rigid motions used to obtain the following pattern from the bent-arrow motif at the far left:

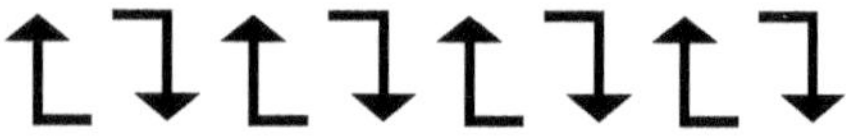

She answered, "rotation and glide–reflection."

(a) Is Lisa's answer correct?

(b) How would you help guide Lisa to the correct answer?

27. A student believes that all rectangles are similar, since all rectangles have four right angles and opposite sides have the same length. How would you respond to this student? In particular, how does the geometrical meaning of "similar" differ from its more common-use meaning?

Thinking Critically

28. A rigid motion takes points A, B, C, and P onto the respective image points A', B', C', and P', where $AB = 3$ cm, $AC = 4$ cm, $AP = 2$ cm, $BC = 2$ cm, $BP = 4$ cm, and $CP = 4$ cm. In each part that follows, use a compass and ruler to draw the smallest set of points that you know must contain the point P' when you start with

(a) only point A'. (*Hint:* The answer is a circle.)

(b) points A' and B'.

(c) points A', B', and C'.

29. (**Writing**) Two successive 90° rotations are taken, first about center O_1 and then about center O_2, where $O_1O_2 = 2$ cm. Describe the basic rigid motion that is equivalent to the successive rotations. Explain carefully, using words and sketches.

30. (**Writing**) Suppose that the 90° rotations in problem 29 are each replaced with a 120° rotation. What basic rigid motion is equivalent to the net outcome of the two rotations? Explain with words and sketches.

31. (a) Find the image $\triangle A'B'C'$ of $\triangle ABC$ under the rigid motion consisting of three consecutive reflections across the concurrent lines m_1, m_2, and m_3.

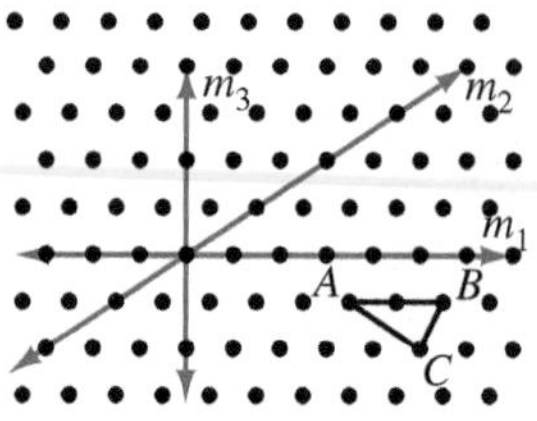

(b) Find a line l so that $\triangle ABC$ is taken onto $\triangle A'B'C'$ by one reflection across l.

32. (a) Find the image $\triangle A'B'C'$ of $\triangle ABC$ under the rigid motion consisting of three consecutive reflections across lines m_1, m_2, and m_3.

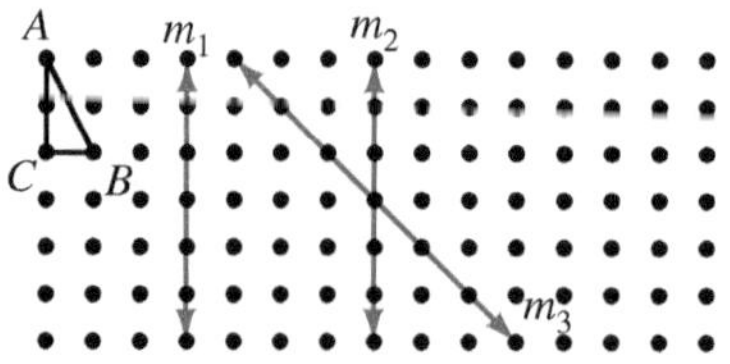

(b) Find the image of $\triangle ABC$ under the glide–reflection whose glide arrow extends from P to P' and whose glide mirror is line l.

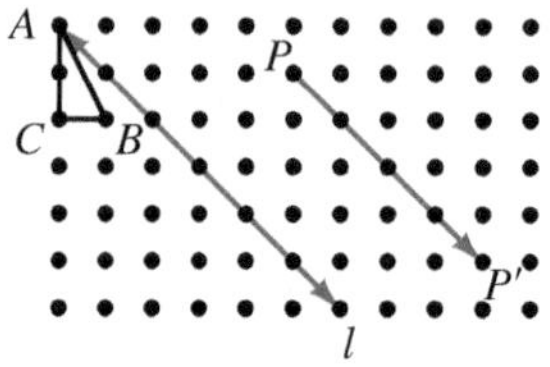

(c) What conclusion can you draw about the two rigid motions described in (a) and (b)?

33. Trace the following figure, where $\triangle ABC$ is congruent to $\triangle A'B'C'$:

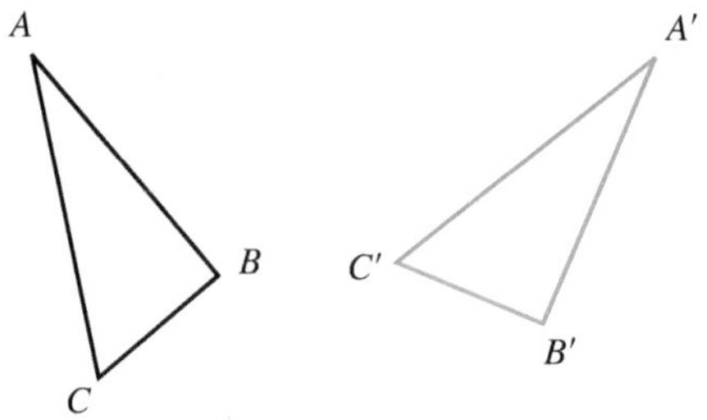

(a) Use a Mira (or other drawing tools) to draw the line m_1 across which A is reflected onto A'. Also, draw the images of B and C under this reflection and label them B_1 and C_1, respectively.

(b) Draw the line m_2 across which B_1 reflects onto B'. What is the image of C_1 across m_2?

(c) Use the lines m_1 and m_2 to describe the basic rigid motion that takes $\triangle ABC$ to $\triangle A'B'C'$.

34. Trace the following figure, where $\triangle ABC$ is congruent to $\triangle A'B'C'$:

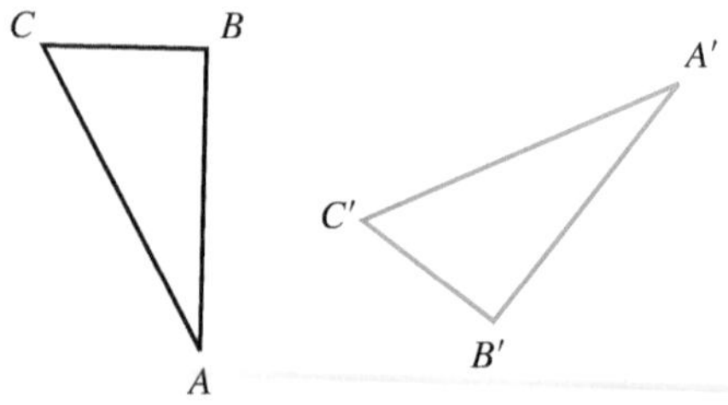

(a) Use a Mira (or other drawing tools) to draw three lines of reflection—m_1, m_2, and m_3—so that

(i) reflection across m_1 takes A onto A' (and B and C are taken to B_1 and C_1, respectively).

(ii) reflection across m_2 takes B_1 onto B' (and C_1 is taken onto C_2).

(iii) reflection across m_3 takes C_2 onto C'.

(b) Describe the type of basic rigid motion that takes $\triangle ABC$ onto $\triangle A'B'C'$.

35. In each of the parts that follow, a complicated sequence of rigid motions is described. Explain how you know what type of basic rigid motion is equivalent to the net outcome of the motion described.

(a) Reflections are taken across 6 lines, and no point is taken back onto its original position.

(b) Reflections are taken across 11 lines, and there are points that are taken back onto their original positions.

(c) Two different glide–reflections are taken in succession, with the net outcome taking some point back onto its original location.

36. Let C and D be two circles on the same side of a mirror. Construct a tangent line to circle C that, after reflection in the mirror, is also tangent to circle D. How many solutions can you find?

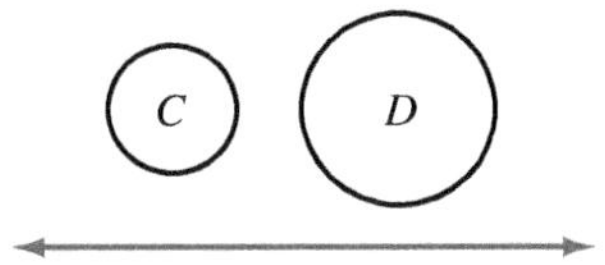

37. A size transformation centered at some point O on line l takes point P onto P'. Explain how to draw (a) the center O of the transformation and (b) the image Q' of the point Q, where Q lies on l.

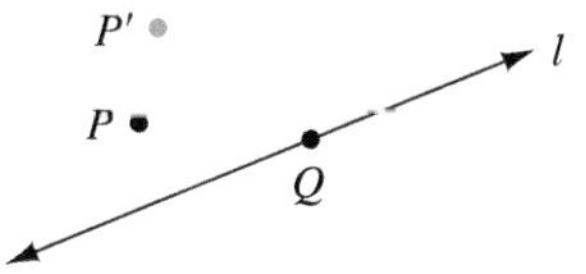

38. A size transformation takes P onto P' and Q onto Q', where the four points P, P', Q, and Q' are collinear as shown here. Redraw the figure and explain **(a)** how to draw the image R' of point R and **(b)** how to locate the center O of the size transformation. (*Hint:* What lines through P' and R' must contain R')?

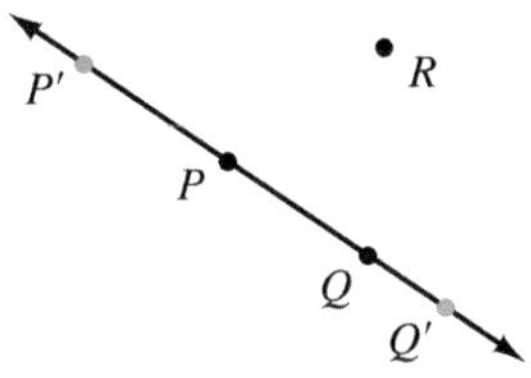

Making Connections

39. A mattress should be turned periodically in order to wear evenly. The mattresses shown here have "flip marks," indicating the axis over which the mattress is flipped. The dashed marks are on the reverse side of the mattress. For example, each mattress is first flipped over the horizontal axis.

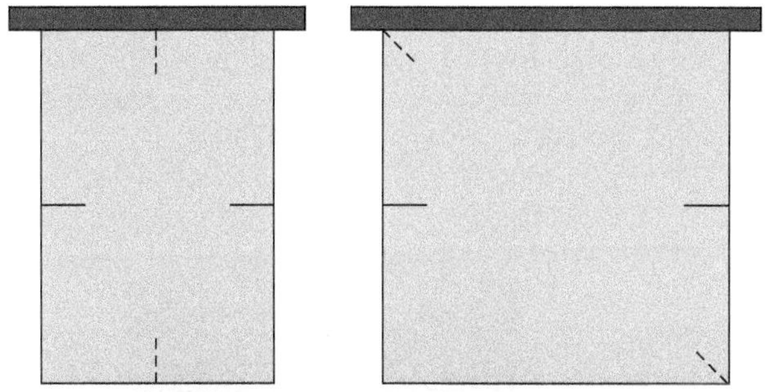

(a) Will the sequence of flips on the left-hand rectangular mattress cycle through all of the positions into which the mattress can be put on the bed? (*Suggestion:* Model the mattress with an index card, with marks on one side indicating a flip from head to foot and marks on the opposite side indicating a side-to-side flip.)

(b) Answer part (a) again, but for the square mattress shown.

40. **Fermat's Principle.** Suppose that a ray of light emanating from point P is reflected from a mirror at point R toward point S. It was known even in ancient times that the incident and reflected rays of light make congruent angles with the line m of the mirror. Pierre de Fermat (1601–1665) proposed an important principle to explain why this is so: *Light follows the path of shortest distance.* According to Fermat's principle, if Q is some point on the mirror other than R, then the distance $PQ + QS$ must exceed the distance $PR + RS$ traveled by the light.

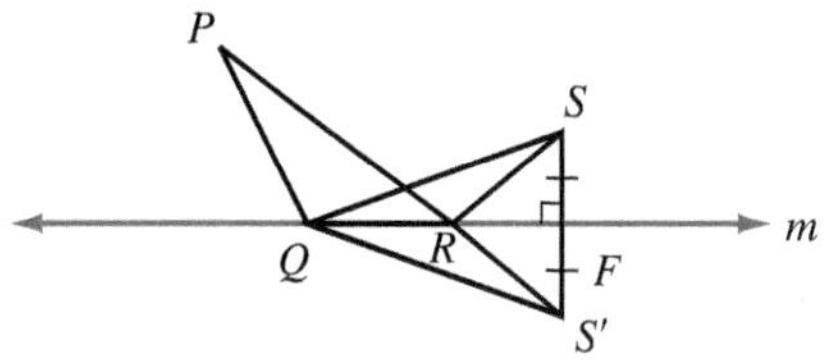

Answer the questions that follow to verify that $PQ + QS > PR + RS$. Let S' be the image of S under reflection across line m.

(a) Why are $RS' = RS$ and $QS' = QS$? (*Hint:* The reflection across m is a rigid motion.)

(b) Why is $PQ + QS' > PS'$?

(c) How does the inequality of part (b) give the desired result that $PQ + QS > PR + RS$?

41. Let P and S be two points on the same side of mirror m. If S' is the point of reflection of point S across m, then the line drawn from P to S' intersects m at the point R of reflection. (See the figure in problem 40.) Suppose P and S are between *two* mirrors m and l as shown next.

(a) Construct a doubly reflected light path $PQRS$ that is reflected off mirror m at Q and then off mirror l at R.

(b) Construct another doubly reflected path $PABS$ that reflects first off mirror l at A and then off mirror m at B.

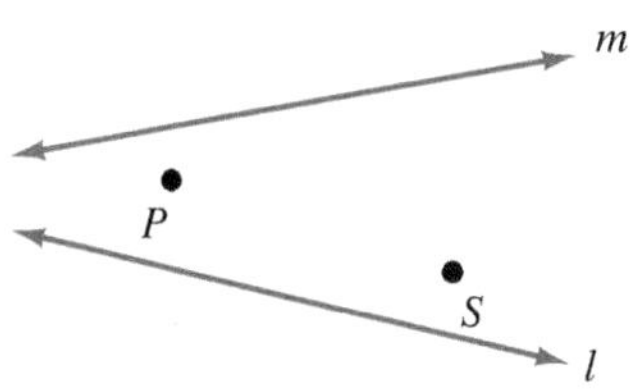

42. A **corner mirror** is formed by placing two mirrors together at a right angle. Explain why looking at yourself in a corner mirror is quite different from seeing yourself in an ordinary mirror. Use sketches to make your ideas clear.

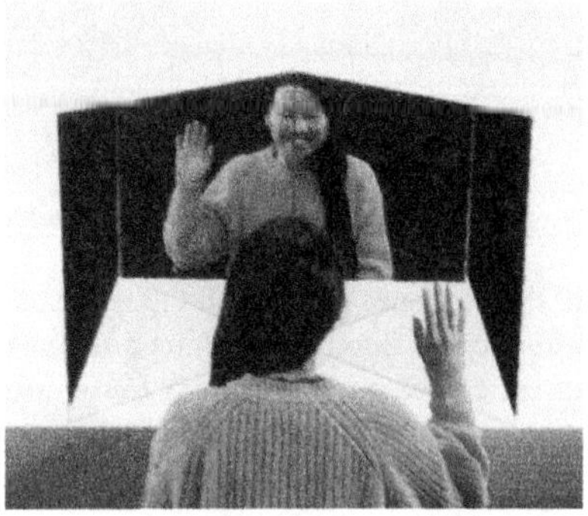

43. A billiard ball is located at a point P along an edge of a rectangular billiard table. Show that there is a billiard shot that strikes the cushions on the other three rails (sides) of the table and then returns to bounce at P.

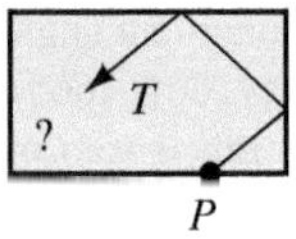

(*Hint:* Suppose the table T is reflected across its sides successively, forming the images T', T'', and T'''.) Explain how a billiard path $PQRSP$ can be found by drawing the line segment $\overline{PP'''}$. Sketch the path $PQRSP$ on the original table T.

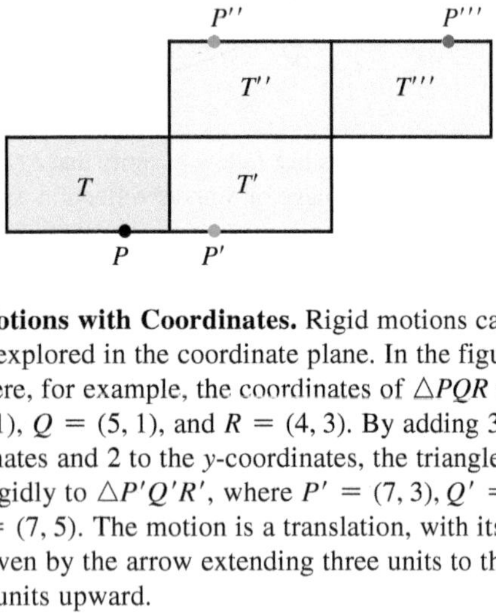

44. Rigid Motions with Coordinates. Rigid motions can be usefully explored in the coordinate plane. In the figure shown here, for example, the coordinates of $\triangle PQR$ are $P = (4, 1)$, $Q = (5, 1)$, and $R = (4, 3)$. By adding 3 to the x-coordinates and 2 to the y-coordinates, the triangle is moved rigidly to $\triangle P'Q'R'$, where $P' = (7, 3)$, $Q' = (8, 3)$, and $R' = (7, 5)$. The motion is a translation, with its slide vector given by the arrow extending three units to the right and two units upward.

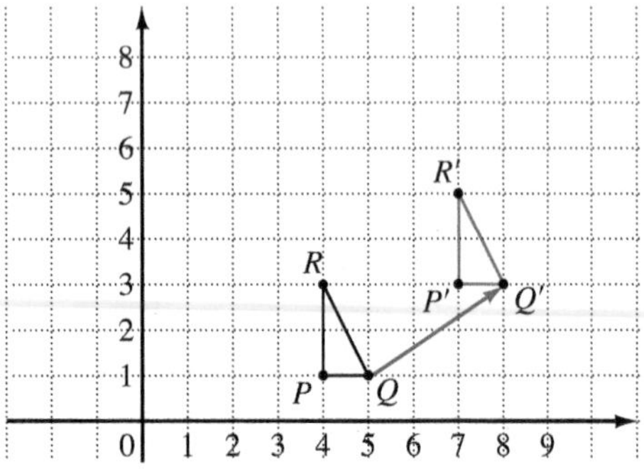

In a similar manner, use a coordinate grid to draw $\triangle PQR$ and its image $\triangle P'Q'R'$ under each of the transformations that follow, always starting with $\triangle PQR$ in its original position given in the preceding diagram. Classify each transformation as a rigid motion.

(a) Subtract 2 from the x-coordinates and add 3 to the y-coordinates.

(b) Replace each y-coordinate with its opposite (that is, with its negative).

(c) Replace each x-coordinate with its opposite.

(d) Replace both the x- and y-coordinates with their opposites.

(e) Exchange the x- and y-coordinates. For example, $P = (4, 1)$ is moved to $P' = (1, 4)$.

(f) If P has the coordinates (x, y), let P' have the coordinates $(-y, x)$. For example, if $P = (4, 1)$ then its image is $P' = (-1, 4)$.

(g) If P has the coordinates (x, y), let P' have the coordinates $(y + 3, x + 3)$. For example, if $P = (4, 1)$ then its image is $P' = (1 + 3, 4 + 3) = (4, 7)$.

45. Size and Similarity Transformations with Coordinates. The coordinates of $\triangle PQR$ are $P = (4, 1)$, $Q = (5, 1)$, and $R = (4, 3)$. By multiplying the x- and y-coordinates by 2, the triangle is moved to $\triangle P'Q'R'$, where $P' = (8, 2)$, $Q' = (10, 2)$, and $R' = (8, 6)$. This size transformation is centered at the origin (0, 0) and has scale factor $k = 2$.

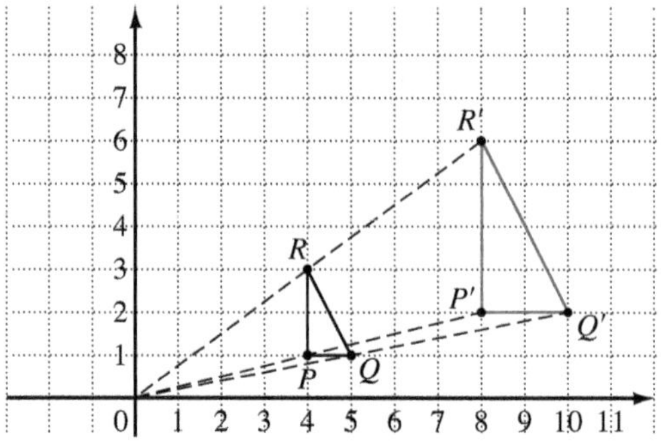

(a) On graph paper, draw the two image triangles obtained by transforming $\triangle PQR$ with size transformations centered at the origin, first with scale factor $k = 3$ and then with scale factor $k = 0.5$.

(b) What combination of a rigid motion and size transformation centered at the origin can map $\triangle PQR$ to the $\triangle STU$ with vertices at the points $S = (1, 2)$, $T = (3, 2)$, and $U = (1, 6)$? Carefully describe how to obtain the coordinates of $\triangle STU$, beginning with the coordinates of $\triangle PQR$.

(c) Why is $\triangle PQR$ similar to $\triangle ABC$, where $A = (-8, -2)$, $B = (-10, -2)$, and $C = (-8, -6)$? Use coordinates to describe the similarity motion that takes triangle PQR to triangle ABC.

Communicating

46. Cutting a Pan of Brownies. Maria made a rectangular pan of brownies that she promised to share equally with her brother Miguel. Once the brownies cool, she intends to make a straight cut across the pan to make two equal portions.

(a) Explain why any straight cut halves the brownies if and only if the cut passes through the center of the pan.

(b) When Maria returned to the kitchen, she discovered that her dad had removed a rectangular brownie for himself from the pan, as shown here.

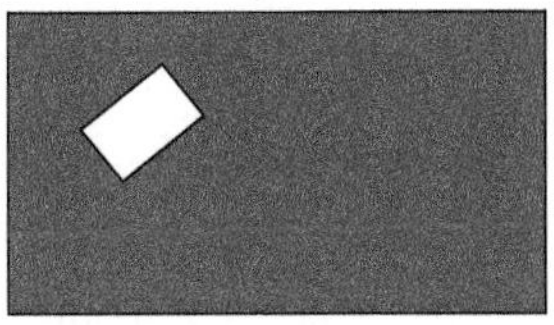

How can Maria share the remainder of the pan of brownies with Miguel by making one straight cut across the pan?

State Assessments

47. (Grade 8)
Figure *PQRST* is translated 3 units to the left and 5 units up to create the figure *P'Q'R'S'T'*. What are the coordinates of the point *T'*?

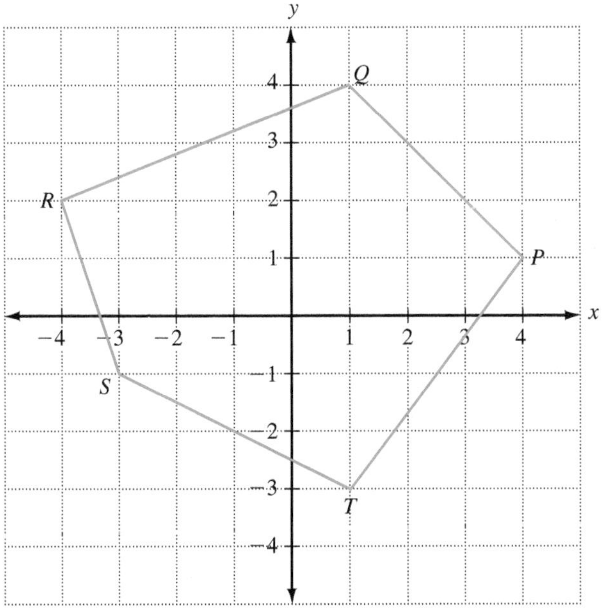

A. (−3, 5) B. (−2, 5) C. (4, −8) D. (−2, 2)

48. (Grade 7)
In the diagram below, a transformation has taken rectangle *KLMN* onto rectangle *WXYZ*. What transformation has been used?

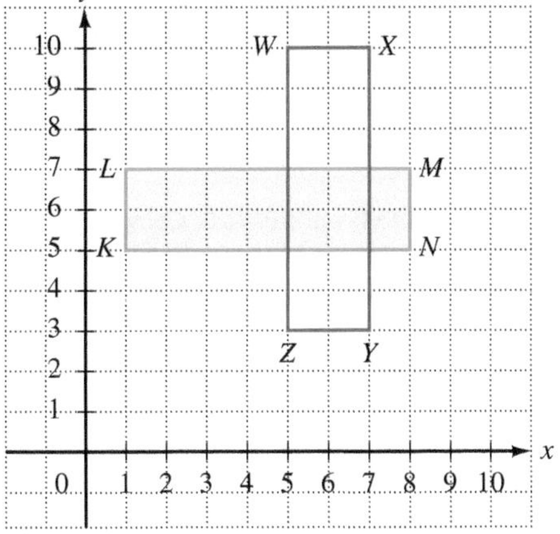

A. A translation 5 units right and 4 units down
B. A translation 5 units right and 3 units up
C. A 90° clockwise rotation about the point (7, 5)
D. A 90° counter clockwise rotation about the point (8, 7)

49. (Grade 8)
Using the origin as the center of dilation, rectangle *KLMN* KLMN is dilated with a scale factor of $\frac{1}{4}$ to form rectangle *K'L'M'N'*.

What is the length of *K'L'*?

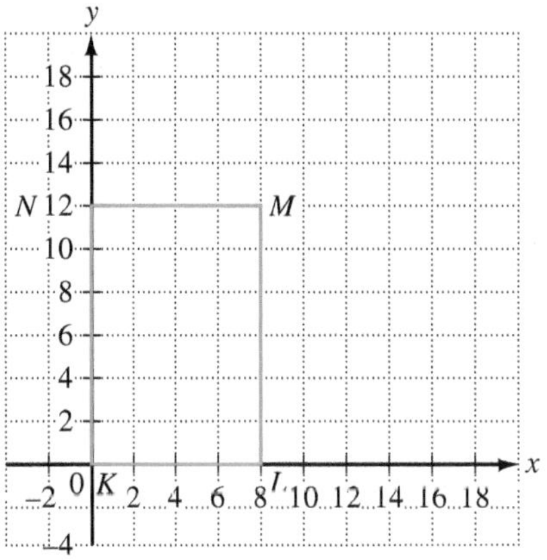

A. 32 units B. 48 units
C. 3 units D. 2 units

50. (Grade 5)
Eugene created the figure below.

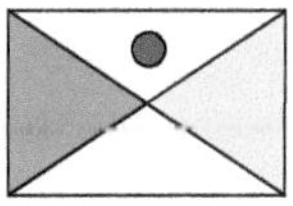

Which of these figures did Eugene get when he rotated his figure clockwise 90°?

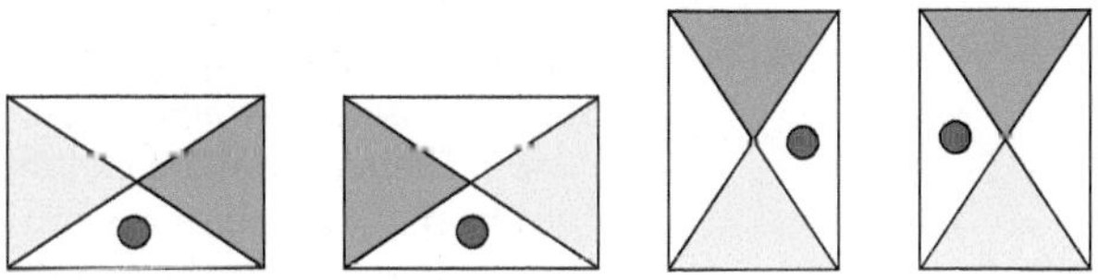

A **B** **C** **D**

51. (Grade 5)
Which of these figures are congruent?

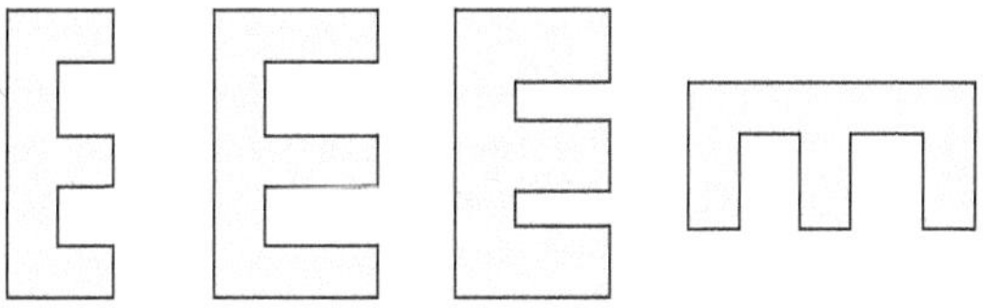

1 2 3 4

A. 1 and 2 B. 1 and 3
C. 2 and 3 D. 2 and 4

11.2 Patterns and Symmetries

Symmetry is a universal principle of organization and form. The circular arc of a rainbow and the hexagonal symmetry of an ice crystal are visible expressions of the symmetry of many of the physical processes of the universe. A seashell and the fanned tail of the peacock are spectacular examples of biological symmetry.

In the human domain, all cultures of the world, even those in prehistoric times, developed a useful intuitive understanding of the basic concepts of symmetry. Decorations on pottery, walls, tools, weapons, musical instruments, and clothing are often highly symmetric. Buildings, temples, tombs, and other structures are usually designed with an eye to symmetry and balance. Music, poetry, and dance frequently incorporate symmetry into their underlying structure.

While people have long had an informal understanding of symmetry, it is only more recently that mathematics has provided a means to a deeper understanding of symmetry and how certain kinds of symmetry can be described and classified. In the classroom, symmetry can draw youngsters to the artistic and aesthetic aspects of mathematics.

What Is Symmetry?

The concept of a rigid motion, which we defined and explored in the last section, makes it possible to give a precise definition of a symmetry of a geometric figure in the plane.

DEFINITION A Symmetry of a Plane Figure

A **symmetry of a plane figure** is any rigid motion of the plane that moves all the points of the figure back to points of the figure.

Thus, all points P of the figure are taken by the symmetry motion to points P' that also are points of the figure. The identity motion is a symmetry of any figure, but of more interest are figures that have symmetries other than the identity. Under a nonidentity symmetry, some points in the figure move to new positions in the figure, even though the figure as a whole appears unchanged by the motion.

The classification theorem of the preceding section tells us that there are just four basic rigid motions. Therefore, any symmetry of a figure is one of these four basic types, and the symmetry properties of a figure can be fully described by listing all of the symmetries of each type.

FROM THE NCTM PRINCIPLES AND STANDARDS

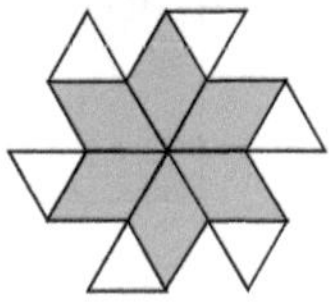

Rotational Symmetry

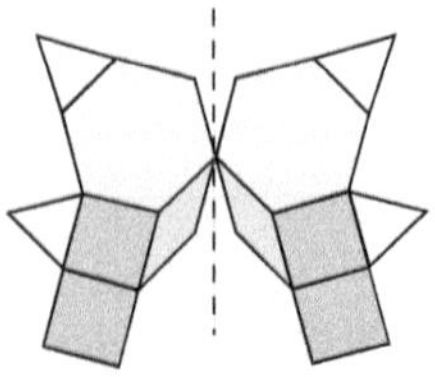

Line Symmetry

Students can naturally use their own physical experiences with shapes to learn about transformations such as slides (translations), turns (rotations), and flips (reflections). They use these movements intuitively when they solve puzzles, turning the pieces, flipping them over, and experimenting with new arrangements. Students using interactive computer programs with shapes often have to choose a motion to solve a puzzle. These actions are explorations with transformations and are an important part of spatial learning. They help students become conscious of the motions and encourage them to predict the results of changing a shape's position or orientation but not its size or shape.

Teachers should guide students to recognize, describe, and informally prove the symmetric characteristics of designs through the materials they supply and the questions they ask. Students can use pattern blocks to create designs with line and rotational symmetry (see figure at left) or use paper cutouts, paper folding, and mirrors to investigate lines of symmetry.

SOURCE: Principles and Standards for School Mathematics by NCTM, *pp. 99–100. Copyright © 2000 by the National Council of Teachers of Mathematics. Reproduced with permission of the National Council of Teachers of Mathematics via Copyright Clearance Center. NCTM does not endorse the content or validity of these alignments.*

Reflection Symmetry

A figure has **reflection symmetry** if a reflection across some line is a symmetry of the figure. The line of reflection is called a **line of symmetry,** or a **mirror line,** of the figure. Each point P of the figure on one side of the line of symmetry is matched to a point P' of the figure on the opposite side of the line of symmetry. Some figures and their lines of symmetry (shown dashed) are displayed in Figure 11.14. A mirror line splits the figure into two mirror-image halves.

Figure 11.14
Plane figures and their lines of symmetry

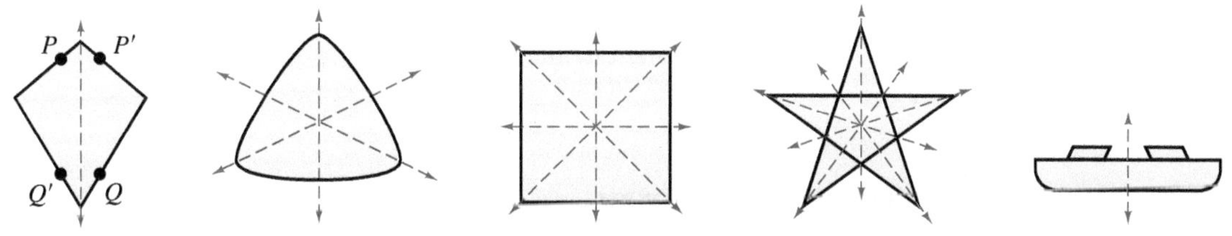

(a) (b) (c) (d) (e)

Reflection symmetry is also called **line symmetry** or **bilateral symmetry.** Like reflection symmetry, bilateral symmetry is used to describe figures in space that have a plane of symmetry. For example, ferries (as suggested in Figure 11.14e) are often bilaterally symmetric across midships to simplify loading and unloading of their cargo of cars and trucks. Infrequent passengers on such ferries can find it very disorienting when the bow and stern are indistinguishable.

EXAMPLE 11.8

Identifying Lines of Symmetry

Identify all lines of symmetry for each of these figures:

(a) (b) (c) (d)

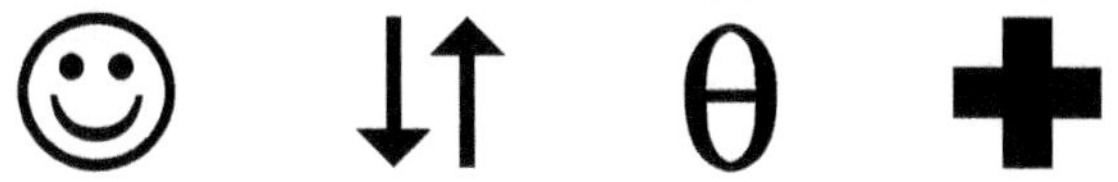

Solution

(a) 1 vertical **(b)** none **(c)** 2: one vertical and one horizontal
(d) 4: one vertical, one horizontal, and the two diagonal lines

Rotation Symmetry

A figure has **rotation symmetry,** or **turn symmetry,** if the figure is superimposed on itself when it is rotated through a certain angle between 0° and 360°. The center of the turn is called the **center of symmetry.** Some examples of figures with rotation symmetry are shown in Figure 11.15.

Figure 11.15
Figures with rotation symmetry

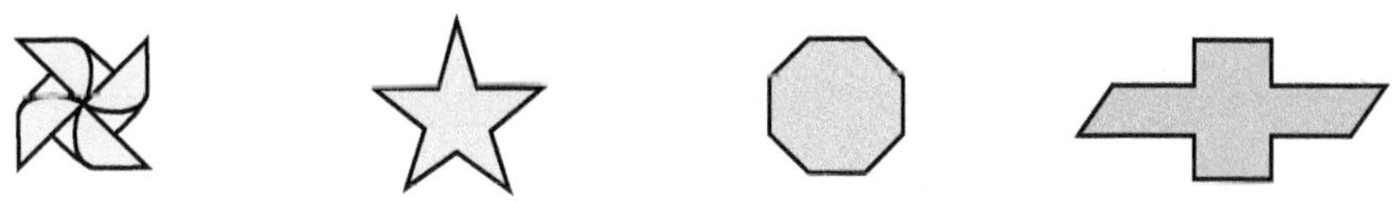

(a) 90° symmetry **(b)** 72° symmetry **(c)** 45° symmetry **(d)** 180° symmetry

A figure with 90° rotation symmetry automatically has 180° and 270° rotation symmetry. For this reason, it is customary to give just the *smallest* positive angle measure that turns the figure onto itself. The only exception is for figures composed of concentric circles, which turn onto themselves after *any* turn about their center. Such figures have **circular symmetry.**

EXAMPLE 11.9

Finding Angles of Rotation Symmetry

Determine the measures of the angles of rotation symmetry of these figures:

(a) (b) (c) (d)

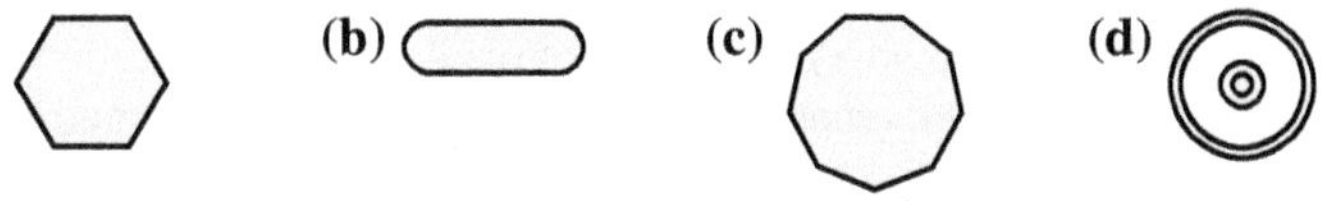

Solution

(a) The smallest positive angle of rotation of a regular hexagon measures 60°, so the turn angles of all of the rotation symmetries are 60°, 120°, 180°, 240°, and 300°.
(b) The only angle of rotation symmetry is 180°.
(c) The smallest amount of turn of a regular 9-gon is $360°/9 = 40°$, so the turn angles are 40°, 80°, 120°, 160°, 200°, 240°, 280°, and 320°.
(d) This figure has circular symmetry.

Point Symmetry

A figure has **point symmetry** if it has 180° rotation symmetry about some point O. Some familiar examples are shown in Figure 11.16. Notice that a half-turn takes the playing card back onto itself, and every point P of the figure has a corresponding point P' of the figure that is directly opposite the turn center O, with $OP = OP'$.

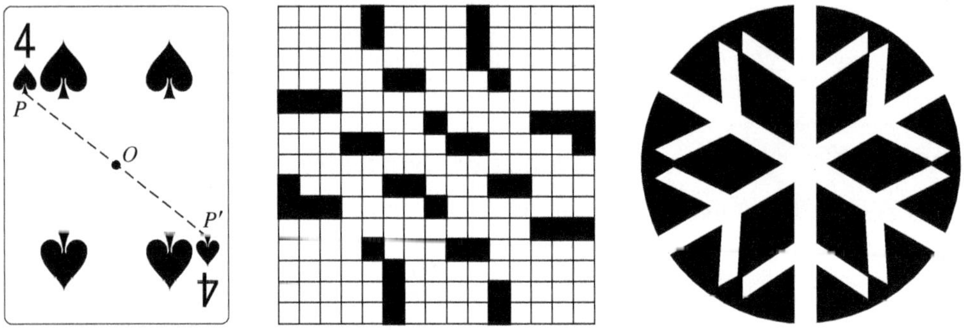

Figure 11.16
The 4 of spades, the crossword puzzle diagram, and the snowflake each have point symmetry, since each has a center point O of half-turn (180°) rotational symmetry. In particular, any point P of the figure corresponds to a point of the figure on the opposite side of O that is at the same distance from O

EXAMPLE 11.10

Identifying Point Symmetry

What letters, in uppercase block form, can be drawn to have point symmetry?

Solution

H, I, N, O, S, X, and **Z.**

The letters **H, I, O,** and **X** also have two perpendicular lines of mirror symmetry. Only **N, S,** and **Z** have just point symmetry.

INTO THE CLASSROOM
Mathematics in Motion

The words *transformation* and *symmetry* often suggest advanced topics best left for gifted middle school students or postponed to the high school curriculum. Quite the opposite is true, however, since there are activities, games, artistic constructions, and problems—all exploring "motion geometry"—that are suitable for students at every grade level. Primary school children can work with paper folding, Miras, pattern blocks, geoboards, or rubber stamps to create and investigate symmetric patterns. For example, each student can create a "half" figure with rubber bands on the upper half of a geoboard. Boards are then exchanged, and the students are challenged to complete a mirror-image figure in the lower half of the geoboard. Older children could replace the geoboard with graph paper or dot paper and investigate rotations and point symmetry, as well as reflections and line symmetry. Geometry software also offers exciting possibilities for investigations in transformation geometry.

Here are three more ideas suggesting how patterns and motions can be approached in the classroom:

- *Follow the leader.* Draw a line with a ruler down a blank sheet of paper. In pairs of students, the "leader" slowly draws a curve, and simultaneously the "follower" draws the reflected curve across the line of symmetry. The students can interchange roles of leader and follower. To explore point symmetry, a prominent dot can be drawn at the center of the sheet. Some students, with a pencil in each hand, might like to attempt a solitaire game.

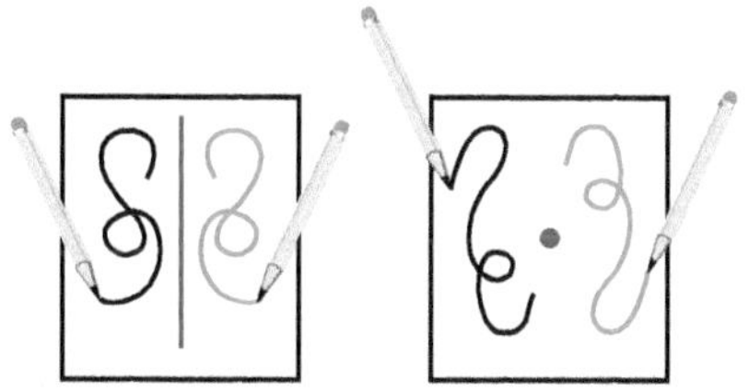

- *Punchy puzzles.* If the square shown on the far left is folded along the dashed lines, then the pattern of holes can be seen to be created with just one punch. How can a square sheet of paper be folded and punched one time only to create the other hole patterns shown?

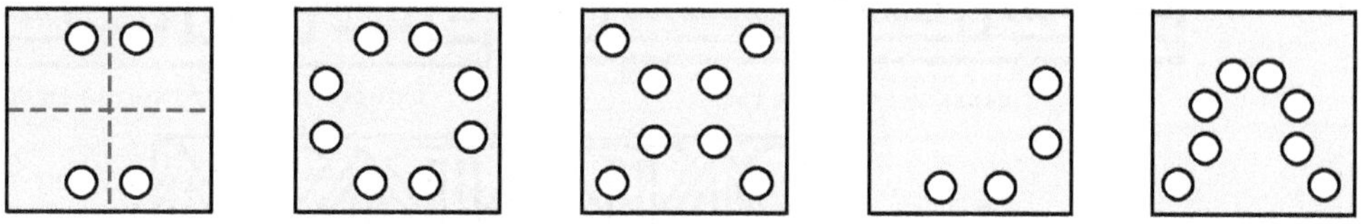

- *Stained-glass window search.* Eight congruent isosceles right triangles, with four of each color, will form a square window. Three windows are shown here, but a reflection and rotation show that the first two windows are really the same. How many different window patterns can be made from the given outline, each with four panes of each of two colors? (You should be able to find 13 distinct patterns, with no two patterns the same, under either a rotation or a reflection.)

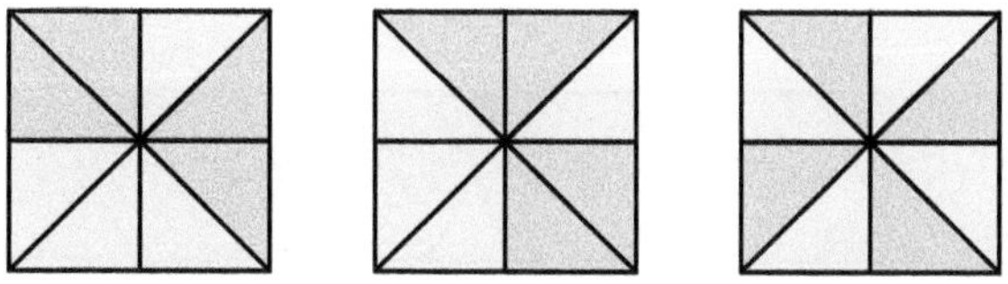

Periodic Patterns: Figures with Translation Symmetries

A **periodic pattern** is a figure with translation symmetry. That is, there is at least one translation that moves the pattern so that it is superimposed on itself. To avoid considering the whole plane, or even just some set of horizontal lines, as a periodic pattern, it is assumed that there is some minimum positive distance required to translate a periodic pattern back onto itself. Thus, a periodic pattern must be an infinite figure (why?) with motifs repeated endlessly at regular intervals. Fragments of periodic patterns are common on wallpaper, decorative brick walls, printed and woven fabrics, ribbons, and friezes (ceiling or façade border decorations in older buildings). Enough of the pattern must be shown to make it clear how to extend the pattern indefinitely.

Two types of periodic patterns in the plane are **border patterns** and **wallpaper patterns.** A border pattern has a repeated motif that has been translated in just one direction to create a strip design also known as a **frieze pattern.** A wallpaper pattern has a motif translated in two nonparallel directions to create an all-over planar design.

Border Patterns and Their Classification

Seven examples of border patterns from a variety of cultures are shown in Figure 11.17.

Some border patterns may have other symmetries in addition to their translation symmetries. However, the possibilities are limited, since any symmetry of a border pattern necessarily takes the infinite strip onto itself. Thus, the only possible rotation symmetry is a half-turn. A careful study has shown that every periodic border pattern has the same symmetries as one of the seven types shown in Figure 11.18 that have been created with pattern blocks.

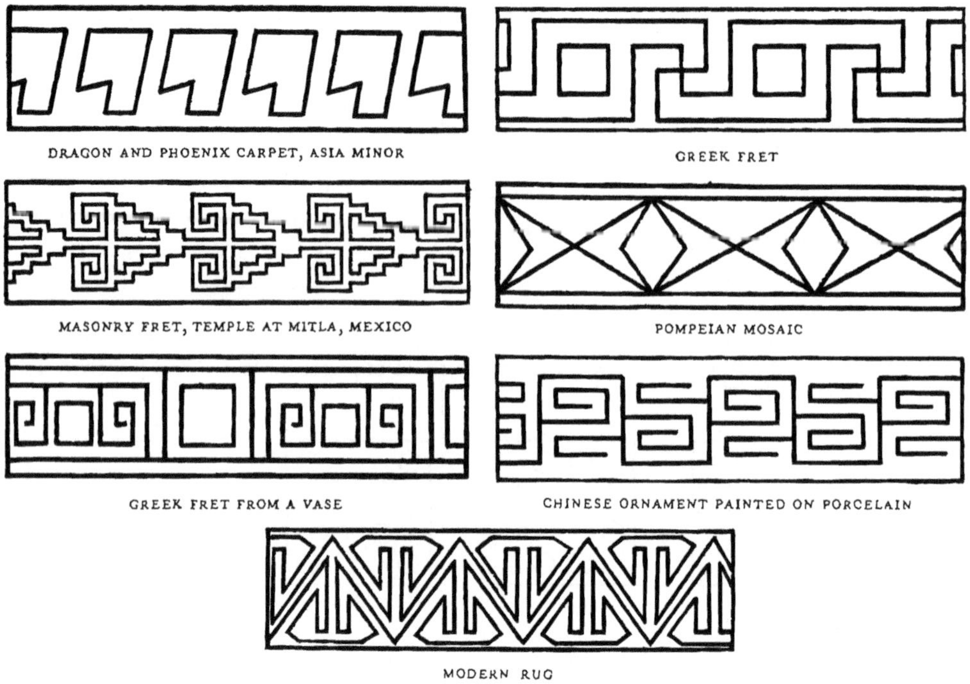

Figure 11.17
Border patterns from around the world

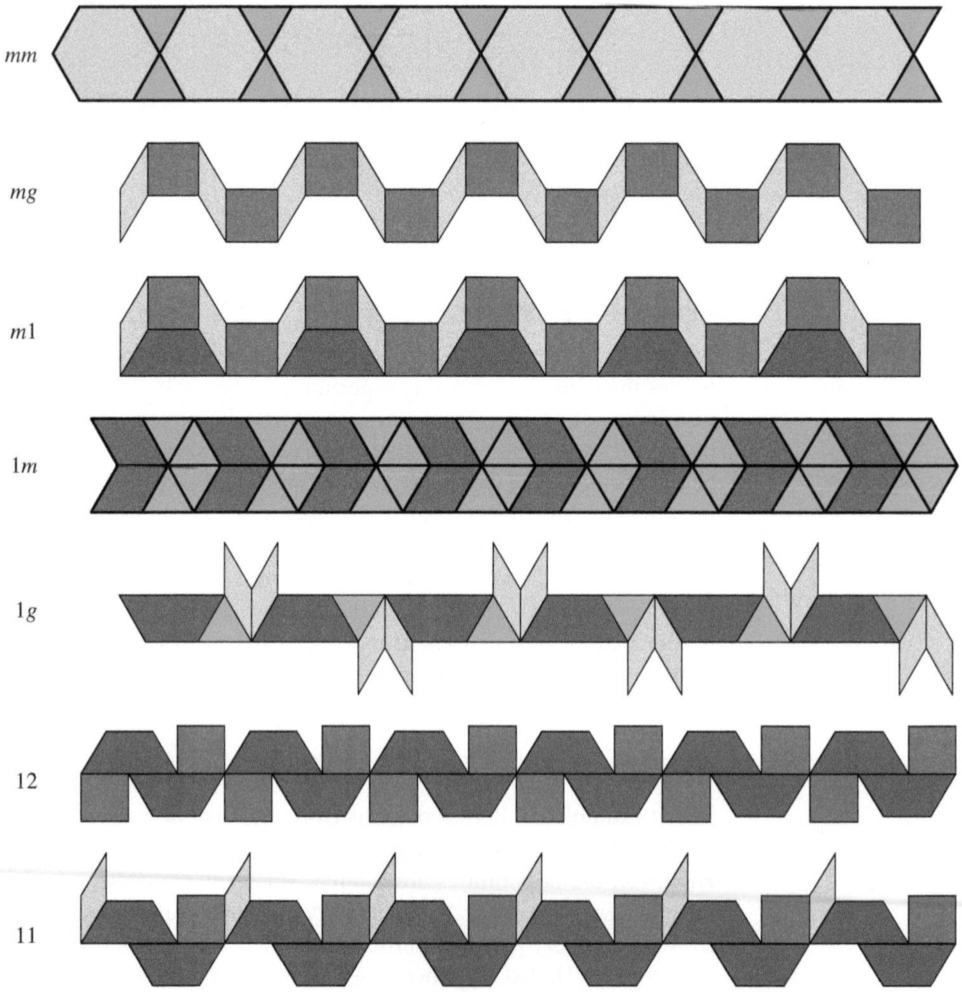

Figure 11.18
The seven symmetry types of border patterns, shown with pattern-block designs

The two-symbol name assigned by the International Crystallographic Union is shown at the left of each pattern in Figure 11.18. To find the classification symbol of any border, follow these steps:

First Symbol: *m*, if there is a vertical line of symmetry
1, otherwise

Second Symbol: *m*, if there is a horizontal line of symmetry
g, if there is glide–reflection symmetry (but no horizontal line of symmetry)
2, if there is half-turn symmetry (but no horizontal line of symmetry or glide–reflection symmetry)
1, otherwise

To check your understanding, cover up the two-symbol name in Figure 11.18 and then follow the preceding directions to see if you obtain the correct classification symbols. With practice, you'll be able to determine the two-symbol name for any border pattern. The symbol "*m*2" will never be encountered, since no border pattern can have such a symmetry.

EXAMPLE 11.11

Classifying Border Patterns

Classify the symmetry type of the following border patterns by assigning the appropriate two-symbol notation:

(a)

(b)

(c)

Solution

It is helpful to turn the patterns upside down or use a mirror or a Mira, since this will help you discover and verify what symmetries are present. A transparency copy of the pattern, if available, is an almost ideal tool to explore and classify patterns of symmetry.

(a) This border has both a vertical and horizontal line of reflection, so the symmetry is of type *mm*.

(b) There is no vertical symmetry line, so the first symbol is 1. There is no horizontal symmetry line, but there is glide–reflection symmetry, so the second symbol is *g*. Altogether, the symmetry is type 1*g*.

(c) There are no lines of reflection, nor is there glide–reflection symmetry. However, there is half-turn symmetry, so the symbol is 12.

SMP 3 Construct viable arguments and critique the reasoning of others.

The classification of border patterns provides excellent reinforcement of the third Standard for Mathematical Practice (SMP 3) from the **Common Core.**

"They make conjectures and build a logical progression of statements to explore the truth of their conjectures. They are able to analyze situations by breaking them into cases, and can recognize and use counterexamples. They justify their conclusions, communicate them to others, and respond to the arguments of others."

The Cooperative Investigation that follows can be adapted for use in the elementary school classroom.

COOPERATIVE INVESTIGATION
Classifying Border Patterns

Materials Needed

For each small group, a paper copy of border patterns (Figure 11.17, for example), an acetate sheet on which the same border patterns have been copied, and a pair of scissors to cut individual patterns from the sheets.

Directions

Each group member selects a pattern to investigate, using both the paper and acetate copies. The two-symbol crystallographic symbol of the pattern is determined by flipping and sliding the acetate copy over the paper copy. Group members then select new patterns to investigate and trade patterns to confirm the classifications of their teammates. The activity ends by having each group, taking turns, report its classification to the other groups.

Wallpaper Patterns

A wallpaper pattern has translation symmetry in two directions. Two examples are shown in Figure 11.19.

Figure 11.19 Examples of wallpaper patterns

The Arabian pattern on the left has centers of both 60° and 120° rotational symmetry. This pattern also has a 180° rotational symmetry at the center of each of the *Z*-shaped black bars. The Egyptian pattern on the right has centers of 90° rotational symmetry. It can be shown that 60°, 90°, 120°, and 180° are the only possible angle measures of rotational symmetry of any wallpaper pattern, a result called the *crystallographic restriction.* This and other restrictions limit the number of symmetry types to be found in a wallpaper pattern. It has been shown that any wallpaper pattern is one of just 17 distinct types.

In the next section, several wallpaper patterns will be created by covering the plane with polygonal tiles. It will also be shown how the straight sides of the tiles can be modified to create tiling patterns in the style of the Dutch graphic artist M. C. Escher.

11.2 Problem Set

Understanding Concepts

1. Carefully trace each figure and draw all of its lines of symmetry.

(a) (b) (c) (d) (e)

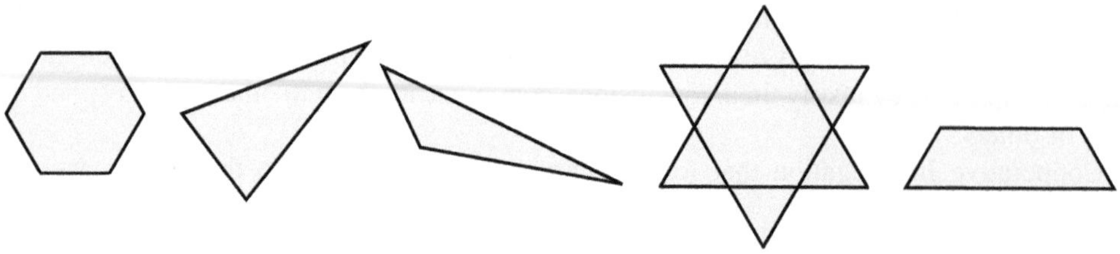

2. Carefully trace each figure and draw all of its lines of symmetry.

(a) (b) (c) (d) (e)

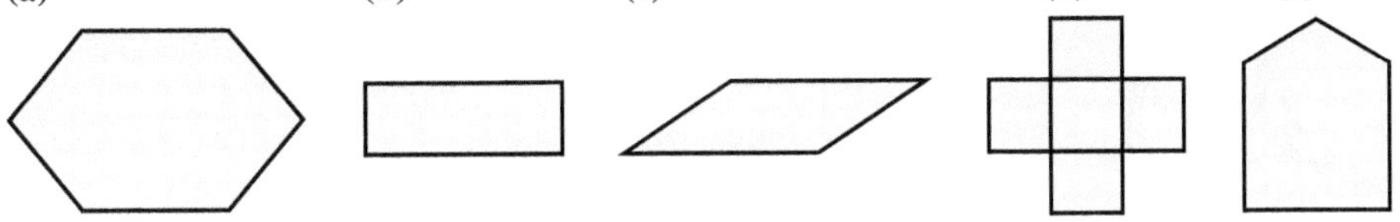

3. Draw polygons with the following symmetries, if possible:
   (a) One line of symmetry, but no rotation symmetry
   (b) Rotation symmetry, but no reflection symmetry
   (c) One line of symmetry and rotation symmetry

4. Describe the most general quadrilateral having the given symmetry property.
   (a) A line of symmetry through a pair of opposite vertices
   (b) A line of symmetry through a pair of midpoints of opposite sides
   (c) Two lines of symmetry, each through a pair of opposite vertices
   (d) Two lines of symmetry, each through a pair of midpoints of opposite sides
   (e) Exactly four lines of symmetry
   (f) A center of 180° rotational symmetry

5. Complete each figure to give it reflection symmetry about line m.

(a)

(b)

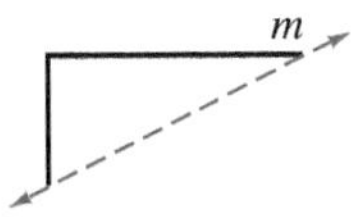

(c)

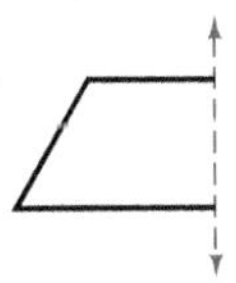

(d)

6. Complete each of these figures to give it point symmetry about point O:

(a)

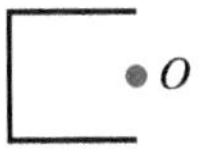

(b)

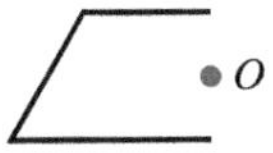

(c)

(d)

7. Copy the figures shown onto graph paper. Then complete each figure to give it reflection symmetry across the dashed line.

(a)

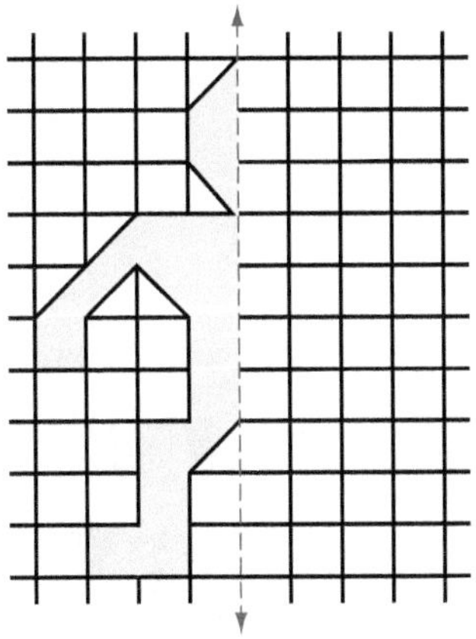

(b)

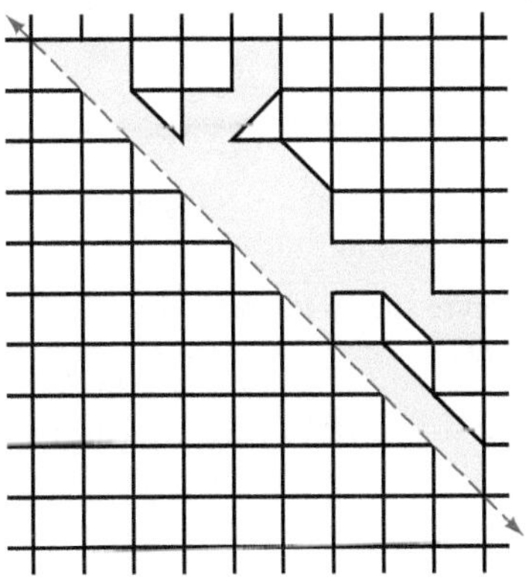

8. A symmetric valentine heart is easy to make: Cut it from a piece of construction paper folded in half once:

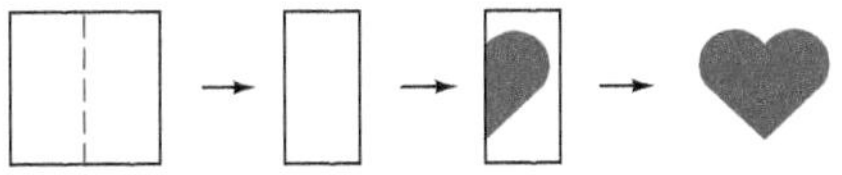

   (a) Suppose the paper is folded in half twice and the same cut is made as shown below. Sketch the shape you obtain when you unfold the cut pattern.

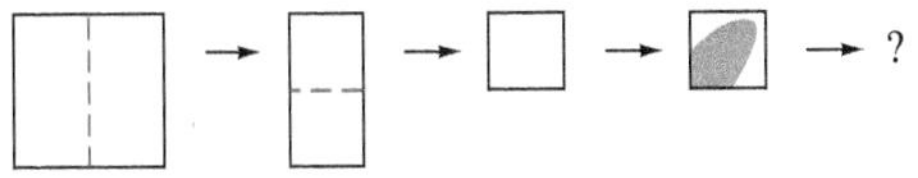

   (b) Describe how to make a sixfold symmetric snowflake by folding and cutting a sheet of paper.

9. Describe all symmetries of each of the following logos:

(a)

(b)

(c)
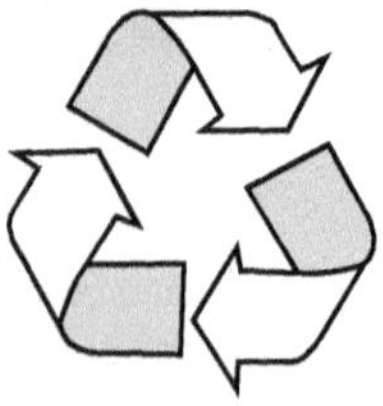

(d)

(e)

10. Describe the symmetries of the wheel covers shown.

(a)

(b)

(c)
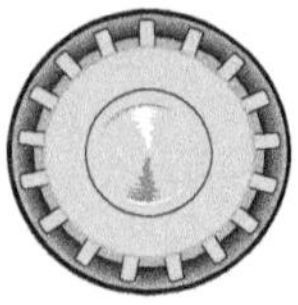

(d)

11. **(a)** Complete the figure shown to give it 90° rotation symmetry about point O.

(b) Repeat part (a), but giving the resulting figure 60° rotation symmetry.

12. Identify the regular n-gons in each part that have the given symmetries.

(a) There are exactly 3 lines of symmetry.
(b) There are exactly 4 lines of symmetry.
(c) There are exactly 19 lines of symmetry.
(d) The polygon has 10° rotation symmetry.
(e) The polygon has both 6° and 15° rotation symmetry.

13. List all the digits from the list **0, 1, 2, 3, 4, 5, 6, 7, 8,** and **9** that have

(a) vertical reflection symmetry.
(b) horizontal reflection symmetry.
(c) vertical and horizontal reflection symmetry.
(d) point symmetry.

Write the digits in the most symmetric way you can.

14. Repeat problem 13, but for the uppercase capital letters, **A, B, . . . , Z,** written as symmetrically as possible.

15. Repeat problem 13 for the lowercase letters, **a, b, . . . , z,** written as symmetrically as possible.

16. Describe all the symmetries of each border pattern, and classify each by the two-symbol notation used in crystallography as in Figure 11.18.

(a) . . . **A A A A A A** . . . .
(b) . . . **B B B B B B** . . . .
(c) . . . **N N N N N N** . . . .

17. Describe all the symmetries of each border pattern, and give its two-symbol classification used by crystallographers as in Figure 11.18.

(a) . . . **H O H O H O** . . .
(b) . . . **M W M W M W** . . .
(c) . . . **9 6 9 6 9 6** . . .

18. Give the two-symbol classification of each of the seven periodic border patterns shown in Figure 11.17.

19. **(a)** Give the two-symbol classification of each of the following four border patterns:

(i) . . . →←→←→← . . .
(ii) . . . ↗↘↗↘↗↘ . . .
(iii) . . . →→→→→ . . .
(iv) . . . ↗↙↗↙↗↙ . . .

(b) Use the arrow motif to create the three types of border patterns not shown.

Into the Classroom

20. **Pattern-Block Symmetries.** Children enjoy creating symmetric designs with pattern blocks. Two simple examples are shown next. This intrinsic interest can be utilized effectively to explore many of the fundamental concepts of symmetry.

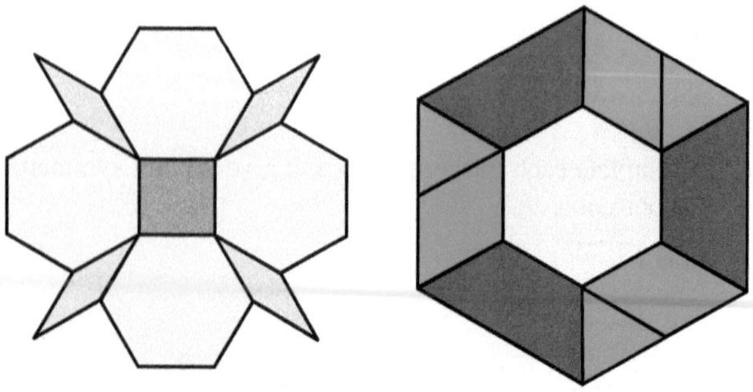

(a) Describe the symmetry of each of the six individual pattern-block shapes (the triangle, square, hexagon, and trapezoid, and the thick and thin rhombuses).

(b) Describe the symmetries of the two designs shown.

(c) Create a design that includes squares and has 60° rotational symmetry.

(d) Create a design that includes squares and has 30° rotational symmetry.

21. (**Writing**) Pattern blocks are well suited to create and classify periodic border patterns. For example, all seven symmetry types are shown in Figure 11.18. Following are periodic borders of type *mg* and 11:

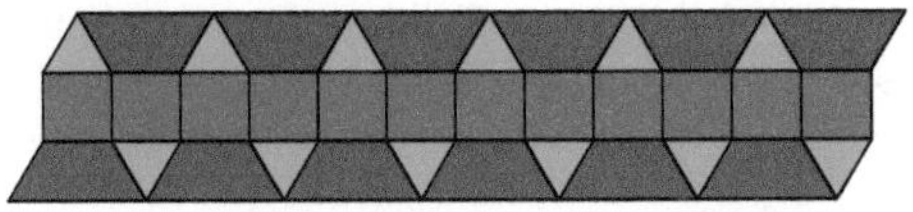

mg

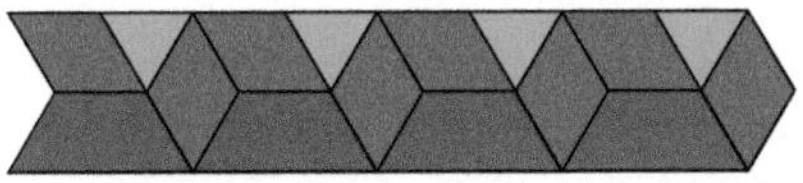

11

(a) Design a lesson that asks children to create pattern-block borders and then examine them for symmetries.

(b) Extend your lesson from part (a) to have children in the upper elementary grades classify the types of symmetry of periodic pattern-block borders. In particular, create examples that exhibit all seven types of symmetry, where each periodic border uses at least three pattern-block shapes.

22. Symmetry Search. The following line grid can be made easily from graph paper:

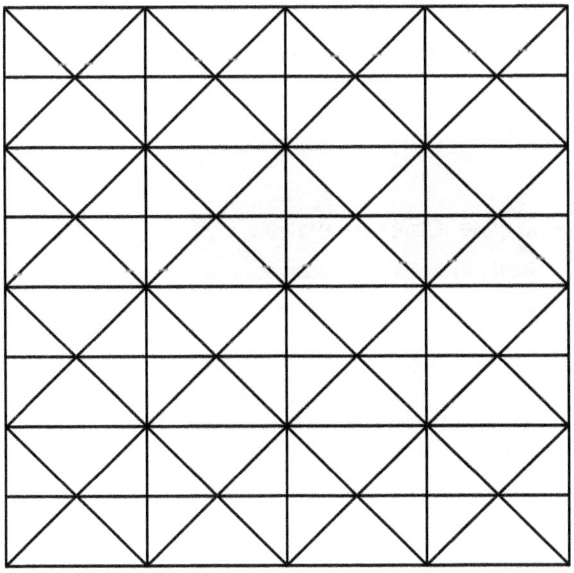

Using only segments in the grid, find polygons with the given symmetries. There are many answers, and you (and youngsters) are free to be creative.

(a) Find an octagon with two lines of symmetry.

(b) Find a quadrilateral with no symmetries.

(c) Find a pentagon with a line of symmetry.

(d) Find a heptagon with a line of symmetry that is neither horizontal nor vertical.

Responding to Students

23. Stacey drew lines of symmetry as shown on the following shapes:

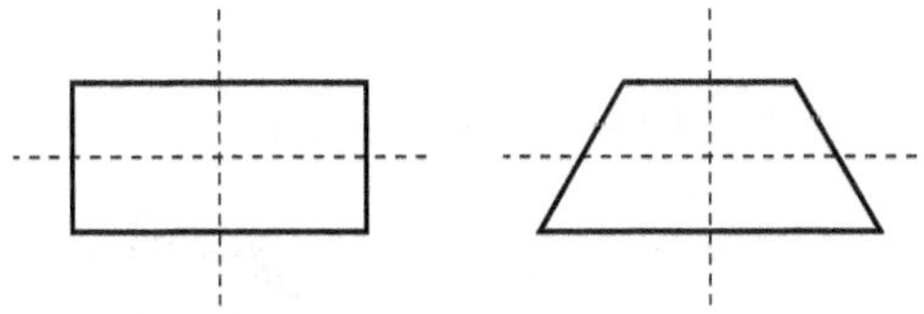

(a) On which shape did she draw the lines of symmetry correctly?

(b) (**Writing**) How would you help explain to her why some lines on the other shape are not lines of symmetry?

24. Kahalah and Ashton agree that **A** and **C** each have a line of symmetry, but they disagree about the letter **B.** How would you help them?

Thinking Critically

25. The Penny Game. Lynn and Kelly are playing a game with just a few simple rules. Each player takes turns placing a penny on a rectangular tabletop. Each new penny must be flat on the table and cannot touch any of the pennies already on the table. The first player unable to put another penny on the table according to these rules is the loser. Lynn, who gets to make the first move, puts the first penny at the center of the table and is confident of a win. What is Lynn's strategy?

26. A **palindrome** is a word, phrase, sentence, or numeral that is read the same either forward or backward. Examples are WOW, NOON, and TOOT.

(a) If a word written in all capital letters has a vertical line of symmetry, why must it be a palindrome?

(b) DAD does not have a line of symmetry. Find another palindrome with no line of symmetry.

(c) What symmetry do you see in the word "pod"?

(d) Find a palindrome with a horizontal line of symmetry.

27. Describe what symmetries you see in these statements:

(a) "Sums are not set as a test on Erasmus."

(b) "Is it odd how asymmetrical is 'symmetry'? 'Symmetry' is asymmetrical. How odd it is."

(c) "Able was I ere I saw Elba." (attributed to Napoleon)

28. Carefully explain why no border pattern has the symbol *m*2. (*Hint:* If a border pattern has a vertical line of symmetry and 180° rotation symmetry, what other symmetry must it also have?)

29. The Maori, the indigenous people of New Zealand, used principles of symmetry to express their belief system. Disregarding the color scheme, classify the following Maori rafter patterns:

(a)

(b)

(c)

(d)

30. Classify the following Inca border patterns:

(a)
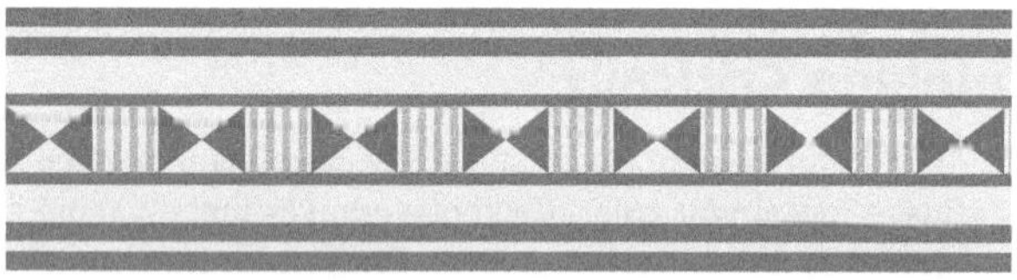

(b)

31. In each strip of rectangles shown, a certain rigid motion applied to the leftmost rectangle takes the figure to the next rectangle. Apply the same motion, but to the second rectangle, to draw the image of the second rectangle in the third rectangle. Continue to use the same motion to fill in the successive rectangles, and then classify the border pattern that is produced.

(a)
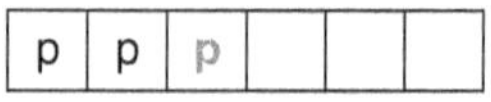

(b)
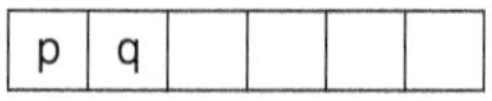

(c)
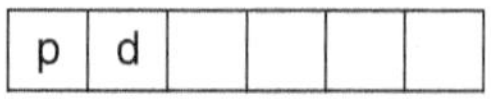

(d)
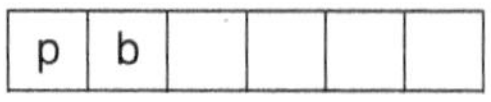

32. These drawings were made by George Pólya for his 1924 paper classifying the 17 wallpaper pattern types:

(a)
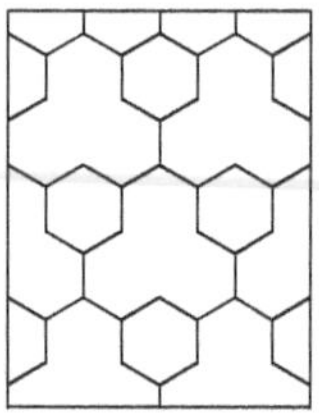

(b)
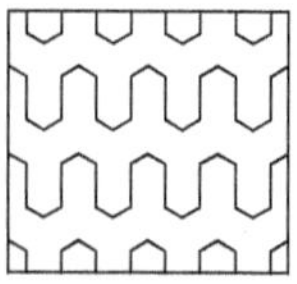

(c)

For each pattern, give

(i) the number of directions of reflection symmetry;

(ii) the number of directions of glide–reflection symmetry;

(iii) the sizes of angles of rotation symmetry.

Making Connections

33. Describe what symmetries, or lack of symmetry, you could see in the following forms and objects:

(a) A pair of scissors **(b)** A T-shirt

(c) A dress shirt **(d)** A golf club

(e) A tennis racket **(f)** A crossword puzzle

34. Describe the symmetry you find in

(a) an addition table.

(b) a multiplication table.

(c) Pascal's triangle.

35. The figures shown result from a famous experiment in physics called the Chladni plate. A square metal plate is supported horizontally at its center, sprinkled with fine dry sand, and then vibrated at different frequencies. The sand migrates to the *nodal lines*, where there is no movement of the plate. In the dark regions between the nodal lines, the plate is in vertical vibrational motion.

(a)

(b)

(c)

(d)

(e)

(f)

Describe the symmetries of each of the six Chladni plates shown. (These Chladni plates were published in 1834 in *Of the Connection of the Physical Sciences*, by Mary Somerville, one of the great female mathematicians of the 19th century.)

36. (**Writing**) Write an illustrated short report entitled "Examples of Symmetry in _____," where the blank is filled in with your choice of topic. For example, you might choose "Sports," "Board Games," "Jewelry," "Musical Forms," "Native American Art," "Corporate Logos," or "Flowers." Use your imagination, and draw on your outside interests and hobbies to be creative. You should include drawings, photocopies, pictures cut from discarded magazines, and so on. Be sure to identify and classify the type of symmetry found in each example.

37. (**Writing**) Go on a symmetry hunt across campus, looking for striking examples of symmetry in buildings, decorative brickwork, sculptures, gardens, or wherever you may find it. Provide photos or drawings of three or four examples that you find especially interesting. Describe and classify the types of symmetry found in your examples. Include a border pattern and a wallpaper pattern.

State Assessments

38. (Grade 5)
Laura came across a broken piece of pottery as shown below. It seemed to be a half of a circular plate with a very symmetrical design. If Laura traced the design on a piece of paper, what transformation can she use to make a drawing of the plate before it was broken?

A. Flip B. Translation
C. Flip and slide D. Rotation

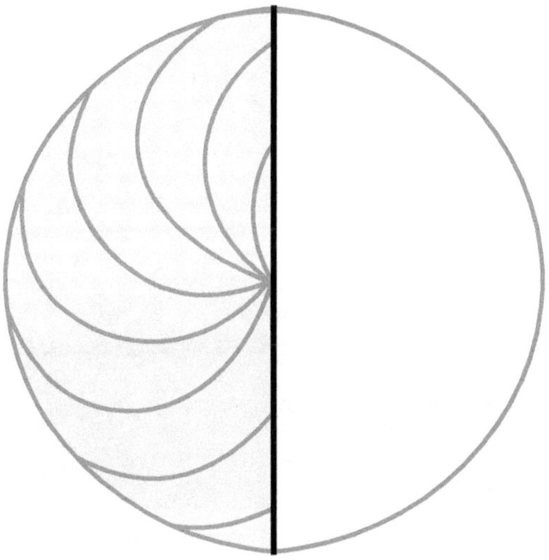

39. (Grade 5)
Jill folded a piece of paper in half twice and then used scissors to cut out a shape that she discarded. When she unfolded the paper it had this shape.

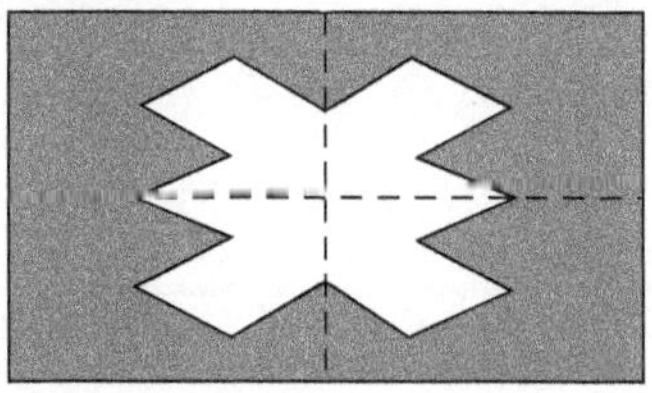

Which of the shapes below shows the paper after it was cut but just before it was unfolded?

A. 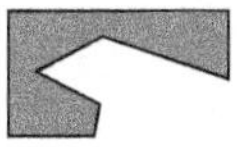 B. 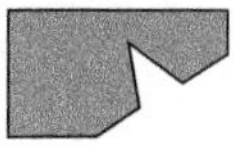 C.  D. 

40. (Grade 8)
The five-sided region at the left has been used as a template to form the border pattern below.

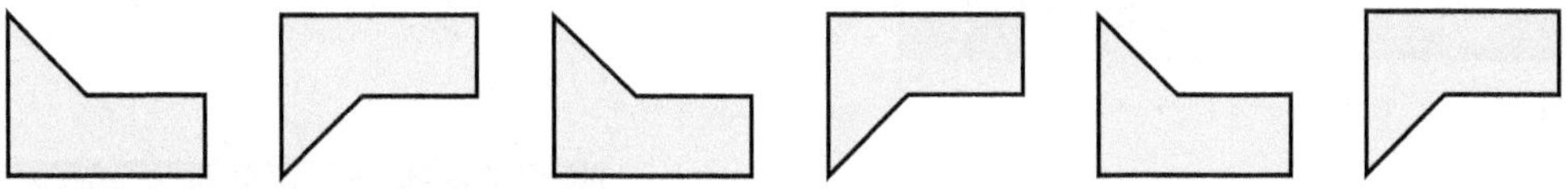

What rigid motion has been used to move the template along the border pattern?

A. reflection B. translation C. rotation D. glide-reflection

11.3 Tilings and Escher-like Designs

This section explores patterns formed by covering the plane with repeated shapes (or motifs) that neither overlap nor leave uncovered gaps. The art of tiling and decorative patterns has a history as old as civilization itself. In virtually every ancient culture, the artisan's choices of color and shape were guided as strongly by aesthetic urges as by structural or functional requirements. Imaginative and intricate patterns decorated baskets, pottery, fabrics, wall coverings, and weapons. Some examples of ornamental patterns from different cultures are shown in Figure 11.20.

Cotton textile hanging, Chimú, Peru. 12th–15th century

Detail of a tiled wall in the Alhambra

Fabric from Africa

Ca'd'Oro Venice Mosaic floor in courtyard

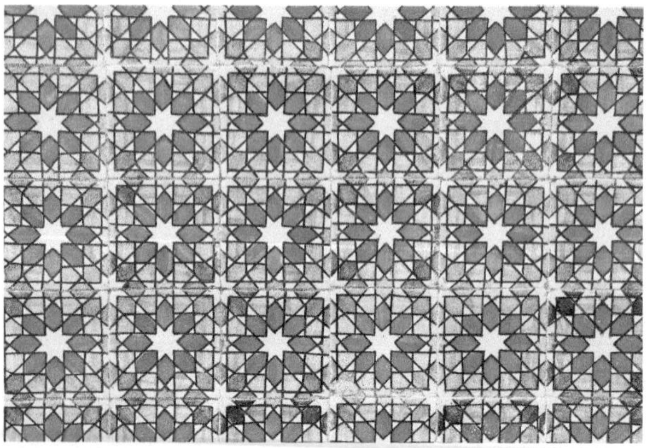

Tiling from Portugal 15th to 16th centuries

Chinese latticework, old window with geometrical and floral patterns

Figure 11.20
Patterns from various cultures

In recent times, the interest in tilings and patterns has gone beyond their decorative value. For example, metallurgists and crystallographers wish to know how atoms can arrange themselves in a periodic array. Similarly, architects hope to know how simple structural components can be systematically combined to create large building complexes, and computer engineers hope to integrate simple circuit patterns into powerful processors called neural networks. The mathematical analysis of tilings and patterns is a response to these contemporary needs. At the same time, the creation and exploration of tilings provides an inherently interesting setting for geometric discovery and problem solving in the elementary and middle school classroom. In particular, children enjoy learning how to create their own periodic drawings in the style of the pioneering Dutch artist M. C. Escher (1898–1972).

Tiles and Tilings

The precise meaning of a tile and a tiling is given in the following definition:

DEFINITION Tiles and Tiling

A simple closed curve, together with its interior, is a **tile.** A set of tiles forms a **tiling** of a figure if the figure is completely covered by the tiles without overlapping any interior points of the tiles.

Tilings are also known as **tessellations,** since the small square tiles in ancient Roman mosaics were called *tesserae* (probably because each tile has four corners and four sides; Greek *tessera*, four) in Latin.

Regular Tilings of the Plane

Each tiling shown in Figure 11.21 is a **regular tiling:** The tiles are congruent regular polygons joined edge to edge.

Figure 11.21
The three regular tilings of the plane

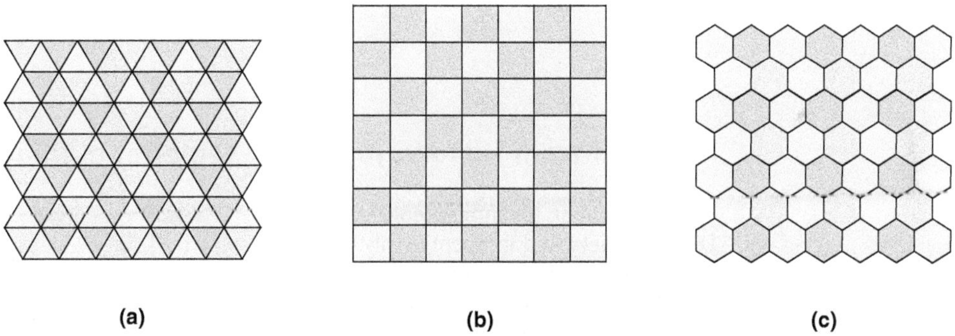
(a)
(b)
(c)

Any arrangement of nonoverlapping polygonal tiles surrounding a common vertex is called a **vertex figure.** Thus, four squares form each vertex figure of the regular square tiling, and three regular hexagons form each vertex figure of the hexagonal tiling. The measures of the interior angles meeting at a vertex figure must add to 360°. For example, in the square tiling, $90° + 90° + 90° + 90° = 360°$.

Suppose we attempt to form a vertex figure with regular pentagons as shown in Figure 11.22. The interior angles of a regular pentagon each measure $(5 - 2) \cdot 180°/5 = 108°$, so three regular pentagons fill in $3 \cdot 108° = 324°$ and leave a 36° gap. However, four regular pentagons create an overlap, since $4 \cdot 108° = 432° > 360°$. Because a vertex figure cannot be formed, no tiling of the plane by regular pentagons is possible.

Figure 11.22
Regular pentagons do not tile the plane

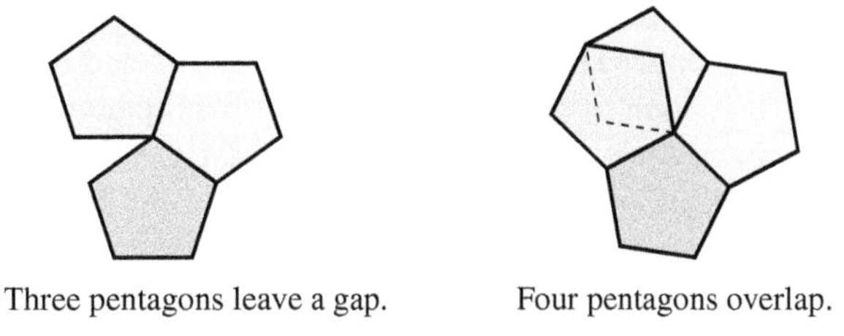
Three pentagons leave a gap.
Four pentagons overlap.

Similarly, since a regular polygon of seven or more sides has an interior angle larger than 120°, three meeting at a vertex must overlap. From all these considerations, we thus have the following theorem:

THEOREM The Regular Tilings of the Plane
There are exactly three regular tilings of the plane: (a) by equilateral triangles, (b) by squares, and (c) by regular hexagons.

Semiregular Tilings of the Plane

An edge-to-edge tiling of the plane with *more than one* type of regular polygon *and* with *identical vertex figures* is called a **semiregular tiling.** It is important to understand the restriction made about the vertex figures: *The same types of polygons must surround each vertex, and they must occur in the same order.* The two vertex figures in Figure 11.23 are not identical, since the two triangles and the two hexagons in the left-hand figure are adjacent, but on the right side the triangles and the hexagons alternate with one another.

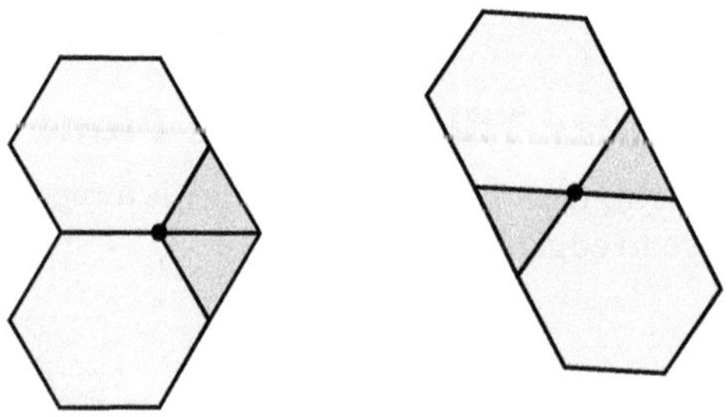

Figure 11.23
There are two distinct types of vertex figures formed by two equilateral triangles and two regular hexagons

Highlight from History: Johannes Kepler and Tiling Patterns

The astronomer Johannes Kepler (1571–1630) is celebrated in scientific history for his identification of the elliptical shape of the orbits of the planets about the sun. Less known is Kepler's contribution to the theory of tiling. Here are some drawings from Kepler's book *Harmonice Mundi,* which he published in 1619.

SOURCE: Kepler, Johannes; HARMONICE MUNDI, 1619.

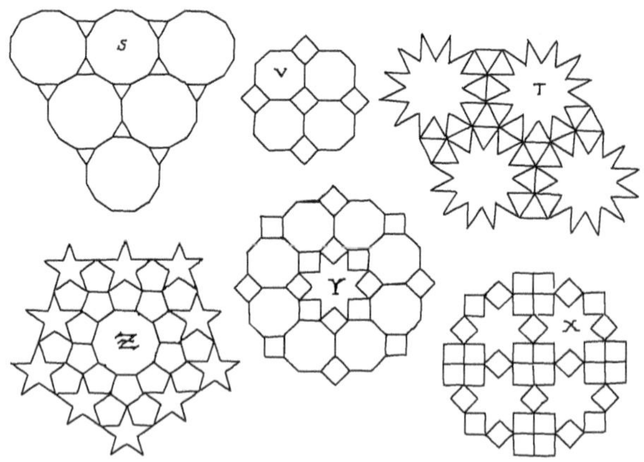

To see if the vertex figures in Figure 11.23 can be extended to form a semiregular tiling, we must check to see if the pattern can be completed to make *all* of the vertex figures match the one shown. It is soon discovered that the left-hand pattern with adjacent triangles cannot be extended. (Try it!) In contrast, the vertex figure at the right with alternating triangles and hexagons extends to a semiregular tiling. You should be able to find that tiling in Figure 11.24.

It can be shown that there are 18 ways to form a vertex figure with regular polygons of two or more types. Only 8 of them extend to a semiregular tiling; these are shown in Figure 11.24.

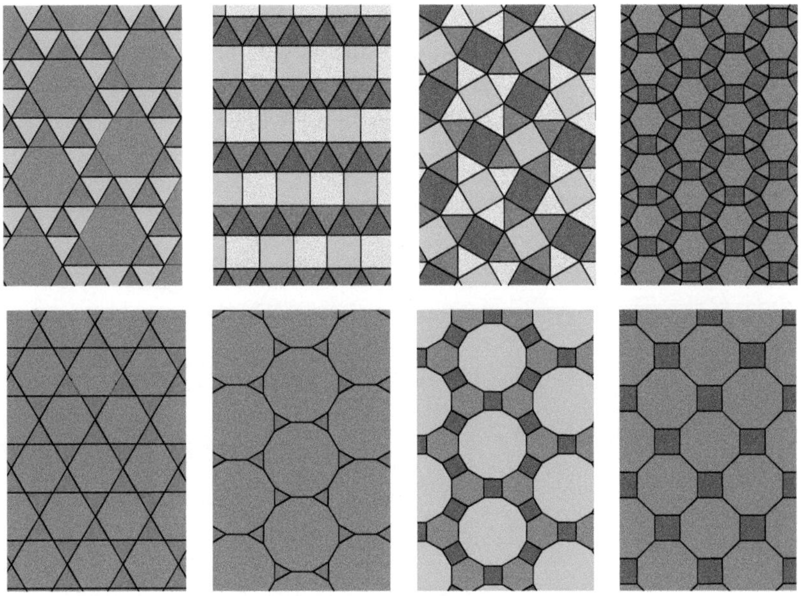

Figure 11.24
The 8 semiregular tilings

Tilings with Irregular Polygons

In the next example, it is helpful to cut tiles from cardboard or heavy card stock. You can then trace around them to see if it is possible to tile the plane with this one shape.

EXAMPLE 11.12

Exploring Tilings with Irregular Polygons

Which of the following polygons tile the plane?

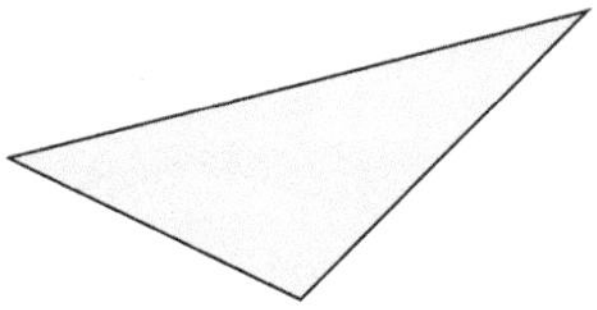

(a) Scalene triangle

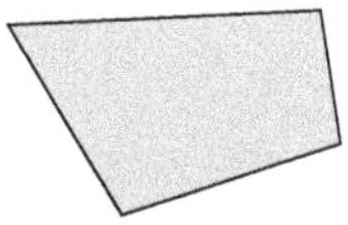

(b) Convex quadrilateral

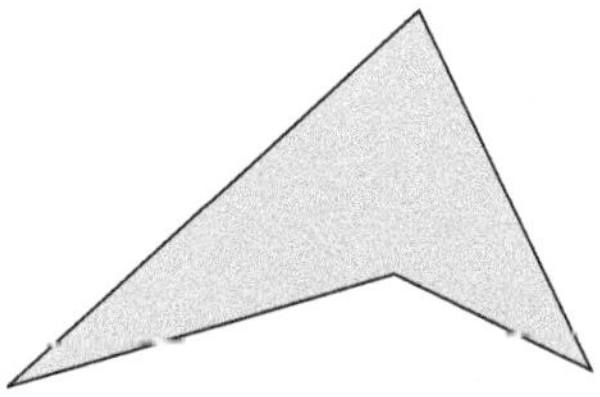

(c) Concave quadrilateral

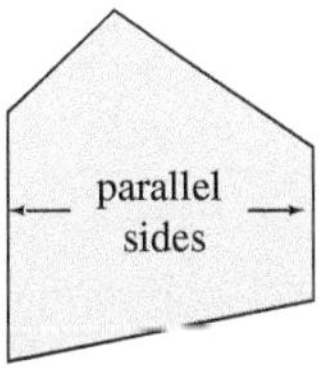

(d) Pentagon with a pair of parallel sides

Solution

(a) If the triangle is turned 180° about the midpoint of one edge, then the two triangles joined at the common edge form a parallelogram. Since parallelograms tile the plane by translations that match opposite edges, it follows that *any triangle will tile the plane.*

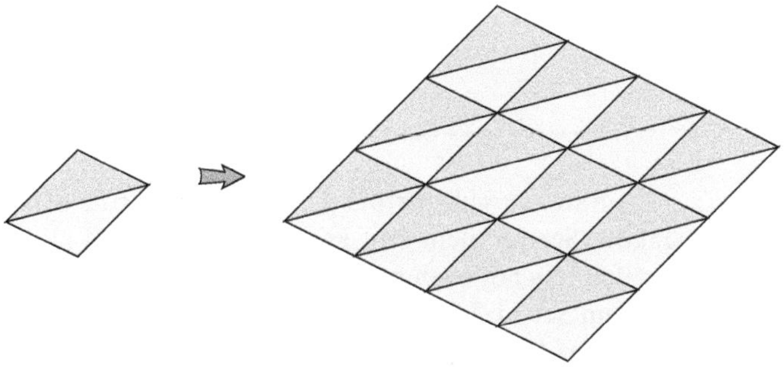

(b) and **(c)** A 180° turn about the midpoint of any side of the quadrilateral forms a convex or concave hexagon with three pairs of opposite congruent edges. Translating the hexagon while matching opposite edges produces the tiling, so *any* quadrilateral, convex or concave, tiles the plane. Each vertex of the tiling is surrounded by angles congruent to the four angles of the quadrilateral, whose measures add up to 360°.

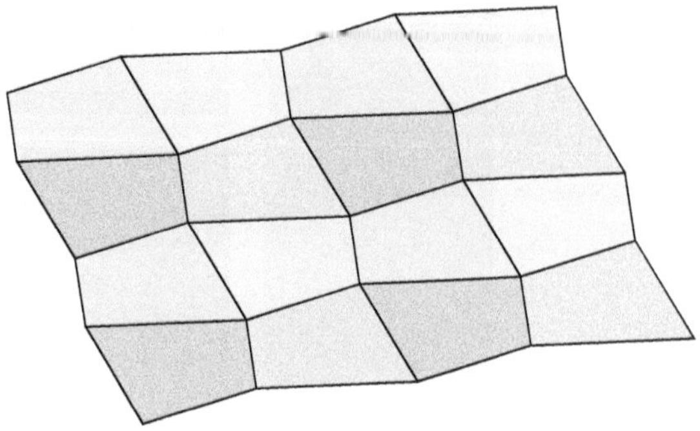

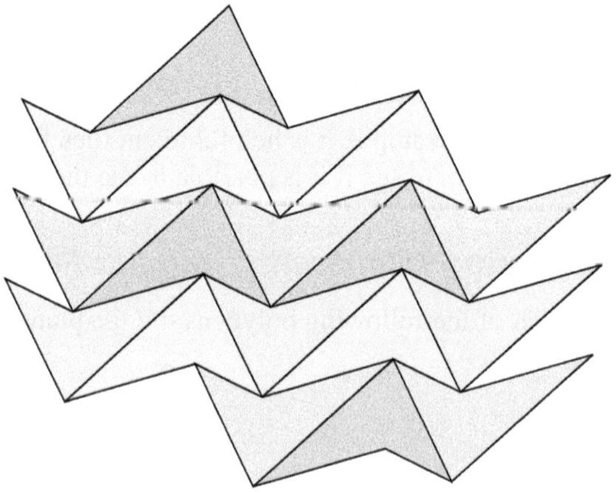

(d) A pentagonal tile with two parallel sides can always tile the plane in the manner shown below. If the parallel edges are congruent, the tiling will be edge to edge, but otherwise there will be a vertex along an edge of each tile.

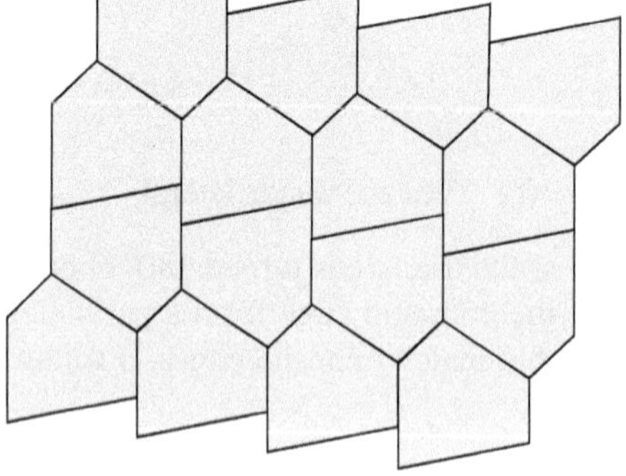

A hexagon will tile the plane if it has a pair of opposite sides that are parallel and of the same length. (See problem 6 of Problem Set 11.3.) If all three pairs of opposite sides are congruent and parallel, it is not even necessary to rotate the tile from one position to any other. (See problem 7 of Problem Set 11.3.) It has been shown that no convex polygon with seven or more sides can tile the plane. The following theorem summarizes these discoveries:

THEOREM Tiling the Plane with Congruent Polygonal Tiles
The plane can be tiled by

- any triangular tile;
- any quadrilateral tile, convex or not;
- certain pentagonal tiles (for example, those with two parallel sides);
- certain hexagonal tiles (for example, those with two opposite parallel sides of the same length).

The plane cannot be tiled by any convex tile with seven or more sides.

Although no convex polygon of seven or more sides can tile the plane, there are many interesting examples of tilings with nonconvex polygons. Figure 11.25 shows a striking example of a spiral tiling by 9-gons (nonagons) created by Heinz Voderberg in 1936.

Figure 11.25 Heinz Voderberg's spiral tiling with nonagons

Escher-like Designs

The Dutch artist Maurits Cornelius Escher (1898–1972) created a large number of artistic tilings. His designs have great appeal to the general public and have also captured the interest of professional geometers. Escher's periodic drawings, such as the birds design shown in Figure 11.26, are often based on modifications of known tilings by polygons.

Figure 11.26 M. C. Escher's birds

To understand how Escher created his birds design, start with a parallelogram as shown at the left of Figure 11.27. Next, replace an edge of the parallelogram with a curve and translate this curve to the opposite side of the parallelogram. Do the same type of modification to the other pair of

parallel sides. This pair of translations creates the outline of the bird motif, which is made recognizable by adding feathers and an eye.

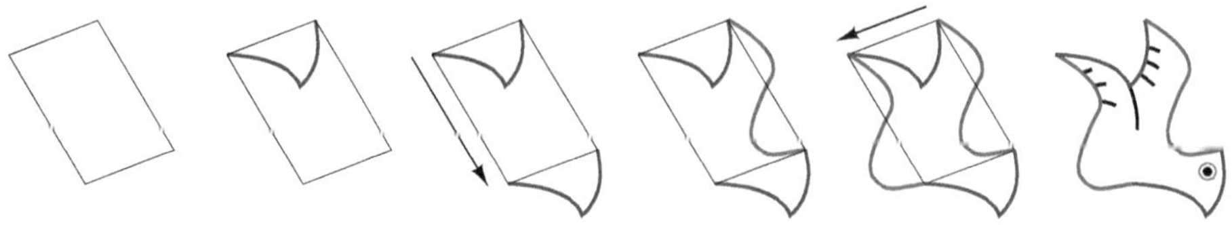

Figure 11.27
Modifying a parallelogram with two translations

Since a parallelogram tiles the plane under translations using the sides of the parallelogram as slide arrows, Figure 11.28 shows both the underlying tiling by parallelograms at the left and the periodic "bird" tiling at the right.

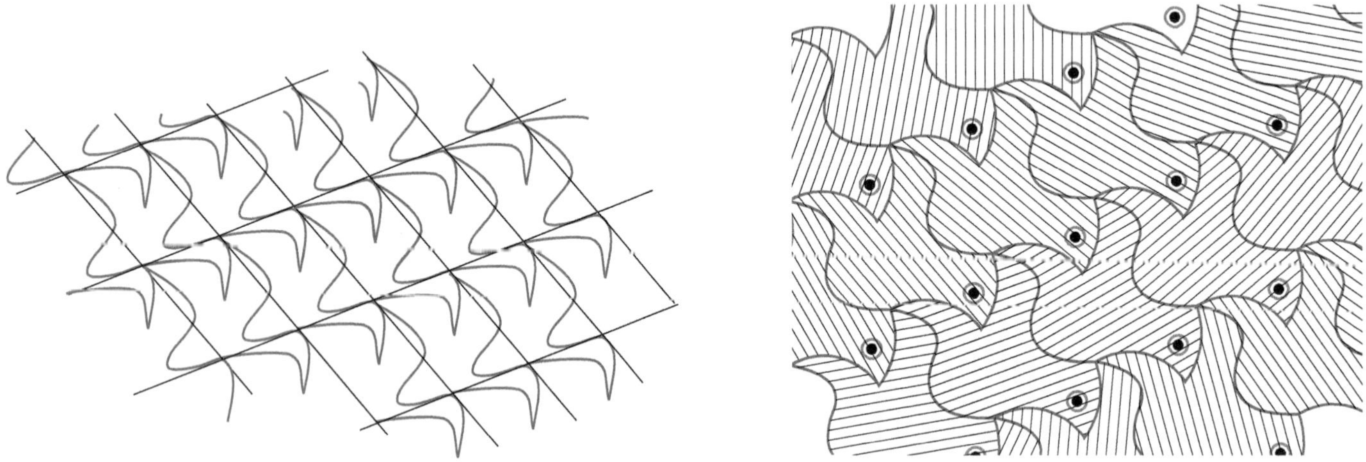

Figure 11.28
A tiling by parallelograms is modified to create a tiling by birds

Similar procedures will transform any polygonal tiling that can be produced by translations to an Escher-like tiling. For example, sixth-grade teacher Nancy Putnam used translations to modify each of the three pairs of opposite parallel congruent sides of a hexagon. Her whale tiling, starting with hexagon *ABCDEF*, is shown in Figure 11.29.

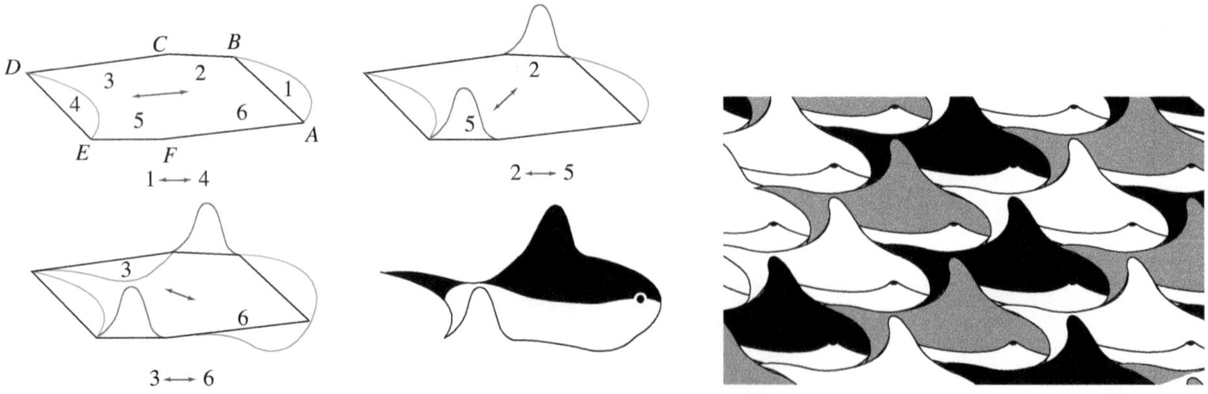

Figure 11.29
Sixth-grade teacher Nancy Putnam modified a hexagon with opposite parallel congruent sides to create an Escher-like tiling

Rotations can also be used to create **Escher-like tiles** with interesting symmetries. Figure 11.30 shows how a lizard tile can be created by modifying a regular hexagon *ABCDEF*. Side $\overline{AB}$ is first modified and then rotated about vertex *B* to modify side $\overline{BC}$. The remaining two pairs of adjacent sides are also modified by rotations, resulting in the outline of the lizard tile shown in the figure, along with its tiling.

Figure 11.30
Modifying a regular hexagon with rotations to create a lizard tiling

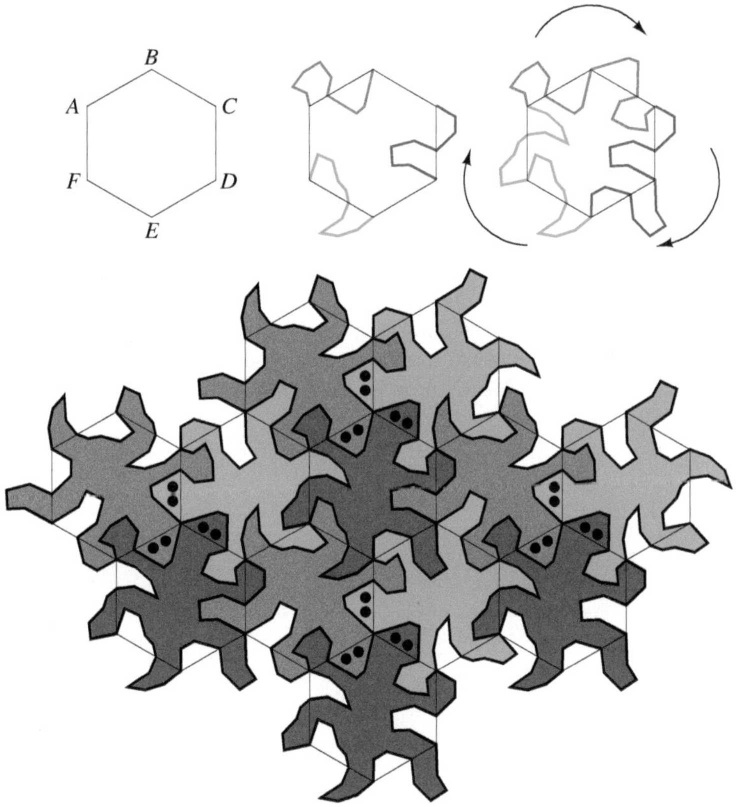

COOPERATIVE INVESTIGATION
Creating an Escher-like Design

Materials Needed

Note cards, 3″ × 5″ (or other card stock), tape, scissors, pencils and colored markers, blank sheets of paper.

Directions

Step 1: Cut a small (say, $2\frac{1}{2}$″-by-3″) rectangle from a note card or card stock.

Step 2: Make an irregular curve joining the corners of one side of the rectangle.

Step 3: Cut out the curve. Translate the cutout piece to the opposite side and tape it in place.

Step 4: Repeat steps 2 and 3 for the remaining two parallel sides of the rectangle, as shown.

Step 5: Do an "inkblot" test. Is your shape a frog, a bird, a face, a ______? Brainstorm with a partner. It may help to rotate your shape or flip it over. Add eyes, mouth, nose, ears, feet, beaks, horns, clothing, scales, fur, and other imaginative details to make your tiling template recognizable and interesting.

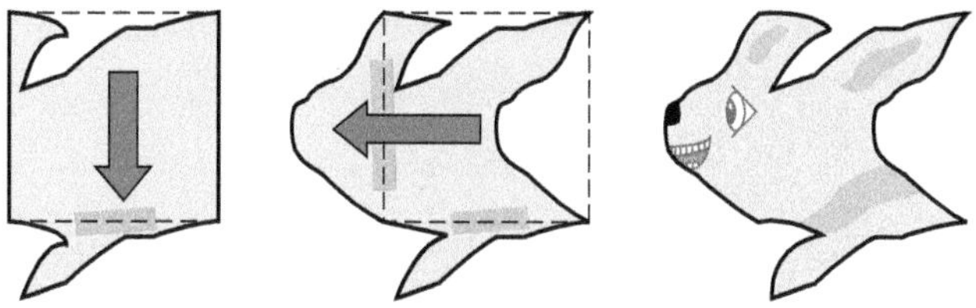

Step 6: Trace around the template on a blank sheet of paper. Translate and trace again, repeating to create at least three rows and three columns of your interlocking tiles to create your Escher-like tiling. Use colored markers to fill in the details, and color adjacent tiles with different colors.

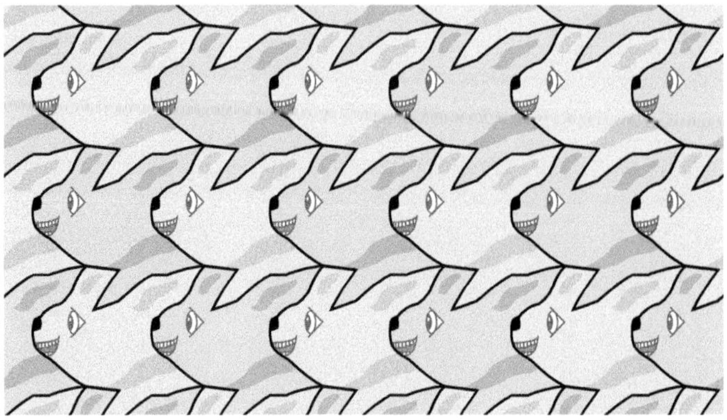

Extensions

Instead of a rectangle, start with any parallelogram (or any hexagon with opposite sides parallel and congruent) and follow the directions just given. It is also possible to adapt this method to create templates based on other tilings of the plane. Several suggestions are described in problems 22 through 25 of Problem Set 11.3.

11.3 Problem Set

Understanding Concepts

1. On dot paper arranged in a square grid, show that the given shape will tile the plane.

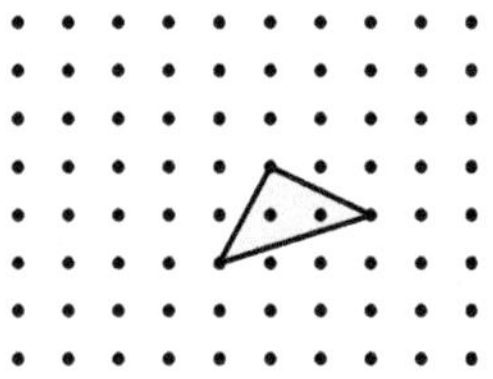

2. Show how to tile the plane with the tile shown.

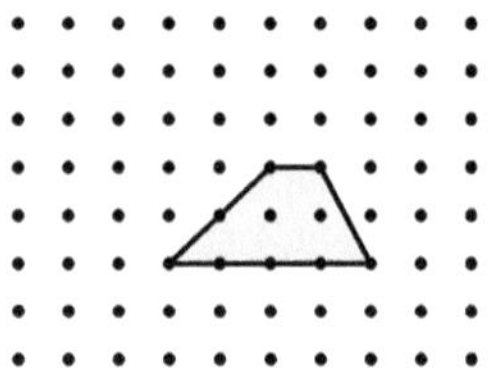

3. On "isometric" dot paper (arranged in a grid of equilateral triangles), show that the given shape will tile the plane.

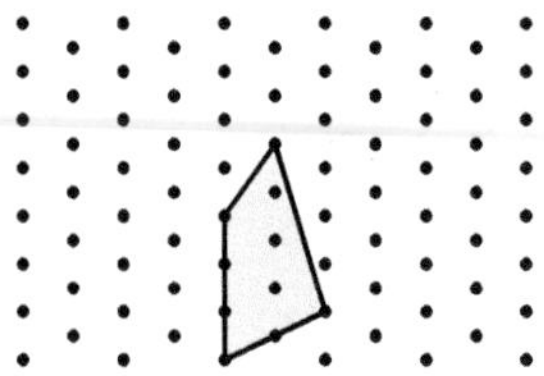

4. Tile the plane with the pentagon shown.

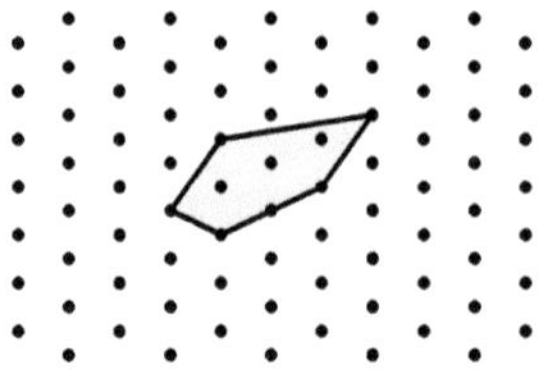

5. Cut a pentagon from a rectangular index card with three straight cuts as shown by the dashed lines. By repeatedly tracing its outline onto a sheet of paper, use the paper pentagon as a pattern to create a tiling of the plane.

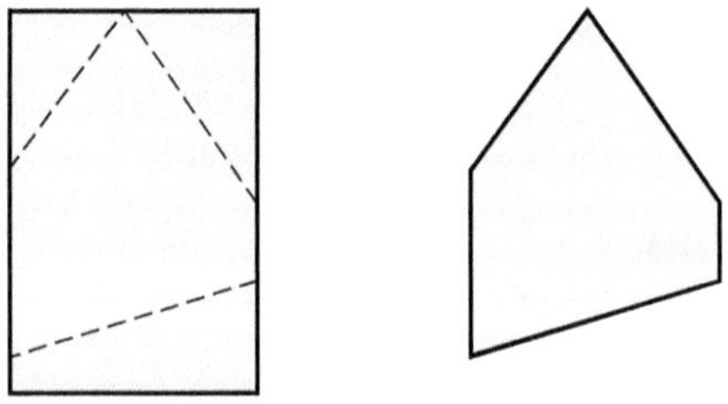

6. Cut a hexagon $ABCDEF$ from a rectangular piece of card stock as shown, using a ruler to ensure that the opposite sides $\overline{AB}$ and $\overline{DE}$ have the same length. No restriction is placed on the position of point C or F. Use the paper template to illustrate that a hexagon with a congruent and parallel pair of opposite sides can tile the plane. (*Suggestion:* Half-turns of the template will be required.)

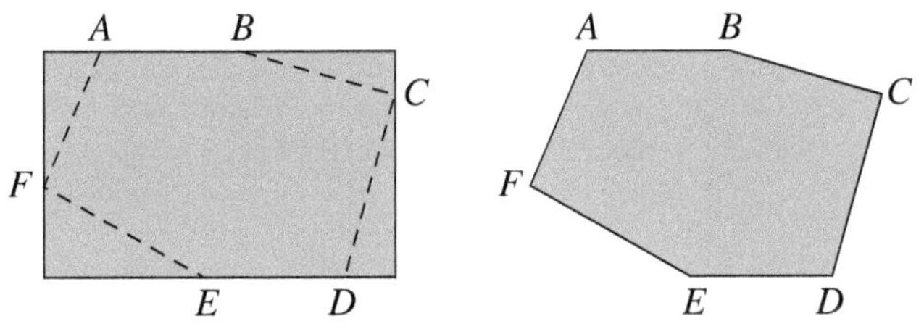

7. Fold a 3″-by-5″ note card in half, and then use scissors to cut (simultaneously) two general quadrilaterals. Rotate one quadrilateral a half-turn and tape the two quadrilaterals together along their corresponding edges to form a hexagonal tile.

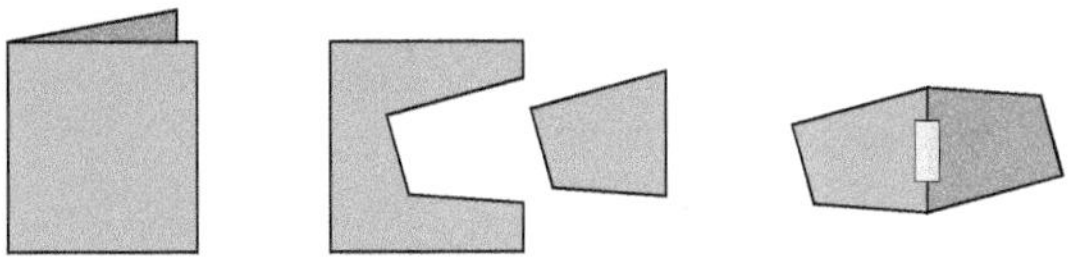

Fold   Cut quadrilaterals   Rotate and tape

Show, using the paper model as a template, that the hexagon tiles the plane. Must the tile be rotated?

8. Hold two index cards tightly together and use scissors to cut a pair of congruent convex quadrilaterals. Separate the quadrilaterals from one another. Cut one of the quadrilaterals along one of the diagonals, and cut the second quadrilateral along the other diagonal. Show that the four triangles T_1, T_2, T_3, and T_4 can be arranged to form a parallelogram and therefore tile the plane with repeated copies of themselves.

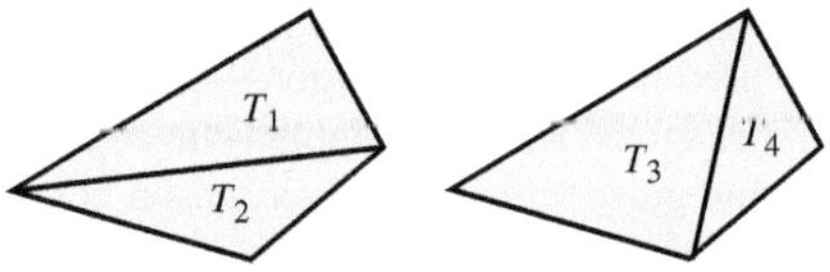

SOURCE: *Data from Quantum Magazine, September/October, 1992.*

9. The vertex figure shown below consists of two adjacent regular hexagons and two equilateral triangles surrounding vertex *V*.

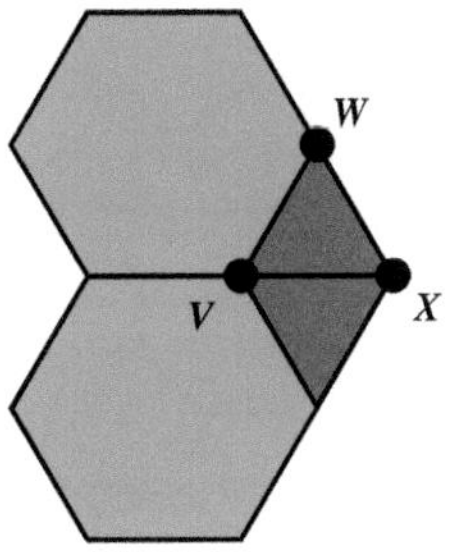

(a) Show that the interior angles of the four polygons add to 360°.

(b) Explain why this vertex figure cannot be extended to form a semiregular tiling of the plane.

10. Answer the questions of problem 9 for this vertex figure of regular polygons.

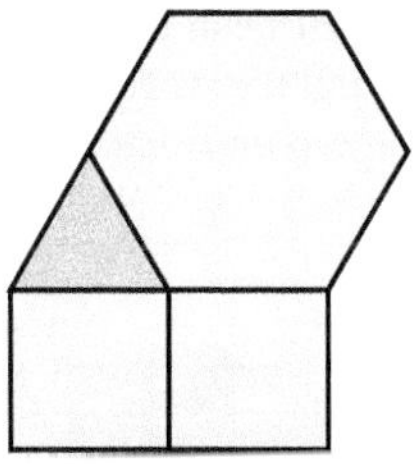

11. Consider the vertex figure formed by a square, a regular pentagon, and a regular 20-gon. Find the measures of the interior angle of each polygon and show that these three measures add up to 360°.

12. Some "letters" of the alphabet will tile the plane. For each letter shown, create an interesting tiling on square dot paper. Look for different patterns that use the same tile.

(a)   (b)   (c)   (d)   (e)

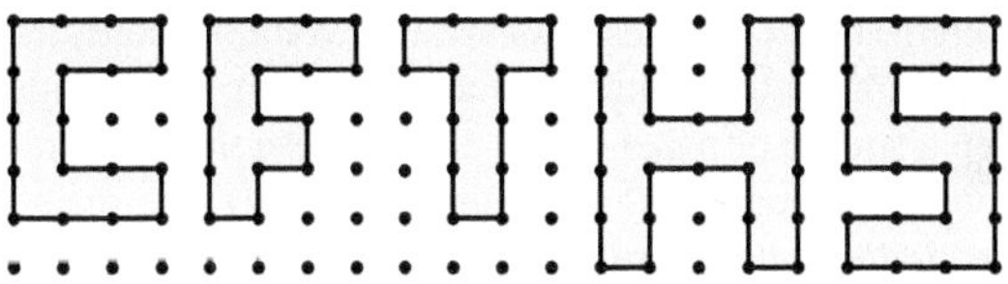

13. A *tetromino* is a tile formed by joining four congruent squares edge to edge, where adjacent squares must share a common edge. Two tetrominoes and two nontetrominoes are shown as follows:

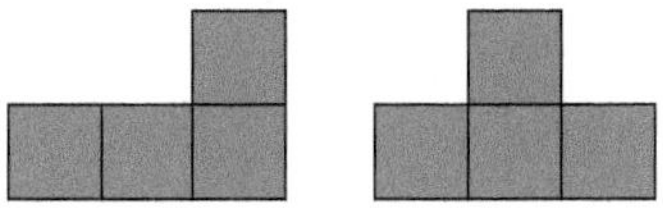

Two tetrominoes

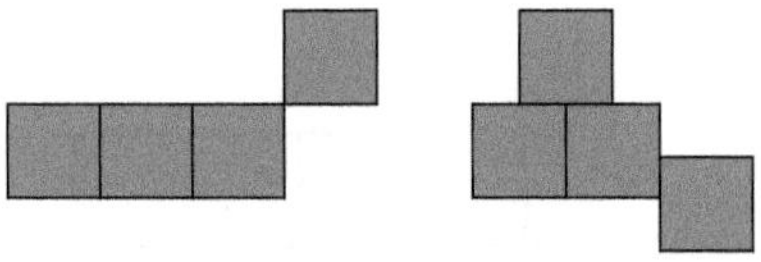

Two nontetrominoes

(a) There are five noncongruent tetrominoes altogether. Find the other three.

(b) Which tetrominoes tile the plane, if one is allowed to use unlimited congruent copies of one tetromino?

(c) The five noncongruent tetrominoes have a total area of 20 square units. Can the five shapes tile a 4-by-5 rectangle? (*Suggestion:* Imagine that the rectangle is colored in a red-and-black checkerboard pattern of unit squares. How many red and how many black squares are covered by each tetromino?)

14. Tiles formed by joining five congruent squares edge to edge are called **pentominoes.** The 12 pentominoes are shown in problem 22 of Problem Set 9.3. Use graph paper (or dot paper) to decide which pentominoes tile the plane.

Into the Classroom

15. (**Writing**) **Patterns in World Cultures.** Children's interest in patterns and symmetries can be heightened by incorporating examples from around the world. For example, the pattern shown here is a pattern on *kente* cloth, a fabric woven by the Ashanti and the Ewe in Togo, West Africa.

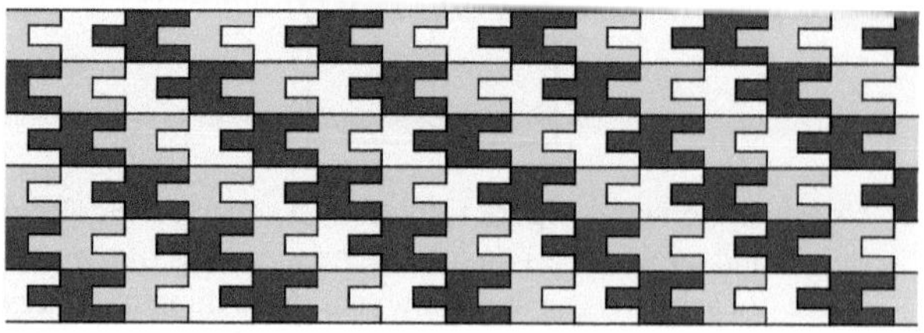

Conduct a search in your library or on the Web to discover four or five examples of wallpaper patterns from a variety of world cultures. Give a careful description of the symmetry properties of each example.

16. (**Writing**) **Connections with Art.** The concepts of symmetry and pattern have close ties with art. For example, consider this assignment in an elementary school classroom: Each student is given a square of paper and is asked to use three colors to decorate each square with a design having 180° symmetry. The decorated squares are then used to tile one or more posters that are hung on the classroom wall.

(a) Carry out the design project just described in your own class.

(b) Create a similar lesson using a different tiling shape with different color and symmetry conditions.

Responding to Students

17. Miguel claims that the following tiling is a regular tiling by squares:

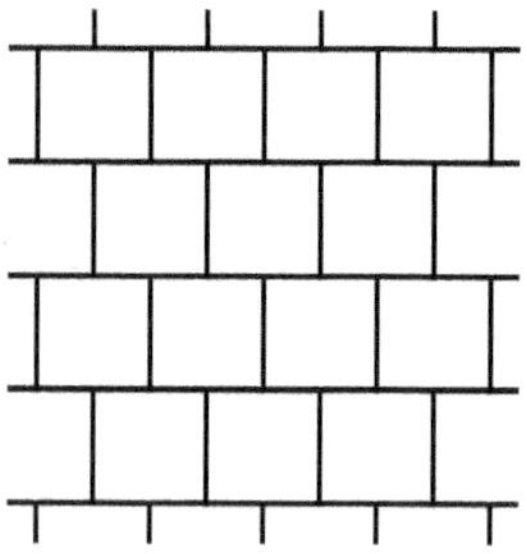

What is Miguel overlooking?

18. Myra cut out a regular pentagon, hexagon, and octagon, and it seems to her that they form a vertex figure such as this:

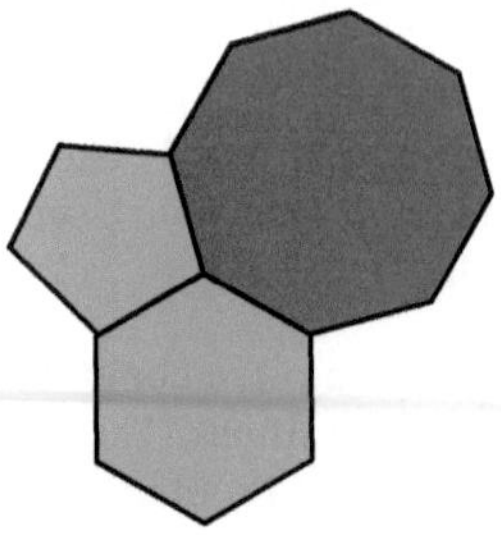

Respond to Myra.

19. Wailea noticed that two adjacent equilateral triangles, a square, and a dodecagon form a vertex figure such as that shown at A. Therefore, she claims that there should be a semiregular tiling in which this vertex figure occurs at every vertex.

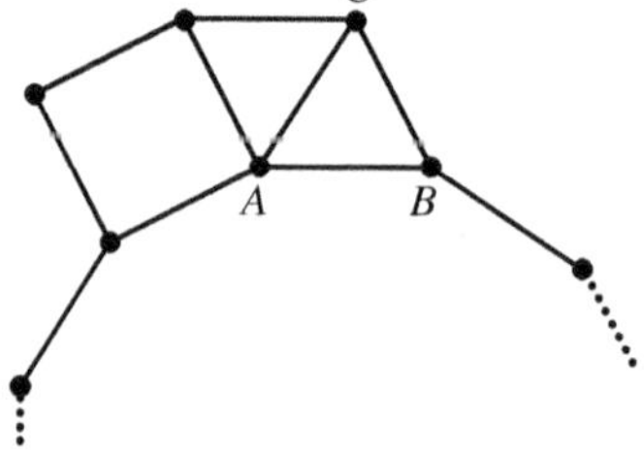

(a) Is Wailea correct that these four polygons create a vertex figure at A?

(b) Explain to Wailea why the vertex figure does not extend to a semiregular tiling.

20. Mason used computer geometry software to make the following figure:

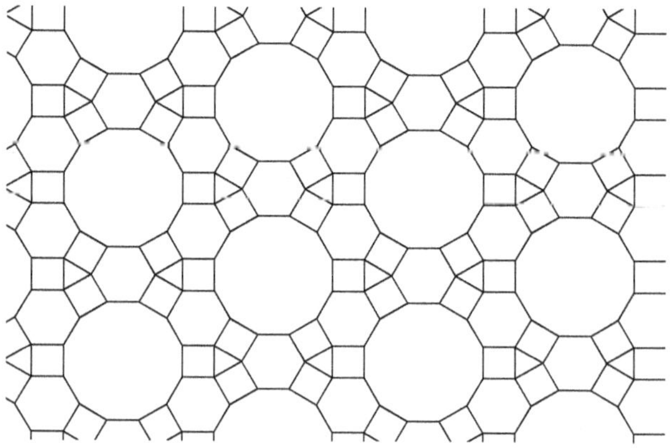

Since he used only regular polygons—equilateral triangles, squares, hexagons, and dodecagons—he believes he has found a new semiregular tiling. What has Mason overlooked?

Thinking Critically

21. Branko Grünbaum and G. C. Shephard (*Tilings and Patterns,* W. H. Freeman and Co., 1987) discovered the tiling shown here in the children's coloring book *Altair Design* (E. Holiday, London: Pantheon, 1970):

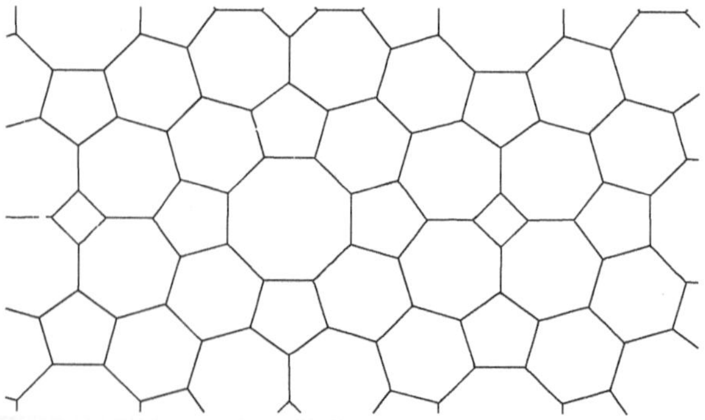

(a) What kinds of polygons appear?

(b) Grünbaum and Shephard claim that this is a "fake" tiling by regular polygons. Explain why.

22. Construct a paper hexagonal tile with each pair of opposite sides parallel and congruent. The template can be cut from a note card following the method described in problem 7. Make cutouts on three adjacent sides. Translate each cutout to the opposite side and tape along the corresponding edges to form a template. (See the instructions in the Cooperative Investigation box "Creating an Escher-like Design.")

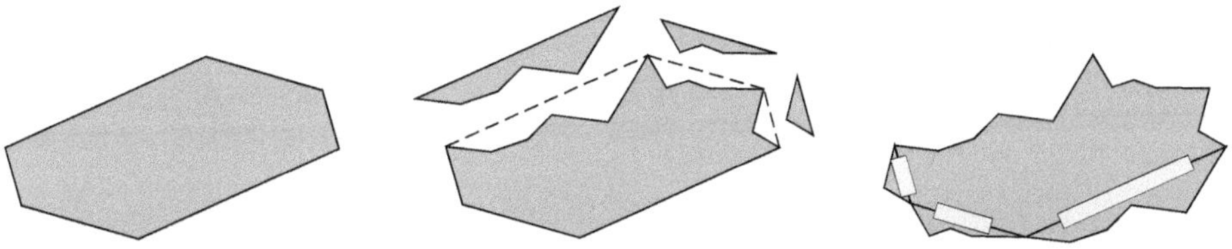

Create a design with your template, adding details such as eyes, mouths, and so forth, to give added interest to the design.

23. Cut an accurate square from a note card. Make cutouts on opposite sides. Rotate each cutout 90° and tape as shown. Use the paper template to create an Escher-like tiling. You will need to use 90° rotations to produce the tiling.

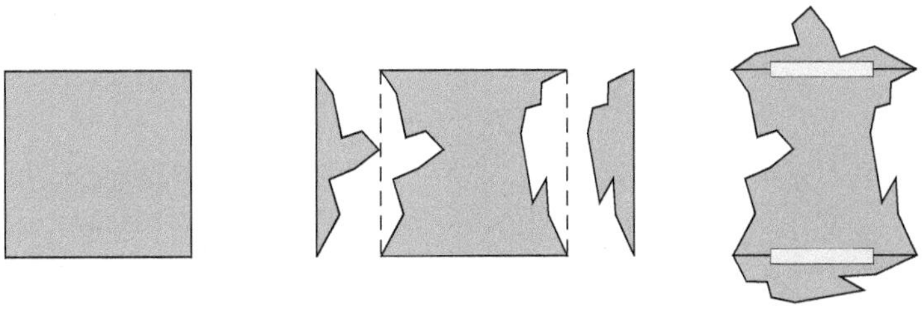

24. Cut an arbitrary triangle from a note card and lightly fold (do not make a heavy crease) one vertex to another to determine the midpoint of the side between the vertices. The side can be modified by making a cutout on one side of the midpoint, rotating the cutout 180° about the midpoint, and taping it in place. The steps to modify one side of a triangle are as follows:

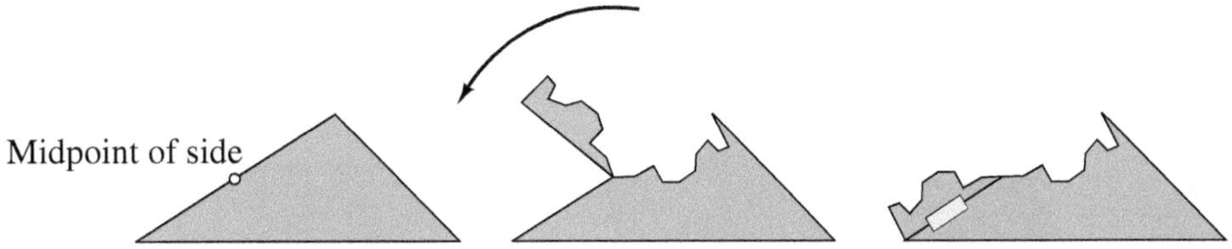

Modify the remaining two sides of the triangle in a similar manner and use the resulting template to create an Escher-like tiling. You will need to use 180° rotations to produce the tiling.

---

25. Cut a convex quadrilateral from a note card. Make midpoint modifications, as described in problem 24, to each of the four sides. Use the resulting template to create an Escher-like design.

26. Suppose a vertex figure of regular polygons includes a regular octagon. Show that the figure must include another octagon and a square.

27. An equilateral triangle and a parallelogram are each examples of "reptiles," short for "repeating tile." In each case, four copies of the tile can be arranged to form a larger, similar shape.

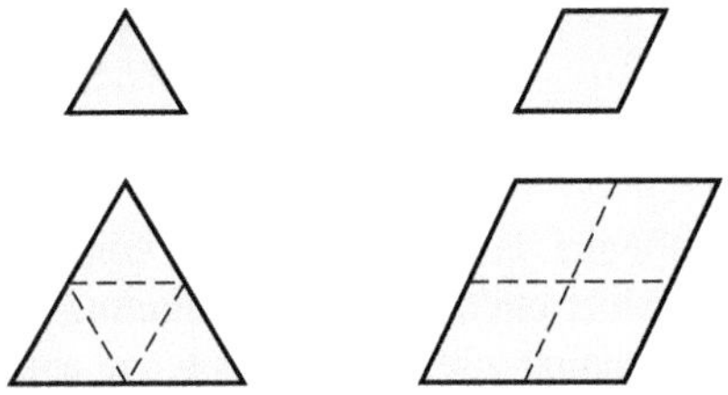

Use square dot paper to show that each of the following shapes is a "reptile":

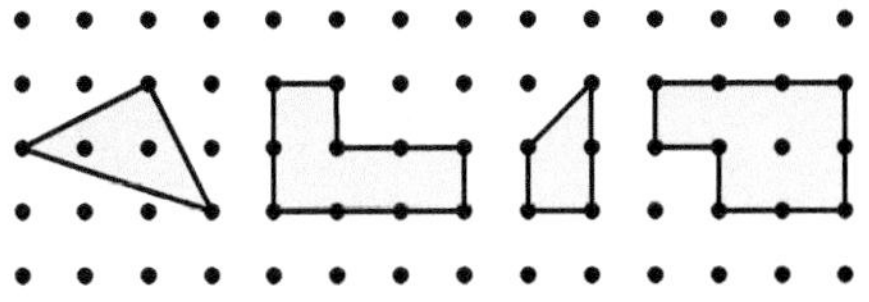

28. A hexiamond is formed from six congruent equilateral triangles. There are 12 different hexiamonds, including the Sphinx, Chevron, and Lobster, shown next. Find the remaining 9 hexiamonds, and see if you can match their shapes to their names: Hexagon, Crook, Crown, Hook, Snake, Yacht, Bar, Signpost, and Butterfly.

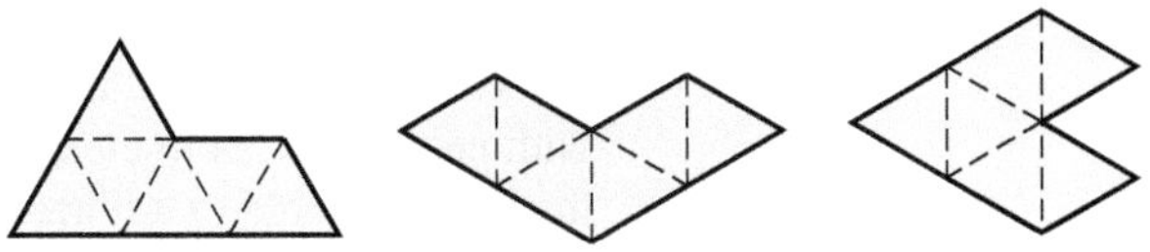

29. (a) Show that the Sphinx is a reptile. (See problems 27 and 28 for the definition of a reptile and a diagram of the Sphinx.)

(b) It requires four copies of the Sphinx to form a second-generation Sphinx. How many copies of the original Sphinx are required to form a third-generation Sphinx? Explain your reasoning and provide a sketch.

(c) Explain why any reptile provides a tiling of the plane.

30. Cut out the seven tangram pieces from a square of card stock, as shown at the left below. The figure at the right shows another way that all seven tangram pieces can be arranged to tile a convex polygon. Use your tangram pieces to discover and sketch all 13 noncongruent convex polygons that can each be tiled using the complete set of seven tangram pieces.

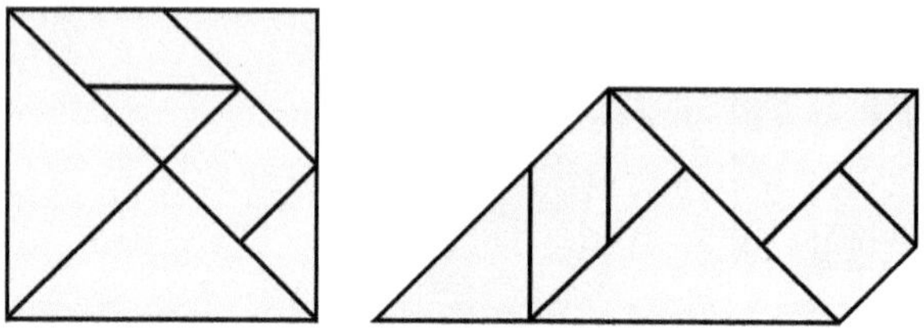

State Assessments

31. (Grade 4)

Naomi has a bucket of pattern blocks with these shapes.

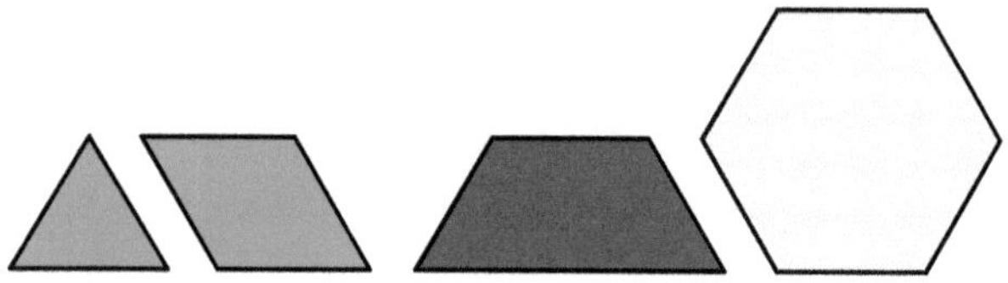

She discovered there were several ways to use a set of these blocks to make this six pointed star.

Which of the following sets of blocks *cannot* be used to make the star?

A. 4 of the red trapezoids
B. 12 of the green triangles
C. 6 of the blue rhombuses
D. 1 yellow hexagon and 6 green triangles

32. (Grade 8)

Which of these shapes does not tile the plane?

A. an equilateral triangle
B. a square
C. a regular pentagon
D. a regular hexagon

33. Provide an analysis of the following problem that was released by a state student assessment office.

Irene is making a tessellation using the shape shown below.

Which of the following tessellations can be made using only a clockwise rotation?

A.

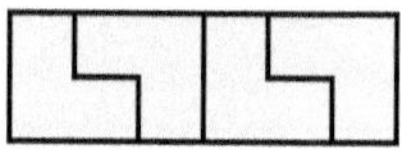

B.

C.

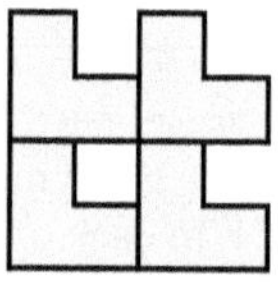

D.

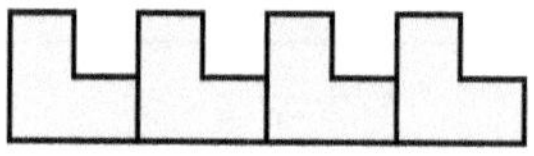

The Chapter in Relation to Future Teachers

This chapter has introduced the basic concepts of transformational geometry, including the rigid and similarity motions that give precise meaning to the fundamental notions of the congruence and similarity of geometric figures. Transformations were then applied to classify patterns according to the type of symmetry they possessed. In the opposite direction, transformations were used to create patterns of interest.

While the chapter has given a thorough and at times technically detailed description of motions and symmetries, the teacher should be aware that much of this material is of appeal and interest even to very young children. For example, a preschooler typically enjoys simple picture puzzles in which a puzzle piece cut from wood must be placed into the matching hole on the puzzle board—and here the child is in fact carrying out a rigid motion and checking for congruence. Similarly, a kindergartner enjoys working with pattern blocks to create an attractive pattern—often, a pattern that exhibits considerable symmetry that is extendable to a tiling of the plane. Somewhat older children enjoy combining mathematics with artistic creativity—and certainly the graphic work originating with Escher has broad appeal to all ages.

Chapter 11 Summary

Section 11.1 Rigid Motions and Similarity Transformations	Page Reference
DEFINITIONS	
• A **transformation** of the plane is a one-to-one correspondence of the points of the plane onto itself.	561
• If point P on a plane corresponds to point P' under a transformation, then P' is the **image** of P and P is the **preimage** of P'.	561
• A **rigid motion,** or **isometry,** is a transformation of the plane that preserves distance: $PQ = P'Q'$ for all points P and Q and their corresponding image points P' and Q'.	561
• The **identity transformation** is the rigid motion for which each point corresponds to itself: $P = P'$ for all points P in the plane.	562
• Two transformations of the plane are **equivalent** if both transformations take each point P onto the same image point P'.	562
• **The four basic rigid motions:** translation (or slide), rotation (or turn), reflection (or flip), and glide–reflection.	562
• A **translation (slide)** is one of the four basic rigid motions of the plane in which all points of the plane are moved the same distance in the same direction.	562
• The **slide arrow,** or **translation vector,** is an arrow that characterizes a translation; the direction of the slide is the direction of the arrow and the distance of the slide is the length of the arrow.	562
• A **rotation (turn)** is one of the four basic rigid motions of the plane in which one point of the plane is held fixed and the remaining points are turned around the center of rotation through the same number of degrees.	563
• The **center of rotation** is the fixed point about which the other points rotate.	563
• The **turn angle** is the number of degrees that the points rotate.	563
• The **turn arrow** is an arrow in the shape of a circular arc drawn to indicate the turn center and the directed angle corresponding to the turn center.	563
• **Reflections (flips)** are one of the four basic rigid motions in which all points P are transformed to the opposite side P', but the same distance away, from a determined line.	564
• The **line of reflection (mirror line)** is the line over which all the points are flipped.	564
• **Glide–reflections** are one of the four basic rigid motions that combine both slides and reflections. The points reflect over a mirror line and slide a fixed distance parallel to the mirror line.	566
• The **glide arrow,** or **glide vector,** is the vector of the glide.	566
• An **orientation-reversing** motion is one that interchanges handedness, such as reflections and glide–reflections. Motions that do not reverse handedness are called **orientation preserving,** such as translations and rotations.	566
• Two figures are **congruent** if and only if one figure is the image of the other under a rigid motion.	572
• **Dilation,** or **size transformation,** with center O and scale factor k takes each point P other than O to the point P' on the ray OP so that $OP = kOP$ and leaves O fixed.	573

• Scale factor k is a **contraction** if $0 < k < 1$ or an **expansion** if $k > 1$.	573
• A **similarity transformation** is a transformation composed of a sequence of dilations and rigid motions.	574
• Two figures are **similar** if and only if there is a similarity transformation that takes one figure onto the other.	574
THEOREMS	
• **Equivalence properties of multiple reflections:**	
A sequence of two reflections across parallel lines is equivalent to a translation.	569
A sequence of two reflections across intersecting lines is equivalent to a rotation.	569
A sequence of three reflections across parallel or concurrent lines is equivalent to a single reflection. Otherwise, a sequence of three reflections is equivalent to a glide–reflection.	570
• **Classification theorem for rigid motions:** Every rigid motion is equivalent to one of the four basic motions: a translation, a rotation, a reflection, or a glide–reflection.	571
• **Distance change under a dilation:** Under a dilation with scale factor k, all distances change by the factor k: $P'Q' = kPQ$.	573

Section 11.2 Patterns and Symmetries	Page Reference
DEFINITIONS	
• A figure has **symmetry** if there is a rigid motion that takes every point of the figure onto an image point that is also in the figure.	582
• A figure has **reflection symmetry** if the reflection over the mirror line, or line of symmetry, takes a figure onto itself.	582
• A figure has **rotation symmetry,** or **turn symmetry,** if there is a rotation about some center O that superimposes the figure onto itself.	583
• Figures have **circular symmetry** if they turn onto themselves for any degree of rotation about some point O.	583
• A figure has **point symmetry** if it has 180° rotation symmetry about some point O.	584
• A **periodic pattern** is a figure with translation symmetry.	585
• A **periodic border pattern** is a periodic pattern with translation symmetry in one direction only.	585
• A **wallpaper pattern** is a planar pattern with translation symmetries in more than one direction.	585
THEOREMS	
• **Classification of border patterns:** Every border pattern has one of seven possible symmetry types (given the symbolic descriptions *mm*, *mg*, *m*1, 1*g*, 12, or 11).	585, 586
• **Classification of wallpaper patterns:** Every wallpaper pattern has one of seventeen possible symmetry types.	585, 586
• **Crystallographic restriction:** A wallpaper pattern can have (smallest) rotation symmetries only of size 60°, 90°, 120°, or 180°.	588

Section 11.3 Tilings and Escher-like Designs	Page Reference
CONCEPTS	
• **Escher-like designs:** Designs resembling the work of M. C. Escher can be created by modifying the straight sides of a polygonal tile to assume more general curves, following rules that preserve some of the tiling properties of the original shape.	599

DEFINITIONS

- A **tile** is a simple closed curve in the plane, together with its interior. 595
- A **tiling,** or **tessellation,** is a covering of a figure or the whole plane with tiles, having no gaps or overlaps. 595
- A **regular tiling of the plane** is a tiling for which the tiles are congruent regular polygons that are joined edge-to-edge. 595
- A **vertex figure** is an arrangement of nonoverlapping polygonal tiles surrounding a common vertex. 595
- A **semiregular tiling of the plane** is an edge-to-edge tiling of the plane with more than one type of regular polygon and with identical vertex figures. 596
- **Tiling with irregular polygons** is tiling with shapes that are not regular polygons or shapes with sides of an equal length. The Dutch artist Escher created a large number of artistic tilings of this type. 596

THEOREMS

- **Regular tilings of the plane:** There are three tilings of the plane, each created with congruent regular polygons placed edge-to-edge: equilateral triangles, squares, or regular hexagons. 596
- **Semiregular tilings of the plane:** There are eight semiregular tilings of the plane. 596
- **Tilings of the plane with a nonregular polygonal tile:** The plane can be tiled with any triangle, any quadrilateral, and certain pentagons and hexagons. No convex polygon with more than six sides can tile the plane. 598

Chapter Review Exercises

Section 11.1

1. Draw the image of $ABCDE$ under the translation that takes A onto A'.

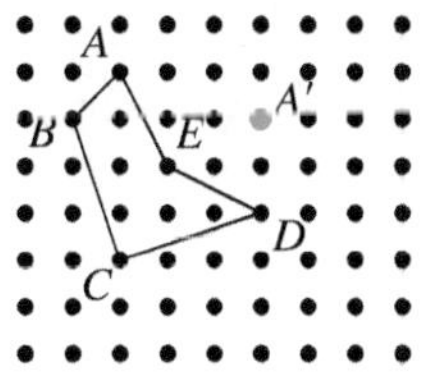

2. Determine the center and turn angle of the rotation that takes A onto A' and B onto B'. Use a protractor, Mira, ruler, or whatever drawing tools you wish.

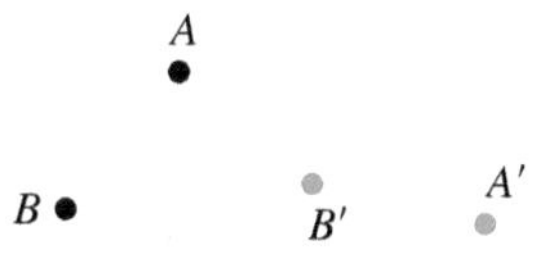

3. Describe the basic rigid motion that takes A, B, and C onto A', B', and C', respectively. Use any drawing tools you wish.

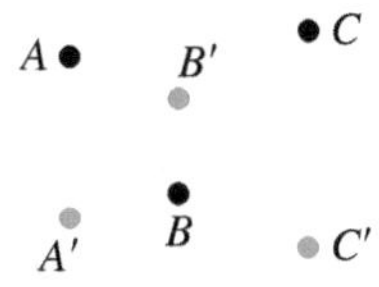

4. A glide–reflection has a horizontal line l as its glide mirror and translates 4 inches to the right. Draw three lines of reflection—m_1, m_2, and m_3—so that successive reflections across m_1, m_2, and m_3 result in a motion equivalent to the glide–reflection.

5. Sketch the image of the square $ABCD$ under each of these transformations:

(a) The dilation centered at O with scale factor 2.

(b) The dilation centered at A with scale factor $1/3$.

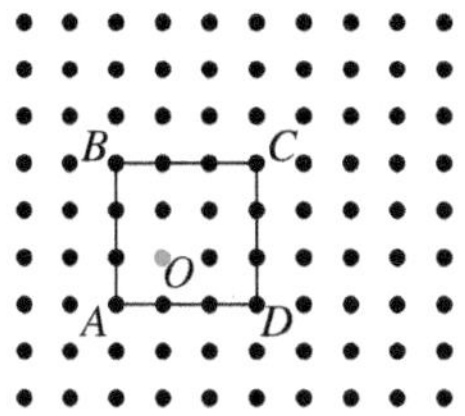

6. Describe the similarity transformation that takes the square *ABCD* onto the square *JKLM*, where *J*, *K*, *L*, and *M* are the midpoints of square *ABCD* as shown.

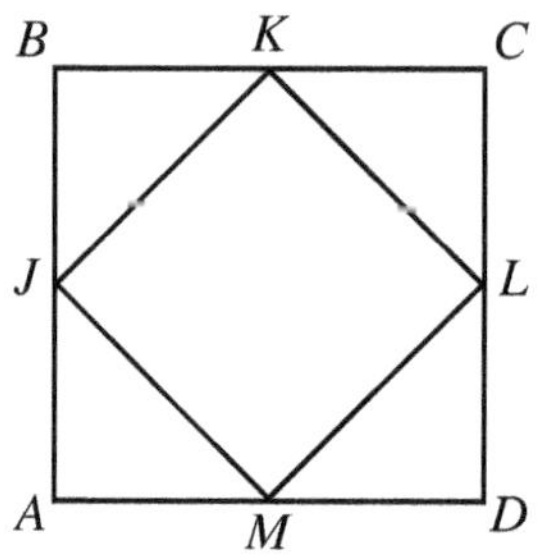

7. Three lines m_1, m_2, and m_3 are shown in each part. What type of rigid motion is equivalent to a sequence of reflections across m_1, m_2, and m_3?

(a)

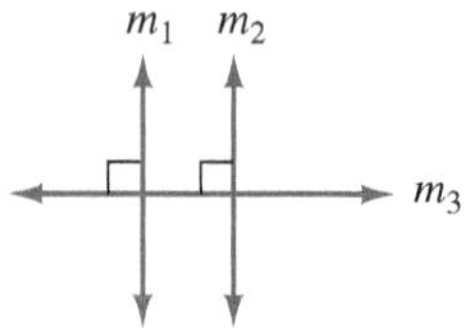

(b) Parallel lines

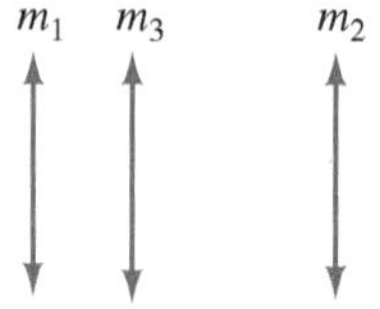

(c)

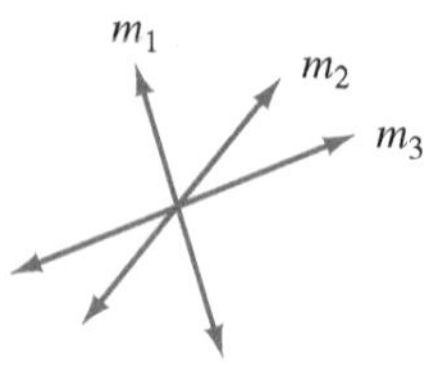

(d)

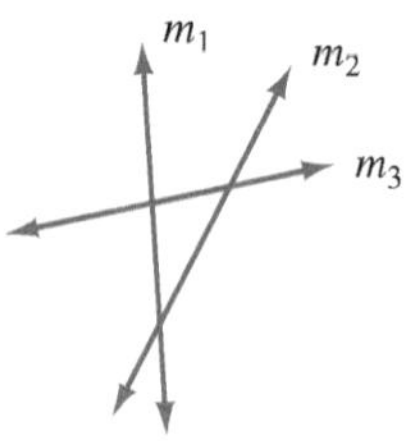

Section 11.2

8. The geometric forms shown are from African art. Describe all lines of symmetry for each figure.

(a)

(b)

(c)

(d)

(e)

(f)

9. What rule is used to separate the letters of the alphabet in the following arrangement?

ABCDE HI K MNO STUVWXYZ
FG J L PRQ

10. What is the common name for these polygons?
    (a) A triangle with no line of symmetry
    (b) A triangle with exactly one line of symmetry
    (c) A triangle with three lines of symmetry
    (d) A kite with two lines of symmetry
    (e) A regular polygon with six lines of symmetry

11. Identify these polygons.
    (a) A triangle with 120° rotational symmetry
    (b) A quadrilateral with 180° rotational symmetry
    (c) A regular polygon with 40° as its smallest angle of rotational symmetry

12. For each of the figures shown in problem 8, give all of the angles of rotation symmetry.

13. Determine the classification symbol of these border patterns.

(a)

FRENCH RENAISSANCE ORNAMENT FROM CASKET

(b)

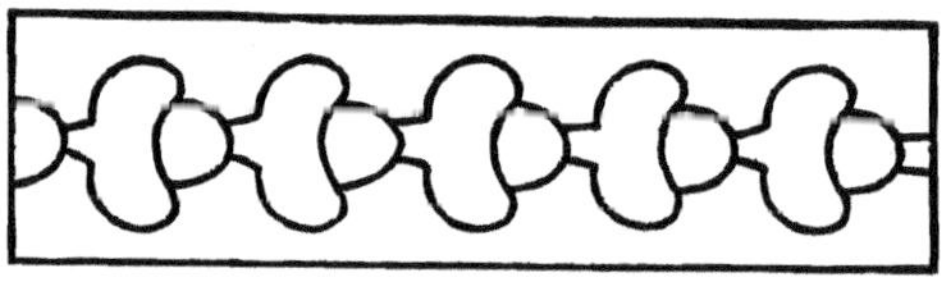

STAINED GLASS, CATHEDRAL OF BOURGES

(c)

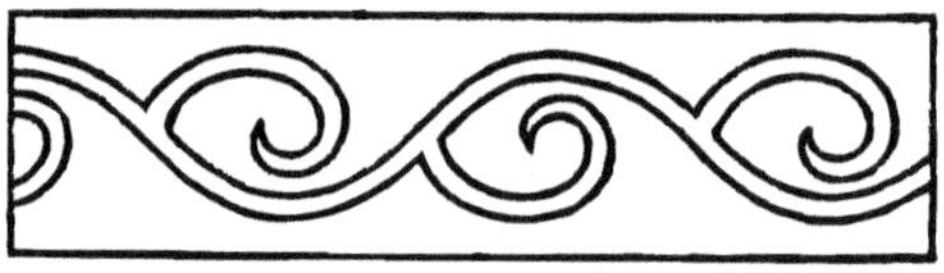

INDIAN PAINTED LACQUER WORK

(d)

MALTESE LACE

(e)

ANCIENT GREEK SCROLL BORDER

(f)

ITALIAN DAMASK OF THE RENAISSANCE

Section 11.3

14. Four regular polygons form a vertex figure in a tiling of the plane. Three of the polygons are a triangle, a square, and a hexagon. What is the fourth polygon?

15. Draw two different vertex figures that each incorporate three equilateral triangles and two squares.

16. Show that the following dart will tile the plane:

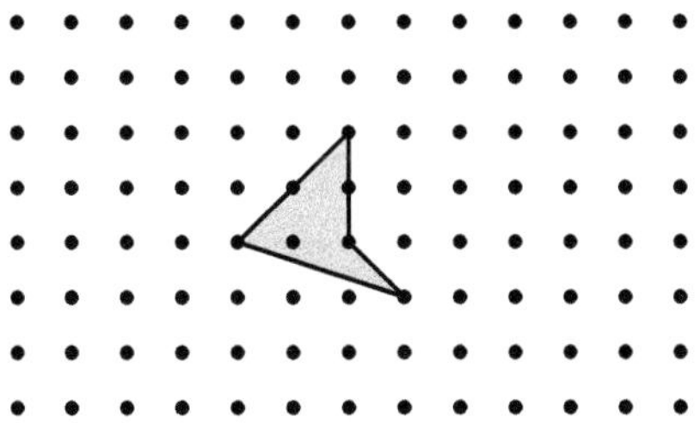

17. Which of the following polygons will tile the plane?

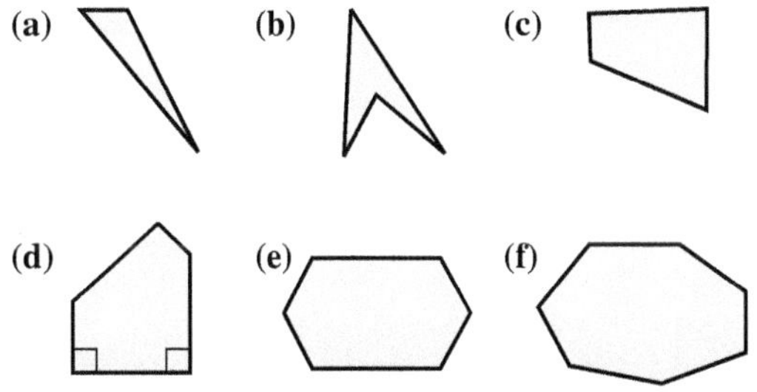

12

Congruence, Constructions, and Similarity

COOPERATIVE INVESTIGATION

Getting Rhombunctious! Folding Paper Polygons

Materials

Rectangular sheets of paper (thin, colorful paper works best, about 4″ by 6″ or 5″ by 7″), rulers, protractors, tape or glue sticks, scissors (optional).

Directions

The goal of this activity is to construct rhombuses and related polygons with paper folding. Some properties of the figures will be investigated by measuring lengths and angles. Unwanted flaps that are not part of the desired final figure can be taped or glued down (or cut off with scissors). It is very important that the folds be made with considerable care. Work in small groups to answer the questions about the properties of the polygons created by your group members.

I. A General Rhombus

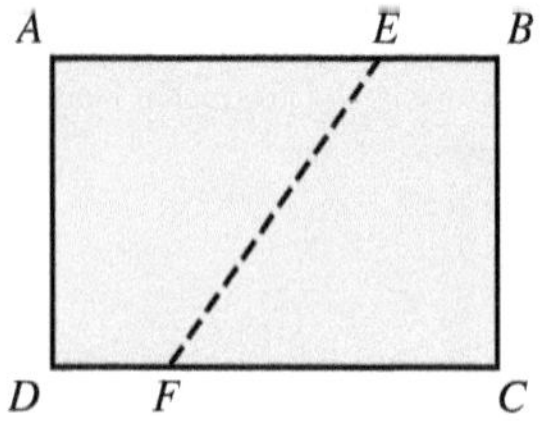

1. Fold C to A to construct segment $\overline{EF}$.

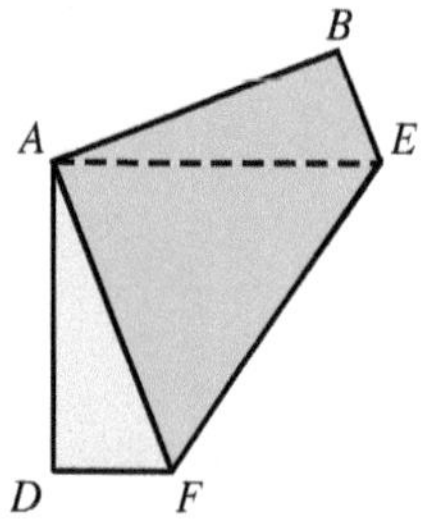

2. Fold up along $\overline{AE}$ and down along $\overline{AF}$.

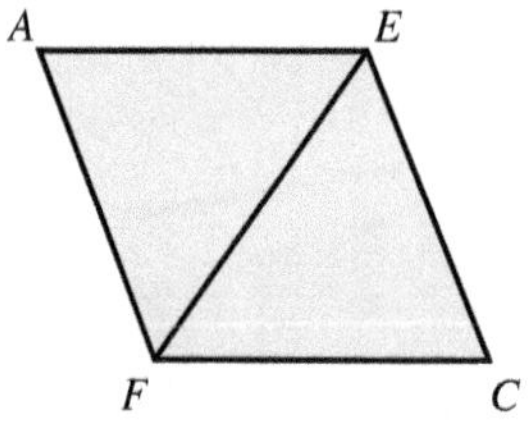

3. Unfold along $\overline{EF}$ to create the rhombus $AECF$ (with the unneeded triangles ACE and ADF taped or glued to the back side).

Questions

1. Use a ruler to measure the sides of $AECF$. Are they equal? How did the folding make this happen?
2. Measure the interior angles of $AECF$. How are they related?
3. Measure $\angle EFC$. Is $\overline{EF}$ an angle bisector?
4. Fold F on E and then unfold to create the segment $\overline{AC}$. At what angle do $\overline{AC}$ and $\overline{EF}$ intersect?

II. A Special Rhombus

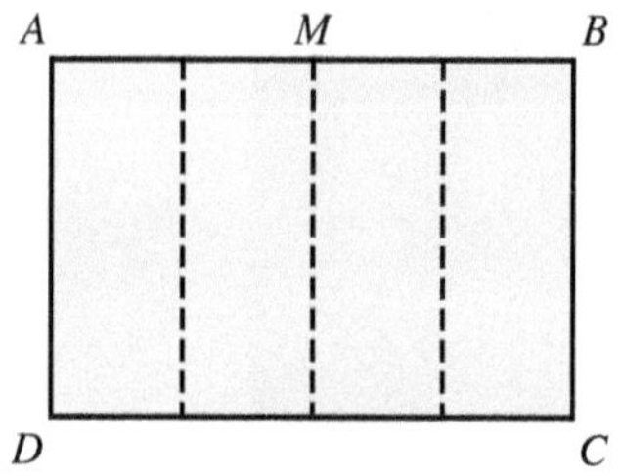

1. Fold a rectangle in half twice, and then unfold.

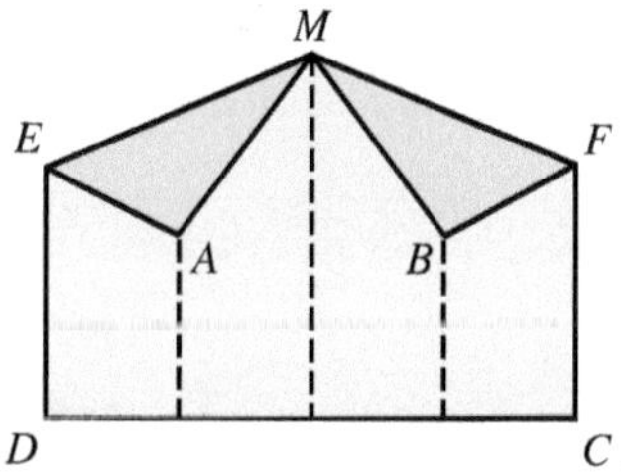

2. Fold A to the 1/4 vertical crease and B to the 3/4 crease, with both folds meeting the midpoint M of side $\overline{AB}$.

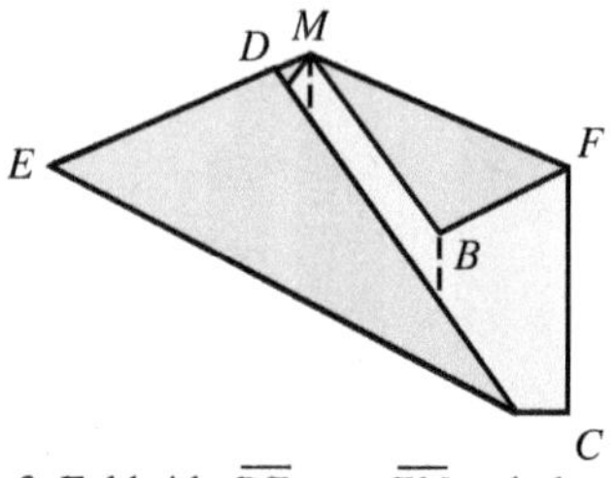

3. Fold side $\overline{DE}$ onto $\overline{EM}$ and glue the fold down.

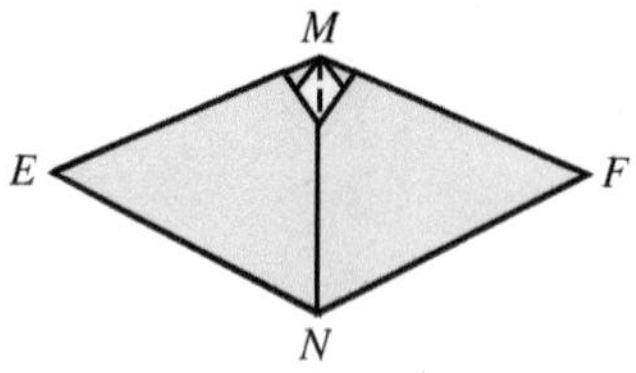

4. Fold side $\overline{CF}$ onto $\overline{MF}$ and glue the fold down.

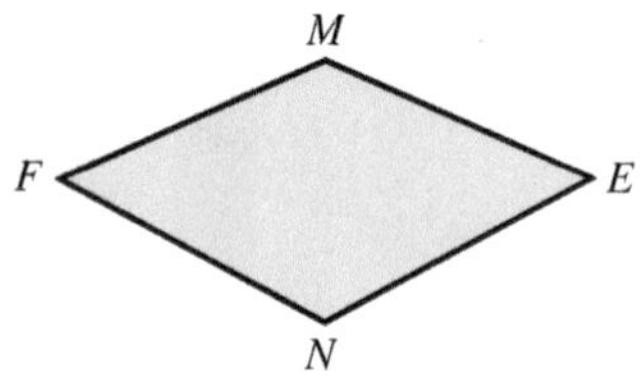

5. Turn the paper over to obtain the rhombus $FMEN$.

Questions

1. What special angles do you find in $FMEN$?
2. If $FMEN$ is folded across $\overline{MN}$, what special triangle is formed?
3. If points F and E are both folded to the midpoint of $\overline{MN}$, what special polygon is constructed?

CHAPTER PREVIEW

Two geometric figures are congruent if they have the same shape and the same size. Therefore, two triangles are congruent if the six measures of one triangle—the three side lengths and the three angle measures—correspond to the six measures of the second triangle. However, we will discover that it is often enough that just three of the six measures are equal. For example, if two triangles have the same lengths of their three sides, then it necessarily follows that the three angle measures are also the same and, therefore, the two triangles are congruent.

Congruence is then put to work constructing figures—perpendicular or parallel lines, angle bisectors, and regular polygons, for example—that have special properties. Unlike a drawing, in which the properties look to hold at least approximately, the properties in a construction can be proved to be exact. In classical Euclidean geometry, only a compass and straightedge (unmarked ruler) are allowed, but sometimes it is more practical to use a ruler (straightedge marked with distances) and a protractor, or a reflective tool such as the Mira, or perhaps even computer software.

Two geometric figures that have the same shape, but not necessarily the same size, are said to be similar. In the concluding section, we will explore simple criteria that guarantee when two triangles are similar. Similar triangles have many practical uses, especially for the indirect measurement of distances such as the height of buildings or trees, the distance across a river, and even the diameter of the sun.

KEY IDEAS

- Congruent triangles
- Properties of congruent triangles: side–side–side (SSS), side–angle–side (SAS), angle–side–angle (ASA), angle–angle–side (AAS)
- Triangle inequality
- Isosceles triangle theorem and its converse
- Thales' theorem
- Compass and straightedge constructions: congruent segment, congruent angle, parallel line, perpendicular line, midpoint and perpendicular bisector of a segment, angle bisector
- Inscribed circle and circumscribed circle of a triangle
- Construction of regular polygons
- Similar triangles
- Properties of similar triangles: angle–angle, side–side–side, side–angle–side
- Applications of similar triangles

12.1 Congruent Triangles

Congruent Line Segments and Their Construction

Since two line segments have the same shape, they are congruent if and only if they have the same length. Construction 1 shows how a compass and straightedge are used to construct a line segment that is congruent to a given segment. Recall that congruence is indicated by the symbol $\cong$.

CONSTRUCTION 1 Construct a Line Segment Congruent to a Given Segment

On a given ray $\overrightarrow{PZ}$, construct a line segment $\overline{PQ}$ that is congruent to a given line segment $\overline{AB}$.

Procedure

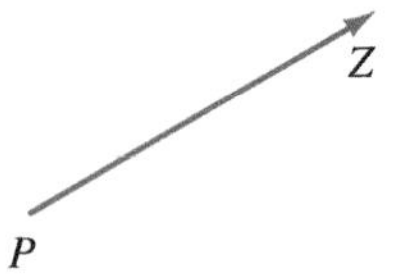

Step 1:
Construct ray $\overline{PZ}$.

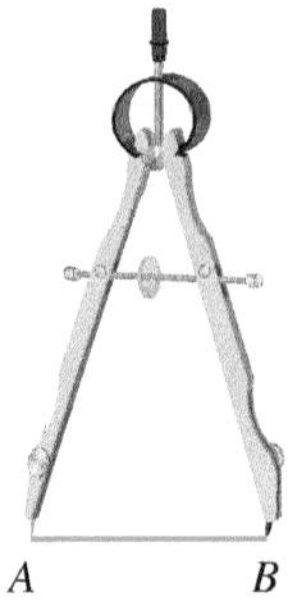

Step 2:
Put the sharp point of the compass at A and the pencil point on B.

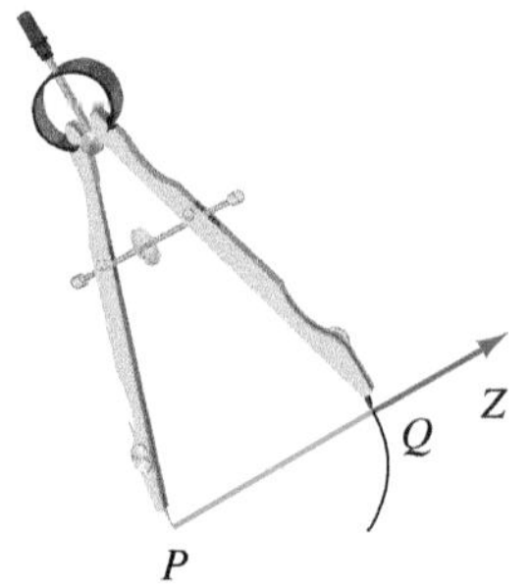

Step 3:
Without changing the opening of the compass, move the point of the compass to P and construct a circular arc that intersects $\overrightarrow{PZ}$. The point Q of intersection determines the segment for which $\overline{PQ} \cong \overline{AB}$.

Corresponding Parts and the Congruence of Triangles

The investigation of congruent triangles will consider these questions:

- What measurements of a triangle completely describe its size and shape?
- Given two triangles, what subsets of measurements are sufficient to decide whether the triangles are congruent to one another?
- Given certain measurements of a triangle ABC, how can a compass and straightedge be used to construct a triangle PQR that is congruent to triangle ABC?

The size and shape of a triangle are described completely if we specify the **six parts of a triangle**, namely, the three sides $\overline{AB}$, $\overline{BC}$, and $\overline{CA}$ and the three angles $\angle A$, $\angle B$, and $\angle C$. A second triangle PQR is congruent to triangle ABC if there is a matching of vertices $A \leftrightarrow P$, $B \leftrightarrow Q$, and $C \leftrightarrow R$ under which *all six* parts of triangle ABC are congruent to the corresponding six parts of triangle PQR. This situation is illustrated in Figure 12.1.

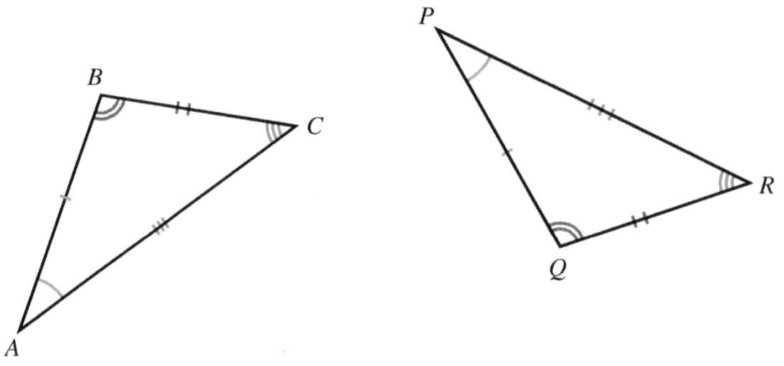

Figure 12.1
Triangles ABC and PQR are congruent under the vertex correspondence $A \leftrightarrow P$, $B \leftrightarrow Q$, and $C \leftrightarrow R$ if and only if $\overline{AB} \cong \overline{PQ}$, $\overline{BC} \cong \overline{QR}$, and $\overline{CA} \cong \overline{RP}$, and $\angle A \cong \angle P$, $\angle B \cong \angle Q$, and $\angle C \cong \angle R$

DEFINITION Congruent Triangles

Two triangles are **congruent** if and only if there is a correspondence of vertices of the triangles such that the corresponding sides and corresponding angles are congruent.

The notation $\triangle ABC \cong \triangle PQR$ is read "Triangle ABC is congruent to triangle PQR." Mathematical notation and symbols must always be read with precision. This is especially true for the symbolic statement $\triangle ABC \cong \triangle PQR$, since it conveys the following information:

- The vertex correspondence is $A \leftrightarrow P$, $B \leftrightarrow Q$, and $C \leftrightarrow R$.
- The corresponding sides are congruent: $\overline{AB} \cong \overline{PQ}$, $\overline{BC} \cong \overline{QR}$, and $\overline{CA} \cong \overline{RP}$.
- The corresponding angles are congruent: $\angle A \cong \angle P$, $\angle B \cong \angle Q$, and $\angle C \cong \angle R$.

The order in which the vertices are listed specifies the vertex correspondence. For the triangles depicted in Figure 12.1, $\triangle ABC \ncong \triangle QRP$ ($\ncong$ is read "is not congruent to"). However, it is correct to say that $\triangle BCA \cong \triangle QRP$.

EXAMPLE 12.1

Exploring the Congruence Relation

Use a ruler and protractor to find the two pairs of congruent triangles among the six triangles shown. State the two congruences in the symbolic form $\triangle$____ $\cong$ $\triangle$____.

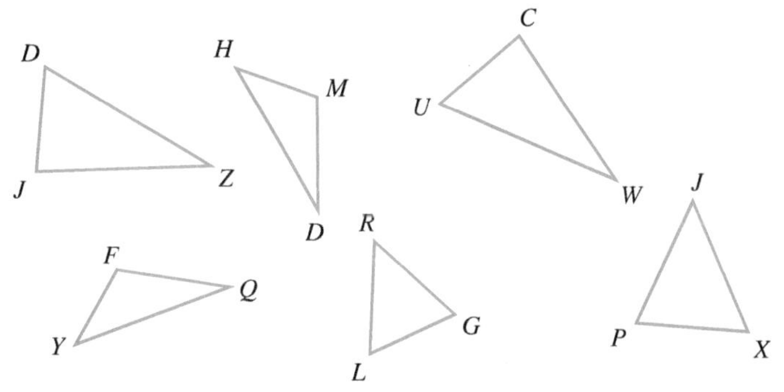

Solution

$\triangle DJZ \cong \triangle UCW$ and $\triangle HMD \cong \triangle YFQ$. The order of the vertices may be permuted in the same way on both sides of a congruence statement. For example, it would also be correct to express the first congruence as $\triangle ZDJ \cong \triangle WUC$.

Suppose that we know just some of the six parts of a triangle ABC. Is the information we have sufficient to construct a triangle PQR that is necessarily congruent to $\triangle ABC$? This question will be explored constructively; that is, we will attempt to use a compass and straightedge to construct a triangle PQR that is congruent to $\triangle ABC$.

The Side–Side–Side (SSS) Property

The next example investigates this construction question: If we have three sides $\overline{AB}$, $\overline{BC}$, and $\overline{AC}$ of a triangle, can a triangle $\overline{PQR}$ be constructed so that $\triangle PQR \cong \triangle ABC$?

EXAMPLE 12.2

Exploring the Side–Side–Side Property

The three sides of triangle ABC are given as shown. Construct a triangle PQR that has sides of the same length as ΔABC.

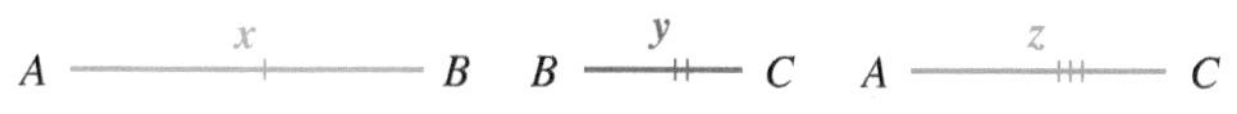

Solution

Step 1: Construct a segment $\overline{PQ}$ of length $x = AB$, using Construction 1.

Step 2: Set the compass to radius $y = BC$, and construct a circle of radius y centered at Q.

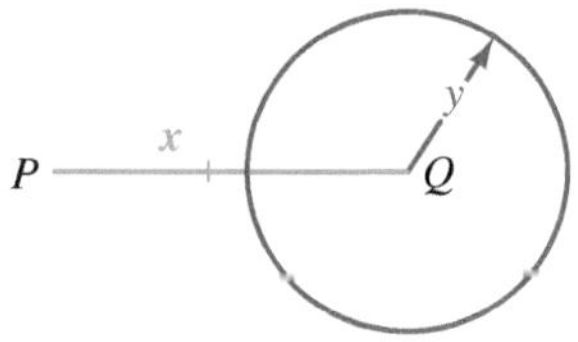

Step 3: Set the compass to radius $z = AC$, and construct a circular arc of radius z centered at P. Let R be either point of intersection with the circle constructed in step 2.

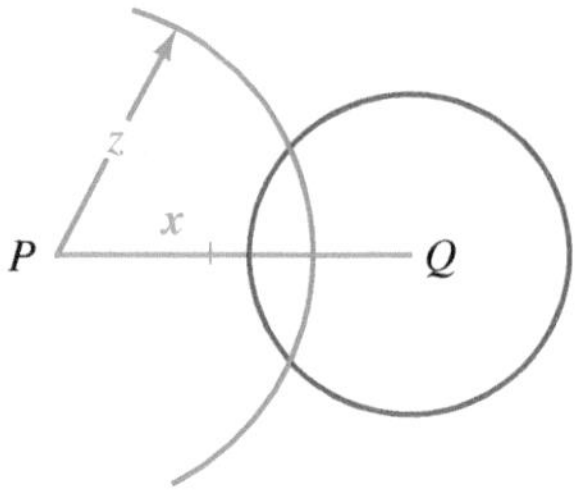

Step 4: Construct the segments $\overline{PR}$ and $\overline{RQ}$.

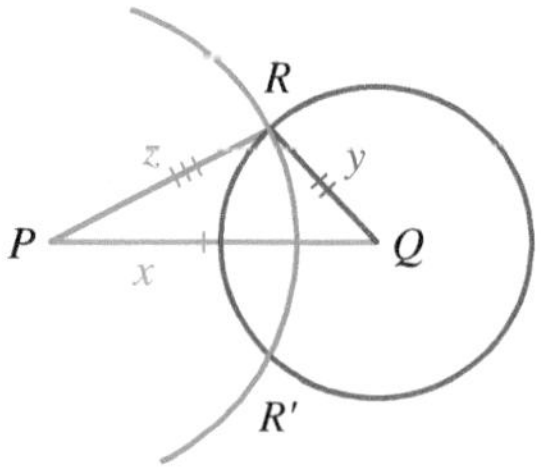

The construction in Example 12.2 shows that the three side lengths uniquely determine the size and shape of $\triangle PQR$. Even if we had chosen the second point of intersection, R', $\triangle PQR'$ would still have been the same size and shape as $\triangle PQR$. Therefore, we have the following basic property:

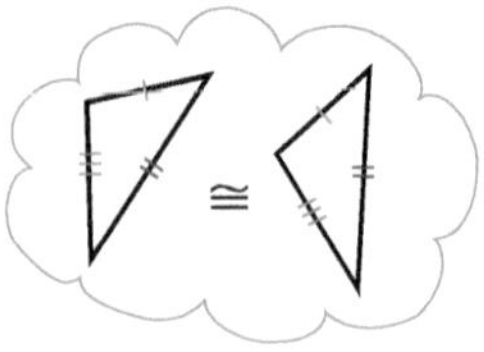

PROPERTY Side–Side–Side (SSS)
If the three sides of one triangle are respectively congruent to the three sides of another triangle, then the two triangles are congruent.

In many formal treatments of Euclidean geometry, the SSS property is adopted as a postulate. That is, SSS is true by assumption, not by proof.

The SSS property is one of the most useful tools available to verify properties about geometric figures. Sometimes, as in the example that follows, it is necessary to create a new triangle in the given figure by constructing a new line segment that is not given in the original figure. This strategy of geometric reasoning is so important that it receives special mention in the seventh Standard of Mathematical Practice of the Common Core.

SMP 7 Look for and make use of structure.

"They [mathematically proficient students] recognize the significance of an existing line in a geometric figure and can use the strategy of drawing an auxiliary line for solving problems."

EXAMPLE 12.3 Using the SSS Property

Let $ABCD$ be a quadrilateral with opposite sides of equal length: $AB = DC$ and $AD = BC$. Show that $ABCD$ is a parallelogram.

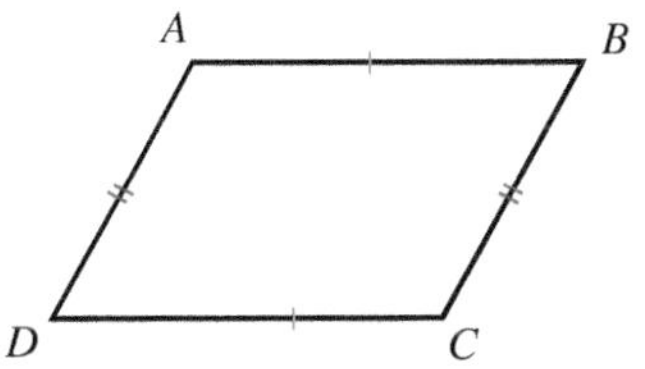

Solution

Construct the diagonal $\overline{AC}$, creating the two triangles ABC and CDA.

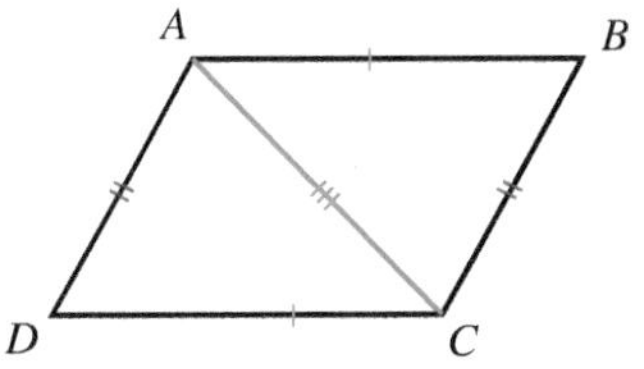

Since the line segment $\overline{AC}$ is congruent to itself, it follows that $\triangle ABC \cong \triangle CDA$ by the SSS property. Thus, the corresponding angles $\angle BAC$ and $\angle DCA$ are congruent. By the alternate-interior-angles theorem of Chapter 9, we conclude that $\overline{AB} \| \overline{DC}$. Similarly, the congruence $\angle BCA \cong \angle DAC$ shows that $\overline{AD} \| \overline{BC}$.

An important consequence of the SSS property is the following construction of an angle congruent to a given angle:

CONSTRUCTION 2 Construct an Angle Congruent to a Given Angle

Construct an angle with vertex Q that is congruent to $\angle D$ as shown.

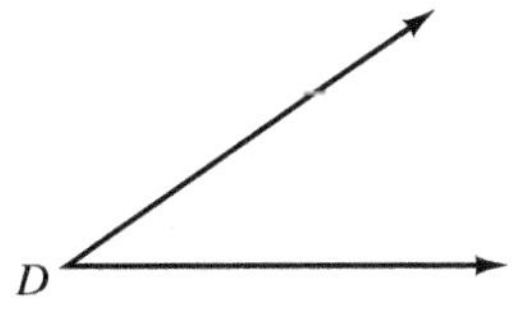

Procedure

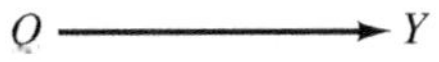 **Step 1:** Use the straightedge to construct a ray $\overrightarrow{QY}$.	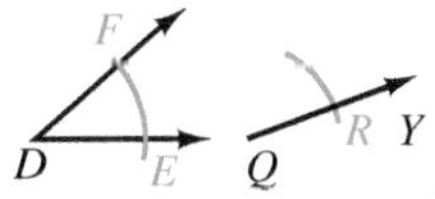 **Step 2:** Construct arcs of the same radius centered at D and at Q; let E and F be the points at which the arc intersects the sides of the given angle, and let R be the point at which the arc intersects $\overrightarrow{QY}$.	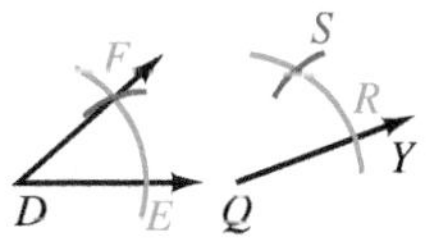 **Step 3:** Place the point of the compass at E and adjust it to construct an arc through F. Construct an arc of the same radius centered at R to locate S.	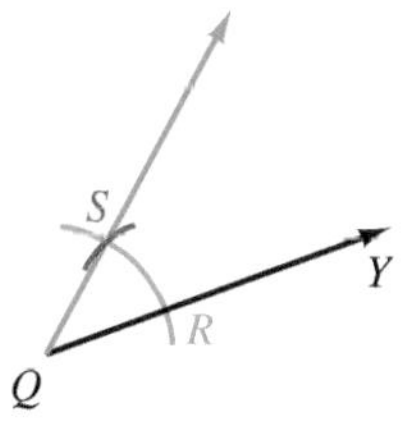 **Step 4:** Use the straightedge to construct a ray from Q through S; then $\angle Q \cong \angle D$.

The construction procedure shows that $DE = QR$, $DF = QS$, and $EF = RS$. Therefore, $\triangle DEF \cong \triangle QRS$ by the SSS property, so the corresponding angles $\angle D$ and $\angle Q$ are congruent.

The Triangle Inequality

The SSS property guarantees that two triangles are congruent if they have corresponding sides of the same length. However, not every triple of given lengths x, y, and z corresponds to a triangle. If we attempt to follow the construction shown in Example 12.2, the circle of radius y must intersect the circle of radius z at two points. If $x \geq y + z$, there are no points of intersection, or, at best, a single point of intersection along the segment of length x. Either way, there is no triangle as seen in Figure 12.2.

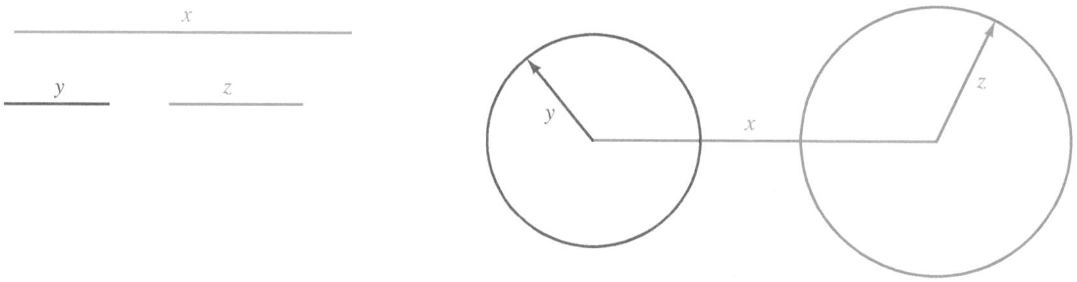

Figure 12.2
If $x \geq y + z$, there is no triangle with sides of length x, y, and z

For a triangle with sides of length x, y, and z to exist, we must have $x < y + z$. This condition is known as the *triangle inequality*.

THEOREM Triangle Inequality
The sum of the lengths of any two sides of a triangle is greater than the length of the third side.

A triangle with sides of length a, b, and c gives rise to three inequalities, as shown in Figure 12.3.

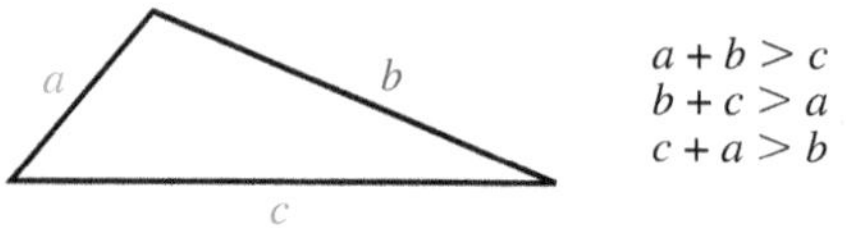

Figure 12.3
The lengths of the sides of any triangle satisfy the triangle inequalities

EXAMPLE 12.4

Applying the Triangle Inequality

The four towns of Abbott, Brownsville, Connell, and Davis are building a new power-generating plant that will serve all four communities. To keep the costs of the power lines at a minimum, the plant is to be located so that the sum of the distances from the plant to the four towns is as small as possible. An engineer recommended locating the plant at point E. A mathematician, seeing that the four towns form a convex quadrilateral $ABCD$ as shown, recommended that the plant be built at point M, the point at which the diagonals of the quadrilateral intersect. Why is location M better than E?

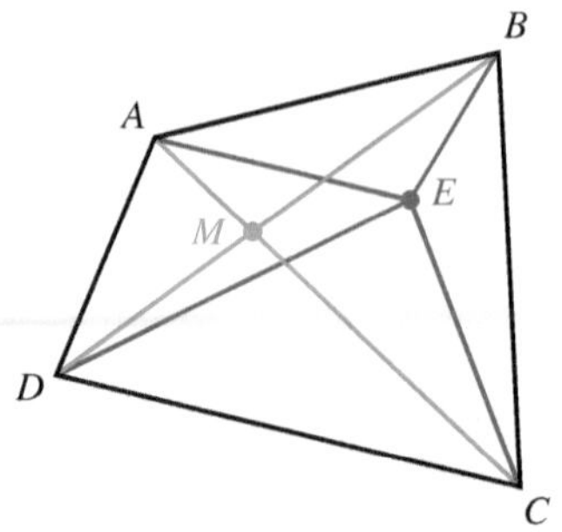

Solution The triangle inequality, applied to $\triangle ACE$, gives $EA + EC > AC$. Since M is on the diagonal $\overline{AC}$, we also see that $AC = MA + MC$, and therefore,

$$EA + EC > MA + MC.$$

If we apply the triangle inequality to $\triangle BDE$, the same reasoning gives us the inequality

$$EB + ED > MB + MD.$$

Adding the two inequalities gives us

$$EA + EB + EC + ED > MA + MB + MC + MD.$$

This inequality shows that the sum of the distances to the towns from point E is greater than the sum of the distances to the towns from point M.

COOPERATIVE INVESTIGATION

Exploring Toothpick Triangles

Materials Needed

Toothpicks of equal length (about 20 per person).

Directions

Three toothpicks, placed end to end, form a triangle in one way—an equilateral triangle. Four toothpicks do not form a triangle. Five and six toothpicks each form just one triangle. Two different (that is, noncongruent) triangles can be formed with seven toothpicks.

Triangles	Number of Toothpicks, n	Number of Triangles, $T(n)$
	3	1
	4	0
	5	1
	6	1
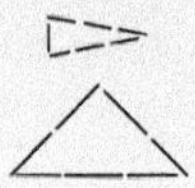	7	2

Explore how many noncongruent triangles you can form with 8, 9, 10, 11, and 12 toothpicks. Extend the table above to include your results.

Questions for Consideration

1. How many isosceles toothpick triangles are there for which the two sides of equal length each use 4 toothpicks?
2. One side of a toothpick triangle uses 3 toothpicks, and a second side uses 5 toothpicks. What are the possible numbers of toothpicks in the third side?
3. If two sides of a toothpick triangle together use 11 toothpicks, what is the largest number of toothpicks that can be used in the third side?
4. Suppose toothpicks form a triangle with p, q, and r toothpicks on its three sides. What can you say about the integer r in terms of the integers p and q?
5. In your table of the number of triangles, suppose $T(n)$ is the number of different toothpick triangles formed from n toothpicks. For odd n, compare $T(n)$ with $T(n + 3)$. For example, compare $T(3)$ with $T(6)$, and compare $T(5)$ with $T(8)$. What pattern do you observe?

The Side–Angle–Side (SAS) Property

The next example explores how to construct a triangle with two given sides and the angle included between the given sides.

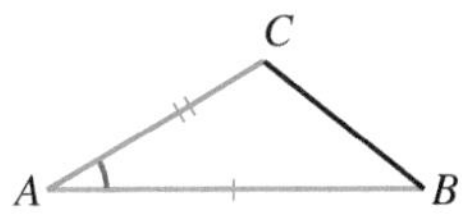

EXAMPLE 12.5

Exploring the Side–Angle–Side Condition

Two sides $\overline{AB}$ and $\overline{AC}$ and the angle $\angle A$ included between these sides are given for $\triangle ABC$ as shown. Show that a triangle PQR can be constructed for which $\triangle PQR \cong \triangle ABC$.

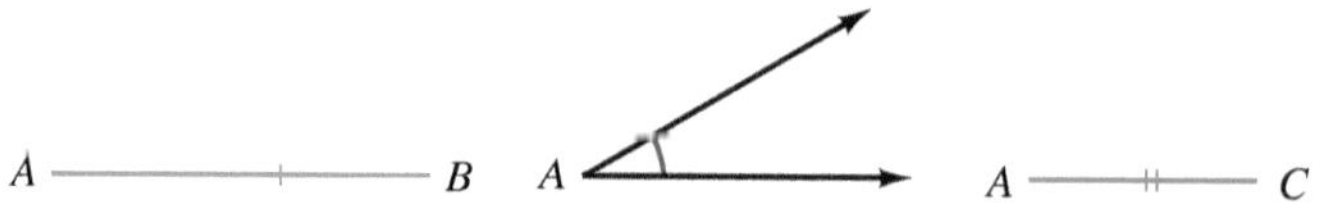

Solution

Step 1: Follow the steps of Construction 2 to construct an angle congruent to $\angle A$; let P denote its vertex.

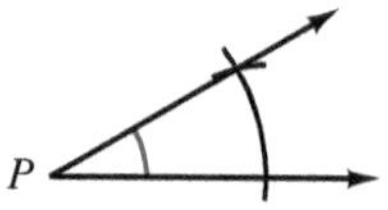

Step 2: Use Construction 1 to construct segments $\overline{PQ}$ and $\overline{PR}$ along the sides of $\angle P$ that are respectively of lengths AB and AC.

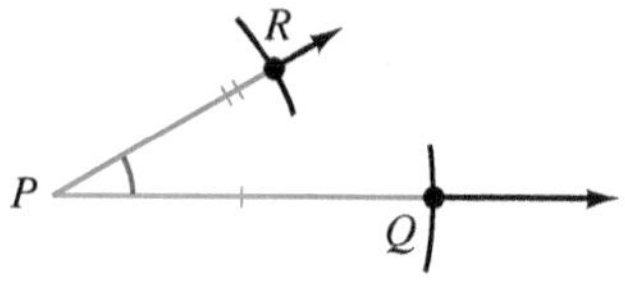

Step 3: Construct segment $\overline{QR}$, completing $\triangle PQR$. This procedure uniquely determines the size and shape of $\triangle PQR$, so there is only one possible triangle whose sides and included angle are congruent to the given ones. Therefore, $\triangle PQR \cong \triangle ABC$.

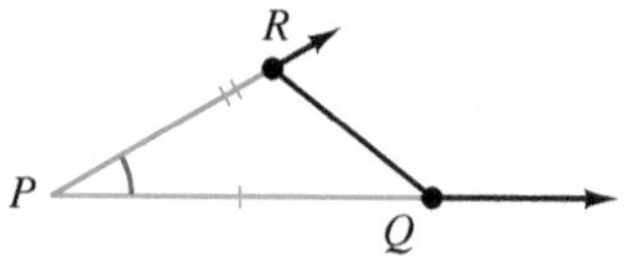

The procedure described in Example 12.5 uniquely determines the size and shape of $\triangle PQR$ when we are given the three parts, side–angle–side, of $\triangle ABC$. The angle has to be the **included angle**, the angle between the given sides. This property is often abbreviated as **SAS (side–angle–side)**.

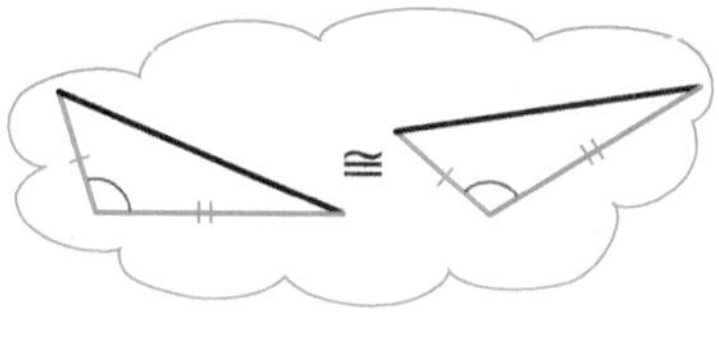

PROPERTY Side–Angle–Side (SAS)

If two sides and the included angle of one triangle are congruent to two sides and the included angle of another triangle, then the two triangles are congruent.

EXAMPLE 12.6

Using the SAS Property

Two line segments $\overline{AB}$ and $\overline{CD}$ intersect at their common midpoint M. Show that $\overline{AD}$ and $\overline{BC}$ are parallel and have the same length.

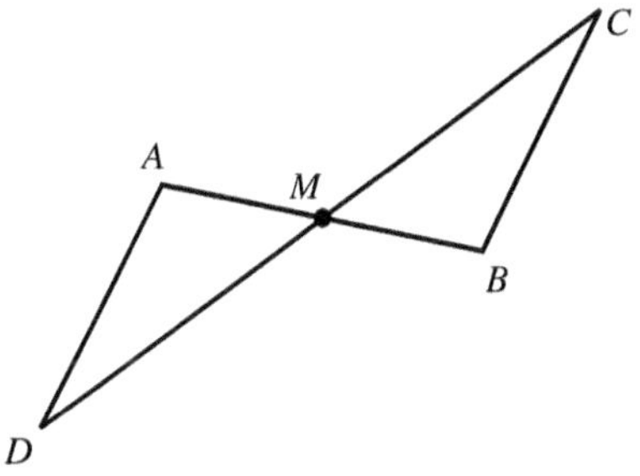

Solution

It is a useful habit to add tick marks and arcs to your drawing to summarize the given information. In this problem, M is the midpoint of $\overline{AB}$, so $AM = BM$. We indicate this equality on the drawing by putting a single tick mark on each of the segments $\overline{AM}$ and $\overline{MB}$. Similarly, $CM = DM$, and we put double tick marks on each of the segments $\overline{CM}$ and $\overline{MD}$. We also use single arcs to indicate the congruence of the vertical angles at M formed by the intersecting segments.

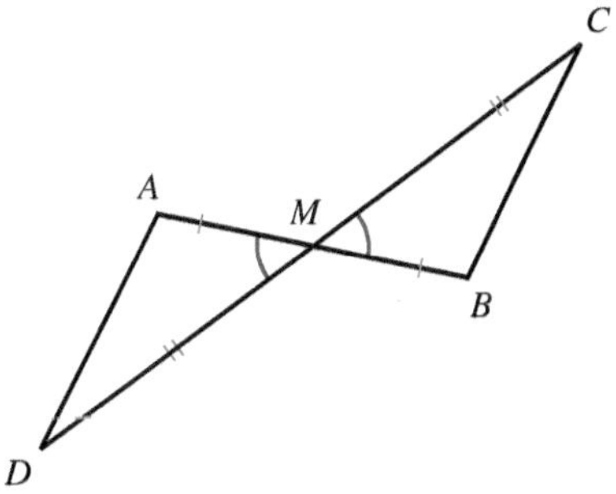

It is now apparent that the SAS property gives us the congruence $\triangle AMD \cong \triangle BMC$. It follows that the corresponding sides $\overline{AD}$ and $\overline{BC}$ are congruent, so that $AD = BC$. We also have $\angle A \cong \angle B$, so the alternate-interior-angles theorem of Chapter 9 guarantees that $\overline{AD}$ and $\overline{CB}$ are parallel.

The following theorem about isosceles triangles is an important consequence of the SAS property:

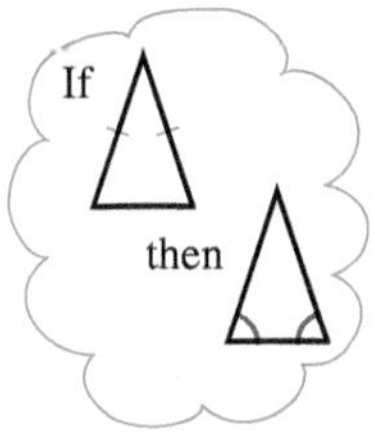

THEOREM Isosceles Triangle Theorem
The angles opposite the congruent sides of an isosceles triangle are congruent.

PROOF Let $\triangle ABC$ be isosceles, with $\overline{AB}$ and $\overline{AC}$ congruent. Consider the vertex correspondence $A \leftrightarrow A$, $B \leftrightarrow C$, and $C \leftrightarrow B$ (which amounts to looking at the same triangle from the back, so to speak!).

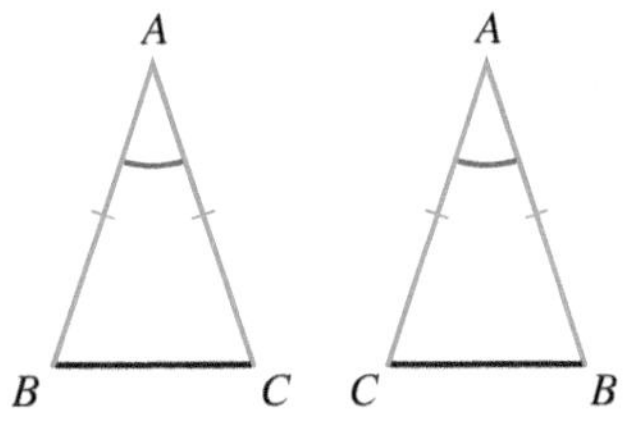

Since $\overline{AB} \cong \overline{AC}$ and $\angle A \cong \angle A$, it follows from the SAS property that $\triangle ABC \cong \triangle ACB$. But then all six corresponding parts of $\triangle ABC$ and $\triangle ACB$ are congruent, including $\angle B \cong \angle C$.

The isosceles triangle theorem has many uses. For example, it gives a simple way to prove **Thales' theorem**.

THEOREM Thales' Theorem

Any triangle ABC inscribed in a semicircle with diameter $\overline{AB}$ has a right angle at point C.

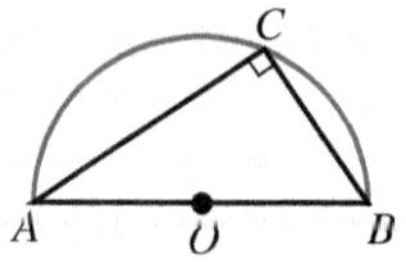

PROOF Once again, it is helpful to draw additional line segments to reveal relationships that would otherwise remain hidden. This time, draw the radius $\overline{OC}$ as shown in the figure. This divides $\triangle ABC$ into two isosceles triangles: $\triangle AOC$ and $\triangle COB$. Both have a pair of sides of length r, the radius of the semicircle. The isosceles triangle theorem tells us that the measures x of the base angles of $\triangle AOC$ are equal. Likewise, the measures y of the base angles of $\triangle COB$ are equal. Since the sum of the measures of the interior angles of $\triangle ABC$ is 180°, we have $x + y + (x + y) = 180°$. This equation tells us that $m(\angle C) = x + y = 90°$.

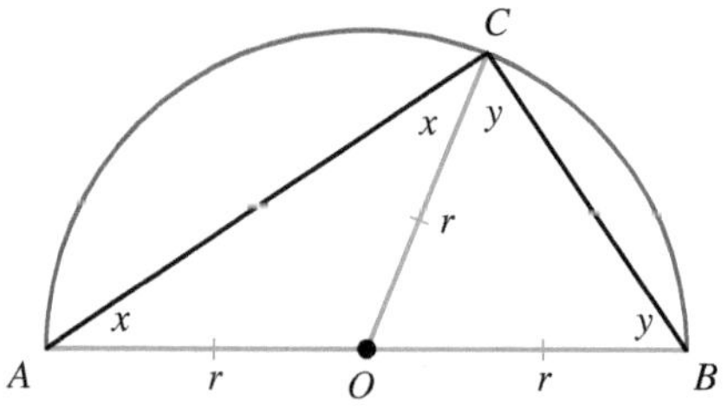

The Angle–Side–Angle (ASA) Property

In the next example, we suppose that two angles of a triangle, and the side included between these angles, are given. Is this information sufficient to construct a congruent triangle?

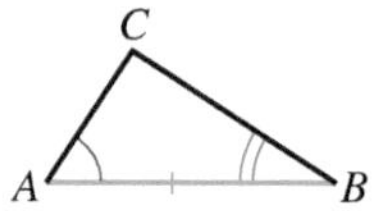

EXAMPLE 12.7

Exploring the Angle–Side–Angle Property

Two angles and their **included side** are given for $\triangle ABC$, as shown. Construct a triangle PQR that is congruent to triangle ABC.

Solution

Step 1: Use Construction 2 to construct $\angle P \cong \angle A$.

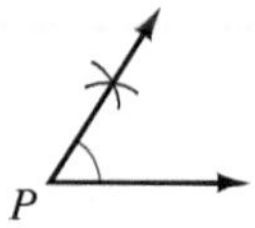

Step 2: Use Construction 1 to construct segment $\overline{PQ}$ on a side of $\angle P$ so that $PQ = AB$.

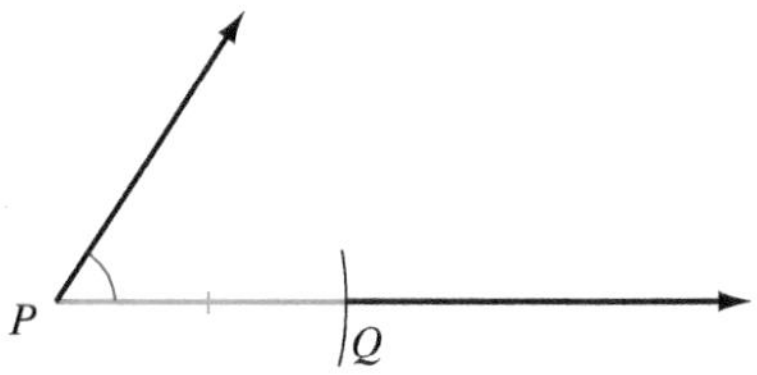

Step 3: Use Construction 2 again to construct an angle congruent to $\angle B$ at vertex Q, with one side containing P and the other side intersecting $\angle P$ to determine point R. The procedure uniquely determines the size and shape of $\triangle PQR$. Therefore, $\triangle PQR \cong \triangle ABC$.

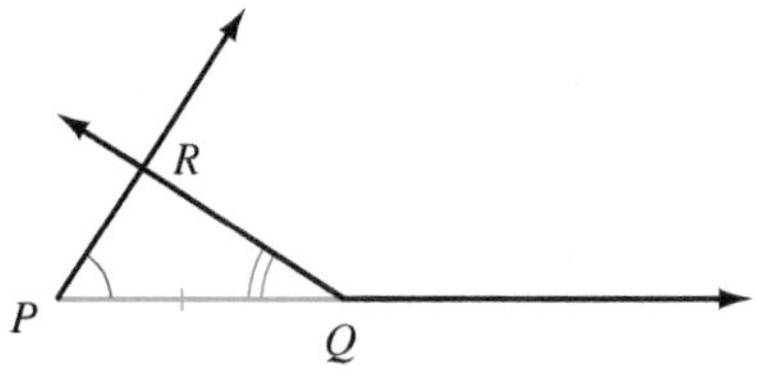

The construction just shown illustrates the **angle–side–angle property,** abbreviated as **ASA.**

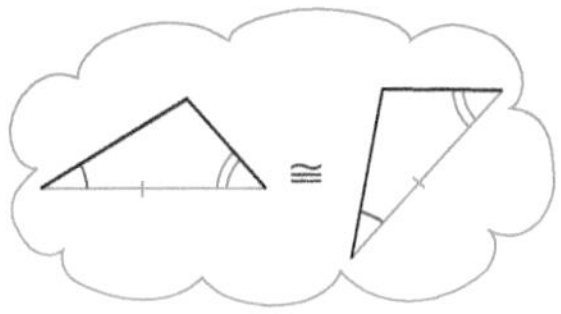

PROPERTY Angle-Side-Angle (ASA)
If two angles and the included side of one triangle are congruent to the two angles and the included side of another triangle, then the two triangles are congruent.

The ASA property allows us to prove that any triangle with two angles of the same measure is isosceles.

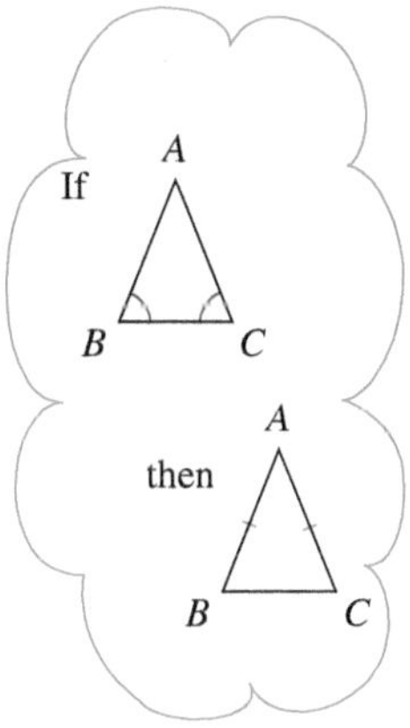

THEOREM Converse of the Isosceles Triangle Theorem
If two angles of a triangle are congruent, then the sides opposite them are congruent; that is, the triangle is isosceles.

PROOF Let $\triangle ABC$ have two congruent angles, say, $\angle B \cong \angle C$. We know that $\overline{BC} \cong \overline{CB}$, since a line segment is congruent to itself. By the ASA property, it follows that $\triangle ABC \cong \triangle ACB$. This means that the corresponding sides of $\triangle ABC$ and $\triangle ACB$ are congruent, so $\overline{AB} \cong \overline{AC}$.

The Angle–Angle–Side (AAS) Property

The side in the ASA theorem is the one included by the two angles. However, if *any* two angles of one triangle are congruent to two angles of a second triangle, then all three pairs of corresponding angles are congruent. This statement follows from the fact that the measures of the three angles of a triangle add up to 180°, so the third angle is uniquely determined by the other two angles. This gives us the **angle–angle–side property,** abbreviated as **AAS.**

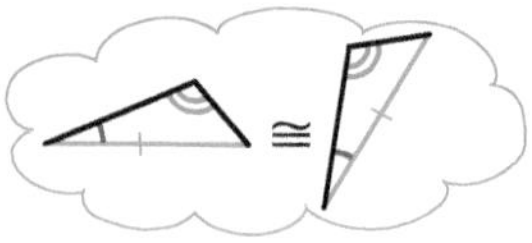

PROPERTY Angle–Angle–Side (AAS)
If two angles and a nonincluded side of one triangle are respectively congruent to two angles and the corresponding nonincluded side of a second triangle, then the two triangles are congruent.

Are There SSA and AAA Congruence Properties?

There is no "SSA" congruence property, since it is possible for two noncongruent triangles to have two pairs of congruent sides and a congruent nonincluded angle. An example is shown in Figure 12.4.

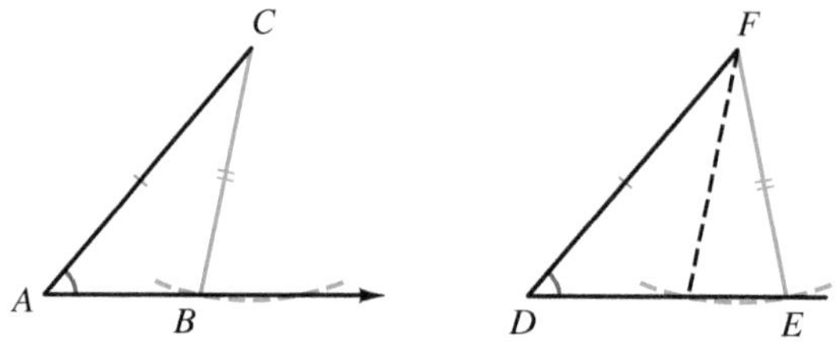

Figure 12.4
Triangles *ABC* and *DEF* are not congruent, even though $AC = DF$, $BC = EF$, and $\angle A \cong \angle D$

Similarly, there is no "AAA" congruence property. For example, Figure 12.5 shows two triangles with three pairs of congruent angles. These triangles are not congruent, since they are of different size. However, the shapes of the two triangles are the same, so they are similar triangles. The properties and applications of similar triangles will be discussed in Section 12.3.

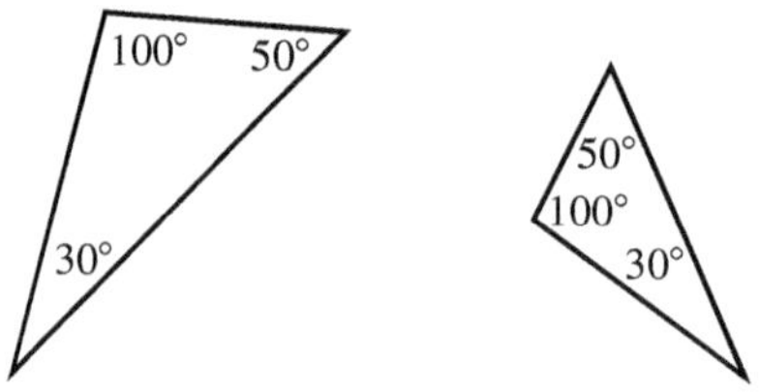

Figure 12.5
The AAA condition guarantees similarity, but not congruence

12.1 Problem Set

Understanding Concepts

1. The two triangles shown are congruent.

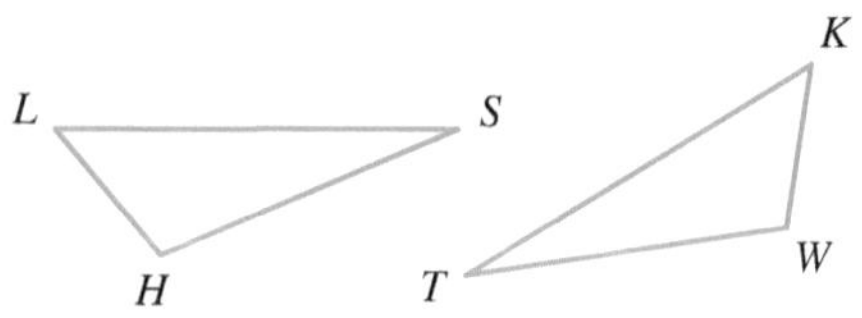

Determine the following:

(a) Corresponding vertices
$L \leftrightarrow$ ____, $H \leftrightarrow$ ____, $S \leftrightarrow$ ____

(b) Corresponding sides
$\overline{LH} \leftrightarrow$ ____, $\overline{HS} \leftrightarrow$ ____, $\overline{SL} \leftrightarrow$ ____

(c) Corresponding angles
$\angle L \leftrightarrow$ ____, $\angle H \leftrightarrow$ ____, $\angle S \leftrightarrow$ ____

(d) $\triangle LHS \cong \triangle$____.

2. Use the congruence symbol ≅ to identify the two pairs of congruent triangles among the six shown here.

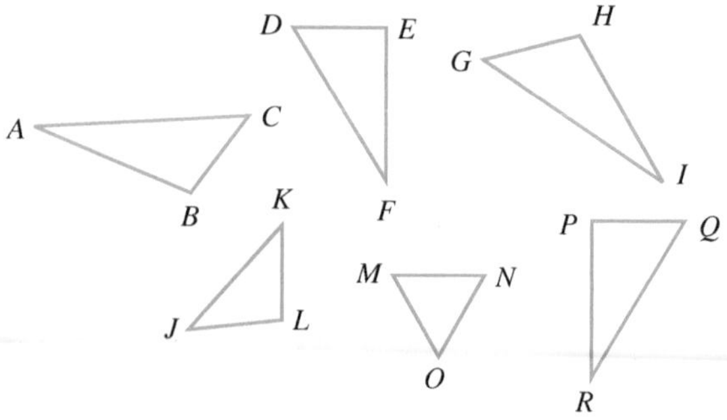

3. Suppose $\triangle JKL \cong \triangle ABC$, where $\triangle ABC$ is as follows:

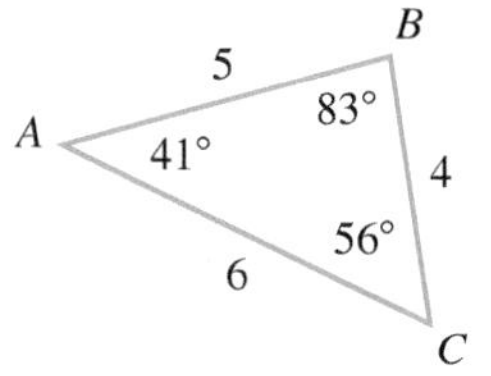

Find the following:

(a) KL **(b)** LJ

(c) $m(\angle L)$ **(d)** $m(\angle J)$

4. (**Writing**) Segments of length x and y are shown, with $x > y$.

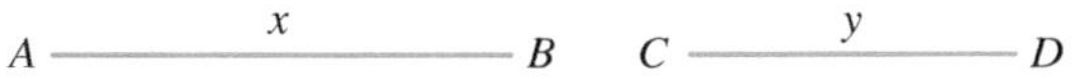

Write step-by-step procedures, using only a straightedge and a compass, to construct the following:

(a) a line segment $\overline{EF}$ of length $x + y$;

(b) a line segment $\overline{GH}$ of length $x - y$.

5. (**Writing**) Use a straightedge and compass to construct a triangle ABC whose sides have the lengths x, y, and z shown. Carefully describe, in pictures and words, the steps in your construction.

6. Can a triangle be constructed with sides of length 2″, 3″, and 6″? If yes, do so. If not, explain why not.

7. Trace the angle $\angle D$ shown. Then use a compass and straightedge to construct an angle $\angle Q$ congruent to $\angle D$. Use a protractor to measure each angle, and report on how closely the measurements of the two angles agree.

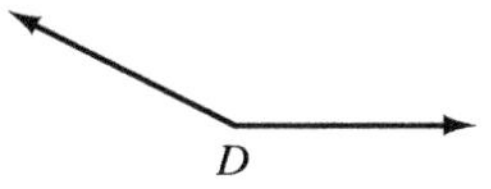

8. (**Writing**) Following are two angles $\angle A$ and $\angle B$:

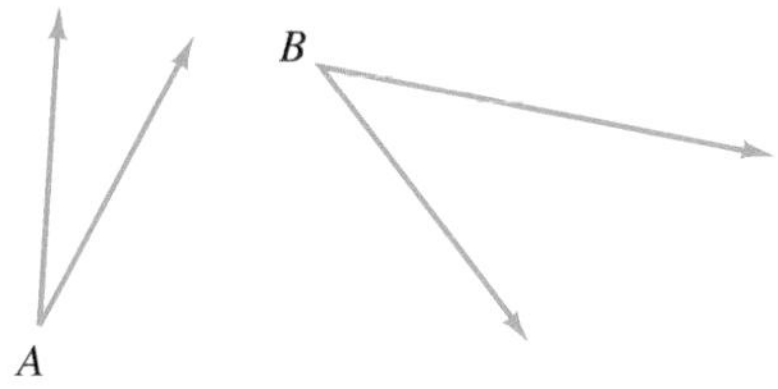

Write step-by-step procedures, using only a straightedge and a compass, to construct

(a) $\angle C$ so that $m(\angle C) = m(\angle A) + m(\angle B)$.

(b) $\angle D$ so that $m(\angle D) = m(\angle B) - m(\angle A)$.

(c) $\angle E$ so that $m(\angle E) + m(\angle A) + m(\angle B) = 180°$.

9. Use a ruler, protractor, and compass to construct, when possible, a triangle with the stated properties. If such a triangle cannot be drawn, explain why. Decide whether there can be two or more noncongruent triangles with the stated properties.

(a) An isosceles triangle with two sides of length 5 cm and an apex angle of measure 28°

(b) An equilateral triangle with sides of length 6 cm

(c) A triangle with sides of length 8 cm, 2 cm, and 5 cm

10. Answer question 9 for these descriptions of triangles.

(a) A triangle with angles measuring 30° and 110° and a nonincluded side of length 5 cm

(b) A right triangle with legs (the sides including the right angle) of length 6 cm and 4 cm

(c) A triangle with sides of length 10 cm and 6 cm and a nonincluded angle of 45°

(d) A triangle with sides of length 5 cm and 3 cm and an angle of 20°

11. Each part that follows shows two triangles, with arcs and tick marks identifying congruent parts. If it is possible to conclude that the triangles are congruent, describe what property or theorem you used. If you cannot be sure that the triangles are congruent, state, "No conclusion possible." The first one is done for you.

(a)

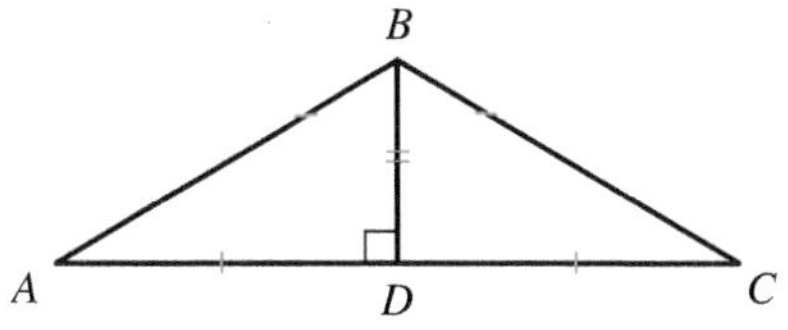

Answer: $\triangle ABD \cong \triangle CBD$ by SAS

(b)

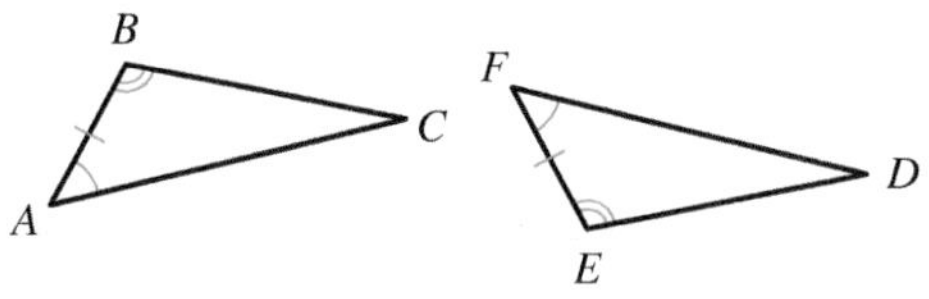

(c)

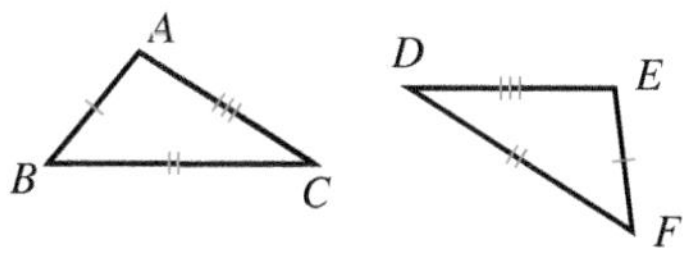

(d)

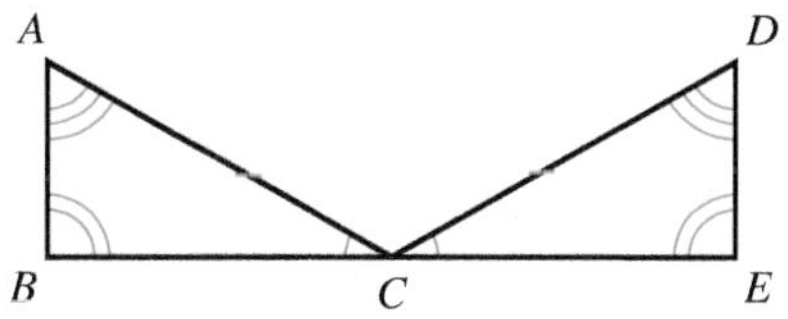

12. Answer question 11 for these diagrams.

(a)

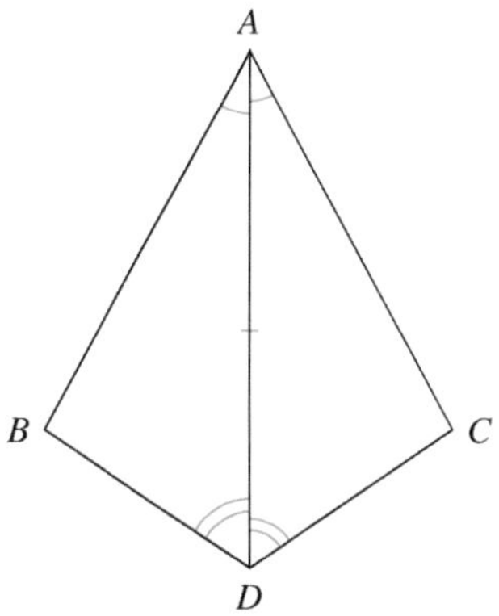

(b)

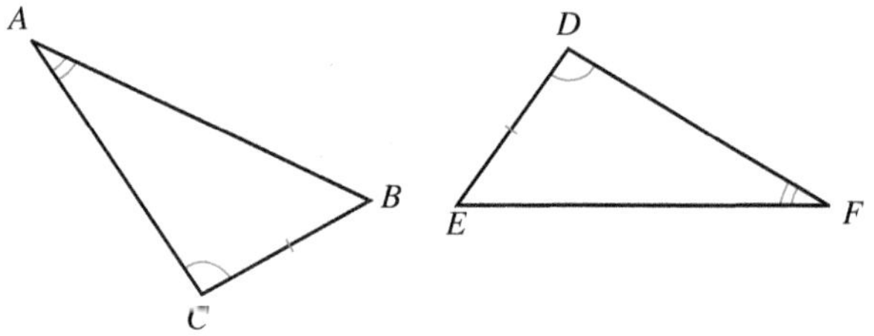

(c)

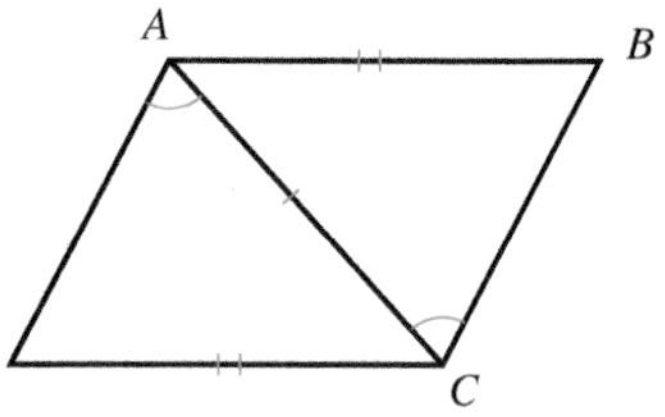

(d)

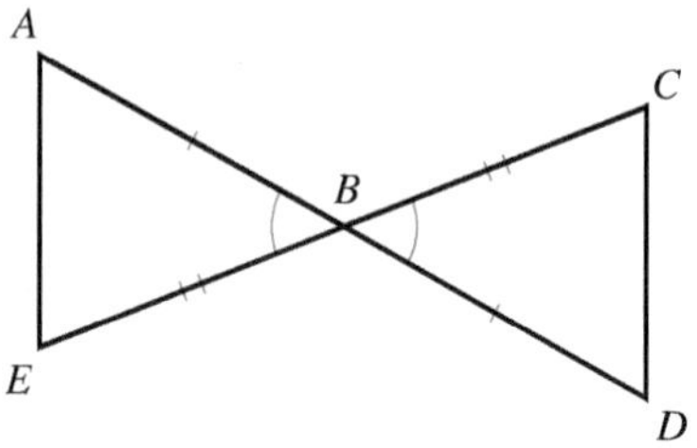

(e)

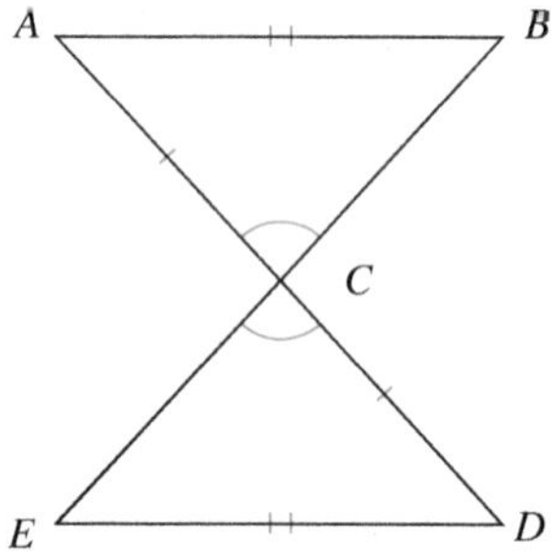

(f)

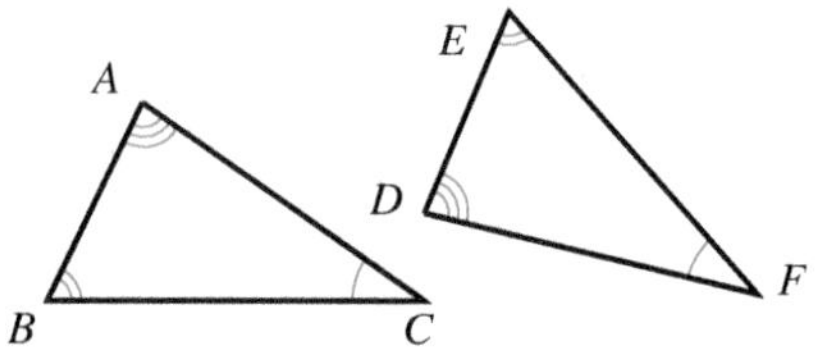

13. In each figure, find a pair of congruent triangles. State what congruence property justifies your conclusion, and express the congruence with the symbol ≅.

(a)

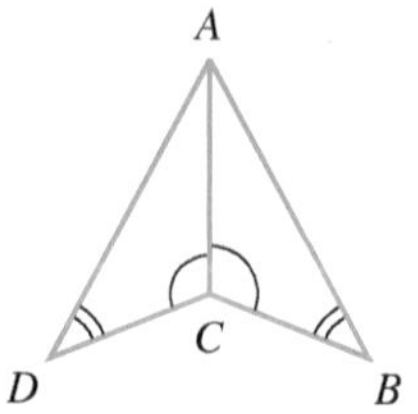

(b)

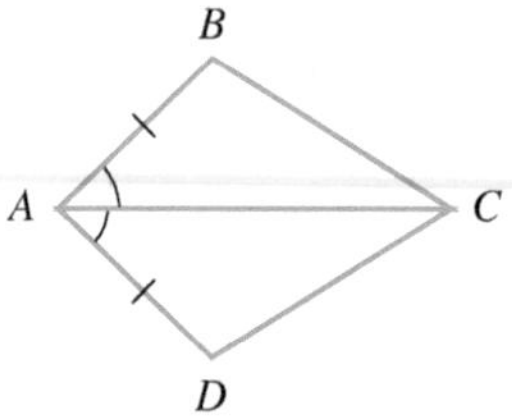

(c)

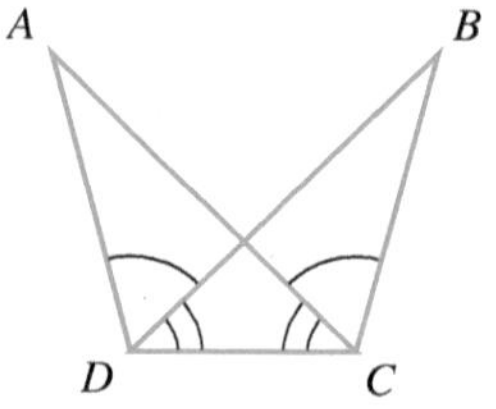

14. In each figure, find a pair of congruent triangles. State the congruence property that justifies your conclusion, and express the congruence with the symbol ≅.

(a)

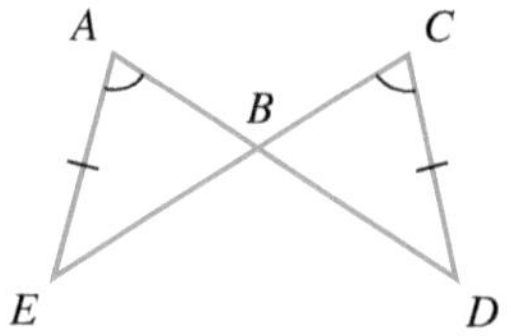

(b)

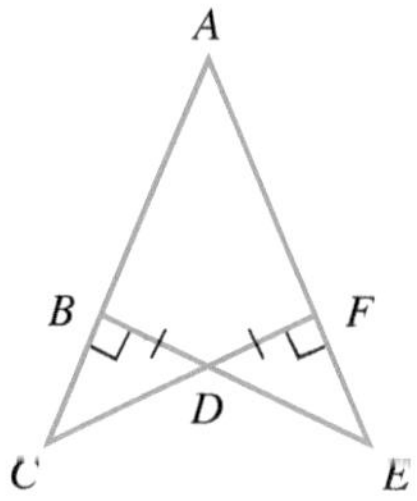

(c)

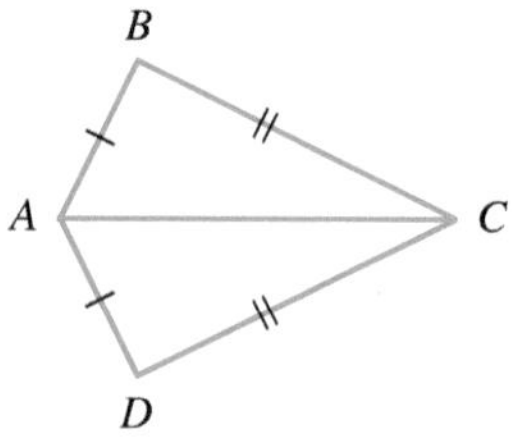

15. Prove that an equilateral triangle is equiangular.

16. Prove that an equiangular triangle is equilateral.

17. Draw an angle $\angle BAC$ and use a protractor to measure the angle. Next, construct an arc centered at A to determine points D and E. Finally, draw arcs of equal radius centered at D and E, denoting their point of intersection as F.

(a) Measure angles 1 and 2. How do they compare with the measure of $\angle BAC$?

(b) Prove that $\triangle AFD$ is congruent to $\triangle AFE$.

(c) Explain why angles 1 and 2 are congruent, using the result of part (b).

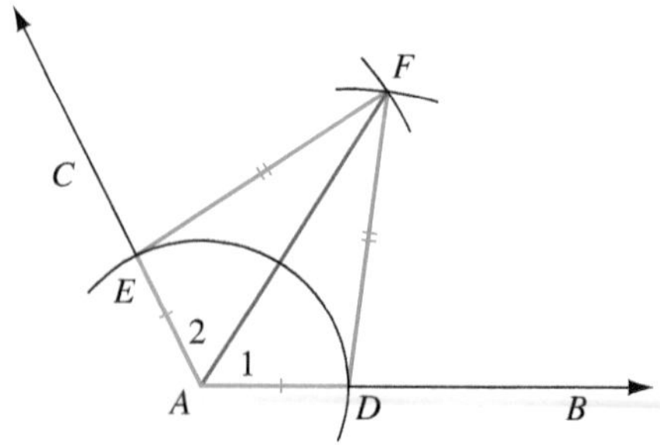

18. Draw a line m and a point P not on the line. Construct an arc centered at P that intersects the line in two points Q and S.

Next, draw two arcs of equal radii centered at Q and S, labeling their intersection as point T. Finally, construct the segment from P to T and let V be its intersection with the line m.

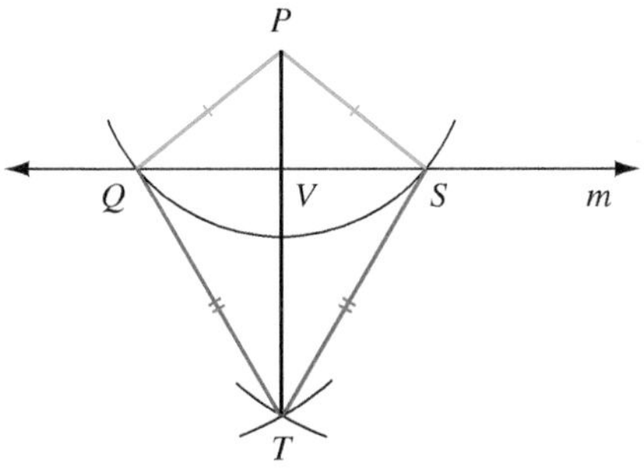

Give reasons that these relationships hold:

(a) $\triangle QPT \cong \triangle SPT$ **(b)** $\angle QPT \cong \angle SPT$

(c) $\triangle QPV \cong \triangle SPV$ **(d)** $\angle QVP$ is a right angle.

Therefore, the construction gives a line $\overleftrightarrow{PT}$ perpendicular to m that passes through point P.

19. Let $ABCD$ be a parallelogram.

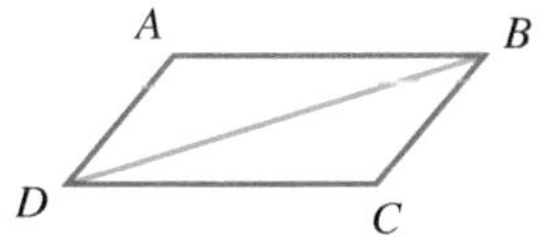

(a) Prove that $\triangle ABD \cong \triangle CDB$. (*Hint:* Use the ASA property.)

(b) Prove that opposite sides of a parallelogram have the same length.

(c) Prove that opposite angles of a parallelogram have the same measure.

20. Let the two diagonals of parallelogram $ABCD$ intersect at point M.

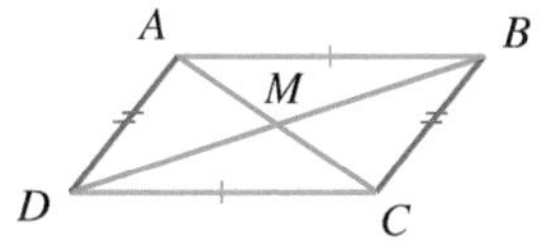

(a) Use the fact that $AB = CD$ (shown in problem 19(b)) to prove that $\triangle ABM \cong \triangle CDM$.

(b) Use part (a) to explain why M is the midpoint of both diagonals of the parallelogram.

21. Let the two diagonals of a rhombus $ABCD$ intersect at M.

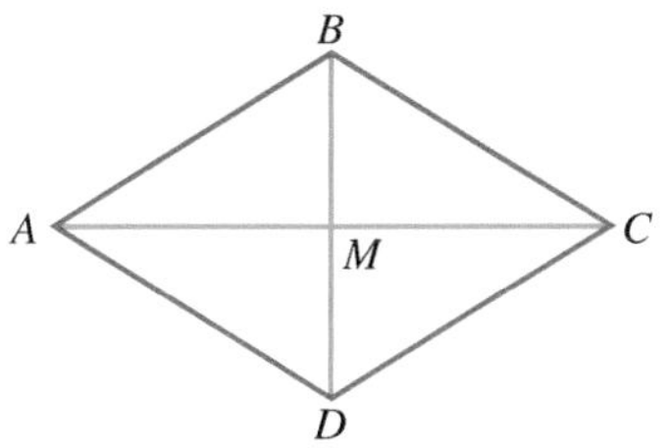

(a) Show that the triangles ABM, CBM, CDM, and ADM are congruent to one another.

(b) Use part (a) to explain why the diagonals of a rhombus bisect the interior angles of the rhombus and intersect at a right angle at M.

22. A triangle has sides of length 4 cm and 9 cm. What can you say about the length of the third side?

23. **(a)** A quadrilateral has sides of length 2 cm, 7 cm, and 5 cm. What inequality does the length of the fourth side satisfy?

(b) Let A, B, C, and D be any four points in the plane. Explain why $AD \leq AB + BC + CD$.

Into the Classroom

24. **(Writing) Copycat Congruence Activity.** Write a detailed lesson plan that expands on the following idea:

> The class is divided into small groups. Each group constructs a triangle on a sheet of paper and uses a ruler and protractor to measure all six parts of the triangle. Next, the measures of three of the parts are written on the back of the sheet, which is taped to the chalkboard with the triangle hidden. Each group then chooses two or three measurements of triangles from other groups. The group's task is to use its measurements to construct congruent triangles cut from sheets of construction paper. Each group then places each triangle over the corresponding triangle on the reverse side of the taped sheets to test the accuracy of its constructions. Finally, each group describes, in written and oral form, how congruence properties were used, or if more than one triangle shape was possible.

Your lesson plan should clearly state the activity's goals, the standards it addresses, the materials required, the directions for carrying out the activity, your assessment of the activity, and any extensions of the activity for further investigation.

Responding to Students

25. Dane has been studying quadrilaterals and wonders whether there is an SSSS congruence property for four-sided polygons. That is, if AB, BC, and DA are given, then the size and shape of the quadrilateral $ABCD$ can be determined. What suggestions would you offer to Dane to investigate this possibility?

26. Valerie believes that there is an SASAS congruence property for quadrilaterals. That is, if the lengths AB, BC, and CD are given, as well as the measures of $\angle B$ and $\angle C$, then the unique size and shape of quadrilateral $ABCD$ can be determined. How would you respond to Valerie?

Thinking Critically

27. Let ABC be a right triangle with hypotenuse $\overline{AB}$ and right angle at vertex C. Explain why the circle centered at the midpoint O of the hypotenuse and passing through point C also passes through points A and B. This gives the following:

Converse of Thales' theorem: The hypotenuse of a right triangle inscribed in a circle is a diameter of the circle.

(*Suggestion:* Imagine that the right triangle ABC is created by constructing the diagonal $\overline{AB}$ of a rectangle $ACBD$.)

28. How can you locate the center of a circular plate with a piece of notebook paper and a ruler? (*Hint:* Use the converse of Thales' theorem stated in problem 27.)

29. Following are two angles and a *nonincluded* side of $\triangle ABC$:

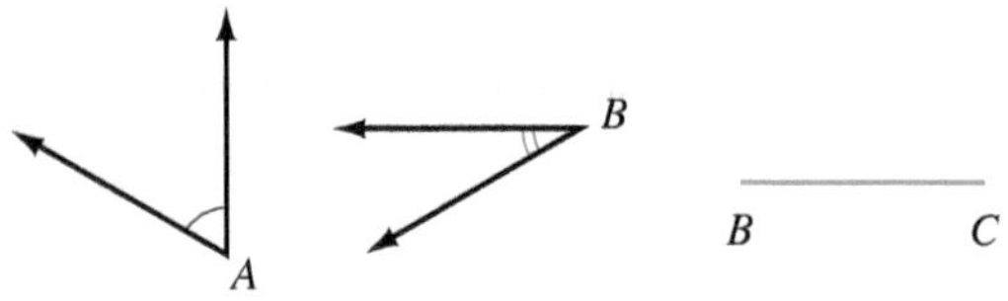

Describe and show the steps of a straightedge-and-compass construction of a triangle PQR that is congruent to $\triangle ABC$.

30. Recall that a trapezoid with a pair of congruent angles adjacent to one of its bases is called an isosceles trapezoid.

(a) Prove that the sides joining the bases of an isosceles trapezoid are congruent. The following figure may help you show that $\overline{AD} \cong \overline{BC}$.

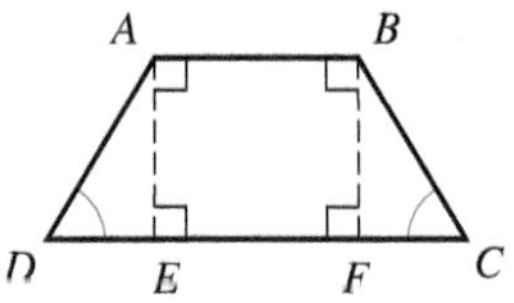

(b) Prove that the diagonals of an isosceles trapezoid are congruent.

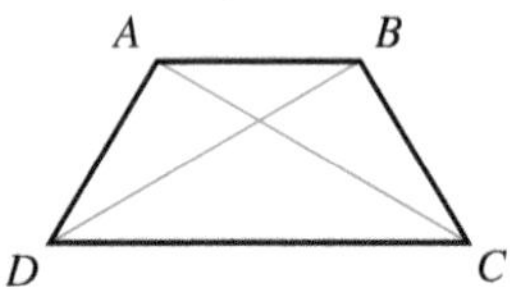

31. Each edge of a tetrahedron is congruent to its opposite edge. For example, $\overline{AB} \cong \overline{CD}$. Prove that the faces of the tetrahedron are congruent to one another.

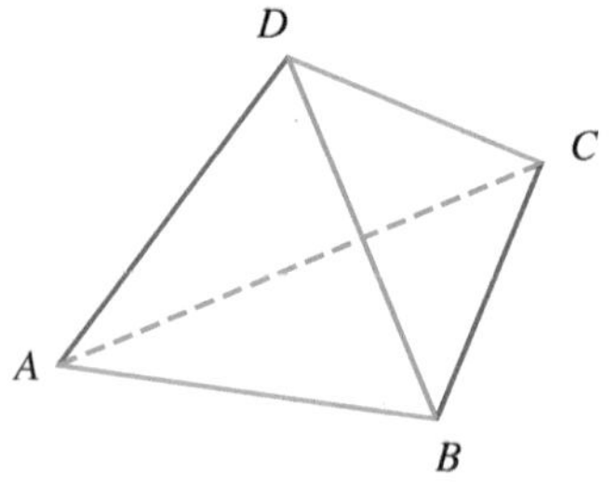

32. (a) Let A and B be two blue points and X and Y be two red points in the plane, where no three of the four points are collinear. Prove that if the blue points are joined to the red points by intersecting line segments, then the sum of their lengths is larger than the total length of the nonintersecting segments.

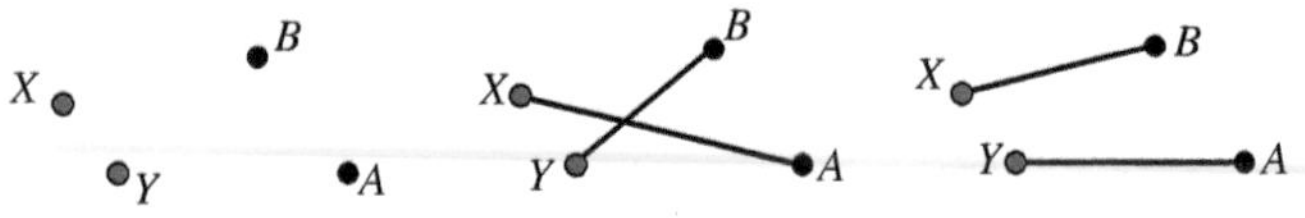

Given point — Segments intersect — Segments do not intersect

[*Suggestion*: Use the triangle inequality.]

(b) Let $\mathcal{B} = \{A, B, \ldots, J\}$ be a set of 10 blue points in the plane and $\mathcal{R} = \{Q, R, \ldots, Z\}$ a set of 10 red points in the plane, where no three of the 20 points are collinear. Explain why the blue points can always be joined to the red points with 10 line segments such that no two of the line segments intersect.

[*Suggestion*: Consider an arrangement with the smallest possible total sum of the lengths of the 10 connecting segments.]

33. Six towns are located at the vertices A, B, C, D, E, and F of a regular hexagon. A power station located at Q would require $QA + QB + QC + QD + QE + QF$ miles of transmission line to serve the six communities. Describe a better location P for the station, and prove that it has the least possible sum of distances to A, B, C, D, E, and F.

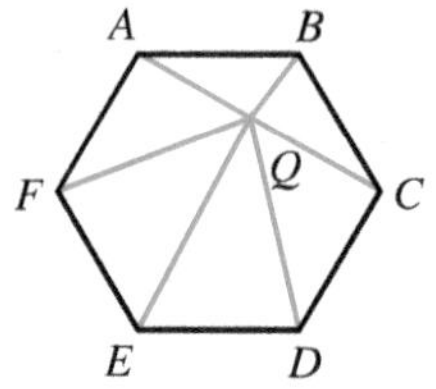

Making Connections

34. A bicycle rack for a car is made from three pieces of metal box tubing. There are bolts at A, B, C, and D that join the tubing and attach the rack to the bumper of the car.

(a) Why is the top of the rack likely to shift sideways?

(b) If a fourth piece of tubing is available, where can it be attached to make the rack rigid? Explain why this works.

(c) Would a rope from A to C make the rack rigid? How about two pieces of rope, from A to C and from B to D?

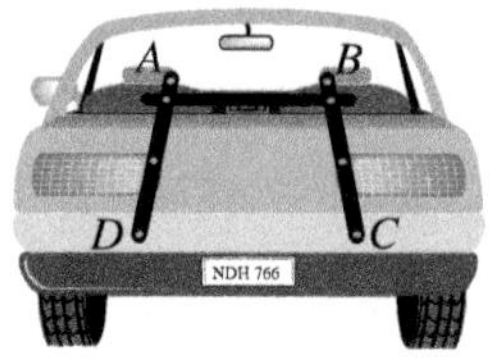

35. Carpenters construct a wall by nailing studs to a top and bottom plate, as shown here:

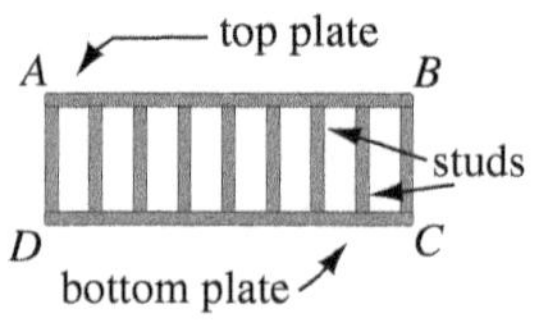

The studs are cut to the same length, and the top and bottom plates are the same length.

(a) If the pieces are properly cut and nailed, is $ABCD$ necessarily rectangular, or are there other shapes the framework can take?

(b) Carpenters frequently "square up" a stud wall by adjusting it so that it has diagonals of equal length. Prove that a parallelogram with congruent diagonals is a rectangle.

(c) Once the wall is "squared up," a diagonal brace is nailed across the frame. Why is this? What geometric principle is involved?

36. The two frameworks shown are constructed with drinking straws and pins. The triangle is a rigid framework by the SSS property, but the quadrilateral is flexible.

Decide if the straw-and-pin frameworks shown next are rigid or flexible. It may be helpful to build the frameworks to check your reasoning.

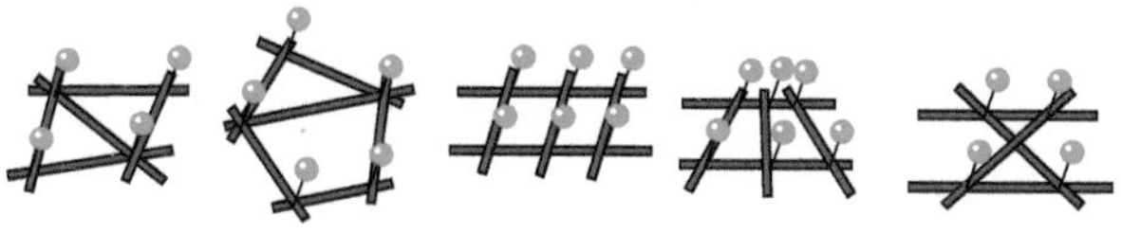

(a) (b) (c) (d) (e)

37. Let A, B, and P be three points on a circle centered at O. In the case that the diameter $\overline{PQ}$ is in the interior of $\angle APB$, as shown below, show that $m\angle AOB = 2m\angle APB$.

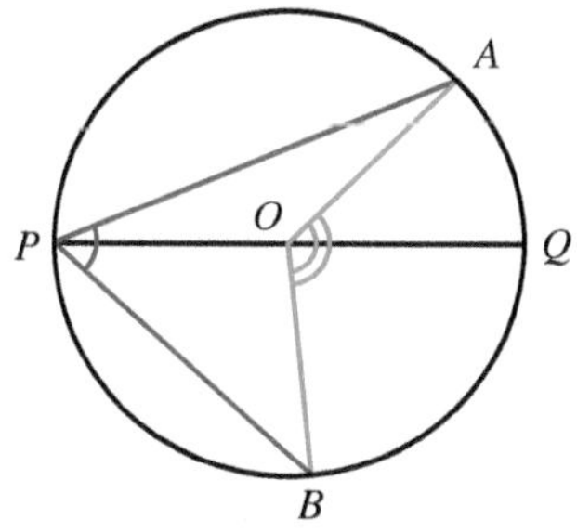

38. Let A, B, and P be three points on a circle centered at O. In the case that the diameter $\overline{PQ}$ is in the exterior of $\angle APB$, as shown below, show that $m\angle AOB = 2m\angle APB$.

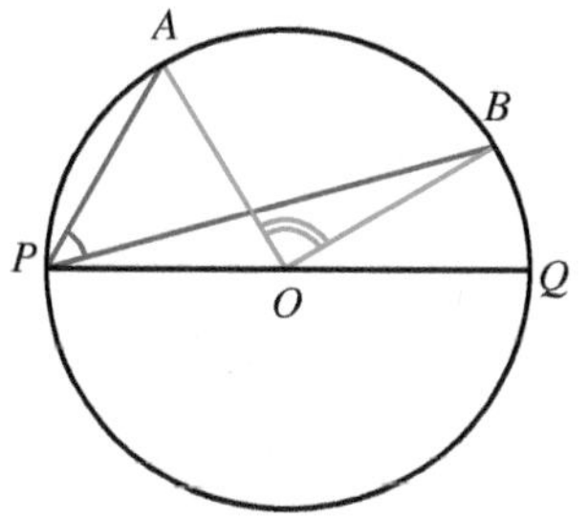

39. Let the convex quadrilateral $ABCD$ be inscribed in a circle. Show that the angles at opposite vertices are supplementary. That is, $m\angle A + m\angle C = 180°$ and $m\angle B + m\angle D = 180°$.

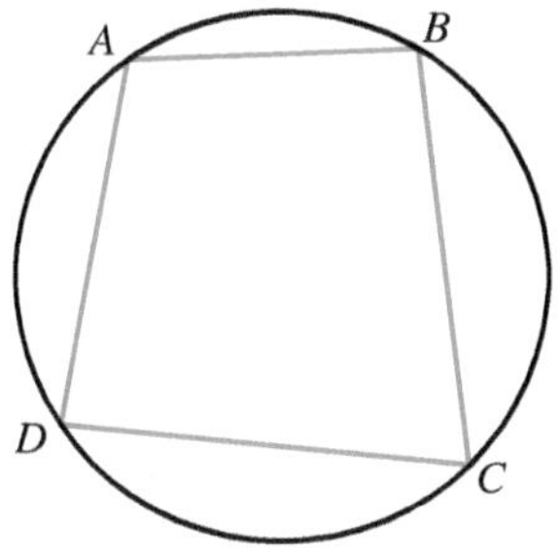

State Assessments

40. (Massachusetts, Grade 5)
Which of the following pairs of quadrilaterals appear to be congruent?

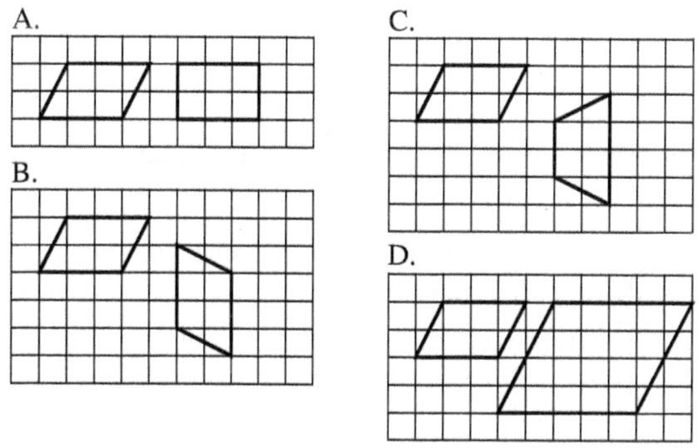

41. (Grade 5)
Your sister's teacher told her that the equilateral triangle shown below is divided into six congruent triangles by the diagonals from each vertex to the midpoint of the opposite side. Your sister isn't clear why this is true. What can you tell her to help her out?

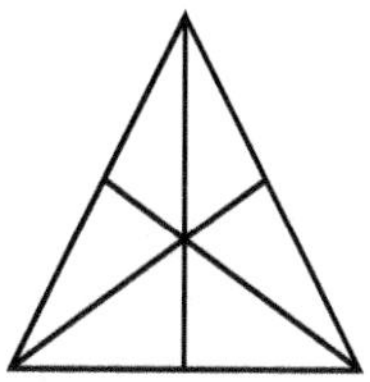

42. (Grade 5)
Which tangram pieces are congruent triangles?

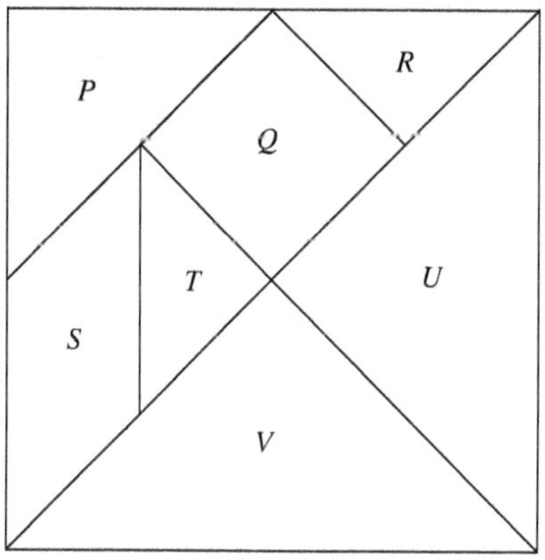

A. P and T
B. P and R
C. T and U
D. U and V
E. P and U

43. (Grade 7)
The two polygons shown below are congruent.

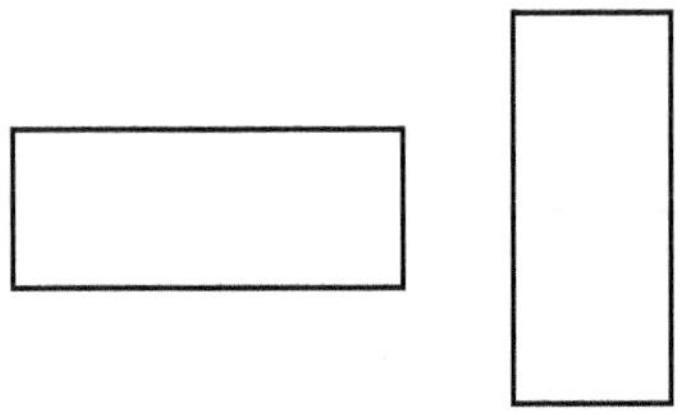

What can be said about the corresponding sides and corresponding angles of the two figures?

A. The corresponding angles have the same measure, but the corresponding sides may have different lengths.

B. The corresponding sides have the same length, but the corresponding angles may have different measures.

C. The corresponding angles have the same measure and the corresponding sides have the same length.

D. The corresponding angles may have different measures and the corresponding sides may have different lengths.

44. (Grade 6)
Which figure does not appear to have a pair of congruent triangles?

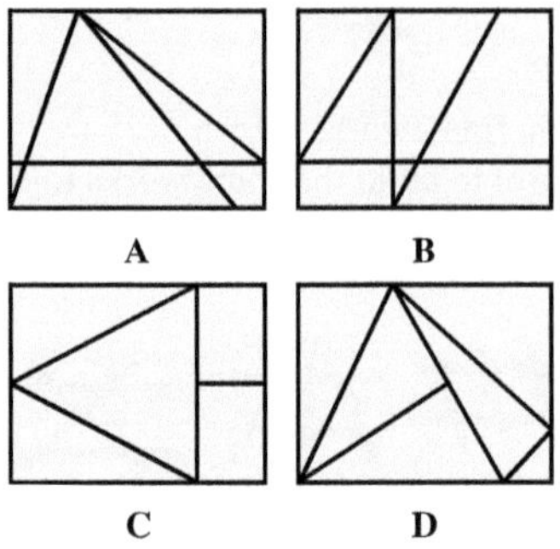
A
B
C
D

12.2 Constructing Geometric Figures

A rhombus is easily constructed by drawing intersecting arcs of circles of the same radius.

In the last section, two basic constructions were described:

- Construction 1 Construct a line segment congruent to a given segment.
- Construction 2 Construct an angle congruent to a given angle.

Since only the straightedge and compass were used, these are examples of Euclidean constructions. In this section, we describe a number of other Euclidean constructions and explore some related applications and theorems. In addition, we investigate constructions with the Mira and by paper folding.

To be certain that a construction results in a figure that has a desired property, a proof of the validity of the construction must be given. For example, Construction 2 of a congruent angle is a consequence of the SSS property. Many constructions can be verified by appealing to the properties of a rhombus listed in Figure 12.6.

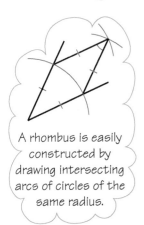

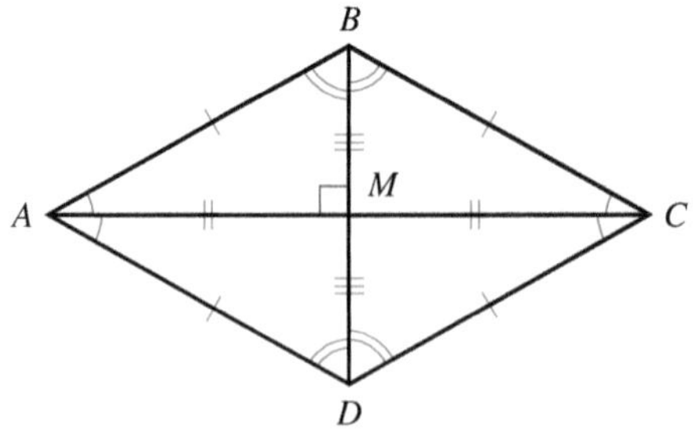
B
M
A
C
D

Figure 12.6
The rhombus *ABCD* has many useful properties:
- The diagonals are angle bisectors
- The diagonals are perpendicular
- The diagonals intersect at their common midpoint *M*
- The sides are all congruent to one another
- The opposite sides are parallel

Constructing Parallel and Perpendicular Lines

If l is a given line and P is a given point not on l. Construction 3 details one way to construct the line k through P that is parallel to line l. In Construction 4, the line m through P that is perpendicular to l is constructed.

CONSTRUCTION 3 Construct a Line Parallel to a Given Line

Given point P and line l as shown, construct a line through P that is parallel to l.

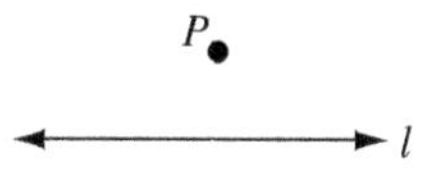

Procedure

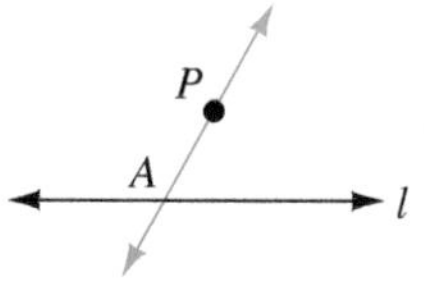	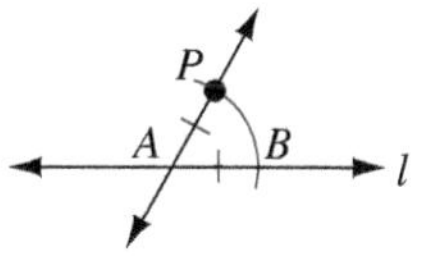	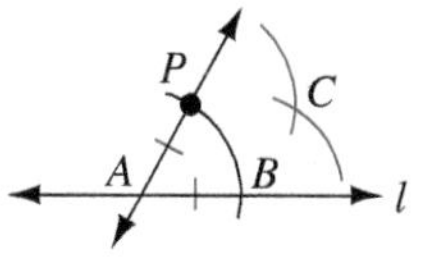	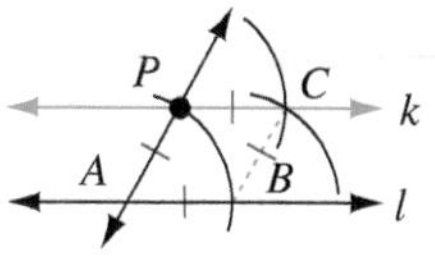
Step 1: Construct any line through P that intersects l at a point labeled A.	**Step 2:** Construct an arc through P centered at A, and let B denote the intersection of the arc with l.	**Step 3:** With the same radius AB, construct arcs with centers at P and B; let C be the intersection of the two arcs.	**Step 4:** Construct the line $k = \overleftrightarrow{PC}$; since $ABCP$ is a rhombus, its opposite sides are parallel, so $k \parallel l$.

CONSTRUCTION 4 Construct a Line Perpendicular to a Given Line Through a Point Not on the Given Line

Given line l and point P not on l as shown, construct a line through P that is perpendicular to l.

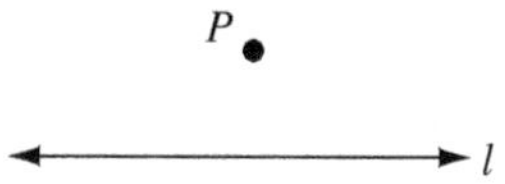

Procedure

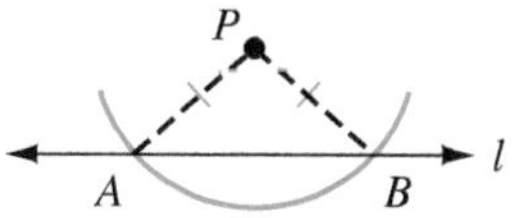	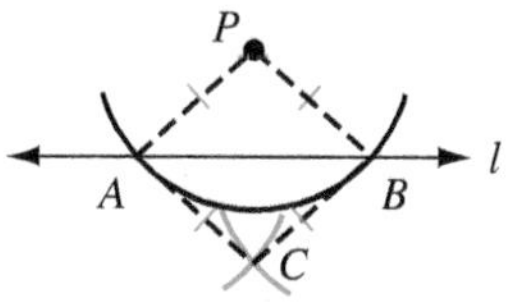	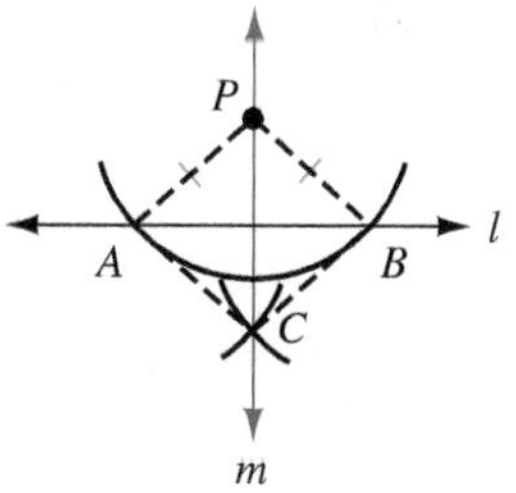
Step 1: Construct an arc at P that intersects l at two points A and B.	**Step 2:** With the compass still at radius AP, construct arcs at A and B and let C be their point of intersection.	**Step 3:** Construct the line $\overleftrightarrow{PC}$; since $\overline{PC}$ is a diagonal of rhombus $ABPC$, it is perpendicular to $\overline{AB}$.

The construction of perpendicular lines has several applications:

- **Nearest point F on a line l from a point P.** The line through P that is perpendicular to line l intersects l at the point F of l that is closest to P (Why? Use the triangle inequality). F is called the **foot** of the perpendicular line from P, and PF is called the **distance from P to l.**
- **Point of reflection P' of P across mirror line l.** The point P' on the perpendicular to l through P for which $PF = P'F$, where F is the point of intersection of l and the perpendicular through P, is called the **point of reflection of P across l.**

- **Altitude of a triangle**. The line through vertex A of triangle ABC that is perpendicular to the opposite side $\overline{BC}$ is called an **altitude** of the triangle. The distance from A to the line containing side $\overline{BC}$ is the **height** of the triangle when $\overline{BC}$ is considered to be the base of the triangle. Often, the word *altitude* is used to mean the height of a triangle.

These applications are illustrated in Figure 12.7.

Figure 12.7 Applications of perpendicular lines

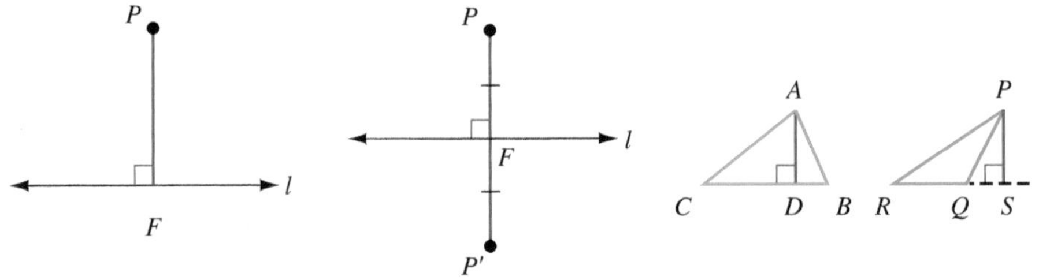

(a) F is the foot of the perpendicular through P.

(b) P' is the reflection of P across l.

(c) $\overline{AD}$ and $\overline{PS}$ are altitudes of triangles ABC and PQR, respectively.

If point P lies on line l, a small modification in the second step of the previous procedure is required to construct the line perpendicular to l at P.

CONSTRUCTION 5 Construct the Line Perpendicular to a Given Line Through a Point on the Given Line

Given line l and point P on l as shown, construct the line through P perpendicular to l.

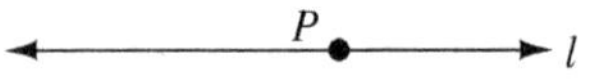

Procedure

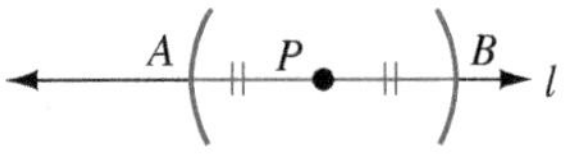

Step 1:
Construct two arcs of equal radius centered at P; let A and B be their points of intersection with l.

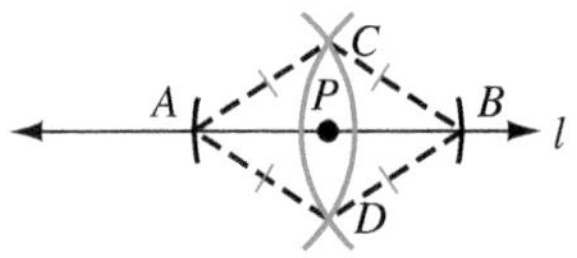

Step 2:
Construct arcs centered at A and B with a radius *greater* than AP; let C and D be their points of intersection.

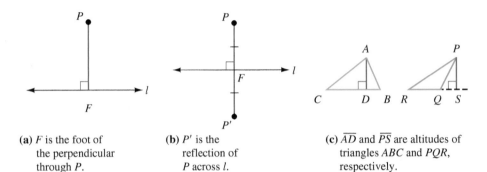

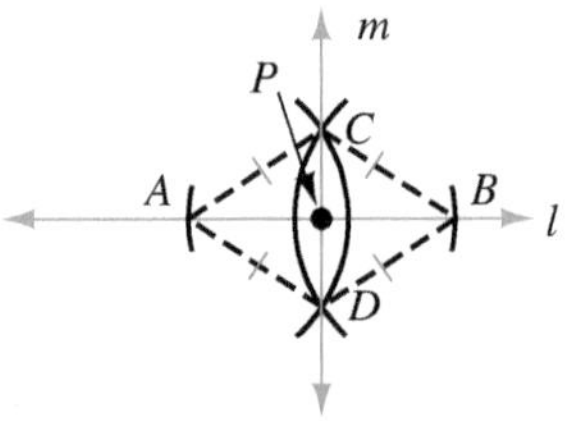

Step 3:
Construct line $m = \overleftrightarrow{CD}$; since $\overline{CD}$ is a diagonal of the rhombus $ADBC$ and P is the midpoint of diagonal $\overline{AB}$, m passes through P and is perpendicular to $l = \overleftrightarrow{AB}$; that is, m is perpendicular to l.

Constructing the Midpoint and Perpendicular Bisector of a Line Segment

The line perpendicular to a segment at its midpoint is called the **perpendicular bisector** of the segment. The following construction is also justified by properties of the rhombus:

CONSTRUCTION 6 Construct the Midpoint and Perpendicular Bisector of a Line Segment

Construct the midpoint and perpendicular bisector of the segment $\overline{AB}$ shown.

Procedure

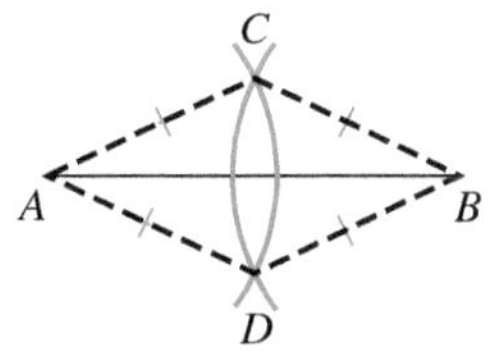

Step 1:
Construct arcs of the same radius centered at A and B and intersecting in points C and D.

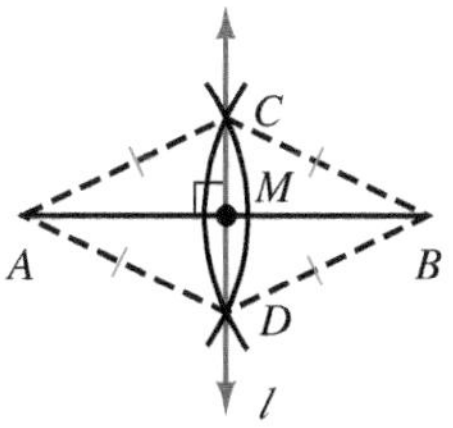

Step 2:
Construct the line $\overleftrightarrow{CD}$. Since $ADBC$ is a rhombus, $\overleftrightarrow{CD}$ intersects $\overline{AB}$ at a right angle at the midpoint M of $\overline{AB}$.

The midpoint of a line segment M is the same distance from A and B. Indeed, every point P of the perpendicular bisector of a segment $\overline{AB}$ is equidistant from the endpoints of the segment. That is, $PA = PB$. We state and prove this property next.

THEOREM Equidistance Property of the Perpendicular Bisector
A point lies on the perpendicular bisector of a line segment if and only if the point is equidistant from the endpoints of the segment.

PROOF Let l be the perpendicular bisector of segment $\overline{AB}$. Thus, l intersects $\overline{AB}$ at right angles at the midpoint M, as shown on the left side of the diagram that follows. Let P be any point on l. By the SAS property, $\triangle PMA \cong \triangle PMB$. This means that the corresponding sides $\overline{PA}$ and $\overline{PB}$ are congruent. Hence, $PA = PB$, as claimed.

The proof of the converse—that if a point P is equidistant from points A and B, then P lies on the perpendicular bisector of $\overline{AB}$—is similar. (See problem 30 of Problem Set 12.2.)

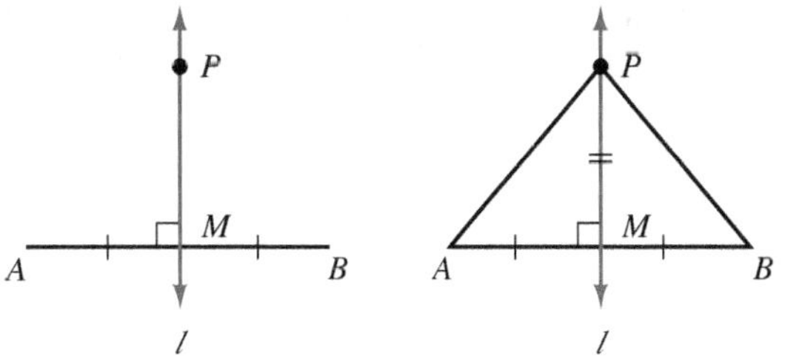

EXAMPLE 12.8

Locating an Airport

The Tri-Cities Airport Authority wishes to locate a new airport to serve its three cities, situated at A, B, and C as shown. If possible, it would like a site P that is the same distance from A, B, and C. How can P be located?

Solution To be equidistant from A and B, P must be on the perpendicular bisector of the segment $\overline{AB}$. Similarly, P must be on the perpendicular bisector of $\overline{BC}$. Since A, B, and C are not collinear, the perpendicular bisectors to $\overline{AB}$ and $\overline{BC}$ are not parallel and we can choose P as their point of intersection. Since $PA = PB$ and $PB = PC$, we have $PA = PC$. Therefore, P is also equidistant from A and C, so P is also on the perpendicular bisector of $\overline{AC}$. Thus, P is equidistant from A, B, and C, as desired.

Point P is the center of a unique circle containing A, B, and C as shown in Figure 12.8. The circle is called the **circumscribed circle** of $\triangle ABC$, and P is called the **circumcenter** of $\triangle ABC$. Frequently, the circumscribing circle is called the **circumcircle** of the triangle.

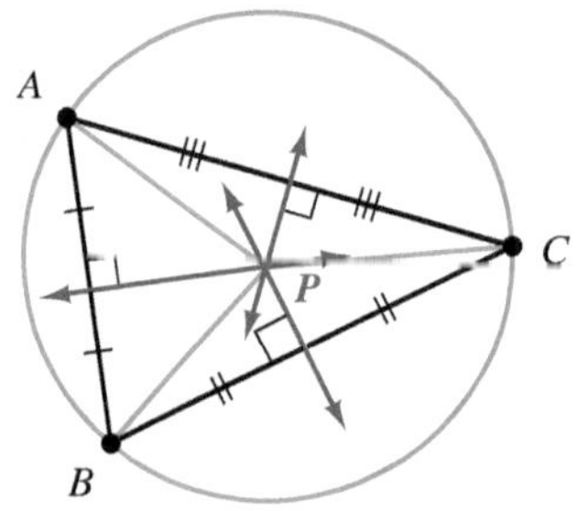

Figure 12.8 The perpendicular bisectors of the sides of $\triangle ABC$ are concurrent at a point P equidistant from A, B, and C. Point P is the center of the circumscribing circle of $\triangle ABC$

THEOREM The Circumscribed Circle of a Triangle

The perpendicular bisectors of a triangle ABC are concurrent at a point P (the *circumcenter* of the triangle) that is the center of a unique circle (the *circumscribed circle*) that passes through the three vertices A, B, and C of the triangle.

COOPERATIVE INVESTIGATION

Finding Special Points of a Triangle with Paper Folding

Materials Needed

Sheets of (thin) white paper, colored markers (black, red, blue, green), ruler.

Directions

First working individually, draw three heavy black lines across a sheet of paper to form a triangle *ABC*, as shown in the leftmost diagram. The lines should be heavy enough to be visible from the back side of the sheet.

Now carry out these folding activities, referring to the following diagrams. Try to be very careful to make accurate, sharp folds with distinct creases.

1. **Point *P*.** Fold and crease the paper to find the circumcenter *P* of triangle *ABC*. Draw a red line along the three creases; label the midpoints of the sides as *K*, *L*, and *M*, and label the intersection of the perpendicular bisectors as *P*. Point *P* is the *circumcenter* of triangle *ABC*. Two of the perpendicular bisectors are shown in the diagram.
2. **Point *H*.** Fold the altitudes of the triangle at each of its vertices. Draw green lines along the creases, and let the point of intersection (which is called the *orthocenter*) be labeled *H*.
3. **Point *G*.** Fold the three medians of the triangle, where a median is the line through a vertex and the midpoint of the opposite side of the triangle. Draw blue lines along the creases, and let *G* be the point of intersection (*G* is known as the *centroid* of the triangle).

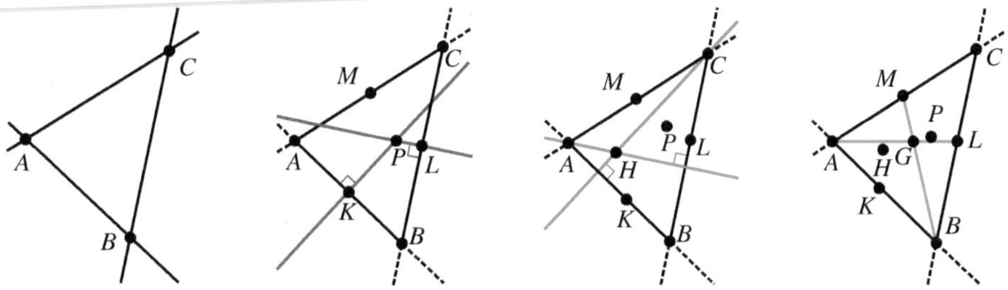

Triangle *ABC* Point *P* Point *H* Point *G*

4. Use your ruler to investigate the points *P*, *H*, and *G*. Discuss your conclusions with the other members of your group. Formulate a conjecture that summarizes your results of the paper-folding investigation.

Constructing the Angle Bisector

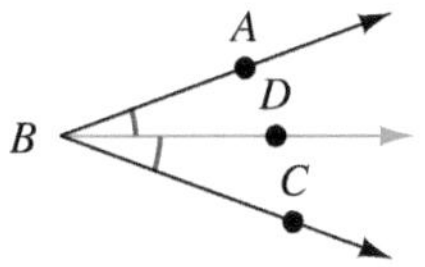

Figure 12.9
$\overrightarrow{BD}$ is the angle bisector of $\angle ABC$ if $\angle ABD \cong \angle DBC$

Given $\angle ABC$ (see Figure 12.9), we wish to construct the ray $\overrightarrow{BD}$ that forms congruent angles with the sides $\overrightarrow{BA}$ and $\overrightarrow{BC}$. If $\angle ABD \cong \angle CBD$, then $\overrightarrow{BD}$ is the **angle bisector** of $\angle ABC$.

Once again, the properties of a rhombus justify the following construction:

CONSTRUCTION 7 Construct the Angle Bisector

Construct the angle bisector of $\angle E$ shown.

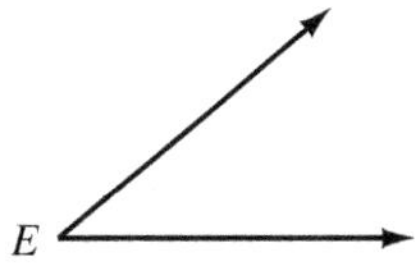

Procedure

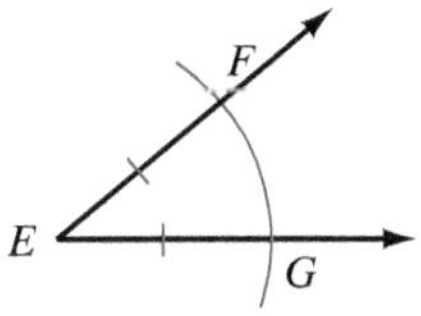 **Step 1:** Construct an arc centered at *E*; let *F* and *G* denote the points at which the arc intersects the sides of $\angle E$.	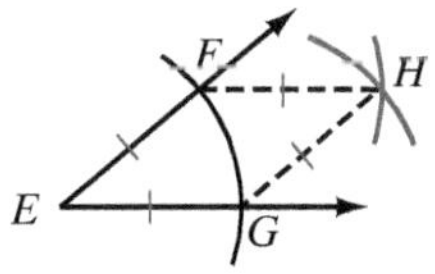 **Step 2:** Construct arcs, centered at *F* and *G*, of radius *EF*. Let *H* be the point of intersection of the two arcs.	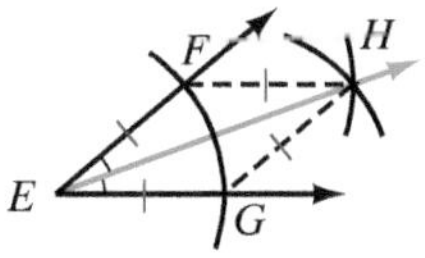 **Step 3:** Construct the ray $\overrightarrow{EH}$. Since the diagonal $\overline{EH}$ forms congruent angles with the sides $\overline{EG}$ and EF of the rhombus *EGHF*, $\overrightarrow{EH}$ is the angle bisector of $\angle E$.

The following theorem gives a condition that states exactly when points lie on the angle bisector:

THEOREM Equidistance Property of the Angle Bisector

A point lies on the bisector of an angle if and only if the point is equidistant from the sides of the angle.

PROOF Let P be any point on the bisector of $\angle A$ as shown. Let F and F' be the feet of the perpendiculars from P to the sides of $\angle A$.

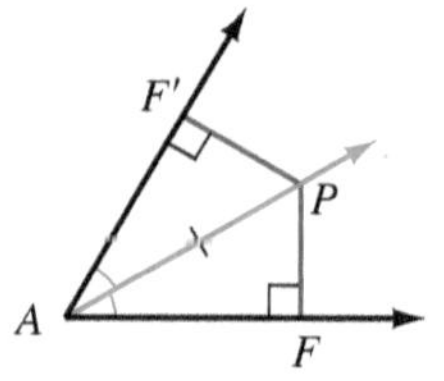

By the AAS congruence property, $\triangle PAF \cong \triangle PAF'$. Therefore, $PF = PF'$. The proof of the converse is similar: It must be shown that if P is equidistant from the sides of $\angle A$, then $\overrightarrow{AP}$ bisects the angle. Details are left to the reader.

By mimicking the solution of the airport problem in Example 12.8, it can be shown that the bisectors of the interior angles of a triangle are concurrent at the point I that is equidistant from three sides of the triangle. (See Figure 12.10.) Point I, called the **incenter** of $\triangle ABC$, is the center of the **inscribed circle**, or **incircle**, of $\triangle ABC$. The inscribed circle is tangent to all three sides of $\triangle ABC$.

To construct the incircle, a perpendicular to a side that passes through I must be constructed. For example, to find the perpendicular to side $\overline{AB}$ containing I, construct the radial segment $\overline{ID}$.

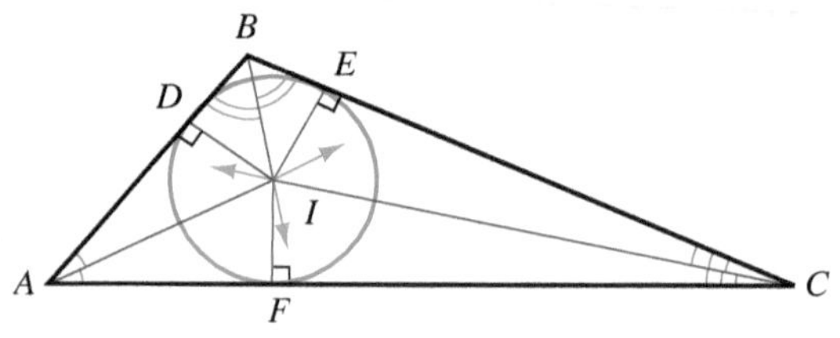

Figure 12.10
The bisectors of the interior angles of a triangle are concurrent at a point *I* equidistant from the sides of the triangle. *I* is the center of the inscribed circle of the triangle, which is tangent to the sides of the triangle at points *D*, *E*, and *F*

THEOREM The Inscribed Circle of a Triangle
The angle bisectors of a triangle ABC are concurrent at a point I (the *incenter* of the triangle) that is the center of a unique circle (the *inscribed circle* or *incircle*) that is tangent to the three sides $\overline{AB}$, $\overline{BC}$, and $\overline{AC}$.

INTO THE CLASSROOM
Constructions in Space

Constructions using a compass and a straightedge produce figures that are confined to a sheet of paper. Interest and excitement can also be generated by constructing figures in space, using sticks, brass fasteners, paper clips, string, cut paper, multilink cubes, polyhedrons, straws, or indeed whatever is available. Such figures can be held and felt, literally giving students a feel for shape. In some cases, the shapes can be bent or flexed to give a dynamic liveliness to figures that would remain of lesser interest when only drawn on a sheet of paper.

Books, pamphlets, and journals published by the National Council of Teachers of Mathematics and other publishing companies provide the teacher with a wide variety of ideas and resources for three-dimensional constructions and related activities. Every teacher will want to gather a personal collection of favorite hands-on constructions suitable for his or her classroom. Here are two suggestions for constructions to do in the classroom:

- **Hinged polygons.** Strips of card stock can be joined with brass fasteners through holes at the ends of each strip. Any triangle is rigid, demonstrating the SSS congruence property. Any quadrilateral, however, is flexible. As the quadrilateral flexes, the sum of the angle measures remains at 360°, as can be checked with a protractor.

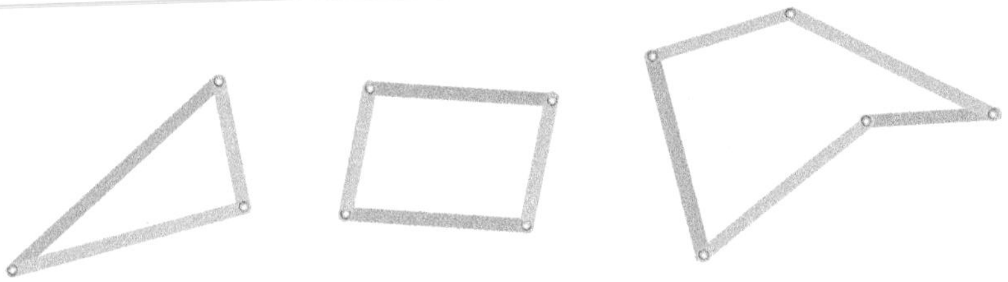

- **Space polygons and polyhedra.** Thin wooden sticks, say, from a "pick-up-sticks" game, or straws can be cut to differing lengths and joined by short pieces of rubber tubing at their endpoints. More than two sticks can meet at a single vertex by inserting a length (or several lengths) of tubing through a hole punched sideways through another section of tubing. It is easy to form quadrilaterals that flex in space, and joining the midpoints of the four sides by elastic bands shows that a parallelogram is always formed. Properties of cubes, tetrahedra, and other polyhedra can also be explored with easily constructed skeletal models.

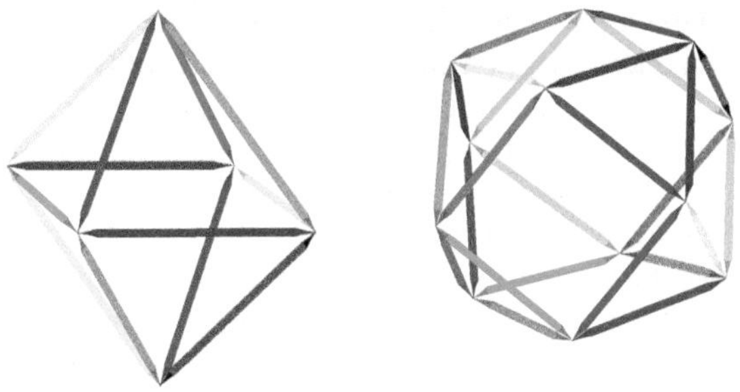

Constructing Regular Polygons

A square is easily constructed with compass and straightedge. For example, construct two perpendicular lines and a circle centered at the point of intersection, as shown in Figure 12.11a.

Figure 12.11 Constructing angle bisectors doubles the number of sides of a regular polygon inscribed in a circle

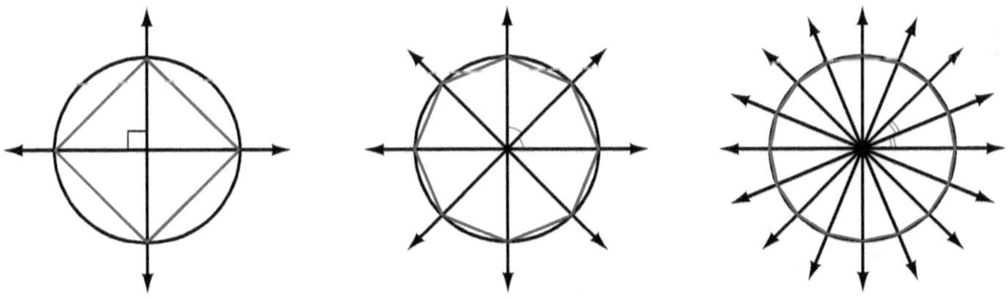

(a) Square **(b)** Regular octagon **(c)** Regular 16-gon

By constructing angle bisectors from the center of a square, the eight vertices of a regular octagon are constructed. Angle bisectors of the octagon's central angles can then be constructed as in Figure 12.11b to yield the regular 16-gon of part (c) of the figure. When the vertices of a polygon all lie on a given circle, the polygon is called an **inscribed polygon**.

A regular hexagon is particularly easy to inscribe in a given circle with a compass and straightedge: Pick any point A on the circumference of the circle centered at O, and then successively strike arcs of radius OA around the circle to locate B, C, D, E, and F. The hexagon $ABCDEF$ is regular, since joining the sides to the center O forms six congruent equilateral triangles: OAB, OBC, . . . , and OFA. On the one hand, as shown in Figure 12.12, connecting every other vertex gives a construction of the inscribed equilateral triangle ACE. On the other hand, constructing angle bisectors of the central angles of the hexagon produces an inscribed regular dodecagon.

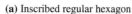

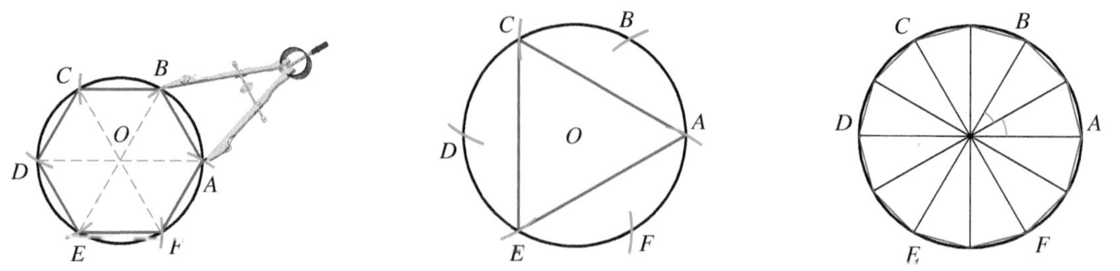

Figure 12.12 The regular 3-, 6-, and 12-gons can be inscribed with compass and straightedge in a given circle

(a) Inscribed regular hexagon **(b)** Inscribed equilateral triangle **(c)** Inscribed regular dodecagon

A compass-and-straightedge construction of a regular pentagon requires more ingenuity. (One method is outlined in problem 33 of Problem Set 12.2 at the end of this section.)

The ancient Greek geometers knew how to construct the regular polygons shown so far. They also knew that the regular 15-gon can be constructed with compass and straightedge. (The construction is outlined in problem 34 in Problem Set 2.2.) For over 2000 years, the only regular polygons known to be constructible with compass and straightedge were the ones contained in Book IV of Euclid's *Elements:* the regular 3-, 4-, 5-, and 15-gons and, by angle bisection, the regular polygons obtained by successively doubling the number of sides.

On March 30, 1796, one month before his 19th birthday, Carl Friedrich Gauss (1777–1855) entered into a notebook his discovery that a number of other regular polygons were constructible, including the 17-gon, the 257-gon, and the 65,537-gon. The numbers 3, 5, 17, 257, and 65,537 are prime numbers of the special form $F_k = 2^{2^k} + 1$, where k is a nonnegative integer. For example, $F_0 = 2^{2^0} + 1 = 2^1 + 1 = 3$, $F_1 = 2^{2^1} + 1 = 2^2 + 1 = 4 + 1 = 5$, and so on. Numbers of this form had been studied earlier by Pierre de Fermat (1601–1665), and prime numbers of the form $2^{2^k} + 1$ are known as **Fermat primes**.

Following is the remarkable theorem of Gauss. The proof of the "only if" part of the theorem is due to Pierre Wantzel (1814–1848).

THEOREM The Gauss-Wantzel Constructibility Theorem

A regular polygon of n sides is constructible with compass and straightedge if and only if n is

1. 4, or
2. a Fermat prime, or
3. a product of distinct Fermat primes, or
4. a power-of-2 multiple of a number that is one of the preceding types 1, 2, or 3.

For example, the regular heptagon is not constructible, since 7 is not a prime number of the form $2^{2^k} + 1$. Nor is a regular nonagon (9-gon) constructible, since 9 has two factors of 3. However, a regular polygon of $1020 = 2^2 \cdot 3 \cdot 5 \cdot 17$ sides is constructible, since its odd prime factors are distinct Fermat primes.

Fermat believed that all numbers of the form $F_k = 2^{2^k} + 1$ were prime, but Euler proved this assertion to be false by showing that $F_5 = 2^{2^5} + 1$ is composite; in fact, $F_5 = 4{,}294{,}967{,}297 = (641) \cdot (6{,}700{,}417)$. Likewise, $F_6, F_7, \ldots, F_{32}$ are now known to be composite. It is generally believed, though not proved, that there are only five Fermat primes, namely, 3, 5, 17, 257, and 65,537.

Highlight from History: Three Impossible Construction Problems

The straightedge allows us to draw a line of indefinite length through any two given points, and the compass* allows us to draw a circle with a given point as its center and passing through any given distinct second point. It then becomes a challenge to find procedures to construct a figure with the use of only these simple tools. Many important contributions to geometry were inspired by attempts to solve the following famous problems, each of which arose in antiquity:

1. *The trisection of an angle:* Divide an arbitrary given angle into three congruent angles.

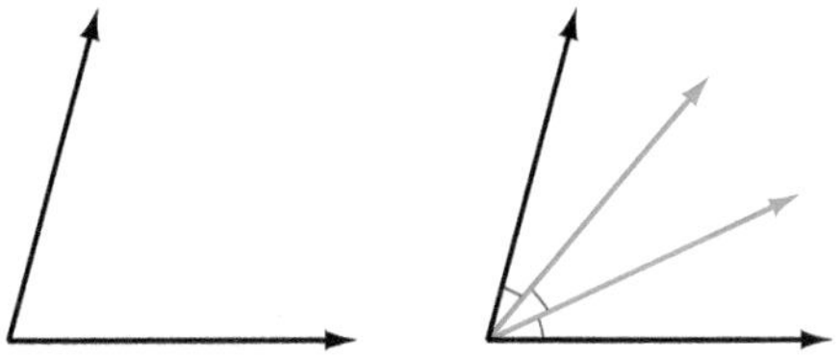

2. *The duplication of the cube:* Given a cube, construct a cube with twice the volume.

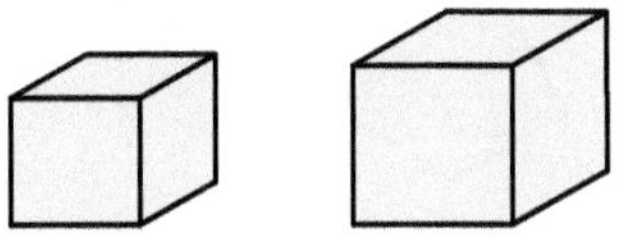

3. *The squaring of the circle:* Given a circle, construct a square of the same area as the circle.

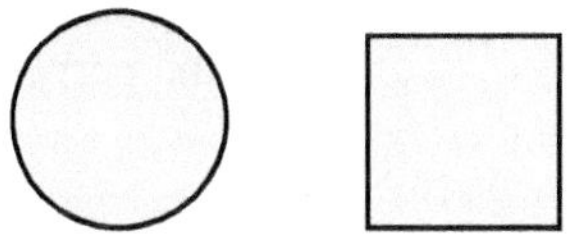

Extensive efforts for over 2000 years failed to solve any of these problems. It was not until the early 1800s that it was shown that these problems were impossible to solve when only the unmarked straightedge and compass were allowed. It is interesting to note that methods of algebra were used to prove the impossibility of these geometric constructions.

The usage "the compass" is common, but some texts and authors still prefer "compasses" or even "a pair of compasses."

Mira™ and Paper-Folding Constructions

Figure 12.13 shows a Mira, which was used in Chapter 11 to construct images under reflection and to investigate lines of symmetry. The Mira can also be used to construct perpendicular lines, midpoints, angle bisectors, and other geometrical objects. With a little practice, constructions with a Mira are quick and yet very accurate.

Figure 12.13
The Mira and its use in three basic constructions

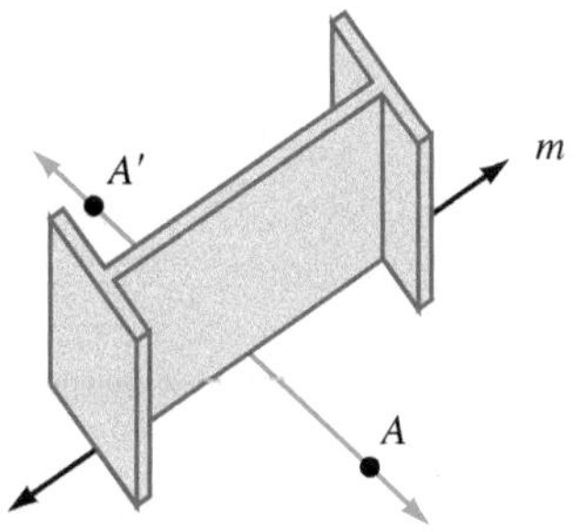

(a) The Mira

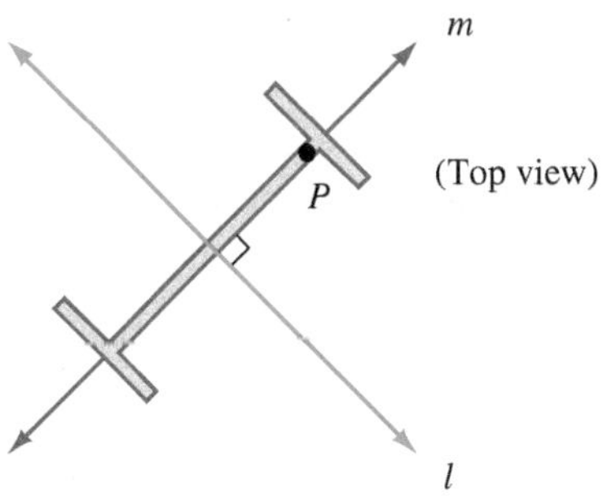

(b) Pivot the Mira about P until line l coincides with its reflection to construct the line m through P that is perpendicular to l.

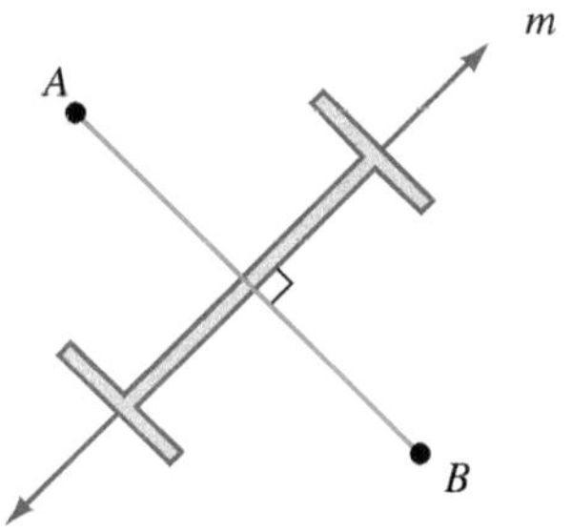

(c) When the reflection of B coincides with A, the drawing edge of the Mira determines the perpendicular bisector of $\overline{AB}$.

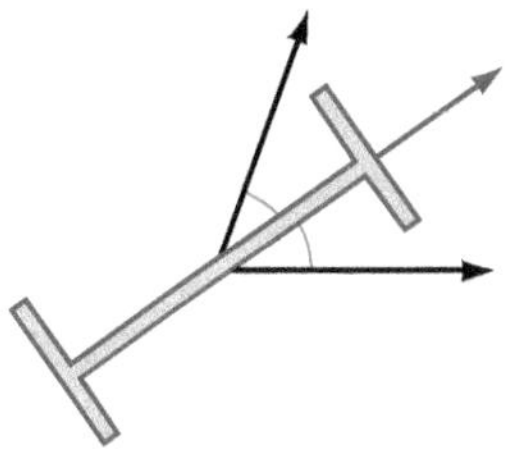

(d) Pivot the Mira about the vertex until the reflection of the near side coincides with the far side to construct the angle bisector.

Mira constructions can usually be converted into equivalent paper-folding procedures, in which the drawing line of the Mira is replaced by the crease line of a fold. It is helpful to draw very dark lines and points so that they can be seen from the reverse side of the paper. A folding construction of the perpendicular bisector is shown in Figure 12.14.

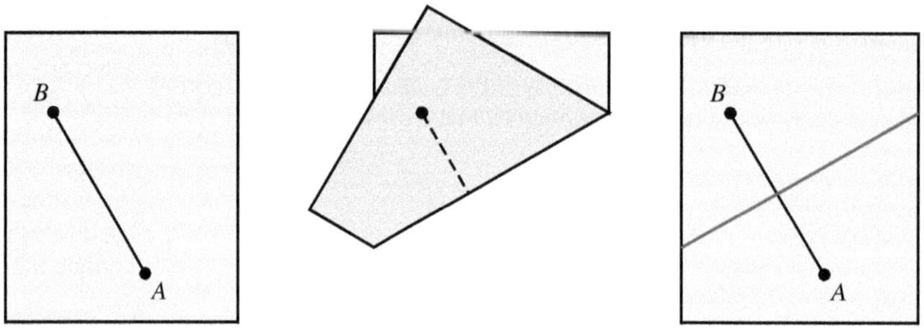

Figure 12.14
Folding point A onto point B forms a crease that is the perpendicular bisector of $\overline{AB}$

Constructions with Geometry Software

Constructions with tangible materials—paper, straws, rubber bands, and the like—continue to be important for children in the earlier grades as they develop their understanding of the principles of geometry. In the upper elementary grades, the traditional Euclidean tools—compass and straightedge —can be used as well as non-Euclidean tools such as the Mira and paper folding. Teachers will also want to encourage their students to take advantage of geometry software, in which figures can first be constructed and then manipulated in a dynamic environment that encourages the exploration of the properties of the figures. There are many geometry programs available such as *Cabri, The Geometer's Sketchpad,* and, more recently, *GeoGebra,* which is available as a free download.

The importance of developing a student's ability to select appropriate tools is highlighted in the fifth Standard of Mathematical Practice (SMP 5) of the Common Core.

SMP 5 Use appropriate tools strategically.

"Mathematically proficient students consider the available tools. . . . These tools might include pencil and paper, concrete models, a ruler, a protractor, . . . or dynamic geometry software. Proficient students are sufficiently familiar with tools appropriate for their grade or course to make sound decisions about when each of these tools might be helpful, recognizing both the insight to be gained and their limitations."

Many constructions that would be tedious and difficult by traditional methods become easy and attractive with geometry software. For example, the tiling with squares and congruent triangles shown in Figure 12.15 is not difficult to construct with computer geometry but would be challenging otherwise.

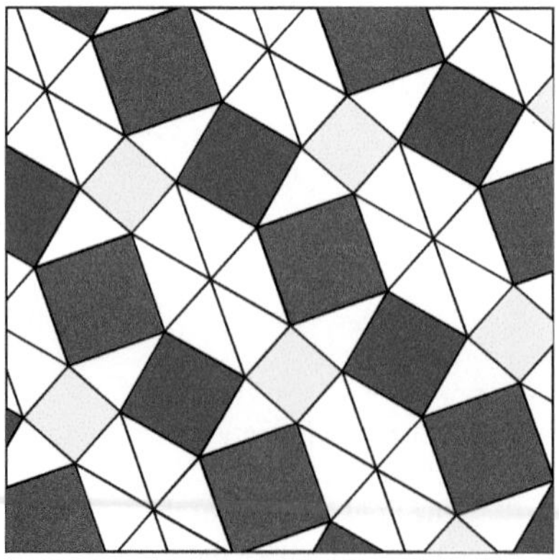

Figure 12.15
A tiling created with geometry software. The triangles are congruent and the sides of the squares are each congruent to a side of the triangles

12.2 Problem Set

Understanding Concepts

1. Let P be a point not on line l. Explain, in diagrams and words, how to construct the point C that is the reflection of point P across line l using straightedge and compass.

2. Let P be a point not on line l. Explain, in diagrams and words, how to construct

(a) the point F on line l that is closest to point P using straightedge and compass. That is, explain how to construct the *foot* F of point P on line l.

(b) the circle centered at P that is tangent to the line l.

3. (**Writing**) Given a segment $\overline{AB}$, describe a straightedge and compass construction of a square $ABCD$. Use words and diagrams to give a step-by-step procedure using one of the constructions of the section as a model. The first step is done for you:

Step 1. Use the straightedge to construct the line $m = \overleftrightarrow{AB}$ through points A and B.

4. (**Writing**) Given a segment $\overline{AB}$, explain step-by-step procedures that describe straightedge and compass constructions of

(a) an equilateral triangle ABC.

(b) a regular hexagon $ABDEFG$ once equilateral triangle ABC has been constructed.

5. Starting with a segment $\overline{AB}$ and a ray $\overrightarrow{AP}$, explain in diagrams and words how to create a compass and straightedge construction of a rhombus $ABCD$, where $\overline{AB}$ is a side of the rhombus and point D is on the ray $\overrightarrow{AP}$.

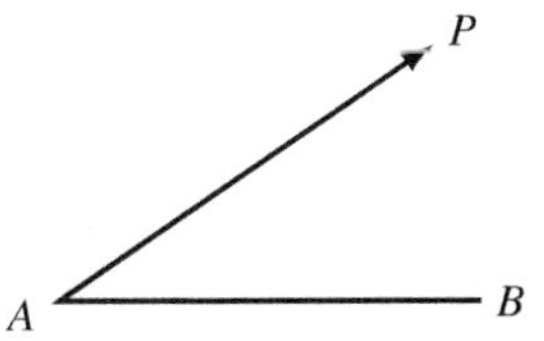

6. Starting with a segment $\overline{AC}$ and a ray $\overrightarrow{AP}$, explain in diagrams and words how to create a compass and straightedge construction of a rhombus $ABCD$, where $\overline{AC}$ is a diagonal of the rhombus and point D is on the ray $\overrightarrow{AP}$.

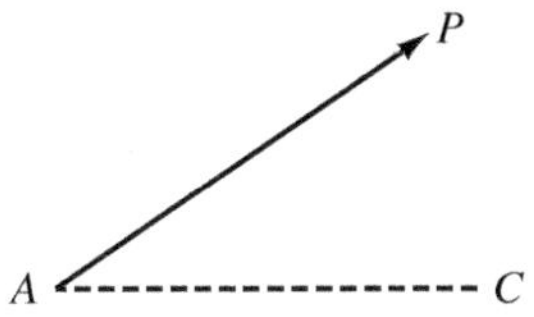

7. Starting with point A and line l not containing A, explain in diagrams and words how to create a compass and straightedge construction of a square $ABCD$ where the points are listed counterclockwise around the square and points B and C are on line l.

8. Starting with point A and line l not containing A (as in problem 7), explain in diagrams and words how to create a compass and straightedge construction of a square $ABCD$ where the points are listed counterclockwise around the square and points B and D are on line l.

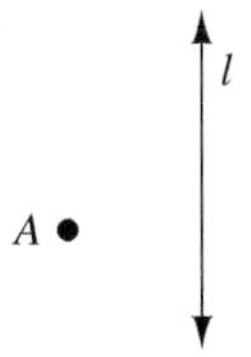

9. Describe, in words and drawings, a Mira construction that gives the line k parallel to a line l and passing through a point P not on l.

10. Answer problem 9 again, but use paper folding instead of the Mira.

11. The drafting triangle and straightedge can be used to construct parallel lines as follows:

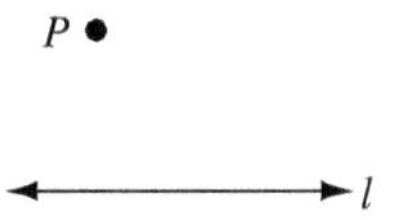

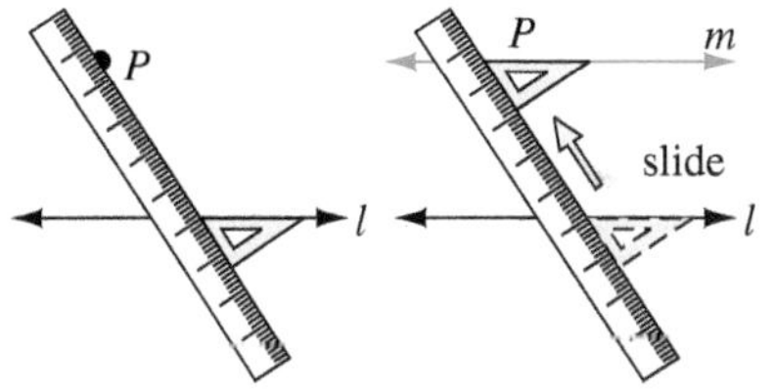

(a) What geometric principle justifies this construction?

(b) Describe, in words and drawings, a procedure using a straightedge and drafting triangle to construct the line m perpendicular to a given line l and passing through a given point P.

12. Use a jar lid (or other handy circular object) to draw three congruent circles through a common point P. Let A, B, and C denote the three points of intersection of pairs of your circles other than P. Now use the jar lid to discover something amazing about the circumscribing circle of $\triangle ABC$.

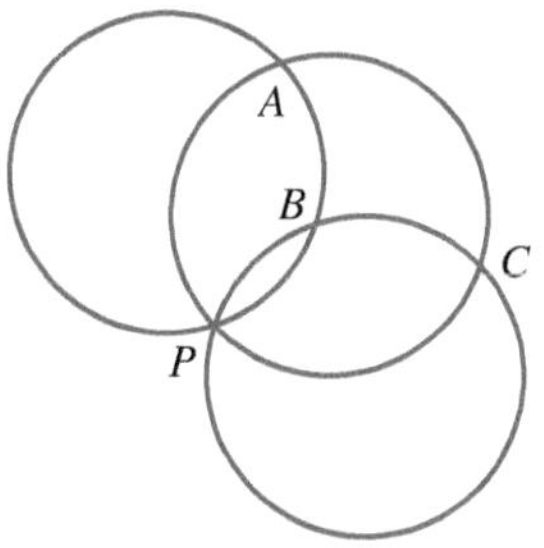

13. Construct an angle ABC and a point T on side $\overrightarrow{BA}$. Show, in words and drawings, how to construct a circle that is tangent to both sides of the angle, with T as one of the points of tangency.

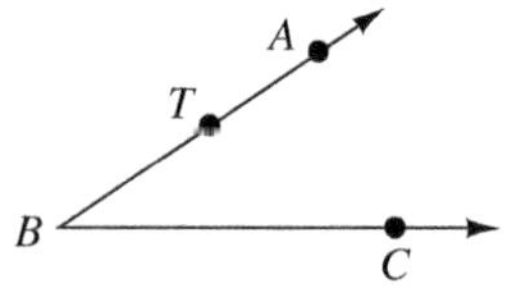

14. First, construct any triangle ABC. Then

(a) construct the incenter I of the triangle.

(b) construct the point D at which the incircle is tangent to side $\overline{AB}$.

(c) construct the incircle of triangle ABC.

(d) explain why the area of triangle ABC is given by the formula

$$\text{area}\,(ABC) = \frac{1}{2}(a + b + c)$$

where the lengths of the sides of the triangle are a, b, and c and r is the radius of the inscribed circle.

15. Construct the circumcenters and circumscribing circles of triangles of the three types shown. Use any tools you wish to draw the perpendicular bisectors of the sides of the triangles.

(a) Acute triangle

(b) Right triangle

(c) Obtuse triangle

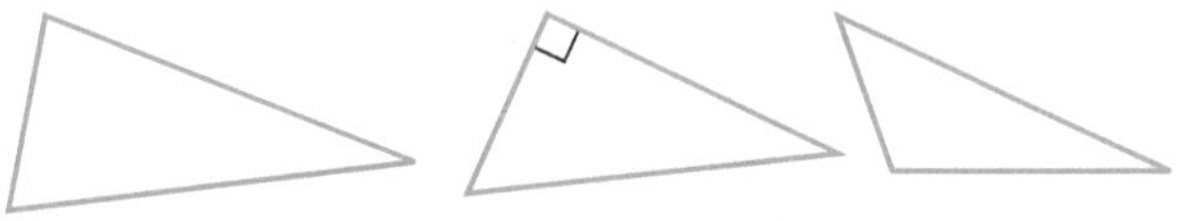

(d) For each case, make a conjecture about where—inside, on, or outside the triangle—the circumcenter will be located.

16. (a) Use any drawing tools you wish to construct the altitudes of the three types of triangles in problem 15.

(b) Make a conjecture about the three altitudes of an acute triangle.

(c) Make a conjecture about the three altitudes of a right triangle.

(d) Make a conjecture about the three lines containing the altitudes of an obtuse triangle.

17. A point P is exterior to a circle centered at Q. Use Thales' theorem to justify why drawing the circle with diameter $\overline{PQ}$ gives a construction of the two lines $\overleftrightarrow{PS}$ and $\overleftrightarrow{PT}$ that are tangent to the circle centered at Q.

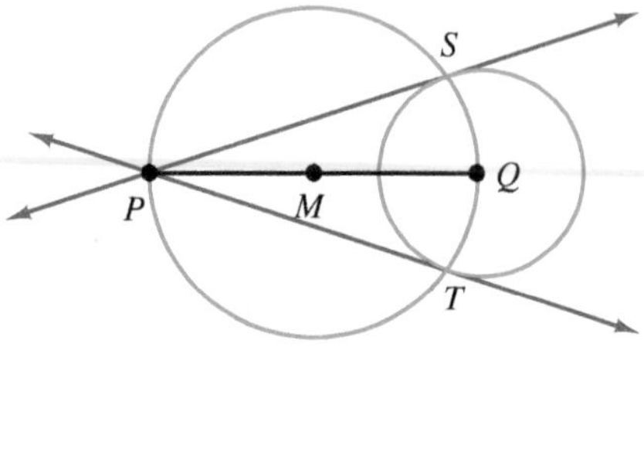

18. Construct a triangle ABC and also its incenter I and the points of tangency T_1 and T_2 of the incircle to sides $\overline{AC}$ and $\overline{BC}$, respectively. Next, construct the line $\overleftrightarrow{AI}$ and the perpendicular to $\overleftrightarrow{AI}$ through point B. What do you find unexpected in your construction?

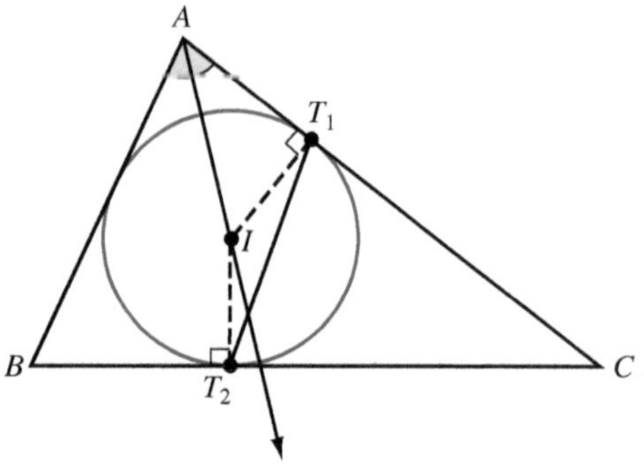

19. Justify the following construction of an equilateral triangle inscribed in a given circle:

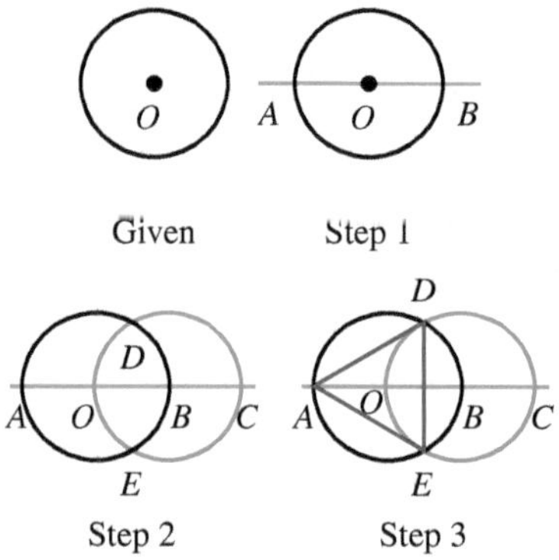

20. A **median** of a triangle is a line segment from a vertex to the midpoint of the opposite side. Using any drawing tools you wish, draw the three medians in each of several triangles. Give a statement that describes what you observe.

21. Reflecting point B to fall on the perpendicular bisector of $\overline{AB}$ shows how a Mira can be used to draw an equilateral triangle ABC on a given side $\overline{AB}$.

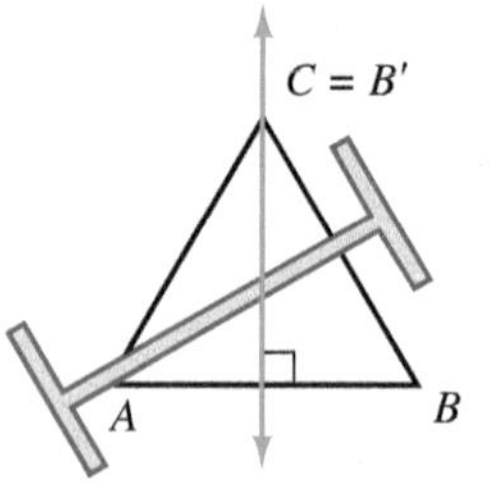

Give careful step-by-step instructions to construct these polygons on a given side with a Mira:

(a) A square

(b) A regular hexagon

22. Let the lower 8½″ edge of an ordinary sheet of paper be the segment $\overline{AB}$. Now describe a way to use paper folding to construct the third vertex C of the equilateral triangle ABC.

23. (a) Prove that the perpendicular bisector of any chord of a circle contains the center of the circle.

(b) Trace partway around a cup or saucer to draw a circular arc. Then explain how to construct the center of the arc. (*Hint:* Use part (a) twice!)

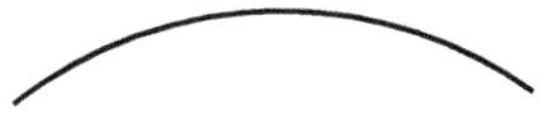

(c) The three congruent circles shown are centered at points A, B, and C on the large circular arc, and the circles at centers A and C contain point B. Where do the dashed lines intersect? Explain why.

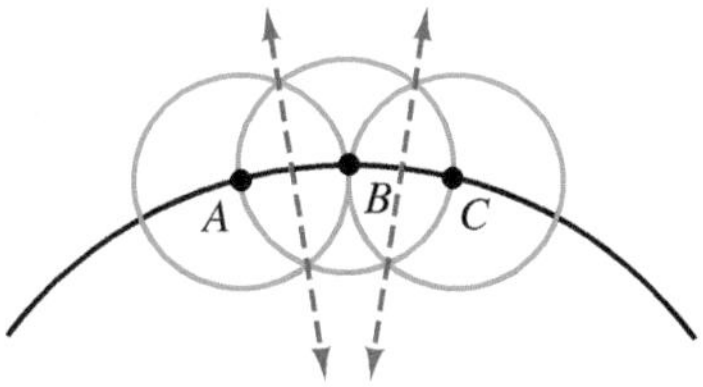

Into the Classroom

24. (**Writing**) Design an activity that investigates which properties of a quadrilateral are determined by how their diagonals intersect. Since students often learn best by doing a constructive activity, use a constructive approach (following an idea of John Van De Walle), where the diagonals are represented with punched strips of card stock (or manila folder) as shown. The holes are regularly spaced (say, 1″ apart), with an odd number of holes to make the middle hole obvious. Two strips are joined by a brass fastener through a hole that is the point of intersection of the two diagonals. For example, the red quadrilateral shown at the right is constructed with diagonals of different length meeting at their midpoints at a right angle.

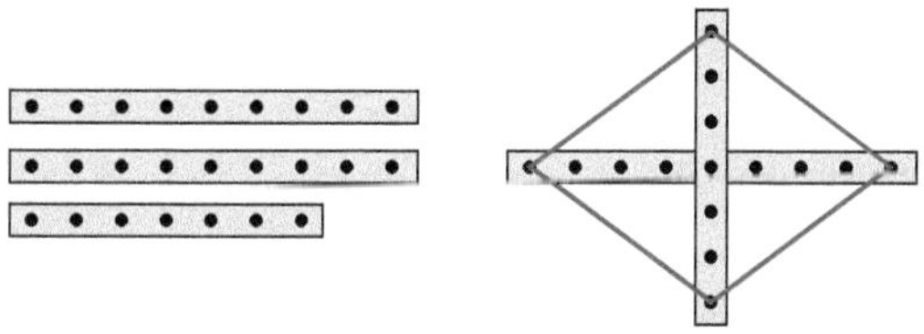

Your activity should investigate how to choose and align diagonals to construct a parallelogram, a rhombus, a square, a kite, and both types of trapezoids—isosceles and nonisosceles.

Responding to Students

25. Waun was asked to construct a regular pentagon and drew the following figure:

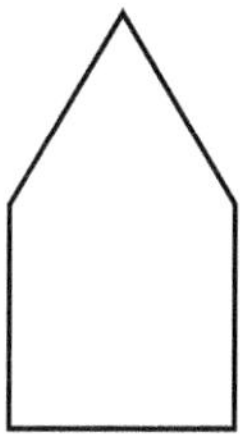

(a) What misunderstanding does Waun have of regular polygons?

(b) What advice can Waun be given about checking if a polygon is regular or not?

26. Kerry has created a new construction for the line through point P and perpendicular to line k, where P is not on k. As shown in the given figure, construct two circles through P centered at two distinct points A and B on k. If Q is the second point of intersection of the circles, then $\overleftrightarrow{PQ}$ is the required perpendicular line. Write a response to Kerry. In particular, if the construction is incorrect, explain why. If it is correct, guide Kerry through a proof that justifies the construction.

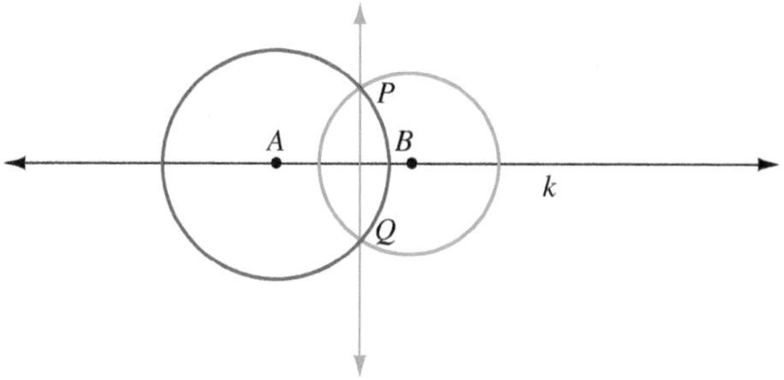

27. Someone claims that trisecting the chord $\overline{BC}$ of an arc centered at A gives a compass-and-straightedge trisection of $\angle A$. How would you respond to this assertion? How well does the method appear to work on an angle of measure 150°? Make a drawing and use a protractor to measure the angles.

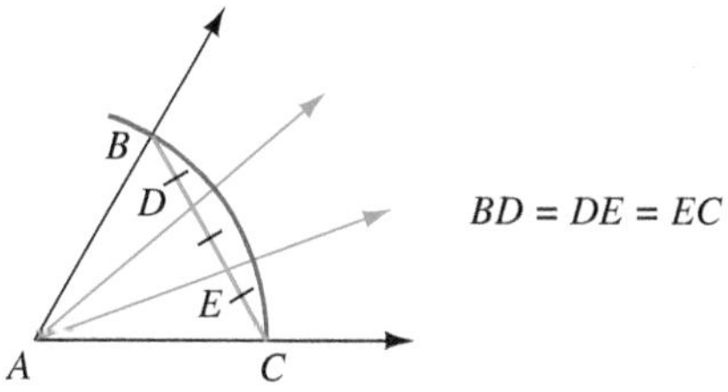

28. Leanne believes that she has constructed a triangle ABQ with two right angles. Her construction is shown here. Draw two circles that intersect at P and Q, and then draw their two diameters $\overline{QR}$ and $\overline{QS}$. Let $\overline{RS}$ intersect the circles at A and B. Then $\angle QBR$ and $\angle QAS$ are both right angles by Thales' theorem. Thus, $\triangle QAB$ has two right angles. How would you respond to Leanne's assertion?

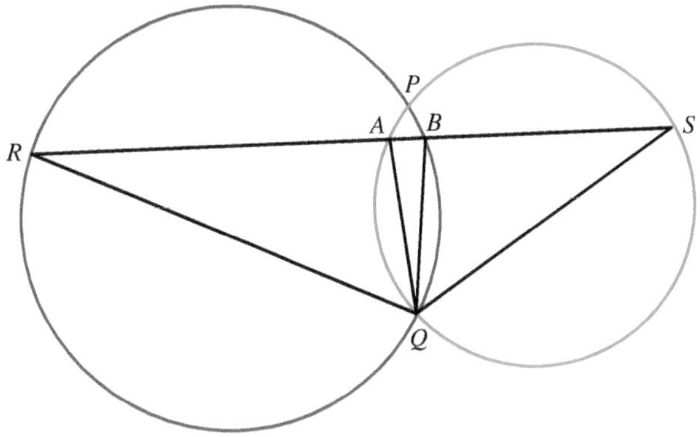

Thinking Critically

29. Trace the following segment $\overline{AB}$ and line l:

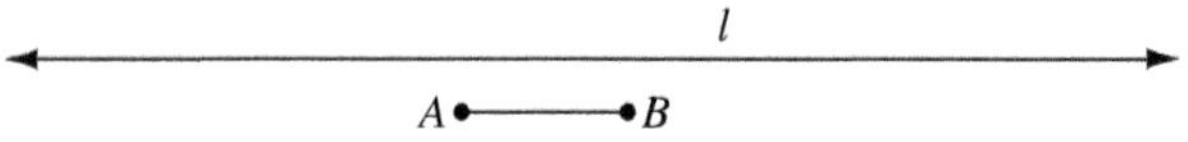

(a) Construct *all* points C on line l for which triangle ABC is isosceles.

(b) Construct *all* points D on line l for which triangle ABD is a right triangle.

30. Complete the proof of the equidistance property of the perpendicular bisector. Do so by showing that if P is equidistant from A and B, then the line $\overleftrightarrow{PM}$ containing P and the midpoint M of $\overline{AB}$ is the perpendicular bisector of $\overline{AB}$.

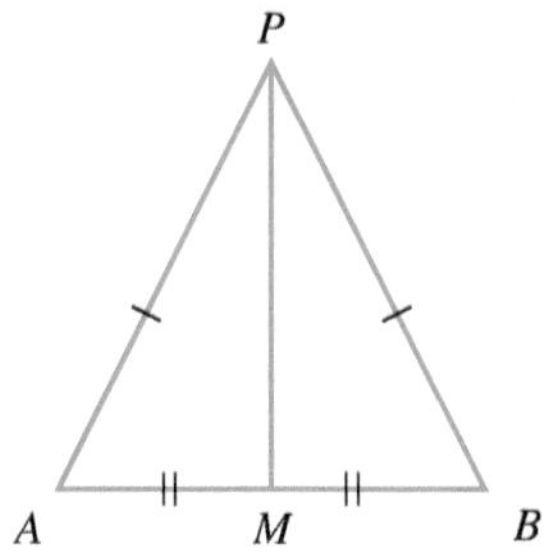

31. Prove that the angle bisectors of a triangle are concurrent, using the equidistance property of the angle bisector. The discussion in Example 12.8 can be used as a model for your proof.

32. An **altitude** of a triangle is a line through a vertex of the triangle that is perpendicular to the line containing the opposite side of the triangle. The altitude through vertex A has been constructed in this figure:

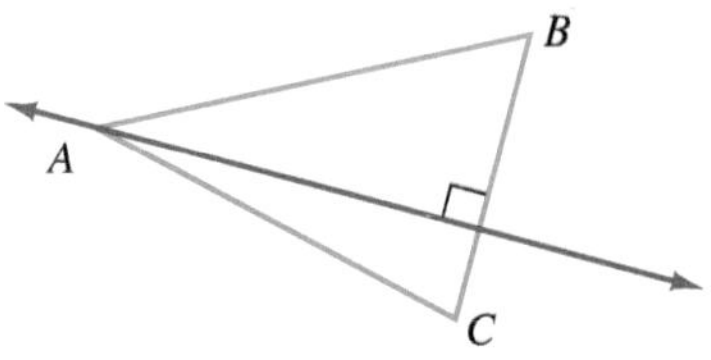

(a) Construct any triangle ABC and all three of its altitudes. What property do you discover about the altitudes of a triangle?

(b) Construct three congruent copies of $\triangle ABC$ on its sides to form $\triangle PQR$, as shown. It becomes apparent that the altitudes of $\triangle ABC$ are simultaneously the perpendicular bisectors of the sides of $\triangle PQR$. Use this fact to explain why the altitudes of any triangle are concurrent. The common point of intersection of altitudes is called the **orthocenter** of the triangle.

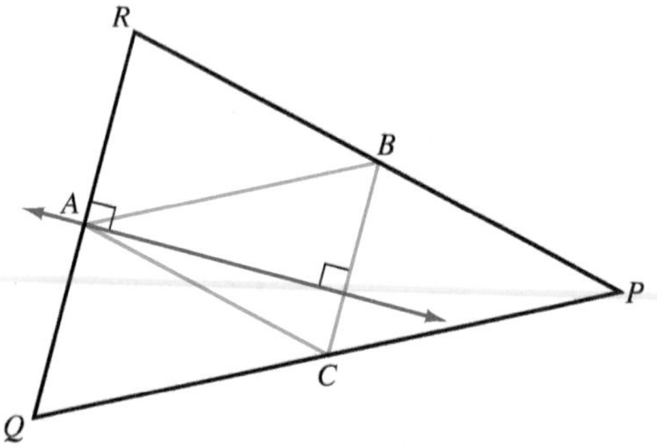

33. (a) Construct a regular pentagon inscribed in a circle by following the steps outlined next.

(b) Use a ruler and protractor to check that *PENTA* is regular.

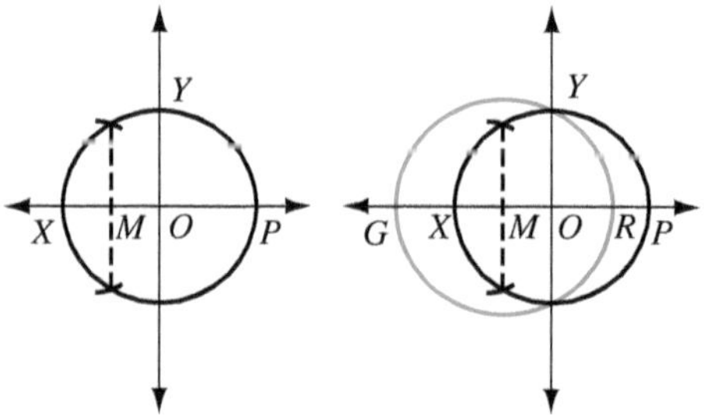

Construct perpendicular diameters to circle, and construct midpoint M of radius $\overline{OX}$.

Construct circle with center M and radius MY; this circle intersects $\overleftrightarrow{XP}$ at G and R.

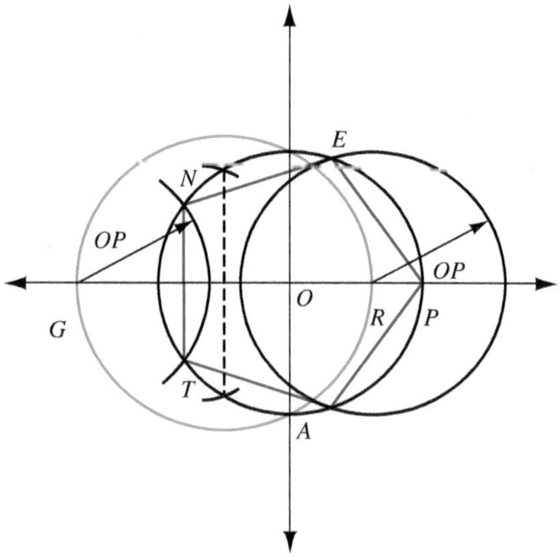

Construct circles at G and R of radius OP to locate the vertices of the regular pentagon *PENTA*.

34. A regular pentagon *PENTA* and an equilateral triangle *PQR* are both inscribed in the same circle centered at O, with P a common vertex. Calculate $m(\angle NOQ)$, and explain why laying segments off of length QN about the circle constructs a regular inscribed 15-gon.

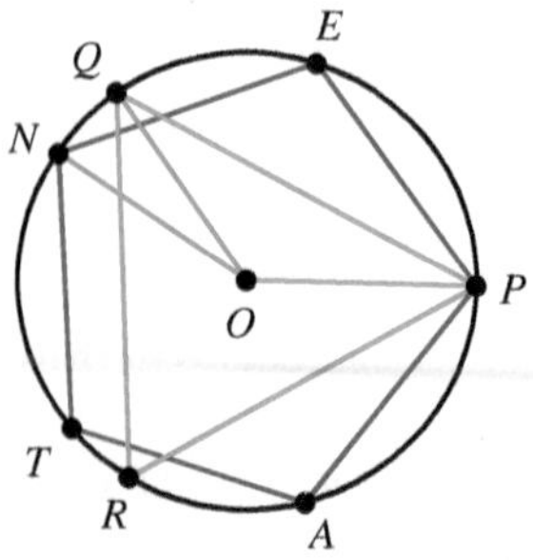

35. **(a)** Construct a heptagon inscribed in a given circle below following the steps outlined:

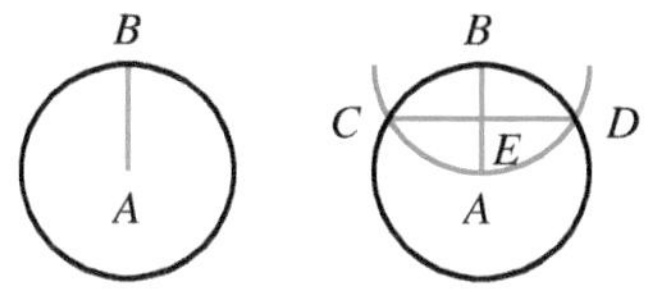

Construct a radius $\overline{AB}$.

Construct an arc at B through A and then draw chord $\overline{CD}$.

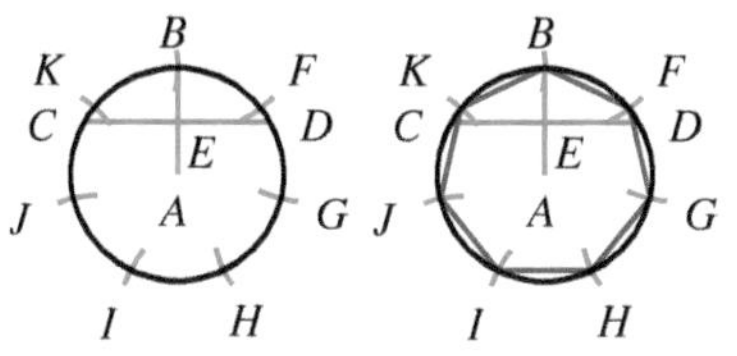

Lay off arcs of radius DE, starting at B.

Construct the heptagon $BFGHIJK$.

(b) Is it possible for $BFGHIJK$ to be a regular heptagon, or is it just a close approximation?

36. Trace around a circular object (such as a CD) and use a ruler to carefully draw the tangent lines at four points T, U, V, and W of the circle to create the circumscribing quadrilateral $ABCD$.

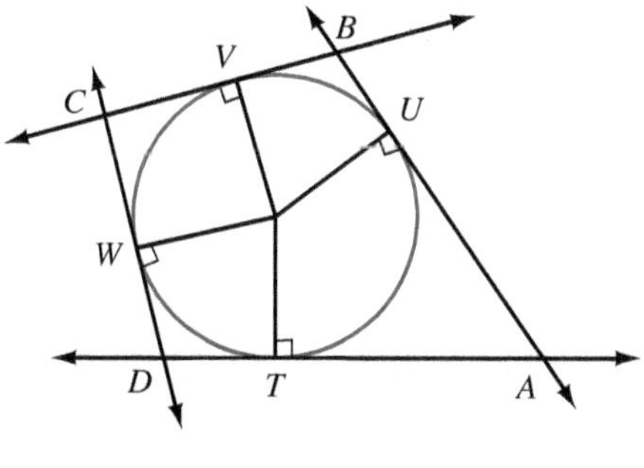

Measure the lengths of the sides AB, BC, CD, and DA to the nearest millimeter and calculate the sums of the lengths $AB + CD$ and $BC + DA$ of opposite sides.

(a) Make a conjecture about the sums of opposite sides of a circumscribing quadrilateral.

(b) Prove your conjecture.

37. List the constructible regular n-gons up to $n = 100$, using the Gauss–Wantzel theorem.

Making Connections

38. To construct a point equidistant from the edges of a board, a carpenter will lay a ruler diagonally across the board, turning the ruler until even numbers touch opposite edges of the board. The point along the ruler with the average value of the end values is marked. For example, with the 2″ and 8″ marks aligned at the edges of the board, the point at the 5″ mark of the ruler is equidistant from the edges of the board. Use congruent triangles to show why the marked point is on the midline of the board.

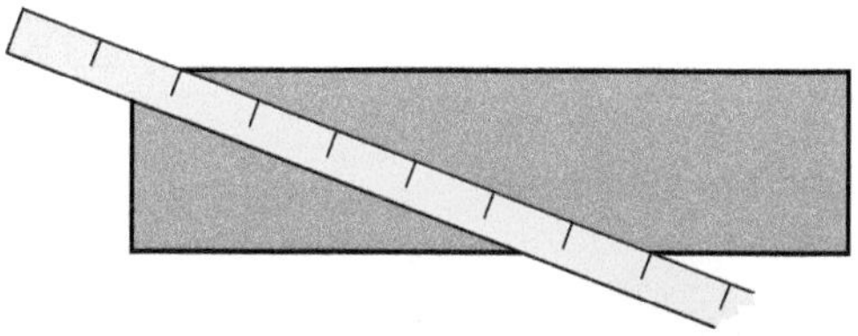

39. Popeye is somewhere off the coast, heading due north. At point A, the lighthouse at C is 35° off his starboard bow. After Popeye has traveled 4 miles further north, the lighthouse is 70° off the bow from point B. How far away is the lighthouse from point B? (*Hint:* What kind of a triangle ABC has been constructed by "doubling the angle," a common navigational technique?)

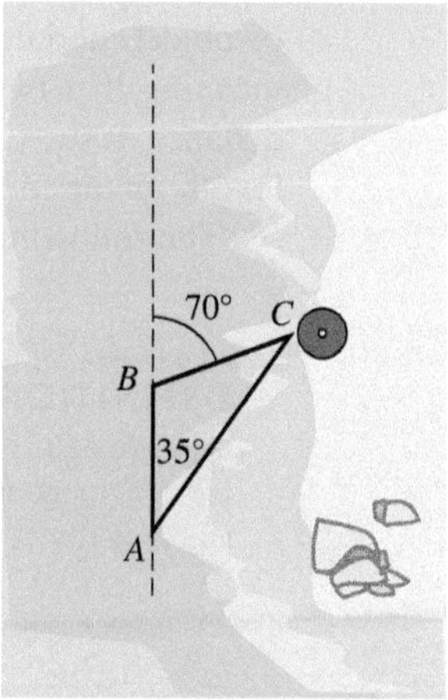

State Assessments

40. (Massachusetts, Grade 4)
I have four sides. Two of my sides are parallel. My other two sides are not parallel. Draw me.

41. (Grade 7)
Three circles are drawn with their centers A, B, and C on the same line, with point B the midpoint of the segment joining A to C. The circles intersect at the points D, E, F and G as shown in the diagram below.

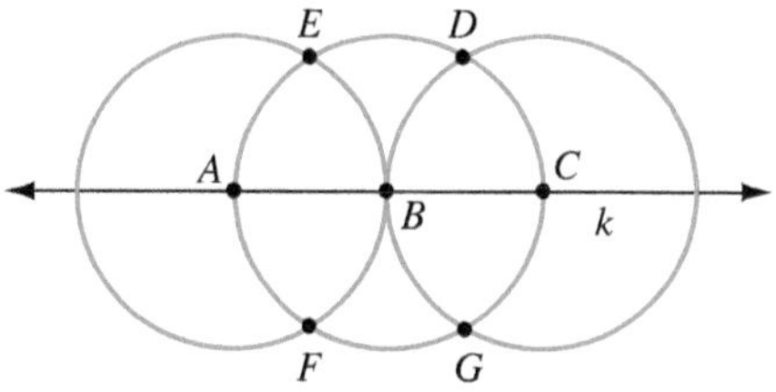

Which of the following polygons cannot be constructed using some of the points $A, B, \ldots, G$?

A. An equilateral triangle
B. A square
C. A trapezoid
D. A regular hexagon

42. (Grade 7)
Loren wishes to construct a square raised garden bed with four boards each 8 feet long. So far, he has joined the boards at their ends to make a quadrilateral. What additional information will ensure the boards form a square?
A. The diagonals are congruent.
B. The diagonals bisect one another
C. Opposite sides are parallel
D. Opposite angles are congruent

12.3 Similar Triangles

Two figures are **similar** if they have the same shape but not necessarily the same size. For example, an overhead projector forms an image on the screen that is similar to the figure in the transparency. The same is true of an enlarged or reduced photocopy. Figures in space can also be similar to one another. Design engineers often build small-scale models of buildings, airplanes, or ships and then make tests and measurements on the model to predict whether or not the design objectives will be met in the full-scale structure.

A map or model will indicate how its size compares with the actual size of the region mapped or object modeled by giving a **scale factor.** For example, a ship model may be scaled at 1:100, meaning that two points at a distance x on the model correspond to points on the real ship at a distance $100x$. Conversely, any length on the actual ship divided by 100 gives the corresponding length on the model.

The following definition gives an exact description of similarity for triangles:

DEFINITION Similar Triangles and the Scale Factor

Triangle ABC is **similar** to triangle DEF, written $\triangle ABC \sim \triangle DEF$, if and only if corresponding angles are congruent and the ratios of lengths of corresponding sides are all equal. That is, $\triangle ABC \sim \triangle DEF$ if and only if $\angle A \cong \angle D$, $\angle B \cong \angle E$, $\angle C \cong \angle F$, and

$$\frac{DE}{AB} = \frac{EF}{BC} = \frac{DF}{AC}.$$

The common ratio of lengths of corresponding sides is called the **scale factor** from $\triangle ABC$ to $\triangle DEF$.

This definition of similarity of triangles agrees with the definition of similarity of general figures given in Chapter 11, since, under a similarity transformation, corresponding side lengths have a common ratio and corresponding angle measures are preserved. However, the approach taken in the current section gives specific criteria for when two triangles are similar to one another. This point of view is often the more useful one for applications. For example, we will use triangle similarity to determine measurements indirectly and to derive relationships between parts of a given figure or pair of figures.

The scale factor from $\triangle DEF$ to $\triangle ABC$ is the reciprocal of the scale factor from $\triangle ABC$ to $\triangle DEF$. For example, if the sides of $\triangle DEF$ are three times the length of the sides of the similar triangle ABC, then the sides of $\triangle ABC$ are one-third the length of the sides of $\triangle DEF$.

Two examples of similar triangles and their scale factors are shown in Figure 12.16.

It is possible to conclude that two triangles are similar even when we have incomplete information about the sides and angles of the triangles. The most commonly used criteria for similarity are the angle–angle (AA), side–side–side (SSS), and side–angle–side (SAS) properties.*

*In some books, these properties are proved on the basis of other assumptions, making the properties theorems. In other texts, the properties are adopted as postulates. In this text, in keeping with an informal treatment of Euclidean geometry, we will refer to the AA, SSS, and SAS criteria for similarity as *properties*.

Figure 12.16
Two pairs of similar triangles

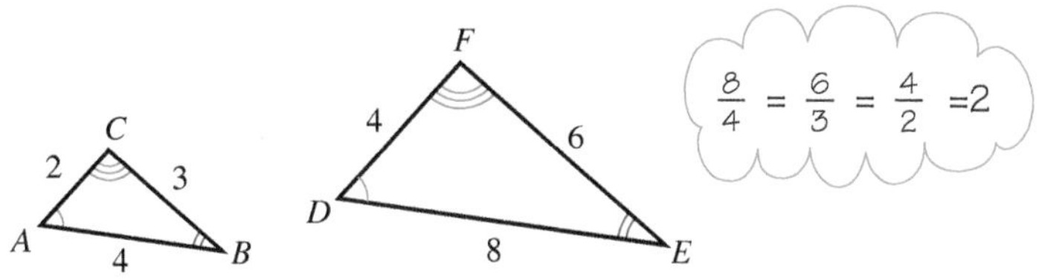

$\triangle ABC \sim \triangle DEF$
Scale factor from $\triangle ABC$ to $\triangle DEF = 2$

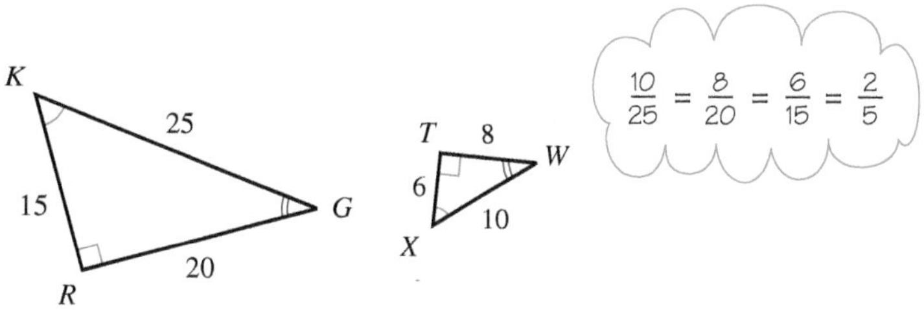

$\triangle KRG \sim \triangle XTW$
Scale factor from $\triangle KRG$ to $\triangle XTW = 2/5$

The Angle–Angle–Angle (AAA) and Angle–Angle (AA) Similarity Properties

In Figure 12.17, $\triangle ABC$ and $\triangle DEF$ have corresponding angles measuring 110°, 40°, and 30°. We see that $\triangle ABC \sim \triangle DEF$. The scale factor can be determined by measuring the lengths of two corresponding sides and forming their ratio. For example, the scale factor is DE/AB.

Figure 12.17
Two triangles with three congruent angles are similar by the AAA similarity property

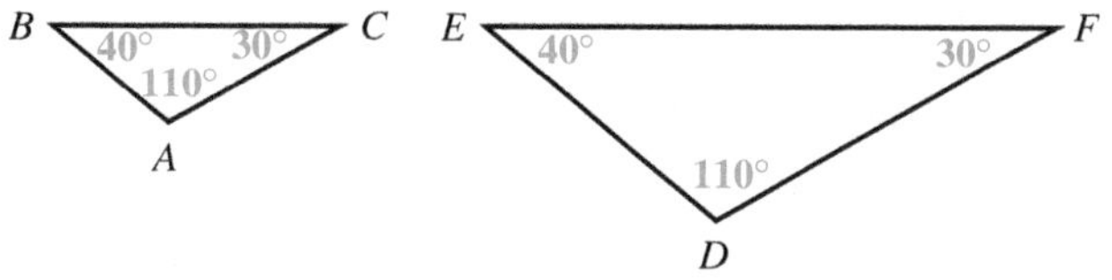

In general, two triangles with three pairs of congruent angles are similar, an observation known as the **angle–angle–angle (AAA) property** of similar triangles. However, all three angles of a triangle are determined once we know two of its angles, since the three measures of the angles add up to 180°. Therefore, the more general property is called the **angle–angle (AA) property of similarity.**

PROPERTY The AA Similarity Property

If two angles of one triangle are respectively congruent to two angles of a second triangle, then the triangles are similar.

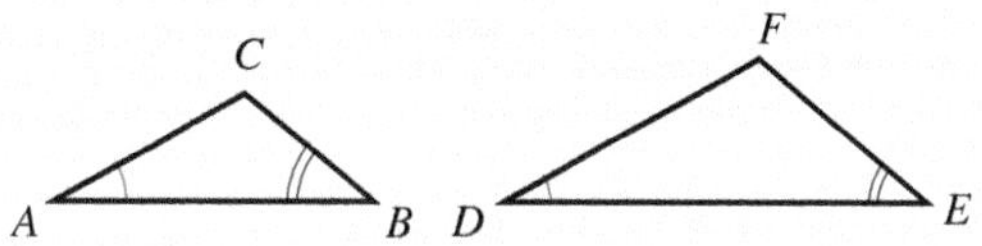

Similar triangles are an important tool for making indirect measurements. The general idea is described by considering two similar triangles with a scale factor that can be determined. If the length of a side of one triangle is easy to measure, this indirectly determines the length of the corresponding side in the second triangle by multiplication by the scale factor. This is an important example of using a **mathematical model**. In the fourth Standard of Mathematical Practice (SMP 4) of the Common Core, we read that

SMP 4 Model with mathematics.

"Mathematically proficient students can apply the mathematics they know to solve problems arising in everyday life, society, and the workplace. . . . They are able to identify important quantities in a practical situation and map their relationships using such tools as diagrams . . . . They can analyze those relationships mathematically to draw conclusions."

EXAMPLE 12.9

Making an Indirect Measurement with Similarity

A tree at point T is in line with a stake at point L when viewed across the river from point N. Use the information in the diagram to measure the width x of the river.

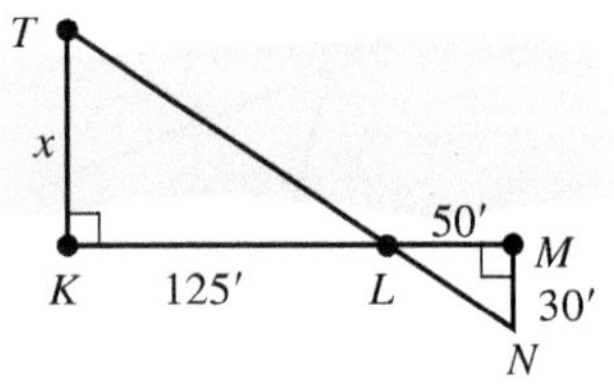

Solution

By the vertical-angles theorem, $\angle TLK \cong \angle NLM$. Also, $\angle K \cong \angle M$, since both are right angles. By the AA similarity property, $\triangle TLK \sim \triangle NLM$. Thus, $\frac{TK}{NM} = \frac{KL}{ML}$, since the ratios of the lengths of corresponding sides are equal. Now, $TK = x$, $NM = 30'$, $KL = 125'$, and $ML = 50'$, so we have the proportion $\frac{x}{30'} = \frac{125'}{50'}$. Therefore, $x = \frac{30' \cdot 125'}{50'} = 75'$. We have thus found the width x of the river by an indirect measurement using similar triangles.

The Side–Side–Side (SSS) Similarity Property

PROPERTY The SSS Similarity Property

If the three sides of one triangle are proportional to the three sides of a second triangle, then the triangles are similar. That is, if $\frac{DE}{AB} = \frac{EF}{BC} = \frac{DF}{AC}$, then $\triangle ABC \sim \triangle DEF$.

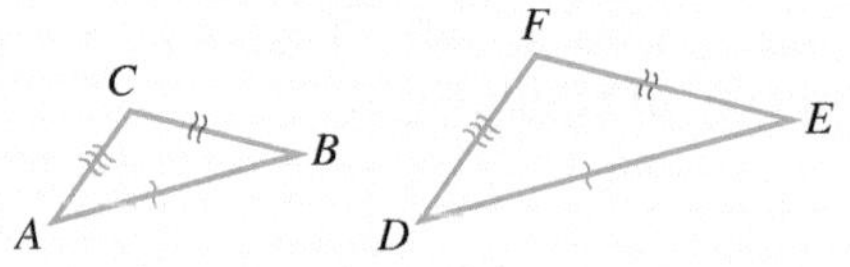

Wavy tick marks are helpful in identifying corresponding proportional sides of similar figures, as shown in the preceding diagram. It is always helpful to convey information in a drawing that clearly reminds you of the properties that are given or have been derived.

EXAMPLE 12.10

Applying the SSS Similarity Property

A contractor wishes to build an L-shaped concrete footing for a brick wall, with a 12-foot leg of the wall meeting a 10-foot leg of the wall at a right angle. The contractor knows that a 3- by 4- by 5-foot triangle has a right angle opposite the 5-foot side. How can the contractor place stakes at points X, Y, and Z to form a right angle at point Y? The wall will be built along string lines stretched from X to Y and from Y to Z.

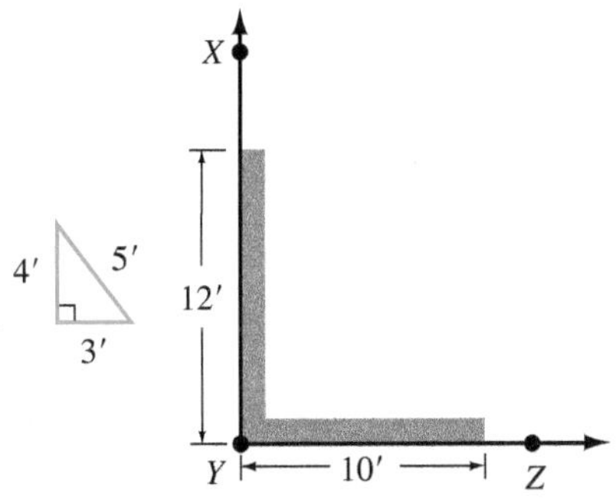

Solution

By the SSS similarity property, the 3–4–5-foot right triangle can be magnified by a convenient scale factor to give an accurate right triangle. A scale factor of 4 gives a 12–16–20-foot right triangle. The contractor can place a stake at *X* that is 16 feet from the corner point *Y*. By using two measuring tapes, a stake is placed at the point *Z* where the 20-foot mark on the tape from *X* crosses the 12-foot mark on the tape from *Y*.

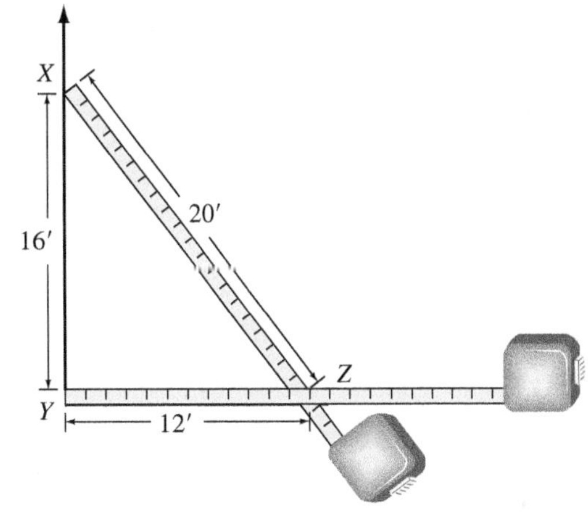

The Side–Angle–Side (SAS) Similarity Property

PROPERTY The SAS Similarity Property

If, in two triangles, the ratios of any two pairs of corresponding sides are equal and the included angles are congruent, then the two triangles are similar. That is, if $\frac{DE}{AB} = \frac{DF}{AC}$ and $\angle A \cong \angle D$, then $\triangle ABC \sim \triangle DEF$.

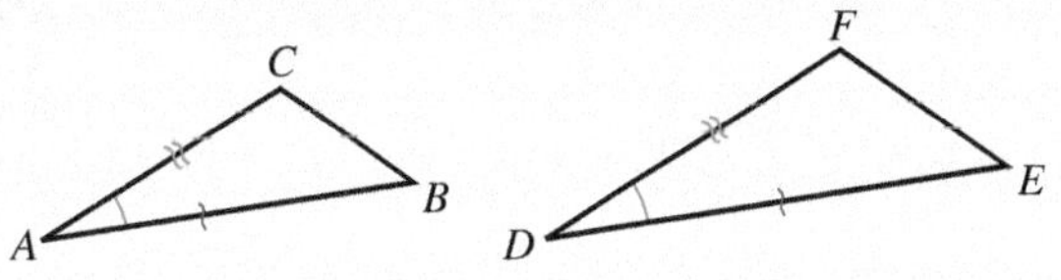

EXAMPLE 12.11

Applying the SAS Similarity Property

The sun is about 93 million miles from Earth, and the distance from Earth to the moon is about 240,000 miles. If the diameter of the moon is 2200 miles, what is the approximate diameter of the sun? (*Hint:* From Earth, the sun and moon appear to have the same diameter.)

Solution

Because the moon (*M*) appears to have the same diameter as the sun (*S*) during an eclipse, they form nearly congruent angles when viewed from Earth (*E*). This illustration is far from a true scale drawing, but it does show that $\triangle EM_1M_2 \sim \triangle ES_1S_2$ by the SAS similarity property. Thus, $\frac{S_1S_2}{M_1M_2} = \frac{ES_1}{EM_1}$, so $S_1S_2 = \frac{ES_1}{EM_1} \cdot M_1M_2 = \frac{93{,}000{,}000}{240{,}000} \cdot 2200 \text{ miles} = 852{,}500 \text{ miles}$. This estimate compares well with the sun's actual diameter of 864,000 miles, given by more accurate methods.

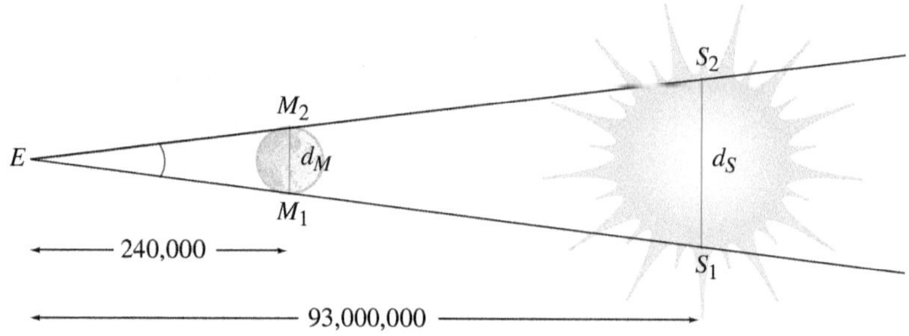

Geometric Problem Solving with Similar Triangles

The examples that follow illustrate how similar triangles can be used to explore properties of geometric figures. These examples all explore figures constructed by joining the midpoints of sides of triangles or quadrilaterals, known as **midpoint figures**.

EXAMPLE 12.12

Exploring the Medial Triangle

If X, Y, and Z are the midpoints of the sides of $\triangle ABC$, then $\triangle XYZ$ is called the **medial triangle** of $\triangle ABC$.

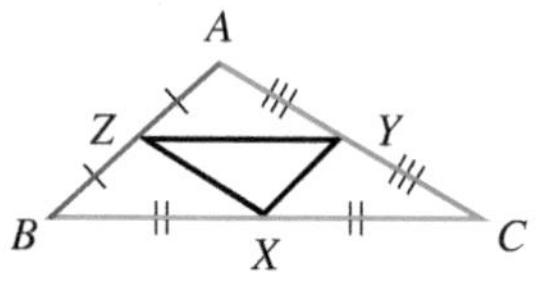

Show that $\triangle ABC \sim \triangle XYZ$, with a scale factor of $\frac{1}{2}$, and that each side of the medial triangle is parallel to the corresponding side of $\triangle ABC$. For example, $\overline{XY} \parallel \overline{AB}$ and $\frac{XY}{AB} = \frac{1}{2}$.

Solution

From the figure shown, it is seen that $\frac{CY}{CA} = \frac{CX}{CB} = \frac{1}{2}$. Therefore, by the SAS similarity property, $\triangle ACB \sim \triangle YCX$, with a scale factor of $\frac{1}{2}$. In particular, $\frac{XY}{BA} = \frac{1}{2}$. Moreover, $\angle CAB \cong \angle CYX$, so $\overline{XY} \parallel \overline{AB}$ by the corresponding-angles property. By the same reasoning, the remaining two sides of $\triangle XYZ$ are also parallel and half of the length of the corresponding sides of $\triangle ABC$. Since the sides of $\triangle XYZ$ are half the length of the corresponding sides of $\triangle ABC$, the two triangles are similar by the SSS similarity property.

EXAMPLE 12.13

Classifying the Midpoint Figure of a Quadrilateral

Two quadrilaterals are shown. It appears that joining successive midpoints of the sides of these quadrilaterals forms a parallelogram. Prove that this is indeed the case.

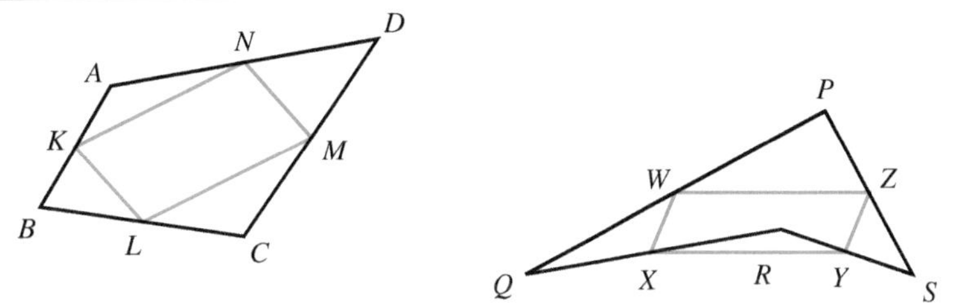

Solution

Let $KLMN$ be the midpoint figure of the quadrilateral $ABCD$. Construct the diagonal $\overline{BD}$ and consider $\triangle ABD$ and $\triangle CBD$. By Example 12.12, $\overline{KN}$ and $\overline{LM}$ are both parallel to $\overline{BD}$ and have length $\frac{BD}{2}$. But then $\overline{KN}$ and $\overline{LM}$ are congruent and parallel segments. The same reasoning shows that $\overline{KL}$ and $\overline{NM}$ are congruent and parallel, since both segments are parallel to $\overline{AC}$ and have length $\frac{AC}{2}$. By definition, $KLMN$ is a parallelogram. The same argument can be adapted to the nonconvex quadrilateral $PQRS$.

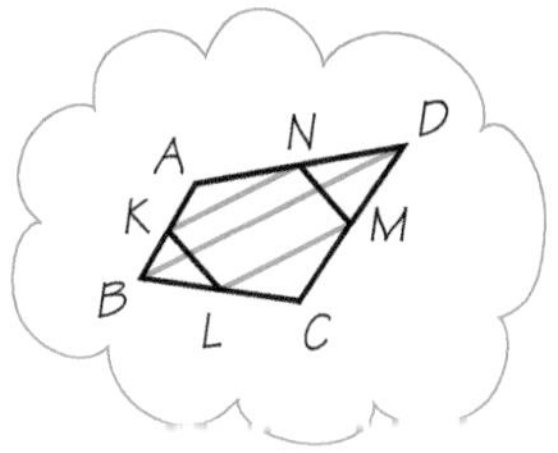

The proof just given is also valid for *space* quadrilaterals, for which the four vertices are not necessarily in the same plane. For example, the quadrilateral $PQRS$ shown in Example 12.13 may be easily visualized as a nonplanar quadrilateral. However, the midpoint quadrilateral $WXYZ$ is a parallelogram, so it lies in a plane. It's interesting to confirm this result with a quadrilateral made of sticks whose midpoints are joined by elastic bands to form the midpoint parallelogram.

EXAMPLE 12.14

Discovering the Centroid of a Triangle

A **median** of a triangle is a line segment joining a vertex to the midpoint of the opposite side. Prove that the three medians of a triangle are concurrent at a point G that is $\frac{2}{3}$ of the distance along each median from the vertex toward the midpoint. The point of concurrence G is the **centroid**, or **center of gravity**, of the triangle.

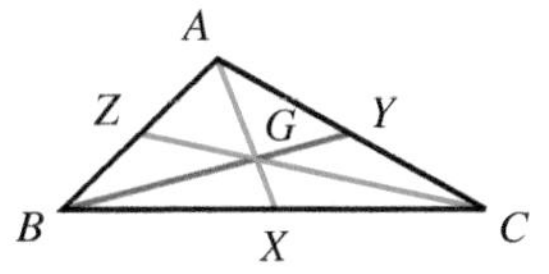

$AG = \frac{2}{3}AX$

$BG = \frac{2}{3}BY$

$CG = \frac{2}{3}CZ$

Solution

Understand the Problem

There are really *two* problems to be solved. First, there is a *distance* problem: We must show that if G is the point where the two medians $\overline{BY}$ and $\overline{CZ}$ intersect, then $BG = \frac{2}{3}BY$, or equivalently, $BG = 2GY$. Second, there is a *concurrence* problem: We must show that if G is the point of intersection of $\overline{BY}$ and $\overline{CZ}$, then the third median, $\overline{AX}$, also passes through G.

Devise a Plan

Since we hope to show that $BG = 2GY$ and $CG = 2GZ$, it may be useful to consider the midpoints M of $\overline{BG}$ and N of $\overline{CG}$. The distance problem for medians $\overline{BY}$ and $\overline{CZ}$ will be solved if it can be shown that M and G trisect $\overline{BY}$ and N and G trisect $\overline{CZ}$. Since Z, M, N, and Y are the successive midpoints of the quadrilateral $ABGC$, we should gain important information by constructing the midpoint figure $ZMNY$, which we know is a parallelogram by Example 12.13:

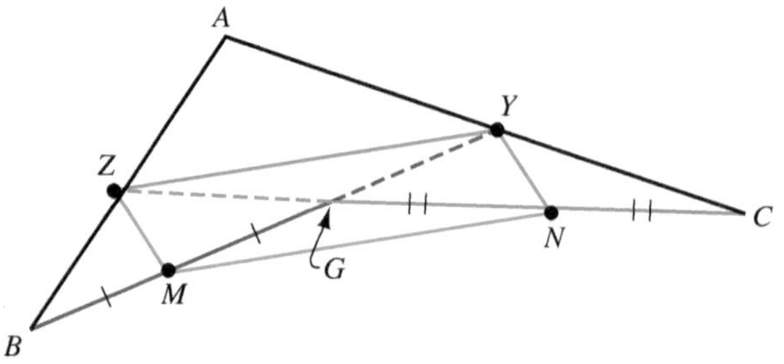

Carry Out the Plan

Because $ZMNY$ is a parallelogram, the point G at which the diagonals intersect is the midpoint of the diagonal $\overline{MY}$ of the parallelogram. Thus, $MG = GY$. But M is the midpoint of $\overline{BG}$, so $BM = MG$. This shows that G is $\frac{2}{3}$ of the distance from B to Y along the median $\overline{BY}$. The same reasoning shows that G and N trisect the median $\overline{CZ}$. The concurrence problem is now solved by symmetry: If G' is the point of intersection of the medians $\overline{BY}$ and $\overline{AX}$, then G' is $\frac{2}{3}$ of the distance from a vertex along either median; therefore, $G = G'$.

Look Back

It is often helpful to review the problem-solving strategies that have been successful. Several strategies used in this example are likely to be helpful with other problems:

- **Divide the problem into simpler parts:** We solved a distance problem and a concurrence problem.
- **Consider a simpler problem first:** G was defined as the intersection of *two* medians, and it was to be shown that G was $\frac{2}{3}$ of the distances along the two medians from the vertices.
- **Use a related result:** The previous example, showing that the midpoints of the sides of any quadrilateral form a parallelogram, was a key idea in the solution.

12.3 Problem Set

Understanding Concepts

1. Which of the given pairs of triangles are similar? If they are similar, explain why, express the similarity with the ~ notation, and give the scale factor.

(a)

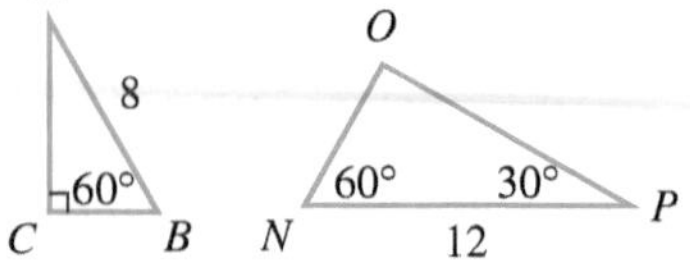

(b)

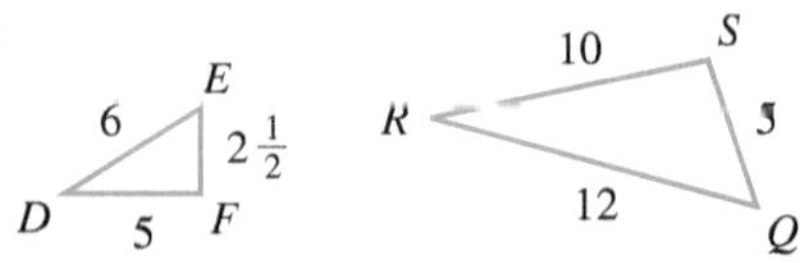

2. Answer the previous problem for these pairs of triangles.

(a)

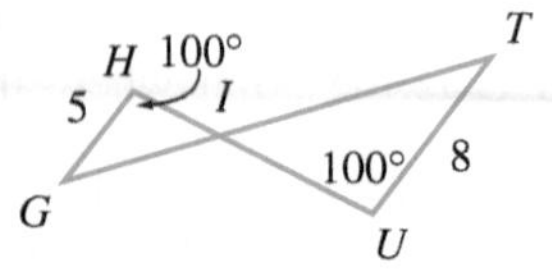

(b)

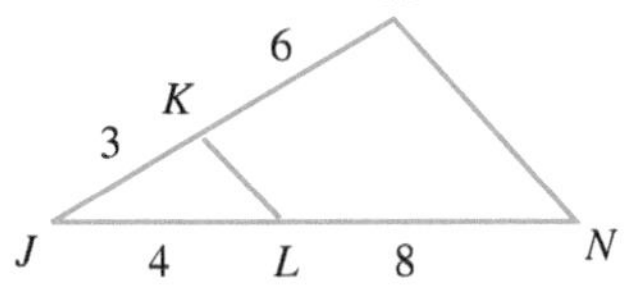

(c)

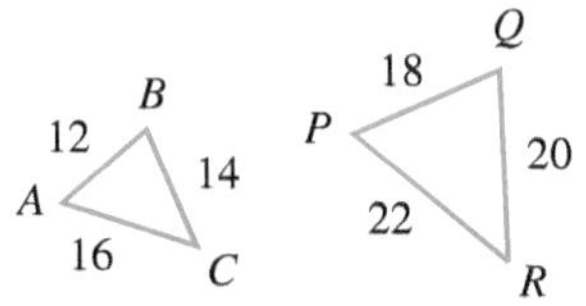

3. Are the figures described next necessarily similar? If so, explain why. If not, draw an example to show why not.
   (a) Any two equilateral triangles
   (b) Any two isosceles triangles
   (c) Any two right triangles having an acute angle of measure 36°

4. Answer problem 3 for these pairs of triangles.
   (a) Any two isosceles right triangles
   (b) Any two congruent triangles
   (c) A triangle with sides of lengths 3 and 4 and an angle of 30° and a triangle with sides of lengths 6 and 8 and an angle of 30°

5. A pair of similar triangles is shown in each part. Find the measures of the segments marked with a letter a, b, c, or d.

   (a)

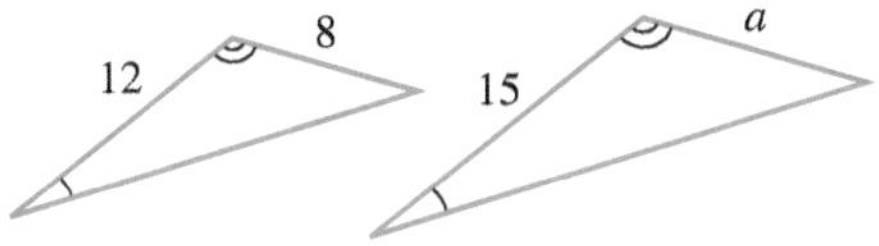

   (b)

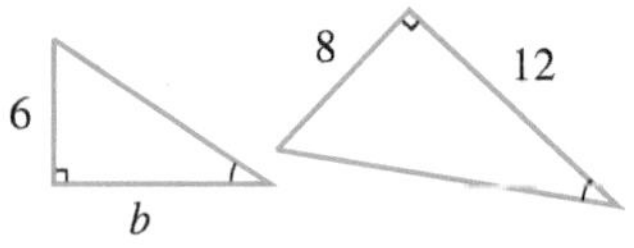

6. Answer problem 5 for these pairs of similar triangles.

   (a)

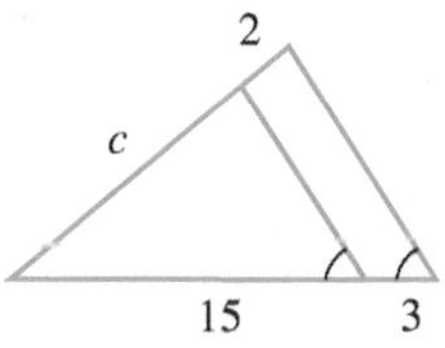

   (b)

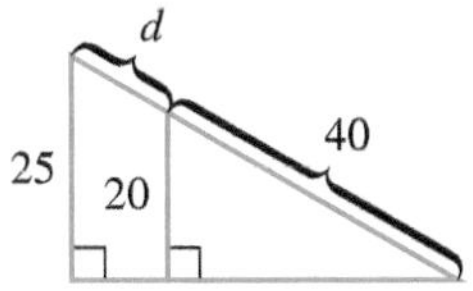

7. (a) A right triangle XYZ with legs of length 5 feet and 12 feet is similar to a triangle ABC of perimeter 3000 feet. Describe triangle ABC.
   (b) Triangle XYZ is also similar to a triangle DEF of area 3000 square feet. Describe triangle DEF.

8. Triangle XYZ is a right triangle with legs of length 3 and 4. Describe *all* triangles that have one side of length 60 and are similar to triangle XYZ.

9. Two convex quadrilaterals $ABCD$ and $EFGH$ have congruent angles at their corresponding vertices: $\angle A \cong \angle E$, $\angle B \cong \angle F$, $\angle C \cong \angle G$, and $\angle D \cong \angle H$. Can you conclude that the two quadrilaterals are similar? Explain.

10. Two convex quadrilaterals have corresponding sides in the same ratio. Are the quadrilaterals necessarily similar? Explain.

11. Suppose $\triangle ABC \sim \triangle DEF$, where only points D and E are shown. Find all possible locations for F and draw the corresponding triangles DEF.

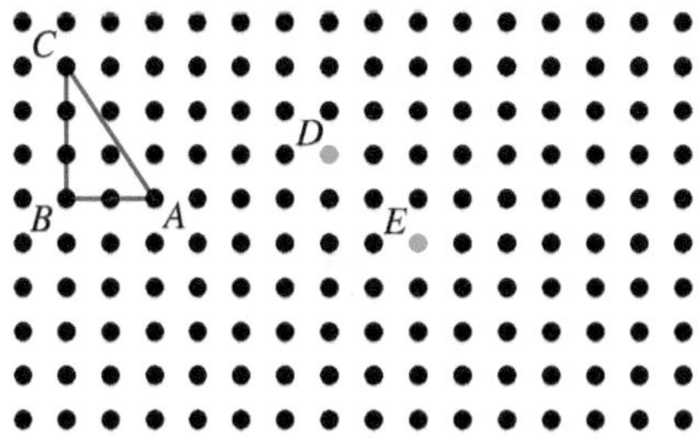

12. The diagonals of the given trapezoid $ABCD$ intersect at E, where $\overline{AB} \parallel \overline{CD}$. Let $x = BE$ and $y = DC$.
   (a) Explain why $\triangle ABE \sim \triangle CDE$.
   (b) Determine x and y.

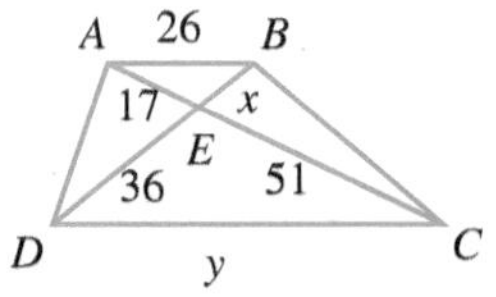

13. Let $\overline{AB}$ and $\overline{CD}$ be parallel, and let $\overline{AD}$ and $\overline{BC}$ intersect at E. Prove that $a \cdot y = x \cdot b$.

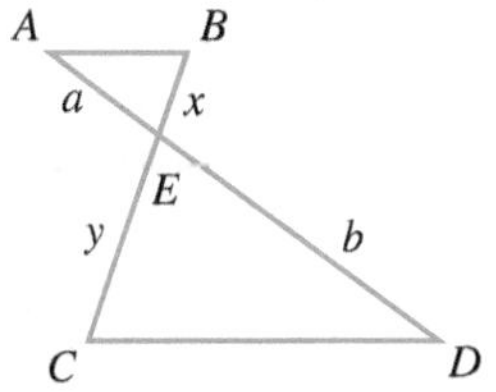

14. $\triangle ABC$ is a right triangle with altitude $\overline{CD}$, as shown.
   (a) Explain why $\triangle ADC \sim \triangle CDB$.
   (b) Find an equation for h, and solve it to show that $h = \sqrt{9 \cdot 25} = 3 \cdot 5 = 15$.

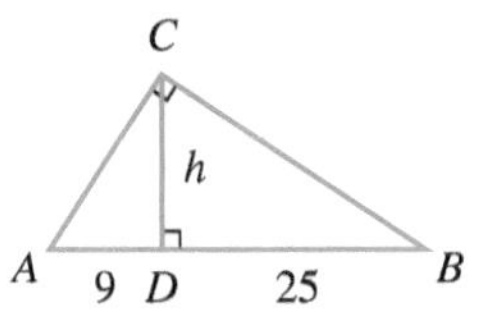

15. An isosceles triangle whose apex angle measures 36°, shown at the left, is sometimes called a **golden triangle:**

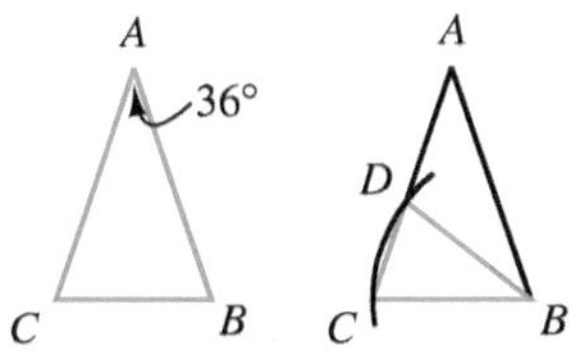

(a) Draw a circular arc centered at B and passing through C. Prove that if D is the point at which the arc intersects $\overline{AC}$, then $\triangle BCD$ is also a golden triangle. (*Hint:* What is $m(\angle C)$?)

(b) Draw an enlarged copy of $\triangle ABC$, with AB at least 3″ long. Construct three more golden triangles CDE, DEF, and EFG, where each contains the next.

16. $\triangle ABC$ is an isosceles triangle with apex C. It has the unusual property that the circular arc centered at A intersects the opposite side at a point D for which $\triangle ABC \sim \triangle BCD$.

(a) Use the preceding property to find the measures of the base and apex angles of $\triangle ABC$.

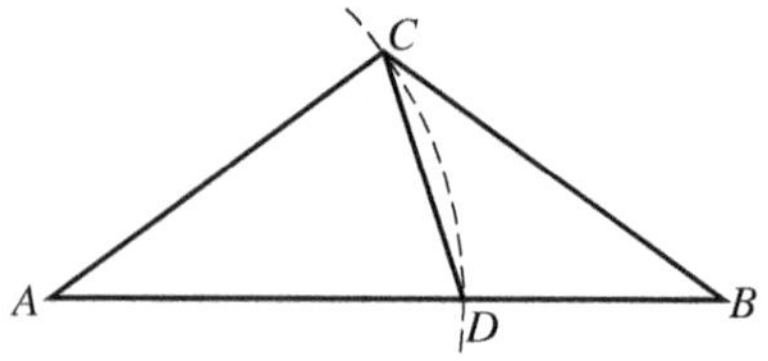

(b) What kind of triangle is $\triangle ACD$? (*Hint:* See problem 15.)

17. Suppose $\triangle ABC \sim \triangle BCA$. What more can you say about $\triangle ABC$?

18. Lined notebook paper provides a convenient way to subdivide a line segment into a specified number of congruent subsegments. The diagram shows how swinging an arc of radius AB subdivides the segment into five congruent parts: $\overline{AX} \cong \overline{XY} \cong \overline{YZ} \cong \overline{ZW} \cong \overline{WB}$.

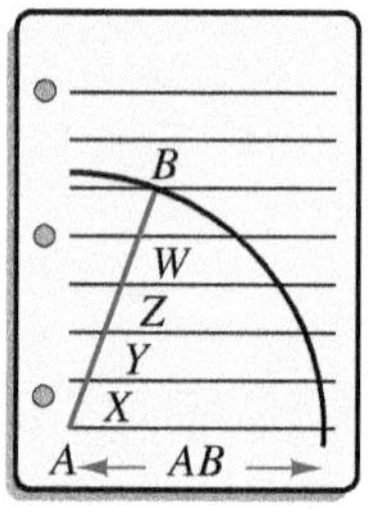

(a) Show how to subdivide a segment $\overline{AB}$ into seven congruent segments.

(b) Carefully explain why and when the procedure you used in (a) is valid.

Into the Classroom

19. The Similarity Matching Game. Staple manila file folders together so that a dozen or so congruent pairs of triangles of various shapes and sizes can be cut out easily. Next, choose one of each congruent pair and cut off a trapezoid from one of its sides, using the side as the longer base of the trapezoid. This converts each congruent pair into a similar pair of triangles. Next, spread out all of the triangles on the classroom floor or perhaps on an overhead, first turning some of the triangles upside down. Now challenge your students to identify the pairs of similar triangles. Have them check their guesses by aligning their chosen pairs of triangles at a vertex to see that the uncovered part of the larger triangle is a trapezoid. Also, have them use rulers to determine the approximate scale factor of each congruent pair of triangles.

20. (**Writing**) Identify a tree, flagpole, clock tower, or other similar prominent object on campus. Students are divided in three- to four-person teams and challenged to measure indirectly the height of the object. Each team submits a report of their method of measurement. Have the teams compare their methods and measurements and decide which team has the most accurate result.

Responding to Students

21. Jackson is having trouble remembering the difference between similar and congruent figures. He asks, "Are all congruent triangles similar?" You think that is a wonderful question with which to begin reflecting on how congruent and similar figures are alike and different. What could you do to help Jackson develop this conceptual understanding?

22. Gloria is recording her findings about the similar triangles shown, and she writes "$\triangle WVU \sim \triangle YXZ$" in her math journal. She then begins to look at corresponding sides and angles. How would you respond to Gloria?

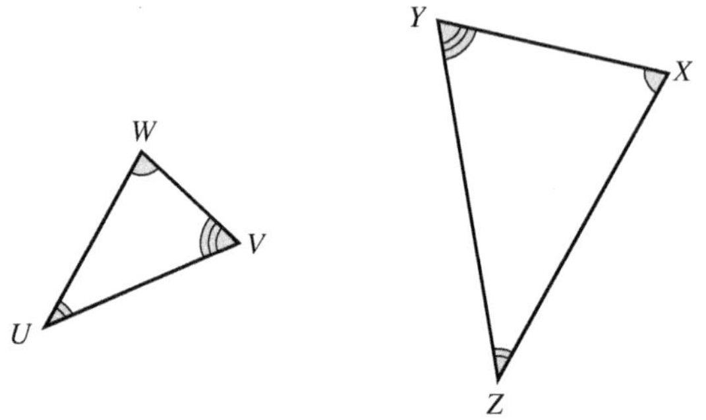

23. Miley thinks that a triangle with sides 5, 9, and 13 is similar to a triangle with sides 10, 14, and 18, since each side is increased by 5 from the one triangle to the other. Reef thinks Miley can't expect to get a similar triangle in this way. After all, says Reef, "If 100 is added to each side, the new triangle would have sides of length 105, 109, and 113 and it would be nearly equilateral, quite unlike the shape of the obtuse 5–9–13 triangle. Going in the other direction, 1 cannot be subtracted from each side, since, by the triangle inequality, no triangle can have sides of length 4, 8, and 12." Respond to Miley and Reef.

24. (**Writing**) After studying similar triangles, your class wonders whether there are analogous similarity properties for quadrilaterals. Respond to the ideas of these students by writing a paragraph to each student.

(a) Megan claims that all rectangles are similar, since every angle in a rectangle is a right angle.

(b) Patrick thinks that there is an SASAS quadrilateral similarity property. That is, if two quadrilaterals $PQRS$ and $WXYZ$ have the relationships

$$\frac{WX}{PQ} = \frac{XY}{QR} = \frac{YZ}{RS}, \angle Q \cong \angle X, \text{ and } \angle R \cong \angle Y,$$

then $PQRS$ is similar to $WXYZ$.

(c) Quincy thinks that there is an SSSS quadrilateral similarity property. That is, if the ratios of the lengths of corresponding sides are all equal, then the two quadrilaterals have the same shape, though not necessarily the same size.

Thinking Critically

25. Let $\overline{CD}$ be the altitude drawn to the hypotenuse of the right triangle ABC:

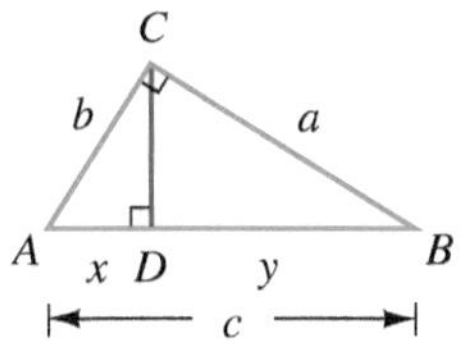

$a = BC$, $b = AC$, $c = AB$
$x = AD$, $y = DB$

(a) Explain why $\triangle ABC \sim \triangle ACD$ and $\triangle ABC \sim \triangle CBD$.

(b) Explain why $\frac{x}{b} = \frac{b}{c}$ and $\frac{y}{a} = \frac{a}{c}$.

(c) Use part (b) to show that $c^2 = a^2 + b^2$. (This gives a proof of the Pythagorean theorem.)

26. Prove that the following square inscribed in a right triangle has sides of length $x = \frac{ab}{a+b}$, where a and b are the lengths of the legs of the triangle:

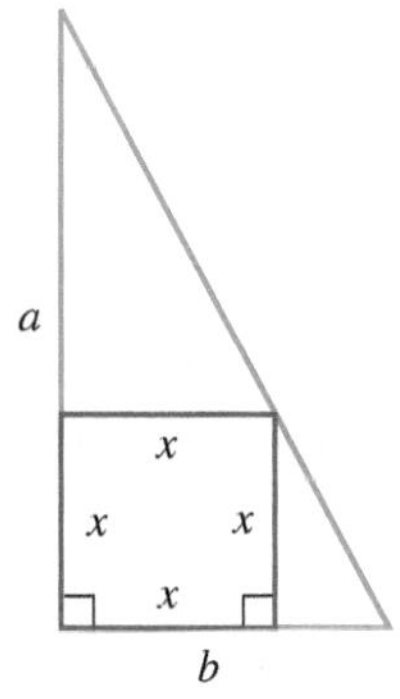

27. The square $ABCD$ has sides of unit length and midpoints at J, K, L, and M. The four segments that join vertices of the square to the midpoints of opposite sides form a smaller, inner square $PQRS$.

(a) Use the Pythagorean theorem to show that

$$DJ = \frac{1}{2}\sqrt{5}.$$

(b) Construct segment $\overline{ST}$ parallel to $\overline{AM}$, where T is on segment $\overline{AP}$. Observe that the segment $\overline{ST}$ has length $\frac{1}{2}$ and creates $\triangle STP$, which is similar to $\triangle DJA$. Use this fact to compute the length PS.

(c) What is the area of the inner square?

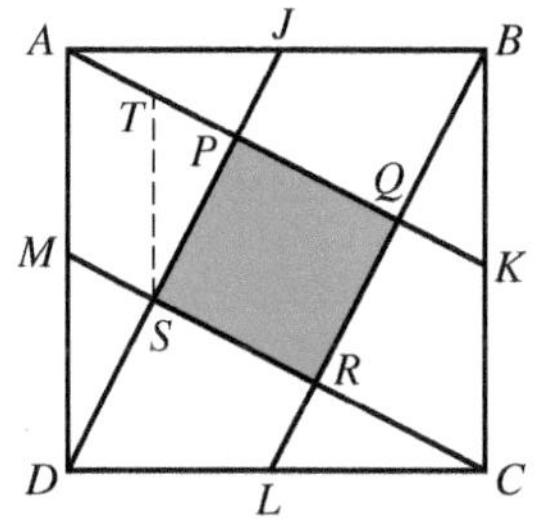

28. The square $ABCD$ has sides of unit length and "one-third" points at J, K, L, and M. Join the vertices of the square to successive "one-third" points to form the smaller, inner square $PQRS$. Follow the steps in problem 27 to show that the inner square formed with "one-third" points has 40% of the area of the larger square $ABCD$.

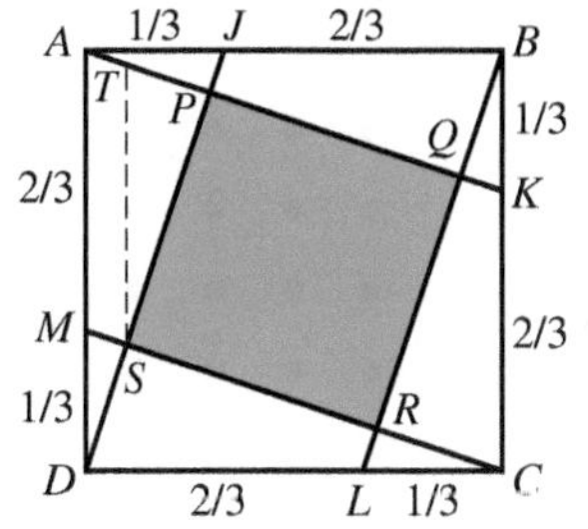

29. Let three arbitrary perpendiculars to the sides of $\triangle ABC$ of the diagram be drawn, meeting in pairs at the points P, Q, and R. Prove that $\triangle PQR \sim \triangle ABC$.

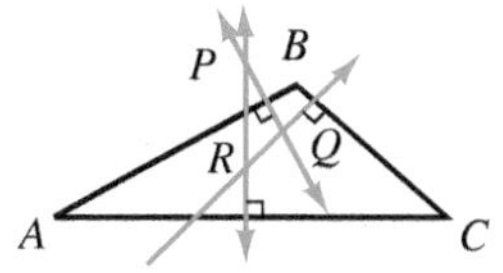

30. Let $PQRS$ be a space quadrilateral, and let W, X, Y, and Z be the midpoints of successive sides. Explain why $\overline{WY}$ and $\overline{XZ}$ intersect at their common midpoint M.

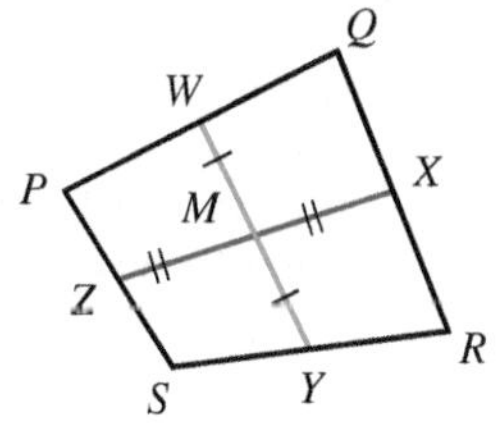

31. Let $ABCD$ be a trapezoid as illustrated, with bases of length $a = AB$ and $b = CD$. Let the diagonals $\overline{AC}$ and $\overline{BD}$ intersect at P, and suppose $\overline{EF}$ is the segment parallel to the bases that passes through P. Show that $EP = FP = \frac{ab}{a+b}$. (Thus, $EF = \frac{2ab}{a+b}$, which is called the **harmonic mean** of a and b.)

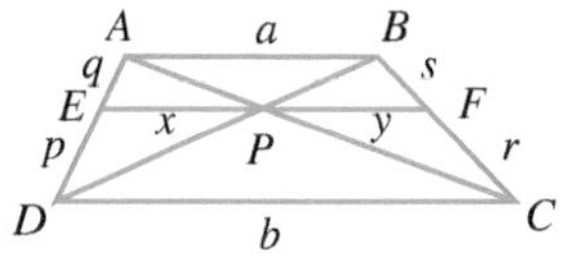

(*Hint:* Explain why $\frac{x}{a} = \frac{p}{(p+q)}$ and $\frac{x}{b} = \frac{q}{(q+p)}$. What happens when these equations are added?)

32. Let M and N be the midpoints of the sides of parallelogram $ABCD$ opposite A as shown. Show that $\overline{AM}$ and $\overline{AN}$ divide the diagonal $\overline{BD}$ into three congruent segments: $BP = PQ = QD$. (*Hint:* Construct $\overline{AC}$, and see Example 12.14; alternatively, notice that $\triangle MBP \sim \triangle ADP$, with scale factor 2).

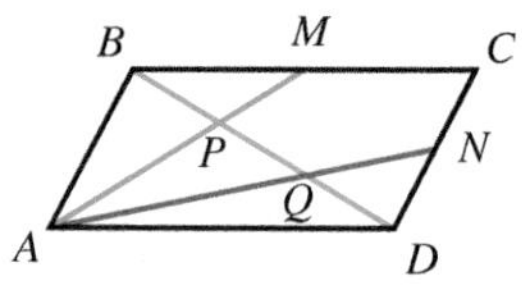

33. Use similarity to find the distances AP, BP, CP, and DP in the figure shown. The smallest squares on the lattice have sides of unit length.

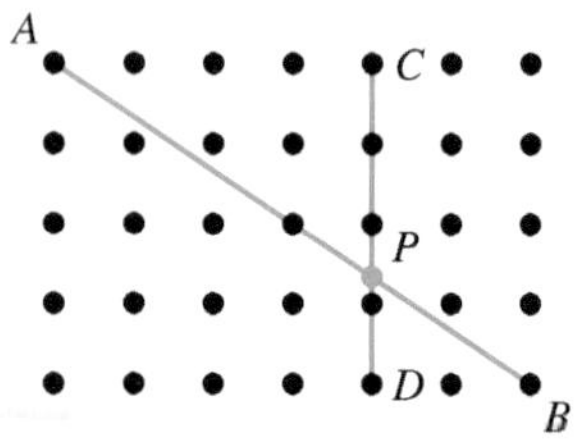

34. (**Writing**)

(a) In Example 12.13, it was shown that joining the successive midpoints of a quadrilateral forms a parallelogram. Explore what you find by *reversing* the construction. That is, given a parallelogram $KLMN$, can you construct a quadrilateral $ABCD$ for which $KLMN$ is the midpoint figure? Is $ABCD$ unique?

(b) Write a report discussing your results in part (a).

Making Connections

35. Mingxi is standing 75 feet from the base of a tree. The shadow from the top of Mingxi's head coincides with the shadow from the top of the tree. If Mingxi is 5′9″ tall and his shadow is 7′ long, how tall is the tree?

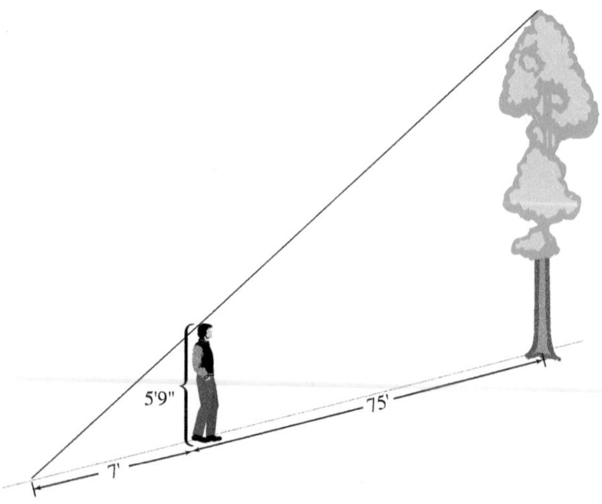

36. Mohini laid a mirror on the level ground 15 feet from the base of a pole, as shown. Standing 4 feet from the mirror, she can see the top of the pole reflected in the mirror. If Mohini is 5′5″ tall, how can she estimate the height of the pole? What must she allow for in her calculation?

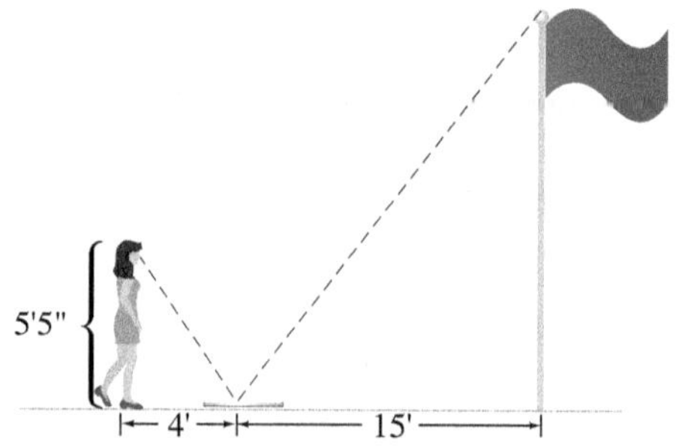

37. By holding a ruler 2 feet in front of her eyes as illustrated in the diagram, Ginny sees that the top and bottom points of a vertical cliff face line up with marks separated by 3.5″ on the ruler. According to the map, Ginny is about a half mile from the cliff. What is the approximate height of the cliff? Recall that a mile is 5280 feet.

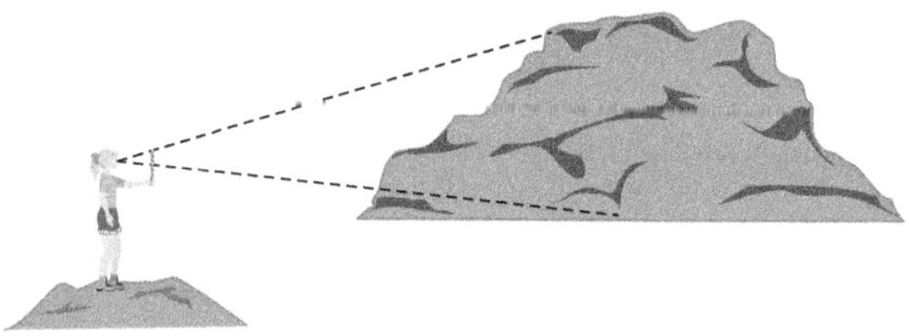

38. A vertical wall 18 feet high casts a shadow 6 feet wide on level ground. If Lisa is 5′3″ tall, how far away from the wall can she stand and still be entirely in the shade?

39. Use similar triangles to solve the following problem posed in 13th-century China: General Tsao stood 2 feet back from the edge of a vertical cliff overlooking the valley below. By holding his staff vertically 2 feet in front of him, he could see that the far bank of the river aligned with the point 1 foot below eye level along his staff and the near bank aligned with the point 3 feet below eye level along the staff. By lowering a rope over the cliff face, he determined that his eyes were 45 feet above the level floor of the valley. How did the general determine the width of the river to know if his army could safely cross to its opposite side? (*Hint:* In the figure shown, the general's eye is at point G, the top of his staff is at point A, and E and F are points on the near and far sides of the river.)

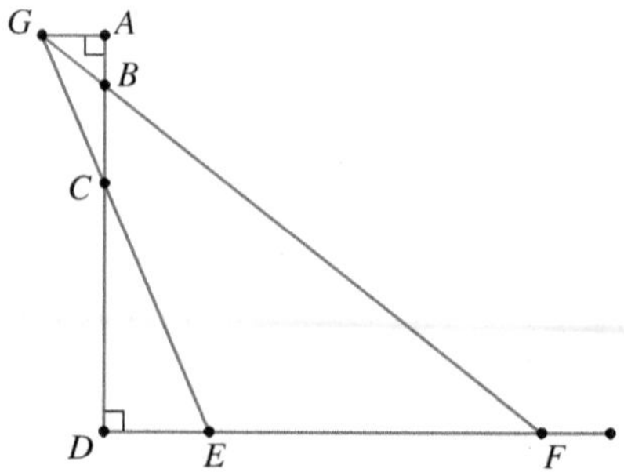

40. A sloping ramp is to be built with vertical supports placed at points B, C, and D as shown. The supports at A and E are 8 and 12 feet high, respectively. How high must the supports be at the points B, C, and D?

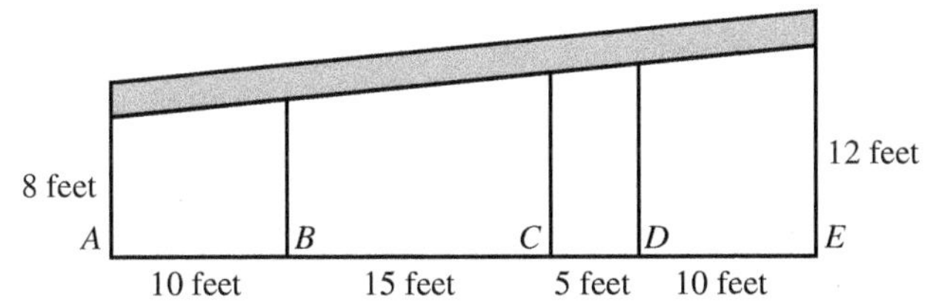

State Assessments

41. (Grade 5)
Triangle PQR is similar to triangle XYZ. What is the length XZ?

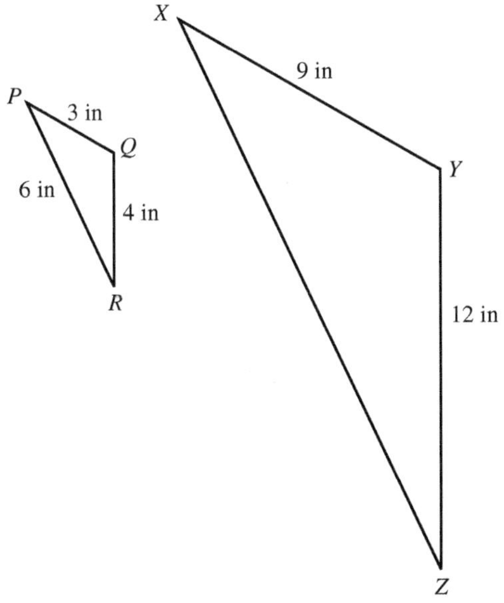

A. 12 in B. 15 in C. 18 in D. 21 in E. 24 in

42. (Massachusetts, Grade 7)
Mr. Liu wants to build a bridge across the creek that runs through his property. He made measurements and drew the map shown here.

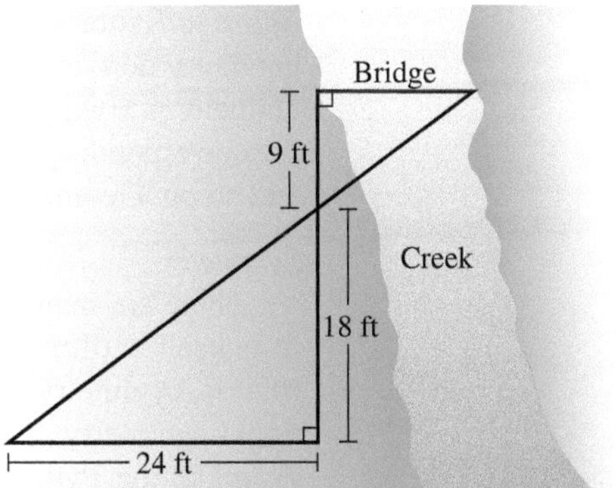

Based on this map, what is the distance across the creek at the place where Mr. Lui wants to put the bridge?

A. 9 feet B. 12 feet
C. 18 feet D. 24 feet

43. (Grade 6)
Triangle UVW is similar to triangle XYZ. Which side of triangle XYZ corresponds to side $\overline{UW}$?

A. $\overline{WU}$ B. $\overline{YZ}$
C. $\overline{XZ}$ D. $\overline{YZ}$

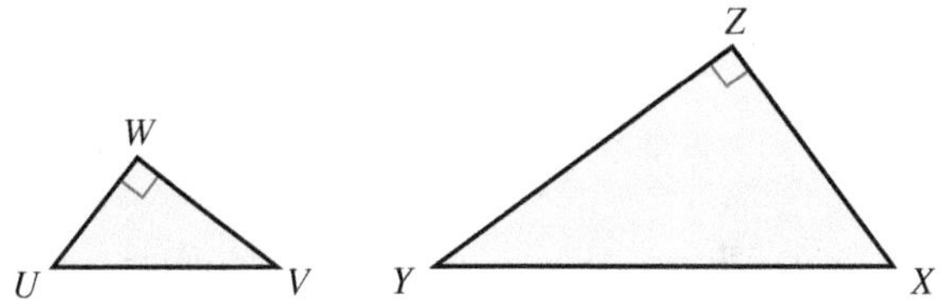

The Chapter in Relation to Future Teachers

This chapter concludes our study of informal geometry. Several themes, which form the basis for the geometry portion of the elementary school curriculum, have cut across these geometry chapters:

- *Congruence*
  The notion of congruence (that is, that one figure is exactly the same size and shape as another) is of great importance. General figures can be shown to be congruent with the use of rigid motions, and triangles can be shown to be congruent by means of angle and side properties.
- *Similarity*
  Similarity embodies the idea that two figures are the same *shape,* but of different size, as if one figure is simply a magnification of the other. Perhaps the most important consequence of similarity is that corresponding lengths in similar figures are proportional, with the scale factor the common ratio of the corresponding lengths. Similarity can be proved by similarity transformations or, in the case of triangles, with angle and side properties.
- *Constructions*
  Traditionally, constructions were restricted to figures and geometric objects that could be created with two Euclidean tools: the straightedge and the compass. In recent times, other construction methods, such as paper folding and the Mira, may be allowed. Figures constructible with one tool may or may not be constructible with a different tool. For example, any angle can

be constructed with the Mira but not with straightedge and compass. Computer geometry programs have also enlarged the set of constructible objects, with fractals as an important example.

- *Invariance*
  Invariance involves the idea that, although many properties of geometric figures change from figure to figure, some, often surprisingly, do not. Indeed, much of the interest in, and utility of, geometry derives from this fact. For example, the ratio of the circumference of any circle to its diameter is always $\pi = 3.14159\ldots$, the volume of any cone is always one-third the volume of the corresponding cylinder, the sum of the exterior angles of any convex polygon is always 360°, and so on. Geometry is replete with remarkable and useful invariances.

- *Symmetry*
  There are many kinds of symmetry in geometry. There is the periodic symmetry that manifests itself in tilings and tessellations; the symmetry of an object as if it were reflected in a mirror (that is, symmetry with respect to a line); the symmetry of an object through a point; and rotational symmetry. There is also a sort of symmetry in many of the formulas of geometry. For example, the Pythagorean expression $c^2 = a^2 + b^2$ is unchanged if a and b are interchanged, and similar symmetries are exhibited by the distance and slope formulas in coordinate geometry.

- *Loci*
  A *locus* is a set of points that satisfies certain conditions. For example, a circle is the locus of all points in a plane that are equidistant from a fixed point O. As another example, the locus of points equidistant from the points A and B is the perpendicular bisector of the segment $\overline{AB}$.

- *Maxima and minima*
  Of all triangles of fixed perimeter, the equilateral triangle has maximum area. Of all rectangles of fixed area, the square has the smallest perimeter. Such questions of maxima and minima often arise in geometry.

- *Limit*
  As an example of the notion of a limit, as n gets larger and larger a regular n-gon more and more closely approximates a circle. Indeed, we would say that the limit of a regular n-gon as n tends to infinity *is* a circle. This notion was used in developing the formula for the area of a circle.

- *Measurement*
  The word *geometry* means, literally, "earth measure." The ability to measure, compare, and communicate information concerning size and amount is basic to geometric thinking and applications of geometry in the real world.

- *Coordinates*
  This theme stresses the idea that geometric objects can be viewed as specified sets of points determined by ordered pairs of numbers in a coordinate plane. This powerful notion makes it possible to use methods of arithmetic and algebra to obtain geometric results.

- *Logical structure*
  As in the rest of mathematics, geometric ideas do not stand alone. Even in informal geometry, it is important to see how some results follow from others and to realize that guessing or conjecturing alone is not enough.

Chapter 12 Summary

Section 12.1 Congruent Triangles	Page Reference
CONCEPTS	
• **Congruent triangles:** Two triangles of the same shape and size.	616
• **Congruent line segments** are segments of the same length.	616

DEFINITIONS

- Two triangles *ABC* and *DEF* are **congruent** if and only if corresponding angles have the same measure and corresponding sides have the same length. That is, $\triangle ABC \cong \triangle DEF$ if and only if $m(\angle A) = m(\angle D)$, $m(\angle B) = m(\angle E)$, $m(\angle C) = m(\angle F)$, $AB = DE$, $BC = EF$, and $CA = FD$. — 616
- The **six parts** of a triangle *ABC* are its three sides $\overline{AB}$, $\overline{BC}$, and $\overline{CA}$ and its three angles $\angle A$, $\angle B$, and $\angle C$. — 616
- Two triangles *ABC* and *DEF* are **congruent triangles** if and only if there is a correspondence of vertices $A \leftrightarrow D$, $B \leftrightarrow E$, $C \leftrightarrow F$ so that the six parts of triangle *ABC* are congruent to the corresponding six parts of triangle *DEF*. In symbols, $\triangle ABC \cong \triangle DEF$ if and only if $AB = DE$, $BC = EF$, $CA = FD$, $m(\angle A) = m(\angle D)$, $m(\angle B) = m(\angle E)$, and $m(\angle C) = m(\angle F)$. — 616

CONSTRUCTIONS

- **Basic compass-and-straightedge constructions:** Construct a congruent copy of a line segment, construct a congruent copy of an angle. — 616

PROPERTIES

- **Congruence properties of triangles:** Two triangles are congruent if they satisfy any one of the following properties: SSS (side–side–side), SAS (side–angle–side), ASA (angle–side–angle), and AAS (angle–angle–side). — 622

THEOREMS

- **Triangle inequality:** The length of any side of a triangle is less than the sum of the lengths of the other two sides. — 620
- **Isosceles triangle theorem and converse:** A triangle has two congruent sides if, and only if, their opposite angles are congruent. — 623
- **Thales' theorem:** A triangle inscribed in a semicircle is a right triangle with the diameter as its hypotenuse. — 624

Section 12.2 Constructing Geometric Figures	Page Reference

CONCEPTS

- **Constructions and drawings:** Geometric figures can be constructed or drawn with compass and straightedge, the Mira, paper folding, and geometry software. — 632
- **Construction versus drawing:** A figure is *constructed* if the defining properties of the figure have been incorporated into the construction: All lengths and angles are exact. A figure is *drawn* if the properties of the figure are only approximate to give the appearance of the desired shape of the figure. — 632

DEFINITIONS

- The **foot of the perpendicular** is the point on a line that is closest to a point not on the line. — 633
- The **altitude of the triangle** is the line through a vertex of a triangle that is perpendicular to the line containing the opposite side of the triangle. — 634
- The **perpendicular bisector** of a line segment is the line perpendicular to the segment at its midpoint. — 634

- **Circumscribing circle,** or **circumcircle** of a triangle, is the unique circle that passes through the three vertices A, B, and C of the triangle. — 636
- The **circumcenter** of a triangle is the center of the circumscribed circle of the triangle. — 636
- An **angle bisector** is a ray that starts at the vertex of an angle and splits the angle into two congruent parts. — 637
- The **inscribed circle,** or **incircle** of a triangle, is the unique circle that is tangent to the three sides of the triangle. — 638
- The **incenter** of a triangle is the center of the inscribed circle. — 638
- The **Fermat primes** are prime numbers of the form $2^{2^k} + 1$, where k is a nonnegative integer. Only five Fermat primes—3, 5, 17, 257, and 65,537—are known to exist. — 640

CONSTRUCTIONS

- **Basic compass-and-straightedge constructions:** Construct a line parallel to a given line through a given point; construct a line perpendicular to a given line through either a point on or away from the given line; construct the midpoint of a line segment and the line perpendicular to the segment through the midpoint; construct the angle bisector of a given angle. — 633, 634, 635, 637

THEOREMS

- **Equidistance property of the perpendicular bisector:** A point lies on the perpendicular bisector of a line segment if and only if the point is equidistant from the endpoints of the segment. — 635
- **Circumscribed circle:** The three perpendicular bisectors of the sides of a triangle are concurrent at a point (the *circumcenter*) that is equidistant to the three vertices of the triangle. That is, the circumcenter is the center of a circle (the *circumcircle*) that passes through the three vertices of the triangle. — 636
- **Equidistance property of the angle bisector:** A point lies on an angle bisector if and only if it is equidistant to the sides of the angle. — 637
- **Inscribed circle:** The three angle bisectors of a triangle are concurrent at a point (the *incenter*) that is the center of a circle (the *incircle*) that is equidistant to the three sides of the triangle. That is, the inscribed circle is tangent to all three sides of the triangle. — 638
- **Gauss–Wantzel theorem:** A regular n-gon has a compass-and-straightedge construction if and only if n is a power of 2 that is 4 or larger or a power-of-2 multiple of a product of distinct Fermat primes. — 640

Section 12.3 Similar Triangles	Page Reference

CONCEPTS

- **Similar triangles** are triangles of the same shape but not necessarily of the same size. Corresponding angles are congruent but corresponding side lengths are magnified or reduced by a common multiplier called the *scale factor*. — 648

DEFINITIONS

- Two triangles ABC and DEF are **similar triangles** if and only if there is a correspondence of vertices $A \leftrightarrow D$, $B \leftrightarrow E$, $C \leftrightarrow F$ so that the corresponding angles are congruent and the corresponding side lengths are proportional. In symbols, $\triangle ABC \sim \triangle DEF$ if and only if $m(\angle A) = m(\angle D)$, $m(\angle B) = m(\angle E)$, $m(\angle C) = m(\angle F)$, $\dfrac{AB}{DE} = \dfrac{BC}{EF} = \dfrac{CA}{FD} = k$, where the constant of proportionality k is the **scale factor.** — 648

• The **midpoint figure** of a given polygon is the polygon whose vertices are the successive midpoints of the given polygon.	652
• The **medial triangle** is the midpoint figure of a given triangle.	652
• A **median** of a triangle is a line segment that joins a vertex to the midpoint of the opposite side of the triangle.	653
PROPERTIES	
• **Similarity properties of triangles:** Two triangles are similar if they have either AA (angle–angle) similarity (two pairs of congruent angles), SSS (side–side–side) similarity (three proportional sides), or SAS (side–angle–side) similarity (two proportional pairs of sides that include congruent angles).	649, 650, 651, 652
THEOREMS	
• The medial triangle of a given triangle is similar to the given triangle, with a scale factor of ½.	652
• The midpoint figure of any quadrilateral is a parallelogram.	652
• The three medians of a triangle are concurrent at a point called the **centroid** of the triangle. The centroid is 2/3 of the distance from a vertex toward the opposite midpoint of the side of the triangle.	653

Chapter Review Exercises

Section 12.1

1. Consider the following two triangles:

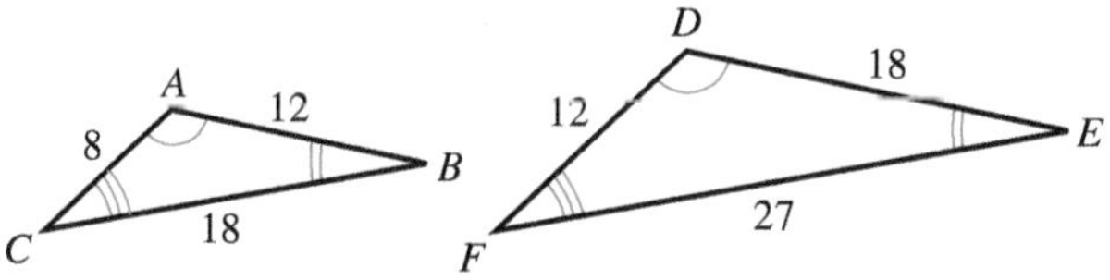

Are the assertions that follow *true* or *false*? Explain your answer.

(a) Five parts of $\triangle ABC$ are congruent to five parts of $\triangle DEF$.

(b) $\triangle ABC$ is congruent to $\triangle DEF$.

2. In each figure, find at least one pair of congruent triangles. Express the congruence by using the $\cong$ symbol, and state why the triangles are congruent.

(a)

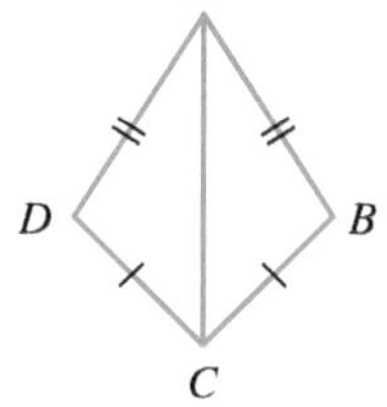

(b)

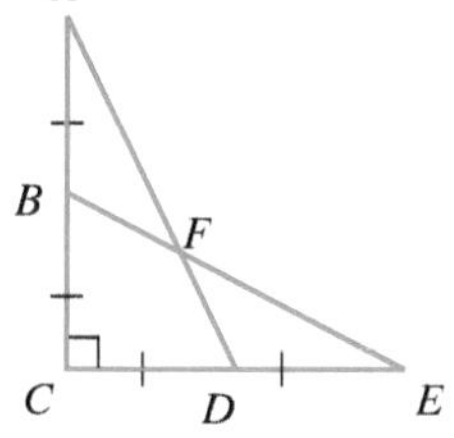

(c)

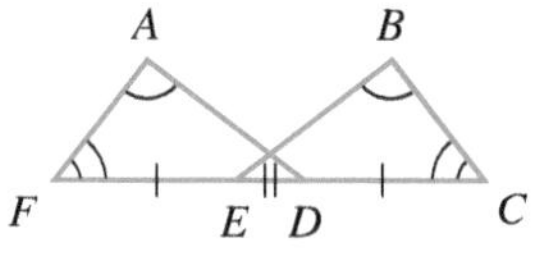

(d)

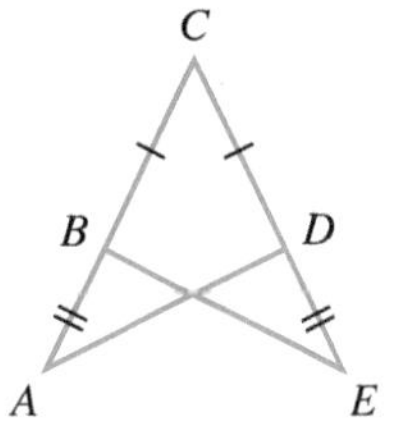

(e)

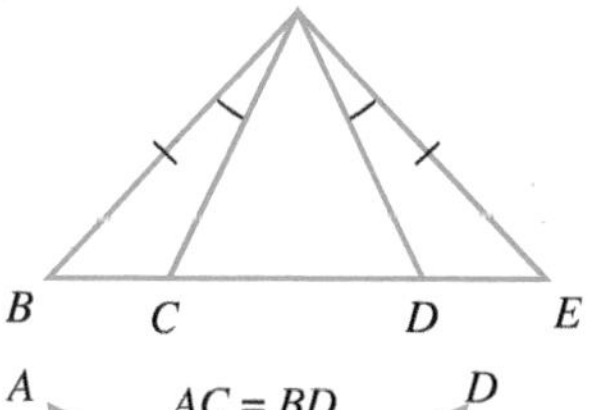

(f)

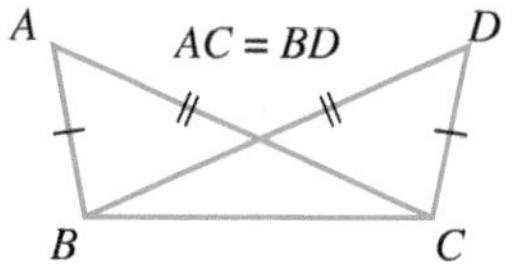

3. Without measuring, fill in the blanks that follow these figures:

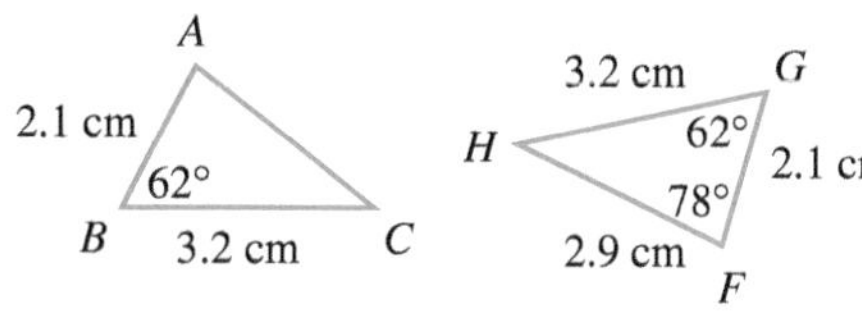

(a) $AC =$ ____

(b) $m(\angle H) =$ ____

(c) $m(\angle A) =$ ____

(d) $m(\angle C) =$ ____

4. Find three pairs of congruent triangles in this figure. Justify your answers.

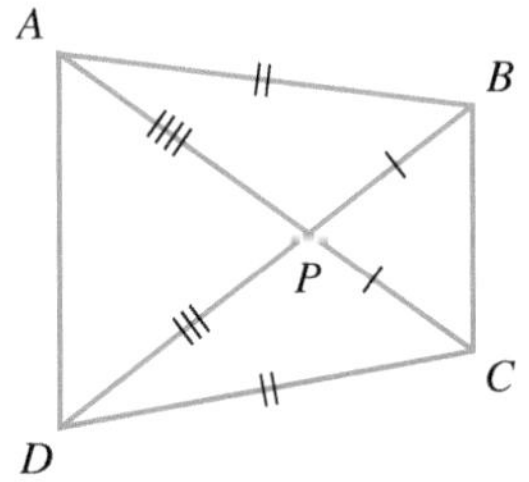

5. Let D be any point of the base $\overline{BC}$ of an isosceles triangle ABC. Locate E on $\overline{AC}$ and F on $\overline{AB}$ so that $EC = BD$ and $BF = DC$. Draw the figure as it is described and then prove that $DE = DF$.

6. Let $ABCD$ be any convex quadrilateral. Show that the sum of the lengths of two of the opposite sides of the quadrilateral is smaller than the sum of the lengths of the two diagonals of the quadrilateral.

Section 12.2

7. Perform the constructions that follow with a compass and a straightedge. Show and describe all of your steps.

(a)

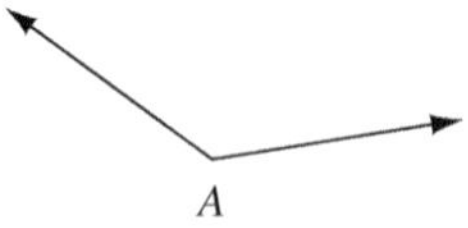

Bisector of $\angle A$

(b)

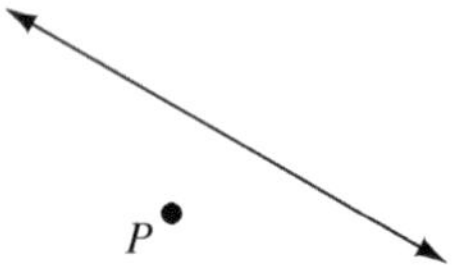

Perpendicular through P

(c)

Perpendicular bisector of $\overline{AB}$

(d)

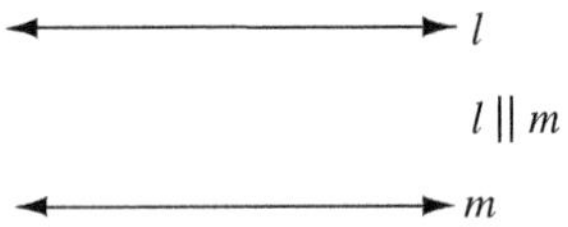

Line k equidistant to l and m

8. Show and describe the position of a Mira that performs each of the constructions of problem 7 in one step.

9. (a) Construct $\triangle ABC$, where $\angle A$, $\overline{AB}$, and $\overline{BC}$ are the parts shown. Is the shape of $\triangle ABC$ uniquely determined?

(b) Is the shape of $\triangle ABC$ uniquely determined if $\angle C$ is obtuse?

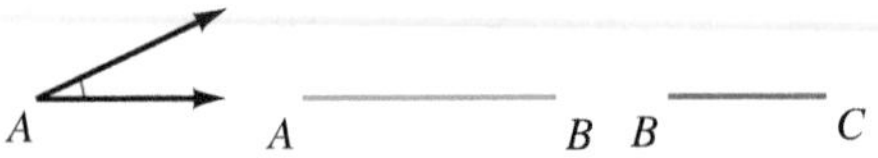

10. Construct a regular hexagon $ABCDEF$, for which diagonal AD is as follows:

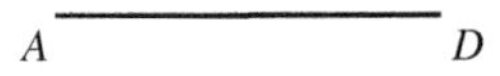

Section 12.3

11. (a) If only a ruler is available, is it possible to determine whether two triangles are similar?

(b) If only a protractor is available, is it possible to determine whether two triangles are similar?

12. Triangle ABC is as shown. Construct a triangle DEF for which $\triangle ABC \sim \triangle DEF$ and $DE = \left(\frac{3}{2}\right)AB$, using a compass and a straightedge.

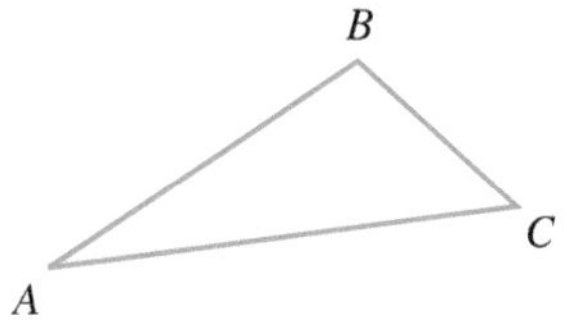

13. Explain why each pair of triangles is similar. State the similarity by using the ~ symbol and give the scale factor.

(a)

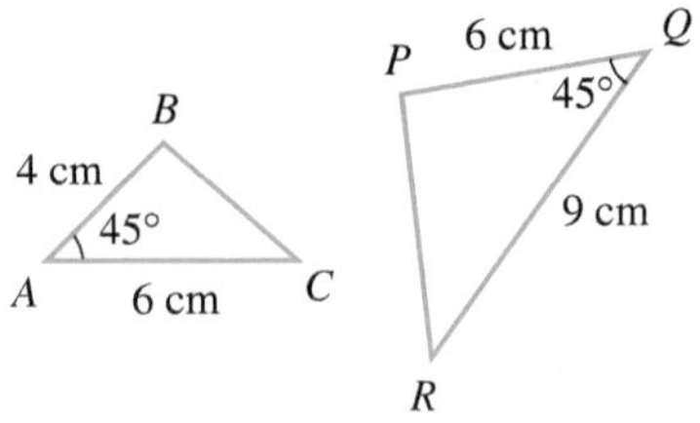

(b)

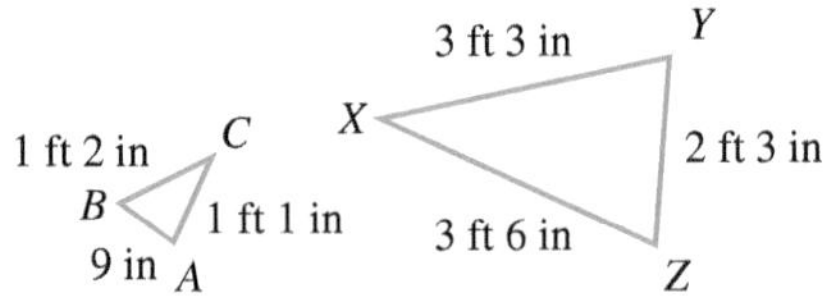

(c)

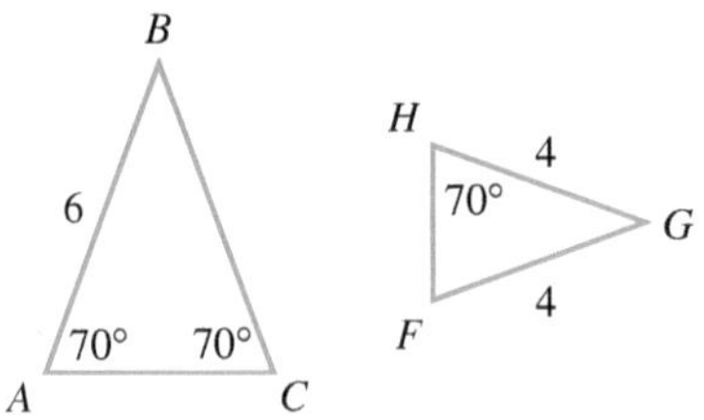

(d)

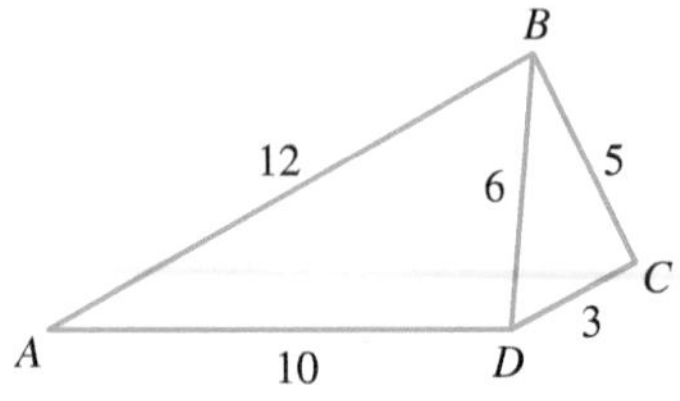

14. Lines k, l, and m are parallel. Find the lengths x and y, using similar triangles.

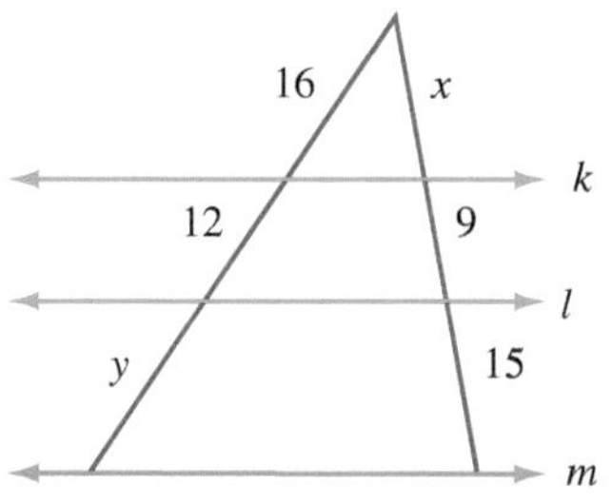

15. In the morning, the shadows cast by the top of a vertical stick 3 feet high and the top of a pyramid were at points S_1 and P_1, respectively, on level ground. That afternoon, because of the motion of the sun in the sky, the points were at S_2 and P_2. If $S_1S_2 = 2$ feet and $P_1P_2 = 270$ feet, what is the height of the pyramid?

16. A person 6 feet tall casts a shadow 7 feet long, and a tree casts a shadow 56 feet long.

(a) What assumption can you make about the sun's rays?

(b) What other assumption can you make to conclude that $\triangle ABC \sim \triangle DEF$?

(c) How tall is the tree?

13 Statistics: The Interpretation of Data

COOPERATIVE INVESTIGATION

How Many Beans in the Bag?

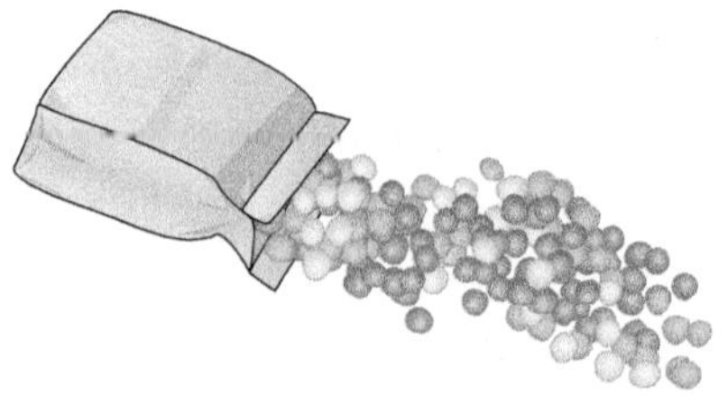

Materials Needed

1. Three or four opaque bags containing about 200 beans each (this number being known only to the instructor), with the same number of beans in each bag.
2. A bright colored marker to go along with each bag.

Directions

1. Divide the class into groups of about ten students each and give each group a bag of beans.
2. The students in each group mark 25 beans with their colored marker and return the marked beans to their bag.
3. The beans in the bag are then thoroughly mixed.
4. Each student, without peeking, removes 5 beans from the bag (without replacement) and notes the number of marked beans obtained.
5. After all the students in a group have made their selection, the collection of all beans selected makes up the sample for the group.
6. Each group determines the fraction of marked beans in its sample (the number of marked beans in the sample divided by the total number of beans in the sample) and uses this figure to estimate the original number of beans in the bag. (Let n denote the number of beans in the bag; then $\frac{25}{n}$ is the fraction of marked beans in the bag.)
7. The students repeat steps 2 through 6, but this time, in step 2, they mark 50 beans and return them to the bag. The students then compare the estimate obtained this time with that obtained the first time and discuss which estimate is likely to be the better one.
8. Finally, the instructor should reveal the actual number of beans in each of the bags and discuss the outcomes with the class.

CHAPTER PREVIEW

What is so important about the understanding of data and the place of data in the world? Statistics is a powerful and subtle science that informs people more thoroughly. Statistics provides major analytical tools and opportunities for jobs in such fields as medicine, the social sciences, education, engineering, the health sciences, business, and many other areas. Most importantly, the use of statistics helps children as well as adults become informed citizens.

How does one learn about statistics? Children who work with data in elementary school get a sense of the presentation, the organization, and the center and spread of their data. Most parents and teachers are surprised that the various ways of picturing data and data analysis are being introduced in elementary schools now. In fact, NCTM standards on data and statistics, which are quoted on page 669, emphasize the development of intuition about sets of data, how to describe them, how to get an idea of what might be called the "center" of the data, and how data "spread." Furthermore, statistics is a part of grades 5 ("make line plots") and 6 ("develop understanding of statistical variability") in the **Common Core** content standards. In addition, the well-thought-out *Guidelines for Assessment and Instruction in Statistics Education (GAISE) Report: A Pre-K–12 Curriculum Framework* (2007) is an influential and accessible text for future teachers.

To show readers that statistics has a real presence in elementary school, we have included many more questions similar to State Assessments in this chapter than in others. Elementary school children need to understand basic statistics in order to build a foundation to prepare them early in their development to be successful in their schooling and career.

This chapter is not a course in statistics—it couldn't be! After all, it isn't possible to learn math, biology, literature, or a social science by reading just one chapter. Instead, this chapter introduces the tools of statistics that appear in elementary and middle school.

KEY IDEAS

- Statistics is a way to organize, present, and represent data through a variety of methods.
- Statistics is a way to make sense of data that have been gathered by computing certain numerical values that help us understand where the **center** of the data is and how the data set spreads out (its **variability**) from its center.
- The various numerical definitions of the center (**mean, median,** and **mode**) of a set give different insights into which measure of the center of the data is most revealing.
- There are various ways to understand the variability of a data set, including **box and whisker plots, quartiles, 5-number summary,** and the concept of the **standard deviation.**
- There is a difference between the population and a sample of the data from the population.
- An approach is given to attempt to predict the center and standard deviation of a population through collecting the standard deviation of a "reasonable" variety of samples. The normal distribution is key to this approach, which is called **statistical inference.**

FROM THE NCTM PRINCIPLES AND STANDARDS

Prior to the middle grades, students should have had experiences collecting, organizing, and representing sets of data. They should be facile both with representational tools (such as tables, line plots, bar graphs, and line graphs) and with measures of center and spread (such as median, mode, and range). They should have had experience using some methods of analyzing information and answering questions, typically about a single population.

In grades 6–8, teachers should build on this base of experience to help students answer more-complex questions, such as those concerning relationships among populations or samples and those about relationships between two variables within one population or sample. Toward this end, new representations should be added to the students' repertoire. Box plots, for example, allow students to compare two or more samples, such as the heights of students in two different classes. Scatterplots allow students to study related pairs of characteristics in one sample, such as height versus arm span among students in one class. In addition, students can use and further develop their emerging understanding of proportionality in various aspects of their study of data and statistics.

Source: Principles and Standards for School Mathematics by NCTM, *p. 249. Copyright © 2000 by the National Council of Teachers of Mathematics. Reproduced with permission of the National Council of Teachers of Mathematics via Copyright Clearance Center. NCTM does not endorse the content or validity of these alignments.*

13.1 Organizing and Representing Data

A collection of data points is called a **data sets** but how can we organize a data set pictorially? We try to organize and represent data so that some of their properties are easier to see. This section covers various kinds of visual representations such as:

- **Dot plots:** summarizing relatively small sets of data—grades in a class, heights of students in a class, birth months of students in a class, and so on.
- **Stem-and-leaf plots:** for essentially the same purposes as dot plots; especially useful in comparing small data sets.
- **Histograms:** summarizing information from large sets of data that can be naturally grouped into intervals.
- **Bar graphs:** conveying information about data with the horizontal scale representing some attributes that are not numerical, such as letter grades.
- **Line graphs:** summarizing trends over time.
- **Pie charts:** representing relative amounts of a whole. Also called **pie graphs** or **circle graphs.**
- **Pictographs:** summarizing relative amounts, trends, and data sets; useful in comparing quantities.

We will start with dot plots and use a specific example.

Dot Plots

The final examination scores for Course ABC are shown in Table 13.1. This table records the data as a simple list of the grades of 35 students.

TABLE 13.1 FINAL EXAMINATION SCORES IN COURSE ABC, SECTION 1

79	78	79	65	95	77	49
91	63	58	78	96	74	68
71	86	91	94	79	69	86
62	78	77	88	67	78	84
69	53	79	75	64	89	77

Although scanning the data gives some idea of how the class did, it is more revealing to organize the data by placing a dot above a number line, as in Figure 13.1. Data depicted in this way are called a **dot plot** or sometimes a **line plot.**

The dot plot makes it possible to see at a glance that the scores range from 49 through 96; that most scores are between 60 and 80, with a large group between 75 and 80; and that the "typical" score is probably about 77 or 78. It also reveals that scores like 49 and 53 are quite atypical, or **outliers.** It is much easier to interpret data arranged on a dot plot than to use "raw," or unorganized, data.

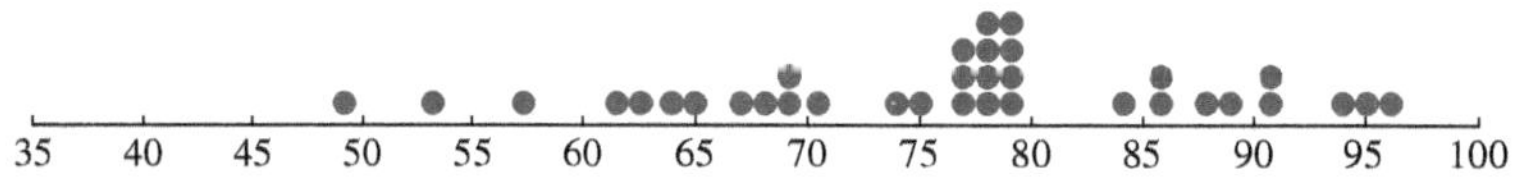

Figure 13.1 Dot plot of the final examination scores in Course ABC, Section 1

Stem-and-Leaf Plots

Stem-and-leaf-plots for displaying data are similar to dot plots and are especially useful for comparing two sets of data. Stem-and-leaf plots are a part of statewide exams, generally around sixth grade.

To draw a stem-and-leaf plot for the data in Table 13.1, we let the tens digits of the scores be the "stems" and let the units digits be the "leaves." The scores 79, 78, and 79 are represented by

The completed plot appears in Figure 13.2.

Figure 13.2 Stem-and-leaf plot of the final examination scores of Course ABC, Section 1

4	9													
5	3	8												
6	2	3	4	5	7	8	9	9						
7	1	4	5	7	7	7	8	8	8	8	9	9	9	9
8	4	6	6	8	9									
9	1	1	4	5	6									

The stem-and-leaf plot gives much the same visual impression as the dot plot and allows a similar interpretation.

To compare two sets of similar data, it is useful to construct stem-and-leaf plots on the same stem in order to develop intuition and make the (visual) comparison easier. Figure 13.3 shows such a plot for final examination scores in Sections 1 and 2 of Course ABC.

Figure 13.3 shows that, although the two classes are comparable, Section 2 had a wider range of scores, with one lower and several higher than those in Section 1. The bottom row of Section 2 shows three scores of 100 percent.

Figure 13.3
Stem-and-leaf plots of the final examination scores in Course ABC, Sections 1 and 2

Section 2		Section 1
3	4	9
9 8 7 5	5	3 8
8 8 5 5 5 5 3 1	6	2 3 4 5 7 8 9 9
5 5 4 4 3 0	7	1 4 5 7 7 7 8 8 8 8 9 9 9 9
9 7 6 4 4 2 0 0 0	8	4 6 6 8 9
6 5 5 0	9	1 1 4 5 6
0 0 0	10	

Histograms

Another common tool for organizing and summarizing data is a *histogram.* A histogram for the data in Figure 13.2 is shown in Figure 13.4. In a **histogram,** scores grouped into intervals indicate the number of data points in each interval by the height of the rectangle constructed on that interval. A histogram must keep the width of the bars consistent and the bars must touch. The vertical rectangular regions in Figure 13.4 show how often the scores occur in each interval. In addition, the histogram gives the data collection a continuous look.

The number of times any particular data value occurs is called its **frequency**. Similarly, the number of data values in any interval is the **frequency of the interval.** Figure 13.4 shows that 14 students have scores in the 70 to 79 range. In other words, the frequency of the interval from 70 to 79 is 14. Note that the horizontal axis indicates data values or ranges of data values (nearly always from lowest to highest), while the vertical axis of a histogram shows frequency.

Figure 13.4
Histogram of the final examination scores in Course ABC, Section 1

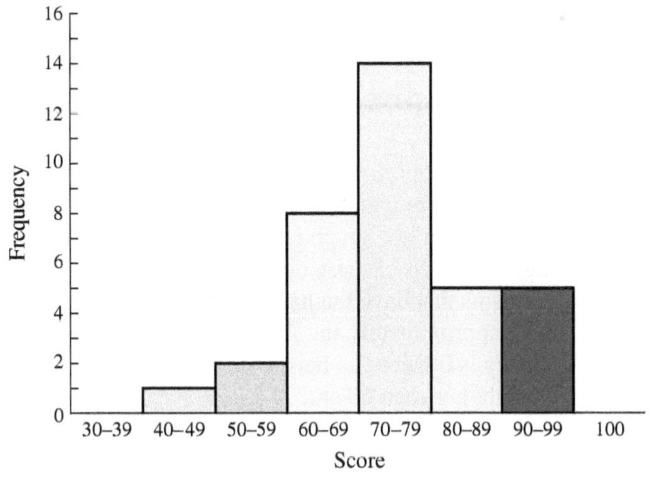

We lose some detail when we use a histogram. In Figure 13.4, for example, the histogram does not show how many students scored 71. However, histograms do allow visualization of large data sets that do not lend themselves to dot or stem-and-leaf plots.

Histograms are useful for both discrete and continuously varying data. We will now give another example in which people's heights vary continuously. You may say that your height is 5′6″ but that is true only to *the nearest inch.* Look at Table 13.2, which gives the heights of 80 boys at Eisenhower High School. The heights were measured to the nearest inch, so the numbers given already represent *grouped data,* for example, a measurement for heights between 65.5 and 66.5 inches. The corresponding histogram is shown in Figure 13.5.

TABLE 13.2 HEIGHTS OF BOYS AT EISENHOWER HIGH SCHOOL, TO THE NEAREST INCH, AND ARRANGED IN INCREASING ORDER

64	67	68	69	69	70	71	72
65	67	68	69	69	70	71	72
66	68	68	69	69	70	71	72
66	68	68	69	69	70	71	72
66	68	68	69	69	70	71	72
67	68	68	69	69	70	71	72
67	68	68	69	70	70	71	73
67	68	69	69	70	70	71	73
67	68	69	69	70	70	71	74
67	68	69	69	70	70	71	74

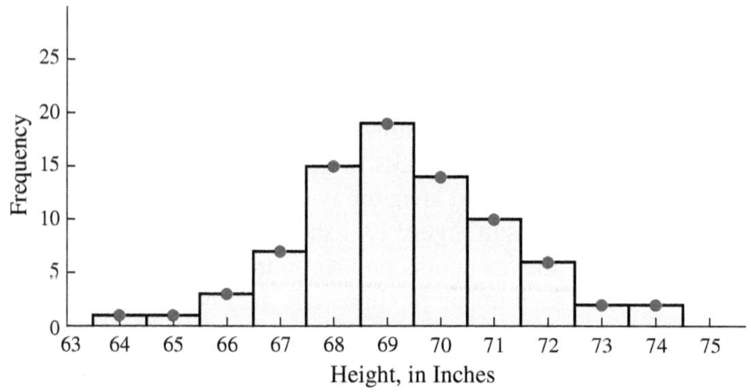

Figure 13.5 Histogram of the heights of boys at Eisenhower High School

Grouping data into classes and displaying the data in a histogram is a useful visualization of the characteristics of the data set. To draw another histogram, choose scales on the two axes in this case (height and frequency in Figure 13.5) that allow all the data to be represented.

Line Graphs

A *line graph* is a useful way to plot data. A **line graph** has a horizontal line (the x-axis) and is usually given in "properties" (such as years, exam scores, or height), and a vertical line (the y-axis) usually shows how frequently something happens or how often something comes about during that year. An example of a line graph is shown in Figure 13.6. First, we look at Figure 13.5 and see the dots in the segment. Each of these dots is the midpoint of the segment. We then have a graph consisting of line segments that have the height of the boys on the x-axis and the frequency of the number of boys who have approximately the same height, in inches, on the y-axis. Note that in this line graph, the frequency is 0 when the height is either 63″ or less or 75″ or more. This means that all of the boys have a height between 63 and 75 inches.

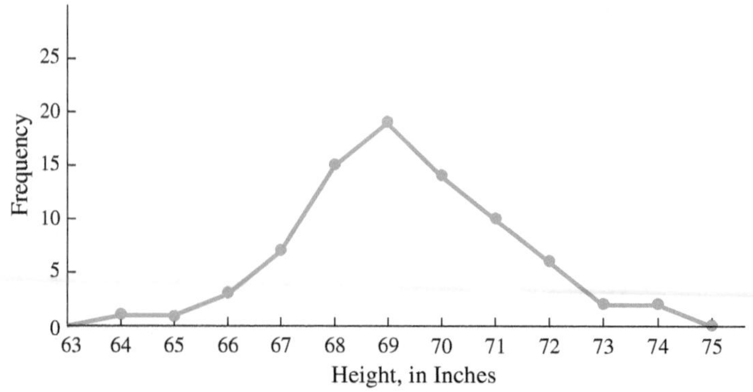

Figure 13.6 Line graph of the heights of boys at Eisenhower High School

Line graphs are particularly effective when they are used to indicate trends over periods of years—trends in the stock market, trends in the consumption of electrical energy, and so on. For example, consider the data in Table 13.3, which gives the salary of teachers in public schools in the United States from 1970 to 2010.

TABLE 13.3 AVERAGE ANNUAL SALARY OF TEACHERS IN PUBLIC ELEMENTARY AND SECONDARY SCHOOLS IN THE UNITED STATES (IN DOLLARS)

Year	1970	1980	1990	2000	2010
Average Salary	$8,626	$15,970	$31,367	$41,807	$55,350

SOURCE: National Education Association, *Estimates of School Statistics*, 1969–70 through 2009–10. (This table was prepared May 2010 by Calvin Long.)

The data points of K–12 teacher salaries can be represented visually by a line graph. In Figure 13.7, the points on the graph are determined by the year (on the x-axis) and the average salary for that year (on the y-axis.) Consecutive points of the table are then connected by straight-line segments.

Figure 13.7
Average annual salary of teachers in public elementary and secondary schools in the United States

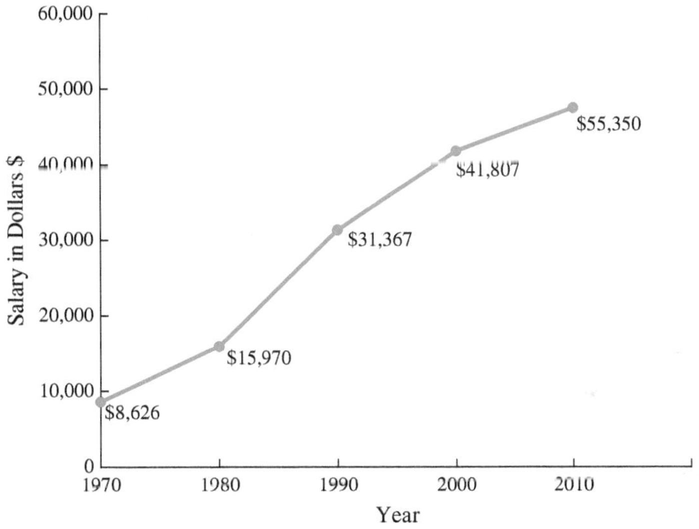

One advantage of a line graph is that it makes it possible to estimate data values not explicitly given otherwise.

EXAMPLE 13.1

Estimating Data Values from a Line Graph

Using the graph in Figure 13.7, estimate the average annual salary of teachers in public elementary and secondary schools in the United States in 1995.

Solution

Understand the Problem

We are asked to estimate the average salary for the 1995 year.

Devise a Plan

Having Figure 13.7 already simplifies our task. The graph suggests that the salary grows steadily each year, and, while the growth is certainly not "straight-line growth" between data points as indicated by the diagram, the straight line joining the data points for 1990 and 2000 surely approximates the actual growth. We look at the midpoint between 1990 and 2000, which is 1995. Thus, the annual salary at 1995 should be the average between the annual salaries at 1990 and 2000.

Carry Out the Plan

We know the midpoint is halfway between 1990 and 2000. Since the annual salaries are $31,367 and $41,807, respectively, the answer for approximating the salary at 1995 is

$$\frac{(31{,}367 + 41{,}807)}{2} = \$36{,}587.$$

Look Back

The solution was achieved by noting that the line graph, which gives the appropriate values of expenditures, suggests that the average salaries increase steadily and that the actual values for intervening years no doubt lie relatively close to the straight-line segments joining the given data points. We notice that the annual salaries (in current dollars) grew well before 2000 and then slowed down. There can be many reasons for this.

Score	Grade
90–100	A
80–89	B
70–79	C
60–69	D
0–59	F

Bar Graphs

Bar graphs, which are similar to histograms, are often used for "categorical" data; that is, the horizontal scale may represent some nonnumerical attribute. For example, consider the final examination scores for Course ABC, Section 1, as listed in Table 13.1. The accompanying table shows that there were 3 Fs, 8 Ds, 14 Cs, 5 Bs, and 5 As. If we indicate grades on the horizontal scale and frequency on the vertical scale, we can construct the bar graph shown in Figure 13.8. In general, the rectangles in a bar graph do not abut, and the horizontal scale may be designated by any attribute—grade in class, year, country, city, and so on. As usual, however, the vertical scale will denote frequency—the number of items in the given class.

Figure 13.8
Bar graph of the final examination grades in Course ABC, Section 1

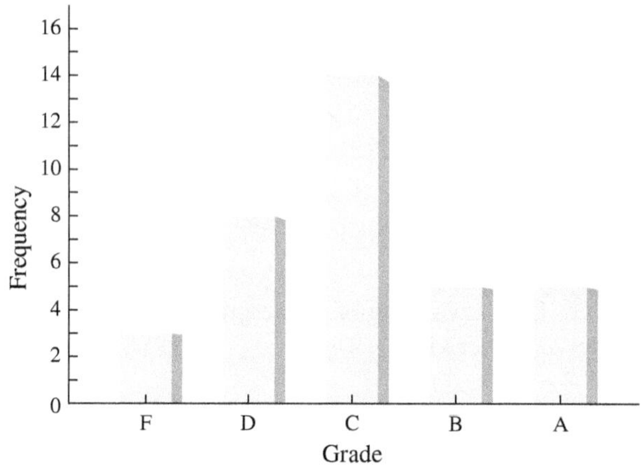

Bar graphs can also help us compare data concerning two or more similar groups of items.

EXAMPLE 13.2

Comparing Grades in Two Classes by Means of a Bar Graph

Draw a bar graph to compare the grades in Course ABC Sections 1 and 2.

Solution

To make the bar graph, draw two adjacent bars (rectangles) for each letter grade to indicate which bars represent which section. If we use the blue bars for Section 1, we have the frequency of the grades given in the first paragraph of the bar graph section. Figure 13.3 on p. 671 tells us the grade distribution of Section 2, which is pictured in pink. We then get the following bar graph. Note that the scales for Section 1 and Section 2 are the same.

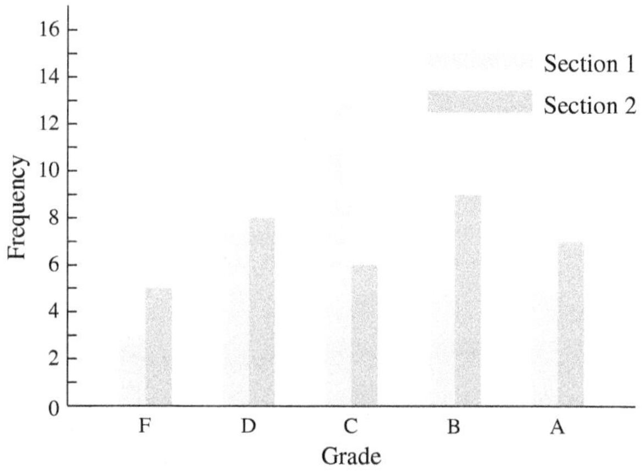

Highlight from History: David Blackwell and Statistics

David Blackwell (1919–2010) discovered a number of brilliant ideas in statistics. Born in Centralia, Illinois, he entered college with the intent to teach elementary school mathematics. He earned his bachelor's degree in mathematics at age 19 and was awarded a Ph.D. in mathematics in 1941 from the University of Illinois at the incredibly young age of 22.

He sought a position at the University of California, Berkeley, but race-based objections prevented his appointment at that time. He taught at Southern University, Clark Atlanta College (now a university) and, in 1944, was appointed full professor at Howard University and head of the Mathematics Department until 1954.

Finally, Dr. Blackwell was hired by the University of California, Berkeley, as a full professor in the newly created Statistics Department in 1955. He was the first African American who was a tenured faculty member at the university.

Dr. Blackwell was a distinguished, longtime member and former vice president of the American Statistical Association, an elected member of the National Academy of Science, and the American Academy of Arts and Sciences. He held 12 honorary doctorates and served as the president of the Institute of Mathematical Statistics, the International Association for Statistics in Physical Sciences, and the Bernoulli Society. At the time of his retirement in 1988, he had changed the course of statistics and had published 80 papers.

Pie Charts

Another pictorial method for conveying information is a **pie chart,** sometimes called a **pie graph** or **circle graph.** Figure 13.9 is an example of a pie chart that gives the percent distribution of employment by aggregate occupational group in 2010. Each of the occupational groups is clearly marked and has a certain percentage of all working people.

The key to the pie chart is understanding that the number of degrees in the angular measure of each part of the chart is its appropriate fraction or percentage of 360°. For example, we know that the occupational group "Education, legal, community service, arts, and media" makes up 11% of those employed because there is an 11 in that slice of pie. Thus, the angular sector (the angle of the pie at the center of this occupational group) for the portion representing "Education, legal, community service, arts, and media" measures 11% of 360°, or

$$0.11 \times 360° = 39.6°,$$

and so on. As shown here, pie charts are most often used to show how the whole is divided.

Occupations that have similar job duties are grouped according to the takes that the workers in those occupations perform. This chart shows the aggregated occupational groups from the 2010 Standard Occupational Classification (SOC) system. For example, the computer, engineering, and science group in this chart includes computer and mathematical occupations; architecture, engineering occupations; and life, physical, and social science occupations.

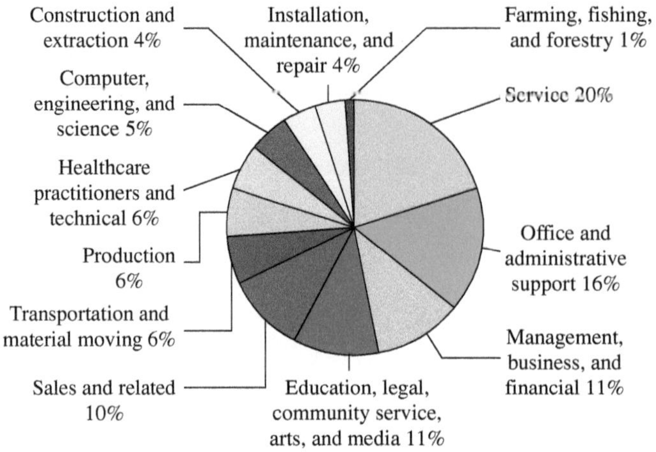

Figure 13.9 Percentage distribution of employment by aggregate occupational group, 2010.

Occupational Outlook Quarterly, Bureau of Labor Statistics.

If the pie chart is drawn in perspective, as if seen from an angle as in Figure 13.10, the central angles are no longer completely accurate. However, the pie chart still gives a good visual understanding of the apportionment of the whole being discussed. Also, in the figure, the pieces of the pie are separated slightly and are in color to produce a more pleasing visual effect.

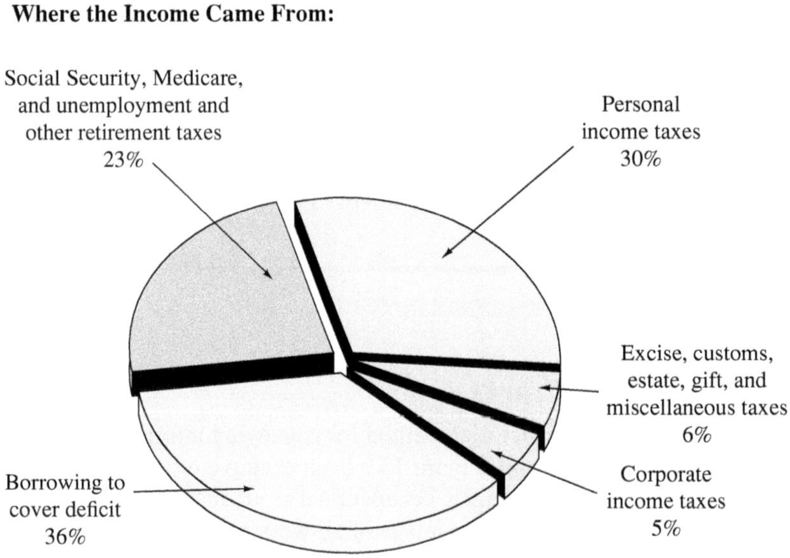

Figure 13.10 Pie chart showing U.S. government sources of revenue for fiscal year 2011.

SOURCE: *2012 IRS Form 1040 instruction booklet*

Pictographs

A **pictograph** uses a set of small figures or icons to represent data and often to represent trends. It is important to be aware that the end-of-year math tests frequently have pictographs in them.

In Figure 13.11, we have a picture of a person at a computer, which presents the **key.** The key, in this case, stands for 10 million computers. For example, in checking the countries, what number of computers are there in Russia? Looking at the pictograph for Russia, there should be about 5.5 times 10 million, or 55 million computers. On the other hand, ask your students what the last "funny little icon" is for Germany. For Germany, it is clear that there is not an exact answer—we can only estimate. Include a table of correct numbers for the ten countries in Figure 13.11.

Figure 13.11
Pictograph of the top ten countries of the world in terms of computer, 2011

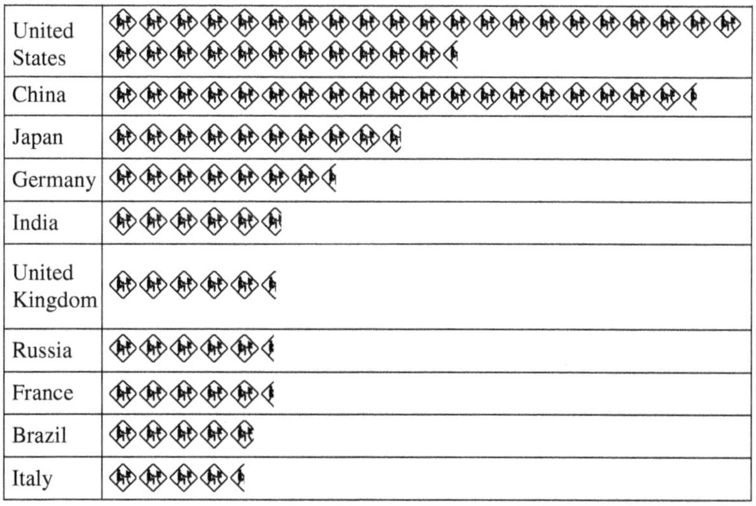

The table for the actual data for the number of computers in the various countries follows:

COUNTRY	UNITED STATES	CHINA	JAPAN	GERMAY	INDIA	UNITED KINGDOM	RUSSIA	FRANCE	BRAZIL	ITALY
Number of Computers (in millions)	310.6	195.1	98.1	71.5	57	54.5	53.5	53.5	48.1	44.7

Data from http://www.c-i-a.com/pr02012012.htm, Calvin Long.

Choosing Good Visualizations

Each of the graphical representations discussed in this chapter is appropriate to summarize and present data so that the reader can visualize frequencies and determine trends. The various representations are more appropriate in some instances than others. A **misleading representation** is a serious distortion with the intent to confuse the reader. For example, the pictograph in Figure 13.12 represents soft drink consumption in Brazil and the United States.

Figure 13.12
Soft drink consumption in Brazil and the United States in 2011 (in liters per person)

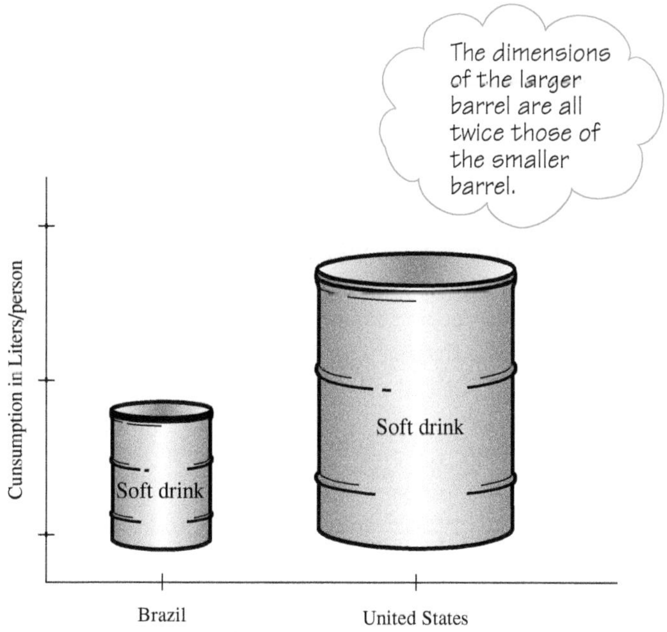

Although the vertical scale accurately indicates that approximately twice as much soft drink per liter was enjoyed in the United States than in Brazil, the pictograph is misleading because the volume of the larger barrel is *8 times* the volume of the smaller barrel. The casual reader is likely to get a badly distorted idea of the relative amount of soft drink consumed in 2011. Of course, that may be precisely what the person who constructed the pictograph intended. Have you ever seen such distortions on television? in the newspaper? in advertisements? Be observant and analytic the next time you see this kind of diagram.

13.1 Problem Set

Understanding Concepts

1. The following scores were obtained on the final examination in an introductory mathematics class of 40 students:

98	80	98	76	79	94	71	45	89	71
62	61	95	77	83	49	65	58	56	89
66	87	74	64	75	58	72	75	48	88
75	51	84	76	95	69	61	69	33	86

(a) After scanning the data, what do you think the "typical" or "average" score is?

(b) Make a dot plot to organize these data.

(c) After looking at the dot plot, what seems to be the "typical" score?

(d) Do you see any scores that seem to be particularly atypical of this data set? Explain.

(e) Write a two- or three-sentence description of the results of the final examination.

2. The students who took the final exam in Problem Solving in Mathematics had the following scores:

42	86	80	90	74	84	86	80	63	92
93	81	95	78	70	41	66	76	87	88
75	88	87	78	89	85	77	87	81	57

(a) Make a dot plot to organize these data.

(b) Make a stem-and-leaf plot of these data.

3. Make a stem-and-leaf plot of the data in problem 1.

4. Make a histogram for the data in problem 1, using the ranges 20–29, 30–39, . . . , 90–99 on the horizontal axis.

5. Make a histogram for the data in problem 2, using the ranges 30–39, 40–49, 50–59, . . . , 90–99 on the horizontal axis.

6. At the same time that heights of the boys at Eisenhower High School were studied, heights of the girls were also studied.

(a) Draw a histogram to summarize the data set that follows, which gives the heights, to the nearest inch, of the 75 girls at Eisenhower High School. Use intervals one unit wide centered at the whole-number values 56, 57, . . . , 74.

57	62	63	64	66
58	62	63	64	66
60	62	63	64	66
60	62	63	65	66
61	62	63	65	66
61	62	63	65	66
61	62	63	65	66
61	63	64	65	66
61	63	64	65	66
61	63	64	65	66
62	63	64	65	67
62	63	64	65	67
62	63	64	65	70
62	63	64	65	70
62	63	64	66	73

(b) Write two or three sentences describing the distribution of the heights of the girls.

7. The scores on the first, second, and third tests given in a class in educational statistics as the term progressed are shown in the accompanying table.

(a) Draw three separate but parallel dot plots for the three sets of scores.

(b) Write a three- or four-sentence analysis of your dot plots suggesting what happened during the term to account for the changing distribution of scores.

First test	92	80	73	74	93	75	76	68	61	76
	83	94	63	74	76	86	82	70	65	74
	83	87	98	77	67	64	87	96	62	64
Second test	52	65	84	91	86	76	73	52	68	79
	88	94	98	84	53	59	63	66	77	81
	94	81	64	56	96	58	64	57	83	87
Third test	97	91	61	67	72	81	63	56	53	59
	43	56	64	78	93	99	84	84	61	56
	73	77	57	46	93	87	93	78	46	87

8. Ms. Smithson earned \$64,000 per year, which she spent as shown in the following table:

Taxes	\$21,000
Rent	\$10,800
Food	\$5,000
Clothes	\$2,000
Car payments	\$4,800
Insurance	\$5,200
Charity	\$7,000
Savings	\$6,000
Misc.	\$2,200

Draw a pie chart to show how Ms. Smithson spent her yearly income.

9. The pie chart of Figure 13.10 shows government sources of revenue for the calendar year 2011. What are the angular sectors for the portion of

(a) personal income taxes?

(b) borrowing to cover deficit?

10. A pie chart showing U.S. government sources of revenue for fiscal year 1991 is given in the 1992 IRS Form instruction booklet.

Where the Income Came From:

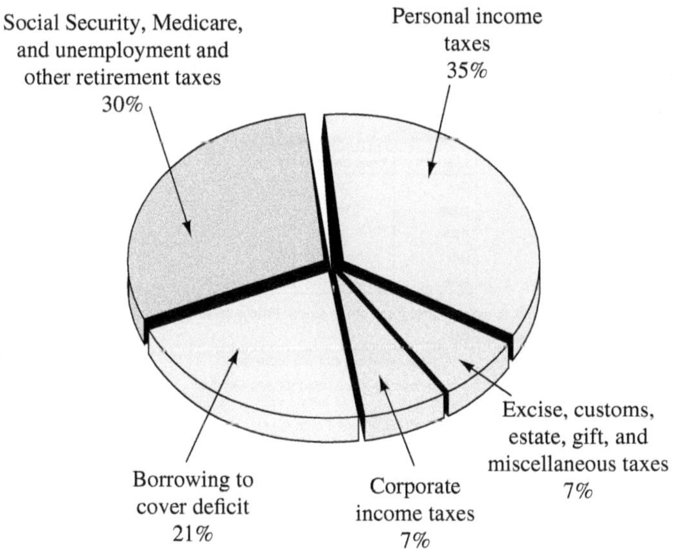

What are the angular sectors for the portion of

(a) personal income taxes?

(b) borrowing to cover deficit?

(c) What do you think the comparisons are between 1991 and 2011? See Figure 13.10.

11. This pie chart indicates how the city of Metropolis allocates its revenues each year:

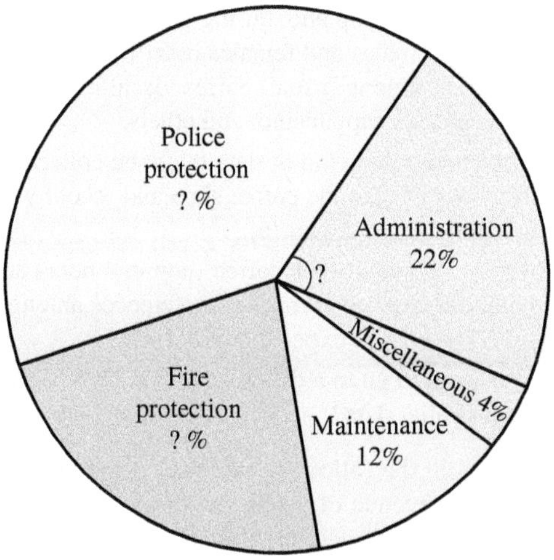

(a) What is the measurement of the central angle of the sector representing administrative expense?

(b) Using a protractor to measure the angle, determine what percent of the city budget goes for police protection.

(c) How does the city's expenditure for maintenance compare with its expenditure for police protection?

(d) How do the expenditures for administration and fire protection compare?

12. This bar graph shows the distribution of grades on the final examination in a class in English literature:

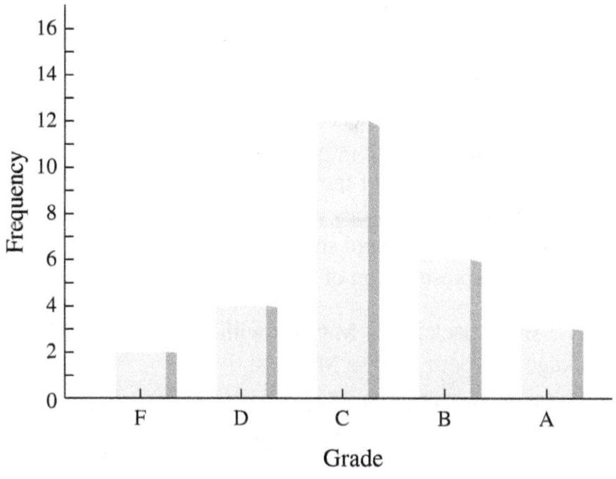

(a) From the bar graph, determine how many students in the class got Cs.

(b) How many more students got Bs than Ds?

(c) What percentage of the students earned As?

13. **(a)** Go to a busy campus parking lot and record the number of cars that are predominately white, black, red, gray, green, and "other."

(b) Make a bar graph to display and summarize your data.

(c) On the basis of part (a), if you were to stand on a busy street corner and watch 200 cars go by, how many would you expect to be predominately white?

(d) If you were on any street corner, how many cars would you expect to be white?

14. (a) Go to a coffee shop and, during the course of an hour, record what males and females order by hearing their preferences among (usual) coffee, decaffeinated coffee, latte, espresso, cappuccino, and others.
(b) Think about what kind of data might be collected at a coffee shop. Go to the coffee shop and record your data in a manner similar to part (a).
(c) If you came back to the coffee shop in 4 hours and again recorded males' and females' preferences among types of coffee, would you expect similar data?
(d) If you were to go to a different coffee shop, would you expect similar data?

15. (a) The data in the following table show the name, year, and revenue of the featured play from 2010 to 2014 at Midtown High School. Construct a bar graph for these data.
(b) From the bar graph, can you give a list, arranged from top to bottom, of the student interest in the plays?

PLAYS AT MIDTOWN HIGH SCHOOL IN A FIVE-YEAR PERIOD

Play	Year	Revenue
Bye Bye Birdie	2010	$4135
Chicago	2011	4572
Singin' in the Rain	2012	4300
West Side Story	2013	8217
Wicked	2014	4045

16. Make a line graph for the histogram of problem 5.

17. The following table shows all employees (in thousands) in education and health services careers. Create a line graph from this table.

All Employees (Thousands) Education and Health Services

Years	2003	2004	2005	2006	2007	2008	2009	2010	2011	2012	2013
All Employees (Thousands)	16,438	16,781	17,177	17,625	18,073	18,605	19,072	19,373	19,705	20,106	20,505

SOURCE OF DATA: U.S. Bureau of Labor Statistics Employment, Hours, and Earnings from the Current Employment Statistics survey (National), Calvin Long.

Into the Classroom

18. Collecting their own data actively engages students in the study of statistical notions, not only heightening student interest but also imparting a sense of meaning and reality to the study of statistics. Name four activities you deem particularly suitable for a class of elementary school students that would involve the collection and representation of data.

19. Buy a small package of M&M's with mixed colors. Open the package and pour out the M&M's.
(a) How many M&M's of each color are in the package?
(b) Make a bar graph of the data from part (a).
(c) Make a pictograph to display the data from part (a).
(d) Would it be reasonable to guess that most packages of M&M's contain about twice as many yellow as green candies?

20. (a) Roll two dice 50 times, and record the number of times (the frequency) you obtained each score.
(b) Draw a bar graph for the data of part (a), showing frequency on the vertical axis and score on the horizontal axis.

21. The accompanying figure is from page 11 of the teacher's edition of *Scott Foresman–Addison Wesley Math Grade 5*, by Randall I. Charles et al., copyright © 2002 Pearson Education, Inc. Reprinted with permission.
(a) The teacher's edition shows a possible correct answer. How many small icons should the plot show for the pine category if the key is such that each icon (small tree symbol) represents ten trees?

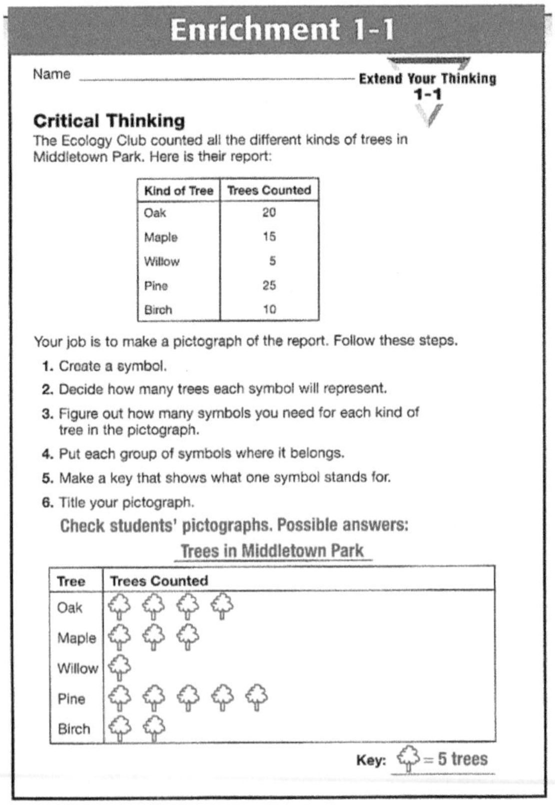

Enrichment 1-1

Name ______________ Extend Your Thinking 1-1

Critical Thinking

The Ecology Club counted all the different kinds of trees in Middletown Park. Here is their report:

Kind of Tree	Trees Counted
Oak	20
Maple	15
Willow	5
Pine	25
Birch	10

Your job is to make a pictograph of the report. Follow these steps.

1. Create a symbol.
2. Decide how many trees each symbol will represent.
3. Figure out how many symbols you need for each kind of tree in the pictograph.
4. Put each group of symbols where it belongs.
5. Make a key that shows what one symbol stands for.
6. Title your pictograph.

Check students' pictographs. Possible answers:

Trees in Middletown Park

(b) Make your own pictograph of these data, with each icon representing two trees.

22. This exercise is best done as a class activity. Our measure of a "yard" was originally the length from the tip of the nose to the fingertip of the outstretched arm of an English king. Working in small groups, use a tape measure to measure this length to the nearest inch for each student in class. Record the data on the chalkboard (and for later use in Problem Set 13.2) in two sets—one set for men and one set for women. Divide the class into several small groups.

(a) Members of one or two small groups each make a dot plot for the data in each of the two sets.

(b) Members of one or two small groups each make a double stem-and-leaf plot for the data in each of the two sets.

(c) Members of one or two small groups each make a histogram for the data in each of the two sets.

(d) Members of one or two small groups each make a frequency polygon for the data in each of the two sets.

(e) Members of one or two small groups each make a pictograph for the data in each of the two sets.

(f) As a class, discuss the various representations of the data. Which ones seem most informative? What conclusions are suggested regarding the length of a "yard"? Discuss briefly.

Responding to Students

23. In this pictograph about books read in September, the box symbol represents two books read:

Fred	▫ ▫ ▫ ▫ ▫
Kristen	▫ ▫ ▫
Nancy	▫ ▫ ▫ ▫
James	▫
Bill	▫ ▫

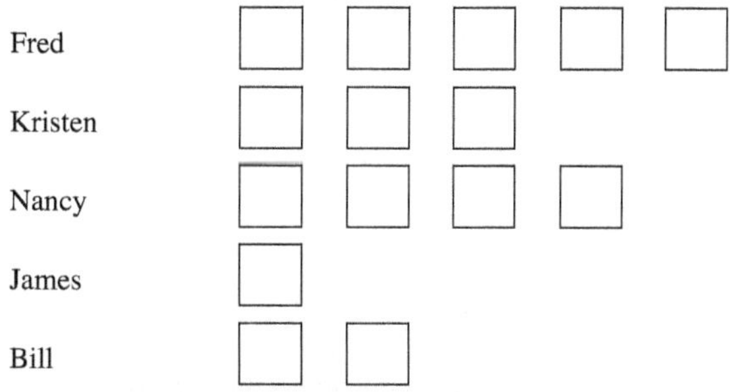

Sarita was asked the following questions about the graph:

1. Who read exactly four more books than James?
2. Who read the most books and how many did he or she read?

For the first question, Sarita studied the graph and answered that Fred read exactly four more books than James. In answering the second question, Sarita said that Fred read the most because he read five books.

(a) For both questions, what is Sarita misinterpreting in regard to the pictograph?

(b) What are the correct answers to the two questions?

(c) How would you guide Sarita to answer similar questions in the future?

24. Jeremy was shown the following graph about the number of fish caught in one week at a local lake:

Fish Caught in One Week

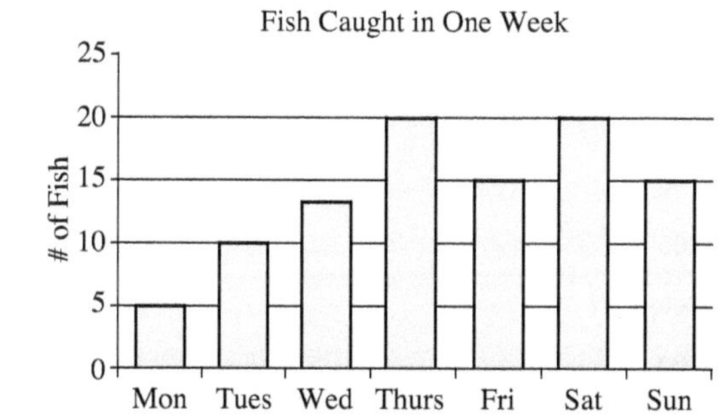

He was asked the following questions:

1. How many more fish were caught on the weekend (Saturday and Sunday) than on Monday?

2. What two days together equal the total number of fish caught on Friday?

After studying the graph:

1. Jeremy said that the answer to question 1 was five, because if you take the amount from Monday and put it with Saturday and Sunday, you get two full bars on the graph.

2. Jeremy answered that Sunday equaled the number of fish on Friday.

(a) How would you help Jeremy understand what was being asked in question 1? How would you guide him toward an answer?

(b) What is Jeremy misinterpreting in regard to question 2? How would you guide him toward an answer?

25. Stacy is working on a science project collecting data on the size of the circumference of her classmates' heads. She wants to represent her data graphically and is trying to decide whether to use a bar graph or a histogram. She says, "Well, aren't they the same?" Help Stacy create a graphic organizer to represent the similarities and differences between the two graphs.

Thinking Critically

Data are often presented in a way that confuses, or even purposely misleads, the viewer.

26. (a) Discuss briefly why the television evening news might show histogram (A) rather than (B) in reporting stock market activity for the last seven days. Is one of these histograms misleading? Why or why not?

(A)
Seven-Day Stock Market Trend

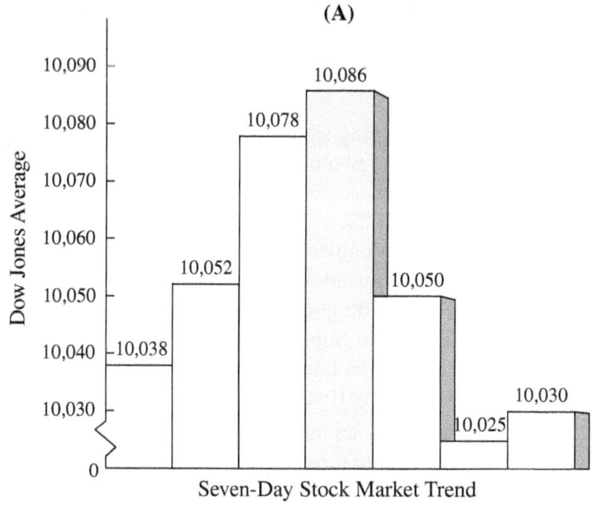

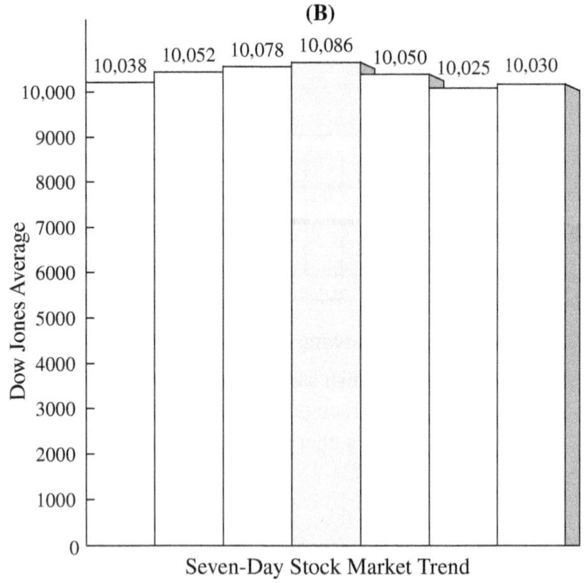

(**b**) What was the percentage drop in the Dow Jones average from the fourth to the fifth day as shown in the preceding histograms? Should an investor worry very much about this 36-point drop in the market?

(**c**) Was the Dow Jones average on day 5 approximately half of what it was on day 4, as suggested by histogram (A)?

27. Longlife Insurance Company printed a brochure with the following pictographs showing the growth in company assets over the ten-year period 2003–2012:

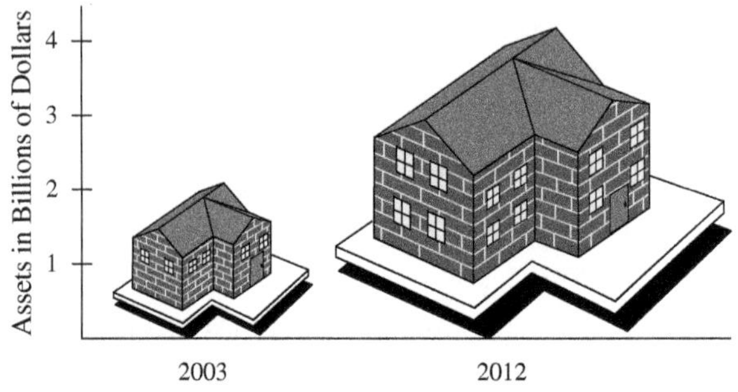

(**a**) Do the pictographs accurately indicate that the assets were \$2 billion in 2003 and \$4 billion in 2012, or might one assume from the pictographs that the assets were actually much greater in 2012 than in 2003? Explain briefly.

(**b**) The larger building shown is just twice the height of the smaller, and the two buildings are similar as geometrical drawings. Do these drawings accurately convey the impression that the assets of Longlife Insurance Company just doubled during the ten-year period? Explain your reasoning. What is the ratio of the volume of the large building to the volume of the small building? (*Suggestion:* Suppose both buildings were rectangular boxes, with the linear dimensions of the second just twice those of the first.)

(**c**) Would it have been more helpful (or honest) to print the actual asset value for each year on the front of each building?

Making Connections

28. The U.S. Department of Commerce, Census Bureau, has a pie chart from March 2011 entitled "Highest level of education attained by persons 25 through 29 years old."

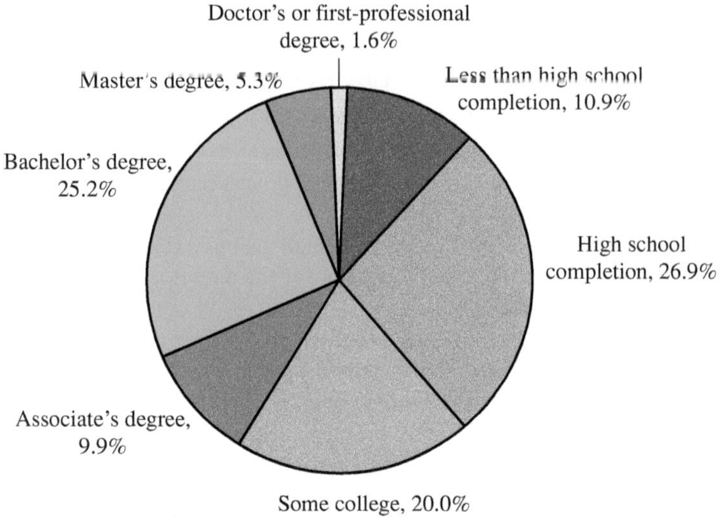

Source: *U.S. Department of Commerce, Census Bureau,* March 2011.

(**a**) What is the total percentage of those in the 25 to 29 year old category who have a degree?

(**b**) What is the total percentage of students who went on to at least some college?

(**c**) What result do you get if you were to add all of the percentages on the pie chart and why is it not 100%.

29. The following graph indicates Buzz Technology's (BT) net income or loss for the years 2010 through 2014 and first three months of 2015.

(**a**) Supposing that BT's income continued to come in at that rate, what would you forecast its net income for 2015 to be?

(**b**) Draw a line graph (including your estimate for 2015) for BT's income for 2010 through 2014.

(**c**) About what percentage increase in net income did BT experience in 2014 over 2012?

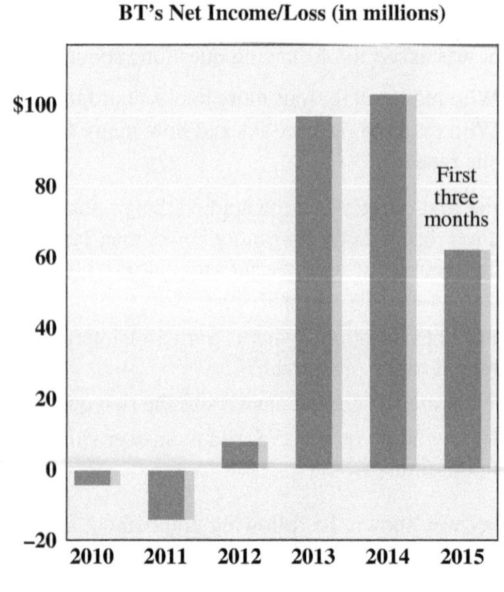

(d) If the assumption in part (a) proved to be correct, what would BT's percentage increase in net income in 2015 over 2013 be?

30. Mathematics has, unfortunately, been viewed traditionally as a "male" subject, which it shouldn't be. Although presently about half of the mathematics undergraduate degrees in the United States are awarded to women, the situation is quite different at the doctoral level, and there are a number of nationwide initiatives that are focused on this issue. The accompanying graphs show the number and percentage of mathematics Ph.D.s earned by women at U.S. schools.

Number of Mathematics Ph.D.s Earned by Women at U.S. Schools

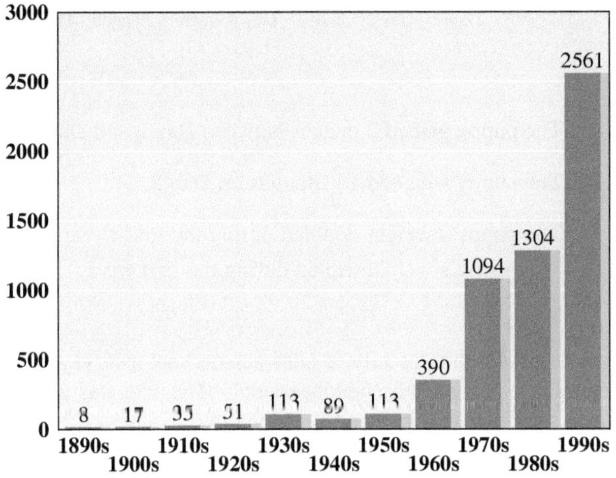

Approximate Percentage of U.S. Mathematics Ph.D.s Earned by Women

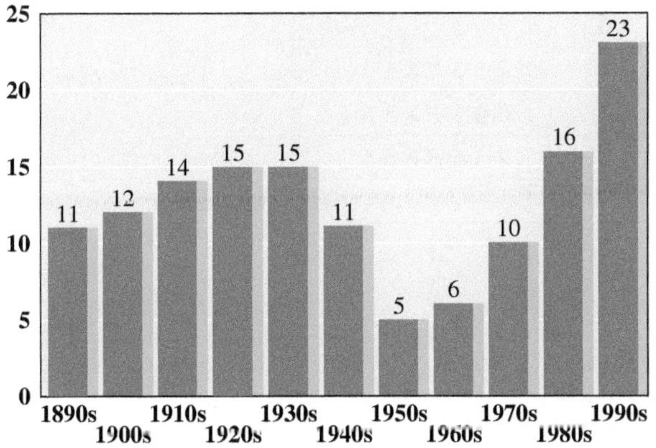

(a) From the graphs, determine how many times as many women earned Ph.D.s in mathematics during the 1990s as during the 1890s.

(b) Determine how many men earned Ph.D.s during the 1890s and during the 1990s.

(c) Determine the ratio of the number of men to the number of women earning Ph.D.s in mathematics during the 1890s and during the 1990s.

31. Consider the SAT scores recorded here:

Student	Verbal Score	Math Score
Dina	502	444
Carlos	590	520
Rosette	585	621
Broz	487	493
Colleen	585	602
Dieter	481	572
Darin	605	599
Luana	547	499

(a) Make a bar graph to summarize the preceding data for verbal scores.

(b) Make a double bar graph to summarize the data, with one of each pair of bars for verbal scores and one for math scores.

32. For fiscal year 2011, federal expenditures were divided as follows:

Social programs—23%

Physical, human, and community development—8%

Net interest on debt—6%

Defense, veterans, and foreign affairs—24%

Social Security, Medicare, and other retirement—37%

Law enforcement and general government—2%

(a) Make a pie chart that reflects these data.

(b) Make a bar graph that reflects these data.

33. Make a line graph that graphically displays these data:

Population of Washington, in Millions	2.10	2.32	2.63	2.83	3.12	3.45	3.80	4.18	4.81	5.76	6.72
Year	1960	1965	1970	1975	1980	1985	1990	1995	2000	2005	2010

State Assessments

Note: There are many SA questions because of the variety and level of depth of these problems.

34. (Grade 3)
This graph shows the results of a classroom vote on favorite ice cream flavors. How many more students voted for vanilla than for chocolate?

Favorite Ice Cream

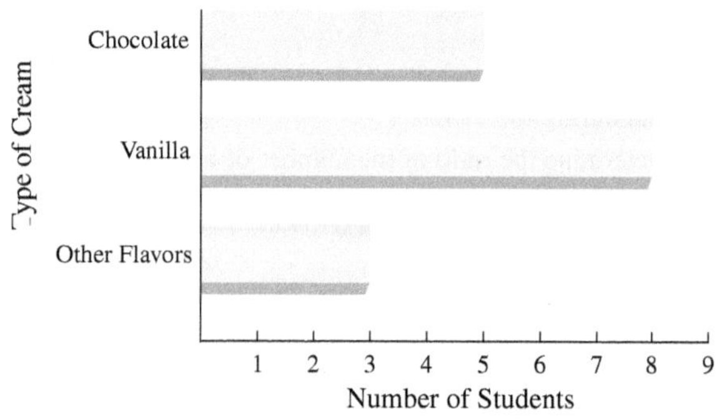

A. 3
B. 4
C. 5
D. 6

35. (Grade 4)
Warren was looking at the books on the self of Mr. Gardiner's fourth grade class. He put a cover on each that said "math" just for fun. He thought about each kind of book to see what kind of book Mr. Gardiner already had. Warren had seen fourteen science books.

Number of Books of Four Types

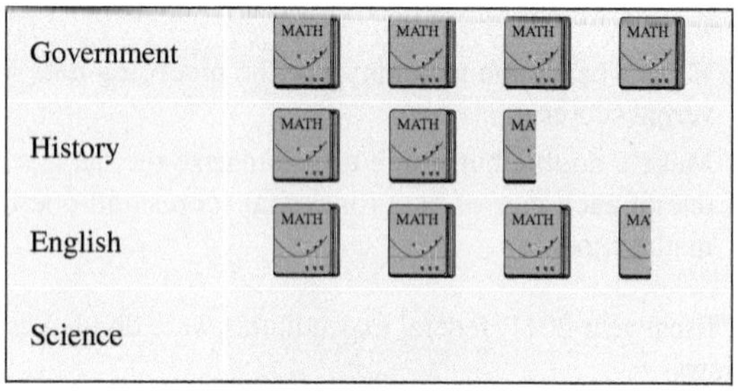

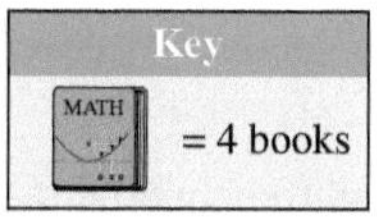

What should the picture graph show for the science book?

A. 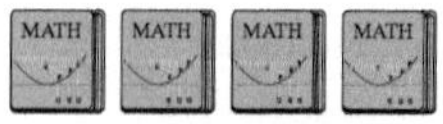

B. 

C. 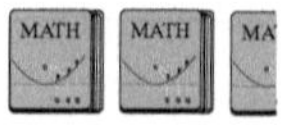

D. 

36. (Massachusetts, Grade 4)
Use the information in the line graph below to answer the question.

The graph shows the weight gain of a puppy during its first week of life. Which is NOT true about the weight of the puppy?

Weight of Puppy During First Week

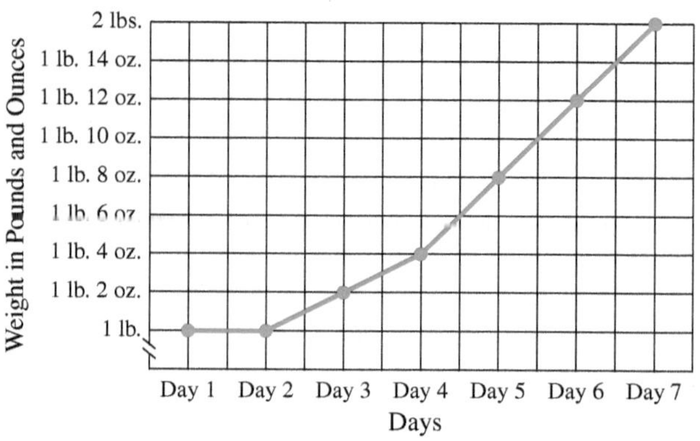

A. The puppy gained 2 ounces between Day 3 and Day 4.

B. The puppy weighed $1\frac{1}{2}$ pounds on Day 5.

C. The puppy's weight doubled during the first week.

D. The puppy's weight tripled during the first week.

37. (Grade 6)
During the month of July, a sales person kept a survey of the number of people who bought sandals. The data was written as a stem-and-leaf plot and is in the table.

Stem	Leaf
0	8
1	2 5 5 5 6 6
2	0 0 1 1 1 5 5 5 5
3	2 5 6 6 6 6 6
4	
5	8

Key
2|5 = 25

Which of the following statements is *true* according to the data in the stem-and-leaf plot?

A. The number of pairs of sandals sold each day was between 0 and 7.

B. The stem-and-leaf plot displays 365 days of sales.

C. The median for the data is 36.

D. The mode for the data is 36.

38. (Grade 6)
Drew has 16 textbooks in his bookstore. This bar graph shows the number of each type of book.

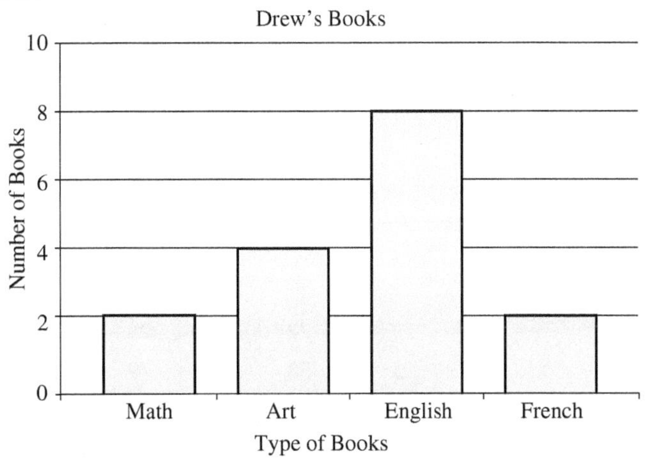

Which circle graph best shows the types of textbooks Drew has in his bookcase?

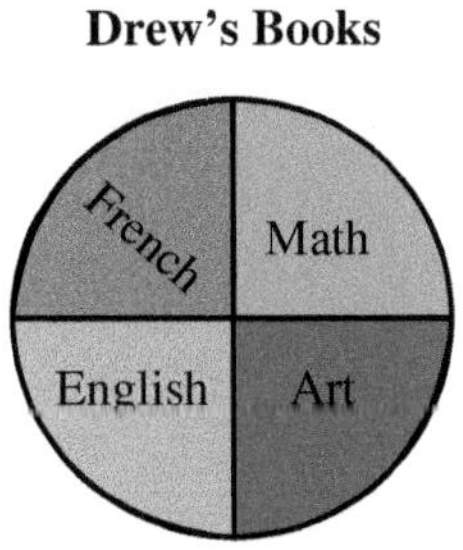

A

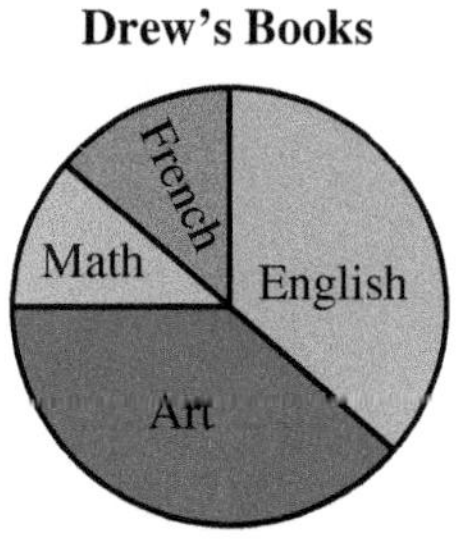

C

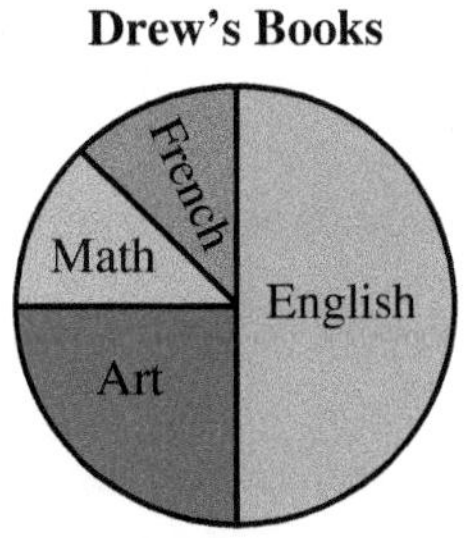

B

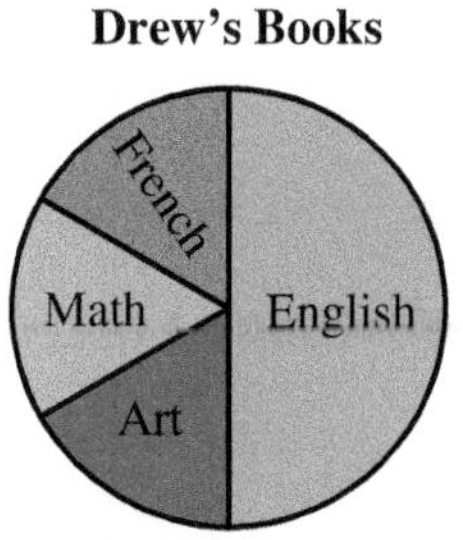

D

13.2 Measuring the Center and Variation of Data

In the previous section, we began to organize and visualize data. Here, we will describe numerically some properties of the data set. We now focus on the *center* and the *spread* of the data set.

Measures of Central Tendency

In statistics, there are three different ways to describe the center of a data set: the *mean* or *average*, the *median*, and the *mode*. These are called measures of **central tendency** of the data set. Consider the following different data sets, R, S, and T, representing grades on tests, and their corresponding dot plots:

	71	79	70	74	46	81	72	77
R	71	77	72	86	79	67	77	61
	76	76	92	79	72	77	63	76

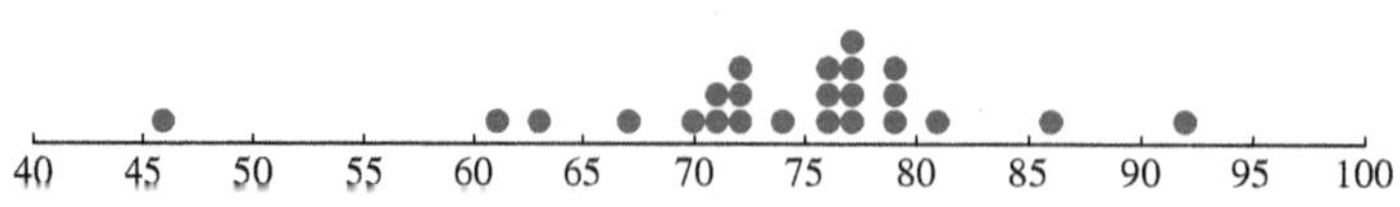

Dot plot for R

	72	77	75	75	67	76	69	76	71
S	68	77	79	82	73	69	76	68	
	69	71	78	72	79	74	73	73	

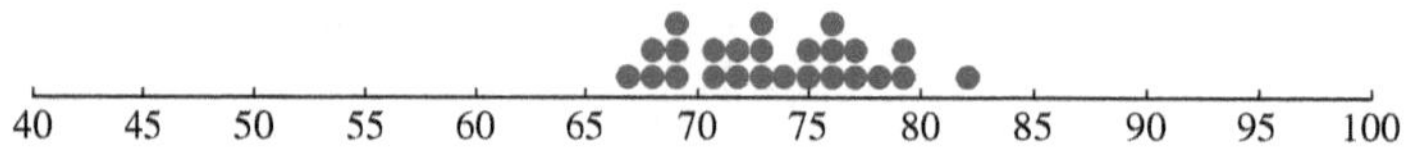

Dot plot for S

	74	84	40	49	91	81	75	
T	79	40	61	70	40	74	85	89
	86	82	40	85	45	96	86	85

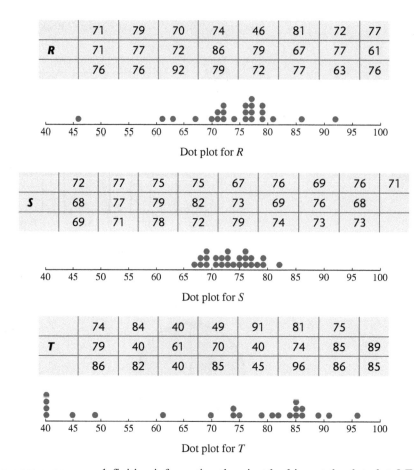

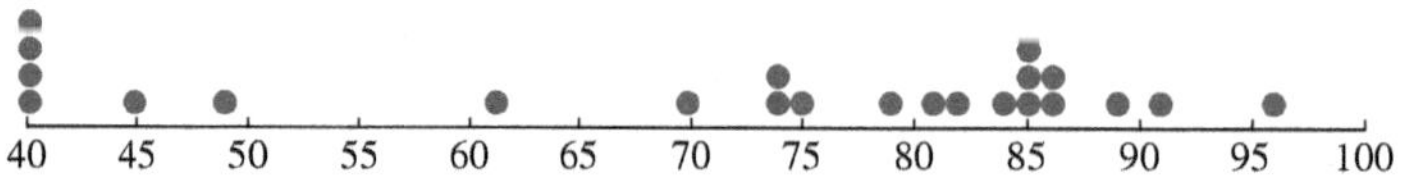

Dot plot for T

What if we want even more definitive information than just looking at the dot plots? For example,

- What is the "typical" grade for the class?
- How did most of the students do?
- Did many of the students perform markedly differently from the bulk of the class?

Let's use dot plots, such as R, S, and T, to help us begin to answer the questions.

1. The scores in data set R seem to cluster about 76, even though they range all the way from 46 to 92 and are generally rather widely spread. Apparently, some of the students did very well, while others did quite poorly. If we had to choose a single grade as typical of the entire class, it would probably be about 76.
2. The scores in data set S are much less spread out than those of R, ranging only from 67 to 82. All the students did reasonably well, with none outstandingly good or outstandingly poor. The grades seem to cluster about 73, a score reasonably typical of the entire class.
3. The scores in data set T are very widely spread, ranging all the way from 40 to 96. Clearly, many students did very poorly, while a substantial number did quite well. It is very difficult to select a single grade as typical. If we ignore the worst grades, we might choose 85 as typical. But there are many scores that differ widely from 85.

These examples illustrate two properties of a data set:

- the typical, or central, value of the data, and
- the dispersion, or spread, of the data about the central value.

Before we identify and quantify these ideas, we will define a center of a data set in three different ways (mean, median, and mode).

The Mean

The dot plots just considered gave us an intuitive, but not very precise, notion of a typical, or central, value of a set of data. One very useful and precisely defined central value is the *mean,* frequently called the *arithmetic mean* or *average*. A simple manipulative device for introducing this notion to students that is quite independent of, and different from, the dot plots just considered is a simple set of blocks. For example, consider the data 7, 5, 7, 3, 8, and 6. Arrange blocks into six stacks containing 7, 5, 7, 3, 8, and 6 blocks, as shown, and ask what the average height of all the stacks is:

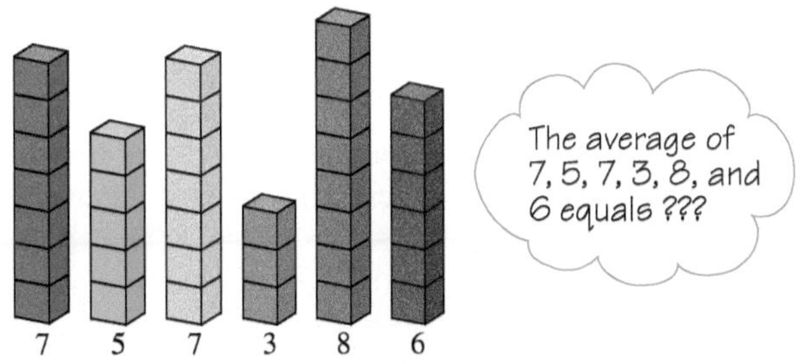
The average of 7, 5, 7, 3, 8, and 6 equals ???
7 5 7 3 8 6

If all the stacks were the same height, the answer would be obvious—it would be their common height. This suggests that we could answer the question by moving blocks from taller stacks to shorter ones in an effort to even them up as shown next. The six stacks each have height 6, suggesting that the average height is 6:

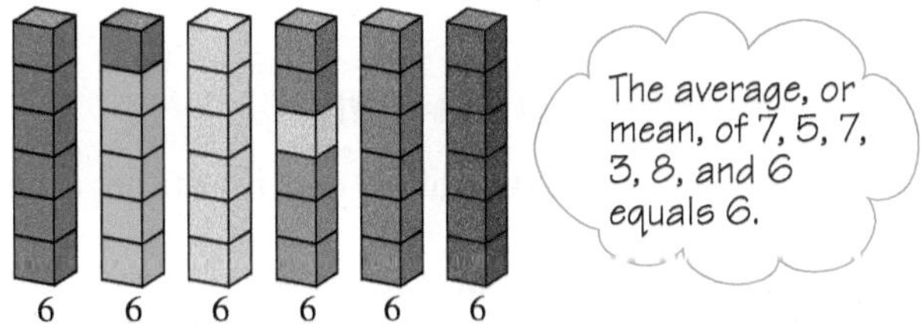
The average, or mean, of 7, 5, 7, 3, 8, and 6 equals 6.
6 6 6 6 6 6

Arithmetically, the number of blocks in the two arrangements has not changed, so

$$6 \cdot 6 = 7 + 5 + 7 + 3 + 8 + 6 \quad \text{and} \quad 6 = \frac{7 + 5 + 7 + 3 + 8 + 6}{6}.$$

Thus, the average height of the original stacks is found by finding the sum of their heights and dividing by the number of stacks. This naturally leads to the following definition:

DEFINITION The Mean of a Set of Data

The **mean,** or **average,** of a collection of values is $\bar{x} = \frac{S}{n}$, where S is the sum of the values and n is the number of values. The symbol $\bar{x}$ should be read as "x bar."

For the data set R, we find the mean by adding all scores and dividing by 24, the number of scores. Thus,

$$\bar{x} = (71 + 79 + 70 + 74 + 46 + 81 + 72 + 77 + 71 + 77 + 72 + 86$$
$$+ 79 + 67 + 77 + 61 + 76 + 76 + 92 + 79 + 72 + 77 + 63 + 76)/24$$

which is approximately 73.8.

This is reasonably close to the score that "looked" average, 76. Actually, 73.8 is smaller than expected, which shows that the mean of a data set is sensitive to atypical data values such as the datum point 46. The mean of the data in R with 46 omitted is 75—very close to our guess. At any rate, for most data sets, data values normally cluster reasonably close to the mean.

Let's consider other frequently used measures of central tendency or central value. For the data sets S and T, we find that the means are, respectively, 73.6 and 71.2. For S, the mean gives a very good estimate of what we intuitively felt was the typical or central value. For T, the mean of 71.2 seems unduly low, and this is again a reflection of the fact that the mean can be strongly affected by the presence of extremely atypical values like 40, 40, 40, 40, 45, 49, and even 61.

Without these values, the mean is a much more acceptable 82.6. At any rate, it is usually the case for most data sets that data values cluster reasonably closely about the mean. This is particularly true if we have criteria for deciding when data values are too atypical so that we may delete them from consideration. We develop such a criterion a little later, but, for now, we consider other frequently used measures of central tendency or central value.

The Median

The **median** of a collection of values is the middle value in the collection when the values are arranged in order of increasing size, or the average of the two middle values in case the number of values is even.

DEFINITION The Median of a Set of Data

Let a collection of n data values be written in order of increasing size. If n is odd, the **median,** denoted by $\hat{x}$, is the middle value in the list. If n is even, $\hat{x}$ is the average of the two middle values. The symbol $\hat{x}$ should be read as "x hat."

EXAMPLE 13.3

Determining a Median

Determine the median of the data in data set R on page 686.

Solution

The scores in R are arranged in order in the line plot of R. Since there are 24 scores, the median is the average of the twelfth and thirteenth scores. We see that $\hat{x} = 76$.

Note that the median in this example closely approximates not only the mean but also our intuitive idea of the middle value of the scores.

It follows from the definition that the median is a data value if the number of data values is odd and is *not* necessarily a data value if the number of values is even. Thus, the median of the nine scores

$$24, 25, 25, 27, 29, 31, 32, 34, 37$$

is 29, the fifth score, while the median of the ten scores

$$42, 42, 43, 44, 44, 46, 47, 47, 47, 49$$

is 45, the average of the two middle scores. The median is not affected by the existence of extremely atypical value unless there are many extreme values.

The Mode

Another value often taken as "typical" of a set of data is the value occurring most frequently. This value is called the **mode.**

DEFINITION A Mode of a Set of Data

A **mode** of a collection of values is a value that occurs the most frequently. If two or more values occur equally often and more frequently than all other values, there are two or more modes.

From the definition, the data set S has three modes, namely, 69, 73, and 76, since each of those numbers appears three times and no data point occurs more than three times. Although the mode is not an important part of modern statistics, it is a part of the curriculum and is on statewide assessment tests of mathematics in grades K–8. The mode is not affected by the existence of extremely atypical values.

EXAMPLE 13.4

Determining Means, Medians, and Modes

Determine the mean, median, and mode for each of the data sets *R*, *S*, and *T* on page 686, and discuss which measures are most representative of the respective data sets.

Solution

Let $\bar{x}_R$, $\hat{x}_R$, $\bar{x}_S$, $\hat{x}_S$, $\bar{x}_T$, and $\hat{x}_T$ denote the means and medians of *R*, *S*, and *T*, respectively. We have already seen that $\bar{x}_R = 73.8$ and that $\hat{x}_R = 76$. Since the mode is the most frequently occurring score, the mode of *R* is 77. In this case, all three measures are reasonably representative of the scores in data set *R*.

Since *S* has 25 scores,

$$\bar{x}_S = (72 + 77 + 75 + 75 + 67 + 76 + 69 + 76 + 71 + 68 + 77 + 79 + 82 + 73 + 69 + 76 + 68 + 69 + 71 + 78 + 72 + 79 + 74 + 73 + 73)/25$$

which is about 73.6.

Also, $\hat{x}_S$ is the middle, or thirteenth, score. Thus, counting on the line plot, we find that $\hat{x}_S = 73$. Finally, *S* has three modes—69, 73, and 76—since each occurs three times and more often than any other score. We observe that the mean, the median, and the middle mode all seem to reasonably represent the set of scores in *S*.

Finally, since *T* contains 23 scores,

$$\bar{x}_T = (74 + 84 + 40 + 49 + 91 + 81 + 75 + 79 + 40 + 61 + 70 + 40 + 74 + 85 + 89 + 86 + 82 + 40 + 85 + 45 + 96 + 86 + 85)/23$$

which is approximating 71.2,

$$\hat{x}_T = 79,$$

and the mode of *T* is 40. As noted earlier, *T* is difficult to characterize. The mean, $\bar{x}_T$, is strongly affected by the several very low scores and so does not seem to represent the data set fairly. Similarly, the mode of 40 is clearly not representative. The median seems to be the most representative value of the set *T*.

EXAMPLE 13.5

Determining an Average

All 12 players on the Uni Hi basketball team played in their 78-to-65 win over Lincoln. Jon Highpockets, Uni Hi's best player, scored 23 points in the game. How many points did each of the other players average?

Solution

Understand the Problem

The problem is to determine averages when we are not explicitly given the data values. We know that Uni Hi scored 78 points, Jon Highpockets scored 23 of the points, and all 12 players on the team played in the game.

Devise a Plan

Since the average score for each of the 11 players other than Jon is the sum of their scores divided by 11, we must determine the sum of their scores.

Carry Out the Plan

Since Uni Hi scored a total of 78 points and Jon scored 23, the total for the rest of the team must have been $78 - 23 = 55$ points. Therefore, the average number of points for these players is $55 \div 11 = 5$.

Look Back

The solution depended on knowing the definition of *average*. The real question was how many points not made by Jon were scored by Uni Hi's team and how many such players there were. Those figures and, hence, the solution were easily obtained by subtraction. Note that we did not need to know (and, in fact, can't figure out) what each player scored.

EXAMPLE 13.6

Determining a Typical Value for a Set of Data

The owner-manager of a factory earned \$850,000 last year. The assistant manager earned \$48,000. Three secretaries earned \$18,000 each, and the other 16 employees each earned \$27,000.

(a) Prepare a dot plot of the salaries of the people deriving their income from the factory.
(b) Compute the mean, median, and mode of the salaries of those working in the factory.
(c) Which is most typical of these salaries—the mean, median, or mode?

Solution

(a) The dot plot is shown here:

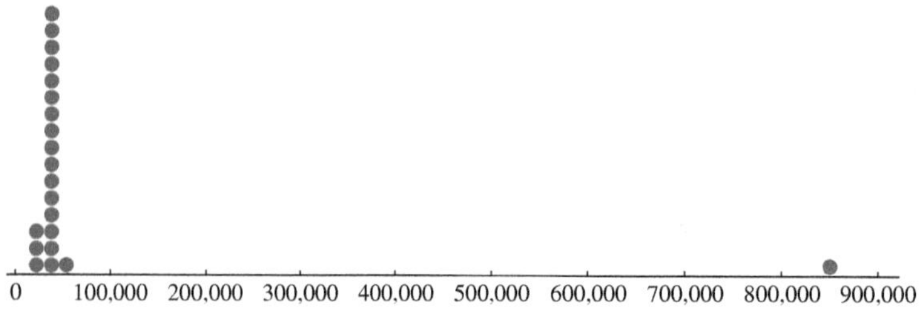

(b) The mean is

$$\bar{x} = \frac{18{,}000 + 18{,}000 + 18{,}000 + 27{,}000 + \cdots + 27{,}000 + 48{,}000 + 850{,}000}{21}$$

which is about \$65,905.
The median, $\hat{x}$, is the 11th in the ordered list of salaries. Thus,

$$\hat{x} = \$27{,}000.$$

The mode is the most frequently occurring salary. Thus,

$$\text{mode} = \$27{,}000.$$

(c) The mean is clearly not typical of the salary most workers at the factory earn. The value of $\bar{x}$ is unduly affected by the huge salary earned by the owner-manager. Here, the median and mode are the same and are more typical of salaries of those deriving their income from the factory, since \$27,000 is the salary of 16 of the 21 people. Here the most typical value is the mode. That the mode and median here are equal is incidental.

While the mean is the most commonly used indicator of the typical value of a data set, the preceding example makes it clear that this choice can be quite misleading. As will be seen in the problem set for this section, it is easy to construct some examples in which the median is the most typical value and in others, it is the mode.

Measures of Variability

The most useful analysis of data would reveal both the center (typical value) and the *spread,* or *variability,* of the data. We now consider how the spread of data is determined and will look at measures of variability. The **range** of the data set could be one way to see how the data points vary. However, the range just takes the largest data point and subtracts the smallest—nothing in between helps us better understand the spread of the data. The range is not a measure of variability, only using the lowest and highest data points. Said another way, there are many different possibilities for the data points between the lowest and highest points. The range certainly tells something about how the data occur, but it is often misleading, particularly if the data set contains a few extremely low or high values that are quite atypical of most of the other values.

Another measure is the **midrange** (or **mid-extreme**), which is the average of the lowest and highest data points. It isn't used in statistics much, but is taught in elementary school. The midrange of the data set R is $\frac{46 + 92}{2} = 69$.

A better understanding of variability uses *quartiles* and, later in this section, the *standard deviation* of a data set. Speaking casually, quartiles divide the data set into four sections, each of which contains, in increasing order, about one-quarter of the data. More precise definitions follow:

DEFINITION Upper and Lower Quartiles

Suppose a set of data is in order of increasing size. Let the number of data values, n, be written as $n = 2r$ when n is even, or $n = 2r + 1$ when n is odd, for some integer r. In either case, the **lower quartile,** denoted by Q_L, is the median of the first r data values. The **upper quartile,** denoted by Q_U, is the median of the last r data values.*

EXAMPLE 13.7

Determining Quartiles

For the four given data sets of this problem, determine the median, $\hat{x}$, and the lower and upper quartiles, Q_L, and Q_U. Note that the resulting values sometimes are and sometimes are not data values.

(a) $A = \{12, 7, 14, 15, 9, 11, 10, 11, 0, 8, 17, 5\}$
(b) $B = \{27, 14, 13, 12, 26, 22, 24, 22, 23, 19, 10, 19, 22\}$
(c) $C = \{16, 22, 20, 15, 12, 14, 16, 14, 21\}$
(d) $D = \{32, 26, 29, 25, 26, 27, 29, 28, 29, 25\}$

Solution

(a) First, we order the values in A to obtain:

$$\begin{array}{cccccccccccc} 0 & 5 & 7 & 8 & 9 & 10 & 11 & 11 & 12 & 14 & 15 & 17 \\ & & & \uparrow & & & \uparrow & & & \uparrow & & \\ & & & Q_L & & & \hat{x} & & & Q_U & & \end{array}$$

Since there are 12 data values, the median is the average of the sixth and seventh values; that is, $\hat{x} = \dfrac{10 + 11}{2} = 10.5$. Since the number of data points is even, we have $2r = 12$, or $r = 6$. The lower quartile, Q_L, is the median of the first six data values: $\dfrac{7 + 8}{2} = 7.5$. Also, $Q_U = \dfrac{12 + 14}{2} = 13$ is the median of the last six data values.

(b) The ordered list for B is as follows:

$$\begin{array}{ccccccccccccc} 10 & 12 & 13 & 14 & 19 & 19 & 22 & 22 & 22 & 23 & 24 & 26 & 27 \\ & & & \uparrow & & & \uparrow & & & \uparrow & & & \\ & & & Q_L & & & \hat{x} & & & Q_U & & & \end{array}$$

The middle data value is $\hat{x} = 22$. The number of data points is odd, so $2r + 1 = 13$, and we have $r = 6$. Thus, $Q_L = \dfrac{13 + 14}{2} = 13.5$ is the median of the first six data values preceding $\hat{x}$. The median of the last six data values is $Q_U = \dfrac{23 + 24}{2} = 23.5$.

(c) The ordered set for C is:

$$\begin{array}{ccccccccc} 12 & 14 & 14 & 15 & 16 & 16 & 20 & 21 & 22 \\ & & \uparrow & & \uparrow & & \uparrow & & \\ & & Q_L & & \hat{x} & & Q_U & & \end{array}$$

*The precise definition of quartiles is not entirely standardized. For example, the lower and upper quartiles for a set of $2r + 1$ data values are often taken as the medians of the first and last $r + 1$ (rather than r) values, respectively, in the ordered list.

Since there are nine data values, $\hat{x} = 16$, the middle data value. Also, $2r + 1 = 9$, so $r = 4$, $Q_L = \dfrac{14 + 14}{2} = 14$ is the median of the first four data values and $Q_U = 20.5$ is the median of the last four data values.

(d) The ordered data set for D is as follows:

$$\begin{array}{cccccccccc} 25 & 25 & 26 & 26 & 27 & 28 & 29 & 29 & 29 & 32 \\ & & \uparrow & & & \uparrow & & \uparrow & & \\ & & Q_L & & & \hat{x} & & Q_U & & \end{array}$$

Since there are ten data points, $\hat{x}$ is the average of the fifth and sixth values, so $\hat{x} = 27.5$. Also, since $2r = 10$, we have $r = 5$. Thus, $Q_L = 26$ is the third of the first five data points and $Q_U = 29$ is the third of the last five data points.

It follows from the definition that approximately 25% of the data values are less than or equal to Q_L, approximately 25% lie between Q_L and $\hat{x}$, approximately 25% lie between $\hat{x}$ and Q_U, and approximately 25% are greater than or equal to Q_U. The difference between Q_L and Q_U, the **interquartile range,** provides a good measure of the spread of the data.

DEFINITION Interquartile Range

The **interquartile range, IQR,** of a data set is the difference between the upper and lower quartiles, or $\text{IQR} = Q_U - Q_L$.

If a data value falls below Q_L by more than $1.5 \cdot \text{IQR}$ or above Q_U by more than $1.5 \cdot \text{IQR}$, it is called an **outlier.** Thus, *outlier* is the term applied to those values we saw earlier that seem to be atypical of the values in a data set.* The study of outliers is an active research area with questions such as whether outliers matter, how they came about, and more.

DEFINITION Outlier

An **outlier** in a set of data is a data value that is *less than* $Q_L - (1.5 \cdot \text{IQR})$ or *greater than* $Q_U + (1.5 \cdot \text{IQR})$.

EXAMPLE 13.8

The Median, the Quartiles, the Interquartile Range, and the Outliers

Determine the median, the quartiles, the interquartile range, and the outliers for data set R on page 686.

Solution

The values in R are ordered in the dot plot on page 686. Since R contains 24 data values, the median is the average of the twelfth and thirteenth data values; that is, $\hat{x} = \dfrac{(76 + 76)}{2} = 76$. Also, $2r = 24$, so $r = 12$ and Q_L and Q_U are the medians of the first and last 12 data values, respectively. Thus, $Q_L = \dfrac{(71 + 71)}{2} = 71$ and $Q_U = \dfrac{(77 + 79)}{2} = 78$. Therefore, the interquartile range is $\text{IQR} = 78 - 71 = 7$. Finally, since

$$Q_L - 1.5 \cdot \text{IQR} = 71 - (1.5 \cdot 7) = 71 - 10.5 = 60.5$$

and

$$Q_U + 1.5 \cdot \text{IQR} = 78 + (1.5 \cdot 7) = 78 + 10.5 = 88.5,$$

it follows that the values 46 and 92 are outliers.

*As with the median, the definitions of interquartile range and outlier are not entirely standardized. The choices made here are among the most common.

Symbolically, if the 24 points shown represent the ordered data values in R, then Q_L, $\hat{x}$, Q_U, the interquartile range, and the outliers are as follows:

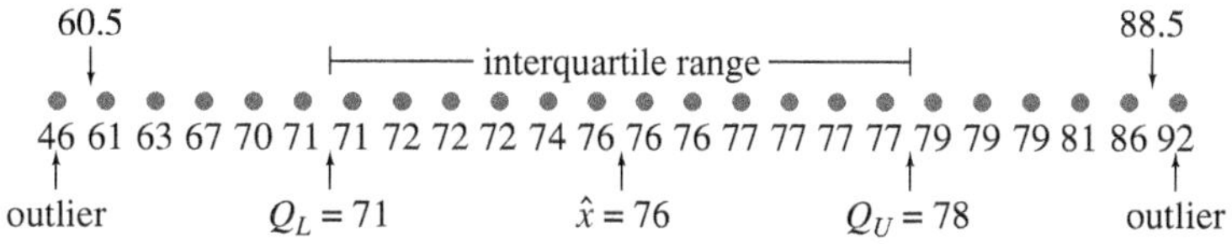

Box Plots

The least and greatest scores, or the *extremes,* along with the lower and upper quartiles and the median, give a concise numerical summary, called the **5-number summary,** of a set of data. For instance, since the median of the data in Example 13.9 is 76 and the **extremes** are 46 and 92, the 5-number summary is 46–71–76–78–92. The 5-number summary can be pictured as

least score–lower quartile–median–upper quartile–highest score.

A *box plot,* often called a *box-and-whisker plot,* gives a graphical visualization of the 5-number summary. Box-and-whisker plots are part of Student Assessments in grade 6. See problem 40.

DEFINITION Box Plot, Box-and-Whisker Plot

A **box plot,** or **box-and-whisker plot,** consists of a central box extending from the lower to the upper quartile, with a line marking the median and with line segments, or whiskers, extending outward from the box to the extremes.

For example, the box plot for Example 13.8 is shown in Figure 13.13.

Figure 13.13
Box plot for data set R

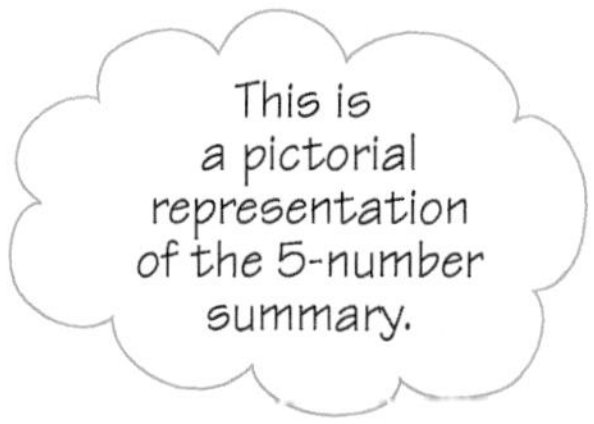

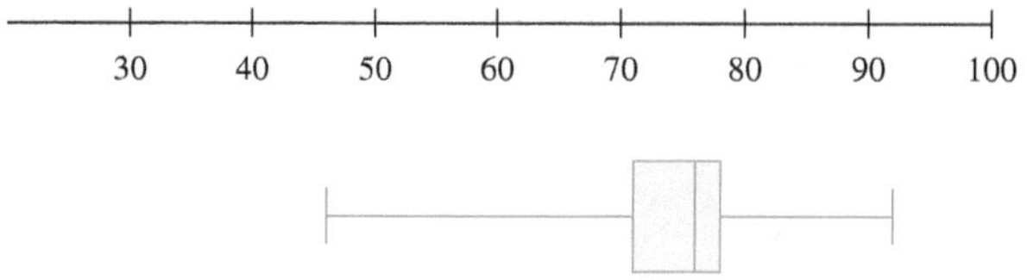

Another advantage of box plots is that we can use them to compare data sets containing widely differing numbers of values, as shown next.

EXAMPLE 13.9

Using Box Plots for Comparisons

The data that follow are the final scores of male and female students in class ABC. Draw box plots to compare the distribution of men's scores with the distribution of women's scores.

Men's scores:	95,	79,	53,	78,	71,	88,	77,	80,	79,	79		
Women's scores:	84,	85,	53,	77,	66,	81,	79,	59,	65,	61,	81,	68,
	68,	80,	76,	87,	85,	74,	92,	76,	70,	85,	55,	79,
	74,	80,	73,	48,	66,	83,	48,	60,	87,	58,	64,	78,
	82,	69,	76,	83,	94,	86,	73,	85,	75,	69,	49,	52,
	59,	68,	65,	75,	31,	69,	73,	56,	95			

Solution

To make the plots, we need the 5-number summaries. First, arrange the scores in order of increasing size:

Men's scores:	53,	71,	77,	78,	79,	79,	79,	80,	88,	95		
Women's scores:	31,	48,	48,	49,	52,	53,	55,	56,	58,	59,	59,	60,
	61,	64,	65,	65,	66,	66,	68,	68,	68,	69,	69,	69,
	70,	73,	73,	73,	74,	74,	75,	75,	76,	76,	76,	77,
	78,	79,	79,	80,	80,	81,	81,	82,	83,	83,	84,	85,
	85,	85,	85,	86,	87,	87,	92,	94,	95			

For the men, the extreme scores are 53 and 95 and the median is 79, the average of the fifth and sixth scores. The lower quartile is 77, the median of the first five scores. The upper quartile is 80, the median of the last five men's scores. Thus, the 5-number summary of the men's scores is

53–77–79–80–95.

Similarly, the 5-number summary of the women's scores is

31–64.5–74–81.5–95.

These summaries give the following box plots:

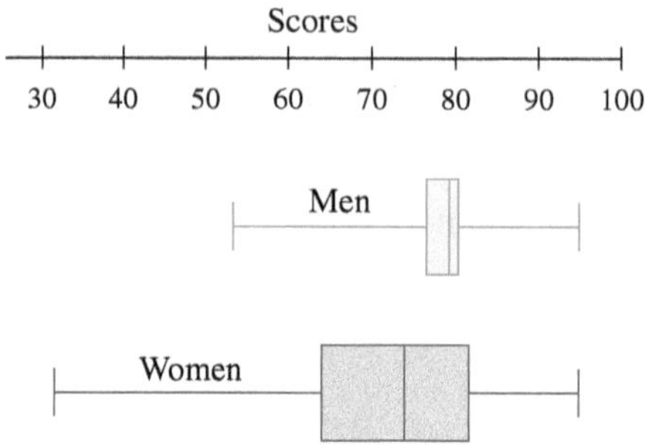

The Standard Deviation

We have already observed that the range is one measure of the spread of a data set. However, it is not a very precise measure, since it depends only on the extreme data values, which may differ markedly from the bulk of the data. This deficiency is largely remedied by the 5-number summary (which includes the IQR) and its visualization by a box plot.

However, just as the mean is an indication of a typical value of a set of data, the **standard deviation** is a measure of the typical deviation of the values from the mean. The idea is to find out first how far *each* of the data points is from the mean [in other words, for example, compute the square of $(\bar{x} - x_1)$ for the first data point] and then do the same for all the others. A way to describe finding the standard deviation casually would be to "square each of the points from the mean, sum them, divide by the number of data points n, and then take a square root after the division by n." Now we will write out the definition formally and then show the procedure for computing the standard deviation.

DEFINITION The Standard Deviation of a Set of Data

Let $x_1, x_2, x_3, \ldots, x_n$ be the values in a set of data and let $\bar{x}$ denote their mean. Then

$$s = \sqrt{\frac{(\bar{x} - x_1)^2 + (\bar{x} - x_2)^2 + \cdots + (\bar{x} - x_n)^2}{n}}$$

is the **standard deviation.***

We now write out the procedure to compute the standard deviation of a data set and then give an example.

PROCEDURE Finding the Standard Deviation of a Data Set with *n* Data Points

To find the way of obtaining the standard deviation, use the following procedure for the data points $\{x_1, x_2, x_3, x_4, \ldots, x_n\}$.

1. First find the mean, $\bar{x}$.
2. For each of the points, find the *square* of the difference between the point and $\bar{x}$. For example, for the fourth point, x_4, find $(\bar{x} - x_4)^2$.

*More advanced books use $n - 1$ rather than n in the denominator. The difference is small, however, so to avoid confusion, we use the definition here.

3. Add all of the squares from Step 2.
4. Divide your answer in Step 3 by n.
5. Take the square root of your answer in Step 4. You will almost always need a calculator to compute the square root.

EXAMPLE 13.10

Computing a Standard Deviation

Compute the mean and standard deviation for this set of data:

35	42	61	29	39

Solution

Step 1: Find the mean of the data.

Since there are five data points, we add the data points and divide by 5 to find the mean:

$$\bar{x} = (35 + 42 + 61 + 29 + 39)/5 = 41.2.$$

Step 2: Set up a table for the data points getting the *square* of the difference between $\bar{x}$ and the point:

		Step 1	Step 2	Value of Square
	Data Value	$\bar{x}$	$(\bar{x} - x_i)^2$	Value of $(\bar{x} - x_i)^2$
x_1	35	41.2	$(41.2 - 35)^2$	38.44
x_2	42	41.2	$(41.2 - 42)^2$	0.64
x_3	61	41.2	$(41.2 - 61)^2$	392.04
x_4	29	41.2	$(41.2 - 29)^2$	148.84
x_5	39	41.2	$(41.2 - 39)^2$	4.84

Step 3: Add all numbers in the last column from Step 2.

The sum is $(38.44 + 0.64 + 392.04 + 148.84 + 4.84) = 584.8$.

Step 4: Divide the answer in Step 3 by the number of data points.

Since there are five data points, we divide the sum of the last column by $n = 5$. Thus, $\frac{584.8}{5} = 116.96$.

Step 5: Compute the standard deviation, s.

The standard deviation is the square root of the answer in Step 4. Thus, $S = \sqrt{\frac{584.8}{5}} = \sqrt{116.96}$, which is approximately 10.815. The answer is the standard deviation is about 10.815.

It turns out that there is an easier formula for calculating standard deviations that, on a calculator, requires only the $\boxed{\sqrt{\ }}$ and $\boxed{x^2}$ keys in addition to the usual keys for arithmetic,

$$s = \sqrt{\frac{x_1^2 + x_2^2 + \cdots + x_n^2}{n} - \bar{x}^2},$$

where $x_1, x_2, \ldots, x_n$ are the data values and $\bar{x}$ is their mean. Many calculators, programs, and other technology make computing standard deviations much simpler.

EXAMPLE 13.11

Alternative Calculation of the Standard Deviation

Calculate the mean and standard deviation of the data set in Example 13.10, using the alternative formula just given.

Solution

The mean, $\bar{x} = 41.2$, is calculated as in Example 13.10. Then

$$s = \sqrt{\frac{35^2 + 42^2 + 61^2 + 29^2 + 39^2}{5} - 41.2^2}, \text{ which is about } 10.8.$$

We will now take a look at three different dot plots and begin our intuition about how much variability data can have by just looking at the plot. Then look again after the standard deviations are computed in Figure 13.14.* If the standard deviation is large, the data are more spread out; if it is small, the data are more concentrated near the mean. This relationship is immediately apparent from the dot plots of data sets R, S, and T discussed earlier. In Figure 13.14, these dot plots are reproduced again, with the addition of the location of the mean as well as the spread of each data set relative to its standard deviation with respect to the mean. A most important fact is that, for most data sets, most data values fall within 1 standard deviation of the mean and almost none lie as far as 3 standard deviations from the mean. Look at the number of dots inside the arrow marked $s = 8.7$ on either side of the mean $\bar{x} = 73.8$. These 19 data points are within 1 standard deviation of the mean $\bar{x} = 73.8$.

Figure 13.14
Dot plots for R, S, and T showing the location of the mean and the spread of the data relative to the standard deviation

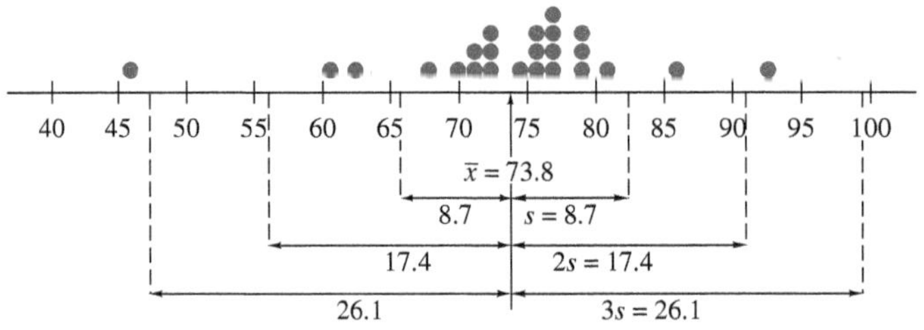

Dot plot for R: All but five data points lie within 1 standard deviation of the mean. Only one data point lies more than 3 standard deviations from the mean.

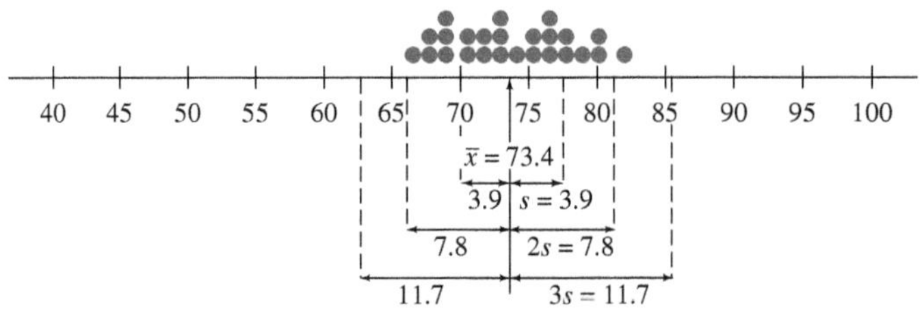

Dot plot for S: All but ten data points lie within 1 standard deviation of the mean. Only one data point lies more than 2 standard deviations from the mean.

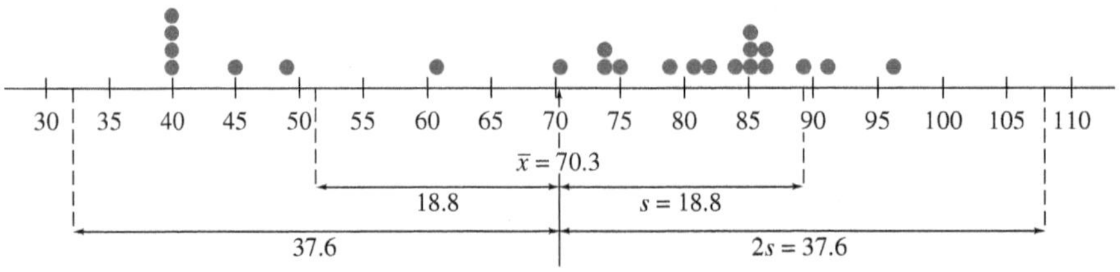

Dot plot for T: All but nine data points lie within 1 standard deviation of the mean. None lies beyond 2 standard deviations of the mean.

*The **variance**, v, is just the square of the standard deviation, s, so the variance is also a good measure of the variability of the data, but the standard deviation is more commonly used. Said algebraically, $v = s^2$.

EXAMPLE 13.12

Determining the Fraction of Data Values Near the Mean

Compute the fraction (expressed as a percentage) of the data values in Example 13.10 that falls

(a) within 1 standard deviation of the mean.
(b) within 2 standard deviations of the mean.

Solution

41.2 − 10.8 = 30.4

41.2 + 10.8 = 52.0

(a) In Example 13.10, $\bar{x} = 41.2$ and s is about 10.8. To be within 1 standard deviation of the mean asks that you find the data points between $(\bar{x} - s)$ and $(\bar{x} + s)$. Since $(\bar{x} - s)$ is approximately 30.4 and $(\bar{x} + s)$ is about 52.0, we are now looking for all data points between 30.4 and 52.0. There are three data points (35, 39, 42) that fall in this range.

$$\frac{3}{5} = .60\ldots, \text{ which is } 60\%.$$

41.2 − 2(10.8) = 19.6

41.2 + 2(10.8) = 62.8

(b) Therefore, data within 2 standard deviations of the mean asks that you look for the data points between $(\bar{x} - 2s)$ and $(\bar{x} + 2s)$. Those data values within 2 standard deviations of the mean must lie between 19.6 and 62.8. This range includes all of the data values, and we have

$$\frac{5}{5} = 1\ldots, \text{ which is } 100\%.$$

13.2 Problem Set

Understanding Concepts

1. Determine the mean, median, and mode for this set of data:

18	27	17	19	21	24	18	15
23	18	17	14	22	19	27	30

2. **(a)** Compute the mean, median, and mode for this set of data:

69	81	77	69	64	85	81	73	79
74	70	78	86	80	71	79	77	70
67	70	79	80	71	67	69	79	81

(b) Draw a dot plot for the data in part (a).

(c) Does either the mean, the median, or the mode seem typical of the data in part (a)?

(d) Might it be reasonable to suspect that the data in part (a) actually come from two essentially different populations (say, daily incomes from two entirely different companies)? Explain your reasoning.

3. **(a)** Compute the quartiles for the data in problem 1.

(b) Give the 5-number summary for the data in problem 1.

(c) Draw a box plot for the data in problem 1.

(d) Determine the interquartile range, IQR, for the data in problem 1.

(e) Identify any outliers in the data of problem 1.

4. **(a)** Compute the quartiles for the data in problem 2.

(b) Give the 5-number summary for the data in problem 2.

(c) Draw a box plot for the data in problem 2.

(d) Determine the interquartile range, IQR, for the data in problem 2.

(e) Identify any outliers in the data of problem 2.

5. What is the midrange of problem 1?

6. What is the midrange of problem 2?

7. **(a)** Draw side-by-side box plots to compare students' performances in class A and class B if the final grades are as shown here:

Class A:	91,	63,	65,	73,	65,	86,
	96,	75,	75,	79,	84,	72,
	80					
Class B:	87,	72,	95,	89,	69,	79,
	56,	64,	66,	67,	89,	47

(b) Briefly compare the performances in the two classes on the basis of the box plots in part (a).

(c) Determine the interquartile range, IQR, for each of the classes in part (a).

(d) Identify any outliers in the classes of part (a).

8. Use the data in problem 1 to do or answer the following:

(a) Compute the mean.

(b) Compute the standard deviation.

9. Use the data in problems 1 and 8 to answer the following:

(a) What percentage of the data is within 1 standard deviation of the mean?

(b) What percentage of the data is within 2 standard deviations of the mean?

(c) What percentage of the data is within 3 standard deviations of the mean?

10. (a) Choose an appropriate scale, and draw a dot plot for this set of measurements of the heights, in centimeters, of 2-year-old ponderosa pine trees:

22.2	23.5	22.5	22.6	23.0	22.8
22.4	22.2	23.0	23.3	23.9	22.7

(b) Compute the mean and standard deviation for the data in part (a).

(c) What percentage of the data is within 1 standard deviation of the mean?

(d) What percentage of the data is within 2 standard deviations of the mean?

(e) What percentage of the data is within 3 standard deviations of the mean?

(f) What is the midrange of the data set?

11. Uniform distribution. A data set has a uniform distribution (or is a **uniform data set**) if every value has exactly the same number of data points. Here is an example: Sandy rolls a five-sided die 10 times and comes up with the data set $D_2 = \{2, 4, 3, 5, 3, 2, 1, 1, 5, 4\}$, so each integer actually occurs twice (and that's why there is a 2 in the subscript of the name of the data set).

(a) What are the mean, mode, and median of D_2?

(b) Suppose that Sandy rolls the die 20 times and each of the integers comes up 4 times. (We'll call that data set D_4.) What are the mean, mode, and median of D_4?

(c) Suppose that Sandy rolls the die 30 times and each of the integers comes up 6 times. (We'll call that data set D_6.) What are the mean, mode, and median of D_6?

(d) What do you notice about the answers to the three parts of the problem?

12. Suppose that D is a data set. We construct a new data set, E, by doubling each element of D. Note that the two sets have the same number of elements. What is the relationship between

(a) the mode of D and the mode of E?

(b) the median of D and the median of E?

(c) the mean of D and the mean of E?

Here is the use of algebra in statistics!

Into the Classroom

Divide the class into groups of three or four students each. Make sure that all members of each small group agree on the answers required of their group.

13. As we have seen before, it often helps to understand a concept if it can be visualized using an appropriate manipulative. Work with about three other students to carry out the following activity:

(a) Suppose you want to demonstrate the idea of the mean of a set of data to fourth graders. Using a set of blocks, form stacks of heights 5, 1, 4, 7, 6, and 7. Now move blocks from higher stacks to lower stacks in an effort to form six stacks, all of the same height. Can this be done? If so, how many blocks are in each stack?

(b) Determine the mean of 5, 1, 4, 7, 6, and 7.

(c) Comparing the results of parts (a) and (b), what do you conclude?

(d) How would you elaborate on your conclusions from the first three parts of this problem to make the idea of the mean clear to your students? Explain carefully.

(e) How would the idea presented in this problem work with these data: 5, 1, 4, 7, 11, and 7? Discuss briefly.

14. Consider the data collected in problem 22 of Problem Set 13.1.

(a) Members of one or two small groups each determine the mode or modes, the mean, and the standard deviation of the data for men.

(b) Members of one or two small groups each determine the mode or modes, the mean, and the standard deviation of the data for women.

(c) Members of one or two small groups each determine the 5-number summaries and draw side-by-side box plots for the two sets of data.

(d) As a class, discuss the results obtained in parts (a), (b), and (c) and decide on a consensus opinion of the most appropriate length of a "yard" and on a reasonable range in which the length might lie, on the basis of the data considered. Do the results differ markedly for men and women?

15. In ancient times, the cubit was taken as the length of the human arm from the tip of the elbow to the tip of the middle finger (generally understood to vary from about 17 to about 21 inches). Working in groups, repeat problem 14, for the cubit rather than the "yard."

16. Working in a group with about three other students, toss seven pennies 30 times and record the number of heads each time.

(a) Determine the mean and standard deviation of the data obtained.

(b) Determine what percent of the data lies within 1 standard deviation of the mean.

(c) What percent of the data differs from the mean by more than 2 standard deviations?

Responding to Students

17. Ms. Chen helped her class gather data on the number of books each student read in one week. Once the data were collected, she helped her students organize the information into the tally chart shown. For homework, Ms. Chen asked her students to find the median and the mode for the data they collected. The next day, Joseph turned in his homework with the answer 4 for the median and the answer 7 for the mode of the set of data.

Number of Books Reported	Number of Students
1	8 tally marks
2	5 tally marks
3	6 tally marks
4	0 tally marks
5	2 tally marks
6	3 tally marks
7	2 tally marks

(a) What mistake did Joseph make when finding the median and the mode for the set of data?

(b) How would you help guide Joseph to find the correct answer?

18. Stefanie was asked to find the mean of the numbers 14, 17, 19, and 26. She gave the answer 76.

(a) What error did Stefanie make when calculating the mean?

(b) How would you help guide Stefanie to calculate the mean of a set of numbers correctly?

19. When Yugi was asked to explain how he found the median for a set of data, he said, "I found the median by crossing off numbers until I had only one left." How would you help guide Yugi so he won't make the same mistake again?

20. Patrick is writing in his journal about the median, mean, mode, and range as measures of central tendency. Are all four of them in fact measures of central tendency?

21. Marilyn talks in class about "the mode" of a data set. Is she making a mistake and, if so, how would you rectify it?

22. Shawn asks why the sum of the differences of the data values from the mean.

$$(\bar{x} - x_1) + (\bar{x} - x_2) + \cdots + (\bar{x} - x_n),$$

isn't used as a measure of variability. How would you respond to Shawn's question?

23. During the classroom discussion resulting from Shawn's question in the previous problem, Leona suggests that the sum of the absolute values of the differences of the data values from the mean be used as a measure of variability since the terms are all positive and thus cannot cancel each other out. Assume that you know the mean is 22.84 in problem 10.

(a) Compute the sum of the absolute values of the differences in problem 10, part (a).

(b) Compute the mean of the absolute values of the differences in problem 10, part (a); that is, divide the sum in part (a) of that problem by 12.

(c) How would you respond to Leona?

Thinking Critically

24. On June 1, 2009, the average age of the 33 employees at Acme Cement was 47 years. On June 1, 2010, three of the staff ages 65, 58, and 62 retired and were replaced by four employees ages 24, 31, 26, and 28. What was the average age of the employees at Acme Cement on June 1, 2010?

25. (a) Compute the mean and standard deviation for these data:

28	34	41	19	17	23

(b) Add 5 to each of the values in part (a) to obtain 33, 39, 46, 24, 22, and 28. Compute the mean and the standard deviation for this new set of values.

(c) What properties of the mean and standard deviation are suggested by parts (a) and (b)?

26. Compute the mean and standard deviation for the data represented by the following two histograms:

(a)

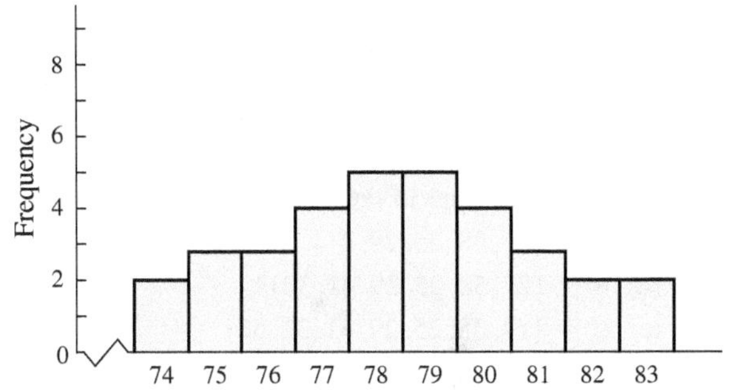

(b)

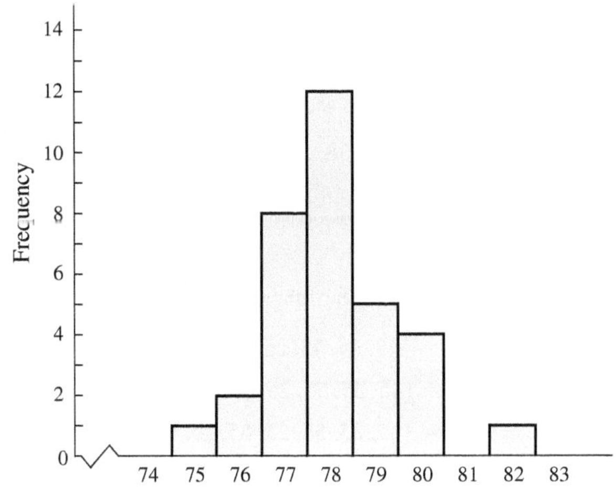

(c) Briefly explain why the standard deviation for the data of part (b) is less than that for the data of part (a).

27. Does the mean, median, or mode seem to be the most typical value for the given set of data? Explain briefly. (*Suggestion:* Draw a line plot.)

42	47	38	16	45	41	16	48	44

28. **(a)** Determine the mean, median, and mode of the data in the dot plot shown.

(b) Does the mean, median, or mode seem to be the most typical of these data? Explain briefly.

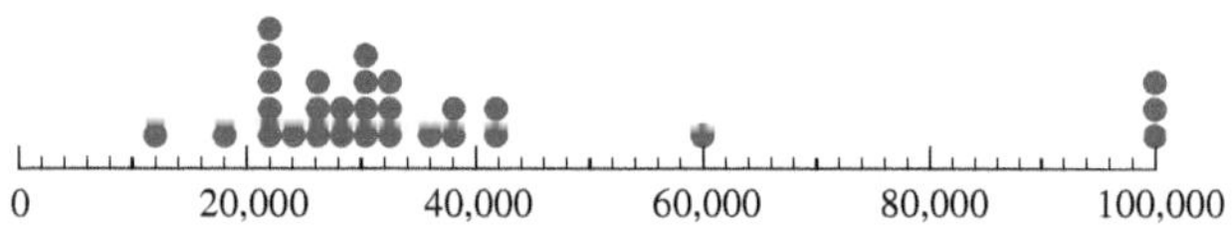

29. Produce sets of data that satisfy these conditions:

(a) mean = median < mode

(b) mean = mode < median

(c) median = mode < mean

30. **(a)** What can you conclude if the standard deviation of a set of data is 0? Explain.

(b) What can be said about the standard deviation of a set of data if the values all lie very near the mean? Explain.

31. A collection of data contains ten values consisting of a mix of 1s, 2s, and 3s.

(a) If $\bar{x} = 3$,what is the data set?

(b) If $\bar{x} = 2$,what are the possibilities for the data set?

(c) If $\bar{x} = 1$,what is the data set?

(d) Could $\bar{x} = 1$ and $s \neq 0$ for this data set? Explain.

32. Compute the means of these collections of data.

(a) $A = \{27, 38, 25, 29, 41\}$

(b) $B = \{27, 38, 25, 29, 41, 32\}$

(c) $C = \{27, 38, 25, 29, 41, 32, 32\}$

(d) $D = \{27, 38, 25, 29, 41, 32, 32, 32, 32, 32, 32\}$

(e) What conclusion is suggested by the calculations in parts (a) through (d)?

(f) Guess the mean of the following set of data, and then do the calculation to see if your guess is correct:

$$E = \{27, 38, 25, 29, 41, 60, 4, 60, 4\}$$

(g) What general result does the calculation in part (f) suggest?

33. **(a)** The mean of each of these collections of data is 45:

$$R = \{45, 35, 55, 25, 65, 20, 70\}$$

$$S = \{45, 35, 55, 25, 65, 20, 70, 45, 45\}$$

$$T = \{45, 35, 55, 25, 65, 20, 70, 80, 10\}$$

Which of R and S has the smaller standard deviation? No computation is needed; justify your response with a single sentence.

(b) Like the means of R and S in part (a), the mean of T is 45. Is the standard deviation for this set the same as that for S? Note that both of these sets have the same number of entries. Explain your conclusion.

34. According to Garrison Keillor, all the children in Lake Wobegon are above average. Is this assertion just a joke, or is there a sense in which it could be true?

35. If the mean of $A = \{a_1, a_2, \ldots, a_{30}\}$ is 45 and the mean of $B = \{b_1, b_2, \ldots, b_{40}\}$ is 65, compute the mean of the combined data set. (*Hint:* The answer is not 55.)

Making Connections

36. **(a)** From the data in the bar graph shown, is it possible to determine the average median income in 2000 for men 25 years old and older with education not exceeding a master's degree? Explain briefly.

Median annual income of persons with income who are 25 years old and over, by highest level of education and sex: 2000

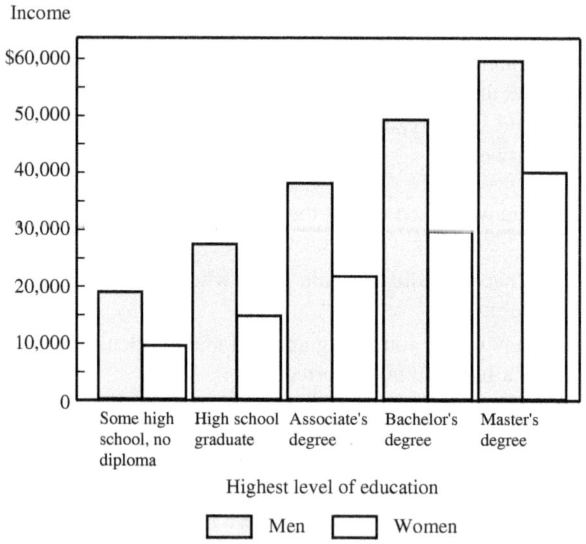

Source: *U.S. Department of Commerce, Bureau of the Census,* Current Population Reports, *Series P-60, "Money Income in the United States: 2000."* United States Department of Commerce.

(b) What legitimate conclusions can you make on the basis of the bar graph shown?

State Assessments

37. (Grade 4)
Ms. Chang said that the class would study statistics and geometry together. There was a bag of 15 squares each of which could be of different side length (in cm.) After picking up a square each, the students came up with a dot plot from the squares. What is the median length of a square in this data?

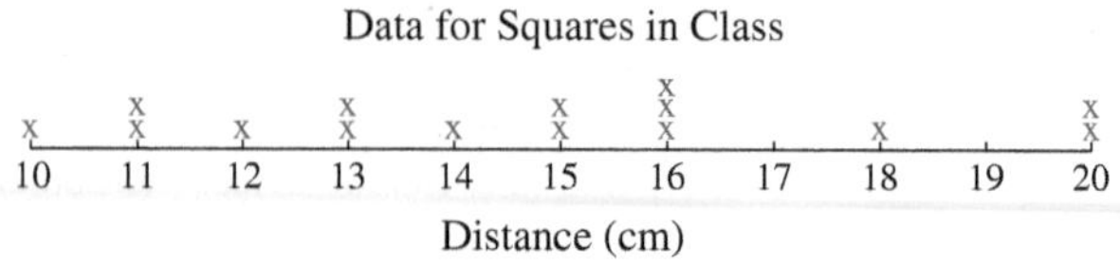

Length of squares (cm)

A. 11 B. 15

C. 16 D. 20

38. (Grade 5)
Hayden picked 143 apples from the 13 apple trees that his family owns. On the average, how many apples did each apple tree produce?

A. 11 B. 12 C. 13

D. 14 E. No such answer

39. (Grade 5)
There were about 102,210 people attending the European Football Championship game. The average price was \$45.75 (in U. S. dollars) for each person. How much money came from the European Football game tickets for all of those who paid to get in?

A. \$450,000 B. \$4,500,000

C. \$45,000,000 D. \$45,000

40. (Grade 6)
What is missing from the box-and-whisker plot?

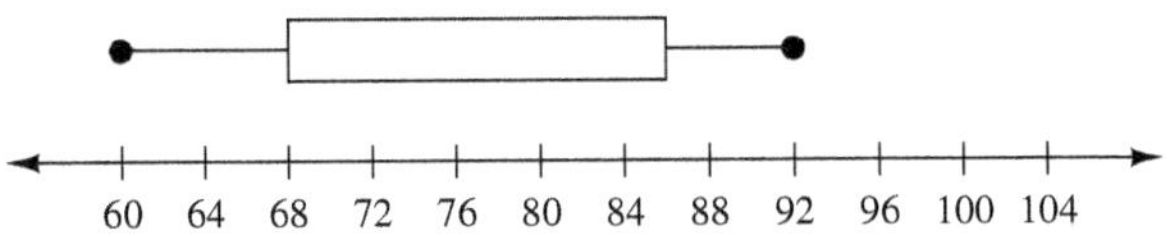

A. Median B. Range

C. Upper quartile D. Lower quartile

41. (Grade 6)
Which statement is *false*?

A. A set of data always has more than one mode.

B. A set of data may have exactly one mode.

C. A set of data may have more than a mode.

D. The mode is the piece of data that occurs most frequently.

42. (Grade 7)
Richard likes to work in a community of older citizens who do drama. He knew that the ages of the actors of the current play were

50, 59, 54, 52, 55, 56, 56,
58, 53, 58, 55, 59, 54, 59,
56, 58, 55, 57, 57, 58.

He thinks that the mode is the most appropriate average to describe his drama group. Which answer is the most reasonable?

A. The mode represents the average fairly.

B. The mode is too low to represent the average. The mean or median is a better choice.

C. The mode is too high to represent the average. The range is a better choice.

D. The mode is too high to represent the average. The mean or median is a better choice.

43. (Grade 9)
Millie entered her horse in a horse show. Her horse got a score of 53. Which measure of data can Millie use to determine if her horse's score got in the top half of all scores at the show?

A. Median B. Mode

C. Mean D. None of the other answers

44. (Grade 6)
Pedro took five tests each of which is worth 100 points. He earned the following scores:

84, 86, 88, 89, 98

What is Pedro's mean (average) score for these five tests?

A. 88 B. 86

C. 89 D. 5

13.3 Statistical Inference

In this section, we give a brief introduction to one aspect of statistics—**statistical inference**—by discussing how one predicts the characteristics of a population by examining the properties of small pieces of it called *samples*. For example, we cannot interview all adults to find out for whom they will vote. On the other hand, is there a way to sample a reasonable number of adults and then predict what the outcome of the voting would most likely be? A first step to this approach is to look at a sufficiently large, unbiased number of samples of the population and use the mean and standard deviation of the samples as a way of getting an idea of what the mean and standard deviation of the population might look like.

The important phrases here are "sufficiently large" and "unbiased." How many samples would it take for us to feel comfortable with our predictions, to what degree are we comfortable with those predictions, and what does it mean for the sample to be taken without bias? There are famous mistakes, such as the prediction by the Gallup poll that Thomas E. Dewey would win the 1948 election when Harry S Truman actually turned out to be the winner. These errors have given insights into how to make such predictions more accurate ("sampling techniques"). The questions posed in this paragraph are deep questions covered in statistics courses, not ones that can be resolved in one chapter of a book. While statistical inference is not a part of the elementary school curriculum, it is important for future teachers to see where the previous two sections (usually called *descriptive statistics*) lead. This section focuses on populations, samples, and distributions.

Populations and Samples

In statistics, a **population** is a particular set of objects about which one desires information. If the desire is to determine the average yearly income of all adults in the United States, the population is the set of *all* adults in the United States. In addition, one must be explicit as to what "adult" means. Other examples of populations are

- all boys in Eisenhower High School in Yakima, Washington,
- all lightbulbs manufactured on a given day by Acme Electric Company, and
- all employees of AT&T,

and so on. One might want to determine

- the average height of boys in Eisenhower High School,
- the average life of lightbulbs produced by Acme Electric Company, or
- the average cost of medical care for employees of AT&T.

Since it is often impractical or impossible to check each member of a population, the idea of statistics is to study a **sample**, which is a subset of the population. Samples can make inferences about the entire population on the basis of the study of the sample.

If the goal of accurate estimation of population characteristics is to be achieved, the population must first be carefully defined. The next step is to sample randomly (which will be discussed shortly). In particular, we must be sure not to favor any one outcome in designing a statistical study. Selection criteria that systematically favor certain outcomes are called **biased.**

Here are two examples of sampling, one of which is unbiased and the other biased: Suppose a study of the heights of boys at Eisenhower High School is desired. Instead of measuring each boy in the school, it is decided to study a sample of just 20 of the boys. If the sample were to be selected by choosing every fourth name out of an alphabetical listing of all 80 boys, would it likely be representative of the entire population? Probably, since there is likely little or no connection between last names and heights. However, a sample consisting of the members of the basketball team is clearly not representative, since basketball players tend to be unusually tall. It turns out that the best approach to sampling is to use a *random sample,* rather than try to find representative samples, and show that the criterion used really is representative of the population. An (unintentional) bias was one of the flaws of the Truman–Dewey election prediction mentioned earlier.

DEFINITION A Random Sample

A **random sample** of size r is a subset of r individuals from the population chosen in such a way that every such subset has an equal chance of being chosen.

For example, suppose an urn contains a mixture of red and white beans and you want to estimate what fraction of the beans is red by selecting a sample of 20 beans. You proceed by mixing the beans thoroughly and then, with your eyes closed, selecting 20 beans. Since each subset of 20 beans has an equal chance of being selected, the sample is indeed random. Note, however, that one must be very careful before asserting that physical mixing is sufficient to ensure randomness. (Have you ever wondered whether a deck of cards was shuffled enough?) These questions are a part of the beautiful subtlety of statistics.

Other schemes also work well. For example, suppose AT&T wishes to study the employees at one of its plants by selecting a random sample of 20 employees and asking them to respond to a questionnaire. One way to obtain a random sample would be to put the names of all the employees on tags, place the tags in a large container, mix the tags thoroughly, and then have someone close his or her eyes and select a sample of 20 tags. The employees whose tags are chosen constitute the random sample.

Another way to obtain the sample is to use a sequence of digits chosen in such a way that each digit is equally likely to be any one of the ten possibilities and the choice of each digit is independent of the choice of every other digit. Such a sequence would be a **random sequence of digits** or **random numbers.** One way to select such a random sequence is to construct a simple spinner with ten 36° sectors numbered 0, 1, 2, 3, 4, 5, 6, 7, 8, and 9, as shown in Figure 13.15. Spinning the spinner repeatedly produces a string of digits. Since the result of each spin is independent of the result of every other spin, each digit is equally likely to be selected and the digit sequence is random.

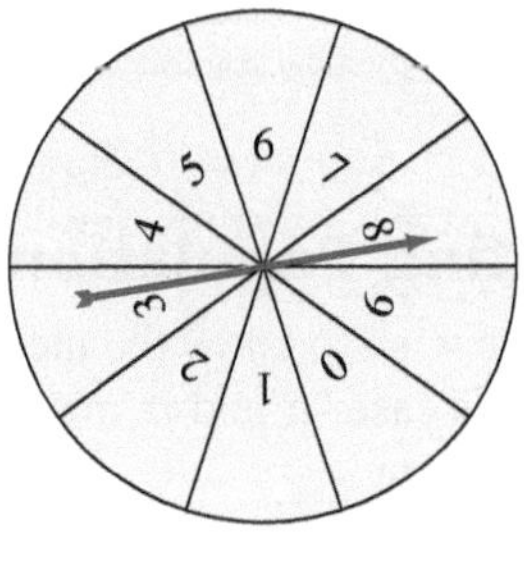

Figure 13.15
A simple random-digit generator

The desired sample of AT&T employees can now be obtained as follows (suppose the plant in question has 9762 employees):

1. Assign each employee a four-digit number from among 0001, 0002, . . . , 9762.
2. Generate a random sequence of four-digit numbers by repeatedly spinning the spinner four times. Since each digit is equally likely to appear on any spin, each four-digit number is equally likely to appear on any four spins. If the spinning process generates 0000, a four-digit number greater than 9762, or any number already obtained, simply ignore it and continue to generate more four-digit numbers. When 20 appropriate numbers have been generated, they can be used to identify the 20 employees to be included in the sample.

Finally, as you might expect, many calculators have built-in statistics routines that will generate **random numbers,** as do most spreadsheet programs.

Population Means and Standard Deviations

Professional statisticians find it helpful to use different symbols for means and standard deviations of populations and of samples. For populations, the mean and standard deviation are denoted by the Greek letters μ and σ (mu and sigma), respectively. Thus, for a population of size N, the population mean and population standard deviation are given by

$$\mu = \frac{x_1 + x_2 + \cdots + x_N}{N} \quad \text{and} \quad \sigma = \sqrt{\frac{(x_1 - \mu)^2 + (x_2 - \mu)^2 + \cdots + (x_N - \mu)^2}{N}},$$

where $x_1, x_2, \ldots, x_N$ are *all* the numbers in the population.

EXAMPLE 13.13

Computing a Population Mean and Standard Deviation

Here is a population that consists of the scores on an exam:

64	65	68	67	59	66	63	66	64	62	65	66	63	66	66	63
63	64	62	67	63	61	60	64	64	63	63	65	64	65	66	63
65	62	63	65	61	64	63	64	62	69	65	65	64	64	63	64
66	65	64	64	63	64	66	67	69	63	65	63	64	64	64	65
67	68	64	62	66	62	64	61	65	62	65	62	65	62	66	63

Use a spreadsheet or other suitable software on your computer or the built-in statistics routine on a suitable calculator to compute μ and σ for this population.

Solution

Since there are 80 numbers in the population,

$$\mu = \frac{64 + 65 + \cdots + 63}{80} \doteq 64.2$$

and

$$\sigma = \sqrt{\frac{(64 - 64.2)^2 + (65 - 64.2)^2 + \cdots + (63 - 64.2)^2}{80}} \doteq 1.9.$$

Estimating Population Means and Standard Deviations

Suppose we wish to know the mean and standard deviation of some large or inaccessible population. Since, in this case, μ and σ may be difficult or impossible to compute, we may estimate them with the sample mean

$$\overline{x} = \frac{x_1 + x_2 + \cdots + x_n}{n}$$

and the sample standard deviation

$$s = \sqrt{\frac{(x_1 - \overline{x})^2 + (x_2 - \overline{x})^2 + \cdots + (x_n - \overline{x})^2}{n}}$$

of a suitably chosen sample $x_1, x_2, \ldots, x_n$ of size n.

EXAMPLE 13.14 Estimating a Population Mean and Standard Deviation

(a) Estimate the mean and standard deviation of the population given in Example 13.13 by using a spinner, as illustrated in Figure 13.15, to select a random sample of size 10.

(b) Compare the results of part (a) with the results obtained in Example 13.13.

Solution

(a) Suppose your spinner generates the digit sequence 5, 5, 2, 9, 1, 0, 4, 5, 3, 1, 2, 4, 1, 9, 4, 6, 6, 9, 1, 7. Using these two at a time, we obtain the following table of two-digit numbers that will tell us which position in the table of Example 13.13 to use:

55	29	10	45	31
24	19	46	69	17

Since all are different, these determine the random sample shown here:

66	64	62	64	66
64	62	64	66	63

The fifty-fifth number in the data set is 66, and so on.

Thus, the mean of the numbers in the sample is

$$\overline{x} = \frac{(66 + 64 + 62 + 64 + 66 + 64 + 62 + 64 + 66 + 63)}{10} = 64.1.$$

The variance is

$$
\begin{aligned}
v &= s^2 \\
&= [(64.1 - 66)^2 + (64.1 - 64)^2 + (64.1 - 62)^2 \\
&\quad + (64.1 - 64)^2 + (64.1 - 66)^2 + (64.1 - 64)^2 \\
&\quad + (64.1 - 62)^2 + (64.1 - 64)^2 + (64.1 - 66)^2 \\
&\quad + (64.1 - 63)^2]/10, \text{ which is about } 2.09
\end{aligned}
$$

and the sample standard deviation is

$$s = \sqrt{v} = \sqrt{2.09}, \text{ which is approximately } 1.45.$$

(b) We observe that $\bar{x}$ and s for the sample are reasonable approximations to μ and σ, respectively, for the population as determined in Example 13.13.

Suppose we were to repeat the preceding example but with a random sample of size 15. For the resulting sample, $\bar{x}$ and s should be slightly better approximations to μ and σ, respectively, from Example 13.13 than the values obtained in Example 13.14. Not surprisingly, it is generally true that larger samples tend to yield better approximations to population characteristics.

Distributions

We return now to the data of Table 13.2 (Page 672), in which the population consisted of all boys in Eisenhower High School. A histogram and line graph of the boys' heights to the nearest inch appear in Figures 13.5 and 13.6, respectively.

Since the heights of the columns in the histogram represent the number, or frequency, of the measurements in each range (63.5–64.5, 64.5–65.5, and so on) and the width of each column is 1, the total area of all the columns in the histogram is 80, the total number of boys in the population.

The **relative frequency** of the measurements in each range in Figure 13.5 is the fraction (expressed as a decimal) of the total number of boys represented in that range. The relative frequency is obtained by taking the frequency distribution in Table 13.2 and dividing it by 80. If the heights of the column in the histogram are determined by relative frequency, the diagram remains the same except for the designation on the vertical scale, as shown in Figure 13.16. Also, since the width of each column is 1, the *area* of each column gives the fraction of the population whose heights fall into that range. Moreover, the area of the first three columns gives the fraction of the population with heights ranging from 63.5 to 66.5 inches, and the total area of the histogram is 1.

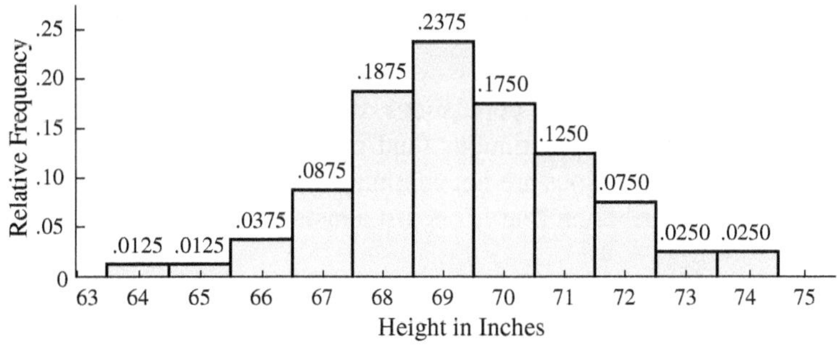

Figure 13.16
Histogram of Figure 13.5 but with the vertical scale denoting relative frequency

As noted earlier, histograms often are representations of grouped data, which make it appear that all the data values in a given range are the same. However, as in the present case, this is frequently not so. The boys' heights are listed to the nearest inch, whereas people's heights actually vary continuously. A truer representation of such data is provided by a line graph, or frequency polygon, as in Figure 13.6. If the vertical scale represented relative frequency rather than frequency, the graph would appear unchanged, as shown in Figure 13.17. Also, since such a diagram can be obtained from a histogram by deleting and adding small triangles of equal area, the area under the **relative-frequency polygon** is still 1 and the area of that portion of the polygon from, say, 63.5 to 66.5 equals the fraction of the population of boys whose heights lie in this range.

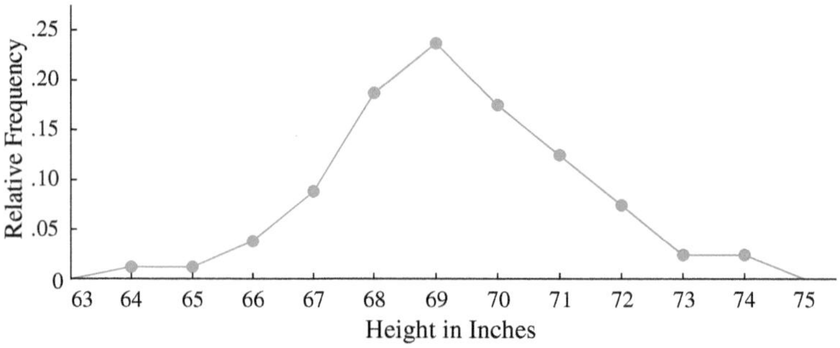

Figure 13.17 Relative-frequency polygon of the heights of boys in Eisenhower High School

In addition, had the measurements been taken more and more closely and the range in the histogram made narrower and narrower, the tops of the columns in the histogram, as well as the corresponding relative-frequency polygon, would have more and more closely approximated a smooth bell-shaped curve called a **normal distribution,** as shown in Figure 13.18. Here also, the area under the curve and above the interval between a and b indicates the relative frequency, or fraction, of the boys measured who have heights between a and b. This fraction indicates the *likelihood,* or *probability,* that a boy chosen at random from the population will have a height in the given range.

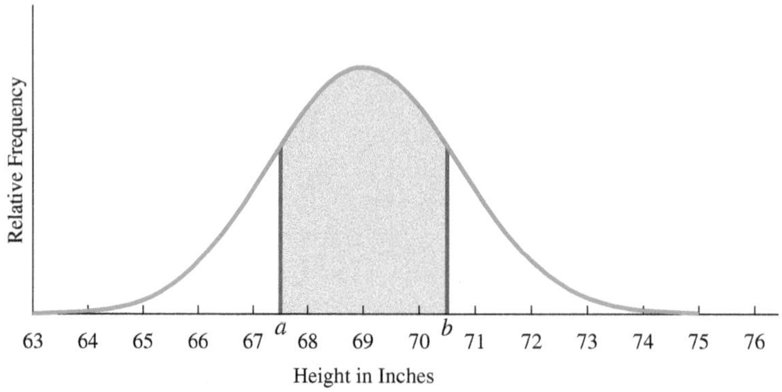

Figure 13.18 Normal distribution of the heights of boys in Eisenhower High School

For many populations, the **distribution** of the measurements of the property being considered will be a continuous (and often normal) curve, as in Figure 13.18. However, in other cases, the observations are not continuous, are not normal, or are discrete. If the observations are discrete, the distribution "curve" is just a histogram.

DEFINITION A Distribution Curve

A curve or histogram that shows the relative frequency of the measurements of a characteristic of a population that lies in any given range is a **distribution curve.** The area under such a curve or histogram is always 1.

Knowing the distribution of a population frequently allows one to say with some precision what the average value is and what percentage of the population lies within different ranges. In particular, the normal distribution has been studied in great detail, and it can be shown that very nearly 68% of the population lies within 1 standard deviation of the mean, very nearly 95% of the population lies within 2 standard deviations of the mean, and very nearly 99.7% (or virtually *all*) of the population lies within 3 standard deviations of the mean, as illustrated in Figure 13.19. Using the language of probability, we would say that the probability that a given data value lies within 1 standard deviation of the mean is 0.68, the probability that a given data value lies within 2 standard deviations of the mean is 0.95, and the probability that any given data value lies within 3 standard deviations of the mean is 0.997 (virtually 100%). Considerations like these are what make it possible for very carefully designed polls and other studies to claim that their results are accurate to within a given tolerance, say, 3%.

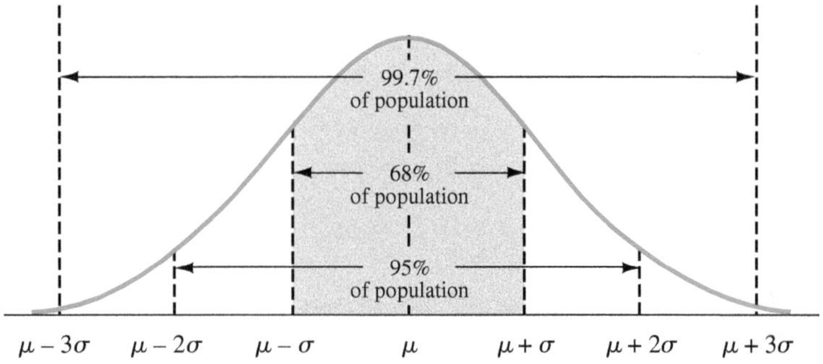

Figure 13.19
Percentage of data within 1, 2, and 3 standard deviations of the mean of a normal distribution

THEOREM The 68–95–99.7 Rule for Normal Distributions
For a population that has a normal distribution, about 68% falls within 1 standard deviation of the mean, about 95% falls within 2 standard deviations of the mean, and about 99.7% falls within 3 standard deviations of the mean.

It turns out that many populations are normally distributed, or approximately so. Thus, the 68–95–99.7 rule is approximately true for these populations and for samples from these populations. For samples, the approximation is increasingly accurate for increasingly larger sample sizes.

z Scores and Percentiles

As noted previously, the normal distribution has been studied deeply and with great care. Some additional facts are as follows:

- The graph of the distribution is symmetric about a vertical line drawn through the mean; that is, if the curve were folded along this line, the two halves of the curve would exactly match each other. Thus, the area under the curve to the left of the mean equals the area under the curve to the right of the mean, and it follows that the mean of the distribution is also its median.
- The maximum height of the curve occurs at the mean, so the mean also equals the mode.
- Since the scale of the vertical axis is relative frequency, the area under the entire curve is 1. Also, the area under the curve over various intervals has been carefully tabulated, thus making such probability statements as the 68–95–99.7 rule possible.

One stratagem that makes it possible to effectively compare two different normal distributions has been the creation of the so-called **standardized form** of the distribution (also called the **z curve**). The way that the standardized form is accomplished is to start with a normal distribution with mean μ and standard deviation σ and change to the variable z. This comparison is accomplished by altering the scale on the horizontal axis by the transformation

$$z = \frac{x - \mu}{\sigma}$$

while maintaining the frequency scale on the vertical axis. This transformation does not materially alter the shape of the distribution, as can be seen in Figure 13.20, which is the standard form of the distribution of Figure 13.19.

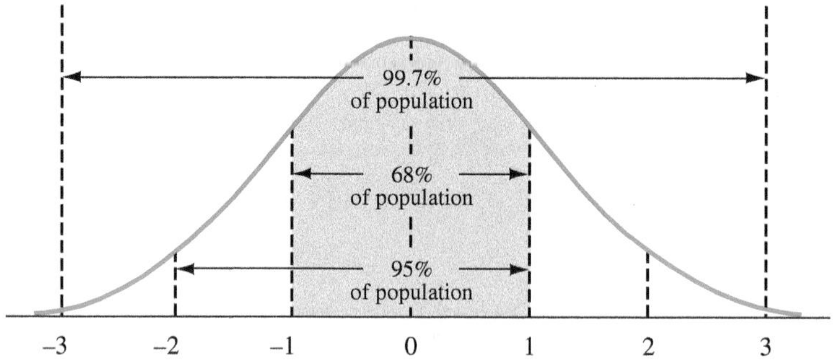

Figure 13.20 Standardized form of a normal distribution, illustrating the 68–95–99.7 rule

Note that *any* normal distribution in standard form will have mean 0 and standard deviation 1. To see this, set $x = \mu$; then $z = \dfrac{\mu - \mu}{\sigma} = 0$. And if we set $x = \mu + \sigma$ (that is, x is 1 standard deviation above the mean), then $z = \dfrac{(\mu + \sigma) - \mu}{\sigma} = \dfrac{\sigma}{\sigma} = 1$. Thus, in standard form, 68% of the population lies between -1 and 1, 95% of the population lies between -2 and 2, and 99.7% of the population lies between -3 and 3, as shown in Figure 13.20.

Normal distributions in standard form may appear tall and skinny or short and fat, depending on their standard deviations (see Figure 13.21), but the foregoing statements remain valid in every case.

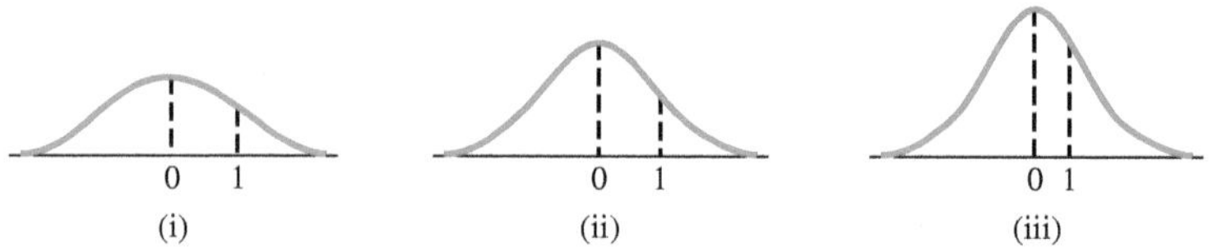

Figure 13.21 Three standard normal distributions with standard deviations $\sigma_1 > \sigma_2 > \sigma_3$ for (i), (ii), and (iii)

The preceding considerations have led to the notion of a *z score,* or *standard score.* If one wants to determine a characteristic of a population that is probably normally distributed (weights of full-term newborn babies, heights of senior boys attending Eisenhower High School, heights of senior girls attending Eisenhower High School, miles per gallon of 2010 Jeep Grand Cherokees, etc.) by a sampling procedure, it is often the case that each observation in the sample is converted to a z score using the following definition:

DEFINITION *z* Score

If x is an observation in a set of data with mean $\bar{x}$ and standard deviation s, the **z score** corresponding to x is given by $z = \dfrac{x - \bar{x}}{s}$.

By using z, it is possible to tell whether an observation is only fair, quite good, or rather poor. For example, a z score of 2 on a national test would be considered quite good, since it is 2 standard deviations above the mean. Indeed, we will see that the z score of 2 lies in the upper 2.5% of the population.

EXAMPLE 13.15

Calculating z Scores

Convert these data to a set of z scores:

66 64 62 64 66 64 62 64 66 63

Solution

$$\bar{x} \doteq \frac{66 + 64 + 62 + 64 + 66 + 64 + 62 + 64 + 66 + 63}{10} = 64.1.$$

Using the alternative formula for calculating the standard deviation given on page 703, we obtain

$$s = \sqrt{\frac{66^2 + 64^2 + 62^2 + 64^2 + 66^2 + 64^2 + 62^2 + 64^2 + 66^2 + 63^2}{10} - 64.1^2}$$

$$\doteq 1.45.$$

Then the different z scores are (since there are only four different data points)

$$\frac{66 - 64.1}{1.45} \doteq 1.31, \qquad \frac{64 - 64.1}{1.45} \doteq -0.07,$$

$$\frac{62 - 64.1}{1.45} \doteq -1.45, \quad \text{and} \quad \frac{63 - 64.1}{1.45} \doteq -0.76.$$

Now consider Example 13.15 further. From Table 13.4, pages 717–718, by indicating the area (and hence the probability) under the portion of a standard normal distribution to the left of a given point, it is possible to determine that 90.49% of the population lies to the left of (below) 1.31 and that only 7.35% of the population lies below -1.45. Considerations like these lead to the notion of a *percentile*.

DEFINITION Percentile

For *any* frequency distribution, the ***r*th percentile** is the number r, with $0 \leq r \leq 100$, such that r% of the data in a data set of a population is less than or equal to r.

EXAMPLE 13.16

Determining a Percentile of a Population

Use Table 13.4, pages 717–718, to determine the percentile corresponding to the z score of -0.76 in a standard normal population.

Solution

Searching down the left column of Table 13.4 on page 717, we come to the row marked -0.7. Moving across this row to the column headed .06, we find .2236, the entry for -0.76. This entry tells us that 22.36% of the population lies to the left of -0.76.

EXAMPLE 13.17

Determining a Percentile of a Sample

Determine the percentile corresponding to 63 in the data set of Example 13.15.

Solution

Since three of the ten scores in the data set are less than or equal to 63, and since $\frac{3}{10} = 0.30 = 30\%$, 63 is at the 30th percentile.

Note that the score of 63 in Example 13.15 gave a z score of -0.76. But the percentiles in Examples 13.16 and 13.17 are different. Why should this be so? The answer is that one percentile number is for a population and the other for a sample. The one for the sample should approximate the one for the population, but that does not mean they should be equal. The population mean and standard deviation were 64.2 and 1.9, respectively, while for the sample, $\bar{x}$ and s were 64.1 and 1.45, respectively. Thus, the population shows more spread, and it is not surprising that the z score -0.76 is further to the left relative to the population as a whole and that there is, therefore, relatively less of the population to its left than to the left of 63 in the sample.

EXAMPLE 13.18

Determining the Percentage of the Population in an Interval

Use Table 13.4 on pages 717–718 to show that 34% of a normally distributed population lies between the z scores -0.44 and 0.44.

Solution

Proceeding as in Example 13.16, we find that 33% of the population has a z score that is less than or equal to -0.44. Similarly, we find that 67% of the population lies to the left of 0.44. Thus, by subtraction, $67\% - 33\% = 34\%$ of the population lies between -0.44 and 0.44, as we were to show.

13.3 Problem Set

Understanding Concepts

1. Describe the population that should be sampled to determine each of the following:
   - **(a)** The percentage of freshmen in U.S. colleges and universities in 2010 who earn baccalaureate degrees within ten years
   - **(b)** The percentage of U.S. college and university football players in 2010 who earn baccalaureate degrees within ten years
   - **(c)** The fraction of people in the United States who feel that they have adequate police protection
   - **(d)** The fraction of people in Los Angeles who feel that they have adequate police protection
   - **(e)** Would you include children in the population you describe in parts (c) and (d)? people in mental institutions? known criminals?

2. Polls are often conducted by telephone. Might such a technique bias the results of the poll? Explain briefly, remembering that there are many cell phones.

3. Suppose a poll is conducted by face-to-face interviews, but the names of the interviewees are selected at random from names listed in the telephone book. Would such a poll yield valid results? Discuss briefly.

4. The registrar at State University wants to determine the percentages of students (a) who live at home, (b) who live in apartments, and (c) who live in dormitories. There are 25,000 students in the university, and the registrar proposes to select a sample of 100 students by choosing every 250th name from the list of all students arranged in alphabetical order.
   - **(a)** What is the population?
   - **(b)** Is the sample random? Explain.

5. In performing a study of college and university faculty attitudes in the United States, investigators first divided the population of all colleges and universities into groups according to size—25,000 or more students, 10,000 to 24,999 students, 3000 to 9999 students, and fewer than 3000 students. Using their judgment, they then chose two schools from each group and asked each school to identify a random sample of 100 of its faculty.
   - **(a)** Was this a good way to obtain a statistically reliable (that is, random) sample of faculty? Why or why not?
   - **(b)** What is the population? Are there four distinct populations? Discuss briefly.

6. The Honorable J. J. Wacaser, United States Representative, recently sent a questionnaire to his constituents to determine their opinion on several bills being considered by the House of Representatives. Discuss how representative of the voters in his district the responses to his poll are likely to be.

7. Discuss how Representative Wacaser (see problem 6) might actually choose a random sample of voters in his district.

8. To determine the average life of lightbulbs it manufactures, a company chooses a sample of the bulbs produced on a given day by selecting every 100th bulb and testing it for failure.
   - **(a)** What is the population?
   - **(b)** Is this a good way to select a sample? Why or why not?
   - **(c)** Is the sample random? Why or why not?

9. Choose a representative sample of 20 students in your college or university, and ask how many hours each person in your sample watches television each week.
   - **(a)** Describe how you chose your sample to ensure that it was representative of your entire student body.

(b) On the basis of your sample, estimate how many hours of television most students on your campus watch each week.

(c) Combine your data with those of all the other students in your class, and determine a revised estimate of the number of hours per week each student in your college or university watches television.

10. Describe and perform a study to determine how many of the students at your college or university have seen the movie *Gone With the Wind.*

11. For a population with a normal distribution with mean 24.5 and standard deviation 2.7,

(a) about 68% of the population lies between what limits?

(b) about 95% of the population lies between what limits?

(c) about 99.7% of the population lies between what limits?

12. Describe two different ways in which a random sample of 100 of the 10,000 students at State University can be obtained.

13. Suppose that only one dentist out of ten actually prefers Whito Toothpaste over all other brands. By taking many random samples of size 10, might it be possible eventually to obtain a sample in which eight out of ten dentists in the sample preferred Whito? Explain.

14. Convert these data sets into sets of z scores:

(a) 17 22 21 19 23 19

(b) 2 7 3 6 5 8

15. (a) Compute the sum of the z scores in problem 14a.

(b) Compute the sum of the z scores in problem 14b.

16. (a) In what percentile is the data value 22 in the data set of problem 14a?

(b) In what percentile is the data value 6 in the data set of problem 14b?

Use Table 13.4 on pages 717–718 to determine the answers to problems 17 and 18.

17. A population is normally distributed.

(a) What percentage of the population lies to the left of the population z score -1.75?

(b) What percentage of the population lies to the left of the population z score 0.26?

18. (a) What percentage of the population lies between the population z scores -1.75 and 1.75?

(b) What percentage of the population lies between the population z scores -0.67 and 0.67?

Teaching Concepts

19. Marita claims that tossing a single die will produce a random sequence of the digits 1, 2, 3, 4, 5, and 6. How would you respond to Marita?

20. After the class discussion of Marita's claim (problem 19), Mark asserts that you could generate a random sequence of the numbers 2, 3, . . . , 12 by repeatedly tossing a pair of dice. How would you respond to Mark?

21. Prompted by the discussions engendered by Marita's and Mark's claims (problems 19 and 20, respectively), Rebecca claims that you could generate a random sequence of two-digit numbers by repeatedly tossing a red die and a green die and recording the result on the red die as the first digit and the result on the green die as the second digit to form a two-digit number. How would you respond to Rebecca?

Thinking Critically

22. A large university was charged with sexual bias in admitting students to graduate school. Admissions were by departments, and the figures are as shown in the accompanying table.

(a) Compute the percentages of men and women applicants the admitted by the school as a whole.

(b) Do the figures in part (a) suggest that sexual bias affected the admission of students?

(c) Compute the percentages of men and women applicants admitted by each department.

(d) Do the figures in part (c) suggest that sexual bias affected admission to the various departments?

	Men		Women	
Department	**Number of Applicants**	**Number Admitted**	**Number of Applicants**	**Number Admitted**
1	373	22	341	24
2	560	353	25	17
3	325	120	593	202
4	191	53	393	94
5	417	138	375	131
6	825	512	108	89
Totals	2691	1198	1835	557

23. Suppose you generate a sequence of 0s and 1s by repeatedly rolling a die and recording a 0 each time an even number comes up and a 1 each time an odd number comes up. Is this a random sequence of 0s and 1s? Explain.

24. A TV ad proclaims that a study shows that eight out of ten dentists surveyed prefer Whito Toothpaste. How could it possibly make such a claim if, in fact, only one dentist out of ten actually prefers Whito?

25. Two sociologists mailed out questionnaires to 20,000 high school biology teachers. On the basis of the 200 responses they received, they claimed that fully 72% of high school biology teachers in the United States believe the biblical account of creation. Is their claim justified by this survey? Explain.

26. In the shoe business, which average of foot sizes is most important—the mean, median, or mode?

27. Consider the data set {6, 11, 10, 8, 12, 8}, where all the data are just 11 less than those in problem 14a. It's as if the

data were drawn from the same distribution moved 11 units to the left.

(a) What would you expect the z scores for this new set of data to be?

(b) Actually compute the z scores for this new set of data.

28. Show that the sum of the z scores for any set of data is equal to 0. (*Suggestion:* Argue from a special case, say, using the data set $\{1, 2, 3\}$, but do not actually compute the z scores; that is, use the strategy "Argue from a special case.")

29. Consider the data set $\{34, 44, 42, 38, 46, 38\}$, where all the data are just twice what they were in problem 14a.

(a) What would you expect the z scores of this data set to be?

(b) Compute the z scores for this data set.

(c) How do you account for the results in part (b)?

Thinking Cooperatively

30. Divide the class into groups of three or four students each and give each group eight pennies. Have each student in each group thoroughly shake and toss the pennies five times and record the number of heads obtained each time. Then have a member of the group record the total number of times zero heads, one head, . . . , eight heads were obtained by the group. Let the population to be studied be the combined set of data obtained by the entire class.

(a) Have each group construct a relative-frequency histogram and polygon (see Figures 13.16 and 13.17, respectively) for the population. Since the polygon shows the distribution of the population, have each group determine a consensus opinion as to whether the distribution approximates a normal distribution.

(b) Have each group compute the population mean and standard deviation.

(c) Have each group determine whether the 68–95–99.7 rule holds for this population.

31. Mai Ling claims that the spinner of Figure 13.15 likely would fail to generate a truly random sequence of digits because, in order to spin it, a person would likely hold it still and start each spin with the pointer in the same (likely horizontal) position each time. But always starting with the pointer in the same position would almost surely cause the spinner to favor some digits over others. How would you respond to Mai Ling?

COOPERATIVE INVESTIGATION

Using Samples to Approximate Characteristics of Populations

The chart shown contains 100 integers (the population) displayed in such a way that they can be represented by a number pair (a, b), where a denotes the row and b the column in which the integer appears. For example, entry (2, 7) is the integer 24, and entry (7, 3) is 26.

	0	1	2	3	4	5	6	7	8	9
0	21	22	20	24	22	29	25	21	27	17
1	25	12	28	21	22	17	28	18	18	26
2	19	17	23	29	19	16	24	24	25	19
3	22	17	26	11	31	19	14	20	23	17
4	26	13	30	26	18	23	37	24	27	28
5	14	15	25	20	24	18	20	30	35	21
6	18	30	22	20	20	23	27	26	33	13
7	24	21	23	26	28	19	28	29	31	23
8	21	27	22	25	21	16	23	27	16	25
9	23	22	24	22	16	15	19	24	25	20

For this investigation, parts (a) through (f) should be executed by each cooperative group of two or three students. Part (g) should involve the entire class.

(a) Use a spinner as in Figure 13.15 to generate five number pairs (a, b) to determine a sample of five numbers from the preceding table. Compute the mean and standard deviation of your five-number sample.
(b) Repeat part (a), but with a sample of size ten.
(c) On the basis of parts (a) and (b), give two estimates for each of the population mean and standard deviation.

(d) Record your means for parts (a) and (b) on the chalkboard.
(e) Determine $\bar{x}_5$, the mean of the means of the samples of size 5, and $\bar{x}_{10}$, the mean of the means of the samples of size 10, from the chalkboard. Also, compute s_5 and s_{10}, the standard deviations of the means of the samples of size 5 and size 10, respectively.
(f) Use the result of part (e) again to estimate the population mean and standard deviation.
(g) The population mean is 22.490 and the population standard deviation is 5.043. Briefly discuss the results of parts (a) through (f) in the context of these two numbers.

The Chapter in Relation to Future Teachers

This chapter is focused on the role of statistics in elementary and middle school. It is meant to prepare future teachers to introduce their students to data collection, representation, and interpretation, as well as to the notions of the center of a data set and measures of its variability. An introduction to statistics that starts in elementary school will be a platform on which your students can build as they become responsible citizens and decision makers. The power of statistics and the interplay between data and the beginning statistics that measure the data are the heart of this chapter.

Chapter 13 Summary

Section 13.1 Organizing and Representing Data	Page Reference
CONCEPTS	
• **Data:** A description, usually in numerical form, of a property (or properties) of a population.	669
• **Dot plot:** A representation of data that uses dots above a number line to represent data values.	669
• **Stem-and-leaf plot:** A representation of data in which the first one or more digits of each data value constitute the stem and the remaining digits constitute the leaves.	670
• **Histogram:** A representation of data in which the height of the bar at a point on the number line represents the number of data points in an interval about the point or the relative frequency of the data in the interval.	671
• **Frequency:** The number of times a data value appears in a data set.	671
• **Frequency of an interval:** The number of times a data value appears in a given interval along a number line.	671
• **Line graph:** A graph formed by joining data points.	672
• **Bar graph:** A histogram-like representation in which the bars are drawn above points on the horizontal axis corresponding to nonnumerical categories.	674
• **Pie chart:** A circle divided into sectors whose sizes (as measured by their angle) correspond to the percentages of a whole.	675
• **Pictograph:** A representation of data in which the frequency is indicated by icons chosen to bear some relationship to the data being represented.	676
• **Misleading representations:** Ways in which representations of data can be organized to confuse the viewer.	677
DEFINITIONS	
• A **data set** is a collection of data that can be organized and represented in various ways.	669
• **Data** is a description, usually in numerical form, of a property of a population.	669
• The **frequency** is the number of times any particular data value occurs.	671
• The **frequency of the interval** is the number of data values in any interval along a number line.	671

Section 13.2 Measuring the Center and Variation of Data	**Page Reference**
CONCEPTS	
• **Central tendency:** Description in a number of ways (mean, median, or mode) of the center of a data set.	685
• **Variability** is the spread of a data set, measured by the range, quartiles, or the standard deviation of the data set.	690
• **Outlier:** A data value less than $\bar{x} - 1.5 \cdot \text{IQR}$ or greater than $\bar{x} + 1.5 \cdot \text{IQR}$.	692
• **Box-and-whisker plot:** A representation summarized by five numbers (the least data value, Q_L, $\bar{x}$, Q_U, and the greatest data value).	693
• **Standard deviation:** A measure of the spread of the data in a sample. The standard formula for the standard deviation s is $s = \sqrt{\dfrac{(x_1 - \bar{x})^2 + (x_2 - \bar{x})^2 + \cdots + (x_n - \bar{x})^2}{n}}$, where $x_1, x_2, \ldots, x_n$ are the n data values of the data set.	694
DEFINITIONS	
• The **mean** is the sum of the data values in a sample, divided by the number of data values.	687
• The **median** is the middle data value for a sample with an odd number of data values, and the average of the two middle values for a sample with an even number of data values. The data values should be in numerical order.	688
• A **mode** is a data value in a sample that appears at least as often as every other data point.	688
• The **range** is the difference between the greatest and least data values in a sample.	690
• The **midrange** is the average of the highest and lowest data points in a sample.	690
• **Quartiles, Q_L and Q_U,** are the points such that 25% of the data set does not exceed Q_L and 25% of the data is not less than Q_U.	691
• The **interquartile range, IQR,** of a data set is the difference between the upper and lower quartiles, or $IQR = Q_L - Q_U$.	692
• An **outlier** is a data value less than $x - 1.5 \cdot$ IQR or greater than $x + 1.5 \cdot$ IQR.	692
• The **extremes** of a data set are the least and greatest scores.	693
• The **standard deviation** is the measure of the spread of data in a sample as given by the standard formula above.	694
• A data set has a **uniform distribution** if every value has exactly the same number of data points.	698
PROCEDURE	
• Using the five-step method to compute the **standard deviation.**	696

Section 13.3 Statistical Inference	**Page Reference**
CONCEPTS	
• **Population:** A particular set of objects about which information is desired.	702
• **Sample:** A subset of the population.	702,703
• **Random sample:** A subset chosen from a population in such a way that every such subset of the same size has an equal chance of being chosen.	702

- **Biased** data are selection criteria that systematically favor certain outcomes. 702
- **Relative frequency:** The frequency of the occurrence of a data value in a sample, expressed as a percentage. 705
- **Relative-frequency histogram:** A histogram in which the heights of the bars indicate relative frequency. 705
- **Distribution:** A curve or histogram showing the relative frequency of the measurements of a characteristic of a population. 706
- **Normal distribution:** A special bell-shaped distribution valid for many populations. 706
- **The 68–95–99.7 rule:** In a normal distribution, 68% of the population lies within 1 standard deviation of the mean, 95% of the population lies within 2 standard deviations of the mean, and 99.7% of the population lies within 3 standard deviations of the mean. 707
- **z score:** If x is a data value in a sample, the corresponding z score to x is $z = \dfrac{x - \bar{x}}{s}$, which standarizes normal distributions. 708

DEFINITIONS

- A **random sequence of digits** is a sequence of digits chosen in such a way that, at each step, each digit has an equal chance of being chosen. 703
- **Random numbers** are numbers formed by successively choosing the appropriate digits from a random sequence of digits. 703
- The **relative frequency** is the frequency of the occurrence of a data value in a sample, expressed as a percentage. 705
- The **relative-frequency histogram** is a histogram in which the heights of the bars indicate relative frequency. 706
- The ***r*th percentile** of a frequency distribution is the number r, with $0 \leq r \leq 100$, such that $r\%$ of the data set of the population is less than or equal to r. 709

Chapter Review Exercises

Section 13.1

1. The following are the numbers of hours of television watched during a given week by the students in Mrs. Karnes's fourth-grade class:

17	8	17	13	16	13	8	9	17	7
8	7	14	14	11	13	11	13	11	17
12	15	11	10	12	13	9	21	19	12

(a) Make a dot plot to organize and display these data.

(b) From the dot plot, estimate the average number of hours per week the students in Mrs. Karnes's class watch television.

2. Make a stem-and-leaf plot to organize and display the data in problem 1.

3. Choosing suitable scales, draw a histogram to summarize and display the data in problem 1.

4. The following are the numbers of hours of television watched during the same week as in problem 1, but by the students in Ms. Stevens's accelerated fourth-grade class:

13	8	9	11	11	12	8	9
11	11	6	8	9	11	11	6
8	9	11	11	6	8	9	11

Prepare a double stem-and-leaf plot to display and compare the number of hours of television watched by Mrs. Karnes's and Ms. Stevens's classes during the given week.

5. (a) Draw a line graph to show the trend in the retail price index of farm products as shown in this table:

1965	1970	1975	1980	1985	1990	1995	2000	2005
35	42	64	88	104	134	168	184	190

(b) Using part (a), estimate the retail price index for farm products in 1972.

(c) Using part (a), estimate the retail price index for farm products in 2010.

6. Find five numbers such that four of the numbers are less than the mean of all five.

7. Draw a pie chart to accurately illustrate how the State Department of Highways spends its budget if the figures are as follows: Administration—12%; New Construction—36%; Repairs—48%; Miscellaneous—4%.

8. (a) Criticize this pictograph, designed to suggest that the administrative expenses for Cold Steel Metal appear to be less than double in 2013 than in 2012 even though, in fact, the administrative expenses did double:

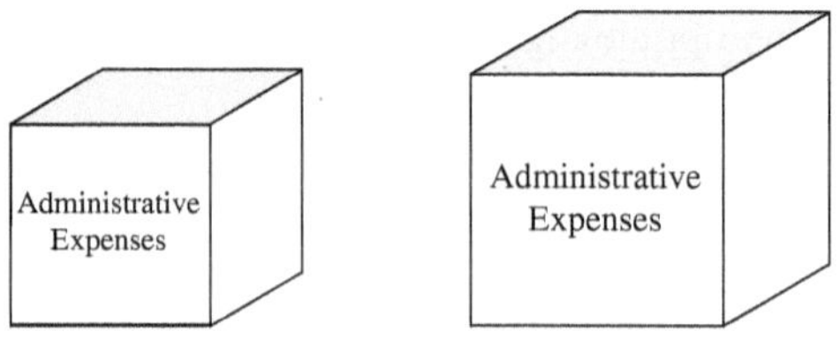

(b) If the administrators are challenged by the stockholders, can they honestly defend the pictograph? (*Hint:* Measure the cubes very carefully with a metric ruler and compute their volumes.)

9. What would you need to know in order to take the following statement seriously? "A survey shows that the average medical doctor in the United States earns $185,000 annually."

Section 13.2

10. Compute the mean, median, mode, and standard deviation for the data in problem 1.

11. (a) Compute the quartiles for the data in problem 1.

(b) Give the 5-number summary for the data in problem 1.

(c) Identify any outliers in the data of problem 1.

(d) Compute the quartiles for the data in problem 4.

(e) Give the 5-number summary for the data in problem 4.

(f) Identify any outliers in the data of problem 4.

(g) Using the same scales, draw side-by-side box plots to compare the data in problems 1 and 4.

12. Compute the mean and standard deviation for the data represented in these two histograms:

(a)

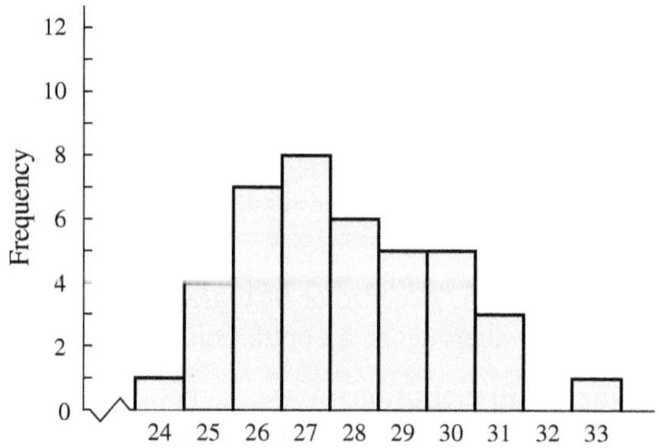

(b)

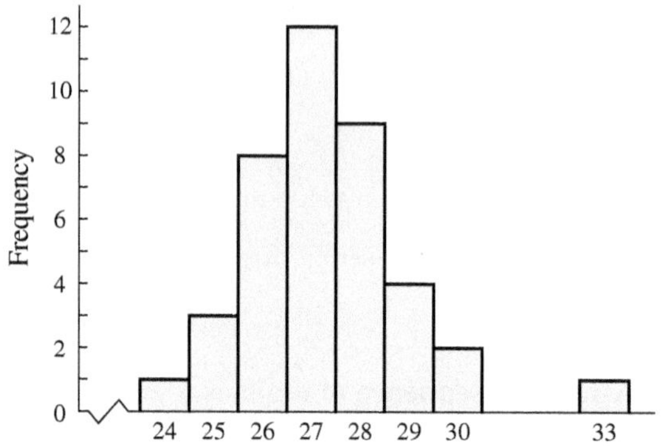

(c) Briefly explain the results of your computations in parts (a) and (b).

13. Three students were absent when the remaining 21 students in the class took a test on which their average score was 77. When the 3 students took the test later on, their scores were 69, 62, and 91. Taking these grades into account, what was the new average of all the test scores?

14. Mr. Renfro's second-period Algebra I class of 27 students averaged 75 on a test, and his fourth-period class of 30 students averaged 78 on the same test. What was the average of all the second- and fourth-period test scores?

Section 13.3

15. In a study of drug use by college students in the United States, the investigators chose a sample of 200 students from State University. Was this an appropriate choice for the study? Explain.

16. Suppose you want to use a sampling procedure to estimate the percentage of people in the United States who are unemployed. How might you describe the population that should be sampled? Should every person residing in the United States be included in the population? Discuss briefly.

17. Discuss briefly the biases that are inherent in samples obtained by voluntary responses to questionnaires like those sent out by members of Congress to their constituents.

18. Determine the z scores for the data set $\{7, 9, 6, 12, 15, 7, 9\}$.

19. What percentile is 12 in problem 18?

20. Use Table 13.4 on pages 717–718 to determine what percentile corresponds to the population z score 1.65.

21. Use Table 13.4 on pages 717–718 to determine what percentage of a standardized normal population lies between the population z scores -0.9 and 0.9.

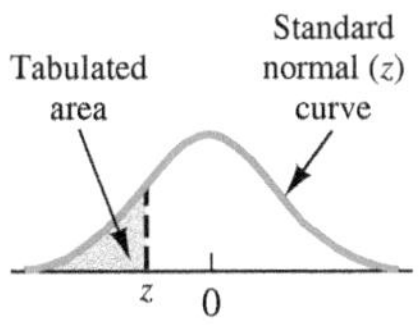

0.1

TABLE 13.4 AREAS TO THE LEFT OF *z* SCORES FOR A STANDARDIZED NORMAL DISTRIBUTION

z*	.00	.01	.02	.03	.04	.05	.06	.07	.08	.09
−3.8	.0001	.0001	.0001	.0001	.0001	.0001	.0001	.0001	.0001	.0000
−3.7	.0001	.0001	.0001	.0001	.0001	.0001	.0001	.0001	.0001	.0001
−3.6	.0002	.0002	.0001	.0001	.0001	.0001	.0001	.0001	.0001	.0001
−3.5	.0002	.0002	.0002	.0002	.0002	.0002	.0002	.0002	.0002	.0002
−3.4	.0003	.0003	.0003	.0003	.0003	.0003	.0003	.0003	.0003	.0002
−3.3	.0005	.0005	.0005	.0004	.0004	.0004	.0004	.0004	.0004	.0003
−3.2	.0007	.0007	.0006	.0006	.0006	.0006	.0006	.0005	.0005	.0005
−3.1	.0010	.0009	.0009	.0009	.0008	.0008	.0008	.0008	.0007	.0007
−3.0	.0013	.0013	.0013	.0012	.0012	.0011	.0011	.0011	.0010	.0010
−2.9	.0019	.0018	.0018	.0017	.0016	.0016	.0015	.0015	.0014	.0014
−2.8	.0026	.0025	.0024	.0023	.0023	.0022	.0021	.0021	.0020	.0019
−2.7	.0035	.0034	.0033	.0032	.0031	.0030	.0029	.0028	.0027	.0026
−2.6	.0047	.0045	.0044	.0043	.0041	.0040	.0039	.0038	.0037	.0036
−2.5	.0062	.0060	.0059	.0057	.0055	.0054	.0052	.0051	.0049	.0048
−2.4	.0082	.0080	.0078	.0075	.0073	.0071	.0069	.0068	.0066	.0064
−2.3	.0107	.0104	.0102	.0099	.0096	.0094	.0091	.0089	.0087	.0084
−2.2	.0139	.0136	.0132	.0129	.0125	.0122	.0119	.0116	.0113	.0110
−2.1	.0179	.0174	.0170	.0166	.0162	.0158	.0154	.0150	.0146	.0143
−2.0	.0228	.0222	.0217	.0212	.0207	.0202	.0197	.0192	.0188	.0183
−1.9	.0287	.0281	.0274	.0268	.0262	.0256	.0250	.0244	.0239	.0233
−1.8	.0359	.0351	.0344	.0336	.0329	.0322	.0314	.0307	.0301	.0294
−1.7	.0446	.0436	.0427	.0418	.0409	.0401	.0392	.0384	.0375	.0367
−1.6	.0548	.0537	.0526	.0516	.0505	.0495	.0485	.0475	.0465	.0455
−1.5	.0668	.0655	.0643	.0630	.0618	.0606	.0594	.0582	.0571	.0559
−1.4	.0808	.0793	.0778	.0764	.0749	.0735	.0721	.0708	.0694	.0681
−1.3	.0968	.0951	.0934	.0918	.0901	.0885	.0869	.0853	.0838	.0823
−1.2	.1151	.1131	.1112	.1093	.1075	.1056	.1038	.1020	.1003	.0985
−1.1	.1357	.1355	.1314	.1292	.1271	.1251	.1230	.1210	.1190	.1170
−1.0	.1587	.1562	.1539	.1515	.1492	.1469	.1446	.1423	.1401	.1379
−0.9	.1841	.1814	.1788	.1762	.1736	.1711	.1685	.1660	.1635	.1611
−0.8	.2119	.2090	.2061	.2033	.2005	.1977	.1949	.1922	.1894	.1867
−0.7	.2420	.2389	.2358	.2327	.2296	.2266	.2236	.2206	.2177	.2148
−0.6	.2743	.2709	.2676	.2643	.2611	.2578	.2546	.2514	.2483	.2451
−0.5	.3085	.3050	.3015	.2981	.2946	.2912	.2877	.2843	.2810	.2776
−0.4	.3446	.3409	.3372	.3336	.3300	.3264	.3228	.3192	.3156	.3121
−0.3	.3821	.3783	.3745	.3707	.3669	.3632	.3594	.3557	.3520	.3483
−0.2	.4207	.4168	.4129	.4090	.4052	.4013	.3974	.3936	.3897	.3859
−0.1	.4602	.4562	.4522	.4483	.4443	.4404	.4364	.4325	.4286	.4247
−0.0	.5000	.4960	.4920	.4880	.4840	.4801	.4761	.4721	.4681	.4641

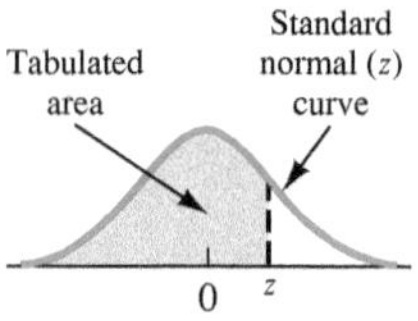

TABLE 13.4 CONTINUED

z*	.00	.01	.02	.03	.04	.05	.06	.07	.08	.09
0.0	.5000	.5040	.5080	.5120	.5160	.5199	.5239	.5279	.5319	.5359
0.1	.5398	.5438	.5478	.5517	.5557	.5596	.5636	.5675	.5714	.5753
0.2	.5793	.5832	.5871	.5910	.5948	.5987	.6026	.6064	.6103	.6141
0.3	.6179	.6217	.6255	.6293	.6331	.6368	.6406	.6443	.6480	.6517
0.4	.6554	.6591	.6628	.6664	.6700	.6736	.6772	.6808	.6844	.6879
0.5	.6915	.6950	.6985	.7019	.7054	.7088	.7123	.7157	.7190	.7224
0.6	.7257	.7291	.7324	.7357	.7389	.7422	.7454	.7486	.7517	.7549
0.7	.7580	.7611	.7641	.7673	.7704	.7734	.7764	.7794	.7823	.7852
0.8	.7881	.7910	.7939	.7967	.7995	.8023	.8051	.8078	.8106	.8133
0.9	.8159	.8186	.8212	.8238	.8264	.8289	.8315	.8340	.8365	.8389
1.0	.8413	.8438	.8461	.8485	.8508	.8531	.8554	.8577	.8599	.8621
1.1	.8643	.8665	.8686	.8708	.8729	.8749	.8770	.8790	.8810	.8830
1.2	.8849	.8869	.8888	.8907	.8925	.8944	.8962	.8980	.8997	.9015
1.3	.9032	.9049	.9066	.9082	.9099	.9155	.9131	.9147	.9162	.9177
1.4	.9192	.9207	.9222	.9236	.9251	.9265	.9279	.9292	.9306	.9319
1.5	.9332	.9345	.9357	.9370	.9382	.9394	.9406	.9418	.9429	.9441
1.6	.9452	.9463	.9474	.9484	.9495	.9505	.9515	.9525	.9535	.9545
1.7	.9554	.9564	.9573	.9582	.9591	.9599	.9608	.9616	.9625	.9633
1.8	.9641	.9649	.9656	.9664	.9671	.9678	.9686	.9693	.9699	.9706
1.9	.9713	.9719	.9726	.9732	.9738	.9744	.9750	.9756	.9761	.9767
2.0	.9772	.9778	.9783	.9788	.9793	.9798	.9803	.9808	.9812	.9817
2.1	.9821	.9826	.9830	.9834	.9838	.9842	.9846	.9850	.9854	.9857
2.2	.9861	.9864	.9868	.9871	.9875	.9878	.9881	.9884	.9887	.9890
2.3	.9893	.9896	.9898	.9901	.9904	.9906	.9909	.9911	.9913	.9916
2.4	.9918	.9920	.9922	.9925	.9927	.9929	.9931	.9932	.9934	.9936
2.5	.9938	.9940	.9941	.9943	.9945	.9946	.9948	.9949	.9951	.9952
2.6	.9953	.9955	.9956	.9957	.9959	.9960	.9961	.9962	.9963	.9964
2.7	.9965	.9966	.9967	.9968	.9969	.9970	.9971	.9972	.9973	.9974
2.8	.9974	.9975	.9976	.9977	.9977	.9978	.9979	.9979	.9980	.9981
2.9	.9981	.9982	.9982	.9983	.9984	.9984	.9985	.9985	.9986	.9986
3.0	.9987	.9987	.9987	.9988	.9988	.9989	.9989	.9989	.9990	.9990
3.1	.9990	.9991	.9991	.9991	.9992	.9992	.9992	.9992	.9993	.9993
3.2	.9993	.9993	.9994	.9994	.9994	.9994	.9994	.9995	.9995	.9995
3.3	.9995	.9995	.9995	.9996	.9996	.9996	.9996	.9996	.9996	.9997
3.4	.9997	.9997	.9997	.9997	.9997	.9997	.9997	.9997	.9997	.9998
3.5	.9998	.9998	.9998	.9998	.9998	.9998	.9998	.9998	.9998	.9998
3.6	.9998	.9998	.9999	.9999	.9999	.9999	.9999	.9999	.9999	.9999
3.7	.9999	.9999	.9999	.9999	.9999	.9999	.9999	.9999	.9999	.9999
3.8	.9999	.9999	.9999	.9999	.9999	.9999	.9999	.9999	.9999	1.0000

14 Probability

Cooperative Investigation
Strings and Loops

Materials

Six pieces of string per student, all pieces the same length (about 7 inches)

Procedure

Students work in pairs.

1. One student twists six lengths of the string into a loose bundle held with one hand. The student's partner then ties six knots, with three knots joining randomly selected pairs of the six strings at the top of the bundle and three other knots joining arbitrary pairs of strings at the bottom of the bundle. When the six knots have been tied, the bundle of string is put on a table.
2. The partners reverse roles and tie six knots in another six-string bundle.
3. The bundles are taken apart to identify what pattern of loops has been created by the six knots. There are three possible loop patterns, where any intertwining of the loops is of no importance:

   T: three small two-string loops;
   M: one medium four-string loop and one small two-string loop;
   L: one large six-string loop.

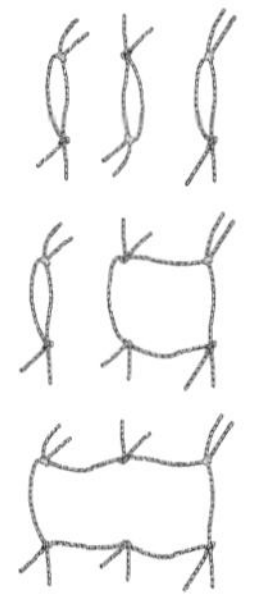

Class Project

Collect the data from all pairs of students and determine the ratios $n(T)/N$, $n(M)/N$, and $n(L)/N$, where $n(T)$, $n(M)$, and $n(L)$ are the respective numbers of times T, M, and L occurred and $N = n(T) + n(M) + n(L)$ is the total number of tied bundles. Which pattern seems most likely to occur? Which seems least likely? Are you surprised?

CHAPTER PREVIEW

Probability is the mathematics of uncertainty, in which the likelihood that a chance event occurs is measured by a number between 0 and 1, where 0 indicates there is no chance of the event occurring, and 1 indicates that the event must certainly occur. Fermat, Pascal, d'Alembert, and others made important contributions to probability theory in the seventeenth century, and the topic has grown in importance ever since.

Probability is obviously an essential tool in engineering—what is the likelihood that a critical part on an airplane may fail? And in medicine—what is the possibility of an allergic reaction to a drug? And in nearly every other discipline of science and technology, an understanding of probability is necessary. Probability also plays a more personal role for members of modern society: "How should I balance risk with an appropriate level of home, car, and life insurance coverage?" "Is it sensible to play the state lottery given my income level?"

Section 1 introduces the basic terminology and principles of probability. In particular, we will discover that there are two ways—experimental and theoretical—that the probability of an outcome is determined. **Experimental probability** is estimated by the number of times an outcome has occurred in past trials, and this number is then used to predict the likelihood that the outcome will occur in the future. **Theoretical probability** is based on considerations such as symmetry.

The calculation of a probability frequently depends on counting the number of ways an event may occur and comparing it to the total number of outcomes. Therefore, Section 2 discusses the basic principles of counting and their implications for probability. Section 3 focuses on two especially important counting principles—permutations and combinations—and illustrates how these principles are used to count arrangements and selections and then determine probabilities.

The concluding Section 4 investigates some additional topics from probability—odds, expected values, geometric probability, and simulations—so that all of the essential topics in probability listed in the Common Core State Standards for Mathematics are covered. For example, a Common Core Content Standard suggests that students "design and use a simulation to generate frequencies for compound events."

KEY IDEAS

- Terminology of probability: sample space, outcome, event, outcome favorable to an event, probability function, equally likely outcomes
- Experimental probability
- Theoretical probability
- Mutually exclusive events
- Complementary events
- Factorials
- The multiplication principle
- Multistage experiments
- Probability trees
- Conditional probability
- Independent events
- Counting permutations
- Combinations
- Odds
- Expected values
- Geometric probability
- Simulations

FROM THE NCTM PRINCIPLES AND STANDARDS

Students in grades 3–5 should begin to learn about probability as a measurement of the likelihood of events. In previous grades, they will have begun to describe events as certain, likely, or impossible, but now they can begin to learn how to quantify likelihood. For instance, what is the likelihood of seeing a commercial when you turn on the television? To estimate probability, students could collect data about the number of minutes of commercials in an hour.

Students should also explore probability through experiments that have only a few outcomes, such as using game spinners with certain portions shaded and considering how likely it is that the spinner will land on a particular color. They should come to understand and use 0 to represent the probability of an impossible event and 1 to represent the probability of a certain event, and they should use common fractions to represent the probability of events that are neither certain nor impossible. Through these experiences, students encounter the idea that although they cannot determine an individual outcome, such as which color the spinner will land on next, they can predict the frequency of various outcomes.

Source: Principles and Standards for School Mathematics by NCTM, *page 181. Copyright © 2000 by the National Council of Teachers of Mathematics. Reproduced with permission of the National Council of Teachers of Mathematics via Copyright Clearance Center. NCTM does not endorse the content or validity of these alignments.*

14.1 The Basics of Probability

Some of the basic terms of probability are defined in this introductory section. These terms are used to describe the basic principles of probability and how these principles are used to calculate the likelihood of chance events.

The Sample Space, Events, and Probability Functions

Many common activities depend on chance outcomes that cannot be predicted with certainty. For example, a move on a board game such as *Monopoly* or *Chutes & Ladders* depends on the roll of a pair of dice, and card games incorporate chance by dealing cards from a well-shuffled deck.

The following terms will be helpful as we begin to quantify the chances that a desirable, or perhaps undesirable, outcome may occur.

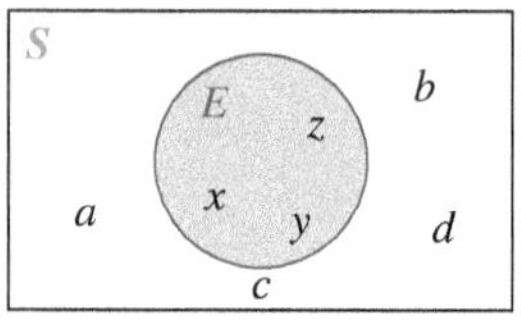

Figure 14.1
A Venn diagram with sample space *S*, event *E*, outcomes *x*, *y*, and *z* favorable to event *E*, and outcomes *a*, *b*, *c*, and *d* unfavorable to *E*

DEFINITIONS The Terminology of Basic Probability

Outcome x: a possible result of one trial of an experiment

Sample space S: the set of all possible outcomes of an experiment

Event E: a subset of outcomes, $E \subseteq S$

Favorable outcome to event E: an outcome x in event E, so that $x \in E$

It is helpful to visualize the sample space, events, and outcomes with a Venn diagram, such as shown in Figure 14.1. The universal set is the set of all of the outcomes in the sample space, and an event is a loop enclosing a subset of the sample space. The diagram shows that outcomes x, y, and z are favorable to event E. However, outcomes a, b, c, and d are unfavorable to event E, since none is a member of E.

EXAMPLE 14.1 Describing a Sample Space and Events with a Venn Diagram

The names of the 12 youngsters at a party are put in a hat, and one name is to be drawn at random to get a special prize.

	Name, age		Name, age		Name, age
a	Abbie, 8	*e*	Ellie, 9	*i*	Isobel, 9
b	Bea, 8	*f*	Frank, 8	*j*	Jason, 8
c	Carl, 9	*g*	Gary, 8	*k*	Karen, 7
d	David, 8	*h*	Hank, 9	*l*	Larry, 10

Illustrate the sample space with a Venn diagram that includes loops for the events E, the set of eight-year-olds; F, the set of nine-year-olds; and G, the set of girls at the party.

Solution

The events E and F are disjoint, so they are indicated with nonoverlapping loops. Because there are girls of both ages 8 and 9, the loop G must overlap with loops E and F. This gives us the following diagram.

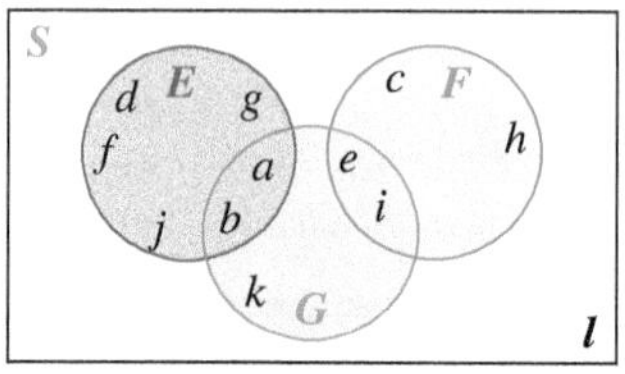

Except for a brief look at geometric probability at the end of this chapter, we will always assume that the sample space is a finite set, say, $S = \{x_1, x_2, \ldots, x_N\}$. Sometimes the number of outcomes in the sample space is obvious, such as $N = 12$ children at the birthday party described in Example 14.1. At other times, the number of outcomes N requires calculation. For example, it will be shown in Section 2 that there are $N = 3{,}628{,}800$ ways for ten people to line up for a photo.

To determine the probability of an event, we first introduce a probability function that measures the likelihood that an outcome x occurs.

DEFINITION Probability Function

A **probability function, P,** is a function defined on the sample space S so that each outcome $x_i \in S$ has a value $P(x_i)$, $0 \leq P(x_i) \leq 1$ and the sum of the probabilities of all the outcomes in the sample space is 1. That is, if $S = \{x_1, x_2, \ldots, x_N\}$, then

$$P(x_1) + P(x_2) + \cdots + P(x_N) = 1.$$

Once the probability function is chosen, the probability $P(E)$ of an event E is the sum of the probabilities of the outcomes favorable to the event.

DEFINITION Probability of an Event

Let $E = \{e_1, e_2, \ldots, e_M\}$ be an event in the sample space S with probability function P. Then the **probability of event E, $P(E)$**, is the sum of the probabilities of the outcomes in S that are favorable to E. That is,

$$P(E) = P(e_1) + P(e_2) + \cdots + P(e_M).$$

The probability function that is appropriate for an experiment is adopted in one of two ways: experimental or theoretical.

Experimental Probability

Experimental probability is based on experience, where the probability of an outcome is determined by how frequently the outcome has occurred in the past.

DEFINITION Experimental Probability

The **experimental probability** of an outcome x is the proportion of times outcome x occurs in a large number of trials of the experiment.

$$P(x) = \frac{\text{number of times outcome } x \text{ occurs}}{\text{number of trials of the experiment}}$$

For example, if a bent coin lands face up 63 times in 100 trials, the experimental probability of tossing a head is $P(H) = \frac{63}{100} = 0.63$, where increasingly accurate determinations of $P(H)$ can be obtained by conducting more experiments. Sometimes statistical data are used to determine an experimental probability. For example, car insurance companies use accident statistics to estimate the likelihood that a driver of ages 16–24 has an accident in a year's time, which is important information for the company in setting the insurance premium appropriately for young drivers.

EXAMPLE 14.2

Finding the Probabilities of Events for an Unfair Number Cube

Repeated rolls of a very asymmetric six-faced number "cube" shown showed that a 1, 3, or 4 each appeared 15% of the time, a 2 or a 5 each appeared 10% of the time, and a 6 appeared 35% of the time.

(a) What is the probability function on the sample space $\{1, 2, 3, 4, 5, 6\}$?

Using this probability function, what is the probability of rolling, in one try,

(b) a 1 or a 6?

(c) an even number?

(d) a prime number?

Solution

(a) The experimental probabilities are $p_1 = p_3 = p_4 = 0.15, p_2 = p_5 = 0.10$, and $p_6 = 0.35$, where p_1 is the probability of a 1, p_2 is the probability of a 2, and so on. Since $0.15 + 0.15 + 0.15 + 0.10 + 0.10 + 0.35 = 1$, the probabilities add to 1.

(b) $P(1 \text{ or } 6) = p_1 + p_6 = 0.15 + 0.35 = 0.5$,

(c) $P(2, 4, \text{ or } 6) = p_2 + p_4 + p_6 = 0.10 + 0.15 + 0.35 = 0.6$,

(d) $P(2, 3, \text{ or } 5) = p_2 + p_3 + p_5 = 0.10 + 0.15 + 0.10 = 0.35$.

The next examples illustrates why there is a close tie between statistics and probability.

EXAMPLE 14.3

Determining Probability from a Histogram

The accompanying table shows the heights of the first 43 presidents of the United States. For example, there has been one president shorter than 5′5″, James Madison at 5′4″. Use the histogram to determine the probability that a randomly selected president is at least as tall as the average height of 5′9″ of the current U.S. male.

Number of presidents	1	0	2	3	5	3	5	6	9	4	4	0	1
From height		5′5″	5′6″	5′7″	5′8″	5′9″	5′10″	5′11″	6′	6′1″	6′2″	6′3″	6′4″
up to	5′5″	5′6″	5′7″	5′8″	5′9″	5′10″	5′11″	6′	6′1″	6′2″	6′3″	6′4″	

Solution

There are $3 + 5 + 6 + 9 + 4 + 4 + 1 = 32$ presidents at least 5′9″ feet tall, for a probability of $\frac{32}{43} \doteq 0.74$.

Cooperative Investigation

Determining Experimental Probabilities

Materials

Each group should have about 25 new pennies, three or four wooden BBQ skewers or other thin dowels matching the side length of a tiled floor, and three or four identical small paper cups.

Directions

Work in small groups of three to four members to determine the experimental probabilities of each of the outcomes described. Data should be collected in a frequency plot from which the experimental probability can be calculated.

Experiment 1

On a smooth, flat table, stand pennies on their edge. Now gently rap the edge of the table so the pennies fall down. What is the experimental probability $P(H)$ of a head? Discuss if there is a surprise and what might account for this outcome.

Experiment 2

Consider the parallel seams between rows of floor tiles. Toss the skewers repeatedly onto the tiled floor and observe if a seam is crossed by the skewer or if it lies between two adjacent seams. Estimate the probability $P(C)$ of a crossing. The Comte de Buffon (1707–1788) showed that the theoretical probability is $2/\pi$. Does the experimental value agree? Combine the data from all of the groups to see if there is a better agreement between the theoretical and experimental probabilities.

Experiment 3

A paper cup tossed onto the floor may land upright, on its side, or upside down. First, have each group member guess the probability of each of these outcomes. Next, conduct trials to determine the experimental probabilities and see which group member had the best intuition.

Theoretical Probability

A theoretical probability function, unlike an experimental probability function, does not depend on past experiments or statistical data. Instead, the function is based on other assumptions. For example, if it is assumed that a coin is not bent and perfectly symmetrical, then tossing the coin should result in heads H or tails T equally often. Therefore, the theoretical probability of landing heads is $P(H) = \frac{1}{2} = 0.5$ and, similarly, $P(T) = \frac{1}{2} = 0.5$.

Most often, theoretical probability depends on assuming that each outcome of the sample space is **equally likely**. In this case, we say we have a **uniform sample space**. For example, each of the outcomes 1, 2, 3, 4, 5, 6 obtained by rolling a number cube (or fair die) is equally likely, so the sample space $S = \{1, 2, 3, 4, 5, 6\}$ is uniform and each outcome has probability 1/6. More generally, we have the following definition.

DEFINITION Theoretical Probability of Equally Likely Outcomes

Suppose that the N outcomes in the sample space $S = \{x_1, x_2, \ldots, x_N\}$ of an experiment are equally likely, so that S is a **uniform sample space.** Then the **theoretical probability,** or **mathematical probability,** of each outcome is the same, $1/N$. That is,

$$P(x_1) = P(x_2) = \cdots = P(x_N) = \frac{1}{N},$$

where $N = n(S)$ is the number of equally likely outcomes.

In Example 14.1, each of the 12 children is given the same chance to win the prize, so each child has the probability $1/12$ of winning.

For an event in a space of equally likely outcomes, the probability of an event is the ratio of the number of outcomes in the event to the total number of outcomes in the sample space. We have the following useful formula.

FORMULA The Probability of an Event in Sample Space of Equally Likely Outcomes

Let E be an event with $n(E)$ outcomes in a sample space S with $n(S)$ equally likely outcomes. Then the probability of the event is

$$P(E) = \frac{n(E)}{n(S)}.$$

EXAMPLE 14.4

Finding the Probabilities of Events for a Number Dodecahedron

In many games, the players roll a dodecahedron with the numbers 1 through 12 on its faces. What is the probability of rolling, in one try,

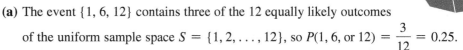

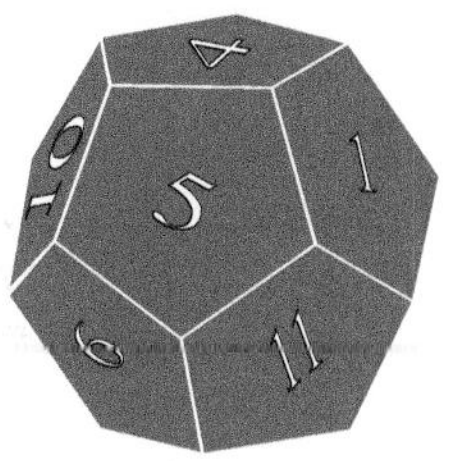

(a) a 1, 6, or 12?
(b) an odd number?
(c) a prime number?

Solution

(a) The event $\{1, 6, 12\}$ contains three of the 12 equally likely outcomes of the uniform sample space $S = \{1, 2, \ldots, 12\}$, so $P(1, 6, \text{ or } 12) = \frac{3}{12} = 0.25$.

(b) Half of the 12 numbers 1 through 12 are odd, so $P(\text{odd}) = \frac{6}{12} = 0.5$.

(c) The primes form the event $\{2, 3, 5, 7, 11\}$, so $P(\text{prime}) = \frac{n(\{2, 3, 5, 7, 11\})}{12} = \frac{5}{12} \doteq 0.42$.

Mutually Exclusive Events

Among the 12 children at the party described in Example 14.1, there are 6 eight-year olds and another 4 nine-year olds. What is the probability that the prize is awarded to one of these children? This is easy to answer: There are 6 + 4 children in the union of the two events, so the probability is $\frac{6+4}{12} = \frac{10}{12}$. In symbols, we have

$$P(\text{eight- or nine-year-old}) = P(E \cup F) = \frac{n(E \cup F)}{12} = \frac{n(E) + n(F)}{12} = \frac{6+4}{12} = \frac{10}{12}.$$

It is important to observe that no child belongs to both events. That is, $n(E \cup F) = n(E) + n(F)$, since $E \cap F = \varnothing$. We will say that events with no outcomes in common are **mutually exclusive.**

DEFINITION Mutually Exclusive Events
Two events E and F are **mutually exclusive** if there is no outcome favorable to both events: $E \cap F = \varnothing$.

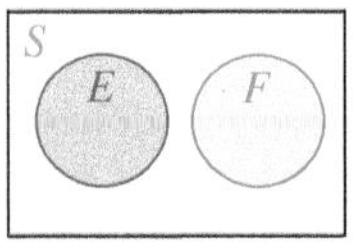

Figure 14.2
Two events E and F with no outcome in common are mutually exclusive

The loops corresponding to mutually exclusive events in the Venn diagram of the sample space can be drawn with no overlap.

The following theorem is evident from Figure 14.2.

THEOREM Probability of Mutually Exclusive Events
If E and F are mutually exclusive events in a sample space S, then

$$P(E \text{ or } F) = P(E \cup F) = P(E) + P(F).$$

If the events in S are equally likely, then

$$P(E \cup F) = \frac{n(E) + n(F)}{n(S)}.$$

It should be noticed that the word "or" corresponds to the union of the sets.

EXAMPLE 14.5

Drawing a Face Card or an Ace

What is the probability that a card drawn from an ordinary deck of 52 cards is either a face card (that is, a jack, queen, or king) or an ace?

Solution

Let E be the set of four aces and F the set of 12 face cards in the deck. Since an ace is not a face card, and a face card is not an ace, E and F are mutually exclusive events. Therefore,

$$P(\text{ace or face card}) = P(E \cup F) = \frac{n(E) + n(F)}{n(S)} = \frac{4 + 12}{52} = \frac{16}{52}.$$

Returning once again to Example 14.1, suppose we asked for the probability that the winner of the prize is either an eight-year-old or a girl. There are 6 eight-year-old children and 5 girls, but the event $E \cup G = \{a, b, d, e, f, g, i, j, k\}$ has 9 members, not 11. This is because $6 + 5$ would count the two eight-year-old girls Abbie and Bea twice rather than once. Since $n(E \cap G) = n(\{a, b\}) = 2$ is the number of children counted twice by $n(E) + n(G) = 6 + 5$, the correct count is obtained by subtracting the number of children counted twice. That is, since E and G are not mutually exclusive, the correct probability is given by

$$P(\text{an eight-year-old or a girl}) = P(E \cup G) = \frac{n(E) + n(G) - n(E \cap G)}{12} = \frac{6 + 5 - 2}{12} = \frac{9}{12}.$$

Thus,

$$P(E \text{ or } G) = P(E) + P(G) - P(E \cap G) = \frac{6}{12} + \frac{5}{12} - \frac{2}{12} = \frac{9}{12}.$$

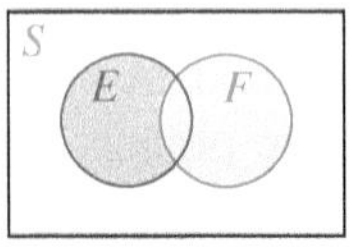

Figure 14.3
If events E and F are not mutually exclusive, then $n(E \cup F) = n(E) + n(F) - n(E \cap F)$ and $P(E \text{ or } F) = P(E) + P(F) - P(E \cap F)$

The same considerations apply to any two events E and F, as illustrated in Figure 14.3. It must be remembered that the word "or" in mathematics means "and/or" to allow the possibility that an outcome is favorable to both events.

For example, since every fourth and every sixth integer in the sample space $\{1, 2, \ldots, 1200\}$ is a multiple of 4 or 6, the probabilities of randomly choosing an integer that is a multiple of 4 or 6 are, respectively, $P(M_4) = \frac{1}{4}$ and $P(M_6) = \frac{1}{6}$. But, since $\text{LCM}(4, 6) = 12$, every twelfth integer is a multiple of both 4 and 6 and, therefore, the probability that an integer is a multiple of both 4 and 6 is $P(M_4 \text{ and } M_6) = \frac{1}{12}$. Thus,

$$P(\text{choosing a multiple of 4 or 6}) = \frac{1}{4} + \frac{1}{6} - \frac{1}{12} = \frac{3 + 2 - 1}{12} = \frac{1}{3}.$$

The formulas for computing the probability of two events that may not be mutually exclusive are highlighted in the following theorem.

THEOREM Probability of Non-Mutually Exclusive Events

If E and F are events in a sample space S, then

$$P(E \text{ or } F) = P(E \cup F) = P(E) + P(F) - P(E \cap F).$$

If the events in S are equally likely, then

$$P(E \cup F) = \frac{n(E) + n(F) - n(E \cap F)}{n(S)}.$$

The formulas in this theorem apply to any pair of events, mutually exclusive or not. If the events are mutually exclusive, then $n(E \cap F) = n(\varnothing) = 0$, and the foregoing formula is the same as given earlier for mutually exclusive events.

EXAMPLE 14.6

Drawing a Face Card or a Heart

What is the probability of drawing either a face card or a heart from an ordinary deck of 52 cards?

Solution

Let F be the set of the 12 face cards and H be the set of 13 hearts in the deck. Since the jack, queen, and king of hearts are three members of both sets, F and H are not mutually exclusive events. The probability of drawing a face card or a heart is, therefore,

$$\begin{aligned} P(\text{face card or heart}) &= P(F \cup H) \\ &= P(F) + P(H) - P(F \cap H) \\ &= \frac{12}{52} + \frac{13}{52} - \frac{3}{52} = \frac{22}{52}. \end{aligned}$$

Complementary Events

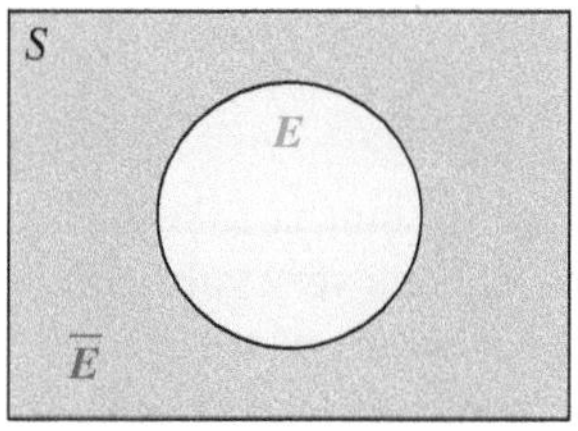

Figure 14.4
The complementary event $\overline{E}$ is the set of outcomes in the sample space that are not favorable to event E

Suppose that E is an event in the sample space S. The set of events *not* favorable to E is the complementary set $\overline{E}$ known as the **complementary event**.

DEFINITION Complementary Event

If E is an event in the sample space S, then the complementary set $\overline{E}$ is the **complementary event**.

In the Venn diagram of the sample space, the outcomes in S that are outside the loop representing event E form the complementary event $\overline{E}$.

Since E and $\overline{E}$ are mutually exclusive events and $S = E \cup \overline{E}$, we have that $1 = P(S) = P(E \cup \overline{E}) = P(E) + P(\overline{E})$. This gives us the two formulas in the following theorem.

THEOREM The Probabilities of Complementary Events
Let E and $\overline{E}$ be complementary events. Then

$$P(E) = 1 - P(\overline{E}) \quad \text{and} \quad P(\overline{E}) = 1 - P(E).$$

The relationship between the probabilities of complementary events is simple, but often there are situations where $P(\overline{E})$ is easier to calculate than $P(E)$, as demonstrated by the next two examples.

EXAMPLE 14.7 Solving "At Least" Problems

A fair coin is tossed ten times. What is the probability that

(a) at least one head appears?
(b) at least two heads appear?

Solution

The first toss has two outcomes, H and T. With two tosses, there are $4 = 2^2$ outcomes, namely HH, HT, TH, and TT. The doubling pattern is clear, and we see that there are $2^{10} = 1024$ equally likely sequences of heads and tails in ten tosses.

(a) If E is the event that at least one head appears, then the complementary event $\overline{E}$ is that no head appears. That is, $\overline{E}$ is the one outcome that all ten tosses are tails. Therefore, $P(\overline{E}) = \frac{1}{1024}$ and

$$P(E) = P(\text{at least one head}) = 1 - P(\overline{E}) = 1 - \frac{1}{1024} = \frac{1023}{1024}.$$

(b) Let F be the event that two or more heads appear in the ten tosses, so that the complementary event $\overline{F}$ is either no head or exactly one head. Thus, event $\overline{F}$ occurs if all ten tosses are tails or if heads land just once on the first, or second, . . . , or tenth toss with the other tosses all tails. Therefore, there are 11 outcomes favorable to $\overline{F}$, so $P(\overline{F}) = \frac{11}{1024}$. By complementary probabilities,

$$P(F) = P(\text{at least two heads}) = 1 - P(\overline{F}) = 1 - \frac{11}{1024} = \frac{1013}{1024}.$$

EXAMPLE 14.8 The Monty Hall Game Show Probability Problem

For many years, Monty Hall was the host of a TV game show in which a contestant is asked to choose one of three doors. Behind two of the doors are trivial, inexpensive prizes, but a very valuable prize is behind the third door. Once the contestant has chosen a door, Monty opens one of the other doors to reveal a trivial prize and the contestant is given the option of either staying with the original door or switching to the other unopened door. What is the contestant's better strategy, switch to the other unopened door or remain with the original choice?

Solution

Many people think it is better to stay with the original choice of door, since once a door is opened to show a trivial prize the probability that the correct door has been chosen seems to jump from $1/3$ to $1/2$. However, let's examine the situation more carefully.

Let $S = \{x, \bar{x}\}$, the sample space of winning strategies, where x is a win by making a "switch" and $\bar{x}$ is a win by the "don't switch" strategy. Clearly, $P(\bar{x}) = \frac{1}{3}$, since the prize is behind one of the three doors. But this means $P(x) = 1 - P(\bar{x}) = 1 - \frac{1}{3} = \frac{2}{3}$. The probability of winning the grand prize is doubled if doors are switched!

Cooperative Investigation
Does It Match?

Materials

Each participant has three 3×5 note cards. One card has a large X on both sides, another card a large O on both sides, and the third card has an X on one side and an O on the opposite side of the card. The Xs and Os should be drawn carefully to look the same.

Directions

Play the following game many times, keeping a record of wins and losses. Player 1 shuffles the three cards, frequently turning the cards over. Holding the cards so the Xs and Os are hidden, player 2 draws a card and can look only at one side. Player 2 then can choose either "same" or "opposite" and wins the game with the correct choice that the symbol on the unseen side of the card is the same or the opposite of the visible symbol. Pairs take turns being player 1 or player 2.

Question

If the visible symbol is an X, the card that was drawn is either the XX card or the XO card. This suggests the probability is 1/2 for "same" and "opposite." Do the data collected from playing the game support this confusion? If not, give a reason why one of the strategies, "same" or "opposite," is better than the other.

14.1 Problem Set

Understanding Concepts

1. (a) List explicitly all the outcomes for the experiment of tossing a penny, a nickel, a dime, and a quarter.
   (b) Determine the probability $P(HHTT)$ of obtaining a head on each of the penny and nickel and a tail on each of the dime and quarter in the experiment of part (a).
   (c) Determine the probability of obtaining two heads and two tails in the experiment of part (a).
   (d) Determine the probability of obtaining at least one head in the experiment of part (a). (*Hint:* Note that the complementary event consists of obtaining four tails.)

2. Describe the sample space for each of these experiments:
   (a) A coin is tossed and a single die is rolled.
   (b) A paper cup is tossed into the air and lands on the floor.
   (c) A card is drawn from an ordinary deck of 52 playing cards.

3. A red and a green die are rolled, with the outcome the pair of numbers on the two dice. For example, (3, 1) is the outcome that the red die shows 3 spots and the green die shows 1 spot.
   (a) Draw a diagram showing the sample space, where each outcome is an ordered pair showing the number of spots on the red and the green die in that order.
   (b) Draw loops in the sample space diagram that show each of these events:

   A = there is a total of 5 spots on the two dice,
   B = there is a total of 9 or more spots on the two dice,
   C = the green die shows a 5.

   (c) Which pairs of events—A and B, B and C, or A and C— are mutually exclusive? For each pair, explain why or why not.

4. Determine the probability of obtaining a total score of 3 or 4 on a single throw of two dice.

5. A family has two children, including at least one boy. What is the probability that the other child is a girl? (*Note:* The answer is *not* one-half.)

6. Acme Auto Rental has three red Fords, four white Fords, and two black Fords. Acme also has six red Hondas, two white Hondas, and five black Hondas. If a car is selected at random for rental to a customer,
   (a) what is the probability that it is a white Ford?
   (b) what is the probability that it is a Ford?
   (c) what is the probability that it is white?
   (d) what is the probability that it is white given that the customer demands a Ford?

7. Five black balls numbered 1, 2, 3, 4, and 5 and seven white balls numbered 1, 2, 3, 4, 5, 6, and 7 are placed in an urn. If one is chosen at random,
   (a) what is the probability it is numbered 1 or 2?
   (b) what is the probability that it is numbered 5 or that it is white?
   (c) what is the probability that it is numbered 5 given that it is white?

8. Mrs. Ricco has seven brown-eyed and two blue-eyed brunettes in her fifth-grade class. She also has eight blue-eyed and three brown-eyed blondes. A child is selected at random.
   (a) What is the probability that the child is a brown-eyed brunette?
   (b) What is the probability that the child has brown eyes or is a brunette?
   (c) What is the probability that the child has brown eyes given that he or she is a brunette?

9. Suppose that you randomly select a two-digit number (that is, one of 00, 01, 02, . . . , 99) from a sequence of random numbers obtained by repeatedly spinning a spinner. What is the probability that the number selected
   (a) is greater than 80?
   (b) is less than 10?
   (c) is a multiple of 3?
   (d) is even or is less than 50?
   (e) is even and is less than 50?
   (f) is even given that it is less than 50?

10. In a certain card game, you are dealt two cards face up. You then bet on whether a third card dealt is between the other two cards. (For example, a 10 is between a 9 and a queen, and so on.) What is the probability of winning your bet if you are dealt
   (a) a 5 and a 7?
   (b) a jack and a queen?
   (c) a pair of 9s?
   (d) a 5 and a queen?

11. In playing draw poker, a flush is a hand with five cards, all in one suit. You are dealt five cards and can throw away any of these and be dealt more cards to replace them. If you are dealt four hearts and a spade, what is the probability that you can discard the spade and be dealt a heart to fill out your flush?

12. If three coins are tossed onto the ground, there can be 0, 1, 2, or all 3 that land heads. Is the sample space $S = \{0, 1, 2, 3\}$ uniform? What probability function is appropriate for this sample space?

13. Two dice are thrown, one an ordinary number cube with its six faces numbered 1 through 6 and the second die a dodecahedron with faces numbered 1 through 12.
   (a) What is $n(S)$, the number of outcomes in the sample space?
   (b) What is the probability of rolling a 2 or an 18?
   (c) What is the probability of rolling a number in the event $\{7, 8, 9, 10, 11, 12, 13\}$?
   (d) What is the probability of rolling a number greater than 4?

14. A bin contains balls numbered 1 through 336. What is the probability that a ball drawn at random is
   (a) a multiple of 6 or 7?
   (b) a multiple of 6 or 8?

15. A standard set of balls used to play pocket billiards contains 16 balls. The balls numbered 1 through 7 are colored balls called "solids" (or "stars") and the balls numbered 9 through 15 are colored by a broad stripe and therefore are known as "stripes." There is also the black eight ball that is not a solid or stripe and the white unnumbered cue ball.
   (a) Draw a Venn diagram showing the events E, the evenly numbered balls, F, the stripes, and G, the solids.

   Use the Venn diagram to determine the probability that a ball chosen at random is
   (b) a solid or a stripe.
   (c) an odd-numbered ball.
   (d) either a solid or an even-numbered ball.

16. The heights of the students in a class are (in inches) 69, 62, 59, 63, 66, 67, 63, 67, 64, 70, 68, 64, 65, 67, 64, 62, 63, 64, 62, and 63. What is the probability that a randomly chosen student
   (a) is less than 5 feet tall?
   (b) is 5 feet, 4 inches tall?
   (c) is more than 5 feet tall?

17. Consider a spinner marked and shaded as indicated here:

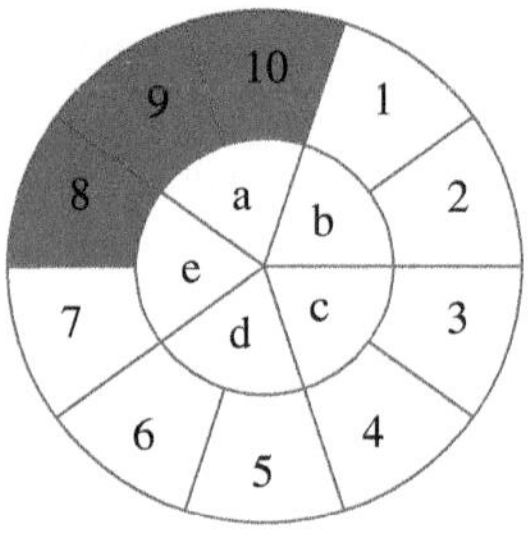

   (a) What is the probability that the spinner lands on the shaded area?
   (b) What is the probability that the spinner lands on region 1 or region 2?
   (c) What is the probability that the spinner lands on region 10 or region 6?
   (d) What is the probability that the spinner lands on region e?
   (e) What is the probability that the spinner lands on region 8 given that it lands on the shaded area?

(f) What is the probability that the spinner lands on a region marked by a vowel given that it lands on an odd-numbered region?

(g) What is the probability that the spinner lands on a region marked by a vowel or an odd number?

Into the Classroom

Students are often interested in first performing an experiment to calculate an experimental probability and then comparing the result obtained with the theoretical probability of the event. For the experiment outlined in the problems that follow, create a lesson plan that describes the materials needed and how to carry out the experiment, either as a class or with small groups. Be sure to give a careful derivation of the mathematical probabilities.

18. Ten note cards are numbered 1 through 10 and placed in a sack. In succession without replacement, two students each draw a card from the sack. Which, if either, of the following is the more likely outcome: The sum of the numbers on the cards is even or the sum is odd?

19. On a two-student team, one student puts either a black or a white marble into a sack without the second student looking. The second student puts a white marble in the sack, shakes the sack, and then draws out one marble. Is the color of the marble drawn helpful in guessing the color of the marble initially placed in the sack? (This is a problem proposed by Lewis Carroll of *Alice in Wonderland* fame in his book *Pillow Problems.*)

Responding to Students

20. **(Writing)** A fifth grader tells you she is certain it will rain sometime over the weekend. It seems that she heard on the weather forecast that there is a 50% chance of rain both Saturday and Sunday, and, of course, 50% + 50% = 100%, a certainty. Write an imagined dialogue with the student to convince her that it may not rain after all.

21. José reads the following problem in his math homework: "Bill and Susan are playing a game with a pair of 1–6 number cubes. They decide that Bill will earn a point if he rolls a multiple of 4 and Susan will earn a point if she rolls an odd number. Are these rules fair or unfair?" After some thought, José answers: fair, because Susan probably won't roll an odd number.

(a) Is José's answer correct?

(b) How would you help correct José's misconceptions of the rules?

22. Mischa and Jorge are playing a game with three sticks as shown here:

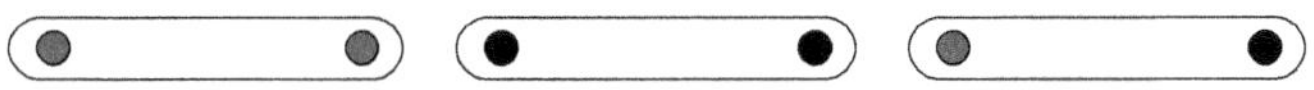

The sticks are put into a paper sack, and then one is drawn halfway out. Jorge sees that the stick has a red spot, so he knows that it is not the stick with two black spots. Since two sticks have a red spot, he claims that the probability that the hidden spot on the stick is black is $\frac{1}{2}$. Mischa doesn't think Jorge is right. Respond to Jorge and Mischa.

Thinking Critically

23. Suppose you are playing a game and are allowed to draw two cards in succession from an ordinary deck of 52 playing cards, where you win the game if either card is a face card. Is your chance of winning better or worse if the card first drawn is replaced in the deck?

24. Let E and F be events in a sample space S. If $P(E) = \frac{1}{2}$, $P(F) = \frac{1}{4}$, and $P(E \text{ and } F) = \frac{1}{8}$, what can be said about $P(E \text{ or } F)$?

25. Let E and F be events in a sample space S. If $P(E) = \frac{1}{2}$, $P(F) = \frac{1}{4}$, and $P(E \text{ or } F) = \frac{1}{2}$, what can be said about $P(E \text{ and } F)$?

26. Let E, F, and G be events in a sample space S, where $P(E) = \frac{1}{11}$, $P(F) = \frac{3}{11}$, and $P(G) = \frac{7}{11}$. If $P(E \text{ and } F \text{ and } G) = 0$, must $P(E \text{ or } F \text{ or } G) = 1$? Explain why or why not.

27. (a) Use a Venn diagram to justify the formula

$$n(E \cup F \cup G) = n(E) + n(F) + n(G) - n(E \cap F) - n(E \cap G) - n(F \cap G) + n(E \cap F \cap G).$$

Hint: Show that the right side of the formula counts each element of the set $E \cup F \cup G$ once by considering these cases: an element that belongs to exactly one set, such as $x \in E \cap \overline{F} \cap \overline{G}$; an element that belongs to exactly two sets, such as $y \in E \cap F \cap \overline{G}$; an element that belongs to all three sets, such as $z \in E \cap F \cap G$.

(b) Explain why

$$P(E \text{ or } F \text{ or } G) = P(E) + P(F) + P(G) - P(E \cap F) - P(E \cap G) - P(F \cap G) + P(E \cap F \cap G).$$

28. A ball is drawn at random from a bin containing balls numbered 1 through 1200. Use the results of the previous problem to determine the probability that the number on the ball is

(a) a multiple of 3, 4, or 5.

(b) a multiple of 4, 5, or 6.

29. Sicherman dice, unlike ordinary dice, have one number cube with faces marked 1, 3, 4, 5, 6, and 8 and a second number cube with faces marked 1, 2, 2, 3, 3, and 4.

(a) Complete this table showing the 36 equally likely outcomes of a roll of Sicherman dice.

	1	3	4	5	6	8
1	(1, 1)	(1, 3)				
2	(2, 1)					
2	(2, 1)					
3						
3						
4						

(b) List the outcomes favorable to the event of rolling a 5 with a pair of Sicherman dice, and give the probability of the event.

(c) What is the most likely value of a roll of a pair of Sicherman dice?

(d) In a two-player game, player A rolls a pair of ordinary dice and player B a pair of Sicherman dice, with the winner determined by the player with the higher roll. Which player, if either, has the advantage?

Making Connections

30. Roulette wheels in the United States have 18 red and 18 black pockets numbered 1 through 36, as well as the 0 and 00 green "house" pockets. In Europe, the roulette wheels are nearly the same but without the 00 house pocket. A bet is made by choosing a subset of pockets. For example, a *Manque* is a bet on 1 through 18, a *Rouge* is a bet on a red pocket, a Dozen is a bet on one of the sequences 1–12, 13–24, or 25–36. For a \$1 bet, the payout (including the \$1 bet) of a win is $36/n$ if n outcomes are favorable to the event bet on.

(a) What is the probability of winning a dollar playing *Manque* if there were no green "house" pockets?

(b) Why do you think a roulette wheel has one or two green "house" pockets?

(c) Which type of roulette wheel better favors the player, American or European?

31. In one unfortunate shipment, 10% of the MP3 players manufactured by Imperfect Electronics had defective switches, 5% had defective batteries, and 2% had both defects. If you purchased an MP3 player from this shipment, what is the probability that your player

(a) has a defective switch or a defective battery?

(b) has a good switch but a defective battery?

(c) has both a good switch and a good battery?

State Assessments

32. (Grade 5)
Lucien, Marcus, and Asad play on the basketball team. They have been keeping track of their success at the free throw line, creating this table.

Name	Free throws made	Free throws missed
Lucien	6	4
Marcus	8	4
Asad	7	2

What is the probability that Marcus makes his next free throw?

A. $\frac{1}{8}$ B. $\frac{1}{2}$ C. $\frac{2}{3}$

D. $\frac{8}{13}$ E. $\frac{9}{13}$

33. (Grade 5)
Each of the spinners shown below is partitioned into six equal sectors.

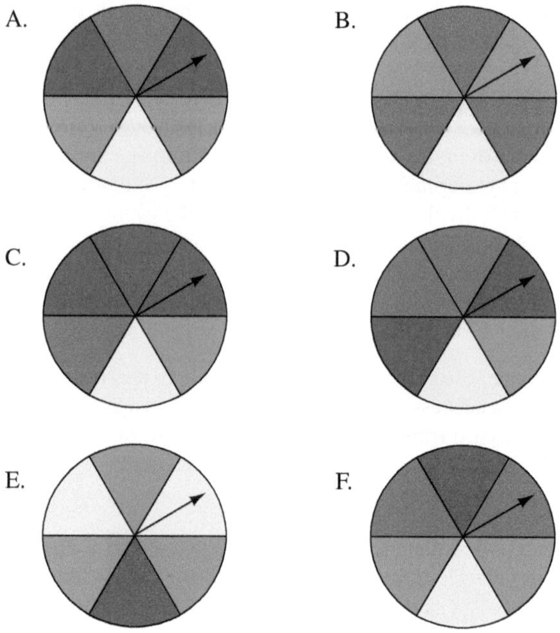

This table shows the outcome of 100 spins of one of the spinners.

Color	Number of spins resulting in this color
Blue	33
Red	34
Yellow	17
Green	16

Which spinner, A, B, C, D, E, or F, was spun 100 times?

34. (Washington State, Grade 4)
Special cakes are baked for May Day in France. A small toy is dropped into the batter for each cake before baking. Whoever gets the piece of cake with the toy in it is "king" or "queen" for the day.

Which cake below would give you the best chance of finding the toy in your piece?

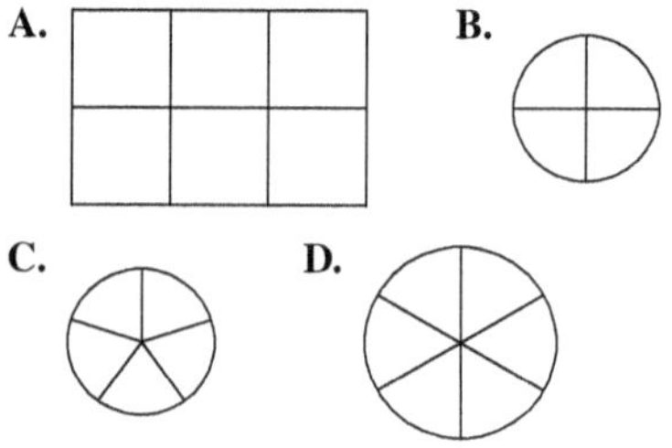

35. (Grade 5)
Mary has a one dollar bill, a five dollar bill, and a ten dollar bill in her purse. She randomly picks out two of the bills to pay for the \$12 paperback she is buying. What is the probability she has enough money with no need to use the remaining bill still in her purse?

A. $\frac{1}{2}$ B. $\frac{2}{3}$ C. $\frac{1}{2}$ D. $\frac{1}{15}$ E. $\frac{1}{3}$

36. (Washington, Grade 8)
Darin played a game with two spinners. The game is played by spinning each spinner and then adding the two resulting numbers. The goal is to spin two numbers that add up to eleven.

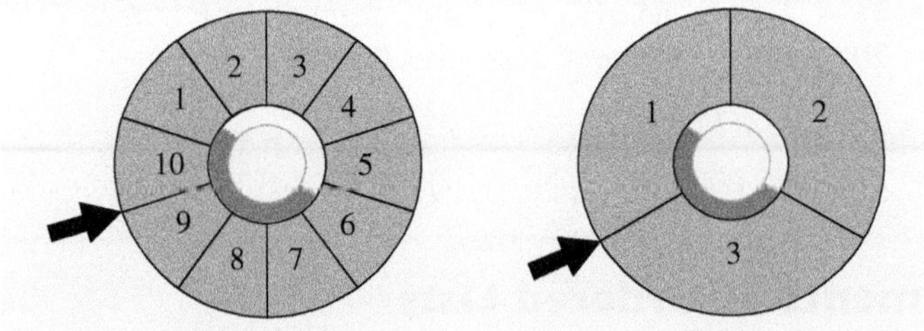

What is the probability that Darin will spin two numbers that add up to eleven?

A. $\frac{1}{3}$ B. $\frac{1}{10}$ C. $\frac{3}{5}$ D. $\frac{3}{10}$

37. (Grade 5)
Darien has a special die with its twelve faces numbered 1, 2, . . . , 12.

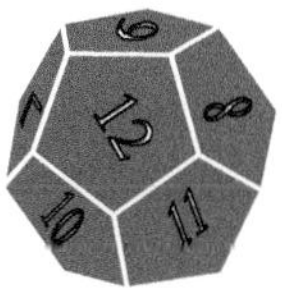

When Darien rolls this die, what type of number is most likely to appear on the upper face?

A. an even number. B. an odd number.
C. a number greater than 5. D. a number less than 9.

14.2 Applications of Counting Principles to Probability

If E is an event in a sample space S of equally likely outcomes, then the probability that an outcome of a trial is favorable to event E is the theoretical probability

$$P(E) = \frac{n(E)}{n(S)}.$$

Therefore, to calculate the probability we must determine the number of outcomes $n(E)$ favorable to E and the number of outcomes $n(S)$ in the sample space. The principles of counting to be discussed in this section will be very helpful.

The Addition Principle of Counting

We already encountered a simple case of the addition principle in the previous section, when it was shown that if E and F are mutually exclusive events in a sample space, then the number of outcomes favorable to either E or F is found by adding the number of outcomes in the events. That is,

$$n(E \text{ or } F) = n(E \cup F) = n(E) + n(F).$$

More generally, we have the following theorem.

THEOREM The Addition Principle of Counting

Let $E, F, \ldots, K$ be events in a sample space, where no outcome belongs to more than one of the events. That is, each pair of events is mutually exclusive. Then

$$n(E \text{ or } F \text{ or} \ldots \text{or } K) = n(E \cup F \cup \cdots \cup K) = n(E) + n(F) + \cdots + n(K).$$

As a consequence, when the terms of the equation of the addition principle of counting are divided by $n(S)$, we obtain the following application to probability.

THEOREM The Addition Principle of Probability

Let $E, F, \ldots, K$ be pairwise mutually exclusive events in a sample space, so that no outcome belongs to more than one of the events. Then the probability that an outcome is favorable to one of the events is

$$P(E \text{ or } F \text{ or} \ldots \text{or } K) = P(E \cup F \cup \cdots \cup K) = P(E) + P(F) + \cdots + P(K).$$

EXAMPLE 14.9

Computing the Probability of Drawing a Face Card, an Ace, or a Deuce

A card is drawn from a well-shuffled deck of 52 cards. What is the probability the card is a face card (jack, queen, or king), an ace, or a deuce (a two)?

Solution

Let F be the event of drawing a face card, A the event of drawing an ace, and T the event of drawing a two. No card from the deck belongs to more than one of the events, so each pair of events is mutually exclusive. Since there are $n(F) = 12$ face cards, $n(A) = 4$ aces, and $n(T) = 4$ twos, there are $n(F \cup A \cup T) = n(F) + n(A) + n(T) = 12 + 4 + 4 = 20$ favorable outcomes. This gives the probability

$$P(F \cup A \cup T) = \frac{n(F) + n(A) + n(T)}{n(S)} = \frac{12 + 4 + 4}{52} = \frac{20}{52} \doteq 0.38.$$

Factorials and Rearrangements of Ordered Lists

Consider this problem.

Three envelopes, addressed with labels A, B, and C, have been randomly stuffed with letters addressed to persons a, b, and c. What is the probability that each letter was placed in the proper envelope?

Since there are only three letters, we can make a list of all ways the letters may have been placed in the envelopes.

A	a	a	b	b	c	c
B	b	c	a	c	a	b
C	c	b	c	a	b	a

The table shows there are six equally likely ways to stuff the envelopes, with just one way all three letters are in the correct envelopes. Therefore, the probability that each letter is in the correct envelope is $1/6$.

Direct counting would be of no use if there were a larger number of letters to be mailed, so we should find a better method to count the number of ways to randomly stuff addressed envelopes. Fortunately, there is a quick way to count: Imagine stuffing envelope A, then stuffing envelope B, and so on, and determining how many ways are available to carry out each step. For three envelopes and letters, there are three steps:

Step 1. Stuff envelope A There are three choices, letter a, or letter b, or letter c.
Step 2. Stuff envelope B Whatever letter was placed in envelope A, there are still two choices of letters to choose from to place in envelope B.
Step 3. Stuff envelope C There is just one letter left, so there is one choice to place a letter in envelope C.

Our counting problem has become much simpler by viewing the outcomes as the result of a **multistep** process. Moreover, the choices made at each step can be depicted with a **possibility tree**, as seen in Figure 14.5. The number of choices at each step are multiplied to show there are $3 \times 2 \times 1 = 6$ arrangements. This tree is drawn sideways, but often a possibility tree is drawn with its branches downward.

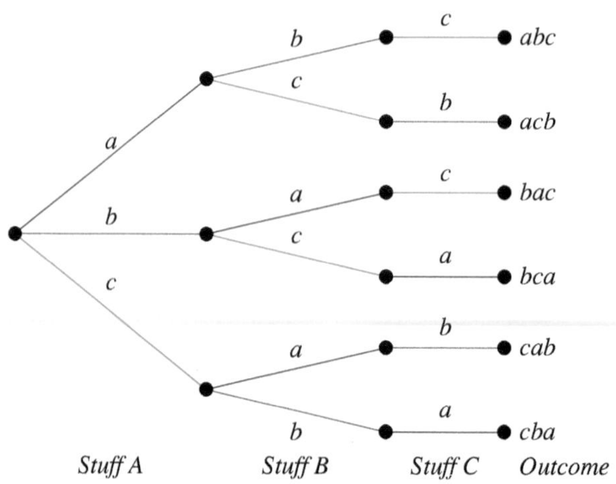

Figure 14.5
This possibility tree shows there are $3 \times 2 \times 1 = 6$ ways to make an ordered list of three items

If there were four addressed letters to be stuffed into four addressed envelope, the same reasoning can be applied, but we now have a four-stage process: There are four choices to stuff the first envelope, then three choices, then two choices, and then one way to stuff the last envelope, giving $4 \times 3 \times 2 \times 1 = 24$ ordered arrangements in all. This type of product is called a **factorial**, written $4! = 4 \times 3 \times 2 \times 1$.

Our reasoning works just as well for counting the number of ways that any number of items can be put into an order. Arrangements into ordered lists are known as **permutations**, so we have the following result.

THEOREM The Number of Permutations of a Set of Objects
The number of permutations of n objects into an ordered list is given by the factorial

$$n! = n \times (n - 1) \times (n - 2) \times \cdots \times 3 \times 2 \times 1.$$

According to this theorem, there are $5! = 5 \times 4 \times 3 \times 2 \times 1 = 120$ ways to place five letters into envelopes, $6! = 6 \times 5 \times 4 \times 3 \times 2 \times 1 = 720$ permutations of six objects, and so on. The number of permutations increases very rapidly, and it can be computed that $10! = 3{,}628{,}800$ and $25!$ is around 1.55×10^{25}.

EXAMPLE 14.10

Determining a Probability with Factorials

Hans is stuffing ten addressed letters $a, b, \ldots, j$ randomly into ten addressed envelopes $A, B, \ldots, J$. What is the probability that

(a) letters A, B, and C each contain the correct letter?
(b) all but letter j has been placed in the correct envelope?

Solution

(a) There are $10! = 3{,}628{,}800$ ways Hans may have stuffed the envelopes, including the $7!$ ways that envelopes A, B, and C are correctly stuffed with letters a, b, and c. This gives the probability

$$P(a, b, c \text{ correct}) = \frac{7!}{10!} = \frac{7 \times 6 \times 5 \times 4 \times 3 \times 2 \times 1}{10 \times 9 \times 8 \times 7 \times 6 \times 5 \times 4 \times 3 \times 2 \times 1} = \frac{1}{10 \times 9 \times 8} = 0.0013\overline{8}.$$

(b) It is impossible that only letter j is in the incorrect envelope so the probability is 0.

The Multiplication Principle of Counting

We have seen that it is often helpful to view the outcome of an experiment or procedure as the result of a sequence of steps. This is known as a **multistage experiment** or a **multistep process.** Since several events are considered along the way, this situation is called a **compound event.**

If the number of options at each step is known and is the same independently of the options chosen at preceding steps, then the number of outcomes is found by multiplying the number of options at each step. This method of counting gives us the following important result.

THEOREM The Multiplication Principle of Counting
Suppose that the outcomes of a multistage experiment or multistep process are the result of choosing one of n_1 options at step 1, choosing one of n_2 options at step 2, and so on, where the number of options at each stage is independent of which choices were made at earlier stages. Then the number of outcomes is the product $n_1 \times n_2 \times \cdots$ of the number of options available at each step.

EXAMPLE 14.11

Calculating the Number of PINs and Passwords

(a) What is the number of PINs? A PIN (personal identification number) is a sequence of four digits.
(b) How many PINs have no repeated digit?
(c) How many case-sensitive passwords have the form of eight symbols that begin with a sequence of four letters and end with a sequence of four digits?

Solution

(a) A PIN is the result of making a sequence of four steps. The first digit can be chosen in ten ways, the second digit can be chosen ten ways independently of which digit was already chosen, and so on. Therefore, there are $10 \times 10 \times 10 \times 10 = 10{,}000$ PINs. This is not surprising, since they can be easily listed: 0000, 0001, 0002, . . . , 9998, 9999.

(b) There are ten digits that can be chosen as the first symbol, which leaves nine digits different from the leading symbol that can be chosen as the second digit of the PIN. Notice that there are always nine choices for the second digit independently of which digit was chosen for the leading digit. Similarly, there are always eight ways to choose the third digit, and always seven ways to choose the fourth digit. This means there are $10 \times 9 \times 8 \times 7 = 5040$ PINs with no repeated digit.

(c) Since capital letters are distinguished from lowercase letters, the first four letter symbols of the password can be chosen in $52 \times 52 \times 52 \times 52 = 7{,}311{,}616$ ways. The last four digits, like PINs, can be chosen in 10,000 ways. Therefore, there are 73,116,160,000 passwords of the form described.

EXAMPLE 14.12 Calculating the Probability of a License Plate Number

Suppose that the license plates in a certain state are always three letters followed by four digits. At a trial, a witness to a robbery identified the license plate of the getaway car as ?CG?49? but couldn't read the first letter or the first and last digits. The license plate of the defendant's car is WCG3495. What is the probability the defendant's car was the getaway car?

Solution

This can be viewed as a three-stage experiment: Determine the missing letter, then the first digit, and finally the last digit. There are 26 letters and 10 digits, so by the multiplication principle, there are $26 \times 10 \times 10 = 2600$ license plates of the form ?CG?49? This means the probability is $1/2600$ that the getaway car is that identified by the witness.

Into the Classroom
Jenifer Martin on the Principles of Counting

The addition and multiplication principles of counting are commonly confused among students. Having hands-on activities to differentiate the two makes it easier for students. To illustrate the addition principle of counting, have students fold a square paper into fourths by folding once horizontally and once vertically. Before unfolding, have students think about how many squares there will be once the paper is unfolded, making sure that they've counted them all. Once they unfold their papers, have students organize their results: For each kind of square, how many are there? There are four small squares, plus one large square, for a total of five squares. You can extend this by asking students to identify the number of quadrilaterals. Students should find 4 small squares + 1 large square + 2 horizontal rectangles + 2 vertical rectangles = 9 quadrilaterals.

To illustrate the multiplication principle of counting, I use paper doll cutouts. Each doll has three dresses (yellow, blue, and gray) and two belts (green and yellow). If a doll is to wear one dress and belt, how many different outfits can you create? Students make every possible outfit, paying attention to the order in which they dress the doll (i.e., picking one dress and trying both belts or picking a belt with each of the three dresses). Having students orate their results one by one iterates the idea that when picking the dress first and trying two belts at a time leads to $2 + 2 + 2 = 2 \times 3 = 6$, but picking the belt first with each of the three dresses results in a pattern of $3 + 3 = 3 \times 2 = 6$. Afterward, students are led into a variety of ways to show their results through a table (which illustrates the Cartesian product), tree diagrams, or organized lists.

The next example is solved by creating the possibility tree that corresponds to a two-stage experiment.

EXAMPLE 14.13

Calculating the Probability of a Two-Stage Experiment

Josie is given an opportunity to draw out two bills at random from a bag containing two one-dollar bills, a five-dollar bill, and a ten-dollar bill. What is the probability that Josie is lucky enough to get a \$15 payout?

Solution

The possibility tree shows the first bill can be withdrawn in four ways and the second bill in three ways, so there are $4 \times 3 = 12$ equally likely outcomes. Of these 12 outcomes, two—drawing a five-dollar bill and then a ten-dollar bill, or the reverse—result in the maximum payout of \$15. This means the probability Josie gets a \$15 payout is $P(15) = \frac{2}{12} = 0.166\ldots.$ The possibility tree also shows that

$$P(11) = P(6) = \frac{4}{12} = 0.333\ldots$$

$$\text{and } P(2) = \frac{2}{12} = 0.166\ldots.$$

It should not be a surprise that $P(2) + P(11) + P(6) + P(15) = \frac{2 + 4 + 4 + 2}{12} = 1.$

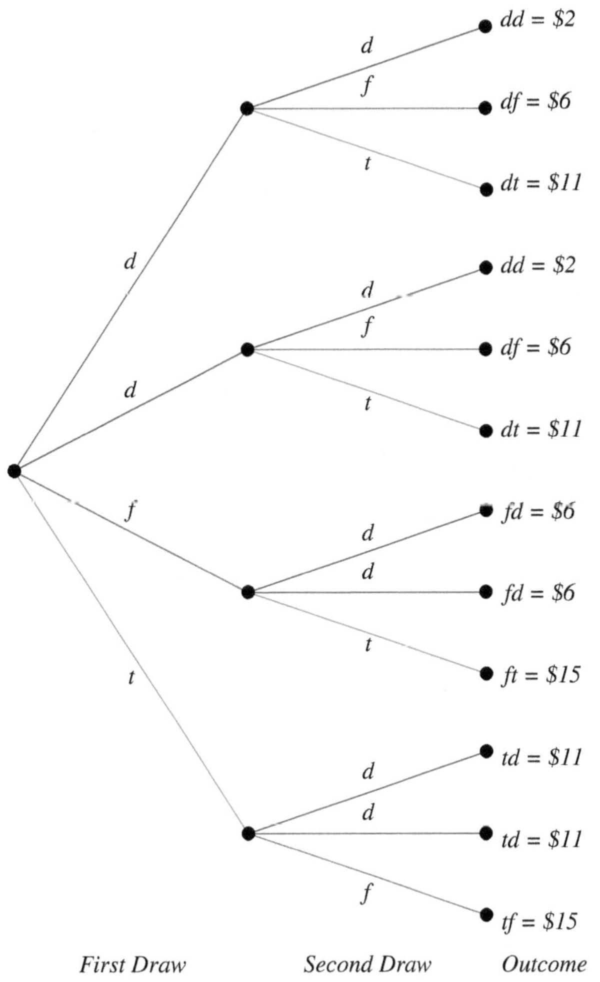

Probability Trees

For complicated multistage experiments, it is often better to replace the possibility tree with a **probability tree.** Each branch of the tree corresponds to the option made at that step, with the probability of the option written along that branch. The probability of outcomes is then found by multiplying the probabilities along the branches of the tree that reach the final outcome.

For example, Figure 14.6 shows a probability tree for Example 14.13. On the first draw, option d has a probability of $1/2$ since there are two dollar bills from among the four bills. On the second draw, option d has a probability of either $1/3$ or $2/3$, depending on whether or not a dollar bill was withdrawn on the first draw.

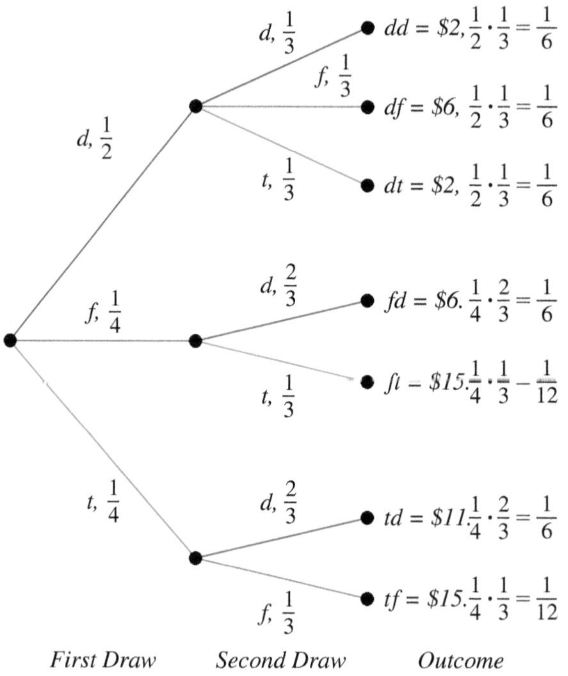

Figure 14.6 A probability tree for drawing two bills randomly from a bag containing two one-dollar bills, a five-dollar bill, and a ten-dollar bill

Conditional Probability

In many probabilistic situations, the probability of one event is modified by knowing if another event has occurred or not. For example, in Example 14.10, suppose it is known that the first bill drawn is the ten-dollar bill. This leaves the five-dollar bill and the two one-dollar bills, so the probability of getting a \$15 payout is $1/3$ with this additional information. We say that $1/3$ is a **conditional probability,** the probability that depends on knowing that another event—drawing a ten-dollar bill first—has already occurred.

DEFINITION Conditional Probability

Let E and F be two events. Then the **conditional probability** $P(E \mid F)$ is the probability that event E occurs given that event F occurs.

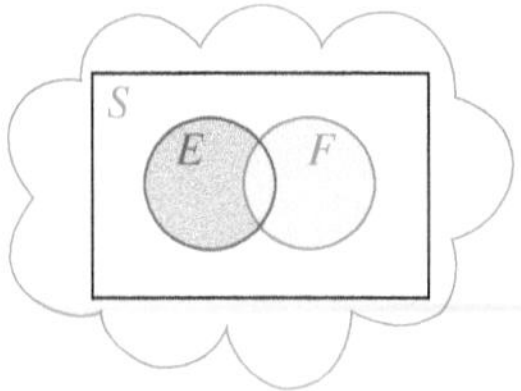

A formula for the conditional probability $P(E \mid F)$ is easy to derive from the Venn diagram of the sample space. If it is assumed that event F occurs, then the only outcomes to be considered are the $n(F)$ outcomes favorable to F. Of these, there are $n(E \cap F)$ outcomes in event E that are also favorable to event E. This gives us the following theorem.

THEOREM The Conditional Probability Formula

Let E and F be events in a uniform sample space. Then the conditional probability of event E, given that event F occurs, is

$$P(E|F) = \frac{n(E \cap F)}{n(F)}.$$

EXAMPLE 14.14

Determining a Conditional Probability in *Monopoly*

Patsy is playing *Monopoly* and does not want to roll a 7, 8, or 9 to avoid landing on properties with expensive rents. She first rolled a 4 with the red die and is now about to roll the green die. What is the probability she avoids trouble?

Solution

Patsy must roll a 1, 2, or 6 with the green die to avoid paying rent, so the probability is $3/6 = 0.5$. But let us see how the conditional probability formula could have been used. There are six ways the red die can land and six ways for the green die to land, so the sample space is the set of 36 equally likely outcomes $\{(1,1), (1, 2), (1, 3), \ldots, (6, 5), (6, 6)\}$. For example, (5, 3) is a roll of 8 with a 5 on the red die and a 3 on the green die. Let E be the event of a safe roll of 2, 3, 4, 5, 6, 10, 11, 12 of the two dice. That is, $E = \{(1, 1), (1, 2), (1, 3), (1, 4), (1, 5), (2, 1), (2, 2), (2, 3), (2, 4), (3, 1), (3, 2), (3, 3), (4, 1), (4, 2), (4, 6), (5, 1), (5, 5), (5, 6), (6, 6)\}$. Next, let F be the event that the red die is a 4, so $F = \{(4, 1), (4, 2), (4, 3), (4, 4), (4, 5), (4, 6)\}$ and $n(F) = 6$. Also, $n(E \cap F) = n(\{(4, 1), (4, 2), (4, 6)\}) = 3$. Therefore,

$$P(E|F) = \frac{n(E \cap F)}{n(F)} = \frac{3}{6} = 0.5.$$

An interesting consequence of the conditional probability formula is obtained by multiplying it by $n(F)$ to obtain $n(E \text{ and } F) = n(E \cap F) = P(E|F)n(F)$. Dividing by $n(S)$, the number of outcomes in the sample space, we get the following result.

THEOREM The Probability of Compound Events

The probability that events E and F both occur is given by

$$P(E \text{ and } F) = P(E \cap F) = P(E|F)P(F).$$

EXAMPLE 14.15

Determining Probabilities of a Spinner

Let the following events be defined for the spinner shown at the right.

R, land on a red sector
B, land on a blue sector
G, land on a blue sector
E, land on an even number
F, land on a prime number

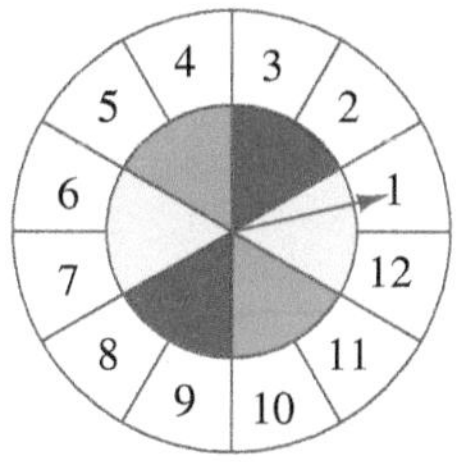

Determine the probabilities of these events.

(a) $P(R \cap F)$
(b) $P(R|G)$
(c) $P(\overline{B} \cap E)$
(d) $P(G \cap F)$
(e) Is it true that $P(B|E)P(E) = P(E|B)P(B)$?

Solution

(a) $P(R \cap F) = P(R|F)P(F) = \frac{2}{5} \cdot \frac{5}{12} = \frac{1}{6}$. Notice that only two of the 12 sectors, 2 and 3, are prime numbers that border a red sector.
(b) $P(R|G) = 0$ since R and G are mutually exclusive.
(c) $P(\overline{B} \cap E) = P(\overline{B}|E)P(E) = \frac{4}{6} \cdot \frac{1}{2} = \frac{1}{3}$. The even numbers 2, 4, 8, and 10 border either a red or green sector.
(d) $P(G \cap F) = P(G|F) \cdot P(F) = \frac{2}{5} \cdot \frac{5}{12} = \frac{1}{6}$. The primes 5 and 11 border green sectors.
(e) $P(B|E)P(E) = \frac{2}{6} \cdot \frac{1}{2} = \frac{1}{6}$ and $P(E|B)P(B) = \frac{2}{4} \cdot \frac{2}{6} = \frac{1}{6}$. This is expected since $P(E \cap F) = P(F \cap E)$.

Part (e) of Example 14.15 illustrates the general property that

$$P(E \cap F) = P(E|F)P(F) = P(F|E)P(E).$$

Independent Events

In many cases, the probability of event E is the same whether or not event F occurs. That is, $P(E|F) = P(E)$. When this is the case, E and F are said to be **independent events** and we have the following result.

THEOREM The Multiplication Property of Independent Events
Let E and F be independent events. Then the probability that an outcome is favorable to both events is the product of the probabilities the outcome is favorable to event E and event F. That is,

$$P(E \text{ and } F) = P(E \cap F) = P(E) \cdot P(F).$$

EXAMPLE 14.16

Flipping a Coin and Rolling a Die

Morry is about to flip a penny and roll a die. What is his probability of getting a head on the coin and an even number on the die?

Solution

Assuming the penny hasn't been glued to the die or there isn't some other reason the events are not independent, the flip of the coin and the roll of the die are independent. The outcome of the coin has no influence on the roll of the die and vice versa. Thus, the event H of flipping a head is independent to the event E of rolling an even number. By the multiplication property of independent events, we see that the probability that Morry gets a head on the coin and an even number on the die is

$$P(H \text{ and } E) = P(H) \cdot P(E) = \frac{1}{2} \cdot \frac{3}{6} = \frac{1}{4}.$$

This result is also evident in the following diagram. There are 12 outcomes in the sample space, of which three outcomes are favorable to $H \cap E$. There is a probability of $1/2$ that the coin flip is in the row of heads and a probability of $3/6$ that the roll of the die is one of the even columns.

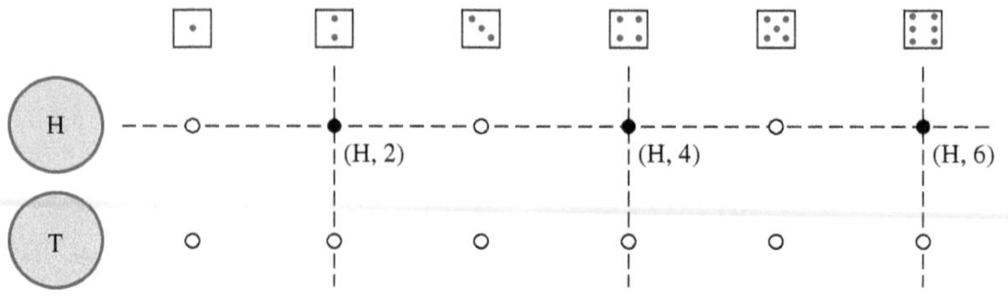

Cooperative Investigation

Compositions and Trains

A *composition* of a positive integer n is an ordered list of one or more positive integers that sum to n. For example, here are the four compositions of $n = 3$:

$$3, 2 + 1, 1 + 2, 1 + 1 + 1.$$

Directions

In groups of two to four, investigate this question:

What is the number of compositions of n?

In particular, answer these questions and perform these tasks:

1. What are the eight compositions of $n = 4$?
2. Fill in the missing entries in this table:

n	1	2	3	4	5	6
Compositions of n			4	8		

3. Consider the four trains of length 3 whose "cars" are shown in the following number-strips diagrams. Is there a connection between trains and compositions? Explain carefully.

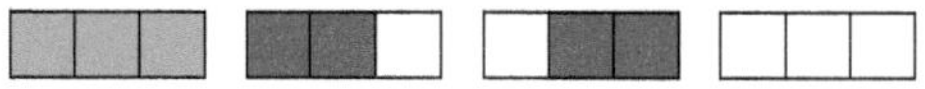

4. Conjecture and prove a formula for the number of compositions of n.

14.2 Problem Set

Understanding Concepts

1. Two dice are thrown. Determine the probability that a score of 4 or 6 is obtained.
2. Two dice are thrown. Determine the probability of obtaining a score of at least 4.
3. A coin is tossed and a die is rolled.
   (a) What is the probability the outcome consists of a head and an even number?
   (b) What is the probability that the outcome is a head or an even number?
4. There are 26 students in Mrs. Pietz's fifth-grade class at the International School of Tokyo. All of the students speak either English or Japanese, and some speak both languages. If 18 of the students can speak English and 14 can speak Japanese,
   (a) how many speak both English and Japanese?
   (b) how many speak English but not Japanese? (*Hint:* Draw a Venn diagram.)
5. All of the 24 students in Mr. Walcott's fourth-grade class at the International School of Tokyo speak either English or Japanese. If 11 of the students speak only English and 9 of the students speak only Japanese, what is the probability that a student chosen at random speaks both languages?
6. In how many ways can you draw a club or a face card from an ordinary deck of playing cards?
7. What is the probability of drawing a red face card or a black ace from an ordinary deck of playing cards?
8. (a) How many four-digit natural numbers can be named with each of the digits 1, 2, 3, 4, 5, and 6 used at most once?
   (b) How many of the numbers in part (a) begin with an odd digit? (*Hint:* Choose the first digit first.)
   (c) How many of the numbers in part (b) end with an odd digit? (*Hint:* Choose the first digit first and the last digit second.)
9. (a) If repetition of digits is not allowed, how many three-digit numbers can be formed from the digits 1, 2, 3, 4, and 5?
   (b) How many of the numbers in part (a) begin with either the digit 2 or the digit 3?
   (c) How many of the numbers in part (a) are even?
   (d) What is the probability that a randomly chosen number in part (a) is odd, and its first digit is even?
10. How many four-letter "words" can be formed from a standard 26-letter alphabet
    (a) if repetition is allowed?
    (b) if repetition is not allowed?
11. How many five-digit numbers can be formed with the first three digits odd and the last two digits even
    (a) if repetition of digits is allowed?
    (b) if repetition of digits is not allowed?
12. (a) If a pair of dice is rolled, in how many ways are the two numbers different?
    (b) If three dice are rolled, in how many ways are at least two of the dice the same?

13. (a) Construct a possibility tree to determine all three-letter code words using only the letters *a*, *b*, and *c* without repetition. For example, *bac* and *cba* are two of the code words.

(b) What is the probability the letter *a* precedes the letter *b* in a three-letter code word chosen at random?

14. Construct a possibility tree to determine the probability of obtaining two heads and two tails when four coins are tossed.

15. Determine the probabilities *u*, *v*, *w*, *x*, *y*, and *z* in the following probability tree:

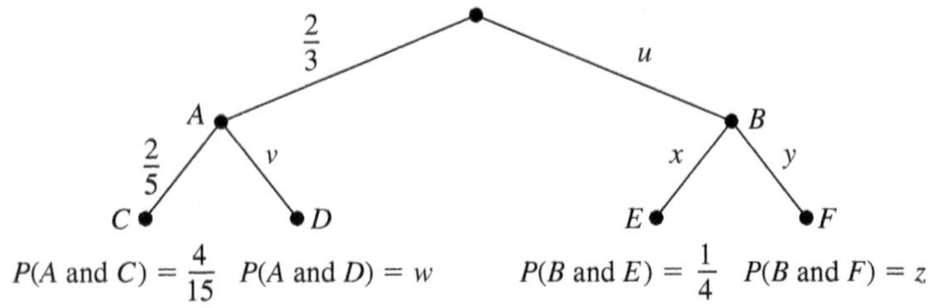

16. A bag contains three white and five black balls. Two balls are drawn without replacement. Construct a probability tree to determine the following probabilities:

(a) Both balls are white.

(b) One ball of each color is drawn.

(c) At least one white ball is drawn.

(d) The second ball is white given that the first ball is white.

(e) The first ball is white given that the second ball is black.

17. (a) How many five-digit numbers are there? (The first digit cannot be 0.)

(b) How many five-digit numbers are even?

(c) How many five-digit numbers have exactly one 0?

(d) How many five-digit numbers have exactly one 7?

Into the Classroom

18. (**Writing**) Children often want to relate classroom concepts to their own lives. For each of the following situations from every-day life, create a minilesson and accompanying problems to illustrate the multiplication principle of counting:

(a) How many ways can you order an ice cream cone? You will need to describe the number of choices of type of cone, flavor, number of scoops, toppings, and so on.

(b) How many ways can you drive from town A to town C, always passing through town B? Note how many roads connect the towns.

(c) How many ways can you and your friends line up at the drinking fountain?

(d) Choose a situation from your own life. Be creative!

Responding to Students

19. (**Writing**) Alejandro knows that, of the first thousand positive integers, 250 are divisible by 4 and 200 are divisible by 5. He then asserts that 450 are divisible by either 4 or 5. Help Alejandro see where his analysis went astray.

20. (**Writing**) Tamara has mix-and-match outfits that include four skirts and six blouses. She claims that she has ten different outfits. How can you help Tamara understand the error in her reasoning?

21. Mr. Brown was introducing counting strategies to his fourth-grade students, preparing them for a unit on probability. He challenged his class to count how many pizzas can be ordered, given a choice of either thick or thin crust and one of four toppings: green peppers, mushrooms, sausage, or pepperoni. How should Mr. Brown respond to the following claims of Ernie and Donna?

(a) Ernie claims that there are six types of pizza that can be ordered.

(b) When Mr. Brown informed the class that there was no pepperoni, Donna claimed that there were now only three types of pizza available.

Thinking Critically

22. If two successive cards are drawn from a standard 52-card deck (without replacement); what is the probability that

(a) the first card is a king and the second card is not a face card?

(b) the first card is a king and the second card is a heart? (Watch out for the king of hearts!)

23. The Chess Club has five freshmen, two juniors, and three seniors.

(a) Make a possibility tree to determine how many slates of officers can fill the two offices of president and treasurer, where the two officers cannot be from the same class.

(b) Answer the problem of part (a) a new way, by considering the three cases where the two offices are filled by a freshman and a junior, a freshman and a senior, and a junior and a senior.

24. A bag contains three red and two green marbles. Two marbles are drawn in succession without replacing the first drawn marble.

(a) Construct the probability tree for the two-stage experiment, showing the probabilities for the four events *rr*, *rg*, *gr*, and *gg*.

(b) Let g^* be the event the first marble drawn is green and $^*r$ be the event the second marble drawn is red. Verify that

$$P(gr) = P(g^* \text{ and } {}^*r) = P({}^*r \mid g^*)P(g^*).$$

(c) Verify the formula

$$P(gr) = P(g^* \text{ and } {}^*r) = P(g^* \mid {}^*r)P({}^*r).$$

25. Two cards are drawn in succession from a 52-card deck. What is the probability both cards are face cards if

(a) the first card drawn is replaced in the deck?

(b) the first card is not replaced?

26. Two dice are "loaded" in such a way that both the 3 and the 4 appear with probability 0.3 and each of the other faces appears with probability 0.1. What is the probability of a 7 being rolled with these unfair dice?

27. Marc's collection of sports cards includes ten baseball, seven basketball, and nine football cards. He let his friend randomly choose two cards as a birthday present. What is the probability the cards are from different sports?

28. A multiple-choice quiz has ten questions, with four choices per question.

(a) How many ways can the quiz be answered?

(b) How many of these ways give an A grade of 90 percent or higher?

29. Rudy has nine loose keys in his pocket, all nearly alike, but only one will open his door. He randomly chooses a key and tries it, but if it is an incorrect key, he randomly draws another key, not replacing any of the keys already tried.

(a) Let $p_1, p_2, \ldots, p_9$ be the probabilities that the 1st, 2nd, . . . , 9th key that Rudy tries is the correct one. Show that the nine probabilities are all the same.

(b) Generalize part (a) by assuming that Rudy has n keys in his pocket, with just one being the correct key. Show that each key tried has the probability $1/n$ of being correct.

30. **(a)** How many ways can a red and a black rook be placed on an 8×8 chessboard?

(b) How many ways can two identical black rooks be placed on an 8×8 board?

(c) How many ways can a red and a black rook be placed on an 8×8 board in different rows and different columns so that they cannot attack one another?

31. A bent coin lands heads up with probability 0.7.

(a) Construct a probability tree for three flips of the unfair coin.

Now use the probability tree to determine these probabilities:

(b) $P(\text{HHH})$

(c) $P(\text{two heads and a tail in any order})$

(d) $P(\text{at least one head appears in the three flips})$

(e) $P(\text{a head and a tail appear among the three flips})$

32. Let R be the event that the arrow on the accompanying spinner touches a red sector, and Y the event the arrow touches a yellow sector. Determine these probabilities:

(a) $P(R \mid Y)$

(b) $P(Y \mid R)$

(c) $P(Y \text{ and } R)$

(d) Verify that $P(Y \text{ and } R) = P(Y \mid R)P(R) = P(R \mid Y)P(Y)$

(e) Are R and Y independent events?

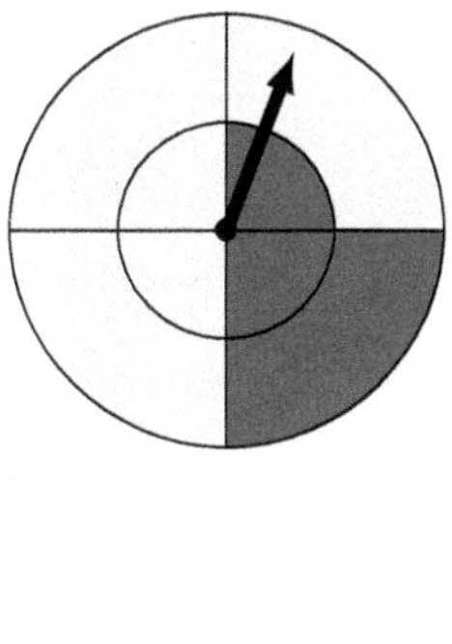

Making Connections

33. An electrician must connect a red, a white, and a black wire to a yellow, a blue, and a green wire in that order. What is the probability a randomly made connection is correct?

34. Suppose that automobile license plates either show three letters followed by three digits or three digits followed by three letters. How many different license plates can be made

(a) if repetition of digits and letters is allowed?

(b) if repetition of digits and letters is not allowed?

35. A four-bit code word is any sequence of four digits, where each digit is either a 0 or a 1. For example, 0100 and 1011 are four-bit code words.

(a) How many different four-bit code words are there?

(b) How many different six-bit code words, such as 001011, are there?

(c) If a vocabulary of 1000 code words is required, how long must the words be?

(d) What is the probability that a randomly selected four-bit code word has no 0 following any 1?

36. The illegal chain letter scam works this way: You receive a letter with a list of (say) 5 names and addresses. You are instructed to send something—say, a dollar—to the person at the top of the list, then remove that name and add your own to the bottom of the list, and send the letter to 10 people not on the list. Your name will be at the bottom of 10 letters but then rises to fourth position on 100 letters, and so on. Eventually, your name is at the top of 100,000 letters, and for a \$1 investment plus small postage and copying costs, you receive \$100,000.

(a) Suppose everyone obeys the instructions, sending their 10 letters within a week. In one week, 10 people have been sent letters, and in two weeks 10 + 100 people have been sent letters. How many people have been sent letters in eight weeks?

(b) Assuming that no one gets a second letter, in how many weeks will all 7 billion people on Earth have a letter?

37. **(a)** Could every person in the United States be given a distinct six-letter identification code using a 26-letter alphabet? A person's code might be YWAWHK.

(b) Would a five-letter code work if both upper- and lowercase letters were used?

(c) If 7 billion people are to be given identification codes using only strings of capital letters, what is the smallest length of the string required?

38. **(a)** Draw the possibility tree for a three-game series between teams A and B where the series ends when a team has two wins.

(b) Draw the possibility tree for a five-game series between teams A and B where the series ends when a team has three wins.

(c) In how many five-game series does the winning team lose the first two games?

(d) In how many five-game series does the winning team lose the first game?

State Assessments

39. (Grade 5)

The Mega Monster Frozen Yogurt Shop offers two cup sizes, five flavors, and two types of sprinkles or none at all. How many ways can an order for yogurt be made by checking three of the boxes on the order slip shown below.

Cup Size	Flavor	Sprinkles
☐ ▽ Small ☐ ▽ Large	☐ Vanilla ☐ Raspberry ☐ Mango ☐ Chocolate ☐ Peach	☐ Chocolate ☐ Rainbow ☐ None

   A. 10   B. 13   C. 20   D. 30

40. (Grade 7)

An ice cream sundae can be ordered with a large or small scoop of ice cream, with one of three toppings—berry, chocolate, or caramel, and with one of three types of sprinkles. How many ways can a sundae be made?

   A. 8   B. 12   C. 16   D. 18   E. 24

41. (Grade 8)

Alek has a bag of jelly beans containing three lemon, four lime, two raspberry, and seven licorice. However, Alek only likes lime jelly beans. What is the probability that, with two successive random draws, Alek is lucky enough to enjoy eating a lime jelly bean each time?

   A. $\dfrac{1}{16}$ B. $\dfrac{1}{15}$ C. $\dfrac{1}{20}$ D. $\dfrac{1}{4}$

14.3 ▷ Permutations and Combinations

Two types of counting situations are encountered so often that they deserve special attention: permutations and combinations. A **permutation** is an *arrangement* of a given number of objects from a specified set into an *ordered list*. A **combination** is a *selection* of a given number of objects from a set to form an *unordered subset* of the objects.

Thus, if we wish to count how many choices can be made, it is very important to ask these questions so that permutations and combinations are not confused:

• *Are the selected objects arranged into an order?*

If so, then we are considering a permutation.

• *Do we need to know only which objects are selected, with the order of the objects making no difference?*

In this situation, we are considering a combination.

To help distinguish permutations from combinations, suppose that we have four textbooks—an algebra book, a biology book, a chemistry book, and a drama book. If two of the books are placed on a shelf, the book on the left can be selected in four ways and the book on the right in three ways. This gives the 4 × 3 = 12 permutations shown in Figure 14.7.

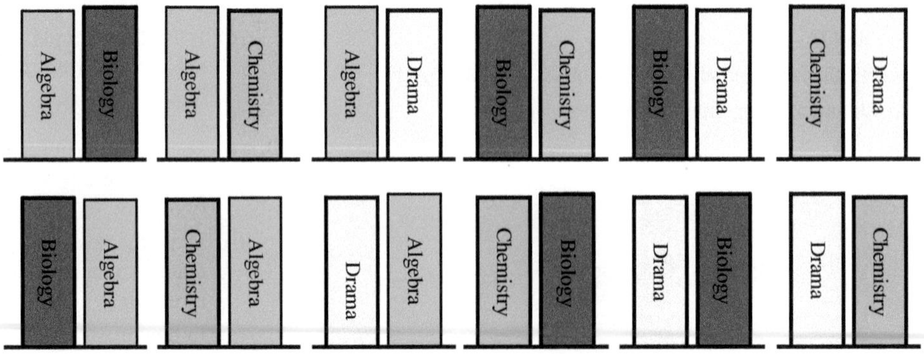

Figure 14.7
Showing the permutations of two books selected from a set of four books

If two books are placed on top of a table with no regard to order, there are six combinations, as shown in Figure 14.8. Each combination corresponds to two permutations of the same books.

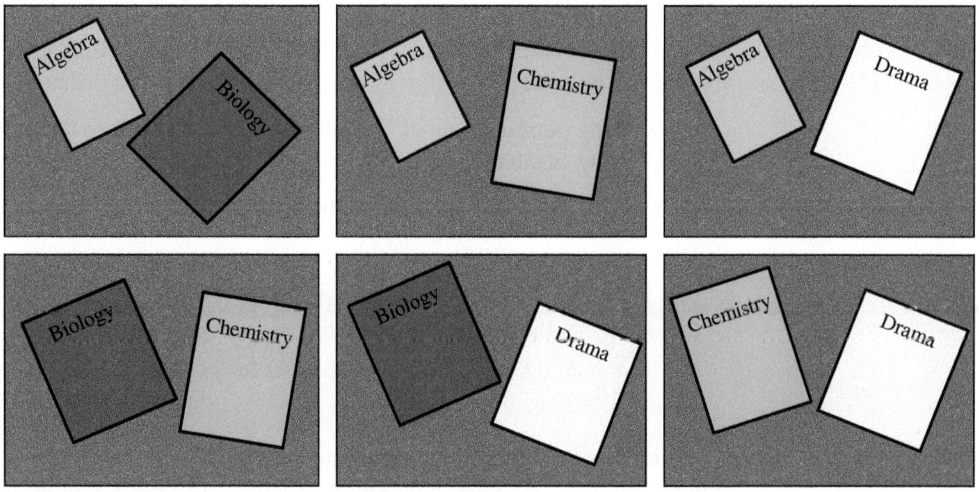

Figure 14.8 Showing the six combinations of two books selected from a set of four books

DEFINITION Permutations and Combinations

Let U be a set of distinct objects.

- An *ordered* sequence of the objects in U is called a **permutation** of the objects of U. If r objects are in the permutation, the sequence is called an ***r*-permutation.**
- An *unordered* selection, or subset, of U is called a **combination** of the objects of U. If r objects are in the combination, the selection is called an ***r*-combination.**

EXAMPLE 14.17 Choosing a Social Committee and Officers for the Math Club

There are five members of the Math Club.

(a) In how many ways can the slate of two officers—a president and a treasurer—be chosen?
(b) In how many ways can the two-person Social Committee be chosen?

Solution

For convenience, suppose the five members of the Math Club form the set $U = \{a, b, c, d, e\}$.

(a) Choosing the officers is the same as choosing a sequence of two members of U, first the president and then the treasurer. That is, we wish to know the number of permutations of five objects taken two at a time. By the multiplication principle of counting, there are five choices for the president, and once the president is chosen, there are four remaining members from which to choose the treasurer. Thus, the number of ways to choose the two officers is 20. As a check, we can make a list of the 20 possible slates of officers, where the first person in each pair is president and the second person is treasurer:

ab	*ac*	*ad*	*ae*	*bc*	*bd*	*be*	*cd*	*ce*	*de*
ba	*ca*	*da*	*ea*	*cb*	*db*	*eb*	*dc*	*ec*	*ed*

(b) Unlike the officers, the members of the Social Committee form an unordered subset of two elements. For example, a committee consisting of a and b is the same committee whether b is chosen first and then a, or a is chosen first and then b. Thus, the number of two-person committees is the number of combinations of five objects taken two at a time. There are ten committees, as shown in the following list:

$\{a, b\}$	$\{a, c\}$	$\{a, d\}$	$\{a, e\}$	$\{b, c\}$	$\{b, d\}$	$\{b, e\}$	$\{c, d\}$	$\{c, e\}$	$\{d, e\}$

In most problems, it is not important to list all of the permutations or all of the combinations. What is of interest is the number of r-permutations or the number of r-combinations when r objects are taken from a set of n different objects. That is, the important questions are these:

- How many permutations of n things taken r at a time are there?
- How many combinations of n things taken r at a time are there?

The following notation will be helpful:

NOTATION ***P*(*n*, *r*) and *C*(*n*, *r*)**

$P(n, r)$ denotes the number of r-permutations from a set of n different objects.

$C(n, r)$ denotes the number of r-combinations from a set of n different objects.

Alternative notations are sometimes used. In particular, nPr may denote $P(n, r)$ and nCr may denote $C(n, r)$. This is the notation most often used by graphing calculators. Sometimes $C(n, r)$ is written $\binom{n}{r}$ and is read as "n choose r," since it gives the number of ways to choose a subset of r objects from a set of n objects.

Formulas for the Number of *r*-Permutations

To derive a formula for $P(26, 5)$, consider counting the number of "code words" that are formed with a sequence of five different capital letters from the 26-letter alphabet. For example, WORDS and VWXYZ are allowable code words since no letter is repeated but LEGAL is not permitted since the letter L appears twice. The order of the letters matters, so the number of code words is the number of 5-permutations from the set of 26 capital letters. Forming a code word is a five-step process, with 26 choices for the first letter of the word, then 25 choices for the second letter, and so on. By the multiplication principle, we conclude there are

$$P(26, 5) = 26 \times 25 \times 24 \times 23 \times 22$$

code words. We see that $P(26, 5)$ is the product of the first five factors of the factorial

$$26! = \underbrace{26 \times 25 \times 24 \times 23 \times 22}_{P(26,\,5)} \times \underbrace{21 \times 20 \times \cdots \times 3 \times 2 \times 1}_{(26\,-\,5)!\,=\,21!}.$$

When this equation is solved for $P(26, 5)$, we see that the number of 5-permutations chosen from the 26-letter alphabet can be written as the quotient of factorials

$$P(26, 5) = \frac{26!}{(26 - 5)!}.$$

The same reasoning applies to counting the number of "code words" that are a sequence of r different letters chosen from an alphabet of n letters. In this way we obtain the formula for the number of r-permutations chosen from a set of n elements.

THEOREM **Formulas for the Number of *r*-Permutations**

Let n and r be natural numbers, with $r \leq n$. Then the number of r-permutations of n is

$$P(n, r) = n(n - 1)(n - 2) \cdots (n - r + 1) = \frac{n!}{(n - r)!}.$$

Since $P(n, n) = n!$ and $P(n, n) = \dfrac{n!}{(n - n)!} = \dfrac{n!}{0!}$, we see it is reasonable to define $0! = 1$.

Here are some calculations of r-permutations. It is helpful to notice that $P(n, r)$ is the product of the first r factors of $n!$.

$P(7, 2) = 7 \cdot 6 = 42 \qquad P(8, 8) = 8! = 40{,}320 \qquad P(13, 4) = 13 \cdot 12 \cdot 11 \cdot 10 = 17{,}160$

Formulas for the Number of r-Combinations

We have just seen that there are $P(26, 5)$ ways to form all of the "code words" consisting of an ordered sequence of five different letters chosen from the alphabet of 26 letters. We can also view forming these words as a two-step process:

Step 1: Choose, without regard to order, the subset of five letters that appear in the word.
Step 2: Order the five letters selected in step 1 to form the code word.

A selection of a subset of five of the 26 letters is a 5-combination from the 26-letter alphabet, since order is not considered. Therefore, there are $C(26, 5)$ ways to complete step 1. Step 2 asks us to form a permutation of all five of the letters selected in step 1, and we know there are 5! permutations. By the multiplication principle, there are $C(26, 5) \cdot 5!$ code words. This gives us the equation $C(26, 5) \cdot 5! = P(26, 5)$. By dividing this equation by 5!, we obtain the formula

$$C(26, 5) = \frac{P(26, 5)}{5!}.$$

We could just as well have counted the number of code words formed with r letters chosen from an alphabet of n letters. This gives us the following theorem.

THEOREM Formulas for the Number of r-Combinations
Let n and r be natural numbers, with $r \leq n$. Then the number of r-combinations of n is

$$C(n, r) = \frac{n(n-1)(n-2)\cdots(n-r+1)}{r!} = \frac{n!}{r!(n-r)!}.$$

Here are some calculations of r-permutations.

$$C(7, 2) = \frac{7 \cdot 6}{2!} = 21, \quad C(8, 8) = \frac{8!}{8!} = 1, \quad C(13, 4) = \frac{13 \cdot 12 \cdot 11 \cdot 10}{4!} = \frac{17{,}160}{24} = 715.$$

The number of r-combinations can be extended from the natural numbers to the larger set of whole numbers by defining $C(n, 0) = 1$ for all $n \geq 0$. It can then be checked that we obtain this table of values of $C(n, r)$, $0 \leq r \leq n$. For example, the entry in row $n = 5$ and column $r = 3$ is

$$C(5, 3) = \frac{5!}{3!(5-3)!} = \frac{5 \cdot 4 \cdot 3}{3!} = \frac{60}{6} = 10.$$

	$r = 0$	1	2	3	4	5
$n = 0$	1					
1	1	1				
2	1	2	1			
3	1	3	3	1		
4	1	4	6	4	1	
5	1	5	10	10	5	1

It may be surprising to discover that this array is Pascal's triangle seen in Chapter 1. An entry in row n and column r is the sum of two entries in the row $n - 1$ and columns r and $r - 1$. For example, $C(5, 3) = 10 = 4 + 6 = C(4, 3) + C(4, 2)$. This could be verified algebraically from the formula for combinations, but it is even simpler to use the definition of the combination numbers and the addition principle of counting. Remember that $C(5, 3)$ counts the number of three-element subsets of the five-element set $\{a, b, c, d, e\}$. These subsets are of two types depending on whether or not the subset contains element a. Those not containing a are the $C(4, 3)$ three-element subsets of $\{b, c, d, e\}$. However, all of the subsets that do contain a are obtained by inserting element a into the $C(4, 2)$ two-element subsets of $\{b, c, d, e\}$.

The reasoning just given can be generalized to count the number of r-element subsets of an n-element set, which gives the following theorem.

THEOREM Pascal's Identity

$$C(n, r) = C(n-1, r) + C(n-1, r-1)$$

Solving Problems with Permutations and Combinations

Counting the number of ways objects can be selected, arranged, or distributed is often the key to the solution of a combinatorial or probabilistic problem. It is important to know if the order of the objects matters or not, so it becomes clear if a permutation or a combination should be used. The examples that follow illustrate how to apply the permutation and combination formulas we have derived.

EXAMPLE 14.18

Forming Committees and Winning a Drawing

The Stamp Club has nine members, including Alicia. The names of the members of the club have been put into a hat.

(a) Four members' names are drawn from the hat to form a refreshment committee. What is the probability Alicia will be a member of the committee?
(b) After the committee is chosen, the names are returned to the hat and three names are drawn in turn. The first name drawn wins third prize, the second name drawn wins second prize, and the third name drawn wins the grand prize. What is the probability that Alicia wins the grand prize before knowing the winners of the second and third prizes? After knowing the winners?
(c) In the drawing described in part (b), what is the probability Alicia wins second or third prize?

Solution

(a) There are $C(9, 4) = \dfrac{9 \cdot 8 \cdot 7 \cdot 6}{4!} = 126$ sets of four names, where a combination is used since the order in which the four names are drawn makes no difference. If Alicia's name is drawn, the other three names are a set of three from the remaining eight names, so there are $C(8, 3) = \dfrac{8 \cdot 7 \cdot 6}{3!} = 56$ drawings for which Alicia is selected. This means that the probability of Alicia being on the committee is $\dfrac{56}{126} \doteq 0.44$.

(b) The order in which the names are drawn makes a difference, so this problem requires permutations. There are $P(9, 3) = 9 \cdot 8 \cdot 7 = 504$ ways to draw three names in order. To win the grand prize, Alicia must be the third name drawn. This means the first two names drawn must be selected from the other eight club members, which can occur in $P(8, 2) = 8 \cdot 7 = 56$ ways. This gives Alicia a probability of $\dfrac{56}{504} \doteq 0.11$ to win the grand prize. This answer should not be a surprise! All nine club members have the same probability to win the grand prize, so the probability is $1/9$. Next suppose that the first two names are drawn. If Alicia wins second or third prize, there is no way for her to win the grand prize and her probability is 0. Otherwise, since seven names are left in the hat, her probability of winning the grand prize is $1/7$. We then have two conditional probabilities:

$$P(\text{Alicia wins grand prize} \mid \text{Alicia won second or third prize}) = 0$$

$$P(\text{Alicia wins grand prize} \mid \text{Alicia did not win second or third prize}) = \frac{1}{7}$$

(c) There are $9 \cdot 8 = 72$ ways to award second and third prize. There are two ways Alicia can win one of the two prizes, with the other prize awarded to one of the other eight club members. This means there are $2 \cdot 8 = 16$ ways that Alicia wins second or third prize, with a probability of $\dfrac{16}{72} = \dfrac{2}{9}$. This can also be seen this way: Alicia's probability of winning either second or third prize is $1/9$, and these are mutually exclusive events so, by the addition principle, the probability of second or third prize is $\dfrac{1}{9} + \dfrac{1}{9} = \dfrac{2}{9}$.

EXAMPLE 14.19

Lining Up Stamp Club Members for a Photo

The two sophomores, four juniors, and three seniors in the Stamp Club lined up randomly for a photo. What is the probability the sophomores, juniors, and seniors are side by side?

Solution

There are $9! = 362{,}880$ ways the nine club members can form a line. The ways to line up with the members of each class together can be viewed as a four-step process:

Step 1: Choose the order in which the class levels appear in the photo in 3! ways.
Step 2: Order the two sophomores in 2! ways.
Step 3: Order the four juniors in 4! ways.
Step 4: Order the three seniors in 3! ways.

The multiplication principle shows there are $3! \cdot 2! \cdot 4! \cdot 3! = 1728$ ways to line up with each class together. This gives the probability $\frac{1728}{362{,}880} = 0.00476\ldots.$

EXAMPLE 14.20

Taking Photos and Forming Committees of Pep Squad Members

Suppose there are six boys and eight girls, all of different heights, on the Pep Squad.

(a) If four members are randomly chosen to line up for a photo, what is the probability that no two boys and no two girls stand side by side?
(b) If a committee of three is chosen by a random drawing of names, what is the probability that all three committee members are boys?
(c) If a committee of four is chosen at random, what is the probability that the committee has two boys and two girls?
(d) If a committee of five is chosen at random, what is the probability there is at most one girl on the committee?

Solution

For each problem, we must describe the sample space S and the event E, then count the number of equally likely outcomes in both S and E, and finally take the ratio of the two counts to give the probability.

(a) The sample space is the number of 4-permutations of the 14 members, so $n(S) = P(14, 4) = 14 \cdot 13 \cdot 12 \cdot 11 = 24{,}024$. There are $P(6, 2) = 6 \cdot 5 = 30$ ways to line up two of the six boys and $P(8, 2) = 8 \cdot 7 = 56$ ways to line up two girls. There are also two allowable arrangements of two boys and two girls with no two boys and no two girls standing side by side: boy–girl–boy–girl and girl–boy–girl–boy. By the multiplication principle, this gives $2 \cdot 30 \cdot 56 = 3360$ boy–girl–boy–girl arrangements. Therefore, the probability is $\frac{3360}{24{,}024} \doteq 0.14$.

(b) The sample space consists of all 3-combinations of the 14 members, so $n(S) = C(14, 3) = \frac{14 \cdot 13 \cdot 12}{3 \cdot 2 \cdot 1} = 364$. Similarly, there are $n(E) = C(6, 3) = \frac{6 \cdot 5 \cdot 4}{3 \cdot 2 \cdot 1} = 20$ ways to choose three of the six boys for the committee. Therefore, the probability is $\frac{20}{364} \doteq 0.055$.

(c) There are $n(S) = C(14, 4) = \frac{14 \cdot 13 \cdot 12 \cdot 11}{4 \cdot 3 \cdot 2 \cdot 1} = 1001$ ways to select four of the 14 squad members to be on the committee. Since there are $C(6, 2) = \frac{6 \cdot 5}{2 \cdot 1} = 15$ ways to choose two of the six boys and $C(8, 2) = \frac{8 \cdot 7}{2 \cdot 1} = 28$ ways to choose two of the eight girls, there are $15 \cdot 28 = 420$ committee choices with two boys and two girls. Therefore, the probability that two boys and two girls are on the randomly selected committee is $\frac{420}{1001} \doteq 0.42$.

(d) There are $n(S) = C(14, 5) = \dfrac{14 \cdot 13 \cdot 12 \cdot 11 \cdot 10}{5 \cdot 4 \cdot 3 \cdot 2 \cdot 1} = 2002$ ways to select a five-member committee. Of these ways, $C(6, 5) = 6$ committees have all boys. Since there are eight ways to select one girl and $C(6, 4) = C(6, 2) = \dfrac{6 \cdot 5}{2 \cdot 1} = 15$ ways to select four of the six boys, there are also $8 \cdot 15 = 120$ committees with one girl and four boys. Altogether, there are $6 + 120 = 126$ committees with at most one girl, so the probability that this occurs with a random choice of the committee is $\dfrac{126}{2002} \doteq 0.06$.

EXAMPLE 14.21

Sharing Birthdays

What is the probability that at least two people in a group celebrate their birthdays on the same day each year?

Solution

Understand the Problem

The size of the group makes a difference. If the "group" has just one person, it is impossible to have a shared birthday and the probability is 0. If the group has 367 people, even if everyone has a different birth date, including February 29, there must still be at least one duplication by the pigeonhole principle so the probability is 1. It seems likely that the probability increases from 0 to 1 as the size of the group increases. It would be of special interest to determine the size of the group when the probability first exceeds 0.5 that at least one birthday is shared.

Devise a Plan

There are many ways a group of people can share birthdays, but far fewer ways in which no two birthdays coincide. This suggests that we determine the probability that no birthdays are shared, and then use complementary probabilities to find the probability there is at least one shared birthday. It may also be assumed for simplicity that we overlook Leap Years, and assume each of 365 days is an equally likely birthday. Although the size of the group will ultimately be a variable n, it is a good idea to first solve a simpler problem. Therefore, we can begin our analysis with a group of five people and then see how to modify our calculations to apply to a group size n.

Carry Out the Plan

Suppose we have a group of five people, and E_5 is the event that no birthdays are shared. Since each person can be born on any of the 365 days of the year, there are 365^5 ways to form a list of five birthdays. If no birthdays are shared, the list must be a 5-permutation selected from 365 days, so there are $P(365, 5) = 365 \cdot 364 \cdot 363 \cdot 362 \cdot 361$ ways no birthday is shared. This gives us the probability $P(E_5) = \dfrac{365 \cdot 364 \cdot 363 \cdot 362 \cdot 361}{365 \cdot 365 \cdot 365 \cdot 365 \cdot 365}$ that no birthdays are shared in a group of five people. The complementary probability is, therefore,

$$P(\overline{E}_5) = 1 - \frac{365 \cdot 364 \cdot 363 \cdot 362 \cdot 361}{365 \cdot 365 \cdot 365 \cdot 365 \cdot 365} = 1 - 0.97\ldots = 0.03\ldots,$$

showing us that it is rare that a birthday is shared in a group of five.

It is easy to see how to modify the preceding calculation so that it applies to a group of a different size. For example, for groups of size $n = 22$, 23, or 30 we have the probabilities

$$P(\overline{E}_{22}) = 1 - \frac{365 \cdot 364 \cdot \cdots \cdot 344}{365^{22}} = 1 - 0.52\ldots = 0.48\ldots$$

$$P(\overline{E}_{23}) = 1 - \frac{365 \cdot 364 \cdot \cdots \cdot 343}{365^{23}} = 1 - 0.49\ldots = 0.51\ldots$$

$$P(\overline{E}_{30}) = 1 - \frac{365 \cdot 364 \cdot \cdots \cdot 336}{365^{30}} = 1 - 0.29\ldots = 0.71\ldots.$$

We see that for a group of 23 people, it is more probable than not that at least two people celebrate their birthday on the same day.

Look Back

Our analysis made two assumptions. First, Leap Days would make little difference since this would result in just a slight lowering of the probability of a shared birthday. It was also assumed that the sample space is uniform, though data show this is not quite the case. The number of births in July through October typically exceeds the number of births in other months, so this clumping slightly increases the probability of a duplicated birthday.

In the following example, several concepts from probability are combined to construct a solution. The alternative notation $\binom{n}{r}$, "n choose r," is used in place of $C(n, r)$.

EXAMPLE 14.22 Combining Concepts: Combinations, Mutually Exclusive Events, and Complementary Events

A hand of five cards is drawn from a standard 52-card deck. What is the probability that both colors—red and black—are represented in the hand? The cards are red if they are hearts or diamonds and are black if they are clubs or spades.

Solution

Let E be the event of all 5-card hands containing at least one red and one black card. The complementary event is, then, $\overline{E} = R \cup B$, where R is the event of all hands with five red cards and B is the event of all hands with five black cards. Since R and B are mutually exclusive events, $P(\overline{E}) = P(R \cup B) = P(R) + P(B)$. There are $\binom{26}{5}$ combinations of five cards taken from the set of 26 red cards. Also, there are $\binom{52}{5}$ five-card hands altogether. Together, they give the probability

$P(R) = \dfrac{\binom{26}{5}}{\binom{52}{5}} = \dfrac{26 \cdot 25 \cdot 24 \cdot 23 \cdot 22}{52 \cdot 51 \cdot 50 \cdot 49 \cdot 48}$. Since there are also 26 black cards, it follows that $P(B) = P(R)$ and $P(\overline{E}) = 2 \cdot \dfrac{26 \cdot 25 \cdot 24 \cdot 23 \cdot 22}{52 \cdot 51 \cdot 50 \cdot 49 \cdot 48} = 0.05\ldots$. Thus, $P(E) = 1 - P(\overline{E}) \doteq 1 - 0.05 = 0.95$.

That is, there is about a 95% probability that a hand of five cards contains both colors.

For our last example of this section, we return to the Cooperative Investigation *Strings and Loops* that opened this chapter. The theoretical probabilities obtained in this example help explain the experimental results of the investigation.

EXAMPLE 14.23 Determining the Theoretical Probabilities of the Strings and Loops Activity

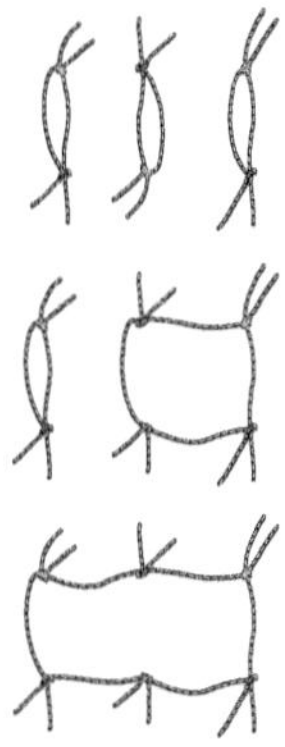

Six pieces of string of equal length are held in a bundle in one person's hand. A second person ties three knots at each end of the bundle of strings, where each knot joins a randomly selected pair of strings. After the six knots are tied, the bundle is examined to see what pattern of loops has been formed. There are three possibilities:

T: three small two-string loops
M: one medium four-string loop and one small two-string loop
L: one large six-string loop

What are the probabilities **(a)** $P(T)$, **(b)** $P(L)$, and **(c)** $P(M)$ of these events?

Solution

We might as well suppose that the first three knots are tied on the same end of the bundle, so that we begin with the pattern

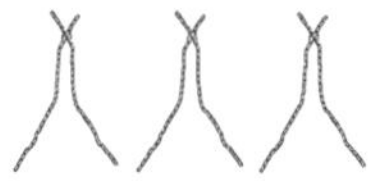

We now consider the compound event in which the next three knots are tied, drawing the probability tree as we proceed.

The Fourth Knot

There are $C(6, 2) = 6 \cdot 5/2 = 15$ ways to choose a pair of loose ends to tie, including three pairs that form a small two-string loop. Thus, the probability of forming a small loop with the fourth knot is $\frac{3}{15} = \frac{1}{5}$. (Another way to see this is to grab any loose end and see that there are five other loose ends, with just one of the five forming a small loop.) By complementary probabilities, the probability that no loop is created is $\frac{4}{5}$. We now see that our probability tree begins this way:

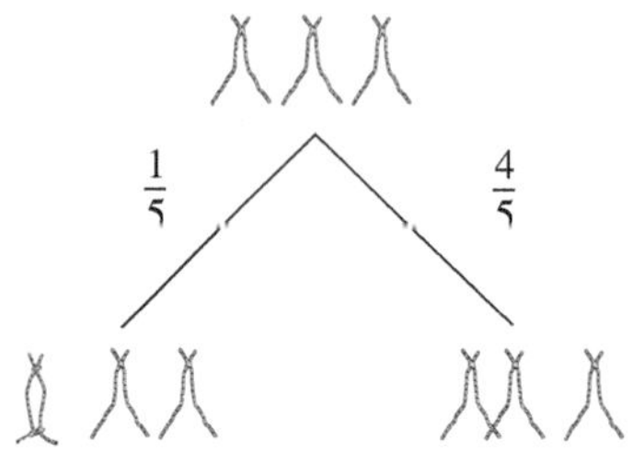

The Fifth Knot

After the fourth knot is tied, there are four loose ends remaining. The next pair to be tied can be selected in $C(4, 2) = 6$ ways. If the left branch of the preceding tree is followed, two of these pairs create a small two-string loop, so there is a probability of $\frac{2}{6} = \frac{1}{3}$ that the fifth knot creates a second small loop. By complementary probability, there is a probability of $\frac{2}{3}$ that no loop is formed. Following the right branch of the tree, we see that there is one pair of loose ends that makes a small two-string loop and one pair that makes a medium four-string loop; evidently, the other four pairs do not make a loop. The respective probabilities to complete the next part of the probability tree are therefore $\frac{1}{6}, \frac{1}{6}$, and $\frac{4}{6} = \frac{2}{3}$. We can now extend the probability tree to account for the fifth knot:

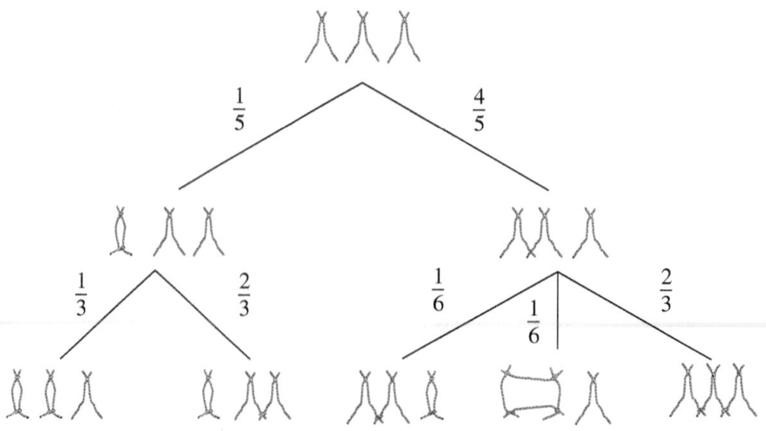

The Sixth Knot

Since only one pair of loose ends remains, the outcome has already been determined once five knots are tied. Thus, we have the following results:

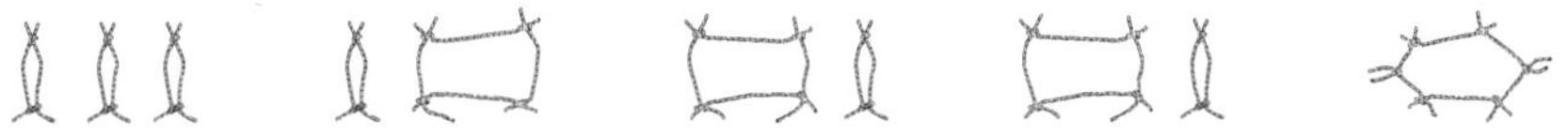

$\frac{1}{5}\cdot\frac{1}{3}=\frac{1}{15}$ $\quad\frac{1}{5}\cdot\frac{2}{3}=\frac{2}{15}$ $\quad\frac{4}{5}\cdot\frac{1}{6}=\frac{2}{15}$ $\quad\frac{4}{5}\cdot\frac{1}{6}=\frac{2}{15}$ $\quad\frac{4}{5}\cdot\frac{2}{3}=\frac{8}{15}$

We see that $P(T) = \frac{1}{15}$, $P(M) = \frac{2}{15} + \frac{2}{15} + \frac{2}{15} = \frac{6}{15}$, and $P(L) = \frac{8}{15}$. Most people are surprised to learn that obtaining three small loops is quite rare and that a single large loop occurs a little more than half the time.

14.3 Problem Set

Understanding Concepts

1. Evaluate these permutation and combination numbers.

(a) $P(7, 3)$ (b) $P(7, 7)$
(c) $P(7, 4)$ (d) $C(7, 3)$
(e) $C(7, 7)$ (f) $C(7, 4)$

2. Evaluate these permutation and combination numbers.

(a) $P(8, 5)$ (b) $P(8, 8)$
(c) $P(8, 3)$ (d) $C(8, 5)$
(e) $C(8, 8)$ (f) $C(8, 3)$

3. For each situation described, first decide whether it is a permutation or a combination, and then answer the question.

(a) A coin is flipped ten times. In how many ways can heads and tails appear equally often?
(b) If Olivia has ten dolls, in how many ways can four of them be displayed in a row on a shelf?
(c) Morrie coaches 12 players on the tennis team. How many ways can he choose five players for a match, where the players must be ranked?
(d) How many different hands of three cards can be drawn from a deck of 15 cards numbered 1 through 15?
(e) How many ways can a hand of five cards be played one at a time?

4. How many ways can a group of 12, including four boys and eight girls, be formed into two six-person volleyball teams

(a) with no restrictions?
(b) so that each team has two of the boys?
(c) so that all of the boys are on the same team?

5. How many ways can a group of seven people go from Allentown to Bovill if one person rides a unicycle, two ride a tandem bike, and the rest walk? (*Careful:* How many ways can two people ride a tandem bike?)

6. The Chess Club has six members. In how many ways

(a) can all six members line up for a picture?
(b) can the club choose a president and a secretary?
(c) can the club choose three members to attend the regional tournament, with no regard to order?

7. There are 14 members of the debate team, including eight girls and six boys.

(a) How many ways can the club members line up for a yearbook photo?
(b) How many ways can they take the yearbook photo with the girls in the front row and the boys in the back row?
(c) How many ways can they randomly choose a slate of three officers, including a president, a secretary, and a treasurer?
(d) How many ways can they form a social committee of four members, including two boys and two girls?

8. In how many ways can a two-scoop cone be purchased at Raskin-Bobbins Thirty-One Flavors Ice Cream Shoppe? Vary the conditions that affect your answer. For example, is a vanilla + chocolate cone different from a chocolate + vanilla cone? Can both scoops be the same flavor?

9. Row $r = 5$ of Pascal's triangle is 1 5 10 10 5 1.

Use Pascal's identity to give the next two rows.

10. Two cards are drawn from a deck of 52 playing cards. What is the probability that

(a) both cards are hearts?
(b) the cards are a pair?

11. Three cards are drawn at random from a deck of 52 playing cards. What is the probability that

(a) all three cards are face cards?
(b) the three cards are of three different suits?

12. A fair coin is flipped ten times. What is the probability that there are five heads and five tails?

13. The six girls and five boys on the debate team lined up at random to walk on stage. What is the probability no two girls are side by side?

14. What is the probability that all six numbers appear in

(a) six rolls of a number cube?

(b) seven rolls of a number cube?

15. What is the probability that a randomly drawn hand of five cards contains all red cards (hearts or diamonds) or all face cards?

16. (a) What is the probability that at least two people in a group of four people have birthdays in the same month?

(b) How many people must be in a group for the probability to first exceed 0.5 that at least two people have a shared birth month?

17. Igor decided to answer each of the ten true or false questions on the quiz by flipping a coin. What is the probability Igor will

(a) get 90% or better?

(b) pass the quiz with a score of 50% or better?

18. Two balls are drawn at random from an urn containing six white and eight red balls.

(a) Use combinations to compute the probability that both balls are white. Recall that

$$C(n, r) = \frac{n(n-1)(n-2)\ \cdots\ (n-r+1)}{r!}.$$

(b) Compute the probability that both balls are red.

19. An urn contains eight red, five white, and six green balls. Four balls are drawn at random.

(a) Compute P(all four are red).

(b) Compute P(exactly two are red and exactly two are green).

(c) Compute P(exactly two are red or exactly two are green).

Into the Classroom

20. **(Writing)** Younger children especially want to check that the counting formulas give correct results by comparing those results with an answer given by direct counting. Design two activities in which groups of four verify that the number of permutations and combinations are indeed given by the respective formulas

$$P(n, r) = n(n-1)\cdots\ (n-r+1) \text{ and } C(n, r) = P(n, r)/r!.$$

(a) **Permutation Activity.** Have each group of four make "photos" of all of the ways in which there are (i) one person, (ii) two persons side by side, and (iii) three persons in a line in the photo. The photo can be an actual digital photo or just a written record of who was in the photo and in what order. After the photos are taken, check that the number of photos in the three cases is matched by the number of permutations given by the formula $P(4, r)$ for $r = 1, 2,$ and 3.

(b) **Combination Activity.** Modify part (a) by placing a loop of string on the floor and having groups of one, two, three, or four stand within the loop. Again, keep a record of all possible ways, and verify that the number of ways is correctly given by the formula for $C(4, r)$ for $r = 1, 2, 3,$ and 4. Comment on why $C(4, 0) = 1$ seems to be reasonable.

Responding to Students

21. The Math Club has four seniors and seven juniors. They wish to form a steering committee of four members, including at least two of the seniors. Being curious, they wondered how many ways the committee could be formed, and three club members presented answers:

- Shelley argued that this could be done by picking two seniors in $C(4, 2)$ ways and then filling out the committee by picking two more people from the nine remaining club members in $C(9, 2)$ ways. Therefore, the committee can be selected in $C(4, 2)C(9, 2)$ ways.
- Murial claimed that there was one committee with all four seniors, $4 \cdot 7$ ways with one junior, and $C(4, 2)C(7, 2)$ ways with an equal number of juniors and seniors. Therefore, there are $1 + 28 + C(4, 2)C(7, 2)$ possible committees.
- Jisoo claims that since there are $C(7, 4)$ committees with all juniors and $4C(7, 3)$ committees with just one senior, then $C(11, 4) - C(7, 4) - 4C(7, 3)$ gives the number of committees with two or more seniors.

Determine which, if any, of the students are correct, and explain what mistakes were made in any incorrect answer.

22. Bai knows that there are two equivalent formulas for $C(n, r)$, namely, $C(n, r) = \dfrac{n(n-1)\cdots\ (n-r+1)}{r!}$ and $C(n, r) = \dfrac{n!}{r!(n-r)!}$. Choosing $r = n$, he finds that $C(n, n) = \dfrac{n!}{n!} = 1$ and $C(n, n) = \dfrac{n!}{n!0!}$. For this to make sense, he observes that we are again led to define $0! = 1$. How do you respond to Bai?

Thinking Critically

23. Consider the set of all five-letter code words without repetition of letters.

(a) What is the probability that a code word begins with the letter a?

(b) What is the probability that, in a code word, c is immediately followed by d?

(c) What is the probability that a code word starts with a vowel and ends with a consonant?

(d) In how many of the original set of five-letter code words are c and d adjacent?

24. A bag contains 10 distinct blue balls $b_1, b_2, \ldots, b_{10}$ and 15 distinct red balls $r_1, r_2, \ldots, r_{15}$.

(a) Explain why there are $C(10 + 15, 2)$ ways to select two balls from the bag. (Order doesn't matter.)

(b) Explain why there are $C(10, 2) + C(15, 2) + 10 \cdot 15$ ways to choose two balls from the bag. [*Suggestion*: Take cases depending on the colors of the chosen balls.]

(c) Parts (a) and (b) show that $C(10 + 15, 2) = C(10, 2) + C(15, 2) + 10 \cdot 15$. Use an argument about a bag with colored balls to show that $C(m + n, 2) = C(m, 2) + C(n, 2) + m \cdot n$ for any natural numbers m and n.

25. Show that $C(m + n, 3) = C(m, 3) + C(n, 3) + mC(n, 2) + nC(m, 2)$ for all natural numbers m and n by giving two answers to the following question:

How many ways can three balls be selected from a bag containing m blue and n red distinct balls?

[*Suggestion*: Modify the combinatorial method of reasoning used in problem 24.]

26. Show that $C(8, 3) + C(8, 4) = C(9, 4)$ by verifying that $\frac{8!}{3!5!} + \frac{8!}{4!4!}$ is equal to $\frac{9!}{4!5!}$. (*Suggestion:* Remember that $5! = 5 \cdot 4!$.)

27. A bag contains eight marbles of assorted colors, of which just one marble is red. Use the symbol $C(n, r)$ to answer these questions:

(a) In how many ways can a subset of any five marbles be chosen?

(b) In how many ways can a subset of five nonred marbles be chosen?

(c) In how many ways can a subset of five marbles be chosen, including the red marble?

(d) What formula expresses the fact that your answer to part (a) is the sum of your answers to parts (b) and (c)?

(e) Create a "marble story" to derive the formula $C(10, 4) = C(9, 4) + C(9, 3)$.

28. There are 11 members of the choir. Six members are to be selected to perform a sextet, and one of these six is to be chosen for the solo part. Assuming that every choir member is equally capable, suppose we wish to know how many sextets with soloist can be selected.

(a) Why is $C(11, 6) \cdot 6$ a correct answer?

(b) Why is $11 \cdot C(10, 5)$ also a correct answer?

(c) Numerically check that $6 \cdot C(11, 6) = 11 \cdot C(10, 5)$.

(d) Invent a "choir story" that shows why $8 \cdot C(14, 8) = 14 \cdot C(13, 7)$.

29. The 15 members of the Chess Club can send eight members to the state meet, including two of the eight named as cocaptains. In how many ways can the eight-person delegation, with the cocaptains, be named?

(a) Why is one correct answer $C(15, 8) \cdot C(8, 2)$?

(b) Why is another correct answer $C(15, 2) \cdot C(13, 6)$?

(c) Algebraically check that $C(15, 8) \cdot C(8, 2) = C(15, 2) \cdot C(13, 6)$.

(d) Invent a "Chess Club" story that shows why $C(19, 7) \cdot C(7, 3) = C(19, 3) \cdot C(16, 4)$.

30. **(Writing)** There are six ways to write 5 as an ordered sum of three positive integers:

$3 + 1 + 1, 1 + 3 + 1, 1 + 1 + 3, 2 + 2 + 1,$
$2 + 1 + 2, 1 + 2 + 2.$

These sums correspond to trains of length five made up of three cars, as shown here:

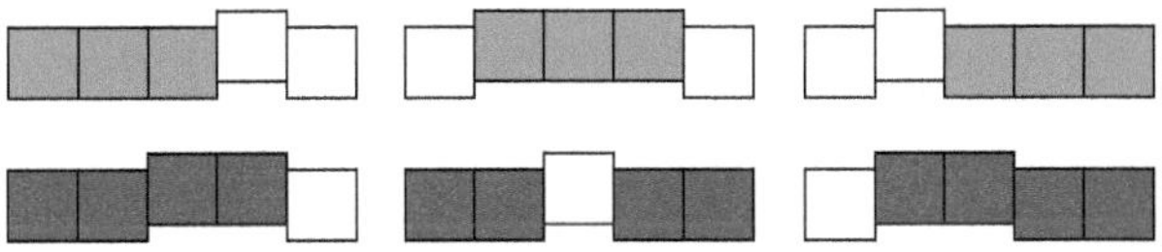

(a) Explain why there are $C(4, 2) = 6$ trains of length 5 with three cars by counting the number of ways to separate a 1×5 rectangle of unit squares into three cars.

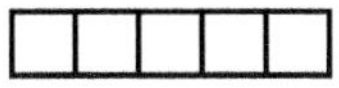

(b) Show all of the ways to write 6 as an ordered sum of four positive integers.

(c) What is the number of ways to write n as an ordered sum of r positive integers?

31. **Permutations with Repetition** Suppose a 10-foot walkway is to be paved with colored tiles, using some permutation of two red, three blue, and five green tiles. Tiles of the same color cannot be distinguished from one another.

(a) Explain why the number of tilings is given by

$$\binom{10}{2}\binom{8}{3}\binom{5}{5}.$$

(b) Use the formulas for the number of r-combinations as ratios of factorials to prove that

$$\binom{10}{2}\binom{8}{3}\binom{5}{5} = \frac{10!}{2!3!5!}.$$

(c) Give a direct explanation why there are $\frac{10!}{2!3!5!}$ tilings of the walkway by first assuming the ten tiles are all different, but then taking into account the same pattern occurs if the tiles of the same color are permuted.

32. How many different signals can be sent up on a flagpole if each signal requires three blue and three yellow flags and the flags are identical except for color?

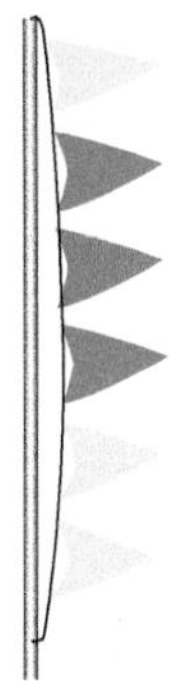

33. (a) How many different arrangements are there of the letters in TOOT?

(b) How many different arrangements are there of the letters in TESTERS?

34. What is the probability that a random permutation of the 11 letters in MISSISSIPPI

(a) correctly spells the name of the state?

(b) contains the two consecutive four-letter string IIII and SSSS?

35. Combinations with Repetition. Jen & Berry's ice cream store offers four flavors: strawberry (S), vanilla (V), chocolate (C), and peppermint (P). A group can order several cones by putting check marks on the order card. For example, the following card indicates an order of 3 strawberry, 0 vanilla, 2 chocolate, and 5 peppermint ice cream cones, for a total of 10 cones ordered:

S	V	C	P
✓✓✓		✓✓	✓✓✓✓✓

This order card suggests how an ice cream order is compactly represented as the list ✓ ✓ ✓ | | ✓ ✓ | ✓ ✓ ✓ ✓ ✓ of 10 checks and 3 bars, where the bars separate the 4 flavors from one another. As another example, the list | ✓ | ✓ ✓ ✓ | ✓ ✓ represents an order of 0 strawberry, 1 vanilla, 3 chocolate, and 2 peppermint ice cream cones, for a total of 6 cones ordered. The question "How many ways can 10 cones be ordered from 4 flavors?" can now be rephrased as "How many ways can 10 checks and 3 bars be arranged?" This question is easy to answer: There are $10 + 4 - 1$ positions, and we can choose the 10 positions for the checks in $C(10 + 4 - 1, 10) = C(13, 10) = C(13, 3) = \frac{13 \cdot 12 \cdot 11}{3 \cdot 2 \cdot 1} = 286$ ways.

(a) Show that 6 cones can be ordered from Jen & Berry's in 84 ways.

(b) Jen & Berry's just introduced two new flavors, making 6 flavors in all. Verify that there are now 1287 ways to place an order for 8 cones.

(c) To meet and beat the competition, Jen & Berry now have 32 flavors of ice cream. How many ways can 4 cones be ordered now?

(d) Carefully explain why there are $C(n + k - 1, n)$ ways to select a total of n objects from k types where repetition is allowed. (That is, there can be any number of objects of each type selected.)

36. Mario donated a total of \$10 to four different charities by randomly putting one-dollar bills in four different collection baskets. What is the probability that

(a) each charity was given at least \$1?

(b) each charity was given at least \$2?

Making Connections

37. (a) Ms. Ruiz has 13 boys and 11 girls in her class. In how many ways can she select a committee to organize a class party if the committee must contain three boys and three girls?

(b) Lourdes, a girl, and Andy, a boy, always fight. How many ways can Mrs. Ruiz select the committee of part (a) if she does not want both Lourdes and Andy on the committee? Note that Lourdes can be on the committee and Andy not on the committee, or vice versa.

38. (a) How many ways can four tennis players divide up to play a set of doubles?

(b) How many ways can five players play doubles with one player sitting out?

(c) How many ways can six basketball players divide up to play three against three?

(d) How many ways can seven basketball players divide up to play three against three with one player sitting out?

39. It is well known that the candidate listed first on a ballot has an advantage in an election. To be equally fair to all candidates, many states require that the ballots be printed in all possible orders with an equal number of ballots in each order. Suppose there are three candidates for governor, four candidates for senate, and five for representative. How many different forms of ballots must be prepared, where all ballots list the races for governor, senator, and representative in that order?

40. California originally operated a 6/49 lottery, meaning that the grand prize went to a player (or players) who picked the same six numbers that were later drawn at random from the set of numbers 1 through 49. In 1990, California went to a 6/53 lottery, but later the state went to a 6/51 lottery.

(a) Find the probability of winning a 6/49 lottery.

(b) Find the probability of winning a 6/53 lottery.

(c) Find the probability of winning a 6/51 lottery.

(d) Why do you think California changed to a 6/53 lottery?

(e) Why do you think the state went to a 6/51 lottery?

41. In the casino game Keno, the player purchases a ticket and marks eight of the "spots" numbered 1 through 80. Every 20 minutes or so, the casino randomly draws 20 balls from a drum of 80 numbered balls. If sufficiently many of the player's spots are among the 20 numbers, the player wins. Usually the player must have five winning spots to receive a prize, and a larger prize is awarded if six, seven, or all eight winning spots are marked.

(a) What is the probability of marking exactly five winning spots?

(b) What is the probability of marking exactly six winning spots?

(c) What is the probability of marking exactly seven winning spots?

(d) What is the probability of marking exactly eight winning spots?

14.4 Odds, Expected Values, Geometric Probability, and Simulations

Our final section takes up topics that complete this chapter's overview of elementary probability.

Odds

When someone speaks of the **odds** in favor of an event E, he or she is comparing the likelihood that the event will happen with the likelihood that it will not happen. Consider an urn containing four blue balls and one yellow ball. If a ball is chosen at random, what are the odds that the ball is blue? Since a blue ball is four times as likely to be selected as a yellow ball, we say that the odds are 4 to 1. The odds are actually the ratio $\frac{4}{1}$, but when quoting odds, one usually writes 4 : 1, which is read "four to one." The latter formulation is the basis for the following definition:

DEFINITION Odds

Let E be an event and let $\overline{E}$ be the complementary event. Then the **odds in favor** of E are $n(E)$ to $n(\overline{E})$, and the **odds against** E are $n(\overline{E})$ to $n(E)$.

EXAMPLE 14.24 The Odds of Rolling a 7 or 11

What are the odds of rolling a 7 or an 11 in one roll of a pair of dice?

Solution

The sample space is the set of outcomes of rolling two dice, say, a red one and a green one. There are six ways for each die to fall, so there are $6 \times 6 = 36$ outcomes in all. For example, "snake eyes" (1, 1) occurs in one way, when each die lands with one spot upward. A roll of 3 occurs in two ways, (1, 2) and (2, 1), with a 1 on the red die and a 2 on the green die or the reverse. There are six ways to roll a 7—(1, 6), (2, 5), (3, 4), (4, 3), (5, 2), and (6, 1)—and two ways to roll an 11—(5, 6) and (6, 5). This means there are eight ways to roll a 7 or an 11, and the other $36 - 8 = 28$ ways do not result in a 7 or an 11. The odds of rolling a 7 or 11 are therefore 8 to 28, or 2:7.

The probability of an event can be determined from the odds in favor of the event, and in the opposite direction, the odds of an event can be calculated from the probability.

EXAMPLE 14.25 Determining Probabilities from Odds

If the odds in favor of event E are 5 to 4, compute $P(E)$ and $P(\overline{E})$.

Solution

Since E and $\overline{E}$ are complementary, $n(S) = n(E) + n(\overline{E})$. Also, since the odds in favor of E are 5 to 4, $n(E) = 5k$ and $n(\overline{E}) = 4k$ for some integer k. Therefore,

$$P(E) = \frac{n(E)}{n(S)} = \frac{n(E)}{n(E) + n(\overline{E})} = \frac{5k}{5k + 4k} = \frac{5}{9}$$

and

$$P(\overline{E}) = \frac{n(\overline{E})}{n(S)} = \frac{n(\overline{E})}{n(E) + n(\overline{E})} = \frac{4k}{5k + 4k} = \frac{4}{9}.$$

In lowest terms, $\frac{n(E)}{n(\overline{E})}$ is $\frac{5}{4}$.

EXAMPLE 14.26 **Odds from Probabilities**

Given $P(E)$, determine the odds in favor of E and the odds against E.

Solution

The odds in favor of E are

$$\frac{n(E)}{n(\bar{E})} = \frac{n(E)/n(S)}{n(\bar{E})/n(S)}$$
$$= \frac{P(E)}{P(\bar{E})}$$
$$= \frac{P(E)}{1 - P(E)}.$$

$P(\bar{E}) = 1 - P(E)$

This last ratio would be expressed as a ratio of integers $\frac{a}{b}$ in lowest terms, and the odds quoted would be stated as a to b.

The odds against E are given by the reciprocal of the ratio giving the odds in favor of E. Thus, the odds against E are

$$\frac{1 - P(E)}{P(E)} = \frac{b}{a}$$

$= \frac{P(\bar{E})}{P(E)}$

and are quoted as b to a.

Expected Value

At a carnival, for a \$4 fee you are offered the chance to play a game that consists of rolling a single die just once. If you play, you win the amount in dollars shown on the die. If you play the game several times, how much would you expect to win? Of course, you may be lucky and win \$6 on each of a series of rolls. However, you *expect* to roll a 6 only about $\frac{1}{6}$ of the time. Since this is so for each of the numbers on the die, you should expect to win, on average, approximately

$$\frac{1}{6}\cdot 1 + \frac{1}{6}\cdot 2 + \frac{1}{6}\cdot 3 + \frac{1}{6}\cdot 4 + \frac{1}{6}\cdot 5 + \frac{1}{6}\cdot 6$$
$$= \frac{1}{6}\cdot(1 + 2 + 3 + 4 + 5 + 6)$$
$$= \frac{1}{6}\cdot 21 = \$3.50$$

per roll. But since it costs you \$4 to play the game, the carnival confidently expects players to *lose* 50¢ per game, on average. Thus, the carnival stands to make a handsome profit if a large number of patrons play the game each night.

The preceding discussion introduces the notion of **expected value.**

DEFINITION **Expected Value of an Experiment**

Let the outcomes of an experiment be a sequence of real numbers (values) $v_1, v_2, \ldots, v_n$, and suppose the outcomes occur with respective probabilities $p_1, p_2, \ldots, p_n$. Then the **expected value** of the experiment is

$$e = v_1p_1 + v_2p_2 + \cdots + v_np_n.$$

EXAMPLE 14.27

Winning at Roulette

An American roulette wheel has 38 compartments around its rim. Two of these are colored green and are numbered 0 and 00. The remaining compartments are numbered from 1 to 36 and are alternately colored black and red. When the wheel is spun, a small ivory ball is rolled in the opposite direction around the rim. When the wheel and the ball slow down, the ball eventually falls into any one of the compartments with equal likelihood if the wheel is fair. One way to play is to bet on whether the ball will fall into a red slot or a black slot. For example, if you bet on red, you win the amount of the bet if the ball lands in a red slot; otherwise, you lose. What is the expected win if you consistently bet $5 on red?

Solution

Since the probability of winning on any given try is $\frac{18}{38}$ and the probability of losing is $\frac{20}{38}$, your expected win is

$$\frac{18}{38}\cdot 5 + \frac{20}{38}\cdot(-5) = \frac{90-100}{38}$$
$$\doteq -0.26.$$

On average, you should expect to lose 26¢ per play. Is it any wonder that casinos consistently make a handsome profit?

In the preceding example, it was pretty clear that you should expect to lose slightly more often than win. This next example is less clear.

EXAMPLE 14.28

Determining the Expected Value of an Unusual Game

Suppose you are offered the opportunity to play a game that consists of a single toss of three coins. It costs you $21 to play the game, and you win $100 if you toss three heads, $20 if you toss two heads and a tail, and nothing if you toss more than one tail. Would you play the game?

Solution

Many people would play the game hoping to "get lucky" and roll HHH frequently. But is this reasonable? What is your expected return? The expected value of the game is

$$\frac{1}{8}\cdot 100 + \frac{3}{8}\cdot 20 + \frac{3}{8}\cdot 0 + \frac{1}{8}\cdot 0 = \frac{160}{8} = \$20.$$

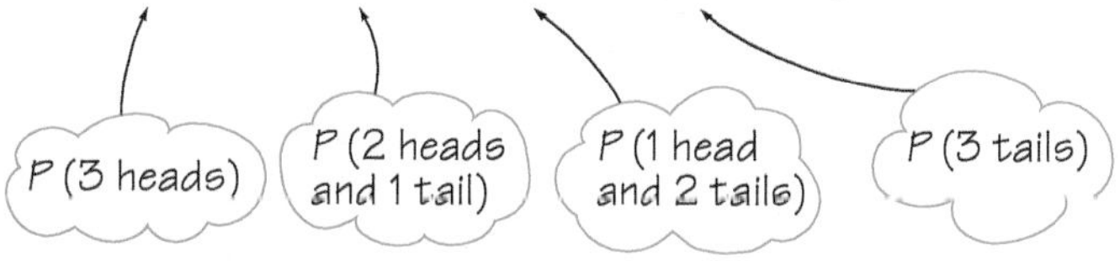

Thus, on average, you should expect to win $1 less than it costs you to play the game each time. Unless the excitement is worth at least $1, you should not play the game.

EXAMPLE 14.29

Expected Winnings at the State Lottery

The State of Washington offers Match 4, in which a player selects four numbers between 1 and 24. Each play costs $2 and has these payouts: $10,000 if all four numbers match, $20 if three numbers match, and $2 if two numbers match. What is the expected value of playing a game of Match 4?

Solution

There are $C(24, 4) = 10{,}626$ equally likely ways the four numbers $\{a, b, c, d\}$ can be selected in any drawing. There is just one way to match all four of these numbers. There are also $C(4, 3) = 4$ ways to choose three matched numbers from the four numbers drawn, with 20 ways to choose the unmatched number, so the number of ways to have exactly three matches is $4 \times 20 = 80$. Finally, there are $C(4, 2) = 6$

ways to choose two numbers to be matched and $C(20, 2) = \frac{20 \times 19}{2} = 190$ ways to choose two nonmatching numbers, showing there are $6 \times 190 = 1140$ ways to match exactly two numbers. The expected amount to be won is therefore, $(\$10{,}000)\frac{1}{10{,}626} + (\$20)\frac{80}{10{,}626} + (\$2)\frac{1140}{10{,}626} \doteq \$1.31$. Since it is certain you have paid \$2 to play the game, you can expect to lose about 69 cents per game on average.

Geometric Probability

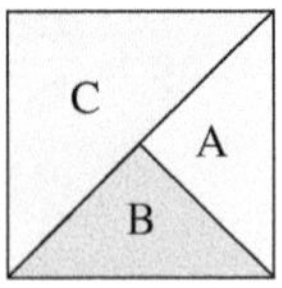

Figure 14.9
Geometric probability is a ratio of areas

Suppose that it is assumed that each point in a region S is equally likely to be chosen. Then the probability of the event corresponding to a subregion E is given by the ratio of areas. That is, the geometric probability is

$$P(E) = \frac{\text{area }(E)}{\text{area }(S)}.$$

An example is illustrated in Figure 14.9, where it is assumed any point of the square S is equally likely. Then $P(A)$ and $P(B)$ are each $1/4$, since regions A and B have areas that are one-quarter of the complete square. Similarly, $P(C) = \frac{1}{2}$ since the area of C is half that of S.

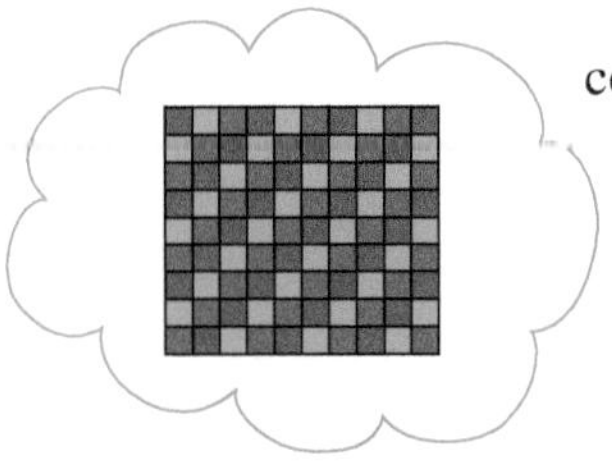

Sometimes, it is the percentage of area that is used to determine the geometric probability. For example, suppose that a huge wall is tiled with small cork squares that are orange, green, and blue. If a dart is thrown to stick in the wall, what is the probability that it has hit a blue tile? Since a third of the region is covered by the blue tiles, it seems reasonable that the probability of hitting a blue tile is $1/3$.

EXAMPLE 14.30

Tossing a Dart

A corkboard wall is covered with the pattern shown here. The large circles have a radius of 2″ and the smaller red circles have radius 1″.

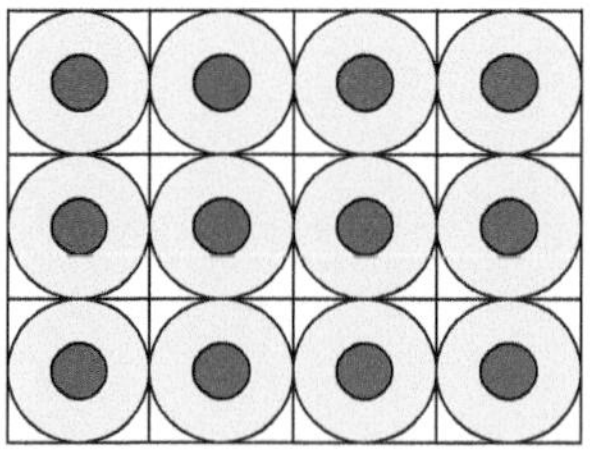

If a dart is thrown randomly at the wall, what is the probability

(a) it hits a red bull's eye?
(b) it hits in the annular yellow ring?
(c) it hits a white region outside any circle?

Solution

(a) The squares have sides of length 4″, so the area is 16 sq. in. The area of a bull's eye is $\pi(1'')^2$, or π sq. in. Therefore, the probability of hitting a bull's eye is $\pi/16$, or about 0.2.
(b) The probability of hitting within a larger circle is $4\pi/16$, so the probability of hitting a yellow ring is $3\pi/16$, or about 0.6.
(c) By complementary probability, the probability of not hitting within a circle is $1 - 4\pi/16$, or about 0.2.

Cooperative Investigation
Estimating π with Geometric Probability

Materials

Darts and a dartboard covered with paper on which the pattern of circles shown is printed.

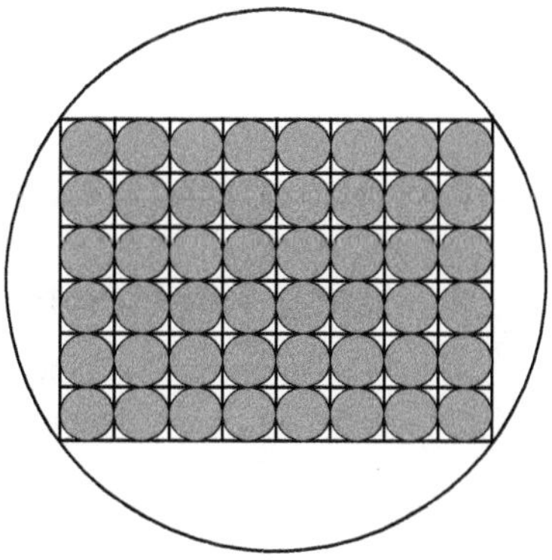

Directions

Repeatedly throw darts and record if the dart landed within some circle or not. Don't count throws that miss the pattern of circles.

Question

Working in small groups, show how to use these data to estimate the value of π.

EXAMPLE 14.31

Determining the Geometric Probability of a Carnival Game

At a carnival, a "double your money" game is played by tossing a quarter onto a large table that has been ruled into a grid of squares of the same size. If your quarter lands entirely within any square, you win back two quarters, but if the coin overlaps a grid line, you lose the quarter you tossed. If a quarter is 2.5 centimeters in diameter and the squares have sides 6 centimeters long, should you play the game?

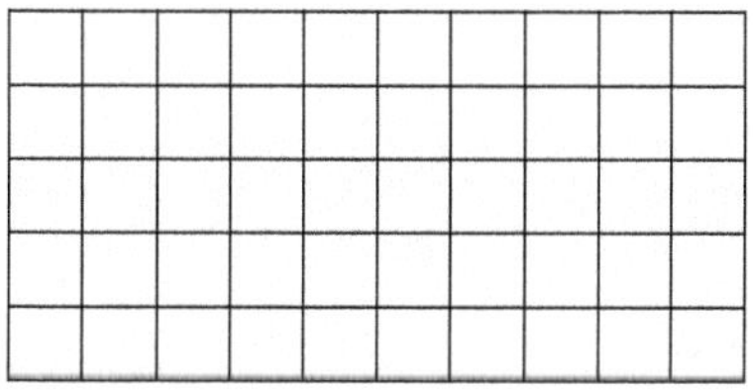

Solution

The sample space S can be considered as the points in a 6- by 6-centimeter square. To win, the center of the quarter must land at least 1.25 centimeters from each side of the large square. That is, the winning region W is a 3.5 by 3.5 centimeter square centered in the larger square, as shown in the diagram at the right. The probability of a win is

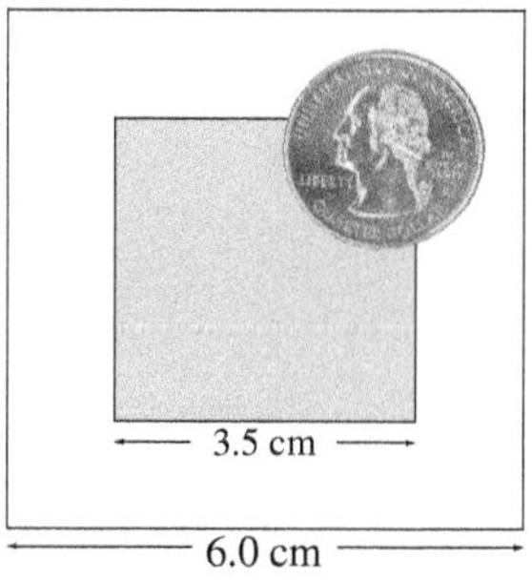

$$P(W) = \frac{\text{area of small square}}{\text{area of large square}} = \frac{3.5^2}{6^2} = \frac{12.25}{36} = \frac{49}{144} \doteq 0.34.$$

We can now compute the expected value of the game, remembering that a win gives us a net gain of \$0.25 (we must subtract the cost of playing from the \$0.50 won) and a loss is −\$0.25:

$$\text{expected value} = \frac{49}{144} \cdot \$0.25 + \frac{95}{144} \cdot (-\$0.25) \doteq -\$0.08.$$

We expect to lose, and the operator to win, about 8¢ per play.

Simulation

Simulation is a method for determining answers to real problems by conducting experiments whose outcomes are analogous to the outcomes of the real problems. Often computer simulations are conducted, since a very large number of trials can be made quickly at a low cost.

Consider, for example, a couple interested in understanding how many boys or girls they might anticipate if they decide to have children. In this case, we assume

- that the birth of either a boy or a girl is equally likely and
- that the sex of one child is completely independent of the sex of any other child.

These assumptions suggest tossing a coin, since the occurrence of a head or a tail is equally likely and what happens on one toss of the coin is completely independent of what happens on any other toss.

EXAMPLE 14.32

Using Simulation to Determine Experimental Probability

Use simulation to determine the experimental probability that a family with three children contains at least one boy and at least one girl.

Solution

Using the preceding assumptions, we can simulate the real problem by repeatedly tossing three coins. Here are the results of such an experiment:

TTH	TTH	HHT	HHT	TTH
TTT	HTT	HTT	HHT	HHT
HHH	HHT	HHT	HHT	HHH
TTT	HHH	HTT	HHT	HHT
HHT	TTH	TTH	HHT	TTH

The experimental probability based on these results is

$$P = \frac{20}{25} = 0.80.$$

Better estimates of probabilities are obtained if the number of simulations is increased. No one really wants to toss a coin thousands of times, but modern technology—graphing calculators and computers—incorporates random number generators that can be programmed to electronically flip coins and gather large amounts of data cheaply and quickly.

For example, the spreadsheet function **RAND()** returns a random decimal number between 0 and 1, so if we multiply by 2 and take the integer part with the function **INT**, we see that the function **INT(2*RAND())** returns a random 0 or 1.

The following table was generated on a spreadsheet to simulate flipping a fair coin 100 times, where 0 and 1 are interpreted as a tail and a head, respectively. As the number of tosses increases, the proportion of heads becomes very close to 0.5.

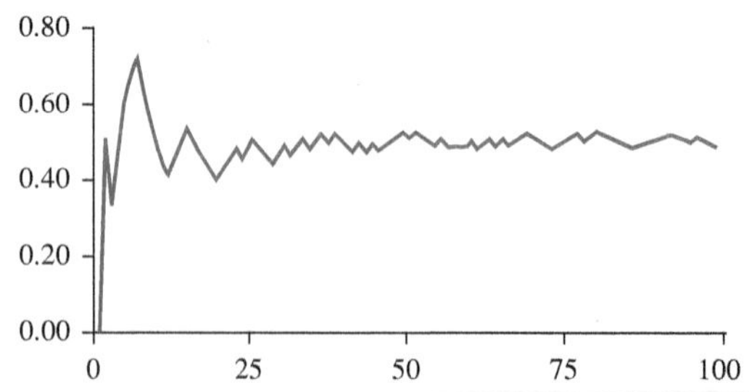

Figure 14.10 Simulating the proportion of heads after n tosses of a coin, where $1 \leq n \leq 100$

The same coin-tossing function is used in the next example to determine probabilities that would be difficult to obtain with a theoretical probability approach.

EXAMPLE 14.33

Determining the Expected Number of Runs in Ten Coin Tosses

Suppose that a "run" is any consecutive string of heads or tails. For example, if a coin is tossed ten times and results in the pattern HTHTHTHHTH, then there are five runs of heads, four runs of tails, and nine runs in all. The sequence HHHHHHHHHH has just one run, the minimum number, and HTHTHTHTHT has ten runs, the maximum. Use simulation to answer these questions.

What is the probability of n runs in ten tosses, where $1 \leq n \leq 10$?
What is the most likely number of runs?
What is the expected number of runs?

Solution

A computer was programmed to perform this experiment 1000 times:

Toss a coin ten times and calculate the number of runs.

The data obtained are shown in the following histogram that gives the frequencies of the number of runs in ten tosses.

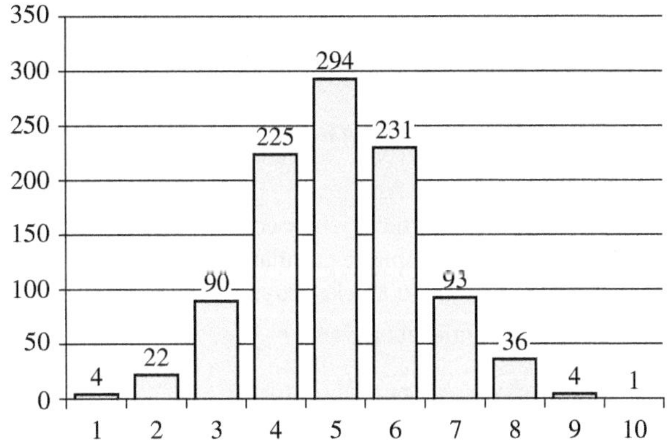

The chart gives experimental probabilities of obtaining n runs for $1 \leq n \leq 10$. For example, the probability of three runs is $90/1000 = 0.09$. The most probable number of runs is five, with an experimental probability of 0.294. The expected number of runs is

$$\frac{4}{1000}\cdot 1 + \frac{22}{1000}\cdot 2 + \frac{90}{1000}\cdot 3 + \frac{225}{1000}\cdot 4 + \frac{294}{1000}\cdot 5 + \frac{231}{1000}\cdot 6$$
$$+ \frac{93}{1000}\cdot 7 + \frac{36}{1000}\cdot 8 + \frac{4}{1000}\cdot 9 + \frac{1}{1000}\cdot 10 = 5.059.$$

14.4 Problem Set

Understanding Concepts

1. Two dice are thrown.
   (a) What are the odds in favor of getting a score of 6?
   (b) What are the odds against getting a 6?
   (c) What is the probability of getting a 6?

2. If $P(A) = \frac{2}{5}$, compute the odds in favor of A resulting from a single trial of an experiment.

3. If $P(A) = \frac{1}{2}$, $P(B) = \frac{1}{3}$, $P(C) = \frac{1}{6}$, and A, B, and C are mutually exclusive, compute the odds in favor of A or C resulting from a single trial of an experiment.

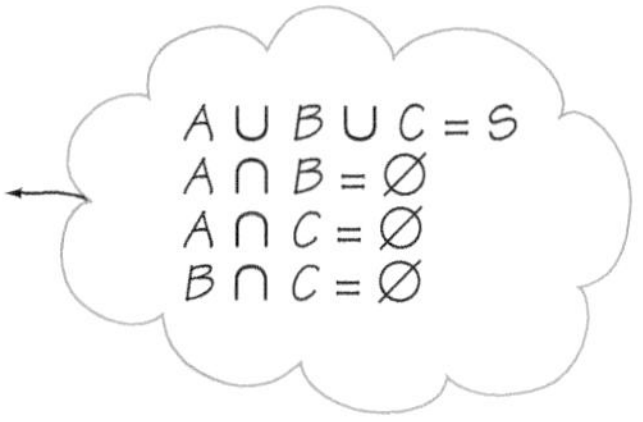

4. What are the odds that a randomly drawn card from a deck of 52 playing cards is
   (a) a heart? (b) an ace? (c) a face card?

5. An experiment has three possible mutually exclusive outcomes, A, B, or C. If the odds of A occurring are 3 to 7 and the odds of B occurring are 1 to 1, what are the odds that C occurs?

6. Compute the expected value of the score obtained by rolling two dice.

7. A game consists of rolling a pair of dice. You win the amounts shown for rolling the score shown in the following table:

Roll	2	3	4	5	6	7	8	9	10	11	12
$ Won	4	6	8	10	20	40	20	10	8	6	4

   Compute the expected value of the game.

8. A game is "fair" if the expected value is zero. Is the following game fair?

   For $1, a player rolls a pair of dice and wins $6 if a 7 is rolled and loses otherwise.

9. The Math Club intends to hold a raffle, in which tickets costing $1 are sold and the winner receives a graphing calculator valued at $100. What is the expected value to a ticket buyer if
   (a) 100 tickets are sold? (b) 200 tickets are sold?

10. What is the expected value of a raffle ticket purchased for $2 if 1000 tickets are sold, the grand prize is valued at $800, and there are five second place prizes each valued at $200?

11. Is roulette, played on an American 38 pocket wheel, a fair game? Assume that a $1 bet wins if the ball lands in any of n chosen pockets, in which case the player is returned $\$36/n$.

12. A point is chosen at random in the large square that circumscribes a circle. What is the probability it is a point in the smaller square that inscribes the circle?

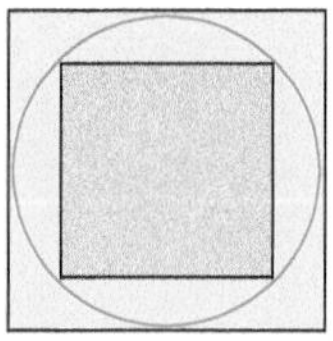

13. A large target is covered with red, yellow, green, and blue hexagons as shown below. If a dart is thrown, what is the probability it lands
   (a) in a red hexagon?
   (b) in a blue or green hexagon?
   (c) outside a yellow hexagon?

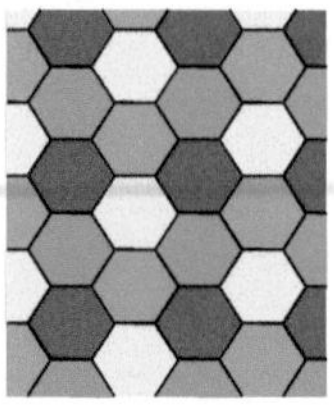

14. A dartboard is marked as shown:

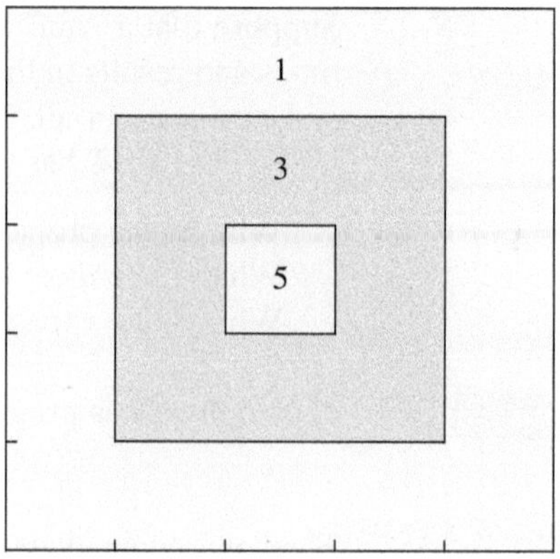

   Josie is good enough that she always hits the dartboard with her darts, but, beyond that, the darts hit in random locations. If a single dart is thrown, compute these probabilities:
   (a) $P(1)$ (b) $P(3)$ (c) $P(5)$
   (d) If Josie wins the number of dollars indicated by the number of the region into which her dart falls, how much is her expected win (the expected value)?
   (e) Suppose it costs Josie $2 each time she throws a dart. Should she play darts as in part (d)? Explain.

15. A $6'' \times 6''$ square has been divided into six regions $A, B, \ldots, F$ by three straight lines through the center of the square as shown below.
   (a) What is the probability that a point chosen at random inside the square is in a blue region?

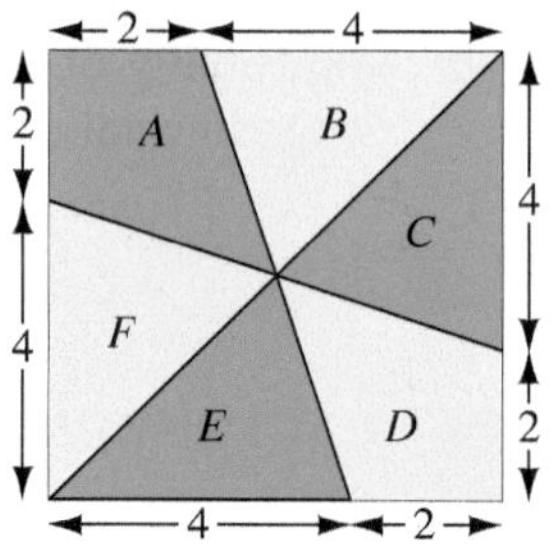

   (b) What region is most likely to contain the point?

16. Assuming it is equally likely that a baby is a girl or a boy, the child's gender can be simulated by the flip of a coin. For example, if two coins are tossed many times, it might be noticed that HH or TT occurs about the same number of times as HT or TH. This suggests that a family with two children will have two children of opposite gender as often as two children of the same gender. Toss four coins 20 times to answer this question.

   In a family with four children not all of the same sex, is it more likely to have two boys and two girls or to have three children of the same sex?

17. Suppose that ten room keys are randomly distributed to ten students in a dorm. What is the probability no one has the correct key? at least one student has the correct key? Answer by a simulation: Number ten index cards 1 through 10, deal them out number side up in a row, and see if the nth card has the number n for any n, $1 \le n \le 10$.

18. A computer is programmed to simulate experiments and to compute experimental probabilities. Match at least one of the computed probabilities with each of the descriptive sentences listed.

(a) $P(A) = 0$ **(b)** $P_e(B) = 0.5$
(c) $P(C) = -0.5$ **(d)** $P_e(D) = 1$
(e) $P(E) = 1.7$ **(f)** $P_e(F) = 0.9$

(i) This event occurred every time.
(ii) There was a bug in the program.
(iii) This event occurred often but not every time.
(iv) This event never occurred.
(v) This event occurred half the time.

Into the Classroom

19. **(a)** **(Writing)** Write up, with careful instructions, graphics, and so on, the following activity:

Work in pairs, with each person first making a paper number cube with blank faces. Each person, unseen by the other, writes a number from 1 through 6 on each face with a pencil, with repetitions and omissions allowed. Each person then rolls his or her cube, keeping it hidden but announcing the outcome of each successive roll. The person continues to roll the cubes until one person announces "stop" and guesses what is written on the faces of the hidden cube. If the guess is correct, that person wins, but if it is incorrect, the opponent is the winner.

(b) Play the game with a partner, and write a report on the game and what principles of probability can be learned and taught.

20. It is often interesting to examine real data to see how experimental probabilities are used in today's society. Use the following table to discuss whether travel is safer by car or by air:

	Deaths per Billion Journeys	Deaths per Billion Kilometers
Car	40	3.1
Air	117	0.05

21. **Classroom Carnival.** Divide the class in groups of four to six students. Each group designs and makes a game such as a dart throw, ring toss, dice game, basket toss, and the like. Each student is then given ten scrip tickets to play at games of their choice, where a ticket is needed to play and a winner is rewarded with some number of tickets. The members of a group take turns running their own game or roaming the class to play games created by other teams. At the end, compare which games and players did well, and discuss why this was so.

Responding to Students

22. Robbie heard on last night's weather forecast that there was a 30% chance of rain. However, it has rained all day long, and Robbie thought it would rain only 30% of the day. How would you help Robbie better understand the meaning of "a 30% chance of rain"?

23. A spinner has three 120° equally sized sectors colored red, green, and yellow, respectively. Maria is certain that her third spin will be yellow, since her first two spins were red and green and she knows that the probability of yellow is $\frac{1}{3}$. What probability concept is Maria overlooking? Help her with a well-written paragraph.

24. Jessie claims that skydiving is safer than getting out of bed, since there were just 49 skydiving deaths in 2012 in the United States compared with 1307 deaths from falling out of bed. Is Jessie's assessment correct?

25. Monica questions whether expected value applies well to lotteries. Respond to Monica's questions:

(a) Isn't it better to play the lottery with fewer players rather than when many players are competing for a large grand prize?

(b) Could it be better to play for \$10 million instead of \$100 million?

Thinking Critically

26. Let E and F be mutually exclusive events in a sample space S. The odds that E occurs is 3:7 and the odds F occurs is 3:2. If it is known either E or F occurred, what are the odds it is E?

27. A couple has decided they would like a child of each sex but no more than four children. What is the expected number of children in the family? Notice that

$$P(3) = P(GGB) + P(BBG)$$
$$= \left(\frac{1}{2}\right)^3 + \left(\frac{1}{2}\right)^3 = \frac{1}{4}$$

is the probability of three children.

28. Two numbers x and y are chosen randomly between 0 and 1, with $x < y$. Equivalently, the numbers correspond to a point $P(x, y)$ in the shaded region shown at the left in the following diagram:

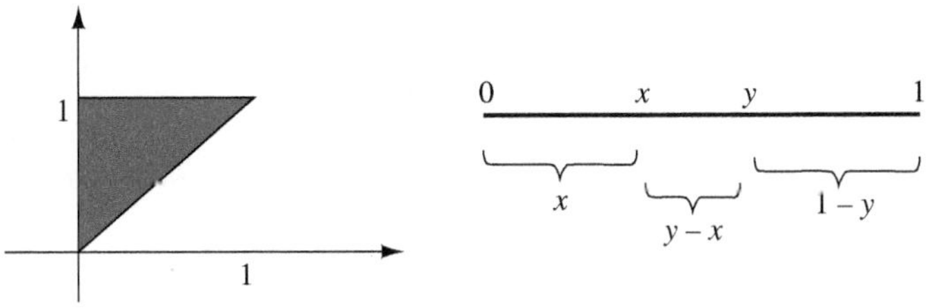

Points x and y divide the unit segment into three segments of length x, $y - x$, and $1 - y$. Carry out the following steps to determine the probability that the three segments form a triangle:

(a) Use the triangle inequalities—for example, $x < (y - x) + (1 - y)$—to show that a triangle can be formed if, and only if, P is to the left of the vertical line $x = \frac{1}{2}$, above the horizontal line $y = \frac{1}{2}$, and below the 45° slanted line with equation $y = x + \frac{1}{2}$.

(b) Graph the region described in part (a) to show that it is a triangle of area $\frac{1}{8}$.

(c) Use parts (a) and (b) to show by geometric probability that the probability that the points x and y in the unit interval form segments from a triangle is $\frac{1}{4}$.

29. Consider a "walk" starting from point A and moving to the right along the edges of the network below, ending at one of the seven points $T, U, V, \ldots Z$. At each intersection point, a die is rolled to determine the direction to walk: upward if a 1 or 2 is rolled, horizontal if a 3 or 4 is rolled, and downward if a 5 or 6 is rolled. For example, a sequence of rolls 122 arrives at point T in three upward steps and the sequence 251 arrives at V. Roll a die to simulate a large number of walks through the network, recording the end point reached.

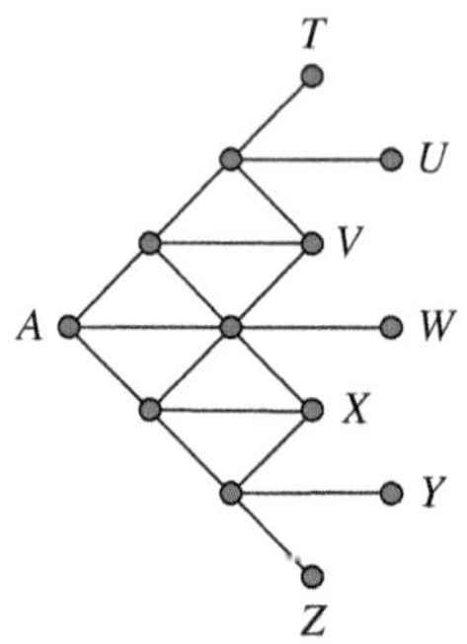

On the basis of your simulation, answer these questions.

(a) What is (are) the most likely ending point(s)?

(b) What is (are) the least likely ending point(s)?

(c) Now compare the results of the simulation with the theoretical probabilities obtained by completing the probability tree started below. It shows that a path reaches points B and C with the respective probabilities $\frac{1}{3}$ and $\frac{1}{3} + \frac{1}{3}\cdot\frac{1}{3} + \frac{1}{3}\cdot\frac{1}{3} = \frac{5}{9}$.

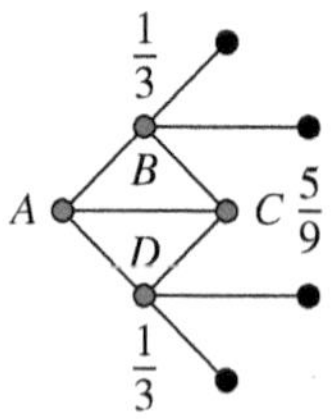

Making Connections

30. In the casino game Keno, 20 balls are drawn at random from a barrel of balls numbered 1 through 80. Six-spot Keno is played by buying a Keno card for \$1 and marking 6 of the 80 numbers. The payouts are:

$1 if 3 numbers match the 20 selected
$8 if 4 numbers match the 20 selected
$50 if 5 numbers match the 20 selected
$1500 if 6 numbers match the 20 selected

There are $C(6, 3)$ ways that exactly 3 of the marked numbers match, and the other 17 balls have numbers chosen in $C(74, 17)$ ways from the 74 numbers that were not marked on the card. The probability of exactly 3 numbers matching is therefore

$$P(3) = \frac{C(6,3)\cdot C(74,17)}{C(80,20)} \doteq 0.12982.$$

By the same reasonings, the probabilities of exactly 4, 5, and 6 matches are

$$P(4) = \frac{C(6,4)\cdot C(74,16)}{C(80,20)} \doteq 0.02854,$$

$$P(5) = \frac{C(6,5)\cdot C(74,15)}{C(80,20)} \doteq 0.00310,$$

$$P(6) = \frac{C(6,6)\cdot C(74,14)}{C(80,20)} \doteq 0.00013.$$

What is the expected payout for your \$1 investment?

State Assessment

31. (Grade 8)
Tracy has a red (R) , a green (G), a blue (B), a white (W), and a purple (P) marble in her bag, and Nina has a red (R), a blue (B), and a white (W) marble in her bag.

(a) Make a list of all of the ways each girl can randomly select a marble from her bag.

(b) What is the probability both marbles have the same color?

(c) What is the probability the two marbles have different colors?

The Chapter in Relation to Future Teachers

This chapter has explored both experimental and theoretical probability. In the lower elementary grades, simple experiments can be performed, with the collected data displayed and analyzed with the strategies of the previous chapter. At the same time, it will be appropriate to introduce some of the terms—*outcome, sample space event, probability*—associated with quantifying the likelihood of a future experiment. Concrete experiences are essential for the continued progress of students in the upper elementary and middle school grades, where counting techniques become increasingly important in the determination of theoretical probabilities.

Chapter 14 Summary

Section 14.1 The Basics of Probability	Page Reference
CONCEPTS	
• **Probability:** The branch of mathematics that quantifies the likelihood of events.	720
• **Experimental probability:** An estimation of probability obtained by conducting repeated trials or by tabulating historical data.	720, 723
• **Theoretical probability:** A determination of probability made by assuming the probability of each outcome.	720, 724
DEFINITIONS	
• An **outcome** x is the result of a single trial of an experiment.	722
• The **sample space, *S*,** is the set of all possible outcomes of an experiment.	722
• An **event, *E*,** is a subset of the sample space.	722
• A **favorable outcome** to event E is an outcome in E.	722
• A **probability function** is a function P on the sample space $S = \{x_1, x_2, \ldots, x_N\}$ that measures the likelihood outcome x_i occurs, where $0 \le P(x) \le 1$ and $P(x_1) + P(x_2) + \cdots + P(x_N) = 1$.	722
• The **probability of event *E*** is the sum of all of the probabilities of outcomes favorable to E.	723
• **Experimental probability**, $P(E)$, is the fraction of the number of times an event occurs over a large number of trials.	723
• **Equally likely outcomes** occur if the N possible outcomes in a sample space S each have the same probability, $1/N$, of occurring.	724
• A sample space $S = \{x_1, x_2, \ldots, x_N\}$ is **uniform** if the outcomes are equally likely with probability $P(x_i) = \dfrac{1}{N}$ for all $x_i \in S$	724
• Events E and F are **mutually exclusive** if no outcome is favorable to both events; that is, $E \cap F = \varnothing$.	725, 726
• A **complementary event, $\overline{E}$,** is the event E does not occur, so $\overline{E} = S - E$.	727
PROPERTIES	
• **Range of probabilities:** The probability of an outcome χ_i satisfies $0 \le P(\chi_i) \le 1$, where $P(\chi_i) = 0$ means that the outcome is impossible and $P(\chi_i) = 1$ means that the outcome is certain to occur.	722
• **Probability of an event *E* in a uniform sample space *S*:** $P(E) = \dfrac{n(E)}{n(S)}$, where $n(E)$ is the number of outcomes favorable to E and $n(S)$ is the number of equally likely outcomes in the sample space.	724
• **Probability of mutually exclusive events:** If A and B are mutually exclusive events, so that $A \cap B = \varnothing$, then $P(A \text{ or } B) = P(A \cup B) = P(A) + P(B)$.	726
• **Probability of two non–mutually exclusive events:** If A and B are any two events, then $P(A \text{ or } B) = P(A \cup B) = P(A) + P(B) - P(A \cap B)$.	727
• **Probability of a complementary event:** The probability of the event $\overline{E} = S - E$ complementary to event E is given by $P(\overline{E}) = 1 - P(E)$.	728

Section 14.2 Applications of Counting Principles to Probability	Page Reference
CONCEPTS	
• A **permutation** of a set of objects is an ordered list of all the objects in the set.	734
• A **factorial** of a whole number n is the product $n! = n \times (n-1) \times (n-2) \times \cdots \times 3 \times 2 \times 1$ when $n \geq 1$ and $0! = 1$.	734
• A **multistage experiment,** or **multistep process,** is an experiment in which all outcomes are the result of choices made in an ordered succession.	734, 735
• A **possibility tree** is a diagram showing the choices that can be made at each step of a multi step process.	734
• A **probability tree** is a diagram showing the probabilities of the choices made at each step of a multistep process.	734
DEFINITIONS	
• The **conditional probability** $P(E\mid F)$ is the probability of event E when it is assumed that event F occurs.	738
• Events E and F are **independent events** if the probability of either event is not dependent on whether or not the other event occurs. That is, $$P(E\mid F) = P(E) \text{ and } P(F\mid E) = P(F).$$	740
THEOREMS	
• **Addition Principle of Counting:** If events $E, F, \ldots, K$ are mutually exclusive, so no outcome belongs to more than one of the events, then the number of outcomes in one of the events is $$n(E \cup F \cup \cdots \cup K) = n(E) + n(F) + \cdots + n(K).$$	733
• **Addition Principle of Probability:** If events $E, F, \ldots, K$ are mutually exclusive, so no outcome belongs to more than one of the events, then the probability that an outcome is favorable to one of the events is $$P(E \cup F \cup \cdots \cup K) = P(E) + P(F) + \cdots + P(K).$$	733
• The **number of permutations of n objects** is the factorial $n! = n \times (n-1) \times (n-2) \times \cdots \times 3 \times 2 \times 1$.	735
• **Conditional Probability Formula:** Let E and F be events in a uniform sample space. Then the conditional probability of event E, given that event F occurs, is $$P(E\mid F) = \frac{n(E \cap F)}{n(F)}.$$	739
• **Probability of Compound Events:** The probability that both events E and F occur is $$P(E \cap F) = P(E\mid F)P(F) = P(F\mid E)P(E).$$	739
• **Probability of Independent Events:** The probability that the independent events E and F both occur is $$P(E \cap F) = P(E)P(F).$$	740

Section 14.3 Permutations and Combinations	Page Reference
CONCEPTS	
• **Permutation:** An ordered arrangement of objects.	744
• **Combination:** An unordered selection of objects that form a subset.	744

DEFINITIONS	
• An ***r*-permutation** is an ordered arrangement of r things taken from a set of n distinct things.	745
• An ***r*-combination** is a subset of r things taken from set of n distinct things.	745
THEOREM	
• **Pascal's identity:** $C(n, r - 1) + C(n, r) = C(n + 1, r)$	748
FORMULAS	
• **Number of *r*-permutations from a set of *n* different objects:** $P(n, r) = n(n - 1) \cdots (n - r + 1) = n!/(n - r)!$	746
• **Number of *r*-combinations from a set of *n* different objects:** $C(n, r) = \binom{n}{r} = n(n - 1) \cdots (n - r + 1)/r! = n!/[r!(n - r)!]$	747

Section 14.4 Odds, Expected Values, Geometric Probability, and Simulations	Page Reference
DEFINITIONS	
• **Odds** are the ratio of the number of times the event can occur to the number of times the event does not occur.	757
• The **expected value** is the average value that can be anticipated if the numerical outcomes of an experiment are $u_1, u_2, \ldots, u_s$ and they occur with respective probabilities $p_1, p_2, \ldots, p_s$. The expected value of the experiment is $e = p_1u_1 + p_2u_2 + \cdots + p_su_s$.	758
• **Geometric probability** is the probability depending on the geometry of an experiment.	760
• **Simulation:** A method of determining answers to real problems by conducting experiments with outcomes analogous to those of the real problem.	761

Chapter Review Exercises

Section 14.1

1. A tetrahedral die has faces marked 1 through 4. An experiment is conducted by rolling the four-faced die and tossing a coin, and seeing which face of the die and side of the coin land downward. Let E be the event the die lands on an even number and H the event the coin lands on its head.

(a) Describe the sample space.

(b) Draw a Venn diagram of the sample space showing the events E and H.

(c) What are the probabilities $P(E)$, $P(H)$, $P(E \cup H)$, and $P(E \cap H)$?

(d) Are E and H mutually exclusive?

2. (a) List explicitly all elements in the sample space if two coins and a die are tossed.

(b) Compute $P(\text{T}, \text{T}, 5)$, the probability of getting two tails on the coins and a 5 on the die in part (a).

3. Compute the probability of obtaining a sum of at most 11 on a single roll of two dice.

4. Two dice are rolled, one an ordinary six-faced cube and the other a dodecahedron with faces numbered 1 through 12. What are the probabilities of rolling

(a) the same two numbers, such as "snake eyes" (two ones) or "box cars" (two sixes)?

(b) no more than a 15?

5. A card is drawn from a deck of 52 playing cards. What is the probability that the card is

(a) a heart or a club?

(b) a diamond or a king?

(c) neither a diamond nor a king?

6. What kind of probability is used when an assertion such as "The probability that penicillin will cure a case of strep throat is 0.9" is made? Explain briefly.

Section 14.2

7. Mischa has four math books and five physics books. How many ways can she

(a) line them up on a shelf?

(b) line them up on a shelf so the math books are together and the physics books are together?

(c) If Mischa randomly puts the books side by side on a shelf, what is the probability that no two of the physics books are together?

8. In Morse code, each letter is coded by a sequence of symbols, where each symbol is either a dot or a dash. For example, a single dot · represents the letter *e*, and the letter *o* is three dashes − − −. The distress signal SOS is transmitted as · · · − − − · · · .

(a) How many letters can be formed if at most three symbols are allowed?

(b) Can all 26 letters be coded if up to four symbols are used?

9. An urn contains four red, five blue, and two green balls. Construct the possibility tree in which two balls are drawn randomly from the urn, with the first ball not replaced. Use the tree to find the

(a) the probability that both balls have the same color, and

(b) the probability that at least one green ball was drawn.

10. The faces of a red cube are marked with one A, two Bs, and 3 Cs. A second, blue cube has faces marked with three As, one B, and two Cs. Construct a probability tree to determine

(a) the probability of rolling a double (both cubes the same).

(b) the probability of rolling an A on at least one of the cubes.

11. Consider the spinner shown here:

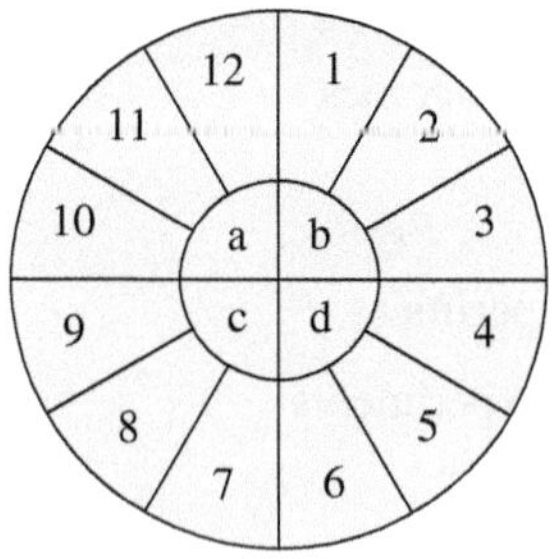

Compute these probabilities:

(a) $P(\text{b and } 8)$ **(b)** $P(\text{b or } 8)$ **(c)** $P(\text{b} \mid 8)$

(d) $P(\text{b and } 2)$ **(e)** $P(\text{b or } 2)$ **(f)** $P(2 \mid \text{b})$

12. A census taker was told by a neighbor that a family of five lived in the next house—two parents and three children. When the census taker visited the house, he was greeted by a girl. What is the probability that the other two children were both boys? Explain briefly.

13. Alex drew a card from a 52 card deck, Betty rolled a number cube, and Carla flipped a coin. What is the probability that

(a) Alex drew a spade, Betty rolled an odd prime, and Carla tossed a head?

(b) Alex drew a spade, or Betty rolled an odd prime, or Carla tossed a head?

Section 14.3

14. How many four-letter code words can be made from the letters *a, b, c, d, e, f,* and *g*

(a) with repetition allowed?

(b) with repetition not allowed?

15. Calculate each of the following:

(a) $7!$ **(b)** $\frac{9!}{6!}$ **(c)** $\frac{8!}{(8-8)!}$

(d) $7 \cdot 6!$ **(e)** $P(8, 5)$ **(f)** $P(8, 8)$

(g) $C(9, 3)$ **(h)** $C(9, 9)$

16. A bicycle lock has a sequence of four cylinders on an axle, where each cylinder can be rotated to one of the digits 0, 1, 2, . . . , 9. Suppose a thief is attempting to steal the bicycle by randomly trying various combinations. What is the probability the thief is successful

(a) in 100 attempts?

(b) in an hour, guessing a new combination each second?

(c) in three hours, trying a new combination each second?

17. An urn contains five yellow, four blue, and eight green marbles.

(a) In how many ways can one select five green marbles?

(b) In how many ways can one select five yellow and five green marbles?

(c) In how many ways can one select five yellow or five green marbles?

18. If you select five marbles from the urn in problem 17, what is the probability that two are yellow given that three are green?

19. An urn contains five white, six red, and four black balls. Two balls are chosen at random.

(a) What is the probability that they are the same color?

(b) What is the probability that both are white?

(c) What is the probability that both are white given that they are the same color?

(d) Verify your answers with a probability tree.

Section 14.4

20. Three coins are tossed.

(a) What are the odds in favor of getting two heads and one tail?

(b) What are the odds in favor of getting three heads?

21. If $P(E) = 0.35$, what are the odds of obtaining E on a single trial of an experiment?

22. (a) If $P(A) = 0.85$, what are the odds in favor of A occurring on any given trial?

(b) If the odds in favor of A are 17 to 8, determine $P(A)$.

23. You play a game where you win the amount shown with the probability shown:

$P(\$5) = 0.50 \quad P(\$10) = 0.25 \quad P(\$20) = 0.10$

(a) What is the expected value of the game?

(b) If it costs you \$10 to play the game of part (a), is it wise to play? Explain.

24. A dart is thrown on a target covered with blue discs touching tangentially in a triangular pattern as shown.

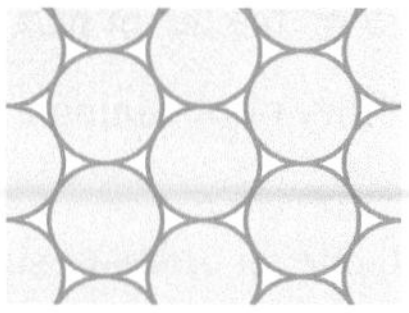

What is the probability of hitting a blue point?

Answers to Odd-Numbered Problems

Chapter 1

Problem Set 1.1 (page 6)

1. **(a)** 21 bikes, 6 trike
(b)

Bikes	Trikes	Bike Wheels	Trike Wheels	Total Wheels
17	10	34	30	64
18	9	36	27	63
19	8	38	24	62
20	7	40	21	61
21	6	42	18	60

(c) Place two wheels next to each seat, and then add a third wheel to as many seats as necessary to make 60 wheels. **(d)** Yes, lift the front wheels of the trikes off the ground. **3.** **(a)** 11 18-cent stamps and 21 29-cent stamps **(b)** Answers will vary.
5. 3 dimes, 3 nickels, 3 pennies (Make an orderly list.)
7. 12 (Work backward from 52.)
9. Answers will vary.
11. **(a)**

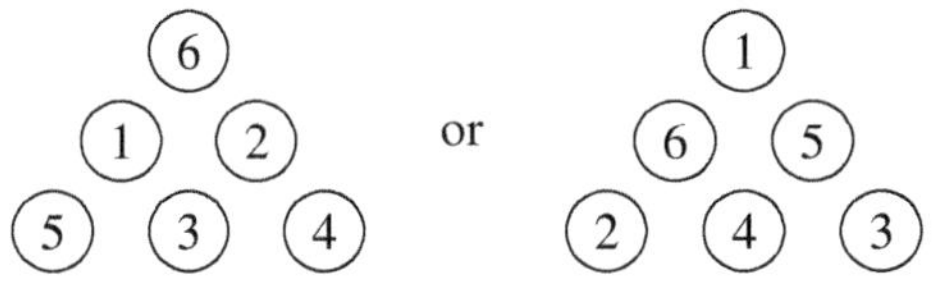

(b) Yes **(c)** Answers will vary. Another alternative is

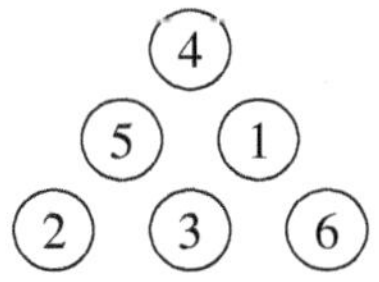

13. **(a)**

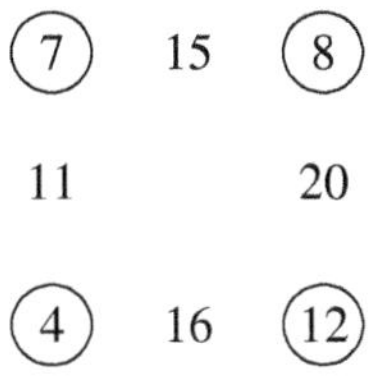

(b)

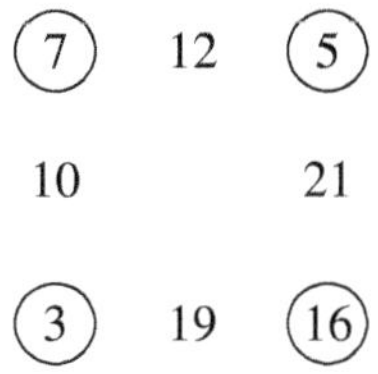

(c)

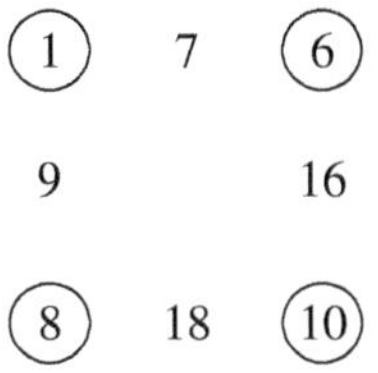

15. **(a)** Each sum is 15. **(b)** Each result is 5. **17.** **(a)** 1, 2, 3, 5, 8, 13, 21 **(b)** 2, 6, 8, 14, 22, 36, 58 **(c)** 3, 5, 8, 13, 21, 34, 55 **(d)** 2, 4, 6, 10, 16, 26 **(e)** 2, 1, 3, 4, 7, 11

Problem Set 1.2 (page 17)

1. **(a)** No. When 10 is multiplied by 5 and 13 is added, the result is 63, not 48. **(b)** Guess and Check, Make a Table, or algebra. **(c)** Using algebra, if x is Nancy's number, then $5x + 13 = 48$ is the answer, $x = 7$.

3.

Guess	Twice the Guess Plus 1	Three Times the Guess Minus 5
4	9	7
5	11	10
6	13	13

The number must be 6.
5. 14; note that $25 + 5 + 5 + 5 = 10 + 10 + 10 + 10 = 40$.
7. 258, 285, 528, 582, 825, 852

9.

Number of Dimes	Number of Nickels	Number of Pennies
2	0	1
1	2	1
1	1	6
1	0	11
0	4	1
0	3	6
0	2	11
0	1	16
0	0	21

11. **(a)** 1, 120 2, 60 3, 40 4, 30 5, 24 6, 20 8, 15 10, 12 **(b)** 10×12 **13.** Jill worked 10 days and Peter worked 15 days. **15.** 9 minutes **17.** 40
19. **(a)** 12, 14, 16, 18, and 20 **(b)** 3-by-3 square
21. **(a)**

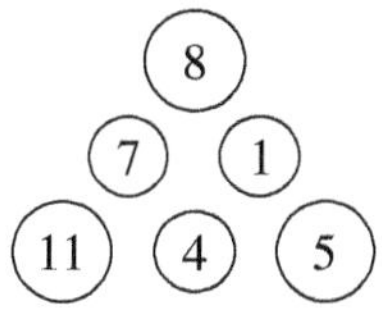

(b)

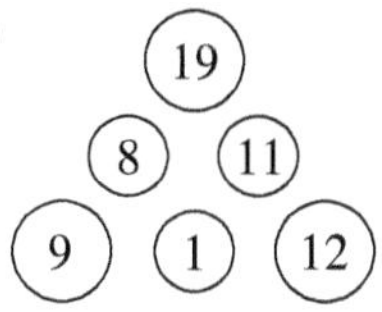

(c)

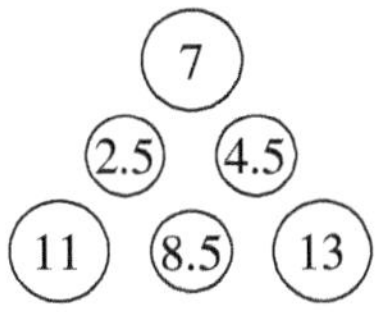

(b)

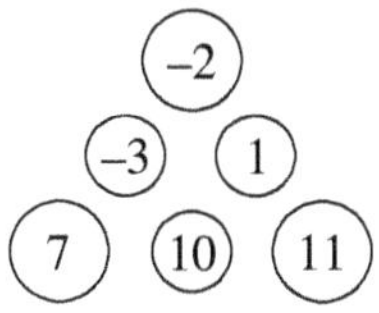

23. **(a)** The organized list has $2 \times 3 = 6$ combinations **(b)** The organized list has $2 \times 3 \times 2 = 12$ combinations.

25. Make a Table. Answer B. There are 20 black chairs

White Chairs	1	1	5	5	9	9	13	13	17	17
Black Chairs	0	4	4	8	8	12	12	16	16	20

Problem Set 1.3 (page 28)

1. (a) 14, 17, 20 **(b)** 3, 5, 7 **(c)** 10, 15, 15 **3. (a)** 16 **(b)** 11 **(c)** 17 **5. (a)** $1 + 2 + 3 + 4 + 5 + 4 + 3 + 2 + 1 = 25$
$1 + 2 + 3 + 4 + 5 + 6 + 5 + 4 + 3 + 2 + 1 = 36$
(b) The sum is $100^2 = 10{,}000$.

7. (a)
• • • •   • • • • •
• • • • •   • • • • • •
• • • • •   • • • • • •

(b) 2, 5, 8, 11, 14, 17 **(c)** 29, 299 **(d)** 34th
9. (a) 86 (b) 78 (c) 73
11. One way is shown:

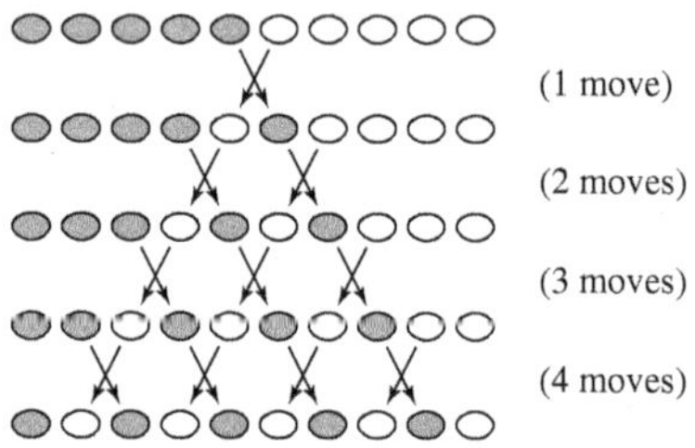

The least number of moves is $t_4 = 1 + 2 + 3 + 4 = 10$.
13. (a) 6, 10 **(b)** 136 **(c)** $\frac{n(n+1)}{2}$ **(d)** Answers will vary.
15. 35 **17.** 5 **19. (a)** 45 **(b)** 55 **(c)** Yes
21. Answers will vary widely. **23.** None. It's a hole!
25. The coins are a penny and a quarter. The problem states that one of the coins is not a quarter. It does not say that neither coin is a quarter. **27.** Try each of the given rules; C is the correct answer.

Problem Set 1.4 (page 39)

1. (a) Each pattern has one more column of dots than the previous one. The next three patterns are shown:

• • • •   • • • • •   • • • • • •
• • • •   • • • • •   • • • • • •

(b) 2, 4, 6, 8, 10, 12 **(c)** Each term is twice the number of the term. Therefore, the 10th term is $2(10) = 20$ and the 100th term is 200. **(d)** The nth even number **(e)** Divide 2402 by 2 and obtain 1201. **3. (a)** Let m be Jackson's number. To triple m and subtract 13 means to compute $3m - 13$. Thus, $3m - 13 = 2$, so $m = 5$ is Jackson's number. **(b)** There are no solutions because the solution of the equation $3n - 13 = 4$ is $n = 17/3$, which is not a whole number. **5.** Each additional car requires 4 more toothpicks, so the formula has the form $t = a + 4c$. At $c = 1, 5 = a + 4$. Therefore, $a = 1$ and the desired formula is $t = 1 + 4c$. **7. (a)** Person A shakes with person B and C. Since B and C have already shaken hands with A, only one shake remains: B with C. Total is $2 + 1 = 3$.
(b) If the people are A, B, C, D, E, F, then A shakes hands with the other five and B has only four people left to shake hands with (C, D, E, and F). Similarly, C has only three handshakes, D has two, E has one, and there is nobody new left for F. Total is $5 + 4 + 3 + 2 + 1 = 15$. **(c)** The logic is the same as (b). The first person shakes hands with the other 199, the second with 198, the third with 197, etc., until we get to the penultimate person, who has only one new hand to shake. $\text{Total} = 199 + 198 + 197 + 196 + \cdots + 2 + 1 = (200)(199)/2 = 19{,}900$, where we have used Gauss insight.
(d) The logic is the same as **(c)**. The first person shakes hands with the other $n - 1$, the second with $n - 2$, the third with $n - 3$, etc., until we get to the penultimate person, who has only one new hand to shake. Thus, the total is

$$(n-1) + (n-2) + (n-3) + \cdots + 2 + 1 = n(n-1)/2$$

where we have used Gauss insight. **9.** There are 100 people in the room. If all of them would shake hands exactly once (problem 7), there would be a total of $(100)(99)/2 = 4950$ handshakes. However, since no husband and wife shake each other's hand, there are 50 fewer handshakes according to the condition of the problem. The answer is $4950 - 50 = 4900$. **11.** Let c be the number of chickens and g the number of goats; $c + g = 100$ and $2c + 4g = 286$ are the conditions of the problem. Thus, $c = 100 - g$ and $200 + 2g = 286$, so $g = 43$ and $c = 57$.
13. (a) In each figure, dots are added to the upper left, upper right, and lower right sides to complete the next larger pentagon:

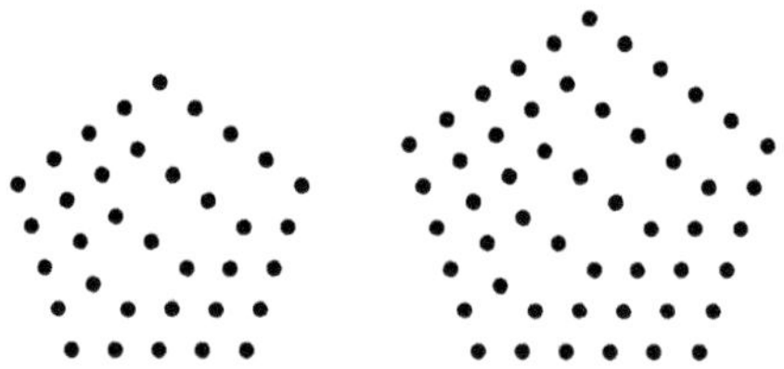

(b) 1, 5, 12, 22, 35, 51, . . .
(c) $1 + 4 + 7 + 10 + 13 = 35$
$1 + 4 + 7 + 10 + 13 + 16 = 51$
(d) 10th term $= 1 + 3(9) = 28$
(e) Use Gauss insight:

$$\begin{aligned} s &= 1 + 4 + 7 + \cdots + 28 \\ s &= 28 + 25 + 22 + \cdots + 1 \\ 2s &= 29 + 29 + 29 + \cdots + 29 \\ \text{Sum} &= \frac{(10)(29)}{2} = 145. \end{aligned}$$

(f) nth term $= 1 + 3(n - 1) = 3n - 2$
(g) Using Gauss insight, we see that there are n terms of $(3n - 1)$. The sum is $\frac{n(3n-1)}{2}$. Therefore, $p_n = \frac{n(3n-1)}{2}$.
15. Following the notation of the example, we have $a + b = 16$, $a + c = 11$, and $b + c = 15$. Subtracting the first two equations gives $b - c = 5$. Adding that to the third equation yields $2b = 20$, so $b = 10$. Working with the first equation then gives $a = 6$, and working with the second producer $c = 5$.
17. Suppose that x denotes the value in the lower small circle. Then the entries in the other small circles are $17 - x$ and $26 - x$, giving the equation $(17 - x) + (26 - x) = 11$. This equation simplifies to $43 - 2x = 11$, or $2x = 43 - 11 = 32$. Therefore, $x = 16$, and the entries in the other two circles are $17 - 16 = 1$ and $26 - 16 = 10$. Alternatively, one can work clockwise to see that the upper-left small circle is $17 - x$ and therefore the

remaining small circle value is $11 - (17 - x) = x - 6$. Then $(x - 6) + x = 26$, or $2x - 6 = 26$. As before, $x = 16$.
19. **(a)** Let x and w be integers that complete the following diagram:

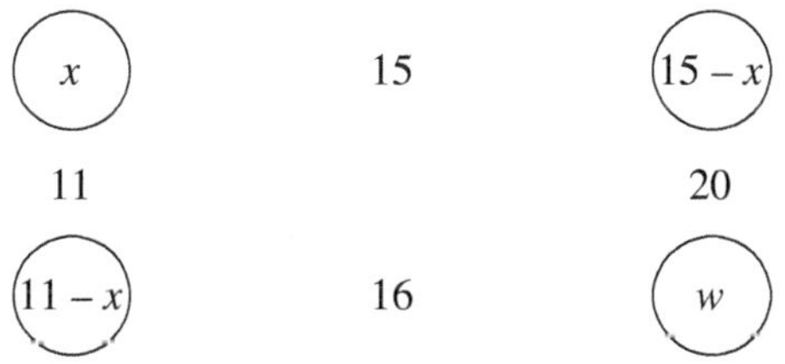

Looking at either $(11 - x) + w = 16$ or $15 - x + w = 20$ gives $w = 5 + x$. Thus, x can be any integer and $w = 5 + x$.
(b) Let x, y, z, w be integers in the circles.

The conditions of the problem are $x + y = 13$, $x + z = 10$, $y + w = 8$, and $w + z = 6$. Subtracting the first two yields $y - z = 3$. Subtracting the last two yields $y - z = 2$, which is impossible. Thus, there are no solutions.

21. **(a)**

n	1	2	3	4	5	6
n^2	1	4	9	16	25	36
$(n + 1)^2$	4	9	16	25	36	49
difference	3	5	7	9	11	13

(b) The difference is $(n + 1)^2 - n^2 = (n^2 + 2n + 1) - n^2 = 2n + 1$. **23.** **(a)** Each of them is divisible by 11. **(b)** The numbers given by the decimal description are $10a + b$ and $10b + a$. Thus, their sum is $10a + b + (10b + a) = 11a + 11b = 11(a + b)$, which is divisible by 11. **25.** Let L and W denote the length and width, respectively, of the rectangle and S denote the length of the sides of the square. Since the rectangle is 3 times as long as it is wide, $L = 3W$. Therefore, the perimeter of the rectangle is $2L + 2W = 6W + 2W = 8W$, and its area is $LW = 3W^2$. The perimeter of the square is $4S$, and its area is S^2. We know that the rectangle and the square have the same perimeter, so $8W = 4S$, or $2W = S$. Also, the area of the square is 4 square feet more than the area of the rectangle, so $S^2 = 3W^2 + 4$. By substitution, $(2W)^2 = 3W^2 + 4$, or $4W^2 = 3W^2 + 4$. Therefore, $W^2 = 4$, and the positive width of the rectangle is $W = 2$. Its length is $L = 3W = 6$. The square has sides of length $S = 2W = 4$. **27.** If x is the number of students and y is the number of adults, then $x + y = 145$ and $3x + 5y = 601$. Solving two equations in two unknowns gives $x = 62$ and $y = 83$.
29. The sum of the two equations is 20 so that the heart is 10.
31. **(a)** Daniel is 10 meters away and Christine is 19 meters.
(b) Daniel has move $10 + 5 + 2.5 = 18.75$ meters and is 1.25 meters from the door. Christine has moved four meters and is 16 meters from the door. Daniel is closer. **(c)** Daniel is correct. He will come as close to the door as he can but will never actually get there. **32.** Let x be the number of rides on Tuesday. Then $2x + 16 = 62$, so $x = 24$. The answer is B.

Problem Set 1.5 (page 47)

1. Yes. The second player can add enough tallies to make a multiple of 5 at each step, forcing the first player to be the one to exceed 30.
3. **(a)** Work backward:

$$39 \times 2 - 78$$
$$78 + 18 = 96$$
$$96 \div 6 = 16$$
$$16 - 7 = 9$$

The input number is 9.
(b) 12 **(c)** Answers will vary. The guess and check method is one possibility.
(d) Let x be the input. After two stages, we have $6(x + 7)$. The output is $[6(x + 7) - 18] \div 2 = \dfrac{6x + 42 - 18}{2} = 3x + 12$.
If the output is 39, as in (a), then $3x + 12 = 39$, so $3x = 27$, or $x = 9$. If the output is 48 for part (b), then the input satisfies $3x + 12 = 48$, so $3x = 36$, or $x = 12$. **5.** **(a)** 18 **(b)** With x being the input and y being the output, the machine stages are formed $x \rightarrow x^2 \rightarrow 6x^2 \rightarrow 6x^2 - 18 \rightarrow 3x^2 - 9$. The answer is $y = 3x^2 - 9$. **7.** **(a)** \$28 **(b)** \$31 **9.** 25. Not all of the information was needed. **11.** 70 or 74. Either answer is possible.
13. The cards must start in the order 4, 5, 3, 6, 2, 7, 1, 8, 0, 9.
15. The following couples are married: Josie and David, Sarah and Will, Taneisha and Floyd, and Kitty and Gus. **17.** To number pages 1 through 9 takes $9 \cdot 1 = 9$ digits. To number pages 10 through 99 takes $90 \cdot 2 = 180$ digits. This leaves $867 - 180 - 9 = 678$ digits to number three-digit pages. Thus, there are $678 \div 3 = 226$ three-digit pages and $226 + 90 + 9 = 325$ pages in the book. **19.** **(a)** 366 **(b)** 731 **21.** Since there are only 10 digits (0, 1, 2, . . . , 9) in any collection of 11 natural numbers, there must be two that have the same units digit. The difference of these two numbers must have a 0 as its units digit and is thus divisible by 10. **23.** If five points are chosen in a square with diagonal of length $\sqrt{2}$, then, by the Pigeonhole Principle, at least two of the points must be in or on the boundary of one of the four smaller squares shown. The farthest these two points can be from each other is $\sqrt{2}/2$ units, if they are on opposite corners of the small square.

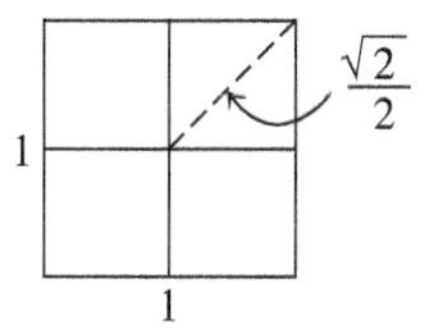

25. If the cups of marbles are arranged as described, each cup will be part of three different groups of three adjacent cups. The sum of all marbles in all groups of three adjacent cups is $3 \cdot (10 \cdot 11/2) = 165$, since each cup of marbles is counted three times. With the marble count of 165 and 10 possible groups of three adjacent cups, by the pigeonhole principle at least one group of three adjacent cups must have 17 or more marbles, since $165 \div 10 = 16.5 > 16$.

27. **(i)** Since there are 20 people at the party, if each person has at least one friend at the party, each must have either 1 or 2 or so on, up to 19, friends at the party. Since $20 > 19$, it follows from the Pigeonhole Principle that at least 2 of the 20 people must have the same number of friends at the party. **(ii)** In this case, 19 people have from 1 to 18 friends at the party. Thus, again by the Pigeonhole Principle, at least 2 people must have the same number of friends at the party. **(iii)** In this case, since at least 2 people have no friends at the party, they have the same number of friends at the party. **29.** 24 **31.** D

Problem Set 1.6 (page 58)

1. **(a)** 9, 98, 987 **(b)** The digits start at 9 and decrease by 1. **(c)** 9876; 98,765; 987,654; 9,876,543; 98,765,432; 987,654,321 **3.** **(a)** 1089, 2178, 3267, 4356, 5445, 6534, 7623, 8712, 9801 **(b)** No. Patterns emerge. **(c)** The first and last products are reversals, as are the second and eighth, and so on. Of course, 5445 is a palindrome and pairs with itself. **5.** The three points P, Q, and R are on a line. **7.** **(a)** There are $F_5 = 5$ arrangements of five logs, supporting the generalization:

However, there are nine arrangements of six logs, instead of eight as suggested by the Fibonacci pattern:

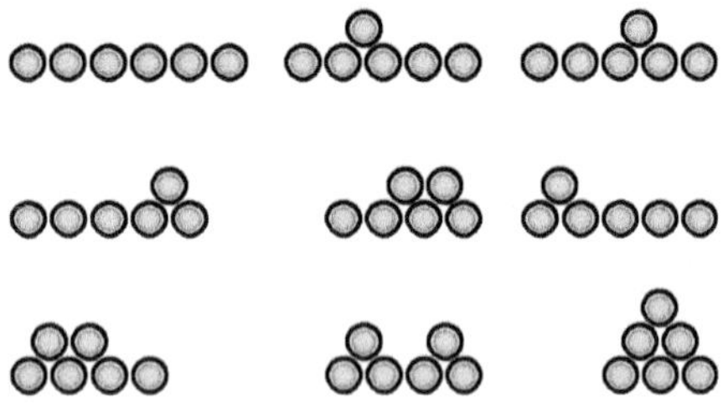

(b) The number of ways to stack n logs at most two layers high is given by the nth Fibonacci number, F_n. **9.** **(a)** Two pennies must be moved. **(b)** Three pennies must be moved:

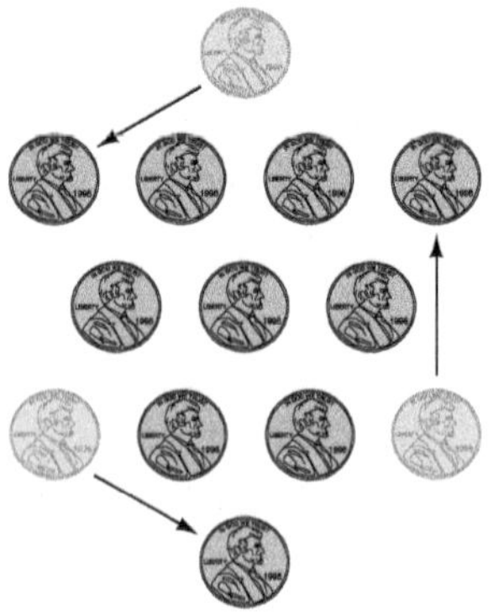

(c) Five pennies must be moved to invert a 15-penny triangle. In general, it can be shown that a triangle with n pennies on a side requires $n/3$ pennies to be moved, where any remainder of the division is dropped. For example, $\frac{10}{3} = 3\frac{1}{3}$, so the triangle of 10 pennies requires 3 moves, and a 15-penny triangle requires $\frac{15}{3} = 5$ pennies to be moved.

11. **(a)**

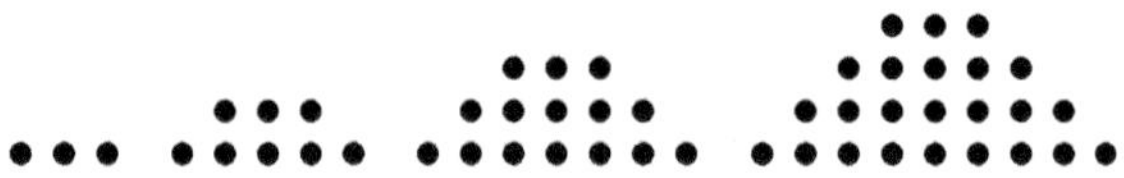

(b)

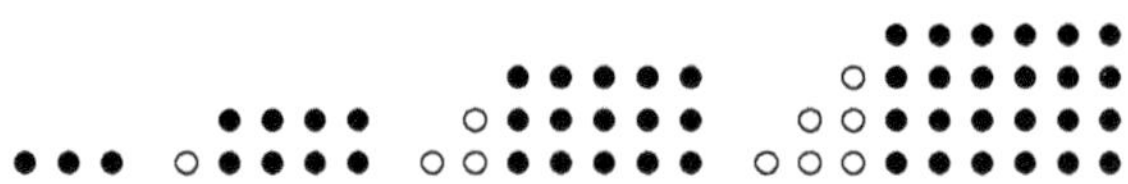

(c) The nth trapezoidal number is $n(n + 2)$, by inductive reasoning. **13.** If n is a multiple of 3, then $n = 3s$ for some whole number s. But then $n^2 = 9s^2$ and so is a multiple of 9. **15.** Assume that n satisfies $2n + 16 = 35$. Since one side of the equation is $2(n + 8)$, an even number (divisible by 2), and the other side, 35, is odd, there can be no solution to this equation. This is a proof by contradiction.

Chapter 1 Review Exercises (page 64)

1. 28 **3.** 9 **5.** 88 square feet **7.** **(a)** Multiply by 5 and then subtract 2. **(b)** Answers will vary. A good strategy is to give Chanty consecutive integers starting with 0.
9. Kimberly is the lawyer and the painter; Terry is the engineer and the doctor; Otis is the teacher and the writer.
11. **(a)** $\frac{n(n+1)}{2} + 1$ **(b)** $\frac{n(n-1)}{2}$ **(c)** n^2
13. **(a)** A one-car train uses 6 toothpicks to form the hexagon. Adding a square + hexagon combination requires an additional 8 toothpicks, so the trains with 1, 3, 5, 7, . . . , cars use 6, 6 + 8, 6 + 8 + 8, 6 + 8 + 8 + 8, . . . toothpicks. In general, a train with $2m + 1$ cars will require $6 + 8m$ toothpicks, where $m = 0, 1, 2, 3, \ldots$. A two-car train uses 9 toothpicks, so trains with 2, 4, 6, 8, . . . cars use 9, 9 + 8, 9 + 8 + 8, 9 + 8 + 8 + 8, . . . toothpicks. In general, a train with $2m + 2$ cars uses $9 + 8m$ toothpicks for $m = 0, 1, 2, 3, \ldots$ **(b)** Since $9 + 8m$ is always an odd number, a train with 102 toothpicks has an odd number of cars, say, $2m + 1$. Then $6 + 8m = 102$, or $8m = 96$. Therefore, $m = 12$, and there are $2(12) + 1 = 25$ cars in the train. **15.** **(a)** 5 **(b)** 9 **(c)** 50
17. **(a)** 4489, 444,889, 44,448,889 **(b)** 44,444,448,888,889 No, as noted earlier, patterns can break down.

19. Since one of any two consecutive integers must be even, $\frac{s(s+1)}{2}$ must be an integer, say q. Thus, $n^2 = 8q + 1$, as was to be shown.

Chapter 2

Problem Set 2.1 (page 75)

1. **(a)** {Arizona, California, Idaho, Oregon, Utah} **(b)** {Maine, Maryland, Massachusetts, Michigan, Minnesota, Mississippi, Missouri, Montana} **(c)** {Arizona}
3. **(a)** {7, 8, 9, 10, 11, 12, 13} **(b)** {9, 11, 13} **(c)** {2, 4, 6, 8, 10, 12, 14, 16, 18, 20} **5.** **(a)** $\{x \in U | 11 \le x \le 14\}$ or $\{x \in U | 10 < x < 15\}$ **(b)** $\{x \in U | x$ is even and $6 \le x \le 16$ and $x \ne 14\}$ **(c)** $\{x \in U | x = 4n$ and $1 \le n \le 5\}$ **(d)** $\{x \in U | x = n^2 + 1$ and $1 \le n \le 4\}$ **7.** **(a)** True. The sets contain the same elements. **(b)** True. Every element in

{6}—namely, 6—is in {6, 7, 8}. **(c)** True. They are the same list of elements. Order doesn't matter.

9. 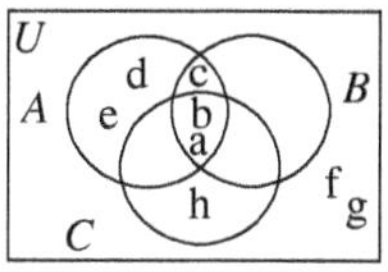

(a) $B \cup C = \{a, b, c, h\}$ **(b)** $A \cap B = \{a, b, c\}$
(c) $B \cap C = \{a, b\}$ **(d)** $A \cup B = \{a, b, c, d, e\}$
(e) $\overline{A} = \{f, g, h\}$ **(f)** $A \cap C = \{a, b\}$
(g) $A \cup (B \cap C) = \{a, b, c, d, e\}$
11. **(a)** {1, 2, 3, 4, 6, 8, 9,12, 16, 18, 24, 36, 48, 72, 144}
(b) {1, 2, 3, 6, 9, 18} **(c)** 18
13. **(a)** 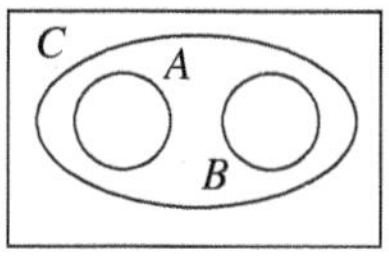

(b) Answers will vary.

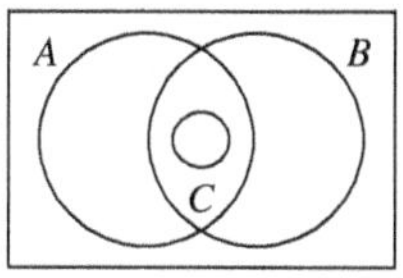

(c) Answers will vary.

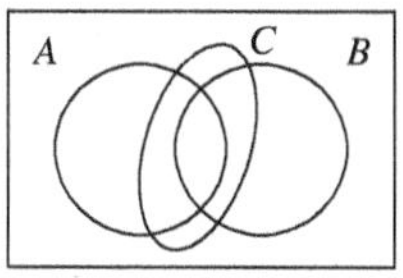

15. It is not always true. For example, if D is any subset of the whole numbers, then $F = \{0, 1\}$ and $G = \{7, 9\}$ is a counterexample.

17. **(a)** 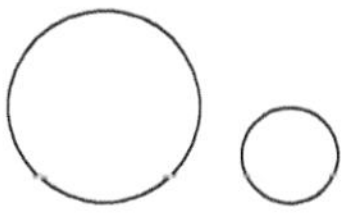 red circles

(b)  large hexagons, 1 red, 1 blue

(c)  triangles and hexagons

(d)  large triangles, 1 red, 1 blue

(e) 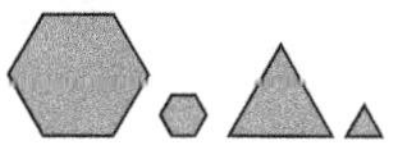 blue figures not circles

(f) ⬡ small red hexagon **19.** **(a)** Answers will vary. One possibility is B = set of students taking piano lessons, C = set of students learning a musical instrument. **(b)** Answers will vary. **(c)** Answers will vary. **21.** Answers will vary.
23. Answers will vary. Following is a possible answer:

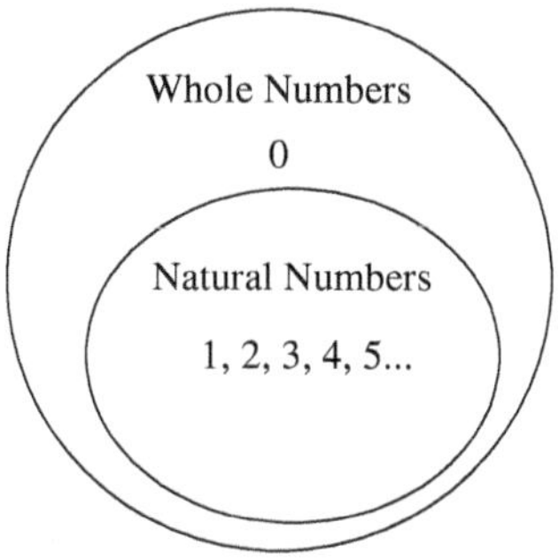

The natural numbers (also called the counting numbers) are a subset of the whole numbers. Zero is a whole number but not a natural number.
25. **(a)** 8 regions

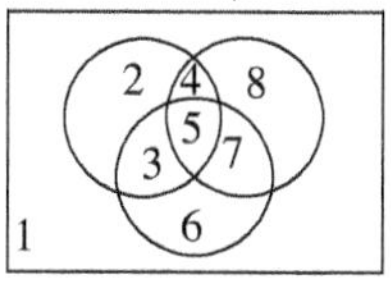

(b) 14 **(c)** $\overline{A} \cap B \cap C \cap \overline{D}$ **(d)** Yes. Each loop contains 8 different regions, and there are 16 regions altogether.

27. 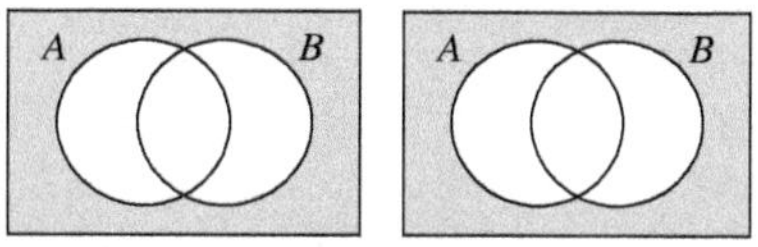

29. **(a)** There are eight subsets: ∅, {P}, {N}, {D}, {P, N}, {P, D}, {N, D}, {P, N, D}. **(b)** There are 16 subsets, ∅, {P}, {N}, {D}, {P, N}, {P, D}, {N, D}, {P, N, D}, {Q}, {P, Q}, {N, Q}, {D, Q}, {P, N, Q}, {P, D, Q}, {N, D, Q}, {P, N, D, Q}. **(c)** Half of the subsets of {P, N, D, Q} contain Q. **(d)** The number of subsets doubles with each additional element, so a set with n elements has 2^n subsets.
31. Answers will vary. One example is shown.

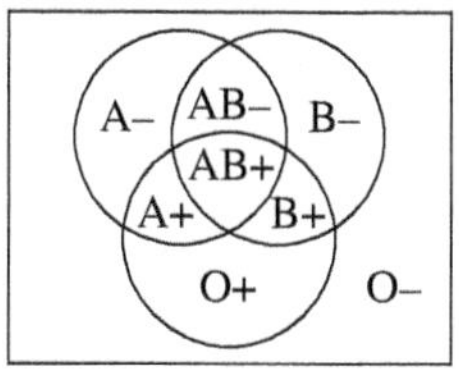

33. $A \cap B = \{6, 12, 18\}$. The elements are those which are divisible by both 2 and 3. **35.** D

Problem Set 2.2 (page 84)

1. **(a)** 13: ordinal first: ordinal **(b)** fourth, second: ordinal; 93: cardinal. In common usage, it represents the number (93) correct out of 100. **3.** **(a)** Equivalent, since there are five letters in the set {A, B, M, N, P} **(b)** Not equivalent **(c)** Equivalent **(d)** Not equivalent **5.** **(a)** $n(A)$ is 0, 1, 2, 3, or 4. **(b)** $n(C)$ is 5, 6, 7, ... **7.** **(a)** $n(A) = 7$ because $A = \{21, 22, 23, 24, 25, 26, 27\}$. **(b)** $n(B) = 0$, since B is the empty set. There is no solution of the equation.
9. **(a)** The correspondence $0 \leftrightarrow 1, 1 \leftrightarrow 2, 2 \leftrightarrow 3, \ldots,$ $w \leftrightarrow w + 1, \ldots$ shows that $W \sim N$. **(b)** The correspondence $1 \leftrightarrow 2, 3 \leftrightarrow 4, \ldots, n \leftrightarrow n + 1, \ldots$ shows that $D \sim E$. **(c)** The correspondence $1 \leftrightarrow 10, 2 \leftrightarrow 100 = 10^2, \ldots$ $n \leftrightarrow 10^n, \ldots$ shows that the sets are equivalent.
11. **(a)** Answers will vary. For example, $Q_1 \leftrightarrow Q_2$, $Q_3 \leftrightarrow Q_4$, and so on.

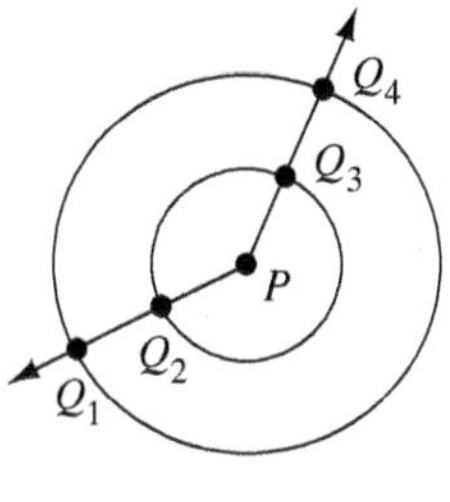

(b) Answers will vary. For example, $Q_1 \leftrightarrow Q_2$, $Q_3 \leftrightarrow Q_4$, and so on.

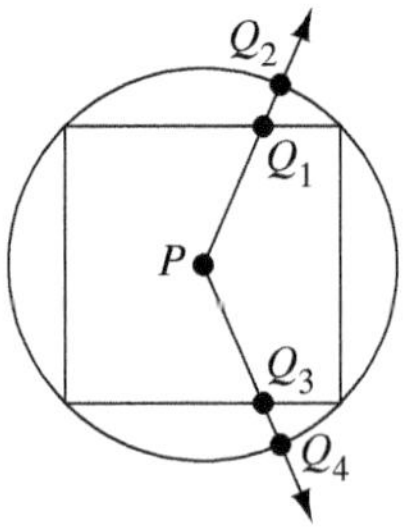

13. (a) True **(b)** False **(c)** True **(d)** True
15. (a) $n(A \cap B) \leq n(A)$. The set $A \cap B$ contains only the elements of A that are also elements of B. Thus, $A \cap B$ cannot have more elements than A. **(b)** $n(A) \leq n(A \cup B)$. The set $A \cup B$ contains all of the elements of the set A *and* any additional elements of B that are not already included. Therefore, $A \cup B$ must have at least as many elements as A. **(c)** Since $n(A \cap B) = n(A \cup B)$ and $A \cap B \subseteq A \subseteq A \cup B$, it follows that $A \cap B = A \cup B$. Thus, $A \cup B$ has no additional elements besides those in $A \cap B$, so neither A nor B has any additional elements. Since $A = A \cap B$ and $B = A \cap B$, it follows that $A = B$. (Caution: This reasoning would not be valid if infinite sets were allowed.) **17.** 200 households **19.** $20 + 5 + 25 = 50$ percent of the students like just one sport, and 5 percent do not like any of the three sports.

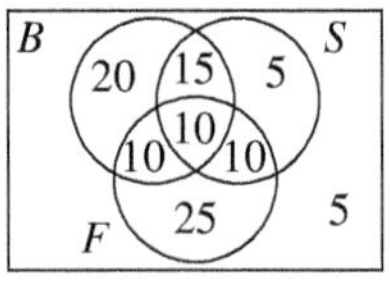

21. (a)

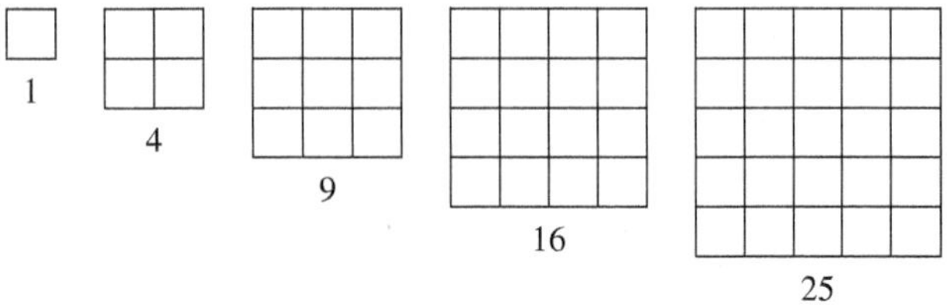

(b)

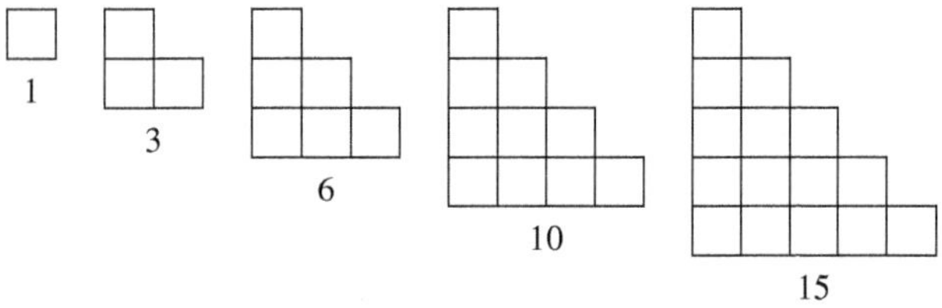

23. Answers will vary. **25.** 400 is the product of two squares, 4 and 100. **27. (a)** The numbers are added in bold. Explanations to Zack will vary.

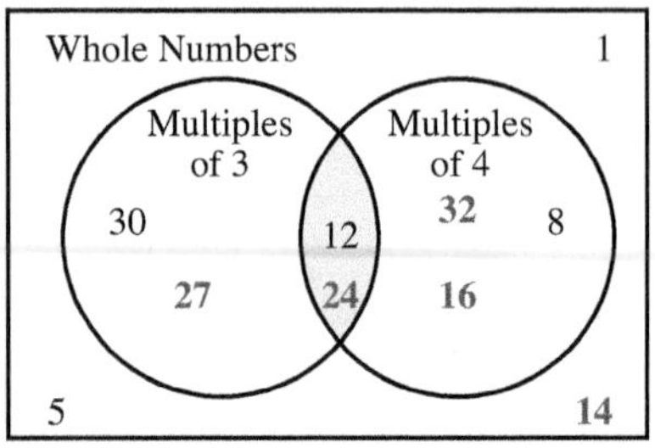

(b) Answers will vary. **29. (a)** Jeff is multiplying the base by the exponent, which is an incorrect procedure. In this case, he is multiplying by three, not taking numbers to the third power.
(b) Use a cube whose sides are of length 1 cm, 2 cm, 3 cm, 4 cm, 5cm, and so on.

31. (a)

Row 0:	1
Row 1:	1 1
Row 2:	1 2 1
Row 3:	1 3 3 1
Row 4:	1 4 6 4 1

(b) The table is the same as Pascal's triangle. Row 5: 1 5 10 10 5 1 and Row 6: 1 6 15 20 15 6 1. **33.** Three students take all three languages, and 26 are not taking any of the three languages.
35. The only way to account for all 60 voters is to start with 3 voters approving all three taxes.

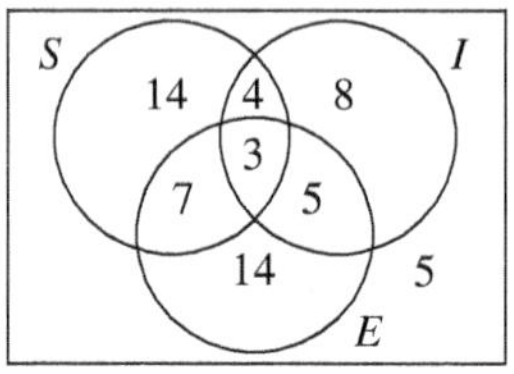

37. (a) All three will meet every $3 \times 4 \times 5 = 60$th day, so there will be 6 days when all three meet.
(b)

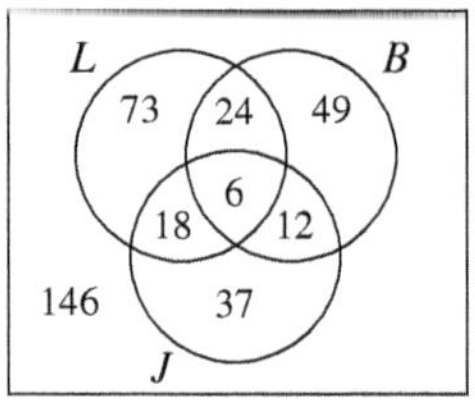

(c) 37 days **(d)** 146 days
39. As the Venn diagram shows, four people have type AB blood.

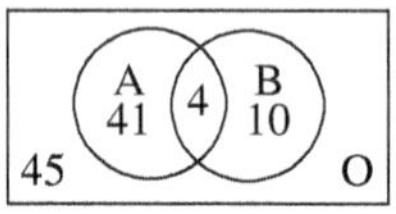

41. A. Seth read a total of $5 + 9 = 14$ and Anna read 16.
43. B. You must add the students taking all three classes and the students only taking French and Spanish ($4 + 8 = 12$). **45.** A

Problem Set 2.3 (page 97)

1. (a) (i) 5 (ii) 5 (iii) 4 **(b)** ii) and (iii)

3. (a)

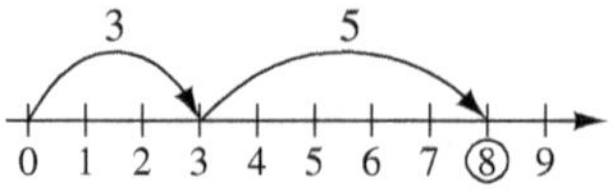

(b)

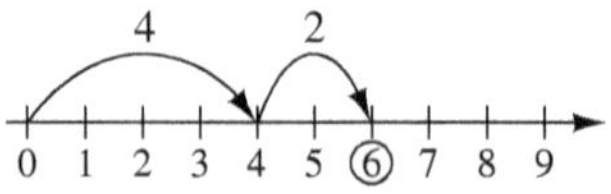

(c)

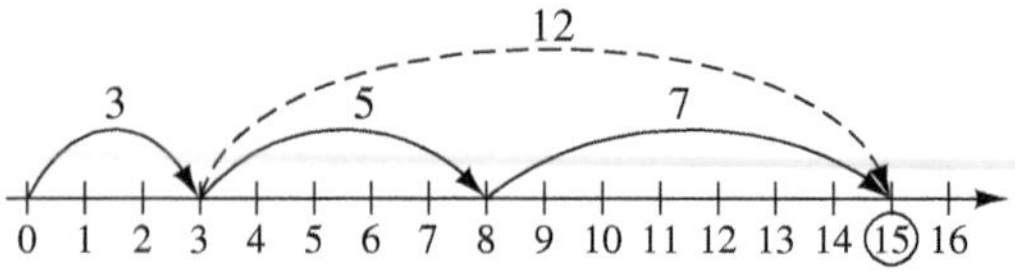

5. (a) 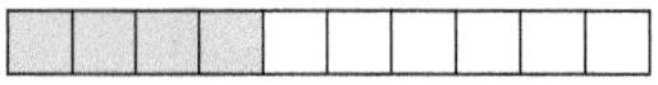

(b)

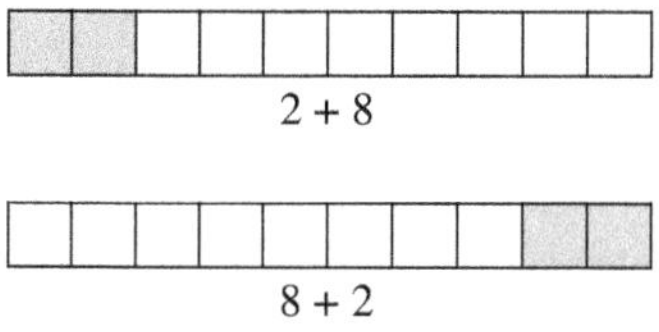

2 + 8

8 + 2

(c)

3 + (2 + 5)   (3 + 2) + 5

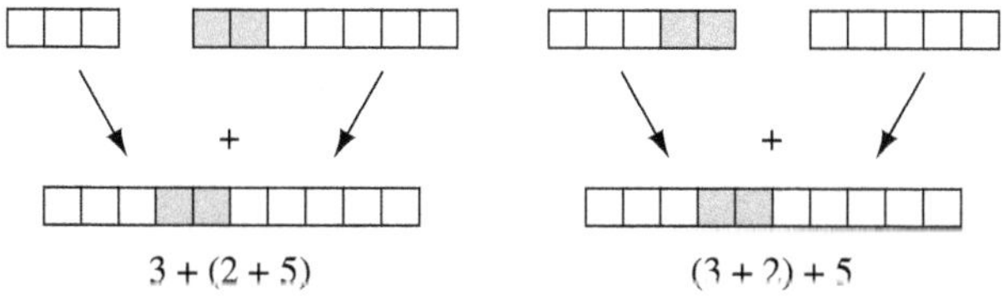

7. Answers will vary. **9. (a)** Commutative property of addition
(b) Closure property **(c)** Additive-identity property of zero
(d) Associative and commutative properties **(e)** Associative and commutative properties
11. (a)

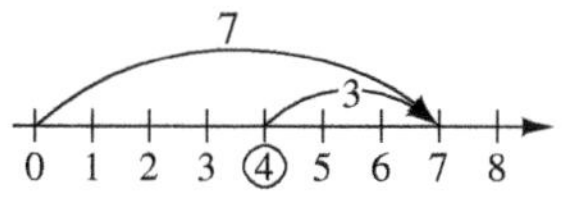

7
3
0 1 2 3 ④ 5 6 7 8

(b)

7
4
0 1 2 ③ 4 5 6 7

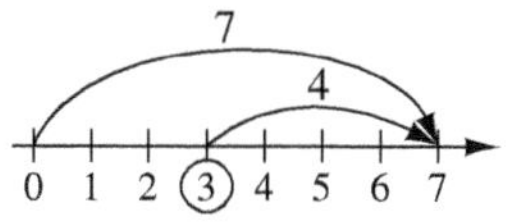

13. (a) $5 + 7 = 12$ $12 - 7 = 5$
          $7 + 5 = 12$ $12 - 5 = 7$
(b) $4 + 8 = 12$ $12 - 8 = 4$
        $8 + 4 = 12$ $12 - 4 = 8$
15. (a) Comparison **(b)** Measurement **17.** Answers will vary.
19. (a) $(8 - 5) - (2 - 1) = 2$ **(b)** $8 - (5 - 2) - 1 = 4$
21. (a)

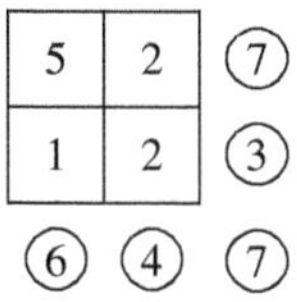

5	2	⑦
1	2	③

⑥ ④ ⑦

(b)

4	2	⑥
5	3	⑧

⑨ ⑤ ⑦

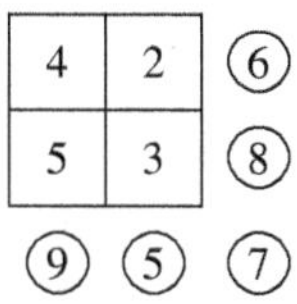

23. Blake has one more marble than before, and Andrea has one fewer than before, so now Blake has two more marbles than Andrea. Use a small number of marbles, say, five each, to demonstrate what happened. After Andrea gives Blake one marble, Blake has six and Andrea has four. Thus, clearly, Blake has two more marbles than Andrea. **25.** Answers will vary. **27.** Answers will vary but should include that, given two of the minuend, subtrahend, and answer, the third can be found. Furthermore, students often struggle when problems are presented in "nonconventional" ways (but this is worth the struggle as it is the beginnings of algebra). Many elementary school students have difficulty when there is a blank in the front of the number sentence.

29. (a)

+	5	4	1	6	9	2	0	8	7	3
3	8	7	4	9	12	5	3	11	10	6
9	14	13	10	15	18	11	9	17	16	12
6	11	10	7	12	15	8	6	14	13	9
4	9	8	5	10	13	6	4	12	11	7
0	5	4	1	6	9	2	0	8	7	3
7	12	11	8	13	16	9	7	15	14	10
5	10	9	6	11	14	7	5	13	12	8
2	7	6	3	8	11	4	2	10	9	5
1	6	5	2	7	10	3	1	9	8	4
8	13	12	9	14	17	10	8	16	15	11

(b) Answers will vary **31. (a)** This student is "subtracting up," which is a common mistake. **(b)** Answers will vary. Here is one: After reminding her about place value, ask her to do a two-digit subtraction, such as $73 - 35$, and reflect on the similarities between what is done for two-digit numbers and what is done for three-digit numbers. That could be followed with $736 - 327 = ?$ as a next-level problem but easier than this one. **33. (a)** Yes, Carmen's method, using the associative property and a good understanding of position value, is correct. **(b)** $71 = 60 + 11$ and $38 = 30 + 8$, so $71 - 38 = 60 - 30 + 11 - 8 = 30 + 3 = 33$. **35.** 53
37.

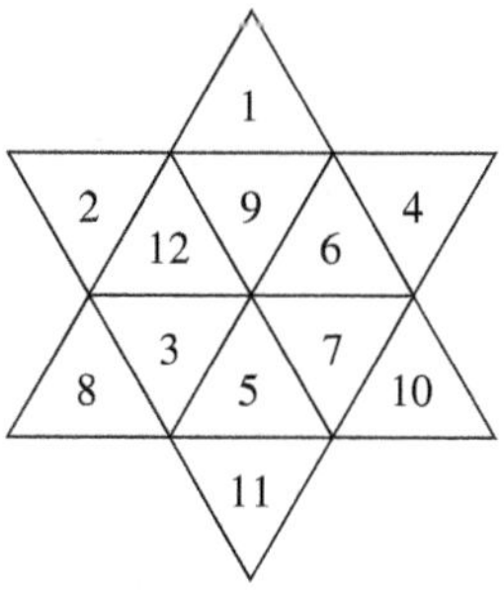

39. (a) $\{2, 3, 4, 5, 6, 7, 8, 9, \ldots\}$ **(b)** $\{0, 1\}$ **(c)** No
(d) Whole numbers that must be in C are all even numbers ≥ 2. Zero and odd numbers may or may not be in C. **41.** It can be shown that Sameer is correct. The easiest way to find a sum is to use the largest Fibonacci number possible as the next summand. For example, $100 = 89 + 8 + 3$. **43.** Statement B **45.** D. The number has to be smaller than 500. **47.** C

Problem Set 2.4 (page 114)

1. (a) $3 \times 5 = 15$, repeated addition **(b)** $6 \times 3 = 18$, number-line model **(c)** $5 \times 3 = 15$, set model **(d)** $3 \times 6 = 18$, number-line model **(e)** $8 \times 4 = 32$, rectangular area model
(f) $3 \times 2 = 6$, multiplication tree model **3. (a)** Answers will vary. **(b)** (i) 36 (ii) 3752 (iii) 286,914 (iv) 336,648
5. (a) Not closed. $2 \times 2 = 4$, and 4 is not in the set.
(b) Closed. $0 \times 0 = 0, 0 \times 1 = 0, 1 \times 1 = 1$, and $1 \times 0 = 0$. All products are in the set. **(c)** Not closed. $2 \times 4 = 8$, and 8 is not in the set. **(d)** Closed. The product of any two even whole numbers is always another even whole number. **(e)** Closed. The product of any two odd whole numbers is always another odd whole number. **(f)** Not closed. $2 \times 2^3 = 2^4$, which is not in the set. **(g)** Closed. $2^m \times 2^n = 2^{m+n}$ for any whole numbers m and n. **(h)** Closed. $7^a \times 7^b = 7^{a+b}$ for any whole numbers a and b. **7. (a)** Commutative property of multiplication
(b) Distributive property of multiplication over addition
(c) Multiplication-by-0 property **9.** Commutative property: $5 \times 3 = 3 \times 5$
11. (a)

2 + 5   2   5

(b)

2 + 5 + 1   2   5   1

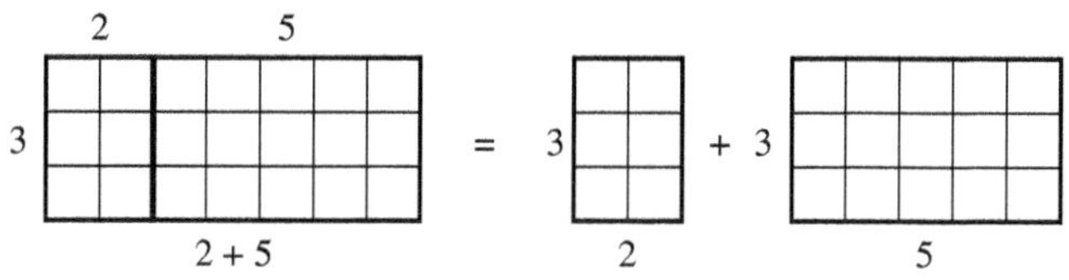

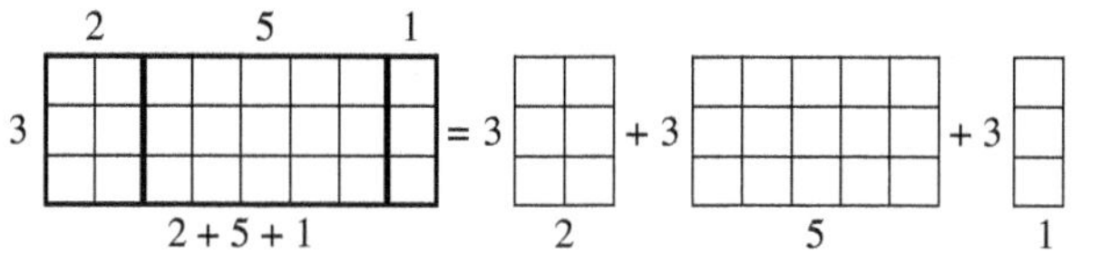

(c)
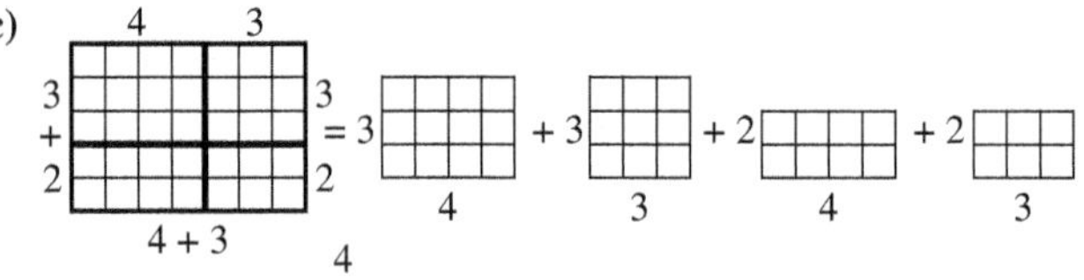

13. The rectangle is $a + b$ by $c + d$, so its area is $(a + b) \times (c + d)$. The rectangle labeled F is a by c, so its area is ac. Similarly, the areas of the other three rectangles are given by ad, bc, and bd. So $(a + b) \times (c + d) = ac + ad + bc + bd$. **15. (a)** Distributive property **(b)** Distributive property **(c)** Distributive property, associative property, and/or multiplicative property of 0

17.
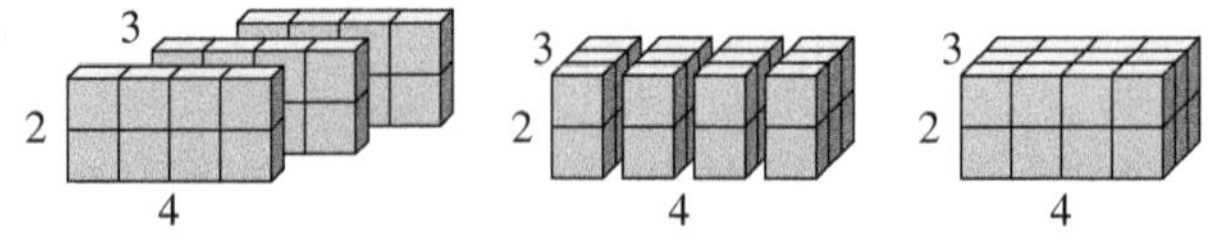

3·(2·4) (3·2)·4 3·2·4

19. (a) $4 \times 8 = 32, 8 \times 4 = 32, 32 \div 8 = 4, 32 \div 4 = 8$ **(b)** $6 \times 5 = 30, 5 \times 6 = 30, 30 \div 5 = 6, 30 \div 6 = 5$ **21. (a)** Repeated subtraction **(b)** Partition **(c)** Missing factor or repeated subtraction **23. (a)** $14 - 7 = 7$. Since $7 \leq 7$, we subtract another 7 to get $7 - 7 = 0$. Thus, there is a remainder of 0. We subtracted twice, so the quotient is 2. **(b)** 7 is already less than 14, so there is no subtraction. The remainder is 7 and, since there were no subtractions, the quotient is 0. **(c)** Answered in each part separately **25. (a)** 29 **(b)** $x = 6$ **27. (a)** 3^{35} **(b)** 3^{10} **(c)** $(yz)^3$ **29. (a)** 2^3 **(b)** 2^5 **(c)** 2^{10} **(d)** 2^{12} **31.** Answers will vary but should include the fact that if we are dividing a by b, then if $a < b$, the process is finished. If a is b or $a > b$, then either $a - b$ is 0 or $a - b > 0$. Thus, in neither case can the remainder be negative, because the process stops before that happens. **33.** Answers will vary. **35.** As listed in the theorem, starting with the second one, "It doesn't matter in which order you multiply two numbers," "it doesn't matter which way you group the terms when multiplying three numbers," "multiplying by 1 never changes the number (and 1 is the only number for which that happens," and "zero times anything is zero." **37.** This student struggles to see the relationship between multiplication and division problems. In addition, students typically have more diffculty when the blank, or box, is at the beginning of the equation. **39.** This student subtracted the number of cupcakes Nelson baked in each pan, instead of dividing to find the number of pans needed to bake all of the cupcakes. This is common, because students understand that they need to do "something" with the numbers, but they aren't sure what. In this case, the student is not thinking of grouping the cupcakes (into pans), which should be a signal that he should divide rather than subtract (which suggests removal). **41.** The equation is $x^2 - 3 = 33$ and, thus, $x = 6$ or -6. Since we are dealing only with whole numbers, Jing's answer is $x = 6$. **43.** The operation is closed, commutative, and associative. The circle is the identity. **45. (a)** "How many tickets must still be sold?" **(b)** "How many cartons will be filled?"; "How many eggs are in the partially filled carton?"
47. 59 **49.** B **51.** A **53.** C

Chapter 2 Review Exercises (page 123)

1. (a) $S = \{4, 9, 16, 25\}$
$P = \{2, 3, 5, 7, 11, 13, 17, 19, 23\}$
$T = \{2, 4, 8, 16\}$
(b) $\overline{P} = \{4, 6, 8, 9, 10, 12, 14, 15, 16, 18, 20, 21, 22, 24, 25\}$
$S \cap T = \{4, 16\}$
$S \cup T = \{2, 4, 8, 9, 16, 25\}$
$S \cap \overline{T} = \{9, 25\}$
3. (a) $\subseteq$ **(b)** $\subset$ **(c)** $\cap$ **(d)** $\cup$
5.

1	4	9	16	25	36	49	64	81	100
↕	↕	↕	↕	↕	↕	↕	↕	↕	↕
a	b	c	d	e	f	g	h	i	j

7.
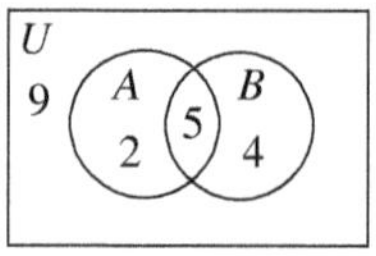

9. (a) Commutative property of addition: $7 + 3 = 3 + 7$
(b) Additive-identity property of 0: $7 + 0 = 7$
11. (a)
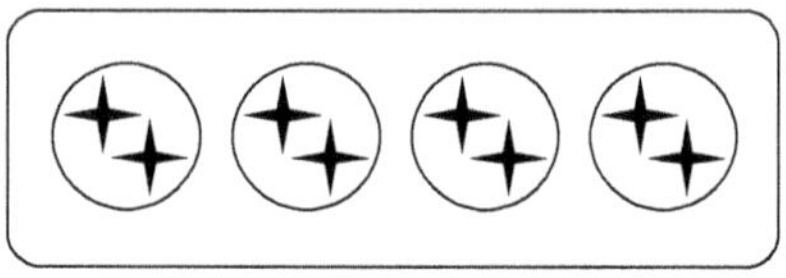

(b)

(c)
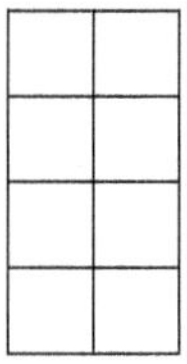

(d)
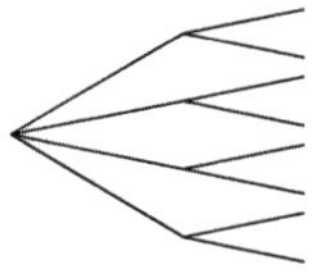

(e)
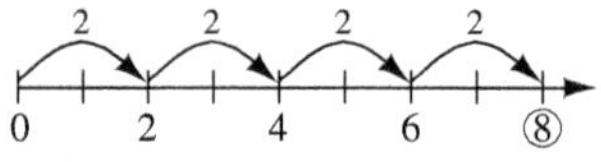

13. $6'' \times 6'' \times 8''$
15. (a)
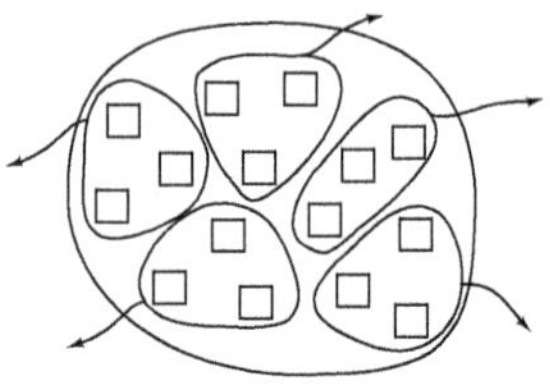

(b)
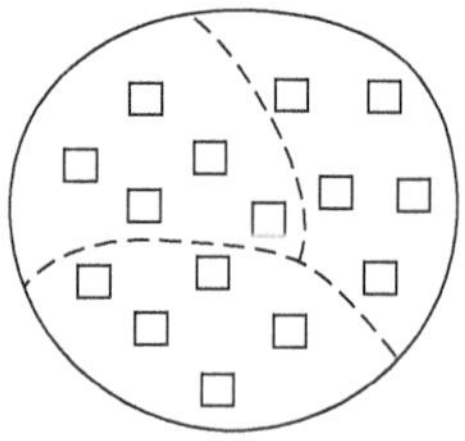

(c)
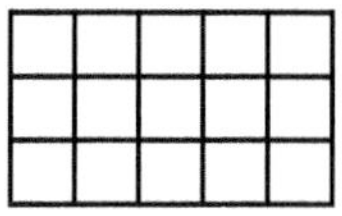

Chapter 3

Problem Set 3.1 (page 135)

1. (a) 2137 **(b)** 1729 **(c)** 697 **(d)** 60 **(e)** 3600 **(f)** 16,920
3. (a)

(b)

(c) 

5. (a) 974 = (1000 − 100) + 70 + 4 = 1000 − 100 + 50 + 20 + 4 = CMLXXIV
(b) 2009 = 2000 + 9 = 2000 + (10 − 1) = MMIX

7. (a) 

(b) 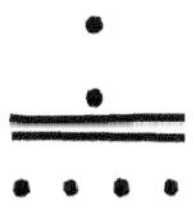

(c) 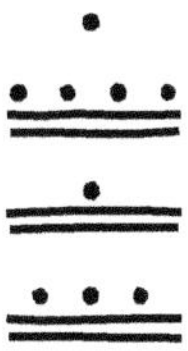

9. MMMCLXXX **11.** MMII, MMIII, MMIV

13. 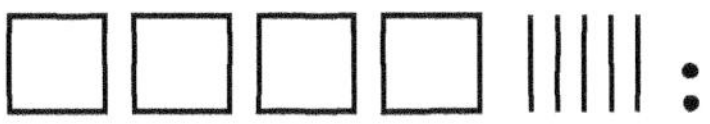

15. 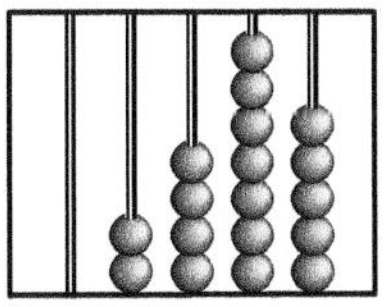

17. First trade 10 of your units for a strip to get 3 mats, 25 strips, and 3 units. Then trade 20 strips for 2 mats to get 5 mats, 5 strips, and 3 units. **19. (a)** One mat must be exchanged for 10 strips, and then 1 strip must be exchanged for 10 units. **(b)** 2 mats, 6 strips, and 8 units **(c)** The borrowing from the hundreds column and the tens columns in this subtraction:

$$\begin{array}{r} \overset{2}{3}\,{}^{1}\overset{9}{\not{0}}\,{}^{1}6 \\ -38 \\ \hline 268 \end{array}$$

21. (a) 21 dollars and 44 dimes, worth \$25.40 **(b)** 4 dimes, worth \$0.40 **(c)** Answers may vary. **23.** Answers will vary. **25.** The problem involves exchanges. Every 60 minutes is 1 hour and every 8 hours is a workday. **27.** Sara wanted to see how the units and the rod were to each other. She figured out that if she cut the rod apart, it would be made out of 10 units. **29.** A **31.** D

Problem Set 3.2 (page 144)

1. (a) 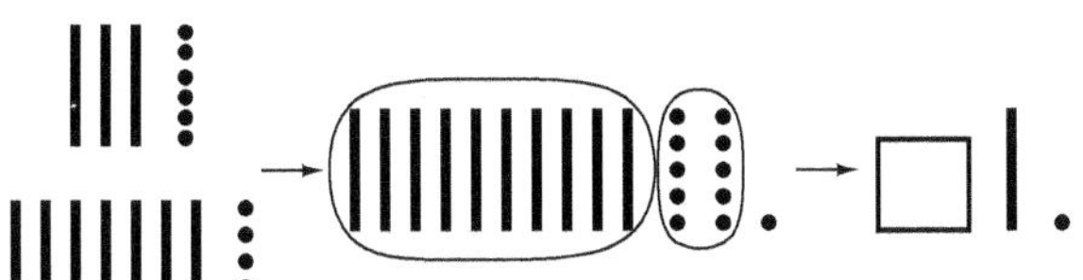

36 + 75 = 111

(b) 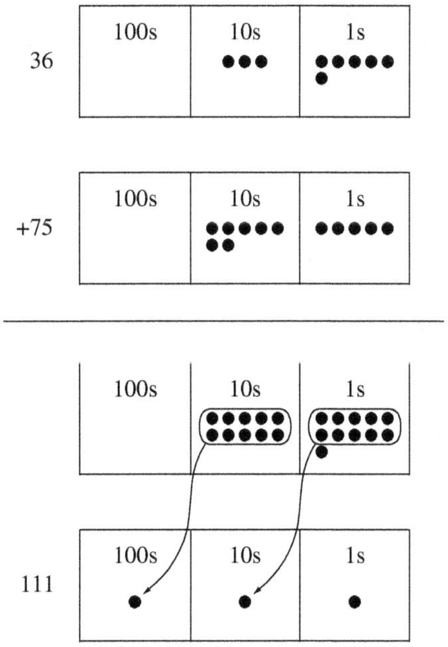

3. (a)
$$\begin{array}{r} 23 \\ +\ 44 \\ \hline 7 \\ 60 \\ \hline 67 \end{array}$$

(b)
$$\begin{array}{r} 57 \\ +\ 84 \\ \hline 11 \\ 130 \\ \hline 141 \end{array}$$

5. (a)
$$\begin{array}{r} 78 \\ -\ 35 \\ \hline 43 \end{array}$$

(b)
$$\begin{array}{rr} 7 & 5 \\ -3 & 8 \\ \hline 6 & (15) \\ -3 & 8 \\ \hline 3 & 7 \end{array}$$

7. In these problems, we must exchange 60 seconds for 1 minute and 60 minutes for 1 hour, or vice versa.

(a)
$$\begin{array}{rl} & 3\text{ hours, } 24\text{ minutes, } 54\text{ seconds} \\ + & 2\text{ hours, } 47\text{ minutes, } 38\text{ seconds} \\ \hline & 5\text{ hours, } 71\text{ minutes, } 92\text{ seconds} \\ = & 5\text{ hours, } 72\text{ minutes, } 32\text{ seconds} \\ = & 6\text{ hours, } 12\text{ minutes, } 32\text{ seconds} \end{array}$$

(b)
$$\begin{array}{rl} & 7\text{ hours, } 56\text{ minutes, } 29\text{ seconds} \\ + & 3\text{ hours, } 27\text{ minutes, } 52\text{ seconds} \\ \hline & 10\text{ hours, } 83\text{ minutes, } 81\text{ seconds} \\ = & 10\text{ hours, } 84\text{ minutes, } 21\text{ seconds} \\ = & 11\text{ hours, } 24\text{ minutes, } 21\text{ seconds} \end{array}$$

(c)
$$\begin{array}{rl} & 5\text{ hours, } 24\text{ minutes, } 54\text{ seconds} \\ - & 2\text{ hours, } 47\text{ minutes, } 38\text{ seconds} \\ \hline & 4\text{ hours, } 84\text{ minutes, } 54\text{ seconds} \\ - & 2\text{ hours, } 47\text{ minutes, } 38\text{ seconds} \\ \hline & 2\text{ hours, } 37\text{ minutes, } 16\text{ seconds} \end{array}$$

(d)
$$\begin{array}{rl} & 7\text{ hours, } 46\text{ minutes, } 29\text{ seconds} \\ - & 3\text{ hours, } 27\text{ minutes, } 52\text{ seconds} \\ \hline & 7\text{ hours, } 45\text{ minutes, } 89\text{ seconds} \\ - & 3\text{ hours, } 27\text{ minutes, } 52\text{ seconds} \\ \hline & 4\text{ hours, } 18\text{ minutes, } 37\text{ seconds} \end{array}$$

9. Using numerals, we have 9 + 4 = 13 ones. Ten of the 1s are exchanged for one 10. 1 + 7 + 8 = 16 tens. Ten of the 10s are exchanged for one 100. **11. (a)** Yes **(b)** Sylvia exchanged 100 ones for 1 hundred and 10 hundreds for 1 thousand. **(c)** No

(d) Sylvia might have carried 10 tens to the tens column. **(e)** Compliment him or her on understanding positional notation so well. **13.** Answers will vary. **15. (a)** One possibility is 8642 and 7531. **(b)** One possibility is 1357 and 2468. **(c)** Answers will vary. **(d)** No. One other solution for (a) could be 8542 and 7631. **17. (a)** Thomas made a common addition error in the ones place, yet was also able to "carry in his head" from the tens to the hundreds place. Perhaps he wasn't paying enough attention in the ones place. **(b)** Annabelle did not carry in either the tens or the hundreds place. **(c)** Xiao made a common addition error of forgetting to add two tens—he just carried the 1 ten and added that to 9 tens. He then wrote down 0 in the tens place and correctly carried 10 tens to form an additional 1 hundreds unit. **19.** One of the 2 tens in 523 was exchanged for 10 ones, and then one of the five 100s was exchanged for 10 tens. **21.** The correct answer is (a) because $273 - (152 + 1) = (273 - 152) - 1 = 121 - 1 = 120$

23. (a)
$$\begin{array}{r} 2437 \\ 281 \\ +3476 \\ \hline 6194 \end{array}$$
(b)
$$\begin{array}{r} 4721 \\ 9012 \\ +7193 \\ \hline 20{,}926 \end{array}$$
(c)
$$\begin{array}{r} 3891 \\ 2493 \\ +5125 \\ \hline 11{,}509 \end{array}$$
(d)
$$\begin{array}{r} 594 \\ 6121 \\ +\ 891 \\ \hline 7606 \end{array}$$

25. The number of scratches in any column gives the number to be exchanged to the next column.

27. (a)
$$\begin{array}{rr} 3' & 8'' \\ 4' & 2'' \\ 6' & 10'' \\ +\ 5' & 11'' \\ \hline 18' & 31'' \\ =\ 20' & 7'' \end{array}$$
(b)
$$\begin{array}{rr} 20' & 7'' \\ -\ 9' & 10'' \\ \hline 19' & 19'' \\ - & 9''\ 10'' \\ \hline 10' & 9'' \end{array}$$

29. H **31.** C **33.** 596,398 327,852

Problem Set 3.3 (page 153)

1. (a)

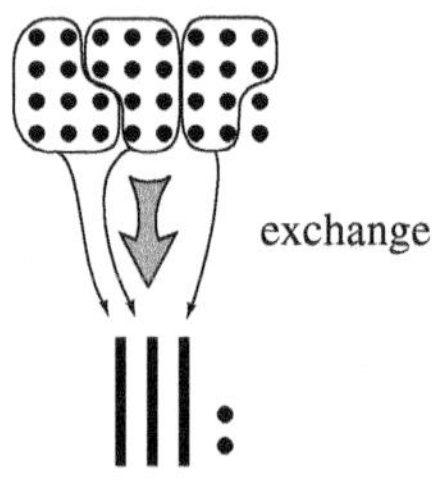

$4 \times 8 = 32$

(b)

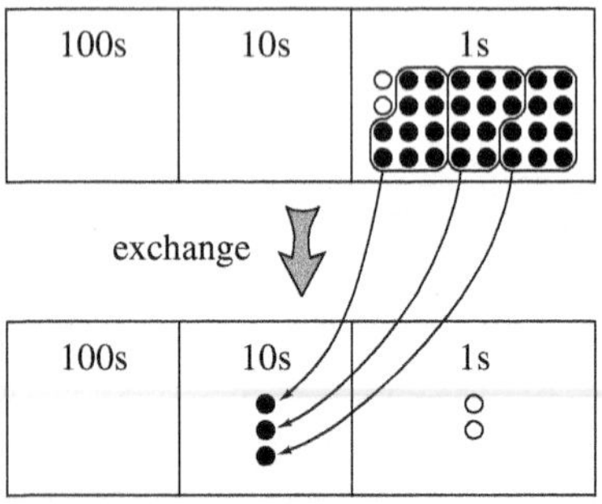

$4 \times 8 = 32$

3. (a) The number of hundreds in $30 \times 70 + 100$; i.e., 200 **(b)** Twenty 10s are being exchanged for two 100s. **5. (a)** Distributive property of multiplication over addition **(b)** Commutative property of multiplication **(c)** Associative property of addition **(d)** Distributive property of multiplication over addition **7.** $437 \div 3 = 145 \text{ R } 2$

9.
$$\begin{array}{r} 241 \\ \times\ 35 \\ \hline 5 \\ 200 \\ 1000 \\ 30 \\ 1200 \\ 6000 \\ \hline 8435 \end{array}$$

11. (a) $871 = 17 \cdot 51 + 4$ **(b)** $21\overline{)72^{9}3}$ with quotient 34R9, so $723 = 21 \cdot 34 + 9$.

13. (a)
$$\begin{array}{r} 14 \\ 4 \\ 10 \\ 213\overline{)3175} \\ 2130 \\ \hline 1045 \\ 852 \\ \hline 193 \end{array}$$
$3175 = 213 \cdot 14 + 193$

(b)
$$\begin{array}{r} 191 \\ 1 \\ 90 \\ 100 \\ 43\overline{)8250} \\ 4300 \\ \hline 3950 \\ 3870 \\ \hline 80 \\ 43 \\ \hline 37 \end{array}$$
The answer is $191 \cdot 43 + 37 = 8250$.

15. (a) $\begin{array}{r} 1\,2\,6\,1 \text{ R } 7 \\ 8\overline{)10^{2}0^{4}9^{1}5} \end{array}$ **(b)** $\begin{array}{r} 1405 \\ 6\overline{)8^{2}432} \end{array}$ with remainder of 2. The quotient is 1405.

17. (a)

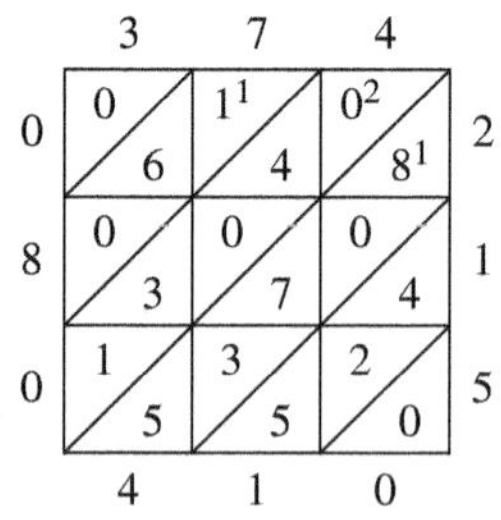

$374 \cdot 215 = 80410$

(b) Answers will vary. **(c)** Answers will vary.

19. (a)
$$\begin{array}{r} 7531 \\ \times\quad 9 \\ \hline 67{,}779 \end{array}$$
(b)
$$\begin{array}{r} 751 \\ \times\quad 93 \\ \hline 69{,}843 \end{array}$$

21. Henry did the problem correctly and with a conceptual understanding. **23. (a)** (a) In this division, the student ignored the remainder in each place value (e.g., 8/3 is 2 rather than 2 R 2 in the hundreds place.) **(b)** (b) In this problem, she understood the need to "bring down the next digit," but instead brought both digits down and then couldn't solve the problem. There is an awareness of the process, but obviously the student is far away from a conceptual understanding of the steps.

(c) In this division problem, the student ignored the zero in the dividend. Oftentimes, students will ignore zeros and assume that no matter where they are, they have "no value." You could also ask the student to estimate 806 ÷ 3 to show that a number close to 28 (28 R 2) can't be correct. **25.** It is correct. Marsha has done the multiplication in the reverse order than is customary.

27. (a) $34 \cdot 54 = (17 \cdot 2) \cdot 54 = 17 \cdot (2 \cdot 54) = 17 \cdot 108$, since 2 evenly divides 34. **(b)** Explanations may vary. **(c)** See the answer to part (a). **(d)** It depends on the base-two representation of the multiplier.

(e)

29	81	11	243
~~14~~	~~162~~	5	486
7	324	~~2~~	~~972~~
3	648	1	1944
1	1296		2673
	2349		

29. (a) 203 minutes **(b)** 3 hours and 23 minutes **31.** Lori's calculator is adding 10 to every asnwer. She must subtract 10 from the displayed answer. **33.** B

Problem Set 3.4 (page 164)

1. (a) 58 **(b)** 66 **(c)** 92 **3. (a)** 142 **(b)** 471 **(c)** 138 **(d)** 192 **5. (a)** 787 **(b)** 637 **(c)** 321 **7. (a)** 800 **(b)** 600 **(c)** 300 **(d)** 500 **9. (a)** 900 **(b)** 900 **(c)** 27,000,000 **(d)** 2000 **11. (a)** 4,340,000 **(b)** 25 million **13. (a)** 52,000 **(b)** 19,000 **(c)** 3000 **15. (a)** 90,000 **(b)** 1,500,000 **(c)** 8,000,000 **(d)** (a) 85,188; (b) 1,635,102; (c) 8,299,179

17. (a) No, rounding to the leftmost digit will give an estimate that is almost 2000 too small, since the last three digits in each number name a number just a bit less than 500. **(b)** Rounding to the nearest hundred to obtain 10,900 **(c)** 10,846 **19.** Answers will vary. **21.** Diley is solving her problems by adding first and then rounding, which defeats the purpose of the estimation process. She should be asked to round off each of the addends and then add for estimation by rounding. **23.** The students will often round the tens digit but leave the ones digit unchanged, as Raphael did.

25. You could ask them which digit is the tens digit. Then, on receiving the answer that it is a 1, say that the two competitors for an answer are 1 or 2 in the tens digit. Since the hundreds digit does not change, the diagram similar to their erroneous one should be

210 220
↖ ↗
217

27. Theresa **29. (a)** 27,451, since the last digit should be 1.

31. (a) Answers will vary, but one possibility is 120, 235, 340, 420. The combined area is about 420,000 square miles. **(b)** 420,905

33. C **35.** About 90 **37.** B

Problem Set 3.5 (page 175)

1. (written as a row) 0, 1, 2, 3, 4, 10, 11, 12, 13, 14, 20, 21, 22, 23, 24, 30, 31, 32, 33, 34, 40, 41, 42, 43, 44, 100 **3.** Only the digits 0 through 5 are used. There are six columns, and in each column the representations have the same rightmost digit and the leftmost digit increases from 0 to 5. Also, the leftmost digits are the same in any row. **5. (a)** 108 **(b)** 254 **(c)** 5 **7.** 14 **9. (a)** 108 **(b)** 413

(c) 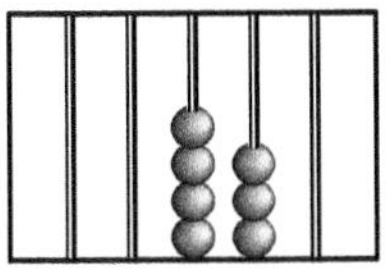

11. (a) 2422_{five} **(b)** 102_{five} **(c)** 10_{five} **(d)** 100_{five}

13. (a) $4 \cdot 5 + 1 + 5 + 4 = 5 \cdot 5 + 5 = 5^2 + 5 = 110_{five}$, which is $25 + 5 + 0 = 30$ in base ten **(b)** $2 \cdot 5^2 + 5 + 3 + 4 \cdot 5^2 + 3 \cdot 5 + 2 = 6 \cdot 5^2 + 4 \cdot 5 + 5 = (5+2)5^2 = 5^3 + 2 \cdot 5^2 = 1200_{five}$. In base ten the answer is 175.

15. (a) $2 \cdot 6 + 3 + 3 \cdot 6 + 5 = 5 \cdot 6 + 6 + 2 = 6^2 + 2 = 102_{six}$. Base-ten answer is 38. **(b)** $(4 \cdot 6^2 + 2 \cdot 6 + 3) + (4 \cdot 6 + 3) = 4 \cdot 6^2 + 6^2 + 6 = 5 \cdot 6^2 = 510_{six}$. Base-ten answer is 186.

17. (a) $\begin{array}{r} 41 \\ -14 \\ \hline 3(6) \\ -14 \\ \hline 22_{five} \end{array}$ Base-ten answer is 12.

(b) $\begin{array}{r} 213 \\ -32 \\ \hline 1(6)3 \\ -32 \\ \hline 131 \end{array}$ Answer is 41 in base ten.

19. (a) $\begin{array}{r} 505 \\ -35 \\ \hline 465 \\ -35 \\ \hline 430_{six} \end{array}$ Answer is 162 in base ten

(b) $\begin{array}{r} 423 \\ -43 \\ \hline 383 \\ -43 \\ \hline 340_{six} \end{array}$ Answer is 132 in base ten.

21. (a) $\begin{array}{r} 23_{five} \\ \times\ 3_{five} \\ \hline 124_{five} \end{array}$

(b) $\begin{array}{r} 432_{five} \\ \times\ 41_{five} \\ \hline 34{,}312_{five} \end{array}$

(c) $\begin{array}{r} 2013_{five} \\ \times\ 23_{five} \\ \hline 101{,}404_{five} \end{array}$

(d) $\begin{array}{r} 13 \\ \times\ 3 \\ \hline 39 \end{array}$; $\begin{array}{r} 117 \\ \times\ 21 \\ \hline 2457 \end{array}$; $\begin{array}{r} 258 \\ \times\ 13 \\ \hline 774 \\ 258 \\ \hline 3354 \end{array}$

23. (a) A positional system to base twenty **(b)** Answers will vary.

Chapter 3 Review Exercises (page 180)

1. (a) 2353 **(b)** 58,331 **(c)** 1998 **3.** Exchange 30 units for 3 strips, and then exchange all 30 strips for 3 mats. The result is 8 mats, 0 strips, and 2 units.

5. (a) $\begin{array}{r} 42 \\ +\ 54 \\ \hline 6 \\ 90 \\ \hline 96 \end{array}$ **(b)** $\begin{array}{r} 47 \\ +\ 35 \\ \hline 12 \\ 70 \\ \hline 82 \end{array}$ **(c)** $\begin{array}{r} 59 \\ +\ 63 \\ \hline 12 \\ 110 \\ \hline 122 \end{array}$

7. **(a)**
```
 357
×  4
  28
 200
1200
1428
```
(b)
```
   642
×   27
    14
   280
  4200
    40
   800
 12000
17,334
```

9. **(a)** $5\overline{)27436}$ = 5487 R 1 **(b)** $8\overline{)39584}$ = 4948 R 0 **11.** 657 rounds to 700, 439 rounds to 400, 1657 rounds to 2000, and 23 rounds to 20. Thus, **(a)** 657 + 439 is approximately 700 + 400 = 1100. The actual sum is 1096. **(b)** 657 − 439 is approximately 700 − 400 = 300. The actual answer is 218. **(c)** $657 \cdot 439$ is approximately $700 \cdot 400 = 280{,}000$. The actual answer is 288,423. **(d)** 1657 ÷ 23 is approximately 2000 ÷ 20 = 100. The actual answer, to the nearest hundredth, is 72.04. **13.** **(a)** 2121_{five} **(b)** $2{,}023{,}221_{five}$

Chapter 4

Problem Set 4.1 (page (194))

1. **(a)**

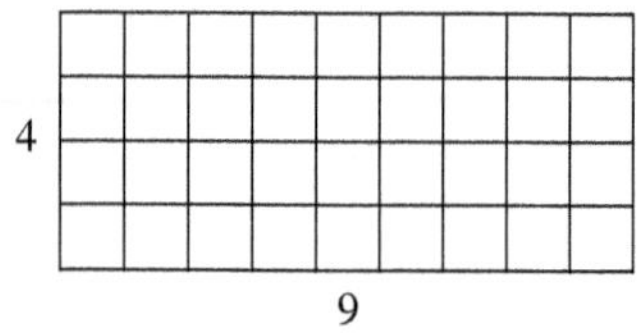

$36 = 4 \cdot 9$

4 divides 36.

(b)

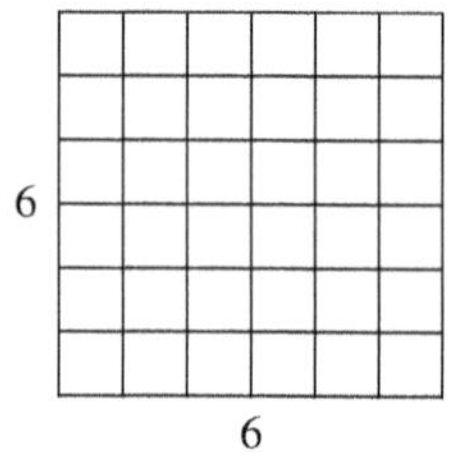

$36 = 6 \cdot 6$

6 divides 36.

3. **(a)** 8, 16, 24, 32, 40, 48, 56, 64, 72, 80 **(b)** 6, 12, 18, 24, 30, 36, 42, 48, 54, 60 **(c)** 24 **5.** Proof by contradiction. Suppose the product *ab*, where at least one of the factors, say *a*, is even. There is then a natural number *k* so that $a = 2k$. But then $ab = (2k)b = 2(kb)$, so 2 is a factor of *ab*, which shows *ab* is even. This is a contradiction, so if *ab* is odd, then both *a* and *b* must be odd. **7.** Let *b* be odd, so $a = 2j + 1$ for some natural number *j*. Then $b^2 = (2j + 1)^2 = 4j^2 + 4j + 1 = 4(j^2 + j) + 1$, so b^2 divided by 4 has a quotient of $j^2 + j$ and a remainder of 1.

9.

Factors of 18	1	2	3	6	9	18
Corresponding Quotients	18	9	6	3	2	1

11. **(a)**
```
   5
 5)25
 2)50
 2)100
 7)700
```
$700 = 2 \cdot 2 \cdot 5 \cdot 5 \cdot 7$

(b)
```
  11
 3)33
 3)99
 2)198
```
$198 = 2 \cdot 3 \cdot 3 \cdot 11$

(c)
```
   3
 5)15
 3)45
 2)90
 5)450
```
$450 = 2 \cdot 3 \cdot 3 \cdot 5 \cdot 5$

(d)
```
  11
 3)33
 2)66
 2)132
 2)264
 2)528
```
$528 = 2 \cdot 2 \cdot 2 \cdot 2 \cdot 3 \cdot 11$

13. **(a)** $136 = 2^3 \cdot 17^1$, $102 = 2^1 \cdot 3^1 \cdot 17^1$
(b) The divisors of 136 are
$2^0 \cdot 17^0 = 1$, $2^1 \cdot 17^0 = 2$, $2^2 \cdot 17^0 = 4$,
$2^3 \cdot 17^0 = 8$, $2^0 \cdot 17^1 = 17$, $2^1 \cdot 17^1 = 34$, and
$2^2 \cdot 17^1 = 68$, $2^3 \cdot 17^1 = 136$.
(c) The divisors of 102 are
$2^0 \cdot 3^0 \cdot 17^0 = 1$, $2^0 \cdot 3^1 \cdot 17^0 = 3$,
$2^0 \cdot 3^0 \cdot 17^1 = 17$, $2^0 \cdot 3^1 \cdot 17^1 = 51$,
$2^1 \cdot 3^0 \cdot 17^0 = 2$, $2^1 \cdot 3^1 \cdot 17^0 = 6$, and
$2^1 \cdot 3^0 \cdot 17^1 = 34$, $2^1 \cdot 3^1 \cdot 17^1 = 102$.
(d) From parts (b) and (c), we see that the greatest common divisor of 136 and 102 is 34.
15. **(a)** $7 \cdot 13$ **(b)** $5^3 \cdot 7^2$ **(c)** $2^3 \cdot 5^3 \cdot 23$ **17.** **(a)** Yes, because $28 = 2^2 \cdot 7^1$, so all the prime factors of 28 appear in *a* and to at least as high a power. **(b)** No. Since 3^2 is a factor of 126, the prime 3 appears to a higher power in 126 than it does in *a*. **(c)** $2^1 \cdot 7^2$ **(d)** The number of factors is $4 \cdot 2 \cdot 3 = 24$ **(e)** The factors, in order, are 1, 2, 3, 4, 6, 7, 8, 12, 14, 21, 24, 28, 42, 49, 56, 84, 98, 147, 168, 196, 294, 392, 588, and 1176. **19.** **(a)** $\sqrt{271} = 16.46\ldots$, so check if any of 2, 3, 5, 7, 11, 13 divide 271. None do, so 271 is prime.
(b) $\sqrt{319} = 17.86\ldots$, so check if any of 2, 3, 5, 7, 11, 13, 17 divide 319. Since 11 is a divisor, 319 is composite. Its prime factorization is $11 \cdot 29$. **(c)** $\sqrt{731} = 27.03\ldots$, so check if any of 2, 3, 5, 7, 11, 13, 17, 19, 23 divide 731. Since 17 is a divisor, 731 is composite. Its prime factorization is $17 \cdot 43$.
(d) $\sqrt{1801} = 42.43\ldots$, so check if any of 2, 3, 5, 7, 11, 13, 17, 19, 23, 29, 31, 37, 41 divide 1801. None do, so 1801 is prime.
21. Show them that $34 = 2 \cdot 17$, yet $17 > \sqrt{34}$.
23. Answers will vary. The prime factorization of *ab* is the product of the prime factorizations of *a* and *b*. This means that if *p* is a prime factor of *ab*, the prime *p* must have also been a factor of at least one of *a* or *b*. This means that *p* is necessarily a factor of *a* or *b*, or possibly both *a* and *b*. **25.** Point out that 90, when divided by 3, has the quotient 30 and 30 is also divisible by 3. Thus, $90 = 3 \times 3 \times 10$, and therefore $90 = 3 \times 3 \times 2 \times 5$ is the correct prime factorization.
27. Point out that 0 is "something," namely, an especially important member of the whole-number system. Since $0 = 0 \times 2$, point out that 0 leaves a 0 remainder when divided by 2, which, by definition, tells us that 0 is an even whole number.
29. **(a)** 1, 3, $3^2 = 9$ **(b)** $5^2 = 25$ with factors 1, 5, and 25; $7^2 = 49$ with factors 1, 7, and 49; $11^2 = 121$ with factors 1, 11, and 121 **(c)** $3^3 = 27$ with factors 1, 3, 9, and 27; $5^3 = 125$ with factors 1, 5, 25, and 125; $7^3 = 343$ with factors 1, 7, 49, and 343
31. Under the given conditions, *p* and *q* must each appear in the prime-factor representation of *n*. Therefore, *p* and *q* both divide *n*.
33. **(a)** The 16-by-13 rectangle has area $16 \cdot 13 = 208$ square units, so it requires 104 dominoes, each of area 2.
(b) Since 208 is not divisible by 3, the rectangle cannot be tiled by triominoes of area 3 square units.
35. **(a)** F_{999} is even, and F_{1000} is odd. **(b)** The Fibonacci number F_n is even if, and only if, *n* is divisible by 3.

37. (a)

n	0	1	2	3	4	5	6	7	8	9	10
P_n	3	0	2	3	2	5	5	7	10	12	17
n	11	12	13	14	15	16	17	18	19	20	21
P_n	22	29	39	51	68	90	119	158	209	277	367
n	22	23	24	25	26	27	28	29	30	31	32
P_n	486	644	853	1130	1497	1983	2627	3480	4610	6107	8090

(b) It appears from the table that n divides P_n if, and only if, n is a prime number. For example, 2 divides $P_2 = 2$, 3 divides $P_3 = 3$, 5 divides $P_5 = 5$, 7 divides $P_7 = 7$, 11 divides $P_{11} = 22$, 13 divides $P_{13} = 39, \ldots,$ 31 divides $P_{31} = 6107 = 31 \cdot 197$. Perrin proved that if n is prime, then n must divide P_n. However, the converse ("only if") conjecture was proved to be incorrect in 1982, when it was shown that the composite number $n = 271,441 = 521^2$ is a factor of $P_{277,441}$. It is the smallest nonprime n that divides its corresponding Perrin number P_n. **39. (a)** The 21-digit string 010001010101010001000 **(b)** There are 35 digits, so we fill in either a 5 × 7 or a 7 × 5 rectangle to get, respectively:

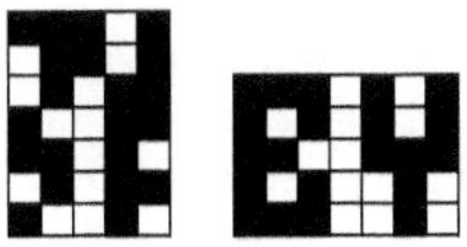

The second rectangle has the desired message, "BY."
(c) Since $96 = 2^5 \cdot 3$, there are 12 rectangles. The product of two prime gives just two rectangles with more than one row and more than one column. **41.** B **43.** A
45. 1 is neither a prime nor a product of primes.
$2 = 2, 3 = 3, 4 = 2^2, 5 = 5, 6 = 2 \cdot 3, 7 = 7,$
$8 = 2^3, 9 = 3^2, 10 = 2 \cdot 5, 11 = 11, 12 = 2^2 \cdot 3, 13 = 13,$
$14 = 2 \cdot 7, 15 = 3 \cdot 5, 16 = 2^4, 17 = 17, 18 = 2 \cdot 3^2,$
$19 = 19, 20 = 2^2 \cdot 5, 21 = 3 \cdot 7, 22 = 2 \cdot 11, 23 = 23,$
$24 = 2^3 \cdot 3, 25 = 5^2, 26 = 2 \cdot 13, 27 = 3^3, 28 = 2^2 \cdot 7,$
$29 = 29, 30 = 2 \cdot 3 \cdot 5, 31 = 31, 32 = 2^5, 33 = 3 \cdot 11,$
$34 = 2 \cdot 17, 35 = 5 \cdot 7, 36 = 2^2 \cdot 3^2, 37 = 37, 38 = 2 \cdot 19,$
$39 = 3 \cdot 13, 40 = 2^3 \cdot 5, 41 = 41, 42 = 2 \cdot 3 \cdot 7, 43 = 43,$
$44 = 2^2 \cdot 11, 45 = 3^2 \cdot 5, 46 = 2 \cdot 23,$
$47 = 47, 48 = 2^4 \cdot 3, 49 = 7^2, 50 = 2 \cdot 5^2$

Problem Set 4.2 (page 203)

1.

	2	3	4	5	6	8	9	10
684	✓	✓	✓	✗	✓	✗	✓	✗
(a) 1950	✓	✓	✗	✓	✓	✗	✗	✓
(b) 2014	✓	✗	✗	✗	✗	✗	✗	✗
(c) 2015	✗	✗	✗	✓	✗	✗	✗	✗
(d) 51,120	✓	✓	✓	✓	✓	✓	✓	✓

3. (a) 0, 2, 4, 6, 8 **(b)** 0, 3, 6, 9 **(c)** 0, 4, 8 **(d)** 0, 5 **(e)** 0, 6 **(f)** 0, 8 **(g)** 0, 9 **(h)** 0 **5. (a)** Make six label boxes (pigeonholes) as follows.

0	1, 9	2, 8	3, 7	4, 6	5

Place the seven numbers into the box according to the last digit of the number. Since there are six boxes, at least one box has two of the numbers. If there are two numbers in the 0 box, or two numbers in the 5 box, both their sum and difference are divisible by 10. Suppose there are two numbers in the 1 box and 9 box. If the last digit of the numbers is the same, either a 1 or a 9, then their difference is divisible by 10. If the two units digits are different, a 1 and a 9, then the sum of the two numbers is divisible by 10. The same reasoning applies to the other three boxes. **(b)** 1, 2, 3, 4, 5, 10, for example **7.** The digits sum to $23 + d$. Therefore, n is divisible by 3 when d is 1, 4, or 7. But n is divisible by 9 when d is 4, so n is divisible by 3 and not 9 when d is 1 or 7. **9.** To be divisible by 12, n must be divisible by both 3 and 4. Therefore, the sum of digits, which is $34 + d + e$ must be a multiple of 3 and de must be divisible by 4. In particular, $d + e$ must be 2, 5, 8, 11, 14, or 17 and e must be even—0, 2, 4, 6, or 8. If $e = 0$, then d is 2 or 8; if $e = 2$, then d is 3 or 9; if $e = 4$, then d is 4; if $e = 6$, then $d = 5$; if $e = 8$, then d is 0 or 6. This gives eight choices of de: 20, 80, 32, 92, 44, 56, 08, and 68.

11. (a)

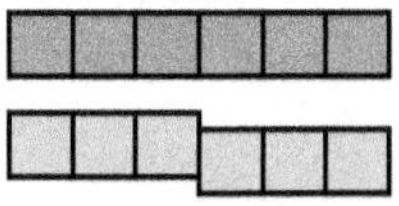

(b)

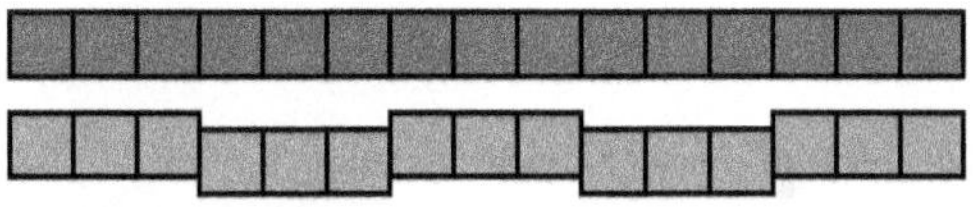

(c)

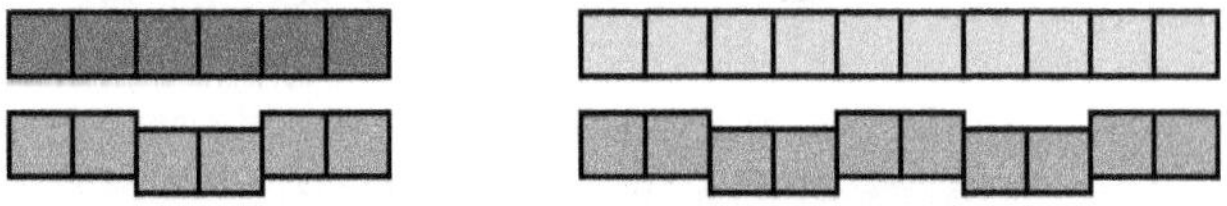

(d)

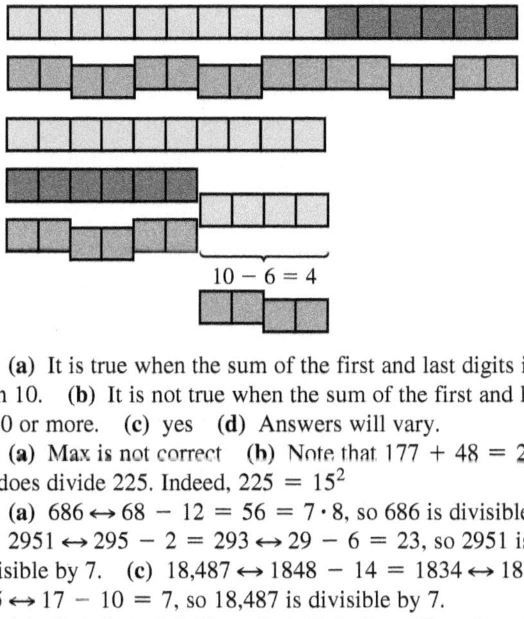

13. (a) It is true when the sum of the first and last digits is less than 10. **(b)** It is not true when the sum of the first and last digits is 10 or more. **(c)** yes **(d)** Answers will vary.
15. (a) Max is not correct **(b)** Note that $177 + 48 = 225$ and 15 does divide 225. Indeed, $225 = 15^2$
17. (a) $686 \leftrightarrow 68 - 12 = 56 = 7 \cdot 8$, so 686 is divisible by 7. **(b)** $2951 \leftrightarrow 295 - 2 = 293 \leftrightarrow 29 - 6 = 23$, so 2951 is not divisible by 7. **(c)** $18{,}487 \leftrightarrow 1848 - 14 = 1834 \leftrightarrow 183 - 8 = 175 \leftrightarrow 17 - 10 = 7$, so 18,487 is divisible by 7.
19. (a) $(2 + 6 + 9 + 2) - (2 + 2 + 4) = 19 - 8 = 11$, so divisible by 11
(b) $(4 + 3 + 0 + 1) - (8 + 6 + 2 + 9) = 8 - 25 = -17$, so not divisible by 11
(c) 7,654,592$(7 + 5 + 5 + 2) + (6 + 4 + 9) = 19 - 19 = 0$, so divisible by 11
(d) 8,352,607$(8 + 5 + 6 + 7) - (3 + 2 + 0) = 26 - 5 = 21$, so not divisible by 11
(e) 718,161,906$(7 + 8 + 6 + 9 + 6) - (1 + 1 + 1 + 0) = 36 - 3 = 33$, so divisible by 11
21. (a) For any palindrome with an even number of digits, the digits in the odd positions are the same as the digits in the even

positions, but with the order reversed. Thus, the difference of the sums of the digits in the even and odd positions is 0, which is divisible by 11. **(b)** Yes **23. (a)** The Fibonacci numbers F_{5k} are divisible by $F_5 = 5$ for every $k = 1, 2, 3, \ldots$. **(b)** The Fibonacci numbers F_{3k} are divisible by $F_3 = 2$ and the Fibonacci numbers F_{4k} are divisible by $F_4 = 3$ for every $k = 1, 2, 3, \ldots$. **(c)** Make and discuss a general conjecture about which Fibonacci numbers divide a Fibonacci number F_n. If n is a prime number, then F_n is only divisible by itself and $F_1 = 1$. If n is a composite number and $n = ij$, then F_n is divisible by F_i. It can be shown that F_i divides F_n if, and only if, i divides n.
25. Each calculation is in error.
(a) $35{,}874 + 7531 + 69{,}450 \leftrightarrow 0 + 7 + 6 = 13 \leftrightarrow 4$ and $113{,}855 \leftrightarrow 5$. The sum should be 112,855.
(b) $8514 + 6854 + 2578 + 6014 \leftrightarrow 0 + 5 + 4 + 2 = 11 \leftrightarrow 2$ and $23{,}860 \leftrightarrow 1$.The sum should be 23,960.
(c) $78{,}962 - 3621 \leftrightarrow 5 - 3 = 2$ and $75{,}331 \leftrightarrow 1$. The difference should be 75,341.
(d) $358 \times 592 \leftrightarrow 7 \times 7 = 49 \leftrightarrow 4$ and $221{,}936 \leftrightarrow 5$. The product should be 211,936.
27. C

Problem Set 4.3 (page 212)

1. (a) 3 **(b)** 2 **(c)** 24 **3. (a)** 216 **(b)** 154 **(c)** 144
5. (a) $\text{GCD}(r, s) = 2^1 \cdot 3^1 \cdot 5^2 = 150$; $\text{LCM}(r, s) = 2^2 \cdot 3^3 \cdot 5^3 = 13{,}500$ **(b)** $\text{GCD}(u, v) = 2^0 \cdot 5^1 \cdot 7^1 \cdot 11^0 = 35$;$\text{LCM}(u, v) = 2^2 \cdot 5^3 \cdot 7^2 \cdot 11^1 = 269{,}500$ **(c)** $\text{GCD}(w, x) = 2^1 \cdot 3^0 \cdot 5^2 \cdot 7^0 = 50$; $\text{LCM}(w, x) = 2^2 \cdot 3^3 \cdot 5^3 \cdot 7^2 = 661{,}500$
7. (a) $\text{GCD}(550, 3500) = 50$; $\text{LCM}(550, 3500) = 1{,}925{,}000/50 = 38{,}500$ **(b)** $\text{GCD}(825, 3915) = 15$;$\text{LCM}(825, 3915) = 3{,}229{,}875/15 = 215{,}325$ **(c)** $\text{GCD}(624{,}1044) = 12$; $\text{LCM}(624, 1044) = 651{,}456/12 = 54{,}288$ **9. (a)** 40 **(b)** 19 **(c)** 59 **11. (b)** For all natural numbers m and n, $\text{GCD}(m, mn) = m$. **(c)** For all natural numbers n, $\text{GCD}(n, 0) = n$. **(d)** For all natural numbers n, $\text{LCM}(n, n) = n$.
(e) For all natural numbers m and n, $\text{LCM}(m, mn) = mn$.
(f) For all natural numbers n, $\text{LCM}(n, 1) = n$.
13. (a) 12 **(b)** 12 **(c)** 18 **(d)** 18 **(e)** The least common multiple of two numbers a and b is the length of the shortest train that can be measured by both the a-train and the b-train.
15. The claim is true if $n \neq 0$. It is false if $n = 0$.
17. Note that mn is certainly a common multiple of m and n, but it is often not the least common multiple. For example, $4 \cdot 6 = 24$ is a common multiple of 4 and 6, but their least common multiple is 12. But use Tawana's comment as a springboard for class discussion by asking, "What must be true if mn is the least common multiple of m and n?" The answer, of course, is that m and n must have no common divisor other than 1, for, in that case,

$$\text{LCM}(m, n) = \frac{mn}{\text{GCD}(m, n)} = \frac{mn}{1} = mn.$$

19. (a) Since $60 = 2^2 \times 3 \times 5$ and $105 = 3 \times 5 \times 7$, it follows that $\text{GCD}(60, 105) = 3 \times 5 = 15$. Thus, the largest square tile possible is 15″ by 15″. **(b)** Since $60 \div 15 = 4$ and $105 \div 15 = 7$, there will be 4 rows of 7 tiles each. Thus, $4 \times 7 = 28$ tiles are required. **21. (a)** The side length of any tiled square must be a multiple of both dimensions of the brick. That is, the side length must be a common multiple of 8″ and 12″. Since $\text{LCM}(8, 12) = (8 \times 12)/\text{GCD}(8, 12) = 96/4 = 24$ inches, the smallest square tiled by 8″ by 12″ bricks is 24″ × 24″. This is tiled with three rows of two bricks each.
(b) Reasoning as in part (a), we conclude that the smallest square that can be tiled with 9″ by 12″ bricks has sides of length $\text{LCM}(9, 12) = (9 \text{ by } 12)/\text{GCD}(9, 12) = 108/3 = 36$ inches. The smallest square is tiled with four rows of three bricks each.
23. (a) $\text{GCD}(a, b, c) = 2^0 \cdot 3^1 \cdot 5^1 \cdot 7^0 = 15$; $\text{LCM}(a, b, c) = 2^2 \cdot 3^3 \cdot 5^3 \cdot 7^1 = 94{,}500$ **(b)** No **(c)** Answers will vary. One possibility is $r = 3, s = 5, t = 7$. **25. (a)** $\text{GCD}(6, 35, 143) = 1$; $\text{LCM}(6, 35, 143) = 30{,}030$ **(b)** Yes **(c)** It must be the case that $\text{GCD}(a, b) = \text{GCD}(a, c) = \text{GCD}(b, c) = 1$.
27. (a) $2^4 \cdot 3^1 \cdot 5^2 \cdot 7^1$ **(b)** $2^3 \cdot 3^2 \cdot 5^1 \cdot 7^1$ **(c)** $2^2 \cdot 5^3 \cdot 7^1 \cdot 11^1$
29. (a) 2 **(b)** 6 **(c)** 11 **(d)** 15
(e) $\frac{3}{4}$ **(f)** $\frac{3}{4}$ **(g)** $\frac{12}{19}$ **(h)** $\frac{21}{25}$
31. (a) $\frac{3}{4} + \frac{1}{6} = \frac{9}{12} + \frac{2}{12} = \frac{11}{12}$
(b) $\frac{7}{10} + \frac{2}{15} = \frac{21}{30} + \frac{4}{30} = \frac{25}{30} = \frac{5 \cdot 5}{5 \cdot 6} = \frac{5}{6}$
(c) $\frac{8}{9} - \frac{4}{15} = \frac{40}{45} - \frac{12}{45} = \frac{28}{45}$ **(d)** $\frac{11}{25} - \frac{3}{10} = \frac{22}{50} - \frac{15}{50} = \frac{7}{50}$
33. 32 **35. (a)** 221 years will pass, since $\text{GCD}(13, 17) = 1$ and $\text{LCM}(13, 17) = 13 \times 17 = 221$. **(b)** In just 36 years, the two species will emerge together, since $\text{GCD}(12, 18) = 6$ and therefore $\text{LCM}(12, 18) = (12 \times 18)/\text{GCD}(12, 18) = 12 \times 18/6 = 36$.
(c) In 45 years, the species will again emerge together, since $\text{GCD}(9, 15) = 3$ and $\text{LCM}(9, 15) = 9 \times 15/\text{GCD}(9, 15) = 9 \times 15/3 = 45$. **37.** B **39. (a)** A muffin only, since 6 divides 30 but 4 does not divide 30. **(b)** Casey is the 12th customer. Since $\text{LCM}(4, 6) = 12$, he was the first to receive both a cookie and a muffin. **(c)** Cookies are given out only on to customers 4, 8, 16, 20, 28, 32, 40, 44, 52, 56, ... These are the multiples of 4 that are not multiples of 6. Since Tom came after Casey, he may be customer 16, 20, 28, 32, 40, 44, 52, 56, ... **(d)** Since $4 \cdot 29 = 116$, there were 116, 117, 118, or 119 customers. None of these are multiples of 6, since $6 \cdot 19 = 114$. Thus, 19 muffins were given away. **41.** D

Chapter 4 Review Exercise (page 219)

1.
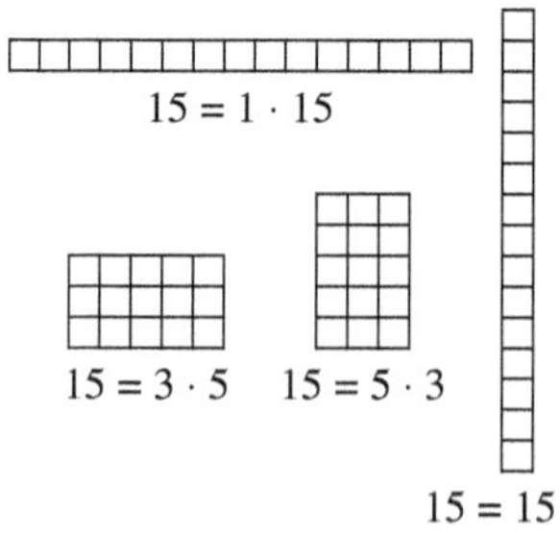

3. (a) $D_{60} = \{1, 2, 3, 4, 5, 6, 10, 12, 15, 20, 30, 60\}$
(b) $D_{72} = \{1, 2, 3, 4, 6, 8, 9, 12, 18, 24, 36, 72\}$
5. (a) $2^5 = 32$ **(b)** $3^5 = 243$
7. (a) Answers will vary; for example, $15 = 3 \cdot 5$; $5 > \sqrt{15}$.
(b) Yes, $3 \leq \sqrt{15}$. **9. (a)** $(1 + 4)(1 + 2) = 15$
(b) 1, 3, 7, 9, 21, 27, 49, 63, 81, 147, 189, 441, 567, 1323, 3969
11. (a) True **(b)** True **(c)** True **(d)** False
13.
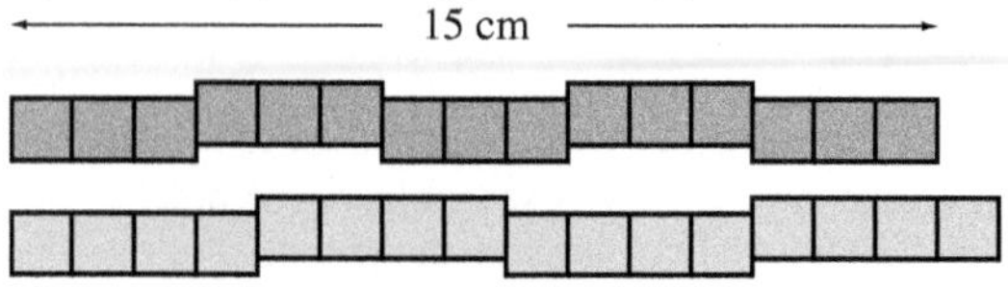

15. **(a)** $D_{63} = \{1, 3, 7, 9, 21, 63\}$ $D_{91} = \{1, 7, 13, 91\}$ and $D_{63} \cap D_{91} = \{1, 7\}$, so $\text{GCD}(91, 63) = 7$.
(b) $M_{63} = \{63, 126, 189, 252, 315, 378, 441, 504, 567, 630, 693, 756, 819, 882, 945, 1008 \ldots.\}$, M91 = {91, 182, 273, 364, 455, 546, 637, 728, 819, 910c}, and $M_{63} \cap M_{91} = \{819, 1638, \ldots\}$, so $\text{LCM}(63, 91) = 819$. **(c)** $7.819 = 5733 = 63{,}91$

17. **(a)**

$$12{,}100\overline{)119{,}790} \quad 9 \text{ R } 10890$$

$$10{,}890\overline{)12{,}100} \quad 1 \text{ R } 1210 \qquad 1210\overline{)10{,}890} \quad 9 \text{ R } 0$$

Thus, $\text{GCD}(119{,}790, 12{,}100) = 1210$.
(b) $\text{LCM}(119{,}790, 12{,}100) = 119{,}790 \cdot 12{,}100/1210 = 1{,}197{,}900$

Chapter 5

Problem Set 5.1 (page 230)

1. **(a)**

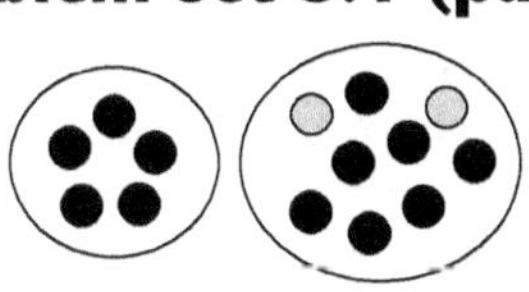

(b)

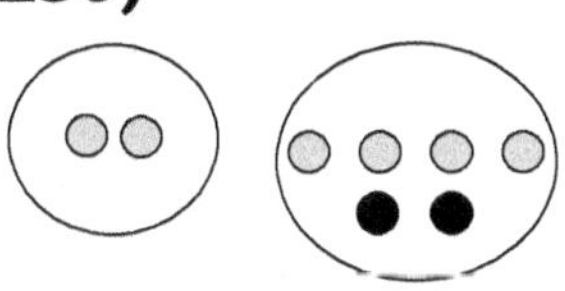

3. **(a)**

(b)

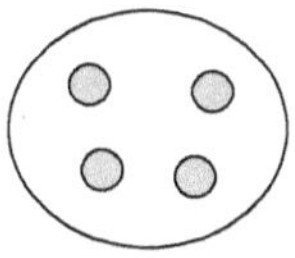

5. **(a)** At mail time, you are delivered a check for \$14.
(b) At mail time, you receive a bill for \$27.
7. **(a)** You are richer by \$17. **(b)** $\$48 + (-31) = 17$
9. **(a)** 4 **(b)** 0 **(c)** 6 **(d)** -5
11. **(a)**

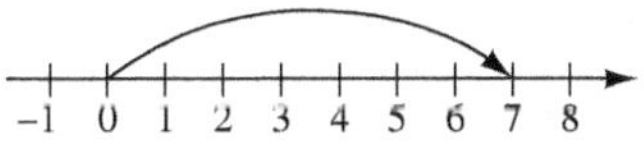

(b)

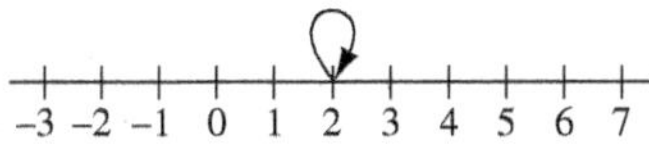

(c)

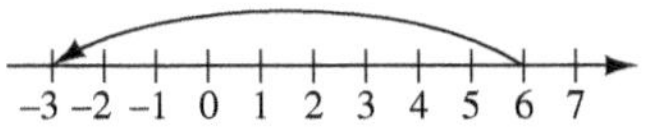

(d)

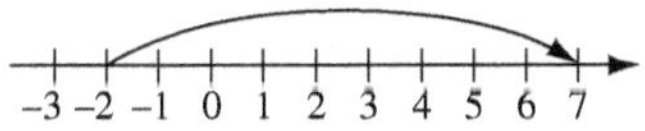

13. **(a)** $13, -13$ **(b)** $1, -1$ **(c)** No values of x **15.** **(a)** Lay down a loop for 4, turn all the markers over to show -4, and turn them all again to show $-(-4)$, or 4, since we finish with the original drop. **(b)** Replace 4 with n in part (a). **17.** Congratulate her, since her claim is true. **19.** **(a)** Four red counters **(b)** Yes; for example, we could have added five red counters and one black counter. **21.** **(a)** $-12, -10, -8, -6, -4, -2, 0, 2, 4, 6, 8, 10, 12$ **(b)** $-11, -9, -7, -5, -3, -1, 1, 3, 5, 7, 9, 11$ **23.** Any integer $-10, -9, \ldots, 9, 10$ **25.** $1, -2, 3, -4, 5$
27. **(a)**

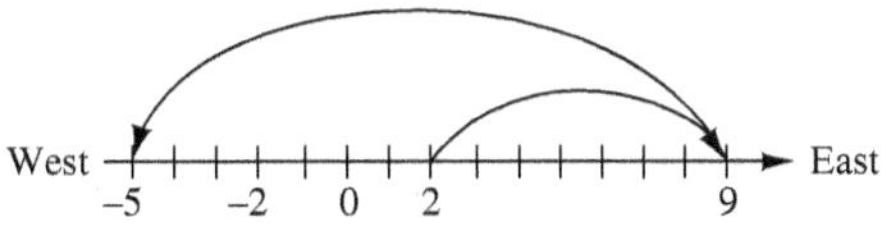

(b) On the first two legs of the trip, the truck drove $|9 - 2| + |(-5) - 9| = |7| + |-14| = 7 + 14 = 21$ blocks.
(c) The gas station is at West Second, so it is $|2 - (-2)| = |4| = 4$ blocks from the warehouse.

29. **(a)** 22nd floor **(b)** 23rd floor **31.** The 47-yard line
33.

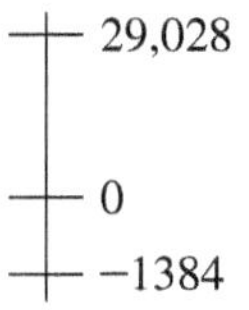

The diagram shows that difference is $|29{,}028 - (-1384)| = 29{,}028 + 1384 = 30{,}412$

Problem Set 5.2 (page 245)

1. **(a)**

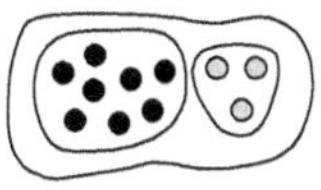

$8 + (-3) = 5$

(b)

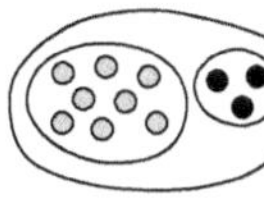

$(-8) + 3 = -5$

(c)

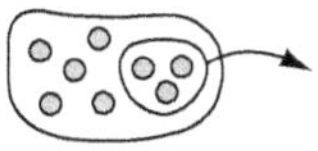

$-8 - (-3) = -5$

(d)

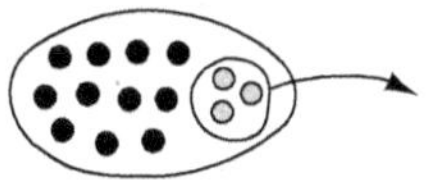

$8 - (-3) = 11$

3. **(a)** At mail time, you receive a bill for \$27 and a bill for \$13; $(-27) - 13) = -40$. **(b)** At mail time, you receive a bill for \$27, and the letter carrier takes away a check for \$13; $(-27) - 13 = -40$. **(c)** The mail carrier brings you a check for \$27 and a check for \$13; $27 + 13 = 40$. **(d)** The mail carrier brings you a check for \$27 and takes away a check for \$13; $27 - 13 = 14$.
5. **(a)**

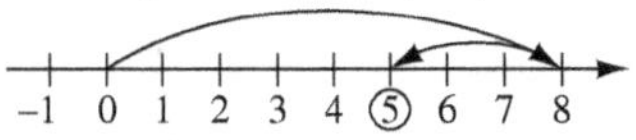

$8 + (-3) = 5$

(b)

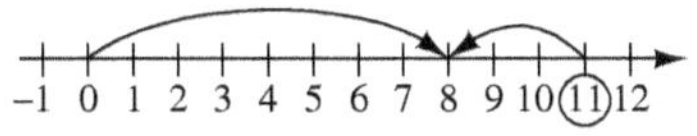

$8 - (-3) = 11$

(c)

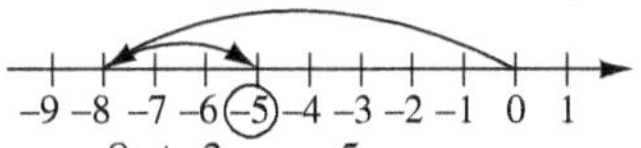

$-8 + 3 = -5$

(d)

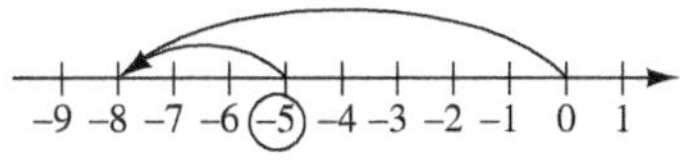

$(-8) - (-3) = -5$

7. **(a)** $13 + (-7)$ **(b)** $13 + 7$ **(c)** $(-13) + (-7)$ **(d)** $-13 + 7$
9. **(a)** 40 **(b)** -12 **(c)** -27 **(d)** -27
11. **(a)** $(-356) + 148 \approx -350 + 150 = -200$
(b) $728 + (-273) \approx 725 - 275 = 700 - 250 = 450$
(c) $298 - (-454) \approx 300 + 450 = 750$
(d) $-827 - 370 \approx -(830 + 370) = -(800 + 400) = -1200$
13. **(a)** 29° **(b)** $2 - (-27) = 29$ **15.** **(a)** More, by \$106 **(b)** $314 - 208 = 106$ **17.** **(a)** $-117 < -24$ **(b)** $0 > -4$ **(c)** $18 > 12$ **19.** $-17, -5, -2, 0, 3, 5, 27$ **21.** Answers will vary. The important point is that negation is a unary operation, in which pressing the button [(−)] takes the opposite, or negative, of the number x currently in the display. The subtraction function [−] is a binary operation that sets up a pending operation in which the currently displayed number x will be the minuend, the next number entered will become the subtrahend y, and pressing [=] (or [ENTER]) calculates the difference $x - y$. **23.** Compliment Melanie on good thinking. **25.** **(a)** Armand is correct. **(b)** $7 - (-3) = 10$ **(c)** One way of subtracting m from n is to add the additive inverse of m to n. That is, $n - m = n + (-m)$. Since the negative of -3 is 3, $7 - (-3) = 7 + 3 = 10$.

27. **(a)** Yes. If $a < b$, then one of the statement $a < b$ and $a = b$ is certainly true. **(b)** Not necessarily; If $a \geq b$, then it is possible that $a = b$, so $a > b$ is false. **(c)** Yes. One of the statements $2 = 2$ and $2 > 2$ is true. **29.** **(a)** 6 **(b)** 6 **(c)** 15 **(d)** 11
31. The distance between a and b on a number line is $|a - b|$.
33.

-1	-6	8	1
4	5	-5	-2
-7	0	2	7
6	3	-3	-4

Magic sum = 2.
35. **(a)** (i) $7 - (-3) = 10, (-3) - 7 = -10$
(ii) $(-2) - (-5) = 3, (-5) - (-2) = -3$ **(b)** No
37. Let $a < b$ so there is a positive integer $c > 0$ for which $a + c = b$. Add n to both sides of this equation to see that $(a + n) + c = b + n$. Therefore, $a + n < b + n$. Each step is reversible, so the if and only if statement has been proved.
39. **(a)** $101 + 3 = 104$ **(b)** $101 + 3 - 5 = 99$ **41.** $3916
43. **(a)** 96, 64, 32, 0, −32, −64, −96, −128 **(b)** The ball is moving upward at 96 feet per second. **(c)** The ball is not moving at that instant. **(d)** The ball is moving downward at 32, 64,96, and 128 feet per second, respectively. **(e)** They are negatives. At any given height, the ball has the same speed on the way up as on the way down. **45.** B

Problem Set 5.3 (page 255)

1. **(a)** 77 **(b)** −77 **(c)** −77 **(d)** 77 **3.** **(a)** 4 **(b)** −4 **(c)** −4 **(d)** 4 **5.** $(-25,753) \cdot (-11) = 283,283$; $283,283 \div (-11) = -25,753$; $283,283 \div (-25,753) = -11$ **7.** **(a)** Richer by $78; $6 \cdot 13 = 78$ **(b)** Poorer by $92; $4 \cdot (-23) = -92$
9. **(a)** $6 \cdot 3 = 18$ **(b)** $4 \cdot (-4) = -16$ **(c)** $(-3) \cdot (-5) = 15$
11. **(a)**

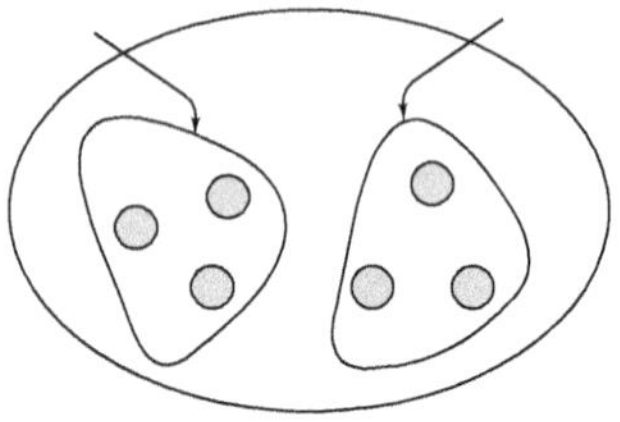

2 × (−3)

(b)

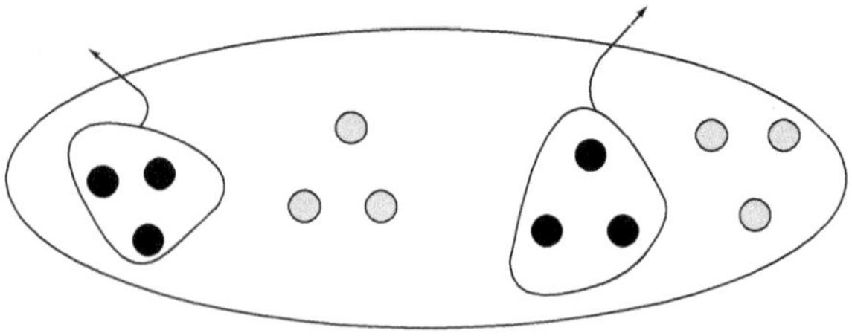

(−2) × 3

(c)

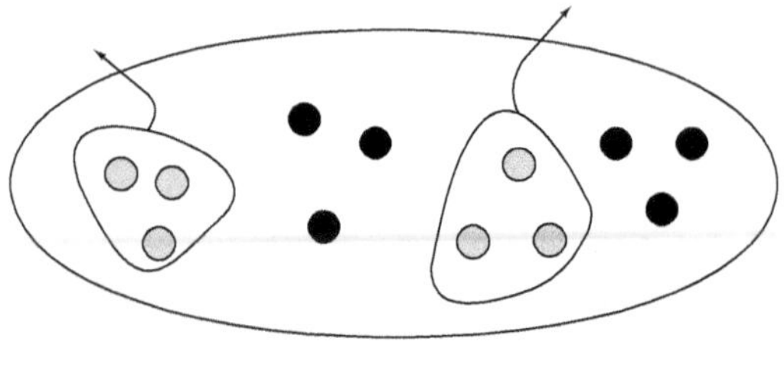

(−2) × (−3)

13.

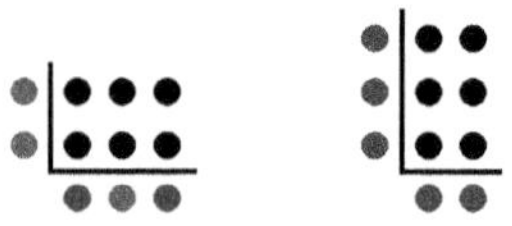

$(-3) \cdot (-2) = 6$ $(-2) \cdot (-3) = 6$
$6 \div (-3) = -2$ $6 \div (-2) = -3$

15. Answers will vary. **17.** Answers will vary. **19.** Kris overlooked the case $n = 0$. When she divided by n, she assumed that n was not zero. **21.** **(a)** Can't tell **(b)** Always nonnegative **(c)** Always positive **(d)** Can't tell **(e)** Always nonpositive
23. Yes. $a(b - c) = a[b + (-c)] = ab + a(-c) = ab - ac$.
25. **(a)** The products are increasing by positive 3 at each step, so the next terms are $0 \cdot (-3) = 0, (-1) \cdot (-3) = 3$, and $(-2) \cdot (-3) = 6$. That is, negative times negative is positive.
(b) The products are decreasing by positive 3 at each step, so the next terms are $(-3) \cdot (0) = 0, (-3) \cdot 1 = -3$, and $(-3) \cdot 2 = -6$. That is, negative times positive is negative.
27.

−6	8	−2
4	0	−4
2	−8	6

The common sum and the middle number must both be 0.

29. **(a)**

$+_{12}$	0	1	2	3	4	5	6	7	8	9	10	11
0	0	1	2	3	4	5	6	7	8	9	10	11
1	1	2	3	4	5	6	7	8	9	10	11	0
2	2	3	4	5	6	7	8	9	10	11	0	1
3	3	4	5	6	7	8	9	10	11	0	1	2
4	4	5	6	7	8	9	10	11	0	1	2	3
5	5	6	7	8	9	10	11	0	1	2	3	4
6	6	7	8	9	10	11	0	1	2	3	4	5
7	7	8	9	10	11	0	1	2	3	4	5	6
8	8	9	10	11	0	1	2	3	4	5	6	7
9	9	10	11	0	1	2	3	4	5	6	7	8
10	10	11	0	1	2	3	4	5	6	7	8	9
11	11	0	1	2	3	4	5	6	7	8	9	10

$\times_{12}$	0	1	2	3	4	5	6	7	8	9	10	11
0	0	0	0	0	0	0	0	0	0	0	0	0
1	0	1	2	3	4	5	6	7	8	9	10	11
2	0	2	4	6	8	10	0	2	4	6	8	10
3	0	3	6	9	0	3	6	9	0	3	6	9
4	0	4	8	0	4	8	0	4	8	0	4	8
5	0	5	10	3	8	1	6	11	4	9	2	7
6	0	6	0	6	0	6	0	6	0	6	0	6
7	0	7	2	9	4	11	6	1	8	3	10	5
8	0	8	4	0	8	4	0	8	4	0	8	4
9	0	9	6	3	0	9	6	3	0	9	6	3
10	0	10	8	6	4	2	0	10	8	6	4	2
11	0	11	10	9	8	7	6	5	4	3	2	1

(b) Clock addition is closed, commutative, and associative; 0 is the additive identity; and every number n has $12 - n$ as its additive inverse. **(c)** I_{12} is a number system with the additive-inverse property. **(d)** Clock multiplication is closed, commutative, and associative, and distributes over clock addition. The number 1 is the multiplicative identity, and $0 \times_{12} n = 0$ for all $n \in I_{12}$. **(e)** No. For example $5 \times_{12} 5 = 1, 7 \times_{12} 7 = 1$, and $11 \times_{12} 11 = 1$.

(f) No. For example $2 \times_{12} 6 = 0, 3 \times_{12} 4 = 0, 3 \times_{12} 8 = 0, 4 \times_{12} 6 = 0$. **31.** **(a)** \$585 **(b)** $15 \cdot 39 = 585$
33. After the first step, all the numbers will be positive and the process will then continue as before. **35.** -10

Chapter 5 Review Exercises (page 260)

1. **(a)** -1 **(b)** 5 **(c)** $-15, -13, -11, \ldots, 11, 13, 15$
3. **(a)** 12 **(b)** -24 **5.** Answers will vary. **(a)** At mail time, you receive a bill for \$114 and a check for \$29. **(b)** The mail carrier brings you a bill for \$19 and a check for \$66.
7. $2 + (-4) = -2$ **9.** **(a)** Poorer by \$23; $45 + (-68) = -23$
(b) Richer by \$113; $45 - (-68) = 113$ **11.** **(a)** -2 **(b)** -22
(c) -32 **(d)** 12 **(e)** 20 **(f)** -4 **13.** **(a)** \$25
(b) $(-12) + 37 = 25$ **15.** **(a)** $3 \cdot 4 = 12$ **(b)** $3 \cdot (-4) = -12$
(c) $(-3) \cdot 4 = -12$ **(d)** $(-3) \cdot (-4) = 12$
17. **(a)** 56 **(b)** -56 **(c)** -56 **(d)** -7 **(e)** -12 **(f)** 12
19. If d divides n, there is an integer c such that $dc = n$. But then $d \cdot (-c) = -n, (-d) \cdot (-c) = dc = n$, and $(-d) \cdot c = -dc = -n$. Thus, d divides $-n$, $-d$ divides n, and $-d$ divides $-n$.

Chapter 6

Problem Set 6.1 (page 276)

1. **(a)** $\frac{1}{6}$ **(b)** $\frac{2}{6}$ **(c)** $\frac{0}{1}$

3. **(a)** 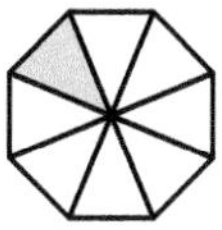

$\frac{1}{8}$

(b) 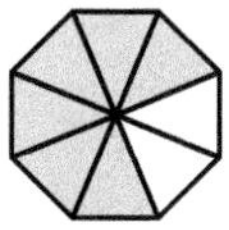

$\frac{6}{8}$

5. **(a)** 4/5 **(b)** 4/5 **(c)** 4/6 **(d)** 2/5 **(e)** 5/6 **(f)** 7/10
(g) 9/14 **7.** **(a)** $\frac{4}{10}$ **(b)** $\frac{3}{10}$ **(c)** $\frac{4}{10}$ **(d)** $\frac{6}{10}$ **(e)** $\frac{2}{10}$

9. **(a)** $\frac{-5}{3}, \frac{2}{3}, \frac{5}{3}$ **(b)** $\frac{-5}{5}, \frac{3}{5}, \frac{7}{5}$

11. **(a)** 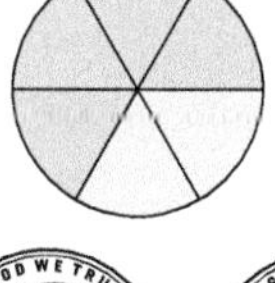

(b) 

(c) 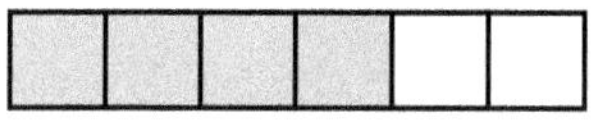

(d) 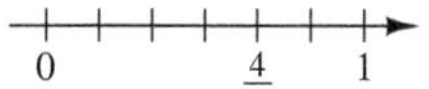

13. **(a)** $\frac{3}{6} = \frac{1}{2}$ **(b)** $\frac{2}{3} = \frac{4}{6}$ **(c)** $\frac{2}{3} = \frac{10}{15}$

15. **(a)** 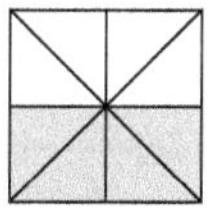 **(b)** 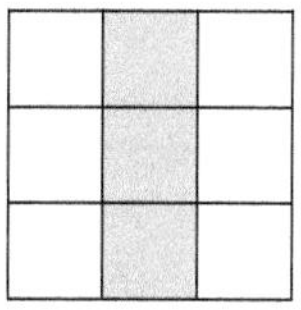 **(c)** 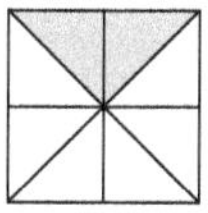

17. **(a)** Equivalent **(b)** Not equivalent **(c)** Not equivalent
(d) Not equivalent **19.** **(a)** Yes **(b)** Yes **(c)** Yes **(d)** No

21. **(a)** $\frac{96}{288} = \frac{2^5 \cdot 3^1}{2^5 \cdot 3^2} = \frac{1}{3}$ **(b)** $\frac{247}{-75} = \frac{13^1 \cdot 19^1}{(-) \cdot 3^1 \cdot 5^2} = \frac{-247}{75}$
(c) $\frac{2520}{378} = \frac{2^3 \cdot 3^2 \cdot 5^1 \cdot 7^1}{2^1 \cdot 3^3 \cdot 7^1} = \frac{20}{3}$ **23.** **(a)** $\frac{9}{24}$ and $\frac{20}{24}$
(b) $\frac{15}{105}, \frac{84}{105}$, and $\frac{70}{105}$ **(c)** $\frac{136}{96}$ and $\frac{21}{96}$ **(d)** $\frac{1}{3}$ and $\frac{4}{3}$

25. **(a)** $\frac{7}{12}, \frac{2}{3}$ **(b)** $\frac{2}{3}, \frac{5}{6}$ **(c)** $\frac{29}{36}, \frac{5}{6}$ **(d)** $\frac{-8}{9}, \frac{-5}{6}$ **(e)** $\frac{2}{3}, \frac{29}{36}, \frac{5}{6}, \frac{8}{9}$

27. Two numbers: $\frac{3}{4}$ and 4 **29.** **(a)** $\frac{1}{2}, \frac{1}{6}, \frac{1}{3}$ **(b)** $\frac{2}{1}, \frac{1}{3}, \frac{2}{3}$
(c) $\frac{3}{1}, \frac{3}{2}, \frac{1}{2}$ **31.** Answers will vary, but typically include a figure of pizzas cut into fourths and sixths to show that $\frac{1}{4}$ of the pizza is larger than $\frac{1}{6}$ of the pizza. **33.** Miley has used excellent reasoning to draw a valid conclusion.

35. **(a)** 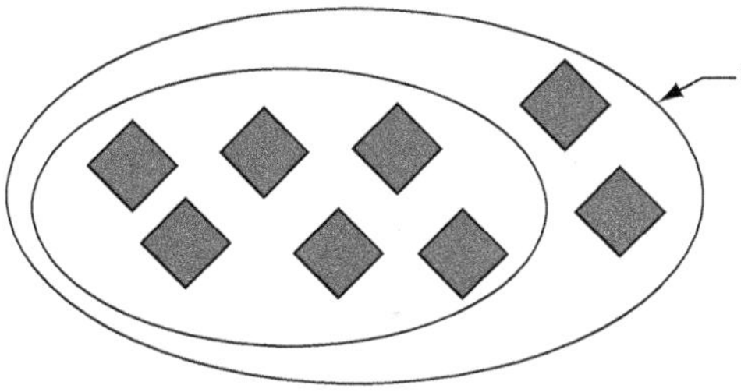

(b) 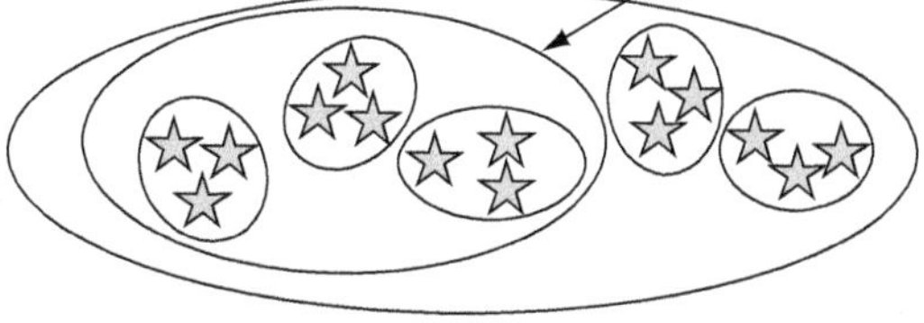

37. **(a)** 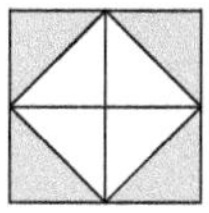 $\frac{4}{8}$ **(b)** 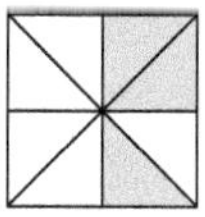 $\frac{3}{8}$ **(c)** 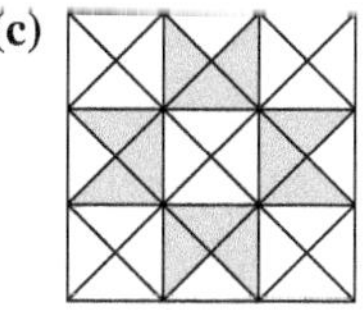 $\frac{12}{36}$

(d) 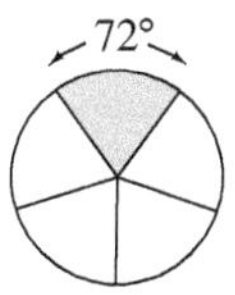

$\frac{1}{5}$ **(e)** 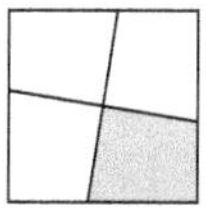 $\frac{1}{4}$

(f) 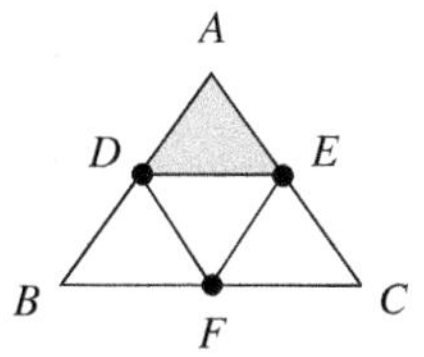

$\frac{1}{4}$ **(g)** 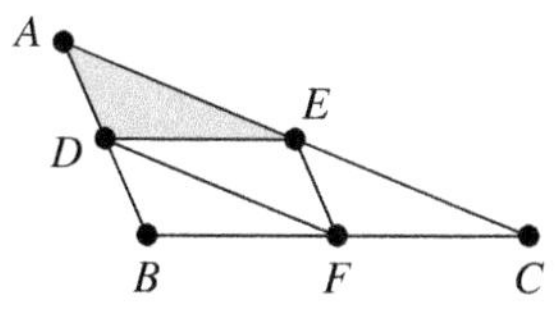

$\frac{1}{4}$

39. Add tick marks to show that the rope was originally 60 feet long.

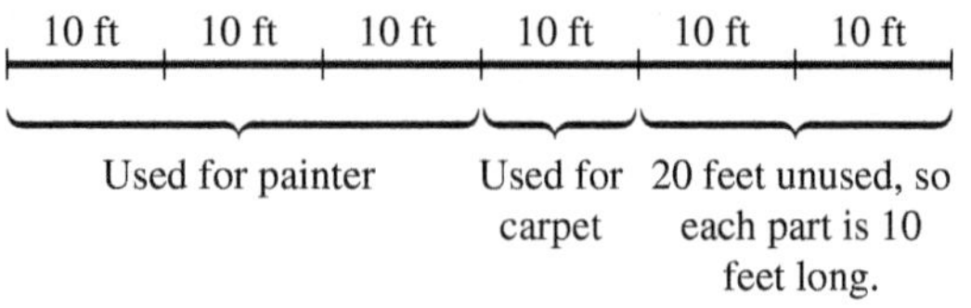

41. Your investigation should suggest that this is a general property of Pascal's triangle. **43. (a)** 18 gal. **(b)** 6 gal. **(c)** 15 gal.
45. (a) $\frac{1}{2}$ **(b)** $\frac{12}{52}$ or $\frac{3}{13}$ **(c)** $\frac{3}{6}$ or $\frac{1}{2}$ **(d)** $\frac{25}{75}$ or $\frac{1}{3}$ **(e)** $\frac{50}{75}$ or $\frac{2}{3}$
47. B **49.** A **51.** B

Problem Set 6.2 (page 288)

1. (a) $\frac{1}{3}+\frac{1}{2}=\frac{5}{6}$ **(b)** $\frac{2}{3}+\frac{1}{2}=\frac{7}{6}$

3. (a)

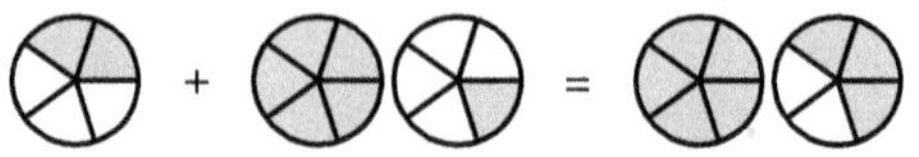

$\frac{2}{5} + \frac{6}{5} = \frac{8}{5}$

(b)

$\frac{1}{4} + \frac{1}{2} = \frac{1}{4} + \frac{2}{4} = \frac{3}{4}$

(c)

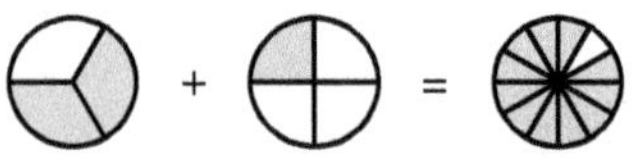

$\frac{2}{3} + \frac{1}{4} = \frac{11}{12}$

5. (a)

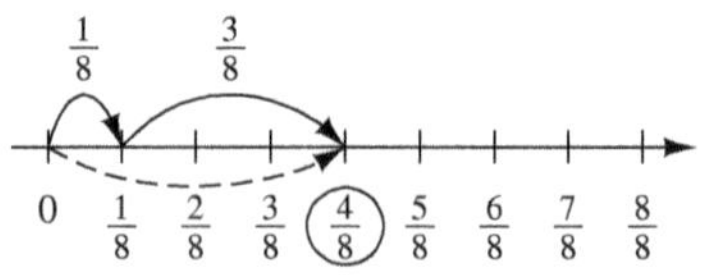

(b)

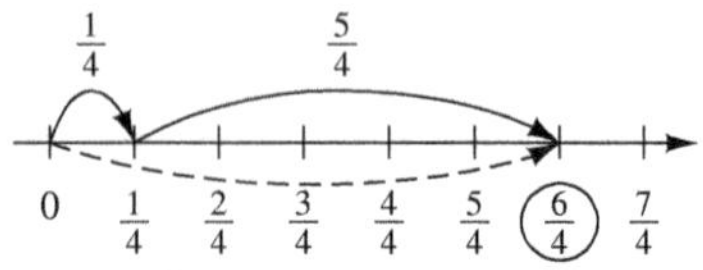

(c)

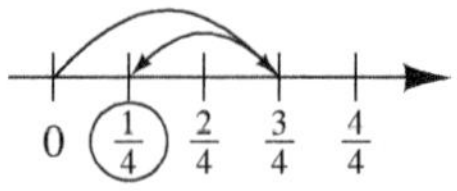

$\frac{3}{4}+\frac{-2}{4}=\frac{1}{4}$

7. (a) $\frac{5}{7}$ **(b)** 2 **(c)** $\frac{5}{6}$ **(d)** $\frac{56}{65}$ **9. (a)** $2\frac{1}{4}$ **(b)** $5\frac{2}{3}$
(c) $4\frac{19}{23}$ **(d)** $-35\frac{71}{100}$ **11. (a)** $\frac{5}{6}-\frac{1}{4}=\frac{7}{12}$
(b) Fraction strip:

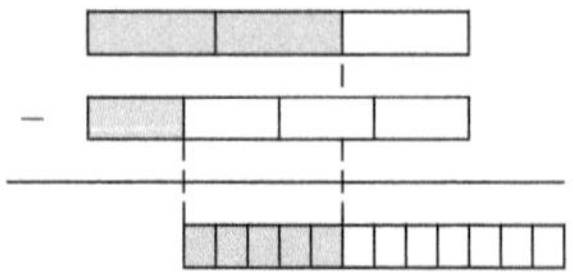

13. (a) $\frac{3}{4}-\frac{1}{3}=\frac{9}{12}-\frac{4}{12}=\frac{5}{12}$
(b)

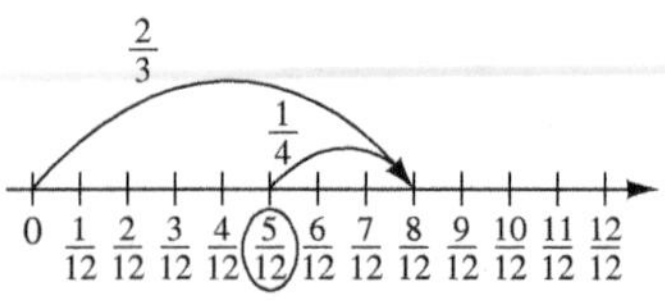

15. (a) $\frac{1}{3}$ **(b)** 0 **(c)** $\frac{625}{642}$ **(d)** $\frac{-107}{250}$
17. (a) $\frac{3}{4}-\frac{2}{3}=\frac{9}{12}-\frac{8}{12}=\frac{1}{12}>0$
(b) $\frac{14}{17}-\frac{4}{5}=\frac{70}{85}-\frac{68}{85}=\frac{2}{85}>0$
(c) $\frac{99}{50}-\frac{19}{10}=\frac{99}{50}-\frac{95}{50}=\frac{4}{50}=\frac{2}{25}>0$
19. Answer will vary. **21.** Answers will vary. Dannea should have eaten half of the remaining pieces, or 3 pieces, not half of the original pizza. **23.** Melanie's two-game average is indeed $\frac{3}{8}$, but she cannot determine this by adding the fractions $\frac{1}{3}$ and $\frac{2}{5}$, since their sum is $\frac{11}{15}$, as John correctly pointed out. When fractions are "added" according to Melanie's rule, it is sometimes known as taking the *mediant* and is written as $\frac{1}{3}\oplus\frac{2}{5}=\frac{3}{8}$.
25. Since $\frac{1}{2}+\frac{1}{3}+\frac{1}{9}=\frac{9+6+2}{18}=\frac{17}{18}$, the will did not call for the distribution of all of the horses.
27. (a) $\frac{1}{10}+6\cdot\frac{1}{70}+6\cdot\frac{1}{21}+3\cdot\frac{1}{14}+3\cdot\frac{11}{105}=1$
(b) $\frac{1}{10}+6\cdot\frac{1}{70}+3\cdot\frac{1}{14}=\frac{2}{5}$ **(c)** $\frac{1}{10}+3\cdot\frac{1}{70}=\frac{1}{7}$
29. (a) It terminates.

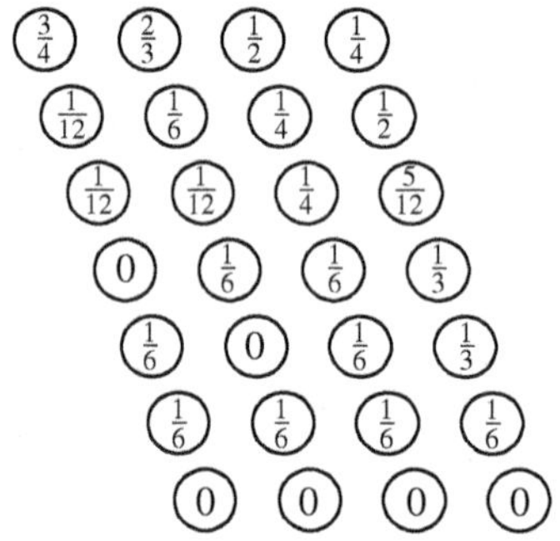

(b) It terminates.

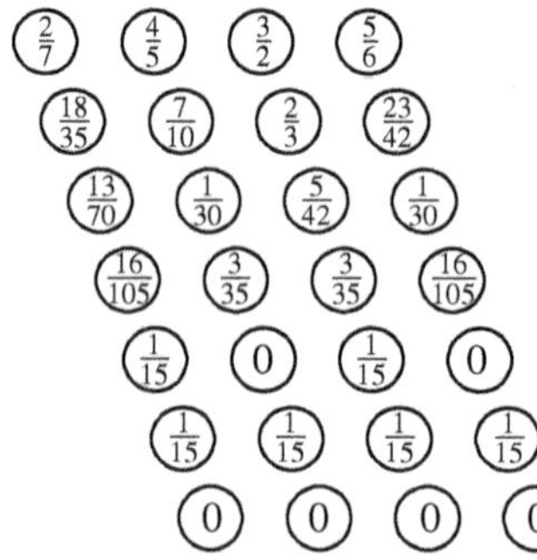

(c) Yes: If the fractions are written with a common denominator, Diffy is played on the whole numbers of the numerators.
31. $3\frac{1}{2}+\frac{5}{8}+\frac{5}{8}=3\frac{1}{2}+\frac{10}{8}=3\frac{1}{2}+\frac{5}{4}=3\frac{2}{4}+1+\frac{1}{4}=4\frac{3}{4}$ inches **33.** $24-10\frac{1}{2}-\frac{1}{16}=24-10-\frac{8}{16}-\frac{1}{16}=14-\frac{9}{16}=13\frac{7}{16}$ **35.** A **37.** D

Problem Set 6.3 (page 302)

1. $4 \times \frac{2}{5} = \frac{8}{5}$

3. **(a)** $4 \cdot \frac{3}{8} = \frac{12}{8}$

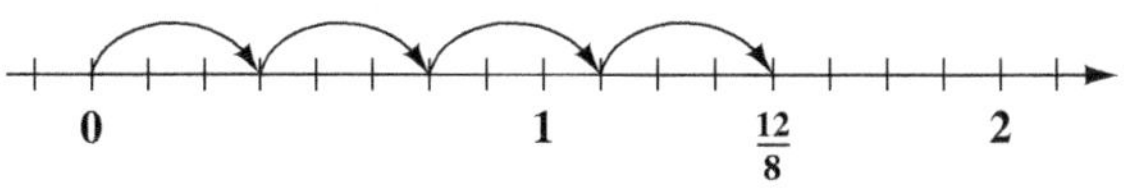

(b) $-4 \cdot \frac{3}{8}$

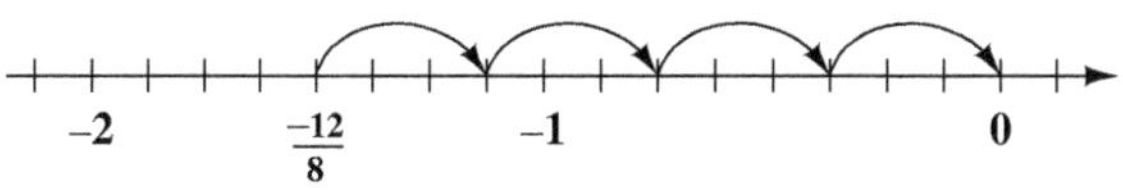

5. $\frac{5}{6} \times 4 = \frac{20}{6} = 3\frac{2}{6}$ **7.** **(a)** $3 \times \frac{5}{2} = \frac{15}{2}$ **(b)** $\frac{2}{3} \times \frac{3}{2} = \frac{6}{6}$ **(c)** $\frac{3}{4} \times 2 = \frac{6}{4}$ **9.** $7\frac{7}{8}$

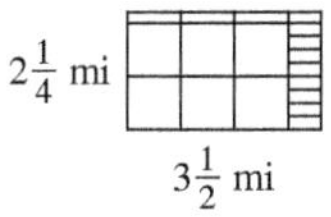

11. **(a)** 8 **(b)** $\frac{1}{5}$ **(c)** 1 **13.** **(a)** $\frac{5}{3}$ **(b)** $\frac{11}{8}$ **(c)** $\frac{2}{7}$ **(d)** 3

15. **(a)** 1 **(b)** $\frac{1}{4}$ **(c)** 1 **17.** **(a)** $\frac{2}{5}x - \frac{3}{4} = \frac{1}{2}$. Add $\frac{3}{4}$ to each side: $\frac{2}{5}x = \frac{1}{2} + \frac{3}{4} = \frac{2}{4} + \frac{3}{4} = \frac{5}{4}$. Multiply both sides by $\frac{5}{2}$: $x = \frac{5}{2} \cdot \frac{2}{5}x = \frac{5}{2} \cdot \frac{5}{4} = \frac{25}{8} = 3\frac{1}{8}$ **(b)** $\frac{2}{3}x + \frac{1}{4} = \frac{3}{2}x$. Subtract $\frac{2x}{3}$ from both sides: $\frac{1}{4} = \frac{3}{2}x - \frac{2}{3}x = \frac{9}{6}x - \frac{4}{6}x = \frac{5}{6}x$. Multiply both sides by $\frac{6}{5}$: $x = \frac{6}{5} \cdot \frac{5}{6}x = \frac{6}{5} \cdot \frac{1}{4} = \frac{6}{20} = \frac{3}{10}$ **19.** $3\frac{3}{5}$ hours

21. Gerry's equation is the same as $x \cdot \frac{5}{3} = 25$.
$x = x \cdot 1 = x \cdot \frac{5}{3} \cdot \frac{3}{5} = 25 \cdot \frac{3}{5} = \frac{25 \cdot 3}{5} = 5 \cdot 3 = 15$ stepping-stones. **23.** Answers will vary. The problem really is asking for $\frac{2}{3} - \frac{1}{4}$ **25.** Answers will vary. One-fourth of a *class* is asked for, not $\frac{1}{4}$ of $\frac{2}{3}$. **27.** Answers will vary. This is a correct example, showing that $3\frac{1}{2}$ recipes can be made.

29. **(a)** Let r, g, and b denote the number of red, green, and blue marbles, respectively. The information given is expressed in the equations $\frac{1}{5}(r + g + b) = r, b = \frac{1}{3}g, r = g - 10$. If the first equation is multiplied by 5 and r is subtracted from each side, then $g + b = 4r$. But $g = 3b$, so $3b + b = 4r$, so $b = r$. Thus, $3r = 3b = g = r + 10$, so $2r = 10$ and $r = 5$. Moreover, $g = 3r = 15$ and $b = r = 5$, so there are $5 + 5 + 15 = 25$ marbles in all.
(b) Let the five loops in the diagram each have the same number of marbles, with the *R* loop containing the red marbles. Then the remaining four loops are one *B* and three *G* loops, making it clear that the number of blue marbles in the *B* loop is $\frac{1}{3}$ of the number of green marbles in the three *G* loops. Since there are 10 fewer red than green marbles, each loop contains 5 marbles, and there are 25 marbles in all: 5 red, 5 blue, and 15 green marbles.

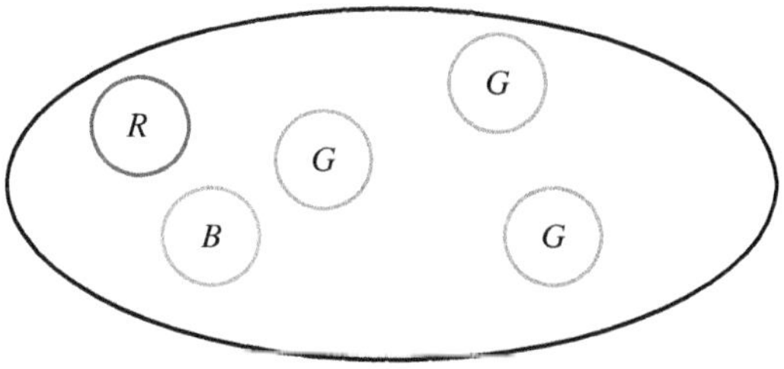

31. Make a cut perpendicular to the cuts already made that is $\frac{1}{4}$ of the width of the cake from a side.

33. **(a)** $\frac{2}{5} \div \frac{3}{4} = \frac{8}{15}$

(b) $\frac{3}{5} \div \frac{5}{6} = \frac{18}{25}$

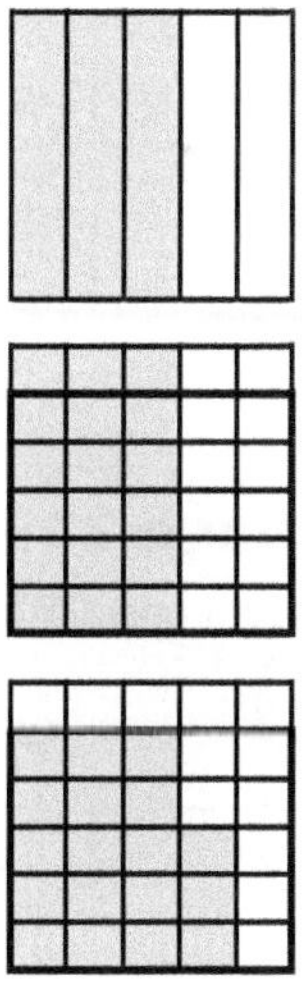

(c) $\frac{7}{8} \div \frac{1}{3} = \frac{21}{8} = 2\frac{5}{8}$

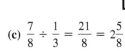

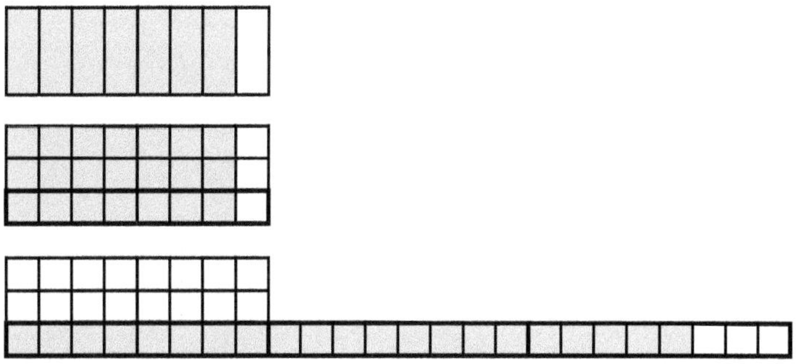

35. **(a)**

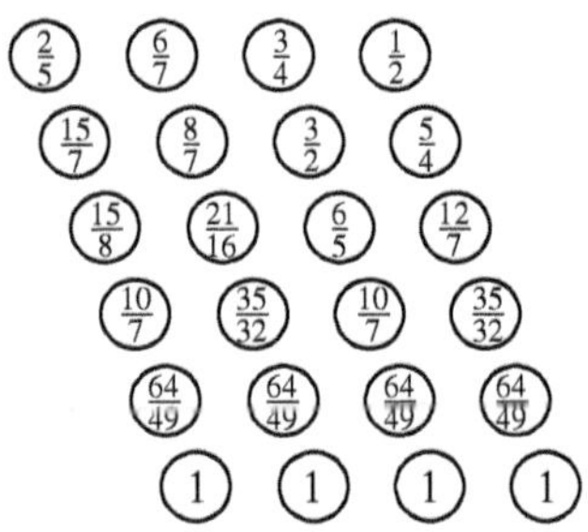

(b)

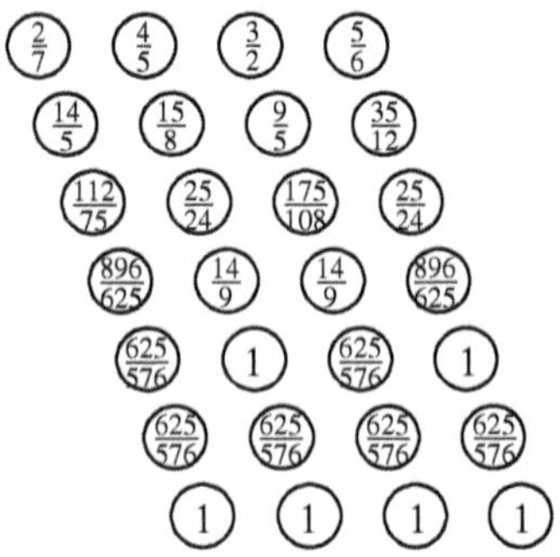

37. 3, 3, 4, 5, 6, 6 **39.** 75 boards will be required, but the board on the end will need to be trimmed.

41. 46 aprons **43.** $2\frac{1}{16}$ cups

45. **(a)** 1 1/2 hours. **(b)** The plants within each of the six loops shown below will take a quarter of an hour, so it will take $6 \cdot \frac{1}{4} = \frac{6}{4} = \frac{3}{2} = 1\frac{1}{2}$ hours to pick all of the plants.

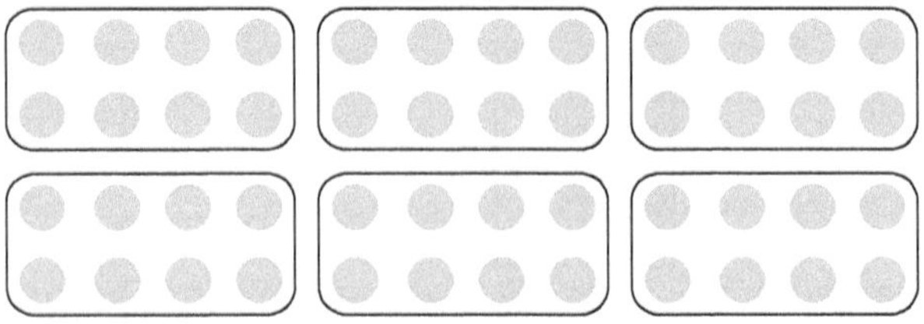

47. B

Problem Set 6.4 (page 313)

1. Commutative and associative properties of addition: $(3 + 2 + 8) + \left(\frac{1}{5} + \frac{2}{5} + \frac{1}{5}\right)$

3. **(a)** $\frac{-4}{5}$

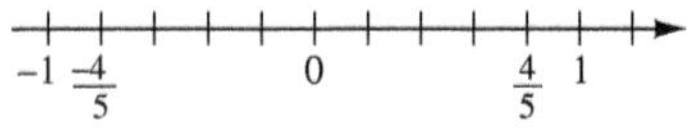

(b) $\frac{3}{2}$

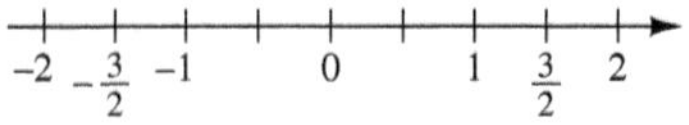

(c) $\frac{8}{3}$

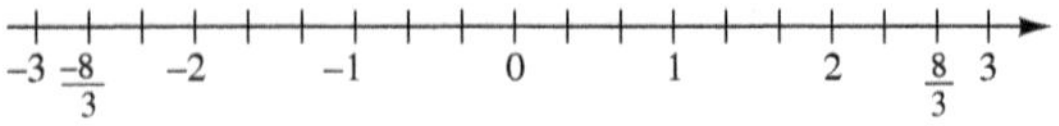

(d) $\frac{-4}{2}$ or -2

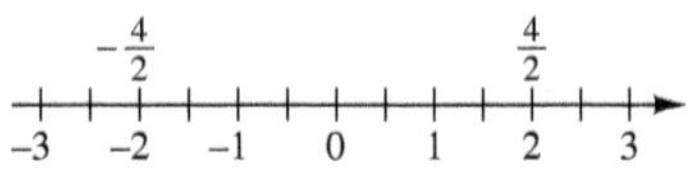

5. **(a)** $\frac{-7}{20}$ **(b)** $\frac{-10}{7}$ **(c)** $\frac{7}{24}$ **7.** **(a)** $\frac{7}{8}$ **(b)** $\frac{-3}{14}$ **(c)** $\frac{1}{2}$

9. **(a)** $\frac{2}{3}$

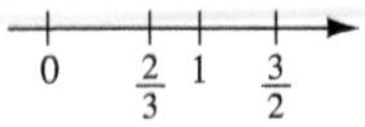

(b) $\frac{-9}{4}$

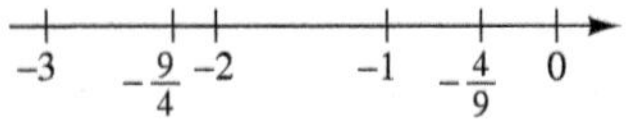

(c) $\frac{-11}{-4}$ or $\frac{11}{4}$

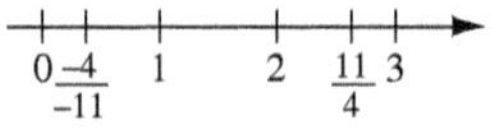

11. **(a)** $\frac{2}{3}$ **(b)** $\frac{1}{3}$ **13.** **(a)** Addition of rational numbers—definition **(b)** Multiplication of rational numbers—definition **(c)** Distributive property of multiplication over addition (for whole numbers) **(d)** Addition of rational numbers—definition **(e)** Multiplication of rational numbers—definition

15. **(a)** $x = \frac{-3}{4}$ **(b)** $x = \frac{1}{8}$ **(c)** $x = \frac{-6}{5}$ **(d)** $x = \frac{-25}{72}$

17. **(a)** $-4 \cdot 4 = -16 < -15 = 5 \cdot (-3)$
(b) $1 \cdot 4 = 4 > -10 = 10 \cdot (-1)$
(c) $(-19) \cdot 3 = -57 > -60 = 60 \cdot (-1)$

19. **(a)** Answers will vary. One answer is $\frac{1}{2}$, since $\frac{4}{9} < \frac{1}{2} < \frac{6}{11}$. **(b)** Answers will vary. One answer is $\frac{2}{19}$, since $\frac{1}{9} = \frac{2}{18} > \frac{2}{19} > \frac{2}{20} = \frac{1}{10}$. **(c)** Answers will vary. Since $\frac{7}{12} = \frac{14}{24} = \frac{28}{48}$ and $\frac{14}{23} = \frac{28}{46}$, one answer is $\frac{28}{47}$. **(d)** Answers will vary. One answer is the mediant: $\frac{36}{145}$. **21.** **(a)** $\frac{1}{4}$ **(b)** $\frac{2}{3}$ **(c)** -1 **(d)** $1\frac{1}{2}$

23. **(a)** $\frac{3}{2}$ **(b)** $\frac{1}{2}$ **(c)** $\frac{3}{5}$ **(d)** $\frac{2}{3}$

25. **(a)** What is his new total acreage? **(b)** How many miles does Janet live from Brian? **(c)** How many square yards of floor space are in the family room? **(d)** How many bottles of ginger ale will Clea fill? **27.** The additional dotted lines partition the unit square into 16 congruent triangles, so each of these has area $\frac{1}{16}$.

Therefore,

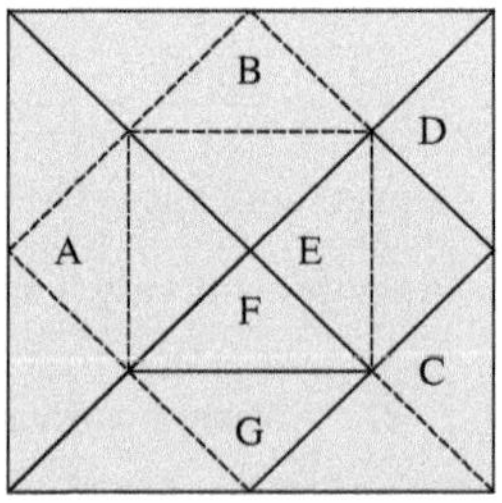

A: $\frac{4}{16} = \frac{1}{4}$, B: $\frac{4}{16} = \frac{1}{4}$, C: $\frac{2}{16} = \frac{1}{8}$, D: $\frac{1}{16}$,
E: $\frac{2}{16} = \frac{1}{8}$, F: $\frac{1}{16}$, G: $\frac{2}{16} = \frac{1}{8}$.

29. It is likely that the student added numerators and added denominators to form the fraction $\frac{4}{8}$, which simplifies to $\frac{1}{2}$. It would be helpful to review the student's understanding of how a common denominator can be obtained by using a pie model with, say, 15 equal sectors. Also, encourage the student to estimate and check for reasonableness of answers. For example, show the student with the pie model that $\frac{3}{5}$ is just a little smaller than $\frac{2}{3}$, so the answer to $\frac{1}{3} + \frac{3}{5}$ can be expected to be very close to $\frac{1}{3} + \frac{2}{3} = 1$. Also, since

$\frac{3}{5}$ is larger than $\frac{1}{2}$, adding $\frac{1}{3}$ to $\frac{3}{5}$ must give an answer larger than $\frac{1}{2}$.

31. $\frac{1386}{97}$ **33.** $\frac{1260}{250} = 5\frac{1}{25}$ **35.** 40 pigeons, 45 holes

37. **(a)** Each column, row, and diagonal adds up to $3x$.

(b)

$\frac{1}{4}$	$\frac{5}{12}$	$\frac{5}{6}$
$\frac{13}{12}$	$\frac{1}{2}$	$-\frac{1}{12}$
$\frac{1}{6}$	$\frac{7}{12}$	$\frac{3}{4}$

(c)

$-\frac{5}{12}$	$\frac{5}{12}$	1
$\frac{7}{4}$	$\frac{1}{3}$	$-\frac{13}{12}$
$-\frac{1}{3}$	$\frac{1}{4}$	$\frac{13}{12}$

$x = \frac{1}{3}, y = \frac{2}{3}, z = \frac{3}{4}$

39. **(a)** If the solar year is $365\frac{1}{4}$ days long, there must be 1 extra day every 4 years. Having leap years on years divisible by 4 accomplishes this. **(b)** $365 + \frac{5}{24} + \frac{48}{24 \cdot 60} + \frac{46}{24 \cdot 60 \cdot 60} = 365\frac{20,926}{24 \cdot 60 \cdot 60}$ days long **(c)** $365\frac{1}{4} - \frac{1}{100} + \frac{1}{400} = 365\frac{97}{400} = 365\frac{20,952}{24 \cdot 60 \cdot 60} = 365\frac{97}{400}$ **(d)** Leap years occur when the year is divisible by 4, but not when the year is also divisible by 100 (except when the year is also divisible by 400). **41.** We will need seven risers of width $5\frac{3''}{4} - \frac{1''}{2} = 5\frac{1''}{4}$. The eighth (top) riser needs to have width $5\frac{3''}{4} - \frac{3''}{4} = 5''$ **43.** Cut the 8-inch side in half and the 10-inch side in thirds to get six $3\frac{1}{3}$-by-4-inch rectangles. **45.** A **47.** D

Chapter 6 Review Exercises (page 322)

1. **(a)** $\frac{2}{4}$ **(b)** $\frac{6}{6}$ **(c)** $\frac{0}{4}$ **(d)** $\frac{5}{3}$ **3.** Answers will vary. Possibilities include $\frac{-6}{8}, \frac{-9}{12}$, and $\frac{-12}{16}$. **5.** **(a)** $\frac{1}{3}$ **(b)** $\frac{4}{33}$ **(c)** $\frac{21}{4}$ **(d)** $\frac{297}{7}$ **7.** **(a)** 36 **(b)** 18

9.

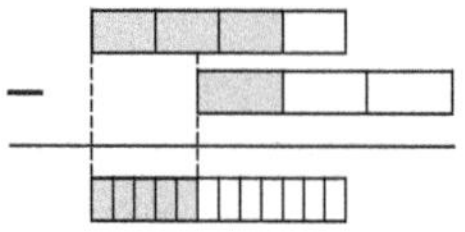

$\frac{3}{4} - \frac{1}{3} = \frac{5}{12}$

11. **(a)** $\frac{1}{8}$ **(b)** $\frac{-7}{9}$ **(c)** 3 **(d)** 1 **13.** $\frac{3}{4}$; this problem corresponds to the sharing (partitive) model of division. **15.** $\frac{57}{10}$ miles $= 5\frac{7}{10}$ miles **17.** **(a)** $x = 2$ **(b)** $x = \frac{-1}{6}$ **(c)** $x = \frac{5}{18}$ **(d)** $x = \frac{9}{16}$ **19.** Write $\frac{5}{6} = \frac{55}{66}$ and $\frac{10}{11} = \frac{60}{66}$. Then $\frac{56}{66}$ and $\frac{57}{66}$ are between $\frac{5}{6}$ and $\frac{10}{11}$. Other answers may be given.

Chapter 7

Problem Set 7.1 (page 337)

1. **(a)** $273.412 = 200 + 70 + 3 + \frac{4}{10} + \frac{1}{100} + \frac{2}{1000}$; $273.412 = 2 \cdot 10^2 + 7 \cdot 10^1 + 3 \cdot 10^0 + 4 \cdot 10^{-1} + 1 \cdot 10^{-2} + 2 \cdot 10^{-3}$ **(b)** $0.000723 = \frac{7}{10,000} + \frac{2}{100,000} + \frac{3}{1,000,000}$; $0.000723 = 7 \cdot 10^{-4} + 2 \cdot 10^{-5} + 3 \cdot 10^{-6}$ **3.** **(a)** 0.21 **(b)** 0.235 **(c)** 0.278 **(d)** 0.302 **5.** 2.45 **7.** **(a)** $\frac{81}{250}$; $250 = 2 \cdot 5^3$ **(b)** $\frac{7}{250}$; $250 = 2 \cdot 5^3$ **9.** **(a)** 0.35 **(b)** 0.4375 **11.** Solution.

(a)

$$\begin{array}{r} 0.875 \\ 8\overline{)7.000} \\ \underline{-6.4} \\ 0.60 \\ \underline{-0.56} \\ 0.040 \\ \underline{-0.040} \\ 0 \end{array}$$

(b)

$$\begin{array}{r} 1.325 \\ 40\overline{)53.000} \\ \underline{-40.0} \\ 13.0 \\ \underline{-12.0} \\ 1.00 \\ \underline{-0.80} \\ 0.200 \\ \underline{-0.200} \\ 0 \end{array}$$

13. **(a)** $\frac{5}{6} = 0.8\overline{3}$ **(b)** $\frac{7}{12} = 0.58\overline{3}$ **15.** **(a)** $\frac{107}{333}$ **(b)** $\frac{1229}{9900}$ **(c)** $\frac{323}{99}$ **17.** **(a)** $\frac{39}{110}$ $x = 0.3\overline{54}$, $10x = 3.\overline{54}$, $1000x = 354.\overline{54}$, $990x = 354.\overline{54} - 3.\overline{54}, = 351$, $x = \frac{351}{990} = \frac{39}{110}$ **(b)** $\frac{313}{60}$ $x = 5.21\overline{6}$, $100x = 521.\overline{6}$, $1000x = 5216.\overline{6}$, $1000x - 100x = 5216.\overline{6} - 521.\overline{6} = 4695$, $x = \frac{4695}{900} = \frac{313}{60}$ **19.** $\frac{1}{999} = 0.\overline{001}$ **21.** Assume that $\sqrt{3}$ is rational. Then there exist integers u and v such that $\sqrt{3} = \frac{u}{v}$ in reduced form. Then $3 = \frac{u^2}{v^2}$ and $3v^2 = u^2$. Hence, 3 is a prime factor of u^2 and thus of u. So $u = 3k$ for some integer k, and $3v^2 = u^2 = (3k)^2 = 9k^2$. This implies that $v^2 = 3k^2$ and that 3 is a factor of v^2 and hence of v. But this means that u and v have a common factor of 3 and $\frac{u}{v}$ is not in reduced form. In view of this contradiction, the assumption that $\sqrt{3}$ is rational must be false. Thus, $\sqrt{3}$ must be irrational. **23.** Answers will vary. It should be possible to find a ruler with fractional markings. **25.** Janeshia failed to realize the place value of the digits 5 and 6 in the number. She wrote 56 as if it were a decimal, instead of a whole number. **27.** **(a)** $\frac{74}{99}$ **(b)** $\frac{7}{9}$ **(c)** $\frac{235}{999}$ **(d)** (i) $\frac{a}{9}$ (ii) $\frac{ab}{99}$ (iii) $\frac{abc}{999}$ **29.** Answers will vary. **(a)** One example is $\sqrt{2} + (3 - \sqrt{2}) = 3$. $\sqrt{2}$ and $(3 - \sqrt{2})$ are both irrational. (See the answer to problem 22.) **(b)** One example is $\sqrt{3} + \sqrt{3} = 2\sqrt{3}$, $\sqrt{3}$ and $2\sqrt{3}$ are both irrational.

31. Answers will vary. For example, $\frac{\sqrt{2}}{2\sqrt{2}} = \frac{1}{2}$, and $\frac{1}{2}$ is rational. **33. (a)** Since $5b^2$ is divisible by 5, its last digit is either 0 or 5. The last digit of a^2 is 0 or 5 only when the last digit of a is 0 or 5. In every case, both a and b are divisible by 5, which contradicts the assumption that the fraction $\frac{a}{b}$ is in lowest terms. **(b)** Irrational **35.** No part of the unit square remains unshaded, and at the nth step an additional shading of area $9/10^n$ is added. **37. (a)** $\frac{1}{11} = 0.\overline{09}$ **(b)** $\frac{15}{22} = 0.6\overline{81}$ **(c)** $\frac{5}{21} = 0.\overline{238095}$ **(d)** $\frac{2}{13} = 0.\overline{153846}$ **39.** C **41.** B

Problem Set 7.2 (page 344)

1. (a) 403.674 **(b)** 339.326 **3. (a)** 174.37 **(b)** 1.7437 **5. (a)** 35, 35.412 **(b)** 27, 27.188 **7. (a)** Estimate $23 + 5 + 1 = 29$. Exact: 28.841 **(b)** Estimate $41.5 - 6.5 + 13 = 35 + 13 = 48$ Exact 48.033 **9.** \$43.51 **11. (a)** 34,796 **(b)** 0.034796 **13. (a)** $a = 4$ **(b)** $b = 6$ **(c)** $c = -3$ **(d)** $d = -6$ **15. (a)** 1.05×10^{-10} **(b)** 5.67×10^{14} **(c)** 1.29×10^{-13} **(d)** 1.20×10^{13} **17. (a)** 0.007, 0.017, $0.01\overline{7}$, 0.027 **(b)** 24.999, 25.312, 25.412, $25.41\overline{2}$ **19.** Answers will vary. **21.** Remind Toni that, in rounding, we are concerned only with the digit immediately to the right of the place to which we are rounding. In this case, the digit is 4 and we round down to 7.2. The basic idea is that this approach guarantees that 7.2447 is closer to 7.2 than to 7.3. Indeed, $|7.2447 - 7.2| = 0.0447$ and $|7.3 - 7.2447| = 0.0553$. **23.** The student is making two fundamental errors. First, no estimated answers are given before commencing with the calculation. Second, the examples show that all of the numbers have been aligned to the right, disregarding the position of the decimal. The position of the decimal apparently is determined by the first addend. The student needs a better understanding of place value in decimal numerals.

25. (b)
21.06   3.21   1.79
24.27   5.00
29.27

(c)
2.374   0.041   5.267
2.415   5.308
7.723

(d)
1   1.414   −0.686
2.414   0.728
3.142

(e) Yes. Part (d) has infinitely many different solutions. **27. (a)** 2.11, 2.321, 2.5531, 2.80841, 3.089251 **(b)** 35.1, −7.02, 1.404, −0.2808, 0.05616 **(c)** 6.01, 3.005, 1.5025, 0.75125, 0.375625

29. (a)
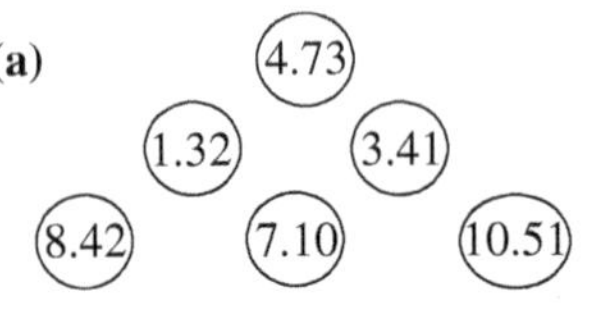

(b)
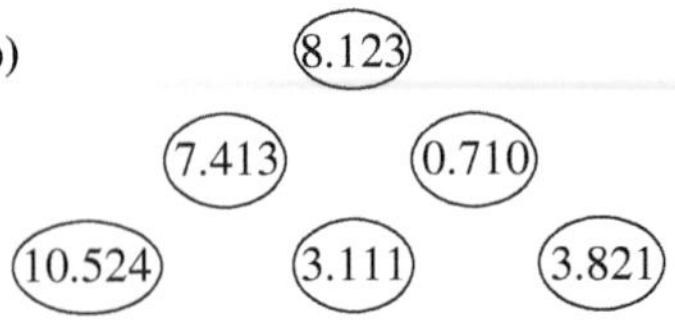

(c)
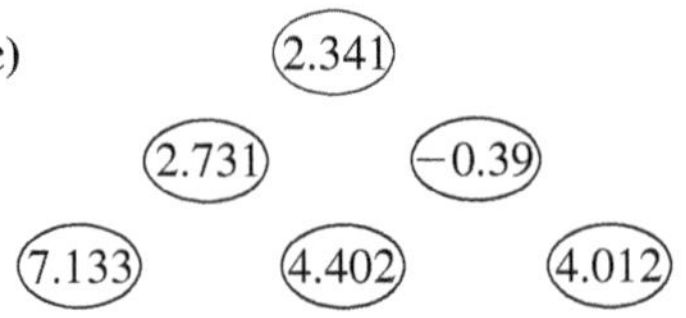

(d)
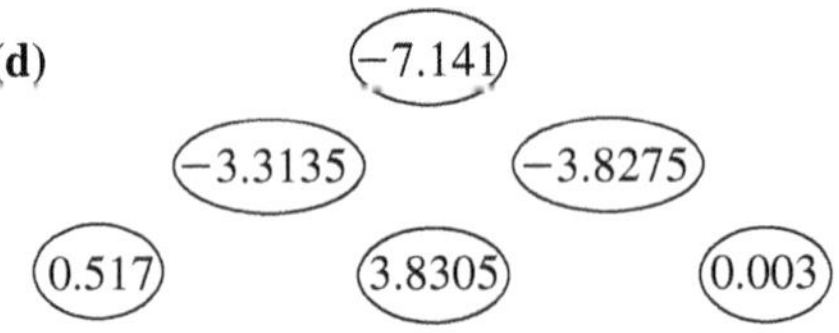

31. \$11.23 **33. (a)** 210.375 in^2 **(b)** 433.125 in^2 **35.** C **37.** B **39.** B **41.** C

Problem Set 7.3 (page 358)

1. (a) $\frac{7}{5}$ **(b)** $\frac{5}{12}$ **(c)** $\frac{7}{12}$ **(d)** $\frac{5}{7}$ **(e)** $\frac{12}{5}$ **(f)** $\frac{12}{7}$ **3. (a)** Yes **(b)** Yes **(c)** No **(d)** Yes **5. (a)** $r = 9$ **(b)** $r = 15$ **(c)** $s = 114.5625$ **7. (a)** \$33.25 **(b)** 5 hr **9.** Since there are 16 ounces in a pound, Brand A is $0.43 \times 16 = 6.88$ per pound, so Brand B is the more expensive. **11.** 89.2857 **13.** 54 miles per hour **15.** The 70¢ difference in the initial cost must be made up by traveling far enough to save 70¢. Since each quarter-mile saves 5¢ and $5 \times 14 = 70$, the break-even distance is 14 quarter-miles, or $3\frac{1}{2}$ miles. **17.** Measure with a metric ruler to determine a, b, c, d, e, f, and g. Then form the ratios $\frac{a}{1}, \frac{b}{2}, \frac{c}{3}, \frac{d}{4}, \frac{e}{5}, \frac{f}{6}$, and $\frac{g}{7}$. Numerically, the ratios are all equal to $\frac{1}{2}$. **19.** Jo correctly multiplied 7×20 to get the weekly usage of 140 gallons. However, she multiplied the weekly rate of 140 by 30 to get the incorrect 4200-gallon monthly rate. The correct answer is $20 \times 30 = 600$ gallons per month. A similar error was made to get a yearly rate of $365 \times 140 = 51{,}100$ gallons when the correct answer is $365 \times 20 = 7300$ gallons per year. **21.** The photo is distorted, since the two rectangles do not have proportional sides: $\frac{4}{6}$ is not the same as $\frac{8}{10}$. Allison needs to first crop her photo, say, by cutting off a half-inch strip from the top and bottom to leave a 4″ by 5″ photo. This can be enlarged by doubling both the vertical and horizontal dimensions to create a distortion-free 8″ by 10″ enlargment. **23. (a)** B **(b)** C **(c)** A **(d)** D **25. (a)** $y = 108$ **(b)** 1 to 4 **(c)** The value of y is multiplied by 4. **27.** 59 to 58 **29. (a)** 32 ounces for 90¢ **(b)** A gallon for \$2.21 **(c)** A 16-ounce box for \$3.85 **31.** 17 to 12 **33.** They all seem to be about equal and to be approximately equal to the golden ratio, $\frac{1 + \sqrt{5}}{2}$. The ratios of measurement approximate the golden ratio, which is the exact ratio. **35.** 13.82 gallons **37.** \$139.93 **39.** $62.9\overline{4}$ ft **41. (a)**

	Cog							
		34	28	23	19	16	13	11
Chainring	24	0.71	0.86	1.04	1.26	1.50	1.85	2.18
	35	1.03	1.25	1.52	1.84	2.19	2.69	3.18
	51	1.50	1.82	2.22	2.68	3.19	3.92	4.64

(b) There are many nearly duplicated ratios. It would be more accurate to say that the bike has 10 speeds. **(c)** Answers will vary. **43.** **(a)** 24 days **(b)** Answers will vary. **45.** B **47.** B

Problem Set 7.4 (page 368)

1. **(a)** 18.75% **(b)** 28% **(c)** 92.5% **(d)** $83.\overline{3}\%$ **3.** **(a)** 19% **(b)** 1.5% **(c)** 215% **(d)** 300% **5.** **(a)** 196 **(b)** 100.8 **(c)** 285.38 **(d)** \$1500 **(e)** 5.4962 **(f)** 8.8725 **7.** **(a)** \$50,400 ÷ 1.05 = \$48,000 **(b)** (12,000 − 11,160) ÷ 12000 = 0.07 = 7% **9.** **(a)** 50% **(b)** 55% (to be 50%, only 5 of the 10 diagonal squares should be blue) **(c)** 30% **(d)** 68% (note there are 8 white squares in each of the four corners)
11. **(a)** $\frac{1}{8} = 0.125 = 12.5\%$ **(b)** $\frac{3}{8} = 0.375 = 37.5\%$ **(c)** $\frac{7}{18} = 0.3888\ldots = 39\%$ **(d)** 50%
13. **(a)** 20 **(b)** 10 **(c)** 8 **(d)** 15 **15.** **(a)** 25% **(b)** $33.\overline{3}\%$ **(c)** 50% **(d)** $66.\overline{6}\%$ **17.** Answers will vary. **19.** Patrick was looking for an additive relationship when he determined his answer, but a look at the proportional (multiplicative relationship) reveals that plant A grew by $\frac{3}{9} \approx 33\%$ compared with plant B, which grew by just $\frac{3}{13} \approx 23\%$. Students need practice in thinking proportionally and using percentage change meaningfully.
21. **(a)** Louis and Arnold prefer to think of saving money instead of how much is left to be paid. **(b)** On the one hand, 35% of \$10 is \$3.50, which is then subtracted from \$10 to leave a sale price of \$6.50. On the other hand, if the price is reduced by 35%, we are still required to pay 65% of \$10, which is \$6.50. **23.** Yoshi has made an error that is often made with percentages. Since $\frac{6}{3} = 2 = 200\%$, it is true that his new pay is 200% of his previous pay. However, it is incorrect to say that the *increase* in his pay is 200%. The increase is \$3, and compared with his previous pay of \$3, the percent increase is $\frac{3}{3} = 1 = 100\%$. Notice that if Yoshi was still paid \$3 a day, he would be making 100% of his previous pay and have a 0% raise.
25. Yes. 5% off parts and 5% off labor results in a discount of 5% on the total job. **27.** **(a)** For the customer, there is no difference. To see why, notice that the purchase price is multiplied by 1.08 to add on the tax and 20% is taken off by multiplying by 0.8. Since 1.08 × 0.8 = 0.8 × 1.08, it makes no difference to the customer if the sales tax is added before or after the coupon discount.
(b) The cashier should redo the calculation. The store should remit only 8% of the *sale* price to the state, not 8% of the regular price.
29. 57.14% **31.** **(a)** 60% **(b)** 0% **33.** \$281.66 **35.** \$17,380 **37.** **(a)** \$3576.80 **(b)** \$15,918.31 **39.** \$11,956.13 **41.** **(a)** 15 **(b)** 11 **(c)** 6 **(d)** 4 **(e)** It seems pretty reasonable. **43.** C **45.** B

Chapter 7 Review Exercises (page 375)

1. **(a)** $2 \cdot 10^2 + 7 \cdot 10^1 + 3 \cdot 10^0 + 4 \cdot 10^{-1} + 2 \cdot 10^{-2} + 5 \cdot 10^{-3}$ **(b)** $3 \cdot 10^{-4} + 5 \cdot 10^{-5} + 4 \cdot 10^{-6}$ **3.** **(a)** 0.48 **(b)** $0.\overline{24}$ **(c)** $0.\overline{63}$
5. **(a)** $\frac{3451}{333}$ **(b)** $\frac{707}{330}$ **7.** **(a)** $\frac{2}{9}$ **(b)** $\frac{4}{11}$ **9.** **(a)** 34.9437 **(b)** 27.999 **(c)** 109.23237 **(d)** 6.0 **11.** $\frac{2}{66}$, 0.33, $\frac{4}{12}$, 0.3334, $\frac{5}{13}$

13. $(3 - \sqrt{2}) + \sqrt{2} = 3$ **15.** **(a)** 928.125 ft^2 **(b)** 8.4375 qts, rounded to 9 qts **17.** 11 to 9 **19.** \$7.88 **21.** $y = \frac{35}{3}$
23. **(a)** 62.5% **(b)** 211.5% **(c)** 1.5% **25.** \$3.53 **27.** 55% **29.** \$3514.98

Chapter 8

Problem Set 8.1 (page 388)

1. **(a)** Constant **(b)** Variable **3.** **(a)** Generalized variables **(b)** Relationship **5.** **(a)** $2x - 3$ **(b)** $\frac{x}{2} + 2$ **(c)** $2x + 6$
7. **(a)** $2q - 10$ **(b)** Assuming that he is successful in his attempt, he will weigh $q - 7/4$ pounds. **(c)** Assuming that he is successful in his attempt, he will weigh $q + \frac{31}{8}$ pounds.
9. B multiplies by 5, C subtracts from 3, D cubes, and E adds 4. Therefore, there are 6 children: A, B, C, D, E, and F. Another correct answer has A and B as in previous sentence, C multiplies by −1 and D adds 3 so this version of a solution has 7 children.
11. **(a)** $y = 8x + 2$ and $y = 3x - 4$ **(b)** $8x + 2 = 3x - 4$ **(c)** $x = -\frac{6}{5}$ **(d)** $y = 8x + 2 = 8 \cdot \left(-\frac{6}{5}\right) + 2 = -\frac{38}{5}$
13. **(a)** Not a function, since element c is associated with more than one element of set B and element d is not associated with any elements of set B. **(b)** A function with range $\{p, q, r\} \subset B$
15. **(a)** Not a function, one point is assigned two points.

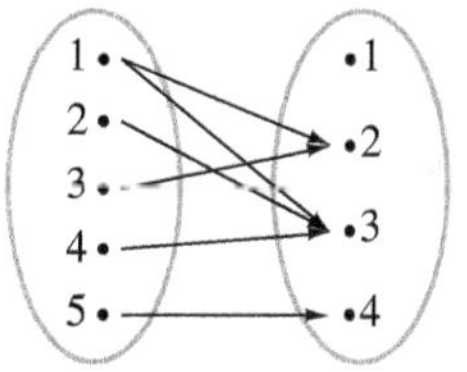

(b) Function, range = {1, 2, 3}

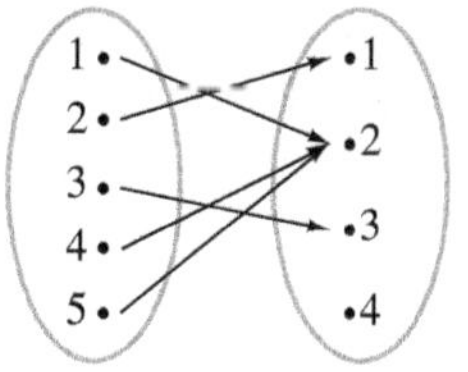

(c) Not a function on A, 3 is not given a point in range.

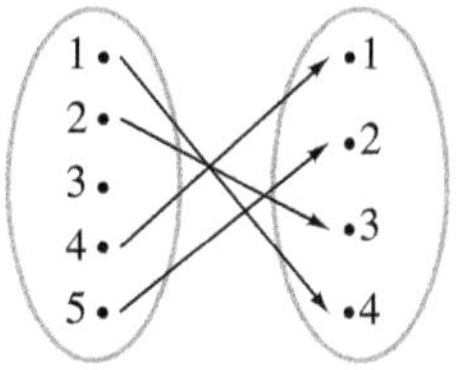

17. **(a)**

x	−2	−1	0	1	2	3
y	−6	−4	−2	0	2	4

(b)

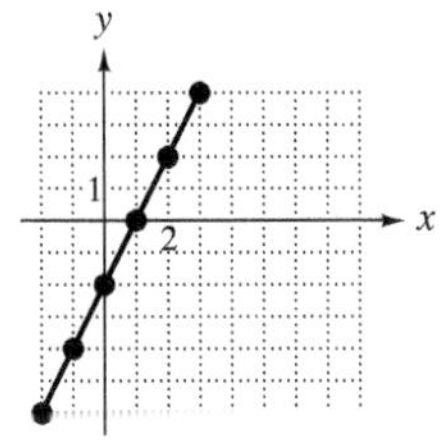

(c) Range = $\{y \mid -6 \le y \le 4\}$ **19. (a)** $h(2) = 3$ **(b)** $h(-2) = 3$ **(c)** $t = 4$ or $t = -4$ **(d)** 52.5824
21. (a) 1:30 **(b)** 1 hr **(c)** 10 mi **(d)** The traffic was heavier coming home from the store, since the return trip took about half an hour and the trip to the store took only about 15 minutes.

23. (a)

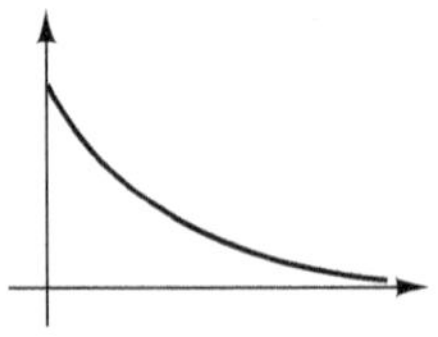

(b)

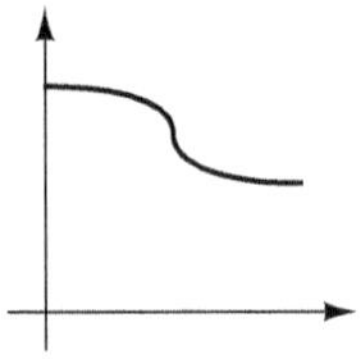

(c)

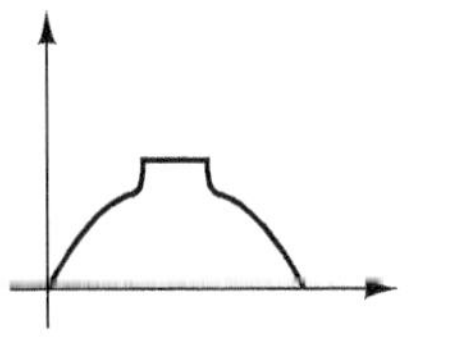

(d)

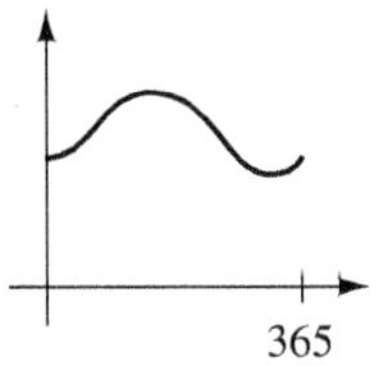

25. (a) Add 5: $y = x + 5$. **(b)** Double and add 2: $y = 2x + 2$ **(c)** Square and add 1: $y = x^2 + 1$ **27.** 6, 8, 10, 12
29. Answers can vary. A useful start is to show that the two expressions do not have the same value when, say, $a = 4$ and $b = 3$. For these values of the variables, $(4 + 3)^2 = 7^2 = 49$, but $4^2 + 3^2 = 16 + 9 = 25$. It can then be shown that $(4 + 3)^2 = (4 + 3)(4 + 3) = 4 \cdot 4 + 4 \cdot 3 + 3 \cdot 4 + 3 \cdot 3 = 4 \cdot 4 + 2 \cdot (4 \cdot 3) + 3 \cdot 3$. Eventually, it can be shown that $(a + b)^2 = (a + b)(a + b) = (a + b)a + (a + b)b = a^2 + ba + ab + b^2 = a^2 + 2ab + b^2$, showing the student how the general formula $(a + b)^2 = a^2 + 2ab + b^2$ is obtained.
31. Answers will vary. **33.** Josiah focused in on the words "9 less" and so thought that he probably needed to subtract. He should be encouraged to plug numbers into the place of the variables to see if his answer was reasonable as a way to remedy this error. The correct answer should be B. **35. (a)** Most students see the first row, and they jump to an answer of "adding the 2 each time." **(b)** Have them model the problem with a function machine. Do it with counters, and have them tell what happened inside the function machine. Then they could try other examples and test their answer.
37. Let p denote the number of pencils and e the number of erasers. Then $15p + 6e = 90$, the total number of cents spent. This is equivalent to $p = \dfrac{30 - 2e}{5} = 6 - \dfrac{2}{5}e$. But p must be a positive whole number, so there are just two choices for e, either 5 or 10. The corresponding values of p are 4 and 2, respectively. Since several more erasers than pencils were purchased, we conclude that Huong bought $e = 10$ erasers and $p = 2$ pencils.
39. (a) All 8 cubes have 3 painted faces. **(b)** The 8 corner cubes have 3 painted faces. Each of the 12 edges of the large cube contains 1 cube with 2 painted faces. Each of the 6 faces of the large cube has 1 cube with 1 painted face. The 1 interior cube has no painted faces. Since $8 + 12 + 6 + 1 = 27 = 3^3$, all of the sugar cubes have been considered. **(c)** The 8 corner cubes have 3 painted faces. Each of the 12 edges of the large cube contains $n - 2$ cubes with 2 painted faces, so $12(n - 2) = 12n - 24$ cubes have 2 painted faces. Each of the 6 faces of the large cube has $(n - 2)^2$ cubes with 1 painted face, so there are $6(n - 2)^2 = 6(n^2 - 4n + 4) = 6n^2 - 24n + 24$ cubes with 1 painted face. There are $(n - 2)^3 = n^3 - 6n^2 + 12n - 8$ interior cubes with no painted faces. **(d)** The expressions, when added, must account for all n^3 sugar cubes. We see that $8 + (12n - 24) + (6n^2 - 24n + 24) + (n^3 - 6n^2 + 12n - 8) = (8 - 8) + (12n - 24n + 12n) + (6n^2 - 6n^2) + n^3 = n^3$.
41. The graphs are straight lines that cross when $t = 4.5$ hr and $y = 45$ mi.

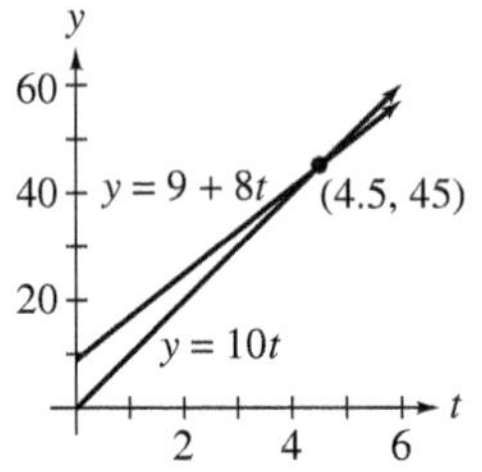

43. For $t \le 40$, Zal's pay is \$18t. For $t > 40$, Zal's pay is \$720 + $(t - 40)$(\$27). **45.** A **47.** Since they each receive the total amount of money and share equally, they each received $\frac{x}{4}$, so A is the correct answer. **49.** A

Problem Set 8.2 (page 406)

1.

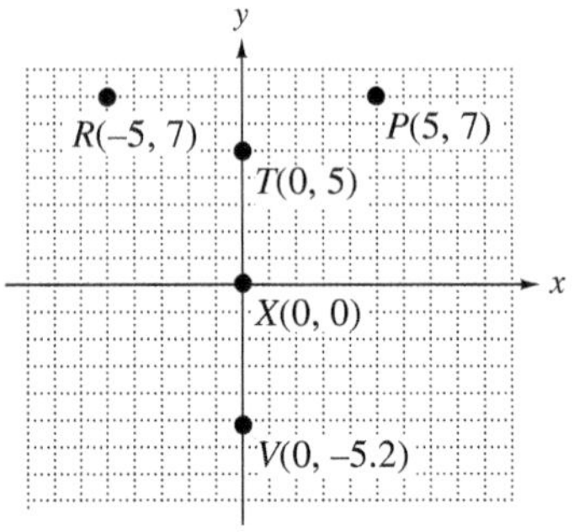

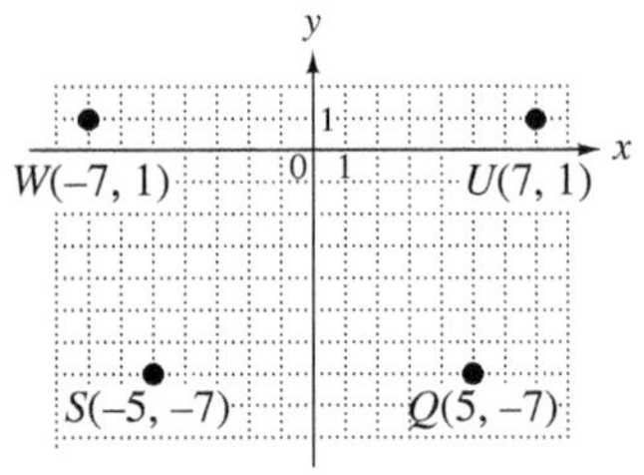

3.

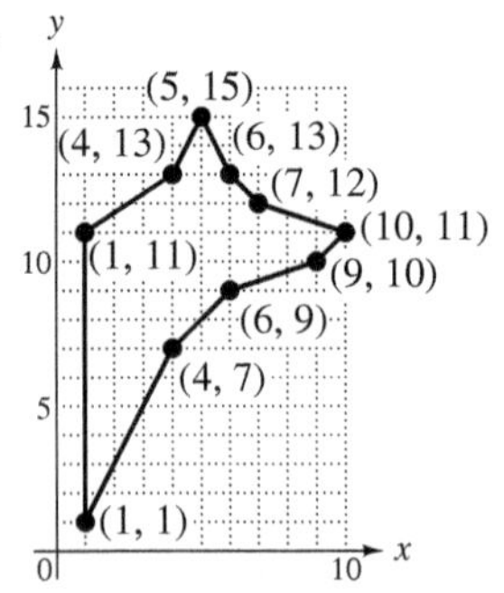

5. (a) $\sqrt{(4 - (-2))^2 + (13 - 5)^2} = \sqrt{36 + 64} = \sqrt{100} = 10$
(b) $\sqrt{(8 - 3)^2 + (8 - (-4))^2} = \sqrt{25 + 144} = \sqrt{169} = 13$
7. (a) 2; upward **(b)** $\dfrac{-11}{0}$ (undefined); vertical **(c)** 2; upward
9. 2 **11.** $a = \dfrac{9}{2}$

13. **(a)**, **(c)**

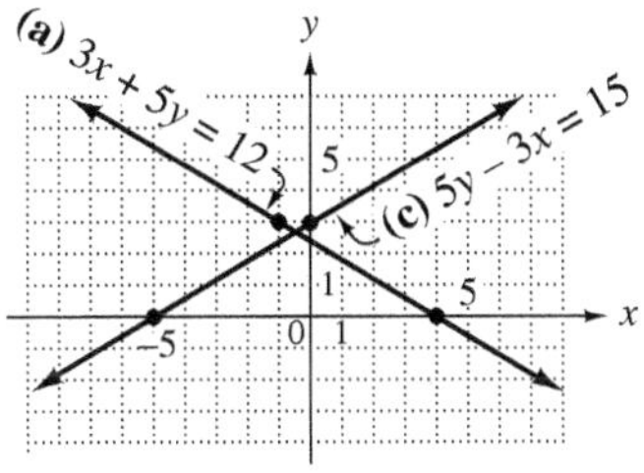

(b), **(d)**

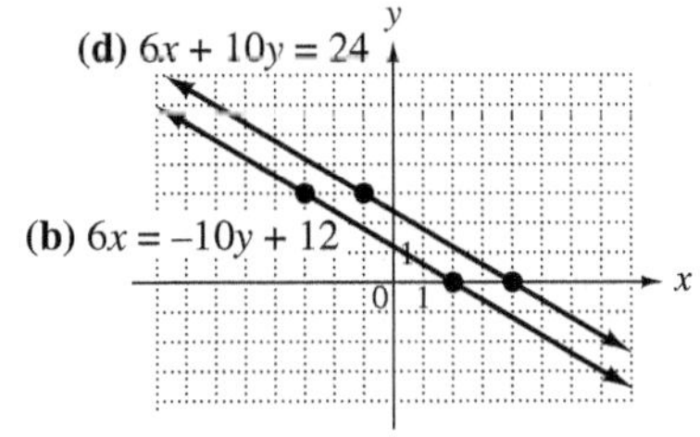

(e) They are the same line.

15. **(a)**

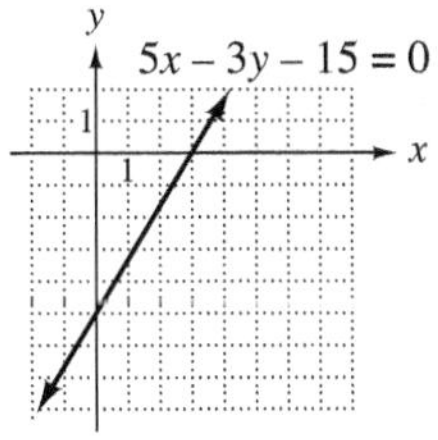

(b) Yes. The function is $y = \frac{5}{3}x - 5$.

(c) Yes. The function is $x = \frac{3}{5}y + 3$.

17. **(a)**

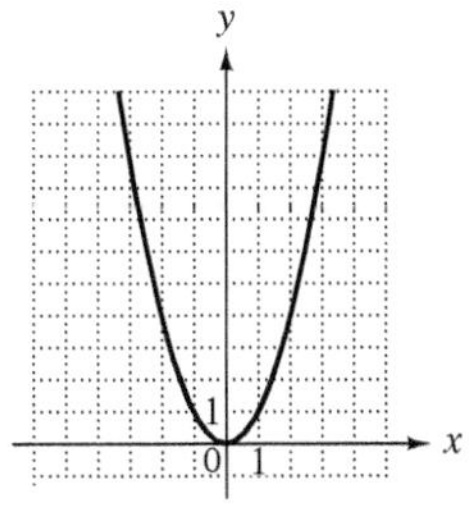

(b)

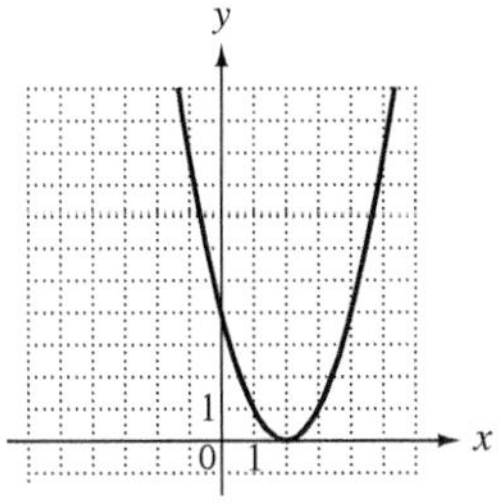

(c)

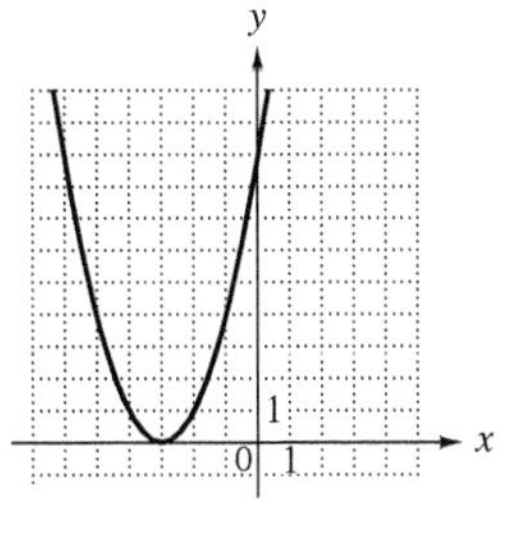

(d) They are horizontal shifts of the same curve.

19. **(a)**

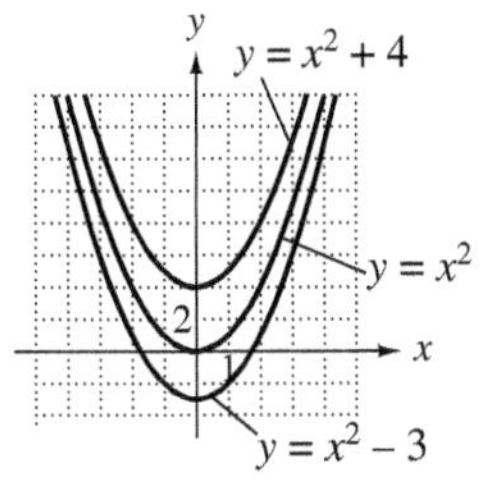

(b) The graphs of $y = x^2 + 4$ and $y = x^2 - 3$ are the same as the graph of $y = x^2$, except that the former is shifted up 4 units and the latter is shifted down 3 units.

21. **(a)** The minimum value of $2x^2 - 4x + 10$ is 8. It occurs when $x = 1$.

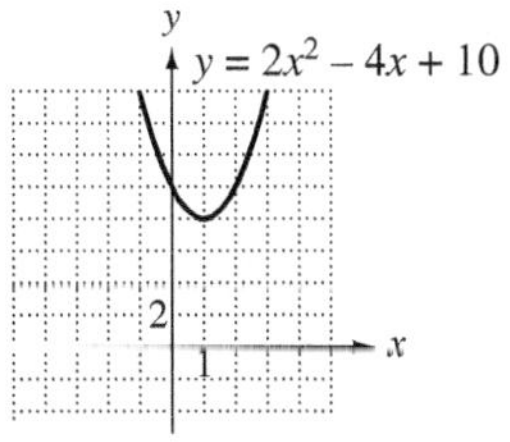

(b) The sketch of the graph shows that $x = 1$ and so the maximum is $2 - 1 + 8 = 9$.

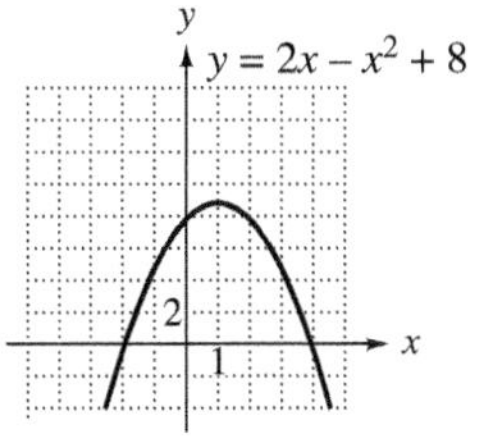

23. Answers will vary. One possibility: Consider squares of size x, and graph the functions $p(x) = 4x$ and $A(x) = x^2$ that respectively give the perimeter and area of the square as functions of x.

25. $r = 10, s = 5$

27.

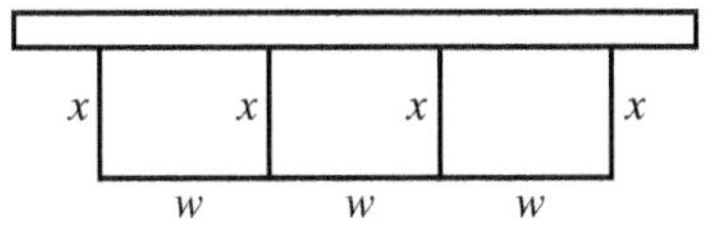

The area is given by $f(x) = (24 - 4x)x$, for $0 \le x \le 6$. The table of values of $f(x)$ for $x = 0, 1, 2, 3, 4, 5$, and 6 suggests that the largest area occurs when $x = 3$, so $w = 4$.

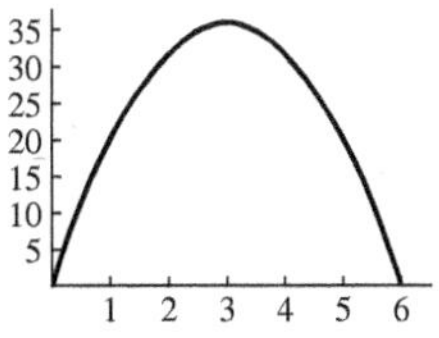

29. Since we are still doubling, but starting with 25 cents, trying a few squares shows that the equation for the value on the nth square is $(.25)\ 2^{n-1}$. The answer again is to take the money on the 64th day. (In fact, the money on a square is larger than a \$1 million on day 23.)

31. **(a)** 30 ft **(b)** about 5% **33.** $m = 2, b = 10$

35. $d = \dfrac{760t}{(60)(60)} \doteq \dfrac{t}{5}$ **37.** $A = (5, 3)$ and $D = (-2, -3)$ so the distance is the square root of $\sqrt{(5 + 2)^2 + (-3-3)^2}$. The distance is $\sqrt{85}$, which is close to 9. Answer is C. **39.** G

Problem Set 8.3 (page 416)

1. **(a)** Answers will vary. For example, (0, 3) and (1.5, 0) **(b)** $m = -2$ **(c)** $y = -2x + 3$. The slope is -2 and the y-intercept is 3. **(d)** Regardless of the point chosen, the slope is still -2. The concept is that the slope is always the same constant.

3. **(a)** $-\dfrac{35}{3}$ **(b)** $\dfrac{24}{5}$

5. (a)

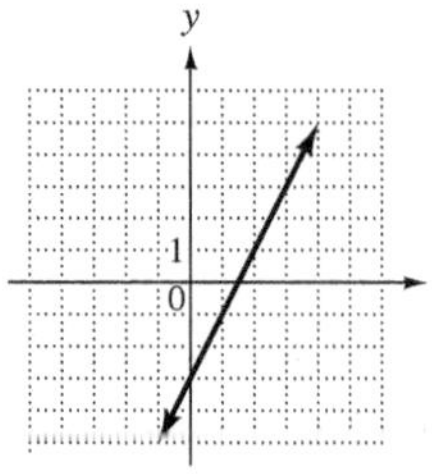

(b)

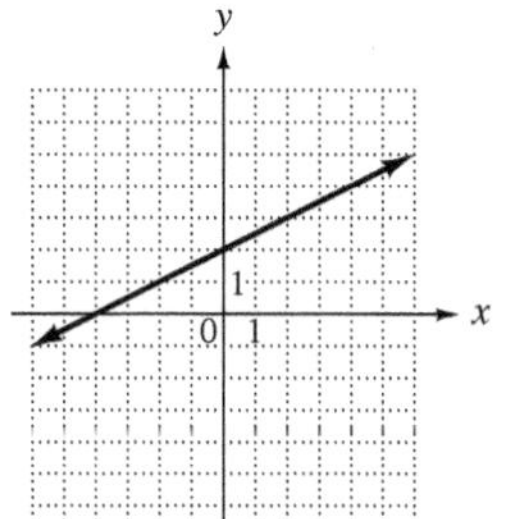

(c)

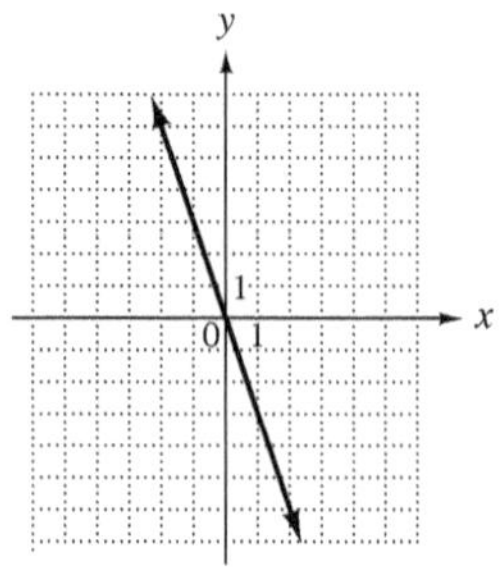

7. (a)

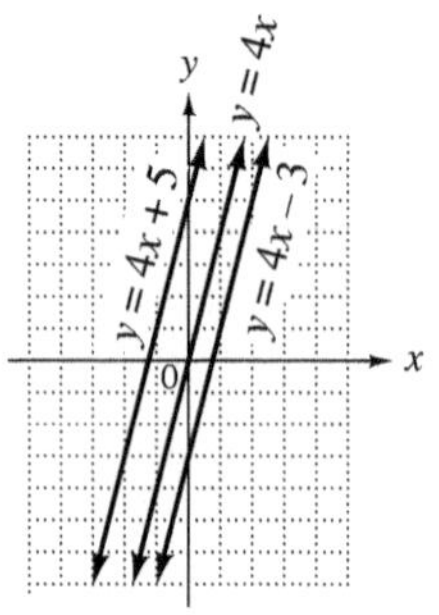

Their slopes are the same, 4.

(b) The lines are parallel and their slopes are all 4.

9. (a) $(RS)^2 = (\sqrt{(7-1)^2 + (10-2)^2})^2 = (\sqrt{36+64})^2 = (\sqrt{100})^2 = 100$

$(RT)^2 = (\sqrt{(5-1)^2 + (-1-2)^2})^2 = (\sqrt{16+9})^2 = (\sqrt{25})^2 = 25.$

$(ST)^2 = (\sqrt{(5-7)^2 + (-1-10)^2})^2 = (\sqrt{4+121})^2 = (\sqrt{125})^2 = 125.$ Since $(RS)^2 + (RT)^2 = 100 + 25 = 125 = (ST)^2$, by the Pythagorean theorem, $\triangle RST$ is a right triangle.

(b)

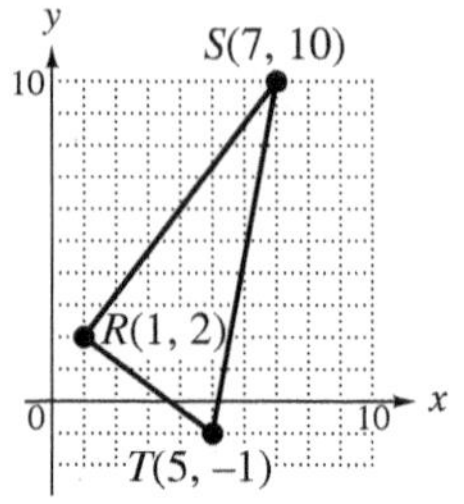

11. (a) $y = \frac{3}{2}x - 7$ **(b)** $y = -\frac{1}{3}x + 3$ **13.** No, since $RS = \sqrt{32}$, which is not equal to the length of the other sides, both of which are $\sqrt{40}$. **15.** The midpoint of $\overline{AB}$ is $\left(\frac{3}{2}, 1\right)$. The slope of $\overline{AB}$ is 6, so the perpendicular of $\overline{AB}$ has slope $-1/6$ and goes through $\left(\frac{3}{2}, 1\right)$. Using the point–slope form gives $y - 1 = -\frac{1}{6}\left(x - \frac{3}{2}\right)$ for the perpendicular bisector.

17. (a) and **(b)** By sketching the circle, is it clear that the answers are (a) $y = -3$ and (b) $y = 3$. **(c)** The lines are parallel but not the same. **19.** Responses vary, but the conclusion is that l is parallel to n. **21.** All are true except (d). **(e)** is true because the two circles can be the same. **23. (a)** A diameter is a line segment that goes through the center and includes two points of the circle. **(b)** A tangent line to a circle is a line that passes through a point on the circle and is perpendicular to the diameter, or radius, at that point.

25. The medians are

$y - b = \frac{b}{a - 2c}(x - a)$, $y = \frac{2b}{2a - c}(x - c)$, and $y = \frac{b}{a + c}x$.

Thus, each pair of medians intersects at the point G. **27. (a)** By trial and error, 10 taxi segments. Alternatively, this is the number of sequences of 3 E(east) and 2 N(north) segments, and there are 10 ways to form such sequences: NNEEE, NENEE, . . . , EEENN.

(b)

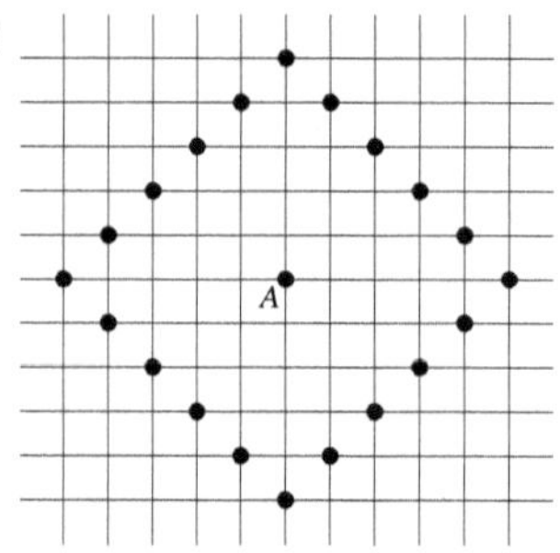

(c) With centered at A, there are six nested taxicab sequres each of which is tilted 45°. The picture of these is below.

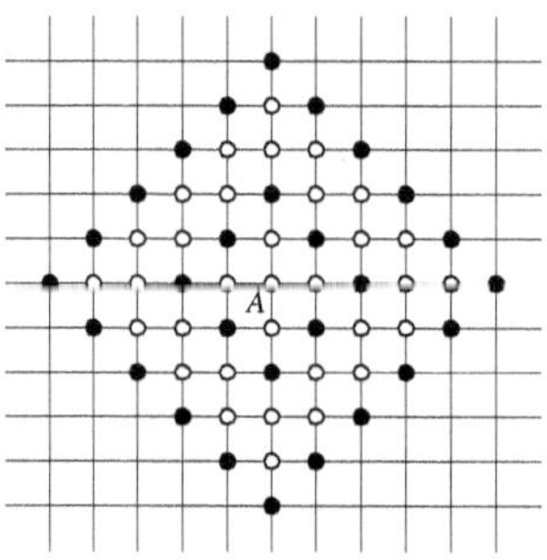

The points a particular distance away form a square oriented with one corner downward. The squares for different distances are nested inside one another. **29.** A

Chapter 8 Review Exercises (page 422)

1. (a) $a + 5$ **(b)** $b < c$ **(c)** $c - b$ **(d)** $\frac{(a + b + c)}{3}$

3. (a) $y = 3x + 2$ **(b)** $y = x(x + 1)$, or $y = x^2 + x$

5. (a)

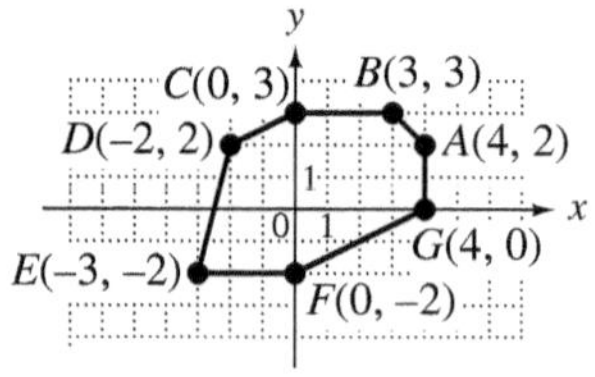

(b) B is in the first quadrant, C is on the positive y-axis, D is in the second quadrant, E is in the third quadrant, and F is on the negative y-axis.

(c) slope $\overline{BC} = \frac{0}{3} = 0$, slope $\overline{CD} = \frac{1}{2}$, slope $\overline{DE} = \frac{4}{1} = 4$, slope $\overline{EF} = \frac{0}{3} = 0$, slope $\overline{FG} = \frac{2}{4} = \frac{1}{2}$, slope $\overline{GA}$ is undefined, $\overline{BC}$ is parallel to $\overline{EF}$, and $\overline{CD}$ is parallel to $\overline{FG}$. **7. (a)** $y = 4 + 2(x - 3)$, using the point–slope form of the equation of a line. In slope–intercept form, the equivalent equation is $y = 4 + 2(x - 3) = 2x - 2$.

(b) $y = -1 + \frac{5 - (-1)}{(-2) - 6}(x - 6) = -1 - \frac{3}{4}(x - 6)$, using the two-point form. In slope–intercept form, the equivalent equation is $y = -\frac{3}{4}x + \frac{7}{2}$. **(c)** $y = 3x - 4$, using the slope–intercept form of

the equation of a line. **9.** The slope of $\overline{ST}$ is $(6 - 0)/(7 - 5) = 3$. The slope of the altitude is $-1/3$, and the altitude must go through $R = (1, 4)$. The answer is the line $y - 4 = -\frac{1}{3}(x - 1)$.
11. **(a)** Values of y are 1, .25, 0.0625, and 0.015625. **(b)** Problem 25 (c) of Example 8.2 is a word problem whose math is seen to be identical with this one once one recognizes that $\$\frac{1}{4} = .25$ cent.

Chapter 9

Problem Set 9.1 (page 440)

1. **(a)** $\overleftrightarrow{AB}$ or $\overleftrightarrow{BA}$ **(b)** $\angle DCE$ or $\angle ECD$
3. **(a)**

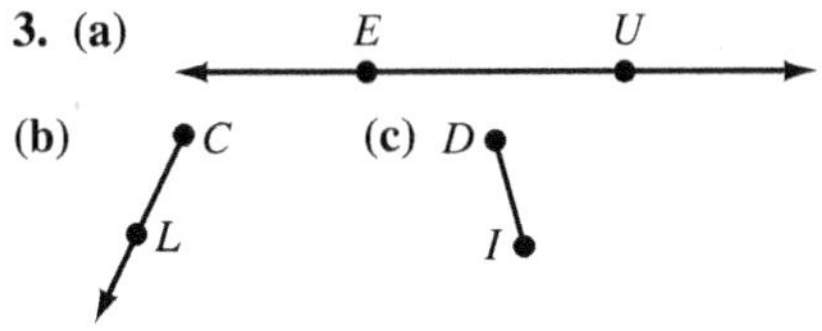

5.

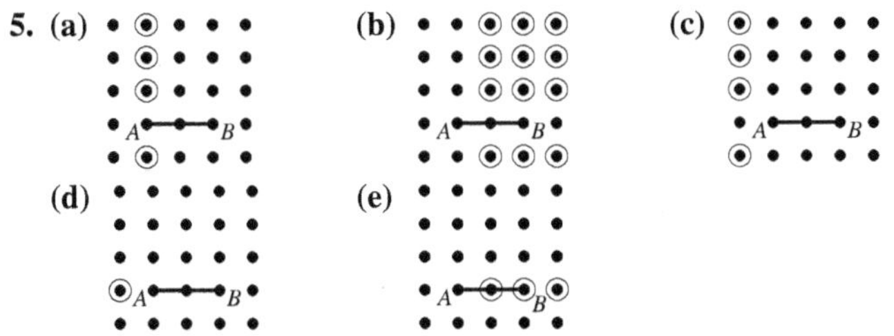

7. P, Q, and R are collinear. **9.** **(a)** $\overleftrightarrow{KL}$ and $\overleftrightarrow{MN}$ are parallel. **(b)** $\overleftrightarrow{LM}$ and $\overleftrightarrow{KN}$ are parallel. **(c)** $\overleftrightarrow{AB}$ is perpendicular to $\overleftrightarrow{KL}$ and $\overleftrightarrow{MN}$. **11.** **(a)** Opposite angles are supplementary: $m(\angle A) + m(\angle C) = 180°$ and $m(\angle B) + m(\angle D) = 180°$. **(b)** In general, $m(\angle P) = m(\angle S)$ and $m(\angle Q) = m(\angle R)$.
13. **(a)** 360° **(b)** 60° **(c)** 12° **(d)** 60° **(e)** 2.5°
15.

Angle	$\angle 1$	$\angle 2$	$\angle 3$	$\angle 4$	$\angle 5$	$\angle 6$
Measure	120°	130°	50°	120°	60°	50°
Angle	$\angle 7$	$\angle 8$	$\angle 9$	$\angle 10$	$\angle 11$	$\angle 12$
Measure	130°	60°	70°	50°	50°	60°

17. **(a)** 40°, since the measures of the interior angles of a triangle add up to 180° **(b)** The measure of an exterior angle is the sum of the measures of its opposite interior angles, so $60° = m(\angle 2) + 20°$. Therefore, $m(\angle 2) = 40°$. **19.** **(a)** $x + x + 30° = 180°$, so $x = 75°$. The interior angles measure 75°, 75°, and 30°. **(b)** $2y + y + 15° = 180°$, so $3y = 165°$, or $y = 55°$. The interior angles measure 55°, 110°, and 15°. **21.** **(a)** No, because an obtuse angle has measure greater than 90° and adding two such measures would exceed 180°, which is the sum of all three interior angle measures for any triangle **(b)** No, because adding together the measures of two right angles and a third angle would exceed 180° **(c)** Each angle must be exactly 60°. **23.** Both $\angle 1$ and $\angle 3$ are complementary to $\angle 2$, so $\angle 1 \cong \angle 3$. **25.** Answers will vary. **27.** Solutions will vary, although asking them to draw two different segments of length one would certainly work. **29.** The confusion at the grade level about parallelism is caused by the segments, as opposed to the lines $\overleftrightarrow{AB}$ and $\overleftrightarrow{CD}$, being examined to see if they intersect. You should talk with Larisa and explain that she should think of lines, rather than the segments, to determine parallelism. You could then ask her whether they intersect "off the paper." **31.** **(a)** Sebastian assumed that all quadrilaterals contain at least one right angle. He also didn't realize that the triangle had a right angle in it (or at least looked like it did.) **(b)** Answers will vary, but Sebastian needs to go back to definitions. **33.** **(a)** 6 lines **(b)** 10 lines **(c)** $C(n, 2) = \frac{n(n-1)}{2}$ lines **35.** The pencil turns through each interior angle of the triangle. Since the pencil faces the opposite direction when it returns to the starting side, it has turned a total of 180°. This demonstrates that the sum of measures of the interior angles of a triangle is 180°.
37. $\triangle BCQ$ is a right triangle, so $m(\angle 2) + m(\angle 3) = 90°$. By the reflection property, $\angle 1 \cong \angle 2$ and $\angle 3 \cong \angle 4$. Therefore, $m(\angle 5) = 180° - m(\angle 1) - m(\angle 2) = 180° - 2m(\angle 2)$ and $m(\angle 6) = 180° - m(\angle 3) - m(\angle 4) = 180° - 2m(\angle 3)$. Adding these equations gives $m(\angle 5) + m(\angle 6) = 360° - 2m(\angle 2) - 2m(\angle 3) = 360° - 2(m(\angle 2) + m(\angle 3)) = 360° - 2\cdot(90°) = 180°$, showing that $\angle 5$ and $\angle 6$ are supplementary. By the alternate-interior-angles theorem, two lines are parallel if, and only if, the interior angles on the *same* side of a transversal are supplementary. Since $\angle 5$ and $\angle 6$ are supplementary, we conclude that $\overleftrightarrow{AB}$ and $\overleftrightarrow{CD}$ are parallel. **39.** $90° - P$
41. Possible answers include the following:

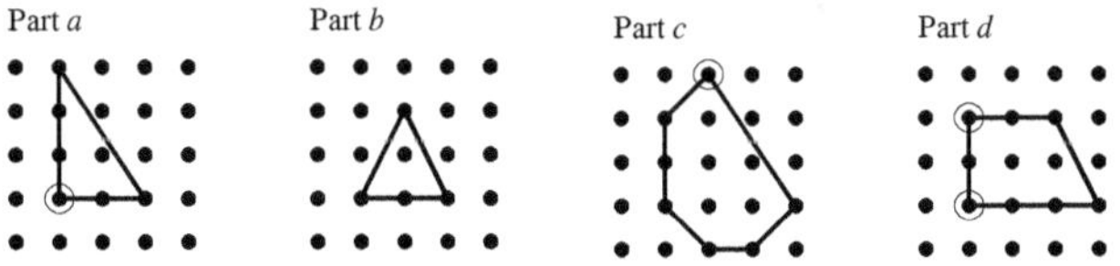

43. Since the lines are parallel, and the angles $m\angle 4 = m\angle 5$ and $m\angle 6 = m\angle 7$, 8 is a right angle so the answer B. **45.** Since $90 + 45 + m\angle X = 180$, the answer is 45° answer B.

Problem Set 9.2 (page 459)

1.

	(a)	(b)	(c)	(d)	(e)	(f)
Simple Curve		✓			✓	✓
Closed Curve		✓	✓	✓		
Polygonal Curve	✓	✓			✓	
Polygon		✓				

3. An example of each figure is given. **(a)**

(b) **(c)** **(d)**

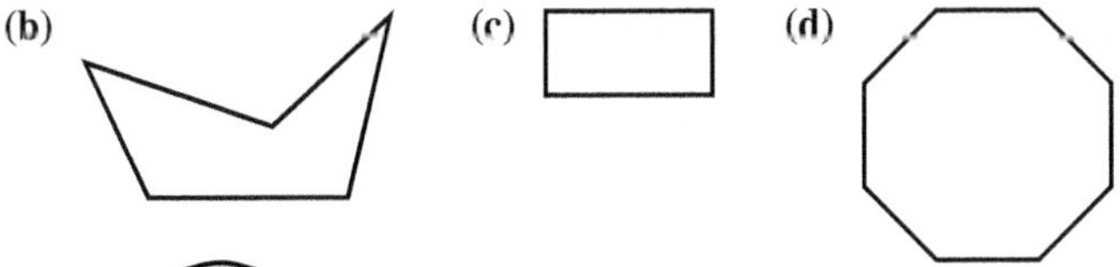

5. **(a)**

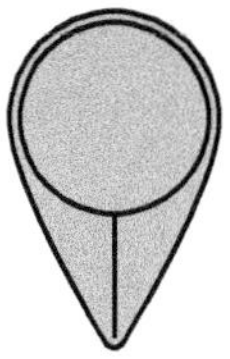

(b)

The shaded region is always convex.

7. **(a)** 6 **(b)** 4 **9.** $2x + 5x + 5x + 5x + 5x + 2x = (6 - 2)(180°)$, or $24x = 720°$, so $x = 30°$. The angles measure 60°, 150°, 150°, 150°, 150°, and 60°. **11.** The sum of the interior angles, since there are 8 sides, is $(8 - 2)180 = 1080$. Thus, $19y = 1080$, so $y = 1080/19$ degrees. The measure of each angle is computed by multiplying this number by the integer in the picture.
13. **(a)** $(6 - 2)(180)° = 720°$ **(b)** $(8 - 2)(180)° = 1080°$

15. (a) 72 **(b)** The average of the interior angles will remain the same. **17. (a)** Yes **(b)** A regular 15-gon **19. (a)** ACEG **(b)** BDFH **(c)** ABDH **(d)** GADE **(e)** EFAD
21.

n	Interior Angle	Exterior Angle	Central Angle
7	$128\frac{4}{7}°$	$51\frac{3}{7}°$	$51\frac{3}{7}°$
8	135°	45°	45°

23. It is helpful to make a sketch to assist in the analysis of the problem.

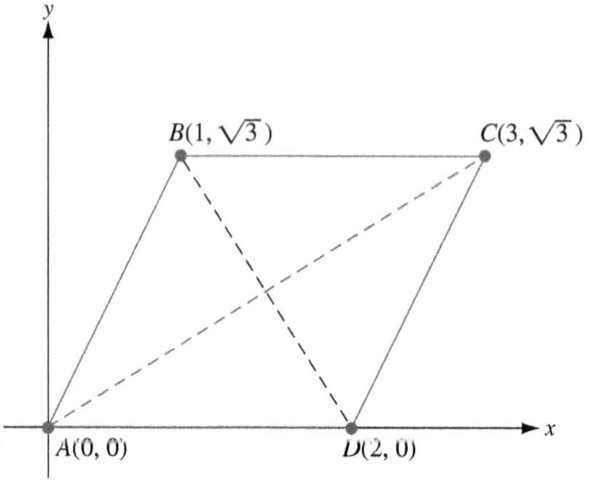

To show that the figure is a rhombus, we must show that the sides have equal lengths. Clearly, $AD = BC = 2$. Moreover, by the distance formula, $AB = \sqrt{(1-0)^2 + (\sqrt{3}-0)^2} = \sqrt{1+3} = 2$ and $DC = \sqrt{(3-2)^2 + (\sqrt{3}-0)^2} = \sqrt{1+3} = 2$.
25. $Q = (a + c, b)$ so the square of the length of $\overline{PR}$ is $(a - c)^2 + b^2$. The square of the length of $\overline{OQ}$ is $(a + c)^2 + b^2$. Thus, the sum of the squares of the length of diagonals is $2(a^2 + b^2 + c^2)$. The sum of the squares of the lengths of the sides of the parallelogram is $(a^2 + b^2) + 2c^2 + a^2 + b^2$, which simplifies to the same expression as the sum of the squares of the lengths of the diagonals. **27. (a)** Equilateral triangle: 60°; square: 90°; regular hexagon: 120° **(b)** Obtuse angles: 120°; acute angles: 60° **(c)** Obtuse angles: 120°; acute angles: 60° **(d)** Obtuse angles: 15°; acute angles: 30°
29. Perhaps JaVonte was thinking that anything larger than a right angle was obtuse and then confused the size of the sides with the size of angles. Talking with JaVonte about the definition of *obtuse angle* and being explicit about the size of the sides having nothing to do with whether a triangle is right, acute, or obtuse would be best.
31. Samuel is right. It is also a square, as well as being a rectangle. A review of the definition will help Samuel to make sure that he masters this subtle point. **33.** Martine is not correct. Here are two different counterexamples: **(a)** a 7-by-24 rectangle and a 15-by-20 rectangle or **(b)** a 3-by-4 rectangle and a 1-by-$\sqrt{24}$ rectangle. It is actually difficult to find integer-valued counterexamples!
35. (a) The boat can drift to any position inside the circle centered at A, where the radius of the circle is the length of the anchor rope. **(b)** The boat is confined to the region in the overlap of circles at points A and C and whose radii are respective lengths of the anchor lines. **37. (a)** $n - 3$ **(b)** $n - 2$ **(c)** Each of the $n - 2$ triangles has 180° as the sum of its three interior angles, so the sum of the interior angles of an n-gon equals $(n - 2) \cdot 180°$.
39. (a) 180°. Tracing the star takes two turns, for a turn angle of 720°. The angle sum at the five points is therefore $5 \cdot 180° - 720° = 180°$. **(b)** 36° **41.** Each new circle creates a new region each time it intersects a previously drawn circle. Since the new circle intersects each of the old circles in two points, this creates the following pattern:

NUMBER OF		NUMBER OF	
Circles	**Regions**	**Circles**	**Regions**
1	2 = 2	6	$22 + 2 \cdot 5 = 32$
2	$2 + 2 \cdot 1 = 4$	7	$32 + 2 \cdot 6 = 44$
3	$4 + 2 \cdot 2 = 8$	8	$44 + 2 \cdot 7 = 58$
4	$8 + 2 \cdot 3 = 14$	9	$58 + 2 \cdot 8 = 74$
5	$14 + 2 \cdot 4 = 22$	10	$74 + 2 \cdot 9 = 92$

43. Proofs will vary. **(a)** Extend side $\overline{AB}$ of the parallelogram, forming the exterior angles that are labeled $\angle a$ and $\angle b$. Then $m(\angle A) + m(\angle a) = 180°$, since $\angle A$ and $\angle a$ are adjacent supplementary angles. But $m(\angle a) = m(\angle B)$ because they are corresponding angles with respect to the parallel sides $\overline{AD}$ and $\overline{BC}$. Therefore, $m(\angle A) + m(\angle B) = 180°$, which shows that $\angle A$ and $\angle B$ are supplementary.

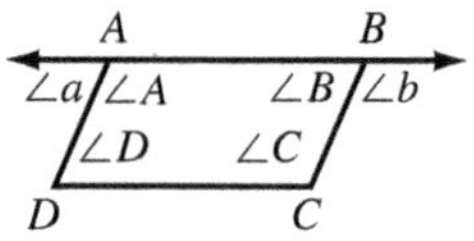

(b) Part (a) shows that any two successive interior angles of a parallelogram are supplementary. Thus, $\angle A$ and $\angle C$ are both supplementary to $\angle B$, so $\angle A \cong \angle C$. Similarly, $\angle B$ and $\angle D$ are both supplementary to $\angle A$, so $\angle B \cong \angle D$. **45.** The square cover could fall through the hole if it were on edge and slightly rotated from its position when in place. **47. (a)** Parallelogram **(b)** Rectangle **(c)** Kite **49.** C, one set of pairs of sides have the same slope **51.** C

Problem Set 9.3 (page 475)

1. (a) Polyhedron **(b)** Not a polyhedron **(c)** Polyhedron
3. (a) Pentagonal prism **(b)** Hexagonal pyramid **(c)** Oblique circular cone **(d)** Triangular prism **5. (a)** 4 **(b)** $\overline{AC}, \overline{AD}, \overline{BC}, \overline{BD}, \overline{CD}, \overline{AB}$ **(c)** A, B, C, D **(d)** $\Delta ABD, \Delta BCD, \Delta ABC, \Delta ACD$
7. (a) Pentagonal right prism **(b)** Oblique hexagonal prism

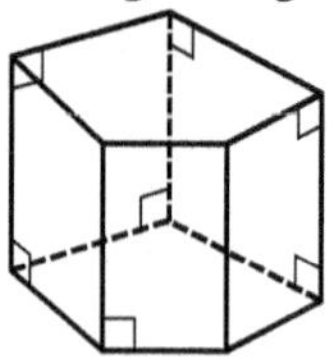

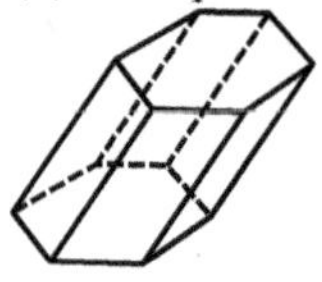

(c) Octahedron

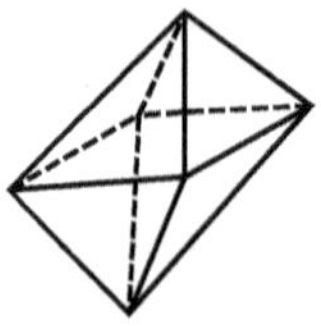

(d) Right circular cone

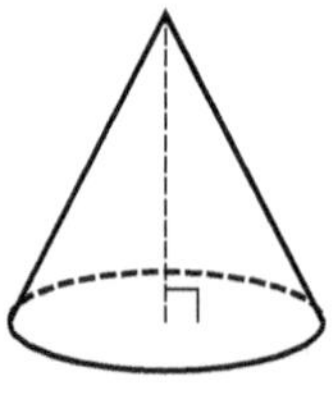

9. (a)

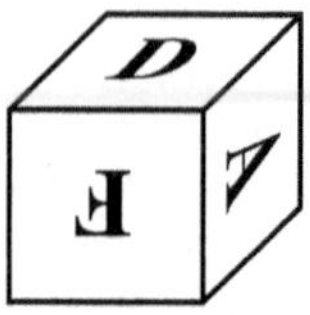

(b)

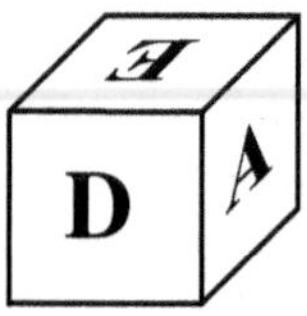

(c)

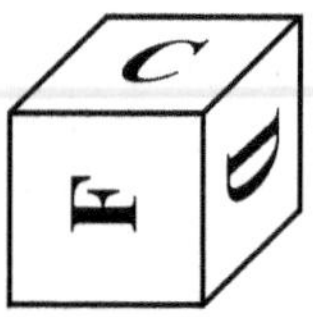

11. Several colorings will work. One such coloring is as follows:

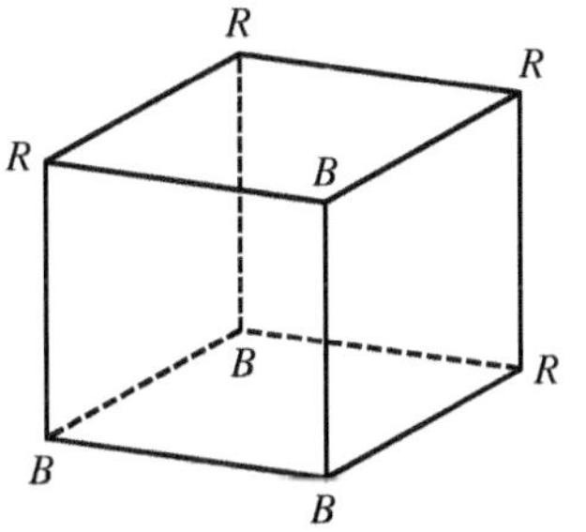

13. **(a)** $F = 7, V = 7$, and $E = 12$, so $V + F = E + 2$, since $7 + 7 = 12 + 2$. **(b)** $F = 10, V = 16$, and $E = 24$, so $V + F = E + 2$, since $16 + 10 = 24 + 2$. **(c)** $F = 20, V = 12$, and $E = 30$, so $V + F = E + 2$, since $12 + 20 = 30 + 2$.
15. **(a)** $F = 10, V = 7$, and $E = 15$, so $V + F = E + 2$, since $7 + 10 = 15 + 2$. **(b)** $F = 40, V = 22$, and $E = 60$, so $V + F = E + 2$, since $22 + 40 = 60 + 2$.
17. Students will see that relaxing one of the hypotheses of a statement may result in the conclusion being wrong. **(a)** $V = 10, F = 12$, and $E = 20$. $V + F = 22 = E + 2$, so Euler's formula holds. **(b)** $V = 12, F = 8$, and $E = 18$. $V + F = 20 = E + 2$, so Euler's formula holds. (c) $V = 12$, $F = 12$, and $E = 24$. $V + F = 24 = E \neq E + 2$. so Euler's formula does not hold. The figure is not a polyhedron, because it has a hole. **19.** Answers will vary. **21.** Responses will vary.
23. **(a)** Only (iv)
(b)

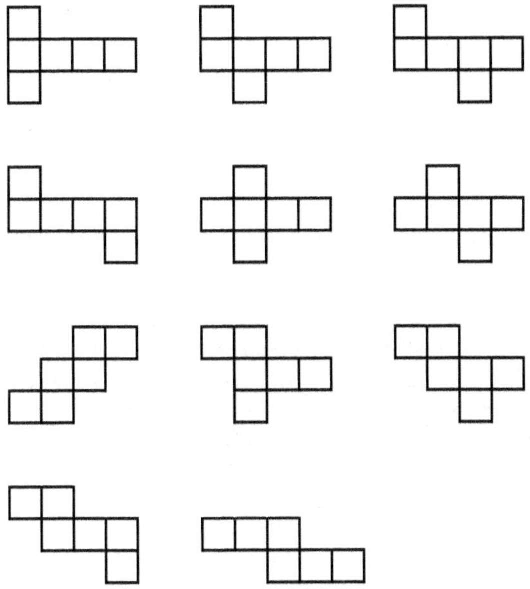

25. **(a)** In her mind, she is possibly picturing the triangle that serves as the "base" or bottom face for the shape. She is then thinking about the triangles that make up the "sides" or other faces and generalizing that, for a geometric solid to have the same number of faces as vertices, it must be the same shape all over the solid figure. When she thinks about the other geometric solids, such as a square pyramid, her mind might think of a square and isn't picturing that four faces or "sides" must come together to form the point or vertex at the top. **(b)** Answers may vary, but Polina's teacher could get out actual geometric solids for her to see and touch. Marking the vertices in some way may help her to see that a shape such as a hexagonal pyramid does have the same number of faces as vertices. Depending on Polina's background, one could point out that, from the Euler formula for polyhedra, the solid must have one less edge then it does vertices (or faces). **27.** **(a)** Tianna is naming each prism by its face rather than its base. The shapes of the bases of a prism are used to name the prism. **(b)** I would guide her by showing that, except for a rectangular prism, there are just two shapes on the prism that are the bases. If she finds just the two shapes that are the same, she can identify the base. The teacher may also offer instruction about the fact that the bases of a prism must be congruent (the same size) and are parallel.

29.

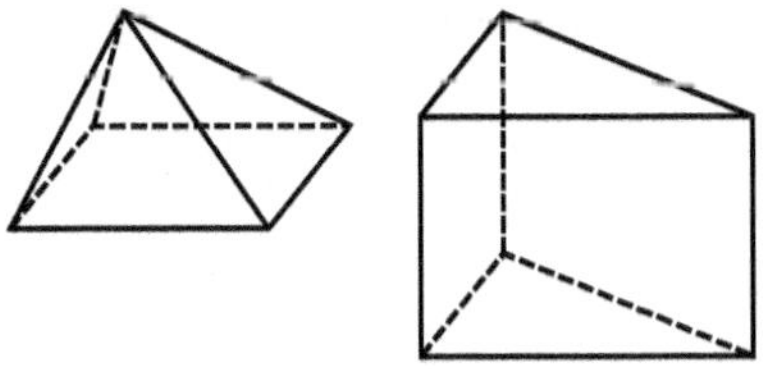

31. **(a)**

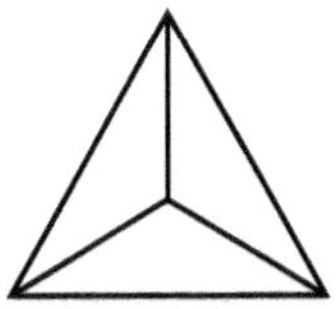

Tetrahedron

(b)

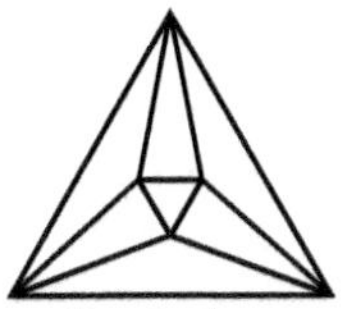

Octahedron

(c)

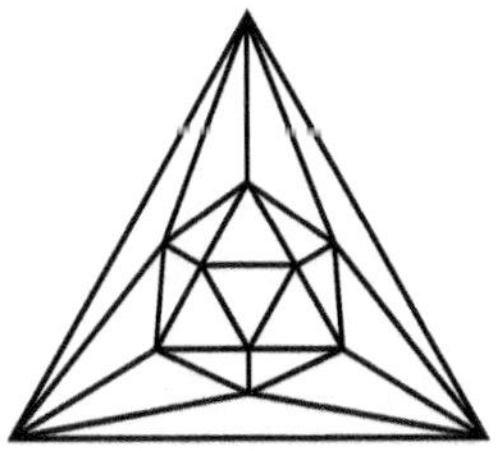

Icosahedron

33. Fire = tetrahedron, earth = cube, air = octahedron, water = icosahedron, universe (world) = dodecahedron **35.** Answers will vary, but include the following: In common with the cube, pyramid A (i) is a polyhedron and (ii) has a square face. In common with the cube, cylinder B (i) is a solid, (ii) has a planar face, and (iii) is a common shape of food containers. In common with the cube, square C (i) has a square face and (ii) has edges meeting at a right angle. **37.** A **39.** **(a)** Common properties include (i) each figure has may polyhedra, (ii) the figures look like they both have a perpendicular to a point on the opposite the base, and (iii) most importantly, the base of each consists of pentagons. **(b)** One is a (right) pyramid and the other is a (right) cylinder and so they are different. Said another way, there is a base for both but one has an opposing base and the other has an opposing apex.

Chapter 9 Review Exercises (page 484)

1. **(a)** $\overleftrightarrow{AC}$ **(b)** $\overline{BD}$ **(c)** AD **(d)** $\angle ABC$ or $\angle CBA$ **(e)** $m(\angle BCD)$, or 90° **(f)** $\overleftrightarrow{DC}$ **3.** **(a)** 143° **(b)** 53° **5.** $x = 45°$, $y = 33°$, $z = 147°$ **7.** **(a)** No, because obtuse angles have a measure greater than 90° and the sum of the three interior angles of a triangle is 180° **(b)** Yes; try angles of 100°, 100°, 100°, and 60°. **(c)** No, because acute angles have a measure less than 90° and the sum of the interior angles of a quadrilateral must be 360° **9.** 360° **11.** Square right prism; triangular pyramid, or tetrahedron; oblique circular cylinder; sphere; hexagonal right prism **13.** **(a)** See Table 9.4. **(b)** $V = 6$, $F = 8$, and $E = 12$. Thus, $V + F = 14$ and $E + 2 = 14$, so Euler's formula holds.

Chapter 10

Problem Set 10.1 (page 502)

1. **(a)** Height, length, thickness, area, diagonal, weight **(b)** Length, diameter of cord **(c)** Height, length, depth **(d)** Length, height, width, weight **3.** **(a)** 6 tga **(b)** 8 tga **(c)** 10 tga **5.** (ii) and (iii), (i), (iv) **7.** **(a)** About $1\frac{1}{3}$ acres **(b)** 77 ares, or 0.77 ha **9.** **(a)** 58.728 kg **(b)** 0.632 g **11.** **(a)** 3.5 kg **(b)** 1200 kg **13.** **(a)** 212 cm **(b)** 50 mm **15.** **(a)** About 28 cm by 22 cm **(b)** About 28 cm by 22 cm **(c)** About 2 cm **(d)** Answers will vary.
17. **(a)** Using Example 10.6, the answer is 20°C. **(b)** Using Example 10.6, 55.4°F. **(c)** It depends on what the children are wearing and if it 0°C or 0°F. **19.** **(a)** Using the two-point form, $y = 1.8x + 32$. **(b)** The variables,x - and y- coordinates as in (a), can be named F instead of x and C instead of y as variables. Compare the equation on p. 501 or Example 10.5. **21.** **(a)** 19 cm **(b)** Answers will vary. A size 8 men's shoe is about 28 cm long. **(c)** Answers will vary. A person 5 feet 7 inches tall is approximately 170 cm tall. **(d)** Answers will vary. The typical door in a house is about 203.5 cm tall and about 75.5 cm wide. **(e)** 91.4 cm **(f)** 110 m, including the end zones **23.** **(a)** Majandra incorrectly associated the measurement of "feet" with the measurement of people only because that is the context in which most children hear that particular measurement discussed. She seems to have a good grasp of the largest and smallest lengths but is having trouble distinguishing the middle amounts. The measurement of yards is a common problem for children in upper elementary grades.
(b) Answers may vary, but one approach would be to take out a ruler and a yardstick and let her physically mark the amounts of 6 feet and 6 yards. This will guide her towards seeing the most reasonable table length. It would be important, however, to make sure that she has some sort of personal measurement reference for the future, since she might not always have access to a ruler or yardstick and will still need to estimate length. The teacher could use arm length, distance between objects, etc., to give Majandra a way to remember these measurement amounts. **25.** **(a)** Corina solved 18 cm = 180 mm and 4 m = 400 cm. **(b)** Corina incorrectly solved the last two. In fact, 40 mm = 4 cm and 3000 mm = 3 m. **(c)** Answers may vary, however, Corina's mistake is common because students are typically taught conversions going from large metric measurements to smaller metric measurements first. They quickly get into a bad habit of multiplying and just "adding zeros" to the end. When metric conversions going from small to large are given, many students just apply the same process of adding zeros and do not give much regard to their value or meaning. Corina needs to be shown with manipulatives what these amounts mean physically. The teacher can then guide her towards the correct answers.
27. **(a)** 8, 16, 32 **(b)** There will be room for one more mouthful.
29.

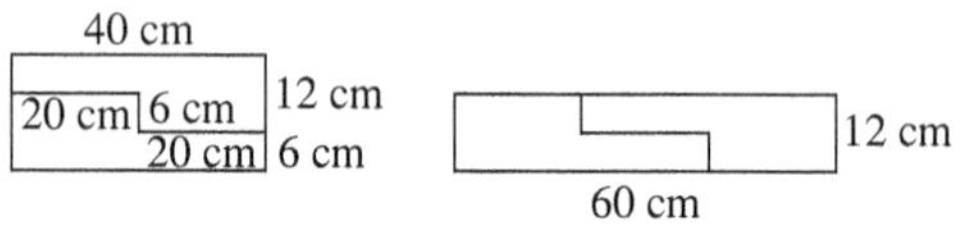

31. $\left(\frac{100\text{ km}}{9\text{ L}}\right)\left(\frac{3.7854\text{ L}}{1\text{ gal}}\right)\left(\frac{1\text{ mi}}{1.6\text{ km}}\right) \doteq 26.3\text{ mi/gal}$

33. **(a)** $(5\text{ gal})\left(\frac{4\text{ qt}}{1\text{ gal}}\right)\left(\frac{32\text{ oz}}{1\text{ qt}}\right) = 640\text{ oz.}$

Since $\frac{640}{80} = 8$, add 8 liquid ounces of concentrate.

(b) Add 80×65 mL = 5200 mL = 5.2 L of water.
35. **(a)** 4917 liters were added. **(b)** 20,089 liters should have been added. **37.** The post is 60 in = 5 ft long, so, at \$2.50 per foot, it costs \$12.50. The tray frame requires four strips, each 2 ft long, or 8 ft altogether. At \$1 per foot, the strips cost \$8.00. The tray bottom is \$6.00. The total cost is \$12.50 + \$8.00 + \$6.00 = \$26.50. **39.** B

Problem Set 10.2 (page 517)

1. About 50 cm^2 **3.** The dodecagon and square have the same area, equal to the sum of the areas of the same subregions in their dissections.

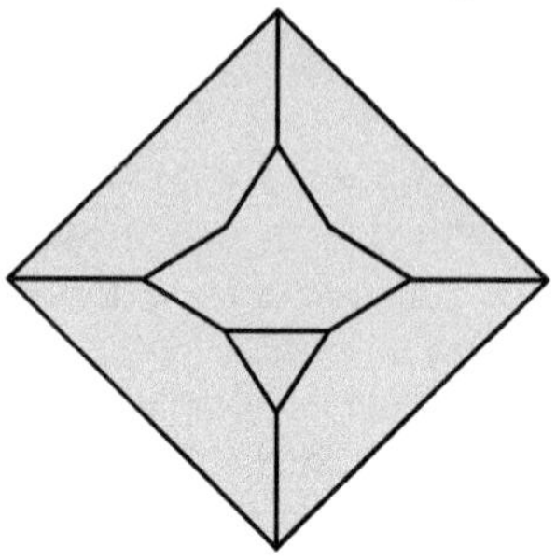

5. **(a)** 12 units **(b)** 8 units **(c)** 4 units

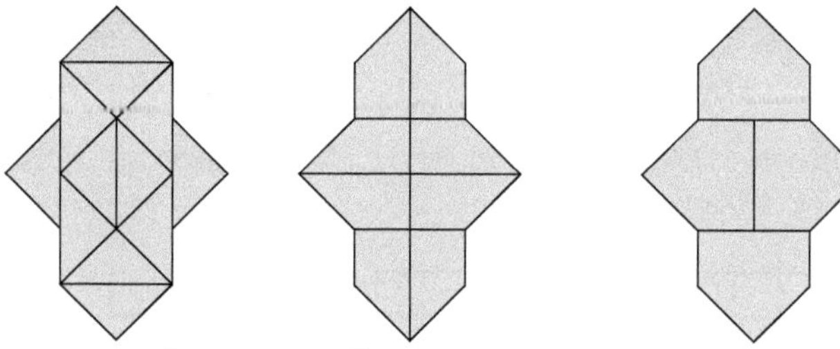

7. **(a)** .007 ft^2 **(b)** 84 km^2
9. $\frac{1}{2} \times 12\text{ m} \times 9\text{ cm} - \frac{1}{2} \times 4\text{ m} \times 3\text{ m} = 48\text{ m}^2$
11. **(a)** If x is the smaller side, then $2(x + (x + 1)) = 20$, so $x = 4.5$ cm. The rectangle is 4.5 cm by 5.5 cm. **(b)** If x is the smaller side, then $(x(x + 1)) = 6$. Factoring, $x^2 + x - 6 = (x + 3)(x - 2) = 0$. Thus, $x = 2$ (since x has to be positive), so the rectangle is 2 cm by 3 cm. **13.** **(a)** 2,369,664 ft^2 **(b)** 470 ha = 4,700,000 m^2 **(c)** 580 ha, 5.8 km^2 **15.** **(a)** 1 cm by 24 cm; 2 cm by 12 cm; 3 cm by 8 cm; 4 cm by 6 cm. The dimensions can also be given in opposite order. **(b)** The 4-cm by 6-cm rectangle has the smallest perimeter, 20 cm. **(c)** The 1-cm by 24-cm rectangle has the largest perimeter, 50 cm.
17. **(a)** 1664.6 m^2, 198.3 m **(b)** 1598.7 in^2, 219 in **(c)** 7.56 cm^2, 16.9 cm **19.** **(a)** The area of each of the triangles is 2 cm^2. The perimeters are computed by use of the Pythagorean theorem to obtain the length of the hypotenuse. **(i)** Hypotenuse is $\sqrt{8}$ so perimeter is 4 cm + $2\sqrt{2}$ cm. **(ii)** Hypotenuse is $\sqrt{17}$ so perimeter is $(5 + \sqrt{17})$ cm. **(iii)** Hypotenuse is $\sqrt{64.25}$ so perimeter is $(8.5 + \sqrt{64.25})$ cm. **(iv)** Hypotenuse is $\sqrt{256.0625}$ so perimeter is $(16.25 + \sqrt{256.0625})$ cm. **(b)** The area of each triangle is the same, but the perimeter is getting longer each time. The triangles are becoming narrower, but longer, depending on which way you view them. **21.** **(a)** 12 square units **(b)** 24 square units **(c)** 4 units **23.** 8 square units **25.** The lanes are staggered π meters apart. **27.** **(a)** Suppose that the radius of the circle is x. Then the area of the circle is πx^2. Since the area of the square is $(2x)^2$, their ratio is $\pi x^2/4x^2 = \pi/4$. Thus, the percentage is about 78.5 percent. **(b)** The perimeter is the length of the curve that bounds the region formed by removing the area inside the circle from the square. Thus, the boundary of the figure consists of two parts: that of the

circle and that of the square. It is the boundary of the very light blue region. Since the radius of the circle is 6 cm, each side of the square is 12 cm. Thus, the answer is $4(12) + 2\pi 6 = 48 + 12\pi$.
29. **(a)** $P = 2(l + w)$. The area is $4 = lw$, so $l = \frac{4}{w}$. Thus, $P = 2\left(\frac{4}{w} + w\right)$. **(b)** Yes, the perimeter will get infinitely large as either w gets closer to zero or w gets very large itself. This Shows that long, skinny rectangles can all have the same area (of 4 cm^2), which is a different type of result than the previous problem.
31. **(a)** Area of the small triangle $\doteq$ 1.7 cm^2, area of the large triangle $\doteq$ 5.1 cm^2, ratio $\doteq$ 3 **(b)** Area of the square $\doteq$ 4 cm^2, area of the arch $\doteq$ 12 cm^2, ratio = 3 **(c)** Area of the pentagon $\doteq$ 6.9 cm^2, area of the arch $\doteq$ 20.7 cm^2, ratio $\doteq$ 3 **(d)** Exactly 3 **(e)** If a polygonal arch is generated by rolling a regular n-gon along a line, the ratio of the area under the polygonal arch to the area of the generating n-gon is exactly 3.
33. **(a)** Starting with the two triangles on top of each other, rotate the top one 180° about the midpoint of any side to form a parallelogram. **(b)** $2 \cdot \text{area (triangle)} = \text{area (parallelogram)} = bh$
35. **(a)** Put two copies of the trapezoid together as shown:

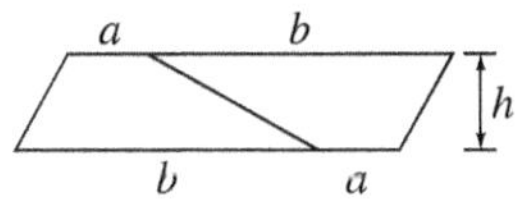

(b) Dissect the trapezoid as shown:

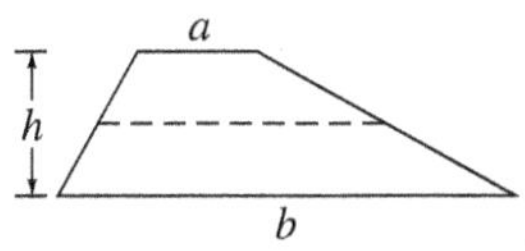

Rotate the bottom half 180° and place as shown:

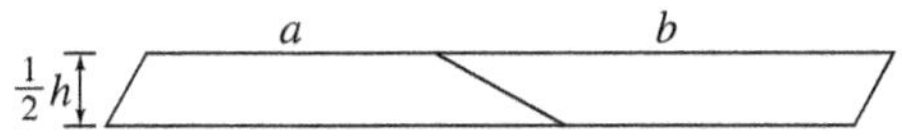

37. Answers will vary in (b) and (c). **(a)** 4 m **(d)** Yes
39. **(a)** 36 inches **(b)** Taneisha received her answer of 24 inches by adding the dimensions of the box: 11 + 7 + 6. **(c)** Answers may vary, but oftentimes when students are presented with a problem that they know how to calculate, such as perimeter, but that is written in a different context (in this case, three dimensions instead of two are given), they don't know what to do with the third number (depth), so they just use it somewhere. Extra numbers that are not needed in problems often confuse students, and they can lose focus on what the task was asking them to do. However, those extra numbers shouldn't be deleted from problems because that is unrealistic. Students should instead be instructed how to highlight what the question is asking them to do and then think about how to apply what they know. Taneisha could be shown a real box and asked to touch the perimeter of the bottom and then apply the given numbers to figure out an answer.
41. March 14 is 3/14 and the first six digits of π are 3.14159. The Math Department can enjoy pi(e). **43.** $P = 6b, A = \frac{3\sqrt{3}}{2}b^2$
45. **(a)** Dissect the vase as shown.

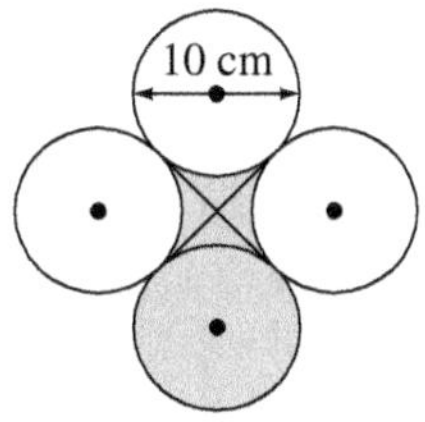

Rearrange the four smaller pieces around the circle as shown.

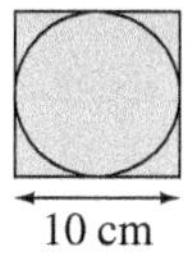

This forms a square with sides of length 10 cm. **(b)** (10 cm) $\cdot$ (10 cm) = 100 cm^2 **(c)** Dissect the vase as shown

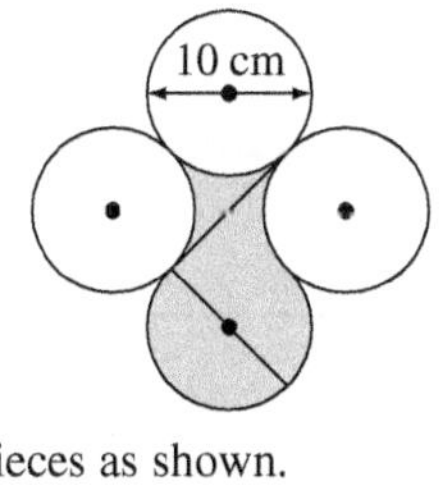

Rearrange the three pieces as shown.

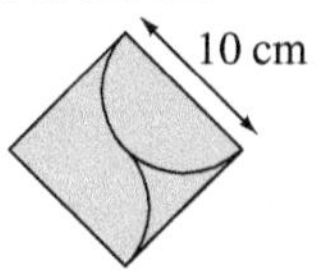

47. Draw a line through point B parallel to $\overline{AC}$. Let D be the intersection of this line with the boundary line containing C. Since the triangles ABC and ADC have the same base AC and the same height, they have the same area. Thus, $\overline{AD}$ is a suitable new boundary line between the two fields. **49.** The areas of the rectangular portions of the sidewalk total 2400 ft^2. The pieces formed with circular areas have a total turning of 360°, so, when placed together, they form a circle with radius 8 ft and area $\pi(8 \text{ ft})^2 = 64\pi \text{ ft}^2$. Total area is $(2400 + 64\pi) \text{ ft}^2$.
51. Since the circle has radius $\frac{9}{2}$, its area is $\pi\left(\frac{9}{2}\right)^2$. This area is approximated by the square of area 8^2, so $\pi\left(\frac{9}{2}\right)^2 \doteq 8^2$. Thus, $\pi \doteq \frac{8^2}{\left(\frac{9}{2}\right)^2} \doteq \left(\frac{16}{9}\right)^2$. **53.** Draw $\overline{AP}$, $\overline{BP}$, and $\overline{CP}$. Then area $(\triangle ABC) = \frac{1}{2}sh = \text{area}(\triangle ABP) + \text{area }(\triangle BPC) + \text{area }(\triangle CPA) = \frac{1}{2}sx + \frac{1}{2}sz + \frac{1}{2}sy = \frac{1}{2}s(x + y + z)$. Therefore, $\frac{1}{2}sh = \frac{1}{2}s(x + y + z)$, and $h = x + y + z$.
Alternative, visual proof:

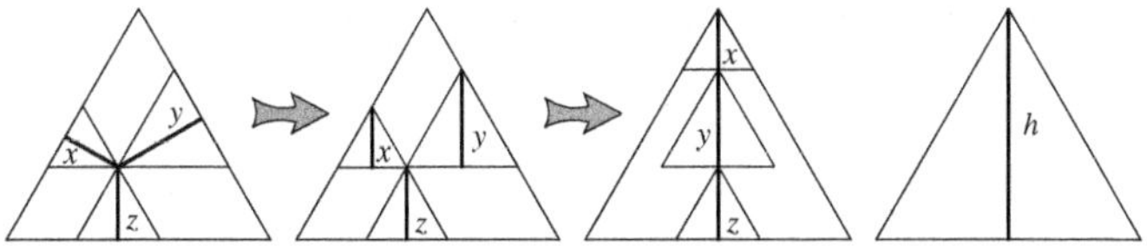

55. The lawn has an area 75 ft × 125 ft = 9375 ft^2. Since 21 in $= \frac{7}{4}$ ft, the lawn area is equivalent to a rectangle 21 in wide and $9375 \div \frac{7}{4} = 5357.14\ldots$. That is, Kelly will walk about 5357 feet, a bit over a mile (1 mile = 5280 ft). **57.** Consider the carpet as a 6-ft by 4-ft rectangle with semicircular ends of radius 2. Then the carpet's area is $(6 \text{ ft})(4 \text{ ft}) + \pi(2 \text{ ft})^2 = 36.57 \text{ ft}^2 \doteq 5266 \text{ in}^2$. The carpet contains about 5266 inches of braid, or about 439 ft. **59.** Since each tile measures $\frac{8}{12} = \frac{2}{3}$ feet on a

side, the dimensions of the kitchen, in tiles, are $10 \div \frac{2}{3} = 15$ and $12 \div \frac{2}{3} = 18$. Therefore, the numbers of tiles needed is $15 \times 18 = 270$. **61.** 90°, since if we view one side of length 300′ as the base, the altitude of the triangle is greatest if the angle is 90° **63.** C **65.** D (rearrange the picture)
67. By looking at the drawing, the lengths of two sides must be from the other sides. Sketching the complete picture below shows all measurements. Starting on the left top side of 1 cm and going clockwise gives the perimeter. The answer is $1 + 7 + 1 + 6 + 2 + 3 + 6 + 3 + 6 + 3 = 38$ cm.

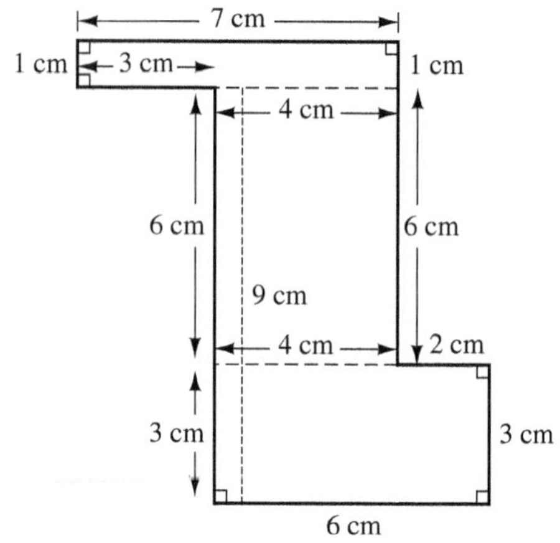

Problem Set 10.3 (page 530)

1. (a) $x^2 = 7^2 + 24^2 = 49 + 576 = 625$, so $x = \sqrt{625} = 25$. **(b)** $x^2 + 8^2 = 17^2$, so $x = \sqrt{225} = 15$. **(c)** $x^2 + 5^2 = 22^2$, so $x = \sqrt{459}$. **3. (a)** $x^2 + (2x)^2 = (25)^2, 5x^2 = 625$, $x = \sqrt{125} = 5\sqrt{5}$ **(b)** Let y be the length of the bottom leg. $y^2 + 7^2 = 12^2$, so $y = \sqrt{95}$. Then $(x + 7)^2 + y^2 = 16^2$, so $x = \sqrt{161} - 7$. **(c)** Let y be the length of the leg of the smaller triangle. $y^2 + 6^2 = 10^2$, so $y = 8$. Then $x^2 = (11 + y)^2 + 6^2 = 397$, so $x = \sqrt{397}$. **5. (a)** $x^2 = 10^2 + 15^2 = 325$, so $x = \sqrt{325}$; $y^2 = x^2 + 7^2 = 325 + 49 = 374$, so $y = \sqrt{374}$. **(b)** $x^2 = 1^2 + 1^2 = 2$, so $x = \sqrt{2}$; $y^2 = x^2 + 1^2 = 2 + 1 = 3$, so $y = \sqrt{3}$. **7. (a)** Height $= \sqrt{15^2 - 9^2}$, so area $= (20)(12) = 240$ square units.
(b) Base $= 2\sqrt{40^2 - 30^2} = 20\sqrt{7}$, so

area $= \frac{1}{2}(20\sqrt{7})(30) = 300 = \sqrt{7}$ square units.

9. The length of the shortcut is $\sqrt{(50 \text{ ft})^2 + (100 \text{ ft})^2} \doteq 112$ ft. The distance along the sidewalk is 50 ft + 100 ft = 150 ft, so the children save about 150 ft − 112 ft = 38 ft. **11. (a)** About 108 miles apart **(b)** About 202 miles apart **13. (a)** $\sqrt{2}$ **(b)** $\frac{2}{\sqrt{3}}$
15. $r^2 + 24^2 = (18 + r)^2$, so $r = 7$.
17. (a) $(14)^2 + (\sqrt{533})^2 = 729 = (27)^2$; yes.
(b) $(7\sqrt{2})^2 + (4\sqrt{7})^2 = 210 \neq 308 = (2\sqrt{77})^2$; no.
(c) $(9.5)^2 + (16.8)^2 = 372.49 = (19.3)^2$; yes.
19.

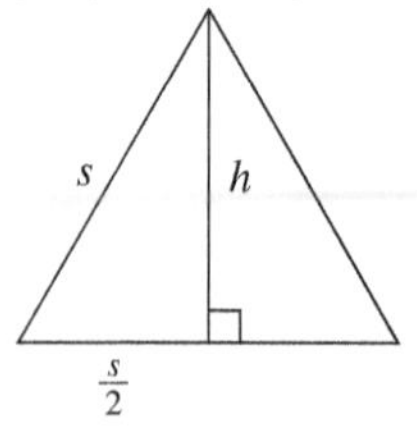

(a) $\left(\frac{s}{2}\right)^2 + (h)^2 = s^2$, by the Pythagorean theorem. Therefore, $h = \frac{\sqrt{3}}{2}s$. **(b)** The area is $\frac{1}{2}sh = \frac{1}{2}(s)\left(\frac{\sqrt{3}}{2}s\right) = \frac{\sqrt{3}}{4}s^2$ square units. **(c)** The area is $6\left(\frac{\sqrt{3}}{4}s^2\right) = \frac{3\sqrt{3}}{2}s^2$ square units.
(d) The inscribed circle has radius $h = \frac{\sqrt{3}}{2}s$, so its area is $\pi\left(\frac{\sqrt{3}}{2}s\right)^2 = \frac{3}{4}\pi s^2$. The circumscribed circle has radius s, so its area is $\pi(s)^2 = \pi s^2$. **21. (a)** The sum of the areas of the two equilateral triangles on the sides of the right triangle equals the area of the equilateral triangle on the hypotenuse. **(b)** Suggest that the student try to discover this relationship by drawing an appropriate diagram and computing the required areas.
23. (a)

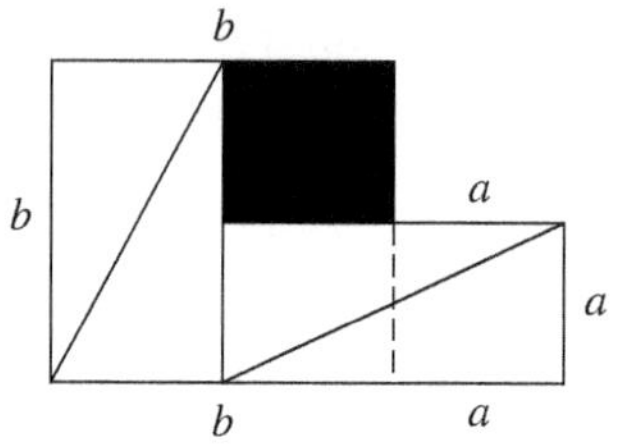

(b) Since the square and double-square are covered by the same five shapes, their areas are equal. The respective areas are c^2 and $a^2 + b^2$, so $c^2 = a^2 + b^2$.
25.

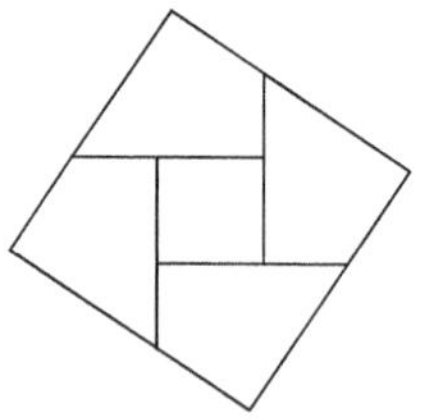

27. Sammy needs to make sure that he realizes that it is one-half the product of the length of the legs which gives the area. The hypotenuse is not involved. There may also be a confusion because the right triangle isn't "standing straight." In addition, he should always use units—in this case, cm²—in his answers. **29.** The diagonal has length 4.
31. π 100 cm² **33.** $90\sqrt{2}$ ft **35.** With lengths of 3, 4, and 5, one still has a right triangle, since $3^2 + 4^2 = 25 = 5^2$.
37. $x \doteq 9.373$, or about $9\frac{3}{8}$ inches **39.** The areas of the two squares are 28 cm² and 11^2 cm². Because of the Pythagorean theorem is the area of the shaded square, $121 - 28 = 93$ cm².
41. B, the hypotenuse of one triangle is the leg of the other. $y = 4$.

Problem Set 10.4 (page 540)

1. (a) Surface area **(b)** Volume **(c)** Volume **(d)** Surface area
3. (a) $\pi \times (10 \text{ m})^2 \times 4 \text{ m} = 400\pi \text{ m}^3 \doteq 1257 \text{ m}^3$
(b) $V = Bh = \pi(3 \text{ ft})^2(7 \text{ ft}) = 63\pi \text{ ft}^3 \doteq 198 \text{ ft}^3$
5. (a) $\frac{1}{3} \times \pi(5 \text{ cm})^2 \times (12 \text{ cm}) = 100\pi \text{ cm}^3 \doteq 314 \text{ cm}^3$
(b) $\frac{1}{2} \cdot \frac{1}{3}(6 \text{ in})^2(14 \text{ in}) = 84\pi \text{ in}^3 \doteq 264 \text{ in}^3$
7. Table 10.4 shows that 2 cups is one-eighth of a gallon, which is the same as 231 in³ divided by 8, or 28.875 in³. A hemisphere of radius r has volume one-half of a sphere of the same radius, so that the volume of a hemisphere is $2\pi r^3/3 = 28.875$.

Solving yields $r^3 = 3(28.875)/2\pi$, so r is the cube root of 13.786 (or about 2.4 inches). **9.** Have Ruth Ann build a pyramid with a rectangular or other polygonal base, using reasonably stiff cardboard. Making careful measurements, she can then determine the area B of the base and height h of the pyramid. She can also compute $V = \frac{1}{3}bh$. Next, she should fill the pyramid with rice, pour it into a suitable rectangular box, and measure the volume. **11. (a)** The circumference of the cone is $\frac{3}{4} \cdot 2 \cdot \pi(4\text{ in}) = 6\pi$ in, so the radius is 3 in. **(b)** $\sqrt{7}$ in **(c)** $v = \frac{1}{3}(\sqrt{7}\text{ in})(\pi(3\text{ in})^2) \doteq$ 25 in^3 **13. (a)** Since the height is 6 cm, the volume is $6\pi r^2$. The table is completed as

Radius in cm	2	4	6	8
Volume in cm^3	24π	96π	216π	384π

(b), (c) Not really. Samuel is correct if the height is fixed (in this case, 6 cm), but if we allow both the radius and the height to change, then the volume may go either up or down as the radius increases. There are many answers, but if the radius goes, for example, from 2 cm to 4 cm but the height decreases from 6 cm to 1 cm, then the volume decreases from 24π to 16π. **15. (a)** Maurice correctly calculated the volume of the figure—it is 48—but it is 48 ft^3, not 48 ft^2. **(b)** Maurice did not realize that the reason he was given only two numbers to multiply was because the area of the base had already been figured out for him. He just multiplied the numbers that were there without thought as to their significance. Maurice then thought that, since he didn't physically multiply three numbers, he shouldn't label it with a "little three."
(c) Answers may vary. Maurice does not seem to fully understand what it means to label a number with "squared" or "cubed." It is important that students be given the correct vocabulary when calculating math so that they learn the concept presented and can internalize what the procedure requires them to do.
17. The respective volumes are

$$V_{\text{cone}} = \frac{1}{3}(\pi r^2)(2r) = \frac{2}{3}\pi r^3,$$

$$V_{\text{sphere}} = \frac{4}{3}\pi r^3, \text{ and}$$

$$V_{\text{cylinder}} = (\pi r^2)(2r) = 2\pi r^3 = \frac{6}{3}\pi r^3,$$

so the volume ratios are 2 to 4 to 6, or 1 to 2 to 3.
19. $V(\text{ring}) = \pi\left(\frac{9}{16}\text{in}\right)^2 \cdot \left(\frac{5}{4}\text{in}\right) - \pi\left(\frac{1}{2}\text{in}\right)^2\left(\frac{5}{4}\text{in}\right) \doteq 0.26\text{ in}^3$, so about 1.56 ounces. **21.** $V(\text{box}) = 160\text{ in}^3$ and $V(\text{tub}) = \pi(3\text{ in})^2(10\text{ in}) \doteq 283\text{ in}^3$. Two boxes is a better buy.
23. (a) $\left(\frac{5}{13}\right)^3(106.75) \doteq 6.07$ pounds
(b) $\frac{1}{3}(177\text{ feet})(45\text{ acres})\left(\frac{43{,}560\text{ ft}^2}{1\text{ acre}}\right) = 115{,}651{,}800\text{ ft}^3$
$(472{,}000{,}000\text{ ft}^3)^{1/3} \doteq 779$ ft **25.** D, Since 2 feet is 24 inches, the new volume is $(24 - 4) \times 10 \times 10 = 2000\text{ in}^3$.

Problem Set 10.5 (page 549)

1. (a) SA $= 2 \cdot \frac{1}{2}(20\text{ cm} + 15\text{ cm})(12\text{ cm}) + (2\text{ cm})(60\text{ cm}) = 540\text{ cm}^2$ **(b)** SA $= 2 \times \frac{1}{2}(6\text{ m} \times 8\text{ m}) + (24\text{ m} \times 5\text{ m}) = 168\text{ m}^2$ **3. (a)** Slant height $= \sqrt{(40\text{ m})^2 + (30\text{ m})^2} = 50$ m; SA $= (60\text{ m})^2 + 4 \cdot \frac{1}{2}(60\text{ m})(50\text{ m}) = 9600\text{ m}^2$ **(b)** Slant height $= \sqrt{(10\text{ ft})^2 - (6\text{ ft})^2} = 8$ ft; SA $= (12\text{ ft})^2 + \frac{1}{2}(4)(12\text{ ft})(8\text{ ft}) = 336\text{ ft}^2$
5. (a) S $= 4\pi(2200\text{ km})^2 = 19{,}360{,}000\pi\text{ km}^2 \doteq 6.08 \times 10^7\text{ km}^2$; $V = \frac{4}{3}\pi(2200\text{ km})^3 \doteq 4.46 \times 10^{10}\text{ km}^3$
(b) SA $= \frac{1}{2}[4\pi(6\text{ cm})^2] + \pi(6\text{ cm})^2 = 108\pi\text{ cm}^2 \doteq 339\text{ cm}^2$; $V = \frac{1}{2}\left[\frac{4}{3}\pi(6\text{ cm})^3\right] = 144\pi\text{ cm}^3 \doteq 452\text{ cm}^3$
7. (a) volume **(b)** volume **(c)** surface area **(d)** surface area
9. $V = \pi(3.25\text{ cm})^2\,11\text{ cm} \doteq 365\text{ cm}^2 = 365\text{ mL} \doteq 12.2$ fl oz
11. Area (sphere) $= 4\pi r^2$ square units. Area (cylinder) $= 2 \cdot \pi \cdot r^2 + 2\pi r \cdot (2r) = 6\pi r^2$ square units. Thus, $4\pi r^2/6\pi r^2 = \frac{2}{3}$.
13. (a) 4, considering area **(b)** One 14″ pizza is nearly the same amount of pizza, but will save \$2.
15.

	I	II	III
Height	6	18	9 cm
Perimeter of Base	10	30	15 cm
Lateral Surface Area	40	360	90 cm^2
Volume	$\frac{80}{27}$	80	10 cm^3

17. (a) 8 liters (doubling the sides increases the volume by a factor of 8) **(b)** $2^{1/3} \times 10\text{ cm} \doteq 12.6$ cm **19.** If r is the radius, then $SA = 2\pi r^2 + (2\pi r)3r = 8\pi r^2 = 392$. $r = \frac{7}{\sqrt{\pi}}$ cm and the height is $\frac{21}{\sqrt{\pi}}$ cm.
21. (a)

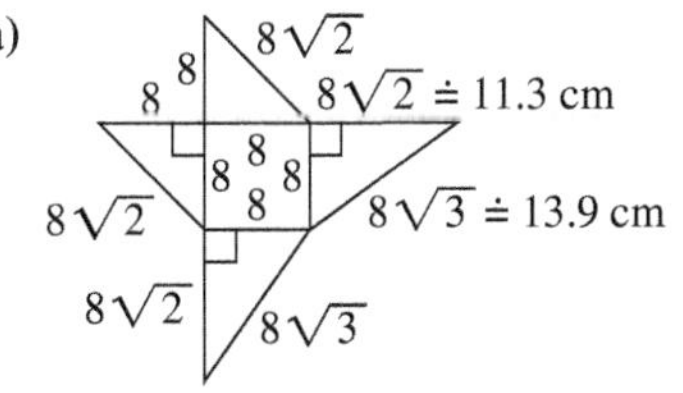

(b) Total area $= 64 + 32 + 32 + 32\sqrt{2} + 32\sqrt{2} = (128 + 64\sqrt{2})\text{ cm}^2 \doteq 218.51\text{ cm}^2$ **(c)** 512 cm^3, $\frac{512}{3}\text{ cm}^3$
23. Jerry was really writing $(4\pi + 24\pi)$, which is the lateral surface area of the cylinder and only the top of the can. It is not unusual for children to compute the surface area by using only those parts of the surface they can actually see.
25. By picking those values of the radius which are less than 1, Inez will see that her assertion is false. (She is correct if $r > 1$).
27. The slant height of the cone is the hypotenuse of a right triangle with legs r and $2r$. That is, the slant height is $s = \sqrt{r^2 + (2r)^2} = \sqrt{5}r$. Thus, the respective surface areas are

$$S_{\text{cone}} = \pi r^2 + \frac{1}{2}(\sqrt{5}r)(2\pi r) = \frac{(1 + \sqrt{5})}{2}2\pi r^2 = 2\tau(\pi r^2),$$

$$S_{\text{sphere}} = 4\pi r^2, \text{ and}$$

$$S_{\text{cylinder}} = 2(\pi r^2) + (2r)(2\pi r) = 6\pi r^2,$$

so the ratio of surface areas is 2τ to 4 to 6, or, equivalently, τ to 2 to 3.

29. **(a)** Each edge is the hypotenuse h of a right triangle with legs of length s. Thus, $h = s\sqrt{2}$. **(b)** The volume of $ACDE = \frac{1}{3}\left(\frac{1}{2}\right)(s)(s)(s) = \frac{1}{6}s^3$. **(c)** Notice that $ACEG$ is formed by cutting four such congruent tetrahedrons, as in part (b), from the cube. Hence, the volume of $ACEG$ is $s^3 - 4 \cdot \frac{1}{6}s^3 = \frac{1}{3}s^3$
(d) The scale factor is $\frac{b}{\sqrt{2}}$, so the volume is $\left(\frac{b}{\sqrt{2}}\right)^3 \frac{1}{3} = \frac{b^3\sqrt{2}}{12}$.
31. E **33.** The top of the can has a disc of $\pi 4^2$ as is the bottom of the can. The side has surface length of $8\pi \times 3$. The total surface area (since there are two discs) is $\pi 4^2 + \pi 4^2 + 24\pi = 56\pi \text{ cm}^2$.
35. A, assuming that the roof is flat.

Chapter 10 Review Exercises (page 556)

1. **(a)** Centimeters **(b)** Millimeters **(c)** Kilometers **(d)** Meters **(e)** Hectares **(f)** Square kilometers **(g)** Milliliters **(h)** Liters **3.** 84 L
5.

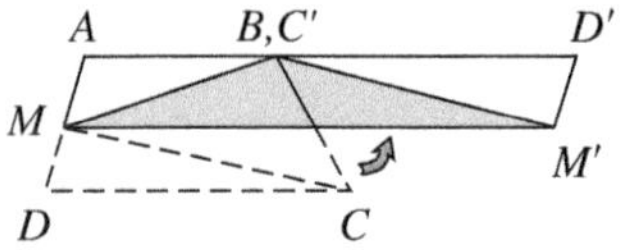

The triangle has half the area of the parallelogram $AD'M'M$, so it also has half the area of the trapezoid $ABCD$.
7. **(a)** $768 \text{ in}^2 = 5\frac{1}{3} \text{ ft}^2$ **(b)** $\frac{1}{2}(8 \text{ m})(9 \text{ m}) + \frac{1}{2}(3 \text{ m}) \cdot (6 \text{ m}) = 45 \text{ m}^2$
(c) $\frac{1}{2}(5 \text{ cm} + 7 \text{ cm})(3 \text{ cm}) = 18 \text{ cm}^2$
9. **(a)** $A = (3 \text{ ft})(4 \text{ ft}) + \frac{1}{2}\pi(1.5 \text{ ft})^2 \doteq 15.5 \text{ ft}^2$, $P = 11 \text{ ft} + \pi(1.5 \text{ ft}) \doteq 15.7 \text{ ft}$ **(b)** $A = \frac{3}{4}\pi(3 \text{ m})^2 \doteq 21.2 \text{ m}^2$, $P = \frac{3}{4} \cdot 2\pi(3 \text{ m}) + 6 \text{ m} \doteq 20.1 \text{ m}$ **11.** $\sqrt{1125} \text{ cm} \doteq 33.5 \text{ cm}$
13. $9 + \sqrt{2} + \sqrt{10} + \sqrt{5} \doteq 15.8$ units
15. $V(\text{sphere}) = \frac{4}{3}\pi(10 \text{ m})^3$ and $V(\text{four cubes}) = 4(10 \text{ m})^3$. Since $\pi > 3$, $\frac{4}{3}\pi > 4$, showing that the sphere has the larger volume.

Chapter 11

Cooperative Investigations (page 560)

1. **(d)** Two lines of symmetry.

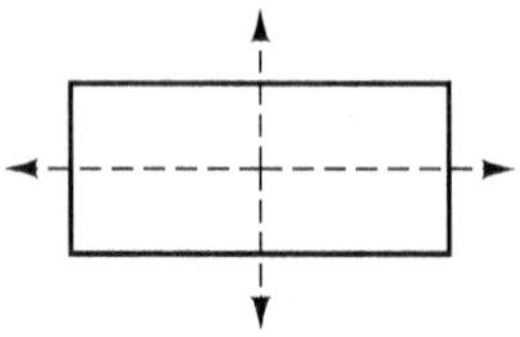

(h) No lines of symmetry.

Problem Set 11.1 (page 575)

1. **(a)** Not a rigid motion; distances between particular cards will change. **(b)** Yes, pieces will be in the same places relative to each other. **(c)** No, distances between particular pieces almost certainly will have changed. **(d)** Yes, the painting has not changed; only its location has. **(e)** No, the distances between bits of dough will have increased.
3.

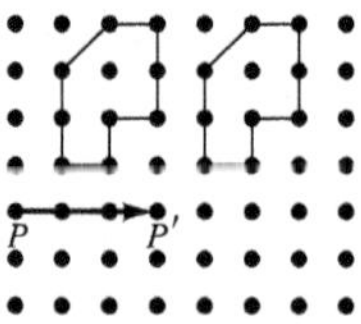

5. **(a)**

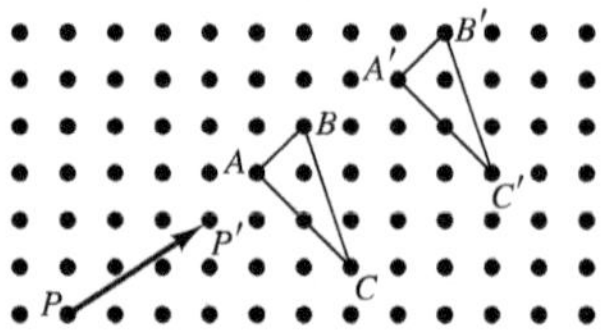

(b) The rigid motion is a translation two units down and three units to the left. This translation does the opposite of the translation that takes ΔABC to $\Delta A'B'C'$.
7.

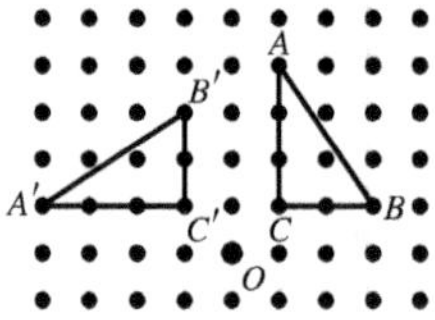

9. **(a)** Center O is the intersection of $\overleftrightarrow{AB}$ and $\overleftrightarrow{A'B'}$. **(b)** 90° (counterclockwise)
(c)

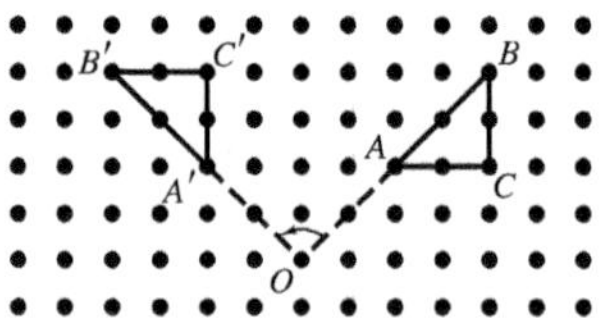

11. The transformation is **(a)** a translation or glide–reflection. **(b)** a rotation about the fixed point by an angle between 0° and 360°. **(c)** a reflection. **(d)** the identity transformation.
13.

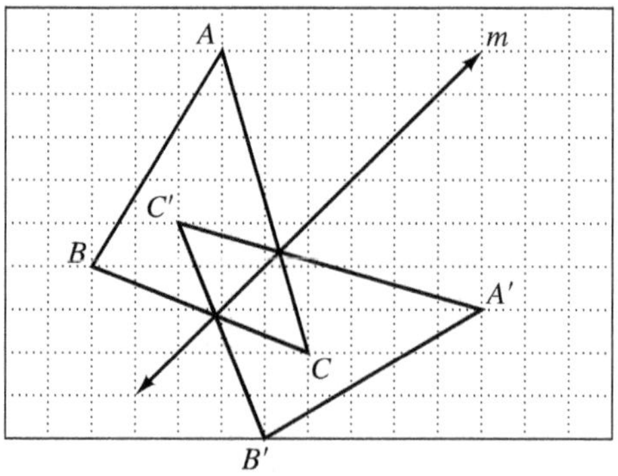

15. **(a)** A is on m. **(b)** A and B are both on m, so $m = \overleftrightarrow{AB}$. **(c)** It takes D to C.
17. **(a)**, **(b)**, **(c)**

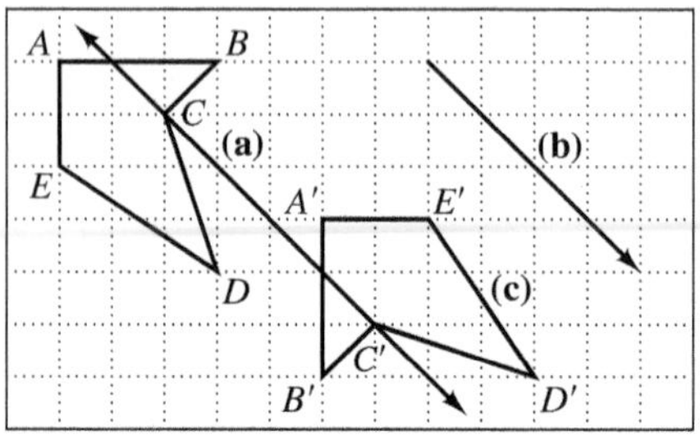

19. **(a)** Translate right eight units. **(b)** j **(c)** l **(d)** Rotate counterclockwise 90° about point R. **(e)** j **(f)** k

21. **(a)** Center P, scale factor $\frac{3}{2}$

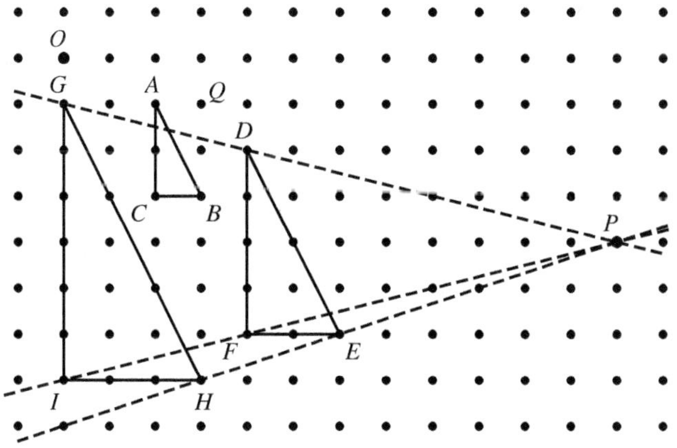

(b) Center Q, scale factor 3 **(c)** The perimeter of ΔDEF is $2(3 + \sqrt{5}) = 6 + 2\sqrt{5}$. The perimeter of ΔGHI is $3(3 + \sqrt{5}) = 9 + 3\sqrt{5}$. **(d)** $12 + 4\sqrt{5}$

(e) Area $\triangle ABC = 1$

Area $\triangle DEF = 4$

Area $\triangle GHI = 9$

The areas are equal to the scale factor squared.

23. Rotate 90° counterclockwise about B. Then perform a size transformation centered at P with scale factor 2. (Other sequences will also work.) See illustration below.

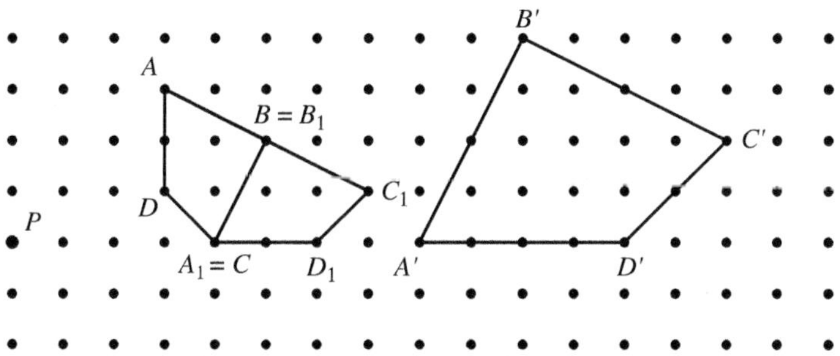

25. **(a)** Ephraim and DeVonte used a 180° rotation to transform their figure. **(b)** Answers will vary. It might help to trace the figure and use a Mira to arrange the pattern blocks into the mirror image. **27.** Under a similarity transformation, all distances are transformed by the same factor and all angle sizes are preserved. Thus, two rectangles are similar if, and only if, they have the same ratio of length to width. In ordinary usage, we might say that two people look similar to one another, but this is not so in the strict mathematical sense of the word. **29.** On a 1-cm-square grid, the two 90° rotations take O_1 to O'_1 and O_2 to O'_2. This motion is equivalent to a 180° rotation about the point O.

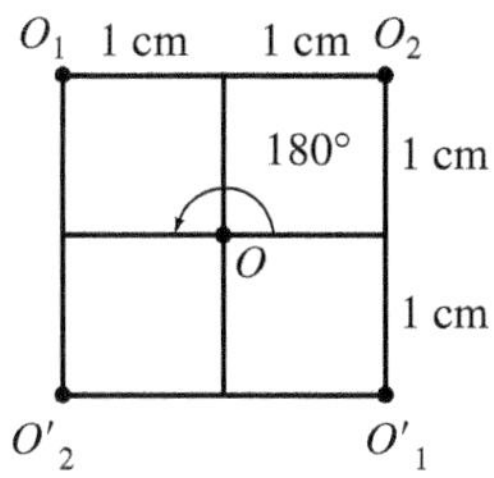

31. **(a), (b)**

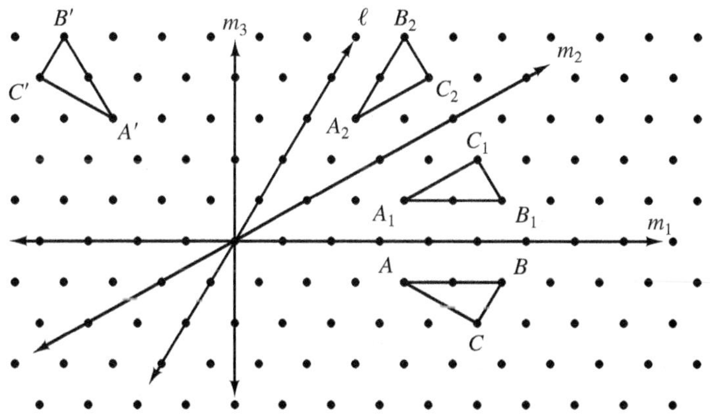

33. **(a), (b)**

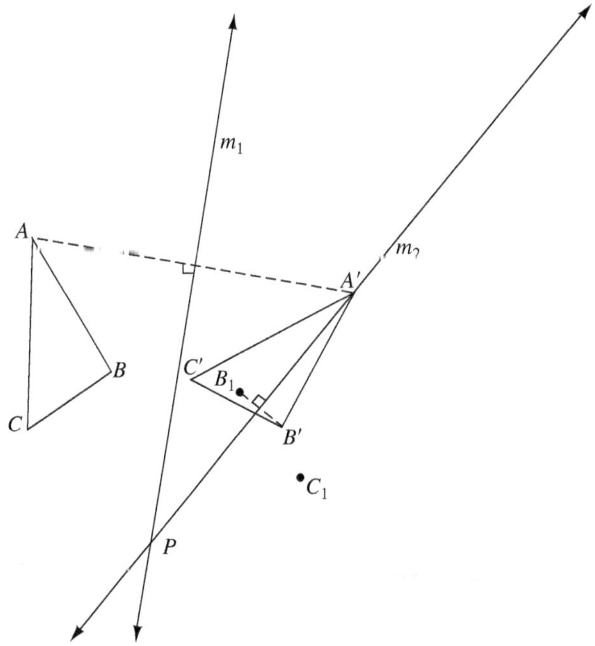

(c) A reflection first across line m_1 and then across line m_2 is equivalent to a rotation about the point P of intersection of m_1 and m_2, through an angle twice the measure x of the directed angle from line m_1 toward line m_2. **35.** **(a)** A translation: Six reflections give an orientation-preserving rigid motion, so it is either a rotation or a translation. Since a rotation has a fixed point (namely, the rotation center), the motion is a translation. **(b)** A reflection: Eleven reflections give an orientation-reversing transformation, so it is either a reflection or a glide–reflection. A glide–reflection leaves no point unmoved, so the motion is a reflection. **(c)** A rotation: Two glide–reflections give an orientation-preserving rigid motion, so it is either a translation or a rotation. Since some point is unmoved, the motion is a rotation. It is possible that the motion is the identity motion, which leaves all points unmoved. **37.** The line PP' passes through O, so constructing this line determines point O. $\overline{PQ}$ and $\overline{P'Q'}$ are parallel, so the line through P' that is parallel to $\overline{PQ}$ will determine Q'.

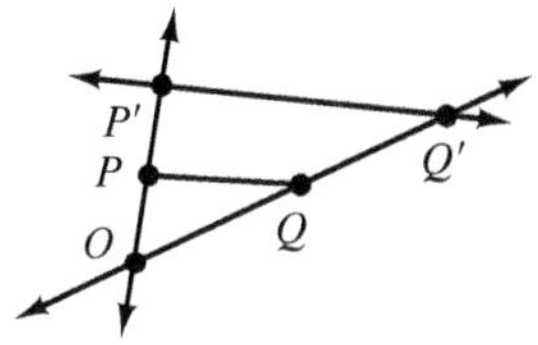

39. (a) Yes **(b)** Yes **41. (a)** First construct the image P' of P across m and the image S' of S across l. Then draw $\overline{P'S'}$ to find Q and R.

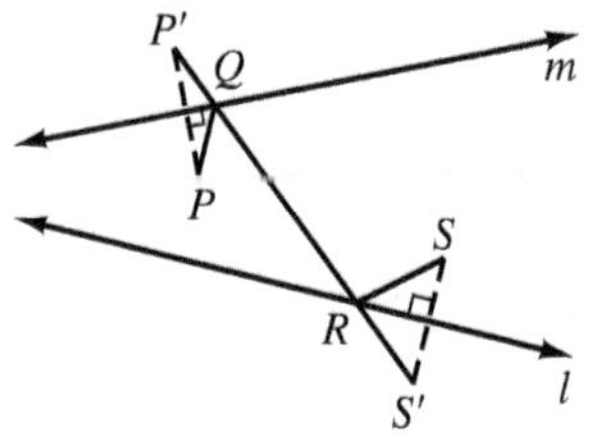

(b) First construct the image P'' of P across l and the image S'' of S across m. Then draw $\overline{P''S''}$ to find A and B.

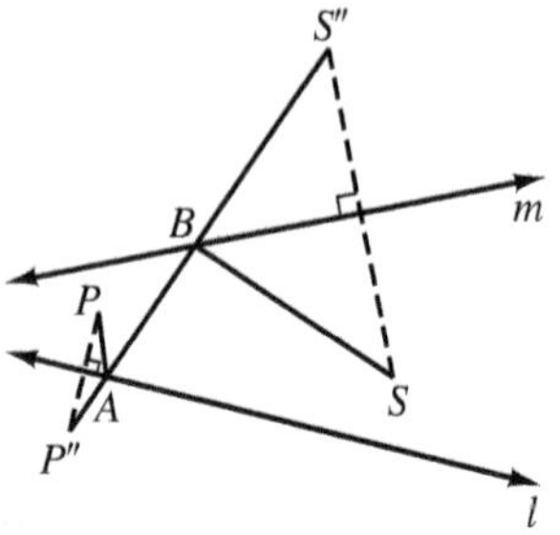

43. Draw $\overline{PP'''}$. Then "fold" the copies of the table back on top of the table to give the billiard path.

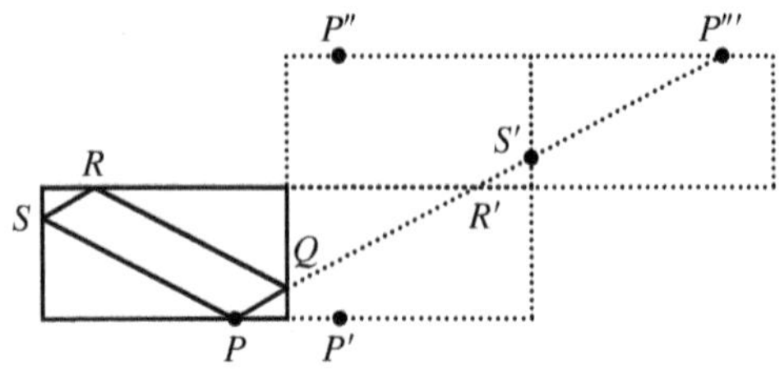

45. (a)

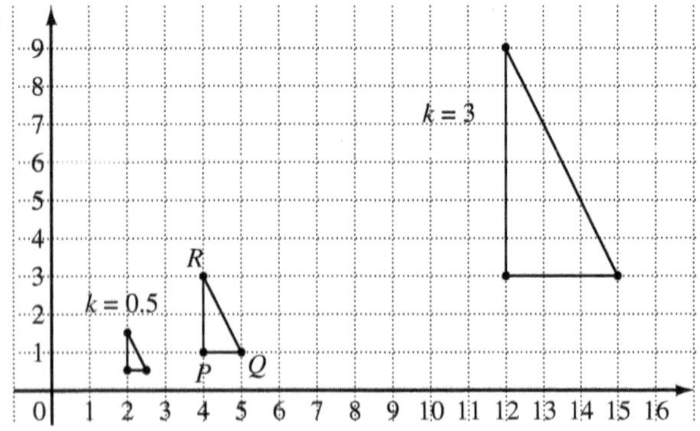

(b) After the size transformation with scale factor 2 that transformed ΔPQR to $\Delta P'Q'R'$, subtract 7 from the x-coordinates of all points to translate $\Delta P'Q'R'$ horizontally to the left by seven units. The combined size transformation and translation take the point (x, y) to the point $(2x - 7, 2y)$. **(c)** First, triangle PQR is dilated with scale factor $k = 2$ to triangle $P'Q'R'$. Then, replacing all coordinates with their negatives gives a half-turn rotation about the origin that moves triangle $P'Q'R'$ to triangle ABC. Altogether, the x- and y-coordinates in the plane are all multiplied by -2.
47. D **49.** D **51.** D

Problem Set 11.2 (page 588)

1. (a)

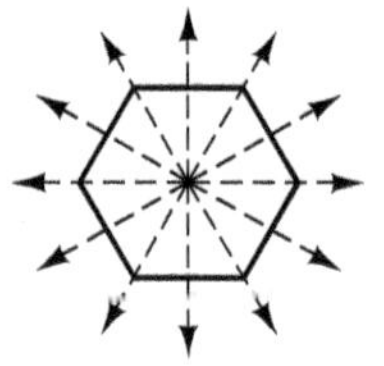

(b)

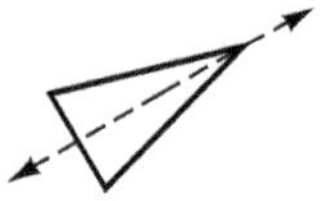

(c)

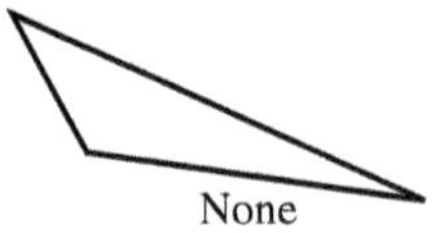

None

(d)

(e)

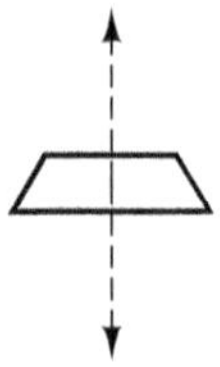

3. (a) Many figures are possible. 
(b) Answers will vary. For example, most parallelograms have no reflection symmetry, but they do have 180° rotation symmetry.
(c) Not possible

5. (a)

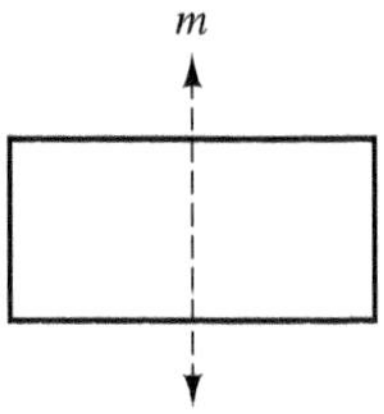

(b)

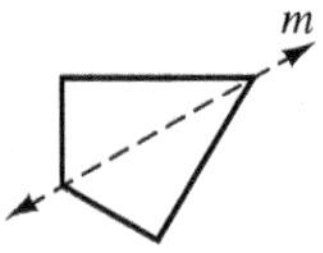

(c)

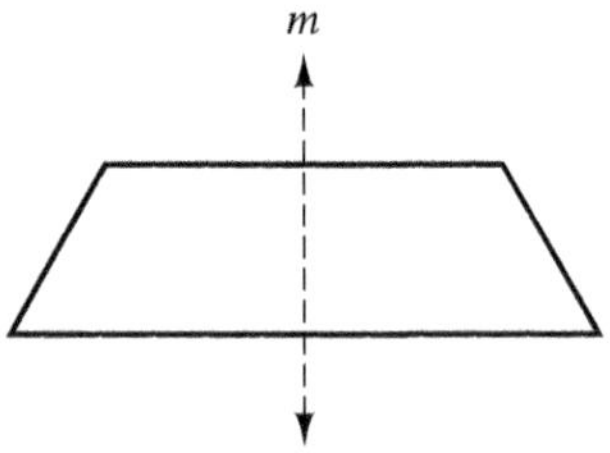

(d)

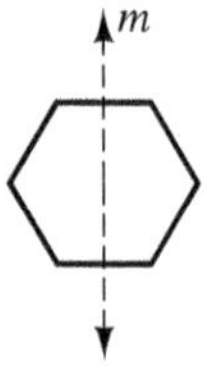

7. (a)

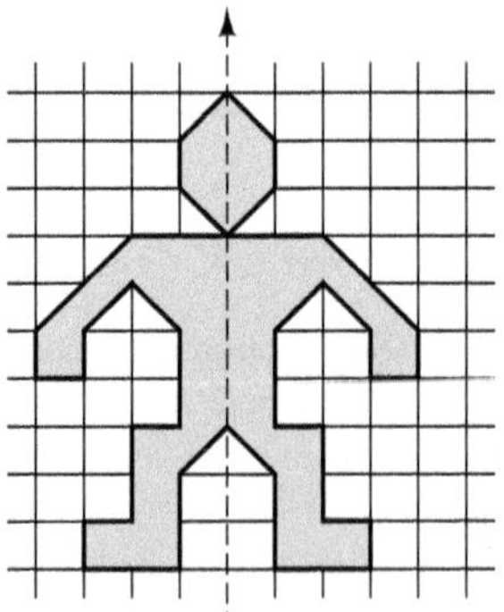

(b)

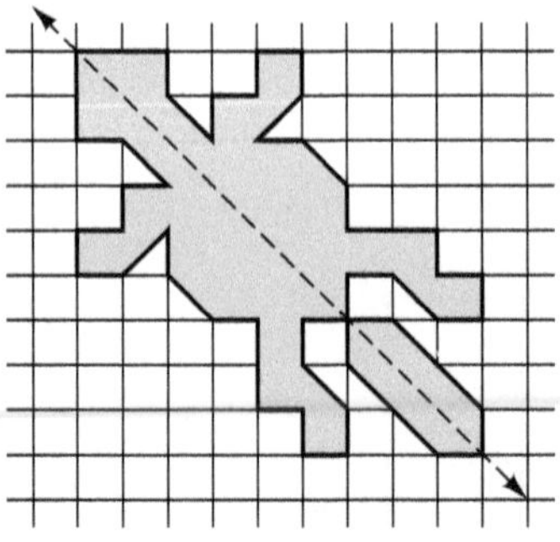

9. (a) One line of symmetry **(b)** Three lines of symmetry and 120° rotation symmetry **(c)** 120° rotational symmetry (no line of symmetry) **(d)** Point symmetry about the center **(e)** Point symmetry about the center

11. (a)

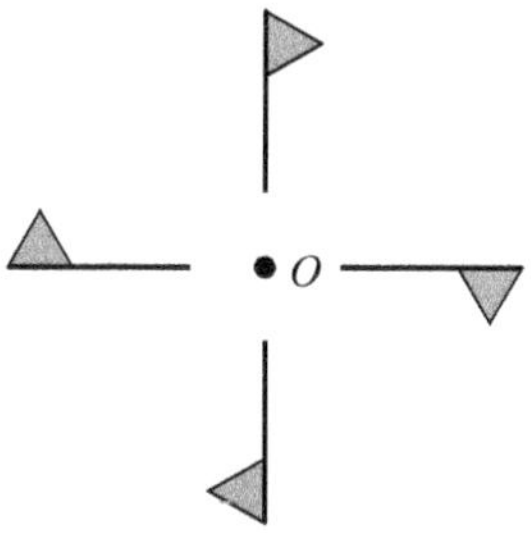

(b)

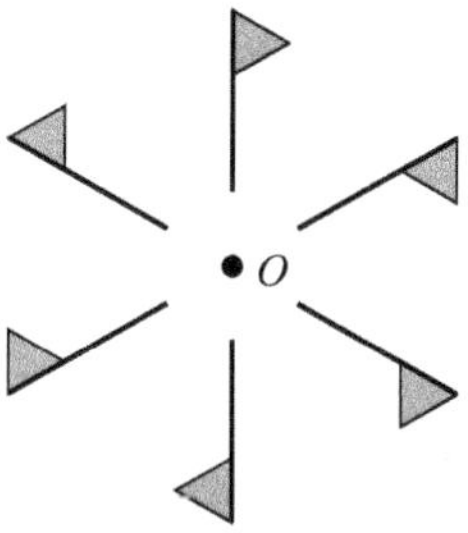

13. **(a)** 0, 8 **(b)** 0, 3, 8 **(c)** 0, 8 **(d)** 0, 8
15. **(a)** **i, l, o, t, u, v, w, x** (b) **c, l, o, x** (c) **l, o, x**
(d) **l, o, s, x, z** **17.** **(a)** There are vertical lines of symmetry through the center of each letter, and there is a horizontal line of symmetry. The symbol type is *mm*. **(b)** Translation symmetry and vertical-line symmetry: *m*1 **(c)** Translation symmetry and half-turn symmetry: 12 **19.** **(a)** (i) *mm* (ii) 1*g* (iii) 1*m* (iv) 12
(b) Remaining patterns are *mg*, *m*1, and 11.

mg ... ↑ ↓ ↑ ↓ ↑ ...

*m*1 ... ↗↖↗↖↗↖ ...

11 ... ↗↗↗↗↗↗ ...

Patterns may vary. **21.** Answers will vary. **23.** **(a)** The lines of symmetry of the nonsquare rectangle are correct. The isosceles trapezoid has a vertical line of symmetry, but no horizontal line of symmetry. **(b)** Use a Mira to compare the reflections over the vertical and horizontal lines, explaining to Stacey why the vertical line is, and the horizontal line is not, a line of symmetry.
25. Lynn can always place a penny at the position that is point symmetric to Kelly's last move. Since Lynn can always make a move, Kelly will be the first player unable to find space for an additional penny on the table. **27.** **(a)** The sequence of letters is the same backward as forward. **(b)** The sequence of words is the same backward as forward. **(c)** The sequence of letters and spaces is the same backward as forward. **29.** **(a)** 1*g* **(b)** 1*m*
(c) *mg* **(d)** *mm*
31. **(a)** **11**

p	p	p	p	p	p

(b) *m*1

p	q	p	q	p	q

(c) 12

p	d	p	d	p	d

(d) 1*g*

p	b	p	b	p	b

33. **(a)** As left-handed people know well, not all scissors are symmetric. **(b)** A T-shirt has bilateral symmetry. **(c)** A man's dress shirt is not quite symmetric, since it buttons right handed.
(d) Most golf clubs are either left-handed or right-handed, so they are not symmetric. Some putters have bilateral symmetry.
(e) Tennis rackets have two planes of bilateral symmetry.
(f) The blacked-out squares usually form a symmetric pattern that may include reflection symmetry and/or 90° rotation symmetry.
35. **(a)** One diagonal line of symmetry; **(b)** two diagonal lines of symmetry and half-turn symmetry; **(c)** (two diagonal lines of symmetry, vertical and horizontal lines of symmetry through the center, and 90° rotation symmetry; **(d)** two lines of diagonal symmetry and half-turn symmetry; **(e)** vertical and horizontal lines of symmetry through the center and half-turn symmetry;
(f) two diagonal lines of symmetry and half-turn symmetry.
37. Answers will vary. **39.** D

Problem Set 11.3 (page 602)

1. Many different tilings can be formed, including the following:

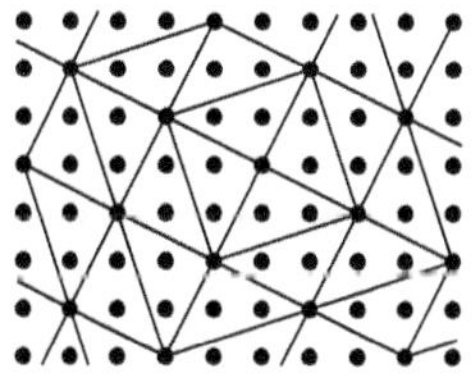

3.

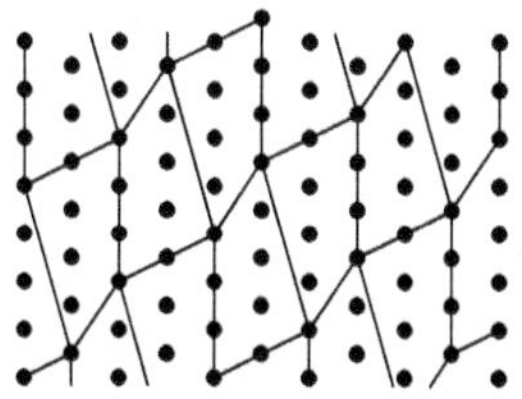

5.

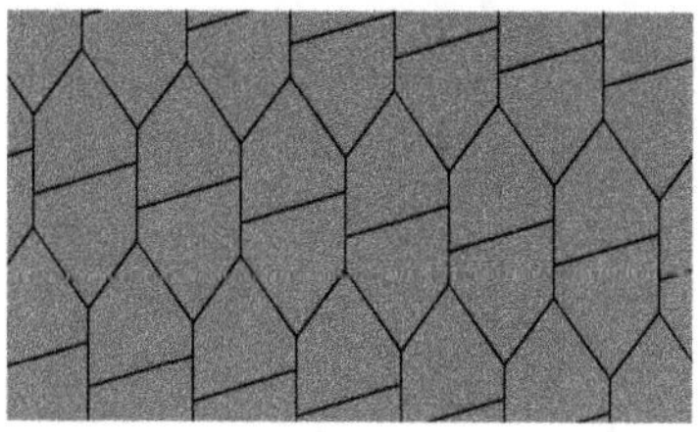

7. No rotation is necessary.

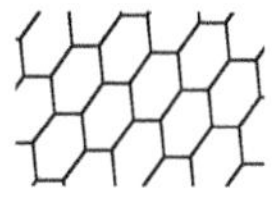

9. **(a)** $60° + 60° + 120° + 120° = 360°$ **(b)** If the vertex figure at *V* is repeated at *W*, the vertex figure at *X* has three triangles, making it different from the figure at *V*. (There are other answers as well. For example, repeating the vertex figure at *X* forces the two hexagons at *W* to be separated by a triangle.)

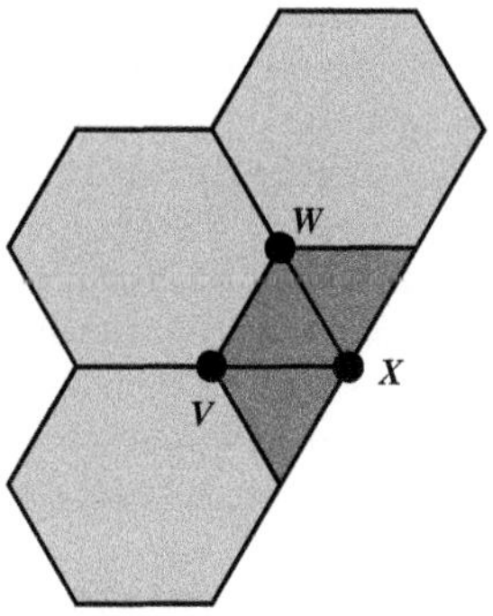

11. The interior angle of a square is 90°, of a regular pentagon is $\frac{(5-2)(180°)}{5} = 108°$, and of a regular 20-gon is $\frac{(20-2)(180°)}{20} = 162°$. Then, $90° + 108° + 162° = 360°$.
13. **(a)**

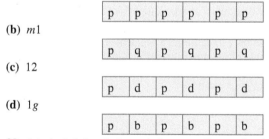

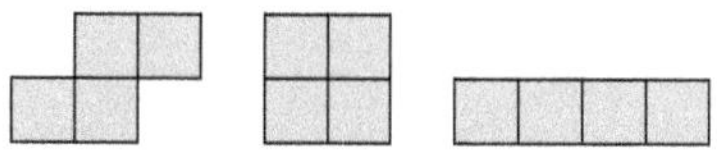

(b) Each tetromino tiles the plane. **(c)** Yes **15.** Answers will vary. **17.** Miguel needs to be reminded that the tiles must meet edge to edge, so that a vertex of any polygon can be placed coincident only with vertices of other polygons. **19.** **(a)** Wailea is correct that the four polygons form a vertex figure, since the sum of the angle measures of the four polygons is $60° + 60° + 90° + 150° = 360°$. **(b)** Remind Wailea that every vertex figure must be identical, so at vertex B there must be an equilateral triangle pointing outward from edge $\overline{BC}$. But this means that three equilateral triangles surround the vertex at C. **21.** **(a)** Squares, pentagons, hexagons, heptagons, octagons **(b)** Many vertex figures that appear in the tiling cannot correspond to regular polygons. For example, regular 5-, 6-, and 8-gons have interior angles of measure 108°, 120°, and 135°. These add up to $108° + 120° + 135° = 363° \neq 360°$. **23.** Designs will vary. **27.** Designs will vary.

27. **(a)**

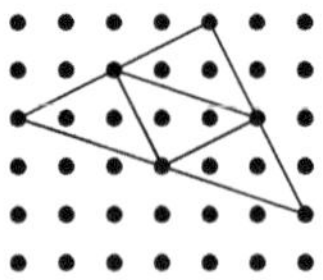

(b)

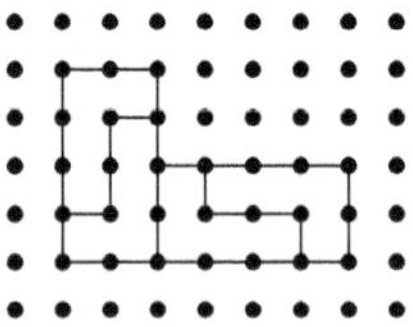

(c)

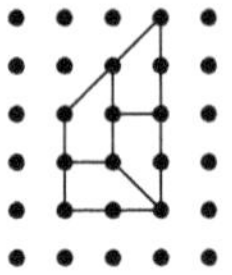

(d)

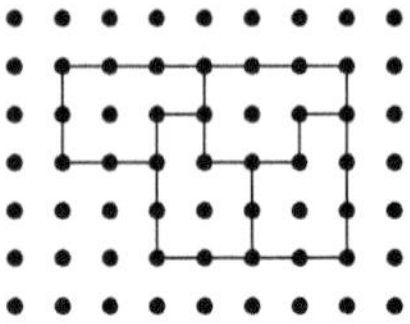

29. **(a)**

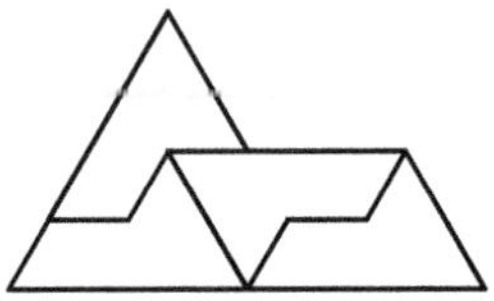

(b) 16:

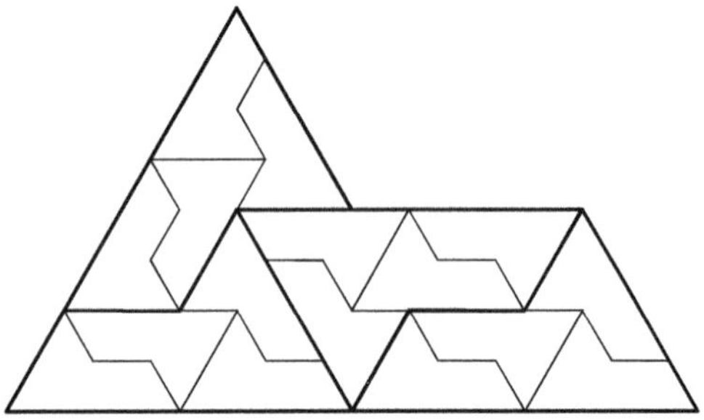

(c) There is no end to the number of generations that can be formed. **31.** A **33.** The criterion of using "only a clockwise rotation" is not clear. Clearly, the rotation must be performed three times, but it is not clear whether the rotation must always be applied with the same center or whether the rotation must be applied to the original figure only or to a previously rotated shape. Although the C tessellation can be obtained with a fixed rotation center by using clockwise rotations of 90°, 180°, and 270°, some students may claim with some justification that the A tessellation is also a correct answer.

Chapter 11 Review Exercises (page 609)

1.

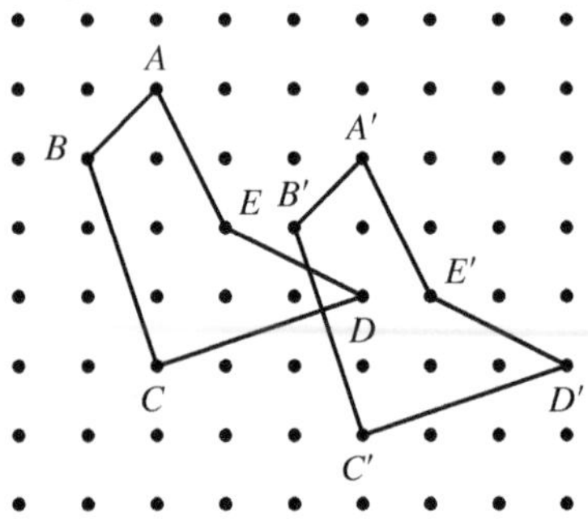

3. Reflection across line l, the perpendicular bisector common to all three segments AA', BB', and CC':

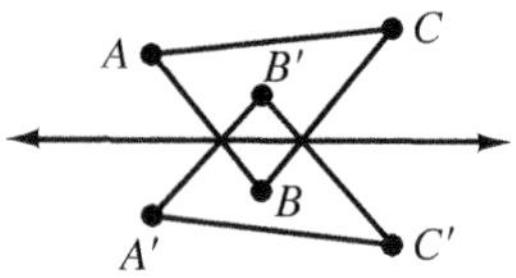

5.

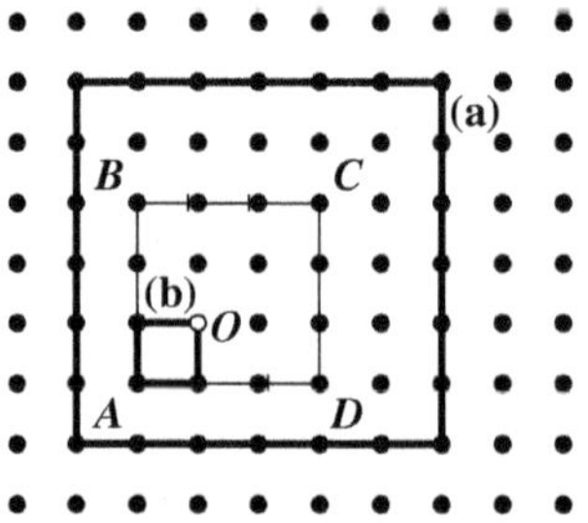

7. **(a)** Glide–reflection **(b)** Reflection **(c)** Reflection **(d)** Glide–reflection **9.** The letters in the upper row have either rotational or mirror symmetry, unlike those in the second row. **11.** **(a)** Equilateral triangle. **(b)** Parallelogram **(c)** A regular 9-gon (enneagon, or nonagon) **13.** **(a)** $m1$. **(b)** $1m$. **(c)** Glide–reflection ($1g$) **(d)** Vertical and horizontal lines of symmetry (mm), and glide and half-turn symmetries **(e)** Half-turn symmetry (12) **(f)** Half-turn symmetry, vertical line of symmetry, glide–reflection (mg)

15.

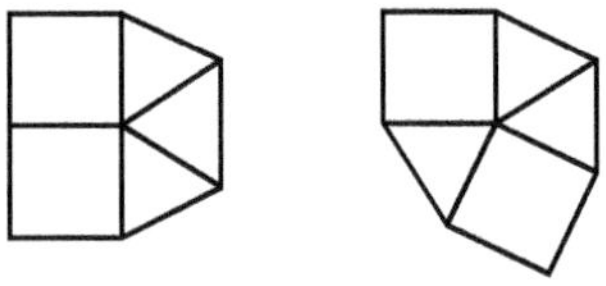

17. (a), (b), (c), (d), and (e) tile the plane, (f) will not

Chapter 12

Problem Set 12.1 (page 626)

1. **(a)** $L \leftrightarrow K, H \leftrightarrow W, S \leftrightarrow T$ **(b)** $\overline{KW}, \overline{WT}, \overline{TK}$ **(c)** $\angle K, \angle W, \angle T$ **(d)** $\triangle KWT$ **3.** **(a)** 4 **(b)** 6 **(c)** 56° **(d)** 41°
5. Use Construction 1 to construct a segment $\overline{AB}$ of length x. Next, construct the circle of radius y centered at B and the circle of radius z centered at A. Finally, let C be one of the points of intersection of the two circles of radii y and z.

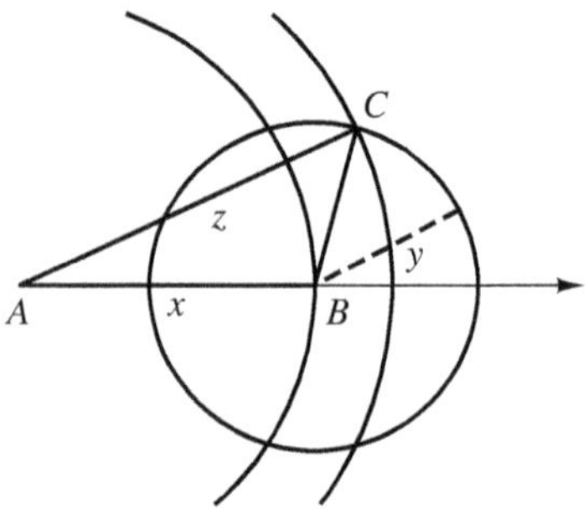

7. Answers will vary but should follow Construction 2 described in the section. **9.** **(a)** One such triangle **(b)** One such triangle **(c)** Impossible by the triangle inequality, since $2 + 5 < 8$
11. **(a)** $\triangle ABD \cong \triangle CBD$ by SAS **(b)** $\triangle ABC \cong \triangle FED$ by ASA **(c)** $\triangle ABC \cong \triangle EFD$ by SSS **(d)** No conclusion possible

13. **(a)** $\triangle ADC \cong \triangle ABC$ by AAS **(b)** $\triangle ABC \cong \triangle ADC$ by SAS **(c)** $\triangle ADC \cong \triangle BCD$ by ASA **15.** Let $\triangle ABC$ be equilateral. Since $AB = BC$, it follows from the isosceles triangle theorem that $\angle A \cong \angle C$. In the same way, since $BC = CA$, it follows that $\angle B \cong \angle A$. Thus, all three angles are congruent. **17.** **(a)** Both $\angle 1$ and $\angle 2$ should measure close to half the measure of $\angle BAC$. **(b)** $\triangle AFD \cong \triangle AFE$ by SSS **(c)** $\angle 1$ and $\angle 2$ are corresponding angles in the triangles shown to be congruent in part (b).
19. **(a)** $\angle ABD \cong \angle CDB$, as alternate interior angles between parallel lines. Likewise, $\angle ADB \cong \angle CBD$, and $DB = BD$. By ASA, $\triangle ABD \cong \triangle CDB$. **(b)** Since $\triangle ABD \cong \triangle CDB$, $\overline{AD} \cong \overline{CB}$ and $\overline{AB} \cong \overline{CD}$. Thus, $AD = CB$ and $AB = CD$. **(c)** Since $\triangle ABD \cong \triangle CDB$, $\angle BAD \cong \angle DCB$ or $m(\angle BAD) = m(\angle DCB)$. $m(\angle ADC) = m(\angle ADB) + m(\angle BDC) = m(\angle CBD) = m(\angle DBA) = m(\angle CBA)$. **21.** **(a)** By problem 20, M is the bisector of each diagonal of the rhombus, so $AM = MC$ and $BM = MD$. Therefore, the four triangles ABM, CBM, CDM, and ADM are congruent by SSS. **(b)** The four angles at M are congruent and sum to 360°. **23.** **(a)** $0 < s < 14$ cm, where s is the length of the fourth side **(b)** $AC \le AB + BC$. Similarly, $AD \le AC + CD$. Combine the two inequalities to get $AD \le AB + BC + CD$ **25.** Dane is mistaken. Have him make a hinged quadrilateral as described in the Into the Classroom box found in Section 12.2. The shape can be changed even though the lengths of the edges remain the same throughout a motion.
27. Since the diagonals of a rectangle intersect at a point the same distance from the vertices of the rectangle (see problem 20b), it is clear that the circle at the midpoint of the hypotenuse through point C passes through all of the vertices of the rectangle.

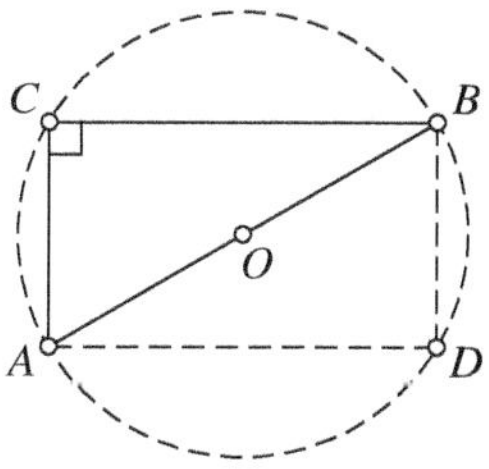

29. Construct a ray $\overrightarrow{QS}$ and use your compass to construct the segment $\overline{QR} \cong \overline{BC}$. At both Q and R, construct parallel rays $\overrightarrow{QT}$ and $\overrightarrow{RU}$ for which $\angle TQR \cong \angle URS \cong \angle B$. Finally, construct the ray $\overrightarrow{RV}$ for which $\angle URV \cong \angle A$. Label P the intersection of $\overrightarrow{QT}$ and $\overrightarrow{RV}$. Thus, $\triangle PQR \cong \triangle ABC$.

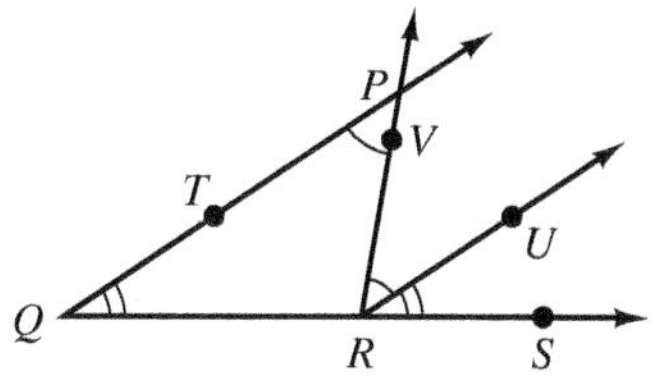

31. If $AB = CD = a$, $BC = AD = b$, and $AC = BD = c$, then each face of the tetrahedron is a triangle with sides of length a, b, and c. By the SSS property, the faces of the tetrahedron are congruent to one another. **33.** The best location for P is the center.
35. **(a)** The framework forms a parallelogram, but not necessarily a rectangle. **(b)** Let $ABCD$ be a parallelogram. By problem 19b, $AB = CD$ and $AD = CB$. $AC = BD$ is given. Then $\triangle ABC \cong \triangle BAD \cong \triangle CDA \cong \triangle DCB$ by the SSS property. Thus, $\angle A \cong \angle B \cong \angle C \cong \angle D$. Since the sum of their measures is 360°, each corner angle is a right angle. **(c)** The brace forms two triangles with fixed side lengths. The two triangles are rigid by the SSS property.
37. $\triangle POA$ is isosceles since $OP = OA$. Therefore, the base angles have the same measure x whose sum $2x$ is the measure of the opposite exterior angle. That is, $m\angle AOQ = 2x = 2m\angle OPA$. Similarly, $m\angle BOQ = 2y = 2m\angle OPB$. Therefore, $m\angle AOB = m\angle AOQ + m\angle BOQ = 2m\angle OPA + 2m\angle OPB = 2m\angle APB$.

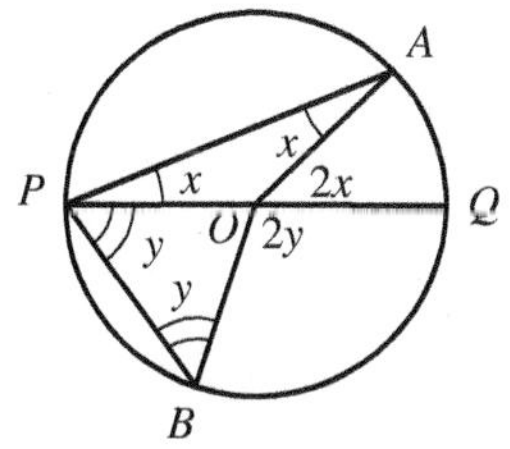

39. Construct the four radii from the center O, so each side $\overline{AB}, \overline{BC}, \overline{CD}, \overline{DA}$ is the base of an isosceles triangle. Let w, x, y, z be the measure of the base angles of the respective triangles, as shown in the diagram below. The sum of the interior angles of the quadrilateral is 360°, so $(w + x) + (x + y) + (y + z) + (z + w) = 360°$. That is, $w + x + y + z = 180°$. But then $m\angle A + m\angle C = (w + x) + (y + z) = 180°$. Similarly, $m\angle B + m\angle D = (x + y) + (z + w) = 180°$.

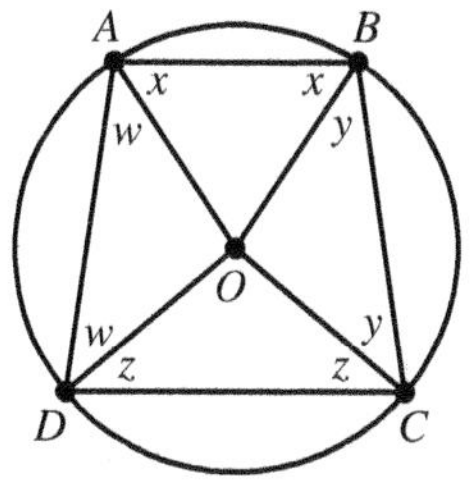

41. Answers will vary but should point out clearly that all eight triangles have the same size and shape. **43.** C

Problem Set 12.2 (page 643)

1. Use Construction 4, since the point C is the point of reflection.
3. Step 2. Use Construction 4 to construct the lines k and l perpendicular to line m that pass through A and B.
Step 3. Construct the circles of radius AB centered at A and B. These circles intersect k and l at the points C and D so that $ABCD$ is the required square.

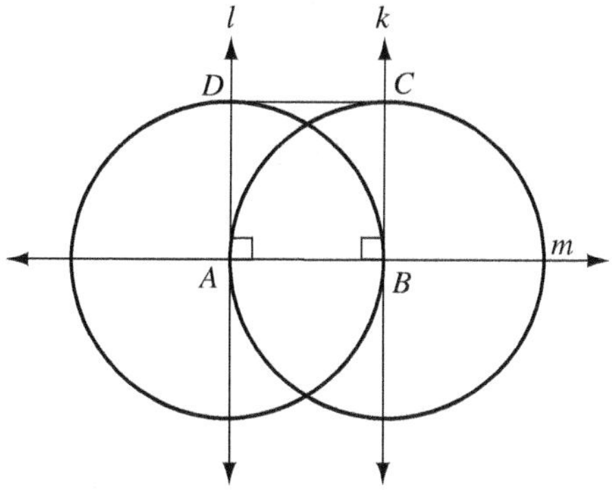

5. Step 1. Construct a circular arc centered at A that passes through point B; let D be the intersection of the arc with the ray $\overrightarrow{AP}$.

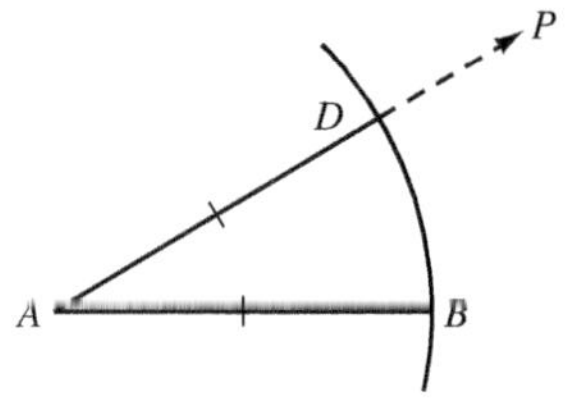

Step 2. Construct a circular arc of radius AB centered at B and D; let C be the point of intersection of these two arcs. Then $ABCD$ is the required rhombus.

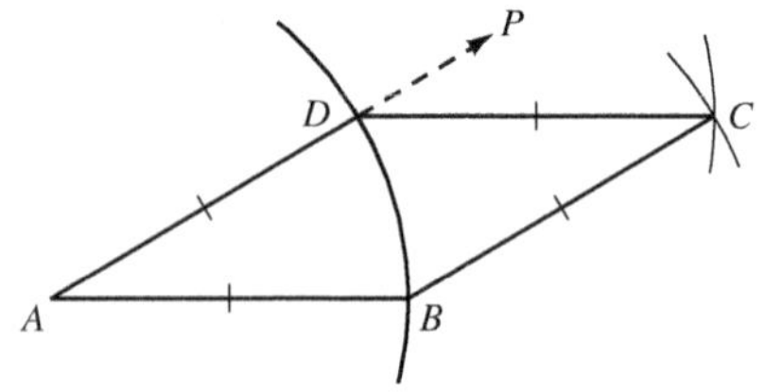

7. Step 1. Use Constructions 3 and 4 to construct lines k and m through A that are respectively perpendicular and parallel to line l. Let k and l intersect at B.

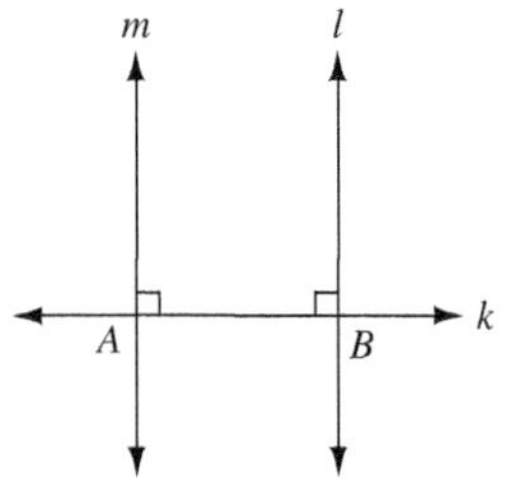

Step 2. Construct the circles of radius AB centered at A and B, and let them intersect l and m at C and D. Then $ABCD$ is the required square.

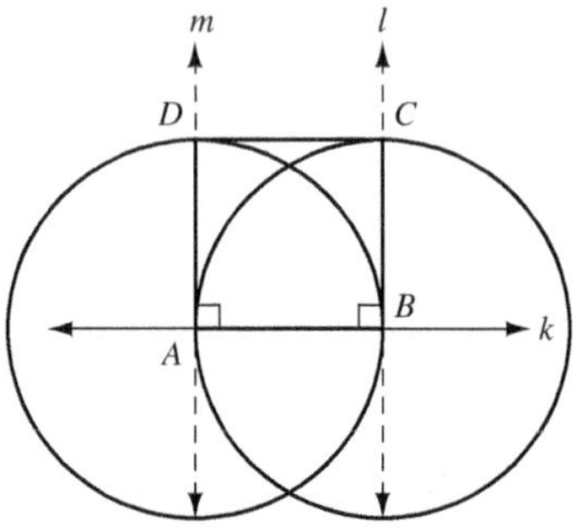

9. Pivot the Mira about P until l coincides with its reflection to construct the line m through P that is perpendicular to l. Pivot the Mira about P until m coincides with its reflection to construct the line k through P that is perpendicular to m. Thus, k is parallel to l. **11. (a)** The corresponding-angles property guarantees that m is parallel to l. **(b)** Align the ruler with the line; slide the drafting triangle, with one leg of the right triangle on the ruler, until the second leg of the triangle meets point P. **13.** Construct the perpendicular to side $\overrightarrow{BA}$ at T and the angle bisector. Let O be their point of intersection. The circle centered at O and passing through T is the desired circle.

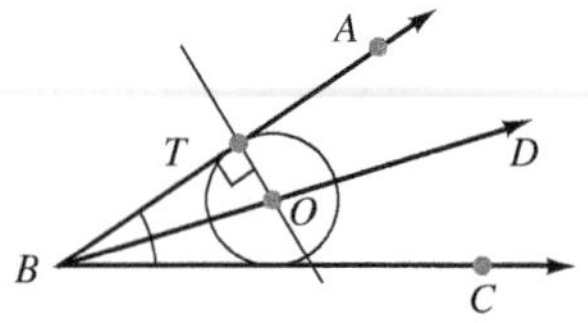

15. (a) The circumcenter will be inside the acute triangle. **(b)** The circumcenter will be at the midpoint of the hypotenuse of a right triangle. **(c)** The circumcenter will be outside an obtuse triangle. **(d)** The circumcenter is inside, on, or outside a triangle if, and only if, the triangle is acute, right, or obtuse, respectively. **17.** Since $\triangle PQS$ is inscribed in the circle with diameter $\overline{PQ}$, it has a right angle at S by Thales' theorem. Thus, $\overline{PS} \perp \overline{SQ}$. Similarly, $\overline{PT} \perp \overline{TQ}$. **19.** By Thales' theorem, $\angle ADB$ is a right angle. We also see that $\triangle ODB$ is an equilateral triangle, since all sides have the length of the radius. Moreover, ODBE is a rhombus, so the side $\overline{DE}$ is a bisector of the 60° angle $\angle ODB$. Thus, $m(\angle ADE) = m(\angle ADB) - m(\angle EDB) = 90° - 30° = 60°$. Similarly, $m(\angle AED) = 60°$. Therefore, all angles of $\triangle ADE$ have measure 60°, so $\triangle ADE$ is equilateral. **21. (a)** Extend $\overline{AB}$ to a longer segment. Erect perpendicular rays to $\overline{AB}$ at both A and B to the same side of $\overline{AB}$. Bisect the right angle at A, and let its intersection with the ray at B determine point C. Erect the perpendicular at C to $\overline{BC}$, and let D be the intersection with the ray constructed at A. Then $ABCD$ is a square erected on the given side $\overline{AB}$. **(b)** Construct the equilateral triangle $\triangle ABO$ with the given side $\overline{AB}$. The reflection of A across $\overline{OB}$ determines the point C. Repeat reflections to determine the remaining vertices D, E, and F to construct the regular hexagon $ABCDEF$. **23. (a)** Suppose the perpendicular bisector of chord $\overline{AB}$ intersects the circle at a point C. Then the circle is the circumscribing circle of $\triangle ABC$. The center of the circumscribing circle is the point of concurrence of the perpendicular bisectors of all three sides of $\triangle ABC$. In particular, the perpendicular bisector of side $\overline{AB}$ contains the center of the circle. **(b)** Construct two nonparallel chords and their respective perpendicular bisectors. By part (a), each bisector contains the center of the circle. Hence, the bisectors intersect at the center. **(c)** The dashed lines are the perpendicular bisectors of chords $\overline{AB}$ and $\overline{BC}$. By part (b), these bisectors intersect at the center of the larger circle.

25. (a) Waun seems to have constructed an equilateral pentagon, since all five sides have the same length. However, not all of the angles are congruent to one another, so Waun's pentagon is not regular. **(b)** Waun needs to check that all sides are congruent and all angles are congruent.

27. Trisecting an angle with compass and straightedge is impossible. In the case of a 150° angle, trisecting the related chord gives angles close to 25°, 100°, and 25°.

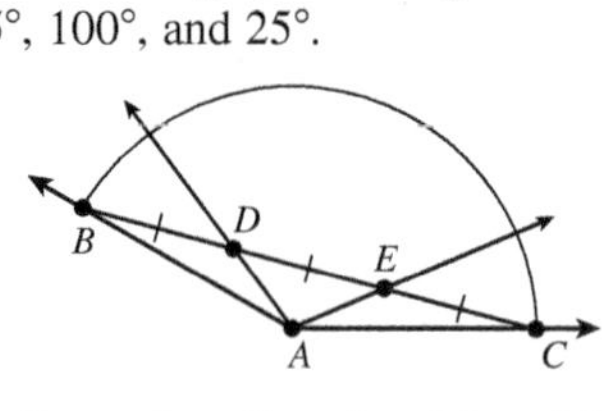

29. (a) There are five points C on line l for which triangle ABC is isosceles. As shown in the accompanying diagram, they are constructed by the circle at A through B, the circle at B through A, and the perpendicular bisector of $\overline{AB}$.

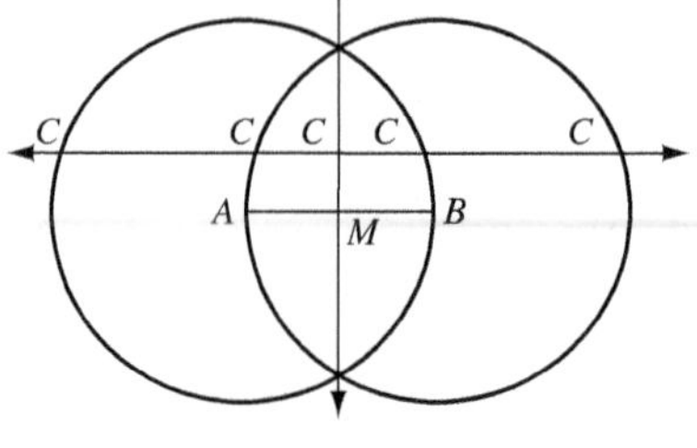

(b) There are four points D on line l for which triangle ABD is a right triangle. As shown in the accompanying diagram, they are constructed by the circle at M, which is the midpoint of $\overline{AB}$, through A, and by the perpendiculars to $\overline{AB}$ at A and B.

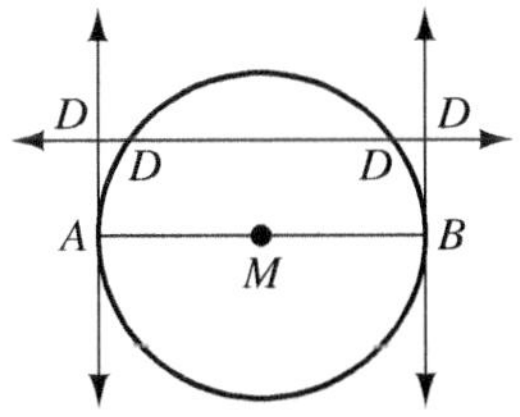

31. Draw $\triangle ABC$ and construct the angle bisectors of $\angle A$ and $\angle B$. Call their point of intersection P. From the equidistance property of the angle bisector, the distance from P to $\overline{AB}$ is the same as the distance from P to $\overline{BC}$, because P is on the angle bisector of $\angle B$. Likewise, the distance from P to $\overline{AB}$ is the same as the distance from P to $\overline{AC}$, because P is on the angle bisector of $\angle A$. Thus, the distance from P to $\overline{BC}$ is the same as the distance from P to $\overline{AC}$, so P lies on the angle bisector of $\angle C$. Therefore, the three angle bisectors are concurrent. **33. (a)** Construction should match the illustration in the text. **(b)** A ruler and protractor will verify that *PENTA* is regular. **35. (a)** Construction should match the illustration in the text. **(b)** *BFGHIJK* is a close approximation, since 7 is not a Fermat prime. **37.** Constructible: 3, 4, 5, 6, 8, 10, 12, 15, 16, 17, 20, 24, 30, 32, 34, 40, 48, 51, 60, 64, 68, 80, 85, 96 **39.** Since the 70° measure of the exterior angle is the sum of measures of the opposite interior angles, one of which is 35°, the interior angle at C also has measure 35°. Therefore, triangle ABC is isosceles, and $AB = BC$, putting the lighthouse at C four miles from B. In general, "doubling the angle" will give a distance that is equal to the distance moved from the initial point where the angle is first observed. **41.** B

Problem Set 12.3 (page 654)

1. (a) First notice that $m(\angle O) = 180° - 60° - 30° = 90°$. Therefore, by the AA similarity property, $\triangle ABC \sim \triangle PNO$. The scale factor from $\triangle ABC$ to $\triangle PNO$ is $\frac{12}{8} = \frac{3}{2}$.

(b) $\frac{QS}{EF} = \frac{SR}{FD} = \frac{RQ}{DE} = 2$, so $\triangle DEF \sim \triangle RQS$ by the SSS similarity property. The scale factor is 2. **3. (a)** Yes, by AA; all angles are the same, 60°. **(b)** No, there are many possible examples. **(c)** Yes, by the AA similarity property

5. (a) $\frac{12}{15} = \frac{8}{a}$, so $a = 10$. **(b)** $\frac{8}{6} = \frac{12}{b}$; $b = 12 \cdot \frac{6}{8} = 9$

7. (a) By the Pythagorean theorem, the hypotenuse of triangle XYZ is $\sqrt{5^2 + 12^2} = \sqrt{169} = 13$, so its perimeter is $5 + 12 + 13 = 30$. Since the perimeter of triangle ABC is 3000 feet, the scale factor is 100. Thus, triangle ABC is a right triangle with legs of lengths 500 and 1200 and a hypotenuse of length 1300. **(b)** The area of triangle XYZ is $5 \times 12/2 = 30$ sq ft and the area of triangle DEF is 100 times as large. Therefore, the scale factor is $\sqrt{100} = 10$, and triangle DEF must be a right triangle with legs of length 50 feet and 120 feet and a hypotenuse of length 130 feet. **9.** No. A square and a nonsquare rectangle are convex quadrilaterals with congruent angles, yet are not similar.

11. There are two choices for F:

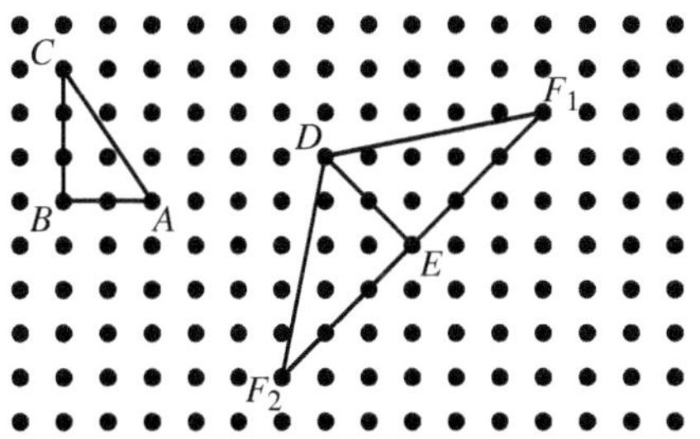

13. $\angle AEB \cong \angle DEC$, since they are vertical angles. Also, since $\overline{AB}$ and $\overline{CD}$ are parallel, $\angle BAE \cong \angle CDE$ by the alternate-interior-angles theorem. By the AA similarity property, $\triangle ABE \sim \triangle DCE$, so $\frac{AE}{DE} = \frac{BE}{CE}$. Therefore, $\frac{a}{b} = \frac{x}{y}$, or $a \cdot y = x \cdot b$. **15. (a)** $m(\angle A) + m(\angle B) + m(\angle C) = 180°$, and $m(\angle B) = m(\angle C)$, since the triangle is isosceles. Therefore, $m(\angle C) = 72°$. In $\triangle BCD$, $\angle C \cong \angle CDB$, so $m(\angle CBD) = 180° - 72° - 72° = 36°$. Thus, $\triangle BCD$ is a golden triangle. **(b)** Answers will vary.

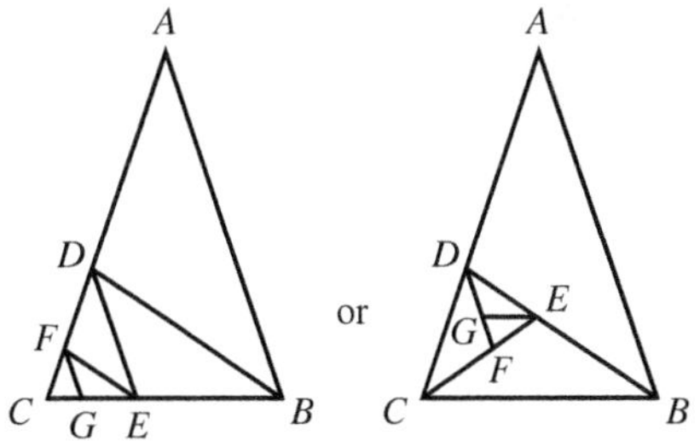

17. All three angles are congruent, so triangle ABC is equilateral. **19.** Answers vary. **21.** This would be a good opportunity to use a Venn diagram to show the attributes of similar and congruent shapes. Provide examples for Jackson to sort, and have him justify his choices on the basis of the relationships common to similar or congruent figures. In particular, all congruent triangles are also similar triangles with a scale factor of 1. However, similar triangles are not congruent if their scale factor is other than 1. **23.** Reef's reasoning is correct and explains why addition won't generally give similar triangles. (The only exception is an equilateral triangle.) Miley needs to understand that multiplication is the key: When each side is to be multiplied by the same number—namely, the scale factor—the two triangles are similar. **25. (a)** Use the AA similarity property. **(b)** Using $\triangle ACD \sim \triangle ABC$, we get $\frac{AD}{AC} = \frac{AC}{AB}$, or $\frac{x}{b} = \frac{b}{c}$. Similarly, $\triangle CBD \sim \triangle ABC$ gives $\frac{BD}{BC} = \frac{CB}{AB}$, or $\frac{y}{a} = \frac{a}{c}$. **(c)** $x = \frac{b^2}{c}$ and $y = \frac{a^2}{c}$. Also, $x + y = c$, so $c = \frac{b^2}{c} + \frac{a^2}{c}$, or $c^2 = a^2 + b^2$. **27. (a)** $DJ = \sqrt{1^2 + \left(\frac{1}{2}\right)^2} = \sqrt{\frac{5}{4}} = \frac{1}{2}\sqrt{5}$ **(b)** $PS = \frac{PS}{AD} = \frac{ST}{DJ} = \frac{1/2}{\sqrt{5}/2} = \frac{1}{\sqrt{5}}$ **(c)** Area$(PQRS) = (PS)^2 = \frac{1}{5}$ **29.** The lines perpendicular to the sides $\overline{AB}$ and $\overline{AC}$ cross at an angle congruent to $\angle A$. That is, $\angle P \cong \angle A$. Similarly, $\angle Q \cong \angle B$, so $\triangle PQR \sim \triangle ABC$ by the AA similarity property.

31. As corresponding angles to parallel lines, $\angle EPD \cong \angle ABD$. Since $\angle EDP = \angle ADB$, it follows from the AA similarity property that $\triangle EPD \sim \triangle ABD$. Thus, $\frac{x}{a} = \frac{p}{p+q}$. A similar argument shows that $\triangle AEP \sim \triangle ADC$, so that $\frac{x}{b} = \frac{q}{p+q}$. Adding these proportions gives $\frac{x}{a} + \frac{x}{b} = \frac{p+q}{p+q} = 1$, so $x = \frac{1}{\frac{1}{a}+\frac{1}{b}} = \frac{ab}{a+b}$. The same procedures show that $y = \frac{ab}{a+b}$. **33.** The right triangles $\triangle ACP$ and $\triangle BDP$ have congruent vertical angles at P. Thus, $\triangle ACP \sim \triangle BDP$ by AA similarity. Since $\frac{AC}{BD} = \frac{4}{2}$, the scale factor is 2. Therefore, $CP = 2DP$. Since $CD = 4$ and $CD = CP + DP$, it follows that $CP = \frac{2}{3}(4) = \frac{8}{3}$ and $DP = \frac{1}{3}(4) = \frac{4}{3}$. By the Pythagorean theorem, $AP = \sqrt{4^2 + (8/3)^2} = \left(\frac{4}{3}\right)\sqrt{13}$, so $BP = \left(\frac{2}{3}\right)\sqrt{13}$. **35.** If Mingxi and the tree are both vertical on flat (not necessarily level) ground, the two triangles are similar by the AA property. Therefore, the height h of the tree satisfies $\frac{h}{5.75'} = \frac{(75+7)'}{7'}$. Solving, we obtain $h = \frac{(5.75')82'}{7'} \doteq 67.36' \doteq 67'4''$. **37.** 385′ **39.** In the figure shown, G is the general's eye and his staff hangs vertically downward from point A. Then, $\triangle GAB \sim \triangle FDB$ and $\triangle GAC \sim \triangle EDC$. Since $AG = 2$ feet, $AB = 1$ foot, and $DB = 44$ feet, it follows that $DF = 44 \cdot 2 = 88$ feet. In the same way, since $AG = 2$ feet, $AC = 3$ feet, and $DC = 42$ feet, it follows that $DE = 42 \cdot \frac{2}{3} = 28$ feet. Therefore, the river is $DF - DE = 88 - 28 = 60$ feet wide.

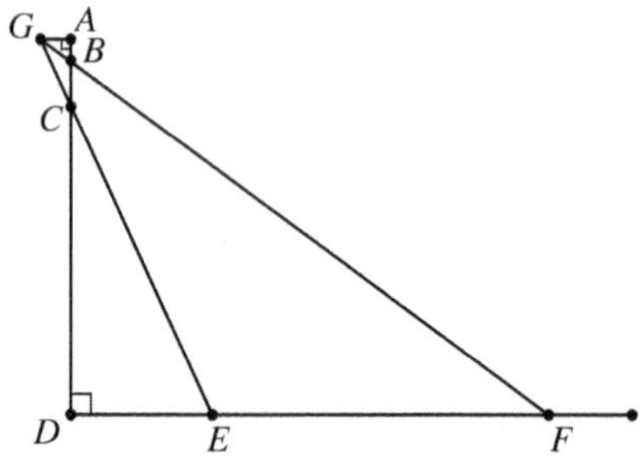

41. C **43.** C

Chapter 12 Review Exercises (page 663)

1. (a) True. Three pairs of angles and two pairs of sides are congruent. **(b)** False. After pairing up congruent angles, we see that the sides with equal lengths are not corresponding sides in the triangles, so the two triangles are not congruent. **3. (a)** 2.9 cm **(b)** 40° **(c)** 78° **(d)** 40° **5.** $\angle B \cong \angle C$, since $\triangle ABC$ is isosceles. By construction, $BF = DC$ and $BD = EC$. Therefore, $\triangle BDF \cong \triangle CED$ by SAS, so $DE = DF$.

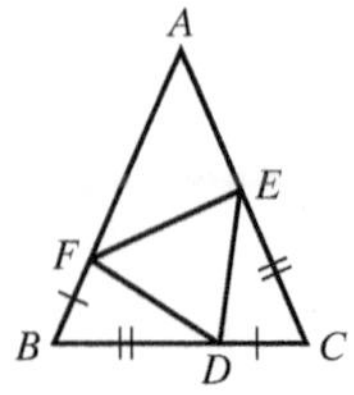

7. (a), (b), (c) Standard constructions as in Section 12.2. **(d)** Draw a circle at any point A on line m, and let it intersect line l at B and C. Construct $\overline{AB}$ and $\overline{AC}$. Draw circles of the same radius at B and C to determine the respective midpoints M and N of $\overline{AB}$ and $\overline{AC}$. Then $k = \overleftrightarrow{MN}$ is the desired line. Alternatively, construct a line perpendicular to l at a point on l. This determines a perpendicular segment between l and m. The perpendicular bisector of the segment is the desired line k.

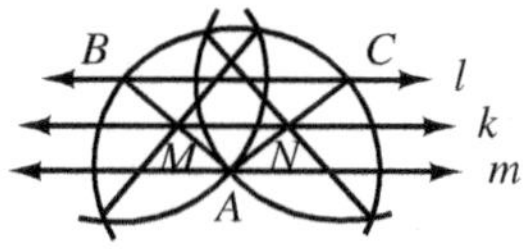

9. (a) Construct $\angle A$, lay off length AB, and draw a circle at B of radius BC. The circle intersects the other ray from A at two points C_1 and C_2, giving two triangles $\triangle ABC_1$ and $\triangle ABC_2$.

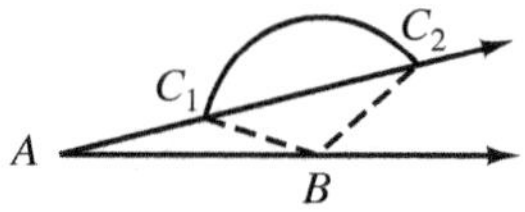

(b) Only $\triangle ABC_1$ has $\angle C = \angle C_1$ obtuse. **11. (a)** Yes, by using the SSS similarity property **(b)** Yes, by using the AA similarity property **13. (a)** $\triangle BAC \sim \triangle PQR$ by SAS similarity. The scale factor is $\frac{6 \text{ cm}}{4 \text{ cm}} = \frac{3}{2}$. **(b)** $\triangle ABC \sim \triangle YZX$ by SSS similarity. The scale factor is $\frac{42''}{14''} = 3$. **(c)** $\triangle ABC \sim \triangle HGF$ by AA similarity. The scale factor is $\frac{4}{6} = \frac{2}{3}$. **(d)** $\triangle ADB \sim \triangle BCD$ by SSS similarity. The scale factor is $\frac{5}{10} = \frac{1}{2}$. **15.** Let K be the top of the stick and Y the top of the pyramid. Then $\triangle KS_1S_2 \sim \triangle YP_1P_2$, with scale factor $\frac{P_1P_2}{S_1S_2} = \frac{270}{2} = 135$. Therefore, the height of the pyramid is 135×3 feet $= 405$ feet.

Chapter 13

Problem Set 13.1 (page 678)

1. (a) Answers will vary. It is difficult to tell with the scores all mixed up.
(b)

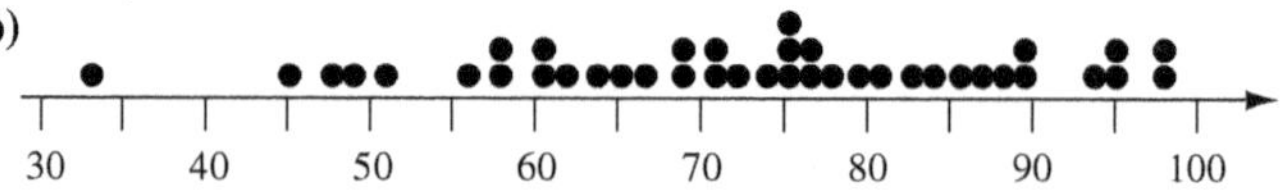

(c) Answers will vary. 75 seems about right. **(d)** The score of 33 seems quite atypical. **(e)** Scores above 90 should receive A grades and scores below 60 an F

3.

3	3
4	5 8 9
5	1 6 8 8
6	1 1 2 4 5 6 9 9
7	1 1 2 4 5 5 5 6 6 7 9
8	0 3 4 6 7 8 9 9
9	4 5 5 8 8

5.

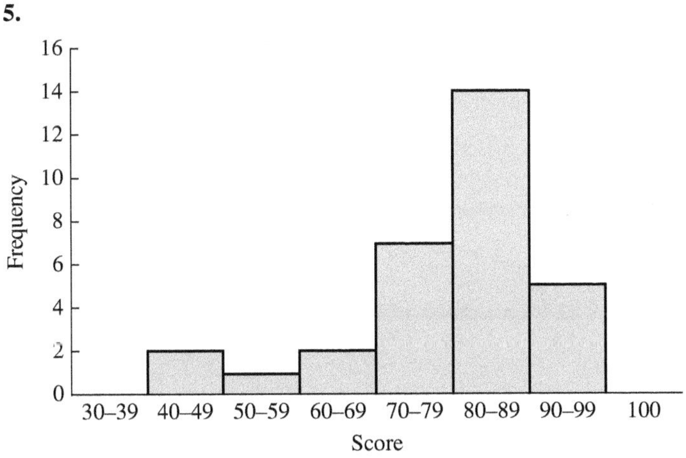

7. (a)

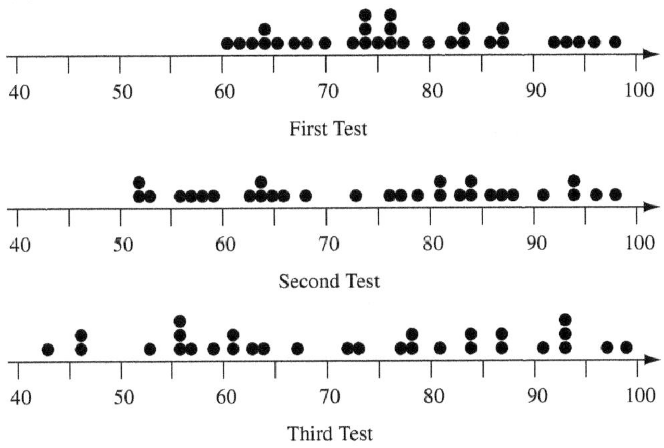

(b) The test scores indicate that the poorer students did increasingly less well as the term progressed. **9. (a)** 108° **(b)** 129.6° **11. (a)** Approximately 79° **(b)** Approximately 39% **(c)** About $\frac{1}{3}$ as much **(d)** About the same amount **13. (c)** Answers will vary. **(d)** It is not really possible to do. If the street corner were close to the original one and there was a lot of traffic on the street, the logic of part (c) might be ok. On the other hand, if you were in a different city, if the street were not very busy, or if the corner were in an area different from that of the original street corner, or at the same corner but at a different time of day, there most likely would be no correlation at all.

15. (a)

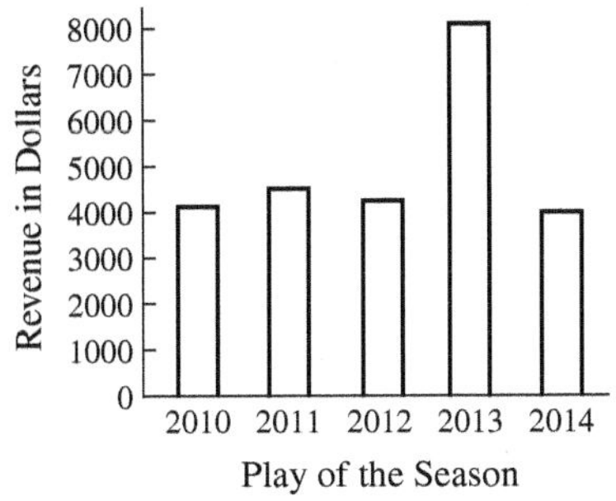

(b) Since student interest may not correlate with revenue, the answer is "no." For example, one can't determine the number of students who bought tickets or whether the price of a ticket changed in different years.

17.

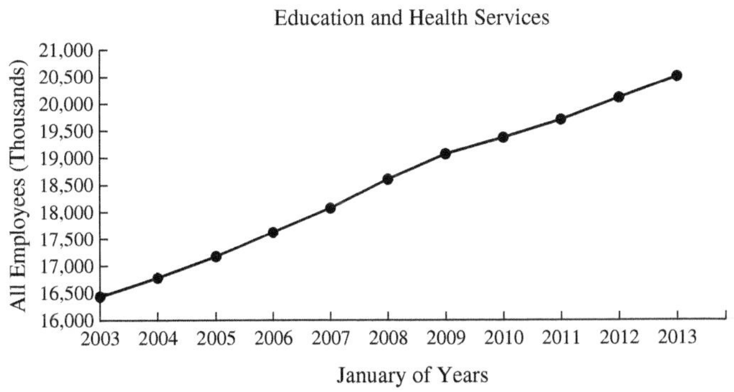

Source: U.S. Bureau of Labor Statistics Employment, Hours, and Earnings from the Current Employment Statistics survey (National), Bureau of Labor Statistics.

19. (a) Answers will vary. **(b)** Bar graphs should represent individual data. **(c)** Answers will vary. **(d)** Answers will vary. **21. (a)** Two-and-a-half icons

(b)

Tree	Trees Counted
Oak	Δ Δ Δ Δ Δ Δ Δ Δ Δ Δ
Maple	Δ Δ Δ Δ Δ Δ Δ ∠
Willow	Δ Δ ∠
Pine	Δ Δ Δ Δ Δ Δ Δ Δ Δ Δ Δ Δ ∠
Birch	Δ Δ Δ Δ Δ

23. (a) Sarita is not using the key on the pictograph. She is assigning a value of 1 to each picture instead of the value of two books. This is how she is arriving at the answer that Fred has read four more books than James. She simply found the person who had four more squares than James. For the second question, she is right that Fred read the most but is again assigning only a value of 1 to each picture and not using the key to interpret the graph. **(b)** For question 1, Kristen has read exactly four more books than James. For question 2, Fred read ten, which is the most. **(c)** Answers may vary. One possible answer is to have Sarita write the number 2 inside each square so that she remembers the value of each picture. Another strategy could be for her to focus on the key and then go to each person in the graph and write down his or her total before ever interpreting the questions. For example, behind Fred, write 10, behind Kristen, write 6, etc. **25.** Following is a possible answer (there could be others; note the ambiguity (done on purpose) of what "circumference of the head" might mean):

Comparing Histograms and Bar Graphs

Histograms	**Bar Graphs**
Organizes and summarizes data	Organizes and summarizes data
Data represented in intervals of numeric data	Usually displays categorical data
Vertical axis indicates frequency	Vertical axis indicates frequency
Usually "continuous data" adjacent to each other (bars must touch)	Can be bars that are separated (do not abut)
Numbers are grouped in a continuous range from left to right	Shows the frequency of each element

27. **(a)** The pictographs are misleading. One would guess the assets had more than doubled. **(b)** The volume for 2012 is eight times the volume for 2003, not double. **(c)** Yes **29.** **(a)** $252,000,000

(b)

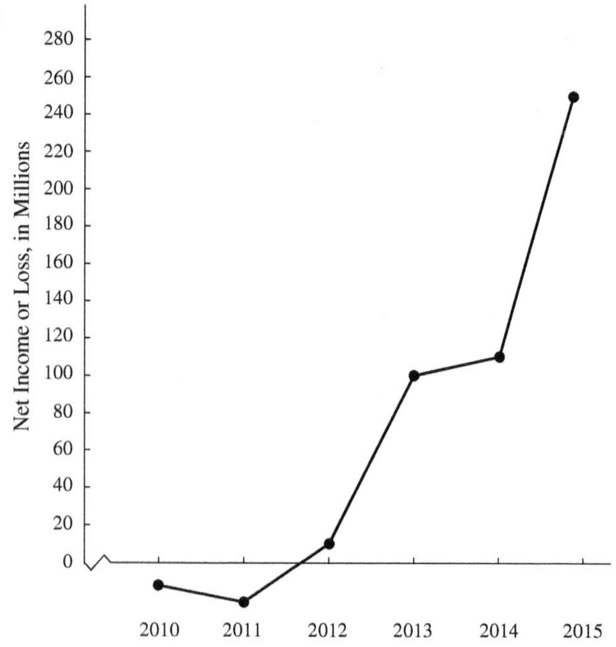

(c) With the net income looking like $105 million for 2014 and $5 million in 2012, the answer comes from $\frac{105 - 5}{5}$ which is 2000%. **(d)** 152%

31. **(a)**

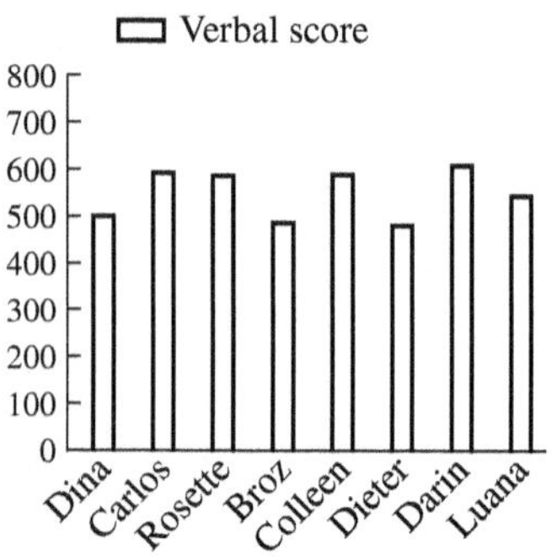

(b)

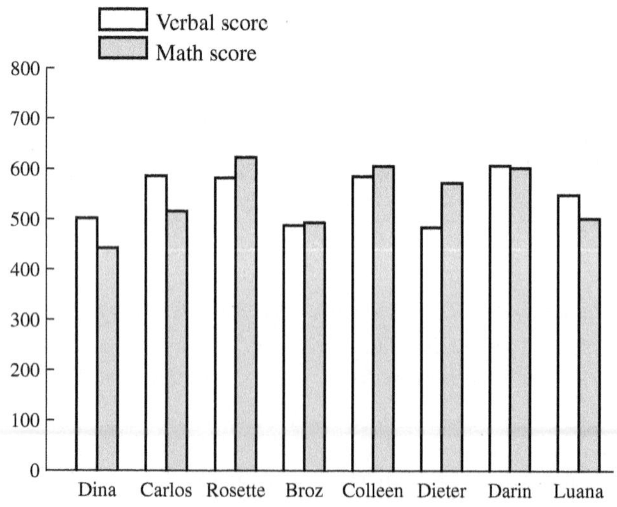

33.

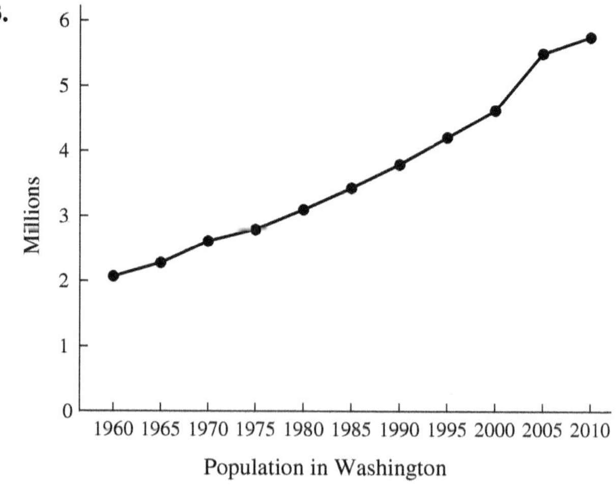

35. B **37.** D

Problem Set 13.2 (page 697)

1. $\bar{x} \doteq 20.6, \hat{x} = 19$, mode $= 18$
3. **(a)** List the values from problem 1 in order: {14, 15, 17, 17, 18, 18, 18, 19, 19, 21, 22, 23, 24, 27, 27, 30}. Q_L is the median for the first eight values: $Q_L = 17.5$, $\hat{x}$ is the median of the entire set: $\hat{x} = 19$. Q_U is the median for the last eight values: $Q_U = 23.5$. **(b)** The 5-number summary is (smallest value–Q_L–$\hat{x}$–Q_U –largest value). For this problem, we list them in order, 14–17.5–19 23.5–30. **(c)** Box-and-whiskers plot

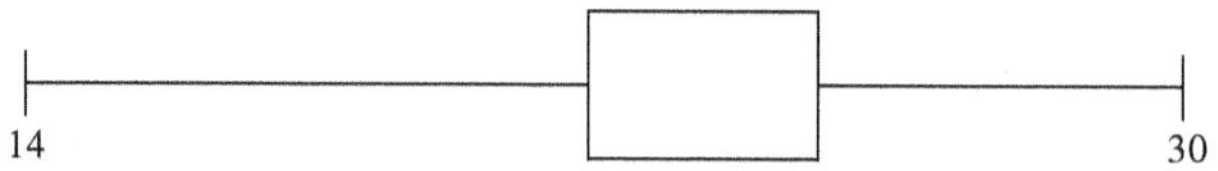

(d) IQR $= Q_U - Q_L = 23.5 - 17.5 = 6$. **(e)** $Q_L - 1.5 \cdot \text{IQR} = 17.5 - (1.5)6 = 8.5$ so no outliers (no data points below14). $Q_U + 1.5 \cdot \text{IQR} = 23.5 + 9 = 32.5$ so no outliers (no data points above 30). **5.** $(14 + 30)/2 = 22$

7. **(a)**

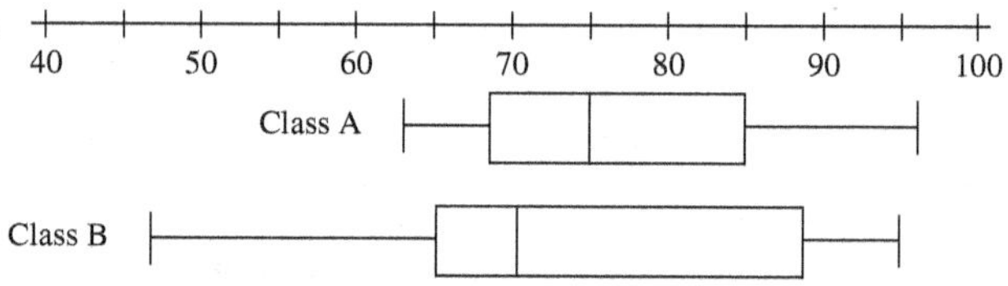

(b) As a whole, class A had higher final grades than class B. **(c)** For class A, IQR $= 16.5$. For class B, IQR $= 23$. **(d)** No data values in class A are less than 43.75, and none is greater than 109.75, so there are no outliers in class A. No data values in class B are less than 30.5, and none is greater than 122.5, so there are no outliers in class B. **9.** **(a)** 69% **(b)** 94% **(c)** 100% **11.** **(a)** The sum of the data points is $2(1) + 2(2) + 2(3) + 2(4) + (2)5 = 30$, so the mean is 3, since there are 10 data points ($n = 10$). Since every integer comes up the same number of times, every integer (between 1 and 5 inclusive) is a mode. The median is 3. **(b)** The sum of the data points is $4(1) + 4(2) + 4(3) + 4(4) + (4)5 = 60$, so the mean is 3, since there are 20 data points ($n = 20$). Since every integer comes up the same number of times, every integer (between 1 and 5 inclusive) is a mode. The median is 3. **(c)** The sum of the data points is $6(1) + 6(2) + 6(3) + 6(4) + (6)5 = 90$, so the mean is 3, since there are 30 data points ($n = 30$). Since every integer comes up the same number of times, every integer (between

1 and 5 inclusive) is a mode. The median is 3. **(d)** Each of the data sets has the same mean, the same modes, and the same median. **13. (a)** Yes, there will be five blocks in each stack. **(b)** $\bar{x} = 5$ **(c)** The height of each of the stacks is the mean of the data set. **(d)** First, have the students consider stacks of equal height. Second, rearrange the stacks to be of unequal heights. In each case, discuss the notion of typical, or average, height. **(e)** The 5, 1, 4, 7, 11, and 7 blocks cannot be arranged into six stacks with an equal number of blocks in each stack. The best that can be done is to have five stacks with six blocks each and one stack with five blocks. Still, this shows that the average value is about 6, perhaps a bit less. **15.** Answers will vary. One set of data produced a female "cubit" of about 17.5 inches and a male "cubit" of about 19.5 inches. **17. (a)** Instead of finding the median for the number of students, Joseph found the median for the number of books reported. When calculating the mode, Joseph looked at the most number of books possible to report, not the most number of books students had read, which was eight. The mode is 1 and median 2.5. **(b)** Ms. Chen could explain the difference between the label (number of books reported) and the outcome (number of students who read that many books). **19.** Presumably, Yugi is crossing off numbers alternately from each side. In that case, Yugi's approach works exactly when there is an odd number of data points. If there is an even number, he must take the average of the last two numbers. Showing him an example using a data set of 4 points would clarify the situation. **21.** Marilyn is making a mistake in talking about "the mode." There can be more than one mode in a data set such as $\{1, 1, 2, 2, 5\}$—which, in fact, has two modes. She should really be using the expression "a mode" unless she is talking about a specific data set. **23. (a)** We know from problem 10 that $x = 22.84$ (approximately). For each of the 12 data values, we subtract the mean from data value and then take the absolute value. The sum of the absolute values is 4.98. **(b)** $4.98 \div 12 = .415$. This result is not the standard deviation but Leona has a different way to look at data spread. **(c)** Compliment Leona on a good idea that avoids the canceling out of the effects of the various terms. **25. (a)** $\bar{x} = 27, s \doteq 8.4$ **(b)** $\bar{x} = 32, s \doteq 8.4$ **(c)** Adding (or subtracting) the same number from each set of data leaves the standard deviation unchanged, and the mean is increased (or decreased) by that number. **27.** $\bar{x} \doteq 37.4, \hat{x} = 42$, mode $= 16$

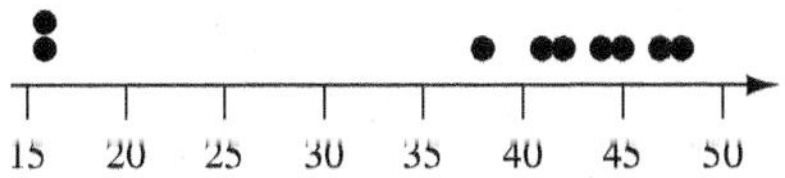

The median of 42 is a likely choice for the typical value. **29.** Many examples satisfy these conditions. **(a)** 5, 7, 10, 14, 14 **(b)** 0, 10, 10, 11, 12, 13, 14 **(c)** 8, 9, 10, 10, 11, 20 **31. (a)** All 3s **(b)**

		1s	2s	3s
Possibilities	1	5	0	5
	2	4	2	4
	3	3	4	3
	4	2	6	2
	5	1	8	1
	6	0	10	0

(c) All 1s **(d)** No, since all data values are equal. **33. (a)** S has the smaller standard deviation. The new data set has two additional values equal to its mean, so its values do cluster more toward its mean. **(b)** No, the standard deviation of set T is greater than that of set S because the data values 45 and 45 have been replaced by 80 and 10, which are farther from the mean. **35.** The total of data values in A is $30 \cdot 45 = 1350$. The total of data values in B is $40 \cdot 65 = 2600$. For the combined data,

$\bar{x} = \dfrac{1350 + 2600}{30 + 40} \doteq 56.4.$ **37.** B **39.** B **41.** A **43.** A

Problem Set 13.3 (page 710)

1. (a) All freshmen in U.S. colleges and universities in 2010 **(b)** All football players in U.S. colleges and universities in 2010 **(c)** All people in the United States **(d)** All people in Los Angeles **(e)** It would be inappropriate to include known criminals and individuals who are in mental institutions. Likely also, young children, and perhaps even teenagers, should not be included in the population. **3.** Probably not. Only those who are listed in the phone book can be selected for an interview, so people without a phone and people with unlisted numbers are not included. Also, face-to-face interviews can create problems such as nonresponse, lying, and interviewer bias. **5. (a)** This is surely a poor sampling procedure. The sample is clearly not random. The selection of the colleges or universities could easily reflect biases of the investigators. The choices of the faculty to be included in the study almost surely also reflect the bias of the administrators of the chosen schools. **(b)** Presumably, the population is all college and university faculty. But the opinions of faculty at large research universities are surely vastly different from those of their colleagues at small liberal arts colleges. Indeed, there are almost surely four distinct populations here. **7.** If the voters are numbered, he can use a random-number generator. Otherwise, since voter lists are essentially random, he might choose every nth person on the list of his constituents, where n is chosen large enough to guarantee a sample of reasonable size. **9.** Some possible answers are as follows: **(a)** Using a numbered list of all registered students, select 20 valid numbers from a random-number table that corresponds to the list of students. **(b)** The mean of the answers obtained should give a reasonable estimate. **(c)** The mean of all the means obtained should give a better estimate than that of part (b). **11. (a)** Between 21.8 and 27.2 **(b)** Between 19.1 and 29.9 **(c)** Between 16.4 and 32.6 **13.** Yes. Just continue taking samples until one finally shows up with eight out of the ten in the sample preferring Whito Toothpaste. **15. (a)** -0.1 **(b)** 0 **17. (a)** $0.04 = 4\%$ to the nearest hundredth **(b)** $0.60 = 60\%$ to the nearest hundredth **19.** Praise her. On any given toss, each digit has an equal chance of coming up. **21.** Rebecca is right if none of the two-digit numbers is to contain 0, 7, 8, or 9 as a digit. Otherwise, she is wrong. **23.** Yes, since all sides of the die are equally likely to come up, all sequences of 0s and 1s are equally likely to appear. **25.** No. The statistic is based on voluntary responses that may or may not represent the opinion of all high school biology teachers. Those mailing back the survey are those who feel strongly about the issue—most likely, those with strong religious beliefs who believe in creationism and those with strongly antireligious views. Those with moderate views would probably be unrepresented. **27. (a)** Since the z scores indicate the location of data points in a data set, in this case they are likely to be the same as in problem 14a. **(b)** $-1.6, 0.9, 0.4, -0.6, 1.4, -0.6$ **29. (a)** Probably twice what they were in problem 14a. **(b)** $-1.6, 0.9, 0.4, -0.6, 1.4, -0.6$ **(c)** The data are all twice as big, but s is also twice as big. Since the z scores are scaled by s, it is not surprising that they remain the same.

31. Compliment Mai Ling on good thinking and suggest that there must be a better way to generate the digits without a bias coming from where the point starts. This is the reason that computer or graphing calculator software is used to get a truly random process.

Chapter 13 Review Exercises (page 715)

1. **(a)**

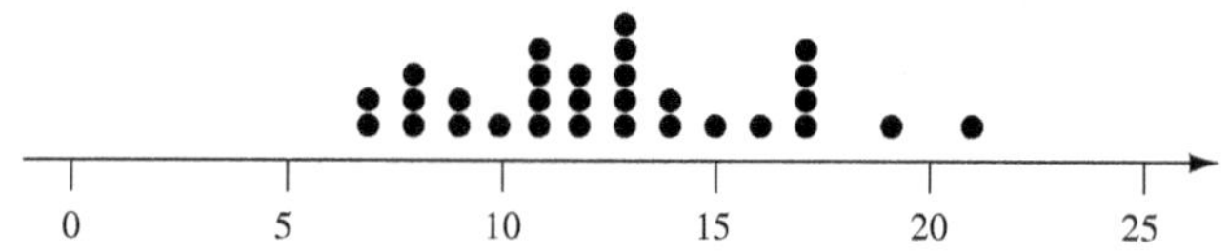

(b) About 13

3.

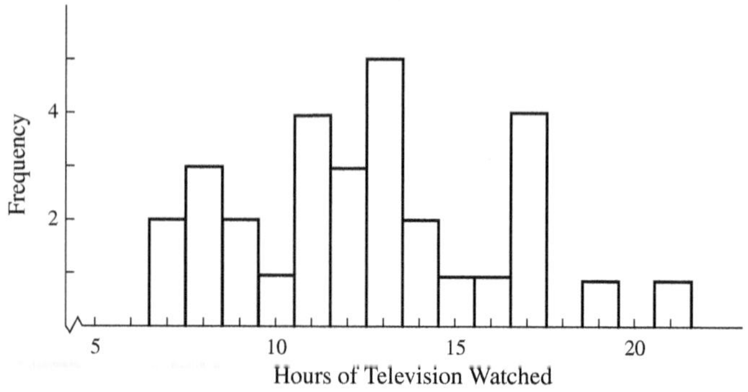

5. **(a)**

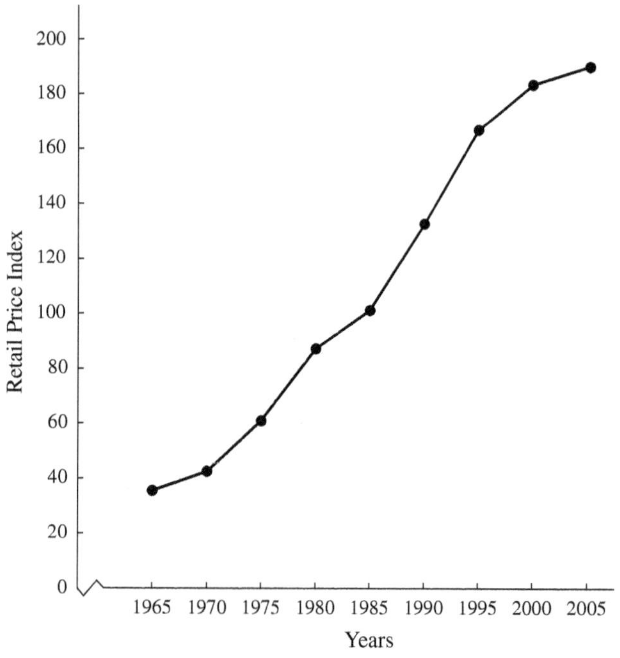

(b) About 48 **(c)** About 200

7.

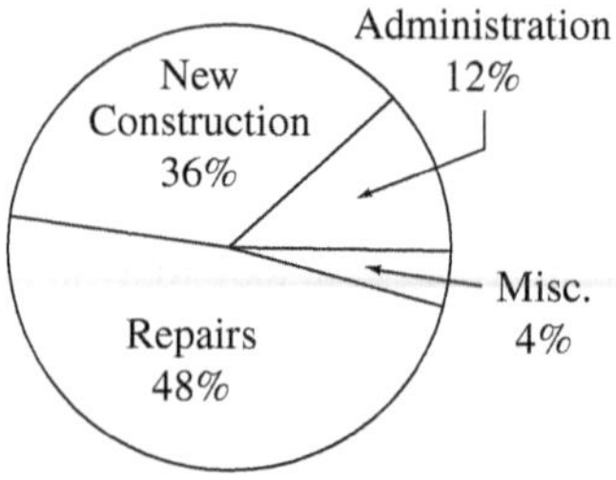

9. How is "medical doctor" defined? Does the term include all specialists? osteopaths? naturopaths? chiropractors? acupuncturists? How was the sampling done to determine the stated average?
11. **(a)** $Q_L = 10, Q_U = 15$ **(b)** 7−10−12.5−15−21
(c) There are no outliers. **(d)** $\bar{x} = 9, Q_L = 8, Q_U = 11$
(e) 6—8—9—11—13 **(f)** There are no outliers.

(g)

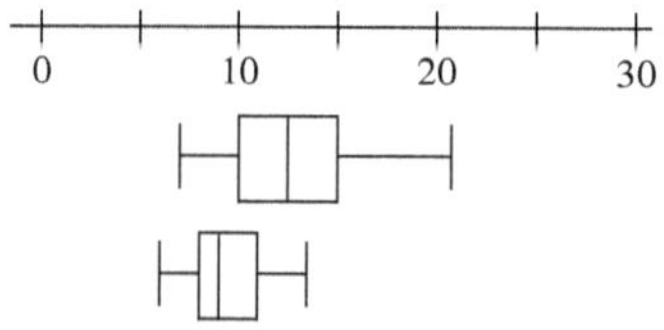

13. There are $21 \cdot 77 = 1617$ points for the 21 students. So there are 1839 points for all 24 students. Thus, the average is $1839 \div 24 \doteq 76.6$.
15. No. The sample represents only the population of students at State University, not university students nationwide.
17. Voluntary responses to mailed questionnaires tend to come primarily from those who feel strongly (either positively or negatively) about an issue or who represent narrow special-interest groups. They are rarely representative of the population as a whole. **19.** Since 12 is the sixth-largest number in the data set, and $\frac{6}{7} = 0.8571$ to the nearest ten-thousandth, 12 is the 85.71th percentile, to the nearest hundredth. **21.** From Table 13.4, the entry for −0.9 is 0.1841 and the entry for 0.9 is 0.8159. Therefore, the desired percentage is 81.59% − 18.41% = 63.18%.

Chapter 14

Problem Set 14.1 (page 729)

1. **(a)** Outcomes listed in order of penny, nickel, dime, and quarter are as follows:

HHHH	THHH	HTHT	TTTH	TTTT
	HTHH	HHTT	TTHT	
	HHTH	TTHH	THTT	
	HHHT	THTH	HTTT	
		THHT		
		HTTH		

(b) $P(\text{HHTT}) = \frac{1}{16}$ **(c)** $P(2 \text{ heads and } 2 \text{ tails}) = \frac{6}{16} = 0.375$

(d) $P(\text{at least one head}) = 1 - \frac{1}{16} = \frac{15}{16}$

3. **(a)** **(b)**

(1, 1)	(1, 2)	(1, 3)	(1, 4)	(1, 5)	(1, 6)
(2, 1)	(2, 2)	(2, 3)	(2, 4)	(2, 5)	(2, 6)
(3, 1)	(3, 2)	(3, 3)	(3, 4)	(3, 5)	(3, 6)
(4, 1)	(4, 2)	(4, 3)	(4, 4)	(4, 5)	(4, 6)
(5, 1)	(5, 2)	(5, 3)	(5, 4)	(5, 5)	(5, 6)
(6, 1)	(6, 2)	(6, 3)	(6, 4)	(6, 5)	(6, 6)

(c) A and B are mutually exclusive, since no outcome belongs to both events. Similarly, A and C are mutually exclusive. B and C are not mutually exclusive, since the outcomes (4, 5), (5, 5) and (6, 5) belong to both events. **5.** There are four possibilities for the two children: BB, BG, GB, and GG. We know that the family has at least one boy. Therefore, the possibilities are BB, BG, and GB. Thus, the probability that the other child is a girl is $\frac{2}{3}$.

7. (a) $\frac{4}{12} \doteq 0.33$ **(b)** $\frac{8}{12} \doteq 0.67$ **(c)** $\frac{1}{7} \doteq 0.14$
9. (a) 0.19 **(b)** 0.1 **(c)** 0.34, since 00 is included **(d)** 0.75 **(e)** 0.25 **(f)** 0.5 **11.** $\frac{9}{47} \doteq 0.19$ **13. (a)** 72 **(b)** 2/72 **(c)** $\frac{42}{72} = \frac{7}{12}$ **(d)** By complementary probabilities, the probability is $1 - 6/72 = 11/12$.
15. (a)

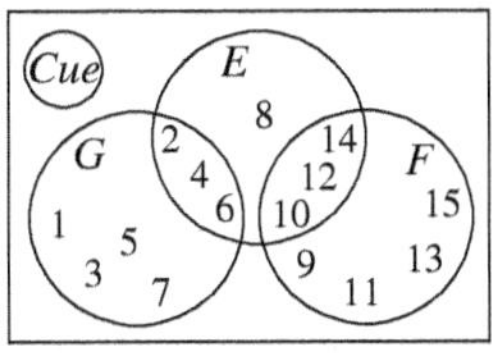

(b) $14/16 = 0.875$ **(c)** $8/16 = 0.5$ **(d)** $11/16 = 0.6875$
17. These answers are determind by ratios of angular measures of appropriate regions. **(a)** $P(\text{shaded area}) = \frac{3}{10} = 0.3$
(b) $P(\text{region 1 or region 2}) = \frac{72°}{360°} = 0.2$
(c) $P(\text{region 10 or region 6}) = \frac{2}{10} = 0.2$
(d) $P(\text{region e}) = \frac{72°}{360°} = 0.2$ **(e)** $P(8 \mid \text{shaded area}) = \frac{1}{3}$
(f) $P(\text{a vowel} \mid \text{odd-numbered region}) = 0.4$
(g) $P(\text{a vowel or odd-numbered region}) = \frac{7}{10} = 0.7$
19. Answers will vary but should indicate why the information is very helpful. Suppose the marble drawn is white. If the initially placed marble is black, there is just one way to draw a white marble. But if the initially placed marble is white, there are two ways to draw a white marble—either the initial or the added marble. That is, there is a $\frac{2}{3}$ probability of drawing a white marble, in which case the initial marble in the sack was also white. Thus, if the marble drawn from the sack is black, it must be the initially placed marble with probability 1, and if the drawn marble is white, the probability is $\frac{2}{3}$ that the initial marble is white. **21. (a)** Jose is incorrect, since there are 18 ways for Susan to get an odd number and only 9 ways for Bill to get a 4, 8, or 12. **(b)** Answers will vary, but it will be helpful to refer to the table of the 36 equally likely outcomes. **23.** The probability is slightly worse, since there is a small probability of drawing the same card twice.
25. $P(E \text{ and } F) = P(E) + P(F) - P(E \text{ or } F) = \frac{1}{2} + \frac{1}{4} - \frac{1}{2} = \frac{1}{4} = P(E)$ so $F \subseteq E$. **27. (a)** If x belongs only to set E, the only non zero term on the right side of the formula that counts x is $n(E)$. If y belongs to E and F but not G, there is a 1 added by each $n(E)$ and $n(F)$ and a 1 removed by $n(E \cap F)$, so y is added once since $1 + 1 - 1 = 1$. $n(E)$. If z belongs to E, F, and G, there is a 1 in each term of the right side of the formula, so y is added once since $1 + 1 + 1 - 1 - 1 - 1 + 1 = 1$. **(b)** Divide each term of the formula of part (a) by $n(S)$.

29. (a)

	1	3	4	5	6	8
1	(1, 1)	(1, 3)	(1, 4)	(1, 5)	(1, 6)	(1, 8)
2	(2, 1)	(2, 3)	(2, 4)	(2, 5)	(2, 6)	(2, 8)
2	(2, 1)	(2, 3)	(2, 4)	(2, 5)	(2, 6)	(2, 8)
3	(3, 1)	(3, 3)	(3, 4)	(3, 5)	(3, 6)	(3, 8)
3	(3, 1)	(3, 3)	(3, 4)	(3, 5)	(3, 6)	(3, 8)
4	(4, 1)	(4, 3)	(4, 4)	(4, 5)	(4, 6)	(4, 8)

(b) A 5 is the event {(1, 4), (2, 3),(2, 3),(4, 1)}, with probability 4/36. **(c)** The most likely roll is a seven, with the favorable outcomes (1, 6),(2, 5),(2, 5),(3, 4),(3, 4), (4, 3) and probability $6/36 = 1/6$. **(d)** The probability of any roll is the same with Sicherman and ordinary dice, so the game is even.
31. (a) 0.13 **(b)** 0.03 **(c)** 0.87 **33.** D **35.** E **37.** D

Problem Set 14.2 (page 741)

1. The events {(1, 3). (2, 2), (3, 1)}and {(1, 5), (2, 4), (3, 3), (4, 2), (5, 1)}are mutually exclusive in the uniform sample space of 36 outcomes, so the probability of rolling a 4 or a 6 is $\frac{3}{36} + \frac{5}{36} = \frac{8}{36} = \frac{2}{9}$. **3.** There are $2 \cdot 6 = 12$ equally likely outcomes. **(a)** {H2, H4, H6}is the event of a head and an even number, so its probability is $3/12 = 1/4$. **(b)** {H1, H2, H3, H4, H5, H6, T2, T4, T6}is the event of a head or an even number, so its probability is $9/12 = 3/4$. **5.** The Venn diagram shows that four students speak both English and Japanese.

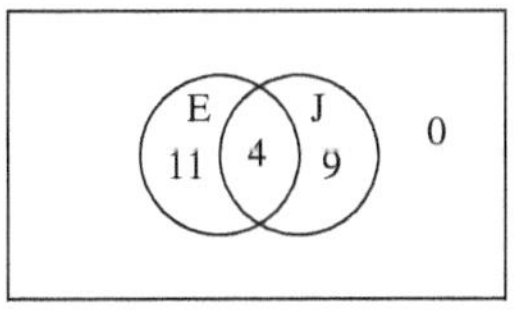

Let x students speak both languages, so $24 = 9 + 11 + x$. Therefore, $x = 4$, so the probability that a student speaks both languages is $4/24 = 1/6$. **7.** There are six red face cards and two black aces. Since the events are mutually exclusive, the probabilities add to give $\frac{6}{52} + \frac{2}{52} = \frac{8}{52} = \frac{2}{13}$.
9. (a) $5 \cdot 4 \cdot 3 = 60$ **(b)** $2 \cdot 4 \cdot 3 = 24$ **(c)** $2 \cdot 4 \cdot 3 = 24$ **(d)** The first digit can be chosen in two ways (a 2 or a 4), the last digit in three ways (a 1, 3, or 5), and the middle digit in three ways from the remaining digits not yet selected. This gives $2 \cdot 3 \cdot 3 = 18$ numbers in the event, with a probability of $\frac{18}{60} = \frac{3}{10}$.
11. (a) $5 \cdot 5 \cdot 5 \cdot 5 \cdot 5 = 3125$ **(b)** $5 \cdot 4 \cdot 3 \cdot 5 \cdot 4 = 1200$
13. (a)

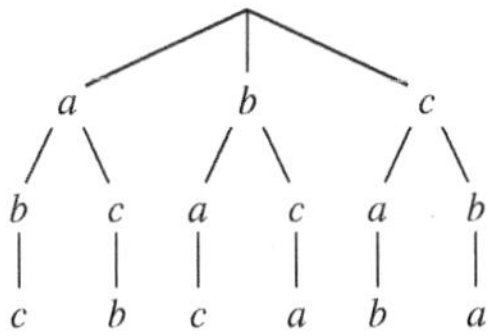

The code words are *abc*, *acb*, *bac*, *bca*, *cab*, and *cba*.

(b) There are three of the $3! = 6$ permutations with a ahead of b, namely, *abc*, *acb*, and *cab*. The probability is $3/6 = 1/2$.

15. $u = 1 - \frac{2}{3} = \frac{1}{3}$,

$v = 1 - \frac{2}{5} = \frac{3}{5}$, $w = \frac{2}{3} \cdot \frac{3}{5} = \frac{2}{5}$, $x = \frac{1}{4} \div \frac{1}{3} = \frac{3}{4}$,

$y = 1 - \frac{3}{4} = \frac{1}{4}$, $z = \frac{1}{3} \cdot \frac{1}{4} = \frac{1}{12}$

17. (a) 90,000, since $9 \cdot 10 \cdot 10 \cdot 10 \cdot 10 = 90{,}000$
(b) 45,000, since $9 \cdot 10 \cdot 10 \cdot 10 \cdot 5 = 45{,}000$
(c) 26,244, since $4 \cdot 9^4 = 26{,}244$ **(d)** $29{,}889 = 6561 + 23{,}328$, since $9^4 = 6561$ begin with a 7 followed by four digits, none of which is a 7, and $8 \cdot 4 \cdot 9^3 = 23{,}328$ have a digit 7 in exactly one of the four places that follows the leading digit, which is neither a 7 nor a 0. **19.** Answers will vary, but all should emphasize that Alejandro is double-counting the numbers that are divisible by both 4 and 5. That is, the 50 numbers 20, 40, . . . , 1000 that are divisible by 20 are counted twice simply by summing $200 + 250$. The correct answer is $450 - 50 = 400$. **21.** Answers will vary, but should **(a)** make it clear to Ernie that the multiplication principle gives the answer 8 (Ernie has apparently added 2 and 4); and **(b)** remind Donna that there are still two crust types available, so the correct answer is 6.

23. (a) Let FP, JP, . . . , FT, and JT denote the events of a slate with a freshman president, junior president, . . . , and junior treasurer. Then the possibility tree shows that there are $10 + 15 + 10 + 6 + 15 + 6 = 62$ possible slates in which the two officers are not from the same class.

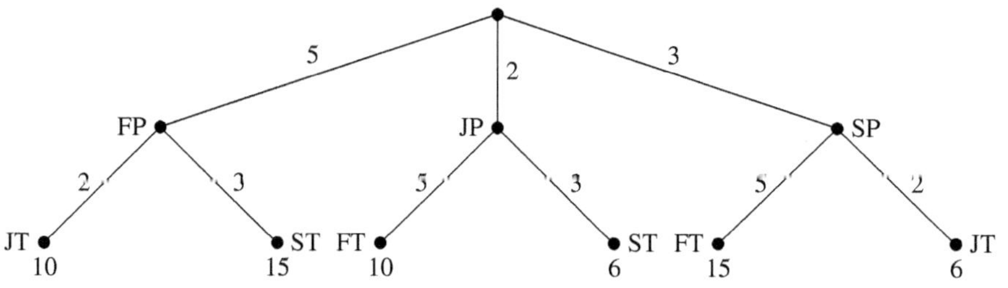

(b) Let $\overline{F}$, $\overline{J}$, and $\overline{S}$ respectively denote the events that no freshman, no junior, and no senior holds an office. To determine $n(\overline{F})$, notice that there are two choices of a junior and three choices of a senior, and two ways to assign them to the offices of president and treasurer. Therefore, $n(\overline{F}) = 2 \cdot 3 \cdot 2 = 12$. Similar reasoning shows that $n(\overline{J}) = 5 \cdot 3 \cdot 2 = 30$ and $n(\overline{S}) = 5 \cdot 2 \cdot 2 = 20$. Thus, since the events are mutually exclusive, there are $12 + 30 + 20 = 62$ different slates of officers in the Chess Club.

25. (a) $\frac{12}{52} \cdot \frac{12}{52} = \frac{9}{169}$ **(b)** $\frac{12}{52} \cdot \frac{11}{51} = \frac{11}{221}$

27. $\frac{10 \cdot 7 + 10 \cdot 9 + 7 \cdot 9}{C(26, 2)} = \frac{223}{325}$

29. (a) Since one of the nine keys is the correct one, $p_1 = \frac{1}{9}$. The first key tried is therefore incorrect with probability $\frac{8}{9}$, and the probability that second key is correct is $\frac{1}{8}$, so, by conditional probabilities $p_2 = \frac{8}{9} \cdot \frac{1}{8} = \frac{1}{9}$. Similarly, $p_3 = \frac{8}{9} \cdot \frac{7}{8} \cdot \frac{1}{7} = \frac{1}{9}$, and so on. **(b)** For n keys, the reasoning followed in part (a) shows that $p_1 = \frac{1}{n}$, $p_2 = \frac{n-1}{n} \cdot \frac{1}{n-1} = \frac{1}{n}$,

$p_3 = \frac{n-1}{n} \cdot \frac{n-2}{n-1} \cdot \frac{1}{n-2} = \frac{1}{n}, \ldots$.

31. (a)

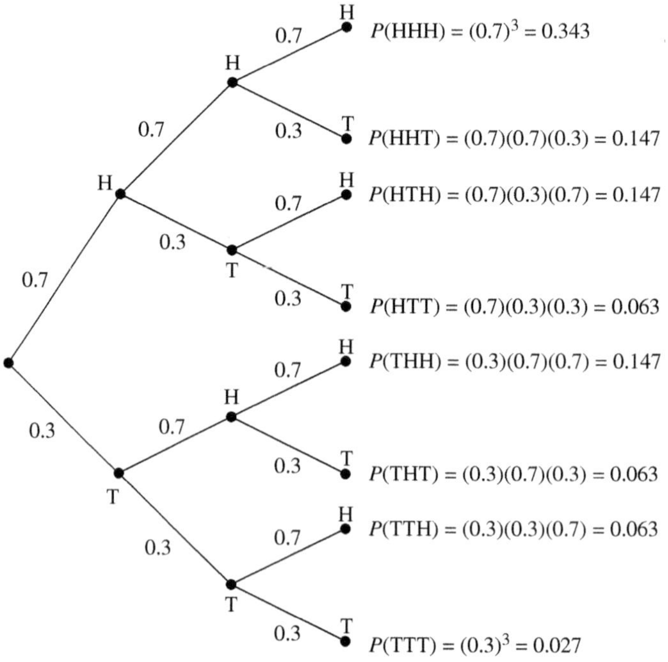

(b) $P(\text{HHH}) = 0.343$
(c) $P(\text{HHT or HTH or THH}) = 0.147 + 0.147 + 0.147 = 0.441$

(d) $P(\text{H appears}) = 1 - P(\text{TTT}) = 1 - 0.027 = 0.973$
(e) $P(\text{H and T appear}) = 1 - P(\text{HHH}) - P(\text{TTT}) = 1 - 0.343 - 0.027 = 0.63$ **33.** $1/3! = 1/6$ **35. (a)** 2^4, or 16, words **(b)** 2^6, or 64, words **(c)** At least 10 bits, since $2^9 = 512$, but $2^{10} = 1024$ **(d)** The five code words 0000, 0001, 0011, 0111, and 1111 have no 0 following a 1, so the probability is $5/16$. **37. (a)** No, since $26^6 = 308{,}915{,}776$ is less than the population of the United States, which reached 310 million in the year 2010. **(b)** Yes, since $52^5 = 380{,}204{,}032$ **(c)** Seven letters are needed since $26^7 = 8{,}031{,}810{,}176$ but $26^6 = 308{,}915{,}776$
39. D **41.** C

Problem Set 14.3 (page 753)

1. (a) $7 \cdot 6 \cdot 5 = 210$ **(b)** $7! = 5040$
(c) $7 \cdot 6 \cdot 5 \cdot 4 = 840$
(d) $(7 \cdot 6 \cdot 5)/(3 \cdot 2 \cdot 1) = 35$ **(e)** $7!/7! = 1$
(f) $7 \cdot 6 \cdot 5 \cdot 4/4 \cdot 3 \cdot 2 \cdot 1 = 35$
3. (a) Combination, $C(10, 5) = 252$ ways **(b)** Permutation, $P(10, 4) = 5040$ ways **(c)** Permutation, $P(12, 5) = 95{,}040$ ways **(d)** Combination, $C(15, 3) = 455$ ways **(e)** Permutation, $5! = 120$ ways **5.** 210 ways, since there are seven ways to choose the unicyclist and $P(6, 2) = 6 \cdot 5 = 30$ ways to choose the front and rear rider for the tandem bike **7. (a)** 14!, around 90 billion **(b)** $8! \cdot 6! = 29{,}030{,}400$ **(c)** $P(14, 3) = 2184$ **(d)** $C(6, 2)\,C(8, 2) = 420$
9. 1 6 15 20 15 6 1
1 7 21 35 35 21 7 1

11. (a) $\dfrac{C(12, 3)}{C(52, 3)} = \dfrac{12 \cdot 11 \cdot 10}{52 \cdot 51 \cdot 50} \doteq 0.01$

(b) $\dfrac{4 \cdot 13^3}{C(52, 3)} = \dfrac{52 \cdot 13^2}{52 \cdot 51 \cdot 50/3!} \doteq 0.398$

13. $6!5!/11! \doteq 0.002$ **15.** The number of ways to have all red cards is $C(26, 5)$ and the numbers of ways to have all face cards is $C(12,5)$. These events are not mutually exclusive, and the number of ways to have five of the six red face cards is $C(6, 5)$. The probability is, therefore,

$$\frac{C(26, 5) + C(12, 5) - C(6, 5)}{C(52, 5)} = \frac{65{,}780 + 792 - 6}{2{,}598{,}960} \doteq 0.026.$$

17. (a) There are 2^{10} ways to answer the quiz. One way is 100% and ten ways are 90%, so the probability of 90% or better is $11/1024$. **(b)** There are $C(10,2) = 45$ ways to score 80%, $C(10, 3) = 120$ ways to score 70%, $C(10, 4) = 210$ ways to score 60%, and $C(10, 5) = 252$ ways to score 50%, so the probability of a passing score is $(1 + 10 + 45 + 120 + 210 + 252)/1024 = 638/1024 = 0.62\ldots$.

19. (a) $P(\text{all 4 red}) = \dfrac{C(8, 4)}{C(19, 4)} = \dfrac{\frac{8 \cdot 7 \cdot 6 \cdot 5}{4 \cdot 3 \cdot 2 \cdot 1}}{\frac{19 \cdot 18 \cdot 17 \cdot 16}{4 \cdot 3 \cdot 2 \cdot 1}} \doteq 0.02$

(b) $P(\text{2 red 2 green}) = \dfrac{C(8, 2) \cdot C(6, 2)}{C(19, 4)} \doteq 0.11$

(c) $P(\text{exactly 2 are red}) = \dfrac{C(8, 2) \cdot C(11, 2)}{C(19, 4)} \doteq 0.40$

$P(\text{exactly 2 are green}) = \dfrac{C(6, 2) \cdot C(13, 2)}{C(19, 4)} \doteq 0.30$

Therefore, $P(\text{exactly 2 are red or exactly 2 are green}) = 0.40 + 0.30 - 0.11 = 0.59$.
21. Both Murial and Jisoo have given correct answers. Murial counted each of the mutually exclusive cases to get $1 + 28 + C(4, 2)\,C(7, 2) = 155$. Jisoo correctly counted the number of complementary ways to get $C(11, 4) - C(7, 4) - 4C(7, 3) = 330 - 35 - 140 = 155$. Shelley's answer of $C(4, 2)C(9, 2) = 216$ is much too high, since her method counts many of the committees more than once. To understand why, suppose the seniors are A, B, C, and D, and the juniors are $a, b, c, \ldots, g$. If A and B are placed on the committee, it can be filled out by also including C and g, resulting in the committee $\{A, B, C, g\}$. Now suppose that A and C are placed on the committee and it is then filled out by including B and g. The same committee $\{A, B, C, g\}$ is formed, but Shelley has counted it more than once.

23. (a) $P(\text{a code word begins with } a) = \dfrac{25 \cdot 24 \cdot 23 \cdot 22}{26 \cdot 25 \cdot 24 \cdot 23 \cdot 22} \doteq 0.04$

(b) $P(\text{a code word } c \text{ followed by } d) = \dfrac{4 \cdot 24 \cdot 23 \cdot 22}{26 \cdot 25 \cdot 24 \cdot 23 \cdot 22} \doteq 0.006$

(c) $P(\text{a code word starts with a vowel and ends with a consonant}) = \dfrac{5 \cdot 21 \cdot 24 \cdot 23 \cdot 22}{26 \cdot 25 \cdot 24 \cdot 23 \cdot 22} \doteq 0.16$ **(d)** $2 \cdot 4 \cdot 24 \cdot 23 \cdot 22 = 97{,}152$

25. Since the order of selection of the three balls does not matter, one answer is given by the combination $C(m + n, 3)$. Taking the respective cases where all three balls are red, all three are blue, one is red and two are blue, and one is blue and two are red gives, altogether, the sum $C(m, 3) + C(n, 3) + mC(n, 2) + nC(m, 2)$. Equating the two expressions then yields the desired formula.
27. (a) $C(8, 5)$ **(b)** $C(7, 5)$, since there are seven nonred marbles from which to choose 5. **(c)** $C(7, 4)$, since there are 7 nonred marbles from which to choose 4. **(d)** $C(8, 5) = C(7, 5) + C(7, 4)$ **(e)** Consider a bag of ten marbles, exactly one red. The number of ways to pick any four marbles is the same as the number of ways to pick four of the nine nonred marbles + the number of ways to pick the red marble and three of the nine nonred marbles.
29. (a) Pick the 8 of the 15 members in $C(15, 8) = 6435$ ways, and pick the two cocaptains from the 8 delegates in $C(8, 2) = 28$ ways. Thus, there are $6435 \cdot 28 = 180{,}180$ ways altogether. **(b)** Pick the two cocaptains in $C(15, 2) = 105$ ways, and then fill out the delegation by choosing 6 of the remaining 13 members in $C(13, 6) = 1716$ ways. This gives $105 \cdot 1716 = 180{,}180$ ways altogether. **(c)** $C(15, 8)C(8, 2) = \dfrac{15!}{8!7!}\dfrac{8!}{2!6!} = \dfrac{15!}{7!2!6!}$ and $C(15, 2)C(13, 6) = \dfrac{15!}{2!13!}\dfrac{13!}{6!7!} = \dfrac{15!}{2!7!6!}$ **(d)** Use 19 members, sending a delegation of 7, including 3 cocaptains.

31. (a) There are $\binom{10}{2}$ ways to choose where the two red tiles are placed along the walkway. From the remaining eight places, there are $\binom{8}{3}$ ways to place the three blue tiles. There are now five empty positions for the five green tiles, and these can be placed in $\binom{5}{5} = 1$ ways. **(b)** $\binom{10}{2}\binom{8}{3}\binom{5}{5} = \dfrac{10!}{2!8!} \cdot \dfrac{8!}{3!5!} \cdot \dfrac{5!}{5!0!} = \dfrac{10!}{2!3!5!}$
(c) There are 10! ways to pave the walkway with ten different tiles. Each color pattern appears the same if the two red tiles are permuted in 2! ways, the three blue tiles are permuted in 3! ways, and the five green tiles are permuted in 5! ways. This means that each color pattern is formed in 2!3!5! ways, so there are $\dfrac{10!}{2!3!5!}$ different color patterns.

33. (a) $\dfrac{4!}{2!2!} = 6$ **(b)** $\dfrac{7!}{2!2!2!1!} = 630$ **35. (a)** $C(6 + 4 - 1, 6) = C(9, 6) = C(9, 3) = \dfrac{9 \cdot 8 \cdot 7}{3 \cdot 2 \cdot 1} = 84$ **(b)** $C(8 + 6 - 1, 8) =$

$C(13, 8) = C(13, 5) = \dfrac{13 \cdot 12 \cdot 11 \cdot 10 \cdot 9}{5 \cdot 4 \cdot 3 \cdot 2 \cdot 1} = 1287$

(c) $C(4 + 32 - 1, 4) = C(35, 4) = \dfrac{35 \cdot 34 \cdot 33 \cdot 32}{4 \cdot 3 \cdot 2 \cdot 1} = 52{,}360$

(d) The number of combinations with repetition is equivalent to the number of ways n ice cream cones can be ordered from k flavors. That is, it is the number of lists of n check marks and $k - 1$ bars that separate one flavor from the next. Therefore, n objects can be chosen from k types, with repetition allowed, in $C(n + k - 1, n)$ ways.

37. (a) $C(13, 3) \cdot C(11, 3) = \dfrac{13 \cdot 12 \cdot 11}{3 \cdot 2 \cdot 1} \cdot \dfrac{11 \cdot 10 \cdot 9}{3 \cdot 2 \cdot 1} = 47{,}190$

(b) $C(10, 2) \cdot C(12, 3) + C(10, 3) \cdot C(12, 3) = 44{,}220$

39. $3!4!5! = 17{,}280$ ways. **41. (a)** $\dfrac{C(8, 5) \cdot C(72, 15)}{C(80, 20)} \doteq 0.0183$

(b) $\dfrac{C(8, 6) \cdot C(72, 14)}{C(80, 20)} \doteq 0.00237$ **(c)** $\dfrac{C(8, 7) \cdot C(72, 13)}{C(80, 20)} \doteq 0.00016$

(d) $\dfrac{C(8, 8) \cdot C(72, 12)}{C(80, 20)} \doteq 0.0000043$

Problem Set 14.4 (page 763)

1. (a) 5:31 or, equivalently, $\dfrac{5}{31}$ **(b)** 31:5 **(c)** $P(6) = \dfrac{5}{36}$

3. $P(A \text{ or } C) = P(A) + P(C) = \dfrac{1}{2} + \dfrac{1}{6} = \dfrac{4}{6} = \dfrac{2}{3}$. Thus, the odds in favor of A or C are $\dfrac{\frac{2}{3}}{1 - \frac{2}{3}} = \dfrac{2}{1}$, or 2:1. **5.** Since $P(A) = 0.3$ and $P(B) = 0.5$, then $P(C) = 1 - (0.3 + 0.5) = 0.2 = \dfrac{1}{5}$. This means the odds that C occurs is 1:4. **7.** $E = \$4 \cdot \dfrac{1}{36} + \$6 \cdot \dfrac{2}{36} +$

$\$8 \cdot \dfrac{3}{36} + \$10 \cdot \dfrac{4}{36} + \$20 \cdot \dfrac{5}{36} + \$40 \cdot \dfrac{6}{36} + \$20 \cdot \dfrac{5}{36} + \$10 \cdot \dfrac{4}{36} +$

$\$8 \cdot \dfrac{3}{36} + \$6 \cdot \dfrac{2}{36} + \$4 \cdot \dfrac{1}{36} = \dfrac{\$600}{36} = \$16.67$ to the nearest penny

9. (a) $(\$100 - \$1)\dfrac{1}{100} + (-\$1)\dfrac{99}{100} = \$0$

(b) $(\$100 - \$1)\dfrac{1}{200} + (-\$1)\dfrac{199}{200} = \$\left(\dfrac{99 - 199}{200}\right) = -\$0.50$

11. The expected value is $\left(\dfrac{36}{n} - 1\right)\left(\dfrac{n}{38}\right) + (-1)\left(\dfrac{38 - n}{38}\right) =$

$\dfrac{36 - n - 38 + n}{38} = -\dfrac{2}{38} \doteq -0.05$. Since the player can expect to lose about 5% of the amount of the bets placed, this game is not fair. **13. (a)** $1/4$ **(b)** $1/2$ **(c)** $3/4$ **15. (a)** Each blue region can be paired with a congruent yellow region, so the blue and yellow areas are equal. Therefore, the probability is $1/2$. **(b)** Each region can be partitioned into two triangles of base 2″ and altitude 3″, so all six regions have the same area. The probability is therefore $1/6$ of a point being in any chosen region. **17.** If sufficiently many trials are conducted, it should be discovered that there is a match about 63% of the time. This means no student is given a correct key with a probability of about 0.37. The probability is nearly the same for any group of more than ten students. **19. (a)** Answers will vary. **(b)** Answers will vary.

21. Answers will vary. **23.** Answers will vary. **25. (a)** With a large number of players, there is an increased chance that more than one winner is declared and the prize must be shared. **(b)** Nearly all of your dreams are more than met with \$10 million! Having a better chance to win \$10 million than \$100 million may be very sensible. **27.** Since

$$P(2) = P(GG) + P(BB) = \left(\frac{1}{2}\right)^2 + \left(\frac{1}{2}\right)^2 = \frac{1}{2}$$
$$P(4) = 1 - (P(2) + P(3)) = 1 - \frac{1}{2} - \frac{1}{4} = \frac{1}{4},$$

the expected value is

$$2P(2) + 3P(3) + 4P(4) = 2 \cdot \frac{1}{2} + 3 \cdot \frac{1}{4} + 4 \cdot \frac{1}{4} = \frac{11}{4} = 2.75.$$

29. (a) The points V and X should each be reached about $1/3$ of the time. **(b)** Points T, U, Y, and Z are rarely reached.
(c)

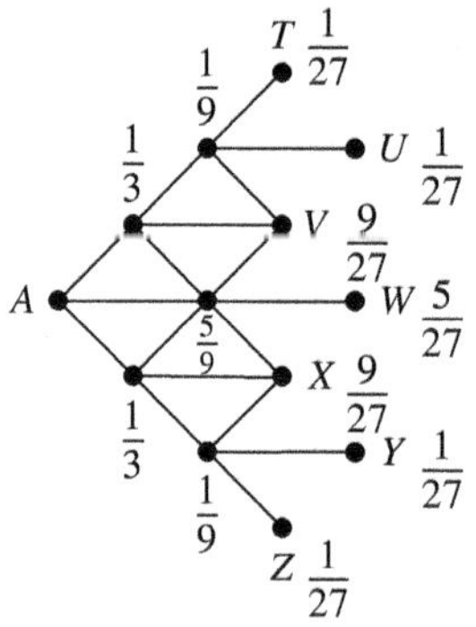

31. (a) 15 ways: RR, GR, BR, WR, PR, RB, GB, BB, WB, PB, RW, GW, BW, WW, PW

(b) $\dfrac{3}{15}$, or $\dfrac{1}{5} = 0.2$ **(c)** 0.8, the complementary probability

Chapter 14 Review Exercises (page 769)

1. (a) $S = \{(1, H), (1, T), (2, H), (2, T), (3, H), (3, T), (4, H), (4, T)\}$.
(b)

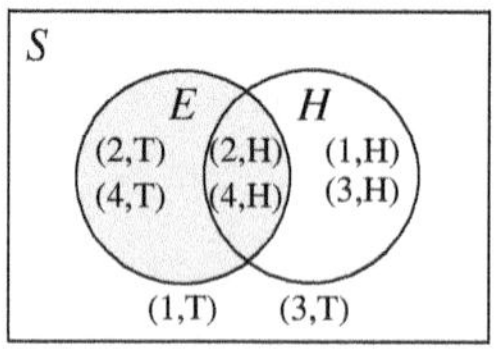

(c) $P(E) = \dfrac{4}{8} = \dfrac{1}{2}$, $P(H) = \dfrac{4}{8} = \dfrac{1}{2}$, $P(E \cup H) = \dfrac{6}{8} = \dfrac{3}{4}$,

$P(E \cap H) = \dfrac{2}{8} = \dfrac{1}{4}$ **(d)** No: $E \cap H = \{(2, H), (4, H)\} \neq \varnothing$

3. $P(\text{sum at most } 11) = 1 - P(\text{sum is } 12) = 1 - \dfrac{1}{36} \doteq 0.97$

5. (a) $\dfrac{13}{52} + \dfrac{13}{52} = \dfrac{1}{2}$ **(b)** $\dfrac{13}{52} + \dfrac{4}{52} - \dfrac{1}{52} = \dfrac{16}{52} = \dfrac{4}{13}$ **(c)** $\dfrac{9}{13}$

7. (a) $9!$ **(b)** $2 \cdot 4! \cdot 5!$ **(c)** $4! \cdot 5!/9!$

9.

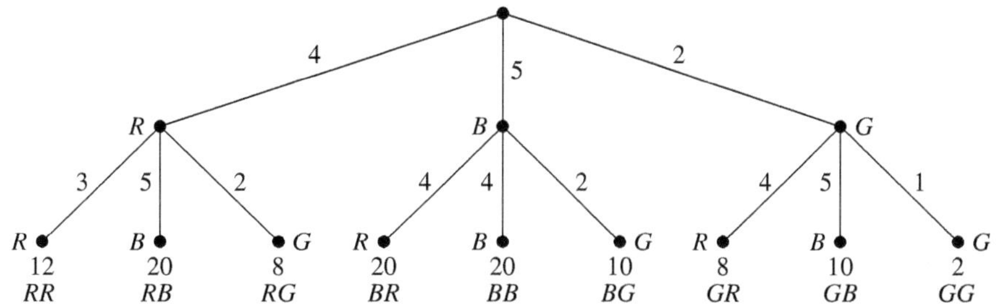

$P(\text{same color } x) = \dfrac{12+20+2}{12+20+8+20+20+10+8+10+2} = \dfrac{34}{110} \doteq 0.31$, $P(\text{at least one green}) = \dfrac{8+10+8+10+2}{110} = \dfrac{38}{110} \doteq 0.35$

11. (a) $P(\text{b and } 8) = 0$ **(b)** $P(\text{b or } 8) = P(\text{b}) + P(8) = \dfrac{90}{360} + \dfrac{30}{360} = \dfrac{1}{3} \doteq 0.33$ **(c)** $P(\text{b} \mid 8) = 0$ **(d)** $P(\text{b and } 2) = \dfrac{1}{12}$ **(e)** $P(\text{b or } 2) = P(\text{b}) + P(2) - P(\text{b and } 2) = \dfrac{90}{360} + \dfrac{30}{360} - \dfrac{30}{360} = 0.25$ **(f)** $P(2 \mid \text{b}) = \dfrac{1}{3}$ **13. (a)** The events are independent so the probabilities multiply to give $P(\text{spade}) \cdot P(3 \text{ or } 5) \cdot P(H) = \dfrac{1}{4} \cdot \dfrac{2}{6} \cdot \dfrac{1}{2} = \dfrac{1}{24}$. **(b)** By complementary probabilities, $1 - P(\text{not spade}) \cdot P(1{,}2{,}4 \text{ or } 6) \cdot P(T) = 1 - \dfrac{3}{4} \cdot \dfrac{4}{6} \cdot \dfrac{1}{2} = 1 - \dfrac{1}{4} = \dfrac{3}{4}$. **15. (a)** 5040 **(b)** 504 **(c)** 40,320 **(d)** 5040 **(e)** 6720 **(f)** 40,320 **(g)** 84 **(h)** 1

17. (a) $C(8, 5) = \dfrac{8 \cdot 7 \cdot 6 \cdot 5 \cdot 4}{5 \cdot 4 \cdot 3 \cdot 2 \cdot 1} = 56$

(b) $C(5, 5) \cdot C(8, 5) = \dfrac{5 \cdot 4 \cdot 3 \cdot 2 \cdot 1}{5 \cdot 4 \cdot 3 \cdot 2 \cdot 1} \cdot \dfrac{8 \cdot 7 \cdot 6 \cdot 5 \cdot 4}{5 \cdot 4 \cdot 3 \cdot 2 \cdot 1} = 56$

(c) $C(5, 5) + C(8, 5) = \dfrac{5 \cdot 4 \cdot 3 \cdot 2 \cdot 1}{5 \cdot 4 \cdot 3 \cdot 2 \cdot 1} + \dfrac{8 \cdot 7 \cdot 6 \cdot 5 \cdot 4}{5 \cdot 4 \cdot 3 \cdot 2 \cdot 1} = 57$

19.

(a) $P(\text{same color}) = \dfrac{2}{21} + \dfrac{1}{7} + \dfrac{2}{35} = \dfrac{10+15+6}{105} = \dfrac{31}{105}$

(b) $P(\text{WW}) = \dfrac{2}{21}$

(c) $P(\text{WW} \mid \text{same color}) = \dfrac{P(\text{WW \& same color})}{P(\text{same color})} = \dfrac{P(\text{WW \& same color})}{P(\text{same color})} = \dfrac{\frac{2}{21}}{\frac{31}{105}} = \dfrac{10}{31}$

(d)

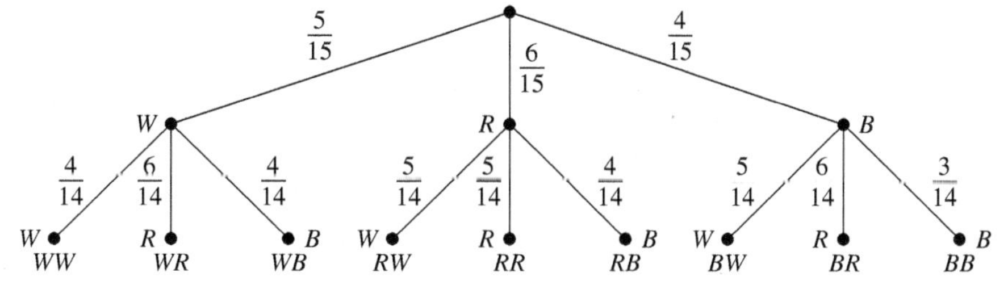

$\dfrac{20}{210} = \dfrac{2}{21}$ $\dfrac{30}{210} = \dfrac{1}{7}$ $\dfrac{20}{210} = \dfrac{2}{21}$ $\dfrac{30}{210} = \dfrac{1}{7}$ $\dfrac{30}{210} = \dfrac{1}{7}$ $\dfrac{24}{210} = \dfrac{4}{35}$ $\dfrac{20}{210} = \dfrac{2}{21}$ $\dfrac{24}{210} = \dfrac{4}{35}$ $\dfrac{12}{210} = \dfrac{2}{35}$

21. 7 to 13

23. (a) E = \$5 · (.50) + \$10 · (.25) + \$20 · (.10) = \$7

(b) No. On average, you expect to lose \$3 per game.

Mathematical Lexicon

Many of the words, prefixes, and suffixes forming the vocabulary of mathematics are derived from words and word roots from Latin, Greek, and other languages. Some of the most common terms are listed below, to serve as an aid to learning and understanding the terminology of mathematics.

acute from Latin *acus* ("needle") by way of *acutus* ("pointed, sharp")
algorithm distortion of Arabic name *al-Khowarazmi* ("the man from Khwarazm"), whose book on the use of Indo-Arabic numeration was translated into Latin as *Liber Algorismi,* meaning "Book of al-Khowarazmi"
angle from Latin *angulus* ("corner, angle")
apex from Latin word meaning "tip, peak"
area Latin *area* ("vacant piece of ground, plot of ground, open court")
associative from Latin *ad* ("to") and *socius* ("partner, companion")
axis from Latin word meaning "axle, pivot"
bi- from Latin prefix derived from *dui-* ("two"); *bi*nary, *bi*nomial, *bi*sect
calculate from Latin *calc* ("chalk, limestone") and diminutive suffix *-ulus* (a *calculus* was a small pebble; *calculare* meant "to use pebbles" = to do arithmetic)
cent- from Latin *centum* ("hundred"); *cent*imeter, per*cent*
circum- from Latin *circum* ("around"); *circum*ference, *circum*scribe
co-, col-, com-, con- from Old Latin *com* ("together with, beside, near"); *com*mutative, *col*linear, *com*plement, *con*gruent
commutative from Latin *co-* ("together with") and *mutare* ("to move")
concurrent from Latin *co-* ("together with") and *currere* ("to run")
conjecture from Latin *co-* ("together with") and *iactus* ("to throw") [conjecture = throw (ideas) together]
cylinder from Greek *kulindros* ("a roller")
de- Latin preposition *de* ("from, down from, away from, out of"); *de*nominator, *de*duction
deca-, deka- from Greek deka- ("ten"); *deca*gon, do*deca*hedron, *deka*meter
deci- from Latin *decimus* ("tenth"); *deci*mal, *deci*meter
diagonal from Greek *dia-* ("through, across") and *gon-* ("angle")
diameter from Greek *dia-* ("through, across") and *metron* ("measure")
digit from Latin *digitus* ("a finger")
distribute from Latin prefix *dis-* ("apart, away") and Latin *tribu* ["a tribe (of Romans)"]
empirical from Latin *empiricus* ("a physician whose art is founded solely on practice")
equal from Latin *æquus* ("even, level")
equilateral from Latin *æquus* ("even, level") and *latus* ("side")
equivalent from Latin *æquus* ("even, level") and *valere* ("to have value")
exponent from Latin *ex* ("away") and *ponent-* = present participial stem of *ponere* ("to put")
figure from Latin *figura* ("shape, form, figure")
fraction from Latin *fractus,* past participle of *frangere* ("to break")
geometry from Greek *geo-* ("earth") and *metron* ("measure")
-gon from Greek *gonia* ("angle, corner"); poly*gon,* penta*gon*
-hedron from Greek *hedra* ("base, seat"); poly*hedron,* tetra*hedron*
hept-, sept- Greek *hept,* from prehistoric Greek *sept,* meaning "seven"; *hept*agon
heuristic from Greek *heuriskein* ("to find, discover")
hex- from Greek *hex,* from prehistoric Greek *sex,* meaning "six"; *hex*agon, *hex*omino
icosahedron from Greek *eikosi* ("twenty") and *hedra* ("bases, seat")
inch from Latin *uncia,* a unit of weight equal to one twelfth of the *libra,* or Roman pound
inscribe from Latin *in* ("in") and *scribere* ("to scratch"); hence, "to write"
integer from Latin *in-* ("not") and Indo-European root *tag-* ("to touch") (an integer is untouched, hence "whole")
inverse from Latin *in* ("in") and *versus,* past participle of *vertere* ("to turn")
isosceles from Greek *isos* ("equal") and *skelos* ("leg")
kilo- from Greek *khiloi* ("thousand"); *kilo*gram, *kilo*meter
lateral from Latin *latus* ("side")
lb. abbreviation for pound, from *libra,* the Roman unit of weight
line from Latin *linum* ("flax") (The Romans made *linea,* linen thread, from flax.)
median from Latin *medius* ("in the middle")
meter from Greek *metron* ("measure, length")
milli- from Latin *mille* ("one thousand")
multiply from Latin *multi* ("many") and Indo-European *pel* ("to fold")
nonagon from Latin *nonus* ("ninth") and Greek *gon* ("angle")
number from Latin *numerus* ("number")
obtuse from Latin *ob* ("against, near, at") and *tusus* ("to strike, to beat") (*obtusus* = beaten down to the point of being dull)
oct- from Greek *octo* ("eight"); *oct*agon, *oct*ahedron
parallel from Greek *para* ("alongside") and *allenon* ("one another")
pent- Greek *pent* ("five"); *pent*agon, *pent*agram, *pent*omino
percent from Latin *per* ("for") and *centum* ("hundred") [percent = for (each) hundred]
peri from Greek *peri* ("around"); *peri*meter
plane from Latin *planus* ("flat")
poly- from Greek *polus* ("many"); *poly*gon, *poly*hedra, *poly*omino
prism from Greek *prisma* ("something that has been sawed")
quadr- Latin *quadr-* ("four"); *quadr*ant, *quadr*ilateral
rectangle from Latin *rectangulus* ("right-angled")
-sect from Latin *sectus,* past participle of *secui* ("to cut"); bi*sect,* inter*sect*
surface from Latin *super* ("over") and *facies* ("form, shape")
symmetric from Greek *sun-* ("together with") and *metron* ("measure")
tetra- Greek *tetra-* ("four"); *tetra*hedron, *tetr*omino
trans- Latin *trans* ("across"); *trans*itive, *trans*lation, *trans*versal
tri- from Latin *tri* ("three"); *tri*angle, *tri*sect
vertex from Latin verb *vertere* ("to turn")
zero from Arabic *çifr* ("empty")

Credits

Photo Credits

Chapter 1

Page 1, Paul Conklin/PhotoEdit. Page 9, AP Images. Page 32, The Image Works.

Chapter 2

Page 67, iStockphoto/Thinkstock. Page 81 (tl), Duane DeTemple. (tm), Duane DeTemple, (tm), Duane DeTemple. (tr), Duane DeTemple. (bl), Cubes–Rectangle, Photos of MathLink® Cubes courtesy of ETA/Cuisenaire®. Copyright © 2010 ETA/Cuisenaire. (bm), Cubes–Prism, Photos of MathLink® Cubes courtesy of ETA/Cuisenaire®. Copyright © 2010 ETA/Cuisenaire. (br), Cubes–Pyramid, Photos of MathLink® Cubes courtesy of ETA/Cuisenaire®. Copyright © 2010 ETA/Cuisenaire. Page 95, Courtesy of Robbin S. Crowell. Page 107, Courtesy of Katie Busbey.

Chapter 3

Page 125, Mila Gligoric/Fotolia. Page 130, Duane DeTemple. Page 169, Courtesy of Tara Morey.

Chapter 4

Page 183, Bloomua/Shutterstock. Page 188, The Image Works. Page 190, Courtesy of Marianne Strayton. Page 196, Science Photo Library/Science Source/Photo Researchers, Inc. Page 200, Pearson Education, Inc. Page 212, Duane DeTemple.

Chapter 5

Page 221, Jim Lozouski/Shutterstock. Page 235, Courtesy of Kristin Hanley. Page 244 (r), Shutterstock. (l), Shutterstock. Page 254, Shutterstock.

Chapter 6

Page 263, David H. Lewis/iStockphoto. Page 284, Courtesy of Ann Clay.

Chapter 7

Page 323, Bonita R. Cheshier/Shutterstock. Page 367, Courtesy of Debbie Goodman.

Chapter 8

Page 377, Gualtiero Boffi/Shutterstock. Page 384, Alamy. Page 395, Library of Congress Prints and Photographs Division [LC-USZ62-1234].

Chapter 9

Page 425, Sakhorn/Shutterstock. Page 428, Cathy Yeulet/Thinkstock. Page 430 (mr), Duane De Temple. Page 430 (tr), Science Source/Photo Researchers, Inc. Page 430 (bl), Science Source/Photo Researchers, Inc. Page 430 (br), Ray Ellis/Science Source/Photo Researchers, Inc. Page 430 (ml), Duane DeTemple. Page 430 (tl), Thinkstock. Page 435 (bl), Superstock. (bm), Alamy. (tl), Beth Anderson/Pearson Education, Inc. (tr), Pearson Education, Inc. (m), Ruler, Beth Anderson/Pearson Education, Inc. (br), Robert Spriggs/Shutterstock. Page 451, Courtesy of Teri M. Rodriguez. Page 457, Alamy. Page 466 (tl), Scott Camazine/Science Source. Page 466 (tm), Science and Society Picture Library/The Image Works. Page 466 (tr), Science and Society Picture Library/The Image Works. Page 466 (m), Scott Sanders/Shutterstock. Page 466 (bl), Addison Wesley/Pearson Education, Inc. Page 466 (br), Addison Wesley/Pearson Education, Inc. Page 472, Courtesy of Ralph Pantozzi. Page 473 (l), Alamy. Page 473 (r), Juergen Berger/Science Source. Page 479, Marcel Clemens/Shutterstock.

Chapter 10

Page 487, The Washington Post/Getty Images. Page 515, NASA. Page 527, Clay tablet. Old Babylonian. Cuneiform inscription in Akkadian. The Trustees of the British Museum [1896.0410.11]/Art Resource, New York. Page 530, Courtesy of Simone Wells-Heard.

Chapter 11

Page 559, Michael Newman/PhotoEdit. Page 580, Beth Anderson/Pearson Education, Inc. Page 592, Duane De Temple. Page 594 (tl), Dukepope/Fotolia. (tr), Javarman/Fotolia. (ml), Richard Laschon/Alamy. (mr), Jonathan/Fotolia. Page 594 (bl), Vdlee/Shutterstock. Page 594 (br), Raywoo/Shutterstock. Page 600, Heinz Voderberg/Collection of the author. Page 600, Nancy Putnam.

Chapter 12

Page 613, Jörg Hackemann/Fotolia.

Chapter 13

Page 667, Zurijeta/Shutterstock. Page 675, Courtesy of University of California, Berkeley.

Chapter 14

Page 719, James W. Porter/Corbis/Glow Images. Page 719, game board, Andriano/Shutterstock. Page 736, Courtesy of Jenifer Martin. Page 759, Aleksandar Bracinac/Shutterstock.

Text Credits

For State Assessment questions used throughout the book:
Washington Standard Student Assessment Questions are reprinted by permission of the Washington State Office of Superintendent of Public Instruction.
Courtesy of the Massachusetts Department of Elementary and Secondary Education. All rights reserved.
California Standard Student Assessment Questions are reprinted by permission of the California Department of Education.

Common Core SMP used throughout the book:
From the Common Core, The Standards for Mathematical Practice. © Copyright 2010. National Governors Association Center for Best Practices and Council of Chief State School Officers. All rights reserved.

Index

N

O

Q

R

T

NCTM Principles and Standards for State Mathematics

Decisions made by teachers, school administrators, and other education professionals about the content and character of school mathematics have important consequences for both students and for society. These decisions should be based on sound professional guidance. *Principles and Standards for School Mathematics* is intended to provide such guidance. The Principles describe particular features of high-quality mathematics education. The Standards describe the mathematical content and processes that students should learn. Together, the Principles and Standards constitute a vision to guide educators as they strive for the continual improvement of mathematics education in classrooms, schools, and educational systems.

You may read the online version of *Principles and Standards for School Mathematics* at http://standards.nctm.org.

Principles for School Mathematics

- ***Equity.*** Excellence in mathematics education requires equity—high expectations and strong support for all students.
- ***Curriculum.*** A curriculum is more than a collection of activities: it must be coherent, focused on important mathematics, and well articulated across the grades.
- ***Teaching.*** Effective mathematics teaching requires understanding what students know and need to learn and then challenging and supporting them to learn it well.
- ***Learning.*** Students must learn mathematics with understanding, actively building new knowledge from experience and prior knowledge.
- ***Assessment.*** Assessment should support the learning of important mathematics and furnish useful information to both teachers and students.
- ***Technology.*** Technology is essential in teaching and learning mathematics; it influences the mathematics that is taught and enhances students' learning.

The Content Standards

Number and Operations Standard

Instructional programs from prekindergarten through grade 12 should enable all students to:

- Understand numbers, ways of representing numbers, relationships among numbers, and number systems.
- Understand meanings of operations and how they relate to one another.
- Compute fluently and make reasonable estimates.

Algebra Standard

Instructional programs from prekindergarten through grade 12 should enable all students to:

- Understand patterns, relations, and functions.
- Represent and analyze mathematical situations and structures using algebraic symbols.
- Use mathematical models to represent and understand quantitative relationships.
- Analyze change in various contexts.

Geometry Standard

Instructional programs from prekindergarten through grade 12 should enable all students to:

- Analyze characteristics and properties of two- and three-dimensional geometric shapes and develop mathematical arguments about geometric relationships.
- Specify locations and describe spatial relationships using coordinate geometry and other representational systems.
- Apply transformations and use symmetry to analyze mathematical situations.
- Use visualization, spatial reasoning, and geometric modeling to solve problems.

Measurement Standard

Instructional programs from prekindergarten through grade 12 should enable all students to:

- Understand measurable attributes of objects and the units, systems, and processes of measurement.
- Apply appropriate techniques, tools, and formulas to determine measurements.

Data Analysis and Probability Standard

Instructional programs from prekindergarten through grade 12 should enable all students to:

- Formulate questions that can be addressed with data and collect, organize, and display relevant data to answer them.
- Select and use appropriate statistical methods to analyze data.
- Develop and evaluate inferences and predictions that are based on data.
- Understand and apply basic concepts of probability.

The Process Standards

Problem Solving Standard

Instructional programs from prekindergarten through grade 12 should enable all students to:

- Build new mathematical knowledge through problem solving.
- Solve problems that arise in mathematics and in other contexts.
- Apply and adapt a variety of appropriate strategies to solve problems.
- Monitor and reflect on the process of mathematical problem solving.

Reasoning and Proof Standard

Instructional programs from prekindergarten through grade 12 should enable all students to:

- Recognize reasoning and proof as fundamental aspects of mathematics.
- Make and investigate mathematical conjectures.
- Develop and evaluate mathematical arguments and proofs.
- Select and use various types of reasoning and methods of proof.

Communication Standard

Instructional programs from prekindergarten through grade 12 should enable all students to:

- Organize and consolidate their mathematical thinking through communication.
- Communicate their mathematical thinking coherently and clearly to peers, teachers, and others.
- Analyze and evaluate the mathematical thinking and strategies of others.
- Use the language of Mathematics to express mathematical ideas precisely.

Connections Standard

Instructional programs from prekindergarten through grade 12 should enable all students to:

- Recognize and use connections among mathematical ideas.
- Understand how mathematical ideas interconnect and build on one another to produce a coherent whole.
- Recognize and apply mathematics in contexts outside of mathematics.

Representation Standard

Instructional programs from prekindergarten through grade 12 should enable all students to:

- Create and use representations to organize, record, and communicate mathematical ideas.
- Select, apply, and translate among mathematical representations to solve problems.
- Use representations to model and interpret physical, social, and mathematical phenomena.

Common Core State Standards for Mathematics

The Common Core State Standards for Mathematics (CCSS-M) ". . . define what students should understand and be able to do in their study of mathematics." The complete statements of the CCSS-M can be obtained at http://www.corestandards.org/Math. The standards are partitioned into two parts, beginning with the *Standards for Mathematical Practice* which are listed below. These are followed by the Standards for *Mathematical Content*, which details the level of mathematical knowledge, skills, and understandings that should be reached according to grade level.

Standards for Mathematical Practice

The Standards for Mathematical Practice describe varieties of expertise that mathematics educators at all levels should seek to develop in their students. These practices rest on important "processes and proficiencies" with longstanding importance in mathematics education. The first of these are the NCTM process standards of problem solving, reasoning and proof, communication, representation, and connections. The second are the strands of mathematical proficiency specified in the National Research Council's report *Adding It Up*: adaptive reasoning, strategic competence, conceptual understanding (comprehension of mathematical concepts, operations and relations), procedural fluency (skill in carrying out procedures flexibly, accurately, efficiently and appropriately), and productive disposition (habitual inclination to see mathematics as sensible, useful, and worthwhile, coupled with a belief in diligence and one's own efficacy).

1. **Make sense of problems and persevere in solving them.** Mathematically proficient students start by explaining to themselves the meaning of a problem and looking for entry points to its solution. They analyze givens, constraints, relationships, and goals. They make conjectures about the form and meaning of the solution and plan a solution pathway rather than simply jumping into a solution attempt. They consider analogous problems, and try special cases and simpler forms of the original problem in order to gain insight into its solution. They monitor and evaluate their progress and change course if necessary. Older students might, depending on the context of the problem, transform algebraic expressions or change the viewing window on their graphing calculator to get the information they need. Mathematically proficient students can explain correspondences between equations, verbal descriptions, tables, and graphs or draw diagrams of important features and relationships, graph data, and search for regularity or trends. Younger students might rely on using concrete objects or pictures to help conceptualize and solve a problem. Mathematically proficient students check their answers to problems using a different method, and they continually ask themselves, "Does this make sense?" They can understand the approaches of others to solving complex problems and identify correspondences between different approaches.

2. **Reason abstractly and quantitatively.** Mathematically proficient students make sense of quantities and their relationships in problem situations. They bring two complementary abilities to bear on problems involving quantitative relationships: the ability to *decontextualize*—to abstract a given situation and represent it symbolically and manipulate the representing symbols as if they have a life of their own, without necessarily attending to their referents—and the ability to *contextualize*, to pause as needed during the manipulation process in order to probe into the referents for the symbols involved. Quantitative reasoning entails habits of creating a coherent representation of the problem at hand; considering the units involved; attending to the meaning of quantities, not just how to compute them; and knowing and flexibly using different properties of operations and objects.

3. Construct viable arguments and critique the reasoning of others. Mathematically proficient students understand and use stated assumptions, definitions, and previously established results in constructing arguments. They make conjectures and build a logical progression of statements to explore the truth of their conjectures. They are able to analyze situations by breaking them into cases, and can recognize and use counterexamples. They justify their conclusions, communicate them to others, and respond to the arguments of others. They reason inductively about data, making plausible arguments that take into account the context from which the data arose. Mathematically proficient students are also able to compare the effectiveness of two plausible arguments, distinguish correct logic or reasoning from that which is flawed, and—if there is a flaw in an argument—explain what it is. Elementary students can construct arguments using concrete referents such as objects, drawings, diagrams, and actions. Such arguments can make sense and be correct, even though they are not generalized or made formal until later grades. Later, students learn to determine domains to which an argument applies. Students at all grades can listen or read the arguments of others, decide whether they make sense, and ask useful questions to clarify or improve the arguments.

4. Model with mathematics. Mathematically proficient students can apply the mathematics they know to solve problems arising in everyday life, society, and the workplace. In early grades, this might be as simple as writing an addition equation to describe a situation. In middle grades, a student might apply proportional reasoning to plan a school event or analyze a problem in the community. By high school, a student might use geometry to solve a design problem or use a function to describe how one quantity of interest depends on another. Mathematically proficient students who can apply what they know are comfortable making assumptions and approximations to simplify a complicated situation, realizing that these may need revision later. They are able to identify important quantities in a practical situation and map their relationships using such tools as diagrams, two-way tables, graphs, flowcharts and formulas. They can analyze those relationships mathematically to draw conclusions. They routinely interpret their mathematical results in the context of the situation and reflect on whether the results make sense, possibly improving the model if it has not served its purpose.

5. Use appropriate tools strategically. Mathematically proficient students consider the available tools when solving a mathematical problem. These tools might include pencil and paper, concrete models, a ruler, a protractor, a calculator, a spreadsheet, a computer algebra system, a statistical package, or dynamic geometry software. Proficient students are sufficiently familiar with tools appropriate for their grade or course to make sound decisions about when each of these tools might be helpful, recognizing both the insight to be gained and their limitations. For example, mathematically proficient high school students analyze graphs of functions and solutions generated using a graphing calculator. They detect possible errors by strategically using estimation and other mathematical knowledge. When making mathematical models, they know that technology can enable them to visualize the results of varying assumptions, explore consequences, and compare predictions with data. Mathematically proficient students at various grade levels are able to identify relevant external mathematical resources, such as digital content located on a website, and use them to pose or solve problems. They are able to use technological tools to explore and deepen their understanding of concepts.

6. Attend to precision. Mathematically proficient students try to communicate precisely to others. They try to use clear definitions in discussion with others and in their own reasoning. They state the meaning of the symbols they choose, including using the equal sign consistently and appropriately. They are careful about specifying units of measure, and labeling axes to clarify the correspondence with quantities in a problem. They calculate accurately and efficiently, express numerical answers with a degree of precision appropriate for the problem context. In the elementary grades, students give carefully formulated explanations to each other. By the time they reach high school they have learned to examine claims and make explicit use of definitions.